# American Electricians' Handbook

## OTHER ELECTRICAL CONSTRUCTION BOOKS OF INTEREST

*Johnson* • Electrical Contracting Business Handbook
*Kolstad* • Rapid Electrical Estimating and Pricing
*Kurtz and Shoemaker* • The Lineman's and Cableman's Handbook
*Kusko* • Emergency/Standby Power Systems
*Linden* • Handbook of Batteries and Fuel Cells
*Lundquist* • On-Line Electrical Troubleshooting
*McPartland* • Handbook of Practical Electrical Design
*McPartland* • McGraw-Hill's National Electrical Code® Handbook
*Ray* • How to Start and Operate an Electrical Contracting Business
*Richter and Schwan* • Practical Electrical Wiring
*Seidman, Mahrous, and Hicks* • Handbook of Electric Power Calculations
*Smeaton* • Switchgear and Control Handbook
*Traister* • Design and Application of Security/Fire Alarm Systems

# American Electricians' Handbook

## Twelfth Edition

**TERRELL CROFT**

Former Editor (Deceased)

**WILFORD I. SUMMERS**

Editor

**McGRAW-HILL, INC.**

New York   St. Louis   San Francisco   Auckland   Bogotá
Caracas   Lisbon   London   Madrid   Mexico   Milan
Montreal   New Delhi   Paris   San Juan   São Paulo
Singapore   Sydney   Tokyo   Toronto

Library of Congress Cataloging-in-Publication Data

Croft, Terrell, date.
   American electricians' handbook / Terrell Croft, former editor,
Wilford I. Summers, editor.—12th ed.
     p.   cm.
  Includes index.
  ISBN 0-07-013933-4
   1. Electric engineering—Handbooks, manuals, etc.  I. Summers,
Wilford I.  II. Title.
TK151.C8  1992                                        91-18395
621.319'24—dc20                                   CIP

ISBN 0-07-013933-4

The sponsoring editor for this book was Harold B. Crawford, the editing supervisor was Stephen M. Smith, and the production supervisor was Donald F. Schmidt. It was set in Caledonia by University Graphics, Inc.

Printed and bound by R. R. Donnelley & Sons Company.

# Contents

*Division 5*   **TRANSFORMERS**

*Division 6*   **SOLID-STATE DEVICES AND CIRCUITS**

*Division 7*   **GENERATORS AND MOTORS**

*Division 8*   **OUTSIDE DISTRIBUTION**

### Division 9    INTERIOR WIRING

### Division 10    ELECTRIC LIGHTING

### *Division 11*   WIRING AND DESIGN TABLES

**Index follows Division 11**

# Preface to the Twelfth Edition

This Twelfth Edition of the *American Electricians' Handbook* has been carefully revised and expanded in accordance with the latest good practice standards for installing electrical conductors, devices, and appliances, and it has been checked and revised to be consistent with the 1990 edition of the **National Electrical Code**® (**NEC**®).* As a result, a great deal of completely new material, including illustrations, has been added.

A few of the most significant advancements discussed in this new edition are revised lamp tables that reflect the energy-saving mood of the nation, new conductor insulation types (such as RHW-2, THHW, and XHHW-2), and expanded ampacities for cable types G and W. Deletion of material concerning many types of asbestos insulations reflects the concern for the safety of workers and the general public. Information on surge protection reflects the expanded uses of electronic equipment and attendant problems with sensitive electronic components. Revisions of many UL and NEMA standards are reflected in new enclosure classifications and receptacle and plug configurations.

Every effort has been made to continue the original purpose of this Handbook, which is to provide a compilation of data and information of the various types of electrical equipment and material presented without advanced mathematics and arranged in as useful a manner as possible for the intelligent selection, installation, maintenance, and operation of equipment.

This editor wishes to express his sincere appreciation to all individuals, organizations, and associations that have provided assistance and illustrations for this edition. A handbook of this nature could not be written without such assistance and suggestions. This editor especially wishes to express appreciation to Charles Schram, Russell Churchill, and Lothar Stern for their major contributions in motors, lighting, and solid-state technology, respectively.

Special appreciation, as always, goes to my wife, Laura, for handling correspondence, proofreading, and typing the manuscript.

Wilford I. "Bill" Summers

*National Electrical Code® and NEC® are Registered Trademarks of the National Fire Protection Association, Inc., Quincy, MA.

# Division 1

# Fundamentals

USEFUL TABLES

## 1.  Natural Trigonometric Functions

| Angle ($\theta$ or lag angle), deg | Sine (or reactive factor) | Cosine (or power factor) | Tangent | Cotangent | Secant | Cosecant | Angle ($\theta$ or lag angle), deg |
|---|---|---|---|---|---|---|---|
| 0  | 0.00000 | 1.00000 | 0.00000 | Infinite | 1.0000 | Infinite | 180 |
| 1  | 0.01774 | 0.99985 | 0.01745 | 57.290 | 1.0001 | 57.299 | 179 |
| 2  | 0 03490 | 0.99939 | 0.03492 | 28.636 | 1.0006 | 28.654 | 178 |
| 3  | 0.05234 | 0.99863 | 0.05241 | 19.081 | 1.0014 | 19.107 | 177 |
| 4  | 0.06976 | 0.99756 | 0.06993 | 14.301 | 1.0024 | 14.335 | 176 |
| 5  | 0.08715 | 0.99619 | 0.08749 | 11.430 | 1.0038 | 11.474 | 175 |
| 6  | 0.10453 | 0.99452 | 0.10510 | 9.5144 | 1.0055 | 9.5668 | 174 |
| 7  | 0.12187 | 0.99255 | 0.12278 | 8.1443 | 1.0075 | 8.2055 | 173 |
| 8  | 0.13917 | 0.99027 | 0.14054 | 7.1154 | 1.0098 | 7.1853 | 172 |
| 9  | 0.15643 | 0.98769 | 0.15838 | 6.3137 | 1.0125 | 6.3924 | 171 |
| 10 | 0.17365 | 0.98481 | 0.17633 | 5.6713 | 1.0154 | 5.7588 | 170 |
| 11 | 0.19081 | 0.98163 | 0.19438 | 5.1445 | 1.0187 | 5.2408 | 169 |
| 12 | 0.20791 | 0.97815 | 0.21256 | 4.7046 | 1.0223 | 4.8097 | 168 |
| 13 | 0.22495 | 0.97437 | 0.23087 | 4.3315 | 1.0263 | 4.4454 | 167 |
| 14 | 0.24192 | 0.97029 | 0.24933 | 4.0108 | 1.0306 | 4.1336 | 166 |
| 15 | 0.25882 | 0.96592 | 0.26795 | 3.7320 | 1.0353 | 3.8637 | 165 |
| 16 | 0.27564 | 0.96126 | 0.28674 | 3.4874 | 1.0403 | 3.6279 | 164 |
| 17 | 0.29237 | 0.95630 | 0.30573 | 3.2708 | 1.0457 | 3.4203 | 163 |
| 18 | 0.30902 | 0.95106 | 0.32492 | 3.0777 | 1.0515 | 3.2361 | 162 |
| 19 | 0.32557 | 0.94552 | 0.34433 | 2.9042 | 1.0576 | 3.0715 | 161 |
| 20 | 0.34203 | 0.93969 | 0.36397 | 2.7475 | 1.0642 | 2.9238 | 160 |
| 21 | 0.35837 | 0.93358 | 0.38386 | 2.6051 | 1.0711 | 2.7904 | 159 |
| 22 | 0.37461 | 0.92718 | 0.40403 | 2.4751 | 1.0785 | 2.6695 | 158 |
| 23 | 0.39073 | 0.92050 | 0.42447 | 2.3558 | 1.0864 | 2.5593 | 157 |
| 24 | 0.40674 | 0.91354 | 0.44523 | 2.2460 | 1.0946 | 2.4586 | 156 |
| 25 | 0.42262 | 0.90631 | 0.46631 | 2 1445 | 1.1034 | 2.3662 | 155 |
| 26 | 0.43837 | 0.89879 | 0.48773 | 2.0503 | 1.1126 | 2.2812 | 154 |
| 27 | 0.45399 | 0.89101 | 0.50952 | 1.9626 | 1.1223 | 2.2027 | 153 |
| 28 | 0.46947 | 0.88295 | 0.53171 | 1.8807 | 1.1326 | 2.1300 | 152 |
| 29 | 0.48481 | 0.87462 | 0.55431 | 1.8040 | 1.1433 | 2.0627 | 151 |
| 30 | 0 50000 | 0.86603 | 0.57735 | 1.7320 | 1.1547 | 2.0000 | 150 |
| 31 | 0.51504 | 0.85717 | 0.60086 | 1.6643 | 1.1666 | 1.9416 | 149 |
| 32 | 0.52992 | 0.84805 | 0.62487 | 1.6003 | 1.1792 | 1.8871 | 148 |
| 33 | 0.54464 | 0.83867 | 0.64941 | 1.5399 | 1.1924 | 1.8361 | 147 |
| 34 | 0.55919 | 0.82904 | 0.67451 | 1.4826 | 1.2062 | 1.7883 | 146 |
| 35 | 0.57358 | 0.81915 | 0.70021 | 1.4281 | 1.2208 | 1.7434 | 145 |
| 36 | 0.58778 | 0.80902 | 0.72654 | 1.3764 | 1.2361 | 1.7013 | 144 |
| 37 | 0.60181 | 0.79863 | 0.75355 | 1.3270 | 1.2521 | 1.6616 | 143 |
| 38 | 0.61566 | 0.78801 | 0.78128 | 1.2799 | 1.2690 | 1.6243 | 142 |
| 39 | 0.62932 | 0.77715 | 0.80978 | 1.2349 | 1.2867 | 1.5890 | 141 |
| 40 | 0.64279 | 0.76604 | 0.83910 | 1.1917 | 1.3054 | 1.5557 | 140 |
| 41 | 0.65606 | 0.75741 | 0.86929 | 1.1504 | 1.3250 | 1.5242 | 139 |
| 42 | 0.66913 | 0.74314 | 0.90040 | 1.1106 | 1.3456 | 1.4945 | 138 |
| 43 | 0 68200 | 0.73135 | 0.93251 | 1.0724 | 1.3673 | 1.4663 | 137 |
| 44 | 0.69466 | 0.71934 | 0.96569 | 1.0355 | 1.3902 | 1.4395 | 136 |
| 45 | 0.70711 | 0.70711 | 1.0000 | 1.0000 | 1.4142 | 1.4142 | 135 |

**Natural Trigonometric Functions**

| Angle ($\theta$ or lag angle), deg | Sine (or reactive factor) | Cosine (or power factor) | Tangent | Cotangent | Secant | Cosecant | Angle ($\theta$ or lag angle), deg |
|---|---|---|---|---|---|---|---|
| 46 | 0.71934 | 0.69466 | 1.0355 | 0.96569 | 1.4395 | 1.3902 | 134 |
| 47 | 0.73135 | 0.68200 | 1.0724 | 0.93251 | 1.4663 | 1.3673 | 133 |
| 48 | 0.74314 | 0.66913 | 1.1106 | 0.90040 | 1.4945 | 1.3456 | 132 |
| 49 | 0.75471 | 0.65606 | 1.1504 | 0.86929 | 1.5242 | 1.3250 | 131 |
| 50 | 0.76604 | 0.64279 | 1.1917 | 0.83910 | 1.5557 | 1.3054 | 130 |
| 51 | 0.77715 | 0.62932 | 1.2349 | 0.80978 | 1.5890 | 1.2867 | 129 |
| 52 | 0.78801 | 0.61566 | 1.2799 | 0.78128 | 1.6243 | 1.2690 | 128 |
| 53 | 0.79863 | 0.60181 | 1.3270 | 0.75355 | 1.6616 | 1.2521 | 127 |
| 54 | 0.80902 | 0.58778 | 1.3764 | 0.72654 | 1.7013 | 1.2361 | 126 |
| 55 | 0.81915 | 0.57358 | 1.4281 | 0.70021 | 1.7434 | 1.2208 | 125 |
| 56 | 0.82904 | 0.55919 | 1.4826 | 0.67451 | 1.7883 | 1.2062 | 124 |
| 57 | 0.83867 | 0.54464 | 1.5399 | 0.64941 | 1.8361 | 1.1922 | 123 |
| 58 | 0.84805 | 0.52992 | 1.6003 | 0.62487 | 1.8871 | 1.1792 | 122 |
| 59 | 0.85717 | 0.51504 | 1.6643 | 0.60086 | 1.9416 | 1.1666 | 121 |
| 60 | 0.86603 | 0.50000 | 1.7320 | 0.57735 | 2.0000 | 1.1547 | 120 |
| 61 | 0.87462 | 0.48481 | 1.8040 | 0.55431 | 2.0627 | 1.1433 | 119 |
| 62 | 0.88295 | 0.46947 | 1.8807 | 0.53171 | 2.1300 | 1.1326 | 118 |
| 63 | 0.89101 | 0.45399 | 1.9626 | 0.50952 | 2.2027 | 1.1223 | 117 |
| 64 | 0.89879 | 0.43837 | 2.0503 | 0.48773 | 2.2812 | 1.1126 | 116 |
| 65 | 0.90631 | 0.42262 | 2.1445 | 0.46631 | 2.3662 | 1.1034 | 115 |
| 66 | 0.91354 | 0.40674 | 2.2460 | 0.44523 | 2.4586 | 1.0946 | 114 |
| 67 | 0.92050 | 0.39073 | 2.3558 | 0.42447 | 2.5593 | 1.0864 | 113 |
| 68 | 0.92718 | 0.37461 | 2.4751 | 0.40403 | 2.6695 | 1.0785 | 112 |
| 69 | 0.93358 | 0.35837 | 2.6051 | 0.38386 | 2.7904 | 1.0711 | 111 |
| 70 | 0.93969 | 0.34202 | 2.7475 | 0.36397 | 2.9238 | 1.0642 | 110 |
| 71 | 0.94552 | 0.32557 | 2.9042 | 0.34433 | 3.0715 | 1.0576 | 109 |
| 72 | 0.95106 | 0.30902 | 3.0777 | 0.32492 | 3.2361 | 1.0515 | 108 |
| 73 | 0.95630 | 0.29237 | 3.2708 | 0.30573 | 3.4203 | 1.0457 | 107 |
| 74 | 0.96126 | 0.27564 | 3.4874 | 0.28647 | 3.6279 | 1.0403 | 106 |
| 75 | 0.96592 | 0.25882 | 3.7320 | 0.26795 | 3.8637 | 1.0353 | 105 |
| 76 | 0.97029 | 0.24192 | 4.0108 | 0.24933 | 4.1336 | 1.0306 | 104 |
| 77 | 0.97437 | 0.22495 | 4.3315 | 0.23087 | 4.4454 | 1.0263 | 103 |
| 78 | 0.97815 | 0.20791 | 4.7046 | 0.21256 | 4.8097 | 1.0223 | 102 |
| 79 | 0.98163 | 0.19081 | 5.1445 | 0.19438 | 5.2408 | 1.0187 | 101 |
| 80 | 0.98481 | 0.17365 | 5.6713 | 0.17633 | 5.7588 | 1.0154 | 100 |
| 81 | 0.98769 | 0.15643 | 6.3137 | 0.15838 | 6.3924 | 1.0125 | 99 |
| 82 | 0.99027 | 0.13917 | 7.1154 | 0.14054 | 7.1853 | 1.0098 | 98 |
| 83 | 0.99255 | 0.12187 | 8.1443 | 0.12278 | 8.2055 | 1.0075 | 97 |
| 84 | 0.99452 | 0.10453 | 9.5144 | 0.10510 | 9.5668 | 1.0055 | 96 |
| 85 | 0.99619 | 0.08715 | 11.430 | 0.08749 | 11.474 | 1.0038 | 95 |
| 86 | 0.99756 | 0.06976 | 14.301 | 0.06993 | 14.335 | 1.0024 | 94 |
| 87 | 0.99863 | 0.05234 | 19.081 | 0.05241 | 19.107 | 1.0014 | 93 |
| 88 | 0.99939 | 0.03490 | 28.634 | 0.03492 | 28.654 | 1.0006 | 92 |
| 89 | 0.99985 | 0.01745 | 57.290 | 0.01745 | 57.299 | 1.0001 | 91 |
| 90 | 1.00000 | 0.00000 | Infinite | 0.00000 | Infinite | 1.0000 | 90 |

## 2. Fractions of Inch Reduced to Decimal Equivalents

| Halves | 4ths | 8ths | 16ths | 32ds | 64ths | Decimal equivalents | Halves | 4ths | 8ths | 16ths | 32ds | 64ths | Decimal equivalents |
|---|---|---|---|---|---|---|---|---|---|---|---|---|---|
| ... | ... | ... | ... | ... | 1/64 | 0.015625 | ... | ... | ... | ... | ... | 33/64 | 0.515625 |
| ... | ... | ... | ... | 1/32 | ... | 0.03125 | ... | ... | ... | ... | 17/32 | ... | 0.53125 |
| ... | ... | ... | ... | ... | 3/64 | 0.046875 | ... | ... | ... | ... | ... | 35/64 | 0.546875 |
| ... | ... | ... | 1/16 | ... | ... | 0.0625 | ... | ... | ... | 9/16 | ... | ... | 0.5625 |
| ... | ... | ... | ... | ... | 5/64 | 0.078125 | ... | ... | ... | ... | ... | 37/64 | 0.578125 |
| ... | ... | ... | ... | 3/32 | ... | 0.09375 | ... | ... | ... | ... | 19/32 | ... | 0.59375 |
| ... | ... | ... | ... | ... | 7/64 | 0.109375 | ... | ... | ... | ... | ... | 39/64 | 0.609375 |
| ... | ... | 1/8 | ... | ... | ... | 0.125 | ... | ... | 5/8 | ... | ... | ... | 0.625 |
| ... | ... | ... | ... | ... | 9/64 | 0.140625 | ... | ... | ... | ... | ... | 41/64 | 0.640625 |
| ... | ... | ... | ... | 5/32 | ... | 0.15625 | ... | ... | ... | ... | 21/32 | ... | 0.65625 |
| ... | ... | ... | ... | ... | 11/64 | 0.171875 | ... | ... | ... | ... | ... | 43/64 | 0.671875 |
| ... | ... | ... | 3/16 | ... | ... | 0.1875 | ... | ... | ... | 11/16 | ... | ... | 0.6875 |
| ... | ... | ... | ... | ... | 13/64 | 0.203125 | ... | ... | ... | ... | ... | 45/64 | 0.703125 |
| ... | ... | ... | ... | 7/32 | ... | 0.21875 | ... | ... | ... | ... | 23/32 | ... | 0.71875 |
| ... | ... | ... | ... | ... | 15/64 | 0.234375 | ... | ... | ... | ... | ... | 47/64 | 0.734375 |
| ... | 1/4 | ... | ... | ... | ... | 0.25 | ... | 3/4 | ... | ... | ... | ... | 0.75 |
| ... | ... | ... | ... | ... | 17/64 | 0.265625 | ... | ... | ... | ... | ... | 49/64 | 0.765625 |
| ... | ... | ... | ... | 9/32 | ... | 0.28125 | ... | ... | ... | ... | 25/32 | ... | 0.78125 |
| ... | ... | ... | ... | ... | 19/64 | 0.296875 | ... | ... | ... | ... | ... | 51/64 | 0.796875 |
| ... | ... | ... | 5/16 | ... | ... | 0.3125 | ... | ... | ... | 13/16 | ... | ... | 0.8125 |
| ... | ... | ... | ... | ... | 21/64 | 0.328125 | ... | ... | ... | ... | ... | 53/64 | 0.828125 |
| ... | ... | ... | ... | 11/32 | ... | 0.34375 | ... | ... | ... | ... | 27/32 | ... | 0.84375 |
| ... | ... | ... | ... | ... | 23/64 | 0.359375 | ... | ... | ... | ... | ... | 55/64 | 0.859375 |
| ... | ... | 3/8 | ... | ... | ... | 0.375 | ... | ... | 7/8 | ... | ... | ... | 0.875 |
| ... | ... | ... | ... | ... | 25/64 | 0.390625 | ... | ... | ... | ... | ... | 57/64 | 0.890625 |
| ... | ... | ... | ... | 13/32 | ... | 0.40625 | ... | ... | ... | ... | 29/32 | ... | 0.90625 |
| ... | ... | ... | ... | ... | 27/64 | 0.421875 | ... | ... | ... | ... | ... | 59/64 | 0.921875 |
| ... | ... | ... | 7/16 | ... | ... | 0.4375 | ... | ... | ... | 15/16 | ... | ... | 0.9375 |
| ... | ... | ... | ... | ... | 29/64 | 0.453125 | ... | ... | ... | ... | ... | 61/64 | 0.953125 |
| ... | ... | ... | ... | 15/32 | ... | 0.46875 | ... | ... | ... | ... | 31/32 | ... | 0.96875 |
| ... | ... | ... | ... | ... | 31/64 | 0.484375 | ... | ... | ... | ... | ... | 63/64 | 0.984375 |
| 1/2 | ... | ... | ... | ... | ... | 0.5 | | | | | | | |

**3. In figuring discounts on electrical equipment,** it is often necessary to apply primary and secondary discounts. By using the values in Table 4, time and labor may be conserved. To find the net price, multiply the list or gross price by the multiplier from the table which corresponds to the discounts.

**example**　The discount on iron conduit may be quoted as 25 and 10 with 2 percent for cash in 10 days. To obtain the actual cost, 25 percent would be deducted from the list price, then 10 percent from that result, and finally 2 percent from the second result. If we assume that the list price of ½-in conduit is $12 per 100 ft, its actual price with the 25, 10, and 2 percent discounts would be

$$\$12.00 - 0.25 \times \$12.00 = \$12.00 - \$3.00 = \$9.00$$
$$\$9.00 - 0.10 \times \$9.00 = \$9.00 - \$0.90 = \$8.10$$
$$\$8.10 - 0.02 \times \$8.10 = \$8.10 - \$0.16 = \$7.94$$

Therefore, the net cost of the conduit would be $7.94 per 100 ft. Now by using the multiplier from Table 4 corresponding to a primary discount of 25 percent and secondary discounts of 10 and 2 percent, which is 0.661,

$$\$12.00 \times 0.661 = \$7.94$$

This is the same result as that obtained by using the longer method.

**4.  Table for Figuring Total Discount Multiplier by Combining Primary and Secondary Discounts**

| Primary discount, per cent | Secondary discounts | | | | | | |
|---|---|---|---|---|---|---|---|
| | 2 % | 5 % | 10 % | 15 % | 5 and 2 % | 10 and 2 % | 10 and 5 % |
| | Multiplier | | | | | | |
| 0 | 0.980 | 0.950 | 0.900 | 0.850 | 0.931 | 0.882 | 0.855 |
| 5 | 0.931 | 0.902 | 0.855 | 0.807 | 0.884 | 0.838 | 0.812 |
| 10 | 0.882 | 0.855 | 0.810 | 0.765 | 0.838 | 0.794 | 0.769 |
| 11 | 0.872 | 0.845 | 0.801 | 0.756 | 0.829 | 0.785 | 0.761 |
| 12 | 0.862 | 0.836 | 0.792 | 0.748 | 0.819 | 0.776 | 0.752 |
| 13 | 0.853 | 0.826 | 0.783 | 0.740 | 0.810 | 0.767 | 0.744 |
| 14 | 0.843 | 0.817 | 0.774 | 0.731 | 0.801 | 0 758 | 0.735 |
| 15 | 0.833 | 0.807 | 0.765 | 0.722 | 0.791 | 0.750 | 0.727 |
| 16 | 0.823 | 0.798 | 0.756 | 0.714 | 0.782 | 0.741 | 0.718 |
| 17 | 0.813 | 0.788 | 0.747 | 0.705 | 0.773 | 0.732 | 0.710 |
| 18 | 0.803 | 0.779 | 0.738 | 0.697 | 0.763 | 0.723 | 0.701 |
| 19 | 0.794 | 0.770 | 0.729 | 0.688 | 0.754 | 0.714 | 0.692 |
| 20 | 0.784 | 0.760 | 0.720 | 0.680 | 0.745 | 0.705 | 0.684 |
| 25 | 0.735 | 0.712 | 0.675 | 0.638 | 0.698 | 0.661 | 0.641 |
| 30 | 0.686 | 0.665 | 0.630 | 0.595 | 0.652 | 0.617 | 0.598 |
| 35 | 0.637 | 0.617 | 0.585 | 0.552 | 0.605 | 0.573 | 0.556 |
| 40 | 0.588 | 0.570 | 0.540 | 0.510 | 0.559 | 0.529 | 0.513 |
| 45 | 0.539 | 0.522 | 0.495 | 0.468 | 0.512 | 0.485 | 0.470 |
| 50 | 0.490 | 0.475 | 0.450 | 0.425 | 0.465 | 0.441 | 0.428 |
| 55 | 0.441 | 0.427 | 0.405 | 0.382 | 0.419 | 0.397 | 0.385 |
| 60 | 0.392 | 0.380 | 0.360 | 0.340 | 0.372 | 0.353 | 0.342 |
| 65 | 0.343 | 0.333 | 0.315 | 0.298 | 0.326 | 0.309 | 0.299 |
| 70 | 0.294 | 0.285 | 0.270 | 0.255 | 0.279 | 0.265 | 0.256 |

**5. Multipliers for Computing Selling Prices Which Will Afford a Given Percentage Profit**

| Percentage profit desired | To obtain selling price, multiply actual cost (invoice cost + freight) by the following value | | Percentage profit desired | To obtain selling price, multiply actual cost (invoice cost + freight) by the following value | |
|---|---|---|---|---|---|
| | When percentage profit is based on cost | When percentage profit is based on selling price | | When percentage profit is based on cost | When percentage profit is based on selling price |
| 5 | 1.05 | 1.053 | 36 | 1.36 | 1.563 |
| 6 | 1.06 | 1.064 | 37 | 1.37 | 1.588 |
| 7 | 1.07 | 1.075 | 38 | 1.38 | 1.613 |
| 8 | 1.08 | 1.087 | 39 | 1.39 | 1.640 |
| 9 | 1.09 | 1.100 | 40 | 1.40 | 1.667 |
| 10 | 1.10 | 1.111 | 41 | 1.41 | 1.695 |
| 11 | 1.11 | 1.124 | 42 | 1.42 | 1.725 |
| 12 | 1.12 | 1.136 | 43 | 1.43 | 1.754 |
| 13 | 1.13 | 1.149 | 45 | 1.45 | 1.818 |
| 14 | 1.14 | 1.163 | 46 | 1.46 | 1.852 |
| 15 | 1.15 | 1.176 | 47 | 1.47 | 1.887 |
| 16 | 1.16 | 1.190 | 48 | 1.48 | 1.923 |
| 17 | 1.17 | 1.204 | 49 | 1.49 | 1.961 |
| 18 | 1.18 | 1.220 | 50 | 1.50 | 2.000 |
| 19 | 1.19 | 1.235 | 52 | 1.52 | 2.084 |
| 20 | 1.20 | 1.250 | 54 | 1.54 | 2.174 |
| 21 | 1.21 | 1.267 | 56 | 1.56 | 2.272 |
| 22 | 1.22 | 1.283 | 58 | 1.58 | 2.381 |
| 23 | 1.23 | 1.299 | 60 | 1.60 | 2.500 |
| 24 | 1.24 | 1.316 | 62 | 1.62 | 2.631 |
| 25 | 1.25 | 1.334 | 64 | 1.64 | 2.778 |
| 26 | 1.26 | 1.352 | 66 | 1.66 | 2.941 |
| 27 | 1.27 | 1.370 | 68 | 1.68 | 3.126 |
| 28 | 1.28 | 1.390 | 70 | 1.70 | 3.333 |
| 29 | 1.29 | 1.409 | 72 | 1.72 | 3.572 |
| 30 | 1.30 | 1.429 | 74 | 1.74 | 3.847 |
| 31 | 1.31 | 1.450 | 76 | 1.76 | 4.168 |
| 32 | 1.32 | 1.471 | 78 | 1.78 | 4.545 |
| 33 | 1.33 | 1.493 | 80 | 1.80 | 5.000 |
| 34 | 1.34 | 1.516 | 90 | 1.90 | 10.000 |
| 35 | 1.35 | 1.539 | 100 | 2.00 | Infinity |

**6. Table Showing Percentage Net Profit**

| Percentage markup above cost | Percentage overhead | | | | | | | |
|---|---|---|---|---|---|---|---|---|
| | 10 % | 12 % | 14 % | 16 % | 18 % | 20 % | 22 % | 24 % |
| | Percentage net profit based on selling price for a given percentage overhead based on gross sales | | | | | | | |
| 10 | −0.90 | −2.90 | −4.90 | −6.90 | −8.90 | −10.90 | −12.90 | −14.90 |
| 15 | 3.05 | 1.05 | −0.95 | −2.95 | −4.95 | −6.95 | −8.95 | −10.95 |
| 20 | 6.67 | 4.67 | 2.67 | 0.67 | −1.33 | −3.33 | −5.33 | −7.33 |
| 25 | 10.00 | 8.00 | 6.00 | 4.00 | 2.00 | 0.00 | −2.00 | −4.00 |
| 30 | 13.08 | 11.08 | 9.08 | 7.08 | 5.08 | 3.08 | 1.08 | −0.92 |
| 33⅓ | 15.00 | 13.00 | 11.00 | 9.00 | 7.00 | 5.00 | 3.00 | 1.00 |
| 35 | 15.93 | 13.93 | 11.93 | 9.93 | 7.93 | 5.93 | 3.93 | 1.93 |
| 40 | 18.57 | 16.57 | 14.57 | 12.57 | 10.57 | 8.57 | 6.57 | 4.57 |
| 45 | 21.00 | 19.00 | 17.00 | 15.00 | 13.00 | 11.00 | 9.00 | 7.00 |
| 50 | 23.33 | 21.33 | 19.33 | 17.33 | 15.33 | 13.33 | 11.33 | 9.33 |
| 55 | 25.50 | 23.50 | 21.50 | 19.50 | 17.50 | 15.50 | 13.50 | 11.50 |
| 60 | 27.50 | 25.50 | 23.50 | 21.50 | 19.50 | 17.50 | 15.50 | 13.50 |
| 65 | 29.40 | 27.40 | 25.40 | 23.40 | 21.40 | 19 40 | 17.40 | 15.40 |
| 70 | 31.18 | 29.18 | 27.18 | 25.18 | 23.18 | 21.18 | 19.18 | 17.18 |
| 75 | 32.85 | 30.85 | 28.85 | 26.85 | 24.85 | 22.85 | 20.85 | 18.85 |
| 80 | 34.45 | 32.45 | 30.45 | 28.45 | 26.45 | 24.45 | 22.45 | 20.45 |
| 85 | 35.95 | 33.95 | 31.95 | 29.95 | 27.95 | 25.95 | 23.95 | 21.95 |
| 90 | 37.37 | 35.37 | 33.37 | 31.37 | 29.37 | 27.37 | 25.37 | 23.37 |
| 95 | 38.72 | 36.72 | 34.72 | 32.72 | 30.72 | 28.72 | 26.72 | 24.72 |
| 100 | 40.00 | 38.00 | 36.00 | 34.00 | 32.00 | 30.00 | 28.00 | 26.00 |

NOTE   Minus (−) values indicate a net loss.

**7. Net profits.** In figuring the net profit of doing business, Table 6 will be found to be very useful. The table may be used in three ways, as explained below.

To Determine the Percentage of Net Profit on Sales That You Are Making   Locate, at the top of one of the vertical columns, your percentage overhead—your "cost of doing business" in percentage of gross sales. Locate, at the extreme left of one of the horizontal columns, your percentage markup. The value at the intersection of these two columns will be the percentage profit which you are making.

   **example**   If your cost of doing business is 18 percent of your gross sales and you mark your goods at 35 percent above cost, your net profit is 7.93 percent of gross sales, obtained by carrying down from the column headed 18 percent and across from the 35 percent markup.

To Determine the Percentage Overhead Cost of Doing Business That Would Yield a Certain Net Profit for a Given Markup Percentage   Locate in the extreme left-hand column the percentage that the selling price is marked above the cost price. Trace horizontally across from this value until the percentage net profit desired is located. At the top of the column in which the desired net profit is located will be found the percentage overhead cost of doing business that will allow this profit to be made.

   **example**   If the markup is 45 percent and the profit desired is 15 percent, an overhead cost of doing business of 16 percent can be allowed, obtained by carrying across from the 45 percent markup to the 15 percent profit and finding that this column is headed by 16 percent overhead.

To Determine the Percentage That Should Be Added to the Cost of Goods to Make a Certain Percentage Net Profit on Sales   Select the vertical column which shows the percentage cost of doing business at its top. Trace down the column until the desired percentage profit is found; from this value trace horizontally to the extreme left-hand column, in which will be found the markup percentage—the percentage to be added to the cost to afford the desired profit.

**example**   It is desired to make a 12 percent net profit when the cost of doing business is 20 percent of gross sales. Select the vertical column with 20 percent at its top. Trace down the column to locate the net profit desired of 12 percent. This will be partway between 11.00 and 13.33. Carrying across to the left from these values gives a required markup between 45 and 50, or approximately 47 percent.

For values which do not appear in the table, approximate results can be obtained by estimation from the closest values in the table. If more accurate results are desired for these intermediate values, the following formulas may be used:

$$P = 100 - h - \frac{10,000}{100 + m} \tag{1}$$

or
$$m = \frac{100(P + h)}{100 - (P + h)} \tag{2}$$

or
$$h = 100 - P - \frac{10,000}{100 + m} \tag{3}$$

where $m$ = percentage markup based on cost of goods, $h$ = percentage overhead based on gross sales, and $P$ = percentage net profit based on selling price.

If you sell your goods at the retail list prices set by the manufacturers, you can use the table by converting the trade discount which you receive to an equivalent percentage markup, according to the following table:

| Manufacturer's discount | Equivalent percentage markup | Manufacturer's discount | Equivalent percentage markup |
|---|---|---|---|
| 10 | 11 | 35 | 54 |
| 15 | 17½ | 40 | 66⅔ |
| 20 | 25 | 45 | 81¾ |
| 25 | 33⅓ | 50 | 100 |
| 30 | 43 | | |

Intermediate values may be calculated from the following formula:

$$m = \frac{100Q}{100 - Q} \tag{4}$$

where $m$ = percentage markup based on cost of goods and $Q$ = manufacturer's discount.

## CONVERSION FACTORS

*(Standard Handbook for Electrical Engineers)*
These factors were calculated with a double-length slide rule and checked with those given by Carl Hering in his "Conversion Tables."

### 8. Length

1 mil  = 0.0254 mm = 0.001 in
1 mm  = 39.37 mils = 0.03937 in
1 cm  = 0.3937 in = 0.0328 ft
1 in  = 25.4 mm = 0.083 ft = 0.0278 yd = 2.54 cm
1 ft  = 304.8 mm = 12 in = 0.333 yd = 0.305 m
1 yd  = 91.44 cm = 36 in = 3 ft = 0.914 m
1 m  = 39.37 in = 3.28 ft = 1.094 yd
1 km  = 3281 ft = 1094 yd = 0.6213 mi
1 mi  = 5280 ft = 1760 yd = 1609 m = 1.609 km

**9. Surface**

1 cmil = 0.7854 mil$^2$ = 0.0005067 mm$^2$ = 0.0000007854 in$^2$
1 mil$^2$ = 1.273 cmil = 0.000645 mm$^2$ = 0.000001 in$^2$
1 mm$^2$ = 1973 cmil = 1550 mil$^2$ = 0.00155 in$^2$
1 cm$^2$ = 197,300 cmil = 0.155 in$^2$ = 0.00108 ft$^2$
1 in$^2$ = 1,273,240 cmil = 6.451 cm$^2$ = 0.0069 ft$^2$
1 ft$^2$ = 929.03 cm$^2$ = 144 in$^2$ = 0.1111 yd$^2$ = 0.0929 m$^2$
1 yd$^2$ = 1296 in$^2$ = 9 ft$^2$ = 0.8361 m$^2$ = 0.000207 acre
1 m$^2$ = 1550 in$^2$ = 10.7 ft$^2$ = 1195 yd$^2$ = 0.000247 acre
1 acre = 43,560 ft$^2$ = 4840 yd$^2$ = 4047 m$^2$ = 0.4047 ha = 0.004047 km$^2$ = 0.001562 mi$^2$
1 mi$^2$ = 27,880,000 ft$^2$ = 3,098,000 yd$^2$ = 2,590,000 m$^2$ = 640 acres = 2.59 km$^2$

**10. Volume**

1 cmil·ft = 0.0000094248 in$^3$
1 cm$^3$ = 0.061 in$^3$ = 0.0021 pt (liquid) = 0.0018 pt (dry)
1 in$^3$ = 16.39 cm$^3$ = 0.0346 pt (liquid) = 0.0298 pt (dry) = 0.0173 qt (liquid) = 0.0148 qt (dry) = 0.0164 L or dm$^3$ = 0.0036 gal = 0.0005787 ft$^3$
1 pt (liquid) = 473.18 cm$^3$ = 28.87 in$^3$
1 pt (dry) = 550.6 cm$^3$ = 33.60 in$^3$
1 qt (liquid) = 946.36 cm$^3$ = 57.75 in$^3$ = 8 gills (liquid) = 2 pt (liquid) = 0.94636 L or dm$^3$ = 0.25 gal
1 liter (L) = 1000 cm$^3$ = 61.025 in$^3$ = 2.1133 pt (liquid) = 1.8162 pt (dry) = 0.908 qt (dry) = 0.2642 gal (liquid) = 0.03531 ft$^3$
1 qt (dry) = 1101 cm$^3$ = 67.20 in$^3$ = 2 pt (dry) = 0.03889 ft$^3$
1 gal = 3785 cm$^3$ = 231 in$^3$ = 32 gills = 8 pt = 4 qt (liquid) = 3.785 L = 0.1337 ft$^3$ = 0.004951 yd$^3$
1 ft$^3$ = 28,317 cm$^3$ = 1728 in$^3$ = 59.84 pt (liquid) = 51.43 pt (dry) = 29.92 qt (liquid) = 28.32 L = 25.71 qt (dry) = 7.48 gal = 0.03704 yd$^3$ = 0.02832 m$^3$ or stere
1 yd$^3$ = 46,656 in$^3$ = 27 ft$^3$ = 0.7646 m$^3$ or stere
1 m$^3$ = 61,023 in$^3$ = 1001 L = 35.31 ft$^3$ = 1.308 yd$^3$

**11. Weight**

1 mg = 0.01543 gr = 0.001 g
1 gr = 64.80 mg = 0.002286 oz (avoirdupois)
1 g = 15.43 gr = 0.03527 oz (avoirdupois) = 0.002205 lb
1 oz (avoirdupois) = 437.5 gr = 28.35 g = 16 drams (avoirdupois) = 0.0625 lb
1 lb = 7000 gr = 453.6 g = 256 drams = 16 oz = 0.4536 kg
1 kg = 15,432 gr = 35.27 oz = 2.205 lb
1 ton (short) = 2000 lb = 907.2 kg = 0.8928 ton (long)
1 ton (long) = 2240 lb = 1.12 tons (short) = 1.016 tons (metric)

**12. Energy**

Torque units should be distinguished from energy units: thus, foot pound and kilogram-meter for energy, and pound-foot and meter-kilogram for torque (see Sec. 67 for further information on torque).

1 ft·lb = 13,560,000 ergs = 1.356 J = 0.3239 g·cal = 0.1383 kg·m = 0.001285 Btu = 0.0003766 Wh = 0.0000005051 hp·h
1 kg·m = 98,060,000 ergs = 9,806 J = 7.233 ft·lb = 2.34 g·cal = 0.009296 Btu = 0.002724 Wh = 0.000003704 hp·h (metric)
1 Btu = 1055 J = 778.1 ft·lb = 252 g·cal = 107.6 kg·m = 0.5555 lb · Celsius heat unit = 0.2930 Wh = 0.252 kg·cal = 0.0003984 hp·h (metric) = 0.0003930 hp·h
1 Wh = 3600 J = 2,655.4 ft·lb = 860 g·cal = 367.1 kg·m = 3.413 Btu = 0.001341 hp·h
1 hp·h = 2,684,000 J = 1,980,000 ft·lb = 273,700 kg·cm = 745.6 Wh
1 kWh = 2,655,000 ft·lb = 367,100 kg·m = 1.36 hp·h (metric) = 1.34 hp·h

### 13. Power

$$1 \text{ g} \cdot \text{cm/s} = 0.00009806 \text{ W}$$

1 ft·lb/min  = 0.02260 W = 0.00003072 hp (metric) = 0.00000303 hp

1 W = 44.26 ft·lb/min = 6.119 kg·m/min = 0.001341 hp

1 hp = 33,000 ft·lb/min = 745.6 W = 550 ft·lb/s = 76.04 kg·m/s = 1.01387 hp (metric)

1 kW = 44,256.7 ft·lb/min = 101.979 kg·m/s = 1.3597 hp (metric) = 1.341 hp = 1000 W

### 14. Resistivity

1 Ω/cmil·ft = 0.7854 Ω/mil²·ft = 0.001662 Ω/mm²·m = 0.0000001657 Ω/cm³ = 0.00000006524 Ω/in³

1 Ω/mil²·ft = 1.273 Ω/cmil·ft = 0.002117 Ω/mm²·m = 0.0000002116 Ω/cm³ = 0.00000008335 Ω/in³

1 Ω/m³ = 15,280,000 Ω/cmil·ft = 12,000,000 Ω/mil²·ft = 25,400 Ω/mm³·m = 2.54 Ω/cm³

### 15. Current density

1 A/in² = 0.7854 A/cmil = 0.155 A/cm² = 1,273,000 cmil/A = 0.000001 A/mil²

1 A/cm² = 6.45 A/in² = 197,000 cmil/A

1000 cmil/A = 1273 A/in²

1000 mil²/A = 1000 A/in²

### 16. Celsius and Fahrenheit Thermometer Scales

| Deg C | Deg F | Deg C | Deg F | Deg C | Deg F | Deg C | Deg F | Deg C | Deg F |
|-------|-------|-------|-------|-------|-------|-------|-------|-------|-------|
| 0 | 32. | 21 | 69.8 | 41 | 105.8 | 61 | 141.8 | 81 | 177.8 |
| 1 | 33.8 | 22 | 71.6 | 42 | 107.6 | 62 | 143.6 | 82 | 179.6 |
| 2 | 35.6 | 23 | 73.4 | 43 | 109.4 | 63 | 145.4 | 83 | 181.4 |
| 3 | 37.4 | 24 | 75.2 | 44 | 111.2 | 64 | 147.2 | 84 | 183.2 |
| 4 | 39.2 | 25 | 77. | 45 | 113. | 65 | 149. | 85 | 185. |
| 5 | 41. | 26 | 78.8 | 46 | 114.8 | 66 | 150.8 | 86 | 186.8 |
| 6 | 42.8 | 27 | 80.6 | 47 | 116.6 | 67 | 152.6 | 87 | 188.6 |
| 7 | 44.6 | 28 | 82.4 | 48 | 118.4 | 68 | 154.4 | 88 | 190.4 |
| 8 | 46.4 | 29 | 84.2 | 49 | 120.2 | 69 | 156.2 | 89 | 192.2 |
| 9 | 48.2 | 30 | 86. | 50 | 122. | 70 | 158. | 90 | 194. |
| 10 | 50. | 31 | 87.8 | 51 | 123.8 | 71 | 159.8 | 91 | 195.8 |
| 11 | 51.8 | 32 | 89.6 | 52 | 125.6 | 72 | 161.6 | 92 | 197.6 |
| 12 | 53.6 | 33 | 91.4 | 53 | 127.4 | 73 | 163.4 | 93 | 199.4 |
| 13 | 55.4 | 34 | 93.2 | 54 | 129.2 | 74 | 165.2 | 94 | 201.2 |
| 14 | 57.2 | 35 | 95. | 55 | 131. | 75 | 167. | 95 | 203. |
| 15 | 59. | 36 | 96.8 | 56 | 132.8 | 76 | 168.8 | 96 | 204.8 |
| 16 | 60.8 | 37 | 98.6 | 57 | 134.6 | 77 | 170.6 | 97 | 206.6 |
| 17 | 62.6 | 38 | 100.4 | 58 | 136.4 | 78 | 172.4 | 98 | 208.4 |
| 18 | 64.4 | 39 | 102.2 | 59 | 138.2 | 79 | 174.2 | 99 | 210.2 |
| 19 | 66.2 | 40 | 104. | 60 | 140. | 80 | 176. | 100 | 212. |
| 20 | 68. | | | | | | | | |

For values not appearing in the table use the following formulas:

$$°C = \tfrac{5}{9} \times (°F - 32) \tag{5}$$

$$°F = (\tfrac{9}{5} \times C°) + 32 \tag{6}$$

### 17.  Greek Alphabet
(Anaconda Wire and Cable Co.)

| Greek letter | Greek name | English equivalent |
|---|---|---|
| A α | Alpha | a |
| B β | Beta | b |
| Γ γ | Gamma | g |
| Δ δ | Delta | d |
| E ε | Epsilon | e |
| Z ζ | Zeta | z |
| H η | Eta | é |
| Θ θ | Theta | th |
| I ι | Iota | i |
| K κ | Kappa | k |
| Λ λ | Lambda | l |
| M μ | Mu | m |
| N ν | Nu | n |
| Ξ ξ | Xi | x |
| O ο | Omicron | ŏ |
| Π π | Pi | p |
| P ρ | Rho | r |
| Σ σ | Sigma | s |
| T τ | Tau | t |
| T υ | Upsilon | u |
| Φ φ | Phi | ph |
| X χ | Chi | ch |
| Ψ ψ | Psi | ps |
| Ω ω | Omega | ō |

## GRAPHICAL ELECTRICAL SYMBOLS

**18. Standard graphical symbols for electrical diagrams** were approved by the American National Standards Institute (ANSI) on October 31, 1975. The complete list of the standardized symbols is given in the ANSI publication *Graphical Symbols for Electrical and Electronic Diagrams*, No. ANSI/IEEE 315–1975. A selected group of these symbols for use in one-line electrical diagrams is given in Secs. 19 and 20 through the courtesy of the Rome Cable Corporation.

### 19.  Graphical Symbols for One-Line Electrical Diagrams
(From American National Standards Institute ANSI/IEEE 315-1975)

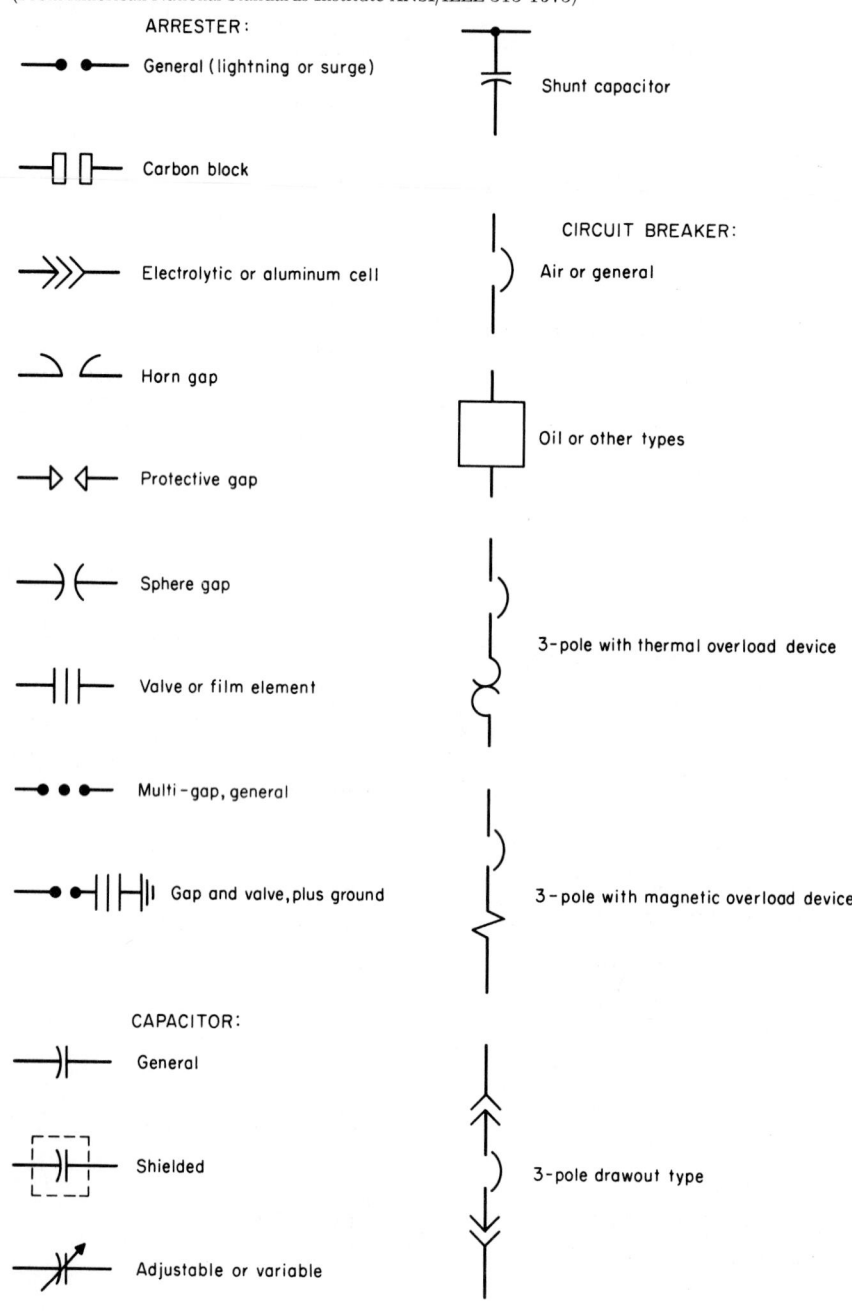

ARRESTER:

General (lightning or surge)

Carbon block

Electrolytic or aluminum cell

Horn gap

Protective gap

Sphere gap

Valve or film element

Multi-gap, general

Gap and valve, plus ground

CAPACITOR:

General

Shielded

Adjustable or variable

Shunt capacitor

CIRCUIT BREAKER:

Air or general

Oil or other types

3-pole with thermal overload device

3-pole with magnetic overload device

3-pole drawout type

## Graphical Symbols

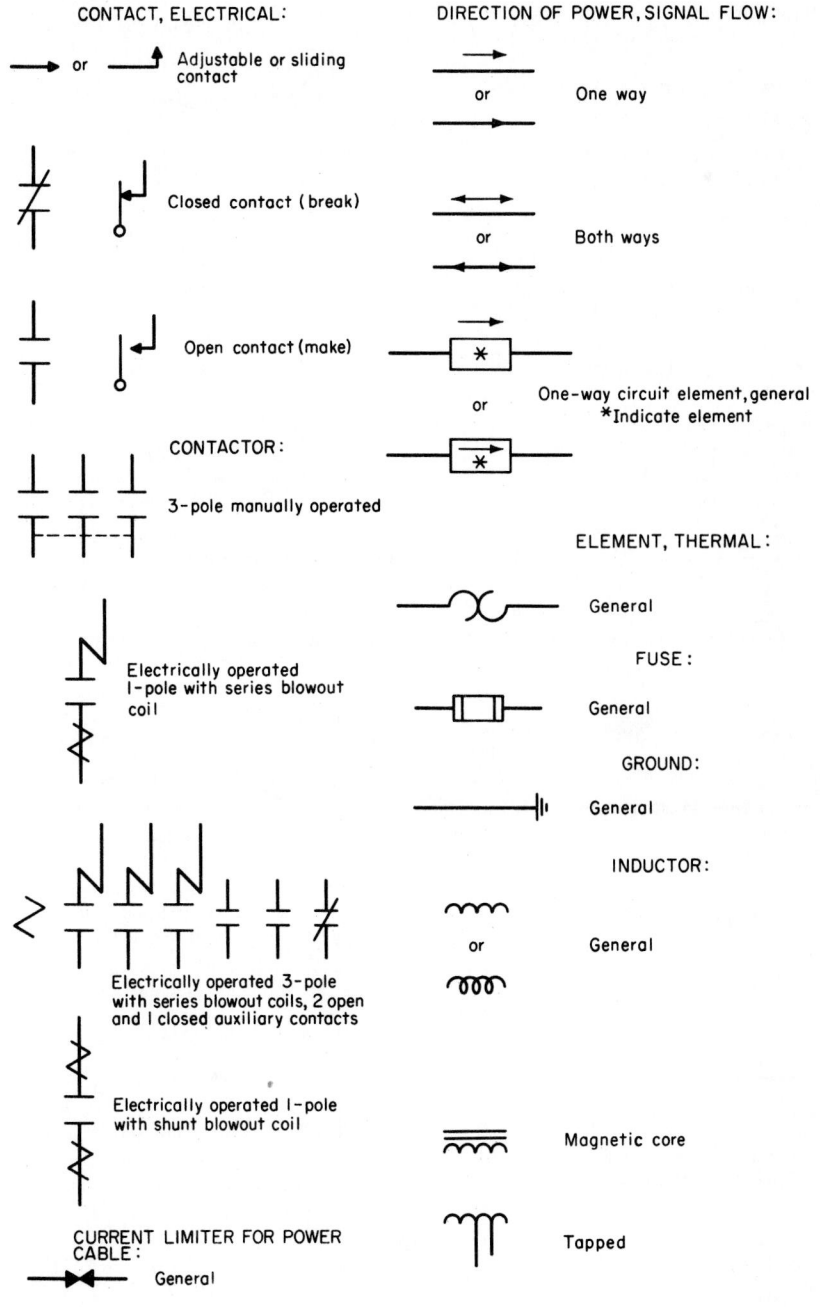

CONTACT, ELECTRICAL:

or        Adjustable or sliding
          contact

Closed contact (break)

Open contact (make)

CONTACTOR:

3-pole manually operated

Electrically operated
1-pole with series blowout
coil

Electrically operated 3-pole
with series blowout coils, 2 open
and 1 closed auxiliary contacts

Electrically operated 1-pole
with shunt blowout coil

CURRENT LIMITER FOR POWER
CABLE:
          General

DIRECTION OF POWER, SIGNAL FLOW:

or        One way

or        Both ways

or        One-way circuit element, general
          *Indicate element

ELEMENT, THERMAL:

          General

FUSE:

          General

GROUND:

          General

INDUCTOR:

or        General

          Magnetic core

          Tapped

**Graphical Symbols** *(Continued)*

Adjustable

Variable

Shunt

RECTIFIER:

Power rectifier

MACHINE, ROTATING:

Basic

Generator, general (Gen)

Motor, general (Mot)

Field, generator or motor

WINDING SYMBOLS:

1-phase

2-phase

3-phase wye (ungrounded)

3-phase wye (grounded)

3-phase delta

RELAY:

Basic (R)

Relay Protective Functions:

Overcurrent

Directional overcurrent

Directional residual overcurrent

Under voltage

Power directional

Balanced current

Differential current

Directional distance

**Graphical Symbols**

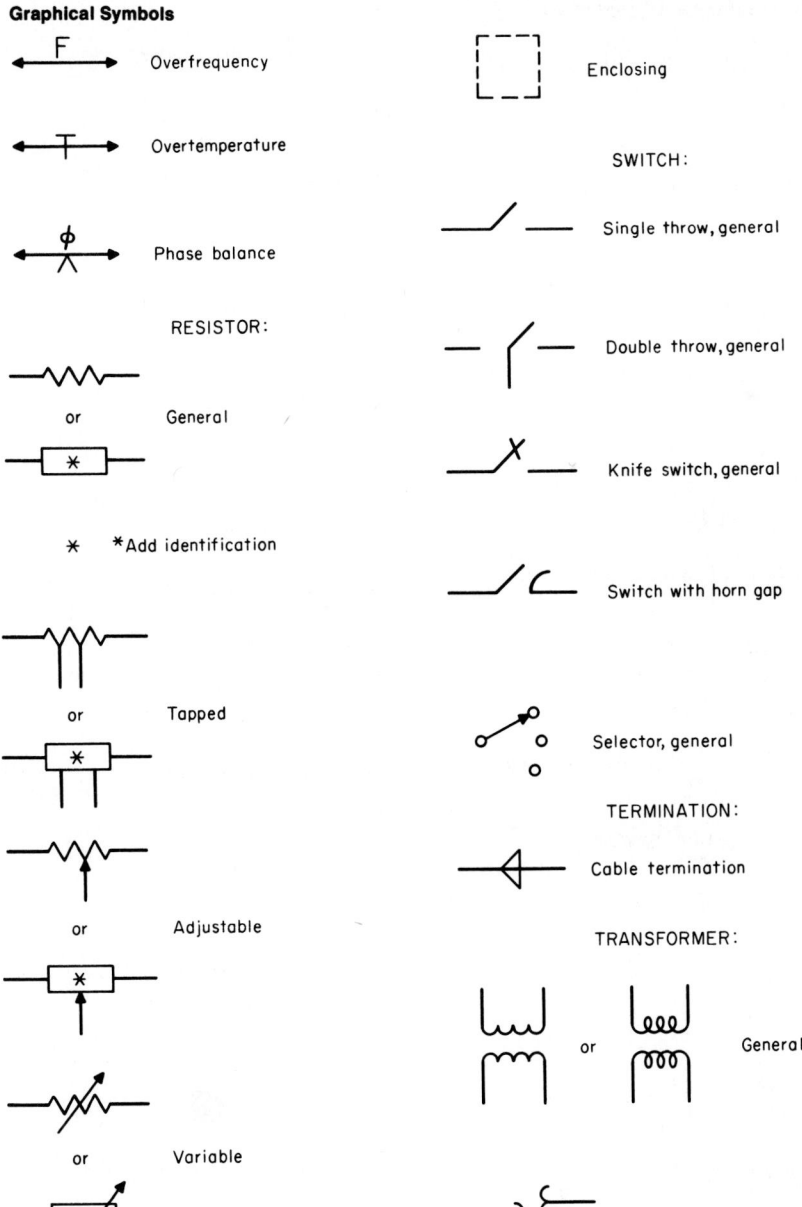

Overfrequency

Overtemperature

Phase balance

RESISTOR:

or   General

*   *Add identification

or   Tapped

or   Adjustable

or   Variable

SHIELDING:

General

Enclosing

SWITCH:

Single throw, general

Double throw, general

Knife switch, general

Switch with horn gap

Selector, general

TERMINATION:

Cable termination

TRANSFORMER:

or   General

With taps, I phase

### Graphical Symbols *(Continued)*

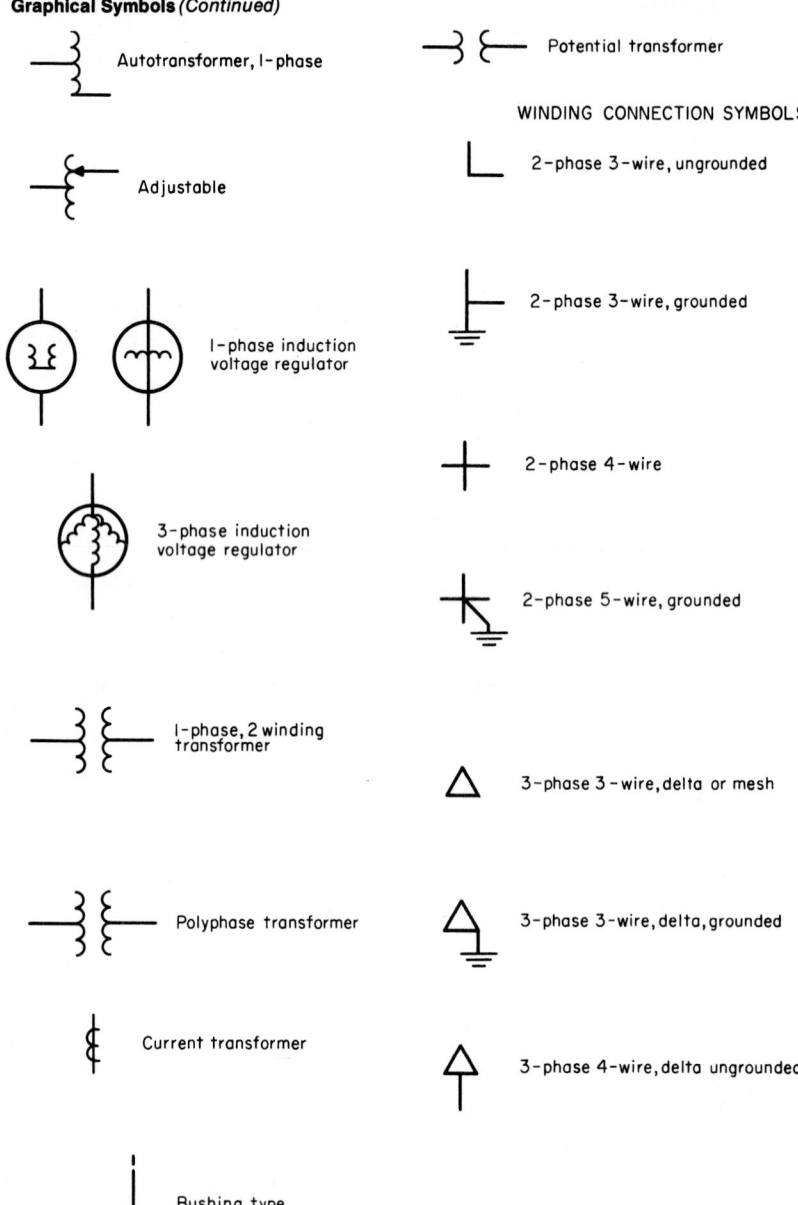

Autotransformer, 1-phase

Adjustable

1-phase induction voltage regulator

3-phase induction voltage regulator

1-phase, 2 winding transformer

Polyphase transformer

Current transformer

Bushing type current transformer

Potential transformer

WINDING CONNECTION SYMBOLS:

2-phase 3-wire, ungrounded

2-phase 3-wire, grounded

2-phase 4-wire

2-phase 5-wire, grounded

3-phase 3-wire, delta or mesh

3-phase 3-wire, delta, grounded

3-phase 4-wire, delta ungrounded

3-phase 4-wire, delta, grounded

## Graphical Symbols

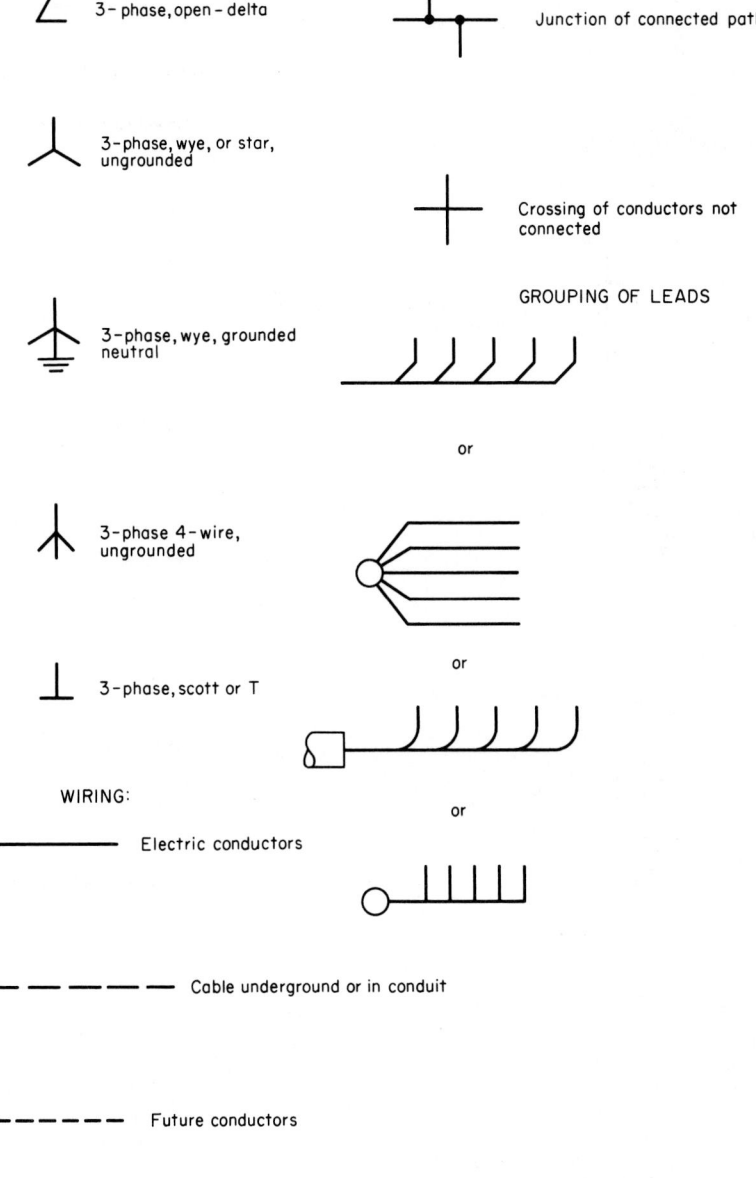

∠  3-phase, open-delta

3-phase, wye, or star, ungrounded

3-phase, wye, grounded neutral

3-phase 4-wire, ungrounded

⊥  3-phase, scott or T

WIRING:

———————  Electric conductors

— — — — — — —  Cable underground or in conduit

— — — — — —  Future conductors

——●——  Spliced conductors or change of type or size

Junction of connected paths

+  Crossing of conductors not connected

GROUPING OF LEADS

or

or

or

## 20.  Graphical Symbols for Meters or Instruments

(From American National Standards Institute ANSI/IEEE 315-1975)

NOTE    The asterisk is not a part of the symbol. Always replace the asterisk by one of the following letter combinations, depending on the function of the meter or instrument, unless some other identification is provided in the circle and explained on the diagram.

| | |
|---|---|
| A | Ammeter |
| AH | Ampere-hour meter |
| CMA | Contact-making (or breaking) ammeter |
| CMC | Contact-making (or breaking) clock |
| CMV | Contact-making (or breaking) voltmeter |
| CRO | Oscilloscope or cathode-ray oscillograph |
| D | Demand meter |
| DB | DB (decibel) meter |
| DBM | DBM (decibels referred to 1 milliwatt) meter |
| DTR | Demand-totalizing relay |
| F | Frequency meter |
| G | Galvanometer |
| GD | Ground detector |
| I | Indicating |
| M | Integrating |
| $\mu$A or UA | Microammeter |
| MA | Milliammeter |
| N | Noise meter |
| OHM | Ohmmeter |
| OP | Oil pressure |
| OSCG | Oscillograph, string |
| PH | Phase meter |
| PI | Position indicator |
| PF | Power-factor meter |
| RD | Recording demand meter |
| REC | Recording |
| RF | Reactive-factor meter |
| S | Synchroscope |
| TLM | Telemeter |
| T | Temperature meter |
| TT | Total time |
| VH | Varhour meter |
| V | Voltmeter |
| VA | Volt-ammeter |
| VAR | Varmeter |
| VI | Volume indicator |
| VOM | Volt-ohm meter |
| VU | Standard volume indicator |
| W | Wattmeter |
| WH | Watthour meter |

## PRINCIPLES OF ELECTRICITY AND MAGNETISM: UNITS

**21. Magnets and magnetism.**  Any body which has the ability to attract iron or steel is called a magnet. The attractive ability of such a body is called magnetism. Certain specimens of iron ore sometimes possess the property when they are taken from the earth. Such natural specimens will attract and hold iron filings and are called natural magnets or lodestones. The attraction for the filings will be greatest at two ends, as illustrated in Fig. 1-1*B*. The two ends that have the greatest attraction for the iron filings are called the *poles* of the magnet. If a natural magnet is suspended by a string from its center so that it is free to turn, it will turn until the axis through its poles is lying north and south. The end or pole which is pointing north is called the *north pole* of the magnet, and the other end or pole, which is pointing south, is called the *south pole*.

It is possible by certain means discussed in Sec. 24 to produce artificial magnets, i.e., to magnetize a piece of iron or steel that did not originally in its natural state possess the property of magnetism. Artificial magnets are of two types, temporary and permanent. *Temporary magnets* are those which will hold their magnetism only as long as the mag-

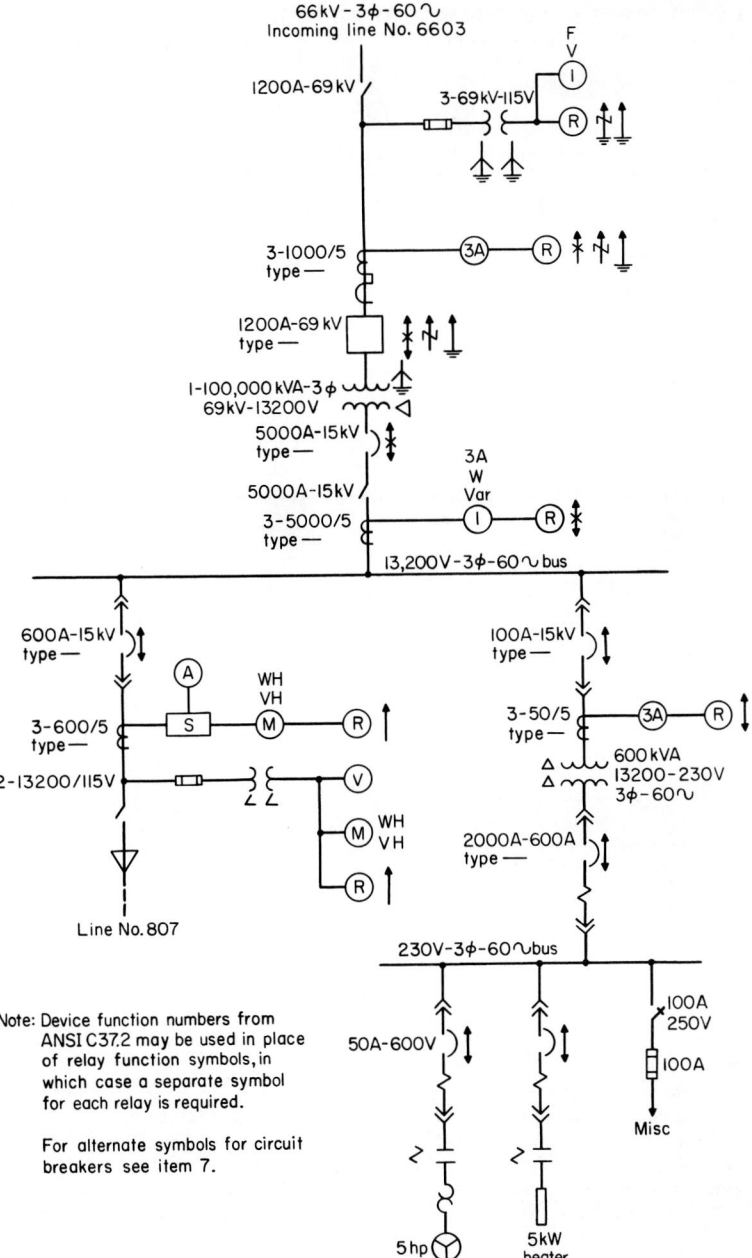

**Fig. 1-1A**   Typical single-line diagram for power equipment. [American National Standards Institute]

netizing force is maintained. *Permanent magnets* are those which will hold their magnetism after the magnetizing force has been removed and will continue to be magnets for a long time unless they are demagnetized by some means such as by being jarred or heated.

Any material that can be magnetized or that is attracted by a magnet is called a magnetic material.

**Fig.    1-1B**  Lodestone    with    iron filings.

**22. Magnetic field.**    The region around a magnet has peculiar properties that exist only as long as the magnet is present. A force exerted upon any piece of magnetic material will be placed in the space near to the magnet. This property or condition of the space around a magnet is called a *magnetic field*. If a magnet is covered with a sheet of paper sprinkled with iron filings, the filings will arrange themselves in definite curves extending from pole to pole, as shown in Fig. 1-2. The direction taken by the filings shows the direction of the magnetic field, i.e., the direction of the force exerted upon a magnetic material placed in the region around the magnet. The presence of this property (magnetic field) around a magnet can be demonstrated by means of a compass needle, a small, light magnet suspended so that it can turn freely. If a compass needle is placed in the region around a magnet, it will turn into a definite position, thereby demonstrating that there is a force acting upon a magnetic material placed in the region around a magnet. The magnitude or strength of the magnetic field (the magnitude of the force exerted upon a magnetic material in the space around a magnet) will differ at different points. The field will be strongest at the poles.

The direction of a magnetic field at any point is the direction in which a force is exerted upon the north pole of a compass needle placed at that point in the field. It will be the direction in which the north-pole end of the axis of the compass needle points.

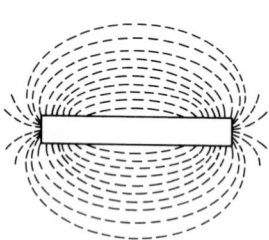

**Fig. 1-2**  Arrangement of iron filings around a magnet.

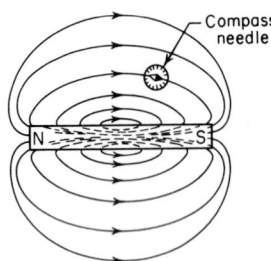

**Fig. 1-3**  Lines of magnetic flux around a bar magnet.

**23. Magnetic flux.**    The magnetic field of a magnet may be depicted by curved lines from the north to the south pole as shown in Fig. 1-3. Arrows placed on the lines, which picture the condition in the space, indicate the direction of the force that would be exerted upon the north pole of a compass needle at that point. The density with which the lines are drawn represents the magnitude of the field, i.e., the magnitude of the force that would be exerted upon a magnetic material placed at that point in the field. These lines, which picture the condition in the space around a magnet, are called *lines of magnetic flux* or simply *magnetic flux*.

**24. Methods of producing artificial magnets.**    A magnetic material can be magnetized to a certain degree by stroking it with a permanent magnet or by placing it in the field of another magnet. Either of these means will normally produce only relatively weak magnets. The general method of magnetizing a material or of producing a magnetic field is to pass an electric current through a coil of wire (see Secs. 85 and 86).

**25. The electron theory states that all matter is made of electricity.**    Matter is anything which has weight and occupies space according to the laws of physics. All matter is made up of molecules, of which there are millions of different kinds. Molecules are made up of atoms, of which there are a limited number. That is, only about 106 elements are now known. All atoms are believed to be composed of electrons, minute particles of negative electricity which normally are held in place in each atom by a positively charged, electrical something which has been named the nucleus.

NOTE The electrons thus interlocked in the atoms are, it is believed, constantly revolving at great speeds in orbits around the positive nuclei. In this, they may be thought of as resembling the eight satellites which rotate about the planet Saturn. In a normal atom the amount of negative electricity of the electrons is neutralized exactly by an equal amount of opposite or positive electricity of the nucleus. Thus a normal atom exhibits no external sign of electrification.

**26. The different kinds of atoms** (the atoms of the different elements) differ only in the number and arrangement of the electrons and in the magnitude of the positive nuclei which compose them. The lighter elements have few electrons; the heavier elements, many. A normal atom of any given element always has the same number of electrons and a positive nucleus of the same magnitude.

examples An atom of hydrogen, the lightest element, it is believed, consists of one electron and a corresponding positive nucleus. An atom of uranium, one of the heaviest elements, has probably 92 electrons and a correspondingly greater positive nucleus.

**27. Thus everything about us, all matter, is composed of electricity.** The electrons can, under certain circumstances, be forced from the atoms. The positive nuclei, except under very special conditions (disruption of the atom), cannot be moved from the atom. Electrical phenomena occur when some of this electricity (electrons) is moved or when the electrical balance which normally obtains within the atoms is disturbed.

**28. Electrons are exceedingly small.** It is estimated that each electron has a diameter of about $\dfrac{1}{12,700,000,000,000}$ in $\left( \dfrac{1}{500,000,000,000} \text{mm} \right)$. An electron weighs about

$$\frac{1}{31,500,000,000,000,000,000,000,000,000,000}\text{oz} \ (9.1 \times 10^{-28} \text{ g}),$$

or about an eighteen hundredth (¹⁄₁₈₄₅) as much as does an atom of hydrogen.

**29. Electricity cannot be generated.** It is evident that there is in the universe a certain definite amount of electricity. Electricity can neither be created nor destroyed. It can, however, be forced to move and thus transmit power or produce electrical phenomena. Electrical energy (not electricity) can be generated (produced from energy of some other form) by forcing electrons to move in certain paths.

**30. An emf** (electromotive force) is the force or pressure, measured in *volts* (V), which makes electrons move or tends to do so. Thus, if an emf is impressed across the two ends of a wire, it will force the electrons of the atoms which compose the wire to move from atom to atom, in the direction of the emf, through the wire, if we assume, of course, that a closed conducting path is provided. A lightning flash is merely a movement of electrons between the atoms of the atmosphere caused by an emf or voltage existing between the clouds and the earth.

**31. An electric current,** measured in *amperes* (A), consists of a movement or flow of electricity. Thus the thing which we call an electric current is merely a shifting of electricity. An electric current could consist of the motion of only negative electricity, of the motion of only positive electricity, or of the motion in opposite directions of both negative and positive electricity. The effects of the current would be the same in all cases. In most cases the current consists of a motion of electrons, negative electricity.

**32. Electric currents may be divided into two general classes,** direct currents and alternating currents.

**33. A direct current** is one which always flows in the same direction.

**34. An alternating current** is one the direction of which is reversed at regular intervals.

**35. Direct currents may also be classified** into (1) continuous currents, which are steady, nonpulsating direct currents; (2) constant currents, which continue to flow for a considerable time in the same direction and with unvarying intensity; and (3) pulsating currents, which are regularly varying continuous currents.

**36. The coulomb (C)** is the name given to the unit quantity of electricity. It is somewhat analogous to our common unit of quantity of water, the gallon (gal). A coulomb of electricity, so calculations show, comprises approximately 6 million million million electrons. However, it is *rate of flow*, which is measured in amperes, that is of importance to the electrician rather than the total quantity of electricity which flows. Hence, the unit coulomb is almost never used directly in practical work.

**37. The ampere** is the name given to the practical unit of rate of flow of electricity; it is analogous to "gallons per minute" or "cubic meters per second" in hydraulics. The ampere represents a rate of flow of 1 C/s. That is, it is equivalent to a flow of 6 million million million electrons per second. It has been internationally agreed (recommended by the Chicago International Electrical Congress of 1893 and legalized by act of Congress in 1894) that the ampere be defined as "that unvarying current, which, when passed through a solution of nitrate of silver in water in accordance with standard specifications, deposits silver at the rate of one thousand one hundred and eighteen millionths (0.001118) of a gram per second."

The ampere also is that unvarying current which when passed through two straight parallel conductors of infinite length and negligible cross section, located at a distance of 1 meter (m) from each other in vacuum, will produce a force between the conductors of 2 × 10⁻⁷ newton (N) per meter of length.

NOTE   The flow of water in a pipe is measured by the quantity of water which flows through it in a second, as 1 gal/s (0.0037 m³/s), 8 gal/s (0.03028 m³/s), etc. Similarly, the flow of electricity in a circuit is measured by the amount of electricity that flows through it in a second, as 1 C/s.

examples   If 2 C flows in a second, the average rate of flow is 2 C/s, and the average current is 2 A. If 20 C flows per second, the current is 20 A, etc.

examples   The current flowing in an ordinary 40-watt (40-W) incandescent Mazda lamp is about ⅓ A. Series street-lighting lamps require from 6.6 to 20 A. The current in a telegraph wire is approximately 0.04 A.

**38. Resistance ($R$ or $r$)** is the name given to the opposition which is offered by the internal structure of the different materials of the earth to the movement of electricity through them, i.e., to the maintenance of an electric current in them. This opposition results in the conversion of electrical energy into heat in accordance with the formula $W = I^2R$, where $W$ = watts, $I$ = intensity of current expressed in amperes, and $R$ = ohms ($\Omega$) of resistance. The electrons of some materials, such as the metals, can be moved from atom to atom within the material with relative ease, i.e., by the application of a small emf. All materials offer some opposition to the maintenance of a current through them, and there is no material in which some current cannot be produced, although it may be minute.

**39. Conductor** is the name given to a material through which it is relatively easy to maintain an electric current.

**40. Insulator** is the name given to a material through which it is very difficult to produce an electric current. Some examples of good insulating materials are glass, mica, and porcelain.

**41. A resistor** is an object having resistance; specifically, a resistor is a conductor inserted in a circuit to introduce resistance. A rheostat is a resistor so arranged that its effective resistance can be varied.

**42. The ohm** is the name given to the practical unit of electrical resistance. A resistance of 1 $\Omega$ is that opposition which will result in electrical energy being converted into heat at the rate of 1 W/A of effective current. In any circuit the rate at which electrical energy is converted into heat is given by the formula $W = I^2R$ of Sec. 38.

It has been agreed that the international ohm shall be represented by the resistance offered to an unvarying electric current by a column of mercury, at the temperature of 0°C, which has a mass of 14.4521 grams (g), a constant cross section, and a length of 106.3 cm.

**43. Impedance ($Z$ or $z$)** is the name given to the total opposition of a circuit or part of a circuit to the passage of an electric current through it, caused by the combined effects of the characteristics of the circuit of resistance, inductance, and capacitance. Impedance is measured in ohms.

**44. Self-inductance** is the phenomenon whereby an emf is induced in a circuit by a change of current in the circuit itself. This emf is always in such a direction that it opposes the change of current which produces it.

Whenever current passes through a conductor, it tends to set up a magnetic field around the conductor. If the current through the conductor changes, the flux produced by it will change. The change in the flux will produce a voltage in the conductor. This voltage is the voltage of self-induction. Since inductance has an effect only when the current in the conductor is changing, inductance will have no effect on a closed dc circuit but will have an effect in ac circuits in which the current is always changing from instant to instant.

**45. Inductance (L)** is defined as the property of a circuit that causes a voltage to be induced in the circuit by a change of current in the circuit. The henry (H) is the unit of inductance. A circuit has an inductance of 1 H when if the current is changed at the rate of 1 A/s, 1 V will be induced in the circuit.

**46. Inductive reactance** ($X_L$) is the name given to the opposition to the flow of changing current due to inductance. It is measured in ohms as resistance is.

**47. Capacitance (C)** is the phenomenon whereby a circuit stores electrical energy. Whenever two conducting materials are separated by an insulating material, they have the ability of storing electrical energy. Such an arrangement of materials (two conductors separated by an insulator) is called a capacitor or condenser. If a source of dc voltage is connected between the two conducting materials of a capacitor, a current will flow for a certain length of time. The current initially will be relatively large but will rapidly diminish to zero. A certain amount of electrical energy will then be stored in the capacitor. If the source of voltage is removed and the conductors of the capacitor are connected to the two ends of a resistor, a current will flow from the capacitor through the resistor for a certain length of time. The current initially will be relatively large but will rapidly diminish to zero. The direction of the current will be opposite to the direction of the current when the capacitor was being charged by the dc source. When the current reaches zero, the capacitor will have dissipated the energy which was stored in it as heat energy in the resistor. The capacitor will then be said to be discharged.

The two conducting materials, often called the plates of the capacitor, will be electrically charged when electrical energy is stored in the capacitor. One plate will have an excess of positive electricity and therefore will be positively charged with a certain number of coulombs of excess positive electricity. The other plate will have an excess of negative electricity and therefore will be negatively charged with an equal number of coulombs of excess negative electricity. When in this state, the capacitor is said to be charged. When a capacitor is charged, a voltage is present between the two conductors, or plates, of the capacitor.

When a capacitor is in a discharged state, no electrical energy is stored in it, and there is no potential difference, no voltage, between its plates. Each plate contains just as much positive as negative electricity, and neither plate has any electric charge.

From the above discussion it is seen that a capacitor has a sustained current only as long as the voltage is changing. A capacitor connected to a dc supply will not have a sustained current. In an ac circuit, the voltage is continually changing from instant to instant. Therefore, when a capacitor is connected to an ac supply, an alternating current continues to flow. The current is first in one direction, charging the capacitor, and then in the opposite direction, discharging the capacitor.

**48. The farad (F)** is the unit of capacitance. It is designated by the symbol F. A circuit or capacitor will have a capacitance of 1 F if when the voltage across it is increased by 1 V, its stored electricity is increased by 1 C. Another definition for a capacitance of 1 F, which results in the same effect, is given below. A circuit or capacitor will have a capacitance of 1 F when if the voltage impressed upon it is changed at the rate of 1 V/s, 1 A of charging current flows.

**49. Capacitive reactance** ($X_c$) is the name given to the opposition to the flow of alternating current due to capacity. It is measured in ohms as resistance and inductive reactance are.

**50. The ohm** is the unit in which all opposition to the maintenance of an electric current is measured. A circuit or part of a circuit has an opposition of 1 $\Omega$ when an emf of 1 V will produce an effective current of 1 A. In any circuit or part of a circuit the current is equal to the emf in volts divided by the total opposition in ohms. Thus,

$$I = \frac{E}{Z} \qquad E = IZ \qquad Z = \frac{E}{I} \qquad (7)$$

Although all opposition to electric current is measured in ohms, the two types, resistive opposition and reactive opposition, are quite different. Resistance results in the loss of electrical energy from the circuit. Reactance results in the interchange of energy between electromagnetic fields and the circuit. It does not result in the loss of energy from the circuit. Capacitive reactance results in the interchange of energy between an electric field and a circuit. Current passing through any type of opposition results in the loss of voltage, or voltage drop.

**51. The volt** is the unit of emf; i.e., it is the unit whereby the tendency to establish and maintain electric currents may be measured. The ampere and the ohm having been arbitrarily defined as previously stated, the volt may now be readily defined:

By international agreement it has been decided that 1 V shall be taken as that emf which will establish a current of 1 A through a resistance of 1 Ω.

**52. Admittance (Y or y)** is the name given to the quantity which is the reciprocal of impedance. It expresses the ease with which an emf can produce a current in an electric circuit. It is measured in a unit called the mho or siemens (S). A circuit or part of a circuit has an admittance of 1 mho when an emf of 1 V will produce an effective current of 1 A. In any circuit or part of a circuit the current is equal to the emf in volts multiplied by the total admittance in mhos. Thus,

$$I = EY \qquad E = \frac{I}{Y} \qquad Y = \frac{I}{E} \tag{8}$$

**53. Conductance (G or g)** is the component of the admittance which results in loss of power from the circuit in the form of heat. It is measured in mhos. For a dc circuit conductance becomes the reciprocal of the resistance.

**54. Susceptance (B or b)** is the component of the admittance which results in no loss of power from the circuit. It is measured in mhos. It does not exist for a dc circuit.

**55. Conductivity.**   The relative ease with which an electric current can be passed through a material is called its percentage conductivity. The conductivity of pure annealed copper is taken as the base, so that pure annealed copper is said to have 100 percent conductivity. Copper of 100 percent conductivity has a resistance of 10.371 Ω/mil·ft at 20°C. The resistance per circular mil-foot of any material can be found by dividing 10.371 by the percentage conductivity of the material.

**56. Work** is the overcoming of mechanical resistance through a certain distance. Work is measured by the product of the mechanical resistance times the space through which it is overcome. It is measured by the product of the moving force times the distance through which the force acts in overcoming the resistance. Work is therefore measured in foot pounds (ft·lb).

> **example**   What work is done if a weight of 6 lb is lifted through a distance of 8 ft?
> **solution**   Work = ft × lb = 8 × 6 = 48 ft·lb

> **example**   If 20 gal of water is pumped to a vertical height of 32 ft, what work has been done?
> **solution**   A gallon of water weighs 8 lb. Therefore,
>
> $$\text{Work} = \text{ft} \times \text{lb} = 32 \times (20 \times 8) = 5120 \text{ ft·lb}$$

> **example**   If the piston in a steam engine travels 1½ ft during a certain interval and the total pressure on the piston is 40,000 lb, what work is done during the interval?
> **solution**   Work = ft × lb = 1.5 × 40,000 = 60,000 ft·lb

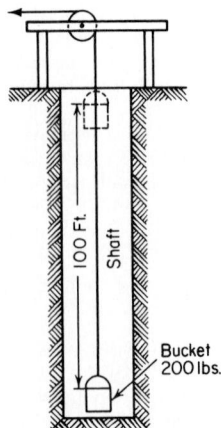

**Fig. 1-4**   Bucket in shaft.

**57. Energy** is capacity for doing work. Any body or medium which is of itself capable of doing work is said to possess energy. A coiled clock spring possesses energy because, in unwinding, it can do work. A moving projectile possesses energy because it can overcome the resistance offered by the air, by armor plate, etc., and thus do work. A charged storage battery possesses energy because it can furnish electrical energy to operate a motor. Energy can be expressed in foot pounds.

**58. Power** is the time rate of doing work. The faster work is done, the greater the power that will be required to do it. For example, if a 10-hp motor can raise a loaded elevator a certain distance in 2 min, a motor of 20 hp (approximately) will be required to raise it the same distance in 1 min.

**59. The horsepower (hp)** is the unit of power and is about equal to the power of a strong horse to do work for a short interval. Numerically, 1 hp is 33,000 ft·lb/min = 550 ft·lb/s = 1,980,000 ft·lb/h. Expressed as a formula,

$$\text{hp} = \frac{L \times W}{33,000 \times t} = \frac{\text{ft·lb/min}}{33,000} \tag{9}$$

where hp = horsepower, $L$ = distance in feet through which $W$ is raised or overcome, $W$ = weight in pounds of the thing lifted or the push or pull in pounds of the force overcome, and $t$ = time in minutes required to move or overcome the weight $W$ through the distance $L$.

**example**  What horsepower is required in raising the load and bucket weighing 200 lb shown in Fig. 1-4 from the bottom to the top of the shaft, a distance of 100 ft, in 2 min?
**solution**  Substitute in the formula

$$\text{hp} = \frac{L \times W}{33,000 \times t} = \frac{100 \times 200}{33,000 \times 2} = 0.3 \text{ hp}$$

**example**  What average horsepower is required while moving the box loaded with stone, in Fig. 1-5, from $A$ to $B$, a distance of 650 ft, in 3 min? A total horizontal pull of 150 is required to move the box.

**solution**  Substitute in the formula

$$\text{hp} = \frac{L \times W}{33,000 \times t} = \frac{650 \times 150}{33,000 \times 3} = 0.98 \text{ hp}$$

**60. Electric power** is the rate of doing electrical work. The unit is the watt or the kilowatt. A kilowatt is 1000 W. Work is being done at the rate of 1 W when a constant current of 1 A is maintained through a resistance by an emf of 1 V.

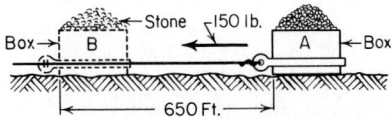

**Fig. 1-5**  Moving loaded box.

**61. Energy of one sort may be transformed into energy of another sort.**  Heat energy in coal may be transformed (with a certain loss) with a boiler, a steam engine, and a generator into electrical energy. The energy possessed by a stream of falling water may be transformed with a waterwheel and a generator into electrical energy. There is a definite numerical relation between different sorts of energy. Thus 1 British thermal unit, or Btu (the unit of heat energy), is equivalent to 778 ft·lb. In electrical units energy is expressed in watthours (Wh) or kilowatthours (kWh).

**62. A kilowatthour** represents the energy expended if work is done for 1 h at the rate of 1 kW.

**63. A horsepower-hour** represents the energy expended if work is done for 1 h at the rate of 1 hp. Therefore 1 hp·h = 60 × 33,000 = 1,980,000 ft·lb.

**64. To reduce horsepower to watts and kilowatts and vice versa.** Since 1 hp = 746 W,

$$\text{hp} = \frac{W}{746} = W \times 0.00134$$

$$W = \text{hp} \times 746 \tag{10}$$

$$\text{hp} = \frac{kW}{0.746} = kW \times 1.34$$

$$kW = \text{hp} \times 0.746 \tag{11}$$

**example**  W = 2460; hp = ?
**solution**  Substitute in the formula

$$\text{hp} = \frac{W}{746} = 2460 \div 746 = 3.3 \text{ hp}$$

**example**  A motor takes 30 kW. What horsepower is it taking?
**solution**  Substitute in the formula

$$\text{hp} = \frac{kW}{0.746} = 30 \div 0.746 = 40.24 \text{ hp}$$

Or instead, using the other formula,

$$\text{hp} = kW \times 1.34 = 30 \times 1.34 = 40.2 \text{ hp}$$

**65. Efficiency** is the name given to the ratio of output to input. No machine gives out as much useful energy or power as is put into it. There are some losses in even the most perfectly constructed machines.

$$\text{Efficiency} = \frac{\text{output}}{\text{input}} \tag{12}$$

Although efficiency is basically a decimal quantity and is so used in making calculations, it is usually expressed as a percentage. Percentage efficiency is equal to the decimal expression of efficiency multiplied by 100. An efficiency of 0.80 expressed as a decimal is an efficiency of 80 percent expressed as a percentage.

Basically efficiency deals with energy. However, when the rate of energy conversion is constant, the values of output and input in terms of power may be used in dealing with efficiency:

$$\text{Input} = \frac{\text{output}}{\text{efficiency}} \tag{13}$$

and
$$\text{Output} = \text{input} \times \text{efficiency} \tag{14}$$

When the formulas are used, output and input must be expressed in the same units and efficiency as a decimal.

**66. Output** is the useful energy or power delivered by a machine, and input is the energy or power supplied to a machine.

**example**  If 45 kW is supplied to a motor and its output is found to be 54.2 hp, what is its efficiency?

**solution**  Since 1 hp = 0.746 kW, 54.2 hp = 54.2 × 0.75 = 40.6 kW. Then substituting in the formula,

$$\text{Efficiency} = \frac{\text{output}}{\text{input}} = \frac{40.6}{45} = 0.90 = 90 \text{ percent}$$

**67. Torque** is the measure of the tendency of a body to rotate. It is the measure of a turning or twisting effort and is usually expressed in pound-feet or in pounds-force at a given radius. Torque is expressed as the product of the force tending to produce rotation times the distance from the center of rotation to the point of application of the force. Thus in Fig. 1-6 there is a torque of 50 × 1 = 50 lb·ft tending to turn the windlass owing to the weight attached to the rope. In the motor of Fig. 1-7 the group of conductors under the north pole produces a combined force of 10 lb. The torque produced by this group of conductors will be 10 × 9/12 = 7.5 lb·ft. The group of conductors under the south pole will similarly produce 10 × 9/12 = 7.5 lb·ft. Both these torques tend to produce rotation in the same direction. Therefore, the total torque produced by the motor will be 7.5 + 7.5 = 15 lb·ft.

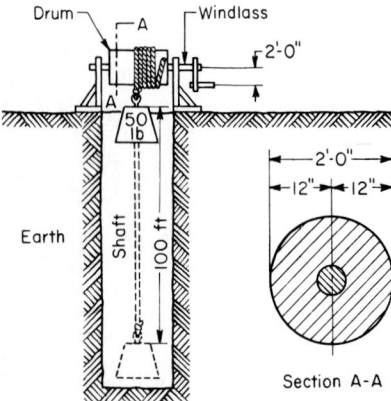

**Fig. 1-6**  Example of work and torque.

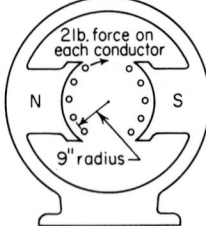

**Fig. 1-7**  Torque produced by the force of the conductors of a motor.

Torque exists even if there is no motion. Thus, in Fig. 1-6 the torque exerted by the weight is 50 lb·ft as long as the weight is supported, whether the drum is moving or standing still. In the motor of Fig. 1-7 a torque of 15 lb·ft is exerted upon the armature, tending to produce rotation whether the motor is revolving or standing still. If there is no rotation, no work can be done; yet there is a torque tending to produce rotation.

**68. Relation between horsepower and torque**

$$\text{hp} = \frac{6.28 \times \text{rpm} \times T}{33,000} \tag{15}$$

or

$$T = \frac{33,000 \text{ hp}}{6.28 \times \text{rpm}} \tag{16}$$

where $T$ = torque in lb ft.

**example**  What is the torque of a 10-hp motor if it delivers a full load at 1150 rpm?

$$T = \frac{33,000 \times 10}{6.28 \times 1150} = 45.7 \text{ lb·ft}$$

**69. Torque and force relations in mechanisms.**  In any mechanism such as the windlass of Fig. 1-6 or a motor belted or geared to a load, if the losses are neglected so that the efficiency is 100 percent, the power output will be equal to the power input. For such a perfect mechanism the following relations will hold true:

$$T_1 \times \text{rpm}_1 = T_2 \times \text{rpm}_2 \tag{17}$$

or

$$\frac{T_1}{T_2} = \frac{\text{rpm}_2}{\text{rpm}_1}$$

where $T_1$ = torque at point 1 in the mechanism, $\text{rpm}_1$ = rpm at point 1 in the mechanism, $T_2$ = torque at point 2 in the mechanism, $\text{rpm}_2$ = rpm at point 2 in the mechanism.

In the windlass of Fig. 1-6 the rpm's for all points in the mechanism are the same. Therefore the torque at any point is equal to the torque at any other point. The torque that must be exerted at the handle of the windlass to raise the weight at a uniform rate of speed would have to be equal to the torque exerted by the weight. (This formula neglects the weight of the rope and the friction.) The force exerted on the handle will be $^{50}\!/_2 = 25$ lb.

The motor of Fig. 1-7 is equipped with a pulley on the shaft for driving a belt. The pulley has a diameter of 6 in. The force exerted upon the belt (the tension of the belt) is determined in the following manner. Since the rpm of the pulley is the same as the rpm of the motor armature, the torque produced by the conductors is the same as the torque exerted by the pulley. The force on the belt is therefore $15/0.5 = 30$ lb.

**70. EMFs may be produced in the following ways:**
1. Electromagnetic induction
2. Thermal action
3. Chemical action
4. Changing electric fields
5. Contact between unlike substances
6. Vibration or heating of crystals

**71. EMFs may be produced by electromagnetic induction in the following three ways:**
1. By moving a conductor across a magnetic field. If the conductor of Fig. 1-8 is moved up or down so as to cut the lines of flux of the magnetic field between the poles of the magnet, an emf will be generated between the two ends of the conductor. This is the method employed for the production of voltage in dc generators (see Sec. 74).

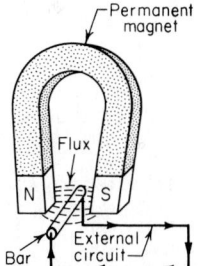

**Fig. 1-8**  EMF generated by magnetic flux.

2. By moving a magnetic field across a conductor. If in Fig. 1-8 the conductor is held stationary and the magnet is moved up or down so that the lines of flux of the magnetic field between the poles of the magnet will cut the conductor, an emf will be generated between the two ends of the conductor. This is the method employed for the production of voltage in most ac generators (see Sec. 144).

3. By changing the strength of the magnetic field linked with a conductor. If in Fig.

1-9 an alternating electric current is passed through winding $A$, it will set up a magnetic field through the iron ring. This magnetic field will be continually changing in strength, owing to the changing magnitude of the current. There will therefore be a continual change in the magnetic flux linked with winding $B$. This change will generate a voltage between the two ends of winding $B$. It is this phenomenon that makes possible the operation of transformers (see Div. 5, "Transformers").

Wherever large quantities of electrical energy are required, the necessary emf is produced by one of the means of electromagnetic induction.

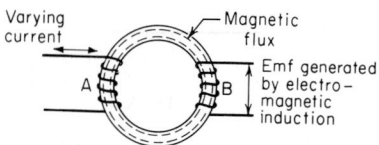

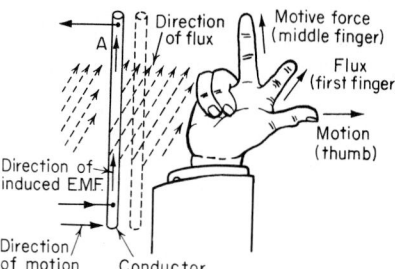

**Fig. 1-9**  Method of producing an emf by electromagnetic induction.

**Fig. 1-10**  Application of right-hand rule.

**72. A hand rule to determine the direction of an induced emf** (see Fig. 1-10). Use the *right hand*. Extend the thumb in the direction of the motion or equivalent motion of the conductor and extend the forefinger in the direction of the magnetic flux. Then the middle finger will point in the direction of the induced emf. (Magnetic flux flows from the north pole to the south pole outside a magnet and from south to north inside the magnet.) This rule can be remembered by associating the sounds of the following word groups: thumb–motion, forefinger–force, and middle finger–motive force. The rule is also known as Fleming's rule.

**73. AC generator (alternator).**   A very simple elementary ac generator is shown in Fig. 1-11. As the conductors are revolved through the magnetic field, a voltage will be produced in each conductor. In view of the series circuit formed by the two conductors, the voltages produced by the conductors will act to send current through the circuit in the same direction. The total voltage between the terminals of the machine will be the sum of the voltages produced by the two conductors at that instant. As long as conductor $A$ is under the north pole and conductor $B$ under the south pole, there will be a terminal voltage which will send current through the external circuit from $C$ to $D$. When the conductors are midway between poles, they will not cut any flux and no voltage will be produced. As conductor $A$ moves under the south pole and conductor $B$ under the north pole, the conductors will again cut flux and voltages will again be produced in the conductors. The direction of these voltages will, however, be just opposite to the direction of the voltages when $A$ was under the north pole and $B$ under the south pole. The terminal voltage of the machine therefore periodically reverses in direction, and the machine is an ac generator.

**74. A dc generator (dynamo)** is shown in Fig. 1-12. The production of voltage in the conductors is exactly the same as for the elementary ac generator of Sec. 73. As the con-

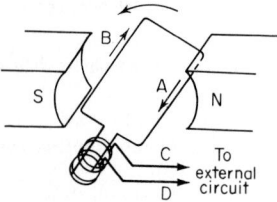

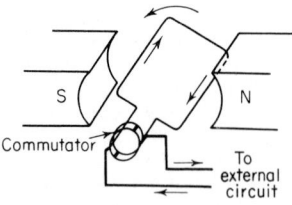

**Fig. 1-11**  Elementary ac generator.

**Fig. 1-12**  Elementary dc generator.

ductors revolve, an alternating voltage is produced in them. In order that the terminal voltage can always act upon the external load in the same direction, some device must be inserted between the conductors and the terminals. This device must reverse the connections of the conductors to the external circuit at
the instant when the voltage of the conductors
is zero and is changing in direction. Such a
device is called a commutator.

**75. The magnitude of the voltage produced by
electromagnetic induction** depends upon the rate
at which the lines of flux are cut by the conductor. Whenever 100 million lines of flux are cut
per second, 1 V is produced. The voltage produced is therefore equal to the number of lines
of flux cut per second divided by 100 million.

**76. Thermal action produces emf's in the follow-
ing ways:**

1. *Seebeck effect.* In a closed circuit consisting of two different metals, an emf will be
produced if the two junctions between the different metals are kept at different temperatures
(see Fig. 1-13). Thermocouples function
through this phenomenon. The magnitude of
the emf produced will depend upon the material of the metals and the difference in temper-

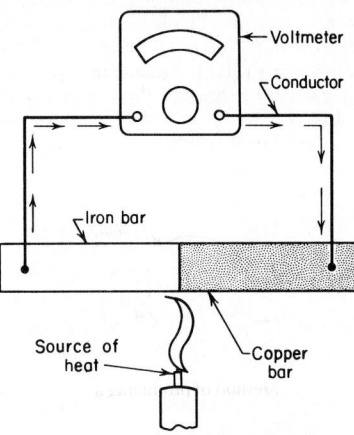

**Fig. 1-13**  EMF generated by heat.

ature between the hot junction and the cold ends. Since only very small voltages can be produced in this way, the method is not applicable when electrical energy of any quantity is required. However, this method is of great practical value for application in tempera-ture measurements.

2. *Peltier effect.* When current passes through the junction between two different metals, energy conversion takes place between energy in heat form and energy in electri-cal form. The action is reversible, and the direction of energy conversion depends upon the direction in which the current passes across the junction. This phenomenon is entirely different and distinct from the conversion of electrical energy into heat caused by the passage of a current through the resistance of the junction of the two materials.

3. *Thomson effect.* When the temperature along a metallic conductor varies in mag-nitude, a very small emf is produced.

**77. EMFs produced by chemical action.**   Certain combinations of chemicals will generate emf's. For instance, if a piece of zinc (Fig. 1-14) and a piece of carbon are immersed in a
solution of sal ammoniac, an emf will be
produced between the zinc and the carbon.
Such a combination is called a battery. If
the key in Fig. 1-14 is closed, an electric
current will flow and the bell will ring. The
voltage of dry cells and storage batteries is
produced in this way by chemical action.

**78. EMFs produced by electric
fields.**   Whenever there is a voltage
between two conductors, certain conditions are produced in the space around and
between the conductors. This condition in
the surrounding space is called an electric
field. Discussion and explanation of this
phenomenon are outside the scope of this

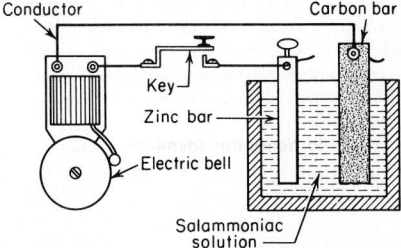

**Fig. 1-14**  EMF generated by chemical action.

book. However, emf's produced by changes in this electric field are of great practical importance, since these changes induce changing voltages in neighboring conductors. Voltages produced by this means in telephone lines from neighboring electric power lines may cause serious interference with communication over the telephone lines.

**79. Contact emf.**   When any two different materials are brought into contact with each other, a very small emf is produced. However, voltages of considerable magnitude may be produced by this phenomenon by the rapid rubbing together of different materials. The rubbing results in rapid change of the contact points between the two materials and thus in an accumulation of the small individual contact emf's into a voltage of considerable magnitude. Practical illustrations of voltages produced by this means are voltages on belts produced by the motion of the belt over the pulleys between automobile bodies and the ground produced by the revolution of the rubber tires over the road. These voltages often are called frictional voltages.

**80. EMFs produced by crystals.** Certain crystals, such as quartz and rochelle salt crystals, have the property of producing very small emf's between opposite faces of the crystals when the crystals are subjected to pressure. Some crystals, such as tourmaline, when heated produce a very small emf between opposite faces. Although emf's produced by these means are of very small magnitude, they are of great practical value in microphones and instruments for the measurement of vibrations in machinery.

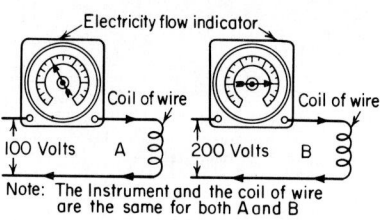

Note: The Instrument and the coil of wire are the same for both A and B

**Fig. 1-15** The effect of increasing voltage.

**81. EMF,** which is measured in volts, causes electricity to flow. A higher voltage is required to force a given current of electricity through a small wire than through a large one. If the voltage impressed on a circuit is increased, the current will be correspondingly increased (see Fig. 1-15).

**82. The distinction between amperes and volts** should be clearly understood. The amperes represent the rate of electricity flow (see Secs. 31 and 37) through a circuit, while the volts represent the tendency causing the flow. There may be a tendency (voltage) and yet no current. If the path of electricity is blocked by an open switch (Fig. 1-16), there will be no

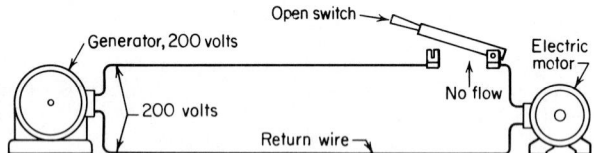

**Fig. 1-16** Electricity flow blocked by an open switch.

current of electricity, though the tendency to produce (voltage) may be high. With a given voltage a greater current of electricity will flow through a large wire than through a small one.

**83. Direction of electric current.** Although in most cases current consists of the actual motion of negative electricity in a certain direction, the conventional direction of a current is the direction in which positive electricity would move to cause the same effects as are produced by the actual motion of electricity. Therefore the direction of current, as it is usually considered, is in the opposite direction to the motion of the electrons.

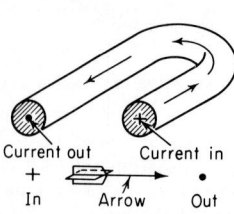

**Fig. 1-17** Symbols indicating direction of current flow.

**84. Symbols for indicating the direction of an emf or currents into or out of the end of a conductor** are shown in Fig. 1-17.

**85. Effects of an electric current.** The two principal effects of an electric current are a heating effect and a magnetic effect.

Whenever an electric current passes through a material, there is a heating effect due to the current. This effect is indicated by the increase in temperature of the material. A certain portion of the electrical energy that is put into the circuit is transformed into heat energy owing to the opposition offered to the flow of current by the resistance of the material. This loss of electrical energy (transferred to heat energy) is associated with a drop in voltage. This drop in voltage is produced by the amount of voltage required to force the current through the resistance. The heating effect with its associated drop of voltage and temperature rise of the material always takes place whenever a current is passed through a circuit. These are also known as $I^2R$ (watts) or $IR$ (drop losses).

There is an association between electricity and magnetism because a magnetic field can be produced by an electric current. In fact, whenever an electric current is passed through a conductor, the current tends to set up a magnetic field around the conductor. The presence of this field can be demonstrated by holding a compass near a wire that is carrying a current (Fig. 1-23). The compass needle will be deflected in a definite direction with the direction of current flow.

**86. Magnetic effect of electric current.**  The magnetic lines of flux (magnetic field) produced by a current passing through a straight wire can be determined by passing the wire through a sheet of paper upon which iron filings are sprinkled, as illustrated in Fig. 1-18. The direction of the lines of flux will be in concentric circles around the axis of the conductor. The field will be strongest close to the conductor.

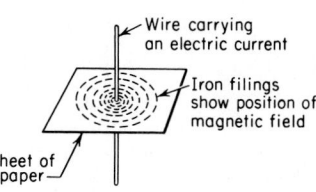

**Fig. 1-18**  Magnetic field around a conductor.

If the current is passed through a coil of wire wound around a piece of iron, as in Fig. 1-19, it will be found that the iron is magnetized in a definite direction. Such a magnet is called an electromagnet. This is the method always employed for producing strong magnets or setting up strong magnetic fields. A coil of wire carrying current will act like a magnet and will produce a magnetic north pole at one end and a magnetic south pole at the other (see Fig. 1-20). The direction of the magnetic field produced will be from the north-pole end around through the space outside of the coil to the south-pole end and then back through the interior of the coil to the north-pole end.

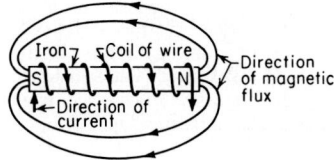

**Fig. 1-19**  Elementary electromagnet.

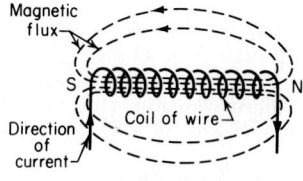

**Fig. 1-20**  Magnetic field from current flowing through a coil of wire.

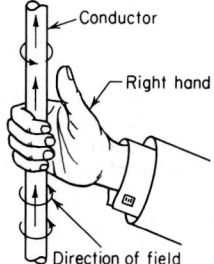

**Fig. 1-21**  Hand rule for direction of field.

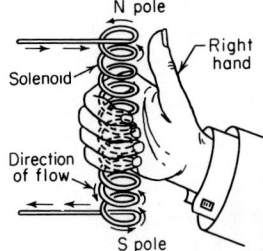

**Fig. 1-22**  Hand rule for determining polarity of a solenoid.

**87. Hand rule for direction of magnetic field about a straight wire** (see Fig. 1-21). If a wire through which electricity is flowing is so grasped with the *right hand* that the thumb points in the direction of electricity flow, the fingers will point in the direction of the magnetic field and vice versa.

**88. Hand rule for polarity of a solenoid or electromagnet** (see Fig. 1-22). If a solenoid or an electromagnet is so grasped with the *right hand* that the fingers point in the direction of the current, the thumb will point in the direction of the magnetic field through the solenoid, i.e., toward the north pole of the solenoid.

**89. Rule for determining direction of current flow with a compass** (see Fig. 1-23). If a compass is placed under a conductor in which electricity is flowing from south to north, the north end of the needle will be deflected to the west. If the compass is placed over the conductor, the north end of the compass will be deflected to the east. If the direction of current

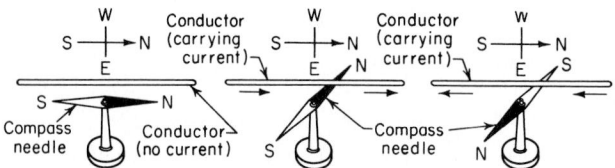

**Fig. 1-23**   Performance of a compass needle near a conductor.

flow in the conductor is reversed, the direction of deflection of the needle will be reversed correspondingly.

**90. The resistances of different materials** vary greatly. Some, such as the metals, conduct electricity very readily and hence are called conductors. Others such as wood or slate are, at least when moist, partial conductors. Still others, such as glass, porcelain, and paraffin, are called insulators because they are practically nonconducting. No material is a perfect conductor, and no material is a perfect insulator (see Sec. 38).

The resistance of a conductor depends not only upon the material of the conductor but also upon the conductor's dimensions and the distribution of the current throughout the cross section of the conductor. The resistance of a given conductor will have its minimum value when the current is uniformly distributed throughout the cross section of the conductor. Uniform current distribution exists in the conductors of most dc circuits. In the conductors of ac circuits the current never is exactly uniformly distributed. The resistance of a circuit to alternating current is always somewhat greater than it is to direct current (see Sec. 122). The amount by which the ac resistance exceeds the dc value depends upon several factors.

Unless otherwise stated, values of resistance should be taken as the resistance for uniform distribution of the current. They are the values to use for dc circuits.

The resistance of materials also depends upon the temperature of the material.

**91. A circular mil** is the area of a circle $\frac{1}{1000}$ in in diameter. A mil is $\frac{1}{1000}$ of an inch (see Fig. 1-24). The areas of electric conductors are usually measured in circular mils. Since the area of any figure varies as the square of its similar dimensions, the area of any circle can be expressed in circular mils by squaring its diameter expressed in thousandths. Thus, since $\frac{3}{8} = \frac{375}{1000} = 0.375$, the area of a circle $\frac{3}{8}$ in in diameter would be $375 \times 375 = 140,625$ cmil. The area of a circle 0.005 in in diameter would be $5 \times 5 = 25$ cmil.

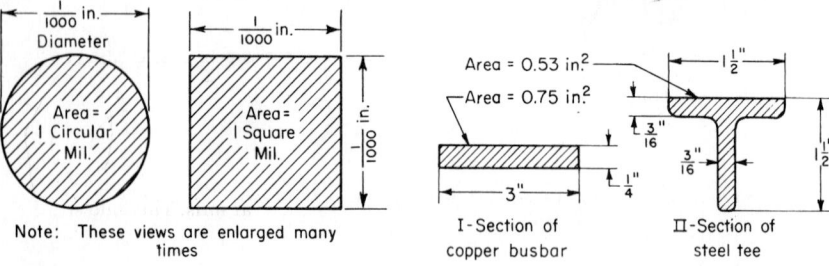

Note: These views are enlarged many times

**Fig. 1-24**   Circular mil and square mil.

**Fig. 1-25**   Conductor sections.

**92. A square mil** is the area of a square having sides $\frac{1}{1000}$ in long (see Fig. 1-24). Areas of rectangular conductors are sometimes measured in square mils. Areas in square mils are obtained by multiplying the length and breadth of the rectangle expressed in thousandths of an inch. Thus, the area of a rectangle $\frac{1}{2}$ in wide and 2 in long would be $500 \times 2000 = 1$ million mils². In actual area, a circular mil is about $\frac{8}{10}$ as great as a square mil.

**93. To reduce square mils or square inches to circular mils, or the reverse,** apply one of the following formulas.

$$\text{Mils}^2 = \text{cmil} \times 0.7854 \tag{18}$$

$$\text{cmil} = \frac{\text{mils}^2}{0.7854} \tag{19}$$

$$\text{cmil} = \frac{\text{in}^2}{0.0000007854} \tag{20}$$

$$\text{In}^2 = \text{cmil} \times 0.0000007854 \tag{21}$$

**example**  The sectional area of the busbar in Fig. 1-25, I, in circular mils is:

cmil = (in²)/0.0000007854 = (3 × ¼) ÷ 0.0000007854 = 0.75 ÷ 0.0000007854 = 955,000 cmil

**example**  The sectional area of the steel $T$ shown in Fig. 1-25, II, in circular mils is:

cmil = in ÷ 0.0000007854 = 0.53 ÷ 0.0000007854 = 674,800 cmil

**94. The circular mil-foot** (cmil·ft) is the unit conductor. A wire having a sectional area of 1 cmil and a length of 1 ft is a circular mil-foot of conductor. The resistance of a circular mil-foot of a metal is sometimes called its *specific resistance* or its *resistivity*. The resistance of a circular mil-foot of copper under different conditions is given in Fig. 1-26. Resistances for other metals and alloys are given in Table 97.

**95. To compute the resistance of a conductor of any common metal or alloy** use the value given for the resistance of a circular mil-foot of the material in Table 97 in the following formula:

$$R = \frac{p \times l}{\text{cmil}} \quad \text{or} \quad \frac{p \times l}{d^2} \tag{22}$$

where $R$ = resistance of the conductor in ohms; $p$ = resistance of 1 cmil·ft of the material composing the conductor, from Table 97; $l$ = length of conductor in feet; $d$ = diameter in mils; and $d^2$ = diameter in mils squared or, what is the same thing, the area of the conductor in circular mils.

The other forms of the formula are

$$p = \frac{d^2 \times R}{l} \qquad l = \frac{d^2 \times R}{p} \qquad d = \sqrt{\frac{p \times l}{R}} \tag{23}$$

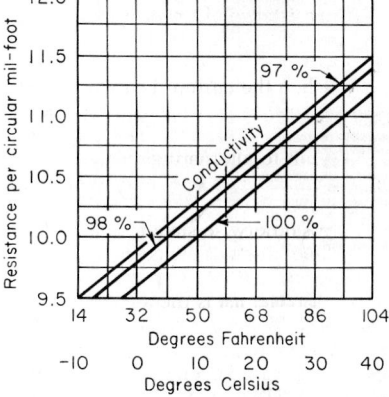

**Fig. 1-26**  Curves showing resistance per circular mil-foot of pure copper at various temperatures and conductivities.

**example**  If we take from Table 97 the resistance of 1 cmil·ft of copper at 23°C (75°F) as 10.5 Ω, what is the resistance of 500 ft of copper wire 0.021 in in diameter?

**solution**  Substituting in the formula,

$$R = \frac{pl}{d^2} = \frac{10.5 \times 500}{21 \times 21} = \frac{5250}{441} = 11.9 \ \Omega$$

**96. An approximate rule for computing the resistance of round copper wire, in ohms per thousand feet,** is to divide 10,500 by the size of the wire in circular mils. This rule should be used only for rough estimating computations.

**example**  Thus, for a 4/0 wire (211,600 cmil), the resistance is approximately $R = 10,500/211,600 = 0.05 \ \Omega/1000$ ft.

## 97. Approximate Specific Resistances and Temperature Coefficients of Metals and Alloys

(*Electrical Engineer's Handbook*, International Textbook Company)

| Metal | $p$ Resistance of 1 cir mil-ft in ohms | | $a$ Average temperature coefficient per degree C between 0 and 100°C | $a$ Average temperature coefficient per degree F between 32 and 212°F | Percentage conductivity | Relative resistance |
|---|---|---|---|---|---|---|
| | 0°C or 32°F | 23.8°C or 75°F | | | | |
| Silver, pure annealed | 8.831 | 9.674 | 0.004000 | 0.002220 | 108.60 | 0.925 |
| Copper, pure annealed | 9.390 | 10.351 | 0.004280 | 0.002380 | 102.10 | 0.980 |
| Copper, annealed | 9.590 | 10.505 | 0.004020 | 0.002230 | 100.00 | 1.000 |
| Copper, hard-drawn | 9.810 | 10.745 | 0.004020 | 0.002230 | 97.80 | 1.022 |
| Gold (99.9 per cent pure) | 13.216 | 14.404 | 0.003770 | 0.002090 | 72.55 | 1.378 |
| Aluminum (99.5 per cent pure) | 15.219 | 16.758 | 0.004230 | 0.002350 | 63.00 | 1.587 |
| (Commercial—97.5 per cent pure) | 16.031 | 17.699 | 0.004350 | 0.002420 | 59.80 | 1.672 |
| Zinc (very pure) | 34.595 | 37.957 | 0.004060 | 0.002260 | 27.72 | 3.608 |
| Iron (approx. pure) | 54.529 | 62.643 | 0.006250 | 0.003470 | 17.50 | 5.714 |
| Iron E.B.B. iron wire | 58.702 | 65.190 | 0.004630 | 0.002570 | 16.20 | 6.173 |
| Platinum (pure) | 65.670 | 71.418 | 0.003669 | 0.002038 | 14.60 | 6.845 |
| Iron, B.B. iron wire | 68.680 | 76.270 | 0.004630 | 0.002570 | 13.50 | 7.407 |
| Nickel | 74.128 | 85.138 | 0.006220 | 0.003460 | 12.94 | 7.726 |
| Tin (pure) | 78.439 | 86.748 | 0.004400 | 0.002450 | 12.22 | 8.184 |
| Steel (wire) | 81.179 | 90.150 | 0.004630 | 0.002570 | 11.60 | 8.621 |
| **Substance** | | | | | | |
| Brass | 43.310 | . . . . | . . . . . . . | . . . . . . . | 22.15 | 4.515 |
| Phosphor bronze | 51.005 | . . . . | 0.000640 | 0.000356 | 18.80 | 5.319 |
| Aluminum bronze | 73.989 | . . . . | 0.001000 | 0.000556 | 12.96 | 7.714 |
| German silver (Cu 50, Zn 35, Ni 15) | 127.800 | . . . . | 0.000400 | 0.000220 | 7.50 | 17.300 |
| Platinoid [Cu, 59, Zn 25.5, Ni 14, W (tungsten) 55] | 251.030 | . . . . | 0.000310 | 0.000172 | 3.82 | 26.180 |
| Manganin (Cu 84, Ni 4, Mn 12) | 280.790 | . . . . | 0.000000 | . . . . . . . | 3.41 | 29.330 |
| Constantan (Cu 58, Ni 41, Mn 1) | {300.77 / 312.80} | . . . . | ±0.000010 | 0.000005 | {3.19 / 3.07} | {31.35 / 32.57} |
| Gray cast iron | 684.000 | . . . . | | | | |

**98. The resistance of conductors that are not circular in section can be computed** by first getting their areas in square inches and then reducing this square-inch value to circular mils as given in Sec. 93. Then proceed with the formula in the preceding paragraph.

**99. Change of resistance with change of temperature.** The resistance of all pure metals increases as they become hot. The resistance of certain alloys is not affected by the temperature. The resistance of a few materials, carbon for instance, decreases as their temperature increases. The proportion that resistance changes per degree rise in temperature is called the temperature coefficient of resistance (see Table 97 for values). Unless otherwise stated, the temperature coefficient is positive; i.e., the resistance increases with increase of temperature. For all pure metals, the coefficient is practically the same and is 0.004 for temperatures in degrees Celsius and 0.0023 in degrees Fahrenheit.

**100. To find the resistance of a conductor at any ordinary temperature,** use this formula:

$$R_h = R_c + [a \times R_c(T_h - T_c)] \qquad \text{or} \qquad T_h - T_c = \frac{R_h - R_c}{a \times R_c} \qquad (24)$$

where $R_h$ = resistance, in ohms, hot; $R_c$ = resistance, in ohms, cold; $T_h$ = temperature of conductor hot, in degrees; $T_c$ = temperature of conductor cold, in degrees; and $a$ = temperature coefficient of the material of the conductor from Table 97. (This is an approximate method, but it is sufficiently accurate for all ordinary work.)

    **example** The resistance of 1 cmil·ft of annealed copper is 9.59 $\Omega$ at 32°F. What will its resistance be at 75°F?

    **solution** From Table 97 the coefficient is 0.00223. Now substitute

$$R_h = R_c + [a \times R_c(T_h - T_c)] = 9.59 + [0.00223 \times 9.59(75 - 32)]$$
$$= 9.59 + [0.00223 \times 9.59 \times 43] = 9.59 + 0.92 = 10.51 \ \Omega, \text{ at } 75°F$$

**101. The temperature rise in a conductor can be determined with the formula of Sec. 100 by measuring hot and cold resistance.** The expression $T_h - T_c$ is the difference between hot and cold temperatures and is therefore the temperature rise or fall.

    **example** The resistance of a set of copper coils measured 20 $\Omega$ at a room temperature of 20°C. After carrying current for some time the resistance measured 20.78 $\Omega$. What was the temperature rise in the coil?

    **solution** The temperature coefficient of copper per degree Celsius from Table 97 is 0.004. Substitute in the formula

$$T_h - T_c = \frac{R_h - R_c}{aR_c} = (20.78 - 20.0) \div (0.004 \times 20) = 0.78 \div 0.08 = 9.75°C$$

Therefore the average temperature rise in the coil was 9¾°C.

**102. Contact resistance** is the resistance at the point of contact of two conductors. When current flows, heat is always developed at such a point. The greater the clamping pressure between the conductors in contact and the greater the area of contact, the less the contact resistance will be. The nature of the surfaces in contact must also be considered. Smooth surfaces have less contact resistance than do rough surfaces. Contacts should always be so designed that, for a given current, the area of contact will be large enough to prevent the contact resistance from being so great as to cause excessive heating. Table 104 indicates safe values.

**103. Ohm's law.** From the preceding sections it is evident that a voltage is required to force a current through a circuit against the resistance caused by the material of the conductor. The relation between current, resistance, and voltage is known as Ohm's law. It is merely a restatement, as applied to electric circuits, of the general law which governs all physical phenomena; i.e., the result produced is directly proportional to the effort or cause and inversely proportional to the opposition. For the phenomenon dealing with current passing through resistance, the emf is the cause, the current is the result or effect, and the resistance is the opposition. Thus,

$$I = \frac{E}{R} = \left( \text{amperes} = \frac{\text{volts}}{\text{ohms}} \right) \qquad \text{(amperes)} \quad (25)$$

$$R = \frac{E}{I} = \left( \text{ohms} = \frac{\text{volts}}{\text{amperes}} \right) \qquad \text{(ohms)} \quad (26)$$

$$E = I \times R = \text{volts} = (\text{amperes} \times \text{ohms}) \qquad \text{(volts)} \quad (27)$$

**104.    Safe Current Densities for Electrical Contacts and for Cross Sections**

| Kind of contact or cross section | Material | Current density | |
|---|---|---|---|
| | | Amp per sq in. | Sq mils per amp |
| Sliding contact (brushes) | Copper brush | 150– 175 | 5,700– 6,700 |
| | Brass gauze brush | 100– 125 | 8,000–10,000 |
| | Carbon brush | 30– 40 | 25,000–33,300 |
| Spring contact (switch blades) | Copper on copper | 60– 80 | 12,500–16,700 |
| | Composition on copper | 50– 60 | 16,700–20,000 |
| | Brass on brass | 40– 50 | 20,000–25,000 |
| Screwed contact | Copper to copper | 150– 200 | 5,000– 6,700 |
| | Composition to copper | 125– 150 | 6,700– 8,000 |
| | Composition to composition | 100– 125 | 8,000–10,000 |
| Clamped contact | Copper to copper | 100– 125 | 8,000–10,000 |
| | Composition to copper | 75– 100 | 10,000–13,000 |
| | Composition to composition | 70– 90 | 11,000–14,000 |
| Fitted contact (taper plugs) | Copper to copper | 125– 175 | 5,700– 8,000 |
| | Composition to copper | 100– 125 | 8,000–10,000 |
| | Composition to composition | 75– 100 | 10,000–13,000 |
| Fitted and screwed contact | Copper to copper | 200– 250 | 4,000– 5,000 |
| | Composition to copper | 175– 200 | 5,000– 5,700 |
| | Composition to composition | 150– 175 | 5,700– 6,700 |
| Cross section | Copper wire | 1,200–2,000 | 500– 800 |
| | Copper wire cable | 1,000–1,600 | 600– 1,000 |
| | Copper rod | 800–1,200 | 800– 1,200 |
| | Composition casting | 500– 700 | 1,400– 2,000 |
| | Brass casting | 300– 400 | 2,500– 3,300 |
| | Brass rod | 575– 750 | 1,300– 1,700 |

where $I$ = the effective current, in amperes, which flows through the resistance in the circuit or in the portion of the circuit under consideration; $R$ = the resistance in ohms of the circuit or the portion of the circuit under consideration; and $E$ = the effective emf in volts required to force the current through the resistance of the circuit or the portion of the circuit under consideration.

**105. Application of Ohm's law.**    Great care must be exercised in the use of Ohm's law. The voltage given by Eq. (27) may or may not be the voltage that must be impressed on the circuit to force the given current through the circuit. When the circuit contains only resistance, the voltage given by Ohm's law is the voltage impressed on the circuit. Ohm's law holds true for all circuits, both direct and alternating, but the voltage obtained by its use is only the voltage required to overcome the resistance of the circuit. Ohm's law cannot be used for a complete motor circuit to determine the resistance of the circuit.

**106. Use Ohm's law for:**

1. Determination of the voltage required to provide a given current through only a resistance

2. Determination of the current which would be produced by a given voltage impressed upon only a resistance

3. Determination of the voltage drop caused by a current passing through a resistance

4. Calculations of complete dc circuits which do not contain any emf other than the impressed voltage

5. Any circuit or portion of a circuit which consists only of resistance

**107. Do not use Ohm's law for:**

1. AC circuits in general

2. Complete motor circuits

3. Complete circuits containing any emf other than the impressed voltage

Refer to Secs. 131 and 132 for calculation of ac circuits.

### 108. Examples of the application of Ohm's law

**example** What will be the current in the dc circuit of Fig. 1-27?

**solution** An entire circuit is shown. It is composed of a dynamo, line wires, and a resistance coil. The emf developed by the dynamo (do not confuse this with the emf impressed by the dynamo on the

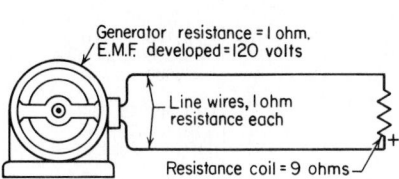

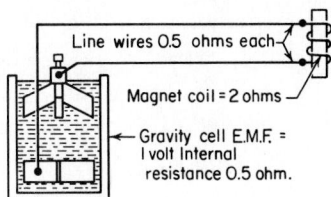

**Fig. 1-27** An entire dynamo circuit.    **Fig. 1-28** An entire battery circuit.

line) is 120 V. The resistance of the entire circuit is the sum of the resistances of dynamo, line wires, and resistance coil. Substituting in the formula,

$$I = \frac{E}{R} = \frac{120}{1 + 1 + 9 + 1} = \frac{120}{12} = 10 \text{ A}$$

**example** What current will flow in the circuit of Fig. 1-28?

**solution** This again is an entire circuit. Substituting in the formula,

$$I = \frac{E}{R} = \frac{1}{0.5 + 0.5 + 2 + 0.5} = \frac{1}{3.5} = 0.28 \text{ A}$$

Note that the internal resistance of the battery must be considered.

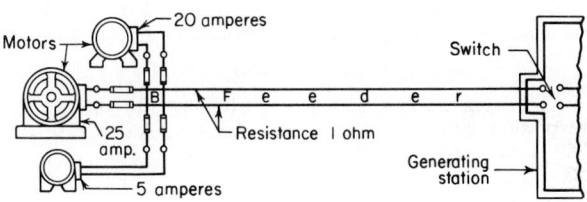

**Fig. 1-29** Feeder to motors.

**example** With 10 A flowing, what will be the voltage drop in the line wires in Fig. 1-29?

**solution** Each has a resistance of 1.0 Ω; hence

$$E = I \times R = 10 \times 2 = 20 \text{ V}$$

**example** What is the resistance of the incandescent lamp of Fig. 1-30A? It is tapped to a 120-V circuit, and the ammeter reads 0.5 A. The branch wires are so short that their resistance can be neglected.

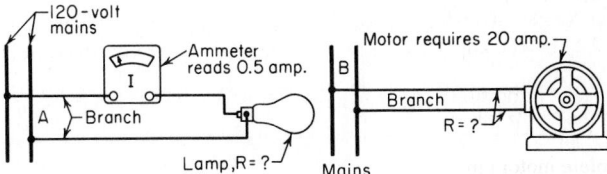

**Fig. 1-30** Portions of circuits.

**solution**  Substitute in the formula

$$R = \frac{E}{I} = \frac{120}{0.5} = 240 \; \Omega$$

**example**  The dc motor of Fig. 1-30B takes 20 A, and the drop in voltage in the branch wires should not exceed 5 V. What is the greatest resistance that can be permitted in the branch conductors?
**solution**  Substitute in the formula

$$R = \frac{E}{I} = \frac{5}{20} = 0.25 \; \Omega$$

This (0.25 Ω) is the resistance of both wires. Each would have a resistance of 0.125 Ω.

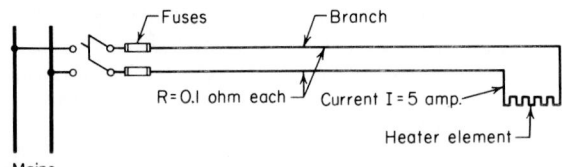

**Fig. 1-31**    Portion of a circuit.

**example**  The electric heater of Fig. 1-31 takes 5 A. The resistance of each branch wire is 0.1 Ω. What will be the drop in volts in each branch wire?
**solution**  Substitute in the formula

$$E = I \times R = 5 \times 0.1 = 0.5 \; V$$

In both branch wires or in the branch circuit the volts lost would be 2 × 0.5 = 1 V.

**example**  Three dc motors (Fig. 1-29) taking respectively 20, 25, and 5 A (these values are stamped on the nameplates of the motors) are located at the end of a feeder having a resistance of 1.0 Ω on each side. What will be the voltage drop in the feeder?
**solution**  Substitute in the formula

$$E = R \times I = (1 + 1) \times (20 + 25 + 5) = 2 \times 50 = 100 \; V$$

**109. DC circuits.**    A constant-current dc circuit has only one characteristic that affects the value of the current, i.e., the resistance offered by the material of which the circuit is made. Therefore, Ohm's law applies to a complete dc circuit or to any portions of the circuit which do not contain any emf other than the impressed voltage. For a motor circuit, therefore, Ohm's law will apply only to the line-conductor portion of the circuit. Ohm's law will give the voltage required to overcome the resistance of the line conductors, the voltage drop of the circuit. The total impressed voltage will be equal to the sum of the voltage at the motor terminals plus the $IR$ voltage of the line conductors. The voltage impressed on the motor will be equal to the impressed line voltage minus the $IR$ voltage drop in the line conductors.

**110. Power in dc circuits** is equal to the product of volts and amperes. Expressing this as a formula,

$$W = I \times E \qquad I = \frac{W}{E} \qquad E = \frac{W}{I} \qquad\qquad (28)$$

and also in circuits where all the energy is converted into heat energy,

$$W = I^2 \times R \qquad W = \frac{E^2}{R} \qquad I = \sqrt{\frac{W}{R}} \qquad E = \sqrt{R \times W} \qquad (29)$$

$$R = \frac{E^2}{W} \qquad R = \frac{W}{I^2} \qquad\qquad (30)$$

where $I$ = current in amperes, $E$ = voltage or emf in volts, $R$ = resistance in ohms, and $W$ = power in watts.

**111.** In applying the above equations be careful that the values of current, voltage, and resistance used in any one problem all apply to the same circuit or to the same portion of a circuit.

**example**   How many watts are consumed by the incandescent lamp in Fig. 1-32?
**solution**   Substitute in the formula

$$W = I \times E = \tfrac{1}{2} \times 120 = 60 \text{ W}$$

**example**   How many watts are taken by the motor of Fig. 1-33? How many kilowatts? How much horsepower?

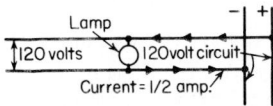

**Fig. 1-32**   Incandescent-lamp branch circuit.       **Fig. 1-33**   Electric motor.

**solution**   Substitute in the formula

$$W = I \times E = 70 \times 220 = 15{,}400 \text{ W}$$

$$kW = \frac{W}{1000} = \frac{15{,}400}{1000} = 15.4 \text{ kW}$$

$$hp = \frac{W}{746} = \frac{15{,}400}{746} = 20.6 \text{ hp}$$

**example**   In the feeder of Fig. 1–34, what power will be lost in the line wires to the motor?

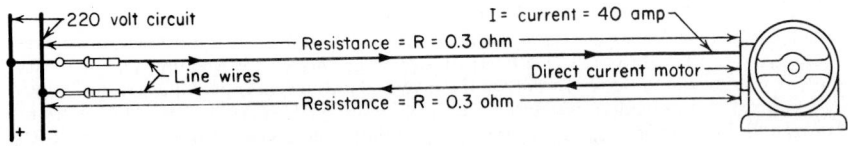

**Fig. 1-34**   Feeder.

**solution**   Substitute in the formula
$$W = I^2 \times R = (40 \times 40) \times (0.3 + 0.3) = 1600 \times 0.6 = 960 \text{ W}$$

**112. Waveform of alternating currents and voltages.**   An alternating current or voltage is one which is reversed at regular intervals. The curve formed by plotting the instantaneous values of the voltage or current against time is called the waveform of the voltage or current. It is best to have the value of an alternating current or voltage vary with time

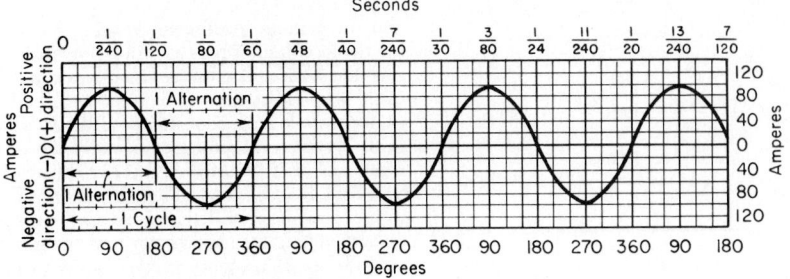

**Fig. 1-35**   Graph of a 60-Hz alternating current having a maximum value of 100 A.

according to what is known as the sine law. The instantaneous values of an alternating current varying according to this law are shown in Fig. 1-35. A voltage or current which varies in this manner is called a sinusoidal voltage or current. Most modern ac generators produce a voltage which is very nearly sinusoidal and is generally so considered. The current in some ac circuits is very nearly sinusoidal and in others may differ considerably from the sine law. In most cases, however, for practical work both alternating voltages and currents can be considered as following the sine law. If it is not satisfactory to do this, the calculations become complicated and discussion is outside the scope of this book.

**113. A cycle** is a complete set of values through which an *alternating current* (see Sec. 34 for an explanation of alternating current) repeatedly passes (see Fig. 1-35). The expres-

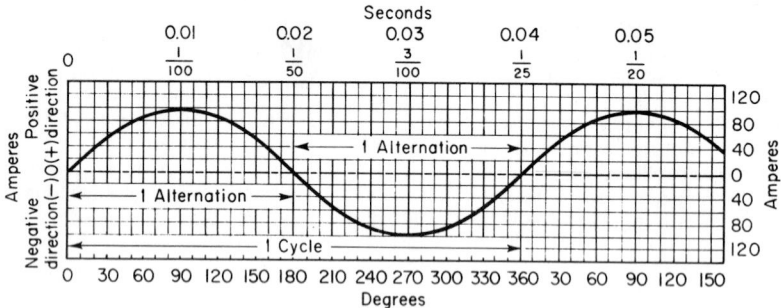

**Fig. 1-36**   Curve of a 25-Hz alternating current.

sion "60 hertz (60 Hz)" means that the current referred to makes 60 complete cycles in a second. It therefore requires $\frac{1}{60}$ s to complete 1 cycle (see Fig. 1-35). With a 25-Hz current, $\frac{1}{25}$ s is required to complete 1 cycle (see Fig. 1-36).

**114. The frequency of an alternating current** is the number of cycles completed in a second. A frequency of 60 Hz is practically standard for lighting and power installations. Most electric power companies generate, transmit, and distribute their power at this frequency. For railroad electrification, 25 Hz is sometimes employed. If 25 Hz is used for lighting, lamps will flicker. Some of the older stations in the United States still generate at 50 Hz. In Europe 50 Hz has been used for light and power and $16\frac{2}{3}$ Hz for railroads.

**115. Electrical degrees.**    The instantaneous values of an ac voltage can be studied in Fig. 1-37, which shows an elementary four-pole generator and the waveform of voltage for one complete revolution of the conductor. When the conductor is in the position marked A, halfway between adjacent poles, it will not cut any flux and therefore no voltage will be induced in the conductor. As the conductor moves from this position, it will start to cut flux and will do so at a greater and greater rate until the conductor is in the position B. At this point, flux will be cut at the maximum rate, and the voltage produced will be maximum. As the conductor moves from position B to position C, it will cut flux at a decreasing rate, until when position C is reached (halfway between poles), the conductor will not cut flux at all. As the conductor moves on from position C, it will start to cut flux, but the direction of the flux is reversed. The direction of the induced voltage will therefore reverse at the instant when the conductor passes through position C. If we pursue the same reasoning for the rest of the revolution, the voltage produced in the conductor will follow the wave-

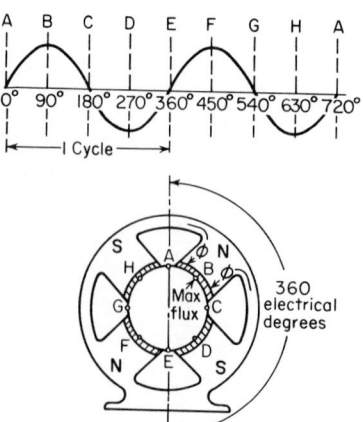

**Fig. 1-37**   Elementary ac generator, illustrating electrical degrees and waveform.

form shown in Fig. 1-37. The arc that the conductor must pass through to generate one complete cycle of voltage is called 360 electrical degrees. In any ac generator there are therefore 360 electrical degrees in the arc between the centerline of one pole and the centerline of the next pole of the same polarity. The number of electrical degrees in one complete revolution for any generator is equal to 360 times the number of pairs of poles for which the machine is wound. One electrical degree is equal to 1 (pair of poles) part of a mechanical degree.

If the elementary generator of Fig. 1-37 produced a 60-Hz voltage, the conductor would have to make 1800 rpm. This would mean that the conductor would make 1 revolution in $^{60}\!/_{1800}$ or $^{1}\!/_{30}$ s. Since the voltage goes through 2 cycles in 1 revolution (see Fig. 1-37), it takes $^{1}\!/_{60}$ s for 1 cycle. This means that in $^{1}\!/_{60}$ s the conductor moves from position $A$ to position $E$.

Refer to Fig. 1-37 and take $A$ position as zero instant of time and as zero electrical degrees. When the conductor has moved from position $A$ to position $B$, it will have passed through $^{360}\!/_{4}$ or 90 electrical degrees, and the elapsed time will be $1/(4 \times 60)$ or $^{1}\!/_{240}$ s. Electrical degrees and time are therefore proportional. Instead of the instantaneous values of the voltage being plotted against time, they can be plotted against electrical degrees. These two methods of plotting such curves are shown in Figs. 1-35 and 1-36.

**116. The word phase,** which is used in ac terminology, refers basically to time. When two alternating currents are in phase, they reach their corresponding zero, maximum, and intermediate values at exactly the same instants. Two in-phase ac quantities are shown in Fig. 1-38. If currents or voltages are not in phase, they reach corresponding values at

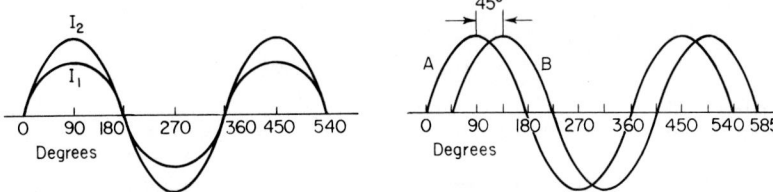

**Fig. 1-38**   Two alternating currents in phase.    **Fig. 1-39**   Two alternating currents out of phase.

different instants of time, as shown in Fig. 1-39. Since electrical degrees are proportional to time, it is standard practice to state the out-of-phase relation of two quantities in electrical degrees. In specifying the phase relation of two quantities, it is necessary to state whether the relationship is leading or lagging. For example, in Fig. 1-39, it is not complete to state that quantity $A$ is 45 degrees out of phase with $B$. The correct statement is either "$A$ leads $B$ by 45 degrees" or "$B$ lags $A$ by 45 degrees." The number of degrees that two quantities are out of phase is the number of electrical degrees that must elapse in the time between the occurrence of a certain instantaneous value of one quantity and the occurrence of the corresponding instantaneous value of the other quantity. For example, in Fig. 1-39 quantity $B$ does not reach its maximum positive value until 45 degrees after quantity $A$ has reached its maximum positive value. Quantity $B$ therefore lags quantity $A$ by 45 degrees. The number of electrical degrees that two quantities are out of phase is called the phase angle.

**117. A three-phase current** consists of three different alternating currents out of phase 120 degrees with each other. A two-phase current consists of two different alternating currents out of phase 90 degrees with each other.

**118. The maximum value of an alternating current or voltage** is the greatest value that it attains. This is an instantaneous value (see Fig. 1-40).

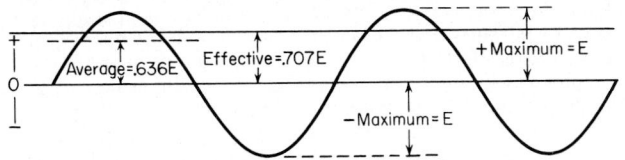

**Fig. 1-40**   Alternating emf values.

**119. The so-called average value** of an alternating current or voltage is the average of the instantaneous values for ½ cycle. Average values are not of much use to practical people.

**120. Effective values** of alternating current and voltage are the ones ordinarily referred to when we speak of alternating quantities. Alternating voltages and currents are constantly changing in value, within a certain range, from instant to instant even if the load is constant. It is not practicable to deal with or indicate with instruments these constantly changing values. The effective value of an alternating current or voltage is the value of a direct current or voltage that would have the same heating effect. A sinusoidal alternating current which has a maximum value of 14.14 A will have the same heating effect as a 10-A direct current. The effective value of the alternating current is therefore 10 A. Practical people deal almost exclusively with effective values. Alternating-current measuring instruments indicate effective values. The relation between effective and maximum values for sinusoidal quantities is given below and illustrated in Fig. 1-40. Effective value is also referred to as rms (root mean square) value.

$$\text{Effective value} = 0.707 \times \text{maximum value} \qquad (31)$$

$$\text{Maximum value} = \frac{\text{effective value}}{0.707} \qquad (32)$$

**example**  What is the effective voltage of a circuit that has a maximum voltage of 156?
**solution**  Substitute in the formula

$$\text{Effective value} = 0.707 \times \text{maximum value} = 0.707 \times 156 = 110 \text{ V}$$

**example**  If a voltmeter on an ac circuit reads 2200, what is the maximum instantaneous voltage?
**solution**  Substitute in the formula

$$\text{Maximum value} = \frac{\text{effective value}}{0.707} = \frac{2200}{0.707} = 3110 \text{ V}$$

**121. Characteristics of ac circuits.** All three of the circuit characteristics, resistance, inductance, and capacitance, may affect the value of the current in an ac circuit.

**122. Resistance of ac circuits.** As mentioned in Sec. 90, the resistance of a circuit to alternating current may be considerably different from its resistance to direct current. This increase in the resistance may be due to two factors. One factor, which takes place inside the conductors, is called skin effect and is discussed in the following sections. The other factor is due to the proximity of magnetic and electrical conducting material to the circuit. The varying magnetic field produced by the alternating current results in electric and magnetic power losses in these neighboring materials. These power losses must be supplied by the power delivered to the electric circuit and therefore result in increasing the effective resistance of the circuit over what it was for direct current. Refer to Sec. 157 for examples of effective resistance to an alternating current.

**123. Skin effect.** When an alternating current flows through a conductor, there is an inductive action whereby the current in the conductor is forced toward its surface. The current density is greater at the surface than at the center, and under certain conditions practically no current may flow along the axis of the conductor. Although skin effect and self-induction both originate from the same magnetic field, they are not otherwise related. Since skin effect increases voltage drop and energy loss, it amounts to an increase in resistance and is so considered. Table 62 of Div. 11 gives values by which dc resistances of conductors must be multiplied to obtain their resistances to alternating currents. Nonconducting cores are sometimes placed in the centers of large cables for alternating currents so that all the metal will be effectively used (see Div. 2 for such conductors).

**124. Skin effect in conductors of magnetic materials** is much greater than in those of nonmagnetic materials owing to the stronger magnetic field that a given current will set up in a magnetic metal.

**125. The effect of inductance in an ac circuit** is that it produces opposition to the flow of the current and tends to make the current lag behind the voltage in time or phase. If a pure inductive circuit (one with only inductance and no resistance) could be built, the current would lag 90 degrees, or ¼ cycle, behind the voltage. In actual circuits containing both resistance and inductance, the current will lag some angle between 0 and 90 degrees behind the voltage. The angle of lag will depend upon relative resistance and inductance.

**126. The opposition to alternating current** which is produced by inductance is called inductive reactance (see Sec. 46). For a sinusoidal current the value of the inductive reactance is

$$X_L = 2\pi f L \tag{33}$$

where $X_L$ = inductive reactance in ohms, $f$ = frequency in hertz, and $L$ = inductance in henrys.

**127. The angle of lag of a current in an inductive circuit** can be calculated from the following equation:

$$\text{Tangent of angle of lag} = \frac{X_L}{R} \tag{34}$$

where $X_L$ = inductive reactance in ohms and $R$ = ac resistance (not dc resistance). Also see method of Sec. 133.

**128. The effect of capacitance in an ac circuit** is that it produces opposition to the flow of the current and tends to make the current lead the voltage. For a pure capacitive circuit (one with only capacitance and no resistance) the current will lead the voltage by 90 degrees, or ¼ cycle. For a circuit with both resistance and capacitance, the current will lead the voltage by some angle between 0 and 90 degrees, depending upon the relative value of the resistance and the capacitance.

**129. The opposition to alternating current** which is produced by capacitance is called capacitive reactance (see Sec. 49). For a circuit with sinusoidal relations the value of the capacitive reactance is

$$X_c = \frac{1}{2\pi f c} \tag{35}$$

where $X_c$ = capacitive reactance in ohms, $f$ = frequency in hertz, and $c$ = capacity in farads.

**130. The angle of lead** of a current in a capacitive circuit can be calculated from the following equation:

$$\text{Tangent of angle of lead} = \frac{X_c}{R} \tag{36}$$

where $X_c$ = capacitive reactance in ohms and $R$ = ac resistance (not dc resistance). Also see method of Sec. 133.

**131. Impedance** is the name given to the total opposition to the flow of alternating current. It is the combined opposition of resistance, inductive reactance, and capacitive reactance. Impedance is measured in ohms and is expressed by the symbol Z. The impedance of a series circuit may be calculated by the following formulas:

For a circuit containing resistance, inductance, and capacitance

$$Z = \sqrt{R^2 + (X_L - X_c)^2} \tag{37}$$

For a circuit containing resistance and inductance

$$Z = \sqrt{R^2 + X_L^2} \tag{38}$$

For a circuit containing resistance and capacitance

$$Z = \sqrt{R^2 + X_c^2} \tag{39}$$

For a circuit containing only resistance

$$Z = R \tag{40}$$

For a circuit containing only inductance

$$Z = X_L \tag{41}$$

For a circuit containing only capacitance

$$Z = X_c \tag{42}$$

where $Z$ = impedance in ohms, $X_L$ = inductive reactance in ohms, and $X_c$ = capacitive reactance in ohms.

**132. Relations between voltage, current, and impedance** are

$$E = IZ \tag{43}$$

$$I = \frac{E}{Z} \tag{44}$$

$$Z = \frac{E}{I} \tag{45}$$

where $E$ = voltage impressed on circuit in volts, $I$ = current in amperes, and $Z$ = impedance in ohms.

The above equations apply to complete ac circuits or any portions of a circuit which do not contain any emf other than the impressed voltage. For a motor circuit, therefore, they will apply only to the line-conductor portion of the circuit. They will give the voltage required to overcome line impedance, the $IZ$ voltage of the circuit.

It is noticed that Ohm's law (Sec. 105) can be used for a complete ac circuit only when the circuit contains only resistance.

**133. The phase angle of a circuit** containing only resistance, inductance, and capacitance can be found from the following formula:

$$\text{Cosine of phase angle} = \frac{R}{Z} \tag{46}$$

where $Z$ = impedance in ohms and $R$ = resistance in ohms.

**134. Power in ac circuits.** The power of an ac circuit is very seldom equal to the direct product of the volts and amperes. To calculate the power of a single-phase ac circuit, the product of the volts and amperes must be multiplied by the power factor (see Sec. 136).

**135. Apparent power** is the term applied to the product of voltage and current in an ac circuit. It is expressed in voltamperes (VA) or in kilovolt-amperes (kVA) or megavolt-amperes (MVA).

**136. Power factor** is the ratio of the true power or watts to the apparent power or voltamperes. The power factor is expressed as a decimal or in percentage. Thus power factors of 0.8 and of 80 percent are the same. In giving the power factor of a circuit, state whether it is leading or lagging. The current is always taken with respect to the voltage. A power factor of 0.75 lagging means that the current lags the voltage. The power factor may have a value anywhere between 0 and 1.0 but can never be greater than 1.0.

**example** In Fig. 1-41, which shows a single-phase circuit, the ammeter $I$ reads 10 A, and the voltmeter $E$ 220 V. The apparent power is the product of volts and amperes, or $IE = 10 \times 220 = 2200$

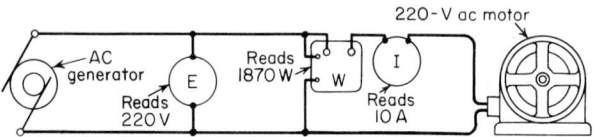

**Fig. 1-41** Example of power factor.

VA. But the wattmeter $W$ reads 1870 W. A wattmeter always indicates real or true power. Therefore the power factor (for a single-phase circuit) is

$$\text{Power factor} = \frac{\text{true watts}}{\text{apparent power}} = \frac{1870}{2200} = 0.85, \text{ or 85 percent}$$

**137. The cosine of the angle of lag or lead is equal to the power factor.** Cosines for different angles can be found in trigonometric tables in handbooks (see Sec. 1). The symbol $\theta$, the Greek letter theta, is often used to designate the angle of lag or lead; hence power factor is sometimes referred to as cos $\theta$ (cosine theta). This means the cosine of the angle $\theta$.

**138. The power factor in a noninductive circuit,** one containing resistance only, is always 1, or 100 percent; i.e., the product of volts and amperes in such a circuit gives true power.

**139. The power factor in a circuit containing inductance or capacitance** may be anything between 0 and 1 (0 and 100 percent), depending on the amount of inductance or capacitance in the circuit.

**140. Typical power factors of various kinds of central-station loads** are as follows:

INCANDESCENT LIGHTING    1.0.

INCANDESCENT LIGHTING WITH SMALL STEP-DOWN TRANSFORMERS    0.95 to 0.98.

INCANDESCENT STREET-LIGHTING SERIES CIRCUITS    0.6 to 0.8.

SODIUM-VAPOR STREET LIGHTING; PARALLEL CIRCUITS    0.8 to 0.85.

SODIUM-VAPOR STREET LIGHTING; SERIES CIRCUITS    0.5 to 0.7.

FLUORESCENT LIGHTING    0.5 to 0.95, depending on type of auxiliary used.

MERCURY-VAPOR LIGHTING    0.5 to 0.95, depending on type of auxiliary used.

SINGLE-PHASE INDUCTION MOTORS; SQUIRREL-CAGE ROTOR    $\frac{1}{20}$ to 1 hp, power factor, 0.55 to 0.75, average 0.68 at rated load; 1 to 10 hp, power factor, 0.75 to 0.86, average 0.82 at rated load.

POLYPHASE INDUCTION MOTORS; SQUIRREL-CAGE ROTOR    1 to 10 hp, power factor, 0.75 to 0.91, average 0.85 at rated load; 10 to 50 hp, power factor, 0.85 to 0.92, average 0.89 at rated load.

POLYPHASE INDUCTION MOTORS; PHASE-WOUND ROTORS    5 to 20 hp, power factor, 0.80 to 0.89, average 0.86 at rated load; 20 to 100 hp, power factor, 0.82 to 0.90, average 0.87 at rated load.

INDUCTION-MOTOR LOADS IN GENERAL    Power factor, 0.60 to 0.85, depending on whether motors are carrying their rated loads.

ROTARY CONVERTERS, COMPOUND-WOUND    Power factor at full load can be adjusted to practically 100 percent. At light loads it will be lagging and at overloads slightly leading.

ROTARY CONVERTERS, SHUNT-WOUND    Power factor can be adjusted to any desired value and will be fairly constant at all loads with the same field rheostat adjustment. Rotary converters, however, should not be operated below 0.95 power factor leading or lagging at full load or overload.

SMALL HEATING APPARATUS    This load has the same characteristics as an incandescent-lighting load. The power factor of the load unit is practically 1, but the distributing transformers will lower it to some extent.

ARC FURNACES    Power factor, 0.80 to 0.90.

INDUCTION FURNACES    Power factor, 0.60 to 0.70.

ELECTRIC-WELDING TRANSFORMERS    Power factor, 0.50 to 0.70.

SYNCHRONOUS MOTORS    Adjustment between 0.80 power factor leading to a power factor of 1. (1) Operating power factors above 0.95 will be obtained only when practically all the load consists of synchronous motors or converters which can be operated at practically a power factor of 1. (2) Power factors of 0.90 to 0.95 can be safely predicted only when the load is entirely incandescent lighting or heating or when a large noninductive load, such as synchronous motors or converters, is used with a smaller proportion of inductive motor load. (3) For the average central-station load, consisting of lighting and motor service, a power factor of 0.80 should be assumed. (4) A power factor of 0.70 should be assumed for a plant having a large proportion of induction motors, fluorescent lighting, electric furnaces, or electric-welding load.

**141. Kilowatts and kilovolt-amperes** (General Electric Company). The term *kilowatt* (kW) indicates the measure of power which is all available for work. Kilovolt-amperes (kVA) indicate the measure of apparent electric power made up of two components, an energy component and a wattless or induction component. Kilowatts indicate real power and kilovolt-amperes apparent power. They are identical only when current and voltage are in phase, i.e., when the power factor is 1. Ammeters and voltmeters indicate total effective current and voltage regardless of the power factor, while a wattmeter indicates the effective product of the instantaneous values of emf and current. A wattmeter, then, indicates real power.

Standard guarantees on ac generators are made on the basis of loads at 80 percent power factor. However, it must not be inferred that a given generator will deliver its rated power output at all power factors. The generator rating in kilowatts will be reduced in proportion to the power factor and probably in a greater ratio if the power factor is very low. The method of rating ac generators by kilovolt-amperes instead of by kilowatts is now in general use.

In discussing an ac load, it is well to state it in terms of kilowatts, power factor, and kilovolt-amperes, thus: 200 kW, 80 percent power factor (250 kVA). This shows that the current in the circuit corresponds to 250 kVA and heats the generator and conductors to that extent but that only 200 kW is available for doing work. An illustration of the distinction between kilowatts and kilovolt-amperes is given in Fig. 1-42.

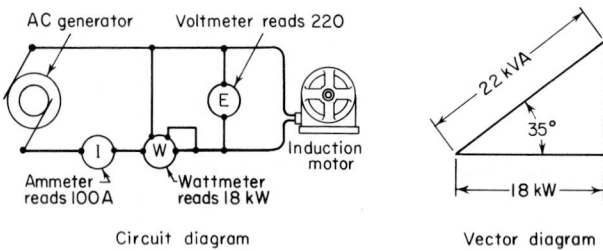

**Fig. 1-42**   Distinction between kilowatts and kilovolt-amperes.

**142. Effects of low power factor.**   It is sometimes considered that the wattless component of a current at low power factor is circulated without an increase of mechanical input over that necessary for actual power requirements. This is inaccurate because internal work or losses due to this extra current are produced and must be supplied by the prime mover. Since these extra losses manifest themselves in heat, the capacity of the machine is reduced. Moreover, wattless components of current heat the line conductors, just as do energy components, and cause losses in them. The loss in any conductor (see Sec. 38) is always

$$W = I^2 R \qquad (47)$$

where $W$ = the loss in watts, $I$ = the current in amperes in the conductor, and $R$ = the resistance in ohms. It requires much larger equipment and conductors to deliver a certain amount of power at a low power factor than at a power factor close to 1.

**143. Correction of low power factor.**   In industrial plants, excessively low power factor is usually due to underloaded induction motors because the power factor of motors is much less at partial loads than at full load. If motors are underloaded, new motors of smaller capacity should be substituted. Power factor can be corrected (1) by installing synchronous motors which, when overexcited, have the property of neutralizing the wattless or reactive components of currents or (2) by connecting static capacitors across the line. See Index: Motor(s), alternating-current—Induction, Application of; Synchronous.

**144. A single-phase alternating emf** will be induced in an armature coil ($S_1$ and $F_1$, Fig. 1-43) which has its sides set, in a generator frame, the same distance apart as are a north and a south magnet that are forced to sweep continuously past the coil sides at a uniform speed. The distance between a north and a south pole is always called 180 electrical degrees. The distance between a north pole and the next north pole is called 360 electrical degrees. In any given generator, the circumferential distance is the same between any two adjacent north and south poles.

**145. In actual ac single-phase generators** there are a number of pairs of north and south poles arranged on a revolving-field structure. A corresponding number of coils, each having its sides set approximately 180 electrical degrees apart, is arranged around the frame. Then the coils are connected in series so that their emf's will be additive. The combination of their emf's is the voltage which is impressed by the generator on the circuit. Some small revolving-armature ac generators have rotating armatures and stationary fields, but their principle is the same as that of the revolving-field machines.

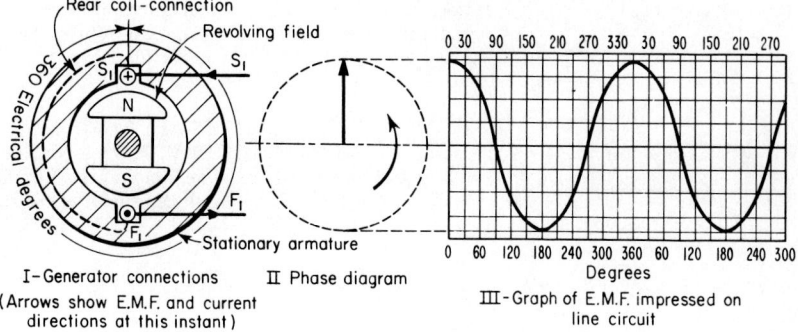

I–Generator connections     II Phase diagram
(Arrows show E.M.F. and current
directions at this instant)

III–Graph of E.M.F. impressed on
line circuit

**Fig. 1-43**   Elementary two-pole single-phase ac generator.

**146. For a single-phase circuit the relations between kilowatts and kilovolt-amperes** are

$$\text{Kilovolt-amperes} = \frac{\text{volts} \times \text{amperes}}{1000} \quad \text{or} \quad \text{kVA} = \frac{E \times I}{1000} \tag{48}$$

$$\text{kW} = \text{kVA} \times \text{power factor} \quad \text{kVA} = \frac{\text{kW}}{\text{power factor}} \tag{49}$$

$$\text{Power factor} = \frac{\text{kW}}{\text{kVA}} \tag{50}$$

For an example see Fig. 1-42.

**147. For a single-phase circuit, the following equations show the relations between power, current, voltage, and power factor:**

$$I = \frac{W}{E \times \text{pf}} \quad E = \frac{W}{I \times \text{pf}} \quad W = E \times I \times \text{pf} \quad \text{pf} = \frac{W}{E \times I} \tag{51}$$

where $I$ = current in amperes, $W$ = power in watts, $E$ = voltage between lines, and pf = power factor.

**examples**   Figures 1-42 and 1-44 show applications of the above equations. The product of volts and amperes $(EI)$ is called voltamperes; see above paragraph.

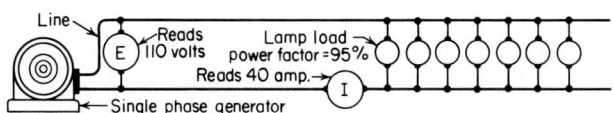

**Fig. 1-44**   A power-factor problem.

**example**   In the circuit of Fig. 1-44, what is the actual load in watts? In kilowatts? Current = 40 A, voltage at load = 110, and power factor of load = 95 percent.
**solution**   Substitute in the formula

$$W = E \times I \times \text{pf} = 110 \times 40 \times 0.95 = 4180 \text{ W}$$

$$\text{kW} = \frac{W}{1000} = \frac{4180}{1000} = 4.18 \text{ kW}$$

**148. A two-phase current** consists of two currents that differ in phase by 90 degrees (see curves of Fig. 1-45). If two sets of coils are arranged on an armature (Fig. 1-45) so that their "starts" $S_1$ and $S_2$ are 90 electrical degrees apart, the emf in one set will attain its maximum value 90 degrees later than that in the other. The emf will force two-phase currents through an external circuit. Instead of being on the same armature, each of the sets of coils may be on different armatures which are so mechanically connected together

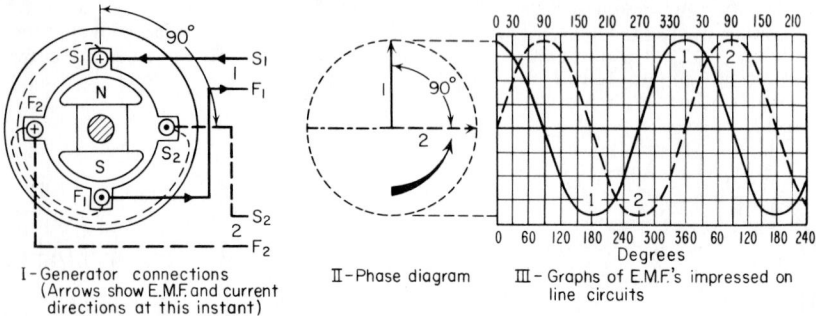

I – Generator connections
(Arrows show E.M.F. and current
directions at this instant)

II – Phase diagram

III – Graphs of E.M.F.'s impressed on
line circuits

**Fig. 1-45**   Two-pole two-phase ac generator illustrating elementary principles.

as to preserve the 90-degree phase relation (see Div. 7, "Generators and Motors," for information on practical machines).

**149. Coil connections for two-phase windings.**   Figure 1-46 shows three methods of connecting two-phase generator coils.

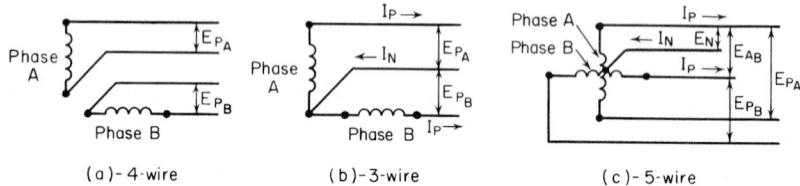

(a)- 4-wire

(b)- 3-wire

(c)- 5-wire

**Fig. 1-46**   Connections for two-phase–generator windings.

**150. Relations of voltage, current, and power that apply to balanced two-phase circuits**

$$I_p = \frac{W}{2E_p \times \text{pf}} \qquad E = \frac{W}{2I_p \times \text{pf}} \qquad W = 2E_p \times I_p \times \text{pf} \qquad \text{pf} = \frac{W}{2E_p \times I_p} \qquad (52)$$

$$\text{Kilovolt-amperes} = \frac{2E_p \times I_p}{1000} \quad \text{or} \quad \text{kVA} = \frac{\text{kW}}{\text{pf}} \tag{53}$$

$$\text{kW} = \text{kVA} \times \text{pf} \tag{54}$$

$$\text{pf} = \frac{\text{kW}}{\text{kVA}} \tag{55}$$

where $I_p$ = phase current in amperes as designated in Fig. 1-46, $E_p$ = phase voltage as designated in Fig. 1-46, $W$ = power in watts, and pf = power factor.

**151. Application of the two-phase system.**   Many years ago certain engineers advocated two-phase generators and distributing systems in preference to three-phase. They then believed that unbalanced load on the phases would have less adverse effect on the performance of the two-phase equipment. Experience has proved that the three-phase system is preferable to and more economical than the two-phase for both transmission and distribution. Two-phase equipment now is seldom purchased except for additions to existing two-phase installations. See Sec. 23 of Div. 3 for relative weights of copper for different systems.

**152. A three-phase current** consists of three alternating currents that differ in phase by 120 degrees, as indicated in Fig. 1-47. If three coils are arranged with their starts, $S_1$, $S_2$, and $S_3$, 120 degrees apart on an armature (Fig. 1-47) and are connected each to an external circuit, a single-phase alternating emf will be impressed by each coil on its own external circuit when the field is rotated at uniform speed. The emf's will differ in phase by 120 degrees and therefore will constitute a three-phase system. The currents in the circuits will be 120 degrees out of phase with each other. Three single-phase generators, if mechanically coupled together so as to maintain a 120-degree phase relation, would produce a three-phase system. Practical three-phase generators usually have more than two poles and consequently have more coils than are indicated in Fig. 1-47. Most alternating-current generators have revolving fields and stationary armatures.

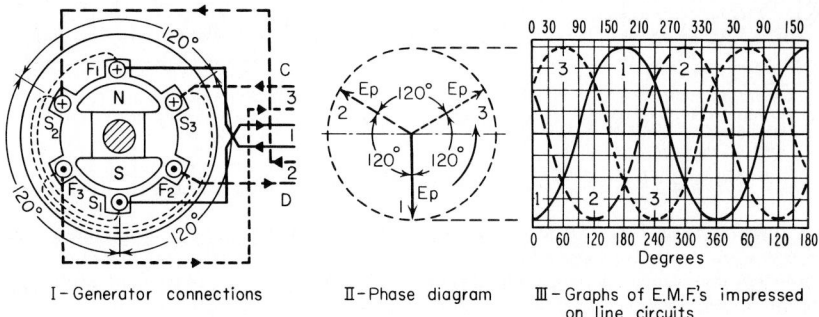

I – Generator connections      II – Phase diagram      III – Graphs of E.M.F.'s impressed on line circuits

**Fig. 1-47**   Two-pole three-phase alternator ensures 120° phase difference between emf's in line wires.

**153. Phase connections.**   Figure 1-48 shows four methods of connecting the phases of a three-phase generator or other apparatus and the external circuits for each. Method I, although it would work, is seldom used for economic reasons given below. It shows the elementary three-phase circuit and illustrates the principle. Each of the three phases would carry a current differing in phase by 120 degrees from the currents in the other two. One common return $N$, as shown at II, can be substituted for the three return wires

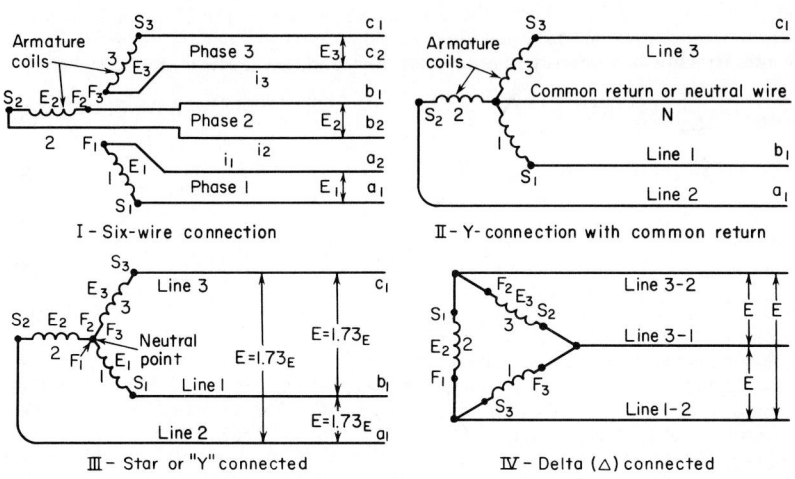

**Fig. 1-48**   Connections for three-phase–generator windings.

of I. Now with a balanced load (one loading each of the phases equally) this return wire would carry no current; hence it may be omitted (star or Y connection of III). In IV is shown the delta connection.

**154. The voltage and current relations in a star- or Y-connected three-phase circuit** are indicated in Fig. 1-49. The armature coils of the generator shown in Fig. 1-49 are 120 electrical

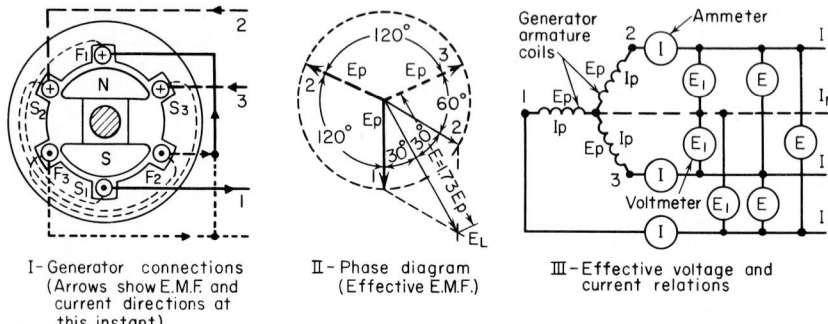

I—Generator connections
(Arrows show E.M.F. and
current directions at
this instant)

II—Phase diagram
(Effective E.M.F.)

III—Effective voltage and
current relations

**Fig. 1-49**   Star- or Y-connected three-phase generator and diagrams.

degrees apart (Fig. 1-47). These emf's will combine as shown in the phase diagram (Fig. 1-49) to produce the voltage $E$ between line wires. The resultant voltage developed by any two of the coils is then equal to $\sqrt{3}$, or 1.73 times the voltage developed in one coil. The following formulas show the relation of voltage and current in the circuit. (All are effective values, and balance is assumed. See Fig. 1-49.)

$$I = I_p \tag{56}$$

$$E = E_p \times \sqrt{3} = E_p \times 1.73 \tag{57}$$

$$E_p = \frac{E}{\sqrt{3}} = \frac{E}{1.73} = E \times 0.577 \qquad \text{or approximately } E_p = 0.58E \tag{58}$$

$$I_N = 0 = \sqrt{(I_p1)^2 + (I_p2)^2 + (I_p3)^2 - (I_p1 \cdot I_p2) - (I_p2 \cdot I_p3) - (I_p3 \cdot I_p1)} \tag{59}$$

where $I$ = amperes per phase in the line, $I_p$ = amperes per phase in each coil, $E$ = volts between phase wires on the line, $E_p$ = volts across each group of armature coils connected across each phase, and $E_1$ = volts between phase wires and neutral. The coils in Fig. 1-49, III, may represent the phase windings of a three-phase generator or transformer, or each coil may represent a transformer or other device, three of which are Y-connected.

**example**   What will be the voltage across line wires of the three-phase circuit for the 120-V Y-connected lamps in Fig. 1-50?

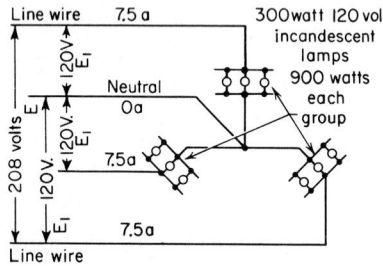

**Fig. 1-50**   Incandescent-lamp groups connected in star or Y.

**solution**   Substitute in the formula $E = E_p\sqrt{3}$ = 120 × 1.73 = 208 V.

**155. Relations for a delta(Δ)-connected three-phase circuit** are shown in Fig. 1-51. When armature coils of a generator (see Diagram I) are connected as indicated, the voltages generated in them are 120 degrees apart. It would appear that the current might flow around through the coils and not into the external circuit, but it is evident from Phase Diagram II that the sum of the effective voltages 1 and 3 generated by two of the coils is equal and opposite to that of the third. Hence, instead of tending to force current around internally, the voltages tend to force current out into the line. The following formulas indicate the relations of the voltages and currents. (All are effective values, and circuit is assumed to be balanced. See Fig. 1-51.)

$$I_L = I_p \times \sqrt{3} = I_p \times 1.73 \tag{60}$$

$$I_p = \frac{I_L}{\sqrt{3}} = I_L \times 0.577 \qquad \text{or approximately } I_p = I_L \times 0.58 \tag{61}$$

$$E = E_p \tag{62}$$

where the symbols have the same meanings as in Fig. 1-51, III.

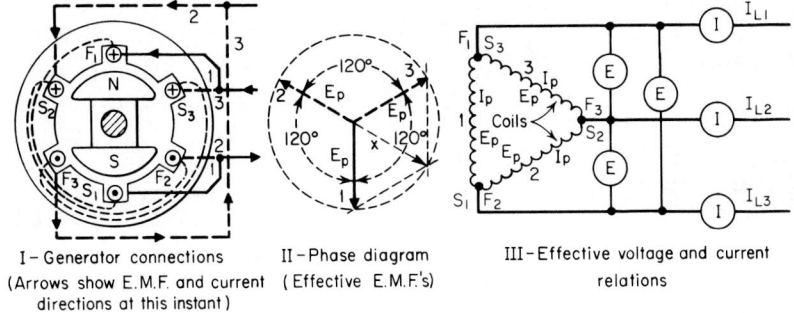

I – Generator connections
(Arrows show E.M.F. and current
directions at this instant)

II – Phase diagram
(Effective E.M.F.'s)

III – Effective voltage and current
relations

**Fig. 1-51** Delta (Δ)-connected three-phase generator and circuit.

NOTE   Each coil (Fig. 1-51) may represent the phase windings of a three-phase transformer or generator, or each coil may represent a transformer or other device, three of which are Δ-connected.

**example**   If each of the coils of the generator in Fig. 1-51 can carry 100 A, what value of current may be drawn from the line wires leading from the machine?
**solution**   Substitute in the formula

$$I_L = I_p \times \sqrt{3} = 100 \times 1.73 = 173 \text{ A}$$

**156. Relations of voltage, current, and power that apply to any balanced three-wire three-phase circuit either delta- or Y-connected.**   Refer to Fig. 1-52 for a key to the letters that appear in the following formulas. For a load with a power factor of 1,

$$I = \frac{W}{E \times \sqrt{3}} = \frac{0.577 \times W}{E} \qquad \text{or approximately} = \frac{0.58 \times W}{E} \tag{63}$$

$$E = \frac{W}{I \times \sqrt{3}} = \frac{0.577 \times W}{I} \qquad \text{or approximately} = \frac{0.58 \times W}{I} \tag{64}$$

$$W = E \times I \times \sqrt{3} = 1.73 \times E \times I \tag{65}$$

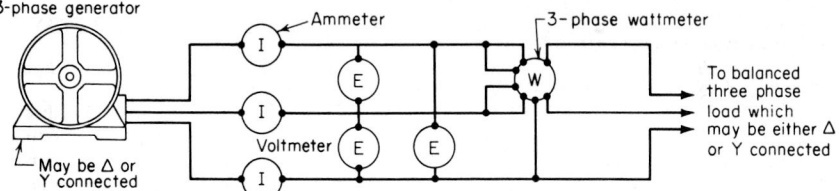

3-phase generator

May be Δ or
Y connected

Ammeter

3-phase wattmeter

Voltmeter

To balanced
three phase
load which
may be either Δ
or Y connected

**Fig. 1-52**   Relations for any Δ- or Y-connected three-phase circuit.

For any load

$$\text{pf} = \frac{W}{1.73 \times I \times E} = \frac{0.577 \times W}{I \times E} \qquad \text{or approximately} = \frac{0.58 \times W}{I \times E} \tag{66}$$

$$E = \frac{W}{\text{pf} \times 1.73 \times I} = \frac{0.577 \times W}{\text{pf} \times I} \qquad \text{or approximately} = \frac{0.58 \times W}{\text{pf} \times I} \tag{67}$$

$$I = \frac{W}{\text{pf} \times 1.73 \times E} = \frac{0.577 \times W}{\text{pf} \times E} \qquad \text{or approximately} = \frac{0.58 \times W}{\text{pf} \times E} \tag{68}$$

$$W = 1.73 \times E \times I \times \text{pf} \tag{69}$$

$$\text{VA} = 1.73 \times E \times I \tag{70}$$

where $I$ = line current, in each of the three wires, in amperes; $W$ = the power transmitted by all three wires in watts; $E$ = voltage across lines; and pf = the power factor of the circuit.

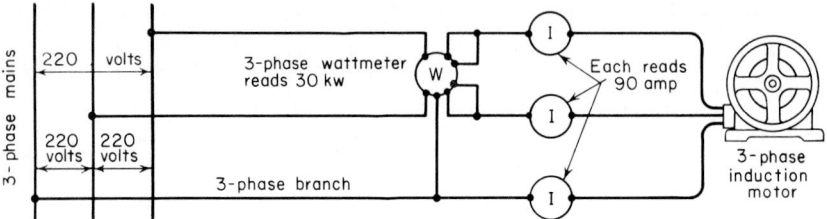

**Fig. 1-53**   Motor on a three-phase circuit.

**example**   What is the power factor in the 220-V circuit to the motor in Fig. 1-53? The three ammeters each indicate 90 A, and the three-phase wattmeter indicates 30 kW (30,000 W).
   **solution**   Substitute in the formula

$$\text{pf} = \frac{0.577 \times W}{I \times E} = 0.577 \times \frac{30,000}{90 \times 220} = \frac{17,310}{19,800} = 0.88 = 88 \text{ percent power factor}$$

**example**   The power factor on the feeder of Fig. 1-54 is known to be 70 percent. The current in
each line is 80 A, and the voltage across each phase is 220. What actual power is being delivered to the
panel?

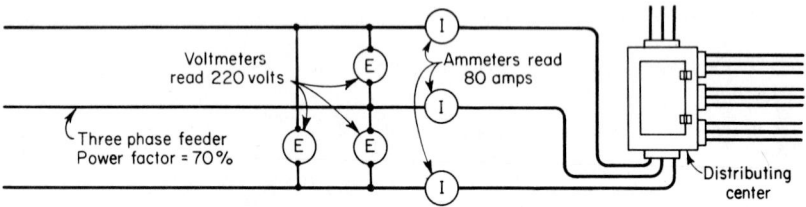

**Fig. 1-54**   Load on a three-phase circuit.

**solution**   Substitute in the formula

$$W = 1.73 \times E \times I \times \text{pf} = 1.73 \times 220 \times 80 \times 0.70 = 21,313.6 \text{ W}$$

$$\text{kW} = \frac{W}{1000} = \frac{21,313.6}{1000} = 21.3 \text{ kW}$$

**examples**   Figure 1-55 shows numerical examples of voltage and current relations in a three-phase
circuit. Note that when a group of three devices or coils is connected in delta, as in the motor, each
device or coil has line voltage impressed on it and must be designed for that voltage. The current in
the line will be 1.73 times the current through the coil. When Y-connected, as on the low-voltage side
of the transformers, each of the three coils need only be designed for 1/1.73, or 0.577 times the line
voltage, and the line current will be the same as the current through each coil. Single-phase loads may
be supplied either from line to neutral as with the 120-V lamps or from two lines as with the 208-V
heater.

**157. The power loss in any circuit traversed by an alternating or a direct current is**

$$W = I^2 \times R \qquad \text{or} \qquad I = \sqrt{\frac{W}{R}} \qquad \text{or} \qquad R = \frac{W}{I^2} \qquad (71)$$

where $W$ = the power lost in heat in the circuit in watts, $I$ = effective current in amperes
in the conductor, and $R$ = resistance of the conductor in ohms. This rule is perfectly
general and applies to all dc circuits and all ac circuits of ordinary voltages and frequencies. The watts power loss $W$ reappears as heat power and heats the conductors. Watts
loss is commonly called $I^2R$ loss.

   **example**   What is the power loss in the incandescent lamp in Fig. 1-56?
   **solution**   Substitute in the formula

$$W = I^2 \times R = (2.2 \times 2.2)98 = 4.84 \times 98 = 474 \text{ W}$$

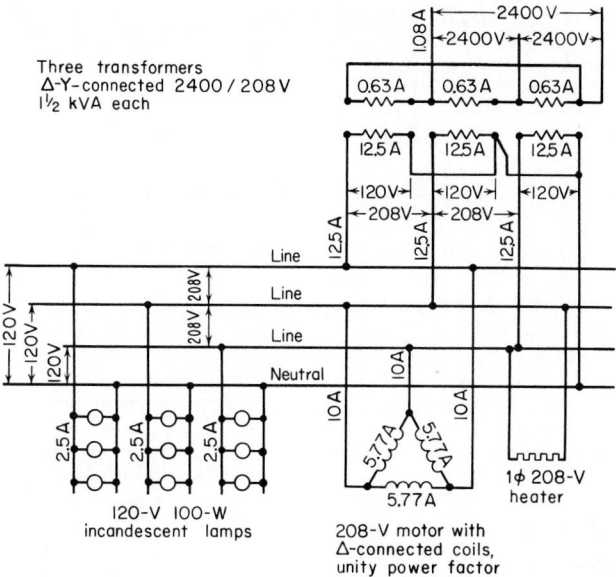

**Fig. 1-55**   Examples of current and voltage relations in three-phase circuits.

**example**   What is the power loss in the inductive winding of Fig. 1-57 with an alternating current of 3 A?

**solution**   Substitute in the formula

$$W = I^2 = R = (3 \times 3)7 = 9 \times 7 = 63 \text{ W}$$

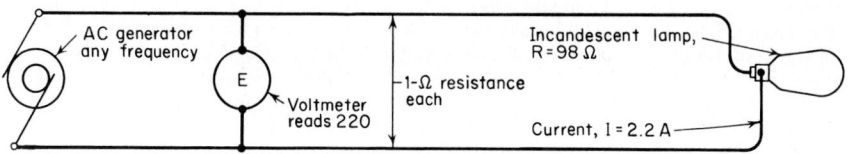

**Fig. 1-56**   Resistance in an ac circuit.

NOTE   In the circuit of Fig. 1-57 the resistance of the coil to direct current is only 5 Ω. The increase in resistance to alternating current over its dc value is due to the power losses produced in the iron core (refer to Sec. 122).

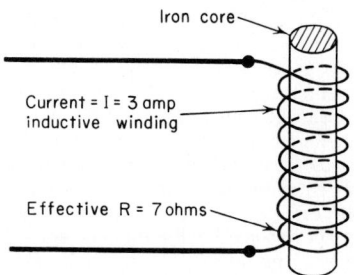

**Fig. 1-57**   Effective resistance in an inductive ac circuit.

**158.    Approximate Sparking Distances, in Air, for Various Mean Effective Voltages between Sharp Needle Points: Sine-Waveform Voltage**

(Locke Insulators, Inc.)

| Voltage | Distance | | Voltage | Distance | |
|---|---|---|---|---|---|
| | In. | Cm | | In. | Cm |
| 1,000 | 0.06 | 0.152 | 20,000 | 1.00 | 2.54 |
| 2,000 | 0.13 | 0.33 | 25,000 | 1.30 | 3.3 |
| 3,000 | 0.16 | 0.41 | 30,000 | 1.63 | 4.1 |
| 4,000 | 0.22 | 0.56 | 35,000 | 2.00 | 5.1 |
| 5,000 | 0.23 | 0.57 | 40,000 | 2.45 | 6.2 |
| 10,000 | 0.47 | 1.19 | 50,000 | 3.55 | 9.0 |
| 15,000 | 0.73 | 1.84 | 100,000 | 9.60 | 24.4 |

NOTE    Above 100,000 V, the gap between needle points is approximately 1 in/10,000 V. Using infinitely sharp needle points up to at least 10,000 V, a graph of the voltage and the corresponding sparking distance would probably result in a straight line passing through the origin. (H. W. Fisher, *Trans. Int. Elect. Cong.*, vol. II, 1904, p. 294.)

**159. Dielectric strength.**    When there is an emf between two conductors that are separated by an insulating material, electrical forces are exerted upon the electrons in the atoms of the insulating material. These forces are trying to pull the electrons out of the atoms. If they succeed in pulling the electrons out, the material ceases to be an insulator and becomes a conductor. When this occurs, it would be said that the insulation had broken down. When the voltage between the conductors is small, the forces exerted in the atoms are not great enough to pull the electrons out of the atoms. As the voltage between the conductors is increased, the forces on the electrons are increased. If the voltage continues to be increased, the forces will eventually become great enough to pull the electrons out of the atoms and the insulation will break down. The voltage at which the material ceases to be an insulator is called the breakdown voltage. Any insulating material can be broken down and cease to be an insulator if the voltage impressed across it is raised high enough. The breakdown voltage of any material depends upon the thickness, condition of the surface, and homogeneity of the material. The breakdown voltage does not increase directly with the thickness of the material; a piece of material 2 in thick will break down at a voltage less than twice the breakdown voltage of a piece 1 in thick. Any irregularities or sharp points on the surface will lower the breakdown voltage.

It is these considerations that govern the thickness of insulation required on conductors, the spacing of bare conductors, distance between live parts and ground, etc.

## MEASURING, TESTING, AND INSTRUMENTS

**160. The magneto test set** is one of the most valuable testing instruments for practical people because of its simplicity and the fact that it is always ready for service. Figure 1-58 shows the circuit and Fig. 1-59 a perspective view of a testing magneto. The apparatus consists of a small hand-operated ac generator in series with a polarized electric bell. Alternating current will ring bells of this type. If the external circuit connected to the terminals of the magneto is closed and the crank of the generator is turned, current will flow and the bell will ring.

The resistance through which magnetos will ring is determined by their design. An ordinary magneto will ring through possibly 20,000 to 40,000 $\Omega$. Electrostatic-capacity effects must be considered when testing with a magneto. When long circuits, such as telephone lines or circuits that are carried in cable for a considerable distance, are tested, the bell of the magneto may ring, owing to capacity, apparently indicating a short circuit, whereas the circuit may be perfectly clear or open. Circuits associated with iron, such as field coils of generators, may have considerable inductance. With highly inductive circuits under test, the magneto may "ring open"; i.e., the bell may not ring at all even if the inductive circuit connected to it is actually closed. In ordinary interior-wiring work the

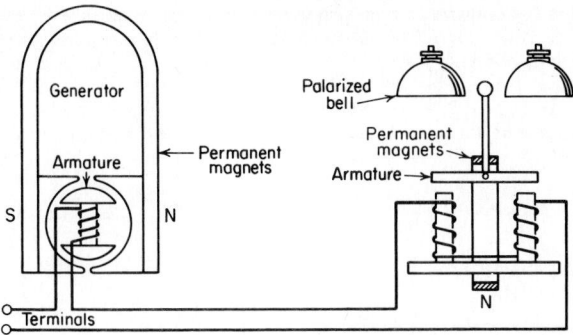

**Fig. 1-58**   Circuits of testing magneto.

effects of capacity and inductance are usually negligible, and the true condition of the circuit will be indicated by the performance of the magneto bell.

**161. A telephone receiver in combination with one or two dry cells constitutes excellent equipment for certain tests.**   A head-telephone receiver (Fig. 1-60) is usually preferable to a receiver of the watchcase type, because it is held on the head by a metal strap, allowing the unrestricted use of both hands. Metal testing clips are soldered to flexible testing cords. The telephone receiver is extremely sensitive and will give a weak click even when the current to it passes through an exceedingly high resistance. In use, one clip is gripped on one conductor of the circuit to be tested, and the other clip is tapped against the other conductor. Prolonged connection should be avoided because it will run down the battery. A vigorous click of the receiver indicates a closed circuit, while a weak click or none at all indicates an open circuit. After practice it is possible to determine approximately the resistance of the circuit under test by the intensity of the receiver click. When the battery and receiver test set are connected to a circuit having some electrostatic capacity, the receiver will give a vigorous click when the clips are first touched to the circuit terminals even if the circuit is open. With successive touchings the click will diminish in intensity if the circuit is open but will not diminish appreciably if the circuit is closed.

**162. The advantages of the telephone receiver over the magneto** for work of certain classes are: (1) The receiver-and-battery outfit costs little. (2) The outfit can be made so compact that it can be carried in a pocket. (3) In making insulation tests with a magneto the circuit may "ring clear"; i.e., the bell will not ring, apparently indicating high insulation resistance,

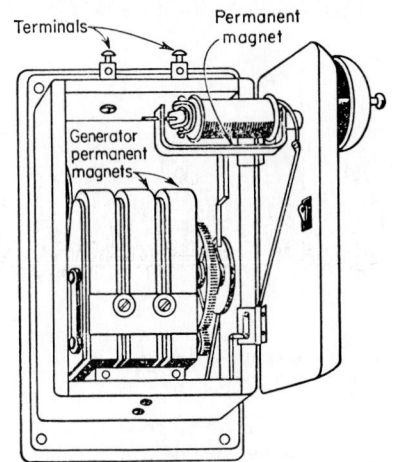

**Fig. 1-59**   Assembly of testing magneto.

whereas the circuit may not be clear, but instead the magneto may be out of order or its local circuit may be open. The indication is negative. With the telephone receiver a slight click is produced even when testing through the highest resistances. The absence of a click usually signifies an open in the testing apparatus itself. Thus the telephone-receiver indication is positive.

**163. A telegraph sounder is sometimes used for testing.**   It is connected in the same way as the telephone receiver of Fig. 1-60 and is adaptable for rough work. When the circuit under test is closed and the flexible-cord clips are touched to the circuit conductors, the sounder clicks. When the circuit is open, there is no click. One feature of the sounder method is that the click is audible at a considerable distance from the instrument.

**164. An electric-bell outfit for testing** is shown in Fig. 1-61. When the free ends for testing are touched to a closed circuit of not too high resistance, the bell rings. If the circuit is open, the bell will not ring. Flexible cord can be used for the testing conductors of the outfit, and testing clips can be provided as in Fig. 1-60.

**165. A test lamp** (Fig. 1-62), consisting merely of a weatherproof rubber-insulated socket into which is screwed an incandescent lamp of the highest voltage rating of the circuits

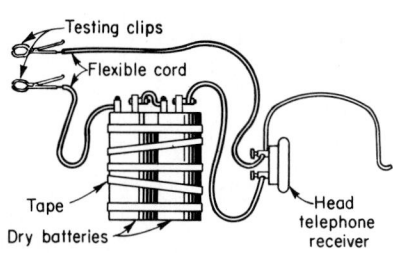

**Fig. 1-60**   Head-telephone and dry-battery testing set.

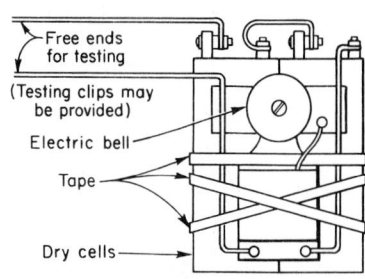

**Fig. 1-61**   Electric-bell–testing outfit.

involved, is very convenient for rough tests on interior-lighting and motor-wiring systems. Porcelain sockets are undesirable because they are so readily broken. Brass sockets should not be used because they may fall across conductors and thereby cause short circuits. Testing clips may be soldered to the ends of the leads which are molded in the socket. Some uses of the testing lamp are given below, and the lamp is very convenient in testing for defective fuses.

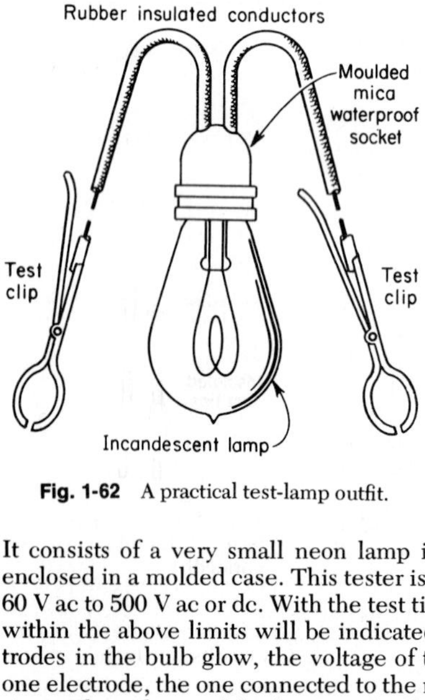

**166. A very convenient and inexpensive test outfit** consists of a pointer which moves along a scale inside an insulated case (Fig. 1-63). It indicates whether the circuit is alternating- or direct-current by vibrating on alternating current, the frequency of the vibrations showing whether the current is 25 or 60 Hz. It also indicates the range of voltage of the circuit up to 550 ac or 600 dc. The sharp points on the ends of the lead wires can be used to pierce insulation for checking insulated leads without destroying the insulation.

**167. Neon-glow lamp testers** provide a very convenient and compact device for determining if a circuit is live, for determining polarity of dc circuits, and for determining if a circuit is alternating- or direct-current. A neon tester for low-voltage work is illustrated in Fig. 1-64.

**Fig. 1-62**   A practical test-lamp outfit.

It consists of a very small neon lamp in series with a 200,000-$\Omega$ protective resistance enclosed in a molded case. This tester is satisfactory for use on circuits of from 90 V dc or 60 V ac to 500 V ac or dc. With the test tips connected to a circuit, the presence of voltage within the above limits will be indicated by the glowing of the neon lamp. If both electrodes in the bulb glow, the voltage of the circuit is alternating. On direct current, only one electrode, the one connected to the negative side of the circuit, will glow. With experience the voltage of the line can be determined approximately by the intensity of the glow. In testing a circuit, it is best first to touch only one side of the line with one test tip, keeping the other test tip free. A glow with this connection will indicate the presence of high voltage. The tester shown is well constructed for safety. The molded case has a voltage breakdown value of 25,000 V and the leads are insulated for 5000 V and provided with insulated test prods.

Glow-tube testers are available for testing for the presence of voltage on high-voltage circuits. In testing high-voltage circuits by this means, no direct connection is made to the circuit. The end of the tester which contains the glow tube is simply held in proximity to

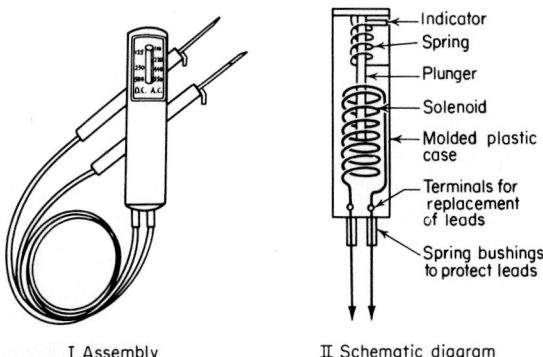

I Assembly          II Schematic diagram

**Fig. 1-63**    Voltage tester, 110 to 600 V. [Square D Co.]

the circuit. The changing electrostatic field will ionize the gas inside the tube and cause it to glow. These testers, or staticscopes, as they are called, will indicate the presence of potential in ac circuits, pulsating dc circuits, x-ray equipment circuits, static from belting, high-frequency circuits, condenser discharges, and automobile ignition. On ac circuits they will give a positive indication on 2000 V and up. Materials which will act as shields to the electrostatic field, such as lead on underground conductors, metal cabinets, and grounded framework, must not be between the tester and the conductors of the circuit being tested. A pocket type of staticscope is shown in Fig. 1-65*b*. Staticscopes are also made in types especially adapted for the testing of overhead lines and station equipment (Fig. 1-65*a*).

**168. A neon lamp tester and fuse puller** is shown in Fig. 1-66. It is made of transparent plastic with the fuse-puller jaws at one end and two hinged prongs at the other end. The test prongs are of a proper size for testing plugs and baseboard receptacles. By using the tester with a screw plug, lamp sockets can be tested without danger of short circuits or of burning out a regular bulb. The hinged prongs can be folded back against the inside of the legs for testing with the regular points. Portable test leads are available. They are sturdily constructed with a plug on one end for engaging the hinged prongs of the tester and with a rubber-insulated clamp and test prod on the other ends of the leads. The leads are 2 ft long and provide a 4-ft spread for test purposes.

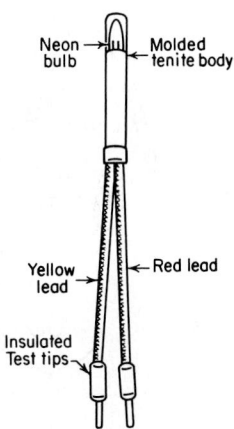

**Fig. 1-64**    Neon-glow lamp tester. [Littelfuse, Incorporated]

**169. Instruments for general-circuit and other industrial testing purposes** are shown in Fig. 1-67. They are applicable for measurements on either dc or ac circuits. Only one instrument is required for measurements of voltage, current, and resistance.

**170. Rules for use of ammeter and voltmeter** (William H. Timbie, *Elements of Electricity*). Place the ammeter in series, always using a short-circuiting switch, where possible, as shown in Fig. 1-68, to prevent injury to the instrument. Place the voltmeter in shunt (Fig. 1-68). Put the + side of the instrument on the + side of the line. Figure 1-68 shows the correct use of an ammeter and a voltmeter to measure the current and the voltage supplied to the motor. The short-circuiting switch $S$ must be opened before the ammeter is read. All the current that enters the motor must then flow through the ammeter and be indicated. The ammeter is of very low resistance (about 0.001 or 0.002 $\Omega$) and does not appreciably cut down the flow of current. The voltmeter is of very high resistance (about 15,000

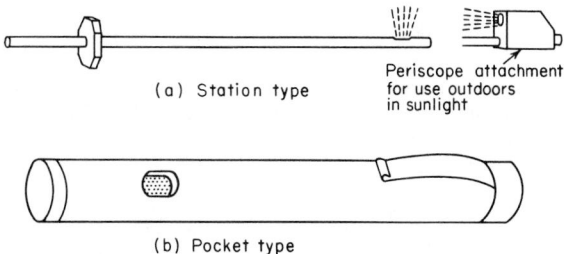

(a) Station type    Periscope attachment
for use outdoors
in sunlight

(b) Pocket type

**Fig. 1-65**   Statiscopes. [Minerallac Electric Co.]

**Fig. 1-66**   Neon-lamp fuse puller and tester. [Martindale Electric Co.]

$\Omega$) and does not allow any appreciable current to flow through it. Yet enough goes through the voltmeter to cause it to indicate the voltage across the terminal $AB$ of the motor. Let us suppose the voltage across the motor to be 110; what would happen if an ammeter of 0.002-$\Omega$ resistance were by mistake placed across $AB$? (Remember that Ohm's law is always in operation.)

**Fig. 1-67A** Hand-held digital multimeter. [Simpson Electric Co.]

**Fig. 1-67B** Volt-ohm-milliammeter. [Simpson Electric Co.]

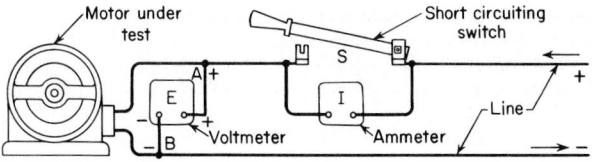

**Fig. 1-68**   Ammeter and voltmeter connections.

**171. All (except electrostatic) ammeters and voltmeters consume power when in use and intro-
duce some error** (William H. Timbie, *Elements of Electricity*). For minimum error (see
Fig. 1-69, I) when measuring a low current and a high voltage, the voltmeter should be
placed around both the ammeter and the apparatus under test.

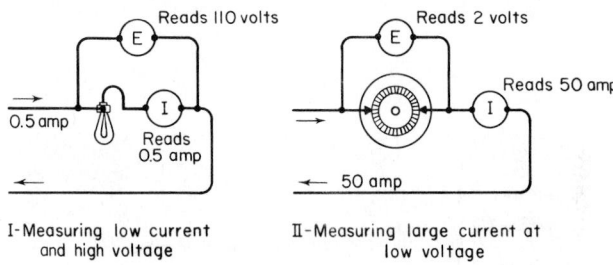

**Fig. 1-69**   Correct methods of connecting instruments.

When measuring the power consumed by a piece of apparatus through which a large
current at low voltage is flowing, the voltmeter should be placed immediately across the
piece of apparatus under test and should not include the ammeter (Fig. 1-69, II).

**172. A method of measuring current with a voltmeter** is shown in Fig. 1-70. If a resistor of
known resistance is connected in series in a circuit and the voltage across the resistor is
measured with a voltmeter, the current can be determined by Ohm's law.

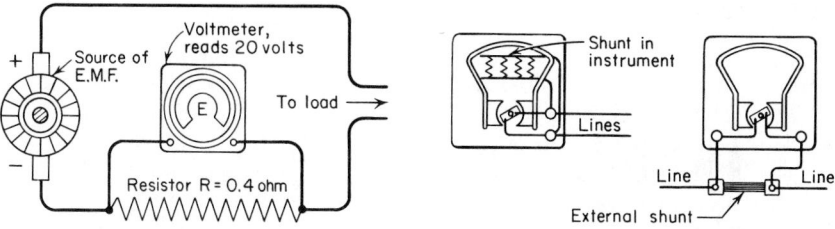

**Fig. 1-70**   Current measurement with voltmeter.        **Fig. 1-71**   Millivoltmeters and shunts.

**example (Fig. 1-70)**   If the drop around an 0.4-Ω resistance in series in a circuit is 20 V, what is the
current in the circuit?
  **solution**   Substitute in the Ohm's-law formula

$$I = \frac{E}{R} = \frac{20}{0.4} = 50 \text{ A}$$

**173. A millivoltmeter is generally used** for making measurements like that of Sec. 172. A
millivoltmeter reads in thousandths of volts, so that a resistor of small resistance can be
used. Ammeters, particularly those for large currents, are often millivoltmeters calibrated
in amperes which are connected around a resistor, in series with the circuit (Fig. 1-71).
The resistor sometimes is in the instrument case and sometimes is inserted in the busbars

of a switchboard (see Fig. 1-71). Resistors of this type are called *shunts* and when furnished by instrument makers are carefully calibrated.

**174. Tong-Test ammeters** provide very useful and convenient test instruments. In many instances when it is desired to measure the current flowing in a cable or other conductor, it is inconvenient, even if it is permissible to break the circuit, to insert an ammeter. The Tong-Test ammeter instantly measures alternating or direct current without opening or

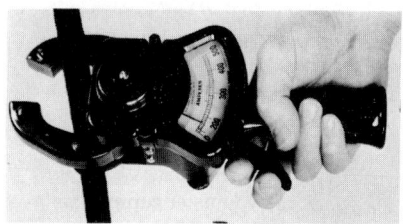 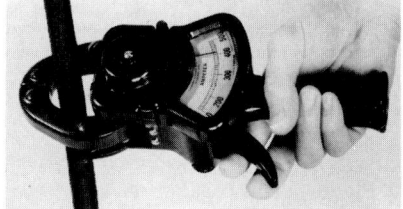

**Fig. 1-72**   Application of Tong-Test ammeter. [Martindale Electric Co.]

interrupting the circuit. Merely encircle the conductor with the tongs as illustrated in Fig. 1-72. As soon as the jaws close, a clear, accurate reading is instantly registered on the scale.

Accurate readings can be taken on ac frequencies from 25 to 400 Hz as well as on direct current.

Readings can be made on either bare or insulated conductors. The jaws of the tongs are insulated, and the Bakelite handle and shield protect the operator from shock. The Tong-Test meter is operated entirely by the magnetic field set up by the current, and it cannot be burned out, as it has no electrical wiring. The tongs are opened by a moderate pressure of one finger on the trigger and are self-closing, requiring only one hand.

When dc measurements are made, the jaws should be opened and closed immediately before the reading is taken to reduce the error caused by hysteresis if the current is rising or falling. If the current is steady, the error can be minimized by reversing the tongs and taking the average of the two readings.

**175. Resistances can be measured** with a voltmeter as indicated in Fig. 1-73. This method is satisfactory for the measurement of resistances whose values range from a few to a few hundred ohms. A resistor of known resistance, a source of direct current, and one voltmeter are required. The same constant current flows through both the known and the unknown resistance. The voltmeter reading $E$ is taken, and then the reading $E_x$. The voltage drops will be proportional to the resistances or

$$\frac{R}{E} = \frac{R_x}{E_x} \quad \text{or} \quad R_x = \frac{E_x \times R}{E} \tag{72}$$

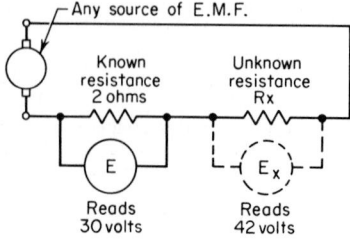

**Fig. 1-73**   Resistance measurement.

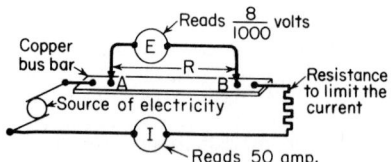

**Fig. 1-74**   Measurement of very low resistance.

**example**   Substituting the values from Fig. 1-72 in the formula,

$$R_x = \frac{E_x \times R}{E} = \frac{42 \times 2}{30} = \frac{84}{30} = 2.8\,\Omega$$

**176. Very small resistances can be measured,** as indicated in Fig. 1-74, with an ammeter and a millivoltmeter. This method is generally satisfactory for resistances having a value of less than 1 $\Omega$ and is convenient for measuring the resistance of busbars, joints between conductors, switch contacts, brush-contact resistance, and other low resistances. As large a current as is feasible should be used. This is another application of Ohm's law.

**example**   What is the resistance of the portion of the busbar between $A$ and $B$, Fig. 1-74?
**solution**   Substitute in Ohm's-law formula

$$R = \frac{E}{I} = \frac{0.008}{50} = 0.00016\,\Omega$$

In the application of Ohm's law as given here, the current through the voltmeter is neglected. Since the resistance of the voltmeter is generally large compared with the resistance to be measured, the error caused by neglecting the voltmeter current is generally negligible.

**177. Insulation resistance is frequently measured** as suggested in Fig. 1-75. A voltmeter of known resistance, preferably of high resistance, and a source of emf (batteries or a gener-

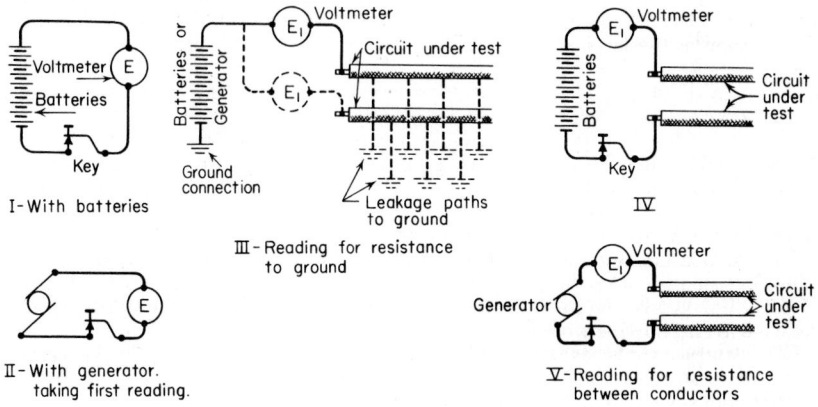

**Fig. 1-75**   Measuring insulating resistance.

ator) are required. First the voltage of the emf source is taken as shown in Diagram I or II. The apparatus is then arranged as shown in Diagram III to measure the resistance from each side of the circuit to ground. In Diagram IV or V are shown the connections for measuring the resistance between conductors. If $E$ = voltage of source, $E_1$ = reading of the voltmeter when connected in series with insulation resistance to be measured, $R_v$ = resistance in ohms of the voltmeter, and $R_x$ = insulation resistance sought, the following formula is used (see Fig. 1-75):

$$R_x = R_v\left(\frac{E}{E_1} - 1\right) \tag{73}$$

**example**   In a certain test (Fig. 1-76) in which a 110-V generator was used as a source of emf and a voltmeter having a resistance of 15,000 $\Omega$ was used to read voltages, the readings indicated in Fig. 1-76 were obtained. What was the insulation resistance to ground of each side of the circuit, and what was the insulation resistance between circuits?
**solution**   For the resistance of conductor 1 (see Fig. 1-76) substitute in the formula

$$R_x = R_v\left(\frac{E}{E_1} - 1\right) = 15,000\left(\frac{100}{5} - 1\right) = 15,000(22 - 1) = 15,000 \times 21$$
$$= 315,000\,\Omega = \text{insulation resistance of conductor 1 to ground}$$

For the resistance of conductor 2 (see Fig. 1-76, III),

$$R_x = R_v\left(\frac{E}{E_1} - 1\right) = 15,000\left(\frac{100}{4} - 1\right) = 15,000(27.5 - 1) = 15,000 \times 26.5$$
$$= 397,500 \ \Omega = \text{insulation resistance of conductor 2 to ground}$$

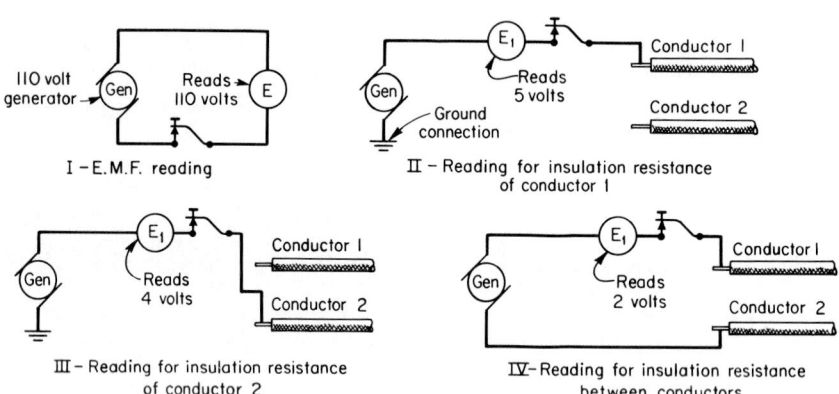

I – E.M.F. reading

II – Reading for insulation resistance of conductor 1

III – Reading for insulation resistance of conductor 2

IV – Reading for insulation resistance between conductors

**Fig. 1-76**  Example of insulation-resistance measurement.

For the insulation resistance between conductors,

$$R_x = R_v\left(\frac{E}{E_1} - 1\right) = 15,000\left(\frac{100}{2} - 1\right) = 15,000(55 - 1) = 15,000 \times 54$$
$$= 810,000 \ \Omega = \text{insulation resistance between conductors 1 and 2}$$

**178. The insulation resistance of a generator** can be determined with a voltmeter of known resistance which is successively connected and read in positions I and II (Fig. 1-77). The formula of Sec. 177 is used. The external circuit connected to the generator should be cut off while the measurements are being taken so that its insulation resistance will not affect the readings.

**179. The insulation resistance of a motor** can be measured with a voltmeter as suggested in Fig. 1-78. The formula of Sec. 177 is used. Unless the external circuit has high insulation resistance, its resistance will affect the result.

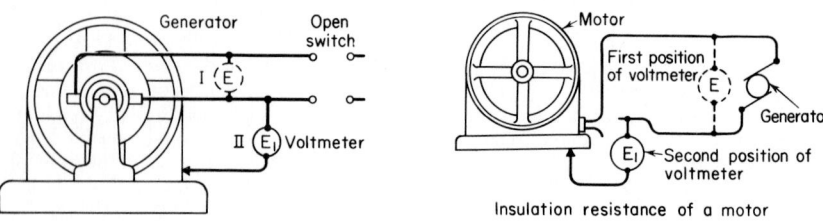

**Fig. 1-77**  Measuring insulation resistance of a generator.

**Fig. 1-78**  Measuring insulation resistance of a motor.

**180. The ohmmeter** is an instrument to measure resistance directly. The one illustrated in Fig. 1-79 has an accuracy of 98 percent. It is especially advantageous for work in the field, since its small size allows the meter to be carried in a pocket. The meter is operated by a No. 2 dry cell, which is mounted inside the meter case.

**181. A resistance tester** (Fig. 1-80) is an instrument frequently used to measure high resistance. It consists of a magneto which is turned by a crank on the side of the case. The scale is calibrated directly in ohms. The resistance to be measured is connected across two terminals. The crank is turned at a moderate speed (about 120 rpm) until the pointer reaches a steady deflection.

**182. Power,** in dc electric circuits or single-phase ac electric circuits with a power factor of 1, can be measured with a voltmeter and an ammeter. For two-wire circuits the power in watts, in accordance with Ohm's law, equals the product of volts times amperes; thus

$$W = I \times E \qquad (74)$$

where $W$ = the power in watts, $I$ = the current in amperes, and $E$ = the emf in volts.

**example** See Sec. 111 for examples of power problems. Although no instruments are shown in these, the principles are the same as if instruments were used.

**example** In Fig. 1-81, I, the power taken by the motor is, substituting in the formula,

$$W = I \times E = 40 \times 220 = 8800 \text{ W}$$

or in kilowatts = 8800/1000 = 8.8 kW.

**example** In Fig. 1-81, II, the power taken by the lamps is

$$W = I \times E = 3 \times 100 = 330 \text{ W}$$

or in kilowatts 330/1000 = 0.33 kW.

**183. Power in single-phase ac circuits** must be measured with a wattmeter unless the power factor is 1. A wattmeter can also be employed for measuring the power of a dc circuit. A wattmeter has two internal coils, namely, a voltage coil and a current coil. The voltage coil is connected across the circuit in which the power is to be

**Fig. 1-79** Ohmmeter. [Weston Instruments, a division of Sangamo Weston, Inc.]

measured, and the current coil is connected in series with the circuit. The wattmeter gives a direct indication of the power of the circuit. Two methods of connecting a wattmeter for

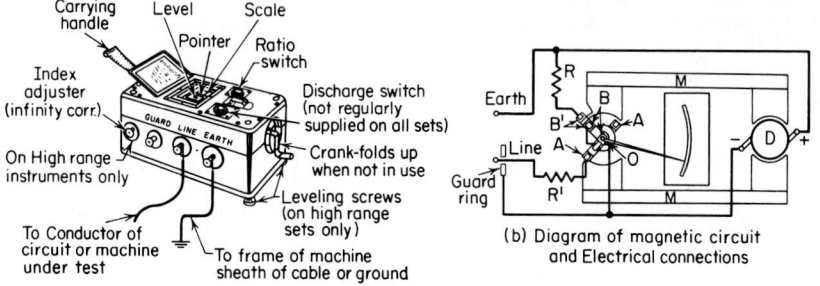

(a) Instrument showing connection of circuit to be tested

(b) Diagram of magnetic circuit and Electrical connections

**Fig. 1-80** Megger®. [Biddle Instruments]

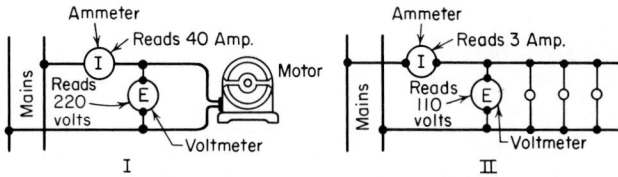

**Fig. 1-81** Power measurements.

measuring the power of a dc circuit or of a single-phase ac circuit are shown in Figs. 1-41 and 1-42. In Fig. 1-41 the voltage coil is connected between the generator and the current coil of the wattmeter, while in Fig. 1-42 the voltage coil is connected between the load and the current coil of the wattmeter. The connection of Fig. 1-41 should be used for loads of small current and that of Fig. 1-42 for loads of large current.

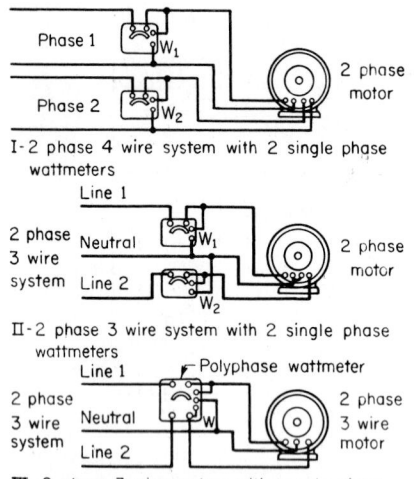

I-2 phase 4 wire system with 2 single phase wattmeters

II-2 phase 3 wire system with 2 single phase wattmeters

III-2 phase 3 wire system with a poly phase wattmeter

**Fig. 1-82**  Measurement of power in two-phase systems.

**184. Power in a two-phase system** can be measured with two single-phase wattmeters connected as shown in Fig. 1-82. Each phase is treated as a separate circuit. One wattmeter reads the power of one phase and the other the power of the second phase. The total power is the arithmetical sum of the two wattmeter readings:

$$\text{Total power} = W_1 + W_2 \qquad (75)$$

A polyphase wattmeter, connected as shown in Fig. 1-82, III, can be used for the measurement of the power of a two-phase system.

**185. Power in three-phase circuits can be measured with wattmeters** by several different methods (see Fig. 1-83). In Diagram I a polyphase wattmeter is shown. An instrument of this type automatically adds the portions of power consumed in each phase and indicates their sum. Instruments made by different manufacturers are arranged differently and must be connected accordingly. Directions accompany each instrument. Diagrams II and III show how the power can be measured, in a balanced circuit, with one wattmeter. One potential lead is connected to the line in which the wattmeter is inserted, and the other potential lead is connected successively to the other two lines. The total power in Diagram II is equal to the sum or difference of the two readings. If resistors are used as indicated in Diagram III, the power can be ascertained without shifting leads. The wattmeter reading of Diagram III multiplied

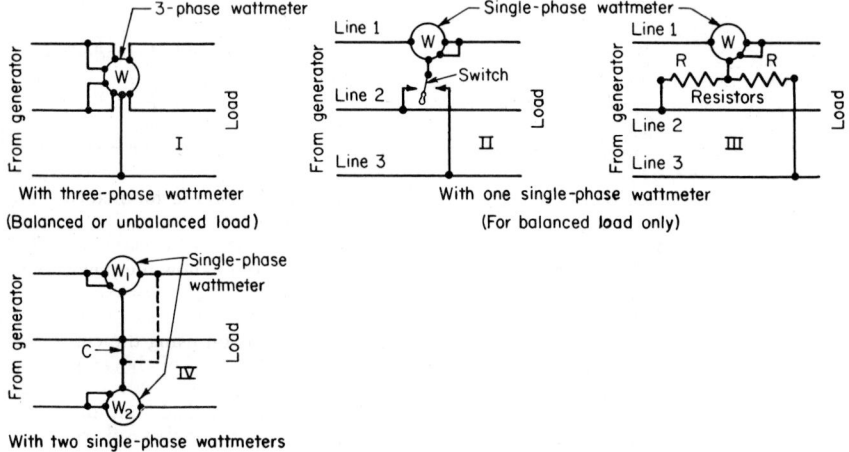

**Fig. 1-83**  Three-phase power measurements with wattmeters.

by 3 will be the true power in a balanced circuit. The resistance of each of the resistors $R$ and $R$ must equal the resistance of the potential or voltage coil of the wattmeter.

With two wattmeters (as in Fig. 1-83, IV) the total power is equal to the sum or difference of the two wattmeter readings. If the power factor is greater than 0.50, the total power is the arithmetical sum of the readings. If it is lower than 0.50, one of the readings is negative, and the power is the arithmetical difference of the two. To ascertain whether

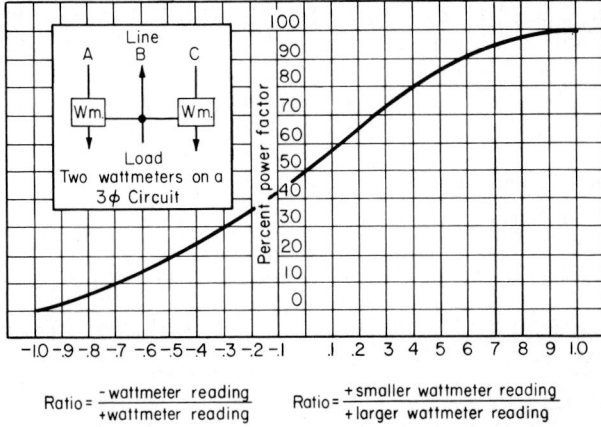

$$\text{Ratio} = \frac{-\text{wattmeter reading}}{+\text{wattmeter reading}} \qquad \text{Ratio} = \frac{+\text{smaller wattmeter reading}}{+\text{larger wattmeter reading}}$$

Directions

Case I :- Both readings positive. Divide smaller by larger reading. Find this ratio on right side of center line above. Follow up the ordinate at this point to its intersection with curve. Opposite this on center line find corresponding percent power factor (above 50 percent)

Case II :- One reading negative. Divide negative by positive reading. Find this ratio on left side of center line above. Follow up the ordinate at this point to its intesection with curve. Opposite this on center line find corresponding percent power factor (below 50 percent)

**Fig. 1-84**   Power-factor curve.

one of the wattmeters is reading negative, temporarily transfer the connection of one of the potential wires (for example $c$ in Diagram IV, as shown by the dotted line) from the middle wire to the outside wire. If its wattmeter reverses, one of the instruments, that of the lesser indication, is reading negatively. The nature of the load usually enables one to judge roughly what the power factor is. With incandescent lamps and fully loaded motors the power factor will be high, but with underloaded and lightly loaded motors it is likely to be low. See Sec. 187 for the method of determining the power factor of three-phase circuits with wattmeters.

**186. The power factor** of a circuit can be determined from readings of voltmeters, ammeters, and wattmeters by use of the formulas of Secs. 146, 147, 150, and 156. The power factor of circuits can also be determined by instruments called power-factor meters, which when properly connected in a circuit read the power factor directly.

**187. The method of determining three-phase power factor with wattmeters** was well described by C. E. Howell in *Electrical World*. It is necessary to know the power factor to connect watthour meters correctly if the wiring is concealed. An abstract follows.

Figure 1-84 shows the power-factor curve for two single-phase meters on a polyphase circuit. It also gives a diagram of connections and instructions on using the curve. The figure should be self-explanatory. Figure 1-85 gives, first, a method of checking results obtained by employing the curve in Fig. 1-84 and a diagram of the connections for obtaining data for the check. The second part of Fig. 1-85 gives a method of determining correct connections for two single-phase meters or one polyphase meter on a three-phase circuit. If this part of Fig. 1-85 is followed, errors in meter connections on three-phase circuits due to the power factor being near 50 percent should be minimal.

The above instructions may be illustrated as follows. A 100-hp three-phase 440-V induction motor was operating on 30 percent full load or 30 hp (29.8 kW) at 60 percent power factor (afterward determined) when an order came through to place a polyphase watthour meter on the installation. Immediately after the meter had been connected the following question was asked: "Should the light element add to or subtract from the heavy element; i.e., is the power factor above or below 50 percent?" As the meter leads were encased in pipe, they could not be traced; therefore the instructions in the second figure pertaining to this point were applied. The connected load of the motor having been thrown off, it was found that one element of the meter gave a negative reading. Sufficient load was then put on to bring the motor to about 80 percent of its full-load rating. Each element of the meter (taken separately) now read positively, but the element which on no load gave a negative reading on 80 percent load read lower than the heavy element. The meter had been correctly connected when installed. Later both methods given above to determine the power factor of a three-phase circuit were applied, and both gave approximately 60 percent power factor (at 30 percent load).

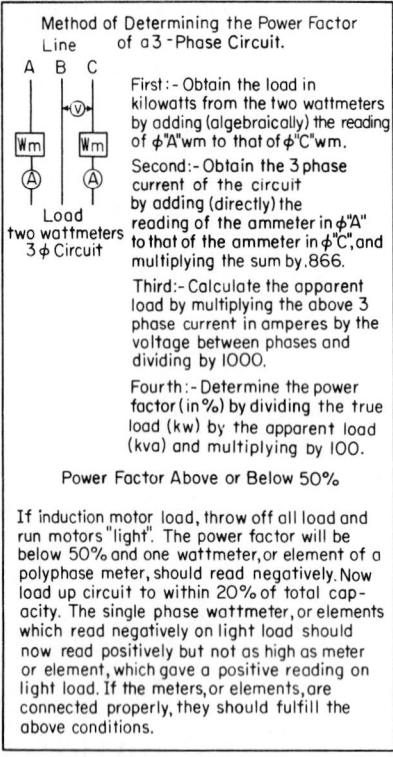

Method of Determining the Power Factor of a 3-Phase Circuit.

First:- Obtain the load in kilowatts from the two wattmeters by adding (algebraically) the reading of φ"A"wm to that of φ"C"wm.

Second:- Obtain the 3 phase current of the circuit by adding (directly) the reading of the ammeter in φ"A" to that of the ammeter in φ"C", and multiplying the sum by .866.

Third:- Calculate the apparent load by multiplying the above 3 phase current in amperes by the voltage between phases and dividing by 1000.

Fourth:- Determine the power factor (in %) by dividing the true load (kw) by the apparent load (kva) and multiplying by 100.

Power Factor Above or Below 50%

If induction motor load, throw off all load and run motors "light". The power factor will be below 50% and one wattmeter, or element of a polyphase meter, should read negatively. Now load up circuit to within 20% of total capacity. The single phase wattmeter, or elements which read negatively on light load should now read positively but not as high as meter or element, which gave a positive reading on light load. If the meters, or elements, are connected properly, they should fulfill the above conditions.

**Fig. 1-85** Chart of instructions for power-factor test.

**188. To read correctly the consumption indicated on the dials of a recording watthour meter** (sometimes, but erroneously, called a recording wattmeter) the directions given below should be followed (*Rules and Regulations* of the Commonwealth Edison Co., Chicago; see Fig. 1-86 for examples).

The pointer on the right-hand dial of a five-dial meter registers $\frac{1}{10}$ kWh or 100 Wh for each division of the dial. A complete revolution of the hand on this dial will move the hand of the second dial one division and register 1 kWh or 1000 Wh. A complete revolution of the hand of the second dial will move the third hand one division and register 10 kWh or 10,000 Wh, and so on.

Accordingly, read the hands from left to right and add two ciphers to the reading of the lowest dial to obtain the reading of the meter in watthours. If there are four dials on the meter, the pointer on the right-hand dial registers 1 kWh or 1000 Wh for each division of the dial, and it is necessary to add three ciphers to the reading on the lowest dial to obtain the reading in watthours, or the meter reads directly in kilowatthours.

Hands should always be read as indicating the figure which they have last passed, not the one to which they are nearest. Thus, if a hand is very close to a figure, whether it has passed this figure or not must be determined from the next lower dial. If the hand of the lower dial has just completed a revolution, the hand of the higher dial has passed the figure, but if the hand of the lower dial has not completed a revolution, the hand of the higher dial has not yet reached the figure even though it may appear to have done so.

When one pointer is on 9, special care must be taken that the pointer on the next higher dial is not read too high, as it will appear to have reached the next number but will not have done so until the hand at 9 has come to 0.

The hands on adjacent dials revolve in opposite directions. Therefore a reading should always be checked after being written down, as it is easy to mistake the direction of the rotation.

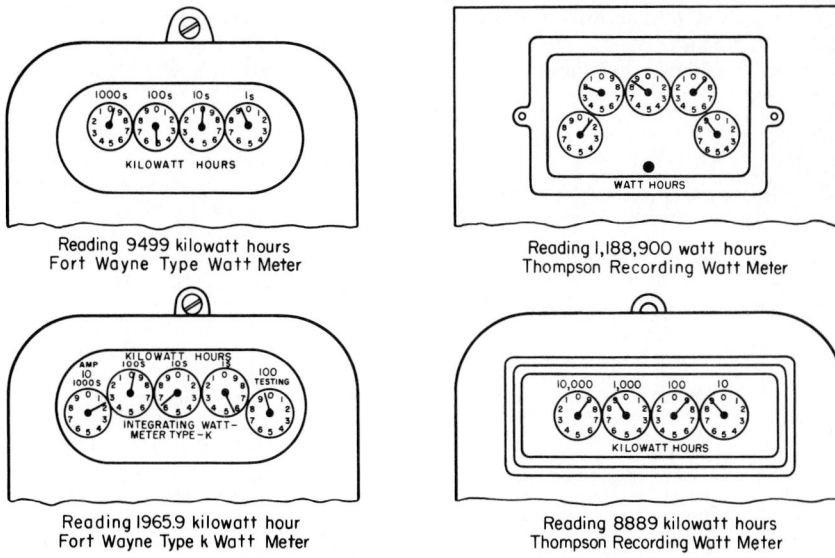

**Fig. 1-86** Examples of watthour-meter readings.

To determine the consumption for a given time, subtract the reading at the beginning of the period from the reading at the end. Always observe if a constant is marked at the bottom of the dial plate. If so, the difference of the readings must be multiplied by this constant to obtain the consumption.

**189. The Wheatstone bridge** is an instrument for measuring medium and high resistances. It is not suitable for measuring resistances of less than 1 $\Omega$. An elementary diagram is shown in Fig. 1-87. $R_1$, $R_2$, and $R$ are adjustable resistances, $R_x$ is the unknown resistance, and $G$ is a delicate galvanometer. A battery supplies emf. It can be shown that if when both keys are pressed, the galvanometer shows no deflection, then

$$\frac{R_1}{R} = \frac{R_2}{R_x} \quad \text{or} \quad R_x = \left(\frac{R_2}{R_1}\right) R \quad (76)$$

**example** If $R_2 = 100\ \Omega$, $R_1 = 10\ \Omega$, and $R = 672$ $\Omega$, what is the value of the unknown resistance?

**solution** Substitute in the formula

$$R_x = \left(\frac{R_2}{R_1}\right) R = \left(\frac{100}{10}\right) 672 = 10 \times 672 = 6720\ \Omega$$

The unknown resistance is 6720 $\Omega$.

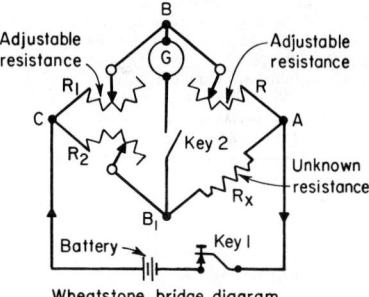

Wheatstone bridge diagram

**Fig. 1-87** Elementary diagram of the Wheatstone bridge.

In commercial bridges, the adjustable resistances $R_2$ and $R_1$ are usually so arranged that the ratio $R_2/R_1$ will be a fraction like ¹⁄₁₀ or ¹⁄₁₀₀ or a number like 10 or 100 so that $R_x$ can be obtained readily by dividing or multiplying $R$ by an easily handled number. $R_1$ and $R_2$ are sometimes called the ratio arms, and $R$ is called the rheostat arm. For most accurate results the resistances $R$, $R_1$, and $R_2$ should be as nearly as possible equal to $R_x$.

**190. A diagram of a commercial bridge of the post-office pattern** is shown in Fig. 1-88. Its principle is similar to that of Fig. 1-87. Brass plugs are used to vary the resistance in the arms $R$, $R_1$, and $R_2$. When a plug is inserted in the opening between two resistance coils, it shunts out the coil. When this bridge is used, the ratio $R_2/R_1$ is arranged by the operator, depending upon the relative value of $R_x$ as compared with $R$. Then $R$ is adjusted until a balance is obtained. When $R_x$ is greater than $R$, the ratio must be 10, 100, or 1000, and when $R_x$ is smaller than $R$, the ratio must be 0.1, 0.01, or 0.001. If $R_1 = R_2$, the value of $R_x$ equals $R$.

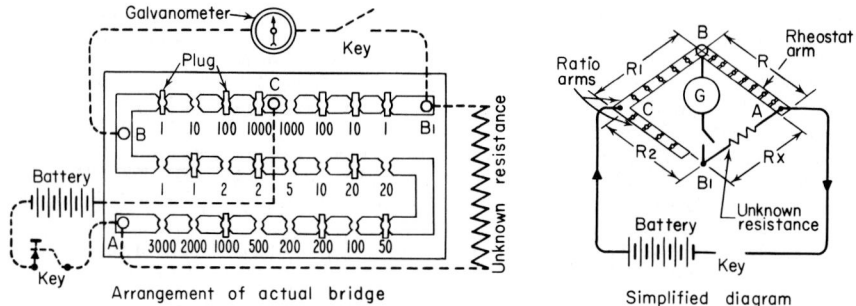

**Fig. 1-88**   Post-office pattern of Wheatstone bridge.

**191. Directions for using a Wheatstone bridge.** (1) Insert the unknown resistance. (2) Make a mental estimation of the probable value of the unknown resistance. If it is not greater than the total resistance in the arm $R$ or smaller than that of any one coil in $R$, then $R_1$ and $R_2$ can be made equal by taking plugs from the proper holes. (3) Take a plug from a coil in $R$ of about the estimated resistance of $R_x$ and press the keys. Note the deflection of the needle, whether it is to the right or to the left. Now unplug a coil in $R$ of about twice the resistance of the first one unplugged. If the needle now deflects in the opposite direction, the value of $R_x$ lies between these two values. If the deflection is in

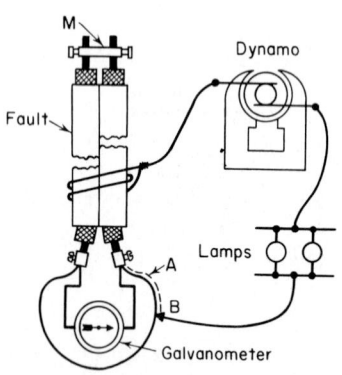

**Fig. 1-89**   Locating a fault in a cable.

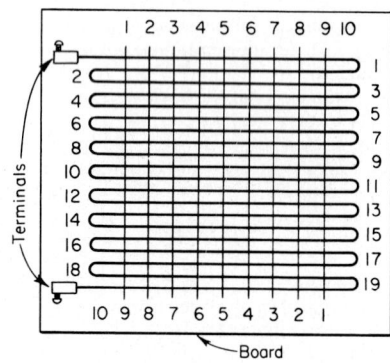

**Fig. 1-90**   Homemade wire bridge.

the same direction, the unplugged resistance in $R$ is too great, and a value of about one-half that originally selected should be tried. Systematically narrow the limits until the best possible balance is obtained. (4) Usually it is impossible to secure an exact balance. When this is the case, proceed as indicated in the following example. Assume that the coil of smallest resistance in the $R$ arm is of 0.1 $\Omega$. With this added, the galvanometer deflects two divisions to the right. The deflection without is three divisions to the left. Therefore a difference of 0.1 $\Omega$ makes a difference of five scale divisions. The resistance that would give no deflection is $\frac{3}{5} \times 0.1 = 0.06$ $\Omega$. (5) Be careful not to allow the metal parts of the bridge plugs to become wet or greasy from the hands. (6) Use a twisting motion when inserting the plugs. Put them in firmly, but do not use enough force to twist off the insulating handles. (7) When closing the keys, close the battery first, and in opening the keys, open the galvanometer key first.

**192. Locating faults in a cable** (Standard Underground Cable Co.). Figure 1-89 shows a simple method, using a dynamo, a galvanometer, and 10 or 15 ft of bare wire. This method is applicable only when both conductors of the cable are of the same size. After making the connections shown, it is necessary only to move the stylus along the bare wire until the galvanometer is not deflected in either direction.

Let $A$ = the length of the wire between the balance point $B$ and the faulty conductor, $C$ = the total length of the wire, and $L$ = the total length of the cable circuit, or twice the length of the cable. Then the distance to the fault = $(A \times L)/C$.

Figure 1-90 shows a simple form of wire bridge which can be used for tests of this kind. The length $A$ can be read directly, and the value of $C$ is 200. A telephone receiver can be used in place of the galvanometer.

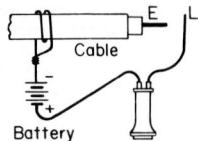

**Fig. 1-91** Test for insulation resistance.

**193. Testing cable insulation with a telephone receiver and battery** (Standard Underground Cable Co.). An extremely simple way to determine whether the insulation resistance of a given wire is faulty is as follows. A telephone receiver and battery are connected as shown in Fig. 1-91. One side of the battery is attached to the lead sheath of the cable or to ground, and the other side to a telephone receiver. A rubber-insulated wire is attached to the other side of the telephone. To test, press the telephone receiver to the ear and touch the wire $L$ to the conductor $E$; a click will always be heard the first time. After keeping both wires in contact for several seconds, break and make the connection once more; if no sound is heard at the instant of reconnection the wire is not faulty. With intervals of time between break and make of 1 s with a battery of 1 V, it can be assumed that no click indicates at least a resistance of 50 megohms. When more battery is used, this number is increased about in proportion to the number of cells. Care must be taken that sounds in the telephone due to induction are not misconstrued for those produced by leaks.

**194. Grounds on series lighting circuits frequently reveal their locations automatically.** If there are two good grounds on a circuit, the lamps connected in the line between the grounds will not burn because the grounds will shunt them out. For example, in Fig. 1-92

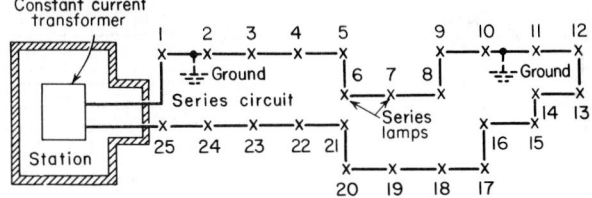

**Fig. 1-92** Effect of two grounds on a series circuit.

with a good ground at 1 and 11, lamps 2 to 10 would be shunted out. Sometimes there may be two grounds on a circuit, but they may not be good enough to shunt out the lamps. (This paragraph and those that follow on testing series circuits are from *Electrical World*.)

The presence of only one ground on a circuit, irrespective of how good it is, will not reveal itself automatically, and the proper operation of the circuit will not be affected by one ground. However, if there is one ground, it constitutes a serious menace to the lives of the station operators and troubleshooters. Furthermore, at any time there may occur another ground that may cause the shunting out of lamps or possibly a fire or destruction of equipment. Hence, it is very desirable to maintain the circuits entirely clear of grounds. It is the practice in all well-maintained stations to test each series circuit for grounds every afternoon. If a ground is discovered, a troubleshooter is sent to locate and clear it before the circuit is thrown into service for the night.

**195. The usual method of testing dead series circuits for grounds** is to disconnect the circuit from all station apparatus and then to connect one terminal of a magneto test set to the circuit and the other to ground. If the bell rings vigorously when the crank is turned, the circuit is grounded. If it does not, the circuit is clear. If the circuit is very long or in cable for a considerable portion of its length, the bell may ring a little even if the circuit is clear of grounds.

**196. The method of locating a ground on a dead series circuit** is illustrated in Fig. 1-93. Disconnect all station apparatus and temporarily ground one side of the circuit as at *B* (Fig. 1-93). Proceed out along the line and connect some testing instrument (a magneto

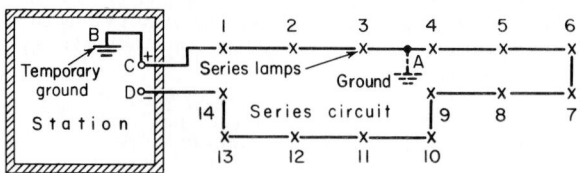

**Fig. 1-93** Locating a ground on a dead series circuit.

test set is most frequently used) in series with the circuit at some point. If when the crank is turned, the magneto bell rings, indicating a closed circuit, the tester is between the station ground and the ground on the circuit. If the magneto rings open, the tester is between the circuit ground and the ungrounded station end of the circuit. If in Fig. 1-93 the test is inserted at lamp 1, 2, or 3, the magneto should ring closed, while if it is inserted at any of the other lamps, it should ring open.

**197. In locating either a ground or an open on a series circuit,** unless the tester has an inkling of the location of the trouble, the tester should proceed first to the middle point of the circuit and there make the first test. This test will indicate on which side of the middle point the trouble is. The tester should then proceed to the middle point of the half of the circuit that shows trouble and there make another test. This will localize the trouble to one-fourth of the circuit. This halving of the sections of the circuit should be continued until the trouble has finally been found.

If there is more than one ground on a series circuit, the trouble is tedious to locate. If the tests made at different points on the circuit are confusing, indicating the existence of several grounds, the best procedure is to open the circuit into several distinct sections and then test each one as a unit, following the methods described in preceding paragraphs.

**198. A ground on a series circuit can sometimes be located with the current from the transformer** by placing a temporary ground on the circuit at the station. For example, if in Fig. 1-93 a temporary ground is connected to terminal *B* and the device that supplies the operating current to the circuit is connected to terminals *C* and *D* and normal operating current is thrown out on the circuit, lamps 1, 2, and 3 will not burn, indicating that the ground is between lamps 3 and 4. This method is attended by some fire risk; hence, it should be used with caution.

**199. A method of locating a ground on a series circuit with a lamp bank** is suggested in Fig. 1-94. A bank of 110-V incandescent lamps, each of the same wattage, is connected in series as indicated, and one end of the bank is permanently grounded. There should be a sufficient number of lamps in the bank so that the sum of the voltages of all of the lamps is at least equal to the voltage impressed on the series circuit by the transformer. For

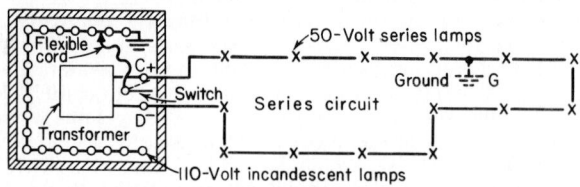

**Fig. 1-94**    Locating a ground on a live series circuit with an incandescent-lamp bank.

instance, if the voltage impressed on the series circuit is 6600, there should be at least sixty 110-V incandescent lamps in the bank (6600 ÷ 110 = 60).

In locating a ground, the flexible cord which is connected to the center point of the double-throw switch is placed successively on different points on the conductor that connects the incandescent lamps in series, the switch being thrown to one or the other of the circuit terminals $C$ or $D$. Move the flexible cord along until the incandescent lamps in the bank, between the point of connection of the cord and the permanent ground, burn at about full brilliancy. When this condition obtains, the voltage impressed across the lamps that are burning fully brilliant is approximately equal to the voltage impressed on the portion of the circuit (to which the switch connects) between the station and the ground. The voltage required across each lamp of the outside circuit being known, the number of lamps between the station and the ground can be readily computed, and the ground is thereby located.

Great care must be exercised in using this method. Practically all series lighting circuits operate at very high voltage. Hence, when the flexible cord is being moved along the lamp bank, the transformer should be entirely disconnected. If it is not, voltage dangerous to life is present.

**example**  Consider Fig. 1-94. There is a ground on the circuit. It is found that two of the incandescent lamps of the bank burn at full brilliancy between the flexible-cord connector and the lamp-bank ground. Since 110-V lamps are used in the bank, the voltage across these two is 220. This means that the voltage on the circuit between points $C$ and $G$ is about 220. Since the series lamps each require about 50 V, there must be 220 ÷ 50 = 4.4, or in round numbers, 4 series lamps between $C$ and the ground $G$. After making a test with the switch point on $C$, it should be thrown over to $D$, and a check test made from the other end of the circuit. The method of figuring is the same in each case.

**200. To locate an open on a series circuit,** ground one end of the circuit at the station as in Fig. 1-93. Then make tests at different points out on the circuit with the magneto connected between line and ground. As long as the magneto bell indicates a closed circuit, the open is on the line side of the tester. When the magneto indicates an open circuit, the open is toward the station from the tester.

**201. The testing out of a concealed wiring system for proper connections** is illustrated in Fig. 1-95. It is assumed that the wires are installed and that the locations of their runs are

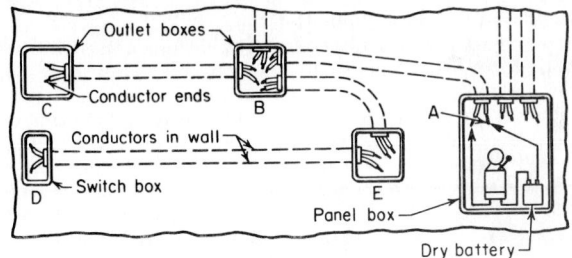

**Fig. 1-95**   Testing out wiring for proper connections.

concealed by the plastering. Only the ends of the conductors are visible at the outlets. It is necessary to identify the conductor ends at each outlet. These tests are usually made with an electric-bell outfit (Fig. 1-61) because the sound of the bell will indicate a closed circuit to the wire tester in a distant room. Hence, one person can test out such a system. In testing out, first skin the ends of all of the conductors and see that none is in contact with any other or with the outlet box. Next, select a pair of conductors ($A$, Fig. 1-95), preferably the pair that serves the group, and connect the bell outfit to the ends of the pair as shown. Then proceed to the outlet ($B$, Fig. 1-95) at which the pair of conductors should terminate and successively touch together the ends of all the wires that terminate in that box until a pair is discovered whose ends touched together ring the bell. This identifies one pair. Tag this pair so that it can be readily found again, and repeat the process on some other pair. Continue until all of the conductors have been identified. (These paragraphs on practical electrical tests are from *Electrical Engineering*.)

**202. The method of testing out the connections for three-way switches** is shown in Fig. 1-96. When finally connected, the circuits should be as shown in Diagram I. It is assumed that the conductors are in place and concealed within walls or ceilings and that only the ends are visible at the outlets, as in Fig. 1-96, II. First, identify the feed conductors and bend back their ends at the outlet box as at $A_3$. Next, twist together, temporarily, the bared ends

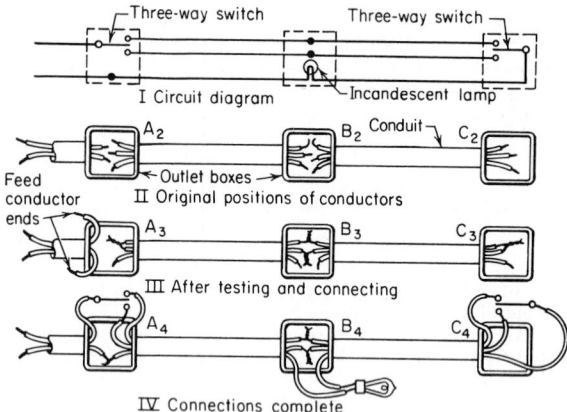

**Fig. 1-96**   Testing out three-way–switch connections.

of any two of the conductors at each of the switch outlets as at $A_3$ and $C_3$. The conductors having their ends thus twisted together will be the switch conductors. Now, at the lamp outlet or outlets identify the short-circuited switch conductors as directed in a preceding paragraph, and connect and solder these switch conductors together as at $B_3$. Connect the remaining conductor ends at the lamp outlets to the lamps $B_4$, connect one of the feed conductors to the center point of the three-way switch $A_4$, and connect the other feed conductor to the lamp wire. The switch conductors are connected to the two points of the switch. At $C_4$ the same procedure is followed.

**203. In testing out a new wiring installation for faults** each branch circuit, main, and feeder should be treated individually. It is usually impracticable to test an installation as a unit, as open switches and loose connections in cutouts may render such a test worthless. If a test is made from the cutout on the two conductors of each individual circuit, the above-mentioned possible elements of uncertainty are eliminated. Test each side of each circuit separately unless the lamps are in position.

**204. Open circuits in multiple wiring installations are usually readily located.** If the lamps are in position and lighting voltage is available, it can be impressed on the circuit. The lamps on the generator side of the open will then burn, while those on the far side will not, which localizes the open. If lighting voltage is not available, all the lamps can be taken out of the sockets, and each of the sides of the circuit can be grounded at the cutout. Then a telephone-and-battery, a bell-and-battery, or a magneto test set can be connected temporarily and successively between one line and ground and between the other line and ground at each outlet on the branch. When the test set indicates an open circuit, the open is between the tester and the ground made at the cutout.

**205. The test for short circuits on a multiple system** is made by temporarily connecting a test set across the terminals of each branch circuit at the cutout. If there is a short circuit on the lines under test, its presence will be immediately evident by the indication of a closed circuit.

**206. The test for continuity of multiple wiring circuits** is made by temporarily connecting a test set across the terminals of each branch cutout and successively short-circuiting, one at a time, the sockets of the branch with a screwdriver, a nail, or other metal object. The test set will then indicate whether the wiring of the circuit is open or closed. If lighting voltage is available and plug cutouts are used, a lamp can be screwed into one socket of the cutout and a plug fuse into the other. Then the tester can proceed from socket to socket and short-circuit each. If the circuit to the socket is continuous, the lamp will light when the socket is short-circuited.

**207. The test for grounds on a multiple wiring installation** is made by temporarily connecting between line and ground a test set of one of the types previously described. If the test set indicates a closed circuit, the line being tested is grounded.

**208. The testing of three-wire circuits to identify the neutral** is effected as suggested in Fig. 1-97. If the neutral is grounded, a test lamp can be connected successively between each

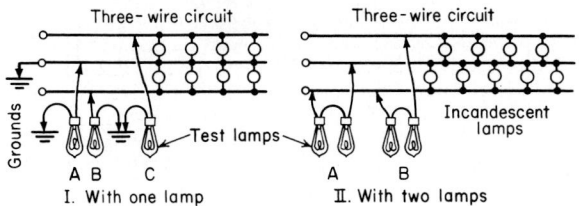

Three-wire circuit    Three-wire circuit

Grounds

Test lamps

Incandescent lamps

A B    C    A    B

I. With one lamp    II. With two lamps

**Fig. 1-97**    Locating neutral wire with a test lamp.

of the three conductors and ground (Fig. 1-97, I). When the ungrounded side of the lamp is touched to the neutral wire, it will not burn, but when touched to either of the outside wires, it will burn. A method that can be used with either a grounded or an ungrounded neutral is illustrated in Fig. 1-97, II. Connect the two test lamps in series successively between one of the line wires and the other two. When connected across the two outer wires, both lamps will burn at full voltage, but when connected between one of the outer wires and neutral, they will burn at only half voltage.

**209. To determine the polarity of dc circuits** hold the two conductors in a glass vessel of water as in Fig. 1-98. It may be necessary to pour a little common salt or acid into the water to render it conducting (pure water is a poor conductor). Bubbles will form only on the negative conductor, indicating the presence of current and the polarity of the circuit. Be careful not to touch the conductor ends together and cause a short circuit (see also Fig. 1-99 and Sec. 167 on neon testers).

**210. The direction of current flow in a dc circuit** can be determined with a compass as described in Sec. 89.

**211. Ground detectors** are desirable on ungrounded systems so that when an accidental ground occurs, it can be remedied before a second ground on another wire causes a short circuit. Control circuits espe-

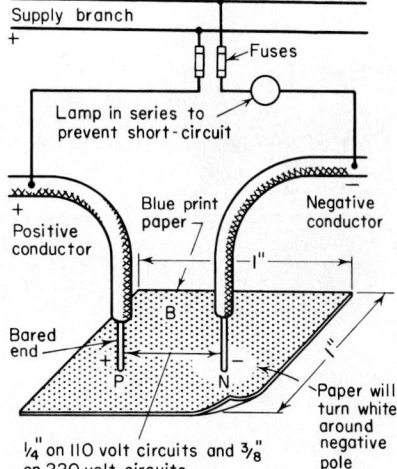

Supply branch

Fuses

Lamp in series to prevent short-circuit

Blue print paper

Negative conductor

Positive conductor

Bored end

Paper will turn white around negative pole

1/4" on 110 volt circuits and 3/8" on 220 volt circuits

Negative conductor

Positive conductor

Glass vessel

Bubbles

**Fig. 1-98**   Determination of polarity with conductor ends in water.

**Fig. 1-99**   Blueprint paper used for testing polarity.

cially should be suitably equipped with ground detectors, as a double ground may cause the closing or tripping of circuits carrying load, with serious consequences.

**212. Ground detectors for two-wire dc circuits.**   Figure 1-100 shows a very good and simple detector for any two-wire low-voltage system. The lamps for the detector should each be of the same wattage and voltage, the voltage being about the same as that of the regular lamps in the plant, and two lamps which, when connected in series, burn with equal brilliancy should be selected. Although somewhat greater sensitiveness can be obtained with low-wattage lamps, such as 25-W lamps, for example, it is believed in general to be better to use lamps of the same wattage as those throughout the plant, because a burned-out or broken detector lamp can then be immediately replaced by a good lamp from the regular stock, thus avoiding the necessity of keeping on hand a few spare special lamps.

The detector lamps, being two in series across the proper voltage for one lamp, burn only dimly. If, however, a ground occurs on any circuit, as at *a*, the current from the positive busbar through lamp 1 divides on reaching *b* instead of all going through lamp 2, as it did when there was no ground. Part now goes down the ground wire and through the ground to *a*, as indicated by the broken line, and thence through the wires to the negative busbar. This reduces the resistance from *b* to the negative busbar, and therefore more current flows through lamp 1 than before, while less current flows through lamp 2. Lamp 1 consequently brightens, and lamp 2 dims. If the ground had occurred at *c* instead of *a*, lamp 2 would have brightened and lamp 1 would have dimmed.

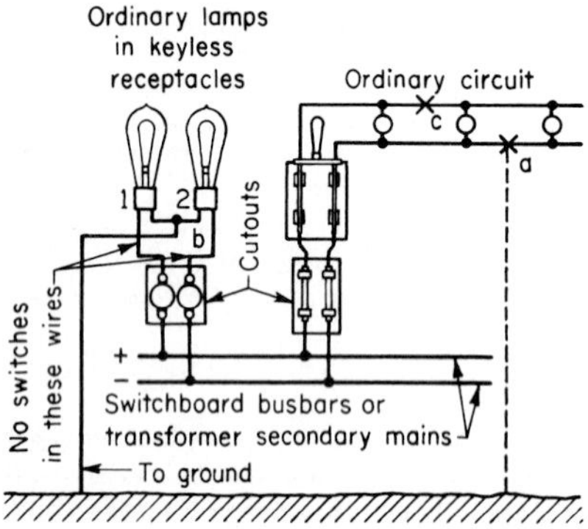

Attention is called to the following points, which are frequently neglected in this form of detector:

1. The lamp receptacles should be keyless, and there should be no switches of any kind in any of the connecting wires, so that the detector will always be in operation. To be of greatest value, the detector must give indications instantly when a ground occurs. The observer should not have to wait until the engineer or electrician remembers to close a switch.

2. The wires should be protected by small fuses where they connect to the busbars. If these fuses are omitted, a short circuit across these wires would either burn up the wires or blow the main generator fuses.

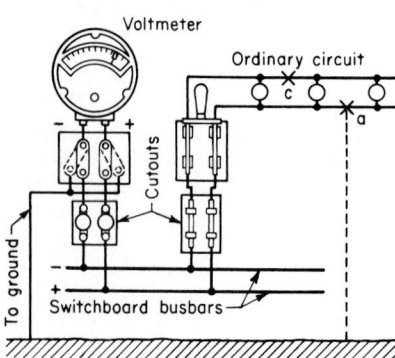

**Fig. 1-100**   Two-lamp ground detector.

3. The lamps should be placed very close together, within 1 or 2 in of each other if possible. The farther apart they are, the harder it is to detect any slight difference in brilliancy between them.

4. The ground wire should be carefully clamped to a pipe which is thoroughly connected to the ground, or some other equally good ground connection should be provided.

**213. On dc control circuits,** a telegraph type of relay, which will sound an alarm, is preferable to the lamps, since sufficient current can flow through the lamps as a result of double grounds to operate some relays and cause circuit breakers to operate.

**214. An ordinary voltmeter can be used as an intermittent ground detector** on dc circuits of any voltage, as shown in Fig. 1-101. The voltmeter ordinarily used to indicate the voltage on the system can, of course, be used for this purpose, the voltmeter switch shown in the cut being arranged to give the different desired connections.

If, for example, the system shown in Fig. 1-101 were of about 100 V, the voltmeter would register 100 when the levers of the switch were on the inside contact points as shown. If, now, the right-hand lever were moved to the outside contact point as shown dotted and there were a ground on the system, as at *a*, current would pass from the positive busbar through the circuit to *a*, thence through the ground to the ground wire, and through the voltmeter to the negative busbar, causing the voltmeter to read something below 100 unless the ground at *a* were practically a perfect connection. In that case the voltmeter reading would be 100. If the positive side of the system were entirely free from grounds, the voltmeter reading would be 0.

**Fig. 1-101**   Voltmeter ground detector.

Let us assume that under these conditions the voltmeter reads 50 and that the resistance of the voltmeter itself is 20,000 Ω. Since with no external resistance, when the voltmeter is connected directly to the busbars, it reads 100, while now it reads 50, the total resistance under the new conditions must be 40,000 Ω. The resistance of the ground at *a* is 40,000 − 20,000 = 20,000 Ω.

If the voltmeter had read only 20, the total resistance would have been $^{100}\!/_{20} \times 20,000$ − 100,000, and the resistance of the ground 100,000 − 20,000 = 80,000 Ω.

**215. Ground detectors for ordinary low-voltage three-phase ac circuits.**  A lamp detector connected as in Fig. 1-102 can be used. The indication is the same as that with the lamp detectors described above. Thus, when a ground comes on one wire, the lamp attached to that wire dims and the other two lamps brighten.

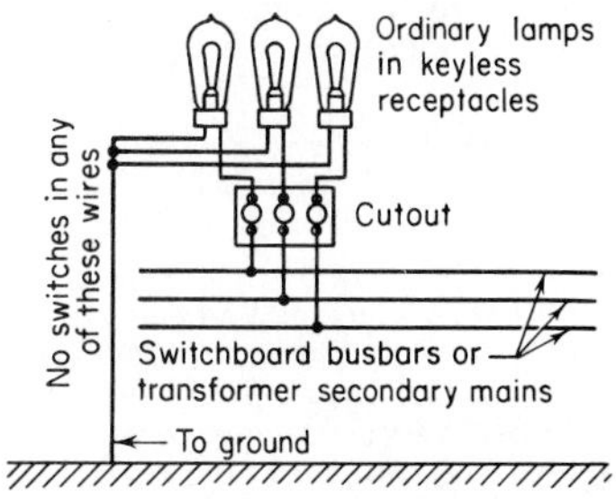

For ordinary two-phase or quarter-phase systems, in which the phases are entirely insulated from each other, the two-lamp detector can be used, with one detector on each phase. There are, however, in this class of wiring several complicated systems, to all of which the lamp-detector principle is applicable, although the exact method of connections differs in each case, so that no general rule can be given.

**216. Electrostatic ground detectors** are used for high-voltage ungrounded ac systems. This instrument for use on three-phase circuits consists of a semispherical vane flexibly supported (Fig. 1-103). Around the circumference of the vane are three stationary curved plates, each being connected through a capacitor to one of the three phase lines. The operation is due to the electrostatic field created between the plates. The vane is grounded, and a ground on any line removes the charge between the plate connected to that line and the ground. This causes the vane to move toward the other two plates against the action of a spring.

**Fig. 1-102**  Three-phase lamp detector.

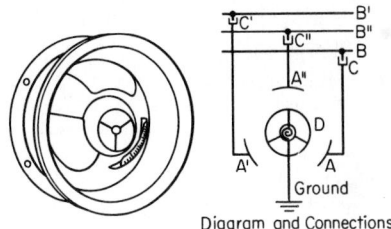

Fig. 1-103  Three-phase electrostatic ground detector. [Westinghouse Electric Corp.]

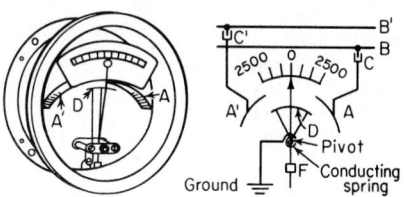

Fig. 1-104  Single-phase electrostatic ground detector. [Westinghouse Electric Corp.]

Thus the vane being off center indicates a ground. The capacitors (Fig. 1-103) insulate the instrument from the high voltage of the line. A single-phase electrostatic ground detector is shown in Fig. 1-104.

**217. A triplex voltmeter,** which is an instrument containing three voltmeter movements with short scales (Fig. 1-105), can also be used on ac systems. Each voltmeter is connected between one line and ground. With no grounds all voltmeters should read zero, but a ground on one line will cause the other two meters to read. The magnitude of the reading will depend on the resistance of the meter and the resistance of the ground, a good ground being indicated by a full line-voltage reading.

**218. Test to determine horsepower of an electric motor.**  By applying the principles outlined elsewhere in this division to a motor under test for output, the power delivered being measured with a Prony brake (Fig. 1-106) may be taken as an example.

**example**  The torque is 10 lb at 3-ft radius, or 30 lb·ft, or 30 lb at 1-ft radius. Since the motor pulley is turning at the rate of 1000 rpm, a point on its circumference travels $2\pi rR = 2 \times 3.14 \times 1 \times$

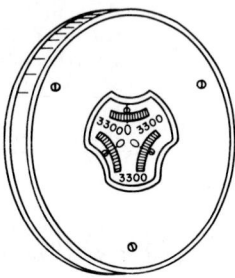

**Fig. 1-105** Electrostatic ground detector (triplex voltmeter type). [General Electric. Co.]

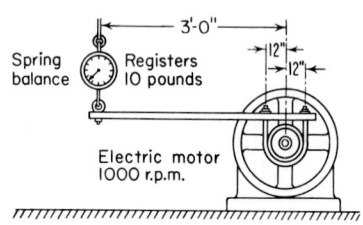

**Fig. 1-106** Horsepower determination with a Prony brake.

1000 = 6280 ft/min. At its circumference the pulley is overcoming a resistance of 30 lb. Therefore it is doing work at the rate of 30 × 6280 = 188,400 ft·lb/min. Since, when work is done at the rate of 33,000 ft·lb/min, a horsepower is developed, the motor is delivering 188,400 ÷ 33,000 = 5.7 hp. It should be noted that though the torque at the circumference of the motor pulley was considered in the example, it is not necessary to take the torque at that point. The torque may be taken at any point if the radius to that point is used instead of the radius of the pulley. The formula for determining the horsepower output of a motor under test with a Prony brake is

$$\text{hp} = 2 \times \pi T S \div 33,000$$

where $\pi = 3.1416$, $T$ = torque in pound-feet, and $S$ is the speed of the motor in rpm. Substituting the values from the above example in this formula,

$$\text{hp} = 2 \times 3.14 \times 30 \times 1000 \div 33,000 = 5.7 \text{ hp}$$

This is the same result secured by the former and longer method. In metric units, work is expressed in kilogram-meters so, conversely torque should be expressed in meter-kilograms or in kilograms at a given radius in meters.

**219. The testing of motors and generators for faults** is treated in Div. 7, "Generators and Motors."

# Properties and Splicing
# of Conductors

## ELECTRICAL CONDUCTING WIRES AND CABLES

**1. Electrical conducting wires and cables** are available in a great variety of types and forms of construction. To cover this one subject completely would require a large volume. The aim of the authors has been, therefore, to include in this division sufficient general information with respect to the materials employed, method of construction, and types available so that the reader can select intelligently the proper cable for a given application. At the end of the section, more detailed tabular information is included for the types of cables that the average worker will use most frequently.

**2. Electric wire and cable terminology**

*Wire.* A slender rod or filament of drawn metal. (This definition restricts the term *wire* to what would ordinarily be understood by the term *solid wire.* In the definition the word *slender* is used in the sense that the length is great in comparison with the diameter. If a wire is covered with insulation, it is properly called an insulated wire, although the term *wire* refers primarily to the metal; nevertheless, when the context shows that the wire is insulated, the term *wire* will be understood to include the insulation.)

*Conductor.* A wire or combination of wires not insulated from one another, suitable for carrying a single electric current. (The term *conductor* does not include a combination of conductors insulated from one another, which would be suitable for carrying several different electric currents. Rolled conductors, such as busbars, are, of course, conductors but are not considered under the terminology given here.)

*Stranded conductor.* A conductor composed of a group of wires or any combination of groups of wires. (The wires in a stranded conductor are usually twisted or braided together.)

*Cable.* (1) A stranded conductor (single-conductor cable) or (2) a combination of conductors insulated from one another (multiconductor cable).

The component conductors of the second kind of cable may be either solid or stranded, and this kind may or may not have a common insulating covering. The first kind of cable is a single conductor, while the second kind is a group of several conductors. The term *cable* is applied by some manufacturers to a solid wire heavily insulated and lead-covered; this usage arises from the manner of the insulation, but such a conductor is not included under this definition of *cable.* The term *cable* is a general one, and in practice it is usually applied only to the larger sizes. A small cable is called a *stranded wire* or a *cord,* both of which are defined below. Cables may be bare or insulated, and insulated cables may be armored with lead or with steel wires or bands.

*Strand.* One of the wires or groups of wires of any stranded conductor.

*Stranded wire.* A group of small wires used as a single wire. (A wire has been defined as a slender rod or filament of drawn metal. If such a filament is subdivided into several smaller filaments or strands and is used as a single wire, it is called stranded wire. There is no sharp dividing line of size between a stranded wire and a cable. If used as a wire, for example in winding inductance coils or magnets, it is called a stranded wire and not a cable. If it is substantially insulated, it is called a cord, defined below.)

*Cord.* A small cable, very flexible and substantially insulated to withstand wear. (There is no sharp dividing line in respect to size between a cord and a cable and likewise no sharp dividing line in respect to the character of insulation between a cord and a stranded wire.

*Concentric strand.* A strand composed of a central core surrounded by one or more layers of helically laid wires or groups of wires.

*Concentric-lay conductor.* A conductor composed of a central core surrounded by one or more layers of helically laid wires. (Ordinarily it is known as a concentric-strand conductor. In the most common type, all the wires are of the same size, and the central core is a single wire.)

*Concentric-lay cable.* A multiconductor cable composed of a central core surrounded by one or more layers of helically laid insulated conductors.

*Rope-lay cable.* A cable composed of a central core surrounded by one or more layers of helically laid groups of wires. (This cable differs from a concentric-lay conductor in that the main strands are themselves stranded and all wires are of the same size.)

*Multiconductor cable.* A combination of two or more conductors insulated from one another. (Specific cables are called *3-conductor cable, 19-conductor cable,* etc.)

*Multiconductor concentric cable.* A cable composed of an insulated central conductor with one or more tubular stranded conductors laid over it concentrically and insulated from one another. (This kind of cable usually has two or three conductors.)

N-*conductor cable.* A combination of N conductors insulated from one another. (It is not intended that the name as given here be actually used. One would instead speak of a 3-conductor cable, a 12-conductor cable, etc. In referring to the general case, one may speak of a multiconductor cable, as in the definition for *cable* above.)

N-*conductor concentric cable.* A cable composed of an insulated central conducting core with tubular stranded conductors laid over it concentrically and separated by layers of insulation. (Usually it is only 2-conductor or 3-conductor. Such conductors are used in carrying alternating currents. The remarks on the expression "N-conductor" in the preceding definition apply here also.)

*Duplex cable.* A cable composed of two insulated single-conductor cables twisted together with or without a common covering.

*Twin cable.* A cable composed of two insulated conductors laid parallel and either attached to each other by the insulation or bound together with a common covering.

*Twin wire.* A cable composed of two small insulated conductors laid parallel and having a common covering.

*Twisted pair.* A cable composed of two small insulated conductors twisted together without a common covering.

*Triplex cable.* A cable composed of three insulated single-conductor cables twisted together.

*Sector cable.* A multiconductor cable in which the cross section of each conductor is approximately the sector of a circle.

*Shielded-type cable.* A cable in which each insulated conductor is enclosed in a conducting envelope so constructed that substantially every point on the surface of the insulation is at ground potential or at some predetermined potential with respect to ground under normal operating conditions.

*Shielded-conductor cable.* A cable in which the insulated conductor or conductors are enclosed in a conducting envelope or envelopes, so constructed that substantially every point on the surface of the insulation is at ground potential or at some predetermined potential with respect to ground.

*Composite conductor.* A conductor consisting of two or more strands of different metals, such as aluminum and steel or copper and steel, assembled and operated in parallel.

*Round conductor.* Either a solid or a stranded conductor of which the cross section is substantially circular.

*Cable filler.* Material used in multiconductor cables to occupy the spaces formed by the assembly of the insulated conductors, thus forming a core of the desired shape.

*Cable sheath.* Protective covering applied to cable.

*Insulation of a cable.* The part of a cable which is relied upon to insulate the conductor from other conductors or conducting parts or from ground.

*Insulation resistance of an insulated conductor.* The resistance offered by a conductor's insulation to an impressed direct voltage, tending to produce a leakage of current through it.

*Lead-covered cable (lead-sheathed cable).* A cable provided with a sheath of lead to exclude moisture and afford mechanical protection.

*Serving of a cable.* A wrapping applied over the core of a cable before the cable is leaded or over the lead if the cable is armored. Materials commonly used for serving are jute, cotton, and duck tape.

*Resistive conductor.* A conductor used primarily because it possesses high electric resistance.

**3. Wire sizes.** The size of wire is usually expressed according to a wire gage, and the different sizes are referred to by gage numbers. Unfortunately several systems of gages have been originated by different manufacturers for their products. However, it has become standard practice in the United States to employ the American wire gage (AWG), also known as the Brown and Sharpe (B&S), to designate copper and aluminum wire and cable used in the electrical industry. The names, abbreviations, and uses of the most important gages employed for the measurement of wires and sheet-metal plates are given in Table 4. A numerical comparison of these gages is given in Table 85. In most cases the larger the gage number, the smaller the size of the wire.

**4.  Names, Abbreviations, and Uses of the Principal Wire and Sheet-Metal Gages**

| Col. no. in Table 85 | Names and abbreviations | | Ordinarily used for measuring |
| --- | --- | --- | --- |
| | Usual | Others | |
| 1 | American wire gage (AWG) | Brown and Sharpe (B&S) | Copper, aluminum, copper-clad aluminum, and other nonferrous wires, rods, and plates. Wall thickness of tubes |
| 2 | Steel wire gage (SWG) | Roebling, American Steel and Wire, Washburn & Moen, National, G. W. Prentiss | Iron and steel wire. Wire nails. Brass and iron escutcheon pins |
| 3 | Birmingham wire gage (BWG) | Stubs iron wire gage, iron wire gage | Galvanized iron and steel wire. Iron and copper rivets. Thickness of wall of nonferrous seamless tubing |
| 4 | Stubs steel wire gage | . . . . . . . . . . . . . . . . . . . . . | Drill rod |
| 5 | British standard wire gage (SWG) | Imperial standard wire gage, standard wire gage, English legal standard | Legal standard wire gage for Canada and Great Britain; used sometimes by American telephone and telegraph companies for bare copper line wire |
| 6 | Steel music wire gage (MWG) | Hammacher, Schlemmer, Felten & Guilleaume | Steel music wire |
| 7 | Manufacturers standard gage (MSG) | . . . . . . . . . . . . . . . . . . . . . | Legal standard for iron and steel plate. Monel metal. Galvanized sheets |
| 8 | American zinc | . . . . . . . . . . . . . . . . . . . . . | Sheet zinc |
| 9 | Birmingham gage (BG) | . . . . . . . . . . . . . . . . . . . . . | Legal standard for iron and steel sheets in Canada and Great Britain |

**5. How to remember the AWG or B&S wire-gage table** *(Westinghouse Diary)*. A wire that is three sizes larger than another wire has half the resistance, twice the weight, and twice the area. A wire that is ten sizes larger than another wire has one-tenth the resistance, ten times the weight, and ten times the area. Number 10 wire is 0.10 (more precisely 0.102) in (2.6 mm) in diameter; it has an area of 10,000 (more precisely 10,380) cmil (5.26 mm²); it has a resistance of 1 Ω/1000 ft (304.8 m) at 20°C (68°F); and it weighs 32 (more precisely 31.4) lb (14.24 kg)/1000 ft (304.8 m).

The weight of 1000 ft (304.8 m) of No. 5 wire is 100 lb (45.36 kg). The relative values of resistance (for decreasing sizes) and of weight and area (for increasing sizes) for consecutive sizes are 0.50, 0.63, 0.80, 1.00, 1.25, 1.60, 2.00. The relative values of the diameters of *alternate* sizes of wire are 0.50, 0.63, 0.80, 1.00, 1.25, 1.60, 2.00. To find resistance, drop one cipher from the number of circular mils; the result is the number of feet per ohm. To find weight, drop four ciphers from the number of circular mils and multiply by the weight of No. 10 wire.

**6. Wire-measuring gages** (Figs. 2-1 and 2-2) are made of steel plate. With the kind shown in Fig. 2-1 the wire being measured is inserted in the slots in the periphery until a slot is found in which the wire just fits. The wire's gage number is indicated opposite the slot. A measuring gage like that of Fig. 2-1 indicates the numbers of one gage or system only. A gage like that of Fig. 2-2 indicates the numbers of four gages but has the disadvantage that, to use it, the end of the wire must be available to be pushed through the slot. The wire is pushed as far toward the small end of the slot as it will go, and its gage number is then indicated opposite the point where the wire stops. The gage of Fig. 2-2 is arranged to indicate gage numbers for the American standard screw gage, English wire gage, and American wire gage, and one scale is divided into thirty-seconds of an inch.

**7. A micrometer** is frequently used to determine the size of a given wire or cable. It provides an accurate means of measuring the diameter of the wire or cable to thousandths of an inch and of estimating to ten-thousandths. After the diameter has been determined, the corresponding gage size may be ascertained by referring to a wire table. The wire to be measured is placed between the thumbscrew and the anvil (Fig. 2-3), and the screw is turned until the wire is lightly held between the screw and the anvil. The screw has 40 threads to the inch, so that one complete turn of the screw in a left-hand direction will open the micrometer ¹⁄₄₀ in. On the edge of the collar is a circular scale divided into 25

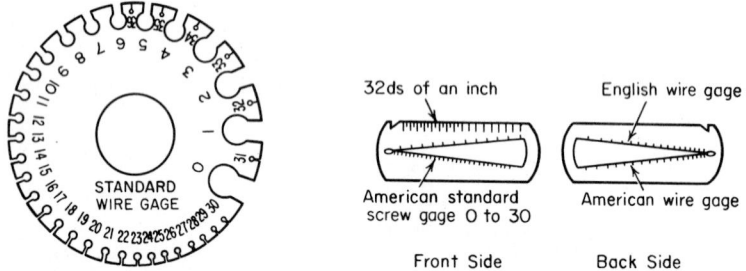

**Fig. 2-1**  Standard wire gage (greatly reduced).    **Fig. 2-2**  Angular wire gage (greatly reduced).

divisions; hence, when the screw is turned through one of these divisions, the micrometer will open ¹⁄₂₅ × ¹⁄₄₀ in = ¹⁄₁₀₀₀ in. The shaft on which the collar turns is marked into tenths of an inch, and each ¹⁄₁₀ is subdivided into four parts. Each of these parts must be equal to ¹⁄₁₀ in × ¹⁄₄ = ¹⁄₄₀ in = 0.025 in. Therefore a complete rotation of the collar or 25 of its divisions will equal one division of the shaft, or 0.025 in.

**8. To read a micrometer** (see Fig. 2-4 and the paragraph above) note the number on the circular scale nearest the index line. This indicates the number of thousandths. Note the number of small divisions uncovered on the shaft scale. Each one of these small divisions indicates 0.025 in (²⁵⁄₁₀₀₀). Add together the number of thousandths indicated on the circular scale and 0.025 times the number of small divisions wholly uncovered on the shaft scale. The sum will be the distance that the jaws are apart. Examples are shown in Fig. 2-4.

**9. Classification of wires or cables.**    Electrical conducting wires and cables may be classified in several different ways depending upon the particular factor of consideration as
follows:

1. According to degree of covering of wire or cable
2. According to material and makeup of electrical conductor
3. According to number of conductors in cable
4. If insulated, according to insulation employed
5. According to protective covering
6. According to service

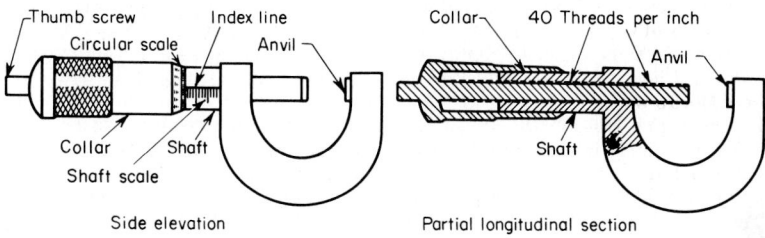

**Fig. 2-3**    A micrometer caliper.

**10. Classification according to degree of covering.**    Probably the broadest classification of
electrical wires and cables is according to the degree of covering employed, as follows:

1. Bare
2. Covered but not insulated
3. Insulated and covered

In many cases *bare* conductors without any covering over the metallic conductors are
employed for electrical circuits. The conductors are supported on insulators so spaced,
depending upon the voltage of the circuit, that the length of the air space between the
conductors will provide sufficient insulation. Bare conductors are employed for most overhead transmission lines. They are also used for certain overhead telephone and telegraph
circuits.

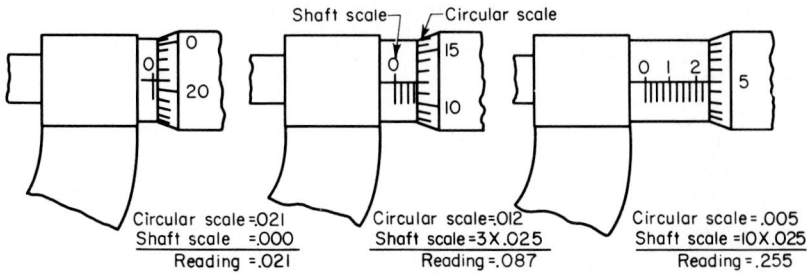

**Fig. 2-4**    A micrometer caliper.

There is a definite distinction between the meanings of the words *bare, covered,* and
*insulated* in electrical-cable terminology:

*Bare.* Designating a conductor without any covering or electrical insulation.

*Covered.* Designating a conductor encased in material whose thickness or composition
is not recognized by the National Electrical Code (NEC) as electrical insulation.

*Insulated.* Designating a conductor encased in material whose thickness or composition is recognized by the Code as electrical insulation.

Insulated wires and cables consist of an electrical conductor covered with some form of
electrical insulation with or without an outer protective covering. With covered wires the
conductor is protected against mechanical injury by some form of covering applied
directly over the conductor. This covering, however, does not provide insulation of the
wet conductor to any great extent and often even becomes a fair conductor itself when

moist. The covering therefore cannot be relied upon to afford consistently any insulation beyond that of the air spacing between the conductors.

Weatherproof wires may be divided into two general classes, those with fibrous coverings and those with homogeneous coverings (Sec. 12).

**11. Weatherproof covered wire with fibrous coverings** is available in two types. Both are designated as URC and are manufactured according to specifications recommended by the Utilities Research Commission. The two types are distinguished by at least one manufacturer as O.K.–URC and Peerless URC. The O.K. is the more common type. The wire is covered with two or three weatherproof cotton braids, which are applied directly over the conductor. The braids are saturated with a black asphaltic compound. The outside surface of the outer braid is coated with mica flake to provide a smooth surface free of tackiness. The Peerless type, a better-grade covering, not only is longer-lived but provides some degree of overall electrical insulation to the conductor. The construction consists of a compacted pad of unspun cotton wrapped helically directly around the conductor and filled with asphalt. This pad provides a homogeneous wall of insulation without the usual braid interstices encountered in ordinary weatherproof covered wire. The asphalt-impregnated pad is covered with a standard outer weatherproof braid. Peerless wire is made in two types, designated as double braid and triple braid. Actually there is only one braid in both types, but the combination of the cotton pad of different thicknesses and the braid gives a covering which is equivalent in thickness to the regular double- or triple-braid O.K. wires. Weatherproof covered wire is available with the conductor material consisting of copper, solid Copperweld, solid bronze, composite copper-Copperweld, all aluminum, or steel-reinforced aluminum.

**12. Weatherproof covered wire with homogeneous covering** is available in two types, neoprene and polyethylene.

The *neoprene type* has a homogeneous covering which consists of a tough and durable vulcanized neoprene compound. This type has the following advantageous characteristics:

1. Highly weather-resistant.
2. Not brittle under extremely low temperatures. It will not migrate under abnormally high temperatures.
3. Preventing normal moisture ingress to conductor, thus permitting maximum continuity of service under storm conditions.
4. Tough, with excellent resistance to mechanical abuse.
5. Light in weight and smooth-surfaced, with adhesion to the conductor. It can be handled and installed without special precautions.

The *polyethylene type* has a homogeneous, seamless thermoplastic covering which is applied by extrusion. Its more important characteristics may be summarized as follows:

1. Highly resistant to weathering, being both weather- and sun-resistant.
2. Physically strong.
3. Resistant to abrasion.
4. Inherently tight on the conductor.
5. Possessing a wide temperature range, $-94°F$ ($-70°C$) to $176°F$ ($80°C$).
6. Light in weight.

**13. Material and makeup of electrical conductors.** The conductor may consist of a single solid wire or a stranded cable made up of several individual bare wires twisted together. Stranded conductors are more flexible than a single solid wire. The flexibility of the cable increases with the number of strands and also depends upon the arrangement of the individual strands in the makeup of the conductor. Solid wires are used for conductors of the smaller sizes and stranded cables for the larger sizes. In the intermediate sizes both solid and stranded conductors are available.

Electrical conducting cables are made of copper, copper-clad aluminum, aluminum, steel, bronze, or a combination of either copper or aluminum with steel. Because of its high conductivity, copper is the material most commonly employed. Practically all insulated wires and cables employ copper or aluminum conductors. Aluminum or steel wires and cables are frequently found to be more economical for certain classes of aerial construction. All-steel conductors are employed for telephone and telegraph work and certain rural distribution lines with light load densities. Steel-copper conductors are employed for long-span river crossings of power lines and for certain rural distribution lines.

A classification of electrical conductors with respect to standard materials and makeups available is given below:

A. Copper
  1. Solid
    *a.* Round
    *b.* Grooved
    *c.* Figure eight
    *d.* Figure nine
    *e.* Square or rectangular
  2. Standard concentric stranded
    *a.* Class AA
    *b.* Class A
    *c.* Class B
    *d.* Class C
    *e.* Class D
  3. Standard rope stranded
    *a.* Class G
    *b.* Class H
  4. Bunched stranded
  5. Bunched and rope stranded
    *a.* Class J
    *b.* Class K
    *c.* Class L
    *d.* Class M
    *e.* Class O
    *f.* Class P
    *g.* Class Q

  6. Annular concentric stranded
  7. Special stranded
  8. Compack stranded
    *a.* Round
    *b.* Sector
    *c.* Segmental
  9. Hollow-core stranded
  10. Tubular segmental (Type HH)
B. Iron or steel
  1. Solid
  2. Stranded
C. Copper-steel
  1. Copperweld
    *a.* Solid
    *b.* Stranded
  2. Copperweld and copper
  3. Copper and steel stranded
D. Aluminum
  1. Solid
  2. Stranded
E. Copper-clad aluminum
  1. Solid
  2. Stranded
F. Aluminum and steel (ACSR)
G. Bronze

Refer to Table 86 for illustrations and applications of different makeups. Refer to Table 87 for a guide to applications of conductor materials.

**14. Copper wire** is made in three grades of hardness, known as hard-drawn, medium-hard-drawn, and soft or annealed. Hard-drawn wire has the greatest tensile strength and the least amount of elongation under stress and is the stiffest and hardest to bend and work. Soft-drawn or annealed wire has the lowest tensile strength and the greatest elongation under stress and is very pliable and easily bent. Medium-hard-drawn wire has characteristics intermediate between those of hard-drawn and soft-drawn wire. The conductivity of copper wires decreases slightly as hardness increases, but there is relatively little difference in the conductivity of the different grades.

Hard-drawn wire is used for long-span transmission lines, trolley contact wires, telephone wires, and other applications for which the highest possible tensile strength is desirable.

Medium-hard-drawn wire is employed for such applications as short-span distribution circuits and trolley feeders, for which slightly lower tensile strength is satisfactory and greater pliability is desired.

Soft-drawn or annealed copper is used for all covered or insulated copper conductors except weatherproof covered cables. Wires with this covering are available in all the three grades of hardness. Bare or weatherproof covered soft wire is used only for short spans.

Copper wire used for rubber-insulated cable must be *tinned* by coating with pure tin to protect the copper against chemical action caused by contact with the rubber.

**15. Makeup of copper conductors.** Copper conductors are manufactured in various forms of makeup as listed in Sec. 13, depending upon their size and application. An understanding of the construction of the different makeups and their applications can be gained from Table 86. The employment of solid or stranded conductors and the choice of the class of stranding depend upon the degree of flexibility desired. Tables 90 to 101 give information on the standard strandings. For very flexible cables for special applications, special strandings are employed. There is no fixed standard for these special strandings, practice varying with different manufacturers. Data for bare copper conductors of the different standard makeups are given in Tables 89 to 101.

**16. Iron or steel electrical conductors.** All-steel conductors for power circuits are made from special low-resistance, high-strength steel stock. They are available in both solid and three-strand construction. Each wire is protected by a heavy galvanized coating of zinc. Since all-steel conductors can be applied economically only for light electrical loads, these conductors are manufactured in only three standard sizes (see Table 103). Stranded conductors are used in preference to solid wires for longer-span construction because their inherent damping capacity reduces the amplitude of vibration in strong lateral winds.

Commercial galvanized-iron wire for telephone and telegraph circuits has been made for many years in three types designated as extra best best (EBB), best best (BB), and steel (see Table 102). These designations are somewhat misleading. Adopted many years ago, they refer to the electrical conductivity of the different grades. All three types are made from high-grade materials and meet the same standard of galvanizing.

Extra best best wire has the best conductivity but the lowest tensile strength. Its weight per mile-ohm is from 4700 to 5000 lb. This wire is uniform in quality, pure, tough, and pliable. It is used largely by commercial telegraph companies, in railway telegraph service, for tie wires, and for signal bonding.

Best best wire has a lower value of conductivity but greater tensile strength. Its weight per mile-ohm is from 5600 to 6000 lb. This grade is used very largely by telephone companies.

Steel is a stiff wire of high tensile strength and low conductivity. It is very difficult to work but is used on short lines which must be erected at low cost and for which conductivity is of little importance. Its weight per mile-ohm is 6500 to 7000 lb.

Additional types have been developed more recently. These newer types along with their special characteristics are given in Table 102.

**17. Copper-steel conductors.** Electrical conductors consisting of a combination of copper and steel are frequently employed for certain types of circuits (see Sec. 13 and Tables 87 and 111). Three general types of construction are listed in Sec. 13 and are illustrated in Table 86.

*Copperweld wire* is composed of a steel core with a copper covering thoroughly welded to it by a molten welding process. This process produces a permanent bond between the two metals which prevents any electrogalvanic action and which will withstand hot rolling, cold drawing, forging, bending, twisting, or sudden temperature changes. This wire is made in three grades: 30 percent conductivity, extra-high strength; 30 percent conductivity, high strength; and 40 percent conductivity, high strength (see Table 110). Solid conductors are made in sizes from No. 12 to No. 4/0 AWG (see Table 112). Concentric-stranded cables are available in sizes with outside diameters from 0.174 to 0.910 in (0.442 to 23.1 mm; see Table 114). Weatherproof covered Copperweld solid wire is made in sizes from No. 12 to No. 2 AWG. Rubber-insulated twisted-pair cables in sizes Nos. 14 and 17 are made for telephone, telegraph, and signal work. Single-conductor rubber-insulated wire may be obtained in any size. Rubber-insulated parallel drop wire is made in one size only, No. 17 AWG.

Composite cables *(Copperweld and copper)* consisting of a combination of certain strands of copper wire with a number of strands of Copperweld wire often are economical for aerial circuits requiring more than average tensile strength combined with liberal conductance (see Table 113 for data).

*Copper and steel cables* consist of a combination of copper and steel wires stranded together. The strands of the different materials are not intended to serve in any dual capacity, and the conductivity of the cable is determined by the sectional area of the copper strands. The size of the cable is designated according to its total sectional area of copper, not according to the total sectional area of the whole cable. For instance, a cable designated as No. 4, consisting of two copper and one steel strands, has a total sectional area of 62,610 cmil (31.7 mm²) and a sectional area of copper of 41,740 cmil (21.1 mm²). The area of the copper, 41,740 cmil, corresponds to the area of a No. 4 wire. The cable is therefore designated as a No. 4 cable, although its total sectional area of copper and steel is between No. 3 and those of No. 2 wire.

These copper and steel cables are available in sizes from No. 2 to No. 12. All sizes except Nos. 10 and 12 consist of two plain hard-drawn copper wires and one extra-galvanized-steel wire. Wire cables of sizes Nos. 10 and 12 consist of one galvanized hard-drawn copper wire and two extra-galvanized-steel wires (see Table 123).

**18. Copper-clad aluminum.**   Copper-clad aluminum is the newest conductor material on the market. A copper-clad aluminum conductor is drawn from copper-clad aluminum rod, the copper being bonded metallurgically to an aluminum core. The copper forms a minimum of 10 percent of the cross-sectional area of the solid conductor or of that of each strand of a stranded conductor.

Although copper-clad aluminum contains only 10 percent of copper by volume (26.8 percent by weight), its electrical performance is equivalent to that of pure copper. It is lighter and easier to handle, and the price advantage, which reflects the value of the copper content, can be as much as 25 percent when copper peaks to one of its periodic highs. Detailed studies by Battelle Laboratories have shown that copper-clad aluminum and copper have the same connection reliability.

Because the electrical industry consumes 60 percent of all copper used in the United States, it is critically affected by copper's fluctuating costs and uncertain supply. Until recently, however, aluminum was the only alternative to copper.

Aluminum, in the more than 40 years since its introduction as an electrical conductor, has significantly penetrated such areas as electric power transmission lines, transformer windings, and telephone communications cables. On the other hand, it has received relatively limited acceptance in nonmetallic-sheathed cable and other small-gage building wires. The reason has been a lack of acceptable means of connecting or terminating aluminum conductors of No. 6 AWG or smaller cross-sectional areas.

Connector manufacturers, the National Electrical Manufacturers Association (NEMA), Underwriters Laboratories (UL), and aluminum companies have devoted much attention to this connection problem. The most significant advance in aluminum termination has been the institution of UL's new requirements and testing procedures for wiring devices for use in branch-circuit–size aluminum conductors. Devices which meet the revised UL requirements are marked CO/ALR and carry that mark on the mounting strap. Only CO/ALR switches and receptacles should be used in aluminum 15- and 20-A branch-circuit wiring.

Copper-clad aluminum is now available to counter the disadvantages of high price and lack of availability of copper and the problems of connection reliability of aluminum. It is a product of a metallurgical material system, i.e., a system in which two or more metals are inseparably bonded in a design that utilizes the benefits of each component metal while minimizing their deficiencies. In copper-clad aluminum conductors, the electrical reliability of copper is combined with the abundant supply, stable price, and light weight of aluminum. Copper-clad aluminum is already being used for building wire, battery cable, magnet wire, and radio-frequency (rf) coaxial cable.

The ampacity (current-carrying capacity) of copper-clad aluminum conductors is the same as that of aluminum conductors. It is required that the wire connectors used with copper-clad aluminum conductors be recognized for use with copper and aluminum conductors and be marked AL-CU, except that No. 12-10 AWG solid copper-clad aluminum conductors may be used with wire-binding screws and in pressure-plate and push-in spring-type connecting mechanisms that are recognized for use for copper conductors. Copper-clad aluminum conductors are suitable for intermixing with copper and aluminum conductors in terminals for splicing connections only when the wire connectors are specifically recognized for such use. Such intermixed connections are limited to dry locations.

Effective January 2, 1987, UL 486C was revised with regard to the performance requirements for wire connectors for use with aluminum and copper-clad aluminum conductors. The effect of this change is that twist-on–type wire connectors, often called "wire nuts," are required to meet new performance requirements if listing is to be continued. As of this writing, no wire nuts have been listed under these requirements. The crimp-type connector (Copalum by AMP) is at present the only UL-listed connector for splicing small CU-to-AL conductors.

**19. Aluminum and aluminum steel-reinforced conductors** often are economical for transmission and rural distribution circuits. Commercial hard-drawn aluminum wire has a conductivity at 20°C of 60.97 percent or a resistance of 17.010 $\Omega$/cmil·ft. Its weight is 0.000000915 ($91.5 \times 10^{-8}$) lb/cmil·ft. An all-aluminum wire for equal conductivity must have a diameter 126 percent and an area 160 percent of that of a copper wire (see Table 115 for comparative values). All-aluminum conductors are available in bare, weatherproof-covered, rubber-insulated, and thermoplastic-insulated constructions. Aluminum steel-reinforced conductors, owing to their high tensile strength, are frequently used for long spans and for high-capacity lines requiring heavy conductors. They consist of a concentric-stranded aluminum cable with a reinforcing steel core. Except in the smaller sizes the steel core is made up of several steel strands. Aluminum steel-reinforced conductors are available either bare or weatherproof-covered (refer to Tables 121 and 122 for data).

**20. Number of conductors.**   Bare or covered (not insulated) cables must, of course, always be single-conductor. Insulated cables are manufactured with one, two, three, or even more conductors per cable. The choice between single- or multiconductor cable is affected by so many factors, which vary with the particular installation, that only general suggestions can be made here. The selection will be influenced by practical field conditions and facilities of installation, cost of cable and enclosure, physical dimensions of the cable and of the available enclosure, electrical load requirements, and voltage and type of power supply system. The use of multiconductor cables usually results in lower cable cost, smaller voltage drop, and more economical utilization of duct space. On the other hand, single-conductor cables are more flexible and are easier to splice and install. For underground transmission and primary distribution applications, single-conductor cable is generally used when the voltage exceeds 35 kV or when the load is 50,000 kVA or more. For all other underground transmission and primary distribution applications, both single-conductor and multiconductor cables are employed. In the past the use of multiconductor cable was most common, but the use of single-conductor cable in this field is rapidly growing. Single-conductor cable is generally preferred for secondary distribution because of the ease in making taps. For interior building wiring, single-conductor cable is most commonly employed in raceway systems.

**21. Cable assembly.**   For the application of electrical conducting wires and cables, it is essential to have a general knowledge of the materials employed in their manufacture and also of the manner of assembling the component parts in the formation of the finished cable. For a bare wire the assembly is, of course, very simple, consisting only of the solid or stranded wire of the conductor. For insulated cables and especially for multiconductor cables the use of many materials and different forms of assembly is involved. The component parts of the more common insulated cables, in the order in which the parts are employed in the manufacture of the cables, are as follows:

A.  Single-conductor cables
   1.  Conductor
   2.  Insulation
   3.  Protective covering
B.  Multiconductor cables
   1.  Without shielding or belting (no insulation around group of conductors)
      *a.*  Conductor
      *b.*  Conductor insulation
      *c.*  Conductor covering
      *d.*  Fillers
      *e.*  Protective covering

   2.  Belted type (insulation around group of conductors)
      *a.*  Conductor
      *b.*  Conductor insulation
      *c.*  Conductor covering
      *d.*  Fillers
      *e.*  Belt insulation
      *f.*  Protective covering
   3.  Shielded type
      *a.*  Conductor
      *b.*  Conductor insulation
      *c.*  Conductor shield
      *d.*  Fillers
      *e.*  Binder tape
      *f.*  Protective covering

**22. Electrical shielding** is often necessary on power cable to confine the dielectric field to the inside of the cable insulation so as to prevent damage from corona or ionization. The shield usually consists of a thin (3-mil, or 0.076-mm) conducting tape of copper or aluminum applied over the insulation of each conductor. The shielding tape sometimes is perforated to reduce power losses due to eddy currents set up in the shield. Sometimes semiconducting tapes consisting of specially treated fibrous tapes or braids are used. These semiconducting tapes are frequently employed for the shielding of aerial cable, since they adhere more closely to the insulation and thus tend to prevent corona.

**23. Fillers** are used in the manufacture of most multiconductor cables. Their purpose is to fill the spaces between the conductors to produce a solid round structure for the complete cable. The materials commonly employed for these fillers are saturated jute, asbestos-base caulk, rubber, and, in some cases, cotton.

**24. Binder tapes** are used in the construction of many multiconductor cables to bind conductors, shields, and fillers together in proper form during the addition of the protective covering. The more common types of binder tapes are rubber-filled cloth tape, combinations of cotton cloth and rubber compounds, steel, and bronze.

**25. Insulation of electrical conductors.** Except for aerial construction, interior exposed wiring on insulators, and special cases of interior-wiring feeder circuits, electrical conductors must be covered with some form of electrical insulation. An ideal insulating material for this purpose should have the following characteristics:

1. Long life
2. Long-time high dielectric strength
3. Resistance to corona and ionization
4. Resistance to high temperature
5. Mechanical flexibility
6. Resistance to moisture
7. Low dielectric loss

It is impossible to find any one material that is best when all these essential characteristics are considered. For instance, impregnated paper has the highest electrical breakdown strength coupled with the longest life of all the materials employed for the insulation of conductors. On the other hand, it is not moisture-resistant, is not so flexible as some other materials, and will not withstand such high temperatures as asbestos. Several different types of insulation are therefore employed. The insulation whose overall characteristics best meet the conditions of service for the particular application should be selected. Tables 76, 128, and 130 of Div. 2 classify the different types of insulation with respect to temperature.

**26. Rubber insulation.** With respect to insulation *rubber* is the word used to designate insulations consisting of compounds of natural rubber or synthetic rubber, or both, combined with such other ingredients as vulcanizing agents, antioxidants, fillers, softeners, and pigments. These natural-rubber and synthetic rubberlike compounds are used more than any other material for the insulation of electrical conductors. They have the desirable characteristics of moisture resistance, ease of handling and termination, and extreme flexibility. On the other hand, they will not withstand such high temperatures or voltage without deterioration as will some other types of insulation. The different ingredients of each compound are combined by the process of vulcanization into a single homogeneous material.

A large variety of rubber compounds are available, with different characteristics that depend upon the particular service conditions for which they have been developed. It is feasible to include here only those in most common use. For applications which require special characteristics experts of the cable manufacturers should be consulted for advice on the best available compound.

In general, the physical and electrical properties of a compound increase with the rubber content. However, rubber alone would not provide suitable insulation. The proper choosing and proportioning of the other ingredients are important in obtaining the desired characteristics.

*Heat-resisting rubber compounds* which will withstand considerably higher temperatures than the Code-grade rubber compound have been developed. Such compounds meeting Underwriters Laboratories specifications are designated as *Type RH and Type RHH insulation.*

*Moisture-resistant rubber compounds* are available for installations in which the wire will be subjected to wet conditions. These compounds are called moisture-resistant and submarine compounds.

*Moisture- and heat-resistant compounds* combine the temperature- and moisture-resistance characteristics of Type RH and the Type RW insulations. They are recognized by the Underwriters Laboratories designation of *Type RHW,* which is approved for use in wet or dry locations at a maximum conductor temperature of 75°C.

*Type SA wire,* called silicone-asbestos wire, is insulated with a silicone rubber compound. The outer covering of this type of wire must consist of heavy glass, asbestos-glass, or asbestos braiding that is impregnated with a heat-, flame-, and moisture-resistant compound.

Many wire and cable manufacturers have a rubber compound for use on general power cables which is of a higher quality than required by the Underwriters Laboratories specifications. These especially high-grade rubber compounds are designated by the individual manufacturer's trade name. Several other nonstandardized compounds, designated by the individual manufacturer's trade name, are available to meet special conditions of service, such as resistance to chemicals, ozone, and corona.

A low-voltage compound called *neoprene*, although not a rubber compound, is recognized by the Underwriters Laboratories as an approved insulation for Type RHW building wire without additional covering over the insulation. It is also widely used in many applications for which its superior mechanical properties are needed, such as self-supporting secondaries for services and service-drop cables, overhead line wires, and motor and appliance lead wires.

**27. Thermoplastic insulation.**   Thermoplastic compounds have been developed for insulations of electrical wires. Those meeting the specifications of the Underwriters Laboratories are designated as *Type TW, THW, THWN, THHN, TA, TBS,* or *MTW*. Type TW, THW, or THWN is moisture-resistant and can be used in wet locations. Type THW or THWN is both moisture- and heat-resistant and is approved for use in wet or dry locations at a maximum conductor temperature of 75°C. Type TBS is thermoplastic insulation with a flame-retardant fibrous outer braid, and it is acceptable only for switchboard wiring. Type TA is a combination thermoplastic-and-asbestos insulation used for switchboard wiring only. Type THWN is a moisture- and heat-resistant insulation that is approved for both wet and dry locations at a maximum conductor temperature of 75°C. It has an outer nylon jacket. Type THHN is similar to Type THWN, except that it is not moisture-resistant and may be used only in dry locations at a maximum conductor temperature of 90°C. Type MTW is moisture-, heat-, and oil-resistant and is restricted to use in machine-tool wiring at a maximum conductor temperature of 60°C in wet locations and 90°C in dry locations.

**28. Thermosetting insulation.**   Three high-quality wires are *Types XHHW, FEP,* and *FEPB*. Type XHHW is a moisture- and heat-resistant, cross-linked, thermosetting polyethylene with an insulating rating of 75°C in wet locations and 90°C in dry locations. Types FEP and FEPB have heat-resistant, fluorinated ethylene propylene insulation and are suitable for use in dry locations at a maximum conductor temperature of 90°C and 200°C for special applications.

**29. Mineral insulation.**   The objective in developing mineral insulation was to provide a wiring material which would be as noncombustible as possible and thus eliminate the hazards of fire resulting from faults or overloads. This insulation consists of highly compressed magnesium oxide and is designated as *Type MI cable*. The outer metallic sheath is copper. A complete description of this cable is given in Sec. 54 of this division and in Div. 9.

**30. Paper insulation.**   Impregnated-paper insulation provides the highest electrical breakdown strength, greatest reliability, and longest life of any of the materials employed for the electrical insulation of conductors. It will safely withstand higher operating temperatures than either rubber or varnished-cambric insulations. On the other hand, it is not moisture-resistant and must always have a covering which will protect the insulation from moisture, such as a lead sheath.

Paper-insulated cables are not so flexible and easy to handle as varnished-cambric or rubber-insulated cables and require greater care and time for the making of splices. They are available in the following types:

1. Solid-type insulation
2. Low-pressure gas-filled
3. Medium-pressure gas-filled
4. Low-pressure oil-filled
5. High-pressure oil-filled (pipe enclosed)
6. High-pressure gas-filled (pipe enclosed)
7. High-pressure gas-filled (self-contained)

The *solid-type insulation* is composed of layers of paper tapes applied helically over the conductor and impregnated with mineral oil (sometimes mixed with resin when so specified). A tightly fitting lead sheath is extruded over the assembled conductors and impregnated insulation. The oil must be heavy enough to prevent bleeding when the cable is cut for splicing and terminations but at the same time must remain semifluid at the lowest operating temperatures. For cables installed as vertical risers, on steep grades, or under high operating temperatures, a heavier oil, designated as a nonmigrating compound, is sometimes employed to prevent the migration of the oil from the high to the low points of the cable. The ordinary solid-type impregnated-paper insulation will give off flammable and explosive gases when exposed to extremely high temperatures. It is common practice to clear cable failures on low-voltage network cables by leaving the power on and burning the fault clear. To reduce the possibility of damage to raceways and manhole structures from consequent explosions, special nonflammable, nonexplosive compounds are sometimes used to impregnate the paper insulation.

*Low-pressure gas-filled cable,* operating under nitrogen gas pressure of 10 to 15 lb/in² (68,947.6 to 103,421.4 Pa), is manufactured and impregnated in the same manner as solid-type cable, but prior to leading it is drained in a nitrogen atmosphere at a temperature somewhat above the maximum allowable operating temperature. To give the gas free access to the insulated conductors, longitudinal gas feed channels are placed in the filler interstices of the three-conductor construction and flutes or another type of channel under the sheath of the single-conductor construction. In the three-conductor construction, two of the three gas feed channels are obtained either by omission of filler material from the interstices or by use of an open helical coil made from steel strip. The third channel is a solid-wall metal tube filled with dry nitrogen gas and sealed off at each end before treatment of the cable length. The end of this solid-wall metal channel is opened at each joint location and ensures positive control of the pressure over the entire cable length by furnishing a bypass path for the gas at low points or dips in the cable run where "slugs" of surplus compound may gradually collect in service in the open-wall channels.

Cable operating records indicate that most service failures are attributable to damage to lead sheaths arising from a variety of causes or to leaks in joint wipes, terminals, or other accessories. An important advantage of low-pressure gas-filled cable, in common with low-pressure oil-filled cable, is that service failures from these causes are practically eliminated because a positive internal pressure is maintained continuously within the sheath, the entrance of moisture is prevented, and operation can continue until it is convenient to make repairs.

*Medium-pressure gas-filled cables* operate under a nitrogen pressure of about 49 lb/in² (337,843 Pa). They are constructed in a manner similar to the low-pressure gas-filled cables, except that a reinforced lead sheath is employed to permit the use of the increased gas pressure.

In *low-pressure oil-filled cables* the paper is impregnated with a relatively thin liquid oil which is fluid at all operating temperatures. The cable is so constructed that channels are provided for longitudinal flow of the oil, and oil reservoirs are furnished at suitable points in the cable installation. A positive pressure of moderate magnitude is thus maintained on the oil at all times, which prevents the formation of voids in the insulation due to changing temperature, stretching, or deformation of the lead sheath. When the cable is heated by load, the oil expands and flows lengthwise through the channels of the cable into the joints and out into the reservoirs. When the cable cools and the oil in the cable contracts, oil is forced back through the channels of the cable from the oil reservoirs. Any damage to the lead sheath will allow moisture to enter a solid-type insulated cable. With oil-filled cable, unless the damage is too severe, the positive internal oil pressure will prevent the entrance of moisture, so that although there will be some loss of oil, operation can continue until it is convenient to make repairs.

*High-pressure oil-filled cables (pipe enclosed)* are insulated and impregnated in the same manner as single-conductor solid-type cables. The necessary number of single-conductor cables are then pulled into a welded steel line pipe which is protected by a high-grade corrosion-protective covering. The enclosing pipe is then filled with oil, which is maintained under constant high pressure. An oil pressure of approximately 200 lb/in² (1,378,951 Pa) prevents the formation of voids in the insulation and is maintained by oil pumps at one or more points on the line.

To provide protection against moisture entrance during shipment, the cable may be shipped with a temporary lead sheath or on a special weathertight reel without a lead covering. In the former case a thin temporary lead sheath is extruded over the skid wires and is stripped from the cable during the pulling operation. When the cable is shipped without a temporary lead sheath, all seams of the weathertight reel are carefully welded.

The outer layer of cable is covered by a "blanket" of material having a very low rate of moisture absorption and diffusion. The edges of the blanket are tightly sealed to the inside surface of the reel flange by a moisture-repellent plastic adhesive and sealing compound. Joints between external mechanical protection, such as lags, sheet-metal layers, etc., and the steel rim of the reel flange are sealed in a similar manner. Each reel is also equipped with a pocket containing a desiccant, and prior to shipment the reel interior is flushed carefully with dry nitrogen gas.

*High-pressure gas-filled cable (pipe enclosed)* is similar to the high-pressure oil-filled type in that the insulated conductors comprising the circuit are installed in a metal pipe but differs from it in that nitrogen gas at a pressure of approximately 200 lb/in² (1,378,951 Pa) is employed as a pressure medium in place of oil.

In the mass-impregnated type, the paper tape is applied to the conductor, and the cable is impregnated in the same manner as the solid type. Shipment is carried out in the same manner as that for high-pressure oil-filled cables, either with a temporary lead sheath or on a weathertight reel without a lead covering.

*High-pressure gas-filled (self-contained) cables* use a lead sheath reinforced with metallic tapes as the pressure-retaining member in place of the steel pipe. In these cables, the permanent lead sheath and the associated reinforcement and protective coverings are applied to the cable prior to shipment from the factory.

A special grade of untreated paper is sometimes employed for insulating magnet wires. It is applied to the conductor in ribbon form as a helix with approximately one-third to one-half lap. This procedure provides a low-cost insulation of constant thickness and slightly higher dielectric strength than that provided by cotton yarn, but it is less sturdy.

**31. Varnished-cambric insulation** has characteristics which in almost every respect are midway between those of rubber and paper. It is more flexible than paper but not so flexible as rubber except for large insulated cables. It is reasonably moisture-resisting, so that it need not always be covered with a lead sheath, but it cannot be operated without such protective covering if it is continuously immersed or is in continuously moist surroundings. With respect to dielectric strength, allowable temperature, and resistance to ionization and corona, it is better than rubber but not so good as impregnated paper. Varnished cambric is not affected by ordinary oils and greases and will withstand hard service.

The term *varnished cambric* is misleading, since the cotton-fabric base of the insulation is not cambric. The correct designation should be *varnished cloth*, but because of long-established custom the material is designated as varnished-cambric insulation. Varnished-cambric–insulated cable consists of conductors helically wrapped with cotton tape which has been previously filled and coated on both sides with insulating varnish. During the wrapping process a heavy nonhardening mineral compound is applied between the tapings to act as a lubricant when the cable is bent and also to fill all spaces so as to prevent ionization and possible capillary absorption of moisture.

**32. Asbestos** provides a truly heat-resisting insulation suitable for use at temperatures beyond the limits allowable for other standard forms of cable insulation. However, it is satisfactory only for low-voltage installations (not more than 8000 V), since it cannot be applied to the conductors so as to give high dielectric strength to the cable insulation.

*Type SA.* Silicone rubber insulated with outer heavy glass, asbestos-glass, or asbestos braid. See Sec. 26.

**33. Cotton yarn** is often employed for the insulation of magnet wire. The wire is covered with one or more wraps of helically wrapped unbleached cotton yarn. When more than one wrap is employed, each wrap is applied in the reverse direction to that of the next inner wrap. The untreated cotton is neither heat-resisting nor impervious to moisture. It is not fully considered to be an insulation, and coils wound with it should be treated or impregnated with an insulating compound by an approved process. The cotton yarn acts as a mechanical separator, holding the conductors apart and providing a medium for the absorption and retention of the impregnating material. Cotton yarn is also used as a protective covering for enamel-insulated magnet wire.

**34. Enamel** is widely used as an insulation for magnet wire. It has excellent resistance to moisture, heat, and oil and possesses high dielectric strength. Enamel wire consists of a comparatively thin, even coating of high-grade organic insulating enamel applied directly to the bare wire. It is made in various types as given in Sec. 82.

**35. Silk yarn** is employed in the same manner as cotton yarn (Sec. 33) for the insulation of magnet wire. Silk coverings have better dielectric characteristics, give a neater appearance, and are mechanically stronger than cotton. Otherwise, all the statements in Sec. 33 respecting cotton-yarn insulations apply to silk.

*Cellulose-acetate (artificial silk) tape or yarn* is used in many applications for the insulation of magnet wire as a substitute for or in combination with natural silk. Artificial silk has somewhat better electrical characteristics and gives a more uniform thickness of insulation. Its strength and abrasion resistance are not so good as those of natural silk.

**36. Fibrous-glass yarn** has been developed for the insulation of conductors. Up to the present time it has been employed only for the insulation of dynamo windings and magnet wire for classes of service requiring operation at high temperatures. The conductors are wound with one or more wraps of alkali-free fibrous-glass yarn.

**37. Protective coverings.** The insulation of most insulated conductors is protected from wear and deterioration due to surrounding conditions by a covering applied over the insulation. Protective coverings are also used on some noninsulated cables such as the weatherproof wire used for distribution purposes (see Secs. 10 and 11). The materials most commonly used for these protective coverings are listed in Sec. 38. No one covering will fulfill all the protective functions that are required for all classes of installations. Each has its particular advantages and limitations and consequently its proper field of application. In many cases a combination of two or more types of covering is required to provide the necessary protection to the conductor and its insulation.

**38. Protective-covering materials and finishes**

I. Nonmetallic
  A. According to material of covering
    1. Fibrous braids
      *a.* Cotton
        (1) Light
        (2) Standard
        (3) Heavy
        (4) Glazed cotton
      *b.* Seine twine or hawser cord
      *c.* Hemp
      *d.* Paper and cotton
      *e.* Jute
      *f.* Asbestos
      *g.* Silk
      *h.* Rayon
      *i.* Fibrous glass
    2. Tapes
      *a.* Rubber-filled cloth tape
      *b.* Combination of cotton cloth and rubber compounds
    3. Woven covers (loom)
    4. Unspun felted cottom
    5. Rubber jackets
    6. Synthetic jackets
    7. Thermoplastic jackets
    8. Jute and asphalt
  B. According to saturant
    1. Asphalt
    2. Paint
    3. Varnish
  C. According to finish
    1. Stearin pitch and mica flake
    2. Paint
    3. Wax
    4. Lacquer
    5. Varnish
II. Metallic
  A. Pure lead sheath
  B. Reinforced lead sheath
  C. Alloy-lead sheath
  D. Flat-band armor
  E. Interlocked armor
  F. Wire armor
  G. Basket-weave armor

**39. Fibrous braids** are used extensively for protective coverings of cables. These braids are woven over the insulation so as to form a continuous covering without joints. The braid is generally saturated (Sec. 45) with some compound to impart resistance to some class of exposure such as moisture, flame, or acid. The outside braid is given one of the finishes described in Sec. 45 in accordance with the application of the cable.

The most common braid is one woven from *light, standard, or heavy cotton yarn.* Such coverings are designated, respectively, as ICEA Classes A, B, and C (ICEA are the initials of the Insulated Cable Engineers Association). The cotton can be furnished in a variety of colors for identification in accordance with an established color code in the industry.

*Glazed-cotton braid* is composed of light cotton yarn treated with a sizing material before fabrication.

*Seine-twine or hawser-cord braid* is composed of cable-laid, hard-twisted cotton yarn, braided to form a heavy, durable covering which will withstand more rough usage than a braid made from common cotton yarn.

*Hemp braid* is woven from strong, durable, long-fibered hemp yarn which will withstand even rougher usage than seine-twine braid.

*Paper-and-cotton braid* is composed of paper twine interwoven with cotton threads.

*Jute braid* is woven from yarn composed of twisted jute fibers.

*Silk and rayon braids* are manufactured in the same manner as glazed-cotton braid from real or artificial silk yarn.

*Fibrous-glass braid* is woven from fine, flexible glass threads and forms a covering resistant to flame, acids, alkalies, and oils.

**40. Fibrous-tape coverings** are frequently used as a part of the protective covering of cables. The material of tape coverings is fabricated into a tape before application to the cable, while the yarn in braid coverings is woven into a fabric during application to the cable. In applying tape coverings, the tape is wrapped helically around the cable, generally with a certain amount of overlapping of adjacent turns. The more common types of fibrous tapes employed in cable manufacture are listed in Sec. 38. Except for duck tape, tape coverings are never used for the outer covering of a cable. They are employed for the covering directly over the insulation of individual conductors and for the inner covering over the assembled conductors of a multiconductor cable. Frequently they are used under the sheath of a lead-sheathed cable. Duck tape made of heavy canvas webbing presaturated with asphalt compound is frequently used over a lead-sheathed cable for protection against corrosion and mechanical injury.

**41. Woven covers** commonly called loom are used for applications requiring exceptional abrasion-resisting qualities. These covers are composed of thick, heavy, long-fibered cotton yarn woven on the cable in a circular loom like that used for fire hose. They are not braids. Although braid coverings also are woven, they are not designated as such.

**42. Unspun felted cotton** is manufactured into a solid felted covering for cables for some special classes of service.

**43. Several types of rubber and synthetic jacket coverings** are available for the protection of insulated cables. These types of coverings do not seem to be standardized. The different manufacturers have their own special compounds designated by individual trade names. These compounds differ from the rubber compounds used for the insulation of cable in that they have been perfected not for insulating qualities but for resistance to abrasion, moisture, oil, gasoline, acids, earth solutions, alkalies, etc. Of course, no one jacket compound will provide protection against all these exposures, as each has its particular qualifications and limitations.

A nonrubber compound with excellent characteristics for a jacket material is called neoprene.

Thermoplastic compounds are becoming more and more popular as jacket materials. Although no one material is perfect in all respects for jacket coverings, thermoplastic compounds possess many of the necessary characteristics for the coverings.

**44. Jute and asphalt coverings** are commonly used as a cushion between the cable insulation and a metallic armor. Frequently they are also employed as corrosion-resisting coverings over a lead sheath or metallic armor. They consist of asphalt-impregnated jute yarn served helically around the cable or of alternate layers of asphalt-impregnated jute yarn serving and asphalt weatherproofing compound.

**45. Saturants and finishes for fibrous coverings.**   Fibrous braids used for covering cables are thoroughly saturated, and the outer surface is finished to provide protection against moisture, flame, weathering, oil, etc. The common materials employed for saturating the braids are listed in Sec. 38*B*, and the common materials employed for finishing the outer surface in Sec. 38*C*.

**46. A pure lead sheath** of uniform thickness tightly applied over an insulated cable by the extrusion process is practically standard covering for use in underground raceways and other wet locations. If the cable will be exposed to special forms of corrosion, electrolytic action, or mechanical strain, one of the other coverings is used in combination with the lead sheath as a protective covering over the sheath.

**47. A reinforced lead sheath** is employed for mechanical strength when internal cable hydrostatic pressures exceed 15 lb/in² (103, 421.4 Pa). This construction consists of a double lead sheath. Around the inner sheath is wrapped a thin tape of hard-drawn copper, bronze, or other elastic metal, preferably nonmagnetic in character. This tape imparts considerable additional strength and elasticity to the sheath. However, it must be protected against wear and corrosion. For this reason a second lead sheath is applied over the tape. Such a finish is recommended for internal pressures up to 30 lb/in² (206,842.7 Pa). It is commonly used on oil-filled cable near the bottom of severe grades or on solid cable at the base of vertical risers and at the bottom of extreme grades where pressure can accumulate.

**48. An alloy-lead sheath** is employed when additional mechanical strength and resistance to crystallization of the sheath are required. The most common alloy is one containing 2 percent tin. Sometimes an antimony-lead alloy is used. An alloy sheath is more resistant to gouging and abrasion during or after installation than a pure lead one.

**49. Flat-band armor** usually consists of two steel tapes applied on the outside of the cable so that the openings between successive turns of the inner tape are covered by the outer tape. Usually these tapes are applied over a jute bedding introduced between them and the lead sheath or cable insulation, and they are frequently finished with an overall layer of asphalted jute as a protection against corrosion. Ordinarily, plain steel tapes are recommended. If corrosion conditions are apt to be severe, galvanized steel or nonferrous materials should be used.

**50. Interlocked armor** consists of a single strip of interlocking metal tape so applied over the insulation of the cable that the cable is always protected throughout its length. The interlocking construction prevents adjacent tapes from being squeezed together during installation or in service, with consequent damage to the edges of the tapes, which sometimes occurs with flat-band steel armor; in addition, it eliminates gaps or open spaces, as there is no break in the continuity of the interlocking strip. The armor is so shaped that its physical strength against mechanical injury to the cable is superior to that of flat-band armor. Its rounded surface tends to deflect blows from shovels or picks and also offers an additional buffer effect to blows that tend to pierce the armor, thereby minimizing damage to the lead sheath and insulation over which it is applied.

For rubber- or varnished-cambric–insulated cables used for interior work the armor is generally applied over a jute bedding which is introduced between the armor and the cable insulation.

For underground installation a lead sheath under the jute bedding of armor is required for paper cables, varnished-cambric cables, or 20 or 25 percent Code-rubber cables.

An overall asphalt-jute finish is sometimes used as additional protection against corrosion, particularly if the cable is to be buried in the earth.

A thermoplastic jacket is often applied over the armor, particularly if corrosive conditions exist, as in chemical plants, paper and steel mills, and similar industrial establishments.

Galvanized steel is the standard material for armor employed in interior building wiring. For other power applications plain steel armor is standard and suitable for most installations, although galvanized armor will give a longer life. Nonmagnetic materials may be furnished for the larger single-conductor cables. Aluminum, copper, or bronze is frequently used when corrosion conditions are apt to be severe.

**51. Wire armor** consists of a layer of round metal wires wound helically and concentrically about the cable. The standard type employs galvanized-steel wires, but if desired, wire armor constructed of nonferrous materials can be obtained. Wire armor is recommended when extreme tensile strength and greatest mechanical protection are necessary. Galvanized-steel construction is stronger than nonferrous. Usually a layer of asphalt-treated jute is introduced between the armor and the sheath to prevent mechanical damage and to minimize electrolytic action. Galvanized armor can be used when corrosion would damage the iron, and further protection in the form of an overall asphalt-jute finish is not uncommon.

**52. Basket-weave armor** is used when light weight and compactness are important. It consists of a braid of metal wire woven directly over the cable as an outer covering. Galvanized-steel, bronze, copper, or aluminum wire is used for the braid, depending upon the service for which the cable is intended. This armor finish is used largely for shipboard wiring.

**53. General-purpose power cables** for underground raceways and interior wiring are available with rubber-, thermoplastic-, varnished-cambric–, paper-, or asbestos-insulated copper or aluminum conductors. The particular type of such an insulation may be any of the types previously listed, depending upon the requirements of the service.

Class B stranding is the adopted standard for copper conductors of these cables, but any of the more flexible standard concentric or rope strandings is available. In addition, varnished-cambric– and paper-insulated copper conductor cable may be obtained with the following construction:

| *Construction* | *Available for* |
|---|---|
| Compack round | Single conductor from No. 1 to 1,000,000. Multiconductor from No. 1 to 1,000,000. Generally employed only for No. 1 |
| Compack sector | Multiconductor No. 1/0 and larger |
| Annular concentric | Single conductor, 750,000 to 5,000,000 cmil |
| Compack segmental | Single conductor, 1,000,000 to 4,000,000 cmil |
| Hollow-core stranded | Single conductor, oil-filled paper cables, No. 2/0 to 2,000,000 cmil |

General-purpose power cables are normally constructed in single- and three-conductor cables for all the types of insulations mentioned. Rubber-insulated wires are also regularly made in two-conductor construction. Four-conductor cables are not standard construction but can be obtained.

One common type of protective covering consists of a rubber-filled cloth tape and an outer cotton braid treated with moisture-resisting, flame-resistant, or moisture- and flame-resistant compounds. Another common protective covering is a neoprene jacket. If the cable will be subjected to rough usage, the outer braid should be made of seine twine, hawser cord, or hemp. If high flame resistance is required, asbestos braid should be used for the protective covering. For underground raceway installation and interior wiring in wet locations, suitable types should be used. If protection against corrosion, electrolysis, or damage due to accidental short-circuiting of open-circuited sheaths is required, the lead sheath is protected by an outer covering of duck tape, reinforced rubber, or jute and asphalt. For cable subjected to special exposures such as acids, alkalies, and gasolines, a suitable rubber or synthetic sheath should be employed (see Sec. 43). Thermoplastic- and latex-insulated wires do not require an outer protective covering. Interlocked metal armor is advantageous as the outer covering for many installations, both indoor and outdoor.

The covered (not insulated) wires described in Sec. 10 are included under the classification of general-purpose power cables.

General-purpose power cables with insulation and covering meeting the minimum requirements of the **National Electrical Code** are frequently called building wires and cables. The **Code** designation letters for these cables according to the type of insulation employed are given in Div. 9. In addition to these type letters, the **Code** uses the following suffix letters:

1. No suffix letter indicates a single insulated conductor.

2. Suffix letter *D* indicates a twin wire with two insulated conductors laid parallel under an outer nonmetallic covering.

3. Suffix letter *M* indicates an assembly of two or more insulated conductors twisted together under a common outer nonmetallic covering.

If no number follows the designating letters of a cable, the cable is for use at not more than 600 V. Cables for use at higher voltages are designated by adding the numerical suffixes in the accompanying table to the letter designations.

The maximum voltages in the table are the operating voltages between phases of single- and two-phase systems and three-phase systems with grounded or ungrounded neutral.

| Numerical suffix | Maximum permissible voltage | Numerical suffix | Maximum permissible voltage |
|---|---|---|---|
| 10 | 1,000 | 40 | 4,000 |
| 20 | 2,000 | 50 | 5,000 |
| 30 | 3,000 | | |

**54. MI (mineral-insulated) cable** was developed to meet the needs for a noncombustible, high-heat–resistant, and water-resistant cable. As described by the **National Electrical Code**, "MI cable [see Fig. 2-5] is a cable in which one or more electrical conductors are insulated with a highly compressed refractory mineral insulation and enclosed in a liquid-tight and gastight metallic tube sheathing." It is available in single-conductor construction in sizes from No. 16 to No. 250 kcmil, in two- and three-conductor construction in sizes from No. 16 to No. 4, in four-conductor construction in sizes from No. 16 to No. 6 and in seven-conductor construction in sizes from No. 16 to No. 10.

The cable is fabricated in the following manner. Magnesium oxide in powder form is molded under great pressure into cylinders approximately 2 in (50.8 mm) in diameter and length, with as many holes through the body of the cylinder as there are conductors in the cable. After molding, the magnesium oxide cylinders are completely dehydrated in electric furnaces. The cylinders are then placed inside a seamless uncoated copper tube whose inside diameter is slightly larger than the diameter of the molded cylinders. Solid uncoated high-conductivity copper rods are then inserted in the holes in the cylinders.

After the ends of the filled copper tube have been sealed, the tube goes through a series of drawing and annealing operations. In the drawing process the compressive forces reduce the magnesium oxide cylinders to powder form and produce a dense homogeneous mass of magnesium oxide which transmits the compressive forces to the conductor rods. This drawing produces, in effect, the equivalent of a solid homogeneous rod with all members—copper tube, copper rods, and magnesium oxide—maintaining the same

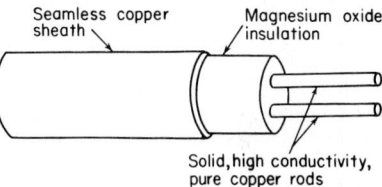

Seamless copper sheath

Magnesium oxide insulation

Solid, high conductivity, pure copper rods

**Fig. 2-5** MI cable construction. [General Cable Co.]

relative position to each other while being proportionately reduced in diameter and cross section and increased in length. (Each component maintains its original proportionate size and relative position throughout the remainder of the process until the desired finished sheath diameter is reached.) The assembly is annealed at intervals during the drawing process to eliminate the effect of work hardening on the metal, and the finished cable is supplied fully annealed.

A departure from the fabrication method just described has been adopted for some constructions. In this method the same starter tube stands vertically while the starter conductors are inserted and jigged into position. Then dried magnesium oxide powder is introduced into the tube and tamped to the desired density. After completion of this vertical filling operation, a drawing and annealing sequence is employed as with the horizontal, plug-type filled-cable assembly.

**55. Nonmetallic-sheathed building cable** is employed for the economical wiring of residences, farm buildings, small stores, etc. It consists of a two-, three-, or four-conductor, rubber- or thermoplastic-insulated cable. When desired, a bare copper ground wire is incorporated in the cable assembly. The ground wire is laid in the interstices between the circuit conductors under the outer braid. Nonmetallic-sheathed cable is manufactured under several trade names. The overall dimensions will differ somewhat with different manufacturers. Type designations are NM and NMC.

**56. Armor-clad building cable** consists of rubber- or thermoplastic-insulated copper wire protected by flexible galvanized-steel armor. It is manufactured with or without a lead sheath under the armor for protection against excessive dampness. The nonleaded type is regularly available in one-, two-, three-, and four-conductor assemblies in sizes from No. 14 to No. 2 and also in size No. 1 for a single conductor only. The leaded type is standard in two- and three-conductor assemblies in sizes from No. 14 to No. 6 and also in size No. 4 for three conductors only. Individual conductors are finished in different colors for identification of any conductor in installation. In installing these cables a short fiber bushing (see Div. 9) is inserted between the conductors and the steel armor wherever the cable is cut. For this reason the cables are frequently called armored bushed cables (ABC). The **National Electrical Code** designation is AC for the nonleaded type and ACL for the leaded type. They are also commonly called BX and BXL.

Armored ground wire consists of a bare copper conductor with a flexible steel armor covering applied directly over the wire.

**57. Metal-clad cable.**   Type MC cable is a factory assembly of one or more conductors, each individually insulated and enclosed in a metallic sheath or interlocking tape or in a smooth or corrugated tube. This category includes wire protected by flexible galvanized-steel armor, aluminum sheathed in a liquidtight and gastight, close-fitting, and continuous aluminum sheath, or copper sheathed in a liquidtight and gastight, close-fitting, and continuous copper or bronze sheath. Type MC cable should not be used in or exposed to corrosive conditions or directly buried in earth or concrete unless it is provided with a metallic sheath which is suitable for the conditions and is protected by materials suitable for the conditions. Metal-clad cable shall be installed as required by Article 334 of the **National Electrical Code.**

**58. Cablebus.**   This is an approved assembly of insulated conductors mounted in a spaced relationship in a ventilated metal-protective supporting structure, including fittings and conductor terminations. Usually assembled at the installation site from components furnished or specified by the manufacturer, cablebus may be used at any voltage or current for which the spaced conductors are rated. Components of this wiring system are described in Div. 9.

**59. Class 2 and Class 3 remote-control, signaling, and power-limited circuits.**   These circuits are the portion of the wiring system between the load side of the overcurrent device or the power-limited supply and all connected equipment. For installation requirements see Article 725 of the NEC.

**60. Service cables.**   Special multiconductor cables have been developed to meet the requirements for connecting the pole line of the street distribution system with the service-entrance equipment on the consumer's premises. The following terminology derived from their application is used for service cables:

1. *Service-drop cables.* Overhead service conductors between pole and building.

2. *Service-entrance cables.* Service conductors which extend along the exterior and enter buildings to the meter, switch, or service equipment.

3. *Combination service-drop and service-entrance cables.* Service conductors continuing without a break from the pole to the meter, switch, or service equipment.

4. *Underground service-entrance cables.* Service conductors installed in conduit or directly in the earth and entering the building to the meter, switch, or service equipment.

Service cable is made in two different types designated by the Underwriters Laboratories. Service-entrance cable is available in sizes No. 12 AWG and larger for copper and No. 10 AWG and larger for aluminum or copper-clad aluminum, with Types RH, RHW, RHH, or XHHW conductors. The cables are classified as follows:

1. *Type SE.* Cable for aboveground installation.

2. *Type USE.* Cable for underground installation including burial directly in the earth. Cable in sizes No. 4/0 AWG and smaller with all conductors insulated is suitable for all underground uses for which Type UF cable is permitted by the **National Electrical Code.** Type USE service-entrance cable is similar in general construction to general power cables for direct burial in the earth.

Service cables are either rubber- or thermoplastic-insulated. They are available in two- and three-conductor assemblies.

**61. Aerial cables.**   Bare, weatherproof (weather-resisting) covered, and rubber-insulated tree wire are all used for overhead pole-line power circuits. Bare wire is used for most power transmission circuits. See Sec. 13 and Table 87 for conductor materials employed and their field of application. Weatherproof (heat-resisting) covered wire is the general standard for distribution and low-voltage transmission circuits. Tree wire is used in distribution circuits when heavy tree growth would make the use of weatherproof wire unsuitable. It consists of a rubber-insulated copper wire protected with an abrasion-resisting covering.

The application of multiconductor aerial power cables is continually increasing. For messenger-type construction, general-purpose power cable with an alloy-lead sheath is commonly employed. To minimize the weight of extremely large cables, it is generally advisable to eliminate the lead sheath. For such installations two types of cable construction have proved satisfactory. One type consists of either paper- or varnished-cambric–insulated cable supplied with a rubber hose jacket and an overall band or interlocked armor of either steel or bronze. The other type has utilized rubber- or varnished-cambric–insulated shielded cable and an overall seine twine or hawser braid saturated with a moisture-resisting compound. A layer of rubberized tape or presaturated duck tape is interposed between the overall shielding band and the outer braid. The rubber insulation should be a corona-resisting compound except for the lower voltages.

Most of the new installations of multiconductor aerial power cables are of the preassembled self-supporting type. These cables consist of one or more (usually three) rubber-insulated conductors bound to a suitable messenger cable by a flat metallic strip applied with open-lay helical wrap. In some cables each insulated conductor is covered by a neoprene jacket. In others the insulation of each conductor is covered with a tough abrasion- and corrosion-resistant aluminum-bronze shielding tape. In a third type of construction each insulated conductor is shielded by means of a tinned-copper tape and then covered with a neoprene jacket.

**62. Power cables for direct burial in the earth without raceways** are available in both metallic and nonmetallic armored types. Cables of different manufacturers are fairly common in general construction but differ in details of assembly and manufacture.

The metallic types consist of a lead-sheathed general-purpose power cable protected with a jute and asphalt covering or a combination covering of a flat or interlocking tape armor together with jute and asphalt. Rubber is generally used for the insulation, but paper or varnished cambric may be employed if desired. The cables are available in one-, two-, and three-conductor assemblies. These metallic types are commonly called *parkway cables.*

The nonmetallic types consist of rubber-insulated conductors protected with some form of nonmetallic covering that will provide a moisture seal, resistance to corrosion, and sufficient mechanical strength. One type of armor assembly consists of a combination of layers of asbestos-base caulk coatings, asbestos braid filled with asbestos-base caulk, and treated fibrous tape, all enclosed in a presaturated heavy duck tape. The outer surface is finished with pitch and mica flake. A common armor assembly is the jute and asphalt covering described in Sec. 44. Rubber and synthetic jackets are also employed for the protective covering of nonmetallic cables for direct burial in the earth. The jackets are frequently enclosed in a saturated duck tape or sisal braid. All the nonmetallic types mentioned are available in one-, two-, and three-conductor assemblies. Some two-conductor cables are made with one central insulated conductor and one noninsulated conductor concentrically stranded over the insulation of the inner conductor.

Thermoplastic-insulated single-conductor and multiconductor Type UF cables are also available for use up to and including 600 V.

**63. Network cables** were developed primarily for underground raceway installation for low-voltage secondary network service. They are single-conductor nonmetallic-sheathed cables suitable for circuits operating at not more than 600 V. These cables are insulated with a special rubber compound covered with a neoprene jacket. Braid-covered or lead-sheathed network cables are also available.

**64. Series street-lighting cables** differ from other power cables in the thickness of the insulation employed on the wires. They are rubber-, varnished-cambric–, or thermoplastic-insulated single-conductor cables, available in shielded or unshielded types. Rubber-insulated cables can be obtained with an outer neoprene jacket or with a lead sheath. Lead-sheathed cables can be obtained with additional coverings of jute or jute and double flat steel-tape armor with jute overall for conditions requiring added protection. Varnished-cambric–insulated cables are always lead-sheathed and may be supplied with any of the additional coverings mentioned above for rubber-insulated leaded cable.

**65. Ornamental-pole and bracket cable** is a two-conductor, rubber-insulated cable specially designed for great flexibility and small overall diameter. It is used for the interior wiring of ornamental lighting poles fed by underground cable and for the exterior wiring of pole-type bracket or mast-arm street-lighting fixtures. For voltages not exceeding 4000, twin-type cable construction, consisting of two rubber-insulated, single-cotton-braid–covered conductors laid parallel and enclosed in overall tape and weatherproof braid, is employed. For voltages greater than 4000, belted construction is employed. This assembly consists of two rubber-insulated and single-cotton-braid–covered conductors laid parallel and enclosed in a close-fitting belt of insulating compound. The belt insulation is covered with a tape and weatherproof braid. Another assembly for belted cable consists of two thermoplastic-insulated conductors laid parallel and enclosed in a close-fitting belt of plastic which serves as both insulation and a jacket.

**66. Borehole or mine-shaft cables** are specially designed to withstand the mechanical strains to which long lengths of vertically suspended cable are subjected. Both rubber and varnished cambric are employed for insulation. The rubber-insulated cables are made both with and without a lead sheath, while the varnished-cambric cables are always lead-sheathed. The cable is protected with a steel-tape or steel-wire armor. Steel-tape armor should not be used unless supports can be provided at intervals not exceeding 50 ft (15.24 m). When the cable must be supported entirely from the top, wire armor must be used. The armor should be designed with sufficient strength to support the entire cable weight with a factor of safety of at least 5 and based on a tensile strength of 70,000 psi (482,632,-990 Pa). The wire armor should be served with bands of steel wire to provide a suitable bond. Conductor-supported cables, instead of the above-described armor-supported ones, are sometimes used for this class of service. The rubber-insulated conductors are protected with a strong loom covering and installed in metallic borehole casings or conduit. Medium-hard-drawn copper is used for the conductors, which entirely support the cable through conductor clamps and strain insulators.

Cables similar to the steel-armor borehole cables are employed for the vertical risers of tall buildings. For nonleaded construction a factor of safety of at least 7 should be used in designing the wire armor, while for leaded construction the minimum factor of safety should be 4.

**67. Submarine cables** for installation underwater in rivers, lakes, harbors, etc., are rubber-insulated power cables protected with a wire armor. The more common types of coverings employed are jute and wire armor; jute, wire armor, and jute; lead, jute, and wire armor; and lead, jute, wire armor, and jute. It is best to consider each submarine installation a separate engineering problem and obtain the recommendations of the cable experts of reliable cable manufacturers.

**68. Control cables** consist of an assembly of several small conductors for carrying the relatively small currents for remote control of motors, circuit breakers, or other power equipment; for relay and metering circuits; for traffic-light control systems; etc. Standard assemblies vary from 1 conductor per cable to 37 conductors per cable. The size of the conductors ranges from No. 14 to No. 9. For most installations a 30 percent rubber compound is used for the insulation of the conductors, although varnished cambric and thermoplastic are also employed. For installations in which the temperature is higher than normal, either a heat-resisting rubber compound or felted asbestos should be employed, the choice depending upon the temperature of the surroundings. Each individual conductor is covered with a color-coded cotton braid for protection of the insulation and identification of circuits. The conductors are assembled with jute fillers and bound together with a rubber-faced or asbestos binder tape.

For rubber-insulated cables the outer covering may be a standard cotton weatherproof braid with a black weatherproof wax finish, a lead sheath, a neoprene jacket, metallic or nonmetallic armor for direct burial in the earth as described for power cable in Sec. 62 of this division, or a braid or jute bedding protected by interlocking steel-tape armor. With rubber-jacket construction, the binder tape is eliminated. For asbestos-insulated cables the outer covering is an asbestos braid saturated with a flame-, heat-, and moisture-resisting compound. For thermoplastic-insulated cables the outer covering is a thermoplastic jacket.

**69. Traffic-control, fire-alarm, and signal cables** are specially designed to meet the conditions encountered in these municipal signal systems. Both rubber- and thermoplastic-insulated cables are used. For rubber-insulated cables various outer coverings are employed, depending upon the installation conditions. These coverings include heavy contton braid, a plain lead sheath, a protected lead sheath, and a neoprene jacket. For thermoplastic-insulated cables the outer covering is a thermoplastic or similar jacket.

**70. Special cables for electric transportation equipment** are manufactured to meet the particular requirements of this class of service with respect to flexibility and resistance to vibration. The conductors are rubber-insulated. The cable covering usually consists of cotton braid or a rubber jacket.

**71. Cables for gas-tube sign and oil-burner ignition systems** should be of a type approved by the Underwriters Laboratories for this service.

The Underwriters kilovolt designations for cables for this service are GTO-5, GTO-10, and GTO-15. In each case the numeral indicates the cable rating in kilovolts. Approved cables made by different manufacturers differ in details of construction. At one time these cables were rubber-insulated. Today most of them are thermoplastic-insulated with an outer thermoplastic jacket.

**72. X-ray cables**   are designed to meet the special conditions encountered in power supply circuits to high-voltage x-ray tubes and similar equipment. As produced by one manufacturer, they consist of a core made up of two rubber-insulated wires and two uninsulated wires. A semiconducting rubber compound is extruded over these conductors and covered with a layer of rubber insulation. Another semiconducting coating is applied over this insulation, and it, in turn, is covered with a close-covering annealed-copper shielding braid. The outer covering may be either cotton braid or a neoprene jacket.

**73. Drive-in-theater cables** are designed for direct installation in the earth under outdoor motion-picture-theater lots to serve sound-distribution networks from the projection booth to speaker posts installed adjacent to the designated automobile-parking stations. Two types of cable are available for this service: two-conductor rubber-insulated cable with an outer neoprene jacket and two-conductor cable with a uni-insulation and jacket of thermoplastic compound.

**74. Portable power cables** are designed for circuits supplying power to portable power equipment such as electric shovels, dredgers, air compressors, and welders. The individual conductors are generally insulated with a thermoset compound. Cables of this type for voltages not exceeding 600 V may be obtained with the following protective coverings: heavy cotton braid, seine-twine braid, loom sheath, thermoset jacket, or jute and wire armor. These coverings are listed in the order of increasing resistance to wear. Only rubber jackets, thermoset jackets, or jute-and-wire–armor coverings are employed for cables for voltages exceeding 600 V. These high-voltage cables may be obtained in the types shown in the accompanying table.

| Type | Covering | Ground wires | Shielding | Max. voltage recommended |
|------|----------|--------------|-----------|--------------------------|
| W | Thermoset jacket | Without | None | 2500 |
| G | Thermoset jacket | With | None | 5000 |
| SH-A | Rubber jacket | Without | On each conductor | |
| SH-B | Rubber jacket | Without | Over cabled conductor | |
| SH-C | Rubber jacket | With | Over cabled conductor | 6000 |
| SH-D | Rubber jacket or jute and wire armor | With | On each conductor | . . .[a] |

[a]No standard above 6000V; consult manufacturers.

The conductors of portable power cables are made especially flexible by the use of Class G or H stranding (see Table 93); for special applications requiring even greater flexibility, special stranding may be employed. Single-conductor cables are regularly available in sizes from No. 8 to 500 kcmil. Two-, three, and four-conductor cables are standard in sizes from No. 8 to No. 4/0.

**75. Fixture wire** is used for the wiring of electrical fixtures. There are 32 approved types, which are listed and described in Table 76. Those wires with a single *F* in their designation may be either solid or stranded. Those with *FF* must be made of wire with especially flexible stranding.

Flexible cords of Types AFC, AFSJ, AFS, and AFPD are multiconductor cords of the AF type of construction and may be employed for the wiring of fixtures. Rubber-covered flexible cords are also used for this purpose.

**77. Flexible cord** is the name applied to wires used to connect portable appliances, small tools, or machinery to the receptacles of the wiring installation or to wire the portable device itself. Flexible cords also are used to wire pendant fixtures from the outlet box to

**76. Fixture Wire[a]**
(Table 402-3, 1990 National Electrical Code)

| Trade name | Type letter | Insulation | Thickness of insulation | | Outer covering | Maximum operating temperature | Application provisions |
|---|---|---|---|---|---|---|---|
| | | | AWG | Mils | | | |
| Heat-resistant rubber-covered fixture wire; solid or 7-strand | RFH-1 | Heat-resistant rubber | 18 | 15 | Nonmetallic | 75°C (167°F) | Fixture wiring; limited to 300 V |
| | RFH-2 | Heat-resistant rubber | 18-16 | 30 | Nonmetallic | 75°C (167°F) | Fixture wiring and as permitted in Secs. 725-16 and 760-16 of the NEC |
| | | Heat-resistant latex rubber | 18-16 | 18 | | | |
| Heat-resistant cross-linked synthetic polymer-insulated fixture wire, solid or stranded | RFHH-2[b] | | 18-16 | 30 | | | |
| | RFHH-3[b] | Cross-linked synthetic polymer | 18-16 | 45 | None or nonmetallic covering | 90°C (194°F) | Fixture wiring and as permitted in Secs. 725-16 and 760-16 of the NEC; multiconductor cable as permitted in Secs. 725-16 and 760-16 |
| Heat-resistant rubber-covered fixture wire; flexible stranding | FFH-2 | Heat-resistant rubber | 18-16 | 30 | Nonmetallic | 75°C (167°F) | Fixture wiring and as permitted in Secs. 725-16 of the NEC |
| | | Heat-resistant latex rubber | 18-16 | 18 | | | |
| Thermoplastic-covered fixture wire; solid or 7-strand | TF[b] | Thermoplastic | 18-16 | 30 | None | 60°C (140°F) | Fixture wiring and as permitted in Secs. 725-16 and 760-16 of the NEC |

| Description | Type | Insulation | AWG | Thickness of moisture-resistant insulation (mils) | Thickness of asbestos (mils) | Braid | Max operating temperature | Application |
|---|---|---|---|---|---|---|---|---|
| Thermoplastic-covered fixture wire; flexible stranding | TFF[b] | Thermoplastic | 18-16 | | 30 | None | 60°C (140°F) | Fixture wiring and as permitted in Sec. 725-16 of the NEC |
| Heat-resistant thermoplastic-covered fixture wire; solid or 7-strand | TFN[b] | Thermoplastic | 18-16 | | 15 | Nylon-jacketed or equivalent | 90°C (194°F) | Fixture wiring and as permitted in Secs. 725-16 and 760-16 of the NEC |
| Heat-resistant thermoplastic-covered fixture wire; flexible stranded | TFFN[b] | Thermoplastic | 18-16 | | 15 | Nylon-jacketed or equivalent | 90°C (194°F) | Fixture wiring and as permitted in Sec. 725-16 of the NEC |
| Asbestos-covered heat-resistant fixture wire | AF | Impregnated asbestos or moisture-resistant insulation and impregnated asbestos | 18-14<br>12-10 | ..<br>20<br>..<br>25 | 30<br>10<br>45<br>20 | None | 150°C (302°F) | Fixture wiring; limited to 300 V and indoor dry location |
| Silicone-insulated fixture wire; solid or 7-strand | SF-1 | Silicone rubber | 18 | | 15 | Nonmetallic | 200°C (392°F) | Fixture wiring; limited to 300 V |
| | SF-2 | Silicone rubber | 18-14 | | 30 | Nonmetallic | 200°C (392°F) | Fixture wiring and as permitted in Secs. 725-16 and 760-16 of the NEC |

**76. Fixture Wire** (*Continued*)

| Trade name | Type letter | Insulation | Thickness of insulation AWG | Thickness of insulation Mils | Outer covering | Maximum operating temperature | Application provisions |
|---|---|---|---|---|---|---|---|
| Silicone-insulated fixture wire; flexible stranding | SFF-1 | Silicone rubber | 18 | 15 | Nonmetallic | 150°C (302°F) | Fixture wiring; limited to 300 V |
| | SFF-2 | Silicone rubber | 18-14 | 30 | Nonmetallic | 150°C (302°F) | Fixture wiring and as permitted in Sec. 725-16 of the NEC |
| Fluorinated ethylene propylene fixture wire; solid or 7-strand | PF | Fluorinated ethylene propylene | 18-14 | 20 | None | 200°C (392°F) | Fixture wiring and as permitted in Sec. 725-16 and 760-16 of the NEC |
| | PGF | | 18-14 | 14 | Glass braid | | |
| Fluorinated ethylene propylene fixture wire; flexible stranding | PFF | Fluorinated ethylene propylene | 18-14 | 20 | None | 150°C (302°F) | Fixture wiring and as permitted in Sec. 725-16 of the NEC |
| | PGFF | | 18-14 | 14 | Glass braid | | |
| Tape-insulated fixture wire; solid or 7-strand | KF-1 | Aromatic polyimide tape | 18-10 | 5.5 | None | 200°C (392°F) | Fixture wiring; limited to 300 V |
| | KF-2 | Aromatic polyimide tape | 18-10 | 8.4 | None | 200°C (392°F) | Fixture wiring and as permitted in Secs. 725-16 and 760-16 of the NEC |
| Tape-insulated fixture wire; flexible stranding | KFF-1 | Aromatic polyimide tape | 18-10 | 5.5 | None | 200°C (392°F) | Fixture wiring; limited to 300 V |
| | KFF-2 | Aromatic polyimide tape | 18-10 | 8.4 | None | 200°C (392°F) | Fixture wiring and as permitted in Sec. 725-16 of the NEC |

| Insulation | Type | Material | AWG | | None | Temp. | Application provisions |
|---|---|---|---|---|---|---|---|
| ECTFE; solid or 7-strand | HF | Ethylene chlorotrifluoro-ethylene | 18-14 | 15 | None | 150°C (302°F) | Fixture wiring and as permitted in Sec. 725-16 of the NEC |
| ECTFE; flexible stranding | HFF | Ethylene chlorotrifluoro-ethylene | 18-14 | 15 | None | 150°C (302°F) | Fixture wiring and as permitted in Sec. 725-16 of the NEC |
| Cross-linked polyolefin insulated fixture wire; solid or 7-strand | XF[b] | Cross-linked polyolefin | 18-14<br>12-10 | 30<br>45 | None | 150°C (302°F) | Fixture wiring; limited to 300 V |
| Cross-linked polyolefin insulated fixture wire; flexible stranded | XFF[b] | Cross-linked polyolefin | 18-14<br>12-10 | 30<br>45 | None | 150°C (302°F) | Fixture wiring; limited to 300 V |
| Modified ETFE; solid or 7-strand | ZF | Modified ethylene tetrafluroethylene | 18-14 | 15 | None | 150°C (302°F) | Fixture wiring and as permitted in Secs. 725-16 and 760-16 of the NEC |
| Flexible stranding | ZFF | Modified ethylene tetrafluoroethylene | 18-14 | 15 | None | 150°C (302°F) | Fixture wiring and as permitted in Sec. 725-16 of the NEC |
| Extruded polytetrafluoroethylene; solid or 7-strand (nickel or nickel-coated copper) | PTF | Extruded polytetrafluoroethylene | 18-14 | rr 20 | None | 250°C (482°F) | Fixture wiring and as permitted in Secs. 725-16 and 760-16 of the NEC (nickel or nickel-coated copper) |
| Extruded polytetrafluoroethylene; flexible stranding (No. 26-36 AWG; silver or nickel-coated copper) | PTFF | Extruded polytetrafluoroethylene | 18-14 | 20 | None | 150°C (302°F) | Fixture wiring and as permitted in Sec. 725-16 of the NEC (silver or nickel-coated copper) |

**76. Fixture Wire** (*Continued*)

| Trade name | Type letter | Insulation | Thickness of insulation | | Outer covering | Maximum operating temperature | Application provisions |
|---|---|---|---|---|---|---|---|
| | | | AWG | Mils | | | |
| Perfluoroalkoxy; solid or 7-strand (nickel or nickel-coated copper) | PAF | Perfluoroalkoxy | 18-14 | 20 | None | 250°C (482°F) | Fixture wiring and as permitted in Secs. 725-16 and 760-16 of the NEC (nickel or nickel-coated copper) |
| Perfluoroalkoxy; flexible stranding | PAFF | Perfluoroalkoxy | 18-14 | 20 | None | 150°C (302°F) | Fixture wiring and as permitted in Sec. 725-16 of the NEC |

[a]Reprinted with permission from NFPA 70-1990, National Electrical Code®, Copyright© 1990, National Fire Protection Association, Quincy, Massachusetts 02269. This reprinted material is not the complete and official position of the NFPA on the referenced subject, which is represented only by the standard in its entirety.
[b]Insulation and outer coverings that meet the requirements of flame-retardant, limited smoke, and are so listed shall be permitted to be designated limited smoke with the suffix /LS after the Code-type designation.

the fixture. Table 130 lists the different types of approved flexible cords with their uses as allowed by National Electrical Code rules.

**78. Radio and TV wires.**   Special types of wires have been developed to meet the needs of the radio and TV industry. They include antenna, lead-in, coil-lead, hookup, and transmission-line wires and microphone cable.

**79. Automotive and aircraft wires and cables** are specially designed to meet the requirements of these two classes of service. They include high-tension ignition cables, primary wires for low-voltage service, and starter cables.

**80. The telephone industry** requires special types of wires and cables. The multiconductor cables for main distribution circuits are dry-paper–insulated and lead-covered. A lead-antimony sheath is generally employed for aerial and underground installation in conduit or ducts. For direct burial in the earth and for some aerial work the lead sheath is protected with jute and steel-tape armor. Specially designed steel-wire–armored cables are used for submarine installation. It is common practice to terminate these paper-insulated cables by splicing to them a short length of textile-insulated lead-covered cable. A common type of this textile-insulated cable employs two layers of silk and one layer of cotton insulation on each conductor, the cable being protected with a pure lead sheath. Rubber-insulated two-conductor cables protected with a weatherproof braid are used for service drops from pole to house. Hard-drawn copper, bronze, and Copperweld conductors are used for these cables.

For circuits exposed to severe moisture and requiring from 2 to 12 conductors, rubber-insulated lead-covered cables are frequently used.

For interior wiring of individual telephone circuits, twisted-pair rubber-insulated cotton-braid–covered cables are employed. There is a trend toward the use of thermoplastic-insulated conductors with thermoplastic or similar outer sheaths for interior work and direct burial in earth.

**81. Cables for shipboard use** should be designed for the special requirements and severe service encountered in installations on board ship. Such cables are generally made in accordance with Institute of Electrical and Electronic Engineers (IEEE) specifications, which are approved by the ICEA.

**82. Magnet wire** is the name applied to single-conductor insulated wires manufactured for the winding of coils for electrical circuits. Round magnet wire is made of soft annealed solid copper wire in sizes from No. 46 to No. 4/0. Square– and rectangular–cross-section magnet wires are also available. Insulations used for magnet wire include baked films (enamel), cotton, silk, nylon yarn, paper, asbestos, fibrous glass, Dacron-glass, or combinations of these materials. The temperature rating of magnet wires depends upon the insulation employed and is based upon the temperature limits stated in Publication No. 1 of the IEEE (refer to Div. 7, Sec. 128). All magnet wire made by a reliable manufacturer will meet certain specifications for physical and electrical characteristics that have been standardized by the National Electrical Manufacturers Association.

ENAMEL COATINGS

*Conventional enamel (generally referred to merely as enamel).* A baked-film insulation of an oleoresinous varnish having good moisture resistance and dielectric properties. Used widely in magnet wire coils of all types. 105°C Class A insulation.

*Formvar enamel.* A baked-film insulation from a varnish based on vinyl acetal resin. Has excellent resistance to heat, abrasion, and solvents, combined with exceptional flexibility and adherence. 105°C Class A insulation.

*Self-bonding Formvar.* A baked-film insulation consisting of Formvar over which is applied a thermoplastic overcoat. When formed into coils and baked or otherwise suitably heat-treated, the turns of wire, because of the thermoplastic overcoat, become bonded together, thus facilitating the fabrication of self-supporting coil assemblies. 105°C Class A insulation.

*Nylon enamel.* A baked-film insulation from a varnish based on polyamide resin which solders readily (self-fluxing) unless it is overbaked. Excellent winding qualities. Not so good as Formvar in electrical properties and moisture resistance. 105°C Class A insulation.

*Formvar and nylon enamel.* A baked-film insulation consisting of Formvar over which is applied a coat of nylon enamel. Combines the advantages of good winding properties of the outer nylon enamel coating with the better electrical properties and moisture resistance of the underlying Formvar enamel coating. 105°C Class A insulation.

*Enamel G.* A baked-film insulation from a varnish based on polyurethane resin. Solders readily (self-fluxing). Good winding and moisture-resistance qualities. Good electrical properties even at high frequencies. Well suited for electronic and communication applications. 120°C, between Class A and Class B insulation.

*Enamel Class B.* A baked-film insulation from a varnish based on polyester resin such as Isonel or comparable materials. Handling properties comparable to Formvar except for somewhat poorer heat shock test. Although stable up to 150°C, it is generally used as Class B (130°C) insulation.

COVERINGS

*Cotton.* Unbleached fine long-staple cotton yarn applied with proper techniques to provide a firm, even layer winding. Rated as 90°C Class O insulation when used alone. When impregnated or immersed in liquid dielectric, it is rated as 105°C Class A insulation.

*Nylon yarn.* High-quality nylon tram supplied as an alternative to silk covering. When used alone, it is rated as 90°C Class O insulation.

*Silk.* The best grade of tram silk is particularly suitable for low-loss radio-frequency coils in which interturn and interlayer capacitances must be kept to a minimum. When used alone, it is rated as 80°C Class O insulation.

*Paper.* 100 percent manila-rope stock or equivalent grade in tape thicknesses of 0.85 to 3.0 mils (0.0216 to 0.0762 mm). When used alone, it is rated as 90°C Class O insulation. When impregnated or immersed in liquid dielectric, it is rated as 105°C Class A insulation.

*Fibrous glass.* Pure, alkali-free, continuous-filament glass yarn is wrapped, saturated, and bonded to the conductor with high-grade insulating varnishes or compounds so applied as to yield a smooth, uniform, homogeneous insulating covering. When impregnated with conventional varnish, it is rated as 130°C Class B insulation. When treated with silicone varnish, it is rated as 180°C Class H insulation.

*Dacron-glass.* A combination insulation consisting of glass and Dacron fibers. Much better abrasion resistance, insulation adhesion, and flexibility than all-glass insulation. When impregnated with standard varnish, it is rated as 130°C Class B insulation. Rating when treated with silicone varnish is not yet established.

**83. In the selection of the proper type of magnet wire** for any application the following factors should be considered:

1. Utilization of winding space
2. Resistance to abrasion
3. Cost
4. Insulating qualities
5. Appearance
6. Operating temperatures

**84. Method of determining the approximate diameter of any cable.** When the approximate overall diameter of a wire or cable not listed in conventional tables is desired, it may be determined by means of one of the following formulas. As a rule, results obtained from these formulas will give minimum outside diameters. To allow for manufacturing tolerances, add 50 mils (0.127 mm) to the outside diameters as obtained from the formulas to get approximate maximum diameters.

For single-conductor cables:

$$OD = d + 2T + 2t_c \tag{1}$$

For multiconductor nonshielded or belted cables:

$$OD = K(d + 2T + 2t_c) + 2t_o \tag{2}$$

For multiconductor belted cable:

$$OD = K(d + 2T) + 2t_B + 2t_o \tag{3}$$

For multiconductor shielded cable:

$$OD = K(d + 2T + 0.032) + 0.020 + 2t_o \tag{4}$$

where OD = outside diameter in inches, $d$ = diameter of bare round conductor in inches, $T$ = thickness of conductor insulation in inches, $t_c$ = thickness of covering over insulation of individual conductors in inches, $t_B$ = thickness of belt insulation in inches, $t_o$ = thickness of overall protective covering in inches, and $K$ = value from following table:

| Conductor | Value of K |
|---|---|
| Two-conductor, round | 2.00 |
| Two-conductor, flat, maximum diameter | 2.00 |
| Two-conductor, flat, minimum diameter | 1.00 |
| Three-conductor, round | 2.16 |
| Three-conductor, ordinary sector | 1.95 |
| Three-conductor, Compack sector | 1.90 |
| Four-conductor, round | 2.41 |
| Four-conductor, ordinary sector | 2.20 |
| Four-conductor, Compack sector | 2.15 |

For any conductor or cable with a known outside diameter, the outside cross-sectional area may be determined by

$$\text{CSA (in}^2) = d^2 \times 0.7854 \tag{5}$$

where $d$ = outside diameter in inches.

## 85.  Comparison of Wire and Sheet-Metal Gage Diameters or Thickness in Inches

| Gage no. | (1) American wire gage (AWG)[a] | (2) Steel wire gage | (3) Birmingham iron wire gage (Stubs) | (4) Stubs steel wire gage | (5) British standard wire gage | (6) Steel music wire gage | (7) Manufacturers standard gage | (8) American zinc | (9) Birmingham |
|---|---|---|---|---|---|---|---|---|---|
| 15/0 | | | | | | | | | 1.0000 |
| 14/0 | | | | | | | | | 0.9583 |
| 13/0 | | | | | | | | | 0.9167 |
| 12/0 | | | | | | | | | 0.8750 |
| 11/0 | | | | | | | | | 0.8333 |
| 10/0 | | | | | | | | | 0.7917 |
| 9/0 | | | | | | | | | 0.7500 |
| 8/0 | | | | | | | 0.5000 | | 0.7083 |
| 7/0 | | 0.4900 | | | 0.5000 | | 0.4687 | | 0.6666 |
| 6/0 | | 0.4615 | | | 0.4640 | | | | 0.6250 |
| 5/0 | | 0.4305 | | | 0.4320 | | 0.4375 | | 0.5883 |
| 4/0 | 0.4600 | 0.3938 | 0.454 | | 0.4000 | | 0.4062 | | 0.5416 |
| 3/0 | 0.4100 | 0.3625 | 0.425 | | 0.3720 | | 0.3750 | | 0.5000 |
| 2/0 | 0.3650 | 0.3310 | 0.380 | | 0.3480 | 0.0087 | 0.3437 | | 0.4452 |
| 1/0 | 0.3250 | 0.3065 | 0.340 | | 0.3240 | 0.0039 | 0.3125 | | 0.3964 |
| 1 | 0.2890 | 0.2830 | 0.300 | 0.227 | 0.3000 | 0.0098 | 0.2812 | 0.002 | 0.3532 |
| 2 | 0.2580 | 0.2625 | 0.284 | 0.219 | 0.2760 | 0.0106 | 0.2656 | 0.004 | 0.3147 |
| 3 | 0.2290 | 0.2437 | 0.259 | 0.212 | 0.2520 | 0.0114 | 0.2500 | 0.006 | 0.2804 |
| 4 | 0.2040 | 0.2253 | 0.238 | 0.207 | 0.2320 | 0.0122 | 0.2344 | 0.008 | 0.2500 |
| 5 | 0.1820 | 0.2070 | 0.220 | 0.204 | 0.2120 | 0.0138 | 0.2187 | 0.010 | 0.2225 |
| 6 | 0.1620 | 0.1920 | 0.203 | 0.201 | 0.1920 | 0.0157 | 0.2035 | 0.012 | 0.1981 |
| 7 | 0.1440 | 0.1770 | 0.180 | 0.199 | 0.1760 | 0.0177 | 0.1875 | 0.014 | 0 1764 |
| 8 | 0.1280 | 0.1620 | 0.165 | 0.197 | 0.1600 | 0.0197 | 0.1719 | 0.016 | 0.1570 |
| 9 | 0.1140 | 0.1483 | 0.148 | 0.194 | 0.1440 | 0.0216 | 0.1562 | 0.018 | 0.1398 |
| 10 | 0.1020 | 0.1350 | 0.134 | 0.191 | 0.1280 | 0 0236 | 0.1406 | 0.020 | 0.1250 |
| 11 | 0.0910 | 0.1205 | 0.120 | 0.188 | 0.1160 | 0.0260 | 0.1250 | 0.024 | 0.1113 |
| 12 | 0.0810 | 0.1055 | 0.109 | 0.185 | 0.1040 | 0.0283 | 0.1094 | 0.028 | 0.0991 |
| 13 | 0.0720 | 0.0915 | 0.095 | 0.182 | 0.0920 | 0.0305 | 0.0937 | 0.032 | 0.0882 |
| 14 | 0.0640 | 0.0800 | 0.083 | 0.180 | 0.0800 | 0.0323 | 0.0821 | 0.036 | 0.0785 |
| 15 | 0.0570 | 0.0720 | 0.072 | 0.178 | 0.0720 | 0.0342 | 0.0703 | 0.040 | 0.0699 |
| 16 | 0.0510 | 0.0625 | 0.065 | 0.175 | 0.0640 | 0.0362 | 0.0625 | 0.045 | 0.0625 |
| 17 | 0.0450 | 0.0540 | 0.058 | 0.172 | 0.0560 | 0.0382 | 0.0562 | 0.050 | 0.0556 |
| 18 | 0.0400 | 0.0475 | 0.049 | 0.168 | 0.0480 | 0.0400 | 0.0500 | 0.055 | 0.0495 |
| 19 | 0.0360 | 0.0410 | 0.042 | 0.164 | 0.0400 | 0.0420 | 0.0437 | 0.060 | 0.0440 |
| 20 | 0.0320 | 0.0348 | 0.035 | 0.161 | 0.0360 | 0.0440 | 0.0375 | 0.070 | 0.0392 |
| 21 | 0.0285 | 0.0317 | 0.032 | 0.157 | 0.0320 | 0.0460 | 0.0344 | 0.080 | 0.0349 |
| 22 | 0.0253 | 0.0286 | 0.028 | 0.155 | 0.0280 | 0.0480 | 0.0312 | 0.090 | 0.0312 |
| 23 | 0.0226 | 0.0258 | 0.025 | 0.153 | 0.0240 | 0.0510 | 0.0281 | 0.100 | 0.0278 |
| 24 | 0.0210 | 0.0230 | 0.022 | 0.151 | 0.0220 | 0.0550 | 0.0250 | 0.125 | 0.0247 |
| 25 | 0.0179 | 0.0204 | 0.020 | 0.148 | 0.0200 | 0.0590 | 0.0219 | 0.250 | 0.0220 |
| 26 | 0.0159 | 0.0181 | 0.018 | 0.146 | 0.0180 | 0.0630 | 0.0187 | 0.375 | 0.0196 |
| 27 | 0.0142 | 0.0173 | 0.016 | 0.143 | 0.0164 | 0.0670 | 0.0172 | 0.500 | 0.0175 |
| 28 | 0.0126 | 0.0162 | 0.014 | 0.139 | 0.0148 | 0.0710 | 0.0156 | 1.000 | 0.0156 |
| 29 | 0.0113 | 0.0150 | 0.013 | 0.134 | 0.0136 | 0.0740 | 0.0141 | | 0.0139 |
| 30 | 0.0100 | 0.0140 | 0.012 | 0.127 | 0.0124 | 0.0780 | 0.0125 | | 0.0123 |
| 31 | 0.0089 | 0.0132 | 0.010 | 0.120 | 0.0116 | 0.0820 | 0.0109 | | 0.0110 |
| 32 | 0.0080 | 0.0128 | 0.009 | 0.115 | 0.0108 | 0.0860 | 0.0101 | | 0.0098 |
| 33 | 0.0071 | 0.0118 | 0.008 | 0.112 | 0.0100 | | 0.0094 | | 0.0087 |
| 34 | 0.0063 | 0.0104 | 0.007 | 0.110 | 0.0092 | | 0.0086 | | 0.0077 |
| 35 | 0.0056 | 0.0095 | 0.005 | 0.108 | 0.0084 | | 0.0078 | | 0.0069 |

[a] The American wire gage sizes have been rounded off to about the usual limits of commercial accuracy.

**85. Comparison of Wire and Sheet-Metal Gage Diameters or Thickness in Inches**

| Gage no. | (1) Ameri-can wire gage (AWG)[a] | (2) Steel wire gage | (3) Bir-ming-ham iron wire (Stubs) | (4) Stubs steel wire gage | (5) British standard wire gage | (6) Steel music wire gage | (7) Manufac-turers standard gage | (8) Ameri-can zinc | (9) Bir-ming-ham |
|---|---|---|---|---|---|---|---|---|---|
| 36 | 0.0050 | 0.0090 | 0.004 | 0.106 | 0.0076 | ...... | 0.0070 | ..... | 0.0061 |
| 37 | 0.0045 | 0.0085 | ..... | 0.103 | 0.0068 | ...... | 0.0066 | ..... | 0.0054 |
| 38 | 0.0040 | 0.0080 | ..... | 0.101 | 0.0060 | ...... | 0.0062 | ..... | 0.0048 |
| 39 | 0.0035 | 0.0075 | ..... | 0.099 | 0.0052 | ...... | ...... | ..... | 0.0043 |
| 40 | 0.0031 | 0.0070 | ..... | 0.097 | 0.0048 | ...... | ...... | ..... | 0.0039 |
| 41 | ...... | 0.0066 | ..... | 0.095 | 0.0044 | ...... | ...... | ..... | 0.0034 |
| 42 | ...... | 0.0062 | ..... | 0.092 | 0.0040 | ...... | ...... | ..... | 0.0031 |
| 43 | ...... | 0.0060 | ..... | 0.088 | 0.0036 | ...... | ...... | ..... | 0.0027 |
| 44 | ...... | 0.0058 | ..... | 0.085 | 0.0032 | ...... | ...... | ..... | 0.0024 |
| 45 | ...... | 0.0055 | ..... | 0.081 | 0.0028 | ...... | ...... | ..... | 0.00215 |
| 46 | ...... | 0.0052 | ..... | 0.079 | 0.0024 | ...... | ...... | ..... | 0.0019 |
| 47 | ...... | 0.0050 | ..... | 0.077 | 0.0020 | ...... | ...... | ..... | 0.0017 |
| 48 | ...... | 0.0048 | ..... | 0.075 | 0.0016 | ...... | ...... | ..... | 0.0015 |
| 49 | ...... | 0.0046 | ..... | 0.072 | 0.0012 | ...... | ...... | ..... | 0.00135 |
| 50 | ...... | 0.0044 | ..... | 0.069 | 0.0010 | ...... | ...... | ..... | 0.0012 |
| 51 | ...... | ...... | ..... | ..... | ...... | ...... | ...... | ..... | 0.0011 |
| 52 | ...... | ...... | ..... | ..... | ...... | ...... | ...... | ..... | 0.00095 |

| Type | Illustration | Application |
|------|-------------|-------------|
| Copper | | |
| Solid, round | | Bare wire in sizes Nos. 4/0 to 45. Covered, noninsulated in sizes Nos. 4/0 to 14. Insulated in sizes Nos. 6 to 14 for power work. Insulated in sizes Nos. 1/0 to 46 for magnet wire |
| Solid, grooved | | Trolley contact wire sizes 350,000 cmil to No. 1/0 |
| Solid, figure eight | | Trolley contact wire sizes 350,000 cmil to No. 1/0 |
| Solid, figure nine | | Trolley contact wire size 400,000 cmil |
| Solid, square | | Magnet wire and windings for electrical equipment |
| Solid, rectangular | | Magnet wire, windings for electrical equipment, busbars |
| Solid, channels | | Busbars |
| Solid, angle | | Busbars |
| Solid, tubes | | Busbars |
| Standard, concentric-stranded | | Bare, covered, and insulated in sizes 5 million cmil to No. 20 |
| Standard, rope-stranded | | Rubber-sheathed cords and cables in sizes 5 million cmil to No. 8 |

## Makeups of Electrical Conductors

| Type | Illustration | Application |
|------|-------------|-------------|
| | Copper | |
| Bunched-stranded | | Flexible cords and fixture wire in sizes Nos. 10 to 22 |
| Annular concentric-stranded with rope core | | Varnished-cambric–insulated and solid-type paper-insulated single-conductor cables in sizes 5 million to 750,000 cmil |
| Compack-stranded, round | | Varnished-cambric–insulated and solid-type paper-insulated cables in sizes 1,000,000 cmil to No. 1 single-conductor and size No. 1 for multiconductor |
| Compack-stranded, sector | | Multiconductor paper-insulated cables, both solid and oil-filled types, in sizes 1,000,000 cmil to No. 1/0 |
| Compack-stranded, segmental | | Varnished-cambric–insulated and solid-type paper-insulated single-conductor cables in sizes 4,000,000 to 1,000,000 cmil |
| Hollow-core, stranded | | Oil-filled paper-insulated single-conductor cables in sizes 2,000,000 cmil to No. 2/0 |
| Tubular-segmental; Type HH | | High-voltage long-distance transmission |
| | Iron or steel | |
| Solid, round | | Bare in sizes Nos. 4, 6, and 8 for power work. Bare in sizes Nos. 4 to 14 for telephone and telegraph work |
| Three-strand | | Bare in sizes Nos. 4, 6, and 8 |
| | Copperweld | |
| Solid, round | | Bare in sizes Nos. 4/0 to 12. Weatherproof covered in sizes Nos. 2 to 12. Rubber-insulated: single conductors in sizes Nos. 4/0 to 12 and two conductors in sizes Nos. 14 and 17 |

**Makeups of Electrical Conductors** (*Continued*)

| Type | Illustration | Application |
|------|:---:|------|
| | Copperweld | |
| Concentric-stranded | | Bare or rubber-covered single-conductor in diameters from 0.910 to 0.174 in |
| | Copperweld and copper | |
| Stranded | | Bare in various combinations of strands in cable sizes from 586,800 cmil to No. 8. Weatherproof covered in sizes Nos. 2 to 8 |
| | Copper and steel | |
| Three-strand | | Bare in sizes Nos. 2 to 12 |
| | All-aluminum | |
| Solid, round | | Bare wire in sizes Nos. 4/0 to 29. Covered, noninsulated in sizes Nos. 4/0 to 12. Insulated in sizes Nos. 6 to 12 for power work. Insulated in sizes Nos. 4 to 26 for magnet wire |
| Stranded | | Bare or weatherproof covered in sizes 1,590,000 cmil to No. 6 |
| | Copper-clad aluminum | |
| Solid, round | | Same as all-aluminum |
| Concentric-stranded | | Same as all-aluminum |
| | Aluminum and steel (ACSR) | |
| Annular aluminum, stranded with steel core | | Bare in sizes 1,590,000 cmil to No. 4. Weatherproof covered in sizes Nos. 4/0 to 8 |
| | Bronze | |
| Solid, round | | Bare or weatherproof covered in sizes Nos. 4 to 14 |

**87.   Application Guide for Electrical Conductor Materials**

Suitable conductor material

| Application | Copper, soft-drawn | Copper, medium-hard-drawn | Copper, hard-drawn | Copperweld | Copperweld and copper | Copper and steel, stranded | All aluminum and copper-clad aluminum | Aluminum and steel (ACSR) | Steel or iron | Bronze |
|---|---|---|---|---|---|---|---|---|---|---|
| **Bare transmission and distribution wires** — Heavy and medium loads — Short and medium spans | | ✓ | ✓ | | | | ✓ | ✓ | | |
| Bare transmission and distribution wires — Heavy and medium loads — Long spans | | | ✓ | | ✓ | | | ✓ | | |
| Bare transmission and distribution wires — Light loads — Short and medium spans | | ✓ | ✓ | | | | ✓ | ✓ | | |
| Bare transmission and distribution wires — Light loads — Long spans | | | ✓ | ✓ | | ✓ | | ✓ | ✓ | |
| **Covered noninsulated transmission and distribution wires** — Heavy loads — Short and medium spans | ✓ | ✓ | | | | | ✓ | | | |
| Covered noninsulated transmission and distribution wires — Heavy loads — Long spans | | ✓ | | | | | | ✓ | | |
| Covered noninsulated transmission and distribution wires — Light loads — Short and medium spans | ✓ | ✓ | | | | | ✓ | | | |
| Covered noninsulated transmission and distribution wires — Light loads — Long spans | | | ✓ | ✓ | ✓ | | | ✓ | | ✓ |
| Insulated cables | ✓ | | | | | | ✓ | | | |
| Telephone and telegraph lines; ground wires | | | ✓ | ✓ | | | | | ✓ | |
| Signal circuits | | | ✓ | ✓ | | | | | | |
| Trolley contact wires | | | ✓ | | | | | | | ✓ |
| Trolley feeders | | ✓ | | | | | | | | |

**88.   Factors for Determining Diameter of Concentric-Stranded Copper Cables**

| Number of strands | 3 | 7 | 12 | 19 | 37 | 61 | 91 | 127 and over |
|---|---|---|---|---|---|---|---|---|
| Factor | 1.244 | 1.134 | 1.199 | 1.147 | 1.151 | 1.152 | 1.153 | 1.154 |

To determine approximate diameter of bare concentric-stranded copper cables, multiply diameter of solid wire of same cross-sectional area by the proper factor from this table.

## 89. Solid Bare Copper Conductors

(Anaconda Wire and Cable Co.)

| Size, AWG | Wire diam, in. | Cross sectional area | | Weight | | Hard-drawn wire | | Medium hard-drawn wire | | Soft or annealed wire | |
|---|---|---|---|---|---|---|---|---|---|---|---|
| | | Cir mils | Sq in. | Per 1,000 ft, lb | Per mile, lb | Min breaking strength, lb | Max resistance per 1,000 ft at 20°C, ohms[b] | Min breaking strength, lb | Max resistance per 1,000 ft at 20°C, ohms[b] | Max breaking strength, lb | Max resistance per 1,000 ft at 20°C, ohms[b] |
| 4/0 | 0.4600 | 211,600 | 0.1662 | 640.5 | 3,382 | 8,143 | 0.05045 | 6,980 | 0.05019 | 5,983 | 0.04993 |
| 3/0 | 0.4096 | 167,800 | 0.1318 | 507.9 | 2,682 | 6,722 | 0.06361 | 5,667 | 0.06329 | 4,745 | 0.06296 |
| 2/0 | 0.3648 | 133,100 | 0.1045 | 402.8 | 2,127 | 5,519 | 0.08021 | 4,599 | 0.07980 | 3,763 | 0.07939 |
| 1/0 | 0.3249 | 105,500 | 0.08289 | 319.5 | 1,687 | 4,517 | 0.1011 | 3,730 | 0.1006 | 2,984 | 0.1001 |
| 1 | 0.2893 | 83,690 | 0.06573 | 253.5 | 1,338 | 3,688 | 0.1287 | 3,024 | 0.1282 | 2,432 | 0.1262 |
| 2 | 0.2576 | 66,370 | 0.05213 | 200.9 | 1,061 | 3,003 | 0.1625 | 2,450 | 0.1617 | 1,929 | 0.1592 |
| 3 | 0.2294 | 52,630 | 0.04134 | 159.3 | 841.2 | 2,439 | 0.2049 | 1,984 | 0.2038 | 1,530 | 0.2007 |
| 4 | 0.2043 | 41,740 | 0.03278 | 126.4 | 667.1 | 1,970 | 0.2584 | 1,584 | 0.2570 | 1,213 | 0.2531 |
| 5 | 0.1819 | 33,100 | 0.02600 | 100.2 | 529.1 | 1,591 | 0.3258 | 1,264 | 0.3241 | 961.9 | 0.3192 |
| 6 | 0.1620 | 26,250 | 0.02062 | 79.46 | 419.6 | 1,280 | 0.4108 | 1,010 | 0.4087 | 762.9 | 0.4025 |
| 7 | 0.1443 | 20,820 | 0.01635 | 63.02 | 332.7 | 1,030 | 0.5181 | 806.6 | 0.5154 | 605.0 | 0.5075 |
| 8 | 0.1285 | 16,510 | 0.01297 | 49.97 | 263.9 | 826.0 | 0.6553 | 643.9 | 0.6499 | 479.8 | 0.6400 |
| 9 | 0.1144 | 13,090 | 0.01028 | 39.63 | 209.3 | 661.2 | 0.8238 | 514.2 | 0.8195 | 380.5 | 0.8070 |
| 10 | 0.1019 | 10,380 | 0.008155 | 31.43 | 165.9 | 529.2 | 1.039 | 401.4 | 1.033 | 314.0 | 1.018 |
| 11 | 0.09074 | 8,234 | 0.006467 | 24.92 | 131.6 | 422.9 | 1.310 | 327.6 | 1.303 | 249.0 | 1.283 |
| 12 | 0.08081 | 6,530 | 0.005129 | 19.77 | 104.4 | 337.0 | 1.652 | 261.6 | 1.643 | 197.5 | 1.618 |
| 13 | 0.07196 | 5,178 | 0.004067 | 15.68 | 82.77 | 268.0 | 2.083 | 208.8 | 2.072 | 156.6 | 2.040 |
| 14 | 0.06408 | 4,107 | 0.003225 | 12.43 | 65.64 | 213.5 | 2.626 | 166.6 | 2.613 | 124.2 | 2.573 |
| 15 | 0.05707 | 3,257 | 0.002558 | 9.858 | 52.05 | 169.8 | 3.312 | 133.0 | 3.295 | 98.48 | 3.244 |
| 16 | 0.05082 | 2,583 | 0.002028 | 7.818 | 41.28 | 135.1 | 4.176 | 106.2 | 4.154 | 78.10 | 4.091 |
| 17 | 0.04526 | 2,048 | 0.001609 | 6.200 | 32.74 | 107.5 | 5.266 | 84.71 | 5.239 | 61.93 | 5.158 |
| 18 | 0.04030 | 1,624 | 0.001276 | 4.917 | 25.96 | 85.47 | 6.640 | 67.61 | 6.606 | 49.12 | 6.505 |
| 19 | 0.03589 | 1,288 | 0.001012 | 3.899 | 20.59 | 67.99 | 8.373 | 53.95 | 8.330 | 38.95 | 8.202 |
| 20 | 0.03196 | 1,022 | 0.0008023 | 3.092 | 16.33 | 54.08 | 10.56 | 43.05 | 10.50 | 30.89 | 10.34 |
| 21 | 0.02846 | 810.1 | 0.0006363 | 2.452 | 12.95 | 43.07 | 13.31 | 34.36 | 13.24 | 24.50 | 13.04 |
| 22 | 0.02535 | 642.5 | 0.0005046 | 1.945 | 10.27 | 34.26 | 16.79 | 27.41 | 16.70 | 19.43 | 16.45 |
| 23 | 0.02257 | 509.5 | 0.0004001 | 1.542 | 8.143 | 27.25 | 21.17 | 21.87 | 21.06 | 15.14 | 20.74 |
| 24 | 0.02010 | 404.0 | 0.0003173 | 1.233 | 6.458 | 21.67 | 26.69 | 17.45 | 26.56 | 12.69 | 26.15 |
| 25 | 0.01790 | 320.4 | 0.0002517 | 0.9699 | 5.121 | 17.26 | 33.66 | 13.92 | 33.49 | 10.07 | 32.97 |

## 89. Solid Bare Copper Conductors (Continued)

| Size, AWG | Wire diam., in. | Cross sectional area | | Weight | | Hard-drawn wire | | Medium hard-drawn wire | | Soft or annealed wire | |
|---|---|---|---|---|---|---|---|---|---|---|---|
| | | Cir mils | Sq in. | Per 1,000 ft, lb | Per mile, lb | Min breaking strength, lb[a] | Max resistance per 1,000 ft at 20°C, ohms[b] | Min breaking strength, lb[a] | Max resistance per 1,000 ft at 20°C, ohms[b] | Max breaking strength, lb[a] | Max resistance per 1,000 ft at 20°C, ohms[b] |
| 26 | 0.01594 | 254.1 | 0.0001996 | 0.7692 | 4.061 | 13.73 | 42.44 | 11.11 | 42.23 | 7.983 | 41.58 |
| 27 | 0.01420 | 201.5 | 0.0001583 | 0.6100 | 3.221 | 10.92 | 53.52 | 8.863 | 53.25 | 6.331 | 52.43 |
| 28 | 0.01264 | 159.8 | 0.0001255 | 0.4837 | 2.554 | 8.698 | 67.49 | 7.070 | 67.14 | 5.021 | 66.11 |
| 29 | 0.01126 | 126.7 | 0.00009954 | 0.3836 | 2.026 | 6.918 | 85.10 | 5.640 | 84.66 | 3.981 | 83.37 |
| 30 | 0.01003 | 100.5 | 0.00007894 | 0.3042 | 1.606 | 5.502 | 107.3 | 4.499 | 106.8 | 3.157 | 105.1 |
| 31 | 0.008928 | 79.70 | 0.00006260 | 0.2413 | 1.274 | 4.376 | 135.3 | 3.589 | 134.6 | 2.504 | 132.6 |
| 32 | 0.007950 | 63.21 | 0.00004964 | 0.1913 | 1.010 | 3.485 | 170.6 | 2.862 | 169.8 | 1.986 | 167.2 |
| 33 | 0.007080 | 50.13 | 0.00003937 | 0.1517 | 0.8011 | 2.772 | 215.2 | 2.283 | 214.1 | 1.575 | 210.8 |
| 34 | 0.006305 | 39.75 | 0.00003122 | 0.1203 | 0.6353 | 2.204 | 271.3 | 1.821 | 269.9 | 1.249 | 265.8 |
| 35 | 0.005615 | 31.52 | 0.00002476 | 0.09542 | 0.5038 | 1.755 | 342.1 | 1.452 | 340.4 | 0.9904 | 335.2 |
| 36 | 0.005000 | 25.00 | 0.00001963 | 0.07567 | 0.3996 | 1.396 | 431.4 | 1.158 | 429.2 | 0.7854 | 422.6 |
| 37 | 0.004453 | 19.83 | 0.00001557 | 0.06001 | 0.3169 | 1.110 | 544.0 | 0.9238 | 541.2 | 0.6228 | 532.9 |
| 38 | 0.003965 | 15.72 | 0.00001235 | 0.04759 | 0.2513 | 0.8829 | 686.0 | 0.7367 | 682.4 | 0.4939 | 672.0 |
| 39 | 0.003531 | 12.47 | 0.000009793 | 0.03774 | 0.1993 | 0.7031 | 865.0 | 0.5876 | 860.5 | 0.3917 | 847.4 |
| 40 | 0.003145 | 9.888 | 0.000007766 | 0.02993 | 0.1580 | 0.5592 | 1,091 | 0.4685 | 1,085 | 0.3106 | 1,069 |
| 41 | 0.002800 | 7.842 | 0.000006159 | 0.02374 | 0.1253 | 0.4434 | 1,375 | 0.3716 | 1,368 | 0.2464 | 1,347 |
| 42 | 0.002494 | 6.219 | 0.000004884 | 0.01882 | 0.09939 | 0.3517 | 1,734 | 0.2947 | 1,725 | 0.1954 | 1,699 |
| 43 | 0.002221 | 4.932 | 0.000003873 | 0.01493 | 0.07882 | 0.2789 | 2,187 | 0.2337 | 2,176 | 0.1549 | 2,142 |
| 44 | 0.001978 | 3.911 | 0.000003072 | 0.01184 | 0.06251 | 0.2212 | 2,758 | 0.1853 | 2,743 | 0.1229 | 2,702 |
| 1 Mil | 0.001000 | 1.000 | 0.000000785 | 0.003027 | 0.01598 | 0.05655 | 10,790 | 0.04738 | 10,730 | 0.03142 | 10,570 |

[a] The breaking strengths are based on ASTM specification requirements, using minimum values for hard- and medium hard-drawn wire, and maximum values for soft wire.

[b] The resistance values in this table are trade maximums and are higher than the average values for commercial wire. The following values for the conductivity of copper were used:

### ASTM Requirements

| | Conductivity, IACS % at 20°C | Resistivity, lb per mile-ohm at 20°C |
|---|---|---|
| Hard-drawn, 0.325 in. and larger.......... | 97.16 | 900.77 |
| Hard-drawn, 0.324 in. and smaller......... | 96.16 | 910.15 |
| Medium hard-drawn, 0.325 in. and larger... | 97.66 | 896.15 |
| Medium hard-drawn, 0.324 in. and smaller.. | 96.66 | 905.44 |
| Soft or annealed.......................... | 98.16 | 891.58 |

## 90.  Stranded Bare Copper Conductors
(Anaconda Wire and Cable Co.)

| Size | | Stranding, class[a] | Cable diam., in. | Weight[b] | | Harddrawn, min breaking strength, lb[c] | Medium harddrawn, min breaking strength, lb[c] | Softdrawn, max breaking strength, lb[c] |
|---|---|---|---|---|---|---|---|---|
| Cir mils | AWG | | | Per 1,000 ft, lb | Per mile, lb | | | |
| 5,000,000 | ... | B | 2.581 | 15,890 | 83,910 | 219,500 | 173,200 | 145,300 |
| 5,000,000 | ... | A | 2.580 | 15,890 | 83,910 | 216,300 | 171,800 | 145,300 |
| 4,500,000 | ... | B | 2.448 | 14,300 | 75,520 | 200,400 | 156,900 | 130,800 |
| 4,500,000 | ... | A | 2.448 | 14,300 | 75,520 | 197,200 | 154,600 | 130,800 |
| 4,000,000 | ... | B | 2.309 | 12,590 | 66,490 | 178,100 | 139,500 | 116,200 |
| 4,000,000 | ... | A | 2.307 | 12,590 | 66,490 | 175,600 | 138,500 | 116,200 |
| 3,500,000 | ... | B | 2.159 | 11,020 | 58,180 | 155,900 | 122,000 | 101,700 |
| 3,500,000 | ... | A | 2.158 | 11,020 | 58,180 | 153,400 | 120,200 | 101,700 |
| 3,000,000 | ... | B | 1.998 | 9,353 | 49,390 | 134,400 | 104,600 | 87,180 |
| 3,000,000 | ... | A | 1.998 | 9,353 | 49,390 | 131,700 | 103,900 | 87,180 |
| 2,500,000 | ... | B | 1.824 | 7,794 | 41,150 | 111,300 | 87,170 | 72,650 |
| 2,500,000 | ... | A | 1.823 | 7,794 | 41,150 | 109,600 | 85,800 | 72,650 |
| 2,000,000 | ... | B | 1.632 | 6,175 | 32,600 | 90,050 | 70,210 | 58,120 |
| 2,000,000 | ... | A | 1.630 | 6,175 | 32,600 | 87,790 | 69,270 | 58,120 |
| 1,750,000 | ... | B | 1.526 | 5,403 | 28,530 | 78,800 | 61,430 | 50,850 |
| 1,750,000 | ... | A | 1.526 | 5,403 | 28,530 | 77,930 | 61,020 | 50,850 |
| 1,500,000 | ... | B | 1.412 | 4,631 | 24,450 | 67,540 | 52,650 | 43,590 |
| 1,500,000 | ... | A | 1.411 | 4,631 | 24,450 | 65,840 | 51,950 | 43,590 |
| 1,250,000 | ... | B | 1.289 | 3,859 | 20,380 | 56,280 | 43,880 | 36,320 |
| 1,250,000 | ... | A | 1.288 | 3,859 | 20,380 | 55,670 | 43,590 | 36,320 |
| 1,000,000 | ... | B, A | 1.152 | 3,088 | 16,300 | 45,030 | 35,100 | 29,060 |
| 1,000,000 | ... | AA | 1.151 | 3,088 | 16,300 | 43,830 | 34,350 | 29,060 |
| 900,000 | ... | B, A | 1.094 | 2,779 | 14,670 | 40,520 | 31,590 | 26,150 |
| 900,000 | ... | AA | 1.092 | 2,779 | 14,670 | 39,510 | 31,170 | 26,150 |
| 800,000 | ... | B, A | 1.031 | 2,470 | 13,040 | 36,360 | 28,270 | 23,250 |
| 800,000 | ... | AA | 1.029 | 2,470 | 13,040 | 35,120 | 27,710 | 23,250 |
| 750,000 | ... | B, A | 0.998 | 2,316 | 12,230 | 34,090 | 26,510 | 21,790 |
| 750,000 | ... | AA | 0.997 | 2,316 | 12,230 | 33,400 | 26,150 | 21,790 |
| 700,000 | ... | B, A | 0.964 | 2,161 | 11,410 | 31,820 | 24,740 | 20,340 |
| 700,000 | ... | AA | 0.963 | 2,161 | 11,410 | 31,170 | 24,410 | 20,340 |
| 600,000 | ... | B | 0.893 | 1,853 | 9,781 | 27,530 | 21,350 | 18,140 |
| 600,000 | ... | A, AA | 0.891 | 1,853 | 9,781 | 27,020 | 21,060 | 17,440 |
| 500,000 | ... | B, A | 0.813 | 1,544 | 8,151 | 22,510 | 17,550 | 14,530 |
| 500,000 | ... | AA | 0.811 | 1,544 | 8,151 | 21,950 | 17,320 | 14,530 |
| 450,000 | ... | B, A | 0.772 | 1,389 | 7,336 | 20,450 | 15,900 | 13,080 |
| 450,000 | ... | AA | 0.770 | 1,389 | 7,336 | 19,750 | 15,590 | 13,080 |
| 400,000 | ... | B | 0.728 | 1,235 | 6,521 | 18,320 | 14,140 | 11,620 |
| 400,000 | ... | A, AA | 0.726 | 1,235 | 6,521 | 17,560 | 13,850 | 11,620 |
| 350,000 | ... | B | 0.681 | 1,081 | 5,706 | 16,060 | 12,450 | 10,580 |
| 350,000 | ... | A | 0.679 | 1,081 | 5,706 | 15,590 | 12,200 | 10,170 |
| 350,000 | ... | AA | 0.710 | 1,081 | 5,706 | 15,140 | 12,020 | 10,170 |
| 300,000 | ... | B | 0.630 | 926.3 | 4,891 | 13,870 | 10,740 | 9,071 |
| 300,000 | ... | A | 0.629 | 926.3 | 4,891 | 13,510 | 10,530 | 8,718 |
| 300,000 | ... | AA | 0.657 | 926.3 | 4,891 | 13,170 | 10,390 | 8,718 |
| 250,000 | ... | B | 0.575 | 771.9 | 4,076 | 11,560 | 8,952 | 7,559 |
| 250,000 | ... | A | 0.574 | 771.9 | 4,076 | 11,360 | 8,836 | 7,265 |
| 250,000 | ... | AA | 0.600 | 771.9 | 4,076 | 11,130 | 8,717 | 7,265 |
| 211,600 | 4/0 | B | 0.528 | 653.3 | 3,450 | 9,617 | 7,479 | 6,149 |
| 211,600 | 4/0 | A, AA | 0.552 | 653.3 | 3,450 | 9,483 | 7,378 | 6,149 |
| 211,600 | 4/0 | A, AA[d] | 0.522 | 653.3 | 3,450 | 9,154 | 7,269 | 6,149 |

## 90.  Stranded Bare Copper Conductors (Continued)

| Size | | Strand- ing, class[a] | Cable diam., in. | Weight[b] | | Hard- drawn, min break- ing strength, lb[c] | Medium hard- drawn, min break- ing strength, lb[c] | Soft- drawn, max breaking strength, lb[c] |
|---|---|---|---|---|---|---|---|---|
| Cir mils | AWG | | | Per 1,000 ft, lb | Per mile, lb | | | |
| 167,800 | 3/0 | B | 0.470 | 518.1 | 2,736 | 7,698 | 5,970 | 5,074 |
| 167,800 | 3/0 | A, AA | 0.492 | 518.1 | 2,736 | 7,556 | 5,890 | 4,876 |
| 167,800 | 3/0 | A, AA[d] | 0.464 | 518.1 | 2,736 | 7,366 | 5,812 | 4,876 |
| 133,100 | 2/0 | B | 0.419 | 410.9 | 2,170 | 6,153 | 4,766 | 4,025 |
| 133,100 | 2/0 | ...... | 0.437 | 410.9 | 2,170 | 6,049 | 4,704 | 3,868 |
| 133,100 | 2/0 | A, AA | 0.414 | 410.9 | 2,170 | 5,927 | 4,641 | 3,868 |
| 105,500 | 1/0 | B | 0.373 | 325.7 | 1,720 | 4,899 | 3,803 | 3,190 |
| 105,500 | 1/0 | ...... | 0.390 | 325.7 | 1,720 | 4,840 | 3,753 | 3,190 |
| 105,500 | 1/0 | A, AA | 0.368 | 325.7 | 1,720 | 4,750 | 3,703 | 3,066 |
| 83,690 | 1 | B | 0.332 | 258.4 | 1,364 | 3,898 | 3,037 | 2,531 |
| 83,690 | 1 | A | 0.328 | 258.4 | 1,364 | 3,804 | 2,958 | 2,432 |
| 83,690 | 1 | AA | 0.360 | 255.9 | 1,351 | 3,620 | 2,875 | 2,432 |
| 66,370 | 2 | B, A | 0.292 | 204.9 | 1,082 | 3,045 | 2,361 | 2,007 |
| 66,370 | 2 | AA | 0.320 | 202.9 | 1,071 | 2,913 | 2,299 | 1,929 |
| 52,630 | 3 | B, A | 0.260 | 162.5 | 858.0 | 2,433 | 1,885 | 1,591 |
| 52,630 | 3 | AA | 0.285 | 160.9 | 849.6 | 2,359 | 1,835 | 1,529 |
| 41,740 | 4 | B, A | 0.232 | 128.9 | 680.5 | 1,938 | 1,505 | 1,262 |
| 41,740 | 4 | AA | 0.254 | 127.6 | 673.8 | 1,879 | 1,465 | 1,213 |
| 33,100 | 5 | B | 0.206 | 102.2 | 539.6 | 1,542 | 1,201 | 1,001 |
| 26,250 | 6 | B | 0.184 | 81.05 | 427.9 | 1,228 | 958.6 | 793.7 |
| 20,820 | 7 | B | 0.164 | 64.28 | 339.4 | 977.2 | 765.3 | 629.6 |
| 16,510 | 8 | B | 0.146 | 50.98 | 269.1 | 777.2 | 610.7 | 499.2 |
| 13,090 | 9 | B | 0.130 | 40.42 | 213.4 | 618.1 | 487.3 | 395.8 |
| 10,380 | 10 | B | 0.116 | 32.05 | 169.2 | 491.6 | 388.9 | 313.9 |
| 6,530 | 12 | B | 0.0915 | 20.16 | 106.5 | 311.1 | 247.7 | 197.5 |
| 4,107 | 14 | B | 0.0726 | 12.68 | 66.95 | 197.1 | 157.7 | 124.2 |
| 2,583 | 16 | B | 0.0576 | 7.975 | 42.11 | 124.7 | 100.4 | 81.15 |
| 1,624 | 18 | B | 0.0456 | 5.014 | 26.47 | 78.98 | 63.89 | 51.02 |
| 1,022 | 20 | B | 0.0363 | 3.155 | 16.66 | 50.06 | 40.69 | 32.11 |

[a] Class of stranding in accordance with ASTM Specification B 8-35T.
[b] Weight increased 2 per cent to allow for cabling, except as follows:

| Cir Mils | Per Cent |
|---|---|
| 5,000,000 and 4,500,000 | 5 |
| 4,000,000 and 3,500,000 | 4 |
| 3,000,000 and 2,500,000 | 3 |

[c] The breaking strengths are based on ASTM specification requirements, using minimum values for hard- and medium hard-drawn cable, and maximum values for soft cable.
[d] Optional construction, see Table 91.

## 91.  Stranding Data for Copper Conductors: Concentric-Stranded; ANSI Standards

| Size | | Class AA | | | Class A | | | Class B | | |
|---|---|---|---|---|---|---|---|---|---|---|
| AWG | Cir mils | Number of strands | Diam of individual strands, mils | Approx over-all diam, in. | Number of strands | Diam of individual strands, mils | Approx over-all diam, in. | Number of strands | Diam of individual strands, mils | Approx over-all diam, in. |
| | 5,000,000 | ... | ..... | ..... | 169 | 172.0 | 2.580 | 217 | 151.8 | 2.581 |
| | 4,500,000 | ... | ..... | ..... | 169 | 163.2 | 2.448 | 217 | 144.0 | 2.448 |
| | 4,000,000 | ... | ..... | ..... | 169 | 153.8 | 2.307 | 217 | 135.8 | 2.309 |
| | 3,500,000 | ... | ..... | ..... | 127 | 166.0 | 2.158 | 169 | 143.8 | 2.159 |
| | 3,000,000 | ... | ..... | ..... | 127 | 153.7 | 1.998 | 169 | 133.2 | 1.998 |
| | 2,500,000 | ... | ..... | ..... | 91 | 165.7 | 1.823 | 127 | 140.3 | 1.824 |
| | 2,000,000 | ... | ..... | ..... | 91 | 148.2 | 1.630 | 127 | 125.5 | 1.631 |
| | 1,900,000 | ... | ..... | ..... | 91 | 144.5 | ..... | 127 | 122.3 | |
| | 1,800,000 | ... | ..... | ..... | 91 | 140.6 | ..... | 127 | 119.1 | |
| | 1,750,000 | ... | ..... | ..... | 91 | 138.7 | 1.526 | 127 | 117.4 | 1.526 |
| | 1,700,000 | ... | ..... | ..... | 91 | 136.7 | ..... | 127 | 115.7 | |
| | 1,600,000 | ... | ..... | ..... | 91 | 132.6 | ..... | 127 | 112.2 | |
| | 1,500,000 | ... | ..... | ..... | 61 | 156.8 | 1.411 | 91 | 128.4 | 1.412 |
| | 1,400,000 | ... | ..... | ..... | 61 | 151.5 | ..... | 91 | 124.0 | |
| | 1,300,000 | ... | ..... | ..... | 61 | 146.0 | ..... | 91 | 119.5 | |
| | 1,250,000 | ... | ..... | ..... | 61 | 143.1 | 1.288 | 91 | 117.2 | 1.289 |
| | 1,200,000 | ... | ..... | ..... | 61 | 140.3 | ..... | 91 | 114.8 | |
| | 1,100,000 | ... | ..... | ..... | 61 | 134.3 | ..... | 91 | 109.9 | |
| | 1,000,000 | 37 | 164.4 | 1.151 | 61 | 128.0 | 1.152 | 61 | 128.0 | 1.152 |
| | 900,000 | 37 | 156.0 | 1.092 | 61 | 121.5 | 1.094 | 61 | 121.5 | 1.094 |
| | 800,000 | 37 | 147.0 | 1.029 | 61 | 114.5 | 1.031 | 61 | 114.5 | 1.031 |
| | 750,000 | 37 | 142.4 | 0.997 | 61 | 110.9 | 0.998 | 61 | 110.9 | 0.998 |
| | 700,000 | 37 | 137.5 | 0.963 | 61 | 107.1 | 0.964 | 61 | 107.1 | 0.964 |
| | 650,000 | 37 | 132.5 | ..... | 61 | 103.2 | ..... | 61 | 103.2 | |
| | 600,000 | 37 | 127.3 | 0.891 | 37 | 127.3 | 0.891 | 61c | 99.2 | 0.893 |
| | 550,000 | 37 | 121.9 | ..... | 37 | 121.9 | ..... | 61d | 95.0 | |
| | 500,000 | 19 | 162.2 | 0.811 | 37 | 116.2 | 0.814 | 37 | 116.2 | 0.814 |
| | 450,000 | 19 | 153.9 | 0.770 | 37 | 110.3 | 0.772 | 37 | 110.3 | 0.772 |
| | 400,000 | 19 | 145.1 | 0.726 | 19 | 145.1 | 0.726 | 37 | 104.0 | 0.728 |
| | 350,000 | 12 | 170.7 | 0.710 | 19 | 135.7 | 0.679 | 37 | 97.3 | 0.681 |
| | 300,000 | 12 | 158.1 | 0.657 | 19 | 125.7 | 0.629 | 37 | 90.0 | 0.630 |
| | 250,000 | 12 | 144.3 | 0.600 | 19 | 114.7 | 0.574 | 37 | 82.2 | 0.575 |
| 4/0 | 211,600 | 7a | 173.9 | 0.522 | 7a | 173.9 | 0.522 | 19 | 105.5 | 0.528 |
| 3/0 | 167,800 | 7b | 154.8 | 0.464 | 7b | 154.8 | 0.464 | 19 | 94.0 | 0.470 |
| 2/0 | 133,100 | 7 | 137.9 | 0.414 | 7 | 137.9 | 0.414 | 19 | 83.7 | 0.418 |
| 1/0 | 105,500 | 7 | 122.8 | 0.368 | 7 | 122.8 | 0.368 | 19 | 74.5 | 0.373 |
| 1 | 83,690 | 3 | 167.0 | 0.360 | 7 | 109.3 | 0.328 | 19 | 66.4 | 0.332 |
| 2 | 66,370 | 3 | 148.7 | 0.320 | 7 | 97.4 | 0.292 | 7 | 97.4 | 0.292 |
| 3 | 52,630 | 3 | 132.5 | 0.285 | 7 | 86.7 | 0.260 | 7 | 86.7 | 0.260 |
| 4 | 41,740 | 3 | 118.0 | 0.254 | 7 | 77.2 | 0.232 | 7 | 77.2 | 0.232 |
| 5 | 33,100 | ... | ..... | ..... | .... | ..... | ..... | 7 | 68.8 | 0.206 |
| 6 | 26,250 | ... | ..... | ..... | .... | ..... | ..... | 7 | 61.2 | 0.184 |
| 7 | 20,820 | ... | ..... | ..... | .... | ..... | ..... | 7 | 54.5 | 0.164 |
| 8 | 16,510 | ... | ..... | ..... | .... | ..... | ..... | 7 | 48.6 | 0.146 |
| 9 | 13,090 | ... | ..... | ..... | .... | ..... | ..... | 7 | 43.2 | 0.130 |
| 10 | 10,380 | ... | ..... | ..... | .... | ..... | ..... | 7 | 38.5 | 0.116 |
| 12 | 6,530 | ... | ..... | ..... | .... | ..... | ..... | 7 | 30.5 | 0.0915 |
| 14 | 4,107 | ... | ..... | ..... | .... | ..... | ..... | 7 | 24.2 | 0.0726 |
| 16 | 2,583 | ... | ..... | ..... | .... | ..... | ..... | 7 | 19.2 | 0.0576 |
| 18 | 1,624 | ... | ..... | ..... | .... | ..... | ..... | 7 | 15.2 | 0.0456 |
| 20 | 1,022 | ... | ..... | ..... | .... | ..... | ..... | 7 | 12.1 | 0.0363 |

a Optional construction for No. 4/0 AWG size in Class AA and Class A is 12 wires of 132.8 mils diameter.
b Optional construction for No. 3/0 AWG size in Class AA and Class A is 12 wires of 118.3 mils diameter.
c Optional construction for 600,000 CM size in Class B is 37 wires of 127.3 mils diameter.
d Optional construction for 550,000 CM size in Class B is 37 wires of 121.9 mils diameter.
The above data are approximate and subject to normal manufacturing tolerances.

## 92.  Stranding Data for Copper Conductors: Concentric-Stranded; ANSI Standards

| Size | | Class C | | | Class D | | |
|---|---|---|---|---|---|---|---|
| AWG | Cir mils | Number of strands | Diam of individual strands, mils | Approx over-all diam, in. | Number of strands | Diam of individual strands, mils | Approx over-all diam, in. |
| | 5,000,000 | 271 | 135.8 | 2.580 | 271 | 135.8 | 2.580 |
| | 4,500,000 | 271 | 128.9 | 2.448 | 271 | 128.9 | 2.448 |
| | 4,000,000 | 271 | 121.5 | 2.307 | 271 | 121.5 | 2.307 |
| | 3,500,000 | 217 | 127.0 | 2.158 | 271 | 113.6 | 2.158 |
| | 3,000,000 | 217 | 117.6 | 1.998 | 271 | 105.2 | 1.998 |
| | 2,500,000 | 169 | 121.6 | 1.824 | 217 | 107.3 | 1.824 |
| | 2,000,000 | 169 | 108.8 | 1.631 | 217 | 96.0 | 1.631 |
| | 1,900,000 | 169 | 106.0 | ..... | 217 | 93.6 | |
| | 1,800,000 | 169 | 103.2 | ..... | 217 | 91.1 | |
| | 1,750,000 | 169 | 101.8 | 1.526 | 217 | 89.8 | 1.526 |
| | 1,700,000 | 169 | 100.3 | ..... | 217 | 88.5 | |
| | 1,600,000 | 169 | 97.3 | ..... | 217 | 85.9 | |
| | 1,500,000 | 127 | 108.7 | 1.414 | 169 | 94.2 | 1.414 |
| | 1,400,000 | 127 | 105.0 | ..... | 169 | 91.0 | |
| | 1,300,000 | 127 | 101.2 | ..... | 169 | 87.7 | |
| | 1,250,000 | 127 | 99.2 | 1.290 | 169 | 86.0 | 1.290 |
| | 1,200,000 | 127 | 97.2 | ..... | 169 | 84.3 | |
| | 1,100,000 | 127 | 93.1 | ..... | 169 | 80.7 | |
| | 1,000,000 | 91 | 104.8 | 1.153 | 127 | 88.7 | 1.154 |
| | 900,000 | 91 | 99.4 | 1.095 | 127 | 84.2 | 1.096 |
| | 800,000 | 91 | 93.8 | 1.032 | 127 | 79.4 | 1.032 |
| | 750,000 | 91 | 90.8 | 0.999 | 127 | 76.8 | 1.000 |
| | 700,000 | 91 | 87.7 | 0.965 | 127 | 74.2 | 0.966 |
| | 650,000 | 91 | 84.5 | ..... | 127 | 71.5 | |
| | 600,000 | 91 | 81.2 | 0.893 | 127 | 68.7 | 0.894 |
| | 550,000 | 91 | 77.7 | ..... | 127 | 65.8 | |
| | 500,000 | 61 | 90.5 | 0.815 | 91 | 74.1 | 0.817 |
| | 450,000 | 61 | 85.9 | 0.773 | 91 | 70.3 | 0.774 |
| | 400,000 | 61 | 81.0 | 0.730 | 91 | 66.3 | 0.731 |
| | 350,000 | 61 | 75.7 | 0.683 | 91 | 62.0 | 0.684 |
| | 300,000 | 61 | 70.1 | 0.632 | 91 | 57.4 | 0.633 |
| | 250,000 | 61 | 64.0 | 0.577 | 91 | 52.4 | 0.578 |
| 4/0 | 211,600 | 37 | 75.6 | 0.529 | 61 | 58.9 | 0.530 |
| 3/0 | 167,800 | 37 | 67.3 | 0.471 | 61 | 52.4 | 0.471 |
| 2/0 | 133,100 | 37 | 60.0 | 0.419 | 61 | 46.7 | 0.419 |
| 1/0 | 105,500 | 37 | 53.4 | 0.376 | 61 | 41.6 | 0.376 |
| 1 | 83,690 | 37 | 47.6 | 0.333 | 61 | 37.0 | 0.333 |
| 2 | 66,370 | 19 | 59.1 | 0.295 | 37 | 42.4 | 0.296 |
| 3 | 52,630 | 19 | 52.6 | 0.261 | 37 | 37.7 | 0.262 |
| 4 | 41,740 | 19 | 46.9 | 0.230 | 37 | 33.6 | 0.231 |
| 5 | 33,100 | 19 | 41.7 | 0.209 | 37 | 29.9 | 0.210 |
| 6 | 26,250 | 19 | 37.2 | 0.186 | 37 | 26.6 | 0.186 |
| 7 | 20,820 | 19 | 33.1 | 0.166 | 37 | 23.7 | 0.166 |
| 8 | 16,510 | 19 | 29.5 | 0.147 | 37 | 21.1 | 0.148 |
| 9 | 13,090 | 19 | 26.2 | 0.131 | 37 | 18.8 | 0.132 |
| 10 | 10,380 | 19 | 23.4 | 0.107 | 37 | 16.7 | 0.117 |
| 12 | 6,530 | 19 | 18.5 | 0.093 | 37 | 13.3 | 0.093 |
| 14 | 4,107 | 19 | 14.7 | 0.074 | 37 | 10.5 | 0.074 |
| 16 | 2,583 | 19 | 11.7 | 0.058 | | | |
| 18 | 1,624 | 19 | 9.2 | 0.046 | | | |
| 20 | 1,022 | 19 | 7.3 | 0.037 | | | |

The above data are approximate and subject to normal manufacturing tolerances.

**93.   Stranding Data for Copper Conductors: Rope-Stranded; ICEA Standards**

| Size, cir mils or AWG | Flexible stranding (Class G) | | | | Extra-flexible stranding (Class H) | | | | Approx over-all diam, in. | Weight per 1,000 ft, net lb |
|---|---|---|---|---|---|---|---|---|---|---|
| | Total number of strands | Diam of individual strands, mils | Number of ropes | Number of strands each rope | Total number of strands | Diam of individual strands, mils | Number of ropes | Number of strands each rope | | |
| 5,000,000 | 1,159 | 65.7 | 61 | 19 | 1,729 | 53.8 | 91 | 19 | 2.959 | 16,200 |
| 4,500,000 | 1,159 | 62.3 | 61 | 19 | 1,729 | 51.0 | 91 | 19 | 2.805 | 14,600 |
| 4,000,000 | 1,159 | 58.7 | 61 | 19 | 1,729 | 48.1 | 91 | 19 | 2.646 | 13,000 |
| 3,500,000 | 1,159 | 55.0 | 61 | 19 | 1,729 | 45.0 | 91 | 19 | 2.475 | 11,400 |
| 3,000,000 | 1,159 | 50.9 | 61 | 19 | 1,729 | 41.7 | 91 | 19 | 2.294 | 9,470 |
| 2,500,000 | 703 | 59.6 | 37 | 19 | 1,159 | 46.4 | 61 | 19 | 2.088 | 8,090 |
| 2,000,000 | 703 | 53.3 | 37 | 19 | 1,159 | 41.5 | 61 | 19 | 1.868 | 6,470 |
| 1,900,000 | 703 | 52.0 | 37 | 19 | 1,159 | 40.5 | 61 | 19 | 1.823 | 6,160 |
| 1,800,000 | 703 | 50.6 | 37 | 19 | 1,159 | 39.4 | 61 | 19 | 1.773 | 5,830 |
| 1,750,000 | 703 | 49.9 | 37 | 19 | 1,159 | 38.9 | 61 | 19 | 1.751 | 5,680 |
| 1,700,000 | 703 | 49.2 | 37 | 19 | 1,159 | 38.3 | 61 | 19 | 1.724 | 5,510 |
| 1,600,000 | 703 | 47.7 | 37 | 19 | 1,159 | 37.2 | 61 | 19 | 1.674 | 5,200 |
| 1,500,000 | 427 | 59.3 | 61 | 7 | 703 | 46.2 | 37 | 19 | 1.617 | 4,860 |
| 1,400,000 | 427 | 57.3 | 61 | 7 | 703 | 44.6 | 37 | 19 | 1.561 | 4,530 |
| 1,300,000 | 427 | 55.2 | 61 | 7 | 703 | 43.0 | 37 | 19 | 1.505 | 4,210 |
| 1,250,000 | 427 | 54.1 | 61 | 7 | 703 | 42.2 | 37 | 19 | 1.477 | 4,060 |
| 1,200,000 | 427 | 53.0 | 61 | 7 | 703 | 41.3 | 37 | 19 | 1.446 | 3,890 |
| 1,100,000 | 427 | 50.8 | 61 | 7 | 703 | 39.6 | 37 | 19 | 1.386 | 3,570 |
| 1,000,000 | 427 | 48.4 | 61 | 7 | 703 | 37.7 | 37 | 19 | 1.320 | 3,240 |
| 900,000 | 427 | 45.9 | 61 | 7 | 703 | 35.8 | 37 | 19 | 1.253 | 2,920 |
| 800,000 | 427 | 43.3 | 61 | 7 | 703 | 33.7 | 37 | 19 | 1.180 | 2,590 |
| 750,000 | 427 | 41.9 | 61 | 7 | 703 | 32.7 | 37 | 19 | 1.145 | 2,440 |
| 700,000 | 427 | 40.5 | 61 | 7 | 703 | 31.6 | 36 | 19 | 1.106 | 2,270 |
| 650,000 | 427 | 39.0 | 61 | 7 | 703 | 30.4 | 37 | 19 | 1.064 | 2,110 |
| 600,000 | 427 | 37.5 | 61 | 7 | 703 | 29.2 | 37 | 19 | 1.022 | 1,940 |
| 550,000 | 427 | 35.9 | 61 | 7 | 703 | 28.0 | 37 | 19 | 0.980 | 1,790 |
| 500,000 | 259 | 43.9 | 37 | 7 | 427 | 34.2 | 61 | 7 | 0.923 | 1,610 |
| 450,000 | 259 | 41.7 | 37 | 7 | 427 | 32.5 | 61 | 7 | 0.878 | 1,460 |
| 400,000 | 259 | 39.3 | 37 | 7 | 427 | 30.6 | 61 | 7 | 0.826 | 1,290 |
| 350,000 | 259 | 36.8 | 37 | 7 | 427 | 28.6 | 61 | 7 | 0.773 | 1,130 |
| 300,000 | 259 | 34.0 | 37 | 7 | 427 | 26.5 | 61 | 7 | 0.716 | 969 |
| 250,000 | 259 | 31.1 | 37 | 7 | 427 | 24.2 | 61 | 7 | 0.653 | 808 |
| 4/0 | 133 | 39.9 | 19 | 7 | 259 | 28.6 | 37 | 7 | 0.602 | 686 |
| 3/0 | 133 | 35.5 | 19 | 7 | 259 | 25.5 | 37 | 7 | 0.536 | 540 |
| 2/0 | 133 | 31.6 | 19 | 7 | 259 | 22.7 | 37 | 7 | 0.477 | 428 |
| 1/0 | 133 | 28.2 | 19 | 7 | 259 | 20.2 | 37 | 7 | 0.424 | 339 |
| 1 | 133 | 25.1 | 19 | 7 | 259 | 18.0 | 37 | 7 | 0.378 | 269 |
| 2 | 49 | 36.8 | 7 | 7 | 133 | 22.3 | 19 | 7 | 0.336 | 213 |
| 3 | 49 | 32.8 | 7 | 7 | 133 | 19.9 | 19 | 7 | 0.299 | 168 |
| 4 | 49 | 29.2 | 7 | 7 | 133 | 17.7 | 19 | 7 | 0.266 | 133 |
| 5 | 49 | 26.0 | 7 | 7 | 133 | 15.8 | 19 | 7 | 0.237 | 106 |
| 6 | 49 | 23.1 | 7 | 7 | 133 | 14.0 | 19 | 7 | 0.210 | 83 |
| 8 | 49 | 18.4 | 7 | 7 | 133 | 11.1 | 19 | 7 | 0.167 | 52 |

AWG size for individual strands are permissible, provided their area is not more than 2 per cent below required area.   Sizes Nos. 2, 3/0, and 4/0 AWG can be supplied in alternate combinations as follows:

| Wire size | No. 2 AWG | No. 3/0 AWG | No. 4/0 AWG |
|---|---|---|---|
| Total number of strands.................... | 259 | 427 | 427 |
| Diameter individual strands (mils)............. | 16.0 | 19.8 | 22.3 |
| Number of ropes.......................... | 37 | 61 | 61 |
| Number of strands each rope................ | 7 | 7 | 7 |

Flexible stranded conductors Nos. 10 to 22 AWG are normally supplied in bunched-stranded form; refer to Tables 94 and 95.

Conforms to all requirements of ICEA Project 46.

The above data are approximate and subject to normal manufacturing tolerances.

**94. Stranding Data for Copper Conductors: Bunched- and Rope-Stranded; ICEA and ASTM Standards**

| Size, AWG or cir mils | Class K | | | | Class M | | | |
|---|---|---|---|---|---|---|---|---|
| | All wires No. 30 AWG | | | | All wires No. 34 AWG | | | |
| | Nominal No. of wires | Suggested construction | Approx over-all diam, in. | Approx weight, lb per 1,000 ft | Nominal No. of wires | Suggested construction | Approx over-all diam, in. | Approx weight, lb per 1,000 ft |
| 20 | 10 | 1 × 10 | 0.038 | 3.2 | 26 | 1 × 26 | 0.038 | 3.2 |
| 18 | 16 | 1 × 16 | 0.048 | 5.0 | 41 | 1 × 41 | 0.048 | 5.0 |
| 16 | 26 | 1 × 26 | 0.060 | 8.0 | 65 | 1 × 65 | 0.060 | 8.0 |
| 14 | 41 | 1 × 41 | 0.078 | 12.8 | 104 | 1 × 104 | 0.078 | 12.8 |
| 12 | 65 | 1 × 65 | 0.101 | 20.3 | 168 | 7 × 24 | 0.101 | 21.0 |
| 10 | 104 | 1 × 104 | 0.126 | 32.5 | 259 | 7 × 37 | 0.126 | 32.5 |
| 9 | 133 | 7 × 19 | 0.150 | 42 | 336 | 7 × 48 | 0.146 | 42 |
| 8 | 168 | 7 × 24 | 0.157 | 53 | 420 | 7 × 60 | 0.162 | 53 |
| 7 | 210 | 7 × 30 | 0.179 | 66 | 532 | 19 × 28 | 0.196 | 67 |
| 6 | 266 | 7 × 38 | 0.210 | 84 | 665 | 19 × 35 | 0.215 | 84 |
| 5 | 336 | 7 × 48 | 0.235 | 106 | 836 | 19 × 44 | 0.240 | 105 |
| 4 | 420 | 7 × 60 | 0.272 | 132 | 1,064 | 19 × 56 | 0.269 | 134 |
| 3 | 532 | 19 × 28 | 0.304 | 169 | 1,323 | 7 × 7 × 27 | 0.305 | 169 |
| 2 | 665 | 19 × 35 | 0.338 | 211 | 1,666 | 7 × 7 × 34 | 0.337 | 212 |
| 1 | 836 | 19 × 44 | 0.397 | 266 | 2,107 | 7 × 7 × 43 | 0.376 | 268 |
| 1/0 | 1,064 | 19 × 56 | 0.451 | 338 | 2,646 | 7 × 7 × 54 | 0.423 | 337 |
| 2/0 | 1,323 | 7 × 7 × 27 | 0.470 | 425 | 3,325 | 19 × 7 × 25 | 0.508 | 427 |
| 3/0 | 1,666 | 7 × 7 × 34 | 0.533 | 535 | 4,256 | 19 × 7 × 32 | 0.576 | 547 |
| 4/0 | 2,107 | 7 × 7 × 43 | 0.627 | 676 | 5,320 | 19 × 7 × 40 | 0.645 | 684 |
| 250,000 | 2,499 | 7 × 7 × 51 | 0.682 | 802 | 6,384 | 19 × 7 × 48 | 0.713 | 821 |
| 300,000 | 2,989 | 7 × 7 × 61 | 0.768 | 960 | 7,581 | 19 × 7 × 57 | 0.768 | 975 |
| 350,000 | 3,458 | 19 × 7 × 26 | 0.809 | 1,120 | 8,806 | 37 × 7 × 34 | 0.825 | 1,130 |
| 400,000 | 3,990 | 19 × 7 × 30 | 0.878 | 1,290 | 10,101 | 37 × 7 × 39 | 0.901 | 1,300 |
| 450,000 | 4,522 | 19 × 7 × 34 | 0.933 | 1,465 | 11,396 | 37 × 7 × 44 | 0.940 | 1,465 |
| 500,000 | 5,054 | 19 × 7 × 38 | 0.988 | 1,635 | 12,691 | 37 × 7 × 49 | 0.997 | 1,630 |
| 550,000 | 5,453 | 19 × 7 × 41 | 1 056 | 1,765 | 13,664 | 61 × 7 × 32 | 1.035 | 1,755 |
| 600,000 | 5,985 | 19 × 7 × 45 | 1.125 | 1,940 | 14,945 | 61 × 7 × 35 | 1.084 | 1,920 |
| 650,000 | 6,517 | 19 × 7 × 49 | 1.166 | 2,110 | 16,226 | 61 × 7 × 38 | 1.133 | 2,085 |
| 700,000 | 6,916 | 19 × 7 × 52 | 1.207 | 2,240 | 17,507 | 61 × 7 × 41 | 1.183 | 2,250 |
| 750,000 | 7,581 | 19 × 7 × 57 | 1.276 | 2,455 | 18,788 | 61 × 7 × 44 | 1.207 | 2,415 |
| 800,000 | 7,980 | 19 × 7 × 60 | 1 305 | 2,585 | 20,069 | 61 × 7 × 47 | 1.256 | 2,580 |
| 900,000 | 9,065 | 37 × 7 × 35 | 1.323 | 2,935 | 22,631 | 61 × 7 × 53 | 1.331 | 2,910 |
| 1,000,000 | 10,101 | 37 × 7 × 39 | 1.419 | 3,270 | 25,193 | 61 × 7 × 59 | 1.404 | 3,240 |

**95.  Stranding Data for Copper Conductors: Bunched-Stranded Ropes; ICEA Standards**

| Nominal size, cir mils or AWG | Class J Min number of No. 30 AWG strands | Class L Min number No. 34 AWG strands | Approx OD, in. | Nominal size, AWG | Class J Min number of No. 30 AWG strands | Class L Min number No. 34 AWG strands | Approx OD, in. |
|---|---|---|---|---|---|---|---|
| 1,000,000 | 9,951 | ...... | 1.52 | 1 | 836 | 2,109 | 0.44 |
| 900,000 | 8,956 | ...... | 1.43 | 2 | 661 | 1,672 | 0.39 |
| 800,000 | 7,961 | ...... | 1.35 | 3 | 524 | 1,326 | 0.36 |
| 750,000 | 7,463 | ...... | 1.31 | 4 | 410 | 1,052 | 0.31 |
| 700,000 | 6,966 | ...... | 1.28 | 5 | 330 | 832 | 0.27 |
| 650,000 | 6,468 | ...... | 1.25 | 6 | 262 | 661 | 0.22 |
| 600,000 | 5,971 | ...... | 1.21 | 7 | 208 | 524 | 0.20 |
| 550,000 | 5,473 | ...... | 1.17 | 8 | 165 | 416 | 0.17 |
| 500,000 | 4,976 | 12,579 | 1.08 | 9 | 131 | 330 | 0.16 |
| 450,000 | 4,478 | 11,321 | 1.00 | 10 | 104 | 262 | 0.12 |
| 400,000 | 3,981 | 10,063 | 0.96 | 12 | 65 | 165 | 0.10 |
| 350,000 | 3,483 | 8,806 | 0.90 | 14 | 41 | 104 | 0.08 |
| 300,000 | 2,986 | 7,557 | 0.82 | 16 | 26 | 65 | 0.06 |
| 250,000 | 2,488 | 6,297 | 0.78 | 18 | 16 | 41 | 0.05 |
| 4/0 | 2,106 | 5,330 | 0.73 | 20 | 10 | 26 | 0.04 |
| 3/0 | 1,670 | 4,227 | 0.61 | | | | |
| 2/0 | 1,325 | 3,353 | 0.55 | | | | |
| 1/0 | 1,050 | 2,658 | 0.49 | | | | |

**96.  Stranding Data for Copper Conductors, Bunch-Stranded: ICEA Special Strandings**

| Nominal size, AWG | Class P Fixture wire, commercial | | Class Q 10,000-cycle heater cord | | Class R Special fixture wire, portable cords, etc. | | Class S Type S cord | | Class T Oscillating fan cord | |
|---|---|---|---|---|---|---|---|---|---|---|
| | Number strands | Size strand, AWG | Number strands | Size strand, AWG | Number strands | Size strand, AWG | Number strands | Size strand, AWG | Number strands | Size strand, AWG |
| 10 | 65 | 28 | ... | ... | 165 | 32 | 104 | 30 | | |
| 12 | 41 | 28 | ... | ... | 104 | 32 | 84 | 31 | | |
| 14 | 26 | 28 | ... | ... | 65 | 32 | 84 | 33 | | |
| 16 | 16 | 28 | 104 | 36 | 41 | 32 | 65 | 34 | | |
| 18 | 10 | 28 | 65 | 36 | 26 | 32 | 41 | 34 | 165 | 40 |
| 20 | 7 | 28 | 41 | 36 | 16 | 32 | 26 | 34 | 104 | 40 |

### 97.   Recommended Stranding Practice for Copper Conductors

| Type of Stranding | Recommended Uses |
|---|---|
| Class AA | For bare cable |
| Class A | For weather-resistant (weatherproof), slow-burning, and slow-burning weather-resistant cables, and for bare cable where greater flexibility than is afforded by Class AA is required |
| Class B | For cable insulated with various materials such as rubber, paper, varnished cloth, etc., and for the cables indicated under Class A where greater flexibility is required |
| Class C and Class K | For cable where greater flexibility is required than is provided by Class B cable |
| Class G | For all rubber-sheathed cords and cables for normal use |
| Class H | For all rubber-sheathed cords and cables where extreme flexibility is required, such as for use on take-up reels, over sheaves, etc. |
| Class J | For use in fixture wire, portable cords, etc. |
| Class L | For use in welding cable, heater cord, Type SJ cord |
| Class P | For commercial fixture wire |
| Class Q | For 10,000-cycle heater cord |
| Class R | For special fixture wire, portable cords, etc. |
| Class S | For Type S cord |
| Class T | For oscillating-fan cord |
| Bunched stranded | For extremely flexible conductors of sizes from No. 10 to No. 22 |

### 98.   Annular, Concentric-Stranded, Rope-Core Conductors: ICEA Standard
(General Cable Co.)

| Nominal size, cir mils | Actual area, cir mils | Approx. rope core size, in. | Number of strands | Diam individual strands, in. | Max over-all diam, in. | Weight per 1,000 ft | |
|---|---|---|---|---|---|---|---|
| | | | | | | Copper only, net lb | Copper and core, net lb |
| 750,000 | 741,735 | 0.375 | 54 | 0.1172 | 1.108 | 2,312 | 2,362 |
| 800,000 | 800,865 | 0.468 | 65 | 0.1110 | 1.164 | 2,497 | 2,575 |
| 900,000 | 906,565 | 0.500 | 66 | 0.1172 | 1.234 | 2,826 | 2,905 |
| 1,000,000 | 1,023,766 | 0.563 | 65 | 0.1255 | 1.346 | 3,192 | 3,304 |
| 1,250,000 | 1,260,020 | 0.750 | 80 | 0.1255 | 1.533 | 3,928 | 4,127 |
| 1,500,000 | 1,512,024 | 1.000 | 96 | 0.1255 | 1.783 | 4,714 | 5,068 |
| 1,750,000 | 1,753,083 | 1.125 | 107 | 0.1280 | 1.923 | 5,466 | 5,914 |
| 2,000,000 | 1,978,387 | 1.3125 | 120 | 0.1284 | 2.114 | 6,168 | 6,778 |
| 2,500,000 | 2,488,320 | 1.500 | 120 | 0.1440 | 2.394 | 7,758 | 8,555 |
| 3,000,000 | 3,044,304 | 1.625 | 116 | 0.1620 | 2.627 | 9,492 | 10,427 |
| 3,500,000 | 3,595,428 | 2.000 | 137 | 0.1620 | 3.007 | 11,319 | 12,735 |
| 4,000,000 | 4,015,332 | 2.250 | 153 | 0.1620 | 3.262 | 12,641 | 14,433 |
| 4,500,000 | 4,408,992 | 2.500 | 168 | 0.1620 | 3.517 | 14,013 | 16,226 |
| 5,000,000 | 4,960,116 | 2.875 | 189 | 0.1620 | 3.897 | 15,765 | 18,691 |

The following tolerances are included in overall diameters:

| Nominal Size, kcmil | Tolerance, In. |
|---|---|
| 750–3,000 | 0.030 |
| 3,001–3,500 | 0.035 |
| 3,501–4,000 | 0.040 |
| 4,001–4,500 | 0.045 |
| 4,501–5,000 | 0.050 |

The following allowances for stranding are included in the metallic weight:

| Nominal Size, kcmil | Tolerance, Percent |
|---|---|
| 750–3,000 | 3 |
| 3,001–4,000 | 4 |
| 4,001–5,000 | 5 |

### 99.  Compack-Stranded Conductors: Segmental
(General Cable Co.)

| Size, cir mils | Number of strands | Approx over-all diam, in. | Size, cir mils | Number of strands | Approx over-all diam, in. |
|---|---|---|---|---|---|
| 1,000,000 | 148 | 1.152 | 2,500,000 | 244 | 1.825 |
| 1,250,000 | 148 | 1.288 | 3,000,000 | 364 | 2.000 |
| 1,500,000 | 148 | 1.412 | 3,500,000 | 364 | 2.159 |
| 1,750,000 | 244 | 1.526 | 4,000,000 | 364 | 2.308 |
| 2,000,000 | 244 | 1.631 | | | |

### 100.  Compack-Stranded Conductors: Round and Sector
(General Cable Co.)

| Size, AWG or cir mils | Round | | 120-degree sector | | 90-degree sector | |
|---|---|---|---|---|---|---|
| | Number of strands | Over-all diam, in. | V-gage depth,[a] in. | Number of strands | V-gage depth,[a] in. | Number of strands |
| 1 | 19 | 0.299 | | | | |
| 1/0 | 19 | 0.336 | 0.288 | 19 | 0.340 | 19 |
| 2/0 | 19 | 0.376 | 0.323 | 19 | 0.382 | 19 |
| 3/0 | 19 | 0.423 | 0.364 | 37 | | |
| 4/0 | 19 | 0.475 | { 0.417 (L) <br> 0.410 (H) } | 37 <br> 37 } | 0.482 | 37 |
| 250,000 | 37 | 0.520 | { 0.455 (L) <br> 0.477 (H) } | 37 <br> 37 } | 0.525 | 37 |
| 300,000 | 37 | 0.570 | { 0.497 (L) <br> 0.490 (H) } | 37 <br> 37 } | 0.572 | 37 |
| 350,000 | 37 | 0.616 | { 0.539 (L) <br> 0.532 (H) } | 37 <br> 37 } | 0.620 | 37 |
| 400,000 | 37 | 0.659 | { 0.572 (L) <br> 0.566 (H) } | 37 <br> 37 | | |
| 500,000 | 37 | 0.736 | { 0.642 (L) <br> 0.635 (H) } | 61 <br> 61 } | 0.740 | 61 |
| 600,000 | 61 | 0.813 | { 0.700 (L) <br> 0.690 (H) } | 61 <br> 61 | | |
| 700,000 | 61 | 0.877 | { 0.754 (L) <br> 0.742 (H) } | 91 <br> 91 | | |
| 750,000 | 61 | 0.908 | { 0.780 (L) <br> 0.767 (H) } | 91 <br> 91 | | |
| 800,000 | 61 | 0.938 | 0.795[b] | 91 | | |
| 1,000,000 | 61 | 1.060 | 0.900[b] | 91 | | |

The above data are approximate and subject to normal manufacturing tolerances.

$L$ and $H$ indicate V-gage depths of the bare conductor for "light" walls (up to and including $13/64$ in.) and "heavy" walls ($15/64$ in. and over), respectively. The shape of the sector and, therefore, the V-gage depth in the larger sizes depend on whether a light or heavy wall of insulation is to be applied.

[a] V-gage depth is the distance from the center point of the exterior arc of the bare sector to the point where the lines of the sides of the conductor if prolonged would intersect. For rough calculation purposes, V-gage depth in inches can be taken as 90 per cent of the square root of the nominal area of the conductor expressed in circular inches.

[b] Representative values; exact values depend on insulation thickness to be used and will be furnished on request.

### 101.  Hollow-Core Stranded Conductors
(General Cable Co.)

| Size, AWG or cir mils | Over-all conductor diam, in. | | Size, AWG or cir mils | Over-all conductor diam, in. | |
|---|---|---|---|---|---|
| | ID of spring core = 0.500 in. | ID of spring core = 0.690 in. | | ID of spring core = 0.500 in. | ID of spring core = 0.690 in. |
| 2/0 | 0.736 | . . . . . | 700,000 | 1.151 | 1.256 |
| 3/0 | 0.768 | 0.924 | 750,000 | 1.180 | 1.286 |
| 4/0 | 0.807 | 0.956 | 800,000 | 1.212 | 1 309 |
| 250,000 | 0.837 | 0.983 | 850,000 | 1.242 | 1.341 |
| 300,000 | 0.880 | 1.017 | 900,000 | 1.261 | 1.365 |
| 350,000 | 0.917 | 1.049 | 1,000,000 | 1.310 | 1.416 |
| 400,000 | 0.953 | 1.082 | 1,250,000 | 1.434 | 1.524 |
| 450,000 | 0.989 | 1.112 | 1,500,000 | 1.547 | 1.635 |
| 500,000 | 1.028 | 1.145 | 1,750,000 | 1.650 | 1.730 |
| 600,000 | 1.084 | 1.201 | 2,000,000 | 1.760 | 1.833 |
| 650,000 | 1.121 | 1.228 | | | |

The above data are approximate and subject to normal manufacturing tolerances.

## 102.  Iron or Steel Telephone and Telegraph Wire

(American Steel and Wire Co.)

| Size, BWG | Diam, in. | Approx weight, lb | | Approx breaking strength, lb | | | Resistance per mile (International Ohms) at 68°F | | |
|---|---|---|---|---|---|---|---|---|---|
| | | Per 1,000 ft | Per mile | EBB | BB | Steel | EBB | BB | Steel |
| 4 | 0.238 | 153 | 811 | 2,028 | 2,271 | 2,433 | 5.98 | 7.15 | 8.32 |
| 6 | 0.203 | 112 | 590 | 1,475 | 1,652 | 1,770 | 8.22 | 9.83 | 11.44 |
| 8 | 0.165 | 74 | 390 | 975 | 1,092 | 1,170 | 12.43 | 14.87 | 17.31 |
| 9 | 0.148 | 60 | 314 | 785 | 879 | 942 | 15.44 | 18.47 | 21.50 |
| 10 | 0.134 | 49 | 258 | 645 | 722 | 774 | 18.79 | 22.48 | 26.16 |
| 11 | 0.120 | 39 | 206 | 515 | 577 | 618 | 23.54 | 28.16 | 32.77 |
| 12 | 0.109 | 32 | 170 | 425 | 476 | 510 | 28.52 | 34.12 | 39.71 |
| 14 | 0.083 | 19 | 99 | 247 | 277 | 297 | 48.98 | 58.59 | 68.18 |

Amertel-85 Wire

| Size, BWG | Nominal diam, in. | Approx weight per mile, lb | Approx coil length, miles | Min breaking strength, lb | Resistance per mile, ohms |
|---|---|---|---|---|---|
| 9 | 0.148 | 314 | ½ | 1462 | 18.47 |
| 10 | 0.134 | 258 | ½ | 1199 | 22.48 |
| 12 | 0.109 | 170 | ½ | 793 | 34.12 |
| 14 | 0.083 | 99 | ½ | 460 | 58.59 |

Amertel-135 Wire

| | |
|---|---|
| Size, BWG............................ | No. 12 |
| Nominal diameter, in................... | 0.109 |
| Minimum breaking strength, lb.......... | 1213 |
| Resistance per mile, ohms............... | 38.23 |
| Approximate weight per mile, lb......... | 170 |
| Approximate weight per coil, lb......... | 150 |
| Approximate length per coil, ft.......... | 4,659 |

Amertel-195 Wire

| | |
|---|---|
| Size, BWG................................ | No. 12 |
| Nominal diameter, in........................ | 0.109 |
| Minimum breaking strength, lb................. | 1800 |
| Resistance per mile (Ohms).................... | 38.8 |
| Approximate weight per mile, lb................ | 170 |
| Approximate weight per coil, lb................ | 160 |
| Minimum weight per coil, lb, approx.......... | 144 |
| Maximum weight per coil, lb, approx.......... | 176 |
| Approximate length per coil, ft.................. | 4,970 |
| Minimum length per coil, ft, approx.......... | 4,470 |
| Maximum length per coil, ft, approx.......... | 5,470 |

### 103.  All-Steel Conductors
(American Steel and Wire Co.)

| Type | Size and diameter | | | Area | | Approx weight | | Breaking strength, lb |
| | Size, BWG | Number and diameter of wires | Diam of conductor, in. | Cir mils | Sq in. | Lb per 1,000 ft | Lb per mile | |
|---|---|---|---|---|---|---|---|---|
| S | 4 | 1 | 0.238 | 56,644 | 0.0445 | 153 | 808 | 5,560 |
| S | 6 | 1 | 0.203 | 41,209 | 0.0324 | 112 | 591 | 4,270 |
| S | 8 | 1 | 0.165 | 27,225 | 0.0214 | 74 | 391 | 2,820 |
| S-3 | 4 | 3/0.138 | 0.297 | 57,132 | 0.0448 | 156 | 823 | 5,560 |
| S-3 | 6 | 3/0.117 | 0.252 | 41,067 | 0.0322 | 112 | 591 | 4,270 |
| S-3 | 8 | 3/0.096 | 0.207 | 27,648 | 0.0217 | 75 | 396 | 2,820 |

Modulus: Type S, 29,000,000; Type S-3, 25,000,000.
Coefficient of expansion: 0.0000066.

**104.  Trolley Wire: Copper and Bronze**
(Anaconda Wire and Cable Co.)

| Nominal size | Cross-sectional area | | | Weight | | Min conductivity, % IACS[a] | D-c resistance or volts drop per amp at 20°C (68°F) | | Min tensile strength, lb per sq in. | Min breaking load, lb | Elongation in 10 in., % |
| | Nominal | Actual | | | | | Ohms or volts per 1,000 ft | Ohms or volts per mile | | | |
| AWG or MCM | MCM | MCM | Sq in. | Lb per 1,000 ft | Lb per mile | | | | | | |
|---|---|---|---|---|---|---|---|---|---|---|---|
| | | | | | | Round 97.16 % Conductivity Hard-drawn Copper | | | | | |
| 1/0 | 105.6 | 105.6 | 0.0829 | 319.5 | 1,687 | 97.16 | 0.1011 | 0.5339 | 54,500 | 4,518 | 2.40 |
| 2/0 | 133.1 | 133.1 | 0.1045 | 402.8 | 2,127 | 97.16 | 0.08021 | 0.4235 | 52,800 | 5,519 | 2.80 |
| 3/0 | 167.8 | 167.8 | 0.1318 | 507.8 | 2,681 | 97.16 | 0.06362 | 0.3359 | 51,000 | 6,720 | 3.25 |
| 4/0 | 211.6 | 211.6 | 0.1662 | 640.5 | 3,382 | 97.16 | 0.05045 | 0.2664 | 49,000 | 8,143 | 3.75 |
| 300 | 300.0 | 300.0 | 0.2356 | 908.0 | 4,794 | 97.16 | 0.03558 | 0.1879 | 46,400 | 10,930 | 4.50 |
| | | | | | | Grooved 97.16 % Conductivity Hard-drawn Copper | | | | | |
| 2/0 | 133.1 | 137.9 | 0.1083 | 417.6 | 2,205 | 97.16 | 0.07741 | 0.4087 | 50,200 | 5,437 | 2.80 |
| 3/0 | 167.8 | 167.3 | 0.1314 | 506.4 | 2,674 | 97.16 | 0.06380 | 0.3369 | 48,500 | 6,373 | 3.25 |
| 4/0 | 211.6 | 212.0 | 0.1665 | 641.9 | 3,389 | 97.16 | 0.05035 | 0.2659 | 46,600 | 7,759 | 3.75 |
| 300 | 300.0 | 299.8 | 0.2355 | 907.6 | 4,792 | 97.16 | 0.03560 | 0.1880 | 44,200 | 10,410 | 4.50 |
| 350 | 350.0 | 351.2 | 0.2758 | 1,063.0 | 5,612 | 97.16 | 0.03040 | 0.1605 | 42,800 | 11,800 | 4.50 |
| | | | | | | Figure-8 97.16 % Conductivity Hard-drawn Copper | | | | | |
| 1/0 | 105.6 | 105.6 | 0.0829 | 319.5 | 1,687 | 97.16 | 0.1011 | 0.5340 | 51,800 | 4,294 | 2.40 |
| 2/0 | 133.1 | 133.1 | 0.1045 | 402.8 | 2,127 | 97.16 | 0.08021 | 0.4325 | 50,200 | 5,246 | 2.80 |
| 3/0 | 167.8 | 167.8 | 0.1318 | 508.0 | 2,682 | 97.16 | 0.06361 | 0.3359 | 48,500 | 6,392 | 3.25 |
| 4/0 | 211.6 | 211.6 | 0.1662 | 640.5 | 3,382 | 97.16 | 0.05044 | 0.2663 | 46,600 | 7,745 | 3.75 |
| 350 | 350.0 | 350.1 | 0.2750 | 1,060.0 | 5,597 | 97.16 | 0.03049 | 0.1610 | 42,800 | 11,770 | 4.50 |
| | | | | | | Figure-9 Deep-section 97.16 % Conductivity Hard-drawn Copper | | | | | |
| 350 | 350.0 | 348.9 | 0.2740 | 1,056.0 | 5,576 | 97.16 | 0.03060 | 0.1616 | 42,800 | 11,730 | 4.50 |
| 400 | 400.0 | 397.2 | 0.3120 | 1,202.0 | 6,347 | 97.16 | 0.02687 | 0.1419 | 41,300 | 12,890 | 4.50 |
| | | | | | | Round 85 % Conductivity Hitenso A Bronze | | | | | |
| 1/0 | 105.6 | 105.6 | 0.0829 | 319.5 | 1,687 | 85 | 0.1156 | 0.6105 | 68,000 | 5,638 | 2.40 |
| 2/0 | 133.1 | 133.1 | 0.1045 | 402.8 | 2,127 | 85 | 0.09172 | 0.4843 | 66,000 | 6,898 | 2.75 |
| 3/0 | 167.8 | 167.8 | 0.1318 | 507.8 | 2,681 | 85 | 0.07275 | 0.3841 | 64,000 | 8,433 | 3.25 |
| 4/0 | 211.6 | 211.6 | 0.1662 | 640.5 | 3,382 | 85 | 0.05768 | 0.3046 | 61,500 | 10,220 | 3.75 |
| 300 | 300.0 | 300.0 | 0.2356 | 908.0 | 4,794 | 85 | 0.04069 | 0.2148 | 58,300 | 13,740 | 4.50 |
| | | | | | | Grooved 85 % Conductivity Hitenso A bronze | | | | | |
| 2/0 | 133.1 | 137.9 | 0.1083 | 417.6 | 2,205 | 85 | 0.08849 | 0.4672 | 66,000 | 7,148 | 2.25 |
| 3/0 | 167.8 | 167.3 | 0.1314 | 506.4 | 2,674 | 85 | 0.07293 | 0.3851 | 64,000 | 8,410 | 2.75 |
| 4/0 | 211.6 | 212.0 | 0.1665 | 641.9 | 3,389 | 85 | 0.05756 | 0.3039 | 61,500 | 10,240 | 3.25 |
| 300 | 300.0 | 299.8 | 0.2355 | 907.6 | 4,792 | 85 | 0.04069 | 0.2149 | 58,300 | 13,730 | 4.00 |
| 350 | 350.0 | 351.2 | 0.2758 | 1,063.0 | 5,612 | 85 | 0.03475 | 0.1835 | 57,000 | 15,720 | 4.00 |

### 104. Trolley Wire: Copper and Bronze (Continued)

| Nominal size AWG or kcmil | Cross-sectional area Nominal kcmil | Actual kcmil | Sq in. | Weight Lb per 1,000 ft | Lb per mile | Min conductivity, % IACS[a] | D-c resistance or volts drop per amp at 20°C (68°F) Ohms or volts per 1,000 ft | Ohms or volts per mile | Min tensile strength, lb per sq in. | Min breaking load, lb | Elongation in 10 in., % |
|---|---|---|---|---|---|---|---|---|---|---|---|
| | | | | | | **Round ASTM Alloy 80 Hitenso BB Bronze** | | | | | |
| 1/0 | 105.6 | 105.6 | 0.0829 | 319.5 | 1,687 | 80 | 0.1228 | 0.6485 | 72,000 | 5,969 | 2.40 |
| 2/0 | 133.1 | 133.1 | 0.1045 | 402.8 | 2,127 | 80 | 0.09742 | 0.5144 | 69,000 | 7,212 | 2.75 |
| 3/0 | 167.8 | 167.8 | 0.1318 | 507.8 | 2,681 | 80 | 0.07727 | 0.4080 | 67,000 | 8,828 | 3.25 |
| 4/0 | 211.6 | 211.6 | 0.1662 | 640.5 | 3,382 | 80 | 0.06127 | 0.3235 | 65,000 | 10,800 | 3.75 |
| 300 | 300.0 | 300.0 | 0.2356 | 908.0 | 4,794 | 80 | 0.04322 | 0.2282 | 61,500 | 14,490 | 4.50 |
| | | | | | | **Grooved ASTM Alloy 80 Hitenso BB Bronze** | | | | | |
| 2/0 | 133.1 | 137.9 | 0.1083 | 417.6 | 2,205 | 80 | 0.09402 | 0.4964 | 69,000 | 7,473 | 2.25 |
| 3/0 | 167.8 | 167.3 | 0.1314 | 506.4 | 2,674 | 80 | 0.07749 | 0.4091 | 67,000 | 8,804 | 2.75 |
| 4/0 | 211.6 | 212.0 | 0.1665 | 641.9 | 3,389 | 80 | 0.06115 | 0.3229 | 65,000 | 10,820 | 3.25 |
| 300 | 300.0 | 299.8 | 0.2355 | 907.6 | 4,792 | 80 | 0.04324 | 0.2283 | 61,500 | 14,480 | 4.00 |
| 350 | 350.0 | 351.2 | 0.2758 | 1,063.0 | 5,612 | 80 | 0.03692 | 0.1949 | 59,500 | 16,410 | 4.00 |
| | | | | | | **Figure-9 Deep-section ASTM Alloy 80 Hitenso BB Bronze** | | | | | |
| 335 | 335.0 | 336.4 | 0.2642 | 1,020.0 | 5,386 | 80 | 0.03854 | 0.2035 | 56,800 | 15,010 | 4.00 |
| | | | | | | **Round ASTM Alloy 65 Trolley Bronze "65"** | | | | | |
| 1/0 | 105.6 | 105.6 | 0.0829 | 319.5 | 1,687 | 65 | 0.1511 | 0.7978 | 68,000 | 5,638 | 2.40 |
| 2/0 | 133.1 | 133.1 | 0.1045 | 402.8 | 2,127 | 65 | 0.1199 | 0.6329 | 65,000 | 6,794 | 2.75 |
| 3/0 | 167.8 | 167.8 | 0.1318 | 507.8 | 2,681 | 65 | 0.09507 | 0.4791 | 63,000 | 8,301 | 3.25 |
| 4/0 | 211.6 | 211.6 | 0.1662 | 640.5 | 3,382 | 65 | 0.07538 | 0.3980 | 61,000 | 10,140 | 3.75 |
| 300 | 300.0 | 300.0 | 0.2356 | 908.0 | 4,794 | 65 | 0.05317 | 0.2808 | 57,800 | 13,620 | 4.50 |
| | | | | | | **Grooved ASTM Alloy 65 Trolley Bronze "65"** | | | | | |
| 2/0 | 133.1 | 137.9 | 0.1083 | 417.6 | 2,205 | 65 | 0.1157 | 0.6110 | 65,000 | 7,040 | 2.25 |
| 3/0 | 167.8 | 167.3 | 0.1314 | 506.4 | 2,674 | 65 | 0.09537 | 0.5036 | 63,000 | 8,278 | 2.75 |
| 4/0 | 211.6 | 212.0 | 0.1665 | 641.9 | 3,389 | 65 | 0.07526 | 0.3974 | 61,000 | 10,160 | 3.25 |
| 300 | 300.0 | 299.8 | 0.2355 | 907.6 | 4,792 | 65 | 0.05321 | 0.2810 | 57,800 | 13,610 | 4.00 |
| 350 | 350.0 | 351.2 | 0.2758 | 1,063.0 | 5,612 | 65 | 0.04544 | 0.2399 | 56,200 | 15,500 | 4.00 |
| | | | | | | **Figure-9 Deep-section ASTM Alloy 65 Trolley Bronze "65"** | | | | | |
| 335 | 335.0 | 336.4 | 0.2642 | 1,020.0 | 5,386 | 65 | 0.04742 | 0.2504 | 54,000 | 14,270 | 4.00 |
| | | | | | | **Round ASTM Alloy 55 Hitenso C Bronze** | | | | | |
| 1/0 | 105.6 | 105.6 | 0.0829 | 319.5 | 1,687 | 55 | 0.1786 | 0.9431 | 76,000 | 6,301 | 2.40 |
| 2/0 | 133.1 | 133.1 | 0.1045 | 402.8 | 2,127 | 55 | 0.1417 | 0.7480 | 73,000 | 7,630 | 2.75 |
| 3/0 | 167.8 | 167.8 | 0.1318 | 507.8 | 2,681 | 55 | 0.1124 | 0.5934 | 71,000 | 9,356 | 3.25 |
| 4/0 | 211.6 | 211.6 | 0.1662 | 640.5 | 3,382 | 55 | 0.08910 | 0.4705 | 69,000 | 11,470 | 3.75 |
| 300 | 300.0 | 300.0 | 0.2356 | 908.0 | 4,700 | 55 | 0.06285 | 0.3319 | 64,800 | 15,270 | 4.50 |

## 104. Trolley Wire: Copper and Bronze

| Nom-inal size | Cross-sectional area | | | Weight | | Min conduc-tivity, % IACS[a] | D-c resistance or volts drop per amp at 20°C (68°F) | | Min tensile strength, lb per sq in. | Min break-ing load, lb | Elonga-tion in 10 in., % |
|---|---|---|---|---|---|---|---|---|---|---|---|
| | Nom-inal | Actual | | | | | Ohms or volts per 1,000 ft | Ohms or volts per mile | | | |
| AWG or kcmil | kcmil | kcmil | Sq in. | Lb per 1,000 ft | Lb per mile | | | | | | |
| Grooved ASTM Alloy 55 Hitenso C Bronze | | | | | | | | | | | |
| 2/0 | 133.1 | 137.9 | 0.1083 | 417.6 | 2,205 | 55 | 0.1368 | 0.7220 | 73,000 | 7,906 | 2.25 |
| 3/0 | 167.8 | 167.3 | 0.1314 | 506.4 | 2,674 | 55 | 0.1127 | 0.5951 | 71,000 | 9,329 | 2.75 |
| 4/0 | 211.6 | 212.0 | 0.1665 | 641.9 | 3,389 | 55 | 0.08895 | 0.4697 | 69,000 | 11,490 | 3.25 |
| 300 | 300.0 | 299.8 | 0.2355 | 907.6 | 4,792 | 55 | 0.06289 | 0.3320 | 64,800 | 15,260 | 4.00 |
| 350 | 350.0 | 351.2 | 0.2758 | 1,063.0 | 5,612 | 55 | 0.05370 | 0.2835 | 62,500 | 17,240 | 4.00 |
| Figure-9 Deep-section ASTM Alloy 55 Hitenso C Bronze | | | | | | | | | | | |
| 335 | 335.0 | 336.4 | 0.2642 | 1,020.0 | 5,386 | 55 | 0.05605 | 0.2959 | 61,500 | 16,250 | 4.00 |
| Round ASTM Alloy 40 Electric Bronze | | | | | | | | | | | |
| 1/0 | 105.6 | 105.6 | 0.0829 | 319.5 | 1,687 | 40 | 0.2456 | 1.297 | 76,000 | 6,301 | 2.40 |
| 2/0 | 133.1 | 133.1 | 0.1045 | 402.8 | 2,127 | 40 | 0.1948 | 1.029 | 73,000 | 7,630 | 2.75 |
| 3/0 | 167.8 | 167.8 | 0.1318 | 507.8 | 2,681 | 40 | 0.1545 | 0.8160 | 71,000 | 9,356 | 3.25 |
| 4/0 | 211.6 | 211.6 | 0.1662 | 640.5 | 3,382 | 40 | 0.1225 | 0.6470 | 69,000 | 11,470 | 3.75 |
| 300 | 300.0 | 300.0 | 0.2356 | 908.0 | 4,700 | 40 | 0.08644 | 0.4564 | 64,800 | 15,270 | 4.50 |
| Grooved ASTM Alloy 40 Electric Bronze | | | | | | | | | | | |
| 2/0 | 133.1 | 137.9 | 0.1083 | 417.6 | 2,205 | 40 | 0.1880 | 0.9928 | 73,000 | 7,906 | 2.25 |
| 3/0 | 167.8 | 167.3 | 0.1314 | 506.4 | 2,674 | 40 | 0.1550 | 0.8183 | 71,000 | 9,329 | 2.75 |
| 4/0 | 211.6 | 212.0 | 0.1665 | 641.9 | 3,389 | 40 | 0.1223 | 0.6458 | 69,000 | 11,490 | 3.25 |
| 300 | 300.0 | 299.8 | 0.2355 | 907.6 | 4,792 | 40 | 0.08647 | 0.4566 | 64,800 | 15,260 | 4.00 |
| 350 | 350.0 | 351.2 | 0.2758 | 1,063.0 | 5,612 | 40 | 0.07384 | 0.3899 | 62,500 | 17,240 | 4.00 |
| Figure-9 Deep-section ASTM Alloy 40 Electric Bronze | | | | | | | | | | | |
| 335 | 335.0 | 336.4 | 0.2642 | 1,020.0 | 5,386 | 40 | 0.07708 | 0.4070 | 61,500 | 16,250 | 4.00 |

[a] These are minimum values of conductivity and usually are exceeded.
These data are approximate and subject to normal manufacturing tolerances.

### 105. Flat Busbar Copper

| Size, in. | | Weight, lb per ft | Area cir mils | Area sq in. |
|---|---|---|---|---|
| 1/16 × | ½ | 0.12 | 39,470 | 0.031 |
| | ¾ | 0.18 | 59,842 | 0.047 |
| | ⅞ | 0.21 | 68,755 | 0.055 |
| | 1 | 0.24 | 80,213 | 0.063 |
| | 1¼ | 0.30 | 99,312 | 0.078 |
| | 1½ | 0.36 | 119,684 | 0.094 |
| ⅛ × | ¾ | 0.36 | 119,684 | 0.094 |
| | 1 | 0.48 | 159,154 | 0.125 |
| | 1¼ | 0.60 | 198,624 | 0.156 |
| | 1½ | 0.72 | 239,368 | 0.188 |
| | 2 | 0.96 | 318,308 | 0.250 |
| | 2½ | 1.20 | 398,523 | 0.313 |
| | 3 | 1.44 | 477,463 | 0.375 |
| | 3½ | 1.68 | 557,677 | 0.438 |
| | 4 | 1.92 | 636,618 | 0.500 |
| | 5 | 2.40 | 795,772 | 0.625 |
| 3/16 × | 1 | 0.72 | 239,368 | 0.188 |
| | 1¼ | 0.90 | 298,402 | 0.234 |
| | 1½ | 1.085 | 358,098 | 0.281 |
| | 1¾ | 1.266 | 417,780 | 0.328 |
| | 2 | 1.44 | 477,993 | 0.375 |
| | 2¼ | 1.62 | 537,140 | 0.422 |
| | 2½ | 1.81 | 596,829 | 0.469 |
| | 2¾ | 1.99 | 656,512 | 0.516 |
| | 3 | 2.17 | 716,196 | 0.563 |
| | 4 | 2.88 | 954,928 | 0.750 |
| | 5 | 3.62 | 1,193,659 | 0.938 |
| | 6 | 4.34 | 1,432,390 | 1.125 |
| ¼ × | 1 | 0.96 | 318,308 | 0.250 |
| | 1½ | 1.44 | 477,463 | 0.375 |
| | 2 | 1.93 | 636,618 | 0.500 |
| | 2½ | 2.41 | 795,772 | 0.625 |
| | 3 | 2.89 | 954,928 | 0.750 |
| | 3½ | 3.38 | 1,114,082 | 0.875 |
| | 4 | 3.86 | 1,273,236 | 1.00 |
| | 4½ | 4.34 | 1,313,024 | 1.13 |
| | 5 | 4.82 | 1,591,545 | 1.25 |
| | 6 | 5.78 | 1,909,857 | 1.50 |
| | 7 | 6.76 | 2,228,164 | 1.75 |
| | 8 | 7.72 | 2,546,473 | 2.00 |
| ⅜ × | 1½ | 2.17 | 716,196 | 0.563 |
| | 2 | 2.89 | 954,928 | 0.750 |
| | 2½ | 3.61 | 1,193,659 | 0.938 |
| | 3 | 4.34 | 1,432,390 | 1.13 |
| | 3½ | 5.06 | 1,671,123 | 1.31 |
| | 4 | 5.79 | 1,909,854 | 1.50 |
| | 4½ | 6.51 | 2,148,586 | 1.69 |
| | 5 | 7.23 | 2,387,318 | 1.88 |
| | 6 | 8.68 | 2,864,781 | 2.25 |
| | 8 | 11.57 | 3,819,708 | 3.00 |
| | 10 | 14.47 | 5,774,636 | 3.75 |
| ½ × | 3 | 5.79 | 1,909,857 | 1.50 |
| | 3½ | 6.75 | 2,228,164 | 1.75 |
| | 4 | 7.72 | 2,546,472 | 2.00 |
| | 4½ | 8.68 | 2,864,780 | 2.25 |
| | 5 | 9.64 | 3,183,090 | 2.50 |
| | 5½ | 10.61 | 3,501,398 | 2.75 |
| | 6 | 11.58 | 3,819,708 | 3.00 |
| | 7 | 12.50 | 4,456,328 | 3.50 |
| | 8 | 15.43 | 5,092,944 | 4.00 |
| | 9 | 17.36 | 5,729,560 | 4.50 |
| | 10 | 19.29 | 6,366,180 | 5.00 |

### 106.  Round Copper Rod

| Diameter | | Approx weight, lb per ft | Cir mils, approx | Cross section, sq in. approx |
|---|---|---|---|---|
| In. | Decimal equivalent | | | |
| 1/8 | 0.125 | 0.047 | 15,600 | 0.01227 |
| 5/32 | 0.1562 | 0.077 | 24,400 | 0.01917 |
| 3/16 | 0.1875 | 0.106 | 35,000 | 0.02761 |
| 1/4 | 0.250 | 0.189 | 62,500 | 0.04909 |
| 5/16 | 0.3125 | 0.296 | 97,800 | 0.07670 |
| 3/8 | 0.375 | 0.426 | 140,000 | 0.11045 |
| 7/16 | 0.4375 | 0.580 | 191,800 | 0.15033 |
| 1/2 | 0.500 | 0.757 | 250,000 | 0.19635 |
| 9/16 | 0.5625 | 0.959 | 316,000 | 0.24850 |
| 5/8 | 0.625 | 1.184 | 390,000 | 0.30680 |
| 3/4 | 0.750 | 1.70 | 562,500 | 0.44179 |
| 7/8 | 0.875 | 2.32 | 765,000 | 0.60132 |
| 15/16 | 0.9375 | 2.68 | 880,000 | 0.69029 |
| 1 | 1.000 | 3.03 | 1,000,000 | 0.78540 |
| 1 1/8 | 1.125 | 3.83 | 1,270,000 | 0.99402 |
| 1 1/4 | 1.250 | 4.73 | 1,560,000 | 1.2272 |
| 1 3/8 | 1.375 | 5.73 | 1,900,000 | 1.4849 |
| 1 1/2 | 1.500 | 6.82 | 2,250,000 | 1.7671 |
| 1 5/8 | 1.625 | 8.04 | 2,650,000 | 2.0739 |
| 1 11/16 | 1.6875 | 8.68 | 2,850,000 | 2.2365 |
| 1 3/4 | 1.750 | 9.28 | 3,070,000 | 2.4053 |
| 1 7/8 | 1.875 | 10.72 | 3,500,000 | 2.7612 |
| 2 | 2.000 | 12.12 | 4,000,000 | 3.1416 |
| 2 1/2 | 2.500 | 19.05 | 6,250,000 | 4.9087 |
| 3 | 3.000 | 27.44 | 9,000,000 | 7.0686 |
| 3 1/2 | 3.500 | 37.23 | 12,250,000 | 9.6211 |

### 107.  Copper Tubing: Standard Iron-Pipe Size

| Size, in. | Approx OD | OD | ID | Wall thickness | Weight, lb per ft | Area, cir mils | Area, sq in. |
|---|---|---|---|---|---|---|---|
| 1/8 | 13/32 | 0.405 | 0.281 | 0.062 | 0.259 | 85,064 | 0.066 |
| 1/4 | 35/64 | 0.540 | 0.375 | 0.082 | 0.459 | 150,000 | 0.117 |
| 3/8 | 11/16 | 0.675 | 0.484 | 0.095 | 0.644 | 221,369 | 0.173 |
| 1/2 | 27/32 | 0.840 | 0.625 | 0.107 | 0.958 | 314,975 | 0.246 |
| 3/4 | 1 1/16 | 1.050 | 0.822 | 0.114 | 1.298 | 426,816 | 0.335 |
| 1 | 1 5/16 | 1.315 | 1.062 | 0.126 | 1.829 | 601,381 | 0.470 |
| 1 1/4 | 1 43/64 | 1.660 | 1.368 | 0.146 | 2.689 | 884,176 | 0.694 |
| 1 1/2 | 1 29/32 | 1.900 | 1.600 | 0.150 | 3.193 | 1,050,000 | 0.824 |
| 2 | 2 3/8 | 2.375 | 2.062 | 0.156 | 4.224 | 1,388,781 | 1.087 |
| 2 1/2 | 2 7/8 | 2.875 | 2.500 | 0.187 | 6.130 | 2,015,625 | 1.579 |
| 3 | 3 1/2 | 3.500 | 3.062 | 0.219 | 8.741 | 2,874,156 | 2.257 |
| 3 1/2 | 4 | 4.000 | 3.500 | 0.250 | 11.41 | 3,750,000 | 2.945 |
| 4 | 4 1/2 | 4.500 | 4.000 | 0.250 | 12.93 | 4,250,000 | 3.337 |

### 108.  Copper Tubing: Extra-Heavy Iron-Pipe Size

| Size, in. | Approx OD | OD | ID | Wall thickness | Weight, lb per ft | Area, cir mils | Area, sq in. |
|---|---|---|---|---|---|---|---|
| ⅛ | 13/32 | 0.405 | 0.205 | 0.100 | 0.371 | 122,000 | 0.095 |
| ¼ | 35/64 | 0.540 | 0.294 | 0.123 | 0.624 | 205,164 | 0.161 |
| ⅜ | 11/16 | 0.675 | 0.421 | 0.127 | 0.847 | 278,384 | 0.218 |
| ½ | 27/32 | 0.840 | 0.542 | 0.149 | 0.253 | 411,836 | 0.323 |
| ¾ | 1 1/16 | 1.050 | 0.736 | 0.157 | 1.706 | 560,804 | 0.440 |
| 1 | 1 5/16 | 1.315 | 0.951 | 0.182 | 2.509 | 824,849 | 0.647 |
| 1¼ | 1 43/64 | 1.660 | 1.272 | 0.194 | 3.460 | 1,137,616 | 0.893 |
| 1½ | 1 29/32 | 1.900 | 1.494 | 0.203 | 4.191 | 1,377,858 | 1.082 |
| 2 | 2⅜ | 2.375 | 1.933 | 0.221 | 5.791 | 1,904,140 | 1.495 |
| 2½ | 2⅞ | 2.875 | 2.315 | 0.280 | 8.839 | 2,906,500 | 2.282 |
| 3 | 3½ | 3.500 | 2.892 | 0.304 | 11.82 | 3,886,320 | 3.052 |
| 3½ | 4 | 4.000 | 3.358 | 0.321 | 14.37 | 4,723,820 | 3.710 |
| 4 | 4½ | 4.500 | 3.818 | 0.341 | 17.25 | 5,672,860 | 4.455 |

### 109.  Copper Tubing: Double-Extra-Heavy Iron-Pipe Size

| Size, in. | Approx OD | OD | ID | Wall thickness | Weight lb per ft | Area, cir mils | Area, sq in. |
|---|---|---|---|---|---|---|---|
| ½ | 27/32 | 0.840 | 0.252 | 0.294 | 1.945 | 642,096 | 0.504 |
| ¾ | 1 1/16 | 1.050 | 0.434 | 0.308 | 2.768 | 914,144 | 0.718 |
| 1 | 1 5/16 | 1.315 | 0.599 | 0.358 | 4.152 | 1,370,449 | 1.076 |
| 1¼ | 1 43/64 | 1.660 | 0.896 | 0.382 | 5.916 | 1,952,784 | 1.534 |
| 1½ | 1 29/32 | 1.900 | 1.100 | 0.400 | 7.271 | 2,410,000 | 1.885 |
| 2 | 2⅜ | 2.375 | 1.503 | 0.436 | 10.246 | 3,381,620 | 2.656 |
| 2½ | 2⅞ | 2.875 | 1.771 | 0.552 | 15.541 | 5,130,300 | 4.028 |
| 3 | 3½ | 3.500 | 2.300 | 0.600 | 21.087 | 6,960,000 | 5.466 |
| 3½ | 4 | 4.000 | 2.728 | 0.636 | 25.930 | 8,568,000 | 6.721 |
| 4 | 4½ | 4.500 | 3.152 | 0.674 | 31.253 | 10,314,080 | 8.101 |

### 110.  General Characteristics of Copperweld Wire

| Grade | Conductivity, % | Density | Weight per cir mil per 1,000 ft, lb | Tensile strength per sq in., 0.162-in. wire, lb | Normal resistivity per cir mil-ft, at 68°F, ohms |
|---|---|---|---|---|---|
| Extra high strength....... | 30 | 8.15 | 0.002775 | 157,000 | 34.57 |
| High strength............ | 30 | 8.15 | 0.002775 | 130,000 | 34.57 |
| High strength............ | 40 | 8.15 | 0.002775 | 118,000 | 25.928 |

Modulus of elasticity (conventional):
        Solid wire............................... 24,000,000 lb per sq in.
        Stranded cable.......................... 23,000,000 lb per sq in.
Coefficient of linear expansion per degree Fahrenheit: 0.0000072.
Thirty per cent conductivity is not recommended in sizes finer than No. 10 AWG.  Finer sizes are supplied in 40 per cent conductivity only.

## 111. Application Guide for Copperweld Products

| Product | Companies and uses | | | | | |
| --- | --- | --- | --- | --- | --- | --- |
| | Power companies | Railroad | Electric railway | Telephone and telegraph | Municipal | Construction mill, etc. |
| Bare wire | Telephone and signal lines<br>Overhead ground wire<br>Grounding wire<br>Lightly loaded lines<br>Tie wires, mousing wire<br>Light guys<br>Remote-control lines<br>Counterpoise wire | Signal, telegraph, and telephone lines<br>Control circuits<br>Overhead ground wire<br>Grounding wire<br>Bond wires, tie wires<br>CTC circuits | Telephone and signal lines<br>Overhead ground<br>Light span wire<br>Light guys | Line wires<br>Guys<br>Antenna wire | Police and fire-alarm circuits | Telephone lines<br>Grounding wire<br>Crane trolley<br>Pipe-insulation wrapping wire |
| Copperweld-copper[a] | Copperweld-copper cables of numerous combinations for long spans, rural distribution, and transmission conductors | Copperweld-copper Catenary messenger Feeder cables for electrification | Copperweld-copper Catenary messenger Feeder cables | | | Copperweld-copper Cables for long spans |
| Weatherproof and plastic-jacketed wire | Telephone and signal lines<br>Remote control | Signal lines, control circuits, tie wires<br>Centralized traffic control circuits | Telephone and signal lines | Tree wire<br>Grounding wire<br>Line wire | Signal and alarm circuits<br>Tie wires | Signal and alarm circuits |
| Rubber- and neoprene-covered | Drop wires for telephone and signal lines | Drop wires for telephone signal lines | Drop wires for telephone lines | Drop wires and block wire | Alarm circuits<br>Drop wires | Telephone and signal lines<br>Drop wires |

**111. Application Guide for Copperweld Products** (*Continued*)

| Product | Companies and uses | | | | | |
|---|---|---|---|---|---|---|
| | Power companies | Railroad | Electric railway | Telephone and telegraph | Municipal | Construction mill, etc. |
| Strand 3, 7, 19 wire, etc. | Guy and messenger Overhead ground wires; long spans; stack guys | Rail bonds Guy and messenger Overhead ground wire Long spans Pull-offs for electrifications | Span wire Guy wire Catenary messenger Pull-offs Overhead ground | Guy and messenger Long spans | Guy and messenger Long spans | Guy and messenger Overhead ground Stack guys Long spans |
| Ground rods and clamps | For all electrical grounding installations in which good earth connections are required | | | | | |
| Anchor rods | For all types of anchors and anchoring installations | | | | | |
| Cable rings and lashing wire | For use with Copperweld messenger strand for all types of aerial cable installations | | | | | |
| Fencing and barbed wire | Whenever high-strength, nonrusting property protection is required; available in many forms of fencing made from Copperweld wire | | | | | |
| Staples, nails | Grounding wire molding, insulator pins, etc. | Grounding wire, insulator pins, trunking, bond wires to ties, etc. Conduit | Grounding wire, insulator pins, etc. | Grounding wire, insulator pins, conduit, etc. | Grounding wire, conduit, etc. | Roofs and building |
| Wall ties | | | | | | Cavity walls |

[a] Copperweld is used by mines, forestries, and game preserves for telephone and signal lines and grounding; by broadcasting stations and individuals for radio antennae, lead-ins, grounding, etc; in the home for clothes lines; and in the construction industry for building and roofing nails and wall ties for masonry cavity-wall construction.

## 112. Solid Copperweld Conductors

| Size, AWG | Diam, in. | Area | | Weight | | Tensile strength, nominal | | | Breaking strength | | |
|---|---|---|---|---|---|---|---|---|---|---|---|
| | | Cir mils | Sq in. | Per 1,000 ft, lb | Per mile, lb | 30% extra high strength, lb per sq in. | 30% high strength, lb per sq in. | 40% high strength, lb per sq in. | 30% extra high strength, lb | 30% high strength, lb | 40% high strength, lb |
| 4/0 | 0.460 | 211,600 | 0.1662 | 587 | 3,100 | ...... | 90,000 | 80,000 | ...... | 14,960 | 13,290 |
| 3/0 | 0.410 | 168,100 | 0.1320 | 466 | 2,460 | ...... | 90,000 | 80,000 | ...... | 11,880 | 10,560 |
| 2/0 | 0.365 | 133,225 | 0.1046 | 370 | 1,954 | ...... | 90,000 | 80,000 | ...... | 9,410 | 8,370 |
| 1/0 | 0.325 | 105,625 | 0.08296 | 293 | 1,547 | ...... | 94,000 | 82,000 | ...... | 7,800 | 6,800 |
| 1 | 0.289 | 83,520 | 0.06560 | 232 | 1,225 | ...... | 102,000 | 90,000 | ...... | 6,690 | 5,900 |
| 2 | 0.258 | 66,565 | 0.05228 | 185 | 977 | 123,000 | 109,000 | 97,000 | 6,430 | 5,700 | 5,070 |
| 3 | 0.229 | 52,440 | 0.04119 | 146 | 771 | 133,500 | 115,000 | 103,000 | 5,490 | 4,740 | 4,240 |
| 4 | 0.204 | 41,615 | 0.03269 | 116 | 612 | 142,500 | 120,000 | 108,000 | 4,660 | 3,920 | 3,530 |
| 5 | 0.182 | 33,125 | 0.02602 | 91.9 | 485 | 150,500 | 125,000 | 113,000 | 3,910 | 3,250 | 2,940 |
| 6 | 0.162 | 26,245 | 0.02061 | 72.8 | 384 | 157,000 | 130,000 | 118,000 | 3,240 | 2,680 | 2,430 |
| 7 | 0.144 | 20,735 | 0.01629 | 57.5 | 304 | 164,000 | 135,000 | 123,000 | 2,670 | 2,200 | 2,000 |
| 8 | 0.128 | 16,385 | 0.01287 | 45.5 | 240 | 170,000 | 140,000 | 128,000 | 2,190 | 1,800 | 1,650 |
| 9 | 0.114 | 12,995 | 0.01021 | 36.1 | 191 | ...... | 145,000 | 133,000 | ...... | 1,480 | 1,360 |
| 10 | 0.102 | 10,404 | 0.00817 | 28.9 | 153 | ...... | 150,000 | 138,000 | ...... | 1,230 | 1,130 |
| 11 | 0.091 | 8,281 | 0.00650 | 23.0 | 121 | ...... | ...... | 138,000 | ...... | ...... | 900 |
| 12 | 0.081 | 6,561 | 0.00515 | 18.2 | 96 | ...... | ...... | 138,000 | ...... | ...... | 710 |

## 113.  Composite Copper-Copperweld Bare Conductors
(Anaconda Wire and Cable Co.)

| Hard-drawn copper equivalent area AWG or kcmil[a] | Type | Diameter, in. | Number and diameter of EHS 30% conductivity copperweld wires, in. | Number and diameter of hard-drawn copper wires, in. | Actual area Cir mils | Actual area Sq in. | Min ultimate strength, lb | Weight Lb per 1,000 ft | Weight Lb per mile | D-c resistance at 20°C (68°F) Ohms per 1,000 ft |
|---|---|---|---|---|---|---|---|---|---|---|
| 350 | E | 0.788 | 7 × .1576 | 12 × .1576 | 471,900 | 0.3706 | 32,420 | 1,403 | 7,409 | 0.03143 |
| 350 | EK | 0.735 | 4 × .1470 | 15 × .1470 | 410,600 | 0.3225 | 23,850 | 1,238 | 6,536 | 0.03143 |
| 350 | V | 0.754 | 3 × .1751 | 9 × .1893 | 414,500 | 0.3255 | 23,480 | 1,246 | 6,578 | 0.03143 |
| 300 | E | 0.729 | 7 × .1459 | 12 × .1459 | 404,400 | 0.3177 | 27,770 | 1,203 | 6,351 | 0.03667 |
| 300 | EK | 0.680 | 4 × .1361 | 15 × .1361 | 351,900 | 0.2764 | 20,960 | 1,061 | 5,602 | 0.03667 |
| 300 | V | 0.698 | 3 × .1621 | 9 × .1752 | 355,100 | 0.2789 | 20,730 | 1,068 | 5,639 | 0.03667 |
| 250 | E | 0.666 | 7 × .1332 | 12 × .1332 | 337,100 | 0.2648 | 23,920 | 1,002 | 5,292 | 0.04400 |
| 250 | EK | 0.621 | 4 × .1242 | 15 × .1242 | 293,100 | 0.2302 | 17,840 | 884.2 | 4,669 | 0.04400 |
| 250 | V | 0.637 | 3 × .1480 | 9 × .1600 | 296,100 | 0.2326 | 17,420 | 889.9 | 4,699 | 0.04400 |
| 4/0 | E | 0.613 | 7 × .1225 | 12 × .1225 | 285,100 | 0.2239 | 20,730 | 848.3 | 4,479 | 0.05199 |
| 4/0 | G | 0.583 | 2 × .1944 | 5 × .1944 | 264,500 | 0.2078 | 15,640 | 789.4 | 4,168 | 0.05199 |
| 4/0 | EK | 0.571 | 4 × .1143 | 15 × .1143 | 248,200 | 0.1950 | 15,370 | 748.4 | 3,951 | 0.05199 |
| 4/0 | V | 0.586 | 3 × .1361 | 9 × .1472 | 250,600 | 0.1968 | 15,000 | 753.2 | 3,977 | 0.05199 |
| 4/0 | F | 0.550 | 1 × .1833 | 6 × .1833 | 235,200 | 0.1847 | 12,290 | 710.2 | 3,750 | 0.05199 |
| 3/0 | E | 0.545 | 7 × .1091 | 12 × .1091 | 226,200 | 0.1776 | 16,800 | 672.7 | 3,552 | 0.06556 |
| 3/0 | J | 0.555 | 3 × .1851 | 4 × .1851 | 239,800 | 0.1884 | 16,170 | 706.7 | 3,732 | 0.06556 |
| 3/0 | G | 0.519 | 2 × .1731 | 5 × .1731 | 209,700 | 0.1647 | 12,860 | 626.0 | 3,305 | 0.06556 |
| 3/0 | EK | 0.509 | 4 × .1018 | 15 × .1018 | 196,900 | 0.1546 | 12,370 | 593.5 | 3,134 | 0.06556 |
| 3/0 | V | 0.522 | 3 × .1212 | 9 × .1311 | 198,800 | 0.1561 | 12,200 | 597.3 | 3,154 | 0.06556 |
| 3/0 | F | 0.490 | 1 × .1632 | 6 × .1632 | 186,400 | 0.1464 | 9,980 | 563.2 | 2,974 | 0.06556 |
| 2/0 | K | 0.534 | 4 × .1780 | 3 × .1780 | 221,800 | 0.1742 | 17,600 | 645.9 | 3,411 | 0.08265 |
| 2/0 | J | 0.494 | 3 × .1648 | 4 × .1648 | 190,100 | 0.1493 | 13,430 | 560.6 | 2,960 | 0.08265 |
| 2/0 | G | 0.463 | 2 × .1542 | 5 × .1542 | 166,400 | 0.1307 | 10,510 | 496.6 | 2,622 | 0.08265 |
| 2/0 | V | 0.465 | 3 × .1080 | 9 × .1167 | 157,600 | 0.1237 | 9,846 | 473.8 | 2,502 | 0.08265 |
| 2/0 | F | 0.436 | 1 × .1454 | 6 × .1454 | 148,000 | 0.1162 | 8,094 | 446.8 | 2,359 | 0.08265 |
| 1/0 | K | 0.475 | 4 × .1585 | 3 × .1585 | 175,900 | 0.1381 | 14,490 | 512.0 | 2,703 | 0.1043 |
| 1/0 | J | 0.440 | 3 × .1467 | 4 × .1467 | 150,600 | 0.1184 | 10,970 | 444.3 | 2,346 | 0.1043 |
| 1/0 | G | 0.412 | 2 × .1373 | 5 × .1373 | 132,000 | 0.1036 | 8,563 | 393.6 | 2,078 | 0.1043 |
| 1/0 | F | 0.388 | 1 × .1294 | 6 × .1294 | 117,200 | 0.09206 | 6,536 | 354.1 | 1,870 | 0.1043 |
| 1 | N | 0.464 | 5 × .1546 | 2 × .1546 | 167,300 | 0.1314 | 15,410 | 481.3 | 2,541 | 0.1315 |
| 1 | K | 0.423 | 4 × .1412 | 3 × .1412 | 139,600 | 0.1096 | 11,900 | 406.2 | 2,144 | 0.1315 |
| 1 | J | 0.392 | 3 × .1307 | 4 × .1307 | 119,600 | 0.09392 | 9,000 | 352.5 | 1,861 | 0.1315 |
| 1 | G | 0.367 | 2 × .1222 | 5 × .1222 | 104,500 | 0.08210 | 6,956 | 312.2 | 1,649 | 0.1315 |
| 1 | F | 0.346 | 1 × .1153 | 6 × .1153 | 93,060 | 0.07309 | 5,266 | 280.9 | 1,483 | 0.1315 |
| 2 | P | 0.462 | 6 × .1540 | 1 × .1540 | 166,000 | 0.1304 | 16,870 | 471.1 | 2,487 | 0.1658 |
| 2 | N | 0.413 | 5 × .1377 | 2 × .1377 | 132,700 | 0.1042 | 12,680 | 381.7 | 2,015 | 0.1658 |
| 2 | K | 0.377 | 4 × .1257 | 3 × .1257 | 110,600 | 0.08687 | 9,730 | 322.1 | 1,701 | 0.1658 |
| 2 | J | 0.349 | 3 × .1164 | 4 × .1164 | 94,840 | 0.07449 | 7,322 | 279.5 | 1,476 | 0.1658 |
| 2 | A | 0.366 | 1 × .1699 | 2 × .1699 | 86,600 | 0.06801 | 5,876 | 256.8 | 1,356 | 0.1658 |
| 2 | G | 0.327 | 2 × .1089 | 5 × .1089 | 83,010 | 0.06520 | 5,626 | 247.6 | 1,307 | 0.1658 |
| 2 | F | 0.308 | 1 × .1026 | 6 × .1026 | 73,690 | 0.05787 | 4,233 | 222.8 | 1,176 | 0.1658 |
| 3 | P | 0.411 | 6 × .1371 | 1 × .1371 | 131,600 | 0.1033 | 13,910 | 373.6 | 1,973 | 0.2090 |
| 3 | N | 0.368 | 5 × .1226 | 2 × .1226 | 105,200 | 0.08264 | 10,390 | 302.7 | 1,598 | 0.2090 |
| 3 | K | 0.336 | 4 × .1120 | 3 × .1120 | 87,810 | 0.06896 | 7,910 | 255.5 | 1,349 | 0.2090 |
| 3 | J | 0.311 | 3 × .1036 | 4 × .1036 | 75,130 | 0.05901 | 5,955 | 221.7 | 1,171 | 0.2090 |
| 3 | A | 0.326 | 1 × .1513 | 2 × .1513 | 68,680 | 0.05394 | 4,810 | 203.6 | 1,075 | 0.2090 |
| 4 | P | 0.366 | 6 × .1221 | 1 × .1221 | 104,400 | 0.08196 | 11,420 | 296.3 | 1,564 | 0.2636 |
| 4 | N | 0.328 | 5 × .1092 | 2 × .1092 | 83,470 | 0.06556 | 8,460 | 240.0 | 1,267 | 0.2636 |
| 4 | D | 0.348 | 2 × .1615 | 1 × .1615 | 78,250 | 0.06145 | 7,340 | 225.5 | 1,191 | 0.2636 |
| 4 | A | 0.290 | 1 × .1347 | 2 × .1347 | 54,430 | 0.4275 | 3,938 | 161.5 | 852 | 0.2636 |

## 113.  Composite Copper-Copperweld Bare Conductors

| Hard-drawn copper equivalent area AWG or kcmil[a] | Type | Diameter, in. | Number and diameter of EHS 30% conductivity copperweld wires, in. | Number and diameter of hard-drawn copper wires, in. | Actual area | | Min ultimate strength, lb | Weight | | D-c resistance at 20°C (68°F) Ohms per 1,000 ft |
|---|---|---|---|---|---|---|---|---|---|---|
| | | | | | Cir mils | Sq in. | | Lb per 1,000 ft | Lb per mile | |
| 5 | P | 0.326 | 6 × .1087 | 1 × .1087 | 82,710 | 0.06496 | 9,311 | 234.9 | 1,240 | 0.3291 |
| 5 | D | 0.310 | 2 × .1438 | 1 × .1438 | 62,040 | 0.04872 | 6,035 | 178.9 | 944.4 | 0.3291 |
| 5 | A | 0.258 | 1 × .1200 | 2 × .1200 | 43,200 | 0.03393 | 3,193 | 128.1 | 676.3 | 0.3291 |
| 6 | D | 0.276 | 2 × .1281 | 1 × .1281 | 49,230 | 0.03866 | 4,942 | 141.8 | 748.9 | 0.4150 |
| 6 | A | 0.230 | 1 × .1068 | 2 × .1068 | 34.220 | 0.02688 | 2,585 | 101.6 | 536.3 | 0.4150 |
| 6 | C | 0.225 | 1 × .1046[b] | 2 × .1046 | 32,820 | 0.02578 | 2,143 | 97.34 | 514.0 | 0.4150 |
| 7 | D | 0.246 | 2 × .1141 | 1 × .1141 | 39,060 | 0.03067 | 4,022 | 112.5 | 594.0 | 0.5232 |
| 7 | A | 0.223 | 1 × .1266 | 2 × .08949 | 32,040 | 0.02517 | 2,754 | 93.66 | 494.6 | 0.5232 |
| 8 | D | 0.219 | 2 × .1016 | 1 × .1016 | 30,970 | 0.02432 | 3,256 | 89.21 | 471.0 | 0.6598 |
| 8 | A | 0.199 | 1 × .1127 | 2 × .07969 | 25,400 | 0.01995 | 2,233 | 74.27 | 392.2 | 0.6598 |
| 8 | C | 0.179 | 1 × .0808[b] | 2 × .08336 | 20,430 | 0.01604 | 1,362 | 60.67 | 320.3 | 0.6598 |
| 9½ | D | 0.174 | 2 × .0808[b] | 1 × .0808 | 19,590 | 0.01539 | 1,743 | 56.46 | 298.1 | 0.9170 |

[a] Hard-drawn copper cable, 97.5 per cent conductivity, IACS, having the same d-c resistance as that of the composite cable after allowing for increases in resistance due to stranding based on Table II, ASTM B229-52.
Manufactured in accordance with ASTM Specification B229-52.
[b] High-strength copperweld, 40 per cent conductivity.
These data are approximate and subject to normal manufacturing tolerances.

## 114.  Stranded Copperweld Conductors

| Nominal diam, in., number of strands, and size, AWG | Diam, in. | Area | | Weight | | Breaking strength | | |
|---|---|---|---|---|---|---|---|---|
| | | Cir mils | Sq in. | Per 1,000 ft, lb | Per mile, lb | 30% extra high strength, lb | 30% high strength, lb | 40% high strength, lb |
| ⅞ (19 No. 5) | 0.910 | 629,375 | 0.4944 | 1,770 | 9,346 | 66,860 | 55,570 | 50,270 |
| 1³⁄₁₆ (19 No. 6) | 0.810 | 498,655 | 0.3916 | 1,403 | 7,408 | 55,400 | 45,880 | 41,550 |
| 2³⁄₃₂ (19 No. 7) | 0.720 | 393,695 | 0.3095 | 1,108 | 5,850 | 45,660 | 37,620 | 34,200 |
| 2¹⁄₃₂ (19 No. 8) | 0.640 | 311,315 | 0.2445 | 877 | 4,630 | 37,450 | 30,780 | 28,210 |
| ⁹⁄₁₆ (19 No. 9) | 0.570 | 246,905 | 0.1940 | 696 | 3,675 | 30,610 | 25,300 | 23,250 |
| ⅝ ( 7 No. 4) | 0.612 | 291,305 | 0.2288 | 820 | 4,330 | 29,360 | 24,700 | 22,240 |
| ⁹⁄₁₆ ( 7 No. 5) | 0.546 | 231,875 | 0.1821 | 650 | 3,432 | 24,630 | 20,480 | 18,500 |
| ½ ( 7 No. 6) | 0.486 | 183,715 | 0.1443 | 515 | 2,720 | 20,410 | 16,880 | 15,300 |
| ⁷⁄₁₆ ( 7 No. 7) | 0.432 | 145,145 | 0.1140 | 407 | 2,149 | 16,820 | 13,860 | 12,600 |
| ⅜ ( 7 No. 8) | 0.384 | 114,695 | 0.0901 | 322 | 1,700 | 13,800 | 11,340 | 10,390 |
| 1¹⁄₃₂ ( 7 No. 9) | 0.342 | 90,965 | 0.0715 | 255 | 1,346 | 11,280 | 9,320 | 8,570 |
| ⁵⁄₁₆ ( 7 No. 10) | 0.306 | 72,828 | 0.0572 | 204 | 1,077 | 9,200 | 7,750 | 7,120 |
| 3 No. 6 | 0.349 | 78,735 | 0.0618 | 220 | 1,162 | 8,260 | 6,830 | 6,200 |
| 3 No. 7 | 0.310 | 62,205 | 0.0489 | 174 | 919 | 6,800 | 5,610 | 5,100 |
| 3 No. 8 | 0.276 | 49,155 | 0.0386 | 138 | 729 | 5,580 | 4,590 | 4,210 |
| 3 No. 9 | 0.246 | 38,985 | 0.0306 | 109 | 576 | 4,560 | 3,770 | 3,470 |
| 3 No. 10 | 0.220 | 31,212 | 0.0245 | 87 | 460 | 3,720 | 3,140 | 2,880 |
| 3 No. 11 | 0.196 | 24,843 | 0.0195 | 69.6 | 367 | ...... | ...... | 2,480 |
| 3 No. 12 | 0.175 | 19,683 | 0.01545 | 55.0 | 290 | ...... | ...... | 2,040 |

To determine copper equivalent of copperweld conductor, multiply circular-mil area by percentage conductivity expressed as a decimal.

### 115.  Comparative Characteristics of Aluminum and Copper Wire

| Characteristic | Commercial hard-drawn aluminum wire | Commercial hard-drawn copper wire | Standard annealed copper wire |
|---|---|---|---|
| Conductivity.............................. | 60.97 % of IACS | 97 % of IACS | 100 % of IACS |
| Resistance per circular mil-foot............... | 17.010 ohms | 10.692 ohms | 10.371 ohms |
| Temperature coefficient of resistance.......... | 0.403 % per °C | 0.393 % per °C | 0.393 % per °C |

IACS stands for the International Annealed Copper Standard, which is the internationally accepted value for the resistivity of annealed copper of 100 per cent conductivity. This value is 10.371 ohms per mil-ft at 20°C and was adopted by the International Electro-Technical Committee (IEC) in 1913.

### 116.  Bare Solid Al Aluminum Hard-Drawn Wire
(ASTM Standard B230-55T)

| Size, AWG | Diameter, in. | Area | | Hard-drawn copper equivalent size, AWG | Net weight, lb per 1,000 ft | Breaking strength, lb | Average tensile strength, lb per sq in. | Nominal d-c resistance, ohms per 1,000 ft 68°F (20°C) |
|---|---|---|---|---|---|---|---|---|
| | | Cir mils | Sq in. | | | | | |
| 8 | 0.1285 | 16,510 | 0.01297 | 10 | 15.20 | 324.2 | 25,000 | 1.030 |
| 7 | 0.1443 | 20,820 | 0.01635 | 9 | 19.16 | 400.7 | 24,500 | 0.8165 |
| 6 | 0.1620 | 26,240 | 0.02061 | 8 | 24.15 | 494.7 | 24,000 | 0.6478 |
| 5 | 0.1819 | 33,090 | 0.02599 | 7 | 30.45 | 623.7 | 24,000 | 0.5138 |
| 4 | 0.2043 | 41,740 | 0.03278 | 6 | 38.41 | 786.8 | 24,000 | 0.4073 |
| 3 | 0.2294 | 52,620 | 0.04133 | 5 | 48.43 | 971.3 | 23,500 | 0.3231 |
| 2 | 0.2576 | 66,360 | 0.05212 | 4 | 61.07 | 1225 | 23,500 | 0.2562 |

**117. Construction Requirements of Concentric-Lay-Stranded Hard-Drawn Aluminum Conductors (ASTM)**

| Conductor size | | Hard-drawn copper equivalent | | Stranding | | | | | | | | | |
| --- | --- | --- | --- | --- | --- | --- | --- | --- | --- | --- | --- | --- | --- |
| | | | | Class AA | | Class A | | Class B | | Class C | | Class D | |
| cmil | AWG | cmil | AWG | Number of wires | Diameter of wire, mils | Number of wires | Diameter of wire, mils | Number of wires | Diameter of wire, mils | Number of wires | Diameter of wire, mils | Number of wires | Diameter of wire, mils |
| 4,000,000 | ... | 2,520,000 | ... | ... | ... | 169 | 153.8 | 217 | 135.8 | 271 | 121.5 | 271 | 121.5 |
| 3,500,000 | ... | 2,200,000 | ... | ... | ... | 127 | 166.0 | 169 | 143.9 | 217 | 127.0 | 271 | 113.6 |
| 3,000,000 | ... | 1,890,000 | ... | ... | ... | 127 | 153.7 | 169 | 133.2 | 217 | 117.6 | 271 | 105.2 |
| 2,500,000 | ... | 1,570,000 | ... | ... | ... | 91 | 165.7 | 127 | 140.3 | 169 | 121.6 | 217 | 107.3 |
| 2,000,000 | ... | 1,260,000 | ... | ... | ... | 91 | 148.2 | 127 | 125.5 | 169 | 108.8 | 217 | 96.0 |
| 1,900,000 | ... | 1,195,000 | ... | ... | ... | 91 | 144.5 | 127 | 122.3 | 169 | 106.0 | 217 | 93.6 |
| 1,800,000 | ... | 1,132,000 | ... | ... | ... | 91 | 140.6 | 127 | 119.1 | 169 | 103.2 | 217 | 91.1 |
| 1,750,000 | ... | 1,101,000 | ... | 61 | 169.4 | 91 | 138.7 | 127 | 117.4 | 169 | 101.8 | 217 | 89.8 |
| 1,700,000 | ... | 1,069,000 | ... | 61 | 166.9 | 91 | 136.7 | 127 | 115.7 | 169 | 100.3 | 217 | 88.5 |
| 1,600,000 | ... | 1,006,000 | ... | 61 | 162.0 | 91 | 132.6 | 127 | 112.2 | 169 | 97.3 | 217 | 85.9 |
| 1,500,000 | ... | 943,000 | ... | 61 | 156.8 | 61 | 156.8 | 91 | 128.4 | 127 | 108.7 | 169 | 94.2 |
| 1,400,000 | ... | 880,000 | ... | 61 | 151.5 | 61 | 151.5 | 91 | 124.0 | 127 | 105.0 | 169 | 91.0 |
| 1,300,000 | ... | 818,000 | ... | 61 | 146.0 | 61 | 146.0 | 91 | 119.5 | 127 | 101.2 | 169 | 87.7 |
| 1,250,000 | ... | 786,000 | ... | 61 | 143.1 | 61 | 143.1 | 91 | 117.2 | 127 | 99.2 | 169 | 86.0 |
| 1,200,000 | ... | 755,000 | ... | 61 | 140.3 | 61 | 140.3 | 91 | 114.8 | 127 | 97.2 | 169 | 84.3 |
| 1,100,000 | ... | 692,000 | ... | 61 | 134.3 | 61 | 134.3 | 91 | 109.9 | 127 | 93.1 | 169 | 80.7 |
| 1,000,000 | ... | 629,000 | ... | 37 | 164.4 | 61 | 128.0 | 61 | 128.0 | 91 | 104.8 | 127 | 88.7 |
| 900,000 | ... | 566,000 | ... | 37 | 156.0 | 61 | 121.5 | 61 | 121.5 | 91 | 99.4 | 127 | 84.2 |
| 800,000 | ... | 503,000 | ... | 37 | 147.0 | 61 | 114.5 | 61 | 114.5 | 91 | 93.8 | 127 | 79.4 |
| 750,000 | ... | 472,000 | ... | 37 | 142.4 | 61 | 110.9 | 61 | 110.9 | 91 | 90.8 | 127 | 76.8 |
| 700,000 | ... | 440,000 | ... | 37 | 137.5 | 61 | 107.1 | 61 | 107.1 | 91 | 87.7 | 127 | 74.2 |

117. Construction Requirements of Concentric-Lay-Stranded Hard-Drawn Aluminum Conductors (Continued)

| Conductor size | | Hard-drawn copper equivalent | | Stranding | | | | | | | | | |
|---|---|---|---|---|---|---|---|---|---|---|---|---|---|
| | | | | Class AA | | Class A | | Class B | | Class C | | Class D | |
| AWG | cmil | cmil | AWG | Number of wires | Diameter of wire, mils | Number of wires | Diameter of wire, mils | Number of wires | Diameter of wire, mils | Number of wires | Diameter of wire, mils | Number of wires | Diameter of wire, mils |
| ... | 650,000 | 409,000 | ... | 37 | 132.5 | 61 | 103.2 | 61 | 103.2 | 91 | 84.5 | 127 | 71.5 |
| ... | 636,000 | 400,000 | ... | 37 | 131.1 | 37 | 131.1 | ... | | ... | | ... | |
| ... | 600,000 | 377,000 | ... | 37 | 127.3 | 37 | 127.3 | 61 | 99.2 | 91 | 81.2 | 127 | 68.7 |
| ... | 550,000 | 346,000 | ... | 37 | 121.9 | 37 | 121.9 | 61 | 95.0 | 91 | 77.7 | 127 | 65.8 |
| ... | 500,000 | 314,000 | ... | 19 | 162.2 | 37 | 116.2 | 37 | 116.2 | 61 | 90.5 | 91 | 74.1 |
| ... | 477,000 | 300,000 | ... | 19 | 158.4 | 37 | 113.5 | ... | | ... | | ... | |
| ... | 450,000 | 283,000 | ... | 19 | 153.9 | 37 | 110.3 | 37 | 110.3 | 61 | 85.9 | 91 | 70.3 |
| ... | 400,000 | 252,000 | ... | 19 | 145.1 | 19 | 145.1 | 37 | 104.0 | 61 | 81.0 | 91 | 66.3 |
| ... | 350,000 | 220,000 | ... | 12 | 170.8 | 19 | 135.7 | 37 | 97.3 | 61 | 75.7 | 91 | 62.0 |
| ... | 336,400 | | 0000 | 12 | 167.4 | 19 | 133.1 | ... | | ... | | ... | |
| ... | 300,000 | 188,700 | ... | 12 | 158.1 | 19 | 125.7 | 37 | 90.0 | 61 | 70.1 | 91 | 57.4 |
| ... | 266,800 | | 000 | 12 | 149.1 | 19 | 118.5 | ... | | ... | | ... | |
| ... | 250,000 | 157,200 | ... | 12 | 144.3 | 19 | 114.7 | 37 | 82.2 | 61 | 64.0 | 91 | 52.4 |
| 0000 | 211,600 | | 00 | 7 | 173.9 | 7 | 173.9 | 19 | 105.5 | 37 | 75.6 | 61 | 58.9 |
| 000 | 167,800 | | 0 | 7 | 154.8 | 7 | 154.8 | 19 | 94.0 | 37 | 67.3 | 61 | 52.4 |
| 00 | 133,100 | | 1 | 7 | 137.9 | 7 | 137.9 | 19 | 83.7 | 37 | 60.0 | 61 | 46.7 |
| 0 | 105,600 | | 2 | 7 | 122.8 | 7 | 122.8 | 19 | 74.5 | 37 | 53.4 | 61 | 41.6 |
| 1 | 83,690 | | 3 | 7 | 109.3 | 7 | 109.3 | 19 | 66.4 | 37 | 47.6 | 61 | 37.0 |
| 2 | 66,360 | | 4 | 7 | 97.4 | 7 | 97.4 | 7 | 97.4 | 19 | 59.1 | 37 | 42.4 |
| 3 | 52,620 | | 5 | ... | | 7 | 86.7 | 7 | 86.7 | 19 | 52.6 | 37 | 37.7 |
| 4 | 41,740 | | 6 | ... | | 7 | 77.2 | 7 | 77.2 | 19 | 46.9 | 37 | 33.6 |
| 5 | 33,090 | | 7 | ... | | 7 | 68.8 | 7 | 68.8 | 19 | 41.7 | 37 | 29.9 |

**118.  Stranded Aluminum Conductor, Bare: Classes AA and A<sup>a</sup> (Hard-Drawn EC-H19)**
(Aluminum Co. of America)

| Conductor size | | Copper equivalent based on equal d-c resistance, Cu 97 % Al 61 % | Stranding | | Cable diam, in. | D-c resistance at 20°C, ohms per 1,000 ft (61 %) | Ultimate strength, lb | Weight per 1,000 ft, lb |
|---|---|---|---|---|---|---|---|---|
| Cir mils or AWG | Square inches | | Class | Number and diam of wires, in. | | | | |
| 6 | 0.0206 | 8 | A | 7 × 0.0612 | 0.184 | 0.6606 | 528 | 24.6 |
| 4 | 0.0328 | 6 | A | 7 × 0.0772 | 0.232 | 0.4155 | 826 | 39.2 |
| 3 | 0.0413 | 5 | A | 7 × 0.0867 | 0.260 | 0.3295 | 1,022 | 49.4 |
| 2 | 0.0521 | 4 | AA, A | 7 × 0.0974 | 0.292 | 0.2613 | 1,266 | 62.3 |
| 1 | 0.0657 | 3 | AA, A | 7 × 0.1094 | 0.328 | 0.2072 | 1,537 | 78.5 |
| 1/0 | 0.0829 | 2 | AA, A | 7 × 0.1228 | 0.368 | 0.1643 | 1,865 | 99.1 |
| 2/0 | 0.1045 | 1 | AA, A | 7 × 0.1379 | 0.414 | 0.1303 | 2,350 | 124.9 |
| 3/0 | 0.1318 | 1/0 | AA, A | 7 × 0.1548 | 0.464 | 0.1033 | 2,845 | 157.5 |
| 4/0 | 0.1662 | 2/0 | AA, A | 7 × 0.1739 | 0.522 | 0.08195 | 3,590 | 198.6 |
| 266,800 | 0.2095 | 3/0 | ..... | 7 × 0.1953 | 0.586 | 0.06500 | 4,525 | 250.4 |
| 266,800 | 0.2095 | 3/0 | A | 19 × 0.1185 | 0.593 | 0.06500 | 4,800 | 250.4 |
| 336,400 | 0.2642 | 4/0 | AA, A | 19 × 0.1331 | 0.666 | 0.05155 | 5,940 | 315.8 |
| 397,500 | 0.3122 | 250,000 | AA, A | 19 × 0.1447 | 0.724 | 0.04363 | 6,880 | 372.5 |
| 477,000 | 0.3746 | 300,000 | AA | 19 × 0.1585 | 0.793 | 0.03636 | 8,090 | 447.8 |
| 477,000 | 0.3746 | 300,000 | A | 37 × 0.1135 | 0.795 | 0.03636 | 8,600 | 447.8 |
| 556,500 | 0.4371 | 350,000 | ..... | 19 × 0.1711 | 0.856 | 0.03116 | 9,440 | 522.4 |
| 556,500 | 0.4371 | 350,000 | AA, A | 37 × 0.1226 | 0.858 | 0.03116 | 9,830 | 522.4 |
| 636,000 | 0.4995 | 400,000 | AA, A | 37 × 0.1311 | 0.918 | 0.02727 | 11,240 | 597.0 |
| 715,500 | 0.5620 | 450,000 | AA | 37 × 0.1391 | 0.974 | 0.02424 | 12,640 | 671.6 |
| 715,500 | 0.5620 | 450,000 | A | 61 × 0.1083 | 0.975 | 0.02424 | 13,150 | 671.6 |
| 795,000 | 0.6244 | 500,000 | AA | 37 × 0.1466 | 1.026 | 0.02181 | 13,770 | 746.3 |
| 795,000 | 0.6244 | 500,000 | A | 61 × 0.1142 | 1.028 | 0.02181 | 14,330 | 746.3 |
| 874,500 | 0.6868 | 550,000 | AA | 37 × 0.1538 | 1.077 | 0.01983 | 14,830 | 820.9 |
| 874,500 | 0.6868 | 550,000 | A | 61 × 0.1198 | 1.078 | 0.01983 | 15,760 | 820.9 |
| 954,000 | 0.7493 | 600,000 | AA | 37 × 0.1606 | 1.124 | 0.01818 | 16,180 | 895.5 |
| 954,000 | 0.7493 | 600,000 | A | 61 × 0.1251 | 1.126 | 0.01818 | 16,860 | 895.5 |
| 1,033,500 | 0.8117 | 650,000 | AA | 37 × 0.1672 | 1.170 | 0.01678 | 17,530 | 970.1 |
| 1,033,500 | 0.8117 | 650,000 | A | 61 × 0.1302 | 1.172 | 0.01678 | 18,260 | 970.1 |
| 1,113,000 | 0.8741 | 700,000 | AA, A | 61 × 0.1351 | 1.216 | 0.01558 | 19,660 | 1,045 |
| 1,192,500 | 0.9366 | 750,000 | AA, A | 61 × 0.1398 | 1.258 | 0.01454 | 21,000 | 1,119 |
| 1,272,000 | 0.999 | 800,000 | AA, A | 61 × 0.1444 | 1.330 | 0.01363 | 22,000 | 1,193 |
| 1,351,500 | 1.062 | 850,000 | AA, A | 61 × 0.1489 | 1.340 | 0.01283 | 23,400 | 1,269 |
| 1,431,000 | 1.124 | 900,000 | AA, A | 61 × 0.1532 | 1.379 | 0.01212 | 24,300 | 1,343 |
| 1,510,500 | 1.186 | 950,000 | AA, A | 61 × 0.1574 | 1.417 | 0.01148 | 25,600 | 1,418 |
| 1,590,000 | 1.249 | 1,000,000 | AA | 61 × 0.1615 | 1.454 | 0.01091 | 27,000 | ʾ,493 |
| 1,590,000 | 1.249 | 1,000,000 | A | 91 × 0.1322 | 1.454 | 0.01091 | 28,100 | ı,493 |

<sup>a</sup> Class AA stranding is usually specified for bare conductors used on overhead lines.  Class A stranding is usually specified for conductors to be covered with weather-resistant (weatherproof) materials and for bare conductors where greater flexibility than afforded by Class AA is required.

**119.    Stranded Aluminum Conductor, Bare: Class B**[a]
(Aluminum Co. of America)
Hard-drawn (EC-H19)—three-quarter hard (EC-H26) intermediate temper (EC-H24)

| Conductor size | | Copper equivalent based upon equal d-c resistance, Cu 97% Al 61% | Stranding, number and diam of wires, in. | Cable diam, in. | D-c resistance at 20°C, ohms per 1,000 ft (61%) | Ultimate strength, lb, EC-H19 | Min ultimate strength, lb, EC-H26 | Min ultimate strength, lb, EC-H24 | Weight per 1,000 ft, lb |
|---|---|---|---|---|---|---|---|---|---|
| Cir mils or AWG | Sq in. | | | | | | | | |
| 6 | 0.0206 | 8 | 7 × 0.0612 | 0.184 | 0.6606 | 528 | 316 | 280 | 24.6 |
| 4 | 0.0328 | 6 | 7 × 0.0772 | 0.232 | 0.4155 | 826 | 500 | 440 | 39.2 |
| 3 | 0.0413 | 5 | 7 × 0.0867 | 0.260 | 0.3295 | 1,022 | 630 | 560 | 49.4 |
| 2 | 0.0521 | 4 | 7 × 0.0974 | 0.292 | 0.2613 | 1,266 | 800 | 700 | 62.3 |
| 1 | 0.0657 | 3 | 19 × 0.0664 | 0.332 | 0.2072 | 1,685 | 1,000 | 890 | 78.5 |
| 1/0 | 0.0829 | 2 | 19 × 0.0745 | 0.373 | 0.1643 | 2,090 | 1,270 | 1,120 | 99.1 |
| 2/0 | 0.1045 | 1 | 19 × 0.0837 | 0.419 | 0.1303 | 2,586 | 1,600 | 1,410 | 124.9 |
| 3/0 | 0.1318 | 1/0 | 19 × 0.0940 | 0.470 | 0.1033 | 3,200 | 2,015 | 1,780 | 157.5 |
| 4/0 | 0.1662 | 2/0 | 19 × 0.1055 | 0.528 | 0.08195 | 3,890 | 2,540 | 2,240 | 198.6 |
| 250,000 | 0.1964 | 157,300 | 37 × 0.0822 | 0.575 | 0.06937 | 4,860 | 3,000 | 2,650 | 234.7 |
| 300,000 | 0.2356 | 188,800 | 37 × 0.0900 | 0.629 | 0.05781 | 5,830 | 3,600 | 3,180 | 281.6 |
| 350,000 | 0.2749 | 220,200 | 37 × 0.097.. | 0.681 | 0.04955 | 6,680 | 4,200 | 3,710 | 328.6 |
| 400,000 | 0.3142 | 251,500 | 37 × 0.1040 | 0.728 | 0.04336 | 7,350 | 4,800 | 4,240 | 375.5 |
| 450,000 | 0.3534 | 283,000 | 37 × 0.1103 | 0.772 | 0.03854 | 8,110 | 5,400 | 4,770 | 422.4 |
| 500,000 | 0.3927 | 314,500 | 37 × 0.1162 | 0.813 | 0.03468 | 9,010 | 6,000 | 5,300 | 469.4 |
| 550,000 | 0.4320 | 346,000 | 61 × 0.0950 | 0.855 | 0.03153 | 10,490 | 6,610 | 5,830 | 516.3 |
| 600,000 | 0.4712 | 377,000 | 61 × 0.0992 | 0.893 | 0.02890 | 11,450 | 7,210 | 6,360 | 563.2 |
| 650,000 | 0.5105 | 409,000 | 61 × 0.1032 | 0.929 | 0.02668 | 11,940 | 7,810 | 6,890 | 610.2 |
| 700,000 | 0.5498 | 440,000 | 61 × 0.1071 | 0.964 | 0.02477 | 12,860 | 8,410 | 7,420 | 657.1 |
| 750,000 | 0.5890 | 472,000 | 61 × 0.1109 | 0.998 | 0.02312 | 13,510 | 9,010 | 7,950 | 704.0 |
| 800,000 | 0.6283 | 503,000 | 61 × 0.1145 | 1.031 | 0.02168 | 14,410 | 9,610 | 8,480 | 751.0 |
| 900,000 | 0.7069 | 566,000 | 61 × 0.1215 | 1.094 | 0.01927 | 15,900 | 10,810 | 9,540 | 844.8 |
| 1,000,000 | 0.7854 | 629,000 | 61 × 0.1280 | 1.152 | 0.01734 | 17,670 | 12,020 | 10,600 | 938.7 |
| 1,100,000 | 0.8639 | 692,000 | 91 × 0.1099 | 1.209 | 0.01576 | 20,210 | 13,220 | 11,660 | 1,033 |
| 1,200,000 | 0.9425 | 755,000 | 91 × 0.1148 | 1.263 | 0.01445 | 21,630 | 14,420 | 12,720 | 1,126 |
| 1,250,000 | 0.9818 | 786,000 | 91 × 0.1172 | 1.289 | 0.01387 | 22,530 | 15,020 | 13,250 | 1,173 |
| 1,300,000 | 1.021 | 818,000 | 91 × 0.1195 | 1.315 | 0.01334 | 23,430 | 15,620 | 13,780 | 1,220 |
| 1,400,000 | 1.100 | 880,000 | 91 × 0.1240 | 1.364 | 0.01239 | 24,750 | 16,830 | 14,850 | 1,314 |
| 1,500,000 | 1.178 | 943,000 | 91 × 0.1284 | 1.412 | 0.01156 | 26,500 | 18,020 | 15,900 | 1.408 |
| 1,600,000 | 1.257 | 1,006,000 | 127 × 0.1122 | 1.459 | 0.01084 | 28,840 | 19,230 | 16,970 | 1,502 |
| 1,700,000 | 1.335 | 1,069,000 | 127 × 0.1157 | 1.504 | 0.01020 | 30,630 | 20,400 | 18,020 | 1,596 |
| 1,750,000 | 1.374 | 1,101,000 | 127 × 0.1174 | 1.526 | 0.00991 | 31,530 | 21,000 | 18,550 | 1.643 |
| 1,800,000 | 1.414 | 1,132,000 | 127 × 0.1191 | 1.548 | 0.00963 | 32,450 | 21,600 | 19,090 | 1,690 |
| 1,900,000 | 1.492 | 1,195,000 | 127 × 0.1223 | 1.590 | 0.00913 | 33,570 | 22,800 | 20,100 | 1,784 |
| 2,000,000 | 1.571 | 1,258,000 | 127 × 0.1255 | 1.632 | 0.00867 | 35,340 | 24,000 | 21,200 | 1,877 |
| 2,500,000 | 1.964 | 1,570,000 | 127 × 0.1403 | 1.824 | 0.00694 | 43,300 | 30,000 | 26,500 | 2,370 |
| 3,000,000 | 2.356 | 1,890,000 | 169 × 0.1332 | 1.998 | 0.00578 | 53,010 | 36,000 | 31,800 | 2,844 |
| 3,500,000 | 2.749 | 2,200,000 | 169 × 0.1439 | 2.158 | 0.00495 | 60,610 | 40,500 | 37,100 | 3,350 |

[a] Class B stranding is usually specified for conductors to be insulated with various materials such as rubber, paper, varnished cloth, etc.

**120.  Stranded Aluminum Conductor, Bare: Class C**[a]

(Aluminum Co. of America)

Hard-drawn (EC-H19)—three-quarter hard (EC-H26)—Intermediate temper (EC-H24)

| Conductor size | | Copper equivalent based upon equal d-c resistance, Cu 97% Al 61% | Stranding, number and diam of wires, in. | Cable diam, in. | D-c resistance at 20°C, ohms per 1,000 ft (61%) | Ultimate strength, lb, EC-H19 | Min ultimate strength, lb, EC-H26 | Min ultimate strength, lb, EC-H24 | Weight per 1,000 ft, lb |
|---|---|---|---|---|---|---|---|---|---|
| Cir mils or AWG | Sq in. | | | | | | | | |
| 2 | 0.0521 | 4 | 19 × 0.0591 | 0.296 | 0.2613 | 1,360 | 800 | 705 | 62.3 |
| 2/0 | 0.1045 | 1 | 37 × 0.0600 | 0.420 | 0.1303 | 2,725 | 1,600 | 1,410 | 124.9 |
| 3/0 | 0.1318 | 1/0 | 37 × 0.0673 | 0.471 | 0.1033 | 3,380 | 2,015 | 1,780 | 157.5 |
| 4/0 | 0.1662 | 2/0 | 37 × 0.0756 | 0.529 | 0.08195 | 4,190 | 2,540 | 2,240 | 198.6 |
| 250,000 | 0.1964 | 157,300 | 61 × 0.0640 | 0.576 | 0.06937 | 5,040 | 3,000 | 2,650 | 234.7 |
| 300,000 | 0.2356 | 188,800 | 61 × 0.0701 | 0.631 | 0.05781 | 5,940 | 3,600 | 3,180 | 281.6 |
| 350,000 | 0.2749 | 220,200 | 61 × 0.0757 | 0.681 | 0.04955 | 6,930 | 4,200 | 3,710 | 328.6 |
| 400,000 | 0.3142 | 251,500 | 61 × 0.0810 | 0.729 | 0.04336 | 7,780 | 4,800 | 4,240 | 375.5 |
| 450,000 | 0.3534 | 283,000 | 61 × 0.0859 | 0.773 | 0.03854 | 8,750 | 5,400 | 4,770 | 422.4 |
| 500,000 | 0.3927 | 314,500 | 61 × 0.0905 | 0.815 | 0.03468 | 9,540 | 6,000 | 5,300 | 469.4 |
| 550,000 | 0.4320 | 346,000 | 91 × 0.0777 | 0.855 | 0.03153 | 10,880 | 6,600 | 5,830 | 516.3 |
| 600,000 | 0.4712 | 377,000 | 91 × 0.0812 | 0.893 | 0.02890 | 11,660 | 7,200 | 6,360 | 563.2 |
| 650,000 | 0.5105 | 409,000 | 91 × 0.0845 | 0.930 | 0.02668 | 12,630 | 7,800 | 6,890 | 610.2 |
| 700,000 | 0.5498 | 440,000 | 91 × 0.0877 | 0.964 | 0.02477 | 13,600 | 8,400 | 7,420 | 657.1 |
| 750,000 | 0.5890 | 472,000 | 91 × 0.0908 | 0.999 | 0.02312 | 14,310 | 9,000 | 7,950 | 704.0 |
| 800,000 | 0.6283 | 503,000 | 91 × 0.0938 | 1.032 | 0.02168 | 15,270 | 9,600 | 8,480 | 751.0 |
| 900,000 | 0.7069 | 566,000 | 91 × 0.0994 | 1.093 | 0.01927 | 17,180 | 10,800 | 9,540 | 844.8 |
| 1,000,000 | 0.7854 | 629,000 | 91 × 0.1048 | 1.153 | 0.01734 | 18,380 | 12,000 | 10,600 | 938.7 |
| 1,100,000 | 0.8639 | 692,000 | 127 × 0.0931 | 1.210 | 0.01576 | 21,000 | 13,200 | 11,660 | 1,033 |
| 1,200,000 | 0.9425 | 755,000 | 127 × 0.0972 | 1.264 | 0.01445 | 22,900 | 14,400 | 12,720 | 1,126 |
| 1,250,000 | 0.9818 | 786,000 | 127 × 0.0992 | 1.290 | 0.01387 | 23,900 | 15,000 | 13,250 | 1,173 |
| 1,300,000 | 1.021 | 818,000 | 127 × 0.1012 | 1.316 | 0.01334 | 23,900 | 15,600 | 13,780 | 1.220 |
| 1,400,000 | 1.100 | 880,000 | 127 × 0.1050 | 1.365 | 0.01239 | 25,700 | 16,800 | 14,850 | 1,314 |
| 1,500,000 | 1.178 | 943,000 | 127 × 0.1087 | 1.413 | 0.01156 | 27,600 | 18,000 | 15,900 | 1,408 |
| 1,600,000 | 1.257 | 1,006,000 | 169 × 0.0973 | 1.460 | 0.01084 | 30,500 | 19,200 | 16,970 | 1,502 |
| 1,700,000 | 1.335 | 1,069,000 | 169 × 0.1003 | 1.505 | 0.01020 | 31,200 | 20,400 | 18,020 | 1,596 |
| 1,750,000 | 1.374 | 1,101,000 | 169 × 0.1018 | 1.527 | 0.00991 | 32,100 | 21,000 | 18,550 | 1,643 |
| 1,800,000 | 1.414 | 1,132,000 | 169 × 0.1032 | 1.548 | 0.00963 | 33,100 | 21,600 | 19,090 | 1,690 |
| 1,900,000 | 1.492 | 1,195,000 | 169 × 0.1060 | 1.590 | 0.00913 | 34,900 | 22,800 | 20,100 | 1,784 |
| 2,000,000 | 1.571 | 1,258,000 | 169 × 0.1088 | 1.632 | 0.00867 | 36,800 | 24,000 | 21,200 | 1,877 |
| 2,500,000 | 1.964 | 1,570,000 | 169 × 0.1216 | 1.824 | 0.00694 | 44,200 | 30,000 | 26,500 | 2,346 |

[a] Class C stranding is specified for conductors where greater flexibility than provided by Class B is required.

**121. Aluminum Conductor, Steel-Reinforced (ACSR) Bare**
(Aluminum Co. of America)

| ACSR Cross section Aluminum Cir mils or AWG | Aluminum Sq in. | Total Sq in. | Copper equivalent based upon equal d-c resistance Cu 97% Al 61% | Stranding Aluminum | Stranding Steel | Diameter Complete cable | Diameter Steel core | D-c resistance at 20°C ohms per 1,000 ft (61%) | Ultimate strength, lb[a] | Weight per 1,000 ft Total | Al | Steel | Weight per mile Total | Al | Steel | % Al | % Steel |
|---|---|---|---|---|---|---|---|---|---|---|---|---|---|---|---|---|---|
| 6 | 0.0206 | 0.0240 | 8 | 6 × 0.0661 | 1 × 0.0661 | 0.198 | 0.0661 | 0.6573 | 1,170 | 36.1 | 24.5 | 11.6 | 190 | 129 | 61 | 67.9 | 32.1 |
| 5 | 0.0260 | 0.0303 | 7 | 6 × 0.0743 | 1 × 0.0743 | 0.223 | 0.0743 | 0.5213 | 1,460 | 45.5 | 30.9 | 14.6 | 240 | 163 | 77 | 67.9 | 32.1 |
| 4 | 0.0328 | 0.0383 | 6 | 6 × 0.0834 | 1 × 0.0834 | 0.250 | 0.0834 | 0.4134 | 1,830 | 57.4 | 39.0 | 18.4 | 303 | 206 | 97 | 67.9 | 32.1 |
| 4 | 0.0328 | 0.0411 | 6 | 7 × 0.0772 | 1 × 0.1029 | 0.257 | 0.1029 | 0.4134 | 2,288 | 67.1 | 39.0 | 28.1 | 354 | 206 | 148 | 58.1 | 41.9 |
| 3 | 0.0413 | 0.0482 | 5 | 6 × 0.0937 | 1 × 0.0937 | 0.281 | 0.0937 | 0.3279 | 2,250 | 72.4 | 49.2 | 23.2 | 382 | 260 | 122 | 67.9 | 32.1 |
| 2 | 0.0521 | 0.0608 | 4 | 6 × 0.1052 | 1 × 0.1052 | 0.316 | 0.1052 | 0.2600 | 2,790 | 91.3 | 62.0 | 29.3 | 482 | 327 | 155 | 67.9 | 32.1 |
| 2 | 0.0521 | 0.0653 | 4 | 7 × 0.0974 | 1 × 0.1299 | 0.325 | 0.1299 | 0.2600 | 3,525 | 106.7 | 62.0 | 44.7 | 563 | 327 | 236 | 58.1 | 41.9 |
| 1 | 0.0657 | 0.0767 | 3 | 6 × 0.1182 | 1 × 0.1182 | 0.355 | 0.1182 | 0.2062 | 3,480 | 115.2 | 78.2 | 37.0 | 608 | 413 | 195 | 67.9 | 32.1 |
| 1/0 | 0.0829 | 0.0967 | 2 | 6 × 0.1327 | 1 × 0.1327 | 0.398 | 0.1327 | 0.1635 | 4,280 | 145.2 | 98.6 | 46.6 | 767 | 521 | 246 | 67.9 | 32.1 |
| 2/0 | 0.1045 | 0.1219 | 1 | 6 × 0.1490 | 1 × 0.1490 | 0.447 | 0.1490 | 0.1297 | 5,345 | 183.1 | 124.3 | 58.8 | 967 | 656 | 311 | 67.9 | 32.1 |
| 3/0 | 0.1318 | 0.1538 | 1/0 | 6 × 0.1672 | 1 × 0.1672 | 0.502 | 0.1672 | 0.1028 | 6,675 | 230.9 | 156.8 | 74.1 | 1,219 | 828 | 391 | 67.9 | 32.1 |
| 4/0 | 0.1662 | 0.1939 | 2/0 | 6 × 0.1878 | 1 × 0.1878 | 0.563 | 0.1878 | 0.08155 | 8,420 | 291.1 | 197.7 | 93.4 | 1,537 | 1,044 | 493 | 67.9 | 32.1 |
| 266,800 | 0.2095 | 0.2211 | 3/0 | 18 × 0.1217 | 1 × 0.1217 | 0.609 | 0.1217 | 0.06500 | 7,100 | 289.7 | 250.4 | 39.3 | 1,530 | 1,322 | 208 | 86.45 | 13.55 |
| 266,800 | 0.2095 | 0.2367 | 3/0 | 6 × 0.2109 | 1 × 0.2109 | 0.633 | 0.2109 | 0.06500 | 9,645 | 343.3 | 250.4 | 91.9 | 1,812 | 1,322 | 485 | 73.2 | 26.8 |
| 266,800 | 0.2095 | 0.2436 | 3/0 | 26 × 0.1013 | 7 × 0.0788 | 0.642 | 0.2364 | 0.06531 | 11,250 | 367.3 | 251.7 | 115.6 | 1,939 | 1,329 | 610 | 68.6 | 31.4 |
| 300,000 | 0.2356 | 0.2740 | 188,700 | 26 × 0.1074 | 7 × 0.0835 | 0.680 | 0.2505 | 0.05809 | 12,650 | 412.9 | 283.0 | 129.9 | 2,180 | 1,494 | 686 | 68.6 | 31.4 |
| 336,400 | 0.2642 | 0.2789 | 4/0 | 18 × 0.1367 | 1 × 0.1367 | 0.684 | 0.1367 | 0.05155 | 8,950 | 365.3 | 315.8 | 49.5 | 1,929 | 1,668 | 261 | 86.45 | 13.55 |
| 336,400 | 0.2642 | 0.3072 | 4/0 | 26 × 0.1138 | 7 × 0.0885 | 0.721 | 0.2655 | 0.05181 | 14,050 | 463.0 | 317.3 | 145.7 | 2,444 | 1,675 | 769 | 68.6 | 31.4 |
| 336,400 | 0.2642 | 0.3259 | 4/0 | 30 × 0.1059 | 7 × 0.1059 | 0.741 | 0.3177 | 0.05193 | 17,040 | 527.1 | 318.1 | 209.0 | 2,783 | 1,679 | 1,104 | 60.35 | 39.65 |
| 397,500 | 0.3122 | 0.3295 | 250,000 | 18 × 0.1486 | 1 × 0.1486 | 0.743 | 0.1486 | 0.04363 | 10,400 | 431.0 | 372.5 | 58.5 | 2,276 | 1,967 | 309 | 86.45 | 13.55 |
| 397,500 | 0.3122 | 0.3630 | 250,000 | 26 × 0.1236 | 7 × 0.0961 | 0.783 | 0.2883 | 0.04384 | 16,190 | 547.2 | 375.0 | 172.2 | 2,889 | 1,980 | 909 | 68.6 | 31.4 |
| 397,500 | 0.3122 | 0.3850 | 250,000 | 30 × 0.1151 | 7 × 0.1151 | 0.806 | 0.3453 | 0.04395 | 19,980 | 622.8 | 375.9 | 246.9 | 3,288 | 1,984 | 1,304 | 60.35 | 39.65 |
| 477,000 | 0.3746 | 0.3954 | 300,000 | 18 × 0.1628 | 1 × 0.1628 | 0.814 | 0.1628 | 0.03636 | 12,300 | 518.0 | 447.8 | 70.2 | 2,735 | 2,364 | 371 | 86.45 | 13.55 |
| 477,000 | 0.3746 | 0.4231 | 300,000 | 24 × 0.1410 | 7 × 0.0940 | 0.846 | 0.2820 | 0.03653 | 17,200 | 614.5 | 450.0 | 164.5 | 3,243 | 2,376 | 867 | 73.25 | 26.75 |

| | | | | | | | | | | | | | | | | | |
|---|---|---|---|---|---|---|---|---|---|---|---|---|---|---|---|---|---|
| 31.4 | 68.6 | 1,091 | 2,376 | 3,467 | 206.6 | 450.0 | 656.6 | 19,430 | 0.03653 | 0.3162 | 0.858 | 7 X 0.1054 | 26 X 0.1355 | 300,000 | 0.4356 | 0.3746 | 477,000 |
| 39.65 | 60.35 | 1,406 | 2,381 | 3,945 | 296.3 | 451.0 | 747.3 | 23,300 | 0.03662 | 0.3783 | 0.883 | 7 X 0.1261 | 30 X 0.1261 | 300,000 | 0.4620 | 0.3746 | 477,000 |
| 13.55 | 86.45 | 433 | 2,756 | 3,189 | 82 | 522 | 604 | 14,300 | 0.03116 | 0.1758 | 0.879 | 1 X 0.1758 | 18 X 0.1758 | 350,000 | 0.4614 | 0.4371 | 556,500 |
| 26.75 | 73.25 | 1,013 | 2,772 | 3,785 | 192 | 525 | 717 | 19,850 | 0.03132 | 0.3045 | 0.914 | 7 X 0.1015 | 24 X 0.1523 | 350,000 | 0.4938 | 0.4371 | 556,500 |
| 31.4 | 68.6 | 1,272 | 2,772 | 4,044 | 241 | 525 | 766 | 22,400 | 0.03132 | 0.341 | 0.927 | 7 X 0.1138 | 26 X 0.1463 | 350,000 | 0.5083 | 0.4371 | 556,500 |
| 39.65 | 60.35 | 1,827 | 2,777 | 4,604 | 346 | 526 | 872 | 27,200 | 0.03139 | 0.409 | 0.953 | 7 X 0.1362 | 30 X 0.1362 | 350,000 | 0.5391 | 0.4371 | 556,500 |
| 26.75 | 73.25 | 1,098 | 3,015 | 4,113 | 208 | 571 | 779 | 21,500 | 0.02880 | 0.318 | 0.953 | 7 X 0.1059 | 24 X 0.1588 | 380,500 | 0.5368 | 0.4752 | 605,000 |
| 31.4 | 68.6 | 1,383 | 3,015 | 4,398 | 262 | 571 | 833 | 24,100 | 0.02880 | 0.356 | 0.966 | 7 X 0.1186 | 26 X 0.1525 | 380,500 | 0.5526 | 0.4752 | 605,000 |
| 39.1 | 60.9 | 1,938 | 3,020 | 4,958 | 367 | 572 | 939 | 30,000 | 0.02888 | 0.426 | 0.994 | 19 X 0.0852 | 30 X 0.1420 | 380,500 | 0.5835 | 0.4752 | 605,000 |
| 26.75 | 73.25 | 1,156 | 3,168 | 4,324 | 219 | 600 | 819 | 22,600 | 0.02740 | 0.326 | 0.977 | 7 X 0.1085 | 24 X 0.1628 | 400,000 | 0.5643 | 0.4995 | 636,000 |
| 26.75 | 73.25 | 1,452 | 3,168 | 4,620 | 275 | 600 | 875 | 25,000 | 0.02740 | 0.365 | 0.990 | 7 X 0.1216 | 26 X 0.1564 | 400,000 | 0.5809 | 0.4995 | 636,000 |
| 39.1 | 60.9 | 2,043 | 3,173 | 5,216 | 387 | 601 | 988 | 31,500 | 0.02747 | 0.437 | 1.019 | 19 X 0.0874 | 30 X 0.1456 | 400,000 | 0.6134 | 0.4995 | 636,600 |
| 39.1 | 60.9 | 1,200 | 3,321 | 4,530 | 230 | 629 | 859 | 23,700 | 0.02614 | 0.333 | 1.000 | 7 X 0.1111 | 24 X 0.1667 | 419,000 | 0.5914 | 0.5235 | 666,600 |
| 26.75 | 73.25 | 1,299 | 3,564 | 4,863 | 246 | 675 | 921 | 26,300 | 0.02436 | 0.345 | 1.036 | 7 X 0.1151 | 54 X 0.1151 | 450,000 | 0.6348 | 0.5620 | 715,500 |
| 31.4 | 68.6 | 1,637 | 3,564 | 5,201 | 310 | 675 | 985 | 28,100 | 0.02436 | 0.387 | 1.051 | 7 X 0.1290 | 26 X 0.1659 | 450,000 | 0.6535 | 0.5620 | 715,500 |
| 39.1 | 60.9 | 2,297 | 3,569 | 5,866 | 435 | 676 | 1,111 | 34,600 | 0.02441 | 0.463 | 1.081 | 19 X 0.0926 | 30 X 0.1544 | 450,000 | 0.6901 | 0.5620 | 715,500 |
| 16.3 | 83.7 | 771 | 3,960 | 4,731 | 146 | 750 | 896 | 22,900 | 0.02192 | 0.266 | 1.063 | 7 X 0.0886 | 45 X 0.1329 | 500,000 | 0.6676 | 0.6244 | 795,000 |
| 26.75 | 73.25 | 1,447 | 3,960 | 5,407 | 274 | 750 | 1,024 | 28,500 | 0.02192 | 0.364 | 1.093 | 7 X 0.1214 | 54 X 0.1214 | 500,000 | 0.7053 | 0.6244 | 795,000 |
| 31.4 | 68.6 | 1,816 | 3,960 | 5,776 | 344 | 750 | 1,094 | 31,200 | 0.02192 | 0.408 | 1.108 | 7 X 0.1360 | 26 X 0.1749 | 500,000 | 0.7261 | 0.6244 | 795,000 |
| 39.1 | 60.9 | 2,550 | 3,971 | 6,521 | 483 | 752 | 1,235 | 38,400 | 0.02197 | 0.489 | 1.140 | 19 X 0.0977 | 30 X 0.1628 | 500,000 | 0.7668 | 0.6244 | 795,000 |
| 26.75 | 73.25 | 1,589 | 4,356 | 5,945 | 301 | 825 | 1,126 | 31,400 | 0.01993 | 0.382 | 1.146 | 7 X 0.1273 | 54 X 0.1273 | 550,000 | 0.7759 | 0.6868 | 874,500 |
| 26.75 | 73.25 | 1,637 | 4,483 | 6,120 | 310 | 849 | 1,159 | 32,300 | 0.01936 | 0.387 | 1.162 | 7 X 0.1291 | 54 X 0.1291 | 900,000 | 0.7985 | 0.7069 | 900,000 |
| 16.3 | 83.7 | 924 | 4,752 | 5,676 | 175 | 900 | 1,075 | 26,900 | 0.01827 | 0.291 | 1.165 | 7 X 0.0971 | 45 X 0.1456 | 600,000 | 0.8011 | 0.7493 | 954,000 |
| 26.75 | 73.25 | 1,737 | 4,752 | 6,489 | 329 | 900 | 1,229 | 34,200 | 0.01827 | 0.399 | 1.196 | 7 X 0.1329 | 54 X 0.1329 | 600,000 | 0.8464 | 0.7493 | 954,000 |
| 16.3 | 83.7 | 1,003 | 5,148 | 6,151 | 190 | 975 | 1,165 | 28,900 | 0.01686 | 0.303 | 1.213 | 7 X 0.1011 | 45 X 0.1516 | 650,000 | 0.8678 | 0.8117 | 1,033,500 |
| 26.75 | 73.25 | 1,880 | 5,148 | 7,028 | 356 | 975 | 1,331 | 37,100 | 0.01686 | 0.415 | 1.246 | 7 X 0.1384 | 54 X 0.1384 | 650,000 | 0.9169 | 0.8117 | 1,033,500 |
| 16.2 | 83.8 | 1,082 | 5,571 | 6,653 | 205 | 1,055 | 1,260 | 30,900 | 0.01573 | 0.315 | 1.259 | 7 X 0.1049 | 45 X 0.1573 | 700,000 | 0.9346 | 0.8741 | 1,113,000 |
| 26.3 | 73.7 | 1,985 | 5,571 | 7,556 | 376 | 1,055 | 1,431 | 40,200 | 0.01573 | 0.431 | 1.293 | 19 X 0.0862 | 54 X 0.1436 | 700,000 | 0.9849 | 0.8741 | 1,113,000 |
| 16.2 | 83.8 | 1,156 | 5,966 | 7,122 | 219 | 1,130 | 1,349 | 33,200 | 0.01469 | 0.326 | 1.302 | 7 X 0.1085 | 45 X 0.1628 | 750,000 | 1.001 | 0.9366 | 1,192,500 |
| 26.3 | 73.7 | 2,128 | 5,966 | 8,094 | 403 | 1,130 | 1,533 | 43,100 | 0.01469 | 0.446 | 1.333 | 19 X 0.0892 | 54 X 0.1486 | 750,000 | 1.0552 | 0.9366 | 1,192,500 |
| 16.2 | 83.8 | 1,235 | 6,368 | 7,603 | 234 | 1,206 | 1,440 | 35,400 | 0.01377 | 0.336 | 1.345 | 7 X 0.1121 | 45 X 0.1681 | 800,000 | 1.068 | 0.9990 | 1,272,000 |
| 26.3 | 73.7 | 2,265 | 6,368 | 8,633 | 429 | 1,206 | 1,635 | 44,800 | 0.01377 | 0.461 | 1.382 | 19 X 0.0921 | 54 X 0.1535 | 800,000 | 1.1256 | 0.9990 | 1,272,000 |
| 16.2 | 83.8 | 1,309 | 6,674 | 8,073 | 248 | 1,281 | 1,529 | 37,600 | 0.01298 | 0.347 | 1.386 | 7 X 0.1151 | 45 X 0.1733 | 850,000 | 1.135 | 1.062 | 1,351,500 |
| 26.3 | 73.7 | 2,408 | 6,674 | 9,172 | 456 | 1,281 | 1,737 | 47,600 | 0.01298 | 0.475 | 1.424 | 19 X 0.0949 | 54 X 0.1582 | 850,000 | 1.1959 | 1.0615 | 1,351,500 |
| 16.2 | 83.8 | 1,389 | 7,165 | 8,554 | 263 | 1,357 | 1,620 | 39,800 | 0.01224 | 0.357 | 1.427 | 7 X 0.1189 | 45 X 0.1783 | 900,000 | 1.202 | 1.124 | 1,431,000 |
| 26.3 | 73.7 | 2,550 | 7,165 | 9,715 | 483 | 1,357 | 1,840 | 50,400 | 0.01224 | 0.489 | 1.465 | 19 X 0.0977 | 54 X 0.1628 | 900,000 | 1.2663 | 1.124 | 1,431,000 |

**121. Aluminum Conductor, Steel-Reinforced (ACSR) Bare** (*Continued*)

| ACSR Cross section | | | Copper equivalent based upon equal d-c resistance Cu 97% Al 61% | Stranding, number and diameter of strands, in. | | Diameter, in. | | D-c resistance at 20°C ohms per 1,000 ft (61%) | Ultimate strength, lb[a] | Weight, lb | | | | | | % of total weight | |
|---|---|---|---|---|---|---|---|---|---|---|---|---|---|---|---|---|---|
| Aluminum | | Total | | Aluminum | Steel | Complete cable | Steel core | | | Per 1,000 ft | | | Per mile | | | | |
| Cir mils or AWG | Sq in. | Sq in. | | | | | | | | Total | Al | Steel | Total | Al | Steel | Al | Steel |
| 1,510,500 | 1.186 | 1.268 | 950,000 | 45 X 0.1832 | 7 X 0.1221 | 1.466 | 0.366 | 0.01159 | 41,600 | 1,709 | 1,432 | 277 | 9,024 | 7,561 | 1,463 | 83.8 | 16.2 |
| 1,510,500 | 1.186 | 1.3366 | 950,000 | 54 X 0.1675 | 19 X 0.1004 | 1.506 | 0.502 | 0.01159 | 53,200 | 1,942 | 1,432 | 510 | 10,254 | 7,561 | 2,693 | 73.7 | 26.3 |
| 1,590,000 | 1.249 | 1.335 | 1,000,000 | 45 X 0.1878 | 7 X 0.1252 | 1.502 | 0.376 | 0.01101 | 43,800 | 1,799 | 1,507 | 292 | 9,499 | 7,957 | 1,542 | 83.8 | 16.2 |
| 1,590,000 | 1.249 | 1.4076 | 1,000,000 | 54 X 0.1716 | 19 X 0.1030 | 1.545 | 0.515 | 0.01101 | 56,000 | 2,044 | 1,507 | 537 | 10,792 | 7,957 | 2,835 | 73.7 | 26.3 |
| 1,780,000 | 1.398 | 1.512 | 1,119,000 | 84 X 0.1456 | 19 X 0.0874 | 1.602 | 0.437 | 0.00984 | 53,600 | 2,074 | 1,687 | 387 | 10,950 | 8,907 | 2,043 | 81.3 | 18.7 |
| 80,000 | 0.0628 | 0.0847 | 50,310 | 8 X 0.1000 | 1 X 0.1670 | 0.367 | 0.1670 | 0.2168 | 5,200 | 149.0 | 75.1 | 73.9 | 787 | 397 | 390 | 50.4 | 49.6 |
| 101,800 | 0.0800 | 0.1266 | 64,160 | 12 X 0.0921 | 7 X 0.0921 | 0.461 | 0.2763 | 0.1712 | 9,860 | 254.1 | 96.0 | 158.1 | 1,342 | 507 | 835 | 37.8 | 62.2 |
| 110,800 | 0.0870 | 0.1378 | 69,700 | 12 X 0.0961 | 7 X 0.0961 | 0.481 | 0.2883 | 0.1573 | 10,730 | 276.6 | 104.5 | 172.1 | 1,460 | 552 | 908 | 37.8 | 62.2 |
| 134,600 | 0.1057 | 0.1674 | 84,600 | 12 X 0.1059 | 7 X 0.1059 | 0.530 | 0.3177 | 0.1295 | 12,920 | 336.0 | 127.0 | 209.0 | 1,774 | 671 | 1,103 | 37.8 | 62.2 |
| 159,000 | 0.1249 | 0.1977 | 100,000 | 12 X 0.1151 | 7 X 0.1151 | 0.576 | 0.3453 | 0.1096 | 15,200 | 396.8 | 150.0 | 246.8 | 2,095 | 792 | 1,303 | 37.8 | 62.2 |
| 176,900 | 0.1389 | 0.2200 | 111,200 | 12 X 0.1214 | 7 X 0.1214 | 0.607 | 0.3642 | 0.09851 | 16,440 | 441.5 | 166.9 | 274.6 | 2,331 | 881 | 1,450 | 37.8 | 62.2 |
| 190,800 | 0.1499 | 0.2373 | 120,000 | 12 X 0.1261 | 7 X 0.1261 | 0.631 | 0.3783 | 0.09134 | 17,730 | 476.3 | 180.0 | 296.3 | 2,515 | 950 | 1,565 | 37.8 | 62.2 |
| 211,300 | 0.1660 | 0.2628 | 132,900 | 12 X 0.1327 | 7 X 0.1327 | 0.663 | 0.3981 | 0.08248 | 19,640 | 527.5 | 199.3 | 328.2 | 2,785 | 1,052 | 1,733 | 37.8 | 62.2 |
| 203,200 | 0.1596 | 0.3020 | 127,800 | 16 X 0.1127 | 19 X 0.0977 | 0.714 | 0.4885 | 0.08576 | 27,500 | 676.7 | 191.7 | 485.0 | 3,573 | 1,012 | 2,561 | 28.3 | 71.7 |

[a] Based on standard-weight zinc-coated steel core wire.

NOTES:

1. An amount not exceeding 10 per cent of the total weight of any one order may be shipped in random lengths, but no piece shorter than 50 per cent of the standard length will be shipped. No random length will be wound on the same reel with a standard length, and all reels will be marked showing number of pieces, length of piece.
2. The actual weight of cable will be held within a tolerance of plus or minus 2 per cent of the weights listed. Invoicing will be based on actual weight.
3. Shipments to each destination will be made to the nearest package specified on each item ordered.

## 122.  Aluminum Conductor, Steel-Reinforced (ACSR) Bare

(Aluminum Co. of America)
With Class B and Class C zinc-coated steel core (B261-55T)

| Size, cir mils or AWG | ACSR Stranding, number and diameter of strands, in. | | Ultimate strength, lb | | | % reduction in strength | |
|---|---|---|---|---|---|---|---|
| | Aluminum | Steel | Standard-weight coating | Class B coating | Class C coating | Class B coating | Class C coating |
| 6 | 6 × 0.0661 | 1 × 0.0661 | 1,170 | 1,170 | 1,150 | 0.0 | 2.0 |
| 5 | 6 × 0.0743 | 1 × 0.0743 | 1,460 | 1,460 | 1,440 | 0.0 | 1.4 |
| 4 | 6 × 0.0834 | 1 × 0.0834 | 1,830 | 1,830 | 1,800 | 0.0 | 1.6 |
| 4 | 7 × 0.0772 | 1 × 0.1029 | 2,288 | 2,245 | 2,205 | 1.9 | 3.6 |
| 3 | 6 × 0.0937 | 1 × 0.0937 | 2,250 | 2,220 | 2,180 | 1.3 | 3.1 |
| 2 | 6 × 0.1052 | 1 × 0.1052 | 2,790 | 2,745 | 2,705 | 1.6 | 3.0 |
| 2 | 7 × 0.0974 | 1 × 0.1299 | 3,525 | 3,385 | 3,255 | 4.0 | 7.7 |
| 1 | 6 × 0.1182 | 1 × 0.1182 | 3,480 | 3,430 | 3,370 | 1.4 | 3.2 |
| 1/0 | 6 × 0.1327 | 1 × 0.1327 | 4,280 | 4,140 | 4,000 | 3.3 | 6.5 |
| 2/0 | 6 × 0.1490 | 1 × 0.1490 | 5,345 | 4,910 | 4,820 | 8.1 | 9.8 |
| 3/0 | 6 × 0.1672 | 1 × 0.1672 | 6,675 | 6,135 | 6,020 | 8.1 | 9.8 |
| 4/0 | 6 × 0.1878 | 1 × 0.1878 | 8,420 | 7,730 | 7,590 | 8.2 | 9.8 |
| 266,800 | 18 × 0.1217 | 1 × 0.1217 | 7,100 | 6,985 | 6,870 | 1.6 | 3.2 |
| 266,800 | 6 × 0.2109 | 7 × 0.0703 | 9,645 | 9,645 | 9,410 | 0.0 | 2.4 |
| 266,800 | 26 × 0.1013 | 7 × 0.0788 | 11,250 | 11,250 | 11,075 | 0.0 | 1.6 |
| 300,000 | 26 × 0.1074 | 7 × 0.083 | 12,650 | 12,650 | 12,460 | 0.0 | 1.5 |
| 336,400 | 18 × 0.1367 | 1 × 0.1367 | 8,950 | 8,810 | 8,660 | 1.6 | 3.2 |
| 336,400 | 26 × 0.1138 | 7 × 0.0885 | 14,050 | 14,050 | 13,830 | 0.0 | 1.6 |
| 336,400 | 30 × 0.1059 | 7 × 0.1059 | 17,040 | 16,740 | 16,430 | 1.8 | 3.6 |
| 397,500 | 18 × 0.1486 | 1 × 0.1486 | 10,400 | 10,050 | 9,960 | 3.4 | 4.2 |
| 397,500 | 26 × 0.1236 | 7 × 0.0961 | 16,190 | 15,930 | 15,680 | 1.6 | 3.2 |
| 397,500 | 30 × 0.1151 | 7 × 0.1151 | 19,980 | 19,600 | 19,240 | 1.9 | 3.7 |
| 477,000 | 18 × 0.1628 | 1 × 0.1628 | 12,300 | 11,800 | 11,700 | 4.1 | 4.9 |
| 477,000 | 24 × 0.1410 | 7 × 0.0940 | 17,200 | 16,940 | 16,700 | 1.5 | 2.9 |
| 477,000 | 26 × 0.1355 | 7 × 0.1054 | 19,430 | 19,130 | 18,820 | 1.5 | 3.1 |
| 477,000 | 30 × 0.1261 | 7 × 0.1261 | 23,300 | 22,500 | 21,600 | 3.4 | 7.3 |
| 556,500 | 18 × 0.1758 | 1 × 0.1758 | 14,300 | 13,770 | 13,650 | 3.7 | 4.5 |
| 556,500 | 24 × 0.1523 | 7 × 0.1015 | 19,850 | 19,560 | 19,280 | 1.5 | 2.9 |
| 556,500 | 26 × 0.1463 | 7 × 0.1138 | 22,400 | 22,100 | 21,700 | 1.3 | 3.1 |
| 556,500 | 30 × 0.1362 | 7 × 0.1362 | 27,200 | 26,200 | 25,200 | 3.7 | 7.4 |
| 605,000 | 24 × 0.1588 | 7 × 0.1059 | 21,500 | 21,250 | 20,950 | 1.2 | 2.6 |
| 605,000 | 26 × 0.1525 | 7 × 0.1186 | 24,100 | 23,800 | 23,400 | 1.2 | 2.9 |
| 605,000 | 30 × 0.1420 | 19 × 0.0852 | 30,000 | 30,000 | 29,500 | 0.0 | 1.7 |
| 636,000 | 24 × 0.1628 | 7 × 0.1085 | 22,600 | 22,400 | 22,000 | 0.9 | 2.7 |
| 636,000 | 26 × 0.1564 | 7 × 0.1216 | 25,000 | 24,200 | 23,400 | 3.2 | 6.4 |
| 636,000 | 30 × 0.1456 | 19 × 0.0874 | 31,500 | 31,500 | 31,000 | 0.0 | 1.6 |
| 666,600 | 24 × 0.1667 | 7 × 0.1111 | 23,700 | 23,400 | 23,100 | 1.3 | 2.5 |
| 715,500 | 54 × 0.1151 | 7 × 0.1151 | 26,300 | 26,000 | 25,600 | 1.1 | 2.7 |
| 715,500 | 26 × 0.1659 | 7 × 0.1290 | 28,100 | 27,200 | 26,300 | 3.2 | 6.4 |
| 715,500 | 30 × 0.1544 | 19 × 0.0926 | 34,600 | 34,000 | 33,300 | 1.7 | 3.8 |
| 795,000 | 45 × 0.1329 | 7 × 0.0886 | 22,900 | 22,900 | 22,700 | 0.0 | 0.9 |
| 795,000 | 54 × 0.1214 | 7 × 0.1214 | 28,500 | 27,700 | 26,900 | 2.8 | 5.6 |
| 795,000 | 26 × 0.1749 | 7 × 0.1360 | 31,200 | 30,200 | 29,200 | 3.2 | 6.4 |
| 795,000 | 30 × 0.1628 | 19 × 0.0977 | 38,400 | 37,700 | 37,000 | 1.8 | 3.6 |

## 122.  Aluminum Conductor, Steel-Reinforced (ACSR) Bare *(Continued)*

| Size, cir mils or AWG | ACSR | | Ultimate strength, lb | | | % reduction in strength | |
|---|---|---|---|---|---|---|---|
| | Stranding, number and diameter of strands, in. | | Standard-weight coating | Class B coating | Class C coating | Class B coating | Class C coating |
| | Aluminum | Steel | | | | | |
| 874,500 | 54 × 0.1273 | 7 × 0.1273 | 31,400 | 30,500 | 29,600 | 2.9 | 5.7 |
| 900,000 | 54 × 0.1291 | 7 × 0.1291 | 32,300 | 31,400 | 30,500 | 2.8 | 5.6 |
| 954,000 | 45 × 0.1456 | 7 × 0.0971 | 26,900 | 26,600 | 26,400 | 1.1 | 1.9 |
| 954,000 | 54 × 0.1329 | 7 × 0.1329 | 34,200 | 33,300 | 32,300 | 2.6 | 5.6 |
| 1,033,500 | 45 × 0.1516 | 7 × 0.1011 | 28,900 | 28,600 | 28,300 | 1.0 | 2.1 |
| 1,033,500 | 54 × 0.1384 | 7 × 0.1384 | 37,100 | 36,100 | 35,000 | 2.7 | 5.7 |
| 1,113,000 | 45 × 0.1573 | 7 × 0.1049 | 30,900 | 30,600 | 30,300 | 1.0 | 2.0 |
| 1,113,000 | 54 × 0.1436 | 19 × 0.0862 | 40,200 | 40,200 | 40,100 | 0.0 | 0.3 |
| 1,192,500 | 45 × 0.1628 | 7 × 0.1121 | 33,200 | 32,800 | 32,500 | 1.2 | 2.1 |
| 1,192,500 | 54 × 0.1486 | 19 × 0.0892 | 43,100 | 43,100 | 43,000 | 0.0 | 0.2 |
| 1,272,000 | 45 × 0.1681 | 7 × 0.1121 | 35,400 | 35,000 | 34,700 | 1.1 | 2.0 |
| 1,272,000 | 54 × 0.1535 | 19 × 0.0921 | 44,800 | 44,200 | 43,600 | 1.3 | 2.7 |
| 1,351,500 | 45 × 0.1733 | 7 × 0.1151 | 37,600 | 37,200 | 36,800 | 1.1 | 2.1 |
| 1,351,500 | 54 × 0.1582 | 19 × 0.0949 | 47,600 | 47,000 | 46,300 | 1.3 | 2.7 |
| 1,431,000 | 45 × 0.1783 | 7 × 0.1189 | 39,800 | 39,400 | 39,000 | 1.0 | 2.0 |
| 1,431,000 | 54 × 0.1628 | 19 × 0.0977 | 50,400 | 49,800 | 49,000 | 1.2 | 2.8 |
| 1,510,500 | 45 × 0.1832 | 7 × 0.1221 | 41,600 | 40,700 | 39,900 | 2.2 | 4.1 |
| 1,510,500 | 54 × 0.1675 | 19 × 0.1004 | 53,200 | 52,500 | 51,800 | 1.3 | 2.6 |
| 1,590,000 | 45 × 0.1878 | 7 × 0.1252 | 43,800 | 42,700 | 41,800 | 2.5 | 4.6 |
| 1,590,000 | 54 × 0.1716 | 19 × 0.1030 | 56,000 | 55,300 | 54,500 | 1.3 | 2.7 |
| 80,000 | 8 × 0.1000 | 1 × 0.1670 | 5,200 | 4,655 | 4,550 | 10.5 | 12.5 |
| 101,800 | 12 × 0.0921 | 7 × 0.0921 | 9,860 | 9,615 | 9,385 | 2.5 | 4.8 |
| 110,800 | 12 × 0.0961 | 7 × 0.0961 | 10,730 | 10,480 | 10,220 | 2.3 | 4.8 |
| 134,600 | 12 × 0.1059 | 7 × 0.1059 | 12,920 | 12,620 | 12,310 | 2.3 | 4.7 |
| 159,000 | 12 × 0.1151 | 7 × 0.1151 | 15,200 | 14,850 | 14,480 | 2.3 | 4.7 |
| 176,900 | 12 × 0.1214 | 7 × 0.1214 | 16,440 | 15,640 | 14,830 | 4.9 | 9.8 |
| 190,800 | 12 × 0.1261 | 7 × 0.1261 | 17,730 | 16,860 | 16,000 | 4.9 | 9.8 |
| 211,300 | 12 × 0.1327 | 7 × 0.1327 | 19,640 | 18,700 | 17,700 | 4.8 | 10.0 |
| 203,200 | 16 × 0.1127 | 19 × 0.0977 | 27,500 | 26,800 | 26,100 | 2.5 | 5.1 |

## 123.  Copper-Steel Conductors
(American Steel and Wire Co.)

| Size, SCP or SCG | Equivalent to hard-drawn copper | | Approx diam, in. | | Modulus of elasticity | Min. breaking strength, lb | Approx weight, lb | | Total cross-sectional area | |
|---|---|---|---|---|---|---|---|---|---|---|
| | Size, AWG | Cir mils | Strand | Each wire | | | Per 1,000 ft | Per mile | Cir mils | Sq in. |
| 2 | 2 | 66,370 | 0.392 | 0.182 | 19,800,000 | 6,378 | 291 | 1,536 | 99,372 | 0.0781 |
| 4 | 4 | 41,740 | 0.310 | 0.144 | 19,800,000 | 4,486 | 182 | 961 | 62,208 | 0.0489 |
| 6 | 6 | 26,250 | 0.248 | 0.115 | 19,800,000 | 3,060 | 116 | 613 | 39,675 | 0.0312 |
| 8 | 8 | 16,510 | 0.196 | 0.091 | 20,500,000 | 2,112 | 73 | 385 | 24,843 | 0.0195 |
| 10[a] | 10 | 10,380 | 0.220 | 0.1019 | 21,700,000 | 3,853 | 88 | 464 | 31,140 | 0.0244 |
| 12[a] | 12 | 6,530 | 0.172 | 0.0808 | 21,700,000 | 2,426 | 55 | 290 | 19,590 | 0.0154 |

Coefficient of linear expansion:
  Sizes 2 to 8—0.0000082 per °F; sizes 10 to 12—0.0000075 per °F.
[a] Made up of one copper and two steel.

## 124. Copper-Clad Aluminum Wire

Cladding Calculations

$A$ = Area or volume percentage of copper-clad aluminum wire.
$W$ = Weight percentage of copper in copper-clad aluminum wire.
$T$ = Radius thickness percentage of copper in copper-clad aluminum wire.

$$W = \frac{8.89A}{6.19A + 270} \times 100$$

$$T = \left( 1 - \sqrt{1 - \frac{A}{100}} \right) \times 100$$

For 10% copper-clad aluminum wire $A = 10$.

Then

$$W = \frac{88.9}{61.9 + 270} \times 100 = 26.785$$

And

$$T = \left( 1 - \sqrt{1 - \frac{1}{10}} \right) \times 100 = 5.13$$

Conductor characteristics

|  | Copper | Copper-aluminum | Aluminum |
|---|---|---|---|
| Density, lb/in³ | 0.323 | 0.121 | 0.098 |
| Density, g/cm³ | 8.91 | 3.34 | 2.71 |
| Resistivity, Ω/cmf | 10.37 | 16.08 | 16.78 |
| Resistivity, microhms/cm | 1.724 | 2.673 | 2.790 |
| Conductivity (IACS %) | 100 | 61–63 | 61 |
| Weight %, copper | 100 | 26.8 | . . . |
| Tensile, 1000 psi, hard | 65.0 | 30.0 | 27.0 |
| Tensile, kg/mm², hard | 45.7 | 21.1 | 19.0 |
| Tensile, 1000 psi, hard | 35.0 | 17.0 | 17.0ᵃ |
| Tensile kg/mm², annealed | 24.6 | 12.0 | 12.0 |
| Specific gravity | 8.91 | 3.34 | 2.71 |

ᵃSemiannealed

Contact resistance vs. pressure
(Two No. 10 AWG wires crossed under pressure load)

| | Contact resistance (milliohms) | | | |
|---|---|---|---|---|
| Loading pressure, g | Copper to copper | Copper-aluminum to copper-aluminum | Aluminum to copper | Aluminum to aluminum |
| 5 | 28.0 | 27.0 | 250.0 | 200.0 |
| 10 | 15.0 | 16.0 | 250.0 | 200.0 |
| 50 | 7.0 | 7.5 | 200.0 | 180.0 |
| 100 | 4.0 | 4.0 | 150.0 | 180.0 |
| 200 | 3.0 | 3.0 | 130.0 | 130.0 |
| 500 | 2.7 | 2.7 | 60.0 | 150.0 |
| 1,000 | 2.5 | 2.0 | 29.0 | 150.0 |
| 5,000 | 2.5 | 2.2 | 20.0 | 160.0 |
| 10,000 | 2.4 | 2.2 | 13.0 | 100.0 |
| 20,000 | 2.5 | 2.3 | 4.5 | 40.0 |
| 50,000 | 2.5 | 2.3 | 3.5 | 3.8 |

## 125.  Copper-Clad Aluminum Wire Guide

| Diameter, mm | Area, mm² | kg/1000 m | | | Ω/1000 m, 20°C | | |
|---|---|---|---|---|---|---|---|
| | | Cu | 10% Cu/Al | 15% Cu/Al | Cu | 10% Cu/Al | 15% Cu/Al |
| 7.95 | 49.5 | 440.0 | 164.8 | 179.5 | 0.3600 | 0.5580 | 0.5470 |
| 2.5 | 4.909 | 43.82 | 16.40 | 17.90 | 3.625 | 5.620 | 5.515 |
| 2.0 | 3.142 | 27.98 | 10.48 | 11.40 | 5.675 | 8.800 | 8.620 |
| 1.95 | 2.986 | 26.62 | 9.98 | 10.88 | 5.951 | 9.230 | 9.045 |
| 1.9 | 2.835 | 25.27 | 9.54 | 10.32 | 6.278 | 9.745 | 9.545 |
| 1.85 | 2.688 | 23.96 | 8.965 | 9.765 | 6.635 | 10.28 | 10.08 |
| 1.8 | 2.545 | 22.68 | 8.498 | 9.250 | 7.010 | 10.86 | 10.63 |
| 1.75 | 2.405 | 21.44 | 8.040 | 8.750 | 7.400 | 11.48 | 11.23 |
| 1.7 | 2.270 | 20.23 | 7.585 | 8.255 | 7.842 | 12.14 | 11.92 |
| 1.65 | 2.138 | 19.06 | 7.145 | 7.765 | 8.348 | 12.92 | 12.68 |
| 1.6 | 2.010 | 17.92 | 6.715 | 7.315 | 8.823 | 13.68 | 13.40 |
| 1.55 | 1.886 | 16.82 | 6.300 | 6.860 | 9.462 | 14.68 | 14.39 |
| 1.5 | 1.767 | 15.75 | 5.900 | 6.425 | 10.40 | 16.11 | 15.80 |
| 1.45 | 1.651 | 14.72 | 5.518 | 6.015 | 10.78 | 16.70 | 16.37 |
| 1.4 | 1.539 | 13.72 | 5.147 | 5.600 | 11.57 | 17.92 | 17.58 |
| 1.35 | 1.431 | 12.76 | 4.777 | 5.205 | 12.42 | 19.28 | 18.89 |
| 1.3 | 1.327 | 11.83 | 4.438 | 4.835 | 13.41 | 20.80 | 20.40 |
| 1.25 | 1.227 | 10.94 | 4.095 | 4.460 | 14.54 | 22.56 | 22.10 |
| 1.20 | 1.130 | 10.08 | 3.775 | 4.115 | 15.75 | 24.40 | 23.95 |
| 1.15 | 1.0387 | 9.257 | 3.465 | 3.778 | 17.15 | 26.59 | 26.05 |
| 1.10 | 0.9503 | 8.470 | 3.175 | 3.458 | 18.73 | 29.02 | 28.46 |
| 1.05 | 0.8659 | 7.717 | 2.895 | 3.145 | 20.58 | 31.83 | 31.22 |
| 1.00 | 0.7854 | 6.990 | 2.620 | 2.858 | 22.62 | 35.05 | 34.42 |
| 0.95 | 0.7088 | 6.317 | 2.362 | 2.578 | 25.14 | 38.97 | 38.20 |
| 0.9 | 0.6362 | 5.670 | 2.122 | 2.316 | 28.04 | 43.50 | 42.65 |
| 0.85 | 0.5675 | 5.057 | 1.892 | 2.062 | 31.35 | 48.55 | 47.62 |
| 0.8 | 0.5027 | 4.470 | 1.675 | 1.825 | 35.44 | 54.95 | 53.95 |
| 0.75 | 0.4481 | 3.937 | 1.473 | 1.605 | 40.35 | 62.50 | 61.25 |
| 0.7 | 0.3848 | 3.430 | 1.281 | 1.400 | 46.34 | 71.85 | 70.45 |
| 0.65 | 0.3318 | 2.957 | 1.108 | 1.207 | 53.49 | 82.90 | 81.35 |
| 0.6 | 0.2827 | 2.520 | 0.942 | 1.026 | 62.95 | 97.55 | 95.65 |
| 0.55 | 0.2376 | 2.115 | 0.792 | 0.862 | 75.00 | 116.1 | 114.0 |
| 0.5 | 0.1963 | 1.747 | 0.655 | 0.712 | 90.59 | 140.5 | 137.8 |
| 0.45 | 0.1590 | 1.416 | 0.540 | 0.578 | 112.0 | 173.5 | 170.2 |
| 0.4 | 0.1257 | 1.118 | 0.419 | 0.456 | 141.1 | 218.5 | 214.2 |
| 0.35 | 0.0962 | 0.8569 | 0.321 | 0.350 | 184.8 | 286.2 | 280.6 |
| 0.3 | 0.0707 | 0.6292 | 0.236 | 0.256 | 251.9 | 390.5 | 382.8 |
| 0.25 | 0.0491 | 0.4382 | 0.164 | 0.179 | 362.5 | 562.0 | 551.5 |
| 0.20 | 0.0314 | 0.2798 | 0.105 | 0.114 | 567.5 | 880.0 | 862.0 |
| 0.15 | 0.0176 | 0.1575 | 0.059 | 0.0643 | 1040.0 | 1611 | 1580.0 |
| 0.12 | 0.0113 | 0.1008 | 0.0378 | 0.0412 | 1575.0 | 2440 | 2395.0 |

**126.   10 Percent Copper-Clad Aluminum Wire Guide**

| AWG | Decimal in | Area, in² | ft/lb Copper | ft/lb 10% Cu/Al | lb/1000 ft Copper | lb/1000 ft 10% Cu/Al | Ω/000 ft, 20°C Copper | Ω/000 ft, 20°C 10% Cu/Al |
|-----|-----------|-----------|--------------|-----------------|-------------------|----------------------|------------------------|--------------------------|
| ... | 0.4375 | 0.1503 | 1.726 | 4.608 | 579.3 | 216.7 | .0542 | .0840 |
| ... | 0.3125 | 0.07666 | 3.384 | 9.035 | 295.4 | 110.5 | .1062 | .1646 |
| 1 | 0.2893 | 0.06573 | 3.947 | 10.54 | 253.3 | 94.7 | .1239 | .1920 |
| 2 | 0.2576 | 0.05212 | 4.978 | 13.29 | 200.9 | 75.1 | .1563 | .2423 |
| 3 | 0.2294 | 0.04133 | 6.278 | 16.76 | 159.3 | 59.6 | .1971 | .3055 |
| 4 | 0.2043 | 0.03278 | 7.915 | 21.13 | 126.3 | 47.2 | .2485 | .3852 |
| 5 | 0.1819 | 0.02599 | 9.984 | 26.66 | 100.2 | 37.5 | .3134 | .4858 |
| 6 | 0.1620 | 0.02061 | 12.59 | 33.62 | 79.94 | 29.9 | .3952 | .6130 |
| 7 | 0.1443 | 0.01635 | 15.87 | 42.37 | 63.03 | 23.6 | .4981 | .7721 |
| 8 | 0.1285 | 0.01297 | 20.01 | 53.43 | 49.98 | 18.7 | .6281 | .9736 |
| 9 | 0.1144 | 0.01028 | 25.24 | 67.39 | 39.62 | 14.8 | .7925 | 1.228 |
| 10 | 0.1019 | 0.008155 | 31.82 | 84.96 | 31.43 | 11.8 | .9988 | 1.548 |
| 11 | 0.0907 | 0.00646 | 40.2 | 107 | 24.9 | 9.31 | 1.26 | 1.95 |
| 12 | 0.0808 | 0.00513 | 50.6 | 135 | 19.8 | 7.40 | 1.59 | 2.46 |
| 13 | 0.0720 | 0.00407 | 63.7 | 170 | 15.7 | 5.87 | 2.00 | 3.10 |
| 14 | 0.0641 | 0.00323 | 80.4 | 215 | 12.4 | 4.64 | 2.52 | 3.91 |
| 15 | 0.0571 | 0.00256 | 101 | 270 | 9.87 | 3.69 | 3.18 | 4.93 |
| 16 | 0.0508 | 0.00203 | 128 | 342 | 7.81 | 2.92 | 4.02 | 6.23 |
| 17 | 0.0453 | 0.00161 | 161 | 430 | 6.21 | 2.32 | 5.05 | 7.83 |
| 18 | 0.0403 | 0.00128 | 203 | 542 | 4.92 | 1.84 | 6.39 | 9.90 |
| 19 | 0.0359 | 0.00101 | 256 | 684 | 3.90 | 1.46 | 8.05 | 12.5 |
| 20 | 0.0320 | 0.000804 | 323 | 862 | 3.10 | 1.16 | 10.1 | 15.7 |
| 21 | 0.0285 | 0.000636 | 407 | 1,087 | 2.46 | .920 | 12.3 | 19.8 |
| 22 | 0.0253 | 0.000503 | 516 | 1,378 | 1.94 | .726 | 16.2 | 25.1 |
| 23 | 0.0226 | 0.000401 | 647 | 1,728 | 1.55 | .580 | 20.3 | 31.5 |
| 24 | 0.0201 | 0.000317 | 818 | 2,184 | 1.22 | .456 | 25.7 | 39.8 |
| 25 | 0.0179 | 0.000252 | 1,030 | 2,750 | .970 | .363 | 32.4 | 50.2 |
| 26 | 0.0159 | 0.000199 | 1,310 | 3,498 | .765 | .286 | 41.0 | 63.6 |
| 27 | 0.0142 | 0.000158 | 1,640 | 4,378 | .610 | .228 | 51.4 | 79.7 |
| 28 | 0.0126 | 0.000125 | 2,080 | 5,554 | .481 | .180 | 65.3 | 101 |
| 29 | 0.0113 | 0.000100 | 2,590 | 6,915 | .387 | .145 | 81.2 | 126 |
| 30 | 0.0100 | 0.0000785 | 3,300 | 8,811 | .303 | .113 | 104 | 161 |
| 31 | 0.0089 | 0.0000622 | 4,170 | 11,134 | .240 | .0896 | 131 | 203 |
| 32 | 0.0080 | 0.0000503 | 5,160 | 13,777 | .194 | .0726 | 162 | 251 |
| 33 | 0.0071 | 0.0000396 | 6,550 | 17,488 | .153 | .0572 | 206 | 319 |
| 34 | 0.0063 | 0.0000312 | 8,320 | 22,214 | .120 | .0449 | 261 | 405 |
| 35 | 0.0056 | 0.0000246 | 10,500 | 28,035 | .0949 | .0355 | 331 | 513 |
| 36 | 0.0050 | 0.0000196 | 13,200 | 35,244 | .0757 | .0283 | 415 | 643 |
| 37 | 0.0045 | 0.0000159 | 16,300 | 43,521 | .0613 | .0229 | 512 | 794 |
| 38 | 0.0040 | 0.0000126 | 20,600 | 55,002 | .0484 | .0181 | 648 | 1000 |
| 39 | 0.0035 | 0.00000962 | 27,000 | 72,090 | .0371 | .0139 | 847 | 1310 |
| 40 | 0.0031 | 0.00000755 | 34,400 | 91,848 | .0291 | .0109 | 1080 | 1670 |

## 127.  12 Percent Copper-Clad Aluminum Wire Guide

| AWG | Decimal in | Area, in² | ft/lb | | lb/1000 ft | | Ω/1000 ft, 20°C | |
|-----|-----------|-----------|-------|-----------|------------|-----------|-----------------|-----------|
|     |           |           | Copper | 12% Cu/Al | Copper | 12% Cu/Al | Copper | 12% Cu/Al |
| ... | 0.4375 | 0.1503 | 1.726 | 4.47 | 579.3 | 223.6 | .0542 | .0834 |
| ... | 0.3125 | 0.07666 | 3.384 | 8.76 | 295.4 | 114.1 | .1062 | .1635 |
| 1 | 0.2893 | 0.06573 | 3.947 | 10.22 | 253.3 | 97.8 | .1239 | .1907 |
| 2 | 0.2576 | 0.05212 | 4.978 | 12.89 | 200.9 | 77.6 | .1563 | .2405 |
| 3 | 0.2294 | 0.04133 | 6.278 | 16.26 | 159.3 | 61.5 | .1971 | .3033 |
| 4 | 0.2043 | 0.03278 | 7.915 | 20.49 | 126.3 | 48.8 | .2485 | .3824 |
| 5 | 0.1819 | 0.02599 | 9.984 | 25.84 | 100.2 | 38.7 | .3134 | .4823 |
| 6 | 0.1620 | 0.02061 | 12.59 | 32.57 | 79.94 | 30.7 | .3952 | .6082 |
| 7 | 0.1443 | 0.01635 | 15.87 | 41.15 | 63.03 | 24.3 | .4981 | .7667 |
| 8 | 0.1285 | 0.01297 | 20.01 | 51.81 | 49.98 | 19.3 | .6281 | .9665 |
| 9 | 0.1144 | 0.01028 | 25.24 | 65.36 | 39.62 | 15.3 | .7925 | 1.219 |
| 10 | 0.1019 | 0.008155 | 31.82 | 82.64 | 31.43 | 12.1 | .9988 | 1.537 |
| 11 | 0.0907 | 0.00646 | 40.2 | 104 | 24.9 | 9.61 | 1.26 | 1.94 |
| 12 | 0.0808 | 0.00513 | 50.6 | 131 | 19.8 | 7.63 | 1.59 | 2.44 |
| 13 | 0.0720 | 0.00407 | 63.7 | 165 | 15.7 | 6.06 | 2.00 | 3.08 |
| 14 | 0.0641 | 0.00323 | 80.4 | 208 | 12.4 | 4.81 | 2.52 | 3.88 |
| 15 | 0.0571 | 0.00256 | 101 | 262 | 9.87 | 3.81 | 3.18 | 4.90 |
| 16 | 0.0508 | 0.00203 | 128 | 331 | 7.81 | 3.02 | 4.02 | 6.18 |
| 17 | 0.0453 | 0.00161 | 161 | 417 | 6.21 | 2.40 | 5.05 | 7.79 |
| 18 | 0.0403 | 0.00128 | 203 | 526 | 4.92 | 1.90 | 6.39 | 9.79 |
| 19 | 0.0359 | 0.00101 | 256 | 667 | 3.90 | 1.50 | 8.05 | 12.4 |
| 20 | 0.0320 | 0.000804 | 323 | 833 | 3.10 | 1.20 | 10.1 | 15.6 |
| 21 | 0.0285 | 0.000636 | 407 | 1,057 | 2.46 | .946 | 12.3 | 19.7 |
| 22 | 0.0253 | 0.000503 | 516 | 1,337 | 1.94 | .748 | 16.2 | 24.9 |
| 23 | 0.0226 | 0.000401 | 647 | 1,675 | 1.55 | .597 | 20.3 | 31.3 |
| 24 | 0.0201 | 0.000317 | 818 | 2,119 | 1.22 | .472 | 25.7 | 39.5 |
| 25 | 0.0179 | 0.000252 | 1,030 | 2,667 | .970 | .375 | 32.4 | 49.7 |
| 26 | 0.0159 | 0.000199 | 1,310 | 3,378 | .765 | .296 | 41.0 | 63.0 |
| 27 | 0.0142 | 0.000158 | 1,640 | 4,255 | .610 | .235 | 51.4 | 79.3 |
| 28 | 0.0126 | 0.000125 | 2,080 | 5,376 | .481 | .186 | 65.3 | 100 |
| 29 | 0.0113 | 0.000100 | 2,590 | 6,711 | .387 | .149 | 81.2 | 125 |
| 30 | 0.0100 | 0.0000785 | 3,300 | 8,547 | .303 | .117 | 104 | 160 |
| 31 | 0.0089 | 0.0000622 | 4,170 | 10,799 | .240 | .0926 | 131 | 202 |
| 32 | 0.0080 | 0.0000503 | 5,160 | 13,369 | .194 | .0748 | 162 | 249 |
| 33 | 0.0071 | 0.0000396 | 6,550 | 16,978 | .153 | .0589 | 206 | 317 |
| 34 | 0.0063 | 0.0000312 | 8,320 | 21,552 | .120 | .0464 | 261 | 402 |
| 35 | 0.0056 | 0.0000246 | 10,500 | 27,322 | .0949 | .0366 | 331 | 510 |
| 36 | 0.0050 | 0.0000196 | 13,200 | 34,247 | .0757 | .0292 | 415 | 640 |
| 37 | 0.0045 | 0.0000159 | 16,300 | 42,194 | .0613 | .0237 | 512 | 788 |
| 38 | 0.0040 | 0.0000126 | 20,600 | 53,476 | .0484 | .0187 | 648 | 995 |
| 39 | 0.0035 | 0.00000962 | 27,000 | 69,930 | .0371 | .0143 | 847 | 1303 |
| 40 | 0.0031 | 0.00000755 | 34,400 | 89,286 | .0291 | .0112 | 1080 | 1660 |

**128. Conductor Application and Insulations[a]**

(Table 310-13, 1990 National Electrical Code)

| Trade name | Type letter | Maximum operating temperature | Application provisions | Insulation | Thickness of insulation — AWG or kcmil | Thickness of insulation — Mils | Outer covering[b] |
|---|---|---|---|---|---|---|---|
| Fluorinated ethylene propylene | FEP or FEPB | 90°C (194°F) / 200°C (392°F) | Dry and damp locations / Dry locations; special applications[b] | Fluorinated ethylene propylene / Fluorinated ethylene propylene | 14–10, 8–2 / 14–8, 6–2 | 20, 30 / 14, 14 | None / Glass braid Asbestos braid |
| Mineral insulation (metal-sheathed) | MI | 90°C (194°F) / 250°C (482°F) | Dry and wet locations / For special application[b] | Magnesium oxide | 16–10, 9–4, 3–250 | 36, 50, 55 | Copper or alloy steel |
| Moisture-, heat-, and oil-resistant thermoplastic | MTW[i] | 60°C (140°F) / 90°C (194°F) | Machine-tool wiring in wet locations as permitted in NFPA Standard No. 79 (see Article 670 of the NEC) / Machine-tool wiring in dry locations as permitted in NFPA Standard No. 79 (see Article 670 of the NEC) | Flame-retardant, moisture-, heat-, and oil-resistant thermoplastic | 22–12, 10, 8, 6, 4–2, 1–4/0, 213–500, 501–1000 | (A) 30, 30, 45, 60, 60, 80, 95, 110 (B) 15, 20, 30, 30, 40, 50, 60, 70 | (A) None (B) Nylon jacket or equivalent |
| Perfluoroalkoxy | PFA | 90°C (194°F) / 200°C (392°F) | Dry and damp locations / Dry locations; special applications[b] | Perfluoroalkoxy | 14–10, 8–2, 1–4/0 | 20, 30, 45 | None |
| Perfluoroalkoxy | PFAH | 350°C (482°F) | Dry locations only; solely for leads within apparatus or within raceways connected to apparatus (nickel or nickel-coated copper only) | Perfluoroalkoxy | 14–10, 8–2, 1–4/0 | 20, 30, 45 | None |

**128. Conductor Application and Insulations<sup>a</sup> (Continued)**

| Trade name | Type letter | Maximum operating temperature | Application provisions | Insulation | Thickness of insulation | | Outer covering[h] |
|---|---|---|---|---|---|---|---|
| | | | | | AWG or kcmil | Mils | |
| Heat-resistant rubber | RH | 75°C (167°F) | Dry and damp locations | Heat-resistant rubber | 14–12[c] / 10 | 30 / 45 | Moisture-resistant, flame-retardant, non-metallic covering[d] |
| Heat-resistant rubber | RHH[i] | 90°C (194°F) | Dry locations | | 8–2 / 1–4/0 / 213–500 / 501–1000 / 1001–2000 | 60 / 80 / 95 / 110 / 125 | |
| Moisture- and heat-resistant rubber | RHW[i,j] | 75°C (167°F) | Dry and wet locations. For over 2000 V, insulation ozone-resistant | Moisture- and heat-resistant rubber | 14–10 / 8–2 / 1–4/0 / 213–500 / 501–2000 / 1001–2000 | 45 / 60 / 80 / 95 / 110 / 125 | Moisture-resistant, flame-retardant, non-metallic covering[d] |
| Moisture- and heat-resistant rubber | RHW-2 | 90°C (194°F) | Dry and wet locations | Moisture- and heat-resistant rubber | 14–10 / 8–2 / 1–4/0 / 213–500 / 501–1000 / 1001–2000 / For 601–2000 V, see Table 310-62 | 45 / 60 / 80 / 95 / 110 / 125 | Moisture-resistant, flame-retardant, non-metallic covering[d] |
| Silicone-asbestos | SA | 90°C (194°F) / 125°C (257°F) | Dry and damp locations / For special application[b] | Silicone rubber | 14–10 / 8–2 / 1–4/0 / 213–500 / 501–1000 / 1001–2000 | 45 / 60 / 80 / 95 / 110 / 125 | Asbestos, glass, or other suitable braid material |
| Synthetic heat-resistant | SIS[i] | 90°C (194°F) | Switchboard wiring only | Heat-resistant rubber | 14–10 / 8 / 6–2 / 1–4/0 | 30 / 45 / 60 / 80 | None |

| Insulation | Trade name | Max. operating temperature | Application provisions | Insulation | AWG or kcmil | Thermoplastic | Asbestos | Outer covering |
|---|---|---|---|---|---|---|---|---|
| Thermoplastic and asbestos | TA | 90°C (194°F) | Switchboard wiring only | Thermoplastic and asbestos | 14–8<br>6–2<br>1–4/0 | 20<br>30<br>40 | 20<br>25<br>30 | Flame-retardant nonmetallic covering |
| Thermoplastic and fibrous outer braid | TBS | 90°C (194°F) | Switchboard wiring only | Thermoplastic | 14–10<br>8<br>6–2<br>1–4/0 | | 30<br>45<br>60<br>80 | Flame-retardant nonmetallic covering |
| Extruded polytetrafluoroethylene | TFE | 250°C (482°F) | Dry locations only; solely for leads within apparatus or within raceways connected to apparatus or as open wiring (nickel or nickel-coated copper only) | Extruded polytetrafluoroethylene | 14–10<br>8–2<br>1–4/0 | | 20<br>30<br>45 | None |
| Heat-resistant thermoplastic | THHN[i] | 90°C (194°F) | Dry and damp locations | Flame-retardant heat-resistant thermoplastic | 14–12<br>10<br>8–6<br>4–2<br>1–4/0<br>250–500<br>501–1000 | | 15<br>20<br>30<br>40<br>50<br>60<br>70 | Nylon jacket or equivalent |
| Moisture- and heat-resistant thermoplastic | THHW | 75°C (167°F)<br>90°C (194°F) | Wet location<br>Dry location | Flame-retardant, moisture- and heat-resistant thermoplastic | 14–10<br>8–2<br>1–4/0<br>213–500<br>501–1000 | | 45<br>60<br>80<br>95<br>110 | None |
| Moisture- and heat-resistant thermoplastic | THW[i,j] | 75°C (167°F)<br>90°C (194°F) | Dry and wet locations<br><br>Special applications within electric discharge lighting equipment: limited to 1000 open-circuit V or less (size 14–8 only as permitted in Sec. 410-31 of the NEC) | Flame-retardant, moisture- and heat-resistant thermoplastic | 14–10<br>8–2<br>1–4/0<br>213–500<br>501–1000<br>1001–2000 | | 45<br>60<br>80<br>95<br>110<br>125 | None |

**128. Conductor Application and Insulations$^a$ (Continued)**

| Trade name | Type letter | Maximum operating temperature | Application provisions | Insulation | Thickness of insulation | | Outer covering$^h$ |
|---|---|---|---|---|---|---|---|
| | | | | | AWG or kcmil | Mils | |
| Moisture- and heat-resistant thermoplastic | THWN$^{i,j}$ | 75°C (167°F) | Dry and wet locations | Flame-retardant, moisture- and heat-resistant thermoplastic | 14–12<br>10<br>8–6<br>4–2<br>1–4/0<br>250–500<br>501–1000 | 15<br>20<br>30<br>40<br>50<br>60<br>70 | Nylon jacket or equivalent |
| Moisture-resistant thermoplastic | TW$^i$ | 60°C (140°F) | Dry and wet locations | Flame-retardant, moisture-resistant thermoplastic | 14–10<br>8<br>6–2<br>1–4/0<br>213–500<br>501–1000<br>1001–2000 | 30<br>45<br>60<br>80<br>95<br>110<br>125 | None |
| Underground feeder and branch-circuit cable, single conductor (for Type UF cable employing more than one conductor see Article 339 of the NEC | UF | 60°C (140°F)<br><br>75°C $_f$ (167°F) | See Article 339 of the NEC | Moisture-resistant<br><br>Moisture- and heat-resistant | 14–10<br>8–2<br>1–4/0 | 60$^e$<br>80$^e$<br>95$^e$ | Integral with insulation |
| Underground service-entrance cable, single conductor (for Type USE cable employing more than one conductor see Article 338 of the NEC) | USE$^j$ | 75°C (167°F) | See Article 338 of the NEC | Heat- and moisture-resistant | 12–10<br>8–2<br>1–4/0<br>213–500<br>501–1000<br>1001–2000 | 45<br>60<br>80<br>95$^g$<br>110<br>125 | Moisture-resistant nonmetallic covering [See Sec. 338-1 (b) of the NEC] |
| Moisture- and heat-resistant cross-linked synthetic polymer | XHHW$^{i,j}$ | 90°C (194°F)<br>75°C (167°F) | Dry and damp locations<br><br>Wet locations | Flame-retardant cross-linked synthetic polymer | 14–10<br>8–2<br>1–4/0<br>213–500<br>501–1000<br>1001–2000 | 30<br>45<br>55<br>65<br>80<br>95 | None |

| Insulation | Type letter | Max operating temperature | Application provisions | Insulation | AWG or kcmil | Thickness of insulation (mils) | Outer covering |
|---|---|---|---|---|---|---|---|
| Moisture- and heat-resistant cross-linked synthetic polymer | XHHW-2 | 90°C 194°F | Dry and wet locations | Flame-retardant cross-linked synthetic polymer | 14–10<br>8–2<br>1–4/0<br>213–500<br>501–1000<br>1001–2000 | 30<br>45<br>55<br>65<br>80<br>95 | None |
| Modified ethylene tetrafluoroethylene | Z | 90°C (194°F)<br>150°C (302°F) | Dry and damp locations<br>Dry locations; special applications[b] | Modified ethylene tetrafluoroethylene | 14–12<br>10<br>8–4<br>3–1<br>1/0–4/0 | 15<br>20<br>25<br>35<br>45 | None |
| Modified ethylene tetrafluoroethylene | ZW[j] | 75°C (167°F)<br>90°C (194°F)<br>150°C (302°F) | Wet locations<br>Dry and damp locations<br>Dry locations; special applications[b] | Modified ethylene tetrafluoroethylene | 14–10<br>8–2 | 30<br>45 | None |

[a] Reprinted with permission from NFPA 70-1990, **National Electrical Code®**, Copyright © 1990, National Fire Protection Association, Quincy, Massachusetts 02269. This reprinted material is not the complete and official position of the NFPA on the referenced subject, which is represented only by the standard in its entirety.

[b] Where environmental conditions require maximum conductor operating temperature above 90°C.

[c] For 14-12 sizes RHH shall be 45 mils thickness insulation.

[d] Some rubber insulations do not require an outer covering.

[e] Includes integral jacket.

[f] For ampacity limitation, see Sec. 339-5 of the NEC.

[g] Insulation thickness shall be permitted to be 80 mils for listed USE conductors that have been subjected to special investigations. The nonmetallic covering over individual rubber-covered conductors of aluminum-sheathed cable and of lead-sheathed or multiconductor cable shall not be required to be flame-retardant. For Type MC cable, see Sec. 334-20 of the NEC. For nonmetallic-sheathed cable, see Sec. 336-25 of the NEC. For Type UF cable, see Sec. 339-1 of the NEC.

[h] Some insulations do not require an outer covering.

[i] Insulation and outer coverings that meet the requirements of flame-retardant, limited smoke, and are so listed shall be permitted to be designated limited smoke with the suffix /LS after the Code-type designation.

[j] Listed wire types designated with the suffix -2 (such as RHW-2) shall be permitted to be used at a continuous 90°C operating temperature wet or dry. Ampacities of those wire types are given in the 90°C column of the appropriate ampacity table.

## 129.    Data for MI Cable
(General Cable Co.)

| Cable section | Conductor size, AWG | Outside diam, mils | | Approx weight, lb per 1,000 ft | Nominal coil length, ft | Cables | Reference Nos. end seal | Threaded gland[a] | Equivalent conduit size (NPT glands) | Current-carrying capacities | |
|---|---|---|---|---|---|---|---|---|---|---|---|
| | | Cable | Conductor | | | | | | | Single isolated cables | Grouped cables[b] |
| One-conductor | | | | | | | | | | | |
|  | 16 | 0.215 | 0.051 | 77 | 1950 | 215/1 | 215/1 | 215 | ⅜ | 25 | 20 |
| | 14 | 0.230 | 0.064 | 89 | 1650 | 230/1 | 230/1 | 230 | ⅜ | 30[c] | 25[c] |
| | 12 | 0.246 | 0.081 | 104 | 1500 | 246/1 | 246/1 | 246 | ⅜ | 40[c] | 30[c] |
| | 10 | 0.277 | 0.102 | 133 | 1075 | 277/1 | 277/1 | 277 | ⅜ | 55[c] | 40[c] |
| | 8 | 0.309 | 0.128 | 170 | 900 | 309/1 | 309/1 | 309 | ½ | 70[c] | 50[c] |
| | 6 | 0.340 | 0.162 | 217 | 750 | 340/1 | 340/1 | 340 | ½ | 100[c] | 70[c] |
| | 4 | 0.402 | 0.204 | 305 | 510 | 402/1 | 402/1 | 402 | ½ | 135[c] | 90[c] |
| | 3 | 0.434 | 0.229 | 361 | 430 | 434/1 | 434/1 | 434 | ¾ | 155[c] | 105[c] |
| | 2 | 0.465 | 0.258 | 425 | 360 | 465/1 | 465/1 | 465 | ¾ | 180[c] | 120[c] |
| | 1 | 0.496 | 0.289 | 498 | 335 | 496/1 | 496/1 | 496 | ¾ | 210[c] | 140[c] |
| | 1/0 | 0.543 | 0.325 | 603 | 285 | 543/1 | 543/1 | 543 | ¾ | 245[c] | 155[c] |
| | 2/0 | 0.590 | 0.365 | 726 | 230 | 590/1 | 590/1 | 590 | 1 | 285[c] | 185[c] |
| | 3/0 | 0.637 | 0.410 | 869 | 205 | 637/1 | 637/1 | 637 | 1 | 330[c] | 210[c] |
| | 4/0 | 0.699 | 0.460 | 1060 | 175 | 699/1 | 699/1 | 699 | 1 | 385[c] | 235[c] |
| Two-conductor | | | | | | | | | | | |
|  | 16 | 0.340 | 0.051 | 166 | 720 | 340/2 | 340/2 | 340 | ½ | 20 | |
| | 14 | 0.371 | 0.064 | 199 | 620 | 371/2 | 371/2 | 371 | ½ | 25[c] | |
| | 12 | 0.402 | 0.081 | 237 | 500 | 402/2 | 402/2 | 402 | ½ | 30[c] | |
| | 10 | 0.449 | 0.102 | 300 | 415 | 449/2 | 449/2 | 449 | ¾ | 40[c] | |
| | 8 | 0.512 | 0.128 | 395 | 305 | 512/2 | 512/2 | 512 | ¾ | 50[c] | |
| | 6 | 0.590 | 0.162 | 534 | 230 | 590/2 | 590/2 | 590 | 1 | 70[c] | |
| | 4 | 0.684 | 0.204 | 734 | 170 | 684/2 | 684/2 | 684 | 1 | 90[c] | |
| Three-conductor | | | | | | | | | | | |
|  | 16 | 0.355 | 0.051 | 184 | 675 | 355/3 | 355/3 | 355 | ½ | 20 | |
| | 14 | 0.387 | 0.064 | 222 | 525 | 387/3 | 387/3 | 387 | ½ | 25[c] | |
| | 12 | 0.434 | 0.081 | 283 | 420 | 434/3 | 434/3 | 434 | ¾ | 30[c] | |
| | 10 | 0.480 | 0.102 | 356 | 355 | 480/3 | 480/3 | 480 | ¾ | 40[c] | |
| | 8 | 0.543 | 0.128 | 470 | 280 | 543/3 | 543/3 | 543 | ¾ | 50[c] | |
| | 6 | 0.621 | 0.162 | 636 | 210 | 621/3 | 621/3 | 621 | 1 | 70[c] | |
| | 4 | 0.730 | 0.204 | 901 | 150 | 730/3 | 730/3 | 730 | 1 | 90[c] | |
| Four-conductor | | | | | | | | | | | |
|  | 16 | 0.387 | 0.051 | 217 | 555 | 387/4 | 387/4 | 387 | ½ | 16 | |
| | 14 | 0.418 | 0.064 | 260 | 490 | 418/4 | 418/4 | 418 | ¾ | 20[c] | |
| | 12 | 0.465 | 0.081 | 329 | 350 | 465/4 | 465/4 | 465 | ¾ | 24[c] | |
| | 10 | 0.527 | 0.102 | 432 | 290 | 527/4 | 527/4 | 527 | ¾ | 32[c] | |
| | 8 | 0.590 | 0.128 | 566 | 225 | 590/4 | 590/4 | 590 | 1 | 40[c] | |
| | 6 | 0.684 | 0.162 | 785 | 170 | 684/4 | 684/4 | 684 | 1 | 56[c] | |
| Seven-conductor | | | | | | | | | | | |
|  | 16 | 0.449 | 0.051 | 294 | 500 | 449/7 | 449/7 | 449 | ¾ | 13 | |
| | 14 | 0.496 | 0.064 | 368 | 410 | 496/7 | 496/7 | 496 | ¾ | 17.5[c] | |
| | 12 | 0.543 | 0.081 | 461 | 325 | 543/7 | 543/7 | 543 | ¾ | 21[c] | |
| | 10 | 0.621 | 0.102 | 622 | 250 | 621/7 | 621/7 | 621 | 1 | 28[c] | |

[a] This number appears on all metal terminal components.
[b] Not more than three cables per group.
[c] National Electrical Code.
The above data are approximate and subject to normal manufacturing tolerances.

**130. Flexible Cords and Cables**[a]
(Table 400-4, 1990 National Electrical Code)

| Trade name | Type letter | Size, AWG | No. of conductors | Insulation | Nominal insulation thickness AWG | Nominal insulation thickness Mils | Braid on each conductor | Outer covering | Use | |
|---|---|---|---|---|---|---|---|---|---|---|
| Asbestos-covered heat-resistant cord | AFC<br>AFPD | 18–10 | 2 or 3 | Impregnated asbestos | 18–14 | 30 | Cotton or rayon | None | Pendant | Dry locations | Not hard usage |
| | AFPO | | 2 | | 12–10 | 45 | None | Cotton, rayon, or saturated asbestos | | | |
| Thermoset-jacketed heat-resistant cord | AFS<br>AFSJ | 18–10<br>18–16 | 2 or 3 | Impregnated asbestos | 18–14 | 30 | None | Thermoset | Portable heaters | Damp locations | Extra-hard usage<br><br>Hard usage |
| Lamp cord | C | 18–10 | 2 or more | Thermoset or thermoplastic | 18–16<br>14–10 | 30<br>45 | Cotton | None | Pendant or portable | Dry locations | Not hard usage |
| Data processing cable | DP: See Notes 2 and 10. | 32 min. | 2 or more | Thermoplastic, thermoset or cross-linked synthetic polymer | 32–27 (50 V)<br>26–23 (50 V)<br>22–20 (50 V)<br>32–16 (300 V)<br>14–10 (300 V)<br>8–2 (300 V) | 8<br>12<br>16<br>20<br>30<br>60 | None | Thermoplastic, thermoset, or cross-linked synthetic polymer | Data processing systems | Dry locations | Power and signaling circuits |

## 130. Flexible Cords and Cables* (Continued)

| Trade name | Type letter | Size, AWG | No. of conductors | Insulation | Nominal insulation thickness | | Braid on each conductor | Outer covering | Use |
|---|---|---|---|---|---|---|---|---|---|
| | | | | | AWG | Mils | | | |
| | E: See Note 6. | | | | 20–16 | 20 | Cotton | Three cotton: outer one flame-retardant and moisture-resistant: See Note 4. | Elevator lighting and control · Nonhazardous locations |
| Elevator cable | See Note 10. | 20–14 | 2 or more | Thermoset | 20–16 | 20 | Flexible nylon jacket | | |
| | | | | | | | Cotton | Three cotton: outer one flame-retardant and moisture-resistant: See Note 4. | Nonhazardous locations |
| Elevator cable | EO: See Note 6. | 20–14 | 2 or more | Thermoset | 14 | 30 | | One cotton and a neoprene jacket: See Note 4. | Elevator lighting and control · Hazardous (classified) locations |
| Elevator cable | ET: See Note 6. | 20–14 | 2 or more | Thermoplastic | 2 0–16 | 20 | Rayon | Three cotton: outer one flame-retardant and moisture-resistant: See Note 4. | Nonhazardous locations |
| | ETLB: See Note 6. | | | | 14 | 30 | None | | |
| | ETO: See Note 6. | | | Thermoplastic | | | None | One cotton and a thermoplastic jacket | Hazardous (classified) locations |
| | ETT: See Note 6. | | | Thermoplastic | | | None | One cotton and a thermoplastic jacket | |

| Trade name | Type letter | Size AWG or kcmil | Number of conductors | Insulation | AWG or kcmil | Nominal insulation thickness (mils) | Braid on each conductor | Outer covering | Use | | Portable, extra-hard usage, and as permitted in Secs. 520-68 and 530-12 of the NEC |
|---|---|---|---|---|---|---|---|---|---|---|---|
| Portable power cable | G | 8–500 kcmil | 2–6 plus grounding conductor(s) | Thermoset | 8–2; 1–4/0; 250–500 kcmil | 60; 80; 95 | | Oil-resistant thermoset | | | |
| Heater cord | HPD | 18–12 | 2, 3, or 4 | Thermoset with asbestos or all thermoset | Thermoset or thermoplastic 18–16; 14–12 | 15; 30 | None | Cotton or rayon | Portable heaters | Dry locations | Not hard usage |
| Parallel heater cord | HPN: See Note 7. | 18–12 | 2 or 3 | Thermosetting | 18–16; 14; 12 | 45; 80; 95 | None | Thermosetting | Portable | Damp locations | Not hard usage |
| Thermoset-jacketed heater cord | HS | 14–12 | | Thermoset with asbestos or all thermoset | 18–16 (thermoset/asbestos); 18–16 (all thermoset) | 15; 30 | None | Cotton and thermoset | Portable or portable heaters | Damp locations | Extra-hard usage |
| | HSJ: See Note 8. | 18–12 | 2, 3, or 4 | | | | | | | | |
| | HSJO: See Note 8. | | | Thermoset with asbestos | 14–12 (thermoset/asbestos) | 30 | | Cotton and oil-resistant compound | | | Hard usage |
| | HSO | 14–12 | | All thermoset | 14–12 (all thermoset) | 45 | None | | | | Extra-hard usage |
| Twisted portable cord | PD | 18–10 | 2 or more | Thermoset or thermoplastic | 14–10 | 45 | Cotton | Cotton or rayon | Pendant or portable | Dry locations | Not hard usage |
| Hard-service cord | S: See Note 5. | 18–2 | 2 or more | Thermoset | 18–16 | 30 | | Thermoset | Pendant or portable | Damp locations | Extra-hard usage |
| | SE: See Note 5. | | | Thermoplastic elastomer | 14–10 | 45 | None | Thermoplastic elastomer | | | |
| | SEO: See Note 5. | | | | 8–2 | 60 | | Oil-resistant thermoplastic elastomer | | | |

## 130. Flexible Cords and Cables<sup>a</sup> (Continued)


| Trade name | Type letter | Size, AWG | No. of conductors | Insulation | Nominal insulation thickness AWG | Nominal insulation thickness Mils | Braid on each conductor | Outer covering | Use Pendant or portable | Use Damp locations | Use Hard usage |
|---|---|---|---|---|---|---|---|---|---|---|---|
| Junior hard-service cord | SJ | 18–10 | 2, 3, 4, or 5 | Thermoset | | | None | Thermoset | | | |
| | SJE | | | Thermo-plastic elastomer | | | | Thermoplastic elastomer | | | |
| | SJEO | | | | 18–12 | 30 | | Oil-resistant thermoplastic elastomer | | | |
| | SJO | | | Thermoset | | | | Oil-resistant thermoset | | | |
| | SJOO | | | Oil-resistant thermoset | | | | Oil-resistant thermoset | | | |
| | SJT | | | Thermo-plastic or thermoset | 10 | 45 | | Thermoplastic | | | |
| | SJTO | | | Thermoset or thermoplastic | | | | Oil-resistant thermoplastic | | | |
| | SJTOO | | | Oil-resistant thermoplastic or thermoset | | | | Oil-resistant thermoplastic | | | |

| | | | | | | | | | | | |
|---|---|---|---|---|---|---|---|---|---|---|---|
| Hard-service cord | SO | 18–2 | 2 or more | Thermoset | 18–16 | 30 | | Oil-resistant thermoset | Pendant or portable | Damp locations | Extra-hard usage |
| | SOO | | | Oil-resistant thermoset | 14–10 | 45 | | Oil-resistant thermoset | | | |
| | | | | | 8–2 | 60 | | | | | |
| All-thermoset parallel cord | SP-1: See Note 7. | 18 | 2 or 3 | Thermoset | 18 | 30 | None | Thermoset | Pendant or portable | Damp locations | Not hard usage |
| | SP-2: See Note 7. | 18–16 | | | 18–16 | 45 | | | | | |
| | SP-3: See Note 7. | 18–10 | | Thermoset | 18–16 | 60 | None | Thermoset | Refrigerators, room air conditioners, and as permitted in Sec. 422-8(d) | Damp locations | Not hard usage |
| | | | | | 14 | 80 | | | | | |
| | | | | | 12 | 95 | | | | | |
| | | | | | 10 | 110 | | | | | |
| All-elastomer (thermoplastic) parallel cord | SPE-1: See Note 7. | 18 | 2 or 3 | Thermoplastic elastomer | 18 | 30 | None | Thermoplastic elastomer | Pendant or portable | Damp locations | Not hard usage |
| | SPE-2: See Note 7. | 18–16 | | | 18–16 | 45 | | | | | |
| | SPE-3: See Note 7. | 18–10 | | Thermoplastic elastomer | 18–16 | 60 | None | Thermoplastic elastomer | Refrigerators, room air conditioners, and as permitted in Sec. 422-8(d) | Damp locations | Not hard usage |
| | | | | | 14 | 80 | | | | | |
| | | | | | 12 | 95 | | | | | |
| | | | | | 10 | 110 | | | | | |

## 130. Flexible Cords and Cables[a] (Continued)

| Trade name | Type letter | Size, AWG | No. of conductors | Insulation | Nominal insulation thickness | | Braid on each conductor | Outer covering | Use | | |
|---|---|---|---|---|---|---|---|---|---|---|---|
| | | | | | AWG | Mils | | | | | |
| All-plastic parallel cord | SPT-1: See Note 7. | 18 | 2 or 3 | Thermoplastic | 18 | 30 | None | Thermoplastic | Pendant or portable | Damp locations | Not hard usage |
| | SPT-2: See Note 7. | 18–16 | | | 18–16 | 45 | | | | | |
| | SPT-3: See Note 7. | 18–10 | | Thermoplastic | 18–16 14 12 10 | 60 80 95 110 | None | Thermoplastic | Refrigerators, room air conditioners, and as permitted in Sec. 422-8(d) | Damp locations | Not hard usage |
| Range or dryer cable | SRD | 10–4 | 3 or 4 | Thermoset | 10–4 | 45 | None | Thermoset | Portable | Damp locations | Ranges or dryers |
| | SRDE | 10–4 | 3 or 4 | Thermoplastic elastomer | | | None | Thermoplastic elastomer | Portable | Damp locations | Ranges or dryers |
| | SRDT | 10–4 | 3 or 4 | Thermoplastic | | | None | Thermoplastic | Portable | Damp locations | Ranges or dryers |

| Type | Trade name | AWG | No. of conductors | Insulation | AWG | | Braid | Outer covering | Use | | |
|---|---|---|---|---|---|---|---|---|---|---|---|
| Hard-service cord | ST: See Note 5.<br>STO: See Note 5.<br>STOO: See Note 5. | 18–2 | 2 or more | Thermoplastic or thermoset<br>Oil-resistant thermoplastic or thermoset | 18–16<br>14–10<br>8–2 | 30<br>45<br>60 | None | Thermoplastic<br>Oil-resistant thermoplastic | Pendant or portable | Damp locations | Extra-hard usage |
| Vacuum-cleaner cord | SV: See Note 7. | 18–17 | 2 or 3 | Thermoset | 18–17 | 15 | None | Thermoset | Pendant or portable | Damp locations | Not hard usage |
| | SVE: See Note 7. | | | Thermoplastic elastomer | | | | Thermoplastic elastomer | | | |
| | SVEO: See Note 7. | | | | | | | Oil-resistant thermoplastic elastomer | | | |
| | SVO | | | Thermoset | | | | Oil-resistant thermoset | | | |
| | SVOO | | | Oil-resistant thermoset | | | | Oil-resistant thermoset | | | |
| | SVT: See Note 7. | 18–17 | 2 or 3 | Thermoset or thermoplastic | 18–17 | 15 | None | Thermplastic | Pendant or portable | Damp location | Not hard usage |
| | SVTO: See Note 7. | | | Thermoset or thermoplastic | | | | Oil-resistant thermoplastic | | | |

## 130. Flexible Cords and Cables<sup>a</sup> (Continued)

| Trade name | Type letter | Size, AWG | No. of conductors | Insulation | Nominal insulation thickness — AWG | Nominal insulation thickness — Mils | Braid on each conductor | Outer covering | Use | | |
|---|---|---|---|---|---|---|---|---|---|---|---|
| | SVTOO | | | Oil-resistant thermoplastic or thermoset | | | | Oil-resistant thermoplastic | | | |
| Parallel tinsel cord | TPT: See Note 3. | 27 | 2 | Thermoplastic | | | None | Thermoplastic | Attached to an appliance | Damp locations | Not hard usage |
| Jacketed tinsel cord | TS: See Note 3. | 27 | 2 | Thermoset | 27 | 15 | None | Thermoset | Attached to an appliance | Damp locations | Not hard usage |
| | TST: See Note 3. | 27 | 2 | Thermoplastic | | | None | Thermoplastic | Attached to an appliance | Damp locations | Not hard usage |
| Portable power cable | W | 8–500 kcmil | 1–6 | Thermoset | 8–2 / 1–4/0 / 250–500 kcmil | 60 / 80 / 95 | | Oil-resistant thermoset | Portable, extra-hard usage, and as permitted in Secs. 520-68 and 530-12 of the NEC | | |

NOTES

1. Except for Types SP-1, SPE-1, SP-2, SPE-2, AFPO, SP-3, SPE-3, SPT-1, SPT-2, SPT-3, HPN, TPT, and 3-conductor parallel versions of SRD, SRDE, SRDT, individual conductors are twisted together.

2. Cables constructed differently than specified herein and listed as component parts of a data processing system shall be permitted.

3. Types TP, TPT, TS, and TST shall be permitted in lengths not exceeding 8 ft when attached directly or by means of a special type of plug to a portable appliance rated at 50 W or less and of such nature that extreme flexibility of the cord is essential.

4. Rubber-filled or varnished-cambric tapes shall be permitted as a substitute for the inner braids.

5. Types S, SO, SE, SEO, SOO, ST, STO, STOO, G, and W shall be permitted for use on theater stages, in garages, and elsewhere where flexible cords are permitted by the NEC.

6. Elevator traveling cables for operating control and signal circuits shall contain nonmetallic fillers as necessary to maintain concentricity. Cables shall have steel supporting members as required for suspension in accordance with Sec. 620-41 of the NEC. In locations subject to excessive moisture or corrosive vapors or gases, supporting members of other materials shall be permitted. Where steel supporting members are used, they shall run straight through the center of the cable assembly and shall not be cabled with the copper strands of any conductor.

In addition to conductors used for control and signaling circuits, Types E, EO, ET, ETLB, and ETT elevator cables shall be permitted to incorporate in the construction one or more No. 20 AWG telephone conductor pairs and/or one or more coaxial cables. The No. 20 AWG conductor pairs may be covered with suitable shielding for telephone, audio, or higher-frequency communication circuits; the coaxial cables consist of a center conductor insulation and shield for use in video or other radio-frequency communication circuits. The insulation of the conductors shall be rubber or thermoplastic of thickness not less than specified for the other conductors of the particular type of cable. Metallic shields shall have their own protective covering. Where used, these components shall be permitted to be incorporated in any layer of the cable assembly but shall not run straight through the center.

7. The third conductor in these cables shall be for grounding purposes only.

8. The individual conductors of all cords except those of heat-resistant cords (Types AFC, AFPD, AFS, and AFSJ) shall have a thermoset or thermoplastic insulation, except that the grounding conductor, where used, shall be in accordance with Sec. 400-23(b) of the NEC. Unvulcanized rubber compounds shall be permitted to be used for all sizes of heater cord Types HSJ and HSJO, and for sizes 18 and 16 AWG Type HPD.

9. Where the voltage between any two conductors exceeds 300 but does not exceed 600, flexible cord of size No. 10 and smaller shall have thermoset or thermoplastic insulation on the individual conductors at least 45 mils in thickness unless Type S, SO, ST, or STO cord is used.

10. Insulation and outer coverings that meet the requirements of flame-retardant, limited smoke, and are so listed shall be permitted to be designated limited smoke with the suffix /LS after the Code-type designation.

### 131.  Types and Symbols of Magnet Wires
(General Cable Co.)

| Type | Symbol | Conductor covering or coating | | |
|---|---|---|---|---|
| | | First | Second | Third |
| **Coatings:** | | | | |
| Single and heavy enamel........ | E, E2 | Enamel | | |
| Single and heavy Formvar...... | R, R2 | Formvar | | |
| Triple Formvar................ | R3 | Formvar | | |
| Quadruple Formvar............ | R4 | Formvar | | |
| Single and heavy Formeze (Type O and Type I or A).......... | RB, RB-2 | Formvar | Bonding material | |
| Triple Formeze (Type II or B).. | RB-3 | Formvar | Bonding material | |
| Quadruple Formeze (Type III or C)....................... | RB-4 | Formvar | Bonding material | |
| Single and heavy Formlon....... | RY, RY-2 | Formvar | Nylon enamel | |
| Single and heavy nylon enamel.. | Y, Y2 | Nylon enamel | | |
| Single and heavy enamel G...... | U, U2 | Enamel G | | |
| Triple enamel G.............. | U3 | Enamel G | | |
| Single and heavy Class B enamel. | K, K2 | Class B enamel | | |
| Single, heavy, and triple Lecton.. | L, L2, L3 | Lecton | | |
| **Coverings:** | | | | |
| Single and double cotton........ | C, C2 | Cotton | Cotton (double only) | |
| Single and double glass bonded.. | GB, G2B or GHB, G2HB | Glass (varnish) | Glass (varnish) (double only) | |
| Single and double dacron-glass... | DG, DG2 | Dacron-glass | Dacron-glass (double only) | |
| Double nylon yarn............ | N2 | Nylon yarn | Nylon yarn | |
| Double silk.................. | S2 | Silk | Silk | |
| Single and double paper........ | P, P2 | Paper | Paper (double only) | |
| Single and double Quinterra-Mylar | QM, QM-2 or QMQ, QMQ-2 | Quinterra-Mylar | Quinterra-Mylar (double only) | |
| **Combination insulations:** | | | | |
| Single and heavy enamel single cotton.................... | EC, E2C | Enamel | Cotton | |
| Single and heavy enamel double cotton.................... | EC2, E2C2 | Enamel | Cotton | Cotton |
| Single and heavy enamel single glass bonded................ | EGB, E2GB | Enamel | Glass (varnish) | |
| Single and heavy enamel double glass bonded | EG2B, E2G2B | Enamel | Glass | Glass (varnish) |
| Single and heavy enamel single paper bonded.............. | EPB, E2PB | Enamel | Paper | |
| Single and heavy enamel double paper...................... | EP2, E2P2 | Enamel | Paper | Paper |
| Enamel nylon yarn............ | EN | Enamel | Nylon yarn | |
| Enamel silk................. | ES | Enamel | Silk | |
| Single and heavy Formvar single cotton.................... | RC, R2C | Formvar | Cotton | |
| Single and heavy Formvar double cotton.................... | RC2, R2C2 | Formvar | Cotton | Cotton |
| Single and heavy Formvar single glass bonded................ | RGB, R2GB | Formvar | Glass (varnish) | |
| Single and heavy Formvar double glass bonded | RG2B, R2G2B | Formvar | Glass | Glass (varnish) |
| Single and heavy nylon enamel single cotton................ | YC, Y2C | Nylon enamel | Cotton | |
| Single and heavy nylon enamel double cotton............... | YC2, Y2C2 | Nylon enamel | Cotton | Cotton |
| Single paper single cotton....... | PC | Paper | Cotton | |
| Double paper single cotton...... | P2C | Paper | Paper | Cotton |

## 132. Data for Round Copper Magnet Wire
(General Cable Co.)

Insulated conductors, max over-all diam, in.

| Size, AWG | Area, cir mils | Single enamel (E) | Heavy enamel (E2) | Single — Formvar (R) Formeze (RB) Formlon (RY) Nylon (Y) Enamel G (U)[a] Class B Enamel (K)[b] Lecton (L)[c] | Heavy — Formvar (R2) Formeze (RB-2) Formlon (RY-2) Nylon (Y2) Enamel G (U2)[a] Class B Enamel (K2)[b] Lecton (L2)[c] | Enamel single cotton (EC) | Enamel bonded paper (EPB) | Enamel single silk (ES) Enamel single nylon (EN) | Enamel single glass (EGB) | Single — Glass (GB) (GHB) Dacron-glass (Dg) | Double — Glass (G2B) (G2HB) Dacron-glass (Dg 2) | Double cotton (C2) |
|---|---|---|---|---|---|---|---|---|---|---|---|---|
| 4/0 | 211,600 | ..... | ..... | ..... | ..... | ..... | ..... | ..... | ..... | ..... | ..... | 0.4806 |
| 3/0 | 167,800 | ..... | ..... | ..... | ..... | ..... | ..... | ..... | ..... | 0.4207 | 0.4247 | 0.4297 |
| 2/0 | 133,100 | ..... | ..... | ..... | ..... | ..... | ..... | ..... | ..... | 0.3754 | 0.3794 | 0.3844 |
| 1/0 | 105,600 | ..... | ..... | ..... | ..... | ..... | ..... | ..... | ..... | 0.3351 | 0.3391 | 0.3441 |
| 1 | 83,690 | ..... | ..... | ..... | ..... | ..... | ..... | ..... | ..... | 0.2992 | 0.3032 | 0.3082 |
| 2 | 66,360 | ..... | ..... | ..... | ..... | ..... | ..... | ..... | ..... | 0.2672 | 0.2712 | 0.2762 |
| 3 | 52,620 | ..... | ..... | ..... | ..... | ..... | ..... | ..... | ..... | 0.2387 | 0.2427 | 0.2477 |
| 4 | 41,740 | ..... | ..... | 0.2092 | 0.2114 | ..... | ..... | ..... | ..... | 0.2133 | 0.2173 | 0.2223 |
| 5 | 33,090 | ..... | ..... | 0.1866 | 0.1886 | ..... | ..... | ..... | ..... | 0.1907 | 0.1947 | 0.1997 |
| 6 | 26,240 | ..... | ..... | 0.1663 | 0.1682 | ..... | ..... | ..... | ..... | 0.1706 | 0.1746 | 0.1776 |
| 7 | 20,820 | ..... | ..... | 0.1484 | 0.1501 | ..... | ..... | ..... | ..... | 0.1527 | 0.1567 | 0.1597 |
| 8 | 16,510 | 0.1324 | 0.1342 | 0.1324 | 0.1342 | 0.1404 | 0.1374 | ..... | 0.1394 | 0.1368 | 0.1408 | 0.1438 |
| 9 | 13,090 | 0.1181 | 0.1198 | 0.1181 | 0.1198 | 0.1251 | 0.1231 | ..... | 0.1251 | 0.1225 | 0.1265 | 0.1275 |
| 10 | 10,380 | 0.1054 | 0.1071 | 0.1054 | 0.1071 | 0.1114 | 0.1104 | ..... | 0.1114 | 0.1089 | 0.1119 | 0.1139 |
| 11 | 8,230 | 0.0941 | 0.0957 | 0.0941 | 0.0957 | 0.0996 | 0.0981 | ..... | 0.1001 | 0.0976 | 0.1006 | 0.1011 |
| 12 | 6,530 | 0.0840 | 0.0855 | 0.0840 | 0.0855 | 0.0895 | 0.0880 | ..... | 0.0900 | 0.0876 | 0.0906 | 0.0911 |
| 13 | 5,180 | 0.0750 | 0.0765 | 0.0750 | 0.0765 | 0.0805 | 0.0790 | ..... | 0.0810 | 0.0787 | 0.0817 | 0.0822 |
| 14 | 4,110 | 0.0670 | 0.0684 | 0.0670 | 0.0684 | 0.0725 | 0.0710 | ..... | 0.0730 | 0.0707 | 0.0737 | 0.0742 |
| 15 | 3,260 | 0.0599 | 0.0613 | 0.0599 | 0.0613 | 0.0654 | 0.0639 | 0.0619 | 0.0659 | 0.0637 | 0.0667 | 0.0672 |
| 16 | 2,580 | 0.0534 | 0.0548 | 0.0534 | 0.0548 | 0.0589 | 0.0574 | 0.0554 | 0.0594 | 0.0573 | 0.0603 | 0.0608 |
| 17 | 2,050 | 0.0478 | 0.0492 | 0.0478 | 0.0492 | 0.0533 | 0.0518 | 0.0498 | 0.0538 | 0.0518 | 0.0548 | 0.0553 |
| 18 | 1,620 | 0.0426 | 0.0440 | 0.0426 | 0.0440 | 0.0481 | 0.0466 | 0.0446 | 0.0486 | 0.0467 | 0.0497 | 0.0502 |
| 19 | 1,290 | 0.0382 | 0.0395 | 0.0382 | 0.0395 | 0.0437 | 0.0422 | 0.0402 | 0.0442 | 0.0423 | 0.0453 | 0.0458 |
| 20 | 1,020 | 0.0341 | 0.0353 | 0.0341 | 0.0353 | 0.0396 | 0.0381 | 0.0361 | 0.0401 | 0.0383 | 0.0413 | 0.0418 |

**132. Data for Round Copper Magnet Wire** (*Continued*)

| Size, AWG | Area, cir mils | Single enamel (E) | Heavy enamel (E2) | Single: Formvar (R) Formeze (RB) Formlon (RY) Nylon (Y) Enamel G(U)[a] Class B Enamel (K)[b] Leton (L)[c] | Heavy: Formvar (R2) Formeze (RB-2) Formlon (RY-2) Nylon (Y2) Enamel G(U2)[a] Class B Enamel (K2)[b] Leton (L2)[c] | Enamel single cotton (EC) | Enamel bonded paper (EPB) | Enamel single silk (ES) Enamel single nylon (EN) | Enamel single glass (EGB) | Single: Glass (GB) (GHB) Dacron-glass (Dg) | Double: Glass (G2B) (G2HB) Dacron-glass (Dg 2) | Double cotton (C2) |
|---|---|---|---|---|---|---|---|---|---|---|---|---|
| 21 | 812 | 0.0306 | 0.0317 | 0.0306 | 0.0317 | 0.0361 | 0.0346 | 0.0326 | 0.0366 | 0.0348 | 0.0378 | 0.0383 |
| 22 | 640 | 0.0273 | 0.0284 | 0.0273 | 0.0284 | 0.0323 | 0.0313 | 0.0293 | 0.0333 | 0.0316 | 0.0346 | 0.0346 |
| 23 | 511 | 0.0244 | 0.0255 | 0.0244 | 0.0255 | 0.0294 | 0.0284 | 0.0264 | 0.0304 | 0.0288 | 0.0318 | 0.0318 |
| 24 | 404 | 0.0218 | 0.0229 | 0.0218 | 0.0229 | 0.0268 | 0.0258 | 0.0238 | 0.0258 | 0.0243 | 0.0263 | 0.0293 |
| 25 | 320 | 0.0195 | 0.0206 | 0.0195 | 0.0206 | 0.0240 | 0.0225 | 0.0215 | 0.0235 | 0.0221 | 0.0241 | 0.0266 |
| 26 | 253 | 0.0174 | 0.0185 | 0.0174 | 0.0185 | 0.0219 | 0.0204 | 0.0194 | 0.0214 | 0.0201 | 0.0221 | 0.0246 |
| 27 | 202 | 0.0156 | 0.0165 | 0.0156 | 0.0165 | 0.0201 | 0.0186 | 0.0176 | 0.0196 | 0.0183 | 0.0203 | 0.0228 |
| 28 | 159 | 0.0139 | 0.0148 | 0.0139 | 0.0148 | 0.0184 | 0.0169 | 0.0159 | 0.0179 | 0.0167 | 0.0187 | 0.0212 |
| 29 | 128 | 0.0126 | 0.0134 | 0.0126 | 0.0134 | 0.0171 | 0.0156 | 0.0146 | ..... | ..... | ..... | 0.0199 |
| 30 | 100 | 0.0112 | 0.0120 | 0.0112 | 0.0120 | 0.0157 | 0.0142 | 0.0132 | ..... | ..... | ..... | 0.0186 |
| 31 | 79.2 | 0.0099 | 0.0107 | 0.0100 | 0.0108 | 0.0144 | 0.0129 | 0.0119 | ..... | ..... | ..... | 0.0175 |
| 32 | 64.0 | 0.0090 | 0.0097 | 0.0091 | 0.0098 | 0.0135 | ..... | 0.0110 | ..... | ..... | ..... | 0.0166 |
| 33 | 50.4 | 0.0080 | 0.0087 | 0.0081 | 0.0088 | 0.0125 | ..... | 0.0100 | ..... | ..... | ..... | 0.0157 |
| 34 | 39.7 | 0.0071 | 0.0077 | 0.0072 | 0.0078 | 0.0116 | ..... | 0.0091 | ..... | ..... | ..... | 0.0149 |
| 35 | 31.4 | 0.0063 | 0.0069 | 0.0064 | 0.0070 | 0.0108 | ..... | 0.0083 | ..... | ..... | ..... | 0.0142 |
| 36 | 25.0 | 0.0057 | 0.0062 | 0.0058 | 0.0063 | 0.0100 | ..... | 0.0077 | ..... | ..... | ..... | 0.0131 |
| 37 | 20.2 | 0.0051 | 0.0056 | 0.0052 | 0.0057 | 0.0094 | ..... | 0.0071 | ..... | ..... | ..... | 0.0126 |
| 38 | 16.0 | 0.0046 | 0.0050 | 0.0047 | 0.0051 | 0.0089 | ..... | 0.0066 | ..... | ..... | ..... | 0.0121 |
| 39 | 12.2 | 0.0040 | 0.0044 | 0.0041 | 0.0045 | 0.0083 | ..... | 0.0060 | ..... | ..... | ..... | 0.0116 |
| 40 | 9.61 | 0.0036 | 0.0039 | 0.0037 | 0.0040 | 0.0079 | ..... | 0.0056 | ..... | ..... | ..... | 0.0112 |
| 41 | 7.84 | 0.0032 | 0.0035 | 0.0033 | 0.0036 | | | | | | | |
| 42 | 6.25 | 0.0029 | 0.0031 | 0.0030 | 0.0032 | | | | | | | |
| 43 | 4.84 | 0.0025 | 0.0028 | 0.0026 | 0.0029 | | | | | | | |
| 44 | 4.00 | 0.0023 | 0.0026 | 0.0024 | 0.0027 | | | | | | | |

The above data are approximate and subject to normal manufacturing tolerances. [a] Single and heavy enamel G (symbols U and U2) available in sizes 10 to 44 AWG inclusive. [b] Single and heavy Class B enamel (symbols K and K2) available in sizes 10 to 44 AWG inclusive. [c] Single and heavy Lecton (symbols L and L2) available in sizes 14 to 24 AWG inclusive.

## 133. Properties of Metals and Alloys for Resistance Wires
(Driver-Harris Co.)

| Material | Specific resistance at 20°C (68°F) | | Temperature coefficient of resistance | | Coefficient of linear expansion | | Specific heat, g.-cal | Thermal conductivity, watts per cm °C | Approx melting point, °C | Tensile strength at 20°C per sq in. | | Specific gravity | Weight, lb per cu in. |
|---|---|---|---|---|---|---|---|---|---|---|---|---|---|
| | Microhm per cu cm | Ohms per cir mil-ft | Temp coeff | Diff in temp, °C | Coeff of exp | Diff in temp, °C | | | | Max | Min | | |
| **Driver-Harris alloys:** | | | | | | | | | | | | | |
| Ohmax | 167 | 1000 | −0.00035 | 20–500 | 0.0000156 | 20–1000 | | | 1,500 | 200,000 | 125,000 | 6.80 | 0.246 |
| Radiohm | 133 | 800 | 0.0007 | 20–500 | 0.0000155 | 20–1000 | | | 1,350 | 175,000 | 90,000 | 7.30 | 0.263 |
| Nichrome | 112 | 675 | 0.00017 | 20–100 | 0.000017 | 20–1000 | 0.107 | 0.136 | 1,400 | 175,000 | 95,000 | 8.247 | 0.2979 |
| Nichrome V | 108 | 650 | 0.00013 | 20–100 | 0.000017 | 20–1000 | 0.104 | 0.149 | 1,400 | 200,000 | 100,000 | 8.412 | 0.3039 |
| 525 Alloy | 100 | 600 | 0.00034 | 20–500 | 0.0000151 | 20–500 | 0.110 | 0.130 | 1,380 | 150,000 | 70,000 | 7.99 | 0.288 |
| Nirex | 98.1 | 590 | 0.000125 | 20–500 | 0.0000161 | 20–750 | 0.109 | 0.136 | 1,388 | 175,000 | 80,000 | 8.55 | 0.3089 |
| Comet | 95 | 570 | 0.00088 | 20–500 | 0.000015 | 20–500 | 0.114 | 0.135 | 1,480 | 160,000 | 75,000 | 8.15 | 0.294 |
| Nilvar | 80.5 | 484 | | | 0.000001 | 20–100 | 0.123 | 0.110 | 1,425 | 150,000 | 70,000 | 8.08 | 0.292 |
| D-H-Nirosta | 73 | 438 | 0.00094 | 20–500 | 0.000020 | 20–1000 | 0.117 | 0.200 | 1,399 | 300,000 | 100,000 | 7.93 | 0.286 |
| 42 Alloy | 66.5 | 400 | 0.0012 | 20–500 | 0.0000053 | 20–400 | | | 1,425 | 150,000 | 70,000 | 8.12 | 0.293 |
| 52 Alloy | 43.2 | 200 | 0.0029 | 20–500 | 0.0000095 | 20–450 | | | 1,425 | 150,000 | 70,000 | 8.247 | 0.2979 |
| Advance | 49 | 294 | 0.0002 | 20–100 | 0.0000149 | 20–100 | 0.094 | 0.218 | 1,210 | 135,000 | 60,000 | 8.9 | 0.321 |
| Manganin | 48.2 | 290 | ±±0.000015 | 15–35 | 0.0000187 | 15–35 | | | 1,020 | 90,000 | 40,000 | 8.192 | 0.296 |
| Lucero | 48.5 | 290 | 0.0010 | 20–250 | 0.0000125 | 20–100 | 0.127 | 0.250 | 1,350 | 150,000 | 70,000 | 8.19 | 0.296 |
| Filmetal D | 41.5 | 250 | 0.00179 | 20–400 | 0.0000143 | 20–500 | | | 1,450 | 175,000 | 95,000 | 8.590 | 0.3103 |
| Midohm | 30 | 180 | 0.00018 | 20–100 | 0.0000175 | 20–500 | | | 1,100 | 100,000 | 50,000 | 8.9 | 0.321 |
| R-63 Alloy | 25 | 150 | 0.0027 | 20–250 | 0.0000152 | 20–500 | 0.126 | 0.385 | 1,425 | 175,000 | 70,000 | 8.72 | 0.315 |
| Hytemco | 20 | 120 | 0.0045 | 20–100 | 0.000015 | 20–1000 | 0.125 | 0.289 | 1,425 | 150,000 | 70,000 | 8.46 | 0.305 |
| Magno | 20 | 120 | 0.0036 | 20–100 | 0.0000143 | 20–500 | 0.127 | 0.272 | 1,435 | 135,000 | 60,000 | 8.750 | 0.316 |
| Manganese nickel | 14 | 85 | 0.0045 | 0–100 | 0.0000146 | 20–500 | 0.129 | 0.615 | 1,435 | 135,000 | 60,000 | 8.813 | 0.3184 |
| Pure nickel | 10 | 60 | 0.0050 | 20–100 | 0.000015 | 20–500 | 0.130 | | 1,450 | 135,000 | 60,000 | 8.9 | 0.321 |
| Lohm | 10 | 60 | 0.00071 | 0–100 | 0.000018 | 20–500 | | 1.29 | 1,100 | 100,000 | 50,000 | 8.9 | 0.321 |
| High brass | 8.3 | 50 | 0.0016 | 0–100 | | | | 1.55 | 905 | 125,000 | 55,000 | 8.53 | 0.308 |
| Low brass | 7.0 | 40 | 0.0071 | 0–100 | | | | | 960 | 85,000 | 43,000 | 8.6 | 0.310 |
| Com. bronze | 4.2 | 25 | 0.0020 | 0–100 | | | | 2.11 | 1,015 | 75,000 | 37,000 | 8.7 | 0.314 |
| **Pure metals:** | | | | | | | | | | | | | |
| Platinum | 10.610 | 63.80 | 0.00398 | 0– | 0.0000089 | 20° | 0.0275 | 0.695 | 1,755 | | | 21.45 | 0.7750 |
| Iron | 9.780 | 58.82 | 0.00726 | 0– | 0.0000117 | 20° | 0.109 | 0.619 | 1,535 | | | 7.86 | 0.2840 |
| Zinc | 5.916 | 35.58 | 0.00347 | 0– | 0.000033 | 20° | 0.0931 | 1.13 | 419.4 | | | 7.14 | 0.2579 |
| Molybdenum | 5.632 | 33.87 | 0.00479 | 0– | 0.000005 | 20° | 0.0647 | 1.46 | 2,620 | | | 10.2 | 0.3685 |
| Tungsten | 5.523 | 33.22 | 0.00524 | 0– | 0.000004 | 20° | 0.0336 | 1.60 | 3,370 | | | 19.3 | 0.6973 |
| Aluminum | 2.670 | 16.06 | 0.00446 | 0– | 0.000024 | 20° | 0.2089 | 2.03 | 660 | | | 2.7 | 0.0975 |
| Gold | 2.350 | 14.13 | 0.00365 | 0– | 0.0000142 | 20° | 0.0316 | 2.96 | 1,063 | | | 19.3 | 0.6973 |
| Copper | 1.724 | 10.37 | 0.00393 | 0– | 0.0000166 | 20° | 0.0951 | 3.88 | 1,083 | | | 8.92 | 0.3223 |
| Silver | 1.622 | 9.755 | 0.00361 | 0– | 0.0000189 | 20° | 0.0559 | 4.19 | 960 | | | 10.5 | 0.3793 |

**134. Electrical resistance wire** is wire that has high resistance to the flow of electric current. It is this higher-resistance characteristic which distinguishes resistance wire from conducting wire. The material used for conducting wires should have as low a resistance as possible. Electrical resistance wire is used for the wiring of rheostats, resistors, heaters, furnaces, electric ranges, etc. Metal alloys are used for the manufacture of resistance wires. The more common alloys employed are nickel-chromium, nickel-copper, nickel-chromium-iron, nickel-iron, and manganese-nickel. In Table 133 are listed the properties of metals and alloys manufactured by the Driver-Harris Co. for resistance wire. Although the table gives the trade names for just one company, the data are fairly representative for the alloys of the other manufacturers.

## CABLE JOINTS AND TERMINAL CONNECTIONS

**135. Cable joints and connections** are an essential part of any electric circuit. It is of utmost importance that they be properly made, since any system is only as strong as its weakest link. The basic requirements of any joint or connection are that it be both mechanically and electrically as strong as the cable with which it is used. High-quality workmanship and materials must be employed so that permanently good electrical contact and insulation (if required) will be ensured. The more common satisfactory methods of making joints and connections in electric cables are discussed in the following sections.

**136. Joints and connections for insulated cables.**   There are two methods of making joints or connections for insulated cables: by means of soldered connections and by means of

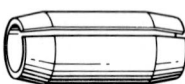

Fig. 2-6 Split copper cable connecting sleeve.

solderless connection devices (see Sec. 188). Soldered connections were formerly the accepted standard, but solderless splicing devices and connectors have gained wide favor for low-voltage work. The use of such devices materially reduces the time required for the making of splices and terminal lug connections. Moreover, if the device is of good design, in addition to the mechanical strength of the connection being fully as great as that of a good soldered connection, the solderless connection has the advantage that the electrical contact will not fail under short circuits or continuous overloads due to the melting of solder.

Soldered joints may be made by means of soldered splices or by a copper sleeve (see Fig. 2-6 and Sec. 145) thoroughly sweated to the cable.

**137. Insulation of joints.**   Every joint in insulated cable must be covered with insulation equivalent in insulating properties to that on the cable itself. Soldered joints in rubber-insulated cables are insulated with rubber tape. The rubber employed is a self-vulcanizing tape often referred to as splicing gum. Tapes made from different types of rubber compound are available. One type is made from normal-aging compound and may be used for cables insulated with **NEC**, intermediate, ASTM 30 percent Class AO, or performance-rubber compounds. A better grade of tape having better aging, heat-resisting, and moisture-resisting properties was developed for Performite-compound–insulated cables. It may be used for high-grade joints in any type of rubber-insulated cables. For insulating joints in cables insulated with an oil-base compound (corona- and ozone-resisting), a corona-resisting tape made from an oil-base compound should be used.

Soldered joints in thermoplastic-insulated wires and cables should be insulated with pressure-sensitive thermoplastic-adhesive tape.

Soldered joints in varnished-cambric– and paper-insulated cables are insulated with varnished-cambric tape.

Joints made with solderless connectors may be insulated as described above for soldered joints, or an insulating cover may be employed.

Refer to Secs. 146 and 147 for more detailed instructions for the application of insulating tapes.

**138. Protection of joints.**   Some form of protection should be supplied over the insulating tape of a joint. For all types of braided cables the joint is protected by applying two layers of friction tape over the insulated joint. Friction tape is made of closely woven cotton fabric treated on both sides with a rubber compound of adhesive character.

Joints in thermoplastic-jacketed cable are protected with pressure-sensitive thermoplastic tape to the same thickness as the cable jacket. Joints in rubber-insulated neoprene-jacketed cable are protected with neoprene tape applied to the same thickness as the cable jacket. These joints are further protected by a covering of anhydrous tape painted with cable paint.

**139. Completed Joints.** Cross-sectional views of typical completed joints are shown in Figs. 2-7, 2-8, 2-12 to 2-15, and 2-20 to 2-25.

**140. Removing protective covering.** For braided cable, strip back the protective covering a sufficient distance from the ends of the cables to be jointed. Trim the edges with a sharp knife, taking great care not to cut the insulation or to leave loose ends to get under taping. For lead-sheathed cable proceed as follows (Rome Cable Corp.).

The length of lead sheath to be removed from single-conductor cables is determined by the creepage distance required for the operating voltage. For multiconductor cable the

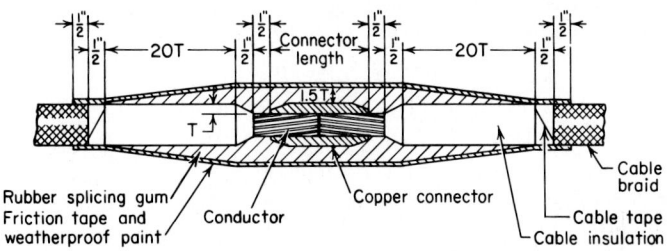

**Fig. 2-7**   Joint in rubber-insulated braided cable.

axial length required for offsetting the conductors is the determining factor. The length of sheath to be removed from each cable end is approximately $1\frac{1}{2}$ in (38.1 mm) less than one-half the total length of the lead sleeve indicated.

Cut halfway through the lead sheath all around at the location determined, and cut the sheath lengthwise to the end. Remove this section of sheath by grasping with pliers and tearing off. This will leave the remaining sheath ends slightly belled.

For belted cable, remove belt, insulation, and fillers to within 1 in (25.4 mm) of the end of the lead sheath.

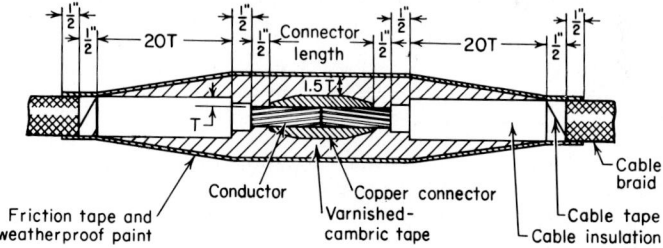

**Fig. 2-8**   Joint in varnished-cambric–insulated braided cable.

For single-conductor shielded cable, remove shielding to within approximately $\frac{1}{4}$ in (6.35 mm) of the end of the lead sheath. For multiconductor cable, remove binder tape and fillers to the end of the lead sheath; remove the shielding tape as far as possible into the crotch.

If the cable is neoprene-jacketed, remove the jacket, underlying tapes, and shielding, if any. Use care not to damage the factory insulation on the cable.

**141. Removing insulation from rubber- or thermoplastic-insulated cables.** Completely remove the insulation for a distance sufficient for making the joint. Cover the bared conductor with a few turns of friction tape so that the conductor will not be nicked. Pencil down the conductor insulation for a distance of $\frac{1}{2}$ in (12.7 mm; see Figs. 2-7 and 2-9). This penciling should be smooth and even and can be done best with a sharp knife, care being taken not to injure the conductor. All the exposed conductor insulation should then be made smooth with fine emery cloth to provide a good surface for the application of the splicing gum. Wrap the tapered ends of the insulation tightly with dry cotton tape, and tie the tape in place. The tape protects the insulation from scorching while the connector is being soldered to the conductors.

**142. Removing insulation from paper- or varnished-cambric–insulated cable** (General Electric Co.). The cable insulation should be removed in steps as indicated in Fig. 2-10.   For paper-insulated cable, the steps should be made by tearing the paper tapes. A convenient method is to hold a loop of fine steel wire (about 0.015 in, or 0.381 mm, in diameter) in place around the insulation at each step in turn, beginning with the step farthest from the end of the cable, and tear the tapes at the wire. A weight of about 1½ lb (0.68 kg) attached to each end of the wire will hold the loop in place while tearing the tape.

For varnished-cambric–insulated cable, the steps should be made by cutting the cloth

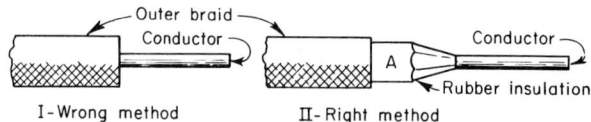

**Fig. 2-9**   Methods of "skinning" wire.

tapes, exercising care to avoid cutting through tapes that will remain on the conductor.

It is recommended that in each case the first step adjacent to the conductor be about ¹⁄₃₂ in (0.79 mm) in height by ½ in (12.7 mm) in length and that the remaining steps be of equal height and spaced about ¾ in (19.1 mm) apart. The number of steps in any case will be the same as that shown in the illustration, but the height of the steps will vary with the thickness of the insulation on the conductor.

The number of tapes to be removed for any given step must be determined by counting the total number of tape layers and measuring the thickness of each. Total insulation usually comprises either paper tapes varying in thickness from 0.004 to 0.008 in (1 to 2 mm) or varnished-cambric tapes about 0.012 in (3 mm) in thickness.

Cotton yarn should be bound into the stepped corners of the insulation and into the corners between insulation and conductor parts (Fig. 2-10). The following procedure

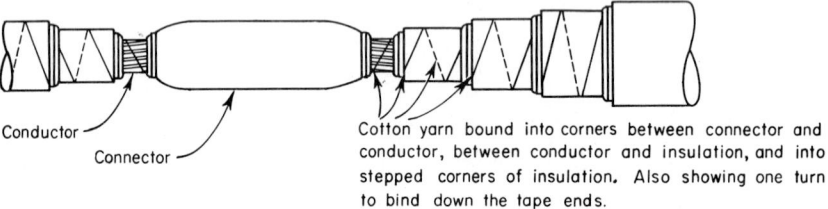

**Fig. 2-10**   Method of stepping and binding insulation.

should be observed. Wash the exposed surfaces of stepped insulation and all metal conductor parts with high-quality Transil oil at a temperature of about 110°C to remove all impurities. Dip the roll of cotton yarn in clean Transil oil at a temperature of 110°C to remove possible moisture. Then wind the yarn in all the stepped corners of the insulation, the corners between the conductor and the insulation, and the corners between the conductor and the connector. In passing from one step of insulation to the next, bind the tape ends down with a turn of yarn. Excessive yarn should be avoided; that is, one turn in the lower steps and two in the higher steps should be sufficient.

**143. Cleaning wire ends.**   Bare wire ends should be scraped bright with the back of a knife blade or rubbed clean with sandpaper or emery cloth to ensure that the solder will adhere readily.

**144.  Split Copper Sleeves for Standard Concentric Round Stranded Conductors**

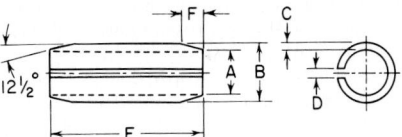

**Fig. 2-11**  Split copper sleeve. [Rome Cable Corp.]

(Dimensions in inches)

| Conductor size, AWG or kcmil | Inside diam A | Outside diam B | Wall thickness C | Slot width D | Over-all length E | Length bevel F |
|---|---|---|---|---|---|---|
| 8 | 0.151 | 0.201 | 0.025 | 0.030 | 1.5 | |
| 7 | 0.169 | 0.225 | 0.028 | 0.030 | 1.5 | |
| 6 | 0.189 | 0.251 | 0.031 | 0.030 | 1.5 | |
| 5 | 0.211 | 0.281 | 0.035 | 0.030 | 1.5 | |
| 4 | 0.237 | 0.315 | 0.039 | 0.030 | 2.0 | |
| 3 | 0.265 | 0.353 | 0.044 | 0.030 | 2.0 | |
| 2 | 0.297 | 0.395 | 0.049 | 0.030 | 2.0 | |
| 1 | 0.337 | 0.449 | 0.056 | 0.070 | 2.0 | 0.077 |
| 1/0 | 0.378 | 0.504 | 0.063 | 0.070 | 2.0 | 0.109 |
| 2/0 | 0.423 | 0.565 | 0.071 | 0.070 | 2.0 | 0.145 |
| 3/0 | 0.475 | 0.635 | 0.080 | 0.070 | 2.0 | 0.185 |
| 4/0 | 0.533 | 0.713 | 0.090 | 0.070 | 2.5 | 0.231 |
| 250 | 0.581 | 0.778 | 0.098 | 0.120 | 2.5 | 0.267 |
| 300 | 0.635 | 0.849 | 0.107 | 0.120 | 2.5 | 0.307 |
| 350 | 0.690 | 0.920 | 0.115 | 0.120 | 2.5 | 0.343 |
| 400 | 0.740 | 0.986 | 0.123 | 0.120 | 3.0 | 0.379 |
| 450 | 0.784 | 1.046 | 0.131 | 0.120 | 3.0 | 0.415 |
| 500 | 0.826 | 1.102 | 0.138 | 0.120 | 3.0 | 0.447 |
| 550 | 0.868 | 1.154 | 0.143 | 0.175 | 3.0 | 0.470 |
| 600 | 0.906 | 1.206 | 0.150 | 0.175 | 3.5 | 0.501 |
| 650 | 0.948 | 1.260 | 0.156 | 0.175 | 3.5 | 0.528 |
| 700 | 0.983 | 1.307 | 0.162 | 0.175 | 3.5 | 0.556 |
| 750 | 1.018 | 1.356 | 0.169 | 0.175 | 3.5 | 0.587 |
| 800 | 1.052 | 1.400 | 0.174 | 0.175 | 4.0 | 0.610 |
| 850 | 1.083 | 1.441 | 0.179 | 0.220 | 4.0 | 0.632 |
| 900 | 1.115 | 1.483 | 0.184 | 0.220 | 4.0 | 0.655 |
| 950 | 1.145 | 1.525 | 0.190 | 0.220 | 4.0 | 0.682 |
| 1000 | 1.175 | 1.565 | 0.195 | 0.220 | 4.5 | 0.704 |
| 1250 | 1.320 | 1.754 | 0.217 | 0.220 | 4.5 | 0.804 |
| 1500 | 1.440 | 1.912 | 0.236 | 0.280 | 5.0 | 0.889 |
| 1750 | 1.560 | 2.074 | 0.257 | 0.280 | 5.5 | 0.984 |
| 2000 | 1.664 | 2.214 | 0.275 | 0.280 | 6.0 | 1.065 |
| 2500 | 1.855 | 2.455 | 0.300 | 0.280 | 6.5 | 1.178 |
| 3000 | 2.033 | 2.683 | 0.325 | 0.300 | 7.5 | 1.178 |

**145. A soldering flux** removes or prevents the formation of an oxide during soldering so that the solder will flow readily and unite firmly the members to be joined. For copper wires the following solution of zinc chloride, which is recommended by Underwriters Laboratories, is good: saturated solution of zinc chloride, 5 parts; alcohol, 4 parts; glycerin, 1 part. Solutions made with acids should be avoided, as more or less corrosion usually occurs in joints made with them. Commercial soldering pastes and sticks give good satisfaction in cleaning joints to be soldered.

**146. Insulation of joints in rubber-insulated cable.** Remove the cotton tape that was wrapped around the tapered ends of the insulation. Clean the rubber insulation with a cloth dampened with high-test gasoline and allow it to dry. The surface over which the splicing tape is to be applied should then be covered with rubber cement, and the solvent allowed to evaporate until it is quite tacky. The rubber splicing tape should then be placed over the joint. Each layer should be applied smoothly and under tension so that there will be no air spaces between the layers. In putting on the first layer, start near the middle of the joint instead of at the end. The diameter of the completed insulated joint should be greater than the overall diameter of the original cable, including the insulation. When a standard split copper connector is used, the thickness of applied insulation over the maximum diameter of the connector should be at least 50 percent greater than the thickness of the insulation on the original cable as indicated in Fig. 2-7. Most splicing tapes are self-vulcanizing, and therefore it is not necessary to vulcanize the joint later.

**147. Insulation of joints in varnished-cambric– or paper-insulated cables.**     Varnished-cambric tape of various widths is used for insulating these joints. A ½-in (12.7-mm) tape is used in the spaces between the cable insulation and the connector end; ¾-in (19.1-mm) tape is used to continue the insulating to the original level of the cable insulation. A 1-in (25.4-mm) tape is used to complete the insulation of the joint or, in the case of joints for three-conductor belted-type cables, the overall of the crotch; 1-in tape is also used either for spacing reinforced conductors or for binding together reinforced conductors in the case of joints for 5-kV belted-type cables.

Before applying the tape, flush all exposed surfaces of the insulation and intervening conductor parts with high-quality Transil oil at 110°C to remove all impurities. Also, during the application of the tape, all surfaces on which the tape is applied and each layer of tape during wrapping should be flushed with a heavy oil compound applied at a temperature of about 20°C with a brush that has been cleaned in high-quality Transil oil at a temperature of approximately 110°C. Tape of ½- or ¾-in (12.7- or 19.1-mm) width should be applied by drawing it tightly in half-lap wrappings; 1-in (25.4-mm) tape should be applied in butt wrappings.

The applied tape insulation should be built up until the thickness of insulation over the maximum diameter of the connector or splice is 50 percent greater than the thickness of the insulation on the original conductor (see Fig. 2-8).

**148. Applying lead sleeve to lead-sheathed cable** (Rome Cable Corp.).    The lead sleeve is slid in place, and the ends are beaten down with a wood tool to fit snugly the lead sheath of the cable. Scrape the wiping surfaces clean and apply stearin flux. Apply paper pasters to limit the length of the wipes. Make the wiped joints by pouring molten solder on the joint. As the solder cools and becomes plastic, work the solder by wiping with a cloth.

Cut and raise V notches in the lead sleeve for filling and venting. Fill the lead sleeve with insulating compound, heated to the recommended temperature and poured into the sleeve through a funnel. The joint is completed by binding the V-notch flaps back in place and sealing them with 50-50 solder.

**149. Cable joints.** The following instructions in Secs. 150 to 153 for the splicing of cables are reproduced here through the courtesy of the Rome Cable Corp.

**150. Straight splice for single-conductor unshielded cables** (Fig. 2-12)

A. RUBBER-INSULATED NEOPRENE-JACKETED CABLE

1. Form the two cables to be joined into their final position, allowing the ends to overlap. Mark the centerline of the joint on both cables, and cut off at this point so that the cables butt squarely together.

2. Strip the factory-applied jacket, insulation, and underlying tape, if any, from both cables for a distance equal to one-half the length of connector plus ¼ in (6.35 mm).

3. Apply compression or solder-type split tinned connector.

4. Pencil the jacketed insulation for a distance of 4 times the overall thickness of jacket and insulation. Apply special cement to the connector, exposed portions of the conductor, the pencils, and the adjacent jacket, allowing the cement to become tacky before proceeding with the joint.

5. Insulate with ozone-resistant rubber splicing tape applied one-half lap to a thickness over the connector of 1½ times the factory-applied insulation and tapering off over the jacket to a point of ½ $L$ from the center.

6. Cover the hand-applied insulation and adjacent jacket for a distance of 1½ in (38.1 mm) with special cement, allowing it to become tacky. Apply neoprene tape over the entire joint to the same thickness as the factory-applied jacket and extending 1½ in beyond the end of the insulating tape.

7. Serve the entire joint with anhydrous tape, extending this serving 1½ in beyond the end of the hand-applied jacket tape.

8. Paint the entire joint with cable paint.

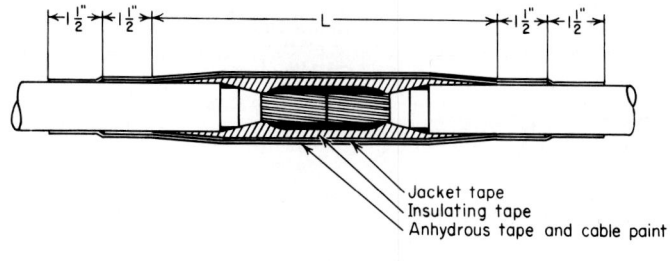

Length of $L$, in

| Conductor size, AWG or kcmil | 600 V | 3000 V | 5000 V |
|---|---|---|---|
| 8–4/0 | 6.5 | 7.5 | 8.5 |
| 250–500 | 7.5 | 8.5 | 9.5 |
| 600–1000 | 8.5 | 9.5 | 10.5 |

**Fig. 2-12**  Straight splice for single-conductor unshielded cable. [Rome Cable Corp.]

B. Rubber-Insulated Thermoplastic-Jacketed Cable

Follow the procedure given in A, substituting the following for step 6:

6. Apply pressure-sensitive thermoplastic tape over the entire joint to the same thickness as the factory-applied jacket and extending 1½ in (38.1 mm) beyond the end of the insulating tape.

C. Thermoplastic-Insulated Thermoplastic-Jacketed Cable

1–2. Follow steps 1 and 2 as in A.

3. Clean the conductors, and apply a compression-type connector with hydraulic press and dies.

4. Pencil the jacketed insulation for a distance of 6 times the overall thickness of jacket and insulation.

5. Insulate with polyethylene-base splicing tape applied one-half lap to a thickness over the connector of 2 times the factory-applied insulation and tapering off over the jacket to a point ½ $L$ from the center.

6. Apply pressure-sensitive thermoplastic tape over the entire joint to the same thickness as the factory-applied jacket and extending 1½ in (38.1 mm) beyond the jacket end.

7. Complete as in steps 7 and 8 of A.

**151. Straight splice for single-conductor shielded cables** (Fig. 2-13)

A. Rubber-Insulated Neoprene-Jacketed Cable

1. Form the two cables to be joined into their final position, allowing the ends to overlap. Mark the centerline of the joint on both cables, and cut off at this point so that the cables butt squarely together.

2. Remove the jacket and underlying tapes, if any, down to the factory-applied shielding from both cables for a distance equal to ½ $L$. Remove the factory-applied shielding and underlying tape, if any, to within ¼ in (6.35 mm) of the end of the jacket or for a distance of ½ $L$.

3. Strip the factory-applied insulation and underlying tape, if any, from both cables for a distance equal to one-half the length of connector plus ¼ in.

4. Apply compression or solder-type split tinned connector.

5. Pencil the factory-applied insulation for a distance equal to 4 times its thickness. Apply special cement to the connector, exposed portions of conductor, the pencils, and the adjacent insulation, allowing the cement to become tacky before proceeding with the joint.

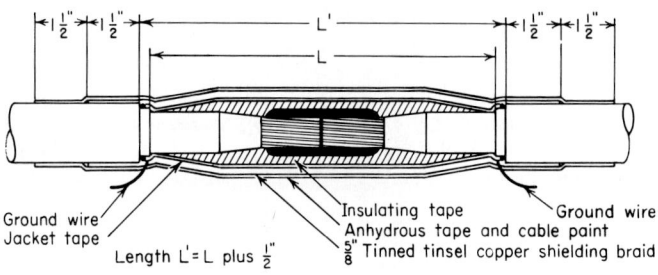

Length L'= L plus $\frac{1}{2}$"

Ground wire  
Jacket tape  
Insulating tape  
Anhydrous tape and cable paint  
$\frac{5}{8}$" Tinned tinsel copper shielding braid  
Ground wire

| Conductor size, AWG or kcmil | Length of L, in | | | | | |
|---|---|---|---|---|---|---|
| | 5000 V | 7500 V | | 15,000 V | | |
| | Grounded neutral or undergrounded neutral | Grounded neutral | Undergrounded neutral | Grounded neutral | Undergrounded neutral | |
| 8–4/0 | 8.5 | 10.0 | 11.0 | 13.5 | 15.5 | |
| 250–500 | 9.5 | 11.0 | 12.0 | 14.5 | 16.5 | |
| 600–1000 | 10.5 | 12.0 | 13.0 | 15.5 | 17.5 | |

**Fig. 2-13**   Straight splice for single-conductor shielded cable. [Rome Cable Corp.]

6. Insulate with ozone-resistant rubber splicing tape applied one-half lap to a thickness over the connector of 1½ times the factory-applied insulation on the cable and tapering off over the cable insulation up to the end of the factory-applied shielding.

7. Cover the hand-applied insulating tape with ⅝-in (15⅞-mm) tinned tinsel copper shielding braid, applied butt edge on the cylindrical portion and one-half lap on the tapered portions. Solder to the factory-applied shielding at both ends, and also apply a light line of solder along the tapered portions to prevent slippage. Use care to do this soldering quickly so as not to damage the shielding braid or insulation. Attach securely and solder the ground wires to the factory-applied shielding at each end of the joint.

8. Cover the shielding braid and the adjacent jacket for a distance of 1½ in (38.1 mm) with special cement, allowing it to become tacky. Apply neoprene tape over the entire joint to the same thickness as the factory-applied jacket and extending 1½ in beyond the jacket end.

9. Serve the entire joint with anhydrous tape, extending this serving 1½ in beyond the end of the hand-applied jacket tape.

10. Paint the entire joint with cable paint.

11. Connect the cable shielding tape ground wires to ground.

B. RUBBER-INSULATED THERMOPLASTIC-JACKETED CABLE  
Follow procedure given in A, substituting the following for step 8:

8. Apply pressure-sensitive thermoplastic tape over the entire joint to the same thickness as the factory-applied jacket and extending 1½ in (38.1 mm) beyond the jacket end.

C. THERMOPLASTIC-INSULATED THERMOPLASTIC-JACKETED CABLE  
1. Follow steps 1, 2, and 3 as in A.

4. Clean the conductors and apply a compression-type connector with hydraulic press and dies.

5. Pencil the factory-applied insulation for a distance equal to 6 times its thickness.

6. Insulate with polyethylene-base splicing tape applied one-half lap to a thickness over the connector of 2 times the factory-applied insulation and tapering off over the cable insulation up to the end of the factory-applied shielding.

7. Cover the hand-applied insulating tape with ⅝-in (15⅞-mm) tinned tinsel copper shielding braid applied as in step 7 of A.

8. Apply pressure-sensitive thermoplastic tape over the entire joint to the same thickness as the factory-applied jacket and extending 1½ in (38.1 mm) beyond the jacket end.

9. Complete as in steps 9, 10, and 11 of A.

**152. Straight splice for three-conductor unshielded cable** (Fig. 2-14)

A. Rubber-Insulated Neoprene-Jacketed Cable

1. Form the two cables to be joined into their final position, allowing the ends to overlap. Mark the centerline of the joint on both cables and cut off at this point so that the cables butt squarely together.

2. Remove the jacket and underlying tape, if any, from both cables for a distance equal to ½ L. Remove the fillers and cut at the ends of the jacket. Bind the conductors of each cable tightly together with a piece of dry cotton tape applied at the end of the jacket. The tape prevents breaking the jacket when spreading the conductors. Spread the conductors radically, and form them into their final position, taking care not to spread them any more than is necessary for insulating. Strip the factory-applied insulation and underlying tape, if any, on each conductor for a distance equal to one-half the length of the connector plus ¼ in (6.35 mm).

3. Apply compression or solder-type split tinned connector.

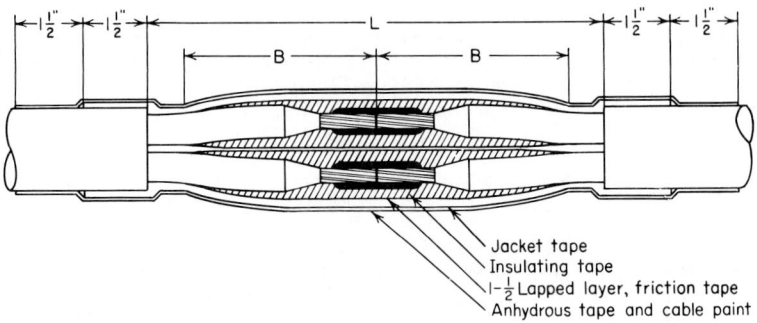

Jacket tape
Insulating tape
1-½ Lapped layer, friction tape
Anhydrous tape and cable paint

| Conductor Size AWG or kcmil | Length of $L$ and $B$, in. | | | | | |
| --- | --- | --- | --- | --- | --- | --- |
| | 600 volts | | 3,000 volts | | 5,000 volts | |
| | $L$ | $B$ | $L$ | $B$ | $L$ | $B$ |
| 8–1 | 8.0 | 3.25 | 9.0 | 3.50 | 10.0 | 4.00 |
| 1/0–4/0 | 9.5 | 3.75 | 10.5 | 4.00 | 11.5 | 4.50 |
| 250–500 | 12.0 | 4.75 | 13.0 | 5.00 | 14.0 | 5.50 |
| 600–1,000 | 15.0 | 6.00 | 16.0 | 6.25 | 17.0 | 6.75 |

**Fig. 2-14**   Straight splice for three-conductor unshielded cable. [Rome Cable Corp.]

4. Pencil the factory-applied insulation on each conductor for a distance equal to 4 times the thickness of the factory-applied insulation. Cover the connectors, the exposed portions of the conductors, the pencils, and the adjacent insulation with special cement, allowing it to become tacky before proceeding with the joint.

5. Insulate with ozone-resistant rubber splicing tape applied one-half lap to a thickness over the connector of 1½ times the factory-applied insulation on each conductor, tapering off over the insulated conductor to a point $B$ from the center.

6. Apply a one-half–lapped layer of friction tape over each insulated conductor, extending it as far as possible into the crotches. Squeeze the conductors together by hand, and bind them in place with a serving of friction tape.

7. Cover the entire joint and adjacent jacket for a distance of 1½ in (38.1 mm) with special cement, allowing it to become tacky. Apply neoprene tape over the entire joint to the same thickness as the factory-applied jacket and extending 1½ in beyond the jacket end.

8. Serve the entire joint with anhydrous tape, extending this serving 1½ in beyond the end of the hand-applied jacket tape.

9. Paint the entire joint with cable paint.

B. RUBBER-INSULATED THERMOPLASTIC-JACKETED CABLE

Follow the procedure given in A, substituting the following for step 7.

7. Apply pressure-sensitive thermoplastic tape over the entire joint to the same thickness as the factory-applied jacket and extending 1½ in (38.1 mm) beyond the jacket end.

C. THERMOPLASTIC-INSULATED THERMOPLASTIC-JACKETED CABLE

1–2. Follow steps 1 and 2 as in A.

3. Clean the conductors, and apply a compression-type connector with hydraulic press and dies.

4. Pencil the factory-applied insulation on each conductor for a distance equal to 6 times its thickness.

5. Insulate with polyethylene-base splicing tape applied one-half lap to a thickness over the connector of 2 times the factory-applied insulation over each conductor, tapering off over the insulated conductor to a point $B$ from the center.

6. Apply a one-half–lapped layer of friction tape over each insulated conductor, extending it as far as possible into the crotches. Squeeze the conductors together by hand, and bind them in place with a serving of friction tape.

7. Apply pressure-sensitive thermoplastic tape over the entire joint to the same thickness as the factory-applied jacket and extending 1½ in (38.1 mm) beyond its end.

8. Complete as in steps 8 and 9 of A.

**153. Straight splice for three-conductor shielded cable** (Fig. 2-15)

A. RUBBER-INSULATED NEOPRENE-JACKETED CABLE

1. Form the two cables to be joined into their final position, allowing the ends to overlap. Mark the centerline on both cables, and cut off at this point so that the cables butt squarely together.

2. Remove the jacket and underlying tape, if any, from both cables for a distance equal to ½ $L$. Remove the fillers and cut at the ends of the jacket. Bind the conductors of each cable tightly together with a piece of dry cotton tape applied at the end of the jacket. This is to prevent breaking the jacket when spreading the conductors. Spread the conductors radially, and form them into their final position, taking care not to spread them any more than is necessary for insulating. Secure the shielding on each conductor at a point about ¼ in (6.35 mm) greater than $B$ from the center of the joint, and strip the shielding and underlying tape, if any, from the point to the end. Strip the factory-applied insulation and underlying tape, if any, on each conductor for a distance equal to one-half the length of the connector plus ¼ in.

3. Apply compression or solder-type split tinned connector.

4. Pencil the factory-applied insulation on each conductor for a distance equal to 4 times the thickness of the factory-applied insulation. Cover the connectors, the exposed portion of the conductors, the pencils, and the adjacent insulation with special cement, allowing it to become tacky before proceeding with the joint.

5. Insulate with ozone-resistant rubber splicing tape applied one-half lap to a thickness over the connector of 1½ times the factory-applied insulation on each conductor, tapering off over the insulated conductor to a point $B$ from the center.

6. Cover the hand-applied insulating tape with ⅝-in (15⅞-mm) tinned tinsel copper shielding braid, applied butt edge on the cylindrical portion and one-half lap on the tapered portions. Solder to the factory-applied shielding at both ends, and also apply a light line of solder along the tapered portions to prevent slippage. Use care to do this soldering quickly so as not to damage the shielding braid or insulation. Attach securely

and solder ground wires to the factory-applied shielding on all three conductors at both ends of the joint. Squeeze the conductors together by hand, and bind them in place with a serving of friction tape.

7. Cover the entire joint and adjacent jacket for a distance of 1½ in (38.1 mm) with special cement, allowing it to become tacky. Apply neoprene tape over the entire joint to

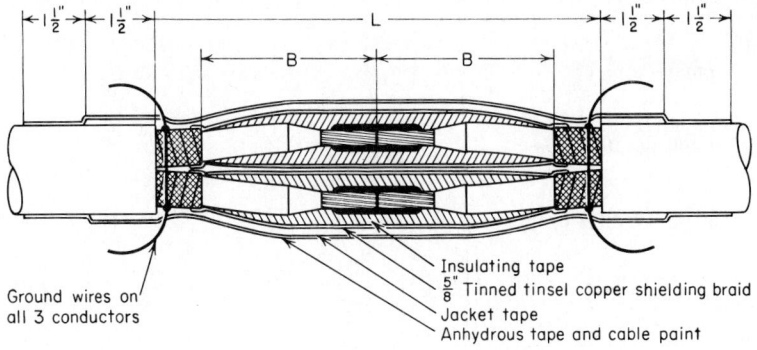

Insulating tape
⅝" Tinned tinsel copper shielding braid
Jacket tape
Anhydrous tape and cable paint

Ground wires on all 3 conductors

| | Length of $L$ and $B$, in | | | | | | | | | |
|---|---|---|---|---|---|---|---|---|---|---|
| Conductor size, AWG or kcmil | 5000 V | | 7500 V grounded neutral | | 7500 V undergrounded neutral | | 15,000 V grounded neutral | | 15,000 V undergrounded neutral | |
| | $L$ | $B$ | $L$ | $B$ | $L$ | $B$ | $L$ | $B$ | $L$ | $B$ |
| 8–1 | 10.5 | 4.25 | 11.5 | 5.00 | 12.5 | 5.50 | 15.0 | 6.75 | 17.5 | 7.75 |
| 1/0–4/0 | 12.0 | 4.50 | 13.0 | 5.25 | 14.0 | 5.75 | 16.0 | 7.00 | 18.5 | 8.00 |
| 250–500 | 14.0 | 4.75 | 15.0 | 5.50 | 16.0 | 6.00 | 18.0 | 7.25 | 20.5 | 8.25 |
| 600–1000 | 17.0 | 5.75 | 18.0 | 6.50 | 19.0 | 7.00 | 21.0 | 8.25 | 23.5 | 9.25 |

**Fig. 2-15**   Straight splice for three-conductor shielded cable. [Rome Cable Corp.]

the same thickness as the factory-applied jacket and extending 1½ in beyond the jacket end.

8. Serve the entire joint with anhydrous tape, extending this serving 1½ in beyond the end of the hand-applied jacket tape.

9. Paint the entire joint with cable paint.

10. Connect the cable shielding tape ground wire to ground.

B. Rubber-Insulated Thermoplastic-Jacketed Cable

Follow the procedure given in A above, substituting the following for step 7.

7. Apply pressure-sensitive thermoplastic tape over the entire joint to the same thickness as the factory-applied jacket and extending 1½ in (38.1 mm) beyond the jacket end.

C. Thermoplastic-Insulated Thermoplastic-Jacketed Cable

1–2. Follow steps 1 and 2 as in A.

3. Clean the conductors and apply a compression-type connector with hydraulic press and dies.

4. Pencil the factory-applied insulation on each conductor for a distance equal to 6 times its thickness.

5. Insulate with polyethylene-base splicing tape applied one-half lap to a thickness over the connector of 2 times the factory-applied insulation over each conductor, tapering off over the insulated conductor to a point $B$ from the center.

6. Cover the hand-applied insulating tape with ⅝-in (15⅞-mm) tinned tinsel copper shielding braid applied as in step 6 of A.

7. Apply pressure-sensitive thermoplastic tape over the entire joint to the same thickness as the factory-applied jacket and extending 1½ in (38.1 mm) beyond the jacket end.

8. Complete as in steps 8, 9, and 10 of A.

**154. Terminating power cables.** The following instructions in Secs. 154 to 162 for terminating power cables are reproduced here through the courtesy of the Rome Cable Corp.

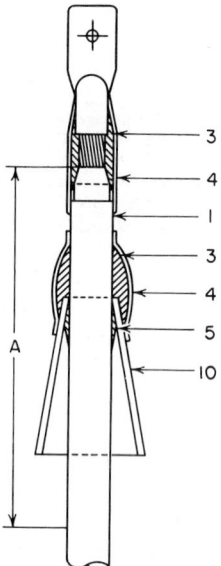

A - Minimum length to
    grounded surface
1 - Jacket
3 - Insulating tape
4 - Anhydrous tape —
    cable paint
5 - Friction tape
10 - Rain shield

|  | Length of $A$, in | | |
|---|---|---|---|
| Conductor size, AWG or kcmil | 600 V | 300 V | 5000 V |
| 8–1000 | 5 | 7 | 10 |

**Fig. 2-16**    Outdoor termination for single-conductor unshielded cable. [Rome Cable Corp.]

GENERAL INSTRUCTIONS    This discussion covers outdoor and indoor terminations for single- and three-conductor unshielded and shielded, rubber- or thermoplastic-insulated cables with neoprene or thermoplastic jackets.

Although outdoor terminations are included for 7500 and 15,000 V, grounded and ungrounded neutral, porcelain potheads are recommended for permanent installations.

In general, cables fall into three classifications: unshielded, leaded, and shielded. Unshielded cables are terminated simply by applying the conductor terminal lug and forming a watertight seal, providing sufficient flashover distance to the nearest ground.

When a leaded or shielded cable is to be terminated, the grounded sheath or shielding must be removed from the cable end to provide sufficient flashover distance. Also, the electric field at the end of the grounded shield is concentrated owing to the divergence of the flux lines toward the thin edge of the shielding and must be relieved, since the high potential gradients may lead to eventual cable failure in this area.

This concentration of electrical stress is relieved by adding a *stress-relief cone* to the exposed insulation surface of the cable, which consists of a double cone of additional insulation with the metallic shielding extended to the middle of the double cone.

Stress-relief cones are recommended for all shielded cables and for 12-kV and higher single-conductor lead-covered cables. (Belling out the lead sheath is usually sufficient for lower-voltage lead-covered cables.)

On rubber-insulated cables, ozone-resistant rubber splicing tape is used as the cone insulation and tinned copper tinsel braid for the cone shielding. On thermoplastic-insulated cables, a polyethylene-base splicing tape is used.

In building the cone, start at the unshielded end and wrap the insulating tape down to the edge of the cable shielding. Continue wrapping back and forth to build up the double cone of proper dimensions. Starting at the middle of the cone, apply the copper tinsel braid to the lower half, making sure that the upper edge is smooth and even. The braid is then tucked under and soldered to the lead sheath or wrapped over and soldered to the cable shielding.

**155. Outdoor termination for single-conductor unshielded cable** (Fig. 2-16)

Rubber- or Thermoplastic-Insulated Neoprene- or Thermoplastic-Jacketed Cable

1. Remove the jacket, factory-applied insulation, and underlying tape, if any, for a distance from the end of the cable equal to the depth of the hole in the lug plus ½ in (12.7 mm) for cable sizes up to 4/0 AWG and 1 in (25.4 mm) for larger cables.

2. Apply a shoulder of friction tape to support the rubber rain shield, when required. This support should be so located that when the rain shield is seated on the support, the bottom will be about 3 in (76.2 mm) from any grounded surface and the distance A as given in the tables. Put the rain shield in place, and apply ozone-resistant rubber splicing tape to form a watertight seal.

3. Apply a compression or solder-type lug on the conductor. A compression-type lug must be used on thermoplastic-insulated cables. Pencil the factory-applied insulation and jacket for a distance equal to 4 times the overall thickness of insulation and jacket. Apply ozone-resistant rubber splicing tape to form a watertight seal.

4. Apply two one-half–lapped layers of anhydrous tape over the tape seals at the lug and upper end of the rain shield. Paint the anhydrous tape with cable paint.

**156. Indoor termination for single-conductor unshielded cable.** Follow the procedure given in Sec. 155, eliminating the rain shield and using the accompanying table for dimension A, which would be the minimum length to grounded surface.

| Conductor size, AWG or kcmil | Length of $A$, in | | |
|:---:|:---:|:---:|:---:|
| | 600 V | 3000 V | 5000 V |
| 8–1000 | 3 | 5 | 8 |

**157. Outdoor termination for single-conductor shielded cable** (Fig. 2-17)

A. Rubber-Insulated Neoprene- or Thermoplastic-Jacketed Cable

1. Remove the jacket, factory-applied shielding, insulation, and underlying tapes, if any, for a distance from the end of the cable equal to the depth of the hole in the lug plus ½ in (12.7 mm) for cable sizes up to 4/0 AWG and 1 in (25.4 mm) for larger cables.

2. Remove all outer coverings down to the factory-applied shielding for a distance A from the end of the factory-applied insulation. Remove the factory-applied shielding and underlying tape to within ¼ in (6.35 mm) of the jacket.

3. The factory-applied shielding must be terminated in a stress cone. Form a stress cone of the proper dimensions, using ozone-resistant rubber splicing tape. Shield the stress cone with ⅝-in (15⅞-mm) tinned tinsel copper shielding braid, terminating the shielding at the point of maximum diameter with a binder of fine copper wire. Solder the braid to the end of the factory-applied shielding and attach the ground wire. Complete

the stress cone with a serving of two one-half–lapped layers of anhydrous tape extending up over the end of the jacket.

4. Seal the rain shield and complete the termination as in steps 2, 3, and 4 of Sec. 155.

B. THERMOPLASTIC-INSULATED THERMOPLASTIC-JACKETED CABLE

Handle in the same way as A except to apply a compression-type lug with hydraulic press and dies. If unpigmented polyethylene-insulated cable is being terminated, it should be covered with two layers of black pressure-sensitive thermoplastic tape applied one-half lap for the distance A prior to application of stress cone and rain shield after removal of shielding tapes.

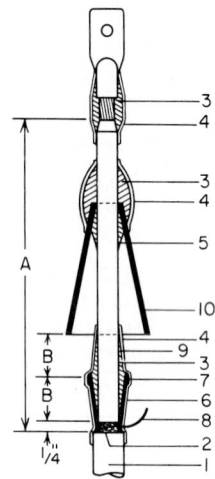

| A- Minimum length | 6- Tinsel shielding braid |
| B- Stress cone length | 7- Binding wire |
| I- Jacket | 8- Ground wire |
| 2- Cable shielding | 9- Stress cone — thickness, same as cable insulation |
| 3- Insulating tape | |
| 4- Anhydrous tape — cable paint. | I0- Rain shield |
| 5- Friction tape | |

| Conductor size, AWG or kcmil | Length of $A$ and $B$, in | | | | | |
| | 5000 V | | 7500 V | | 15,000 V | |
| | $A$ | $B$ | $A$ | $B$ | $A$ | $B$ |
| 8–1000 | 14 | 2 | 17 | 2.5 | 24 | 3.5 |

**Fig. 2-17**   Outdoor termination for single-conductor shielded cable. [Rome Cable Corp.]

**158. Indoor termination for single-conductor shielded cable.**   Follow the procedure given in Sec. 157, eliminating the rain shield and using the accompanying table for dimensions $A$ and $B$.

| Conductor size, AWG or kcmil | Length of *A* and *B*, in. | | | | | |
|---|---|---|---|---|---|---|
| | 5,000 volts | | 7,500 volts | | 15,000 volts | |
| | *A* | *B* | *A* | *B* | *A* | *B* |
| 8–1,000 | 10 | 2 | 13 | 2.5 | 20 | 3.5 |

**159. Outdoor termination for three-conductor unshielded cable** (Fig. 2-18)

A. RUBBER-INSULATED NEOPRENE- OR THERMOPLASTIC-JACKETED CABLE

1. Remove the jacket and underlying tape, if any, for the distance *A* as measured from the center conductor of the termination. (NOTE: Allowance must be made for forming the outer conductors into position.) Remove the fillers and cut at the end of the jacket. Bind the conductors tightly together with a piece of dry cotton tape applied at the end of the jacket before the conductors are spread apart. Spread the conductors radially, and form them into their final position, taking care not to spread them any more than is necessary to permit applying the rain shields and allowing sufficient clearance (1 in, or 25.4 mm/kV)

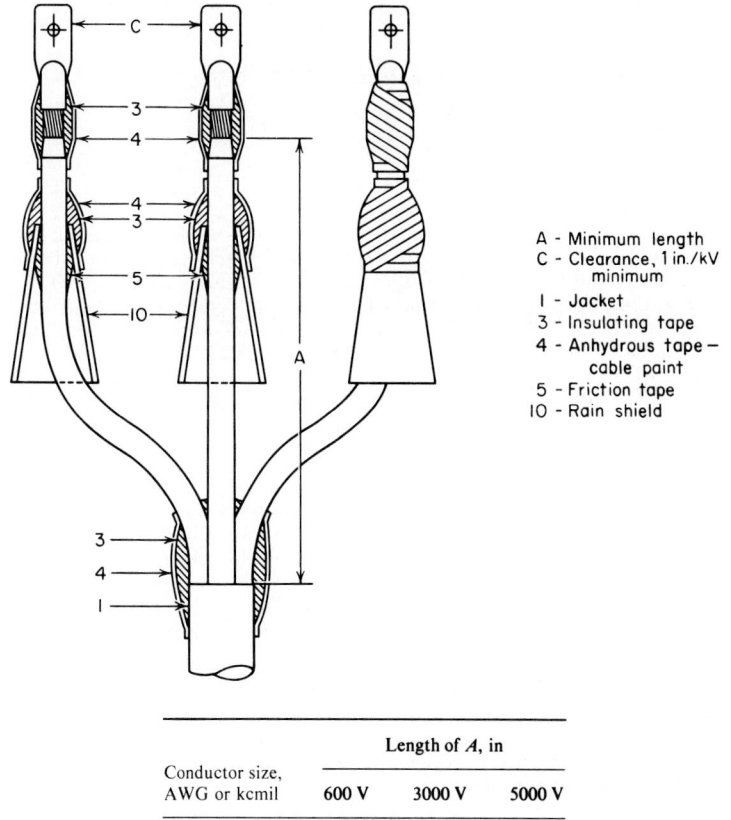

A - Minimum length
C - Clearance, 1 in./kV minimum
1 - Jacket
3 - Insulating tape
4 - Anhydrous tape – cable paint
5 - Friction tape
10 - Rain shield

| Conductor size, AWG or kcmil | Length of *A*, in | | |
|---|---|---|---|
| | 600 V | 3000 V | 5000 V |
| 8–1000 | 8 | 10 | 13 |

**Fig. 2-18**  Outdoor termination for three-conductor unshielded cable. [Rome Cable Corp.]

between lugs. Bare the end of each conductor for a length equal to the depth of the hole in the lug plus ½ in (12.7 mm) for cable sizes up to 4/0 AWG and 1 in for larger cables.

2. Terminate each conductor as described in steps 2, 3, and 4 of Sec. 155.

3. Form a watertight seal with ozone-resistant rubber splicing tape at the end of the jacket. Apply two one-half–lapped layers of anhydrous tape over the tape seal, and paint with cable paint.

B. THERMOPLASTIC-INSULATED THERMOPLASTIC-JACKETED CABLE

Handle in the same way as A except to apply a compression-type lug with hydraulic press and dies. If unpigmented polyethylene-insulated cable is being terminated, it should be covered with two layers of black pressure-sensitive thermoplastic tape applied one-half lap for the distance A prior to application of stress cone and rain shield.

**160. Indoor termination for three-conductor unshielded cable.**    Follow the procedure given in Sec. 159, eliminating the rain shield and using the accompanying table for dimension A.

| Conductor size, AWG or kcmil | Length of $A$, in. | | |
|:---:|:---:|:---:|:---:|
| | 600 volts | 3,000 volts | 5,000 volts |
| 8–1,000 | 8 | 10 | 12 |

**161. Outdoor termination for three-conductor shielded cable** (Fig. 2-19)

A. RUBBER-INSULATED NEOPRENE- OR THERMOPLASTIC-JACKETED CABLE

1. Remove the jacket and underlying tape, if any, for the distance A as measured for the center conductor of the termination. (NOTE: Allowance must be made for forming the outer conductors into position.) Remove the fillers and cut at the end of the jacket. Bind the conductors tightly together with a piece of dry cotton tape applied at the end of the jacket before the conductors are spread apart. Spread the conductors radially, and form them into their final position, taking care not to spread them any more than is necessary to permit applying the rain shields and allowing sufficient clearance (1 in, or 25.4 mm/kV) between lugs. Bare the end of each conductor for a length equal to the depth of the hole in the lug plus ½ in (12.7 mm) for cable sizes up to 4/0 AWG and 1 in for larger cables.

2. Terminate each conductor as described in steps 2, 3, and 4 of Sec. 157. (NOTE: Remove shielding and underlying tapes to within 6 in, or 152.4 mm, of the jacket.)

3. Form a watertight seal with ozone-resistant rubber splicing tape at the end of the jacket. Apply two one-half–lapped layers of anhydrous tape over the tape seal and paint with cable paint.

B. THERMOPLASTIC-INSULATED THERMOPLASTIC-JACKETED CABLE

Handle in the same way as A except to apply a compression-type lug with hydraulic press and dies. If unpigmented polyethylene-insulated cable is being terminated, it should be covered with two layers of black pressure-sensitive thermoplastic tape applied one-half lap for the distance A prior to application of stress cone and rain shield.

**162. Indoor termination for three-conductor shielded cable.**    Follow the procedure given in Sec. 161, eliminating the rain shield and using the accompanying table for dimensions A and B.

| Conductor size, AWG or kcmil | Length of $A$ and $B$, in. | | | |
|:---:|:---:|:---:|:---:|:---:|
| | 5,000 volts | | 7,500 volts | |
| | $A$ | $B$ | $A$ | $B$ |
| 8–1,000 | 16 | 2 | 19 | 2.5 |

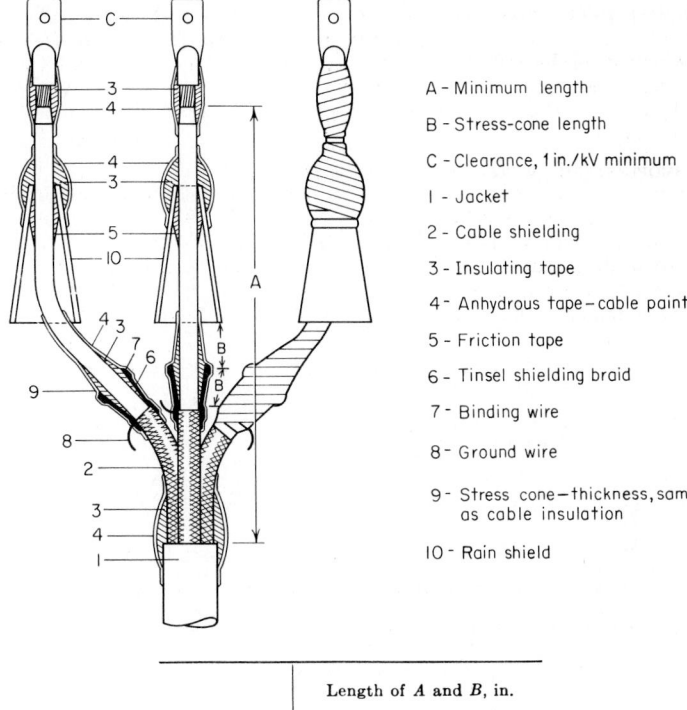

A – Minimum length

B – Stress-cone length

C – Clearance, 1 in./kV minimum

1 – Jacket

2 – Cable shielding

3 – Insulating tape

4 – Anhydrous tape – cable paint

5 – Friction tape

6 – Tinsel shielding braid

7 – Binding wire

8 – Ground wire

9 – Stress cone – thickness, same as cable insulation

10 – Rain shield

| Conductor size, AWG or kcmil | Length of $A$ and $B$, in. | | | |
|---|---|---|---|---|
| | 5,000 volts | | 7,500 volts | |
| | $A$ | $B$ | $A$ | $B$ |
| 8–1,000 | 20 | 2 | 22 | 2.5 |

**Fig. 2-19**   Outdoor termination for three-conductor shielded cable. [Rome Cable Corp.]

**163. Arc proofing** (Rome Cable Corp.).    Underground systems require arc proofing if primary cables or a combination of primary and secondary cables are contained in a manhole. If a primary cable fails even though fast cutoff devices are used, fire protection is necessary to prevent damaging the adjacent cables.

1. Make up splices when required, and form cables into their final position in close triangular configuration, binding them together with a 12-in (304.8-mm) serving of ⅛-in (3.175-mm) tarred marlin twine.

2. Apply a butted layer of heavy wetted asbestos tape 3 in (76.2 mm) wide on the exposed bound-together cables, tying the starting and finishing ends of each roll with tarred marlin twine. Wrap a second layer of wetted asbestos tape so that the butted joints will occur midway between the butted joints of the first layer, tying the starting and finishing ends with tarred marlin twine.

3. On unshielded cables, a helical wrap of No. 6 AWG solid copper wire should be applied over the asbestos tape with a 2-in (50.8-mm) lay in the opposite direction to the lay of the asbestos tapes. The starting and finishing ends should have three close-wrapped turns with sufficient overlength for connecting to ground.

4. Apply a ⅜-in (9.525-mm) layer of asbestos cement over the wet asbestos tape extending into the duct mouths.

5. Ground the ends of the No. 6 AWG copper drain wire.

**164. Portable-cable joints.** The following instructions in Secs. 165 to 177 for making joints in all-rubber portable cables have been reproduced here through the courtesy of the General Electric Co.

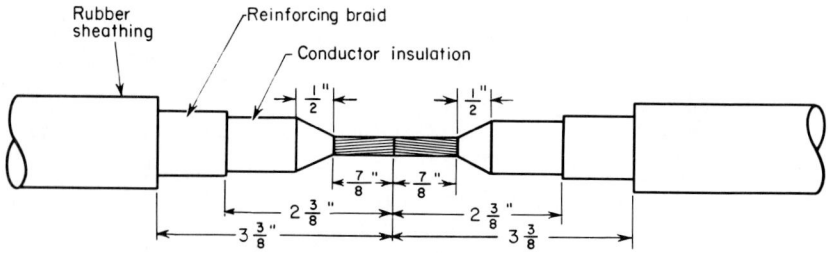

(a) - Rubber sheathing, reinforcing braid, and conductor
insulation removed, and conductor insulation penciled

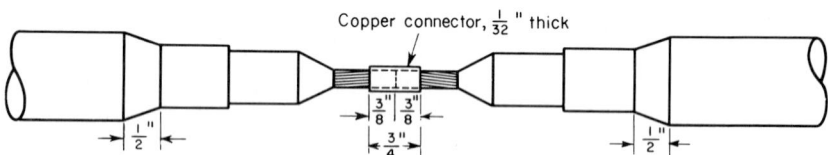

(b) - Copper connector assembled and rubber sheathing penciled

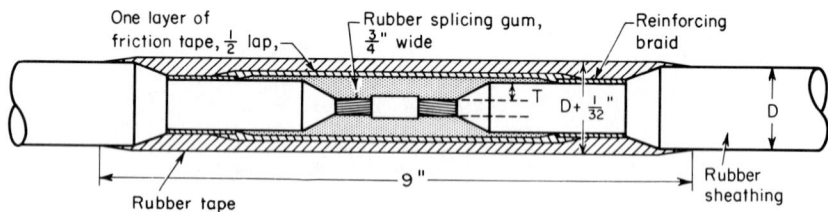

(c) - Splice completed with splicing gum, friction tape, and rubber
tape applied. The joint should now be vulcanized

| Conductor size, AWG | Thickness of insulation, $T$, in. | Over-all diam of cable, $D$, in. | Conductor size, AWG | Thickness of insulation, $T$, in. | Over-all diam of cable, $D$, in. |
|---|---|---|---|---|---|
| 8 | 3/64 | 0.44 | 3 | 5/64 | 0.64 |
| 6 | 3/64 | 0.50 | 2 | 5/64 | 0.67 |
| 5 | 3/64 | 0.53 | 1 | 5/64 | 0.73 |
| 4 | 3/64 | 0.58 | 0 | 5/64 | 0.82 |

**Fig. 2-20**  Single-conductor soldered joint in an all-rubber portable cable. [General Cable Co.]

**165. A typical single-conductor portable-cable joint** is shown in Fig. 2-20. The following directions for making this kind of joint refer to that figure.

1. Bring the cables in position for splicing, with their ends overlapping a little, and cut the cable ends off squarely.

2. Remove 3⅜ in (85.725 mm) of rubber sheathing from the end of each cable. Remove the reinforcing braid for a distance of 2⅜ in (60.325 mm), and remove the conductor insulation for a distance of ⅞ in (22.225 mm).

3. Pencil down the conductor insulation for a distance of ½ in (12.7 mm). This penciling should be smooth and even and can be done best with a sharp knife, taking care not to injure the conductor. All the exposed conductor insulation should then be made smooth with fine emery cloth to provide a good surface for the later application of the splicing gum.

4. Wrap the tapered ends of the insulation tightly with dry cotton tape and tie the tape in place. The tape protects the insulation from scorching while the connector is being soldered to the conductors.

5. Assemble the connector in place over the conductor joint. The connector should be made of soft-drawn copper strap, ¾ in (19.05 mm) wide and ½₂ in (0.79 mm) thick and long enough to wrap around the conductor without overlapping. Apply a coating of non-acid soldering flux over the surface of the conductor and then slide the connector over the conductor; let the connector slot come at the top, and place it so that the two conductors butt at the center of the connector. Pinch the connector with pliers so that it closes into maximum contact with the conductor. Solder the conductor fast to the connector by pouring hot solder over the joint until it is thoroughly heated and the solder runs freely through the slot and into the strands of the cable. Solder that collects on the bottom of the connector should be brought to the top to fill the slot. Before the solder sets, wipe the connector smooth and clean with a piece of cloth. After cooling, any sharp projections should be smoothed off with a file and fine emery cloth.

6. Remove the cotton tape that was wrapped around the tapered ends of the insulation. Clean the rubber insulation with a cloth dampened with high-test gasoline, and allow it to dry. Then apply a coating of rubber cement over the entire surface on which the splicing gum is to be applied, and allow the solvent to evaporate before applying any tape.

7. Apply the splicing gum and build it up over the rubber insulation to a diameter of about ¹⁄₁₆ in (1.58 mm) larger than the rubber insulation on the cable, as shown in Fig. 2-20.

8. Apply one layer of friction tape half-lapped over the splicing gum and over the ends of the reinforcing braid, as shown in Fig. 2-20.

9. Pencil the rubber sheathing as shown, and clean the tapered surface and the outside surface of the cable for a distance of at least ¾ in (19.05 mm) at each end with fine emery cloth and high-test gasoline; allow to dry.

10. Cover the entire surface of the joint with a coating of rubber cement. Be sure to cover all the surface over which the final wrapping is to be applied.

11. After the solvent has evaporated, apply the rubber tape in half-lapping; build it up to a diameter of about ¹⁄₁₆ in (1.58 mm) greater than the diameter of the cable, and let it extend over the surface of the rubber sheathing as indicated in Fig. 2-20, so that the overall length of the joint will be 9 in (228.6 mm).

12. The joint should now be vulcanized in a suitable mold. The inside diameter of the mold should be about the same as the diameter of the cable. If a steam vulcanizer is used, a steam pressure of about 30 lb (206,843 Pa) should be applied for 15 to 30 min, depending on the size of the mold and the joint that is being vulcanized.

13. To complete the process, cover the joint with black wax and wipe it smooth.

**166. Typical examples of multiconductor portable-cable joints** are shown in Figs. 2-21 and 2-22. In each case, sufficient rubber sheathing and fillers are removed from the cable ends to permit staggering the individual splices as shown in Fig. 2-23 in order to obtain a flexible joint of approximately the same outside diameter as the cable. The individual conductors are spliced in the same way as described for single-conductor cable. The individual joints and the ends of the rubber-filled tape, if present, are bound with one layer of friction tape, half-lapped, as shown in the illustrations.

To maintain the overall length of the joint within a given dimension it is necessary that the first splice be at a minimum distance from the end of the rubber sheathing. This dimension is given in the data accompanying the illustrations. The distance between the individual splices.is approximately 3 in (76.2 mm).

In splicing Type G cable, exercise care to avoid damaging the ground wires. These wires should be bent out of the way until the individual conductors have been insulated.

In splicing Type SH, Class A cable, each of the single conductors must be completely shielded; i.e., the copper-mesh shielding must be carried across the joint and soldered, as shown in Fig. 2-22. The cable-shielding braid is removed ½ in (12.7 mm) farther than the rubber-filled tape on each end of each conductor. The remaining braid is then opened and turned back ½ in, care being taken not to loosen the braid farther than this point; such loosening is prevented by applying and soldering at this point three or four turns of tightly drawn 0.0126-in (0.32-mm) tinned-copper wire. The ¾- by 0.015-in (19.05- by 0.381-mm) flat copper braid is then applied half-lapped over the unshielded portion of the insulation. The ½ in of cable-shielding braid, which was just turned back, is drawn over the ¾-in-wide shielding braid, and the two shields are bound together by continuing the tightly drawn butt wraps of 0.0126-in tinned-copper wire for a total distance of ⅞ in (22.225 mm). The wire should then be soldered to the braid, using a nonacid soldering flux; this is necessary because there must be a positive electric contact across the joint. In applying

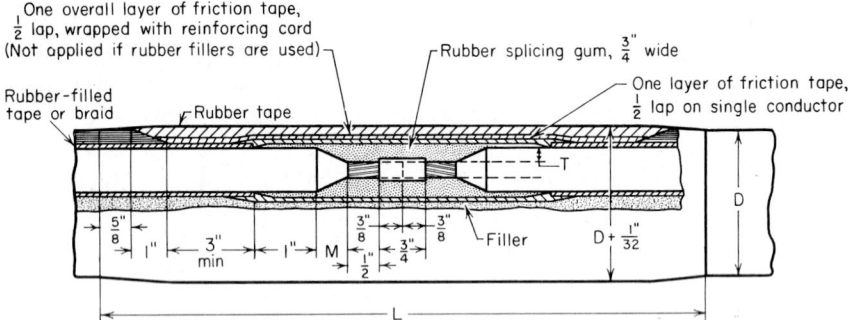

**Fig. 2-21** Joint for multiconductor Type W and Type G portable cable and two-conductor flat cable. [General Electric Co.]

the shield, be sure that no gaps or openings are left, since an opening in the shielding of one conductor might lead to corona.

Bring the single conductors together with approximately the same twist as the rest of the cable.

In the case of Type G cable, the ground wires should now be cut to their proper length. They should be connected in the same manner as the cable conductors, and all sharp

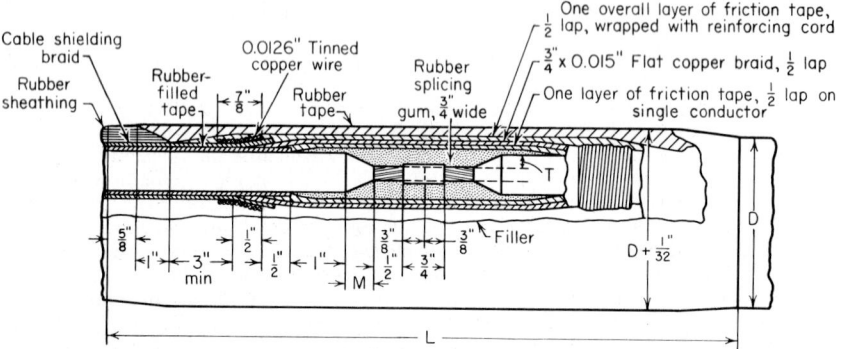

**Fig. 2-22** Joint for Type SH, Class A portable cable. [General Electric Co.]

projections removed. The bare wire must be wrapped with friction tape, half-lapped. The ground wire should then be placed in the interstices.

In joining the shields of Type SH, Classes C and D cable, follow in general the same instructions as given for Type SH, Class A, and in joining the ground wires follow the instructions for Type G cable.

Returning to the instructions for all types, fill the interstices and make the cable its proper shape by means of a suitable filler. The type of filler used depends on the makeup of the cable being spliced. If rubber filler is used in the cable, rubber filler should be used in the joint. If jute fillers are used in the cable, jute fillers should be used in the joint.

Rubber filler is applied by placing a coating of rubber cement over the entire surface of the individual conductors and filler space and, when the solvent has evaporated, by laying strips of splicing tape longitudinally in the interstices until these are filled completely.

When jute fillers are used, the individual conductors, the jute fillers, and the ground wires, if any, should be covered completely with one layer of half-lapped friction tape. The surface of the friction tape should then be wrapped with reinforcing cord. Apply the cord in tightly drawn wraps spaced ½ in (12.7 mm) apart, having a pitch of approximately 2 in (50.8 mm). The joint should be wrapped first with the pitch in one direction and then with the pitch in the opposite direction, thereby obtaining the equivalent of a reinforcing braid. Tie the cord firmly at both ends.

The rubber sheathing should then be tapered and cleaned as described for single-con-

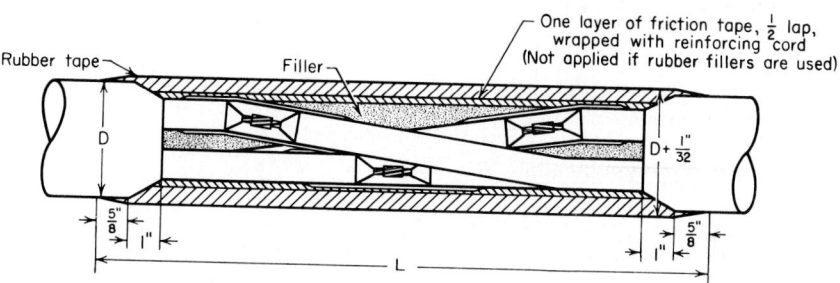

**Fig. 2-23** Joint for 600-V three-conductor portable Type W cable, showing the staggering of individual splices. [General Electric Co.]

ductor cable. All the surface that is to be covered with splicing gum should first be covered with a coating of rubber cement. When the solvent has evaporated, wrap the surface with splicing tape to the diameter shown in the illustrations.

The joint is now ready to be vulcanized. The same vulcanizing process is used for both shielded and nonshielded cables. Preheat the vulcanizer before inserting the joint, and vulcanize according to instructions furnished with the particular vulcanizing equipment. If a steam vulcanizer is used, a steam pressure of about 30 lb (206,843 Pa) should be applied for 30 to 60 min, depending on the size of the mold and the joint that is being vulcanized. It is sometimes necessary to apply a dressing of thin paper to the mold to prevent sticking of the rubber. After vulcanizing, the joint should be covered with black wax and wiped smooth (see the instructions for vulcanizing).

If the voltage is 2500 or above, a feeder cable that contains provision for a ground connection should be used. The power-line end must be grounded to a suitable permanent ground, and the end at the apparatus must be grounded securely through a bolted connection to the apparatus frame.

Included in this instruction is one method for making the conductor splice. Other methods can be used successfully. Most operating companies have their own ideas as to how these joints should be made. One suggestion is added: the copper splice should be made as smooth as possible and as close to the original diameter as practical.

**167. Nonsoldered joints** (for types of cable used in mines). These instructions apply to types of all-rubber portable cable which are widely used in coal and other mines on such equipment as gathering-reel locomotives, cutters, loaders, and drills. The types covered are single-conductor, two-conductor flat (with the conductors side by side), and two-conductor concentric (with an inner and outer conductor).

This method of jointing eliminates solder, and although it is thus particularly applicable to work in a mine where the explosive hazard is high, it can, of course, be used for any job (600 V and below) for which the omission of solder seems to be justified or desirable. This type of connector is useful for emergency repairs because it is simply and easily installed. (Refer to tables on page 2-118.)

**168. Nonsoldered joint for single-conductor portable cable** (refer to Fig. 2-24)

1. Remove the outer jacket and insulation for a distance of approximately 5 in (127 mm) from each cable end.

2. Overlap the two conductor ends, as shown in the drawing, and install two Newberry clamps at the ends of the conductors.

## Dimensions, in Inches, of Multiconductor Cable Joints
See Figs. 2-21, 2-22, and 2-23

### Types W and G, 600 V, two-, three-, and four-conductor

| Conductor size, AWG | $T$ | Two-conductor | | | Three-conductor | | | Four-conductor | | |
|---|---|---|---|---|---|---|---|---|---|---|
| | | $D$ | $L$ | $M$ | $D$ | $L$ | $M$ | $D$ | $L$ | $M$ |
| 8 | $\frac{9}{64}$ | 0.75 | | | 0.78 | | | 0.84 | | |
| 6 | $\frac{9}{64}$ | 0.91 | | | 0.95 | | | 1.03 | | |
| 4 | $\frac{9}{64}$ | 1.06 | | | 1.12 | | | 1.27 | | |
| 3 | $\frac{9}{64}$ | 1.12 | 18 | ½ | 1.19 | 21 | ½ | 1.36 | 24 | ½ |
| 2 | $\frac{9}{64}$ | 1.19 | | | 1.31 | | | 1.42 | | |
| 1 | $\frac{9}{64}$ | 1.41 | | | 1.50 | | | 1.62 | | |
| 0 | $\frac{9}{64}$ | 1.55 | | | 1.62 | | | 1.84 | | |

### Type G, 5000 V, three-conductor

| Conductor size, AWG | | 5000 volts | | |
|---|---|---|---|---|
| | $T$ | $D$ | $L$ | $M$ |
| 8 | | | | |
| 6 | $\frac{10}{64}$ | 1.66 | | |
| 4 | $\frac{10}{64}$ | 1.80 | | |
| 3 | $\frac{10}{64}$ | 1.86 | 24 | 1 |
| 2 | $\frac{10}{64}$ | 1.94 | | |
| 1 | $\frac{10}{64}$ | 2.03 | | |
| 0 | $\frac{10}{64}$ | 2.14 | | |

### Type SH, 5000 V, three-conductor

| Conductor size, AWG | | 5000 volts | | |
|---|---|---|---|---|
| | $T$ | $D$ | $L$ | $M$ |
| 8 | | | | |
| 6 | | 2.00 | | |
| 4 | | 2.20 | | |
| 3 | $13\frac{1}{64}$ | 2.24 | 26 | 1 |
| 2 | | 2.33 | | |
| 1 | | 2.44 | | |
| 0 | | 2.53 | | |

### Two-conductor, flat, 600 V

| Conductor size, AWG | | Two-conductor | | |
|---|---|---|---|---|
| | $T$ | $D$ | $L$ | $M$ |
| 8 | $\frac{9}{64}$ | 0.60 × 0.81 | | |
| 6 | $\frac{9}{64}$ | 0.75 × 0.98 | | |
| 4 | $\frac{9}{64}$ | 0.84 × 1.09 | 18 | ½ |
| 3 | $\frac{9}{64}$ | 0.88 × 1.16 | | |
| 2 | $\frac{9}{64}$ | 0.88 × 1.29 | | |
| 1 | $\frac{9}{64}$ | 1.13 × 1.43 | | |

3. Taper the insulation on each side of the joint, leaving ½ in (12.7 mm) of bare conductor between the end of the taper and the edge of the connector.

4. Remove the outer jacket and reinforcing braid to the final dimensions.

5. Thoroughly clean the surface to which splicing gum is to be applied and cover with a coating of rubber cement. Allow the solvent to evaporate.

6. Apply splicing gum to the dimensions as shown in Fig. 2-24.

7. Cover the splicing gum with one layer of friction tape, half-lapped and finally wrapped with reinforcing cord.

8. Taper the ends of the outer rubber jacket and clean the entire surface of the joint.

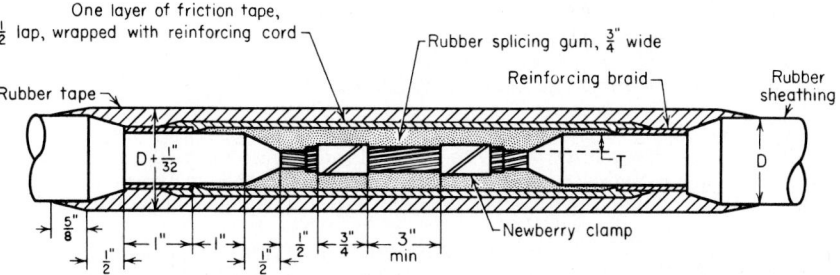

**Fig. 2-24**   Nonsoldered single-conductor cable joint. [General Electric Co.]

9. Apply a coating of rubber cement over the entire joint. When the solvent has evaporated, apply the outer jacket with wrappings of splicing gum.

10. Vulcanize the joint.

**169. Nonsoldered joint for two-conductor (flat) portable cable** (refer to Fig. 2-25)

1. Remove the outer jacket to the dimensions shown in the drawing, making an allowance for an approximately 5-in (127-mm) overlap of the conductors.

2. When the cable is supplied with jute fillers, the fillers should be laid back over the cable and reassembled later.

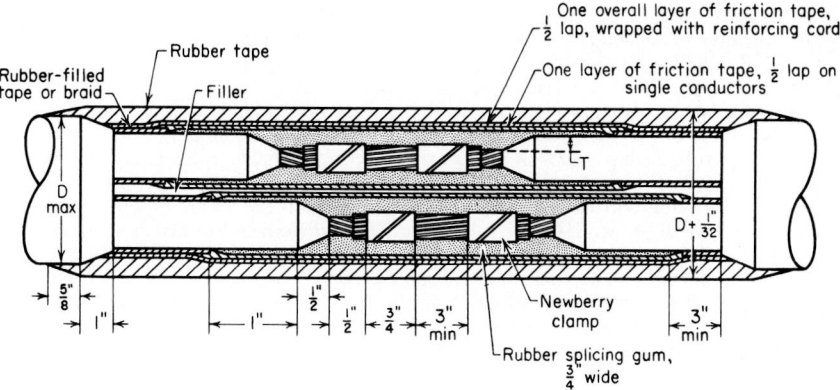

**Fig. 2-25**   Nonsoldered two-conductor (flat) cable joint. [General Electric Co.]

3. Remove sufficient insulation from each of the four cable ends to allow installation of the connectors.

4. The conductors are then connected by overlapping the two ends as shown in the drawing, and a Newberry clamp is installed at each of the conductor ends. This clamp has been used very successfully around mines for a number of years.

5. Remove the outer tape or braid to the dimensions shown.

6. Taper the conductor insulation on each side of the connectors as shown in the drawings.

7. Thoroughly clean the surface over which splicing gum is applied, and apply a

coating of rubber cement. After the solvent has evaporated, wrap splicing gum to a diameter slightly greater than that on the original cable.

8. Cover each insulated conductor with a layer of friction tape half-lapped.

9. If the original cable is supplied with jute or rope fillers, they should be replaced in the interstices of the joint. If rubber fillers were used in the original cable, they should be replaced by rubber fillers (see standard instructions for splicing portable cables in Sec. 166).

10. After the fillers have been applied, cover the insulated conductors and fillers with a layer of friction tape half-lapped and wrap with reinforcing cord (see the instructions for the splicing of standard portable cables).

11. Taper the ends of the overall jacket, thoroughly clean the entire surface which is to be covered by rubber tape, and apply a coat of rubber cement.

12. After the solvent has evaporated, apply the outer jacket with wrappings of rubber tape to the dimensions that are shown on the drawing.

13. Vulcanize the joint.

**170. Vulcanizers** are easy to operate and can be used very successfully by persons with little experience. These instructions should enable anyone to operate the average steam vulcanizer without trouble. Steam vulcanizers are generally shipped without water in the jacket. The user should make sure that the water is placed in the unit to the proper level before the vulcanizer is employed.

Most units are equipped with a relief or safety valve. This valve is set to "pop" at a given pressure. It can be adjusted by removing the top and tightening the screw to increase pressure or loosening the screw to lower the pressure. The pressure control can be set to open at from 30 to 65 lb (206,843 to 448,159 Pa) pressure and to close at from 25 to 60 lb (172,369 to 413,685 Pa). This control can generally be adjusted by tightening or loosening the control-spring tension in the same way as the relief valve is adjusted.

The heaters should always be covered with water to prevent their burning out. If no steam is escaping anywhere, a filling of water should last a considerable time even when used daily. The water should normally be checked about once a week. The vulcanizer should be placed in a practically level position while the current is connected; otherwise the water may drain away from the immersion heaters.

Most steam vulcanizers are operated with electric-heating units, but any form of heat can be used, provided it can be carefully controlled so that a constant heat is applied to the joint to be vulcanized. Gas furnaces and blowtorches have been successfully used.

**171. General instructions for operation of vulcanizers.** Three factors are essential for the successful vulcanizing of rubber-jacketed cables: time, termperature, and cleanliness.

**172. The time required to vulcanize rubber** is approximately 20 min; however, not every vulcanizing job can be done in this time. First, the mass of rubber, cable, and copper conductors must be brought up to temperature before the cure begins.

The outer rubber that comes in direct contact with the mold begins to cure before the inner mass, but the work must be left in the vulcanizer long enough to cure all the rubber. Therefore, the time will vary considerably, depending on the volume to be heated and the temperature of the work before it is put into the vulcanizer. For instance, in winter a cable brought into a shop or repaired outdoors will require more time than if vulcanized in the hot summer. Also, a large cable will require more time than a small one. With these factors in mind, it is readily seen that a definite time schedule cannot be set forth.

In most cases the time will be between 30 and 45 min, although there are conditions that require 1 h or even more and others that require only 20 min.

The exact length of time for curing is not known. In the first place, there is no positive schedule to follow. After a few jobs have been done, the time can be estimated. It is best to leave the first few jobs in slightly longer than might seem necessary to be sure of a full cure. Then reduce the time as you gain experience. An approximate schedule for a novice is as follows:

| Cable Size, In | Time, Min |
| --- | --- |
| ¾ and smaller | 20 |
| ¾–1¼ | 25 |
| 1¼–1¾ | 30 |
| 1¾–2¼ | 35 |
| 2¼–3 | 40 |
| 3 –3½ | 50 |

This schedule is an estimate for complete splices. When only small patches are made, the time will be somewhat less, as the small volume of new rubber and the adjacent old rubber are quickly brought up to temperature and may be cured before the mass of cable has been fully heated.

Care must be taken not to undercure a joint because undercured rubber deforms easily and therefore makes a poor joint from the standpoint of abrasion. A simple method of determining an undercure is to cut in two a piece of the surplus fin; if the fin will stick together where it was cut, the joint was undercured. Because the fin is thin, it will always cure first. Therefore, the fin might be overcured, but the joint itself might still be undercured. This does not necessarily mean an unsatisfactory job but is mentioned simply as an aid to the operator.

**173. Temperature for vulcanizing.**    The controls on vulcanizers are set to open at from 30 to 65 lb (206,843 to 448,159 Pa) steam pressure and close at from 25 to 60 lb (172,769 to 413,685 Pa). This range under ordinary conditions provides a sufficiently accurate temperature for curing the rubber.

About the only condition that might require any change would be subzero weather during which the vulcanizer itself is exposed. In such cases the radiation is so severe that it may be necessary to increase the pressure by adjusting the control. Since cables are usually repaired indoors, however, extremely cold weather is seldom a factor.

**174. Cleanliness in vulcanizing.**    The important reason for vulcanizing electric-cable splices is to make them watertight and give them added strength. Obviously, it is necessary for the applied rubber to mold into a solid mass and to stick to the old cable jacket.

One of the chief aids in obtaining these results is keeping the work clean. Oil or grease is known to be detrimental to rubber, and dust or dirt on the surface of the rubber being applied is likely to prevent its molding together. When new rubber is applied to the old cable jacket, it must be absolutely clean to obtain adhesion.

**175. How to patch a jacket.**    The amount of work to be done depends on the extent of the injury. For instance, in many cases a small cut can be patched simply, whereas a severe injury may require the removal of the portion of jacket that is injured and replacing this portion with new rubber. In any case, thoroughly inspect the conductor insulation at the injury to be sure it is not also injured. If it is, the jacket should be removed for a sufficient length to repair the conductor insulation properly. It may be necessary to make a complete splice.

To vulcanize new rubber to old rubber, it is necessary to roughen the old rubber and scrape or grind off any old scale or coating at the place where the new rubber is expected to stick. The old rubber should also be prepared for 1 in (25.4 mm) around the injured spot.

When the jacket must be removed, it will be found easier to clean and roughen the cable for about 2 in (50.8 mm) each way beyond the portion to be removed before cutting it out.

Two kinds of rubber are used in the repair of electric cables: jacket rubber and insulating rubber. Insulating rubber is used only for conductor insulation. Jacket rubber is used for all outside work.

When new rubber is applied to the old, proceed as follows:

1. Clean and roughen old rubber as described previously.

2. With naphtha or benzol, thoroughly clean the areas where the rubber is to be applied and allow them to dry.

3. Brush rubber cement over the surfaces to which the new rubber is expected to stick, as well as an inch or so beyond. Allow the solvent to evaporate.

4. Apply new rubber, which is supplied in tape form. If the rubber is wrapped tightly and evenly, all air is excluded, and the new rubber need not be built up much larger than the mold. Only a little experience is necessary to demonstrate how much rubber to apply.

5. The repair is now ready to vulcanize. Naturally a suitable mold must be on hand. Molds are made in several types and sizes to fit the particular joint in question.

The proper mold should be hot when ready to use. Swab or brush sparingly an application of mold dressing in the mold before putting in the job.

Shut the mold to within $\frac{1}{6}$ in (4.23 mm) of being closed, depending on the size of the cable, and allow to set for about 5 min to permit the rubber to soften; then close the mold. When this procedure is followed, there is a better chance of filling out low places due to unevenness in wrapping. Furthermore, when a complete splice is made, there may be a distortion of core assembly if full pressure is applied before the rubber is allowed to soften.

6. After curing, remove from the mold, cut off surplus fins, apply wax, and the job is finished. The joint should be allowed to cool somewhat before it is removed from the mold.

**176. Temporary vulcanizing equipment.** Frequently it is necessary to vulcanize a cable joint in the field, and no equipment is available. To cope with this condition, temporary vulcanizing equipment can be constructed.

An open-top vat of suitable size should be obtained. This vat should be large enough to allow the entire joint to be submerged in heated liquid in the vat when the spliced cable is placed directly over the vat, allowing the cable joint to dip into the vat.

Paraffin is generally used in the vat for the vulcanizing medium, but any wax with a high enough boiling point could be used. In no case should oil be used.

After the joint has been constructed in the usual manner, it should be placed in a suitable mold. There are a number of different molds that can be easily constructed.

1. Split a piece of pipe which has approximately the same inside diameter as the completed splice. The two pieces of pipe should be held tightly on the joint by wrappings of wire or metal tape or by standard U clamps.

2. In a metal block, drill a hole of approximately the same diameter as the splice. Split the block in half, and place it over the joint. The two halves of the block should be held tightly together by bolts or acceptable clamps.

If possible, the molds should be heated up to about 275°F (135°C) before they are placed over the joint. The inside of the molds should be coated with an approved mold dressing to obtain the best and most workmanlike finished splice.

**177. The procedure to follow in vulcanizing a joint with the equipment of Sec. 176 is as follows:**

1. Make up the joint in the usual manner.

2. Preheat the molds if possible up to 275°F (135°C), coat them with a mold dressing, and clamp them over the joint.

3. In the meantime, the paraffin in the vat should be heated and held to from 275 to 300°F (149°C) by satisfactory heating equipment. Blowtorches or gas furnaces can be used for this purpose, as can any available heating equipment which will hold these temperatures satisfactorily.

4. Place the joint in the heated paraffin, bearing in mind that the entire joint and mold are submerged in the liquid.

5. Hold the joint in this medium for a period of time depending on the size of joint involved and the thickness of rubber to be vulcanized, keeping the temperature of the paraffin as nearly constant as possible. The approximate time required to vulcanize the joint is listed in the steam-vulcanizer instructions. The exact time for curing a splice can best be obtained by experience with the particular equipment used.

6. When the splice has been cured, it should be removed from the vat and allowed to cool for a time before removing the mold.

It is natural that perfect results cannot be obtained with this temporary equipment, but a number of splices have been vulcanized by this method, and all have operated satisfactorily when properly made.

The mold dressing generally used is a saturated solution of soap chips.

**178. Terminal connections of conductors** should be made by means of solder lugs or solderless connectors, except that No. 8 or smaller conductors may be connected to binding posts by clamps or screws, provided the terminal plates of the binding parts have upturned lugs. The upturned plates are necessary so that the wires will not be forced out of place as the screw or nut is tightened.

**179. The correct method of "making up" a lead wire around a binding post** is shown in Fig. 2-26. First, an eye, of such internal diameter that it will slip over the post, is bent with pliers in the bared-and-cleaned end of the lead wire, as illustrated at I. Then the eye is dropped down over the post (III) in such a position that rotation of the bolt or nut in tightening will tend to wrap the eye end around the post rather than to unwrap it. That is, the eye should wrap around the post in a right-handed direction, in the same direction as that in which the nut rotates while being turned on. If the eye is laid on left-handed as at II, it will unwrap and open while the nut is being turned tightly down on it.

**180. Soldering wires in terminal lugs.** If many lugs are to be soldered, a convenient and timesaving method of making the connections is to melt a pot of solder over a plumber's furnace, heat the lug in the solder, pour the solder in the hole in the lug, and then plunge the bared end of the conductor into it, as shown in Fig. 2-27. The insides of the holes of many commercial lugs are "tinned" so that the solder adheres to them readily. The bared end of the conductor should also first be tinned. This may be done as follows. The end of the wire is carefully scraped with a knife or with a piece of fine sandpaper (sandpaper is best because it cannot nick the wire) and then smeared with soldering flux and thrust into the solder pot. If a soldering stick is used, the wire must be heated in the solder before the stick compound will melt and adhere. It requires only a short time to tin the wire end in the pot.

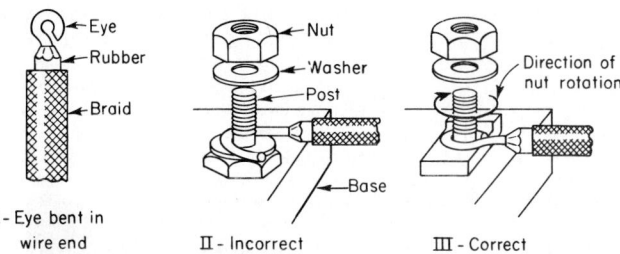

I- Eye bent in wire end          II - Incorrect          III - Correct

**Fig. 2-26**   How a wire should be "made up" on a binding post.

NOTE Immediately after the tinned end is pushed into the hold in the lug, the lug should be soused with a piece of wet waste to cool it rapidly. Scrape or file off any shreds or globules of solder that adhere to the exposed surfaces of the lug and brighten it with fine sandpaper if necessary.

**181. In skinning a wire end which is to be soldered into a lug,** the insulation should be cut back just far enough so that it will abut against the shoulder of the lug, as suggested in Fig. 2-28, I. The appearance is very unsightly and indicates careless work if there is a gap between the shoulder and the insulation, as at II. If because of some mishap a connection has the appearance of II, a partial correction can be made by filling the gap with servings of tape, as shown at III. Tape of standard ¾-in (19.1-mm) width should be torn into strips about ¼ in (6.35 mm) wide before applying.

**182. Only enough molten solder should be poured into the hole in the lug to fill it almost to the brim** when the conductor is in position. If too much is poured in, it will be squeezed out by the wire and will flow over the lug. It must then be removed at a sacrifice of time.

**183. To ensure proper adhesion between wire, solder, and lug** the temperature of all three must be above the melting point of solder when they are brought together in soldering. If this condition is not fulfilled, only a good friction fit of all three parts will result. Such a friction fit does not afford a good electrical connection. To secure maximum mechanical strength and electrical conductivity, it is essential that the solder be maintained at the melting point until it has thoroughly permeated the interstices of the conductor.

**184. The wire terminal and lug should be held in the molten solder until they acquire the temperature of the solder.** To prevent adhesion of solder to the outside of the lug a light oil or possibly soft soap should be applied to its outer surfaces. Be careful to see that no oil is permitted to reach the inside of the lug. When holding the bared ends of heavy conductors in the solder pot, it is advisable to wrap the insulation with a rag previously wrung out in cold water to prevent, insofar as possible, the melting of the insulating compound and the consequent smearing of the terminal. Such smearing will not impair the effectiveness of a properly made joint, though it will detract from the appearance of the finished job.

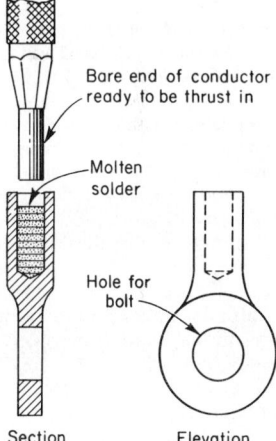

Section          Elevation

**Fig. 2-27**   Soldering wire in lug.

**185. Another method of soldering wires in lugs is to heat the lug with a blowtorch flame.**    When the lug is sufficiently hot, wire solder is fed into the hole. The solder melts and the bared conductor end is then thrust into it, as described above. However, the use of a blowtorch in this way should be avoided if possible, as it blackens the exposed surfaces of the lug. Cleaning with fine sandpaper is then necessary, and it requires considerable time.

**186. Some suggestions for handling blowtorches.**    Only the very best grade of gasoline should be used, and it must be clean and be kept in a clean can; otherwise the burner will become clogged. Never try to fill a torch from a big can. A pint or quart receptacle should be used for this purpose. If this is done, the torch can be held in one hand and filled with the other without danger of overfilling or spilling. The torch should be a little more than two-thirds full, so that there will be room for sufficient air to prevent the necessity of frequent repumping to maintain the pressure. See that the filler plug is closed tight to prevent the escape of air from tank. The fiber washer under the plug must be replaced when worn out. Common washing soap rubbed into thread and joints will stop leaks. The pump should be in good working order; a few drops of lubricating oil well rubbed in will soften the pump washer. Do not turn the needle valve too tight, as there is danger of enlarging the orifice of the burner. See that the burner is sufficiently heated when starting. One filling of the drip cup is generally sufficient if the flame is shielded from draft while heating the burner; if it is not, fill the cup again and light the gasoline as before. A long or yellow flame or raw gasoline shooting from the burner shows that the burner is not hot enough to properly generate gas. When a gasoline torch is used, as much as 90 percent of its heat may be dissipated without doing any work whatever. In performing most blowtorch operations, a great part of this heat may be readily saved by making a shield of sheet iron or asbestos to direct the heat to the object to be heated.

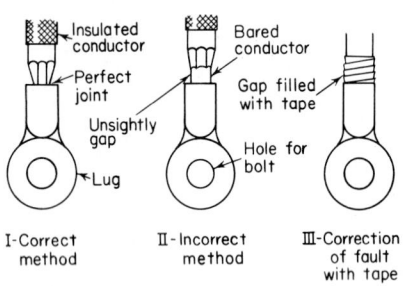

I-Correct method    II-Incorrect method    III-Correction of fault with tape

**Fig. 2-28**    Finished connections.

**187. Wire connectors.**    The selection of proper wire connectors for various sizes, combinations, and types of conductors is an extremely important factor in a sound electrical installation.

Wire connectors are generally classified as either thermal or pressure types. Thermal connectors include those in which heat is applied to form soldered, brazed, or welded joints and terminals. Soldered joints and lugs for copper conductors have been used for many years and, except for service-wire and ground connections, are still permitted. However, soldering is rarely used in modern installations because of the greater dependability and installation ease of present-day solderless pressure connectors. And although it is possible to solder aluminum conductors, this is not recommended unless the installer is fully familiar with the special techniques involved. On the other hand, welded aluminum connections in circular-mil sizes are very satisfactory and provide the best possible joint for aluminum conductors. At the same time, this process is generally restricted to the types of installations in which the cost of the necessary equipment and skilled labor can be justified.

There is wide use of grounding grids or busing on large construction jobs, and grounding connections must be permanent in the interests of safety. For joining bare copper conductors together or to reinforcing rods, ground rods, or steel surfaces, a mold-type welding process is highly recommended (exothermic welding). This process makes use of a mold and starting and mixing powder. After conductors, sleeves, and/or lugs have been placed in the mold, a flint gun is used to ignite the starting powder. This forms a liquid copper, which fuses conductors into a solid mass and welds a permanent electrical connection. Offered by several manufacturers, the mold-type welding process can be performed easily without special training.

**188. Solderless connectors.**    A solderless pressure connector is a device which establishes the connection between two or more conductors or between one or more conductors and a terminal by means of mechanical pressure and without the use of solder. This broad definition covers most solderless connectors and terminal lugs.

**189. Screw-on pigtail connectors.** The common screw-on pigtail connector (see Fig. 2-29) consists of an insulated cap made of plastic, Bakelite, porcelain, or nylon and an internal threaded core with or without a metallic coil spring. Such connectors require no hand tools and are simply twisted onto appropriate combinations of bared conductors. Connector types consisting of metallic spring coils (copper-coated or steel) are listed by Underwriters Laboratories as pressure cable connectors, which means that they can be used for connecting branch-circuit conductors No. 14 and larger according to listed sizes and combinations. Sometimes these connectors can also be used as fixture-splicing connectors, in which case they will be dual-listed by Underwriters Laboratories.

Connectors without metallic coil springs are regarded as fixture-type splicing connec-

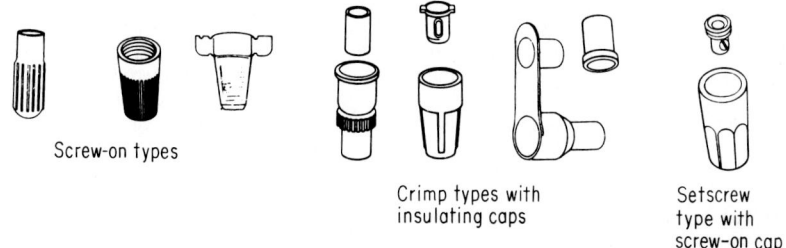

Screw-on types

Crimp types with
insulating caps

Setscrew
type with
screw-on cap

**Fig. 2-29** Typical solderless pigtail connectors. [*Electrical Construction and Maintenance*]

tors for joining fixture wires or branch-circuit wires to fixture wires in a combination of sizes from No. 18 to No. 10 within the listed capacities of the connector. They are not acceptable for general use in branch-circuit wiring.

Most screw-on connectors are designed for copper-to-copper wire connections. Effective January 2, 1987, UL 486C was revised with regard to performance requirements and no twist-on wire connectors have subsequently been listed for terminating No. 12-10 AWG aluminum or copper-clad aluminum conductors. The crimp-type connector (Copalum by AMP) is at present the only UL-listed connector for splicing small copper-to-aluminum conductors.

With screw-on–type connectors, solid conductors are inserted parallel into the connector. Then the connector is twisted to form a locking action on the conductors and provide a dependable connection. Stranded wires in the smaller sizes are twisted together before the connector is installed. And in all cases, proper-size connectors should be selected to accommodate the size and combination of conductors involved in each splice.

Maximum voltage ratings for screw-on connectors are 600 V (1000 V inside a fixture or sign), although some types cannot be used on circuits of more than 300 V. Also, temperature ratings will vary according to the type of insulating cap. Data for voltage and temperature ratings of the various connectors can be obtained from individual manufacturers' catalogs.

**190. Bolted-type pressure connectors** include those for making pigtail, straight, T, or terminal connections. Such connectors depend on the applied force of bolts or screws to produce the clamping and contact pressures between conductors and the connector. Figure 2-30 shows typical examples.

Setscrew types          Straight coupling types

**Fig. 2-30** Typical coupling-type solderless pressure connectors. [*Electrical Construction and Maintenance*]

For making pigtail connections in wire combinations up to No. 10, setscrew connectors with separate insulated caps, which thread onto the connectors after securing the wires, are readily available. These connectors are used in branch-circuit wiring, fixture hanging, and equipment hookups. The only tool required is a screwdriver, and the connectors can be reused. As such, they are especially recommended in installations where wiring connections will be changed frequently. Various setscrew connectors with thread-on insulated caps are suitable as pressure wire connectors or fixture-splicing connectors according to individual manufacturers' listings (see Fig. 2-29).

Other types of bolted connectors are available for practically any desired wire combinations, sizes, or arrangements. Such connectors are constructed of copper, bronze, or alloys of similar metals for use with copper conductors. Connectors for joining aluminum wires together are constructed of aluminum, tin-plated silicon bronze, or tin-plated copper-alloy materials.

If copper or copper-clad aluminum conductors will be connected to aluminum conductors, select connectors which are designed for this purpose and which provide bimetal or tinned spacers in other than dry locations. Bimetal or tinned spacers permit separation of aluminum and copper conductors, thus preventing the possibility of galvanic corrosion. Careful compliance with manufacturers' recommendations as to cleaning aluminum conductors, applying connector aids, and properly positioning the aluminum and copper conductors will provide a satisfactory joint.

**191. Run and tap connectors.**   Many forms of bolted connectors, such as those shown in Fig. 2-31, are designed for making branch connections from main conductors. In such instances, minimum and maximum wire sizes are listed for both main and branch (tap) ranges. Such ranges in wire sizes are listed for each connector and for satisfactory connections should be carefully observed. Split-bolt, clamp-type, gutter-tap, and parallel cable-tap connectors are a few of the common types used for splicing branch conductors to main conductors. Except for the parallel cable-tap type, these connectors must be taped after the wires have been connected. Some parallel cable-tap connectors feature a two-piece molded-plastic cover that fits over the connector and covers uninsulated parts. This eliminates the time required for taping, and makes it possible to remove the connector quickly if this ability is a factor to be considered.

**192. Bolted-type lugs.**   Wires are fastened to bolted-type lugs by one or more bolts, depending on design and ampere ratings. Two- to four-barrel lugs are used to terminate multiple conductors.

**193. Terminal blocks.**   If numerous connections are necessary, as in complex control circuitry, consideration should be given to the use of terminal blocks and terminal-block kits

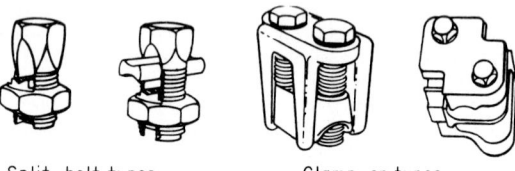

Split-bolt types          Clamp-on types

**Fig. 2-31**  Typical run and tap solderless pressure connectors. [*Electrical Construction and Maintenance*]

made available by a number of manufacturers. Such assemblies contain bolted-type terminal blocks. The terminal blocks can be obtained in fully assembled or individual blocks, and the terminals are available for a wide range of wire sizes. With these units connections can be made quickly, and all terminals can be readily identified, thus simplifying troubleshooting or later changes in circuitry. And the terminal blocks can be installed in standard pull boxes or auxiliary gutters.

**194. Compression-type connectors and lugs** include those in which hand, pneumatic, or hydraulic tools indent or crimp tubelike sleeves which hold one or more conductors. The crimping action changes the size and shape of the connector; and a properly crimped joint deforms the conductor strands enough to provide good electrical conductivity and mechanical strength. With the smaller sizes of solid wires, some indenter splice caps require that the wires be twisted together before being placed in the cap and crimped, whereas others are intended to be inserted parallel. The distinction will vary with the type of indentations, and the manufacturer's recommendations on the correct method of making the splice should be followed. See Figs. 2-29 and 2-34.

One of the disadvantages of indenter or crimp connectors is that special tools are required to make the joint. However, their use can be justified by the assurance of a dependable joint and low cost of connectors.

**195. Connector designations and markings.** Connectors are designed for copper, copper-clad aluminum, or aluminum conductors, or for all of them. Connectors with no markings are generally suitable for copper conductors only. If a connector bears the marking AL, it is designed solely for the connection of aluminum conductors. Connectors marked AL-CU are for use with either copper, copper-clad aluminum, or aluminum conductors.

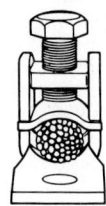

Straight coupling connectors marked AL-CU are ideal if aluminum conductors are run to equipment with terminals designed for copper conductors. In such cases, a short piece of copper wire extends from the equipment terminals to one side of the straight coupling connector, and the aluminum conductor is connected to the opposite side. On the other hand, terminals in many new meter sockets, switches, and panelboards are designed for connection to either copper, copper-clad aluminum, or aluminum wires. If this is the case, a statement to that effect will appear on the inside of the enclosure, and no markings will be found on the terminals.

**Fig. 2-32** Solderless connectors of the pressure-washer type. [Electrical Division of H. K. Porter Company, Inc.]

**196. Splice kits for circuits over 600 V.** For electrical connections in circuits rated above 600 V, proper insulation is as important as providing the actual continuity of the current path. Splices and terminals must be made carefully in full conformity with instructions of connector manufacturers.

Several manufacturers feature splicing kits for making high-voltage splices. These kits

  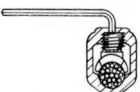    

(a)-Nat. Elec. Prod. Corp.    (b)-Thomas and Betts Co.    (c)-Thomas and Betts Co.    (d)-Trumbull Elec. Mfg. Co.

**Fig. 2-33**   Solderless lugs of pressure-washer type.

vary according to the circuit voltage and type of cable (interlocked armor, neoprene-jacketed with or without shielding, or lead-covered) for the particular conductor insulations.

A common high-voltage connecting device is the pothead, which is used to connect cables to equipment or to other circuits. A pothead is a sealed terminal which provides connection to the conductor in the cable, furnishes moisture proofing for the conductor's insulation, and seals in cable-impregnating oil. Stress cones are also available in kit forms.

**197. Splices in covered aerial conductors** are made in the same manner as splices in bare aerial conductors of the same material. The splice may be covered simply with friction tape as shown in Fig. 2-35, but it is preferable to apply rubber tape covered with two layers of friction tape as shown in Fig. 2-36.

**198. Splices in steel or copper-and-steel stranded cables** are generally made with seamless solderless splicing sleeves. Steel sleeves are used for all-steel wires, and either copper or steel sleeves for the combination cables.

**Fig. 2-34** Typical compression connectors, crimp or indenter types. [*Electrical Construction and Maintenance*]

**199. Splices in Copperweld or Copperweld-and-copper cables** are made with seamless solderless splicing sleeves. Two sleeves should be used for making each splice in extra-high-strength cable.

**200. Splices in all-aluminum cable or in aluminum cable, steel-reinforced,** are made by means of twisted solderless oval sleeves or solderless compression sleeves. The former joints are made with oval single-tube sleeves of seamless aluminum. For all-aluminum cables, one sleeve is sufficient for each joint. For aluminum cable, steel-reinforced, two sleeves should be used for each joint, except for Nos. 8 and 7 cables, for which one sleeve is sufficient. Two twisting wrenches (Fig. 2-37) are employed for twisting the sleeve and cable into the finished joint. The wrenches should be rotated in a clockwise direction when facing the mouth of the sleeve. Before making the joint, be sure that the inside of the sleeves and the ends of the cable are perfectly clean and free from dirt and grease. The following methods of making the joints are recommended by the Aluminum Co. of America.

1. For aluminum cable, steel-reinforced (ACSR), sizes 4/0 to 1/0, give each sleeve 4½ complete twists distributed as shown in Fig. 2-38. This requires setting one wrench at the inner end of each sleeve and setting the other wrench 3 times progressing toward the outer end of each sleeve. Make these in the order shown in the illustration. At the ends of the joint the wrench should not be less than ¼ in (6.35 mm) from the end of the sleeve.

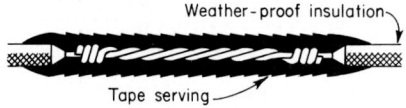

**Fig. 2-35**   Western Union splice insulated with tape.

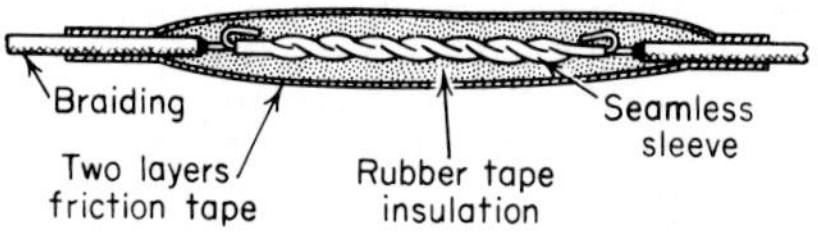

**Fig. 2-36**   Splice in covered conductor.

2. For aluminum cable, steel-reinforced (ACSR), sizes 1 to 6, give each sleeve four complete uniform twists, as shown in Fig. 2-39. This requires one setting of the twisting wrenches at the ends of each sleeve. The wrenches should not be less than ¼ in (6.35 mm) from the end of the sleeve.

3. For twisting joints, consisting of a single sleeve, as used on all-aluminum cables and sizes 7 and 8 ACSR, give the middle third as many complete twists as it will stand (from 2 to 3½ depending upon the size of cable and length of sleeve). Then give each end third length one complete twist.

**201. Compression joints on all-aluminum cable** are made by compressing a seamless aluminum sleeve (Fig. 2-40) over the butted ends of the two cable lengths. A portable hydraulic compressor and dies are required for making these joints. The following method of making a joint is recommended by the Aluminum Co. of America.

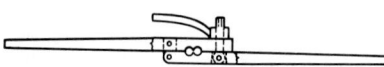

**Fig. 2-37**   Twisting wrench for applying aluminum twisting sleeves. [Aluminum Co. of America]

1. Impregnate the cable ends thoroughly with red-lead paint (red lead and linseed oil) by dipping them in a bucket of paint a little deeper than one-half the length of the joint.

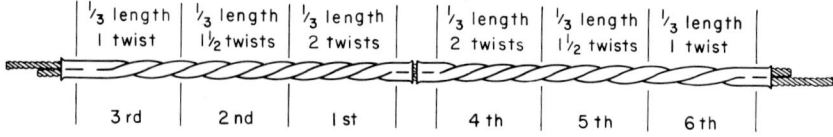

**Fig. 2-38**   Joint for ACSR cable sizes 4/0 to 1/0. [Aluminum Co. of America]

2. Slip the cable ends into the joint, taking care that they meet exactly at the middle.

3. Compress the joint, starting at the center and working toward the ends, allowing dies always to overlap the previous position.

For old or blackened cable, each strand should be scraped or sandpapered clean. The bore of the joint is large enough to insert the end of the cable easily after the strands have been opened up, cleaned, and rearranged.

In making a repair joint between a piece of new cable and a piece of old cable, the most convenient method is first to run the end of the old cable clear through the joint, then to open up and clean the end of each strand for a distance equal to one-half the length of the joint. Replace the strands, impregnate with red-lead paint, and pull the cable back until the end is exactly in the middle of the joint. Then insert the new cable at the outer end of the joint after first removing grease from the strands and impregnating with red-lead paint.

**202. Compression joints for most aluminum cable, steel-reinforced,** consist as illustrated in Fig. 2-41 of a steel compression sleeve (item 1) on the steel core and an aluminum compression sleeve (item 2) on the complete cable. The overall compression sleeve is fitted with two holes through which a heavy red-lead filler is injected. The holes are sealed with aluminum plugs (item 3). The following method of making a joint is recommended by the Aluminum Co. of America.

1. Before applying the joint, see that its bore and the ends of cable to be jointed are free from grease and dirt, i.e., thoroughly clean.

2. Slip the aluminum compression joint over one cable end and back out of the way along the cable.

3. Using a hacksaw, cut off the aluminum strands, exposing the steel core for a distance of a little more than half the length of the steel compression joint. Use care not to

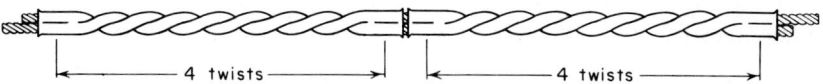

— 4 twists —          — 4 twists —

**Fig. 2-39**   Joint for ACSR cable sizes 1 to 6. [Aluminum Co. of America]

nick the steel core with the saw. Before doing this, serve the cable with wire just back of the cut.

4. Insert the steel core into the steel compression joint, bringing the ends exactly to the center.

5. Compress the steel compression joint, beginning at the center and working out toward the ends, allowing dies always to overlap the previous position.

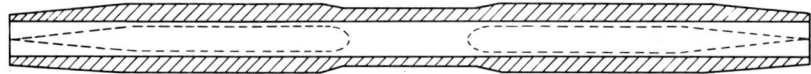

**Fig. 2-40**   Seamless aluminum sleeve for compression joint in all-aluminum cables. [Aluminum Co. of America]

6. Remove serving from the cable and slip the aluminum compression joint up over the steel compression joint. Center the aluminum joint by sighting the ends of the steel joint through the filler holes in the aluminum joint.

7. Using a caulking gun (¼-in, or 6.35-mm, round nozzle, maximum), inject heavy red lead (approximately 93 percent red lead, 7 percent linseed oil by weight) through

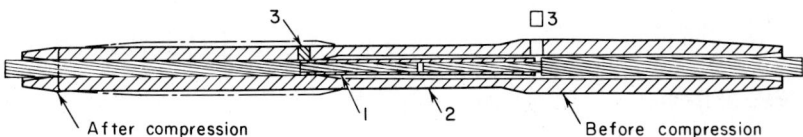

3          ☐3

After compression          Before compression

**Fig. 2-41**   Compression joint for ACSR cable. [Aluminum Co. of America]

both holes provided in the aluminum joint until the space between the aluminum joint and the steel joint is completely filled. This can be observed through the filling holes.

8. Insert the aluminum plugs in the filler holes and hammer them firmly in place. They will be completely locked during compression.

9. Compress the aluminum compression joint, starting at the center and working toward the ends, allowing dies always to overlap the previous position.

Some types of aluminum cable, steel-reinforced, have such a large percentage of steel that the steel compression sleeve of the above joint has a larger diameter than the complete cable. This necessitates the use of the joint shown in Fig. 2-42. One end of the overall aluminum sleeve is counterbored so that it can be slipped over the steel sleeve.

The space between the counterbore and the cable is filled with a third aluminum sleeve (item 4). The procedure in making such a joint is in general the same as previously described. It is self-evident, however, that the aluminum joint (item 2) must be slipped over the cable with that end which has the smaller bore first and also that the sleeve (item 3) must be slipped on the other end of the cable—both before the steel joint (item 1) is applied. After the steel joint is applied and before the aluminum joint is slipped into place, the sleeve is slipped along until it is flush with the ends of the aluminum strands. Otherwise the procedure is the same as previously described for making compression joints in aluminum cable, steel-reinforced.

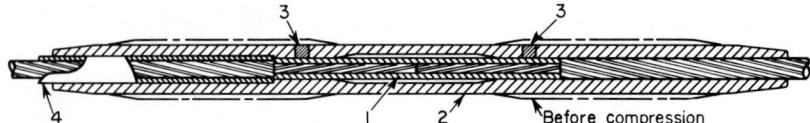

**Fig. 2-42**   Compression joint for ACSR cable with large steel core. [Aluminum Co. of America]

## ALUMINUM-BUILDING-WIRE INSTALLATION PRACTICES

### 203. Aluminum building wire (The Aluminum Association, Inc.)

A. Basic Installation Techniques

1. *Stripping insulation.* Never ring a conductor when stripping insulation. One way to avoid ringing is to pencil or whittle the insulation (Fig. 2-43). Another method is to skin the insulation back from the cut end of the conductor and then cut outward (Fig. 2-44). For the smaller sizes of wire the insulation may be removed quickly and easily by means of a wire stripper (Fig. 2-45). Be sure to match the size of notch on the stripper to the size of the wire.

2. *Splicing.* To splice two lengths of aluminum conductor use an aluminum compression-type splicing connector (Fig. 2-46). Manufacturers supply connectors filled with joint compound to seal the splice against moisture and other contaminants.

First, strip the insulation from the end of each conductor by one of the methods described. Strip back far enough so that the conductor will go fully into the connector, but also make sure that the insulation fits closely to the connector.

Next, insert the stripped end of the conductor into the connection as far as it will go. Then apply the crimping tool designed for the type of connector you are using and crimp

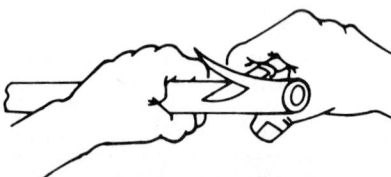

**Fig. 2-43**   Never ring a cable: ringing may lead to a break. Remove insulation as you would sharpen a pencil.

**Fig. 2-44**   Another safe method of removing insulation from a conductor is to peel the insulation back and then cut outward.

fully in accordance with the manufacturer's instructions (Fig. 2-47). Be sure to select the correct crimping tool die for the size of conductor being spliced.

Finally, wipe off any excess joint compound and apply insulating tape according to the usual procedure (Fig. 2-48 and accompanying tables).

3. *Pulling conductors in conduit or electrical metallic tubing.* The following procedures are applicable to conduit of all types including aluminum. (*a*) Run a "fish" line through the conduit, either by attaching the line to a piston-type device propelled through the conduit by compressed air or by pushing a round flexible speedometer-type steel wire through the conduit. Polyethylene fish tapes may be used for shorter runs of up to about 100 ft (30.5 m). (*b*) Attach a cleanout brush to the fish line, and behind it attach the pull line; then pull both through the conduit by means of the fish line. (*c*) Attach the pull line

to the conductor or conductors. A basket grip over the insulation may be used provided it is of a type that will not score the conduit during the pull (Fig. 2-49). A bare steel basket grip should not be used. (*d*) If conductors are pulled with a rope, stagger the conductor ends and anchor in position with tape to provide maximum flexibility around bends (Fig. 2-50). (*e*) Try to feed conductors into the conduit end closest to the first bend to reduce

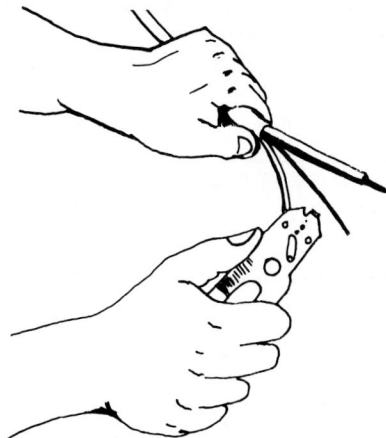

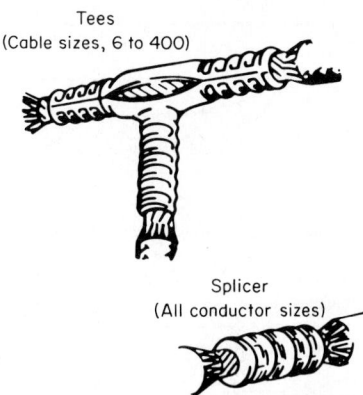

Compression Connectors

Tees
(Cable sizes, 6 to 400)

Splicer
(All conductor sizes)

**Fig. 2-45** Use a wire stripper for removing insulation from smaller wire sizes. Match stripper notch to the wire size.

**Fig. 2-46** Splice aluminum wires with a plated aluminum compression-type splicing connector. The compound in the connector seals the joint against moisture.

pulling tension. (*f*) Use pulling equipment with adequate power available to make a steady pull on the cables without jerking. (*g*) Use a pulling compound compatible with the conductor insulation as the conductors are fed into the conduit to reduce the coefficient of friction and the required pulling tension. (*h*) For single conductors on reels, stagger reels one behind another while feeding in conduit in order to maintain equal pulling tensions and prevent conductors from crossing over and jamming in the conduit. (*i*) Whenever possible, pull conductors downward to allow gravity to assist in pulling with reduced tension (Fig. 2-51). (*j*) Arrange pulls for large conductors so that maximum pulling tension on conductors does not exceed $0.006 \times N \times$ cmil, where $N$ = the number of conductors pulled and cmil = the circular-mil area for each conductor. (*k*) When conductor ends are prepared for pulling, be sure not to nick the stranded aluminum conductor during insulation removal. Damaged strands can reduce the pulling tension capabilities of the conductor. To avoid this, pencil the insulation for removal, as described above; do not ring-cut the insulation. (*l*) Adhere to all **NEC** requirements.

4. *Pulling conductors in trays.* If aluminum cable is to be pulled in trays or cable racks, take the following precautions, plus those applicable to conduit. (*a*) If pulling attachments are used on the conductors, cover them with rubberlike or plastic tapes to prevent scoring of the trays and installa-

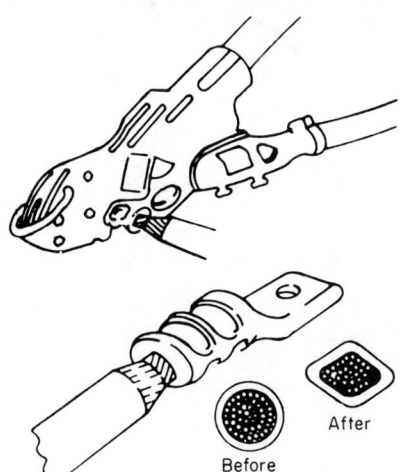

After

Before

**Fig. 2-47** The crimping tool and its die must be correct for the conductor size being terminated to make a satisfactory joint. Apply enough pressure to distort all strands fully and make required number of crimps.

tion sheaves during a conductor pull (Fig. 2-52). (*b*) Use lightweight (aluminum) large-radius sheaves around bends and smaller sheaves on the straight sections of cable-support trays to facilitate cable installations and to reduce required pulling tensions. (*c*) Maintain minimum bending radii on all cable turns, bends, and drops (see Table 203F below). (*d*)

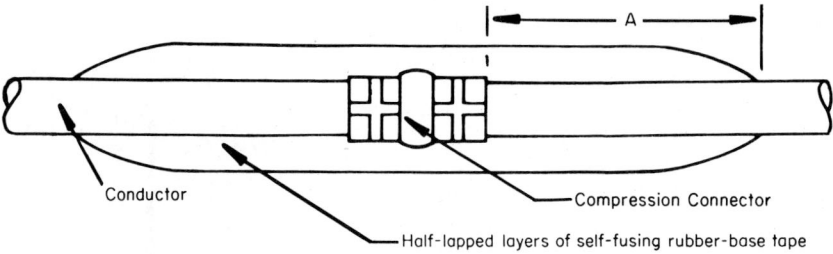

**Cross-Linked Polyethylene Cables**

| Insulation thickness, in | *A*, in | Minimum no. of tape (half-lapped layers) |
|---|---|---|
| .055 | 2½ | 2 |
| .065 | 2¾ | 2 |
| .075 | 2¾ | 3 |
| .085 | 3 | 3 |
| .095 | 3 | 4 |
| .105 | 3¼ | 4 |
| .110 | 3¾ | 4 |

**Rubber-Butyl-Neoprene Cables**

| Insulation thickness, in | Jacket thickness, in | *A*, in | Minimum no. of tape (half-lapped layers) |
|---|---|---|---|
| ⁵⁄₆₄ | .030 | 3¼ | 4 |
| ⁵⁄₆₄ | .045 | 3¾ | 5 |
| ⁵⁄₆₄ | .065 | 4 | 5 |

**Fig. 2-48**    Apply insulating tape according to the usual procedure, making sure to protect the joint from physical damage.

If cables are anchored on trays, be sure that straps or other cable-anchoring devices do not cut into the insulation. (*e*) Conductors in the same circuit should be grouped together, but whenever possible maintain spacing between conductors in different circuits to obtain optimum current-carrying capacity and prevent hot spots. (*f*) Be sure that tray supports are capable of handling the maximum weight of conductors and planned conductor additions.

**Fig. 2-49**    Use an insulated basket grip to attach a pull line to the conductors. A bare steel grip should not be used because it will score the inside of the conduit.

**Fig. 2-50**    When a rope pull must be used, skin the cable ends and stagger them after locking with tape. This procedure will hold the tie to a minimum cross section.

If permanent bends are made at terminations in aluminum building wires, the following are the minimum recommended bending radii (see Table 203F) based on the overall diameter (OD) of insulated conductor (inches):

| | |
|---|---|
| 1.000 and less | 4 × OD |
| 1.001 to 2.000 | 5 × OD |
| 2.001 and over | 6 × OD |

It is recommended that such bends be made before the terminal is applied to minimize electrical-contact distortion.

5. *Junction boxes.* Junction boxes are used when abrupt changes in direction of conduit runs are made, to provide for expansion and contraction of the conductors upon heating and cooling, and when tap connections are to be made (Fig. 2-53). Tap connectors are described in Subsec. B.

6. *Joint compound.* Connections in larger-size aluminum conductors are made with the aid of joint compound, as indicated above under splicing instructions. The compound is applied to the bare conductor after it has been wire-brushed. A coating of joint compound remains on the surface of the conductor, preventing the oxide from re-forming when it is broken in the installation operation. Manufacturers usually supply compression connectors already filled with joint compound to seal the connection from moisture and other contaminants. The compound must be added to connections made with mechanical connectors.

If current density is not high, successful long-lived connections may be made in aluminum conductors in sizes used in branch circuits and in residential wiring without joint compound, although it is a preferred practice to use the compound on all connections.

The user is cautioned that two classes of joint compounds are available and that each has its own applications. The type used with connections contains small metallic particles that help penetrate the aluminum oxide. The second type of compound does not contain these particles and should be used only for flat-surface, lug-to-lug, and lug-to-bus connections.

B. CONNECTORS   Only all-aluminum pressure-type connectors marked AL-CU to indicate that they have been tested and are listed by Underwriters Laboratories for aluminum, copper, or aluminum-to-copper connections interchangeably should be used. The connec-

**Fig. 2-51** Plan pulls so that the cable moves downward. This will speed the work and reduce the strain on the cables and pull line.

tors are usually plated to avoid the formation of oxide and to resist corrosion. Copper-clad aluminum conductors in feeder sizes also require the use of UL-listed AL-CU connectors.

Pressure connectors are of two basic types, a mechanical screw type and a compression

**Fig. 2-52** When bare steel basket grips and similar pull-line-to-cable devices must be used, cover them with plastic tape or rubberlike shields.

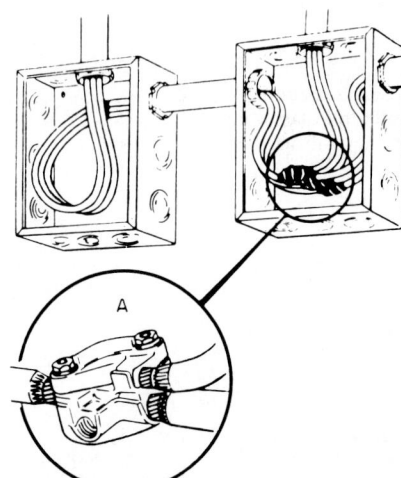

**Fig. 2-53** Use proper connectors and sufficient insulation, and leave enough slack for possible wiring changes and cable movement. The connector allows a tap without cutting the feeder (A).

type applied with a tool and die. Both are designed to apply sufficient pressure to shatter the brittle aluminum oxide from the strand surfaces and provide low-resistance metal-to-metal contact. Both basic types work reliably with aluminum, although many contractors prefer compression connectors because they are less susceptible to installation error.

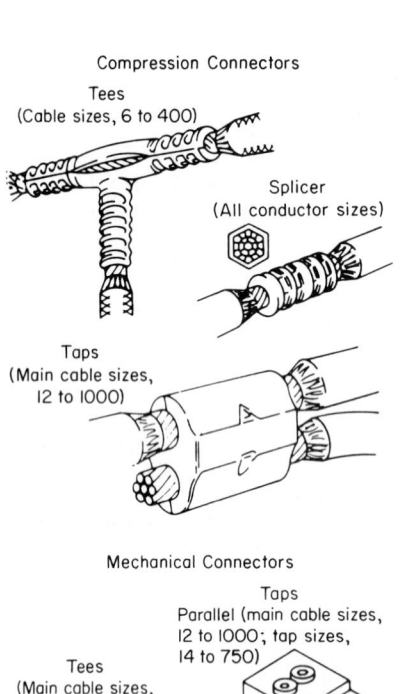

Compression Connectors

Tees
(Cable sizes, 6 to 400)

Splicer
(All conductor sizes)

Taps
(Main cable sizes,
12 to 1000)

Mechanical Connectors

Taps
Parallel (main cable sizes,
12 to 1000; tap sizes,
14 to 750)

Tees
(Main cable sizes,
1/10 to 500; tap sizes,
14 to 1/0)

Crossover, Tee,
or Parallel Tap

(Main and tap
cable sizes,
8 to 500)

**Fig. 2-54** Connectors have been designed and manufactured for every conceivable contingency. Make sure the connector is Underwriters Laboratories–listed for aluminum.

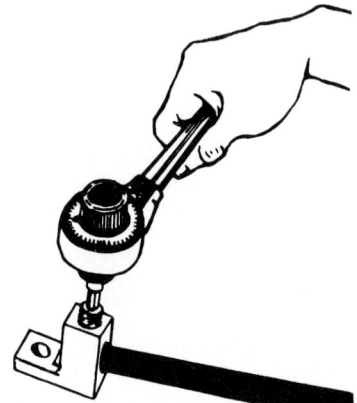

**Fig. 2-55** Proper torque is important: overtightening may sever the wires or break the fitting; undertightening will lead to overheating and failure.

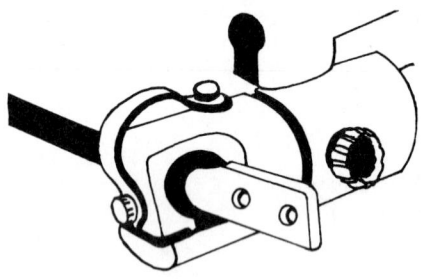

Mechanical Terminals

**Fig. 2-56** The crimping tool must be fully closed. Failure to close the tool will lead to an unsatisfactory and weak joint.

**Fig. 2-57** UL standard 486C was revised with regard to performance requirements and no twist-on connectors have subsequently been listed for use with aluminum or copper-clad aluminum conductors.

Effective January 2, 1987, UL standard 486C, covering connectors for use with aluminum wire, was revised. It is likely that some mechanical connectors now UL-listed for use with aluminum will not pass the more stringent requirements to be employed. Meanwhile, installers are cautioned to avoid mechanical pressure connectors with too wide a range of wire sizes because the screw may not adequately engage the strands of the smaller conductors. Connectors with copper bodies or steel bolts should also be avoided, as they may not give a reliable connection. Installers are also advised to obtain from conductor manufacturers recommendations concerning specific connectors for use with their products.

Connectors for every conceivable need are available. Some typical connectors are shown in Fig. 2-54. Whichever type you use, follow the manufacturer's instructions carefully.

There are available copper-bodied connectors that carry the AL-CU designation and are tin-plated and resemble aluminum. They should not be used with aluminum conductors larger than No. 6. If in doubt, file the edge of the connector to see if copper shows through.

In making connections, first strip the insulation as instructed in Subsec. A1. Then apply joint compound if it is not already contained in the connector. If the connector is a mechanical screw type, apply the manufacturer's recommended torque (Fig. 2-55). If it is a compression type, crimp it as recommended by the manufacturer (Fig. 2-56). Be sure to select the correct size of die and close the tool completely for full compression. Wipe off any excess compound. Then tape the joint as instructed in Subsec. A2, or apply the insulating enclosure that comes with some types of connectors.

Connections in branch circuits may be made with approved pigtail connectors of the crimp-on type (Fig. 2-70). No joint compound is required on such connections if load currents are low because of a limitation to small wire size (No. 10 AWG maximum), although its use is preferred because it generally improves performance.

C. TERMINATIONS  Plated aluminum-bodied terminal lugs are used to connect aluminum conductors to transformers, switches, busbars, motors, and other equipment. Like connectors, they are of two basic types, a mechanical screw type and a compression type applied with tool and die. Some typical terminal lugs are shown in Fig. 2-58. They are applied to the conductor in the same manner as described above under "Connectors."

Insist that all equipment be furnished with all-aluminum terminals that carry the UL AL-CU label. Terminals that are copper-bodied and tin-plated should not be used with aluminum conductors larger than No. 6.

When all components (bus, studs, lugs) are aluminum, only aluminum bolts should be

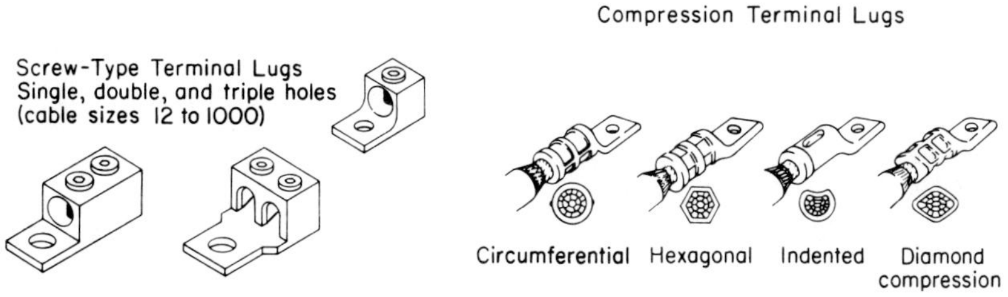

**Fig. 2-58**  Typical plated aluminum terminal lugs come in a variety of styles and are applied in same manner as standard connectors.

used to make the connections. The following procedures should be used. (*a*) Aluminum bolts should be anodized-alloy 2024-T4 and conform to ANSI B18.2.1 specifications and ASTM B211 or B221 chemical and mechanical property limits. (*b*) Nuts should be aluminum-alloy 6061-T6 or 6262-T9 and conform to ANSI B18.2.2. (*c*) Washers should be flat aluminum-alloy aluminum-clad 2024-T4, Type A plain, standard wide series, conforming to ANSI B27.2. SAE or narrow-series washers should not be used. (*d*) Hardware should be assembled as shown in Fig. 2-59. (*e*) All hardware should be lubricated with a suitable joint compound before tightening. (*f*) Bolts should be tightened to the manufacturer's recommended torque. In the absence of such recommendations, torque values listed in NEMA CC-1 should be used. See Table C-1.

If you are adding to an existing installation containing copper bus or studs or if it is impossible to obtain the required equipment with aluminum terminations, use a steel bolt with a Belleville spring washer to allow for the differing rates of thermal expansion of the materials. The following procedures should be used. (*a*) The steel bolt should be plated or galvanized, medium-carbon-steel–heat-treated, quenched, and tempered equal to ASTM A325 or SAE Grade 5. (*b*) Nuts should be heavy semifinished hexagon, conforming to ANSI B18.2.2, the thread to be unified coarse series (UNC), Class 2B. (*c*) Flat washers should be steel, Type A plain standard wide series, conforming to ANSI B27.2. SAE or narrow-series washers should not be used. (*d*) Belleville conical spring washers

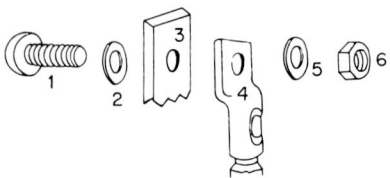

1. Aluminum bolt
2. Aluminum washer
3. Aluminum bus
4. Aluminum lug
5. Aluminum washer
6. Aluminum nut

**Fig. 2-59**  When all the components are aluminum, aluminum hardware should be used and installed as shown.

**TABLE C-1**

| Diameter of bolt, in | Nominal torque | |
|---|---|---|
| | ft/lb | in/lb |
| ⅜ (L) | 19 | 228 |
| ½ (L) | 40 | 480 |
| ⅝ (L) | 55 | 660 |

NOTE   L = lubricated.

**TABLE C-2**

| Bolt size, in | Belleville | | Nominal load, lb, to flatten |
|---|---|---|---|
| | Diameter, in | Thickness, in | |
| ¼ | 11⁄16 | 0.050 | 800 |
| 5⁄16 | 13⁄16 | 0.060 | 1000 |
| ⅜ | 15⁄16 | 0.070 | 1400 |
| ½ | 1 3⁄16 | 0.085 | 2700 |
| ⅝ | 1½ | 0.100 | 4000 |

NOTE   Material is hardened steel; finish, cadmium plate.

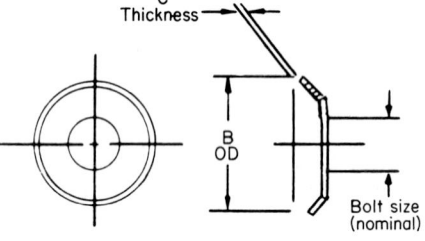

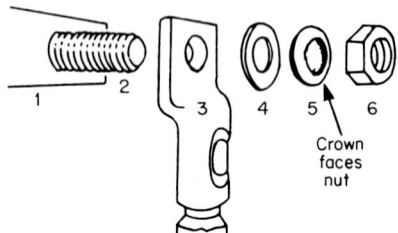

1. Aluminum or copper bus
2. Steel or copper stud
3. Aluminum lug
4. Steel flat washer
5. Steel Belleville
6. Steel nut

**Fig. 2-60**  A Belleville spring washer is used to make an aluminum-to-copper or steel joint. Note that the crown of the washer should be under the nut.

come in sizes for use with bolts ranging in sizes indicated in Table C-2. (*e*) Hardware should be assembled as shown in Fig. 2-60. (*f*) All hardware should be lubricated with a suitable joint compound before tightening. (*g*) Bolts should be tightened sufficiently to flatten the spring washer and be left in that position.

With equipment having terminals that will accommodate only copper conductors, a "gutter" splice may be used to connect the aluminum conductor. The aluminum conductor is spliced to a short length of copper conductor, and the copper-conductor stub is then connected to the equipment terminal (Fig. 2-61). An AL-CU compression-type connector is used to make the splice.

Two-hole lugs should be used for size 4/0 and larger conductors, and each bolt connecting a lug to a terminal or bus should not carry current exceeding the value in Table C-3.

In connecting large aluminum conductors (500,000 cmil, or 253.35 mm², and up) to heavy equipment having copper terminal studs or pads, or both, large compression-type lugs, preferably with two holes, should be used (Fig. 2-62). With other than aluminum bolts, Belleville spring washers and heavy flat washers in consecutive arrangement as shown in Fig. 2-62 must be used. If aluminum bolts and nuts are used, only the heavy washer bearing on the aluminum lug is necessary.

Wiring devices on branch circuits having ratings not exceeding 35 A usually are furnished with binding-screw terminals.

The installation of aluminum nonmetallic sheathed cable (Type NM) for branch circuits in buildings is the same as for residential wiring, which is covered in Subsec. 203E.

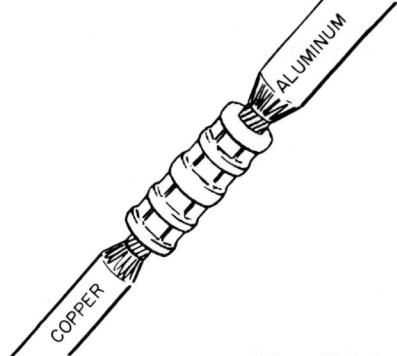

**Fig. 2-61**  A gutter splice is used when the terminal lugs are not removable and are approved for copper-cable connection only.

Figures 2-62 to 2-68 show typical connections of aluminum conductors to equipment terminals.

### TABLE C-3.  Bolt Loading

| Bolt Size, In | Current Capacity, A |
|:---:|:---:|
| ⅜ | 225 |
| ½ | 300 |
| ⅝ | 375 |
| ¾ | 450 |

#### D. SERVICE ENTRANCES

1. *Industrial.* With the aluminum service-drop cable in general use today, the connection of the service cable to aluminum service-entrance conductors requires only simple splices made as described in Subsec. A2, "Splicing." Connection of aluminum service-entrance conductors is made in the same manner when copper service-drop cable is still in use. In that case be sure to use compatible aluminum-bodied connectors.

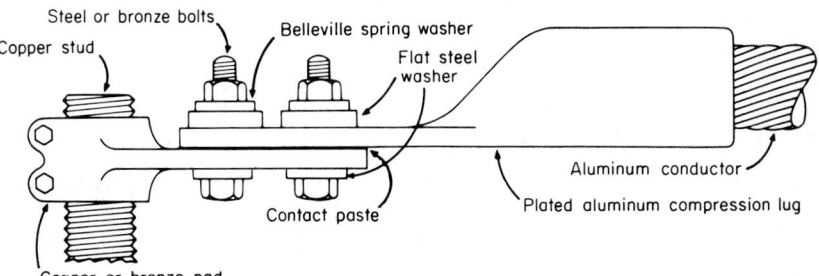

**Fig. 2-62**  Method of connecting large aluminum conductors to equipment studs or terminal pads made of copper. If the bolts are made of aluminum, it is not necessary to provide the Belleville spring washer.

Underground service cable also is largely aluminum today. This is true of both small- and medium-size services at secondary voltages as well as of large installations with transformers or unit substations supplied at primary voltages. A typical medium-size service entrance is shown in Fig. 2-69.

2. *Residential.* The connection to service-drop cable in residential services is made in the same manner as in small industrial or commercial service entrances. Type SE alu-

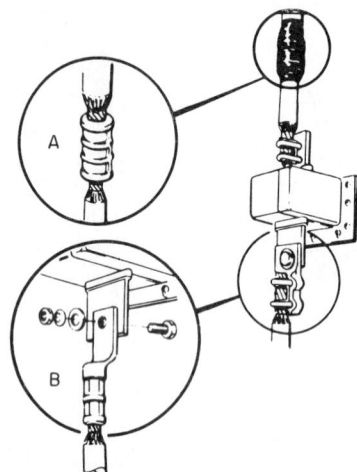

**Fig. 2-63** Where possible, current transformer terminals should be replaced with a compression type (B). If they cannot be removed, a section of copper cable should be spliced to aluminum (A).

**Fig. 2-64** Copper power transformer terminals require a short copper stub spliced to aluminum cable.

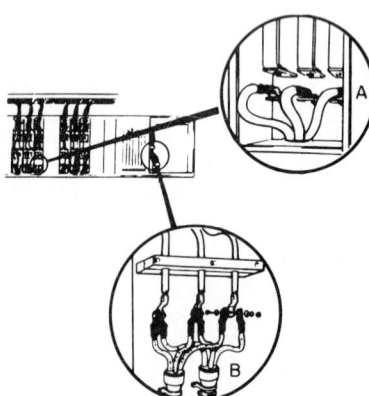

**Fig. 2-65** When connecting aluminum conductors to a unit substation with copper bus, use compression-type aluminum lugs attached with a Belleville washer (A). Copper primary leads on the transformer are connected to aluminum feeders in aluminum connectors and bolted back to back with Belleville washers (B).

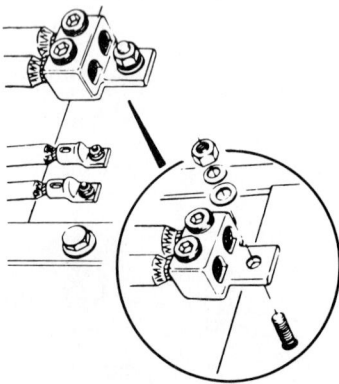

**Fig. 2-66** Copper lug connections on switchgear are replaced with equivalent aluminum connectors by using double-barrel lugs. A Belleville washer is used with copper or steel studs.

minum cable is used for this purpose and is connected to the utility service drop with splicing connectors, as described in Subsec. A2.

The size of service-entrance cable to be used depends on the load to be served. The 1990 NEC requires at least 100-A service for all single-family residences with six or more two-wire branch circuits. The **NEC** requires that service-entrance cable be not smaller than No. 6 AWG except in certain light-load installations. If initial loads are computed to

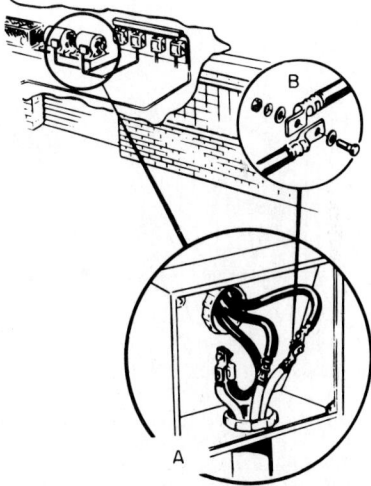

**Fig. 2-67** Copper terminations on switch connections may be replaced with aluminum by using smaller multiple aluminum conductors to reduce the voltage drop and increase the current-carrying capacity.

**Fig. 2-68** Three methods of making motor connections are shown in A. All terminals are preferably aluminum. If the bolt is steel or copper, a Belleville washer is necessary (B).

be 10 kW or more, the service must have an ampacity of not less than 100 A, which requires as a minimum No. 2 AWG aluminum SE cable. In any event, the size of service-entrance cable should be chosen in accordance with **NEC** requirements and the installation made in the manner required by local codes. In most areas local code bodies have adopted the **NEC** for electrical installations.

Meter boxes and service panels are generally available with aluminum AL-CU terminals, and aluminum service-entrance cable may be connected directly to these.

With equipment having terminals suitable only for copper cable, a gutter splice may be used to connect the aluminum cable, as described in Subsec. 203C, "Terminations." However, splices are prohibited in service-entrance conductors except under special conditions. See NEC, Sec. 230-46, 1990.

E. Circuit-Size Wiring   Because of reports of field failures originating at receptacle outlets wired with aluminum and a subsequent investigation indicating the apparent cause of many to be insufficient tightening of binding screws at the time of installation, UL revised the requirements for solid aluminum conductors in November 1971. UL also instituted new requirements and new testing procedures for wiring devices for use with circuit-size aluminum conductors.

One of the basic requirements for both wire and devices is a critical heating test of 500 4-h cycles with high overload currents. The new tests were designed to simulate field conditions with marginal workmanship. A 20-A device is tested on cyclic loads of 53 A, with binding screws tightened to a low torque of only 6 lb·in. No. 12 solid aluminum conductor is tested at 40 A (267 percent of rated load), with binding screws tightened to 6 lb·in.

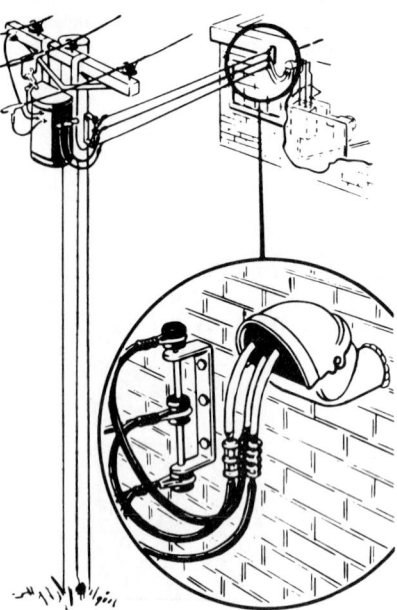

**Fig. 2-69** Most service-drop cables are now aluminum; they may be quickly field-spliced to aluminum service-entrance cable with a variety of compression connectors. A drip loop should be provided. If either the service-drop cable or the service-entrance cable is copper, the aluminum should always be higher than the copper.

Solid aluminum-alloy conductors must meet these requirements before receiving UL listing as "insulated aluminum wire." Devices which pass the revised UL requirements are rated CO/ALR and carry that marking on the mounting strap. Field reports confirm the reliability of the new, more flexible alloy wire and the CO/ALR switches and receptacles. Only these products should be used in aluminum 15- and 20-A branch-circuit wiring.

PIGTAILING METHOD Pigtailing, either field- or factory-wired, as illustrated in Fig. 2-70, is recognized by the NEC. As of the writing of this handbook, listed connectors only in the tool-applied crimp-type construction are available for pigtailing to 15- and 20-A wiring devices.

The correct method of terminating aluminum wire at a binding screw (Fig. 2-71) is as follows. (a) Wrap the freshly stripped end of the wire two-thirds to three-fourths of the distance around the wire-binding screw post as shown in Fig. 2-71A. The loop is made so that the rotation of the screw in tightening will tend to wrap the wire around the post rather than to unwrap it. (b) Tighten the screw until the wire is snugly in contact with the underside of the screw head and with the contact plate on the wiring device as shown in Fig. 2-71B. (c) Tighten the screw an additional half

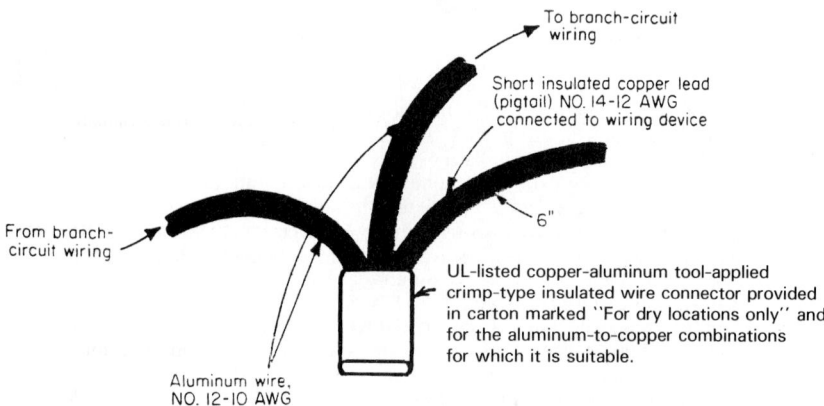

To branch-circuit wiring

Short insulated copper lead (pigtail) NO. 14-12 AWG connected to wiring device

From branch-circuit wiring

6"

Aluminum wire, NO. 12-10 AWG

UL-listed copper-aluminum tool-applied crimp-type insulated wire connector provided in carton marked "For dry locations only" and for the aluminum-to-copper combinations for which it is suitable.

Note: Detailed installation instructions provided with pressure wire connectors must be followed.

**Fig. 2-70** Pigtailing copper-to-aluminum conductors.

turn, thereby providing a firm connection. If a torque screwdriver is used, tighten to 12 lb·in (see Fig. 2-71C). (d) Position the wires behind the wiring device so as to decrease the likelihood of the terminal screws' loosening when the device is positioned into the outlet box.

Figure 2-72 illustrates incorrect methods for connection that should not be used.

In the following types of devices the terminals should not be connected directly to aluminum conductors but may be used with UL-labeled copper or copper-clad conduc-

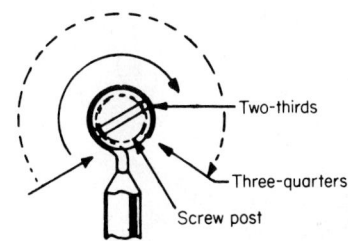

Step A: Strip and wrap wire

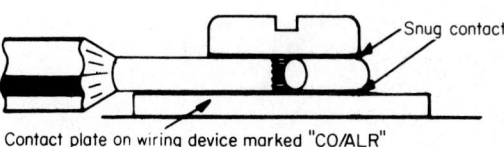

Step B: Tighten screw to full contact

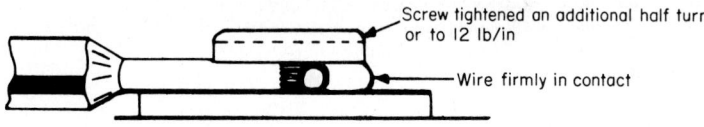

Step C: Complete connection

**Fig. 2-71**   Correct method of terminating aluminum wire at wire-binding screw terminals of receptacles and snap switches. [Underwriters Laboratories Inc.]

tors: receptacles and snap switches marked "AL-CU"; receptacles and snap switches having no conductor marking; receptacles and snap switches having back-wired terminals or screwless terminals of the push-in type.

The new CO/ALR marking applies only to 15- and 20-A wiring devices such as receptacles and switches. All other devices such as circuit breakers, fusible protection, and oven or dryer receptacles continue to be marked AL-CU when approved for use with aluminum.

You may be called upon to inspect homes wired with aluminum between 1965 and 1972 to determine whether or not the installation is functioning satisfactorily. If examination discloses overheating or loose connections, it is advisable to replace the older receptacles with new devices carrying the CO/ALR marking. The UL-listed crimp-on compression tool (Copalum by AMP) may be used. See Fig. 2-70.

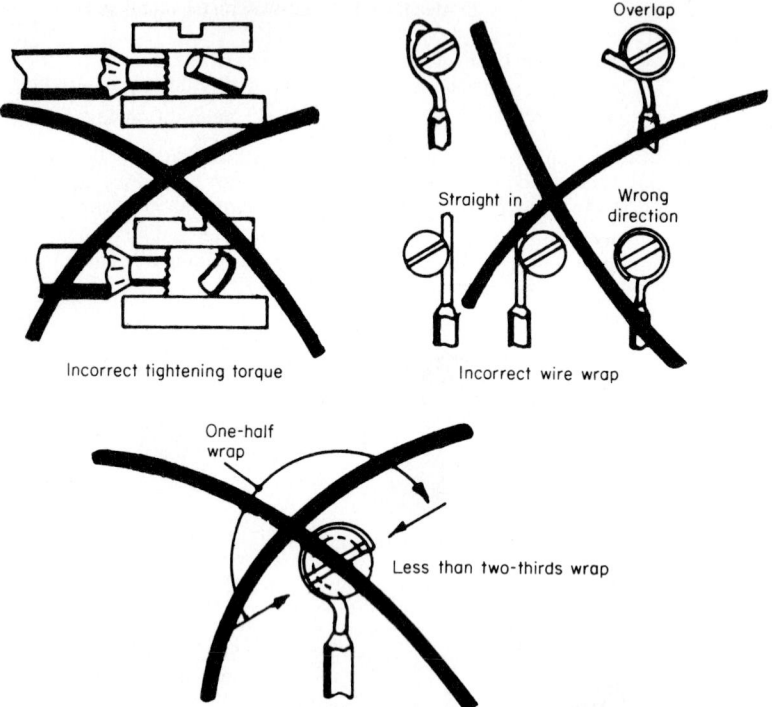

**Fig. 2-72** Incorrect methods of terminating aluminum wire at wire-binding screw terminals of receptacles and snap switches. [Underwriters Laboratories Inc.]

### 203F. Bending Radii for Cable Not in Conduit or on Sheaves or While under Tension

Power cables without metallic shielding or armor

The minimum recommended bending radii as multiples of the overall cable diameter given in the following tabulation are for both single and multiconductor cable with or without lead sheath and without metallic shielding or armor.

| Thickness of conductor insulation, 64ths in | Minimum bending radius as multiple of cable diameter; overall diameter of cable, in | | |
|---|---|---|---|
| | 1.00 and less | 1.01 to 2.00 | 2.00 and over |
| 10 and less | 4 | 5 | 6 |
| 11 to 20 | 5 | 6 | 7 |
| 21 and over | . . . | 7 | 8 |

Power cables with metallic shielding or armor

*a. Interlocked armored cables.* The minimum recommended bending radius for all interlocked armored cables is in accordance with table but not less than 7 times the overall diameter of the cable except as noted below (*c*) for shielded cable.

*b. Flat-tape– and wire-armored cables.* The minimum recommended bending radius for all flat-tape–armored and all wire-armored cables is 12 times the overall diameter of the cable.

*c. Shielded cables.* For all cables having metallic shielding tapes the minimum recommended bending radius is 12 times the overall diameter of the completed cable.

*d. Wire-shielded cables.* Wire-shielded cables should have the same bending radius as power cables without metallic shielding tape.

NOTE   Reprinted in whole or in part from ICEA S-66-524, NEMA WC-7.

**TERMINATION AND SPLICE KITS**

**204. All-tape and tape-cast terminations and splices** (3M Company). The 3M system for splicing and terminating consists in part of a series of engineered drawings for splicing and terminating power cables. These data sheets contain data for splicing and terminating solid dielectric shielded power cable, as shown in Fig. 2-73. The designs include all-tape and tape-cast (tape-resin) in-line and T splices for 5- to 25-kV cable, all-resin in-line and T splices for 5- to 8-kV cable, and all-tape terminations for 5- to 25-kV cable, copper or aluminum conductors from No. 8 AWG to 2500 kcmil. Each engineered drawing includes comprehensive, step-by-step installation instructions and complete bills of materials and dimensions.

I. Metallic-shield type.        II. Wire-shield type.

**Fig. 2-73** Shielded cable. [3M Co.]

The application of these data sheets is for splicing and terminating 5- to 25-kV (up to 8-kV for all-resin splices) metallic and wire shielded-type power cables from No. 8 AWG to 2500 kcmil copper or aluminum conductors on the following solid cables: polyethylene (high- and low-density), cross-linked polyethylene (XLP), ethylene propylene rubber (EPR), and butyl rubber.

Splices and terminations for shielded cable operating from 5000 to 25,000 V (to 8000 V for all-resin splices) shall be made by using tape and/or materials in accordance with drawings engineered by the manufacturer of the splice material. The terminations, as constructed according to these designs, shall meet all the design dielectric requirements of the IEEE 48-1975 standard for potheads. The splices, as constructed according to these designs, shall meet the requirements of the IEEE standard for power-cable joints. All splices and terminations constructed according to these designs must be able to be hypotted (submitted to a high-voltage test) according to ICEA standards S-66-524 and S-19-81.

Splices and terminations for metallic shielded-type cables, operating from 5000 to 25,-000 V (to 8000 V for all-resin splices), shall be made by using materials in accordance with drawings engineered by the manufacturer of the splice material, such as the 2047 splice and termination drawing series available from the 3M Company. In-line splices shall be made according to engineering drawing 2047-A-1, 2047-K-4, or 2047-X-1; T splices shall be made according to engineering drawing 2047-C-2, 2047-M-2, or 2047-Y-1; and standard terminations shall be made according to engineering drawing 2047-B-2. In areas where space is limited, terminations shall be made according to engineering drawing 2047-B-16.

The tape and tape-cast 3M systems for splicing and terminating designs can be used on cables with a rated operating temperature of 90°C and an emergency overload rating of 130°C.

All termination designs meet the design dielectric requirements of the IEEE 48-1975 standard for potheads. All splice designs meet the requirements of the IEEE standard for power-cable joints. All splices and terminations made according to these designs may be dc-hypotted according to ICEA standards S-66-524 and S-19-81.

**205. In-line splicing kits for 15-kV and 25-kV URD cable** (3M Company: see Fig. 2-74). Specifically engineered for direct-burial URD applications, Quick-Splice accommodates a wide variety of cable types, both 15-kV and 25-kV, for use with nominal 175- to 280-mil (4.45- to 7.11-mm) polyethylene and cross-linked polyethylene (XLP) or ethylene propylene rubber (EPR), insulated concentric neutral (URD) cable, aluminum or copper, stranded or solid conductor. Its application range is as follows: 15 kV, from No. 2 AWG to 750 kcmil; 25 kV, from No. 2 AWG to 500 kcmil.

INSTALLATION

1. Prepare cables as you would any splice. Apply one half-lapped layer of vinyl tape over the semiconductive jacket of the cables.

2. Apply silicone grease over cable insulation and vinyl tape. Then slide splice end

caps onto each cable until they are butted against the concentric neutral wires. Slide the splice body onto the cable.

3. Crimp the connector over the conductors. Use conventional crimping tools; no special tools or techniques are required.

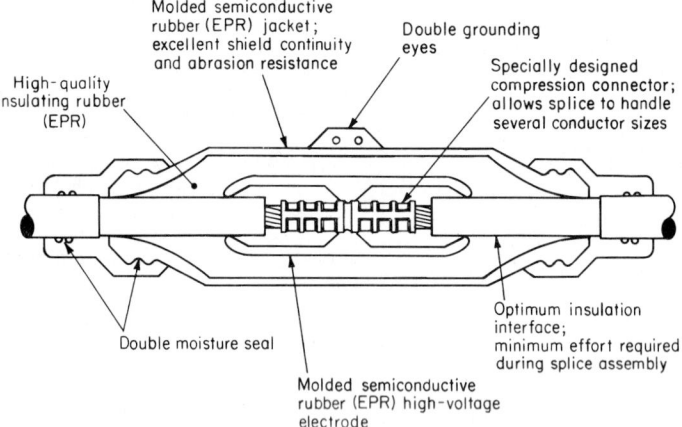

**Fig. 2-74**    3M Quick-Splice in-line URD splice kit. [3M Co.]

4. Slide the splice body into its final position over the connector. Leave a small area of insulation exposed at both ends of the splice body. Remove the applied vinyl tape.

5. Slide end caps onto the splice body. Return concentric neutral wires to the edge of the splice body and connect to the ground eyes.

**206. Resin splicing kits**  (3M Company). A full line of splicing kits utilizing a resin sealing compound in a voltage range of 600, 1000, and 5 kV is available.

A. MULTIMOLD SPLICING KIT   (see Fig. 2-75) Odd-size and odd-shaped splices can be difficult to insulate, especially when they must be moistureproof for underground appli-

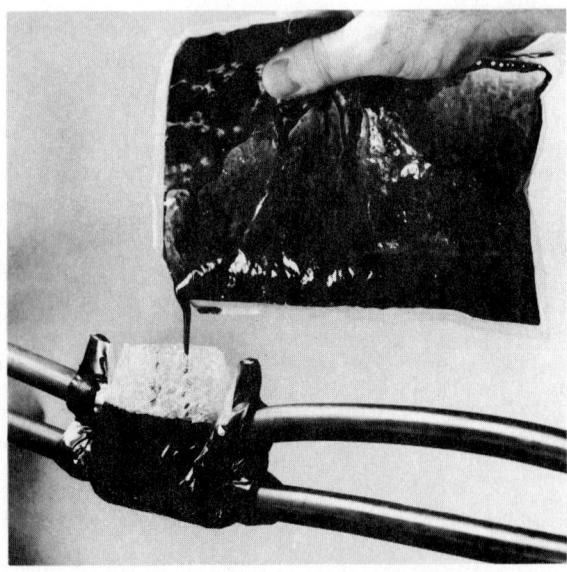

**Fig. 2-75**    Multimold splice kit. [3M Co.]

cations. Versatile multimold splicing kits are designed to handle almost any type of splice configuration up to 1000 V either aboveground or buried. Each kit contains a wraparound mold body of porous webbing on a thin, tough polyester film to ensure the proper thickness of resin around the connection, sealing putty strips on the mold body to form a resin-tight envelope for the resin, and Scotchcast brand electrical insulating resin No. 4, packed in a Unipak brand container to ensure convenience and proper mixing ratios.

Use multimold splicing kits for any odd-shaped and odd-size splices, as in secondary distribution, plant grounds, parking-lot and airport-runway lighting, electrical sprinkling systems, sheath repair, remodeling wiring, sealing anodes, leads, and cathodic protection. The kits are for use on cables operating at 1000 V and under.

B. Y AND IN-LINE SPLICING KIT   (see Fig. 2-76) Each kit is packaged to simplify the job and contains all the materials necessary to make a splice. The contents include two-piece snap-together mold bodies, tape for sealing ends, two funnels, and insulating and sealing compound Scotchcast brand No. 4 resin in a Unipak container. These field-splicing kits conform in strength and electrical properties to the cable. To use them, snap the transparent mold bodies over the prepared cable, mix the resin in the Unipak container, and pour. When the resin cures, in about 30 min, a moisturetight seal and a perfect electrically

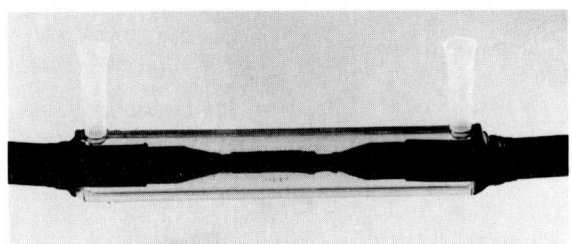

**Fig. 2-76**   Y and in-line splice kit. [3M Co.]

insulated splice are complete. Scotchcast splicing kits can be used for overhead, underground, or direct-burial applications.

**207. Porcelain termination kits** (3M Company; see Fig. 2-77).   It is possible to make a custom pothead in the field with a Scotchcast porcelain termination kit. Such a kit will terminate No. 2 AWG to 750 kcmil shielded or concentric neutral cable from 5 to 25 kV (see Fig. 2-77 and accompanying tables).

High-quality wet-process porcelain insulators assure maximum reliability for a non-tracking surface. They are sky-blue to blend against the horizon.

3M has designed a sealed metal topcap which not only assures a permanent seal but also eliminates the necessity for special internal or external connectors on copper conductors. A pressure contact is made on the conductor by a pad which is driven against the conductor by a setscrew. Load-cycling tests prove that this pressure contact will prevent overheating of the pothead. The entire contact is sealed when the poured elastomeric compound has hardened and the metal sealing plug is installed. The aerial wire connection is made in the same manner and has a quick disconnect feature. Special termination lugs are necessary for aluminum conductors.

To accommodate a wide range of cable sizes and handle shielded and concentric neutral cables, an elastomeric insulation is molded around the cable on the job. The elastomeric compound cures to a solid void-free elastomer in minutes; no outside heat is required.

The termination kits are provided with a single-tongue mounting ring which is made of a conductive chromate-coated casting and is an integral part of the termination. The ring can be attached to a standard crossarm bracket.

Special termination lugs which adapt the kits for use on cables with aluminum conductors have been designed. Conventional crimping tools are used.

**208. Molded-rubber termination kits for No. 2 AWG to 500 kcmil** (3M Company). Each 3M Quick-Term MT-series kit will handle a wide range of cable and conductor sizes. The kits are lightweight and are available in skirted or tubular silicone rubber insulators. They permit free-hanging applications or standard mounting with optional cable support.

The skirted insulators are designed to protect terminations on 15-kV concentric neutral cable in contaminated environments (see Fig. 2-78, I). The tubular insulators are for indoor use, pad mounts, switchgear, or mildly contaminated environments (see Fig. 2-78, II).

3M stress cones are designed to slide easily over cable insulation. A fold-back base permits visual alignment and sealing. Skirted and tubular insulators come prestretched over a collapsible core. They position quickly, then shrink tightly in place when the core is unwound.

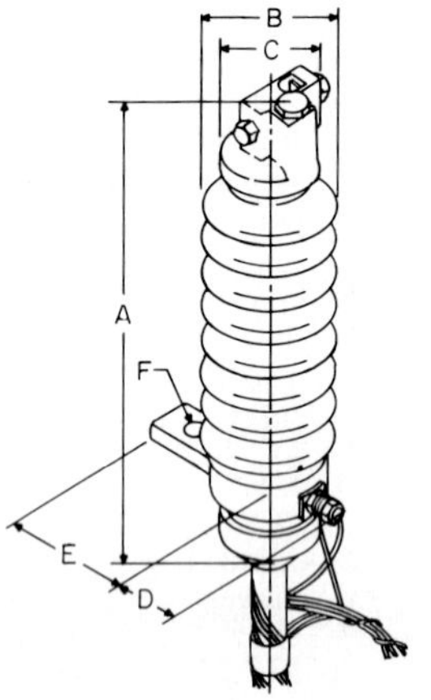


For use on concentric neutral (URD) and synthetic insulated shielded cable. The terminations meet the design dielectric requirements of the IEEE No. 48 standard for potheads.

**Porcelain Termination Kits**

| Dimensions | 5903, in | 5904, in | 5905, in | 5906, in | 5912, in | 5913, in |
|---|---|---|---|---|---|---|
| A | 12¼ | 14¼ | 12¾ | 14¾ | 12¾ | 14¾ |
| B | 3⅜ | 3⅜ | 4½ | 4½ | 4½ | 4½ |
| C | 2⅜ | 2⅜ | 3⅛ | 3⅛ | 3⅛ | 3⅛ |
| D | 2⅜ | 2⅜ | 2⅝ | 2⅝ | 2⅝ | 2⅝ |
| E | 3⅝ | 3⅝ | 4 | 4 | 4 | 4 |
| F | 0.530 minimum | 0.530 minimum | 0.530 minimum | 0.530 minimum | 0.530 minimum | 0.530 minimum |
| Dry arcing distance | 6¼ | 8⅜ | 6¼ | 8⅜ | 6¼ | 8⅜ |
| Leakage distance | 11½ | 15½ | 13½ | 17½ | 13½ | 17½ |

*Ratings*

| | | |
|---|---|---|
| 5903 | 15 kV | 110 bil[a] |
| 5904 | 25 kV | 150 bil |
| 5905 | 15 kV | 110 bil |
| 5906 | 25 kV | 150 bil |
| 5912 | 15 kV | 110 bil |
| 5913 | 25 kV | 150 bil |

[a]Basic impulse insulation level.

**Fig. 2-77** Porcelain termination kit. [3M Co.]

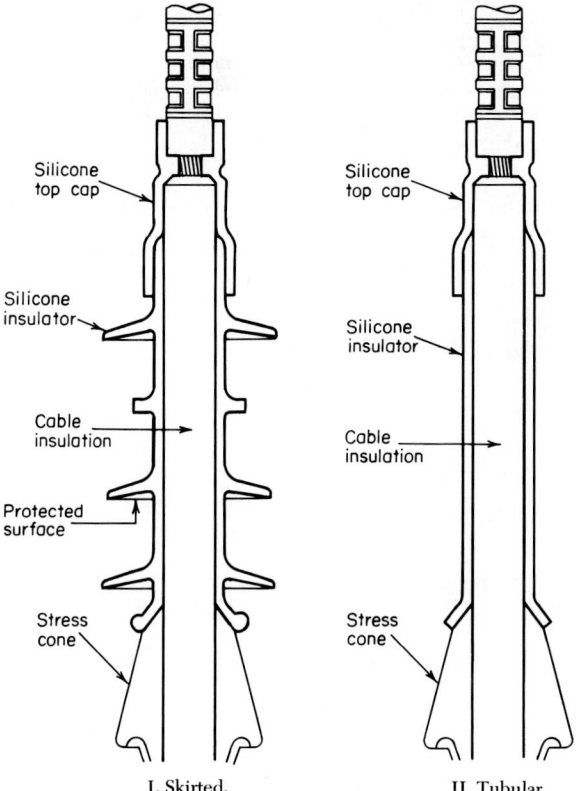

I. Skirted.                    II. Tubular.

**Fig. 2-78**    Molded-rubber termination kits. [3M Co.]

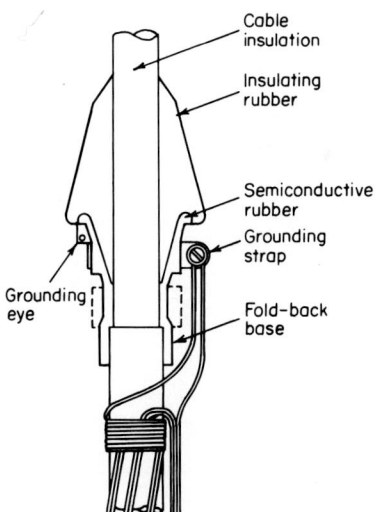

**Fig 2-79**    Stress cone for rubber termination kit. [3M Co.]

INSTALLATION

1. Lubricate the prepared cable and the bore of the stress cone with the silicone grease provided. The specially formulated rubbers slide easily over the cable insulation, and live action ensures a tight interface fit during load cycling.

2. The base of the stress cone comes rolled back to help assure proper alignment with cable-shield cutoffs, thus providing a close, void-free fit between the cable shield and the base of the stress cone even if the cable jacket is not cut off squarely. See Fig. 2-79.

3. The base of the stress cone unrolls down over the cable shield to form a tight-fitting, weatherproof seal-and-jacket continuity even if the cable is in a bent position. The ground wire is fastened as the last step in the installation of the stress cone.

Division **3**

# Circuits and Circuit Calculations

## TYPES OF CIRCUITS

**1. A series circuit** is one in which all components are connected in tandem as in Figs. 3-1 and 3-2. The current at every point of a series circuit is the same. Series circuits find their most important commercial application in series street lighting. They are seldom if ever used in the United States for the transmission of power.

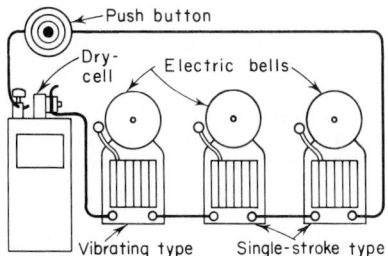

**Fig. 3-1**  Series electric-bell circuit.

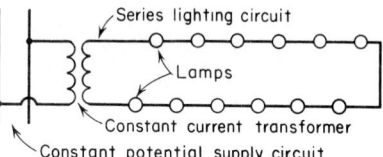

**Fig. 3-2**  Series street-lighting circuit.

**2. Multiple, parallel, or shunt circuits** are those in which the components are so arranged that the current divides between them (Figs. 3-3 and 3-4). Commercially, the distinction between multiple and series circuits is that in series lighting circuits the current is maintained constant and the generated emf varies with the load whereas in multiple circuits the current through the generator varies with the load and the generator emf is maintained practically constant.

**3. Adding receivers in parallel** on multiple circuits is really equivalent to increasing the cross section of the imaginary conductor formed by all the receivers in parallel between the + and the − sides of the circuit.

**4. The distribution of current in a multiple circuit** is shown in Fig. 3-5. Motors, heating devices, or other equipment requiring electricity for their operation could be substituted for the incandescent lamps if the proper current values were substituted for

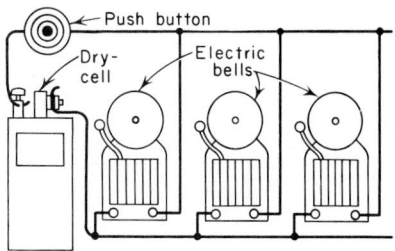

**Fig. 3-3**  Electric bells in parallel.

**3-1**

those shown. Note that the current in the main conductors decreases toward the end of the run and that the current supplied by the source, the generator, is equal to the sum of the currents required by all the components. The voltage at the end of the run is less than that at the generator.

**5. A multiple-series or parallel-series circuit** consists of a number of minor circuits in series with each other and with several of these series circuits then connected in parallel, as shown in Figs. 3-6 and 3-7. Incandescent lamps and motors used in railway work are

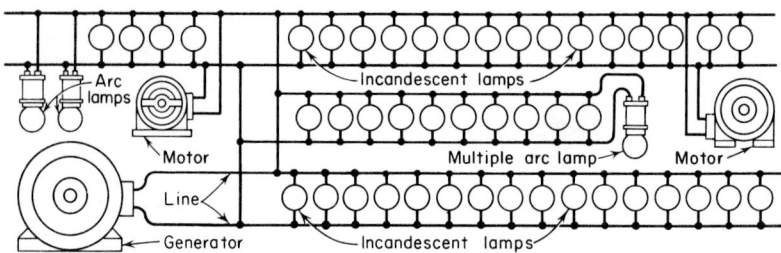

**Fig. 3-4** A multiple circuit for light and power.

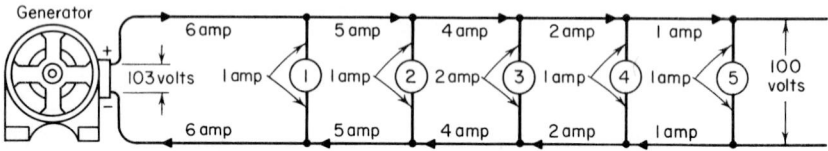

**Fig. 3-5** Multiple-circuit currents.

sometimes arranged in this way. For example, on a railway car operating from a 600-V third rail, the 120-V lamps are connected five in series (see Fig. 3-6). The motors on railway cars which operate from a 3000-V dc overhead trolley wire are usually connected two in series so that the voltage per motor will be only 1500 V (see Fig. 3-7).

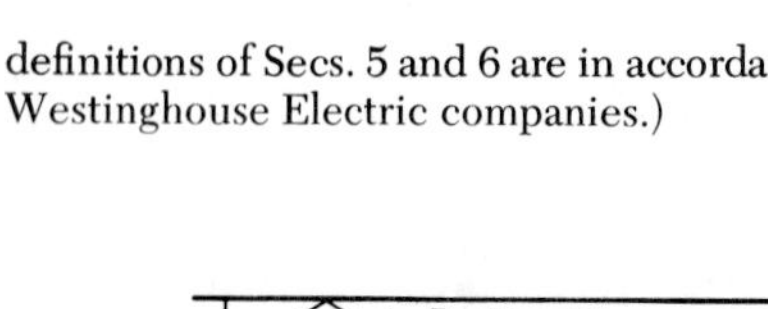

**6. A series-multiple or series-parallel circuit** is one in which first a number of minor circuits are connected in parallel and then several of the parallel-connected minor circuits are connected in series across a source of emf as in Fig. 3-8. This method of connection is seldom used. (There appears to be a difference of opinion as to what constitutes a series-multiple and what a multiple-series circuit. The definitions of Secs. 5 and 6 are in accordance with the practice of the General Electric and Westinghouse Electric companies.)

**Fig. 3-6** Multiple-series circuit supplying lamps on a railway car.

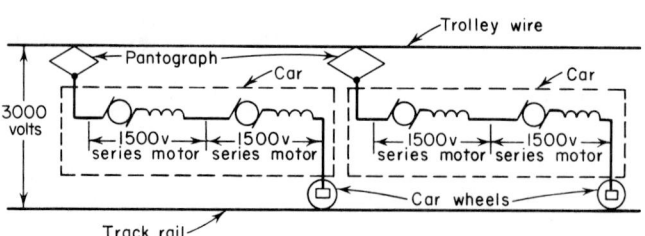

**Fig. 3-7** Multiple-series circuit supplying motors on railway cars.

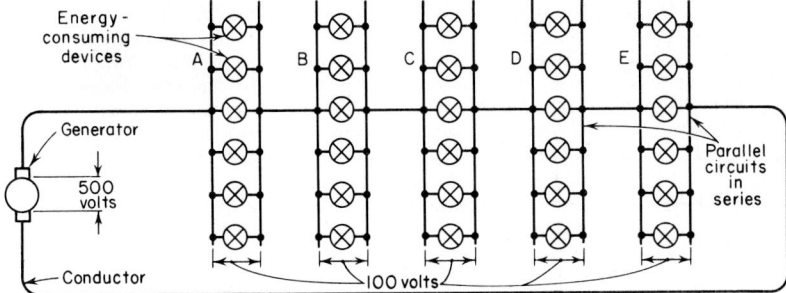

**Fig. 3-8**   A series-parallel or series-multiple circuit.

**7. A divided circuit** (Fig. 3-9) is really one form of multiple or parallel circuit. The distinction between the two sorts appears to be that, as ordinarily used, the term *divided* refers to an isolated group of a few conductors in parallel rather than to a group of a large number of conductors in parallel.

**8. The joint resistance of a number of conductors in parallel** can be computed with the following formula. There should be as many terms in the denominator of the formula as there are conductors in parallel.

$$R = \cfrac{1}{\cfrac{1}{r_1} + \cfrac{1}{r_2} + \cfrac{1}{r_3} + \cfrac{1}{r_4}, \text{etc.}} \qquad (1)$$

**example**   What is the joint resistance of the conductors in the divided circuit shown in Fig. 3-9? In other words, what is the resistance from $A$ to $B$?

**solution**   Substitute in the formula

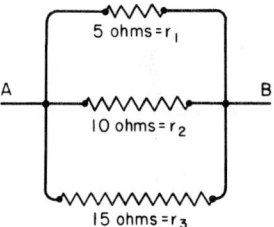

**Fig. 3-9**   A divided circuit.

$$R = \cfrac{1}{\cfrac{1}{r_1} + \cfrac{1}{r_2} + \cfrac{1}{r_3}} = \cfrac{1}{\cfrac{1}{5} + \cfrac{1}{10} + \cfrac{1}{15}} = \cfrac{1}{\cfrac{6}{30} + \cfrac{3}{30} + \cfrac{2}{30}} = \cfrac{1}{\cfrac{11}{30}} = 1 \times \frac{30}{11} = 2.73\,\Omega$$

**9. Multiple circuits** are employed for the distribution of electrical energy for all lighting and power work with the exception of street lighting, for which series circuits frequently are used.

**10. The complete electric system** from the generating plant to the particular piece of equipment utilizing the electrical energy for heating, lighting, power, etc., may be divided into three main parts. These are the transmission system, the distribution system, and the interior wiring in the building where the energy is utilized. Two simple complete electric systems illustrating the component parts are shown in Figs. 3-10 and 3-11. All

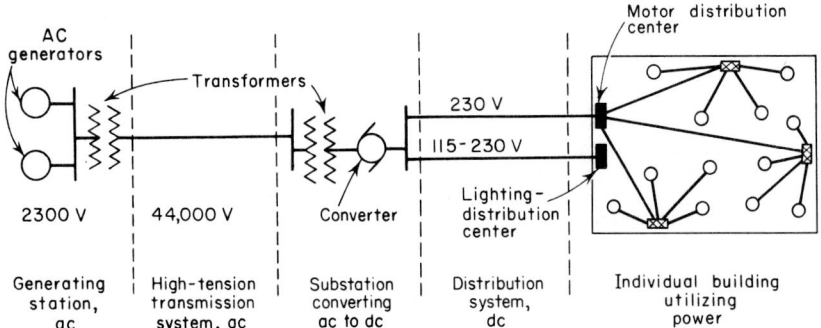

**Fig. 3-10**   Simple complete electric system, generating and transmitting alternating current, distributing direct current.

wires of a circuit are represented for simplicity by a single line. Voltages commonly used for the different parts of the system are indicated in the figures. In the system shown in Fig. 3-10 the electricity is generated as alternating current at 2300 V. At the generating station it is stepped up to 44,000 V and transmitted at this voltage over the transmission system to the substation. In the substation the energy is converted by means of a rotary converter to direct current. From the substation the energy is distributed over two distribution systems to the different buildings. One of these distribution systems is two-wire, 230-V direct current for the motor load, and the other is three-wire, 115- to 230-V direct current supplying the energy for lighting. In the system of Fig. 3-11 the energy is generated, transmitted, and distributed as alternating current throughout. This system is more common.

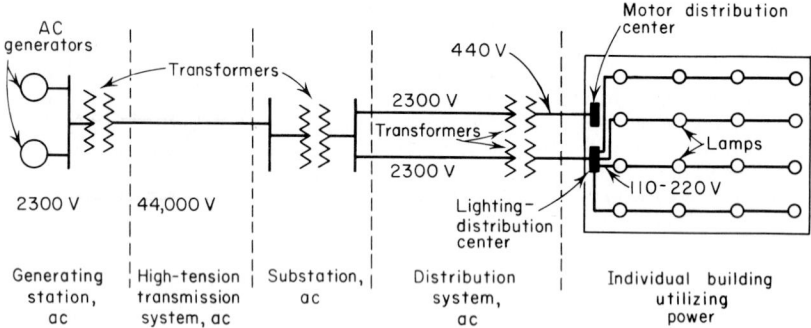

**Fig. 3-11**  Simple complete electric system, generating, transmitting, and distributing alternating current.

**11. A feeder** or feeder circuit (Figs. 3-12 and 3-13) is a set of conductors in a distributing system extending from the original source of energy in the installation to a distributing center and having no other circuits connected to it between the source and the center. The source may be a generating station or a substation or, in the case of building or house wiring, a connection to the service conductors from the street (see Figs. 3-12 and 3-13).

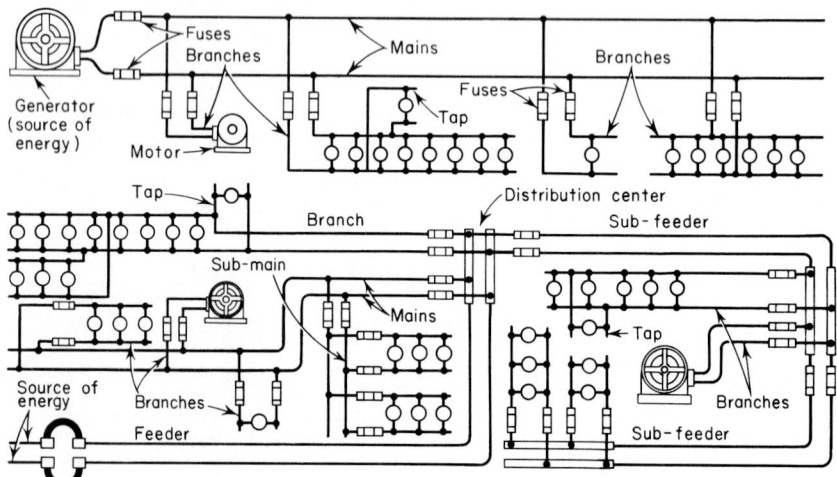

**Fig. 3-12**  Diagram illustrating circuit nomenclature.

**12. A subfeeder** is an extension, fed through a cutout, of a feeder or of another subfeeder from one distributing center to another and having no other circuit connected to it between the two distributing centers.

**13. A main** (Figs. 3-12 and 3-13) is any supply circuit to which other energy-consuming circuits (submains, branches, or services) are connected through automatic cutouts (fuses or circuit breakers) at different points along its length, which is of the same size of wire for its entire length, and which has no cutouts in series with it for its entire length. If a main is supplied by a feeder, the main is frequently of smaller wire than the feeder which serves it. An energy-utilizing device is never connected directly to a main, a cutout always being interposed between the device and the main.

**14. A submain** (Fig. 3-12) is a subsidiary main, fed through a cutout from a main or from another submain, to which branch circuits or services are connected through cutouts. A submain is usually of smaller wire than the main or other submain which serves it.

**15. A branch or branch circuit** (Fig. 3-12) is a set of conductors feeding, through an over-current device (from a distribution center, main, or submain) to which one or more energy-consuming devices are connected directly, i.e., without the interposition of additional overcurrent devices. The only overcurrent device associated with a branch is the one through which the branch is fed at the main, submain, or distribution center.

**16. A tap or tap circuit** (Fig. 3-12) is a circuit which serves a single energy-utilizing device and is connected directly to a branch without the interposition of a cutout.

**17. A distributing or distribution center** in an electrical-energy distribution system is the location at which a feeder, subfeeder, or main connects to a number of subordinate circuits which it serves. The switches and automatic overcurrent devices for the control and protection of the subcircuits are usually grouped at the distributing center. In interior-wiring parlance, a distributing center is often an arrangement or group of fittings by which two or more minor circuits are connected at a common location to another larger circuit. A panelboard with circuit breakers or a group of fuses is a distribution center (see Fig. 3-12).

**18. A service** is the conductors and equipment for delivering energy from the electricity supply system to the wiring system of the premises served.

A *service cable* is the service conductors made up in the form of cable.

*Service conductors* are the supply conductors which extend from the street main or

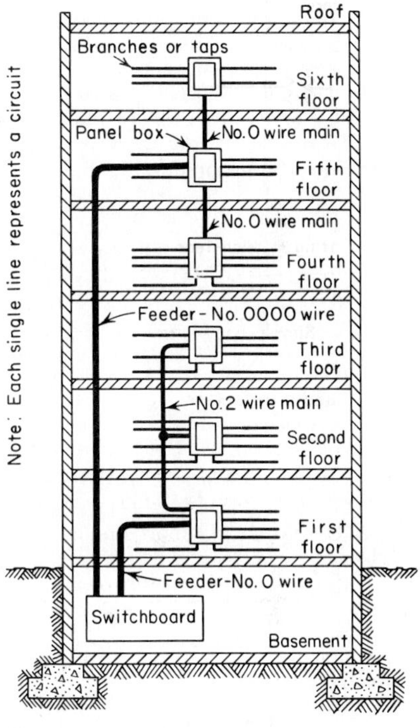

**Fig. 3-13** Examples of feeders, mains, and branches.

from transformers to the service equipment of the premises supplied.

A *service drop* is the overhead service conductors from the last pole or other aerial support to and including the splices, if any, connecting to the service-entrance conductors at the building or other structure.

A *service lateral* is that portion of underground service conductors between the street main, including any risers at a pole or other structure or from transformers, and the first point of connection to tbe service-entrance conductors in a terminal box inside or outside the building wall. If there is no terminal box or meter or other enclosure with adequate space, the point of connection shall be considered to be the point of entrance of the service conductors into the building.

*Service-entrance conductors (overhead system)* are the service conductors between the terminals of the service equipment and a point usually outside the building, clear of building walls, where joined by a tap or splice to the service drop.

*Service-entrance conductors (underground system)* are the service conductors between the terminals of the service equipment and the point of connection to the service lateral.

If service equipment is located outside the building walls, there may be no service-entrance conductors or they may be entirely outside the building.

*Service equipment* is the necessary equipment, usually consisting of a circuit breaker or switch and fuses and their accessories, located near the point of entrance of supply conductors to a building or other structure or an otherwise-defined area, that is intended to constitute the main control and means of cutoff of the supply.

## ELECTRICAL SYSTEMS

**19. Electrical systems.** Several systems can be used for distributing electrical energy over the various arrangements of circuits previously described. They are as follows:

1. Direct-current two-wire (Fig. 3-14)
2. Direct-current three-wire (Fig. 3-15)
3. Single-phase two-wire (Fig. 3-14)
4. Single-phase three-wire (Fig. 3-15)
5. Two-phase four-wire (Fig. 3-17)
6. Two-phase three-wire (Fig. 3-18)
7. Two-phase five-wire (Fig. 3-19)
8. Three-phase three-wire (Fig. 3-20)
9. Three-phase four-wire (Fig. 3-21)

It is generally not good practice to supply lamps and motors from the same circuit (unless the motors are of small, fractional-horsepower size) for the following reasons.

1. When the motors are started, the large starting current causes a voltage drop on the feeder, which will make the lights grow dim or blink.

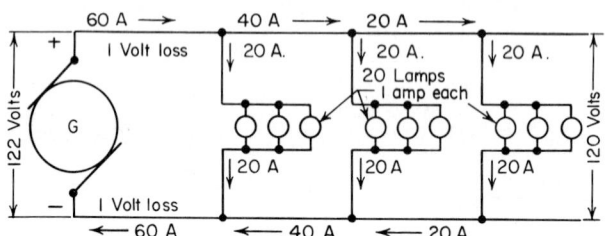

**Fig. 3-14** Two-wire system (direct-current or single-phase).

2. Overloads and short circuits are more common on motor circuits and should not be the cause of putting the lights out.

3. Lamps for satisfactory service must operate within closer voltage limits than motors, and therefore the circuits must be designed for less voltage drop than is allowable on motor circuits.

4. Frequently it is more economical to operate motors on a higher voltage than that of the lighting circuit.

In installations in which busway wiring is employed, it is often very satisfactory to have a common motor and lighting bus for the feeders and mains. If these systems are properly planned with generous allowance in the size of the buses, they will overcome the above criticisms of the common system and provide a very economical installation.

In the illustrations of the various systems, motors and lamps are shown connected to the same circuits simply to indicate the manner in which the different loads would be connected to that type of circuit. The reader is cautioned against forming the impression that they should be connected to the same circuit except in special cases. In considering any one of the illustrations, if one were dealing with a motor circuit, the lamps should be eliminated from the figure, or if one were dealing with a lighting circuit, the motors should be eliminated.

**20. The three-wire dc or single-phase ac system is used because it saves copper** (see Figs. 3-15 and 3-22). Incandescent lamps for 110 to 120 V are more economical than those for higher or lower voltages. A system of any consequence operating at 110 V would require very large conductors to maintain the line drop within reasonable limits. With the three-wire system, a low voltage, say, 110, is impressed on the receivers while one twice as great, say, 220, is used for distribution. Since the weight of conductors for a given power loss varies inversely as the square of the voltage, it is evident that a considerable saving is possible with the three-wire system. In the United States the three-wire system is of most importance as applied to 110/220-V lighting systems.

**21. The principle of the three-wire dc or single-phase ac system is illustrated** in Fig. 3-22. Incandescent lamps for 110 V could be connected two in series across 220 V as shown at I, and although each lamp would operate at 110 V, the energy to the group would be transmitted at 220 V and the outside conductor could, with equal loss, be one-fourth the

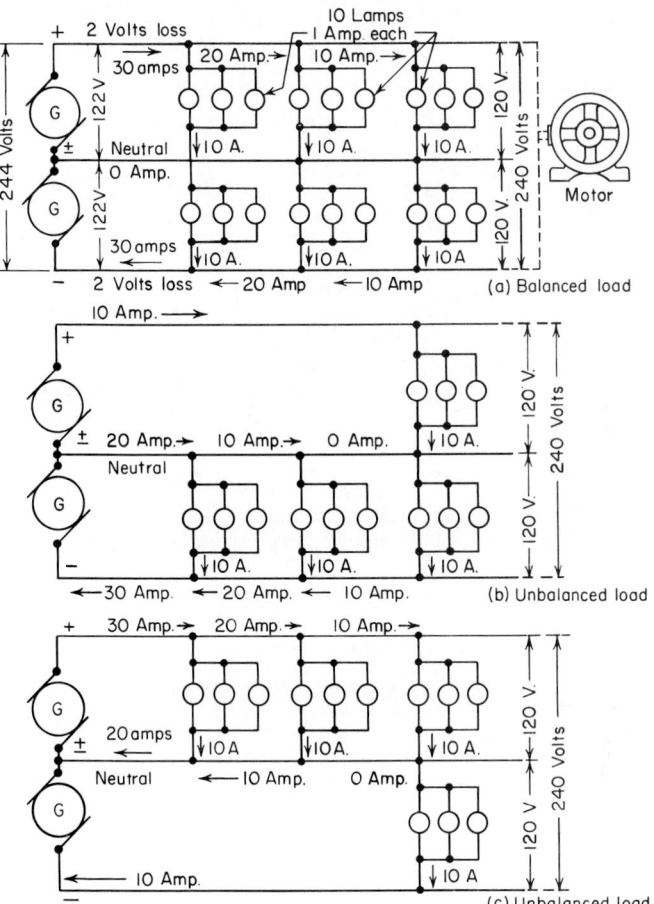

**Fig. 3-15**   Three-wire system (direct-current or single-phase).

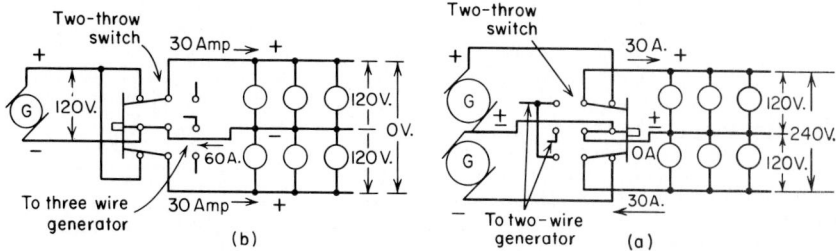

**Fig. 3-16**   Three-wire convertible system (direct-current or-single phase): (a) connections when operating three-wire; (b) connections when operating two-wire. In the case of arc lamps and Cooper-Hewitt lamps, it is important that they all be connected on the side of the circuit which does not reverse polarity when operating two-wire.

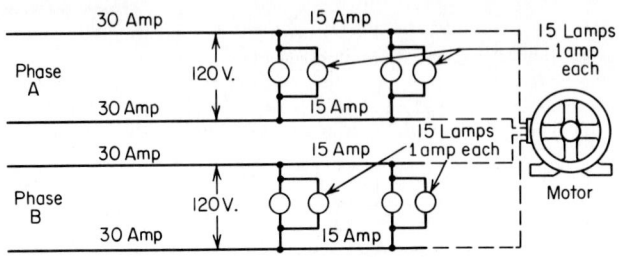

**Fig. 3-17**  Two-phase four-wire system.

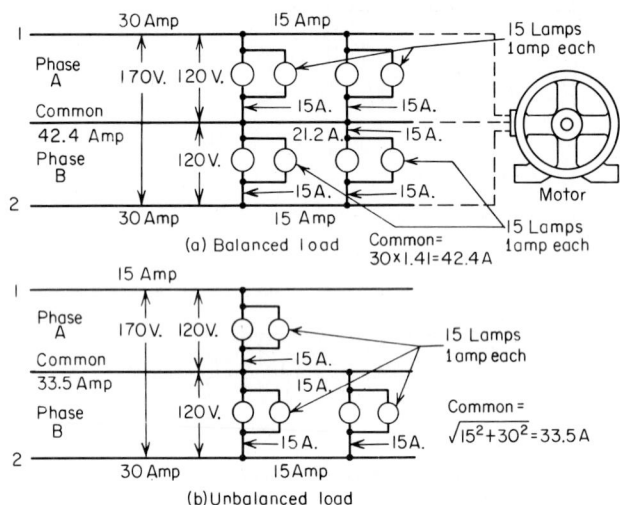

(a) Balanced load

Common=
30×1.41=42.4 A

(b) Unbalanced load

**Fig. 3-18**  Two-phase three-wire system.

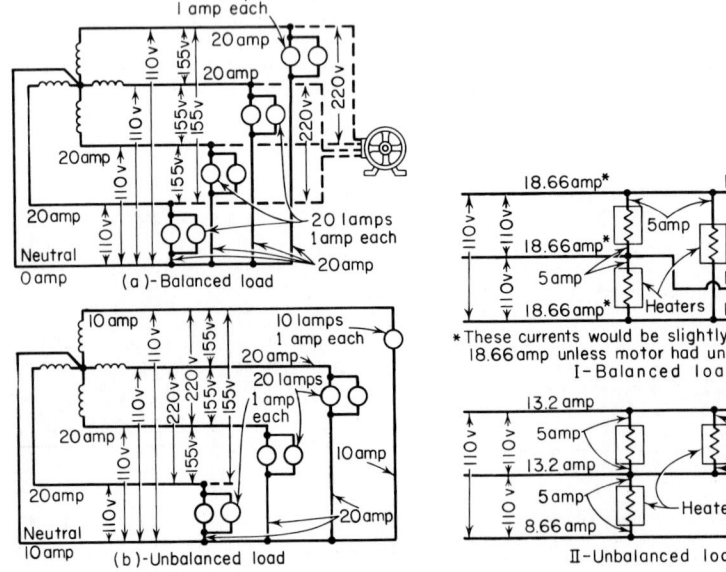

(a)-Balanced load

(b)-Unbalanced load

**Fig. 3-19**  Two-phase five-wire system.

*These currents would be slightly less than
18.66 amp unless motor had unity power factor
I—Balanced load

II—Unbalanced load

**Fig. 3-20**  Three-phase three-wire system.

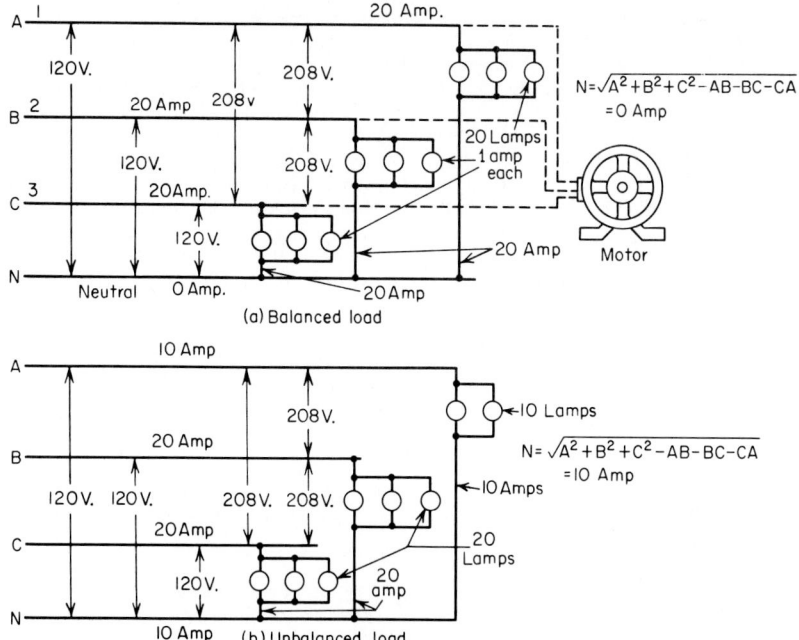

**Fig. 3-21**   Three-phase four-wire Y system.

size that would be necessary if the energy were transmitted at 110 V. This arrangement (Fig. 3-22, I), although it would operate, is not commercially feasible because each lamp of each pair of lamps in series must be of the same size, and if one lamp goes out, its partner is also extinguished. These disadvantages might be partially corrected by running a third wire, as at Fig. 3-22, II. Then one lamp might be turned off and the others would burn, and a single lamp might be added to either side of the system between the third wire and either of the outside wires. But unless the total resistance of all the lamps connected to one side was practically equal to that of all the lamps connected to the other side, the voltage across one side would be higher than that across the other. On the high side the lamps would burn bright and on the low side dim. Obviously, it is not feasible in practice to arrange or balance the sides so that they will have the same resistance. Hence

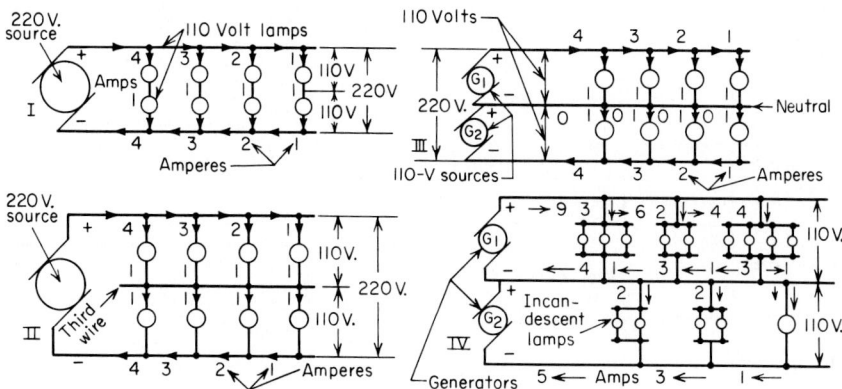

**Fig. 3-22**   Elements of the three-wire system.

practicable three-wire systems must use some other method whereby the electricity will be transmitted at, say, 220 V and the pressure across the lamps will be, say, 110 V.

**22. Commercial three-wire dc or single-phase ac systems** (Fig. 3-22, III and IV) consist of two outer conductors, having (for lighting installations) a pressure of 220 V impressed across them and a neutral wire so connected to sources of voltage that the pressure between it and either of the outside wires is 110 V. In Fig. 3-22, III, generators are the sources of voltage. The neutral wire joins at the point where the generators are connected together. When the system is perfectly balanced, the neutral wire carries no current and the system is in effect a 220-V system. Perfect balance seldom obtains in practice. When the balance is not perfect, the neutral wire conveys a current equal to the difference between the current taken by one side and that taken by the other side. Note from Fig. 3-22, IV, that the current in different parts of the neutral wire may be different and that it is

| System | Connections | Voltage relations | Relative weights of copper in percentage based on voltage drop | Relative weights of copper in percentage based on carrying capacity |
|---|---|---|---|---|
| Two-wire, d-c or single-phase | Fig. 14 | 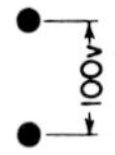 | 100 | 100 |
| Three-wire, d-c or single-phase | Fig. 15 | 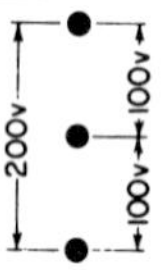 | With neutral same size as outside conductors 37.5 <br><br> With neutral 0.70 the size of the outside conductors 34 | With neutral same size as outside conductors 75 <br><br> With neutral 0.70 the size of the outside conductors 67.5 |
| Three-wire, convertible, d-c or single-phase | Fig. 16 | 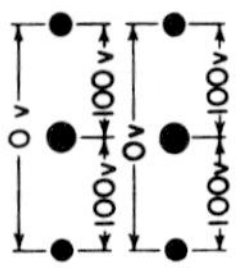 | Neutral double size of outside conductors 100 | Neutral double size of outside conductors 100 |
| Two-phase, four-wire | Fig. 17 | 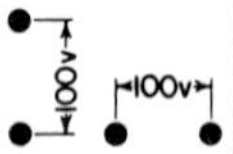 | 100 | 100 |
| Two-phase, three-wire | Fig. 18 | 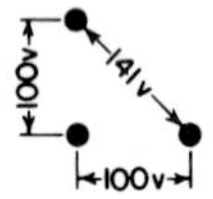 | Common 1.41 times as large as outside conductors 73 | Common 1.41 times as large as outside conductors 85 |
| Two-phase, five-wire | Fig. 19 |  | With common 1.41 times as large as outside conductors 34 <br> With common same size as outside conductors 31.3 | With common 1.41 times as large as outside conductors 67 <br> With common same size as outside conductors 62.5 |
| Three-phase, three-wire delta | Fig. 20 | 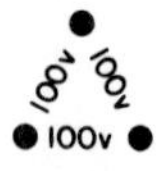 | 75 | 87 |
| Three-phase, four-wire wye | Fig. 21 | 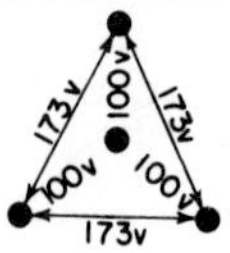 | With neutral same size as outside conductors 33.3 <br> With neutral 0.70 as large as outside conductors 31 | With neutral same size as outside conductors 67 <br> With neutral 0.70 as large as outside conductors 62 |

**Fig. 3-23**   Copper economics of different electric systems.

not necessarily in the same direction in all parts of the neutral wire. Each incandescent lamp in Fig. 3-22, IV, is assumed to take 1 A, and the small figures indicate the currents in different parts of the circuit.

**23. Comparisons of systems.** The relative weights of copper required for the different systems are given in Fig. 3-23. The weight of the copper required for the conductors of a two-wire dc circuit is assumed for convenience to be 100 percent. The values given in the column headed "Relative weights of copper in percentage based on voltage drop" are true for direct current, if we assume for all systems equal voltages on the lamps or other receivers, equal amounts of power transmitted, equal voltage drops, and balanced loads. For ac circuits the values are true for circuits in which the effect of inductance is negligible. For ac circuits in which the effect of inductance is not negligible, the amount of copper to produce the same voltage drop would be somewhat larger than the values shown in the table. The values given in the last column are based on the assumption that the carrying capacity of a wire is directly proportional to its cross-sectional area. Since the larger the wire, the smaller the allowable carrying capacity per unit of cross-sectional area, the percentages for systems of more than two wires would be somewhat less than shown. The actual saving in dollars gained by one of the systems that employs more than two wires will not be so great as that indicated in the table, owing to the fact that the total cost for a single large wire is less than for two small wires containing the same total amount of copper.

**24. Application of alternating current and of direct current.** AC systems have become almost universally standard in the United States for the transmission, distribution, and utilization of electrical energy. This is true because of the flexibility of ac systems in economically stepping up or down the voltage without the use of rotating equipment. Some vestiges of old Edison dc systems still exist both for city distribution and in old office buildings served by isolated plants.

When special characteristics of the load, such as electroplating or adjustable-speed motors, make the utilization of direct current either necessary or advisable, the alternating current can be converted to direct current by means of motor-generator sets, rectifiers, or, in special cases, converters.

**25.  Standard DC Voltages and Their Applications**

| Voltages | | Application |
|---|---|---|
| Generators and energy-delivering apparatus | Motors and energy-utilization apparatus | |
| 125ᵃ | 115 | Used for multiple-circuit lighting.   Usually obtained from a 115/230-volt, three-wire system |
| 125ᵃ–250ᵃ 575–600ᵃ | $\left\{ \begin{array}{c} 110\ -220 \\ 115^a-230^a \\ 550^a \end{array} \right\}$ | Direct-current motors |
| 600ᵃ | . . . . . . . . . . . | Urban and interurban electric railways |
| 1,200 1,500 | $\left. \begin{array}{c} . . . . . . . . . . . . \\ . . . . . . . . . . . . \end{array} \right\}$ | Interurban railways |
| 3,000 | . . . . . . . . . . . | Trunk-line railways |

ᵃ Electric Power Club standard voltage ratings.

## 26.  Standard AC Voltages and Their Applications

| Voltage | | Application |
|---|---|---|
| Generators and energy-delivering apparatus | Motors and energy-utilization apparatus | |
| 120/208 | 110<br>110 single-phase motors<br>110–120 lamps and appliances | Single-phase, used for small motors, lighting, and appliances, usually obtained from a 120/240-volt, three-wire, single-phase system or a three-phase, four-wire system |
| 240<br>480<br>600 | 220<br>440<br>550 | Usually three-phase three-wire, used for distribution for power for polyphase motors up to possibly 50 to 60 hp sizes.    220 volts used occasionally for lamps and heaters.    265/460 volt, four-wire, three-phase system is often used (265 volts for fluorescent lamps and 460 volts for supplying motors) |
| 2,400 | 2,200 | Single phase for primary distribution in residential districts; three-phase, three-wire for polyphase motors greater than about 60 or 100 hp, feeders for large industrial plants |
| 2,400/4,160 | 2,300/4,000 | For three-phase four-wire distribution in moderately heavily loaded districts |
| 6,900<br>11,500<br>13,800<br>18,000 | 6,600<br>11,000<br>13,200 | Highest voltages for which generators or motors can, ordinarily, be effectively designed.  Distribution systems for large cities; for power transmission over distances of a few miles.  For high-voltage distribution feeders in industrial plants and large office buildings |
| 22,000<br>33,000<br>44,000<br>66,000<br>88,000<br>110,000<br>132,000 | The voltages higher than  13,200  are used for transmission only and not for generation or utilization. | For power transmission overhead and underground, for distances up to about 125 miles, selection being roughly on the basis of 1,000 volts per mile of transmission distance |
| 154,000<br>220,000<br>275,000 | . . . . . . . . . . . . . . | For long-distance power transmission over aerial lines |

NOTE   The voltages up to 13,800 have been standardized by the American National Standards Institute, the National Electrical Manufacturers Association, and the Institute of Electrical and Electronic Engineers. The voltages from 22,000 to 154,000 were standardized by the Edison Electric Association and are in common use by power companies. The 220,000- and 275,000-V systems have been used on a few transmission lines.

**27. Selection of a frequency.**   Two frequencies now are standard in the United States: 25 and 60 Hz. A frequency of 50 Hz has been used considerably in Europe and in some installations in the United States. All other things being equal, 25 Hz would seem at first sight preferable because there is less inductive effect with it than with a higher frequency. It therefore follows that the inherent voltage regulation of a 25-Hz system is better than that of a 60-Hz system and also that the 25-Hz system is a trifle more efficient. For transmission distances of less than a few miles neither of these factors is of much consequence one way or the other. Alternating current at 25 Hz is not particularly well adapted for electric lighting because a flickering due to the filament cooling down every half cycle while the current is at a low value is very noticeable and causes eyestrain. At 60 Hz the time of low current values is so short that no flickering is noticeable.

Many years ago 25 Hz was considered necessary for the satisfactory operation of rotary converters, so most dc electric traction systems have used a 25-Hz transmission system. All dc railroad electrifications have also used 25 Hz because of the higher power factor and better commutation of the 25-Hz series motors. Again, it is not economically feasible to build large slow-speed motors for main drives in steel mills and cement plants at 60 Hz, so these types of plants have often adopted 25 Hz. Transformers and most other apparatus, except the motors indicated, are cheaper for 60 Hz, and delivery time is shorter.

Since 60 Hz is best for most installations, it has become practically the standard frequency in the United States, and probably more than 95 percent of the equipment sold is for 60 Hz. Most of the 50-Hz installations have been or are being changed over to 60 Hz.

Higher frequencies than the standard 60-Hz distribution frequency are sometimes used for special applications, such as the operation of high-speed motors and fluorescent lamps (refer to Divs. 7 and 10). These higher application frequencies are obtained from the 60-Hz supply by means of conversion equipment.

**28. Effect of increased voltage and decreased voltage on generators, lines, transformers, meters, lamps, and motors.**   Sometimes it is necessary or desirable to adopt, for some existing installation, a voltage either higher or lower than that which has been in use. The table below indicates the general effects of such a change. The voltage impressed on incandescent lamps should be between 110 and 120. In new installations it is well to adopt a high lamp voltage, say, about 120. The operating voltage of an old installation may be gradually increased sometimes as much as 10 V without adverse results (*Electrical World*).

| *Effect of Increased Voltage on* | *Effect of Decreased Voltage on* |
|---|---|
| Generators | Generators |
| Increase in excitation. | Decrease in excitation. |
| Will the exciter voltage be sufficient to produce the required excitation at full load or at partial load of low power factor? | Is the resistance of the field rheostat sufficient to maintain the lower voltage at no load? |
| Will the fields overheat at the increased excitation required? | Decrease in core loss. |
| Increase in core loss. | Increase in armature copper loss for the same kilovolt-ampere output. |
| Will the iron overheat? | Will this be offset by the reduction in field and core losses? |
| Decrease in armature copper loss for same kilovolt-ampere output. | |
| Will this offset the additional field and core loss? | |
| Lines | Lines |
| Increase in transmission radius for the same kVA output. | Decrease in transmission radius for the same kVA output |
| Transformers | Transformers |
| For the same kVA output, increase in core losses, decrease in copper losses, somewhat lower all-day efficiency. | For the same kVA output, decrease in core losses, increase in copper losses, somewhat higher all-day efficiency. |
| Meters | Meters |
| Effect negligible. | Effect negligible. |
| Lamps | Lamps |
| No effect after the new voltage lamps are installed. | No effect after the new voltage lamps are installed. |
| Motors | Motors |
| Reduced slip. | Increased slip. |
| Increased torque. | Decreased torque. |
| Increased efficiency. | Decreased efficiency. |
| Decreased power factor of induction motors. | Increased power factor of induction motors. |

**29. Common distribution systems** (electrical systems design). Basically distribution systems are classified according to the voltage level used to carry the power either directly to the branch circuits or to load-center transformers or substations at which feeders to branch circuits originate. The following are the most common types of distribution systems based on voltage.

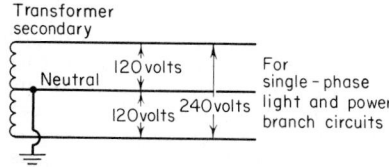

**Fig. 3-24**   Single-phase three-wire distribution system.

1. 120/240-V three-wire single-phase combination light and power distribution to lighting and appliance branch-circuit panelboards and to power panels (Fig. 3-24). This type of system is restricted to applications in which the total load is small and is primarily lighting. Stores, small schools, and other small commercial occupancies use this system. In most cases of small commercial buildings, the use of 208Y/120-V three-phase distribution offers greater economy because of the higher operating efficiency of three-phase circuits. In cases in which 120/240-V distribution is used as the basic distribution, the service to the premises is made at that voltage. Of course, 120/240-V distribution is frequently an effective and economical system for lighting-subfeeder distribution in electrical systems which use a higher-voltage basic distribution system with load-center step-down to utilization voltages for local and incidental lighting and receptacle circuits.

2. 208Y/120-V three-phase four-wire distribution (Fig. 3-25a) is the most common type of system used in commercial buildings, in some institutional occupancies, and in some industrial shops with limited electrical loads. This system offers substantial economy over the 120/240-V system in the amount of copper conductor required to carry a given amount of power to a load. It is a combination light- and power-distribution system, providing 120 V phase-to-neutral for lighting and single-phase loads and 208 V phase-to-phase for single- or three-phase motor or other power loads. This distribution system is used as the basic distribution in occupancies in which the service to the building is of the same voltage. It is also the most common subdistribution system for lighting and receptacle circuits in occupancies using higher-voltage distribution to load centers.

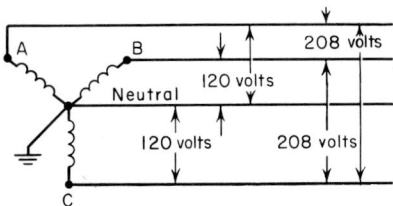

For three-phase power circuits and single-phase light and power branch circuit

(a)- Three-phase, 4 wire wye (or star) with grounded neutral rated 120/208 volts

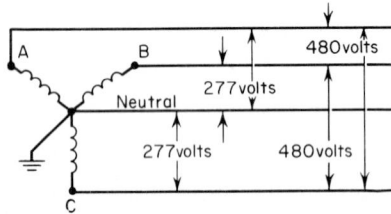

For three-phase power circuits and lighting circuits using 227 volts ballasts.

120 volt lighting and receptacle loads are fed from this system through single-phase transformers rated 480/120-240volts or three phase transformers rated 480/120-208 volts

(b)-Three phase, 4-wire wye (or star) with grounded neutral rated 480Y/277 or 460Y/265, depending upon voltage spread under local conditions

**Fig. 3-25**   Three-phase Y distribution systems.

3. 240-V three-phase three-wire distribution (Fig. 3-26) is a common system for power loads in commercial and industrial buildings. In such cases, service to the premises is made at 240 V, three-phase. Feeders carry the power to panelboards or wireways supplying branch circuits for motor loads. Lighting loads are usually handled by a sep-

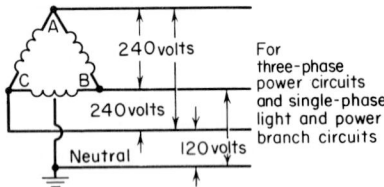

**Fig.   3-26**   Three-phase   delta   distribution system.

arate single-phase service to the building. This system offers economic application when the power load is large compared with the lighting load. In some 240-V three-phase three-wire systems, a grounded center tap on one of the phases is used as a neutral to provide 120 V for lighting and receptacle circuits.

4. 480-V three-phase three-wire distribution is commonly used in commercial and industrial buildings with substantial motor loads. Service to the building can be made at this voltage, and the 480-V feeders carried to motor loads and to step-down transformers for lighting and receptacle circuits. In many cases, 480-V feeders will be derived from load-center substations within the building and carried to motor loads or power panels.

5. 480Y/277-V three-phase four-wire distribution (Fig. 3-25b) has become an important system for use in commercial buildings and industrial buildings. In office and other commercial buildings, the 480-V three-phase four-wire feeders are carried to each floor, where 480 V, three-phase, is tapped to a power panel or to motors; general area fluorescent lighting using 277-V ballasts is connected between each phase leg and the neutral; and 208Y/120-V three-phase four-wire circuits are derived from step-down transformers for local lighting, appliance, and receptacle circuits. Application of this system offers an economic advantage over a 208Y/120-V system when less than about half of the load devices require 120- or 208-V power. When the 480Y system can be used, it will cost less than the 208Y/120-V system owing to copper savings through the use of smaller sizes of conductors and the lower cost of system elements because of lower current capacities. If

the required amount of 120- or 280-V power is more than half of the total load in a building, the cost of the step-down transformers to supply these circuits will offset the savings in the 480-V circuiting. The 480Y system is more advantageous in high-rise or other large-area buildings than in small buildings.

6. 2400-V three-phase distribution is an industrial-type system used to feed heavy motor loads directly and motor and lighting loads through load-center substations and lighting transformers.

7. 4160/2400-V three-phase four-wire distribution with a grounded neutral is a more common industrial system than the above 2400-V, delta-connected system. This system is widely used to supply load-center substations in which the voltage is stepped down to 480 to feed motors and lighting transformers for 120/240- and/or 208Y/120-V circuits. It can also be used in distribution to substations stepping down the voltage directly to 208Y/120.

8. 4800-V three-phase distribution is a delta-connected industrial system for feeding 480-V substations supplying motors and lighting transformers.

9. 7200-V three-phase distribution is another industrial system used with substations for stepping down voltage to lower levels for power and lighting.

10. 13.2Y/7.2-kV (or 13.8-kV) three-phase four-wire distribution is a modern, widely used distribution system for large industrial plants. Power at this voltage is delivered to substations which step down the voltage to 480 for motor loads and which supply 480/120/240- or 480/208Y/120-V transformers for lighting. Or 480Y/277-V substations can be used to supply motor loads and 277-V fluorescent or mercury-vapor lighting for office and industrial areas. Lighting transformers are then used to supply 120-V circuits for lighting and convenience receptacles.

The voltage values given for these distribution systems are, of course, subject to the usual variation or spreads due to distance of transmission and distribution, local conditions of utility supply, and settings of transformer taps. In addition to the distribution voltages given, other systems may operate at 6.6, 8.3, 11, and 12 kV. Of the high-voltage (over 600 V) distribution systems, 4160 and 13,200 V are the most common and represent good design selection and economy of application for most cases. In many areas, delta-connected supplies have been changed to four-wire Y systems with consequent increase in power-handling capacity as a result of increased phase-to-phase voltage. The trend today is toward the use of the 13-kV systems instead of other high-voltage distribution systems for large industrial plants. To a limited extent, high-voltage distribution finds application in large commercial buildings. The most recent trend in distribution in office and other multifloor commercial buildings is toward distribution at 480Y/277 V, three-phase four-wire, with grounded neutral.

## CIRCUIT CALCULATIONS

**30. Five factors should be considered** in determining the size of wire for the distribution of electricity. A wire should be of such size that (a) the current will not heat it to a temperature that would ruin the insulation or cause a fire, (b) it will have sufficient mechanical strength so that it will not be broken under the ordinary strains to which it is reasonable to assume that it will be subjected, (c) it will not be so large as to exceed the limitations required for its economical installation, (d) it will carry the electricity to the point at which it will be used without an excessive drop or loss of voltage, (e) the cost of energy lost—the $I^2R$ loss—due to the voltage overcoming the resistance will not be excessive (refer to Sec. 82). A conductor may satisfy any one of the five conditions and not satisfy the four others.

**31. Safe current-carrying capacity** should always be considered in designing circuits. When current passes through a conductor, some of the electrical energy is converted into heat energy, the amount of energy thus converted being equal to $I^2R$ (see Div. 1). This heat energy raises the temperature of the conductor and its insulation and covering above that of the surrounding medium and is dissipated into the atmosphere through thermal conduction of the metal of the conductor, conductor insulation and covering, and surrounding materials (such as conduit, ducts, earth, etc.) and through convection of the surrounding air over these materials. It is not good practice to operate bare conductors at a temperature in excess of from 70 to 80°C, since trouble is apt to occur at joints and connections when they are operated at higher temperatures. A curve for determining current capacities of bare copper conductors is given in Sec. 15 of Div. 11. For covered and insulated cables the maximum allowable temperature of the conductor is limited by the maximum temperature that will not be harmful to the insulating and covering materials (refer to Div. 2 for maximum safe temperatures for different types of insulation). The allowable safe current-carrying capacity of an insulated conductor will therefore depend upon the type of conductor insulation and covering, the conditions of installation (whether in conduit, exposed to air, or buried in earth, and the number of conductors grouped in close proximity to each other), and the temperature of the surrounding atmosphere or earth (ambient temperature). A curve for determining the carrying capacity of weatherproof copper conductors when installed outside buildings is given in Sec. 16 of Div. 11. Curves for aluminum cable, steel-reinforced, are given in Sec. 26 of Div. 11, and a table for parkway cables buried directly in the ground is given in Sec. 27 of Div. 11. It would require more space than is available in this book to include the necessary tables for allowable carrying capacities of cables installed in underground raceway systems. Complete tables can be obtained from the various cable manufacturers or from the Insulated Cable Engineers Association (ICEA).

The maximum allowable safe current-carrying capacities of wires for interior wiring are specified by the **National Electrical Code** (see Tables 17 to 23 of Div. 11). The maximum current that a wire will have to carry should never exceed the allowable safe carrying capacity for the size of wire, type of insulation employed on the wire, and method of installation as given in these tables except in the case of motor circuits. Motors draw a current at the instant of starting that is much greater than the normal full-load running current. Owing to the fact that this large starting current lasts for only a short time, the **Code** does not require that the carrying capacity of the wires for motor circuits be as great as the starting current. For branch circuits supplying continuous-duty motors the size of the wire must be sufficient to carry at least 125 percent of the full-load rated current of the motor. The wires of branch circuits supplying motors in classes of service having short-time duty must have carrying capacities as large as the average load currents required by the motors. In the majority of cases, the average load currents required by motors of this class will not exceed the percentages of the full-load rated currents given in Table 12 of Div. 11. The wires between the slip rings of wound-rotor induction motors and the secondary controller must have a carrying capacity of at least 125 percent of the full-load secondary current of the motor. The wires between the secondary controller and the resistors must have a carrying capacity that is not less than the percentages of the full-load secondary currents given in Table 13 of Div. 11. The size of wire for motor feeders or mains must be sufficient to carry at least the maximum-demand running current of the motors supplied by the circuit. The method of computing the maximum-demand running current is given in Sec. 54.

The ampacities which are listed in Tables 18 to 23 of Div. 11 are based upon room temperatures of 30°C. If the room will have a temperature greater than this, the ampacities listed in the table should be multiplied by the proper correction factor as given in Tables 18, 19, 21, and 22 of Div. 11.

**32. Mechanical strength of wires.**    Wires should be of sufficient size so that their mechanical strength will be great enough to withstand the strains of installation and of the service to which they will be subjected. For general overhead distribution mains, no wire of soft-drawn copper smaller than No. 6 should be used. If medium- or hard-drawn copper is used, no wire smaller than No. 8 should be employed. For overhead outside wiring on private premises, no wire smaller than No. 10 should be used for spans up to 50 ft and no wire smaller than No. 8 for longer spans. Service drops from an overhead distribution main to the service-entrance conductors must not be smaller than No. 10 if of soft-drawn copper or smaller than No. 12 if of medium- or hard-drawn copper. The service-entrance conductors to a building shall generally not be smaller than No. 8 copper or No. 6 aluminum or copper-clad aluminum:

1. For installations consisting of not more than two two-wire branch circuits they shall not be smaller than No. 8 copper or No. 6 aluminum or copper-clad aluminum.

2. By special permission because of limitations of supply source or load requirements they shall not be smaller than No. 8 copper or No. 6 aluminum or copper-clad aluminum.

3. For installations to supply only limited loads of a single branch circuit, such as small polyphase power, controlled water heaters, and the like, they shall not be smaller

than the conductors of the branch circuit and in no case smaller than No. 12 copper or No. 10 aluminum or copper-clad aluminum.

One-family residences with an initial load of 10 kVA or six or more two-wire branch circuits must have a 100-A service.

In the interior wiring of buildings no wire smaller than No. 14 can be used except for the following special cases:

1. For fixture wiring and flexible cords, No. 16 up to 100 ft and No. 18 up to 50 ft for fixture wire can be used. Also, tinsel cords or cords having equivalent characteristics of smaller size may be approved for use with specific appliances.

2. Number 18 wire can be used for stationary motors, rated 1 hp or less, for the conductors between the motor and an approved terminal enclosure.

3. For elevator, dumbwaiter, and escalator wiring, except for conductors which form an integral part of control equipment, the minimum allowable size of conductors is as follows:

*a. Traveling cables.*
  (1) For lighting circuits, No. 14, except that No. 20 or larger conductors may be used in parallel provided the carrying capacity is equivalent to at least that of No. 14 wire
  (2) Operating control and signal circuits, No. 20

*b. Other wiring.* For all operating control and signal circuits, No. 24.

4. In the wiring for cranes and hoists No. 16 wire may be used for some circuits under certain conditions (see Div. 9).

5. For industrial machinery wiring the following exceptions are allowed:

*a.* Conductors to moving parts: copper conductors for control purposes to continuously moving parts may be No. 16 if all such conductors are insulated for the maximum voltage of any conductor in the cable or tubing.

*b.* Conductors to electronic and precision devices: copper conductors to electronic and precision devices may be No. 22, except that if pulled into raceways they shall not be smaller than No. 18.

6. For remote-control limited-power and signaling circuits, Nos. 18 and 16 conductors may be used provided they are installed in a raceway or a cable approved for the purpose or in listed flexible cords (Class 1 circuits).

The **National Electrical Code** gives detailed rules for these cases.

**33. There are certain limitations in the size of wires** that it is feasible to employ to satisfy construction and installation requirements. For convenience in installation, it is seldom advisable to use larger than 3-in conduit. When all the wires of a circuit are run in the same conduit, this limits the size of the conductors to 500,000 cmil (253.35 mm). If only one wire were run in a conduit, the size of wire could be considerably larger. It is seldom wise or convenient to install conductors of larger size than 1,000,000 cmil (506.7 mm). A 2½-in conduit is about as large as can usually be installed between floors and ceilings. In such cases, therefore, the size of wire is limited to 300,000 cmil (152.01 mm) if two or three wires are placed in the same conduit. For open wiring, wires larger than 1,000,000 cmil are cumbersome to handle.

**34. Percentage line drop, or voltage loss,** can be figured as either a percentage of the voltage required at the receiver or as a percentage of the voltage impressed by the generator or other energy source on the line. For instance, in Fig. 3-27, the voltage impressed on the receivers (lamps and motor) is 220. The line loss is 11 V; hence the pressure impressed on the line is 220 + 11 = 231 V. The voltage loss as a percentage of the voltage

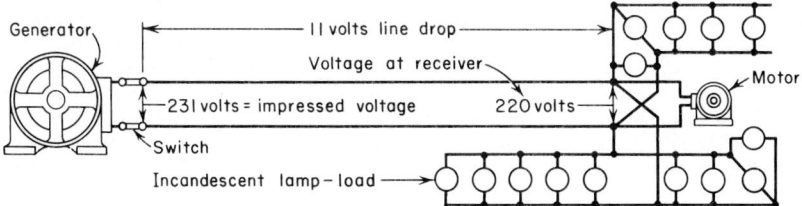

**Fig. 3-27**   Percentage line drop.

at the receiver is 11/220 = 0.05 = 5 percent. The voltage loss as a percentage of the voltage impressed on the line is 11/231 = 0.048 = 4.8 percent. In practical work the percentage loss or drop is usually taken as a percentage of the voltage required at the receivers, because this is the most convenient and direct method. In this book the term *percentage drop* refers to a percentage of the voltage required at the receivers unless otherwise noted.

**35. Allowable voltage drops.**  The conductors of an interior-wiring system should be of sufficient size so that the voltage drop from the point of service entrance to the equipment utilizing the energy, such as lamps or motors, is not excessive. If the voltage drop is too great, the operating conditions will not be satisfactory, owing to one or more of the following causes:

1.   Shortened life of lamps caused by overvoltage under light-load conditions
2.   Unsatisfactory illumination under maximum-load conditions
3.   Unsatisfactory motor speed, torque, or temperature

No definite standards have been adopted for the maximum allowable voltage drop. The National Electrical Code recommends that the voltage drop from the point of service entrance to the final distribution point should not be greater than 3 percent for power, lighting, or heating loads.

For *lighting loads* the voltage drop from the point of service entrance to any lamp should not exceed 5 percent. If it is possible without employing excessively large conductors, it is good practice to keep this total voltage drop within 3 percent. The voltage drop in any branch lighting circuit or in any lighting feeder should not be greater than 2 percent.

For *power loads* the voltage drop from the point of service entrance to any motor or heating unit should not exceed 5 percent. The distribution of this voltage drop among the different parts of the system will depend upon the layout of the wiring installation. For systems operating at 208 V or higher, it is generally good practice to limit the voltage drop in branch circuits to 1 percent and in feeders to 4 percent. Only in exceptional cases should the voltage drop in a motor branch circuit exceed 2 percent.

Values of voltage drop that will be in accordance with good practice for different systems are given in Sec. 52 of Div. 11.

**36. Allowance for growth in planning circuits.**  In installing the wiring for electric power and lighting installations, the demands that may be placed upon the system in the future should be carefully considered. It requires a relatively small additional expense, when installing a system, to put in wires somewhat larger than are required by the present lighting equipment. On the other hand, if the wires installed are of just sufficient size to accommodate the present load and it is necessary to increase the load at some future date, the cost of removing the old installation and installing a new one is considerable.

Standards of illumination have been continually increasing, and it is reasonable to assume from the trend of practice that the general level of illuminating intensities employed will be still further increased. Both employees and employers are appreciating more and more the benefits to be gained from high intensities of illumination.

It is therefore good practice to limit the initial loading of branch lighting circuits to 50 percent of their maximum allowable load. Such limitation of the initial loading will allow the load on branch lighting circuits in most commercial and industrial installations to be increased to 160 percent of their initial loading. It would appear that the loading could be increased to 200 percent of the initial loading, but this is not true, since the National Electrical Code requires that the total load connected to branch lighting and appliance branch circuits shall not exceed 80 percent of their rating if in normal operation the load will continue for 3 h or more, as in store lighting and similar loads. In the following sections dealing with the calculation of the loads on lighting and appliance circuits, the term *future load* means the load that would be placed on the circuit if it were loaded to 80 percent of its rating.

The layout and equipment in industrial and commercial buildings are continually being changed. These changes frequently involve increased load requirements. Therefore, no definite rules can be laid down for the proper allowance that should be made for growth, since conditions vary so widely for different installations. Each installation requires careful study with respect to probable future load requirements. The actual connected loads on the circuit should be increased by a sufficient amount so that the size of circuit will be ample to accommodate whatever loads it is estimated may be connected to the circuit in the future. For good general practice, it is recommended that the allowance for growth be at least 50 percent of the initial load.

**37. Calculation of load in amperes on two-wire branch lighting and appliance circuits.**  The actual load in amperes on a two-wire branch lighting circuit is equal to the sum of the watt ratings of the individual loads connected to the circuit divided by the voltage of the circuit. Each heavy-duty lampholder outlet for lighting other than general illumination should be figured at 600 VA. Each plug receptacle outlet connected to the circuit should be figured at at least 180 VA per outlet. If the load to be connected to a plug outlet is known to be more than 180 VA, its load should be considered as the actual known wattage that it will supply.

For show-window lighting a load of not less than 200 VA for each linear foot of show window, measured horizontally along its base, may be allowed in lieu of the 180 VA per outlet. If fixed multioutlet assemblies are supplied by the circuit, each 5 ft (1.524 m) or fraction thereof of each separate and continuous length of the assembly should be considered as a load of 180 VA, except in locations in which a number of appliances are likely to be used simultaneously. In that case each 1 ft (0.3048 m) or fraction thereof should be considered as a load of 180 VA. In computing the load of lighting units that employ ballasts, transformers, or autotransformers, such as fluorescent-lighting units, the load must be based on the voltampere load or current of the complete unit and not simply the wattage of the lamp. Refer to Sec. 129 of Div. 10 for information on these loads. For incandescent-lighting units the voltampere loads will be the same as the rated watts of the lamps. For branch lighting circuits it is sufficiently accurate to assume that the total voltampere load is equal to the direct numerical summation of the individual voltampere loads. This is also true for the summation of lighting loads in terms of amperes.

As stated in Sec. 36, the future load on branch lighting and appliance circuits in commercial and industrial installations is taken as 80 percent of the rating of the branch circuit.

The standard sizes of lamps are given in Table 1 of Div. 11 and more complete data in Div. 10.

**example**  A two-wire, 120-V, 20-A branch lighting circuit supplies five 200-W incandescent lamps. Determine the actual connected and probable future load.

**solution**  Present connected load $= \dfrac{5 \times 200}{120} = 8.3$ A

$$\text{Future load} = 0.80 \times 20 = 16 \text{ A}$$

**example**  A two-wire, 120-V, 20-A branch lighting circuit supplies four 150-W and three 200-W incandescent lamps. Determine the actual connected and probable future load.

**solution**  Present connected load $= \dfrac{(4 \times 150) + (3 \times 200)}{120} = 10$ A

$$\text{Future load} = 0.80 \times 20 = 16 \text{ A}$$

**38. Calculation of load in amperes on three-wire branch lighting and appliance circuits.**  The actual and future loads in amperes on either side of a three-wire branch lighting circuit can be computed in the same manner as for two-wire circuits in Sec. 37, the voltage between the neutral and either outside wire and the loads connected between the neutral and the one outside wire under consideration being used.

**example**  A three-wire, 120/240-V, 15-A branch lighting circuit supplies the loads in the following diagram. Determine the actual connected and probable future loads.

**solution**     Present connected load 1 to $N = \dfrac{4 \times 200}{120} = 6.7$ A

Present connected load 2 to $N = \dfrac{5 \times 150}{120} = 6.3$ A

Future load 1 to $N = 0.80 \times 15 = 12$ A

Future load 2 to $N = 0.80 \times 15 = 12$ A

**39. Calculation of load in amperes on branch electric-range circuits.**  For electric ranges and other cooking appliances of 1¾-kW rating or less, the load on the branch circuit is equal to the ratings of the appliances. For ranges or cooking appliances of ratings greater than 1¾ kW the watt load on the branch can be computed in accordance with Table 2 of Div. 11.

For two-wire branches, the load in amperes is equal to the watt load, as determined above, divided by the voltage of the circuit. For three-wire branches, the load in amperes on each outside wire of the circuit is equal to the watt load, as determined above, divided by the voltage between outside wires. The load in amperes on the neutral wire can be taken as 70 percent of the load on an outside wire.

**40. Calculation of lighting and appliance load on two-wire dc or single-phase distribution panels.**  The actual connected load is equal to the sum of the actual connected loads of all the branch circuits fed from the panel. Each appliance circuit in the panel should be figured at at least 10 A for a circuit rated 15 A and at least 15 A for a circuit rated 20 A. The future load on the panel is figured in the same manner as the actual load, the future loads of the branch circuits being used instead of the actual loads and each appliance or spare circuit being figured at at least 15 A. If an appliance circuit is of a type allowing more than 15 A, the circuit should be figured at its maximum allowable amperage as given in Div. 9.

**example**  Determine the actual connected and probable future loads on a two-wire 120-V lighting panel supplying the following circuits:

| Circuit number and rating | Present load, amp | Future load, amp |
|---|---|---|
| 1–15 amp lighting | 10 | 12 |
| 2–15 amp lighting | 7 | 12 |
| 3–20 amp lighting | 9 | 16 |
| 4–15 amp appliance | 10 | 15 |
| 5–15 amp lighting | 8 | 12 |
| 6–20 amp appliance | 15 | 20 |
| 7–15 amp spare | | 15 |
| 8–15 amp spare | | 15 |
| | 59 | 117 |

**solution**                  Present connected load on panel = 59 A
                                  Probable future load = 117 A

**41. Calculation of lighting and appliance load on three-wire dc or single-phase distribution panels.**  The actual connected load on each side of the panel is computed in the same manner as for two-wire panels in Sec. 40, using the voltage between the neutral and one outside wire and making separate calculations for each line wire. If the circuits have been well laid out, there should not be much difference between the loads on the two sides of the panel.

**example**  A three-wire 120/240-V lighting panel supplies the following circuits. All branch circuits are two-wire, 120 V. Determine the actual connected and the probable future loads. All circuits are rated 15 A.

| Branch circuits between wire 1 and neutral | | | Branch circuits between wire 2 and neutral | | |
|---|---|---|---|---|---|
| Circuit | Present load, amp | Future load, amp | Circuit | Present load, amp | Future load, amp |
| 1 | 7 | 12 | 1 | 8 | 12 |
| 2 | 8 | 12 | 2 | 7 | 12 |
| 3 | 8 | 12 | 3 | 8 | 12 |
| 4 | 7 | 12 | 4 (plug) | 10 | 15 |
| 5 (plug) | 10 | 15 | 5 (spare) | | 15 |
| 6 (spare) | | 15 | 6 (spare) | | 15 |
| | 40 | 78 | | 33 | 81 |

**solution**  The present connected load would be taken as 40 A. The future load would be taken as 81 A.

**42. Calculation of lighting and appliance load on two-phase three-wire distribution panels.**  The loads for the two-line wires are computed as in Sec. 41. The actual and future loads on the common wire of the panel are taken as equal to 1.41 times the actual and future loads, respectively, of the more heavily loaded side of the panel.

**example**  A three-wire two-phase 120-V lighting panel supplies the same circuits on the two sides of the panel as given in the example under the three-wire single-phase panel. Determine the actual connected and probable future loads.
**solution**  The actual connected load on each outside wire of the panel would be taken as 40 A. The future load on each outside wire of the panel would be taken as 81 A.
The actual connected load on the neutral would be 1.41 × 40 = 56.4 A.
The future load on the neutral would be 1.41 × 81 = 114.2 A.

**43. Calculation of lighting and appliance load on four-wire three-phase distribution panels.**  The actual connected load on any one bus except the neutral bus is computed as in Sec. 41. The load on each of the four buses usually is taken as being equal to that of the most heavily loaded outside bus.

**44. Fundamentals of load calculations on mains and feeders.**  In some cases the carrying capacity of feeders or mains need not be so great as the total ampere load of all the equipment supplied by the circuit. This is due to the fact that all the apparatus may not be operating at the same time or, if operating, may not all be taking full-load current from the line at the same time. A demand factor is employed to determine the maximum current that it is estimated the circuit will ever be required to carry. A demand factor is the ratio between the maximum current that the circuit will ever have to carry and the total load connected to the circuit. The maximum-demand current of a circuit, therefore, is equal to the total connected load times the proper demand factor (see Table 4, Div. 11, for lighting demand factors; Table 6, Div. 11, for maximum-demand factors for motor circuits; Table 2, Div. 11, for demand factors for household electric ranges, wall-mounted ovens, counter-mounted cooking units, and other household cooking appliances; and Table 3, Div. 11, for demand factors for household electric clothes dryers). If four or more fixed appliances, in addition to electric ranges, air-conditioning equipment, or space-heating equipment, are connected to the same feeder or main in a single or multifamily dwelling, a demand factor of 75 percent may be applied to the fixed-appliance load, not including the electric ranges. The computed load of a feeder supplying fixed electrical space-heating equipment shall be the total connected load on all branch circuits, except that if reduced loading of the conductors results from units operating on duty cycle, intermittently, or from all units not operating at one time, the authority enforcing this Code may grant permission for feeder conductors to be of a capacity less than 100 percent, provided the conductors are of sufficient capacity for this load.

**45. Calculation of load on two-wire dc or single-phase ac lighting and appliance feeders or mains.**  The ampere or watt load is equal to the sum of the maximum-demand lighting and portable-appliance load, the maximum-demand fixed-appliance load, and the maximum-demand electric-range load. The present or future maximum-demand lighting and portable-appliance load is equal to the sum of the actual or future lighting and portable-appliance loads fed from the circuit times the proper demand factor as determined from Table 4, Div. 11. The maximum-demand fixed-appliance load is figured in accordance with Sec. 44. The maximum-demand electric-range load is figured in accordance with Table 2, Div. 11.

In one-family dwellings, in individual apartments of multifamily dwellings having provisions for cooking by tenants, and in each hotel suite having cooking facilities or a serving pantry, a feeder load of not less than 1500 VA for each two-wire 20-A branch circuit shall be included for small appliances (portable appliances supplied from receptacles of 15- to 20-A rating) in pantry and breakfast room, dining room, kitchen, and laundry. This is a National Electrical Code requirement.

**example**  An apartment in a multifamily dwelling has a total floor area of 700 ft². It is not equipped with an electric range. Determine the maximum-demand present load for a two-wire 120-V feeder supplying the apartment.

solution

$$\text{General lighting load from Table 4, Div. 11} = 700 \times 3 = 2100 \text{ VA}$$
$$\text{Small-appliance load, two circuits, each at 1500 W} = 3000 \text{ VA}$$
$$\text{Total computed load} = 5100 \text{ VA}$$

Application of demand factor:

$$\text{3000 VA at 100\%} = 3000 \text{ VA}$$
$$\text{2100 VA at 35\%} = \underline{\phantom{0}735 \text{ VA}}$$
$$\text{Maximum-demand load} = 3735 \text{ VA}$$
$$\text{Maximum-demand current} = \frac{3735}{120} = 31.1 \text{ A}$$

**46. Calculation of load on three-wire dc or single-phase ac lighting and appliance feeders or mains.** The load on either outside wire is equal to the sum of the maximum-demand lighting and portable-appliance load, the maximum-demand fixed-appliance load, the maximum-demand electric-range load, and the maximum-demand household electric-clothes-dryer load fed from that wire. The present or future maximum-demand lighting and portable-appliance load is equal to the sum of the actual or future lighting and portable-appliance loads of all the panels fed from that wire times the proper demand factor as determined from Table 4, Div. 11. The maximum-demand fixed-appliance load is equal to the sum of all the fixed-appliance load fed from that wire times the proper demand factor as given in Sec. 44. The maximum-demand current for the electric-range load is equal to the maximum-demand watt range load as determined in accordance with Table 2, Div. 11, divided by the voltage between outside wires of the circuit. Refer to Sec. 45 for required allowance for small appliances in dwelling occupancies. The maximum-demand current for the household electric-clothes-dryer load is equal to the maximum-demand watt clothes-dryer load as determined in accordance with Table 3, Div. 11, divided by the voltage between outside wires of the circuit. In adding branch-circuit loads for space heating and air cooling in dwelling occupancies, the smaller of the two loads may be omitted from the total if it is unlikely that both of the loads will be served simultaneously. The air-cooling (air-conditioning) load is a motor load. Therefore for the calculation of load on circuits of this character refer to Secs. 53, 54, and 55.

The load on the neutral wire is the maximum unbalance of the load on the two sides of the system. The maximum unbalanced load is the sum of the maximum-demand lighting and portable-appliance load, the maximum-demand fixed-appliance load, and the maximum-demand electric-range load for the more heavily loaded side of the circuit. The maximum-demand household electric-range load is considered as 70 percent of the electric-range load on the outside conductors as determined from Table 2 of Div. 11. If the maximum unbalanced load for the neutral as determined from the above procedure is more than 200 A, then a further demand factor of 70 percent can be applied to that portion of the unbalanced load in excess of 200 A. There shall be no reduction of the neutral capacity for that portion of the load which consists of electric discharge lighting.

The National Electrical Code allows the following optional method of calculation of load for a one-family residence or apartment unit. For each dwelling unit served by a 120/240-V, three-wire, 100-A or larger service when the total load is supplied by one feeder or one set of service-entrance conductors, the following percentages may be used in lieu of the method of determining feeder (and service) loads detailed previously in this section.

**Optional Calculation for Dwelling-Unit Load in Kilovolt-Amperes (Table 220-30, NEC)**

Largest of the following five selections:

1. 100 percent of the nameplate rating(s) of the air conditioning and cooling, including heat-pump compressors.

2. 100 percent of the nameplate rating(s) of electric thermal storage and other heating systems where the usual load is expected to be continuous at the full nameplate value. Systems qualifying under this selection shall not be figured under any other selection in this table.

3. 65 percent of the nameplate rating(s) of the central electric space heating including integral supplemental heating in heat pumps.

4. 65 percent of the nameplate rating(s) of electric space heating if less than four separately controlled units.

5. 40 percent of the nameplate rating(s) of electric space heating of four or more separately controlled units.

Plus 100 percent of the first kVA of all other load and 40 percent of the remainder of all other load.

All other load shall include 1500 VA for each 20-A appliance outlet circuit; lighting and portable appliances at 3 W/ft²; all fastened-in-place appliances (ranges, wall-mounted ovens, and counter-mounted cooking units) at nameplate-rated load (kVA for motors and other low–power-factor loads).

The following example is given in the **National Electrical Code** for explanation of the optional method of load calculation for a one-family dwelling.

**example**  The dwelling has a floor area of 1500 ft² exclusive of unoccupied cellar, unfinished attic, and open porches. It has a 12-kW range, a 2.5-kW water heater, a 1.2-kW dishwasher, 9 kW of separately controlled electric space heating installed in five rooms, a 5-kW clothes dryer, and a 6-A, 230-V room air-conditioning unit.

**solution**  The air conditioner is $6 \times 230 \div 1.38$ kVA. This is less than the connected load of 9 kVA of space heating; therefore the air-conditioner load need not be included in the service calculation.

| | |
|---|---:|
| 1500 ft² at 3 VA | 4.5 kVA |
| Two 20-A appliance outlet circuits at 1500 VA each | 3.0 kVA |
| Laundry circuit | 1.5 kVA |
| Range (at nameplate rating) | 12.0 kVA |
| Water heater | 2.5 kVA |
| Dishwasher | 1.2 kVA |
| Space heating | 9.0 kVA |
| Clothes dryer | 5.0 kVA |
| | 38.7 kVA |

$$\text{First 10 kVA at } 100\% = 10.00 \text{ kVA}$$
$$\text{Remainder at } 40\% (28.7 \text{ kVA} \times 0.4) = 11.48 \text{ kVA}$$
$$\text{Calculated load for service size} = 21.48 \text{ kVA} = 21,480 \text{ VA}$$
$$21,480 \div 240 = 90 \text{ A}$$

Therefore this dwelling may be served by a 100-A service.

**47. Calculation of load on four-wire two-phase lighting and appliance feeders or mains.**  The ampere loads on either phase can be computed as in Sec. 45, each phase being considered a separate single-phase circuit.

**48. Calculation of load on three-wire two-phase lighting and appliance feeders or mains.**  For the two-line wires proceed as in Sec. 46.

The load on the neutral wire will be 1.41 times the load for the neutral as determined in accordance with the procedure of Sec. 46. If the load on the neutral as determined in this manner is more than 200 A, no further demand factor can be applied to that portion of the load for this type of system.

**example**  Determine the present and future loads for a three-wire two-phase feeder in an industrial building of 9000 ft² supplying the following loads:

| Panels connected between wire 1 and common wire | | | Panels connected between wire 2 and common wire | | |
|---|---|---|---|---|---|
| Panel | Present load, amp | Future load, amp | Panel | Present load, amp | Future load, amp |
| *A* | 50.8 | 65.2 | *A* | 47.8 | 61.3 |
| *B* | 43.7 | 55.6 | *B* | 52.7 | 67.9 |
| *C* | 41.4 | 54.7 | *C* | 36.8 | 45.9 |
| | 135.9 | 175.5 | | 137.3 | 175.1 |

**solution**  From Table 4, Div. 11, the demand factor is 1.0.
Present load on wire 1 = 135.9 × 1.0 = 135.9 A.
Future load on wire 1 = 175.5 × 1.0 = 175.5 A.
Present load on wire 2 = 137.3 × 1.0 = 137.3 A.
Future load on wire 2 = 175.1 × 1.0 = 175.1 A.
Present load on common wire = 137.3 × 1.41 = 193.6 A.
Future load on common wire 175.5 × 1.41 = 248 A.

**49. Calculation of load on five-wire two-phase lighting and appliance feeders or mains.**  For the four line wires proceed as for the line wires in Sec. 46.

The following example shows how to compute the maximum demand for the four-line conductors and the common neutral of a typical five-wire two-phase system:

**example**  The actual connected loads on the four outside wires of a five-wire two-phase lighting feeder are 100.5, 97.8, 95.4, and 100.0 A. The feeder is supplying the load in a factory building. Determine the present load for the neutral wire.

**solution**  From Table 4, Div. 11, the demand factor is 1.0. The load on the neutral = (100.5 × 1.0)1.41 = 142 A.

**50. Calculation of load on four-wire three-phase lighting and appliance feeders or mains.**  The ampere load on any outside wire is computed as in Sec. 46, except for the calculation of the maximum-demand current for the electric-range load. The maximum-demand current for the range load should be determined from the following formula:

$$\text{Maximum-demand range current} = \frac{\substack{\text{watt load as determined from Table 2, Div.}\\ \text{11, for twice the number of ranges con-}\\ \text{nected between any two outside wires}}}{\substack{2 \times \text{voltage between an outside wire and}\\ \text{neutral}}} \qquad (2)$$

For the ampere load of the neutral wire, proceed as for the neutral in Sec. 46.

**example**  Thirty ranges rated at 12 kW each are supplied by a three-phase four-wire 120/208-V feeder, 10 ranges on each phase. Determine the maximum-demand current for each outside wire.

**solution**  As there are 20 ranges connected to each ungrounded conductor, the load should be calculated on the basis of 20 ranges (or in case of unbalance, twice the maximum number between any two-phase wires), since diversity applies only to the number of ranges connected to adjacent phases and not to the total.

The current in any one conductor will be one-half the total watt load of two adjacent phases divided by the line-to-neutral voltage. In this case, 20 ranges, from Table 2, Div. 11, will have a total watt load of 35,000 W for two phases; therefore, the current in the feeder conductor would be

$$17,500 \div 120 = 146 \text{ A}$$

**51. Determining size of lighting and appliance distribution panel buses.**  The required ampere capacity of the buses in lighting and appliance distribution panels can be calculated in the same manner as given in the preceding sections for the calculation of the ampere loads on feeders and mains.

**52. Minimum allowable load to use for any lighting feeder or main.**  The National Electrical Code specifies definite rules for computing the minimum loads for a given area. Feeders or mains must be of sufficient size to carry safely at least the amperes required by these rules regardless of the actual connected load. The areas employed in applying the rules should be gross floor areas as determined from the outside dimensions of the building and the number of floors. Floor areas of open porches, garages in connection with dwelling occupancies, and unfinished and unused spaces in dwellings, unless adapted for future use, need not be included.

The minimum allowable watt load for the feeder or main supplying power to any area for lighting and appliances must be equal to the sum of the minimum allowable general-lighting load, the portable-appliance load, the fixed-appliance load (other than ranges), fixed electrical space-heating equipment load, and the electric-range load. The minimum allowable general-lighting load is equal to the area times the minimum allowable watts per square foot for that type of area times the proper demand factor. The minimum allowable watts per square foot and the allowable demand factors for different types of areas are given in Table 5 of Div.11. The portable-appliance load requirements are given in Table 5 of Div. 11. If in normal operation the load on the installation will continue for 3 h or more, as in store lighting, the minimum allowable load, as determined above, must be increased by 25 percent. The fixed-appliance load is equal to the summation of the watt rating of the actual fixed appliances installed in the area if the number of such appliances is four or less. For electric ranges, and other cooking appliances of 1¾-kW rating or less, a load equal to the summation of the watt ratings of the ranges or appliances installed must be included. For electric ranges and other cooking appliances rated more than 1¾ kW, the minimum allowable load can be computed for Table 2 of Div. 11. If a number of ranges are supplied by a three-phase four-wire feeder, the minimum allowable watt load should be computed from the following formula:

$$\text{Minimum allowable range watt load} = 3 \times \frac{\begin{array}{c}\text{watt load as determined from Table}\\ \text{2 of Div. 11 for twice the number of}\\ \text{ranges connected between any two}\\ \text{outside wires}\end{array}}{2} \quad (3)$$

The minimum allowable load for motor circuits is identical with the values obtained from the instructions of Secs. 53 and 54.

The minimum allowable load for circuits supplying both motors and other loads is equal to the motor load plus the minimum allowable load of other types. After the minimum allowable voltampere load has been determined, the corresponding ampere load can be determined from the following formulas and instructions, depending upon the type of electric system employed:

$$I = \frac{\text{minimum allowable voltamperes}}{KE} \quad (4)$$

where $K$ = 1 for two-wire dc or two-wire single-phase ac
   = 1.73 for three-wire three-phase ac
   = 2 for three-wire dc, three-wire single-phase ac, three-wire two-phase ac, or four-wire two-phase ac
   = 3 for four-wire three-phase ac
   = 4 for five-wire two-phase ac
   $E$ = voltage between outside wire and neutral if the system has a neutral; otherwise the voltage between any two line wires
   $I$ = current in any line wire except the neutral

For neutral for three-wire dc, three-wire single-phase ac, or four-wire three-phase, follow instructions given in Sec. 46.

For neutral for three-wire or five-wire two-phase, follow instructions given in Secs. 48 and 49.

**example**   Determine the minimum allowable loading of a feeder supplying a multifamily dwelling having an area of 30,800 ft$^2$ with 44 apartments. There is no electric-range load.

**solution**   From Table 5, Div. 11, 3.0 VA must be allowed for each square foot plus 3000 VA for appliances for each apartment. The demand factor from Table 5 of Div. 11 is 1.0 for the first 3000 VA, 0.35 for the next 117,000 VA, and 0.25 for all load in excess of 120,000 VA.

| | |
|---|---:|
| Lighting load = 3 × 30,800 | = 92,400 |
| Appliance load = 44 × 3000 | = 132,000 |
| Total load based on area | = 224,400 VA |

Minimum allowable voltampere load = 3000 + (117,000 × 0.35) + (224,400 − 120,000)0.25
   = 70,050 VA

For a two-wire 120-V single-phase or dc system:

$$I = \frac{70,050}{120} = 584 \text{ A}$$

For a three-wire 120/240-V single-phase or dc system:

$$I \text{ (outside wires)} = \frac{70,050}{2 \times 120} = 292 \text{ A}$$

$$I \text{ (neutral)} = 200 + (92 \times 0.70) = 264.4 \text{ A}$$

For a four-wire 115-V two-phase system:

$$I = \frac{70,050}{2 \times 115} = 304.6 \text{ A}$$

For a three-wire 115-V two-phase system:

$$I \text{ (outside wires)} = \frac{70,050}{2 \times 115} = 304.6 \text{ A}$$

$$1.41 \times 304.6 = 429.5 \text{ A}$$

$$(\text{common wire}) = 429.5 \text{ A}$$

For a five-wire 115/230-V two-phase system:

$$I \text{ (outside wires)} = \frac{70,050}{4 \times 115} = 152.3 \text{ A}$$

$$I \text{ (neutral)} = 1.41 \times 152.3 = 214.7 \text{ A}$$

For a four-wire 208Y/120-V three-phase system:

$$I \text{ (outside wires)} = \frac{70,050}{3 \times 120} = 194.6 \text{ A}$$

$$I \text{ (neutral)} = I \text{ (outside wire)} = 121 \text{ A}$$

**example**  Determine the minimum allowable loading for a three-wire 120/240-V feeder supplying an apartment house having a total floor area of 32,000 ft$^2$ with 40 apartments. One-half of the apartments are equipped with electric ranges of 10 kW each.

**solution**  From Table 5, Div. 11, 3 VA must be allowed for each square foot plus 3000 VA for appliances for each apartment. The demand factor for the lighting and appliance load from Table 5 of Div. 11 is 1.0 for the first 3000 VA, 0.35 for the next 117,000 VA, and 0.25 for all load in excess of 120,000 VA. From Table 2 of Div. 11, the maximum demand for the range load is 35 kW.

| | |
|---|---:|
| Lighting load = 3 × 32,000 | = 96,000 |
| Appliance load = 40 × 3000 | = 120,000 |
| Total lighting and appliance load | = 216,000 VA |
| Minimum allowable lighting and appliance load = 3000 | |
| +(117,000 × 0.35) + (216,000 − 120,000)0.25 = | 67,950 VA |
| Range load | = 35,000 VA |
| Total minimum allowable voltampere load | = 102,950 VA |

$$I \text{ (outside wires)} = \frac{102,950}{2 \times 120} = 429 \text{ A}$$

Neutral feeder:

$$\text{Lighting and small-appliance load} = 67,950 \text{ VA}$$
$$\text{Range load, 35,000 W at 70\%} = 24,500 \text{ VA}$$
$$\text{Computed load (neutral)} = 92,450 \text{ VA}$$
$$92,450 \div 240 = 385 \text{ A}$$

Further demand factor:

$$200 \text{ A at 100\%} = 200 \text{ A}$$
$$185 \text{ A at 70\%} = 130 \text{ A}$$
$$\text{Maximum-demand neutral wire} = 330 \text{ A}$$

**example**  Determine the minimum allowable general-lighting load for a feeder supplying an office building having an area of 100,000 ft$^2$.

**solution**  From Table 5, Div. 11, 3½ VA plus 1 VA for receptacle loads must be allowed per square foot. From Table 5, Div. 11, the demand factor is 1.0.

$$\text{Voltampere load based on area} = 100,000 \times 4.5 = 450,000 \text{ VA}$$

For a four-wire 120/208-V three-phase system:

$$I \text{ (outside wires)} = \frac{450,000}{3 \times 120} = 1250 \text{ A}$$

$$I \text{ (neutral)} = 200 + (1250 - 200)0.70$$

$$= 935 \text{ A}$$

**53. Calculation of load on motor branch circuits.**  For continuous-duty motors the load on the branch circuit feeding a single motor is taken as 125 percent of the full-load rated current of the motor. The load on a branch circuit supplying a motor in a class of service having short-time duty depends upon the character of the loading. In the majority of cases the load need not be greater than the percentages of the full-load rated currents given in Table 12 of Div. 11. The average full-load rated currents of the different types and sizes of motors are given in Tables 7 to 10 of Div. 11.

The load on conductors connecting the secondary of a wound-rotor, polyphase induction motor to its controller is taken as 125 percent of the full-load secondary current of the motor for continuous-duty motors and not less than the percentages given in Table 12 of Div. 11 of the full-load secondary current for short-time–duty motors. The loads on the conductors connecting the controller with the secondary resistors must be taken at not less than the proper percentage, as given in Table 13 of Div. 11, of the full-load secondary current. The value of the full-load secondary current, if not marked on the nameplate of the motor, should be obtained from the manufacturer of the motor, since its value will depend upon the design of the motor. For preliminary studies the secondary current may be taken to be equal to the motor full-load line current for values of full-load current up to about 20 A. For larger motors the secondary current is usually less than the full-load line current. For a motor with a full-load line current of about 50 to 70 A, the secondary current is about two-thirds of that value; for a full-load line current of about 80 to 120 A, the secondary current is about one-half of that value; for a full-load line current of about 150 to 250 A, the secondary current is about one-third the full-load line value.

**54. Calculation of load on motor feeders and mains.** Two values of load should be computed for motor feeders or mains: one the maximum-demand starting current and the other the maximum-demand running current. The maximum-demand running current can be used in determining the size of wire required to carry the current safely. Although this meets the requirements of the National Electrical Code, many authorities consider it better practice to use the maximum-demand starting current in determining the size of wire according to carrying capacity, since then the circuit can be protected against both straight overload and short circuit. When the maximum-demand running current is used for determining the size of wire according to carrying capacity, the circuit can be protected only against short circuit or very heavy overloads and not against overloads of moderate severity. The maximum-demand running current is used for computing the voltage drop of the circuit. The maximum-demand starting current is employed in determining the proper size of protective equipment for the circuit. Values of the average starting currents of motors are given in Tables 42 and 43 of Div. 11.

The National Electrical Manufacturers Association (NEMA) adopted in 1940 a standard of identifying code letters that must be marked by manufacturers on motor nameplates to indicate the motor kilovolt-ampere input with locked rotor. These code letters with their classification are given in Table 42 of Div. 11. In determining the starting current to employ for circuit calculations, use values from Table 43 of Div. 11.

In many installations in which the number of motors exceeds five, all the motors will not be running at full load at the same time. It is general practice, therefore, on such feeders or mains to use a maximum-demand factor so that the estimated maximum-demand current is less than the sum of the full-load rated currents of all the motors fed from the circuit. Values of maximum-demand factors that have been found satisfactory for ordinary installation are given in Table 6 of Div. 11. It should be remembered, in applying these demand factors, that they are average values satisfactory for ordinary installations. Before they are used, a careful study of the operating conditions of the plant should be made to determine whether the conditions of instantaneous loading of the motors will come within the average conditions or whether there are special requirements that will submit the feeders to a greater loading than will be taken care of by these factors.

Permission for the use of a demand factor must be obtained from the authority enforcing the code. Some authorities recommend that no demand factor be used in determining the size of circuit to install so that the additional current capacity, thus allowed in the circuit, will give some spare capacity for growth. Refer to Sec. 36 for further discussion on allowance for growth.

The following formulas give the methods of computing the maximum-demand starting and running currents for motor feeders and mains:

$$\text{Starting current} = \text{starting current of largest motor} + \left( \text{demand factor} \times \begin{array}{c} \text{sum of full-load rated currents of all the motors except largest} \end{array} \right) \qquad (5)$$

$$\text{Running current} = 1.25 \times \text{full-load current of largest motor} + \left( \text{demand factor} \times \begin{array}{c} \text{sum of full-load rated currents of all the motors except largest} \end{array} \right) \qquad (6)$$

If a number of motors of equal horsepower rating are the largest in the group supplied by the circuit, one of the these motors should be taken as the largest motor for the calculation of the load on the circuit.

If two or more motors must be started at the same time, it will generally be necessary to increase the load on the circuit above the values obtained from Eqs. (5) and (6).

**example**   Determine the load on a 220-V feeder supplying two 10-hp and two 15-hp motors. All the motors are the three-phase, squirrel-cage induction, normal-starting-current type started at reduced voltage.

**solution**   From Table 10 of Div. 11, the full-load current of a 10-hp 220-V squirrel-cage induction three-phase motor is 28 A and the full-load current of a 15-hp motor is 42 A. The starting current of a 15-hp squirrel-cage induction motor started at reduced voltage from Table 43 of Div. 11 is 200 percent of the full-load current.

$$\text{Maximum-demand starting current} = (2.00 \times 42) + 42 + (2 \times 28)$$
$$= 84 + 42 + 56$$
$$= 182 \text{ A}$$
$$\text{Maximum-demand running current} = (1.25 \times 42) + 42 + (2 \times 28)$$
$$= 150 \text{ A}$$

**example**   Determine the load on a 440-V feeder supplying three 5-hp motors, two 10-hp motors, and three 15-hp motors. All the motors are of the low-starting-current, squirrel-cage type, three-phase.

**solution**   From Table 10 of Div. 11 the full-load current of a 5-hp three-phase low-starting-current squirrel-cage motor is 7.6 A; of a 10-hp motor, 14 A; and of a 15-hp motor, 21 A. From Table 6 of Div. 11 the maximum-demand factor is 0.75. Permission has been obtained for the use of this factor.

$$\text{Maximum-demand starting current} = (2.5 \times 21) + 0.75[(3 \times 7.6) + (2 \times 14) + (2 \times 21)]$$
$$= 52.5 + 0.75(22.8 + 28 + 42)$$
$$= 122 \text{ A}$$
$$\text{Maximum-demand running current} = (1.25 \times 21) + 0.75[(3 \times 7.6) + (2 \times 14) + (2 + 21)]$$
$$= 26.25 + 0.75 (22.8 + 28 + 42)$$
$$= 95 \text{ A}$$

**example**   Determine the load on a 230-V dc feeder supplying three 3-hp motors, five 5-hp motors, and one 20-hp motor.

**solution**   From Table 7 of Div. 11 the full-load current of a 3-hp 230-V dc motor is 12.2A; of a 5-hp motor, 20 A; of a 20-hp motor, 72 A. The starting current of a dc motor from Table 43 of Div. 11 is 150 percent of its full-load current. From Table 6 of Div. 11 the maximum-demand factor for nine motors is 0.75. Permission has been obtained for the use of this factor.

$$\text{Maximum-demand starting current} = (1.50 \times 72) + 0.75[(3 \times 12.2) + (5 \times 20)]$$
$$= 108 + 0.75(36.4 + 100)$$
$$= 108 + 102 = 210 \text{ A}$$
$$\text{Maximum-demand running current} = (1.25 \times 72) + 0.75[(3 \times 12.2) + (5 \times 20)]$$
$$= 90 + 0.75(36.4 + 100)$$
$$= 90 + (0.75 \times 136.4)$$
$$= 90 + 102 = 192 \text{ A}$$

**55. Load on combined motor and lighting circuits.**   If both motors and lamps are fed from the same circuit, proceed as follows in determining the total load of the circuit:
   1. Determine the lighting and appliance load.
   2. Determine the motor load.
   3. Total load is equal to the sum of the lighting and motor loads.

**56. General considerations in computing voltage drop.**   When calculations are made for the voltage drop in a circuit, the future current (Sec. 36) should be used for lighting and appliance circuits, the maximum running current with allowance for growth (Secs. 36 and 54) for motor feeders or mains, and the full-load current for motor branch circuits.

The resistance employed in the calculations should be the value corresponding to the operating temperature of the conductors. The following rules with respect to the values of operating temperature to employ in voltage-drop calculations represent good practice for wires that are loaded between 50 and 100 percent of their allowable carrying capacity.

   1. Use 50 to 60°C for wires insulated with Code or moisture-resistant rubber compounds, synthetic rubberlike compounds, or thermoplastic compounds; Type TW wire.

   2. Use 70°C for wires insulated with heat-resistant rubber compounds, varnished cambric, paper, thermoplastic, slow-burning compounds, or weatherproof compounds; Types RH, RHH, RHW, THW, THWN, and MI.

   3. Use 100°C for all wires not included in Pars. 1 and 2. For wires loaded less than 50 percent of their allowable carrying capacity the temperature should be reduced from 15 to 20° below the above values. The resistance of copper wire varies somewhat with the

method of drawing the wire. As a general rule the voltage drop of a circuit will depend upon several variables, the value of which cannot possibly be determined accurately. It is not practical, therefore, to spend too much time in accurately determining the value of the resistance of the wires. An average value of 98 percent conductivity for copper wires is generally satisfactory. Copper of 98 percent conductivity has a resistivity of approximately 10.6 $\Omega$/cmil · ft at 20°C, 11.2 $\Omega$ at 30°, 11.6 at 40°, 11.8 at 50°, 12.3 at 60°, and 12.7 at 70°. Tables of resistance of wires are given in Div. 11.

The length of the circuit should be taken as the distance along the circuit from the supply end to the load center. If a load is distributed along the circuit, the total current does not flow the complete length of the circuit. Therefore, if the actual length of the circuit were used in computing the voltage drop, the drop determined would be greater than the drop that would actually occur. The load center of a circuit is that point in the circuit where, if the load were concentrated at that point, the drop would be the same as the voltage drop to the farthest load in the actual circuit.

**57. The load center of a circuit** can be determined in the following manner. Multiply each load by its distance from the supply end of the circuit. Add these products for all the loads fed from the circuit and divide this sum by the sum of the individual loads. The result thus obtained is the distance from the supply end of the circuit to the load center. It is this length that should be employed in computing the voltage drop of the circuit (see solution of example in Fig. 3-28).

The load center of a group of receivers symmetrically arranged (Fig. 3-29) and all of the same output will be in the middle of the group. Always take the distance along the circuit as $L$, Fig. 3-29.

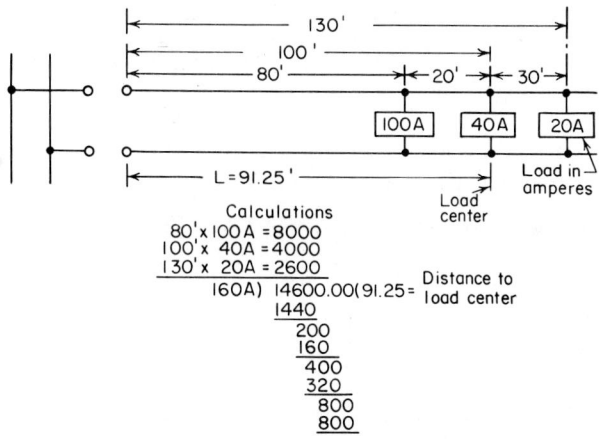

**Fig. 3-28**  Method of computing location of load center.

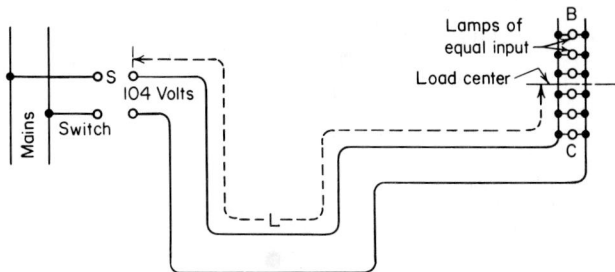

**Fig. 3-29**  Location of load center.

**58. Calculation of voltage drop in dc two-wire circuits by means of formula.**  The voltage drop can be calculated by means of either Eq. (7) or Eq. (8), and the size of wire to produce a given voltage drop can be calculated by Eq. (9).

$$VD = 2R \times L \times I \tag{7}$$

where $VD$ = drop in volts in the circuit, $R$ = resistance of wire in ohms per foot (values of resistance for copper or aluminum conductors can be obtained from Table 61 of Div. 11), $I$ = current in amperes, and $L$ = length one way of circuit in feet, or

$$VD = \frac{2K \times L \times I}{cmil} \tag{8}$$

where $VD$ = drop in volts in the circuit; $K$ = resistivity of material of conductor in ohms per circular mil-foot (values of $K$ for copper conductors can be obtained from Sec. 56; sufficient accuracy will be obtained for most cases by using the following more approximate values: 12 for circuits loaded between 50 and 100 percent of their allowable carrying capacity and 11 for circuits loaded at less than 50 percent of their allowable carrying capacity); $I$ = current in amperes; $L$ = length one way of circuit in feet; and cmil = area of conductor in circular mils. $K$ = 18 for aluminum conductors.

$$cmil = \frac{2K \times I \times L}{VD} \tag{9}$$

**59. Determination of voltage drop in dc two-wire circuits by means of chart.**  A graph for computing the voltage drop in circuits (Sec. 53 of Div. 11) was originally proposed by R. W. Stovel and N. A. Carle in *The Electric Journal* for June 1908. It is based on a resistivity of $10.7 \ \Omega/\text{cmil} \cdot \text{ft}$ of copper wire. The chart will therefore give satisfactory results for circuits loaded at less than 50 percent of their allowable carrying capacity. For most circuits which are loaded to between 50 and 100 percent of their allowable carrying capacity, the voltage drop obtained from the chart should be increased by 10 percent. The length of the circuit that should be employed in using the chart is the distance from the supply end to the load center. The voltage drop as read from the chart is the total drop in both wires. To determine the voltage drop of a circuit from the chart, proceed as follows:

1.  Start on the lower left-hand scale with the current for the circuit.
2.  Follow this point vertically upward to the diagonal line for the size of wire of the circuit.
3.  From this intersection proceed horizontally to the right to the diagonal line representing the length of the circuit one way.
4.  At this intersection drop vertically to the voltage-drop scale.
5.  The reading on the voltage-drop scale will be the drops in both wires of the circuit for lightly loaded circuits as discussed above.
6.  For normally loaded circuits the voltage drop will be equal to 1.1 times the value read from the chart.

**example**  What will be the voltage drop of the motor feeder for the last problem of Sec. 54 if the length of circuit is 150 ft and a No. 4/0 Type RH wire is employed?

**solution**  The maximum-demand running current of 195 A should be used in determining voltage drop. From Table 61 of Div. 11 a No. 4/0 wire has 211,600 cmil and from Table 18 of Div. 11 has a carrying capacity of 230 A.

Using the formula,

$$VD \times \frac{24 \times I \times L}{cmil} = \frac{24 \times 195 \times 150}{211,600} = 3.32 \text{ V drop}$$

Using the chart, start at the bottom of the chart on the left-hand side at 195 A and follow this point vertically upward until it intersects the diagonal line marked No. 0000. From this intersection proceed horizontally to the right until the diagonal line marked 150 is reached. At this intersection drop vertically downward to the voltage-drop scale, and read 3.0 V. Since the wire is loaded to more than 50 percent of its capacity, the voltage drop equals $3.0 \times 1.1 = 3.3 \text{ V}$.

**60. Determination of voltage drop in dc three-wire circuits.**  Either the formulas of Sec. 58 or the chart of Sec. 53, Div. 11, can be used for determining the voltage drop of dc three-wire circuits. In either case the current used should be that of the more heavily loaded outside wire and the size of conductor that of the outside conductors. The drop thus obtained from either the formula or the chart will be the drop between the outside wires. What is desired is the drop across each receiver or the drop between an outside wire and the neutral wire. The approximate voltage drop to each receiver will be one-half of the value determined from either the formula or the chart. This value will be correct for a balanced load. When the load is unbalanced, the current in the neutral wire will cause an additional voltage drop which will not have the same effect on the two sides of the circuit. Since in a well-laid-out system the load will be nearly balanced, the method given is satisfactory for most cases.

   **example**  Determine the voltage drop on a three-wire lighting feeder 100 ft long. The actual load on one side is 147 A, and on the other side 158 A. The future load on one side is 170 A, and on the other side 175 A. A No. 4/0 Type TW wire is used for the outside copper conductors.

   **solution**  The future load on the more heavily loaded side should be used in determining voltage drop. A No. 4/0 Type TW wire from Table 61, Div. 11, has 211,600 cmil, and from Table 18, Div. 11, a carrying capacity of 195 A.

   Using the formula,

$$VD \times \frac{24 \times I \times L}{\text{cmil}} = \frac{24 \times 175 \times 100}{211,600} = 1.98 \text{ V}$$

Using the chart, start at the bottom of the chart on the left-hand side at 175 A and follow this point vertically upward until it intersects the diagonal line marked No. 0000. From this intersection follow horizontally to the right until the diagonal line marked 100 is reached. At this intersection drop vertically downward to the voltage-drop scale, and read 1.8 V drop. The voltage drop will be equal to 1.8 × 1.1 = 1.98.

   The drop across each receiver is one-half of the drop determined by the formula or chart. The drop across each receiver is, therefore, 0.99 V determined from either formula or chart.

   **61. The voltage drop in ac circuits** is affected by several factors that have no effect in dc circuits. These factors are (1) power factor of the load, (2) inductance of the circuit, (3) capacity of the circuit, and (4) increased resistance of the circuit to alternating current.

   As discussed in Div. 1, when alternating current flows in a circuit, the resistance is increased owing to skin effect. This increase in resistance for most circuits is so small that it need not be considered. Unless the size of the conductor is greater than 750,000 cmil (380 mm²) for 25-Hz circuits or 300,000 cmil (152 mm²) for 60-Hz circuits, skin effect can be neglected and the resistance of the wire taken in the same way as for direct current.

   As discussed in Div. 1, the phenomenon of inductance causes a voltage which opposes the flow of current in the circuit to be induced in an ac circuit. Inductance therefore offers opposition to the flow of alternating current in a circuit. This opposition is called the inductive reactance of the circuit and is represented by the symbol $X_L$. The value of the inductive reactance of a circuit depends upon the size of the wire, the distance between the wires of the circuit, the frequency of the current flowing in the circuit, the material of the conductor, and the presence of any magnetic material in proximity to the circuit. The voltage drop due to inductance produced in a circuit is equal to the current times the inductive reactance. For small-size conductors the effect of inductance is so small that it can be neglected (see Sec. 54 of Div. 11).

   The two conductors of any circuit with the insulation between them produce a capacitor. They therefore introduce capacity into the circuit. The effect of capacity in an ac circuit is to offer opposition to the flow of current. This opposition is called capacity reactance and is denoted by the symbol $X_c$. The voltage drop in a circuit due to capacity is equal to the current times the capacity reactance. The effect of capacity upon the voltage drop of the circuit is so small except for high-voltage long-distance transmission lines that it is neglected.

   The total voltage drop in an ac circuit due to the resistance and reactance is affected by the power factor of the load connected to the circuit.

   **62. Summary of factors that must be considered in calculating drop for ac circuits.**  Skin effect can usually be neglected unless the size of wires is greater than 300,000 cmil (152 mm²) for 60-Hz circuits or 750,000 cmil (380 mm²) for 25-Hz circuits.

   The effect of capacity can be neglected except for long-distance transmission lines.

   The effect of inductance can be neglected unless the size of wire exceeds the values given in Table 54 of Div. 11.

   **63. Line or circuit reactance can be reduced in three ways.**  One way is to diminish the distance between wires. The extent to which this can be carried is limited, in the case of a pole line, to the least distance at which the wires are safe from swinging together in the middle of a span. In inside wiring (knob or tube work), it is limited by the spacings required by the National Electrical Code. In conduit work, nothing can be done about reducing the distance between wires. Another way of reducing the reactance is to increase the size of the conductors, but it is not possible to secure much reduction by this method unless the size is increased by an excessive amount. The third way of reducing reactance is to divide the load into a greater number of circuits. Voltage drop in lines due to inductive reactance is best diminished (Mershon) by subdividing the copper or by bringing the conductors closer together. It is little affected by changing the size of the conductors.

**64. Determining power factors of feeders or mains.**   Although the values of power factors obtained by the following method are approximate, they are accurate enough for most circuit calculations. To determine the power factor of a circuit supplying several motors proceed as follows:

1. Multiply the horsepower of each motor by its power factor at 75 percent of rated load.

2. Add these products for all the motors fed from the line.

3. The approximate power factor of the circuit will equal the sum obtained in Par. 2 divided by the total horsepower connected to the circuit.

Approximate power factors of different types of loads are given in Table 56 of Div. 11.

**example**   Determine the power factor for a feeder supplying two 5-hp motors, five 10-hp motors, and one 50-hp motor.

**solution**

$$\text{Hp of motor} \times \text{power factor} = \text{product of hp and pf}$$
(From Table 79, Div. 11)

| | | | | |
|---:|:--:|:--:|:--:|---:|
| 5 | × | 0.83 | = | 4.15 |
| 5 | × | 0.83 | = | 4.15 |
| 10 | × | 0.86 | = | 8.6 |
| 10 | × | 0.86 | = | 8.6 |
| 10 | × | 0.86 | = | 8.6 |
| 10 | × | 0.86 | = | 8.6 |
| 10 | × | 0.86 | = | 8.6 |
| 50 | × | 0.89 | = | 44.5 |
| 110 total connected hp | | | | 95.8 |

$$\text{Approximate power factor of circuit} = \frac{\text{sum of products of hp and pf}}{\text{total connected hp}}$$

$$= \frac{95.8}{110} = 87.1$$

**65. Calculation of voltage drop in two-wire single-phase circuits when effect of inductance can be neglected.**   These circuits can be calculated in the same way as dc circuits, by means of either the formulas of Sec. 59 or the chart of Sec. 53 of Div. 11.

**66. Calculation of voltage drop in three-wire single-phase circuits when effect of inductance can be neglected.**   Circuits of this type can be calculated in exactly the same manner as three-wire dc circuits (see Sec. 60).

**67. Calculation of voltage drop in four-wire two-phase circuits when effect of inductance can be neglected.**   A four-wire two-phase circuit can be considered as two separate single-phase circuits, and the voltage drop computed in the same manner as given for two-wire dc circuits in either Sec. 58 or Sec. 59.

**68. Calculation of voltage drop in three-wire two-phase circuits when effect of inductance can be neglected.**   The voltage drop in the common wire of three-wire two-phase systems somewhat unbalances the voltages of the two phases, and therefore the voltage drop of the two phases is not the same. An exact method of the calculation of the voltage drop of these circuits is too complicated for the scope of this book. The following method will give the approximate voltage drop on the phase having the greater voltage drop. It is accurate enough for most interior-wiring calculations. Proceed as follows:

1. Determine the voltage drop in one outside wire. It will be equal to one-half of the voltage drop determined by means of either the formulas of Sec. 58 or the chart of Sec. 53 in Div. 11.

2. Determine the voltage drop in the common wire. It will be equal to one-half of

the voltage drop determined by means of either the formulas of Sec. 58 or the chart of Sec. 53 in Div. 11, the size of wire and current for the common wire being used.

3. The total voltage drop is taken as equal to the drop in one outside wire plus 0.8 times the voltage drop in the common wire.

**example** Determine the voltage drop of a three-wire two-phase 60-Hz lighting feeder which is 200 ft long and installed in conduit. The current in each outside wire is 130 A. A No. 2/0 wire is used for the outside wires, and a No. 4/0 for the common wire. All wires are Type TW.

**solution** Since the largest wire used is No. 4/0 from Table 54 of Div. 11, the effect of inductance can be neglected. From Table 61 of Div. 11 a No. 2/0 wire has 133,100 cmil, and a No. 4/0 wire 211,600 cmil. From Table 18 of Div. 11 all wires are loaded to more than 50 percent of their carrying capacity.

$$\text{Drop in two outside wires} = \frac{24 \times I \times L}{\text{cmil}} = \frac{24 \times 130 \times 200}{133,100} = 4.7 \text{ V}$$

$$\text{Drop in one outside wire} = \frac{4.7}{2} = 2.35 \text{ V}$$

$$\text{Current in common wire} = 1.41 \times 130 = 183.5 \text{ A}$$

$$\text{Drop in common wire} = \frac{24 \times I \times L}{2 \times \text{cmil}} = \frac{24 \times 183.5 \times 200}{2 \times 211,600} = 2.08 \text{ V}$$

$$\text{Total voltage drop} = 2.35 + (0.8 \times 2.08) = 2.35 + 1.66$$
$$= 4.01 \text{ V drop}$$

**69. Calculation of voltage drop in five-wire two-phase circuits when effect of inductance can be neglected.** In a well-laid-out system the load will be very nearly balanced under normal conditions. Therefore, there will be practically no current in the neutral wire. For lighting loads supplied by this system the lamps are connected between the neutral wire and the respective outside wires. The drop to any lamp therefore will be equal to the drop in one outside wire. The drop in one outside wire will be equal to one-half of the drop determined by either the formulas of Sec. 58 or the chart of Sec. 53 of Div. 11. The drop to motors fed by this system will be equal to the drop in two outside wires and, therefore, to the drop determined by either the formulas or the chart.

**70. Calculation of voltage drop in three-wire three-phase circuits when effect of inductance can be neglected.** The voltage drop in these circuits will be equal to 0.866 times the voltage drop of a two-wire dc circuit carrying the same current as the three-phase circuit. This drop may be determined from either the formulas of Sec. 58 or the chart of Sec. 53 of Div. 11.

**example** Determine the voltage drop of a three-wire three-phase 60-Hz motor feeder which is 150 ft long. Three No. 2 Type TW wires are installed in conduit. The current is 85 A.

**solution** Since the size of wire is No. 2 from Table 54 of Div. 11, the effect of inductance can be neglected. From Table 61 of Div. 11 a No. 2 wire has 66,370 cmil, and from Table 18 of Div. 11 a carrying capacity of 95 A.

$$\text{Drop in a two-wire dc circuit carrying 85 A over a No. 2 wire} = \frac{24 \times I \times L}{\text{cmil}}$$

$$= \frac{24 \times 85 \times 150}{66,370} = 4.61 \text{ V}$$

Therefore $\qquad$ Drop of the three-phase system $= 0.866 \times 4.61$
$$= 4.0 \text{ V}$$

**71. Calculation of voltage drop in four-wire three-phase circuits when effect of inductance can be neglected.** In a well-laid-out system the load will be very nearly balanced under normal load conditions. Therefore, there will be practically no current in the neutral wire. For lighting loads supplied by this system the lamps are connected between the neutral wire and the respective outside wires. The drop to any lamp, therefore, is equal to the drop in one outside wire. The drop in one outside wire will be equal to one-half of the drop determined by either the formulas of Sec. 64 or the chart of Sec. 53 of Div. 11. The drop to motors fed by this system is the drop between any two outside wires. This drop is equal to 0.866 times the voltage drop of a two-wire dc system carrying the same current as the three-phase system. (Use either the formulas of Sec. 58 or the chart of Sec. 53 of Div. 11 for determining the dc drop.)

**72. Calculation of circuits when effect of inductance cannot be neglected.**  Two methods are given in the following paragraphs for calculating the voltage drop of circuits when the effect of inductance is so great that it cannot be neglected (see Sec. 54 of Div. 11 for rules indicating when inductance must be considered). In one of these methods the voltage drop in the actual ac circuit is calculated by mulitplying the drop of a dc circuit by a factor called the drop factor. This drop factor is the ratio between the actual drop and the drop that would occur if there were no inductance. The drop factor is affected by the size of wire, spacing of the wires of the circuit, the frequency of the current, and the power factor of the load. Values of drop factors for various conditions in which concentric stranded copper conductors are used are given in Table 59 of Div. 11.

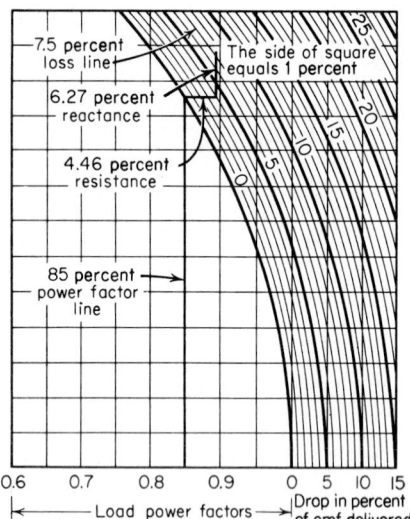

The other method given for computing the voltage drop of circuits when the effect of inductance cannot be neglected makes use of a diagram called the Mershon diagram (Fig. 3-30).

Under ordinary conditions of use both of these methods will give results of about the same degree of accuracy. If the Mershon diagram is used carefully, it will give the more accurate results of the two. But since in the calculation of most circuits several factors are based on assumption, the greater accuracy of the Mershon diagram is of questionable value. The drop-factor method is certainly accurate enough for the

**Fig. 3-30**  Illustrating the application of the Mershon diagram for computing a single-phase circuit. The side of each small square equals 1 percent; percentage resistance is measured horizontally and percentage reactance vertically.

calculation of all interior-wiring circuits, and the authors believe that it is the more easily applied method of the two. For circuits employing some type of conductor material or construction other than standard copper, the Mershon diagram must be used, since no data are given for drop factors for conductors of other types.

**73. Drop-factor method of calculating voltage drop when effect of inductance cannot be neglected.**  Proceed as follows:

1. Determine from either Sec. 58 or Sec. 59 the voltage drop of a two-wire dc system carrying the same current and using the same size of wire.

2. Determine the ratio of reactance to resistance for the size of wire, spacing of wires, and frequency of the circuit considered (see Table 57 or 58 of Div. 11).

3. Determine the power factor of the circuit (see Sec. 64 and Table 56 of Div. 11).

4. Determine the drop factor for the circuit corresponding to the determined ratio of reactance to resistance and the power factor of the circuit from Table 59 of Div. 11.

5. The voltage drop of the circuit then is determined as follows, depending upon the electric system:

For two-wire single-phase systems:

$$\text{Voltage drop} = \text{drop factor} \times \text{dc voltage drop} \qquad (10)$$

For three-wire single-phase systems:

$$\text{Voltage drop between outside wires} = \text{drop factor} \times \text{dc voltage drop} \qquad (11)$$

Voltage drop between outside wire and neutral (lamps)
$$= \tfrac{1}{2}\,(\text{drop factor} \times \text{dc voltage drop}) \qquad (12)$$

For four-wire two-phase systems, each phase is considered as a separate two-wire single-phase system:

$$\text{Voltage drop} = \text{drop factor} \times \text{dc voltage drop} \qquad (13)$$

For three-wire system tapped from four-wire three-phase system:

$$\text{Voltage drop} = 0.75(\text{drop factor} \times \text{dc voltage drop}) \qquad (14)$$

For three-wire two-phase systems, two dc drops must first be determined: one for a two-wire dc circuit carrying the same current and with the same size of wire as the outside conductors of the two-phase system, and the other for a two-wire dc circuit carrying the same current and with the same size of wire as the common wire of the two-phase system. Then

$$\begin{array}{l}\text{Voltage drop of two-} \\ \text{phase three-wire system}\end{array} = \dfrac{\dfrac{\text{drop}}{\text{factor}} \times \begin{array}{c}\text{dc voltage drop for current} \\ \text{and size of outside wire}\end{array}}{2}$$

$$+ \left( 0.8 \times \dfrac{\dfrac{\text{drop}}{\text{factor}} \times \begin{array}{c}\text{dc voltage drop for current and} \\ \text{size of wire of common wire}\end{array}}{2} \right) \quad (15)$$

For five-wire two-phase systems:

$$\text{Voltage drop for lamps} = \dfrac{\text{drop factor} \times \text{dc voltage drop}}{2} \qquad (16)$$

$$\text{Voltage drop for motors} = \text{drop factor} \times \text{dc voltage drop} \qquad (17)$$

For three-wire three-phase systems:

$$\text{Voltage drop} = 0.866 \times \text{drop factor} \times \text{dc voltage drop} \qquad (18)$$

For four-wire three-phase systems:

$$\text{Voltage drop for lamps} = \dfrac{\text{drop factor} \times \text{dc voltage drop}}{2} \qquad (19)$$

$$\text{Voltage drop for motors} = 0.866 \times \text{drop factor} \times \text{dc voltage drop} \qquad (20)$$

**example**   Determine the voltage drop of a two-wire single-phase 60-Hz motor circuit carrying a current of 250 A for a distance of 150 ft. A 300,000-cmil Type RH wire is used, installed in conduit. The power factor of the circuit is 80 percent.

**solution**   From Table 54 of Div. 11, since the size of wire is 300,000 cmil, the effect of inductance cannot be neglected. The carrying capacity from Table 18 of Div. 11 is 285 A.

The drop of a two-wire dc circuit carrying 250 A over a 300,000-cmil wire for 150 ft is $\dfrac{24 \times I \times L}{\text{cmil}} = \dfrac{24 \times 250 \times 150}{300,000} = 3.0$ V.

From Table 57 of Div. 11 the ratio of reactance to resistance for a 300,000-cmil wire installed in conduit on a 60-Hz system is 1.01.

From Table 59 of Div. 11 the drop factor for a ratio of reactance to resistance of 1.01 and a power factor of 80 percent is 1.40.

Voltage drop of the actual circuit = drop factor × dc drop = 1.40 × 3.0 = 4.2 V.

**example**   Determine the voltage drop of the above circuit if it were a three-wire single-phase one.

**solution**   The voltage drop between outside wires would be equal to the drop factor times the drop of a two-wire dc system carrying the same current = 1.40 × 3.0 = 4.2 V.

Voltage drop across lamps if connected to the circuit would be the drop between an outside wire and the neutral = $\dfrac{1.40 \times 3.0}{2} = 2.1$ V.

**example**   Determine the voltage drop of the above circuit if it were three-wire tapped from a four-wire three-phase system.

**solution**   Voltage drop across lamps = 0.75 × 1.4 × 3.0 = 3.15 V.

**example**   Determine the voltage drop of the preceding circuit if it were a four-wire two-phase system.

**solution**   Voltage drop = drop factor × dc drop

$$= 1.40 \times 3.0 = 4.2 \text{ V}$$

**example**   Determine the voltage drop of the preceding circuit if it were a three-wire two-phase one. The 250 A is the current in the outside wires.

**solution**   The current in the common wire = $1.41 \times 250 = 352$ A. Use a 500,000-cmil wire for the common.

The voltage drop for a two-wire dc system carrying 250 A over a 300,000-cmil wire for 150 ft =
$$\frac{24 \times I \times L}{300,000} = \frac{24 \times 250 \times 150}{300,000} = 3.0 \text{ V.}$$

The voltage drop for a two-wire dc circuit carrying 352 A for 150 ft over a 500,000-cmil wire =
$$\frac{24 \times I \times L}{500,000} = \frac{24 \times 352 \times 150}{500,000} = 2.54 \text{ V.}$$

From preceding problems the drop factor for the outside conductors is 1.40

From Table 57 of Div. 11 the ratio of reactance to resistance for a 500,000-cmil wire installed in conduit on a 60-Hz system is 1.75.

From Table 59 of Div. 11 the drop factor for a ratio of reactance to resistance of 1.75 and a power factor of 80 percent is 1.88.

$$\text{Voltage drop of actual circuit} = \frac{\begin{array}{c}\text{dc voltage drop}\\\text{for current and}\\\frac{\text{drop}}{\text{factor}} \times \begin{array}{c}\text{size of wire of}\\\text{outside wires}\end{array}\end{array}}{2} + 0.8 \times \frac{\begin{array}{c}\text{dc voltage drop}\\\text{for current and}\\\frac{\text{drop}}{\text{factor}} \times \begin{array}{c}\text{size of wire of}\\\text{common wire}\end{array}\end{array}}{2}$$

$$= \frac{1.40 \times 3.0}{2} + \frac{0.8 \times 1.88 \times 2.54}{2} = 2.1 + 1.91$$

$$= 4.01 \text{ V}$$

**example**   Determine the voltage drop of the preceding circuit if it were a five-wire two-phase system.

**solution**   Voltage drop for motors = drop factor × dc voltage drop

$$= 1.40 \times 3.0 = 4.2 \text{ V}$$

If any lamps were connected to the circuit, the voltage drop across each lamp

$$= \frac{\text{drop factor} \times \text{dc voltage drop}}{2}$$

$$= \frac{1.40 \times 3.0}{2} = 2.1 \text{ V}$$

**example**   Determine the voltage drop of the preceding circuit if it were a three-wire three-phase system.

**solution**

$$\text{Voltage drop} = 0.866 \times \text{drop factor} \times \text{dc voltage drop}$$
$$= 0.866 \times 1.40 \times 3.0 = 3.64 \text{ V}$$

**example**   Determine the voltage drop of the preceding circuit if it were a four-wire three-phase system.

**solution**

$$\text{Voltage drop for motors} = 0.866 \times \text{drop factor} \times \text{dc voltage drop}$$
$$= 0.866 \times 1.40 \times 3.0 = 3.64 \text{ V}$$

If any lamps were connected to the circuit, the voltage drop across each lamp

$$= \frac{\text{drop factor} \times \text{dc voltage drop}}{2}$$

$$= \frac{1.40 \times 3.0}{2} = 2.1 \text{ V}$$

**74. Calculation of voltage drop of two-wire single-phase circuits by Mershon diagram when effect of inductance cannot be neglected.**   Proceed as follows:

1. Determine the ac resistance of the total length of wire. This will be equal to the resistance per foot of wire of the size and material employed multiplied by 2 times the length of circuit one way in feet. Care should be exercised in taking values of resistance from the tables to observe in what units the resistance is given. Some values are given in $\Omega/1000$ ft, and others in $\Omega/\text{mi}$. For copper conductors use Table 61 or 62 of Div. 11. The ac resistance will be equal to the dc resistance at the working temperature of the wire times the skin-effect correction factor.

2. Determine the resistance volts drop. This drop will be equal to the resistance of the total length of wire as has been determined in step 1 and multiplied by the current.

3. Determine the inductive reactance of the total length of wire. This will be equal to the reactance per foot of wire for the size, material, and spacing of wires multiplied by 2 times the length of circuit one way in feet. Care should be exercised in taking values of reactance from the tables to observe in what units the reactance is given. Some values are given in $\Omega/1000$ ft, and others in $\Omega/\text{mi}$.

4. Determine the reactance volts drop. This drop will be equal to the reactance of the total length of wire as has been determined in step 3 and multiplied by the current.

5. Find what percentage the resistance volts drop is of the voltage delivered at the end of the line.

6. Find what percentage the reactance volts drop is of the voltage delivered at the end of the line.

7. Determine the power factor of the circuit (see Sec. 64 of this division and Table 56 of Div. 11).

8. On the Mershon diagram (Fig. 3-30) start at the point where the vertical line corresponding to the power factor of the circuit intersects the smallest circle.

9. From this point lay off horizontally to the right the percentage of resistance volts drop. From the point thus obtained lay off vertically upward the percentage of reactance volts drop.

10. The circle upon which the last point obtained falls will give the voltage drop in percentage of the voltage at the end of the line.

**75. Calculations of voltage drop by Mershon diagram for any circuit when effect of inductance cannot be neglected.**    First determine, as outlined in Sec. 74, the voltage drop for a two-wire single-phase circuit of the same size of wire carrying the same current as the actual circuit. Then proceed as follows for the particular system involved.

Three-wire single-phase:

$$\text{Voltage drop between outside wires} = \text{drop as read from diagram} \qquad (21)$$

Voltage drop between outside wire and neutral (lamps)

$$= \frac{\text{drop as read from diagram}}{2} \qquad (22)$$

Three-wire tapped from three-phase four-wire:

Voltage drop between outside wire and neutral (lamps)

$$= 0.75(\text{drop as read from diagram}) \qquad (23)$$

Four-wire two-phase systems: Each phase is considered as a separate two-wire single-phase system.

Three-wire two-phase systems: Determine two drops from the diagram, one for a circuit with the size of wire and current of outside wires of a two-phase system, and the other for a circuit with size of wire and current of common wire.

Then

$$\text{Voltage drop} = \frac{\begin{array}{c}\text{drop for circuit with}\\ \text{size of wire and current}\\ \text{of outside wires}\end{array}}{2} + .08 \times \frac{\begin{array}{c}\text{drop for circuit with}\\ \text{size of wire and current}\\ \text{of common wire}\end{array}}{2} \qquad (24)$$

Five-wire two-phase systems:

$$\text{Voltage drop for lamps} = \frac{\text{drop from diagram}}{2} \qquad (25)$$

$$\text{Voltage drop for motors} = \text{drop as read from diagram} \qquad (26)$$

Three-wire three-phase systems:

$$\text{Voltage drop} = 0.866 \times \text{drop as read from diagram} \qquad (27)$$

Four-wire three-phase systems:

$$\text{Voltage drop for lamps} = \frac{\text{drop from diagram}}{2} \qquad (28)$$

$$\text{Voltage drop for motor} = 0.866 \times \text{drop as read from diagram} \qquad (29)$$

**example**   Determine the voltage drop for the circuit of the example in Sec. 74 for the different systems.

**solution**   By referring to the problem in Sec. 74, the voltage drop of the circuit for a two-wire single-phase system is 8.25 V.

For a three-wire single-phase system:

$$\text{Voltage drop for motors} = \text{drop as read from diagram}$$
$$= 8.25 \text{ V}$$

$$\text{Voltage drop for lamps} = \frac{\text{drop as read from diagram}}{2} = \frac{8.25}{2} = 4.13$$

For a four-wire two-phase system:

$$\text{Voltage drop} = \text{drop as read from diagram} = 8.25 \text{ V}$$

For a three-wire two-phase system:

Use must be made of the Mershon diagram to determine another voltage drop. The drop obtained in Sec. 74 is for a two-wire single-phase circuit carrying the same current as the outside wires of the three-wire two-phase circuit and with the same size of wire as the outside wires. Another voltage drop for a two-wire single-phase circuit carrying the same current as the common wire of the three-wire two-phase circuit and with the size of wire of the common must be determined.

$$\text{Current in common wire} = 1.41 \times 200 = 282 \text{ A}$$

Assume that a 400,000-cmil wire is used for common wire (as large a size would not be required by carrying capacity).

If the Mershon diagram is applied, a two-wire single-phase circuit of 85 percent pf carrying 282 A over 400,000-cmil wires spaced 6 in apart will give a 7.3 percent volts drop, or $\dfrac{7.3 \times 110}{100} = 8.03$ V.

$$\text{Voltage drop for three-wire two-phase circuit} = \frac{8.25}{2} + \left( 0.8 \times \frac{8.03}{2} \right)$$

$$= 4.13 + 3.21$$
$$= 7.34$$

For a five-wire two-phase system:

$$\text{Voltage drop for lamps} = \frac{\text{drop from diagram}}{2} = \frac{8.25}{2} = 4.13$$

$$\text{Voltage drop for motors} = \text{drop as read from diagram}$$
$$= 8.25 \text{ V}$$

For a three-wire three-phase system:

$$\text{Voltage drop} = 0.866 \times \text{drop as read from diagram}$$
$$= 0.866 \times 8.25 = 7.15 \text{ V}$$

For a four-wire three-phase system:

$$\text{Voltage drop for lamps} = \frac{\text{drop from diagram}}{2} = \frac{8.25}{2}$$
$$= 4.13 \text{ V}$$

$$\text{Voltage drop for motors} = 0.866 \times \text{drop as read from diagram}$$
$$= 0.866 \times 8.25 = 7.15 \text{ V}$$

### 76. How to proceed in determining the proper size of wire for a circuit

1. Determine the ampere load on the circuit.

For branch lighting and appliance circuits refer to Secs. 36 to 38.

For two-wire dc or single-phase ac lighting and appliance mains or feeders refer to Secs. 40, 44, and 45.

For three-wire dc or single-phase ac lighting and appliance mains or feeders refer to Secs. 41, 44, and 46.

For four-wire two-phase lighting and appliance mains or feeders refer to Secs. 44 and 47.

For three-wire two-phase lighting and appliance mains or feeders refer to Secs. 42, 44, and 48.

For five-wire two-phase lighting and appliance mains or feeders refer to Secs. 44 and 49.

For four-wire three-phase lighting and appliance mains or feeders refer to Secs. 43, 44, and 50.

For circuits supplying electric ranges refer to Secs. 39 and 45.

For branch motor circuits refer to Sec. 53.

For motor feeders or mains refer to Sec. 54.

2. For branch lighting or appliance circuits refer to Div. 9 and determine if outlets and equipment supplied by circuits meet the National Electrical Code requirements.

3. For lighting feeders or mains determine the minimum allowable load (refer to Sec. 52).

4. Select from Secs. 17 to 23 of Div. 11 the size of wire that will carry safely the current with the type of insulation employed on the wires (see also Secs. 30 and 31).

For 15-A branch lighting circuits, although No. 14 wire will safely carry the current, it is best not to use smaller than No. 12 wire. If the length of the branch lighting circuit to the load center is over 75 ft, wire at least as large as No. 10 should be used.

A branch circuit supplying an electric range with a rating of 8¾ kW or more must not be smaller than No. 8 except for the neutral conductor, which must not be smaller than No. 10.

For lighting feeders or mains use the estimated future load current unless it is less than the minimum allowable load, in which case the minimum allowable load should be used.

For continuous-duty branch motor circuits use 125 percent of the full-load rated current of the motor.

For branch motor circuits supplying short-time–duty motors, use the proper percentage of the rated current as given in Table 12 of Div. 11.

For motor feeders or mains use the maximum-demand running current with allowance for growth.

For the determination of the required carrying capacity of the neutral wire refer as follows:

For three-wire dc or single-phase systems refer to Sec. 46.

For three-wire two-phase systems refer to Sec. 48.

For five-wire two-phase systems refer to Sec. 49.

For four-wire three-phase systems refer to Sec. 46.

The neutral conductor of a three-wire circuit supplying a household electric range, a wall-mounted oven, or a counter-mounted cooking unit with a maximum demand of 8¾ kW or more is allowed to have a carrying capacity of only 70 percent of the outside wires provided that the neutral is not smaller than No. 10.

5. Determine whether the size of wire required according to carrying capacity is sufficient for mechanical strength (refer to Sec. 32).

Be sure that the size of wire required according to carrying capacity is not too large for satisfactory installation (refer to Sec. 33).

6. Find the distance to the load center of the circuit (refer to Sec. 57).

7. Decide what voltage drop is allowable in the circuit (refer to Sec. 35, and Table 52 of Div. 11).

8. Determine what the voltage drop in the circuit will be in using the size of wire required according to carrying capacity.

If the circuit is an ac one, before calculating the voltage drop, determine from Sec. 62 and Sec. 54 of Div. 11 whether or not the effects of inductance and skin effect can be neglected.

For branch lighting circuits use the future-load current unless the maximum allowable load for the type of branch circuit as given in Div. 9 is less than this value. In that case, use the maximum allowable load.

For lighting feeders or mains use the future-load current unless it is smaller than the minimum allowable load, in which case the minimum allowable load should be used.

For branch motor circuits use the full-load current of the motor.

For motor feeders or mains use the maximum-demand running current with proper allowance for growth.

For the length of the circuit use the distance from the source of supply to the load center, measured along the circuit.

For methods of calculating voltage drop refer to the following sections:

| | |
|---|---|
| Direct-current two-wire | Sec. 58 or 59 |
| Direct-current three-wire | Sec. 60 |
| Alternating-current when inductance can be neglected: | |
|     Single-phase two-wire | Sec. 65 |
|     Single-phase three-wire | Sec. 66 |
|     Two-phase four-wire | Sec. 67 |
|     Two-phase three-wire | Sec. 68 |
|     Two-phase five-wire | Sec. 69 |
|     Three-phase three-wire | Sec. 70 |
|     Three-phase four-wire | Sec. 71 |
| Alternating-current when the effect of inductance cannot be neglected | Sec. 73 |
| | or |
| | Secs. 74 and 75 |

9. If the voltage drop determined in step 8 is greater than the allowable amount, take the next larger conductor and determine the voltage drop it will give. Proceed in this way until the conductors are large enough to keep the voltage drop within the required amount. For ac circuits, this would sometimes require too great an increase in the size of the conductors. In such cases it is best to replan the system so as to divide the load among a greater number of circuits. For systems employing a common or neutral wire, when the size of wire must be increased to keep the voltage drop within the allowable amount, it is best to increase the size of the neutral in the same proportion as the outside wires are increased.

**example**  Determine the proper size of wire to use for a two-wire single-phase 15-A 60-Hz 115-V branch lighting circuit supplying six 150-W lamps. The length of the circuit to the load center is 50 ft. The wires are to be installed in conduit and have rubber insulation.

**solution**

$$\text{Actual connected load} = \frac{6 \times 150}{115} = 7.8 \text{ A}$$

$$\text{Future load (Sec. 37)} = 0.80 \times 15 = 12 \text{ A}$$

The future load of 12 A will be used in selecting the size of wire according to both carrying capacity and voltage drop.

From item 4 of this section at least a No. 14 wire is required by the **Code**, but it is best to use not smaller than No. 12 wire.

From Table 52 of Div. 11 a voltage drop of 2.3 V would be allowable.

Determine the voltage drop (Sec. 58 or 59) for a No. 12 wire carrying 12 A 50 ft.

$$\text{Voltage drop with No. 12 wire} = \frac{24 \times 12 \times 50}{6530} = 2.21 \text{ V}$$

Since this drop is less than the allowable amount, two No. 12 wires should be used.

**example**  A two-wire 15-A single-phase 60-Hz 115-V branch lighting circuit supplies seven 150-W lamps. The length of the circuit to the load center is 100 ft. The wires are to be installed in conduit and to be rubber-insulated. Determine the proper size of wire to use.

**solution**

$$\text{Actual connected load} = \frac{7 \times 150}{115} = 9.13 \text{ A}$$

$$\text{Future load} = 0.80 \times 15 = 12 \text{ A}$$

The future load of 12 A will be used in selecting the size of wire according to both carrying capacity and voltage drop.

From item 4 of this section at least a No. 14 wire is required, but it will be best to use not smaller than No. 10 wire when the circuit is over 100 ft long.

From Table 52 of Div. 11 a voltage drop of 2.3 V would be allowable.

From Table 54 of Div. 11 the effect of inductance can be neglected for a lighting circuit using No. 10 wire on a 60-Hz system.

$$\text{Voltage drop with No. 10 wire} = \frac{24 \times 12 \times 100}{10,380} = 2.77 \text{ V}$$

Since this voltage drop is greater than the allowable amount, take the next larger size of wire, No. 8, and check for voltage drop.

$$\text{Voltage drop with No. 8 wire} = \frac{24 \times 12 \times 100}{16,510} = 1.74 \text{ V}$$

Since this voltage drop is less than the allowable amount, two No. 8 wires should be used.

   **example**   Determine the proper size of wire for the three-wire 115/230-V dc main of the first example under Sec. 46. The length of the circuit is 125 ft. The wires are to be thermoplastic-covered, Type TW, and installed in conduit. The installation has only a main and branch circuits, no feeder.
   **solution**   From Sec. 46,

Present load on more heavily loaded outside wire = 145 A
Future load on more heavily loaded outside wire = 242 A
From Sec. 52 and Table 5 of Div. 11 minimum allowable load = 5.0 × 10,000 × 1.0
= 50,000 W

$$\text{Minimum allowable current in outside conductor} = \frac{50,000}{2 \times 115} = 217.4 \text{ A}$$

Since the future load is greater than the minimum allowable load, the future load on the more heavily loaded side of 242 A will be used in determining the size of wire according to both carrying capacity and voltage drop.
   From Table 18 of Div. 11 a 350,000-cmil wire is required for the outside conductors to carry 242 A.
   Since the future load on the neutral is 229.4 A, then from Table 18 of Div. 11, a 300,000-cmil wire is required to carry 229.4 A.
   From Table 52 of Div. 11 a voltage drop of 2.3 V is allowable on the lamps.

$$\text{Voltage drop across lamps (Sec. 60)} = \frac{1}{2} \times \frac{24 \times 242 \times 125}{350,000} = 1.04 \text{ V}$$

Since this voltage drop is less than the allowable amount, the circuit should have outside wires of 350,000 cmil and a neutral of 300,000 cmil.

   **example**   Determine the proper size of wire to use for a branch motor circuit supplying a 10-hp 440-V three-phase squirrel-cage induction motor on continuous duty. The motor is of the normal-starting-current, normal-torque type started with reduced voltage. It is a 60-Hz system. The length of the circuit is 50 ft. The wires are to be thermoplastic-covered, Type TW, installed in conduit.
   **solution**   From Table 10 of Div. 11 the full-load current = 14 A.

$$125 \text{ percent of the full-load current} = 14 \times 1.25 = 17.5 \text{ A}$$

From Table 18 of Div. 11 a No. 12 wire is required to carry 17.5 A.
From Table 52 of Div. 11, 4.4 V drop is allowable.
From Table 54 of Div. 11 the effect of inductance can be neglected for a 60-Hz motor circuit using No. 12 wire.
The voltage drop on a two-wire dc circuit carrying 14 A (full-load current) over a No. 12 wire for 50 ft = $\dfrac{24 \times 14 \times 50}{6530}$ = 2.58 V.

$$\text{Voltage drop of actual circuit} = 0.866 \times 2.58 = 2.24 \text{ V}$$

Since this drop is less than the allowable amount, three No. 12 wires should be used.

   **example**   Determine the proper size of wire to use for a branch motor circuit supplying a series-wound dc motor on a 30-min rating varying duty. The motor is a 15-hp 230-V machine. The length of the circuit is 100 ft. The wires are Type RH and installed in conduit.
   **solution**   From Table 7 of Div. 11 the full-load current is 55 A.
   Since this is a short-time–duty service, by referring to Table 12 of Div. 11, use 150 percent of the full-load current in determining size of wire according to carrying capacity.
   From Table 18 of Div. 11 a No. 4 wire is required to carry (1.50 × 55) = 83 A when Type RH wire is used.

From Sec. 52 of Div. 11, 4.4 V drop is allowable.
In determining voltage drop, use full-load current.

$$\text{Voltage drop} = \frac{24 \times 55 \times 100}{41,740} = 3.16 \text{ V}$$

Since this drop is less than the allowable amount, two No. 4 wires would be used.

**example**   Determine the proper size of wire to use for a three-wire, three-phase 440-V, 60-Hz feeder supplying five 10-hp motors, six 15-hp motors, and two 50-hp motors. All the motors are of the normal-torque, normal-starting-current, squirrel-cage type, started with reduced voltage. It is a 60-Hz circuit, 200 ft long. The wires are installed open with a regular flat spacing of 6 in. Type RHH insulation is employed on the wires. Permission to use a demand factor of 0.6 has been obtained from the officials enforcing the Code.

**solution**   From Table 10 of Div. 11 the full-load current of a 10-hp motor of this type is 14 A; of a 15-hp motor, 21 A; and of a 50-hp motor, 65 A.

By using the formula of Sec. 54,

Starting current of 50-hp motor (Sec. 43 of Div. 11) = 2.0 × 65 = 130 A

Maximum-demand starting current of feeder

= [(5 × 14) + (6 × 21) + (1 × 65)] × 0.6 + 130 = (261 × 0.6) + 130 = 286 A

Maximum-demand running current of feeder

= [(5 × 14) + (6 × 21) + (1 × 65)]0.6 + (65 × 1.25) = (261 × 0.6) + 81 = 238 A

From Table 18 of Div. 11 a No. 1/0 wire is required to carry 238 A (maximum-demand running current) if Type RHH (90°C) insulation is employed.
From Sec. 52 of Div. 11, 17.6 V drop is allowable.

Effective spacing of wires (Sec. 55, Div. 11) = 1.26 × 6 = 7.56

From Sec. 54 of Div. 11 the effect of inductance cannot be neglected.
Determine the voltage drop according to the method given in Sec. 73, using the maximum-demand running current.
Voltage drop of a two-wire dc circuit carrying 238 A 200 ft over No. 1/0 wire = (24 × 238 × 200)/105,560 = 10.8 V.
Ratio of reactance to resistance for No. 1/0 wires spaced 8 in apart on a 60-Hz circuit (Table 57 of Div. 11) = 0.9.
Power factor of circuit (Sec. 64 and Table 56, Div. 11):

$$
\begin{array}{rl}
5 \times 10 \times 0.86 = & 43.0 \\
6 \times 15 \times 0.86 = & 77.4 \\
2 \times 50 \times 0.89 = & \underline{89.0} \\
& 209.4
\end{array}
$$

$$\text{pf} = \frac{209.4}{240} = 87.5$$

Drop factor of circuit from Table 59 of Div. 11 for a ratio of reactance to resistance of 0.9 and a power factor of 0.875 = 1.34.
Voltage drop of circuit = 0.866 × 1.34 × 10.8 = 12.3 V.
Since this drop is less than the allowable amount, use three No. 1/0 wires.

**77. In determining circuit lengths from drawings or blueprints** a long piece of tough paper divided into the same measure as the drawing (see Fig. 3-31) can be effectively used in scaling distances. Always allow for rises or drops for wall outlets. The rotometer (Fig. 3-

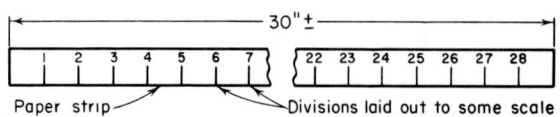

**Fig. 3-31**   Paper scale for measuring circuit lengths.

32) is a convenient tool for scaling distances. The little wheel is run over the course of the circuit. The pointer indicates feet directly for drawings of certain scales. For other scales the dial reading must be multiplied by a constant to obtain actual lengths.

**78. Indicating loads on diagrams.**   In computing the ampere loads on circuits, if drawings are available, it is good practice to note the amperes of the loads on the drawings alongside the circuits. It is good practice to indicate both the present and the estimated future loads, one colored pencil being used for the present load and another colored pencil for the future loads.

**79. Use of tables in circuit calculations.**   If tables that will give the values of loads, voltage drops, etc. are laid out when circuit calculations are made, much time can be saved in checking the size of circuits or in making changes in the system. A typical table is given in Sec. 80.

**80.  Branch Lighting Circuits**

| Panel | Branch circuit | Present load, amp | Future load, amp | Size wire according to carrying capacity | Length, ft | Volts drop | Size wire to use | Remarks |
|-------|----------------|-------------------|------------------|------------------------------------------|------------|------------|------------------|---------|
| *A* | 1 | 10.0 | 12 | No. 14 | 50 | 3.50 | No. 10 | No. 10 used to keep drop to 1.39 |
| ..... | 2 | 7.8 | 12 | No. 14 | 25 | 1.75 | No. 14 | |
| ..... | 3 | 9.1 | 12 | No. 14 | 30 | 2.10 | No. 12 | To keep voltage drop down |
| ..... | 4 | 8.0 | 12 | No. 14 | 20 | 1.40 | No. 14 | |
| *B* | etc. | | 48 | | | | | |

**81. The question of energy loss in a circuit should not be slighted in circuit calculations.**   It is well known that in overcoming resistance electrical energy is wasted, and as it costs money to develop or buy electrical energy, it is evident that in any commercial system such waste must be kept to a minimum. This can be done by decreasing the resistance of the conductors or, what amounts to the same thing, by increasing the size of the conductors. Inasmuch as this is also an expensive matter, care must be exercised that the additional sum added to the expenditure in copper (and conduit if used) is not so excessive as to more than counterbalance the cost of the energy continually saved.

**Fig. 3-32**   A rotometer.

**82. Conductor economy** in interior-wiring installations should always be considered a matter subordinate to the National Electrical Code and recommended voltage-drop requirements. Obviously, any conductor selected for a specific installation must fulfill the requirements of mechanical strength, ample carrying capacity, and recommended voltage drop. Frequently one of these three considerations will definitely determine the size of the conductor; however, a calculation may show that the resistance or $I^2R$ (power) loss is excessive. Then it may be desirable to use a larger size of conductor than would otherwise be necessary.

**83. Annual charges** may be considered, in connection with the economical selection of a conductor size, as being made up of two items, *resistance-loss charges* and *investment charges.* Resistance-loss charges depend upon the resistance, the current, and the unit cost of energy and can be decreased by an increase of conductor size. This, however, calls for a greater investment with corresponding larger investment charges. A conductor should, for maximum economy, be selected of such a size that the total annual charge will be a minimum. In Fig. 3-33 the effect of a variation of conductor size on resistance-loss charge, investment charge, and total annual charge is shown graphically for wires not installed in raceways. The interest charges on the conductor increase directly with its cross-sectional area (curve *A*, Fig. 3-33). If the wires were installed in conduit, the cost of the conduit and fittings should be added to the conductor cost. The resistance-loss charges decrease inversely as the cross-sectional area of the conductor (curve *B*, Fig. 3-33). The total annual charge (curve *C*, Fig. 3-33), the sum of curves *A* and *B*, is at its minimum value directly over the point where curves *A* and *B* intersect. That is, the conductor size which will have the least total annual cost is the one for which the annual interest charge equals the annual resistance-loss charge. This proposition was demonstrated by Lord Kelvin.

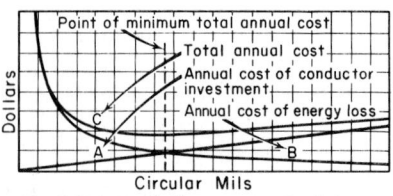

**Fig. 3-33** Graph illustrating Kelvin's law.

**84. Kelvin's law** was deduced in 1881 by Sir William Thomson (Lord Kelvin). The original law was modified into the following more exact form by Gisbert Kapp: "The most economical area of conductor is that for which the annual cost of energy wasted is equal to the interest on that portion of the capital outlay which can be considered proportional to the weight of copper used." On the basis of the above law it can be shown that

$$\text{cmil} = 55,867 \times I \times \sqrt{\frac{C_e}{C_c \times A}} \qquad (30)$$

where cmil = area, in circular mils, of the most economical conductor; $I$ = the mean annual current; $C_e$ = cost of energy per kilowatthour, in dollars; $C_c$ = cost of copper, in dollars, per pound installed; and $A$ = annual charge, in percentage, of the cost of the conductor. This can ordinarily be assumed to be about 10 percent. Note that the length of the circuit is not a factor in the equation. The above equation is best adapted to the solution of cases in which the conductors are not insulated. The method of the following paragraph should always be used to check it.

**85. Method of determining economical conductor size.** This method involves the preparation of a table, as shown below, showing energy cost and interest cost of an arbitrarily chosen length of conductor for a number of sizes of conductor. The cost per unit length of conductors (preferably 1000 ft, or 305 m) can be obtained from the manufacturers' wire lists. For conduit installations the corresponding cost of conduit and fittings should be added to the cost of conductors. The cost of energy lost can be computed from the equations of Sec. 86 or from Sec. 87. By use of a comparative table, as given below, costs are effectively shown and a decision can be speedily reached. This method can be used as a check upon the Kelvin's-law equation. In the table the least total annual cost is $15.69. Hence a 4/0 conductor is, for this case, the one of maximum economy.

| Size of rubber-covered conductor | No. 3/0 wire | No. 4/0 wire | 250,000– cir-mil wire | 300,000– cir-mil wire |
|---|---|---|---|---|
| Cost of 400 ft of conductor............................................. | $100.00 | $120.00 | $160.00 | $200.00 |
| Annual charges on above cost at 10 per cent..................... | $ 10.00 | $ 12.00 | $ 16.00 | $ 20.00 |
| Cost of energy lost in conductor at 2 cts per kwhr........... | $  5.94 | $  3.69 | $  2.64 | $  2.27 |
| Total annual cost of conductor....................................... | $ 15.94 | $ 15.69 | $ 18.64 | $ 22.27 |

[1]For derivation see *The Electrical Engineer's Pocketbook* (International Textbook Company). See also A. V. Abbot, *Transmission of Electrical Energy;* Louis Bell, *Electric Power Transmission;* Alfred E. Still, *Overhead Electric Power Transmission.*

**86. Cost of energy lost.**  If we take 12 Ω as the resistance of a circular mil-foot of commercial copper wire at 50°C, the power loss in any conductor can be found thus:

$$P = \frac{12 \times I^2 \times L}{\text{cmil}} \tag{31}$$

where $P$ = power lost in the conductor in watts, $I$ = the mean annual current in amperes in the conductor, $L$ = length of the conductor in feet, and cmil = area of the conductor in circular mils.

The cost of energy lost can be computed from the following formula:

$$\text{Cost of energy per year in dollars} = \frac{N \times P \times 8760 \times R}{1000 \times 100} \tag{32}$$

where $N$ = number of conductors; $P$ = power loss in one conductor; and $R$ = cost of electric energy in cents per kilowatthour. In the formula, the 8760 is the number of hours in a year, the 1000 is the factor to change watts to kilowatts, and the 100 changes cents to dollars.

**87.   Cost of Energy Lost for 500 H of Operation with Cost of Electrical Energy at 2 Cents per Kilowatthour**

| Wire size, AWG | Length of run, ft | Load, amp | | | | | | | | | | | | | |
|---|---|---|---|---|---|---|---|---|---|---|---|---|---|---|---|
| | | 6 | 10 | 15 | 20 | 25 | 30 | 35 | 40 | 50 | 60 | 70 | 80 | 90 | 100 |
| 14 | 50 | $0.10 | $0.26 | $0.60 | | | | | | | | | | | |
| | 100 | 0.19 | 0.52 | 1.20 | | | | | | | | | | | |
| | 200 | 0.38 | 1.04 | 2.40 | | | | | | | | | | | |
| | 300 | 0.57 | 1.06 | 3.60 | | | | | | | | | | | |
| 12 | 50 | 0.06 | 0.16 | 0.36 | $0.64 | | | | | | | | | | |
| | 100 | 0.12 | 0.32 | 0.72 | 1.28 | | | | | | | | | | |
| | 200 | 0.24 | 0.64 | 1.44 | 2.56 | | | | | | | | | | |
| | 300 | 0.36 | 0.96 | 2.16 | 3.84 | | | | | | | | | | |
| 10 | 50 | 0.04 | 0.10 | 0.22 | 0.40 | $0.62 | $0.90 | | | | | | | | |
| | 100 | 0.08 | 0.20 | 0.45 | 0.80 | 1.24 | 1.80 | | | | | | | | |
| | 200 | 0.16 | 0.40 | 0.90 | 1.60 | 2.48 | 3.60 | | | | | | | | |
| | 300 | 0.24 | 0.60 | 1.35 | 2.40 | 3.72 | 5.40 | | | | | | | | |
| 8 | 50 | 0.03 | 0.06 | 0.15 | 0.25 | 0.40 | 0.60 | $0.77 | $1.06 | | | | | | |
| | 100 | 0.05 | 0.12 | 0.30 | 0.50 | 0.80 | 1.20 | 1.54 | 2.12 | | | | | | |
| | 200 | 0.09 | 0.24 | 0.60 | 1.00 | 1.60 | 2.40 | 3.08 | 4.24 | | | | | | |
| | 300 | 0.14 | 0.36 | 0.90 | 1.50 | 2.40 | 3.60 | 4.62 | 6.36 | | | | | | |
| 6 | 50 | 0.02 | 0.04 | 0.09 | 0.16 | 0.25 | 0.36 | 0.49 | 0.64 | $1.00 | $1.44 | | | | |
| | 100 | 0.03 | 0.08 | 0.18 | 0.32 | 0.50 | 0.72 | 0.98 | 1.28 | 2.00 | 2.88 | | | | |
| | 200 | 0.06 | 0.16 | 0.36 | 0.64 | 1.00 | 1.44 | 1.76 | 2.56 | 4.00 | 5.76 | | | | |
| | 300 | 0.09 | 0.24 | 0.54 | 0.96 | 1.50 | 2.16 | 2.64 | 3.84 | 6.00 | 8.64 | | | | |
| 4 | 50 | 0.01 | 0.03 | 0.06 | 0.08 | 0.15 | 0.22 | 0.32 | 0.40 | 0.62 | 0.90 | $1.22 | $1.60 | | |
| | 100 | 0.02 | 0.05 | 0.12 | 0.16 | 0.30 | 0.44 | 0.64 | 0.80 | 1.24 | 1.80 | 2.44 | 3.20 | | |
| | 200 | 0.04 | 0.10 | 0.24 | 0.32 | 0.60 | 0.88 | 1.28 | 1.60 | 2.48 | 3.60 | 4.88 | 6.40 | | |
| | 300 | 0.05 | 0.15 | 0.36 | 0.48 | 0.90 | 1.32 | 1.92 | 2.40 | 3.72 | 5.40 | 7.32 | 9.60 | | |
| 2 | 50 | ..... | 0.02 | 0.04 | 0.06 | 0.10 | 0.13 | 0.18 | 0.24 | 0.40 | 0.60 | 0.77 | 1.00 | $1.26 | $1.60 |
| | 100 | 0.01 | 0.03 | 0.07 | 0.12 | 0.20 | 0.25 | 0.36 | 0.48 | 0.80 | 1.20 | 1.54 | 2.00 | 2.52 | 3.20 |
| | 200 | 0.02 | 0.06 | 0.14 | 0.24 | 0.40 | 0.50 | 0.72 | 0.56 | 1.60 | 2.40 | 3.08 | 4.00 | 5.04 | 6.40 |
| | 300 | 0.04 | 0.10 | 0.21 | 0.36 | 0.60 | 0.75 | 1.08 | 1.44 | 2.40 | 3.60 | 4.62 | 6.00 | 7.56 | 9.60 |

$$\text{Cost of energy lost per year in dollars} = \frac{H \times R \times K}{500 \times 2} \tag{33}$$

where $H$ = hours of operation of circuit per year; $R$ = cost of electrical energy in cents per kilowatthour; $K$ = dollars as obtained from table for size of wire, current, and length of run. The table and formula can be used for dc or single-phase ac circuits. If the load on a three-wire dc or single-phase ac circuit is nearly balanced, the formula will give satisfactory results if the current is taken as the current in the more heavily loaded wire.

**88. Factors for determining the mean annual current.** To ascertain the mean annual current for substitution in the Kelvin's-law equation (30) or in Eq. (31) of Sec. 86 multiply the maximum current by the ratio applying to the conditions under consideration, which is given in the column headed "Factor" in the following table. The table is calculated on the basis 24 h × 365 days = 8670 h per year.

**example** If a maximum current of 1000 A ($I$) flows three-fourths of the time, or 6570 h per year, and a current of 750 A ($\frac{3}{4}I$) flows one-fourth of the time, or 2190 h per year, the factor 0.944 would be used. That is, $0.944 \times 1000$ A $= 944$ A $=$ mean annual current for substitution in Kelvin's-law equation.

| Proportion of maximum current $I$ carried | | | | Factor |
|---|---|---|---|---|
| $\frac{1}{4}I$ | $\frac{1}{2}I$ | $\frac{3}{4}I$ | $I$ | |
| 0 | 0 | 0 | 1 | 1.000 |
| 0 | 0 | $\frac{1}{4}$ | $\frac{3}{4}$ | 0.944 |
| 0 | $\frac{1}{4}$ | 0 | $\frac{3}{4}$ | 0.901 |
| 0 | 0 | $\frac{1}{2}$ | $\frac{1}{2}$ | 0.844 |
| 0 | 0 | 0 | $\frac{3}{4}$ | 0.866 |
| $\frac{1}{4}$ | 0 | 0 | $\frac{3}{4}$ | 0.875 |
| 0 | $\frac{1}{4}$ | $\frac{1}{4}$ | $\frac{1}{2}$ | 0.838 |
| 0 | 0 | $\frac{3}{4}$ | $\frac{1}{4}$ | 0.820 |
| $\frac{1}{4}$ | 0 | $\frac{1}{4}$ | $\frac{1}{2}$ | 0.810 |
| 0 | $\frac{1}{2}$ | 0 | $\frac{1}{2}$ | 0.790 |
| 0 | $\frac{1}{4}$ | $\frac{1}{2}$ | $\frac{1}{4}$ | 0.771 |
| $\frac{1}{4}$ | $\frac{1}{4}$ | 0 | $\frac{1}{2}$ | 0.760 |
| $\frac{1}{4}$ | 0 | $\frac{1}{2}$ | $\frac{1}{4}$ | 0.744 |
| $\frac{1}{2}$ | 0 | 0 | $\frac{1}{2}$ | 0.729 |
| 0 | $\frac{1}{2}$ | $\frac{1}{4}$ | $\frac{1}{4}$ | 0.718 |
| 0 | 0 | 0 | $\frac{1}{2}$ | 0.707 |
| $\frac{1}{4}$ | $\frac{1}{4}$ | $\frac{1}{4}$ | $\frac{1}{4}$ | 0.685 |
| 0 | $\frac{3}{4}$ | 0 | $\frac{1}{4}$ | 0.661 |
| $\frac{1}{2}$ | 0 | $\frac{1}{4}$ | $\frac{1}{4}$ | 0.650 |
| $\frac{1}{4}$ | $\frac{1}{2}$ | 0 | $\frac{1}{4}$ | 0.611 |
| $\frac{1}{2}$ | $\frac{1}{4}$ | 0 | $\frac{1}{4}$ | 0.586 |
| $\frac{3}{4}$ | 0 | 0 | $\frac{1}{4}$ | 0.545 |
| 0 | 0 | 0 | $\frac{1}{4}$ | 0.500 |

Proportion of time current is carried

Division **4**

# General Electrical Equipment and Batteries

## INTRODUCTION

**1. Introduction.** In this division the miscellaneous equipment, materials, and devices which are employed for electrical installations are discussed. Certain items which are used for specific types of installations have not been included in this division; an explanation of these devices or materials will be found in Div. 8, "Outside Distribution," or in Div. 9, "Interior Wiring." Since there is so much material to cover for generators, motors, and transformers, a separate division (Div. 7) has been devoted to generators and motors and another (Div. 5) to transformers.

The aim in the preparation of Div. 4 has been to present the different types of equipment, materials, and devices which are available, with sufficient explanation of their characteristics, installation, and maintenance so that proper equipment can be intelligently selected, installed, and maintained in good condition for satisfactory service. Much useful information that is not ordinarily readily available at short notice has been included.

## SWITCHES

**2. A switch** is a device for making, breaking, or changing connections in an electric circuit under the conditions of load for which it is rated. It is not designed for interruption of a circuit under short-circuit conditions. Refer to Sec. 61 for a discussion of the difference between circuit breakers and switches.

**3. Switches** may be classified in several different ways:

I. According to number of poles
  A. Single-pole
  B. Two- or double-pole
  C. Three- or triple-pole
    1. Standard
    2. Solid-neutral
  D. Four-pole: standard
  E. Four-pole: solid-neutral
  F. Five-pole: standard
  G. Five-pole: solid-neutral
II. According to number of closed positions
  A. Single-throw
  B. Double-throw
III. According to type of contact
  A. Knife-blade
  B. Butt-contact
    1. Single-line
    2. Multiple-line
    3. Surface
  C. Mercury
IV. According to number of breaks
  A. Single-break
  B. Double-break
V. According to method of insulation
  A. Air-break
  B. Oil-immersed
VI. According to method of operation
  A. Operating force
    1. Manual
    2. Magnetic
    3. Motor or solenoid
  B. Mechanism
    1. Lever
    2. Dial
    3. Drum
    4. Snap
      a. Tumbler
      b. Rotary
      c. Pushbutton
VII. According to speed of operation
  A. Quick-break
  B. Quick-make
  C. Slow-break
VIII. According to enclosure
  A. Open
  B. Enclosed
    1. General-purpose
    2. Driptight

  3. Weather-resisting
  4. Watertight
  5. Dusttight
  6. Submersible
  7. Hazardous locations
  8. Bureau of Mines
IX. According to protection provided to circuits or apparatus
X. According to type of service
  A. Power switches
    1. General-purpose
      a. Open
      b. Safety
    2. Disconnecting
    3. Motor-circuit
    4. Motor-starting
    5. Field switches
    6. Service-entrance
    7. Bolted-pressure-contact
  B. Wiring switches
    1. According to method of installation
      a. Flush-mounting
        (1) Standard
        (2) Interchangeable
        (3) Combination
      b. Surface
      c. Pendant
      d. Through-cord
      e. Door
      f. Canopy
      g. Appliance
      h. Special exposures
    2. According to function
      a. General-purpose
        (1) Single-pole
        (2) Double-pole
      b. Three-way
      c. Four-way
      d. Three- or four-pole
      e. Multiple-circuit or electrolier
      f. Momentary-contact
      g. Heater
      h. Dimmer
  C. Control switches
  D. Instrument switches
  E. Miscellaneous types

**4. A pole of a switch** is that part of a switch which is used to make or break a connection and which is electrically insulated from other contact-making or -breaking parts. A single-pole switch will make or break connections in only one conductor or leg of a circuit; a two-pole switch, in two legs, etc. Schematic diagrams illustrating the meaning of the number of poles are shown in Fig. 4-1.

For some installations in systems having a grounded neutral wire, it is always desirable to break the neutral connection as the circuit passes through a switch. In these cases, a regular switch with a number of poles one less than the number of conductors of the circuit can be used. The neutral conductor is then carried directly to a neutral bar in the switch. Another installation can be made by the use of a *solid-neutral switch* (Fig. 4-2).

The number of switch blades or contacts for a solid-neutral switch is one less than the number of conductors of the circuit, but a neutral strap is located between two of the switch blades or contacts to carry the neutral leg solidly through the switch without breaking the circuit of the neutral conductor. A switch designated as a three-pole solid-neutral switch will accommodate a three-conductor circuit and provides two switched poles and one solid-neutral strap. A solid-neutral switch is shown in Fig. 4-2, II. Figure 4-15 shows

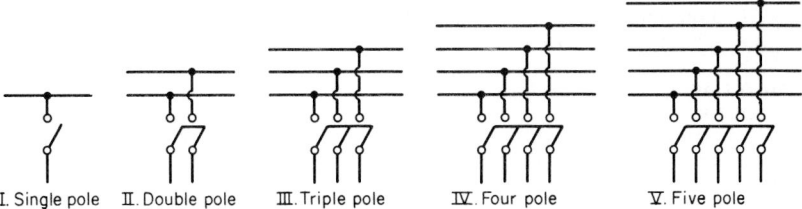

I. Single pole    II. Double pole    III. Triple pole    IV. Four pole    V. Five pole

**Fig. 4-1**    Meaning of poles.

a three-pole switching-neutral fused switch which satisfies National Electrical Code requirements.

**5. A single-throw switch** (Fig. 4-3, I) will make a closed circuit only when the switch is thrown in one position. A *double-throw switch* (Fig. 4-3, II) will make a closed circuit when thrown in either of two positions. Special *multiple-throw switches* can be made to meet the requirements of conditions which require that a switch make a closed circuit when thrown in more than two positions. Examples of multiple-throw switches are the dial and drum switches employed for control and instrument work.

**6. Switch contacts.** In the construction of switches, different methods are employed for making contact between the movable and stationary parts of the poles. *Knife-blade switches* consist of some form of movable copper blade which makes contact by being forced between forked contact jaws, as illustrated in Fig. 4-4.

The construction of the contact members of *butt-contact switches* varies widely with different manufacturers. In all cases, however, the contact is formed by pressing the movable member against the stationary contact. (The two contacts are butted together when closed.) The two members are so constructed that there will be a stiff spring

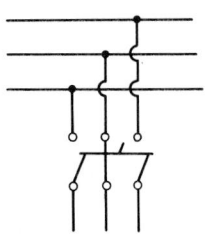

I Schematic diagram of three-
pole solid-neutral switch

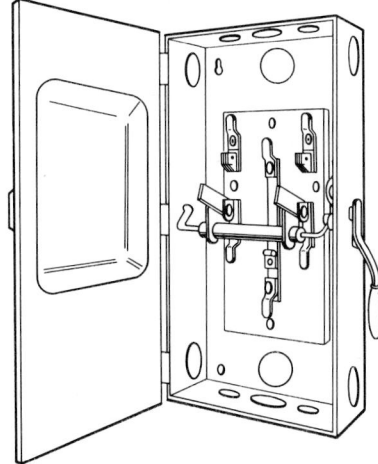

II Three pole, unfused switch

**Fig. 4-2**    Solid-neutral switch and connections.

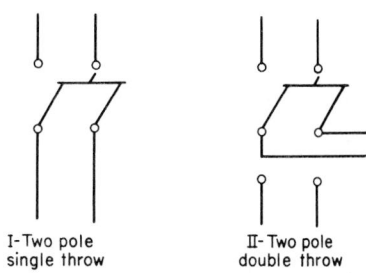

I-Two pole          II-Two pole
single throw        double throw

**Fig.    4-3**    Single-    and    double-throw switches.

action between them to ensure a good contact when in the closed position. The contact formed between the two members may be of the *single-line, multiple-line,* or *surface* type. Single-line contact is obtained by making the surface of the contacting members curved so that, when closed, the members are in contact only along a single line instead of over the entire surface. With multiple-line contacts the two members are not in contact over their entire surfaces, contact being made along several lines. The multiple lines of contact are obtained by making the movable member of laminated (brush) or cylindrical construction. In butt-contact switches of the surface type, the contact members, when in the closed position, are butted together so that they are in contact over as much of the

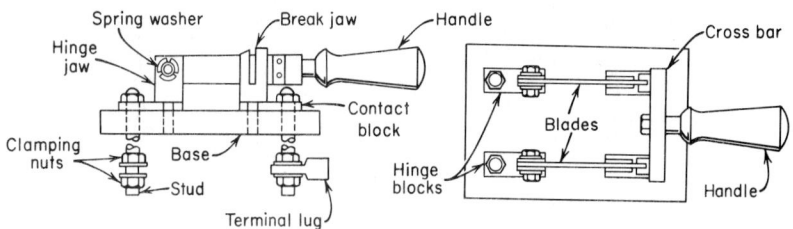

**Fig. 4-4**   Knife-blade switch parts.

surface as possible. The disadvantage of the surface type of contact is that it is very difficult to obtain a good contact over the entire surface. Typical butt-contact constructions are illustrated in Fig. 4-5.

One of the essential functions of any switch is to maintain a good low-resistance contact in the closed position. If the contact is poor, there will be considerable resistance at this point. The high contact resistance will result in overheating of the switch and possible opening of the circuit due to the functioning of circuit-overload protective devices. Switches are generally constructed so that the contacts will be closed under pressure with a sliding or rolling action to maintain clean, low-resistance contacts. The contact surfaces are frequently silver-plated to prevent oxidation and reduce contact resistance.

**7. Mercury contact switches** are sometimes used for low-capacity circuits in control and branch-circuit wiring. They consist of a glass tube containing mercury and two stationary contact members located in opposite ends of the tube. The tube is mounted in a supporting case so that the position of its axis can be changed by an operating handle. When the handle is in the off position, the axis of the tube is tilted so that the mercury is at one end of the tube and makes contact with only one of the stationary contacts. When the handle is thrown to the closed position, it rotates the tube, causing the mercury to move in the tube in a manner that closes the circuit from the stationary contact at one end of the tube through the mercury to the stationary contact at the other end of the tube. Refer to Secs. 30 and 31 for further information on wiring switches for interior-wiring work.

**8. Both knife-blade and butt-contact switches are made in single- and double-break types.** A single-break switch, in opening, breaks each pole of the circuit at only one point. A double-break switch, in opening, breaks each pole of the circuit at two points. Breaking the circuit in two places doubles the quenching effect of the switch in suppressing the arc. This reduction in the arcing at the contacts prolongs the life of the contacts by reducing the formation of destructive beads and pitted areas.

**9. Insulation of switches.** The great majority of switches are designed to interrupt the circuit in the surrounding air. The live parts must be spaced far enough apart so that the air space between them will have sufficient insulating ability for the voltage which exists between the adjacent parts. Oil switches (Fig. 4-6) have their contact parts immersed in oil so that the circuit is interrupted under oil.

**10. Switches can be operated** by hand (Fig. 4-7), by electromagnets (Fig. 4-8), by motors (Fig. 4-8), or by means of solenoids (Fig. 4-8). Most power switches are operated by means of simple lever action (Fig. 4-4) or by means of a lever attached to the switch through a toggle mechanism. Dial switches consist of a movable contact mounted on a rotatable arm and of several fixed contacts arranged in circular form. As the arm is rotated, the movable contact makes connection with the successive fixed contacts, one at a time. This type of switch is used for instrument, control, storage-battery end-cell, and transformer tap-changing switches. Drum switches consist of a set of contact segments mounted on a central movable drum and a set of stationary contact fingers (Fig. 4-9). Drum switches are employed for motor-starting, control, and instrument switches.

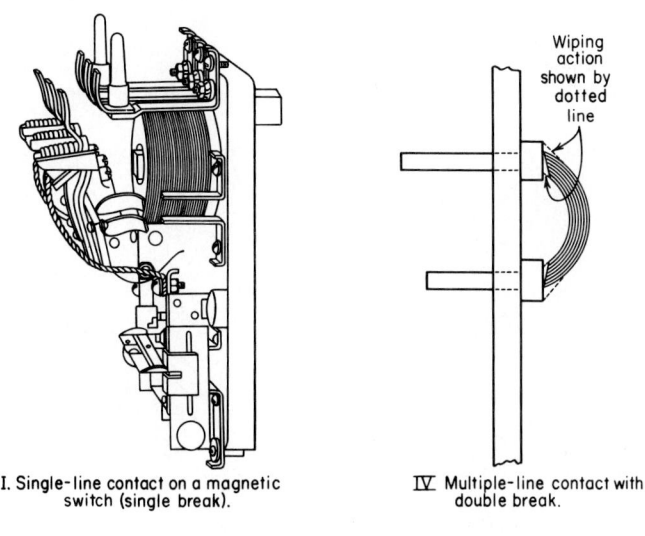

I. Single-line contact on a magnetic switch (single break).

Ⅳ. Multiple-line contact with double break.

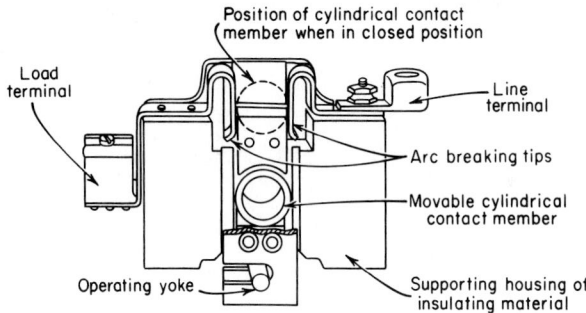

Ⅱ Single-line contact with double break (switch in open position).

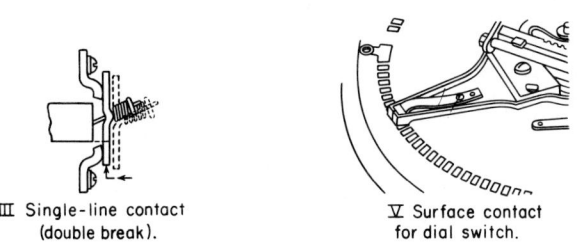

Ⅲ Single-line contact (double break).

Ⅴ Surface contact for dial switch.

**Fig. 4-5** Butt-contact constructions.

*Magnetic switches* consist of switch contacts operated by means of an electromagnet. They are used extensively as the switching element for motor controllers. They also find application in general wiring work when it is desired to control a circuit at some point remote from the switch. See Secs. 36 and 37.

*Snap switches* are small-capacity switches in which the circuit is made or broken with a quick motion independent of the speed of operation of the switch by the operator. In the rotary type of snap switch (Fig. 4-10), the switch blades are given a rotary motion by means of the handle or button through a spring-and-cam mechanism. The blades of a pushbutton switch (Fig. 4-11) are operated by a rocking action imparted to them by the push button through a spring-and-cam mechanism. The tumbler or toggle switch (Fig. 4-12) is operated in a manner similar to the pushbutton switch except that the blades are actuated by a lever instead of a button.

**11. Speed of operation of switches.**    With an ordinary manual switch the speed of closing or opening a switch depends upon the operator. If current conditions are such that an arc may be drawn, attachments which make it impossible to open the switch slowly and thus

I. Small-capacity three-pole double-throw switch. [General Electric Co.]

II. Medium-voltage industrial type. [Westinghouse Electric Corp.]

III.    High-voltage    industrial type. [General Electric Co.]

**Fig. 4-6**    Oil switches.

draw a dangerous arc should be provided. Many switches are provided with quick-make as well as quick-break features. A quick-break attachment for an open knife switch is shown in Fig. 4-13. When the switch is opened, the auxiliary blade is held in the switch jaws by friction until the main blade has been withdrawn. Withdrawal of the main blade

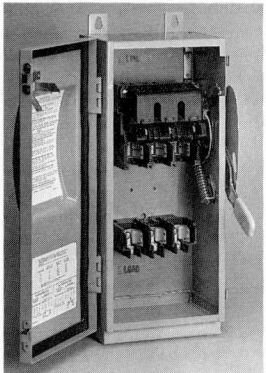

I. Heavy-duty lever-operated safety switch. [Crouse-Hinds Co.]

II. Lever and toggle mechanism. [Westinghouse Electric Corp.]

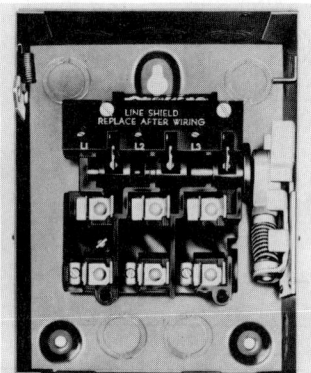

III. Rotary motion operated by lever outside box. [General Electric Co.]

**Fig. 4-7**  Operating mechanisms for manual switches.

increases the spring tension so that it suddenly jerks the auxiliary blade out of the jaws and quickly breaks the circuit. In switches of the enclosed safety type, the operating handle is frequently connected through linkages and a spring so that after the handle has been moved a certain distance, the spring quickly completes the closing or opening action, giving the switch quick-break or -make features.

**12. Enclosure of switches.**  Switches may be classified as open or enclosed, depending upon whether their current-carrying parts are exposed or enclosed in a protecting box or casing. All switches should be of the enclosed type (externally operated) unless they are mounted on switchboards or panelboards. Even in these cases it is generally better practice to use switchboards and panelboards of the dead-front–construction type. In no case should open switches be employed where they may be operated by unqualified persons.

Switches can be obtained with enclosures to meet the requirements of different classes of exposures. The standard (National Electrical Manufacturers Association) types of non-ventilated switch enclosures are as follows:

I. Solenoid-operating mechanism. [Westinghouse Electric Corp.]

II. Motor-operating mechanism. [Westinghouse Electric Corp.]

**Fig. 4-8**    Electrically operated switches.

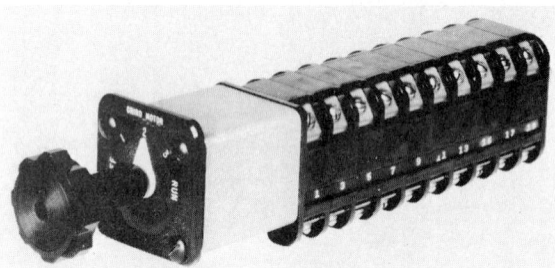

**Fig. 4-9**    Drum-type control switch. [General Electric Co.]

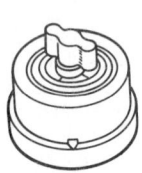

**Fig. 4-10** Rotary snap switch: surface-mounting type.

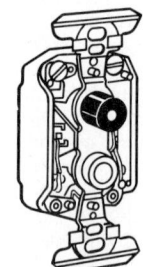

**Fig. 4-11** Pushbutton switch: flush-mounting type, old style.

TYPE 1    Type 1 enclosures are intended for indoor use primarily to provide a degree of protection against contact with the enclosed equipment, in locations where unusual service conditions do not exist. Type 1 enclosures may be of the nonventilated or the ventilated type.

TYPE 3R    Type 3R enclosures are intended for outdoor use primarily to provide a degree of protection against falling rain, but also to remain undamaged by the formation of ice on the enclosure. They are not intended to provide protection against conditions

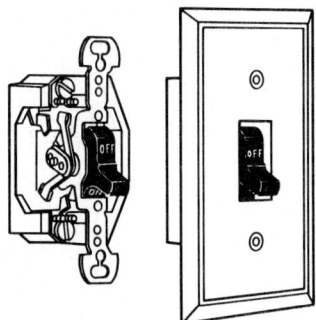

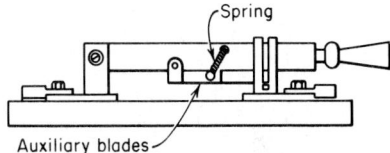

Spring

Auxiliary blades

**Fig. 4-12**  Toggle or tumbler switch: flush-mounting ac, dc type.

**Fig. 4-13**  A quick-break switch.

such as dust, internal condensation, or internal icing. Type 3R enclosures may be of the nonventilated or the ventilated type.

TYPE 4    Type 4 enclosures are intended for indoor or outdoor use primarily to provide a degree of protection against windblown dust and rain, splashing water, and hose directed water; and to be undamaged by the formation of ice on the enclosure. They are not intended to provide protection against conditions such as internal condensation or internal icing.

TYPE 7    Type 7 enclosures are intended for use indoors in locations classified as Class I, Group A, B, C, or D, as defined in the National Electrical Code. The letters A, B, C, and D indicate the gas or vapor atmospheres in the hazardous location. The enclosures are marked with the appropriate class and group(s) for which they have been qualified. These enclosures are capable of withstanding the pressures resulting from an internal explosion of the specified gases and containing such an explosion sufficiently that an explosive gas-air mixture existing in the atmosphere surrounding the enclosure will not be ignited. Enclosed heat-generating devices do not cause external surfaces to reach temperatures capable of igniting explosive gas-air mixtures in the surrounding atmosphere.

TYPE 9    Type 9 enclosures are intended for indoor use in locations classified as Class II, Group E or G, as defined in the National Electrical Code. The letters E and G indicate the dust atmospheres in the hazardous location. The enclosures are marked with the appropriate class and group(s) for which they have been qualified. These enclosures are capable of preventing the entrance of dust. Enclosed heat-generating devices do not cause external surfaces to reach temperatures capable of igniting or discoloring dusts on the enclosure or igniting dust-air mixtures in the surrounding atmosphere.

For complete listings of NEMA enclosures, see Secs. 7-253 and 7-253A.

**13. Overcurrent protection** for circuits or apparatus is frequently incorporated with a switch. For the protection of circuits the protective device incorporated with the switch consists of fuses. For the protection of equipment the protective device may be any one of the types listed in Sec. 46 (I, II, IV, and V). Standard fused switches are equipped with fuse clips for accommodating Class H cartridge fuses or, in some cases, Type S screw-base plug fuses if the switch rating is 30 A.

Fused switches used on systems with a grounded neutral may have one less pole than the number of conductors of the circuit, or a solid-neutral or switching-neutral switch may be employed. In the first case, the neutral leg is not carried through the switch, but the neutral conductor is simply carried through the switch box in the space around the switch. A solid-neutral switch (Fig. 4-14) has one less blade or contact than the number of conductors of the circuit, with a neutral strap located in the space between two of the switch blades or contacts to carry the neutral leg solidly through the switch without breaking the circuit of the neutral conductor. No fuse is located in the neutral leg. A switching-neutral switch (Fig. 4-15) has a blade or contact for each conductor of the cir-

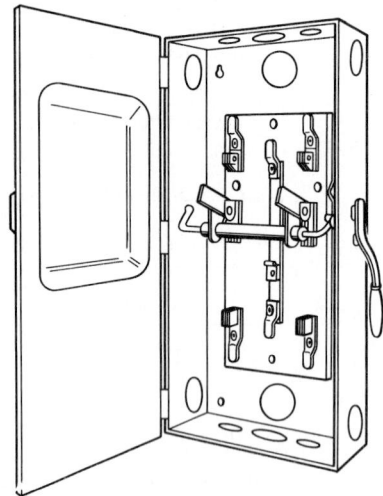

**Fig. 4-14**   Three-pole solid-neutral fused switch.

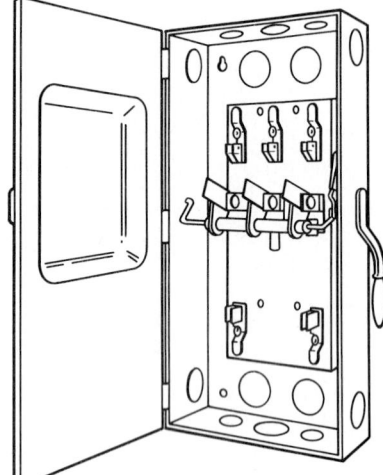

**Fig.    4-15**   Three-pole    switching-neutral fused switch.

cuit. The blade or contact for the neutral conductor is connected by a copper strap to its terminal without any fuse clips. Such a switch opens the neutral conductor leg and all other circuit conductors when the switch is opened and satisfies NEC rules.

**14. General-purpose or general-use power switches** are intended for use in general-distribution and branch circuits. They are rated in amperes and are capable of interrupting their rated current at their rated voltage. These switches may be of the open or enclosed externally operated type. The enclosed switches are generally referred to as safety switches.

General-purpose or general-use switches are NEMA-designated as Type HD, ND, or LD switches, depending upon the severity of service for which they are designed. *Type HD* (heavy-duty) switches, the most ruggedly constructed of the three types, are designed for the heaviest-duty service with respect to both frequency of operation and current capacity for handling heavy overloads. *Type ND* (normal-duty) switches are of lighter construction than Type HD and are designed for service conditions in which frequent operation is not required and overload conditions are not severe. *Type LD* (light-duty) switches are of light construction and are designed primarily for entrance service and general-purpose use when the load is light and operation very infrequent.

**15. Open switches** are constructed with their current-carrying parts exposed. Most open switches are of the plain lever knife-blade type. Knife-blade switches in very heavy current-carrying capacities are hard to operate owing to the large rubbing-contact surfaces required. For this reason open switches of very large capacity are usually of the brush-butt-contact type (Fig. 4-5, IV).

Most open lever-type knife-blade switches are of the single-break type; they are made in a variety of forms. They can be obtained in single-, double-, triple-, and four-pole types, for either front or rear connection (see Figs. 4-16 and 4-17). They may be of the fused or unfused type and either single- or double-throw.

Open switches are used principally for mounting on live-front switchboards.

**16. The names of knife-switch parts** are given in Fig. 4-4. The contact between the break jaws and the blade should be carefully inspected, as it is at this point that knife switches are most apt to give trouble by overheating. The contact between the hinge jaws and the blade seldom limits the capacity of a switch, because it is under pressure from the hinge bolt and the spring washers. The capacity of a switch is determined by its temperature rise.

About 1000 A/in² (645 mm²) of copper section and 50 to 75 A/in² of sliding-contact surface are usually allowed in designing switches.

A switch that will carry, possibly, 1000 A with a 20°C temperature rise will carry possibly 2000 A with about a 60°C rise. The radiation of heat from the switch increases more rapidly than does the rise in temperature, and as the heat generated varies as the square of the current, it is evident that the temperature rise will be somewhat less than proportional to the square of the current.

With a given current, a switch will break about double the voltage with alternating current that it will with direct current. This is due to the fact that an alternating current decreases to a zero value during each cycle. The NEC recognizes this and specifies that the spacings for 250 V dc are also approved for 500 V ac.

The voltage drop from contact block to hinge block of a good switch should not exceed about 12 mV with full-load current.

**17. Safety switches** consist of a switch mounted inside a sheet-metal or cast-iron box and

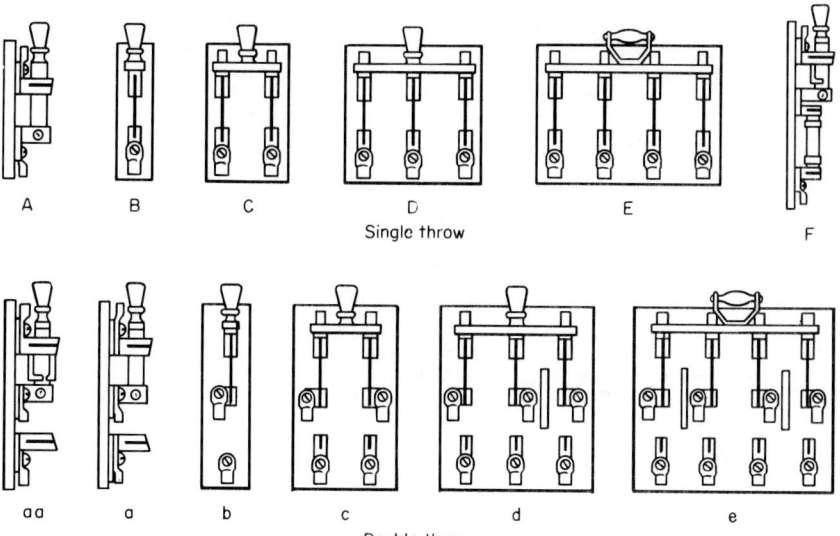

**Fig. 4-16**   Front-connected knife-blade switches. [*Electric System Handbook*]

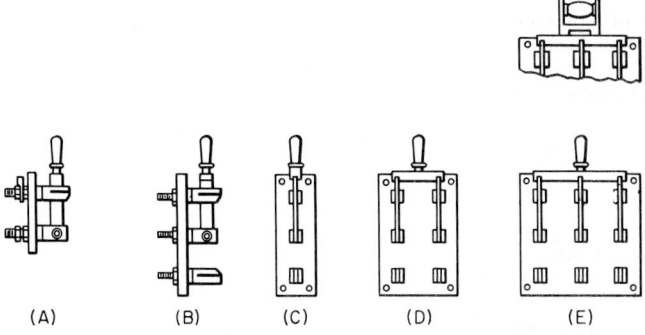

**Fig. 4-17**   Rear-connected knife-blade switches. [*Electric System Handbook*]

operated from outside the box by a handle connected to the switch mechanism. Various types are available with contacts of either the single- or the double-break type employing either knife-blade or butt-contact construction. The operating handle may be located on the side of the switch box, in the center of the front cover, or in the front side of the box (see Fig. 4-18).

Certain safety switches are so arranged that their doors cannot be opened when the switch is closed. Others possess this feature and the additional one that absolutely no live metal parts are exposed when the switch and door are open.

I. Handle located in middle.
[General Electric Co.]

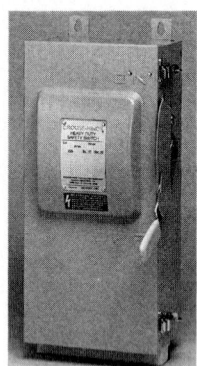

II. Handle located on side
[Crouse-Hinds Co.]

III. Handle located on front side.
[Eaton Corp., Cutler-Hammer Products]

**Fig. 4-18**  Safety switches with different types of operating-handle arrangement.

Safety switches are made in two-, three-, four-, and five-pole assemblies either fused or unfused and in single- or double-throw types. The three-, four-, and five-pole switches may be of the standard, solid-neutral, or switching-neutral type. Safety switches are made in HD, ND, and LD types of construction. They can be obtained with or without a quick-break mechanism. Many switches are constructed to give quick-make action in closing the switch as well as quick-break action in opening. Typical constructions for Class HD knife-blade safety switches are shown in Figs. 4-19 and 4-20. Butt-contact design (Fig. 4-21) generally results in a more compact construction with smaller overall dimensions. A very compact type of rotary knife-blade switch (Fig. 4-22) is made, however, in 30- to 100-A sizes.

**18. To make a contact between switch blade and jaws,** considerable skill is required. After a switch has been assembled, the jaws are first bent into correct position either by hand

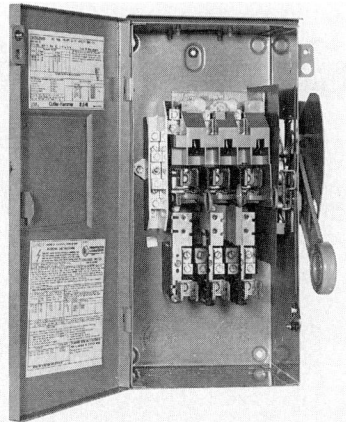

**Fig. 4-19** Fusible safety switch in raintight enclosure. [Eaton Corp., Cutler-Hammer Products]

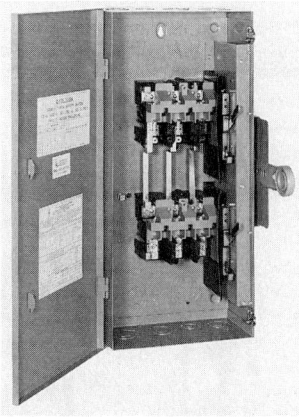

**Fig. 4-20** Nonfusible double-throw safety switch. [Eaton Corp., Cutler-Hammer Products]

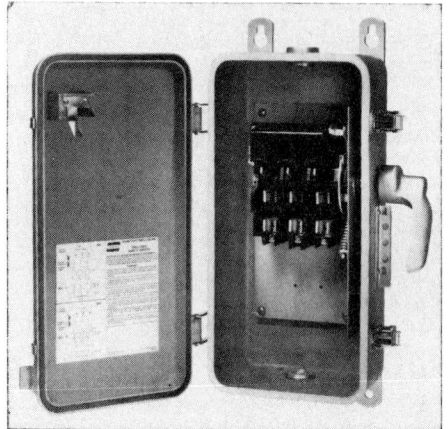

**Fig. 4-21** Side-operated heavy-duty raintight safety switch. [Crouse-Hinds Co.]

**Fig. 4-22** 400-A, 600-V visible-blade safety switch. [Square D Co.]

or by driving a block of wood against the distorted portion with a hammer. Then they are ground in with Vaseline and fine (FF) pumice. Often the fit of a switch is reasonably good at the start, and merely working the blade in and out of the jaws by hand will grind them in. Before the grinding process is started, the portion of the blade that wipes the jaws should be daubed with the Vaseline-and-pumice compound. The abrasive not only grinds in the fit but wears off the lacquer, which if it remained, might cause a bad contact. The surplus compound should be removed with a rag.

**19. A test for good blade contact** can be made by trying to insert a "feeler," a leaf of very thin steel, mica, or paper, between the jaws and the blade at the corners and sides. About 0.001 to 0.004 in (0.0254 to 0.1016 mm) is about the right thickness for a feeler. An excellent feeler can be made by hammering down to a knife edge the edges of a strip of very thin metal possibly 4 in (101.6 mm) long and ¾ in (19.05 mm) wide. If the feeler slips in at any point, it is evident that the fit is poor at that point and the contact bad. Proper forming of the jaw will correct the difficulty. In some cases switches have been made to carry, without excessive temperature rise, currents 50 percent greater than their normal ratings merely by carefully fitting their jaws to their blades.

**20. The standard sizes of general-purpose power switches** are listed in Table 46, Div. 11.

**21. High-capacity switches.** There are three basic types of fused high-capacity, load-break switches in ratings from 800 to 6000 A (ac). These switches are shown in Fig. 4-24.

I. Fused type. [Westinghouse Electric Corp.]    II. Fused type. [General Electric Co.]

**Fig. 4-23**  Open-type knife-blade disconnecting switches.

1. Bolted-pressure-contact switches contain a pressure mechanism which firmly bolts the blades to the contacts when the switch handle is moved to the closed position. Auxiliary springs maintain an initial pressure during opening and closing of the blades to prevent arcing and pitting of the main contacts. Class L fuses are bolted to fuse terminals.

2. Circuit-breaker-type switches are actually circuit breakers with the overload trips removed, and bolt-on Class L current-limiting fuses are used as the overcurrent protection.

3. Fused circuit breakers include all the mechanical features of conventional large-capacity circuit breakers and, in addition, have replaceable Class L fuses which function to increase the fault-current–interrupting ability. They are intended to be used in the same manner as other circuit breakers and are mounted in enclosures with hinged doors or covers over the accessible fuses. They are rated at 600 V or less.

Fused circuit breakers are classified in two categories, Classes 1 and 2.

*a.* Class 1 fused circuit breakers meet all the performance requirements of large branch-circuit and service circuit breakers. The Class L fuses function only to extend the fault-current–interrupting rating beyond the short-circuit test requirement applicable.

*b.* Class 2 fused circuit breakers use fuses so coordinated that they function at currents below those specified in short-circuit test requirements. Except for this feature of short-circuit operation, Class 2 fused circuit breakers meet all requirements applicable to large branch-circuit and service circuit breakers and, in addition, are required to clear circuits up to and including 25 times their ampere rating, and circuits of 1000 A or less regardless of ampere rating, without causing the operation of any fuses which are part of the device. Class 2 devices are limited to constructions which are designed to accommodate and coordinate with Class L fuses.

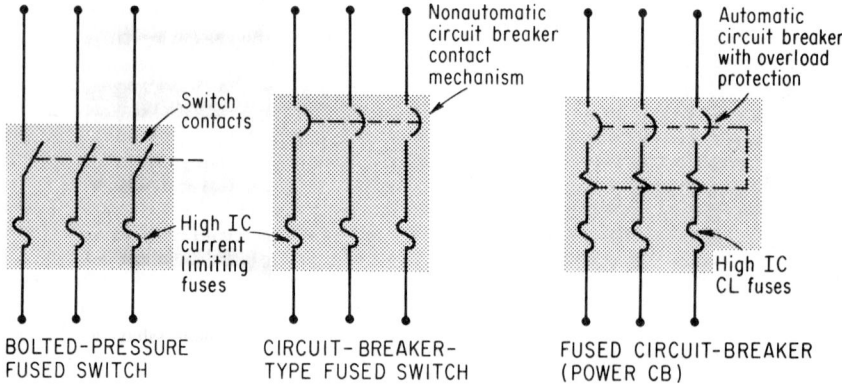

**Fig. 4-24**  High-capacity switches (800 to 6000 A).

All the high-capacity switches described in Sec. 21 use Class L current-limiting fuses with interrupting-current (IC) ratings of 100,000 or 200,000 A rms, and they are designed for ac circuits. Typical switch ratings are 240, 480, 500, and 600 V. Since the NEC prohibits the use of fuses in parallel (unless factory-assembled and approved for the purpose), one of these switch types will be required if fuse ratings larger than 600 A are needed for a given service or circuit.

All these high-capacity switches incorporate a manual means of disconnection. In addition, shunt-trip coils can be provided to permit electrical operation of switches. With an electrically operated switch ground-fault interrupters can be added to open the main circuit in the event of line-to-ground faults. Such devices are described in Sec. 74 and shown in Fig. 4-104.

**22. Isolating switches** are intended simply to isolate circuits from their source of power. They are not intended to be used to interrupt the current of the circuit and are to be operated only when the circuit has been opened by some other means. The principal application of these switches is to isolate equipment or parts of an electric circuit for inspection or repair. They are commonly used on circuits rated at more than 600 V. The switches are made in a great variety of open knife-blade types to meet different requirements of both indoor and outdoor service. Since they are generally located out of arm's reach, they are provided with an eye on the free end of the blade for operation with a hook rod. Typical switches of this type are shown in Fig. 4-23. Enclosed safety industrial disconnecting switches are shown in Figs. 4-25 and 4-26.

**23. Motor-circuit switches** are used to interrupt current in motor circuits. They are rated in horsepower and must be capable of interrupting at their rated voltage the maximum operating overload current of a motor of the same horsepower as the switch rating. The maximum operating overload current is taken as 6 times the rated full-load motor current for ac motors and 4 times the rated full-load motor current for dc motors. Motor-circuit switches are of the same construction as the general-purpose safety switches discussed in Sec. 17. See Table 48, Div. 11, for horsepower ratings of fused switches. The switches are rated in amperes for general use and are given a horsepower rating for motor-circuit application.

**Fig. 4-25** Safety enclosed fusible disconnecting switch. [General Electric Co.]

**Fig. 4-26** Manual-magnetic safety disconnecting switch with motor-circuit switch and control transformer. [Eaton Corp., Cutler-Hammer Products]

**24. Motor-starting switches** are employed to start motors which may be started directly across the line. Standard motor-circuit switches may be used for this purpose, but it is generally better practice to employ a switch designed specifically for starting duty. These switches are made in manual and magnetic types. They generally are provided with some form of motor-overload protective device. For a discussion of motor-starting switches, refer to across-the-line starting equipment in Div. 7. When the control of a dc motor is mounted on a switchboard, a multiple-contact open knife switch (Fig. 4-27) is frequently employed to cut out the starting resistance. These switches may be provided with a ratchet device on the hand lever so that they cannot be moved too rapidly from step to step.

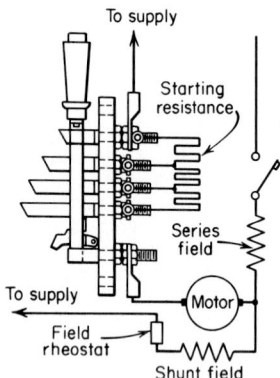

**Fig. 4-27** Multiple-contact open knife-blade switch for motor starting. [*Electric System Handbook*]

**25. Field switches** are used to interrupt the field current of generators and synchronous motors. These switches (Fig. 4-28) are provided with an auxiliary contact, an extra blade attached to one of the main blades, and auxiliary quick-break blades attached to each main blade. In operating the switch to disconnect the field from its source, the auxiliary quick-break blades do not break contact with their jaws until after the short auxiliary blade has made contact with the extra jaws. Thus the field circuit is shorted through the discharge resistance before the circuit is disconnected from its source. Since the field circuit is highly inductive, if the circuit were suddenly broken, a high induced voltage which might puncture the insulation of the field coils would be produced. The field switch of Fig. 4-28 is for mounting on the front of the switchboard or panel. A preferable arrangement is to employ a switch mounted on the rear of the board, as shown in Fig. 4-29. There are electrically operated field switches which allow the field switch to be mounted near its machine while its operation is controlled from the switchboard.

**26. Service or entrance switches.** The service equipment to a building consists of the necessary circuit breaker or switch and fuses and their accessories, which are located near the point of entrance of the supply conductors. It constitutes the main control and means of cutoff for the supply to that building. In large industrial plants comprising several buildings, the service equipment for each building will generally consist simply of a fused

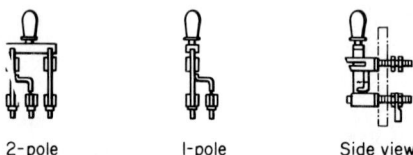

2-pole          I-pole          Side view

I. Field switch for mounting on the front of a switchboard.

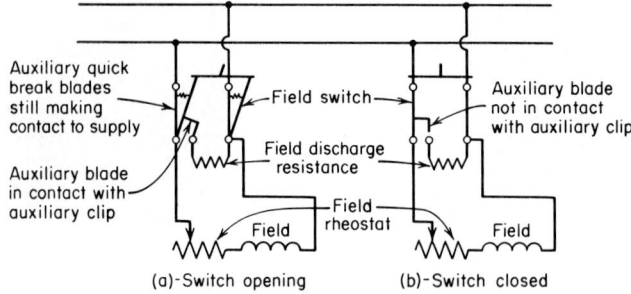

(a)-Switch opening          (b)-Switch closed

II. Schematic diagram showing functioning of a switch.

**Fig. 4-28** Field switch.

safety switch or an industrial circuit breaker. For single buildings supplied directly from a utility company's lines, the service equipment must include a switch, fuses, and a meter or a circuit breaker and a meter. The meter is furnished by the utility, but the rest of the equipment must generally be supplied by the consumer. Service or entrance switches are specifically designed to meet these entrance conditions so that the meter and fused switch can be mounted in a convenient, neat, and compact manner. One end of many switch boxes is provided with knockouts or hubs for conduit connection (Fig. 4-32). This allows the box to be connected by conduit to a separate meter trough either for a single meter or for several meters, as shown in Fig. 4-30. Combination units consisting of the switching member and a meter socket (Fig. 4-33) are available in different types of design.

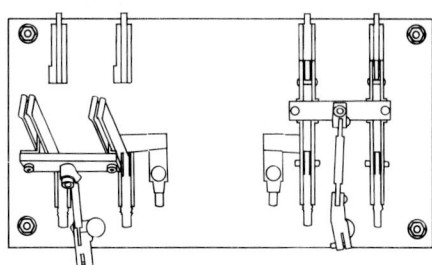

I. Switch open at left, closed at right.

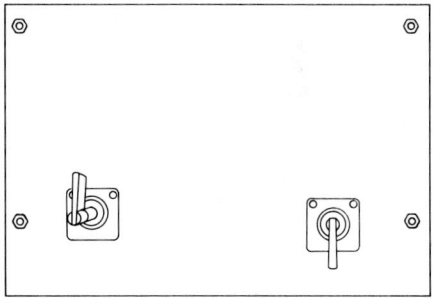

II. Operating handles.

**Fig. 4-29**   Field switch for mounting on the rear of a switchboard with front-operating handle.

**27. Service or entrance switches are made** in a great variety of types to meet the desired sequence between meter, switch, and fuses (refer to Div. 9) and the other requirements of individual utility companies. It has been impossible to introduce any standardization into the types of service switch owing to the divergence of opinion among utility representatives. Some of these switches simply include the entrance switching device (Figs. 4-30, 4-31, and 4-32) and overload protection for the service. Others have in addition a switching device and overload protection for an electric-range circuit. Still others, which combine the service-entrance equipment with overload protection for branch circuits, frequently are called load and service centers (Figs. 4-34 and 4-35). There are available types which include test links mounted on the main switch base

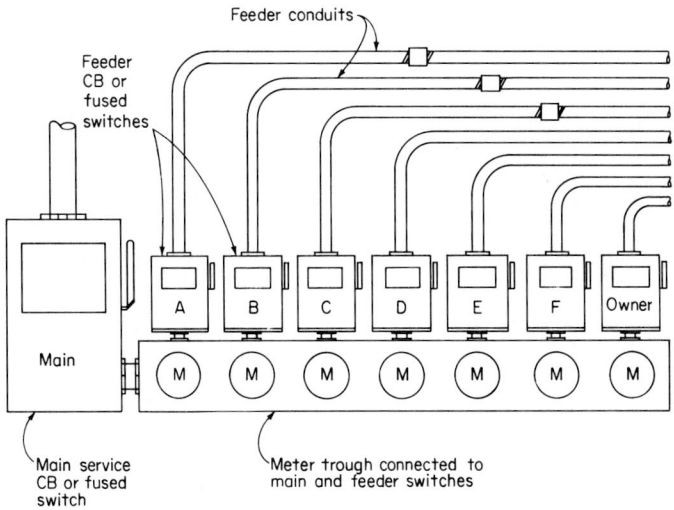

**Fig. 4-30**   Installation of service switches connected with conduit to a separate meter trough.

so that the utility company can test the meters without disconnecting the load and interrupting the customer's service. In some types the main fuses are accessible to the customer, while in others they may be sealed by the utility company so that they are not accessible to the customer.

Service switches may be divided broadly into three types, depending upon the type of main switching device: fuse-puller switches (Fig. 4-31), circuit breakers (Figs. 4-32, 4-33, 4-34, and 4-35), and the standard type of safety switch (Fig. 4-30). When the main switching device is a circuit breaker, the equipment generally is of the load-center type, each branch being provided with a circuit breaker. This type has the advantage that it furnishes a means of switching as well as overload protection for each branch. The other types of load center provide only overload protection for the branches with no means of switching.

**Fig. 4-31** Fuse-puller type of service switch. [Crouse-Hinds Co.]

**Fig. 4-32** Circuit-breaker type of service switch. [Crouse-Hinds Co.]

**Fig. 4-33** Raintight combination meter socket circuit-breaker type of service switch. [Crouse-Hinds Co.]

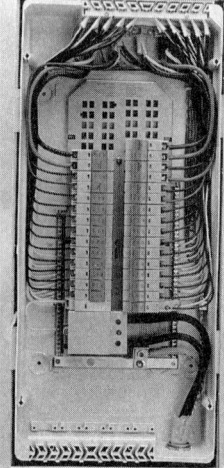

**Fig. 4-34** Trilliant® home power system constructed of fire-retardant Noryl resin. [Square D Co.]

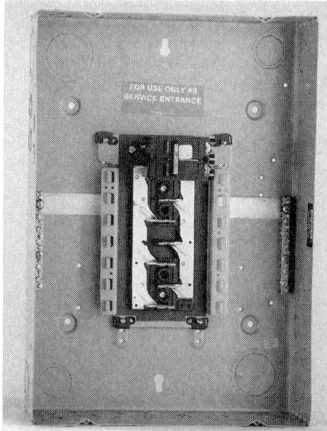

**Fig. 4-35.** Combination circuit-breaker–type service switch and load center with a split neutral bus. [Crouse-Hinds Co.]

**28. Service-entrance equipment** utilizes the detachable socket type of meter. The meter is provided with prongs that fit into clips or jaws provided in a meter-mounting trough. Some of these troughs are wired on the job, and others are factory-wired by the manufacturer. They are made in indoor and outdoor weatherproof types. Examples of troughs and installations of this type are shown in Figs. 4-33, 4-36, and 4-37.

**Fig. 4-36** Raintight socket meter trough for a single meter. [Crouse-Hinds Co.]

**Fig. 4-37** Multiple-meter-socket meter stack and disconnect. [Crouse-Hinds Co.]

**29. Service switches are available in raintight enclosures** so that the service-entrance equipment can be located on the outside of the building. Typical equipment is shown in Fig. 4-33.

**30. Wiring switches,** as designated in this book, include all the relatively small switches that are employed in interior-wiring installations for the control of branch circuits and individual lamps or appliances. Except for mercury switches, the switching mechanism consists of a snap action operated by one of the methods given in Sec. 10. Wiring switches are made in many types to meet the requirements for different conditions of mounting.

The *flush* ac, dc switch (Figs. 4-38 and 4-39) consists of a switch mechanism mounted in a porcelain or composition molded case, the complete unit being designed to be mounted inside a switch box, outlet box, or conduit fitting so that the operating handle or button is the only portion of the switch which extends beyond the cover of the box or conduit fitting. The contacts may be of the knife-blade or the mercury type (Sec. 7). Those with knife blades are available with toggle (tumbler), pushbutton, or locking types of

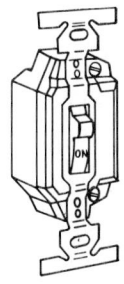
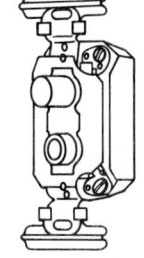
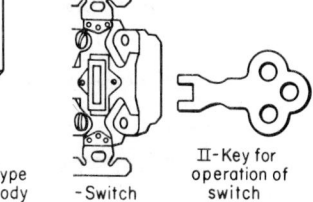

I - Tumbler type with porcelain body       II - Tumbler type with composition body       III - Push button type with porcelain body       - Switch       II - Key for operation of switch

**Fig. 4-38** Standard line of flush ac, dc snap switches.

**Fig. 4-39** Locking type of tumbler flush switch, standard-line construction.

operating mechanisms. Mercury switches are almost completely silent in operation, making practically no noise on the make or break. They are used in installations where the click of a switch might prove disturbing, as in hospitals and nurseries, and they are rated as ac snap switches as well as ac, dc types. Locking switches do not have a protruding operating handle but are operated by a key inserted in a slot in the face of the switch mounting.

Flush-mounted switches may be classified into subtypes depending upon the physical size and mounting of the switch. The *standard line* of switches (Figs. 4-38 and 4-39) is made with more generous proportions and is the more rugged of the two types. It should be used for heavy-duty installations and in locations where frequent operation is required. The overall dimensions range from 1 to 2 in (25.4 to 50.8 mm) in depth, $1\frac{7}{16}$ to 2 in (36.5 to 50.8 mm) in width, and $2\frac{3}{8}$ to $3\frac{3}{16}$ in (60.3 to 80.9 mm) in length. Most of these switches are supplied with two sets of supporting screw holes, spaced from center to center $3\frac{9}{32}$ and $2\frac{3}{8}$ in (83.3 and 60.3 mm), respectively. They are made in sizes up to 30 A capacity. Switches of standard-line dimensions are made in regular tumbler, pushbutton, locking, and mercury types.

The *ac* snap switches (Fig. 4-40) represent a major development in the wiring-device field. These switches are quiet in operation and have a long life because of their silver-alloy butt contacts. Unlike ac, dc switches, the ac snap switch can be used up to its full ampere rating for incandescent-lamp and fluorescent-lamp loads and any other resistive or inductive loads except motors. Motor loads are limited to 80 percent of the ac switch rating.

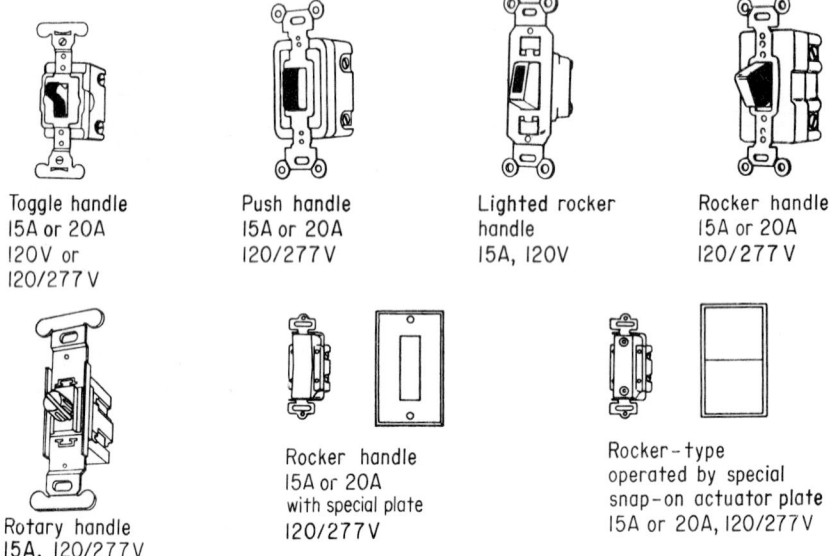

| Toggle handle<br>15A or 20A<br>120V or<br>120/277V | Push handle<br>15A or 20A<br>120/277V | Lighted rocker<br>handle<br>15A, 120V | Rocker handle<br>15A or 20A<br>120/277V |

Rotary handle
15A, 120/277V

Rocker handle
15A or 20A
with special plate
120/277V

Rocker-type
operated by special
snap-on actuator plate
15A or 20A, 120/277V

**Fig. 4-40**   Typical types and ratings of modern ac snap switches.

Although the ac snap switch is limited to ac circuits, it is obvious that such switches can be used on most installations because ac supplies represent more than 98 percent of the systems in the United States. Ratings of ac snap switches are 15 or 20 A at 120 or 120/277 V. These switches are widely used for local control of 277-V fluorescent luminaires, which are supplied by 480Y/277-V branch circuits.

As shown in Fig. 4-40, ac snap switches are available in a wide variety of shapes and styles to blend with modern decor. The switches are available in single-pole, double-pole, three-way, four-way, and momentary-contact types. In addition, lighted (to indicate an off position) or pilot-light (to indicate an on position) operating handles are available in these switches. Terminals are either screw-type or pressure-locking push-in types. Only the screw terminals are suitable for the connection of aluminum conductors. See Sec. 3 of Div. 9 for definitions of various types of snap switches.

The modern *dimmer snap switch* is an electronic device which provides full-range dimming of incandescent lamps from off to full brightness. These devices are rated from 600 to 1000 W and are wired in the same manner as the standard single-pole switches. Similar devices are available for use with special rapid-start ballasts for the dimming of fluorescent lamps.

The *interchangeable line* of switches (Figs. 4-41 and 4-42) is constructed in a much more compact manner with less generous proportions. As many as three of these switches can be installed in the space required for a single standard-line switch. The largest size of switch made in this line is 20 A. Three switches or a combination of three units of switches, receptacles, and pilot lights can be assembled on a single supporting strap yoke and mounted in a single-gang outlet box or switch box (see Figs. 4-41 and 4-42).

*Surface switches* (Figs. 4-43 and 4-44) are designed for mounting on the surface of a wall, ceiling, or box so that all or practically all of the switch body extends beyond the surface upon which it is mounted. The contacts are of the knife-blade type operated by tumbler or rotary action. A line of outlet fittings for surface mounting is available for use with exposed nonmetallic-sheathed–cable wiring (refer to Sec. 203, Div. 9).

Pendant switches (Fig. 4-45) are used for the control of lamps or other devices which are mounted overhead out of reach from persons standing on the floor. These switches are supported at the end of a pendant two-conductor cord.

Through-cord switches (Fig. 4-46) are designed to be inserted in a portable cord for control of the cord circuit.

Door switches (Fig. 4-47) are designed to control circuits through the opening and closing of doors. They are mounted in boxes located in the doorjamb. Two types are available, one constructed so that the switch is closed when the door is open and the other so that the switch is closed when the door is closed.

Canopy switches (Fig. 4-48) are small, compact switches for mounting in the canopies of lighting fixtures for the control of the lamps directly at the fixture. They can be obtained in pull-chain or toggle-switch types.

Appliance switches are made in a variety of types for mounting in the appliance enclosure.

Rotary surface switches mounted in special enclosures to meet the conditions of special exposures, such as weather, water, or explosive atmospheres, can be obtained.

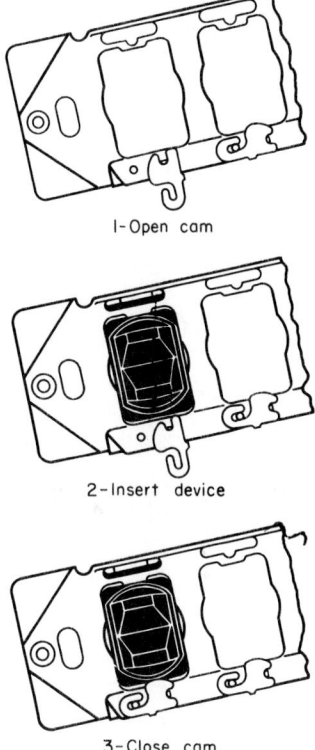

1-Open cam

2-Insert device

3-Close cam

**Fig. 4-41** Method of assembling units and strap.

**31. Wiring switches are available in different types to perform different functions in the wiring system.** The regular switches designated simply as *single-pole* or *double-pole* are for the general-purpose use of opening and closing circuits for the control of lamps or other devices from a single point. The single-pole switches (Fig. 4-49, I and II) break the circuit of only one side of the line, while the double-pole switch (Fig. 4-49, III) breaks the circuit of both sides.

*Three-way switches* are used if it is desired to control lights from two different points, such as a hall light to be controlled from the lower hall and the upper hall. The lights can be turned on or off from either position. Three-way switches when properly connected (see Sec. 32) break the circuit of only one side of the line; they do not isolate the circuit from the supply. The method of controlling a lamp from two positions with two three-way switches is shown in Fig. 4-50.

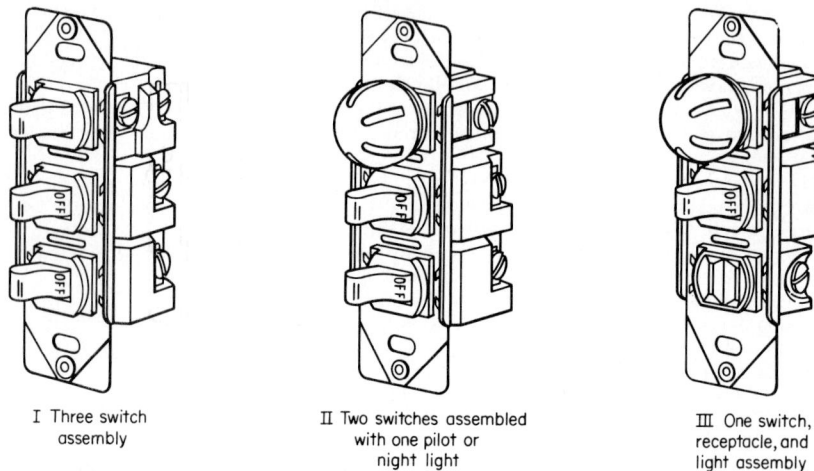

I  Three switch          II Two switches assembled       III  One switch,
   assembly                 with one pilot or                 receptacle, and
                            night light                       light assembly

**Fig. 4-42**   Typical assemblies of interchangeable wiring devices.

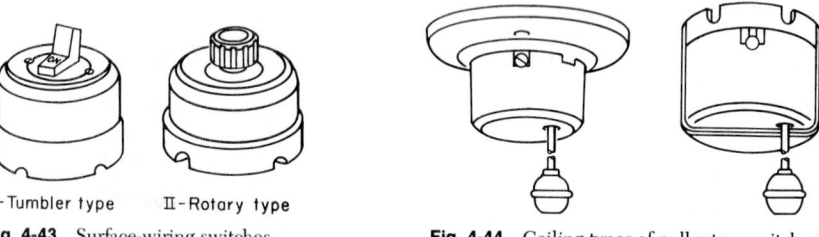

I-Tumbler type    II-Rotary type

**Fig. 4-43**   Surface-wiring switches.      **Fig. 4-44**   Ceiling types of pull rotary switches.

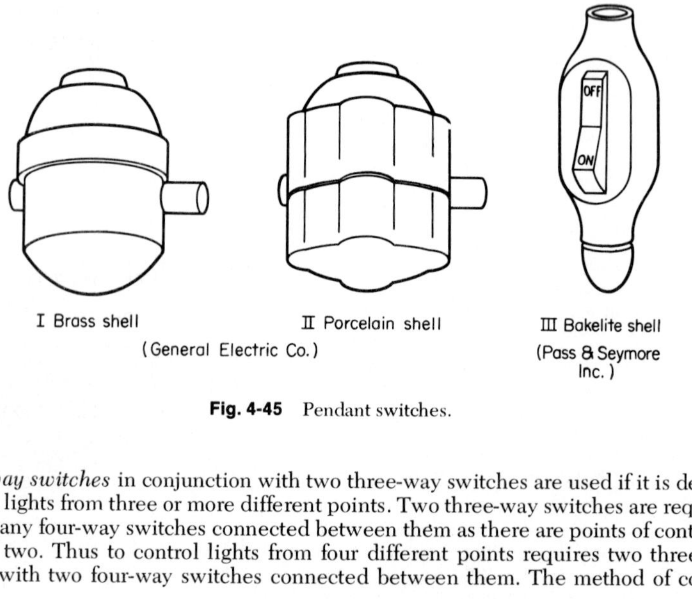

I  Brass shell           II  Porcelain shell          III  Bakelite shell
(General Electric Co.)                            (Pass & Seymore
                                                      Inc.)

**Fig. 4-45**   Pendant switches.

*Four-way switches* in conjunction with two three-way switches are used if it is desired to control lights from three or more different points. Two three-way switches are required with as many four-way switches connected between them as there are points of control in excess of two. Thus to control lights from four different points requires two three-way switches with two four-way switches connected between them. The method of control

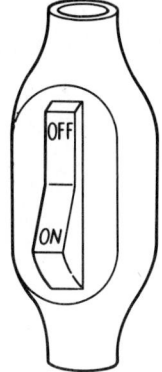

**Fig. 4-46** Through-cord switch. [Pass & Seymour, Inc.]

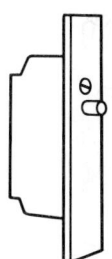

**Fig. 4-47** Door switch. [General Electric Co.]

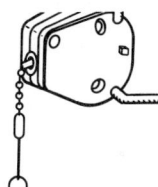 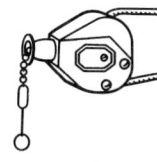  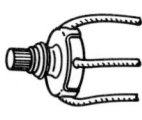

**Fig. 4-48** Types of canopy switches. [General Electric Co.]

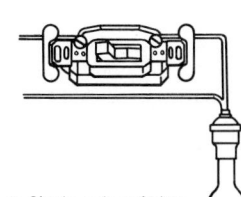 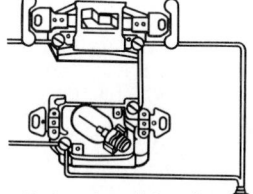 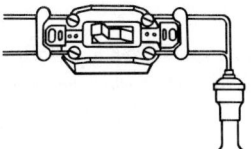

I–Single-pole switches

For controlling lights from one point-breaking one side of line

II–Single-pole switch and pilot lamp receptacle

To be used where it is desirable to indicate if lights are "ON" or "OFF", e.g., a cellar light controlled from the kitchen with pilot light in the kitchen

III–Double-pole switches

For controlling lights from one point-breaking both sides of line

**Fig. 4-49** Connections and functions of single- and double-pole switches.

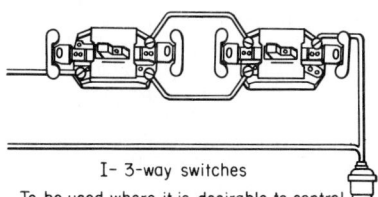 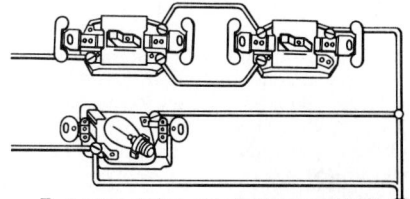

I– 3-way switches

To be used where it is desirable to control lights from two different points, e.g., a hall light to be controlled from the lower hall and upper hall or sleeping room.

II– 3-way switches and pilot lamp receptacle

To be used where it is desirable to control lights from two different points, and to indicate if lights are "on" or "off", e.g., a garage light controlled from the kitchen or garage with pilot light in the kitchen.

**Fig. 4-50** Connections and functions of three-way switches.

from three different locations is shown in Fig. 4-51. Four-way switches when properly connected break the circuit of only one side of the line.

*Momentary-contact switches* are used if it is desired to close or open a circuit for only a short length of time. The switch is provided with a spring so that it will return to its original position as soon as the handle or button is released. These switches are available in standard-line tumbler and pushbutton types of flush switches and in pushbutton surface types.

*Heater switches* are designed for the control of circuits for electric ranges and electric heating devices. They are made in single- and double-pole types for the control of single, double, or triple heats.

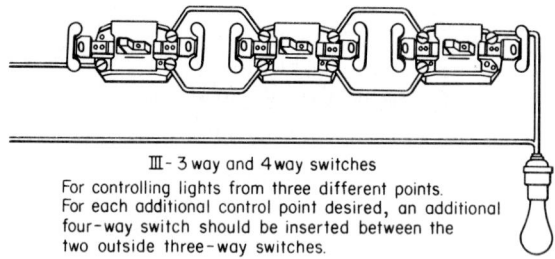

III- 3 way and 4 way switches
For controlling lights from three different points.
For each additional control point desired, an additional four-way switch should be inserted between the two outside three-way switches.

**Fig. 4-51**   Connections and functions of three-way and four-way switches.

**32. Caution should be exercised in connecting** three-way and four-way switches, since it is possible to connect these switches with both sides of the circuit connected to each switch. This is an improper connection which should never be used. The switch is not designed for this purpose and in case of failure of the switch mechanism a short circuit may result. The acceptable methods of connection are shown in Fig. 4-58. Refer to Sec. 40 for additional information on connection of these switches.

**33. The rating of snap switches** for general wiring installation must be in accordance with the requirements of the National Electrical Code. For this purpose the Code classifies loads into three types: (1) noninductive loads other than tungsten-filament lamps, (2) tungsten-filament lamps, and (3) inductive loads.

Snap switches shall be used within their ratings. AC general-use snap switches are suitable only for use on alternating-current circuits to control resistive and inductive loads, including electric discharge lamps, not exceeding the ampere rating of the switch at the voltage involved; tungsten-filament lamp loads not exceeding the ampere rating of the switch at 120 V; and motor loads not exceeding 80 percent of the ampere rating of the switch at its rated voltage.

AC-DC general-use snap switches are suitable for use on either ac or dc circuits to control resistive loads not exceeding the ampere rating of the switch at the voltage applied and inductive loads not exceeding 50 percent of the ampere rating of the switch at the applied voltage. Switches rated in horsepower are suitable to control motor loads within their rating at the voltage applied and tungsten-filament lamp loads not exceeding the ampere rating of the switch at the applied voltage if T-rated.

The T rating of an ac, dc switch is its ability to control satisfactorily a tungsten-filament lamp load of that rating. The peculiar current characteristic of a tungsten-filament lamp over other types of circuit loading is that the resistance of the tungsten filament to the passage of current is extremely low when the lamp filament is cold. Thus, when the circuit is closed, there is a heavy inrush current. As the temperature of the lamp filament increases, the current decreases and very soon reaches its normal value. Therefore, switches used with these loads must be capable of satisfactorily handling the inrush current, which will be considerably greater than the normal hot operating current of the circuit.

The common types of inductive loads encountered in general wiring installations are fluorescent lamps and mercury-vapor lamps. Switches used on signs and outline lighting must conform with the Code requirements given in Article 600 of the Code.

**34. Control switches** are used for controlling from a switchboard the operation of electrically operated equipment such as switches, circuit breakers, rheostats, and prime-mover governors. They are generally of the drum or rotary type. Two representative control switches are shown in Figs. 4-52 and 4-53. Pilot lamps are generally used in connection with control switches to indicate to the operator the position of the controlled equipment.

**35. Instrument switches** are employed to change the connections of meters from one circuit to another. The most common type is of rotary construction, as shown in Fig. 4-54. The plug-and-receptacle type of switch is also employed to connect a common

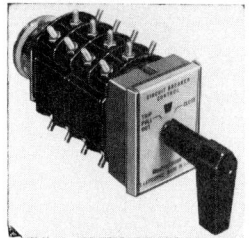 

**Fig. 4-52**   Rotary type of control switch. [Westinghouse Electric Corp.]

voltmeter, ammeter, or synchroscope to various circuits. One plug is used for each set of common receptacles so that it is impossible to make cross connections. Examples of plug-and-receptacle equipment are shown in Figs. 4-55 and 4-56.

**36. Low-voltage relay control systems** are widely used in installations which require multiple-switch control of outlets. To a large degree such systems replace conventional three-way and four-way switch control, particularly when a large number of multiple-switch locations are required in a given installation. Figure 4-57 shows three basic types of low-voltage relay switching systems. The system shown in Fig. 4-57, 1, operates on a 24-V system, and the switching relay, located at the controlled outlet, contains an on-off latching-type double coil wired to one or more single-pole, double-throw, momentary-contact switches. Pressing to the on position of any switch operates the on coil and closes the single-pole contact which completes the circuit to the connected load. After the on button is released, the current does not flow through the coil, but the single-pole line contacts remain in the closed position because they are mechanically latched. Pressing to the off position of any of the paralleled switches energizes the off coil in the relay, and the single-pole line contacts are opened, disconnecting the load; the contacts remain mechanically latched in the open position until the on switch position is operated again.

**Fig. 4-53**   Bypass contact control switch. [Westinghouse Electric Corp.]

**Fig. 4-54**   Rotary type of instrument switch with cover removed. [Westinghouse Electric Corp.]

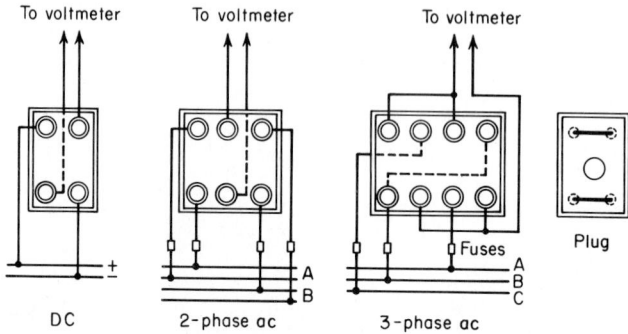

**Fig. 4-55**   Voltmeter receptacles and plug. [*Electric System Handbook*]

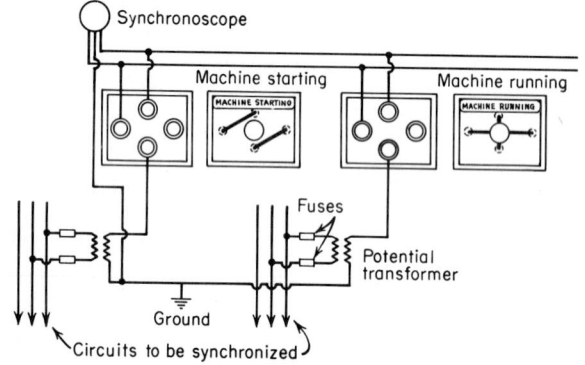

**Fig. 4-56**   Synchronizing plugs and receptacles. [*Electric System Handbook*]

In Fig. 4-57, 2, a two-wire latching relay is operated at 12 or 24 V by one or more single-pole, single-throw, momentary-contact switches (placed in parallel). Thus, every time that a momentary-contact switch is actuated, the relay either closes or opens the circuit and latches in either position like a pull-chain type of fixture.

The system shown in Fig. 4-57, 3, uses a compact transformer relay in a single unit and is located at each controlled outlet. This unit operates on a thermal principle and trips the single-pole 115-V contacts into the on or off position according to the operation of the single-pole, double-throw, momentary-contact switches. Although this transformer-relay system is no longer manufactured, it is described so that the reader will understand the principle of operation because numerous systems of this type are still in existence.

The system in Fig. 4-57, 1, is the most popular low-voltage relay system and is available from several manufacturers. This system lends itself to a *master* and *submaster* control so that many circuits can be operated at once through a motorized master unit. Special relays and switches are available to provide pilot-light indication at switches to show when the controlled load is in the on position. In most cases the relay is located at the outlet to which the controlled load is connected.

The system in Fig. 4-57, 2, is available from only one manufacturer. Generally, the relays are in a large, centrally located junction box or panel, and the 115-V circuits extend from this central point to each controlled outlet. This system can also be converted to a master system, and pilot-light indicators are available.

Full details, diagrams, and suggested layouts are available from manufacturers of these systems in the form of wiring handbooks. Switch ratings are 10 to 15 A at 120 or 277 V.

**37. Magnetic switches** in ratings up to several hundred amperes are available in either magnetically held or mechanically held types. Voltage ratings are 600 V or less.

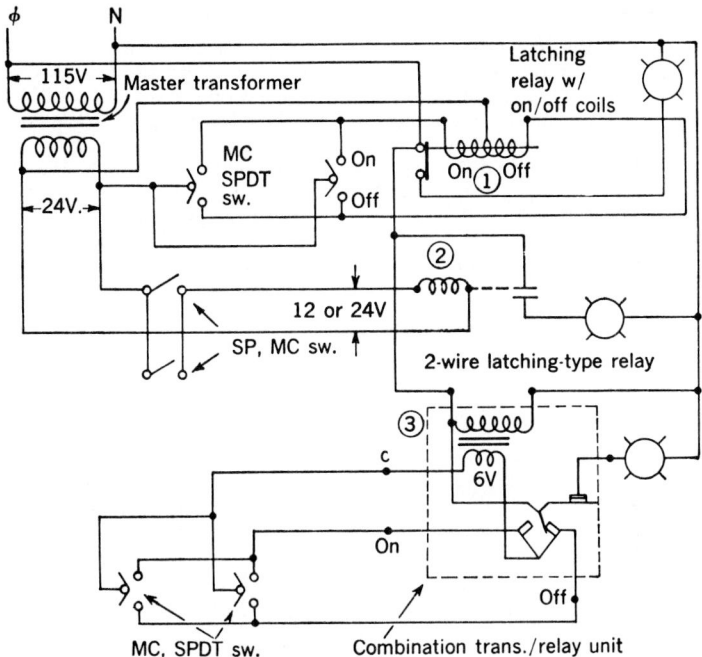

**Fig. 4-57**  Three types of low-voltage relay switching systems.

A magnetically operated switch is an on-off switch in which the opening and closing operations are effected by energizing an electromagnetic mechanism. Such switches or contactors are used to provide automatic and/or remote switching of feeders, branch circuits, and individual loads.

The basic switch is operated by energizing its coil to close the contacts, and it is held in the closed position by maintaining current through the coil.

A variation of the magnetically held contactor is the mechanically held contactor, in which the operating coil is momentarily energized to close the contacts and momentarily energized to open the contacts. Because of the definite switch action with mechanically maintained open and closed positions, such a magnetic contactor is commonly distinguished from the magnetically held type by calling it a *remote switch,* a mechanical switch which can be operated by means of a control circuit to a remote switch.

A remote switch permits the operation of the contact assembly from one or more distant control stations. In such applications the switch is placed in the circuit which it is to control. This may be in the middle of a split-bus panelboard, where the switch controls one section of bus. It may be adjacent to a panel, where the switch controls a branch circuit or feeder from the panel. Or it may be placed on a column, where it switches lighting circuits for a given area. Then, in each case, control conductors are run in cable or conduit from the switch enclosure to one or more control points at which pilot devices provide for operation of the remote switch. The pilot devices may be push buttons, toggle switches, or an automatic device, such as a time switch.

A basic contactor application for a full panel control might involve locating the contactor at widely spaced lighting panelboards supplying outdoor lighting, with all the control circuits being brought to pilot switches at a common point of control. For the control of individual circuits supplying lighting loads, contactors may be located near the panelboard or near the load.

**38. Miscellaneous types of switches** are available for special applications. Knife-blade switches of reduced dimensions are available for low-voltage circuits such as telephone, signaling, and battery circuits.

## INSTALLATION OF SWITCHES

**39. A switch or a circuit breaker must not be installed** so that it will disconnect a grounded conductor of a circuit unless the switching mechanism simultaneously disconnects all wires of the circuit or is so arranged that the grounded conductor cannot be disconnected until the ungrounded conductor or conductors have first been disconnected.

**40. Three-way and four-way switches** are in reality single-pole switches and therefore must be installed so that all switching is done only in the ungrounded conductor. When the wiring for these switches is enclosed in a metal enclosure, the wiring must be so installed that at every point the same enclosure includes wires of both polarities. Standard and special methods of wiring such switches are illustrated in Figs. 4-58 and 4-59. Refer to Secs. 31 and 32 for additional information on three- and four-way switches.

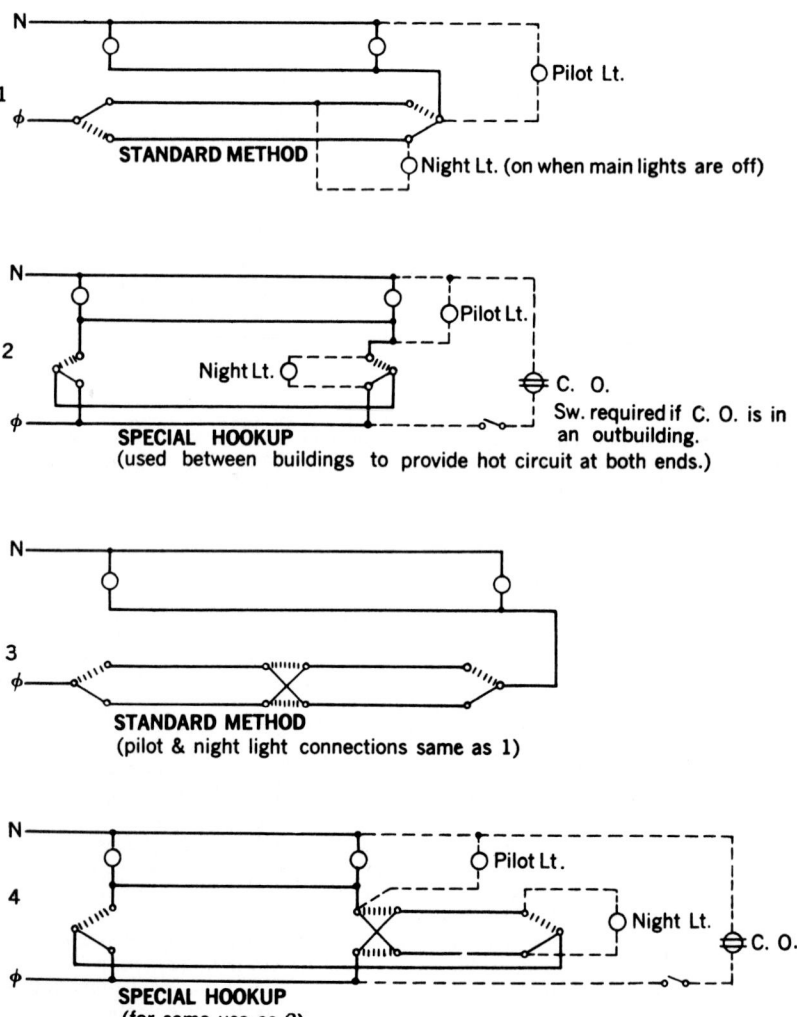

**Fig. 4-58**  Standard and special hookups with three- and four-way switches and connections of night (neon) lights and pilot lights.

**41. Position of mounting.** Single-throw knife switches shall be mounted so that gravity will tend to open and not to close them, or when approved for the use in the inverted position, they shall be provided with a locking device that will ensure that the blades remain in the open position when so set (Figs. 4-60 and 4-61). Double-throw switches can be mounted so that the throw is either vertical or horizontal. If mounted with the throw vertical, a switch must be provided with a locking device which will hold the blades in the open position when so set.

**42. In connecting knife switches** in a circuit, the switch should be wired so that the blades will be dead when the switch is open (Figs. 4-62 and 4-63).

A fused switch should always be wired so that the fuses will be dead when the switch is open.

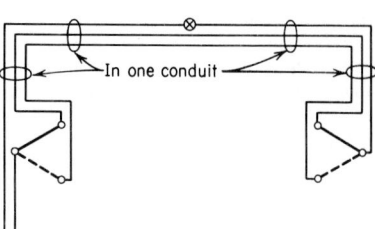

**Fig. 4-59** How wires of opposite polarity must be enclosed in the same conduit in the wiring of three-way switches.

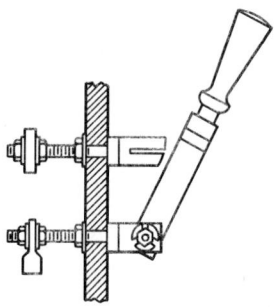

**Fig. 4-60** Single-throw switch mounted vertically so that gravity will not tend to close it.

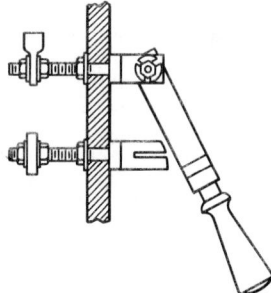

**Fig. 4-61** Wrong method of mounting a single-throw switch. In this position gravity tends to close the switch.

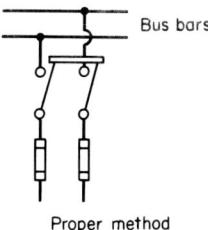

Proper method
Switch blades dead when
Switch is open

**Fig. 4-62** Approved method of connecting a knife switch.

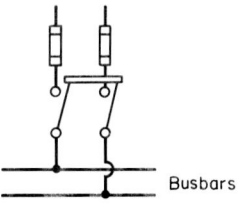

Switch blades alive when
switch is open. This connection
should be avoided.

**Fig. 4-63** Unapproved method of connecting a knife switch.

## PROTECTIVE DEVICES

**43. Various forms of protective devices** are employed to protect electrical circuits and equipment from injury under abnormal conditions. These devices may be classified as follows according to the type of protection which they provide:

1. Overload protection
2. Underload protection

3. Undervoltage protection
4. Overvoltage protection
5. Reverse-current or power protection
6. Reverse phase rotation
7. Lightning and surge protection
8. Ground-fault protection

**44. Relays and releases.** Protection for the various types of abnormal conditions is in many cases provided by relays or release devices which actuate the opening of circuit breakers or switches. If protection is afforded by a simple release device, the abnormal condition causes the functioning of the release device, which acts directly upon a holding catch of the switch or breaker mechanism. With relay protection the abnormal condition causes the functioning of the relay, which opens or closes an auxiliary electric circuit, which in turn trips the switch or breaker mechanism.

Release devices may be of thermal or magnetic types. Thermal releases can be employed only for overload protection. Their operation depends upon the deflection of a bimetallic element as it is heated by the current of the circuit. Magnetic releases consist of a solenoid acting upon an iron plunger or armature. For current protection the coil of the solenoid is connected in series with the circuit, while for voltage protection the coil is connected across the circuit which is to be protected.

Relays may be of thermal, magnetic, or induction types. The principles of operation of the thermal and magnetic types are the same as those for releases of the same type. Induction relays operate upon the same general principle as induction motors and therefore are applicable only to ac systems.

**45. Time characteristics.** Relays and release devices may be classified according to the time that will elapse between the occurrence of the abnormal condition and the opening of the circuit. There are three general types: the instantaneous or high-speed, the definite-time, and the inverse-time. In the instantaneous type, time delay is purposely omitted from its action so that the circuit is opened almost instantaneously upon the occurrence of the abnormal condition for which the release or relay is set. A definite-time relay or release is purposely so designed that there is a definite-time delay in its action. The abnormal condition must exist for a definite time before the device will function to open the circuit. The length of time of the disturbance required to cause the device to function is independent of the magnitude of the disturbance. An inverse-time relay or release is purposely so designed that there is a time delay in its action. The length of time of the disturbance required to cause the device to function is dependent upon the magnitude of the disturbance. The greater the magnitude of the disturbance, the quicker will the device function to open the circuit. All thermal relays and releases, owing to the principle of their construction, provide inverse-time protection.

**46. Overcurrent protective devices** are used to protect wires, wiring fittings and devices, and apparatus against excessive currents. The several types of overcurrent protective devices may be divided into the following groups:

| *Means of Protection* | *For Protection of* |
|---|---|
| I. Fuses | |
| A. Low-voltage fuses | |
| 1. Ordinary Edison-base plug fuses | Replacement in existing installations in which there is no evidence of tampering or overfusing for branch incandescent-lighting and heating-appliance circuits |
| 2. Dual-element plug fuses<br>   *a.* Standard screw-base type<br>   *b.* Type S | Any type of branch circuit and small motors |
| 3. Ordinary one-time cartridge fuses | Mains or feeders in which faults occur very infrequently |
| 4. Time-delay renewable cartridge fuses | Mains or feeders in which faults occur frequently |
| 5. Dual-element cartridge fuses | Most mains or feeders, motor or appliance branch circuits, and overload on motors |
| 6. Silver-sand fuses<br>   *a.* High-interrupting type | Mains and feeders in which high interrupting capacity is necessary |
|    *b.* Current-limiting type | Mains and feeders in which magnitude of fault current must be limited |

    *B.* High-voltage fuses

  II. Thermal cutouts                      Motors

 III. Circut breakers actuated by

    *A.* Thermal overcurrent release device    Circuits

    *B.* Thermal relay                  Circuits

    *C.* Magnetic release device        Circuits and apparatus

    *D.* Magnetic relay                Circuits and apparatus

    *E.* Combined thermal and magnetic   Circuits and apparatus
       release

    *F.* Induction relays              Circuits and apparatus

 IV. Manual switches actuated by thermal   Motors
    releases

  V. Magnetic switches actuated by

    *A.* Thermal relays              Motors

    *B.* Magnetic relays             Motors

 VI. Ground-fault interrupters        Circuits, persons, and apparatus

    With fuses and thermal cutouts, in case of overcurrent the circuit is opened inside the device itself. These two overcurrent protective devices must be capable, therefore, of interrupting the line current and of extinguishing the arc thus formed. Relays or overload release devices themselves do not interrupt the line current of the circuit which they are protecting. An overload release device, when operated by a current greater than that for which it is set, trips the holding catch of the switch mechanism and thus allows the switch mechanism to open. The current is interrupted by the switch mechanism and not by the overload release device. An overload relay, instead of acting directly upon the holding catch of the switch mechanism, opens an auxiliary circuit. This auxiliary circuit may be the operating coil of a magnetic switch or the shunt-trip coil of a circuit breaker. As with the overload release devices, the relay does not interrupt the main-line current itself. The main circuit is opened by the contacts of the switch mechanism.

    **47. Fuses and circuit breakers** are used for the overcurrent protection of both circuits and apparatus. Thermal cutouts, thermal overload release devices, and thermal relays are used for the protection of motors against overloads. They should not be used for the protection of circuits unless they are specifically designed and approved for that purpose. Thermal overload protective devices have admirable characteristics for the protection of motors against overloads, but they should not be relied upon for protection against short circuits unless they are specifically designed and approved for that purpose. Unless they are so designed, in the case of a short circuit there is danger of the excessive current destroying the thermal element before it has actuated the switch mechanism, with the consequent failure of opening the circuit. Circuit breakers actuated by thermal overload release devices and designed for protection against short circuits have been developed and are widely used for the protection of circuits of small and medium current-carrying capacity. When thermal overload protective devices are employed to protect motors, fuses of the proper rating or circuit breakers with the proper setting should always be located at some point back in the line ahead of the thermal protective device.

    **48. Short-circuit calculations.** When considering the use of circuit breakers and fuses, the amount of short-circuit current available at the *line terminals* of such devices must be known so that an overcurrent device with the proper interrupting-capacity (IC) rating can be selected.

    Ordinary plug fuses and Class H cartridge fuses, in general, are tested for short circuits not exceeding 10,000 A. Most common-type small-frame-size molded-case circuit breakers have IC ratings of 5000 A, although the present trend is toward ratings of 10,000 A. Larger breakers have IC ratings of 10,000, 15,000, 25,000, and up to 100,000 A, according to types and voltage ratings. The IC ratings for high-interrupting-capacity and/or current-limiting fuses are shown in Fig. 4-65.

    Under normal conditions a circuit draws current according to the load impedance and the applied voltage. When a solid short circuit occurs, there will be an abnormally high flow of fault current until the overcurrent device opens. If the short-circuit current is higher than the interrupting rating of the affected overcurrent device, the device may be completely destroyed. In addition, its destruction may cause considerable damage to other equipment.

    There is no easy method of calculating available short circuits at various points of a given installation because numerous factors and terms must be clearly understood before such calculations can be made accurately. To explain these factors would require a complete book. See Secs. 49 through 49S from the NEMA standard on determination of short-circuit currents in low-voltage systems.

A conservative rule-of-thumb formula which can be used to determine the short-circuit current at the load terminals of a supply transformer would be $I_{sc} = 100\%/\%Z_t \times I_s$, where $I_{sc}$ = maximum secondary short-circuit current based on transformer impedance and assuming a primary source of *infinite capacity*, $\%Z_t$ = transformer impedance, and $I_s$ = full-load secondary current.

**example**  Assume a 75-kVA transformer with a 5 percent impedance and a 120/240-V secondary. The full-load secondary current $I_s = 75,000/240 = 312$ A; then $I_{sc} = 100\%/5\% \times 312 = 6240$ A. Thus, the maximum short-circuit current is 20 times the full-load secondary current for this transformer. If the transformer impedance were 2½ percent, the short-circuit current would be 40 times the full-load secondary current, or 12,480 A. Thus, the transformer impedance is a major factor in determining the maximum secondary short-circuit current. Equally important is the available primary current, which is almost always less than what would be considered an "infinite" capacity. For a given transformer, the lower the available primary current, the lower the secondary short-circuit current, which identifies the previously described formula as quite conservative.

Another major point is that the maximum secondary short-circuit current determined by the previous formula is applicable directly at the secondary terminals of the trans-

**48A. Diameters of Wires of Various Materials That Will Be Fused by a Current of a Given Strength**

(Charles E. Knox, *Electric Light Wiring;* derived from tables of W. H. Preece)

| Current, A | Copper | | Aluminum | | Nickel silver | | Iron | |
|---|---|---|---|---|---|---|---|---|
| | Diameter, in | Nearest AWG gage | Diameter, in | Nearest AWG gage | Diameter, in | Nearest AWG gage | Diameter, in | Nearest AWG gage |
| 1 | 0.0021 | 43 | 0.0026 | 41 | 0.0033 | 39 | 0.0047 | 37 |
| 2 | 0.0034 | 39 | 0.0041 | 38 | 0.0053 | 35 | 0.0074 | 33 |
| 3 | 0.0044 | 37 | 0.0054 | 35 | 0.0069 | 33 | 0.0097 | 30 |
| 4 | 0.0053 | 35 | 0.0065 | 34 | 0.0084 | 31 | 0.0117 | 29 |
| 5 | 0.0062 | 34 | 0.0076 | 32 | 0.0097 | 30 | 0.0136 | 27 |
| 10 | 0.0098 | 30 | 0.0120 | 28 | 0.0154 | 26 | 0.0216 | 24 |
| 15 | 0.0129 | 28 | 0.0158 | 26 | 0.0202 | 24 | 0.0283 | 21 |
| 20 | 0.0156 | 26 | 0.0191 | 24 | 0.0245 | 22 | 0.0343 | 19 |
| 25 | 0.0181 | 25 | 0.0222 | 23 | 0.0284 | 21 | 0.0398 | 18 |
| 30 | 0.0205 | 24 | 0.0250 | 22 | 0.0320 | 20 | 0.0450 | 17 |
| 35 | 0.0227 | 23 | 0.0277 | 21 | 0.0356 | 19 | 0.0498 | 16 |
| 40 | 0.0248 | 22 | 0.0303 | 20 | 0.0388 | 18 | 0.0545 | 15 |
| 45 | 0.0268 | 21 | 0.0328 | 20 | 0.0420 | 18 | 0.0589 | 15 |
| 50 | 0.0288 | 21 | 0.0352 | 19 | 0.0450 | 17 | 0.0632 | 14 |
| 60 | 0.0325 | 20 | 0.0397 | 18 | 0.0509 | 16 | 0.0714 | 13 |
| 70 | 0.0360 | 19 | 0.0440 | 17 | 0.0564 | 15 | 0.0791 | 12 |
| 80 | 0.0394 | 18 | 0.0481 | 16 | 0.0616 | 14 | 0.0864 | 12 |
| 90 | 0.0426 | 18 | 0.0520 | 16 | 0.0667 | 14 | 0.0935 | 11 |
| 100 | 0.0457 | 17 | 0.0558 | 15 | 0.0715 | 13 | 0.1003 | 10 |
| 120 | 0.0516 | 16 | 0.0630 | 14 | 0.0808 | 12 | 0.1113 | 9 |
| 140 | 0.0572 | 15 | 0.0698 | 14 | 0.0895 | 11 | 0.1255 | 8 |
| 160 | 0.0625 | 14 | 0.0763 | 13 | 0.0978 | 10 | 0.1372 | 7 |
| 180 | 0.0676 | 14 | 0.0826 | 12 | 0.1058 | 10 | 0.1484 | 7 |
| 200 | 0.0725 | 13 | 0.0886 | 11 | 0.1135 | 9 | 0.1592 | 6 |
| 225 | 0.0784 | 12 | 0.0958 | 10 | 0.1228 | 8 | 0.1722 | 5 |
| 250 | 0.0841 | 12 | 0.1028 | 10 | 0.1317 | 8 | 0.1848 | 5 |
| 275 | 0.0897 | 11 | 0.1095 | 9 | 0.1404 | 7 | 0.1969 | 4 |
| 300 | 0.0950 | 11 | 0.1161 | 9 | 0.1487 | 7 | 0.2086 | 4 |

former. The short-circuit current will be reduced at all downstream points according to the line impedance of the secondary supply conductors. As a result, a fairly long run of conductors from a pole transformer to the service switch would appreciably reduce the maximum short-circuit current available at this point from that shown in the previous example.

It should also be emphasized that certain equipment, such as motors or capacitors, in operation at the time a fault occurs will *add* current to the fault and *increase* the short-circuit current.

**49. Determination of short-circuit currents in low-voltage systems (NEMA).** To determine the interrupting-capacity requirements for low-voltage air circuit breakers, it is necessary to obtain certain information, at the point of application, on the short-circuit condition that exists in every part of a distribution system.

PART A: THREE-PHASE SYSTEMS   The circuit diagram shown in Fig. 4-64 includes the factors which are usually considered in the calculation of short-circuit currents. The calculations were made for a system, as shown in the diagram, for various transformer kilovolt-ampere (kVA) ratings and secondary voltages in combination with common sizes and various lengths of cable. The calculations were based on those covered in AIEE (now IEEE) Transaction Paper No. 55-442, and the results are shown in Tables 49C through 49J.

With reference to Fig. 4-64, the following basic characteristics were used in the study:

1. Primary-source availability is assumed to be 1000 MVA at the primary of the transformer, with a source circuit $X/R$ ratio of 25.

2. Transformer kilovolt-ampere sizes are as shown in Table 49A.

3. There are distribution voltages of 208, 240, 480, and 600 V, three-phase, 60 Hz.

4. Tables 49C through 49J are based on the transformer impedances shown in Table 49A. Impedances of transformers rated 150 kVA through 500 kVA are the minimum specified by standards. The impedance of

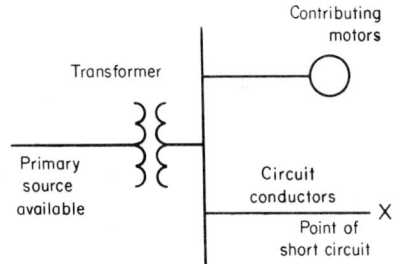

**Fig. 4-64**   Circuit investigated. [National Electrical Manufacturers Association]

transformers rated 750 kVA through 2000 kVA are 5.75 percent of the nominal, minus the 7.5 percent tolerance; $X/R$ ratios are not specified by NEMA standards and have been selected from data furnished by manufacturers.

5. Motor contribution: for system voltages of 120 and 120/208 V, it is usual to assume that the connected load consists of 50 percent lighting and 50 percent motor load. This corresponds to an equivalent symmetrical contribution of twice the full connected load. For system voltages of 240 to 600 V, it is usual to assume that the connected load consists of 100 percent motor load and, in the absence of exact information, that 25 percent of the motors are synchronous motors and 75 percent are induction motors. This corresponds to an equivalent symmetrical contribution of 4 times the full connected load. Short-circuit calculations have been based on a motor-impedance or $X/R$ ratio of 6.

6. The feeder conductors chosen are in the size normally used for standard frame sizes of low-voltage air circuit breakers. Calculations are based on three-phase conductors at 75°C in magnetic ducts. Tables 49C through 49J are calculations made for copper conductors having the characteristics shown in Table 49B. The characteristics of aluminum conductors, when compared on the basis of equivalent ampacity, are so similar to those of copper conductors that separate calculations for aluminum conductors are not included. In general, the impedances of aluminum conductors are slightly higher than those of equivalent copper conductors, and the available short-circuit–current amperes are slightly lower.

7. The available short-circuit current can be read directly when the transformer kilovolt-amperes, wire size, and distance from the transformer are known simply by referring to the corresponding values listed in Tables 49C through 49J.

PART B: SINGLE-PHASE, THREE-WIRE, 120/240-V SYSTEMS   The circuit diagram shown in Fig. 4-65 includes the factors which are usually considered in the calculation of short-circuit currents. The calculations were made for a system, as shown in the diagram, for the various transformer kilovolt-ampere ratings given in Table 49K in combination with common sizes and various lengths of cables. Cable impedances are shown in Table 49L.

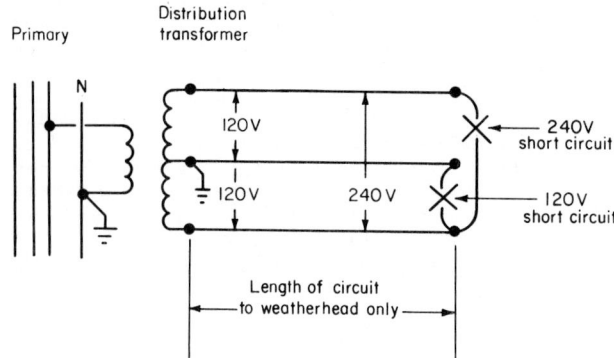

**Fig. 4-65**   Circuit investigated. [National Electrical Manufacturers Association]

With reference to Fig. 4-65, the following basic characteristics were used in the study:

1. Primary source available is assumed to be a line-to-ground transformer primary connection near a primary substation at which the planned maximum future available short-circuit duty corresponds to a 20,000-A symmetrical rms current on a 34,500-V system, with an assumed source-circuit $X/R$ ratio of 25.

2. Transformer kilovolt-ampere sizes are as shown in Table 49K.

3. Tables 49M through 49S are based on the transformer impedances shown in Table 49K, which in turn is based on typical distribution transformer data, from which the impedances are the smallest of those listed for transformers with at least 7200/12470Y-V primary voltage ratings and are based on the full 240-V secondary-winding rated kilovolt-amperes. $R$ and $X$ for a 120-V half-winding short circuit, based on full-rated kilovolt-amperes, are calculated as 1.5 and 1.2 times the tabulated full-winding $R$ and $X$, respectively.

### 49A.   Transformer Impedances
(Based on NEMA standards publication for secondary unit substations, Publication No. 210-1970)

| Three-phase kVA | Percent impedance | | | $X/R$ ratio | Short-circuit table |
|---|---|---|---|---|---|
| | $R$ | $X$ | $Z$ | | |
| 150 | 1.109 | 1.664 | 2.0 | 1.5 | A-3 |
| 225 | 1.109 | 1.664 | 2.0 | 1.5 | A-4 |
| 300 | 1.236 | 4.327 | 4.5 | 3.5 | A-5 |
| 500 | 1.131 | 4.356 | 4.5 | 3.85 | A-6 |
| 750 | 0.9685 | 5.230 | 5.32 | 5.4 | A-7 |
| 1000 | 0.9270 | 5.237 | 5.32 | 5.65 | A-8 |
| 1500 | 0.8604 | 5.249 | 5.32 | 6.10 | A-9 |
| 2000 | 0.7968 | 5.259 | 5.32 | 6.60 | A-10 |

NOTE   The data in Table 49A are based on liquid-filled self-cooled transformers. Dry-type transformers have, in general, higher reactances, and the results of this study should also be conservative for systems in which they are used.

4. The conductors chosen are in the size normally used for 120/240-V services. Calculations are based on triplexed conductors having the characteristics shown in Table 49L.

5. The available short-circuit current can be read directly when the transformer kilovolt-amperes, wire size, and distance from the transformer are known simply by referring to the corresponding values listed in Tables 49M through 49S.

These tables are intended primarily for estimating the short-circuit currents available at residential locations and therefore do not include the motor contribution, which is considered to be minimal and not a significant factor.

**49B.    Copper and Aluminum Feeder or Branch-Circuit Impedances**

| Probable design ampacity | Copper feeders | | | | Aluminum feeders | | | |
|---|---|---|---|---|---|---|---|---|
| | Size, AWG or MCM | Ampacity | $R$ | $X$ | Size, AWG or MCM | Ampacity | $R$ | $X$ |
| 15 | 14 | 15 | 3.065 | 0.0423 | 12 | 15 | 3.185 | 0.0396 |
| 20 | 12 | 20 | 1.932 | 0.0396 | 10 | 25 | 2.000 | 0.0371 |
| 30 | 10 | 30 | 1.214 | 0.0371 | 8 | 30 | 1.257 | 0.0389 |
| 40 | 8 | 40 | 0.7638 | 0.0389 | 6 | 40 | 0.8071 | 0.0401 |
| 60 | 6 | 65 | 0.4890 | 0.0401 | 4 | 65 | 0.5077 | 0.0378 |
| 75 | 4 | 85 | 0.3089 | 0.0378 | 2 | 90 | 0.3185 | 0.0359 |
| 100 | 2 | 115 | 0.1952 | 0.0359 | 1/0 | 120 | 0.2012 | 0.0350 |
| 125 | 1 | 130 | 0.1554 | 0.0360 | 2/0 | 135 | 0.1593 | 0.0342 |
| 150 | 1/0 | 150 | 0.1241 | 0.0350 | 3/0 | 155 | 0.1270 | 0.0334 |
| 175 | 2/0 | 175 | 0.09963 | 0.0342 | 4/0 | 180 | 0.1011 | 0.0326 |
| 200 | 3/0 | 200 | 0.07963 | 0.0334 | 250 | 205 | 0.08648 | 0.0328 |
| 400 | Two 3/0 | 400 | 0.03981 | 0.0167 | Two 250 | 410 | 0.04324 | 0.0164 |
| 225 | 4/0 | 230 | 0.06374 | 0.0326 | 300 | 230 | 0.07207 | 0.0321 |
| 250 | 250 | 255 | 0.05449 | 0.0328 | 350 | 250 | 0.06229 | 0.0315 |
| 500 | Two 250 | 510 | 0.02724 | 0.0164 | Two 350 | 500 | 0.03114 | 0.01575 |
| 800 | Three 300 | 855 | 0.01531 | 0.0107 | Three 400 | 810 | 0.01835 | 0.01033 |
| 300 | 350 | 310 | 0.03966 | 0.0315 | 500 | 310 | 0.04493 | 0.0301 |
| 600 | Two 350 | 620 | 0.01983 | 0.01575 | Two 500 | 620 | 0.02246 | 0.01505 |
| 900 | Three 350 | 930 | 0.01322 | 0.0105 | Three 500 | 930 | 0.01498 | 0.01003 |
| 1000 | Three 400 | 1005 | 0.01181 | 0.01033 | Three 600 | 1020 | 0.01272 | 0.00997 |
| 350 | 500 | 380 | 0.02911 | 0.0301 | 750 | 385 | 0.03165 | 0.0290 |
| 700 | Two 500 | 760 | 0.01455 | 0.01505 | Two 750 | 770 | 0.01582 | 0.0145 |
| 450 | 750 | 475 | 0.02078 | 0.0290 | 1250 | 485 | 0.02160 | 0.0273 |
| 1600 | Four 750 | 1900 | 0.00519 | 0.00725 | Four 1250 | 1940 | 0.00540 | 0.006825 |

NOTE  Impedances are for three-phase conductors in magnetic conduit. Ampacities given are for Type TW for wire sizes through No. 8 AWG for copper and No. 6 AWG for aluminum, and Type THW or RHW for larger sizes. Units of $R$ and $X$ are in $\Omega/1000$ ft at 60 Hz.

**49C.   Symmetrical Three-Phase Bolted Short-Circuit Currents in Amperes at Various Distances from a Liquid-Filled Transformer** (Rated 150 kVA; $Z$ = 2.00 percent)

| Transformer secondary voltage | Distance, ft | Wire size of copper conductors | | | | | | | | | | |
|---|---|---|---|---|---|---|---|---|---|---|---|---|
| | | 14 AWG | 12 AWG | 10 AWG | 8 AWG | 6 AWG | 4 AWG | 2 AWG | 1 AWG | 1/0 AWG | 2/0 AWG | 3/0 AWG |
| 208 | 0 | 21451 | 21451 | 21451 | 21451 | 21451 | 21451 | 21451 | 21451 | 21451 | 21451 | 21451 |
| | 5 | 6324 | 8834 | 11634 | 14271 | 16345 | 17932 | 19012 | 19390 | 19704 | 19951 | 21055 |
| | 10 | 3526 | 5241 | 7512 | 10166 | 12754 | 15105 | 16911 | 17583 | 18149 | 18601 | 18979 |
| | 25 | 1503 | 2325 | 3549 | 5260 | 7378 | 9894 | 12397 | 13499 | 14490 | 15332 | 16065 |
| | 50 | 767 | 1202 | 1874 | 2877 | 4247 | 6125 | 8375 | 9531 | 10664 | 11706 | 12677 |
| | 100 | 387 | 611 | 963 | 1503 | 2281 | 3435 | 5000 | 5906 | 6871 | 7840 | 8819 |
| | 200 | 194 | 308 | 487 | 768 | 1181 | 1821 | 2747 | 3325 | 3976 | 4675 | 5432 |
| | 500 | 78 | 123 | 196 | 311 | 482 | 754 | 1164 | 1432 | 1746 | 2101 | 2507 |
| | 1000 | 39 | 62 | 98 | 156 | 243 | 381 | 593 | 734 | 901 | 1093 | 1319 |
| | 2000 | 19 | 31 | 49 | 73 | 121 | 191 | 299 | 371 | 458 | 558 | 677 |
| | 5000 | 7 | 12 | 19 | 31 | 48 | 77 | 120 | 149 | 185 | 226 | 275 |
| 240 | 0 | 19257 | 19257 | 19257 | 19257 | 19257 | 19257 | 19257 | 19257 | 19257 | 19257 | 19257 |
| | 5 | 6883 | 9328 | 11845 | 14021 | 15615 | 16779 | 17549 | 17815 | 18035 | 18208 | 18350 |
| | 10 | 3953 | 5772 | 8058 | 10540 | 12769 | 14650 | 16013 | 16505 | 16914 | 17239 | 17508 |
| | 25 | 1716 | 2636 | 3975 | 5785 | 7908 | 10262 | 12428 | 13328 | 14116 | 14769 | 15328 |
| | 50 | 881 | 1375 | 2132 | 3241 | 4716 | 6652 | 8833 | 9892 | 10896 | 11789 | 12597 |
| | 100 | 446 | 702 | 1103 | 1715 | 2582 | 3842 | 5488 | 6406 | 7356 | 8281 | 9187 |
| | 200 | 224 | 354 | 561 | 881 | 1351 | 2068 | 3090 | 3712 | 4403 | 5128 | 5896 |
| | 500 | 90 | 142 | 226 | 358 | 555 | 864 | 1329 | 1630 | 1981 | 2372 | 2815 |
| | 1000 | 45 | 71 | 113 | 180 | 279 | 438 | 681 | 841 | 1031 | 1248 | 1501 |
| | 2000 | 22 | 35 | 56 | 90 | 140 | 221 | 344 | 427 | 526 | 640 | 775 |
| | 5000 | 9 | 14 | 22 | 36 | 56 | 88 | 138 | 172 | 213 | 260 | 316 |
| 480 | 0 | 9628 | 9628 | 9628 | 9628 | 9628 | 9628 | 9628 | 9628 | 9628 | 9628 | 9628 |
| | 5 | 7068 | 7922 | 8520 | 8906 | 9145 | 9306 | 9408 | 9442 | 9471 | 9493 | 9512 |
| | 10 | 5317 | 6521 | 7510 | 8218 | 8677 | 8991 | 9192 | 9260 | 9317 | 9361 | 9397 |
| | 25 | 2909 | 4054 | 5319 | 6497 | 7412 | 8105 | 8574 | 8737 | 8873 | 8979 | 9067 |
| | 50 | 1625 | 2413 | 3452 | 4656 | 5818 | 6863 | 7656 | 7949 | 8195 | 8391 | 8555 |
| | 100 | 858 | 1318 | 1987 | 2892 | 3954 | 5131 | 6214 | 6664 | 7058 | 7384 | 7664 |
| | 200 | 440 | 687 | 1066 | 1620 | 2358 | 3326 | 4416 | 4946 | 5448 | 5894 | 6298 |
| | 500 | 178 | 282 | 444 | 693 | 1052 | 1583 | 2300 | 2714 | 3154 | 3593 | 4035 |
| | 1000 | 89 | 142 | 225 | 354 | 545 | 839 | 1266 | 1531 | 1830 | 2150 | 2495 |
| | 2000 | 45 | 71 | 113 | 179 | 277 | 432 | 664 | 815 | 990 | 1186 | 1407 |
| | 5000 | 18 | 28 | 45 | 72 | 112 | 176 | 273 | 338 | 415 | 504 | 608 |
| 600 | 0 | 7703 | 7703 | 7703 | 7703 | 7703 | 7703 | 7703 | 7703 | 7703 | 7703 | 7703 |
| | 5 | 6326 | 6809 | 7130 | 7331 | 7454 | 7537 | 7589 | 7607 | 7622 | 7633 | 7643 |
| | 10 | 5196 | 5994 | 6579 | 6967 | 7210 | 7373 | 7477 | 7513 | 7542 | 7565 | 7584 |
| | 25 | 3216 | 4227 | 5192 | 5971 | 6514 | 6900 | 7151 | 7237 | 7308 | 7364 | 7410 |
| | 50 | 1909 | 2732 | 3709 | 4697 | 5523 | 6182 | 6642 | 6805 | 6940 | 7046 | 7135 |
| | 100 | 1040 | 1568 | 2294 | 3191 | 4126 | 5031 | 5762 | 6042 | 6279 | 6470 | 6631 |
| | 200 | 542 | 839 | 1281 | 1900 | 2665 | 3573 | 4474 | 4869 | 5224 | 5524 | 5785 |
| | 500 | 222 | 349 | 547 | 845 | 1262 | 1852 | 2592 | 2988 | 3388 | 3766 | 4126 |
| | 1000 | 112 | 176 | 279 | 437 | 667 | 1015 | 1499 | 1787 | 2100 | 2422 | 2755 |
| | 2000 | 56 | 89 | 141 | 222 | 343 | 531 | 808 | 983 | 1183 | 1401 | 1642 |
| | 5000 | 22 | 35 | 56 | 89 | 139 | 218 | 338 | 417 | 510 | 616 | 738 |

| Two 3/0 AWG | 4/0 AWG | 250 MCM | Two 250 MCM | Three 300 MCM | 350 MCM | Two 350 MCM | Three 350 MCM | Three 400 MCM | 500 MCM | Two 500 MCM | 750 MCM | Four 750 MCM |
|---|---|---|---|---|---|---|---|---|---|---|---|---|
| 21451 | 21451 | 21451 | 21451 | 21451 | 21451 | 21451 | 21451 | 21451 | 21451 | 21451 | 21451 | 21451 |
| 20787 | 20320 | 20407 | 20918 | 21125 | 20569 | 21002 | 21150 | 21166 | 20692 | 21065 | 20790 | 21282 |
| 20155 | 19287 | 19450 | 20407 | 20808 | 19754 | 20569 | 20856 | 20889 | 19984 | 20692 | 20168 | 21115 |
| 18433 | 16680 | 17015 | 19001 | 19908 | 17639 | 19368 | 20019 | 20096 | 18120 | 19648 | 18506 | 20631 |
| 16065 | 13533 | 14022 | 17015 | 18558 | 14945 | 17639 | 18757 | 18896 | 15675 | 18120 | 16272 | 19871 |
| 12677 | 9748 | 10313 | 14022 | 16319 | 11418 | 14945 | 16643 | 16871 | 12332 | 15675 | 13106 | 18506 |
| 8819 | 6203 | 6705 | 10313 | 13107 | 7732 | 11418 | 13555 | 13874 | 8636 | 12332 | 9434 | 16272 |
| 4551 | 2948 | 3254 | 5701 | 8186 | 3915 | 6654 | 8667 | 9021 | 4540 | 7508 | 5125 | 11944 |
| 2507 | 1570 | 1749 | 3254 | 5014 | 2145 | 3915 | 5397 | 5685 | 2534 | 4540 | 2910 | 8275 |
| 1319 | 911 | 908 | 1749 | 2820 | 1126 | 2145 | 3071 | 3264 | 1345 | 2534 | 1560 | 5125 |
| 544 | 331 | 371 | 732 | 1218 | 464 | 910 | 1338 | 1432 | 558 | 1089 | 652 | 2392 |
| 19257 | 19257 | 19257 | 19257 | 19257 | 19257 | 19257 | 19257 | 19257 | 19257 | 19257 | 19257 | 19257 |
| 18795 | 18465 | 18525 | 18885 | 19030 | 18639 | 18943 | 19047 | 19058 | 18724 | 18987 | 18792 | 19139 |
| 18350 | 17727 | 17841 | 18525 | 18808 | 18057 | 18639 | 18841 | 18864 | 18220 | 18724 | 18349 | 19022 |
| 17110 | 15790 | 16038 | 17516 | 18169 | 16500 | 17778 | 18247 | 18301 | 16854 | 17977 | 17137 | 18680 |
| 15328 | 13293 | 13681 | 16038 | 17188 | 14409 | 16500 | 17332 | 17432 | 14977 | 16854 | 15437 | 18135 |
| 12597 | 10022 | 10516 | 13681 | 15497 | 11468 | 14409 | 15742 | 15913 | 12239 | 14977 | 12880 | 17137 |
| 9187 | 6658 | 7142 | 10516 | 12916 | 8115 | 11468 | 13279 | 13536 | 8950 | 12239 | 9674 | 15437 |
| 4991 | 3290 | 3614 | 6147 | 8554 | 4305 | 7075 | 8995 | 9316 | 4947 | 7888 | 5537 | 11895 |
| 2815 | 1780 | 1977 | 3614 | 5445 | 2411 | 4305 | 5827 | 6113 | 2832 | 4947 | 3233 | 8603 |
| 1501 | 927 | 1036 | 1977 | 3145 | 1281 | 2411 | 3412 | 3616 | 1526 | 2832 | 1764 | 5537 |
| 624 | 380 | 427 | 837 | 1385 | 532 | 1038 | 1519 | 1623 | 640 | 1240 | 746 | 2676 |
| 9628 | 9628 | 9628 | 9628 | 9628 | 9628 | 9628 | 9628 | 9628 | 9628 | 9628 | 9628 | 9628 |
| 9570 | 9527 | 9535 | 9581 | 9600 | 9549 | 9589 | 9602 | 9603 | 9560 | 9594 | 9569 | 9613 |
| 9512 | 9427 | 9442 | 9535 | 9571 | 9471 | 9549 | 9575 | 9578 | 9493 | 9560 | 9511 | 9599 |
| 9341 | 9138 | 9175 | 9397 | 9487 | 9245 | 9433 | 9497 | 9504 | 9298 | 9460 | 9340 | 9554 |
| 9067 | 8688 | 8758 | 9175 | 9349 | 8889 | 9245 | 9369 | 9384 | 8988 | 9298 | 9067 | 9482 |
| 8555 | 7895 | 8019 | 8758 | 9084 | 8250 | 8889 | 9123 | 9150 | 8427 | 8988 | 8568 | 9340 |
| 7664 | 6646 | 6840 | 8019 | 8594 | 7204 | 8250 | 8666 | 8716 | 7488 | 8427 | 7718 | 9067 |
| 5771 | 4453 | 4706 | 6366 | 7382 | 5199 | 6772 | 7523 | 7623 | 5605 | 7092 | 5947 | 8339 |
| 4035 | 2846 | 3073 | 4706 | 5956 | 3537 | 5199 | 6154 | 6295 | 3944 | 5605 | 4301 | 7353 |
| 2495 | 1645 | 1807 | 3073 | 4277 | 2152 | 3537 | 4497 | 4658 | 2473 | 3944 | 2768 | 5947 |
| 1155 | 723 | 805 | 1497 | 2302 | 988 | 1799 | 2476 | 2607 | 1166 | 2084 | 1338 | 3778 |
| 7703 | 7703 | 7703 | 7703 | 7703 | 7703 | 7702 | 7703 | 7703 | 7703 | 7703 | 7703 | 7703 |
| 7673 | 7650 | 7654 | 7678 | 7688 | 7662 | 7682 | 7689 | 7690 | 7668 | 7685 | 7672 | 7695 |
| 7643 | 7599 | 7607 | 7654 | 7673 | 7622 | 7662 | 7675 | 7677 | 7633 | 7668 | 7642 | 7687 |
| 7554 | 7447 | 7467 | 7583 | 7630 | 7503 | 7602 | 7635 | 7639 | 7531 | 7616 | 7553 | 7665 |
| 7410 | 7206 | 7244 | 7467 | 7558 | 7314 | 7503 | 7569 | 7576 | 7367 | 7531 | 7409 | 7627 |
| 7135 | 6762 | 6831 | 7244 | 7419 | 6960 | 7314 | 7439 | 7454 | 7059 | 7367 | 7137 | 7553 |
| 6631 | 6003 | 6122 | 6831 | 7153 | 6343 | 6960 | 7192 | 7220 | 6513 | 7059 | 6650 | 7409 |
| 5430 | 4450 | 4638 | 5816 | 6454 | 4996 | 6072 | 6537 | 6595 | 5282 | 6270 | 5517 | 7009 |
| 4126 | 3079 | 3280 | 4638 | 5537 | 3677 | 4996 | 5667 | 5759 | 4012 | 5282 | 4297 | 6430 |
| 2755 | 1891 | 2056 | 3280 | 4295 | 2398 | 3677 | 4464 | 4586 | 2705 | 4012 | 2979 | 5517 |
| 1364 | 872 | 966 | 1731 | 2549 | 1170 | 2042 | 2714 | 2835 | 1366 | 2326 | 1551 | 3869 |

**49D.   Symmetrical Three-Phase Bolted Short-Circuit Currents in Amperes at Various Distances from a Liquid-Filled Transformer** (Rated 225 kVA; $Z$ = 2.00 percent)

| Transformer secondary voltage | Distance, ft | Wire size of copper conductors | | | | | | | | | | |
|---|---|---|---|---|---|---|---|---|---|---|---|---|
| | | 14 AWG | 12 AWG | 10 AWG | 8 AWG | 6 AWG | 4 AWG | 2 AWG | 1 AWG | 1/0 AWG | 2/0 AWG | 3/0 AWG |
| 208 | 0 | 32080 | 32080 | 32080 | 32080 | 32080 | 32080 | 32080 | 32080 | 32080 | 32080 | 32080 |
| | 5 | 6800 | 9890 | 13734 | 17861 | 21527 | 24599 | 26817 | 27617 | 28284 | 28812 | 29251 |
| | 10 | 3655 | 5553 | 8235 | 11683 | 15469 | 19355 | 22669 | 23982 | 25110 | 26029 | 26807 |
| | 25 | 1525 | 2378 | 3681 | 5584 | 8098 | 11371 | 15022 | 16781 | 18443 | 19916 | 21247 |
| | 50 | 773 | 1216 | 1908 | 2963 | 4454 | 6609 | 9405 | 10953 | 12550 | 14097 | 15608 |
| | 100 | 389 | 614 | 971 | 1525 | 2335 | 3571 | 5322 | 6385 | 7561 | 8791 | 10089 |
| | 200 | 195 | 309 | 490 | 773 | 1195 | 1856 | 2836 | 3463 | 4186 | 4981 | 5870 |
| | 500 | 78 | 124 | 197 | 312 | 485 | 760 | 1179 | 1456 | 1784 | 2158 | 2593 |
| | 1000 | 39 | 62 | 98 | 156 | 243 | 382 | 597 | 740 | 911 | 1108 | 1342 |
| | 2000 | 19 | 31 | 49 | 78 | 122 | 192 | 300 | 373 | 460 | 562 | 683 |
| | 5000 | 7 | 12 | 19 | 31 | 48 | 77 | 120 | 150 | 185 | 226 | 276 |
| 240 | 0 | 28802 | 28802 | 28802 | 28802 | 28802 | 28802 | 28802 | 28802 | 28802 | 28802 | 28802 |
| | 5 | 7547 | 10727 | 14434 | 18098 | 21097 | 23454 | 25081 | 25654 | 26128 | 26502 | 26812 |
| | 10 | 4138 | 6214 | 9040 | 12476 | 15982 | 19307 | 21945 | 22943 | 23787 | 24464 | 25031 |
| | 25 | 1747 | 2712 | 4166 | 6240 | 8880 | 12136 | 15518 | 17054 | 18458 | 19665 | 20728 |
| | 50 | 888 | 1395 | 2182 | 3366 | 5010 | 7316 | 10172 | 11682 | 13191 | 14604 | 15942 |
| | 100 | 448 | 707 | 1115 | 1747 | 2661 | 4036 | 5935 | 7057 | 8271 | 9508 | 10779 |
| | 200 | 225 | 356 | 564 | 889 | 1371 | 2120 | 3217 | 3908 | 4695 | 5549 | 6485 |
| | 500 | 90 | 143 | 227 | 359 | 558 | 873 | 1351 | 1665 | 2035 | 2453 | 2937 |
| | 1000 | 45 | 71 | 113 | 180 | 280 | 440 | 686 | 850 | 1045 | 1270 | 1534 |
| | 2000 | 22 | 35 | 56 | 90 | 140 | 221 | 346 | 429 | 530 | 646 | 784 |
| | 5000 | 9 | 14 | 22 | 36 | 56 | 88 | 139 | 172 | 213 | 261 | 318 |
| 480 | 0 | 14401 | 14401 | 14401 | 14401 | 14401 | 14401 | 14401 | 14401 | 14401 | 14401 | 14401 |
| | 5 | 9130 | 10746 | 11973 | 12805 | 13330 | 13685 | 13911 | 13987 | 14051 | 14101 | 14142 |
| | 10 | 6288 | 8184 | 9955 | 11355 | 12317 | 12996 | 13436 | 13587 | 13711 | 13809 | 13890 |
| | 25 | 3133 | 4549 | 6300 | 8161 | 9796 | 11151 | 12120 | 12467 | 12756 | 12984 | 13174 |
| | 50 | 1685 | 2559 | 3791 | 5366 | 7082 | 8826 | 10295 | 10872 | 11367 | 11767 | 12106 |
| | 100 | 873 | 1356 | 2083 | 3120 | 4440 | 6068 | 7759 | 8527 | 9229 | 9832 | 10364 |
| | 200 | 444 | 697 | 1091 | 1683 | 2505 | 3658 | 5086 | 5841 | 6595 | 7302 | 7971 |
| | 500 | 179 | 283 | 448 | 704 | 1077 | 1646 | 2451 | 2939 | 3477 | 4038 | 4628 |
| | 1000 | 90 | 142 | 226 | 357 | 551 | 856 | 1308 | 1596 | 1928 | 2293 | 2700 |
| | 2000 | 45 | 71 | 113 | 179 | 279 | 436 | 675 | 832 | 1017 | 1226 | 1468 |
| | 5000 | 18 | 28 | 45 | 72 | 112 | 176 | 275 | 341 | 420 | 511 | 619 |
| 600 | 0 | 11520 | 11520 | 11520 | 11520 | 11520 | 11520 | 11520 | 11520 | 11520 | 11520 | 11520 |
| | 5 | 8572 | 9563 | 10251 | 10695 | 10968 | 11152 | 11268 | 11308 | 11340 | 11366 | 11387 |
| | 10 | 6510 | 7933 | 9087 | 9904 | 10431 | 10791 | 11021 | 11099 | 11164 | 11214 | 11256 |
| | 25 | 3602 | 4994 | 6511 | 7903 | 8970 | 9773 | 10312 | 10499 | 10654 | 10776 | 10877 |
| | 50 | 2022 | 2994 | 4264 | 5718 | 7102 | 8329 | 9251 | 9590 | 9874 | 10099 | 10288 |
| | 100 | 1070 | 1641 | 2469 | 3580 | 4870 | 6283 | 7564 | 8091 | 8552 | 8931 | 9256 |
| | 200 | 550 | 858 | 1328 | 2016 | 2924 | 4107 | 5424 | 6057 | 6654 | 7182 | 7657 |
| | 500 | 223 | 352 | 554 | 865 | 1311 | 1969 | 2852 | 3358 | 3894 | 4426 | 4959 |
| | 1000 | 112 | 177 | 281 | 442 | 680 | 1047 | 1576 | 1904 | 2272 | 2665 | 3088 |
| | 2000 | 56 | 89 | 141 | 223 | 346 | 539 | 329 | 1016 | 1233 | 1476 | 1750 |
| | 5000 | 22 | 35 | 56 | 90 | 140 | 219 | 342 | 422 | 519 | 629 | 758 |

| Two 3/0 AWG | 4/0 AWG | 250 MCM | Two 250 MCM | Three 300 MCM | 350 MCM | Two 350 MCM | Three 350 MCM | Three 400 MCM | 500 MCM | Two 500 MCM | 750 MCM | Four 750 MCM |
|---|---|---|---|---|---|---|---|---|---|---|---|---|
| 32080 | 32080 | 32080 | 32080 | 32080 | 32080 | 32080 | 32080 | 32080 | 32080 | 32080 | 32080 | 32080 |
| 30614 | 29607 | 29795 | 30900 | 31356 | 30146 | 31084 | 31409 | 31447 | 30412 | 31224 | 30622 | 31702 |
| 29251 | 27445 | 27786 | 29795 | 30662 | 28421 | 30146 | 30765 | 30837 | 28906 | 30412 | 29292 | 31334 |
| 25711 | 22390 | 23028 | 26871 | 28740 | 24224 | 27628 | 28975 | 29139 | 25157 | 28206 | 25913 | 30279 |
| 21247 | 16995 | 17816 | 23028 | 25995 | 19393 | 24224 | 26396 | 26677 | 20670 | 25157 | 21734 | 28669 |
| 15608 | 11374 | 12188 | 17816 | 21776 | 13820 | 19393 | 22374 | 22798 | 15219 | 20670 | 16431 | 25913 |
| 10089 | 6804 | 7432 | 12188 | 16374 | 8751 | 13820 | 17102 | 17629 | 9953 | 15219 | 11042 | 21734 |
| 4851 | 3073 | 3412 | 6215 | 9326 | 4157 | 7392 | 9973 | 10455 | 4878 | 8483 | 5565 | 14645 |
| 2593 | 1604 | 1793 | 3412 | 5414 | 2215 | 4157 | 5870 | 6218 | 2636 | 4878 | 3046 | 9486 |
| 1342 | 820 | 919 | 1793 | 2940 | 1145 | 2215 | 3217 | 3432 | 1373 | 2626 | 1599 | 5565 |
| 548 | 332 | 373 | 739 | 1239 | 467 | 922 | 1365 | 1463 | 563 | 1108 | 659 | 2484 |
| | | | | | | | | | | | | |
| 28802 | 28802 | 28802 | 28802 | 28802 | 28802 | 28802 | 28802 | 28802 | 28802 | 28802 | 28802 | 28802 |
| 27780 | 27062 | 27193 | 27977 | 28297 | 27440 | 28105 | 28334 | 28360 | 27627 | 28202 | 27775 | 28538 |
| 26812 | 25493 | 25737 | 27193 | 27808 | 26194 | 27880 | 27931 | 26542 | 27627 | 26818 | 28279 |
| 24212 | 21623 | 22114 | 25061 | 26430 | 23030 | 25611 | 26597 | 26714 | 23736 | 26030 | 24305 | 27529 |
| 20728 | 17136 | 17824 | 22114 | 24395 | 19130 | 23030 | 24691 | 24899 | 20168 | 23736 | 21021 | 26364 |
| 15942 | 12003 | 12758 | 17824 | 21101 | 14245 | 19130 | 21570 | 21901 | 15489 | 20168 | 16548 | 24305 |
| 10779 | 7450 | 8085 | 12758 | 16552 | 9398 | 14245 | 17175 | 17621 | 10566 | 15489 | 11605 | 21021 |
| 5401 | 3465 | 3833 | 6827 | 9979 | 4635 | 8026 | 10603 | 11065 | 5400 | 9115 | 6120 | 14956 |
| 2937 | 1829 | 2040 | 3833 | 5979 | 2510 | 4635 | 6454 | 6813 | 2974 | 5400 | 3423 | 10097 |
| 1534 | 940 | 1053 | 2040 | 3313 | 1309 | 2510 | 3615 | 3848 | 1566 | 2974 | 1819 | 6120 |
| 630 | 382 | 429 | 848 | 1416 | 537 | 1056 | 1557 | 1668 | 647 | 1266 | 756 | 2805 |
| | | | | | | | | | | | | |
| 14401 | 14401 | 14401 | 14401 | 14401 | 14401 | 14401 | 14401 | 14401 | 14401 | 14401 | 14401 | 14401 |
| 14270 | 14175 | 14192 | 14296 | 14337 | 14225 | 14312 | 14342 | 14345 | 14249 | 14325 | 14269 | 14368 |
| 14142 | 13955 | 13988 | 14192 | 14274 | 14052 | 14225 | 14283 | 14289 | 14101 | 14249 | 14139 | 14334 |
| 13766 | 13327 | 13408 | 13889 | 14086 | 13559 | 13968 | 14109 | 14125 | 13673 | 14028 | 13764 | 14236 |
| 13174 | 12383 | 12530 | 13408 | 13784 | 12805 | 13559 | 13829 | 13860 | 13015 | 13673 | 13182 | 14075 |
| 12106 | 10811 | 11057 | 12530 | 13215 | 11515 | 12805 | 13298 | 13357 | 11868 | 13015 | 12152 | 13764 |
| 10364 | 8568 | 8912 | 11057 | 12197 | 9565 | 11515 | 12345 | 12449 | 10084 | 11868 | 10510 | 13182 |
| 7127 | 5210 | 5577 | 8113 | 9876 | 6311 | 8813 | 10139 | 10326 | 6937 | 9377 | 7478 | 11696 |
| 4628 | 3127 | 3413 | 5577 | 7463 | 4013 | 6311 | 7788 | 8022 | 4557 | 6937 | 5048 | 9845 |
| 2700 | 1732 | 1916 | 3413 | 4989 | 2317 | 4013 | 5301 | 5532 | 2700 | 4557 | 3060 | 7478 |
| 1195 | 739 | 826 | 1571 | 2489 | 1021 | 1913 | 2698 | 2856 | 1214 | 2243 | 1402 | 4343 |
| | | | | | | | | | | | | |
| 11520 | 11520 | 11520 | 11520 | 11520 | 11520 | 11520 | 11520 | 11520 | 11520 | 11520 | 11520 | 11520 |
| 11454 | 11404 | 11413 | 11467 | 11488 | 11430 | 11475 | 11490 | 11492 | 11443 | 11481 | 11453 | 11503 |
| 11387 | 11290 | 11307 | 11413 | 11455 | 11341 | 11430 | 11460 | 11463 | 11366 | 11443 | 11386 | 11486 |
| 11192 | 10958 | 11001 | 11255 | 11358 | 11081 | 11297 | 11370 | 11379 | 11141 | 11328 | 11189 | 11436 |
| 10877 | 10440 | 10521 | 11001 | 11200 | 10671 | 11081 | 11224 | 11240 | 10786 | 11141 | 10877 | 11352 |
| 10288 | 9523 | 9666 | 10521 | 10896 | 9933 | 10671 | 10941 | 10972 | 10138 | 10786 | 10301 | 11189 |
| 9256 | 8065 | 8291 | 9666 | 10331 | 8716 | 9933 | 10414 | 10472 | 9047 | 10138 | 9314 | 10877 |
| 7034 | 5460 | 5762 | 7734 | 8924 | 6348 | 8210 | 9089 | 9204 | 6830 | 8584 | 7235 | 10035 |
| 4959 | 3515 | 3791 | 5762 | 7249 | 4353 | 6348 | 7482 | 7648 | 4842 | 6830 | 5272 | 8889 |
| 3088 | 2042 | 2241 | 3791 | 5245 | 2665 | 4353 | 5509 | 5701 | 3057 | 4842 | 3417 | 7235 |
| 1437 | 902 | 1004 | 1860 | 2849 | 1229 | 2231 | 3061 | 3221 | 1450 | 2581 | 1662 | 4642 |

**49E. Symmetrical Three-Phase Bolted Short-Circuit Currents In Amperes at Various Distances from a Liquid-Filled Transformer** (Rated 300 kVA; $Z$ = 4.50 percent)

| Transformer secondary voltage | Distance, ft | Wire size of copper conductors | | | | | | | | | | |
|---|---|---|---|---|---|---|---|---|---|---|---|---|
| | | 14 AWG | 12 AWG | 10 AWG | 8 AWG | 6 AWG | 4 AWG | 2 AWG | 1 AWG | 1/0 AWG | 2/0 AWG | 3/0 AWG |
| 208 | 0 | 20041 | 20041 | 20041 | 20041 | 20041 | 20041 | 20041 | 20041 | 20041 | 20041 | 20041 |
| | 5 | 6694 | 9431 | 12376 | 14910 | 16656 | 17822 | 18517 | 18736 | 18915 | 19050 | 19160 |
| | 10 | 3658 | 5509 | 7985 | 10824 | 13418 | 15537 | 16953 | 17418 | 17794 | 18078 | 18306 |
| | 25 | 1529 | 2385 | 3680 | 5520 | 7806 | 10459 | 12923 | 13911 | 14755 | 15425 | 15973 |
| | 50 | 774 | 1219 | 1913 | 2963 | 4419 | 6430 | 8804 | 9975 | 11085 | 12058 | 12915 |
| | 100 | 389 | 615 | 973 | 1528 | 2335 | 3547 | 5200 | 6150 | 7152 | 8134 | 9096 |
| | 200 | 195 | 309 | 490 | 775 | 1196 | 1854 | 2816 | 3416 | 4093 | 4815 | 5588 |
| | 500 | 78 | 124 | 197 | 312 | 485 | 760 | 1177 | 1450 | 1772 | 2134 | 2549 |
| | 1000 | 39 | 62 | 98 | 156 | 243 | 383 | 597 | 739 | 908 | 1103 | 1331 |
| | 2000 | 19 | 31 | 49 | 78 | 122 | 192 | 300 | 373 | 460 | 560 | 680 |
| | 5000 | 7 | 12 | 19 | 31 | 48 | 77 | 120 | 150 | 185 | 226 | 275 |
| 240 | 0 | 18805 | 18805 | 18805 | 18805 | 18805 | 18805 | 18805 | 18805 | 18805 | 18805 | 18805 |
| | 5 | 7404 | 10132 | 12822 | 14930 | 16283 | 17154 | 17665 | 17824 | 17956 | 18056 | 18137 |
| | 10 | 4148 | 6157 | 8711 | 11420 | 13681 | 15391 | 16476 | 16823 | 17103 | 17315 | 17484 |
| | 25 | 1755 | 2725 | 4169 | 6160 | 8507 | 11041 | 13202 | 14017 | 14697 | 15225 | 15651 |
| | 50 | 891 | 1400 | 2191 | 3371 | 4970 | 7091 | 9444 | 10536 | 11536 | 12382 | 13106 |
| | 100 | 449 | 709 | 1119 | 1753 | 2664 | 4010 | 5787 | 6770 | 7778 | 8734 | 9641 |
| | 200 | 225 | 356 | 565 | 891 | 1374 | 2120 | 3195 | 3854 | 4586 | 5349 | 6148 |
| | 500 | 90 | 143 | 227 | 360 | 558 | 874 | 1350 | 1660 | 2022 | 2426 | 2884 |
| | 1000 | 45 | 71 | 113 | 180 | 280 | 441 | 686 | 849 | 1043 | 1264 | 1522 |
| | 2000 | 22 | 35 | 56 | 90 | 140 | 221 | 346 | 429 | 529 | 644 | 781 |
| | 5000 | 9 | 14 | 22 | 36 | 56 | 88 | 139 | 172 | 213 | 261 | 317 |
| 480 | 0 | 9402 | 9402 | 9402 | 9402 | 9402 | 9402 | 9402 | 9402 | 9402 | 9402 | 9402 |
| | 5 | 7549 | 8282 | 8728 | 8981 | 9121 | 9210 | 9265 | 9282 | 9297 | 9309 | 9318 |
| | 10 | 5783 | 7023 | 7930 | 8492 | 8809 | 9007 | 9123 | 9160 | 9191 | 9215 | 9235 |
| | 25 | 3106 | 4387 | 5775 | 6975 | 7803 | 8356 | 8684 | 8787 | 8871 | 8935 | 8986 |
| | 50 | 1692 | 2553 | 3709 | 5043 | 6268 | 7272 | 7943 | 8163 | 8341 | 8475 | 8582 |
| | 100 | 877 | 1362 | 2084 | 3080 | 4253 | 5520 | 6601 | 7008 | 7348 | 7612 | 7825 |
| | 200 | 445 | 700 | 1095 | 1685 | 2485 | 3545 | 4722 | 5268 | 5768 | 6191 | 6553 |
| | 500 | 179 | 284 | 449 | 706 | 1079 | 1641 | 2410 | 2853 | 3321 | 3781 | 4232 |
| | 1000 | 90 | 142 | 226 | 357 | 552 | 857 | 1302 | 1581 | 1895 | 2231 | 2591 |
| | 2000 | 45 | 71 | 113 | 180 | 279 | 437 | 675 | 830 | 1011 | 1213 | 1442 |
| | 5000 | 18 | 28 | 45 | 72 | 112 | 176 | 275 | 341 | 419 | 509 | 615 |
| 600 | 0 | 7522 | 7522 | 7522 | 7522 | 7522 | 7522 | 7522 | 7522 | 7522 | 7522 | 7522 |
| | 5 | 6621 | 6984 | 7195 | 7314 | 7381 | 7425 | 7452 | 7460 | 7468 | 7474 | 7479 |
| | 10 | 5602 | 6339 | 6809 | 7079 | 7229 | 7323 | 7380 | 7398 | 7414 | 7426 | 7436 |
| | 25 | 3481 | 4594 | 5588 | 6299 | 6729 | 7001 | 7160 | 7209 | 7250 | 7282 | 7308 |
| | 50 | 2019 | 2936 | 4024 | 5075 | 5872 | 6436 | 6780 | 6889 | 6977 | 7044 | 7097 |
| | 100 | 1075 | 1644 | 2445 | 3443 | 4456 | 5373 | 6034 | 6259 | 6441 | 6579 | 6690 |
| | 200 | 552 | 862 | 1333 | 2005 | 2849 | 3842 | 4777 | 5155 | 5480 | 5738 | 5949 |
| | 500 | 224 | 353 | 556 | 868 | 1310 | 1946 | 2749 | 3172 | 3591 | 3975 | 4325 |
| | 1000 | 112 | 178 | 281 | 444 | 682 | 1046 | 1559 | 1866 | 2199 | 2538 | 2883 |
| | 2000 | 56 | 89 | 141 | 224 | 347 | 540 | 827 | 1009 | 1219 | 1447 | 1697 |
| | 5000 | 22 | 35 | 56 | 90 | 140 | 220 | 341 | 422 | 517 | 625 | 750 |

| Two 3/0 AWG | 4/0 AWG | 250 MCM | Two 250 MCM | Three 300 MCM | 350 MCM | Two 350 MCM | Three 350 MCM | Three 400 MCM | 500 MCM | Two 500 MCM | 750 MCM | Four 750 MCM |
|---|---|---|---|---|---|---|---|---|---|---|---|---|
| 20041 | 20041 | 20041 | 20041 | 20041 | 20041 | 20041 | 20041 | 20041 | 20041 | 20041 | 20041 | 20041 |
| 19598 | 19247 | 19288 | 19660 | 19802 | 19374 | 19703 | 19815 | 19824 | 19442 | 19738 | 19495 | 19902 |
| 19160 | 18485 | 18568 | 19288 | 19568 | 18741 | 19374 | 19593 | 19611 | 18874 | 19442 | 18977 | 19765 |
| 17892 | 16406 | 16617 | 18221 | 18887 | 17025 | 18436 | 18949 | 18992 | 17332 | 18601 | 17568 | 19363 |
| 15973 | 13627 | 13997 | 16617 | 17825 | 14692 | 17025 | 17946 | 18030 | 15215 | 17332 | 15616 | 18728 |
| 12915 | 9971 | 10471 | 13997 | 15960 | 11422 | 14692 | 16181 | 16335 | 12165 | 15215 | 12748 | 17568 |
| 9096 | 6358 | 6842 | 10471 | 13080 | 7809 | 11422 | 13430 | 13675 | 8620 | 12165 | 9292 | 15616 |
| 4669 | 2996 | 3301 | 5814 | 8302 | 3954 | 6727 | 8740 | 9055 | 4557 | 7512 | 5099 | 11668 |
| 2549 | 1585 | 1764 | 3301 | 5088 | 2160 | 3954 | 5457 | 5732 | 2543 | 4557 | 2904 | 8176 |
| 1331 | 815 | 912 | 1764 | 2849 | 1130 | 2160 | 3098 | 3287 | 1348 | 2543 | 1559 | 5099 |
| 546 | 331 | 372 | 734 | 1224 | 465 | 913 | 1344 | 1437 | 559 | 1091 | 652 | 2389 |
| 18805 | 18805 | 18805 | 18805 | 18805 | 18805 | 18805 | 18805 | 18805 | 18805 | 18805 | 18805 | 18805 |
| 18470 | 18202 | 18232 | 18516 | 18624 | 18297 | 18548 | 18633 | 18640 | 18348 | 18574 | 18388 | 18699 |
| 18137 | 17617 | 17679 | 18232 | 18445 | 17809 | 18297 | 18464 | 18477 | 17910 | 18348 | 17988 | 18594 |
| 17164 | 15986 | 16145 | 17410 | 17923 | 16460 | 17572 | 17969 | 18002 | 16698 | 17697 | 16880 | 18287 |
| 15651 | 13694 | 13992 | 16145 | 17097 | 14554 | 16460 | 17189 | 17253 | 14975 | 16698 | 15296 | 17794 |
| 13106 | 10442 | 10884 | 13992 | 15608 | 11718 | 14554 | 15780 | 15901 | 12358 | 14975 | 12855 | 16880 |
| 9641 | 6924 | 7398 | 10884 | 13193 | 8331 | 11718 | 13483 | 13686 | 9093 | 12358 | 9712 | 15296 |
| 5183 | 3372 | 3700 | 6354 | 8822 | 4392 | 7259 | 9230 | 9522 | 5019 | 8018 | 5572 | 11897 |
| 2884 | 1806 | 2005 | 3700 | 5594 | 2442 | 4392 | 5969 | 6245 | 2860 | 5019 | 3249 | 8647 |
| 1522 | 934 | 1045 | 2005 | 3203 | 1291 | 2442 | 3471 | 3673 | 1535 | 2860 | 1769 | 5572 |
| 628 | 381 | 428 | 842 | 1397 | 534 | 1044 | 1532 | 1636 | 641 | 1246 | 747 | 2687 |
| 9402 | 9402 | 9402 | 9402 | 9402 | 9402 | 9402 | 9402 | 9402 | 9402 | 9402 | 9402 | 9402 |
| 9360 | 9326 | 9330 | 9366 | 9380 | 9338 | 9370 | 9381 | 9382 | 9344 | 9373 | 9349 | 9389 |
| 9318 | 9251 | 9258 | 9330 | 9357 | 9274 | 9338 | 9359 | 9361 | 9287 | 9344 | 9297 | 9376 |
| 9193 | 9027 | 9046 | 9222 | 9289 | 9086 | 9242 | 9295 | 9299 | 9118 | 9258 | 9143 | 9336 |
| 8986 | 8666 | 8705 | 9046 | 9178 | 8786 | 9086 | 9190 | 9198 | 8848 | 9118 | 8897 | 9271 |
| 8582 | 7993 | 8072 | 8705 | 8961 | 8230 | 8786 | 8984 | 9001 | 8349 | 8848 | 8440 | 9143 |
| 7825 | 6847 | 6996 | 8072 | 8548 | 7277 | 8230 | 8594 | 8626 | 7487 | 8349 | 7648 | 8897 |
| 6030 | 4644 | 4878 | 6541 | 7469 | 5326 | 6869 | 7573 | 7645 | 5675 | 7115 | 5948 | 8228 |
| 4232 | 2951 | 3177 | 4878 | 6107 | 3629 | 5326 | 6272 | 6388 | 4009 | 5675 | 4323 | 7303 |
| 2591 | 1686 | 1850 | 3177 | 4411 | 2196 | 3629 | 4615 | 4761 | 2509 | 4009 | 2786 | 5948 |
| 1179 | 732 | 815 | 1528 | 2359 | 998 | 1831 | 2531 | 2659 | 1176 | 2112 | 1343 | 3801 |
| 7522 | 7522 | 7522 | 7522 | 7522 | 7522 | 7522 | 7522 | 7522 | 7522 | 7522 | 7522 | 7522 |
| 7500 | 7483 | 7485 | 7503 | 7510 | 7489 | 7505 | 7511 | 7511 | 7492 | 7507 | 7495 | 7515 |
| 7479 | 7444 | 7447 | 7485 | 7498 | 7456 | 7489 | 7500 | 7501 | 7462 | 7492 | 7468 | 7508 |
| 7414 | 7328 | 7338 | 7429 | 7464 | 7358 | 7439 | 7467 | 7469 | 7375 | 7448 | 7388 | 7488 |
| 7308 | 7139 | 7159 | 7338 | 7406 | 7200 | 7358 | 7412 | 7417 | 7233 | 7375 | 7258 | 7454 |
| 7097 | 6776 | 6817 | 7159 | 7293 | 6899 | 7200 | 7305 | 7314 | 6963 | 7233 | 7011 | 7388 |
| 6690 | 6116 | 6197 | 6817 | 7074 | 6354 | 6899 | 7098 | 7115 | 6472 | 6963 | 6562 | 7258 |
| 5618 | 4626 | 4787 | 5918 | 6471 | 5089 | 6108 | 6528 | 6568 | 5318 | 6249 | 5494 | 6894 |
| 4325 | 3208 | 3401 | 4787 | 5631 | 3773 | 5089 | 5731 | 5801 | 4070 | 5318 | 4307 | 6358 |
| 2883 | 1953 | 2119 | 3401 | 4419 | 2457 | 3773 | 4566 | 4669 | 2749 | 4070 | 2997 | 5494 |
| 1404 | 887 | 982 | 1778 | 2624 | 1188 | 2088 | 2784 | 2899 | 1382 | 2362 | 1560 | 3885 |

**49F.  Symmetrical Three-Phase Bolted Short-Circuit Currents in Amperes at Various Distances from a Liquid-Filled Transformer** (Rated 500 kVA; $Z$ = 4.50 percent)

| Transformer secondary voltage | Distance, ft | 14 AWG | 12 AWG | 10 AWG | 8 AWG | 6 AWG | 4 AWG | 2 AWG | 1 AWG | 1/0 AWG | 2/0 AWG | 3/0 AWG |
|---|---|---|---|---|---|---|---|---|---|---|---|---|
| | | | | | Wire size of copper conductors | | | | | | | |
| 208 | 0 | 33273 | 33273 | 33273 | 33273 | 33273 | 33273 | 33273 | 33273 | 33273 | 33273 | 33273 |
| | 5 | 7226 | 10756 | 15274 | 20103 | 24153 | 27215 | 29148 | 29762 | 30258 | 30629 | 30926 |
| | 10 | 3780 | 5839 | 8844 | 12831 | 17252 | 21654 | 25100 | 26326 | 27329 | 28093 | 28704 |
| | 25 | 1546 | 2432 | 3807 | 5868 | 8669 | 12408 | 16580 | 18525 | 20308 | 21815 | 23105 |
| | 50 | 778 | 1230 | 1942 | 3044 | 4633 | 6987 | 10108 | 11842 | 13624 | 15319 | 16929 |
| | 100 | 390 | 618 | 980 | 1547 | 2385 | 3685 | 5561 | 6714 | 7997 | 9339 | 10745 |
| | 200 | 195 | 309 | 492 | 779 | 1208 | 1887 | 2906 | 3563 | 4326 | 5169 | 6113 |
| | 500 | 78 | 124 | 197 | 313 | 487 | 765 | 1191 | 1474 | 1810 | 2194 | 2643 |
| | 1000 | 39 | 62 | 98 | 156 | 244 | 384 | 600 | 745 | 918 | 1118 | 1356 |
| | 2000 | 19 | 31 | 49 | 78 | 122 | 192 | 301 | 374 | 462 | 564 | 686 |
| | 5000 | 7 | 12 | 19 | 31 | 48 | 77 | 120 | 150 | 185 | 227 | 276 |
| 240 | 0 | 31235 | 31235 | 31235 | 31235 | 31235 | 31235 | 31235 | 31235 | 31235 | 31235 | 31235 |
| | 5 | 8157 | 11926 | 16459 | 20894 | 24287 | 26683 | 28134 | 28586 | 28953 | 29228 | 29448 |
| | 10 | 4323 | 6628 | 9901 | 14031 | 18280 | 22161 | 24966 | 25917 | 26686 | 27266 | 27728 |
| | 25 | 1779 | 2791 | 4353 | 6654 | 9692 | 13552 | 17560 | 19312 | 20862 | 22130 | 23187 |
| | 50 | 897 | 1416 | 2232 | 3487 | 5275 | 7864 | 11162 | 12916 | 14660 | 16262 | 17732 |
| | 100 | 450 | 712 | 1129 | 1779 | 2736 | 4205 | 6289 | 7539 | 8905 | 10299 | 11722 |
| | 200 | 225 | 357 | 567 | 898 | 1391 | 2167 | 3322 | 4060 | 4908 | 5833 | 6852 |
| | 500 | 90 | 143 | 227 | 361 | 561 | 881 | 1370 | 1692 | 2075 | 2510 | 3015 |
| | 1000 | 45 | 71 | 113 | 180 | 281 | 442 | 691 | 857 | 1056 | 1285 | 1555 |
| | 2000 | 22 | 35 | 57 | 90 | 141 | 222 | 347 | 431 | 532 | 650 | 790 |
| | 5000 | 9 | 14 | 22 | 36 | 56 | 88 | 139 | 173 | 214 | 261 | 318 |
| 480 | 0 | 15617 | 15617 | 15617 | 15617 | 15617 | 15617 | 15617 | 15617 | 15617 | 15617 | 15617 |
| | 5 | 10584 | 12443 | 13702 | 14442 | 14849 | 15101 | 15250 | 15296 | 15336 | 15367 | 15392 |
| | 10 | 7094 | 9428 | 11537 | 13052 | 13966 | 14536 | 14865 | 14966 | 15051 | 15115 | 15168 |
| | 25 | 3344 | 4987 | 7100 | 9374 | 11291 | 12746 | 13664 | 13955 | 14189 | 14365 | 14505 |
| | 50 | 1747 | 2700 | 4096 | 5957 | 8031 | 10108 | 11741 | 12323 | 12799 | 13162 | 13451 |
| | 100 | 889 | 1395 | 2176 | 3327 | 4846 | 6776 | 8780 | 9656 | 10431 | 11065 | 11593 |
| | 200 | 448 | 708 | 1116 | 1743 | 2637 | 3932 | 5581 | 6458 | 7330 | 8131 | 8866 |
| | 500 | 180 | 285 | 452 | 714 | 1102 | 1703 | 2573 | 3108 | 3705 | 4330 | 4986 |
| | 1000 | 90 | 143 | 227 | 359 | 558 | 871 | 1342 | 1647 | 2000 | 2391 | 2829 |
| | 2000 | 45 | 71 | 113 | 180 | 280 | 440 | 685 | 846 | 1037 | 1255 | 1507 |
| | 5000 | 18 | 28 | 45 | 72 | 112 | 177 | 277 | 343 | 424 | 516 | 626 |
| 600 | 0 | 12494 | 12494 | 12494 | 12494 | 12494 | 12494 | 12494 | 12494 | 12494 | 12494 | 12494 |
| | 5 | 9935 | 10958 | 11578 | 11925 | 12115 | 12235 | 12307 | 12330 | 12350 | 12366 | 12378 |
| | 10 | 7501 | 9204 | 10467 | 11249 | 11687 | 11957 | 12115 | 12164 | 12205 | 12237 | 12263 |
| | 25 | 3947 | 5628 | 7491 | 9139 | 10290 | 11058 | 11511 | 11651 | 11765 | 11851 | 11920 |
| | 50 | 2132 | 3233 | 4735 | 6505 | 8167 | 9553 | 10484 | 10788 | 11032 | 11216 | 11362 |
| | 100 | 1101 | 1713 | 2632 | 3915 | 5455 | 7152 | 8628 | 9190 | 9659 | 10023 | 10316 |
| | 200 | 558 | 878 | 1376 | 2124 | 3147 | 4525 | 6083 | 6818 | 7495 | 8071 | 8565 |
| | 500 | 224 | 355 | 563 | 885 | 1356 | 2069 | 3054 | 3627 | 4238 | 4842 | 5440 |
| | 1000 | 112 | 178 | 283 | 448 | 692 | 1075 | 1639 | 1993 | 2395 | 2827 | 3292 |
| | 2000 | 56 | 89 | 142 | 225 | 349 | 547 | 846 | 1042 | 1270 | 1527 | 1818 |
| | 5000 | 22 | 35 | 56 | 90 | 140 | 221 | 345 | 427 | 525 | 638 | 771 |

| Two 3/0 AWG | 4/0 AWG | 250 MCM | Two 250 MCM | Three 300 MCM | 350 MCM | Two 350 MCM | Three 350 MCM | Three 400 MCM | 500 MCM | Two 500 MCM | 750 MCM | Four 750 MCM |
|---|---|---|---|---|---|---|---|---|---|---|---|---|
| 33273 | 33273 | 33273 | 33273 | 33273 | 33273 | 33273 | 33273 | 33273 | 33273 | 33273 | 33273 | 33273 |
| 32089 | 31160 | 31266 | 32252 | 32631 | 31493 | 32365 | 32664 | 32687 | 31670 | 32455 | 31806 | 32895 |
| 30926 | 29181 | 29404 | 31266 | 32006 | 29853 | 31493 | 32071 | 32117 | 30194 | 31670 | 30455 | 32524 |
| 27656 | 24152 | 24679 | 28528 | 30224 | 25674 | 29082 | 30385 | 30498 | 26417 | 29500 | 26983 | 31458 |
| 23105 | 18350 | 19135 | 24679 | 27562 | 20614 | 25674 | 27867 | 28079 | 21747 | 26417 | 22622 | 29820 |
| 16929 | 12115 | 12951 | 19135 | 23247 | 14598 | 20614 | 23762 | 24121 | 15944 | 21747 | 17035 | 26983 |
| 10745 | 7101 | 7754 | 12951 | 17457 | 9113 | 14598 | 18154 | 18648 | 10316 | 15944 | 11356 | 22622 |
| 5019 | 3138 | 3485 | 6446 | 9769 | 4247 | 7659 | 10428 | 10914 | 4976 | 8757 | 5655 | 15148 |
| 2643 | 1622 | 1813 | 3485 | 5577 | 2242 | 4247 | 6044 | 6398 | 2666 | 4976 | 3074 | 9726 |
| 1356 | 824 | 925 | 1813 | 2991 | 1152 | 2242 | 3272 | 3490 | 1382 | 2666 | 1607 | 5655 |
| 550 | 333 | 374 | 743 | 1249 | 468 | 927 | 1375 | 1474 | 565 | 1113 | 660 | 2503 |
| 31235 | 31235 | 31235 | 31235 | 31235 | 31235 | 31235 | 31235 | 31235 | 31235 | 31235 | 31235 | 31235 |
| 30336 | 29622 | 29701 | 30457 | 30746 | 29872 | 30542 | 30771 | 30788 | 30005 | 30610 | 30109 | 30946 |
| 29448 | 28087 | 28253 | 29701 | 30268 | 28594 | 29872 | 30316 | 30351 | 28855 | 30005 | 29056 | 30662 |
| 26902 | 24031 | 24445 | 27562 | 28889 | 25235 | 27986 | 29010 | 29097 | 25825 | 28308 | 26273 | 29839 |
| 23187 | 18993 | 19667 | 24445 | 26782 | 20930 | 25235 | 27017 | 27182 | 21885 | 25825 | 22615 | 28554 |
| 17732 | 13070 | 13868 | 19667 | 23223 | 15415 | 20930 | 23641 | 23932 | 16649 | 21885 | 17631 | 26273 |
| 11722 | 7900 | 8579 | 13868 | 18116 | 9968 | 15415 | 18731 | 19162 | 11168 | 16649 | 12185 | 22615 |
| 5659 | 3567 | 3951 | 7186 | 10656 | 4785 | 8453 | 11312 | 11789 | 5572 | 9575 | 6294 | 15867 |
| 3015 | 1858 | 2074 | 3951 | 6241 | 2555 | 4785 | 6737 | 7111 | 3028 | 5572 | 3479 | 10546 |
| 1555 | 948 | 1062 | 2074 | 3396 | 1321 | 2555 | 3708 | 3947 | 1581 | 3028 | 1835 | 6294 |
| 634 | 383 | 431 | 854 | 1431 | 539 | 1064 | 1575 | 1687 | 649 | 1276 | 759 | 2843 |
| 15617 | 15617 | 15617 | 15617 | 15617 | 15617 | 15617 | 15617 | 15617 | 15617 | 15617 | 15617 | 15617 |
| 15505 | 15413 | 15422 | 15519 | 15556 | 15443 | 15530 | 15559 | 15561 | 15460 | 15538 | 15473 | 15581 |
| 15392 | 15210 | 15228 | 15422 | 15495 | 15271 | 15443 | 15501 | 15505 | 15305 | 15460 | 15531 | 15545 |
| 15056 | 14615 | 14665 | 15133 | 15313 | 14772 | 15186 | 15328 | 15339 | 14855 | 15228 | 14919 | 15437 |
| 14505 | 13676 | 13781 | 14665 | 15016 | 13993 | 14772 | 15046 | 15068 | 14154 | 14855 | 14277 | 15261 |
| 13451 | 12015 | 12222 | 13781 | 14444 | 12617 | 13993 | 14505 | 14548 | 12912 | 14154 | 13136 | 14919 |
| 11593 | 9496 | 9833 | 12222 | 13391 | 10465 | 12617 | 13508 | 13591 | 10942 | 12912 | 11307 | 14277 |
| 7883 | 5627 | 6018 | 8922 | 10863 | 6790 | 9620 | 11106 | 11276 | 7421 | 10153 | 7933 | 12629 |
| 4986 | 3289 | 3593 | 6018 | 8133 | 4226 | 6790 | 8461 | 8693 | 4787 | 7421 | 5273 | 10565 |
| 2829 | 1783 | 1975 | 3593 | 5328 | 2392 | 4226 | 5656 | 5894 | 2786 | 4787 | 3147 | 7933 |
| 1221 | 749 | 837 | 1611 | 2581 | 1036 | 1964 | 2798 | 2963 | 1232 | 2303 | 1421 | 4512 |
| 12494 | 12494 | 12494 | 12494 | 12494 | 12494 | 12494 | 12494 | 12494 | 12494 | 12494 | 12494 | 12494 |
| 12436 | 12389 | 12393 | 12443 | 12462 | 12404 | 12449 | 12464 | 12465 | 12413 | 12453 | 12420 | 12475 |
| 12378 | 12284 | 12294 | 12393 | 12431 | 12315 | 12404 | 12434 | 12436 | 12333 | 12413 | 12346 | 12457 |
| 12206 | 11975 | 12000 | 12244 | 12337 | 12055 | 12271 | 12345 | 12350 | 12098 | 12293 | 12131 | 12401 |
| 11920 | 11476 | 11528 | 12000 | 12183 | 11638 | 12055 | 12198 | 12210 | 11722 | 12098 | 11787 | 12310 |
| 11362 | 10545 | 10653 | 11528 | 11882 | 10867 | 11638 | 11913 | 11935 | 11028 | 11722 | 11151 | 12131 |
| 10316 | 8968 | 9171 | 10653 | 11310 | 9554 | 10867 | 11372 | 11415 | 9839 | 11028 | 10056 | 11787 |
| 7854 | 5988 | 6303 | 8549 | 9819 | 6902 | 8994 | 9961 | 10059 | 7371 | 9328 | 7737 | 10857 |
| 5440 | 3760 | 4055 | 6303 | 7958 | 4649 | 6902 | 8181 | 8337 | 5149 | 7371 | 5563 | 9582 |
| 3292 | 2129 | 2340 | 4055 | 5680 | 2785 | 4649 | 5952 | 6146 | 3190 | 5149 | 3548 | 7737 |
| 1484 | 919 | 1024 | 1928 | 2995 | 1256 | 2317 | 3217 | 3383 | 1482 | 2677 | 1695 | 4874 |

**49G. Symmetrical Three-Phase Bolted Short-Circuit Currents in Amperes at Various Distances from a Liquid-Filled Transformer** (Rated 750 kVA; $Z$ = 5.32 percent)

| Transformer secondary voltage | Distance, ft | Wire size of copper conductors | | | | | | | | | | |
|---|---|---|---|---|---|---|---|---|---|---|---|---|
| | | 14 AWG | 12 AWG | 10 AWG | 8 AWG | 6 AWG | 4 AWG | 2 AWG | 1 AWG | 1/0 AWG | 2/0 AWG | 3/0 AWG |
| 208 | 0 | 42763 | 42763 | 42763 | 42763 | 42763 | 42763 | 42763 | 42763 | 42763 | 42763 | 42763 |
| | 5 | 7457 | 11346 | 16672 | 22935 | 28735 | 33434 | 36471 | 37425 | 38183 | 38742 | 39179 |
| | 10 | 3834 | 5982 | 9220 | 13758 | 19209 | 25172 | 30243 | 32119 | 33661 | 34834 | 35762 |
| | 25 | 1555 | 2453 | 3864 | 6019 | 9048 | 13311 | 18451 | 21013 | 23450 | 25577 | 27436 |
| | 50 | 780 | 1235 | 1956 | 3081 | 4727 | 7229 | 10695 | 12712 | 14855 | 16970 | 19047 |
| | 100 | 391 | 619 | 983 | 1556 | 2408 | 3745 | 5717 | 6958 | 8368 | 9880 | 11507 |
| | 200 | 195 | 310 | 493 | 781 | 1214 | 1902 | 2945 | 3626 | 4425 | 5320 | 6337 |
| | 500 | 78 | 124 | 197 | 313 | 488 | 767 | 1197 | 1484 | 1826 | 2219 | 2682 |
| | 1000 | 39 | 62 | 98 | 156 | 244 | 384 | 601 | 747 | 922 | 1125 | 1365 |
| | 2000 | 19 | 31 | 49 | 78 | 122 | 192 | 301 | 375 | 463 | 566 | 689 |
| | 5000 | 7 | 12 | 19 | 31 | 48 | 77 | 120 | 150 | 185 | 227 | 277 |
| 240 | 0 | 40669 | 40669 | 40669 | 40669 | 40669 | 40669 | 40669 | 40669 | 40669 | 40669 | 40669 |
| | 5 | 8488 | 12757 | 18342 | 24437 | 29592 | 33446 | 35805 | 36527 | 37103 | 37526 | 37859 |
| | 10 | 4401 | 6834 | 10436 | 15307 | 20823 | 26394 | 30743 | 32263 | 33491 | 34410 | 35132 |
| | 25 | 1791 | 2822 | 4434 | 6870 | 10227 | 14787 | 19970 | 22406 | 24643 | 26526 | 28124 |
| | 50 | 900 | 1423 | 2252 | 3538 | 5408 | 8206 | 11973 | 14092 | 16284 | 18379 | 20372 |
| | 100 | 451 | 714 | 1133 | 1791 | 2768 | 4290 | 6508 | 7882 | 9420 | 11039 | 12745 |
| | 200 | 225 | 357 | 568 | 901 | 1398 | 2187 | 3377 | 4149 | 5047 | 6045 | 7165 |
| | 500 | 90 | 143 | 227 | 361 | 562 | 884 | 1378 | 1707 | 2098 | 2546 | 3070 |
| | 1000 | 45 | 71 | 114 | 180 | 281 | 443 | 693 | 861 | 1062 | 1294 | 1570 |
| | 2000 | 22 | 35 | 57 | 90 | 141 | 222 | 348 | 432 | 534 | 652 | 793 |
| | 5000 | 9 | 14 | 22 | 36 | 56 | 88 | 139 | 173 | 214 | 262 | 319 |
| 480 | 0 | 20334 | 20334 | 20334 | 20334 | 20334 | 20334 | 20334 | 20334 | 20334 | 20334 | 20334 |
| | 5 | 12388 | 15230 | 17289 | 18515 | 19172 | 19562 | 19782 | 19848 | 19905 | 19948 | 19983 |
| | 10 | 7731 | 10761 | 13823 | 16229 | 17734 | 18663 | 19181 | 19334 | 19461 | 19555 | 19631 |
| | 25 | 3448 | 5256 | 7748 | 10708 | 13487 | 15766 | 17252 | 17721 | 18093 | 18367 | 18582 |
| | 50 | 1771 | 2765 | 4267 | 6383 | 8946 | 11782 | 14225 | 15137 | 15888 | 16461 | 16915 |
| | 100 | 895 | 1411 | 2217 | 3435 | 5113 | 7393 | 9985 | 11203 | 12321 | 13263 | 14062 |
| | 200 | 450 | 711 | 1126 | 1769 | 2704 | 4103 | 5986 | 7046 | 8142 | 9189 | 10186 |
| | 500 | 180 | 285 | 454 | 718 | 1112 | 1730 | 2643 | 3220 | 3876 | 4581 | 5342 |
| | 1000 | 90 | 143 | 227 | 360 | 560 | 878 | 1360 | 1675 | 2045 | 2461 | 2933 |
| | 2000 | 45 | 71 | 113 | 180 | 281 | 442 | 689 | 853 | 1049 | 1273 | 1535 |
| | 5000 | 18 | 28 | 45 | 72 | 112 | 177 | 277 | 345 | 425 | 519 | 630 |
| 600 | 0 | 16267 | 16267 | 16267 | 16267 | 16267 | 16267 | 16267 | 16267 | 16267 | 16267 | 16267 |
| | 5 | 12149 | 13822 | 14853 | 15414 | 15708 | 15887 | 15990 | 16021 | 16049 | 16070 | 16088 |
| | 10 | 8548 | 11013 | 13017 | 14311 | 15029 | 15456 | 15694 | 15765 | 15825 | 15870 | 15907 |
| | 25 | 4155 | 6129 | 8544 | 10932 | 12749 | 14003 | 14737 | 14957 | 15133 | 15264 | 15367 |
| | 50 | 2182 | 3364 | 5065 | 7245 | 9520 | 11602 | 13085 | 13575 | 13966 | 14255 | 14482 |
| | 100 | 1112 | 1744 | 2715 | 4127 | 5942 | 8134 | 10236 | 11087 | 11811 | 12379 | 12837 |
| | 200 | 561 | 885 | 1395 | 2176 | 3279 | 4845 | 6761 | 7730 | 8662 | 9482 | 10206 |
| | 500 | 225 | 356 | 565 | 893 | 1376 | 2123 | 3190 | 3835 | 4543 | 5270 | 6013 |
| | 1000 | 112 | 178 | 284 | 449 | 697 | 1088 | 1673 | 2049 | 2481 | 2955 | 3479 |
| | 2000 | 56 | 89 | 142 | 225 | 350 | 550 | 855 | 1056 | 1293 | 1561 | 1870 |
| | 5000 | 22 | 35 | 57 | 90 | 140 | 221 | 346 | 429 | 529 | 644 | 780 |

| Two 3/0 AWG | 4/0 AWG | 250 MCM | Two 250 MCM | Three 300 MCM | 350 MCM | Two 350 MCM | Three 350 MCM | Three 400 MCM | 500 MCM | Two 500 MCM | 750 MCM | Four 750 MCM |
|---|---|---|---|---|---|---|---|---|---|---|---|---|
| 42763 | 42763 | 42763 | 42763 | 42763 | 42763 | 42763 | 42763 | 42763 | 42763 | 42763 | 42763 | 42763 |
| 40964 | 39517 | 39664 | 41187 | 41766 | 39986 | 41344 | 41810 | 41843 | 40236 | 41469 | 40428 | 42157 |
| 39179 | 36475 | 36798 | 39664 | 40794 | 37452 | 39986 | 40885 | 40950 | 37943 | 40236 | 38313 | 41566 |
| 34162 | 28963 | 29733 | 35463 | 38039 | 31178 | 36274 | 38270 | 38433 | 32244 | 36877 | 33042 | 39880 |
| 27436 | 20935 | 22000 | 29733 | 33990 | 24021 | 31178 | 34433 | 34742 | 25576 | 32244 | 26770 | 37330 |
| 19047 | 13137 | 14159 | 22000 | 27679 | 16204 | 24021 | 28412 | 28921 | 17907 | 25576 | 19297 | 33042 |
| 11507 | 7423 | 8154 | 14159 | 19812 | 9700 | 16204 | 20734 | 21390 | 11096 | 17907 | 12319 | 26770 |
| 5167 | 3196 | 3561 | 6716 | 10438 | 4367 | 8065 | 11208 | 11781 | 5148 | 9310 | 5883 | 16913 |
| 2682 | 1637 | 1833 | 3561 | 5782 | 2274 | 4367 | 6292 | 6681 | 2714 | 5148 | 3140 | 10424 |
| 1365 | 828 | 930 | 1833 | 3047 | 1161 | 2274 | 3342 | 3571 | 1394 | 2714 | 1625 | 5883 |
| 552 | 333 | 375 | 746 | 1258 | 470 | 932 | 1387 | 1488 | 567 | 1122 | 663 | 2547 |
| 40669 | 40669 | 40669 | 40669 | 40669 | 40669 | 40669 | 40669 | 40669 | 40669 | 40669 | 40669 | 40669 |
| 39262 | 38117 | 38228 | 39432 | 39887 | 38478 | 39554 | 39921 | 39947 | 38674 | 39652 | 38824 | 40194 |
| 37859 | 35685 | 35932 | 38228 | 39120 | 36443 | 38478 | 39191 | 39242 | 36831 | 38674 | 37124 | 39728 |
| 33830 | 29406 | 30036 | 34845 | 36924 | 31226 | 35484 | 37104 | 37232 | 32103 | 35963 | 32757 | 38286 |
| 28124 | 22130 | 23090 | 30036 | 33613 | 24895 | 31226 | 33968 | 34215 | 26259 | 32103 | 27296 | 36324 |
| 20372 | 14412 | 15430 | 23090 | 28212 | 17433 | 24895 | 28831 | 29261 | 19061 | 26259 | 20365 | 32757 |
| 12745 | 8344 | 9125 | 15430 | 20981 | 10753 | 17433 | 21830 | 22429 | 12192 | 19061 | 13430 | 27296 |
| 5867 | 3650 | 4058 | 7560 | 11548 | 4953 | 9005 | 12341 | 12925 | 5811 | 10314 | 6608 | 18048 |
| 3070 | 1879 | 2102 | 4058 | 6526 | 2601 | 4953 | 7080 | 7499 | 3096 | 5811 | 3572 | 11465 |
| 1570 | 953 | 1070 | 2102 | 3477 | 1333 | 2601 | 3807 | 4062 | 1600 | 3096 | 1861 | 6608 |
| 636 | 384 | 432 | 859 | 1445 | 541 | 1072 | 1592 | 1707 | 652 | 1288 | 763 | 2905 |
| 20334 | 20334 | 20334 | 20334 | 20334 | 20334 | 20334 | 20334 | 20334 | 20334 | 20334 | 20334 | 20334 |
| 20179 | 20012 | 20023 | 20178 | 20236 | 20053 | 20193 | 20240 | 20243 | 20078 | 20205 | 20097 | 20275 |
| 19983 | 19691 | 19716 | 20023 | 20138 | 19777 | 20053 | 20146 | 20153 | 19826 | 20078 | 19864 | 20215 |
| 19455 | 18748 | 18819 | 19564 | 19847 | 18977 | 19640 | 19868 | 19884 | 19099 | 19702 | 19193 | 20038 |
| 18582 | 17264 | 17422 | 18819 | 19372 | 17742 | 18977 | 19416 | 19448 | 17981 | 19099 | 18162 | 19749 |
| 16915 | 14703 | 15018 | 17422 | 18462 | 15613 | 17742 | 18552 | 18616 | 16051 | 17981 | 16378 | 19193 |
| 14062 | 11065 | 11545 | 15018 | 16806 | 12447 | 15613 | 16984 | 17107 | 13129 | 16051 | 13648 | 18162 |
| 8889 | 6108 | 6589 | 10296 | 13014 | 7555 | 11264 | 13368 | 13614 | 8362 | 12011 | 9024 | 15604 |
| 5342 | 3438 | 3780 | 6589 | 9260 | 4502 | 7555 | 9699 | 10012 | 5157 | 8362 | 5732 | 12585 |
| 2933 | 1825 | 2029 | 3780 | 5774 | 2476 | 4502 | 6170 | 6462 | 2905 | 5157 | 3304 | 9024 |
| 1239 | 756 | 847 | 1646 | 2677 | 1051 | 2020 | 2914 | 3096 | 1254 | 2383 | 1452 | 4844 |
| 16267 | 16267 | 16267 | 16267 | 16267 | 16267 | 16267 | 16267 | 16267 | 16267 | 16267 | 16267 | 16267 |
| 16178 | 16102 | 16108 | 16188 | 16217 | 16123 | 16195 | 16219 | 16221 | 16135 | 16201 | 16145 | 16237 |
| 16088 | 15937 | 15949 | 16108 | 16167 | 15980 | 16123 | 16171 | 16174 | 16005 | 16135 | 16025 | 16206 |
| 15817 | 15448 | 15482 | 15871 | 16017 | 15561 | 15909 | 16028 | 16036 | 15624 | 15941 | 15672 | 16115 |
| 15367 | 14656 | 14732 | 15482 | 15770 | 14895 | 15561 | 15792 | 15808 | 15020 | 15624 | 15115 | 15965 |
| 14482 | 13192 | 13357 | 14732 | 15289 | 13681 | 14895 | 15335 | 15367 | 13922 | 15020 | 14102 | 15672 |
| 12837 | 10804 | 11108 | 13357 | 14381 | 11677 | 13681 | 14473 | 14538 | 12098 | 13922 | 12412 | 15115 |
| 9193 | 6718 | 7134 | 10202 | 12077 | 7937 | 10853 | 12288 | 12434 | 8573 | 11338 | 9072 | 13641 |
| 6013 | 4019 | 4369 | 7134 | 9371 | 5086 | 7937 | 9688 | 9910 | 5703 | 8573 | 6223 | 11703 |
| 3479 | 2206 | 2436 | 4369 | 6348 | 2932 | 5086 | 6699 | 6952 | 3391 | 5703 | 3804 | 9072 |
| 1518 | 933 | 1042 | 1993 | 3162 | 1284 | 2417 | 3416 | 3607 | 1523 | 2817 | 1751 | 5372 |

**49H.  Symmetrical Three-Phase Bolted Short-Circuit Currents in Amperes at Various Distances from a Liquid-Filled Transformer** (Rated 1000 kVA; $Z$ = 5.32 percent)

| Transformer secondary voltage | Distance, ft | Wire size of copper conductors | | | | | | | | | | |
|---|---|---|---|---|---|---|---|---|---|---|---|---|
| | | 14 AWG | 12 AWG | 10 AWG | 8 AWG | 6 AWG | 4 AWG | 2 AWG | 1 AWG | 1/0 AWG | 2/0 AWG | 3/0 AWG |
| 208 | 0 | 56787 | 56787 | 56787 | 56787 | 56787 | 56787 | 56787 | 56787 | 56787 | 56787 | 56787 |
| | 5 | 7574 | 11683 | 17607 | 25222 | 33185 | 40485 | 45680 | 47393 | 48758 | 49767 | 50555 |
| | 10 | 3859 | 6053 | 9425 | 14341 | 20669 | 28330 | 35692 | 38675 | 41211 | 43199 | 44798 |
| | 25 | 1558 | 2462 | 3891 | 6097 | 9266 | 13910 | 19914 | 23137 | 26359 | 29322 | 32034 |
| | 50 | 781 | 1237 | 1962 | 3098 | 4775 | 7368 | 11078 | 13326 | 15793 | 18325 | 20919 |
| | 100 | 391 | 620 | 985 | 1560 | 2419 | 3776 | 5808 | 7111 | 8615 | 10263 | 12086 |
| | 200 | 195 | 310 | 493 | 782 | 1216 | 1909 | 2966 | 3663 | 4485 | 5418 | 6492 |
| | 500 | 78 | 124 | 197 | 313 | 488 | 768 | 1201 | 1489 | 1835 | 2235 | 2707 |
| | 1000 | 39 | 62 | 98 | 156 | 244 | 385 | 602 | 748 | 924 | 1128 | 1372 |
| | 2000 | 19 | 31 | 49 | 78 | 122 | 192 | 301 | 375 | 464 | 567 | 690 |
| | 5000 | 7 | 12 | 19 | 31 | 48 | 77 | 120 | 150 | 186 | 227 | 277 |
| 240 | 0 | 54026 | 54026 | 54026 | 54026 | 54026 | 54026 | 54026 | 54026 | 54026 | 54026 | 54026 |
| | 5 | 8661 | 13248 | 19654 | 27426 | 34917 | 41210 | 45379 | 46701 | 47750 | 48521 | 49123 |
| | 10 | 4437 | 6937 | 10735 | 16138 | 22793 | 30327 | 36981 | 39507 | 41599 | 43201 | 44471 |
| | 25 | 1796 | 2835 | 4472 | 6982 | 10541 | 15625 | 21908 | 25119 | 28222 | 30973 | 33410 |
| | 50 | 901 | 1426 | 2260 | 3563 | 5477 | 8405 | 12510 | 14934 | 17539 | 20143 | 22736 |
| | 100 | 451 | 715 | 1135 | 1797 | 2784 | 4336 | 6638 | 8097 | 9764 | 11565 | 13523 |
| | 200 | 225 | 358 | 569 | 902 | 1402 | 2198 | 3408 | 4200 | 5133 | 6181 | 7379 |
| | 500 | 90 | 143 | 227 | 361 | 563 | 886 | 1383 | 1715 | 2111 | 2568 | 3106 |
| | 1000 | 45 | 71 | 114 | 181 | 282 | 444 | 694 | 863 | 1065 | 1299 | 1578 |
| | 2000 | 22 | 35 | 57 | 90 | 141 | 222 | 348 | 432 | 534 | 653 | 795 |
| | 5000 | 9 | 14 | 22 | 36 | 56 | 89 | 139 | 173 | 214 | 262 | 319 |
| 480 | 0 | 27013 | 27013 | 27013 | 27013 | 27013 | 27013 | 27013 | 27013 | 27013 | 27013 | 27013 |
| | 5 | 13890 | 18024 | 21437 | 23662 | 24899 | 25632 | 26038 | 26158 | 26260 | 26336 | 26399 |
| | 10 | 8134 | 11771 | 15927 | 19678 | 22293 | 23995 | 24960 | 25245 | 25476 | 25647 | 25783 |
| | 25 | 3500 | 5405 | 8164 | 11739 | 15522 | 19037 | 21568 | 22409 | 23079 | 23575 | 23962 |
| | 50 | 1782 | 2796 | 4358 | 6642 | 9601 | 13217 | 16736 | 18177 | 19407 | 20375 | 21156 |
| | 100 | 898 | 1417 | 2236 | 3491 | 5270 | 7812 | 10954 | 12559 | 14111 | 15486 | 16705 |
| | 200 | 450 | 713 | 1130 | 1781 | 2738 | 4202 | 6255 | 7467 | 8769 | 10071 | 11368 |
| | 500 | 180 | 286 | 454 | 720 | 1117 | 1744 | 2684 | 3288 | 3986 | 4753 | 5603 |
| | 1000 | 90 | 143 | 227 | 361 | 561 | 881 | 1370 | 1692 | 2072 | 2504 | 3003 |
| | 2000 | 45 | 71 | 113 | 180 | 281 | 443 | 691 | 857 | 1055 | 1284 | 1553 |
| | 5000 | 18 | 28 | 45 | 72 | 112 | 177 | 278 | 345 | 426 | 521 | 633 |
| 600 | 0 | 21610 | 21610 | 21610 | 21610 | 21610 | 21610 | 21610 | 21610 | 21610 | 21610 | 21610 |
| | 5 | 14355 | 17121 | 18986 | 20047 | 20604 | 20935 | 21123 | 21179 | 21228 | 21265 | 21296 |
| | 10 | 9333 | 12661 | 15769 | 18021 | 19352 | 20153 | 20595 | 20725 | 20834 | 20915 | 20981 |
| | 25 | 4269 | 6448 | 9344 | 12585 | 15401 | 17551 | 18881 | 19289 | 19613 | 19851 | 20037 |
| | 50 | 2206 | 3432 | 5261 | 7766 | 10661 | 13658 | 16056 | 16905 | 17592 | 18108 | 18513 |
| | 100 | 1117 | 1759 | 2756 | 4249 | 6264 | 8903 | 11737 | 12999 | 14123 | 15041 | 15802 |
| | 200 | 562 | 888 | 1404 | 2202 | 3352 | 5048 | 7266 | 8471 | 9687 | 10815 | 11858 |
| | 500 | 225 | 357 | 567 | 896 | 1386 | 2152 | 3272 | 3970 | 4755 | 5587 | 6470 |
| | 1000 | 112 | 178 | 284 | 450 | 699 | 1095 | 1693 | 2081 | 2535 | 3040 | 3610 |
| | 2000 | 56 | 89 | 142 | 225 | 351 | 552 | 860 | 1063 | 1306 | 1582 | 1905 |
| | 5000 | 22 | 35 | 57 | 90 | 140 | 221 | 347 | 430 | 531 | 648 | 786 |

| Two 3/0 AWG | 4/0 AWG | 250 MCM | Two 250 MCM | Three 300 MCM | 350 MCM | Two 350 MCM | Three 350 MCM | Three 400 MCM | 500 MCM | Two 500 MCM | 750 MCM | Four 750 MCM |
|---|---|---|---|---|---|---|---|---|---|---|---|---|
| 56787 | 56787 | 56787 | 56787 | 56787 | 56787 | 56787 | 56787 | 56787 | 56787 | 56787 | 56787 | 56787 |
| 53648 | 51159 | 51422 | 54043 | 55048 | 51987 | 54317 | 55125 | 55181 | 52419 | 54534 | 52747 | 53726 |
| 50555 | 46037 | 46611 | 51422 | 53367 | 47737 | 51987 | 53525 | 53638 | 48573 | 52419 | 49198 | 54700 |
| 42202 | 34346 | 35562 | 44430 | 48695 | 37835 | 45808 | 49094 | 49374 | 39527 | 46825 | 40797 | 51817 |
| 32034 | 23379 | 24830 | 35562 | 42123 | 27635 | 37835 | 42859 | 43369 | 29854 | 39527 | 31593 | 47583 |
| 20919 | 13967 | 15186 | 24830 | 32650 | 17683 | 27635 | 33758 | 34532 | 19835 | 29854 | 21642 | 40797 |
| 12086 | 7658 | 8460 | 15186 | 22087 | 10183 | 17683 | 23311 | 24193 | 11780 | 19835 | 13216 | 31593 |
| 5267 | 3236 | 3614 | 6918 | 10984 | 4457 | 8391 | 11868 | 12532 | 5286 | 9782 | 6076 | 18676 |
| 2707 | 1647 | 1847 | 3614 | 5937 | 2298 | 4457 | 6485 | 6907 | 2751 | 5286 | 3194 | 11055 |
| 1372 | 831 | 933 | 1847 | 3089 | 1167 | 2298 | 3395 | 3633 | 1404 | 2751 | 1639 | 6076 |
| 553 | 334 | 375 | 748 | 1265 | 471 | 936 | 1396 | 1499 | 568 | 1128 | 666 | 2582 |
| 54026 | 54026 | 54026 | 54026 | 54026 | 54026 | 54026 | 54026 | 54026 | 54026 | 54026 | 54026 | 54026 |
| 51566 | 49586 | 49786 | 51868 | 52659 | 50224 | 52080 | 52719 | 52763 | 50563 | 52250 | 50822 | 53192 |
| 49123 | 45447 | 45891 | 49786 | 51329 | 46780 | 50224 | 51451 | 51540 | 47445 | 50563 | 47944 | 52381 |
| 42313 | 35431 | 36461 | 44087 | 47575 | 38388 | 45188 | 47889 | 48110 | 39811 | 46005 | 40874 | 50074 |
| 33410 | 25123 | 26489 | 36461 | 42117 | 29092 | 38388 | 42716 | 43132 | 31107 | 39811 | 32662 | 46614 |
| 22736 | 15503 | 16757 | 26489 | 33792 | 19286 | 29092 | 34759 | 35431 | 21413 | 31107 | 23164 | 40874 |
| 13523 | 8665 | 9537 | 16757 | 23753 | 11391 | 19286 | 24924 | 25760 | 13079 | 21413 | 14570 | 32662 |
| 6006 | 3705 | 4132 | 7835 | 12269 | 5077 | 9441 | 13202 | 13898 | 5997 | 10935 | 6867 | 20198 |
| 3106 | 1893 | 2121 | 4132 | 6737 | 2634 | 5077 | 7341 | 7803 | 3147 | 5997 | 3646 | 12280 |
| 1578 | 957 | 1075 | 2121 | 3534 | 1342 | 2634 | 3879 | 4147 | 1613 | 3147 | 1880 | 6867 |
| 637 | 385 | 433 | 862 | 1455 | 543 | 1077 | 1604 | 1722 | 655 | 1297 | 766 | 2953 |
| 27013 | 27013 | 27013 | 27013 | 27013 | 27013 | 27013 | 27013 | 27013 | 27013 | 27013 | 27013 | 27013 |
| 26706 | 26449 | 26469 | 26740 | 26840 | 26521 | 26765 | 26848 | 26853 | 26563 | 26786 | 26596 | 26908 |
| 26399 | 25889 | 25934 | 26469 | 26669 | 26040 | 26521 | 26684 | 26694 | 26125 | 26563 | 26190 | 26803 |
| 25475 | 24259 | 24388 | 25670 | 26161 | 24665 | 25804 | 26199 | 26226 | 24876 | 25910 | 25037 | 26493 |
| 23962 | 21762 | 22043 | 24388 | 25339 | 22594 | 24665 | 25416 | 25471 | 23002 | 24876 | 23307 | 25991 |
| 21156 | 17715 | 18230 | 22043 | 23787 | 19194 | 22594 | 23944 | 24055 | 19905 | 23002 | 20437 | 25037 |
| 16705 | 12561 | 13244 | 18230 | 21058 | 14546 | 19194 | 21358 | 21566 | 15553 | 19905 | 16331 | 23307 |
| 9738 | 6482 | 7054 | 11592 | 15315 | 8228 | 12928 | 15848 | 16221 | 9243 | 13989 | 10099 | 19238 |
| 5603 | 3544 | 3917 | 7054 | 10299 | 4720 | 8228 | 10879 | 11298 | 5467 | 9243 | 6140 | 14822 |
| 3003 | 1852 | 2066 | 3917 | 6134 | 2538 | 4720 | 6601 | 6949 | 2998 | 5467 | 3433 | 10099 |
| 1250 | 760 | 853 | 1670 | 2747 | 1061 | 2061 | 3001 | 3197 | 1271 | 2445 | 1476 | 5130 |
| 21610 | 21610 | 21610 | 21610 | 21610 | 21610 | 21610 | 21610 | 21610 | 21610 | 21610 | 21610 | 21610 |
| 21453 | 21321 | 21331 | 21470 | 21522 | 21357 | 21483 | 21526 | 21528 | 21379 | 21494 | 21396 | 21556 |
| 21296 | 21033 | 21054 | 21331 | 21434 | 21108 | 21357 | 21441 | 21447 | 21151 | 21379 | 21185 | 21502 |
| 20823 | 20182 | 20243 | 20917 | 21171 | 20382 | 20985 | 21191 | 21205 | 20491 | 21039 | 20574 | 21343 |
| 20037 | 18823 | 18961 | 20243 | 20742 | 19246 | 20382 | 20781 | 20809 | 19461 | 20491 | 19623 | 21081 |
| 18513 | 16402 | 16689 | 18961 | 19913 | 17239 | 19246 | 19993 | 20050 | 17643 | 19461 | 17945 | 20574 |
| 15802 | 12754 | 13230 | 16689 | 18379 | 14119 | 17239 | 18539 | 18650 | 14783 | 17643 | 15282 | 19623 |
| 10465 | 7339 | 7874 | 11929 | 14704 | 8931 | 12907 | 15044 | 15280 | 9796 | 13650 | 10492 | 17201 |
| 6470 | 4213 | 4614 | 7874 | 10806 | 5454 | 8931 | 11263 | 11585 | 6201 | 9796 | 6847 | 14214 |
| 3610 | 2258 | 2506 | 4614 | 6933 | 3045 | 5454 | 7377 | 7700 | 3555 | 6201 | 4024 | 10492 |
| 1540 | 941 | 1053 | 2038 | 3289 | 1305 | 2492 | 3572 | 3787 | 1554 | 2928 | 1795 | 5828 |

**49I. Symmetrical Three-Phase Bolted Short-Circuit Currents in Amperes at Various Distances from a Liquid-Filled Transformer** (Rated 1500 kVA; $Z$ = 5.32 percent)

| Transformer secondary voltage | Distance, ft | Wire size of copper conductors | | | | | | | | | | |
|---|---|---|---|---|---|---|---|---|---|---|---|---|
| | | 14 AWG | 12 AWG | 10 AWG | 8 AWG | 6 AWG | 4 AWG | 2 AWG | 1 AWG | 1/0 AWG | 2/0 AWG | 3/0 AWG |
| 208 | 0 | 84472 | 84472 | 84472 | 84472 | 84472 | 84472 | 84472 | 84472 | 84472 | 84472 | 84472 |
| | 5 | 7680 | 11990 | 18493 | 27611 | 38542 | 50427 | 60415 | 64063 | 67042 | 69290 | 71056 |
| | 10 | 3882 | 6118 | 9612 | 14888 | 22136 | 31909 | 42859 | 47925 | 52524 | 56353 | 59567 |
| | 25 | 1561 | 2471 | 3916 | 6168 | 9469 | 14488 | 21442 | 25483 | 29770 | 33990 | 38117 |
| | 50 | 782 | 1239 | 1968 | 3114 | 4820 | 7499 | 11452 | 13941 | 16765 | 19792 | 23046 |
| | 100 | 391 | 620 | 986 | 1563 | 2429 | 3806 | 5896 | 7260 | 8859 | 10652 | 12689 |
| | 200 | 195 | 310 | 493 | 783 | 1219 | 1916 | 2987 | 3698 | 4545 | 5515 | 6648 |
| | 500 | 78 | 124 | 197 | 313 | 488 | 769 | 1204 | 1495 | 1845 | 2250 | 2732 |
| | 1000 | 39 | 62 | 98 | 156 | 244 | 385 | 603 | 750 | 926 | 1132 | 1378 |
| | 2000 | 19 | 31 | 49 | 78 | 122 | 192 | 302 | 375 | 464 | 568 | 692 |
| | 5000 | 7 | 12 | 19 | 31 | 48 | 77 | 120 | 150 | 186 | 227 | 277 |
| 240 | 0 | 80426 | 80426 | 80426 | 80426 | 80426 | 80426 | 80426 | 80426 | 80426 | 80426 | 80426 |
| | 5 | 8816 | 13699 | 20933 | 30715 | 41754 | 52821 | 61356 | 64300 | 66664 | 68421 | 69793 |
| | 10 | 4470 | 7031 | 11008 | 16930 | 24849 | 35040 | 45679 | 50290 | 54334 | 57590 | 60257 |
| | 25 | 1800 | 2848 | 4507 | 7086 | 10834 | 16449 | 24003 | 28244 | 32622 | 36796 | 40748 |
| | 50 | 902 | 1429 | 2268 | 3586 | 5542 | 8593 | 13039 | 15793 | 18874 | 22113 | 25520 |
| | 100 | 451 | 715 | 1137 | 1802 | 2798 | 4378 | 6762 | 8307 | 10107 | 12104 | 14347 |
| | 200 | 225 | 358 | 569 | 903 | 1405 | 2208 | 3437 | 4251 | 5216 | 6317 | 7596 |
| | 500 | 90 | 143 | 228 | 362 | 563 | 887 | 1388 | 1722 | 2124 | 2589 | 3141 |
| | 1000 | 45 | 71 | 114 | 181 | 282 | 444 | 696 | 865 | 1068 | 1305 | 1587 |
| | 2000 | 22 | 35 | 57 | 90 | 141 | 222 | 348 | 433 | 535 | 655 | 798 |
| | 5000 | 9 | 14 | 22 | 36 | 56 | 89 | 139 | 173 | 214 | 262 | 320 |
| 480 | 0 | 40213 | 40213 | 40213 | 40213 | 40213 | 40213 | 40213 | 40213 | 40213 | 40213 | 40213 |
| | 5 | 15517 | 21588 | 27679 | 32400 | 35305 | 37072 | 38042 | 38325 | 38559 | 38732 | 38871 |
| | 10 | 8510 | 12799 | 18403 | 24488 | 29580 | 33335 | 35597 | 36278 | 36820 | 37217 | 37527 |
| | 25 | 3547 | 5541 | 8557 | 12806 | 17945 | 23602 | 28424 | 30204 | 31661 | 32765 | 33633 |
| | 50 | 1792 | 2825 | 4440 | 6884 | 10254 | 14828 | 20007 | 22427 | 24638 | 26489 | 28049 |
| | 100 | 900 | 1424 | 2253 | 3543 | 5417 | 8224 | 12001 | 14122 | 16311 | 18398 | 20374 |
| | 200 | 451 | 714 | 1134 | 1793 | 2771 | 4296 | 6519 | 7896 | 9437 | 11056 | 12760 |
| | 500 | 180 | 286 | 455 | 721 | 1121 | 1757 | 2723 | 3354 | 4095 | 4927 | 5873 |
| | 1000 | 90 | 143 | 227 | 361 | 562 | 884 | 1379 | 1708 | 2099 | 2548 | 3072 |
| | 2000 | 45 | 71 | 114 | 181 | 281 | 443 | 694 | 861 | 1062 | 1294 | 1570 |
| | 5000 | 18 | 28 | 45 | 72 | 112 | 177 | 278 | 346 | 427 | 522 | 636 |
| 600 | 0 | 32170 | 32170 | 32170 | 32170 | 32170 | 32170 | 32170 | 32170 | 32170 | 32170 | 32170 |
| | 5 | 17150 | 22054 | 25991 | 28491 | 29855 | 30655 | 31095 | 31224 | 31334 | 31417 | 31484 |
| | 10 | 10128 | 14583 | 19578 | 23964 | 26939 | 28839 | 29899 | 30210 | 30462 | 30648 | 30795 |
| | 25 | 4373 | 6746 | 10162 | 14537 | 19077 | 23192 | 26080 | 27022 | 27770 | 28320 | 28747 |
| | 50 | 2227 | 3494 | 5442 | 8278 | 11921 | 16306 | 20479 | 22152 | 23568 | 24670 | 25553 |
| | 100 | 1122 | 1772 | 2794 | 4360 | 6574 | 9717 | 13553 | 15486 | 17336 | 18959 | 20381 |
| | 200 | 563 | 891 | 1412 | 2226 | 3421 | 5244 | 7788 | 9279 | 10872 | 12454 | 14014 |
| | 500 | 225 | 357 | 568 | 900 | 1396 | 2179 | 3351 | 4103 | 4970 | 5919 | 6967 |
| | 1000 | 112 | 179 | 284 | 451 | 702 | 1101 | 1712 | 2113 | 2588 | 3125 | 3745 |
| | 2000 | 56 | 89 | 142 | 226 | 352 | 553 | 864 | 1071 | 1319 | 1604 | 1939 |
| | 5000 | 22 | 35 | 57 | 90 | 141 | 222 | 347 | 432 | 533 | 651 | 791 |

| Two 3/0 AWG | 4/0 AWG | 250 MCM | Two 250 MCM | Three 300 MCM | 350 MCM | Two 350 MCM | Three 350 MCM | Three 400 MCM | 500 MCM | Two 500 MCM | 750 MCM | Four 750 MCM |
|---|---|---|---|---|---|---|---|---|---|---|---|---|
| 84472 | 84472 | 84472 | 84472 | 84472 | 84472 | 84472 | 84472 | 84472 | 84472 | 84472 | 84472 | 84472 |
| 77645 | 72404 | 73006 | 78536 | 80697 | 74234 | 79133 | 80863 | 80984 | 75155 | 79597 | 75846 | 82157 |
| 71056 | 62121 | 63356 | 73006 | 77107 | 65698 | 74234 | 77451 | 77694 | 67417 | 75155 | 68691 | 79952 |
| 54766 | 41853 | 43947 | 59221 | 67522 | 47908 | 62002 | 68365 | 68951 | 50936 | 64046 | 53246 | 73933 |
| 38117 | 26298 | 28329 | 43947 | 55151 | 32383 | 47908 | 56571 | 57555 | 35744 | 50936 | 38472 | 65563 |
| 23046 | 14859 | 16320 | 28329 | 39579 | 19402 | 32383 | 41394 | 42680 | 22177 | 35744 | 24600 | 53246 |
| 12689 | 7899 | 8777 | 16320 | 24842 | 10702 | 19402 | 26515 | 27743 | 12541 | 22177 | 14242 | 38472 |
| 5367 | 3276 | 3668 | 7125 | 11570 | 4550 | 8736 | 12588 | 13364 | 5429 | 10296 | 6280 | 20823 |
| 2732 | 1657 | 1861 | 3668 | 6097 | 2322 | 4550 | 6687 | 7144 | 2789 | 5429 | 3249 | 11760 |
| 1378 | 833 | 937 | 1861 | 3130 | 1173 | 2322 | 3448 | 3696 | 1414 | 2789 | 1653 | 6280 |
| 554 | 334 | 76 | 750 | 1272 | 472 | 940 | 1405 | 1509 | 570 | 1134 | 668 | 2617 |
| 80426 | 80426 | 80426 | 80426 | 80426 | 80426 | 80426 | 80426 | 80426 | 80426 | 80426 | 80426 | 80426 |
| 75054 | 70839 | 71297 | 75735 | 77446 | 72259 | 76199 | 77576 | 77671 | 72988 | 76564 | 73537 | 78599 |
| 69793 | 62342 | 63324 | 71297 | 74586 | 65215 | 72259 | 74854 | 75045 | 66608 | 72988 | 67642 | 76844 |
| 56099 | 44217 | 46099 | 59798 | 66769 | 49628 | 62085 | 67443 | 67912 | 52283 | 63764 | 54284 | 71978 |
| 40748 | 28843 | 30862 | 46099 | 56189 | 34827 | 49628 | 57387 | 58215 | 38035 | 52283 | 40589 | 65009 |
| 25520 | 16703 | 18262 | 30862 | 41896 | 21505 | 34827 | 43564 | 44737 | 24365 | 38035 | 26814 | 54284 |
| 14347 | 8997 | 9972 | 18262 | 27234 | 12089 | 21505 | 28914 | 30135 | 14084 | 24365 | 15905 | 40589 |
| 6145 | 3761 | 4206 | 8120 | 13058 | 5204 | 9909 | 14163 | 15000 | 6193 | 11621 | 7143 | 22899 |
| 3141 | 1907 | 2140 | 4206 | 6957 | 2668 | 5204 | 7616 | 8125 | 3200 | 6193 | 3722 | 13209 |
| 1587 | 960 | 1079 | 2140 | 3591 | 1350 | 2668 | 3952 | 4234 | 1626 | 3200 | 1900 | 7143 |
| 639 | 385 | 434 | 865 | 1464 | 544 | 1083 | 1616 | 1736 | 657 | 1305 | 770 | 3002 |
| 40213 | 40213 | 40213 | 40213 | 40213 | 40213 | 40213 | 40213 | 40213 | 40213 | 40213 | 40213 | 40213 |
| 39544 | 3982 | 39026 | 39616 | 39835 | 39138 | 39671 | 39850 | 39862 | 39229 | 39716 | 39299 | 39981 |
| 38871 | 37767 | 37867 | 39026 | 39460 | 38099 | 39138 | 39492 | 39516 | 38282 | 39229 | 38422 | 39751 |
| 36859 | 34297 | 34594 | 37299 | 38359 | 35197 | 37593 | 38442 | 38501 | 35650 | 37822 | 35989 | 39077 |
| 33633 | 29294 | 29899 | 34594 | 36601 | 31042 | 35197 | 36770 | 36890 | 31882 | 35650 | 32504 | 37996 |
| 28049 | 22108 | 23049 | 29899 | 33384 | 24814 | 31042 | 33721 | 33956 | 26141 | 31882 | 27142 | 35989 |
| 20374 | 14421 | 15431 | 23049 | 28094 | 17413 | 24814 | 28693 | 29107 | 19017 | 26141 | 20294 | 32504 |
| 10699 | 6883 | 7565 | 13184 | 18502 | 9006 | 15101 | 19368 | 19984 | 10309 | 16697 | 11449 | 25047 |
| 5873 | 3652 | 4060 | 7565 | 11550 | 4954 | 9006 | 12338 | 12918 | 5810 | 10309 | 6604 | 17997 |
| 3072 | 1880 | 2103 | 4060 | 6529 | 2602 | 4954 | 7081 | 7500 | 3096 | 5810 | 3571 | 11449 |
| 1261 | 765 | 859 | 1694 | 2818 | 1072 | 2102 | 3092 | 3304 | 1288 | 2510 | 1501 | 5448 |
| 32170 | 32170 | 32170 | 32170 | 32170 | 32170 | 32170 | 32170 | 32170 | 32170 | 32170 | 32170 | 32170 |
| 31828 | 31539 | 31560 | 31864 | 31976 | 31617 | 31892 | 31984 | 31990 | 31663 | 31915 | 31699 | 32051 |
| 31484 | 30911 | 30959 | 31560 | 31784 | 31075 | 31617 | 31800 | 31812 | 31168 | 31663 | 31240 | 31933 |
| 30451 | 29074 | 29215 | 30662 | 31213 | 29520 | 30808 | 31254 | 31284 | 29754 | 30925 | 29931 | 31583 |
| 28747 | 26233 | 26545 | 29215 | 30286 | 27161 | 29520 | 30371 | 30432 | 27617 | 29754 | 27957 | 31015 |
| 25553 | 21549 | 22138 | 26545 | 28528 | 23239 | 27161 | 28702 | 28824 | 24049 | 27617 | 24651 | 29931 |
| 20381 | 15439 | 16246 | 22138 | 25405 | 17776 | 23239 | 25742 | 25976 | 18952 | 24049 | 19853 | 27957 |
| 12039 | 8046 | 8744 | 14262 | 18698 | 10170 | 15848 | 19318 | 19751 | 11394 | 17097 | 12418 | 23258 |
| 6967 | 4415 | 4876 | 8744 | 12692 | 5866 | 10170 | 13386 | 13886 | 6781 | 11394 | 7601 | 18070 |
| 3745 | 2312 | 2577 | 4876 | 7610 | 3163 | 5866 | 8180 | 8603 | 3733 | 6781 | 4269 | 12418 |
| 1562 | 950 | 1065 | 2084 | 3423 | 1325 | 2570 | 3738 | 3981 | 1586 | 3047 | 1841 | 6362 |

**49J.   Symmetrical Three-Phase Bolted Short-Circuit Currents in Amperes at Various Distances from a Liquid-Filled Transformer** (Rated 2000 kVA; $Z$ = 5.32 percent)

| Transformer secondary voltage | Distance, ft | Wire size of copper conductors | | | | | | | | | | |
|---|---|---|---|---|---|---|---|---|---|---|---|---|
| | | 14 AWG | 12 AWG | 10 AWG | 8 AWG | 6 AWG | 4 AWG | 2 AWG | 1 AWG | 1/0 AWG | 2/0 AWG | 3/0 AWG |
| 208 | 0 | 111710 | 111710 | 111710 | 111710 | 111710 | 111710 | 111710 | 111710 | 111710 | 111710 | 1117 |
| | 5 | 7729 | 12130 | 18903 | 28785 | 41492 | 56805 | 71359 | 77175 | 82082 | 85890 | 889 |
| | 10 | 3893 | 6147 | 9698 | 15141 | 22844 | 33797 | 47182 | 53926 | 60373 | 66018 | 709 |
| | 25 | 1563 | 2475 | 3927 | 6202 | 9565 | 14765 | 22210 | 26715 | 31654 | 36713 | 418 |
| | 50 | 782 | 1240 | 1970 | 3121 | 4842 | 7561 | 11633 | 14245 | 17259 | 20560 | 242 |
| | 100 | 391 | 620 | 987 | 1565 | 2434 | 3821 | 5938 | 7332 | 8980 | 10847 | 129 |
| | 200 | 195 | 310 | 493 | 784 | 1220 | 1920 | 2998 | 3716 | 4575 | 5563 | 67 |
| | 500 | 78 | 124 | 197 | 313 | 489 | 770 | 1205 | 1498 | 1849 | 2258 | 27 |
| | 1000 | 39 | 62 | 98 | 156 | 244 | 385 | 603 | 750 | 928 | 1134 | 13 |
| | 2000 | 19 | 31 | 49 | 78 | 122 | 192 | 302 | 375 | 464 | 568 | 6 |
| | 5000 | 7 | 12 | 19 | 31 | 48 | 77 | 120 | 150 | 186 | 227 | 2 |
| 240 | 0 | 106436 | 106436 | 106436 | 106436 | 106436 | 106436 | 106436 | 106436 | 106436 | 106436 | 1064 |
| | 5 | 8887 | 13904 | 21532 | 32380 | 45735 | 60749 | 73845 | 78747 | 82777 | 85835 | 882 |
| | 10 | 4485 | 7074 | 11133 | 17299 | 25864 | 37638 | 51226 | 57685 | 63640 | 68667 | 729 |
| | 25 | 1802 | 2854 | 4524 | 7134 | 10972 | 16845 | 25081 | 29936 | 35145 | 40340 | 454 |
| | 50 | 902 | 1430 | 2272 | 3597 | 5573 | 8682 | 13297 | 16222 | 19561 | 23167 | 270 |
| | 100 | 451 | 716 | 1138 | 1805 | 2806 | 4399 | 6822 | 8410 | 10277 | 12377 | 147 |
| | 200 | 225 | 358 | 569 | 904 | 1407 | 2213 | 3452 | 4275 | 5258 | 6385 | 77 |
| | 500 | 90 | 143 | 228 | 362 | 564 | 888 | 1390 | 1726 | 2131 | 2600 | 31 |
| | 1000 | 45 | 71 | 114 | 181 | 282 | 444 | 696 | 866 | 1070 | 1307 | 15 |
| | 2000 | 22 | 35 | 57 | 90 | 141 | 222 | 348 | 433 | 536 | 655 | 7 |
| | 5000 | 9 | 14 | 22 | 36 | 56 | 89 | 139 | 173 | 214 | 262 | 3 |
| 480 | 0 | 53218 | 53218 | 53218 | 53218 | 53218 | 53218 | 53218 | 53218 | 53218 | 53218 | 532 |
| | 5 | 16329 | 23631 | 31927 | 39318 | 44376 | 47607 | 49403 | 49923 | 50346 | 50656 | 509 |
| | 10 | 8682 | 13292 | 19728 | 27509 | 34937 | 41086 | 45082 | 46327 | 47310 | 48026 | 485 |
| | 25 | 3568 | 5603 | 8738 | 13327 | 19266 | 26497 | 33468 | 36286 | 38676 | 40541 | 420 |
| | 50 | 1797 | 2838 | 4478 | 6996 | 10568 | 15671 | 21960 | 25158 | 28233 | 30943 | 333 |
| | 100 | 901 | 1427 | 2262 | 3567 | 5486 | 8422 | 12540 | 14968 | 17572 | 20170 | 227 |
| | 200 | 451 | 715 | 1136 | 1798 | 2786 | 4341 | 6648 | 8111 | 9780 | 11583 | 135 |
| | 500 | 180 | 286 | 455 | 722 | 1123 | 1764 | 2742 | 3387 | 4149 | 5014 | 60 |
| | 1000 | 90 | 143 | 227 | 361 | 563 | 886 | 1384 | 1715 | 2112 | 2569 | 31 |
| | 2000 | 45 | 71 | 114 | 181 | 282 | 444 | 695 | 863 | 1065 | 1300 | 15 |
| | 5000 | 18 | 28 | 45 | 72 | 112 | 177 | 278 | 346 | 428 | 523 | 6 |
| 600 | 0 | 42574 | 42574 | 42574 | 42574 | 42574 | 42574 | 42574 | 42574 | 42574 | 42574 | 425 |
| | 5 | 18739 | 25386 | 31517 | 35877 | 38399 | 39889 | 40698 | 40932 | 41128 | 41273 | 413 |
| | 10 | 10506 | 15603 | 21945 | 28331 | 33240 | 36622 | 38571 | 39144 | 39603 | 39937 | 401 |
| | 25 | 4420 | 6882 | 10556 | 15589 | 21381 | 27327 | 32010 | 33642 | 34955 | 35931 | 366 |
| | 50 | 2237 | 3523 | 5524 | 8521 | 12568 | 17859 | 23501 | 25990 | 28191 | 29975 | 314 |
| | 100 | 1124 | 1778 | 2812 | 4412 | 6720 | 10125 | 14569 | 16977 | 19397 | 21634 | 236 |
| | 200 | 563 | 892 | 1416 | 2238 | 3453 | 5337 | 8049 | 9700 | 11520 | 13398 | 153 |
| | 500 | 225 | 358 | 569 | 901 | 1400 | 2192 | 3390 | 4168 | 5077 | 6089 | 72 |
| | 1000 | 112 | 179 | 284 | 451 | 703 | 1105 | 1721 | 2129 | 2614 | 3168 | 38 |
| | 2000 | 56 | 89 | 142 | 226 | 352 | 554 | 866 | 1075 | 1325 | 1614 | 19 |
| | 5000 | 22 | 35 | 57 | 90 | 141 | 222 | 348 | 432 | 534 | 652 | 7 |

| Two 3/0 AWG | 4/0 AWG | 250 MCM | Two 250 MCM | Three 300 MCM | 350 MCM | Two 350 MCM | Three 350 MCM | Three 400 MCM | 500 MCM | Two 500 MCM | 750 MCM | Four 750 MCM |
|---|---|---|---|---|---|---|---|---|---|---|---|---|
| 111710 | 111710 | 111710 | 111710 | 111710 | 111710 | 111710 | 111710 | 111710 | 111710 | 111710 | 111710 | 111710 |
| 99969 | 91260 | 92321 | 101546 | 105226 | 94419 | 102576 | 105512 | 105717 | 95972 | 103366 | 97124 | 107714 |
| 88927 | 74999 | 77026 | 92321 | 99159 | 80818 | 94419 | 99749 | 100163 | 83599 | 95972 | 85656 | 103964 |
| 63965 | 46751 | 59608 | 70814 | 83587 | 55110 | 75185 | 84981 | 85944 | 59432 | 78417 | 62790 | 94002 |
| 41876 | 27959 | 30384 | 49608 | 65045 | 35339 | 55110 | 67189 | 68684 | 39588 | 59432 | 43134 | 80820 |
| 24206 | 15329 | 16930 | 30384 | 44127 | 20367 | 35339 | 46539 | 48271 | 23544 | 39588 | 26388 | 62790 |
| 12998 | 8022 | 8941 | 16930 | 26436 | 10977 | 20367 | 28414 | 29885 | 12953 | 23544 | 14813 | 43134 |
| 5418 | 3296 | 3695 | 7232 | 11879 | 4597 | 8916 | 12974 | 13815 | 5503 | 10570 | 6387 | 22081 |
| 2745 | 1662 | 1867 | 3695 | 6179 | 2334 | 4597 | 6791 | 7267 | 2808 | 5503 | 3277 | 12144 |
| 1381 | 835 | 939 | 1867 | 3151 | 1176 | 2334 | 3474 | 3728 | 1419 | 2808 | 1660 | 6387 |
| 554 | 334 | 376 | 752 | 1275 | 472 | 942 | 1409 | 1514 | 571 | 1137 | 669 | 2636 |
| 106436 | 106436 | 106436 | 106436 | 106436 | 106436 | 106436 | 106436 | 106436 | 106436 | 106436 | 106436 | 106436 |
| 97163 | 90075 | 90893 | 98370 | 101300 | 92551 | 99174 | 101524 | 101685 | 93790 | 99798 | 94713 | 103273 |
| 88240 | 76343 | 78004 | 90893 | 96433 | 81138 | 92551 | 96895 | 97222 | 83433 | 93790 | 85128 | 100272 |
| 66655 | 50210 | 52891 | 72557 | 83553 | 57983 | 76254 | 84686 | 85470 | 61901 | 78969 | 64897 | 92138 |
| 45490 | 31027 | 33520 | 52891 | 67278 | 38530 | 57983 | 69145 | 70441 | 42724 | 61901 | 46153 | 80979 |
| 27079 | 17344 | 19086 | 33520 | 47434 | 22780 | 38530 | 49735 | 51373 | 26135 | 42724 | 29085 | 64897 |
| 14774 | 9167 | 10197 | 19086 | 29310 | 12463 | 22780 | 31355 | 32862 | 14639 | 26135 | 16662 | 46153 |
| 6215 | 3788 | 4244 | 8267 | 13479 | 5270 | 10156 | 14684 | 15606 | 6295 | 11993 | 7290 | 24523 |
| 3158 | 1914 | 2150 | 4244 | 7070 | 2684 | 5270 | 7759 | 8294 | 3226 | 6295 | 3761 | 13724 |
| 1592 | 962 | 1082 | 2150 | 3621 | 1355 | 2684 | 3989 | 4278 | 1633 | 3226 | 1910 | 7290 |
| 639 | 386 | 434 | 866 | 1469 | 545 | 1086 | 1623 | 1744 | 658 | 1310 | 771 | 3028 |
| 53218 | 53218 | 53218 | 53218 | 53218 | 53218 | 53218 | 53218 | 53218 | 53218 | 53218 | 53218 | 53218 |
| 52063 | 51093 | 51170 | 52187 | 52564 | 51363 | 52280 | 52591 | 52610 | 51517 | 52357 | 51636 | 52815 |
| 50901 | 49007 | 49185 | 51170 | 51918 | 49587 | 51363 | 51972 | 52012 | 49899 | 51517 | 50136 | 52417 |
| 47437 | 43180 | 43705 | 48220 | 50028 | 44738 | 48729 | 50170 | 50271 | 45502 | 49119 | 46069 | 51254 |
| 42032 | 35289 | 36278 | 43705 | 47053 | 38127 | 44738 | 47344 | 47548 | 39484 | 45502 | 40489 | 49414 |
| 33327 | 25105 | 26445 | 36278 | 41776 | 28991 | 38127 | 42343 | 42735 | 30950 | 39484 | 40489 | 46069 |
| 22745 | 15513 | 16760 | 26445 | 33639 | 19265 | 28991 | 34572 | 35220 | 21362 | 30950 | 23076 | 40489 |
| 11219 | 7094 | 7839 | 14113 | 20580 | 9442 | 16446 | 21725 | 22550 | 10928 | 18454 | 12261 | 29479 |
| 6011 | 3707 | 4133 | 7839 | 12271 | 5078 | 9442 | 13199 | 13890 | 5996 | 10928 | 6862 | 20136 |
| 3107 | 1894 | 2122 | 4133 | 6739 | 2635 | 5078 | 7342 | 7803 | 3147 | 5996 | 3645 | 12261 |
| 1267 | 767 | 862 | 1706 | 2855 | 1078 | 2123 | 3138 | 3359 | 1297 | 2543 | 1514 | 5622 |
| 42574 | 42574 | 42574 | 42574 | 42574 | 42574 | 42574 | 42574 | 42574 | 42547 | 42574 | 42574 | 42574 |
| 41984 | 41484 | 41521 | 42045 | 42239 | 41617 | 42092 | 42252 | 42262 | 41696 | 42131 | 41757 | 42367 |
| 41391 | 40403 | 40486 | 41521 | 40686 | 41617 | 41933 | 41953 |  | 40845 | 41696 | 40966 | 42162 |
| 39605 | 37269 | 37521 | 39976 | 40923 | 38049 | 40230 | 40993 | 41045 | 38448 | 40429 | 38747 | 41557 |
| 36691 | 32589 | 33131 | 37521 | 39338 | 34170 | 38049 | 39485 | 39589 | 34932 | 38448 | 35406 | 40581 |
| 31442 | 25441 | 26362 | 33131 | 36388 | 28081 | 34170 | 36686 | 36895 | 29356 | 34932 | 30307 | 38747 |
| 23688 | 17168 | 18256 | 26362 | 31358 | 20361 | 28081 | 31914 | 33298 | 22026 | 29356 | 23326 | 35496 |
| 12955 | 8433 | 9233 | 15745 | 21562 | 10905 | 17833 | 22455 | 23083 | 12387 | 19532 | 13661 | 28215 |
| 7229 | 4520 | 5015 | 9233 | 13862 | 6090 | 10905 | 14740 | 15380 | 7108 | 12387 | 8040 | 20888 |
| 3813 | 2339 | 2613 | 5015 | 7988 | 3225 | 6090 | 8637 | 9126 | 3827 | 7108 | 4402 | 13661 |
| 1573 | 954 | 1071 | 2108 | 3493 | 1336 | 2610 | 3826 | 4083 | 1603 | 3109 | 1865 | 6665 |

### 49K.  Transformer Impedances

(Based on NEMA standards publication for secondary unit substations, Publication No. 210-1970)

| Single-phase kVA | Percent impedance | | | Short-circuit table |
|---|---|---|---|---|
| | R | X | Z | |
| 25 | 1.3 | 1.4 | 1.9 | A-13 |
| 37.5 | 1.0 | 1.4 | 1.7 | A-14 |
| 50 | 1.0 | 1.4 | 1.7 | A-15 |
| 75 | 1.0 | 1.5 | 1.8 | A-16 |
| 100 | 0.9 | 1.6 | 1.8 | A-17 |
| 167 | 0.9 | 1.7 | 1.9 | A-18 |
| 250 | 0.8 | 2.9 | 3.0 | A-19 |

### 49L.  Aluminum Conductor Impedances

Impedances are for triplexed conductors as shown in Fig. 4-66, which are identical to each other except that the neutral (ground) conductor is uninsulated. Conductor sizes, insulation thicknesses, and three-conductor ampacities are taken from the National E'ectrical Code as specified for Type TW for aluminum wire sizes up through 4/0 AWG and Types RHV and THW for larger conductors. $R$ and $X$ values are given per conductor at 50°C total temperature, no metallic enclosure, $\Omega/1000$ ft at 60 Hz.

| Probable design ampacity | Conductor data | | Line to line | | Line to neutral | |
|---|---|---|---|---|---|---|
| | Size, AWG or MCM | Ampacity | R | X | R | X |
| 30 | 6 | 40 | 0.741 | 0.0349 | 0.741 | 0.0298 |
| 50 | 4 | 55 | 0.466 | 0.0329 | 0.466 | 0.0286 |
| 100 | 2 | 100 | 0.292 | 0.0312 | 0.292 | 0.0276 |
| 110 | 1 | 110 | 0.232 | 0.0314 | 0.232 | 0.0273 |
| 125 | 1/0 | 125 | 0.185 | 0.0305 | 0.185 | 0.0267 |
| 150 | 2/0 | 150 | 0.146 | 0.0297 | 0.146 | 0.0263 |
| 200 | 4/0 | 200 | 0.0920 | 0.0284 | 0.0920 | 0.0256 |
| 225 | 300 | 230 | 0.0650 | 0.0280 | 0.0650 | 0.0252 |
| 250 | 350 | 250 | 0.0557 | 0.0277 | 0.0557 | 0.0250 |
| 400 | Two 4/0 | 400 | 0.0460 | 0.0142 | 0.0460 | 0.0128 |
| 500 | Two 350 | 500 | 0.02785 | 0.01385 | 0.02785 | 0.0125 |

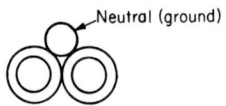

**Fig. 4-66**  Aluminum conductor impedance for triplexed conductors. [National Electrical Manufacturers Association]

**49M. Symmetrical Single-Phase Short-Circuit Currents in Amperes at Various Distances from a Distribution Transformer**
(Rated 120/240 V, three-wire, 25 kVA; Z = 1.91 percent)

| Transformer secondary voltage | Distance, ft | Wire size of aluminum conductors | | | | | | | | | | |
|---|---|---|---|---|---|---|---|---|---|---|---|---|
| | | 6 AWG | 4 AWG | 2 AWG | 1 AWG | 1/0 AWG | 2/0 AWG | 4/0 AWG | 300 MCM | 350 MCM | Two 4/0 MCM | Two 350 MCM |
| 240, line to line | 0 | 5438 | 5438 | 5438 | 5438 | 5438 | 5438 | 5438 | 5438 | 5438 | 5438 | 5438 |
| | 5 | 4830 | 5037 | 5174 | 5222 | 5260 | 5292 | 5337 | 5360 | 5367 | 5387 | 5403 |
| | 10 | 4310 | 4674 | 4927 | 5017 | 5090 | 5152 | 5239 | 5283 | 5298 | 5337 | 5367 |
| | 25 | 3191 | 3796 | 4283 | 4471 | 4628 | 4765 | 4963 | 5065 | 5101 | 5191 | 5264 |
| | 50 | 2180 | 2841 | 3478 | 3753 | 3997 | 4218 | 4555 | 4735 | 4800 | 4963 | 5101 |
| | 75 | 1643 | 2251 | 2907 | 3215 | 3501 | 3772 | 4204 | 4444 | 4532 | 4751 | 4946 |
| | 100 | 1315 | 1857 | 2488 | 2803 | 3107 | 3404 | 3899 | 4184 | 4290 | 4555 | 4800 |
| | 125 | 1095 | 1579 | 2171 | 2481 | 2788 | 3097 | 3632 | 3952 | 4072 | 4373 | 4662 |
| | 150 | 937 | 1371 | 1923 | 2222 | 2525 | 2838 | 3398 | 3743 | 3875 | 4204 | 4532 |
| | 175 | 819 | 1212 | 1725 | 2011 | 2306 | 2618 | 3191 | 3554 | 3695 | 4046 | 4408 |
| | 200 | 728 | 1085 | 1563 | 1835 | 2121 | 2428 | 3006 | 3383 | 3531 | 3899 | 4290 |
| | 500 | 310 | 479 | 731 | 890 | 1071 | 1286 | 1761 | 2132 | 2293 | 2693 | 3241 |
| | 1000 | 158 | 248 | 386 | 477 | 584 | 716 | 1035 | 1313 | 1442 | 1761 | 2293 |
| 120, line to ground | 0 | 8080 | 8080 | 8080 | 8080 | 8080 | 8080 | 8080 | 8080 | 8080 | 8080 | 8080 |
| | 5 | 5670 | 6390 | 6927 | 7128 | 7292 | 7432 | 7634 | 7737 | 7773 | 7851 | 7924 |
| | 10 | 4282 | 5221 | 6023 | 6349 | 6625 | 6869 | 7230 | 7421 | 7488 | 7634 | 7773 |
| | 25 | 2421 | 3306 | 4268 | 4728 | 5156 | 5564 | 6225 | 6604 | 6743 | 7042 | 7353 |
| | 50 | 1390 | 2027 | 2835 | 3277 | 3725 | 4191 | 5034 | 5570 | 5776 | 6225 | 6743 |
| | 75 | 974 | 1457 | 2114 | 2496 | 2903 | 3348 | 4215 | 4810 | 5048 | 5569 | 6223 |
| | 100 | 749 | 1136 | 1683 | 2013 | 2374 | 2783 | 3621 | 4229 | 4481 | 5034 | 5776 |
| | 125 | 608 | 931 | 1397 | 1685 | 2007 | 2379 | 3171 | 3772 | 4027 | 4589 | 5388 |
| | 150 | 512 | 788 | 1194 | 1449 | 1737 | 2076 | 2820 | 3403 | 3656 | 4215 | 5048 |
| | 175 | 442 | 683 | 1042 | 1270 | 1531 | 1842 | 2538 | 3100 | 3347 | 3896 | 4748 |
| | 200 | 389 | 603 | 925 | 1131 | 1369 | 1654 | 2307 | 2845 | 3086 | 3621 | 4481 |
| | 500 | 159 | 250 | 392 | 487 | 601 | 743 | 1099 | 1431 | 1591 | 1951 | 2669 |
| | 1000 | 80 | 126 | 200 | 250 | 310 | 387 | 586 | 781 | 879 | 1099 | 1591 |

**49N. Symmetrical Single-Phase Short-Circuit Currents in Amperes at Various Distances from a Distribution Transformer**
(Rated 120/240 V, three-wire, 37.5 kVA; Z = 1.72 percent)

| Transformer secondary voltage | Distance, ft | Wire size of aluminum conductors | | | | | | | | | | |
|---|---|---|---|---|---|---|---|---|---|---|---|---|
| | | 6 AWG | 4 AWG | 2 AWG | 1 AWG | 1/0 AWG | 2/0 AWG | 4/0 AWG | 300 MCM | 350 MCM | Two 4/0 MCM | Two 350 MCM |
| 240, line to line | 0 | 9040 | 9040 | 9040 | 9040 | 9040 | 9040 | 9040 | 9040 | 9040 | 9040 | 9040 |
| | 5 | 7579 | 8071 | 8398 | 8511 | 8602 | 8678 | 8785 | 8838 | 8856 | 8911 | 8947 |
| | 10 | 6390 | 7213 | 7804 | 8015 | 8188 | 8335 | 8540 | 8643 | 8679 | 8785 | 8856 |
| | 25 | 4181 | 5316 | 6332 | 6744 | 7099 | 7410 | 7866 | 8100 | 8183 | 8422 | 8593 |
| | 50 | 2582 | 3593 | 4707 | 5235 | 5725 | 6188 | 6919 | 7317 | 7461 | 7866 | 8183 |
| | 75 | 1855 | 2687 | 3706 | 4235 | 4755 | 5274 | 6152 | 6659 | 6847 | 7367 | 7807 |
| | 100 | 1445 | 2139 | 3043 | 3540 | 4049 | 4578 | 5524 | 6101 | 6320 | 6919 | 7461 |
| | 125 | 1182 | 1775 | 2576 | 3035 | 3518 | 4036 | 5006 | 5625 | 5865 | 6515 | 7142 |
| | 150 | 1000 | 1515 | 2231 | 2652 | 3106 | 3604 | 4571 | 5214 | 5468 | 6152 | 6847 |
| | 175 | 867 | 1322 | 1967 | 2354 | 2778 | 3252 | 4203 | 4856 | 5119 | 5823 | 6574 |
| | 200 | 764 | 1172 | 1758 | 2115 | 2511 | 2961 | 3887 | 4543 | 4811 | 5524 | 6320 |
| | 500 | 316 | 495 | 769 | 948 | 1159 | 1417 | 2026 | 2540 | 2773 | 3376 | 4291 |
| | 1000 | 159 | 252 | 396 | 493 | 608 | 755 | 1120 | 1456 | 1617 | 2026 | 2773 |
| 120, line to ground | 0 | 13830 | 13830 | 13830 | 13830 | 13830 | 13830 | 13830 | 13830 | 13830 | 13830 | 13830 |
| | 5 | 8099 | 9654 | 10904 | 11389 | 11792 | 12139 | 12641 | 12900 | 12991 | 13214 | 13399 |
| | 10 | 5513 | 7206 | 8843 | 9557 | 10184 | 10752 | 11613 | 12076 | 12241 | 12641 | 12991 |
| | 25 | 2755 | 3983 | 5497 | 6297 | 7089 | 7887 | 9266 | 10097 | 10408 | 11152 | 11895 |
| | 50 | 1492 | 2256 | 3318 | 3948 | 4627 | 5378 | 6863 | 7887 | 8297 | 9266 | 10408 |
| | 75 | 1021 | 1570 | 2367 | 2862 | 3417 | 4059 | 5424 | 6451 | 6881 | 7895 | 9238 |
| | 100 | 776 | 1203 | 1838 | 2242 | 2704 | 3253 | 4476 | 5449 | 5872 | 6863 | 8297 |
| | 125 | 626 | 975 | 1502 | 1842 | 2236 | 2712 | 3807 | 4713 | 5117 | 6062 | 7525 |
| | 150 | 524 | 820 | 1269 | 1562 | 1906 | 2325 | 3310 | 4151 | 4533 | 5424 | 6881 |
| | 175 | 451 | 707 | 1099 | 1357 | 1660 | 2034 | 2927 | 3708 | 4068 | 4906 | 6337 |
| | 200 | 396 | 621 | 969 | 1198 | 1470 | 1807 | 2624 | 3349 | 3688 | 4476 | 5872 |
| | 500 | 160 | 253 | 400 | 499 | 619 | 772 | 1166 | 1547 | 1737 | 2172 | 3108 |
| | 1000 | 80 | 127 | 202 | 253 | 315 | 395 | 605 | 815 | 922 | 1166 | 1737 |

## 490. Symmetrical Single-Phase Short-Circuit Currents in Amperes at Various Distances from a Distribution Transformer

(Rated 120/240 V, three-wire, 50 kVA; Z = 1.72 percent)

| Transformer secondary voltage | Distance, ft. | Wire size of aluminum conductors | | | | | | | | | | |
|---|---|---|---|---|---|---|---|---|---|---|---|---|
| | | 6 AWG | 4 AWG | 2 AWG | 1 AWG | 1/0 AWG | 2/0 AWG | 4/0 AWG | 300 MCM | 350 MCM | Two 4/0 MCM | Two 350 MCM |
| 240, line to line | 0 | 12035 | 12035 | 12035 | 12035 | 12035 | 12035 | 12035 | 12035 | 12035 | 12035 | 12035 |
| | 5 | 9525 | 10351 | 10912 | 11108 | 11267 | 11400 | 11587 | 11679 | 11712 | 11808 | 11872 |
| | 10 | 7658 | 8936 | 9905 | 10262 | 10556 | 10807 | 11162 | 11341 | 11404 | 11587 | 11712 |
| | 25 | 4632 | 6127 | 7587 | 8216 | 8773 | 9273 | 10025 | 10418 | 10559 | 10959 | 11255 |
| | 50 | 2731 | 3913 | 5320 | 6032 | 6721 | 7396 | 8512 | 9145 | 9378 | 10025 | 10559 |
| | 75 | 1926 | 2852 | 4056 | 4719 | 5397 | 6102 | 7361 | 8128 | 8419 | 9216 | 9937 |
| | 100 | 1486 | 2239 | 3267 | 3860 | 4491 | 5173 | 6466 | 7303 | 7629 | 8512 | 9378 |
| | 125 | 1209 | 1841 | 2730 | 3261 | 3838 | 4481 | 5756 | 6623 | 6969 | 7899 | 8874 |
| | 150 | 1019 | 1562 | 2344 | 2820 | 3347 | 3948 | 5181 | 6054 | 6410 | 7361 | 8419 |
| | 175 | 881 | 1357 | 2052 | 2483 | 2966 | 3525 | 4707 | 5572 | 5932 | 6886 | 8006 |
| | 200 | 775 | 1199 | 1824 | 2217 | 2662 | 3183 | 4311 | 5160 | 5519 | 6466 | 7629 |
| | 500 | 318 | 499 | 781 | 967 | 1188 | 1462 | 2128 | 2714 | 2987 | 3686 | 4841 |
| | 1000 | 160 | 253 | 399 | 497 | 616 | 767 | 1149 | 1510 | 1686 | 2128 | 2987 |
| 120, line to ground | 0 | 18421 | 18421 | 18421 | 18421 | 18421 | 18421 | 18421 | 18421 | 18421 | 18421 | 18421 |
| | 5 | 9359 | 11584 | 13503 | 14281 | 14939 | 15515 | 16361 | 16803 | 16960 | 17340 | 17663 |
| | 10 | 6027 | 8169 | 10416 | 11457 | 12403 | 13283 | 14662 | 15424 | 15698 | 16361 | 16960 |
| | 25 | 2866 | 4235 | 6021 | 7017 | 8040 | 9113 | 11071 | 12314 | 12791 | 13925 | 15131 |
| | 50 | 1522 | 2329 | 3490 | 4202 | 4992 | 5895 | 7776 | 9155 | 9726 | 11071 | 12791 |
| | 75 | 1035 | 1604 | 2451 | 2989 | 3605 | 4337 | 5967 | 7263 | 7826 | 9147 | 11056 |
| | 100 | 784 | 1223 | 1887 | 2318 | 2819 | 3426 | 4834 | 6012 | 6541 | 7776 | 9726 |
| | 125 | 631 | 988 | 1534 | 1892 | 2313 | 2830 | 4060 | 5125 | 5615 | 6755 | 8675 |
| | 150 | 528 | 829 | 1292 | 1598 | 1961 | 2410 | 3498 | 4465 | 4917 | 5967 | 7826 |
| | 175 | 454 | 713 | 1116 | 1383 | 1701 | 2098 | 3072 | 3955 | 4373 | 5342 | 7127 |
| | 200 | 398 | 626 | 982 | 1219 | 1502 | 1858 | 2739 | 3549 | 3937 | 4834 | 6541 |
| | 500 | 160 | 254 | 402 | 503 | 625 | 781 | 1187 | 1588 | 1789 | 2249 | 3281 |
| | 1000 | 80 | 127 | 202 | 254 | 316 | 397 | 610 | 826 | 936 | 1187 | 1789 |

**49P. Symmetrical Single-Phase Short-Circuit Currents in Amperes at Various Distances from a Distribution Transformer**
(Rated 120/240 V, three-wire, 75 kVA; Z = 1.80 percent)

| Transformer secondary voltage | Distance, ft | Wire size of aluminum conductors | | | | | | | | | | |
|---|---|---|---|---|---|---|---|---|---|---|---|---|
| | | 6 AWG | 4 AWG | 2 AWG | 1 AWG | 1/0 AWG | 2/0 AWG | 4/0 AWG | 300 MCM | 350 MCM | Two 4/0 MCM | Two 350 MCM |
| 240, line to line | 0 | 17180 | 17180 | 17180 | 17180 | 17180 | 17180 | 17180 | 17180 | 17180 | 17180 | 17180 |
| | 5 | 12439 | 13943 | 15002 | 15377 | 15682 | 15939 | 16298 | 16477 | 16540 | 16731 | 16855 |
| | 10 | 9346 | 11428 | 13133 | 13786 | 14335 | 14805 | 15479 | 15818 | 15938 | 16298 | 16540 |
| | 25 | 5138 | 7116 | 9261 | 10260 | 11180 | 12037 | 13371 | 14086 | 14344 | 15093 | 15652 |
| | 50 | 2885 | 4262 | 6039 | 7007 | 7990 | 9001 | 10781 | 11846 | 12246 | 13371 | 14344 |
| | 75 | 1999 | 3024 | 4442 | 5271 | 6157 | 7123 | 8974 | 10181 | 10652 | 11954 | 13219 |
| | 100 | 1528 | 2340 | 3505 | 4211 | 4990 | 5872 | 7661 | 8907 | 9410 | 10781 | 12246 |
| | 125 | 1237 | 1908 | 2891 | 3502 | 4189 | 4986 | 6672 | 7907 | 8419 | 9801 | 11397 |
| | 150 | 1038 | 1610 | 2459 | 2995 | 3607 | 4328 | 5903 | 7103 | 7611 | 8974 | 10652 |
| | 175 | 895 | 1392 | 2139 | 2616 | 3166 | 3822 | 5290 | 6444 | 6942 | 8268 | 9995 |
| | 200 | 786 | 1226 | 1892 | 2321 | 2820 | 3421 | 4791 | 5894 | 6379 | 7661 | 9410 |
| | 500 | 319 | 504 | 792 | 985 | 1216 | 1507 | 2233 | 2898 | 3216 | 4027 | 5486 |
| | 1000 | 160 | 254 | 402 | 502 | 624 | 779 | 1179 | 1564 | 1756 | 2233 | 3216 |
| 120, line to ground | 0 | 26505 | 26505 | 26505 | 26505 | 26505 | 26505 | 26505 | 26505 | 26505 | 26505 | 26505 |
| | 5 | 10941 | 14226 | 17346 | 18688 | 19857 | 20905 | 22478 | 23315 | 23612 | 24358 | 24984 |
| | 10 | 6605 | 9330 | 12482 | 14056 | 15553 | 17004 | 19383 | 20747 | 21246 | 22478 | 23612 |
| | 25 | 2982 | 4504 | 6612 | 7856 | 9190 | 10657 | 13522 | 15469 | 16240 | 18106 | 20221 |
| | 50 | 1553 | 2405 | 3671 | 4475 | 5392 | 6478 | 8880 | 10769 | 11582 | 13522 | 16240 |
| | 75 | 1049 | 1639 | 2536 | 3121 | 3805 | 4637 | 6585 | 8233 | 8976 | 10734 | 13532 |
| | 100 | 792 | 1243 | 1937 | 2395 | 2937 | 3608 | 5226 | 6656 | 7320 | 8880 | 11582 |
| | 125 | 636 | 1001 | 1566 | 1943 | 2391 | 2952 | 4330 | 5583 | 6177 | 7565 | 10116 |
| | 150 | 532 | 838 | 1315 | 1634 | 2016 | 2497 | 3696 | 4807 | 5341 | 6585 | 8976 |
| | 175 | 456 | 720 | 1133 | 1410 | 1743 | 2163 | 3223 | 4220 | 4704 | 5828 | 8065 |
| | 200 | 400 | 632 | 995 | 1240 | 1534 | 1908 | 2857 | 3760 | 4203 | 5226 | 7320 |
| | 500 | 161 | 255 | 404 | 506 | 630 | 790 | 1209 | 1628 | 1842 | 2328 | 3463 |
| | 1000 | 80 | 128 | 203 | 255 | 318 | 399 | 616 | 836 | 951 | 1209 | 1842 |

**49Q. Symmetrical Single-Phase Short-Circuit Currents in Amperes at Various Distances from a Distribution Transformer**
(Rated 120/240 V, three-wire, 100 kVA; Z = 1.84 percent)

| Transformer secondary voltage | Distance, ft | Wire size of aluminum conductors | | | | | | | | | | |
|---|---|---|---|---|---|---|---|---|---|---|---|---|
| | | 6 AWG | 4 AWG | 2 AWG | 1 AWG | 1/0 AWG | 2/0 AWG | 4/0 AWG | 300 MCM | 350 MCM | Two 4/0 MCM | Two 350 MCM |
| 240, line to line | 0 | 22423 | 22423 | 22423 | 22423 | 22423 | 22423 | 22423 | 22423 | 22423 | 22423 | 22423 |
| | 5 | 15087 | 17370 | 19013 | 19596 | 20071 | 20468 | 21021 | 21293 | 21388 | 21709 | 21896 |
| | 10 | 10716 | 13611 | 16131 | 17125 | 17966 | 18691 | 19728 | 20245 | 20427 | 21021 | 21388 |
| | 25 | 5495 | 7858 | 10618 | 11979 | 13272 | 14509 | 16480 | 17549 | 17935 | 19122 | 19973 |
| | 50 | 2987 | 4501 | 6560 | 7736 | 8974 | 10292 | 12721 | 14228 | 14801 | 16480 | 17935 |
| | 75 | 2046 | 3139 | 4709 | 5661 | 6710 | 7892 | 10269 | 11895 | 12542 | 14389 | 16233 |
| | 100 | 1555 | 2407 | 3665 | 4451 | 5341 | 6376 | 8578 | 10190 | 10856 | 12721 | 14801 |
| | 125 | 1254 | 1951 | 2998 | 3664 | 4430 | 5340 | 7352 | 8899 | 9557 | 11374 | 13585 |
| | 150 | 1051 | 1640 | 2535 | 3112 | 3782 | 4590 | 6426 | 7891 | 8528 | 10269 | 12542 |
| | 175 | 904 | 1415 | 2196 | 2704 | 3298 | 4023 | 5703 | 7084 | 7696 | 9351 | 11641 |
| | 200 | 793 | 1244 | 1936 | 2390 | 2924 | 3579 | 5125 | 6424 | 7009 | 8578 | 10856 |
| | 500 | 321 | 507 | 800 | 997 | 1235 | 1536 | 2301 | 3018 | 3366 | 4258 | 5943 |
| | 1000 | 161 | 255 | 404 | 505 | 628 | 786 | 1197 | 1598 | 1799 | 2301 | 3366 |
| 120, line to ground | 0 | 35187 | 35187 | 35187 | 35187 | 35187 | 35187 | 35187 | 35187 | 35187 | 35187 | 35187 |
| | 5 | 12176 | 16458 | 20845 | 22823 | 24588 | 26195 | 28640 | 29947 | 30411 | 31661 | 32651 |
| | 10 | 7015 | 10204 | 14156 | 16244 | 18301 | 20359 | 23854 | 25906 | 26662 | 28640 | 30411 |
| | 25 | 3058 | 4687 | 7031 | 8467 | 10054 | 11857 | 15557 | 18197 | 19268 | 21958 | 25087 |
| | 50 | 1573 | 2455 | 3792 | 4660 | 5669 | 6891 | 9702 | 12021 | 13047 | 15557 | 19268 |
| | 75 | 1058 | 1662 | 2593 | 3209 | 3938 | 4842 | 7022 | 8941 | 9830 | 11968 | 15575 |
| | 100 | 797 | 1256 | 1969 | 2446 | 3016 | 3729 | 5496 | 7110 | 7876 | 9702 | 13047 |
| | 125 | 640 | 1009 | 1587 | 1976 | 2443 | 3032 | 4513 | 5898 | 6568 | 8150 | 11216 |
| | 150 | 534 | 843 | 1329 | 1657 | 2052 | 2554 | 3827 | 5038 | 5631 | 7022 | 9830 |
| | 175 | 458 | 724 | 1144 | 1427 | 1770 | 2206 | 3322 | 4396 | 4927 | 6166 | 8746 |
| | 200 | 401 | 635 | 1003 | 1253 | 1555 | 1941 | 2935 | 3899 | 4379 | 5496 | 7876 |
| | 500 | 161 | 255 | 406 | 508 | 634 | 795 | 1222 | 1653 | 1875 | 2380 | 3582 |
| | 1000 | 80 | 128 | 203 | 255 | 319 | 401 | 619 | 843 | 959 | 1222 | 1875 |

## 49R. Symmetrical Single-Phase Short-Circuit Currents in Amperes at Various Distances from a Distribution Transformer

(Rated 120/240 V, three-wire, 167 kVA; Z = 1.92 percent)

| Transformer secondary voltage | Distance, ft | Wire size of aluminum conductors | | | | | | | | | | |
|---|---|---|---|---|---|---|---|---|---|---|---|---|
| | | 6 AWG | 4 AWG | 2 AWG | 1 AWG | 1/0 AWG | 2/0 AWG | 4/0 AWG | 300 MCM | 350 MCM | Two 4/0 MCM | Two 350 MCM |
| 240, line to line | 0 | 35475 | 35475 | 35475 | 35475 | 35475 | 35475 | 35475 | 35475 | 35475 | 35475 | 35475 |
| | 5 | 19525 | 23936 | 27475 | 28798 | 29894 | 30820 | 32120 | 32758 | 32982 | 33750 | 34194 |
| | 10 | 12537 | 17015 | 21545 | 23522 | 25273 | 26841 | 29158 | 30337 | 30753 | 32120 | 32982 |
| | 25 | 5867 | 8728 | 12461 | 14503 | 16580 | 18711 | 22417 | 24581 | 25382 | 27824 | 29729 |
| | 50 | 3085 | 4745 | 7148 | 8616 | 10242 | 12087 | 15831 | 18410 | 19441 | 22417 | 25382 |
| | 75 | 2090 | 3250 | 4988 | 6091 | 7356 | 8852 | 12131 | 14617 | 15664 | 18603 | 22048 |
| | 100 | 1580 | 2470 | 3826 | 4704 | 5728 | 6966 | 9803 | 12086 | 13084 | 15831 | 19441 |
| | 125 | 1270 | 1992 | 3102 | 3830 | 4686 | 5738 | 8214 | 10289 | 11221 | 13746 | 17359 |
| | 150 | 1061 | 1669 | 2608 | 3228 | 3964 | 4875 | 7064 | 8951 | 9815 | 12131 | 15664 |
| | 175 | 912 | 1435 | 2250 | 2790 | 3434 | 4237 | 6194 | 7918 | 8719 | 10847 | 14262 |
| | 200 | 799 | 1259 | 1978 | 2456 | 3029 | 3746 | 5514 | 7097 | 7841 | 9803 | 13084 |
| | 500 | 322 | 509 | 806 | 1008 | 1252 | 1565 | 2372 | 3152 | 3540 | 4519 | 6525 |
| | 1000 | 161 | 255 | 406 | 508 | 633 | 793 | 1215 | 1634 | 1847 | 2372 | 3540 |
| 120, line to ground | 0 | 56066 | 56066 | 56066 | 56066 | 56066 | 56066 | 56066 | 56066 | 56066 | 56066 | 56066 |
| | 5 | 13594 | 19398 | 26197 | 29608 | 32847 | 35970 | 41023 | 43855 | 44878 | 47654 | 49947 |
| | 10 | 7424 | 11163 | 16241 | 19169 | 22245 | 25538 | 31671 | 35595 | 37096 | 41023 | 44878 |
| | 25 | 3128 | 4862 | 7461 | 9124 | 11033 | 13305 | 18364 | 22348 | 24058 | 28331 | 34076 |
| | 50 | 1591 | 2500 | 2908 | 4842 | 5950 | 7328 | 10673 | 13641 | 15021 | 18364 | 24058 |
| | 75 | 1066 | 1682 | 2645 | 3292 | 4068 | 5048 | 7504 | 9790 | 10891 | 13513 | 18512 |
| | 100 | 802 | 1267 | 1999 | 2493 | 3090 | 3849 | 5782 | 7629 | 8536 | 10673 | 15021 |
| | 125 | 642 | 1016 | 1606 | 2007 | 2491 | 3110 | 4702 | 6248 | 7017 | 8813 | 12629 |
| | 150 | 536 | 848 | 1343 | 1679 | 2086 | 2608 | 3962 | 5289 | 5955 | 7504 | 10891 |
| | 175 | 460 | 728 | 1153 | 1443 | 1795 | 2246 | 3423 | 4586 | 5173 | 6532 | 9572 |
| | 200 | 402 | 638 | 1011 | 1265 | 1574 | 1972 | 3012 | 4047 | 4572 | 5782 | 8536 |
| | 500 | 161 | 256 | 407 | 510 | 637 | 800 | 1235 | 1679 | 1908 | 2430 | 3709 |
| | 1000 | 80 | 128 | 204 | 256 | 319 | 402 | 623 | 850 | 968 | 1235 | 1908 |

**49S. Symmetrical Single-Phase Short-Circuit Currents in Amperes at Various Distances from a Distribution Transformer**
(Rated 120/240 V, three-wire, 250 kVA; Z = 3.01 percent)

| Transformer secondary voltage | Distance, ft | Wire size of aluminum conductors | | | | | | | | | | |
|---|---|---|---|---|---|---|---|---|---|---|---|---|
| | | 6 AWG | 4 AWG | 2 AWG | 1 AWG | 1/0 AWG | 2/0 AWG | 4/0 AWG | 300 MCM | 350 MCM | Two 4/0 MCM | Two 350 MCM |
| 240, line to line | 0 | 33936 | 33936 | 33936 | 33936 | 33936 | 33936 | 33936 | 33936 | 33936 | 33936 | 33936 |
| | 5 | 20488 | 24812 | 27957 | 29027 | 29875 | 30557 | 31453 | 31861 | 32002 | 32694 | 32956 |
| | 10 | 13123 | 17851 | 22416 | 24284 | 25866 | 27209 | 29044 | 29890 | 30178 | 31453 | 32002 |
| | 25 | 6025 | 9043 | 12998 | 15137 | 17278 | 19416 | 22926 | 24797 | 25454 | 27898 | 29312 |
| | 50 | 3131 | 4849 | 7361 | 8902 | 10610 | 12535 | 16348 | 18834 | 19786 | 22926 | 25454 |
| | 75 | 2111 | 3300 | 5098 | 6246 | 7566 | 9128 | 12514 | 14996 | 16010 | 19168 | 22322 |
| | 100 | 1592 | 2500 | 3892 | 4800 | 5861 | 7149 | 10084 | 12395 | 13382 | 16348 | 19786 |
| | 125 | 1278 | 2011 | 3146 | 3894 | 4778 | 5866 | 8425 | 10537 | 11468 | 14195 | 17717 |
| | 150 | 1067 | 1682 | 2640 | 3275 | 4031 | 4970 | 7227 | 9152 | 10021 | 12514 | 16010 |
| | 175 | 916 | 1446 | 2273 | 2825 | 3485 | 4310 | 6323 | 8038 | 8891 | 11174 | 14585 |
| | 200 | 802 | 1267 | 1996 | 2484 | 3069 | 3804 | 5618 | 7234 | 7987 | 10084 | 13382 |
| | 500 | 322 | 511 | 809 | 1012 | 1259 | 1575 | 2394 | 3184 | 3576 | 4591 | 6632 |
| | 1000 | 161 | 256 | 406 | 509 | 634 | 796 | 1221 | 1643 | 1857 | 2394 | 3576 |
| 120, line to ground | 0 | 55685 | 55685 | 55685 | 55685 | 55685 | 55685 | 55685 | 55685 | 55685 | 55685 | 55685 |
| | 5 | 14229 | 20560 | 27961 | 31591 | 34948 | 38060 | 42761 | 45166 | 45989 | 48862 | 50560 |
| | 10 | 7626 | 11596 | 17073 | 20246 | 23568 | 27078 | 33371 | 37129 | 38492 | 42791 | 45989 |
| | 25 | 3165 | 4950 | 7657 | 9407 | 11428 | 13843 | 19205 | 23338 | 25064 | 29868 | 35458 |
| | 50 | 1600 | 2523 | 3963 | 4925 | 6072 | 7504 | 10998 | 14090 | 15512 | 19205 | 25064 |
| | 75 | 1071 | 1693 | 2671 | 3331 | 4126 | 5134 | 7672 | 10037 | 11171 | 14012 | 19203 |
| | 100 | 804 | 1273 | 2014 | 2516 | 3124 | 3899 | 5884 | 7784 | 8715 | 10998 | 15512 |
| | 125 | 644 | 1020 | 1616 | 2021 | 2513 | 3143 | 4771 | 6354 | 7140 | 9042 | 12994 |
| | 150 | 537 | 851 | 1349 | 1689 | 2102 | 2632 | 4011 | 5367 | 6046 | 7672 | 11171 |
| | 175 | 460 | 730 | 1158 | 1450 | 1806 | 2264 | 3459 | 4644 | 5242 | 6661 | 9792 |
| | 200 | 403 | 639 | 1015 | 1271 | 1583 | 1986 | 3041 | 4093 | 4626 | 5884 | 8715 |
| | 500 | 161 | 256 | 407 | 511 | 638 | 803 | 1240 | 1687 | 1918 | 2449 | 3746 |
| | 1000 | 80 | 128 | 204 | 256 | 320 | 403 | 624 | 852 | 971 | 1240 | 1918 |

**50. Low-voltage enclosed fuses** (600-V or less) may be classified as follows:

A. Type of enclosure and contact construction
   1. Plug
   2. Cartridge
      a. Ferrule contact: Class CC, G, H, J, K, R, or T
      b. Knife-blade: Class H, J, K, L, R, or T
B. Renewability
   1. One-time
   2. Renewable
C. Time overload present for fuse to blow
   1. Ordinary type; very little time delay
   2. Time delay

The basic features of the different types of enclosures and contact construction are shown in Figs. 4-67 and 4-68. Plug fuses are made in sizes up to and including 30 A. Class CC, G, H, J, and K cartridge fuses with ferrule contacts are made in sizes up to and including 60 A. Class H, J, K, R, and T cartridge fuses with knife-blade contacts are made in sizes from 70 up to and including 600 A. Above 600 A, Class L current-limiting fuses are available in ratings up to 6000 A.

One-time fuses cannot be reused after they have blown from the occurrence of a fault. Renewable fuses are designed so that the fusible element can be replaced, after it has blown, by a new element. Thus the fuse cases can be used over and over, and the fuse expense is thereby reduced in service in which the blowing of fuses is frequent and short-circuit currents are less than 10,000 A.

There is a certain lapse of time for all fuses between the occurrence of an overload and the opening of the circuit by rupture of the fusible element. For the ordinary type (the original one-time fuse) this time lapse is very short. The ordinary plug fuse, for example, opens the circuit in approximately 3 s at 200 percent load. The time lapse required for a fuse to blow can be controlled through proper design. According to the standards of the Underwriters Laboratories, any fuse with an opening time greater than 12 s at 200 percent load is characterized as a time-delay (time-lag) fuse. Such fuses are marked with the letter D, time-delay or dual-element. Other classes require 8 to 10 s at 500 percent.

Low-voltage fuses rated 600 A or less are of standard ratings and dimensions as established by the National Electrical Code and the Underwriters Laboratories. The standard dimensions for Class H cartridge fuses are given in Sec. 52. The dimensions for fuses rated above 600 A are covered by UL 198C, high-interrupting-capacity fuses, current-limiting type (Type L), and UL 198H, Class T fuses.

Fuses are made in several types as listed previously. Certain types are definitely limited in their application, some can be used for nearly any application, and others are designed for special applications for which their characteristics are required so that proper protection may be realized and system coordination maintained.

**51. Underwriters Laboratories standards on fuses** require that all fuses meet the following requirements:

1. The standard on fuses requires that the fuses be of certain dimensions within specified tolerances and meet the following tests:

2. They must withstand heavy short circuits. Failure to do this would result in a very great fire hazard, as such failure would mean that the fuse would explode or belch fire.

3. They must operate at a reasonably low temperature. Failure to do this would make a hazard of the fuse.

4. They must carry rated current. Failure to do this would result in premature burn-out, causing unnecessary expense and annoyance. All fuses are required to carry a 10 percent overload indefinitely when tested in the open. When fuses are installed in enclosures, as is usually the case, they carry less current. Hence the 10 percent overload requirement makes sure that they carry rated current in actual practice.

5. They must blow promptly at an overload. Failure to do this would result in repair bills amounting to many times the cost of the fuse. The requirements are that fuses blow at a 35 percent overload when tested in the open so that they will blow at a 25 percent overload when installed in an enclosure, as they usually are. The blowing time permitted on 0- to 60-A fuses is 1 h and on larger fuses 2 h.

**52. Dimensions of Underwriters Laboratories–Listed Cartridge-Enclosed Class H Fuses**

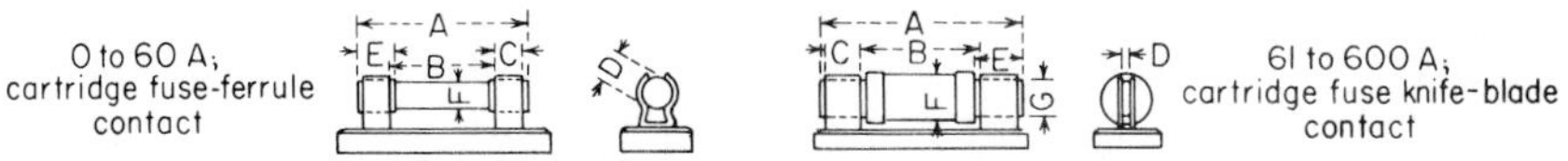


**Fig. 4-67** Class H and K standard fuses and holders.

| | | *A* | *B* | *C* | *D* | *E* | *G* |
|---|---|---|---|---|---|---|---|
| **Voltage** | Rated capacity, A | **Length over terminals, in** | **Distance between contact clips, in** | **Width of contact clips, in** | Diameter of ferrules or thickness of terminal blades, in | Minimum length **of ferrules or of terminal blades outside of tube, in** | **Width of terminal blades, in** |
| **0–250** | **0–30** | **2** | **1** | **½** | **9/16** | **½** | |
| | **31–60** | **3** | **1¾** | **⅝** | **13/16** | **⅝** | |
| | **61–100** | **5⅞** | **4** | **⅞** | **⅛** | **1** | **¾** |
| | **101–200** | **7⅛** | **4½** | **1¼** | **3/16** | **1⅜** | **1⅛** |
| | **201–400** | **8⅝** | **5** | **1¾** | **¼** | **1⅞** | **1⅝** |
| | **401–600** | **10⅜** | **6** | **2⅛** | **¼** | **2¼** | **2** |
| **251–600** | **0–30** | **5** | **4** | **½** | **13/16** | **½** | |
| | **31–60** | **5½** | **4¼** | **⅝** | **1 1/16** | **⅝** | |
| | **61–100** | **7⅞** | **6** | **⅞** | **⅛** | **1** | **¾** |
| | **101–200** | **9⅝** | **7** | **1¼** | **3/16** | **1⅜** | **1⅛** |
| | **201–400** | **11⅝** | **8** | **1¾** | **¼** | **1⅞** | **1⅝** |
| | **401–600** | **13⅜** | **9** | **2⅛** | **¼** | **2¼** | **2** |

**53. Cartridge-fuse types.** Cartridge fuses come in a wide range of types, sizes, and ratings. Various classes are designated by NEMA standards and Underwriters Laboratories (UL) standards. In broad terms, these fuses are classified as one-time, renewable, dual-element, current-limiting, or high-interrupting-capacity.

Present NEMA and UL standards indicate that standard (so-called National Electrical Code type) cartridge fuses of one-time or renewable types are designated as Class H. Such fuses are generally classified at an interrupting-capacity rating of 10,000 A. Refer to Sec. 52 for dimensions of 250- or 600-V fuseholders.

Cartridge-fuse classifications, based on existing UL requirements at IC ratings above 10,000 rms symmetrical amperes, are Class J, L, R, T, G, CC, or K. These fuses are high-interrupting-capacity or current-limiting types. They are shown in Figs. 4-68*A* and 4-68*B*. The term *high-interrupting-capacity fuse* indicates a fuse interrupting rating at some value above 10,000 to 200,000 rms symmetrical amperes, depending upon the particular fuse. A *current-limiting fuse* is a fuse which safely interrupts all available currents within its interrupting rating and limits the peak let-through current $I_p$ and the total amperes squared-seconds $I^2t$ to a specified degree. UL states that *current-limiting* indicates that a fuse, when tested on a circuit capable of delivering a specific short-circuit current (rms symmetrical amperes) at rated voltage, will start to melt within 90 electrical degrees and will clear the circuit within 180 electrical degrees (½ cycle).

1. Class J and L Fuses Both the Class J and the Class L fuses are current-limiting, high-interrupting-capacity types. The interrupting rating is 200,000 rms symmetrical amperes, and it is marked on the label of each Class J or L fuse.

Class J fuse dimensions are different from those for standard Class H cartridge fuses of the same voltage rating and ampere classification. As such, they require special fuseholders that will not accept non-current-limiting fuses. This arrangement complies with the last sentence of Sec. 240-60(b) of the National Electrical Code, which reads: "Fuseholders for current-limiting fuses shall not permit insertion of fuses which are not current limiting."

Class J fuses of 60 A or less are ferrule type, and from 61 to 600 A they have slots in the fuse blades to permit bolted or knife-blade connections to fuseholders.

Class L fuses are divided into several different amperage classifications; and fuse-blade mounting holes, the number of which varies according to fuse sizes, permit bolted connection to fuseholders.

UL standards list specific dimensions for Class J fuses and recommend specific dimensions for Class L fuses. Both Class J and Class L fuses, covered in these UL rules, closely parallel the existing NEMA standard (FU-1), "Low-Voltage Cartridge Fuses," which includes construction and test recommendations for Class J and Class L fuses.

2. CLASS G FUSES   Any 300-V cartridge fuse intended to be used on circuits exceeding 10,000 rms symmetrical amperes has been given a designation by UL as Class G. These 300-V cartridge fuses successfully passed UL-supervised tests at 100,000 rms symmetrical amperes. Figure 4-68A shows the dimension of these fuses.

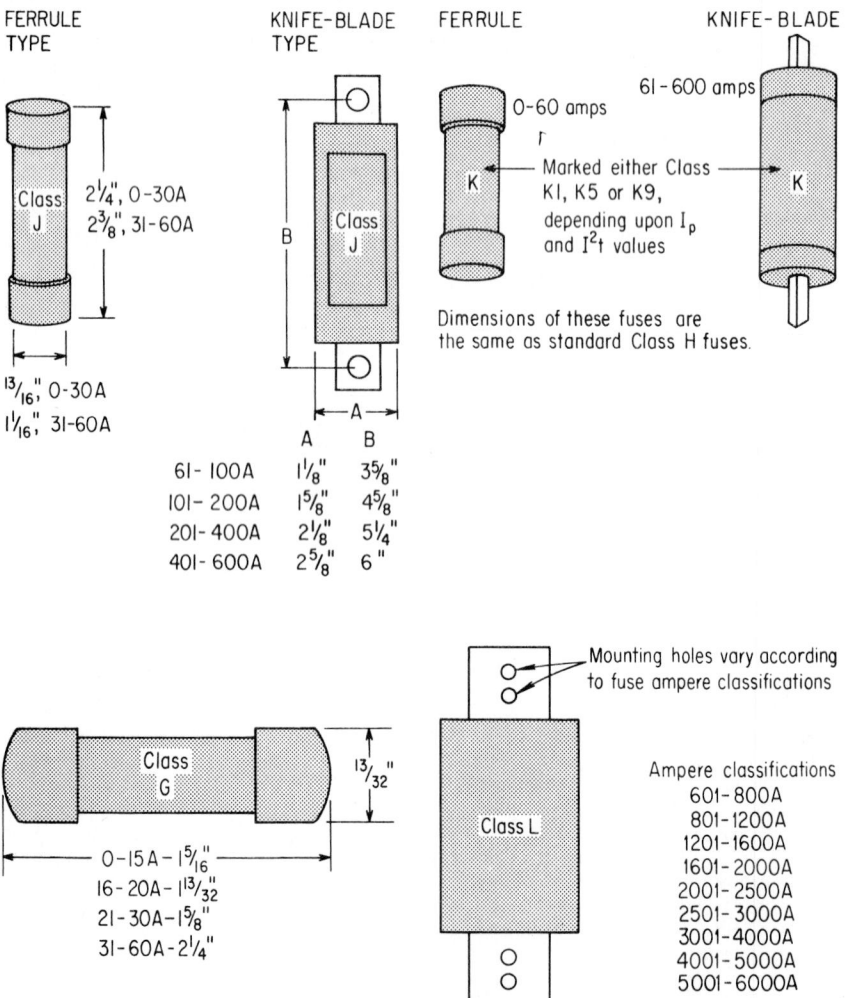

**Fig. 4-68A**   Ampere classifications of Class CC, G, H, K, J, L, and T high-interrupting-capacity or current-limiting cartridge fuses.

3. CLASS K FUSES  These fuses have interrupting ratings from 50,000 to 200,000 rms symmetrical amperes at various peak let-through currents $I_p$ and maximum let-through energy conditions $I^2t$. Initially these fuses were divided into three groups, K1, K5, and K9. These groups indicate the degree of $I^2t$ and $I_p$ values. Class K1 fuses provide more current limitation than do K5 or K9 fuses. UL can verify proposed available current ratings of switches or motor starters, based on the use of a specific K-type fuse (K1, K5, etc.). K9 fuses have virtually disappeared from the marketplace.

4. CLASS R FUSES  These fuses are cartridge-type with a high interrupting capacity and are current-limiting with a built-in rejection feature that prohibits the installations of standard Class H fuses in Class R fuseholders. Class R fuses may be inserted in fuseholders of the standard Class H design. The fuses have passed the UL tests at 200,000 rms symmetrical amperes and have a current-limiting range for the individual case sizes. Class R fuses are divided into two subclasses according to the peak let-through current and total amperes-squared–seconds characteristics. These subclasses are identified by the designations RK1 and RK5 and are physically interchangeable. Fig. 4-68$B$ shows these fuses.

5. CLASS T FUSES  These fuses are cartridge-type with a high interrupting capacity and are current-limiting. The fuses are intended for use with fuseholders which will properly receive Class T fuses but will now receive fuses of any other type. The fuses have passed UL tests of 200,000 rms symmetrical amperes with peak let-through currents and total amperes-squared–seconds characteristics specified for the individual case sizes. Fig. 4-68$A$ shows these fuses. There are two voltage classes, 300 and 600.

All presently UL-listed Class K fuses have the same dimensions as conventional Class H 250-V or 600-V, 0- to 600-A fuses (Fig. 4-67). Because of this interchangeable feature,

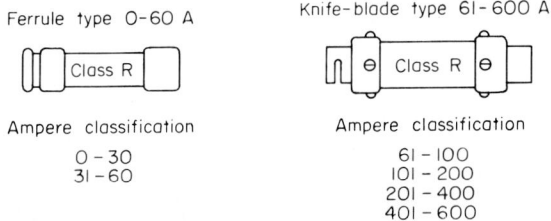

Ferrule type  0–60 A

Knife-blade type  61–600 A

Ampere classification

0 – 30
31 – 60

Ampere classification

61 – 100
101 – 200
201 – 400
401 – 600

1. Interrupting rating 200,000 A rms symmetrical.
2. 250 or 600 V ac.
3. Current–limiting.
4. Dimensions are the same as standard Class H fuses.
5. Fuseholders with a Class R rejection feature will not accept Class H fuses.

Class T fuses: ferrule type

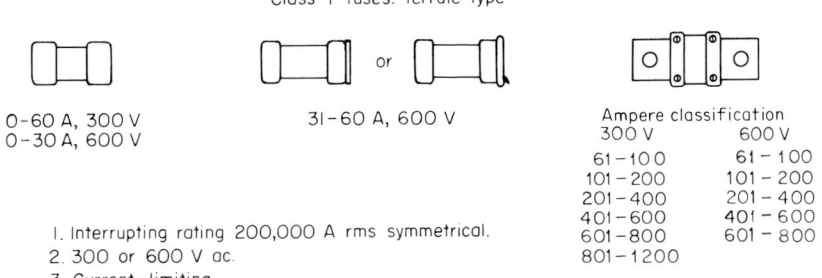

0–60 A, 300 V
0–30 A, 600 V

31–60 A, 600 V  or

Ampere classification

| 300 V | 600 V |
|---|---|
| 61–100 | 61 – 100 |
| 101–200 | 101 – 200 |
| 201–400 | 201 – 400 |
| 401–600 | 401 – 600 |
| 601–800 | 601 – 800 |
| 801–1200 | |

1. Interrupting rating 200,000 A rms symmetrical.
2. 300 or 600 V ac.
3. Current–limiting.
4. Class T fuseholders will not receive fuses of any other type.
5. For 300 V use fuse clip or bolted-in mounting for knife-blade type, and bolted-in mounting for 600 V knife-blade type.
6. 300-V fuses may be used on 277/480-V systems where the voltage of the supply does not exceed 300. The same is true for Class G (300-V) fuses.

**Fig. 4-68$B$**  Class R and T current-limiting fuses.

Class K fuses are not labeled "current-limiting" even though the $I_p$ and $I^2t$ values of K1 fuses compare closely with those for Class J fuses.

**54. The ordinary one-time cartridge fuse** (Fig. 4-69) is the oldest type of cartridge fuse in common use today. It consists of a tube of vulcanized fiber, paper, or some similar material within which the fuse element is mounted. The fuse terminals are connected to contact pieces at the ends of the tube. An insulating porous powder resembling chalk surrounds the fuse and fills or nearly fills the tube. When the fuse blows, the powdered material quenches the arc. The ordinary type of cartridge fuse is available in ratings from 1 up to and including 600 A and for use on circuits with maximum voltages of 250 or 600 V, respectively. These fuses have very little time delay, and their use therefore is limited to short-circuit protection on circuits in which faults occur infrequently. They are known as Class H fuses and have an interrupting-capacity rating of 10,000 A in UL listings.

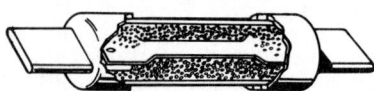

**Fig. 4-69** Ordinary one-time fuse. [Bussmann Division of Cooper Industries Inc.]

**55. Renewable cartridge fuses** (Figs. 4-70 to 4-72) are designed so that the fusible element can be readily replaced when blown by a new element. They are available in ratings from 3 up to and including 600 A and for use on circuits with maximum voltages of 250 or 600 V, respectively. Most renewable cartridge fuses are of the time-delay type. The time delay is accomplished through special link construction that combines parts with a heavy cross-sectional area with parts having a very reduced cross-sectional area. As an example, consider the superlag fuse shown in Fig. 4-72. In this fuse, heavy lag plates are attached to the center of the link. The principal blowing parts of the fuse are near the terminals. The heavy lag plates keep the center of the link relatively cool. An extra reduced section is provided in the center of these plates so that they do not increase the current-carrying capacity of the link too much. As a result, the lag plates serve only to help the terminals conduct away and temporarily store some of the heat generated in the weak spots so that it takes a longer time to get the weak spots heated sufficiently to blow. Thus a time lag far in excess of that of other types of fuses is obtained.

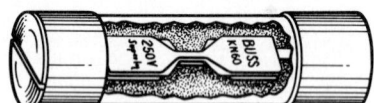

**Fig. 4-70** Ferrule-contact time-lag renewable cartridge fuse. [Bussmann Division of Cooper Industries Inc.]

**Fig. 4-71** Knife-blade-contact time-delay renewable cartridge fuse. [Reliance Fuse Division of Reliance Electric Co.]

On light overloads the center weak spot of the superlag link will sometimes burn out. This happens because the rise in temperature of the entire strip is so gradual that the heavy metal terminals of the fuse can conduct the heat away from the end weak spots fast enough to keep them relatively cool. Whether the two end weak spots or the center weak spot of the bus superlag fuse blows first, the superior time lag obtained through the use of the lag plates is not affected.

When a short circuit occurs, the two end weak spots melt instantaneously, and because of the well-known fact that two arcs in series cannot maintain themselves as long as a single arc on the same voltage, much less metal is vaporized.

The thick lag plates also serve to reduce the amount of metal vaporized. First, their cooling effect at ordinary overloads makes it possible to reduce the amount of metal in the weak spots. Second, the thick lag plates in the center of the link serve to keep cooler the entire mass of metal in the center, and therefore the arc that follows the short circuit cannot so readily melt

**Fig. 4-72** Cutaway view of time-lag fuse of the renewable type. [Superlag fuse of the Bussmann Division of Cooper Industries Inc.]

this center section. As a consequence, less metal is vaporized on short-circuit blows than with any other type of fuse. This means that less pressure is created in the fuse, and the danger of a fuse rupturing or exploding or belching fire on short circuits therefore is considerably reduced.

The time-delay renewable fuse is used to take advantage of lower replacement costs for the protection of mains and feeders in which faults occur frequently.

**56. Dual-element Class K cartridge fuses** combine a thermal element for protection against overloads up to approximately 800 percent of their rating and a fuse link element for protection against heavier overloads and short circuits. They are available in ratings of $\frac{1}{10}$ up to and including 600 A and for use on circuits with maximum voltages of 250 or 600 V respectively.

A typical dual-element fuse is shown in Fig. 4-73. It consists of two copper links $A$ and $C$ located in the two end sections of the container and a heavy copper center strap $B$. The center strap $B$ is held in position by a soldered connection to a spring. The soldered connection acts as the thermal element, while the copper links in the end sections provide the short-circuit element. Moderate overloads will not blow the short-circuit fuse element but will open the circuit through the action of the thermal element. The center strap $B$ is heated by the light overloads. When $B$ is heated up to the melting point of the solder, the spring is released and the thermal-cutout portion opens. The heat coil

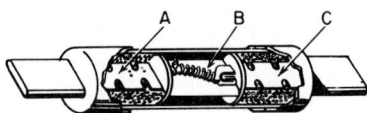

**Fig. 4-73** Nonrenewable and dual-element cartridge fuse. [Fusetron of the Bussmann Division of Cooper Industries Inc.]

and center strap are quite heavy, so that it takes considerable time to raise the temperature of the strap sufficiently to melt the solder. This feature together with the time required to melt the solder gives the fuse its time-lag characteristics. On heavy overloads the fuse link blows immediately, before the thermal-cutout portion has time to function.

Dual-element fuses provide highly desirable protection for most circuits, including mains, feeders, and lighting and appliance branch circuits, as well as motor branch circuits and the overload protection of the motor itself. Their long time lag prevents useless shutdowns caused by ordinary fuses or breakers opening on motor-starting currents or other harmless overloads. They will hold even if all motors on a circuit start at one time; yet they protect against short circuits with all the speed of an ordinary fuse.

Dual-element fuses have a lower resistance than any ordinary fuses. Hence switches and panelboards will operate at a much cooler temperature with dual-element fuses than with ordinary fuses. This prevents damage and the needless blowing of fuses so often caused by excessive heating.

Dual-element fuses also give a new kind of protection to switches and panelboards: thermal protection. The thermal cutout in a dual-element fuse will open whenever its temperature reaches approximately 280° (138°C). Thus if poor contact heat develops from any cause, the dual-element fuse cuts off the current before damaging temperatures can be reached.

On motor installations dual-element fuses are particularly advantageous. On normal installations, because of their long time lag, a fuse of a size of about 100 to 125 percent of ampere rating can be installed in a disconnect switch or branch-circuit panel. When so used, dual-element fuses give motor-running protection as safe and dependable as that furnished by the most expensive devices made.

Double protection can be given to motors already protected by other devices simply by replacing the fuses used for short-circuit protection with dual-element fuses of motor-running–protection size. Then if such other devices fail for any reason, the dual-element fuses will open to protect the motor against any dangerous overload or single-phasing condition.

On new installations, dual-element fuses permit use of proper-size switches and panels instead of oversize ones. With ordinary fuses, switches and panels must be oversized because fuses much larger than the operating load must be used to hold the starting current. But dual-element fuses hold starting currents; therefore proper-size switches and panels to fit the load can be installed. Their use often solves the problem of finding space for the switch or panel and generally saves money as well.

On present installations, by replacing oversize fuses with dual-element fuses switches or panels can be loaded near their capacity. A larger motor or additional motors can often be installed without the trouble of changing the switch or panel.

Refer to Table 48, Div. 11, for increased horsepower ratings of fused switches when time-delay fuses are used as the branch-circuit protection for motors.

**57. Small-dimension fuses** are available for the protection of instruments and other special apparatus. Fuses for instrument protection generally are enclosed in clear glass containing tubes.

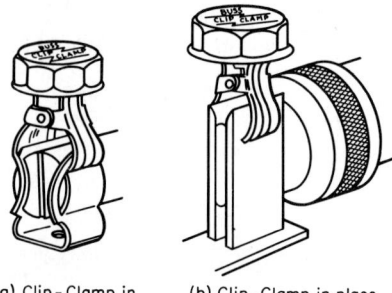

**58. Fuse accessories** which are advantageous in the use of fuses for special applications are available.

Clip-clamps (Fig. 4-74) which will ensure good contact between the fuse terminals of cartridge fuses and the fuse clips are available.

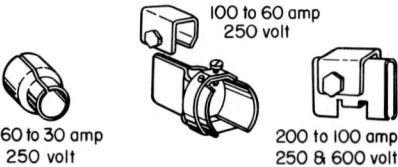

(a) Clip-Clamp in place on 0 to 60 amp fuse clip.

(b) Clip-Clamp in place on 70 to 600 amp fuse clip.

60 to 30 amp 250 volt

100 to 60 amp 250 volt

200 to 100 amp 250 & 600 volt

**Fig. 4-74** Clip-clamps for cartridge fuses. [Bussmann Division of Cooper Industries Inc.]

**Fig. 4-75** Fuse reducers. [Bussmann Division of Cooper Industries Inc.]

Fuse reducers (Fig. 4-75) make it possible to use cartridge fuses of a smaller size than that for which the fuse clips are intended. They are available for the following reductions:

| For 250-volt Fuses, Amp | For 600-volt Fuses, Amp |
|---|---|
| 60– 30 | 60– 30 |
| 100– 30 | 100– 30 |
| 100– 60 | 100– 60 |
| 200– 60 | 200– 60 |
| 200–100 | 200–100 |
| 400–200 | 400–200 |
| 600–400 | 600–400 |

**59. Plug fuses** are available in three types (refer to Fig. 4-76): (*a*) the ordinary type, (*b*) the time-delay type with standard screw base, and (*c*) the Type S (tamper-resisting) type. All plug fuses are of the one-time type. The standard sizes are 15, 20, 25, and 30 A. Sizes of 1, 2, 3, 5, 6, 8, and 10 A are also regularly available. Fuses of 15-A capacity or less are provided with a hexagonal window, and those of greater capacity with a round window. Plug fuses should be used only in circuits not exceeding 125 V or in circuits of a system having a grounded neutral and no conductor at more than 150 V to ground. Plug fuses installed in residences should be of the time-delay type on circuits of 20 A or less.

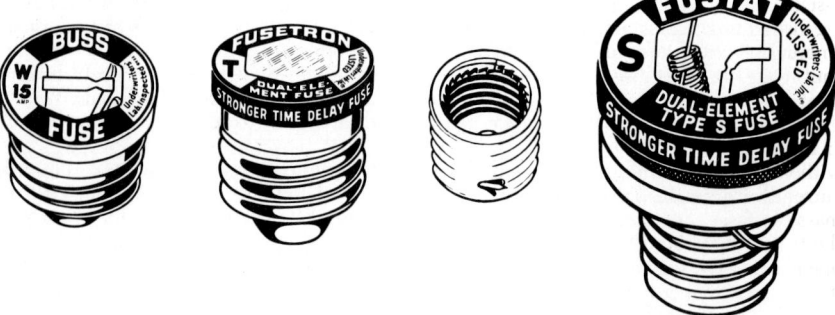

**Fig. 4-76** Types of plug fuses. [Bussmann Division of Cooper Industries Inc.]

The *ordinary plug fuse* (Fig. 4-77) consists of a wire or strip of fusible alloy mounted in a porcelain container. The container is fitted with a screw base corresponding to the standard medium lamp-base dimensions and threads. The top of the container is transparent to make the fuse link visible. Since the ordinary plug fuse has very little time delay (it opens in approximately 3 s at 200 percent load), it is subject to blowing on harmless transient overloads. This practically limits the satisfactory use of the ordinary plug fuse to incandescent-lighting or heating-appliance circuits.

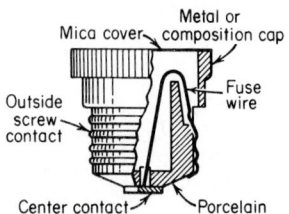

**Fig. 4-77**  Edison-base type of plug fuse.

*Plug fuses of the time-delay type* are available in both the Edison-base type (Fig. 4-78) and the Type S (tamper-resisting type; Fig. 4-79). Edison-base plug fuses are approved only as a replacement item in existing installations in which there has been no evidence of overfusing or tampering.

A *dual-element, time-delay type* of standard screw-base plug fuse is shown in Figs. 4-80 and 4-81. This fuse consists of a fuse link to which a thermal cutout is added. The internal construction of the screw-base type is shown in Fig. 4-80. An overload causes the thermal cutout to heat up, and if the overload is continued long enough, the solder in the thermal cutout softens to permit the spring to pull out the end of the fuse link, thus opening the circuit. Because it takes some time to melt solder even with a heavy current, the thermal cutout cannot open quickly, and the fuse link is heavy enough so that it will not open quickly on motor-starting currents. Hence the time-lag fuse will not open on motor-starting currents that last only a short while. When a short circuit or an overload of approximately 800 percent occurs, the fuse link opens in exactly the same manner as an ordinary fuse.

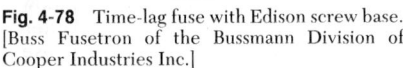
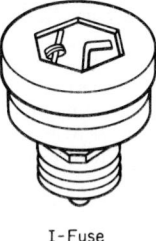

I-Fuse          II-Adapter

**Fig. 4-78**  Time-lag fuse with Edison screw base. [Buss Fusetron of the Bussmann Division of Cooper Industries Inc.]

**Fig. 4-79**  Time-lag fuse with tamperproof Type S screw base. [Buss Fustat of the Bussmann Division of Cooper Industries Inc.]

A time-lag fuse designed especially for the protection of small motors is shown in Fig. 4-81. It is intended to be mounted directly on the motor. When the receptacle is properly installed, the heat from the motor windings is conducted to the thermal cutout in the time-lag fuse at point 1. Should the windings reach a prohibitive temperature with just a normal amount of current flowing, the time-lag fuse will open at point 1, and the motor will be shut down. The opening of the time-lag fuse under these conditions is primarily due to the heat of the motor windings rather than to the current flowing through the fuse. Thus the motor is automatically protected against an excessive rise of the ambient temperature or failure of air circulation or any other condition causing excessive heating of the motor without a corresponding rise in the flow of current. If, on the other hand, the flow of current is so high that the insulation in the motor will be injured or destroyed before the mass of the motor is heated sufficiently to operate the time-lag fuse, as mentioned above, the thermal cutout is heated by the excessive flow of current through the heat coil and opens at point 2. The flow of current is stopped no matter how cold the exterior of the motor may be. If the motor does not properly come up to speed owing to low voltage, a tight belt, dry bearings, or any other cause, the time-lag fuse will open at point 1 or 2. If the motor should be stalled, the excessive flow of current would cause the time-lag fuse to open at point 2. If a short or ground should occur, the excessive current would cause the fuse to open in the fuse link, point 3. However, this method generally has been replaced with thermal devices similar to those described in Sec. 68.

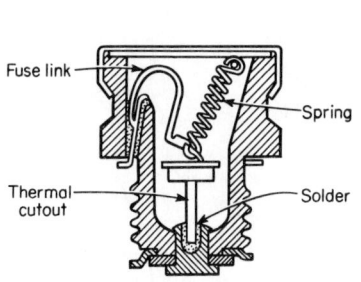

**Fig. 4-80**    Construction of a time-lag fuse of the screw-base type. [Bussmann Division of Cooper Industries Inc.]

**Fig. 4-81**    Time-lag fuse for protection of small motors. [Bussmann Division of Cooper Industries Inc.]

The *tamper-resisting type of plug fuse* (Fig. 4-79) is known as the Type S fuse. It is identical with the standard screw-base type of time-delay plug fuse except for the construction of the base. The Type S fuse has a different-size base which is screwed into an adapter. The complete assembly is then screwed into a regular Edison-base fuseholder. Adapters are made in sizes from 1 to 30 A. The 15-A adapter will take fuses of any size from 7 to 15 A. The 20-A adapter will take only a 20-A fuse, and the 30-A adapter will accommodate a 20-, 25-, or 30-A fuse. Type S fuses make safe protection remain safe, since (1) once the correct adapter has been installed, an oversize fuse cannot be inserted, and (2) bridging and tampering are practically impossible. They give the best fuse protection for ordinary lighting, appliance, and small motor branch circuits, and they are required when plug fuses are to be used as overcurrent protection in new installation.

**60. High-voltage fuses.**    Fuses for circuits having a voltage greater than 600 are specially constructed so that they will be safe for the interruption of current under such voltages. Two types for protection of power circuits are shown in Figs. 4-82 and 4-83. The fuse of Fig. 4-82 is of the oil-fuse-cutout type. The fuse link shown at II has a section of fusible alloy at the bottom. The entire link except the top laminated-metal terminals is enclosed

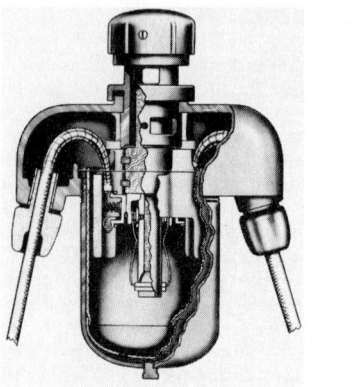

I. Cutaway view of assembly.            II. Fuse link of cutout.

**Fig. 4-82**    D&W oil fuse cutout. [General Electric Co.]

in flat tubing of insulating material. The link is mounted in a casing and immersed in oil. The fuse of Fig. 4-83 consists of a tube lined with boric acid, which provides a source of deionization for extinguishing the arc. A high-voltage fuse for protecting a potential transformer is shown in Fig. 4-84.

I. RBA expulsion-type fuse and disconnect mounting.

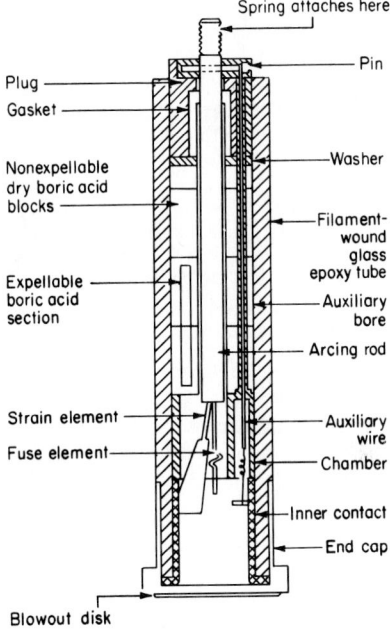

II. Cross-section view of RBA expulsion-type power-fuse refill showing construction details.

**Fig. 4-83**    High-voltage power fuse of the boric acid type. [Westinghouse Electric Corp.]

**61. Circuit breakers** are of the two basic types, air circuit breakers (frequently called switchgear) and molded- (insulated-) case breakers. They are switching devices designed to open a circuit automatically at a predetermined overcurrent without damage to themselves when applied within their ratings. Circuit breakers can be closed and opened manually or electrically. Additionally, they have inherent short-circuit ratings to distinguish them from switches, which have limited current-interruption ability.

A circuit-breaker assembly essentially includes (1) an operating mechanism, (2) a contact-and-arc extinguisher system, and (3) an intelligence or current-detection scheme.

A.  OVERLOAD RESPONSE    Early swtichgear circuit breakers utilized air dashpots, but modern designs use solid-state circuitry and current transformers. To provide a trip response for overload current, molded-case circuit breakers employ (1) bimetals, (2) hydraulic dashpots, or (3) electronic circuits. State-of-the-art circuit breakers use current transformers and solid-state circuitry to provide inverse-time delay to overloads. Conventional

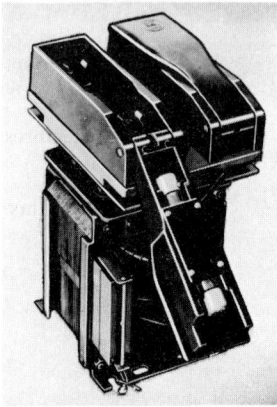

**Fig. 4-84** High-voltage fuse mounted on potential transformer. [Westinghouse Electric Corp.]

molded-case breakers use bimetals or a cylindrical dashpot having a dynamic core which moves within a fluid.

1. Bimetal designs include two bonded metallic strips having dissimilar thermal-expansion rates. Heat generated by the current causes the bimetal to bend against a trip bar, which unlatches the spring mechanism and opens the contacts as shown in Fig. 4-85. The time-current curve (or inverse-time delay) of a typical 100-A molded-case breaker is also shown in Fig. 4-85; it illustrates the rapid decrease of trip time with increasing overload.

2. Dashpots provide inverse delay to overloads by means of a movable slug inside a hermetically sealed tube (Fig. 4-86). A bias spring positions the slug (core) away from the pole face until an overload through the coil attracts the core upward, which enhances the magnetic field, pulling the armature down to unlatch the breaker. The core does not move until minor overloads overcome the bias-spring force. The speed of the core motion varies with increasing current, but tripping due to short-circuit current occurs instantaneously owing solely to the magnetomotive force of the coil.

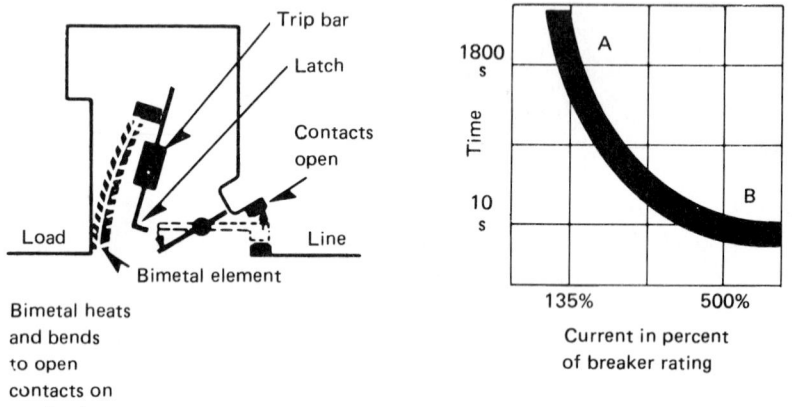

**Fig. 4-85** Principle of thermal trip action. [Westinghouse Electric Corp.]

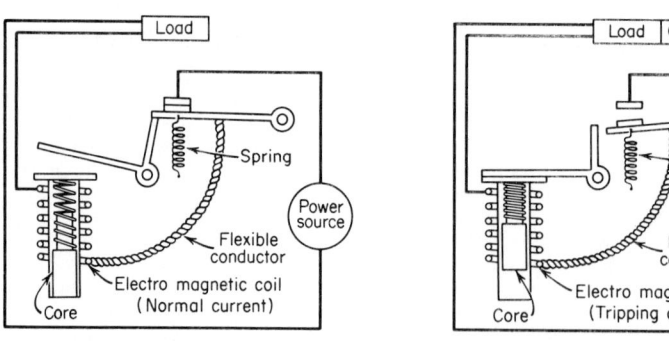

**Fig. 4-86** Principle of magnetic breaker.

B. Magnetic (Instantaneous) Trip Action   Electromagnets in series with the load current effect tripping without delay, typically within 1 cycle for molded-case breakers (Fig. 4-87, I). Fault current passing through the breaker causes the electromagnet to effect unlatching and open the contacts. The breaker trips at a predetermined pickup point $A$ in Fig. 4-87, III.

C. Thermal-magnetic breakers combine the thermal and instantaneous features described above and employ both bimetals and electromagnets to provide a trip curve (Fig. 4-87, III). Conventional 150-A and smaller frames have fixed instantaneous trip levels (Fig. 4-87, I), but larger frames have adjustable trip levels to suit user requirements. Figure 4-87, III depicts this adjustment feature, which is usually from 3 to 10 times the nominal trip unit rating.

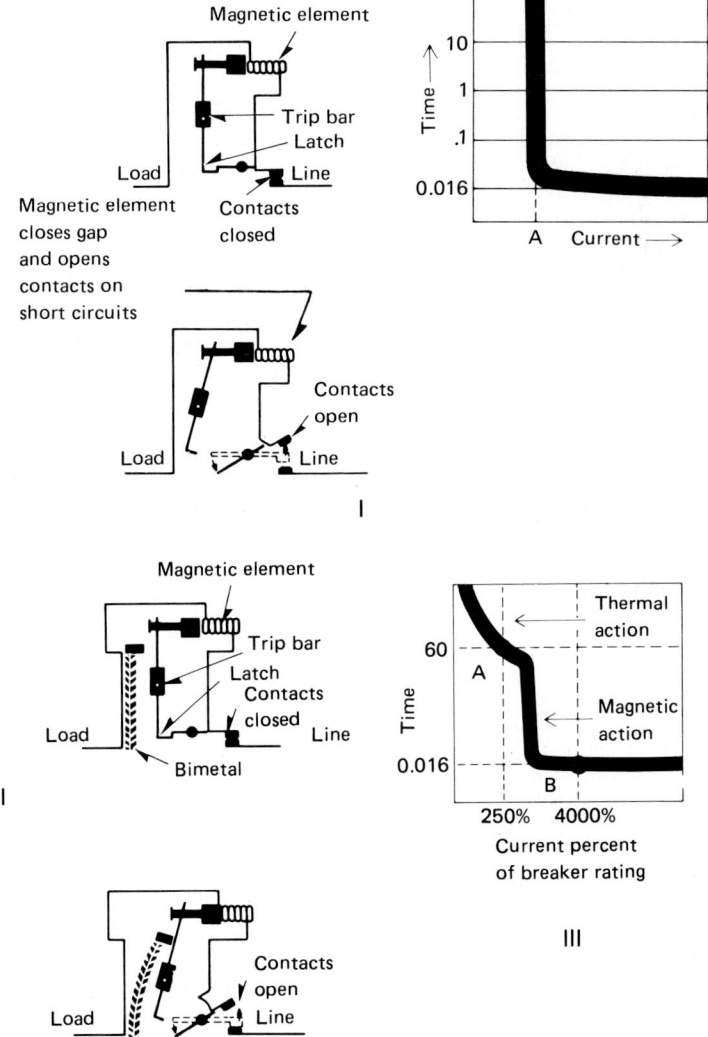

**Fig. 4-87**   Magnetic and thermal-magnetic breakers. I. Magnetic (instantaneous) trip action. II. Thermal-magnetic action. III. Typical time-current curve for a 100-A thermal-magnetic breaker. [Westinghouse Electric Corp.]

The National Electrical Code (NEC) defines thermal-magnetic designs as time-limit, and they satisfy the large majority of circuit-breaker applications. They are sensitive to temperature changes to protect cable and equipment, allow longer overloads under low-temperature conditions, and trip instantly during short circuits.

D. TEMPERATURE CONSIDERATION (AMBIENT COMPENSATION)  The thermal elements of molded-case breakers can be modified to compensate for ambient-temperature variations and minimize the need to uprate them in lower temperatures and derate them for high ambients. An additional bimetal is used in the thermal element to coordinate with the primary bimetal and counteract the effects of varying ambient temperature (Fig. 4-88). This provides a more constant current rating and reduces nuisance tripping for installations having a circuit breaker at a location and ambient temperature different from those of the circuit conductor.

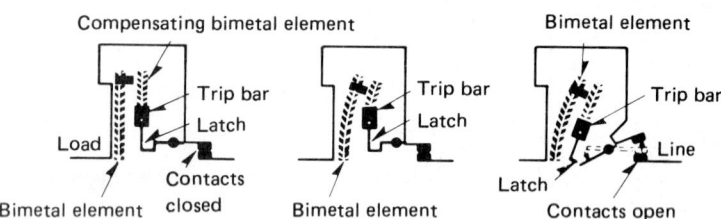

**Fig. 4-88**   Ambient-compensating breaker. [Westinghouse Electric Corp.]

**62. Current-limiting fused circuit breakers.**  There are available both switchgear and molded-case circuit breakers which combine special current-limiting fuses to provide compact devices having high fault interruption, typically 200 kA at 600 V ac. Their prime advantage lies in avoiding useless melting of fuses due to overloads and moderate short circuits by utilizing the breakers for opening and having the fuses operate only during major faults. Additionally, these designs have interlocks to prevent single-phase operation. Refer to Fig. 4-89, I and II, for a typical molded-case design and a schematic of the trip action.

**63. Current-limiting circuit breakers.**  Molded-case circuit breakers are also available with special contact designs to provide current limiting without fuses. They are listed by Underwriters Laboratories and represent state-of-the-art construction. They effect rapid arc extinction, minimize potential circuit problems, and protect conventional downstream devices. Figure 4-90 illustrates one pole of a typical design.

**64. Class CTL devices.**  Certain types of plug-in molded-case breakers are designed to satisfy Sec. 384-15 of the National Electrical Code, which requires that the number of circuit devices (fuses or breakers) not exceed the number of circuits for which a panelboard is designed (Fig. 4-91). Consequently, the circuit breakers have unique dimensions and mounting configurations to comply with the maximum number of poles for which the panelboard is designed.

**64A. Electronic circuit breakers.**  State-of-the-art designs of both molded-case and switchgear types of circuit breakers have replaced conventional thermal-magnetic and dashpot trips with current transformers, flux-transfer shunt trips, and solid-state circuitry which are an integral part of the breaker construction (Fig. 4-92). The result is increased accuracy, reliability, and repeatability that were never available with previous breakers. Typically, electronic-circuit-breaker designs have the same physical dimensions and nominal ratings as the conventional breakers they supersede (Fig. 4-93).

The electronic-type units use current transformers to reduce the current to the correct ratio for input to the printed-circuit board (Fig. 4-94). Dual transformation provides an efficient intelligence circuit with low power requirements. Under prolonged overloads or short-circuit conditions, the circuit board generates an output to a low-power flux-transfer device which trips the circuit breaker.

Solid-state circuitry allows users to fashion the trip curve to their exact needs, in that *both* current magnitude *and* time are adjustable with the turn of a dial. Furthermore,

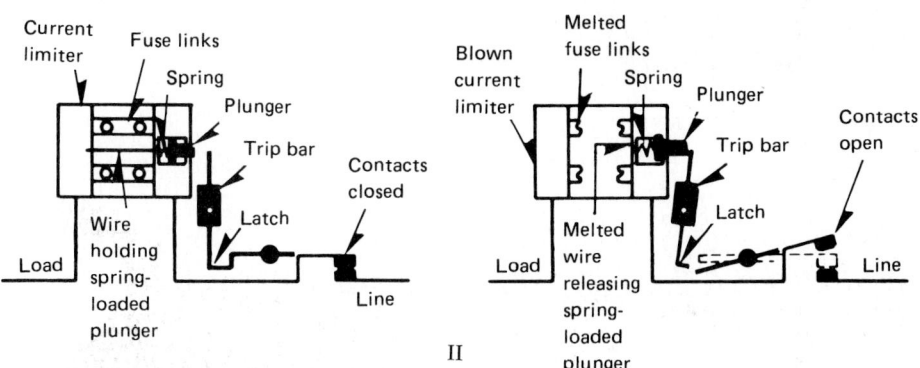

**Fig. 4-89**  High-interrupting-capacity breakers. I. High-interrupting-capacity breaker by use of current-limiting devices. II. Current-limiter action. [Westinghouse Electric Corp.]

latest designs have inherent equipment ground-fault protection, enabling users to dial both pickup and trip time on overloads, short circuits, and ground-fault currents.

**65. Manual or magnetic switches** provided with thermal or magnetic overload relays are commonly used for the overload protection of motors. These switches are discussed in Div. 7.

**66. Motor burnouts** usually are caused by currents which exceed the motor current rating. In the thermal overload relay, excessive current is translated into a temperature increase. The elevated temperature actuates the relay to trip the motor from the line.

A heater element in the relay is connected in series with the motor. Tripping is usually accomplished by the relay opening the motor-control circuit. The trip point (expressed in amperes) is determined by the heater rating.

Heaters of various current ratings are interchangeable within the specific product lines of any given manufacturer. All heaters, also called current elements, perform the same function; they interpret an increase in motor current as a temperature increase in the overload relay.

**67. Types of thermal overload relays.**  The two main types of thermal overload relays in use today are the bimetallic and the eutectic-alloy types. Bimetallic types (see Fig. 4-95)

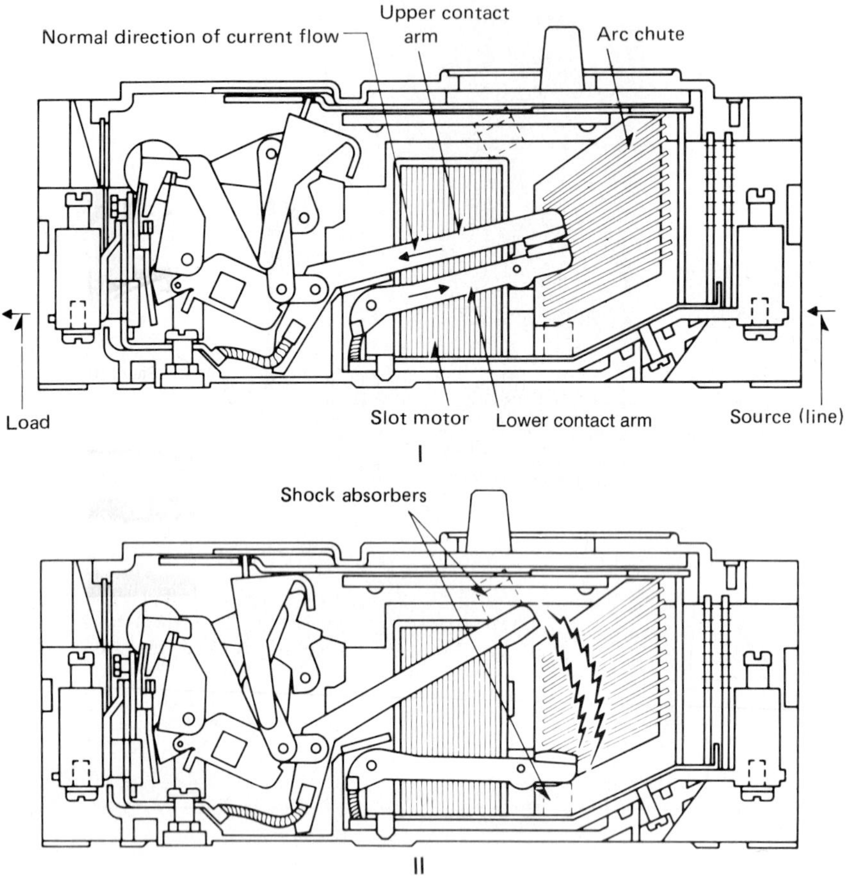

Upper contact arm
Normal direction of current flow
Arc chute
Load
Slot motor
Lower contact arm
Source (line)
I
Shock absorbers
II

**Fig. 4-90** Contact action during fault interruption. I. Contacts closed: on position. II. Contacts open: position during high-level fault interruption. [Westinghouse Electric Corp.]

CLASS CTL DEVICE
ON
OFF
15A
N

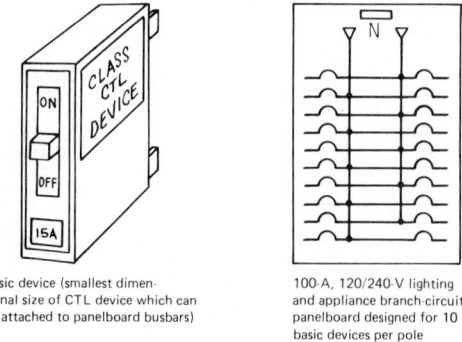

Basic device (smallest dimensional size of CTL device which can be attached to panelboard busbars)

100-A, 120/240-V lighting and appliance branch-circuit panelboard designed for 10 basic devices per pole

**Fig. 4-91** Class CTL insert-type overcurrent devices and panel assemblies limit the number of circuits in lighting panelboards.

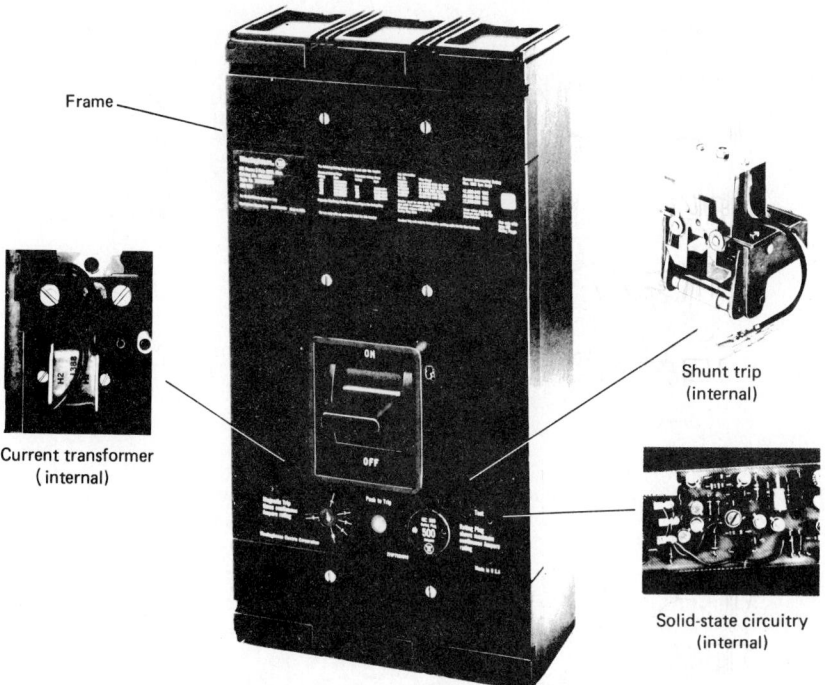

Frame

Current transformer
( internal)

Shunt trip
(internal)

Solid-state circuitry
(internal)

**Fig. 4-92**   Solid-state circuit breaker. [Westinghouse Electric Corp.]

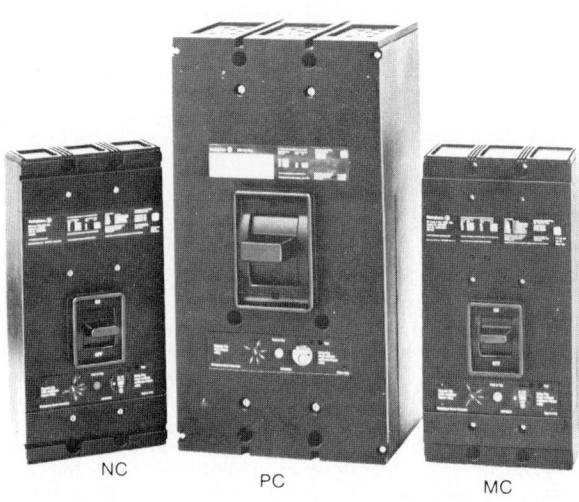

NC

PC

MC

**Fig. 4-93**   Solid-state Seltronic circuit breakers. [Westinghouse Electric Corp.]

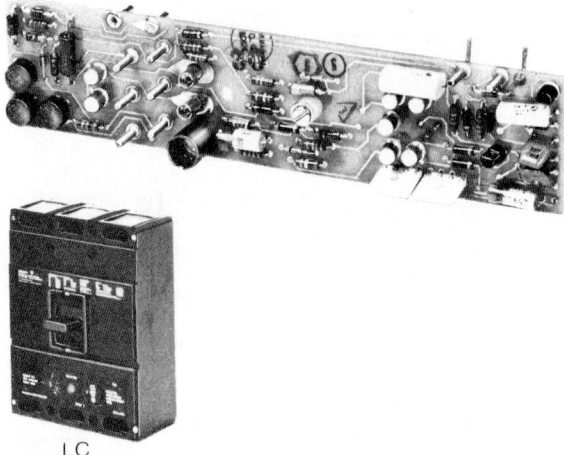

**Fig. 4-94**   Printed-circuit board used in Seltronic. [Westinghouse Electric Corp.]

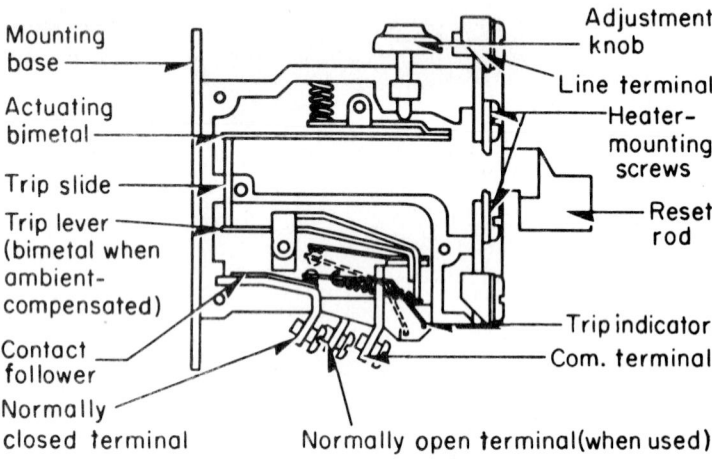

**Fig. 4-95**   Bimetallic overload relay. [Westinghouse Electric Corp.]

operate much in the manner of a common bimetallic-element thermostat. Current passing through the heater heats the bimetallic strip, causing it to deflect. In deflecting, the strip opens a normally closed contact which is wired into the motor's control circuit. Operation of a bimetallic-type overload relay is illustrated in Fig. 4-96.

Eutectic-alloy relays are commonly called solder-pot overload relays. Operation of a solder-pot relay is illustrated in Fig. 4-97. In such a relay, the contacts are kept closed by a spring-loaded pawl engaging a ratchet. When the motor current is below the value of the heater rating, the ratchet shaft is secured by a pool of hardened solder (eutectic-alloy). As the current rises to a value determined by the heater rating, the solder melts, permitting the ratchet to slip and disengage the pawl.

Both types of thermal overload relays have specific advantages. The bimetallic type is considered more versatile than the solder-pot variety. For applications requiring automatic reset, ambient compensation, and adjustment between incremental heater ratings, the bimetallic type must be used. On the other hand, the solder-pot type performs better

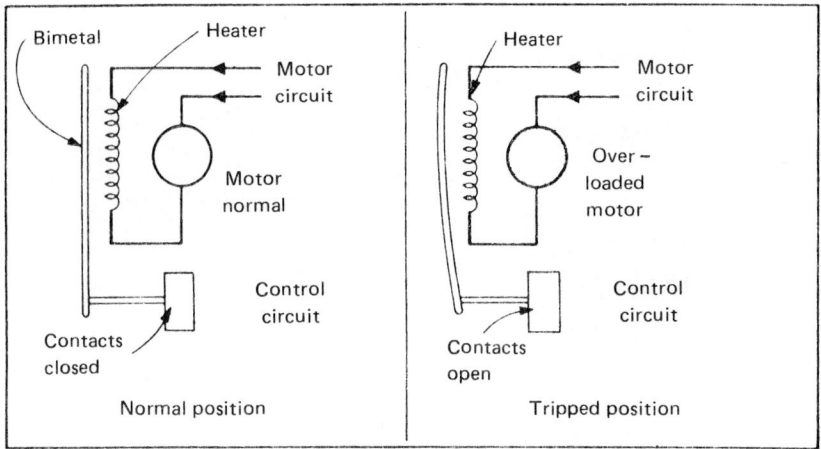

**Fig. 4-96**   Operation of bimetallic overload relay. [Westinghouse Electric Corp.]

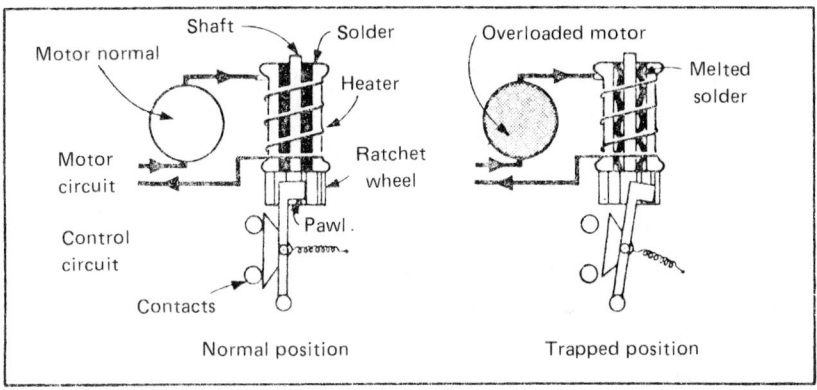

**Fig. 4-97**   Operation of a eutectic-alloy (solder-pot) overload relay.

when abnormal physical shock and vibration are encountered and when higher control-circuit ratings are required. Both are available in single-pole and three-pole configurations (Fig. 4-98).

In most applications, however, the critical factor is not the choice of solder-pot versus bimetallic overload relay. The essential consideration is proper selection of the heater element.

All thermal overload relays have a class designation that specifies the maximum number of seconds in which an overload relay will trip when carrying a current equal to 600 percent of the heater-element rating. Present industry standards recognize Class 10, 20, and 30 overload relays. They require that Class 10 and 30 relays be marked with their class designation. Class 20 overload relays need not be marked.

Class 10 overload relays should be used for hermetic refrigerant motor-compressors, submersible pumps, and similar applications in which the motor can only tolerate stalled (locked) rotor conditions for less than 10 s. Class 30 overload relays may be required for motors driving high-inertia loads, such as ball mills, reciprocating pumps, loaded conveyors, etc. Most motors will be properly protected by a Class 20 overload relay selected in accordance with the control manufacturer's instructions.

**68. A thermal protector** is another form of heat-sensitive element that is authorized for running overcurrent protection. One or more thermal protectors are included in the

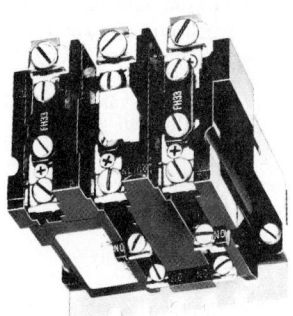
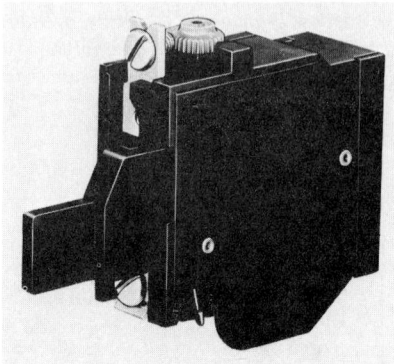

I. Block type.                                    II. Single pole.

**Fig. 4-98**   Thermal overload relays. [Westinghouse Electric Corp.]

motor windings at the time that the motor is manufactured. They respond to the temperature at their location by opening a circuit. Thermal protectors may open the power circuit within the motor itself or may open a control circuit which deenergizes a motor controller, thus removing power from the motor. Thermal protectors are rated in motor horsepower and operate in response to currents that are high multiples of a nominal motor full-load current (140 to 170 percent).

**69. Magnetic relays and magnetic overload release devices** consist of a solenoid acting upon an iron plunger or armature (Fig. 4-95). The coil of the solenoid is connected in series with the circuit to be protected or frequently in ac circuits to the secondary of a current transformer which has its primary connected in series with the main circuit. The current at which the relay will function can be regulated by adjusting the height of the plunger or the position of the movable armature. Instantaneous-type relays and overload release devices function instantaneously upon the occurrence of an overload of any value of current above that for which the device is set.

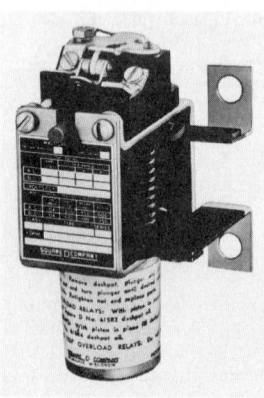

By the attachment of an air bellows or an oil dashpot to the plunger the device can be given inverse- or definite-time characteristics. With the inverse-time type, the time that an overload must exist before the device functions is inversely proportional to the severity of the overload. A slight overload must continue for a considerable time before the device will function, while a heavy overload will cause the device to function almost instantaneously. With the definite-time type, the device does not function until the overload has persisted for a certain definite length of time, depending upon the setting of the device. The elapsed time between the occurrence of the overload and the functioning of the device does not depend upon the severity of the overload. The essential difference between a magnetic relay and a magnetic overload release device is the same as for thermal relays and thermal overload release devices. The overload release device acts directly upon the catch of the switch mechanism, while a relay acts upon an auxiliary circuit, which in turn trips the switch mechanism.

**Fig. 4-99**   Magnetic inverse-time overload relay. [Square D Co.]

Magnetic relays are used in conjunction with circuit breakers and magnetic switches for overload protection. Magnetic overload release devices are employed in conjunction with circuit breakers only.

An inverse-time magnetic overload relay, which is used with magnetic contactors in motor controllers, is shown in Fig. 4-99.

A modified type of magnetic overload release, called a hydraulic-magnetic release, is used on some circuit breakers. A breaker equipped with this type of release is shown in Fig. 4-100. The hydraulic-magnetic tripping element of this circuit breaker is simply a solenoid coil wound around a hermetically sealed nonmagnetic cylinder containing a spring-loaded movable iron core and a silicone fluid. When a current up to and including the rated current is passing through the coil, the iron core stays at the end of the cylinder opposite the tripping armature, as shown in Fig. 4-101, I.

When the iron core reaches the opposite end of the cylinder, the magnetic field has increased in intensity to a point to attract the armature. In turn, the armature releases the

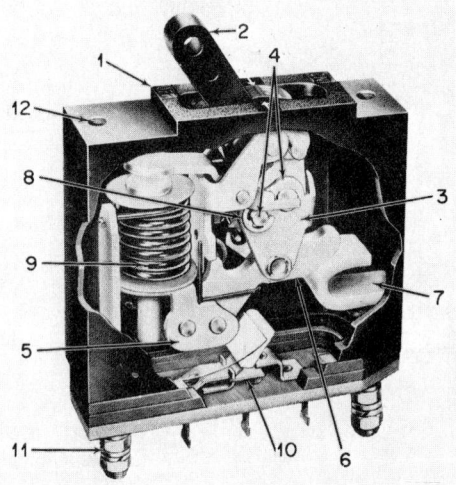

1. Moldarta case.
2. Handle indication. Operational positions of the handle are clearly marked "on" and "off." After a fault clearance, service is restarted by turning handle back to on position.
3. Trip-free toggle mechanism.
4. Free bearing surfaces. These are of dissimilar metals to prevent sticking and bearing wear.
5. Corrosion resistance. All nonhardened ferrous parts are cadmium-plated with a chromate dip which provides maximum resistance to corrosion.
6. Self-cleaning contact. Moving on a sliding pivot point, the contact arm causes a wiping motion of the contacts whenever they are opened or closed. This action assures low contact resistance and long contact life.
7. High-speed arc blowout. When the circuit is broken, a coil-formed stationary contact creates a magnetic field, forcing the arc into the arcing chamber. Arcing takes place on special surfaces, not on contact surfaces used for normal contact.
8. Accurate protection.
9. Hydraulic-magnetic trip unit.
10. Auxiliary switch. A single-pole double-throw auxiliary switch is available as an optional feature when required. Contacts are mechanically actuated by a toggle mechanism but are electrically isolated from the breaker circuit. Auxiliary switches are rated 5 A, 125 V; 1 A, 50 V dc, noninductive.
11. Rear-connected terminals.
12. Threaded inserts. To ensure simplicity of mounting threaded brass inserts are provided to prevent stripping during installation.

**Fig. 4-100**    Hydraulic-magnetic circuit breaker. [Westinghouse Electric Corp.]

tripping mechanism, and the circuit breaker opens the circuit (see Fig. 4-101, III). On short circuits, the magnetic flux produced in the coil alone is strong enough to attract the armature regardless of core position, and the circuit interruption is instantaneous.

**70. A combined thermal and magnetic overload release** is used on some circuit breakers. A breaker with this combined release is shown in Fig. 4-102. The inverse-time element of the thermal release provides overload protection for overloads of ordinary magnitude without causing circuit interruption on harmless overloads. The magnetic instantaneous trip provides for split-second tripping on short circuit.

**71. Induction-type overload relays** are seldom used for ordinary interior-wiring work. They are widely used by central-station companies for the protection of circuits and apparatus. A detailed discussion of these relays is outside the scope of this book. Induction relays operate upon the general principle of induction motors and therefore are applicable only to ac systems. They can be obtained with instantaneous-, inverse-, or definite-time characteristics.

**72. Comparison of the time characteristics of overcurrent protective devices.** Typical time-current characteristics of the different overcurrent protective devices are illustrated in Fig. 4-103. The curve shown for thermal and magnetic devices is typical. All such devices

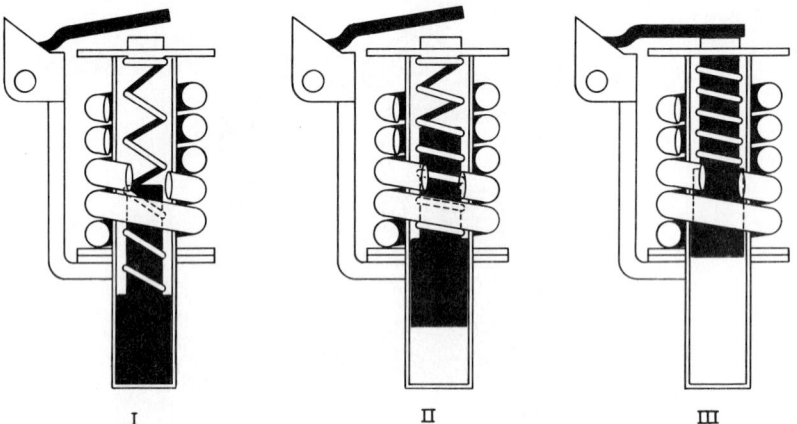

I          II          III

**Fig. 4-101** Hydraulic-magnetic overload release. [Westinghouse Electric Corp.]

1. Thermal time-delay trip.
2. Magnetic instantaneous trip.
3. Over-center toggle mechanism.
4. Silver-alloy contacts.
5. High contact pressure.
6. Common tripper bar.
7. Arc chute.
8. Solderless pressure wire connectors.
9. Three-position handle: indicates "on," "automatically tripped," or "off."
10. Totally enclosed case.

**Fig. 4-102** Combined thermal-and-magnetic–trip molded-case circuit breaker rated 400 A, 600 V ac, 250 V dc. [Square D Co.]

will have a characteristic of the form shown, although different devices, depending upon design, will vary from that shown in the time of opening the circuit for the various overloads. Time-lag fuses have characteristics similar to those shown for the thermal and magnetic devices. The curve labeled "fuse" is for the ordinary type of fuse, not for time-lag ones. It will be noticed that fuses of the regular type have a considerable time-delay characteristic. They will not blow immediately on an overcurrent unless the overcurrent is approximately 200 percent or greater than the rating of the fuse. Fuses will carry indefinitely a current of 80 percent of their rating. The inverse-time overload protective devices of the thermal, magnetic, and time-lag fuse type have a greater time-delay characteristic than regular fuses and will therefore allow much greater overloads for a short period without interrupting the circuit. These inverse-time characteristics make these overload protective devices ideal for the overload protection of motors. They will allow the heavy starting current to pass but will protect the motor against sustained overloads that would be harmful to it. They will also permit the motor to operate with intermittent peak overloads of short duration. If ordinary fuses are used for the overload protection of induction motors, the fuses must be shorted during the starting period, for a fuse of proper rating to protect the motor would not allow the starting current to pass. Although ordinary fuses will permit a motor to operate with moderate intermittent peak overloads, they are not nearly so flexible in this respect as the thermal and magnetic protective devices. Devices with inverse-time characteristics are also advantageous for protection of circuits, since they will prevent interruption of the circuit due to short overloads, which, although of high magnitude, would be harmless to the circuit.

**73. Protection other than overload protection** can be provided by means of release devices or relays employed in connection with circuit breakers or magnetic switches.

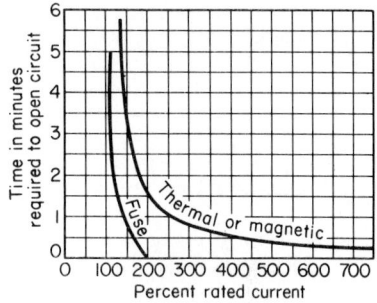

**Fig. 4-103**   Time-current characteristics of overcurrent protective devices.

Undervoltage protection may be provided by relays or releases actuating hand-operated circuit breakers or by magnetic switches. With magnetic switches, undervoltage protection is generally provided through the operating coil of the switch. When the voltage drops below a certain value, the current of the operating coil is so reduced that it will no longer hold the switch in the closed position.

Underload, overvoltage, or reverse-current, power, or phase-rotation protection can be provided by means of relays employed in connection with circuit breakers or magnetic switches.

**74. Ground-fault interrupters.** Another form of overcurrent protection is the ground-fault interrupter. Other types of overcurrent protection do not provide reliable protection from line-to-ground faults because of widely varying ground-path impedances, including the arc itself in arcing ground faults. As a result, conventional overcurrent devices may see the arcing ground fault only as a load current. And the continuation of the arcing ground fault will eventually cause a fire or damage equipment.

The basic principle of a ground-fault interrupter is to sense current leakage to ground and then cause an interconnected overcurrent device to trip open the faulted circuit. Two basic types of ground-fault interrupters are shown in Fig. 4-104. In the first type (Fig. 4-104, 1) the interrupter is designed for *life protection*, which means that it must cause the circuit to open at very low magnitudes of current (about 5 mA). The heart of such an interrupter is a toroidal coil through which all circuit conductors pass. The interrupter senses any unbalance in the circuit, which would be a leakage to ground on one of the circuit conductors (but not on the other). After this unbalance is detected, a differential transformer and solid-state circuitry supply sufficient current to operate a coil of a special circuit breaker, and the circuit is opened within milliseconds (as fast as $\frac{25}{1000}$ s) after the ground fault occurs. This type of ground-fault interrupter is widely used with 120-V electrical circuits supplying swimming-pool equipment and bathrooms, garages and outdoor receptacles of dwelling units, and any other applications for which protection from line-to-ground shock hazards is deemed necessary.

The second type of ground-fault interrupter is designed for *equipment protection*. Figure 4-104, 2, shows the basic function of this system. With this type of interrupter an electrically operated main or feeder switch or circuit breaker is operated through current and time-delay relays, which are energized when a current transformer detects a ground fault flowing through a jumper from the metal enclosure to the insulated circuit neutral conductor. These units are adjustable so that the value of ground-fault current flowing in the circuit can be selected for a given installation before the switch or circuit breaker will

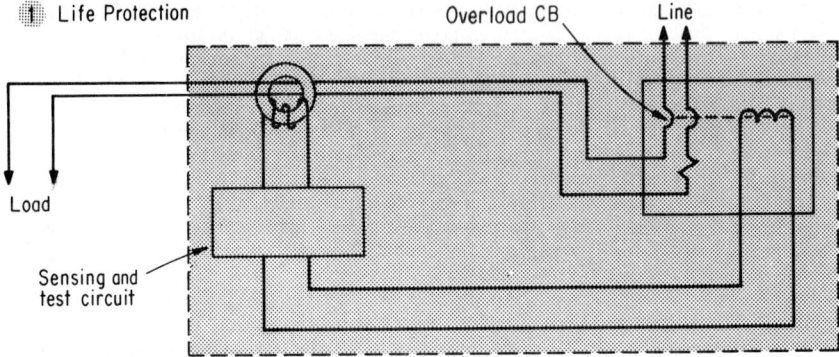

Interrupter senses unbalance in circuit when hot wire becomes grounded. Then a semiconductor differential CB causes circuit to open. Unit opens in less than 25/1000 s when a leakage of 5 mA or more occurs.

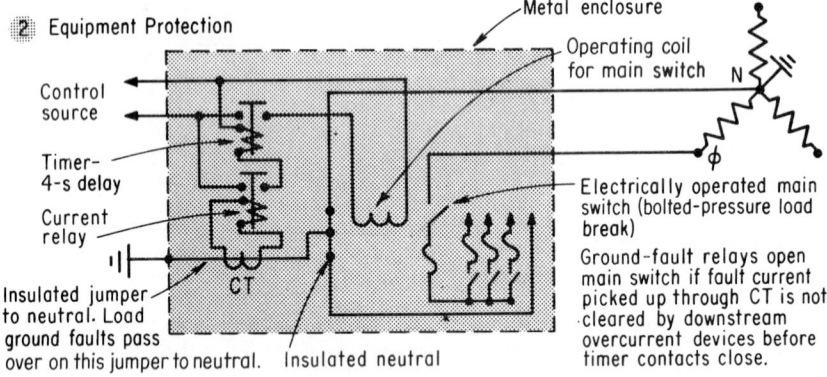

**Fig. 4-104**  Two classifications of ground-fault circuit interrupters.

trip open. Usually such interrupter devices are set at specific current ratings to allow downstream overcurrent devices to open when grounds occur on the load side of them. Thus, the main switch or circuit breaker opens only when downstream overcurrent devices fail to open under ground-fault conditions.

Several manufacturers can provide both types of ground-fault interrupters described in this section, and these interrupters are highly recommended as a means of optimum protection for life and property.

Recently a large industrial corporation installed a system of low-voltage ground-fault protection in its research and development laboratories to protect against shock. The system encompasses all laboratory receptacles in which portable equipment of either 110 or 220 V may be applied. The ground-fault equipment is mounted in a lighting-panel enclosure which is only slightly larger than the standard size. The arrangement is shown in Fig. 4-105.

The ground-fault system consists of five basic units: (A) ground relay, (B) power unit, (C) zero-sequence-coupling transformer, and (D) 100-A main circuit breaker with a 120-V shunt-trip device mounted in the same enclosure. It functions as follows:

All incoming (three-phase and neutral) conductors pass through the window of the zero-sequence-coupling transformer. The secondary of this transformer is connected to the ground relay. The power unit provides dc control voltage to the sensing element of the relay and the main breaker shunt trip for tripping from a 120-V ac source.

Under normal conditions, the load current flowing in the cables creates a magnetic flux in the zero-sequence-coupling transformer that characteristically balances exactly, thereby canceling the flux. No signal is produced. When insulation fails or *when a person becomes a line-to-ground conductor,* leakage current reaches a predetermined level anywhere on the downstream side of the zero-sequence-coupling transformer. This produces a signal to the trigger element, which, in effect, closes the circuit to the shunt trip instantly, thereby tripping the main circuit breaker and isolating the fault.

Section 230-95 of the 1984 NEC requires ground-fault protection on services. The main provision of this paragraph is that ground-fault protection must be applied as follows:

1. On grounded Y circuits
2. 150 V to ground or higher
3. Up to 600 V phase to phase
4. Minimum setting, 1200 A or lower
5. Services rated 1000 A or higher

There are a number of fine-print notes in this code paragraph. The first has to do with application of ground protection on fused switches. This application is all right if the fuse speed and the switch interrupting ratings are coordinated to provide safe interruption for all fault magnitudes. Another note indicates that ground-fault protection is desirable on many circuits rated less than 1000 A. The provisions of Sec. 230-95 require that to obtain coordination for faults on feeders and branches it is usually desirable and sometimes necessary that ground-fault protection be included on the feeders as well as on the main service.

Though not stated in Sec. 230-95, ground-fault protection is very useful on circuits of 240 V and 208 V and all grounded delta circuits. Uncleared ground faults can be equally damaging on these applications, which are not covered at present in the NEC. It should be evident that in large installations in which the incoming service is high-voltage, for

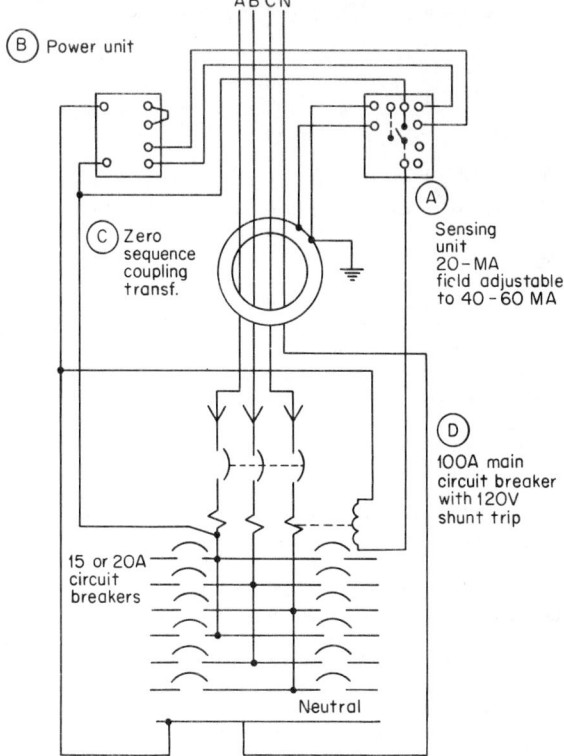

**Fig. 4-105**  Ground-fault interrupter in panelboard enclosure. [Federal Pacific Electric Co.]

example, 13,800 V, and the distribution system consists of several step-down transformers to perhaps 480 V or less, the need for ground-fault protection on all these individual main 480-V feeder circuits is just as important as if the main service to the facility were 480 V.

GROUND-SHIELD SSYTEM   Modern ground-fault-protection (GFP) systems operate on the principle of *zero-sequence detection* or *core balance*, as it is more commonly known. This basic approach has been made practical by the advent of solid-state relays, with their inherent low burden characteristics. A GFP system consists of a special-design CT, or ground sensor, and a solid-state relay. The Y-connected 3- to 4-W transformer source is solidly grounded, and three single-phase loads are connected to it. The ground sensors are available with small or large window configurations, designed to enclose all phase and neutral bus or cable conductors but not the ground conductor (see Fig. 4-106).

**Fig. 4-106**   Family of ground-shield current sensors. [BBC Brown Boveri, Inc.]

The sensors will respond only to ground-fault currents. Balanced or unbalanced load currents and two-phase or three-phase short circuits not involving a ground return will have no effect on the sensor.

Application of a ground-shield system is simple and direct. One sensor and one relay are used with any type of circuit. The sensor is selected by physical size; the relay, by sensitivity range and speed of operation. A minimum pickup setting on the relay offers maximum system protection but at a possible sacrifice in selectivity, depending on the downstream-equipment characteristics. The application does not require a special insulated enclosure or similar complexities. It should be noted that the system neutral must be grounded on the load side of the ground sensor, that is, at downstream panelboards.

OPERATIONAL TESTS   Section 230-95(c) of the **NEC** requires a performance test when a ground-fault protection system is first installed. It is not necessary to schedule periodic maintenance and testing of the system. However, if tests are desired to confirm the proper functioning of the system, one of the following procedures can be used. With draw-out relays, tests can be accomplished by a simple push-to-test-button operation (see Fig. 4-107).

1. Mounted in switchgear.

*a.* After deenergizing the main, set the ground-fault relay to minimum amperes.

*b.* Loop a test coil of approximately No. 14 wire through the sensor window. Special multiturn test cables are available (see Fig. 4-108).

*c.* Apply enough test amperes so that the ampere turns exceed the relay ampere setting. The relay will trip the breaker. Immediately return the test current to zero.

2. Bench tests (without circuit breaker).

*a.* Set the ground-fault relay to minimum amperes.

*b.* Connect the relay and sensor as shown in Fig. 4-109 (typical).

*c.* Apply enough test amperes so that the ampere turns exceed the relay ampere setting. The auxiliary relay will pick up when ground-fault relay operates. Immediately return the test current to zero.

*Precaution.* The relay output circuit will be damaged unless a normally open auxiliary switch opens the trip circuit after the trip operation.

**Fig. 4-107**  Push-to-test ground-fault relay. [BBC Brown Boveri, Inc.]

**Fig. 4-108**  Special multiturn cable. [BBC Brown Boveri, Inc.]

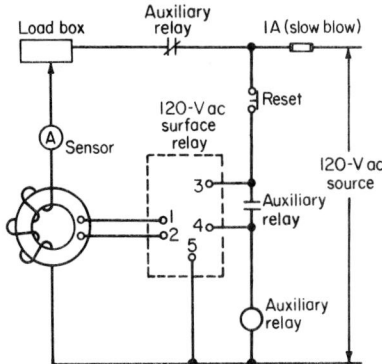

**Fig. 4-109**    Bench test. [BBC Brown Boveri, Inc.]

**75. Lightning protection.**    High voltages may be electrostatically induced in outdoor electric lines during lightning storms. These voltages produce traveling surges of high voltage which travel along the lines and into electrical equipment, whether it is located outdoors or inside buildings. The high induced voltages would puncture the insulation of equipment and be dangerous to life. Lightning arresters are used to limit these voltages to a safe value and provide a path to ground for the dissipation of the energy of the surge. To provide this protection satisfactorily, lightning arresters must fulfill the following functions:

1. They must not allow the passage of current to ground as long as the voltage is normal.

2. When the voltage rises to a definite amount above normal, they must provide a path to ground for dissipation of the surge energy without further rise in voltage of the circuit.

3. As soon as the voltage has been reduced below the setting of the arrester, the arrester must stop the flow of current to ground and reseal itself so as to insulate the conductor from ground.

4. Arresters must not be injured by the discharge and must be capable of automatically repeating their action as frequently as is required.

Lightning arresters should be provided on all overhead systems. For the small consumer who purchases power from a public utility, this protection is supplied by the utility on its distribution circuits. For a large consumer it may be advantageous to install lightning arresters between the public-utility lines and the consumer's substation. All communication circuits entering a building must be provided with a protector which will give protection against abnormal voltage to ground caused not only by lightning but by accidental contact between communication circuits and power circuits. This protection is usually provided by the communication company which is supplying the service. All signal circuits which are run outdoors for any part of their length or which are installed so that there is any possibility of accidental contact between the signal circuit and power conductors must be provided with a protector.

The National Electrical Code requires radio equipment to be protected against lightning as follows. In receiving stations, each conductor of a lead-in from an outdoor antenna shall be provided with a listed antenna discharge unit, except that when the lead-in conductors are enclosed in a continuous metallic shield, the lightning arrester may be installed to protect the shield or may be omitted if the shield is permanently and effectively grounded. Lightning (surge) arresters shall be located outside the building or inside the building between the point of entrance of the lead-in and the radio set or transformers and as near as practicable to the entrance of the conductors into the building. Antenna discharge units shall not be located near combustible material or in a hazardous location.

Grounding conductors shall be not smaller than No. 10 copper or No. 8 aluminum or No. 17 copper-clad steel or bronze.

**76. Surge (lightning) arresters** are made in many different forms. A discussion of the construction, application, and installation of the various types is outside the scope of this book. Typical types for power work are illustrated in Fig. 4-110, and types for signal and communication work in Fig. 4-111.

Surge arresters can be installed indoors or outdoors. When located indoors, they must be well away from other equipment, passageways, and combustible material or parts of

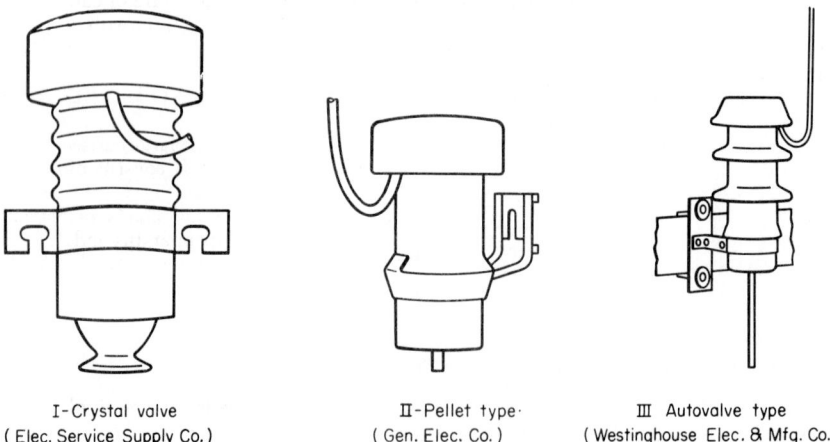

I-Crystal valve          II-Pellet type·          III  Autovalve type
( Elec. Service Supply Co.)      ( Gen. Elec. Co. )      ( Westinghouse Elec. & Mfg. Co.)

**Fig. 4-110**   Surge arresters of the power type.

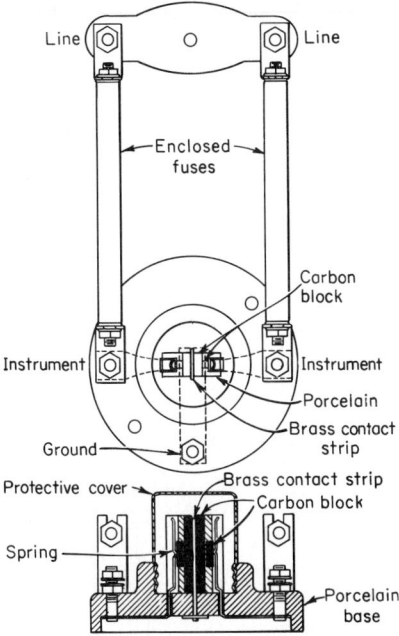

I. Standard telephone-line protector.          II. Arrester for signal circuits.
                                                 [General Electric Co.]

**Fig. 4-111**   Surge arresters of the low-voltage type.

the building, and if they contain oil, they must be enclosed in vaults of construction similar to that required for oil-filled transformers (see Div. 5). If arresters which contain oil are located outdoors, provision must be made to drain away any accumulation of oil. This may be done by properly constructed ditches and drains, or the oil may be absorbed and the danger of spreading removed by paving the yard with cinders or other absorbent material to a depth of several inches.

An arrester must not be located in a hazardous location (see Div. 9). When surge arresters are used, an arrester should be installed for each ungrounded circuit wire. The National Electrical Code requires that the connections between an arrester and a line wire or bus and between the arrester and ground shall be of copper wire or cable or the equivalent. Except for secondary services these connections shall not be smaller than No. 6. The connections shall be made as short and as straight as practicable, avoiding as far as possible all bends and runs, especially sharp bends.

Proper grounding of lightning arresters is of utmost importance. Refer to Div. 9 for discussion of grounding rules and practice.

**77. Surge protective equipment** is used in connection with lightning (surge) arresters and in large power systems to limit the value of short-circuit current. It consists of coils of wire connected in series with the circuit for the purpose of introducing inductance. Coils used with surge arresters are called choke coils (Fig. 4-112). The inductance of the coils tends to shunt the high-frequency lightning surge away from the apparatus and through the arrester to ground. Those used for limiting short-circuit currents are called reactors. A typical construction of a reactor is shown in Fig. 4-113.

## SWITCHBOARDS AND SWITCHGEAR

**78. A switchboard** as defined in the National Electrical Code is a large single panel, frame, or assembly of panels on which are mounted, on the face or back or both, switches, overcurrent and other protective devices, buses, and, usually, instruments. Switchboards

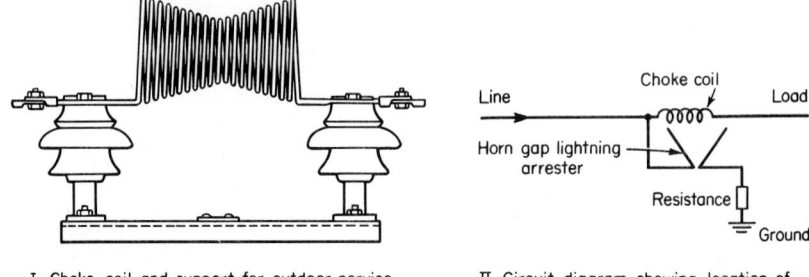

I  Choke coil and support for outdoor service

II  Circuit diagram showing location of choke coil

**Fig. 4-112**  Choke coil for use with surge arrester. [*Electric System Handbook*]

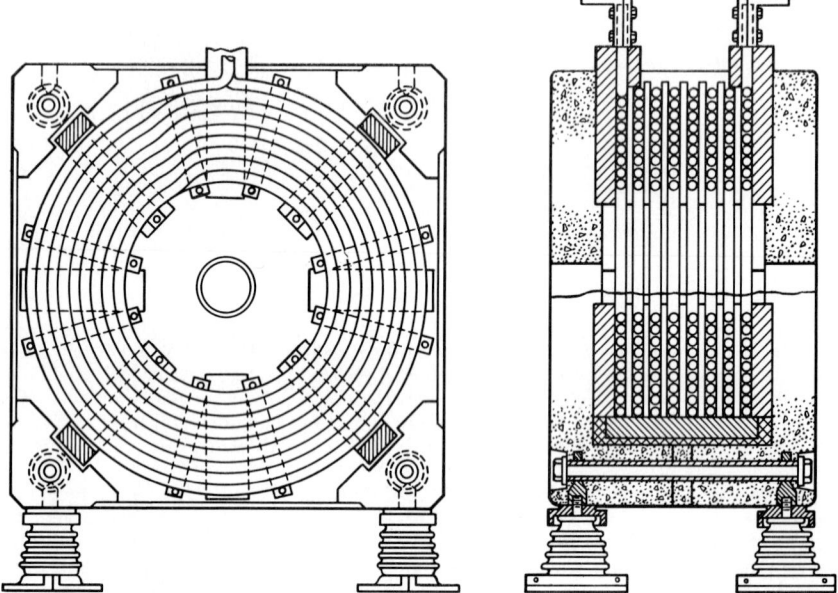

**Fig. 4-113**  Current-limiting reactor, series-wound with single insulated cable—horizontal axis. [*Electric System Handbook*]

are generally accessible from the rear as well as from the front and are not intended to be installed in cabinets.

**79. The types of switchboards,** classified by basic features of construction, are as follows:

1. Live-front vertical panels
2. Dead-front boards
3. Safety enclosed boards (metal-clad)

Live-front switchboards (Fig. 4-114) have the current-carrying parts of switching equipment mounted on the exposed face or front of the panels. They are not as a rule employed in new boards when the voltage exceeds 600 V.

Large-capacity equipment with resultant increases in weight makes enclosed metal-clad switchboards more desirable from a structural and safety point of view. If high short-circuit currents are available, it is much safer to confine short circuits to metal-clad equipment for the protection of personnel and buildings.

On existing live-front switchboards, obsolete equipment and panels can be replaced with more modern components. In drilling new holes in slate or marble a standard steel twist drill may be used if the heel of the drill is ground off, thus providing the drill with a sharper angle. Slate panels are drilled dry, whereas marble panels must be drilled with the use of water. It is best, however, to use modern masonry drills when drilling such panels.

Avoid excessive pressure when drilling through panels. Abandoned holes can be filled with cement or plaster.

Dead-front switchboards (Fig. 4-115) are the more modern type and are employed for systems of all voltages. With this type no live parts are mounted on the front of the board.

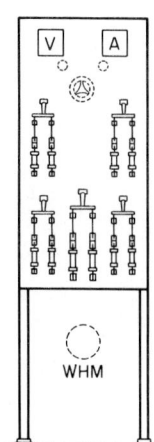

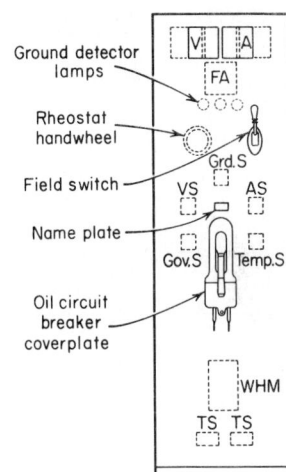

**Fig. 4-114**   Live-front switchboard panel.        **Fig. 4-115**   Dead-front switchboard panel.

Safety enclosed boards are used for most new installations. Common terms used to designate equipment of this type are *metal-enclosed switchgear* and *metal-clad switchgear*. Most safety enclosed boards are of the unit or sectional type. They consist of a combination of the desired number and type of standardized unit sections. Each section is a standard factory-assembled combination of a formed steel panel and apparatus mounted on a steel framework.

Safety enclosed switchgear of the truck or draw-out type (Figs. 4-116 and 4-117) consists of a steel enclosure mounted on an angle-iron framework. All apparatus which requires attention or inspection is mounted on a truck or draw-out structure. The equipment for each circuit is mounted on either a separate truck, which can easily be rolled out from the rest of the structure, or a metal structure supported on guides so that it can easily be slid in and out from the enclosure (Fig. 4-117).

**80. Safety enclosed switchgear** may be classified with respect to purpose of application as follows:

1. General medium- or high-voltage switchgear
2. Primary unit substations
3. Rectifier unit substations
4. Secondary unit substations or power centers
5. General low-voltage switchgear
6. Low-voltage distribution switchboards
7. Motor-control-center switchboards

Medium-voltage switchgear (Fig. 4-118) provides for the required control and metering equipment for generators, transformer supply circuits, feeders, large motors, etc., for systems with voltage up to 15,000 V. A large variety of standardized units is available.

Primary unit substations of the metal-clad type (Figs. 4-119 and 4-120) consist of one or more transformers mechanically and electrically connected to and coordinated with one

**Fig. 4-116** Draw-out type of safety enclosed circuit-breakers switchboard. [I-T-E Electrical Products Division of Siemens]

**Fig. 4-117** Operation of draw-out type of circuit-breaker switchboard. [I-T-E Electrical Products Division of Siemens]

**Fig. 4-118**   Medium-voltage metal-enclosed switchgear. [Westinghouse Electric Corp.]

**Fig. 4-119**   Installed power-distributing center for a brewery in the northeastern United States.

or more feeder or motor-control sections. A variety of combinations is available in both outdoor and indoor construction.

Rectifier unit substations provide the transforming, rectifying, switching, metering, and control equipment necessary for the conversion of three-phase ac power at any of the conventional voltages to direct current.

Secondary unit substations (Fig. 4-121) permit the distribution of power in industrial plants or other large buildings at the higher, more economical voltages and the transformation of the power to the desired utilization voltage at points near the load. They usually

**Fig. 4-120**   Outdoor metal-enclosed switchgear. [Westinghouse Electric Corp.]

**Fig. 4-121**   Secondary unit power center. [Westinghouse Electric Corp.]

consist of three basic types of components: high-voltage (incoming) section, transformer section, and low-voltage (outgoing) sections.

General low-voltage switchgear (Figs. 4-122 and 4-123) provides for the required control and metering equipment for generators, incoming circuits from transformers, feeders, large motors, etc., for systems with voltage not greater than 600 V. A large variety of standardized units is available.

**Fig. 4-122** General-purpose low-voltage switchgear assembly. [Westinghouse Electric Corp.]

**Fig. 4-123** Typical low-voltage distribution switchgear assembly. [I-T-E Electrical Products Division of Siemens]

Low-voltage distribution switchboards may be of the so-called building type or of the multipurpose type. The building type is specifically designed for the control of low-voltage distribution circuits (600 V and below) in offices, hospitals, and commercial types of buildings. A circuit-breaker type is shown in Fig. 4-124 and a fusible-switch type in Fig. 4-125. Multipurpose distribution switchboards are for general use in commercial or industrial applications. Typical assemblies are shown in Fig. 4-126, which also indicates the major features of a well-designed modern switchboard.

In general, low-voltage distribution switchboards fall into three distinct NEMA classifications encompassing specific structural features, equipment arrangement, and electrical characteristics, which are noted as follows. Common to all classes are total metal enclosure (except bottom), top or bottom gutter, and vertical sections electrically connected.

CLASS I    Maximum current limited to 2000 A; group-mounted branches; wall-supported; front accessibility only for line and load connections; flush back.

CLASS II    No maximum current requirement; group-mounted branches; branches front-accessible only; main bus rear-accessible; self-supported.

CLASS III    No maximum current requirement; individually mounted branches; branches and mains rear-accessible; self-supported.

Motor-control centers (Fig. 4-127) provide a compact centralized location for full-voltage motor-control equipment. Each compartment contains a magnetic starter with a circuit breaker or a fused switch.

**81. Materials employed for the panels of switchboards** are slate, marble, ebony-asbestos, and steel. At one time slate was the material principally used for switchboard panels, but it has been almost entirely replaced by ebony-asbestos for live-front boards and by steel for dead-front and safety enclosed ones. Marble is used on live-front boards sometimes for its higher insulation qualities over slate but more often for its better appearance.

The supporting frame for panels may be of angle iron or pipe.

**Fig. 4-124**   Building-type distribution switchboard. [Westinghouse Electric Corp.]

**Fig. 4-125**  Power-control-center switchboard. [General Electric Co.]

**Fig. 4-126**  A modern switchboard with a main disconnect switch and feeder circuit breakers. [Square D Co.]

**82. Location of switchboards.** Switchboards of the safety enclosed type may be placed in any nonhazardous location without restriction. Switchboards with exposed live parts must be located in permanently dry locations and only where they will be under competent supervision and accessible only to qualified persons. The rear of a switchboard may be made accessible only to qualified persons by a metal grillwork enclosure, entrance to which is by means of a locked door. Switchboards should be located so as to reduce to a minimum the probability of communicating fire to adjacent easily ignitable material. A clearance of at least 3 ft (0.91m) must be left between the top of a board and a nonfireproof ceiling unless an adequate fireproof shield is provided between the board and the ceiling or the switchboard is totally enclosed. Refer to Sec. 110-16 of the National Electrical Code for required clearances around switchboards and control centers.

**Fig. 4-127**  Motor-control-center switchboard. [Nelson Electric]

**83. Spacings required by the National Electrical Code.** Except at switches and circuit breakers, at least the distance between bare metal parts, busbars, etc., shown in the accompanying table shall be maintained. It should be noted that these distances are the minimum allowable, and it is recommended that greater distances be adopted wherever conditions permit.

| | Opposite polarity when mounted on the same surface, in | Opposite polarity when held free in air, in | Live parts to ground, in[a] |
|---|---|---|---|
| Not over 125 V | ¾ | ½ | ½ |
| Not over 250 V | 1¼ | ¾ | ½ |
| Not over 600 V | 2 | 1 | 1 |

[a] For spacing between live parts and doors of cabinets, see Sec. 373-11(a) of the NEC.

At switches, enclosed fuses, etc., parts of the same polarity may be placed as close together as convenience in handling will allow unless close proximity causes excessive heating.

**84. Busbar Spacings, Minimum Distance between Parts of Opposite Polarity and Minimum Distance between Live Parts and Ground for Different Voltages** (Electrical Engineer's Equipment Co.)

| Voltage | Distance between centers of buses, in | | Minimum distance between opposite live parts, in | | Minimum distance between live parts and ground, in | |
|---|---|---|---|---|---|---|
| | A | B | A | B | A | B |
| 250 | 1½ – | 2½ | 1 – 2 | | ¾– 1½ | |
| 600 | 2 – | 3 | 1½– 2½ | | 1 – 2 | |
| 1,100 | 4 – | 5 | 2½– 3½ | | 1½– 2½ | |
| 2,300 | 5 – | 6½ | 2¾– 4 | | 2 – 2¾ | |
| 4,000 | 6 – | 7½ | 3 – 4½ | | 2¼– 3 | |
| 6,600 | 7 – | 8 | 3½– 4½ | | 2½– 3 | |
| 7,500 | 8 – | 9 | 4 – 4½ | | 2¾– 3¼ | |
| 9,000 | 9 – | 10 | 4¼– 4½ | | 3 – 3½ | |
| 11,000 | 9 – | 11 | 4½– 4¾ | | 3¼– 3¾ | |
| 13,200 | 9 – | 12 | 4¾– 5 | | 3½– 4¼ | |
| 15,000 | 9 – | 14 | 5 – 5½ | | 3¾– 4½ | |

**84. Busbar Spacings, Minimum Distance between Parts of Opposite Polarity and Minimum Distance between Live Parts and Ground for Different Voltages**

| Voltage | Distance between centers of buses, in | | Minimum distance between opposite live parts, in | | Minimum distance between live parts and ground, in | |
|---|---|---|---|---|---|---|
| | *A* | *B* | *A* | *B* | *A* | *B* |
| 16,500 | 10 | – 14 | 5½– 6 | | 4½– 5 | |
| 18,000 | 11 | – 14 | 6 – 7 | | 5 – 6 | |
| 22,000 | 12 | – 15 | 7½– 9 | | 6 – 7 | |
| 26,000 | 14 | – 16 | 10 –12 | | 8 – 9 | |
| 35,000 | 18 | – 22 | 12 –15 | | 10 –12 | |
| 45,000 | 22 | – 27 | 16 –18 | | 13½–15 | |
| 56,000 | 28 | – 31 | 17½–19 | | 16 –17½ | |
| 66,000 | 34 | – 38 | 22 –24 | | 18½–23 | |
| 75,000 | 36 | – 42 | 26 –30 | | 25 –27½ | |
| 90,000 | 46 | – 54 | 32 –35 | | 27 –29 | |
| 104,000 | 54 | – 60 | 34½–39 | | 28½–32 | |
| 110,000 | 60 | – 72 | 38 –41 | | 33 –36 | |
| 122,000 | 66 | – 78 | 42 –47 | | 35½–39 | |
| 134,000 | 74 | – 84 | 48½–56 | | 39 –41 | |
| 148,000 | 82 | – 96 | 59 –67 | | 45 –50 | |
| 160,000 | 88 | –104 | 70 –84 | | 53 –62 | |

NOTE  The distances given in the *A* columns are based on a safety factor of 3.5 between live parts of opposite polarity and safety factor of 3 between live parts and ground. *B* column shows good practice for the larger plants. Tubular buses should be used on all buses above 22,000 V. Bus support porcelains should have a wet test of 2 times the voltage used.

## PANELBOARDS

**85. A panelboard** as defined by the National Electrical Code is a single panel or a group of panel units designed for assembly in the form of a single panel, including buses and with or without switches and/or automatic overcurrent protective devices for the control of light, heat, or power circuits, designed to be placed in a cabinet or cutout box placed in or against a wall or partition and accessible only from the front. Panelboards provide a compact and convenient method of grouping circuit switching and protective devices at some common point.

Panelboards may be of either the flush or the surface type (Fig. 4-128). The flush type is used with concealed-wiring installations and has the advantage of not taking up space in the room by extending beyond the surface of the wall. Surface-type boxes are used for installations employing exposed wiring. The boxes are generally constructed of sheet steel, which must be not less than No. 16 Manufacturers standard gage in thickness. The steel must be galvanized or be covered with some other protective coating to prevent corrosion.

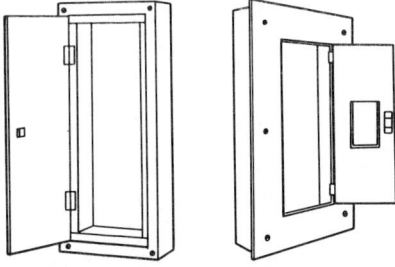

I–Surface mounting type    II–Flush mounting type

**Fig. 4-128**    Panel boxes.

Gutters are provided around the panelboards in cabinets to allow sufficient space for wiring (Figs. 4-128 and 4-129). The Code requires that all cabinets which contain connections to more than eight conductors be provided with back or side wiring spaces. These wiring spaces must be separated from the panelboard or other devices in the cabinet by partitions so that they will be separate closed compartments, unless all wires are led from the cabinet at points directly opposite their terminal connections to the panelboard. The minimum width of gutters required if conductors are deflected upon entering or leaving cabinets is given in the table on page 4-98.

**Minimum Bending Space, In**
(Table 373-6(a), 1990 NEC)[a]

| AWG or circular-mil size of wire | Wires per terminal | | | | |
|---|---|---|---|---|---|
| | 1 | 2 | 3 | 4 | 5 |
| 14–10 | Not specified | | | | |
| 8–6 | 1–½ | | | | |
| 4–3 | 2 | | | | |
| 2 | 2½ | | | | |
| 1 | 3 | | | | |
| 1/0–2/0 | 3½ | 5 | 7 | | |
| 3/0–4/0 | 4 | 6 | 8 | | |
| 250 kcmil | 4½ | 6 | 8 | 10 | |
| 300–350 kcmil | 5 | 8 | 10 | 12 | |
| 400–500 kcmil | 6 | 8 | 10 | 12 | 14 |
| 600–700 kcmil | 8 | 10 | 12 | 14 | 16 |
| 750–900 kcmil | 8 | 12 | 14 | 16 | 18 |
| 1,000–1,250 kcmil | 10 | | | | |
| 1,500–2,000 kcmil | 12 | | | | |

[a]Reprinted with permission from NFPA 70-1990, National Electrical Code®, Copyright © 1990, National Fire Protection Association, Quincy, Massachusetts 02269. This reprinted material is not the complete and official position of the NFPA on the referenced subject, which is represented only by the standard in its entirety.

NOTE   Bending space at terminals shall be measured in a straight line from the end of the lug or wire connector (in the direction that the wire leaves the terminal) to the wall, barrier, or obstruction.

A panelboard consists of a set of copper or plated-aluminum busbars, called mains, from which provision is made for tapping off several circuits through overcurrent protective devices and/or switching mechanisms. The provisions for taps generally are built up from unit sections assembled to form the complete board. This plan allows the manufacturers to build a few standard types of unit sections which can be assembled in a great variety of combinations to meet the varying requirements of different installations. This results in economy of manufacture and in the greatest flexibility of possible combinations to meet all requirements. Typical panelboard construction is illustrated in Fig. 4-129. This figure shows several types of unit sections in the same panelboard to illustrate the flexibility of assembly. Similar panelboards are designed for plug-in or bolt-on circuit breakers or fuse assemblies.

All panelboards have ampere ratings, and such ratings are shown on the nameplate of each panelboard along with the voltage rating. The panelboard ampere rating is the ampacity of the busbars to which the branch overcurrent units are connected.

Main circuit breakers, fused pullouts, or fused switches can be provided in a panelboard. A panelboard without integral main overcurrent protection is called *mains only,* which means that the panel has only main lugs. Such a panelboard is shown in Fig. 4-131.

If a panelboard supplies wiring installed by nonmetallic methods or any other system which will include equipment grounding conductors in a raceway or cable, a grounding terminal bar must be installed to terminate all such grounding wires. The grounding terminal bar must be bonded to the cabinet. It can be bonded to the *neutral* bar of a panelboard only if the panel is used as service equipment.

To provide a high degree of selectivity and flexibility most panelboards today are designed for insert-type fusible or circuit-breaker assemblies. These assemblies are plugged into or bolted onto "receiver" panel interior busbars. With such an arrangement the installer or designer can select the proper size of overcurrent devices for circuits or

Table 373-6(b), 1990 NEC
**Minimum Wire Bending Space at Terminals for Sec. 373-6(b)(2), in[a]**

| | Wires per terminal | | | |
| Wire size | 1 | 2 | 3 | 4 or more |
|---|---|---|---|---|
| 14–10 | Not specified | | | |
| 8 | 1½ | | | |
| 6 | 2 | | | |
| 4 | 3 | | | |
| 3 | 3 | | | |
| 2 | 3½ | | | |
| 1 | 4½ | | | |
| 1/0 | 5½ | 5½ | 7 | |
| 2/0 | 6 | 6 | 7½ | |
| 3/0 | 6½ (½) | 6½ (½) | 8 | |
| 4/0 | 7 (1) | 7½ (1½) | 8½ (½) | |
| 250 | 8½ (2) | 8½ (2) | 9 (1) | 10 |
| 300 | 10 (3) | 10 (2) | 11 (1) | 12 |
| 350 | 12 (3) | 12 (3) | 13 (3) | 14 (2) |
| 400 | 13 (3) | 13 (3) | 14 (3) | 15 (3) |
| 500 | 14 (3) | 14 (3) | 15 (3) | 16 (3) |
| 600 | 15 (3) | 16 (3) | 18 (3) | 19 (3) |
| 700 | 16 (3) | 18 (3) | 20 (3) | 22 (3) |
| 750 | 17 (3) | 19 (3) | 22 (3) | 24 (3) |
| 800 | 18 | 20 | 22 | 24 |
| 900 | 19 | 22 | 24 | 24 |
| 1000 | 20 | | | |
| 1250 | 22 | | | |
| 1500 | 24 | | | |
| 1750 | 24 | | | |
| 2000 | 24 | | | |

[a]Reprinted with permission from NFPA 70-1990, National Electrical Code®, Copyright © 1990, National Fire Protection Association, Quincy, Massachusetts 02269. This reprinted material is not the complete and official position of the NFPA on the referenced subject which is represented only by the standard in its entirety.

NOTE   Bending space at terminals shall be measured in a straight line from the end of the lug or wire connector in a direction perpendicular to the enclosure wall. For removable and lay-in wire terminals intended for only one wire, bending space shall be permitted to be reduced by the number of inches shown in parentheses.

feeders. Overcurrent-device assemblies are available in single-, two-, and three-pole units.

In selecting a panelboard for any installation it is good design to provide enough space so that additional overcurrent devices may be inserted at a later date.

It is also wise to use surface-mounted panelboards in areas where surface-type wiring is acceptable. This will enable new raceways or cables to be added with less difficulty.

If installations require concealed wiring, spare conduits should be stubbed to accessible areas, such as above the lift-out ceiling-panel construction in common use today.

Every panelboard should include a circuit legend with a neatly typed and accurate description of the circuits supplied by each overcurrent device. Circuit identification is a National Electrical Code requirement and is extremely useful when troubleshooting, repairs, or additional wiring is required.

1. Galvanized sheet-steel box
2. Panel adjusting screws
3. Galvanized channel irons
4. Removable frame barrier separating wiring gutter from panelboard, simplifying panel adjustments
5. Molded Bakelite end section covering main lugs or neutral bar
6. "Plug fuse only" unit section
7. Branch-circuit terminal screws
8. Single-fusing snap-switch, plug-fuse type of unit section
9. Molded Bakelite section plates
10. Double-fusing, two-pole snap-switch, plug-fuse type of unit section
11. Index numbers for identification of circuits
12. Single-fusing snap-switch, cartridge-fuse type of unit section
13. Wiring gutter
14. Double-fusing, two-pole snap-switch, cartridge-fuse type of unit section
15. Snap switches *entirely* removable from front, each enclosed with an individual Bakelite cover
16. Standard knockout arrangement, usually satisfactory for most installations

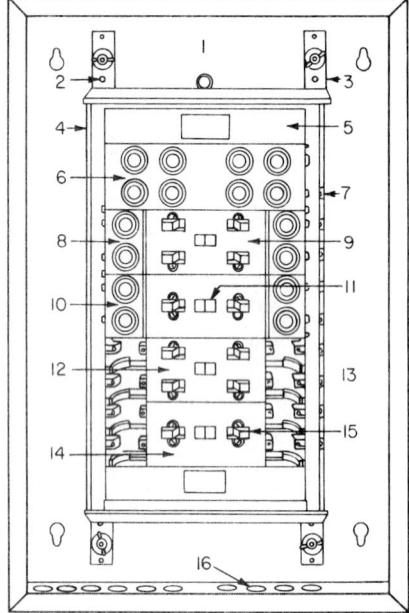

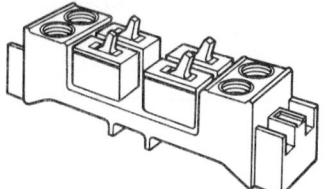

I. Unit section of tumbler switches and fuse-plug holders.

II. Panelboard assembly, illustrating flexibility of construction.

**Fig. 4-129**   Typical panelboard construction. [General Electric Co.]

Many panelboards and attached overcurrent devices contain terminals which are suitable for both copper and aluminum wire terminations. In such cases an AL-CU marking will appear in the panelboard and/or on the overcurrent assemblies. If there is no marking as to what type of conductor material may be attached to terminals, it can be assumed that only copper conductors are suitable. In such cases aluminum wire can still be used by splicing a short length of copper wire to the aluminum connector with a suitable connector, especially in the case of branch-circuit connections.

**86. Lighting and appliance branch-circuit panelboards.**   In solving all installation problems with panelboards the first consideration is to determine whether the panelboard will be considered a lighting and appliance branch-circuit type. The reason for this is that National Electrical Code rules are much stricter for lighting and appliance branch-circuit panelboards than for other types.

The Code defines a lighting and appliance branch-circuit panelboard as one having *more than* 10 percent of its overcurrent devices rated 30 A or less, for which neutral connections are provided. For example, if any panelboard with less than 10 overcurrent devices contains *one* overcurrent device rated at 30 A for which neutral connections are provided, it would be considered a lighting and appliance branch-circuit panelboard (1 ÷ 9 = 11%).

In another example, panelboards that supply loads without any neutral connections are not considered lighting and appliance branch-circuit types whether or not the overcurrent devices are 30 A or less.

When it is determined that a panelboard is a lighting and appliance branch-circuit type, the following **National Electrical Code** rules apply:

1. Individual protection, consisting of not more than two main circuit breakers or sets of fuses having a combined rating not greater than that of the panelboard, is required on the supply side. This main protection may be contained within the panelboard (as shown in Fig. 4-130) or in a separate enclosure ahead of it. There are two exceptions to this Code rule.

**Fig. 4-130** Circuit-breaker type of lighting panelboard with main breaker. [General Electric Co.]

**Fig. 4-131** A 480Y/277-V circuit-breaker lighting panelboard without an integral main breaker. [Square D Co.]

*a.* Individual protection is not required when the panelboard feeder has overcurrent protection not greater than that of the panelboard.

**example**   Two 400-A panelboards can be connected to the same feeder if the feeder *overcurrent device* is rated or set at 400 A or less.

*b.* Individual protection is not required where such existing panelboards are used as service equipment in supplying an individual residential occupancy.

**example**   Take a split-bus panelboard in which the line section contains three to six circuit breakers or fuses, none of which are rated 20 A or less. In such an arrangement one of the main overcurrent devices supplies the second part of the panel, which contains 15- or 20-A branch-circuit devices. The other main overcurrent devices (over 20 A) supply feeders or major appliances such as cooking equipment, clothes dryers, water heaters, or space-conditioning equipment (refer to Fig. 4-132). This arrangement is permitted only for existing panelboards in existing individual residential occupancies.

2. A lighting and appliance branch-circuit panelboard is limited to not over 42 overcurrent devices (excluding the main overcurrent devices) in any one cabinet or cutout box (refer to Fig. 4-133). In enumerating such devices a single-pole circuit breaker is counted as one overcurrent device; a two-pole circuit breaker, as two overcurrent devices; and a three-pole circuit breaker, as three overcurrent devices.

In addition, such panelboards shall be provided with physical means to prevent the installation of more overcurrent devices than that number for which the panelboard was designed, rated, and approved. This rule concerns a circuit-limitation concept and the use of Class CTL overcurrent devices, which are discussed in Sec. 64 and shown in Figs. 4-94 and 4-134.

The lighting panelboard shown in Fig. 4-130 is a circuit-breaker type with a main 200-

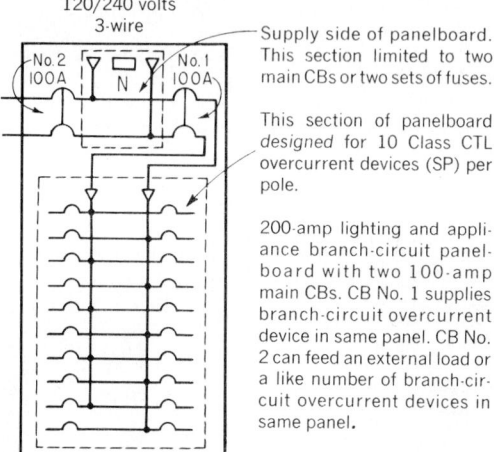

120/240 volts
3-wire

No. 2
100 A

No. 1
100 A

N

Supply side of panelboard. This section limited to two main CBs or two sets of fuses.

This section of panelboard *designed* for 10 Class CTL overcurrent devices (SP) per pole.

200-amp lighting and appliance branch-circuit panelboard with two 100-amp main CBs. CB No. 1 supplies branch-circuit overcurrent device in same panel. CB No. 2 can feed an external load or a like number of branch-circuit overcurrent devices in same panel.

**Fig. 4-132**  Typical arrangements of a split-bus lighting panelboard which has two main circuit breakers and Class CTL branch breakers.

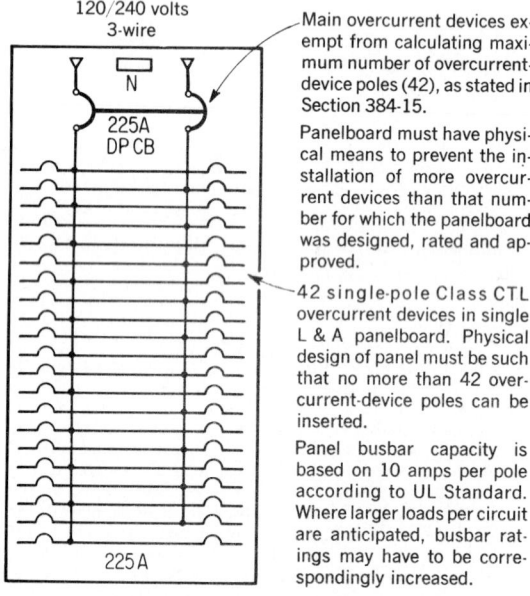

120/240 volts
3-wire

N

225 A
DP CB

225 A

Main overcurrent devices exempt from calculating maximum number of overcurrent-device poles (42), as stated in Section 384-15.

Panelboard must have physical means to prevent the installation of more overcurrent devices than that number for which the panelboard was designed, rated and approved.

42 single-pole Class CTL overcurrent devices in single L & A panelboard. Physical design of panel must be such that no more than 42 overcurrent-device poles can be inserted.

Panel busbar capacity is based on 10 amps per pole according to UL Standard. Where larger loads per circuit are anticipated, busbar ratings may have to be correspondingly increased.

**Fig. 4-133**  Typical arrangement which shows National Electrical Code rules for lighting panelboards.

A circuit breaker and thirty-two 20-A single-pole breakers. This panel is used for a four-wire three-phase grounded-neutral system. The main breaker is three-pole.

Other National Electrical Code provisions which apply to *all* types of panelboards are:

1. Panelboards equipped with snap switches rated 30 A or less, as shown in Fig. 4-129, shall have overcurrent protection not in excess of 200 A. Circuit breakers are not considered snap switches.

2. Panelboards having switches on the load side of any type of fuse shall not be

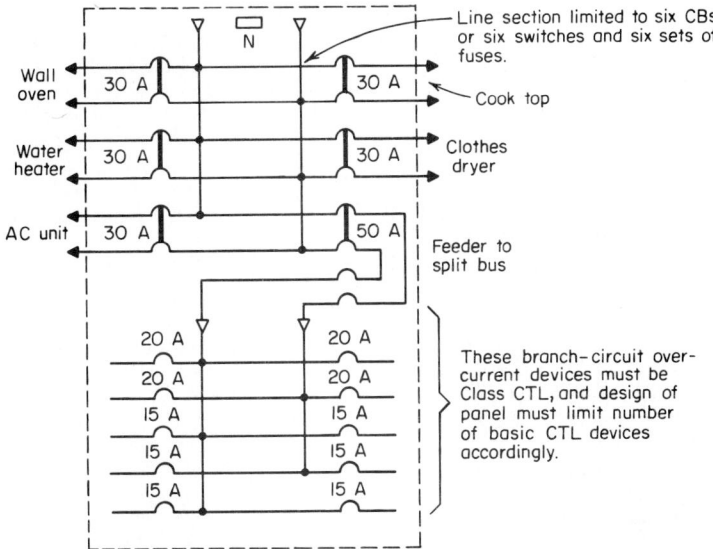

**Fig. 4-134**   Suitable arrangement for an existing 200-A lighting panelboard used as service equipment for an individual residential occupancy.

installed except for use as service equipment. In Fig. 4-129 the snap switch is on the line side of the plug fuses and satisfies this Code rule.

3. The total load on any overcurrent device located in a panelboard shall not exceed 80 percent of its rating if in normal operation the load will be continuous (3 h or more), unless the assembly including the overcurrent device is approved for continuous duty at 100 percent of its rating.

**87. Service-equipment panelboards.**   For loads up to 800 A, panelboards which contain six or fewer main fused switches, fused pullouts, or circuit breakers are available. These panels constitute service equipment and frequently contain split buses which supply branch-circuit or feeder overcurrent devices installed in the same enclosure (see Figs. 4-132 and 4-134).

**88. Feeder distribution panels** generally contain circuit overcurrent devices rated at more than 30 A to protect *subfeeders* that extend to smaller branch-circuit panelboards.

**89. Power-distribution panels** are similar to the feeder-distribution type.   They have bus-bars normally rated up to 1200 A at 600 V or less and contain control and overcurrent devices sized to match connected motor or other power circuit loads. Generally, the devices are three-phase (see Figs. 4-135, 4-136, and 4-137).

Figure 4-136 shows an Underwriters Laboratories 200,000-A short-circuit rating of the panelboard by the use of Class R fuse clips to reject the use of non-current-limiting fuses.

Special panelboards containing relays and contactors can be obtained and installed when remote control of specific equipment is specified. A thorough knowledge of all available types of panelboards facilitates selection and installation of the proper unit.

**90. Location and installation.**   All panelboards should be located as near as possible to the loads that they supply and control. Mounting heights should be such that the distance from the floor or working level to the top of the uppermost overcurrent device is not more than 6½ ft (1.98 m) and the distance from the lowest overcurrent device to the floor is not less than 6 in (152.4 mm).

Panels to be installed in wet or damp locations, in dust-laden areas, outdoors, or in any hazardous area must be of the type approved for use in such locations.

If qualified persons only are to have access to panelboards, install panel trims with locking-type catches on the doors.

There are several precautions that can simplify panelboard installation, improve workmanship, and produce a neater job. Among them are:

**Fig. 4-135** Circuit-breaker type of power-distribution panelboard. [Square D Co.]

**Fig.   4-136** Power   panelboard   with fused-switch units. [I-T-E Electrical Products Division of Siemens]

**Fig. 4-137**    Converti-fuse power panelboard; fuse-puller switching units. [General Electric Co.]

1. Coordinate panelboard cabinet installation with the raceway system, and maintain proper alignment.

2. Align floor-slab conduit stub-ups with openings in the bottom of the cabinets. Maintain alignment during a concrete pour by using interlocking conduit spacer caps or a wood or metal template. This is exceptionally important when the cabinet is to be surface-mounted.

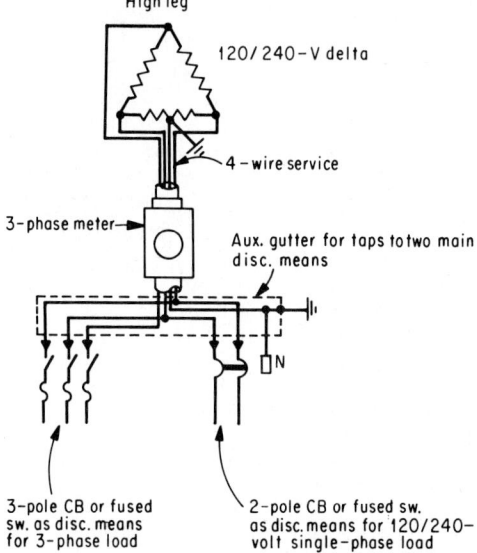

High leg

120/240-V delta

4-wire service

3-phase meter

Aux. gutter for taps to two main disc. means

N

3-pole CB or fused sw. as disc. means for 3-phase load

2-pole CB or fused sw. as disc. means for 120/240-volt single-phase load

**Fig. 4-138**  An effective way of separating single- and three-phase loads of four-wire delta supplies.

3. Cabinet knockouts should be properly sized and spaced to match feeder and branch-circuit raceway layouts. Most often the standard prestamped single and/or multiple-concentric cabinet knockouts are used. On some installations, however, it is best to order blank enclosures and cut the required knockouts at the jobsite.

4. If cabinets are to be installed flush in a wall, provide temporary anchoring facilities to position the cabinet securely until the wall is constructed. Enclosures supported only by conduit attachments often are knocked out of alignment.

5. Surface-mounted panelboard enclosures should be securely fastened to the wall or other structural surface. In the case of large, heavy panelboards, the addition of supplementary pedestals or "legs" (structural-steel or pipe sections with floor flanges) will increase rigidity and help distribute the weight.

6. If a number of panelboard enclosures are to be mounted side by side, the addition of a suitably sized auxiliary gutter, wiring trough, or pull box in between, above, or below the grouped cabinets will facilitate cable pulling and circuit installation. In many cases, it will eliminate the need for oversize cabinets to accommodate the required conductors. Section 373-8 of the National Electrical Code definitely prohibits the use of enclosures for switches and overcurrent devices (and this includes panelboard cabinets) as a feed-through or splicing gutter for conductors unless the enclosure has been designed to provide adequate space for this.

**GENERAL WIRING MATERIALS AND DEVICES**

**91. Insulators** of various types are employed for interior-wiring work when the conductors are not installed in raceways. The more common types may be classified as knobs, cleats, tubes, crane insulators, and rack insulators. The accompanying sections and illustrations give data on many standard types. Insulators for outdoor and underground installations are discussed in Div. 8.

Split knobs are shown in Figs. 4-139 and 4-140. Those of Fig. 4-139 are factory-assembled with a supporting nail or screw. Various standard types of solid-knob insulators are shown in Fig. 4-141. Standard types of rack insulators are shown in Fig. 4-142. Standard

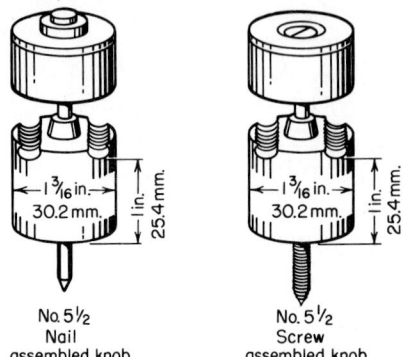

No. 5½
Nail
assembled knob

No. 5½
Screw
assembled knob

**Fig. 4-139**  Assembled wiring knobs for No. 12 to No. 14 wire.

wire cleats are shown in Figs. 4-143 and 4-144. Crane insulators are shown in Fig. 4-145. All these insulators are made of porcelain.

**92.  Wire Table for Single-Wire Cleats** [a]
(Knox Porcelain Corp.)

| AWG or cmil Size of Wire Received | Catalog Numbers |
|---|---|
| 14–8 | 1 |
| 6–4 | 1½–2 |
| ⅛–⅜ | 2½ |
| ⅜–⅝ | 3 |
| ⅝–300 MCM | 3½ |
| 300–500 MCM | 4 |

[a]See Fig. 4-144.

**93. Use of screws or nails with split knobs.**    Nails hold better than screws in certain woods. The breaking of knobs at the time of putting them up with screws is not the only source of trouble, for the binding tension applied often acts to crack the knob a considerable time after it has been put in place. It is an objectionable practice of many wire installers in putting up knobs with screws to drive the screws in nearly all the way with a hammer, giving them only a couple of turns with a screwdriver to tighten them. The principal argument in favor of the use of nails is the resulting great saving of the installers' time as compared with the time required for putting in screws. The insulating value of the two constructions is practically the same.

**94. Insulated racks** are often convenient for supporting cables in open-wiring installations. They consist (Fig. 4-146) of porcelain insulators mounted in an iron base and clamped together with an iron top for dc work and a brass top for ac work. They are made in assemblies for one, two, three, and four cables and can be supplied with insulator openings of inside diameters ranging from ³⁄₁₆ to 3 in (4.8 to 76.2 mm). The diameters vary by ¹⁄₁₆ in (1.6 mm) from ³⁄₁₆- to ⅝-in (4.8- to 15.9-mm) sizes and by ⅛ in (3.2 mm) from ⅝- to 3-in (3.2- to 76.2-mm) sizes.

**95. Universal insulator supports** (Fig. 4-147) are malleable-iron clamps fitted with cup-pointed, core-hardened steel setscrews for securing porcelain and glass insulators to exposed steel framework.

**96. Porcelain tubes** are made in three types: standard solid tube, split tube, and floor tube. The standard lengths, as indicated by the $L$ dimension in Fig. 4-148, are ½, 1, 1½, 2, 2½, 3, 4, 5, 6, 8, 10, 12, 14, 16, 18, 20, 22, and 24 in (12.7, 25.4, 38.1, 50.8, 63.5, 76.2, 101.6, 127.0, 152.4, 203.2, 254.0, 304.8, 355.6, 406.4, 457.2, 508.0, 558.8, and 609.6 mm). The diameters are given in Sec. 97.

**98. Receptacles and caps or plugs** are wiring devices for the purpose of providing a convenient means of attaching cord-and-plug–connected appliances or equipment. Such

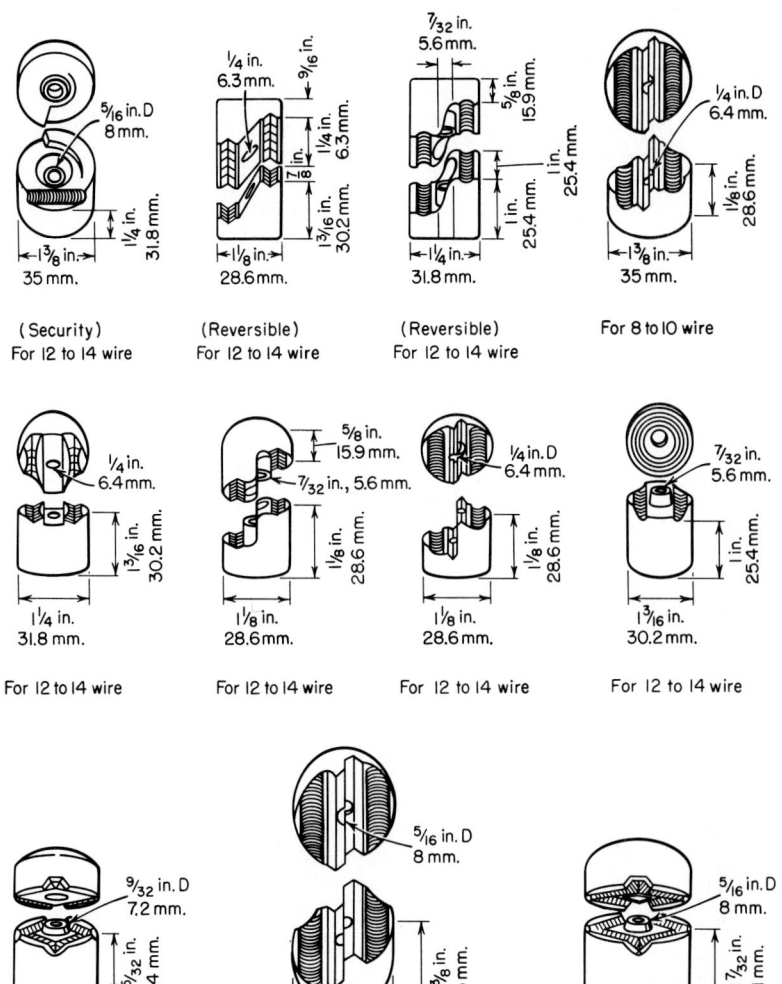

**Fig. 4-140**  Split knobs.

### 97.  Tube Diameters

| Diameters | Tube size (see Sec. 96) | | | | | | | | | | | | | |
|---|---|---|---|---|---|---|---|---|---|---|---|---|---|---|
| | 5/16 | 3/8 | 1/2 | 5/8 | 3/4 | 1 | 1 1/4 | 1 1/2 | 1 3/4 | 2 | 2 1/4 | 2 1/2 | 2 3/4 | 3 |
| Outside diameter, in | 9/16 | 11/16 | 13/16 | 15/16 | 1 3/16 | 1 7/16 | 1 13/16 | 2 3/16 | 2 9/16 | 2 15/16 | 3 5/16 | 3 11/16 | 4 1/4 | 4 1/2 |
| Outside diameter, mm | 14.3 | 17.5 | 20.6 | 23.8 | 30.2 | 36.5 | 46.1 | 55.6 | 65.1 | 74.6 | 84.1 | 93.7 | 108 | 114 |
| Inside diameter, in | 5/16 | 3/8 | 1/2 | 5/8 | 3/4 | 1 | 1 1/4 | 1 1/2 | 1 3/4 | 2 | 2 1/4 | 2 1/2 | 2 3/4 | 3 |
| Inside diameter, mm | 7.9 | 9.5 | 12.7 | 15.9 | 19.1 | 25.4 | 31.7 | 38.1 | 44.5 | 50.8 | 57.1 | 63.5 | 69.8 | 76.2 |

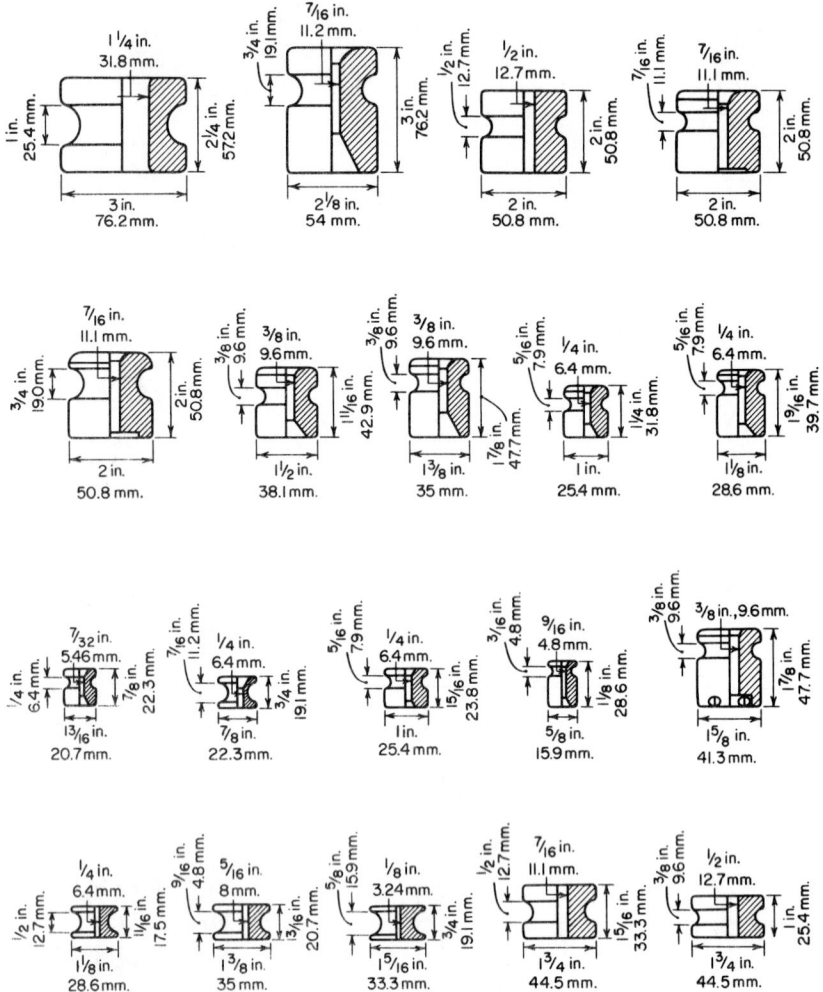

**Fig. 4-141**   Standard types of solid-knob insulators.

devices are available in a wide variety of voltage (600 V or less) and ampere ratings (10 to 400 A). The two broad classifications are grounding and nongrounding types.

The National Electrical Code requires all 15- and 20-A receptacles to be grounding types, and nongrounding-type receptacles can be used only for replacement purposes. Because millions of nongrounding-type receptacles were installed before the Code rule which required grounding types, it was obvious that nongrounding types would have to continue to be manufactured for replacement purposes. Also, because many existing receptacle outlets are not grounded, the Code states that grounding-type receptacles shall not replace nongrounding-type receptacles unless the receptacle outlets are properly grounded. The reason for this rule is to prevent a false sense of security that would exist if a grounding-type receptacle were attached to an existing outlet with no ground connection to the grounding terminal of the receptacle. Ground-fault circuit-interrupter (GFCI) receptacles may be used to replace nongrounding receptacles without the necessity of providing grounding conductors, and the existing nongrounding-type receptacles shall be permitted to be replaced with grounding-type receptacles where supplied through a GFCI receptacle.

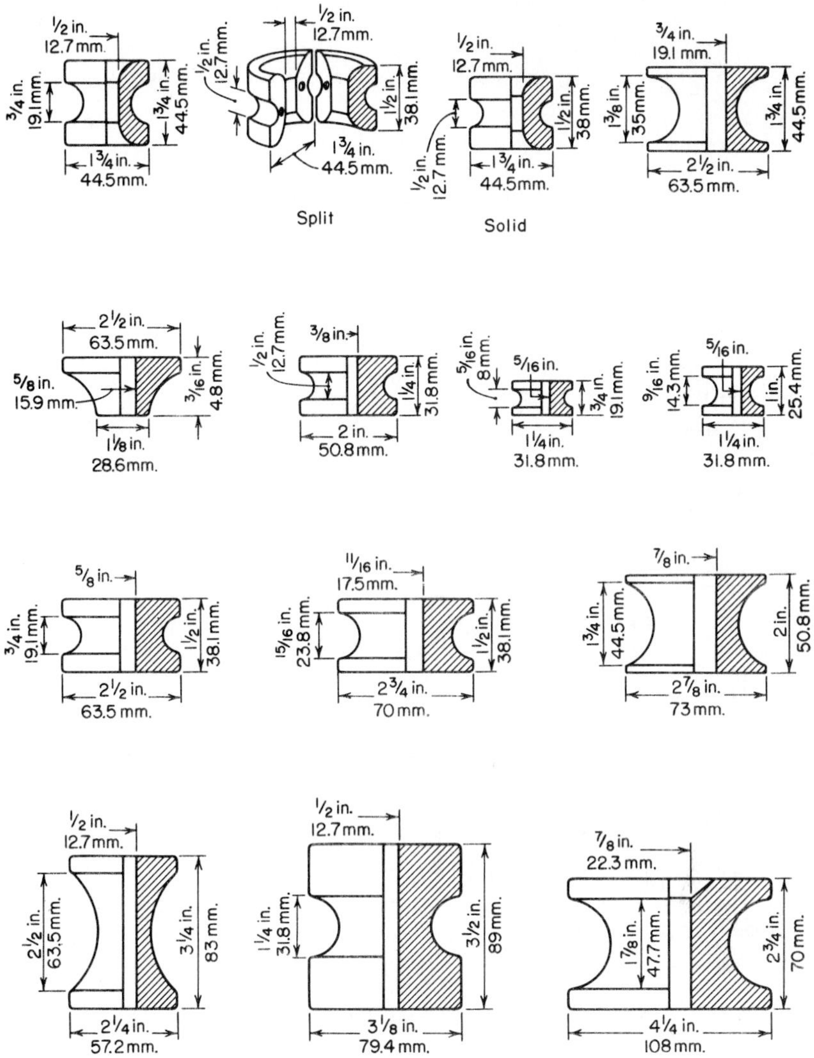

Split

Solid

**Fig. 4-142**    Standard types of rack insulators.

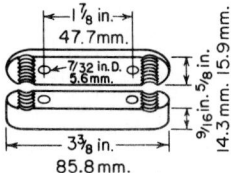

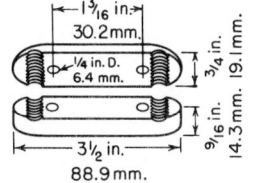

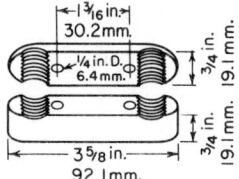

**Fig. 4-143**    Two-wire cleats.

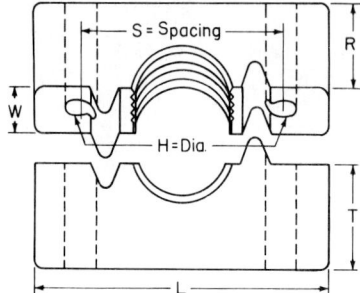

**Fig. 4-144**    Single-wire cleats (see Sec. 92).

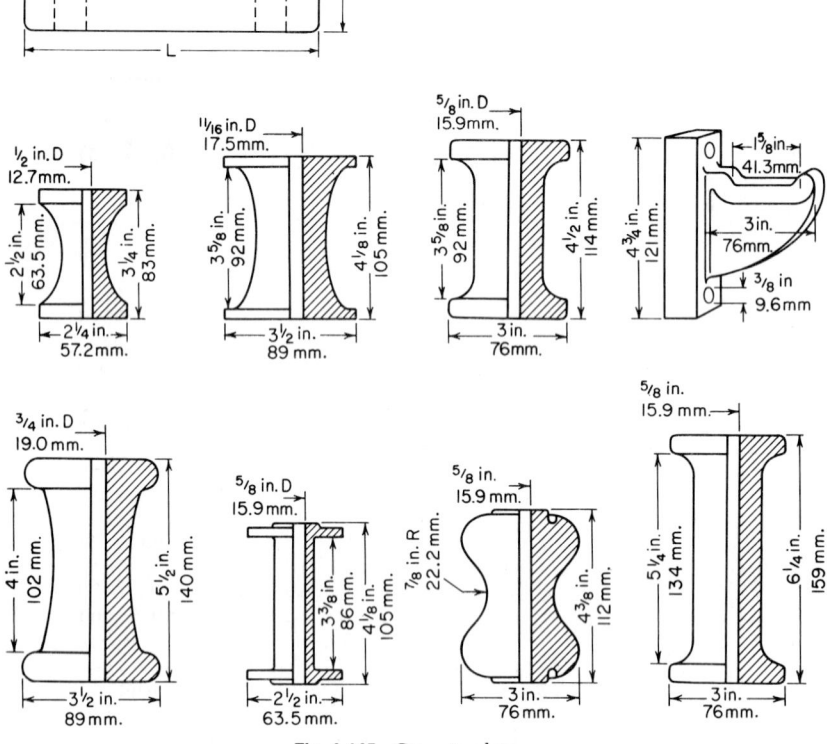

**Fig. 4-145**    Crane insulators.

**Fig. 4-146**    Insulated cable rack. [Westinghouse Electric Corp.]

I- Showing No. 502 support with No. 3½ insulator. Support is tapped standard for No. 24-16 thread machine screw.

II- Showing No. 502 support with attachment for type A No. 2 B.&D. cleat Support is tapped standard for No. 24-16 thread machine screw.

**Fig. 4-147**    Universal insulator supports. [Steel City Division of Midland-Ross Corp.]

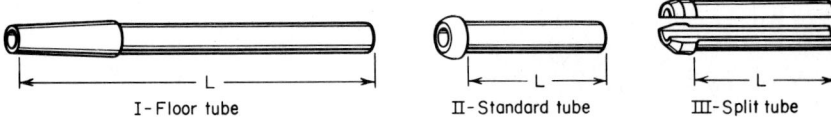

I-Floor tube          II-Standard tube          III-Split tube

**Fig. 4-148**   Porcelain insulating tubes.

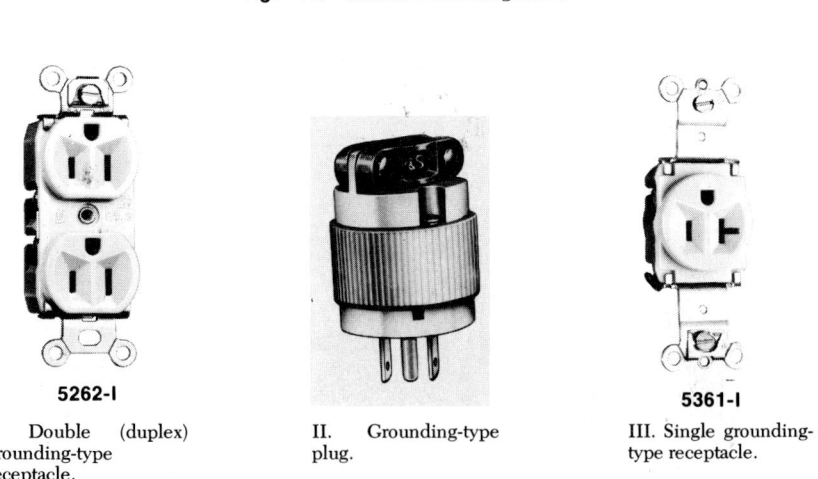

5262-I

5361-I

I.   Double   (duplex)   grounding-type receptacle.

II.   Grounding-type plug.

III. Single grounding-type receptacle.

**Fig. 4-149A**   Grounding-type receptacles and plugs, 125 V, 15 A. [Pass & Seymour, Inc.]

Another significant development concerning receptacles is a National Electrical Code rule which states: "*Grounding-type* receptacles shall be installed only on circuits of the *voltage class and current* for which they are rated." Figures 4-150 and 4-151 list the various receptacles and caps for general-purpose nonlocking and locking types.

The nonlocking-type plug-and-receptacle configurations shown in Fig. 4-150 include both grounding and nongrounding types. Current ratings are 15 to 60 A, and voltage ratings range from 125 to 600 V and two-pole, two-wire to four-pole, five-wire.

Figure 4-151 shows the receptacle-and-plug configurations for locking types, in ratings of 15, 20, and 30 A at 125 to 600 V, two-pole, two-wire to four-pole, five-wire.

Figures 4-150 and 4-151 are the configurations adopted by ANSI and NEMA; these configurations satisfy the National Electrical Code rule which calls for specific voltage and current ratings for all grounding-type receptacles. Also, these configurations are designed so that lower-voltage–rated receptacle caps cannot be inserted into receptacles supplied at higher voltages. This plan also achieves a high degree of standardization so that manufacturers of cord-and-plug–equipped apparatus can provide standard plug caps for such equipment and be reasonably assured that installers will provide a matching receptacle for a given current and voltage class.

Another significant advantage of the NEMA configurations is the wide selection of receptacles to allow different configurations on the same installation if several different voltages, classes of current (ac or dc), or different frequencies are used. In such cases, maintenance personnel usually install plug caps to match specific receptacle configurations.

**99. Single- or double-wipe contacts and terminal connections.**   In many smaller ratings, such as 15 A, 125 V, receptacles are available with single-wipe or double-wipe contacts to receive plug caps. The single-wipe contact is in contact on only one side of the blade of an inserted plug cap. The other side of the plug-cap blade is wedged against a non-current-carrying part of the receptacle. With double-wipe contacts *both sides* of a plug-cap blade are in contact with the receiving part of the receptacle when inserted, and this ensures better contact. Better-grade receptacles have double-wipe contacts, and they

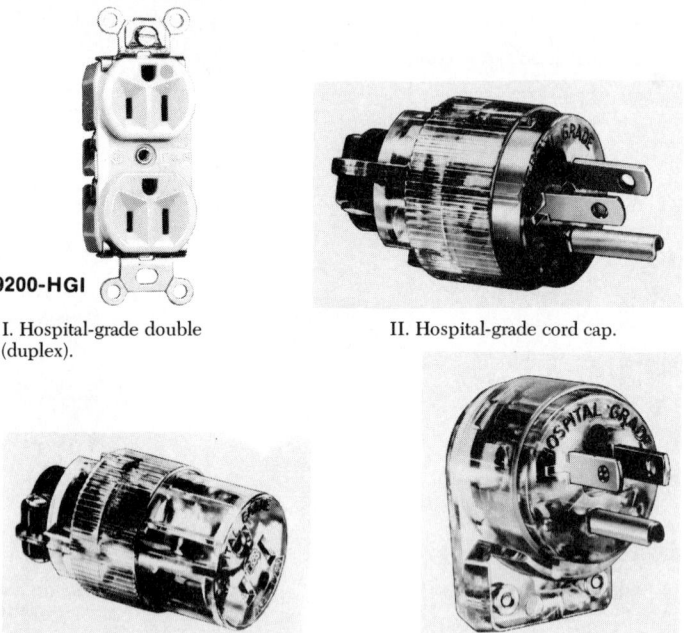

**9200-HGI**

I. Hospital-grade double (duplex).

II. Hospital-grade cord cap.

III. Hospital-grade female cord cap.

IV. Angle hospital-grade male cord cap.

**Fig. 4-149B** Hospital-grade grounding-type receptacles and plugs. [Pass & Seymour, Inc.]

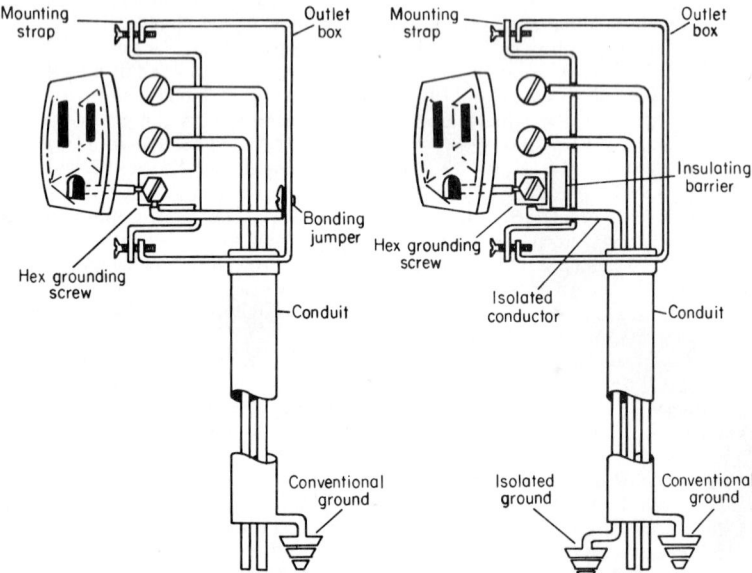

**Fig. 4-149C** A standard receptacle and an isolating receptacle with the grounding conductor isolated from the yoke of the receptacle. [Pass & Seymour, Inc.]

should be used whenever possible to avoid early failures or loose connections when plug caps are inserted frequently.

In general, smaller receptacles (15 or 20 A) use binding screws, recessed pressure-locking terminals, or a combination of both as a means of connecting supply wires. With binding screws, supply wires are skinned and wrapped around the screw in a *clockwise* direction. Since the screw is tightened in a clockwise position, this ensures a secure connection. Never fasten a wire around a binding screw in a counterclockwise direction because a poor connection will result when the binding screw is tightened.

The pressure-locking terminals are recessed inside the receptacles. The supply wires are simply skinned about ½ in (12.7 mm) and are then pushed through a hole (usually in the rear of the receptacle) into the pressure-locking contact. As the wire is inserted, the spring-held contact separates, and when the wire is pushed all the way in, it is firmly gripped.

Some receptacles contain binding screws and pressure-locking terminals so that an installer can select either method of connection. Another reason is that such receptacles can be used with either copper or aluminum wire. Since spring-contact pressure-locking terminals are not suitable for aluminum wire connections, the binding-screw method can be used. It should be noted that binding screws are considered acceptable for fastening aluminum wires (No. 12 or No. 10 solid) when approved for the purpose and marked CO/ALR.

**Fig. 4-149D** Class A ground-fault circuit interrupter to feed receptacles or a swimming pool. [Pass & Seymour, Inc.]

Receptacles 30 A or larger generally have *setscrew* terminals, and in such cases the terminals will be marked AL-CU if suitable for the connection of aluminum conductors. If there is no such marking, it can be assumed that only copper wire connections are permitted.

A class of receptacles has been developed for hospital use. These devices are subjected to more severe damage tests than conventional receptacles. They are identified by the marking "hospital grade" and by a green dot on the face of the receptacle. These receptacles are for use in nonhazardous locations in health care facilities. The see-through construction of the plugs permits visual inspection of the connections.

Figure 149C depicts a method of isolating the equipment-grounding conductor from the yoke of the receptacle. Electronic equipment frequently picks up radio-frequency interference and transient signals through the conventionally grounded system. The complicated circuitry of medical and communication systems is particularly susceptible to transient signals of a very low magnitude. The NEC· permits the grounding terminals of receptacles to be isolated from the yoke so that a conductor may be run directly to the grounding-electrode connection in the service equipment. Receptacles with an isolated ground shall be appropriately identified.

A GFCI to feed receptacles is shown in Fig. 149D. Ground-fault circuit interrupters for receptacles are required on construction sites; in bathrooms, attached garages, crawl spaces, unfinished basements, and outdoor receptacles of dwellings; and within 6 ft of kitchen sinks. Hotels and motels are also required to have a GFCI-protected receptacle in each bathroom. Trip units are set at 5 mA and will clear as fast as $^{25}/_{1000}$ s. These GFCI receptacles are equipped with a reset button if the unit should trip and a test button that should be used monthly to ensure proper operation of the GFCI. The receptacles may be obtained with a feed-through feature to protect other downstream receptacles. Should a leakage current of 5 mA occur from either phase conductor to ground, the shunt trip will energize and disconnect the receptacle from the circuit. However, phase-to-phase leakage will not affect the GFCI receptacle trip unit.

**100. Split-bus receptacles.**   In the 125-V, 15- or 20-A sizes, parallel-blade–type duplex receptacles are available with break-off jumpers or links so that one or both sides of the receptacles can be separated to avoid feed-through connections. With such receptacles it is common practice to break the connection on only the side of the receptacle that contains the screw connections for the ungrounded circuit conductors. With this connection separated, each half of the duplex receptacle can be connected to a different circuit or switch,

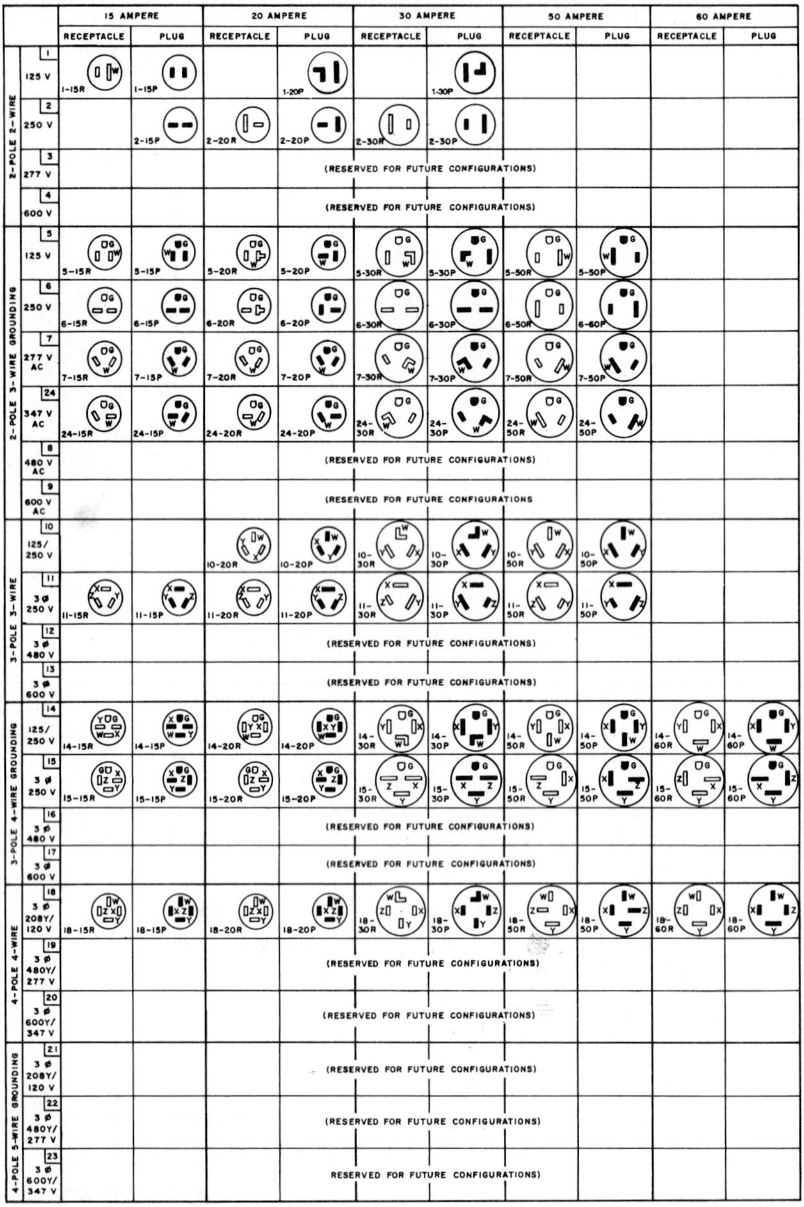

**Fig. 4-150** National Electrical Manufacturers Association configurations for general-purpose non-locking plugs and receptacles. [This figure is reproduced by permission from NEMA WD 6-1988, "General-Purpose Wiring Devices."]

or one-half of the receptacle can be controlled by a switch while the other half remains energized at 120 V continuously.

The break-off link on the other side is broken only when two separate circuits are required, particularly when these circuits are *not* three-wire, 120/240 V. Some receptacles with pressure-locking terminals have break-off features also.

**101. Interchangeable line.** Receptacles for use in the interchangeable line are available

**Fig. 4-151** National Electrical Manufacturers Association configurations for locking-type plugs and receptacles. [This figure is reproduced by permission from NEMA WD 6-1988, "Specific-Purpose Wiring Devices."]

in 15- and 20-A two-pole, two-wire and two-pole, three-wire. One such type rated at 15-A 125-V two-pole, two-wire is shown in Fig. 4-42, along with other interchangeable devices. For new installations such a receptacle would have to be a two-pole, three-wire, grounding type.

**102. Special four-pole, five-wire receptacles.**   Harvey Hubbell, Inc., has developed a line of five-wire receptacles and caps, rated at 30 A, 250 V ac, dc, or 600 V ac, which retain the advantages of locking-type receptacles. These receptacles are available in eight noninterchangeable center-pin and slot configurations as illustrated in Fig. 4-152. This permits a user to designate one configuration for one voltage, amperage, and/or frequency; a different configuration for a different voltage, amperage, and/or frequency; and so on. Such a system prevents the connection of equipment into the wrong source of power.

**Fig. 4-152** Noninterchangeable center-pin and slot configurations permit the use of eight differently rated circuits. [Harvey Hubbell, Inc.]

**103. Heavy-duty industrial receptacles.** Figure 4-153 shows typical heavy-duty industrial receptacles which are available in ratings up to 400 A at 600 V ac. The receptacles shown in Fig. 4-153 are five-pole, four-wire and are rated at 60 A, 250 V dc, 600 V ac. One receptacle contains a spring door which closes the plug opening when the receptacle cap is removed, and the other receptacle has a threaded cover which is screwed in place when the receptacle is not in use. Such covers are recommended in many industrial applications to prevent the entry of water, dirt, or dust.

**104. Range and clothes-dryer receptacles.** Figure 4-154 shows three-pole flush- and surface-mounted range receptacles rated at 250 V ac, 50 A. This crowfoot configuration is commonly used for connecting cord-and-plug–equipped 115/230-V household electric ranges in which the neutral conductor is used to ground the frame of the range. A similar receptacle, rated 30 A, is used for the connection of cord-and-plug–equipped electric clothes dryers of 5.5 kW or less at 115/230 or 120/208 V. The major difference is that the neutral pole is an L shape.

**105. Special receptacle housings** are available for receptacles installed outdoors, in hazardous locations, or for other applications in which receptacles must be protected from weather elements, contaminants, or hazardous concentrations of flammable gases, liquids, or dust.

**106. A box must be installed at each outlet, switch, or junction point** for all wiring installations except for exposed wiring on insulators or cable. An outlet is any point in a wiring system where current is taken for supplying fixtures, lamps, heaters, or any other current-consuming equipment. The boxes must be constructed of metal except for open wiring on insulators, concealed knob-and-tube work, or other nonmetallic wiring methods. In these installations boxes made of insulating material can be used (see Div. 9). Boxes used in interior-wiring work may be classified as outlet, utility, sectional switch, and floor boxes.

**Fig. 4-153**   Heavy-duty industrial receptacles. [Crouse Hinds Co.]

Utility boxes are designed principally for exposed wiring. Conduit fittings (see Div. 9) are used at the outlets for many exposed-conduit wiring installations. Sectionalized device boxes are used in concealed wiring if they are not embedded in masonry structure. Floor boxes are used for making outlets in floors from conduit wiring systems. Boxes are made of No. 10 to No. 14 gage sheet steel. The surfaces are galvanized. Galvanized boxes will resist corrosion and preserve the electrical conductivity of the raceway system. The different types of boxes are illustrated in Sec. 113 with the dimensions in which they are most commonly available.

**107. Box accessories** consist of support brackets or ears, studs, cable clamps, and covers. Many boxes are equipped with some form of mounting bracket or with mounting ears. Some of the available types of supporting brackets are shown in Table 111. Standard mounting ears for the support of device boxes are shown in Table 117 along with other methods of supporting sectional device boxes. Many boxes are provided with fixture studs for the support of a fixture which receives its electrical supply through the outlet box. Boxes are often

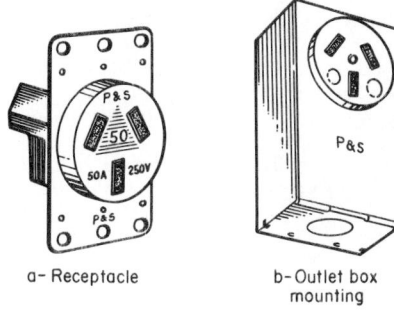

a– Receptacle          b– Outlet box mounting

**Fig. 4-154**    50-A range outlet. [Pass & Seymour, Inc.]

equipped with clamps for attaching armored cable or nonmetallic-sheathed cable to the box. Different types of clamps are shown in Table 112.

**108. Covers for outlet boxes** are made in a variety of types. Some of the types available are shown in Table 114.

**109. Outlet boxes** are made in round, square, octagonal, and oblong shapes. Round boxes should never be used when a conduit must enter the box through the side, as it is difficult to make a good connection with a locknut or bushing on a rounded surface.

The *round* and *octagonal* boxes are used for ceiling outlets in all types of concealed wiring installed in buildings of all types of construction. The octagonal boxes are also used for wall-bracket lighting outlets. Both round and octagonal boxes are made with various combinations of knockouts in sides and bottom and with built-in clamps for armored cable or nonmetallic-sheathed cable. They can be obtained with built-in ⅜- or ½-in fixture studs or with holes for mounting a separate fixture stud. Various types of covers are available for these boxes as shown in Sec. 114. Some are of the flat type, while others have the central portion of the cover raised so that additional interior space is made available in the box. Octagonal boxes constructed especially for concrete work have the bottom or back plate, as it is called, detachable from the sides. The back plate and sides are provided with conduit knockouts. The back plate may be provided with ⅜- or ½-in built-in fixture studs or with holes for mounting a separate fixture stud.

*Square* and *oblong* boxes are used principally for sidewall outlets in conduit wiring when these are embedded in masonry or installed in brick or tile walls. Sometimes they also are used with conduit wiring in other types of building construction. The square boxes are used sometimes for ceiling outlets. The oblong boxes, which frequently are called gang boxes, are used if more internal space is required than is provided by the square boxes. The term *gang* refers to the number of standard-line wiring devices which the box will accommodate. A three-gang box will accommodate three standard wiring devices mounted side by side. They are made with various combinations of knockouts in sides and bottom. The square boxes may be provided with ⅜- or ½-in built-in fixture studs or with holes for mounting separate fixture studs. Various types of covers are available for the square boxes shown in Sec. 114. Covers are available with openings and supporting lugs for the accommodation of flush wiring devices. The opening is then covered with a standard flush plate. The portion of the box cover around the central opening is raised so that the box is set back a slight amount in the wall and the cover brings the surface of the wiring device flush with the wall surface. The box and the box cover are plastered over a sufficient amount so that the remaining openings will be completely covered with the flush plate. Covers are available with the central portion raised by different amounts to accommodate different thicknesses of plaster. The covers for the oblong boxes are always of the type just described. When the square boxes are used in exposed wiring, a flush plate would not give a neat appearance, and an outlet-box cover which covers the entire opening of the box therefore is used. Openings in the cover accommodate switch handles or receptacles.

**110. Utility boxes** are designed for use in exposed-conduit wiring installations to produce a neat-appearing job. They are provided with knockouts in bottom and sides. Data on covers are given in Sec. 114.

**110A. Sectional device boxes** are employed in concealed-wiring installations which are not embedded in masonry. They are constructed with removable sides so that any number of boxes can be ganged side by side to provide compactly for several wiring devices at one location. No covers are required, since the boxes are designed to accommodate the standard flush wiring devices with standard flush plates. They are made in types with conduit knockouts and with clamps for armored cable, nonmetallic-sheathed cable, and loom. Most boxes are 3 in long and 2 in wide.

**110B. Extension rings** (Fig. 4-155) are available for octagonal and square boxes to increase the depth of the boxes. The standard depths in which they are available for the different boxes are given in Sec. 113.

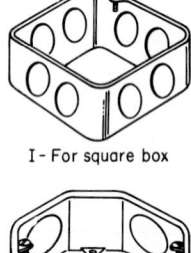

I - For square box

II-For octagonal box

**Fig. 4-155**   Outlet-box extension rings.

## 111.  Bracket Identification[a]

| Bracket designation | | Description and application |
|---|---|---|
| B | | For mounting on the face of stud.  Can be offset from studding to clear door frame or snugged up against studding as desired |
| F | | A sturdy, simple flat bracket for easy side mounting.  Exceptionally rigid and easy to install.  Hugs close to studding |
| A | | Angle bracket snugs against both face and side of studding.  When nails are driven both ways, it becomes the strongest mounting available.  Stands up under the strain of marking box openings in dry wall |
| O | | Side-mount bracket mounts on face and side of studding.  Locating lugs engage face of studding for normal plaster thickness.  Lugs are easily removed so box can be mounted on side of studding in various positions to accommodate all wall thicknesses |
| D | | The D bracket is an integral part of the box, affording great strength and rigidity.  These boxes are extremely versatile—easily adapted for ⅜, ½ and ⅝-in. wall thickness.  Wall thickness is marked on box |
| E | | Face-mounting bracket on end of device box for horizontal mounting of switches or receptacles |
| T | | Exceptionally strong twin mounting bracket for oversize 4- by 2-in. device boxes.  Can be nailed from front or side or both.  Box stands out from stud for door frame clearance |

[a] Courtesy of General Electric Co.

## 112.    Clamp Identification[a]

| Clamp designation | | For use with | Description and application |
|---|---|---|---|
| C | | BX armored cable | For outlet and bar hanger boxes.  Two-way entrance design permits cable to enter through side or back of box.  Well suited for ceiling work |
| K | | Nonmetallic and BX armored cable | For outlet boxes and device boxes.  Provides visual inspection of antishort bushings.  Holds bushings tightly against armor stop |
| D | | Nonmetallic cable | For outlet boxes and device boxes.  Two-way entrance design permits cable to enter through side or back of box.  Compresses when tightened; grips cables firmly |
| L | | Nonmetallic cable | For outlet boxes and device boxes.  Compresses when tightened; grips cable firmly.  Particularly adaptable for side-entry installations |
| No. 9 | | Nonmetallic cable or loom | For ceiling pan installations |
| No. 10 | | Nonmetallic cable | For ceiling pan installations.  Tapped for outside clamp screw.  Compresses when tightened; grips cable firmly |
| No. 12 | | BX armored cable | For shallow pans.  Tapped hole for outside clamp screw |
| BD | | Nonmetallic cable | For beveled-corner device boxes.  Holds cable with even pressure |

[a] Courtesy of General Electric Co.

## 113.  Data on Wiring Boxes
(Courtesy of General Electric Co.)

| | Sketch No. | Depth, in. | Accessories | | | Knockouts | | | |
|---|---|---|---|---|---|---|---|---|---|
| | | | | | | Sides | | Bottom | |
| | | | Stud | Ears | Clamps | Conduit | Cable $2\frac{1}{32}$ in. | Conduit | Cable $2\frac{1}{32}$ in. |
| Sketch 1 | 1 | $1\frac{1}{2}$ | ... | ... | ... | Four $\frac{1}{2}''$ | ... | One $\frac{1}{2}''$ | |
| | | $1\frac{1}{2}$ | ... | ... | ... | Four $\frac{3}{4}''$ | ... | One $\frac{1}{2}''$ | |
| **$3\frac{1}{4}$-in. Extension Rings** | | | | | | | | | |
| Sketch 2 | 2 | $1\frac{1}{2}$ | ... | ... | ... | Four $\frac{1}{2}''$ | | | |
| | | $1\frac{1}{2}$ | ... | ... | ... | Four $\frac{3}{4}''$ | | | |
| **$3\frac{1}{2}$-in. Octagon Boxes** | | | | | | | | | |
| Sketch 3 | | $1\frac{1}{2}$ | ... | ... | 2-C | Two $\frac{1}{2}''$ | 4 | One $\frac{1}{2}''$ | 4 |
| | | $1\frac{1}{2}$ | ... | ... | 2-L | Two $\frac{1}{2}''$ | 4 | One $\frac{1}{2}''$ | 4 |
| | 3 | $1\frac{1}{2}$ | ... | Yes | 2-L | Two $\frac{1}{2}''$ | 4 | One $\frac{1}{2}''$ | 4 |

113. Data on Wiring Boxes (*Continued*)

| | Sketch No. | Depth, in. | Accessories | | | | Knockouts | | | |
|---|---|---|---|---|---|---|---|---|---|---|
| | | | | | | | Sides | | Bottom | |
| | | | Brkt. | Stud | Ears | Clamps | Conduit | Cable $^{23}/_{32}$ in. | Conduit | Cable $^{23}/_{32}$ in. |
| 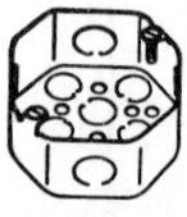 Sketch 4 | 4 | 1½ | ... | ... | ... | ... | Four ½″ | ... | Five ½″ | |
| | | 1½ | ... | ... | ... | ... | Four ¾″ | ... | Three ½″ Two ¾″ | |
| | | 1½ | ... | ... | ... | ... | Two ½″ Two ¾″ | ... | Three ½″ Two ¾″ | |
|  Sketch 5 | 5 | 1½ | ... | ... | ... | 2-C | Two ½″ | 4 | One ½″ | 4 |
| | | 1½ | ... | ... | Yes | 2-C | Two ½″ | 4 | One ½″ | 4 |
| | | 1½ | ... | Yes | ... | 2-C | Two ½″ | 4 | ......... | 4 |
| | | 1½ | ... | ... | ... | 2-L | Two ½″ | 4 | One ½″ | 4 |
| 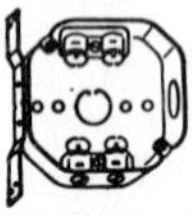 Sketch 6 | 6 | 1½ | F | No | No | 2-L | Two ½″ | 4 | One ½″ | 4 |
| | | 1½ | ... | Yes | ... | 2-D | Two ½″ | 4 | ......... | 4 |
| | | 1½ | ... | Yes | Yes | 2-D | Two ½″ | 4 | ......... | 4 |
|  Sketch 7 | 7 | 1½ | .. | ... | ... | 2-D | Two ½″ | 4 | One ½″ | 4 |
| | | 1½ | ... | ... | Yes | 2-D | Two ½″ | 4 | One ½″ | 4 |
| 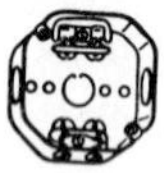 Sketch 8 | 8 | 1½ | ... | ... | ... | 2-K | Two ½″ | 4 | One ½″ | 4 |
| | | 1½ | ... | Yes | ... | 2-K | Two ½″ | 4 | ......... | 4 |
| | | 2⅛ | ... | ... | ... | ... | Four ½″ | ... | Five ½″ | |
| | | 2⅛ | ... | ... | ... | ... | Four ¾″ | ... | Three ½″ Two ¾″ | |
| | | 2⅛ | ... | ... | ... | ... | Four 1″ | ... | Three ½″ Two ¾″ | |
| | | 2⅛ | ... | ... | ... | ... | Two ½″ Two ¾″ | ... | Three ½″ Two ¾″ | |
| **4-in. Octagon Extension Rings** | | | | | | | | | | |
| 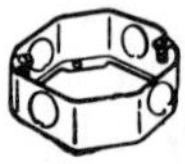 Sketch 9 | 9 | 1½ | ... | ... | ... | ... | Four ½″ | | | |
| | | 1½ | ... | ... | ... | ... | Four ¾″ | | | |
| | | 1½ | ... | ... | ... | ... | Two ½″ Two ¾″ | | | |
| 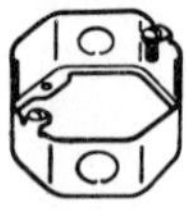 Sketch 10 | 10 | 2⅛ | ... | ... | ... | ... | Four ½″ | | | |
| | | 2⅛ | ... | ... | ... | ... | Four ¾″ | | | |
| | | 2⅛ | ... | ... | ... | ... | Four 1″ | | | |
| | | 2⅛ | ... | ... | ... | ... | Two ½″ Two ¾″ | | | |

## 113. Data on Wiring Boxes

|  | Sketch No. | Depth, in. | Clamps | Brkt. | Knockouts Sides Conduit | Knockouts Sides Cable 21/32 in. | Knockouts Bottom Conduit | Knockouts Bottom Cable 21/32 in. |
|---|---|---|---|---|---|---|---|---|
|  Sketch 11 |  | 1¼ | ... | ... | Twelve ½" | ... | Five ½" | |
|  |  | 1¼ | ... | B | Twelve ½" | ... | Five ½" | |
|  | 11 | 1½ | ... | ... | Twelve ½" | ... | Five ½" | |
|  |  | 1½ | ... | ... | Eight ¾" | ... | Three ½" Two ¾" | |
|  |  | 1½ | ... | ... | Eight ½" Four ¾" | ... | Three ½" Two ¾" | |
| Sketch 12 | 12 | 1⅜ | ... | B | Twelve ½" | ... | Five ½" | |
|  |  | 1½ | ... | B | Six ¾" | ... | Three ½" Two ¾" | |
|  |  | 1½ | ... | B | Six ½" Three ¾" | ... | Three ½" Two ¾" | |
|  |  | 1½ | ... | A | Nine ½" | ... | Five ½" | |
|  |  | 1½ | ... | F | Nine ½" | ... | Five ½" | |
|  |  | 1½ | ... | O | Nine ½" | ... | Five ½" | |
| 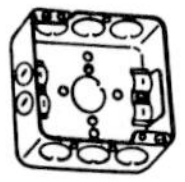  |  | 1½ | ... | O | Six ¾" | ... | Three ½" Two ¾" | |
| Sketch 13 |  | 1½ | • | ... | Twelve ½" | | | |
|  |  | 1½ | 2-C | ... | Six ½" | 4 | One ½" | 4 |
|  |  | 1½ | 2-C | B | Six ½" | 4 | One ½" | 4 |
|  |  | 1½ | 2-C | F | Three ½" | 4 | One ½" | 4 |
|  |  | 1½ | 2-C | O | Three ½" | 4 | One ½" | 4 |
|  | 13 | 1½ | 2-L | ... | Six ½" | 4 | One ½" | 4 |
|  |  | 1½ | 2-L | B | Six ½" | 4 | One ½" | 4 |
|  |  | 1½ | 2-D | ... | Six ½" | 4 | One ½" | 4 |
|  |  | 1½ | 2-D | A | Three ½" | 4 | One ½" | 4 |
|  |  | 1½ | 2-D | B | Six ½" | 4 | One ½" | 4 |
|  |  | 1½ | 2-D | F | Three ½" | 4 | One ½" | 4 |
|  |  | 1½ | 2-K | ... | Six ½" | 4 | One ½" | 4 |
| Sketch 14 | 14 | 1½ | 2-K | B | Six ½" | 4 | One ½" | 4 |
|  | 15 | 2⅛ | ... | ... | Twelve ½" | ... | Five ½" | |
| 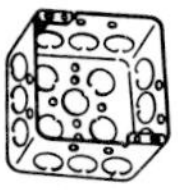  |  | 2⅛ | ... | ... | Eight ¾" | ... | Three ½" Two ¾" | |
| |  | 2⅛ | ... | ... | Eight 1" | ... | Three ½" Two ¾" | |
|  |  | 2⅛ | ... | ... | Four 1¼" | ... | Three ½" Two ¾" | |
| Sketch 15 |  | 2⅛ | ... | ... | Eight ½" Four ¾" | ... | Three ½" Two ¾" | |

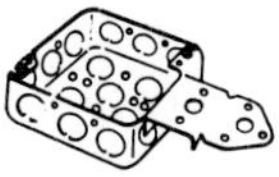

### 4-in. Square Extension Rings

|  | Sketch No. | Depth, in. | Clamps | Brkt. | Sides Conduit | | | |
|---|---|---|---|---|---|---|---|---|
|  |  | 1½ | ... | ... | Twelve ½" | | | |
|  |  | 1½ | ... | ... | Eight ¾" | | | |
| Sketch 16 | 16 | 1½ | ... | ... | Eight ½" Four ¾" | | | |
| |  | 2⅛ | ... | ... | Twelve ½" | | | |
| 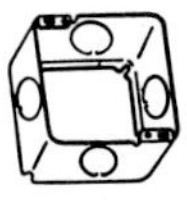  |  | 2⅛ | ... | ... | Eight ¾" | | | |
|  | 17 | 2⅛ | ... | ... | Eight 1" | | | |
|  |  | 2⅛ | ... | ... | Four 1¼" | | | |
| Sketch 17 |  | 2⅛ | ... | ... | Eight ½" Four ¾" | | | |

**113.  Data on Wiring Boxes** (*Continued*)

| | Sketch No. | Depth, in. | Knockouts | |
|---|---|---|---|---|
| | | | Sides | Bottom |
| 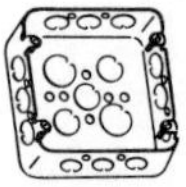 | 18 | 1½ | Twelve ½" | Three ½", two ¾" |
| | | 1½ | Eight ¾" | Three ½", two ¾" |
| | | 1½ | Eight ½", four ¾" | Three ½", two ¾" |
| Sketch 18 | 19 | 2⅛ | Twelve ½" | Three ½", two ¾" |
|  | | 2⅛ | Eight ¾" | Three ½", two ¾" |
| | | 2⅛ | Eight 1" | Three ½", two ¾" |
| | | 2⅛ | Four 1¼" | Three ½", two ¾" |
| | | 2⅛ | Eight ½", four ¾" | Three ½", two ¾" |
| Sketch 19 | | | | |

4¹¹⁄₁₆-in. Square Extension Rings

| | Sketch No. | Depth, in. | Sides | |
|---|---|---|---|---|
| 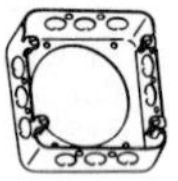 | 20 | 1½ | Twelve ½" | |
| | | 1½ | Eight ¾" | |
| | | 1½ | Eight ½", four ¾" | |
| Sketch 20 | | 2⅛ | Twelve ½" | |
| | | 2⅛ | Eight ¾" | |
| | | 2⅛ | Eight 1" | |
| 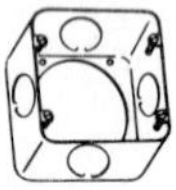 | 21 | 2⅛ | Four 1¼" | |
| | | 2⅛ | Eight ½", four ¾" | |
| Sketch 21 | | | | |

3½-in. Round—½-in. Deep

| | Sketch No. | Accessories | | | Knockouts | |
|---|---|---|---|---|---|---|
| | | Studs | Ears | Clamps | Conduit | Loam or cable ²¹⁄₃₂ in. |
| 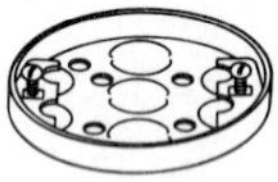 Sketch 22 | 22 | No | Yes | Two No. 9 | Three ½" | 4 |

3½-in. Round—¾-in. Deep

| | Sketch No. | Studs | Ears | Clamps | Conduit | Loam or cable ²¹⁄₃₂ in. |
|---|---|---|---|---|---|---|
|  | 23 | Yes | Yes | Two No. 10 | ......... | 4 |
| Sketch 23 | | No | Yes | Two No. 10 | One 1½" | 4 |
| 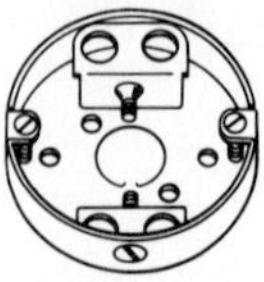 | | Yes | Yes | Two No. 12 | ......... | 4 |
| Sketch 24 | 24 | No | Yes | Two No. 12 | One ½" | 4 |

4-in. Round—½-in. Deep

| | Sketch No. | Studs | Ears | Clamps | Conduit | Loam or cable ²¹⁄₃₂ in. |
|---|---|---|---|---|---|---|
|  Sketch 25 | 25 | No | Yes | No | Five ½" | 1 |

## 113.  Data on Wiring Boxes

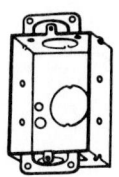 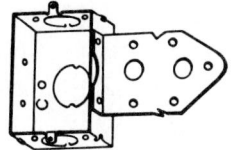  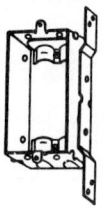 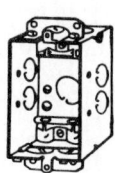

| Sketch 26 | Sketch 27 | Sketch 28 | Sketch 29 | Sketch 30 |
|-----------|-----------|-----------|-----------|-----------|

| Sketch No. | Depth, in. | Accessories | | | Knockouts | | | | | |
|---|---|---|---|---|---|---|---|---|---|---|
| | | | | | Each end | | Each side | | Bottom | |
| | | Brkt. | Clamps | Ears | Conduit | Cable $2\frac{1}{32}$ in. | Conduit | Cable $2\frac{1}{32}$ in. | Conduit | Cable $2\frac{1}{32}$ in. |
| 26 | 1½ | No | No | Yes | One ½″ | ........ | ...... | ..... | One ½″ | |
| | 1½ | F | No | No | One ½″ | ........ | ...... | ..... | One ½″ | |
| 27 | 1½ | B | No | No | One ½″ | ........ | ...... | ..... | One ½″ | |
| | 1½ | No | 1-L | Yes | One ½″ | Corner K.O.'s for N-M cable 1 end | | | | |
| 28 | 1½ | No | 2-L | Yes | ...... | Corner K.O.'s for N-M cable | | | | |
| 29 | 1½ | F | 2-L | No | ...... | Corner K.O.'s for N-M cable | | | | |
| | 1½ | B | 2-L | No | ...... | Corner K.O.'s for N-M cable | | | | |
| | 2 | No | No | No | One ½″ | ........ | Two ½″ | .... | One ½″ | |
| | 2 | No | No | Yes | One ½″ | ........ | Two ½″ | .... | One ½″ | |
| | 2 | F | No | No | One ½″ | ........ | Two ½″ | .... | One ½″ | |
| | 2 | A | No | No | One ½″ | ........ | Two ½″ in 1 side | .... | One ½″ | |
| | 2 | B | No | No | One ½″ | ........ | Two ½″ | .... | One ½″ | |
| 30 | 2 | No | 2-K | Yes | ...... | 2 | ...... | 2 | One ½″ | |
| | 2 | No | 2-L | Yes | ...... | 2 | ...... | 2 | One ½″ | |

**113. Data on Wiring Boxes** (*Continued*)

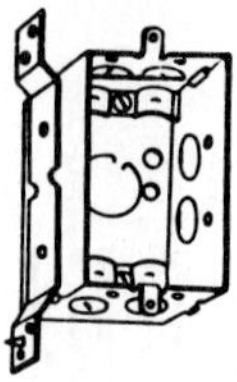

Sketch 31

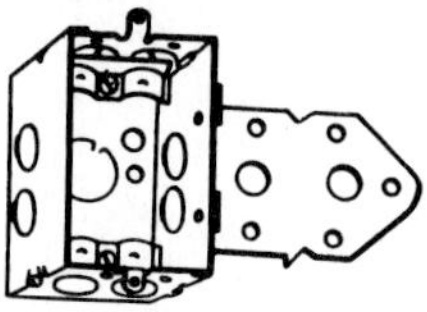

Sketch 32

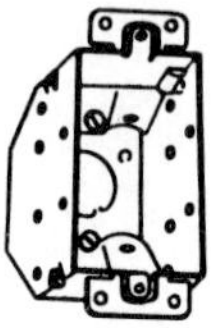

Sketch 33

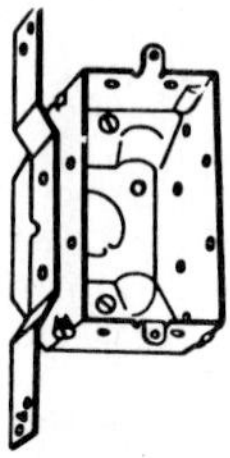

Sketch 34

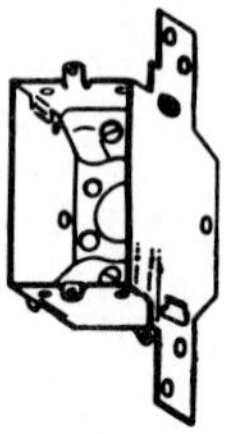

Sketch 35

| Sketch No. | Depth, in. | Accessories | | | Knockouts | | | | | |
|---|---|---|---|---|---|---|---|---|---|---|
| | | | | | Each end | | Each side | | Bottom | |
| | | Brkt. | Clamps | Ears | Conduit | Cable 21/32 in. | Conduit | Cable 21/32 in. | Conduit | Cable 21/32 in. |
| | 2 | F | 2-K | No | ...... | 2 | ...... | 2 in 1 side | One ½″ | |
| 31 | 2 | F | 2-L | No | ...... | 2 | ...... | 2 in 1 side | One ½″ | |
| | 2 | A | 2-K | No | ...... | 2 | ...... | 2 in 1 side | One ½″ | |
| | 2 | A | 2-L | No | ...... | 2 | ...... | 2 in 1 side | One ½″ | |
| | 2 | B | 2-K | No | ...... | 2 | ...... | 2 | One ½″ | |
| 32 | 2 | B | 2-L | No | ...... | 2 | ...... | 2 | One ½″ | |
| | 2¼ | No | 2-BD | Yes | ...... | Ea. bevel corner 2 | ...... | ..... | One ½″ | |
| | 2¼[a] | No | No | No | ...... | Ea. bevel corner 2 | ...... | ..... | One ½″ | |
| | 2¼ | No | No | Yes | ...... | Ea. bevel corner 2 | | | | |
| | 2¼[a] | No | 2-BD | No | ...... | Ea. bevel corner 2 | ...... | ..... | One ½″ | |
| 33 | 2¼ | No | 2-BD | Yes (single screw) | ...... | Ea. bevel corner 2 | ...... | ..... | One ½″ | |
| | 2¼ | D | No | No | ...... | Ea. bevel corner 2 | ...... | ..... | One ½″ | |
| 34 | 2¼ | F | 2-BD | No | ...... | Ea. bevel corner 2 | ...... | ..... | One ½″ | |
| 35 | 2¼ | D | 2-BD | No | ...... | Ea. bevel corner 2 | ...... | ..... | One ½″ | |
| | 2¼ | A | 2-BD | No | ...... | Ea. bevel corner 2 | ...... | ..... | One ½″ | |

[a] Leveling bumps and straight through nail holes.

### 113. Data on Wiring Boxes

3-in. Long by 2-in. Wide Device Boxes (*Continued*)

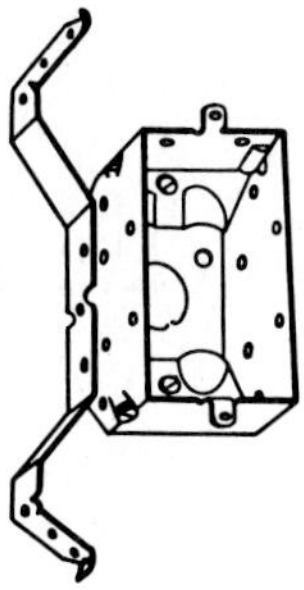

Sketch 36

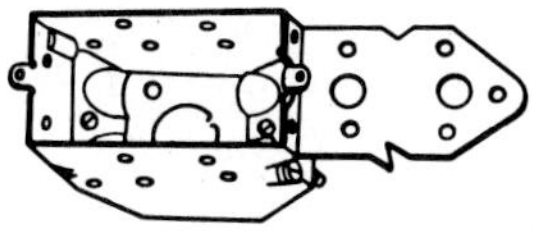

Sketch 37

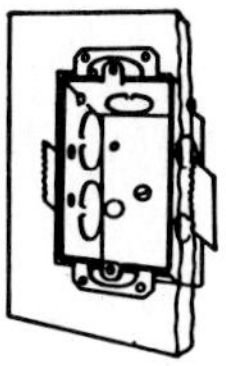

Sketch 38

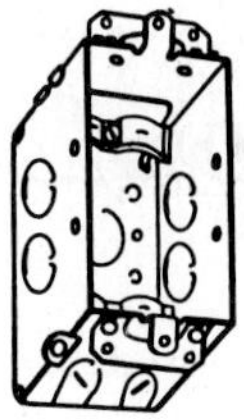

Sketch 39

| Sketch No. | Depth, in. | Accessories | | | Knockouts | | | | | |
|---|---|---|---|---|---|---|---|---|---|---|
| | | | | | Each end | | Each side | | Bottom | |
| | | Brkt. | Clamps | Ears | Conduit | Cable $^{21}/_{32}$ in. | Conduit | Cable $^{21}/_{32}$ in. | Conduit | Cable $^{21}/_{32}$ in. |
| 36 | $2\frac{1}{4}$ | O | 2-BD | No | ...... | Ea. bevel corner 2 | ...... | ..... | One ½" | |
| | $2\frac{1}{4}$ | B | 2-BD | No | ...... | Ea. bevel corner 2 | ...... | ..... | One ½" | |
| 37 | $2\frac{1}{4}$ | E | 2-BD | No | ...... | Ea. bevel corner 2 | ...... | ..... | One ½" | |
| | $2\frac{1}{2}$ | No | No | No | One ½" | ........ | Two ½" | ..... | One ½" | |
| | $2\frac{1}{2}$ | No | No | Yes | One ½" | ........ | Two ½" | ..... | One ½" | |
| | $2\frac{1}{2}$ | F | No | No | One ½" | ........ | Two ½" in 1 side | ..... | One ½" | |
| | $2\frac{1}{2}$ | A | No | No | One ½" | ........ | Two ½" in 1 side | ..... | One ½" | |
| | $2\frac{1}{2}$ | B | No | No | One ½" | ........ | Two ½" | ..... | One ½" | |
| | $2\frac{1}{2}$ | O | No | No | One ½" | ........ | Two ½" in 1 side | ..... | One ½" | |
| | $2\frac{1}{2}$ | E | No | No | One ½" | ........ | Two ½" | ..... | One ½" | |
| 38 | $2\frac{1}{2}$ | SP970 | No | Yes | One ½" | ........ | Two ½" | | | |
| | $2\frac{1}{2}$ [a] | No | 2-K | No | ...... | 2 | ...... | 2 | One ½" | |
| | $2\frac{1}{2}$ [a] | No | 2-L | No | ...... | 2 | ...... | 2 | One ½" | |
| | $2\frac{1}{2}$ | No | 2-K | Yes | ...... | 2 | ...... | 2 | One ½" | |
| 39 | $2\frac{1}{2}$ | No | 2-L | Yes | ...... | 2 | ...... | 2 | One ½" | |
| | $2\frac{1}{2}$ | F | 2-K | No | ...... | 2 | ...... | 2 in 1 side | One ½" | |
| | $2\frac{1}{2}$ | F | 2-L | No | ...... | 2 | ...... | 2 in 1 side | One ½" | |

[a] Leveling bumps and straight through nail holes.

## 113. Data on Wiring Boxes (*Continued*)

**3-in. Long by 2-in. Wide Device Boxes (*Continued*)**

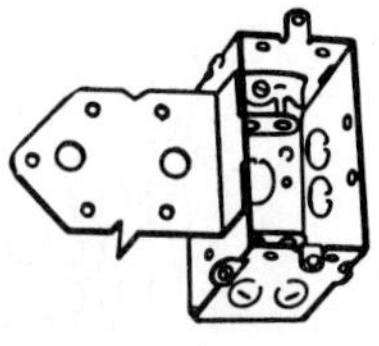

Sketch 40

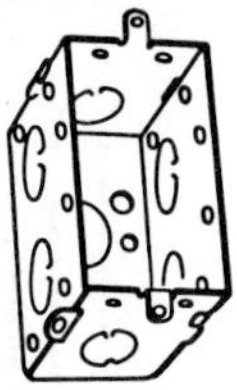

Sketch 41

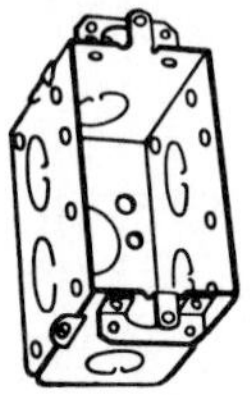

Sketch 42

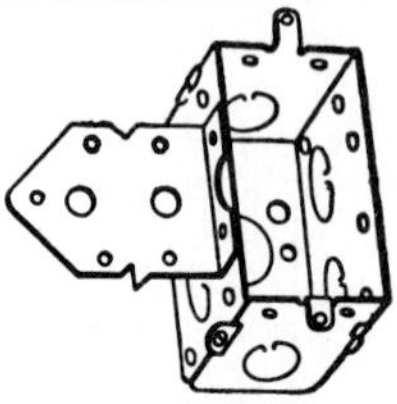

Sketch 43

| Sketch No. | Depth, in. | Accessories | | | Knockouts | | | | | |
|---|---|---|---|---|---|---|---|---|---|---|
| | | | | | Each end | | Each side | | Bottom | |
| | | Brkt. | Clamps | Ears | Conduit | Cable $\frac{21}{32}$ in. | Conduit | Cable $\frac{21}{32}$ in. | Conduit | Cable $\frac{21}{32}$ in. |
| | 2½ | A | 2-K | No | ...... | 2 | ...... | 2 in 1 side | One ½″ | |
| | 2½ | A | 2-L | No | ...... | 2 | ...... | 2 in 1 side | One ½″ | |
| 40 | 2½ | B | 2-K | No | ...... | 2 | ...... | 2 | One ½″ | |
| | 2½ | B | 2-L | No | ...... | 2 | ...... | 2 | One ½″ | |
| | 2½ | O | 2-K | No | ...... | 2 | ...... | 2 in 1 side | One ½″ | |
| | 2½ | O | 2-L | No | ...... | 2 | ...... | 2 in 1 side | One ½″ | |
| | 2½ | E | 2-K | No | ...... | 2 | ...... | 2 | One ½″ | |
| | 2½ | E | 2-L | No | ...... | 2 | ...... | 2 | One ½″ | |
| | 2½[a] | No | 2-D | No | ...... | 2 | ...... | 2 | One ½″ | 4 |
| | 2½ | No | 2-D | Yes | ...... | 2 | ...... | 2 | One ½″ | 4 |
| | 2½ | O | 2-D | No | ...... | 2 | ...... | 2 in 1 side | One ½″ | 4 |
| 41 | 2¾ | No | No | No | One ½″ | ........ | Two ½″ | ..... | One ½″ | |
| | 2¾ | No | No | No | One ¾″ | ........ | Two ½″ | ..... | One ½″ | |
| 42 | 2¾ | No | No | Yes | One ½″ | ........ | Two ½″ | ..... | One ½″ | |
| | 2¾ | No | No | Yes | One ¾″ | ........ | Two ½″ | ..... | One ½″ | |
| | 2¾ | A | No | No | One ¾″ | ........ | Two ½″ in 1 side | ..... | One ½″ | |
| 43 | 2¾ | B | No | No | One ½″ | ........ | Two ½″ | ..... | One ½″ | |
| | 2¾ | O | No | No | One ¾″ | ........ | Two ½″ in 1 side | ..... | One ½″ | |

[a] Leveling bumps and straight through nail holes.

**113.  Data on Wiring Boxes**

3-in. Long by 2-in. Wide Device Boxes (*Continued*)

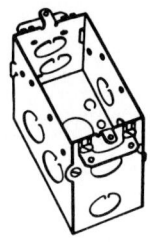
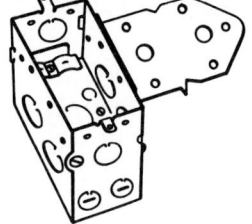
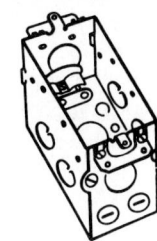

Sketch 44                     Sketch 45                     Sketch 46

| Sketch No. | Depth, in. | Accessories | | | Knockouts | | | | | |
| --- | --- | --- | --- | --- | --- | --- | --- | --- | --- | --- |
| | | | | | Each end | | Each side | | Bottom | |
| | | Brkt. | Clamps | Ears | Conduit | Cable 2¹⁵⁄₃₂ in. | Conduit | Cable 2¹⁵⁄₃₂ in. | Conduit | Cable 2¹⁵⁄₃₂ in. |
| 44 | 3½ | No | No | Yes | Two ½″ | ........ | Two ½″ | ..... | One ½″ | |
| | 3½ | No | No | Yes | One ¾″ | ........ | Two ¾″ | ..... | One ½″ | |
| | 3½ | B | 2-K | No | One ½″ | 2 | Two ½″ | ..... | One ½″ | |
| 45 | 3½ | B | 2-L | No | One ½″ | 2 | Two ½″ | ..... | One ½″ | |
| 46 | 3½ | No | 2-K | Yes | One ½″ | 2 | Two ½″ | ..... | One ½″ | |
| | 3½ | No | 2-L | Yes | One ½″ | 2 | Two ½″ | ..... | One ½″ | |

4-in. Long by 2-in. Wide Device Boxes

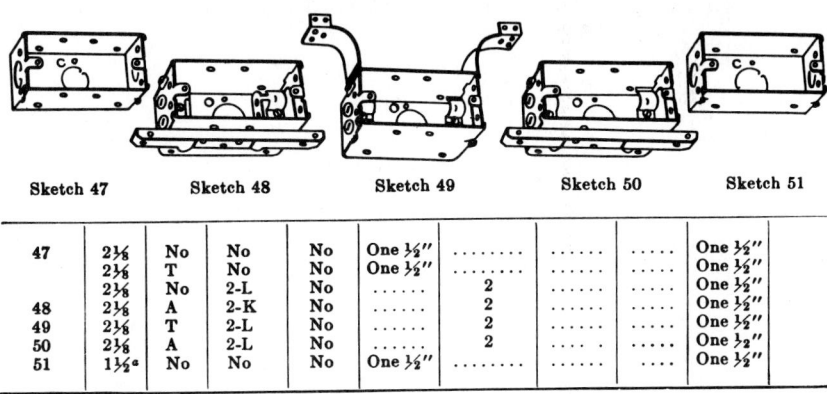

Sketch 47        Sketch 48        Sketch 49        Sketch 50        Sketch 51

| 47 | 2⅛ | No | No | No | One ½″ | ........ | ...... | ..... | One ½″ | |
| --- | --- | --- | --- | --- | --- | --- | --- | --- | --- | --- |
| | 2⅛ | T | No | No | One ½″ | ........ | ...... | ..... | One ½″ | |
| | 2⅛ | No | 2-L | No | ...... | 2 | ...... | ..... | One ½″ | |
| 48 | 2⅛ | A | 2-K | No | ...... | 2 | ...... | ..... | One ½″ | |
| 49 | 2⅛ | T | 2-L | No | ...... | 2 | ...... | ..... | One ½″ | |
| 50 | 2⅛ | A | 2-L | No | ...... | 2 | ...... | ..... | One ½″ | |
| 51 | 1½ᵃ | No | No | No | One ½″ | ........ | ...... | .... | One ½″ | |

ᵃ Straight through nail holes.

### 113.    Data on Wiring Boxes (Continued)

Utility Boxes
3¾ by 1½ by 1½ In.

| | Sketch No. | Bracket | Knockouts | | |
|---|---|---|---|---|---|
| | | | Each end | Each side | Bottom |
| 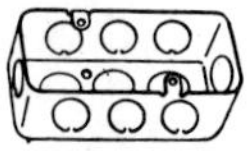 Sketch 52 | 52 | None | One ½" | Three ½" | Three ½" |

4 by 2⅛ by 1½ In.

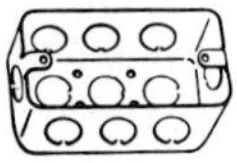

| | Sketch No. | Bracket | Each end | Each side | Bottom |
|---|---|---|---|---|---|
| Sketch 53 | 53 | None | One ½" | Three ½" | Three ½" |
| | | B | One ½" | Three ½" one side | Three ½" |
| 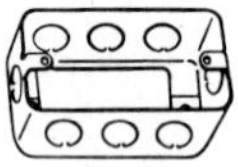 Sketch 54 | 54 | None | One ½" | Three ½" | |

4 by 2⅛ by 1⅞ In.

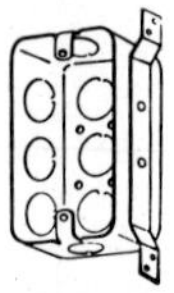

| | Sketch No. | Bracket | Each end | Each side | Bottom |
|---|---|---|---|---|---|
| Sketch 55 | | None | One ½" | Three ½" | Three ½" |
| | | A | One ½" | Three ½" one side | Three ½" |
| 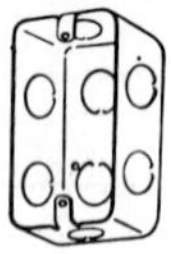 | 55 | F | One ½" | Three ½" one side | Three ½" |
| Sketch 56 | 56 | None | One ¾" | Two ¾" | Two ¾" |

4 by 2⅛ by 2⅛ In.

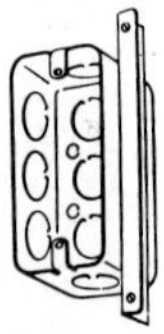

| | Sketch No. | Bracket | Each end | Each side | Bottom |
|---|---|---|---|---|---|
| | | None | One ½" | Three ½" | Three ½" |
| Sketch 57 | 57 | A | One ½" | Three ½" one side | Three ½" |
| | 58 | B | One ½" | Three ½" one side | Three ½" |
| | | F | One ½" | Three ½" one side | Three ½" |
| 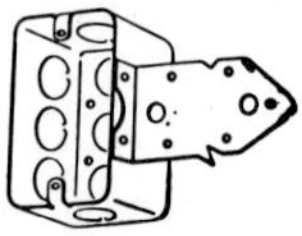 | | None | One ¾" | Two ¾" | Two ¾" |
| | | B | One ¾" | Two ¾" one side | Two ¾" |
| Sketch 58 | | O | One ¾" | Two ¾" one side | Two ¾" |

## 113. Data on Wiring Boxes

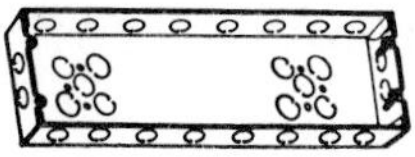

| Knockouts | | | | | | No. of gangs | Length outside, in. |
|---|---|---|---|---|---|---|---|
| Sides | | Ends | | Bottoms | | | |
| $\frac{1}{2}$ in. | $\frac{3}{4}$ in. | $\frac{1}{2}$ in. | $\frac{3}{4}$ in. | $\frac{1}{2}$ in. | $\frac{3}{4}$ in. | | |
| 8 | ... | 4 | ... | 3 | 2 | 2 | $6\frac{13}{16}$ |
| | 8 | ... | 4 | 3 | 2 | 2 | $6\frac{13}{16}$ |
| 4 | 4 | 2 | 2 | 3 | 2 | 2 | $6\frac{13}{16}$ |
| 10 | ... | 4 | ... | 6 | 4 | 3 | $8\frac{5}{8}$ |
| | 10 | ... | 4 | 6 | 4 | 3 | $8\frac{5}{8}$ |
| 6 | 4 | 2 | 2 | 6 | 4 | 3 | $8\frac{5}{8}$ |
| 12 | ... | 4 | ... | 6 | 4 | 4 | $10\frac{1}{2}$ |
| | 12 | ... | 4 | 6 | 4 | 4 | $10\frac{1}{2}$ |
| 6 | 6 | 2 | 2 | 6 | 4 | 4 | $10\frac{1}{2}$ |
| 14 | ... | 4 | ... | 6 | 4 | 5 | $12\frac{1}{4}$ |
| | 14 | ... | 4 | 6 | 4 | 5 | $12\frac{1}{4}$ |
| 8 | 6 | 2 | 2 | 6 | 4 | 5 | $12\frac{1}{4}$ |
| 16 | ... | 4 | ... | 6 | 4 | 6 | $14\frac{1}{16}$ |
| | 16 | ... | 4 | 6 | 4 | 6 | $14\frac{1}{16}$ |
| 8 | 8 | 2 | 2 | 6 | 4 | 6 | $14\frac{1}{16}$ |
| 18 | ... | 4 | ... | 6 | 4 | 7 | $15\frac{15}{16}$ |
| | 18 | ... | 4 | 6 | 4 | 7 | $15\frac{15}{16}$ |
| 10 | 8 | 2 | 2 | 6 | 4 | 7 | $15\frac{15}{16}$ |
| 20 | ... | 4 | ... | 6 | 4 | 8 | $17\frac{5}{8}$ |
| | 20 | ... | 4 | 6 | 4 | 8 | $17\frac{5}{8}$ |
| 10 | 10 | 2 | 2 | 6 | 4 | 8 | $17\frac{5}{8}$ |
| 22 | ... | 4 | ... | 6 | 4 | 9 | $19\frac{1}{2}$ |
| | 22 | ... | 4 | 6 | 4 | 9 | $19\frac{1}{2}$ |
| 12 | 10 | 2 | 2 | 6 | 4 | 9 | $19\frac{1}{2}$ |

3-gang box will take both 3 and 4 gang covers. 4-gang box will take both 4 and 5 gang covers. 5-gang box will take both 5 and 6 gang covers. 6-gang box will take both 6 and 7 gang covers. 7-gang box will take both 7 and 8 gang covers. 8-gang box will take both 8 and 9 gang covers. 9 gang box will take both 9 and 10 gang covers.

**Concrete Boxes**

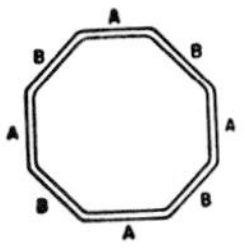


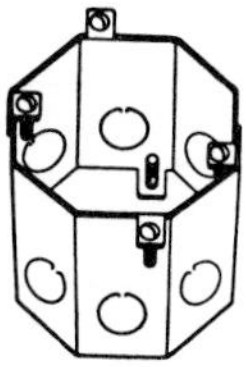

| Depth, in. | Knockouts | |
|---|---|---|
| | A, in. | B, in. |
| 2 | $\frac{1}{2}$ | $\frac{3}{4}$ |
| $2\frac{1}{2}$ | $\frac{1}{2}$ | $\frac{3}{4}$ |
| 3 | $\frac{1}{2}$ | $\frac{3}{4}$ |
| 3 | $\frac{3}{4}$ | 1 |
| $3\frac{1}{2}$ | $\frac{1}{2}$ | $\frac{3}{4}$ |
| $3\frac{1}{2}$ | $\frac{3}{4}$ | 1 |
| 4 | $\frac{1}{2}$ | $\frac{3}{4}$ |
| 4 | $\frac{3}{4}$ | 1 |
| 5 | $\frac{1}{2}$ | $\frac{3}{4}$ |
| 6 | $\frac{1}{2}$ | $\frac{3}{4}$ |

**Concrete Box Plates**

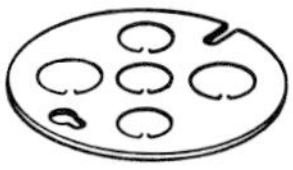

Three $\frac{1}{2}$″ K.O., two $\frac{3}{4}$″ K.O., no stud

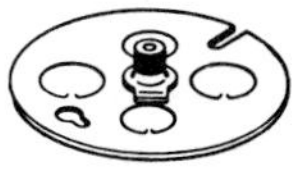

Two $\frac{1}{2}$″ K.O., two $\frac{3}{4}$″ K.O., $\frac{3}{8}$″ stud

## 114.  Box Covers

| Illustration | Description | Illustration | Description |
|---|---|---|---|
| **For 3¼-in. Round and Octagonal Boxes** | | | Flat with ½-in. knockout in center |
| | Raised, closed; ⅜ in. deep | | |
| | Flat, closed | | Flat with slots for surface devices |
| | Raised, with 1²¹⁄₃₂-in. keyed opening for Federal sign receptacle.  ⅜ in. deep | | Raised, closed; ⅝ in. deep |
| | Flat, with 1⁹⁄₁₆-in. opening and screw holes on 1¾-in. centers for Benjamin sign receptacle | | Raised with ½-in. knockout in center, ⅝ in. deep |
| | Flat, with ½-in. knockout in center | | Raised with ⅜-in. eyelet; ⅝ in. deep |
| | Raised, with ½-in. knockout in center; ⅜-in. deep | | Raised with 1½-in. grooved opening; ⅝ in. deep |
| | Raised, with 1½-in. opening for sign receptacles notched for protruding lug on porcelain; ⅜ in. deep | | Raised with plain 1½-in. opening; ⅝ in. deep |
| | Raised, with 1½-in. diameter opening and bent tongue to fit notches in new standard sign receptacles; ⅜ in. deep | **For 4-in. Round and Octagonal Boxes** | |
| | Raised, with ²⁷⁄₆₄-in. metal eyelet for drop cord; ⅜ in. deep | | Raised, closed, ⅝ in. deep |
| | Flat, with slots for surface devices; opening 1¹⁵⁄₁₆ in., screw centers 1⁵⁄₁₆ in. to 1¹³⁄₁₆ in. | | Flat, closed |
| | Cover with pigtail receptacle | | Flat, with ½-in. knockout in center |
| | Cover with terminal receptacle | | Raised, with ½-in. knockout in center; ⅝ in. deep |
| **For 3½-in. Round and Octagonal Boxes** | | | Raised, with 2¾-in. opening; ⅝ in. deep |
| | Flat, closed | | |

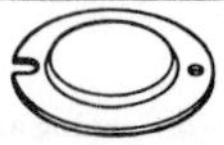

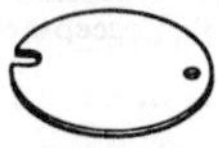

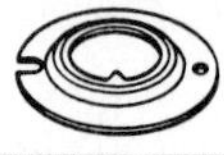

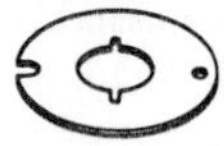

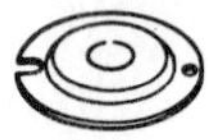

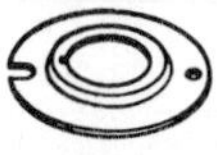

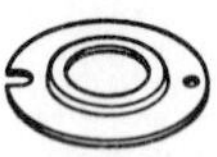

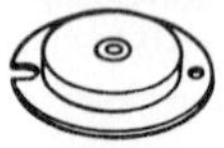

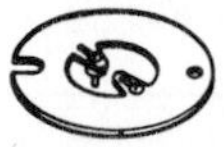

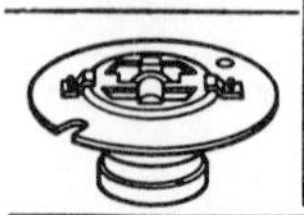

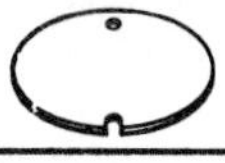

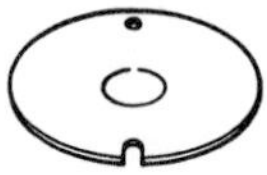

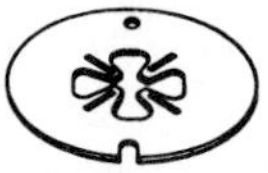

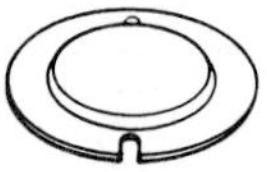

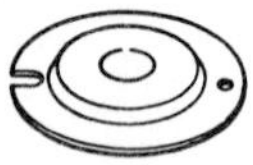

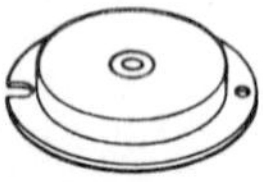

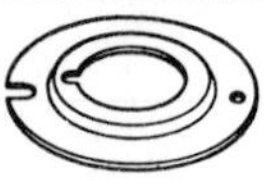

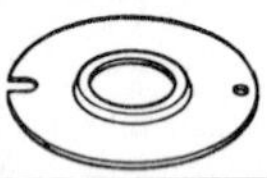

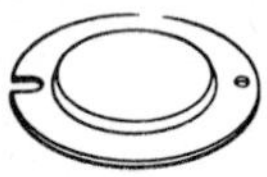

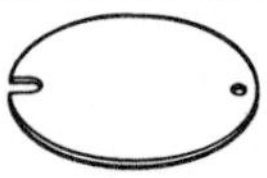

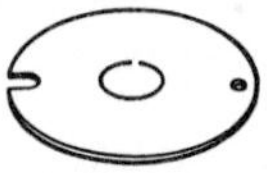

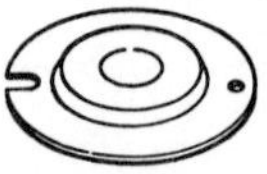

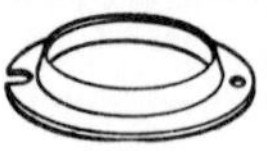

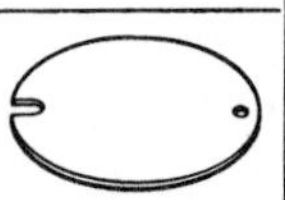

## 114. Box Covers

| Illustration | Description |
|---|---|
| 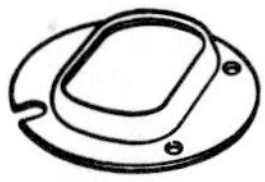 | Raised, ⅝ in. high, for one flush device; also suitable for bracket outlet |
| 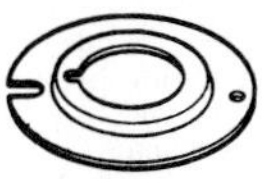 | Raised, ⅝ in. high with 1½ in. diameter opening for sign receptacles notched for protruding lug on porcelain |
| 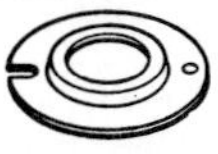 | Raised ⅝-in., with 1½-in. diameter opening and bent tongue to fit notches in new standard sign receptacles |
| 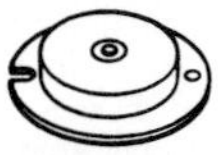 | Raised ⅝ in., with $^{27}/_{64}$-in. metal eyelet for drop cord |
| 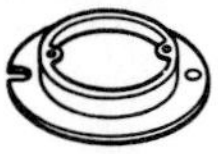 | Raised ⅝ in., 2¾ in. opening, $1^{1}/_{16}$ in. deep lugs tapped 8-32 on 2¾-in. centers |
| 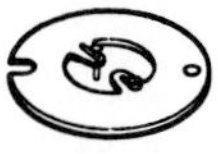 | Flat, with slots for surface devices; opening $1^{15}/_{16}$ in., screw centers $1^{5}/_{16}$ to $1^{13}/_{16}$ in.; two ⅞-in. × 6-32 screws |
| 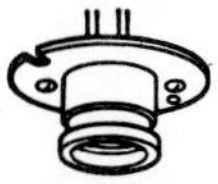 | Cover with pigtail receptacle |
| 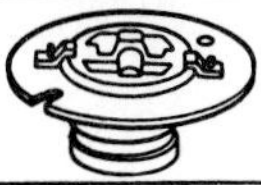 | Cover with terminal receptacle |
| | **For 4-in. Square Boxes** |
| 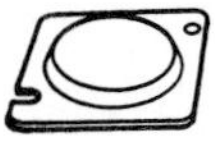 | Raised, closed, ⅝ in. deep |
| 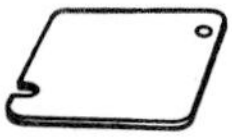 | Flat closed cover |
| 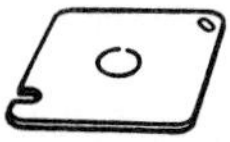 | Flat with ½-in. knockout in center |
| 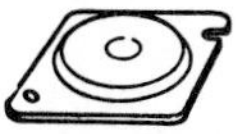 | Raised, ⅝ in. deep with ½-in. knockout in center |
| 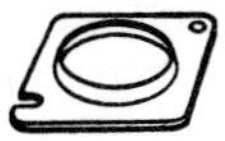 | Raised with 2¾-in. opening; ⅝ in. deep |
| 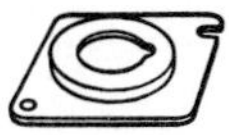 | Raised with 1½-in. opening for sign receptacles, notched for protruding lug on porcelain; ⅝ in. deep |
| 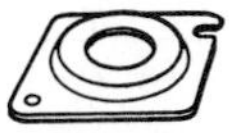 | Raised with 1½-in. opening and bent tongue to fit notches in new standard sign receptacle; ⅝ in. deep |
| 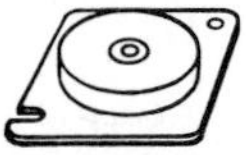 | Raised with $^{27}/_{64}$-in. metal eyelet for drop cord; ⅝ in. deep |
| 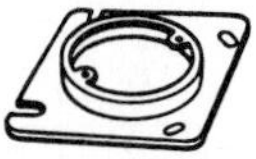 | Raised with 2¾-in. opening, ⅝ in. deep; lugs tapped 8-32 on 2¾-in. centers |
| 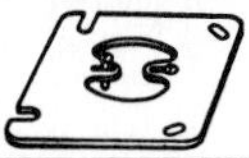 | Flat, with slots for surface devices, opening $1^{15}/_{16}$ in., screw centers $1^{5}/_{16}$ and $1^{13}/_{16}$ in. |
| | **For 4-in. Square Boxes to Accommodate Flush Wiring Devices with Flush Plate Covers** |
| 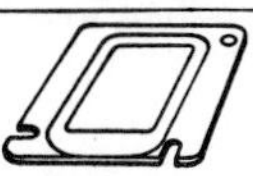 | Raised ¼ in. for one-gang plate |
| 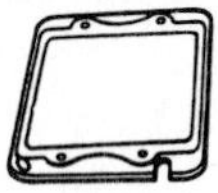 | Raised ¼ in. for two-gang plate |
| 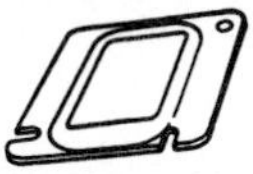 | Raised ½ in. for one-gang plate |
| 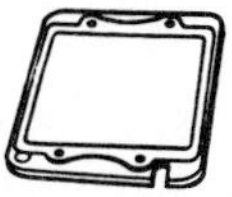 | Raised ½ in. for two-gang plate |
| 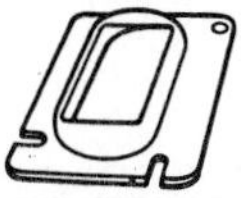 | Raised ¾ in. for one-gang plate |
| 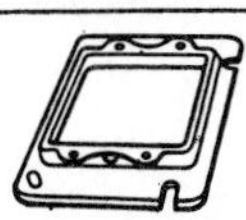 | Raised ¾ in. for two-gang plate |
| 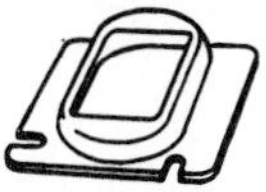 | Raised 1 in. for one-gang plate |

**114.  Box Covers** (*Continued*)

| Illustration | Description | Illustration | Description |
|---|---|---|---|
|  | Raised 1 in. for two-gang plate |  | Raised ½ in.; for one push-button switch and one single flush receptacle |
|  | Raised 1¼ in. for one-gang plate |  | Raised ½ in.; for one square-handle toggle switch and one single flush receptacle |
|  | Raised 1¼ in. for two-gang plate |  | Raised ½ in.; for one push-button switch and one duplex receptacle |
| **For 4-in. Square Boxes for Accommodating Flush Wiring Devices in Exposed Wiring** |  |  | Raised ½-in.; for one square-handle toggle switch and one duplex receptacle |
|  | Raised ½ in.; for one push-button switch | **For 4¹¹⁄₁₆-in. Square Boxes** |  |
|  | Raised ½ in.; for one square-handle toggle switch |  | Raised, closed; ⅝ in. deep |
|  | Raised ½ in.; for one single flush receptacle |  | Flat, closed |
|  | Raised ½ in.; for one duplex receptacle |  | Flat with ½-in. knockout in center |
|  | Raised ½ in.; for two push-button switches |  | Raised, with ½-in. knockout in center; ⅝ in. deep |
|  | Raised ½ in.; for two square-handle toggle switches |  | Raised, with 2¾-in. opening; ⅝ in. deep |
|  | Raised ½ in.; for two single flush receptacles |  | Raised, with 2⁷⁄₆₄-in. metal eyelet for drop cord; ⅝ in. deep |
|  | Raised ½ in.; for two duplex receptacles |  | Raised, with 2¾-in. opening, ⅝-in. deep; lugs tapped 8-32 on 2¾-in. centers |

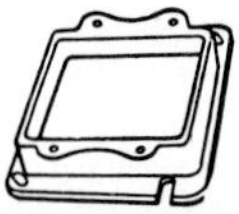

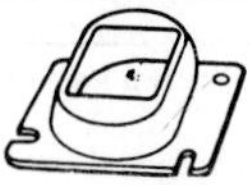

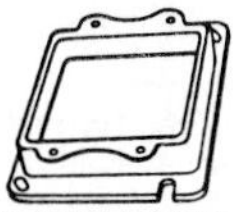

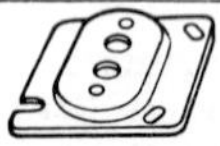

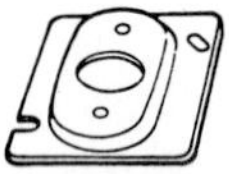

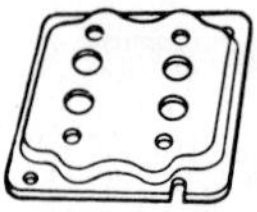

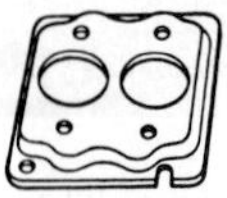

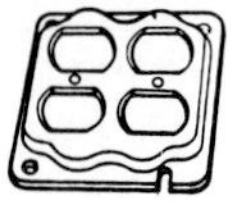

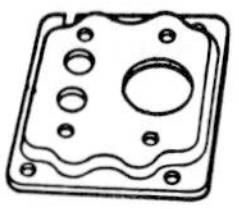

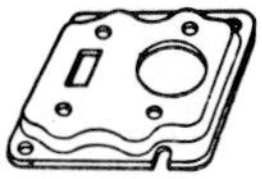

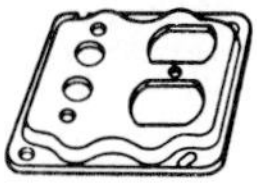

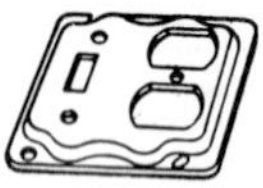

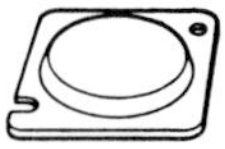

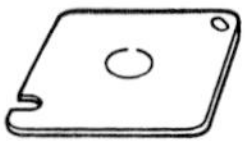

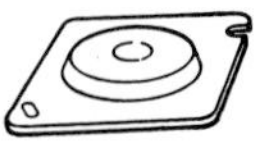

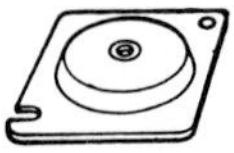

## 114. Box Covers

| Illustration | Description | Illustration | Description |
|---|---|---|---|
| For 4¹¹⁄₁₆-in. Square Boxes for Accommodating Flush Wiring Devices and Flush Plate Covers | | For Utility Boxes | |
| | Raised ¾ in. for one-gang plate | | Blank |
| | Raised 1 in. for one-gang plate | | For single push-button switch |
| | Raised 1¼ in. for one-gang plate | | For standard duplex receptacle |
| | Raised ¾ in. for two-gang plate | | For standard square-handle toggle switch |
| For Oblong Gang Boxes for Accommodating Flush Wiring Devices and Flush Plate Covers | | | For single T-slot and Edison-base receptacle |
| | Available for two-, three-, four-, five-, six-, seven-, eight-, and nine-gang flush plates | | Cover with knockouts for interchangeable devices |

**115. Nonmetallic boxes.** The material used in the construction of nonmetallic boxes is porcelain, polyvinyl chloride (PVC), fiberglass, or Bakelite. Except for surface-type wiring with nonmetallic-sheathed cable, porcelain boxes are not used because they are easily broken.

Polyvinyl chloride boxes are widely used with rigid nonmetallic conduits, and the boxes are shaped similarly to the more common metal types. Refer to Div. 9 for uses of these boxes with nonmetallic conduits.

The boxes shown in Fig. 4-156 have no cable clamps. This fact permits one more wire in the box in determining the maximum number according to Sec. 370-6(a) of the National Electrical Code. Without cable clamps nonmetallic-sheathed cables must be supported within 8 in of nonmetallic boxes. Ceiling boxes are required to have clamps. Another important advantage is that no grounding connection is required to the box as with metallic boxes. When nonmetallic cables, which contain grounding wires, enter nonmetallic boxes, they are spliced together with an approved solderless pigtail connector. If a grounding-type receptacle is attached to a nonmetallic box, the pigtail connection should include a short lead for connection to the grounding terminal on the receptacle.

**Fig. 4-156** Plastic boxes for use with nonmetallic-sheathed cable. [Slater Electric Inc.]

If a metallic switch plate is to be attached to a flush switch, switches are available with a grounding terminal. Then the grounding conductor of the supply cable is connected to this terminal, and the metal frame of the switch is grounded. This, in turn, will ground the attached metal switch plate. Such switches should be used if the switch attached to a nonmetallic box is located within 5 ft (1.5 m) horizontally or 8 ft (2.4 m) vertically of a grounded object (water pipes, concrete floors, etc.).

**116. Flush plates** are covers which are used with flush wiring devices in concealed-wiring installations to produce a neat covering of the outlet. They are supported by screws turning into tapped holes in mounting ears or straps on sectional boxes or on covers for square or oblong boxes as described in Sec. 114. Holes in the plates accommodate the handles of flush switches or plugs for flush receptacles. The plates are made in single-, two-, and three-gang sizes with various types and combinations of openings to meet almost any combination of flush wiring devices. Flush plates made of Bakelite, brass, enameled metal, or stainless steel can be obtained. The standard finish for brass plates is brush brass. Plates can be obtained with almost any special finish to meet the requirements of any interior. Some of these special finishes are listed below:

FLUSH PLATE FINISHES

| | | |
|---|---|---|
| Aluminum (spray) | Bronze, polished | Gilt, rich |
| Black lacquer | Statuary (light) | Gunmetal |
| Bower-Barff lacquer | Brown lacquer (imitation Bakelite) | Ivory enamel |
| | | Nickel, dull |
| Brass, sand blast, antique | Cadmium, brushed | Polished |
| Sand blast, brush | Polished | Silver, butler's (brushed) |
| Flemish | Chromuim, dull | Oxidized |
| Lemon | Polished | Polished |
| Oxidized | Copper, antique | Satin |
| Polished | Brush | Verd-antique lacquer |
| Bronze, antique | Mottled | White enamel |
| Brush | Oxidized | Various colors |
| Japanese (dark) | Polished | Wood grain |

Flush plates are made for the standard and interchangeable lines of wiring devices. Typical plates are shown in Figs. 4-157 and 4-158. Some manufacturers offer engraved plates to identify standard or special equipment.

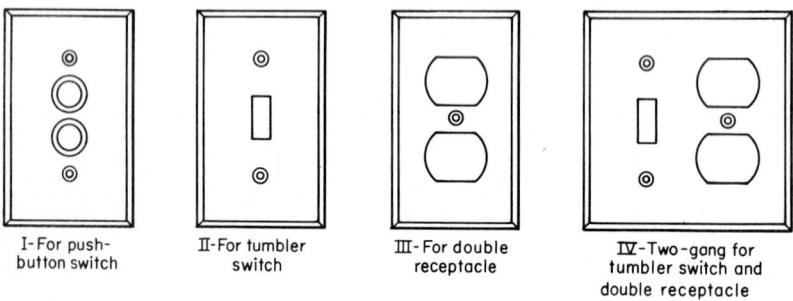

I-For push-button switch     II-For tumbler switch     III-For double receptacle     IV-Two-gang for tumbler switch and double receptacle

**Fig. 4-157**   Flush plates for standard-line wiring devices.

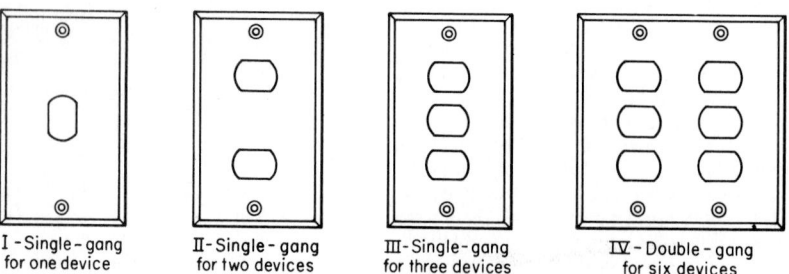

I - Single - gang for one device     II- Single - gang for two devices     III- Single-gang for three devices     IV - Double - gang for six devices

**Fig. 4-158**   Flush plates for interchangeable-line wiring devices.

**117. Types of Mounting Features for Sectional Boxes**

| Name | Illustration | Description |
|---|---|---|
| Standard mounting ears | | Box equipped with two ears, one secured to each end of box with two 8-32 large fillister-head screws. Ears adjustable and reversible. Set of extra holes (not tapped) in each side for changing position of ears to sides if desired (8-32 tap required) |
| Rectangular mounting bracket without lath support | | For plaster or wallboard jobs, with holes for gripping plaster and with aligning-cleat prongs and four nail holes. Bracket is perfectly flat so it does not interfere with wallboard. Bracket is welded to box |
| Extended mounting ears | Ears on ends for vertical mounting in plaster with lath support. | For mounting in plaster or baseboard. Each ear fastened to box with two 8-32 fillister-head screws. Can be located on sides or ends. Tapped mounting holes supplied regularly only on ends. Aligning cleat prongs and adequate nail holes for rigid fastening |
| Box supports and lath holders | | For mounting in plaster walls. Supports available in lengths of $16\frac{1}{2}$, $18\frac{1}{2}$, $20\frac{1}{2}$, $22\frac{1}{2}$, $24\frac{1}{2}$, and $26\frac{1}{2}$ in. |
| Rectangular mounting bracket with lath support | | Same as item 2 with lath supports welded to box for wood-lath or wallboard jobs. Back lip of support projects beyond front lip providing adequate support behind wall board and still retaining channel for supporting end of wood lath front and back |
| Extended mounting ears | Ears on sides for horizontal mounting in baseboards.  Ears on sides for horizontal mounting in plaster.  Ears on ends for vertical mounting in plaster without lath support. | For mounting in plaster or baseboard. Each ear fastened to box with two 8-32 fillister-head screws. Can be located on sides or ends. Tapped mounting holes supplied regularly only on ends. Aligning cleat prongs and adequate nail holes for rigid fastening |

**118. Depth of outlet boxes.**  The National Electrical Code specifies that outlet boxes intended to enclose flush devices must have an internal depth of at least $^{15}/_{16}$ in (23.8 mm) for concealed work except that if the installation of such a box is impracticable, a box of not less that $^1/_2$-in (12.7-mm) internal depth may be used. Outlet boxes with conduits entering the side should be $2^1/_8$ in (54 mm) deep for installation in lath and plaster. Boxes $1^1/_2$ in (38.1 mm) deep will generally be satisfactory for brickwork. Shallow boxes of $^1/_2$- or $^3/_4$-in (12.7- or 19.1-mm) depth are sometimes used on terra-cotta ceilings with the box set on 'he surface of the tile and embedded in the plaster. The conduits are brought into the ɔack of the box.

**119. Metal straps for supporting outlet boxes** can be obtained in various types with or without attached fixture studs. Several representative types are described below.

A

For shallow boxes in new work or for holding boxes to concrete forms. With boxes of a depth of $^1/_2$ in (12.7 mm), in which the bar is nailed to joists or studding, the edge of the box will be flush with ordinary plaster. It will fit any box having a $^1/_2$-in knockout. It is made in 18- and 24-in (457- and 609-mm) lengths.

B

For boxes $1^1/_2$ in (38.1 mm) deep without switch covers or plaster rings; offset brings the box edge flush with the plaster. It will fit any box having a $^1/_2$-in (12.7-mm) knockout. The length of the bar is $19^1/_2$ in (749 mm). Offset is $1^1/_{16}$ in (27 mm) deep.

C

For boxes $1^1/_2$ in (38.1 mm) deep with switch covers or plaster rings; offset brings covers $^5/_8$ in (15.9 mm) high, flush with plaster. It will fit any box having a $^1/_2$-in (12.7-mm) knockout. The length of the bar is $19^1/_2$ in (749 mm). Offset is $1^{11}/_{16}$ in (43 mm) deep.

D

For mounting shallow boxes or plates in buildings already plastered. Make a small hole about $1^1/_2$ in (38.1 mm) in diameter; push bar all the way into the hole, long end first as shown in cut; hold stud in one hand and pull wire with the other until the bar is centered across the hole. It will fit any box having a $^1/_2$-in (12.7-mm) knockout. The length of the bar is 12 in (305 mm).

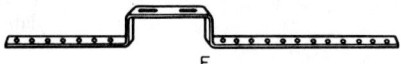

E

For boxes $1^1/_2$ in (38.1 mm) deep with covers and integral studs or without studs. Offset has slots for stove bolts to hold the box and is the right depth to bring $^5/_8$-in (15.9-mm) covers flush with the plaster. The length of the bar is 21 in (533 mm). Offset is $1^{11}/_{16}$ in (43 mm) deep.

**120. Wallboard hanger** (General Electric Co.). Instructions for supporting a switch box with a special wallboard hanger are given below and illustrated in Fig. 4-159.

For installing switch boxes in old work where plaster, wallboard, plasterboard, or similar construction is present:

    1. Cut hole the exact size of the switch box.
    2. Assemble the switch box and hanger. Only partially tighten the bolt.

3. Push assembly into the wall hole until the sides of the hanger spring free on the inside wall.

4. Tighten by screwing the bolt on the inside of the box.

Inside and outside pressure against the wall give the box a rigid installation.

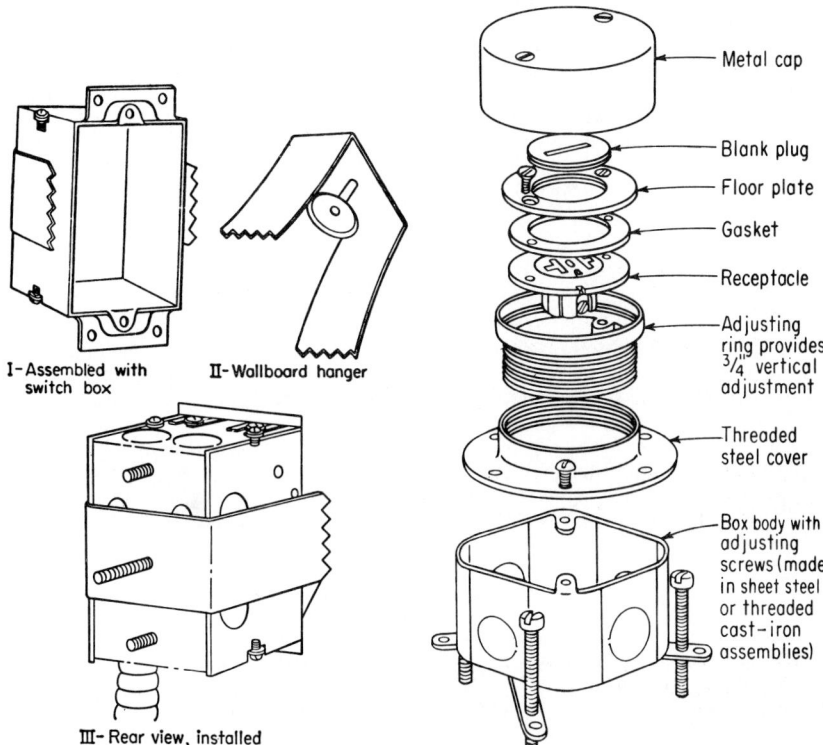

I—Assembled with switch box

II—Wallboard hanger

III— Rear view, installed

Metal cap

Blank plug

Floor plate

Gasket

Receptacle

Adjusting ring provides $3/4''$ vertical adjustment

Threaded steel cover

Box body with adjusting screws (made in sheet steel or threaded cast-iron assemblies)

**Fig. 4-159**  Wallboard switch hanger. [General Electric Co.]

**Fig. 4-160**  Adjustable floor box with flush grounding-type receptacle. [Steel City Division of Midland-Ross Corp.]

**121. Floor boxes** (Fig. 4-160) are made in adjustable and nonadjustable types to provide outlets from concealed conduits embedded in floors. The adjustable ones allow the top to be adjusted in height and angle to suit the floor conditions. They may be obtained in single- and multiple-outlet units.

**122. Door boxes** (Fig. 4-161) are for the installation of door switches in the jamb of the door. They are made in different types to accommodate the different types of door switches.

**123. Pull boxes** are described in Div. 9.

**124. Lampholders** consist of an assembly for supporting lamps, comprising the lamp socket, protective and insulating covering, and means for attachment to some form of support. The term *socket* is sometimes incorrectly used in place of lampholder. Each lampholder must have a lamp socket, but the socket is only the part of the lampholder which engages with the base or terminal of the lamp. Lampholders are made in a great variety of types to fulfill the requirements of all classes of installations and types of lamp bases.

**125. Brass-shell–type interchangeable lampholders** are

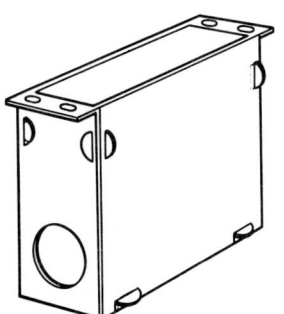

**Fig. 4-161**  Type of door box.

the old standard type for incandescent lamps. They consist of two fundamental parts, a body (Fig. 4-162, III) and a cap (Fig. 4-162, I) or base (Fig. 4-162, II). The two parts are held together by means of a fluted catch which allows them to be easily separated or assembled. Any cap or base can be used with any body. The body consists of a socket mounted on a porcelain or composition support and encased in an outside brass shell.

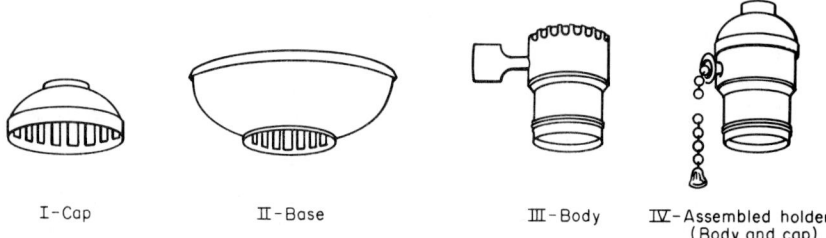

I–Cap    II–Base    III–Body    IV–Assembled holder (Body and cap)

**Fig. 4-162**   Fluted-catch interchangeable lampholders. [General Electric Co.]

The socket is insulated from the shell by means of a fiber casing. When the porcelain or composition portion of the body simply forms a support and assembly for the socket and electric-circuit connecting screws, it is known as a keyless body. If the body contains a small switch for turning the lamp on or off, it may be of the key, pull-chain, or push type (Fig. 4-163). Caps are made with standard inside pipe thread for connection to 1/8-, 1/4-, 3/8-, or 1/2-in pipe or with 3/8-in male thread. Caps are also constructed with porcelain- or composition-bushed holes or with cord-lamp grip for pendant support on a drop cord. The cord holes are 0.406 in (10.31 mm) in diameter, and the cord grips will accommodate cords from 0.375 to 0.5000 in (from 9.525 to 12.7 mm) in diameter. Bases are available for mounting directly on outlet boxes and for exposed wiring of either the cleat-mounting or the concealed-base type.

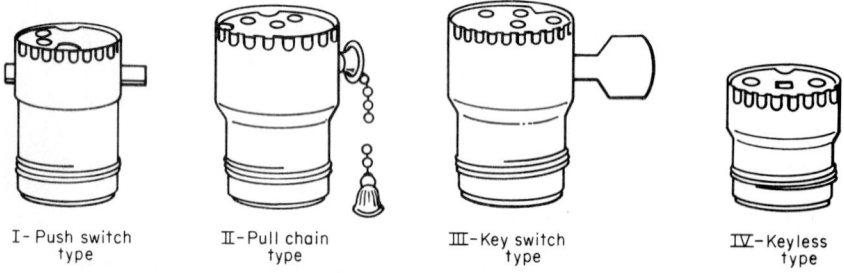

I– Push switch type    II–Pull chain type    III–Key switch type    IV–Keyless type

**Fig. 4-163**   Fluted-catch lampholder bodies.

**126. Threaded-catch interchangeable lampholders** for incandescent lamps (Fig. 4-164) are a newer development. They consist of a body fastened by means of a threaded ring to a cap or base. The bodies, caps, and bases may have outer brass shells or may be made of composition. The same types of bodies, caps, and bases are made as for the fluted-catch ones of Sec. 125. The line is interchangeable in that any cap or base will fit any body.

**127. The porcelain snap-catch interchangeable line of lampholders** for incandescent lamps (Fig. 4-165) is used in installations subjected to moisture, acid fumes, or other corroding influences. The bodies are fitted with two bayonet hooks, which engage two flexible phosphor-bronze contacts in the cap or base. The two parts are thus securely locked together when assembled but permit easy disassembly without affecting the wiring. The bayonet hooks and bronze contact catches provide the electrical connection between cap and body as well as supplying the means of mechanically supporting the body from the cap. The same general types of caps, bases, and bodies are included in this line as for the fluted-catch line. A porcelain line for the same type of applications is made with the cap and body fastened together by means of two screws.

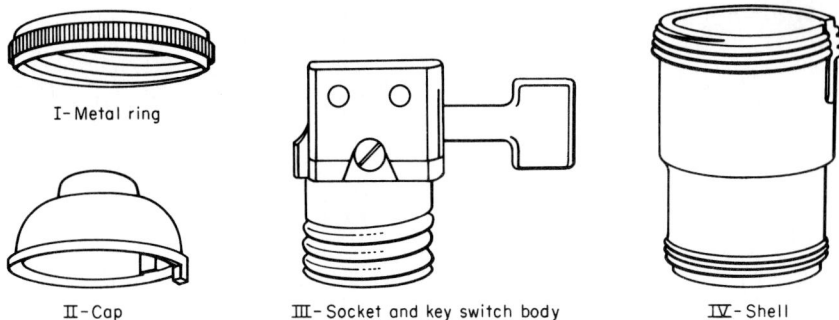

I–Metal ring

II–Cap                    III–Socket and key switch body            IV–Shell

**Fig. 4-164** Threaded-catch interchangeable lampholders. The threaded metal ring (I), knurled to provide a grip, holds the cap and shell together. The cap (II) cannot pull loose from the shell, and no amount of vibration will loosen the threaded ring when it is properly assembled. A lug in the cap fits in the shell slot, preventing rotation between cap and shell. After the cap has been fitted to the shell, the flange on the threaded ring fits over the flange on the cap, and the ring is securely threaded to the shell.

**128. Noninterchangeable lampholders** are made in a great variety of types to meet the requirements for lampholders for show-window, showcase, cove, trough, or reflector lighting; candelabra or intermediate-base lamps; Lumiline or fluorescent lamps; mounting on outlet boxes; mounting in fixture canopies; lamp outlets in exposed wiring on insulators; and sign lighting, etc. Most of these holders are made of porcelain, but plastic materials are used to a limited extent. A few of the more common types are shown in Fig. 4-166.

**129. Holders are available for surface mounting** for use primarily in installations wired with nonmetallic-sheathed cable (Figs. 4-167 and 4-168). For dry locations these holders have cases of all-plastic construction. Holders for excessively moist or dusty locations have porcelain bases and plastic tops. Both types are made in keyless and pull-chain constructions.

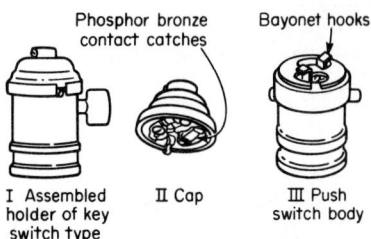

Phosphor bronze contact catches          Bayonet hooks

I Assembled holder of key switch type          II Cap          III Push switch body

**Fig. 4-165** Porcelain snap-catch interchangeable lampholders.

**130. Weatherproof lampholders** (Fig. 4-169) with one-piece porcelain, composition, or rubber cases are made with sealed-in leads for wet locations.

**131. Shade holders.** Many lampholders have the body supplied with threads or grooves for engaging shade holders. Shade holders are made with openings of 2¼, 3¼, and 4 in (57.15, 82.55, and 101.6 mm) diameter for accommodating shades. Shades may be held to the shade holder with a wire spring or by means of screws (Fig. 4-170).

**132. Insulating socket bushings** must be used where a cord enters a socket to protect it against abrasion and grounding against the shell. The most popular bushings are of hard

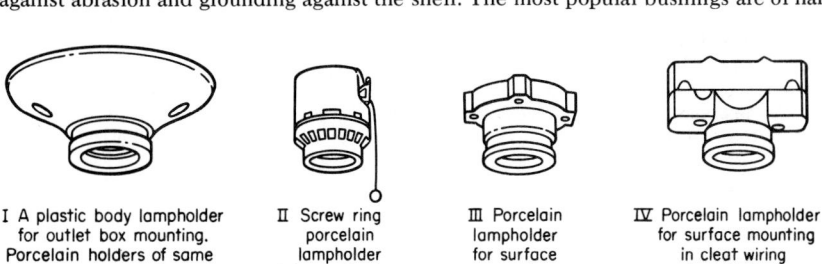

I A plastic body lampholder for outlet box mounting. Porcelain holders of same type more commonly used.

II Screw ring porcelain lampholder for mounting in fixture canopies

III Porcelain lampholder for surface mounting

IV Porcelain lampholder for surface mounting in cleat wiring installation

**Fig. 4-166** Types of interchangeable lampholders.

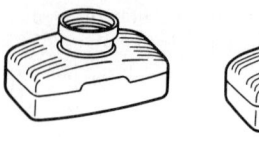

**Fig. 4-167**   All-plastic lampholders for use with nonmetallic-sheathed cable.

**Fig. 4-168**   Porcelain-base lampholders for use with nonmetallic-sheathed cable.

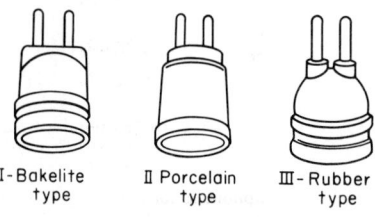

I-Bakelite type    II Porcelain type    III-Rubber type

**Fig. 4-169**   Weatherproof lampholders. [General Electric Co.]

rubber or of a compound resembling it. Patented bushings which automatically grip the cord by a wedging action can be purchased.

Most lampholders of the pendant type are fitted with an approved bushing constructed as an integral part of the lampholder cap.

**133. Rosettes** are devices for supporting and connecting to the circuit the cord and sockets of flexible drop cords. They are made in different types for open wiring and for attaching to outlet boxes or moldings. Rosettes similar to the one shown in Fig. 4-171 are used for open-wiring work, and ones similar to that of Fig. 4-172 for attachment to outlet boxes. The drop cord passes through the hole in the center and is attached to connections inside the body of the rosette. The connections should be relieved of any strain by making a knot in the wires just inside the rosette body. Fused rosettes were at one time employed, but they are no longer allowed by the National Electrical Code rules.

An all-plastic rosette for use with nonmetallic-sheathed cable is shown in Fig. 4-173.

**134. Lampholders for fluorescent lamps** are available in various types to meet all the usual conditions of installation. Combination lampholders and starter sockets also are available. Typical constructions are shown in Fig. 4-174.

## POWER CAPACITORS

**135. Power capacitors** are capacitors with relatively large values of capacity which are used on power-distribution systems or in industrial plants for improving the power factor.

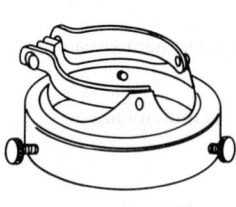

I-Clamp type for lampholders with grooves

II- Threaded type for lampholders with threads. Supporting screws for shade

III- Threaded type with wire spring support for shade

**Fig. 4-170**   Shade holders.

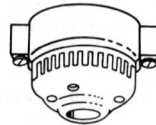

**Fig. 4-171**   Cleat-base rosette for open wiring.

**Fig. 4-172**   Rosette for attachment to outlet box.

Since many power companies include low–power-factor penalties, kilovolt-ampere demand rates, or power-factor bonuses in their rate schedules, it is often economical for industrial consumers to install capacitors for power-factor improvement. These capacitors are connected across the line and neutralize the effect of lagging power-factor loads, thus reducing the current for a given kilowatt load. The amount of reactive kilovolt-amperes of capacitors required to raise the power factor to any given value can easily be determined from the chart of Fig. 4-175.

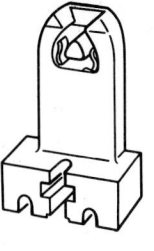
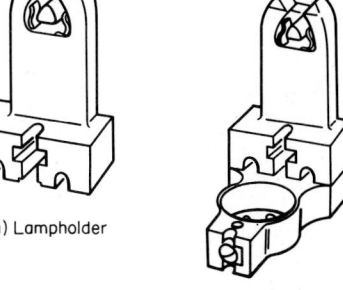

(a) Lampholder

(b) Combination lamp-
holder and starter
socket.

**Fig. 4-173** All-plastic rosette for use with non-metallic-sheathed cable.

**Fig. 4-174** Lampholders for fluorescent lamps. [Harvey Hubbell, Inc.]

The best point to connect capacitors to the circuit depends upon cost considerations. Relatively small capacitor units, A in Fig. 4-176, can be connected at the individual loads, or the total capacitor kilovolt-amperes can be grouped at one point and connected to the main bus. Both of these methods are shown schematically in Fig. 4-176. Greater power-factor corrective effect for a given total capacitor kilovolt-amperes will result with the capacitors located directly at each individual load, since the current is thereby reduced all the way from the load to the source. The first cost of an installation of individual capacitors will be greater, however, than that for one unit of the same total kilovolt-amperes located at a central point. The greater saving in operating expense due to individual capacitors must be weighed against their increased first cost.

**136. A complete capacitor bank** is made up of the necessary standard units to give the desired kilovolt-amperes, connected in parallel with each other. Each unit consists of one or more cells enclosed in a hermetically sealed steel box (Fig. 4-177). The cells are aluminum-foil, paper- or film-insulated capacitors impregnated with insulating fluid. Standard units can be obtained with the cells internally connected for single-, two-, or three-phase operation. The units may be of enclosed- or rack-type construction. The enclosed construction is made primarily for individual motor applications (Fig. 4-181). It consists of a dusttight steel conduit box (Fig. 4-178) mounted on top of the hermetically sealed capacitor. The conduit box contains the terminals, discharge resistor, and fuses if desired (see Fig. 4-182). Rack-type banks consist of the required number of hermetically sealed capacitors with exposed terminals, supported on a steel rack. The complete structure may be provided with enclosing screens, dusttight steel enclosing cases, or a steel enclosing cabinet for outdoor installation. Each capacitor unit of the bank is individually fused, and the entire bank is provided with discharge resistors or coils inside the enclosure (see Fig. 4-183). Typical assemblies for wall and ceiling mounting are shown in Figs. 4-179 and 4-180.

**137. Drainage of stored charge.** When capacitors are disconnected from the supply, they are generally in a charged state. Considerable energy is stored in a capacitor under this condition, and there is a voltage present between its terminals. If the capacitor were left in this charged state, a person servicing the equipment might receive a dangerous shock or the equipment might be damaged by an accidental short circuit. Therefore, all capacitors must be provided with a means of draining the stored charge (see Figs. 4-182 and 4-183). The National Electrical Code requires that the drainage equipment shall be so designed that it will discharge the capacitor to 50 V or less within 1 min after the capacitor has been disconnected from the source of supply in capacitors rated 600 V or less and in 5 min in capacitors rated more than 600 V. The discharge equipment may consist of resistors or inductive coils which are permanently connected to the terminals of the capacitor bank. If the discharge circuit is not permanently connected to the terminals of the capacitor bank, automatic means must be provided for connecting the capacitor to the discharge circuit on the removal of the supply voltage. When the capacitors are connected directly to other equipment without a switch or overcurrent device being interposed, a means shall be provided to reduce residual voltage to 50 V or less within 5 min of disconnection (Fig. 4-181).

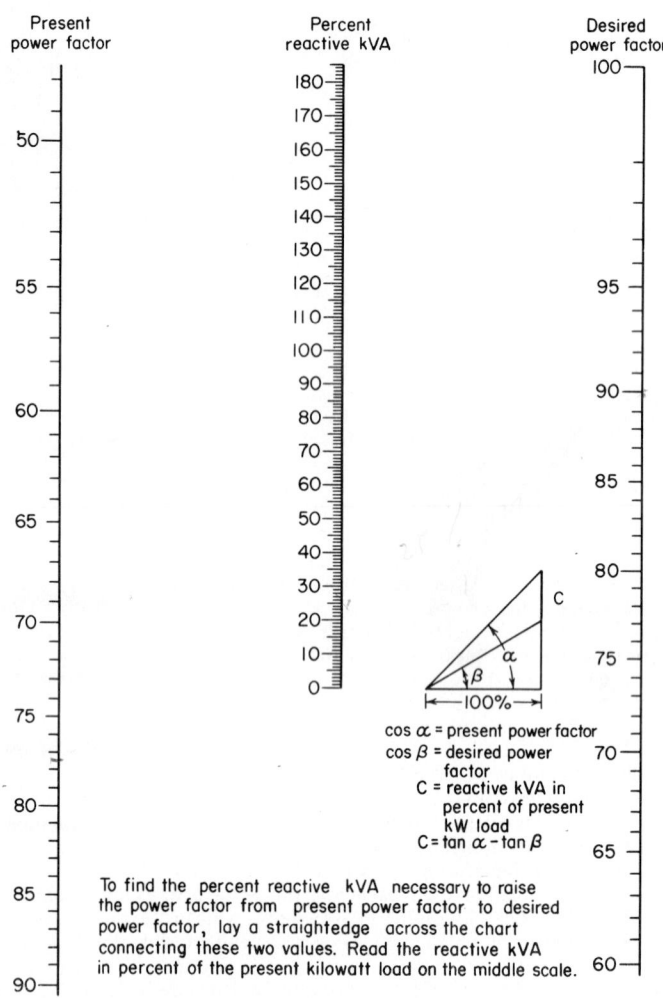

To find the percent reactive kVA necessary to raise the power factor from present power factor to desired power factor, lay a straightedge across the chart connecting these two values. Read the reactive kVA in percent of the present kilowatt load on the middle scale.

**Fig. 4-175**  Chart for use in determining the percentage of reactive kilovolt-amperes required to raise the power factor to a desired value.

### 138. The National Electrical Code rules for size of conductors, overcurrent protection, and disconnecting means are given below:

1. CAPACITOR RATING   The maximum capacitor ratings that are permitted for use with NEMA-type three-phase 60-Hz Classification B motors are presented in Table 11 of Div. 11.

2. CAPACITOR CIRCUITS    Capacitor circuits shall conform to the following:

*a. Conductor ratings.* The current-carrying capacity of capacitor circuit conductors shall be not less than 135 percent of the rated current of the capacitor. The current-carrying capacity of conductors which connect a capacitor to the terminals of a motor or to motor-circuit conductors shall be not less than one-

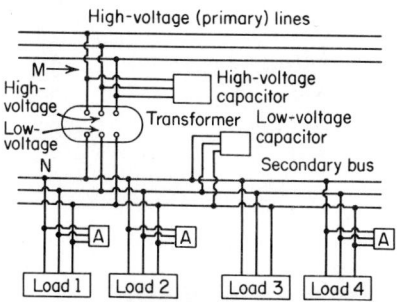

**Fig. 4-176**    Location of capacitors in electric system. [General Electric Co.]

**Fig.    4-177**    Three-phase    capacitor    unit enclosed in a hermetically sealed steel box. [Westinghouse Electric Corp.]

third of the carrying capacity of the motor-circuit conductors but not less than 135 percent of the rated current of the capacitor.

*b. Overcurrent protection*

(1) An overcurrent device shall be provided in each ungrounded conductor.

*Exception: A separate overcurrent device is not required on the load side of a motor-running overload protective device.*

(2) The rating or setting of the overcurrent device shall be as low as practicable.

*c. Disconnecting means*

(1) A disconnecting means shall be provided in each ungrounded conductor.

*Exception: A separate disconnecting means is not required for a capacitor connected on the load side of a motor-overload protective device.*

(2) The disconnecting device shall open all ungrounded conductors simultaneously.

(3) The disconnecting device may be used for disconnecting the capacitor from the line as a regular operating procedure.

(4) The continuous current-carrying capacity of the disconnecting device shall be not less than 135 percent of the rated current of the capacitor.

3. RATING OR SETTING OF THE MOTOR OVER-CURRENT DEVICE    If a motor installation includes a capacitor connected on the load side of the motor-overload device, the rating or setting of the motor-overload device shall be determined in accordance with Sec. 430 of Div. 7 except that, instead of using the full-load rated current of the motor as provided in that section, a lower value corresponding to the improved power factor of the motor circuit shall be used. Section 437 of Div. 7 applies with respect to the rating of the motor-circuit conductors.

4. GROUNDING    Capacitor cases shall be grounded in accordance with Div. 9.

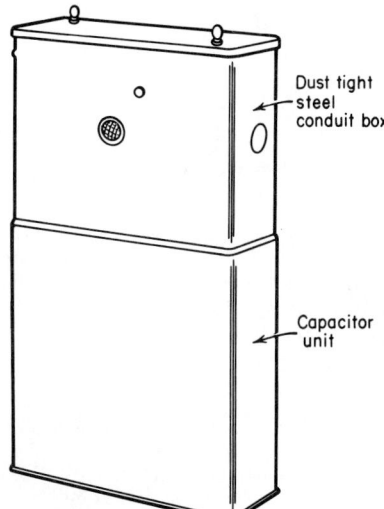

**Fig. 4-178**    Enclosed capacitor unit.

**Fig. 4-179**   Typical wall-mounting bracket capacitors. [Westinghouse Electric Corp.]

**Fig. 4-180**   Typical small rack capacitor bank for wall or ceiling mounting. [Western Electric Co.]

5. Enclosing and Guarding   Capacitors containing more than 3 gal (11.36 L) of flammable liquid shall be enclosed in vaults or outdoor fenced enclosures. Capacitors shall be enclosed, located, or guarded so that persons cannot come into accidental contact or bring conducting materials into accidental contact with exposed energized parts, terminals, or buses associated with them except that no additional guarding is required for enclosures accessible only to authorized and qualified persons. For isolation by elevation, see Sec. 37 of Div. 11.

6. Marking   Each capacitor shall be provided with a nameplate giving the maker's name, rated voltage, frequency, kilovar or amperes, number of phases, and, if filled with a combustible liquid, the amount of liquid in gallons. When filled with a nonflammable liquid, the nameplate shall so state. The nameplate shall indicate whether or not a capacitor unit has a discharge device inside the case.

**139. Capacitors which are constructed with a liquid which will not burn** can be placed at any convenient location provided that they are not exposed to mechanical injury. If exposed to combustible dust or flyings or if located in the vicinity of easily ignitable material, they must be enclosed in dusttight metal enclosures. Capacitors constructed with a liquid which will burn must be intalled in vaults unless each container does not contain more than 3 gal (11.36 L) of liquid. Vaults, when required, should be constructed to meet the same requirements as for transformers (see Div. 5).

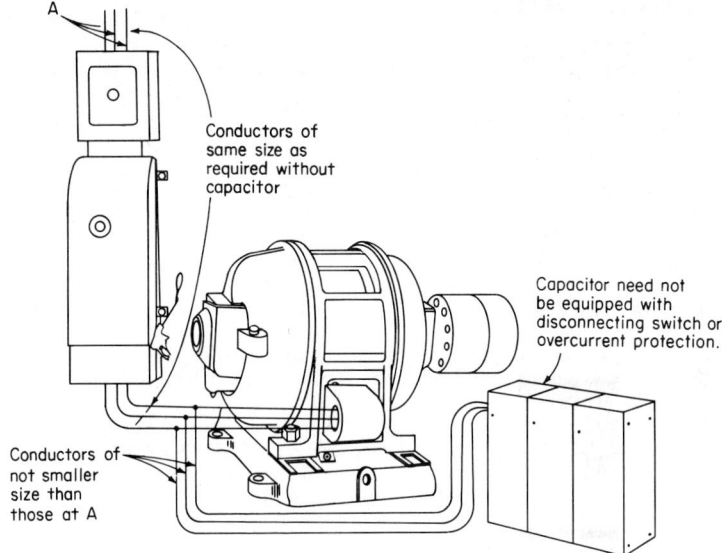

**Fig. 4-181**   Schematic diagram of connections when an enclosed capacitor is installed at motor terminals. [General Electric Co.]

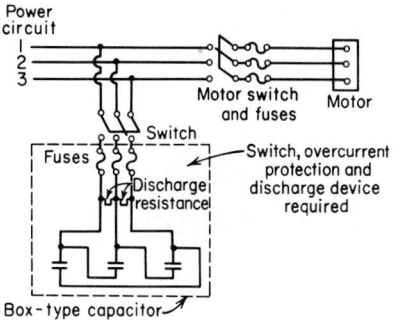

**Fig. 4-182** Connection diagram of enclosed capacitor unit, with fuses, installed on the line side of a motor switch.

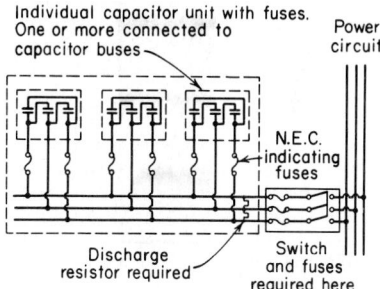

**Fig. 4-183** Connection diagram of rack-type three-phase 230-, 460-, or 575-V capacitor installed on power circuit.

## BATTERIES: GENERAL

**140. An electric battery** is a device for producing an emf by chemical means. When such a source of emf is connected to a closed electric circuit, chemical energy is transformed into electrical energy. An emf will be produced by chemical means whenever two dissimilar solid conductors are immersed in a conducting liquid. The solid conductors are called electrodes, and the conducting liquid is called the electrolyte. Such a combination of chemicals resulting in the production of an emf is called a voltaic cell. A battery may consist of a single cell or a combination of cells. The voltage of a cell depends upon the material of the electrodes and the electrolyte and is independent of the dimensions of the cell. The current and power capacity of a cell are, however, directly dependent upon the dimensions of the cell and the weight of active material in the electrodes. Although there are an infinite number of different combinations of electrodes and electrolytes which will produce a voltaic cell, there are only a limited number of combinations which are practicable.

**141. Classifications of batteries.**   For practical purposes, batteries may be classified as primary and secondary. A *primary battery* is used only for discharge (conversion of chemical energy into electrical energy). As such a battery is discharged, the material of one of the electrodes goes into solution in the electrolyte. The electrode is thus consumed, and the character of the electrolyte is altered so that with primary batteries it is necessary to renew from time to time both the electrode which goes into solution and the electrolyte. A *secondary battery* is alternately discharged and charged. As a battery discharges, the electrodes and electrolyte undergo chemical changes. After a secondary battery has been discharged, the electrodes and electrolyte can be restored to their original charged condition by passing a current through the battery in the reverse direction from that of discharge. In charging a battery, electrical energy is transformed into chemical energy. *Secondary batteries* are generally called *storage batteries.*

**142. The internal resistance of batteries** is the resistance offered to the flow of current inside the battery due to its electrodes and electrolyte. Owing to this internal resistance, the voltage at the terminals of a battery is less when it is discharging than when it is on open circuit. The internal resistance of a battery will change with the condition of discharge. As the battery discharges, the internal resistance increases, so that the terminal voltage decreases as the battery discharges. The terminal voltage of a battery is thus dependent upon the rate of discharge and the length of discharge. The voltage of a battery will decrease rather slowly as the battery discharges until nearly all the active material of the electrode has gone into solution. The battery is then discharged, and the voltage will drop rapidly to a very low value with any further attempt to discharge.


**PRIMARY BATTERIES**

**143. The standard Daniell cell** is a primary cell that has an emf which is practically 1 V when delivering a constant current. There are many forms of Daniell cell, each of which is particularly adapted to certain services, but all have very nearly the same emf (1.07 + V). The emf is not changed appreciably by the degree of concentration of the solutions, by the temperature, by the resistance, or by the purity of the zinc or copper, etc. In short, it makes a very good rough-and-ready standard.

A very good model is that used by the British Post Office. The jar is made with two compartments, one containing a porous cup immersed in water, in which are placed a copper plate and crystals of copper sulfate, the other containing a zinc plate and a 50 percent saturated solution of zinc sulfate. The zinc plate is fastened so as to be just clear of the solution, and a pencil of zinc is placed in the bottom. When in use, the porous cup is placed in the second compartment, thus raising the level of the zinc solution so as to immerse the zinc. Under working conditions the emf is about 1.07 V; when new, it is about 1.079 V.

**144. The gravity-type primary cell,** which is used in telegraph work, is suitable for closed-circuit work but should not be used for applications in which it is apt to stand for a long time on open circuit.

**145. In setting up the gravity cell,** place the copper electrode (−) in the bottom of the jar and pour in about 3 lb (1.36 kg) of copper sulfate crystals. Next place the zinc electrode (+) and fill with water to cover the zinc; to the water add a tablespoonful of sulfuric acid. Cover the electrolyte with a layer of pure mineral oil, which should be free from naphtha or acid and have a flash point about 400°F (204°C). If the oil is not used, the creeping can be stopped by dipping the edge of the jar in hot paraffin. When the cell is thus set up, it should be short-circuited for a day or two to form zinc sulfate which will protect the zinc electrode; this preliminary run also reduces the internal resistance. The temperature of the cells should be kept above 70°F (21°C), since the resistance increases rapidly with a decrease in temperature.

The internal resistance of the gravity cell is ordinarily from 2 to 3 Ω. A blue color in the bottom of the cell denotes a good condition, but a brown color shows that the zinc is deteriorating. When renewing the copper sulfate, it is best to empty the cell and set it up with a completely new electrolyte. The blue line, which marks the boundary between the copper sulfate and the zinc sulfate, should stand about halfway between the electrodes. If it comes too close to the zinc, some of the copper sulfate can be siphoned out, or the cell can be short-circuited so as to produce more zinc sulfate. If the blue line goes too low, some water and crystals of copper sulfate should be added.

**146. The Fuller cell** is well adapted to telephone work or any intermittent work. It can stand on open circuit for several months at a time without any appreciable deterioration.

**147. The Fuller cell is set up as follows:** Mix the electrolyte by adding 6 oz (0.17 kg) of potassium bichromate and 17 oz (0.48 kg) of sulfuric acid to 56 oz (1.59 kg) of soft water; pour this mixture into a suitable glass jar. Into a suitable porous cup put 1 teaspoonful of mercury and 2 teaspoonfuls of salt; place the cup and a zinc electrode in the glass jar, and fill to within 2 in (50.8 mm) of the top with soft water. Put on the cover, insert a carbon electrode, and the cell is ready for use.

The color of the solution is orange when in working order. The resistance varies from 0.5 to 4 $\Omega$, depending upon the condition and dimensions of the porous cup and upon the concentration of the solution.

**148. The Lalande cell,** frequently called the caustic soda cell, is suitable for either open- or closed-circuit work. The mechanical construction of this cell is especially good. The positive pole is a plate of compressed oxide of copper, the surfaces of which are reduced to metallic copper to improve the conductivity. This form of plate also acts as a depolarizer. The negative pole is of pure zinc amalgamated throughout by adding mercury when the casting is made. The electrolyte is a solution of caustic soda. The top of the solution is covered with a heavy mineral oil to prevent the solution from evaporating.

The emf of all types and sizes of caustic soda cells initially is approximately 0.90 V per cell. The voltage on closed circuit will depend upon the rate of discharge and the size of cell. The larger sizes, having lower internal resistance, will, of course, have a higher voltage for a given rate of discharge than the smaller cells under similar conditions. For ordinary purposes, however, it is safe to figure a mean effective voltage of approximately 0.67 V per cell at normal temperature.

**149. Primary batteries** are manufactured by the Primary Battery Division of Thomas A. Edison Industries in three types: the renewable carbon type, the nonrenewable Carbonaire battery, and the renewable copper oxide type.

These batteries are an ideal power supply for many low-voltage dc services for which the utmost freedom from power interruptions is extremely important. Their inherent operating advantages make them particularly desirable for the following applications in which batteries are used for direct operation or as an emergency standby power supply:

1. Navigation aids (river, channel, pier, and obstruction lights)

2. Alarm systems (municipal, police, and fire-alarm installations; fire and burglar alarms in factories, schools, office buildings, banks, public institutions, and private properties)

3. Annunciator systems

4. Electric fences

5. Elevator signals

6. Farm radios

7. Laboratory services and apparatus (industrial, school, and scientific)

8. Mine signal and communication systems

9. Railway signal and communication service

10. Telegraph service (local sounder and main-line circuits)

11. Telephone service (operation of transmitters on magneto switchboards, intercommunicating systems, interrupters or pole changers)

12. Time-clock systems (program, self-winding, etc., in schools, factories, and public buildings)

These primary batteries offer the following advantages:

1. Unlike storage cells, they provide a self-contained power supply requiring no charging. They can be applied anywhere, whether or not commercial power is accessible, and are therefore ideal for operating isolated installations.

2. They consistently deliver rated ampere-hour capacity at satisfactory operating voltage in continuous or intermittent service.

3. They require little attention or maintenance other than occasional inspection from the time they are placed in service until they are exhausted.

4. They eliminate the cost of providing and maintaining battery-charging facilities.

5. They are easy to install without experienced help. Transporting batteries to and from a charging station is unnecessary.

6. In the renewable-type cells, there are accurate visual indicator panels which warn of approaching and complete exhaustion of rated capacity. This feature makes it possible to see and make sure that ample battery capacity is always available—a decided advantage in preventing service failures due to exhausted batteries.

**150. The Edison Carbonaire battery is** a nonrenewable primary battery of the air-depolarized, add-water type, sealed in a single molded hard-rubber case. The battery is shipped dry and is made ready for use merely by adding a small quantity of water.

The add-water construction has a distinct advantage in that the battery is kept inert and free from internal deterioration during storage.

During its service life, the Carbonaire requires no maintenance other than infrequent inspections of the solution height and the addition of a small amount of water if the solution has fallen below the recommended operating level.

There are no visual indications of exhaustion in the Carbonaire, and its service life must be calculated on the basis of the known current consumption of the connected apparatus.

Carbonaire batteries (Fig. 4-184A) are manufactured in four types: the 2-S-J-1, the 3-S-J-1, the Crystalaire, Type F, and the Type Y. The Carbonaire batteries shown in Fig. 4-184B have largely replaced the hard-rubber-case types shown in Fig. 4-184A.

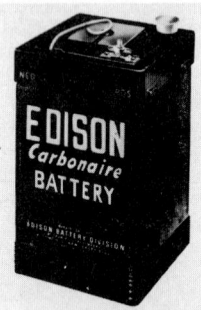

I. Type 2-S-J-1 Carbonaire battery.    II. Type 3-S-J-1 Carbonaire battery.    III. Type Y.

**Fig. 4-184A**    Edison Carbonaire batteries. [McGraw-Edison Co., Power Systems Division]

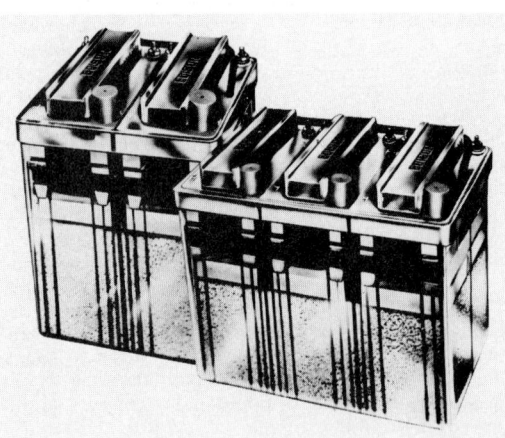

**Fig. 4-184B**    Edison Types ST-22 and ST-33 renewable Carbonaire primary batteries. [McGraw-Edison Co., Power Systems Division]

The technical details and nominal service ratings for the Carbonaire batteries in Fig. 4-184A are given in the accompanying table. They may be applied to many services for which a large ampere-hour reserve of dependable low-voltage current is needed and for which their simplicity of handling, setting up, and maintenance may be used to good advantage.

| Characteristics | 2-S-J-1 | 2-M-J-1 | 3-S-J-1 | Y |
|---|---|---|---|---|
| Over-all dimensions, in. | | | | |
|   Length.................................... | 8⁵⁄₁₆ | 8⁵⁄₁₆ | 12 | 8¼ |
|   Width..................................... | 7⅜ | 7⅜ | 7⅜ | 8¼ |
|   Height.................................... | 9½ | 9½ | 9½ | 13½ |
| Shipping weight, lb...................... | 24 | 24 | 35 | 33½ |
| Service weight, lb........................ | 29¼ | 29¼ | 45 | 40½ |
| Nominal capacity, amp-hrs................. | 1000 | 2000 | 1000 | 2500 |
| Nominal voltage.......................... | 2.4–2.5 | 1.2–1.25 | 3.6–3.75 | 1.2–1.25 |
| Nominal current, amp..................... | 0.25 | 0.50 | 0.25 | 1.00 |

When the battery is filled, it will get warm and then cool off, which causes the solution to rise a little and then to recede. This is normal, and no more water need be added on this account.

There is no advantage in maintaining the solution level at any exact point during the battery life except that if it falls below 2 in (50.8 mm) from the top of the filler, the level should be brought up to 1½ in (38.1 mm) by adding water. The top of the battery, especially the central carbons, should be kept clean and dry.

If the battery is used outdoors, it should be protected in a well-ventilated housing. Where temperatures are extreme, the housing should be partially sunk in the ground but constructed of material not permeable to dampness. Ventilation provides air for the battery to breathe and dissipates dampness.

Under adverse climatic conditions, there may be a minor wetness, or salt formation, around the battery top. This may be wiped off.

In connection with battery housings, it should be noted that since the Carbonaire requires oxygen for its operation, it should never be installed in completely airtight or watertight housings.

The batteries may be stored for any reasonable length of time, and they will remain inert and free from internal deterioration. This is due to the fact that they are shipped dry with the openings sealed at the factory.

## STORAGE BATTERIES: GENERAL

**151. A storage battery** is a device which can be used repeatedly for storing energy at one time in the form of chemical energy for use at another time in the form of electrical energy. It consists of two kinds of plates bearing the necessary electrochemically active materials immersed in a proper solution. The solution is called the electrolyte. Charging a battery consists of connecting the two terminals to a dc supply of proper polarity for a sufficient length of time. Electrical energy is delivered by the dc supply to the battery, in which it produces certain chemical reactions so that the energy is converted into chemical energy. If a charged battery has its two terminals connected through a closed external electric circuit, the active materials of the plate will react chemically with the electrolyte, producing a flow of current in the circuit. This conversion of chemical energy into electrical energy is called discharging the battery. Charging is the process of putting energy into the battery (delivering energy to the battery), while discharging is the process of taking energy from the battery (battery delivering energy to external electric circuit).

**152. Storage cell** is the name given to the fundamental unit of any storage battery. It consists of one positive plate or a group of positive plates electrically connected together, one negative plate or a group of negative plates electrically connected together, separators, electrolyte, and a suitable container. A storage battery may consist of a single cell or a group of cells electrically interconnected.

**153. Types of storage batteries.** Storage batteries may be classified in two ways as follows:

  1. According to fundamental type of service for which they are suitable
    *a.* Stationary
    *b.* Portable

2. According to fundamental materials of construction
   a. Lead-acid
   b. Nickel-iron-alkaline
   c. Nickel-cadmium

Stationary batteries are those designed for service in a permanent location.

Portable batteries are those designed for service requiring the transportation of the batteries during service.

**154. Storage-battery terminology.** Most of the following definitions of terms are taken from the standards of the Institute of Electrical and Electronic Engineers.

*Active materials.* Materials of plates reacting chemically to produce electrical energy during the discharge. The active materials of storage cells are restored to their original composition, in the charged condition, by oxidation or reduction processes produced by the charging current.

*Grid.* A metallic framework for conducting the electric current and supporting the active material.[1]

*Positive plate.* The grid and active material from which the current flows to the external circuit when the battery is discharging.

*Negative plate.* The grid and active material to which the current flows from the external circuit when the battery is discharging.

*Electrolyte.* An aqueous solution of sulfuric acid used in lead cells and of certain hydroxides used in nickel-iron-alkaline cells.

*Separator.* A device for preventing metallic contact between the plates of opposite polarity within the cell.

*Group.* Assembly of a set of plates of the same polarity for one cell.

*Element.* The positive and negative groups with separators assembled for a cell.

*Couple.* The element of a cell containing two plates, one positive and one negative. This term is also applied to a positive and negative plate connected together as one unit for installation in adjacent cells.

*Jar.* The container for the element and electrolyte of a cell. Specifically, a jar for lead-acid cells is usually of hard-rubber composition or glass, but for nickel-iron-alkaline cells it is a nickel-plated–steel container frequently referred to as a can.

*Tank.* A lead container, supported by wood, for the element and electrolyte of a cell. This is restricted to some relatively large types of cells.

*Case.* A container for several cells. Specifically, wood cases are containers for cells in individual jars; rubber or composition cases are provided with compartments for the cells.

*Tray.* A support or container for one or more cells.

*Terminal posts.* The points of the cell or battery to which the external circuit is connected.

*End cells.* The cells of a battery which may be cut in or out of the circuit for the purpose of adjusting the battery voltage.

*Pilot cell.* A selected cell whose temperature, voltage, and specific gravity of electrolyte are assumed to indicate the condition of the entire battery.

*Ampere-hour capacity.* The number of ampere-hours which can be delivered by a cell or battery under specified conditions as to temperature, rate of discharge, and final voltage.

*Energy density.* The watthours per pound (Wh/lb) of battery weight.

*Power density.* The watts per pound (W/lb) of battery weight.

*Watthour capacity.* The number of watthours which can be delivered by a cell or battery under specified conditions as to the temperature, the rate of discharge, and the final voltage.

*Time rate.* The rate in amperes at which a battery will be fully discharged in a specified time, under specified conditions of temperature and final voltage, as, for example, the 8-h rate or the 20-min rate.

*Open-circuit voltage.* The voltage of a cell or battery at its terminals when no current is flowing. For the purpose of measurement, the small current required for the operation of a voltmeter is usually negligible.

*Closed-circuit voltage.* The voltage at the terminals of a cell or battery when current is flowing.

---

[1]In certain types of batteries the active material is enclosed in containers which are held in place by the grid.

*Average voltage.* The average value of the voltage during the period of charge or discharge. It is conveniently obtained from the time integral of the voltage curve.

*Initial voltage.* The voltage of a cell or battery at the beginning of a charge or discharge. It is usually taken after the current has been flowing for a sufficient period of time for the rate of change of voltage to become practically constant.

*Final voltage.* The prescribed voltage upon reaching which the discharge is considered complete. The final voltage is usually chosen so that the useful capacity of the cell is realized. Final voltages vary with the type of battery, the rate of the discharge, temperature, and the service in which the battery is used.

*Polarity.* An electrical condition determining the direction in which current tends to flow. By common usage the discharge current is said to flow from the positive or peroxide plate through the external circuit. In a nickel-iron-alkaline battery the positive plate is that containing nickel peroxide.

*Charge.* The conversion of electrical energy into chemical energy within the cell or battery. This consists of the restoration of the active materials by passing a unidirectional current through the cell or battery in the opposite direction to that of the discharge. A cell or battery which is said to be charged is understood to be fully charged.

*Charging rate.* The current expressed in amperes at which a battery is charged.

*Constant-current charge.* A charge in which the current is maintained at constant value. For some types of lead batteries this may involve two rates called the starting and the finishing rates.

*Constant-voltage charge.* A charge in which the voltage at the terminals of the battery is held at a constant value. A modified constant-voltage system is usually one in which the voltage of the charging circuit is held substantially constant but in which a fixed resistance is inserted in the battery circuit, producing a rising-voltage characteristic at the battery terminals as the charge progresses. This term is also applied to other methods for producing automatically a similar characteristic.

*Boost charge.* A partial charge, usually at a high rate for a short period.

*Equalizing charge.* An extended charge given to a battery to ensure the complete restoration of the active materials in all the plates of all the cells.

*Trickle charge.* A continuous charge at low rate approximately equal to the internal losses and suitable to maintain the battery in a fully charged condition. This term is also applied to very low rates of charge suitable not only for compensating for internal losses but for restoring intermittent discharges of small amounts delivered from time to time to the load circuit.

*Finishing rate.* The rate of charge expressed in amperes to which the charging current for some types of lead batteries is reduced near the end of charge to prevent excessive gassing and temperature rise.

*Discharge.* The conversion of the chemical energy of the battery into electrical energy.

*Reversal.* Change in normal polarity of a storage cell.

*Local action or self-discharge.* The internal loss of charge which goes on continuously within a cell regardless of connections to an external circuit.

*Floating.* A method of operation in which a constant voltage is applied to the battery terminals sufficient to maintain an approximately constant state of charge.

*Specific gravity of electrolyte.* The electrolyte of lead-acid batteries increases in concentration to a fixed maximum value during charge and decreases during discharge. The concentration is usually expressed as the specific gravity of the solution. This variation of the specific gravity of the solution affords an approximate indication of the state of charge of the battery.

The specific gravity of the electrolyte in nickel-iron-alkaline batteries does not change appreciably during charge or discharge and therefore does not indicate the state of charge. The specific gravities, however, are an indication of the electrochemical usefulness of the electrolyte.

*Gassing.* The evolution of oxygen or hydrogen or both.

*Efficiency.* The ratio of the output of a cell or battery to the input required to restore the initial state of charge under specified conditions of temperature, current rate, and final voltage.

*Ampere-hour efficiency (electrochemical efficiency).* The ratio of the ampere-hours output to the ampere-hours of the recharge.

*Volt efficiency.* The ratio of the average voltage during the discharge to the average voltage during the recharge.

*Watthour efficiency (energy efficiency).* The ratio of the watthours output to the watthours of the recharge.

*General data.* Batteries are usually rated in terms of the number of ampere-hours which they are capable of delivering when fully charged and under specified conditions as to temperature, rate of discharge, and final voltage. For different classes of service, different time rates (see definition of *time rate*) are frequently used. For comparing the capacities of batteries of different size but of the same general design, it is customary to use the same time rate, and a comparison based on the different lengths of time they will discharge at the same rate is not recommended, as it is misleading.

*Misrating.* A battery which fails to deliver its rated capacity on the third successive measured cycle of charge and discharge under specified current rates, temperature of electrolyte, specific gravity, and final voltage shall be considered to be improperly rated.

## LEAD-ACID STORAGE BATTERIES

**155. Lead-acid batteries.**[1] In the charged condition the active materials consist of lead peroxide on the positive plate and sponge lead on the negative plate. The electrolyte is a mixture of sulfuric acid and water. The strength of the electrolyte is measured in terms of specific gravity, which is the ratio of the weight of a given volume of electrolyte to an equal volume of water. Concentrated sulfuric acid has a specific gravity of about 1.830; water has a specific gravity of 1.000. The acid and water are mixed in a proportion to give the specific gravity desired. For example, an acid manufacturer to supply electrolyte of 1.210 gravity will mix roughly about 1 part of concentrated acid to 4 parts of water.

In a fully charged battery all the active material of the positive plates is lead peroxide, and that of the negative plates is pure sponge lead. In a fully charged battery all the acid is in the electrolyte, and the specific gravity is at its maximum value. The active material of both the positive and the negative plates is porous, so that it has absorption qualities similar to those of a sponge, and the pores are therefore filled with some of the battery solution. As the battery discharges, the acid, which is in the pores of the plates, separates from the electrolyte, forming a chemical combination with the active material, changing it to lead sulfate. As the discharge continues, additional acid is drawn or diffused from the electrolyte into the pores of the plates and further sulfate is formed. It can be readily understood that as this process continues, the specific gravity of the electrolyte will gradually decrease, because the proportion of acid is decreasing. On charge the reverse action takes place: the acid in the sulfated active material is driven out and back into the electrolyte. This return of the acid to the electrolyte increases the specific gravity, so that it will continue to rise until all the acid is driven out of the plates and back into the electrolyte. After all the acid has been driven back into the electrolyte, further charging will not raise the specific gravity any higher, as all the acid in the cells is in the electrolyte, and the battery is said to be fully charged. The material of the positives is again lead peroxide, and that of the negatives is spongy lead; the specific gravity is maximum.

Practically speaking, on discharge the plates absorb acid, and the specific gravity of the electrolyte decreases. On the charge the plates return the absorbed acid to the electrolyte, and the specific gravity increases. Figure 4-185, showing a cell charged, discharging, discharged, and charging, illustrates clearly the chemical action which takes place on charge and discharge:

When a cell is fully charged [Fig. 4-185(1) of the diagram], the negative plate is lead sponge, Pb; the positive plate is lead peroxide, $PbO_2$; and the specific gravity of the electrolyte (sulfuric acid, $H_2SO_4$, and water, $H_2O$) is at its maximum. Chemical energy is stored in the cell in this condition.

When a cell is put on discharge [Fig. 4-185(2)], the $H_2SO_4$ of the acid is divided into $H_2$ and $SO_4$. The $H_2$ passes in the direction of the current to the positive plate and combines with some of the oxygen of the lead peroxide and forms $H_2O$; the $SO_4$ combines with the liberated Pb of the positive plate to form lead sulfate. The $SO_4$ also forms lead sulfate at the negative or lead sponge (Pb) plate. As the discharge progresses, both plates finally contain considerable lead sulfate, $PbSO_4$ [see Fig. 4-185(3)]. The water formed has diluted the acid, lowering the specific gravity of the electrolyte. When the plates are

---

[1] A large portion of the following information in this section on lead-acid batteries has been taken from the literature of the Electric Storage Battery Co.

entirely sulfated, current will cease, since the plates are then identical, and any active electric cell requires two dissimilar plates in electrolyte. In common practice, however, the discharge is always stopped before the plates have become entirely reduced to lead sulfate.

During charge [Fig. 4-185(4)], the lead sulfate, $PbSO_4$, on the positive plate is converted into lead peroxide, $PbO_2$, while the lead sulfate on the negative plate is converted into sponge lead, Pb, and the electrolyte gradually becomes stronger as the $SO_4$ from the plates combines with hydrogen from the water to form acid, $H_2SO_4$, until no more sulfate remains and all the acid has been returned to the electrolyte. It will then be of the same strength as before the discharge, and the same acid will be ready to be used over again during the next discharge.

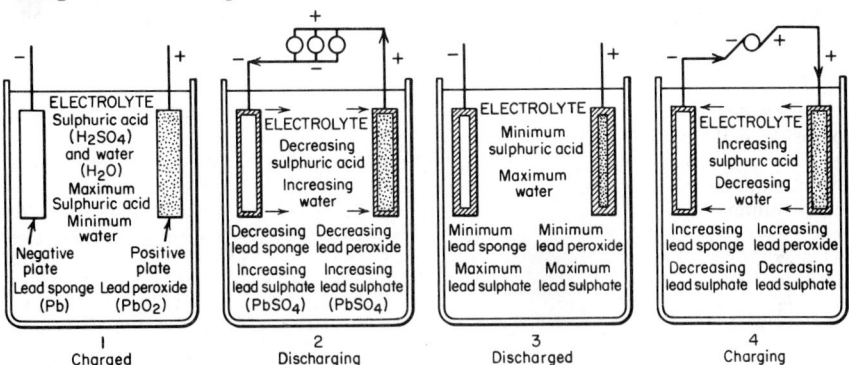

**Fig. 4-185**  Chemical action in a cell on cycle of discharge and charge.

**156. Change in electrolyte during charge and discharge.** During discharge, as stated in Sec. 155, some of the acid leaves the electrolyte and combines with the plates. During charge, the acid which has been absorbed by the plates is driven back into the electrolyte. When all the acid has been driven back into the electrolyte, the battery is fully charged.

When all the acid has been driven into the electrolyte, the electrolyte is stronger than when some of the acid is in the plates. This strength is measured in terms of specific gravity. The specific gravity of the electrolyte, therefore, changes as the battery charges or discharges. It falls on discharge and rises on charge. It is, then, an excellent indication of the state of charge of a battery, being greatest when the cell is fully charged and least when discharged.

The difference between the full-charge and discharge values of the gravity depends upon the type of cell under consideration. For example, for the type of cell in use in starting work in automobiles, the full-charge gravity in temperate climates is 1.280; discharged, 1.150.

The specific gravity of the electrolyte is readily determined by means of a float called a hydrometer. With a high specific gravity, the hydrometer, or float, does not sink so far in the electrolyte as it does when the specific gravity is low (see Fig. 4-186).

**157. Voltage characteristics.** The voltage of each cell is approximately 2 V on an open circuit but is higher than this when the battery is being charged and lower when being discharged. The nominal voltage of a battery is, therefore, the number of cells multiplied by 2.

The voltage at any time on discharge or charge depends upon several factors, such as the current rate, the state of charge or discharge, and the temperature. No general averages to cover all conditions can therefore be given. In usual 6- to 8-h discharge service, the average cell voltage (Sec. 154) during discharge is roughly 1.95 V with a final voltage of about 1.75 V. As soon as the cell is put on charge, its voltage rises to about 2.15 V and then increases during charge until at the end it is between 2.4 and 2.7, depending upon local conditions. The average voltage during the entire charge is usually considered to be 2.33 V.

Typical voltage-discharge characteristics are given in Figs. 4-187 and 4-188. The effect of the discharge rate upon the voltage characteristics is shown in Fig. 4-189.

**158. High discharge rates (amperes) are often confused with overdischarge (too many**

**ampere-hours taken out).** A lead-acid battery of the type sold under the trade names Exide, Chloride, and Ironclad can be discharged, without injury to the plates, at any rate of current that it will deliver. The maximum permissible rate of discharge is limited only by the current-carrying ability of the wiring, motor, or other apparatus to which the battery is connected or by the current-carrying ability of the cell terminals and connectors and not by the plates themselves.

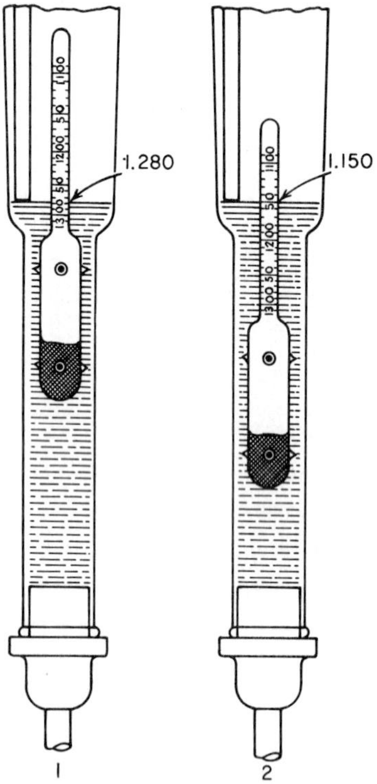

**159. Rating of lead-acid batteries.** All batteries are given a normal (ampere-hour) capacity rating based on a certain time rate of discharge under specified conditions of temperatures and final voltage. For example, a certain battery may have an 8-h rating of 1000 Ah discharging to 1.75 V per cell at a temperature of 77°F (24°C).

The ampere-hour capacity of a battery depends upon the amount of available active material in the plates that can be reached by the electrolyte, the amount of sulfuric acid in the electrolyte, the rate of discharge, and the allowable safe limit of discharge. To produce an ampere-hour of electricity on discharge requires the combination of a certain amount of sponge lead (negative active material) and a certain amount of peroxide of lead (positive active material) with a certain amount of sulfuric acid from the electrolyte. The allowable rating therefore depends upon the construction of the plates and the number of plates that are constructed in parallel on each side of the battery.

**160. Effect of discharge rates on discharge capacity of lead-acid batteries.** The useful ampere-hour capacity of a storage battery depends on the rate of discharge and is greater for a long low rate or an intermittent rate than for a short high rate. Any particular battery is given a so-called normal rate. This so-called normal rate is not the capacity obtainable under all conditions. Each different rate of discharge governs the available capacity that can be obtained from the battery.

**Fig. 4-186**  Use of hydrometer in determining specific gravity of electrolyte.

When a battery is discharged continuously at a constant rate, its available ampere-hour capacity is a function of the rate of discharge, the available capacity being lower at the higher rates, as is shown in Fig. 4-190. The reduction in available capacity of high-rate continuous discharge is due to depletion of the acid in the pores of the plates. The depletion is due to the fact that at high rates of discharge the acid in the pores of the plates combines with the active material and is withdrawn from the solution more rapidly than it can be replenished by diffusion from the free electrolyte in the cell. It is this limitation of available acid in the pores of the plates that limits the capacity at high continuous rates

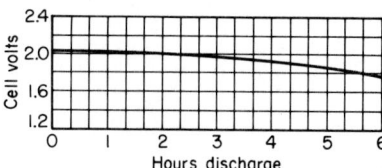

**Fig. 4-187** Type MVM Exide-Ironclad discharge characteristics at the 6-h rate. [Electric Storage Battery Co.]

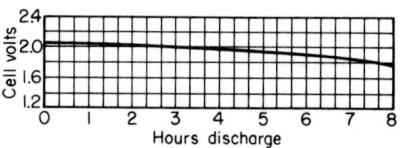

**Fig. 4-188** Type MVM Exide-Ironclad discharge characteristics at the 8-h rate. [Electric Storage Battery Co.]

of discharge, rather than any limitation due to the plates themselves. This explains the fact that after a battery has been exhausted at a high discharge rate, the balance of its normal capacity can be obtained by continuing the discharge at lower rates or by allowing the battery to recuperate while standing on open circuit for a time and then continuing the high-rate discharge. For this reason also, as stated in Sec. 158, it is impossible to damage the plates by overdischarge at high rates, as the voltage of the battery will drop below a usable value before the active material in the plates has been discharged to the danger point.

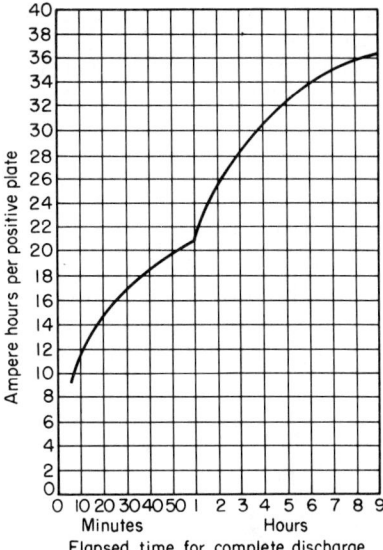

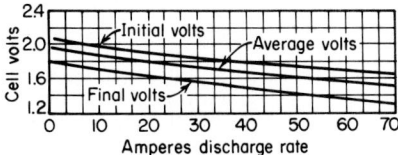

**Fig. 4-189**  Type MVM Exide-Ironclad initial, average, and final volts at various discharge rates. [Electric Storage Battery Co.]

**Fig. 4-190**  Curve illustrating effect of discharge rate upon available ampere-hour capacity of lead-acid storage batteries. [Electric Storage Battery Co.]

Figure 4-190 shows the available capacity per positive plate of an Exide-Ironclad battery when discharged continuously, in varying lengths of time.

If a battery is discharged intermittently, it is evident that during periods of rest between discharges diffusion will continue, thus renewing the strength of acid in the pores of plates and increasing the available capacity corresponding to the discharge rate. The rate of diffusion depends on the difference in strength of the acid in the pores and that outside the plates, and if this difference is great, the diffusion is rapid at first but decreases as the two strengths become more nearly equal. It follows from this reasoning that the reduction in available capacity due to high rates of discharge largely disappears when the discharge is intermittent.

If the total elapsed time during which discharges are made is greater than 6 h, and if the discharges are distributed throughout that time so that there is time for this diffusion to take place, the full 6-h capacity of the battery will be available regardless of the rates at which the discharges are taken.

An illustration of the effect of acid diffusion is given in Fig. 4-191, which shows a test made on a cell of the type used in submarine boats by the United States government.

This cell has a rated capacity of 3000 A for 1 h. The curve shows it to have been discharged at that rate for 58 min, at which time it was practically exhausted at that rate, and if the discharge had been continued at that rate, the voltage would have fallen rapidly. The rate of discharge was, however, reduced to 1350 A and continued for 45 min, then to 910 A for 30 min, then to 525 A for 1 h and 20 min, and finally to 300 A for 3 h and 15 min. At each reduction in the rate of discharge the rate of acid absorption by the plates was reduced, thus allowing diffusion to strengthen the acid in the pores of the plate and so permitting the discharge to be continued. In each case the voltage at the lower discharge rate was higher than the voltage at the preceding higher rate. During the 300-A discharge it will be noted that the voltage continued to rise for over an hour after the start at that rate, showing that the acid from the free electrolyte was entering the pores of the plates faster than it was being absorbed.

The curve showing ampere-hour output shows that 6000 Ah was delivered by the cell and that after the battery had been exhausted at the high rate of discharge, as much more energy was still available at lower rates.

**161. Discharge limits of lead-acid batteries.**  In an emergency, little if any permanent harm will result if the battery is discharged to the full amount that it will give, provided that it is promptly recharged.

It has already been pointed out that the drop in specific gravity of electrolyte should not exceed a certain definite value, which varies according to the type of cell and will be furnished on application to the manufacturer.

The danger of harm from overdischarge may be illustrated by a comparison between the active material of the plates and the action which goes on when some of the electrolyte

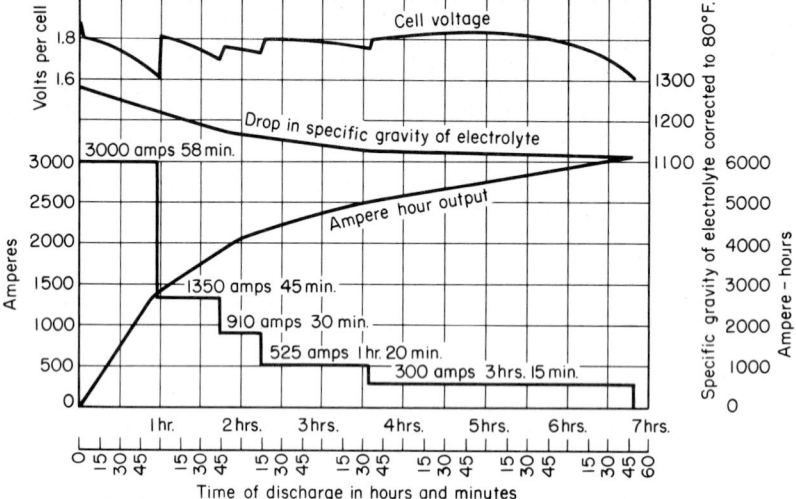

**Fig. 4-191**   Curves showing effect of acid diffusion in lead-acid storage batteries on intermittent discharge. [Electric Storage Battery Co.]

is allowed to act on copper wiring terminals of a battery. It is noticeable that a comparatively large amount of copper sulfate is formed when only a small quantity of the metal is eaten away by the acid. In the same manner, when the acid combines with the lead in the active material, the resulting lead sulfate occupies more space than the active material from which it is formed. The active material of all battery plates is porous, and this expansion of the sulfated material is accommodated by reduction in the size of pores in the active material. All battery plates are designed to accommodate a certain amount of this expansion of the active material during sulfation, and in batteries of the type under consideration this is limited to the amount represented by a certain specific-gravity reading.

Further discharge, even if it can be obtained at a satisfactory voltage, results in an excessive expansion, which so closes the pores in the active material that it becomes increasingly difficult to recharge the battery properly after an excessive discharge, and unless a proper recharge is given, the battery is likely to deteriorate.

**162. Charging and charging rates for lead-acid batteries.**   A battery must, of course, be charged with direct current, and the current must be connected to the battery so that it will go through it in the proper direction. The positive pole of the charging source must be connected to the positive terminal of the battery, and the negative of the charging source to the negative of the battery.

While a battery is being charged, the amount of sulfate in the plates decreases, and the ability of the plate to give up the acid becomes reduced; in other words, during the early part of a charge the plates can give up the acid at a rapid rate, as there is a large amount of sulfate available. Therefore a battery that is considerably discharged can be charged at a high rate, but as the charge approaches completion, currents at a high rate cannot be utilized, and if high rates are maintained, only a portion of the current is used to withdraw acid from the plates, and the balance of the current acts to decompose the water in the electrolyte into oxygen and hydrogen, which are given off in the form of gas. Gassing of the battery, therefore, at any time shows whether or not the charging rate is too high. Consequently, when the cells are gassing on charge the rate of charge should be reduced so as not to waste the current. Furthermore, the action of the bubbles of gas in escaping from the pores of the plates and in "boiling" to the top of the electrolyte has a tendency to wash and wear the active material away from the plates, particularly the positive.

It is a well-known fact that batteries wear out. This wear shows itself to the eye principally in the positive plate, the active material of which softens with use, and were it not for this unavoidable fact, the life of batteries would be very much longer than at present. As the active material of the plate softens with use, there is a tendency for the softened material on the surface of the plate to fall to the bottom of the jar in the form of sediment. The action of the gas in escaping from the pores of the plates and the little whirlpools created in the electrolyte when the bubbles of gas boil to the surface hasten this shedding of material and shorten the life of the battery.

Excessive gassing, therefore, should be avoided if the best life of the battery is to be obtained. A small amount of gassing at low rates and for a short time, at the completion of a charge, is not objectionable, but violent gassing having the appearance of boiling should be avoided.

**163. Charging methods for lead-acid batteries.**   In practice, charging methods vary with the type of service. For example, in propelling electric vehicles, the battery is discharged over a period of time and then recharged. The rate in amperes to use for recharge depends upon the time available and the type of cell. The lower the rate of charge, the longer will be the time required. The shorter the time available, the higher the rate must be to recharge, provided the rate is not higher than recommended for the type of cell. In any event, the rate must always be low at the end of charge when gassing begins. This is known as the finish charge rate.

Batteries charged by this cycle method of discharge and charge can have their charging equipment arranged and designed to provide a taper charge rate automatically and inherently, i.e., high at the start of charge, when a high rate can be utilized, and low at the end of charge when gassing begins. In addition, the equipment can be arranged to stop the charge at the proper time, and all without any manual attention whatever.

Batteries, used in starting, lighting, and ignition work on automobiles equipped with a generator for charging are charged whenever the engine is running at ordinary speeds. At very low speeds or when the engine is not running, the battery supplies current for lights, ignition, and cranking. In automobile service, the rate must be sufficient to keep the battery charged and yet not overcharge it.

Batteries used for reserve emergency, standby, or voltage regulation are kept fully charged by a trickle current or floating charge. A constant voltage of appropriate value impressed across the battery terminals is sufficient for proper charging.

Regardless of the charging method, the rate in amperes must not cause excessive gassing; neither should the cell temperature rise above $110°F$ ($43°C$).

**164. Equalizing charge for lead-acid batteries.**   Wherever practicable, batteries are given an equalizing charge at regular intervals, for example, weekly for batteries used in propelling vehicles and other services in which the battery is discharged considerably and then recharged, monthly in floating service.

In cycle service an equalizing charge is a continuation of the regular charge at a low rate. In floating service the equalizing charge is obtained by raising the voltage across the battery to increase the current into the battery. The equalizing charge continues until all cells gas freely and until it is certain by taking voltage and gravity readings that *all* cells are fully charged.

As has already been pointed out, the object of charging is to withdraw all acid from the plates. In practice, the regular charges in cycle service are not always given long enough to withdraw all the acid completely from the plates. In fact, this is not necessary, provided the acid is completely withdrawn regularly by giving an equalizing charge. If this is not done and some sulfate is allowed to remain in the plates for a considerable time, it will gradually increase, the pores of the plate will become clogged, and the battery will lose capacity and deteriorate in such a way that it becomes increasingly difficult to restore it to its normal condition. To carry frequent (such as daily) regular charges to the full extent would involve an unnecessary amount of charging and gassing, which is not desirable.

A battery is not fully charged until all the acid has been driven out of the plates by charging. To charge a battery fully, do not try to charge to a fixed or definite gravity, but charge until the specific gravity or voltage stops rising.

**165. Effect of temperatures upon lead-acid batteries.**   The cell temperature should not exceed 110°F (43°C). The effect of high temperature is primarily to shorten the life of the wood separators which are installed between the positive and negative plates of some types of cells.

There is always a tendency for wood in contact with sulfuric acid to become carbonized. This tendency is greatly increased at temperatures above 110°F. If a battery is operated regularly under such conditions, it will probably be necessary to renew the wood separators before the battery itself is worn out.

In the materials used in any commercial storage battery, there are also present some impurities which cause very slight action in the cell even when it is not in active operation. At high temperatures, these internal losses are increased, as is evidenced by the fact that a battery placed in storage will not lose its charge seriously over a period of, say, 6 months if kept in a cool place. If kept in a temperature around 100°F (38°C), it will lose much more of its charge in, say, 3 months.

Low temperature temporarily decreases both the discharge voltage and the ampere-hour capacity which can be taken out of the battery. The battery acts as if it were numbed by the cold and unable to make the same effort as at normal temperature. The effect of

**Fig. 4-192**   Automobile type of lead-acid storage battery. [Electric Storage Battery Co.]

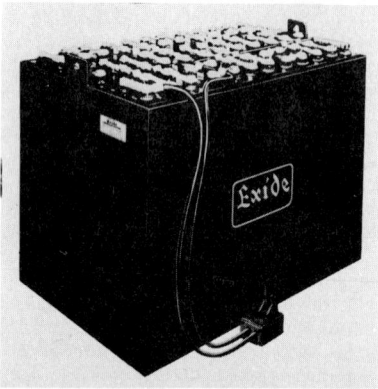

**Fig. 4-193**   Ironclad type of lead-acid storage battery. [Electric Storage Battery Co.]

cold is only temporary, the battery returning to its normal state upon its return to normal temperature even without charge. There is no danger of the electrolyte freezing in a fully charged cell, but this is likely to occur in an overdischarged battery or in one that has had water added without subsequent charging.

**166. Types of lead-acid storage batteries.**   Lead-acid storage batteries may be roughly divided into three groups, depending upon the class of service for which they are intended. Those for automobile starting, lighting, and ignition work consist of cells assembled in a single hard-rubber container (Fig. 4-192). Those for electric-vehicle applications consist of ironclad-type cells assembled in a suitable container (Fig. 4-193). Those for stationary service consist of cells enclosed in glass jars (Fig. 4-194).

**167. Lead-acid batteries are easily kept in good condition** for a long time of trouble-free service if the following maintenance instructions are observed:

1. Keep battery clean outside (Sec. 166).
2. Add water at regular intervals (Secs. 169 and 170).
3. Maintain battery in a healthy state of charge (Secs. 172 and 173).
4. Keep written records (Sec. 174).

**168. Cleanliness**

1. Keep the battery, its connections, and surrounding parts clean and dry by wiping with a dry rag, but do not

**Fig. 4-194**   Lead-acid storage battery in enclosing glass jar. [Electric Storage Battery Co.]

remove the grease from the seal nuts. Keep the vent plugs in place and tight, and make sure their gas-escape holes are open. If electrolyte is spilled or if any parts are damp with acid, apply a solution of ammonia or of baking soda in the proportions of 1 lb (0.45 kg) of soda to 1 gal (3.8 l) of water, then rinse with water and dry; do not allow solution to get into cells. If this treatment is given two or three times a year and the battery kept clean between times by regular washings with water or blowing off with an air jet, the life and service of the battery in general and trays or supports in particular will be increased considerably. Before hosing a battery without removing it from the compartment, consult equipment manufacturer for permission. When washing high-voltage batteries, open the connections at several places to avoid possible shocks.

2. If the terminals or connections show any tendency to corrode, scrape the corroded surface clean, wash it with soda solution or with ammonia solution, rinse with water, and coat it thinly with Vaseline or No-Ox-Id grease. No corrosion will occur unless electrolyte is spilled and allowed to remain.

3. Soda solution or ammonia will neutralize the effect of acid on clothing, cement, etc.

### 169. Adding water

1. During operation, water must be regularly added to each cell. Do not allow the surface of the electrolyte to get below the level specified by the manufacturer. Keep it above this point by removing the vent plugs regularly from all the cells and adding sufficient approved water to each cell as often as necessary. Do not fill so high that electrolyte will be lost through the vent plugs, eventually resulting in ruined cells. Less harm will result in allowing the level to get a little low than in adding water too high. After filling, be sure to replace and securely tighten the plugs. The intervals at which water must be added depend largely on the operating schedule, but it should not be necessary to add water more often than once a week; otherwise the battery is being given too great a charge.

2. In cold weather the time to add water is just before a charge, so that gassing (bubbling of the electrolyte resulting from charging) will ensure thorough mixing and any danger of the water freezing be avoided.

3. Electrolyte loses some of its water by the charging of the battery and some by evaporation, but its acid is never lost in this manner; therefore, it will not be necessary to add new electrolyte unless some should get outside the cell through carelessness or by adding too much water so that the container is too full.

4. Nothing but water is required to be added to storage batteries. Never add any special powders, solutions, or jellies. A great many special powder solutions or jellies are injurious, having a corrosive or rotting action on the battery plates, reducing the voltage and capacity of the cells.

5. All the cells in the battery should take the same amount of water. If one cell takes more than the others, examine it for leakage.

6. Keep a written record of the amount of water that is being added from time to time.

### 170. Kind of water

1. The quality of water to add is distilled (not merely boiled) or other approved water. By approved water is meant that of which the battery manufacturers have analyzed a sample and found safe for their batteries. The local source of water is usually suitable, but before using it the battery manufacturers should be consulted. Most companies will do this without charge for users of their batteries. Transportation charges should be prepaid, and the sample marked for identification.

2. If water is drawn from a tap or spigot, it should be allowed to run a few moments before it is used, to remove pipe accumulations. Water should not be transported or stored in a vessel of any metal except lead. Glass, earthenware, rubber, plastic, or wooden receptacles that have not been used for any other purpose are satisfactory.

### 171. Discharge limits

1. In an emergency, little if any permanent harm will result if the battery is discharged to the full amount that it will give, provided that it is promptly and fully recharged.

2. If an ampere-hour meter is in use, the discharge should be stopped and the battery promptly recharged before or upon reaching its capacity limit in ampere-hours, as given by the manufacturers. The ampere-hour meter will run slow if the discharge is at high rates or if the meter is not calibrated at proper intervals and will then show less discharge than the battery actually gives. This fact should not be overlooked.

3. The specific gravity of the electrolyte falls on discharge and is therefore an indication of the amount of discharge. The difference between the full-charge and discharge values of the gravity depends on the type of cell. The manufacturer of the battery should be consulted for the proper values for any particular cell.

### 172. Hydrometer readings: specific gravity

1. If the specific-gravity or hydrometer reading is known, one can tell that the battery is fully charged or the amount by which it is discharged. With all cells connected in series, the gravity reading of one cell, known as a pilot cell, will indicate the state of discharge or charge of the whole battery.

2. The specific gravity is easily determined by allowing a hydrometer to float in the electrolyte. When the specific gravity is high, the hydrometer will not sink so far into the electrolyte as when the specific gravity is low (see Fig. 4-186).

3. To take a reading, insert the nozzle of the hydrometer syringe (Fig. 4-195) into the cell, squeeze the bulb, and then slowly release it, drawing up just enough electrolyte from the cell to float the hydrometer freely. With the syringe held vertically, the reading on the stem of the hydrometer at the surface of the liquid is the gravity reading of the electrolyte. After the testing is completed, always return the electrolyte to the cell from which it was taken.

4. Both temperature and level of electrolyte affect the specific-gravity reading somewhat, and it is therefore desirable to record the temperature and level of the electrolyte at the same time its gravity reading is taken. A gravity reading should not be taken immediately after adding water, as the reading will give a false indication of the specific gravity. Allow a day or so for the water to mix with the electrolyte by gassing (bubbling) of the electrolyte resulting from charging or floating the battery.

5. After every 50-odd gravity readings of a pilot cell, a different cell should be used as a pilot to avoid lowering of its gravity due to possible loss of a small amount of electrolyte each time the gravity is read.

6. Hydrometer syringes are available for mounting through the vent plug of a pilot cell (Fig. 4-196). Since they are continuously in place, no dripping is experienced and the pilot cell need not be changed after every 50-odd readings.

### 173. Full-charge specific gravity

1. The proper specific gravity of the electrolyte with the cells fully charged and with the electrolyte at the proper level, as specified by the maker, will vary somewhat with different types of batteries. The manufacturer should be consulted for the proper value.

2. The specific gravity of a new battery is adjusted at the factory and will not require adjusting during the life of the battery unless electrolyte is actually lost out of the battery. If, however, electrolyte is lost, it should be replaced with electrolyte of about the same specific gravity as in the surrounding cells.

3. The full-charge specific gravity will decrease in value as the battery ages. No definite value can be given, but this decrease is very small: not over a few points per year at the most. This change is mentioned so that it will be understood.

4. Before adjusting low gravity, first make sure charging will not raise gravity. To do this, continue an equalizing charge until the specific gravity shows no rise and then for 3 h more. Never make a gravity adjustment on a cell which does not gas when it is being charged.

5. To adjust low gravity, first have ready sulfuric acid of specific gravity between 1.265 and 1.300, sufficiently pure for storage-battery use. Add this instead of water when restoring level until the gravity at the end of an equalizing charge is normal. Then stop adding acid and return to the use of water. A quicker method, but one requiring more work and acid, is to withdraw some of the low-gravity electrolyte from the cell and at once replace it with this new electrolyte. Do not allow a cell to stand partly empty. The amount to withdraw will have to be determined by trial, as it depends upon the gravities of both the old and the new electrolyte. Charge until all cells have been gassing for 1 h. Then, if the gravity is not normal, repeat adjustment until it is.

6. To adjust high gravity, remove some of the electrolyte and replace with water until the gravity at the end of an equalizing charge is normal.

### 174. Readings: written records

1. To facilitate following the operation of the battery, it is advisable to record the specific gravity and voltage of each cell at intervals. The relative state of charge should be the same each time readings for record are taken: for instance, on a manually cycled battery, at the end of a charge; on a floated battery, 10 to 15 min after starting the monthly equalizing charge.

2. Cell voltage readings should be taken while the charging current is being maintained and not after it has been reduced or interrupted. During these readings the battery voltage or the charging current into the battery should be kept constant. Cell gravities should be taken 10 or 15 min after charge has been completed and not while cells are gassing heavily.

3. The individual cell voltages, read to the hundredth of a volt, should be recorded once a month, in which case three or four times a year will be sufficient for recording the cell gravities. Otherwise cell gravities should be recorded monthly.

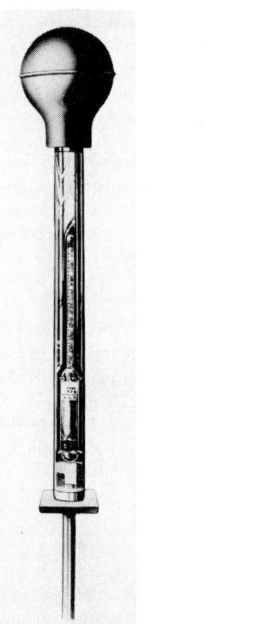     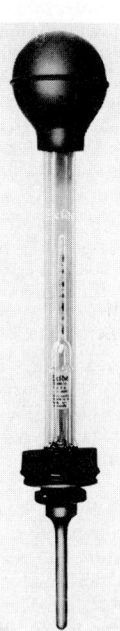

**Fig. 4-195**   Hydrometer syringe. [Electric Storage Battery Co.]

**Fig. 4-196**   Vent-plug hydrometer syringe. [Electric Storage Battery Co.]

4. Review the monthly cell readings and compare with those for the previous month promptly. Plotting the readings saves time in reviewing and comparing. Prompt action upon indication of trouble may save time and expense later (see Sec.175).

### 175. Trouble

1. The chief indications of trouble in a cell are falling off in gravity or voltage relative to the rest of the cells and lack of gassing on equalizing charge.

2. If a battery seems to be in trouble, the first thing to do is to give it an equalizing charge (Sec. 164). Then take a gravity reading of each cell. If all the cells gas evenly on the equalizing charge and the gravity of them all goes above a certain value as specified by the manufacturer, then all the battery needed was the charge. Before making an adjustment, determine whether the jar is cracked by adding water to the proper height and allowing the cell or jar to stand several hours, noting whether level falls. If a jar is cracked, change it. Never make a gravity adjustment on a cell which does not gas. If a cell will not gas on the equalizing charging, investigate for impurities or inspect it for short circuits. For the latter, remove the elements from the jar and examine the separators carefully to make sure that none is broken or damaged, thus causing a short circuit. Also examine plates to see that they are in good condition, and note the height of sediment in the bottom of the jar. Remove any collection of "moss" on the top or edges of the plates. Handle elements very carefully so that plates will not be broken from the straps. Replace damaged separators.

### 176. Impurities.

Impurities in the electrolyte will cause a cell to work irregularly. If it is known that any impurity has got into a cell, it should be removed at once. In case removal is delayed and any considerable amount of foreign matter becomes dissolved in the electrolyte, this solution should be replaced with new solution immediately, thoroughly flushing the cell with water before putting in the new electrolyte. If in doubt as to whether the electrolyte contains impurities, submit a sample for test.

**177. Sediment.** The sediment which collects underneath the plates need cause no alarm unless it deposits too rapidly, in which case there is something wrong with the way the battery is operated. In a new battery there is always a thin layer at the start. As the battery wears, the sediment becomes higher, but for batteries which are floated, the plates usually wear out before the sediment space is filled.

**178. Putting battery into storage**

1. If the use of the battery is to be temporarily discontinued, give it a charge until all the cells gas and add water to the cells during this charge so that the gassing will ensure thorough mixing and prevent freezing in cold weather. Add enough water to raise the level of the electrolyte to the proper level. After the charge has been completed, remove all fuses to prevent the use of the battery during the idle period. Make sure all vent plugs are in place.

2. At certain periods the battery should be reconnected, water added, and the battery charged. These periods are every 2 months in climates averaging 70 to 80 °F (21 to 27°C) and every 6 months in climates averaging 40°F (4.4°C).

**179. Putting battery into commission again.** Add water, if needed, and give a charge until the gravity of the electrolyte has ceased rising over a period of 3 h.

### NICKEL-IRON-ALKALINE BATTERIES

**180. The Edison storage battery** is the result of an effort to avoid many of the disadvantages of the lead–sulfuric acid combination and is a radical departure from it in every detail of construction. The positive plate consists of hollow, perforated, sheet-steel tubes filled with alternate layers of nickel hydrate and metallic nickel. The hydrate is the active material, and the metal, which is made in the form of miscroscopically thin flakes, is added to provide good conductivity between the walls of the tube and the remotest active material. The negative plate is made up of perforated, flat, sheet-steel boxes or pockets loaded with iron oxide and a small amount of mercury oxide, the latter also for the sake of conductivity.The grids which support these tubes and pockets are punchings of sheet steel. The cell terminals and container are likewise of steel, and all metallic parts are heavily nickel-plated. The electrolyte is a 21 percent solution of caustic potash containing also a small amount of lithium hydrate. All separators and insulating parts are made of rubber. The details of the construction of the Edison storage battery are shown in Fig. 4-197.

The current used in charging causes an oxidation of the positive plate and a reduction of the negative, and these operations on discharge are reversed. The electrolyte acts merely as a medium and does not enter into combination with any of the active material as it does in the acid battery. Its specific gravity remains practically constant throughout the complete cycle of charge and discharge. The charge and discharge curves are shown in Fig. 4-198.

The chief characteristics of the battery are ruggedness, due to its solid, steel construction; low weight, due to its stronger and lighter supporting metal; long life, due to the complete reversibility of the chemical reactions and the absence of shedding active material; and low cost of maintenance, due to freedom from the diseases, such as sulfation, so commonly met with in storage-battery practice and from the necessity of internal cleaning and plate renewals. The arguments against it are high first cost and high internal resistance. The importance of these must, of course, be weighed with the advantages and the resultant considered in each proposed installation. The battery has attained its chief prominence in vehicle propulsion, but its characteristics also recommend it for many other purposes.

**181. Voltage characteristics.** The voltage of each cell is approximately 1.5 V on open circuit, but it is higher than this when the battery is being charged and lower when being discharged. The voltage at any time depends upon state of charge or discharge, the temperature, and the density of the electrolyte. The average discharge voltage is approximately 1.2 V per cell. Typical voltage charge and discharge curves are shown in Fig. 4-198. Cells are generally discharged until the voltage drops from 1.0 to 0.9 V per cell. Further discharge is generally not satisfactory, since the voltage drops very rapidly if the discharge is continued past this point. The maximum voltage during charge will be

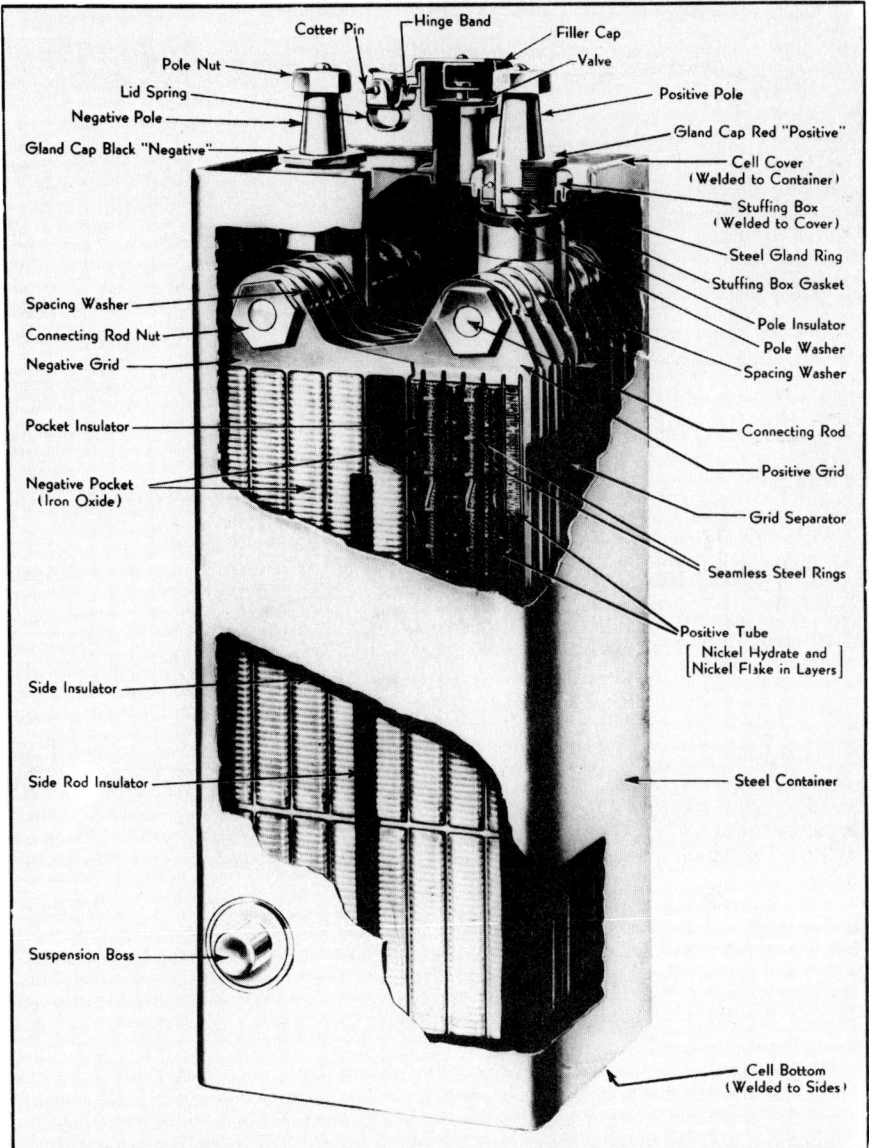

**Fig. 4-197**  Edison nickel-iron-alkaline storage cell with container cut away to show construction detail. [Electric Storage Battery Co.]

between 1.80 and 1.90 V per cell. The voltage for any degree of discharge is affected slightly by the rate of discharge, as shown in Fig. 4-199.

**182. Rating of nickel-iron-alkaline batteries.**  All batteries are given a normal ampere-hour capacity rating based on a certain rate of discharge to a final voltage of 1.0 V per cell. Some current ratings are based on a 5-h continuous discharge rate, and others on a 3⅓-h continuous rate.

The ampere-hour capacity for discharging continuously at a constant rate to a final voltage of 1.0 V per cell will be affected by the rate of discharge (Fig. 4-199). The higher the rate of discharge, the lower the capacity of the battery. The effect upon the capacity will

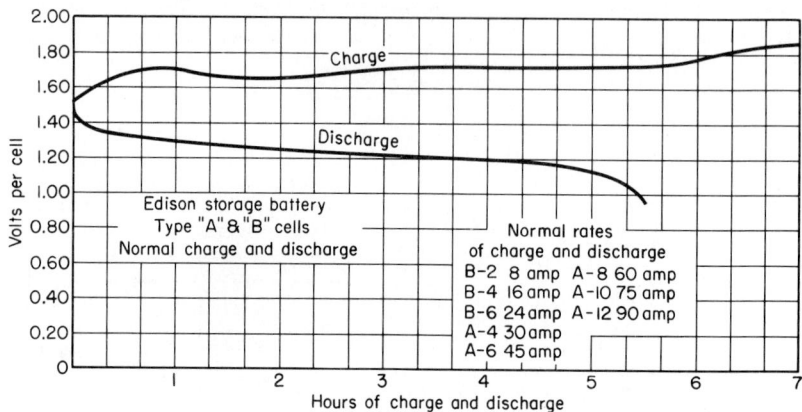

**Fig. 4-198**    Charge and discharge curves of Edison battery.

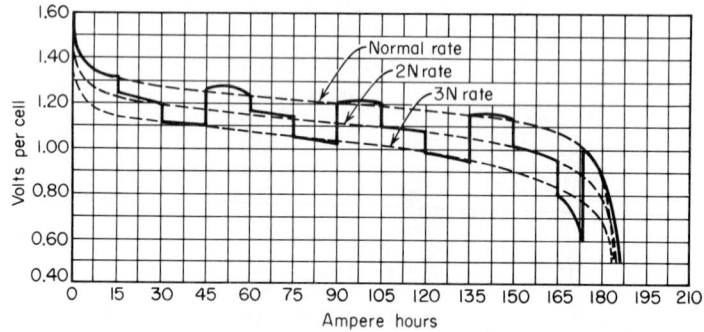

**Fig. 4-199**    Discharging at high rates: a four-cell battery, normal rate 30 A, discharged successfully at 30, 60, and 90 A, with comparison of continuous discharge at each of these rates. [Electric Storage Battery Co.]

not be so great, however, as it is for lead-acid batteries. After a high rate of discharge, the balance of the normal capacity of the battery can be obtained by continuing the discharge at a lower rate. If the battery is discharged intermittently, the normal capacity can be obtained unless the total elapsed time of discharge is excessively short. These batteries are very rugged and withstand very severe service. They are not harmed by occasional short-circuit discharges.

**183. Charging.**    The best method of charging Edison batteries is by the average-constant-current method. The battery is connected in series with an adjustable resistance to a constant-potential dc supply, as shown in Fig. 4-200. The positive terminal of the battery should be connected to the positive terminal of the supply. The maximum voltage available at the batteries should be at least 1.85 times the number of cells in series. Throughout the charging period the rheostat should be periodically adjusted so that the average current will be maintained at its normal rated value. At each adjustment the current should be set a few amperes above the rated value so that by the time the next adjustment is made the current will not have dropped much below normal. The battery should be charged until a maximum voltage has been reached and maintained for 30 min. If an ampere-hour meter is used with the battery in charge and discharge, satisfactory charge is generally obtained by charging the battery at its rated current until the ampere-hours of charge is 125 percent of the ampere-hours of discharge. The condition of charge cannot be judged by means of specific-gravity readings. Automatic equipment for the control of the current during charge is being used to a greater extent.

Before starting to charge, see that the solution is at the proper level. If the solution is low, bring it to the proper level by adding pure distilled water as instructed in Sec. 169.

If the battery is in a compartment, open the covers before starting a charge. If necessary, and if full capacity is not required, a battery can be taken off charge at any time and used.

**184. Effect of temperature.** Do not charge in a hot place or allow the temperature of the solution to exceed 115°F (46°C) on charge. High temperatures during charge or discharge will shorten the life of any kind of battery. Better efficiency is obtained if cells are charged at a temperature of 80 to 90°F (27 to 32°C). If the temperature of the solution exceeds 115°F while charging, allow the cells to cool before continuing the charge.

**185. Charging at high rates.** In an emergency, when time for a normal charge is not available, charging can be done at any higher rates than normal, provided there is no frothing and the temperature does not rise above 115°F (46°C).

**186. Charging at low rates.** If the discharge requirements are such that a low constant rate is used or if there is an intermittent rate of such value that for a given time period a low average rate will be had, then a charge rate of less than normal can be used, provided that this charge rate is approximately 120 percent of the constant or average discharge rate. The term *low discharge rate* is to be construed to mean a constant or average rate of less than 80 percent of normal.

When charging at low rates, it must be thoroughly understood that the required ampere-hour input must be put in and that, therefore, the time periods of charge must be correspondingly increased over that necessary to charge a cell fully at normal rate.

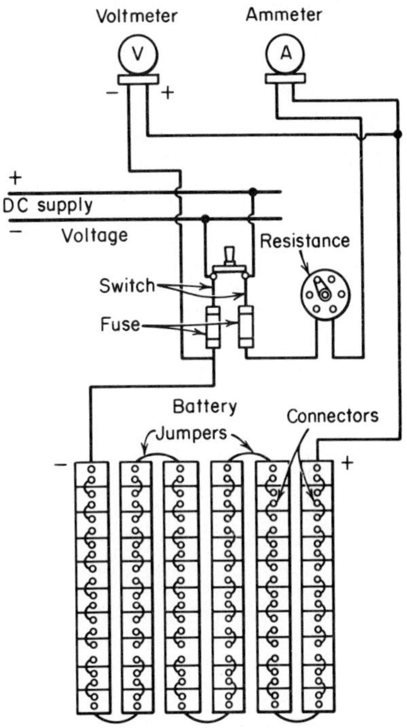

**187. Boosting charge.** An Edison battery can

**Fig. 4-200** Correct connections for charging a battery at constant current through a variable resistor. [Electric Storage Battery Co.]

be boosted, i.e., given a supplementary charge, at high rates during brief periods of idleness, thereby materially adding to the available capacity. The principal limiting feature is that the temperature of the solution in the cells nearest the center or the warmest part of the battery does not exceed 115°F (46°C). A battery can be boosted whether it is entirely discharged or only partially discharged. The object of the boost is to supplement the remaining charge so that additional work can be done without waiting for regular charge.

The following table gives figures that can be used under average conditions, but values that will not cause excessive heating must be determined in each case by experience.

5 min at 5 times normal
15 min at 4 times normal
30 min at 3 times normal
60 min at 2 times normal

Frothing at the filler opening is an indication that the boosting has been carried too far (if the solution is at the proper height), and the boosting should be discontinued at once.

**188. Overcharging** is to be interpreted as charging a battery at the normal rate for periods in excess of the specified hours normal charge. For A, B, C, and N types, 12 h is considered an overcharge when the previous discharge has been taken only to an average of approximately 0.5 V per cell, and 15 h is considered an overcharge when the previous discharge has been taken to zero voltage and short-circuited. For G and L types these overcharge periods are, respectively, 8 and 10 h.

Overcharging in conjunction with proper discharges is used to compensate for either lack of work of battery or change of solution. In cases in which batteries have become sluggish owing to lack of work, as when a battery is seldom totally discharged in regular service, the battery should be periodically completely discharged to zero at normal rate and then short-circuited for 1 or 2 h. Follow with a regular overcharge.

With new Edison batteries better capacities will result if they are given plenty of work. It is, therefore, advisable to give new batteries additional work every 2 weeks for the first 2 months and every 2 months thereafter for 6 months. This should consist of a complete discharge to zero at normal rate with a short circuit of at least 2 h followed by an overcharge.

When an Edison battery does not give satisfactory capacity on discharge at rates several times normal, it is considered sluggish. This sluggishness or low capacity may result from persistent low-rate discharging, frequent low-rate charging, long stands, seldom discharging completely, or a weak solution. With the exception of the last named the primary cause is lack of work. Edison batteries thrive on work; therefore, the proper procedure is to discharge the battery completely at normal rate to zero and then short-circuit it for 2 h. Follow this by a overcharge. If the condition is rather pronounced, the cycle should be repeated. In the worst cases results can usually be obtained by several repetitions of the above. Ordinarily, this method will restore underworked batteries.

Completely discharge batteries before starting overcharges. Test for height of solution, and bring the solution to the proper height. Tests of solution height should be made before and after the battery has been completely discharged at normal rate, followed by a short circuit of at least 2 h and then an overcharge.

**189.  Nickel-Iron-Alkaline Batteries**
(Electric Storage Battery Co.)

| Cell type | Rating[a] | | Weight,[b] lb per cell | |
|---|---|---|---|---|
| | Ampere-hour capacity | Normal rate, amp | Standard | High type |
| N2 | 11¼ | 2¼ | 1.94 | |
| L20 | 12½ | 3¾ | 1.88 | |
| L30 | 18¾ | 5⅝ | 2.64 | |
| L40 | 25 | 7½ | 3.24 | |
| B1, B1 *H*[c] | 18¾ | 3¾ | 5.3 | 6.7 |
| B2, B2 *H*[c] | 37½ | 7½ | 6.0 | 7.2 |
| B4, B4 *H*[c] | 75 | 15 | 9.5 | 10.9 |
| B6, B6 *H*[c] | 112½ | 22½ | 13.0 | 15.2 |
| A4, A4 *H*[c] | 150 | 30 | 16.5 | 19.3 |
| A5, A5 *H*[c] | 188 | 37½ | 19.6 | 22.5 |
| A6, A6 *H*[c] | 225 | 45 | 22.4 | 25.5 |
| A7, A7 *H*[c] | 263 | 52½ | 25.8 | 28.6 |
| A8, A8 *H*[c] | 300 | 60 | 31.2 | 35.9 |
| A10, A10 *H*[c] | 375 | 75 | 38.1 | 43.8 |
| A12, A12 *H*[c] | 450 | 90 | 47.3 | 53.5 |
| A14, A14 *H*[c] | 525 | 105 | 56.6 | 60.8 |
| A16, A16 *H*[c] | 600 | 120 | 62.9 | 68.0 |
| A20 *H*[c] | 750 | 150 | ..... | 84.8 |
| A24 *H*[c] | 900 | 180 | ..... | 102.5 |
| G4, G4 *H*[c] | 100 | 30 | 12.5 | 19.0 |
| G6, G6 *H*[c] | 150 | 45 | 17.5 | 21.1 |
| G7, G7 *H*[c] | 175 | 52½ | 20.8 | 23.0 |
| G9, G9 *H*[c] | 225 | 67½ | 24.0 | 27.5 |
| G11, G11 *H* | 275 | 82½ | 30.4 | 36.2 |
| G14, G14 *H* | 350 | 105 | 37.2 | 44.5 |
| G18, G18 *H* | 450 | 135 | 52.4 | 56.8 |
| G22 *H*[c] | 550 | 165 | ..... | 70.0 |
| C4 | 225 | 45 | 24.3 | |
| C5 | 281 | 56¼ | 29.3 | |
| C6 | 338 | 67½ | 34.4 | |
| C7 | 394 | 78¾ | 39.8 | |
| C8 | 450 | 90 | 45.5 | |
| C10 | 563 | 112½ | 61.3 | |
| C12 | 675 | 135 | 71.0 | |
| D6 | 450 | 90 | 45.6 | |
| D8 | 600 | 120 | 60.0 | |
| D10 | 750 | 150 | 77.3 | |
| D12 | 900 | 180 | 91.6 | |

[a] Ratings are on basis of 5-hr rate for A, B, C, and N Type cells and 3⅓-hr rate for G and L Type, with average of 1.2 volts per cell and final of 1.0 volt per cell.

[b] Weights are for completely assembled cells, including trays, connectors, etc.

[c] The letter *H* indicates high-type cells; these cells have the same characteristics as the standard-type cells but are built higher so as to hold more electrolyte and are used in installations where frequent flushing is not convenient.

**190. Maintenance of nickel-iron-alkaline batteries.**   The attention required by this battery is of the simplest character. It is chiefly important that the electrolyte be replenished from time to time with distilled water so that the plates will be entirely immersed and the outside of the cells be kept clean and dry, for if this is not done, leakage of current will occur with consequent corrosion of containers by electrolysis. Perhaps once or twice during the total useful life of the cell, the electrolyte may need renewal.

**191. Cleaning.**   The cells, trays, and battery compartment must be kept dry, and care must be taken that dirt and other foreign substances do not collect at the bottom or between the cells.

Dirt and dampness are likely to cause current leakage, which may result in serious injury to the cells.

If protection of cell tops from moisture is required, they should be given a light coat of rosin or liquid Esbaline, this material being applied to the cover of the cell and sparingly to the outside of the filling aperture, care being taken not to get any great quantity on the lid hinge. Esbaline can be applied best with a small paintbrush, care being taken not to get any on the inside of lugs or on cell poles.

Rosin Esbaline must be applied warmed to approximately 170 to 190°F (77 to 88°C) and thinned to good paint consistency with benzine, etc.

A wet-steam jet or even an air blast will be found most satisfactory for cleaning but must not be used on cells while they are in the compartments. It has been found that a pressure of 70 lb with a 1-in rubber steam hose about 10 ft long into which has been inserted a piece of iron pipe about 12 in long with an orifice ⅛ in in diameter will give wet steam with a velocity to clean the battery satisfactorily. (This orifice can be made by plugging one end of an iron pipe and drilling out with a ⅛-in drill.) When removing encrustations from the tops of cells, do not allow them to fall between or into the cells. Before reassembling, make sure that all poles, connectors, and jumper lugs are clean. Also cells, trays, and compartments must be dry before the battery is replaced.

Occasionally, cells and trays after being cleaned should be recoated with Esbalite, an alkaliproof insulating paint put out by the Electric Storage Battery Co.

The cells should be thoroughly cleaned of all grease, dirt, dried salts, and paint blisters or flakes and be perfectly free from all moisture. Painting may be done with a brush or, if the quantity is large, by dripping.

When cleaning and recoating with any cell coating be careful not to allow any of the materials to get into the cells.

**192. Water or flushing.** Do not allow the level of the solution to drop below the tops of the plates. Never fill higher than the proper level. If filled too high, solution will be forced out during charge.

For replenishing solution in Edison cells during operation use only pure distilled water or water which has been tested and approved by the Electric Storage Battery Co. Although pure distilled water is recommended for use in storage batteries generally, there are certain places in the United States where the local water supply is of such purity that it can be satisfactorily used. It is extremely important that no other water than pure distilled water be used unless it has received the approval of the Electric Storage Battery Co. after test at the factory's laboratories. The use of impure water will result in a slow poisoning of the electrolyte by an accumulation of impurities, the effect of which may not appear within a few months but will become apparent within the course of several years.

When solution has been spilled, use standard refill solution, which has a specific gravity of approximately 1.215 at 60°F (15.5°C).

Battery compartments and trays must be kept dry and clean at all times; so take care when filling cells not to spill water over and around the cells and not to exceed specified height.

Test for height of solution before placing battery on charge. Do not test for solution height while battery is charging; the gassing during charge creates a false level.

A reasonably heavy-walled glass tube about 8 in (203 mm) long and of not less than ³⁄₁₆ in (4.8 mm) inside diameter with ends cut straight and fused enough to round the edges can be used as illustrated in Fig. 4-201 to find the level of electrolyte above plate tops. A short length of tightly fitting rubber tube forced over one end and projecting about ⅛ in (3.2 mm) will prove a very good finger grip. Insert the tube until the tops of the plates are touched; close the upper end with the finger and withdraw the tube. The height of the liquid in the tube should be as specified by the manufacturer.

**193. Specific gravity of nickel-iron-alkaline batteries.** The density, or specific-gravity, reading of the electrolyte of an Edison cell has no value in determining the state of charge or discharge, as the specific gravity does not change during the charging or discharging of the cell to any marked extent. The small changes ordinarily observed are due either to large changes in temperature or to loss of water from the electrolyte by evaporation or electrolysis in operating the cell. Therefore it is not necessary to take frequent readings to determine the specific gravity. The only time it is necessary to obtain a specific-gravity reading is to determine when a change of electrolyte would be advantageous. When making such readings, certain fundamental conditions must be observed. A suitable hydrometer must be obtained and used in accordance with the rules laid down in Sec. 172 on hydrometer readings. The glass container must be clean and must not contain acid or other impurities, as these tend to give lower readings than the true specific gravity.

Do not take a specific-gravity reading when the cell is charging, as the bubbles of gas contained in the electrolyte will cause a lower reading than the true one. Readings taken when the temperature of the electrolyte is either very high or very low will give results that will vary with the temperature. The specific gravities quoted in this section are for 60°F (15.5°C). Temperatures very much higher than this will give lower gravity readings, while temperatures very much lower than this will give higher readings.

Specific-gravity readings taken immediately after watering the cell are of no value, as the water has had no chance to be thoroughly mixed with the electrolyte and the resulting readings will be low. The specific gravity should be taken when the electrolyte is at the proper height after a complete charge. It is best to allow the cells to stand for a short period after the completion of the charge to permit free bubbles of gas to dissipate before taking readings. Corrections for temperature should be made by adding 0.0025 for each 10° above 60°F (each 5.5° above 15.5°C) to the observed reading or subtracting 0.0025 for every 10° below 60°F.

**Fig. 4-201**  Quick method of determining height of solution. [Electric Storage Battery Co.]

After taking a specific-gravity reading, return the solution to the same cell from which it was taken. Otherwise the gravity of the solution in the cell to which it is added will be increased and the gravity of the solution in the cell from which it was taken will be decreased owing to the addition of water being made necessary.

The potash electrolyte in Edison cells has a normal specific gravity of approximately 1.200 at 60°F when at the normal level and thoroughly mixed by charging and when a sample is taken at least ½ h after charge to allow for dissipation of gas.

**194. Solution renewal.**  Throughout the total useful life of the cell the electrolyte gradually weakens and may need renewal once or twice, depending on the severity of service, and in some cases, when maintenance and operation have been poor or when contamination has been allowed by the use of impure water, etc., a third time might be necessary.

The low-limit specific gravity beyond which it is inadvisable to run an electrolyte is 1.160. Operation at lower specific gravity than 1.160 should not be allowed, since such operation will produce sluggishness, loss of capacity, and rapid breakdown on severe service. Should the specific gravity be above 1.160 and at the same time sluggishness and loss of capacity be evident, do not immediately renew the electrolyte until it is found that an electrolyte sample sent to the factory shows a prohibitive accumulation of impurities and until the following treatment fails to produce marked improvement:

1. Discharge at normal rate to zero voltage. (When the current can no longer be kept up, either reverse the battery on the line with sufficient resistance in series or connect in series with another more nearly charged battery; then continue the discharge.) It is of prime importance that the rate be kept at normal throughout.

2. Short-circuit the battery in groups of not more than about five cells each for at least 2 h.

3. Charge at normal rate for 15 h for A, B, C, and N types and 10 h for G and L types.

4. Discharge at normal rate to approximately 1.0 to 0.9 V per cell.

5. Charge at normal rate for 7 h for A, B, C, and N types and 4¾ h for G and L types.

6. Discharge at normal rate to approximately 1.0 to 0.9 V per cell.

If the cells do not respond noticeably to this treatment, there is probably very marked contamination of the electrolyte, and the result of the analysis of a representative sample sent to the manufacturer's factory in accordance with instructions will show this. Therefore, regardless of the specific gravity, the electrolyte should be replaced as follows:

1. When previous electrolyte has reached approximately the low limit of 1.160, the new solution should be standard renewal.

2. When previous electrolyte is 1.190 or above, the new solution should be standard refill.

It is always advisable, in case the battery exhibits trouble of any sort, to communicate all details to the manufacturers or their representative, so that immediate advice may be obtained.

Do not use any other solution than Edison electrolyte. Do not pour out old solution until you have received new solution and are ready to use it. Never allow cells to stand empty. State type and number of cells when ordering Edison electrolyte for renewal.

When ready to renew solution, first completely discharge the battery at normal rate to zero and short-circuit for 2 h or more. This is done to protect the elements. Then empty cells completely. It is not necessary to shake or rinse cells, and under no circumstances should cells be filled with water.

Immediately after emptying each cell, pour in new solution. Do not allow to stand empty. Fill to exactly the proper height. For this purpose use a clean glass or enamelware funnel. A plain iron funnel can be used if it has no soldered seams, but do not by any means use one of tinned or galvanized iron. A clean rubber tube can be used to siphon the solution directly from the container to the cell. If the tube is new, it should be thoroughly soaked in electrolyte for a couple of hours in such a position as to retain the solution. This is to remove thoroughly any impurities on the rubber. Fill to exactly the proper height, for if cells are filled too full when renewing the solution and allowed to remain that way, the specific gravity of the electrolyte will be too high when the level of the solution returns to proper height. This condition may lead to serious results and can easily be avoided by reasonable care.

Do not attempt to put in all solution received, as an excess is allowed to make up for any loss due to spilling. It may be necessary to add some more electrolyte after cells have stood a little time, as some electrolyte may be absorbed by the plates.

The specific gravity of the Edison electrolyte for renewal as shipped is about 1.250, but this will quickly fall to normal when put into a battery, owing to mixture with the old, weak solution remaining in the plates.

Do not attempt to use the electric filler for refilling cells. It was not designed for this purpose and will not work.

When the new electrolyte is in and the battery is again connected for service, give it an overcharge at the normal rate as outlined under Sec. 188, "Overcharging."

### 195. Cautions in operation of nickel-iron-alkaline batteries

1. Never put lead battery acid into an Edison battery or use utensils that have been used with acid; you may ruin the battery.

2. Never bring a lighted match or other open flame near a battery.

3. Never lay a tool or any piece of metal on a battery.

4. Always keep the filter caps closed except when necessary to have them open for filling.

5. Keep batteries clean and dry externally.

6. Edison electrolyte is injurious to the skin or clothing and must be handled carefully. Solution spilled on the person should be immediately washed away with plenty of water.

### 196. Laying up nickel-iron-alkaline batteries.

If the battery is to be laid up for any length of time, be sure that the plates are covered to the proper height by solution or electrolyte. The battery should be stored in a dry place. Do not leave it in a damp place, as damage to the containers may result from electrolysis. Never let the battery stand unfilled.

Edison batteries are easy to lay up. Merely discharge to zero voltage and short-circuit. They can be left standing idle indefinitely in this condition without injury.

New Edison cells have received sufficient cycles of charge and discharge before shipment to give considerable capacity above that rated. However, Edison cells will increase still further in capacity if thoroughly worked in. Therefore it is best, if new cells are to stand for some time before being put into commission, to discharge them to zero at normal rate and short-circuit at least 6 h.

When putting the battery back into commission, go over each cell and see that all poles and connections are in good condition as for a new cell. See that the plates are properly covered with electrolyte and then charged as here instructed. First, if not already discharged, discharge cells to zero at normal rate and short-circuit. Follow this by an overcharge at normal rate, and then discharge at normal rate. Then charge at normal rate for normal hours of charge. If the battery shows signs of sluggishness, repeat the overcharge and carry the discharge down to zero until cells are fully active; then give regular charge.

## NICKEL-CADMIUM BATTERIES

**197. Nickel-cadmium batteries** are a relatively new addition in the United States to the storage-battery family. These batteries consist of an interleaved assembly of positive and negative plates (Fig. 4-202).

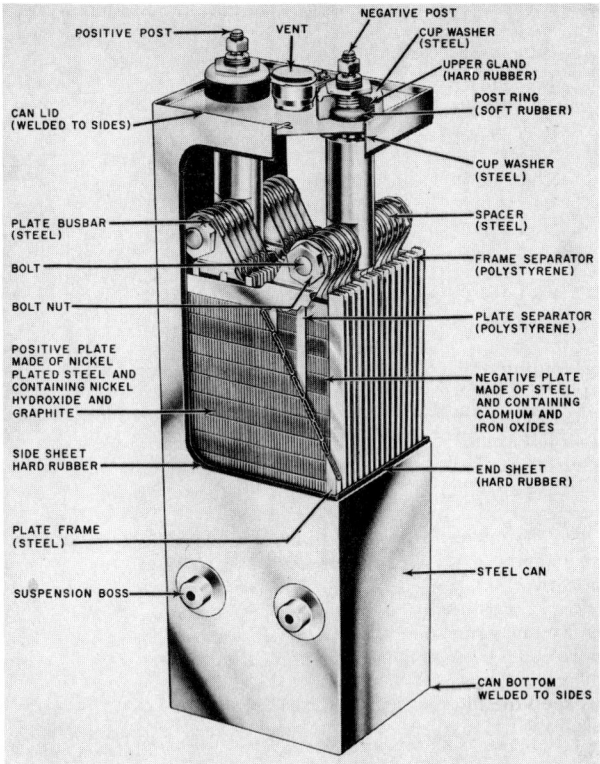

**Fig. 4-202**   Cutaway view of a nickel-cadmium battery. [Nicad Division of Gould–National Batteries, Inc.]

Cell containers are made of nickel-plated steel. Active materials, nickel hydroxide (positive) and cadmium oxide (negative), are encased in finely perforated steel pockets. Plates consist of rows of pockets crimped together and locked into steel frames under many tons of pressure.

Positive and negative plates are welded or bolted to heavy steel busbars. Plate groups are interleaved and are separated by thin plastic rods.

The alkaline electrolyte is a solution of potassium hydroxide.

The active materials are converted on charging and discharging in accordance with the chemical reaction, which may be written as follows:

$$2Ni(OH)_3 + CD \rightleftharpoons 2Ni(OH)_2 + Cd(OH)_2$$

On discharge of the battery the reaction proceeds in the forward direction, while on charge the reaction is reversed. The system stores chemical energy when the battery is charged, and this chemical energy is converted back to electrical energy on discharge. Reversibility of this reaction under extreme environmental conditions is one of the outstanding properties of the nickel-cadmium electrochemical system.

During charge or discharge of a nickel-cadmium cell there is practically no change in the specific gravity of the electrolyte. The sole function of the electrolyte is to act as a conductor for the transfer of hydroxide ions from one electrode to the other, depending on whether the cell is being charged or discharged.

### 198.  Data for Nickel-Cadmium Batteries
(Nicad Division of Gould–National Batteries, Inc.)

| Cell type No. | Amp-hr capacity at 8-hr rate | Width of tray, in. over all | Height of tray, in. over all | Length of tray containing, in. over all | | | | Weight per cell, lb | |
|---|---|---|---|---|---|---|---|---|---|
| | | | | 2 cells | 3 cells | 4 cells | 5 cells | Net | Ship. |
| EBZ17 | 80 | 6 | 16 | 13¼ | 20⅛ | 25½ | 32¼ | 21.0 | 24.0 |
| EBZ19 | 90 | 6½ | 16 | 13¼ | 20⅛ | 25½ | 32¼ | 23.2 | 26.3 |
| EBZ22 | 105 | 7¼ | 16 | 13¼ | 20⅛ | 25½ | 32¼ | 26.5 | 30.7 |
| EBZ25 | 120 | 8 | 16 | 13¼ | 20⅛ | 25½ | 32¼ | 29.7 | 33.5 |
| EBZ28 | 135 | 8¾ | 16 | 13¼ | 20⅛ | 25½ | 32¼ | 32.8 | 37.4 |
| EBZ31 | 150 | 9½ | 16 | 13¼ | 20⅛ | 25½ | 32¼ | 36.1 | 41.7 |
| ERX23 | 165 | 7½ | 18 | 15¼ | 23⅛ | 29⅝ | 37¼ | 36.1 | 41.7 |
| ERX25 | 180 | 8 | 18 | 15¼ | 23⅛ | 29⅝ | 37¼ | 38.7 | 45.6 |
| ERX27 | 195 | 8½ | 18 | 15¼ | 23⅛ | 29⅝ | 37¼ | 41.3 | 49.2 |
| ERX29 | 210 | 9 | 18 | 15¼ | 23⅛ | 29⅝ | 37¼ | 43.8 | 52.1 |
| ERX33 | 240 | 10 | 18 | 15¼ | 23⅛ | 29⅝ | 37¼ | 49.0 | 57.8 |
| ERX37 | 270 | 11 | 18 | 15¼ | 23⅛ | 29⅝ | 37¼ | 54.0 | 63.8 |

### 199. Characteristics
*Nominal voltage per cell.* 1.2 V (a 6-V battery consists of 5 cells).
*Temperature range.* −60 to +200°F (−51 to +93°C).
*Maximum discharge current.* Up to 25 times rated ampere-hour capacity.
*Capacity at −60°F (−51°C).* Up to 90 percent at low rates.
*Internal resistance very low.* 0.001 Ω for 10-Ah–type cell.
*Full charge.* By constant potential in 1 h at 1.55 V per cell.
*Full charge.* By constant potential in 6 h at 1.43 V per cell.
*Charge, trickle.* 1.35 V per cell to maintain charged battery.
*Gassing discharge.* None.
*Gassing charge.* Virtually none below 1.47 V per cell.
*Vibration resistance.* Excellent.
*Shock.* 80 g.
*Altitude.* Pressure-release valve opens 25 psi (172,369 Pa) above ambient.
*Cycle life.* No known limit.
*Storage life.* No known limit in any state of charge.
*Charge retention.* Up to 70 percent after 1 year at room temperature.
*Orientation.* Any position on discharge.

**200. Advantages of nickel-cadmium batteries** as stated by Nicad Division of Gould–National Batteries, Inc.:

*High surge currents.* Good voltage maintenance under extremely high current discharge conditions.

*Constant-voltage source.* Close voltage regulation during discharge (see Figs. 4-203 and 4-204).

*Rapid-charge acceptance.* Can be completely recharged at high rates without damage.

*Long life.* Designed to give exceptionally long life under cycle and float service.

*Extreme-temperature operation.* Normal and high-rate discharge and charge possible at temperatures from −40 to +165°F (−40 to +74°C).

*Excellent charge retention.* Charged cells filled with electrolyte will retain approximately 70 percent of their charge after 1 year of idle storage at normal temperature.

*Storageability.* Can be laid up in any state of charge for long periods without attention or fear of deterioration.

*Discharge in any position.* Can be discharged in any position.

*Easy maintenance.* Negligible loss of water during service. Records of specific gravity are not necessary.

*Vibration and shock resistance.* Capable of withstanding up to 50 *g* shock and severe vibration.

*Alkaline electrolyte.* The potassium hydroxide electrolyte does not give off corrosive fumes on charge or discharge.

*Pressure sealing.* The smaller cells are normally provided with pressure-sensitive sealed vents.

*Economy.* Low cost per year due to long life. Very low maintenance costs. Low-cost, high-rate performance.

*Small size and light weight.* Nicad sintered plate batteries are smaller and lighter than conventional batteries under high-current-drain conditions.

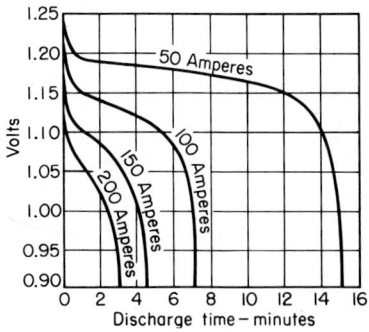

**Fig. 4-203**  Typical discharge characteristics for a high-rate nickel-cadmium cell. [Nicad Division of Gould–National Batteries, Inc.]

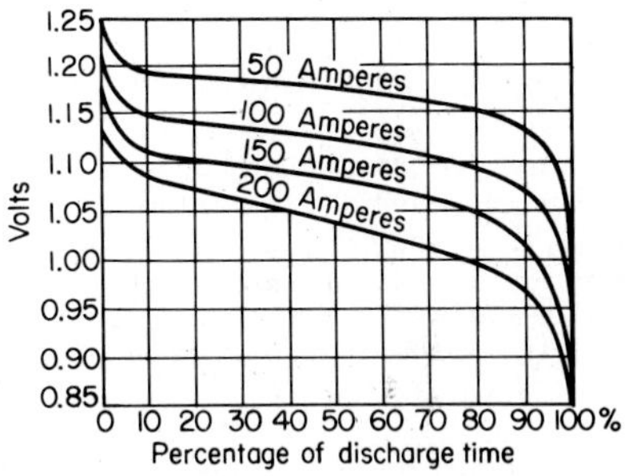

**Fig. 4-204**  Discharge-voltage characteristics of a nickel-cadmium cell. [Nicad Division of Gould–National Batteries, Inc.]

**201. Installation and care of nickel-cadmium batteries.**  The following instructions for the installation and care of nickel-cadmium batteries are those recommended by the Nicad Division of the Gould–National Batteries, Inc.

**202. Important precautions**

1.  Maintain the battery compartment, cells, and trays in a clean and dry condition. Dirt and moisture will cause self-discharge, corrosion, and eventual leakage of cell containers.

2.  Check the electrolyte level *before* adding distilled water. Maintain correct electrolyte level, but do not add water in excess of the recommended maximum level. Do not spill water or electrolyte on the cells or trays.

3.  The electrolyte is an alkaline solution of caustic potash–*not* sulfuric acid, as used in lead batteries.

4.  Sulfuric acid or traces thereof will rapidly ruin the Nicad battery by corroding its steel plates and cell containers. Therefore, use only the hydrometer and level test tube furnished with the battery.

5.  Do not discharge at normal rates below 1.10 V per cell. Repeated overdischarges will damage the battery.

6.  Always keep the vent caps closed except when checking the electrolyte or adding water.

7.  Never examine the cells with an open flame. Keep tools and other metal objects away from the battery.

8.  Apply boric acid solution if electrolyte is splashed on person or clothing.

**203. Installing the battery.**  Place the battery in a clean, dry room located so that it can be easily inspected and watered. On mobile equipment such as vehicles or ships it must be securely tied down.

Avoid placing the battery in a hot location (above 125°F, or 52°C) or where it will be exposed to corrosive gases or fumes.

It is not good practice to install Nicad batteries and lead-acid batteries in the same room unless there is ample ventilation to carry away fumes from the lead-acid batteries.

All battery rooms and compartments must have good ventilation and drainage, and at the same time cinders, road dirt, soot, dust, rain, snow, and seawater must be kept out.

Accumulations of dirt and moisture on cell tops and particularly between the cells will cause corrosion and leakage of cell containers.

Holes or gratings which permit ready entry of dust, water, etc., must be closed up. However, four small drainage holes (maximum diameter of holes $\frac{1}{4}$ in, or 6.35 mm) should be provided in the bottom, one at each corner of the compartment.

Compartments which have previously housed lead-acid batteries must be washed out, neutralized with ammonia or washing-soda solution, allowed to dry thoroughly, and then painted with Nicadvar asphalt paint. Wood liners must be removed and replaced by new ones. Ample space should be provided above the battery for inserting the electrolyte-level test tube and the hydrometer into the cells.

If the battery is to be serviced through the top of a compartment, allow a minimum clearance of 2 in (50.8 mm) between the top of the battery and the underside of the access cover.

If the battery is to be serviced from the side, the clearance between the battery and the ceiling of the compartment should be at least 8 in (203.2 mm).

Small stationary batteries can be placed directly on a clean, dry floor or on a suitable wall shelf. Large batteries should be placed on racks.

Check the electrolyte level and the specific gravity and be sure that the battery is thoroughly clean and dry before placing it in position.

Nothing should be placed or allowed to lodge in the open spaces between or underneath the cells. These air gaps serve as electrical insulation between the cells and are an essential part of the battery. For this reason they must be maintained open, clean, and unobstructed at all times.

Variation of specific gravity may be found between individual cells. A maximum variation of plus or minus 0.005 point is permissible. For example, if normal specific gravity called for is 1.210, a minimum of 1.205 and a maximum of 1.215 specific gravity is satisfactory.

The cell containers must not be grounded, and the bottoms of the cells must not be allowed to rest on any object. Battery trays should never be stacked directly on top of one another, nor should water or spilled electrolyte be allowed to accumulate under the battery.

All cables leading to the battery posts should be fitted with nickel-plated cable lugs. Do not use bare copper cable lugs or connectors. After connecting cable lugs to battery, cover lugs with Nicad petroleum jelly No. 32982.

All wiring to the battery should be properly spaced and firmly secured to prevent any chance of a short circuit.

Wires or cables should never be allowed to rest on top of the cells.

Never connect to the battery a device or instrument that might cause an unnecessary constant drain, however small, as it will ultimately discharge the battery when left standing on open circuit. Voltmeters, for instance, should be connected to the battery only by means of a normally open pushbutton switch.

If either side of a high-voltage battery is grounded, it will expose personnel and battery to hazardous conditions. The shock hazard to personnel is obvious. In such cases additional high-resistance grounds in any part of the circuit will cause trouble and possibly a serious short circuit. For example, an inductive surge from breaker operations can cause a slight ground to become larger and rapidly discharge the battery. When checking battery voltage, also take readings between each terminal of the battery and ground for possible leakage, as ground indicator lights may not show slight grounds.

Most storage batteries consist of a single series of cells of the same type and ampere-hour capacity. If the battery consists of more than one tray, be sure that the negative end terminal of each tray is connected to the positive end terminal of the following tray. Cells wrongly connected will receive a reverse charge and will be damaged if the condition is allowed to continue.

On completing the installation, make sure that no loose objects such as screws or tools have been accidentally left in the battery compartment.

Check and tighten all cell-post nuts, as loose electrical connections will heat up and can cause sparking.

Verify that every person who is going to take care of the battery has a copy of these instructions.

Mount the instruction card accompanying the battery in a conspicuous position for future reference.

**204. Charging.**   The positive terminal post of every cell is identified by a plus mark on the cell container. When charging, always connect the positive terminal of the battery to the positive lead from the charger. Only direct current can be used for charging storage batteries. If only alternating current is available, a rectifier or a motor generator is necessary to convert the alternating into direct current. Information concerning various types of charging equipment is available from manufacturers.

When a battery is first placed in service, the charging rate may vary somewhat before it is stabilized. During the first week or two the adjustment should be checked every few days. A reasonable amount of overcharging, particularly at low- or trickle-charge rates, has a beneficial effect on Nicad batteries. When in doubt at any time as to the state of charge, it is advisable to overcharge rather than to undercharge.

Variations in line voltage due to local conditions can be large enough to throw charge voltages off normal. A normal setting during the day may rise to a point of overcharge and excessive charging at night and on weekends. If such a condition exists, it is desirable to install a constant-voltage transformer ahead of the charger.

The open-circuit voltage is the voltage of a storage battery when standing idle, i.e., when it is neither on charge nor on discharge. All types of storage batteries standing idle for a period of time have an open-circuit voltage independent of the state of charge. It therefore cannot be used to indicate the state of charge of a battery.

. The open-circuit voltage of a Nicad cell is approximately 1.30 V. Nicad batteries possess the characteristic common to all storage batteries that their voltage rises throughout the charge. Hence when the battery voltage ceases to rise and the charge current remains steady, it is an indication that the battery is fully charged. The voltage of a fully charged battery, still on charge, will depend upon the magnitude of the charge current; the heavier the current, the higher the battery voltage.

The accuracy of voltmeters and ammeters is of considerable importance. Temperature changes and even slight vibration can change their adjustment. Good storage-battery maintenance requires that meters, particularly panel-mounted ones, be checked periodically to an accuracy of 0.5 percent.

Several methods of charging batteries are described below.

**205. Charging engine-starting batteries.**   The generator voltage regulator should be set so as to hold the battery voltage between 1.45 and 1.50 times the number of cells in the battery. Readings should be taken at the battery terminal posts with the voltage regulator at proper working temperature, with its cover in place, and only after the battery voltage has ceased to rise.

If the engine is started infrequently, as in emergency standby services, or if the engine is operated only for very short periods at a time (insufficient to keep the battery in a fully charged condition), it is recommended that the battery be maintained on constant trickle charge, preferably from a dry disk or plate, metallic rectifier, at a voltage equal to 1.40 to 1.45 times the number of cells in the battery. At this charge voltage the battery may require watering as often as every 6 to 9 months.

**206. Trickle charging or floating.**   Fully charged batteries floated across the line should be maintained at a voltage equal to 1.40 times the number of cells in the battery. If the discharges, although momentary, are relatively heavy and frequent, the voltage should be raised to 1.45 times the number of cells in the battery to ensure that the total input will exceed the total output over a period of time. Otherwise, the battery will become slowly discharged and require so-called equalization or overcharges from time to time to bring it back to a fully charged condition.

The recommended battery voltages will hold the individual cell voltages below 1.47 V. Above 1.47 V the cells will begin to gas and hence consume water. Variations in voltage up to a maximum of 0.05 V between the individual cells of a battery on float should be disregarded, as they are of no importance. One of the most valuable features of the Nicad battery is that its floating voltage is not critical, provided, of course, that it is high enough to compensate for the loads imposed on the battery from time to time. Floating at above 1.47 V per cell will not harm the battery, but its water consumption will be considerably increased, requiring additional attention.

Trickle charging is actually preservation charging and should be used only to keep a charged battery in a fully charged condition. It cannot be used as a substitute for normal charging of a discharged battery. Trickle chargers should be equipped with a variable resistance and a voltmeter of suitable range and accuracy. Information on suitable equipment for trickle and high-rate charging of Nicad batteries will be furnished by the manufacturer upon request.

**207. Constant-current charging.**  This method consists of charging at a constant current, not for a definite length of time or to a definite end voltage, but until the battery voltage ceases to rise, indicating that the battery is fully charged.

The length of time required to charge the battery by this method depends on the magnitude of the charge current and the state of charge of the battery at the time it is put on charge.

It is usual to insert a variable resistance of suitable size between the dc line and the battery and to reduce the resistance from time to time by hand, during the charge, so as to hold the charge current reasonably constant.

A dc line voltage of 1.40 times the number of cells in the battery is necessary at the beginning of the charge and of 1.85 times the number of cells in the battery at the end of the charge. In theory, practically any rate of charge can be employed, provided that the electrolyte temperature, which is the limiting factor, is not allowed to exceed 145°F (63°C).

Too high a charge current, however, particularly if the electrolyte is above the maximum permissible level, may cause the electrolyte to be forced out of the cell vents.

When charged by the constant-current method at the normal (7-h) charge rate, Nicad batteries will commence to gas after about 4½ h, i.e., when the battery voltage has risen to about 1.47 V per cell. Therefore, if a period of more than 7 h between discharges is available, it is desirable to charge at a lower rate than normal. This will reduce gassing and consequently the amount of water required by the battery. For example, a 100-Ah battery has a "normal" charge rate of 20 A (for 7 h) but may conveniently be charged at 14 A for 10 h or 10 A for 14 h. The specific gravity of the electrolyte remains practically constant during charge and discharge; therefore, specific-gravity readings are not necessary.

**208. Ascertaining state of charge.**  Open-circuit voltage readings (no current passing into or being delivered by the battery) *cannot* be used as an indication of the state of charge of any storage battery.

The density of the electrolyte of the Nicad battery does not change appreciably on charge or discharge, and specific-gravity measurements, therefore, do *not* indicate its state of charge at any time.

To determine the state of charge of a partially charged battery it becomes necessary to take simultaneous current and voltage readings. There are several ways of doing this, and for services in which it is necessary to determine the state of charge frequently a satisfactory method can generally be worked out.

For certain applications involving heavy rate discharges of short duration, such as switch tripping, it has been found advantageous to install a voltmeter and a fixed resistance near the battery to provide an artificial load equal in value to the normal load. Voltage readings obtained while the battery is connected momentarily to the artificial load indicate the ability of the battery to carry the normal load.

Information regarding methods and equipment can be obtained from manufacturers of nickel-cadmium batteries.

**209. Discharging.**  Heavy discharges (such as engine starting) will not damage the Nicad battery. Do not, however, discharge the battery below 1.10 V per cell at from 3- to 10-h rates or below 1.20 V per cell at lower current rates. Overdischarging at low rates regularly continued below these end voltages will damage the battery and is an indication that the battery is too small for its work.

**210. Maintenance.**  Keep the cells and trays clean and dry externally at all times. Moisture and dirt allowed to accumulate on top of and particularly between the cells will permit stray intercell currents, resulting in corrosion through electrolysis of the cell containers. For this reason any water or electrolyte spilled on the cells or the trays must be wiped off. Use compressed air or, better still, low-pressure steam to clean cells and trays. Do not allow dirt to enter the vents when cleaning cells. After cleaning, regrease the cell tops and connectors with Nicad petroleum jelly No. 32982 to protect the metal.

Batteries that are charged at high rates may gas heavily toward the end of charge, giving off minute quantities of potassium hydroxide. This combines with carbon dioxide in the air, forming potassium carbonate, which deposits as a noncorrosive, inert white powder on the cell tops and connectors. Potassium carbonate is electrically conductive when damp and if allowed to build up can cause current leakage and possibly discharge the battery. Any accumulation should be removed with a brush or damp cloth.

Keep all vent caps closed to prevent air from entering the cells. Open caps only to check the electrolyte. Caps must be closed when the battery is charging.

Always check and service only one cell at a time.

Never place or drop any metal articles, such as post nuts, cable lugs, or tools, on or between the cells. These will cause heavy short circuits which may damage the cell containers.

Never permit sparks, open flame, or lighted cigarettes near a storage battery. All storage batteries when gassing give off a highly explosive mixture of hydrogen and oxygen. A nonmetallic flashlight is desirable for battery inspection. Keep all connections tight.

Use only spirit thermometers when taking temperature readings. Ordinary mercury thermometers may break. Mercury running into the cell between its plates will cause sparking and explosions.

Always keep the plates covered with electrolyte. Serious damage can be caused by exposing the tops of the plates to the air. If electrolyte has been spilled accidentally from the battery, proceed as described in Sec. 213.

**211. Damage from impurities.**    Impurities of all kinds must be kept out of the cells, as they have a harmful effect and can eventually ruin the battery.

Even a trace of sulfuric acid can ruin a Nicad battery by attacking and corroding its steel plates and cell containers. To prevent contamination never use any tools or utensils such as hydrometers, funnels, rubber hoses, battery fillers, etc., which have been used at any time for servicing lead-acid batteries.

Any vegetable oil or grease accidentally introduced into the cells will cause them to froth on charge.

**212. Electrolyte.**    The electrolyte (solution) in Nicad batteries is alkaline and consists of specially purified caustic potash (KOH, potassium hydroxide) dissolved in distilled water. The specific gravity of the electrolyte does not change with the state of charge but remains practically constant on charge and discharge.

The use of other than Nicad electrolyte can damage the battery. Ordinary commercial grades of caustic potash should never be used, as they are not sufficiently pure.

Nicad refill or renewal electrolyte of proper specific gravity is available in nonreturnable containers holding 5, 10, 15, 20, and 130 lb (2.3, 4.5, 6.8, 9, and 59 kg). Use an enamelware or glass pitcher and funnel for filling cells with electrolyte or water. Earthenware, hard-rubber, and plastic utensils are also suitable.

Refill electrolyte is used to replace electrolyte accidentally lost in transit or otherwise, while renewal electrolyte is used when changing the electrolyte, as described in Sec. 216.

If electrolyte should be lost by accident from any of the cells, replace the lost quantity with Nicad refill electrolyte. If refill electrolyte is not available, take the battery out of service and add enough water to cover the plates (so as to prevent damage to the plates by exposing them to the air) and procure renewal electrolyte. Upon its arrival empty out all the old electrolyte and fill the cells with renewal electrolyte. Charge the battery and check that the specific gravity is correct and uniform in all the cells before putting the battery back into service.

**213. Method of adjusting electrolyte specific gravity.**    Cells having lost their electrolyte in shipping or by accident should be filled immediately to the proper level with Nicad refill electrolyte, and no adjustment of electrolyte specific gravity will be needed. If, however, they have been filled with water pending the arrival of refill electrolyte, the following electrolyte adjustment treatment will be necessary:

Dump the weak electrolyte-water mixture from the cells and fill with refill electrolyte. Cells which have lost electrolyte in varying amounts need adjustment of electrolyte specific gravity. Adjustment should be made during charge and toward the end of the charge when the cells are gassing freely.

If electrolyte is too strong, add distilled water. If electrolyte is too weak, add special 1.400-specific-gravity electrolyte supplied by Nicad on request. Three adjustments may have to be made before the normal specific-gravity reading is obtained. Allow 30 min between each adjustment during charge.

The amount of distilled water or special 1.400-specific-gravity electrolyte required can be found only by trial. For example, one cell may need only half a syringe full to bring the electrolyte to normal specific gravity. Yet another cell may need several syringes full, depending on the size of the cell and the strength of the electrolyte in the cell.

Each time that electrolyte is withdrawn from a cell during this adjustment, it should be replaced with an equal amount of special electrolyte or water to maintain correct electrolyte level above the plate tops.

Once adjusted to the correct specific gravity at maximum level, the cell will need only distilled water to keep the electrolyte at maximum level.

The above-described method of adjusting the electrolyte specific gravity may seem slow and cumbersome but is necessary to assure good performance and to maintain the capacity of the cells.

The electrolyte will readily absorb carbon dioxide from the air to form potassium carbonate, which has the effect of temporarily lowering the capacity of the battery. Electrolyte must therefore be stored in airtight containers. Cell vent caps should be kept closed at all times except when adding water or checking the electrolyte, and this should always be done as quickly as possible, opening only one vent cap at a time.

When handling electrolyte wear goggles and rubber gloves; avoid splashes. The electrolyte is injurious to skin and clothing and must therefore always be handled carefully. Particularly guard the eyes! A generous quantity of concentrated boric acid solution (5 oz of boric acid powder to each quart, or 0.14 kg/0.95 L, of water) should be kept handy in a bottle or open bowl for neutralizing any accidental splashes on person or clothing. Use an eyecup for eye injuries.

Boric acid powder will dissolve in warm water within 1 h. With cold water allow 24 h. Do not use the boric acid solution on cells or trays.

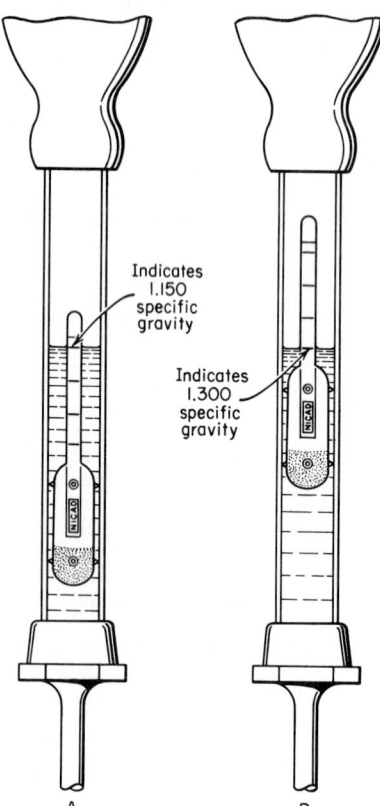

Indicates 1.150 specific gravity

Indicates 1.300 specific gravity

A          B

**Fig. 4-205**    Application of hydrometer. [Nicad Division of Gould–National Batteries, Inc.]

**214. Checking the electrolyte.** Storage batteries normally lose water through natural evaporation and particularly when gassing freely on charge. While there are no corrosive or obnoxious gases given off by Nicad batteries, traces of the potassium hydroxide are lost with the gas, resulting in a gradual lowering of the specific gravity of the electrolyte over the years.

The level of the electrolyte as well as its specific gravity must therefore be checked periodically, as serious damage will be done to the plates if the electrolyte level falls below the top of the plates or the specific gravity is less than the minimum value stated on the wall card.

The electrolyte level is determined by inserting the $\frac{3}{16}$-in- (4.8-mm-) bore plastic tube shipped with the battery through the vent until it rests on top of the plates, then placing the finger tightly over the end and withdrawing the tube for inspection. Be sure to return the electrolyte in the tube to the cell from which it was withdrawn.

A hydrometer (Fig 4-205) is used to check the specific gravity of the electrolyte, which should be within the range specified on the wall card. The illustrations show the positions of the float in electrolyte which is too weak (Fig. 4-205B) and too strong (Fig. 4-205A).

Use only a Nicad hydrometer to take specific-gravity readings. First rest the tip of the nozzle firmly on top of the plates in the cell; then squeeze and release the bulb. This method will prevent Celoil (see Sec. 217) from being drawn up into the barrel. Draw up sufficient solution to permit the float to move freely, and then tap the glass barrel of the hydrometer gently with the finger to prevent the float from giving a false reading by sticking to the barrel wall.

The maximum level of the electrolyte is halfway between the tops of the plates and inside the cell covers (do not include vent height). At this level and down to the minimum level of ½ in (12.7 mm) above the plate tops, the specific gravity of the electrolyte should be within the range specified on the wall card. These figures apply to normal temperatures. When extreme temperatures prevail and the observed hydrometer readings are outside these limits, it will be necessary to apply temperature- and electrolyte-volume-correction factors as described below.

To arrive at the true specific gravity of the electrolyte at 72°F (22°C), the temperature used as a base for purposes of calculation, observe and make a record of:

    1. The electrolyte specific gravity
    2. The electrolyte temperature
    3. The electrolyte level above the plates

If the electrolyte temperature is above 72°F, add to the specific gravity reading 0.001 for every 4° above 72°F (every 2.2° above 22°C).

If the temperature is below 72°F, subtract 0.001 from the specific-gravity reading for every 4° below 72°F.

For every ¼ in (6.35 mm) of electrolyte above the top of the plates add 0.005 to the specific-gravity reading.

The following examples show typical calculations:

    **example 1**  Hydrometer reads 1.190. Electrolyte temperature is 96°F; add $0.001 \times (96 - 72) \div 4$ = 0.006. Specific gravity, corrected for temperature only, is 1.196. Electrolyte level is 1¾ in; add $0.005 \times 1¾ \div ¼ = 0.035$. Specific gravity, corrected for both temperature and volume, is 1.231.

    **example 2**  Hydrometer reads 1.190. Electrolyte temperature 40°F; subtract $0.001 \times (72 - 40) \div 4$ = 0.008. Specific gravity, corrected for temperature only, is 1.182. Electrolyte level is ¾ in; add $0.005 \times ¾ \div ¼ = 0.015$. Specific gravity, corrected for both temperature and volume, is 1.197.

A Nicad spirit thermometer part No. 91791 is necessary to apply proper temperature-correction factors to the hydrometer readings.

Gas bubbles in sample of electrolyte withdrawn from cells which are gassing must be allowed to disappear; otherwise false readings will be obtained.

Specific-gravity readings should not be taken on cells to which water or special 1.400-specific-gravity electrolyte has just been added, but deferred until they have had time to mix properly with the electrolyte during charge.

Always return the sample of electrolyte to the cell from which it was taken. After use wash out the hydrometer thoroughly with water to remove all traces of electrolyte, as any electrolyte allowed to remain in the hydrometer will absorb carbon dioxide from the air to form a thin coating on the float, which will cause false readings.

**215. Adding water.**  Always check the electrolyte level before adding any water to the cells. For maximum permissible level of the electrolyte refer to Sec. 214 or the wall card packed with the battery. Do not overfill.

Nicad batteries use very little water, particularly when they are on float or trickle charge. Batteries on float at a voltage equal to 1.40 times the number of cells in the battery will usually be found to require watering less than once a year.

If the cells are overfilled, the electrolyte will be forced out of the vents on charge and saturate the trays, causing electrolysis between the cells, corrosion of the cell containers, and troublesome grounds in the electrical circuit. Overfilling will also dilute the electrolyte to such an extent that the specific gravity will become too weak and the plates will be damaged.

Maintain the proper electrolyte level by the periodic addition of distilled water; do not add electrolyte. Do not use so-called distilled water for storage batteries, as it generally contains small amounts of sulfuric acid through being stored in carboys having contained sulfuric acid intended for use in lead batteries.

Some drinking water may be usable, but it is always advisable to forward to the manufacturer a laboratory analysis or a sample of the water to be used. Transportation should be prepaid. The minimum quantity of water to be sent is 2 qt (1.9 L), and the minimum quantity of electrolyte is 1 qt (0.95 L) for proper analysis. Ship in thoroughly cleaned glass or plastic bottles. Tag for easy identification. Do not send by mail. Send by express service. Label electrolyte "corrosive liquid."

Although the quantity of any impurities introduced into the battery each time the battery is watered may be insignificant, the cumulative destructive effect over a number of years will be considerable. Distilled water is so inexpensive and in most cases so easily procured that its use is easily justified in terms of battery life. Stills for producing distilled water at very low cost are available from a number of reliable manufacturers. Store distilled water in clean, airtight glass or plastic bottles or jars.

**216. Change of electrolyte.**  Batteries which are charged at high rates for long periods and hence gas freely, with consequent loss of water and potassium hydroxide, may require change of electrolyte. Nicad batteries seldom if ever require change of electrolyte. However, it is desirable to check the specific gravity about once a year. When the specific gravity of the electrolyte, at a height of ½ in (12.7 mm) above the tops of the plates, has fallen to the minimum specified, further operation of the battery will cause a rapid reduction in its life. At this point the battery should be discharged at the 7-h discharge rate to a voltage of 0.5 to 0.8 V per cell and have its electrolyte changed.

Remove the cells from their trays, and working on only one tray at a time, turn them upside down and empty out the electrolyte. Do not rinse the cells with water or electrolyte. Do not allow the cells to stand empty more than 30 min, as exposure of the plates to the air will damage the capacity. Fill immediately to the maximum height, i.e., halfway between the tops of the plates and inside the cell covers, with Nicad renewal electrolyte. The renewal electrolyte should, of course, always be procured in advance and be available before starting to empty the cells.

Avoid splashes. Protect the eyes! Wear goggles and rubber gloves.

While the above operations are performed, the trays or cells should not be stacked on top of one another, nor should the cell posts be allowed to touch anything except insulation. Any chance of short-circuiting the cells should be avoided. The trays should be examined for damage, and any broken parts replaced. Note that the trays are so designed that there is a clear space between the cells and the surface on which the trays rest.

After being thoroughly cleaned, preferably by steam or compressed air, the cells and the trays should be allowed to dry thoroughly and then painted or, preferably, dipped in Nicadvar, a specially prepared corrosion-resisting asphalt-base paint. After the battery has been reassembled, it should be given a 14-h charge at the 7-h charge rate before being put back into service.

**217. Celoil.**  Nicad batteries normally contain a ¼-in (6.35-mm) layer of Celoil floating on top of the electrolyte, which retards the natural evaporation of water from the electrolyte.

Celoil is a pure, acid-free, and nonsaponifying oil specially prepared for use in Nicad cells. Do not use anything but Nicad Celoil.

**218. Laying up.**  When a Nicad battery is to be laid up for a few months, make sure that the specific gravity of the electrolyte is within the specified range and that the level is at least ½ in (12.7 mm) above the tops of the plates.

Nicad batteries can be taken out of service in any state and left idle for years without deterioration.

Remove all the intertray connectors to break any stray currents, and store the battery in a cool place free from dust and moisture.

Regrease the cell tops to protect the metals with Nicad petroleum jelly No. 32982.

If the battery has been stored for some time, give it a freshening charge before placing it back in service. Charge battery for 7 h at normal rate or for 14 h at half normal rate.

**219. Returning cells to factory.**  If a cell or battery appears to be abnormal in any respect or if an accident has occurred, it may be necessary to return it to the factory for test and repair.

Before a cell or battery is returned, a complete description of the unusual condition should be reported to the service department. If in the opinion of the company the return is indicated, full shipping instructions will be sent with return-material authorization.

Return shipments cannot be accepted without the company's authorization.

In case a cell has to be returned to the factory, it should be discharged at the 7-h rate to 0.5 to 0.8 V. The electrolyte must be left in the cell.

Tighten the post nuts, and drive a tapered rubber or wooden plug of suitable size into the vent hole. Turn the cell upside down. Leave it standing in this position for 1 h and then check that no electrolyte has leaked out.

Pack the cell, surrounded by a large quantity of liquid-absorbing material, such as sawdust, in a suitable container and ship prepaid in accordance with instructions which will be furnished by Nicad.

### INSTALLATION OF STORAGE BATTERIES

**220. In installing storage batteries,** follow the rules of the National Electrical Code, which are given below.

**221. Scope.** The provisions of this section shall apply to all stationary installations of storage batteries.

**Fig. 4-206** Alkali-type battery with cells mounted in an insulating tray. [Electric Storage Battery Co.]

**222. Definition of nominal voltage.** The nominal battery voltage shall be calculated on the basis of 2.0 V per cell for the lead-acid type and 1.2 V per cell for the alkali type.

**223. Wiring and apparatus supplied from batteries.** Wiring, appliances, and apparatus supplied from storage batteries shall be subject to the requirements of the Code applying to wiring, appliances, and apparatus operating at the same voltage, except as otherwise provided for communication systems in Article 800 of the Code.

**224. Insulation of batteries of not over 250 V.** The provisions of this section shall apply to storage batteries having the cells so connected as to operate at a nominal battery voltage not exceeding 250 V.

1. *Vented lead-acid batteries.* Cells and multicompartment batteries with covers sealed to containers of nonconductive, heat-resistant material shall not require additional insulating support.

2. *Vented alkaline-type batteries.* Cells with covers sealed to jars of nonconductive, heat-resistant material shall require no additional insulation support. Cells in jars of conductive material shall be installed in trays of nonconductive material with not more than 20 cells (24 V) in the series circuit in any one tray (Fig. 4-206).

3. *Rubber jars.* Cells in rubber or composition containers shall require no additional insulating support if the total nominal voltage of all cells in series does not exceed 150. If the total voltage exceeds 150, batteries shall be sectionalized into groups of 150 V or less, and each group shall have the individual cells installed in trays or on racks.

4. *Sealed cells or batteries.* Sealed cells and multicompartment sealed batteries constructed of nonconductive, heat-resistant material shall not require additional insulating support. Batteries constructed of a conducting container shall have insulating support if a voltage is present between the container and ground.

**225. Insulation of batteries of over 250 V.** The provisions of Sec. 224 shall apply to storage batteries having the cells so connected as to operate at a nominal voltage exceeding 250 V, and in addition the provisions of this section shall also apply to such batteries.

1. Cells shall be installed in groups having a total nominal voltage of not over 250 V on any rack. Insulation, which can be air, shall be provided between racks and shall have a minimum separation between live battery parts of opposite polarity of 2 in (50.8 mm) for battery voltages not exceeding 600 V.

**Fig. 4-207** Installation of lead-acid batteries in sealed glass jars mounted on a three-tier rack. [Electric Storage Battery Co.]

**226. Racks and trays.** Racks and trays shall conform to the following:

1. Racks Racks, as required in this section, refer to frames designed to support cells or trays (see Fig. 4-207). They shall be substantial and be made of any of the following:

*a*. Metal so treated as to be resistant to deteriorating action by the electrolyte and provided with nonconducting members directly supporting the cells or with continuous insulating material other than paint or conducting members

*b*. Other construction such as fiberglass or other suitable nonmetallic materials

2. Trays Trays are frames, such as crates or shallow boxes usually of wood or other nonconductive material, so constructed or treated as to be resistant to deteriorating action by the electrolyte.

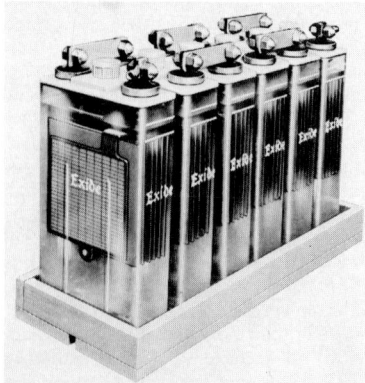

**Fig. 4-208** Lead-acid battery of cells in sealed glass jars mounted in a supporting tray. [Electric Storage Battery Co.]

**227. Battery locations.** Battery locations shall conform to the following:

1. VENTILATION Provisions shall be made for sufficient diffusion and ventilation of the gases from the battery to prevent the accumulation of an explosive mixture.

2. LIVE PARTS Guarding of live parts shall comply with Sec. 110-17 of the **NEC**.

**228. Vents shall conform to the following:**

1. VENTED CELLS Each vented cell shall be equipped with a flame arrester designed to prevent destruction of the cell due to ignition of gases within the cell by an external spark or flame under normal operating conditions.

2. SEALED CELLS Sealed batteries or cells shall be equipped with a pressure-release vent to prevent excessive accumulation of gas pressure, or they shall be designed to prevent scatter of cell parts in the event of a cell explosion.

**229. Danger from gas with storage batteries.** The hydrogen and oxygen given off the battery during charge, when unmixed with a large amount of air, form a combination that will explode violently if ignited by an open flame or an electric spark. If the battery is in a compartment, open or ventilate the compartment while the battery is being charged so that these gases may become mixed with air. Do not bring exposed flame, match, cigar, etc., near the battery when charging or shortly afterward.

## GENERAL CONSTRUCTION MATERIALS

**230. Wire nails** are formed from wire of the same diameter as the shank of the nail is to be. The wire from which nails are made, hence the nail diameter, is measured by the steel-wire gage (see Table 85, Div. 2), which is the same as the Washburn & Moen gage and is used by practically all nail manufacturers, though it is sometimes given a different name.

The size of nails is designated by the *penny system*, which originated in England. Two explanations are offered as to how this curious designation came about. One is that the sixpenny, fourpenny, tenpenny, etc., nails derived their names from the fact that 100 cost sixpence, fourpence, etc. The other explanation, which is more probable, is that 1000 tenpenny nails, for instance, weighed 10 lb (4.5 kg). The ancient as well as the modern abbreviation for penny is *d*, which is the first letter of the Roman coin denarius; the same abbreviation in early history was used for the English pound in weight. At any rate, the penny has persisted as a term in the nail industry.

Ordinary nails are made of steel wire with a natural bright-steel finish. The following special coatings, finishes, and heat treatments can be obtained at some additional cost:

| | | |
|---|---|---|
| Galvanized (hot-process) | Brass-plated | Blued |
| Galvanized (electro) | Cadmium-plated | Annealed |
| Cement-coated (regular) | Nickel-plated | Oil-quench–hardened |
| Cement-coated (clear) | Chromium-plated | Japanned |
| Coppered | Painted | Parkerized |
| Tinned | Acid-etched | |

Nails made of copper, aluminum, brass, or stainless-steel wire are also available.

Nails are made in several different forms to meet the requirements of different applications. The more common types with dimensional data are given in the following sections.

**231.  Dimensions of Common Nails and Brads**
(American Steel and Wire Co.)

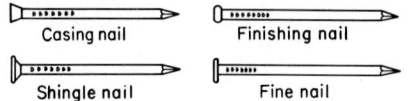

**Fig. 4-209**   Casing, finishing, shingle, and fine nails.

| Size | Length, in | Steel-wire gage | Approximate no. per lb | Diameter in decimals, in | Approximate diameter, in | Nearest AWG gage |
|------|-----------|-----------------|------------------------|--------------------------|--------------------------|------------------|
| 2d | 1 | 15 | 876 | 0.0720 | ⁵⁄₆₄ | 13 |
| 3d | 1¼ | 14 | 568 | 0.0800 | ⁵⁄₆₄ | 12 |
| 4d | 1½ | 12½ | 316 | 0.0985 | ⁷⁄₆₄ | 10 |
| 5d | 1¾ | 12½ | 271 | 0.0985 | ⁷⁄₆₄ | 10 |
| 6d | 2 | 11½ | 181 | 0.1130 | ⁷⁄₆₄ | 9 |
| 7d | 2¼ | 11½ | 161 | 0.1130 | ⁷⁄₆₄ | 9 |
| 8d | 2½ | 10¼ | 106 | 0.1314 | ⅛ | 8 |
| 9d | 2¾ | 10¼ | 96 | 0.1314 | ⅛ | 8 |
| 10d | 3 | 9 | 69 | 0.1483 | ⁹⁄₆₄ | 7 |
| 12d | 3¼ | 9 | 63 | 0.1483 | ⁹⁄₆₄ | 7 |
| 16d | 3½ | 8 | 49 | 0.1620 | ⁵⁄₃₂ | 6 |
| 20d | 4 | 6 | 31 | 0.1920 | ³⁄₁₆ | 6 |
| 30d | 4½ | 5 | 24 | 0.2070 | ¹³⁄₆₄ | 4 |
| 40d | 5 | 4 | 18 | 0.2253 | ⁷⁄₃₂ | 3 |
| 50d | 5½ | 3 | 14 | 0.2437 | ¼ | 2 |
| 60d | 6 | 2 | 11 | 0.2625 | ¹⁷⁄₆₄ | 2 |

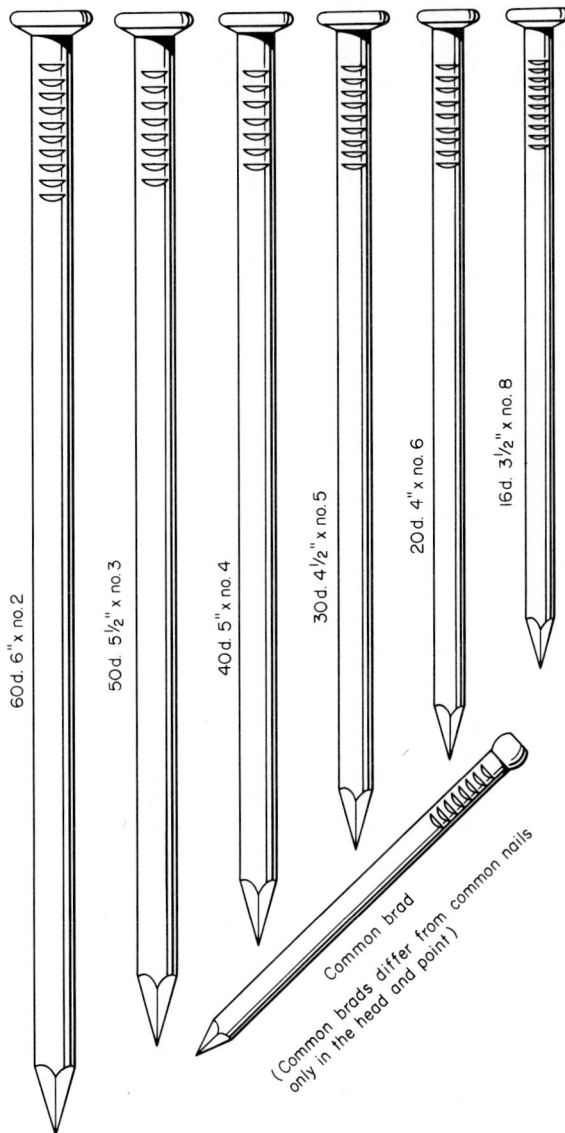

**Fig. 4-210**    Common nails (actual size).

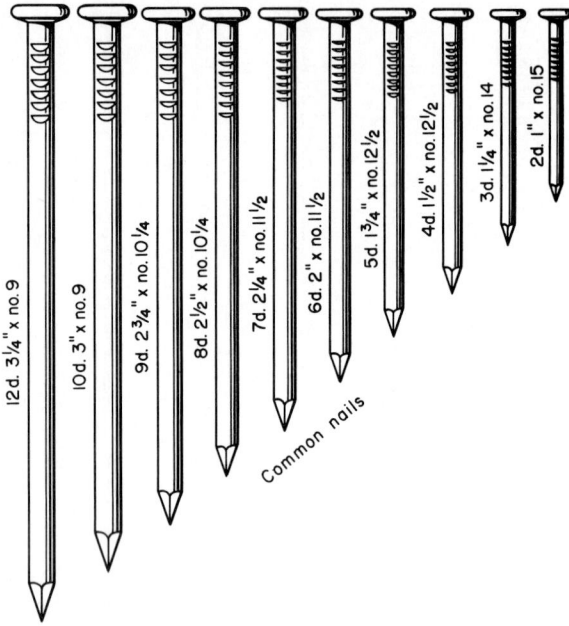

**Fig. 4-211**   Common nails (continued).

## 232.   Dimensions of Casing, Finishing, Shingle, and Fine Nails

(American Steel and Wire Co.)

Nail diameters are measured by the steel-wire gage (see Div. 2, Table 85). Equivalent AWG gage numbers and fractional-inch equivalents are given in Fig. 4-209.

| Size | Length, in. | Casing | | Finishing | | Shingle | | Fine | |
|------|------|------|------|------|------|------|------|------|------|
| | | Gage | Approx no. per lb | Gage | Approx no. per lb | Gage | Approx no. per lb | Gage | Approx no. per lb |
| 2d | 1 | 15½ | 1,010 | 16½ | 1,351 | .... | ... | 16½ | 1,351 |
| 3d | 1¼ | 14½ | 635 | 15½ | 807 | 13 | 429 | 15ᵇ | 778 |
| 3½d | 1⅜ | .... | ..... | .... | ..... | 12½ | 345 | | |
| 4d | 1½ | 14 | 473 | 15 | 584 | 12 | 274 | 14 | 473 |
| 5d | 1¾ | 14 | 406 | 15 | 500 | 12 | 235 | | |
| 6d | 2 | 12½ | 236 | 13 | 309 | 12 | 204 | | |
| 7d | 2¼ | 12½ | 210 | 13 | 238 | 11 | 139 | | |
| 8d | 2½ | 11½ | 145 | 12½ | 189 | 11 | 125 | | |
| 9d | 2¾ | 11½ | 132 | 12½ | 172 | 11 | 114 | | |
| 10d | 3 | 10½ | 94 | 11½ | 121 | 10 | 83 | | |
| 12d | 3¼ | 10½ | 87 | 11½ | 113 | | | | |
| 16d | 3½ | 10 | 71 | 11 | 90 | | | | |
| 20d | 4 | 9 | 52 | 10 | 62 | | | | |
| 30d | 4½ | 9 | 46 | | | | | | |
| 40d | 5 | 8 | 35 | | | | | | |
| 2dᵃ | 1 | .... | ..... | .... | ..... | .... | ... | 17 | 1,560 |
| 3dᵃ | 1⅛ | .... | ..... | .... | ..... | .... | ... | 16 | 1,015 |

ᵃ These sizes are called Extra Fine.
ᵇ This nail is only 1⅛ in. long.

**233. Dimensions of wood screws.** Roundheaded wood screws do not measure full-length but are from 1/16 to 3/16 in short. For example, a No. 4 by 1/2-in roundheaded wood screw measures aobut 7/16 in long under the head, and a No. 20 by 2-in screw measures about 1 7/8 in under the head.

**234. Wood screws.** Diameters are measured by the American Screw Co.'s gage and range in size from No. 0 to No. 30. They range in length from 1/4 to 6 in. The increase in length is by eighths of an inch up to 1 in, then by quarters of an inch up to 3 in, and by half inches up to 5 in. Manufacturers' standards vary, but generally the threaded portion is approximately seven-tenths of the total length. There is no standard number of threads per inch for the products of all manufacturers. The gage and diameter refer to the unthreaded portion of the screw. The screws are available in three types of head (Fig. 4-212): flat, round, and oval. The oval head (see illustration of machine screws, Fig. 4-213)

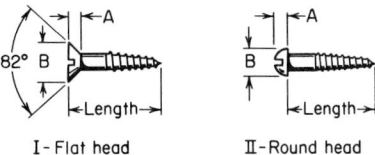

I- Flat head        II-Round head

**Fig. 4-212**   Wood screws.

| Screw gage no. | Diameter, in | | | Flat head | | Round head | | | Clearance drill | | Greatest length obtainable, in |
|---|---|---|---|---|---|---|---|---|---|---|---|
| | In decimals | In fractions | Nearest AWG gage | A | B | A | B | Counterbore for head | No. | Diam, in | |
| 0 | 0.05784 | 1/16 − | 15 | 1/16 | 7/64+ | .... | .... | .... | ..... | ...... | 3/8 |
| 1 | 0.07100 | 5/64 − | 14 | 1/16 | 9/64 − | .... | .... | .... | ..... | ...... | 1/2 |
| 2 | 0.08416 | 5/64 + | 12 | 1/16 | 5/32+ | 1/16 | 13/64 | 7/32 | 44 | 0.086 | 7/8 |
| 3 | 0.09732 | 3/32+ | 11 | 1/16 | 3/16 | 5/64 | 13/64 | .... | .... | ...... | 1 1/4 |
| 4 | 0.11048 | 7/64+ | 9 | 1/16 | 7/32 − | 5/64 | 13/64 | 7/32 | 33 | 0.113 | 1 1/2 |
| 5 | 0.12364 | 1/8 − | 8 | 1/16 | 15/64+ | 5/32 | 15/64 | .... | .... | ...... | 2 1/2 |
| 6 | 0.13680 | 9/64 − | 7 | 5/64 | 17/64+ | 3/32 | 1/4 | 17/64 | 28 | 0.1415 | 3 |
| 7 | 0.14996 | 5/32 − | 7 | 3/32 | 19/64 − | 7/64 | 9/32 | .... | .... | ...... | 3 |
| 8 | 0.16312 | 5/32+ | 6 | 7/64 | 19/64+ | 7/64 | 9/32 | 5/16 | 18 | 0.1695 | 4 |
| 9 | 0.17628 | 11/64+ | 5 | 7/64 | 11/32+ | 1/8 | 21/64 | .... | .... | ...... | 4 |
| 10 | 0.18944 | 3/16+ | 5 | 7/64 | 3/8 − | 1/8 | 11/32 | 23/64 | 10 | 0.1935 | 4 |
| 11 | 0.20260 | 13/64 − | 4 | 1/8 | 25/64+ | 9/64 | 3/8 | .... | ..... | ...... | 4 |
| 12 | 0.21576 | 7/32+ | 4 | 1/8 | 27/64 | 5/32 | 25/64 | 13/32 | 7/32 | 0.2188 | 6 |
| 13 | 0.22892 | 15/64 − | 3 | 1/8 | 29/64 | 5/32 | 27/64 | .... | .... | ...... | 6 |
| 14 | 0.24208 | 1/4 − | 3 | 9/64 | 15/32+ | 5/32 | 29/64 | 29/64 | 1/4 | 0.250 | 6 |
| 15 | 0.25524 | 1/4 + | 2 | 9/64 | 1/2 | 11/64 | 29/64 | .... | .... | ...... | 6 |
| 16 | 0.26840 | 17/64+ | 2 | 5/32 | 17/32 − | 11/64 | 31/64 | .... | .... | ...... | 6 |
| 17 | 0.28156 | 9/32 | 1 | 5/32 | 35/64+ | 11/64 | 1/2 | .... | .... | ...... | 6 |
| 18 | 0.29472 | 19/64 − | 1 | 11/64 | 37/64 | 3/16 | 17/32 | 35/64 | ..... | 0.302 | 6 |
| 19 | 0.30788 | 5/16 − | 0 | 3/16 | 39/64 − | .... | 17/32 | .... | .... | ...... | 6 |
| 20 | 0.32104 | 21/64 − | 0 | 13/64 | 5/8 + | 13/64 | 9/16 | 37/64 | ..... | 0.323 | 6 |
| 21 | 0.33420 | 21/64+ | 0 | 13/64 | 21/32 | 13/64 | 19/32 | .... | ..... | ...... | 6 |
| 22 | 0.34736 | 11/32+ | 0 | 13/64 | 11/16 − | 7/32 | 5/8 | .... | .... | ...... | 6 |
| 23 | 0.36052 | 23/64+ | 2/0 | 7/32 | 45/64+ | 7/32 | 41/64 | .... | .... | ...... | 6 |
| 24 | 0.37368 | 3/8 − | 2/0 | 7/32 | 47/64 | 15/64 | 21/32 | 43/64 | ..... | 0.377 | 6 |
| 25 | 0.38684 | 25/64 − | 3/0 | 7/32 | 49/64 − | 15/64 | 11/16 | .... | .... | ...... | 6 |
| 26 | 0.40000 | 13/32 − | 3/0 | 15/64 | 25/32+ | 1/4 | 45/64 | .... | .... | ...... | 6 |
| 27 | 0.41316 | 13/32+ | 3/0 | 15/64 | 13/16 | 1/4 | 23/32 | .... | .... | ...... | 6 |
| 28 | 0.42632 | 27/64+ | 3/0 | 1/4 | 27/32 | 1/4 | 47/64 | .... | .... | ...... | 6 |
| 29 | 0.43948 | 7/16+ | 4/0 | 1/4 | 55/64+ | 17/64 | 3/4 | .... | .... | ...... | 6 |
| 30 | 0.45264 | 29/64 | 4/0 | 17/64 | 57/64 | 9/32 | 25/32 | .... | .... | ...... | 6 |

(Note in Counterbore column: Depth of hole = height of head A)

is used for especially high-grade work such as cabinetwork, which must present a very neat appearance.

The ordinary materials used for making wood screws are steel and brass. The flat-headed screws are regularly available in steel, with a natural bright finish or galvanized, and in brass. The roundheaded screws are made of steel, with a blued or nickel-plated finish, and of brass. The oval-headed screws are regularly made of brass with a nickel-plated finish.

**235. Machine screws.**    Diameters are designated by the American Screw Co.'s gage and range in size from No. 2 to No. 34. They range in length from ⅛ to 4 in. The increase in length is by sixteenths of an inch up to 1 in, by eighths of an inch up to 2 in, and by quarters of an inch up to 4 in. The threaded portion extends over either all or nearly all the total length. Machine screws are designated by their gage number and number of threads per inch. A No. 10-24 screw has a diameter corresponding to size No. 10 of the American screw gage and has 24 threads per inch. The gage number refers to the diameter over the thread. The screws are available with heads of four different types: flat, round, oval, and fillister (Fig. 4-213). The length of flatheaded screws is measured overall, the length of roundheaded ones includes about one-half of the head, the length of oval-headed screws includes the countersinks, and the length of fillister-headed ones is measured from the rim of the head.

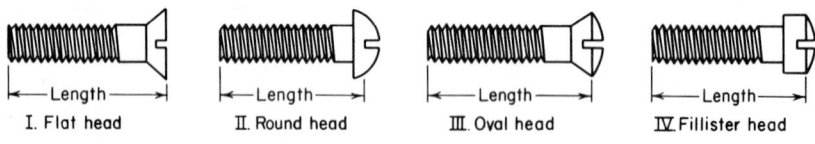

I. Flat head          II. Round head          III. Oval head          IV. Fillister head

**Fig. 4-213**   Machine screws.

Standard machine screws are made of iron, aluminum, plastic, nylon, or brass. The iron screws may be provided with any one of the following finishes: natural bright, blued, nickel, brass- or silver-plated, silver- or copper-oxidized, various bronzes, japanned, lacquered, coppered, tinned, galvanized, Bower-Barff, gunmetal.

**236.   Data for Machine Screws**

| Screw gage No. | Diam, in. | Standard threads per in. | Screw gage No. | Diam, in. | Standard threads per in. |
|---|---|---|---|---|---|
| 2 | 0.08416 | 48, 56, or 64 | 12 | 0.21576 | 20 or 24 |
| 3 | 0.09732 | 48 or 56 | 14 | 0.24208 | 18, 20, or 24 |
| 4 | 0.11048 | } 32, 36, or 40 | 16 | 0.26840 | } 16, 18, or 20 |
| 5 | 0.12364 | | 18 | 0.29472 | |
| 6 | 0.13680 | 30, 32, or 36 | 20 | 0.32104 | 16 or 18 |
| 7 | 0.14996 | 30 or 32 | 24 | 0.37368 | 14, 16, or 18 |
| 8 | 0.16312 | 30, 32, or 36 | 30 | 0.45264 | 14 or 16 |
| 9 | 0.17628 | } 24, 30, or 32 | 34 | 0.50528 | 13 |
| 10 | 0.18944 | | | | |

**237. Machine bolts and nuts.**    Machine bolts are designated by the diameter of the shank of the bolt. They range in length from 1½ to 30 in. The increase in length is by half inches up to 8 in and by inches up to 30 in. Bolts are regularly furnished with one nut. The standard number of threads per inch for the bolts and nuts is as follows:

| Diameter . . . . . . . . . . . . . . . | ¼ | ⁵⁄₁₆ | ⅜ | ⁷⁄₁₆ | ½ | ⁹⁄₁₆ | ⅝ | ¹¹⁄₁₆ | ¾ | ⅞ | 1 |
|---|---|---|---|---|---|---|---|---|---|---|---|
| Number of threads per in . . | 28 | 24 | 24 | 20 | 20 | 18 | 18 | 16 | 16 | 14 | 14 |

Bolts are made with either square or hexagonal heads and may be furnished with nuts of the corresponding shape. They are made of iron and may be finished with a natural black or galvanized coating.

**238. Dimensions and Strengths of Machine Bolts and Nuts**

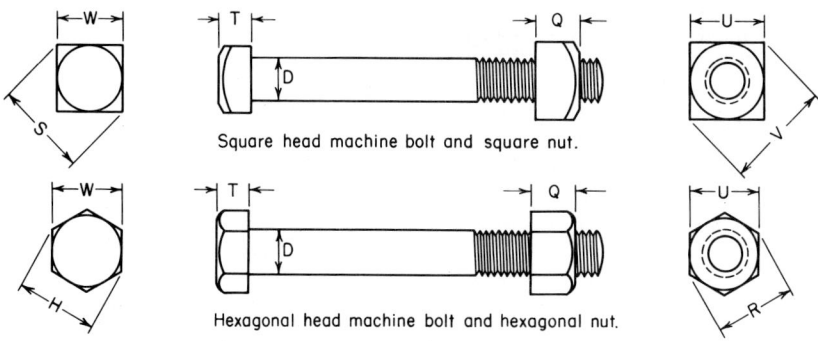

Square head machine bolt and square nut.

Hexagonal head machine bolt and hexagonal nut.

**Fig. 4-214**  Machine bolts and nuts.

| D | T | W | S | H | Q | U | R | V | Safe tensile strength of bolt at root of thread at 6,000 lb per sq in. | Safe shearing strength at 7,500 lb per sq in. | |
|---|---|---|---|---|---|---|---|---|---|---|---|
| | | | | | | | | | | Full bolt | Root of thread |
| $\frac{1}{4}$ | $\frac{3}{16}$ | $\frac{3}{8}$ | $1\frac{7}{32}$ | $\frac{7}{16}$ | $\frac{3}{16}$ | $\frac{7}{16}$ | $\frac{1}{2}$ | $\frac{5}{8}$ | 160 | 370 | 200 |
| $\frac{5}{16}$ | $1\frac{5}{64}$ | $1\frac{5}{32}$ | $2\frac{1}{32}$ | $3\frac{5}{64}$ | $\frac{1}{4}$ | $1\frac{7}{32}$ | $\frac{5}{8}$ | $\frac{3}{4}$ | 270 | 575 | 340 |
| $\frac{3}{8}$ | $\frac{9}{32}$ | $\frac{9}{16}$ | $5\frac{1}{64}$ | $2\frac{1}{32}$ | $\frac{5}{16}$ | $4\frac{7}{64}$ | $5\frac{7}{64}$ | | 410 | 830 | 510 |
| $\frac{7}{16}$ | $2\frac{1}{64}$ | $2\frac{1}{32}$ | $1\frac{5}{16}$ | $\frac{3}{4}$ | $\frac{3}{8}$ | $2\frac{3}{32}$ | $2\frac{7}{32}$ | $1\frac{5}{64}$ | 560 | 1,130 | 700 |
| $\frac{1}{2}$ | $\frac{3}{8}$ | $\frac{3}{4}$ | $1\frac{1}{16}$ | $2\frac{7}{32}$ | $\frac{7}{16}$ | $1\frac{3}{16}$ | $6\frac{1}{64}$ | $1\frac{5}{32}$ | 760 | 1,470 | 940 |
| $\frac{5}{8}$ | $1\frac{5}{32}$ | $1\frac{5}{16}$ | $1\frac{21}{64}$ | $1\frac{5}{64}$ | $\frac{9}{16}$ | 1 | $1\frac{5}{32}$ | $1\frac{27}{64}$ | 1,210 | 2,300 | 1,510 |
| $\frac{3}{4}$ | $1\frac{9}{32}$ | $1\frac{1}{8}$ | $1\frac{19}{32}$ | $1\frac{19}{64}$ | $\frac{11}{16}$ | $1\frac{3}{16}$ | $1\frac{3}{8}$ | $1\frac{43}{64}$ | 1,810 | 3,310 | 2,260 |
| $\frac{7}{8}$ | $2\frac{1}{32}$ | $1\frac{5}{16}$ | $1\frac{55}{64}$ | $1\frac{1}{2}$ | $\frac{13}{16}$ | $1\frac{3}{8}$ | $1\frac{9}{32}$ | $1\frac{61}{64}$ | 2,520 | 4,510 | 3,140 |
| 1 | $\frac{3}{4}$ | $1\frac{1}{2}$ | $1\frac{1}{2}$ | $1\frac{47}{64}$ | $\frac{15}{16}$ | $1\frac{9}{16}$ | $1\frac{13}{16}$ | $2\frac{7}{32}$ | 3,300 | 5,890 | 4,130 |
| $1\frac{1}{8}$ | $2\frac{7}{32}$ | $1\frac{11}{16}$ | $2\frac{23}{64}$ | $1\frac{31}{64}$ | $1\frac{1}{8}$ | $1\frac{13}{16}$ | $2\frac{3}{32}$ | $2\frac{9}{16}$ | 4,160 | 7,450 | 5,200 |
| $1\frac{1}{4}$ | $1\frac{5}{16}$ | $1\frac{7}{8}$ | $2\frac{21}{32}$ | $2\frac{5}{32}$ | $1\frac{1}{4}$ | 2 | $2\frac{5}{16}$ | $2\frac{27}{32}$ | 5,350 | 9,200 | 6,670 |
| $1\frac{3}{8}$ | $1\frac{1}{32}$ | $2\frac{1}{16}$ | $2\frac{59}{64}$ | $2\frac{25}{64}$ | $1\frac{3}{8}$ | $2\frac{3}{16}$ | $2\frac{17}{32}$ | $3\frac{3}{32}$ | 6,340 | 11,130 | 7,900 |
| $1\frac{1}{2}$ | $1\frac{1}{8}$ | $2\frac{1}{4}$ | $3\frac{3}{16}$ | $2\frac{19}{32}$ | $1\frac{1}{2}$ | $2\frac{3}{8}$ | $2\frac{47}{64}$ | $3\frac{3}{8}$ | 7,770 | 13,225 | 9,700 |
| $1\frac{5}{8}$ | $1\frac{7}{32}$ | $2\frac{7}{16}$ | $3\frac{3}{4}$ | $2\frac{13}{16}$ | $1\frac{5}{8}$ | $2\frac{9}{16}$ | $2\frac{31}{32}$ | $3\frac{5}{8}$ | 9,090 | 15,550 | 11,360 |
| $1\frac{3}{4}$ | $1\frac{5}{16}$ | $2\frac{5}{8}$ | $3\frac{23}{32}$ | $3\frac{1}{32}$ | $1\frac{3}{4}$ | $2\frac{3}{4}$ | $3\frac{3}{16}$ | $3\frac{57}{64}$ | 10,470 | 18,000 | 13,080 |
| $1\frac{7}{8}$ | $1\frac{13}{32}$ | $2\frac{13}{16}$ | $3\frac{63}{64}$ | $3\frac{1}{4}$ | $1\frac{7}{8}$ | $2\frac{15}{16}$ | $2\frac{25}{64}$ | $4\frac{5}{32}$ | 12,300 | 20,700 | 15,370 |
| 2 | $1\frac{1}{2}$ | 3 | $4\frac{1}{4}$ | $3\frac{15}{32}$ | 2 | $3\frac{1}{8}$ | $3\frac{5}{8}$ | $4\frac{27}{64}$ | 13,800 | 23,560 | 17,250 |

**239. Data on carriage bolts** are given in Div. 8.

**240. Toggle bolts,** which are used for fastening raceways and electrical devices to hollow tile or plaster-on-metal-lath surfaces, are of two general types. The screw type (Fig. 4-215) is the most frequently used but has the disadvantage that if it is ever necessary to remove the screw entirely, the toggle is lost within the wall. If the object fastened must be removed and replaced, a nut-type toggle bolt (Figs. 4-216 and 4-217) can be used. When the type of Fig. 4-216 is used, it is usually necessary, after the device is in place, to cut off the part of the bolt that extends so as to give a neat appearance. The so-called plumbers' toggle bolt (Fig. 4-217) has a removable hexagonal cap so that the device can be inserted in the wall before the object to be fastened is slipped over the bolt. Then, on putting the cap in place, the whole bolt is backed into the wall, hiding the surplus thread from view. Cone-headed toggles (Fig. 4-218) are used principally for the erection of metal raceway and have the advantage that the toggle head will readily pass through the hole in the raceway backing. A spring-in type of toggle bolt is shown in Fig. 4-219. The wings are tempered spring steel having a cam action against the saddle which throws them to the open position. Toggle bolts are made in several diameters and lengths

**241. Lag screws** are discussed in Div. 8.

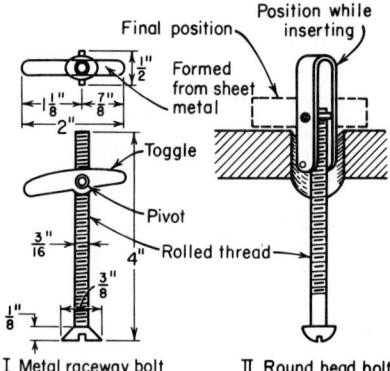

Final position ~ Position while inserting

I Metal raceway bolt          II Round head bolt

**Fig. 4-215**   Screw-type toggle bolts.

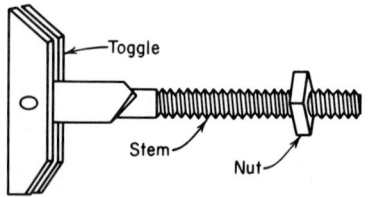

**Fig. 4-216**   Nut-type toggle bolt.

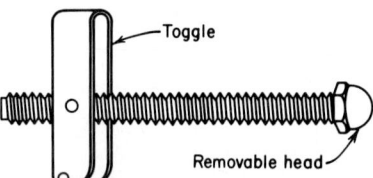

**Fig. 217**   Plumbers' toggle bolt.

**242. Expansion sleeves, shields, and bolts** are used for attaching equipment and devices to stonelike materials such as concrete, brick, tile, etc. They are made in a variety of types and consist of a lead sleeve or malleable-iron expansion shield which is inserted in a hole cut in the masonry. The equipment to be supported is attached to the sleeve or shield. As the screw is turned into the sleeve or shield, it expands the body of the sleeve or shield against the sides of the hole in the masonry so that it grips the masonry with considerable holding power. These devices are called by various trade names such as expansive anchors, expansion sleeves, expansion screws, or expansion bolts and anchors. Data and illustrations of some common types are given in the following sections.

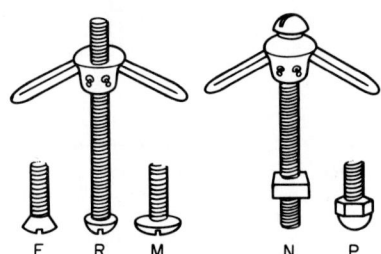

F. With flathead screw.

R. With roundhead screw.

M. Mushroom head, furnished in ⅛-in diameter only.

N. Reverse R or F screw, and add nut on any size.

P. Plumbers' toggle; reverse R or F screw and add cap nut on any size.

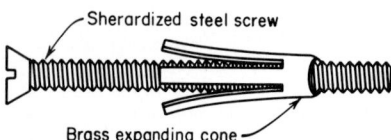

**Fig. 4-218**   Toggle bolt for metal raceway.

**Fig. 4-219**   Spring-in toggle bolt.

**243. Diamond multisize screw anchors**   These are designed to accommodate in one anchor several diameters of wood screws. The purpose is to reduce the number of anchors required to accommodate all sizes of screws. They are made in several lengths covering the majority of uses and are put up in boxes of 100.

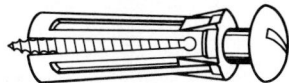

**Fig. 4-220**

| For wood screws, No. | Length, in. | Size of drill, in. | Weight, lb per 100 |
|---|---|---|---|
| 6– 8 | $\frac{3}{4}$ | $\frac{1}{4}$ | 1 |
| 6– 8 | $1\frac{1}{2}$ | $\frac{1}{4}$ | $1\frac{3}{4}$ |
| 10–14 | $\frac{3}{4}$ | $\frac{5}{16}$ | $1\frac{1}{2}$ |
| 10–14 | 1 | $\frac{5}{16}$ | 2 |
| 10–14 | $1\frac{1}{2}$ | $\frac{5}{16}$ | $2\frac{3}{4}$ |
| 16–18 | 1 | $\frac{3}{8}$ | 3 |
| 16–18 | $1\frac{1}{2}$ | $\frac{3}{8}$ | $4\frac{1}{4}$ |
| 20–24 | $1\frac{3}{4}$ | $\frac{7}{16}$ | $5\frac{1}{4}$ |

**244. Diamond caulking anchors for use with machine screws.**   Caulking anchors are put up in standard packages of 100 for sizes No. 10-24 and smaller and in packages of 50 for sizes No. 12-24 and larger. One caulking tool is in each box. All suggested safe loads are based on ideal conditions.

**Fig. 4-221**

| Diam of bolt or screw | Hole size, in. | | Suggested safe load, lb | Weight, lb per 100 |
|---|---|---|---|---|
| | Diam | Depth | | |
| 6–32 | $\frac{1}{4}$ | $\frac{3}{8}$ | 80 | 1 |
| 8–32 | $\frac{5}{16}$ | $\frac{1}{2}$ | 90 | $1\frac{1}{2}$ |
| 10–24 | $\frac{3}{8}$ | $\frac{5}{8}$ | 175 | 2 |
| 12–24 | $\frac{7}{16}$ | $\frac{3}{4}$ | 320 | $3\frac{1}{2}$ |
| $\frac{1}{4}$–20 | $\frac{1}{2}$ | $\frac{7}{8}$ | 400 | $4\frac{1}{2}$ |
| $\frac{5}{16}$ | $\frac{5}{8}$ | 1 | 480 | 11 |
| $\frac{3}{8}$ | $\frac{3}{4}$ | $1\frac{1}{4}$ | 720 | 16 |
| $\frac{7}{16}$ | $\frac{7}{8}$ | $1\frac{1}{2}$ | 950 | 24 |
| $\frac{1}{2}$ | $\frac{7}{8}$ | $1\frac{1}{2}$ | 1,000 | 24 |
| $\frac{5}{8}$ | 1 | 2 | 1,250 | 41 |

**245. Ackerman-Johnson expansive screw anchors.**   The Ackerman-Johnson expansive screw anchor consists of a hard, biconoidal, internally threaded nut within a lead-compo-

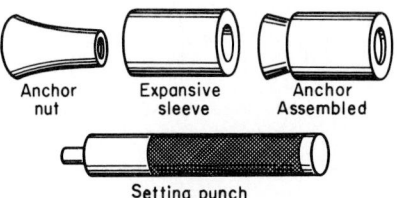

Anchor nut     Expansive sleeve     Anchor Assembled

Setting punch

**Fig. 4-222**

sition expansive sleeve. It is used for attaching fixtures or wiring to any stonelike material such as concrete, brick, tile, etc.

METHOD OF INSTALLING ANCHOR   When installed, the sleeve is driven farther toward the base of the tapered nut, being thereby expanded and swaged tightly against the sides of the hole, giving perfect holding contact throughout the length and circumference of the anchor.

To determine the anchor size best suited for the work, first select the screw or bolt to be preferred and then anchors of corresponding size number; for example, for work demanding No. 10-24 screws, specify No. 10-24 anchor; if ⅜-in bolts are required, use No. ⅜-in anchor.

A setting punch is included with every package of 50 or 1100 anchors.

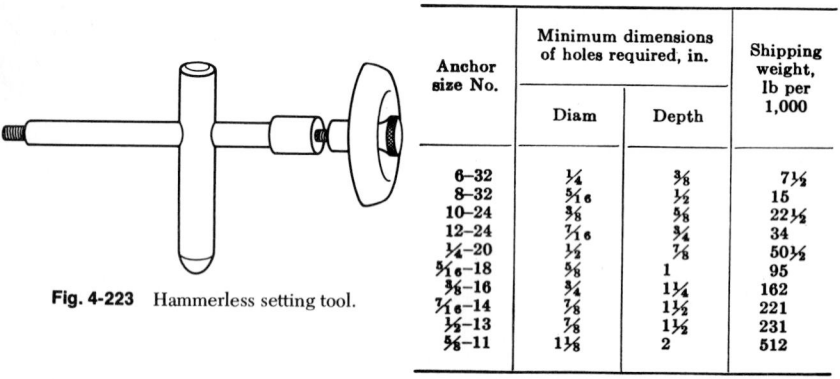

| Anchor size No. | Minimum dimensions of holes required, in. | | Shipping weight, lb per 1,000 |
|---|---|---|---|
| | Diam | Depth | |
| 6–32 | ¼ | ⅜ | 7½ |
| 8–32 | 5/16 | ½ | 15 |
| 10–24 | ⅜ | ⅝ | 22½ |
| 12–24 | 7/16 | ¾ | 34 |
| ¼–20 | ½ | ⅞ | 50½ |
| 5/16–18 | ⅝ | 1 | 95 |
| ⅜–16 | ¾ | 1¼ | 162 |
| 7/16–14 | ⅞ | 1½ | 221 |
| ½–13 | ⅞ | 1½ | 231 |
| ⅝–11 | 1⅛ | 2 | 512 |

**Fig. 4-223**   Hammerless setting tool.

HAMMERLESS SETTING TOOL   Anchors are set perfectly, in tile or other thin materials in which the hole extends through or has a weak bottom, by means of the hammerless setting tool (see Fig. 4-223).

The stud is screwed into the anchor, which is then inserted in position; turning the wheel moves the threaded stud rearward, thus drawing the anchor nut into its ductile sleeve, expanding the latter to any degree required for safe anchorage. The tool is furnished in sizes to use with the following sizes of anchor: 8-32, 10-24, 12-24, and ¼-20.

**246. Diamond lag-screw shield.**   The lag-screw shield is made of malleable iron in two wedge-shaped halves. The outer wedge slides forward when the screw is turned, creating a powerful expansion. The shield is galvanized by the Diamond hot-dip process.

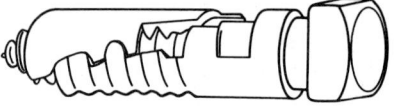

**Fig. 4-224**

| Diam screw, in. | Outside diam and drill required, in. | Standard package | Long standard pattern | | Short standard pattern | |
|---|---|---|---|---|---|---|
| | | | Length shield, in. | Shipping weight, lb | Length shield, in. | Shipping weight, lb |
| ¼ | ½ | 100 | 1½ | 4½ | 1 | 3 |
| 5/16 | ½ | 100 | 1¾ | 5 | 1¼ | 3 · |
| ⅜ | ⅝ | 100 | 2¼ | 10 | 1¾ | 7½ |
| ½ | ¾ | 100 | 3 | 16 | 2 | 10 |
| ⅝ | ⅞ | 100 | 3½ | 21 | | |
| ¾ | 1 | 50 | 3½ | 24 | | |

**247. Interlocking Keystone Expansion Shields** (For use with machine bolts or machine screws)

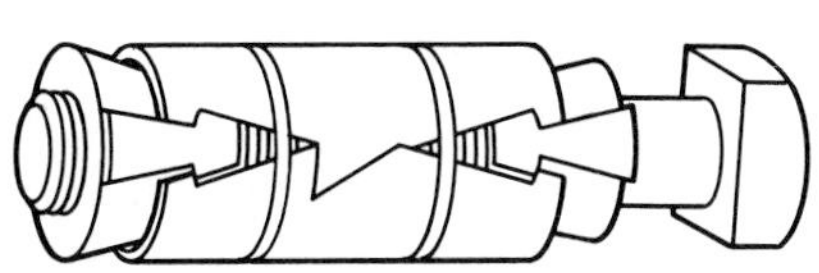

| Diam, in. | Length, in. | Outside diam and drill required, in. | Weight, lb per 100 | Diam, in. | Length, in. | Outside diam and drill required, in. | Weight, lb per 100 |
|---|---|---|---|---|---|---|---|

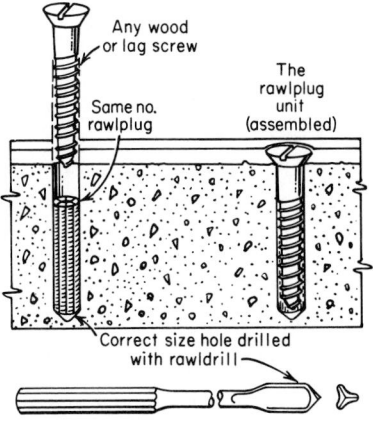

**Fig. 4-225**  Double. 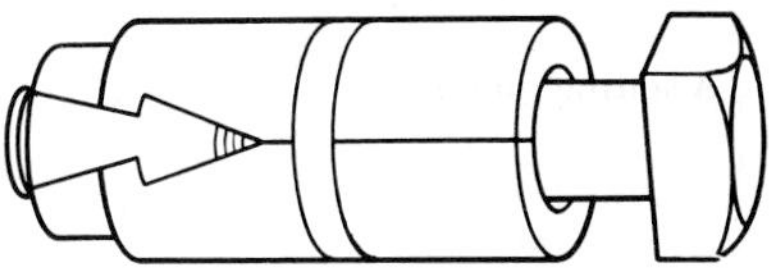        **Fig. 4-226**  Single.

| Diam, in. | Length, in. | Outside diam and drill required, in. | Weight, lb per 100 | Diam, in. | Length, in. | Outside diam and drill required, in. | Weight, lb per 100 |
|---|---|---|---|---|---|---|---|
| ¼ | 1½ | ½ | 6 | ½ | 1 5/16 | ½ | 6 |
| 5/16 | 1¾ | 9/16 | 7 | 5/16 | 1½ | 9/16 | 6 |
| ⅜ | 2 | 11/16 | 11 | ⅜ | 1⅝ | 11/16 | 10 |
| 7/16 | 2½ | ⅞ | 16 | ½ | 1⅞ | ⅞ | 16 |
| ½ | 2½ | ⅞ | 20 | ⅝ | 2 | 1 | 20 |
| ⅝ | 2⅞ | 1 | 29 | ¾ | 2¾ | 1 3/16 | 46 |
| ¾ | 3¼ | 1 ⅛ | 52 | | | | |
| ⅞ | 4 | 1 ½ | 88 | | | | |
| 1 | 4¼ | 1 ⅝ | 114 | | | | |
| 1¼ | 6 | 2 ⅛ | 300 | | | | |
| 1½ | 7½ | 2 ½ | 450 | | | | |

**248. Rawlplugs** (Fig. 4-227) are made from stiffened strands of jute fiber. The fiber strands are compressed together by a patented process. Once in place, a Rawlplug can never crumble or pulp. It is unaffected by moisture or change in temperature.

The screw entering the Rawlplug automatically threads it, which permits removal and replacement of the screw as often as desired without stripping the threads so formed. The screw cannot be withdrawn by direct or indirect pulling.

Rawlplugs require only a small hole. This makes installation easy and assures a neat job. The Rawlplug is invisible when in position, as the diameter of the hole and Rawlplug is smaller than the head of the screw, giving the screw the appearance of being screwed into the masonry itself.

Owing to their composition, Rawlplugs resist and absorb shocks. They cannot work loose, slip, sheer, or lose their viselike grip. When properly inserted, the Rawlplug should

**Fig. 4-227**  Installation of Rawlplugs.

develop the entire strength of the screw or of the surrounding material, such as concrete, brick, stone, slate, marble, plaster, glass, hard rubber, gypsum block, wallboard, terracotta, composition walls and floors, wood, and metal.

Rawlplugs are made in the sizes listed below in lengths of ⅝, ¾, 1, 1¼, 1½, 2, 2½, 3, and 3½ in.

| Rawlplug No. | Use with screw No. | Diam drill required, in. | Rawldrill No. (not diam) | Use with tool holder No. | Size twist drill, in. |
|---|---|---|---|---|---|
| | | Rawlplug, Rawldrill, and Tool-holder Size Data | | | |
| 6 | 5–6 | ⁵⁄₃₂ | 6 | 14 | ⁵⁄₃₂ |
| 8 | 7–8 | ¹¹⁄₆₄ | 8 | 14 | No. 15 |
| 10 | 9–10 | ³⁄₁₆ | 10 | 14 | No. 10 |
| 12 | 11–12 | ¼ | 12 | 14 | ¼ |
| 14 | 14ᵃ | ⁹⁄₃₂ | 14 | 14 | ⁹⁄₃₂ |
| 16 | 16ᵃ | ⁵⁄₁₆ | 16 | 20 | ⁵⁄₁₆ |
| 20 | 20ᵇ | ⅜ | 20 | 20 | ⅜ |
| | | Lag-screw Sizesᶜ | | | |
| ⅜ | ⅜ | ⁷⁄₁₆ | | | |
| ⁷⁄₁₆ | ⁷⁄₁₆ | ½ | | | |
| ½ | ½ | ⅝ | | | |
| ⅝ | ⅝ | ¾ | | | |

ᵃ Or ¼-in. lag screw.
ᵇ Or ⁵⁄₁₆-in. lag screw.
ᶜ For ¼-in. lag screw use No. 14 or No. 16 Rawlplug.  For ⁵⁄₁₆-in. lag screw use No. 20.

**249. Rawl-Drives** (Fig. 4-228) consist of a high-grade heat-treated steel pin with a flat head and split expanded central portion which, when driven in, grips the side of the hole with a spring action. They are easily and quickly installed, since they drive like a nail into a drilled hole no larger than the diameter of the pin itself. Rawl-Drives are for use only in solid masonry of brick, concrete, or stone. They should not be used in soft or brittle materials, such as plaster, wood, composition wood, glass, tile, etc. The manufacturer specifies that they have the following holding powers:

**Holding Power of Rawl-Drives**

| Size Rawl-drive, in. | Average direct pull, lb | |
|---|---|---|
| | 1-2-4 concrete | Common brick |
| ³⁄₁₆ × 1¼ | 1,285 | 634 |
| ¼ × 1½ | 2,050 | 1,183 |
| ⁵⁄₁₆ × 1½ | 3,500 | 1,833 |
| ⅜ × 2 | 5,010 | 2,150 |
| ½ × 3 | 6,015 | 3,700 |

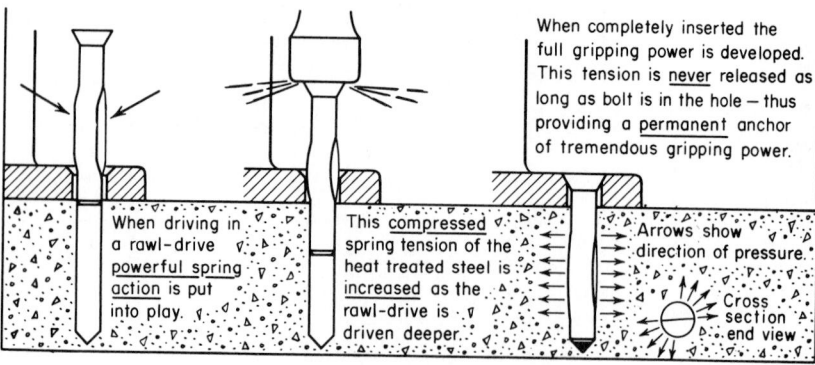

When completely inserted the full gripping power is developed. This tension is never released as long as bolt is in the hole — thus providing a permanent anchor of tremendous gripping power.

When driving in a rawl-drive powerful spring action is put into play.

This compressed spring tension of the heat treated steel is increased as the rawl-drive is driven deeper.

Arrows show direction of pressure.

Cross section end view

**Fig. 4-228**  Principle of Rawl-Drive.

**250.  Data for Rawl-Drives**

| Diam of bolt and drill, in. | Length of bolt, in. | Use Rawldrill No. or size | Packed[a] in boxes of | Approx weight, lb per 100 bolts |
|---|---|---|---|---|
| 3/16 | 1 | No. 10 | 100 | 1¾ |
| 3/16 | 1¼ | No. 10 | 100 | 2 |
| 1/4 | 1¼ | No. 12 | 100 | 3 |
| 1/4 | 1½ | No. 12 | 100 | 3½ |
| 1/4 | 2 | No. 12 | 100 | 4 |
| 5/16 | 1½ | No. 16 | 50 | 6 |
| 5/16 | 1½ | No. 16 | 50 | 7 |
| 3/8 | 2 | No. 20 | 25 | 10 |
| 3/8 | 2½ | No. 20 | 25 | 12 |
| 3/8 | 3 | No. 20 | 25 | 13 |
| 3/8 | 3½ | No. 20 | 25 | 15 |
| 1/2 | 3 | ½ in. | 25 | 25 |
| 1/2 | 3½ | ½ in. | 25 | 28 |
| 1/2 | 4 | ½ in. | 25 | 30 |

[a] A standard package consists of four boxes of the same size, length and type.

**251.  Anchor Selection Chart by Type of Material**
(The Rawlplug Co., Inc.)

| | Rawl-plug | Saber-tooth | Multi-calk | Calk-in | Rawl-drive | H/S drop-in | Double | Lag shield | Nailin | Spring-wing | Rawly |
|---|---|---|---|---|---|---|---|---|---|---|---|
| Brick | | • | • | • | • | • | | • | • | | |
| Concrete | | • | • | • | • | • | • | • | • | | |
| Concrete block | | | | | | • | | | • | • | • |
| Cinder block | | | | | | • | | | | • | • |
| Stone | | • | • | • | • | • | • | | • | | |
| Marble | | • | | • | | • | | | • | | |
| Building tile | | | | | | • | | | | • | • |
| Ceramic tile | | | | | | • | | | | • | • |
| Terrazzo | | • | | • | | • | | | • | | |
| Terra cotta | | | | | | • | | | | • | • |
| Plaster | | | | | | | | | | • | • |
| Dry wall | | | | | | | | | | • | • |
| Slate | | | | | | • | | | | | • |
| Stucco | | | | | | | | | | | |
| Glass | | | | | | | | | | | |

(Rawl-plug column label, spanning Stone through Terra cotta: "Use in any masonry material")

**252. Powder-actuated tools.**    A highly efficient method of attaching fasteners or materials to concrete or steel is the powder-actuated tool and fastening system. This integrated system (Fig. 4-229) permits fastening without drilling in one timesaving operation. It consists of a tool, a powder charge, and a fastener; and it fastens wood and steel to concrete of any compressive strength and to structural steel up to 1 in (25.4 mm) thick. A line of more than 100 drive pins, eye pins, and threaded studs meets the requirements for most jobs.

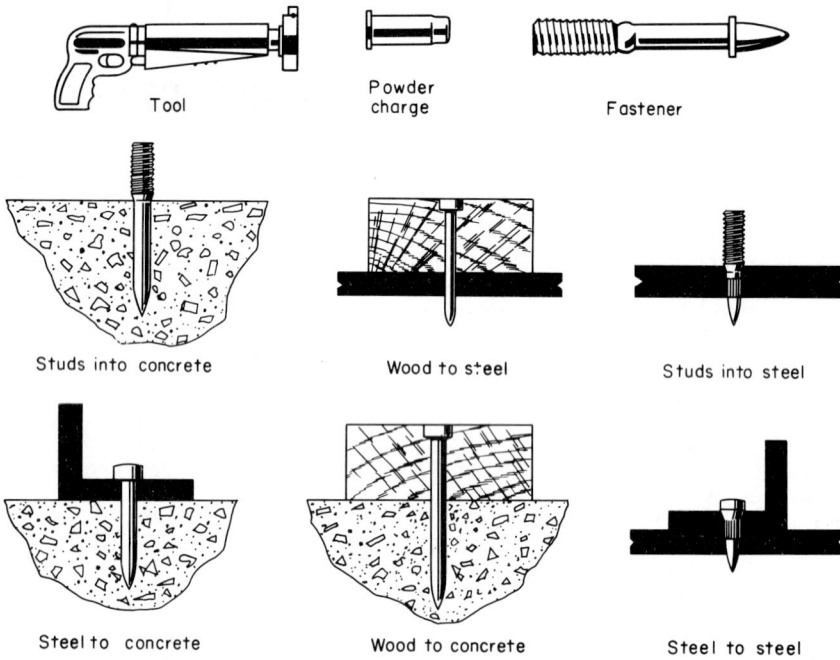

**Fig. 4-229**    A basic powder-actuated tool system. [Ramset, Olin Corp.]

There are several major manufacturers of powder-actuated tool systems, and it is suggested that users be fully familiar with these tools and fastening materials. These manufacturers provide guidebooks on how to use the systems efficiently and safely. In some states and local areas mechanics who use power-actuated tools must be certified or licensed to indicate that they completely understand the use or proper applications of these systems.

**Fig. 4-230**    Ram-Master semiautomatic power-load feed assembly. [Ramset, Olin Corp.]

Division **5**

# Transformers

## CONSTRUCTION, TYPES, AND CHARACTERISTICS

**1. A transformer** is an apparatus for converting electrical power in an ac system at one voltage or current into electrical power at some other voltage or current without the use of rotating parts.

**2. A constant-potential transformer** (Fig. 5-1) consists essentially of three parts: the primary coil which carries the alternating current from the supply lines, the core of magnetic material in which is produced an alternating magnetic flux, and the secondary coil in which is generated an emf by the change of magnetism in the core which it surrounds. Sometimes the transformer may have only one winding, which will serve the dual purpose of primary and secondary coils.

The high-tension winding is composed of many turns of relatively fine copper wire, well insulated to withstand the voltage impressed on it. The low-tension winding is composed of relatively few turns of heavy copper wire capable of carrying considerable current at a low voltage.

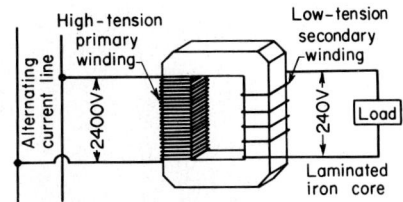

**Fig. 5-1**  The elementary transformer.

**3. Transformer terminology.**  The *primary winding* is the winding of the transformer which is connected to the source of power. It may be either the high- or the low-voltage winding, depending upon the application of the transformer.

The *secondary winding* is the winding of the transformer which delivers power to the load. It may be either the high- or the low-voltage winding, depending upon the application of the transformer.

The *core* is the magnetic circuit upon which the windings are wound.

The *high-tension winding* is the one which is rated for the higher voltage.

**Fig. 5-2**   Assembly of transformer-core laminations.

**Fig. 5-3**   Transformer assembly. [I-T-E Electrical Products Division of Siemens]

The *low-tension winding* is the one which is rated for the lower voltage.

A *step-up transformer* is a constant-potential transformer so connected that the delivered voltage is greater than the supplied voltage.

A *step-down transformer* is one so connected that the delivered voltage is less than that supplied; the actual transformer may be the same in one case as in the other, the terms *step-up* and *step-down* relating merely to the application of the apparatus.

**4. Transformer cores.** Until recently all transformer cores were made up of stacks of sheet-steel punchings firmly clamped together. One method of assembly and clamping of the sheets is shown in Figs. 5-2 and 5-3. Sometimes the laminations are coated with a thin varnish to reduce eddy-current losses. When the laminations are not coated with varnish, a sheet of insulating paper is inserted between laminations at regular intervals.

A new type of core construction consists of a continuous strip of silicon steel which is wound in a tight spiral around the insulated coils and firmly held by spot welding at the end. The core and windings of such a transformer are shown in Fig. 5-4. This type of construction reduces the cost of manufacture and reduces the power loss in the core due to eddy currents.

**Fig. 5-4** Assembled core and coils for wound-core distribution transformer. [I-T-E Electrical Products Division of Siemens]

**5. Classification of transformers**

1. According to method of cooling
   a. Self–air-cooled (dry type)
   b. Air-blast–cooled (dry type)
   c. Liquid-immersed, self-cooled
   d. Oil-immersed, combination self-cooled and air-blast
   e. Oil-immersed, water-cooled
   f. Oil-immersed, forced-oil–cooled
   g. Oil-immersed, combination self-cooled and water-cooled
2. According to insulation between windings
   a. Windings insulated from each other
   b. Autotransformers
3. According to number of phases
   a. Single-phase
   b. Polyphase
4. According to method of mounting
   a. Pole and platform
   b. Subway
   c. Vault
   d. Special

5. According to purpose
   a. Constant-potential
   b. Varying-potential
   c. Current
   d. Constant-current
6. According to service
   a. Large power
   b. Distribution
   c. Small power
   d. Sign lighting
   e. Control and signaling
   f. Gaseous-discharge lamp transformers
   g. Bell ringing
   h. Instrument
   i. Constant-current
   j. Series transformers for street lighting

**6. Cooling of transformers.** A certain amount of the electrical energy delivered to a transformer is transformed into heat energy because of the resistance of its windings and the hysteresis and eddy currents in the iron core. Means must be provided for removing this heat energy from the transformer and dissipating it into the surrounding air. If this were not done in a satisfactory manner, the transformer would operate at an excessively high temperature, which would destroy or harm the insulation of the transformer. The different methods of cooling employed are listed in Sec. 5 and described below.

I. Ventilated dry type with openings to aid circulation of air. [I-T-E Electrical Products Division of Siemens]

II. With solid-metal casing. [General Electric Co.]

**Fig. 5-5**   Dry-type transformers.

In *self–air-cooled* transformers (Fig. 5-5) the windings are simply surrounded by air at atmospheric pressure. The heat is removed by natural convection of the surrounding air and by radiation from the different parts of the transformer structure. Air cooling has long been employed for transformers of very small capacity. The development of satisfactory coil insulation materials, such as porcelain, mica, glass, and asbestos, which will withstand higher temperatures than the more common insulating materials has made possible the application of air cooling to transformers of large capacity. Except in the smaller sizes,

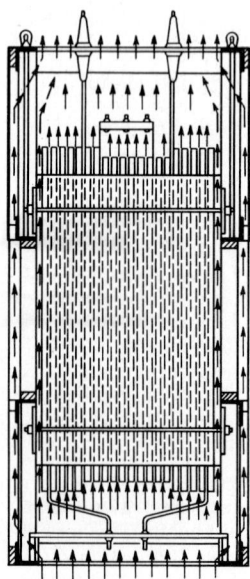

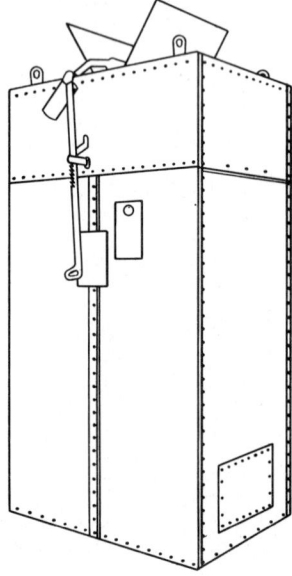

**Fig. 5-6**   Longitudinal section of air-blast transformer, showing direction of air currents.

**Fig. 5-7**   Exterior view of 1330-kVA air-blast transformer, showing the sheet-steel casing.

the sheet-metal enclosure is provided with louvers or gratings to allow free circulation of the air over and through the windings. Self–air-cooled transformers are commonly called dry-type transformers. The present-day use of self–air-cooled transformers has been extended to units of at least 3000-kVA capacity at 15,000 V.

In *air-blast-cooled transformers* (Figs. 5-6 and 5-7), the core and windings are enclosed in a metal enclosure through which air is circulated by means of a blower. This method has been used for large power transformers in ratings up to 15,000 kVA with voltages not exceeding 35,000.

In *liquid-immersed, self-cooled transformers*, the core and windings are immersed in some insulating liquid and enclosed in a metal tank. The liquid, in addition to providing some of the required insulation between the windings, carries the heat from the core and windings to the surface of the tank. The heat is then removed into the surrounding atmosphere by natural convection of the sur-

**Fig. 5-8** Small distribution transformer with smooth tank. [Moloney Electric Co.]

rounding air and by radiation from the tank. In the smaller sizes the tanks have a smooth surface (Fig. 5-8). In larger sizes tanks are corrugated or finned or have external tubes (Fig. 5-10), and in very large units the tanks must be supplied with external radiators (Fig. 5-11), through which the oil circulates by natural convection owing to differences in temperature in the liquid. This method can be employed for units of any size or voltage rating, although large-capacity units become rather expensive and bulky. The common liquid employed is an insulating oil. Nonflammable and nonexplosive liquids have been developed for use as a cooling and insulating medium for electrical equipment. These liquids are used in transformers where their nonflammable and nonexplosive qualities warrant their additional expense. The use of such a liquid is particularly advantageous for trans-

**Fig. 5-9** Medium-size transformer with corrugated tank. [Moloney Electric Co.]

**Fig. 5-10** Tubular-type tank as used on medium-size power transformers. [Moloney Electric Co.]

formers installed in buildings, since the transformer can then be installed in general areas without the use of a fireproof vault enclosure. Transformers insulated with such a liquid are designated as askarel-insulated transformers. Askarel has been banned by the Environmental Protection Agency and is being replaced in many instances by a listed less flammable nonpropagating liquid. See Sec. 450-23 of the National Electrical Code.

Large oil-immersed transformers are frequently cooled by means of a combination of *self-cooling* and *air blast* (Fig. 5-12). The construction of the transformer is in general the same as for those which are oil-immersed and self-cooled, with the addition of a motor-

**Fig. 5-11**    Large self-cooled transformer with radiators. [Moloney Electric Co.]

driven blower or blowers mounted integrally with the transformer tank. The blowers provide a forced circulation of air up through the radiators to supplement the natural convection air currents. The blower motors are generally automatically controlled by means of a thermostat. When the oil temperature reaches a certain value, the thermostat closes the motor circuit. After the temperature has been reduced to a definite value, the thermostat opens the motor circuit, shutting off the fans.

*Gas-vapor transformers* (Fig. 5-13) are sometimes employed for large units. The transformer is insulated with a quantity of gas necessary for start-up, along with a vaporizable liquid which provides insulation and cooling during operation. During operation, a pump delivers a stream of this liquid from the sump to the upper part of the core and coils. Here, under low pressure, the liquid is evenly distributed over the core structure and all current-carrying parts. Enough of the liquid evaporates to absorb heat from the core and coils. The hot vapors, being heavy, flow downward and enter the cooler-tube headers near the bottom of the tank. Normally, the bottom of the cooling tubes will be hotter than the top. In the cooling tubes, the vapors condense back into liquid, thus releasing their heat to the cooler-tube surfaces. The liquid returns to the pump by gravity, for recirculation.

**7. The oil used in transformers** (*Standard Handbook for Electrical Engineers*) performs two important functions. It serves to insulate the various coils from each other and from the core, and it conducts the heat from the coils and core to some cooler surfaces, where it is either dissipated in the surrounding air or transferred to some cooling medium. It is evident that the oil should be free from any conducting material, it should be sufficiently

I. Liquid-filled transformer with standard and optional accessories.

II. Fan cooling is temperature-activated by a liquid temperature gage.

**Fig. 5-12** Liquid-filled transformer that is self-cooled, forced-air. [I-T-E Electrical Products Division of Siemens]

thin to circulate rapidly when subjected to differences of temperature at different places, and it should not be ignitable until its temperature has been raised to a very high value. Although numerous kinds of oils have been tried in transformers, at the present time mineral oil is used almost exclusively. This oil is obtained by fractional distillation of petroleum unmixed with any other substances and without subsequent chemical treatment. A good grade of transformer oil should show very little evaporation at 100°C, and it should not give off gases at such a rate as to produce an explosive mixture with the surrounding air at a temperature below 180°C. It should not contain moisture, acid, alkali, or sulfur compounds.

It has been shown by C. E. Skinner that the deteriorating effect of moisture on the insulating qualities of an oil is very marked; moisture to the extent of 0.06 percent reduces

**Fig. 5-13**   Gas-vapor substation-type transformer. [Westinghouse Electric Corp.]

the dielectric strength of the oil to about 50 percent of the value when it is free from moisture, but there is very little further decrease in the dielectric strength with an increase in the amount of moisture in the oil.

Dry oil will stand an emf of 25,000 V between two 0.5-in (12.7-mm) knobs separated by 0.15 in (3.8 mm). The presence of moisture can be detected by thrusting a red-hot nail in the oil; if the oil "crackles," water is present. Moisture can be removed by raising the temperature slightly above the boiling point of water, but the time consumed (several days) is excessive. The oil is subsequently passed through a dry-sand filter to remove any traces of lime or other foreign materials.

**8. The insulating value of the oil, in oil-immersed transformers** *(Standard Handbook),* is depended on very largely to help insulate the transformer; this is done by providing liberal oil ducts between coils and between groups of coils, in addition to the solid insulation. The oil ducts thus serve the double purpose of insulating and cooling the windings.

Since the oil is a very important part of the insulation, every effort is made in modern transformers to preserve both its insulating and its cooling qualities. Oxidation and moisture are the chief causes of deterioration. Oil takes into solution about 15 percent by volume of whatever gas is in contact with it. In the open-type transformer, oil rapidly darkens, owing to the effects of oxygen in solution in the oil and the oxygen in contact with the top surface of the hot oil.

1. EXPANSION TANK (OR CONSERVATOR)   One of the first devices used to reduce oxidation

was the expansion tank (or conservator), which consisted of a small tank mounted above and connected with the main tank by means of a constricted connection so that the small tank could act as a reservoir to take up the expansion and contraction of the oil due to temperature changes and reduce the oil surface exposed to air.

2. INERTAIRE TRANSFORMER  This transformer has the space above the oil in the tank filled with a cushion of inert gas which is mostly nitrogen. The nitrogen atmosphere is initially blown in from a cylinder of compressed nitrogen and is thereafter maintained by passing the inbreathing air through materials which remove the moisture and the oxygen, permitting dry nitrogen to pass into the case. A breathing regulator, which consists of a mercury U tube with unequal legs, allows inbreathing of nitrogen when the pressure in the case is only slightly below atmospheric but prevents outbreathing unless the pressure in the case becomes 5 psi (34,474 Pa) higher than atmospheric pressure. The elimination of oxygen from within the transformer case eliminates the oxidation of the oil and prevents fire and secondary explosion within the case.

**9. Insulation between windings.**  The great majority of transformers are constructed with two or more windings which are electrically insulated from each other. In some cases a single winding is employed, parts of the winding functioning as both primary and secondary. These transformers are called autotransformers. They are frequently used when the voltage ratio is small. Autotransformers should never be used for high voltage ratios, as the low-voltage winding is not insulated from the high-voltage one, so that in case of trouble it would be dangerous to both life and equipment. Refer to Sec. 32 for further discussion.

**10. Transformer insulation.**  The type of insulation used in dry-type–transformer design and construction has a definite bearing on the size and operating temperature of the unit. Currently four classes of insulation, each having a separate NEMA specification and temperature limit, are being used. A look at these will facilitate selection of the proper unit to meet prescribed installation and operating conditions.

1. CLASS A TRANSFORMERS  When properly applied and loaded in an ambient not over 40°C, these transformers will operate at not more than a 55°C temperature rise on the winding. These units can be used as control-type transformers when higher temperatures might affect other temperature-sensitive devices in the enclosure or as distribution transformers in locations (textile mills, sawmills, etc.) where combustible flyings might be present in the surrounding atmosphere.

2. CLASS B TRANSFORMERS  These units have a higher-temperature insulating system and are physically smaller and about half the weight of Class A units of corresponding rated capacities. When properly loaded to rated kilovolt-amperes and installed in an ambient not over 40°C, Class B units will operate at a maximum 80°C rise on the winding. For years dry-type distribution transformers have been of the Class B type.

3. CLASS F TRANSFORMERS  These units also have a high-temperature insulating system and, when properly loaded and applied in an ambient not over 40°C, will operate at no more than 115°C rise on the winding. The units are smaller in size than similarly rated Class B units and currently are available from a number of manufacturers in ratings of 25 kVA and lower, both single- and three-phase design. One manufacturer designs in-wall, flush-mounted dry-type transformers as Class F units.

4. CLASS H TRANSFORMERS  These units are insulated with a high-temperature system of glass, silicone, and asbestos components and are probably the most compact ones available. When properly loaded and applied in an ambient not over 40°C, Class H transformers will operate at a maximum 150°C rise on the winding. This class of insulation is used primarily in designs in which the core and coil are completely enclosed in a ventilated housing. Generally, this stipulation covers units with ratings of 30 kVA and larger. Some experts recommend that the hottest spot on the metal enclosure be limited to a maximum rise of 40°C above a 40°C ambient.

It should be noted that Class B insulation is being replaced with Class F or H insulation in transformers of recent design.

Another significant factor which concerns all dry-type transformers is that they should never be overloaded. The way to avoid this is to size the primary or secondary overcurrent device as close as possible to the full-load primary or secondary current for other than motor loads. If close overcurrent protection has not been provided, loads should be checked periodically. Overloading a transformer causes excessive temperature, which, in turn, produces overheating. This results in rapid deterioration of the insulation and will cause complete failure of the transformer coils.

**11. Transformers are built in both single- and polyphase units.**  A polyphase transformer consists of separate insulated electric windings for the different phases, wound upon a single core structure, certain portions of which are common to the different phases.

THREE-PHASE TRANSFORMERS (*Standard Handbook*)  Although there are numerous possible arrangements of the coils and cores in constructing a polyphase transformer, it can be stated that a polyphase transformer generally consists of several one-phase transformers with separate electric circuits but having certain magnetic circuits in common. A three-phase transformer is illustrated in Fig. 5-14, together with the component one-phase transformer. It will be observed that a three-phase transformer requires 3 times as much copper as the one-phase component transformer but less than 3 times as much iron. Thus in comparison with three individual transformers the three-phase unit is somewhat lighter and more efficient. Each component transformer operates as though the others were not

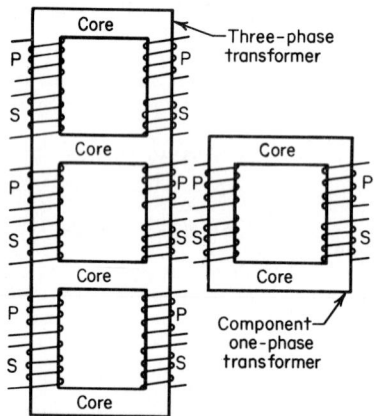

**Fig. 5-14**  Three-phase core-type transformer.

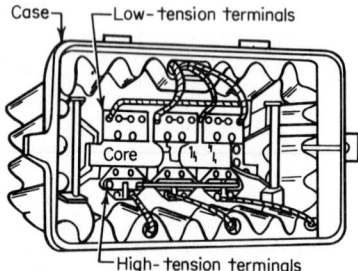

**Fig. 5-15**  Interior view of a three-phase transformer.

present, the flux of one transformer combining with that of an adjacent transformer to produce a resultant flux exactly equal to that of each one alone. Figure 5-15 shows the interior of a Westinghouse three-phase transformer.

**12. Application of three-phase transformers** (A. D. Fishel).  For central stations of medium size, three-phase transformers are rarely superior to single-phase except when the large sizes can be applied, in which case the transformers are normally installed in substations or central stations. The chief reason for this is the nonflexibility of a three-phase transformer. It is usually purchased for a particular size and type of load, and if that load should be changed, the transformer, representing a comparatively heavy investment, remains on the hands of the central station, whereas a single-phase transformer of one-third the size could usually be adapted for some other service.

This feature becomes of less importance as the central station increases its size, and three-phase transformers for purely power service are now being used by a considerable number of the large central stations in the United States. The three-phase transformer costs less to install, and the connections are simpler, points that are of importance in connection with outdoor installations. The fact that a failure of a three-phase transformer would interrupt service more than the failure of one single-phase transformer in a bank of three is of little importance because of the comparatively few failures of modern transformers. On the other hand, especially for 2200-V service, the single-phase transformer has been carried to a high degree of perfection and is manufactured in much larger quantities, so that better performance is usual and in some cases initial cost is lower. Three-phase distribution transformers are used extensively in underground city network service on account of the smaller space required by them in the manhole, their higher efficiency, and their lower initial cost. For overhead service for pole or platform mounting, three single-phase units are more common on account of the ease of handling and mounting the smaller-sized units.

**13. Methods of mounting.** Transformers are constructed with different types of metal enclosing structures to meet the requirements of different conditions of installation. One type of enclosure (Figs. 5-8 and 5-19) is designed for *mounting on poles,* either directly or with hanger irons, for use in overhead distribution work. Another type of enclosure, called the *platform type* (Figs. 5-10, 5-11, and 5-12), is suitable for installations in which the transformer stands upon its own base. It can be mounted on any flat horizontal surface having sufficient mechanical strength, such as a floor or a platform between poles. *Subway transformers* have watertight tanks which are designed primarily for underground installations when the transformer may be completely submerged in water. *Vault transformers* also have watertight enclosures so that they will not be injured by total submersion, but they are not designed to operate satisfactorily under such conditions. The vault transformers are intended for operation in underground vaults in which the transformer would not be required to operate for any considerable length of time while submerged. Small transformers for power and special application as listed in Sec. 5 (6c to 6j) are designed with special types of mounting to meet the requirements of installation for these types of service.

**14. Purpose of transformers.** Transformers can be classified as in Sec. 5 (5) according to the purpose of transformation for which they are employed.

The function of a *constant-potential* transformer is to change the voltage of a system. It is designed to operate with its primary connected across a constant-potential supply and to provide a secondary voltage which is substantially constant from no load to full load. It is the ordinary, common type of transformer. The currents of both primary and secondary vary with the load supplied by the transformer. A *varying-potential* transformer is also intended for changing the voltage of a system but is so designed that when operated with its primary connected to a constant-potential supply the secondary voltage will vary widely with the load. Such transformers are necessary for the operation of many gaseous-discharge lamps.

A *current* transformer is one which is designed for changing the current of a system. The primary winding of such a transformer is connected in series with the circuit of which it is desired to change the current. The voltage of both primary and secondary will change with the value of the current of the system. Such transformers are used for instrument transformers and in some series street-lighting installations. *Constant-current* transformers are designed for supplying a constant value of secondary current regardless of the load on the transformer. The primary is connected to a constant-potential source. The secondary voltage varies proportionally with the load, while the secondary current remains constant. The primary current and kilovolt-amperes will be constant for all loads, but the kilowatt input and power factor will vary with the load.

**15. The most important application of constant-potential transformers** is raising the voltage of an electric transmission circuit so that energy can be transmitted for considerable distances with small voltage drop and small energy loss and then lowering the voltage for safe usage by motors, lights, and appliances.

**16. Theory of operation of the constant-potential transformer** (see Fig. 5-1). It has been shown in Div. 1 that turns of wire wound on an iron core have self-induction. When an alternating voltage is applied to such turns, there flows through them a current that generates a countervoltage or emf that opposes the applied voltage. From formulas the transformer designer can compute just how many turns are necessary for a transformer of a given size so that it will generate a countervoltage equal to the applied voltage. So, in designing the primary winding of the transformer of Fig. 5-1, the designer would select such a number of turns for the primary winding that the countervoltage generated by it would be nearly 2400 V. Hence, when the primary winding is connected to a 2400-V circuit, it generates a countervoltage of practically 2400, and no appreciable current flows. A small current, the exciting current, just enough to magnetize the core, does flow, but it is so small that it can be disregarded in this discussion.

Since the primary and secondary windings are on the same core, the magnetic flux generated by the magnetizing or exciting current flowing in the primary winding also cuts the turns of the secondary winding and generates an emf in them. This emf will be, in accordance with a well-known law, opposite in direction to that impressed on the primary. If the secondary circuit is open, no current can flow in it, but if it is closed, a certain current, proportional to the impedance of the secondary circuit, will flow. This current, because of the direction of the emf generated in the secondary, will be in such a direction that the magnetic flux produced in the core by it will oppose the flux due to the primary winding. It will therefore decrease the effective or resultant flux in the core by a small amount which will decrease the counter-emf of the primary winding and permit more current to flow into the primary winding. As noted elsewhere, the ratio of the number of turns in the primary winding to the number of turns in the secondary winding determines the ratio of the primary to the secondary voltage.

If the voltage impressed on the transformer is maintained constant, the voltage of the secondary will be nearly constant also. When more current flows in the secondary, there will be a corresponding increase in primary current. As the load on a transformer increases, the impressed voltage remaining constant, there is actually a slight drop from the no-load voltage of the secondary, due to certain inherent characteristics of the transformer, but in a properly designed device this drop will be very small. Although the construction and elementary theory of the transformer are very simple, a theoretical explanation of all the phenomena involved in its operation is very complicated. Only the principal features have been described. Some minor, though very important, considerations that would complicate the explanation have not been treated.

**17. Transformer ratios.**   The voltage ratio of a constant-potential transformer, i.e., the ratio of primary to secondary voltage, depends primarily upon the ratio of the primary to the secondary turns. The voltage ratio will vary slightly with the amount and power factor of the load. For general work the voltage ratio can be taken as equal to the turn ratio of the windings.

The current ratio of a constant-potential transformer will be approximately equal to the inverse ratio of the turns in the two windings. For example, for transforming or "stepping down" from 2400 to 120 V the ratio of the turns in the windings will be 20:1. The currents in the primary and the secondary windings will be, very closely, inversely proportional to the ratio of the primary and secondary voltages because, if the small losses of transformation are disregarded, the power put into a transformer will equal the power delivered by it. For example, for a transformer with windings having a ratio of 20:1, if its secondary winding delivers 100 A at 50 V, the input to its primary winding must receive almost exactly 5 A at 1000 V. The input and output are each practically equal, and each would equal almost exactly 5000 W.

**18. The regulation of a transformer** is the change in secondary voltage from no load to full load. It is generally expressed as a percentage of the full-load secondary voltage:

$$\text{Percent regulation} = \frac{\text{no-load secondary voltage–full-load secondary voltage}}{\text{full-load secondary voltage}} \times 100 \quad (1)$$

The regulation depends upon the design of the transformer and the power factor of the load. Although with a noninductive load such as incandescent lamps the regulation of transformers is within about 3 percent, with an inductive load the drop in potential between no load and full load increases to possibly about 5 percent. If the motor load is large and fluctuating and close lamp regulation is important, it is desirable to use separate transformers for the motors.

**19. The efficiency of a transformer** is, as with any other device, the ratio of the output to input or, in other words, the ratio of the output to the output plus the losses. As a formula it can be expressed thus:

$$\text{Efficiency} = \frac{\text{output}}{\text{input}} = \frac{\text{output}}{\text{output} + \text{copper loss} + \text{iron loss}} \quad (2)$$

Average efficiencies of transformers are given in Secs. 43 to 57.

**20. The copper loss of a transformer** is determined by the resistances of the high-tension and low-tension windings and of the leads. It is equal to the sum of the watts of $I^2R$ losses in these components at the load for which it is desired to compute the efficiency.

**21. The iron loss of a transformer** is equal to the sum of the losses in the iron core. These losses consist of eddy- or Foucault-current losses and hysteresis losses. Eddy-current losses are due to currents generated by the alternating flux circulating within each lamination composing the core, and they are minimized by using thin laminations and by insulating adjacent laminations with insulating varnish. Hysteresis losses are due to the power required to reverse the magnetism of the iron core at each alternation and are determined by the amount and the grade of iron used for the laminations for the core.

**22. Transformer ratings.**   Transformers are rated at their kilovolt-ampere (kVA) outputs. If the load to be supplied by a transformer is at 100 percent power factor (pf), the kilowatt (kW) output will be the same as the kilovolt-ampere output. If the load has a lesser power factor, the kilowatt output will be less than the kilovolt-ampere output proportionally as the load power factor is less than 100 percent.

> **example**   A transformer having a full-load rating of 100 kVA will safely carry 100 kW if the 100 kW is at 100 percent pf, 90 kW at 90 percent pf, or 80 kW at 80 percent pf.

Transformers are generally rated on the kilovolt-ampere load which the transformer can safely carry continuously without exceeding a temperature rise of 55°C for Class A insulation or 80°C for Class B insulation when maintaining rated secondary voltage at rated frequency and when operating with an ambient temperature of 40°C. (Ambient temperature is the temperature of the surrounding atmosphere.) The actual temperature of any part of the transformer is the sum of the temperature rise of that part plus the ambient temperature. See Sec. 10 for an explanation of transformer insulation classifications.

The usual service conditions under which a transformer should be able to carry its rated load are:

1. At rated secondary voltage or not in excess of 105 percent of rated value
2. At rated frequency
3. Temperature of the surrounding cooling air at no time exceeding 40°C and average temperature of the surrounding cooling air during any 24-h period not exceeding 30°C
4. Altitude not in excess of 3300 ft
5. If water-cooled, temperature of cooling water not exceeding 30°C and average temperature of cooling water during any 24-h period not exceeding 25°C

GUIDE FOR LOADING OIL-IMMERSED DISTRIBUTION AND POWER TRANSFORMERS

**23. The following sections on guidance for loading of transformers** are excerpts from the American National Standards Institute's publication Appendix 057.92. This publication is not a part of the ANSI standard but is, as stated, a guide for the use of transformers.

The actual output which a transformer can deliver at any time in service without undue deterioration of the insulation may be more or less than the rated output, depending upon the ambient temperature and other attendant operating conditions.

**24. Loading on basis of test temperature rise.**   For each degree Celsius in excess of 2° that the test temperature rise is below the standard temperature rise specified in the standard, the transformer load may be increased above rated kilovolt-amperes by the percentages shown in col. 3 of the accompanying table. Making use of this factor gives the kilovolt-amperes that the transformer can deliver with a 55°C temperature rise. The leeway of 2° provides for a negative tolerance in the measurement of temperature rise.

### Loading on Basis of Ambient Temperature

| Type of cooling | Per cent of rated kva | |
|---|---|---|
| | Decrease load for higher temperature | Increase load for lower temperature |
| Self-cooled............ | 1.5 | 1.0 |
| Water-cooled............. | 1.5 | 1.0 |
| Forced-air-cooled.......... | 1.0 | 0.75 |
| Forced-oil-cooled......... | 1.0 | 0.75 |

Some transformers are designed to have the difference between hottest-spot and average copper temperatures greater than the nominal allowance of 10°C. This will result in a temperature rise for average copper of less than 55°C, but the hottest-spot copper temperature rise may be at the limiting value of 65°C. Such transformers should not be loaded above rating as outlined under this heading. The manufacturer should be consulted to give information as to design of hottest-spot allowances.

**25. Loading on basis of ambient temperature.**   For each degree Celsius that the average temperature of the cooling medium is above or below 30°C for air or 25°C for water, a transformer can be loaded for any period of time below or above its kilovolt-ampere rating as specified in the table of Sec. 24. Average temperature should be for periods of time not exceeding 24 h with maximum temperatures not more than 10°C greater than average temperatures for air and 5°C for water. On the basis used in this guide for calculating loss of life, life expectancy will be approximately the same as if the transformer had been operated at rated kilovolt-amperes and standard ambient temperatures over that period.

The use of transformers in cooling air above 50°C or below 0°C or with cooling water above 35°C is not covered by the table of Sec. 24 and should be taken up with the manufacturer.

**26. Loading on basis of load factor.**   When the load factor for a period of time not exceeding 24 h is below 100 percent, the maximum loading of a transformer during that period may be increased above rated kilovolt-amperes by the percentages shown in the accompanying table for each percent that the load factor is below 100 percent. On the basis used in this guide for calculating loss of life, life expectancy will be approximately the same as if the transformer had been operated at rated kilovolt-amperes during that period.

**Loading on Basis of Load Factor**

| Type of cooling | Increase in per cent of rated kva | Maximum per cent increase[a] |
|---|---|---|
| Self-cooled................. | 0.5 | 25 |
| Water-cooled.............. | 0.5 | 25 |
| Forced-air-cooled........... | 0.4 | 20 |
| Forced-oil-cooled........... | 0.4 | 20 |

[a] Corresponds to 50 per cent load factor.

**27. Loading on basis of short-time loads above rating.**   When short-time loads above rating occur not more than once in any 24-h period, the maximum loading of a transformer during that period can be increased conservatively above rated kilovolt-amperes, as given in Table 28. On the basis used in this guide for calculating loss of life, life expectancy will be approximately the same as if the transformer had been operated at rated kilovolt-amperes during that period.

**28.   Daily Overloads to Give Normal Life Expectancy**

| Time, hr | Times rated kilovolt-amperes | | | | | | | | |
|---|---|---|---|---|---|---|---|---|---|
| | Self-cooled and water-cooled transformers | | | Forced-air-cooled transformers rated 133 % or less of self-cooled rating | | | Forced-air-cooled transformers rated more than 133 % of self-cooled rating and all forced-oil-cooled | | |
| Initial load, %[a] | 90 | 70 | 50 | 90 | 70 | 50 | 90 | 70 | 50 |
| ½ | 1.59 | 1.77 | 1.89 | 1.45 | 1.58 | 1.68 | 1.36 | 1.47 | 1.50 |
| 1 | 1.40 | 1.54 | 1.60 | 1.31 | 1.38 | 1.50 | 1.24 | 1.31 | 1.34 |
| 2 | 1.24 | 1.33 | 1.37 | 1.19 | 1.23 | 1.26 | 1.14 | 1.18 | 1.21 |
| 4 | 1.12 | 1.17 | 1.19 | 1.11 | 1.13 | 1.15 | 1.09 | 1.10 | 1.10 |
| 8 | 1.06 | 1.08 | 1.08 | 1.06 | 1.07 | 1.07 | 1.05 | 1.06 | 1.06 |

[a] Percentages fix the load which is assumed to exist before the short-time load is applied.   Use either average load for 2 hr previous to load above rating or average load for 24 hr (less overload period), whichever is greater.

Ambient temperature assumed for this table is 30°C for air and 25°C for water.

As the loads may be applied once every 24 hr, and as there is some evidence that the rate of insulation deterioration at about 100°C doubles with less than 8°C increase in insulation temperature, the values have been based on 4 rather than on 8°C.

**29. Effect of various factors existing at one time.**   When two or more of the following factors affecting loading for normal life expectancy exist at one time, the effects are cumulative and the increase in loads due to each can be added to secure the maximum suggested load (each increase must be based on rated kilovolt-amperes):

1. Loading on basis of test temperature rise.
2. Loading on basis of ambient temperature.
3. Either loading on basis of load factor or loading on basis of short-time overloads. Do not use both.

**30.  Capacities of Transformers for Induction Motors**
(General Electric Co.)

| Size of motor, hp | Kva per transformer | | |
|---|---|---|---|
| | Two single-phase transformers | Three single-phase transformers | One three-phase transformer |
| 1 | 0.6 | 0.6 | |
| 2 | 1.5 | 1.0 | |
| 3 | 2.0 | 1.5 | 2.0 |
| 5 | 3.0 | 2.0 | 3.0 |
| 7½ | 4.0 | 3.0 | 5.0 |
| | | | 7.5 |
| 10 | 5.0 | 4.0 | 10.0 |
| 15 | 7.5 | 5.0 | 15.0 |
| 20 | 10.0 | 7.5 | 20.0 |
| 30 | 15.0 | 10.0 | 30.0 |
| 50 | 25.0 | 15.0 | 50.0 |
| 75 | 40.0 | 25.0 | 75.0 |
| 100 | 50.0 | 30.0 | 100.0 |

**31. Capacities of transformers for operating motors** (General Electric Co.).   For the larger motors the capacity of the transformers in kilovolt-amperes should equal the output of the motor in horsepower. Thus a 50-hp motor requires 50 kVA in transformers. Small motors should be supplied with a somewhat larger transformer capacity, especially if, as is desirable, they are expected to run most of the time near full load or even at slight overload. Transformers of less capacity than those noted in Table 30 should not be used even when a motor is to be run at only partial load.

**32. The autotransformer** *(Standard Handbook).*   The most efficient and effective method of operating a stationary transformer (when the ratio of transformation is not too large) is as an autotransformer, i.e., with certain portions of the windings used simultaneously as the primary and the secondary circuits. The electrical circuits of a one-phase autotransformer (sometimes called a compensator or a balance coil) are indicated in Fig. 5-16. The autotransformer has only one coil, a certain portion of which is used for both the high-tension and the low-tension windings. The number of turns of this coil is the same as would be required if it were used exclusively for the high-tension winding and a separate additional coil were used for the low-tension winding. Moreover, when the ratio of transformation is 2:1 or 1:2, the amount of copper in the one coil is exactly the same whether it is used as an autotransformer or as a high-tension coil of a two-coil transformer of the same rating. Not only is less copper required for an autotransformer than for a two-coil transformer, but less iron is needed to surround the copper.

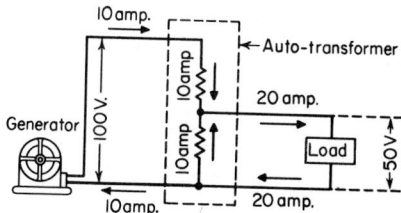

**Fig. 5-16**   Electric circuits of a 1-kVA single-phase autotransformer.

Referring again to Fig. 5-16, note that the one-coil transformer is designed for 10 A throughout and for a total emf of 100 V. The voltage per turn is uniform throughout, so that to obtain 50 V it is necessary merely to select any two points on the continuous winding such that one-half of the total number of turns is included between them. The load current of 20 A (required for 1000 W at 50 V) is opposed by the superposed 10 A of primary current, so that even in this section of the coil the resultant current is only 10 A.

If an ordinary two-coil transformer had been used, the circuits would have been as noted in Fig. 5-17, while the required constructive material would have been approximately as indicated in Fig. 5-18, I. With respect to its constructive material, a 1-kW 2:1-ratio autotransformer is the equivalent of a 1:1-ratio 0.5-kW two-coil transformer as shown in Fig. 5-18, II. The latter transformer requires about 14 lb (6.35 kg) of copper and 28 lb (12.7 kg) of iron as compared with about 22 lb (10 kg) of copper and 34 lb (15.4 kg) of iron for the transformer of Fig. 5-18, I. Moreover, the losses of the autotransformer are correspondingly less than those of a two-coil transformer. Refer to Sec. 117 for limitations in the allowable use of autotransformers.

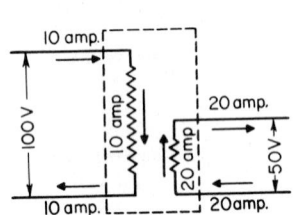

**Fig. 5-17** Electric circuits of a 1-kVA single-phase two-coil transformer equivalent to the autotransformer of Fig. 5-16.

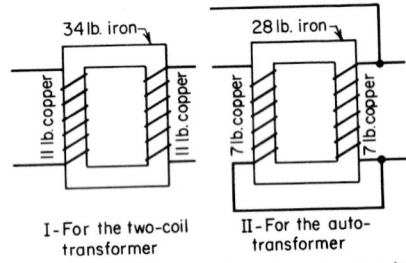

I - For the two-coil transformer    II - For the auto-transformer

**Fig. 5-18** Comparison of constructive material required for a two-coil transformer and for an autotransformer.

**33. Constant-potential transformers** for the transformation of a large amount of power, more than 500 kVA, are called power transformers. Transformers for general constant-potential power transformation, whose rating is 500 kVA or less, are called distribution transformers. All the methods of cooling are employed for power transformers. The choice depends upon which will result in the best overall economy including first cost, operating expense, and space occupied. Distribution transformers generally are liquid-immersed, self-cooled. Power and distribution transformers are normally of the standard type with the windings insulated from each other, although those with autotransformer construction can be obtained for special applications in which the voltage ratio is small. Power transformers are always of the platform type. Distribution transformers are made with tanks for pole and platform mounting and with tanks of the subway and vault types. The tanks of the platform type of transformers of 50-kVA capacity and smaller are equipped with lugs or brackets (Fig. 5-19) for direct pole mounting or for the attachment of hanger irons for crossarm pole mounting.

**34. Network transformers** are distribution transformers specially constructed and equipped with attached auxiliaries such as junction boxes and switches for disconnecting and grounding the high-voltage cable to meet the requirements of transformers for supplying low-voltage networks.

**35. Self-protected distribution transformers** are equipped with lightning and overload protective equipment built integrally with the transformers. They are made in two types: completely self-protecting (CSP), i.e., having both overload and lightning protection, and with only surge protection (SP). The connections of the lightning protective equipment can be made in different ways to satisfy any desired grounding practice. A sectional view of such a CSP transformer is shown in Fig. 5-20.

**36. Voltage taps.** It is frequently desirable on transformers to obtain a voltage ratio which is slightly different from the standard ratio obtained with the complete winding. Most transformers are provided with two 5 percent taps on the high-voltage winding for this purpose. The taps are for reduced voltage and allow the rated secondary voltage to be obtained when the supply voltage is below the rated value. Many transformers are

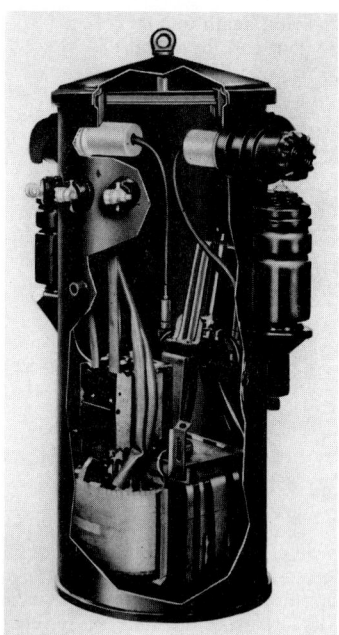

**Fig. 5-19** Distribution transformer for pole mounting. [Marcus Transformer Co.]

**Fig. 5-20** Sectional view of CSP transformer. [Westinghouse Electric Corp.]

equipped with a ratio adjuster (Fig. 5-21). The ratio adjuster provides a means of changing taps on the transformer simply by turning an operating handle without having to take down and remake connections. The diagram of connections of a ratio adjuster is shown in Fig. 5-22. The transformer of the illustration is provided with two primary coils instead of one, as would be the more general practice. The transformer should be disconnected from the line before the taps are changed with the ratio adjuster, as the device is not intended for operation under load.

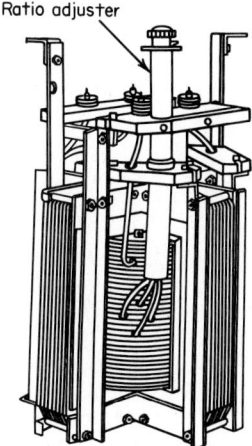

Ratio adjuster

**Fig. 5-21**    Core and coils for a 6600-V distribution transformer showing ratio adjuster. [General Electric Co.]

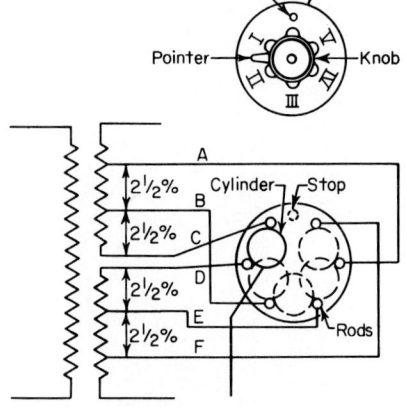

| Position | Connecting | % of Winding |
|----------|-----------|--------------|
| I | C to D | 100 |
| II | B to D | 97.5 |
| III | B to E | 95 |
| IV | A to E | 92.5 |
| V | A to F | 90 |

**Fig. 5-22**  Diagram of ratio-adjuster connections to transformer with four 2½ percent taps. [General Electric Co.]

The standard taps for distribution transformers are given in Secs. 38 and 39. A rated–kilovolt-ampere tap is a tap through which the transformer can deliver its rated kilovolt-ampere output without exceeding its rated temperature rise. A reduced–kilovolt-ampere tap is a tap through which the transformer can deliver only a kilovolt-ampere output less than its rated value without exceeding its rated temperature rise.

Large power transformers frequently are equipped with load-ratio control equipment by means of which the connections to taps of the transformer can be changed under load.

### 37.  Preferred Ratings of Transformers
(kVA ratings)

| Single-phase | | | | Three-phase | | | |
|---|---|---|---|---|---|---|---|
| 3 | 100 | 2,500 | 20,000 | 9 | 300 | 5,000 | 37,500 |
| 5 | 167 | 3,333 | 25,000 | 15 | 500 | 7,500 | 50,000 |
| 10 | 250 | 5,000 | 33,333 | 30 | 750 | 10,000 | 60,000 |
| 15 | 333 | 6,667 | | 45 | 1,000 | 12,000 | 75,000 |
| 25 | 500 | 8,333 | | 75 | 1,500 | 15,000 | 100,000 |
| 37½ | 833 | 10,000 | | 112½ | 2,000 | 20,000 | |
| 50 | 1,250 | 12,500 | | 150 | 2,500 | 25,000 | |
| 75 | 1,667 | 16,667 | | 225 | 3,750 | 33,333 | |

## 38. Standard Ratings for Single-Phase Transformers
(American National Standards Institute)

| | Transformer high-voltage | | Taps | | Standard kva ratings for low-voltage ratings of (1) | | | | | | | |
|---|---|---|---|---|---|---|---|---|---|---|---|---|
| Preferred nominal system voltage | Rating (9) | BIL, kv | Above | Below | 120/240 (2) | 240/480 (2) | 240 × 480 (3) | 600 | 2,400 or 4,800 | 2,520 or 5,040 | 6,900 | 7,200 or 7,560 or 7,980 |
| 480 | 480(8) | 30 | None | 2–5 % | 3–100 | | | | | | | |
| 600 | 600(8) | 30 | None | 2–5 % | 3–100 | | | | | | | |
| | | | None | None | 3–50 | | | | | | | |
| 2,400 or 2,400/4,160Y | 2,400 | 60 | None | 4–2½ % | 10–167 | | | | | | | |
| | | | 2–2½ % | 2–2½ % | | 10–100 | — | 25 | | | | |
| | | | | | | | — | 50 | | | | |
| | | | | | | | | 100 | | | | |
| | | | | | | | 167 | 167 | | | | |
| | | | | | 250–500 | | 250–500 | 333–500 | | | | |
| 2,400/4,160Y | 4,160 | 75 | None | 4–2½ % | 3–100 | | | | | | | |
| | | | 2–2½ % | 2–2½ % | | 10–100 | | | | | | |
| 4,800 or 4,800/8,320Y | 4,800 | 75 | None | None | 3–167 | | | | | | | |
| | | | 2–2½ % | 2–2½ % | | 10–100 | — | 25 | | | | |
| | | | | | | | — | 50 | | | | |
| | | | | | | | | 100 | | | | |
| | | | | | | | 167 | 167 | | | | |
| | | | | | 250–500 | | 250–500 | 333–500 | | | | |
| 2,400 or 2,400/4,160Y or 4,800 or 4,800/8,320Y | 2,400 × 4,800 | 75 | None | None | 3–100 | | | | | | | |

**Standard Ratings for Single-Phase Transformers** (*Continued*)

| Preferred nominal system voltage | Transformer high-voltage Rating (9) | BIL, kv | Taps Above | Taps Below | 120/240 (2) | 240/480 (2) | 240 × 480 (3) | 600 | 2,400 or 4,800 | 2,520 or 5,040 | 6,900 | 7,200 or 7,560 or 7,980 |
|---|---|---|---|---|---|---|---|---|---|---|---|---|
| | | | | | Standard kva ratings for low-voltage ratings of (1) | | | | | | | |
| 7,200 or 7,200/12,470Y | 7,200 | 95 | None<br>None<br>2-2½% | None<br>4-2½%<br>2-2½% | 3-50<br>3-167<br>—<br>—<br>250-500<br>— | 10-100 | —<br>—<br>167<br>250-500<br>— | 25<br>50<br>100<br>167<br>333-500<br>— | 50<br>100<br>167-500 | 333-500 | | |
| 7,200/12,470Y | 12,470GrdY<br>7,200<br>(4) | 95 | None<br>None | None<br>4-2½% | 3-25<br>3-50<br>3-25<br>3-50 | | | | | | | |
| 7,620/13,200Y | 7,620 | 95 | None<br>2-2½% | None<br>2-2½% | 3-167<br>—<br>—<br>250-500 | 10-100 | —<br>—<br>167<br>250-500 | 25<br>50<br>100<br>167<br>333-500 | | | | |
| 7,620/13,200Y | 13,200GrdY<br>7,620<br>(4) | 95 | None<br>2-2½% | None<br>2-2½% | 3-5<br>3-5<br>3-25<br>3-50 | | | | | | | |

| System (V) | Rated (V) | BIL (kV) | | | 5–167 | 10–100 | 167 / 250–500 | | 50/100 ; 167–500 | 333–500 | | 167 / 333–500 |
|---|---|---|---|---|---|---|---|---|---|---|---|---|
| 12,000 | 12,000 | 95 | None<br>2–2½% | 4–2½%<br>2–2½% | — | 10–100 | 167<br>250–500 | 25<br>50<br>100<br>167<br>333–500 | 50<br>100 | 167–500<br>333–500 | | |
| 13,200 | 13,200 | 95 | 2–2½% | 2–2½% | —<br>—<br>250–500<br>— | 10–100 | —<br>—<br>167<br>250–500<br>— | 167<br>333–500<br>— | 167–500 | 333–500 | | |
| 14,400 | 13,800 | 95 | 14,400/14,100 | 13,500/13,200 | 37½–167<br>— | 10–100 | —<br>167<br>250–500<br>— | 167<br>333–500<br>— | 167–500 | 333–500 | | |
| 13,200 or 14,400 | 14,400(5) | 95 | None | 13,800/13,200<br>12,870/12,540(6) | 250–500<br>— | 10–100 | 167<br>250–500<br>— | 167<br>333–500<br>— | 50<br>100 | 333–500 | | |
| 23,000 | 22,900 | 150 | 24,100/23,500 | 22,300/21,700 | 167<br>250–500 | 25<br>50<br>100 | —<br>167<br>333–500<br>— | 100<br>167<br>333–500<br>— | 100–167<br>333–500 | | —<br>333–500 | 167<br>333–500 |
| 34,500 | 34,400 | 200 | 38,200/35,300 | 33,500/32,600 | 25<br>100 | 25<br>50<br>100 | 167<br>333–500<br>— | 167<br>333–500<br>— | 167<br>333–500 | | —<br>333–500 | 167<br>333–500 |
| 46,000 | 43,800 | 250 | 46,200/45,000 | 42,600/41,400 | 50<br>100 | 50<br>100 | 167<br>333–500<br>— | 167<br>333–500<br>— | 167<br>333–500 | | —<br>333–500 | 167<br>333–500 |
| 69,000 | 67,000 | 350 | 70,600/68,800 | 65,200/63,400 | 100 | 100 | 167<br>333–500<br>— | —<br>— | 167<br>333–500 | | —<br>333–500 | 167<br>333–500 |

## 38. Standard Ratings for Single-Phase Transformers (Continued)

(1) Standard kVA ratings are 3, 5, 10, 15, 25, 37½, 50, 75, 100, 167, 250, 333, and 500. Kilovolt-ampere ratings separated by a dash (—) indicate that all intervening standard ratings are included. Groupings by kVA ratings are as given in original ANSI publication.
(2) Low-voltage rating of 120/240 or 240/480 V is suitable for series, multiple, or three-wire service.
(3) Low-voltage rating 240 × 480 V or high-voltage rating, 2400 × 4800 V is suitable for series or multiple service but not for three-wire service.
(4) One high-voltage bushing only.
(5) Transformers in this class are designed to cover in one rating the general voltage range 13,000 – 15,000 V.
(6) The lowest voltage shall be reduced kVA rating. All others shall be rated at kVA.
(7) Suitable only when system ground conditions permit the use of 18-kV arresters.
(8) Not equipped with tap changers.
(9) Transformers of the following ratings will also be available:

| Transformer high-voltage rating | Basic impulse level | High-voltage taps | | Standard kva ratings for low-voltage ratings of (1) | | |
| --- | --- | --- | --- | --- | --- | --- |
| | | Above | Below | 120/240(2) | 240/480(2) | 240 × 480(3) |
| 16,340 | 95 | 17,200/16,770 | 15,910/15,480 | 5-25<br>50<br>100 | 10-100<br>—<br>— | 167<br>333-500 |
| 24,940GrdY/14,400(4) (7) | 125 | None | 13,800/13,200<br>12,870/12,540(6) | 3-25<br>3-50 | | |
| 14,400/24,940GrdY(7) | 125 | None | 13,800/13,200<br>12,870/12,540(6) | 3-50 | 10-50 | |

## 39. Standard Ratings for Three-Phase Transformers
(American National Standards Institute)

| Preferred nominal system voltage | Transformer high voltage | | Taps | | Standard kva ratings for low-voltage ratings of:[a,c] | | | | | |
|---|---|---|---|---|---|---|---|---|---|---|
| | Rating[a] | Basic impulse level (BIL), kv | Above | Below | 208Y/120 | 240[b] or 480 | 240 × 480[b] | 240 × 480 | 2,400 or 4,160Y/ 2,400 or 4,800 | 12,470Y/ 7,200 or 13,200Y/ 7,620 |
| 2,400 | 2,400 | 45 | None | None | 9-75 | 9-45 | 75 | | | |
| | | | None | 4-2½% | 112½-150 225-500 | | | 225-500 | | |
| | | | 2-2½% | 2-2½% | | | 112½-150 — | | | |
| | 4,160Y/2,400 | 60 | None | None | 9-75 | 9-45 | 75 | | | |
| | | | 2-2½% | 4-2½% | 112½-150 225-500 | | | 225-500 | | |
| 2,400/4,160Y | 4,160Y | 60 | None | None | | 9-45 | 75 | | | |
| | | | 2-2½% | 2-2½% | | | 112½-150 — | | | |
| | 4,160 | 60 | 1-5% | 1-5% | | 9-45 | 75 | | | |
| | | | None | 2-5% | | | 112½-150 — | 225-500 | | |
| | | | None | 4-2½% | 9-75 112½-150 225-500 | 9-45 | 75 | | | |
| 4,800 | 4,800 | 60 | None | 4-2½% | 150 300-500 | | 75 | | | |
| | | | 2-2½% | 2-2½% | | | 150 — | | | |
| | 8,320Y/4,800 | 75 | None | None | 9-75 | 9-45 | | 300-500 | | |
| | | | None | 4-2½% | 150 300-500 | | | | | |

NOTE. Footnotes appear at end of table.

# Standard Ratings for Three-Phase Transformers (Continued)

| Preferred nominal system voltage | Transformer high voltage | | Taps | | Standard kva ratings for low-voltage ratings of:[a,c] | | | | | |
| | Rating[a] | Basic impulse level (BIL), kv | Above | Below | 208Y/120 | 240[b] or 480 | 240 × 480[b] | 240 × 480 | 2,400 or 4,160Y / 2,400 or 4,800 | 12,470Y / 7,200 or 13,200Y / 7,620 |
|---|---|---|---|---|---|---|---|---|---|---|
| 4,800/8,320Y | 8,320Y | 75 | None | None | — | 9–45 | 75 | | | |
| | | | 2–2½% | 2–2½% | | — | 150 | 300–500 | | |
| 7,200 | 7,200 | 75 | None | 2–5% | 9–75 | | | | | |
| | | | None | 4–2½% | 112½–150 225–500 | | | | | |
| | | | 1–5% | 1–5% | — | 9–45 | 75 | | | |
| | | | 2–2½% | 2–2½% | | — | 112½–150 | 225–500 | | |
| 12,000 | 12,000 | 95 | None | 2–5% | 15–75 | | | | | |
| | | | None | 4–2½% | 112½–150 225–500 | | | | | |
| | | | 1–5% | 1–5% | — | 15–45 | 75 | | 150 | |
| | | | 2–2½% | 2–2½% | | — | 112½–150 | 225–500 | 300–500 | |
| 7,200/12,470Y | 12,470Y/7,200 | 95 | None | 2–5% | 15–75 | | | | | |
| | | | None | 4–2½% | 112½–150 225–500 | | | | | |
| | 12,470Y | | 1–5% | 1–5% | — | 15–45 | 75 | | | |
| | | | 2–2½% | 2–2½% | | — | 112½–150 | 225–500 | | |

| Nominal voltage | Rated voltage | BIL, kv | Rated voltage | Taps | Rated voltage | Taps | kva[a] | | | | | |
|---|---|---|---|---|---|---|---|---|---|---|---|---|
| 7,620/13,200Y | 13,200Y/7,620 | 95 | | 1-5% | | 2-2½% | 15-75 | 15-45 | 75 | 112½-150 / 225-500 | | |
| | 13,200Y | 95 | | 1-5% | | 2-2½% | | | | 225,500 | | |
| 13,200 | 13,200 | 95 | None | 2-2½% | | 4-2½%ᵇ | 15-75 | 15-45 | 75 | 112½-150 / 225-500 | 150 / 300-500 | 150 / 300-500 |
| 14,400 | 13,800 | 95 | 14,400/14,100 | 2-2½% | 13,500/13,200 | 2-2½% | 15-75 | 15-45 | 75 | 112½-150 / 225-500 | 150 / 300-500 | 150 / 300-500 |
| 23,000 | 22,900 | 150 | 24,100/23,500 | | 22,300/21,700 | | | | | 150 / 300-500 | 150 / 300-500 | 150 / 300-500 |
| 34,500 | 34,400 | 200 | 36,200/35,300 | | 33,500/32,600 | | | | | 300-500 | 300-500 | 300-500 |
| 46,000 | 43,800 | 250 | 46,200/45,000 | | 42,600/41,400 | | | | | 300-500 | 300-500 | 300-500 |
| 69,000 | 67,000 | 350 | 70,600/68,800 | | 65,200/63,400 | | | | | 500 | 500 | 500 |

ᵃ Standard kva ratings are 9, 15, 30, 45, 75, 112½, 150, 225, 300, 500. Kilovolt-ampere ratings separated by a dash (—) indicate that all intervening ratings are included.

ᵇ A 120-volt reduced kva tap is provided.

ᶜ All transformers are delta-connected unless otherwise specified.

**40. Standard Accessory Equipment for Single-Phase Transformers**
(American National Standards Institute)

| Accessory | 12.47GrdY/7.2 13.2GrdY/7.62 24.94GrdY/14.4 — 120/240 — 3 to 25 | 12.47GrdY/7.2 13.2GrdY/7.62 24.94GrdY/14.4 — 120/240 — 3 to 25 | — 37½ to 50 | 5 and below — 600 and below — 3 to 25 | — 37½ to 50 | — 75 to 100 | — 167 — 167 | 7.2 to 15 14.4/24.94GrdY/16.34 — 5,000 and below — 3 to 25 | — 37½ to 50 | — 75 to 100 | 600 and below — 167 | 22.9 to 67 — 600 and below — 25,50,100 | — 167 | — 250 to 500 | 7.2 to 15 — 167 | 2.4 to 67 — 250 to 500 | Above 600 7.2 to 15 — 167 | 22.9 to 67 — 100 | 22.9 to 67 — 167 | 7.2 to 67 — 250 to 500 |
|---|---|---|---|---|---|---|---|---|---|---|---|---|---|---|---|---|---|---|---|---|
| Support lug, Type (see Figs. 23, 24, 25) | A | A | B | A | B | B | C | A | B | B | C |  |  |  |  |  |  |  |  |  |
| Combination oil-drain plug and sampling device |  |  |  | X | X | X | X | X | X | X | X | X | X | X | X | X | X | X | X | X |
| Combination oil drain, filter-press connection, and sampling valve | X |  |  |  |  | X | X |  | X | X | X |  | X | X | X | X | X | X | X | X |
| Liquid-level marking |  | X | X | X | X | X | X | X | X | X | X | X | X | X | X | X | X | X | X | X |
| Magnetic liquid-level gage |  |  |  |  |  |  |  |  |  |  |  |  |  |  | X |  | X | X | X | X |
| Dial-type thermometer |  |  |  |  |  |  |  |  |  |  |  |  |  |  | X |  | X | X | X | X |
| Upper filter-press connection | X |  |  |  |  | X | X |  | X | X | X |  | X | X | X | X | X | X | X | X |
| Upper filter-press valve |  |  |  |  |  |  |  |  |  |  |  |  |  |  | X |  | X | X | X | X |
| Tap changer, internal operation | X |  | X | X | X | X | X | X | X | X | X | X | X | X | X | X | X | X | X | X |
| Tap changer, external operation |  |  |  |  |  |  |  |  |  |  |  |  |  |  |  |  | X | X | X | X |
| Pressure-vacuum gage provision |  |  |  |  |  |  |  |  |  |  |  |  |  |  |  |  | X | X | X | X |
| Jacking provision |  |  |  |  |  |  |  |  |  |  |  |  |  |  |  |  |  |  |  |  |
| Rolling provision |  |  |  |  |  |  |  |  |  |  |  |  |  |  |  |  |  |  |  |  |
| Handhole in cover |  |  |  |  |  |  |  |  |  |  |  |  |  |  |  |  |  |  |  |  |
| Lifting lugs | X | X | X | X | X | X | X | X | X | X | X | X | X | X | X | X | X | X | X | X |
| Tank grounding provision | X | X | X | X | X | X | X | X | X | X | X | X | X | X | X | X | X | X | X | X |
| Tank grounding connector | X | X | X | X | X | X | X | X | X | X | X | X | X | X | X | X | X | X | X | X |
| Low-voltage grounding connection | X | X | X | X | X | X | X | X | X | X | X | X | X | X | X | X | X | X | X | X |
| Low-voltage grounding provision |  |  |  |  |  |  |  |  |  |  |  |  |  |  |  |  |  |  |  |  |
| High-voltage bushing terminals | X | X | X | X | X | X | X | X | X | X | X | X | X | X | X | X | X | X | X | X |
| Low-voltage bushing terminals | X | X | X | X | X | X | X | X | X | X | X | X | X | X | X | X | X | X | X | X |
| Low-voltage bushing arrangement | X | X | X | X | X | X | X | X | X | X | X | X | X | X | X | X | X | X | X | X |
| Nameplate location | X | X | X | X | X | X | X | X | X | X | X | X | X | X | X | X | X | X | X | X |
| Nameplate extension | A | A | A | A | A | A | A | A | A | A | A | A | A | A | A | A | A | A | A | A |
| Nameplate, type | X | X | X | X | X | X | X | X | X | X | X | X | B | B | B | B | B | B | B | B |
| Stenciled rating | X | X | X | X | X | X | X | X | X | X | X | X | X | X | X | X | X | X | X | X |

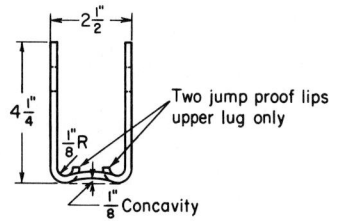

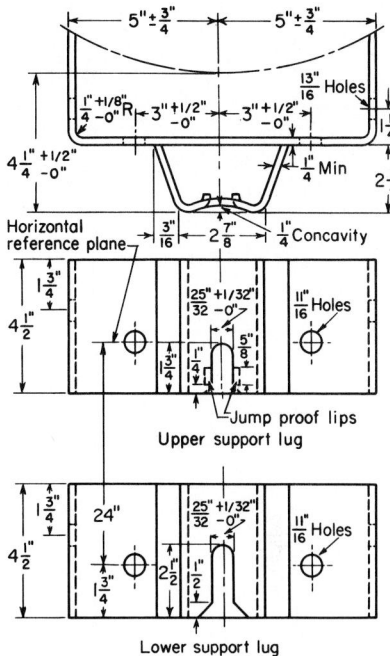

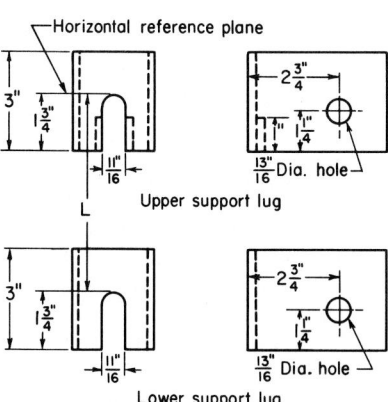

Upper support lug

Lower support lug

NOTES

1. Both support lugs No. 10 gage MSG (minimum) steel.

2. May be flanged for welding to tank.

3. External side surfaces to be flat and parallel.

4. Jumpproof lips omitted on lower lug.

5. Support lugs are spaced at ¾ in less than spacing of pole bolts for ease in mounting.

6. Tolerance for bolt-slot dimensions shall be ±¹⁄₆₄ in. Tolerance for all other dimensions shall be ±¹⁄₁₆ in.

7. Support lugs to be used with ⅝-in pole bolts.

8. For $L$ dimension, see ANSI Table 12-27.001$a$ and $b$.

**Fig. 5-23** Transformer support lug A. [American National Standards Institute]

Upper support lug

Lower support lug

NOTES

1. Support lugs may be flanged for welding to tank.

2. Jumpproof lips omitted on lower support lug.

3. Tolerance, except where indicated otherwise, shall be ±¹⁄₁₆ in.

4. Support lugs to be used with ¾-in pole bolts.

**Fig. 5-24** Transformer support lug B. [American National Standards Institute]

NOTES

1. Support lugs attached to transformer and intended for bolting to adapter plates for direct pole mounting or to conventional crossarm hangers.

2. Slots or holes shall be suitable for ⅝-in bolts.

3. Support-lug faces are to be in one plane.

4. For $L$ dimension see ANSI Table 12-27.001$a$ and $b$.

5. The dimensions shown must be maintained to obtain a standard mounting and are not intended to show details of construction.

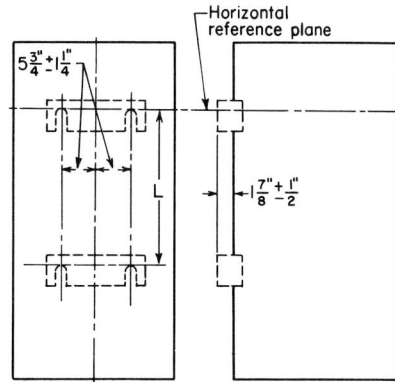

**Fig. 5-25** Transformer support lug C. [American National Standards Institute]

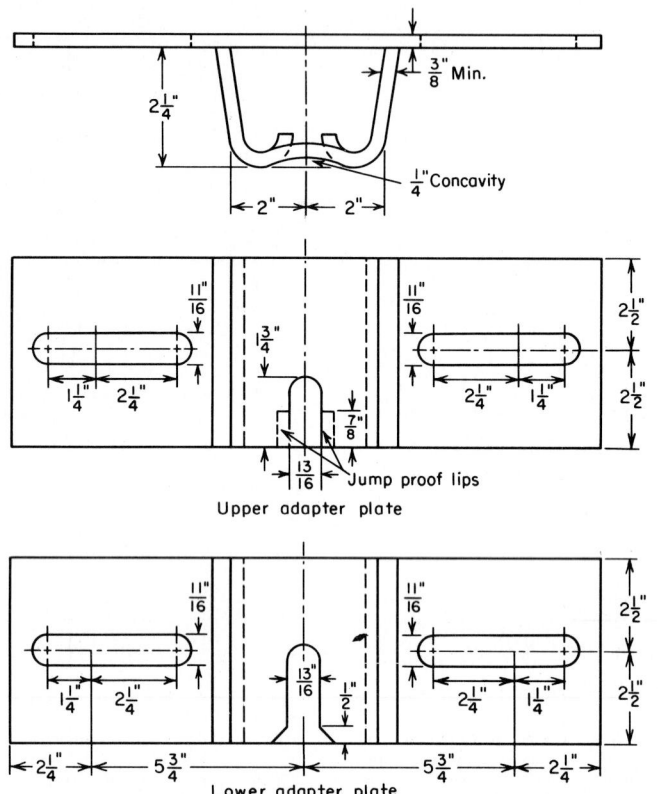

**Upper adapter plate**

**Lower adapter plate**

NOTES

1. Upper and lower adapter plates are identical, except that bolt slot is ¾ in longer and jumpproof lips are omitted on lower plate and bottom of bolt slot is chamfered.

2. Use ⅝-in bolts to bolt adapter plates to transformer support lugs. Use ¾-in bolts to bolt adapter plates to pole.

3. For ease of inserting pole bolts in adapter-plate slots and for tolerance in boring pole-bolt holes, the distance between the tops of bolt slots is ¾ in less than bolt spacing, and lower edges of slots are chamfered on lower adapter plate.

4. Tolerances, except where indicated otherwise, shall be ±¹⁄₁₆ in, except that adapter-plate bolt slot tolerances shall be ±¹⁄₆₄ in.

5. Adapter plates, nuts, and bolts shall be hot-dip–galvanized.

6. For use with 75–112½ kVA three-phase, 15 kV and below.

**Fig. 5-26**  Type C adapter plates for direct pole mounting of transformers. [American National Standards Institute]

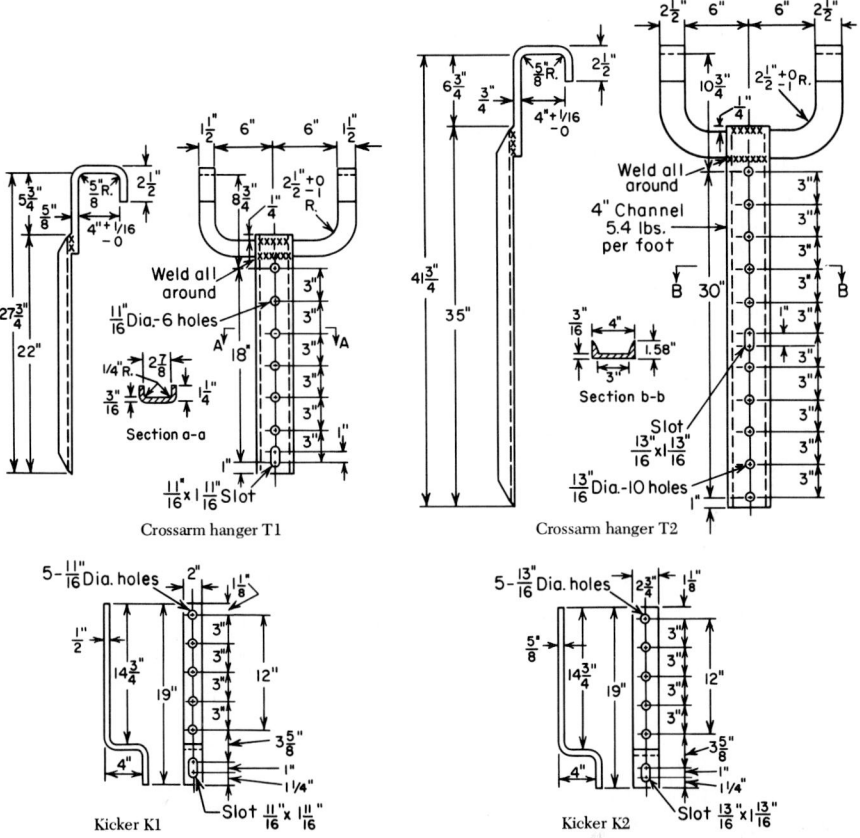

| Hanger crossarm size | Kicker size | Number of ⅝-in. bolts and nuts | | Number of ¾-in. bolts and nuts | | kva range of use for HV 15 kv and below (4) | |
|---|---|---|---|---|---|---|---|
| | | Hanger | Kicker | Hanger | Kicker | Single phase | Three phase |
| T1 | K1 | 2 | 1 | .... | ... | 3–25 | 9 |
| T2 (1) | K2 | 2 (2) | ... | 2 (3) | 1 | 37½–50 | 15–45 |

(1) Both sizes of bolts are included.
(2) For use on 15 kva, three phase.
(3) For use on 37½–50 kva, single phase, and 30–45 kva, three phase.
(4) Including 24.94GrdY/14.4 kv, 14.4/24.94GrdY kv, 16.34 kv.

NOTE

Tolerances, except where indicated otherwise, shall be ±⅟₁₆ in, except that bolt hole and slot tolerances shall be ±⅟₆₄ in.

All bolts and nuts to be square-headed NC threads.
⅝-in bolts, 1¾ in long, and threaded within ³⁄₁₆ in or less of bolt head.
¾-in bolts, 2¼ in long, and threaded within ⅜ in or less of bolt head.
All T-crossarm hangers, kickers, nuts, and bolts shall be hot-dip–galvanized.

**Fig. 5-27**  T-crossarm hangers and kickers for the mounting of transformers. [American National Standards Institute]

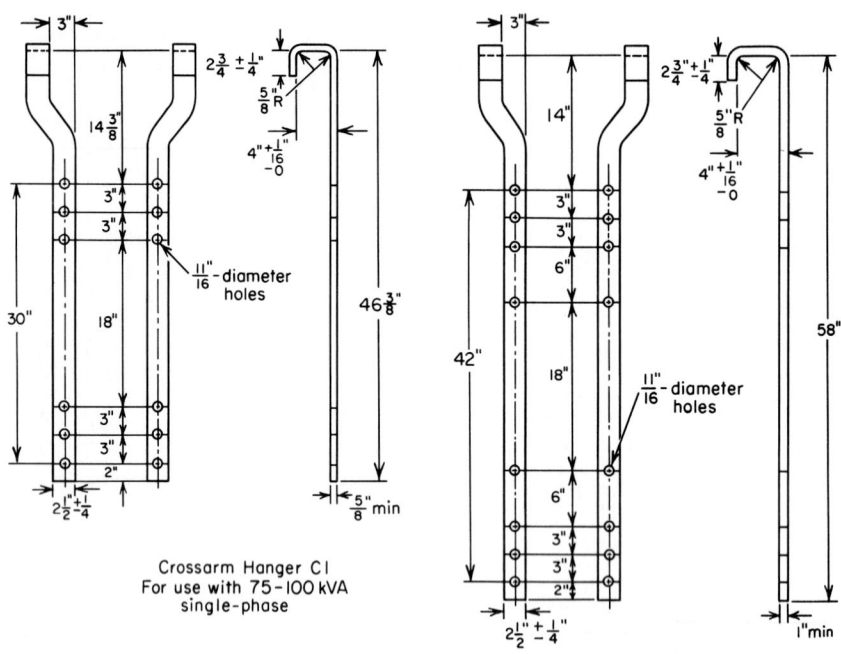

Crossarm Hanger C1
For use with 75–100 kVA
single-phase

Crossarm Hanger C3
For use with 75–150 kVA three-phase
and 167 kVA single-phase

NOTES

1. Crossarm hangers shall be hot-dip–galvanized.

2. C1- and C3-crossarm hangers each to be equipped with four square-head bolts and square nuts having NC threads as follows:

  C1: ⅝-in bolts, 2 in long, threaded to within 3⁄16 in or less of bolt head.

  C3: ⅝-in bolts, 2½ in long, threaded to within ⅜ in or less of bolt head.

3. Tolerances, except where indicated otherwise, shall be ±1⁄16 in, except that bolt hole and slot tolerance shall be ±1⁄64 in.

**Fig. 5-28**   Crossarm hangers. [American National Standards Institute]

## 41. Standard Accessory Equipment for Three-Phase Transformers
(American National Standards Institute)

| Item No. | Paragraph reference | High-voltage, kv → Low-voltage, volts → Kva | 4.8 and below / 8.32Y and below — 600 and below | | | | 7.2 to 13.8 / 12.47Y to 13.2Y — 5,000 and below | | | | 22.9 — 600 and below | All — 600 and below | 22.9 — Above 600 | All — Above 600 |
|---|---|---|---|---|---|---|---|---|---|---|---|---|---|---|
| | | | 9 to 15 | 30 to 45 | 75 to 112½ | 150 | 9 to 15 | 30 to 45 | 75 to 112½ | 150 | 150 | 225 to 500 | 150 | 300 to 500 |
| 1 | 12-27.950 | Support lug, type | A | B | C | C | A | B | C | C | | | | |
| 3 | 12-27.661 | Combination oil-drain plug and sampling device | | X | X | | | X | X | | | | | |
| 4 | 12-27.662 | Combination oil drain, filter-press connection, and sampling valve | | | | X | | | | X | X | X | X | X |
| 5 | 12-27.621 | Liquid-level marking | X | X | X | X | X | X | X | X | | | | |
| 6 | 12-27.622 | Magnetic liquid-level gage | | | | | | | | | X | X | X | X |
| 7 | 12-27.630 | Dial-type thermometer | | | | | | | | | | X | | X |
| 8 | 12-27.663 | Upper filter-press connection | | | | | | | | | X | X | X | X |
| 9 | 12-27.611 | Tap changer, internal operation | X | X | X | X | X | X | X | X | | | | |
| 10 | 12-27.612 | Tap changer, external operation | | | | | | | | | X | X | X | X |
| 11 | 12-27.650 | Pressure-vacuum gage provision | | | | | | | | | | X | | X |
| 12 | 12-27.671 | Jacking provision | | | | | | | | | | X | | X |
| 13 | 12-27.672 | Rolling provision | | | | | | | | | | X | | X |
| 14 | 12-27.932 | Handhole in cover | | | | X | X | X | X | X | | | | |
| 15 | 12-27.933 | Handhole in cover | | | | | | | | | X | X | X | X |
| 16 | 12-27.673 | Lifting lugs | X | X | X | X | X | X | X | X | X | X | X | X |
| 17 | 12-27.991 | Tank grounding provision | X | X | X | X | X | X | X | X | | | | |
| 18 | 12-27.992 | Tank grounding provision | | | | | | | | | X | X | X | X |
| 21 | 12-27.995 | Low-voltage grounding provision | X | X | X | X | X | X | X | X | | | | |
| 22 | 12-27.114 | High-voltage bushing terminals | X | X | X | X | X | X | X | X | X | X | X | X |
| 23 | 12-27.115 | Low-voltage bushing terminals | X | X | X | X | X | X | X | X | X | X | X | X |
| 24 | 12-27.152 | Low-voltage bushing arrangement | X | X | X | X | X | X | X | X | X | X | X | X |
| 25 | 12-27.751.1 | Nameplate location | X | X | | | X | X | | | | | | |
| 26 | 12-27.751.2 | Nameplate location | | | X | X | | | X | X | | | | |
| 27 | 12-27.751.3 | Nameplate location | | | | | | | | | X | X | X | X |
| 29 | 12-27.755 | Nameplate extension | X | X | X | X | X | X | X | X | | | | |
| 30 | 12-27.752 | Nameplate, type | A | A | A | A | A | A | A | A | B | B | B | B |
| 31 | 12-27.762 | Stenciled rating | X | X | X | X | X | X | X | X | | | | |

NOTE   Paragraph references (col. 2) are to ANSI publication.

**42. Average data for distribution transformers** are given in Secs. 43 to 57. The tables give approximate information for transformers with ratings on the low-voltage side of 120/240, 240/480, 600, or 2400 V.

### 43.  Performance of 60-Hz Single-Phase Oil-Filled Distribution Transformers
(Standard voltages of 600 or 2400 V on high-voltage side)

| Kva | No load loss | Total loss at 75°C | Efficiencies at 75°C | | | | Regulation at 75°C | | % im-pedance at 75°C |
|---|---|---|---|---|---|---|---|---|---|
| | | | Full | ¾ | ½ | ¼ | 100 % pf | 80 % pf | |
| 1½ | 20 | 66 | 95.8 | 96.1 | 95.9 | 94.2 | 3.1 | 3.6 | 3.6 |
| 3 | 27 | 98 | 96.9 | 97.1 | 97.1 | 96.0 | 2.4 | 2.6 | 2.5 |
| 5 | 41 | 134 | 97.4 | 97.6 | 97.5 | 96.4 | 1.9 | 2.6 | 2.5 |
| 7½ | 55 | 184 | 97.6 | 97.8 | 97.7 | 96.8 | 1.8 | 2.3 | 2.3 |
| 10 | 59 | 241 | 97.7 | 97.9 | 97.9 | 97.3 | 1.9 | 2.3 | 2.3 |
| 15 | 80 | 333 | 97.8 | 98.0 | 98.1 | 97.5 | 1.7 | 2.2 | 2.2 |
| 25 | 120 | 496 | 98.1 | 98.3 | 98.3 | 97.7 | 1.6 | 2.7 | 2.9 |
| 37½ | 154 | 649 | 98.3 | 98.5 | 98.5 | 98.1 | 1.4 | 2.6 | 2.8 |
| 50 | 190 | 795 | 98.4 | 98.6 | 98.6 | 98.2 | 1.3 | 2.3 | 2.5 |
| 75 | 288 | 1,194 | 98.4 | 98 6 | 98.6 | 98.1 | 1.3 | 3.1 | 3.6 |
| 100 | 357 | 1,553 | 98.5 | 98.6 | 98.7 | 98.2 | 1.3 | 3.2 | 3.8 |
| 150 | 550 | 2,292 | 98.5 | 98.7 | 98.7 | 98.3 | 1.3 | 3.8 | 4.8 |
| 200 | 800 | 2,905 | 98.6 | 98.7 | 98.7 | 98.2 | 1.2 | 3.7 | 4.8 |
| 250 | 1,115 | 3,815 | 98.5 | 98.6 | 98.6 | 98.0 | 1.2 | 3.7 | 4.8 |
| 333 | 1,310 | 4,665 | 98.6 | 98.7 | 98.7 | 98.2 | 1.2 | 3.7 | 4.8 |
| 500 | 1,675 | 6,325 | 98.8 | 98.9 | 98.9 | 98.5 | 1.1 | 3.6 | 4.8 |

### 44.  Performance of 60-Hz Single-Phase Oil-Filled Distribution Transformers
(Standard voltages of 7200 V on high-voltage side)

| Kva | No load loss | Total loss at 75°C | Efficiencies at 75°C | | | | Regulation at 75°C | | % im-pedance at 75°C |
|---|---|---|---|---|---|---|---|---|---|
| | | | Full | ¾ | ½ | ¼ | 100 % pf | 80 % pf | |
| 1½ | 22 | 65 | 95.9 | 96.1 | 95.8 | 93.8 | 2.9 | 2.9 | 3.0 |
| 3 | 32 | 106 | 96.6 | 96.9 | 96.7 | 95.4 | 2.6 | 3.1 | 3.0 |
| 5 | 51 | 164 | 96.8 | 97.1 | 96.9 | 95.6 | 2.3 | 2.5 | 2.6 |
| 7½ | 59 | 229 | 97.0 | 97.3 | 97.3 | 96.4 | 2.3 | 3.0 | 3.0 |
| 10 | 71 | 288 | 97.2 | 97.5 | 97.6 | 96.8 | 2.3 | 3.0 | 3.0 |
| 15 | 97 | 364 | 97.6 | 97.9 | 97.9 | 97.1 | 1.9 | 3.1 | 3.3 |
| 25 | 144 | 544 | 97.9 | 98.1 | 98.1 | 97.4 | 1.7 | 3.1 | 3.4 |
| 37½ | 208 | 725 | 98.1 | 98.3 | 98.2 | 97.5 | 1.5 | 2.9 | 3.2 |
| 50 | 270 | 940 | 98.2 | 98.3 | 98.3 | 97.6 | 1.4 | 3.1 | 3.6 |
| 75 | 370 | 1,310 | 98.3 | 98.4 | 98.4 | 97.8 | 1.4 | 4.0 | 5.1 |
| 100 | 470 | 1,700 | 98.3 | 98.5 | 98.4 | 97.8 | 1.4 | 3.8 | 4.8 |
| 150 | 725 | 2,425 | 98.4 | 98.5 | 98.5 | 97.8 | 1.3 | 3.9 | 5.0 |
| 200 | 915 | 3,180 | 98.4 | 98.6 | 98.5 | 97.9 | 1.3 | 3.9 | 5.0 |
| 250 | 1,115 | 3,815 | 98.5 | 98.6 | 98.6 | 98.0 | 1.3 | 3.9 | 5.0 |
| 333 | 1,310 | 4,665 | 98.6 | 98.7 | 98.7 | 98.2 | 1.2 | 3.8 | 5.0 |
| 500 | 1,675 | 6,325 | 98.8 | 98.9 | 98.9 | 98.5 | 1.1 | 3.8 | 5.0 |

### 45.  Performance of 60-Hz Single-Phase Oil-Filled Distribution Transformers
(Standard voltage of 13,200 V on high-voltage side)

| Kva | No load loss | Total loss at 75°C | Efficiencies at 75°C | | | | Regulation at 75°C | | % impedance at 75°C |
|---|---|---|---|---|---|---|---|---|---|
| | | | Full | ¾ | ½ | ¼ | 100 % pf | 80 % pf | |
| 2½ | 44 | 113 | 95.7 | 95.8 | 95.3 | 92.8 | 2.7 | 3.1 | 3.1 |
| 5 | 58 | 180 | 96.5 | 96.8 | 96.6 | 95.0 | 2.5 | 2.9 | 2.9 |
| 10 | 92 | 288 | 97.2 | 97.4 | 97.2 | 96.0 | 2.0 | 2.8 | 2.8 |
| 15 | 122 | 396 | 97.4 | 97.6 | 97.5 | 96.4 | 1.9 | 2.9 | 3.0 |
| 25 | 170 | 550 | 97.9 | 98.0 | 97.9 | 97.0 | 1.6 | 3.0 | 3.3 |
| 37½ | 230 | 748 | 98.1 | 98.2 | 98.1 | 97.3 | 1.5 | 3.0 | 3.3 |
| 50 | 300 | 958 | 98.1 | 98.3 | 98.2 | 97.4 | 1.4 | 2.9 | 3.2 |
| 75 | 415 | 1,375 | 98.2 | 98.3 | 98.3 | 97.5 | 1.4 | 4.0 | 5.0 |
| 100 | 530 | 1,755 | 98.3 | 98.4 | 98.3 | 97.6 | 1.4 | 4.0 | 5.1 |
| 150 | 750 | 2,520 | 98.3 | 98.5 | 98.4 | 97.8 | 1.3 | 3.9 | 5.0 |
| 200 | 950 | 3,240 | 98.4 | 98.5 | 98.5 | 97.9 | 1.3 | 3.9 | 5.0 |
| 250 | 1,115 | 3,815 | 98.5 | 98.6 | 98.6 | 98.0 | 1.3 | 3.9 | 5.0 |
| 333 | 1,310 | 4,665 | 98.6 | 98.7 | 98.7 | 98.2 | 1.2 | 3.8 | 5.0 |
| 500 | 1,675 | 6,325 | 98.8 | 98.9 | 98.9 | 98.5 | 1.1 | 3.8 | 5.0 |

### 46.  Performance of 60-Hz Single-Phase Oil-Filled Distribution Transformers
(Standard voltage of 22,000 V on high-voltage side)

| Kva | No load loss | Total loss at 75°C | Efficiencies at 75°C | | | | Regulation at 75°C | | % impedance at 75°C |
|---|---|---|---|---|---|---|---|---|---|
| | | | Full | ¾ | ½ | ¼ | 100 % pf | 80 % pf | |
| 10 | 135 | 420 | 96.0 | 96.2 | 96.0 | 94.2 | 3.0 | 4.9 | 5.2 |
| 15 | 175 | 530 | 96.6 | 96.8 | 96.6 | 95.0 | 2.5 | 4.7 | 5.2 |
| 25 | 230 | 740 | 97.1 | 97.3 | 97.2 | 96.0 | 2.2 | 4.6 | 5.2 |
| 37½ | 300 | 985 | 97.4 | 97.6 | 97.5 | 96.4 | 2.0 | 4.5 | 5.2 |
| 50 | 375 | 1,225 | 97.6 | 97.8 | 97.7 | 96.7 | 1.9 | 4.4 | 5.2 |
| 75 | 500 | 1,630 | 97.9 | 98.0 | 98.0 | 97.0 | 1.7 | 4.2 | 5.2 |
| 100 | 610 | 1,995 | 98.1 | 98.2 | 98.1 | 97.3 | 1.7 | 4.2 | 5.2 |
| 150 | 825 | 2,715 | 98.2 | 98.3 | 98.3 | 97.5 | 1.4 | 4.1 | 5.2 |
| 200 | 1,015 | 3,430 | 98.3 | 98.4 | 98.4 | 97.7 | 1.4 | 4.1 | 5.2 |
| 250 | 1,200 | 3,930 | 98.5 | 98.6 | 98.5 | 97.8 | 1.3 | 4.0 | 5.2 |
| 333 | 1,450 | 4,810 | 98.6 | 98.7 | 98.6 | 98.0 | 1.2 | 4.0 | 5.2 |
| 500 | 1,950 | 6,570 | 98.8 | 98.8 | 98.8 | 98.2 | 1.1 | 3.9 | 5.2 |

**47.  Performance of 60-Hz Single-Phase Oil-Filled Distribution Transformers**
(Standard voltage of 44,000 V on high-voltage side)

| Kva | No load loss | Total loss at 75°C | Efficiencies at 75°C | | | | Regulation at 75°C | | % impedance at 75°C |
|---|---|---|---|---|---|---|---|---|---|
| | | | Full | ¾ | ½ | ¼ | 100 % pf | 80 % pf | |
| 25 | 310 | 858 | 96.7 | 96.8 | 96.5 | 94.8 | 2.4 | 5.0 | 5.7 |
| 37½ | 400 | 1,122 | 97.1 | 97.2 | 97.0 | 95.5 | 2.1 | 4.8 | 5.7 |
| 50 | 490 | 1,370 | 97.3 | 97.4 | 97.2 | 95.8 | 2.0 | 4.8 | 5.7 |
| 75 | 630 | 1,810 | 97.6 | 97.7 | 97.6 | 96.4 | 1.8 | 4.6 | 5.7 |
| 100 | 770 | 2,225 | 97.8 | 97.9 | 97.8 | 96.7 | 1.7 | 4.6 | 5.7 |
| 150 | 1,010 | 3,003 | 98.0 | 98.1 | 98.0 | 97.1 | 1.5 | 4.5 | 5.7 |
| 200 | 1,230 | 3,743 | 98.2 | 98.3 | 98.2 | 97.3 | 1.5 | 4.4 | 5.7 |
| 250 | 1,420 | 4,405 | 98.3 | 98.4 | 98.3 | 97.5 | 1.4 | 4.4 | 5.7 |
| 333 | 1,695 | 5,395 | 98.4 | 98.5 | 98.5 | 97.7 | 1.3 | 4.4 | 5.7 |
| 500 | 2,200 | 7,030 | 98.6 | 98.7 | 98.7 | 98.0 | 1.2 | 4.3 | 5.7 |

**48.  Performance of 60-Hz Single-Phase Oil-Filled Distribution Transformers**
(Standard voltage of 66,000 V on high-voltage side)

| Kva | No load loss | Total loss at 75°C | Efficiencies at 75°C | | | | Regulation at 75°C | | % impedance at 75°C |
|---|---|---|---|---|---|---|---|---|---|
| | | | Full | ¾ | ½ | ¼ | 100 % pf | 80 % pf | |
| 50 | 640 | 1,515 | 97.1 | 97.1 | 96.7 | 94.7 | 2.0 | 5.3 | 6.5 |
| 75 | 820 | 1,950 | 97.5 | 97.5 | 97.1 | 95.5 | 1.8 | 5.1 | 6.5 |
| 100 | 970 | 2,385 | 97.7 | 97.7 | 97.4 | 95.9 | 1.7 | 5.1 | 6.5 |
| 150 | 1,220 | 3,160 | 97.9 | 98.0 | 97.8 | 96.5 | 1.5 | 5.0 | 6.5 |
| 200 | 1,440 | 3,900 | 98.1 | 98.1 | 98.0 | 96.9 | 1.5 | 4.9 | 6.5 |
| 250 | 1,630 | 4,615 | 98.2 | 98.3 | 98.1 | 97.2 | 1.4 | 4.9 | 6.5 |
| 333 | 1,875 | 5,635 | 98.3 | 98.4 | 98.3 | 97.5 | 1.4 | 4.9 | 6.5 |
| 500 | 2,400 | 7,380 | 98.5 | 98.6 | 98.6 | 97.9 | 1.3 | 4.8 | 6.5 |

**49.  Performance of 60-Hz Three-Phase Oil-Filled Distribution Transformers**
(Standard voltage of 2400 V on high-voltage side)

| Kva | No load loss | Total loss at 75°C | Efficiencies at 75°C | | | | Regulation at 75°C | | % impedance at 75°C |
|---|---|---|---|---|---|---|---|---|---|
| | | | Full | ¾ | ½ | ¼ | 100 % pf | 80 % pf | |
| 10 | 90 | 345 | 96.7 | 97.0 | 97.0 | 96.0 | 2.6 | 3.4 | 3.4 |
| 15 | 113 | 443 | 97.1 | 97.4 | 97.5 | 96.5 | 2.3 | 3.3 | 3.4 |
| 25 | 162 | 650 | 97.5 | 97.7 | 97.8 | 97.0 | 2.0 | 3.3 | 3.4 |
| 37½ | 214 | 890 | 97.7 | 97.9 | 98.0 | 97.3 | 1.9 | 3.9 | 4.3 |
| 50 | 275 | 1,117 | 97.8 | 98.0 | 98.1 | 97.4 | 1.9 | 4.2 | 4.9 |
| 75 | 370 | 1,525 | 98.0 | 98.2 | 98.2 | 97.7 | 1.7 | 3.8 | 4.4 |
| 100 | 455 | 1,860 | 98.2 | 98.4 | 98.4 | 97.9 | 1.5 | 3.8 | 4.6 |
| 150 | 590 | 2,445 | 98.4 | 98.6 | 98.6 | 98.1 | 1.4 | 3.4 | 4.1 |
| 200 | 785 | 3,093 | 98.5 | 98.6 | 98.6 | 98.2 | 1.3 | 3.4 | 4.2 |
| 300 | 1,100 | 4,350 | 98.6 | 98.7 | 98.7 | 98.3 | 1.2 | 3.7 | 4.8 |
| 450 | 1,550 | 6,125 | 98.4 | 98.8 | 98.8 | 98.6 | 1.2 | 3.7 | 4.8 |

**50.  Performance of 60-Hz Three-Phase Oil-Filled Distribution Transformers**
(Standard voltage of 4800 V on high-voltage side)

| Kva | No load loss | Total loss at 75°C | Efficiencies at 75°C | | | | Regulation at 75°C | | % impedance at 75°C |
|---|---|---|---|---|---|---|---|---|---|
| | | | Full | ¾ | ½ | ¼ | 100 % pf | 80 % pf | |
| 10 | 94 | 354 | 96.6 | 96.9 | 96.9 | 95.8 | 2.7 | 3.7 | 3.7 |
| 15 | 120 | 465 | 97.0 | 97.3 | 97.3 | 96.3 | 2.4 | 3.7 | 3.8 |
| 25 | 170 | 685 | 97.3 | 97.6 | 97.7 | 96.9 | 2.2 | 3.4 | 3.5 |
| 37½ | 225 | 935 | 97.6 | 97.8 | 97.9 | 97.2 | 2.0 | 3.7 | 4.0 |
| 50 | 275 | 1,165 | 97.7 | 98.0 | 98.0 | 97.4 | 1.9 | 3.6 | 4.0 |
| 75 | 370 | 1,585 | 97.9 | 98.2 | 98.2 | 97.7 | 1.8 | 3.5 | 4.0 |
| 100 | 455 | 1,955 | 98.1 | 98.3 | 98.3 | 97.8 | 1.6 | 3.5 | 4.0 |
| 150 | 615 | 2,615 | 98.3 | 98.5 | 98.5 | 98.0 | 1.4 | 3.4 | 4.0 |
| 200 | 785 | 3,275 | 98.4 | 98.6 | 98.6 | 98.1 | 1.4 | 3.4 | 4.0 |
| 300 | 1,100 | 4,350 | 98.6 | 98.7 | 98.7 | 98.3 | 1.2 | 3.7 | 4.8 |
| 450 | 1,550 | 6,125 | 98.7 | 98.8 | 98.8 | 98.4 | 1.2 | 3.7 | 4.8 |

**51.  Performance of 60-Hz Three-Phase Oil-Filled Distribution Transformers**
(Standard voltage of 7200 V on high-voltage side)

| Kva | No load loss | Total loss at 75°C | Efficiencies at 75°C | | | | Regulation at 75°C | | % impedance at 75°C |
|---|---|---|---|---|---|---|---|---|---|
| | | | Full | ¾ | ½ | ¼ | 100 % pf | 80 % pf | |
| 10 | 100 | 385 | 96.3 | 96.7 | 96.7 | 95.5 | 2.9 | 2.9 | 3.0 |
| 15 | 132 | 512 | 96.7 | 97.0 | 97.0 | 96.0 | 2.6 | 3.3 | 3.3 |
| 25 | 190 | 735 | 97.2 | 97.4 | 97.5 | 96.5 | 2.3 | 3.4 | 3.5 |
| 37½ | 260 | 1,000 | 97.4 | 97.7 | 97.7 | 96.9 | 2.1 | 3.4 | 3.5 |
| 50 | 315 | 1,205 | 97.7 | 97.9 | 97.9 | 97.1 | 1.9 | 3.1 | 3.2 |
| 75 | 440 | 1,650 | 97.8 | 98.0 | 98.0 | 97.3 | 1.7 | 3.1 | 3.4 |
| 100 | 560 | 2,050 | 98.0 | 98.2 | 98.1 | 97.4 | 1.6 | 3 4 | 3.9 |
| 150 | 730 | 2,760 | 98.2 | 98.4 | 98.4 | 97.8 | 1.5 | 3.5 | 4.2 |
| 200 | 850 | 3,370 | 98.4 | 98.5 | 98.5 | 98.0 | 1.4 | 3.5 | 4.2 |
| 300 | 1,150 | 4,470 | 98.5 | 98.7 | 98.7 | 98.2 | 1.3 | 3 9 | 5.0 |
| 450 | 1,600 | 6,300 | 98.6 | 98.8 | 98.8 | 98.3 | 1.2 | 3.9 | 5.0 |

**52.  Performance of 60-Hz Three-Phase Oil-Filled Distribution Transformers**
(Standard voltage of 12,000 V on high-voltage side)

| Kva | No load loss | Total loss at 75°C | Efficiencies at 75°C | | | | Regulation at 75°C | | % impedance at 75°C |
|---|---|---|---|---|---|---|---|---|---|
| | | | Full | ¾ | ½ | ¼ | 100 % pf | 80 % pf | |
| 10 | 110 | 410 | 95.1 | 96.4 | 96.4 | 95.1 | 3.1 | 4.7 | 4.8 |
| 15 | 145 | 545 | 96.5 | 96.8 | 96.8 | 95.7 | 2.8 | 4.7 | 5.0 |
| 25 | 207 | 775 | 97.0 | 97.3 | 97.3 | 96.2 | 2.4 | 3.7 | 4.0 |
| 37½ | 275 | 1,035 | 97.3 | 97.6 | 97.6 | 96.7 | 2.2 | 3.7 | 4.0 |
| 50 | 345 | 1,275 | 97.5 | 97.7 | 97.7 | 96.9 | 2.0 | 3.9 | 4.3 |
| 75 | 475 | 1,745 | 97.7 | 97.9 | 97.9 | 97.1 | 1.9 | 3.9 | 4.6 |
| 100 | 600 | 2,160 | 97.9 | 98.1 | 98.1 | 97.3 | 1.8 | 4.6 | 5.7 |
| 150 | 765 | 2,895 | 98.1 | 98.3 | 98.3 | 97.6 | 1.5 | 3.7 | 4.4 |
| 200 | 890 | 3,540 | 98.3 | 98.4 | 98.4 | 97.9 | 1.4 | 3.6 | 4.4 |
| 300 | 1,225 | 4,795 | 98.4 | 98.6 | 98.6 | 98.1 | 1.4 | 3.9 | 5.0 |
| 450 | 1,650 | 6,690 | 98.5 | 98.7 | 98.7 | 98.3 | 1.3 | 3.9 | 5.0 |

**53.  Performance of 60-Hz Three-Phase Oil-Filled Distribution Transformers**
(Standard voltage of 13,200 V on high-voltage side)

| Kva | No load loss | Total loss at 75°C | Efficiencies at 75°C | | | | Regulation at 75°C | | % impedance at 75°C |
|---|---|---|---|---|---|---|---|---|---|
| | | | Full | ¾ | ½ | ¼ | 100 % pf | 80 % pf | |
| 15 | 165 | 565 | 96.4 | 96.7 | 96.6 | 95.2 | 2.8 | 4.6 | 4.8 |
| 25 | 230 | 800 | 96.9 | 97.2 | 97.1 | 96.0 | 2.4 | 3.6 | 3.6 |
| 37½ | 305 | 1,065 | 97.3 | 97.5 | 97.4 | 96.4 | 2.2 | 4.1 | 4.5 |
| 50 | 375 | 1,305 | 97.5 | 97.7 | 97.6 | 96.6 | 2.0 | 4.0 | 4.5 |
| 75 | 515 | 1,785 | 97.7 | 97.9 | 97.8 | 96.9 | 1.8 | 3.2 | 3.5 |
| 100 | 630 | 2,190 | 97.9 | 98.0 | 98.0 | 97.2 | 1.7 | 4.5 | 5.5 |
| 150 | 790 | 2,920 | 98.1 | 98.3 | 98.2 | 97.6 | 1.6 | 3.8 | 4.5 |
| 200 | 930 | 3,580 | 98.3 | 98.4 | 98.4 | 97.8 | 1.5 | 3.7 | 4.5 |
| 300 | 1,225 | 4,795 | 98.4 | 98.6 | 98.6 | 98.1 | 1.4 | 3.9 | 5.0 |
| 450 | 1,650 | 6,690 | 98.5 | 98.7 | 98.7 | 98.3 | 1.3 | 3.9 | 5.0 |

**54.  Performance of 60-Hz Three-Phase Oil-Filled Distribution Transformers**
(Standard voltage of 22,000 V on high-voltage side)

| Kva | No load loss | Total loss at 75°C | Efficiencies at 75°C | | | | Regulation at 75°C | | % impedance at 75°C |
|---|---|---|---|---|---|---|---|---|---|
| | | | Full | ¾ | ½ | ¼ | 100 % pf | 80 % pf | |
| 37½ | 330 | 1,250 | 96.8 | 97.1 | 97.1 | 96.1 | 2.6 | 4.8 | 5.2 |
| 50 | 385 | 1,510 | 97.1 | 97.4 | 97.4 | 96.5 | 2.4 | 4.7 | 5.2 |
| 75 | 525 | 1,945 | 97.4 | 97.7 | 97.7 | 96.8 | 2.1 | 4.5 | 5.2 |
| 100 | 640 | 2,365 | 97.7 | 97.9 | 97.9 | 97.1 | 1.9 | 4.4 | 5.2 |
| 150 | 800 | 3,085 | 98.0 | 98.2 | 98.2 | 97.6 | 1.7 | 4.3 | 5.2 |
| 200 | 985 | 3,815 | 98.1 | 98.3 | 98.3 | 97.7 | 1.6 | 4.2 | 5.2 |
| 300 | 1,355 | 5,320 | 98.3 | 98.4 | 98.5 | 97.9 | 1.5 | 4.2 | 5.2 |
| 450 | 1,850 | 7,100 | 98.4 | 98.6 | 98.6 | 98.1 | 1.3 | 4.1 | 5.2 |

**55.  Performance of 60-Hz Three-Phase Oil-Filled Distribution Transformers**
(Standard voltage of 44,000 V on high-voltage side)

| Kva | No load loss | Total loss at 75°C | Efficiencies at 75°C | | | | Regulation at 75°C | | % impedance at 75°C |
|---|---|---|---|---|---|---|---|---|---|
| | | | Full | ¾ | ½ | ¼ | 100 % pf | 80 % pf | |
| 75 | 675 | 2,230 | 97.1 | 97.3 | 97.2 | 96.0 | 2.3 | 4.9 | 5.7 |
| 100 | 840 | 2,635 | 97.4 | 97.6 | 97.5 | 96.3 | 2.0 | 4.7 | 5.7 |
| 150 | 1,100 | 3,485 | 97.7 | 97.9 | 97.8 | 96.8 | 1.8 | 4.6 | 5.7 |
| 200 | 1,350 | 4,290 | 97.9 | 98.0 | 98.0 | 97.0 | 1.7 | 4.6 | 5.7 |
| 300 | 1,765 | 5,830 | 98.1 | 98.2 | 98.2 | 97.4 | 1.6 | 4.5 | 5.7 |
| 450 | 2,340 | 7,865 | 98.3 | 98.4 | 98.4 | 97.7 | 1.4 | 4.4 | 5.7 |

**56.  Performance of 60-Hz Three-Phase Oil-Filled Distribution Transformers**
(Standard voltage of 66,000 V on high-voltage side)

| Kva | No load loss | Total loss at 75°C | Efficiencies at 75°C | | | | Regulation at 75°C | | % impedance a. 75°C |
|---|---|---|---|---|---|---|---|---|---|
| | | | Full | ¾ | ½ | ¼ | 100 % pf | 80 % pf | |
| 150 | 1,450 | 3,970 | 97.4 | 97.5 | 97.3 | 95.9 | 1.9 | 5.2 | 6.5 |
| 200 | 1,700 | 4,800 | 97.7 | 97.8 | 97.6 | 96.4 | 1.8 | 5.2 | 6.5 |
| 300 | 2,200 | 6,340 | 97.9 | 98.0 | 97.9 | 96.8 | 1.6 | 5.0 | 6.5 |
| 450 | 2,850 | 8,420 | 98.2 | 98.3 | 98.1 | 97.2 | 1.5 | 4.9 | 6.5 |

**57.  Performance of 60-Hz Single-Phase Air-Cooled Transformers**
(Standard voltage of 2400 V on high-voltage side)

| Kva | No load loss | Total loss at 75°C | Efficiencies at 75°C | | | | Regulation at 75°C | | % impedance at 75°C |
|---|---|---|---|---|---|---|---|---|---|
| | | | Full | ¾ | ½ | ¼ | 100 % pf | 80 % pf | |
| 150 | 755 | 2,265 | 98.51 | 98.60 | 98.51 | 97.78 | 1.2 | 3.8 | 5.0 |
| 200 | 925 | 2,890 | 98.58 | 98.66 | 98.60 | 97.95 | 1.1 | 3.8 | 5.0 |
| 250 | 1,080 | 3,495 | 98.62 | 98.72 | 98.67 | 98.07 | 1.1 | 3.8 | 5.0 |
| 333 | 1,325 | 4,460 | 98.68 | 98.78 | 98.75 | 98.20 | 1.1 | 3.6 | 5.0 |
| 500 | 1,760 | 6,315 | 98.75 | 98.86 | 98.85 | 98.39 | 1.1 | 3.7 | 5.0 |

**58. Small power transformers** (Figs. 5-5 and 5-29) are constant-potential, self–air-cooled transformers for industrial purposes. The windings are completely enclosed in metal casings. The transformers are available in single- and three-phase construction of conventional or autotransformer types with cases designed for open or conduit wiring. Single-phase units are available in sizes up to 50 kVA, while three-phase units are made as large as 150 kVA. Primary voltage ratings range from 125 to 2400. Standard secondary voltages are 115/230 or 230/460.

**59. Small power transformers** are available for taking power from motor disconnecting switches for localized lighting at the machine. Types are available for operating 115-, 64-, 32-, or 6-V lamps from 115-, 230-, 460-, or 575-V power circuits. Standard sizes are 75, 150, 225, and 300 W. Another type of small power transformer is the one designed for domestic oil-burner ignition (see Fig. 5-30).

**60. Sign-lighting transformers** are single-phase, constant-potential, self–air-cooled transformers for stepping down from 115 or 230 V to the proper voltage for low-voltage incandescent lamps in signs. They are made in sizes up to 5 kVA with the windings completely enclosed in metal casings.

**Fig. 5-29**   Small power transformers, self–air-cooled. [General Electric Co.]

**Fig. 5-30**   Replacement-type domestic oil-burner ignition transformers. [General Electric Co.]

**61. Control and signal transformers** are constant-potential, self–air-cooled transformers for the purpose of supplying the proper voltage for control circuits of electrically operated switches or other equipment and for signal circuits. They may be of the open type with no protective casing over the windings or of the enclosed type with a metal casing over the windings. Some of the available mounting types are shown in Figs. 5-31 and 5-32.

**62. Transformers for many gaseous-discharge lamps** must be of the varying-voltage type. These transformers are designed so that the secondary voltage drops rapidly with load. This voltage characteristic is accomplished by means of a three-winding transformer with magnetic leakage paths between the primary and each secondary, as shown in Fig. 5-33. As the load increases, the leakage flux of the primary produces a greater voltage drop in the primary winding, thereby reducing the secondary voltage. Some of these transformers are equipped with a capacitor mounted inside the same enclosing case for power-factor improvement. The transformers are of the self–air-cooled type. Various mounting types for luminous-tube work are shown in Fig. 5-34.

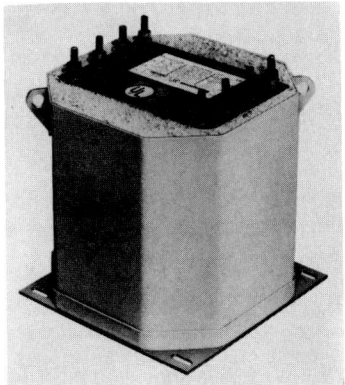

I. Enclosed type with flexible terminal leads.

II. Open type with lug terminals.

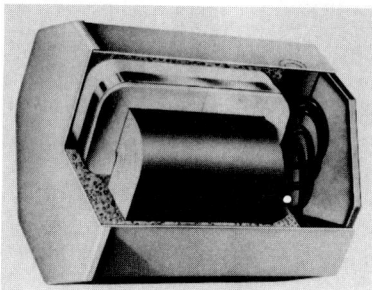

III. Enclosed type with flexible terminal leads, encapsulated core and coil.

**Fig. 5-31**   Typical control transformers. [Westinghouse Electric Corp.]

**Fig. 5-32**   Control transformer. [General Electric Co.]

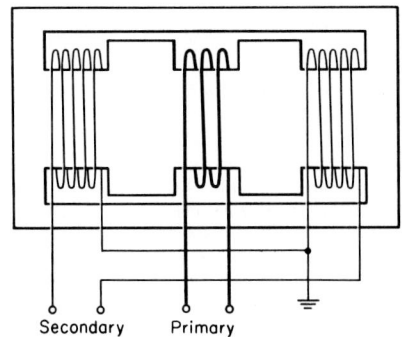

**Fig. 5-33**   Cross section of a typical neon-sign transformer, showing windings and leakage paths in magnetic circuit.

**Fig. 5-34**   Luminous-tube transformers rated 3000 to 15,000 V, 30 to 120 mA, showing low- and high-voltage terminals. [General Electric Co.]

**63. Bell-ringing transformers** are especially designed small-capacity constant-potential transformers for applications such as the operation of doorbells, buzzers, door openers, and annunciators. They are self–air-cooled units enclosed in metal cases. They can be obtained for primary voltages of 120 or 240 with a single secondary voltage of 10 or with three secondary voltages of 6, 12, and 18.

**64. Instrument transformers** are employed for the purpose of stepping down the voltage or current of a circuit for the operation of instruments, such as ammeters, voltmeters, wattmeters, etc., and relays for various protective purposes. It would be practically impossible to build delicate devices such as instruments and relays so that they would have satisfactory insulation for high voltages. Frequently such devices while energized must be accessible to operators, so that it is necessary to have the devices insulated from the high-voltage circuit. Instrument transformers are of two types: potential and current. They are made in various types, depending upon the voltage of the circuits, requirements of installations, and accuracy required. Instrument transformers will always introduce a certain amount of error into the readings obtained on instruments supplied by them. The transformers are not rated on a thermal basis of the load which they can safely carry but upon an accuracy basis. The voltampere rating is the load which the transformer will carry without exceeding the specified accuracy limit. If accuracy can be sacrificed, the

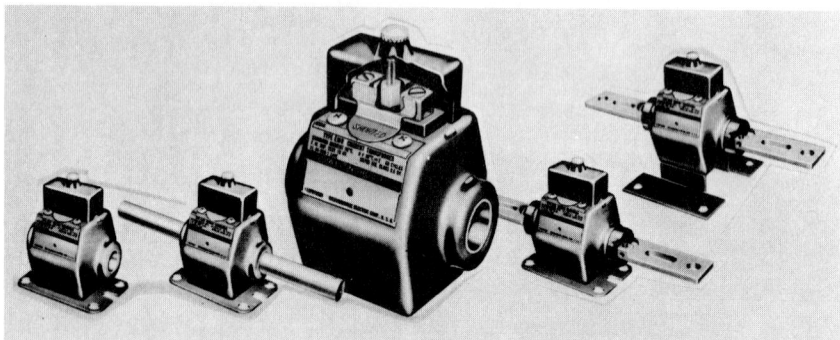

**Fig. 5-35**   Typical current instrument transformers. [Westinghouse Electric Corp.]

transformers can be used for loads considerably above their rated values. Instrument transformers are made in indoor, outdoor, and portable types for test purposes. Potential transformers have a normal secondary voltage of 120 V. Current transformers have a normal secondary current of 5 A. Typical instrument transformers are shown in Figs. 5-35 and 5-36.

**65. The current transformer** (*Standard Handbook*), considered electrically and without any reference to the change in its design to accomplish its specific duty, differs from the shunt or potential transformer merely in the method of use. The latter transformer is ordinarily supplied with a constant impressed voltage, the load being changed by varying the impedance (load) of the total secondary circuit, while the total impedance of the second-

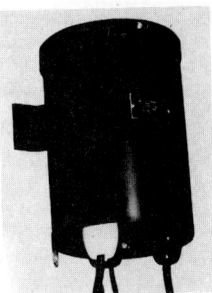

I. Relatively low voltage, unfused; dry, indoor type.

II. Relatively low voltage, fused; dry, indoor type.

III. Relatively low voltage, unfused; dry, outdoor type.

IV. Medium and high voltage, fused; oil-filled, indoor type.

V. Medium and high voltage, fused; oil-filled, both indoor and outdoor type.

**Fig. 5-36**    Typical potential transformers. [General Electric Co.]

ary circuit of the former transformer is normally held constant, and the change in load is due to a simultaneous change in the primary current and emf. In the potential transformer the actual ratio of the primary to the secondary current is of minor importance, while every effort is made to design the apparatus so that the ratio of the secondary power to the primary power is as nearly unity as possible. In the design of a series transformer no thought whatsoever is given to the ratio of the primary and secondary watts, but attention is concentrated on the endeavor to obtain a definite ratio of secondary to primary amperes.

The electric and magnetic circuits of a current transformer can conveniently be represented by the diagram shown in Fig. 5-37, in which it is used for reducing the line current to a value suitable for measurement by a low-reading ammeter, which can be thoroughly insulated from the main circuit. If the small transformer loss is neglected, the current through the ammeter $A$ is, since the ratio of turns $M = 2:5$, equal to two-fifths of the line current $I_L$.

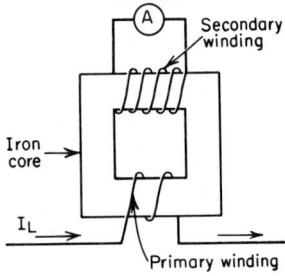

**Fig. 5-37**    Elementary series transformer.

**66. Application of the current instrument transformer.**  When an alternating current is so large that to connect measuring or operating instruments directly in the circuit would be impracticable or when the voltage is so high that to do so would be unsafe, the current transformer provides a means of reproducing the effect of the primary current on a scale suited to the instrument and of insulating the instrument from the main circuit. It is a special development of the transformer principle in which a constant ratio of primary to secondary current is the important consideration instead of the usual constant ratio of primary to secondary voltage.

Current transformers are used with ac ammeters, wattmeters, power-factor meters, watthour meters, compensators, protective and regulating relays, and the trip coils of circuit breakers. It is a standard practice in the United States to design current transformers (regardless of their capacity or ratios of transformation) to supply a rated secondary current of 5 A.

    **example**  A 600-A current transformer has a ratio of 120:1; i.e., when a current of 600 A flows in the primary circuit, $600 \div 120 = 5$ A will flow in the secondary circuit.

Measuring instruments for use with these transformers are so designed and are provided with scales such that they give a normal scale deflection when 5 A flows through them.

**67. It is unsafe to open the secondary circuit of a current transformer when there is any current in the primary.**  When the secondary circuit is closed, the current in this circuit creates a magnetomotive force (mmf) which is in opposition to the mmf of the primary current, and the core flux is thereby limited to the value necessary to generate in the secondary coil an emf sufficient to produce therein a current only slightly less than the primary current in magnetizing effect. When the secondary is open, there is no opposing mmf for limiting the core flux, which may reach a high value. Thus even a small value of primary current produces an excessive value of core flux and a correspondingly large secondary emf. The secondary voltage under these conditions reaches a value which may both damage the insulation and prove dangerous to life. Absolutely no harm can come from short-circuiting the secondary terminals of the current transformer, and this method is used when it is necessary to insert or disconnect instruments in the secondary circuit.

**68. The constant-current transformer** (*Standard Handbook*).  The operation of low-voltage lamps in parallel on a constant-potential system necessitates a prohibitive expenditure for conducting material when the area to be lighted is extensive and the lamps are widely separated. For such service it is the common practice to operate the lamps, which are connected in series, with a constant current. The constant-current transformer is a special form of transformer which converts alternating current at a constant potential to a constant (alternating) current with a voltage varying with the load. It consists of a primary coil upon which the constant voltage is impressed, a secondary coil (or coils) movable with respect to the primary, and a core of low magnetic reluctance. It depends for its regulation upon the magnetic leakage between the primary and secondary coils.

Consider first the primary coil; with the constant emf impressed upon this coil the total magnetism within the coil will be practically constant under all conditions. The emf generated in the secondary will depend upon the strength of the field which it surrounds. In all types of stationary transformers the secondary current is opposite in general time direction to the primary, so that there is not only a repulsive thrust between the two coils but also a considerable tendency for the magnetic lines from the primary to be forced out into space without penetrating the secondary. In the ordinary constant-potential transformer the repelling action between the two currents is prevented from producing motion of the coils by the rigid mechanical construction, while the proximity of the primary and secondary coils limits the magnetic leakage.

In the constant-current transformer, however, the repelling action is utilized to adjust the relative positions of the primary and secondary coils; when the coils are widely separated, the paths for the leakage lines are increased and the lines which the secondary surrounds are fewer than when the coils are quite close together. The counterweights mechanically attached to the movable coil (or coils) are so arranged that when the desired current exists in the secondary coil (independent of its position along the core), the weights are just balanced. An increase in the current increases the repulsion and causes the coils to separate. With any current less than normal, the repelling force diminishes, and the primary and secondary coils approach each other, thereby restoring the current to normal. The primary can be wound for any reasonable potential (say, as high as 10,000 V), while the secondary can be wound for the voltage required for operating the number of lamps in the circuit—from 15 to 200 or more lamps.

**69. Mechanical construction of the constant-current transformer.** The magnetic circuit of a constant-current transformer is usually of the *shell* type, the three limbs being placed vertically. In small sizes (Fig. 5-38) one of the coils is arranged in a fixed position, while the other is movable. In some of the larger sizes there are two fixed primary coils and two movable secondary coils, while in others both the primary and secondary coils are movable. In any event the gravitational action on the movable coil or the gravitational action of one movable coil against another to which it is mechanically interconnected is counterbalanced accurately with an excess or deficiency just equal to the repulsive thrust of the primary and secondary coils at the desired load current. By the use of cam mechanisms for the counterweights or of eccentrically placed extra weights, the excess force of the counterweights can be arranged to be equal to the variable repulsive thrust corresponding to a constant value of current in the coils at all positions of the movable coils. In fact, the transformer can be adjusted to regulate for a current of constant value at all loads or for one which either increases or decreases with increase of loads, while both the real value of the load current and its rate of change with the variation in load can be adjusted at will. To prevent any "hunting" action of the movable coils each transformer is sometimes equipped with a dashpot (see Fig. 5-38).

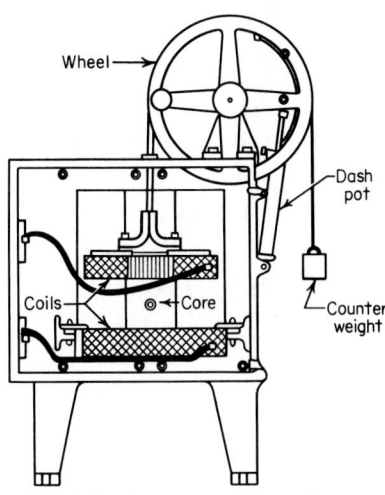

**Fig. 5-38**   Constant-current transformer.

**70. Commercial constant-current transformers** are built for natural air cooling or for immersion in oil. Oil has proved an excellent medium for insulation, cooling, and lubrication. This type of transformer is extensively used for series street-lighting service with either gaseous-discharge or incandescent lamps. The efficiency of a constant-current transformer is high, being about 96 percent at full load for a 100-lamp transformer. The power factor, which depends upon the magnetic leakage, is low at all loads; it reaches from 75 to 80 percent at full load and decreases therefrom in almost direct proportion to the decrease in load.

Constant-current transformers are made in outdoor-pole–mounting, subway-mounting, and indoor-station types. A primary voltage of 2400 is standard. The secondary current may be 6.6 or 20 A. The standard ratings are 5, 10, 15, 20, 25, 30, 35, 40, 50, 60, and 70 kVA.

**71. Series transformers** are for use with series incandescent street lamps for operation of lamps at a different value of current from that of the main series circuit or for isolation of a lamp of the same current value as the main circuit from the main circuit, which operates at a high voltage. The principle of operation is the same as for instrument current transformers. The primary is connected in series with the main series circuit. The lamp or lamps are connected to the secondary. These transformers are made in single-lamp types for operating one 6.6-, 15-, or 20-A series lamp from a 6.6-A main circuit; in two-lamp series type for operating two 6.6-, 15-, or 20-A series lamps connected in series to the secondary; in two-lamp multiple type for operating two 6.6-, 15-, or 20-A series lamps connected to two separate secondaries; in group series type for operating a group of 6.6-A series lamps connected in series to the secondary; and in single-lamp type for operating a single 115-V lamp from a series 6.6-A main circuit. The group series type is made in ratings of 0.25, 0.5, 1.0, 2, 3, 4, 5, 7.5, and 9.0 kVA. The single-lamp type for a 115-V lamp is made in sizes to accommodate one lamp of any watt size from 40 to 1000 W. These are made in types for vault or subway mounting, for mounting in bases of ornamental lighting poles, and for pole mounting (Figs. 5-39 and 5-40).

**72. The induction regulator** (Fig. 5-41) is a special type of transformer, built like an induction motor with a coil-wound secondary, which is used for varying the voltage delivered to a synchronous converter or ac feeder system. In comparison with a variable-ratio trans-

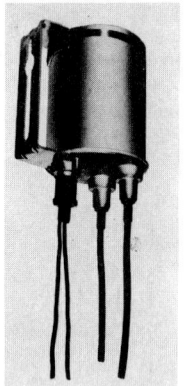 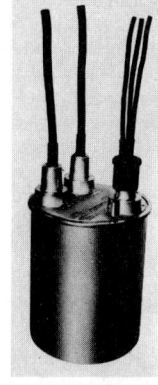

I. Aerial type.    II. Pole-base type.

**Fig. 5-39**  Series transformers for supplying a single lamp from series street-lighting circuits. [General Electric Co.]

former it possesses the advantage of being operated without opening the circuit and without short-circuiting any transformer coil. The primary of the induction regulator is subjected to the constant voltage of the supply system, the delivered voltage obtained from the secondary winding being varied by rotating the primary structure through a certain number of degrees with reference to the secondary structure. The primary structure is normally stationary, although it is movable either automatically or by hand for the purpose of varying the secondary voltage.

**73. The step-by-step potential regulator** is merely a stationary transformer provided with a large number of secondary taps and equipped with a switching mechanism for joining any desired pair of these taps to the delivery circuit, according to the emf required. A diagram of the circuits of a regulator of this type is shown in Fig. 5-42. In comparison with the induction type of regulator the step-by-step type is less noisy in operation, requires less magnetizing current, and is more rapid in action. However, it provides only a limited number of voltage steps and may give trouble from arcing at the switch contacts.

I. Small.    II. Medium.    III. Large.

**Fig. 5-40**  Series transformers for operation of a group of lamps from series street-lighting circuits. [General Electric Co.]

## CONNECTIONS: POLARITY

**74. Polarity of transformers** (Line Material Co.).   The polarity of a transformer is simply an indication of direction of flow of current from a terminal at any one instant. The idea is quite similar to the polarity marking on a battery.

As you face the high-voltage side of a transformer, the high-voltage terminal on your right is always marked $H_1$ and the other high-voltage terminal is marked $H_2$. This is an established standard.

By definition, the polarity is additive if when you connect the adjacent high-voltage and low-voltage terminals (Fig. 5-43a) and excite the transformer, a voltmeter between the other two adjacent terminals reads the sum of the high-voltage and low-voltage winding voltages. For additive polarity, the low-voltage terminal on your right when facing the low-voltage side should then be marked $X_1$ and the other low-voltage terminal $X_2$.

For subtractive polarity, the voltmeter in Fig. 5-43b reads the difference between the two winding voltages. In other words, the voltages subtract. In the case of subtractive polarity, the low-voltage terminal on your left when facing the low-voltage side is marked $X_1$.

In making transformer connections, particularly bank connections, polarity of individual transformers must be checked. In making such connections it is necessary to

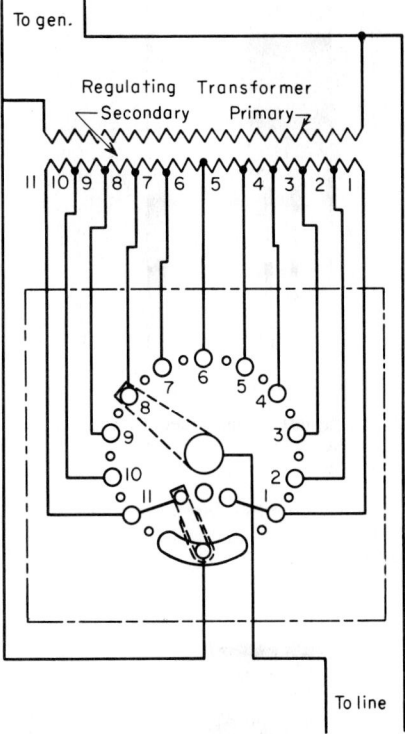

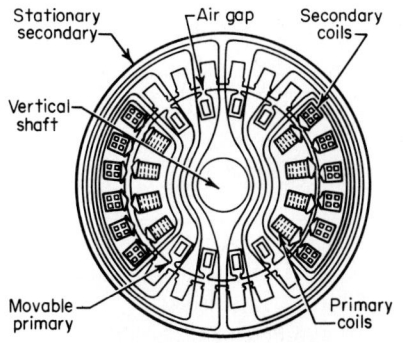

**Fig. 5-41** Section through a single-phase induction-type potential regulator.

**Fig. 5-42** Single-phase step-by-step potential regulator.

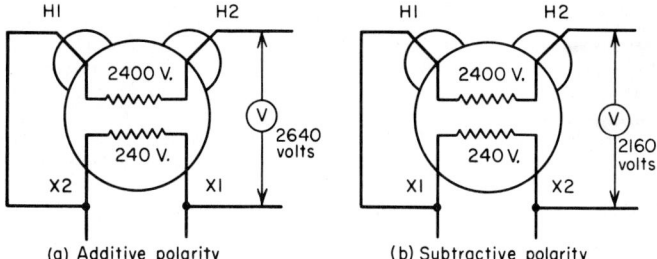

**Fig. 5-43** Transformer polarity. [Line Material Co.]

remember that all $H_1$ terminals are of the same polarity and all $X_1$ terminals are of the same polarity. Thus, if you were connecting two single-phase transformers in parallel, you should connect the two $H_1$ terminals together, then the two $H_2$ terminals together, the two $X_1$ terminals together, and the two $X_2$ terminals together. By following this procedure, you can satisfactorily parallel transformers regardless of whether they are both of the same polarity or one is of additive and one is of subtractive polarity.

**75. Tests for polarity of single-phase transformers.** If a standard transformer of known correct polarity and of the same ratio and voltage as the transformer to be tested is available, the following simple method can be used. Connect together (Fig. 5-44, I) the high-tension and the low-tension leads as if for parallel operation, inserting a fuse in one of the secondary leads. If both transformers are of the same polarity, no current will flow in the low-tension windings and the fuse will not blow. If the transformers are of opposite polarities, the low-tension windings will short-circuit each other and the fuse will blow. The fuse should be sufficiently small so that there can be no possibility of injuring the transformers.

A method of testing for polarity of single-phase transformers with a voltmeter is shown in Fig. 5-44, II. Connect the transformer as shown. Make, successively, voltmeter readings $V$, $V_1$, and $V_2$. If the transformer has what is called additive polarity, $V + V_1$ will equal $V_2$. If the transformer has what is called subtractive polarity, $V_2$ will equal $V - V_1$. For example, in Fig. 5-44, II, the transformer has additive polarity, and the primary line voltage $V$ (2200 V) plus the secondary transformer voltage $V_1$ (220 V) equals the voltage between $A$ and $B$, or 2420 V. With subtractive polarity the voltmeter $V_2$ would read (2200 − 220), or 1980 V.

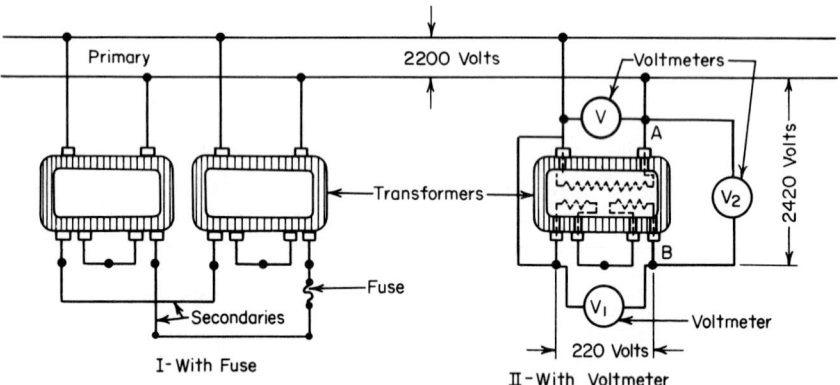

**Fig. 5-44**   Testing transformers for polarity.

## SINGLE-PHASE CONNECTIONS

**76. Connections for standard distribution transformers** are shown in Figs. 5-45 and 5-46. Distribution transformers of medium and small capacity are almost invariably arranged with two secondary coils, which can be connected in series or parallel with each other so that the same transformer can be used to supply either of two different secondary voltages or a three-wire system. In transformers of the 120/240-V type their secondary windings can be so connected as to deliver 120 or 240 V or for a 120/240-V three-wire circuit. The connections of the secondary coils are made, either by splicing the secondary leads or with connectors, outside the transformer case. Some transformers are constructed with two primary coils as shown in Fig. 5-45. If the necessary primary-coil connections are made, they can be used on primary circuits of either of two voltages. Standard practice with respect to number of coils and voltage ratings is given in Secs. 38 and 39.

**77. Transformer connections for three-wire secondary service** are shown in Figs. 5-45, 5-46, and 5-47. In the arrangement in Figs. 5-45 and 5-46, one transformer only is used. Its secondary windings are connected in series, and a tap is made to the point of connection between the two windings, providing 240 V between the two outside wires and 120 V on each of the side circuits. The transformer should have a capacity equal to the load to be supplied, and the three-wire circuits should be carefully balanced. If the three-wire circuits are decidedly unbalanced, the transformer should have a capacity equal to twice the load on the more heavily loaded of the two side circuits.

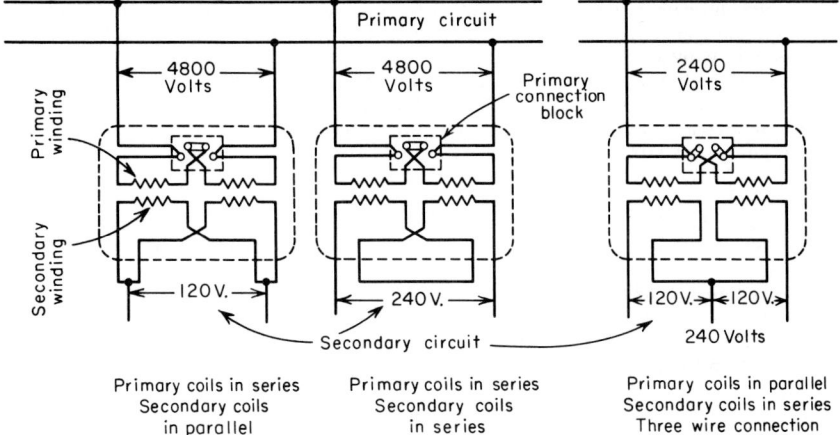

**Fig. 5-45**  Connections of standard distributing transformers.

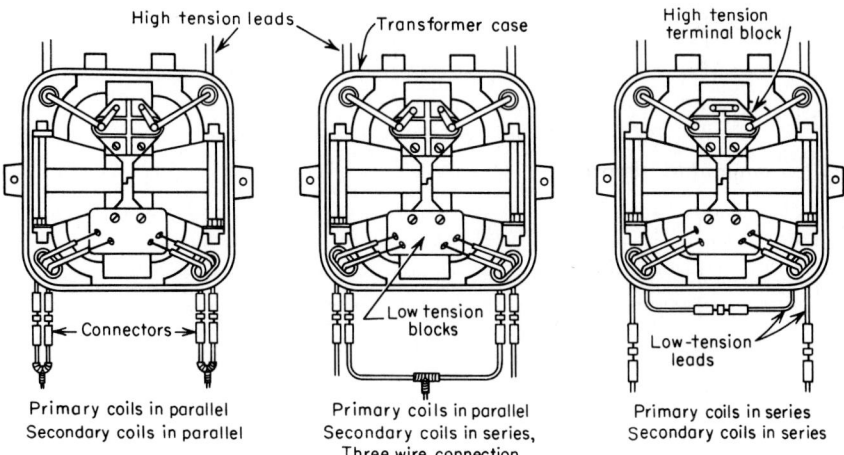

**Fig. 5-46**  Method of interconnecting transformer secondaries with connectors.

In Fig. 5-47, two transformers are shown connected to serve a three-wire circuit. The three-wire load should be balanced as nearly as possible, and if it is very nearly balanced, each transformer should have a capacity equal to one-half of the total load. If the load is badly unbalanced, each transformer should have a capacity equal to the load on its side of the circuit. See "Parallel Operation" below.

## TWO-PHASE CONNECTIONS

**78. Transformers connected to four-wire two-phase circuits** are shown in Fig. 5-48. As a rule, two-phase primary lines are four-wire as shown, and to such a four-wire line the transformers are connected to each of the side circuits, as if each side circuit were a single-phase circuit not having any connection with the other. The total load should be so divided between the phases that the loads on each will be as nearly equal as possible. Each transformer should be designed for line voltage and will carry line current. Each transformer should have a kilovolt-ampere capacity equal to one-half of the kilovolt-ampere load that is served by the two transformers.

**79. Transformers connected to three-wire two-phase circuits** are shown in Fig. 5-49. The current in the center line wire *AA* for balanced load is 1.41 times the current in either of the outer wires. Each transformer has line voltage impressed on it and carries one-half the total load. A General Electric Co. publication comments thus: "Considerable unbalancing of voltage at the end of a transmission line or cable is experienced with the three-wire, two-phase system owing to the mutual induction between phases. Where the power factor

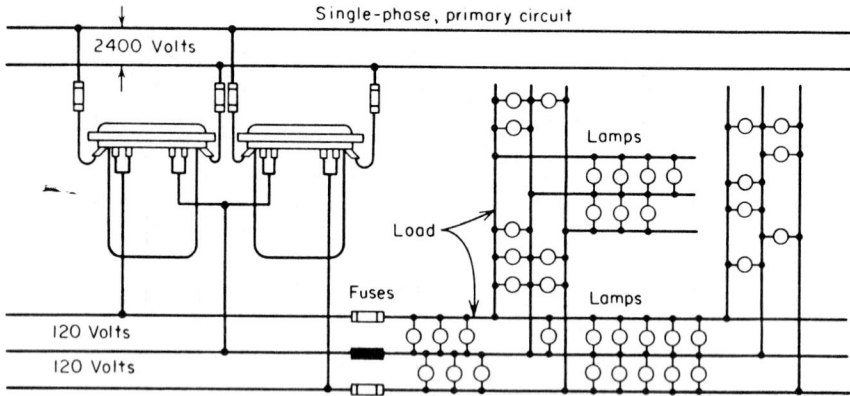

**Fig. 5-47**    Two transformers serving a three-wire circuit.

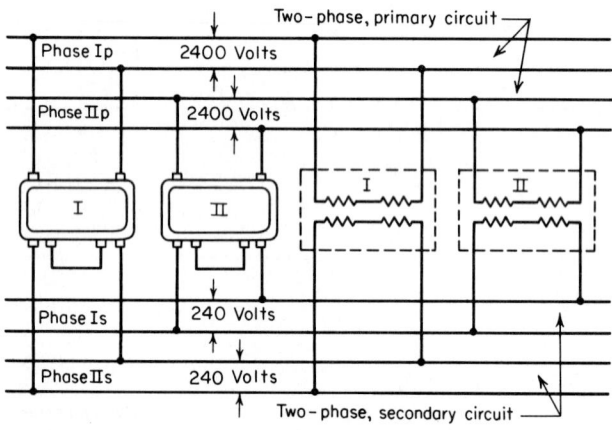

**Fig. 5-48**    Transformers with two-phase to two-phase four-wire connection.

is low, a still worse regulation is obtained, making satisfactory operation difficult. Very few systems now operate on this plan, and practically all of them could be improved by the use of some other system."

**80. Mixed connections are sometimes made with two-phase transformers,** as shown in Fig. 5-50. With improper connections such as those shown, difficulty will be experienced in the operation of motors, and they may not run at all.

### THREE-PHASE CONNECTIONS

**81. Comparison of one three-phase transformer with a group of single-phase transformers** (*Standard Handbook*) that can be employed for obtaining the same service has been summed up by J. S. Peck as follows. The advantages of the three-phase transformer are (1) lower cost, (2) higher efficiency, (3) less floor space and less weight, (4) simplification in outside wiring, and (5) reduced transportation charges and reduced cost of installation. The disadvantages of the three-phase transformer are (1) greater cost of spare units, (2) greater derangement of service in the event of breakdown, (3) greater cost of repair, (4) reduced capacity obtainable in self-cooling units, and (5) greater difficulties in bringing

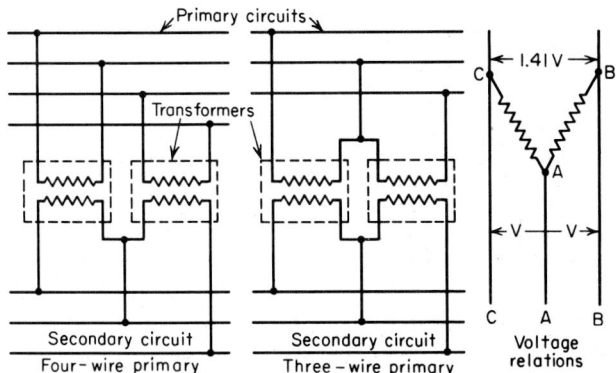

**Fig. 5-49**  Connections for transformers on three-wire two-phase circuits.

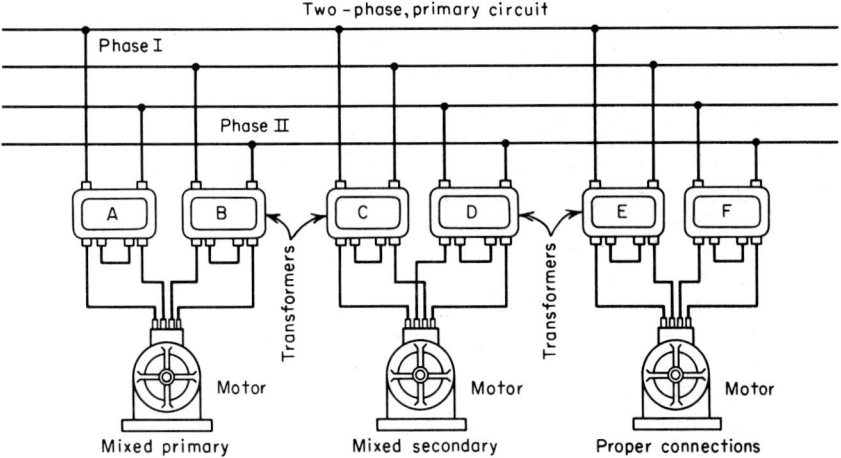

**Fig. 5-50**  Correct and incorrect connections for transformers serving two-phase motors.

out taps for a large number of voltages. It is considered that the three-phase transformer has certain real and positive advantages over the one-phase type, while its disadvantages are chiefly those which result in the event of breakdown—an abnormal condition which occurs at rarer and rarer intervals as the art of transformer design and manufacture advances.

**82. Connections of transformers for the transformation of three-phase power** are given in Figs. 5-51 to 5-58 inclusive. In all these figures the polarity of the transformer units is additive. If the transformers had subtractive polarity, the interconnections between transformers would be made in exactly the same manner as shown in the figures insofar as terminal marking is concerned. The interconnections for subtractive polarity would be different with respect to the relative location of terminals. The connections shown in these figures result in standard angular displacements.

Much of the material in the following sections has been taken from a publication of the Line Material Company.

**83. Transformers with both primary and secondary coils delta (Δ)-connected** are shown in Fig. 5-51a. All three of the transformers are connected in series in a closed circuit, and each line wire is connected to the connection between two of the transformers. The voltage imposed on either the primary or the secondary of the transformer is the primary or the secondary line voltage, respectively. The current in either winding = line current ÷ $\sqrt{3}$, or line current × 0.58. The kilovolt-ampere capacity of each transformer should be

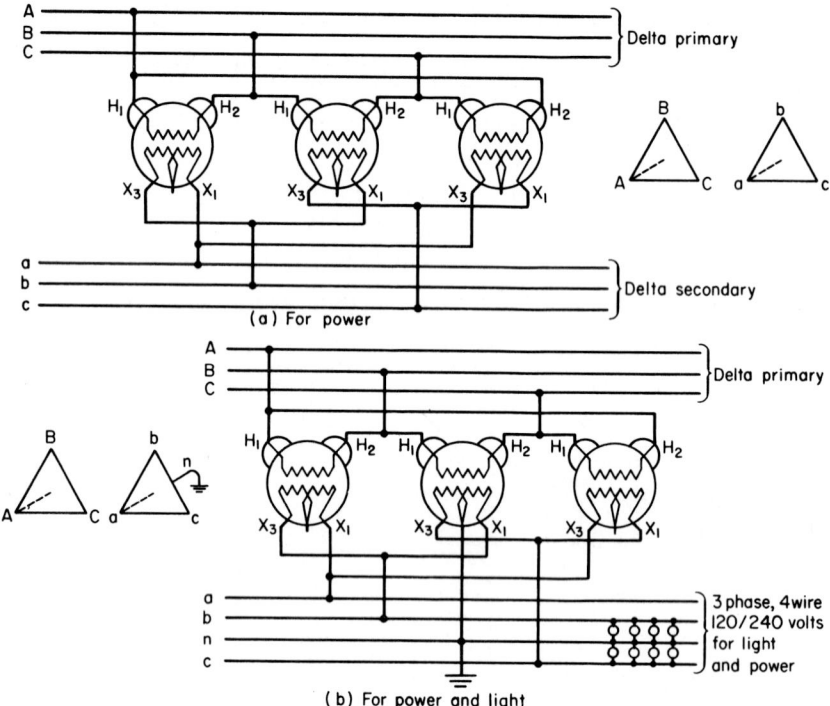

**Fig. 5-51**   Transformers delta-connected on both primary and secondary.

equal to one-third the total kilovolt-amperes of the load to be served. The total kilovolt-ampere load transmitted by a balanced three-phase line = $1.73IE$ (where $I$ is the line current in each line wire and $E$ is the voltage between wires). Therefore, the kilovolt-ampere capacity of each transformer should be $1.73IE ÷ 3 = 0.58IE$. This type of three-phase transformation has been the one most commonly used in the past. The ungrounded primary system will continue to supply power even though one of the lines is grounded owing to a fault. If one of the transformers in the bank should fail, secondary power can be supplied from the two remaining units by changing to the open-delta connection. Thus this type of connection is ideal from the standpoint of service continuity.

When light and power are to be supplied from the same bank of transformers, the mid-tap of the secondary of one of the transformers is grounded and connected to the fourth wire of the three-phase secondary system as shown in Fig. 5-51b. The lighting load is then divided between the two hot wires of this same transformer, the grounded wire being common to both branches.

**84. Transformers with both primary and secondary coils, star-connected,** from a three-wire primary circuit are shown in Fig. 5-52. The current in each transformer winding is the same as the line current, and the voltage imposed on each winding = line voltage ÷ $\sqrt{3}$ = line voltage × 0.58. The kilovolt-ampere capacity of each transformer should be equal to one-third the total kilovolt-amperes of the load to be served. The total kilovolt-ampere load transmitted by a balanced three-phase line = $1.73IE$ (where $I$ is the line current in each line wire and $E$ is the voltage between wires). Therefore, the kilovolt-ampere capacity of each transformer should be $1.73IE \div 3 = 0.58IE$.

It is necessary that the primary neutral be available when this connection is used, and the neutrals of the primary system and of the bank are tied together as shown. If the three-

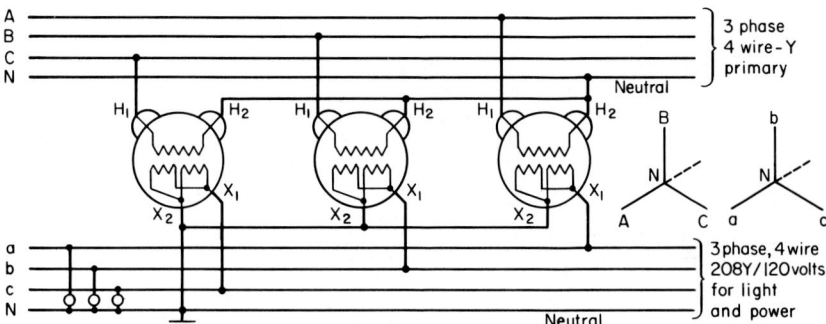

**Fig. 5-52**    Transformers star-connected on both primary and secondary (four-wire primary circuit).

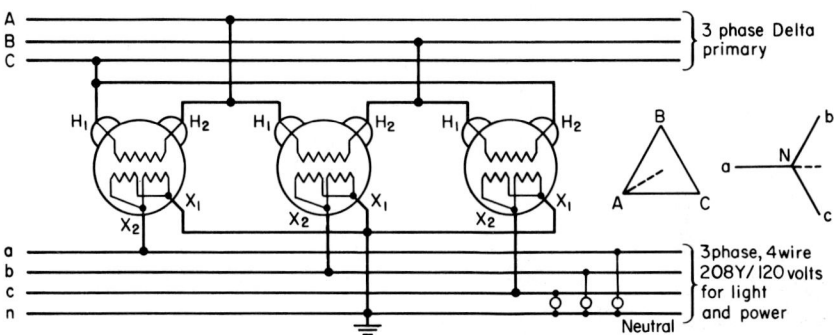

**Fig. 5-53**    Transformers delta-connected primary and star-connected secondary.

phase load is unbalanced, part of the load current flows in the primary neutral. Also the third-harmonic component of the transformer exciting current flows in the primary neutral. For these reasons, it is necessary that the neutrals be tied together as shown. If this tie were omitted, the line to neutral voltages on the secondary would be very unstable. That is, if the load on one phase were heavier than on the other two, the voltage on this phase would drop excessively and the voltage on the other two phases would rise. Also, large third-harmonic voltages would appear between lines and neutral, both in the transformers and in the secondary system, in addition to the 60-Hz component of voltage. This means that for a given value of rms voltage, the peak voltage would be much higher than for a pure 60-Hz voltage. This overstresses the insulation both in the transformers and in all apparatus connected to the secondaries.

**85. Transformers delta-connected to the primary circuit and star-connected to the secondary circuit** are shown in Fig. 5-53. Any group of transformers can have either their primary or their secondary coils connected in either star or delta. With the primary delta-connected and the secondary star-connected as shown, the secondary voltage will be 1.73 times what it would be if it were delta-connected. The neutral of the secondary three-phase system is grounded. The single-phase loads are connected between the different phase wires and neutral while the three-phase power loads are connected to the three-phase wires. Thus, 120 V is supplied to the lighting loads and 208 V to the power load. Advantages of this type of bank are the fact that the single-phase load can be balanced on the three phases in each bank by itself and the fact that the secondaries of different banks can be tied together.

**86. Transformers star-connected primary and delta-connected secondary** are shown in Fig. 5-54. This is the reverse of the grouping described in Sec. 85, and the secondary voltage will be only 0.58 times as great as if both secondary and primary were star-connected.

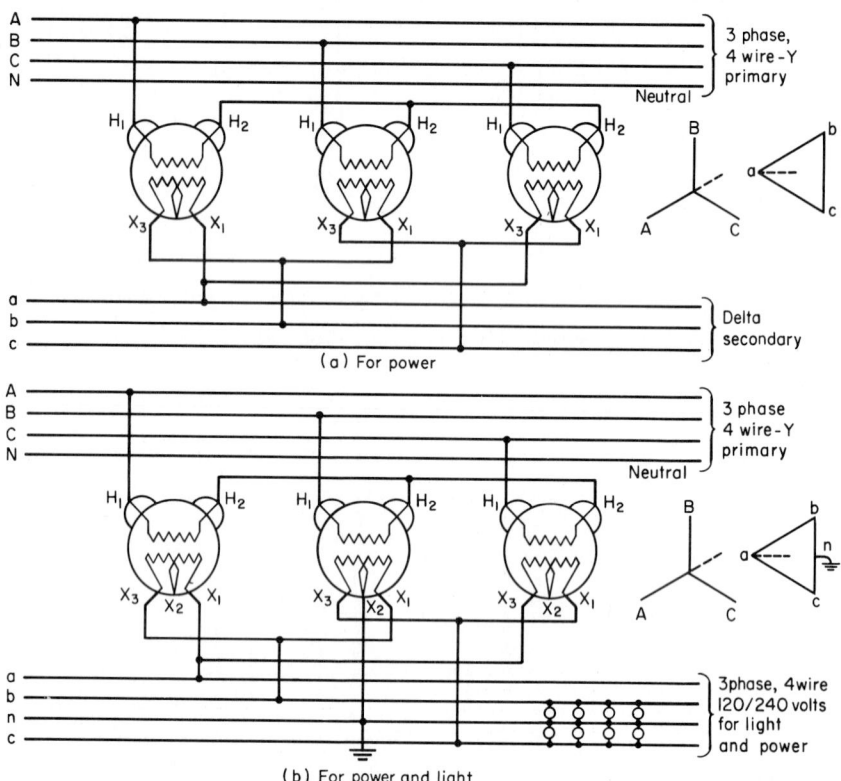

**Fig. 5-54**    Transformers star-connected primary and delta-connected secondary.

The present tendency in utilities is to replace the 2400 delta system with the 2400/4160Y-V three-phase four-wire system. This change in effect raises the distribution voltage from 2400 to 4160 without any major changes in connected equipment. The same transformers that were previously connected between lines on the 2400-V delta system are now connected between lines and neutral on the new 2400/4160Y-V system. Three-phase banks that had previously been connected delta-delta are now connected Y-delta as shown.

When service for both light and power is to be supplied, the Y-delta bank takes the form shown in Fig. 5-54b.

If one unit of a Y-delta bank goes bad, service can be maintained by means of the connection of Fig. 5-55. In the regular Y-delta bank with three units, the neutral of the primaries of the transformers is not ordinarily tied in with the neutral of the primary

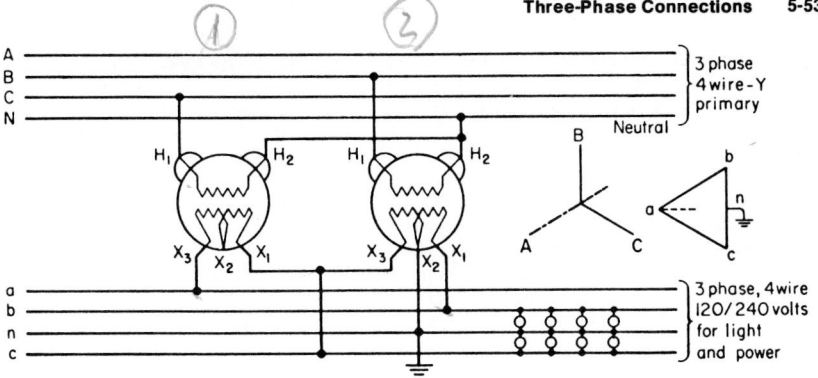

**Fig. 5-55** Y-delta connection with one unit missing.

system. In fact, this bank can be used even when the primary neutral is not available. In the bank with two units, however, it is necessary to connect to the neutral as shown. The main disadvantage of this hookup is the fact that full-load current flows in the neutral even though the three-phase load may be balanced. In addition to maintaining service in an emergency, this type of bank is satisfactory if the main part of the load is lighting and the three-phase load is small.

**87. The three-phase V, or open-delta, connection** (Fig. 5-56). Line voltage is impressed on each transformer, and line current flows in each transformer coil. This method is often used for motors but has the objection that if one of the transformers becomes inoperative, the three-phase circuit served will be fed by only one transformer and hence will be inoperative.

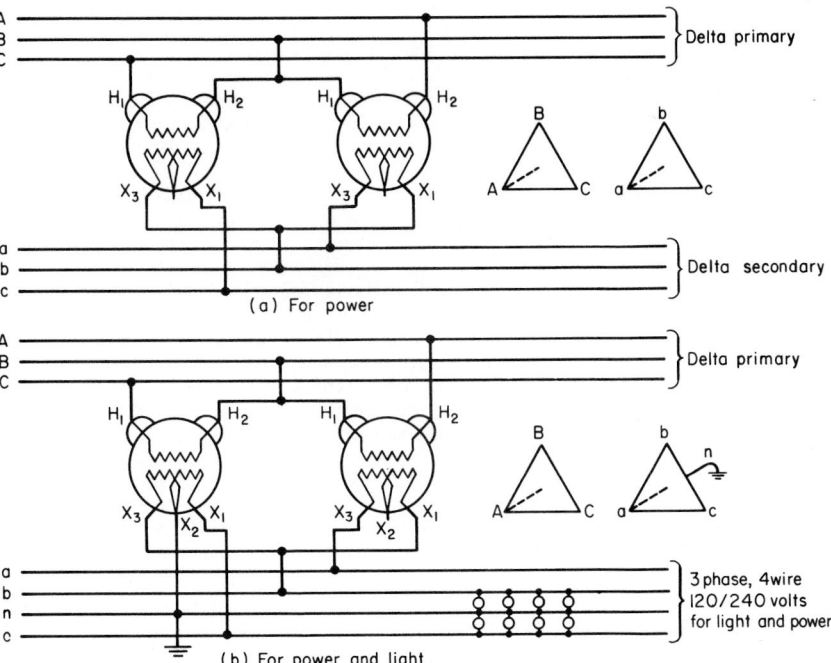

**Fig. 5-56** Three-phase V, or open-delta, connection of two transformers.

The combined capacity of two transformers (Harry B. Gear and Paul F. Williams, *Electric Central Station Distributing Systems*, D. Van Nostrand Co.) connected by this method and serving a given load should be 15.5 percent greater than the combined capacity of three transformers delta- or star-connected and serving the same load. For instance, if three 5-kVA transformers (total capacity, 15 kVA) are required for a certain installation and they are replaced by two 7½-kVA transformers (total capacity, 15 kVA), the two transformers will be overloaded by 15.5 percent at a full load of 15 kW at 100 percent pf.

For example, assume that in a three-transformer installation the current in the secondary line is 17.3 A. This imposes a load of 10 A on the transformer secondary coils. At 200 V this is 2 kW per transformer, or 6 kW in all. If two 3-kW transformers are put in to replace the three 2-kW units, the capacity of the secondary coils would be 15 A. But, as above noted, with the open-delta connection the current in the secondary coil is the same as the current in the line, and the 15-A winding must carry 17.3 A, or 15.5 percent overload.

In the grouping of Fig. 5-56, to reverse the direction of a motor served by the group, interchange any two of the primary phase wires or reverse any two of the secondary wires.

In some cases two transformers for open-delta grouping of proper aggregate capacity to serve a given load will be cheaper than three transformers for star or delta grouping to serve the same load, but this is not always the case.

The open-delta connection can be used in an emergency if one of the transformers in a delta-delta bank fails. This type of bank is also used to supply power to a three-phase load which is temporarily light but which is expected to grow. When the load increases to a point at which the two transformers in the bank are overloaded, an increase in capacity of 1.732 times can be obtained by adding another unit of the same size and using the delta-delta connection.

When the secondary circuits are to supply both light and power, the open-delta bank takes the form shown in Fig. 5-56b. In addition to the applications listed above for the open-delta bank for power, this type of bank is used when there is a large single-phase load and only a small three-phase load. In this case, the two transformers would be of different kVA sizes, the one across which the lighting load is connected being the larger.

**88. Disadvantages of the V, or open-delta, connection** (*Standard Handbook*). For normal operation not only must each of the V-connected transformers be larger than each of the delta-connected transformers, but the two transformers must have a combined rating 15.5 percent greater than the three transformers. This fact taken alone does not represent a disadvantage of the V connection, because the two larger transformers are exactly equal in constructive material and operating efficiency to the three smaller transformers. The real objection to the V connection for serious work resides in the tendency for the local impedance of the transformers to produce an enormous unbalance of the secondary voltages and of the primary currents. In spite of this disadvantage (which is really of little consequence in 2200-V primary distribution work) many V-connected groupings are in satisfactory operation.

**89. Comparison between the delta, the star, and the open-delta methods of connection** (*Standard Handbook*). The choice between the methods would be governed largely by the service requirements. When the three transformers are delta-connected, one can be removed without interrupting the performance of the circuit, the two remaining transformers in a manner acting in series to carry the load of the missing transformer. The desire to obtain immunity from a shutdown due to the disabling of one transformer has led to the extensive use of the delta connection of transformers, especially on the low-potential delivery side. It is to be noted that if one transformer is crippled, the other two will be subjected to greatly increased losses.

Thus, if three delta-connected transformers are equally loaded until each carries 100 A, there will be 173 A in each external circuit wire. If one transformer is now removed and 173 A continues to be supplied to each external circuit wire, each of the remaining transformers must carry 173 A, since it is now in series with an external circuit. Therefore, each transformer must now show 3 times as much copper loss as when all three transformers were active, or the total copper loss is now increased to a value of 6 relative to its former value of 3. An open-delta installation is made frequently when considerable future increase in load is expected. The increase can be accommodated by adding the third transformer to the bank at a later date and thus increasing the capacity of the load that can be carried by about 75 percent.

A change from delta to Y in the secondary circuit alters the ratio of the transmission emf to the receiver emf from 1 to $\sqrt{3}$. On account of this fact, when the emf of the transmission circuit is so high that the successful insulation of transformer coils becomes of constructive and pecuniary importance, the three-phase line sides of the transformers are connected in "star" and the neutral is grounded. The windings of most transformers operating on systems of 100,000 V or more are star-connected.

See also Sec. 88 regarding properties of an open-delta–connected group.

**90. Comparative cost of transformers for different grouping for three-phase service.** The accompanying table shows the costs of the single-phase transformers, of proper capacities for either a delta or an open-delta grouping, and of a three-phase transformer to serve a 75-kVA installation. The relative costs will be the same for the present date.

| Method of connection or grouping | Number of transformers required | Capacity of each transformer, kVA | Aggregate capacity | Cost per transformer | Aggregate cost |
|---|---|---|---|---|---|
| Delta (Δ) | 3 | 25 | 75 | \$ 500 | \$1500 |
| Open delta (V)[a] | 2 | 50 | 100 | 800 | 1600 |
| Three-phase transformer | 1 | 75 | 75 | 1400 | 1400 |

[a]The theoretical aggregate capacity of two single-phase transformers for open-delta grouping for a 75-kVA three-phase load (see Sec. 87) would be $75 \times 1.15 = 86.3$ kVA or $86.3 \div 2 = 43.2$ kVA per transformer. The nearest commercial capacity to 43.2 kVA is 50 kVA, which gives an aggregate capacity of 100 kVA.

**91. Nonstandard angular displacement.** All the transformer connections shown in Figs. 5-51 to 5-56 inclusive are made to give the standard angular displacement or vector relation between the primary and secondary voltage systems as defined by the different standards publications including those of the ANSI. These standard angular displacements are 0° for delta-delta or Y-Y connected banks and 30° for delta-Y or Y-delta banks.

Angular displacement becomes important when two or more three-phase banks are interconnected into the same secondary system or when three-phase banks are paralleled. In such cases it is necessary that all the three-phase banks have the same displacement.

If subtractive-polarity transformers are used in the connections illustrated in Figs. 5-51 to 5-56, it is found that the secondary connections are much simplified from those shown for the additive-polarity units. The additive-polarity connections for standard angular displacement are somewhat complicated, particularly in cases with a delta-connected secondary, by the crossed secondary interconnections between units.

For this reason simplified bank connections which give nonstandard angular displacement between the primary and secondary systems are sometimes used with additive-polarity units. The diagrams of Figs. 5-57 and 5-58 cover these simplified bank connections for three additive-polarity units for the more common three-phase connections with the delta-connected secondary.

## SPECIAL TRANSFORMER CONNECTIONS

**92. Transformers connected for transforming from three-phase to two-phase or the reverse** are illustrated in Fig. 5-59, which shows what is known as the Scott connection. The transformers required are special, and each has a lead brought out from the middle point of the high-tension winding, and a special voltage tap giving 86.6 percent of the high-tension winding is arranged. Usually two transformers just alike are purchased so that they will be interchangeable. These special transformers can be purchased from any of the large manufacturers. Those shown are Westinghouse transformers. Two standard single-phase transformers for such service should have an aggregate kilovolt-ampere rating 15½ percent greater than their group or nominal rating.

**93. Explanation of the transformation from three-phase to two-phase** (*Standard Handbook*). Assume the simple case of a total power of 30,000 W, 1.0 pf, at 100 V, three-phase to be transformed (without loss) to 30,000 W, 100 V, two-phase (see Fig. 5-60). If the load is balanced on the two-phase side, there will be 15,000 W per phase, or 150 A at 100 V. Since the three-phase power is represented as $\sqrt{3}IE = 30,000$, where $I$ is the current per line wire and $E$ is the emf between line wires, $I$ must equal 173.2 A, because $E$ has been taken as 100 V.

As shown in Fig. 5-60, the three-phase coils of one transformer must be designed for 100 V and 173.2 A, while the three-phase coils of the other transformer must be designed for 86.6 V and 173.2 A. The current through the coil $CD$ divides equally. A part (86.6 A) goes through $DA$ and an equal part (86.6 A) passes differentially through $DB$; thus the

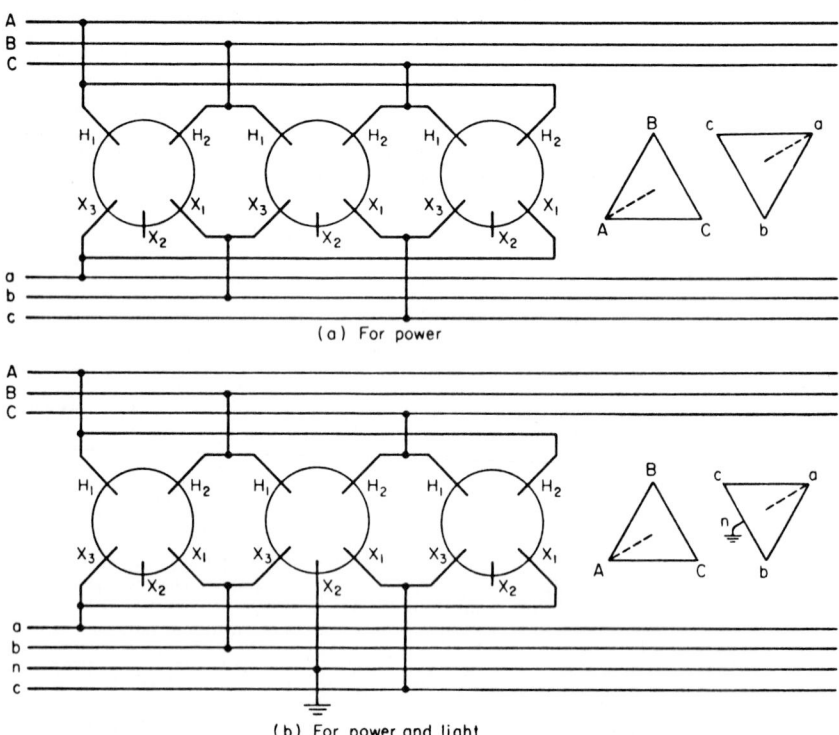

(a) For power

(b) For power and light

**Fig. 5-57**  Delta-delta connections with nonstandard angular displacements.

mmf of these two currents has a resultant of zero, and it has no effect upon the core flux insofar as transformer $T'$ is concerned. The coil $ADB$ carries a total value of current of 173.2 A throughout all its turns, but the current in one half is 60 time degrees out of phase with that in the other half. That is to say, the 173.2 A in one half is made up of a load current of 150 A, in leading time quadrature with which is 86.6 A, while that in the other half is made up of a load current of 150 A, in lagging time quadrature with which is a superposed current of 86.6 A. The magnetizing effect of the 173.2 A is, therefore, 150 A, and the current in the two-phase side of transformer $T'$ is 150 A. In the $T$ transformer the mmf of 173.2 A in 86.6 percent turns is just equal to that of 150 A in 100 percent turns; these two currents are directly in time-phase opposition, and the apparatus operates in all respects like a one-phase transformer. The phase relations and the relative values of the several components of currents are shown in the vector diagram of Fig. 5-60.

**94. Kilovolt-ampere ratings of transformers for Scott connection (three-phase to two-phase) for serving a given horsepower load.**  The accompanying table gives the ratings recommended by the Westinghouse Electric Corp. for transformers serving squirrel-cage

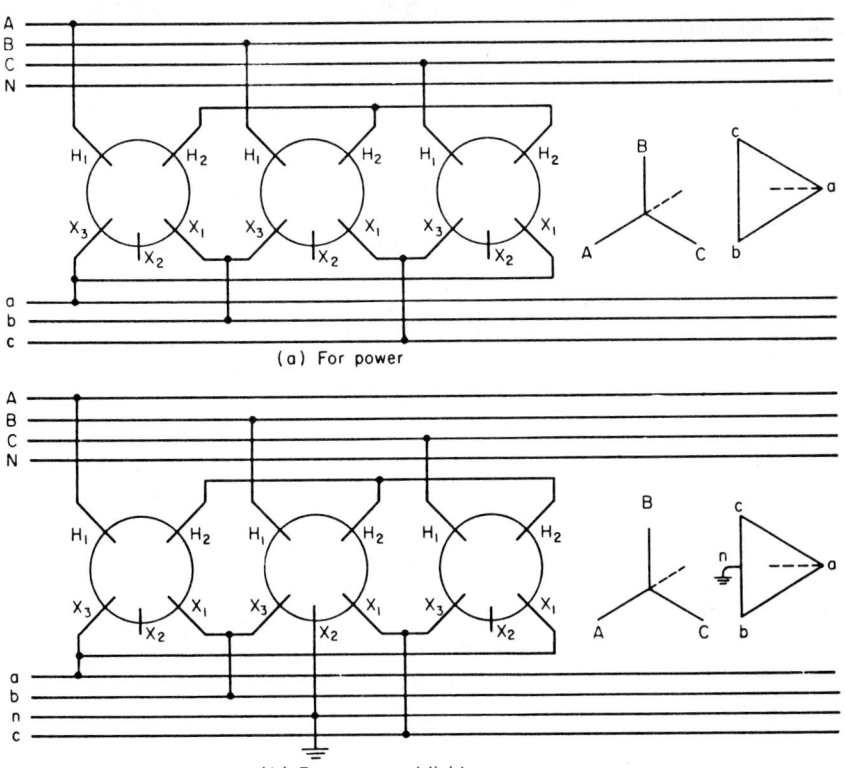

(a) For power

(b) For power and light

**Fig. 5-58** Y-delta connections with nonstandard angular displacements.

Three-phase, primary circuit

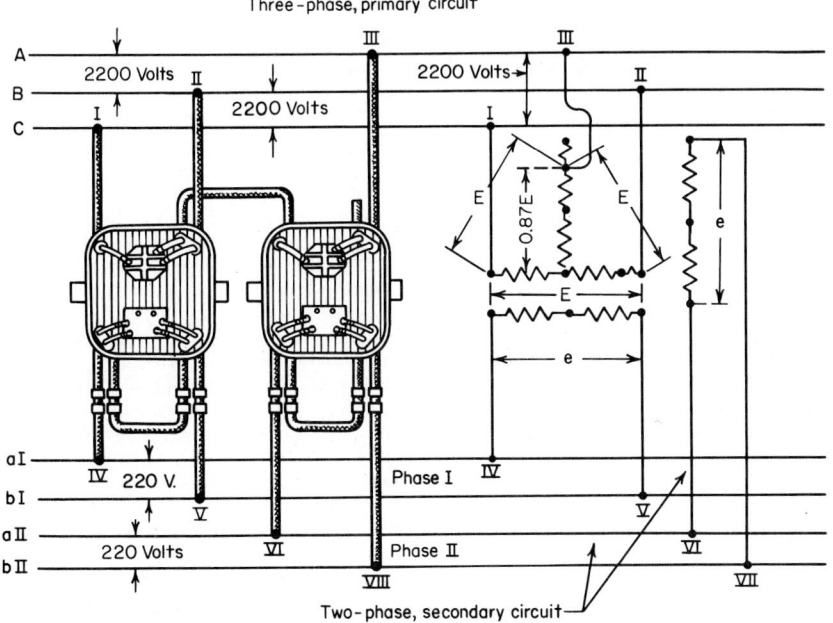

**Fig. 5-59** Transformers connected for three-phase to two-phase transformation.

induction motors and indicates the efficiency of the installation. The temperature guarantee with performances as shown is a 50°C rise.

| Hp of motor | Number of transformers | Kva capacity of each transformer | Total kva load imposed on the group of two transformers when motor is operating at full load | Efficiency of bank of transformers with full load on motor |
|---|---|---|---|---|
| ½ | 2 | ½ | 0.75 | 92.8 |
| 1 | 2 | ½ | 1.35 | 94.5 |
| 2 | 2 | 1 | 2.40 | 95.2 |
| 3 | 2 | 1½ | 3.4 | 95.9 |
| 5 | 2 | 2½ | 5.5 | 96.4 |
| 7½ | 2 | 4 | 8.1 | 97.0 |
| 10 | 2 | 5 | 10.7 | 97.2 |
| 15 | 2 | 7½ | 15.7 | 97.4 |
| 20 | 2 | 10 | 20.9 | 97.6 |
| 30 | 2 | 15 | 31.5 | 97.8 |
| 40 | 2 | 20 | 42.0 | 98.0 |
| 50 | 2 | 25 | 51.0 | 98.0 |
| 75 | 2 | 37½ | 77.0 | 98.3 |

$$\text{kVA on each transformer} = \frac{\text{hp} \times 0.746 \times 0.59}{(\text{efficiency} \times \text{pf of motor})}$$

$$\text{Efficiency} = \frac{\text{kW on transformer}}{\text{kW transformer loss} + \text{kW on transformer}}$$

**95. T-connected transformers for transforming from three-phase to three-phase** *(Standard Handbook).*   A method of employing two transformers in three-phase transformation which practically overcomes the disadvantages of the V connection and possesses considerable merit is found in the T connection. As indicated in Fig. 5-61, one transformer is connected across between two of the line wires, while the other is joined between the third line wire and the middle point of the first transformer. The current in the primary coil of each transformer is the same in value as that in the primary coil of the other, and the secondary currents in the two transformers are likewise equal in value. The voltage impressed across one transformer is only 86.6 percent of that across the other, so that if each transformer is designed especially for its work, one will have a rating of $EI$ and the other a rating of $0.866EI$, where $I$ is the current in each line wire and $E$ is the emf between lines. The combined rating will therefore be 1.866, as compared with 1.732$EI$ for three one-phase transformers connected either Δ or Y, or with 2.0$EI$ for two V-connected transformers.

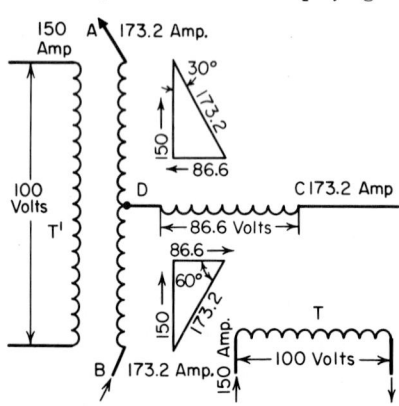

**Fig. 5-60**   Three-phase to two-phase transformation.

**96. Requisites for transformers for T connection.**   The two transformers should possess the same ratio of primary to secondary turns, and a tap should be brought out from the central point of one of the transformers. It is not essential that the former transformer be designed for exactly 86.6 percent of the voltage of the latter; the normal voltage of one can be 90 percent of the other without producing detrimental results. Moreover, transformers designed for the same normal emf and intended for V connection can be T-connected with considerable improvement in service.

**97. In comparing the T connection with the Δ or the Y connection** (*Standard Handbook*), it is to be noted that each connection accomplishes the transformation without sensible distortion of phase relations. The T connection allows the neutral point to be reached equally as well as does the Y connection. The Δ connection, however, is the only one capable of transforming in emergencies with one disabled transformer. With reference to its ability to maintain balanced phase relations, the T connection is much better than the V connection. The aggregate kilovolt-ampere rating of two T-connected transformers should be 15½ percent greater than the nominal kilovolt-ampere rating of the group.

**98. Booster transformers** (Harry B. Gear and Paul F. Williams, *Electric Central Station Distributing Systems*). Ordinary distributing transformers applied as illustrated (Fig. 5-62) are used when it is necessary to raise, by a fixed percentage, the voltage delivered by a line, as it is when transformer ratios do not give quite the right voltage or when line drop is excessive. A booster raises the voltage of any primary circuit in which it may be inserted by the amount of the secondary voltage of the booster (see Fig. 5-62).

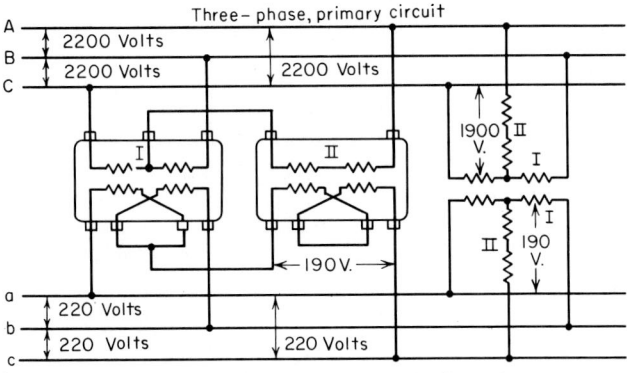

**Fig. 5-61** Transformers, T-connected for three-phase transformation.

**examples** On a long single-phase 2080-V lighting branch so heavily loaded that the pressure drops more than the amount for which the normal regulation of the feeder will compensate, a 110-V transformer inserted in the line as a booster will raise the pressure of the primary branch on the load side of the booster by 110 V. This raises the secondary pressure 5.5 percent on all the transformers beyond the booster.

With 440-V service supplied by star-connected 230-V transformers, a 10 percent booster in each phase raises the normal pressure of 230/400 V to 253/440 V.

The connections for a simple booster are shown in Fig. 5-62, I, the line pressure being raised from 2080 to 2184 V, or 5 percent. The connection at II is that for an augmented booster in which the line pressure is raised from 2080 to 2190 V, because the primary of the booster is connected across the line on the far side and the booster is boosted as well as the line. This gives an increase of 5.5 percent in the line pressure.

Figure 5-62, III, shows a 10 percent simple booster and IV an augmented 11.1 percent booster.

The transformers shown in Fig. 5-62 have a 10:1 or 20:1 ratio, and the percentages shown apply only to transformers of this ratio. If boosters having a ratio of 2080 to 115/230 are used, the percentages are increased by about 10 percent. Figure 5-62, I, would then become 5.5 percent; II, 6.05 percent; III, 11.1 percent; and IV, 12.2 percent.

**99. The proper connection of the secondary for a booster or bucking transformer** must usually be determined by trial for a transformer of any given type, but once it has been determined, the same connection can be used for any transformer of the same type. The connections shown in Figs. 5-62 and 5-63 are correct for transformers of the principal makers. Boost and buck transformers are also used in circuits of 600 V and less. See Sec. 109.

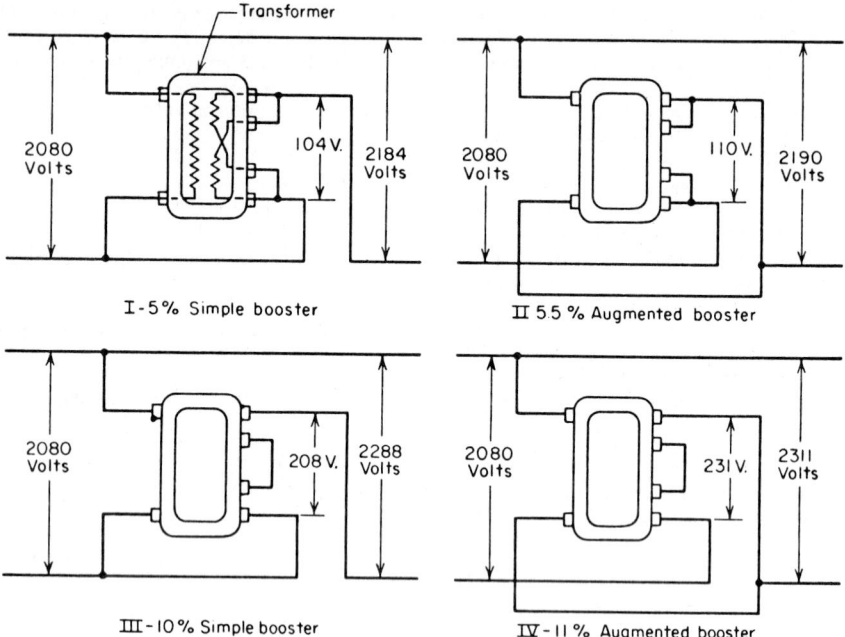

I - 5% Simple booster

II 5.5% Augmented booster

III - 10% Simple booster

IV - 11% Augmented booster

**Fig. 5-62**   Boosting transformers.

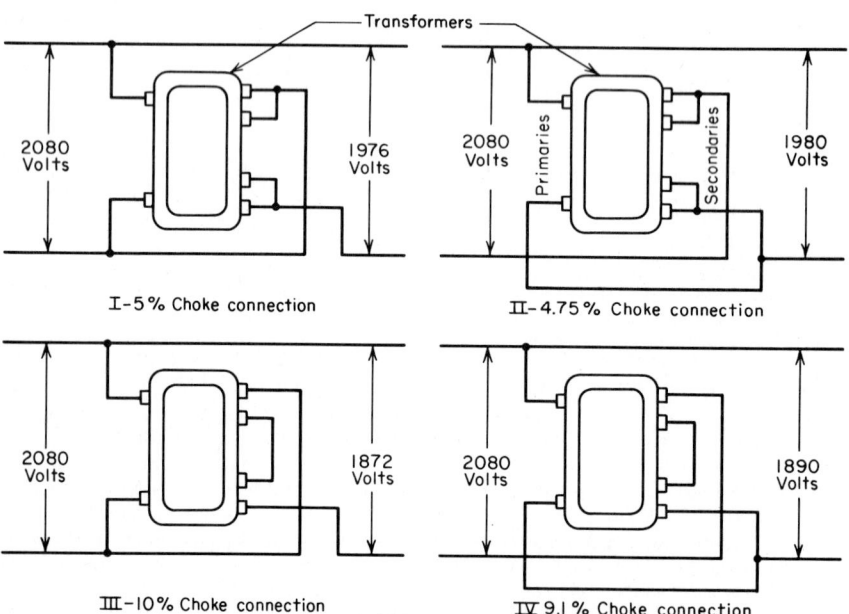

I - 5% Choke connection

II - 4.75% Choke connection

III - 10% Choke connection

IV 9.1% Choke connection

**Fig. 5-63**   Bucking (choking) transformers.

**100. Boosters are connected in a two-phase circuit** in a manner similar to that shown in Fig. 5-62 for a single-phase circuit. In three-wire two-phase feeders the boosters (secondary windings) are cut into the outer wires, and the primary windings are connected between the middle and the outside wires.

**101. Booster transformers in three-phase circuits are connected** as shown in Fig. 5-64 (*Electric Central Station Distributing Systems*). The insertion in any phase wire of the booster voltage affects two phases. The boosting and bucking effects, with transformers of various ratios, with the boosting transformers used in one, two, or three phases, are expressed in percentage of the primary voltage in the tables of Sec. 102.

**102. Voltage boosting and bucking effect of transformers connected in three-phase circuits** (*Electric Central Station Distributing Systems*). Values in the body of the tables are the percentages that the voltages will be increased or decreased, respectively, by the insertion of booster or bucking transformers, of different ratios, in one, two, or three of the phase wires. Transformers are connected for

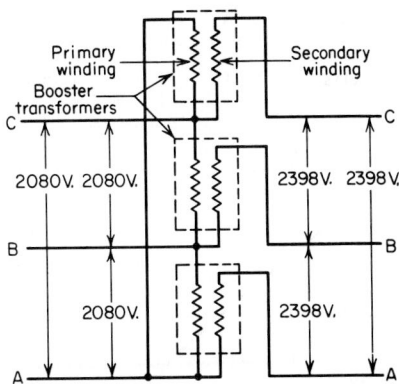

**Fig. 5-64**    Booster transformers in a three-wire, three-phase circuit.

boosting as shown in Fig. 5-64. *AB, BC,* and *CA* refer to the three phases of Fig. 5-64.

BOOSTING

| Ratios | 10:1 | | | 20:1 | | | 9:1 | | | 18:1 | | |
|---|---|---|---|---|---|---|---|---|---|---|---|---|
| Booster in | *AB* | *BC* | *CA* | *AB* | *BC* | *CA* | *AB* | *BC* | *CA* | *AB* | *BC* | *CA* |
| *A* phase | 10.0 | 10.0 | 5.3 | 5.00 | 0.00 | 2.65 | 11.0 | 0.0 | 5.8 | 5.5 | 0.00 | 2.9 |
| *A* and *B* | 15.3 | 10.0 | 5.3 | 7.65 | 5.00 | 2.65 | 16.8 | 5.5 | 5.8 | 8.4 | 2.75 | 2.9 |
| *A, B,* and *C* | 15.3 | 15.3 | 15.3 | 7.65 | 7.65 | 7.65 | 16.8 | 16.8 | 16.8 | 8.4 | 8.40 | 8.4 |

BUCKING

| | | | | | | | | | | | | |
|---|---|---|---|---|---|---|---|---|---|---|---|---|
| *A* phase | 10.0 | 0.0 | 4.6 | 5.0 | 0.0 | 2.3 | 11.00 | 0.00 | 5.06 | 5.5 | 0.00 | 2.53 |
| *A* and *B* | 14.6 | 10.0 | 4.6 | 7.3 | 5.0 | 2.3 | 16.06 | 11.00 | 5.06 | 8.3 | 5.50 | 2.53 |
| *A, B,* and *C* | 14.6 | 14.6 | 14.6 | 7.3 | 7.3 | 7.3 | 16.06 | 16.06 | 16.08 | 8.03 | 8.03 | 8.03 |

**103. Bucking transformers** (*Electric Central Station Distributing Systems*). When the secondary is connected in reverse order, the transformer becomes a "choke," reducing the line pressure instead of raising it.

   **examples**  A 5 percent choke connection is shown in Fig. 5-63, I, a 4.75 percent choke in II, a 10 percent choke in III, and a 9.1 percent choke in IV.
   The transformers shown in Fig. 5-63 have a ratio of 10:1 or 20:1, and the percentages shown are only for transformers of that ratio. If bucking transformers having a ratio of 2080 to 115/230 are used, the choking percentages would be changed to the following values: Fig. 5-63, I, 5.5 percent; II, 5.24 percent; III, 11 percent; and IV, 10 percent.

**104. The capacity of a transformer that is to be used as a booster** (Fig. 5-65) will be determined by (1) the current *I* which will flow in the ac line and (2) the voltage *E* by which the emf on the line will be boosted. If, for example, the current in the line is 20 A and it is desired to raise the voltage (Fig. 5-65) by 110 V, then required capacity of the boosting transformer = 20 A × 110 V = 2200 VA = 2.2 kVA. Hence, for these conditions it would be necessary to use at I a standard 2½-kVA transformer. The secondary coils of this transformer would be connected in multiple so that they would develop 110 V.

If it is desired to boost the voltage by 220 V, then 20 A × 220 V = 4400 VA = 4.4 kVA. Hence, theoretically a 4.4-kVA transformer would be required. Note that in any case the secondary winding of the boosting transformer must be capable of carrying the primary line current (20 A, in Fig. 5-65). Furthermore, the secondary coils of the boosting transformer must be so designed and connected that they will produce at their terminals the voltage by which it is desired to boost the emf which is impressed on the line by the generator in the station.

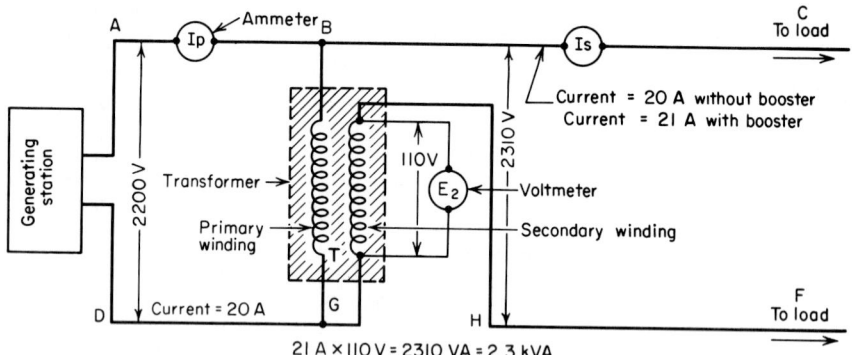

**Fig. 5-65**   Example of computing capacity required for a booster transformer.

**105. As a practical proposition it is usually desirable to select for the boost voltage one of the voltages for which the secondaries of distributing transformers are regularly designed** or for which they can be connected. That is, the boosting voltage should, ordinarily, be either 110, 220, or 440. If a boosting voltage other than one of these standard voltages is necessary, it will then probably be necessary to obtain a specially designed transformer or a transformer which has taps on the secondary coils.

**106. In every case the line current will be increased somewhat by the boosting transformer.** If the load remains constant, the increase in that portion of the circuit *HFCB* (Fig. 5-65) will be in the ratio by which the voltage impressed on this portion of the circuit has been boosted. For example, if the current in the ac line *HFCB* of Fig. 5-65 was 20 A before the boosting transformer was inserted, the current after its insertion would be computed in this way:

$$\frac{I_2}{I_1} = \frac{E_2}{E_1} \tag{3}$$

$$I_2 = \frac{E_2 \times I_1}{E_1} \text{ A} \tag{4}$$

where $I_1$ = the line current, in amperes, before inserting the booster transformer; $E_1$ = emf, in volts, across *BH* (Fig. 5-65) before inserting the booster transformer; $I_2$ = current, in amperes, in *HFCB* after the booster is inserted; and $E_2$ = emf, in volts, across *BH* after the booster is inserted.

   **example**  If we substitute the values from Fig. 5-65 in Eq. (4), the current in *HFCB* after inserting the booster = $I_2$ = ($E_2 \times I_1$) ÷ $E_1$ = (2310 × 20) ÷ 2200 = 21 A. After inserting the booster, the current $I_P$ in *ABCD* will be 21 A + the current carried by the primary winding. The value of 20 A in Sec. 104 is not strictly correct, but this relatively small difference will not appreciably affect the kilo-volt-ampere rating of the booster transformer which would be used.

**107. There are certain precautions that should be observed in the installation of boosters** *(Electric Central Station Distributing Systems)* to protect them from injury. The booster secondary is in series with the line, and current is drawn through its primary windings in proportion to the load on the line. If the primary of the booster is opened while the secondary is carrying the line current, the booster acts as a choke coil in the main circuit. This causes a large drop of pressure in the booster, imposing upon its secondary windings a difference of potential of 2 to 5 times normal. Under these conditions the insulation of a 2000-V transformer can be subjected to a pressure of 10,000 to 20,000 V or more, depending upon the load carried by the main circuit at the time.

If a fuse is used in the primary, its blowing creates the above condition, and the arc holds across the terminals of the fuse block until it burns itself clear. It has often been observed that if boosters have been "protected" by fuses in this way, the transformer has burned out shortly after the blowing of its primary fuses if not at the time.

Also, in using ordinary static transformers as boosters, it should be remembered that the service is different from that for which the transformer is designed. If this feature is neglected, the insulation may be subjected to excessive voltages or dielectric stresses, thus causing breakdown. The standard distributing transformer is designed to withstand a pressure of about 10,000 V between the high-voltage coils and the case, while the low-voltage coils (the 220-V coils) are designed to withstand a test pressure of 4000 V between coils, core, and case for a period of 1 min.

When a 2200–110/220-V transformer is used as a booster, the low-voltage coils are subjected to a continuous pressure of about 2300 V above ground. This sustained voltage will, if the case is grounded, be likely to destroy the insulation of the low-voltage coils. Therefore the transformer case should, if feasible, be insulated from the ground, although this may involve life hazard. When the transformer is installed on a pole, the insulation is automatically provided, since the pole acts as the insulating medium.

**108. Booster cutout** (Gear and Williams).   In connecting or disconnecting a booster, the main line should be opened before putting it in or out. If service on the line cannot be interrupted or if it is desired to switch the booster in or out at certain times, it can be done with a series arc cutout as in Fig. 5-66. The operation of the cutout simultaneously opens the primary and short-circuits the secondary of the booster. The switch must be of a type having a positive action so that arcing will not damage its contacts at the moment when the secondary is short-circuited. The arc cutout must have sufficient carrying capacity to carry the main-line current when the booster is shunted out, and standard series cutouts should not be used if the line current is likely to exceed 20 to 25 A.

When the augmented booster is used, the terminals of the primary winding of the transformer which goes to the cutout should be connected to the terminal of the cutout which is shown as not being in use in Fig. 5-66.

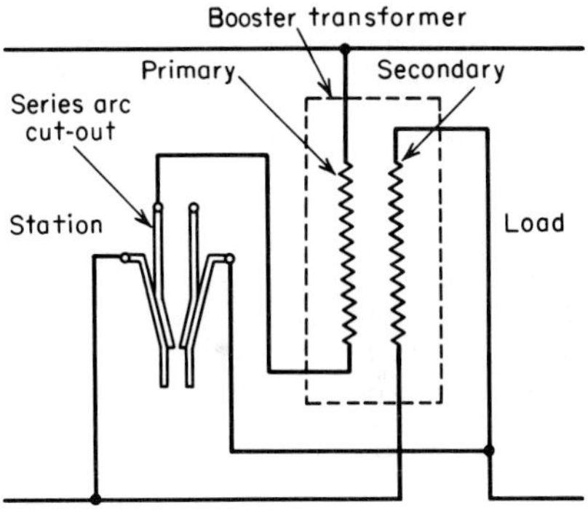

Fig. 5-66   Series arc cutout for a booster.

**109. Boost-buck transformers in low-voltage circuits.**   For circuits of 600 V or less, the most common application for a boost-buck transformer is boosting 208 to 230 or 240 V and vice versa for industrial and commercial air-conditioning systems, boosting 110 to 120 V, and 240 to 277 V for lighting systems, and voltage correction for heating systems and all types of induction motors.

In such cases a buck-boost transformer raises or lowers a supply line voltage by as much as 20 percent. A typical transformer has two primary (input) windings, both rated at either 120 or 240 V, and two secondary (output) windings, both rated at 12, 16, or 24 V. With this arrangement the primary and secondary windings can be connected together so that the electrical characteristics are changed from those of an insulating transformer to those of a boosting or bucking autotransformer.

Connections are quite similar to those shown in Figs. 5-62 and 5-63, and complete literature on boost and buck transformers can be obtained from Acme Electric Corp.

## CONNECTIONS FOR THREE-PHASE TRANSFORMERS

**110. Methods of connecting the windings of three-phase transformers** (*Standard Handbook*).   The windings of each component transformer are connected to the external cir-

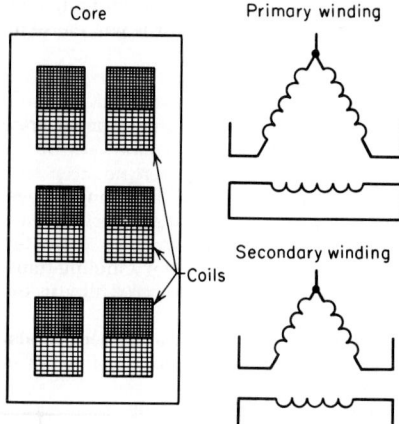

Core    Primary winding

Secondary winding

Coils

**Fig. 5-67**  Operating a damaged three-phase transformer.

cuits just as though this component were a one-phase unit; i.e., the primaries can be connected either Y or delta. Moreover, the relative advantages of the Y connection and the delta connection are quite the same with one three-phase transformer as with three one-phase transformers. The delta connection is advantageous in some cases, in that if the windings of one phase become damaged by short circuiting, grounding, or any other defect, it is possible to operate with the other phase windings V-connected.

**111. In operating a damaged three-phase transformer on two coils,** it is necessary to separate the damaged transformer windings electrically from the other coils, as indicated in Fig. 5-67. The high-potential winding of the damaged phase should be short-circuited upon itself, and the corresponding low-potential winding should also be short-circuited upon itself. The winding thus short-circuited will choke down the flux passing through the portion of the core surrounded by the windings without producing in any portion of the winding a current greater than a small fraction of the current which would normally exist in such portion at full load.

## PARALLEL OPERATION

**112. Parallel operation of transformers.**  Transformers will operate satisfactorily in parallel (Fig. 5-68), i.e., with their high- and low-tension windings respectively connected directly to the same circuits, provided that they have (1) the same ratio of transformation, (2) the same voltage ratings, and (3) approximately the same regulation. If the low-tension voltages are different, the transformer having the highest voltages will circulate current to those of lower voltage and cause a continuous loss. If transformers connected in parallel do not have the same regulation, they will not share the total load in proportion to their ratings. The greater share of the load will be taken by the transformer having the best regulation.

In connecting large transformers in parallel, especially when one of the windings is for

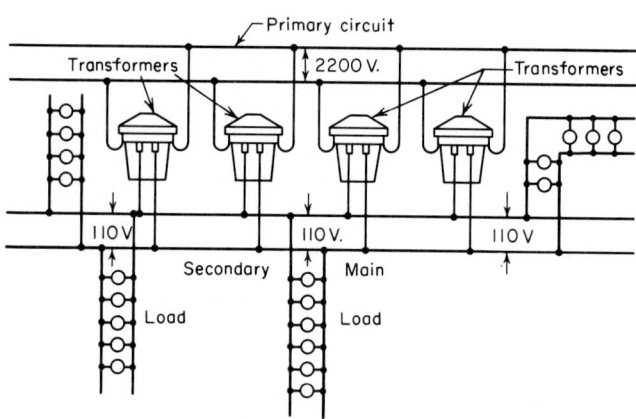

**Fig. 5-68**  Transformers banked or operating in parallel.

a comparatively low voltage, it is necessary that the resistance of the joints and interconnecting leads does not vary materially for the different transformers, or it will cause an unequal division of load.

**113. With transformers of the same general voltage and capacity characteristics connected in multiple in a secondary network** (A. D. Fishel, *Distributing Transformers*), little trouble will be encountered, as the impedance of the line between two transformers on separate poles spaced about 100 or 200 ft (30.5 or 61 m) apart will normally neutralize any difference in the transformer impedances. When transformers operated in multiple are placed on the same pole, the question of equal sharing of the load may be of some importance. The standard transformers of reliable manufacturers do not differ very widely in impedance characteristics, however, and it is usually practicable to operate transformers of the various standard types in parallel. Often the commercial desirability of paralleling transformers of different sizes will overbalance the undesirability of some inequality in the sharing of the load which might result.

**114. A method of forcing equal division of load between transformers having considerably different impedance characteristics** (A. D. Fishel, *Distributing Transformers*) is shown in Fig. 5-69. Standard 2200-V distributing transformers are usually provided with arrangements for the series-parallel connecting of both the high-tension and the low-tension windings, and therefore the connections shown can be used. As the high-tension windings are in series, the currents in the primary windings will be the same; hence the transformers will be equally loaded.

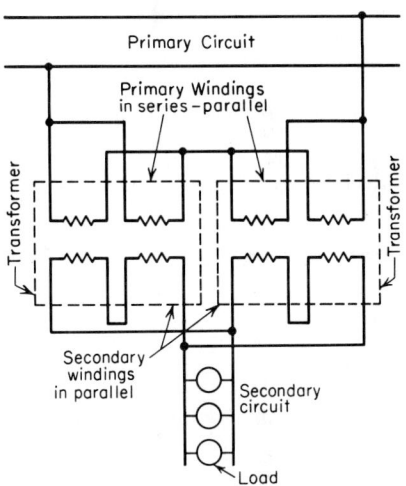

**Fig. 5-69**   Connection for forcing parallel operation of transformers which have different impedance characteristics.

**115. Some three-phase transformers can be paralleled, and some cannot** (W. M. McConahey, *Electric Journal*). A transformer having its coils connected in delta on both high-tension and low-tension sides cannot be made to parallel with one connected either in delta on the high-tension and star on the low-tension or in star on the high-tension and delta on the low-tension side. However, a transformer connected in delta on the high-tension and in star on the low-tension side can be made to parallel with transformers (having their coils joined in accordance with certain schemes) connected in star on the high-tension side and in delta on the low-tension side. Some three-phase transformers cannot be made to parallel (without changing the internal connection arrangement of their coils) with others using the same type of connections for the two windings. For example, a transformer connected delta to delta may have its coils so interconnected that it will not parallel with another transformer, connected delta to delta. By changing the internal connections between the coils, however, it will be possible to bring out the terminals in such a way that parallel operation can be obtained.

**116. How to determine whether or not three-phase transformers will operate in parallel.**   If the transformers are available, connect them as indicated in Fig. 5-70, leaving two leads on one of the transformers unjoined. Test with a voltmeter across the unjoined leads. If there is no voltage between $E$ and $e$ or between $F$ and $f$ of transformer 2, the polarities of the transformers are the same, the connections can be completed, and the transformers put in service.

If a voltage difference is found between $E$ and $e$ or between $F$ and $f$ or between both, the polarities of the transformers are not the same. Then connect transformer lead $A$ successively to mains 1, 2, and 3 and at each connection test with the voltmeter between $e$ and $f$ and the legs of the main to which lead $A$ is not connected. If with any trial connection the voltmeter readings between $f$ and $e$ and either of the two legs is found to be zero, the transformer will operate with leads $f$ and $e$ connected to those two legs. If no system of connections can be found that will satisfy this condition, the transformer will not operate in parallel without changes in its internal connections, and it may be that it will not operate in parallel at all (see Sec. 115).

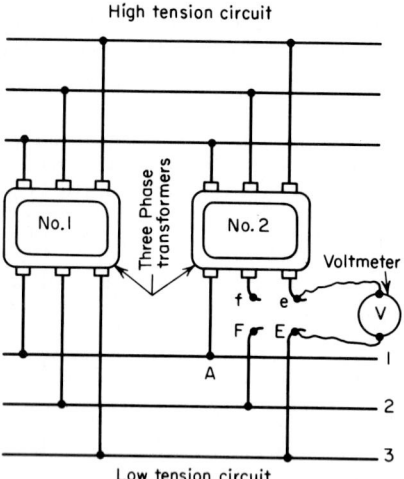

**Fig. 5-70** Testing three-phase transformers for parallel operation.

## CONNECTIONS AND APPLICATIONS OF AUTOTRANSFORMERS

**117. Use of autotransformers.** The National Electrical Code allows the use of autotransformers for supplying interior-wiring systems under certain conditions.

A branch circuit may be supplied through an autotransformer only if the system supplied has an identified grounded conductor which is solidly connected to a similar identified grounded conductor of the system supplying the autotransformer.

Autotransformers may be used as part of the ballast for lighting units. In these applications an autotransformer which raises the voltage to more than 300 V shall be supplied only by a grounded system.

Autotransformers may be used as starting compensators for ac motors. The compensator supplies a reduced voltage to the motor circuit while the machine is accelerating from rest. Ordinarily each autotransformer used for this purpose is provided with several taps so that a number of low voltages can be obtained.

**118. A starting-compensator arrangement for a two-phase induction motor** (*Standard Handbook*) is shown in Fig. 5-71. There are two autotransformers, the two separate phase lines being connected to the ends of the separate autotransformer windings. During the starting period the motor is connected between two of the ends and two intermediate taps. Figure 5-72 shows a starting-compensator arrangement for a three-phase induction motor. The three autotransformer windings are Y-connected, and low-voltage points are permanently selected along each leg of the Y. It is not necessary to employ three autotransformers for

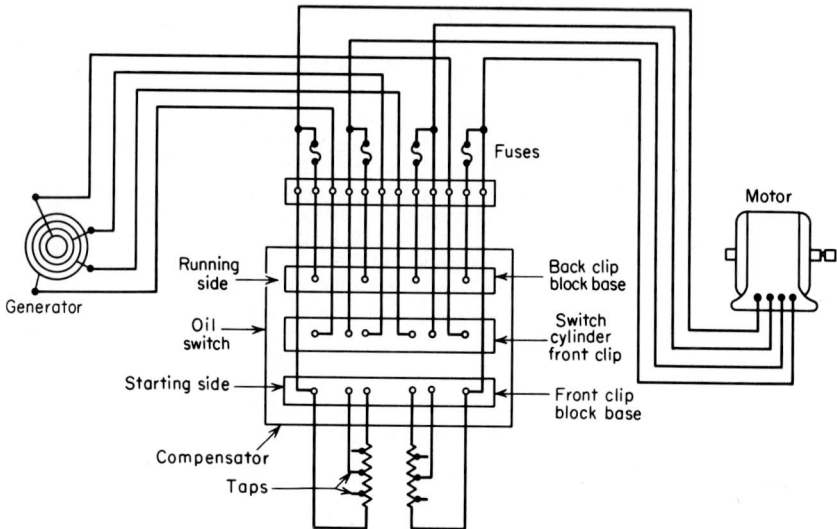

**Fig. 5-71** Two-phase starting compensator.

starting a three-phase motor; two V-connected autotransformers are quite satisfactory for this purpose. Figure 5-73 shows two V-connected autotransformers for starting a three-phase induction motor and operating it at four different voltages.

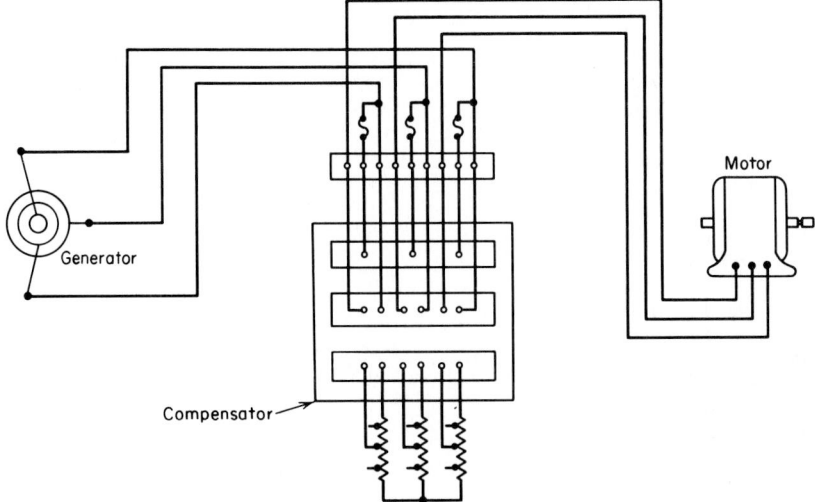

**Fig. 5-72**   Three-phase starting compensator.

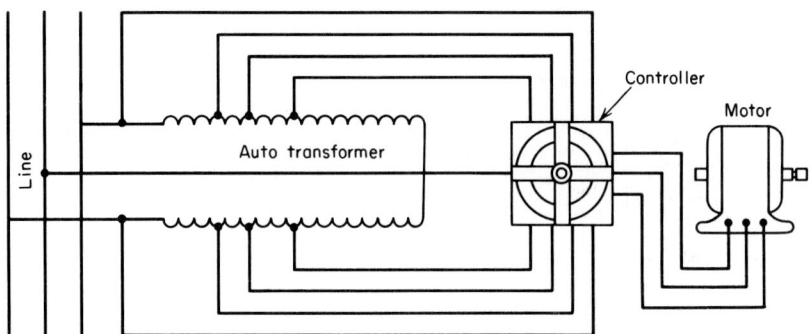

**Fig. 5-73**   An autotransformer three-phase starting compensator.

**119. The coils of a three-phase transformer can be connected for operation as an autotransformer** equally as well as can those of a one-phase transformer. The interconnections would ordinarily be by the Y method, although the delta method or a combination of the Y and delta methods may be used. Figure 5-74 represents a Y connection for autotransformer operation.

## INSTALLATION, CARE, AND OPERATION

**120. The successful operation of transformers** (Allis-Chalmers Corporation) is dependent upon proper installation and operation as much as upon proper design and manufacture. Although a transformer requires less care than almost any other type of electrical apparatus, neglect of certain fundamental requirements may lead to serious trouble, if not loss of the transformer.

**121. The following sections (122 to 141) give the requirements of the** National Electrical Code **for the installation of transformers.**   For further information see Article 450 of the NEC.

**122. Application.**   Article 450 of the NEC applies to the installation of all transformers except (1) current transformers, (2) dry-type transformers which constitute a component part of other apparatus and which conform to the requirements for such apparatus, (3) transformers which are an integral part of an x-ray, electrostatic-coating, or high-frequency apparatus, (4) transformers used with Class 2 or Class 3 power-limited circuits which shall conform to Article 725 of the Code, (5) transformers for sign and outline lighting which shall conform to Article 600 of the Code, (6) transformers for electric discharge lighting which shall conform to Article 410 of the Code, (7) transformers used for power-limited fire-protective signaling circuits which shall comply with Article 760, and (8) liquid-filled or dry-type transformers used for research, development, or testing laboratories.

Article 450 applies to the installation of transformers in hazardous locations except as modified by Article 500 of the Code.

**123. Location.** Transformers and transformer vaults shall be readily accessible to qualified personnel for inspection and maintenance. The location of oil-insulated transformers and transformer vaults is covered in Secs. 132, 133, and 137; dry-type transformers, in Sec. 130; and askarel-insulated transformers, in Sec. 131.

**124. Overcurrent protection.** Overcurrent protection shall conform to the following requirements. As used in this section, the word *transformer* means a transformer or a polyphase bank of two or three single-phase transformers operating as a unit.

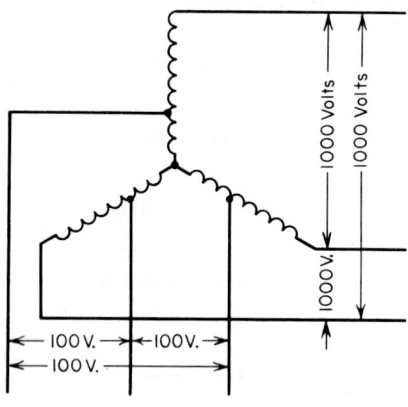

**Fig. 5-74**  Y-connected three-phase autotransformer.

A. PROTECTION ON PRIMARY SIDE, OVER 600 V, IN SUPERVISED LOCATIONS   The continuous-current rating of fuses shall not exceed 250 percent of the rated primary current, or if circuit breakers are used, they shall be set at not more than 300 percent of the rated primary current of the transformers. Exceptions to this general protection requirement are as follows: (1) If 250 percent of the rated primary current does not correspond to a standard rating of fuses, the next standard rated fuse may be used. (2) An overcurrent device is not required at the transformer if the primary circuit has fuses rated at 250 percent or circuit breakers rated at 300 percent. (3) Transformers may be protected on the primary and secondary side per table below.

**Transformers over 600 Volts Having Overcurrent Protection on the Primary and Secondary Sides**[a]
(Table 450-3(a) (1), 1990 National Electrical Code)

| | Maximum overcurrent device | | | | |
|---|---|---|---|---|---|
| | Primary | | Secondary | | |
| | Over 600 V | | Over 600 V | | 600 V or below |
| Transformer rated impedance | Circuit-breaker setting | Fuse rating | Circuit-breaker setting | Fuse rating | Circuit-breaker setting or fuse rating |
| Not more than 6% | 600% | 300% | 300% | 250% | 125%[b] |
| More than 6% and not more than 10% | 400% | 300% | 250% | 225% | 125%[b] |

[a]Reprinted with permission from NFPA 70-1990, National Electrical Code®, Copyright © 1990, National Fire Protection Association, Quincy, Massachusetts 02269. This reprinted material is not the complete and official position of the NFPA on the referenced subject, which is represented only by the standard in its entirety.

[b]In supervised installations where conditions of maintenance and supervision assure that only qualified persons will monitor and service the installation, 250% overcurrent protection shall be permitted.

B. Protection on Primary Side, over 600 V, and on Secondary Side    Primary overcurrent protection may be increased to the values shown in the accompanying table if the transformer is equipped with coordinated thermal overload protection by the manufacturer or secondary protection is provided as shown in the table.

C. Protection on Primary Side, 600 V or Less    An individual overcurrent device shall be rated or set at not more than 125 percent of the rated primary current of the transformer. Exceptions to this general protection requirement are as follows: (1) If the rated primary current is 9 A or more and 125 percent of this current does not correspond to a standard rating of fuse or nonadjustable circuit breaker, the next higher standard rating may be used; if the rated primary current is less than 9 A, the primary overcurrent device may be rated or set at 167 percent; or if the rated primary current is less than 2 A, the primary overcurrent device may be set at no more than 300 percent. (2) An individual overcurrent device is not required at the transformer if the primary overcurrent device provides the protection. (3) Primary and secondary protection is provided.

D. Protection on Primary Side, 600 V or Less, and on Secondary Side    If the secondary overcurrent protection is set at 125 percent, an overcurrent device shall be provided in the primary feeder or at the transformer rated or set at not more than 250 percent of the rated primary current. When provided with a manufacturer-installed coordinated thermal overload protection that interrupts the primary current, an overcurrent device shall be installed in the primary feeder or at the transformer set at not more than 6 times the rated current of the transformer for transformers having not more than 6 percent impedance and at not more than 4 times the rated current of the transformer for transformers having more than 6 but not more than 10 percent impedance. Exceptions to these general protection requirements are as follows: (1) If the rated secondary current of the transformer is 9 A or more and 125 percent of this current does not correspond to standard ratings of fuses or nonadjustable circuit breakers, the next higher rating may be used. (2) If the rated secondary current is less than 9 A, the secondary overcurrent device may be set at not more than 167 percent of the rated secondary current.

E. Potential (Voltage) Transformers    When potential transformers are installed indoors or enclosed, they shall be protected with primary fuses.

**125. Secondary ties.**    As used in this section, the word *transformer* means a transformer or a bank of transformers operating as a unit. A secondary tie is a circuit operating at 600 V or less between phases which connects two power sources or power supply points, such as the secondaries of two transformers. The tie may consist of one or more conductors per phase.

A. Tie Circuits    Tie circuits shall be provided at each end with overcurrent protection *according to the ampacity of the conductors,* except under the conditions described in Subpars. A1 and A2 of this section, in which cases the overcurrent protection may be in accordance with Subpar. A3 of this section.

1. *Loads at transformer supply points only.* If all loads are connected at the transformer supply points at each end of the tie and overcurrent protection is not provided, the rated ampacity of the tie shall be not less than 67 percent of the rated secondary current of the largest transformer connected to the secondary tie system.

2. *Loads connected between transformer supply points.* If a load is connected to the tie at any point between transformer supply points and overcurrent protection is not provided, the rated ampacity of the tie shall be not less than 100 percent of the rated secondary current of the largest transformer connected to the secondary tie system, except as otherwise provided in Subpar. A4.

3. *Tie-circuit protection.* Under the conditions described in Subpars. A1 and A2 of this section, both ends of each tie conductor shall be equipped with a protective device which will open at a predetermined temperature of the tie conductor under short-circuit conditions. This protection shall consist of one of the following: (1) a fusible-link cable connector, terminal, or lug, commonly known as a limiter, being of a size corresponding with that of the conductor and of construction and characteristics, according to the operating voltage and the type of insulation on the tie conductors, or (2) automatic circuit breakers actuated by devices having comparable current-time characteristics.

4. *Interconnection of phase conductors between transformer supply points.* If the tie consists of more than one conductor per phase, the conductors of each phase shall be interconnected to establish a load supply point, and the protection specified in Subpar. A3 shall be provided in each tie conductor at this point, except as follows:

Loads can be connected to the individual conductors of a paralleled conductor tie without interconnecting the conductors of each phase and without the protection specified in Subpar. A3 at load connection points, provided the tie conductors of each phase have a combined capacity not less than 133 percent of the rated secondary current of the largest transformer connected to the secondary tie system; the total load of such taps does not exceed the rated secondary current of the largest transformer, and the loads are equally divided on each phase and on the individual conductors of each phase as far as practicable.

5. *Tie-circuit control.* If the operating voltage exceeds 150 V to ground, secondary ties provided with limiters shall have a switch at each end which when open will deenergize the associated tie conductors and limiters. The current rating of the switch shall be not less than the rated current of the conductors connected to the switch. It shall be capable of opening its rated current, and it shall be constructed so that it will not open under the magnetic forces resulting from short-circuit current.

B. OVERCURRENT PROTECTION FOR SECONDARY CIRCUITS   When secondary ties are used, an overcurrent device rated or set at not more than 250 percent of the rated secondary current of the transformers shall be provided in the secondary connections of each transformer, and in addition an automatic circuit breaker actuated by a reverse-current relay set to open the circuit at not more than the rated secondary current of the transformer shall be provided in the secondary connection of each transformer.

**126. Parallel operation.**   Transformers can be operated in parallel and switched as a unit if the overcurrent protection for each transformer meets the requirements of Sec. 124.

**127. Guarding.**   Transformers shall be guarded as follows:

A. MECHANICAL PROTECTION   Appropriate provisions shall be made to minimize the possibility of damage to transformers from external causes if the transformers are located where they are exposed to physical damage.

B. CASE OR ENCLOSURE   Dry-type transformers shall be provided with a noncombustible moisture-resistant case or enclosure which will afford reasonable protection against the accidental insertion of foreign objects.

C. EXPOSED LIVE PARTS   The transformer installation shall conform with the Code provisions for guarding live parts.

D. VOLTAGE WARNING   The operating voltage of exposed live parts of transformer installations shall be indicated by signs or visible markings on the equipment or structures.

**128. Grounding.**   Exposed non-current-carrying metal parts of transformer installations including fences, guards, etc., shall be grounded when required under the conditions and in the manner prescribed by the Code for electrical equipment and other exposed metal parts.

**129. Marking.**   Each transformer shall be provided with a nameplate giving the name of the manufacturer, rated kilovolt-amperes, frequency, primary and secondary voltage, amount and kind of insulating liquid, if any, and the impedance if the transformer rating is 25 kVA or larger. The nameplate of each dry-type transformer shall include the temperature class of the insulation system.

### PROVISIONS FOR DIFFERENT TYPES OF TRANSFORMERS

**130. Dry-type transformers installed indoors.**   Transformers rated 112½ kVA or less shall have a separation of at least 12 in (305 mm) from combustible material unless separated therefrom by a fire-resistant heat-insulating barrier or unless of a rating not exceeding 600 V and completely enclosed except for ventilating openings.

Transformers with a rating of more than 112½ kVA shall be installed in a transformer-room of fire-resistant construction unless they are constructed with Class B (80°C rise) or higher rating and are separated from combustible material by not less than 6 ft (1.83 m) horizontally and 12 ft (3.66 m) vertically or are separated therefrom by a fire-resistant heat-insulating barrier; or are constructed with 80°C rise or higher ratings and completely enclosed except for ventilating openings.

Transformers rated more than 35,000 V shall be installed in a vault.

**131. Askarel-insulated transformers installed indoors.**   Askarel-insulated transformers rated in excess of 25 kVA shall be furnished with a pressure-relief vent. If installed in a poorly ventilated place, they shall be furnished with a means of absorbing any gases generated by arcing inside the case, or the pressure-relief vent shall be connected to a chimney or flue which will carry such gases outside the building. Askarel-insulated transformers rated more than 35,000 V shall be installed in a vault. The Environmental Protection Agency has found polychlorinated biphenyls (PCBs) to be hazardous to health, and askarels will be phased out of use.

**132. Oil-insulated transformers installed indoors.** Oil-insulated transformers shall be installed in a vault constructed as specified in Article 450 of the Code except as follows:

A. NOT OVER 112½-KVA TOTAL CAPACITY The provisions for transformer vaults specified below apply except that the vault may be constructed of reinforced concrete not less than 4 in (102 mm) thick.

B. NOT OVER 600 V A vault is not required, provided suitable arrangements are made when necessary to prevent a transformer oil fire from igniting other materials and the total transformer capacity in one location does not exceed 10 kVA in a section of the building classified as combustible or 75 kVA in a section classified as fire-resistant.

C. ELECTRIC-FURNACE TRANSFORMERS Transformers of a total rating not exceeding 75 kVA may be installed without a vault in a building or room of fire-resistant construction if arrangements necessary to prevent a transformer oil fire from spreading to other combustible material are provided.

D. DETACHED BUILDINGS Transformers may be installed in a building that does not conform with Code provisions for transformer vaults provided that neither the building nor its contents present a fire hazard to any other property and that the building is used only in supplying electric service and the interior is accessible only to qualified persons.

E. PORTABLE AND SURFACE MINING EQUIPMENT (such as electric excavators) Oil-insulated transformers may be used without a vault if (1) provisions are made to drain leaking fluid to the ground, (2) safe egress is provided for personnel, and (3) personnel are protected with at least a ¼-in (6.35-mm) steel barrier.

**133. Oil-insulated transformers installed outdoors.** Combustible material, combustible buildings and parts of buildings, fire escapes, and door and window openings shall be safeguarded from fires originating in oil-insulated transformers installed on roofs, or attached or adjacent to a building or combustible material. Space separations, fire-resistant barriers, automatic water-spray systems, and enclosures which confine the oil of a ruptured transformer are recognized safeguards. One or more of these safeguards shall be applied according to the degree of hazard involved when the transformer installation presents a fire hazard. Oil enclosures may consist of fire-resistant dikes, curbed areas or basins, or trenches filled with coarse crushed stone. Oil enclosures shall be provided with trapped drains if the exposure and quantity of oil are such that removal of oil is important.

**134. Listed less flammable liquid-insulated transformers.** A listed less flammable liquid is one which, when subjected to a source of ignition, may burn but the flame of which will not spread from the source of ignition. These liquids were developed to replace askarel, which has been banned by the Environmental Protection Agency. Transformers insulated with a listed less flammable liquid, having a fire point not less than 300°C, may be installed indoors without a vault in Type I or Type II building construction provided there is a liquid-containment area and no combustibles are stored. Outdoor transformers may be installed attached to, adjacent to, or on the roof of Type I or Type II buildings. Indoor transformers rated over 35,000 V shall be installed in a vault.

**PROVISIONS FOR TRANSFORMER VAULTS**

**135. Walls, roofs, and floors.** The walls and roofs of vaults shall be constructed of reinforced concrete, brick, load-bearing tile, concrete block, or other fire-resistive constructions which have adequate structural strength for the conditions and a minimum fire resistance of 3 h according to ASTM Standard 119-83, *Fire Tests of Building Construction and Materials* (NFPA No. 251-1985). The floors of vaults in contact with the earth shall be of concrete not less than 4 in (102 mm) thick, but when the vault is constructed with a vacant space or other stories below it, the floor shall have adequate structural strength for the load imposed thereon and a minimum fire resistance of 3 h. If transformers are protected with an automatic sprinkler, a water spray, halon, or carbon dioxide, construction may have a 1-h rating.

**136. Doorways.** Vault doorways shall be protected as follows:

A. TYPE OF DOOR Each doorway leading into a vault from the building shall be provided with a tight-fitting door of a type approved for openings in Class A situations as defined in the National Fire Protection Association *Standard for the Installation of Fire Doors and Windows*, No. 80, with a minimum fire rating of 3 h. The authority enforcing the Code may require such a door for an exterior-wall opening when conditions warrant. If transformers are protected with an automatic sprinkler, a water spray, halon, or carbon dioxide, construction may have a 1-h rating.

B. Sills   A door sill or curb of sufficient height to confine within the vault the oil from the largest transformer shall be provided, and in no case shall the height be less than 4 in (102 mm).

C. Locks   Entrance doors shall be equipped with locks, and doors shall be kept locked, access being allowed only to qualified persons. Personnel doors shall swing out and be equipped with panic bars, pressure plates, or other devices that are normally latched but open under simple pressure.

**137. Ventilation.**   Whenever practicable, vaults shall be located where they can be ventilated to the outside air without using flues or ducts. The ventilation shall be adequate to prevent a transformer temperature rise in excess of the transformer rating as prescribed in ANSI/IEEE C57. 12.00-1987, *General Requirements for Liquid-Immersed Distribution, Power and Regulating Transformers* and ANSI/IEEE C57.12.01-1979, *Dry-Type Distribution and Power Transformers.*

Transformers with ventilating openings shall be installed so that the ventilating openings are not blocked by walls or other obstructions.

**138. Ventilation openings.**   When required by Sec. 137, openings for ventilation shall be provided in accordance with the following:

A. Location   Ventilation openings shall be located as far as possible from doors, windows, fire escapes, and combustible material.

B. Arrangement   Vaults ventilated by natural circulation of air may have roughly half the total area of openings required for ventilation in one or more openings near the floor and the rest in one or more openings in the roof or in the sidewalls near the roof, or all the ventilation area may be provided in one or more openings in or near the roof.

C. Size   In vaults ventilated to an outdoor area without using ducts or flues, the combined net area of all ventilating openings after deducting the area occupied by screens, gratings, or louvers shall be not less than 3 in$^2$ (1936 mm$^2$)/1 kVA of transformer capacity in service, except that the net area shall be not less than 1 ft$^2$ (0.093 m$^2$) for any capacity under 50 kVA.

D. Covering   Ventilation openings shall be covered with durable gratings, screens, or louvers, according to the treatment required to avoid unsafe conditions.

E. Dampers   All ventilation openings to the indoors are to be provided with automatic-closing dampers that operate in response to fire. Such dampers shall have a fire rating of not less than 1½ h.

F. Ducts   Ventilating ducts shall be constructed of fire-resistant material.

**139. Drainage.**   If practicable, vaults containing more than 100-kVA transformer capacity shall be provided with a drain or other means which will carry off any accumulation of oil or water that may collect in the vault unless local conditions make this impracticable. The floor shall be pitched to the drain where provided.

**140. Water pipes and accessories.**   Any pipe or duct systems foreign to the electrical installation shall not enter or pass through a transformer vault. Piping or other facilities provided for fire protection or for water-cooled transformers are not deemed to be foreign to the electrical installation.

**141. Storage in vaults.**   Materials shall not be stored in transformer vaults.

**142. The following instructions** (Secs. 143 to 166) for procedure in installing, caring for, and operating transformers have been taken from National Electrical Manufacturers Association Publication 37–46 and from *Transformer Installation Book 5019A* of the Allis-Chalmers Corporation. Except as noted, the authority is the NEMA publication.

**143. Location.**   Accessibility, ventilation, and ease of inspection should be given careful consideration in locating transformers. *Water-cooled transformers* depend almost entirely upon the flow of water through the cooling coils for carrying away heat, so that the temperature of the surrounding air has little effect upon that of the transformers. For this reason air circulation is of minor importance, and water-cooled transformers can be located in any convenient place without regard to ventilation.

**144. Self-cooled (dry-type) transformers** depend upon surrounding air for carrying away their heat. For this reason care must be taken to provide adequate ventilation. For indoor installation the room in which the transformers are placed must be well ventilated so that heated air can escape readily and be replaced by cool air from outside. Inlet openings should be near floor level and distributed to be most effective. The outlet or outlets should be as high above the apparatus as construction of the building will permit. The number and size of air outlets required will depend on their distance above the transformer and on the efficiency and load cycle of the apparatus. In general, about 20 ft² (1.86 m²) of outlet opening or openings should be provided for each 1000 kVA of transformer capacity. Air inlets should be provided with the same total area as the outlets. If the transformer must operate for considerable periods at continuous full load, areas of the inlet and outlet openings should be increased to about 60 ft² (5.57 m²)/1000 kVA of transformer capacity.

**145. Storage.** When a transformer can be set up immediately in its permanent location and filled with oil, it is advisable to do so even though it will not be put into service for some time. If this is not convenient, it should be stored in a dry place having no rapid or radical temperature changes and, if possible, filled with dry transformer oil. The transformer should not be stored or operated in the presence of corrosive gases such as chlorine. If an indoor transformer is stored outdoors, it should be thoroughly covered to keep out rain.

**146. Handling.** When a transformer is being lifted, the lifting cables must be held apart by a spreader to avoid bending the lifting studs or other parts of the structure.

If a transformer cannot be handled by a crane, it may be skidded or moved on rollers, but care must be taken not to damage the base or tip the transformer over. A transformer should never be lifted or moved by placing jacks or tackle under the drain valve, cooling-coil outlets, radiator connections, or other attachments. When rollers are used under large transformers, skids must be employed to distribute the stress over the base.

When working about a transformer, particular care must be taken in handling all tools and other loose articles, since anything metallic dropped among the windings and allowed to remain there may cause a breakdown.

**147. Inspection preliminary to installation.** Transformers are in first-class operating condition when shipped by the manufacturer; i.e., they have been thoroughly tested for defects and are perfectly dry. When received, they should be examined before removing the shipment, and if any injury is evident or any indication of rough handling is visible, a claim should be filed at once and the manufacturer notified.

Moisture may condense on any metal if the metal is colder than the air, and if present, it lowers the dielectric strength and may cause a failure of the transformer. Therefore, if transformers or oil drums are brought into a room warmer than they are, they should be allowed to stand before opening until there is no condensation on the outside and they are thoroughly dry.

Before installation, each individual transformer should be thoroughly examined for indications of moisture and inspected for breakage, injury, or displacement of parts during shipment. In addition, all accessible nuts, bolts, and studs should be tightened if necessary. Before being placed in service, transformers having a plurality of voltage connections should be carefully checked to ensure that they are connected for operation at the required voltage and on the proper tap.

It is standard practice to ship transformers connected for their maximum voltage.

If transformers are water-cooled, the cooling coils should be tested for leaks at a pressure of 80 to 100 psi (551,581 to 689,476 Pa). Water, oil, or preferably air may be used in the coil for obtaining the pressure. The coil must be outside the tank, i.e., away from the winding insulation, if water is used for the pressure test. When pressure is obtained, the supply should be disconnected, and after 1 h it should be determined whether any fall in pressure is due to a leak in the coil or to a leak in the fittings at the ends of the coil.

**148. Transformers shipped filled with oil.** Each transformer shipped filled with oil should be inspected to see whether there is any condition indicating the entrance of moisture during shipment.

If the transformer is received in damaged condition, so that water or other foreign material has had a chance to enter the tank, the transformer should be emptied of oil and treated as though not shipped in oil, and in no case may drying be omitted.

In all cases samples of oil should be taken from the bottom and tested. The dielectric strength of the oil when shipped is at least 22 kV between 1-in (25.4-mm) disks spaced 0.1 in (2.54 mm) apart. A new transformer should not be put into service with oil which tests below this value.

**149. Transformers shipped assembled without oil.** Each transformer shipped assembled but not filled with oil should be carefully inspected for damage in shipment. A thorough inspection can be made only by removing core and coils from the tank. All dirt should be wiped off and parts examined for breakage or other injuries. All conductors and terminals should be examined to check their proper condition and position. The coil and core clamps should be tightened and, if necessary.

The tank should be inspected and, if necessary, cleaned.

When a transformer is shipped assembled but not filled with oil, moisture may be absorbed during transportation. For this reason it is good practice to dry out all such transformers, especially transformers above 7500 V, before putting them into service.

**150. Transformers shipped disassembled.** Only very large transformers are shipped in this way, and special instructions covering features incident to this method of shipping are supplied by the manufacturer. These instructions should be carefully followed.

**151. Drying core and coils.** There are a number of approved methods of drying out transformer core and coils, any one of which will be satisfactory if carefully performed. However, too much stress cannot be laid upon the fact that if the drying is carelessly or improperly performed, great damage may result to the transformer insulation through overheating.

The methods in use may be broadly divided into two classes:

1. Drying with the core and coils in the tank with oil.
2. Drying with the oil removed. The core and coils may or may not be removed from the tank.

**152. Drying with oil in tank.** Under the first class, the moisture is driven off by sending current through the winding while immersed in oil, with the top of the tank open to the air or with some other arrangement made for adequate ventilation. The current necessary for this class can be secured by the short-circuit method.

This method consists in heating the windings and oil up to a high temperature for a limited time under short circuit with a partial load on the windings, the high oil temperature being obtained by blanketing the tank (or by reducing the flow of water for water-cooled transformers).

When a transformer is short-circuited in this manner, only a fraction of the normal voltage should be applied to one winding. When this method is used, the load current and maximum top-oil temperature should be in accordance with the following tabulation.

During the drying run, ventilation additional to that ordinarily provided should be maintained by slightly raising the manhole cover and protecting the opening from the weather. With good ventilation, the moisture, as it is driven off in the form of vapor, will escape to the outside atmosphere, and no condensation of moisture will take place on the underside of the cover or elsewhere, provided these parts are lagged with heat-insulating material to prevent condensation of moisture within.

The accompanying table shows the short-circuit current in percentage of full-load current which can be used for this method of drying transformers, with the corresponding maximum allowable top-oil temperature in degrees Celsius. Less than 5 percent of normal voltage will usually be required to circulate the current in the windings.

| Self-cooled transformers | Water-cooled transformers | Maximum top-oil temperatures, °C |
|:---:|:---:|:---:|
| Short-circuit amp in percentage of full load | | |
| 50 | 50 | 85 |
| 75 | 60 | 80 |
| 85 | 75 | 75 |

These temperature limits and loads must be adhered to strictly to obtain the desired results without danger to the transformers.

It should be noted that the higher allowable temperatures go with the smaller loads; i.e., more blanketing or less water will be required for the smaller loads than for the higher to bring the oil temperature up to the point shown in the table.

**152A. When to discontinue drying.** Drying should be continued until oil from the top and bottom of the tank tests 22 kV or higher for seven consecutive tests taken 4 h apart with the oil maintained at maximum temperature for the load held and without filtering. The testing of the oil for dielectric strength should be made between parallel 1-in- (25.4-mm-) diameter disks, spaced ¹⁄₁₀ in (2.54 mm) apart. All ventilating openings should then be closed, the transformer kept at the same temperature for another 24 h without filtering the oil, and as before the oil should be tested at 4-h intervals. A decrease in the dielectric strength of the oil indicates that moisture is still passing from the transformer into the oil, and drying should be continued. The temperature of the oil samples when tested should preferably be at room temperature and not in excess of 40°C.

Unless constant or increasing dielectric strength as shown by these tests indicates that drying is complete, the ventilators should be opened, the oil filtered, and the drying process continued.

After the short-circuit run has been discontinued, the transformer should be operated for 24 h at approximately two-thirds voltage and at the same high temperature, similar tests of oil samples should be made, and the oil filtered if necessary. After a satisfactory two-thirds-voltage test, full voltage should be applied for 24 h and the same tests repeated. Water-cooled transformers may require some water to hold the top-oil temperature within the 85°C limit during this test.

**153. Drying with oil removed.** Typical of the second class, i.e., drying with the oil removed, are the three following methods:
1. By internal heat
2. By external heat
3. By internal and external heat

1. BY INTERNAL HEAT    For this method alternating current is required. The transformer should be placed in its tank without the oil and with the cover left off to allow free circulation of air. Either winding can be short-circuited, and sufficient voltage should be impressed across the other winding to circulate enough current through the coils to maintain the temperature at from 75 to 80°C. About one-fifth of normal full-rated current is generally sufficient to do this. The impressed voltage necessary to circulate this current varies within wide limits among different transformers but will generally be approximately 0.5 to 1.5 percent of normal voltage at normal frequency.

The end terminals of the winding, not taps, must be used so that current will circulate through the total winding. The amount of current can be controlled by a rheostat in series with the exciting winding. Proper precautions should be taken to protect the operator from dangerous voltage.

This method of drying out is superficial and slow and should be used only with small transformers and then only when local conditions prohibit the use of one of the other methods.

2. BY EXTERNAL HEAT    The transformer should be placed in a box with holes in the top and near the bottom to allow air circulation. The clearance between the sides of the transformer and the box should be small so that most of the heated air will pass up through the ventilating ducts among the coils and not around the sides. The heat should be applied at the bottom of the box. With some types of transformers it is better to distribute the heat evenly around the lower coils.

The best way to obtain the heat is from grid resistors, using either alternating or direct current. The temperature limits of ingoing air are 85 to 90°C. The transformer must be carefully protected against direct radiation from the heaters. Care must also be taken to see that there is no flammable material near the heaters, and to this end it is advisable to line the wooden box completely with asbestos. Also, when forced air is used, suitable baffles should be placed between heater and inlet to the transformer enclosure.

Instead of the heater being placed inside the box containing the transformer, it can be placed outside, the heat being carried into the bottom of the box through a suitable pipe. When this plan is followed, the heat may be generated by the direct combustion of gas, coal, or wood, provided that none of the products of combustion are allowed to enter the box containing the transformer. Heating by combustion is not advocated except when electric current is not available.

This method, although effective, requires a much longer time than Method 3.

3. BY INTERNAL AND EXTERNAL HEAT    This is a combination of Methods 1 and 2. The transformer should be placed in a box and external heat applied as in 2, and current circulated through the windings as in 1. The current should, of course, be considerably less than when no external heat is applied.

This method is used occasionally when direct current only is available, a certain amount of current being passed through the high-voltage winding only, as the cross-sectional area of the low-voltage conductor is generally too large for it to be heated with an economical amount of direct current. The use of direct current for drying out is not recommended except when alternating current cannot be obtained. When this method of drying is used, the temperature should be measured by the increase-in-resistance method.

Method 3 requires technically skilled supervision.

**153A. Time required for drying.** There is no definite length of time for drying. Up to 3 weeks may be required, depending upon the condition of the transformer, the size, the voltage, and the method of drying used.

**153B. Insulation resistance.** The measurement or determination of insulation resistance is of value in determining the course of drying only when the transformer is without oil. If the initial insulation resistance is measured at ordinary temperatures, it may be high although the insulation is not dry, but as the transformer is heated up, it will drop rapidly.

As the drying proceeds at a constant temperature, the insulation resistance will generally increase gradually until toward the end of the drying period, when the increase will become more rapid. Sometimes the resistance will rise and fall through a short range one or more times before reaching a steady high point. This is caused by moisture in the interior parts of the insulation working its way out through the outer portions which were dried at first.

As the temperature varies, the insulation resistance also varies greatly; therefore the temperature should be kept nearly constant, and the resistance measurements should all be taken at as nearly the same temperature as possible. The insulation resistance in megohms varies inversely with the temperature, and, for a 10°C change of temperature, the megohms change by a ratio of 2:1. Measurements should be taken every 2 h during the drying period.

**153C. Resistance curve.** A curve of the insulation-resistance measurements should be plotted with time as abscissa and resistance as ordinate. By observation, the knee of the curve (the point where the insulation resistance begins to increase more rapidly) can be determined, and the run should continue until the resistance is constant for 12 h.

The Allis-Chalmers Corporation further recommends that the insulation resistance should be as great as or greater than that given by the following formula and that the megohm resistance should be taken with all windings grounded except the winding being tested.

Megohms at 85°C in air or at 40°C in oil are equal to

$$\frac{kV \times 30}{\sqrt{kVA/cycles}}$$

where kV = kilovolts of winding involved and kVA = kilovolt-amperes of transformer per phase.

The megohms vary inversely with the temperature, and for a 10°C change of temperature the megohms change in the ratio of 2:1. For example,

$$\text{Megohms at } 85°C = \ \ 300$$
$$\text{Megohms at } 75°C = \ \ 600$$
$$\text{Megohms at } 65°C = 1200$$
$$\text{Megohms at } 55°C = 2400$$

**154. Precautions to be observed in drying without oil.** As the drying temperature approaches the point at which fibrous materials deteriorate, great care must be taken to see that there are no points at which the temperature exceeds 85°C. Several thermometers should be used, and they should be placed well in among the coils near the top and screened from air currents. Ventilating ducts offer particularly good places in which to place some of the thermometers. As the temperature rises rapidly at first, the thermometers must be read at intervals of about ½ h. To keep the transformer at a constant temperature for insulation-resistance measurements, one thermometer should be placed where it can be read without removing it or changing its position. The other thermometers should be shifted about until the hottest points are found and should remain at these points throughout the drying period. Whenever possible, the temperature should be checked by the increase-in-resistance method.

**155. Caution in drying out.** It is well to have a chemical fire extinguisher or a supply of sand at hand for use in case of necessity.

It is not safe to attempt the drying out of transformers without giving them constant attention.

**156. Sampling and testing of oil.**    The sample container should be a large-mouthed glass bottle. All bottles should be cleaned and dried with gasoline before being used. A cork stopper should be used.

The sample for dielectric tests should be at least 16 oz (0.473 L); if other tests are to be made, 1 qt (32 oz, or 0.946 L).

Test samples should be taken only after the oil has settled for some time, varying from 8 h for a barrel to several days for a large transformer. Cold oil is much slower in settling and may hardly settle at all. Oil samples from the transformer should be taken from the oil-sampling valve at the bottom of the tank. Oil samples from a barrel should be taken from the bottom of the drum. A brass or glass "thief" can be conveniently used for this purpose. The same method should be used for cleaning the thief as is used for cleaning the container.

When samples of oil are drawn from the bottom of the transformer or large tank, sufficient oil must first be drawn off to make sure that the sample will comprise oil from the bottom of the container and not from the oil stored in the sampling pipe. A glass receptacle is desirable so that if water is present, it can be readily observed. If water is found, an investigation of the cause should be made and a remedy applied. If water is not present in sufficient quantity to settle out, the oil may still contain considerable moisture in a suspended state. It should, therefore, be tested for dielectric strength.

For *testing oil for dielectric strength,* some standard device for oil testing should be used. The standard oil-testing spark gap has disk terminals 1 in (25.4 mm) in diameter spaced 0.1 in (2.54 mm) apart. The testing cup should be cleaned thoroughly, to remove any particles of cotton fiber, and rinsed out with a portion of the oil to be tested.

The spark-gap receptacle should be filled with oil, both oil and spark gap being at room temperature or approximately 25°C. After filling the receptacle, allow ½ to 1 min for air bubbles to escape before applying voltage.

The rate of increase in voltage should be about 3000 V/s. Five breakdowns should be made on each filling, and then the receptacle emptied and refilled with fresh oil from the original sample. The average voltage of 15 tests (5 tests on each of three fillings) is usually taken as the dielectric strength of the oil. It is recommended that the test be continued until the mean of the averages of at least three fillings is consistent.

The dielectric strength of oil when shipped is at least 22 kV tested in the standard gap. If the dielectric strength of the oil in a transformer in service tests at less than 17,500 V, it should be filtered. New oil of less than the standard dielectric strength should not be put in a transformer.

**157. Drying oil and filling transformer.**    In removing moisture from transformer oil, it is preferable to filter from one tank and discharge into another, although if necessary oil may be drawn from the bottom of a tank and discharged at the top. When there is much water in the oil, it should be allowed to settle and then be drawn off and treated separately.

Before the transformer is filled with oil, all accessories, such as valves, gages, thermometers, plugs, etc., must be fitted to the transformer and made oiltight. The threads should be filled in accordance with instructions from the manufacturer before putting them in place. The transformers must be thoroughly cleaned.

Metal hose must be used for filling instead of rubber hose, because oil dissolves the sulfur found in rubber and may cause trouble if the sulfur attacks the copper.

The oil used should be clean, dry oil of the grade recommended by the manufacturer.

The use of a filter press is recommended, and if one is not available, some precaution should be taken to strain the oil before putting it in the transformer.

After filling the transformer, the oil should be allowed to settle at least 12 h, and then samples taken from the bottom should again be tested before voltage is applied to the transformer.

It is very important that the surfaces of the oil when cold (25°C) be at the oil level indicated by the mark on the oil gage. When the transformer is not in service, the oil level must never be allowed to fall to a point at which it does not show in the gage. When it is necessary to replenish the oil, care must be taken to see that no moisture finds its way into the tank. As the oil heats up with the transformer under load, it will expand and rise to a higher level.

**158. Putting into service.**   When the voltage is first applied to the transformer, it should, if possible, be brought up slowly to its full value so that any wrong connection or other trouble can be discovered before damage results. After full voltage has been applied successfully, the transformer should preferably be operated in that way for a short period without load. It should be kept under observation during this time and also during the first few hours that it delivers load. After 4 or 5 days' service it is advisable to test the oil again for moisture.

If the transformer is *water-cooled,* the main water valve should be opened as soon as the oil temperature reaches 45°C. If there are two or more sets of cooling coils in parallel, the valves of all sections should be adjusted for equal rates of flow. This can be estimated by feeling the weight of the discharge streams from the different sections. It can be determined best, however, by noting the difference in temperature between ingoing and outgoing water from each section. A careful measure should be taken of the total amount of water flowing through all sections, and the total rate of flow should be adjusted to a value not less than that specified.

**159. Care.**   The idea that a transformer in service needs no attention may lead to serious results. Careful inspection is essential, and the directions given in this section should be followed.

In spite of all precautions, moisture may be absorbed by the transformer if it is of the open type, and during the first few days of operation it is well to inspect the inside of the manhole cover for moisture. If sufficient moisture has condensed to drip from the cover, the transformer should be taken out of service and dried. The oil should be tested and dried if necessary.

Closed-type transformers should have their oil tested at top and bottom after the first few days of operation to make sure that no moisture is being given off from the transformer into the oil.

*Samples of oil* from all transformers should be drawn and tested at least once every 6 months or according to the manufacturer's recommendations.

During the first month of service of transformers having a potential of 40,000 V or over, samples of oil should be drawn each week from the bottom of the tank and tested.

If at any time the oil should test below 17,500 V, it should be filtered.

Closed-type transformers when properly dried out and installed will need thorough inspection only infrequently, i.e., only when there are specific indications of trouble. Other types of transformers should be taken out of service periodically for a thorough inspection. The inside of the cover and the tank above the oil should be regularly inspected to see that they are clean, dry, and free from moisture and that the thermometer bulb is clean. If an appreciable amount of dirt or sediment is found inside the case, it is best to untank the transformer and remove the oil from the tank. The transformer and the tank should then be cleaned thoroughly and the oil filtered and tested. In cleaning, only dry cloths or waste should be used. Care should be taken to see that all nuts are tight and that all parts are in their proper places. If the transformer is water-cooled, the cooling coils should be cleaned thoroughly. The transformer and the oil should be replaced in the tank, and when the cover is put on, all cracks and openings should be closed tightly.

In the case of water-cooled transformers the rate of flow should be checked from time to time, and if it is found to have diminished, the cause should be looked for and remedied. The most frequent cause of clogging of cooling coils is the presence of air in the water, resulting in the formation of a scaly oxide.

**160. Removing scale from cooling coils.**   Scale and sediment can be removed from a cooling coil without removing the coil from the tank. Both inlet and outlet pipes should be disconnected from the water system and temporarily piped to a point a number of feet away from the transformer where the coil can be filled and emptied safely. Especial care must be taken to prevent any acid, dirt, or water from getting into the transformer.

All the water should be blown or siphoned from the cooling coils, which should then be filled with a solution of hydrochloric (muriatic) acid (specific gravity, 1.10). Equal parts of commercially pure concentrated hydrochloric acid and water will give this specific gravity.

It may be found necessary to force this solution into the cooling coils. When this is done, one end of the coil should be partially restricted, so that the solution will not be wasted when the coil is full. After the solution has stood in the coil about 1 h, the coil should be flushed out thoroughly with clean water. If all the scale is not removed the first time, the operation should be repeated until the coil is clean, using new solution each time. The number of times it is necessary to repeat the process will depend on the condition of the coil, though ordinarily one or two fillings will be sufficient.

As the chemical action which takes place may be very violent and may often force acid, sediment, etc., from both ends of the coil, it is well to leave both ends partially open to prevent abnormal pressure.

**161. Idle cooling coils.** When a water-cooled transformer is idle and exposed to freezing temperatures, the water must be blown out of the cooling coil. In addition to blowing out the water, the cooling coils should be dried by forcing heated air through them. If it is not convenient to do this, the coil should be filled with transformer oil.

**162. Stopping oil leaks** (Allis-Chalmers Corporation). Oil leaks at gasketed joints can often be stopped by tightening the bolts, but if this is not effective, new gaskets must be installed. Special cork is available for this purpose. An adhesive such as shellac, Bakelite varnish, or other suitable material should be used.

Oil leaks at welds can be stopped by peening, soldering, or welding. Peening is often effective for small stains, and soldering alone or a combination of peening and soldering will usually be effective with larger leaks. For this purpose a hard solder having a high melting point (about 365°F, or 185°C) can be used.

Leaks which cannot be stopped by any of these methods must be welded. When transformer tanks are welded with the core and coils in place, the oil should not be removed.

**163. General operation.** An artificially cooled transformer should not be run continuously, even at no load, without the cooling medium. Therefore, it is essential to maintain a proper circulation in the cooling system.

If the water circulation in a water-cooled transformer is stopped for any reason, the load should be immediately reduced as much as possible and close watch kept of the temperature of the transformer. When the oil at the top of the tank reaches 80°C, the transformer must be cut out of service at once. This temperature should be recognized as an absolute limit and must not be exceeded. It should be held only during an emergency period of short duration.

Nearly all cooling water will in time cause scale or sediment to form in the cooling coil. The time required to clog up the cooling coils depends on the nature and amount of foreign matter in the water. The clogging materially decreases the efficiency of the coil and is indicated by a high oil temperature and a decreased flow of water, load condition and water pressure remaining the same.

**164. Temperature.** Thermometers should be read daily or more often. If, at rated load or less, the oil temperature reaches 80°C for an oil-immersed, self-cooled transformer or an oil-immersed, forced-air–cooled transformer or 65°C for an oil-immersed, water-cooled transformer, it is advisable to check operating conditions.

Oil-immersed, self-cooled transformers or oil-immersed, forced-air–cooled transformers should not be operated for long periods of time at oil temperatures in excess of 80°C on account of the increased rate of deterioration of the insulations. The oil temperature in these transformers should not be allowed to exceed 90°C even for short periods of time.

If the oil temperature in oil-immersed, water-cooled transformers should exceed 65°C at rated load or less, the cooling coils need cleaning, an insufficient amount of cooling water is being used, or the temperature of the cooling water is higher than 25°C. The oil temperature in oil-immersed, water-cooled transformers should not be allowed to exceed 75°C even for short periods of time. A lower oil temperature is recommended for oil-immersed, water-cooled transformers on account of the greater difference between the temperatures of the windings and of the oil than in oil-immersed, self-cooled transformers.

Regardless of oil temperatures as indicated by thermometers, transformers should not be operated continuously at overloads in excess of 1 percent for each degree that the ambient is below 30°C for air and 25°C for water. In no case should the overload exceed 30 percent for self-cooled transformers and 25 percent for water-cooled transformers unless stipulated by the specification or contract. During overloads the transformer should be watched with special care.

Moisture may get into an open-type transformer owing to the fact that as oil is heated and cooled, it expands and contracts, causing air to be expelled from and drawn into the transformer. If the air which enters the transformer is at the same time cooled by contact with the cover to below its dew point, moisture will condense.

It is therefore good practice to operate transformers at several degrees above air temperatures at all times. This will largely prevent condensation.

**165. Recommendations relative to pole-mounted transformers.** In pole mounting, convenient lugs or eyebolts are provided on the side of the case to which the rope lifting the transformer may be attached. It will be found convenient to fasten the hanger irons to the case before the transformer is raised to the crossarm. The transformer can then be raised up to and slightly above the crossarm, and the hooks on the hanger irons can then be made to engage the crossarm by lowering the transformer.

Pole-mounted transformers may be filled with oil either before or after mounting, as desired. It is sometimes necessary to add oil a short time after the transformer has been installed, owing to the fact that the insulation will absorb a certain amount of oil. It may be found necessary to replenish the oil from time to time during actual operation so that the normal oil level is kept constant. When the transformer oil is replenished, care should be taken that no moisture finds its way inside the case.

Great care should be exercised in putting on the cover. If the gasket is not properly in place or the cover not securely bolted to the case, moisture in the form of snow or rain may be driven into the transformer tank.

It is very important that the surface of the oil when cold (25°C) be at the oil level indicated on the inside of the tank or on the oil gage.

The following practice is recommended for the care of pole-mounted distribution transformers in service:

1. The oil level should be inspected once every year, and enough oil added to bring the level up to the mark inside the tank or on the oil gage.

2. Periodically the condition of the oil should be inspected, and if necessary the oil should be removed and replaced with good clean oil.

3. A periodic check of the load should be made to ensure that the transformer is not being overloaded.

**166. Care and operation of transformers immersed in nonflammable, nonexplosive liquids** are, in general, the same as for those which are oil-immersed. The liquid may be dehydrated and filtered by means of special equipment designed for the purpose. Regular oil-filtering equipment cannot be used. The same great care with respect to moisture must be taken with these transformers as with the regular types. The transformer cover should not be taken off in a manhole.

Precautions should be taken in handling the liquid, as it has an irritating effect upon the skin, more so to some persons than to others. Especially the eyes, nose, and lips are affected when coming in contact with the liquid, and safety precautions must be observed when handling it.

When a person handles the liquid or works on a transformer filled with the liquid, an application of castor oil is recommended for the eyes and castor oil or cold cream for the nose and lips. In case the liquid comes in contact with the skin, the part should be thoroughly washed and cleaned.

A person should be careful about looking down into the transformer case, as the liquid gives off a gas at 80°C which causes irritation to eyelids, nostrils, and lips. When a person works around one of these transformers, the hands should be coated with grease and the eyes, nose, and lips protected as advised above.

## THE NOISE PROBLEM

**167. Transformer noise.** Transformers create noise when energized. Actually, the noise is a characteristic hum that is generated by vibrations in the laminated core structure. This hum has a fundamental frequency about twice that of the applied frequency. Its relative loudness to the ear is a function of transformer design and construction characteristics, the ambient noise level of the area where it is located, and the manner in which it is installed. In some applications, such as factories, transformer noise is masked by the higher level of surrounding noise, and it poses no problem. In low-noise-level or quiet areas (hospitals, schools, libraries, apartment buildings, office interiors, churches, etc.) transformer hum can be quite objectionable and cause serious concern. In fact, this became such an application problem that manufacturers improved transformer design and construction so that they now provide low-noise-level standard units and give them a definite sound rating in decibels.

**168. Selection.**   The decibel (dB), a unit of measure of sound level, has become a key factor in the selection and installation of transformers in areas where noise considerations are important. The following table of average ambient noise levels in common installation areas is the starting point:

| Area | Average Sound Level, dB |
|---|---|
| Average home | 30–45 |
| Retail store | 45–55 |
| Office area (without machines) | 45–70 |
| Office area (with machines) | 50–75 |
| Average factory | 75–95 |

To facilitate comparison, transformer manufacturers publish the sound-level ratings in decibels of their units as determined by prescribed standard test procedures.

**169. Typical ratings are listed in the following table:**

| Transformer Rating, kVA | Average Sound Level, dB |
|---|---|
| 0–  9 | 40 |
| 10–  50 | 45 |
| 51–150 | 50 |
| 151–300 | 55 |
| 301–500 | 60 |

The tables in Secs. 168 and 169 comprise a handy reference when audible-noise considerations are a factor in transformer applications. It is a simple matter of comparing transformer noise with ambient noise. Knowing the average ambient sound level of the installation area, one can select a transformer with the proper sound rating (from the table or manufacturer's data). A basic selection rule is to choose a transformer with a decibel rating lower than the average decibel level of the area in question. Otherwise, the installed results might be disappointing.

One point must be kept in mind. The sound level of a specific transformer, measured and recorded under established test conditions at the factory, may be quite different after the unit has been installed. Because of location, mechanical and/or acoustical conditions, and mounting methods, the audible sound level may be considerably higher. The ultimate objective must be to keep the job-installed noise level of the unit below the area ambient noise level.

**170. Carefully check the sound-level ratings** of standard transformers, particularly when applications are to be made in areas of medium and low ambient levels. If ambient noise will be considerably higher than the transformer decibel ratings in the table in Sec. 169, there should be no problem. If the spread between transformer and ambient noise is not so great, careful attention must be given to installation. Often the combination of a sufficiently quiet standard unit and recognized noise-attenuating mounting methods will resolve the problem. However, if audible noise in specific areas is especially critical, it may be necessary to order special quiet-type transformers from the factory.

**171. Noise attenuation.**   The manner in which a transformer is located and installed has a definite bearing on the overall audible noise level in the area. If the unit is mounted in such a way that the noise vibrations are transmitted to a large vibrating surface such as sheet metal or wood, the sound can be amplified to a higher level than the unit's decibel rating. Similar amplification occurs when the unit is located in the corner of a room or close to large reflecting surfaces which act like a megaphone and give the transformer hum an acoustical level boost. In some cases of poor mounting, a special quiet-type transformer can produce a louder sound output than that of a carefully mounted standard transformer.

Some experts claim that improper location and installation can increase transformer sound levels from 10 to 20 dB. Such a condition might easily become intolerable because a 3-dB increase in sound level reportedly has the effect of almost doubling the sound volume as detected by the human ear.

**172. There are a number of basic installation precautions and mounting techniques** which, if carefully noted and followed, will minimize the audible sound level of energized transformers. Some of the major considerations are:

1. PROPER LOCATION   This is the first consideration in a low-sound-level installation. To keep within or below prescribed area decibel limits:

   *a.* Keep the transformer as far as possible from the area in which its noise would be most objectionable.

   *b.* Avoid mounting the unit in a room corner up near the ceiling. Three-sided corners act as megaphones and amplify the sound.

   *c.* Avoid installations in narrow halls and corridors or in corners of stairwells. The transformer sound reflected from the walls can become additive to the primary sound of the transformer and cause additional decibel buildup.

   *d.* If feasible, experimental temporary operation and positioning of a free-standing transformer in a room or area will quickly indicate the best location and orientation of the unit.

   *e.* If necessary, cover the walls of the transformer room with acoustical dampening material (fiberglass, acoustical tile, or similar absorbent materials), to reduce propagation of transformer noise from the room to any adjacent areas. It should be noted, however, that such material has a major effect on the high harmonics of transformer noise but little, if any, effect on the fundamental hum. Although there are special sound-insulating materials available for the 120-cycle frequency range, their present form and cost make them impractical for the above application.

2. TRANSFORMER MOUNTING   Mounting methods play an important role in control and reduction of the audible sound coming from the unit. The prime objective is to isolate the noise, that is, to prevent its mechanical transmission to the supporting structure and connected raceway system. This can be accomplished with one or a combination of the following installation techniques:

   *a.* Use solid mounting when the transformer can be secured to a heavy, solid mass which cannot vibrate audibly. Such would be reinforced concrete for floor or wall.

   *b.* For installation on a structural frame, wall, ceiling, or column, use the flexible-mounting technique, employing special vibration isolating pads called *flexible mounts* or vibration dampeners. There must be no solid-metal contact between the transformer and supporting surface; otherwise the vibration of the pads would be "short-circuited." These external pads are furnished and installed by the installer. Internal vibration dampeners between the core-coil assembly and enclosure are or can be furnished by the manufacturer.

   *c.* Use flexible connections between the raceway system and the transformer enclosure to prevent transmission of noise vibrations from the enclosure to the raceway system, panels, and other mechanical parts. Flexible metal conduit and liquidtight flexible metal conduit are frequently used for these relatively short coupling sections.

## POLE AND PLATFORM MOUNTING

**173. Mounting distributing transformers.**   Units of the smaller capacities are supported either directly on the poles or on crossarms in accordance with instructions furnished by their manufacturers. Refer to Sec. 40 and Figs. 5-23 to 5-28 for details of transformer mounting lugs, adapter plates, crossarm hangers, etc.

With some of the newer transformer designs single transformers up to 167 kVA can be mounted satisfactorily directly on a pole (Fig. 5-75). Three 50-kVA transformers of the proper design can be safely supported from the crossarm on a single pole. In crossarm mounting of transformers Gear and Williams recommend that for transformers of capacities larger than 20 kVA double crossarms should be used at the top, as the top arms carry most of the weight. "Where the installation consists of three 15-kVA or larger transformers it is advisable to use a larger-sized crossarm than the standard. An arm having a cross section of 4 by 5½ in has been found ample for installations aggregating 90 to 100 kVA." Gear and Williams state: "Where a large amount of power is needed which requires a number of 50-kVA units which cannot be conveniently installed inside the building, they can be safely and conveniently installed on a platform between two or more poles," as shown in Fig. 5-77. A platform for supporting three 50-kVA units can be built by bolting in gains, between two poles, two 3- by 10-in planks and nailing to them a floor of 2-in plank.

**Fig. 5-75** Overhead-type conventional 25-kVA transformer with two high-voltage cover bushings. [Cooper Power Systems]

**Fig. 5-76** Overhead-type single-phase conventional transformer with two high-voltage sidewall bushings. [Cooper Power Systems]

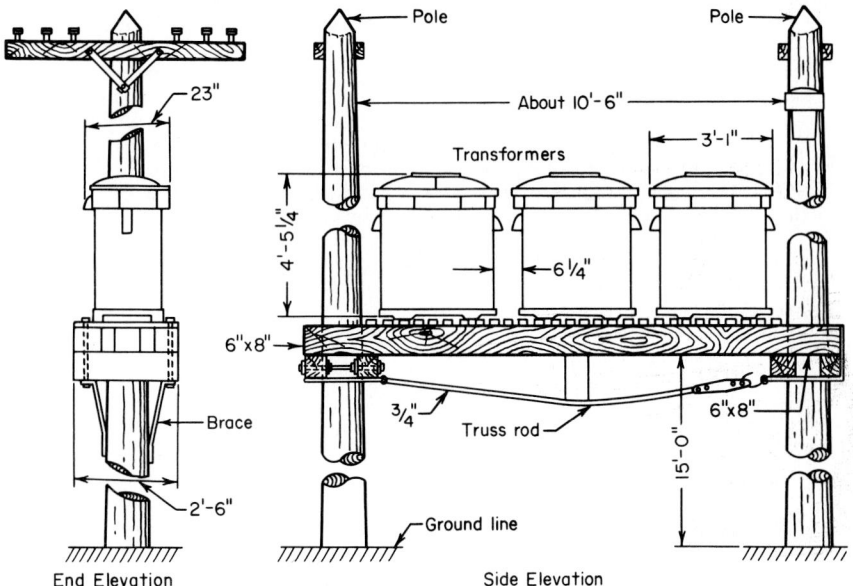

**Fig. 5-77** Platform for large transformers.

**174. Methods of crossarm and platform mounting of transformers.** The methods of mounting transformers described and illustrated in the following paragraphs were taken from *Report of the Committee on Overhead Line Construction* of the Pennsylvania Electric Association. The practices outlined were those followed by the Allegheny County Light Co. The methods provide ample clearances for persons climbing the poles and assure that the wiring will remain in place and not give trouble from short circuits. Platforms are recommended for supporting the larger transformers because of the accessibility for repairs or replacements that they provide.

**175. Crossarm method of mounting single transformers of from 1- to 4-kVA capacity.** The transformer should be supported by the iron hangers furnished by the manufacturer and hung at the central point on the crossarm and not out on the arm away from the pole. At the bottom of the hanger a section of an arm, not longer than the diameter of the pole,

should be fastened to the pole with two lag bolts. The transformer can be hung on the bottom arm, if one is in place and supports lines, provided this arm is in the second gain or a lower one. The primary mains feeding the transformer should be on an upper arm.

In installations in which the transformers are more than 4 ft (1.2 m) below the arm supporting the primary mains, it is advisable to mount Western Union pins horizontally in the line arms. On these pins the primary wires can be tied to maintain them rigid. Iron pins should also be mounted in the transformer arm to take the stress imposed on the primary conductors by the fuse terminal screws.

**Fig. 5-78** Pole-mounted 25-kVA distribution transformer protected with conventional surge arrester and separate current-limiting fuse. [Cooper Power Systems]

**176. Method of mounting transformers of from 5- to 10-kVA capacity** (Fig. 5-79). The same rules should be followed as outlined in the preceding paragraph with the following additions. The transformers, on account of their increased weight and dimensions, should not be hung on a line arm. A specially placed arm should be used underneath existing arms and other apparatus. In addition to using the regular hangers which accompany transformers, a pair of iron braces 24 by 2 by ¼ in (610 by 51 by 6.35 mm) should be placed between the transformer lugs and the hanger with the hanger bolts passing through one of the holes in the braces. These braces are to be run in an upward direction and fastened to the pole with a standard through bolt (see Fig. 5-79). If the arm weakens or entirely rots away, these two braces are of sufficient strength to support the transformer and permit crossarm replacement.

**177. Method of mounting two 5-kVA or two 10-kVA transformers** (Fig. 5-80). Construction similar to that described above should be used except that a standard arm on which the hanger irons can rest should be placed at the bottom. Also only one special brace (24 by 2 by ¼ in) per transformer should be placed between the lug and hanger iron next to the pole.

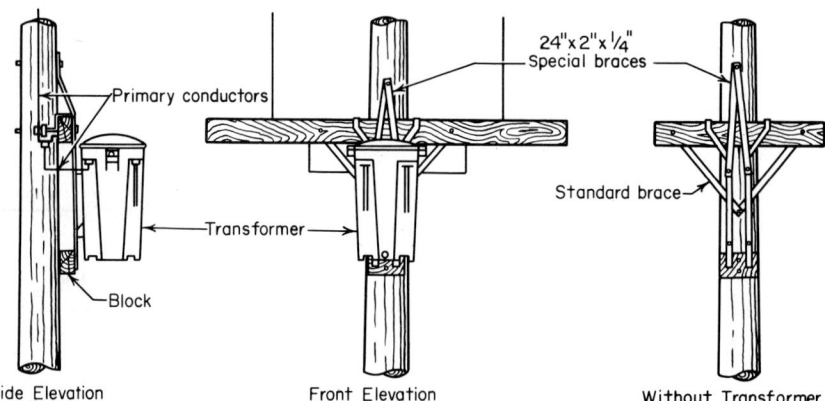

Side Elevation          Front Elevation          Without Transformer

**Fig. 5-79**    Method of supporting a 5- to 10-kVA transformer.

**178. Method of mounting three 5-kVA transformers** (Fig. 5-81).  The construction should be similar to that outlined in the preceding paragraphs, except that the special braces supporting the outside transformers are 33 in (838 mm) between centers of holes. It is also advisable to place an additional crossarm on the rear side of the pole. This arm braces the front arm and provides a place where fuse blocks can be mounted.

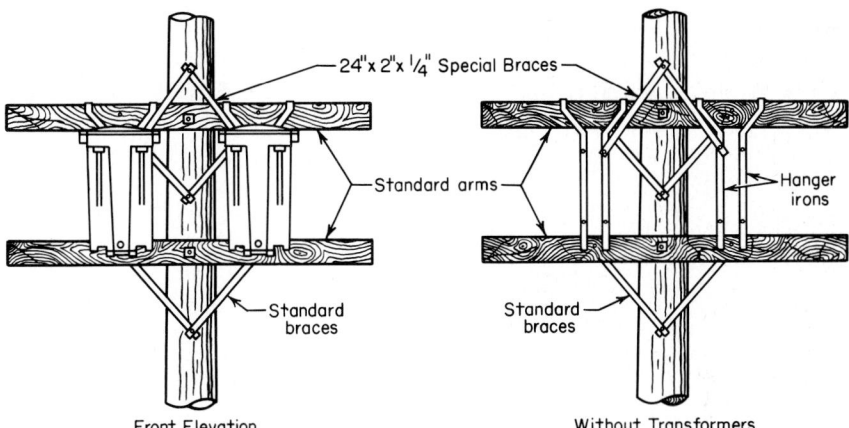

Fig. 5-80    Method of supporting two 5-kVA or two 10-kVA transformers.

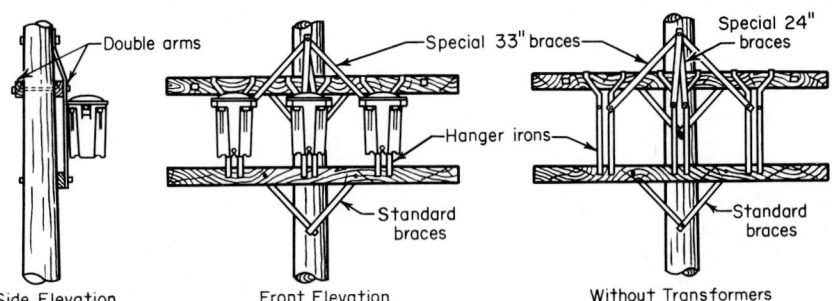

Fig. 5-81    Method of supporting three 5-kVA transformers.

**179. Method of mounting three 10-kVA transformers** (Figs. 5-82 and 5-83).  The construction is similar to that for three 5-kVA transformers with the following addition. The top arm supporting the transformers should be reinforced over its entire length with a piece of 5- by 3-in (127- by 76-mm) angle iron, which should be placed with the 3-in (76-mm) leg on the top of the arm.

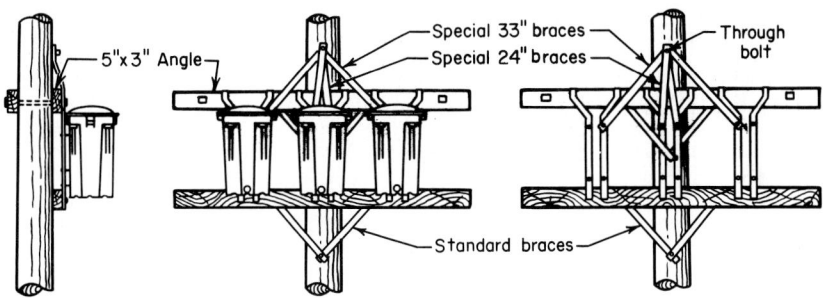

Fig. 5-82    Method of supporting three 10-kVA transformers.

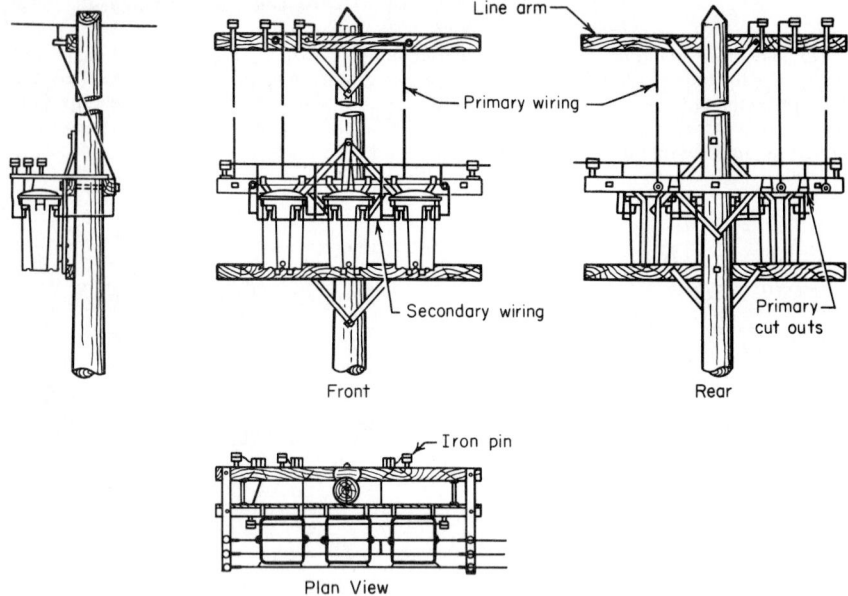

**Fig. 5-83** Wiring for three 5-kVA or for three 10-kVA transformers.

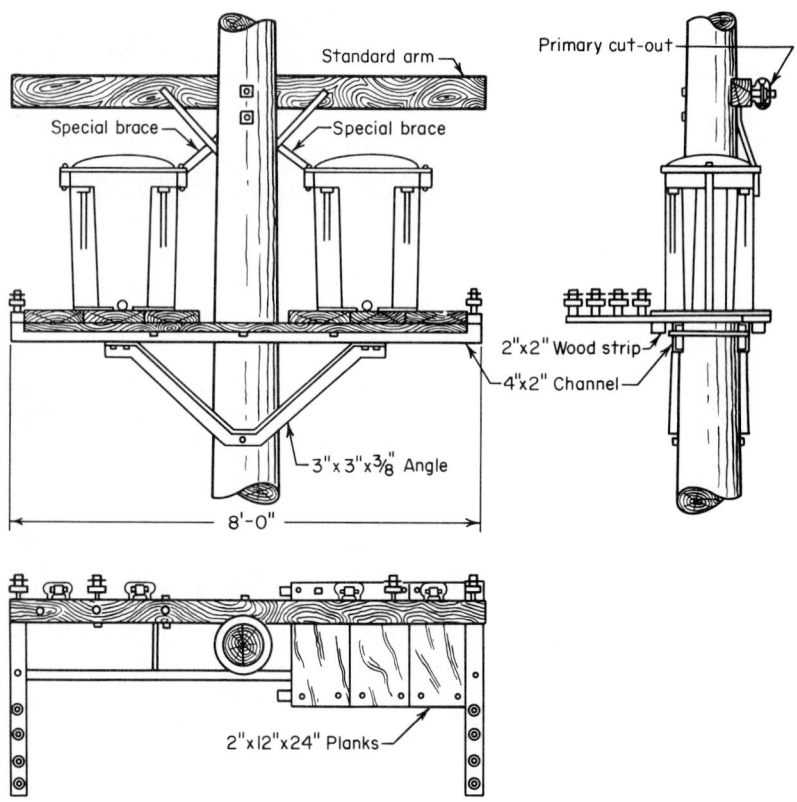

**Fig. 5-84** Single-pole platform for two 20-kVA, two 30-kVA, or one 50-kVA transformer.

**180. Method of mounting two 20-kVA, two 30-kVA, or one 50-kVA transformer** (Fig. 5-84).   For transformers of these capacities a single-pole platform is recommended. The beams for the platform should be 4-in (101-mm) by 6-lb (2.7-kg) channel iron 8 ft (2.4 m) long. The braces used are a single piece of angle iron 3 by 3 by ⅜ in (76.2 by 76.2 by 9.5 mm) bent in a V shape. Pine or oak planks 2 by 12 by 24 in (51 by 305 by 610 mm) are to be laid

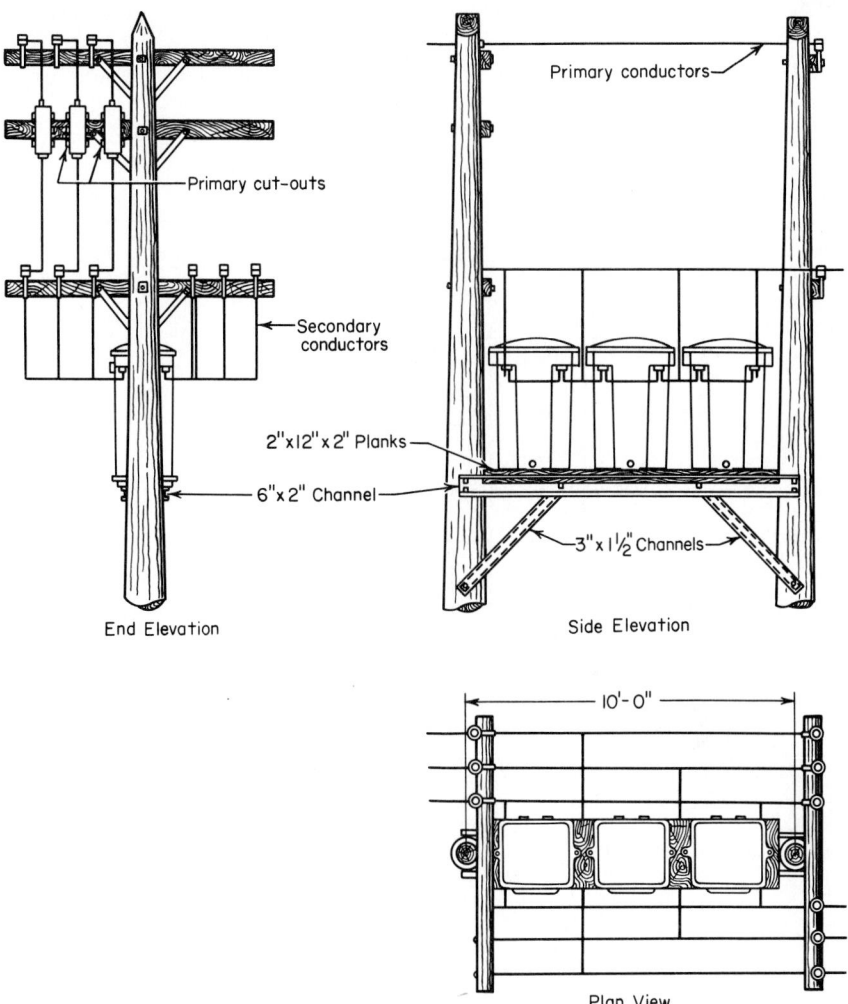

End Elevation                    Side Elevation

Plan View

**Fig. 5-85**  Double-pole platform for three 20-kVA, three 30-kVA, two 50-kVA, or three 50-kVA transformers.

across the channel irons for the transformers to rest upon. The wooden platform is to be held together by a 2- by 2-in (51- by 51-mm) strip of wood running on the outside of the channel iron, to which the planks are secured by 4-in wood screws or 20-penny nails.

**181. Method of mounting three 20-kVA, three 30-kVA, two 50-kVA, or three 50-kVA transformers** (Fig. 5-85).   The poles should be spaced 10 ft (3 m) apart on centers. The main channel irons are 6 in (152 mm) by 10 lb (4.5 kg) by 10 ft 6 in (3.2 m) overall. Braces are of 3-in (76-mm) by 4-lb (1.8-kg) channel.

## SATURABLE-CORE REACTOR

**182. Saturable-core reactor.**   In many control circuits it is desirable to be able to control the value of the inductive reactance. This can often be accomplished advantageously by means of a device called a saturable-core reactor. A saturable-core reactor consists of a magnetic core associated with dc and ac windings. A common form of construction is shown in Fig. 5-86. When there is no current in the dc winding, the inductive reactance of the ac winding will be high, since there will be a high rate of change of flux with

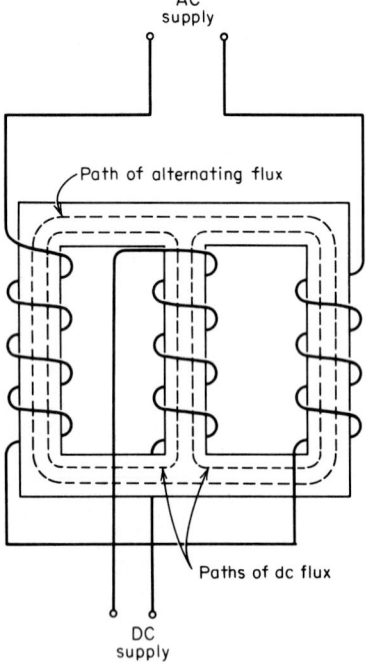

**Fig. 5-86**   Saturable-core reactor.

current. If sufficient direct current is passed through the dc winding to saturate the magnetic core, then the changing current in the ac winding will produce only very small changes in the flux of the core, and the inductive reactance of the ac winding will be very small. If the current of the dc winding is adjusted, the inductive reactance of the ac winding can be varied smoothly from practically zero to its maximum value. It might appear that a saturable-core reactor could be constructed with one ac and one dc winding on a common core. This simple construction would not be satisfactory, since the variations of the flux produced by the alternating flux would induce large voltages of changing magnitude in the dc winding. To be satisfactory, the reactor must be so constructed that the flux linked with the dc winding is not affected by the alternating current in the ac winding. This can be accomplished by the construction of Fig. 5-86. The currents in the two ac coils tend to produce flux in opposite directions through the central core on which the dc winding is wound. Therefore, the currents in the ac windings neutralize each other with respect to the production of flux linking with the dc winding.

Division **6**

# Solid-State Devices and Circuits

## FUNDAMENTALS OF ELECTRONICS

**1. From vacuum tubes to transistors.** The *science of electronics* was founded on the ability of the electron tube (most importantly, the vacuum tube) to generate, amplify, and control an electric signal to accomplish a wide variety of functions. During its half-century dominance of the electronics art, the vacuum tube fostered the invention of radio, television, and radar, of factory automation and computers, and indeed of all the phenomena that we associate with electronics even today. Yet, with the invention of the transistor by Bell Laboratories in 1948, the tube was headed for extinction. Today, except for a few special functions, it no longer occupies an important place in electronic technology.

In place of the electron tube, the transistor has emerged as the cornerstone of modern electronics. Based on the theory of electron conduction in a solid crystalline material rather than in a vacuum or a gaseous environment, the transistor not only can perform virtually all the functions formerly associated with the tube, it can do them faster, more cheaply, and more reliably. Moreover, it occupies an infinitesimally small amount of space and, unlike the tube, requires no power-wasting filament that ties large equipment to the commercial power lines.

The dramatic effects of these advantages are most readily evident in the field of computers. The most advanced computer in the vacuum-tube era cost millions of preinflation dollars; consisted of an entire roomful of equipment utilizing tens of thousands of vacuum tubes; required a power plant just to heat the filaments and an air-conditioning plant to keep the computer cool; was difficult to keep in operation owing to vacuum-tube failures; and had less capability than today's simple calculator, which fits into a vest pocket, requires only an occasional battery replacement, and is thrown away at the first sign of trouble because of its low replacement cost.

But while the transistor is the cornerstone of today's electronics technology, it is not, in itself, the means for accomplishing the exponential expansion of electronic capabilities since the mid-1970s. That distinction is reserved for the invention and development of integrated circuits.

**2. Integrated circuits.** There are two types of integrated circuits (IC) in use today: monolithic and hybrid.

A monolithic IC consists of a semiconductor substrate, called a chip or die, in which all necessary active and passive components are formed and interconnected to perform a desired circuit function (Fig. 6-1*a*). The chip is then encased in a suitable package, which may subsequently be interconnected on a printed-circuit board with other monolithic ICs and/or discrete components to form circuits of greater scope and complexity.

A hybrid integrated circuit (Fig. 6-1b) consists of an insulating substrate with deposited passive elements (resistors and capacitors) and metallized interconnect patterns, to which

(a)

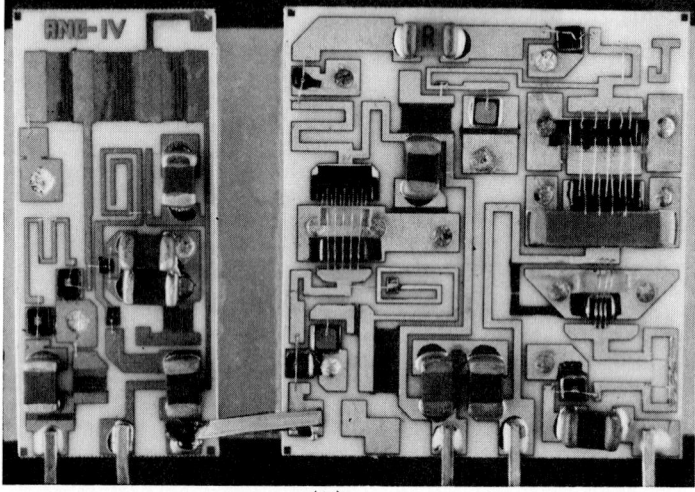

(b)

**Fig. 6-1**   (a) A modern monolithic microprocessor integrated circuit contains upward of 70,000 transistors on a tiny silicon die. (b) An example of a hybrid circuit utilizes power transistor chips and integrated circuits with deposited passive elements on an insulating substrate in a radio-frequency amplifier module. [Motorola Inc.]

active elements are added in uncased chip form either by wire bonding or by other die-attachment techniques. The active elements may be discrete devices or monolithic integrated circuits. Hybrid circuits can be considerably smaller than circuits of equivalent complexity made by interconnecting the same components in encapsulated form on a more conventional printed-circuit board. However, they do not have most of the major attributes of monolithic ICs and are utilized principally where size reduction is required but where the required performance cannot be built into a monolithic structure.

### 3. Characteristics of monolithic integrated circuits

SIZE AND WEIGHT REDUCTION    Monolithic integrated circuits consist of a number of transistors and associated resistors fabricated and interconnected within the same monolithic piece of solid semiconductor material. Since these functional elements are fashioned by a slight change in the atomic structure of the material rather than by the addition of actual physical elements, the dimensions of these elements are so small that it is possible today to fabricate and interconnect tens of thousands of individual components within a piece of material no larger than the fingernail of a small child. And researchers are already anticipating circuits of million-transistor complexity within a single monolithic building block of similar dimensions.

Obviously, it is the size and weight reduction of integrated circuits, compared with equivalent vacuum-tube circuits (or even those made from individual discrete transistors), that has put the power of a full-scale computer into a pocket or purse. But size and weight reduction have long ceased to be the principal objectives of integrated circuits. Though vitally important in certain applications such as medical electronics and space apparatus, the physical characteristics of equipment are dictated more by the dimensions of the peripherals—the display and keyboard of calculators and computers, the loudspeakers and picture tubes of radio and TV, etc.—than they are by the associated electronics. Indeed, the size and weight advantages of integrated circuits would be of limited importance if they were not paced by corresponding reductions in cost and the improvements in component reliability and circuit performance.

COST REDUCTION    The cost of the active elements associated with transistors, particularly transistors within complex integrated circuits, borders on the insignificant. This is so because semiconductor devices are made by batch processing, in which individual manufacturing processes operate not on a single device but simultaneously on a batch of devices ranging from hundreds of integrated circuits to thousands of discrete transistors. The basic cost of such devices therefore reflects primarily the cost of packaging, testing, and marketing such devices rather than the raw materials or manufacture of the basic element. As a result, discrete small-signal transistors can be purchased for only pennies (in large quantities), and complex ICs housing up to thousands of fully interconnected transistors are available for only a few dollars. Each transistor serves the function of an erstwhile vacuum tube, whose individual cost equaled or exceeded that of many of today's complex ICs. It is clear therefore that an attempt to duplicate the functions of today's very-large-scale integrated circuits (VLSI circuits) before the era of semiconductor technology would have priced the resulting equipment out of the market.

IMPROVED RELIABILITY    Semiconductor electronics is based on the movement of free electrical particles (electrons and holes) within a loop of solid materials. Once such particles have been introduced into the material during manufacture, their movement can be controlled but the particles themselves cannot be destroyed. The basic semiconductor device therefore has no inherent wear-out mechanism. While the packaging of semiconductor devices does introduce some potential failure modes and while such devices can be damaged through overvoltage and excessive heat, modern manufacturing technology and circuit design techniques have reduced such failures almost to the vanishing point. It is this attribute of semiconductors that makes possible today's highly complex electronic equipment utilizing millions of transistors.

IMPROVED PERFORMANCE    While transistors have replaced vacuum tubes in most applications, there are a few application areas in which tubes still dominate. Primarily, these are the areas of very high power and very high (microwave) frequencies. Since ICs are based on transistor technology, it is clear that they have not yet penetrated these applications to a significant degree. In all other applications, particularly those involving complex circuitry utilizing many transistors, the IC not only represents the most cost-effective component but also provides the best performance.

This performance advantage is the result of the close spacing between transistors within the monolithic substrate of the IC, which greatly reduces signal-propagation delay within a large system. Propagation-delay time is composed of two constituents: (1) the time required for a circuit stage (transistor) to perform its operation and (2) the time required for the signal to travel to the succeeding stage or stages. In the early days of transistors, the reaction time of the transistor was relatively slow so that interstage travel time was not a significant factor in total propagation delay. Today's transistors, however, can operate at subnanosecond speeds. Thus, even if we assume that a signal travels along its conductive path between transistors with the speed of the light (approximately 1 ft/ns), signal travel time through large systems composed of discrete transistors can become a significant part of the total delay. Within ICs, the spacing between transistors can be held to a few micrometers, eliminating travel time as a major factor of overall operating speed for most electronic applications.

## DISCRETE SOLID-STATE COMPONENTS

Integrated circuits have penetrated all digital electronic equipment. They are still limited, however, in the amount of power they can produce, and their uses in linear applications have not yet been fully exploited. Discrete solid-state components therefore are still very much in evidence, and there is no indication that their use rate will diminish significantly in the immediate future.

Among the most prevalent discrete components are transistors themselves (both signal transistors and power transistors), as well as zener diodes and rectifiers, thyristors, and a growing family of transducers such as light-, temperature-, and pressure-activated devices. Most of the nontransistor components had counterparts in the electron-tube era, but they are subject to the same benefits of smaller size, improved performance, greater reliability, and lower cost when converted to the solid state.

**4. Fundamentals of solid-state devices.**  Solid-state devices, whether discrete or IC, are fabricated from semiconductor materials, so named because their conductivity lies some-

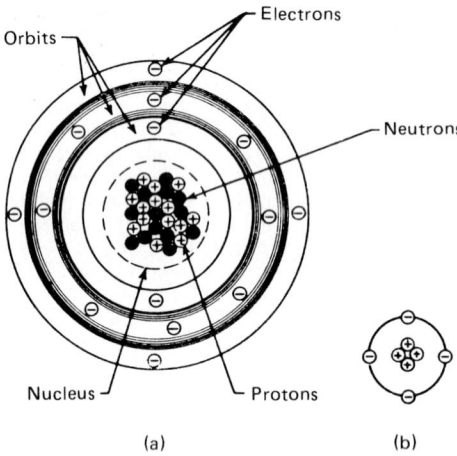

(a)                                    (b)

**Fig. 6-2**  (a) A diagram of a silicon atom shows a nucleus of 14 protons and 14 neutrons and three orbital rings containing a total of 14 electrons. (b) The electrical equivalent of the silicon atom.

where between that of a conductive material and that of an insulator. While a variety of semiconductor materials (germanium, selenium, copper oxide, etc.) have been used to fabricate solid-state devices in the past and still others (notably gallium arsenide) are coming into limited use for special purposes, the most common material in current use is silicon. This material is used for both discrete components and integrated circuits. A basic understanding of its properties therefore is essential to an understanding of today's solid-state devices.

THE SILICON ATOM    From an electrical standpoint, the silicon atom (Fig. 6-2) consists of a nucleus containing 14 protons (each with an electrical charge of $+1$) and three orbital rings housing a total of 14 electrons (each with a charge of $-1$), which electrically neutralize the protons of the nucleus. The outermost ring contains 4 electrons. Since the electrons in the two inner rings are tightly bound to the nucleus, they contribute nothing to the electrical properties of the atom and therefore can be disregarded. The 4 outer-ring electrons, however, are loosely bound to their neutralizing protons and can easily be dislodged from their orbits by some external force.

THE CRYSTAL STRUCTURE    The atoms of silicon tend to arrange themselves into a crystal structure. With proper processing, the material can be of single-crystal form, in which all the atoms are arranged in a continuous, well-ordered lattice (Fig. 6-3) whereby each of the four outer-ring electrons associated with one atom forms a bond with one of the electrons of a neighboring atom so that every electron of every atom is tightly bound within the crystal lattice. Thus, while the four outer electrons of an individual silicon atom can easily be removed by an external force, once they have been bound in the lattice, they are very difficult to dislodge. Intrinsic (pure) single-crystal silicon therefore behaves like an insulator at very low temperatures.

ELECTRONS AND HOLES    As temperature is raised, however, thermal energy is imparted to the lattice structure, causing it to vibrate. When this occurs, electrons at random points throughout the crystal gain enough energy to break their bond and become free to wander through the lattice. Wherever such a bond is broken, a hole is left in the lattice. And because a hole is caused by an electron leaving the orbit of a particular atom, that atom becomes a positive ion with an electrical charge of $+1$.

At any given temperature above absolute zero, there will be a given number of electrons that gain enough energy to break the covalent bond. The higher the temperature, the greater the number of broken bonds; hence the greater the number of electron-pole pairs scattered throughout the crystal. These electrons and holes can be regarded as charge carriers which are free to move through the crystal lattice. The electrons represent negative charges; the holes, positive charges.

Covalent bonds

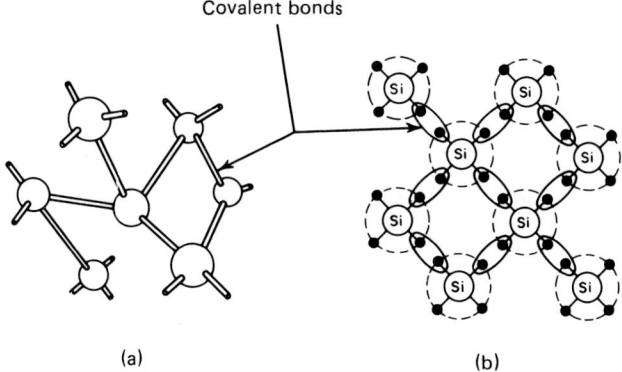

(a)                                      (b)

**Fig. 6-3**  Symbolic diagram (a) and schematic representation (b) of a pure silicon lattice structure. Covalent bonds between adjacent atoms firmly fix all electrons in their respective orbits, rendering the structure nonconductive at very low temperatures.

Once liberated from its covalent bond, an electron moves at random through the lattice until it happens to land in the vicinity of a hole in the structure. Now the hole, having a positive charge, will have an attracting influence on the negative electron, which, when sufficiently close, will jump into the hole. As a result, both the hole and the electron will disappear as free charges within the crystal.

A hole, too, is free to move through the crystal lattice. When a hole is formed, it will affect an attracting force on electrons in covalent bonds in the immediate vicinity because of its positive charge. Since, in a vibrating lattice, these bonds are none too tight, the energy gained by an electron in an adjacent bond, coupled with the attracting force of the hole, will cause the electron to break its bond in order to fill the hole. Note that the hole itself has not disappeared; it has merely moved to an adjacent atom. In this way, the hole, too, can move at random throughout the crystal. Not until the hole meets a free electron will recombination take place and cause both hole and electron to disappear as free charges.

For every hole-electron pair that is generated at one point in the lattice, there will be recombination of a free hole and electron at another point; and for every recombination, another hole-electron pair will be generated elsewhere. At any given temperature, therefore, there will always be a predictable number of carriers, both positive and negative, within the lattice, and the higher the temperature, the greater the number of free charges in the material. But it is interesting to compare the number of free charges with the total number of atoms in the silicon material. Intrinsic silicon has $5 \times 10^{22}$ atoms per cubic centimeter. At room temperature (25°C), approximately $1.5 \times 10^{10}$ of these atoms are ionized. Therefore, there are only one hole and one free electron for approximately every 3 trillion atoms in the crystal structure.

As the temperature is increased, the number of free carriers increases proportionally and the silicon's electrical resistivity is decreased proportionally.

**5. Impurity concentrations in silicon.**    The only major electrical characteristic of intrinsic silicon is a temperature-sensitive resistance which (except, perhaps, as a temperature-sensing element) is undesirable in any actual use. To utilize the properties of a semiconductor material, it is necessary to introduce additional charge carriers deliberately into the intrinsic semiconductor lattice. This is done by introducing impurity atoms into the structure. If the number of charge carriers introduced by these impurity atoms is much larger than the number of charge carriers resulting from thermal agitation of the lattice, the temperature sensitivity of the material is reduced to practical limits.

Two types of impurity atoms are of primary importance: those with three electrons in their outer rings, such as boron; and those with five electrons in their outer rings, such as phosphorus.

P-TYPE MATERIAL    If an atom of boron is substituted for a silicon atom in the lattice structure, the electrons of the boron atom can form covalent bonds with only three adjacent silicon atoms. Since the fourth silicon atom of the crystal structure is not symmetrically bound in the lattice, a defect exists in the lattice structure. This defect takes on the characteristics of a hole. This hole, by virtue of capturing an electron from another adjacent silicon atom, will move through the lattice in the form of a positive charge.

Impurity atoms having three electons in the outer ring are called *acceptor atoms*, because such impurities will accept electrons from the silicon lattice. Silicon doped with acceptor atoms is called *p*-type silicon, indicating that the resulting charge carriers are holes having a positive charge.

Although the thermally generated holes in heavily doped *p*-type material have a negligible effect on the characteristics of the material, their number being small in comparison with the impurity-caused holes at normal temperatures, the thermally generated electrons do play an important part in semiconductor devices. Because their number is small compared with the number of holes, free electrons in *p*-type materials are called *minority carriers*. The holes, for obvious reasons, are called *majority carriers*.

Now if an electrical force, e.g., a battery, is applied across a *p*-type material, the holes will appear to drift toward the negative terminal of the battery and the free electrons will drift toward the positive terminal. Because the number of majority carriers predominates, conduction in doped semiconductor material is majority-carrier conduction or, in this case, *p*-type conduction.

N-TYPE MATERIAL    If an intrinsic piece of silicon is doped with an impurity atom having five electrons in its outer ring, only four of the impurity atom's electrons can form covalent bonds with the silicon atoms; the fifth, being very loosely bound in the lattice, is free to detach itself from the impurity atom and wander through the lattice as a negative charge. In this case, however, the free electron is not balanced by a corresponding hole because the crystal lattice is unbroken. Under this condition, the material has an excess of free negative-charge carriers and is called *n*-type material. Still, the material remains electrically neutral because the detached electron leaves a positive phosphorus ion in its wake. Impurity atoms that produce *n*-type material are called *donor atoms*.

The *n*-type material behaves very similarly to the *p*-type material just described. The majority carriers in this case are, of course, electrons. The minority carriers are holes generated by thermal excitation.

**6. P-n junctions and their characteristics.**    When two oppositely doped materials are fused to form a *p-n* junction, free electrons from the *n* side and free holes from the *p* side of the junction migrate across the junction to combine with the oppositely charged carriers in the regions near the junction. Since both materials are originally electrically neutral, this migration and subsequent combination create a small electrical potential across the junction, the *p* side becoming slightly negative and the *n* side becoming slightly positive (Fig. 6-4a). This "space charge" in the vicinity of the junction quickly reaches an equilibrium condition in which the acquired negative charge on the *p* side prohibits any further migration of electrons from the *n* region and the acquired positive charge on the *n* side repels any further incursion of holes from the *p* side.

When an external voltage is applied across the junction, the equilibrium condition is altered. With the voltage connected as shown by the battery in Fig. 6-4b (positive to the

$p$ side, negative to the $n$ side), some of the excess electrons that originally migrated from $n$ to $p$ are drawn out of the material and the voltage barrier at the junction is reduced. As a result, majority carriers from both sides can move freely and continuously across the junction. Under this condition, the junction is *forward-biased* and represents little resistance to the flow of current.

Conversely, if the potential of the battery is reversed (negative to the $p$ side, positive to the $n$ side; Fig. 6-4c), electrons from the battery enter the $p$ region and fill in more of the holes in that region. This increases the potential across the junction and prevents the

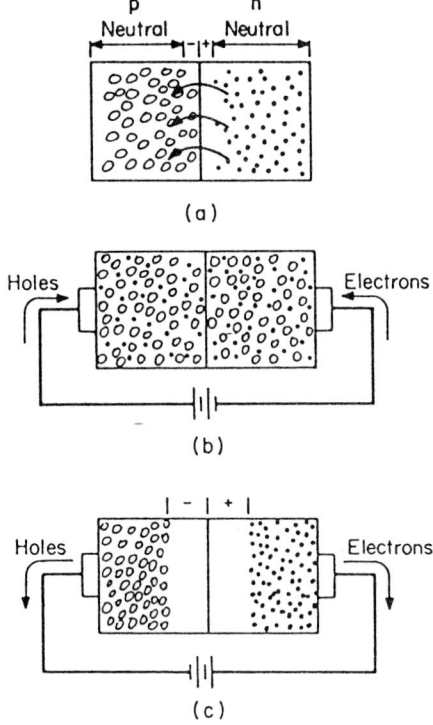

**Fig. 6-4** Junction characteristics. (*a*) An unbiased junction. (*b*) A forward-biased junction permits nearly unrestricted current flow. (*c*) A reverse-biased junction represents high resistance to majority carriers.

migration of majority carriers across the junction. The junction is then *reverse-biased* and represents a very high resistance to the flow of current.

But while a reverse-biased junction is a high resistance to majority carriers, it is a very low resistance to minority carriers. These can cross the junction quite easily and form a *leakage current*. Since the number of minority carriers in doped semiconductor material is quite small, the leakage current is correspondingly small.

RECTIFICATION    From the foregoing, it is evident that a semiconductor junction has the properties of a rectifier, permitting a current flow when biased in one direction and prohibiting a current flow (except for leakage) when biased in the other direction. The voltage-current characteristics of such a junction are shown in Fig. 6-5. Here it is seen that, when biased in the forward direction, the current rises quickly as the applied voltage is raised. The bend in the curve at low voltage is due to the fact that the externally applied voltage must neutralize the space charge before full current is permitted to flow. Thereafter, however, the current flow is directly proportional to the applied voltage, being limited only by the bulk resistivity of the material itself.

Biased in the reverse direction, the current is seen to be extremely small (leakage) until a breakdown point is reached. At that point, current flow increases greatly for any small increase in voltage. Unless current is limited by an external resistance, the junction can be destroyed.

CAPACITANCE EFFECT    A capacitor is formed when two conductors are separated by an insulator (dielectric) across which a difference of potential can exist. For a *p-n* junction, with no applied external voltage, the region in the immediate vicinity of the junction is depleted of free current-carrying charges. This is so because any mobile electrons near the junction in the *n* region have migrated across the junction to combine with the mobile holes in the *p* region. With the junction area thus depleted of free-charge carriers, the region can be considered to be an insulator. This region, however, is bounded at each

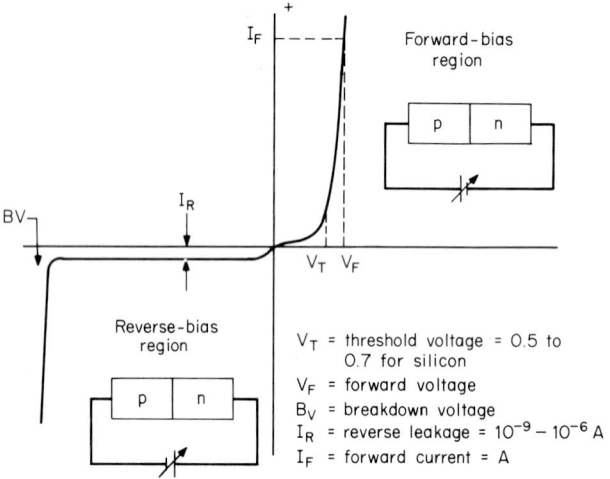

**Fig. 6-5**   Junction current-versus-voltage characteristics. Note that the reverse bias is measured in microamperes and is virtually independent of voltage until breakdown $V_B$ is reached.

end by a conductive region of *p*- and *n*-type material in which there are numerous free charges (holes and electrons, respectively), so that the structure fits the definition of a capacitor.

The value of the capacitance is a function of the area of the junction (die size), the resistivity of the material (determined primarily by the doping level of the high-resistivity side of the structure), and the voltage applied across the junction. For a reverse-biased junction, this capacitance decreases as the reverse voltage is increased. This is so because an increase in reverse bias increases the width of the depletion area near the junction, thereby widening the dielectric region between the two conductive regions of the structure.

The capacitance associated with a *p-n* junction represents a limiting factor on the frequency response of semiconductor devices. However, in some devices, such as varactor diodes, it is utilized for constructive purposes.

**7. Practical applications for *p-n* junctions.**  A simple *p-n* junction forms the nucleus of a number of practical semiconductor components. Among the most prevalent are signal and rectifier diodes, zener and reference diodes, and varactor diodes.

RECTIFIER DIODES    The characteristics of a *p-n* junction are utilized most extensively in the half-wave rectifier diode to convert an ac voltage (or current) into pulsating direct current (Fig. 6-6). This characteristic is used in power supplies, demodulator circuits of AM radios, signal clippers, square-wave generators, and a variety of other applications. Many of these functions, particularly those associated with small-signal applications, are currently being combined with other circuit functions in integrated circuits, so that the use of such diodes in discrete form is diminishing. The primary exception is the rectifier diode for power supplies, which is required to pass relatively high currents and is not yet replaceable with ICs. The selection of such rectifiers, in both capabilities and packaging, continues to expand.

The use of silicon rectifiers in the most popular power supply configurations, as well as the more important circuit characteristics, is shown in Fig. 6-7.

SCHOTTKY RECTIFIERS   When a metal is brought into contact with a highly doped (low-resistivity) semiconductor material, an ohmic contact is formed. The resulting junction is highly conductive, exhibiting linear voltage-current relationships with little series resis-

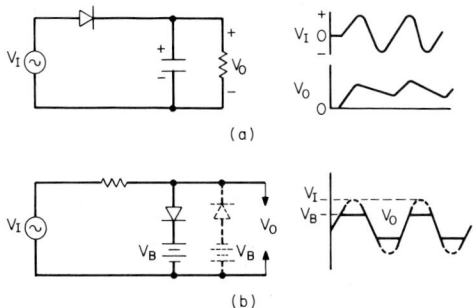

(a)

(b)

**Fig. 6-6**  Schematic diagrams and waveforms of diodes used as (a) rectifiers and (b) clippers.

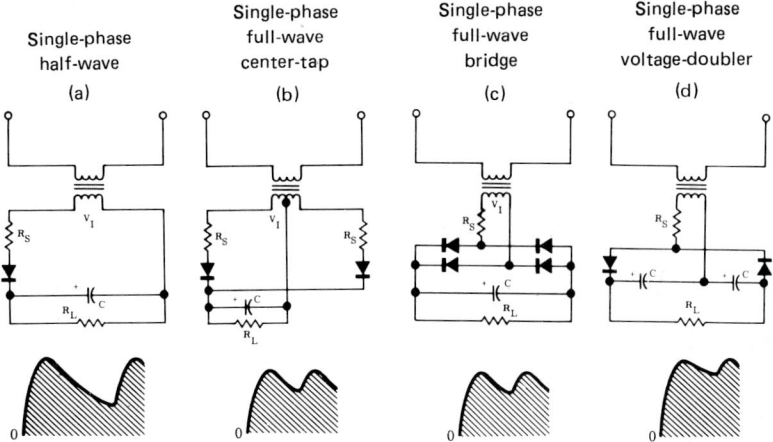

**Fig. 6-7**  Practical power supply circuits and their characteristics.

| Characteristics | (a) Single-phase, half-wave | (b) Single-phase, full-wave, center-tap | (c) Single-phase, full-wave, bridge | (d) Single-phase, full-wave voltage doubler |
|---|---|---|---|---|
| $V_i$ | $0.910\ V_o$ | $0.825\ V_o$ | $0.805\ V_o$ | $0.552\ V_o$ |
| Peak inverse voltage per rectifier | $2.56\ V_o$ | $2.34\ V_o$ | $1.14\ V_o$ | $1.56\ V_o$ |
| Ripple | $0.12\ V_o$ | $0.06\ V_o$ | $0.06\ V_o$ | $0.09\ V_o$ |
| $I_p$ per rectifier | $7.80\ I_o$ | $4.75\ I_o$ | $4.75\ I_o$ | $3.00\ I_o$ |
| $I_{rms}$ per rectifier | $2.50\ I_o$ | $1.63\ I_o$ | $1.33\ I_o$ | $1.10\ I_o$ |
| Secondary VA | $2.35\ P_o$ | $2.16\ P_o$ | $2.16\ P_o$ | $1.32\ P_o$ |
| Primary VA | $2.35\ P_o$ | $3.05\ P_o$ | $2.16\ P_o$ | $1.72\ P_o$ |

Condition: $R_{L(min)} = 50\ R_s$.

tance. Such junctions are used for the connection of metal leads to seimiconductor devices. However, when a metal is brought into contact with a lighly doped (high-resistivity) semiconductor material, the resulting junction exhibits rectifying properties similar to those of a *p-n* junction. Alternatively known as Schottky rectifiers (diodes), surface-barrier diodes, and hot-carrier diodes, such structures are commonly employed as VHF-UHF mixers, as which they offer faster response time and lower noise than other types of devices, and as high-speed rectifiers in power supply circuits, for which they have a number of advantages over standard *p-n* junction rectifiers. Among these advantages are:

1. Virtually nonexistent offset voltage at low current and substantially lower forward voltage drop than comparable silicon rectifiers at high currents (Fig. 6-8)

2. Substantial improvement in rectification efficiency

3. Significantly higher switching speed owing to the absence of storage time as a speed-limiting parameter

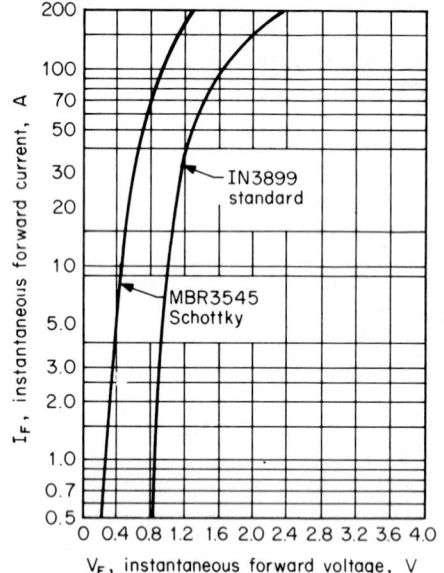

**Fig. 6-8** Typical offset voltage and forward voltage drop of standard and Schottky rectifiers with similar current ratings.

This combination of features makes Schottky rectifiers highly useful in low-voltage, high-current power supplies, in switching power supplies, and in switching circuits associated with integrated circuits.

The primary disadvantage of a Schottky rectifier is its substantially higher reverse-bias leakage current, which may be of an order of magnitude greater than that of *p-n* junctions. Total power dissipation with symmetrical signals, however, favors the Schottky diode because of its low voltage drop in the on condition.

The lower turn-on potential of Schottky diodes is used to increase the switching speed of conventional saturated-logic circuits by keeping the switching transistor out of deep saturation while in the on condition. This circumstance increases the turn-off speed by eliminating storage time as a factor of turn-off time. This is accomplished by connecting a reverse-biased Schottky diode across the collector-base junction of the transistor. As a positive-going base signal drives the transistor toward saturation, resulting in a collector voltage lower than the base voltage (tending to forward-bias the collector-base junction), the lower turn-on potential of the Schottky diode causes the diode to conduct before the normally reverse-biased collector-base junction of the transistor goes into conduction. Thus, the collector-base forward voltage is clamped by the low on voltage of the diode to a value that is less than required to cause reverse conduction in the collector-base junction.

ZENER DIODES  Operated as a rectifier, a *p-n* junction utilizes the forward-conduction–reverse-blocking action of the structure. As a zener diode, it employs the "breakdown" region of the diode.

Figure 6-9 shows that, for an applied reverse voltage, the diode current is very small (leakage current) until the breakdown point is reached, at which point the reverse current increases very rapidly. This breakdown voltage is usually called zener voltage ($V_z$). A zener diode differs from a rectifier diode in that the breakdown-voltage point $V_z$ is very precisely controlled during design and manufacture.

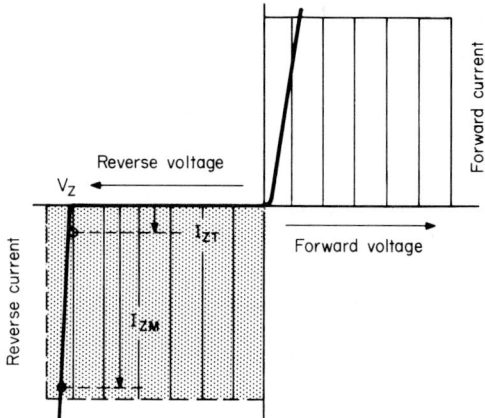

**Fig. 6-9**  Typical zener-diode characteristics and major parameters. $V_Z$ = zener breakdown voltage; $I_{ZT}$ = zener test current at which $V_Z$ is specified; $I_{ZM}$ = maximum permissible zener current.

When used as a zener diode, the device is operated in the breakdown region, where it acts as a constant-voltage device and can be utilized as a voltage regulator in power supplies. The simple circuit in Figure 6-10 illustrates this application. Here, when the applied unregulated dc potential is low (below the breakdown point of the diode), the diode current is very small and the device acts essentially as an open circuit. Therefore, its effect on the load circuit is virtually zero. If the applied dc voltage exceeds $V_z$, however, the junction breaks down and its voltage drop remains essentially constant. Therefore, the voltage available to the load remains approximately constant and equal to the zener voltage regardless of variations in source voltage or load-current requirements. Of course, if the applied voltage should drop below the point where diode breakdown is maintained or if the load resistance $R_L$ were to decrease to the point where it could no

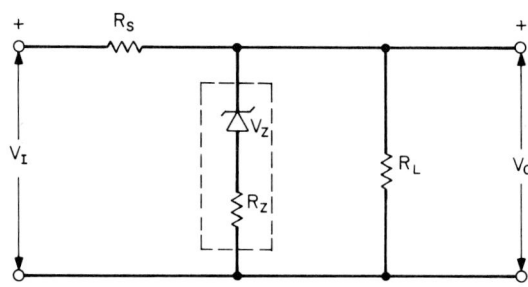

**Fig. 6-10**  Simple power supply voltage regulator.

longer sustain the voltage required for zener breakdown, the diode would come out of conduction and voltage regulation would be lost. With proper design, the value of the applied dc voltage and the value of the series resistance $R_s$ are chosen so that the voltage drop across $R_L$ in the absence of the zener diode would be greater than $V_z$ under any anticipated load-resistance variations.

Zener diodes are also used for a variety of other applications, including clipper circuits (by replacing the batteries utilized in the circuits of Fig. 6-6), overvoltage-protection circuits (similar to the voltage-regulator circuit of Fig. 6-10), and surge-protection circuits for all sorts of electrical equipment as exemplified by the circuit of Fig. 6-11, where

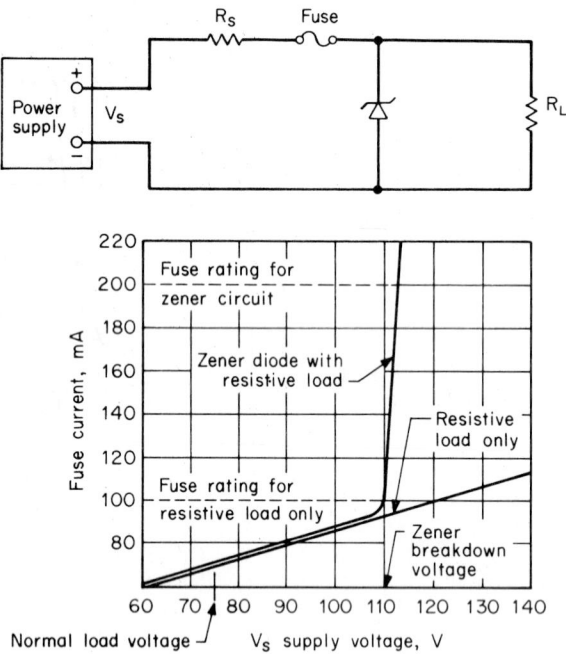

**Fig. 6-11** Zener-diode protection permits the use of fuses with higher current ratings by generating a sudden burst of current at the predetermined fault point.

the zener is used to protect against excessive fuse burnout due to prolonged periods of operation near the melting point of the fuse. The zener provides a sudden burst of current at voltages beyond $V_z$ so that a more tolerant fuse may be selected.

Commercial zener diodes are available with zener voltages ranging from about 2.4 to 200 V and with power ratings from milliwatts to 50 W or more so that a great many power supply regulation requirements can be satisfied. Manufacturers' data sheets usually provide a minimum current ($I_{ZK}$) which will assure reasonable regulation, as well as a maximum current ($I_{ZM}$) beyond which the diode is subject to damage.

REFERENCE DIODES   The forward and reverse characteristics of a typical zener diode are affected by temperature. This is shown in the graph of Fig. 6-12, where it is seen that a change in temperature from 25 to 100°C causes a decrease in voltage of approximately 150 mV in the forward-biased region of the diode. This corresponds to a negative temperature coefficient (TC). In the zener region, the voltage increases as temperature is increased, resulting in a positive TC.

For many applications this change in zener voltage with temperature can be tolerated. Critical applications, however, require temperature-compensated diodes, or reference diodes, whose breakdown voltage remains constant under varying temperature conditions.

A reference diode can be formed by combining one or more forward-biased diodes with a reverse-biased diode, as shown in Fig. 6-13. Here it is seen that the increase in voltage across the reverse-biased zener junction is counterbalanced by the decrease in voltage across the forward-biased junction at 7.5 mA. By judicious selection of diode junctions and proper operation, excellent temperature compensation can be achieved. It is important, however, to operate the TC device at a current level specified in the data sheet. Otherwise, voltage compensation is not obtained because the $\Delta V$'s at different current levels are not necessarily equal.

THE TRANSISTOR    The transistor is the most versatile component in the semiconductor family. Its most important characteristic is current and/or voltage amplification. Because of this capability it is the heart of most electronic circuits involving signal amplification and switching.

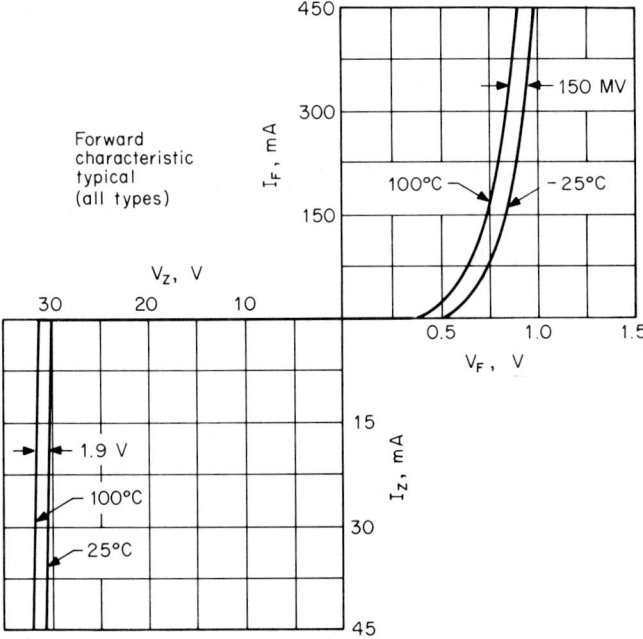

**Fig. 6-12**   Typical forward and reverse diode characteristics as a function of temperature.

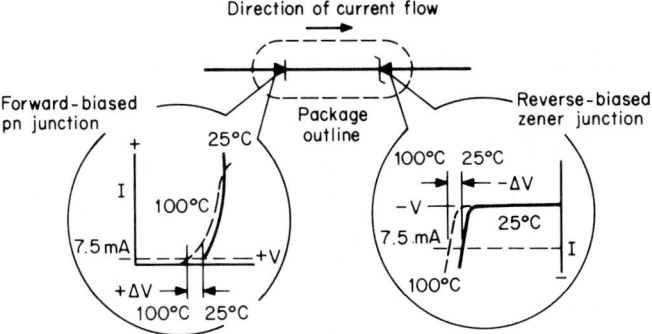

**Fig. 6-13**   The temperature effect on zener breakdown voltage can be compensated for by a series-connected forward-biased diode with an equal temperature change of opposite polarity.

There are two basic kinds of transistors: *unipolar transistors,* usually referred to as field-effect transistors (FETs), and *bipolar transistors.* The FET was first described in the early 1930s but was not exploited commercially at that time. The bipolar transistor was invented in the late 1940s and rapidly became a commercial product. The FET was finally developed as a commercial product early in the 1960s, long after the bipolar transistor had established itself. FETs, especially MOSFETs, are now widely used in the front ends of low-noise radio and TV equipment, and they are the dominant devices in digital integrated circuits.

**8. Bipolar transistors.**   The basic element of any bipolar transistor is the *p-n* junction. When the junction (diode) is forward-biased (*p* layer positive, *n* layer negative), current through the device increases rapidly as the voltage is increased. Under reverse bias, only a very small leakage current can flow until the reverse voltage becomes high enough to cause breakdown.

A bipolar transistor is formed by sandwiching a very thin layer of *n*-doped material between two layers of *p*-type material (*p-n-p*) or a thin layer of *p*-type material between two *n*-doped layers (*n-p-n*). The characteristics of such a structure are varied by varying the geometries and resistivities of the three layers.

The three layers of a bipolar transistor (Fig. 6-14) are the *emitter,* the *base,* and the *collector.* The emitter represents the current source, where the current carriers originate. The base is the control element, and the collector is the element through which the current carriers are transferred to an external circuit.

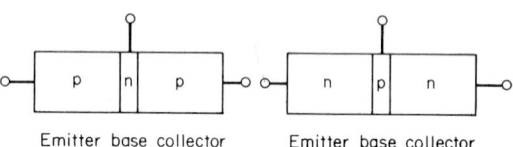

Emitter  base  collector        Emitter  base  collector

**Fig. 6-14**   Physical representation of bipolar transistors.

The main difference between a *p-n-p* and an *n-p-n* transistor is that the former operates with a negative voltage on the collector while the latter operates with a positive collector voltage (with respect to the emitter). This makes it possible to have complementary circuits that often provide improved performance over single-ended circuits, which can be implemented with only one type of transistor.

Transistor action can be understood by analyzing the current flow through an *n-p-n* transistor under the influence of externally applied voltages (Fig. 6-15). For the normal case, the external voltages are applied so that the emitter junction is forward-biased and the collector junction is reverse-biased. By examining first the collector junction with the emitter open (*a*), it is obvious that only minority carriers from the base and collector can cross this junction. If the number of minority carriers in these regions is assumed to be relatively small, the total current across this junction is correspondingly small and is referred to as leakage current ($I_{CBO}$).

If the collector circuit is opened and a forward bias is applied to the emitter-base junction (*b*), majority carriers from the emitter and the base (electrons and holes, respectively) find it quite easy to cross this junction, causing a heavy base current.

With both junctions properly biased (*c*), if the base region were quite thick, the electrons from the emitter would combine with abundantly available holes in the base and cause an equivalent and compensating flow of current out of the base lead. In that case, the current in the collector would not be greatly affected by the emitter-injected carriers and would remain at its leakage-current value while the base current would be quite high.

However, for a very thin base region, the *lifetime* of the electrons injected into the base (before recombination takes place) is long enough so that most of them can traverse the entire base region toward the collector. Since these electrons represent minority carriers in the base region, the reverse-biased collector-base junction is actually forward-biased for the minority carriers, and the electrons will be swept across the junction into the collector region and from there into the collector power source. Therefore, the collector current equals the leakage current plus most of the injected current, while the base current is very small.

The above action gives rise to the most commonly used expressions for transistor current gain, beta ($\beta$), which is the ratio $I_C/I_B$ and is called the common-emitter current gain. Beta can be quite high since, for a very thin, high-resistivity base region, almost all the injected carriers diffuse into the collector, with very few exiting via the base lead.

*Circuit configurations.* There are three basic transistor circuit configurations: common-emitter, common-base, and common-collector. They differ principally in the manner in which the signal is applied to the transistor and where the load is attached. Figure 6-16 shows these basic circuits. Since the common-emitter circuit is by far the most prevalent, data sheets normally characterize the transistor in terms of this circuit.

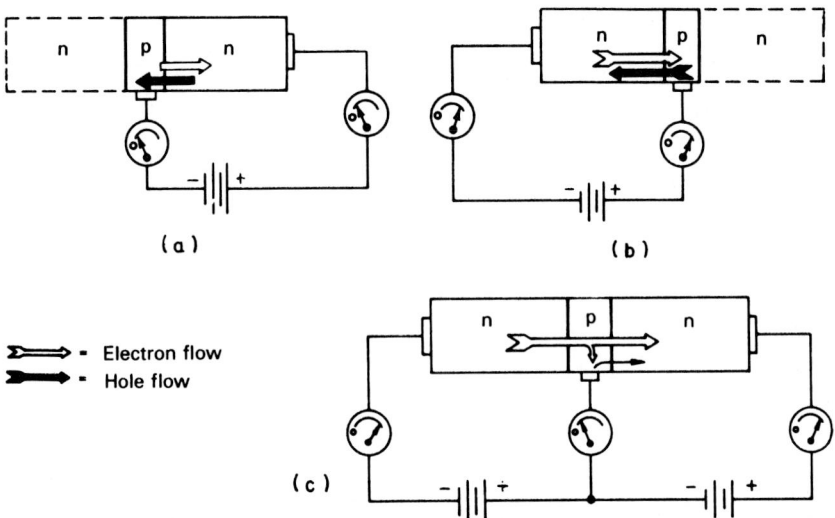

**Fig. 6-15**   Transistor action. With the collector junction reverse-biased (*a*), only minority carriers can cross the junction, and the resulting current flow is small. A forward-biased emitter junction (*b*) permits a heavy flow of majority carriers. With both junctions properly biased (*c*), emitter-injected electrons cross the thin base region and flow out of the collector lead.

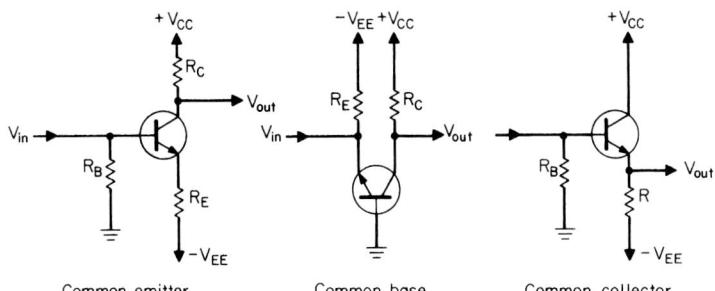

**Fig. 6-16**   Basic transistor circuit configurations and their relative characteristics.

| Circuit | Current gain | Voltage gain | Input resistance | Output resistance |
|---|---|---|---|---|
| Common emitter | High | High | Low | High |
| Common base | <1 | High | Very low | Very high |
| Common collector | High | <1 | High | Low |

BIPOLAR-TRANSISTOR CHARACTERISTICS    Figure 6-17 shows the characteristics of a typical small-signal transistor in a common-emitter circuit. The input circuit, curve $a$, shows that a base-emitter voltage $V_{BE}$ of less than approximately 0.5 V (for silicon) causes virtually no emitter current $I_E$ to flow, thereby keeping the transistor cut off. Above 0.5 V, $I_E$ rises sharply, limited only by the ohmic resistance of the emitter region. Since the latter is very small, a very small rise in $V_{BE}$, beyond the threshold voltage $V_T$, causes a large injection of emitter current into the base and from there into the collector, with a small portion flowing out of the base lead.

A typical plot of the division of emitter current between base and collector is shown in Fig. 6-17$b$. The numbers vary considerably for different transistors, but the shape of the curve remains similar. This shows that the $\beta$ of the transistor can vary considerably over the range of the curve and that, for large signals, distortion can be quite high.

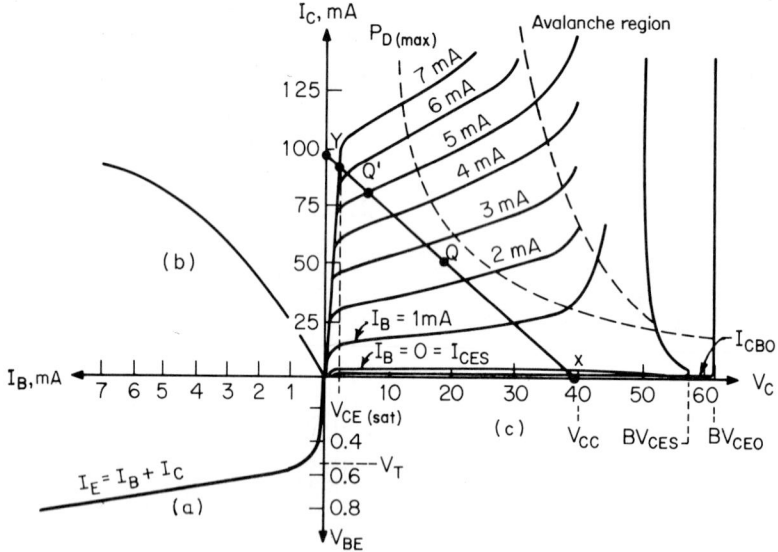

**Fig. 6-17**    Typical input-output characteristic curves of a bipolar transistor in a common-emitter circuit. ($a$) Emitter current versus base voltage (input characteristics). ($b$) Collector current versus base current. ($c$) Collector current versus collector voltage for various values of base current (output characteristics).

The collector curves in Fig. 6-17$c$ are the most significant. They can be used to determine the relationship between the input and output signals of a transistor.

When the transistor is cut off ($I_B = 0$), a residual current $I_{CES}$ flows in the collector circuit. This leakage current can be reduced somewhat (but not to zero) by applying reverse bias to the emitter junction. The limiting value, with reverse bias applied to the emitter-base junction, is $I_{CBO}$. This is equivalent to the collector-base leakage current when the emitter is open-circuited.

Collector current increases rapidly as the base is energized. The maximum $I_C$, $[I_{C(max)}]$, is that which would damage the internal transistor structure. This value of collector current is given as a maximum rating on data sheets. Thus, the output current could range from $I_{C(max)}$ to $I_{CES}$ (or $I_{CBO}$ in the event of reverse bias), but in practice $I_C$ is limited to a value far less than $I_{C(max)}$.

POWER DISSIPATION    Power dissipation $P_D$ is the product of $V_{CE}$ and $I_C$, and it causes the collector junction to heat up. Beyond a critical junction temperature $T_{J(max)}$, the device could be damaged. Thus the $P_{D(max)}$ rating of a transistor limits the maximum $V_{CE}$ and $I_C$ that can be applied simultaneously. The locus of a $P_{D(max)}$ rating is the parabolic curve on the $V_{CE}/I_C$ plot of Fig. 6-17. A load line, $XQY$, must be chosen to maintain steady-state operation to the left of the locus.

SWITCHING ACTION    From the above it is evident that when the voltage applied between base and emitter is less than approximately 0.5 V, there is only a very small current flow through the transistor. Between collector and emitter, therefore, the transistor represents a very high resistance, or open switch. When $V_{BE}$ is raised significantly above the threshold voltage, the internal resistance of the transistor drops to a very small value so that the device approximates a closed switch. As shown in Fig. 6-17, it takes a change of only 7 mA of $I_B$ along the load line $XQY$ to cause $V_{CE}$ to cause the voltage across this particular transistor to change from $V_{CC}$ to near zero, that is, from an open switch to a closed switch. This switch action is accomplished by a $V_{BE}$ of only a few tenths of a volt.

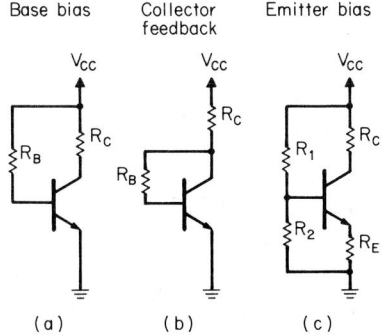

**Fig. 6-18**    Conventional common-emitter bias circuits. The following table gives approximate characteristic expressions:

| $I_C$ | $\beta \dfrac{V_{CC}}{R_B}$ | $\dfrac{V_{CC}}{R_C + R_B/\beta}$ | $\dfrac{R_2}{R_1 + R_2} \cdot \times \dfrac{V_{CC}}{R_C}$ |
|---|---|---|---|
| $S$ | $1$ | $\dfrac{1}{1 + \beta R_C/R_B}$ for $R_B = \beta R_C$ $S = 0.5$ | $\dfrac{1}{1 + \beta R_E \dfrac{R_1 R_2}{R_1 + R_2}}$ |
| To make $I_C = \dfrac{I_{C(\text{sat})}}{2}$ | $R_B = 2\beta R_C$ | $R_B = \beta R_C$ | $\dfrac{R_1}{R_2} = 1 + 2\dfrac{R_C}{R_E}$ |

BIASING    When operated as an amplifier, the transistor must first be biased to some quiescent value of collector current, so that both positive- and negative-going input-voltage excursions will cause corresponding changes in output voltage and current. The ideal bias point is represented by $Q$ on the load line because this permits approximately equal excursions in $I_C$ and $V_{CE}$ without signal clipping. The bias point is established by a quiescent base current that results in a dc collector voltage of approximately $V_{CC}/2$.

Several circuits are used for establishing the bias point. Among the most familiar are those in Fig. 6-18. The basic performance difference is in the bias-point stability. At point $Q$ on the load line in Fig. 6-17, the transistor has a beta ($I_C/I_B$) of approximately 20. If a transistor with a beta of 40 were substituted (simulated by dividing all $I_B$ values by 2) and if $I_B$ were held by the bias circuit to 2.5 mA as before, the operating point would move up the load line to point $Q'$, a much higher value of $I_C$. As a result, considerable distortion would occur for high-value input signals.

The bias-point stability factor $S$ is defined as the percent change in $I_C$ for a percent change in $\beta$, or $\Delta I_C/I_C = S \, \Delta\beta/\beta$. If a percent change of $\beta$ causes a corresponding percent change in $I_C$ (the least desirable condition), then $S = 1$. If $I_C$ is independent of $\beta$ (corresponding to a zero change in $I_C$ when $\beta$ is varied), then $S = 0$. The formulas accompanying

Fig. 6-18 give $I_C$ and $S$ as functions of $\beta$ and assign values for $S$ under specific operating conditions. The bias arrangement in Fig. 6-18$c$, using emitter degeneration, is preferred because, by proper choice of resistor values, the effect of $\beta$ on $I_C$ can be made almost negligible.

**9. Field-effect transistors (FETs).**   There are two types of field-effect transistors: the junction field-effect transistor (JFET) and the metal-oxide–semiconductor field-effect transistor (MOSFET). Of these, the MOSFET is by far the more popular and forms the basis for most digital large-scale and very-large-scale integrated circuits (LSI and VLSI). JFETs, because of their low-noise advantages, are utilized extensively in TV and radio front ends as well as in input circuitry for integrated operational amplifiers, etc.

JUNCTION FIELD-EFFECT TRANSISTORS   In its simplest form, a JFET starts with a bar of $n$- or $p$-doped silicon which behaves as a resistor (Fig. 6-19$a$). By convention, the terminal into which current is injected is called the *source;* the other terminal is the *drain.* Current flow between source and drain is dependent on the drain-source voltage and on the resistivity of the material.

In Fig. 6-19$b$, $p$-type regions have been diffused into an $n$-type substrate, leaving an $n$ channel between source and drain. (A complementary $p$ channel is made by reversing the dopants in all regions of the structure.) The $p$ regions control the current flow between source and drain and are called *gates.*

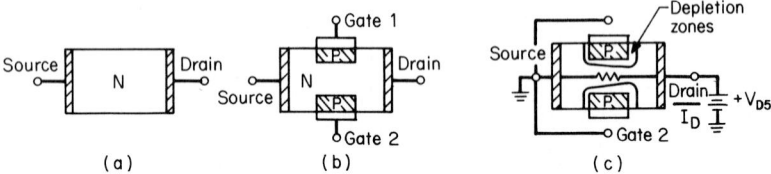

**Fig. 6-19**   Development of junction field-effect transistors.

When the gates are connected to the source and a positive voltage is applied between drain and source (Fig. 6-19$c$), current flow in the channel creates a voltage gradient along the length of the channel. This voltage reverse-biases the gate-substrate junctions and creates the usual depletion layer in the vicinity of the junction. The depletion layer, being devoid of mobile charges, is in effect an insulating region which cannot be penetrated by the channel current. This reduces the channel width, thereby increasing its resistance.

The overall effect is shown in Fig. 6-20$a$, where an increase in $V_{DS}$ causes a corresponding increase in drain current $I_D$, but the rate of change of current tapers off as the increasing voltage drop in the channel increases the spread of the depletion region into the channel. At $V_P$ (pinch-off voltage) an equilibrium condition is reached whereby an increase in $V_{DS}$ generates a neutralizing increase in channel resistance, and drain current remains essentially constant until the breakdown condition is reached.

By applying a reverse-bias voltage between the gates and the substrate (Fig. 6-20$b$), the channel width is reduced by the gate voltage and the maximum drain current is lowered. Further increases in gate voltage correspondingly reduce the channel width and produce successive reductions in maximum drain current.

It will be noted that current flow in an FET is made of majority carriers only and that, unlike in a bipolar transistor, the current does not cross a junction. Hence, an FET is sometimes called a *unipolar transistor.*

A more practical structure for a junction FET is illustrated in Fig. 6-21. In this single-ended geometry, the $p$-type substrate takes the place of gate 2 in Fig. 6-20. An $n$-

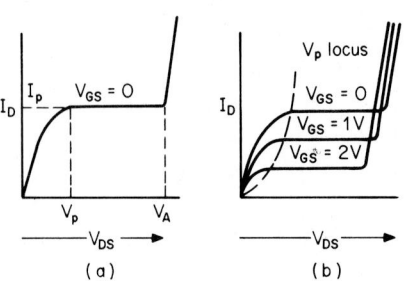

**Fig. 6-20**   Drain-current characteristics.

type layer is epitaxially grown on top of this substrate, and a $p$-type gate diffusion into the epitaxial layer forms the second gate and creates the channel.

MOS FIELD-EFFECT TRANSISTORS   The MOSFET operates on a somewhat different control mechanism. Figure 6-22 shows the development. Two separate low-resistivity $n$-type regions (source and drain) are diffused into a high-resistivity substrate, as shown in Fig. 6-22$a$. Next, the surface of the structure is covered with an insulating oxide layer (Fig. 6-22$b$), and holes are cut into the layer, allowing metallic contact to source and drain. Then, a metal area is overlaid on the oxide, covering the entire region between source and drain, and simultaneously metal contacts to drain and source are made as shown in Fig. 6-22$c$. The metal area between source and drain is the gate terminal. Note that there is no physical penetration of the gate metal through the oxide into the substrate. Since drain and source are isolated by the opposite-polarity substrate, the structure is analogous to two diodes connected back to back (Fig. 6-22$d$). If a voltage were applied between source and drain (in the absence of a gate voltage), the resulting current flow would be very small because one of the back-to-back diodes would be reverse-biased regardless of the polarity of the applied voltage.

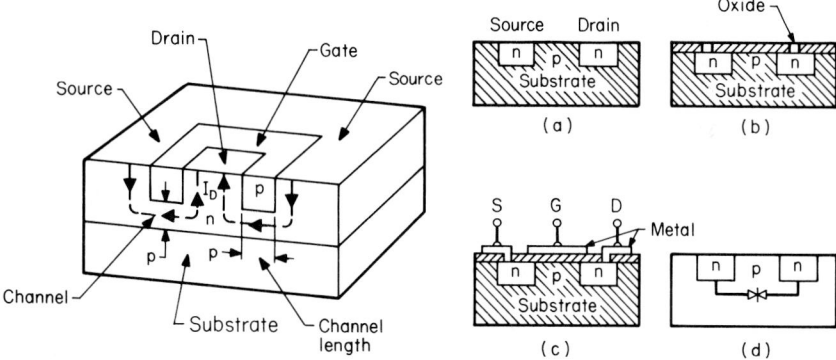

**Fig. 6-21**   Cross section of a JFET structure.

**Fig. 6-22**   Development of an MOSFET.

The metal gate area, in conjunction with the insulating oxide layer and the semiconductor layer underneath, forms a capacitor. The metal area is the top plate; the substrate material, the bottom plate. When a positive potential is applied to the gate metal, the positive charge at the metal side of the capacitor induces a corresponding negative charge at the semiconductor side. As the positive charge at the gate is increased, the negative charge "induced" in the semiconductor increases until the region beneath the oxide is changed from $p$-type to $n$-type (Fig. 6-23), and current can flow between drain and source through the induced channel. In other words, drain-current flow is "enhanced" by the gate potential. Thus drain-current flow can be modulated by the gate voltage; i.e., channel resistance is inversely proportional to gate voltage. (The $n$-channel structure may be changed to a $p$-channel device by reversing conductivity of the semiconductor regions.)

The FET just described is called an enhancement-type MOSFET. A depletion-type MOSFET can be made by diffusing a low-resistivity $n$ channel into the space between source and drain so that considerable drain current will flow when the gate potential is at zero volts. In this manner,

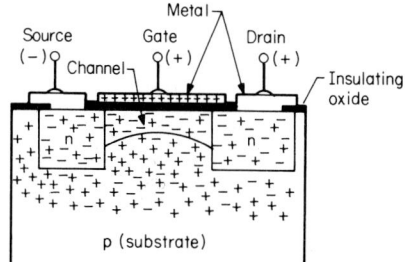

**Fig. 6-23**   A typical MOSFET operation. A positive voltage applied to the gate converts the $p$-type material underneath the gate insulation to $n$-type material and creates a conductive channel between source and drain.

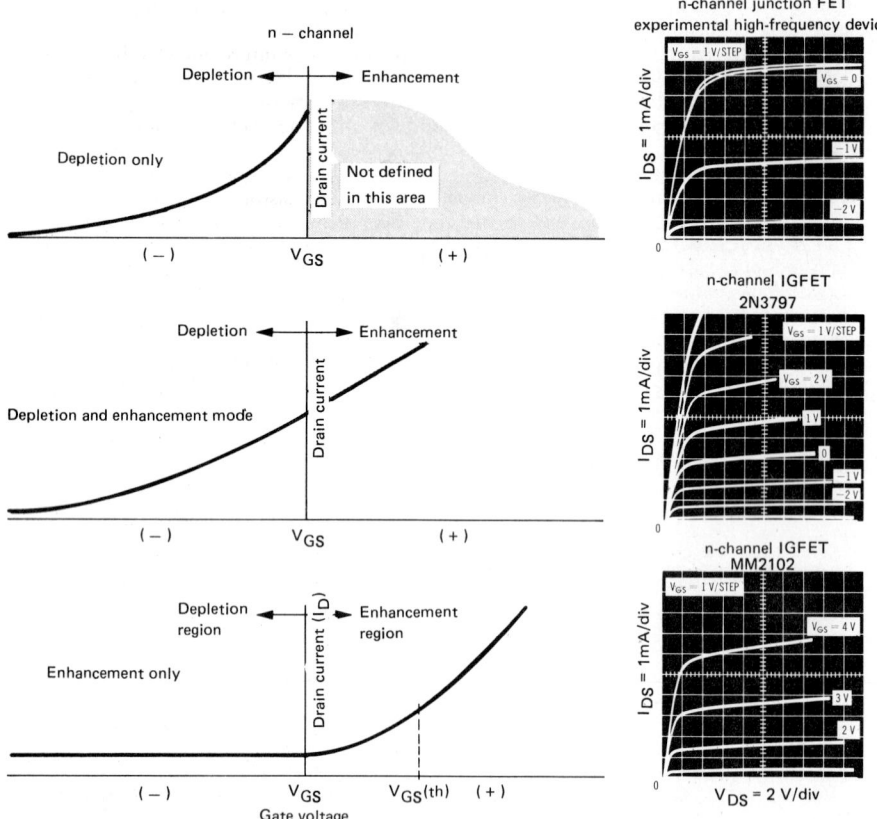

**Fig. 6-24** Gate-voltage-transfer characteristics and associated scope traces for different types of FETs.

the MOSFET can be made to exhibit *depletion* characteristics by applying a negative voltage to the gate.

For a moderate-resistivity induced channel, the FET will exhibit both enhancement and depletion characteristics. Typical performance curves for such structures are illustrated in Fig. 6-24.

COMPARISON OF SMALL-SIGNAL BIPOLAR AND MOS TRANSISTORS    The MOSFET enjoys several clear-cut advantages over its bipolar counterpart. First, it has an extremely high input resistance, making it more adaptable to many circuit applications. Second, it is a low-noise device and is often preferred for front-end communications circuits. Third, it is extremely compatible with integrated-circuit processing, particularly for LSI and VLSI circuits, for which its smaller geometry, lower power consumption, and simpler processing make it the preferred basic device for many applications.

On the other hand, bipolar transistors are capable of operating at higher speeds and higher power. In discrete form, they are generally available with a wider range of specifications and often at considerably lower cost for applications for which both types of devices are suitable.

POWER MOSFETS    While MOSFETs for small-signal applications have been widely used for some years, their performance in power circuits has been limited. The limiting factor of the conventional small-signal MOSFET structure has been the lateral channel, which has an inherently low current-carrying capacity and a relatively high on resistance coupled with a low reverse breakdown voltage.

The disadvantages of the lateral FET from a power standpoint were overcome with the development of a structure that permits the control of a vertical current flow by means of a gate field. The structure utilizes thousands of "source" sites interconnected in parallel on a single die (Fig. 6-25). A common drain at the bottom of the die minimizes the geometry of the structure to make such a multicell architecture practical. Moreover, it permits very short channel lengths to reduce the on resistance of each cell to a minimum. The overall result is a device which allows current and voltage ratings to meet virtually any desired electronic application while maintaining the higher speed and circuit-simplification advantages of MOSFETs in comparison with bipolar transistors.

The power MOSFET has made great strides, but emphasis on cost reduction is continuing. Improvements of present devices over the early double-diffused MOS structure have been significant, and prices are now approaching those associated with bipolar power transistors for medium voltage and current applications.

**10. Thyristors.**  Thyristors and their trigger devices can take numerous forms, but they share these characteristics:

1. They are "open circuits," capable of withstanding rated voltage until triggered.

2. They become low-impedance current paths when triggered and remain so, even after the trigger source has been removed, until current through that path is interrupted or is reduced below a minimum "holding" level.

The regenerative action which holds a thyristor in the on state is due to multiple layers of opposite $p$- and $n$-doped silicon which result in the two-transistor-equivalent circuit structure shown in Fig. 6-26. Here, part of the current through transistor 2 is injected back into transistor 1 to supplement the trigger current and sustain conduction when the trigger is removed. This characteristic, coupled with the thyristor's low on resistance, makes it possible to control a portion of each cycle of an ac power waveform into a load, for low-dissipation "dimming" or motor-speed–control applications, to switch capacitive discharge currents precisely in electronic pilot ignition systems, and to fulfill innumerable other control purposes.

There are a number of semiconductor devices under the title of thyristor; the most important of these are silicon-controlled rectifiers (SCRs) and triacs.

SCRs    As a rectifier, the SCR conducts current in only one direction. But unlike a two-layer rectifier, which begins conduction when its anode becomes slightly positive with respect to its cathode, the SCR will remain nonconductive even in the forward direction until the anode voltage exceeds a certain minimum value called the *forward breakover voltage* ($V_{DRM}$). Moreover, the value of $V_{DRM}$ can be varied through the injection of a small current into a third, or *gate*, element, which governs the amplitude of the anode voltage needed to cause conduction, or firing.

The current-voltage relationship of an SCR is shown in Fig. 6-27. With the gate terminal open or shorted to the cathode and an external voltage applied only between anode and cathode, the reverse-bias characteristics (anode negative with respect to cathode) are identical to those of a conventional $p$-$n$ junction, or rectifier. Under forward-bias condi-

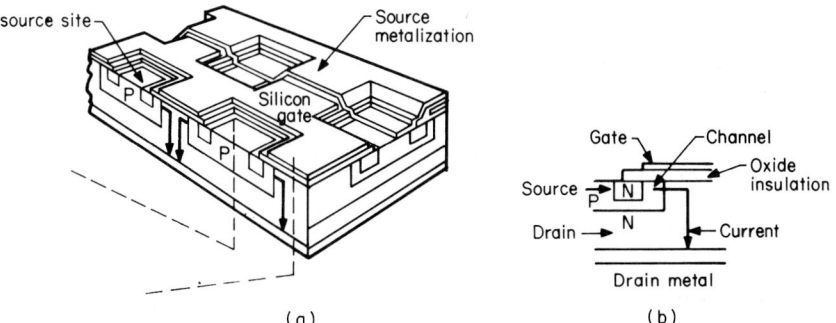

(a)                    (b)

**Fig. 6-25**  A power MOS transistor. The cross section (*a*) illustrates overall interaction of source sites, while (*b*) shows the details of a single transistor element in the structure.

tions, however, current flow does not begin (except for a small leakage current) until $V_{DRM}$ is reached. At that point current flow begins to increase very rapidly until, at the *forward-breakover-current point*, a switchback effect takes place and the voltage across the device suddenly drops to a very low value. At that point, the voltage-current relationship becomes similar to that of a conventional forward-biased diode.

It is important to recognize the difference between the reverse-bias breakdown and the forward-bias-breakover effect. In the former, the voltage across the junction remains at its breakdown value so that power dissipation in the device is high and care must be taken to keep the maximum current at a relatively low value to prevent device damage. In the forward breakover region, owing to the switchback effect, the voltage across the device is very low, and it can carry large amounts of current without exceeding the power-dissipation rating of the device. Hence, the device with its gate open (or shorted to the cath-

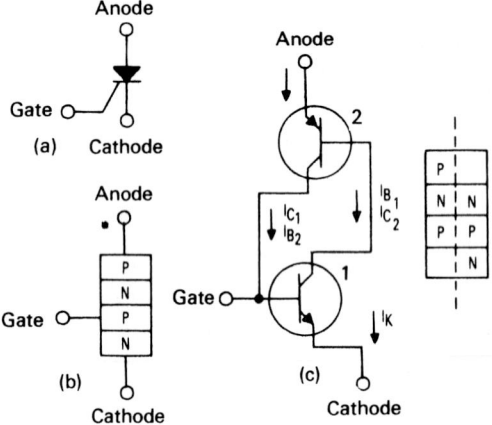

**Fig. 6-26**  Two-transistor analogy of an SCR. (*a*) Schematic symbol of the SCR. (*b*) A *p-n-p-n* structure represented by a schematic symbol. (*c*) A two-transistor model of an SCR.

ode) acts like a voltage-operated switch. The switch is turned on when the applied anode voltage exceeds the breakover voltage and is turned off when the anode current through the device is reduced to the *holding current* $I_H$, at which point the device returns to its high-resistance state and the switch is turned off.

The switching action of the SCR is greatly enhanced by employing its gate characteristics. When a current is injected into the gate, the breakover voltage of the device is reduced. As this current is increased, the breakover point becomes less and less until the

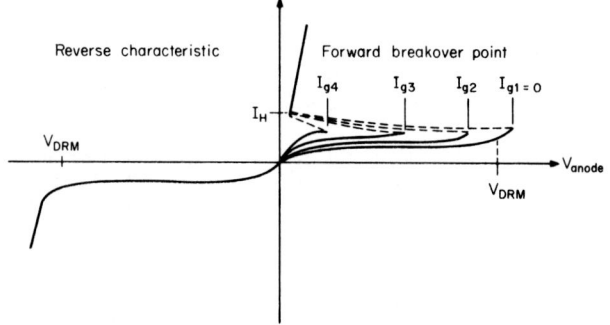

**Fig. 6-27**  Thyristor characteristics illustrating breakover as a function of gate current.

barrier to forward conduction is removed entirely and the device acts as a forward-biased rectifier. The amount of gate current required to reduce breakover voltage from its open-gate value (open switch) to virtually zero (closed switch) is extremely small, so that a small amount of gate current can control a large amount of anode current. In this respect, the SCR acts very much like a sensitive relay, in which a small current through the relay coil can control a much larger current through the relay contact circuit.

A simplified SCR circuit is shown in Fig. 6-28, where the SCR is connected in series with the load $R_L$ and an applied voltage $V_{ac}$. $V_{ac}$ is less than the breakover voltage of the SCR, so that the SCR is an open circuit until a gate signal is applied. If a gate-signal pulse

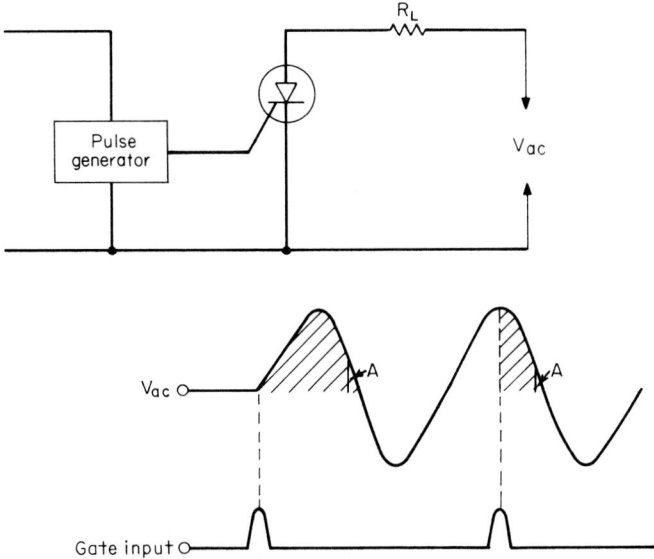

**Fig. 6-28** A simple control circuit indicates the amount of current control available with SCRs.

is applied at the beginning of the positive-going cycle of $V_{ac}$, the SCR is turned on and the current through the load follows the positive half cycle of the applied voltage until that voltage approaches the point $A$ where it can no longer sustain a current above the holding-current value. The SCR is then turned off.

It must be noted that the SCR can be turned on only during the positive-going portion of the applied voltage. No amount of gate current can turn it on during the negative portion of the cycle, so that a single SCR acts like a triggered half-wave rectifier.

Various techniques can be used to permit SCRs to function during both halves of an applied ac voltage. This can be accomplished by connecting two SCRs in an inverse-parallel configuration so that one device is turned on during the positive-going half cycle and the other during the negative portion (Fig. 6-29a). Another is to utilize a triac in place of an SCR (Fig. 6-29b).

TRIACS  A triac is a device which operates exactly like an SCR except that it can be triggered on during both the positive-going and the negative-going half cycles of an applied voltage. Thus, it acts like a triggered full-wave rectifier or, as indicated by its schematic symbol, like a pair of inverse-parallel–connected SCRs. In application, the triac offers circuit simplification and a reduction in component requirements where full-wave power control is required.

A triac may be triggered into conduction by either a positive- or a negative-going gate signal, but sensitivity varies considerably. The recommendations given in manufacturers' data sheets should be followed for best performance.

GATE-TURNOFF THYRISTORS (GTOs)  A gate-turnoff thyristor (GTO) combines the most desirable characteristics of SCRs and bipolar power transistors. Like a transistor, the GTO

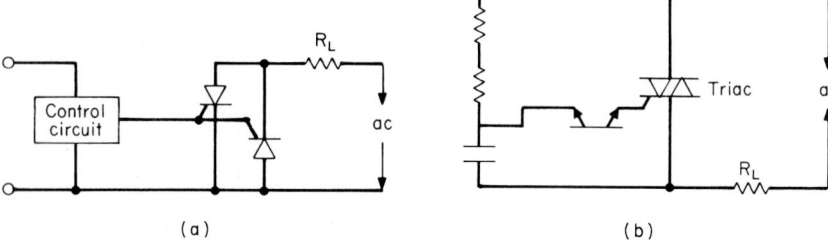

**Fig. 6-29** Full wave power control achieved by two inverse-parallel–connected SCRs (*a*) and through the use of a triac to replace the SCRs (*b*).

can be switched on and off by a low-power gate drive signal; like an SCR, it can pass high forward currents when turned on and block high forward voltages when turned off. But unlike the power transistor, which requires a constant drive source to keep it turned on, the GTO is turned on by a momentary pulse and remains in the on condition until the anode voltage is reduced below the level that sustains a holding current or until it is turned off by a reverse-polarity gate-turnoff pulse. In this way, it differs, too, from more conventional thyristors (SCRs, triacs), which cannot be turned off by a gate signal. Present devices have voltage ratings up to 1400 V and current ratings on the order of 20 A. Gate current trigger requirements are 300 mA. Devices will withstand surge currents as high as 200 A.

TRIGGERS FOR THYRISTORS    The most common form of power control using SCRs (or triacs) is phase control of alternating current. Power control is achieved by controlling the timing of the gate trigger pulse. If the trigger pulse is applied at the beginning of the positive-going anode cycle, the thyristor is turned on for virtually the entire positive half of the cycle, allowing the maximum load current to flow. As the trigger pulse is delayed, correspondingly less of the positive-going anode voltage waveform is applied to the load. Figure 6-28 explains visually how this phase control of alternating current is achieved.

There are some instances when a thyristor can be adequately triggered by the slowly rising voltage or current from the 60-Hz line without a separate trigger device, e.g., the simple motor-speed–control circuit in Fig. 6-30. Here, the setting of $R_3$ determines the point of the applied-voltage cycle at which the gate voltage will be high enough to trigger the SCR. The circuit offers control over almost the entire half of the positive portion of the applied voltage. Much better control can be achieved, however, by employing a separate trigger device to provide reliable turn-on pulses timed to trigger the SCR at any desired point during the cycle.

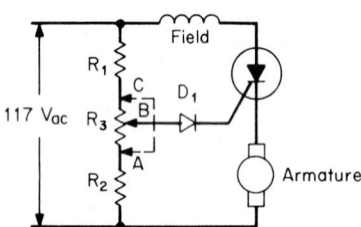

**Fig. 6-30** Half-wave motor-speed–control circuit. The thyristor is triggered directly from the 60-Hz line.

The most common trigger circuit is a relaxation oscillator (Fig. 6-31) using one of a variety of semiconductor switching devices. As shown, when the voltage on the capacitor rises to a point where the electronic switch turns on, the discharge of the capacitor produces a trigger pulse to the load. The charge time of the capacitor can be varied by the setting of $R_T$, so that pulse spacing can be varied at will.

A number of trigger devices serve the purpose of the electronic switch. Among them are unijunction transistors (UJTs), three- and four-layer diodes, and sidacs.

*Unijunction transistors.* The UJT is a three-terminal device having only a single semiconductor junction. Its basic structure and equivalent circuit are shown in Fig. 6-32*a* and *b*.

In the physical structure, when a positive voltage is applied between base 2 ($B_2$) and base 1 ($B_1$) and the emitter is shorted to $B_1$, the current through the device sets up an internal voltage drop. The voltage at the *p-n* junction, being slightly positive, reverse-

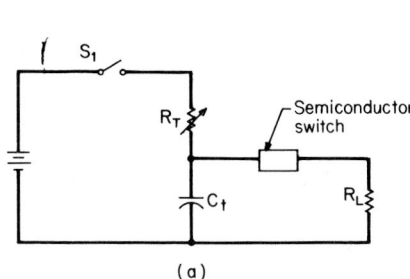

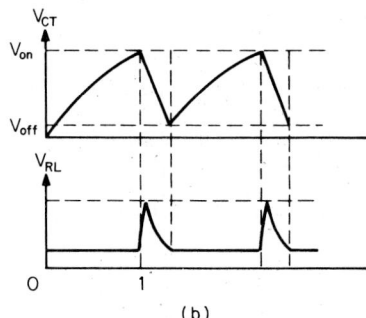

(a)

(b)

**Fig. 6-31**  Basic relaxation-oscillator circuit and associated waveforms.

biases the diode junction, and there is a small reverse-bias (leakage) current through the diode. The diode represents a high resistance in parallel with $R_{b1}$ and has no effect on the internal voltage distribution.

If a positive-going voltage is applied between the emitter and $B_1$, a point will be reached where the voltage overcomes the internal reverse-bias voltage on the $p$-$n$ junction and the junction becomes forward-biased, representing a very small resistance in parallel with $R_{b1}$. At that point, the voltage at point $A$ drops to that of a forward-biased junction.

The emitter $I$-$V$ characteristic curve of a UJT is shown in Fig. 6-33. To an external circuit, it appears that the UJT has a negative resistance to applied emitter voltage between $V_P$, the diode turn-on point (peak-point voltage), and $V_V$, the point where the diode becomes fully conductive and its voltage drops to its forward-biased value (valley point).

A relaxation oscillator utilizing a UJT (Fig. 6-34) provides an output across $R_L$ after capacitor $C_t$ has charged up through $R_t$ to reach the peak-point voltage of the UJT. Subsequently, $C_t$ discharges through $R_L$ and the UJT to generate a positive pulse across $R_L$. The frequency of the pulsed output can be varied by varying the resistance of $R_t$, which permits triggering the thyristor load at any desired point of the positive-going anode cycle.

Unijunction transistors are highly stable devices for general-purpose trigger applications and as pulse generators and timing circuits at frequencies ranging from 1 Hz to 1 MHz.

A number of other trigger devices are available. All take advantage of the phenomenon of negative-resistance characteristics to produce switching action. The $V$-$I$ characteristic curves of some of the more popular devices are illustrated in Fig. 6-35.

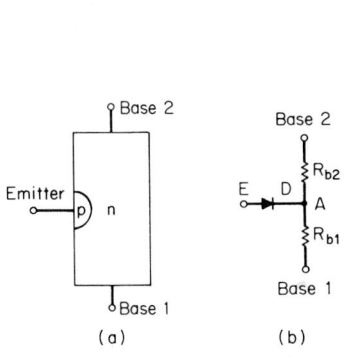

(a)                    (b)

**Fig. 6-32**  A unijunction transistor: basic construction (a) and equivalent electrical circuit (b).

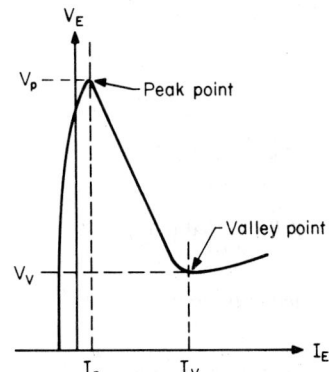

**Fig. 6-33**  $I$-$V$ characteristics of a unijunction-transistor input (emitter-$B_1$) circuit.

*Programmable unijunction transistors (PUT).* Programmable unijunction transistors are similar to UJTs except that their basic characteristics are programmable (adjustable) by means of external voltage dividers. This ability stabilizes circuit performance for variations in device parameters. The general operating frequency range is from 0.01 Hz to 10 kHz, making PUTs suitable for long-duration timer circuits.

*Bilateral triggers: DIACs.* Specifically designed as low-cost triggers in line-operated triac control circuits such as light dimmers, motor controls, and temperature controls, these devices exhibit symmetrical bidirectional characteristics with defined switching voltage and current points. The negative-resistance region extends over the full range of operating current, so that the concept of holding current is not applicable. Switching voltages on the order of 35 V or greater are achievable.

*Silicon bidirectional switches (SBS).* These devices are actually integrated circuits as opposed to the four layer discrete structures of other thyristors. They offer greater accuracy than conventional triggers, the difference between positive and negative breakover voltage being less than 0.5 V. A third lead, designated the gate, permits variations in characteristics such as breakover voltage. Switching voltages up to 10 V are achievable.

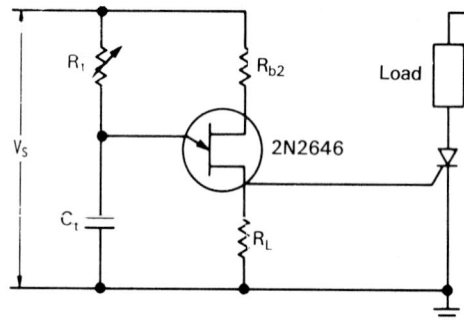

**Fig. 6-34** Basic relaxation-oscillator circuit using a UJT trigger device.

*Sidacs.* These components extend trigger capabilities to significantly higher voltages and currents than are achievable with other devices. With voltage ratings up to 300 V and a steady-state rms current rating of 1 A, they can replace triacs in numerous low-power applications.

**11. Optoelectronics devices.** The field of optoelectronics deals mainly with the phenomena of converting light into electric current and, conversely, turning electric current into light. Phototubes, followed by solid-state cadmium sulfide photocells, have been early and successful examples of light-to-current conversion, while tungsten-filament and gas-filled bulbs have been serving current-to-light conversion for many years. Their applications, however, have been restricted to direct current and low frequencies. Semiconductor light sources (LEDs and lasers) and detectors have brought about significant changes in optoelectronic applications.

Light sources basically consist of injection laser diodes (ILDs) and light-emitting diodes (LEDs). Both are semiconductor chips made of gallium arsenide (GaAs) or related III-V compounds.

Detectors consist primarily of silicon *p-n* junctions that perform the conversion function, but these may be combined (on the same chip) with other semiconductor devices such as transistors or integrated circuits to improve sensitivity and resulting output.

LIGHT SOURCES One basic characteristic of semiconductor *p-n* junctions is their sensitivity to light. Specifically, in the cutoff region the leakage current of such a diode increases with light intensity, while in the avalanche breakdown region it actually emits light. A diode therefore can be operated both as a light detector and as a light source.

The important characteristics of light sources are *spectral width,* the portion of the spectrum covered by the emitted light; *peak wavelength,* the wavelength at which maximum output is achieved; *emission pattern,* the "spread" of the light beam as it leaves the emission point; *power output,* the amount of light emitted by the device at its peak wavelength;

**Basic thryristors**

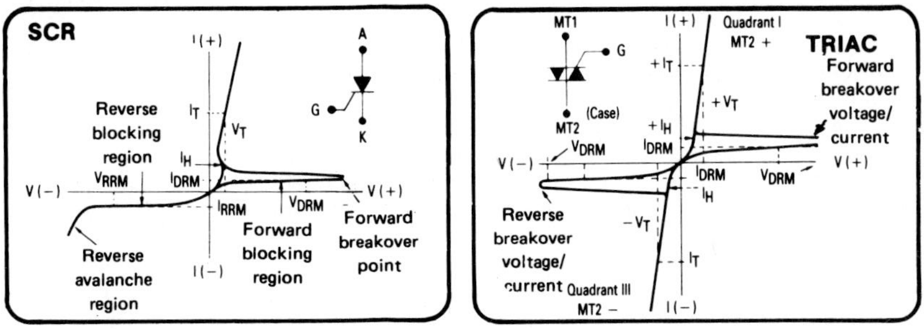

**Trigger devices**

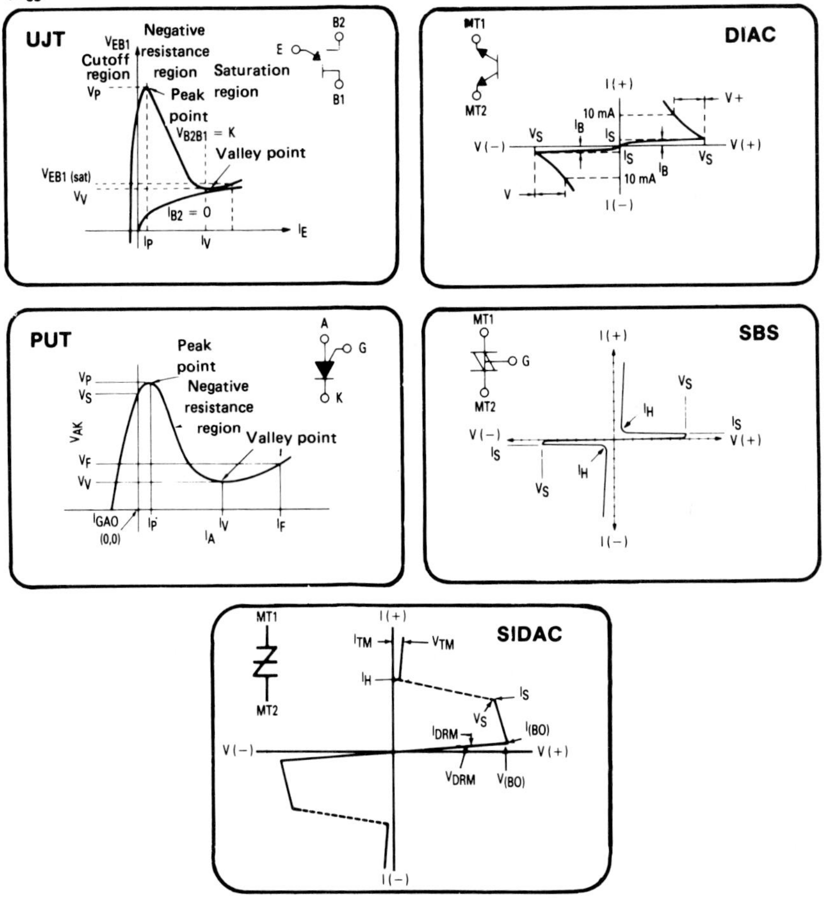

**Fig. 6-35** Voltage-current characteristics of various popular thyristor devices.

*speed,* usually determined by the time it takes for the output to rise from 10 to 90 percent of peak power in response to an applied signal pulse; and *reliability and lifetime,* the anticipated mean time between failures of a component.

In electronic applications, the ILD is usually considered for long-distance communications in the fiber-optic transmission of CATV and telephone signals. It has a very narrow optical beam (Fig. 6-36) and can launch a greater amount of power into small-diameter fiber cores than an LED. And, because of its significantly higher speed, it has greater frequency-modulation rates. The LED is normally preferred for short-distance communications and for other electronic functions such as optocoupling (isolation) and light-control applications. This preference is due to its substantially lower cost, its spectral match with a wide variety of detectors, its comparatively long life, and its general compatibility with most short-distance communications requirements.

The spectral response of an LED depends on the material used in its manufacture. Gallium arsenide is used primarily for LEDs with outputs in the infrared region of the spec-

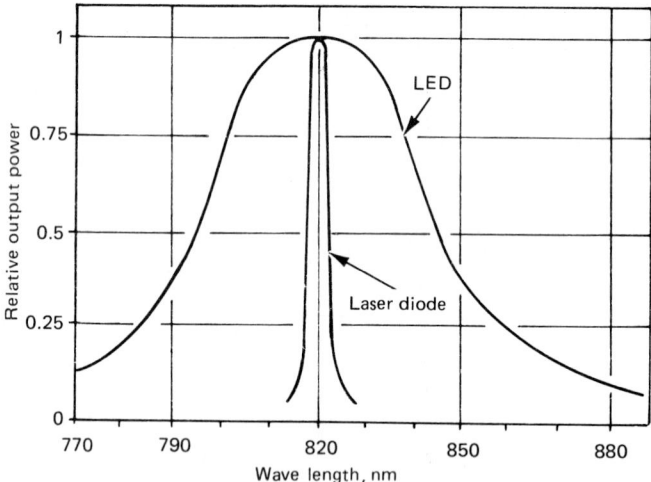

**Fig. 6-36**    Typical spectral width of laser diodes and LEDs.

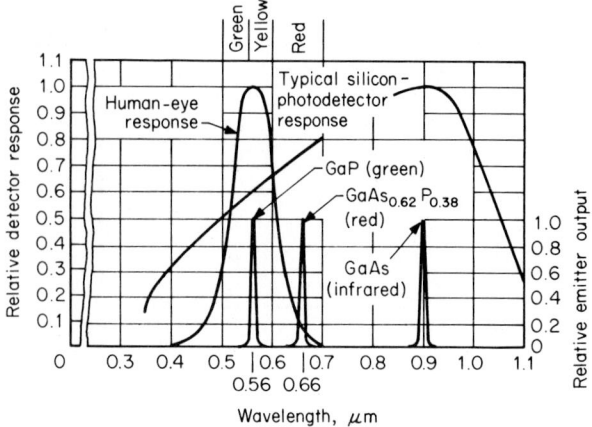

**Fig. 6-37**    Spectral output of various LED materials and spectral response of two common detectors, the human eye and silicon photodetectors.

trum between 900 and 1000 nm. Gallium arsenide phosphide produces a visible red light at about 660 nm, and gallium phosphide yields green at about 560 nm.

Visible-light diodes are widely used in panel lights, circuit-condition indicators, light modulators, displays, and the like. Infrared emitters are employed in fiber-optics communications, in card and tape readers, for shaft and position encoders, and in other applications requiring the use of photodetectors in place of the human eye because, as shown in Fig. 6-37, the detector response is much greater to infrared than to the visible-light spectrum.

Packaging has a substantial effect on the characteristics of an LED. It affects not only the power output but the spatial response as well. Figure 6-38 shows variation of output power and light intensity to be linear functions of current for typical LEDs. One limiting

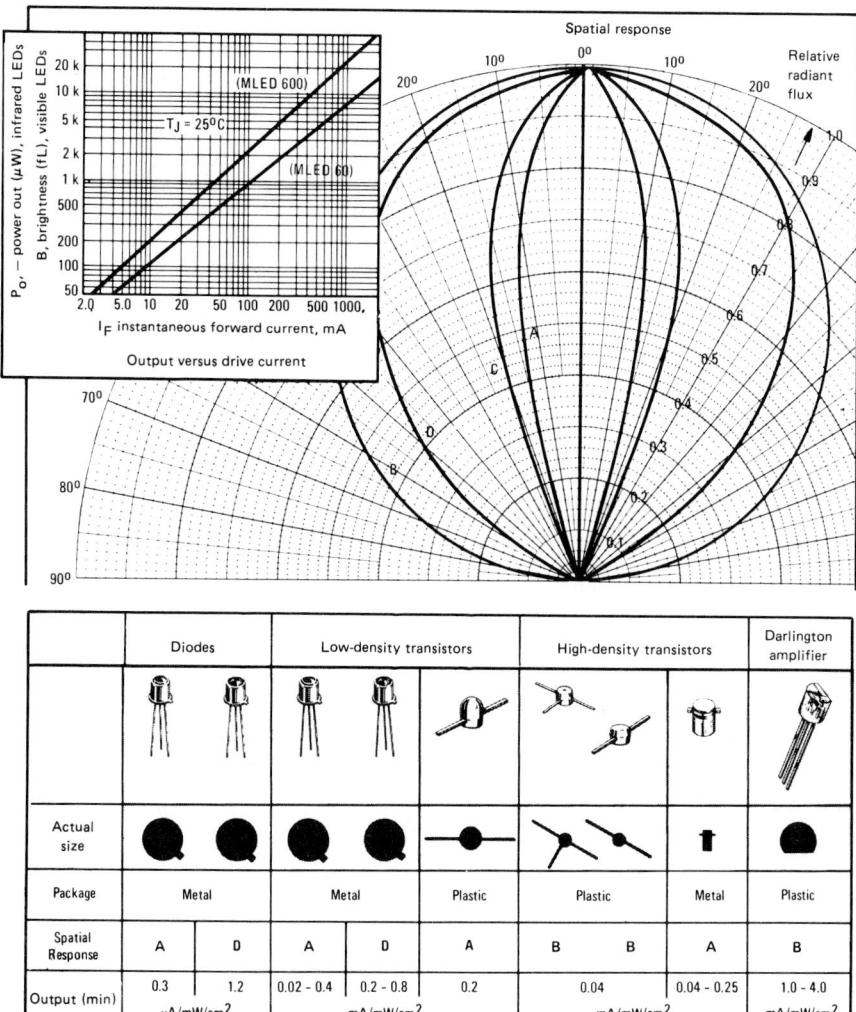

| | Diodes | | Low-density transistors | | | High-density transistors | | | Darlington amplifier |
|---|---|---|---|---|---|---|---|---|---|
| Package | Metal | | Metal | | Plastic | Plastic | | Metal | Plastic |
| Spatial Response | A | D | A | D | A | B | B | A | B |
| Output (min) | 0.3 | 1.2 | 0.02 - 0.4 | 0.2 - 0.8 | 0.2 | 0.04 | | 0.04 - 0.25 | 1.0 - 4.0 |
| | μA/mW/cm² | | mA/mW/cm² | | | mA/mW/cm² | | | mA/mW/cm² |

**Fig. 6-38** Typical characteristics of commonly available LEDs. Graph, upper left, shows typical light output as a function of drive current. Spatial-response curves are functions of package and lens designs.

value is the current limit set by the package in which the die is housed. Spatial response is a function of the placement of the die within the package (with respect to the lens) and of the shape of the lens.

Spatial distribution is also a function of the basic chip design. A simple $p$-$n$ junction yields a lambertian emission pattern which radiates in all directions. While this is desirable for some applications, it is undesirable in others. A number of techniques have been developed to control the emission pattern of the junction. Some of the prevalent chip structures and packaging methods are illustrated in Fig. 6-39.

PHOTODETECTORS   All semiconductor materials are light-sensitive. An increase of light within a specific range of frequencies, like an increase in temperature, imparts energy to the crystal lattice, causing some of the electrons to break their covalent bonds and become free. In a homogeneous semiconductor material or in a forward-biased $p$-$n$ junction, the number of light-generated carriers is relatively small compared with the normally available free carriers and add little to the overall current flow. In a reverse-biased junction, however, the increase in current due to light-induced carriers becomes significant.

PHOTODIODES   If a near-intrinsic layer of silicon is interspersed between oppositely doped layers, a *PIN diode* is formed. The *I* (intrinsic) layer, being nearly devoid of free carriers, adds significantly to the width of the depletion region, making the structure far more light-sensitive than the ordinary $p$-$n$ junction.

If the reverse bias on a $p$-$n$ junction is increased to a point near the reverse breakdown point, the electric field within the depletion region will be high enough to give any light-liberated carrier enough energy to liberate additional carriers through collision with other atoms in the lattice. This avalanche effect actually multiplies the current generated by the impinging light and greatly increases the sensitivity of the diode.

*Phototransistors.* When one of the junctions of a transistor, e.g., the normally reverse-biased collector-base junction, is exposed to light, a phototransistor is created. Such a device adds gain to a photodiode and therefore increases sensitivity considerably. It pays for this, however, with a decrease in response time. Whereas the response time of diodes is measured in nanoseconds, that of phototransistors is specified in microseconds. Even greater gain and correspondingly slower response time are achieved with photodarlington devices.

Virtually any semiconductor device can be made photosensitive. Thus, manufacturers are now providing phototriacs and photo SCRs in order to reduce the component count in light-sensitive equipment requiring such devices.

With the advent of integrated circuits, optodetectors are being combined with complete preamplifiers to provide an amplified low-impedance output that is far less noise-sensitive than that of a discrete detecting device.

OPTOCOUPLERS   An excellent example of optoutilization to replace conventional electronic components is the optical coupler. Consisting of an infrared-emitting diode coupled to a phototransistor in a single package, the device advantageously replaces such components as interstage transformers and relays as well as coupling and feedback networks.

The diagrams in Fig. 6-40 show such a unit being used as a linear-signal coupler and as a pulse coupler. In the former mode a constant current supplied to the emitter biases this diode, and since the output (infrared) is directly proportional to the diode current, any increase or decrease of diode bias current resulting from an applied modulation input causes corresponding variations in light output. These are coupled to the detector, which provides an equivalent linear current output. Devices with current transfer ratios $(I_F/I_C)$ ranging from 2 to 1000 percent (depending on detector gain) are currently available.

As pulse couplers, the devices are equally interesting. The principal operating difference is that in the switching mode they require no bias current. The detector is either off (with a dark current of typically less than 20 nA) or on (with a maximum continuous forward current of around 50 mA).

Used as an electronic relay, the coupler is fast (much faster than a mechanical relay), and it has no contacts to bounce, pit, or corrode. It is small and insensitive to vibration, and, in contrast to other forms of electronic relays, its output is completely isolated from its input.

Couplers of this type could be produced with an almost infinite variety of gain, sensitivity, and output current. Currently they are available with transistor output, Darlington

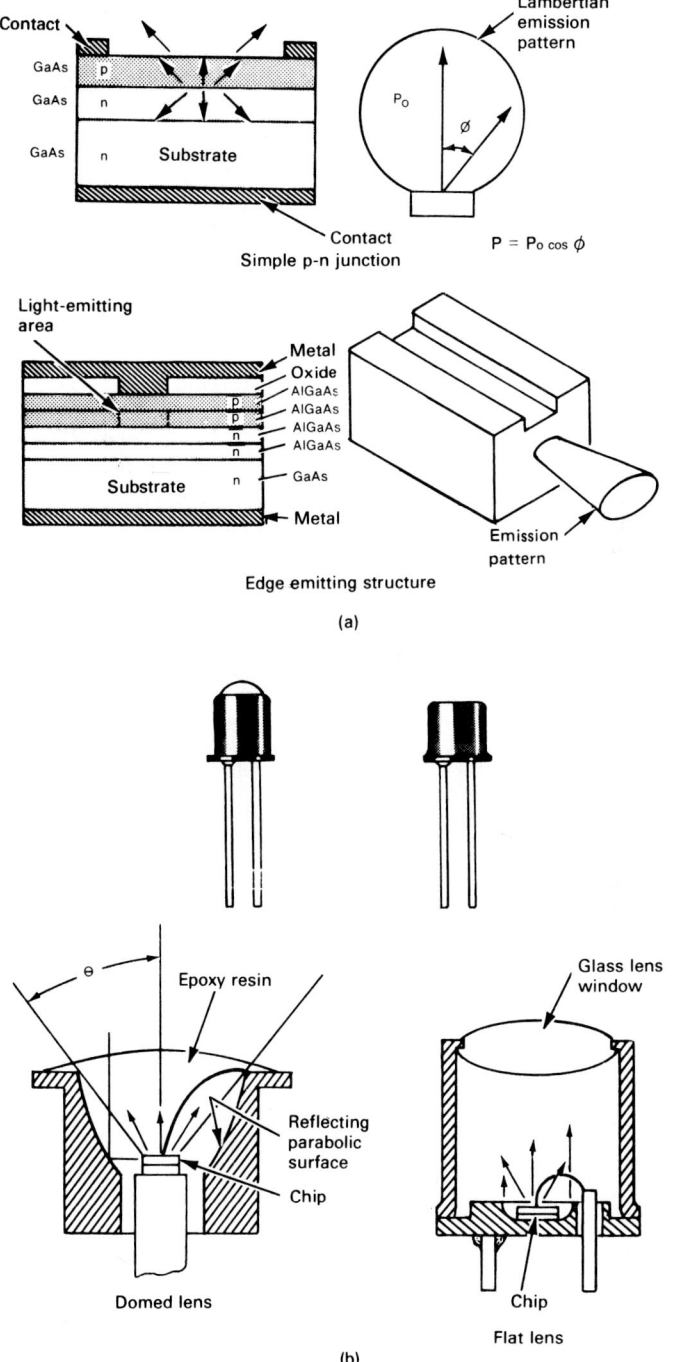

**Fig. 6-39** Chip design (*a*) and packaging methods (*b*) affect the radiation pattern of LEDs.

output, SCR output, triac driver output, and Schmitt trigger output. They are even being combined with on-chip amplifiers. This is a case of matching design to a required application, and the couplers clearly lend themselves ideally to custom fabrication when end-use quantities are high enough to permit economical production.

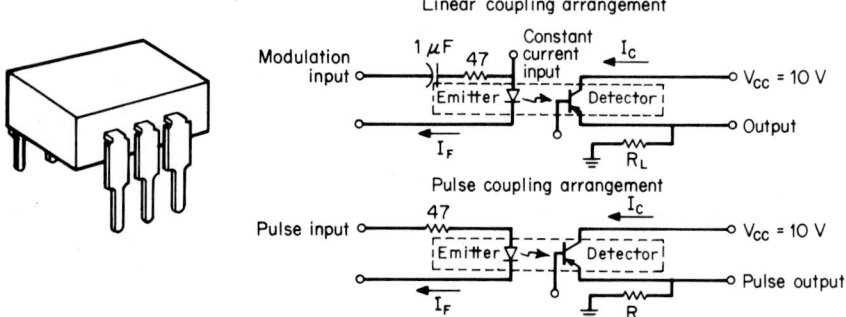

**Fig. 6-40** An optoelectronic coupler consists of an LED and a photodetector with similar spectral-response characteristics coupled to each other in a single package. It permits efficient transfer of linear or pulse signals between circuits where impedance matching is difficult.

**12. Varactor diodes.** Normally, a *p-n* junction (diode) is operated in the forward conduction region, as a rectifier, or in the reverse breakdown (avalanche) region, as a zener. There is, however, a third region, the region between forward conduction and reverse breakdown (Fig. 6-41), in which the junction is nonconductive and acts as a capacitor. This junction capacitance is an undesirable parasitic in most applications but becomes the mechanism whereby the junction can act as a variable-capacitance (tuning) diode or frequency multiplier.

An unbiased semiconductor junction simulates a slightly charged capacitor, with the depletion region representing the dielectric and the *n*- and *p*-type regions adjacent to the junction representing the two conductive plates. If an external voltage is connected across the diode so as to reinforce the contact potential (reverse bias), the depletion-layer width increases, resulting in a decrease in capacitance. Conversely, if an external forward bias is applied, the depletion region narrows and capacitance increases. However, if the

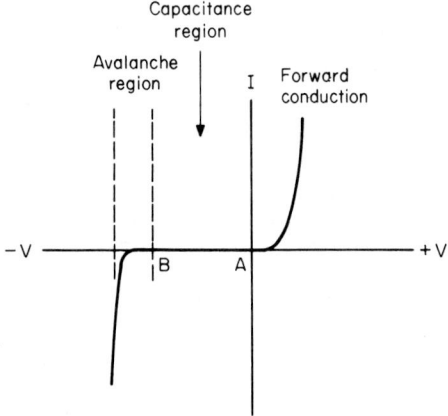

**Fig. 6-41** The three regions of a junction *I-V* characteristic curve are used for various applications: the forward conduction region for rectification, the capacitance region for voltage-variable capacitance diodes, and the avalanche region for zener diodes.

forward voltage is made large enough to overcome the contact potential, forward conduction occurs and the capacitance effect is destroyed. By applying a variable dc voltage of a magnitude between points $A$ and $B$ of Fig. 6-41, the capacitance of the junction can be varied to cover the tuning requirements of radio and television receivers.

Three parameters are of particular importance with varactor diodes: nominal capacitance, capacitance ratio, and $Q$. The nominal capacitance $C_t$ is normally specified on data sheets at a particular value of reverse-bias voltage. This serves as a reference point for comparing the capacitance values of various varactor diodes. Capacitance ratio is the ratio $C_{max}/C_{min}$ for a specified reverse-bias voltage range. $Q$ is the figure of merit that defines the quality of the device as a capacitor. As a reference, mechanical tuning capacitors that are being largely replaced by varactor tuning diodes often have $Q$'s on the order of 1000 or greater. Tuning-diode $Q$'s are generally considerably lower but still are adequate to achieve the desired performance.

The capacitance change over a specific voltage range for a commercial line of tuning diodes is shown in Fig. 6-42.

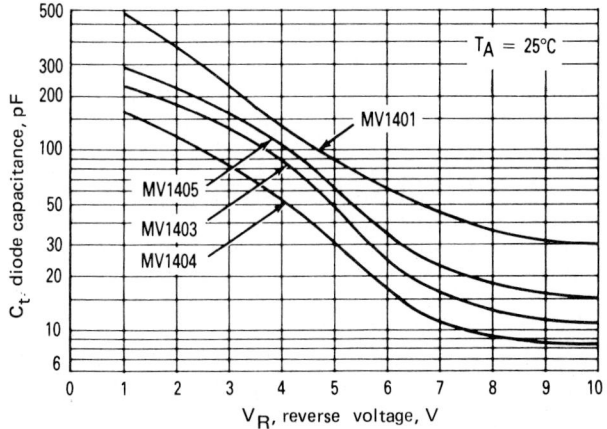

**Fig. 6-42** Typical capacitance versus reverse-voltage variations of one commercial line of hyperabrupt junction diodes.

Varactors can be applied to most sections of a receiver or transmitter where variable capacitors are required. Usable frequencies range to several thousand megahertz, and nominal capacitance values range from about 6 to 500 pF or greater. Capacitance ratios run from approximately 15, for AM broadcast-band tuning diodes, to as low as 2.5 for diodes used for general tuning and control purposes.

Varactor diodes are useful for a variety of radio-frequency applications, including electronic tuning, harmonic generation, and parametric amplification.

**13. Transducers (sensors).**   Transducers are representative of a class of device that converts one form of energy into another. Sensors are often defined as the basic element within a transducer in which the actual conversion takes place. This may be followed by additional circuitry that provides signal conditioning and amplification to accomplish an end result.

Semiconductor devices, owing to their low cost and high reliability, are being adopted in increasing numbers to accomplish the basic sensing functions. Optoelectronic devices in the form of light emitters and detectors (Sec. 11) are examples of semiconductor applications for sensing purposes. Other categories include temperature sensors and pressure sensors.

PRESSURE SENSORS   One form of pressure sensor employs the piezoresistive effect in silicon to convert a change in pressure into a change in electric current. A commercial structure of this type is illustrated in Fig. 6-43. Here a cavity, or chamber, is etched into one side of a bar of silicon, leaving a thin, flexible bridge of silicon that acts as a diaphragm.

A resistive element is diffused into this diaphragm at a point where it is most susceptible to stress when the diaphragm is flexed. A pair of contacts is placed at a point of the resistor to measure the transverse voltage at that point when a current flows through the resistor.

The resistance value is subject to change as the resistor is stressed as a result of pressure being applied to the diaphragm (piezoresistive effect). This causes a change in voltage at the points of contact that is available for measurement or control purposes by means of the package pins.

The device can measure absolute pressure with respect to a vacuum or differential pressure with respect either to atmospheric pressure or to two different pressures applied to opposite sides of the diaphragm. For absolute-pressure measurements, the chamber is sealed off in an evacuated atmosphere to create a vacuum on the underside of the diaphragm. This flexes the diaphragm and generates a reference offset voltage equal to 14.5 psi (1 atm). When an external pressure is applied to the opposite side of the diaphragm, the output voltage varies linearly above or below the reference point, depending on the polarity of the external pressure, whether positive or negative (Fig. 6-44).

For differential-pressure measurements, the chamber is left unsealed so that an external positive (downward) pressure applied to the "pressure" side of the diaphragm acts against existing atmospheric pressure at the underside to create a differential voltage. Alternatively, the difference between two external pressures can be measured by applying these simultaneously to opposite sides of the diaphragm, through pressure ports associated with the package (Fig. 6-45).

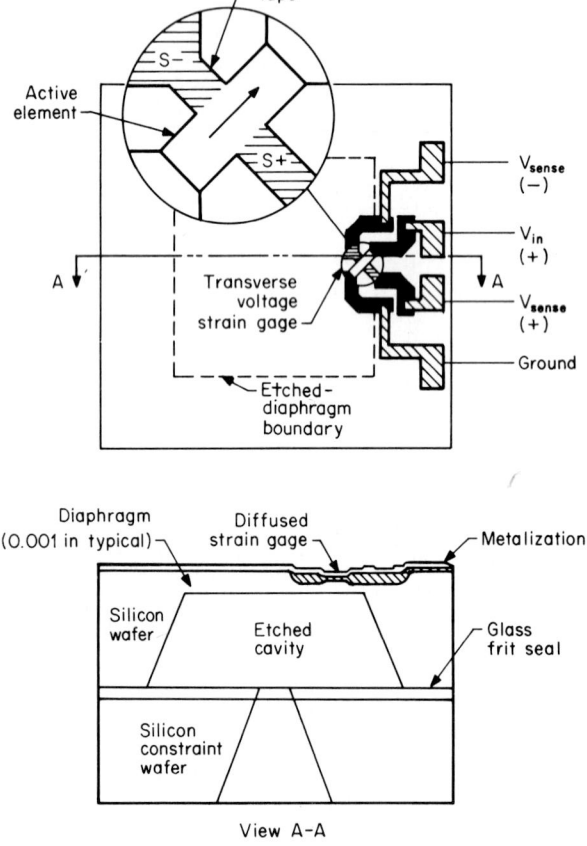

**Fig. 6-43**   Top view and cross section of a commercial pressure-sensor chip. [Motorola Inc.]

While piezoresistive techniques in silicon provide a very accurate and linear output, they are temperature-sensitive and generate an initial offset voltage. Both effects can readily be compensated for with external circuitry, but integrated-circuit technology has been employed to put temperature compensating and calibrating directly on the chip during the fabrication process. The compensating thin-film resistor network is laser-trimmed during the computer-controlled manufacturing process to provide good temperature stability and to eliminate the offset (Fig. 6-46). The more advanced devices feature on-chip amplification to increase the output-voltage span from millivolts to volts.

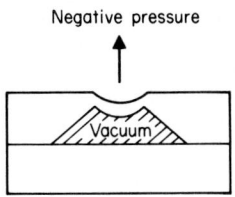

Negative pressure

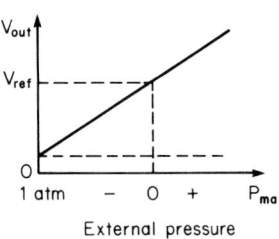

External pressure

**Fig.     6-44**  Absolute-pressure measurement.

## INTEGRATED CIRCUITS

The evolution of integrated circuits has initiated tremendous innovations in the conception, design, and manufacture of electronic equipment. The individual components such as resistors, capacitors, and inductors, which once were the building blocks of all electronic equipment, are being relegated to relatively minor peripheral functions. Circuit designs are being implemented without the benefit of capacitors and inductors, which are difficult to combine with active components on a practically sized silicon chip. Even the resistor is losing its battle against active devices, such as transistors, which can often serve the resistive function more compatibly and in much less space. Digital techniques are replacing analog functions in communications circuitry. The computer is replacing the time-honored "breadboard" for developing, testing, and evaluating new circuit and system designs; and in the factory the microscope is replacing the soldering iron as the basic manufacturing tool in many instances.

But there still remain a large number of applications for which ICs are not yet adaptable. Integrated circuits are not able to produce high power output; they are uncomfortable in VHF-UHF applications, and they are not particularly well suited to very simple electronic equipment that might have special requirements outside the realm of standard

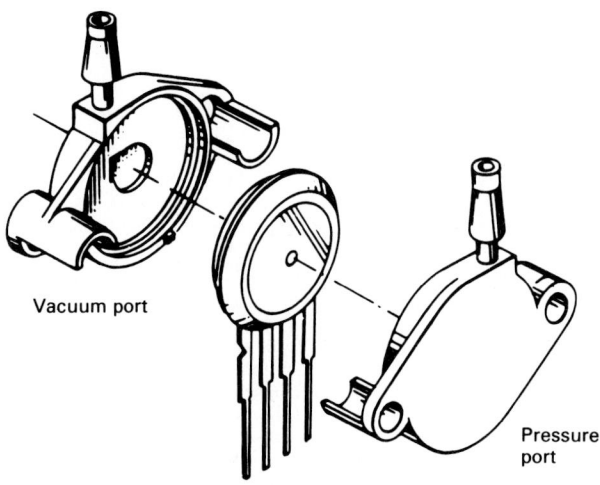

Vacuum port

Pressure port

**Fig. 6-45**  Packaged pressure sensor with ports for the application of external pressure.

off-the-shelf ICs. Indeed, while ICs are expanding into ever larger subsystem functions, they will continue to need discrete-component supplements for some time to come.

All integrated circuits have a common goal of reducing equipment size and cost and improving reliability and performance. There are many different paths toward achieving these goals. The more important of these are listed in Table 1.

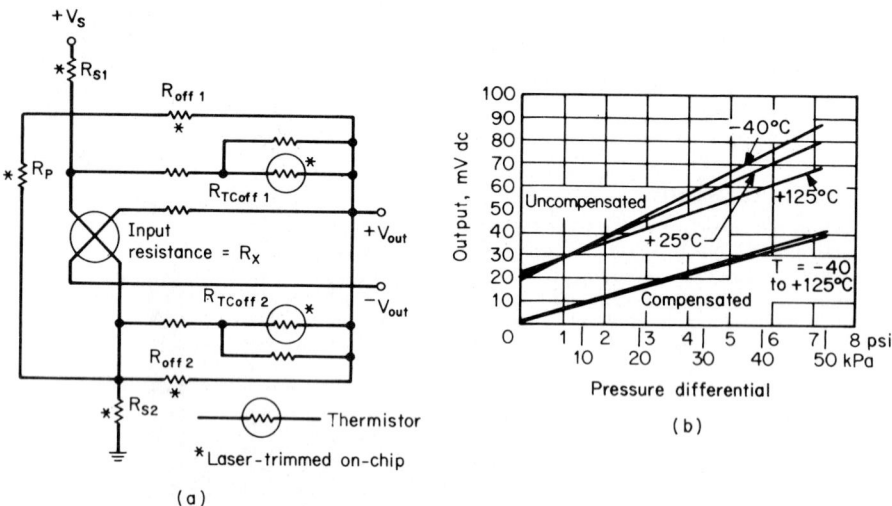

(a)

(b)

**Fig. 6-46** Schematic diagram of a fully compensated monolithic pressure sensor (*a*) and output characteristics of compensated and uncompensated devices (*b*).

**Table 1  Integrated-Circuit Classifications**

| Technology | Applications | Complexity | Selection |
|---|---|---|---|
| Monolithic | Digital | SSI | Standard |
| Bipolar | Linear | MSI | Semicustom |
| MOS | | LSI | Custom |
| Hybrid | | VLSI | |
| Thin film | | | |
| Thick film | | | |

### 14. Integrated-circuit technology

MONOLITHIC CIRCUITS  Monolithic circuits are those in which all associated components are fabricated and interconnected on or within a single chip of silicon. Processes employed are basically separated into *bipolar,* using the two-junction transistor as its basic element, and *MOS,* using the unipolar field-effect structure as its building block.

Bipolar processing generally yields greater speed and permits the design of higher-voltage and higher-power circuits. MOS processing, including the widely implemented complementary MOS (CMOS) structure, offers greater component density and much lower internal power dissipation. Therefore, bipolar circuits are widely used in linear circuits, in small-scale-integration (SSI) and medium-scale-integration (MSI) logic circuits, where they are so well entrenched that they cannot be economically replaced, and in circuits where state-of-the-art speed is required.

MOS circuits, particularly CMOS circuits, are utilized primarily in digital LSI and VLSI designs, where their small geometries and extremely low power requirements permit implementation of circuit complexities that cannot readily be achieved by other means.

Progress is being made in combining bipolar and MOS technologies, each to its own advantage, within a single chip, and some such circuits are being marketed. Generally, however, the two technologies are being used independently.

HYBRID CIRCUITS    Hybrid circuits utilize a combination of monolithic integrated circuits and discrete components attached to and interconnected by a thin- or thick-film conductive pattern deposited on an insulating glass or ceramic substrate (Fig. 6-47). They are used basically for size and weight reduction of complex circuitry whose component complement does not permit monolithic fabrication. High-power circuits and radio-frequency circuits which require capacitors and inductors are typical candidates for thin-film construction techniques.

**Fig. 6-47**    Hybrid UHF power amplifier module capable of 2.2-W output at frequencies up to 900 MHz. It is designed for portable cellular radio applications.

The advantages of thin-film circuits over conventionally wired circuits using printed-circuit boards lie in the fact that most of the passive elements of a complete circuit, such as resistors and capacitors, can be fabricated directly and simultaneously on the insulating substrate. Similarly, all interconnections can be made at one time, often in conjunction with one of the deposition steps of the passive elements. Then the active components (or monolithic ICs) are added separately and interconnected with the previously deposited pattern on the substrate. To save space, active elements are often utilized in chip form and wire-bond techniques are employed for subsequent interconnections. This practice, however, introduces manufacturing and testing complications. The recent proliferation of surface-mounted packages for both discrete active components and ICs (Fig. 6-48) greatly simplifies hybrid-circuit manufacturing with little penalty on space utilization.

DIGITAL CIRCUITS    Digital circuits are those dealing exclusively with the technology of *ones* and *zeros*. Their active elements, predominantly transistors, are used as switches which are either on or off. These circuits are used as digital logic elements in computer and control applications. They are composed primarily of basic building blocks, called gates, which are interconnected in a variety of ways to form more complex functions such as flip-flops, counters, and a myriad of other circuit entities. In all instances, however, the basic circuit is driven by a signal whose voltage level is either high or low (one or zero), causing the circuit to be turned on or off and the output voltage to be either high or low, or true or false, depending on the terminology associated with a particular application. Generally, digital circuits, no matter how complex, are operated at low voltages, normally ranging between 3 and 10 V.

LINEAR CIRCUITS    Linear circuits are not as clear-cut in definition. Generally, they encompass analog functions, such as amplifiers, whose output is *proportional* to the *level* of the input signal. Also included in the linear category are analog-to-digital (A/D) and digital-to-analog (D/A) converters, comparators, voltage regulators, and even interface circuits whose functions are more digital than analog in nature.

In recent years, the proliferation of circuits and the increasing complexity which permits the combination of digital and linear functions in a single chip have made these classifications virtually obsolete. The present trend is toward definitions that classify circuits in terms of their primary intended applications. Categories such as logic circuits, microcomputer circuits, voice-data circuits, analog circuits, power-conversion circuits, etc., are becoming more commonplace. While such definitions also have limitations and while there is as yet no standardization of such classifications, they serve as the basic subsections for the following discussions. These discussions, moreover, will be limited primarily to standard circuits, available off the shelf from IC manufacturers and distributors as the basic building blocks for system designs. A brief discussion of the growing trend toward semicustom and custom approaches is included.

**Fig. 6-48** Tiny surface-mounted packages now replace unencapsulated chips in hybrid-circuit designs because of their simpler installation and testing.

**15. Integrated-circuit complexity.** Integrated circuits in general and logic circuits in particular are classified within four basic categories: small-scale integration (SSI), medium-scale integration (MSI), large-scale integration (LSI), and very-large-scale integration (VLSI). Initially, each of these terms encompassed a degree of chip complexity involving a specified number of equivalent gate circuits (gate equivalents). The rapid progress in IC technology has rendered this practice virtually obsolete, since the VLSI circuit of only a decade ago would hardly be classified as more than MSI today. Nevertheless, the terms still apply on a relative basis, although the judgment of where one classification ends and the other begins is largely a matter of personal interpretation.

**16. Integrated-circuit selection**

Standard versus Custom Circuits    From the very beginning of the technology, the development and production of state-of-the-art ICs has been a time-consuming and expensive endeavor. For IC manufacturers, each new circuit represented a gamble on whether the device would sell in large enough quantities to permit recovery of the development cost and, subsequently, to produce a profit. In the earliest days, the odds were good because there were enough universally required circuits to assure large-scale acceptance. Gates and flip-flops for digital equipment, operational amplifiers, and voltage regulators in the linear field were certain to find ready markets. IC manufacturers found it easy to identify such circuits and rapidly built up inventories of a large variety of "standard" off-the-shelf devices that were sold to all comers.

Even then, however, some equipment manufacturers opted to have special proprietary circuits designed on a custom basis if their volume requirements were large enough to amortize development costs. In many instances, these custom circuits, after an agreed-upon time, became standard circuits offered on the open market.

As the technology advanced, it became more difficult to identify increasingly complex circuits that would find a large general market. Accordingly, the development of standard circuits has become a greater gamble for the manufacturer. And as the demand for ever-increasing device complexity has increased, the number of original-equipment manufacturers (OEMs) able to afford the cost of designing such proprietary VSLI circuits is dwindling. As a result, an intermediary technology has emerged that permits the development of very complex semicustom circuits, utilizing a library of predefined basic circuits.

Called applications-specific integrated circuits (ASICs), these devices sacrifice some custom-design flexibility, by being constrained to the basic-circuit library, but are much less costly and less time-consuming to produce.

ASIC technology is applicable largely to circuits so complex that they can be implemented only through computer-aided–design (CAD) techniques. Currently, two types of ASIC techniques have reached production status: gate arrays and standard cells.

GATE ARRAYS   The most complex digital logic circuits can be designed through the interconnection of basic gate circuits in various unique arrangements and sequences. IC manufacturers take advantage of this fact by producing complete wafers full of identical basic gate circuits. Then they provide a computer-stored program for interconnecting the gates into circuits of increasing complexity. During the design cycle, the system designer formulates a complex design by using the library of subcircuits issued by the IC manufacturer. The computer-stored program is then utilized to develop an interconnect system that converts the network of gates on the wafer into an exact duplication of the complex design. Use of prefabricated gate arrays and a proven CAD program saves months of time compared with custom design of VLSI circuits.

MACROCELL ARRAYS   Macrocell arrays are similar to gate arrays, except that the prediffused wafers do not contain prewired basic gates but rather a pattern of islands each of which contains a number of prediffused transistors and resistors (Fig. 6-49). The IC manufacturer issues a library of predefined functions called *macros* that the computer program can produce from the discrete parts within each island and subsequently interconnect into a custom IC. In theory, the marcrocells—flip-flops, counters, registers, and a wide variety of additional digital building blocks—can be designed more efficiently from discrete components than from basic gates, yielding smaller geometries (equivalent to lower cost) and improved performance. As with gate arrays, the designer develops VLSI circuits with the macros contained in the library, and the computer develops the necessary mask patterns that convert the prediffused macrocell wafers into functional VLSI circuits.

COMMERCIAL IMPLEMENTATION   The efficiency of the gate-array–macrocell-array concepts depends upon the availability of arrays that closely match the needs of the designer. For example, a prediffused array with fewer on-chip gates (or islands) than required for the envisioned VLSI circuit would increase the number of IC packages; an array with too many on-chip gates would result in unused wafer space. IC manufacturers therefore offer a variety of array sizes (Table 2) that permit a close match between need and availability. Moreover, such arrays are available for a variety of technologies (bipolar and MOS) to offer a number of performance options.

STANDARD CELLS   Standard-cell concepts go one step further toward true custom designs than do gate or macrocell arrays. These too start with a computer-stored catalog of basic circuits (standard cells) that form the building blocks for VLSI designs. Rather than using prediffused wafers of basic cells, with each cell requiring an equal amount of chip space, a standard cell is designed to provide maximum efficiency with a layout requiring a minimum amount of space. The computer then provides the necessary mask patterns for the fabrication of the complete circuit, not just the interconnect patterns for the prediffused arrays. The use of custom-designed standard cells, compared with equivalent cells made from array components, can save a significant amount of chip space (Fig. 6-50) and produce end results that closely rival those of true custom circuits. Yet, because these end circuits are fashioned from predefined building blocks with the computer doing much of the design, development time is substantially reduced.

## INTEGRATED-CIRCUIT LOGIC FUNCTIONS

Logic circuits are implemented with either bipolar or MOS technologies. Bipolar logic circuits encompass a variety of different basic "families," each with its unique advantages and limitations. These various families are divided into two generic categories: saturated logic and nonsaturated logic. In saturated-logic designs, the transistor (acting as a switch) is driven from the cutoff condition to the saturated condition by the input signal. In nonsaturated logic, the transistor is not switched from full off to full on but swings above and below a specified reference level. Thus nonsaturated logic avoids some of the inherent transistor time delays associated with saturated-mode switching but pays for the resulting improvement in switching speed with increased power consumption.

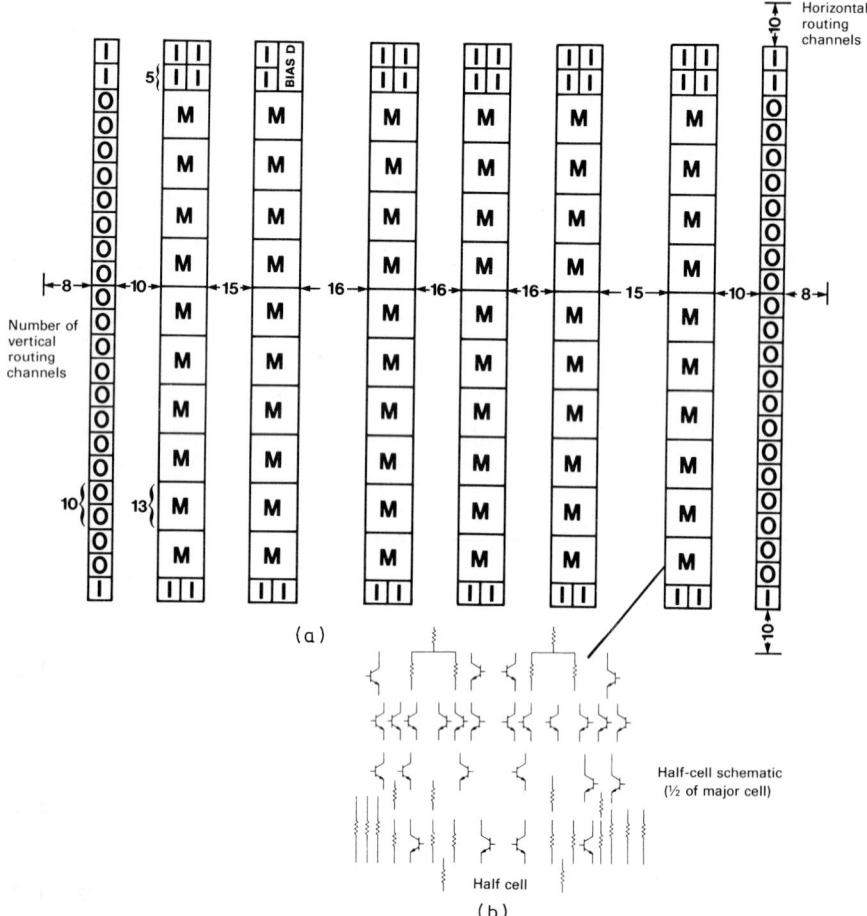

**Fig. 6-49**  A typical macrocell-array chip layout (*a*) illustrates cell placement and routing channels. Component content (*b*) represents one-half of the complement within each major cell. *M* = major cell; *I* = input cell; *O* = output cell.

Integrated logic circuits have gone through a number of iterations. The early forms, carryovers from discrete designs, were bipolar resistor-transistor-logic (RTL) and diode-transistor-logic (DTL) circuit designs, all utilizing saturated-logic forms. These have been largely replaced with a variety of transistor-transistor-logic (TTL) designs utilizing bipolar technology, with bipolar nonsaturated emitter-coupled-logic (ECL) designs where maximum speed is required, and with CMOS (complementary MOS) designs using MOS technology where low power consumption is a basic requirement. In general, SSI and MSI circuits are being implemented with bipolar and CMOS technologies, both of which offer a wide selection of standard circuits as building blocks. LSI and VLSI designs utilize CMOS technology to an ever-increasing degree.

**Table 2  Macrocell Arrays[a]**

| Technology: | ECL | | | TTL | | | ECL-TTL |
|---|---|---|---|---|---|---|---|
| Bipolar array number: | MCA 600 ECL | MCA 1200 ECL | MCA 2500 ECL | MCA 500 ALS | MCA 1300 ALS | MCA 2800 ALS | MCA 2900 ETL |
| Gate equivalent | 652 | 1192 | 2472 | 533 | 1280 | 2720 | 2958 |
| Major or primary cells | 24 | 48 | 110 | 24 | 60 | 130 | 130 |
| Macrocell components | 67 | 106 | 178 | 77 | 140 | 250 | 272 |
| I/O ports | | | | | | | |
|   Input only | 28 | 34 | 52 | 30 | 36 | ... | 30 TTL, 72 ECL |
|   Output only | ... | ... | ... | 15 | ... | ... | ... |
|   Uncommitted | 18 | 26 | 68 | 12 | 40 | 120 | 40 TTL, 48 ECL |
| Maximum gate delay, ns | 1.2 | 1.2 | 0.5 | 3.0 | 3.0 | 1.1 | 1.1 |
| Maximum toggle frequency, MHz | 160 | 160 | 400 | 80 | 80 | 125 | 125 |
| Power dissipation, W (typical) | 2.2 | 4.0 | 6.5 | 1.0 | 1.4 | 2.5 | 4 |

| Technology | 3-μm silicon-gate HCMOS | | | | 2-μm silicon-gate HCMOS | | | | |
|---|---|---|---|---|---|---|---|---|---|
| CMOS array number | HCA 6306 | HCA 6312 | HCA 6324 | HCA 6348 | HCA 6206 | HCA 6212 | HCA 6225 | HCA 6236 | HCA 6248 |
| Gate equivalent | 648 | 1200 | 2295 | 4860 | 648 | 1200 | 2430 | 3699 | 4860 |
| Major or primary cells | 216 | 400 | 765 | 1620 | 216 | 400 | 810 | 1233 | 1620 |
| Macrocell components | 104 | 104 | 104 | 104 | 104 | 104 | 104 | 104 | 104 |
| I/O ports | | | | | | | | | |
|   Input only | 2 + 1 or 0 + 1 | 17 + 1 | 58 + 1 | 53 + 1 | 2 + 1 or 0 + 1 | 17 + 1 | 8 + 1 | 0 + 1 | 53 + 1 |
|   Output only | ... | ... | ... | ... | ... | ... | ... | ... | ... |
|   Uncommitted | 35 | 42 | 56 | 54 | 35 | 42 | 80 | 95 | 54 |
| Maximum gate delay, ns | | 2.5 typical | | | | | 1.9 typical | | |
| Maximum toggle frequency, MHz | | 70 typical | | | | | 85 typical | | |
| Power dissipation, W (typical) | Dependent on the number of outputs switched simultaneously | | | | Dependent on the number of outputs switched simultaneously | | | | |

[a] These arrays run the gamut of technologies and offer a variety of cell configurations. Comparative speed-power specifications are given. [Motorola Inc.]

**17. Transistor switching characteristics.** While the performance of a logic family is judged on a number of characteristics, the most important of these is switching speed. Switching speed defines the time it takes for a transistor, or gate, to change its output from one state to another in response to a change in the input signal.

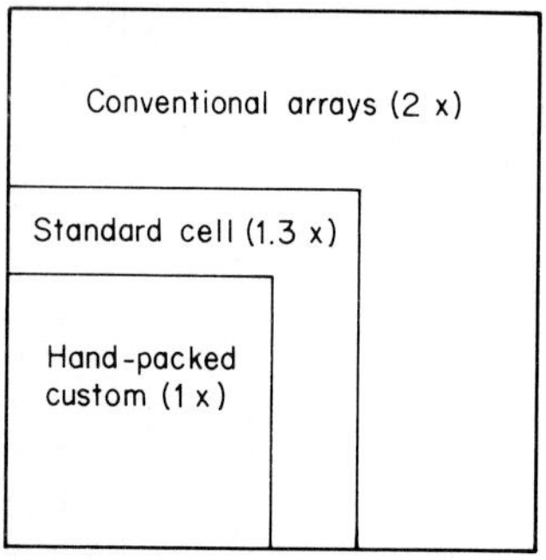

Figure 6-51 shows a typical switching transistor and its associated input and output waveforms. In a "perfect" switch, the output current would follow the input-voltage waveform exactly. The practical circuit involves a number of delays. Note that collector current does not begin to flow until time $t_1$, some time after $t_0$, where the base voltage rises to its maximum value. The difference between $t_0$ and $t_1$ is the *turn-on delay time* $(t_d)$, caused by the need to charge or discharge the transistor interelectrode capacitances ($C_{BE}$ and $C_{CB}$) before current flow can begin.

At $t_1$ collector current begins to flow but cannot rise instantly to its maximum value because of the finite time required for charge carriers to traverse the base region of the transistor between the emitter and the collector. The time required for the current to rise from 10 to 90 percent of its maximum value (from $t_1$ to $t_2$) is called *rise time* $(t_r)$.

**Fig. 6-50** Relative comparison of die size between full-custom and semicustom circuits.

When the transistor is driven into saturation, both the emitter and the collector junctions are forward-biased and dump minority carriers (electrons, in the case of an *n-p-n* transistor) into the base region. When the base voltage suddenly drops to zero, at $t_3$, these carriers must be ejected through the collector circuit. Hence, current continues to flow until all these carriers have been pulled out of the base region. The time difference between $t_3$ (the drop of the base signal to zero) and the point where collector current begins to drop at $t_4$ is called *storage time* $(t_s)$. Even then, collector current cannot drop instantly to zero because the interelectrode capacitances cannot discharge instantly. The gradual decay of collector current between $t_4$ and $t_5$ is the *fall time* $(t_f)$.

From the above, it is clear that a practical transistor switch involves both a turn-on delay time $(t_{on})$ and a turn-off delay time $(t_{off})$. Of these, $t_{off}$ is by far the greater and is significantly influenced by $t_s$, which in turn results from driving the transistor into saturation.

THE GATE FUNCTION  The basic circuit of all logic elements is the gate circuit. Virtually all building blocks associated with digital equipment, flip-flops, counters, shift registers, etc., can be made by interconnecting a series of gate circuits. Hence, the complexity of

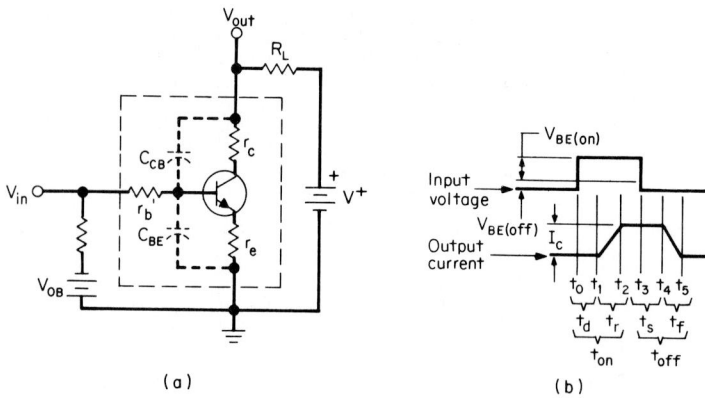

**Fig. 6-51** A typical saturated-mode switching circuit (*a*) and its input and output waveforms (*b*).

any given logic circuit is specified by the number of equivalent individual gate functions it encompasses. The electrical characteristics of a basic gate circuit within a given logic family (TTL, ECL, CMOS, etc.) are often used for comparing the capabilities of various logic families.

A gate is simply an electrical or electronic switching circuit, with two or more inputs, through which power is applied to a load.

The mechanical equivalent (Fig. 6-52a) shows an OR gate consisting of two switches (relays) connected in such a way that power is applied to the lamp (load) when either switch A or switch B is closed. Figure 6-52b shows an AND gate which applies power to the load only when switch A and switch B are closed.

In Fig. 6-52c, an additional relay is connected to the OR circuit so that power is applied to the load only when switch A and switch B are not closed. This is a NOT OR, or NOR, gate.

Similarly, an AND circuit connected in such a way that the load is powered only when switch A and switch B are not closed simultaneously (Fig. 6-52d) is a NAND gate.

The additional relays perform the function of an *inverter;* i.e., they invert a normally high (or 1) output of an OR or AND gate into a low (or 0) output, and vice versa. Hence, an OR or AND gate can be converted to the NOR or NAND function by running the outputs through an inverter.

In Fig. 6-53, the switching relays of Fig. 6-52 have been replaced by diodes which serve similar functions. In the OR gate (a), the output is high (or 1) when the input to diode A or diode B is high. In the two-input AND gate (b), the output is high only when both diode inputs are high. If either input were low, the corresponding diode would conduct and the output would be equal to the very low (0) voltage drop of the conducting diode.

When a common-emitter transistor circuit is added to the above circuits (Fig. 6-54), the transistor automatically converts a low-level input into a high-level output (and a high-level input into a low-level output), thereby acting as an inverter. Thus, a transistor added to a diode AND gate results in a NAND gate, and a transistor added to a diode OR gate results in a NOR gate.

**18. Saturated-logic gates.**    In many switching-circuit families the transistors are driven from deep-in cutoff, where the collector voltage is equal approximately to the supply voltage, to a high level of conduction (saturation), where the collector junction becomes forward-biased, resulting in a very low collector voltage, $V_{CE(sat)}$, that remains relatively constant even if the base-drive voltage is further increased. In this latter condition, virtually all the power supply voltage is dropped across the collector load resistor—a limiting condition that prevents any further increase in collector current even if the charge injected into the base is increased by a further rise in base voltage. This is called saturated-mode switching.

DIODE-TRANSISTOR LOGIC (DTL)    The basic DTL gate circuit is the DTL NAND gate shown in Fig. 6-55. Note that it uses the NAND circuit of Fig. 6-54 plus an offset-voltage diode circuit. The offset-voltage diode circuit, being returned to a negative bias voltage through resistor $R_3$, improves noise immunity and transistor turn-off time.

TRANSISTOR-TRANSISTOR LOGIC (TTL)    The DTL circuit just described was the basis of a more advanced logic line currently in use. Known as transistor-transistor logic (TTL), this logic is considerably faster than other forms of saturated logic.

The basic diagram of a typical TTL gate circuit is shown in Fig. 6-56. The principal difference between DTL and TTL is that the latter uses the collector-base junctions of a multiple-emitter transistor $Q_1$ as the input diodes and the collector-base junction as the offset diode. This, however, has a considerable effect on the operation, as follows:

When the input voltage to one or more of the $Q_1$ emitters is very low, say, zero, the base-emitter junction is forward-biased, through $R_1$, to a high positive voltage. Current flow through this junction, therefore, establishes approximately 0.7 V at the base. The collector of $Q_1$ is returned to ground through the base-emitter junctions of $Q_2$ and $Q_3$. Thus, the collector-base junction is also forward-biased, and it would appear that there should be a current flow through the emitter-base junctions of $Q_2$ and $Q_3$. This cannot occur, however, because it would require a $Q_1$ base voltage of at least three junction drops $[V_{BE(Q3)} + V_{BE(Q2)} + V_{CB(Q1)}]$, or 2.1 V, to turn both $Q_2$ and $Q_3$ fully on. Therefore, $Q_2$ and $Q_3$ remain cut off and the output voltage $V_{out}$ is equal to $V^+$ (minus a small leakage drop).

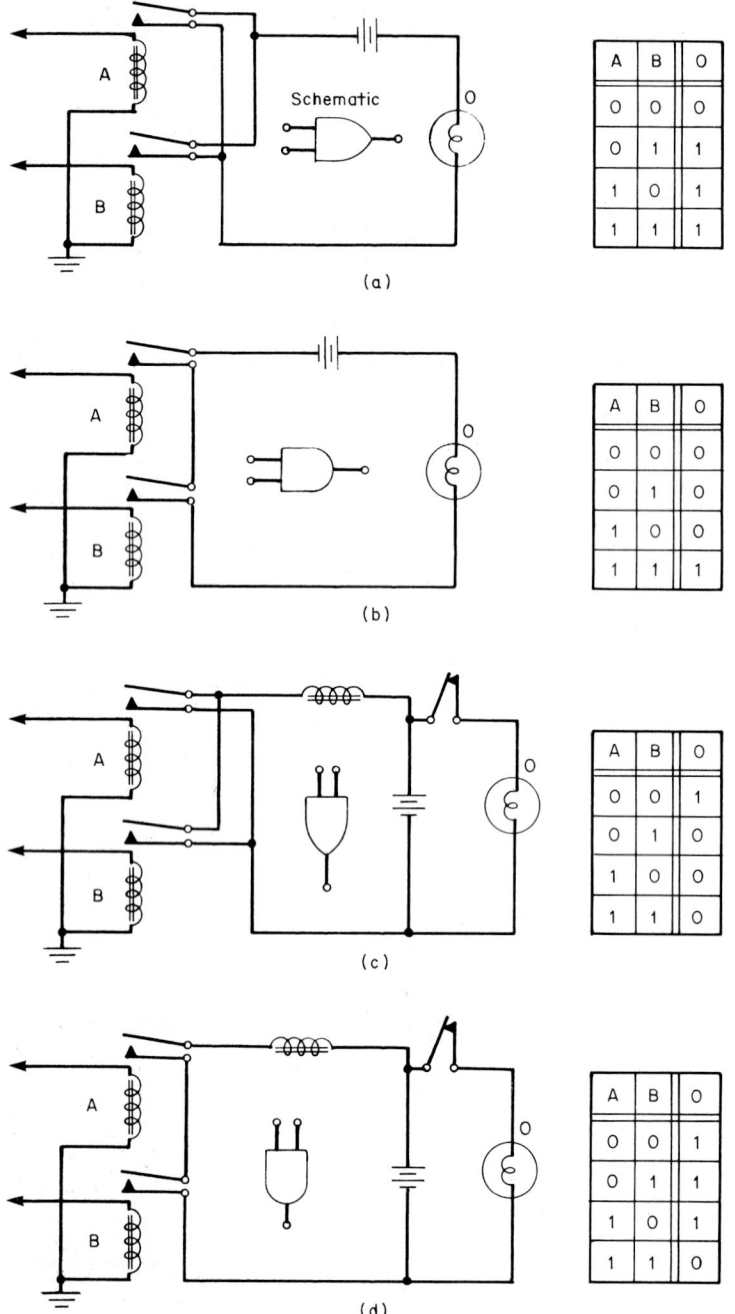

**Fig. 6-52** Basic mechanical gate functions. (*a*) Mechanical equivalent of a two-input OR gate and its representative functional truth table which illustrates the status of the output *o* as a function of the status of inputs *A* and *B*. (*b*) Mechanical AND gate and its accompanying truth table. (*c*) NOR gate and its truth table. (*d*) NAND gate and its truth table.

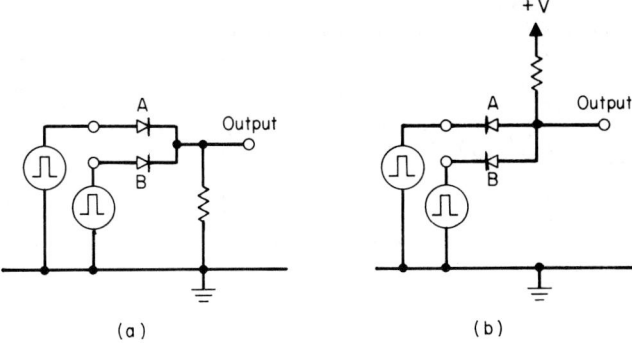

**Fig. 6-53** Diode logic OR (*a*) and AND (*b*) gates.

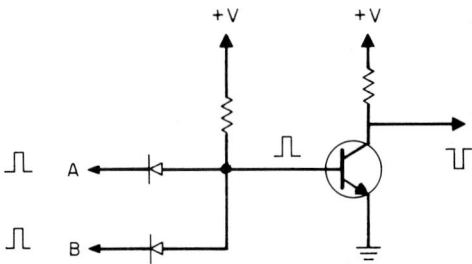

**Fig. 6-54** Basic NAND function.

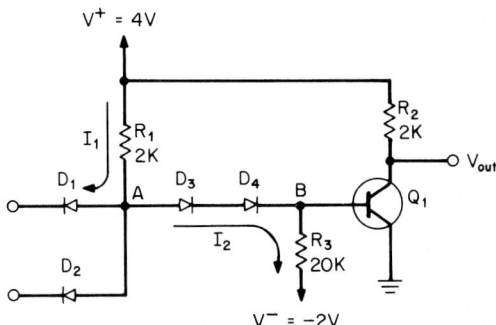

**Fig. 6-55** Basic DTL NAND gate. Operation: With a low-state voltage, say, $V_{CE(sat)} = 0.1$ V, applied to input diode $D_1$ or $D_2$, current $I_1$ will flow through the diode, causing the voltage at point $A$ to equal one diode drop, or approximately 0.7 V. At the same time, current $I_2$ will flow through diodes $D_3$ and $D_4$, causing the voltage at point $B$ to be two diode drops less than that at point $A$, or approximately $-0.7$ V. With a negative voltage thus applied to the base of transistor $Q_1$, the transistor will be cut off and its collector voltage $V_{out}$ is equal to $V^+$, or 4 V (neglecting leakage current). If we now apply a high-state voltage, say, 4 V, to both input diodes, they will be reverse-biased and become nonconductive. The base of $Q_1$ therefore "sees" a relatively high positive voltage, and the transistor will be driven into saturation. Its output voltage then is $V_{CE(sat)}$. Therefore, $V_{out}$ swings from $V_{CE(sat)}$ to nearly $V^+$.

Now if the input voltage is increased to approximately 0.6 V, the $Q_1$ base voltage rises to 1.3 V, and the collector junction of $Q_1$ and the base junction of $Q_2$ will be brought to the threshold of conduction through $R_2$. As input voltage increases still further to about 1.3 V, $Q_1$ base voltage rises to 2.0 V, the collector and $Q_2$ base junctions become fully conductive, and the voltage across $R_2$ rises to 0.6 V. This brings $Q_3$ to the threshold of conduction, so that any additional increase in input voltage causes $Q_3$ to turn on and the output voltage drops to $V_{CE(sat)}$.

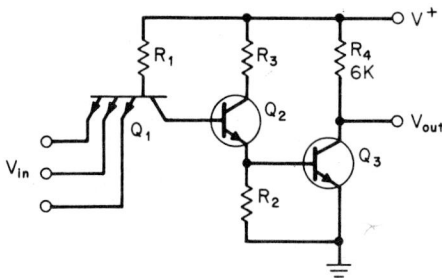

**Fig. 6-56**  Basic TTL gate circuit.

For most of the operating cycle the multiple-emitter input transistor acts as though it were a pair of back-to-back diodes. During one small part of the turn-off operation it does act like a transistor, thereby improving circuit speed. Specifically, when the input is in the high state and both $Q_2$ and $Q_3$ are conducting, the voltage at the base of $Q_1$ is 2.1 V while the base voltage of $Q_2$ is 1.4 V. Thus, $Q_2$ is in saturation, and a heavy charge is built up in its base region. Now the input suddenly drops to zero and the $Q_1$ base voltage drops to 0.7 V (the forward voltage drop of the input diode). If the $Q_1$ collector-base junction were an ordinary diode, it would be reverse-biased and $Q_2$ would be suddenly cut off. As a result, the stored charge in the $Q_2$ base would have to be dissipated by recombination, and $Q_2$ would continue to conduct until equilibrium was restored. Storage time, therefore, would be relatively long.

With the transistor, however, for the conditions existing at $Q_2$ turn-off, the emitter of $Q_1$ quickly drops to nearly zero volts. The base drops to 0.7 V, but the collector remains near 1.4 V because of the charge in the base of $Q_2$. These are the proper biases for transistor action and cause a sudden surge of $Q_1$ collector current that quickly clears $Q_2$ of its stored charge. Thus, storage time of $Q_2$ is drastically reduced.

**19. Nonsaturated-logic gates.**  See Secs. 20, 21, and 22.

**20. Schottky TTL.**  The basic TTL family has been substantially improved by a number of variations involving the use of Schottky diodes to keep the transistors out of saturation during turn-on. Since the switching time delay associated with transistor *storage time* is the result of driving the device into saturation, nonsaturated logic can considerably improve switching speed.

Schottky TTL simply involves the use of a Schottky diode between collector and base of a transistor, as shown in Fig. 6-57. Since this configuration can be achieved during the fabrication of the transistor without the use of a separate diode, it adds little to the device cost, but it has a great effect on performance.

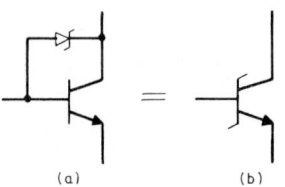

**Fig.  6-57**  A  representative Schottky-clamped transistor (*a*) is normally indicated by the schematic symbol in (*b*).

Specifically, since the turn-on voltage for the diode is considerably less than that for a conventional junction diode (in this case, the collector-base junction of the transistor), diode conduction occurs before the $V_{CE(sat)}$ region of the transistor is reached. Therefore, the collector voltage will be clamped at a value that prevents the collector-base junction from becoming forward-biased, and storage-time delay during turn-off is avoided (see Secs. 17 to 23).

By raising the minimum achievable collector voltage

somewhat, the maximum output voltage of the transistor is correspondingly reduced. This lowers the noise margin slightly, but not enough to be seriously detrimental for most applications.

*Noise margin* can be simply described with the aid of the basic RTL inverter circuit and transfer characteristics in Fig. 6-58. From the transfer characteristics it is seen that when the input to the transistor is low (say, 0 to 0.6 V in this instance), the output voltage is high and current is delivered to the load. As the input voltage rises to approximately 0.6 V, the transistor begins to conduct and its output rapidly drops toward $V_{CE(sat)}$ (approximately 0.1 V). For any input voltage greater than 0.9 V, the output remains at $V_{CE(sat)}$. The output voltage swing therefore is from 0.1 to 1.65 V.

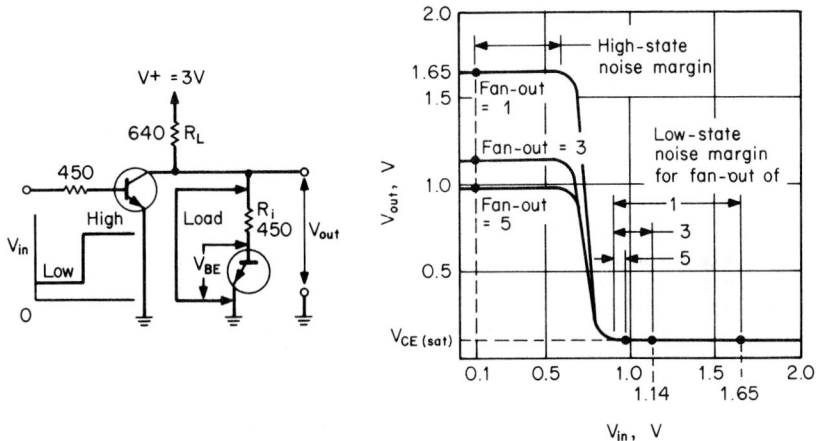

**Fig. 6-58**  A simple RTL inverter circuit is used here to demonstrate the concept of noise immunity and the fanout effect on noise immunity.

Under low-input conditions, if a noise pulse of 0.5 V or higher were to appear at the input of the gate, the noise voltage would add to the 0.1-V input voltage and the gate would tend to change its output from high to low. Therefore, the gate is immune to noise signals up to 0.5 V, or it has a high-state noise margin of 0.5 V.

The RTL circuit is used in the above example because it permits a simple discussion of noise immunity (it is no longer a popular logic format). It also offers a simple demonstration of the effect of fanout (FO) on noise immunity. In this instance, if the output voltage were applied simultaneously to five succeeding stages, the high-state voltage would drop to approximately 1 V and the low-state noise immunity would approach zero.

*Schottky TTL logic families* now include a variety of circuit innovations. All are based on the familiar 54/74 TTL series, which for years has been the dominant logic family for general-purpose applications. Only within the last few years has the Schottky LSTTL series made significant inroads, replacing the original 7400 series with lower power requirements and higher speed. The latest in this progressive logic series are the advanced low-power Schottky (ALSTTL) family and the FAST® Schottky TTL series.

The *low-power Schottky (LSTTL) family* combines an improvement in current and power reduction over the standard 7400 TTL by a factor of 5. This is accomplished by advanced processing and by using Schottky diode clamping to prevent saturation.

The *advanced low-power Schottky TTL family (ALSTTL)* provides a 50 percent power reduction compared with the standard 54/74 LSTTL and yet offers improved circuit performance over the standard LS owing to a state-of-the-art oxide-isolated process (MOSAIC). ALS also differs from LS in that p-n-p transistors on the input stage are utilized to lower input currents and raise thresholds.

The *FAST Schottky TTL family* provides a 75 to 80 percent power reduction compared with the standard Schottky 54/74S TTL and yet offers a 20 to 40 percent improvement in circuit performance over the standard Schottky owing to the MOSAIC process. Also, FAST circuits contain additional circuitry to provide a flatter power-frequency curve. The input configuration of FAST uses a lower input current, which translates into higher fanout.

Speed-power comparisons of these three families are indicated in Table 3.

**Table 3  Speed-Power Characteristics for Schottky TTL Logic[a]**

| Characteristic | Symbol | LS | ALS | FAST | Units |
|---|---|---|---|---|---|
| Quiescent supply current or gate | $I_G$ | 0.4 | 0.2 | 1.1 | mA |
| Power or gate (quiescent) | $P_G$ | 2.0 | 1.0 | 5.5 | mW |
| Propagation delay | $t_p$ | 9.0 | 5.0 | 3.7 | ns |
| Speed-power product | ... | 18 | 5.0 | 19.2 | pJ |
| Clock frequency (D-F/F) | $f_{max}$ | 33 | 35 | 125 | MHz |
| Clock frequency (counter) | $f_{max}$ | 40 | 45 | 125 | MHz |

[a]All typical ratings.

**21. Emitter-coupled logic (ECL).** Emitter-coupled logic is the fastest form of logic currently available. It is nonsaturated logic which is extremely flexible, with both OR and NOR logic outputs available from the same basic gate configuration. ECL utilizes a pair of input transistors, one of which is in a conductive state while the other is nonconductive. Switching is accomplished by means of a signal appearing across a common-emitter resistor. The basic gate circuit of an emitter-coupled logic circuit is shown in Fig. 6-59. It consists principally of a switching circuit, followed by an emitter-follower output circuit. The switching circuit consists of transistors $Q_1$ and $Q_2$, which are connected in a differential amplifier configuration. Operation is as follows:

Assume for a moment that all input transistors $Q_1$ are cut off owing to logical 0 being applied to their gates. $Q_2$ has its base connected to a fixed stable bias source $V_{BB}$ of $-1.15$ V, which causes $Q_2$ to conduct heavily since its emitter is connected, through $R_E$, to $-5.2$ V. Under these conditions the voltage drop across the base-emitter junction of $Q_2$ is approximately 0.75 V, and the current through $Q_2$ is

$$I_{Q2} = [V_{EE} + V_{BE(Q2)} + V_{BB(Q2)}]/RE$$
$$= (-5.2 + 0.75 + 1.15)/1.24$$
$$= -2.66 \text{ mA}$$

and $\qquad V_{C2} = I_{Q2}R_{C2} \approx -0.8$ V

In the on condition, therefore, $Q_2$ is not in the saturated mode because the voltage across the collector-base junction is still positive ($V_C - V_B$), keeping this junction reverse-biased.

Transistor $Q_3$, seeing $-0.8$ V on its base compared with $-5.2$ V as an emitter source, goes into heavy conduction, yielding the OR output of

$$V_{OR} = V_{C(Q2)} - V_{BE(Q3)} = -(0.8 + 0.75) = -1.55 \text{ V}$$

which represents the output-low condition.

At the same time, the base of $Q_4$, being connected to the collector of $Q_1$ (which is nonconductive), sees a voltage of zero, causing this transistor to conduct even more heavily. Its output voltage is

$$V_{NOR} = V_{C(Q1)} - V_{BE(Q4)} = -0.75 \text{ V}$$

which represents the output-high condition indicative of the NOR function.

If a positive-going signal is now applied to one or more of the $Q_1$ transistors, the transistors begin to conduct, tending to increase the emitter voltage of $Q_1/Q_2$. In turn, this reduces the forward bias on $Q_2$, causing a nearly equivalent decrease in $Q_2$ collector current. Thus, it is clear that emitter current remains essentially constant but is switched from $Q_2$ to $Q_1$. At the same time, the output-voltage levels of $Q_3$ and $Q_4$ reverse from their previous states, which again is consistent with their respective OR/NOR functions. The transfer characteristics illustrated in the graph of Fig. 6-59b indicate the following:

1. When an ECL gate is driven by another ECL gate, it provides two output signals: one identical to the input, the other equal to the inverted input.

2. The fixed logic levels are −0.75 (logical 1) and 1.55 (logical 0) V, resulting in a difference of 800 mV between the on state and the off state.

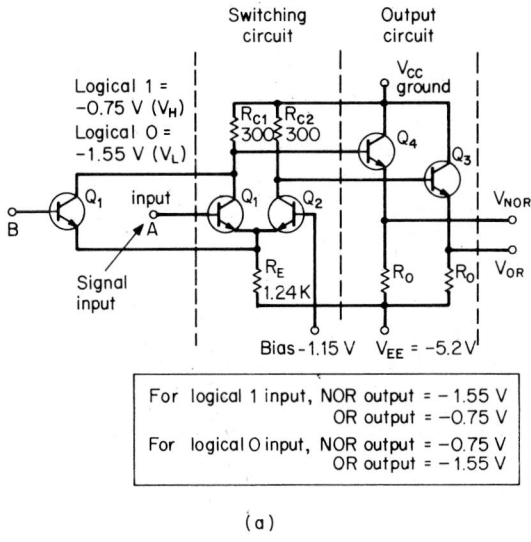

(a)

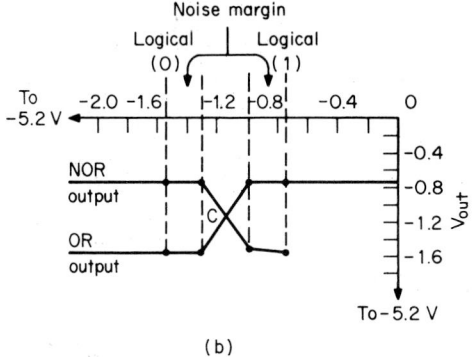

(b)

**Fig. 6-59**    (a) Basic ECL gate circuit. (b) Typical transfer characteristic curve.

3. The transistors are never driven into saturation; therefore, storage time is eliminated.

4. The noise margin is approximately 250 mV.

5. The total current flow in the circuit remains relatively constant, thereby maintaining a relatively constant drain on the power supply and preventing internally generated noise spikes.

The circuit, furthermore, has a very high input impedance and a very low output impedance, which makes high fanout possible. Thus, ECL not only is the highest speed logic available but provides other features that make it highly desirable for advanced systems. Its principal detriment is its limited (800-mV) logic swing, which makes it somewhat more sensitive to externally generated noise than some other logic forms. In addition, power consumption is considerably higher than for other logic forms.

Like TTL, emitter-coupled logic also comprises a number of families with progressively improved performance. Three families are in popular use: ECL 10K, ECL 10KH, and ECL III.

The ECL 10K series has become the industry standard for high-speed applications. To make the circuits comparatively easy to use, edge speed (rise and fall times) are deliberately slowed to 2.0 ns while the important propagation delay is held to 2.0 ns. The slow edge speed permits use of wire-wrap and standard printed-circuit lines; however, the circuits are specified to drive transmission lines for optimum performance.

The newer ECL 10KH family features 100 percent improvement in propagation delay and clock speeds while maintaining a power supply current equal to that of the ECL 10K. This ECL family is voltage-compensated, which allows guaranteed dc and switching parameters over a $\pm 5$ percent power supply range. Noise margins of ECL 10KH are 75 percent better than those of the ECL 10K series. ECL 10KH is compatible with ECL 10K and ECL III, a key element in allowing users to enhance existing systems by increasing the speed in critical timing areas.

ECL III, with its 1-ns gate propagation delays and greater than 1-GHz flip-flop toggle rates, is the industry speed leader. The 1-ns rise and fall times require a transmission-line environment for all but the smallest systems. For this reason, all circuit outputs are designed to drive transmission lines, and all output logic levels are specified when driving 50-$\Omega$ loads. Because of ECL III's fast edge speeds, multilayer boards are recommended above 200 MHz. ECL III's popularity is with high-speed test and communications equipment.

Speed-power comparisons for Motorola ECL families are given in Table 4.

**Table 4   ECL Family Comparisons**

|  |  | MECL 10K | | |
| --- | --- | --- | --- | --- |
| Parameter | MECL 10KH | 10,100 series 10,500 series | 10,200 series 10,600 series | MECL III |
| Gate propagation delay | 1.0 ns | 2 ns | 1.5 ns | 1ns |
| Output edge speed | 1.2 ns | 2.2 ns | 2.0 ns | 1ns |
| Flip-flop toggle speed | 250 MHz min | 125 MHz min | 200 MHz min | 300–500 MHz min |
| Gate power | 25 mW | 25 mW | 25 mW | 60mW |
| Speed-power product | 25 pJ | 50 pJ | 37 pJ | 60pJ |

**22. CMOS logic.**   While individual $p$-channel and $n$-channel MOS devices have certain performance advantages over bipolar structures and while a $p$-channel logic line did gain some popularity for a short time, these advantages were not sufficient to displace bipolar logic. Not until processing technology permitted *complementary* MOS cells to become space-competitive with individual bipolar cells did MOS become a serious challenge. Now, CMOS logic families promise to become the dominant logic form over the next few years, particularly in the areas of general-purpose logic, where speed is not an overriding consideration, and in VLSI circuitry, where high chip density demands the lowest possible dissipation.

Some of the features of CMOS are attributable directly to the basic MOS transistor structure, while others are the result of (or are enhanced by) the complementary symmetry configuration. For example, the MOS transistor inherently has a very high input resistance, thereby eliminating dc fanout restrictions and providing low power dissipation. The CMOS configuration reduces power dissipation even more and increases speed, noise immunity, and logic swing.

A simple $p$-channel MOSFET inverter and its transfer characteristics are shown in Fig. 6-60a. [The use of a second (fixed-bias) MOSFET in place of a conventional load resistor is particularly beneficial since a resistor would require far more chip area. Thus MOS is cost-effective by increasing the permissible number of circuits on a wafer.]

As indicated by the transfer characteristic curve, for an input voltage $V_{in}$ between ground and $V_{to}$, transistor $Q_s$ is off, so the output voltage $V_o$ approaches the $V^-$ state. As the input voltage approaches threshold voltage $V_{to}$, $Q_s$ begins to conduct, and for further increases of $V_{in}$ the output voltage is reduced. Note, however, that for a fixed MOSFET load resistance $V_o$ can never reach zero because the resistance of $Q_s$ never reduces to zero regardless of the value of $V_{in}$. In fact, its on resistance is substantially higher than that of bipolar transistors. Therefore, the total output-voltage swing is always less than the supply voltage $V^-$.

The complementary configuration (Fig. 6-60b) results in far more satisfactory performance. In this connection, the signal is applied simultaneously and in phase to both transistors so that when the signal value is zero $Q_1$ is off and $Q_2$ is on. Under this condition, the output voltage is very nearly the full supply voltage. When the gate voltage goes high (positive), transistor $Q_1$ is turned on while $Q_2$ is turned off. This causes $V_o$ to go virtually to zero because the current flowing through $Q_1$ is the leakage current of $Q_2$, which is very, very low. (The resistance of an MOSFET in cutoff is approximately 5000 M$\Omega$, resulting in a leakage current of less than 1 nA.)

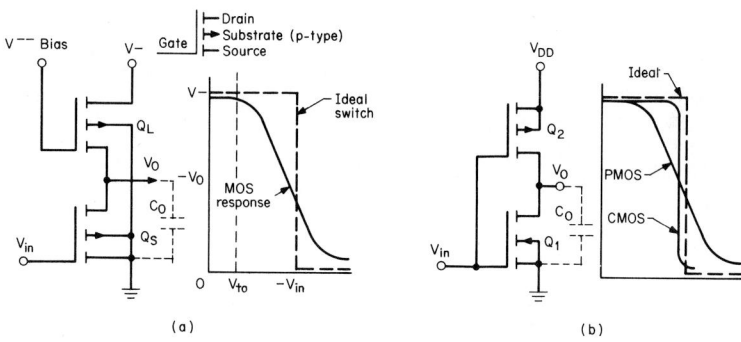

**Fig. 6-60**   Inverter circuits and their response characteristics. (*a*) Typical PMOS inverter circuit using a fixed-bias MOSFET as a load. Transfer characteristics compare response with a "perfect" switch. (*b*) Typical CMOS inverter circuit with *p*- and *n*-channel MOSFETs. Characteristics closely approach an "ideal" switch.

The transfer characteristic curve for the complementary circuit is shown in comparison with that of a single-ended circuit. Observe that the slope of the complementary circuit curve is much steeper in the transition region, thus providing much greater noise immunity. This is caused by the input signal acting on both transistors in opposition, turning one device on and the other off.

In addition, it is evident that the circuit conducts current only during a very short time after turn-off (Fig. 6-61), as required to charge capacitor $C_o$ through the load (the load is assumed to be another CMOS inverter circuit). This reduces power dissipation of the structures to such a low value that thousands of them can be fabricated within a tiny chip of silicon without special heat-sinking techniques.

Progressive improvements in technology have resulted in the current availability of two CMOS logic families: a standard metal-gate family and a high-speed silicon-gate family. In silicon-gate CMOS, the metal (aluminum) gate associated with standard CMOS is replaced by a heavily doped silicon layer which serves a similar function but has the following advantages:

1. It reduces the threshold voltage required to turn the transistor on.

2. It provides self-alignment of the gate electrode, thereby simplifying manufacture and improving fabrication accuracy.

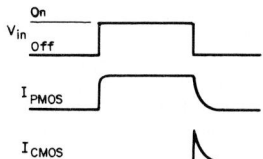

**Fig. 6-61**   A comparison of current flow in a PMOS and a CMOS inverter during a full on-off cycle of $V_{in}$.

3. It reduces gate-to-drain capacitance, thereby increasing circuit speed.

4. It permits improved space utilization, thereby increasing packing density.

General comparisons of speed and power requirements for the two CMOS families are indicated in Table 5.

**Table 5    Speed-Power Characteristics for CMOS Logic[a]**

| Characteristic | Symbol | Standard CMOS | High-speed CMOS | Units |
|---|---|---|---|---|
| Quiescent supply current or gate | $I_G$ | 0.0001 | 0.0003 | mA |
| Power or gate (quiescent) | $P_G$ | 0.0006 | 0.001 | mW |
| Propagation delay | $t_p$ | 125 | 8.0 | ns |
| Speed-power product | ... | 0.075 | 0.01 | pJ |
| Clock frequency (D-F/F) | $f_{max}$ | 4.0 | 40 | MHz |
| Clock frequency (counter) | $f_{max}$ | 5.0 | 40 | MHz |

[a]All typical ratings.

**23. Comparison of logic-family characteristics.**   The variety of established and newly introduced logic families challenges the system designer with choosing the best available technology for each design. Each family offers distinct advantages and limitations.

The three most often used characteristics for determining family selection are propagation delay, operating frequency, and power consumption. For the logic families described here, these characteristics are displayed and compared in the graphs of Fig. 6-62. Graph $a$ illustrates the tradeoffs between power dissipation and propagation delay at low operating frequencies. As frequency increases, the changes in power-dissipation characteristics of the various families vary considerably, as shown in Graph $b$. These factors, plus the ever-changing impact of economics, must be considered in the selection of the most suitable logic line.

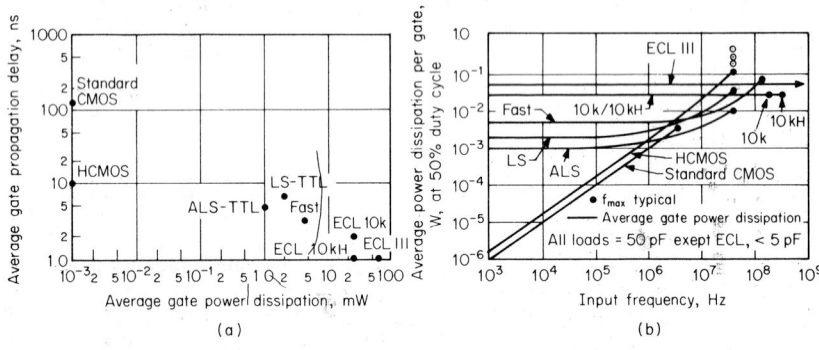

(a)                                                            (b)

**Fig. 6-62**   A comparison of major characteristic of various logic families. ($a$) General comparison of speed-power characteristics. ($b$) Variations in power dissipation as a function of operation frequency.

LINEAR-CIRCUIT FAMILIES   Linear (analog) integrated circuits are as pervasive and as unique as the markets they serve. They do not lend themselves readily to classification by building-block families, as do digital circuits for which standardized basic circuits can be used repetitively in various combinations to build highly complex end systems. Linear circuits tend to be complete functional circuits applicable to a wide variety of specific unique functions. And as circuit and chip complexity increases, this situation will become even more prevalent.

There are, however, a number of linear classifications that do encompass the family approach. Chief among these are operational amplifiers and voltage regulators. These represent basic circuits utilized in many different pieces of equipment, each of which requires some variation in performance specifications.

**24. Operational amplifiers.**   Throughout history, there has been a continuous goal to develop the "ideal" device for a particular class of application: one that would satisfy all necessary requirements, thereby eliminating the need for any other product. In the field of signal amplifiers, the operational amplifier comes close to approaching that ideal.

Originally, the operational amplifier was designed principally for performing arithmetic operations. In such applications, high amplifier gain was not of critical importance, but absolute stability and accuracy were. These were achieved through the design of a high-gain amplifier that permitted the application of a large amount of negative feedback to provide the necessary accuracy.

This is still the basic design criterion for operational amplifiers. But while virtually any operational amplifier will serve a tremendous number of amplifier functions, the ideal device still remains to be invented. Hence, while the basic operational amplifier exhibits tremendous versatility, the operational-amplifier classification probably contains more type numbers than any other single applications category.

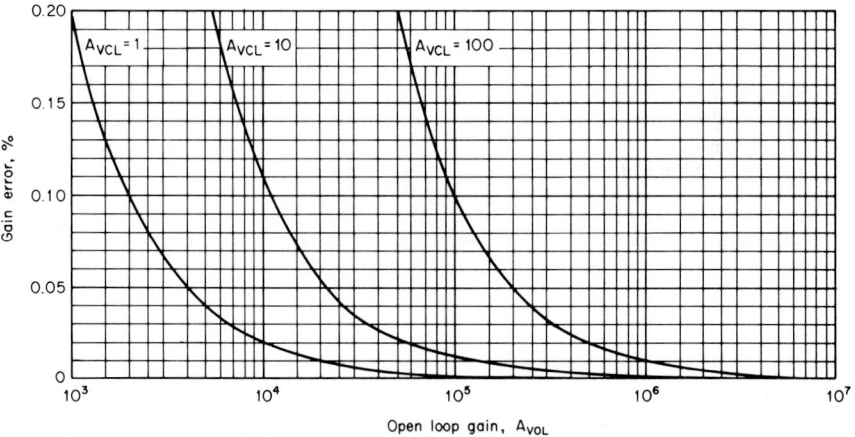

**Fig. 6-63**   When operational amplifiers are used at modest closed-loop gains, closed-loop–gain error can be held to negligible levels.

Basically, the operational amplifier is a multistage low-level amplifier with an open-loop gain ($A_{VOL}$, before the application of negative feedback) ranging from a low of, say, 25,000 (25 V/mV) to 1 million or more. With such high gain levels readily available, further gain increases are apt to be by-products of improvements in other parameters rather than specific design goals. This is evident from Fig. 6-63, which shows the percent of amplifier gain error as a function of open-loop gain. It is clear that for a reasonable closed-loop–gain requirement, say, 100 or less, an open-loop gain of $1 \times 10^6$ will permit sufficient feedback to hold gain error to negligible levels.

Since integrated circuits are intolerant of capacitors, analog functions are achieved with dc amplifier configurations. This usually involves *differential* amplifier input circuits which provide stable, direct-coupled amplification with very high circuit gain.

The block diagram for a typical operational amplifier is shown in Fig. 6-64a. Most integrated-circuit operational amplifiers follow this format. The primary differences are those of the circuits within each of the blocks that give the amplifier its ultimate performance specifications.

The first stage of an operational amplifier is a differential amplifier that provides most of the circuit gain. It is desirable to have high gain in this section so that any imperfections in succeeding stages (offset voltage, etc.) have little or no effect on the output. The first stage, too, should employ a current source at the common-emitter node for good common-mode rejection.

The second stage does not require a current source in the emitter because common-mode rejection and other matching-dependent characteristics of the total amplifier are determined primarily by the specifications of the first stage. It is needed primarily to provide additional gain. Its input resistance should be relatively high to prevent excessive loading of the first stage. Therefore, an emitter-follower or Darlington-amplifier stage is often employed.

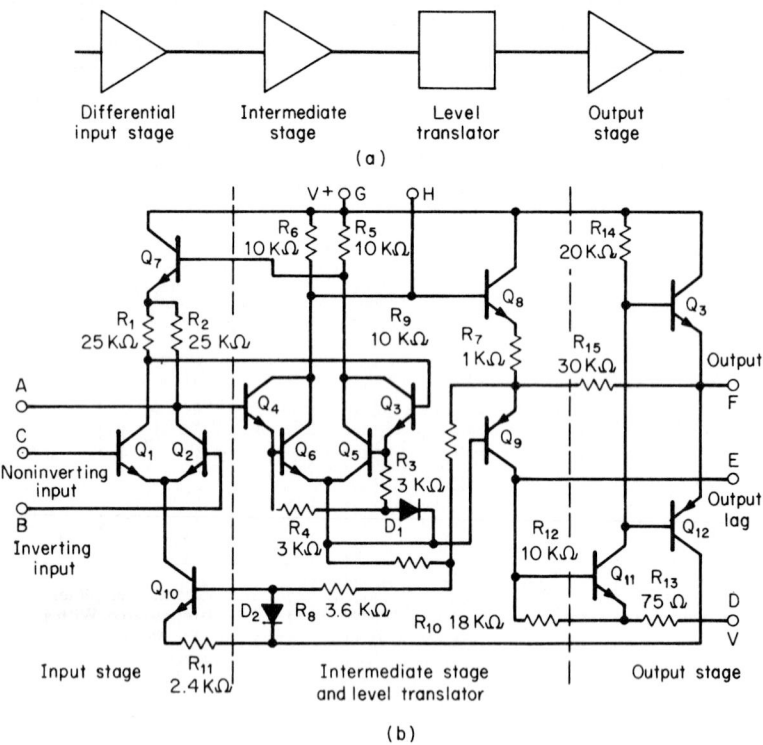

**Fig. 6-64** A block diagram of a typical operational amplifier (*a*) and a schematic diagram of an MC1709 circuit (*b*).

Since a single-ended output is normally employed for the second stage, it follows that a dc voltage is present at its output. In a direct-coupled system, this dc level is propagated through the amplifier chain so that the amplifier output voltage would have a dc component in addition to a desired ac output signal. Therefore, some means of level translation is employed between the second and final stages. By eliminating the dc level at the final stage, the output voltage will vary about a zero reference level, thus preventing any undesired dc current in the load and also increasing the permissible output-voltage swing.

DIFFERENTIAL AMPLIFIER   The differential amplifier is the ideal circuit for direct coupling because of its versatility, its excellent stability, and its high immunity to interfering signals.

From the standpoint of stability, the circuit can be made virtually insensitive to temperature changes, which often cause excessive drift in other configurations. It is versatile in that it may be adapted for applications requiring floating inputs and outputs, as in the case of some sense amplifiers, or for applications in which grounded inputs and/or outputs are more desirable. In both cases, it exhibits the same drift-free interference-rejection capabilities.

To illustrate these capabilities, examine the schematic diagram of a simple differential amplifier shown in Fig. 6-65. Note that the two transistors with their respective collector resistors form a bridge which, if the transistor and resistor characteristics are identical, is perfectly balanced. Thus, the voltage across the output terminals is zero. If we now apply a differential-mode input signal and if $R_{i1}$ equals $R_{i2}$, so that the input voltages are equal in amplitude but opposite in phase, there will be a difference in voltage between the two output terminals which is proportional to the gain of the transistors.

If a common-mode input signal is applied (caused by line pickup or other interference), the input signals to each transistor will be equal in amplitude and in phase. The bridge will remain balanced, and the voltage between the output terminals will remain zero. Thus, the circuit provides high gain for differential-mode signals and no output at all for common-mode signals.

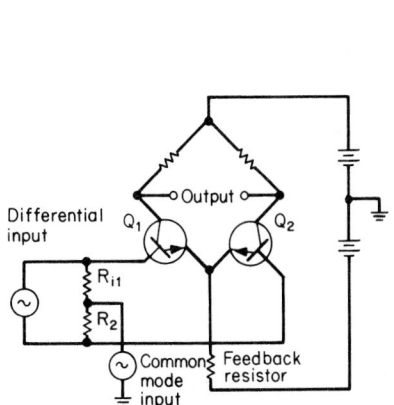

**Fig. 6-65**  A basic differential-amplifier circuit.

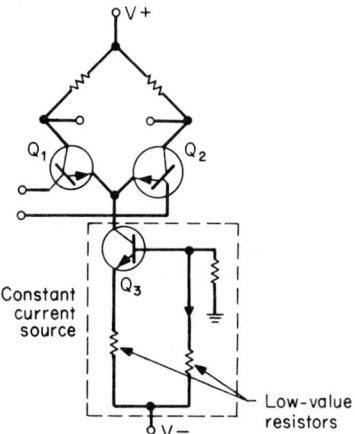

**Fig. 6-66**  A differential amplifier operating from a constant-current source. With a fixed voltage at the base of $Q_3$, this transistor provides a constant current regardless of load because its collector current is relatively independent of collector voltage.

In applications in which output must be taken between one output terminal and ground, the common-mode signal will produce some output voltage, although there will be a very substantial reduction of gain for the common-mode signal as compared with the gain for the differential-mode input.

If, for example, a common-mode signal causes the current through the transistors to rise, the voltage drop across the common-emitter resistor rises also. This represents degeneration, and the gain of the transistors is very low. For differential-mode signals, the current through one transistor rises while that through the other transistor drops by an equal amount. Thus, the current through the emitter resistor remains constant and no degeneration occurs. The gain for the differential-mode signal, therefore, with matched conditions assumed, is unaffected whereas that for the common-mode signal is substantially reduced.

A modified version of the basic differential amplifier, using a constant-current source circuit in place of an emitter resistor (Fig. 6-66), represents the most common basic amplifier not only for operational-amplifier input stages but for virtually all low-power amplification purposes.

DARLINGTON CIRCUIT  In many applications for which an amplifier is used with a high-impedance signal source, it is desirable to have a higher input resistance than is available with conventional bipolar transistor circuits. In such cases, today's technology offers field-effect transistors and Darlington (bipolar) transistor configurations. The Darlington circuit, as used as the intermediate amplifier stage in Fig. 6-64, is shown in its simplistic form in Fig. 6-67.

The input resistance of a transistor is a function of its emitter resistance $r_e$ and its $\beta$ such that $R_{in} = \beta r_e$. In the Darlington circuit shown, the input resistance of $Q_2$ becomes the emitter resistance (approximately) of $Q_1$, and the input resistance of the circuit is

$$R_{in} = \beta r_{e(Q1)}$$

$$\text{but} \quad r_{e(Q1)} = \beta r_{e(Q2)}$$

$$\therefore R_{in} = \beta_{(Q1)} \times \beta_{(Q2)} r_{e(Q2)}$$

$$= \beta^2 r_{e(Q2)} \text{ for transistors with equal betas}$$

Thus the input resistance of the Darlington transistor pair is extremely high, and its loading effect on the driving source is negligible. The total $\beta$ of the Darlington is also the square of the individual betas so that extremely high current gain can be obtained.

On the other hand, the ac input resistance of the pair is dependent on the input capacitance, which now shunts a much higher input resistance. Therefore, the frequency response curve of a Darlington drops off very rapidly as frequency is increased. In operational amplifiers, which are primarily intended for low-frequency applications, this is not a serious limitation, particularly since the high open-loop gain of such devices permits large amounts of negative feedback to compensate.

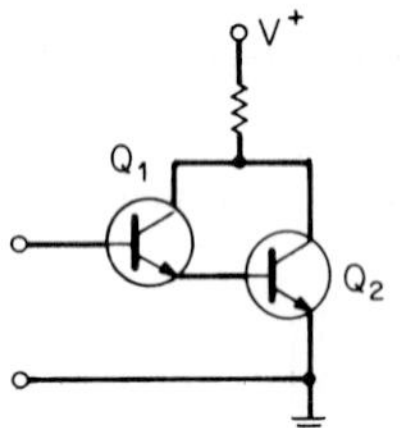

**Fig. 6-67** A simplified circuit of a Darlington transistor pair.

OPERATIONAL-AMPLIFIER SPECIFICATIONS   Operational-amplifier specifications, as listed on manufacturers' data sheets, cover a wide range of characteristics. Among the most important are the following:

*Open-loop voltage gain* ($A_{VOL}$). This is normally specified in volts (output) per millivolts (input). This specification defines the maximum available voltage gain, without feedback, and gives an indication of the eventual accuracy that can be obtained when feedback is applied (see Fig. 6-63).

*Common-mode rejection ratio (CMRR)*. CMRR is defined as the ratio of common-mode input voltage to differential-mode input voltage that will yield the same differential-mode output voltage. With well-designed monolithic integrated circuits, common-mode rejection ratios can be as high as 100 dB.

*Input offset voltage* ($V_{IO}$). Although IC technology permits close matching of adjacent transistors, perfect matching is not achievable. A slight mismatch in gain between the two differential-amplifier transistors will result in an undesirable dc voltage across the differential output terminals. $V_{IO}$ is defined as the offsetting base-emitter voltage required for equal emitter currents in the two transistors.

*Input offset current* ($I_{IO}$). $I_{IO}$ is the difference in base currents required to produce equal emitter currents in differential-amplifier transistors.

*Input bias current* ($I_{IB}$). $I_{IB}$ refers to the amount of current flowing in input terminals under no-signal conditions. When an amplifier is required to respond to extremely small input signals, the input bias current limits the response. Today's standard operational amplifiers have normal $I_{IB}$'s on the order of only a few nanoamperes, and even this can be substantially reduced through the use of FET input devices to the picoampere range.

*Slew rate (SR)*. Slew rate refers to the maximum time rate of change of closed-loop–amplifier output voltage. It is determined by applying a step-function input signal and measuring the slope of the output pulse, as shown in Fig. 6-68a. Obviously, if an applied signal voltage varies more rapidly than the slew rate of the amplifier, the amplifier cannot respond exactly and the output signal is distorted. This can occur on a sine-wave signal, as illustrated in Fig. 6-68b, as well as on a pulsed signal.

Slew rate is specified in terms of the maximum signal-voltage change per microsecond that can be tolerated without signal distortion.

*Other parameters*. Among other parameters normally specified for operational amplifiers are *power supply voltage*, which indicates the maximum output voltage obtainable; *bandwidth*, which indicates frequency response; and *temperature coefficient* of the input offset voltage, which describes the effect of temperature on operational-amplifier operation.

**25. Power supply circuits.** The power supply, which converts ac to dc operating voltages, is the one circuit that is universally required for all ac line-operated electronic equipment. In addition, the sensitivity of active components to changes in operating voltage and the potentially large variations in applied ac source voltage and load currents demand that power supplies for most equipment be well regulated in order to maintain stable operation under these variable conditions. Voltage-regulator circuits therefore are important functions for most power supplies.

Voltage regulators can be designed with discrete components to match the specific requirements of any equipment. However, the prevailing need for such circuits has resulted in a large variety of monolithic IC regulators to approximate most ideal requirements closely at a small fraction of the cost of equivalent discrete circuits.

**26. Voltage regulators.** A voltage regulator operates on the principle that a variable resistance in series or in shunt with the load resistance of an unregulated power supply can be used to compensate for any variation in output voltage or load resistance to keep the voltage across the load at a "constant" value.

The simplest shunt voltage regulator (Fig. 6-69a) is that of a zener diode in series with a voltage-dropping resistor $R_s$. Operated in the breakdown region, the zener acts much like a voltage-activated variable resistor, in that the voltage across its terminals tends to remain constant while the current through the device varies in accordance with an applied voltage. Thus, in the simple circuit of Fig. 6-69a, if the input voltage to the circuit $V_I$ were to increase (for a constant load resistance) or if the load resistance were to increase (for a constant input voltage), the output voltage $V_O$ would tend to increase. As a result, the current through the zener would increase, and this increased current, flow-

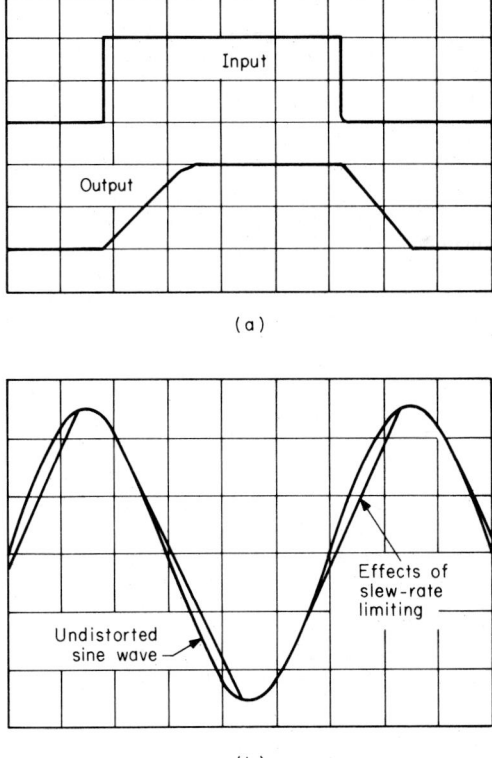

**Fig. 6-68** The effect of slew-rate limiting on a pulse input (a) and on a sine wave (b).

ing through $R_s$, counteracts the initial increase in $V_O$, thereby keeping this voltage close to its original value.

An improved regulator circuit is the series regulator of Fig. 6-69$b$. Here, the series resistance $R_s$ of the previous circuit is replaced with the collector-emitter resistance of a power transistor. The collector-base feedback resistor, in conjunction with the zener diode, varies the bias on the transistor, thereby causing a change in its internal resistance to compensate for any change in output voltage.

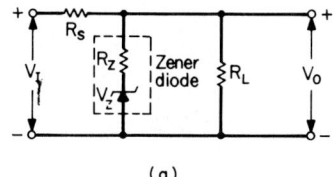

(a)

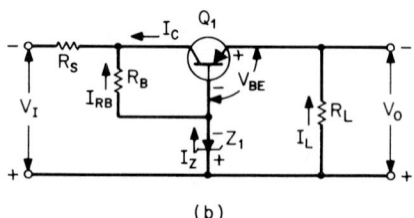

(b)

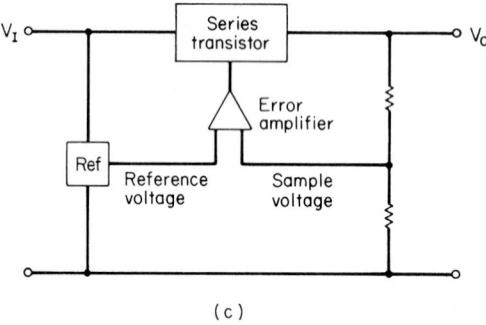

(c)

**Fig. 6-69**   Progression of voltage regulators. The circuit in $c$ forms the basis of today's more complex regulator ICs. ($a$) Basic zener shunt regulator. The internal zener resistance $R_z$ is very small compared with $R_s$ and $R_L$. ($b$) A simple series-pass regulator improves efficiency and performance. ($c$) An improved series regulator uses the gain of a differential amplifier to increase regulator sensitivity and provide additional functions.

A further improvement in performance results from the circuit configuration in Fig. 6-69$c$, in which a differential amplifier (or operational amplifier) is used to control the resistance of the series-connected transistor. The differential amplifier continuously samples the output voltage, compares it with a fixed reference voltage, and produces an "error voltage" to control the bias on the series transistor. Because the differential amplifier has considerable gain, even very small changes in output voltage will be detected and compensated for. The circuit in Fig. 6-69$c$ forms the basis of most IC voltage regulators in use today.

IC voltage regulators are available in a variety of configurations to meet the specific needs of the system designer. These configurations include the following:

1. Three-terminal fixed-output regulators with either positive or negative output
2. Three-terminal adjustable-output regulators with positive or negative output
3. Floating regulators for increased voltage range
4. Tracking regulators with a combination positive and negative output voltage
5. Switching regulators for use in switching power supplies.

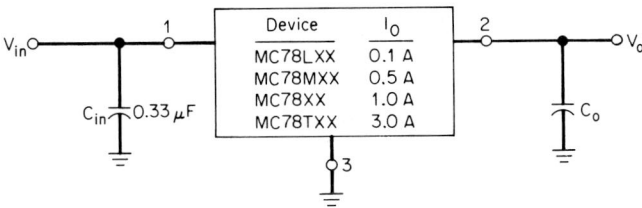

**Fig. 6-70** Basic circuit configurations for positive fixed-output three-terminal regulators. $C_{in}$: Required if regulator is located more than a few (about 2 to 4) inches away from input supply capacitor; for long input leads to regulator up to 1 $\mu$F may be needed for $C_{in}$. $C_{in}$: Should be a high-frequency type of capacitor. $C_0$: Improves transient response. XX: These two digits of the type number indicate nominal output voltage. Available voltages are 5, 6, 8, 12, 15, 18, 20, and 24 V.

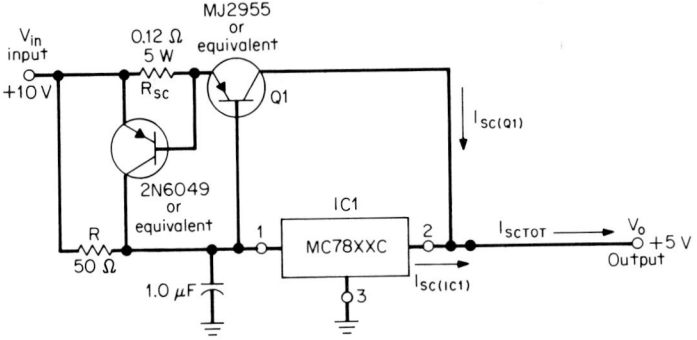

**Fig. 6-71** A current-boost configuration for positive three-terminal regulators. $R$: Used to divert IC-regulator bias current; determines at what output-current level $Q_1$ begins conducting $0 < R \le \frac{V_{BEON(Q1)}}{I_{BIAS(IC1)}}$. $R_{SC} \approx \frac{6 \text{ V}}{I_{SC(Q1)}}$. $I_{SCTOT} = I_{SC(Q1)} + I_{SC(IC1)}$. Values shown are for a 5-V, 5-A regulator using an MC7805CK on a 2.5°C/W heat sink and $Q_1$ on a 1°C/W heat sink for $T_A$ up to 70°C. [Motorola Inc.]

THREE-TERMINAL FIXED-OUTPUT REGULATORS Such regulators are available with a variety of fixed-output voltages in a range from approximately 2 to 40 V. Permissible output currents run the gamut from 100 mA to approximately 3 A, and both positive and negative output regulators are readily obtainable. The diagram in Fig. 6-70 illustrates the simple application and the range of voltage and current available for one commercial line of regulators, but even this wide variety of capabilities is expandable through the addition of external components. Paralleling the regulator with an external series-pass transistor (Fig. 6-71) can increase current-handling capacity; adding an external preregulator (Fig. 6-72) permits operation with higher input voltage than the maximum of 35 to 40 V for which this series of regulators was designed.

The simplicity of obtaining the exact output voltage required with an adjustable three-terminal output-voltage regulator is shown in Fig. 6-73. As in the previously used series, maximum output voltage is approximately 40 V, but any value from 1.2 V to the maximum limit can be obtained through the adjustment of $R_2$. Current values from 100 mA to 3 A are available.

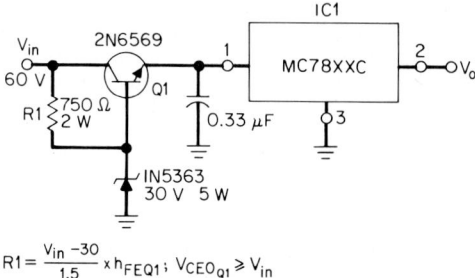

**Fig. 6-72** A preregulator for input voltages above the specified maximum input voltage of the regulator. Values shown are for $V_{in} = 60$ V. $Q_1$ should be mounted on a 2°C/W heat sink for operation at $T_A$ up to +70°C. $IC_1$ should be appropriately heat-sinked for the package type used. [Motorola Inc.]

FLOATING REGULATORS   When wide flexibility in output voltage and current is required, a floating regulator can be employed. In the configuration illustrated in Fig. 6-74 the regulator is isolated from the main power supply by a dedicated power source of its own. Its output-voltage and -current capabilities are limited only by the choice of external series-pass transistors.

TRACKING REGULATORS   Applications requiring a dual polarity (+ and −) power supply, such as operational-amplifier circuits, can be served with tracking regulators. These consist of two regulators, interconnected so that the positive and negative regulators have the same output-voltage levels, which track each other in the event of parameter changes.

The circuit of Fig. 6-75 uses an MC1568 or

**Fig. 6-73**  An adjustable three-terminal–regulator circuit.

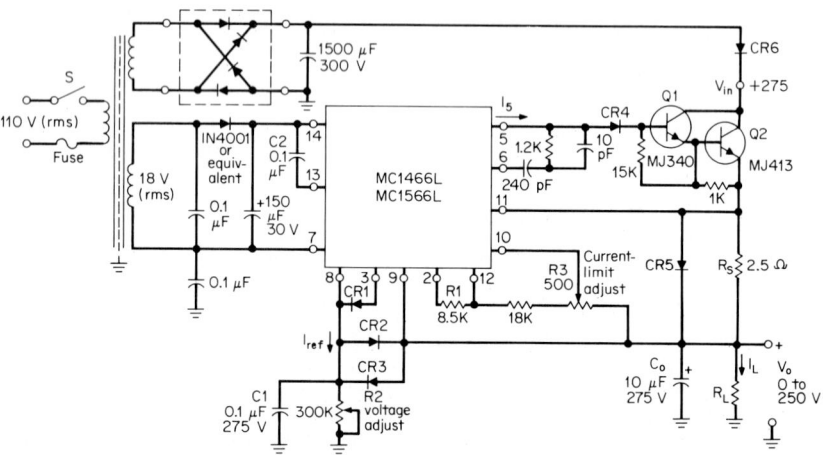

**Fig. 6-74**  An MC1566, MC1466 floating-regulator configuration. For constant-voltage operation, output voltage $V_o$ is given by $V_o = (I_{ref})(R_2)$, where $R_2$ is the resistance from pin 8 to ground and $I_{ref}$ is the output current of pin 3. The recommended value of $I_{ref}$ is 1.0 mA dc. Resistor $R_1$ sets the value of $I_{ref}$ at $8.5/R_1$, where $R_1$ is the resistance between pins 2 and 12. For constant-current operation, (a) select $R_s$ for a 250-mV drop at the maximum desired regulated output current $I_{max}$; and (b) adjust potentiometer $R_3$ to set constant-current output at the desired value between zero and $I_{max}$. Values shown are for a 0- to 250-V, 100-mA regulator using an MC1486L with $Q_1$ and $Q_2$ mounted on a 1°C/W heat sink for $T_A \le 70$°C. [Motorola Inc.]

MC1468 monolithic dual regulator for this purpose. Its outputs are set internally for $\pm 15$ V, but an external adjustment can change both outputs simultaneously from 8.0 to 20 V through the use of two balancing resistors.

Alternatively, tracking regulators can be developed by interconnecting two separate regulators, one positive and the other negative, in such a way that the output of one drives the input of the other.

SWITCHING REGULATORS Switching power supplies, or "switchers," are making rapid inroads into the power supply market. Compared with linear supplies, they have significant advantages in efficiency, size, and weight. They are, however, more complex and in the past have been considered principally for high-power applications. With the advent of IC regulators and improvement and cost reductions of other solid-state components, the cost-performance tradeoffs are no longer severe even for medium- and low-power requirements. As anticipated, therefore, switching power supplies are rapidly increasing their market share.

The basic circuit configuration of a typical flyback switcher is shown in Fig. 6-76. In this off-line circuit, the ac line voltage is rectified by a bridge rectifier and filtered by capacitor $C_1$. The resulting dc voltage is then "chopped" by high-frequency square-wave pulses applied to the transistor in series with the primary-transformer winding. The

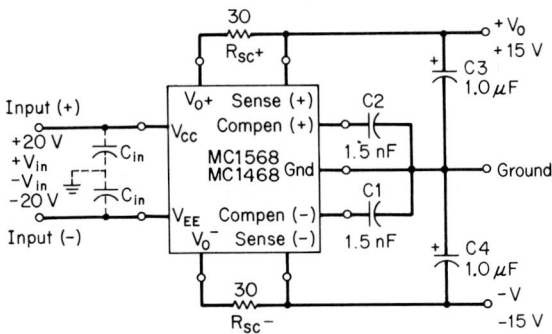

**Fig. 6-75** A dual-tracking regulator provides equal but opposite polarity outputs when both positive and negative voltages are required.

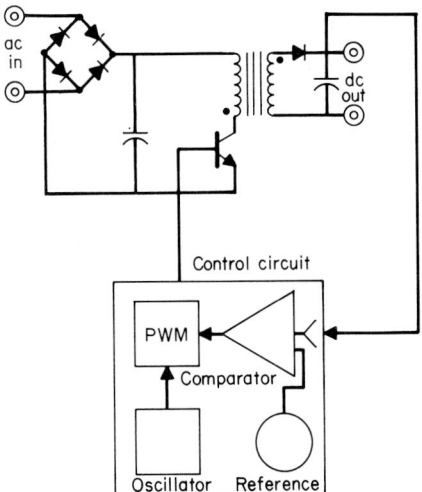

**Fig. 6-76** A representative block diagram of a typical flyback switching power supply.

resulting square wave is then passed through a high-frequency transformer. This provides the required step up or step down to produce the desired dc output after a second rectification and filtering process.

The output voltage is regulated by varying the width of the pulses applied to the chopper transistor. This is done through a pulse-width–modulator control circuit consisting of a triangular-wave generator, a pulse-width modulator, a dc reference source, and a comparator. The comparator senses the dc output voltage of the supply and compares it with the reference source. Any deviation of the output voltage from its design value will cause a control voltage from the comparator to provide a compensating pulse-width compression or expansion by the pulse-width modulator.

The control circuits are quite complex, as indicated by the block diagram of a typical unit in Fig. 6-77. Monolithic IC processing, however, has reduced the cost to a point at which switching power supplies are cost-competitive with series-regulated supplies for many applications. This is particularly true in view of the cost savings accruing from lower-cost transformer and filtering requirements.

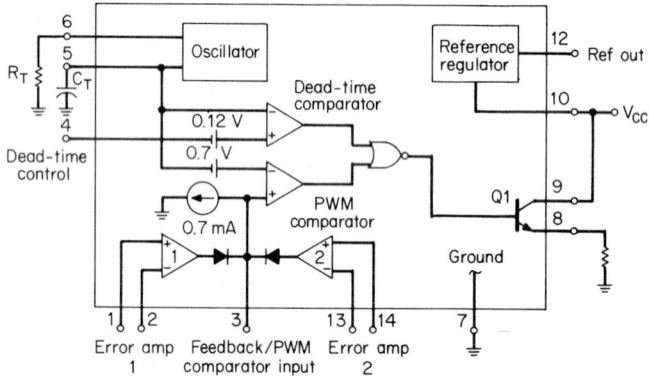

**Fig. 6-77**   A block diagram of an MC34060 pulse-width–modulation control circuit incorporating the primary building blocks for the control of a switching power supply. [Motorola Inc.]

### MICROCOMPUTERS

Computer concepts in the 1940s were credited with revolutionizing the scientific and engineering fields by replacing people power with electronic power for computational purposes. In the 1950s, by using vacuum tubes, large computer installations performed rather primitive routines and proved the value of these concepts in actual practice. In the early 1960s, widespread use of the transistor and its attending improvement in reliability, performance, and size expanded the influence of the computer to include business and process control applications. In the 1970s, integrated-circuit technology, with its attending cost reductions, spawned the microprocessor, which broadened computer applications to affect virtually every human endeavor and practice. From the laboratory to the factory floor, from the office to the warehouse, from the kitchen to the automobile, the invasion of the microprocessor has created dramatic changes in equipment architecture and its use.

Computers, whether mainframe, mini, or micro, all have essentially the same basic organization; that is, they all consist of a central processing unit (CPU) which performs the basic arithmetic and logic operations in response to a sequential series of instructions (program), a memory section which stores the program, and an input/output section (I/O), which communicates with external equipment. The principal difference between the three classes of computers is in the word length that can be processed and in the memory capacity, being largest for mainframes and smallest for microcomputers. The larger the word size and the internal memory, the faster the operation of the computer and the larger and more powerful the program which can be sorted and executed.

Additionally, mainframe computers and minicomputers are usually made with bipolar components (TTL or ECL), which are inherently faster than the MOS technology employed for microprocessor units (MPUs), the CPUs associated with microcomputers. But the distinction between the various computer classifications is blurring as the latest microcomputers exceed the capabilities of today's minicomputers and approach those of existing mainframes. That is, microcomputers utilizing the latest 32-bit MPUs can serve applications formerly in the domain of the more powerful minis and mainframes, depending on the amount of memory and other peripheral circuitry used in association. It is likely that in the future the distinction between computers will be based more on the amount of circuitry crammed into a computer's architecture than on the inherent capability of the technology employed.

The pervasiveness of the microcomputer, however, depends not so much on the upper limit of its speed and processing power as it does on its minuscule size and its incredibly

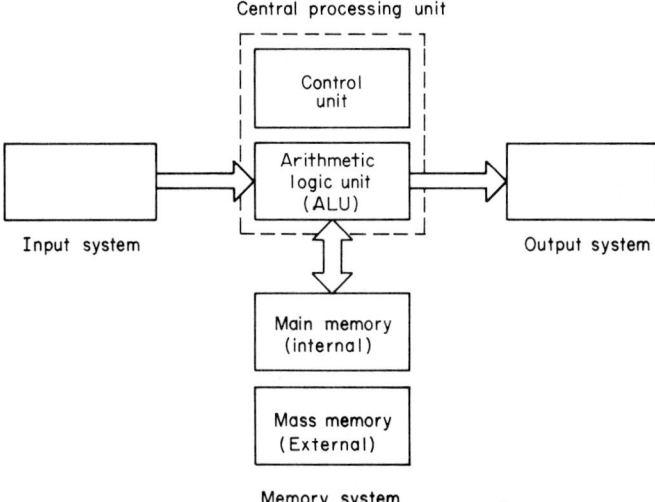

**Fig. 6-78**   Basic components of a typical computer.

low cost. It is this combination that makes a microcomputer suitable not only as a general-purpose machine which must be reprogrammed for each specific end use but also as a preprogrammed and dedicated "engine" that loses its identity as a computer and becomes an electronic ignition system, a smart cash register with built-in inventory control, a control system for home appliances, or one of countless other machines in everyday use.

**27. Computer architecture.**   The diagram in Fig. 6-78 shows the basic components of a typical computer. The heart of the system is the central processing unit (CPU), which consists of an arithmetic-logic unit (ALU) and a control unit. In a microcomputer, these two entities are normally processed on a single chip called a microprocessor unit (MPU). Together, these two sections provide all the action within the computer: the control unit "fetches" the instructions and the data from the memory section in the proper sequence and feeds these to the ALU; the ALU performs the actual arithmetic and logic operations in accordance with the instructions it receives from the control unit and passes the results on to the output section. The output section converts the results into the corresponding electrical stimulus needed to activate the peripheral equipment associated with the computer. This equipment may be a printer, a display terminal, a control mechanism such as a motor or relay, a modem (for transmission to a remote location over telephone lines), or any other electronic or electromechanical system.

The memory system represents the brain of the computer. The main memory stores all the instruction (program) to be run and all the data which are to be manipulated. In a dedicated microcomputer, the memory section is invisible to the user. It consists of an electronic read-only memory (ROM), which stores the code sequence for the particular program that is to be run. In a general-purpose computer, much of the internal memory is random-access (RAM), often called a read-write memory because it can be programmed and reprogrammed at will. With such computers, the specific program to be run must first be loaded into the RAM from an external source through the input section.

The main-memory system shown in the diagram represents that portion of the system which is an integral part of the microcomputer itself. In a microcomputer dedicated to one specific application, this is all the memory needed. General-purpose computers operate in conjunction with associated mass memory, which is external to the computer. Such memories are in the form of disks, tapes, and cassettes which represent permanent storage for the various programs that the computer may be called upon to run at different times. These programs must be transferred from the external memory to the computer's internal main memory before they can be used effectively.

The input section converts the output of a signal source into a format acceptable to the microprocessor and memory. The signals from such input peripherals as the keyboard, disk and tape readers, etc., may already be in the proper format. Other potential input systems such as pressure and temperature sensors, modems, and light or presence detectors have output signals that are incompatible with either the computer memory or the MPU. These must first be converted into the proper format by the input section.

| 1 | 0 | 1 | 1 | 0 | 0 | 1 | 0 |
|---|---|---|---|---|---|---|---|

**Fig. 6-79** An 8-bit computer word (byte) can be implemented by eight parallel-connected transistors, or memory cells, each programmed to either a high level (1) or a low-level (0) state. The word can represent a letter, a number, a symbol, or an instruction code.

**28. Computer language.** The language of the modern computer is *binary*. That is, all data and all instructions processed by the computer consist of a series of binary digits (bits) representing either a zero (0) or a one (1).

A zero is normally represented by a low voltage level, while a one is represented by a high voltage level. The actual voltage value representing a 1 or a 0 depends upon the type of logic employed.

The basic computer word consists of eight bits, called a byte (Fig. 6-79). A single byte can represent decimal numbers from 0 to 256 ($2 \times 10^8$); 2 bytes (16 bits) can represent numbers up to 65,536 ($2 \times 10^{16}$), etc. Similarly, a 2-byte (16-bit) word can encompass over 65,000 memory addresses or program instructions.

Within the computer, information is processed in blocks of parallel bits (bytes or multiple bytes) which are transferred from one section to another along parallel paths called *buses*. A microcomputer generally has three distinct buses: a data bus, an address bus, and a control bus (Fig. 6-80).

The data bus carries the data words between the ALU and the main memory. It is therefore bidirectional and usually has as many lines as the number of data bits which the ALU is able to process in one operation. Thus, an 8-bit MPU has an 8-bit data bus; a 16-bit machine has a 16-bit data bus, etc. An 8-bit machine could be used to process a 16-bit data word, but it would require two cycles of operation and would be considerably slower than a 16-bit machine.

The address bus is used by the control unit to fetch data and instructions from specific locations in the main memory and transport it to the ALU for processing. The most popular 8-bit machines utilize 16-bit address buses to access in excess of 65,000 memory address locations.

The control bus accesses all portions of the microcomputer. It controls the sequence of events that carries out a particular instruction, activating the read and write lines of the memory, as needed, initiating the transfer of instructions and data to various parts of the CPU in the proper sequence, and generally controlling the entire sequence of program execution.

**29. Microcomputer hierarchy.** The first major component of the microcomputer era was a PMOS 4-bit microprocessor introduced by Intel Corp. in 1971. This limited-capability chip was soon joined by 8-bit NMOS MPUs from a number of manufacturers. As processing technology improved to permit increased component density, processor power increased first to 16-bit and, most recently, to 32-bit data manipulation on a single chip of silicon. The latter utilize CMOS technology, which, owing to its extremely low power requirements, permits the operation of tens of thousands of transistors in such a small area without demanding extraordinary measures of heat dissipation.

While microprocessor power was expanding from 4 to 32 bits, microcomputer-chip architecture was expanding in other directions as well. Recall that a microprocessor consists of a single chip housing the ALU and control-unit functions. A complete microcomputer requires memory and I/O circuitry as well. Understandably, as increasing component densities facilitated the increase of on-chip componentry, designers began to utilize

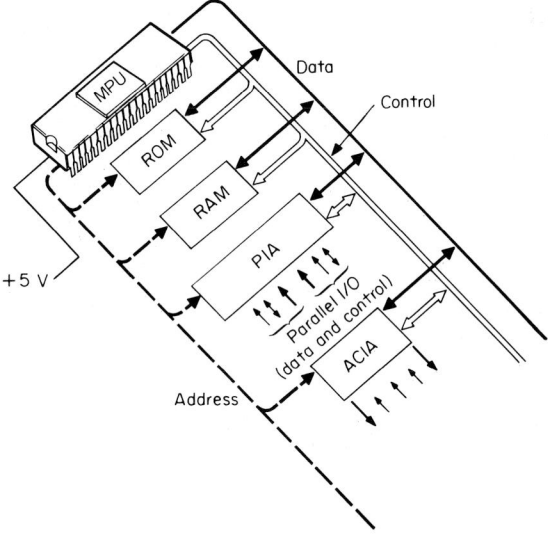

**Fig. 6-80**   A typical MPU bus system uses three separate interconnect paths to carry signals between the MPU and its peripheral chips. PIA = peripheral interface adapter; ACIA = asynchronous communications interface adapter.

the excess chip space to place memory and I/O functions on the same chip with the MPU. This resulted in a diversified series of components called microcomputer units (MCUs), which provide virtually complete microcomputer capabilities on a single chip. The repertoire of MCUs is necessarily large because of the many different applications in which microcomputers are employed. These varying applications require different amounts of memory and, in many instances, different I/O functions. Thus, with a basic MPU structure as a core, semiconductor manufacturers are offering MCUs that are tailor-made for different end-use functions by surrounding the MPU with differing complements of on-chip peripheral circuits (Table 6). This capability is becoming so commonplace that some manufacturers are including microprocessor cores and associated peripheral choices as basic cells to be used in the design of semicustom (ASIC) microcomputers.

**30. Product availability.**   The proliferation of microcomputer applications has spawned a large array of component building blocks with which to implement microcomputer systems. Fundamental, still, for low-level applications are the 8-bit MPUs with their wide assortments of associated memory and I/O chips. These have recently been joined by increasing numbers of 8-bit MCUs which can be selected for dedicated applications without the need for additional peripheral chips except, perhaps, for interface circuits to match an MCU to the voltage and current requirements of specific input and output equipment.

For more sophisticated computer applications, 16-bit microprocessors have become commonplace, and 32-bit MPUs are in the early stages of implementation. These circuits are so complex that manufacturers have not yet combined them with memory and I/O peripherals on the same chip, but associated peripheral chips to match the characteristics of the MPUs are being introduced in considerable numbers. These peripherals are often as complex as the MPUs themselves and greatly reduce design time as well as the size of the resulting equipment.

Implementation of microcomputers in a wide range of personal computers as well as dedicated machines is further aided by the availability of increasing numbers of board-level products. These predesigned boards greatly speed up the equipment design cycle. They are being offered not only by the chip manufacturers themselves but also by other manufacturers. Regardless of manufacturer, the boards are designed in families that work together to form complete systems. They offer a choice of basic single-board–microcomputer modules as well as additional peripheral modules that provide a variety of memory capacity and interface choices.

**Table 6   On-Chip Peripheral Variations Available with a Basic MC6805 MPU Core**

| Features / Suffix | HCMOS MC68HC05 | CMOS MC146805 | | | |
|---|---|---|---|---|---|
| | C4 | E2 | F2 | G2 | H2 |
| Number of pins | 40 | 40 | 28 | 40 | 40 |
| RAM (bytes) | 176 | 112 | 64 | 112 | 112 |
| User ROM (bytes) | 4160 | 0 | 1089 | 2106 | 2106 |
| I/O lines, bidirectional | 24 | 16 | 16 | 32 | 24 |
| I/O lines, unidirectional | 0 | 4 | 0 | 4 | 0 |
| Timer (bits) | 16 | 8 | 8 | 8 | 8 |
| Special features | SPI, SCI | Bus expander | . . . | . . . | COP, synthetic audio output tone generator |
| EPROM version | . . . | . . . | MC1468705F2 | MC1468705G2 | |

NOTE   SPI = serial peripheral interface; SCI = serial communications interface; COP = computer-operating-properly reset timer which acts as a watchdog to reset the CPU automatically if it is not reset by a program sequence within a given time.

Board families are defined in terms of the bus structure they support; that is, all boards within a given bus family, say, the VMEbus family, have identical pin assignments so that they will be properly interconnected when plugged into a backplane. It is possible, therefore, to purchase VMEbus boards (modules) from different vendors and be assured that they will be compatibly interconnected when plugged into a VMEbus chassis. The large variety of modules available for some bus structures has given rise to the open-system philosophy which permits rapid assembly of complete functional systems with tailor-made capabilities by choosing the appropriate modules and plugging them into the slots (sockets) of a prewired chassis (Fig. 6-81).

### ELECTRONIC MEMORIES

While MCUs normally have on-chip memory sections with a memory capability deemed adequate (by the manufacturer) for the intended application, the MPUs implement their memory sections with peripheral chips. Since all microcomputers require some sort of memory, these devices are the most prevalent of the microprocessor peripherals.

There are many different types of memories. They are classified in a variety of ways, as shown in Table 7. These memories are electronic memories associated with the actual operation of the computer, not with the permanent mass storage of operating programs or large files of data. The mass-storage memories, such as disks, tapes, and cassettes, are separate from the microcomputers and are used only as sources of information to be entered into the computer when required. They are *nonvolatile*, meaning that the information they contain is "permanent" (unless deliberately removed), and they store large amounts of data in a serial fashion. This means that access time is relatively slow.

On the other hand, the so-called main memory, or operating memory, within the computer is electronic in nature, and information is stored in a matrix of cells that are accessed in a parallel manner, so that information transfer to and from such memories is relatively fast.

As with logic circuits, memory characteristics are determined by the technology employed. Bipolar ECL and TTL memories are the fastest, NMOS memories have the highest chip density and therefore the highest storage capacity, and CMOS memories have the lowest power dissipation.

**Fig. 6-81**   A prewired chassis and board assortment simplifies microcomputer design and assembly.

**31. Memory organization.**  An electronic memory consists of a number of storage units called *cells*. Each cell is a transistor circuit that can be charged to either of two binary states: a high (voltage) level or a low level. A high level is designated as 1, and the low level is designated as 0. In practice, the cells are operated in groups called *words* which store information in binary coded form. The content of a memory word can be the binary code for a number, a letter, or an instruction for manipulating data.

A memory is specified by two numbers that indicate the total number of cells (bits) in the memory and the manner in which these cells are grouped. A $1024 \times 1$-bit memory (commonly called 1K memory) contains 1024 cells, each of which is separately addressable and programamble. A $128 \times 8$-bit memory also contains 1024 cells. Each one of these is also individually programmable, but each is accessible only with a group of seven other cells with which it is permanently associated.

**Table 7   Memory Descriptors**[a]

| | |
|---|---|
| By function | RAM, ROM, PROM, EPROM, EEPROM |
| By capacity (number of bits) | 1K (1000), 16K, 64K, 256K, 1M (1,000,000) |
| By organization | $4K \times 1$, $4K \times 4$, $2K \times 8$, etc. |
| By technology | NMOS, CMOS, bipolar (TTL, ECL) |
| By operation | Dynamic, static (RAM only) |

[a]A representative description might be: a $256K \times 1$-bit NMOS dynamic RAM.

**32. Memory capacity.**   Memory chips are available in a variety of sizes and cell organizations. Capacity has been increasing rapidly as computer power and processing technology have increased.

Early memories had a capacity of only 128 bits, or 16 bytes. The subsequent 1K-bit memory was considered quite a breakthrough but was quickly replaced by 4K, 16K, and 64K memories. Today, 265K-bit memories are available and 1M (megabit) devices are being introduced. These large-capacity memories are fabricated with NMOS technology, with smaller capacity units being available in CMOS and bipolar processing.

**33. Memory functions.**   Functionally, electronic memories are divided into two basic classifications: random-access memories (RAMs) and read-only memories (ROMs). RAMs are *volatile* memories whose content disappears when power to the memory is interrupted.

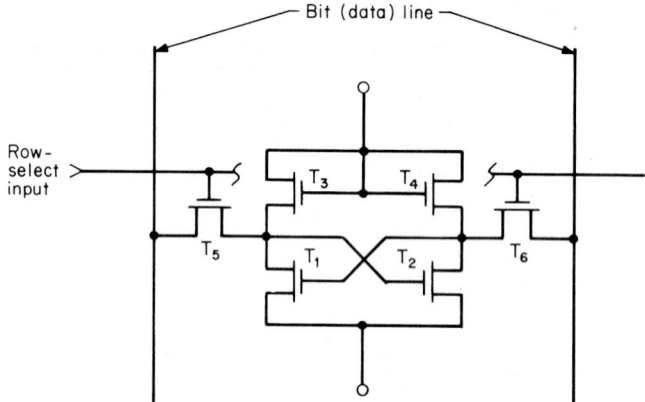

**Fig. 6-82**   A typical six-transistor static MOS memory cell. $T_1$ and $T_2$ comprise a flip-flop, with $T_3$ and $T_4$ representing load resistors (note the fixed bias on these transistors). $T_5$ and $T_6$ are isolation transistors which, when turned on by a row-select signal, cause the state of the flip-flop to be reflected on the bit (data) lines.

ROMs are *nonvolatile*, in that they retain their content even when power is shut off. Hence, RAMs are used for the actual manipulation of a program or the storing of a program on a temporary basis, while ROMs are employed for storing permanent data such as microprograms, look-up tables, and display graphics.

The addresses of both RAM and ROM are accessible in a random fashion, but RAMs are read-write memories whose contents can be changed at will, whereas ROMs are read-only memories. In general, ROMs are preprogrammed during manufacture with the custom program specified by the customer. There are, however, a number of variations— PROMs, EPROMs, EEPROMs—which are custom-programmable by the user.

RANDOM-ACCESS MEMORIES (RAMs)   There are two types of RAMs: static and dynamic. In a static RAM, the individual cells are flip-flops which, once set to either 1 or 0, will retain their setting indefinitely, until either deliberately reset or until power is removed. The basic cell (Fig. 6-82) consists of six transistors. $T_1$ and $T_2$ represent the actual flip-flop, $T_3$ and $T_4$ function as load resistors for the flip-flop, and $T_5$ and $T_6$ are isolation transistors that isolate the cell from the access lines (bit lines) until the cell is ready to be addressed.

The basic cell of a dynamic MOS RAM (Fig. 6-83) utilizes only a single transistor and therefore requires considerably less chip space. Data storage is accomplished by charging the associated capacitor for a 1 level or discharging it for a 0 level. In the diagram, the capacitor is charged when both $X$ and $Y$ lines are made high and discharged when the $X$ line is high and the $Y$ line is low. Unfortunately, the small capacitor will not retain its charge very long. Therefore, dynamic RAMs must be refreshed (the charged cells must be recharged) periodically (say, every millisecond) in order to maintain their information. The refreshing circuitry adds substantial complexity to the circuit. On the other hand, dynamic RAMs dissipate a relatively small amount of power. For this reason and because of the smaller space requirements, they are preferred for memories with a large number of cells (large capacity).

Read-Only Memories (ROMs, PROMs, EPROMs, EEPROMs)

*ROMs.* Read-only memories are an integral part of all microcomputers. They store the operating routines of the system (microprograms) and provide conversion data, display codes, and other information that is periodically accessed by the computer but is not subject to change.

A ROM consists basically of a series of OR gates addressed by a decoder circuit (Fig. 6-84). The truth table associated with the diagram indicates the state of the four output lines, $b_0$ to $b_3$, as a function of the input signals $A$ and $B$. If, for example, the input signals are both HIGH, then output lines $b_1$ and $b_3$ are HIGH because the output of AND gate 11 in the decoder is HIGH. Similarly, if $B$ is HIGH and $A$ is LOW, output lines $b_0$, $b_1$, and $b_3$ will be HIGH because the output of decoder gate 10 is HIGH. Output $b_2$ will be HIGH only when the output of NAND gate 01 is HIGH, which occurs only if $A$ is HIGH and $B$ is LOW. Thus, for any combination of input signals, the output can be determined by the number and placement of the diodes that comprise the OR gates.

The simple circuit described is a 4 × 4 ROM consisting of 4 words, with 4 bits per word. If the number of output lines were extended to 8, the resultant ROM would be

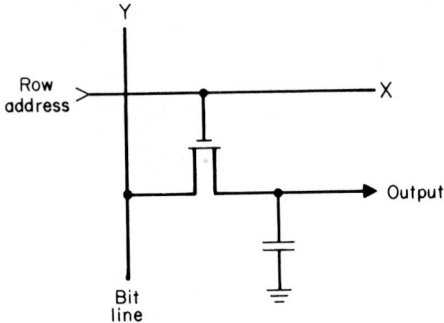

**Fig. 6-83**  A single-transistor dynamic RAM cell. Additional circuitry is needed to refresh all cells in the memory on a regular, periodic schedule.

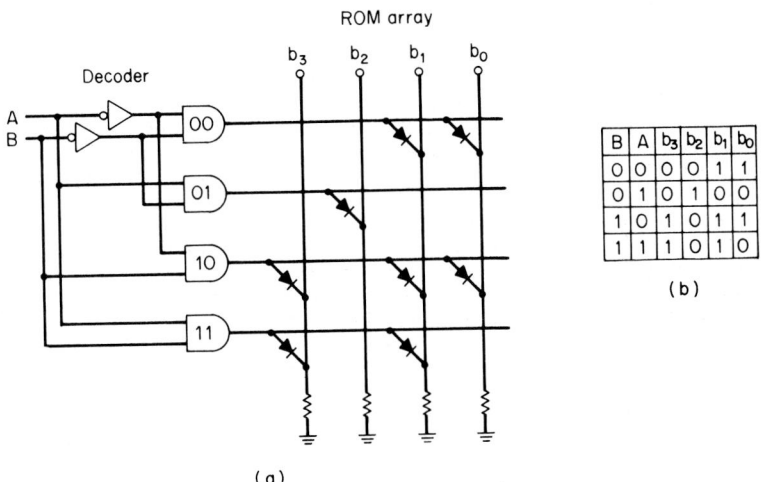

| B | A | $b_3$ | $b_2$ | $b_1$ | $b_0$ |
|---|---|---|---|---|---|
| 0 | 0 | 0 | 0 | 1 | 1 |
| 0 | 1 | 0 | 1 | 0 | 0 |
| 1 | 0 | 1 | 0 | 1 | 1 |
| 1 | 1 | 1 | 0 | 1 | 0 |

(b)

(a)

**Fig. 6-84**  (a) Basic configuration of a typical read-only memory. (b) Truth table corresponding to the simple circuit.

$4 \times 8$. An increase in the capacity of the decoder will provide ROMs with thousands of different 8-bit output combinations.

ROMs are available as factory mask-programmable devices which are fabricated by the manufacturer to any set of custom patterns specified by the user. They can also be obtained as standard preprogrammed devices containing various popular functions such as character generators, code converters, look-up tables, etc.

*PROMs.* The ROM of Fig. 6-84 can be converted into a unit that is user-programmable (PROM) simply by placing diodes in all possible locations within the matrix and fabricating a fusible link in series with each diode (Fig. 6-85). The user can burn out any link, using a PROM programmer, in order to create any special set of codes desired. Of course, this can be done only once, so that this type of PROM is sometimes called a one-time programmable ROM.

*Erasable programmable ROMs (EPROMs).* The ideal memory is a nonvolatile one which maintains its program even after power is removed from the system (as with a standard mask-programmed or one-time field-programmed PROM discussed above), yet one

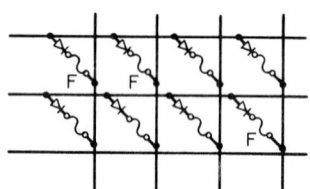

**Fig. 6-85** Fusible links in series with each diode can be "blown" in a PROM programmer to create any desired pattern arrangement.

**Fig. 6-86** Ultraviolet erasable read-only memory (UV EPROM).

whose program can be altered at will (as with a standard RAM). Development efforts in that direction are bearing fruit, and several iterations of programmable ROMs have been introduced. These do not yet have the reprogramming speed and ease of the conventional RAM, but they are enjoying considerable popularity for a number of applications, and they are indicative of the rapidly expanding technology.

One of the earliest practical ROMs that was repeatedly erasable and reprogrammable in the field was labeled EPROM. The information stored in the memory cells could be erased by subjecting the device to ultraviolet light, which entered the package through a quartz window at the top (Fig. 6-86). While the EPROM can be erased and reprogrammed repeatedly, it must be removed from the system and bulk-erased by exposing the entire unit to a high-intensity ultraviolet-light source. There are a number of suitable ultraviolet programmers, but the erase time is on the order of half an hour, which falls far short of meeting the criteria for the ideal memory. Moreover, the EPROM can be unintentionally erased by sunlight and fluorescent light and requires special shielding after programming. Despite these limitations, the UV EPROM still enjoys considerable use, and its principles of operation are fundamental to even the more advanced devices currently emerging.

The EPROM is basically an MOS device whose structure is illustrated in Fig. 6-87. It is characterized by two stacked polysilicon gates which are separated from each other and from the substrate of the cell by an insulating layer of silicon dioxide. The gate closest to the substrate is a floating gate which is not connected in the circuit. The upper gate is the control gate, which is used to energize the cell. Under the initial condition, the cell acts as a conventional MOS transistor so that when a positive voltage is applied to the control gate and the drain, a channel is created between source and drain, permitting a drain-current flow.

If the drain voltage and the control-gate voltage are raised to a high level (say, 30 V), some of the electrons in the channel gain enough energy to cross the silicon dioxide barrier and impinge upon the floating gate, where they become trapped. This raises the turn-on threshold level of the MOS transistor so that it will not be turned on when normal operating voltages are applied. The cell then is said to be charged, representing an open circuit and yielding a high output level (1) when connected in a conventional circuit.

The charge on the floating gate can be removed by the application of light to the structure. Sunlight or fluorescent light will erase the charge but would require an excessive amount of time. A strong dose of ultraviolet light will bulk-erase the entire memory in about 30 min.

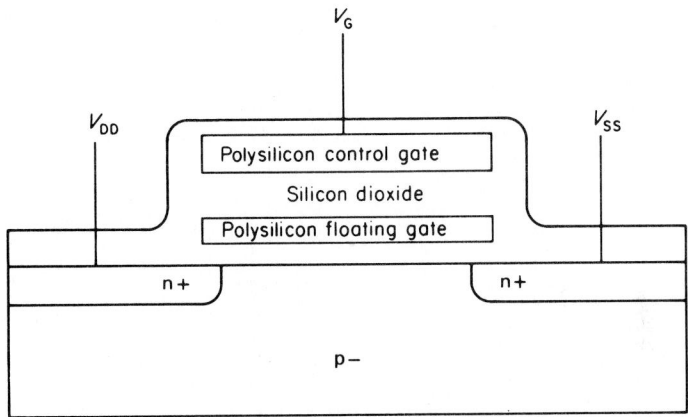

**Fig. 6-87** A typical EPROM-device structure.

*Electrically erasable programmable ROMs (EEPROMs).* The EEPROM is very similar in theory and structure to the EPROM. The principal difference is in the thickness of the oxide between the floating gate and the substrate. Whereas the thickness of this oxide layer for the EPROM is on the order of 1000 Å, that of the EEPROM is on the order of 100 Å or less. As a result of this thinner oxide, the floating gate can be charged *and discharged* by a voltage applied to the control gate. Write and erase times for typical EEPROMs with present technology are on the order of 10 ms. While the EEPROM is not competitive with RAM speeds, its in-circuit program-erase capability makes it suitable for a great many applications. Devices of this type are currently available with 64K-bit capacity. They are also included as on-chip peripheral circuits for a number of MCUs to expand MCU versatility.

**34. Accessing the memory.** The total number of cells in a single memory chip can range from several hundred on the low end to 1 million (1 megabit) with current state-of-the-art technology. The cells are arranged in rows and columns (Fig. 6-88), with the exact location of each cell specified by its row and column number. There are two basic organizational schemes for accessing the memory: bit organization and word organization.

BIT ORGANIZATION In a bit-organized memory chip (e.g., 1K × 1, 64K × 1, etc.), each cell is addressed in conjunction with similarly located cells on other chips of the same capacity and organization. In Fig. 6-89, for example, eight identical chips are used to store information in byte-sized (8-bit) words. Each chip stores

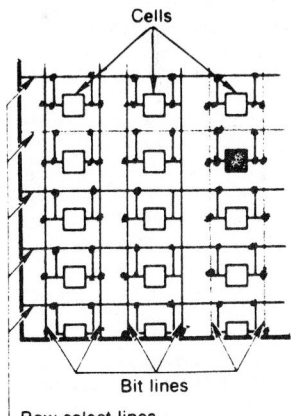

**Fig. 6-88** A semiconductor memory is a rectilinear pattern of digital circuits, called cells, each of which can hold a high-level (logic-1) or low-level (logic-0) voltage charge. Each cell is accessed for programming or reading simply by energizing the proper column and row.

one of the 8 bits, and all chips are addressed simultaneously for programming or reading.

WORD ORGANIZATION    A word-organized memory chip is one in which a number of columns, corresponding to the number of bits in the word, are addressed simultaneously. In this manner, a complete word can be stored in a single chip (Fig. 6-90).

**35. Memory circuitry.** A memory array receives a number of signals from the microprocessor. The primary ones are as follows:

1. The address signal which activates the rows and columns on each chip that represent the cells associated with a given address

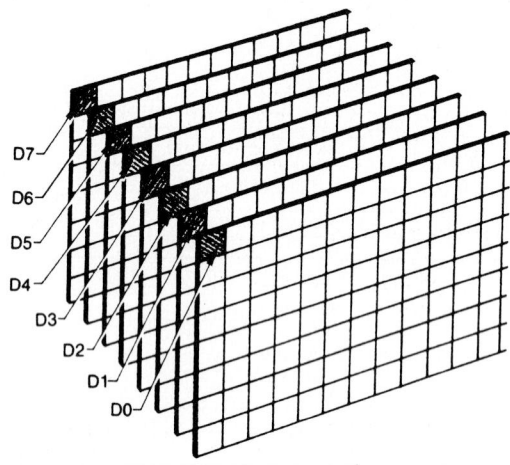

**Fig. 6-89**  Bit-organized memory requires multiple memory chips, with each chip storing 1 bit of a complete word.

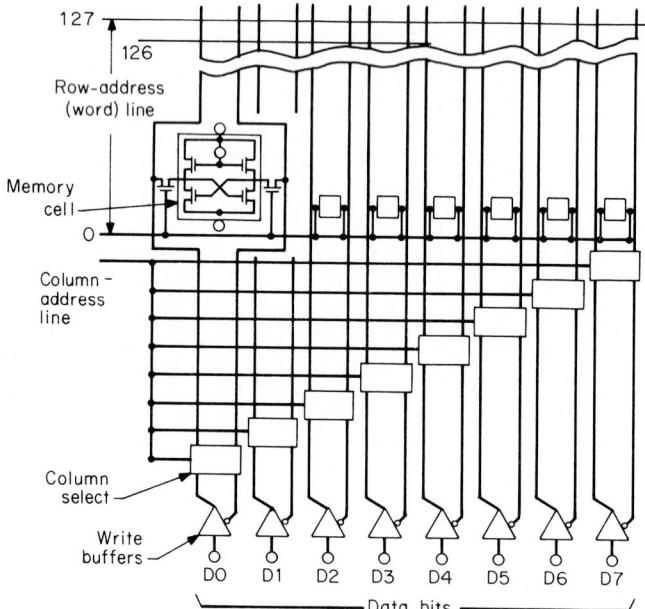

**Fig. 6-90**  A 1K (128 × 8) byte-organized chip addresses all bits associated with a complete word on a single chip.

2. The binary word representing the data to be entered into the addressed cells in the case of a programming step

3. A read-write signal generated by the processor which determines whether information is to be entered into the memory or extracted from it

The circuitry required to respond to these signals and generate the desired results resides directly on the memory chip along with the array of memory cells. Typical operation can be described with the aid of diagrams utilized in a representative MOS static memory chip.

THE MEMORY CELL Figure 6-91 is a conventional MOS flip-flop in which $T_1$ and $T_2$ are cross-coupled inverters and $T_3$ and $T_4$ represent their respective load resistors. A positive pulse applied to point $A$ appears at the gate of $T_2$, turning this transistor on. As a result, the voltage at point $B$ goes to 0 (ground), and since this voltage is applied to the gate of $T_1$, the latter is turned off. Voltage at $A$ therefore goes high and remains high even when the external signal is removed. In this condition, the cell is considered to store a 1.

To reprogram the cell to contain a 0, a positive pulse is applied to point $B$, turning $T_1$ on and $T_2$ off. Again, this condition remains stable until it is deliberately changed by another positive signal to point $A$. Clearly, the cell is programmable.

ROW ADDRESS In practice, a cell is programmed by means of a two-rail bit-line system (Fig. 6-92), in which a signal and its complement are simultaneously applied to points $A$ and $B$. To be applied to the cell, however, isolation transistors $T_5$ and $T_6$ must be turned on. Programming is accomplished by applying the desired programming signals on the bit lines and then momentarily turning on $T_5$ and $T_6$ with a pulse applied simultaneously to their gates. Once programmed, a cell is read by briefly turning $T_5$ and $T_6$ on without having an external signal on the bit lines.

COLUMN ADDRESS In Fig. 6-93, the column-select circuitry and a write buffer have been added to Fig. 6-90. Note that the write buffer generates the programming signal and its complement, to be applied to the bit lines. To access the bit lines, however, transistors $T_7$ and $T_8$ must be turned on. The address code defines which of the columns is to be activated. In Fig. 6-93, each of the columns can be separately addressed, as required for a single-bit organization. In Fig. 6-90, all eight of the column-select circuits are activated simultaneously, as required for a byte-organized orientation.

READ-WRITE CIRCUIT In Fig. 6-94, a read buffer circuit has been added to the previous diagram, permitting the data on the bit lines to be applied to the data bus for transmittal back to the MPU. Note, however, that the data line that carries the data from the memory

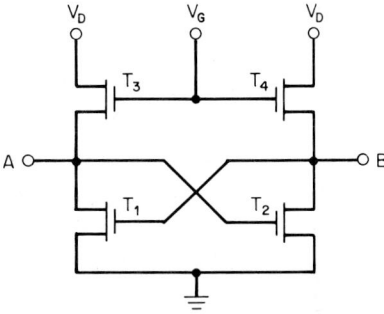

**Fig. 6-91** A typical MOS static memory cell.

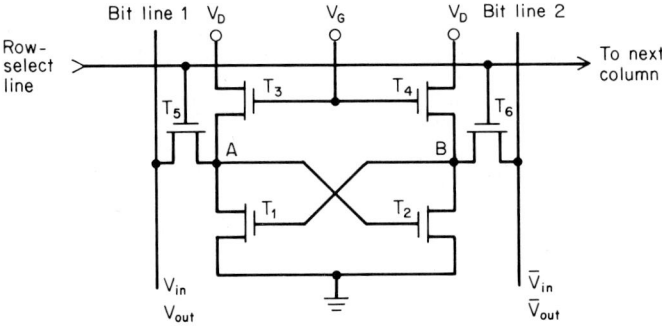

**Fig. 6-92** A cell is programmed or read by a signal on the row-select line which causes the cell to be connected to its bit lines through $T_5$ and $T_6$.

to the MPU is the same as the line that carries the programming data from the MPU to the memory. To avoid undesirable feedback, the read buffer must be disabled during a write operation. This is accomplished with a three-state read buffer circuit as in Fig. 6-94*b*.

THREE-STATE BUFFER    A three-state buffer is a circuit whose output can be high, low, or open-circuited. A high output at bit line 1 turns $T_4$ on and $T_1$ off. If $T_2$ and $T_3$ are turned

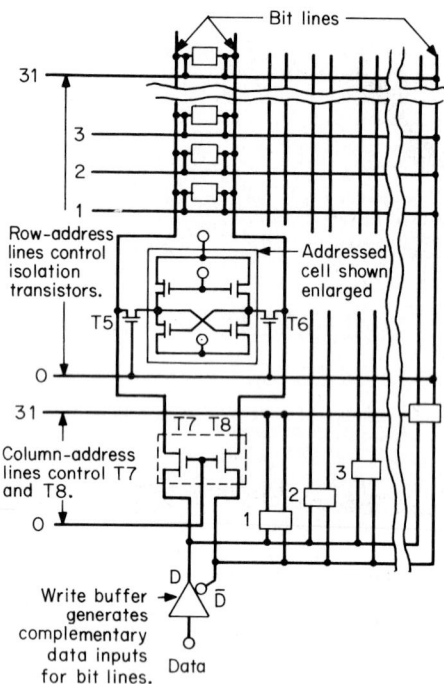

**Fig. 6-93**    Column-address circuitry. Data can be entered into the memory only when transistors $T_7$ and $T_8$ are turned on.

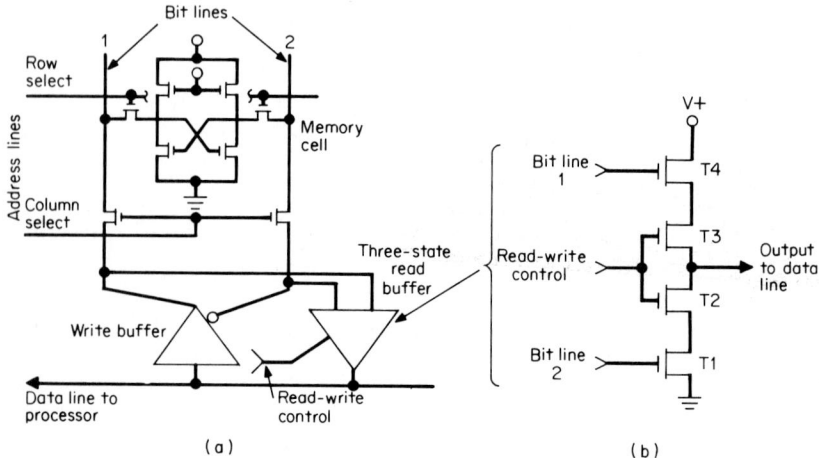

(a)

(b)

**Fig. 6-94**    (*a*) Connection of read buffer to memory column. (*b*) Three-state read-buffer schematic diagram.

on by a separate high-level signal on their gates, then the output line is connected to the positive-voltage source through $T_3$ and $T_4$, and the output is high. A high output at bit line turns on $T_1$, turns off $T_4$, and presents a low signal to the output. But if $T_2$ and $T_3$ are turned off, then regardless of the state of $T_1$ and $T_4$ the output is isolated from the inputs because of the very high resistance of $T_2$ and $T_3$.

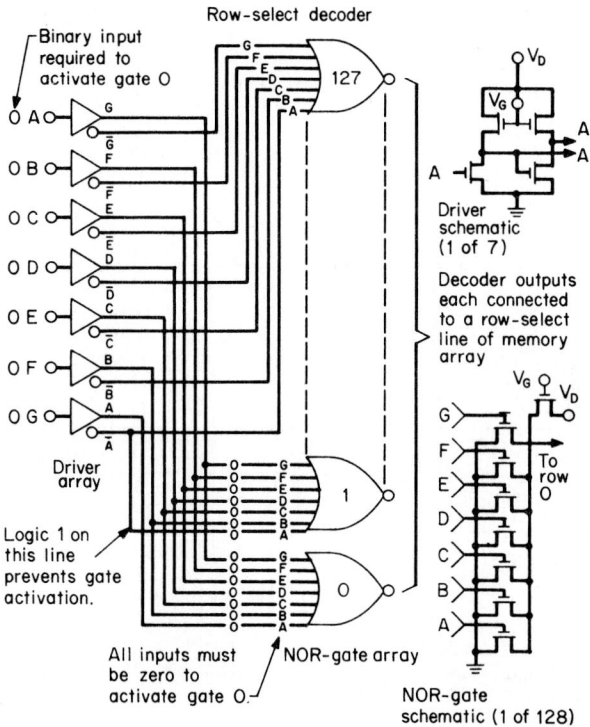

**Fig. 6-95**  Example of an address decoder for a single-column 128 × 8-bit memory. The output from each NOR gate enables one of the 128 different rows in the memory matrix.

$T_2$ and $T_3$ are turned off during writing operations and turned on at all other times. Thus, the cell is always connected to the data line through the read buffer except during a write cycle, during which the read buffer is in the open-circuit state.

**36. Address decoding.**    The number of unique bit patterns required to address all rows in a single-column, word-organized memory equals $2^n$, where $n$ is the number of rows. Thus, a small 128 × 8-bit memory, for example, requires 128, or $2^7$, unique address patterns, which can be generated with a 7-bit binary code. These 128 unique address patterns can be applied to a single-address decoder that has 128 output lines. For each address, the decoder provides a high-level signal on a different output line. There are many different decoder circuit configurations. The circuit shown in Fig. 6-95 is for illustration and does not necessarily represent any specific design.

DECODER CIRCUIT    The *decoder circuit* consists of 128 seven-input NOR gates having a common driver circuit. Each NOR gate is activated (puts out a high-level signal) only when all seven inputs are low. The driver circuit, in turn, is configured so that each of the 128 possible input patterns activates a different NOR gate.

Each NOR gate consists of seven parallel-connected transistors (switches) in series with a single fixed-bias transistor (resistor). When all seven parallel-connected transistors are turned off by a low-level input to each, the output is high and the row of cells associated with that gate is activated.

To provide the required all-zero signals to the gate to be activated, the 7-bit energizing signal is first sent through double-inverter driver stages, one for each bit. The output of each driver then makes available both the signal (A, B, C, . . .) and its complement ($\overline{A}$, $\overline{B}$, $\overline{C}$, . . .).

For example, if gate 0 (row 0) is to be activated by the first address pattern (0000000), then gate-0 input terminals are connected to outputs (A, B, C, D, E, F, G) of the driver. To activate gate 1 by address No. 2 (0000001), gate-1 terminals are connected to driver outputs $\overline{A}$, B, C, D, E, F, G. The last gate in the system (No. 127) has all its inputs connected to the inverted outputs of the driver. It is possible, therefore, to connect the driver outputs to each gate in such a manner that each gate receives an all-zero input for a different combination of 1s and 0s at the decoder input.

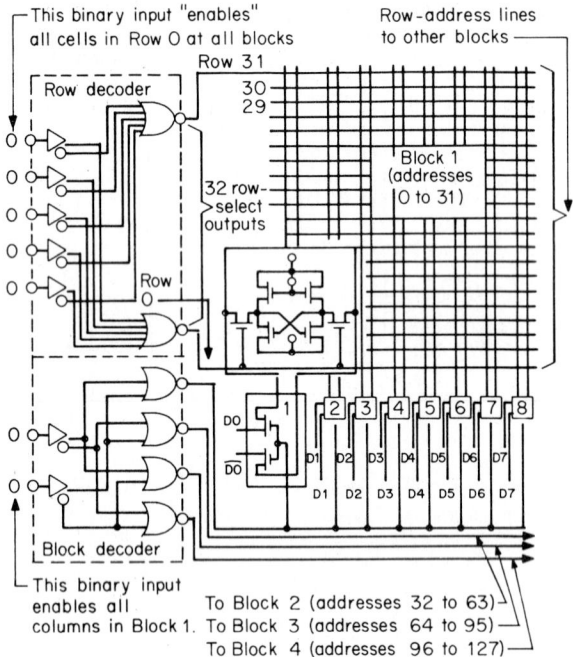

**Fig. 6-96**    A row-and-block–select decoder.

Most memories, instead of being arranged in one continuous sequence of rows, are arranged physically and electrically in different blocks. A $128 \times 8$-bit memory, for example, could contain four blocks, each with 32 rows and eight columns. Here, the 8-bit data signal could be applied via a data bus to all eight columns of each block simultaneously. However, this arrangement needs separate decoding circuits to activate not only the desired row but the desired eight-column block as well. Accordingly, a 2-bit block-select decoder (Fig. 6-96) is needed for four-pattern ($2^2$) selectivity. However, since the row-select decoder need now decode only 32 lines ($2^5$) rather than 128, a 5-bit decoder is sufficient for row selection. Thus, again, a 7-bit decoder handles the entire job.

The only difference between this circuit and the previous one is that four gates of the decoder are used to select the blocks while the row-select function is handled by 32 five-input NOR gates. The same binary address pattern that accesses a memory address in a one-block system can access the same address in a four-block organization and other organizations as well.

NOTE    Much of the information provided in this division is based on the work of technical and marketing specialists of the Motorola Semiconductor Sector. The author gratefully acknowledges the kind permission of Motorola for the use of material from its various data sheets, application notes, and other documents.

## PRODUCT RELIABILITY

**37. Semiconductor reliability.** In today's marketplace, quality and reliability are paramount to the success of a product. Quality is defined here as the ability of a device to operate within its specifications when received by the user; reliability refers to its capability to operate within specifications over long periods of time and under the most adverse environmental conditions to be encountered in practice.

Semiconductors are inherently extremely reliable. Properly designed and used, they can be expected to outlive the equipment in which they are employed. Yet they are subject to a wide range of failure mechanisms that must be anticipated and circumvented during manufacture in order to achieve this inherent feature.

To ensure high reliability, the government has developed a stringent series of tests and screens involving all phases of manufacturing for military applications. The manufacturing techniques required to meet these qualifications have been adopted in some measure by manufacturers of products intended for commercial applications so that, except for very special requirements, there are no longer significant variations in inherent reliability between commercial and military products.

BASIC MILITARY QUALIFICATIONS TESTS   The basic MIL-qualification tests, called JAN (Joint Army-Navy), are divided into groups to satisfy three objectives:

1. Group A—verifies that form, fit, and function conform to design specifications.
2. Group B—assures manufacturing integrity and verifies reliability in ground support applications.
3. Group C—provides evidence of long-term reliability under harsh environmental conditions where severe mechanical and environmental stresses exist.

Examples of the types of tests included in each group are given below.

*Group A Tests*

*Visual/mechanical.* Consists of a sampling of devices that is examined to determine whether they meet the applicable materials, design, construction, marking, and workmanship standards.

*DC tests.* Verify the major voltage, current, and other dc parameters of the device first under normal (25°C) temperature conditions, then at an appropriate high and/or low temperature limit to confirm satisfactory performance over the entire temperature range.

*AC tests.* Check parameters associated with dynamic operating conditions, such as capacitances, noise figure, switching time, etc.

*Safe operating area (SOA) tests.* Limited to power transistors, these tests corroborate the power limits over which the devices are designed to operate.

*Current surge.* Applies principally to diodes, e.g., rectifiers.

*Selected tests.* Assigned to specific devices for unique specifications that do not fit a general specification.

Thus the complete spectrum of electrical device characteristics is accounted for, and successful completion of this test sequence provides assurance that the devices are capable of operating, at least initially, in accordance with their design.

*Group B Tests*

This sequence of tests includes screens that are intended to verify that the devices are mechanically sound and that they can be expected to operate satisfactorily over time and under adverse operating conditions. Since a number of these screens involve stress factors that could result in ultimate performance degradation, the electrical parameters expected to be affected are tested before and after the applied screen to ascertain that the performance change remains within prescribed limits.

*Solderability.* Determines the solderability of wires, lugs, tabs, and all types of terminals that are normally joined by a soldering operation. The procedure includes an accelerated aging test that simulates a minimum of 2 to 10 years natural aging under a combination of various storage conditions that have different deleterious effects.

*Resistance to solvents.* Verifies that the markings will not become illegible when the component is subjected to various solvents employed by equipment manufacturers.

*Thermal shock.* Simulates transferring equipment from a heated shelter to the outside environment in an arctic area. While temperature limits may vary with specified conditions, the test usually involves exposure of the components to temperatures of −55°C and +125°C, or the maximum rated specification of the component.

*Surge current test.* Subjects devices under test to high forward current stress conditions to determine the ability of the chip and the contacts to withstand current surges.

*Hermetic seal.* Determines the effectiveness of the seal of semiconductors with internal cavities. There are two types of required tests—fine leak and gross leak.

*Steady-state operation life.* Designed to weed out components that might be subject to "infant mortality" on the proven premise that if a semiconductor device is likely to fail, it will do so during the early hours of normal operation.

*Intermittent operation life.* Similar to the steady-state operation life test, above, except that the devices are subject to sudden sequential on-off periods to provide added stress.

*Blocking life.* Performed on rectifier diodes only, this test is normally run at an elevated temperature of 150°C for 340 h, with the primary blocking junction reverse-biased. The reverse-bias voltage is 80 to 85 percent of the rated voltage of the device.

*Decap internal visual.* Verifies conformance to the original specifications by "opening" samples of a completed lot and comparing the internal structure with the qualified design report.

*Scanning electron microscope (SEM) inspection.* Verifies the quality of the metallization on the semiconductor die by checking for defects such as voids, separations, notches, cracks, depressions, or tunnels and other processing-related faults.

*Bond strength.* Verifies the integrity of wire (or clip) bonds within the package by applying stresses that will cause the maximum potential for wire breaks and associated die failures.

*Thermal resistance tests.* Determine the efficiency of the chip-to-header interface in transferring heat from the chip to the header.

*High-temperature (nonoperating) life.* Verifies that the devices will not degrade after storage at maximum temperature for 340 h.

*Group C Tests*

Whereas previous test groups deal with the design and manufacturing integrity of semiconductors, this group of tests adds additional assurances that the products will withstand severe environmental stresses over prolonged time periods.

*Thermal shock (glass strain).* Exposes the devices to sudden extreme changes in temperature by alternately immersing them in hot and cold liquids ranging from 0 to 100°C and, subsequently, testing them for hermeticity and electrical parameters to determine if any failures occurred.

*Terminal strength.* Checks the capabilities of the device leads, welds, and seals for their ability to withstand pulls and bends and other physical stresses.

*Moisture resistance.* An accelerated test method of evaluating the resistance of the device to the deteriorative effects of high humidity and heat typical of tropical conditions.

*Mechanical shock.* Consists of mounting the devices in a fixture that is subsequently dropped repeatedly from a predetermined height to produce a high-impact force on the various mechanical parts and bonds.

*Vibration.* Simulates the use of devices in the field, when mounted on a jeep, truck, or airplane, in order to confirm the integrity of the package and the wire bonds within the package.

*Constant acceleration.* Designed to detect structural and mechanical weaknesses not detected in other shock and vibration tests.

*Salt atmosphere.* An accelerated laboratory corrosion test simulating the effect of shipboard or seacoast atmospheres on devices by subjecting them to a controlled salt atmosphere fog stream to detect subsequent flaking, pitting, or corrosion that will interfere with device application.

The above tests are required for all standard military (MIL-Qualified) semiconductors and are described in detail in documents MIL-S-19500 and MIL-STD-750 maintained by the Department of the Navy. In addition, many of them are employed routinely during the manufacturing process of commercial products as a safeguard against processing faults and to provide a continual check of end-product quality and reliability.

**38. Semiconductor packages.**  Semiconductor devices are available in a seemingly endless variety of packages. Some of these variations are cost-related, others are function-related, and still others are preference-related.

Cost-related selections normally include metal and ceramic packages on the high end of the scale, and plastic packages on the low end. Plastic packages are dominant in commercial equipment, while metal and ceramic packages are usually mandated for military and high-reliability applications, or where power requirements are beyond the capabilities of plastic packages.

For this discussion, function-related packages relate to sizes and shapes dictated by the functions contained on the chip. Thus, discrete components may utilize from two to four pins, while integrated circuit packages may require hundreds of pins to accommodate the I/O requirements of the enclosed function.

Preference-related packages are those that offer the same or similar functions in different package configurations to accommodate design preferences. Typical choices include such variations as leaded or surface-mount packages, in-line or dual in-line pin configurations or special structures such as press-fit and stud packages.

Integrated circuits, typically, are housed in dual in-line packages (Fig. 6-97a) to accommodate a large number of leads in an acceptable amount of space. Large-scale integrated

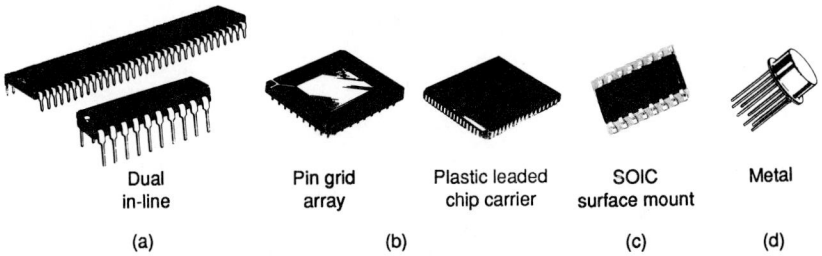

| Dual in-line | Pin grid array | Plastic leaded chip carrier | SOIC surface mount | Metal |
|:---:|:---:|:---:|:---:|:---:|
| (a) | (b) | (c) | (d) | |

**Fig. 6-97**   Typical integrated circuit package configurations.

circuits, whose I/O requirements cannot easily be satisfied even with dual in-line structures are accommodated by special VLSI packages such as pin grid arrays (PGAs) and leaded chip carriers (LCCs) (Fig. 6-97b). And for equipment where available space is at a premium, equivalent space-saving small outline integrated circuit (SOIC) packages for surface-mount applications (Fig. 6-97c) are gaining in popularity and availability. While integrated circuits for military and hi-rel applications are usually made of ceramic, a few metal packages with relatively high pin counts are also available (Fig. 6-97d).

The selection of discrete component packages is equally prolific. Higher-power devices come in standard JEDEC-registered (TO-numbered) metal and plastic housings (Fig. 6-98a) and a wide variety of package variations carrying a manufacturer's proprietary case numbers. Often, these offer similar specifications in a number of different package choices. Small-signal components are equally endowed. Typical examples of JEDEC-registered three-leaded transistors and two-terminal diodes are shown in Fig. 6-98b. Packages housing multiple transistors and diodes take the form of conventional dual in-line integrated circuits, with the number of leads depending on the number of devices contained.

Mounting Considerations for Power Packages[1]   Heat is a principal enemy of semiconductor devices. Except for lead-mounted parts used at low currents, a heat exchange is required to prevent junction temperatures from exceeding their rated limits, thereby risking device failure. Proper mounting of power devices, during manufacture and eventual substitution or replacement, is a prerequisite for reliable operation. This necessitates attention to the following areas:

1. Mounting surface preparation
2. Application of thermal compounds
3. Electrical insulation
4. Assembly fastening
5. Lead bending and soldering

Intimate thermal contact between the package and the heat sink is essential for proper heat dissipation. Rough or scratched heat sink surfaces, excessively large mounting holes, and oxidized or unclean surfaces all can have deleterious effects on heat transfer, thereby contributing to potential device degradation in high-power circuits. Significantly improved contact can be achieved through the use of thermal joint compounds ("grease"), which fill air voids between all mating surfaces. Satisfactory joint compounds can reduce the thermal resistivity of such voids from $1200°C \cdot in/W$ to approximately $60°C \cdot in/W$. This compares much more favorably to the $0.10°C \cdot in/W$ for copper film.

---

[1]Abstracted from Motorola Application Note AN1040.

Since power semiconductors normally have their collectors or anodes connected to the case, many applications require that the case be electrically insulated from circuit ground. This can be accomplished most effectively by insulating the entire heat sink–semiconductor assembly from ground rather than using an insulator between the semiconductor package and the heat sink. Where this is not possible, in cases where the chas-

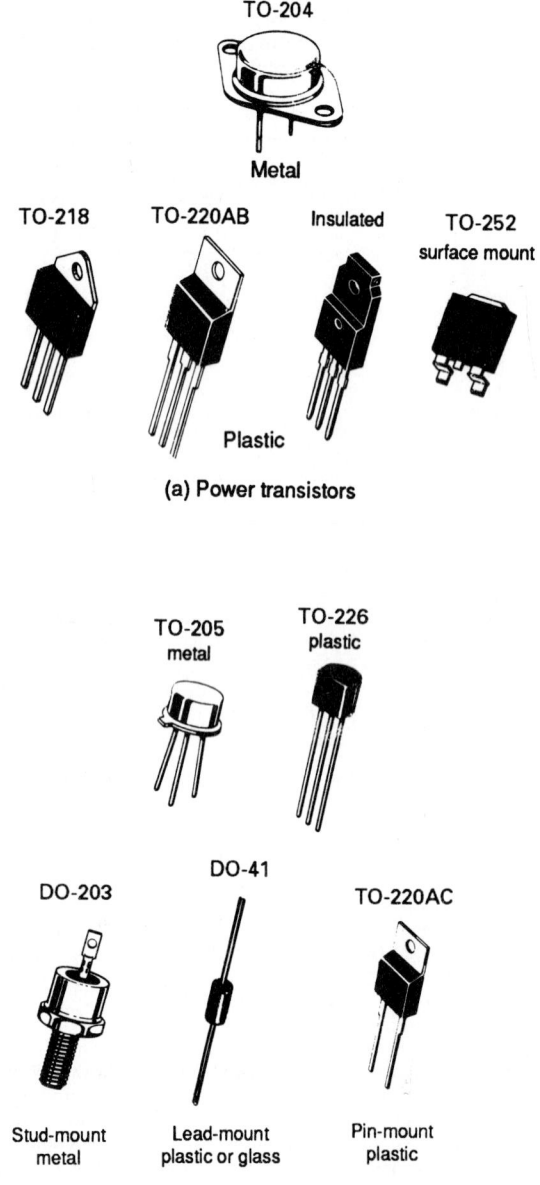

(a) Power transistors

(b) Small-signal transistors and diodes

**Fig. 6-98**  Representative sampling of packages used for discrete devices.

sis serves as the heat sink, or where the heat sink is common to several devices, insulating (mica or silicone rubber) washers can be used between the package and the heat sink.

When electrical insulation is required prior to heat sinking, the need for a thermal grease becomes more pronounced since there are now two isolating surfaces between the package and the heat sink. Data obtained by Thermalloy, Inc. (Fig. 6-99) show relative thermal resistance values between different types of insulating materials and also for various mounting screw torque values. It is quite evident that the use of grease is much more effective in reducing thermal resistance than an increase in the amount of force used to mate the two surfaces. Indeed, excessive tightening of the mounting screw can cause

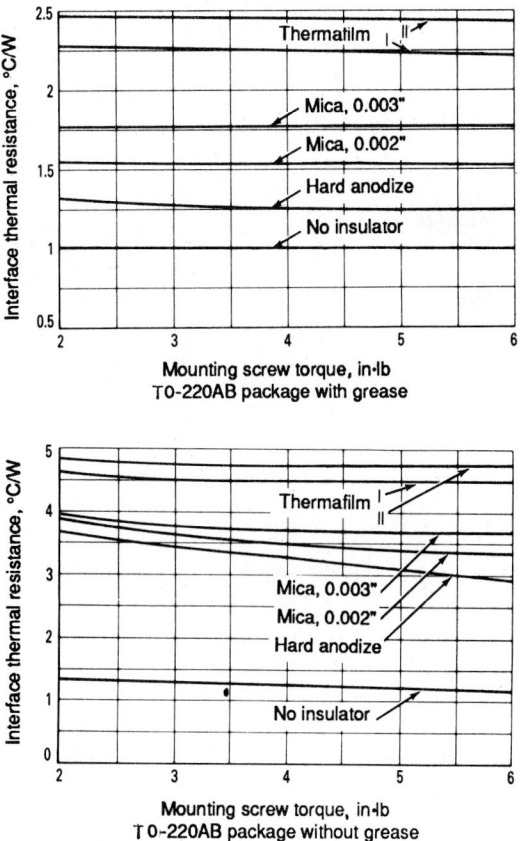

**Fig. 6-99** Interface thermal resistance for various types of insulating materials as a function of mounting screw torque—with and without the use of grease.

potential problems, particularly if the mounting hole is too large, by pulling the package into the hole (Fig. 6-100). The resulting buckling not only reduces surface contact, but can also cause cracking of the die in the package. Die rupture can also occur when the leads of a device must be bent in order to fit into a fixture or socket.

Typical examples of mounting several types of semiconductor packages are illustrated in Fig. 6-101.

To reduce insulation problems, some of the more recent power devices are being produced in insulated packages. These newer packages, housing multiple-chip and monolithic integrated power circuits as well as discrete components, come in two basic forms.

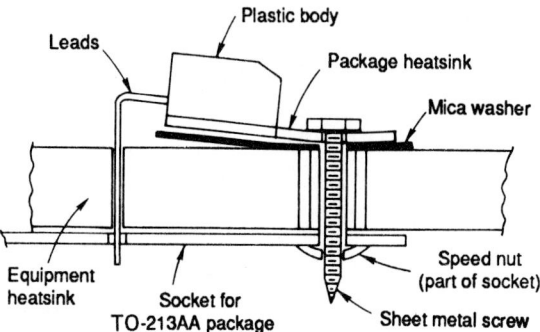

**Fig. 6-100**   Exaggerated example of improperly mounted semi-conductor package.

The first (Fig. 6-102*a*) uses a mounting plate that is insulated from the chip and can be fastened directly to a grounded chassis or heat sink. The second (Fig. 6-102*b*) utilizes a thermal conductive plastic overmold to cover the metal mounting base of the basic package.

**39. Transient suppression[1].**   A not-infrequent cause for semiconductor problems is the incidence of transients, which can stress the devices beyond their electrical design limits.

Some transients are internally generated as a result of inductive switching, commutation voltage spikes, etc. These are easily suppressed because their energy content is known and predictable. Others, however, may be created outside the circuit and coupled into it. These are more difficult to anticipate, and their suppression is often beyond the control of the circuit designer. Such transients include lightning-generated spikes and substation problems, but can also be caused by switching of parallel loads within the same branch of a power distribution system. Effective transient suppression requires that the impulse energy within the transient be dissipated in an added transient suppressor at a low enough voltage so that the capabilities of the circuit or associated components will not be exceeded.

Several types of transient suppressors have been in widespread use. Among these are carbon block spark gaps, gas tubes, selenium rectifiers, metal oxide varactors, and zener diodes. Of these, zener diodes, especially those designed specifically for surge suppression, e.g., the 1N6267–1N6303 series, have the most favorable set of characteristics.

There are two characteristics of primary importance for surge suppressors. The first, the zener breakdown voltage rating, must be equal to or lower than the maximum voltage of the equipment or circuits it is to protect. The second is the maximum power rating of the suppressor, beyond which it may itself be damaged. The latter must take into account both the maximum peak voltage and current ratings of the device, as well as the voltage excursions and pulse width of the transients.

Typical waveshapes for indoor and outdoor locations (Fig. 6-103) are defined in IEEE standard 28, ANSI standard C62.1, and can be considered as realistic representations. Peak ac power line voltage transients in indoor locations are usually limited to approximately 6.0 kV due to the spark-over spacing between conductors used in standard wiring practices. In outdoor locations, such as electrical service entrances, power line connections to additional buildings, etc., no such limitations exist.

The amplitude of randomly induced voltage transients and their energy content are difficult to define, but data from surge counters and other sources (Fig. 6-104) have provided some insight. In this plot, data for low exposure were taken from locations with little load switching or lightning activity; medium exposure is considered as areas with severe switching transients and frequent lightning activity; high-exposure systems are those supplied by long unprotected overhead lines with high spark-over clearances.

---

[1]Abstracted from Motorola Application Note AN843.

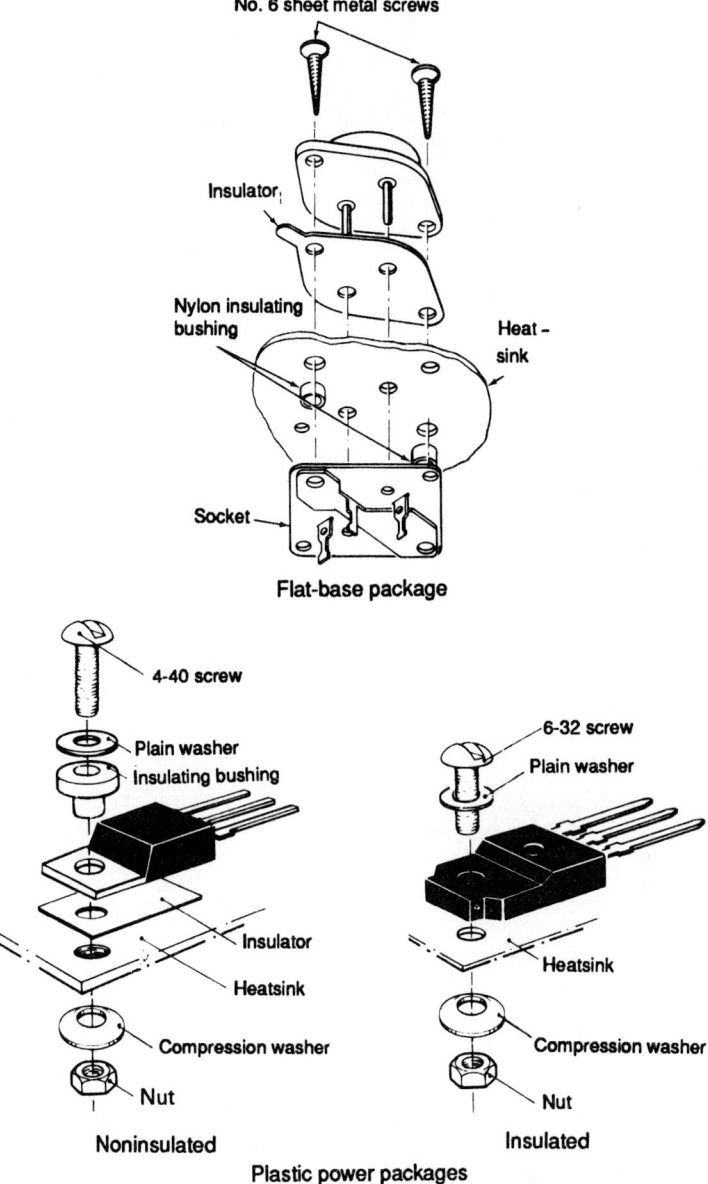

**Fig. 6-101** Typical methods for mounting various types of power device packages.

Surge characteristics deemed representative of various locations within a building are given in Table 8.

A safe approximation of the peak power contained in a transient can be obtained by considering its shape to be that of a rectangular pulse with the same peak power. For example, an exponential discharge with a time constant of 1.0 ms can be approximated with a rectangular 1.0-ms pulse with the same peak power as the transient.

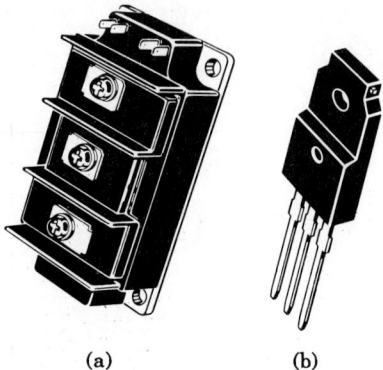

(a)                              (b)

**Fig. 6-102**  Two commercial types of power packages requiring no external insulation.

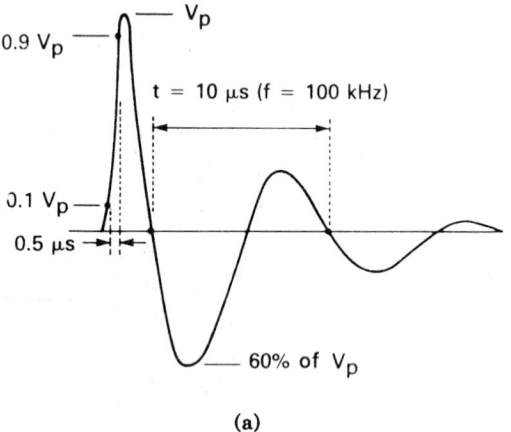

(a)

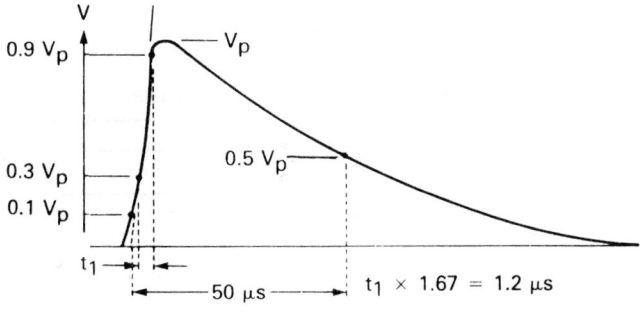

Open-Circuit Voltage Waveform

(b)

**Fig. 6-103**  Typical (a) indoor and (b) outdoor transient waveshapes.

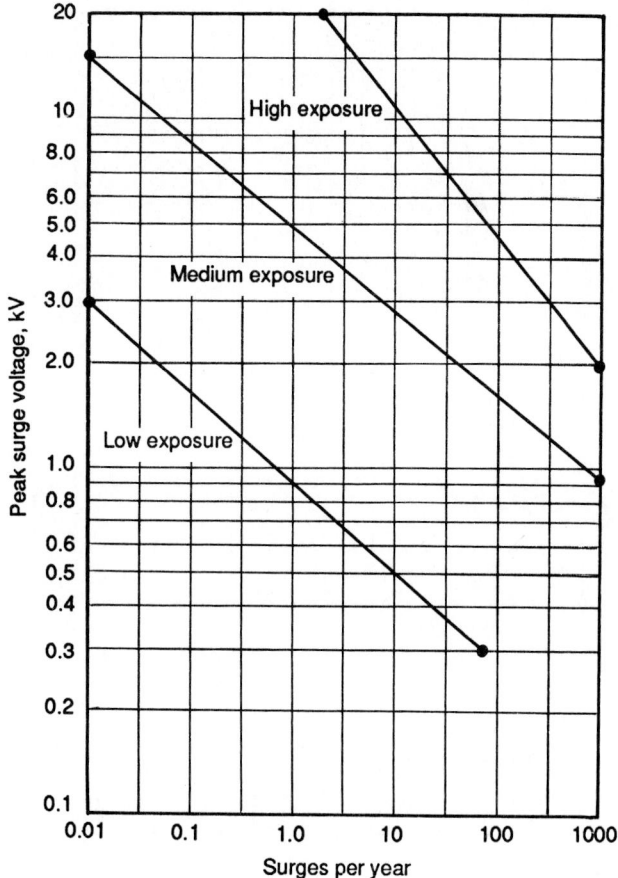

**Fig. 6-104** Peak surge voltages versus surges per year. [EIA Paper 587.1/F, May 1979, p. 10]

**Table 8   Surge Voltages and Currents Anticipated at Various Locations**

| Location | Surge voltage | Surge current | Energy (joules[a]) dissipated in suppressor with clamp voltage of | | |
| | | | 250 V | 500 V | 1000 V |
|---|---|---|---|---|---|
| 1 | 6.0 kV | 200 A | 0.4 | 0.8 | 1.6 |
| 2 | 6.0 kV | 500 A | 1.0 | 2.0 | 4.0 |
| 3 | 10 kV or more | 10 kA or more | 20 | 40 | 80 |

[a]Joules = power × time.
1. Outlets and circuits a long distance from electrical service entrance.
2. Major bus lines and circuits a short distance from service entrance.
3. Electrical service entrance and outdoor locations.

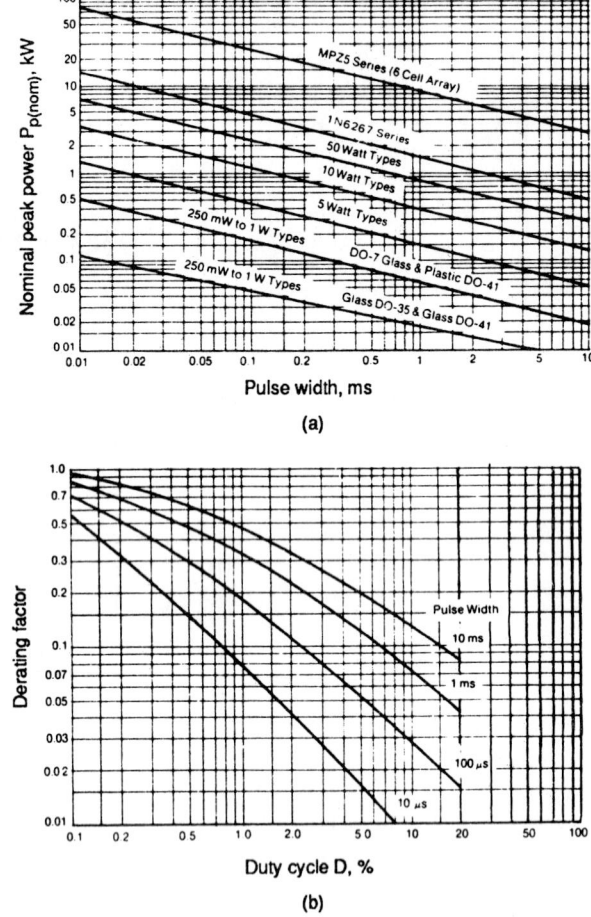

**Fig. 6-105** (a) Peak power ratings of typical zener diodes as a function of pulse width. (b) Power derating as a function of duty cycle.

Because pulse width determines the heating effect of the transient, the peak power dissipation ratings of typical zener diodes are also a function of pulse width, as shown in Fig. 6-105a. And when the transient is expected to be repetitive, peak power must be further derated as in Fig. 6-105b.

# Generators and Motors

## PRINCIPLES, CHARACTERISTICS, AND MANAGEMENT OF
## DC GENERATORS (DYNAMOS)

**1. Direct-current generators** impress on the line a direct or continuous emf, one that is always in the same direction. Commercial dc generators have commutators, which distinguish them from ac generators. The function of a commutator and the elementary ideas of generation of emf and commutation are discussed in Div. 1. Additional information about commutation as applied to dc motors, which in general is true for dc generators, is given below.

**2. Excitation of generator fields.** To generate an emf, conductors must cut a magnetic field which in commercial machines must be relatively strong. A permanent magnet can be used for producing such a field in a generator of small output, such as a telephone magneto or the magneto of an insulation tester, but in generators for light and power the field is produced by electromagnets, which may be excited by the machine itself or be separately excited from another source.

Self-excited machines may be of the series, shunt, or compound type, depending upon the manner of connecting the field winding to the armature. In the series type of machine, the field winding (the winding which produces the magnetic field) is connected in series with the armature winding. In the shunt type, the field winding is connected in parallel, shunt, with the armature winding. Compound machines have two field windings on each pole. One of these windings is connected in series with the armature winding, and the other is connected in parallel or shunt with the armature winding.

**3. Armature windings** of dc machines may be of the *lap* or the *wave* type. The difference in the two types is in the manner of connecting the armature coils to the commutator. A coil is the portion of the armature winding between successive connections to the commutator. In the lap type of winding (see Fig. 7-1) the two ends of a coil are connected to adjacent commutator segments. In the wave type of winding (see Fig. 7-2) the two ends of a coil are connected to commutator segments that are displaced from each other by approximately 360 electrical degrees.

The type of armature winding employed affects the voltage and current capacity of the machine but has no effect upon the power capacity. This is due to the fact that the number of parallel paths between armature terminals is affected by the type of winding. For a wave-wound machine there are always two paths in parallel in the armature winding between armature terminals. For a lap-wound machine there are as many parallel paths in the armature winding as there are pairs of poles on the machine. For the same number

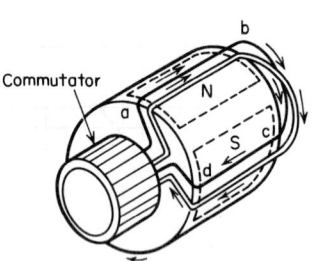

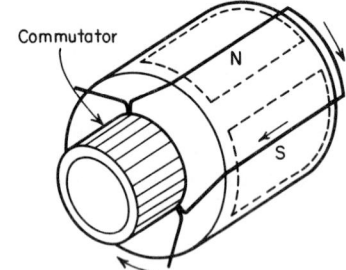

**Fig. 7-1**  Two coils of a four-pole lap-wound armature.

**Fig. 7-2**  Coil on a four-pole wave-wound armature.

and size of armature conductors, a machine when wave-connected would generate a voltage that would equal the voltage generated when lap-connected times the number of pairs of poles. But the current capacity would be decreased in the same proportion that the voltage was increased. The current capacity of a machine when wave-connected is therefore equal to the capacity when lap-connected divided by the number of pairs of poles.

**4. The value of the voltage generated by a dc machine** depends upon the armature winding, the speed, and the field current. For a given machine, therefore, the voltage generated can be controlled by adjusting either the speed or the field current. Since generators are

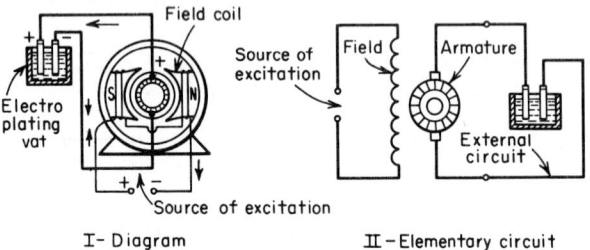

I– Diagram                  II – Elementary circuit

**Fig. 7-3**  Separately excited generator.

usually operated at a constant speed, the voltage must be controlled by adjusting the field current.

**5. Separately excited dc generators** are used for electroplating and for other electrolytic work for which the polarity of a machine must not be reversed. Self-excited machines may change their polarities. The essential diagrams are shown in Fig. 7-3. The fields can be excited from any dc constant-potential source, such as a storage battery, or from a rectifier connected to an ac supply.

The field magnets can be wound for any voltage because they have no electric connection with the armature. With a constant field excitation, the voltage will drop slightly from no load to full load because of armature drop and armature reaction.

Separate excitation is advantageous when the voltage generated by the machine is not suitable for field excitation. This is true for especially low- or high-voltage machines.

**6. Series-wound generators** have their armature winding, field coils, and external circuit connected in series with each other so that the same current flows through all parts of the circuit (see Fig. 7-4). If a series generator is operated at no load (external circuit open), there will be no current through the field coils, and the only magnetic flux present in the machine will be that due to the residual magnetism which has been retained by the poles from previous operation. Therefore, the no-load voltage of a series generator will be only a few volts produced by cutting the residual flux. If the external circuit is closed and the current increased, the voltage will increase with the increase in current until the magnetic circuit becomes saturated. With any further increases of load the voltage will decrease.

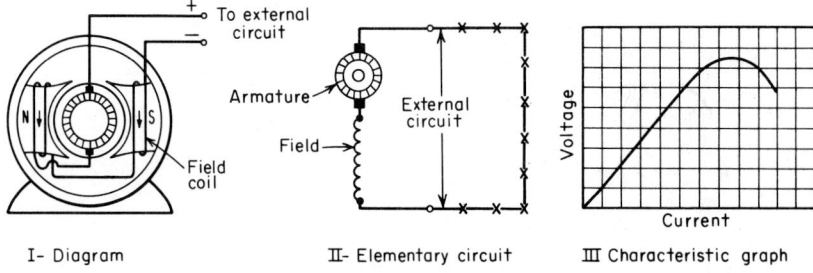

I- Diagram      II- Elementary circuit      III Characteristic graph

**Fig. 7-4**   Series generator.

Series generators have been used sometimes in street-railway service. They have been connected in series with long trolley feeders supplying sections of the system distant from the supply point in order to boost the voltage. However, power rectifiers have replaced dc generators for most installations of this type.

**7. The shunt-wound generator** is shown diagrammatically in Fig. 7-5, I and II. A small part of the total current, the exciting current, is shunted through the fields. The exciting current varies from possibly 5 percent of the total current in small machines to 1 percent in large ones. The exciting current is determined by the voltage at the brushes and the

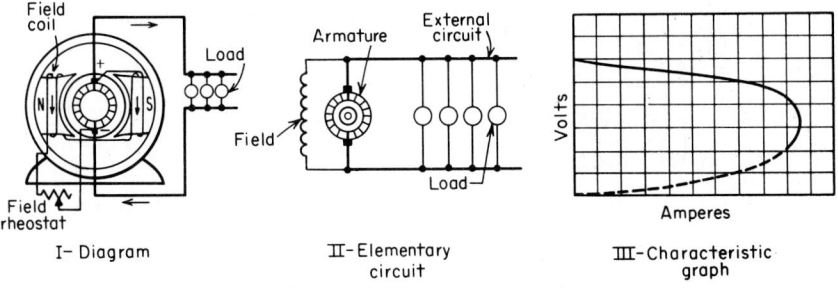

I- Diagram      II-Elementary circuit      III-Characteristic graph

**Fig. 7-5**   Shunt generator.

resistance of the field winding. Residual magnetism in the field cores permits a shunt generator to "build up." This small amount of magnetism that is retained in the field cores induces a voltage in the armature (William H. Timbie, *Elements of Electricity*). This voltage sends a slight current through the field coils, which increases the magnetization. Thus, the induced voltage in the armature is increased. This in turn increases the current in the fields, which still further increases the magnetization, and so on, until the normal voltage of the machine is reached and conditions are stable. This "building-up" action is the same for any self-excited generator and often requires 20 to 30 s.

If a shunt generator (Timbie) runs at a constant speed, as more and more current is drawn from the generator, the voltage across the brushes falls slightly. This fall is due to the fact that more and more of the generated voltage is required to force the increasing current through the windings of the armature; i.e., the armature $IR$ drop increases. This leaves a smaller part of the total emf for brush emf, and when the brush voltage falls, there is a slight decrease in the field current, which is determined by the brush voltage. This and armature reactions cause the total emf to drop a little, which still further lowers the brush potential. These causes combine to lower the voltage gradually, especially at heavy overloads. The curve in Fig. 7-5, III, shows these characteristics. For small loads the curve is nearly horizontal, but at heavy overloads it shows a decided drop. The point at which the voltage of a commercial machine drops off rapidly is beyond the operating range and is of importance only for short-circuit conditions.

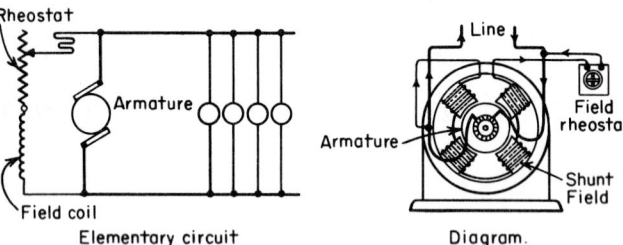

Elementary circuit                     Diagram.

**Fig. 7-6**   Shunt-wound generator with a rheostat.

The voltage of a shunt machine can be kept fairly constant by providing extra resistance in the field circuit (see Fig. 7-6), which may be cut out as the brush potential falls. This will allow more current to flow through the field coils and increase the number of magnetic lines set up in the magnetic circuit. If the speed is kept constant, the armature conductors cut through the stronger magnetic field at the same speed and thus induce a greater emf and restore the brush potential to its former value. This resistance can be cut out either automatically or by hand (see Sec. 41 of Div. 1).

Shunt-wound generators give a fairly constant voltage, even with varying loads, and can be used for any system which incorporates constant-potential loads. They will operate well in parallel because the voltage of the machines decreases as the load increases. Shunt

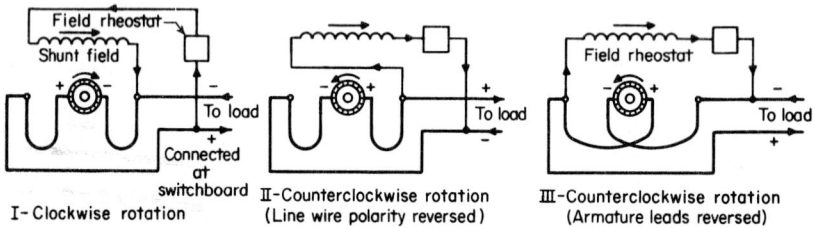

I- Clockwise rotation          II-Counterclockwise rotation          III-Counterclockwise rotation
                               (Line wire polarity reversed)          (Armature leads reversed)

**Fig. 7-7**   Changing the rotation direction of a shunt machine.

generators running in parallel will divide the load well between themselves if the machines have similar characteristics.

The necessary change in connections when reversing the direction of rotation of a shunt-wound machine is indicated in Fig. 7-7. Rotation is *clockwise* when, facing the commutator end of a machine, the rotation is in the direction of the hands of a clock. *Counterclockwise* rotation is the reverse. When changing the direction of rotation, do not reverse the direction of current through the field windings. If the direction is reversed, the magnetism developed by the windings on starting will oppose the residual magnetism and the machine will not "build up."

**8. Parallel operation of shunt generators.**   As suggested in Sec. 7, shunt-wound generators will in general operate very well in parallel and will divide the load well if the machines have similar characteristics. If the machines do not have similar characteristics, one machine will take more than its share of the load and may tend to drive the other as a motor. When this machine is running as a motor, its direction of rotation will be the same as when it was generating; hence the operator must watch the ammeters closely for an indication of this trouble. Shunt generators are now seldom installed. Figure 7-8 shows the connections for shunt generators that are to be operated in parallel.

**9. The compound-wound generator** is shown diagrammatically in Fig. 7-9, I. If a series winding is added in a proper manner to a shunt generator (Fig. 7-5), the two windings will tend to maintain a constant voltage as the load increases. The magnetization due to the series windings increases as the line current increases, thus tending to increase the

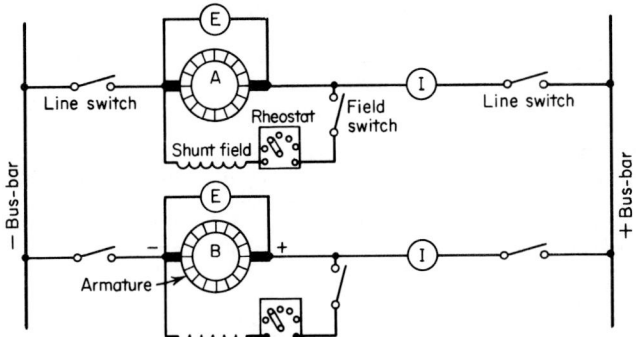

**Fig. 7-8**    Connections for shunt generators for parallel operation.

generated voltage. The drop of voltage at the brushes that occurs in a shunt generator can thus be compensated for.

The series winding must be connected in such a manner that its current will aid that of the shunt-field winding in producing magnetic flux. With this proper connection, the machine is said to be cumulatively connected. If the series field is connected in the reverse manner, so that the series field tends to produce flux in the opposite direction to

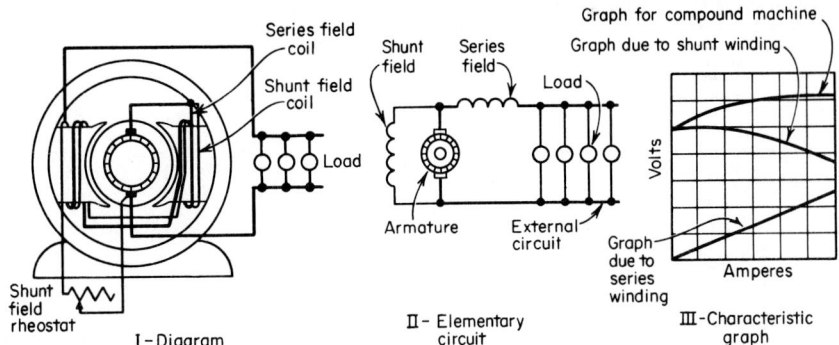

**Fig. 7-9**    Compound generator.

that produced by the shunt field, the machine is said to be differentially connected. Differential connection of compound machines is used in very special cases.

**10. A flat-compounded generator** is one having its series coils so proportioned that the voltage remains practically constant at all loads from 0 to 1¼ full load.

**11. An overcompounded generator** has its series windings so proportioned that its full-load voltage is greater than its no-load voltage. Overcompounding is necessary when it is desirable to maintain a practically constant voltage at some point out on the line distant from the generator. It compensates for line drop. The characteristic curve (Fig. 7-9, III) indicates how the terminal voltage of a compound-wound machine is due to the action of both shunt and series windings. Generators are usually overcompounded, so that the full-load voltage is from 5 to 10 percent greater than the no-load voltage.

Although compound-wound generators are usually provided with a field rheostat, it is not intended for regulating voltage, as the rheostat of a shunt-wound machine is. It is provided to permit initial adjustment of voltage and to compensate for changes of the resistance of the shunt winding caused by heating. With a compound-wound generator, the voltage having been once adjusted, the series coils automatically strengthen the magnetic field as the load increases. For dc power work, compound-wound generators are used almost universally when rectifiers are not employed.

**11A. An undercompounded generator** is one with a relatively weak series-field winding, so that the voltage decreases with increased load.

**12. If a compound-wound generator is short-circuited,** the field strength due to the series windings will be greatly increased but the field due to the shunt winding will lose its strength. For the instant or so that the shunt magnetization is diminishing, a heavy current will flow. If the shunt magnetization constitutes a considerable proportion of the total magnetization, the current will decrease after the heavy rush and little harm will be done if the armature has successfully withstood the heavy rush. However, if the series magnetization is quite strong in proportion to the shunt, their combined effect may so magnetize the fields that the armature will be burned out.

**13. A short-shunt compound-wound generator** has its shunt field connected directly across the brushes (see Fig. 7-9, II). Generators are usually connected in this way because this arrangement tends to maintain the shunt-field current more nearly constant on variable loads, as the drop in the series winding does not directly affect the voltage on the shunt field.

**14. A long-shunt generator** has its shunt-field winding connected across the terminals of the generator (see Fig. 7-10).

**15. Nearly all commercial dc generators have more than two poles.** A two-pole machine is a bipolar machine; one having more than two poles is a multipolar machine. Figure 7-11 shows the connections for a four-pole compound-wound machine. Diagrams for machines having more poles would be similar. In multipolar machines there is usually one set of brushes for each pair of poles, but with wave-wound armatures, such as those used for railway motors, one set of brushes may suffice for a multipolar machine. The connections of different makes of machines vary in detail; since manufacturers will always furnish complete diagrams, no attempt will be made to give them here. The directions of the field windings on generator frames

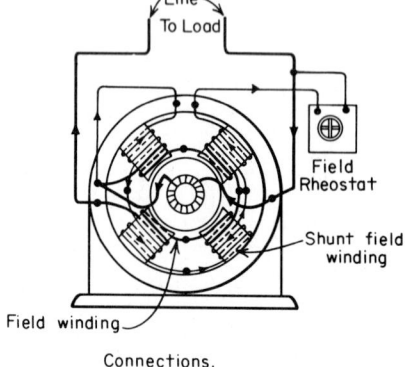

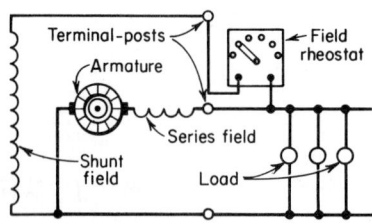

**Fig. 7-10**  Long-shunt compound-wound generator.

**Fig. 7-11**  Four-pole compound-wound generator.

are given in Fig. 7-12. The directions of the windings on machines having more than four poles are similar in general to those of the four-pole machines.

**16. A series shunt for a compound generator** consists of a low-resistance connection across the terminals of the series field (see Figs. 7-13 and 7-14) through which the compounding effect of the series winding can be regulated by shunting more or less of the armature current around the series coils. The shunting resistance may be in the form of grids, on large machines, or of ribbon resistors. In the latter case it is usually insulated and folded into small compass.

**17. Parallel operation of compound-wound generators** is readily effected if the machines are of the same make and voltage or are designed with similar electrical characteristics (Westinghouse Electric Corp.). The only change that is usually required is the addition of an equalizer connection between machines. If the generators have different compounding ratios, it may be necessary to adjust the series-field shunts to obtain uniform conditions.

**18. An equalizer, or equalizer connection,** connects two or more generators operating in parallel at a point where the armature and series-field leads join (see Fig. 7-13), thus connecting the armatures in multiple and the series coils in multiple so that the load will divide between the generators in proportion to their capacities. The arrangement of connections to a switchboard (Westinghouse Electric Corp.) is illustrated in Fig. 7-14. Consider, for example, two overcompound-wound machines operating in parallel without an equalizer. If for some reason there is a light increase in the speed of one machine, it will

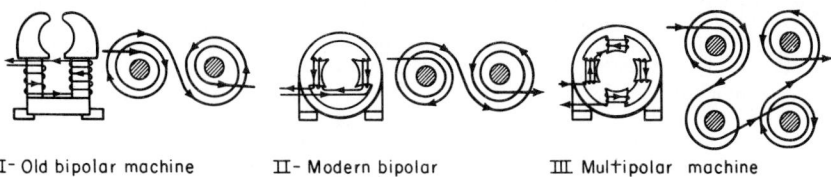

I- Old bipolar machine        II- Modern bipolar        III. Multipolar machine
                                     machine

**Fig. 7-12**   Directions of field windings on generator frames.

take more than its share of load. The increased current flowing through its series field will strengthen the magnetism, raise the voltage, and cause the machine to carry a still greater amount until it carries the entire load. When equalizers are used, the current flowing through each series coil is inversely proportional to the resistance of the series-coil circuit and is independent of the load on any machine; consequently, an increase of voltage on one machine builds up the voltage of the other at the same time, so that the first machine cannot take all the load but will continue to share it in proper proportion with the other generators.

**19. Operation of a shunt and a compound dynamo in parallel** is not successful because the compound machine will take more

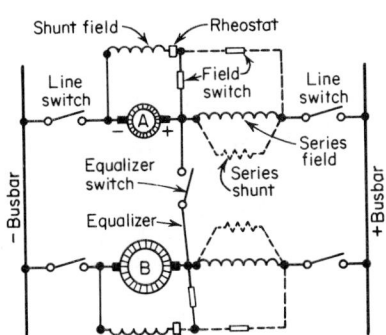

**Fig. 7-13**   Elementary connections for parallel operation of compound-wound generators.

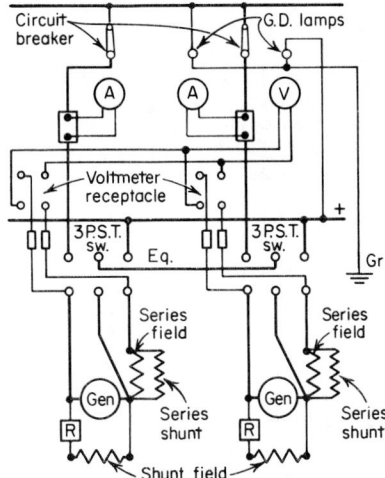

**Fig. 7-14**   Connections of two compound-wound generators to a switchboard.

than its share of the load unless the shunt-machine field rheostat is adjusted at each change in load.

**20. Connecting leads for compound generators.**   See that all the cables for machines of equal capacity that lead from the series fields of the various machines to the busbars are of equal resistance. This means that if the machines are at different distances from the switchboard, different sizes of wire should be used or resistance inserted in the low-resistance leads.

With generators of small capacity the equalizer is usually carried to the switchboard, as suggested in Figs. 7-14 and 7-15, but with larger ones it is carried under the floor directly between the machines (Fig. 7-16). The positive and the equalizer switches of each machine may be mounted side by side on a pedestal near the generator (Fig. 7-16). The difference in potential between the two switches is only that due to the small drop in the

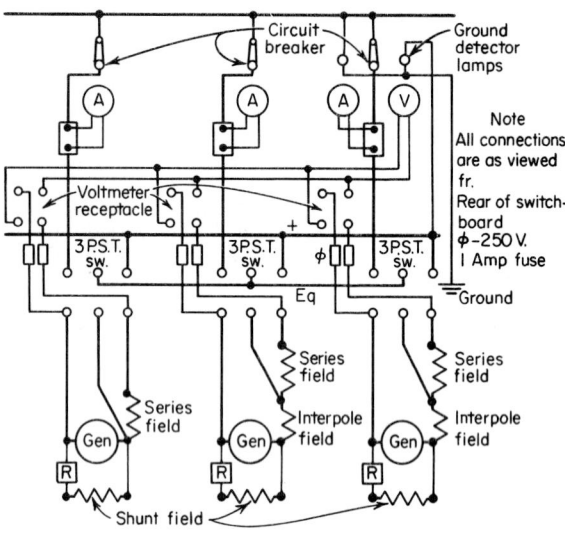

**Fig. 7-15**  Connections of two dc commutating-pole generators in parallel with one generator without commutating poles.

series coil. The positive busbar is carried under the floor near the machines, permitting leads of minimum length. Leads of equal length should be used for generators of equal capacities. If the capacities are unequal (see Sec. 24), it may be necessary to loop the leads (see Fig. 7-16).

**21. Ammeters and circuit breakers for compound generators** should, as in Fig. 7-14, always be inserted in the lead not containing the compound winding. If the ammeters are put in the compound-winding lead, the current indications will be inaccurate because current from either side of the machine can flow through either the equalizer or the compound-winding lead.

**22. Starting a shunt- or compound-wound generator.**  (1) See that there is enough oil in the bearings, that the oil rings are working, and that all field resistance is cut in. (2) Start the prime mover slowly and permit it to come up to speed. See that the oil rings are working. (3) When machine is up to normal speed, cut out field resistance until the voltage of the machine is normal or equal to or a trifle above that on the busbars. (4) Throw on the load. If three separate switches are used, as in Fig. 7-13, close the equalizer switch first, the series-coil line switch second, and the other line switch third. If a three-pole switch is used, as in Fig. 7-14, all three poles

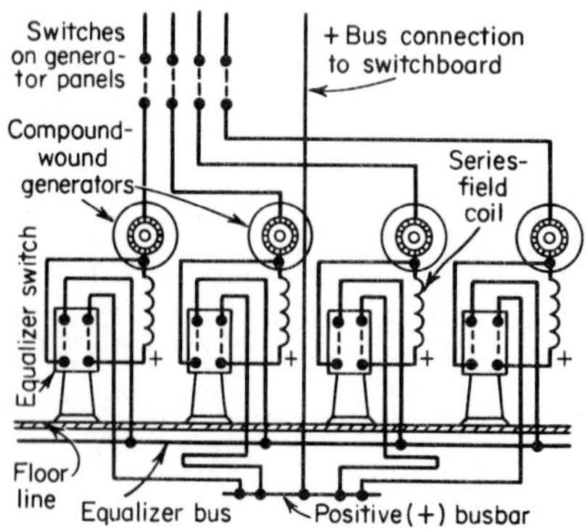

**Fig. 7-16**  Equalizer carried directly between machines.

are, of course, closed at the same time. (5) Watch the voltmeter and ammeter and adjust the field rheostat until the machine takes its share of the load. A machine generating the higher voltage will take more than its share of the load and if its voltage is too high, it will run the other machine as a motor.

**23. Shutting down a shunt- or compound-wound generator operating in parallel with others.** (1) Reduce the load on the machine as much as possible by cutting resistance into the shunt-field circuit with the field rheostat. (2) Throw off the load by opening the circuit breaker if one is used; otherwise open the main generator switches. (3) Shut down the driving machine. (4) Wipe off all oil and dirt, clean the machine, and put it in good order for the next run. Turn all resistance in the field rheostat. Open the main switch.

**24. Adjusting the division of load between two compound-wound generators.** First adjust the series shunts of both machines so that, as nearly as possible, the voltages of both will be the same at one-fourth, one-half, three-fourths, and full load. Then connect the machines in parallel, as suggested in Fig. 7-13, for trial. If, upon loading, one machine takes more than its share of the load (amperes), increase the resistance of the path through its series-field coil path until the load divides between the machines proportionally to their capacities. Only a small increase in resistance is usually needed. The increase can be provided by inserting a longer conductor between the generator and the busbar, or iron or nickel silver washers can be inserted under a connection lug. Inasmuch as adjustment of the series-coil shunt affects both machines when the machines are connected in parallel, nothing can be accomplished through making such an adjustment.

**25. Commutating-pole dc generators.** Generators which do not have commutating poles (Westinghouse Electric Corp.) and which operate under severe overloads and over a wide speed range are apt to spark under the brushes at the extreme overloads and at the higher speeds. This happens because the field due to the armature current distorts the main field to such an extent that the coils being commutated under the brush are no longer in a magnetic field of the proper direction and strength. To overcome this, *commutating poles* are placed between the main poles (see Fig. 7-17). These poles introduce a magnetic field of such direction and strength as to maintain the magnetic field, at the point where the coils are commutated, at the proper strength for good commutation. Commutating poles are sometimes called *interpoles*. Commutating poles is the preferable term.

The winding on the commutating poles is connected in series with the armature so that the strength of the corrective field increases and decreases with the load. The adjustment and operation of commutating-pole generators are not materially different from those of non-commutating-pole machines.

When the brush position of a commutating-pole machine has once been properly fixed,

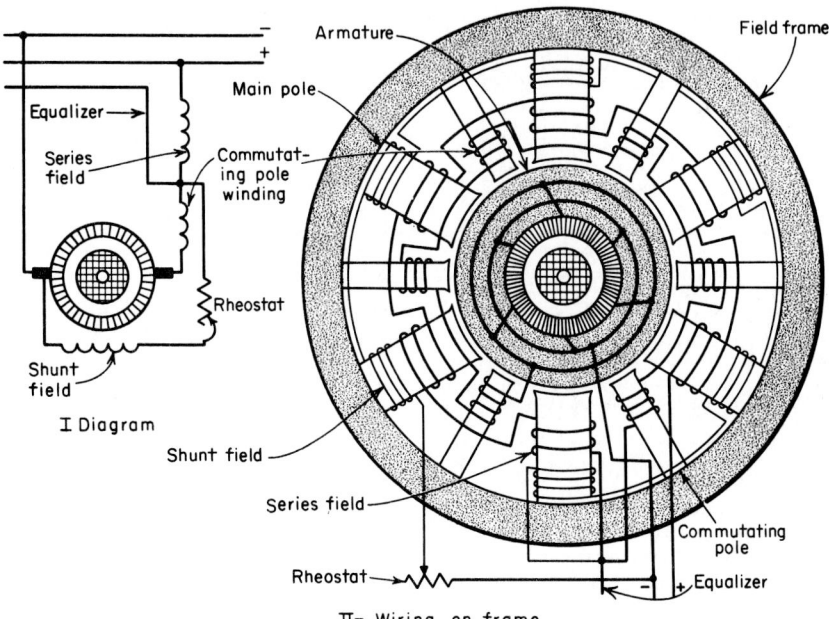

**Fig. 7-17**  Compound-wound commutating-pole (interpole) machine.

no shifting is afterward required or should be made, and most commutating-pole generators are shipped without any shifting device. An arrangement for clamping the brush-
holder rings securely to the field frame is provided.

In commutating-pole apparatus, accurate adjustment of the brush position is necessary.
The correct brush position is on the no-load neutral point, which is located by the manufacturer. A template is furnished with each machine, or some other provision is made so
that the brush location can be determined in the field. If the brushes are given a backward
lead on a commutating-pole generator, the machine will overcompound and will not commutate properly. With a forward lead of the brushes, a generator will undercompound and
will not commutate properly.

**26. The object in using a commutating pole** is to produce within the armature coil under
commutation an emf of the proper value and direction to reverse the current in the coil
while it is yet under the brush—a result that is essential to perfect commutation. The
variation in the flux distribution in the air gap of a commercial dc machine of the ordinary
shunt-wound type, at no load and under full load, is shown in Fig. 7-18. Consider now the
value and position of the flux in the coil under the brush when the machine is operating
at full load. The motion of the armature through this flux causes the generation within the
coil of an emf, and the sign of this emf is such as to tend to cause the current in the coil to
continue in the direction which it had before the coil reached the brush; hence it opposes
the desired reversal of the current before the coil leaves the brush.

There is an additional detrimental influence which tends to retard the rapid reversal of
the current even when all other influences are absent. This influence is due to the local
magnetizing effect of the current in the coil under the brush. This effect causes the lines
of force which surround the conductor to change in value with the fluctuations of the
current as it tends to be reversed. As a result, an emf which opposes the change in the
value of the current is generated in the coil. This reactive emf is in the same direction as
that due to the cutting of the flux by the coil under the brush and is likewise proportional
to the speed.

It will be apparent that even if the field distortion were completely neutralized, the
detrimental reactive emf would remain. The improved and practically perfect commutation of the commutating-pole machine is due to the fact that the flux, which is superposed
locally upon the main field, not only counterbalances the undesirable main flux cut by the

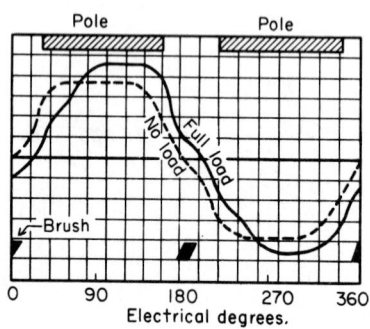

**Fig. 7-18**  Distribution of magnetic flux at no
load and at full load, without commutating poles.

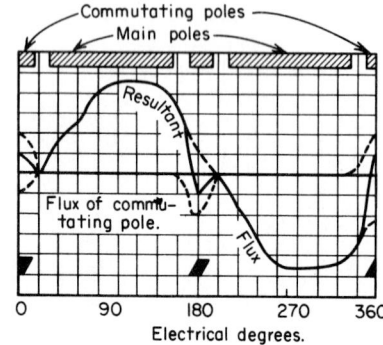

**Fig. 7-19**  Distribution of magnetic flux at full
load, with and without commutating poles.

coil under the brush but causes an emf sufficient to equal and oppose the reactive emf to
be generated within the coil. This effect will be appreciated from a study of Fig. 7-19,
which represents the distorted flux of the motor of the usual design, as shown in Fig. 7-
18, and indicates the results to be expected when the flux due to the auxiliary or commutating pole is given the relatively proper value.

This desirable effect is the more pronounced the weaker the main field. The commutation voltage, if correct for a low speed, is correct for a high speed; and with the increase
of load-current and main-field distortion there is a proportional increase of the counter-
magnetizing field produced in the coil under the brush, up to the point of magnetic satu-

ration of the auxiliary pole. Sparkless operation is ensured for all operating ranges of both speed and load.

**27. The action of the magnetic flux in a commutating-pole generator** is illustrated in Fig. 7-20. The direction of the main-field flux is shown by the dashed line. The direction of the armature magnetization is shown by the dotted lines. The direction of the flux in the commutating pole is shown by the full line. It is evident that the commutating-pole flux is in a direction opposite to that of the armature flux, and as the commutating-pole coil is more powerful at the commutating point in its magnetizing action than the armature coils, the flux of the armature coils is neutralized. With a less powerful magnetizing force from the commutating pole than from the armature at the commutating point, the armature would overpower the commutating pole and reverse the direction of the flux, which would result in a bad commutating condition.

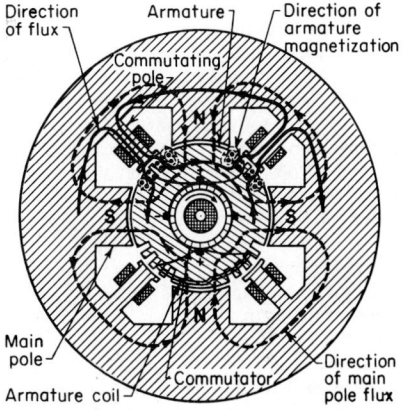

**Fig. 7-20**   Distribution of flux in a commutating-pole generator.

**28. Determining the neutral point of a motor or generator** (Fig. 7-21). Two copper-wire contact points, or contactors, $C$ are inserted in an insulating block and are allowed to extend through it about $\frac{1}{16}$ in (1.6 mm). The distance between the centers of the points should be equal to the width of one commutator bar. The contactors are connected to a millivoltmeter $V$, which preferably should be of the differential type. Place both points on the commutator while the machine is being rotated with the brushes lifted from the commutator and the shunt field is being excited. While shifting the points around the periphery of the commutator, hold the block so that an imaginary line connecting the two contact points will be perpendicular to the axis of the commutator.

If a differential voltmeter is used, its needle will indicate either to the right or to the left of the zero point until the contacts are exactly over the neutral position; then the voltmeter will read zero. The brushes can now be shifted so that an imaginary line, par-

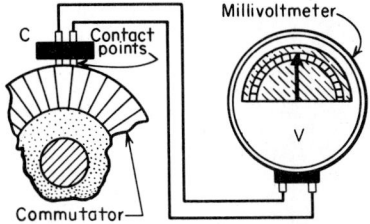

**Fig. 7-21**   Connections for determination of the neutral point.

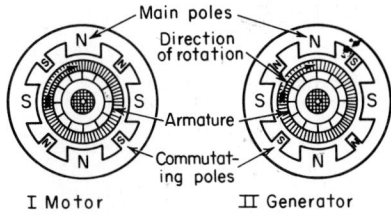

**Fig. 7-22A**   Polarity of commutating poles (clockwise rotation).

allel to the axis of the commutator and bisecting the bearing surface of the brush, will coincide with a point equidistant between the two contact points.

**29. Determining the proper polarity of the commutating poles.** For a motor, proceeding from pole to pole around the frame in the direction of armature rotation, each commutating pole should have the same polarity as the main pole which just precedes it (Fig. 7-22A, I). For a generator, proceeding from pole to pole around the frame in the direction of armature rotation, each commutating pole should have a polarity opposite to that of the main pole which just precedes it (Fig. 7-22A, II).

**30. Commutating-pole machines will run in parallel** with each other and with non-commutating-pole machines provided correct connections are made (see illustrations). The series-field windings on commutating-pole machines are usually less powerful than on non-commutating-pole machines, and particular attention should therefore be paid to getting the proper drop in accordance with instructions of Sec. 24. A connection diagram is shown in Fig. 7-15.

**30A. Compensating field windings.**   The commutating poles of a dc motor or generator do not neutralize the effect of the armature current upon the flux of the machine. They simply produce a local flux at the commutating location of the armature conductors. For machines that must operate under very severe conditions of overload and speed range, for satisfactory operation means must be provided to neutralize the tendency of the armature current to distort and change the flux of the machine.

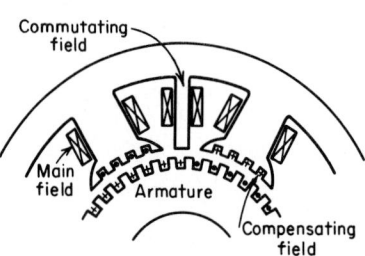

**Fig. 7-22B**   Section of a dc machine showing the location of a compensating field winding.

This neutralization is produced by means of an additional winding called a *compensating field winding*. A compensating field winding consists of coils embedded in slots in the pole faces of the machine, as shown in Figs. 7-22B and 7-22C. The compensating winding is connected in series with the armature in such a manner that the current through the individual conductors of the compensating field winding will be opposite to the direction of the corresponding armature conductors, as shown in Fig. 7-22B.

**31. Three-wire dc generators** are ordinary dc generators with the modifications and additions described below. They are usually wound for 125/250-V, three-wire circuits. In the case of commercial three-wire generators (Westinghouse Electric Corp.), four equidistant taps are made in the armature winding, and each pair of taps diametrically opposite each other is connected through a balance coil. The balance coil may be external (Fig. 7-23) or wound within the armature. The middle points of the two balance coils are connected, and this junction constitutes the neutral point to which the third, or neutral, wire of the system is connected. A constant voltage is maintained between the neutral and outside wires which, within narrow limits, is one-half the generator voltage. The generator shaft is extended at the commutator end

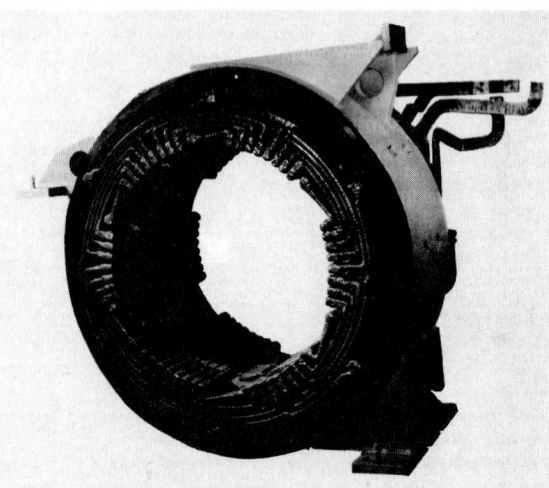

**Fig. 7-22C**   Frame and field structure of a dc motor showing compensating windings in the faces of the main poles. [Westinghouse Electric Corp.]

for the collector rings. Four collector brushes and brush holders are used in addition to the regular dc brushes and brush holders.

**32. The series coils of compound-wound three-wire generators are divided into halves** (see Fig. 7-23), one of which is connected to the positive and one to the negative side. This is done to obtain compounding on either side of the system when operating on an unbalanced load. To understand this procedure, consider a generator with the series field on the negative side only and with most of the load on the positive side of the system. The current flows from the positive brush through the load and back along the neutral wire without passing through the series field. The generator is then operating as an ordinary shunt machine. If most of the load is on the negative side, the current flows out the neutral wire and back through the series fields, boosting the voltage by the maximum amount. Such operation is evidently not satisfactory, and so divided series fields are provided.

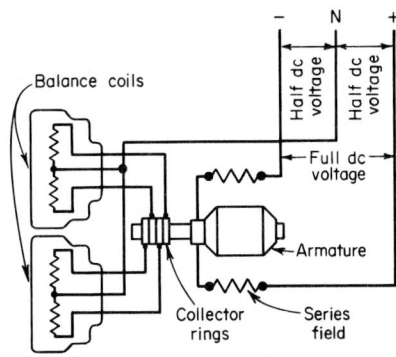

**Fig. 7-23** Connections for a three-wire generator.

**33. Switchboard connections for three-wire generators.** Figure 7-24 is a diagrammatical representation of switchboard connections for two three-wire generators operated in multiple (Westinghouse publication). Two ammeters indicate the unbalanced load. The positive lead and equalizer are controlled by a double-pole circuit breaker, as are the negative lead and equalizer. Note that both the positive and the negative equalizer connections as well as both the positive and the negative leads are run to the circuit breakers as well as to the main switches on the switchboard. This must be done in all cases. Otherwise, when two or more machines are running in multiple and the breaker comes out, opening the main circuit to one of them but not breaking its equalizer leads, its ammeter is left connected to the equalizer busbars and current is fed into it from the other machines through the equalizer leads, either driving it as a motor or destroying the armature winding (see also Figs. 7-25 and 7-26).

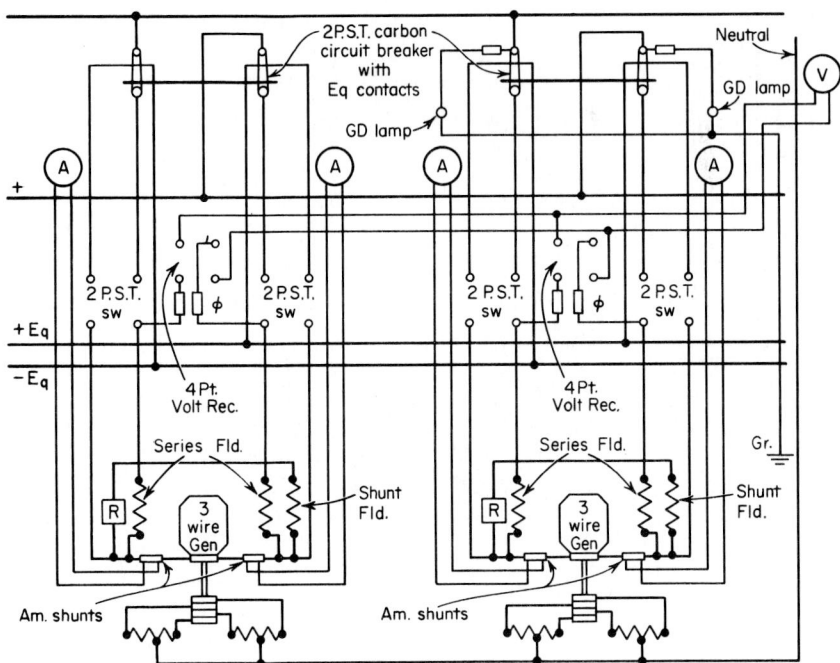

**Fig. 7-24** Connections of two three-wire dc generators operating in parallel, 125/250 V.

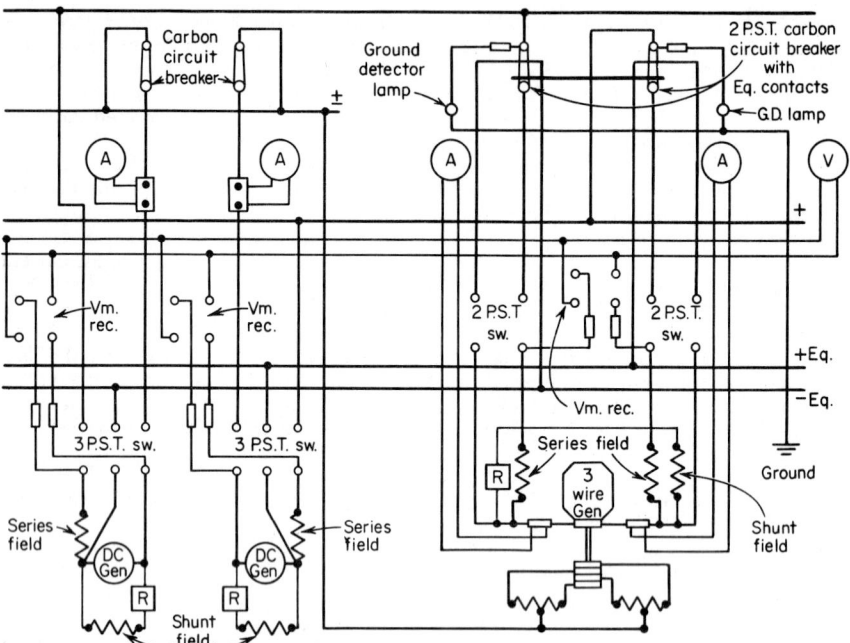

**Fig. 7-25** Connections of a three-wire dc generator, 125/250 V, in parallel with two two-wire generators, 125 V.

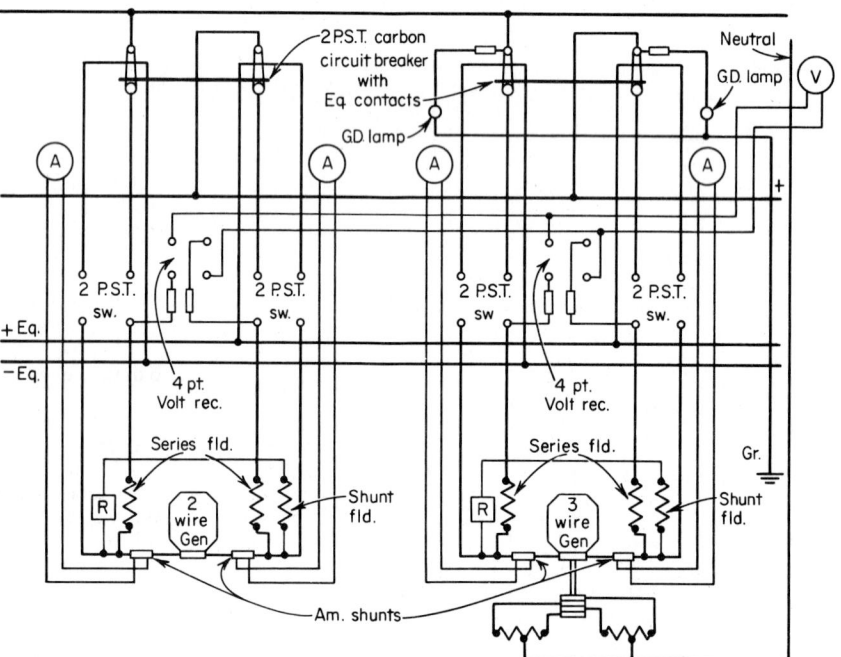

**Fig. 7-26** Connections of a three-wire dc generator, 125/250 V, in parallel with a two-wire generator, 250 V.

**34. As there are two series fields, two equalizer buses are required when several three-wire machines are installed** (see Fig. 7-24) and are to be operated in parallel. The two equalizers serve to distribute the load equally between the machines and to prevent crosscurrents due to differences in voltage on the different generators. Because of the equalizer connections, two small terminal boards, one for each side of the generator, are supplied. Arrangement is also made for ammeter shunts on the terminal boards.

An ammeter shunt is mounted directly on each of the contact boards of the machine. The total current output of the machine can thereby be read at the switchboard. Because the shunts are at the machine, there is no chance for current to leak across between generator switchboard leads without causing a reading on the ammeters. Two ammeters must be provided for reading the current in the outside wires. It is important that the current be measured on both sides of the system, for with an ammeter on one side of the system only it is possible for a large unmeasured current to flow in the other side with disastrous results.

**35. Wires connecting the balance coils to a three-wire generator** must be short and of low resistance. Any considerable resistance in these wires will affect the voltage regulation. The unbalanced current flows along these connections; consequently, if they have much resistance, the resulting drop in voltage reduces the voltage on the heavily loaded side.

Switches are not ordinarily placed in the circuits connecting the four collector rings to the balance coils. When necessary, the coils can be disconnected from the generator by raising the brushes from the collector rings. Switching arrangements often make it necessary to run the balance-coil connections to the switchboard and back, requiring heavy leads to keep the drop low, or if heavy leads are not used, poor regulation may result. The balance coils are so constructed that there is very little likelihood of anything happening to them that will not be taken care of by the main circuit breakers. Complete switchboard connection diagrams are given in Figs. 7-24, 7-25, and 7-26.

**36. Commutating-pole three-wire generators.** In three-wire generators, connections are so made that one-half of the commutating-pole winding is in the positive side and the other half is in the negative side. This ensures proper action of the commutating pole at an unbalanced load (see Figs. 7-24, 7-25, and 7-26 and the text accompanying them).

**37. Three-wire dc generators can be operated in parallel with each other** (Westinghouse publication) and in parallel with other machines on the three-wire system (see Figs. 7-24, 7-25, and 7-26). When a three-wire, 250-V generator is operated in multiple with two-wire, 125-V generators, the series fields of the two two-wire generators must be connected, one on the positive side and one on the negative side of the system, and an equalizer must be run to each machine. Similarly, when a three-wire, 250-V generator is operated in multiple with a 250-V two-wire generator, the series field of the 250-V two-wire generator must be divided and one-half connected to each outside wire. The method of doing this is to disconnect the connectors between the series-field coils and to reconnect these coils so that all the $N$ pole fields will be in series on one side of the three-wire system and all the $S$ pole fields in series on the other side of the system.

**38. Testing for polarity.** When a machine that is to operate in parallel with others is connected to the busbars for the first time, it should be tested for polarity. The + lead of the machine should connect to the + busbar and the − lead to the − busbar (Fig. 7-27, I). The machine to be tested should be brought up to normal voltage but not connected to the bars. The test can be made with two lamps (Fig. 7-27, II), each lamp of the voltage of the circuit. Each is temporarily connected between a machine terminal and the bus terminal of the main switch. If the lamps do not burn, the polarity of the new machine is correct, but if they burn brightly, its polarity is incorrect and should be reversed. A voltmeter can be used (Fig. 7-27, III). A temporary connection is made across one pair of outside terminals, and the voltmeter is connected across the other pair. No deflection or a small deflection indicates correct polarity. (Test with voltmeter leads one way and then reverse them, as indicated by the dotted lines.) A full-scale deflection indicates incorrect polarity. Use a voltmeter having a voltage range equal to twice the voltage on the busbars.

**39. Third-brush generators** were often used on automobiles to provide the necessary electric power for the charging of the storage battery and the operation of lights. If an ordinary generator were employed for this purpose, the voltage would vary over a wide range as the speed of the car changed. The voltage would vary nearly proportionally to the speed. This, of course, would not be satisfactory either for the proper charging of the battery or for operation of the car lights. The third-brush generator is a special shunt generator with the field winding connected between one of the main brushes and the auxiliary, or third, brush (see Fig. 7-28). As the speed of the automobile increased, thereby increasing the speed of the generator, the voltage tended to increase. This increase in voltage increased the current delivered by the generator. But the increase in current so changed the magnetic-flux distribution in the machine that the voltage between the third brush and main brush A was reduced. This reduced the field current and therefore the flux of the machine and tended to bring the main voltage between brushes A and C back to its former value. This action did not maintain an absolutely constant voltage for different speeds, but the voltage was held within certain limits so that an excessive voltage which would over-

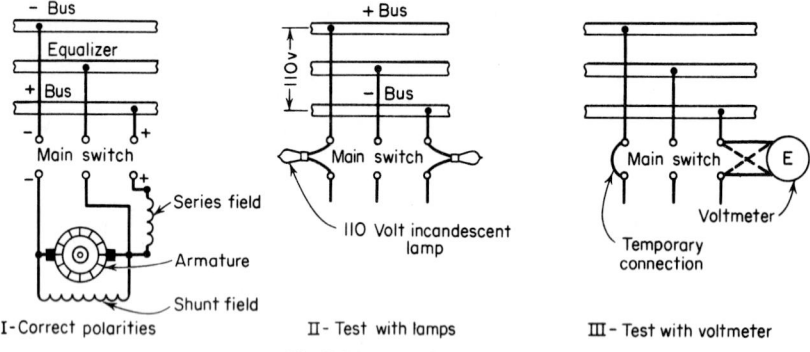

**Fig. 7-27** Tests for polarity.

charge the battery and shorten the life of the lamps was not developed at high speeds. It should be noted that present-day automobiles use an ac generator (alternator) with rectifier diodes to provide a 12-V dc supply.

**40. Diverter-pole generators** are a special type of dc generator (developed by the Electric Products Co.) that have particular advantages for the charging of batteries by the floating

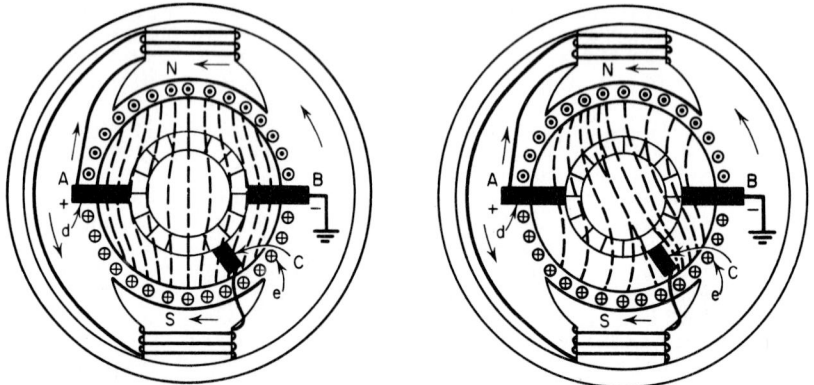

**Fig. 7-28** Third-brush generator. [Charles L. Dawes, "Direct Currents," *A Course in Electrical Engineering*]

method. The machine is constructed with additional pole pieces (the diverter poles) located midway between the main poles in the same manner as commutating poles. Each main pole is connected by a magnetic bridge to one diverter pole. The windings of the main poles are connected in shunt with the armature winding, and the windings of the diverter poles in series with the armature winding. The construction, connections, and load-voltage characteristics for such a generator are shown in Fig. 7-29. These generators

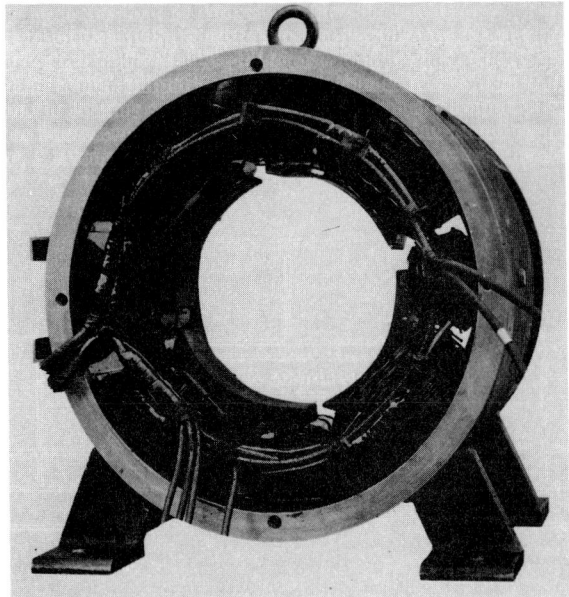

I. Frame and field-winding construction.

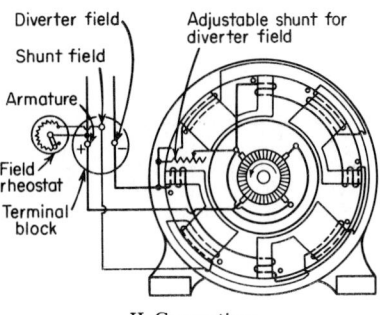

II. Connections.

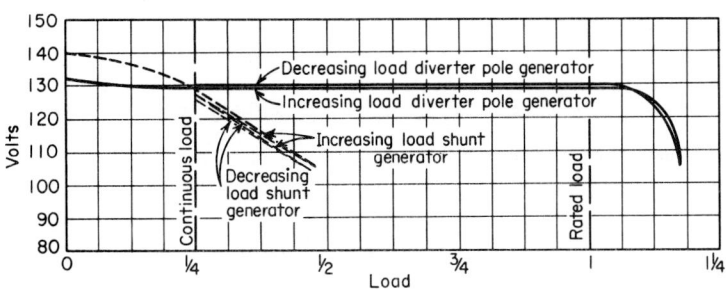

III. Load-voltage characteristics.

**Fig. 7-29**    Diverter-pole generator. [Electric Products Co.]

will produce an almost constant terminal voltage from no load to 110 percent of rated load. Above 110 percent of rated load the voltage drops very rapidly.

At no load, a part of the magnetic flux resulting from the shunt coil on the main pole piece is diverted and does not pass through the armature. As the load increases, the series winding on the diverter pole rediverts this flux to the armature and provides a commutating field.

If the shunt and series windings are properly proportioned, the flux in the armature varies with the load to compensate for the *IR* drop in the generator and for speed changes of the driving motor.

A flat voltage curve is obtained, since the necessary magnetic changes produced by the series winding take place only in the diverter pole, the flux from the main pole remaining constant. The flux densities in the diverter pole are kept low, so that the magnetic changes which occur in this part of the magnetic circuit take place on the straight portion of the magnetization curve, thus eliminating most of the curvature from the voltage characteristic. By correct adjustment of the diverter-pole winding by means of an adjustable shunt, a very straight, flat curve is obtained with only a very slight rise on approaching zero load.

At some value of the load current the ampere-turns on the diverter pole will equal those on the main pole, and at this time the magnetic flux leaking across the bridge to the diverter pole is all rediverted across the air gap; hence there is no further leakage flux for an increased current to be rediverted to the armature. When the load is increased beyond this point, the increased ampere-turns on the diverter pole combine with the armature cross-magnetizing force to send magnetic flux in the reverse direction across the leakage bridge, which tends to demagnetize the main pole and reduce the generator voltage. Good commutation is assured, as the diverter pole provides a commutating field of the correct direction for improving commutation, and this field varies with the current output just as in a commutating-pole generator.

**41. Control generators** are specially designed dc generators which are employed for the precise automatic control of dc motors. They are used to perform a wide variety of functions, such as:

1. Controlling and regulating speed, voltage, current, or power accurately over a wide range

2. Controlling tension and torque to maintain product uniformity in winding, drawing, and similar operations

3. Speeding up acceleration or deceleration to increase the production of high-inertia machines, etc.

Small motors can be supplied directly with power from the control generator. In the majority of applications, however, the capacity of the control generator is not sufficient to supply the motor directly, and the control generator energizes the field of a main generator which supplies the motor with power. A control generator must possess the characteristics of moderately fast response and low power consumption from the activating control circuit. The application of control generators for motor control is discussed in greater detail in Sec. 141.

**41A. The amplidyne** is a control generator manufactured by the General Electric Co. Its construction, as shown in Fig. 7-30A, differs from that of the conventional dc generator in the following ways:

1. Separately excited field windings (up to four individual windings) with only a small number of ampere-turns in each winding. The field windings are called *control fields* or *control windings*.

2. Brushes (*A* and *B* in Fig. 7-30A) located in the conventional neutral position which are short-circuited on each other.

3. Pole structure split in two.

4. The addition of a second set of brushes (*C* and *D* in Fig. 7-30A) 90 electrical degrees from the conventional neutral position. These brushes are the load brushes of the machine and are connected to the output terminals of the machine.

5. A compensating field winding wound around the pole structure of the machine instead of being located in slots on the face of the poles. The compensating field winding is connected in series with the load-brush circuit.

The amplidyne functions in the following manner. When a control-field winding or windings are energized, a magnetic flux will be produced in paths *G* and *H* of Fig. 7-30A. The rotation of the armature through this flux will generate voltage in the armature conductors. However, because of the small number of ampere-turns in the control-field windings, the voltage generated from brush $B$ to brush $A$ will produce only a reasonable value of current in the short-circuited armature. The direction of this short-circuited armature current will be as shown by the designations inside the conductors in Fig. 7-30A. The

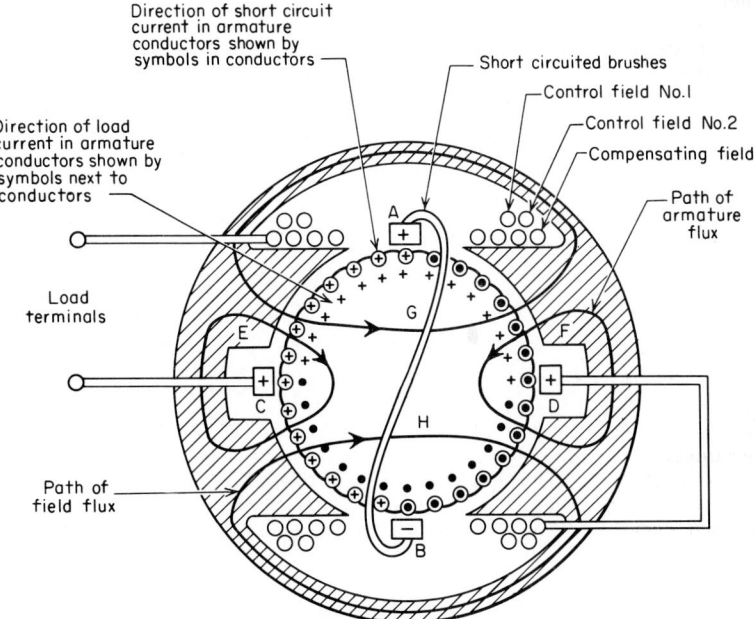

**Fig. 7-30A**   Cross section of an amplidyne generator.

short-circuited armature current will produce magnetic flux in the paths $E$ and $F$. Voltages will be induced in the armature conductors by this flux $E$ and $F$. These conductor voltages will produce no net voltage from brush $B$ to brush $A$, since the voltages in the conductors from $B$ to $C$ and $B$ to $D$ will just neutralize the voltages produced in the conductors from $C$ to $A$ and $D$ to $A$, respectively. A voltage will be produced, however, from brush $C$ to brush $D$. Since brushes $C$ and $D$ are connected to the terminals of the machine, an output voltage will be produced whenever a control-field winding is energized. When brushes $C$ and $D$ are connected to an external load circuit, current will flow through the armature conductors because of the load current in the directions designated in Fig. 7-30A by the markings alongside each conductor. The actual current through each individual armature conductor will be the combination of the armature short-circuit current in paths $B$ to $A$ and the armature load current in paths $C$ to $D$. The load current in the armature should not materially affect the flux of the machine. Therefore, a compensating field winding is connected in series with the load. This winding is so located and designed that its ampere-turns will practically neutralize the ampere-turns produced by the armature load current. The functioning of the machine is summarized as follows:

1. Excitation of a control-field winding produces flux $G$ and $H$.
2. Flux $G$ and $H$ produces no net voltage between brushes $C$ and $D$ but does produce net voltage in armature paths $BCA$ and $BDA$.
3. Voltage in paths $BCA$ and $BDA$ produces current in the armature because of the short circuit $AB$.
4. Short-circuit current in the armature produces flux $E$ and $F$.
5. Flux $E$ and $F$ produces no net voltage between brushes $A$ and $B$ but does produce net voltage between brushes $C$ and $D$. This is the load or output voltage of the machine.
6. Voltage $CD$ produces load current in the armature and external circuit.

7. Ampere-turns (flux-producing tendency) of the load armature current are neutralized by ampere-turns of compensating field winding.

Thus the control-field current controls the short-circuit armature current, which in turn controls the output voltage. Since the control-field winding is designed with a very small watt capacity, a very small change of input power to a control-field winding will produce a large change in output power.

The symbolic representation of an amplidyne generator as used in wiring diagrams is shown in Fig. 7-30B.

**41B. Rototrol** is the trade name of the control generator made by the Westinghouse Electric Corp. The Rototrol is essentially a small dc generator that is similar in electrical and

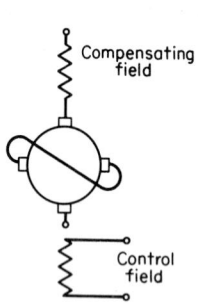

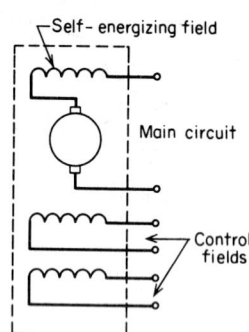

**Fig. 7-30B**   Basic symbol for an amplidyne.          **Fig. 7-30C**   Basic symbol for the Rototrol.

mechanical construction to a standard dc generator of equal size, except that the Rototrol is provided with a number of field windings. The voltage output of the machine depends entirely upon the control and interaction of these field windings. The Rototrol is provided with a self-energizing field winding and two or more control-field windings (see Fig. 7-30C.) The self-energizing field winding is generally connected in series with the Rototrol armature and furnishes the necessary excitation for the normal operation of the machine. All the other field windings are called control-field windings. The control-field windings are used to measure and compare standard and actual values representative of the quantity to be regulated. One of the control-field windings is called a *pattern-field winding*, and the others are called *pilot-field windings*. The pattern field is the control field, which is separately excited from an independent source and is used as a calibration, or standard of comparison. The pilot fields, the rest of the control-field windings, measure directly or indirectly the quantity to be regulated.

**42. The efficiency of a dc generator** increases with the load up to a certain point. Figure 7-31 indicates typical efficiency performances. Large-capacity machines have higher full-load efficiencies than small ones. Generators should be of such capacity that insofar as possible they will operate at loads in the neighborhood of their normal ratings.

**43. Approximate data on standard compound-wound dc commutating-pole generators.** The efficiency of a generator depends on its

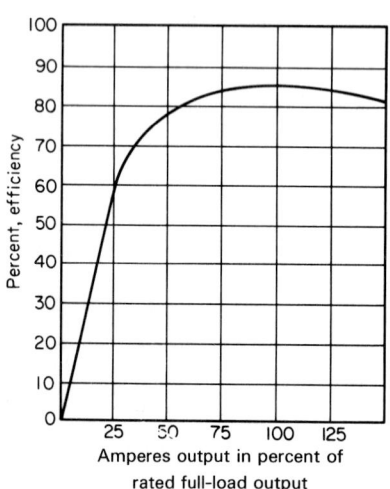

**Fig. 7-31**   Efficiency graph of a 12 kW compound-wound dc generator.

design and, to a certain extent, on its speed and voltage. Average values that are fairly representative of modern practice are given in the following table.

| Capacity, kw | Output current, amp | | | Efficiency, per cent | | |
|---|---|---|---|---|---|---|
| | 125 volts | 250 volts | 500 volts | ½ load | ¾ load | Full load |
| 5 | 40 | 20 | 10 | 77.0 | 81.0 | 82.5 |
| 10 | 80 | 40 | 20 | 82.0 | 85.0 | 86.0 |
| 15 | 120 | 60 | 30 | 82.5 | 86.5 | 86.5 |
| 20 | 160 | 80 | 40 | 84.0 | 86.5 | 87.5 |
| 25 | 200 | 100 | 50 | 85.0 | 88.0 | 89.0 |
| 35 | 280 | 140 | 70 | 87.0 | 89.0 | 89.5 |
| 50 | 400 | 200 | 100 | 88.0 | 89.5 | 90.5 |
| 60 | 480 | 240 | 120 | 88.5 | 90.5 | 91.0 |
| 75 | 600 | 300 | 150 | 88.5 | 90.5 | 91.0 |
| 90 | 720 | 360 | 180 | 88.5 | 90.5 | 91.0 |
| 100 | 800 | 400 | 200 | 89.0 | 90.5 | 91.0 |
| 125 | 1,000 | 500 | 250 | 90.5 | 91.0 | 91.0 |
| 150 | 1,200 | 600 | 300 | 90.5 | 91.3 | 91.5 |
| 200 | 1,600 | 800 | 400 | 91.0 | 91.5 | 92.0 |
| 300 | 2,400 | 1,200 | 600 | 91.3 | 91.8 | 92.0 |
| 400 | 3,200 | 1,600 | 800 | 91.8 | 92.3 | 92.5 |
| 500 | 4,000 | 2,000 | 1,000 | 91.8 | 92.2 | 92.5 |
| 750 | 6,000 | 3,000 | 1,500 | 92.0 | 92.3 | 92.5 |
| 1,000 | 8,000 | 4,000 | 2,000 | 92.5 | 93.0 | 93.5 |

**44. Rating of dc generators.** The standard methods of rating generators are as follows.

CONTINUOUS RATING A generator given a continuous power output rating will carry its rated load continuously in an ambient temperature of 40°C without exceeding a specified rise in temperature. Some large generators (larger than 1 kW/rpm) designed for metal-rolling-mill service (other than reversing hot-mill service) will carry 115 percent of their rated load continuously without injury to themselves provided service conditions are normal. The factor of 1.15 is known as a *service factor*.

OVERLOAD A general-use industrial generator will carry 150 percent of its rated current for 1 min without injury to itself. Some large generators (larger than 1 kW/rpm) have additional overload-time capabilities.

**45. Brushes: their adjustment and care** *(Westinghouse Instruction Book).* Brushes on a dc generator should be on or near the no-load neutral point of the commutator. This neutral point on most standard non-commutating-pole generators is in line with the center of the pole, and the brushes should be set a little in advance of this neutral point. The brushes of non-commutating-pole generators should be given a slight forward lead in the direction of rotation of the armature. Motor brushes should be set somewhat back of the neutral point, the backward lead in this case being approximately equal to the forward lead on generators. The exact position in either case is that which gives the best commutation at normal voltage for all loads. In no case should the brushes be set far enough from the neutral point to cause dangerous sparking at no load. For commutating-pole machines it is essential that the brushes be located at the neutral point.

The ends of all brushes should be fitted to the commutator so that they make good contact over

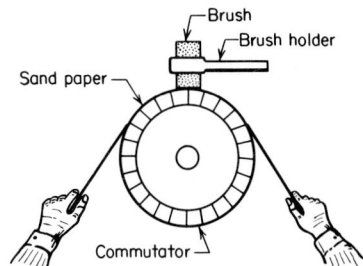

**Fig. 7-32** Sandpapering brushes.

their entire bearing faces. This can be most easily accomplished after the brush holders have been adjusted and the brushes inserted as follows. Lift a set of brushes sufficiently to permit a sheet of sandpaper to be inserted. Draw the sandpaper in one direction only, preferably in the direction of rotation, under the brushes (Fig. 7-32), being careful to keep the ends of the paper as close to the commutator surface as possible and thus avoid rounding the edges of the brushes. Treat each set of brushes similarly in turn. Start with coarse and finish with fine sandpaper. With copper-plated brushes, bevel their edges slightly so that the copper will not touch the commutator.

**46. Operating Instructions.**    Do not lubricate the commutator with oil; a piece of muslin moistened with Vaseline can be used to clean and lubricate the commutator.

Emery is a conductor and should not be used in fitting brushes or cleaning the commutator; use sandpaper or sandstone (Sec. 77), and do not use it on the commutator too frequently.

Do not use greater brush tension than necessary; tension greater than 2 lb/in$^2$ (13,790 Pa) is seldom required. When replacing brushes, use the quality and size originally supplied with the machine and fit them to the commutator with sandpaper before use (Sec. 45). Do not open generator-field circuits quickly; open the switch slowly, permitting the arc to extinguish gradually, which should take about 5 s.

**47. Direct-Current Generator and Motor Defects**
(From *Machinery*, by special permission)

| | | | No. | Description |
|---|---|---|---|---|
| Sparking at the brushes | Faults of — Brushes | Not set diametrically opposite | 1 | A. Should have been set properly at first, by counting bars, or by measurement on the commutator |
| | | | 2 | B. Can be done if necessary while running; move rocker until brush on one side sparks least, then adjust other brushes so they do not spark |
| | | Not set at neutral points | | Move rocker back and forth slowly until sparking stops |
| | | Not properly trimmed | 3 | A. Brushes should be properly trimmed before starting. If there are two or more brushes one may be removed and retrimmed |
| | | Not in line | 4 | B. Clean with alcohol or ether, then grind and reset carefully (see 1, 4, 38) |
| | | | | Adjust each brush until bearing is on line and square on commutator bar, bearing evenly the whole width (see 12 A) |
| | | Not in good contact | 5 | A. Clean commutator of oil and grit. See that brushes touch |
| | | | | B. Adjust tension screws and springs to secure light, firm and even contact (see 38 B) |
| | Commutator | Rough; worn in grooves or ridges; out of round | 6 | A. Grind with fine sandpaper on curved block, and polish with crocus cloth. Never use emery in any form |
| | | | 7 | B. If too bad to grind down turn off true in a lathe or preferably in its own bearings, with a light tool and rest, a light cut; running slowly<br>NOTE Armature should have ⅟₆₄ to ⅛-in (0.159 to 3.175 mm) end motion when running, to wear commutator evenly and smoothly (see 31) |
| | | High bars | 8 | Set "high bar" down carefully with mallet or block of wood, then clamp end nuts tightly, or file, grind, or turn true. A high bar may cause singing (see line 38) |
| | | Low bars | 9 | Grind or turn commutator true to the surface of the low bars |
| | | Weak magnetic field | 10 | A. Broken circuit } repair if external, rewind if internal } in field coils<br>B. Short circuit<br>C. Machine not properly wound, or without proper amount of iron—no remedy but to rebuild it |
| | Armature | Short-circuited coils | 11 | A. Remove copper dust, solder, or other metallic contact between commutator bars<br>B. See that clamping rings are perfectly free, and insulated from commutator bars; no copper dust, carbonized oil, etc., to cause an electrical leak<br>C. Test for cross connection or short circuit, and if such is found rewind armature to correct<br>D. See that brush holders are perfectly insulated. No copper dust, carbon dust, oil or dust, to cause an electrical leak (see 1, 2, 60) |
| Armature faults | | Broken coils | 12 | A. Bridge the break temporarily by staggering the brushes until machine can be shut down (to save bad sparking) and then repair<br>B. Shut down machine if possible, and repair<br>C. If coil is broken inside, rewinding is the only sure remedy. May be temporarily repaired by connecting to next coil, across mica<br>D. Solder commutator lugs together, or put in a "jumper," and cut out, and leave open the broken coil. Be careful not to short-circuit a good coil in doing this (see 11) |
| | | Cross connections | 13 | Cross connections may have same effect as short circuit, treat as such (see 11)<br>Each coil should test complete without cross and no ground |

**47. Direct-Current Generator and Motor Defects** (*Continued*)

| | | | | Defect | Remedy | No. |
|---|---|---|---|---|---|---|
| Sparking at the brushes | Excessive current in armature | Generator | | Excessive load | A. Reduce number of lamps and load | 14 |
| | | | | Ground and leak from short circuit on line | B. Test out, locate, and repair | |
| | | | | Dead short circuit on line | C. NOTE. Dead short circuit will or should blow safety fuse. Shut down, locate fault, and repair before starting again and put in a new fuse | |
| | | Motor | | Excessive voltage | D. Use proper current only and with proper rheostat, controller, and switch | |
| | | | | Excessive amperes on constant-current circuit | E. See that controller, etc., are suitable with ample resistance | |
| | | | | Friction | F. See that there is no undue friction or mechanical resistance anywhere (see 3 B and 35, 36) | |
| | | | | Too great load on pulley | G. Reduce load on motor to its rated capacity or less | |
| Heating of parts | Armature | | | Overloaded | Overload. Too many amperes, lights, or too much power being taken from machine (see 11, 12, 13, 14) | 15 |
| | | | | Short circuit | Short-circuited, generally dirt, etc., at commutator bars (see 11, 12, 13, 14) | 16 |
| | | | | Broken circuit | Broken circuit often caused by a loose or broken band (see 11, 12, 13, 14) | 17 |
| | | | | Cross connection | Cross connection. Often caused by a loose coil abrading on another coil or core (see 11, 12, 13, 14) | 18 |
| | | | | Moisture in coils | Dry out by gentle heat. May be done by sending a small current through, or causing machine to generate a small current itself, by running slowly | 19 |
| | | | | Eddy currents in core | Iron of armature hotter than coils after a run. Faulty construction. Core should be made of finely laminated insulated sheets. No remedy but to rebuild | 20 |
| | | | | Friction | Hot boxes or journals may affect armature (see 23, 33) | 21 |
| | Field coils | | | Excessive current | Shunt — A. Decrease voltage at terminals by reducing speed. Increase field resistance by winding on more wire, or finer wire or putting resistance in series with fields | 22 |
| | | | | | Series — B. Decrease current through fields by shunt, removing some of field winding or rewind with coarser wire | |
| | | | | | NOTE. Excessive current may be from a short circuit or from moisture in coils causing a leakage (see 10, 24) | |
| | | | | Eddy currents | Pole pieces hotter than coils after short run, due to faulty construction, or fluctuating current; if latter, regulate, and steady current | 23 |
| | | | | Moisture in coils | Coils show less than normal resistance, may cause short circuit or body contact to iron of dynamo. Dry out as in 19 (see also 22, note) | 24 |

| | | | |
|---|---|---|---|
| **Heating of parts** — **Bearings** | 25 | Not sufficient or poor oil | *A.* See that plenty of good mineral oil, filtered clean and free from grit, feeds but be careful that it does not get on commutator or brush holder (see 11)<br>*B.* Cylinder oil or vaseline can be used if necessary to complete run, mixed with sulfur or white lead, or hydrate of potash. Then clean up and put in good order |
| | 26 | Dirt or grit in bearings | *A.* Wash out grit with oil while running, then clean up and put in order. Be careful about flooding commutator and brush holder<br>*B.* Remove caps and clean and polish journals and bearings perfectly, then replace. See that all parts are free and lubricate well<br>*C.* When shut down, if hot, remove bearings and let them cool naturally; then clean, scrape and polish, and assemble; see that all parts are free, and lubricate well |
| | 27 | Rough journals or bearings | Smooth and polish in a lathe, removing all burrs, scratches, tool marks, etc., and rebabbitt old boxes and fit new ones |
| | 28 29 | Journals too tight in bearings; bent shaft | Slacken cap bolts, put in liners, and retighten till run is over, then scrape, ream, etc., as may be needed or bend or turn true in lathe or grinder. Possibly a new box or shaft will be needed |
| | 30 | Bearings out of line | Loosen bearing bolts, line up and block, until armature is in center of pole pieces, ream out dowel and bolt-holes and secure in new position |
| | 31 | End pressure of pulley hub or shaft collars | *A.* See that foundation is level and armature has free end motion<br>*B.* If there is no end motion, file or turn ends of boxes or shoulders on shaft to provide end motion<br>*C.* Then line up shaft and belt, so that there is no end thrust on shaft, but so that the armature plays freely endways when running |
| **Noises** | 32 | Belt too tight | *A.* Reduce load so that belt may be loosened and yet not slip. Avoid vertical belts if possible<br>*B.* Choose larger pulleys, wider and longer belts with slack side on top. Vibrating and flapping belts cause winking lamps |
| | 33 | Armature out of center of pole pieces | *A.* Bearings may be worn out and need replacing, throwing armature out of center (see 36)<br>*B.* Center armature in polar space, and adjust bearings to suit (see 30)<br>*C.* File out polar space to give equal space all round<br>*D.* Spring pole away from armature; this may be difficult or impossible in large machines |
| | 34 | Armature or pulley out of balance | Faulty construction, armature and pulley should have been balanced when made. May be helped by balancing on knife-edges now |
| | 35 | Armature strikes or rubs pole pieces | *A.* Bend or press down any projecting wires, and secure with tie bands<br>*B.* File out pole pieces where armature strikes (see 30, 33) |
| | 36 | Collars or shoulders on shaft strike or rub box | Bearings may be loose or worn out. Perhaps new bearings are needed (see 30, 31) |

**47. Direct-Current Generator and Motor Defects** (*Continued*)

| | | Defect | Remedy | No. |
|---|---|---|---|---|
| **Noises** | | Loose bolt connection or screws | See that all bolts and screws are tight, and examine daily to keep them so | 37 |
| | | Brushes sing or hiss | A. Apply stearic acid (adamantine) candle, vaseline, or cylinder oil to commutator and wipe off; only a trace should be applied<br>B. Move brushes in and out of holder to get a firm smooth, gentle pressure, free from hum or buzz (see 3, 6, 7, 8, 9, 31) | 38 |
| | | Flapping of belt | Use an endless belt if possible; if a laced belt must be used, have square ends neatly laced | 39 |
| | | Slipping of belt from overload | Tighten belt or reduce load (see 32) | 40 |
| | | Humming of armature lugs or teeth | A. Slope end of pole piece so that armature does not pass edges all at once<br>B. Decrease magnetism of field, or increase magnetic capacity of tooth | 41 |
| **Speed** | | Engine fails to regulate with varying load | Adjust governor of engine to regulate properly, from no load to full load, or get a better engine | 42 |
| | | Series motor, too much current, and runs away | A. Series motor on constant current: (1) put in a shunt and regulate to proper current; (2) use regulator or governor to control magnetism of field for varying load<br>B. Series motor on constant potential: (1) insert resistance and reduce current; (2) use a proper regulator or controlling switch; (3) change to automatic speed-regulating motor | 43 |
| | Runs too fast — Shunt motor | Field rheostat not properly set<br>Not proper current<br>Motor not properly proportioned | A. Adjust field rheostat to control motor<br>B. Use current of proper voltage and no other, with a proper rheostat<br>C. Get a better motor, one properly designed for the work | 44 |
| | Runs too slow | 45, Engine fails to regulate. 46, Overload. 47, Short-circuit in armature. 48, Striking or rubbing of armature. 49, Friction. 50, Weak magnetic field | 45, same as 42; 46 see 14 A; 47, short circuit in armature (see 11); 48 rubbing armature (see 35); 49, friction (see 3 B); 50, weak magnetic field (see 10) | 45 to 50 |
| **Motor** | Stop or fail to start — Circuit open | Great overload (see 14 F and G)<br>Excessive friction (see 25, 33, 35) | Open switch and find and repair trouble. Keep switch open and rheostat "off" to see if everything is right<br>Shunt motor on constant-potential circuit, fuse may blow or armature burn out | 51<br>52 |
| | | Fuse melted or switch open<br>Broken wire or connection<br>Brushes not in contact<br>Current fails or is shut off at station | A. Find and repair trouble after opening switch, then put in fuse (see 14 C)<br>B. Open switch, find and repair trouble (see 12)<br>C. Open switch and adjust (see 5)<br>D. Open switch and return starting-box lever to off position, wait for current | 53 |
| | | Short circuit of field<br>Short circuit of armature | Test for and repair if possible. Examine insulation of binding posts and brush holders<br>Poor insulation, dirt, oil, and copper, or carbon dust often result in a short circuit | 54<br>56 |
| | | Short circuit of switch. Runs backward. Wrong connections | Connect correctly as per diagram; if no diagram is at hand, reverse connections to brushes or others until direction of rotation is satisfactory | 57 |

| Dynamo or generator | | Remedy | Test |
|---|---|---|---|
| Reversed residual magnetism | Reversed current through field coils | *A.* Use current from another machine or a battery through field in proper direction to correct fault. Test polarity with a compass | 58 |
| | Reversed connections | *B.* If connections or winding are not known, try one way and test; if not correct reverse connections, try again and test | |
| | Earth's magnetism | *C.* Connect as per diagram for desired rotation, see that connections to shunt and series coils are properly made (see 57) | |
| | Proximity of another dynamo | | |
| | Brushes not in right position (see 1, 2, 3) | *D.* Shift brushes until they operate better (see 1, 2, 3) | |
| Too weak residual magnetism | | Same as 58 *A* | 59 |
| Short circuit in machine | | See 11, 54, 56 | 60 |
| Short circuit in external circuit | | A lamp socket, etc., may be short-circuited or grounded, and prevent building up shunt or compound machines. Find and remedy before closing switch (see 54, 56) | 61 |
| Field coils opposed to each other | | Reverse connections of one of field coils and test. Find polarity with compass; if necessary try 58 *A, C, D.* If necessary reverse connections and recharge in opposite directions | 62 |
| Open circuit | Broken wire | *A.* Search out and repair (see 12) | 63 |
| | Faulty connections | *B.* Search out and repair (see 37) | |
| | Brushes not in contact | *C.* Search out and repair (see 5) | |
| | Safety fuses melted or broken | *D.* Search out and repair (see 53 *A*) | |
| | Switch open | *E.* Search out and repair (see 53 *D*) | |
| | External circuit open | *F.* Search out and repair with dynamo switch open until repairs are completed | |
| Too great load on dynamo | | Reduce load to pilot lamp on shunt and incandescent machines; after voltage is obtained close switches in succession slowly, and regulate voltage (see 14 *A* and 65) | 64 |
| Too great resistance in field rheostat | | Bring up to voltage gradually with rheostat, and watch pilot lamp; regulate carefully | 65 |

**48. When starting, a generator may fail to excite itself** (*Westinghouse Instruction Book*). This may occur even when the generator operated perfectly during the preceding run. It will generally be found that this trouble is caused by a loose connection or a break in the field circuit, by poor contact at the brushes due to a dirty commutator or perhaps to a loss of residual magnetism, or by the incorrect position of the brushes. Examine all connections, try a temporarily increased pressure on the brushes, and look for a broken or burned-out resistance coil in the rheostat. An open circuit in the field winding can some-times be traced with the aid of a magneto bell, but this is not an infallible test, as some magnetos will not ring through a circuit of such high resistance and reactance even though it is intact. If no open circuit is found in the rheostat or in the field winding, the trouble is probably in the armature. But if nothing is wrong with the connections or the winding, it may be necessary to excite the field from another generator or some other outside source.

Calling the generator we desire to excite 1 and the other machine from which current is to be taken 2, we should follow this procedure. Open all switches and remove all brushes from generator 1; connect the positive brush holder of generator 1 with the posi-tive brush holder of generator 2; also connect the negative holders of the machines together (it is desirable to complete the circuit through a switch having a fuse of about 5 A capacity in series). Close the switch. If the generator in trouble connects to busbars fed by other generators, the same result can be effected by insulating the brushes of the machine in trouble from their commutator and closing the main switch (see Fig. 7-33A.) If the shunt winding of generator 1 is all right, its field will show considerable magnetism. If possible, reduce the voltage of generator 2 before opening the exciting circuit; then break the connections. If this cannot be done, throw in all the rheostat resistance of gen-erator 1; then open the switch very slowly, lengthening the arc which will be formed until it breaks.

A simple means of getting a compound-wound machine to pick up is to short-circuit it through a fuse having approximately the current capacity of the generator (see Fig. 7-33B). If sufficient current to melt this fuse is not generated, there is something wrong with the armature, either a short circuit or an open circuit. If, however, the fuse has blown, make one more attempt to get the machine to excite itself. If it does not pick up, some-thing is wrong with the shunt winding or connections.

If a new machine refuses to excite and the connections seem to be all right, reverse the connections of the shunt field; i.e., connect the wire which leads from the positive brush to the negative brush and the wire which leads from the negative brush to the positive brush. If this change of connections does no good, change back and locate the fault as previously suggested.

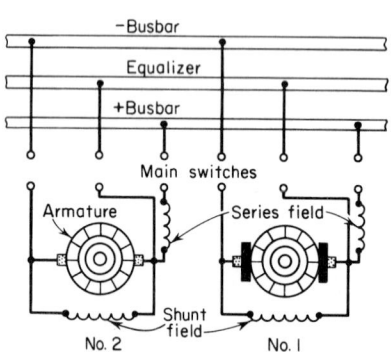

**Fig. 7-33A**  Exciting a generator.

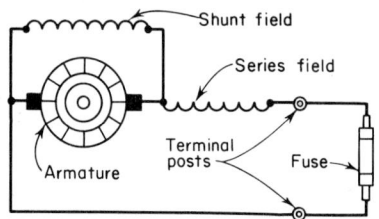

**Fig. 7-33B** Another method of exciting a generator.

**49. The proper connections for a shunt motor** are shown in Fig. 7-34. The field $B$ is con-nected as shown, so that when switch $D$ is closed, it becomes excited before the armature circuit through switch $E$ is closed. Thus when the motor armature has current admitted to it through switch $E$ and starting resistance box $A$, the field is already on and the full torque of the motor is obtained. The torque of a motor is equal to the product of a constant, the flux per pole, the ampere-turns on the armature, and the number of poles. Hence, if the full field is not on the motor at starting, full torque will not be obtained.

**50. If a motor will not start when the starting box is operated** and when current is flowing in the armature, investigate to see if field flux is present. This can be done by holding a piece of iron, such as a key, against the pole piece. If the flux exists, the key will be drawn strongly against the pole piece; if there is no flux, there will be practically no attraction.

**51. Reversed field-spool connection.** In some cases a manufacturer may have shipped a motor with one or more field spools reversed. Then no torque or, perhaps, very weak torque will be noticed. Under such conditions a trial with an iron key will indicate the presence of field magnetism, yet the weakness or total absence of torque will be present, and a trial of polarity should be made.

**52. Running in the wrong direction.** Sometimes a motor when set up and started will run in the wrong direction. The only change necessary is to reverse the field connection. Thus

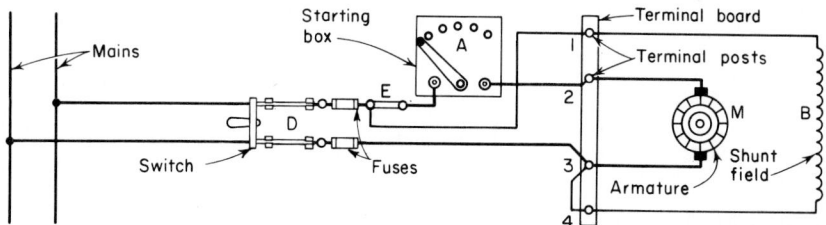

**Fig. 7-34**    Control-apparatus connections for a shunt motor.

Fig. 7-35, I, shows the connection for one direction of rotation, and Fig. 7-35, II, that for the other. Note that in Fig. 7-35, I, the brushes are shifted backward against the direction of rotation. For the opposite rotation, a backward lead, as shown in Fig. 7-35, II, must be chosen.

**53. Testing polarity of field.** This can be done in two ways. The first way is to use a compass, bringing it near the various poles and noting the direction of the deflection of the needle. Since in all motors the poles alternate in magnetic polarity, in one pole the magnetism coming out and in the next going in, a certain end of a compass needle will point toward one pole and away from the next when conditions are normal. If, however, two adjacent poles show similar magnetism, the trouble is located and the offending spool should be reversed. This should be done end for end, not by turning the axis. The latter operation does not change the direction of magnetism, while the former does.

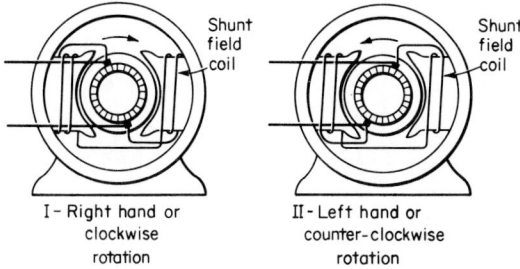

I – Right hand or
clockwise
rotation

II – Left hand or
counter-clockwise
rotation

**Fig. 7-35**    Connections for shunt-wound motors.

Direction of magnetism is determined by the following rule. When looking at the face of an electromagnet (such as the field spool of a motor), a pole will be north if the current is flowing around it in a direction opposite to the motion of the hands of a watch (Fig. 7-36) and south if in the same direction as the motion of the hands of a watch (see also the rules outlined in Div. 1).

Another method of determining whether the magnetism of the poles is correct is to use two ordinary nails, their lengths depending upon the distance between pole tips. The point of one nail should touch one pole tip, the point of the other nail should touch the other pole tip, and the heads of nails should touch each other.

When the current flows around the field spools, the polarity between any poles is properly related if the nails placed as suggested stick together by the magnetism. If there is no tendency to stick, the polarity of the two poles is alike and therefore wrong.

**54. Open field circuit.** If, on closing the field switch, no magnetism is obtained by trial with an iron key as suggested above, there is an open circuit within one of the spools or in the wires leading to these spools. The open circuit can be located by shunting out one spool at a time and allowing current to flow through the rest until the defective spool is discovered. On a two-pole motor try first one spool and then the other. For a very short time, say, 10 min, double voltage can be carried on a spool. On a motor having four or more poles, three spools can always be left in circuit during the open-circuit investigations.

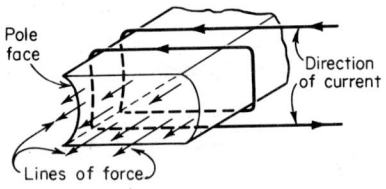

Pole face

Direction of current

Lines of force

**Fig. 7-36** Direction of magnetism and current about a pole.

**55. A method of locating an open-circuited field coil** is illustrated in Fig. 7-37A. Connect one terminal of the voltmeter to one side of the field-coil circuit, and with the bared end of a wire or a contactor successively touch the junctions of the field-coil leads around the frame. When the open coil is bridged, the voltmeter will show a full deflection. Another way is to connect the field-coil circuit terminals to a source of voltage. Connect the voltmeter successively across each coil as indicated by the dotted lines in Fig. 7-37A. There will be no deflection on the voltmeter until the open coil is bridged, when the full voltage of the circuit will be indicated.

**56. A grounded field coil can be located** (Fig. 7-37B) by connecting a source of voltage to the machine terminals, having first raised the brushes from the commutator, if it is a dc machine. Connect one terminal of the voltmeter to the frame and the other to a lead with a bared end. Tap exposed parts of the field circuit with the bared end of the lead. The voltmeter deflection will be least near the grounded coil.

**57. Heating of field coils** (*Westinghouse Instruction Book*). Field coils may heat from any of the following causes: (1) too low speed, (2) too high voltage, (3) too great forward or backward lead of brushes, (4) partial short circuit of one coil, and (5) overload.

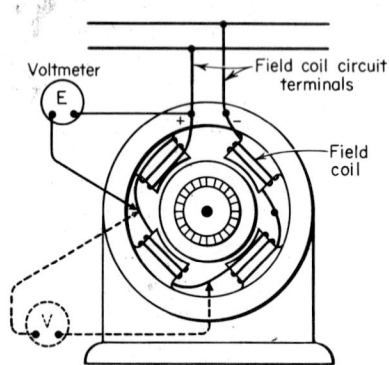

Voltmeter

Field coil circuit terminals

Field coil

**Fig. 7-37A** Locating field-coil troubles.

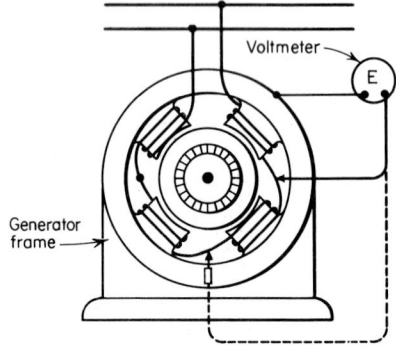

Voltmeter

Generator frame

**Fig. 7-37B** Locating a grounded coil.

**58. Direct-current armatures can be tested for common troubles** with the arrangement of Fig. 7-38. Terminals *b* and *c* are clamped to the commutator at points displaced 180 electrical degrees from each other and connected with a source of steady current through an adjustable resistance and an ammeter. For a two-pole machine the terminals *b* and *c* will be located at opposite sides of the commutator. The terminals of a low-reading voltmeter (a galvanometer can often be used) are connected to two bare metal points, which are separated, by a distance equal to the width of one commutator segment plus the width of one mica strip, by an insulating block *j*. In use, the current is adjusted to produce a convenient deflection of the voltmeter when each of the points rests on an adjacent bar.

The points are moved around the commutator and bridged across the insulation between every two bars. If the voltmeter deflection is the same for every pair of bars, there is no trouble in the armature.

**59. Sparking due to an open armature circuit.** A cause of a sparking commutator is an open circuit in the winding, either in the armature body or, more often, at the point where the lead from the armature winding is soldered to the commutator. In the latter case resoldering is a ready remedy. If, however, the point of open circuit cannot be located, the bars can be bridged over on the commutator itself by fastening with solder or otherwise a strip of copper around the segments which indicate the break.

The indication of this trouble is very apparent, for if an open circuit exists, the long, heavy spark which accompanies it soon eats away the mica between the two segments on each side of the break. This shows positively where to bridge over. An open circuit also shows itself, when the machine is running, by the viciousness of the spark. It is unlike any other kind of commutator sparking, being heavy, long, and destructive in its action.

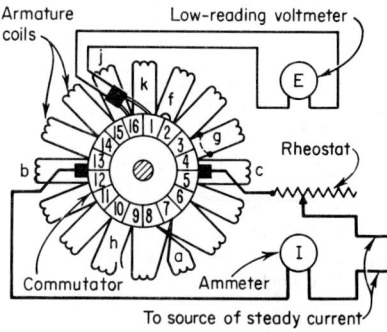

**Fig. 7-38**   Method of testing an armature.

**60. A poor connection between a bar and coil leads** will cause a considerable deflection of the voltmeter (Fig. 7-38) when one of the points rests on the bar in trouble and the other on either of the adjacent bars.

**61. An open-circuited coil** such as $h$, Fig. 7-38, will prevent the flow of current through its half of the armature. There will be no deflection on that half of the armature until the "open" is bridged. Then the voltage of the testing circuit will be indicated.

**62. Tests for open armature circuits.** Another method (Fig. 7-39A) is to apply to the commutator, at two opposite points, a low voltage, say from a battery or a dynamo with its voltage kept low. Place an ammeter in circuit and clean the surface of the commutator so that it is bright and smooth.

The terminal ends leading the current into and out of the commutator should be small, so that each rests only on a single segment (Fig. 7-39A). Note the ammeter reading and

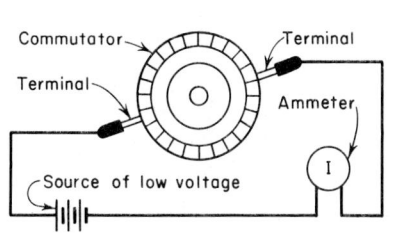

**Fig. 7-39A**   Testing for an armature open circuit with an ammeter.

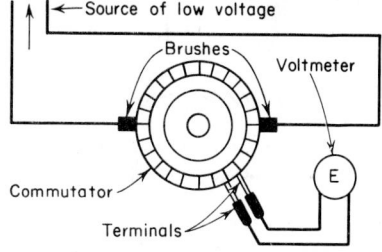

**Fig. 7-39B**   Testing for an armature open circuit with a voltmeter.

rotate the armature slowly. At the point where the open circuit exists, the ammeter needle will go to zero if the leads to the commutator bar have become entirely open-circuited. This occurs because the segment is attached to the winding through the commutator leads.

If the armature does not show the above symptoms, try connecting a low-reading voltmeter or a galvanometer to two adjacent segments while the current is passing through the armature as described from some external low-voltage source (Fig. 7-39B). Note the deflection. Pass from segment to segment in this manner, recording the drop between the successive pairs of bars. This drop, if the current is held constant from the external source, should be the same between each pair of adjacent segments. If any pair shows a higher drop than the others near it, a higher-resistance connection exists there, perhaps causing sparking and biting of commutator insulation, to a less degree, to be sure, than with an actual open circuit but perhaps enough to cause the trouble requiring the investigation.

**63. The test for armature short circuits** can be made as indicated in Fig. 7-39B. Called a bar-to-bar test, it is most valuable in locating faults in armatures. This is the method to use if a short circuit from one segment to another is suspected. When the section in which the short circuit or partial short circuit exists comes under the contacts, a low or perhaps no deflection is shown on the galvanometer or voltmeter, thus locating the defective place. Such short circuits, if they occur when running, owing to defective insulation, burn out the coil short-circuited. When the coil passes through the active field in front of the pole piece, an immense current is induced in it, causing destruction of the insulation. When this occurs, the coil should be open-circuited and bridged over, as suggested above, until a new coil can be inserted.

**64. If two bars or a coil is short-circuited** as at $f$ or $g$, Fig. 7-38, respectively, there will be little or no voltmeter deflection when the two bars connecting to the "short circuit" are bridged by the points.

**65. A grounded armature coil** can be detected in the same manner as indicated in Fig. 7-39A for a field coil. Impress a low voltage on the terminals clamped to the commutator. Ground one side of the voltmeter on the shaft or spider, and touch a lead connected to the other side to all the bars in succession. The minimum deflection will obtain when the bars connecting to the grounded coil are touched.

**66. Crossed coil leads** as at $a$, Fig. 7-38, are indicated by a twice-normal deflection when the points bridge the bars 6 and 7 to which the crossed coils should rightly connect. The crossing of the coil leads connects two coils in series between the adjacent commutator segments and hence causes a twice-normal drop.

**67. Reversed armature coil.** Instead of the armature winding progressing uniformly around from bar to bar of the commutator, at some point a coil may be connected backward. Such a reversed coil often causes bad sparking. One way to locate such a trouble is to pass a current through the armature at opposite points on the commutator. Then with a compass explore around the armature the direction of magnetism from slot to slot. If a coil is reversed when the compass comes before it, the needle will reverse, giving a very definite indication of the improperly connected coil.

**68. Heating of armature** (*Westinghouse Instruction Book*). Excessive heating of the armature may develop from any of the following causes: (1) too great a load, (2) a partial short circuit of two coils heating the two particular coils affected, and (3) short circuits or grounds on armature or commutator.

**69. Hot armature coils.** Sometimes when a new machine is started, local heating occurs in the armature, following the exact shape of the armature coil. This may occur because, in receiving the final turning off, the commutator bars were bridged with copper from one segment to another by the action of the turning tool. An examination of the commutator surface will reveal this bridging. When it is removed, satisfactory operation will ensue if the trouble has not gone too far and seriously injured the insulation of the coil.

**70. Care of commutators.** Commutators should be kept smooth by the occasional use of No. 00 sandpaper. A small quantity of Vaseline should be used as a lubricant. The lubricant should be applied to high-voltage generators with a piece of cloth attached to the end of a dry stick. If the commutator gets out of true, it should be turned down (refer to Sec. 77). Inspect the commutator surface carefully to see that the copper has not been burned over from segment to segment in the mica, and remove with a scraper any particles of copper which may be embedded in the mica. Keep oil away from the mica end rings of the commutator, as oily mica will soon burn out and ground the machine.

**71. Process of commutation and correction of glowing and pitting.** The path of the current is as shown in Fig. 7-40. A is the carbon brush; $C$, $C'$, $C''$ are the commutator segments; $B$, $B'$, $B''$ are the windings of the armature. At the position shown, coil $B$ is short-circuited by the carbon, the current passing into the face of the brush and out again as shown by the dotted line. This local current may be many times larger than the normal flow of current and is the one that causes pitting.

With perfect commutation and no sparking or glowing, an emf should be created in the short-circuited coil under the brush by means of the flux from the pole tip away from which the armature is revolving. This emf should be just large enough to reverse the current within the short-circuited coil and to render it equal to the current in the winding proper. Since on one side of the brush the current is in one direction and on the other side in the other direction, the act of commutation beneath the brush is to reverse this current and bring it up to the correct amount in the opposite direction.

With copper brushes this reversal of current must be very accurately effected. With carbon brushes there is a much smaller tendency to spark; hence they will stand a certain inexactness in commutation adjustment. Experiments indicate that the carbon can resist as much as 3 V creating current in the wrong direction and still not spark or glow. This is the property that has caused the use of carbon brushes instead of copper on most apparatus. When, however, this potential, induced in the wrong direction, rises above 3 V during the passage of the armature coil underneath the brush, trouble from sparking and glowing occurs.

This is the reason that the brushes in a motor are pulled backward as far as possible at no load, so that the coil short-circuited by the brush can enter the fringe of flux from the pole tip, thus creating the proper reversal of current while the coil is passing under the brush. Since adjacent poles are opposite in polarity, only one pole can provide the proper flux direction for this reversal. In a motor it is always the pole behind the brush, and thus the brush requires a backward lead. In a generator it is the pole ahead of the brush in the direction of rotation. Hence generators require a forward lead.

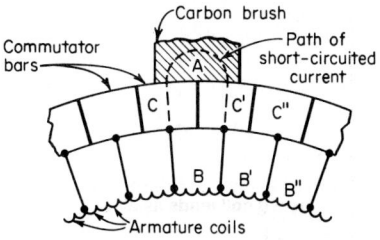

**Fig. 7-40** Armature coil short-circuited when commutating.

If a motor gives trouble from glowing and pitting, the cause is probably this induced current, and the remedy is, first, to see that the lead of the brushes brings them in the most satisfactory position. If no change of lead or brush position which will eliminate the trouble can be found, the width of the brush must be changed. The wider the brush, the longer the coil suffers a short circuit, as described. Conversely, the narrower the brush, the sooner the current must be reversed. There is, therefore, a width of brush which best satisfies both conditions.

Usually, however, when glowing occurs, the cause is too wide a brush, and often serious trouble from this cause can be entirely eliminated by varying the width of the brush, perhaps only by ⅛ in (3.175 mm).

**72. Sparking due to a rough commutator.** First, the commutator surface may not be perfectly smooth after receiving its last turnoff. The work may have been poorly done by the manufacturer, with the result that the commutator surface, instead of being left smooth, is somewhat rough. The result, especially with high-speed commutators, is that the brush does not make first-class contact with the commutator surface. It may chatter with attending noise, and thus the operation of many motors (especially those of high voltage) will be attended by sparking. As a result, the commutator surface, instead of becoming bright and smooth with time, becomes rough and dull or raw. Under these conditions the brushes do not make good contact, and the heat generated even under proper commutator conditions, owing to the resistance of brush contact, is multiplied several times, with a consequent increase of the commutator's temperature. In addition, the friction of brush contact (which should give a coefficient of 0.2) is, with a rough commutator, much higher than it should be. This high friction tends to increase the temperature.

**73. Heating of a commutator** (*Westinghouse Instruction Book*) may develop from any of the following causes: (1) overload, (2) sparking at the brushes, (3) too high brush pressure, and (4) lack of lubrication on the commutator.

**74. Hot commutator.** All this trouble (Sec. 72) is cumulative. The result is that the temperature will finally rise to a point at which the solder in the commutator will melt, perhaps short-circuiting or open-circuiting the winding. A commutator will stand very slight sparking, but if sparking is noticeable and if it continues for long periods of time, trouble is apt to result. If the load is usually very light on a motor and full load or overload is infrequent, a smoothing of the commutator occurs during the light-load period, averting trouble. This is the reason that certain railway motors, which sometimes show sparking under their normal hour-rating load, give satisfaction in commutation. The coasting of the car smooths the imperceptible damage done by the sparking during the heavy load.

**75. Loose commutator segments.**    A further and more serious cause of sparking and commutator trouble is the fact that the commutator may not be "settled" when shipped by the manufacturer. A commutator is made of many parts (Fig. 7-41) insulated one from another and bound together by mechanical clamping arrangements. The segments themselves are held on each end by a clamp ring, which must be insulated from them and should hold each segment individually from any movement relative to another.

Since the clamp must touch and hold down all segments, a failure to do so in any case results in a loose bar, which moves relatively to the next bar and causes roughness and thus sparking, with all its cumulative troubles. The roughness of commutators due to poor turning or to poor design is shown uniformly over the surface of the commutator on which brushes rest. A roughness due to a high or loose bar is shown by local trouble near the bad bar and its corresponding bars around the commutator. The jump of the brush occurs at the high bar and causes the sparking (see also Secs. 76 and 78).

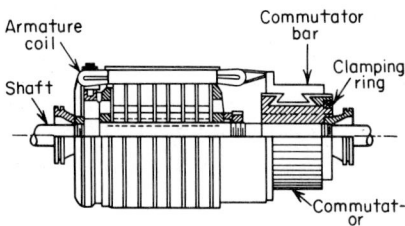

**Fig. 7-41**    Section of a dc motor armature.

**76. Blackening of a commutator.**    Sparking due to a loose or high bar causes a local blackening instead of a uniform blackening, which occurs in the case of poor design or a poor commutator surface resulting from poor turning. Also, if the speed of the commutator is low enough, a spark will occur when the bad segment passes the brush. At ordinary speeds or in cases of several loose bars, the appearance of the sparking will not differ from that due to poor design or poor turning. In such a case the commutator surface must be examined to identify the cause.

It must be remembered that the slightest movement of a bar, especially with higher-voltage and high-commutator-speed machines, may cause the trouble. A splendidly designed motor may operate very poorly because of a commutator fault.

**77. Grinding commutator** (Westinghouse Electric Corp.).    When a commutator must be resurfaced, the resurfacing should always be done with a grinding rig, whether the surface is to be ground concentric or to remove high bars or flat spots. A handstone should never be used on a commutator to obtain a true surface because it simply follows the irregularities in the surface and in some cases may even exaggerate them. The grinding rig consists of an abrasive stone set up similarly to a lathe tool in a rigging or carriage which may be moved back and forth in an axial direction and may be equipped with a radial feed. It should be supported very rigidly so that the stone is subjected to a very minimum of vibration. In large dc equipment, such a rigging can be mounted on a brush arm by removing the brush holders on that arm. In some cases, to obtain maximum rigidity it may even be desirable to brace the brush-holder bracket arm while grinding. It is also possible by removing the brush rigging to hold the grinder on parallels supported from the bedplate.

Grinding should be done when the machine is running in its own bearings and at rated speed in the case of a constant-speed machine. If grinding is done at a low speed, any slight unbalance will cause the commutator to run eccentric at rated speed.

Great care must be exercised to prevent copper and stone dust from entering the windings. The grinding rig should be equipped with a vacuum-cleaner arrangement, fitted over the stone to catch all dust. If a suction system is not available, the necks of the commutator and the front end windings should be protected by pasting heavy paper over them or by covering them with a cloth hood properly applied.

A simple, effective, and inexpensive device for collecting dust during the grinding of commutators and collectors is illustrated in Fig. 7-42. It can be made from three small pieces of equipment: (1) a 4-ft (1.2-m) length of 1½-in rubber hose, (2) a vacuum bag from a household vacutm cleaner, and (3) a small ejector which can either be purchased or be improvised from a 1½-in Y pipe fitting.

At one end of the rubber hose cut away, in the form of a long arc, the bottom and about half of the sidewalls to fit the radius of the commutator or ring to be ground. Cut a hole in the hose in the middle of the remaining arc portion to fit the grinding stone. Fit the hose to the stone and commutator or ring so as to form an enclosure, or shoe, around the grinding surface to collect the dust. Attach the other end of the hose to the intake of the ejector.

Attach the vacuum bag to the exhaust of the ejector. Use compressed air from a shop air line to produce the suction.

The stones used in grinding commutators may be classed as rough, medium, and fine. The rough stone has a grit of about 80 mesh and is used only when a very large amount of copper is to be removed. It should be employed very seldom, because if sufficient copper is to be removed to warrant its use, it would be better to take a cut off the surface in a lathe. The medium stone has a grit of about 120 mesh and is used for the bulk of the grinding work, the fine stone being used only to obtain a fine finish. The fine stone should have a grit of about 200 mesh.

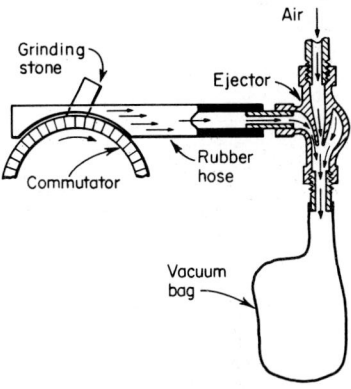

After grinding, all commutator slots should be cleaned out thoroughly and the edges of the bars beveled. Beveling accomplishes two things. It removes the burrs caused by the stone's dragging copper over the slots and eliminates the sharp edge at the entering side of the bar under a brush. The bevel on the bars is achieved with a special beveling tool; it should be about $\frac{1}{32}$ in (0.8 mm) chamfer at 45° for medium thickness of bars. For thinner or wider bars, the beveling can be changed accordingly.

Practically all up-to-date machines have under-cut mica. This undercutting should be kept $\frac{1}{16}$ in (1.6 mm) deep $\pm\frac{1}{64}$ in (0.4 mm). If it is apparent that enough copper is going to be removed by grinding so that the undercutting will be shallow, the commutator should be reundercut before grinding. This is done by means of a small circu-

**Fig. 7-42**  Method of commutator grinding, showing dust collector. [Westinghouse Electric Corp.]

lar high-speed saw about 0.003 in (0.076 mm) thicker than the nominal thickness of the mica. In undercutting, great care must be taken to see that a thin sliver of mica is not left against one side of the slot. Sometimes such a sliver must be removed by scraping by hand.

After grinding, undercutting the mica, and beveling the edges of the bars, the commutator surface should be polished while operating at rated speed. Aloxite or sandpaper (never emery cloth or paper) should first be used, as it will remove the burrs due to beveling. After a very fine grade of sandpaper is used, a high polish can be obtained by burnishing the commutator with dense felt or canvas. The surface can be even further improved if a small amount of light oil is applied to the canvas during the polishing.

**78. Loose commutator clamp rings.**  First, draw the clamps of the commutator down firmly so that when the commutator is at normal temperature, the clamping rings cannot be screwed down farther without excessive effort. This is necessary so that all the bars may have direct pressure from the clamp, rendering impossible any movement up or down. Second, smooth the surface of the commutator.

To get the clamps down firmly, run the motor. If roughness appears, shut the motor down at a convenient time, and while it is hot, tighten the clamping rings. If the tightening bolts can be screwed up somewhat, the machine should again be put in service for at least 4 h. At the end of this time, shut it down again and make another trial on the tightening bolts. If the tightening bolts cannot be screwed farther, the commutator should be surfaced, either by turning with a tool or by grinding. If the clamps are down tight and the surface of the commutator has been properly smoothed, there will be no further trouble.

**79. Slotting commutators** (Alan Bennett, *American Machinist*, Sept. 26, 1912). There seems to be a prevalent idea that slotting should cure all commutator troubles irrespective of their causes. This is not true, but slotting is a cure for certain specific troubles. If the peripheral speed of the commutator is so slow that the dirt which may collect in the slots between commutator bars will not be thrown out by centrifugal force, slotting may aggravate rather than correct commutation difficulties (see Sec. 83).

**80. The principal reason for slotting commutators** is to relieve them of high mica, i.e., mica that projects above the surface. High mica is generally due to one of two causes: either the mica is too hard and does not wear down at an equal rate with the copper, or the commutator does not hold the mica securely between the segments, allowing it to work out by the combined action of centrifugal force and the heating and cooling of the commutator.

It is evident that a commutator with a surface made irregular by projecting mica rotating at high speed under a brush must impart to the brush a vibratory action and thus impair the close contact that should exist between brush and commutator. The result is that sparking takes place more or less violently, depending on the condition of the commutator surface and the rate of speed.

This condition generally manifests itself after the machine has been running for some time and often will account for the development of sparking which did not occur at the time of installation. Often a case of this kind is aggravated by increasing the brush tension, causing a still faster rate of wear of copper over mica, with attendant increased heating of the commutator.

**81. What is accomplished by slotting.**    A harder brush can at times be used, with the idea of grinding off the mica and bringing it down to the commutator surface. Instead of the trouble being cured, the commutator will, in most cases, assume the raw appearance of being freshly·sandpapered instead of the glossy surface it should have, and both brush and commutator will wear rapidly.

This condition can be restored to normal and the commutator kept to a true surface by slotting, after which, with proper care and the use of proper brushes, commutator troubles will generally cease, provided the electrical design of the machine is not at fault. Even then there are cases that may be benefited to a certain extent by slotting, by reason of the good brush contact obtained. Most cases that show improvement are ones in which the trouble is not inherent in the design of the machine but is due to mechanical causes.

With a slotted commutator it is possible to use a brush of fine grain and soft texture because there is not the same tendency to wear away the brush as with an unslotted commutator. The commutator will then take on the much-desired polish that is generally not possible with the harder brush. The life of both brush and commutator will be increased, and friction and consequent heating will be reduced. These advantages will effect a saving that will more than offset the cost of slotting.

**82. Various methods of slotting.**    A variety of slotting devices are on the market (Figs. 7-43 to 7-47). Some operate with the armature swung between the centers of a lathe; others use a special tool in a shaper, with the armature secured to its bed. Still others are operated by hand, with the armature resting on blocks. In all cases the full width of the mica should be removed and the resulting slot carefully cleaned of burrs and rough edges. The slotting need not be carried deeply in the commutator: $\frac{1}{16}$ in (1.6 mm) is generally considered sufficient (see also Sec. 77).

**83. A slotted commutator should have proper and frequent care,** as there is a chance of small particles of copper being dragged across from bar to bar and for dirt, oil, and carbon dust to accumulate in the slots and short-circuit the commutator.

**84. High mica in commutators.**    Some motors, under certain conditions, roughen their commutators after a short term of service although there seems to be no excessive sparking under or at the edges of the brushes. Roughening may occur even though the commutator has been well settled. The commutator acts as if the mica used between bars to insulate the various segments from each other had protruded upward, causing roughness and excessive sparking.

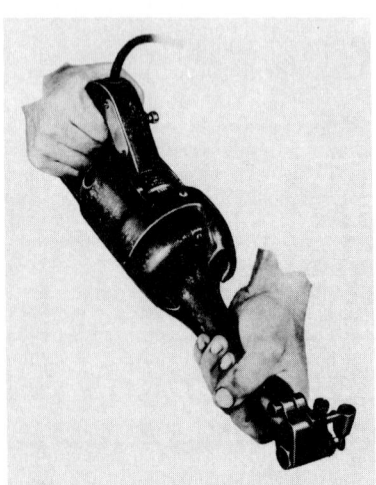

**Fig. 7-43**   Hand-type mica miller. [The Martindale Electric Co.]

Actual raising of the mica is a very rare occurrence, and if it occurs, it does so at certain spots and is easily and positively identified. An actual uniform protruding of mica all over a commutator, as described, is practically an unknown phenomenon. What does occur is an eating away of the copper surface of the commutator, leaving the high mica between the bars. A good machine will not spark enough to cause this condition. A poor machine will.

The phenomenon is easily identified, as the commutator surface looks raw all over instead of smooth and bright with a good brown gloss. If the condition is allowed to continue, a general roughness appears, accompanied by sparking, until finally the sparking and heating will so increase that the machine may flash over from brush to brush, blowing the fuses or opening the circuit breakers. The trouble is aggravated if the motor operates continuously under heavy load. If there are periods of light load, the commutator has an opportunity to be smoothed down by the brushes. This condition is appreciated by railway-motor designers. A railway motor coasts a considerable portion of the time. Thus the commutator is smoothed, neutralizing the roughening that occurs under load.

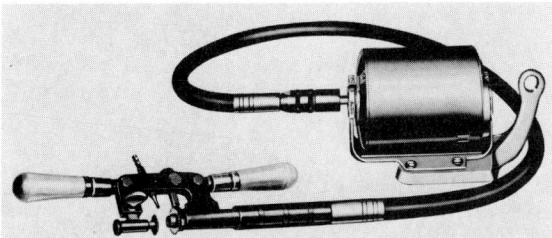

**Fig. 7-44**  Suspension-type flexible-shaft undercutter. [The Martindale Electric Co.]

**85. Remedying a roughened high-mica commutator.**    (1) Use the commutator on work for which the load is somewhat intermittent, (2) replace it altogether, or (3) slot the commutator. Then, as there no longer are two different materials to wear down or to be worn away by sparking, an unequal surface will not result. The mica need be cut down only $\frac{1}{16}$ in (1.6 mm), and a narrow, sharp chisel will do the work satisfactorily. No trouble will result from short-circuiting in this case, since centrifugal force keeps the slots clean. Some manufacturers ship machines with slotted commutators.

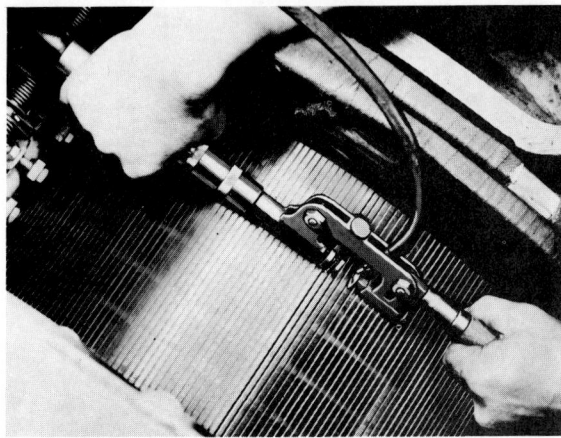

**Fig. 7-45**  Flexible-shaft undercutter in use on a large generator. [The Martindale Electric Co.]

**86. Brush troubles.**    When there is an excessive drop in speed from no load to full load, the position of the brushes on the commutator (Sec. 45) should first be investigated. No brush position that causes sparking should be chosen. The following sections outline brush troubles and their remedies.

**87. Sparking of the brushes** may be due to one of the following causes (*Westinghouse Instruction Book;* see also Table 47): (1) The machine may be overloaded. (2) The brushes may not be set exactly at the point of commutation—a position where there is no perceptible sparking can always be found, and at this point the brushes should be set and secured. (3) The brushes may be wedged in the holders. (4) The brushes may not be fitted to the circumference of the commutator. (5) The brushes may not bear on the commutator with sufficient pressure. (6) The brushes may be burned on the ends. (7) The commutator may be rough; if so, it should be smoothed off. (8) A commutator bar may be loose or may project above the others. (9) The commutator may be dirty, oily, or worn out. (10) The carbon in the brushes may be unsuitable. (11) The brushes may not be spaced equally

**Fig. 7-46**    Bench-type undercutter for a horizontal commutator. [The Martindale Electric Co.]

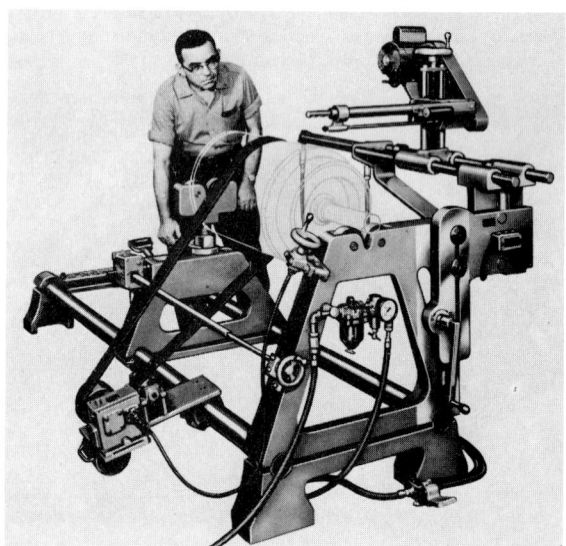

**Fig. 7-47**    Shop-type undercutter. [The Martindale Electric Co.]

around the periphery of the commutator. (12) Some brushes may have extra pressure and may be taking more than their share of the current. (13) Mica may be high. (14) The brushes may be vibrating. (15) Brush toes may not be in line.

These are the more common causes, but sparking may be due to an open circuit or a loose connection in the armature. This trouble is indicated by a bright spark which appears to pass completely around the commutator and can be recognized by the scarring of the commutator at the point of open circuit. If a lead from the armature winding to the commutator becomes loose or is broken, it will draw a bright spark as the break passes the brush position. This trouble can be readily located, as the insulation on each side of the disconnected bar will be more or less pitted. The commutator should run smoothly and true, with a dark, glossy surface.

**88. Glowing and pitting of carbon brushes.** This condition may be due to either of two causes: poor design or a wrong position of the brushes on the commutator. The error of design may be only in the choice of width of carbon brush used. The pitting is due to glowing. If the glowing is at the edge of the carbon, it is plainly visible and easily located. It may, however, occur underneath the carbon, so that only with difficulty can it be seen. Such glowin g pits the carbon face by heat disintegration. With some machines three-fourths of the brush face may be eaten away, and the pits may be ¼ to ½ in (6.35 to 12.7 mm) deep when discovered. A usual (incorrect) decision is that the current per square inch of contact is too great, the calculation being made by dividing the *line amperes* by *the square-inch cross section of either the positive or the negative brushes.* If this calculation gives a value under 45 or 50, it is certain that the cause of the trouble has not been judged correctly.

The real cause of the glowing is, to be sure, excessive current through the carbon, but this is not the line current if the calculation, as stated, shows a brush-face density below 50 A/in² (7.75 A/cm²). It is a local current caused by the short circuiting of two or more segments of the commutator when the brush rests upon them. The usual overlap of a carbon brush is about two segments, and while these two segments are under the brush, the armature coils connected to them are short-circuited. If the design of the machine is such that the coil so short-circuited encloses stray flux from the pole tip, this flux will create in the short-circuited coil a current perhaps many times larger than the brush is capable of carrying, with the result that glowing and pitting occur.

**89. Chattering of brushes** is sometimes experienced on dc machines. Under certain conditions it may become so prominent as not only to be of annoyance but also actually to break the carbons. An examination of the commutator will reveal no roughness, the surface being, perhaps, perfectly smooth and bright. This trouble occurs principally with the type of brush holder which has a box guide for the carbon. The spring which forces the brush into contact rests on the carbon, which has fairly free play in the box guide. Chattering usually occurs with high-speed commutators, running at 4000 to 5000 ft/min (20.3 to 25.4 m/s) peripheral speed.

Such brush holders are necessary on commutators which, like those on engine-driven machines, may run out of true owing to the shaft play in the bearings caused by the reciprocating motion of the engine. The clamped type of holder is usually free from bad chattering but rocks on a commutator that runs out, causing poor contact and perhaps sparking.

Lubricating the commutator causes the chattering to disappear immediately, but

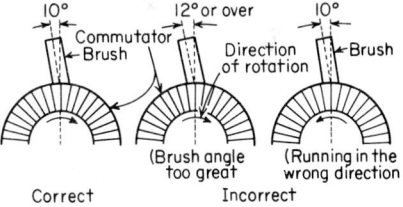

**Fig. 7-48**   Methods of setting brushes.

there is no commutator compound which gives a lubricating effect lasting over possibly ½ h. Thus it is not practical to lubricate often enough to prevent the chattering. There will be no chattering if the angle of the brush with the radial line, passing through the center of the carbon and the center of the commutator, is less than 10° and if the carbon trails instead of leads on the commutator. Figure 7-48 shows the correct setting which will stop all serious chattering, together with two incorrect settings which may give trouble.

**90. Low speed.** The fault may be in the winding of the armature or field, in which case a remedy is difficult. Considerable range of speed can be obtained by the choice of brush position on the commutator. For many motors a speed variation of 15 percent can be obtained, without sparking, by brush shift. Therefore, if the discrepancy of speed is within this amount, the brushes should be moved to counteract it. A backward shift of the brush gives increased speed, and a forward shift decreased speed. At any brush position, however, there must be practically no sparking. A first-class motor should run at full load within 4 percent (up or down) of the nameplate speed if the voltage is as specified on the nameplate. The speed at no load should not be more than 5 percent higher than this; and the speed at full load, hot, should not be over 5 percent greater than the speed at full load, cold.

**91. Bearing troubles of dc motors and generators.**   See Sec. 236.

**92. Sporadic motor sparking** has been known to be due to irregular short circuits on the line which were caused by wind blowing the line wires together.

## PRINCIPLES, CHARACTERISTICS, AND MANAGEMENT OF AC GENERATORS (ALTERNATORS)

**93. Types of ac generators.**   The different types of ac generators (alternators), classified according to the method of producing the voltage, are listed below:
1.  Synchronous alternators
    *a.* Revolving field
    *b.* Revolving armature
2.  Induction alternators
    *a.* Stator winding
    *b.* Rotor winding
3.  Inductor alternators

**94. Synchronous ac generators are discussed in an elementary way** in Secs. 73, 115, 144, 145, 148, and 152 of Div. 1. These generators may be constructed with either the armature or the field structure as the revolving member. Small generators up to 50 kW are commonly made with the revolving-armature construction. Practically all other synchronous alternators employ the revolving-field construction. The required magnetic field is produced by dc electromagnets, which are excited by a small dc generator or exciter. The fundamental construction and connections for a revolving-field alternator are shown in Fig. 7-49.

**95. The emf in a synchronous alternator is generated as suggested in Fig. 7-50.**   As each field coil, $D$ for instance, sweeps past the armature coils, the lines of flux from the field coil cut the armature coils. As coil $D$ passes from $A$ to $C$, an alternating emf represented by the curve $ABC$ will be generated in the armature. It should be understood that in commercial alternators the armature coils are set in slots and arranged differently from those shown in Fig. 7-50, which only illustrate a principle.

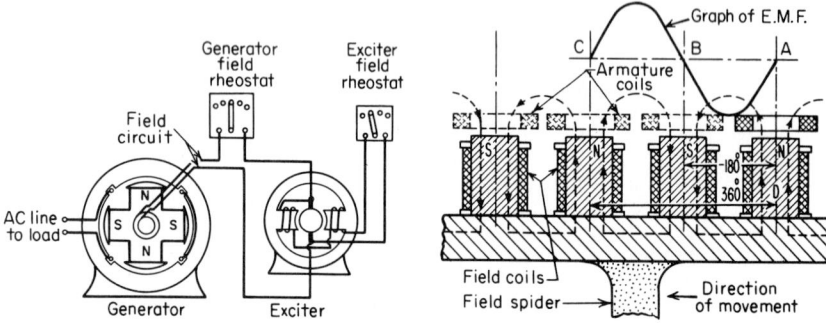

**Fig. 7-49**  Elementary diagram of an ac generator and exciter.

**Fig. 7-50**  Armature and field structure developed.

The value of the voltage generated by a given synchronous alternator depends upon the speed and direct field current. Since the speed must be held constant to maintain the proper frequency, the voltage must be controlled by adjustment of the field current.

**96. The speed and number of poles of an alternator determine the frequency which it generates.**

$$f = \frac{p \times \text{rpm}}{120} \quad \text{or} \quad p = \frac{120f}{\text{rpm}} \quad \text{or} \quad \text{rpm} = \frac{120f}{p} \tag{1}$$

where $f$ = frequency in hertz (cycles per second), rpm = revolutions per minute of rotor, and $p$ = the number of field poles.

**example** What is the frequency of a two-pole alternator running at 3600 rpm?
**solution** Substitute in the formula

$$f = \frac{p \times \text{rpm}}{120} = \frac{2 \times 3600}{120} = \frac{7200}{120} = 60 \text{ Hz}$$

**example** How many poles has a 25-Hz alternator running to 500 rpm?
**solution** Substitute in the formula

$$p = \frac{120f}{\text{rpm}} = \frac{120 \times 25}{500} = \frac{3000}{500} = 6 \text{ poles}$$

## 97. Synchronous Speeds: AC Generators and Motors

*Application to Generators.* The table shows the speeds at which the rotor of an alternator which has a given number of field poles must turn to generate voltage at given frequencies.

*Application to Motors.* The table indicates the synchronous speed of the rotary magnetic field of an induction motor having a given number of poles and taking current at a given frequency.

The table also shows the speeds of synchronous motors having a given number of field poles and taking currents at given frequencies.

| Number of poles | RPM when frequency is | | | | | | | | | | | |
|---|---|---|---|---|---|---|---|---|---|---|---|---|
| | 25 | 30 | 33⅓ | 40 | 50 | 60 | 66⅔ | 80 | 100 | 120 | 125 | 133⅓ |
| 2 | 1,500 | 1,800 | 2,000 | 2,400 | 3,000 | 3,600 | 4,000 | 4,800 | 6,000 | 7,200 | 7,500 | 8,000 |
| 4 | 750 | 900 | 1,000 | 1,200 | 1,500 | 1,800 | 2,000 | 2,400 | 3,000 | 3,600 | 3,750 | 4,000 |
| 6 | 500 | 600 | 667 | 800 | 1,000 | 1,200 | 1,333 | 1,600 | 2,000 | 2,400 | 2,500 | 2,667 |
| 8 | 375 | 450 | 500 | 600 | 750 | 900 | 1,000 | 1,200 | 1,500 | 1,800 | 1,875 | 2,000 |
| 10 | 300 | 360 | 400 | 480 | 600 | 720 | 800 | 960 | 1,200 | 1,440 | 1,500 | 1,600 |
| 12 | 250 | 300 | 333 | 400 | 500 | 600 | 667 | 800 | 1,000 | 1,200 | 1,250 | 1,333 |
| 14 | 214 | 257 | 286 | 343 | 428 | 514 | 571 | 686 | 857 | 1,029 | 1,071 | 1,143 |
| 16 | 188 | 225 | 250 | 300 | 375 | 450 | 500 | 600 | 750 | 900 | 938 | 1,000 |
| 18 | 167 | 200 | 222 | 267 | 333 | 400 | 444 | 533 | 667 | 800 | 833 | 889 |
| 20 | 150 | 180 | 200 | 240 | 300 | 360 | 400 | 480 | 600 | 720 | 750 | 800 |
| 22 | 136 | 164 | 182 | 217 | 273 | 327 | 364 | 436 | 545 | 655 | 682 | 720 |
| 24 | 125 | 150 | 167 | 200 | 250 | 300 | 333 | 400 | 500 | 600 | 625 | 667 |
| 26 | 115 | 138 | 154 | 185 | 231 | 280 | 308 | 370 | 461 | 554 | 577 | 615 |
| 28 | 107 | 128 | 143 | 171 | 214 | 257 | 286 | 343 | 429 | 514 | 536 | 571 |
| 30 | 100 | 120 | 133 | 160 | 200 | 240 | 267 | 320 | 400 | 480 | 500 | 533 |
| 32 | 94 | 113 | 125 | 150 | 188 | 225 | 250 | 300 | 375 | 450 | 487 | 500 |
| 36 | 83 | 100 | 111 | 133 | 166 | 200 | 222 | 266 | 333 | 400 | 417 | 444 |
| 44 | 79 | 82 | 91 | 109 | 136 | 164 | 182 | 218 | 273 | 327 | 341 | 363 |
| 48 | 63 | 75 | 83 | 100 | 125 | 150 | 167 | 200 | 250 | 300 | 312 | 333 |
| 54 | 56 | 66 | 74 | 90 | 111 | 133 | 148 | 178 | 222 | 266 | 278 | 296 |
| 60 | 50 | 60 | 67 | 80 | 100 | 120 | 133 | 160 | 200 | 240 | 250 | 266 |
| 68 | 44 | 53 | 59 | 71 | 88 | 106 | 118 | 141 | 176 | 212 | 221 | 235 |
| 72 | 42 | 50 | 55 | 67 | 83 | 100 | 111 | 133 | 166 | 200 | 208 | 222 |
| 96 | 31 | 38 | 42 | 50 | 64 | 75 | 82 | 100 | 125 | 150 | 156 | 167 |
| 100 | 30 | 36 | 40 | 48 | 60 | 72 | 80 | 96 | 120 | 120 | 150 | 160 |

**98. Single-phase alternator.** The circumferential distance from the centerline of one pole to the centerline of the next pole of the same polarity constitutes 360 electrical degrees. See Fig. 7-50, which shows how a single-phase emf is generated. Figure 7-49 is a diagrammatic illustration of a single-phase alternator, and Fig. 7-51 shows diagrammatically two different kinds of single-phase windings. Single-phase alternators are seldom made now except for emergency or standby power.

**99. Two-phase alternator.** In a generator of the type indicated in Fig. 7-52, the centers of the two component coils I and II are situated 90 electrical degrees apart, and the single-phase emf's generated in coils I and II by the passage of the field system past them differ in phase by 90 degrees. This property has given rise to the term *quarter-phase* for this type of machine, but it is more frequently called a two-phase machine. The emf in coil I is zero when that in coil II is a maximum, and vice versa. The curves of emf in coils I and II can be plotted as indicated in Fig. 7-53. Figure 7-54 shows two methods of connecting the armature windings of two-phase alternators. The armature coils can be arranged in one or more slots per pole per phase, as suggested diagrammatically in Fig. 7-55. In commercial machines the windings are almost always arranged in more than one slot per pole. (See Div. 1 for further information about two-phase currents.)

**100. Three-phase alternator coils** are arranged as illustrated in Fig. 7-58, and the curves of instantaneous emf are displaced from one another by 120 electrical degrees as indicated in Fig. 7-57. These curves also represent the emf's for the winding shown diagrammatically by coils I, II, and III in Fig. 7-56. Here three coils are distributed (60 electrical degrees apart) over a pole

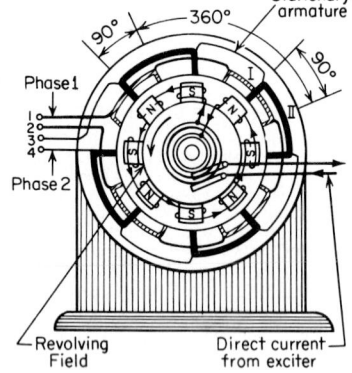

Fig. 7-52   Two-phase alternator.

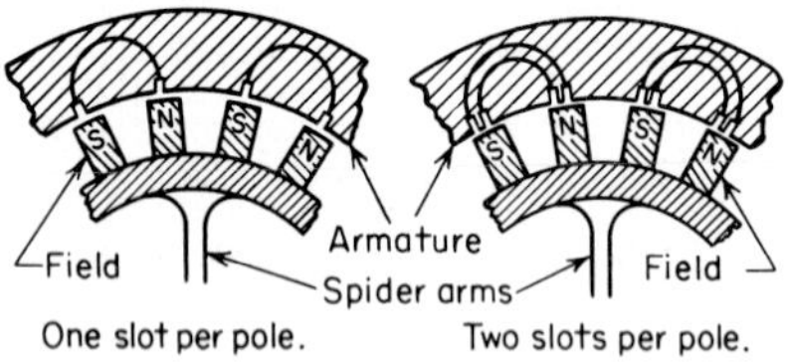

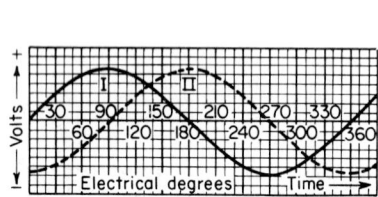

**Fig. 7-51**   Single-phase armature windings.

pitch, and the phase displacement between the emf's is 60 degrees. However, if in connecting the coils the middle coil is connected in the reverse sense from the other two, the result will be three voltages 120 electrical degrees apart, as shown in Fig. 7-57.

The two methods of connecting three-phase armature windings are shown in Fig. 7-59. These methods are discussed in greater detail in Div. 1. Armature windings can be arranged in one or more slots per pole per phase (Fig. 7-60). The Y method is almost always used to connect three-phase generators.

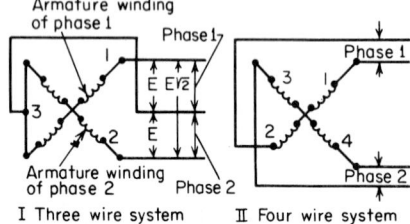

**Fig. 7-53**   Graph of two-phase current.

**Fig. 7-54**   Methods of connecting two-phase generator armature windings.

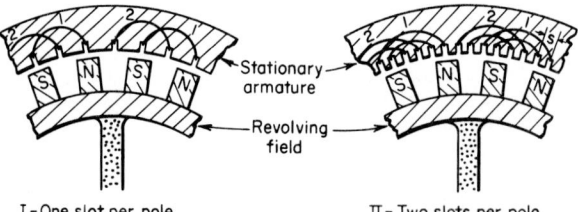

**Fig. 7-55**   Two-phase armature windings.

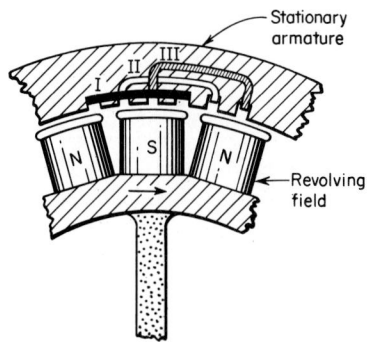

**Fig. 7-56**    Six-phase grouping.

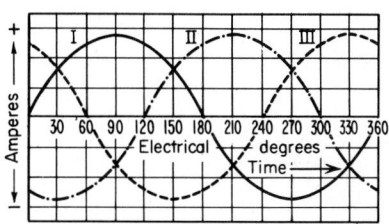

**Fig. 7-57**    Graph of three-phase currents.

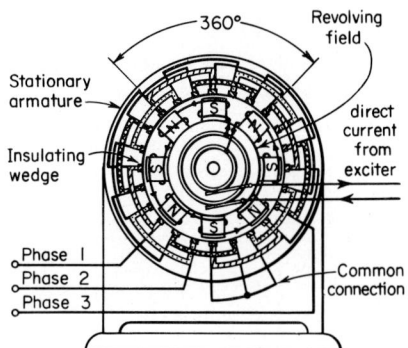

**Fig. 7-58**    Three-phase Y-connected alternator.

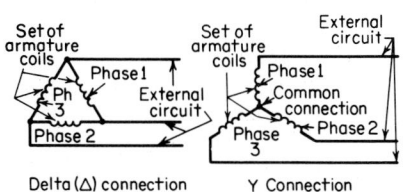

**Fig. 7-59**    Methods of connecting three-phase armature coils.

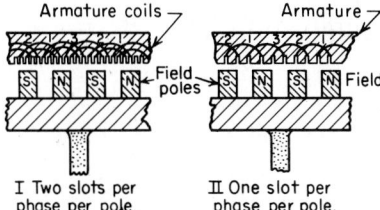

**Fig. 7-60**    Three-phase armature windings.

## 101.  Approximate Performance Values of AC Generators

(208Y/120, 240, 480Y/277, 480, 600, and 2400 volts, three-phase)

It should be understood that values will vary somewhat with speed and other conditions. Those given are general and approximate only and do not apply to any particular manufacturer's line. A slow-speed machine is assumed to be one turning at from 100 to 200 rpm; a medium-speed machine, one turning at from 200 to 300 rpm; and a high-speed machine, one turning at from 300 to 1200 rpm. In the table, S indicates slow speed, M medium speed, and H high speed.

| Output, kva | | Current | | | | Efficiency | | | Exciter capacity required |
|---|---|---|---|---|---|---|---|---|---|
| | | Three-phase | | | | ½ load | ¾ load | Full load | |
| | | 208Y/120, 240 volts | 480Y/277, 480 volts | 600 volts | 2,400 volts | | | | |
| 50 | S<br>M<br>H | 120.3 | 60.1 | 48.0 | 12.0 | 85.5[a]<br><br>86.6 | 88.0[a]<br><br>89.8 | 89.0[a]<br><br>90.8 | 7.0<br>2.0 |
| 75 | S<br>M<br>H | 180.4 | 90.2 | 72.2 | 18.0 | 88.0[a]<br><br>87.1 | 90.0[a]<br><br>89.7 | 91.3[a]<br><br>90.8 | 8.0<br>3.0 |
| 100 | S<br>M<br>H | 240.6 | 120.3 | 96.2 | 24.1 | 89.0[a]<br><br>87.7 | 91.0[a]<br><br>90.2 | 92.0[a]<br><br>91.3 | 9.0<br>3.0 |
| 125 | S<br>M<br>H | 301.0 | 150.0 | 120.0 | 30.1 | 91.0[a]<br><br>90.1 | 92.0[a]<br><br>91.7 | 92.5[a]<br><br>92.7 | 9.0<br>5.0 |
| 150 | S<br>M<br>H | 360.8 | 180.4 | 144.3 | 36.1 | 90.5[a]<br>91.0<br>90.2 | 91.7[a]<br>92.0<br>91.8 | 92.2[a]<br>93.0[a]<br>92.8 | 14.0<br>9.0<br>4.5 |
| 200 | S<br>M<br>H | 481.1 | 241.6 | 192.4 | 48.1 | 90.7[a]<br>91.0<br>90.1 | 92.3[a]<br>93.0[a]<br>92.7 | 93.4[a]<br>93.5[a]<br>93.5 | 12.0<br>11.0<br>6.0 |
| 300 | S<br>M<br>H | 723.0 | 362.0 | 289.0 | 72.0 | 91.0[a]<br>92.0[a]<br>89.2 | 93.0[a]<br>93.5[a]<br>92.1 | 93.5[a]<br>94.2[a]<br>93.2 | 20.0<br>15.0<br>12.0 |
| 400 | S<br>M<br>H | 962.0 | 481.0 | 385.0 | 96.2 | 92.0[a]<br>92.0[a]<br>90.2 | 93.0[a]<br>94.0[a]<br>92.3 | 94.0[a]<br>94.5[a]<br>93.8 | 23.0<br>14.0<br>12.0 |
| 500 | S<br>M<br>H | 1,203.0 | 602.0 | 481.0 | 120.0 | 92.5[a]<br>91.8<br>90.8 | 94.0[a]<br>93.5<br>93.5 | 94.5[a]<br>94.4<br>94.5 | 23.0<br>16.0<br>13.0 |
| 600 | S<br>M<br>H | 1,450.0 | 722.0 | 578.0 | 144.0 | 92.5[a]<br>92.4<br>90.0 | 94.0[a]<br>94.1<br>92.4 | 94.5[a]<br>94.8<br>93.8 | 28.0<br>22.0<br>20.0 |
| 700 | S<br>M<br>H | 1,690.0 | 841.0 | 673.0 | 168.0 | 93.0[a]<br>91.8<br>90.0 | 94.0[a]<br>94.1<br>92.5 | 94.6[a]<br>95.0<br>94.0 | 35.0<br>24.0<br>20.0 |
| 800 | S<br>M<br>H | 1,930.0 | 977.0 | 773.0 | 193.0 | 92.8[a]<br>92.1<br>91.5 | 94.5[a]<br>94.0<br>93.0 | 95.3[a]<br>95.0<br>94.0 | 32.0<br>23.0<br>17.0 |
| 1,000 | S<br>M<br>H | 2,406.0 | 1,203.0 | 962.0 | 241.0 | 93.0[a]<br>92.3<br>92.5 | 94.0[a]<br>94.2<br>94.0 | 94.8[a]<br>95.0<br>94.6 | 35.0<br>29.0<br>25.0 |
| 1,250 | S<br>M<br>H | 3,000.0 | 1,500.0 | 1,200.0 | 300.0 | 93.5[a]<br>92.5<br>92.0 | 94.5[a]<br>94.6<br>94.2 | 95.7[a]<br>95.5<br>95.3 | 38.0<br>30.0<br>26.0 |
| 1,500 | S<br>M<br>H | 3,640.0 | 1,804.0 | 1,443.0 | 361.0 | 93.6[a]<br>92.2<br>93.0 | 94.7[a]<br>94.4<br>95.1 | 95.4[a]<br>95.5<br>95.9 | 42.0<br>38.0<br>22.0 |
| 2,000 | S<br>M<br>H | 4,850.0 | 2,420.0 | 1,924.0 | 481.0 | 94.0[a]<br>92.6<br>92.3 | 95.0[a]<br>94.8<br>94.7 | 95.8[a]<br>95.8<br>95.7 | 50.0<br>42.0<br>38.0 |

[a]Engine-type machines: efficiencies do not include friction of bearings.

**102. Exciters for ac generators** are compound-wound dc generators, flat-compounded and rated at 125 V for the smaller sizes of generators and at 250 V for the larger sizes. Systems of excitation in general use are:

1.  Individual exciter for each generator unit
    *a.*  Direct-connected to the alternator shaft
    *b.*  Belt-connected to the alternator shaft
    *c.*  Brushless with integral rectifiers
2.  Exciter-bus system supplied by one of the following combinations:
    *a.*  Induction-motor–driven exciters and steam-driven exciters
    *b.*  Induction-motor–driven exciters and hydraulic-turbine–driven exciters
    *c.*  Built-in rectifiers, solid-state

The individual exciter unit (Fig. 7-61) has the advantage of rapid response at the time of system short circuits or rapid fluctuations in load. It also has a high efficiency due to

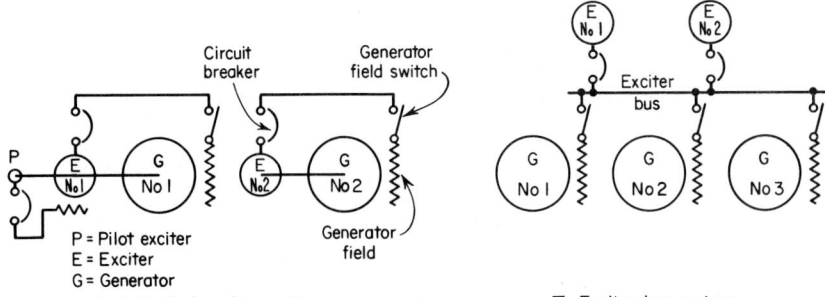

**Fig. 7-61**    Systems of excitation for ac generators.

being driven by the highly efficient prime mover of the generator and to eliminating the loss in a generator field rheostat, since the generator voltage is controlled through varia- tion of the field current of the exciter. The direct connection to the generator shaft is used for 1200-, 1800-, and 3600-rpm alternators. The belt connection is ordinarily used for slower-speed alternators so that a cheaper, higher-speed exciter (usually 1800 rpm) can be used.

The exciter-bus system (Fig. 7-61) has the advantage of not crippling the alternator because of trouble with the exciter and of keeping closer voltage regulation when the prime mover driving the alternator is subject to speed variations. With this system at least two of the exciters should be steam- or hydraulic-driven to ensure ability to start the plant up after a shutdown.

Since failure of the excitation power will produce a failure of the ac generation and a possible shutdown of the station, the exciter system should be as reliable as possible. With the exciter-bus system there should be at least one or two spare exciter units. In some plants, for extreme reliability a storage battery is floated on the exciter bus to ensure continuity of the dc excitation. Induction motors should be used for driving the exciters, because they can be started rapidly and will not fall out of step during voltage fluctuations caused by short circuits on the system.

With the individual-exciter system, pilot exciters are frequently used to supply the field of the main exciter. This procedure adds to the rapidity of response to voltage fluctuations.

**103. Synchronizing.**  Two or more ac generators will not operate satisfactorily in parallel unless (1) their voltages, as registered by a voltmeter, are the same, (2) their frequencies are the same, and (3) their voltages are in phase. If the machines are not in phase, even if their indicated voltages and their frequencies are the same, the voltage of one will, at given instants, be different from that of the other and there will be an interchange of current between the machines. When two or more generators all satisfy the three listed requirements, they are in synchronism. Synchronizing is the operation of getting machines into synchronism. Incandescent lamps or instruments are, as described below, used to indicate when machines are in synchronism.

**104. Synchronizing a single-phase circuit with lamps.** The elementary principle involved in determining synchronism is indicated in Fig. 7-62. If the voltage and frequency of generators $A$ and $B$ are the same and the machines are in phase, point $a$ will be at the same potential as point $a'$ at every instant. Hence the lamps between $a$ and $a'$ will not light as long as the three conditions are satisfied. As long as the conditions are not satisfied, there will be a fluctuating crosscurrent from $a$ to $a'$ and a constant fluctuation of the brilliancy of the incandescent lamps. When the lamps become dark and remain so, the generators are in synchronism and may be thrown together. Had the connection at $a'$ been made to the $b'$ generator lead, the lamps would be bright when the generators were

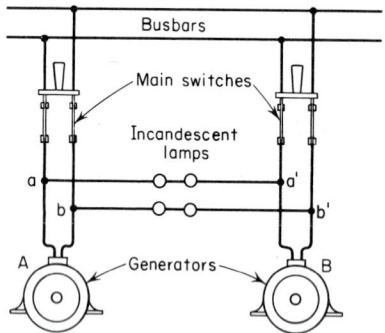

**Fig. 7-62**  Circuits for synchronizing with lamps.  **Fig. 7-63**  Circuits for synchronizing high-voltage circuits with lamps.

in synchronism, but for reasons outlined in Sec. 107 the connection shown, which provides the dark-lamp method of synchronizing, is preferred. The same conditions occur in the $b$—$b'$ set of lamps as in the $a$—$a'$ set. A voltmeter of proper rating can be substituted for the lamps.

If the voltage generated is so high that it is not desirable to connect a sufficient number of lamps in series for it, a single lamp fed through voltage transformers can be used for synchronizing, as suggested in Fig. 7-63.

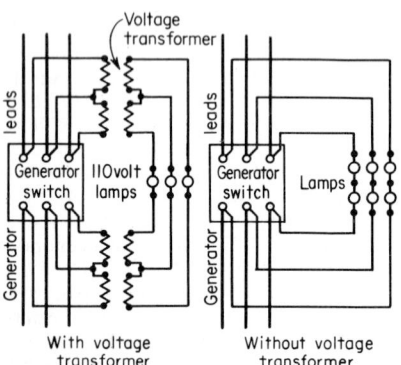

**Fig. 7-64**  Connections for phasing out three-phase circuits.

**105. Phasing out three-phase circuits.** Before connecting the leads from a polyphase generator that is to operate in parallel with others to the generator switch, the circuits must be *phased out;* i.e., the leads must be so arranged that when the generator switch is thrown, each lead from the generator will connect to the corresponding lead of the other generator. If this is not arranged, there may be considerable damage owing to an interchange of current when the two machines are in parallel. After once phasing out it is necessary to synchronize only one phase of the machine with the corresponding phase of the other machine.

Connections for phasing out three-phase circuits are shown in Fig. 7-64. If voltage transformers are not used, the sum of the voltages of the lamps in each line should be approximately the same as the voltage of the circuits. On 460-V circuits, two 230-V or four 115-V lamps should be used in each phasing-out lead.

To phase out, run the two machines at about synchronous speed. If all the lamps do not become bright and dark together, interchange any two of the main leads on one side of the switch, leaving the lamps connected to the same switch terminals, after which all the lamps should fluctuate together, indicating that the connections are correct. The machines are in phase when all the lamps are dark.

**106. The synchronizing connections for three-phase generators** are shown in Fig. 7-65. A synchronizing plug can be used instead of the single-pole synchronizing switch shown. The illustration indicates the connections used when machines are to be synchronized to a bus. If only two machines are to be synchronized, the connections are the same as shown in Fig. 7-65 except that the bus transformer and the corresponding lamp are omitted and one plug is required instead of two.

**107. Synchronizing dark or light.** *Synchronizing dark* appears to be the preferable method. All the connections shown are for synchronizing dark. When the lamps are dark,

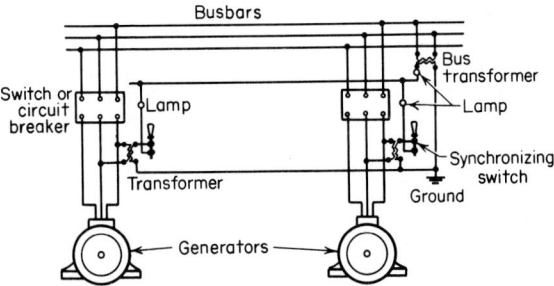

**Fig. 7-65**   Connections for synchronizing three-phase circuits when transformers are required.

the machines are in phase, and it is necessary to close the switch when the pulsation is the slowest obtainable or ceases altogether, i.e., at or just before the middle of the longest dark period.

Should a filament break, the synchronizing lamps would remain dark and thus apparently indicate synchronism and possibly cause an accident. Therefore it is considered desirable by some authorities to reverse the synchronizing circuit connections and thereby to *synchronize light.* Synchronizing light eliminates the danger due to the breaking of a filament but has the disadvantage that the time of greatest brilliancy is difficult to determine. The light period is relatively long compared with the dark period, so that synchronizing light is usually considered more difficult, and if the synchronizing-light method did not eliminate the danger due to filament breakage, it would never be used.

The probability of a filament's breaking when synchronism approaches and the machines are not in phase is remote. If breakage occurs at any other time in the operation, it will be noticed. As a protection against accidents due to breakage, two synchronizing lamps should always be placed in multiple.

**108. The number of lamps to use in a group to indicate synchronism** is determined by the voltage of the generators. With high-voltage circuits it is not feasible to use a sufficient number of lamps, so a transformer that has a secondary voltage of 115 V is employed (see the diagrams). The greatest voltage impressed on the lamps is double that of one generator or the secondary voltage of one transformer. Thus the maximum voltage on the lamps when two 230-V generators are being synchronized is 460 V. The dark period can be shortened by impressing on the lamps a voltage higher than their normal. For two 230-V machines, for example, three 115-V lamps might be used, but the life of the lamps would be greatly reduced.

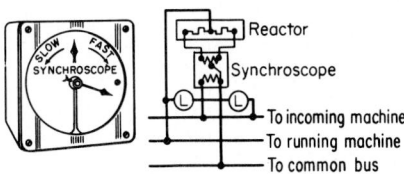

**Fig. 7-66**   Synchroscope and wiring diagram.

**109. A synchroscope** (see Fig. 7-66) is an instrument that indicates the difference in phase and frequency between two alternators. It shows whether the machine to be synchronized is running fast or slow. The pointer rotates clockwise if the machine is running fast and counterclockwise if it is running slow. When the pointer remains stationary pointing upward, the machines are in synchronism. It is quite common to use two synchronizing lamps in addition to a synchroscope, so that one system is a check against the failure of the other. Should the lamps remain dark any longer than a few seconds, the operator should look for trouble in either the lamps or the connections. After the main switches have been thrown, thus connecting the machines in parallel, the machines will hold themselves in synchronism.

**110. Although for successful parallel operation** ac generators need not be of the same type, output, and speed, it is universally conceded that the question of waveshape is important, since if the waves are of different shapes, crosscurrents will always be present. Similar waveshapes are more readily obtained with machines of similar type. Satisfactory parallel operation, the previously mentioned conditions being fulfilled, consists in obtaining

    1.  Correct division of the load among the machines
    2.  Freedom from hunting

**111. Division of load.**    Machines with similar characteristics tend to divide the common load proportionally to the ratings of the machines. Such a proportional load division may be disturbed if the steam supply to the engines is defective or variable from any cause. The steam supply is regulated by the engine governors, and defects in one or more of these governors will cause poor load division. It is essential that the governors of all the engines have similar speed-regulation characteristics so that a sudden change in the load will cause the same amount of regulation on each engine. Correct load division is therefore essentially a problem for the engine governors.

Varying the voltage of an alternator running in parallel with others by adjusting its field rheostat will not vary the load on it as with a dc generator. To increase the energy delivered by an alternator the prime mover must be caused to do more work. An engine should be given more steam or a waterwheel more water.

**112. Adjustment of field current.**    When the rheostats of two alternators running in parallel at normal speed are not adjusted to give a proper excitation, a crosscurrent will flow between the armatures. The intensity of this current depends only upon the difference in the field currents and the impedances of the armature windings. It may vary over a wide range, from a minimum of zero when both field currents are normal to more than full-load current when they differ greatly. The effect of this crosscurrent is to increase the temperature of the armatures and, consequently, to decrease the allowable useful output of the generators. It is important that the rheostats be adjusted to reduce it to a minimum. This crosscurrent registers on the ammeters of both generators and usually increases both readings. The sum of the ammeter readings will be minimum when the idle current or crosscurrent is zero.

In general, the proper field current for a machine running in parallel with others is that which it would have if running alone and delivering its load at the same voltage. To determine the proper position of the rheostats trial adjustments must be made after the alternators have been paralleled until the position at which the sum of the ammeter readings is minimum is found.

To illustrate this method let us consider two similar alternators $A$ and $B$ (Fig. 7-67) operating in parallel. When the generator field rheostats of both are properly adjusted, no crosscurrents will flow through the armatures, and the main ammeters will show equal readings if each machine is receiving the same amount of power from its prime mover. If the rheostat of $A$ is partly cut in to reduce its field current, a crosscurrent lagging in $B$ and leading in $A$ will flow between the armatures. The effect of this crosscurrent will be to strengthen $A$'s magnetization and weaken $B$'s until they are approximately equal. The resultant emf of the system will thereby be lowered.

On the other hand, if the rheostat of $B$ is partly cut out to increase its field current, a crosscurrent leading in $A$ and lagging in $B$ will flow between the armatures, strengthening $A$'s magnetization and weakening $B$'s magnetization until they are again equal. The resultant emf of the system will thereby be raised. A crosscurrent of the same character is therefore produced by decreasing one field current or increasing the other; i.e., in both cases it will lead in the first machine and lag in the second. The emf of the system will, however, be decreased in one case and increased in the other. It is obvious that by simultaneously adjusting the two rheostats the strength of the crosscurrent can be varied considerably and the emf of the system maintained constant.

For the first trial adjustment, cut in $A$'s rheostat several notches and cut out $B$'s by the same amount so as not to vary the emf of the system. If this reduces the sum of the main

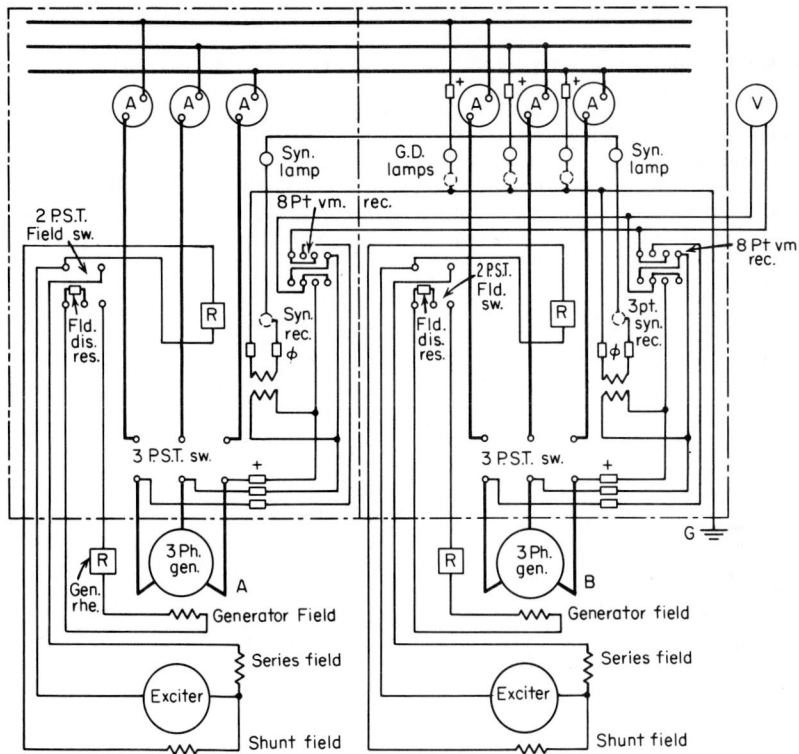

**Fig. 7-67**   Two three-phase alternators of similar characteristics operating in parallel.

ammeter readings, continue the adjustment in the same direction until the result is minimum. After this point has been reached, a further adjustment of the rheostat in either direction will increase the ammeter readings. If the first adjustment increases the sum of the ammeter readings, it is being made in the wrong direction; in this case move the rheostats back to the original positions, and then cut out $A$'s rheostat and cut in $B$'s. If both adjustments increase the sum of the ammeter readings, the original positions of the rheostats are the proper ones.

In making these adjustments it may be difficult to locate the exact points at which the crosscurrent is minimum, as it may be possible to move the rheostats over a considerable range when near the correct positions without materially changing the ammeter readings. When the adjustment is carried this far, it is close enough for practical operation. If the generators are provided with power-factor meters, the same result can be obtained by adjusting the field currents until all the power-factor meters read the same while maintaining the voltage constant.

**113. Starting a single alternator.**   (1) See that there is plenty of oil in the bearings, that the oil rings are free to turn, and that all switches are open. (2) Start the exciter and adjust for normal voltage. Start the generator slowly. See that the oil rings are turning. (3) Permit the machine to reach normal speed. Turn the generator field rheostat so that all its resistance is in the field circuit. Close the field switch. (4) Adjust the rheostat of the exciter for the normal exciting voltage. Slowly increase the alternator voltage to normal by cutting out the resistance of the field rheostat. (5) Close the main switch.

**114. Starting an alternator to run in parallel with others.**   (1) Bring the exciter and generator to speed as described in Sec. 113. Adjust the exciter voltage and close the field switch, the generator field resistance being all in. (2) Adjust the generator field resistance so that the generator voltage will be the same as the busbar voltage. (3) Synchronize, as outlined above. Close the main switch. (4) Adjust the field rheostat until crosscurrents are at a minimum (see Sec. 112), and adjust the governors of the prime movers so that the load will be properly distributed between the operating units in proportion to their capacities. (5) Adjust the governor of the machine which is to take the greatest load so that it will admit more steam.

**115. Cutting out a generator which is running in parallel with others** (*Westinghouse Instruction Book*). (1) Preferably cut down the driving power until it is just sufficient to run the generator at no load. This will reduce the load on the generator. (2) Adjust the resistance in the field curcuit until the armature current is at a minimum. (3) Open the main switch.

*Caution.* The field circuit of a generator which is to be disconnected from the busbars must not be opened before the main switch has been opened, for if the field circuit is opened first, a heavy current will flow between the armatures.

**116. Induction generators** have the same construction as induction motors (see Sec. 144). A revolving magnetic field is produced by the stator currents in exactly the same manner as in an induction motor (see Sec. 145). Induction generators are made in two types as listed in Sec. 93. The type in which the stator winding is the source of voltage is used to produce voltages of ordinary power frequencies, such as 25, 50, or 60 Hz. The type in which the rotor winding is the source of voltage is used to produce voltages of higher frequencies than ordinary power frequencies, such as 90, 100, 175, 180, or 400 Hz. These higher frequencies are often required for the operation of high-speed portable tools and machines, 400-Hz lighting, and large electronic computers.

Induction generators, in which the stator windings are the source of voltage, have their stator windings connected to the electric system which is to receive the power and their rotor windings short-circuited as shown in Fig. 7-68. The machine cannot function as a generator until a revolving magnetic field has been produced in the machine. The current which produces the rotating magnetic field must therefore be supplied to the stator winding from a source external to the machine. Therefore, an induction generator of this type must be operated in parallel with a synchronous generator. Such an induction generator is, in effect, an induction motor which is driven at a speed above the speed of its rotating magnetic field.

Suppose that an induction motor which has a slip of 5 percent at full load when operating as a motor is driven at a speed 5 percent greater than the speed of the rotating magnetic field. The flux cut by the rotor conductors will be practically the same as when it was operating as a motor. But the direction of motion of the conductors relative to the flux will be reversed. Hence the machine will become a generator and deliver power to the line approximately equal to its full-load motor rating. The frequency of the current supplied by an induction generator will always be the same as that of the synchronous generator with which it is in parallel. The induction generator tends to supply a leading current, just as does a condenser. The power factor at which it operates is determined by the slip and the design and does not depend upon the load.

Induction generators in which the rotor is the source of voltage are used for the generators in motor-generator frequency-changer sets. The stator windings are connected to a 25-, 50-, or 60-Hz main power supply, and the rotor supplies the power to the higher-frequency circuits as shown in Fig. 7-69. If a 60-Hz current is passed into the primary

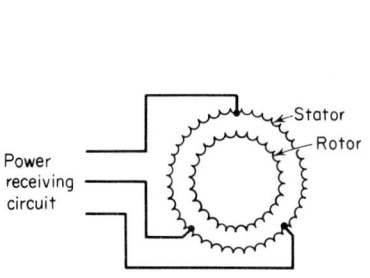

**Fig. 7-68** Induction-generator connections.

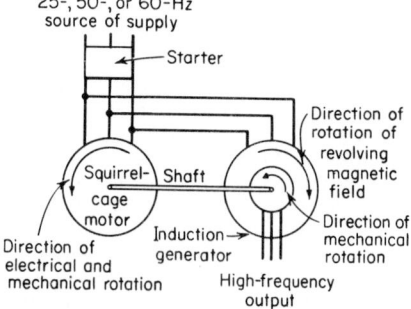

**Fig. 7-69** Schematic diagram of induction frequency-changer set.

winding of a standard wound-rotor motor and the motor is operated at synchronous or no-load speed, practically no voltage is generated in the secondary. If, however, the rotor is held stationary, a 60-Hz voltage can be obtained from the secondary. If the rotor is revolved in the opposite direction to that in which it would revolve as a motor, a voltage of a frequency higher than 60 Hz is generated. The high frequency depends upon the speed and number of poles in the generator and can be calculated from the following formula:

$$\text{High frequency} = \frac{\text{poles} \times \text{rpm}}{120} + \text{line frequency} \tag{2}$$

The voltage delivered by the induction frequency changer depends upon the design of the primary and secondary windings, the speed of the set, and the applied primary voltage.

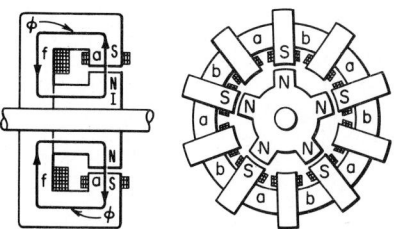

**Fig. 7-70**   An inductor alternator. This type of construction has been used for relatively small alternators employed in radio transmission and gives as high as 200,000 Hz. [Charles H. Sanderson, ed., *Electric System Handbook*]

**117. Inductor alternators** are employed to produce voltages of frequencies between 500 and 10,000 Hz for supplying the power to induction furnaces for the melting and heating of steel and alloys. The rotating element carries no electrical windings but consists simply of a toothed steel member (Fig. 7-70). The stationary member supports two sets of windings, the field windings and the armature coils. As the toothed member revolves, it varies the reluctance of the magnetic paths and thereby varies the flux produced by the current in the field windings. The flux also links with the armature coils, and its variation induces a voltage in these coils.

## PRINCIPLES, CHARACTERISTICS, AND MANAGEMENT OF ELECTRIC MOTORS

**118. Types of electric motors.**   Electric motors are manufactured in a number of different types. They may be divided into three main groups, depending upon the type of electric system from which they are designed to operate: dc, single-phase ac, and polyphase ac. There are several types of motors in each one of these groups, constructed so that they produce different starting and running characteristics. The principal types of electric motors follow:

Direct-current
  Shunt-wound
    Straight shunt-wound
    Stabilized shunt-wound
  Series-wound
  Compound-wound
  Permanent magnet
Polyphase alternating-current
  Induction
    Squirrel-cage
      Normal-torque, normal-starting-current
      Normal-torque, low-starting-current
      High-torque, low-starting-current
      Low-torque, low-starting-current

High-resistance–rotor
Automatic-start
Multispeed
Wound-rotor
Commutator, brush-shifting
Synchronous
Direct-current excited
Permanent-magnet
Reluctance
Single-phase alternating-current
Repulsion
Induction
Shading-pole–starting
Inductively split-phase–starting
Capacitor-type
Capacitor start
Permanent-split capacitor
Two-value capacitor
Repulsion-start, induction-run
Repulsion-induction
Series
Universal
Series-wound
Compensated series-wound

**119. Speed classification of motors.** Each electric motor possesses an inherent speed characteristic by which it can be classified in one of several groups. The following classification of speed characteristics is that adopted by the National Electrical Manufacturers Association (NEMA).

1. A *constant-speed motor* is one in which the speed of normal operation is constant or practically constant; for example, a synchronous motor, an induction motor with small slip, or a direct-current shunt-wound motor.

2. An *adjustable-speed motor* is one in which the speed can be varied gradually over a considerable range but when once adjusted remains practically unaffected by the load, such as a direct-current shunt-wound motor with field resistance control designed for a considerable range of speed adjustment.

3. A *multispeed motor* is one which can be operated at any one of two or more definite speeds, each being practically independent of the load; for example, a direct-current motor with two armature windings or an induction motor with windings capable of various pole groupings. In the case of multispeed permanent-split capacitor and shaded-pole motors, the speeds are dependent upon the load.

4. A *varying-speed motor* is one in which the speed varies with the load, ordinarily decreasing when the load increases, such as a series-wound or repulsion motor.

5. An *adjustable varying-speed motor* is one in which the speed can be adjusted gradually but, when once adjusted for a given load, will vary in considerable degree with change in load, such as a direct-current compound-wound motor adjusted by field control or a wound-rotor induction motor with rheostatic speed control.

6. The *base speed of an adjustable-speed motor* is the lowest-rated speed obtained at rated load and rated voltage at the temperature rise specified in the rating.

**120. Service classification of motors.** Electric motors are classified into two groups, depending upon the type of service for which they are designed. *General-purpose motors* are those motors designed for general use without restriction to a particular application. They meet certain specifications as standardized by NEMA. A *definite-purpose motor* is one which is designed in standard ratings and with standard operating characteristics for use under service conditions other than usual or for use on a particular type of application. A *special-purpose motor* is one with special operating characteristics or special mechanical construction, or both, which is designed for a particular application and which does not meet the definition of a general-purpose or a definite-purpose motor.

**121. Brake motors** are motors equipped with electrically controlled brakes as an integral part of the motor assembly. The brake motor manufactured by one company consists of one or more rotating steel disks splined on a pinion on the motor shaft, with stationary friction linings on each side of each disk. A helical spring in the center applies pressure to provide the required braking, and two, three, or four magnets, depending on the rating of the brake, supply force to compress the spring and release the brake while the motor is running.

When power is applied to the motor, the brake is immediately energized, since the brake leads are connected directly to the motor leads in the conduit box. The current energizes the magnets, which pull the armature plate toward the end plate. This action removes the pressure on the revolving disks and allows them to move freely between the friction linings, releasing the brake. Since the rotating disks are separated from the friction surfaces at all times except during actual braking, the motor delivers full rated horsepower at the output shaft.

When the motor is disconnected from the power supply, the magnets are immediately deenergized, and the spring pushes the armature plate away from the adjustable plate toward the motor. This applies braking pressure on the surfaces between the revolving disks and the friction linings, bringing the motor to a quick, smooth stop. This inherent smooth action, free from hammer blow, keeps stresses at the minimum in the brake and in belts, cables, gears, or chains through which the motor drives.

**122. Gear motors** are motors equipped with a built-in reduction gear as an integral part of the motor assembly. The motor itself is generally a 60-Hz, nominal 1750-rpm machine. Output-shaft speeds between 4 and 1430 rpm are available. These motors can be obtained equipped with almost any type of general-purpose polyphase induction, single-phase, or dc motor.

Loads of the same horsepower and speed rating will require different gear sizes, depending upon the type of load. Therefore, time of operation and frequency and severity of shock must be determined to select the proper gear motor for a specific application. To assist engineers in their selection, the American Gear Manufacturers Association (AGMA) has defined three classes of service, according to the degree to which all these variables are present:

*Class I.* Steady loads not exceeding the normal rating of the motor on 8-h-per-day service or moderate shock loads if service is intermittent.

*Class II.* Steady loads not exceeding the normal rating of the motor on 24-h-per-day service or moderate shock loads running 8 h per day.

*Class III.* Moderate shock loads on 24-h-per-day service or heavy shock loads running 8 h per day.

Classes of service for a great number of applications are given in Sec. 123.

**123.   Application Classification: Typical Gear-Motor Applications Grouped According to Normal Character of Load**

(General Electric Co.)

| Application | Load classification | Gear-motor class | |
|---|---|---|---|
| | | 8–10 hr per day | 24 hr per day |
| **Agitators:** | | | |
| Pure liquids............................................ | Uniform | I | II |
| Liquids and solids.................................... | Moderate shock | II | II |
| Liquids—variable density.......................... | Moderate shock | II | II |
| **Blowers:** | | | |
| Centrifugal............................................ | Uniform | I | II |
| Lobe...................................................... | Moderate shock | II | II |
| Vane...................................................... | Uniform | I | II |
| **Brewing and distilling:** | | | |
| Bottling machinery................................... | Uniform | I | II |
| Brew kettles—continuous duty.................... | Uniform | — | II |
| Cookers—continuous duty.......................... | Uniform | — | II |
| Mash tube—continuous duty....................... | Uniform | — | II |
| Scale hopper—frequent starts..................... | Moderate shock | II | II |
| **Can-filling machines**.................................... | Uniform | I | II |
| **Cane knives**................................................ | Moderate shock | II | II |
| **Car dumpers**.............................................. | Heavy shock | III | — |
| **Car pullers**................................................ | Moderate shock | II | — |
| **Clarifiers**.................................................. | Uniform | I | II |
| **Classifiers**................................................. | Moderate shock | II | II |
| **Clay-working machinery:** | | | |
| Brick press............................................. | Heavy shock | III | III |
| Briquette machine.................................... | Heavy shock | III | III |
| Clay-working machinery............................. | Moderate shock | II | II |
| Pug mill................................................. | Moderate shock | II | II |
| **Compressors:** | | | |
| Centrifugal............................................ | Uniform | I | II |
| Lobe...................................................... | Moderate shock | II | II |
| Reciprocating:ᵃ | | | |
| Multicylinder......................................... | Moderate shock | II | II |
| Single Cylinder...................................... | Heavy shock | III | III |
| **Conveyors—uniformly loaded or fed:** | | | |
| Apron.................................................... | Uniform | I | II |
| Assembly................................................ | Uniform | I | II |
| Belt...................................................... | Uniform | I | II |
| Bucket................................................... | Uniform | I | II |
| Chain..................................................... | Uniform | I | II |
| Flight.................................................... | Uniform | I | II |
| Oven...................................................... | Uniform | I | II |
| Screw.................................................... | Uniform | I | II |
| **Conveyors—heavy duty not uniformly fed:** | | | |
| Apron.................................................... | Moderate shock | II | II |
| Assembly................................................ | Moderate shock | II | II |
| Belt...................................................... | Moderate shock | II | II |
| Bucket................................................... | Moderate shock | II | II |
| Chain..................................................... | Moderate shock | II | II |
| Flight.................................................... | Moderate shock | II | II |

For footnote references see end of table.

**123.   Application Classification: Typical Gear-Motor Applications Grouped According to Normal Character of Load**

| Application | Load classification | Gear-motor class | |
|---|---|---|---|
| | | 8–10 hr per day | 24 hr per day |
| Live roll[b] | | b | b |
| Oven | Moderate shock | II | II |
| Reciprocating | Heavy shock | III | III |
| Screw | Moderate shock | II | II |
| Shaker | Heavy shock | III | III |
| **Cranes and hoists:** | | | |
| Main hoists: | | | |
|    Heavy duty | Heavy shock | III | III |
|    Medium duty | Moderate shock | II | II |
| Reversing | Moderate shock | II | II |
| Skip hoists | Moderate shock | II | II |
| Travel motion | Moderate shock | II | II |
| Trolley motion | Moderate shock | II | II |
| **Crushers:** | | | |
| Ore | Heavy shock | III | III |
| Stone | Heavy shock | III | III |
| **Dredges:** | | | |
| Cable reels | Moderate shock | II | |
| Conveyors | Moderate shock | II | II |
| Cutter head drives | Heavy shock | III | III |
| Jig drives | Heavy shock | III | III |
| Maneuvering winches | Moderate shock | II | |
| Pumps | Moderate shock | II | II |
| Screen drive | Heavy shock | III | III |
| Stackers | Moderate shock | II | II |
| Utility winches | Moderate shock | II | |
| **Elevators:** | | | |
| Bucket—uniform load | Uniform | I | II |
| Bucket—heavy load | Moderate shock | II | II |
| Bucket—continuous | Uniform | I | II |
| Centrifugal discharge | Uniform | I | II |
| Escalators | ............ | I | I |
| Freight | ............ | II | II |
| Gravity discharge | Uniform | I | II |
| Man lifts | | | |
|    Passenger[b] | ............ | b | b |
|    Service-hand lift | ............ | III | |
| **Fans:** | | | |
| Centrifugal | Uniform | I | II |
| Cooling Towers: | | | |
|    Induced draft | ............ | II | II |
|    Forced draft[b] | ............ | b | b |
| Induced draft | Moderate shock | II | II |
| Large (mine, etc.) | Moderate shock | II | II |
| Large industrial | Uniform | I | II |
| Light (small diameter) | Uniform | I | II |
| **Feeders:** | | | |
| Apron | Moderate | II | II |
| Belt | Moderate | II | II |
| Disk | Uniform | I | II |
| Reciprocating | Heavy shock | III | III |
| Screw | Moderate shock | II | II |
| **Food industry:** | | | |
| Beet slicer | Moderate shock | II | II |
| Cereal cooker | Uniform | I | II |

For footnote references see end of table.

**123.  Application Classification: Typical Gear-Motor Applications Grouped According to Normal Character of Load** (*Continued*)

| Application | Load classification | Gear-motor class | |
|---|---|---|---|
| | | 8–10 hr per day | 24 hr per day |
| Dough mixer | Moderate shock | II | II |
| Meat grinders | Moderate shock | II | II |
| Generators (not welding) | Uniform | I | II |
| Hammer mills | Heavy shock | III | III |
| Laundry washers, reversing | Moderate shock | II | II |
| Laundry tumblers | Moderate shock | II | II |
| **Line shafts:** | | | |
| Driving processing equipment | Moderate shock | II | II |
| Other line shafts | Uniform | I | II |
| **Machine tools:** | | | |
| Bending roll | .............. | — | II |
| Notching press—belt driven | Uniform | I | II |
| Plate planer | Heavy shock | III | III |
| Punch press—gear driven | Heavy shock | III | III |
| Tapping machines | .............. | — | III |
| Other Machine Tools: | | | |
| Main drives | Moderate shock | II | II |
| Auxiliary drives | Uniform | I | II |
| **Metal mills:** | | | |
| Draw bench—carriage | .............. | III | III |
| Draw bench—main drive | Uniform | II | III |
| Forming machines | Heavy shock | III | III |
| Pinch dryer and scrubber rolls, reversing[b] | .............. | b | b |
| Slitters[a] | Moderate shock | II | II |
| Table conveyors: | | | |
| Nonreversing | Moderate shock | II | II |
| Reversing[a] | .............. | | |
| Wire drawing and flattening machine | Moderate shock | II | II |
| Wire winding machine | .............. | — | II |
| **Mills, rotary type:** | | | |
| Ball[a] | Moderate shock | II | II |
| Cement kilns[a] | .............. | — | II |
| Driers and coolers | Moderate shock | II | II |
| Kilns | Moderate shock | II | II |
| Pebble[a] | Moderate shock | II | II |
| Rod | Heavy shock | III | III |
| Tumbling barrels | Heavy shock | III | III |
| **Mixers:** | | | |
| Concrete mixers, continuous | Moderate shock | II | II |
| Concrete mixers, intermittent | Moderate shock | I | |
| Constant density | Uniform | I | II |
| Variable density | Moderate shock | II | II |
| **Oil industry:** | | | |
| Chillers | Moderate shock | II | II |
| Oil-well pumping[b] | .............. | b | b |
| Paraffin filter press | Moderate shock | II | II |
| Rotary kilns | Moderate shock | II | II |
| **Paper mills:** | | | |
| Agitators (mixers) | Moderate shock | II | II |
| Barker auxiliaries, hydraulic | .............. | — | III |
| Barker, mechanical | .............. | — | III |
| Barking drum | Moderate shock | II | II |
| Beater and pulper | Moderate shock | — | II |
| Bleacher | Uniform | I | II |

For footnote references see end of table.

**123.  Application Classification: Typical Gear-Motor Applications Grouped According to Normal Character of Load**

| Application | Load classification | Gear-motor class | |
|---|---|---|---|
| | | 8–10 hr per day | 24 hr per day |
| Calenders[a] | Moderate shock | — | II |
| Calenders—super | Heavy shock | — | III |
| Converting machines, except cutters and platers | ............ | — | II |
| Conveyors | Uniform | — | II |
| Couch | Moderate shock | — | II |
| Cutters, platers | ............ | — | III |
| Cylinders | Moderate shock | — | II |
| Driers[a] | Moderate shock | — | II |
| Felt stretcher | Moderate shock | — | II |
| Felt whipper | ............ | — | III |
| Jordans | Heavy shock | — | III |
| Log haul | ............ | — | III |
| Presses[a] | Uniform | — | II |
| Pulp machines | ............ | — | II |
| Reel | ............ | — | II |
| Stock chests[a] | Moderate shock | — | II |
| Suction roll[a] | Uniform | — | II |
| Washers and thickeners | ............ | — | II |
| Winders | Uniform | — | II |
| Printing presses | Uniform | I | II |
| **Pullers:** | | | |
| Barge haul | ............ | II | III |
| **Pumps:** | | | |
| Centrifugal | Uniform | I | II |
| Proportioning[a] | Moderate shock | II | II |
| Reciprocating: | | | |
|   Single acting, 3 or more cylinders | Moderate shock | II | II |
|   Double acting, 2 or more cylinders | Moderate shock | II | II |
|   Single acting, 1 or 2 cylinders[b] | ............ | b | b |
|   Double acting, single cylinder[b] | ............ | b | b |
| Rotary: | | | |
|   Gear type | Uniform | I | II |
|   Lobe, vane | Uniform | I | II |
| **Rubber industry:** | | | |
| Mixer | Heavy shock | III | III |
| Rubber calendar[a] | Moderate shock | II | II |
| Rubber mill (2 or more)[a] | Moderate shock | II | II |
| Sheeter[a] | Moderate shock | II | II |
| Tire-building machines | ............ | II | II |
| Tire and tube press openers | ............ | I | I |
| Tubers and strainers | Moderate shock | II | II |
| **Sewage-disposal equipment:** | | | |
| Bar screens | Uniform | I | II |
| Chemical feeders | Uniform | I | II |
| Collectors, circuline or straightline | Uniform | I | II |
| Dewatering screws | Moderate shock | II | II |
| Grit collectors | Uniform | I | II |
| Scum breakers | Moderate shock | II | II |
| Slow or rapid mixers | Moderate shock | II | II |
| Sludge collectors | Uniform | I | II |
| Thickeners | Moderate shock | II | II |
| Vacuum filters | Moderate shock | II | II |
| **Screens:** | | | |
| Air washing | Uniform | I | II |
| Rotary—stone or gravel | Moderate shock | II | II |
| Traveling water intake | Uniform | I | II |

For footnote references see end of table.

**123.   Application Classification: Typical Gear-Motor Applications Grouped According to Normal Character of Load** (*Continued*)

| Application | Load classification | Gear-motor class | |
|---|---|---|---|
| | | 8–10 hr per day | 24 hr per day |
| Slab pushers.......................................... | Moderate shock | II | II |
| Steering gear.......................................... | Moderate shock | II | II |
| Stokers.............................................. | Uniform | I | II |
| Textile industry: | | | |
| Batchers........................................... | Moderate shock | II | II |
| Calenders.......................................... | Moderate shock | II | II |
| Card machines*..................................... | Moderate shock | II | II |
| Cloth finishing machines (Washers, pads, tenters, driers, calenders, etc).............................................. | Moderate shock | II | II |
| Dry cans........................................... | Moderate shock | II | II |
| Dyeing machinery................................... | Moderate shock | II | II |
| Looms.............................................. | Moderate shock | II | II |
| Mangles............................................ | Moderate shock | II | II |
| Nappers............................................ | Moderate shock | II | II |
| Range drives....................................... | ............. | *b* | *b* |
| Soapers............................................ | Moderate shock | II | II |
| Spinners........................................... | Moderate shock | II | II |
| Tenter frames...................................... | Moderate shock | II | II |
| Winders (other than batchers)....................... | Moderate shock | II | II |
| Yarn preparatory machines (cards, spinners, slashers, etc.).... | Moderate shock | II | II |
| Windlass............................................ | Moderate shock | II | II |

*Classes listed are minimum, and normal conditions are assumed. In view of varying load conditions, it is suggested that these applications be carefully reviewed before final selection is made.
*b*Refer to company.

**124. Direction of mounting of motors.** The standard type of motor is designed to be mounted with its shaft horizontal. Motors can be obtained, however, for mounting with their shafts vertical, but such motors are more expensive and require more careful attention.

**125. Bearings for motors** may be of the sleeve or the ball-bearing type. Ball-bearing construction is of two kinds: (1) ball bearings with provision for in-service lubrication and (2) prelubricated ball bearings which require no service lubrication. Ball bearings generally are used for small motors up to 30 hp. The type to use is largely one of individual preference, since both types when properly designed and maintained will give satisfactory service. Most manufacturers are prepared to furnish from regular stock with either sleeve or ball bearings any type of motor up to 200 hp except some totally enclosed motors.

For especially quiet operation, sleeve bearings are preferable.

The application of grease-packed ball bearings is particularly advantageous under the following conditions:

1. When the motor frame does not remain in a stationary position after installation
2. If the motor is located in an inaccessible place
3. For many totally enclosed motor applications
4. For motors with vertically mounted shafts
5. For high speeds
6. For heavy-thrust loads

The cost of ball-bearing–equipped motors is slightly higher than that of sleeve-bearing motors.

**126. Types of motor enclosure.** The various types of electric motors can be obtained with constructions which offer different degrees of enclosure and protection to the operating parts and windings. The different standard types as explained and defined by NEMA are as follows:

OPEN

*General-purpose.* Ventilating openings permit the passage of external cooling air over and around the windings of the machine.

*Dripproof.* Ventilating openings are so constructed that successful operation is not interfered with when drops of liquid or solid particles strike or enter the enclosure at any angle from 0 to 15° downward from the vertical.

*Splashproof.* Ventilating openings are so constructed that successful operation is not interfered with when drops of liquid or solid particles strike or enter the enclosure at any angle not greater than 100° downward from the vertical.

*Guarded.* Openings giving direct access to live metal or rotating parts (except smooth rotating surfaces) are limited in size by the structural parts or by screens, baffles, grills, expanded metal, or other means, to prevent accidental contact with hazardous parts. Openings giving access to such live or rotating parts shall not permit the passage of a cylindrical rod 0.75 in (19.05 mm) in diameter.

*Semiguarded.* Some of the ventilating openings, usually in the top half, are guarded as in the case of a *guarded* machine, but the others are left open.

*Dripproof guarded.* This type of dripproof machine has ventilating openings as in a *guarded* machine.

*Externally ventilated.* Designating a machine that is ventilated by a separate motor-driven blower mounted on the machine enclosure. Mechanical protection may be as defined above. This machine is sometimes known as a *blower-ventilated* or *force-ventilated* machine.

*Pipe-ventilated.* Openings for the admission of ventilating air are so arranged that inlet ducts or pipes can be connected to them. Open pipe-ventilated machines may be self-ventilated (air circulated by means integral with the machine) or force-ventilated (air circulated by means external to, and not part of, the machine).

*Weather-protected.* Type I: Ventilation passages are so constructed as to minimize the entrance of rain, snow, and airborne particles to the electric parts, and having its ventilated openings constructed as to prevent the passage of a cylindrical rod 0.75 in (19.05 mm) in diameter. Type II: In addition to the enclosure described for a Type I machine, ventilating passages at both intake and discharge are so arranged that high-velocity air and airborne particles blown into the machine by storms or high winds can be discharged without entering the internal ventilating passages leading directly to the electric parts.

The normal path of the ventilating air that enters the electric parts of the machine is so arranged by baffling or separate housings as to provide at least three abrupt changes in direction, none of which is less than 90°. In addition, an area of low velocity not exceeding 600 ft/min (183 m/min) is provided in the intake air path to minimize the possibility of moisture or dirt being carried into the electric parts.

*Encapsulated windings.* An ac squirrel-cage machine having random windings filled with an insulating resin which also forms a protective coating. This type of machine is intended for exposure to more-severe environmental conditions than the usual varnish treatment can withstand.

*Sealed windings.* An ac squirrel-cage machine making use of form-wound coils and having an insulation system which, through the use of materials, processes, or a combination of materials and processes, results in a sealing of the windings and connections against contaminants. This type of machine is intended for exposure to more-severe environmental conditions than the usual insulation system can withstand.

TOTALLY ENCLOSED

*Nonventilated.* Not equipped for cooling by means external to the enclosing parts.

*Fan-cooled.* Equipped for exterior cooling by means of a fan or fans, integral with the machine but external to the enclosing parts.

*Fan-cooled guarded.* All openings giving direct access to the fan are limited in size by design of the structural parts or by screens, grills, expanded metal, etc., to prevent accidental contact with the fan. Such openings do not permit the passage of a rod 0.75 in (19.05 mm) in diameter.

*Explosionproof.* Designed and constructed to withstand an explosion of a specified gas or vapor which may occur within it and to prevent the ignition of the specified gas or vapor surrounding the machine by sparks, flashes, or explosions of the specified gas or vapor which may occur within the machine casing.

*Dust-ignition-proof.* Designed and constructed in a manner which will exclude ignitable amounts of dust or amounts which might affect performance or rating, and which will not permit arcs, sparks, or heat otherwise generated or liberated inside of the enclosure to cause ignition of exterior accumulations or atmospheric suspensions of a specific dust on or in the vicinity of the enclosure.

*Pipe-ventilated.* Openings are so arranged that when inlet and outlet ducts or pipes are connected to them there is no free exchange of the internal air and the air outside the case. Totally enclosed pipe-ventilated machines may be ventilated as described for other pipe-ventilated machines.

*Water-cooled.* Cooled by circulating water; the water or water conductors coming in direct contact with the machine parts.

*Water-air-cooled.* Cooled by circulating air which, in turn, is cooled by circulating water. It is provided with a water-cooled heat exchanger for cooling the internal air and a fan or fans, integral with the rotor shaft or separate, for circulating the internal air.

*Air-to-air–cooled.* Cooled by circulating the internal air through a heat exchanger which, in turn, is cooled by circulating internal air. It is provided with an air-to-air heat exchanger for cooling the internal air and a fan or fans, integral with the rotor shaft or separate, but external to the enclosing part or parts, for circulating the external air.

**127. Torques of motors.** The *locked-rotor or static torque* of a motor is the minimum torque which it will develop at rest for all angular positions of the rotor, with rated voltage applied at rated frequency.

The *pull-up torque* of a motor is the minimum torque developed by the motor during the period of acceleration from rest to the speed at which breakdown torque occurs. For motors which do not have a definite breakdown torque the pull-up torque is the minimum torque developed up to rated speed.

The *breakdown torque* of a motor is the maximum torque that the motor will develop with rated voltage applied at rated frequency without an abrupt drop in speed.

The *pull-in torque* of a synchronous motor is the maximum constant torque under which the motor will pull its connected inertia load into synchronism at rated voltage and frequency.

The *pullout torque* of a synchronous motor is the maximum sustained torque which the motor will develop at synchronous speed with rated voltage applied at rated frequency with normal excitation.

The *full-load torque* of a motor is the torque necessary to produce its rated horsepower at full-load speed.

**128. The rating of an electric motor** includes the service classification (Sec. 120), voltage, full-load current, speed, number of phases and frequency if alternating-current, and full-load horsepower. The horsepower rating that is stamped on the motor nameplate by the manufacturer is the horsepower load which the motor will carry without injury to any part of the motor.

In any motor a certain amount of the input energy is lost in the machine. This energy is dissipated in the form of heat, which raises the temperature of the motor. As the load on a motor is increased, the losses increase, and therefore the temperatures of the different parts of the machine increase with the load. If too great a load is put on the motor, the motor will become excessively hot, causing probable injury to the insulation of the windings.

Motor and generator insulation systems are divided into classes according to the thermal endurance of the system for temperature-rating purposes. Four classes of insulating systems are used in motors and generators, namely, Classes A, B, F, and H. These classes have been established in accordance with IEEE standards. The insulation class is one which by experience or accepted test can be shown to have suitable thermal endurance when operated at the limiting temperature specified in the NEMA temperature-rise standard for the machine under consideration. *Experience* means successful operation for a long time under actual operating conditions of machines designed with temperature rise at or near the temperature-rating limit. *Accepted test* means a test on a system or model system which simulates the electrical, thermal, and mechanical stresses occurring in service.

Depending on the type of motor, winding type, motor service factor, etc., the insulation classes are generally based on maximum internal winding temperatures measured by the change-of-resistance method using maximum actual anticipated normal winding temperatures of 105 to 115°C for Class A systems, 130 to 140°C for Class B systems, 155 to 160°C for Class F systems, and 180°C for Class H systems.

In general, maximum ambient temperature is 40°C for any motor unless that motor is specifically designed for a different figure, in which case the correct ambient should be stamped elsewhere on the nameplate.

General-purpose open-type motors are usually rated for continuous operation in a 40°C ambient temperature under normal service conditions. These conditions include, in addition to the 40°C ambient temperature, operation at an altitude no greater than 3300 ft (1006 m) above sea level, applied voltage within 10 percent of rated value, system frequency within 5 percent of rated value, combined variation of both voltage and frequency within 10 percent of rated values, unobstructed or unhampered ventilation, and proper mounting and mechanical connection to the driven load. Motors may be provided with various classes of insulation as shown in the motor insulation list above.

**129. Starting currents of motors.**    In stating the current taken by a motor, specify just what current is meant, as starting currents are given in several different ways. In some cases the starting current pulled from the line is not the same as the current flowing into the motor itself at the instant of starting. If the rotating member of a motor is held stationary (locked rotor) and voltage is applied to the machine, a certain current would be indicated. This current is the first inrush current that will occur at starting. If the starting current is measured with a well-damped ammeter under actual starting conditions (the rotor is free to revolve), the current indicated on the ammeter will be less than the first inrush current. This is true because the ammeter does not have a chance to indicate the peak current before the motor has started to revolve and the current has decreased somewhat. It is standard practice to take the free-rotor values of starting currents as 75 percent of the locked-rotor values.

The *locked-rotor current of a motor* is the current taken directly by the motor with the revolving part held stationary.

The *locked-rotor current of a motor and its starter* is the current taken from the line with the revolving part held stationary.

The *free-rotor current of a motor* is the maximum current taken directly by the motor during the starting period as indicated by a well-damped ammeter.

The *free-rotor current of a motor and its starter* is the maximum current drawn from the line during the starting period as indicated by a well-damped ammeter.

The *average starting current* of a motor is the average value of the starting current that must be considered in selecting fuses or circuit breakers. Fuses with a rating equal to the average starting current of a motor will allow the motor to start without rupturing the fuses. Owing to the time-lag characteristic of fuses, the peak value of the starting current will not rupture fuses of a rating considerably less than the peak value of the starting current.

**130. Permissible motor-starting currents.**    The operating conditions of an electric power system are affected by the currents taken by motors under starting conditions. Excessive power fluctuations will cause large changes in the voltage of the system and excessive voltage drops under the peak-current conditions. These power and voltage fluctuations may produce unsatisfactory operating conditions for one or more of the following reasons:

1. Actuation of undervoltage protective or release devices on motor controllers, causing the shutdown of motors

2. Stalling of induction motors that are operating close to full load or slightly over full load

3. Objectionable dips in the light output of lamps and flickering of lamps

4. In the case of small private generating plants, danger of overloading of generators

Each electric power company has definite rules for permissible starting currents for motors supplied from its system. At present there is no standardization of these rules, each company having its own set of rules. Some of these rules limit the starting current on a basis of an allowable percentage of full-load current of the motor. With one of the newer rules, called the *increment method*, the maximum starting current of any motor must not be greater than a certain number of amperes per kilowatt of the customer's maximum-demand load.

**131. The speed regulation of a motor** is the percentage drop in speed between no load and full load based on the full-load speed.

$$\text{Percent speed regulation} = \frac{\text{no-load speed} - \text{full-load speed}}{\text{full-load speed}} \times 100 \qquad (3)$$

## DIRECT-CURRENT MOTORS

**132. Direct-current motors.** All dc motors must receive their excitation from some outside source of supply. Therefore they are always separately excited. The interconnection of the field and armature windings can be made, however, in one of the three different ways employed for self-excited dc generators. A dc motor may be designed, therefore, to operate with its field connected in parallel or in series with its armature, producing, respectively, a shunt or a series motor. If the machine is provided with two field windings, one connected shunt and the other series, it is a compound motor. The speed of a shunt motor connected directly to the line is nearly constant from no load to full load, while the speed of a series motor drops off rapidly as the load is increased. If a series motor were operated at no load with normal voltage, it would attain a dangerous speed, so high in most cases that it would throw itself apart by centrifugal force. A series motor should never be operated at no load unless there is sufficient external resistance connected in series with the motor to limit its speed to a safe value. For this reason a belt drive should never be used with a series motor.

Standard compound motors are designed to operate with the shunt and series fields connected so as to aid each other (cumulative). Their operating characteristics therefore are a compromise between those of shunt and series motors. Their speed drops off considerably as the load is increased, but not nearly so much as with series motors. A compound motor will not run away under no load. The amount of change of speed from no load to full load depends upon the strength of the series field. The stronger the series field, the greater the change in speed from no load to full load. Special compound motors with their field windings connected to buck each other (differential motors) are used sometimes for special applications. If the series field is of just the proper strength, a differential compound motor will operate at a more constant speed than a shunt motor, but it tends to be unstable under overload conditions.

Many shunt motors have a weak series-field winding in addition to the shunt-field winding. The series-field winding is connected cumulatively and consists of only a very few turns. The purpose of the series-field winding is to counteract partially the effect of the armature current upon the speed of the motor. The armature current tends to reduce the strength of the magnetic field of the motor. Therefore, as the load on the motor is increased, the decrease in the strength of the field may cause a rise in the speed. A motor with a rising-speed characteristic is unstable. A few series-field turns will sufficiently counteract the demagnetizing effect of the armature current and stabilize the speed characteristic of the motor. These motors, although actually cumulative compound motors with a very weak series field, are called stabilized shunt motors.

Typical characteristics for shunt, series, and compound motors are shown in Figs. 7-71, 7-72, and 7-73.

The speed regulation due to variations in load from rated load to no load of standard dc motors which meet NEMA requirements will not exceed the following values:

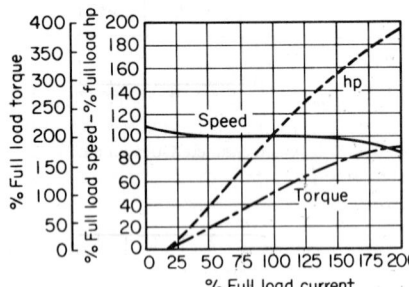

**Fig. 7-71** Typical characteristics for a shunt motor.

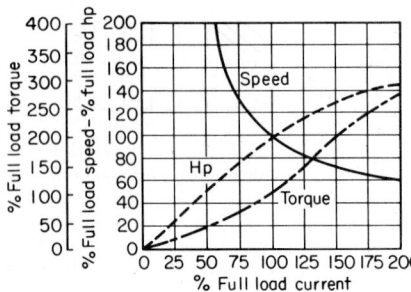

**Fig. 7-72** Typical characteristics for a series motor.

| hp | Shunt-wound regulation, % | Compound-wound | |
|---|---|---|---|
| | | Speed, rpm | Regulation, % |
| 1/20–1/8, inclusive | . . . | 1725 | 30 |
| 1/20–1/8, inclusive | . . . | 1140 | 35 |
| 1/6–1/3, inclusive | . . . | 1140 | 30 |
| 1/6–1/3, inclusive | . . . | 1725 | 25 |
| 1/2–3/4, inclusive | . . . | 1725 | 22 |
| 1/2 | . . . | 1140 | 25 |
| Less than 3 | 25 | . . . | . . . |
| 3–50, inclusive | 20 | . . . | . . . |
| 51–100, inclusive | 15 | . . . | . . . |
| 101 and larger | 10 | . . . | . . . |

**133. Torque characteristics of dc motors.**   The torque developed by any one of the three types of dc motors, either at starting or while running, depends upon the product of the magnetic flux and the armature current. Since in a shunt motor the flux is nearly constant at all times, the torque varies nearly directly with the value of the armature current. In a series motor the armature current flows directly through the field winding, so that the flux is not constant but varies almost directly with the value of the armature current. The torque of a series motor, therefore, varies approximately as the square of the armature current. Doubling the armature current will make the torque 4 times as great. The cumulative compound motor has a torque characteristic that is a compromise between that of the series motor and that of the shunt motor. The torque will increase faster than the increase in armature current but not so fast as the square of the armature current. The way in which the torque varies with the armature current for a cumulative compound motor depends upon the relative strength of the series and shunt fields. The curves of Fig. 7-74 indicate the relation between torque and armature current for the different types of dc motors.

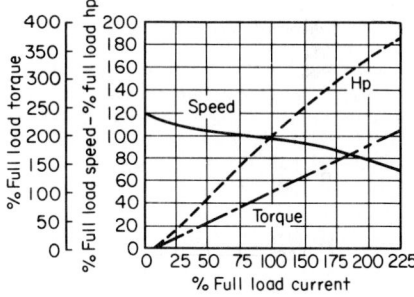

**Fig. 7-73**   Typical characteristics for a compound motor.

The maximum value of starting torque that it is satisfactorily possible for any dc motor to produce is limited only by the commutating conditions. These usually limit the current to approximately 200 percent of full-load current for a shunt motor and 250 percent of full-load current for a series or compound motor. The corresponding torques are approximately 200 percent of full-load torque for a shunt motor, 300 percent for a compound motor, and 400 percent for a series motor.

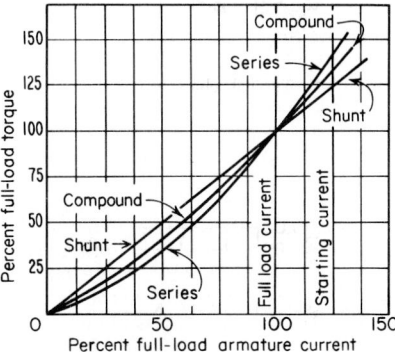

**Fig. 7-74**   Torque characteristics for dc motors.

There is no value of pullout torque for a dc motor. If the load on a dc motor were continually increased, the motor would gradually slow down as the load increased until finally the load would become so great that the motor would gradually stop. Of course, it would never be feasible in practice to overload the machine to such an extent that it would stop, and long before such a load was reached the fuses in the circuit would blow.

**134. Method of starting dc motors** (EC&M Division, Square D Co.).   A dc motor of any capacity, when its armature is at rest, will offer a very low resistance to the flow of current, and an excessive and perhaps destructive current would flow through it if it were connected directly across the supply mains while at rest. Consider a motor adapted to a normal full-load current of 100 A and having a resistance of 0.25 $\Omega$; if this motor were connected across a 250-V circuit, a current of 1000 A would flow through its armature; in other words, it would be overloaded 900 percent, with consequent danger to its windings and also to the driven machine. For the same motor, with a rheostat having a resistance of 2.25 $\Omega$ inserted in the motor circuit, at the time of starting the total resistance to the flow of current would be the resistance of the motor (0.25 $\Omega$) plus the resistance of the rheostat (2.25 $\Omega$), or a total of 2.5 $\Omega$. Under these conditions exactly full-load current, or 100 A, would flow through the motor, and neither the motor nor the driven machine would be overstrained in starting. This indicates the necessity of a rheostat for limiting the flow of current in starting the motor from rest.

An electric motor is simply an inverted generator or dynamo. Consequently when its armature begins to revolve, a voltage is generated within its windings just as a voltage is generated in the windings of a generator when driven by a prime mover. This voltage generated within the moving armature of a motor opposes the voltage of the circuit from which the motor is supplied and hence is known as a *counter-emf.* The net voltage tending to force current through the armature of a motor when the motor is running is, therefore, the line voltage minus the counter-emf.

For the motor cited above, when the armature reaches such a speed that a voltage of 125 is generated within its windings, the effective voltage will be 250 minus 125, or 125 V, and therefore the resistance of the rheostat may be reduced to 1 $\Omega$ without the full-load current of the motor being exceeded. As the armature further increases its speed, the resistance of the rheostat may be further reduced until when the motor has almost reached full speed, all the rheostat may be cut out and the counter-emf generated by the motor will almost equal the voltage supplied by the line, so that an excessive current cannot flow through the armature.

It is general practice to provide a rheostat for starting a dc electric motor except for small fractional-horsepower motors of ¼ hp or smaller. In special cases, when the motor is started very infrequently and the starting load is light, larger motors can be started by throwing them directly across the line. Motors as large as 5 hp have been successfully started in this way, but the manufacturer should be consulted whenever one contemplates starting a motor larger than 1 hp by this method. The resistor providing the starting resistance is divided into sections and is so arranged that the entire length or maximum resistance of the rheostat is in circuit with the motor at the instant of starting and that the effective length of the resistance conductor, and hence its resistance, can be reduced as the motor comes up to speed.

In cutting out the resistance of a starting rheostat care must be used not to cut it out too rapidly. If the resistance is cut out more rapidly than the armature can speed up, a sufficient counter-emf will not be generated to oppose the flow of current properly and the current will be excessive.

The standard starting rheostat, when thrown into the first position, limits the armature current to about 150 percent of the full-load current. This current will produce a starting torque of about 150 percent of full-load torque for a shunt motor, of about 175 percent for a standard cumulative compound motor, and of about 225 percent for a series machine. If greater starting torque is required, it can be obtained by advancing the starter to the second or third point. The allowable starting torque that can be obtained in this way is limited by commutating conditions to about 200 percent of full-load torque for a shunt motor, to 300 percent for a compound motor, and to 400 percent for a series motor. The currents corresponding to these maximum allowable torques are approximately 200 percent of full-load current for a shunt motor, 250 percent of full-load current for a cumulative compound motor, and 250 percent of full-load current for a series motor.

**135. Speed control of dc motors.**   The speed of dc motors for a given load can be controlled by the following methods:

    1. Armature control
        *a.* Variable resistance in series with armature
        *b.* Variable-voltage generator supplying armature (Ward-Leonard system)
        *c.* Variable voltage supplied to armature by controlled electronic rectifier
        *d.* Resistance shunted across armature

2. Field control
   *a.* Shunt-field rheostat
   *b.* Variable-voltage generator supplying shunt field
   *c.* Variable voltage supplied to shunt field by controlled electronic rectifier
   *d.* Resistance shunted across series field
3. Combination of armature and field control

**136. Armature speed control.**  Any dc motor may have its speed reduced below normal by armature control. With this method, the speed is controlled by varying the voltage impressed upon the armature, while the voltage of the field is held constant. For a constant-torque load, the speed will vary in approximate proportion to the voltage impressed on the armature. The allowable horsepower output will decrease proportionally to the decrease in speed.

Armature speed control can be obtained by inserting an external variable resistance in series with the armature circuit. The location of such a resistance for the different types of dc motors is shown in Fig. 7-75. This method is easy to apply and requires only simple and relatively inexpensive equipment, and gradual control of the speed down to practically zero speed can be obtained. Armature series resistance control, however, has the following objections:

1. *Bulk of rheostat.* This may not be very objectionable if only a few motors are so controlled, but for larger numbers the extra space becomes a factor, and in many cases it is difficult to find sufficient room near the motor.

2. *Inefficiency of the system.* The same amount of power is supplied at all speeds, but at low speeds only a small part is converted into useful work, the balance being wasted in the rheostat as heat.

3. *Poor speed regulation with varying loads.* Since the impressed voltage at the armature terminals is equal to the line voltage minus the resistance drop in the rheostat ($V_t = V - I_aR_x$), any change in the current drawn by the motor produces a change in the terminal voltage, the counter-emf, and therefore the speed.

Armature series resistance control is the type of speed control most commonly employed for series motors. Series motors requiring speed control are used chiefly for intermittent duty, and the total loss of energy in the control resistance therefore is not so great as it would be for continuous duty. Generally, armature series resistance control is not employed for shunt and compound motors unless a wider speed range than can be obtained satisfactorily by field control is desired or unless very low speeds are required.

Armature speed control obtained by a variable-voltage generator (Ward-Leonard system) requires an individual motor-generator set for each motor whose speed is to be controlled. Although this method is applicable to any type of dc motor, it is most commonly used with shunt or compound motors. A schematic diagram is given in Fig. 7-76. The Ward-Leonard system gives very close control of the speed over a wide range but has the disadvantage of requiring three full-size machines for each motor drive. For automatic drive control with dc motors, a variation of the Ward-Leonard system can be used, employing control generators (see Secs. 41, 41A, 41B, and 141).

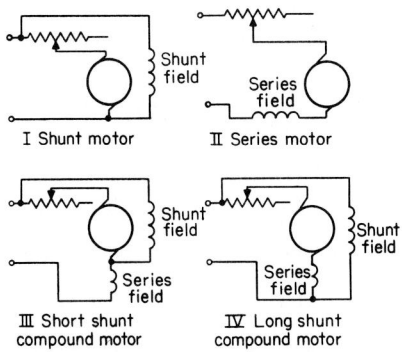

**Fig. 7-75** Speed control with armature resistance.

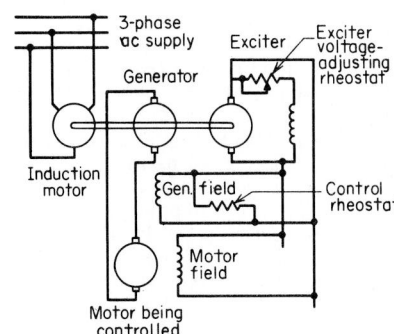

**Fig. 7-76** Ward-Leonard system of speed control.

The variable voltage for armature control can be obtained from a controlled electronic rectifier. Electronic armature speed control gives close control of the speed over a wide range. Electronic control of dc motor speed can be adapted for automatic drive control (see Sec. 140).

Resistance shunted across the armature is useful for controlling the speed of dc motors to very low values at light loads. It is used only in conjunction with armature series resistance control. Connections for this method are shown in Fig. 7-77.

**137. Field speed control.**   The speed of dc motors can be controlled by varying the field current of the motor while the armature voltage is held constant. Reducing the field-current will reduce the flux and consequently increase the speed of the motor. Field control gives close control of the speed over a limited range. The speed of any shunt motor can be increased by approximately 25 percent above normal by field control. If a greater speed range than this is required, a shunt motor designed for adjustable-speed service should be used. A speed range of 4:1 is the practical limit for field control with an adjustable-speed shunt motor. The simplest method of obtaining field speed control for shunt or compound motors is by means of a field rheostat connected in series with the shunt field. The proper location of a field rheostat in motor circuits is shown in Fig. 7-78. This method is the one

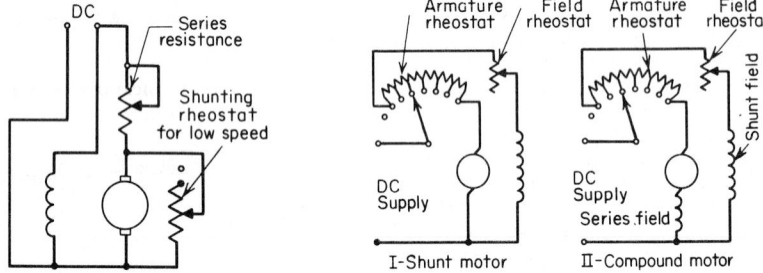

**Fig. 7-77**   Speed control using a shunt across the armature.

**Fig. 7-78**   Control of speed with a field rheostat.

most commonly used for speed control of shunt and compound motors. It is very easy to apply, requires simple and inexpensive equipment that occupies only a very small space, does not consume any appreciable amount of power in the control equipment, and gives good speed regulation with varying loads.

Field control of shunt or compound motors can be obtained by energizing the shunt field from a variable-voltage generator. This method requires an individual motor-generator set for each motor to be controlled and has no advantage over the simple field rheostat method except for automatic drive control of small motors requiring only a limited speed range. When this method is employed in automatic drive control, the shunt field of the motor is excited by a control generator (see Secs. 41, 41A, and 41B).

Field control of shunt and compound motors is sometimes obtained by supplying the shunt field from a controlled electronic rectifier. The fundamental principles and connections of controlled rectifiers are the same as those discussed for rectifiers employed for electronic armature control. Electronic field control generally is employed only in combination with electronic armature control when a very wide speed range is required.

Field control of series motors is obtained by shunting a resistance across the series field. If a resistance or rheostat is placed in parallel with the field of a series-wound motor, the speed will be increased instead of decreased at a given load. This procedure is known as shunting the field of the motor. This shunt would not be applied until the motor had been brought up to normal full speed by cutting out the starting resistance. With a shunted field a motor drives a load at a speed higher than normal and therefore requires a correspondingly increased current. More current is required to produce a given torque. The allowable torque at speeds above normal is therefore less than the rated full-load torque. With this type of speed control the motor is practically a constant-horsepower machine, the allowable rating being almost the same for all speeds.

**138. A combination of armature and field control** is employed when a wide speed range is required. The following combinations are frequently employed:

1. Armature resistance combined with a shunt-field rheostat
2. Ward-Leonard system combined with a shunt-field rheostat
3. Electronic armature control combined with electronic field control

A common combination used for controlling the speed of standard constant-speed motors is to employ armature control for reducing the speed to 50 percent below normal and field control for increasing the speed 25 percent above normal.

**139. Automatic drive control (servo system)** is the automatic control of a motor to meet specified requirements of the driven equipment or material. A few of the more common requirements that are automatically fulfilled by automatic drive control are:

1. Production of uniform acceleration and deceleration
2. Maintenance of constant speed regardless of load and control of speed over a wide speed range
3. Production of constant torque over a wide speed range

The essential elements of an electric drive-control system are shown in the block diagram of Fig. 7-79. An electric drive-control system must have the ability (1) to pick up a signal whenever there is a deviation from the required performance of a driven machine or process, (2) to convert this variation signal into a proportional electrical signal, (3) to compare this electrical signal with the desired standard, (4) to have the deviation from the standard activate a correction in the electrical input to the drive motor, and (5) to provide antihunting characteristics.

Requirements 1 and 2 are fulfilled by the conversion element (pickup) of Fig. 7-79. The conversion element may take a variety of forms. It may be a pilot generator coupled to the drive motor or driven machine, which will pick up any variation in speed and translate this speed variation into a proportional voltage variation. It may be a resistor in series with the armature of the drive motor. The voltage across such a resistor will act as a pickup for any variation in the current delivered to the motor. Such a pickup can be used for constant-torque control of a motor. Phototubes are used as the pickup device in many cases. A phototube circuit will pick up and convert into a voltage any deviation from the proper position of a machine tool or of a driven object. The pickup device may be a set of synchros, which will pick up and convert into a signal voltage any deviation from a desired angular position (see Sec. 177).

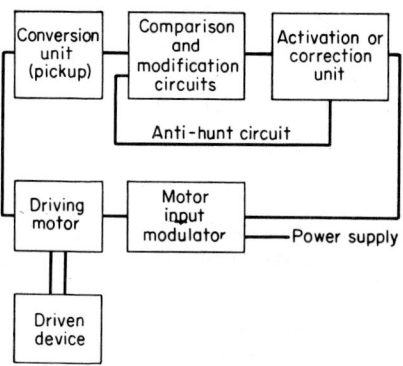

**Fig. 7-79**  Elements for automatic drive control.

Requirement 3 of the control system is fulfilled by the comparison circuit of Fig. 7-79. For example, consider the conditions for automatic speed control of a motor. The voltage of the conversion element (a pilot generator) is connected in series with a standard voltage, and any variation in the voltage of the pilot generator results in a variation of current in the comparison circuit. This variation in current will be controlled in magnitude and direction by the variation in the voltage of the pilot generator. The output of the comparison circuit may not be great enough to produce proper functioning of the activation circuit. In such cases, a modification circuit, which will consist of an electronic amplifying circuit, must be employed. Often it is desired that the controlled machine be influenced by some factor or factors in addition to the main controlling function. Such requirements are fulfilled by modification circuits.

Requirement 4 of the control system is fulfilled by the activation circuit of Fig. 7-79. The output of the comparison and modification circuit must be connected to an activation circuit, which will control the input circuit to the drive motor so as to correct the deviation of the driven device. For example, the output of the comparison circuit may act upon the control circuit of a thyristor and in this manner control the input to the drive motor.

Requirement 5 of the control system is fulfilled through coupling or feedback circuits between comparison and activation circuits. The purpose of these antihunt circuits is to prevent overshooting of the correction and to prevent serious oscillations about the desired controlled condition. Although these antihunt circuits are in many cases essential, a discussion of them is beyond the scope of this book.

In the majority of dc drive-control applications, the requirements are met through the use of dc motors controlled by electronic or electrodynamic control systems. In both these control systems, dc motor performance is controlled through (1) control of input to the motor armature, (2) control of input to the motor field, or (3) combination control of input to the motor armature and field. In the electronic control systems, the motor input is adjusted by means of thyristor rectifiers. In electrodynamic control systems, the motor input is adjusted by means of control generators (see Sec. 41).

**140. In the electronic control of dc motors,** either the armature or the field, or both, are supplied through electronic rectifiers. The output of the rectifiers is automatically controlled in accordance with the requirements of the driven machine or process.

**141. Automatic electrodynamic drive control** is frequently called a *servo system*. In electrodynamic control of dc motors, the input to the motor is automatically controlled in accordance with the requirements of the load by means of a control generator. Small motors may be supplied with power directly from the control generator. When the capacity of the control generator is not sufficient to supply the motor directly, the control generator energizes the field of a main generator and thus controls the input to the motor. Electrodynamic control circuits for constant-speed control of a dc motor are shown in Figs. 7-80 and 7-81. The circuit of Fig. 7-80 employs an amplidyne control generator, and that of Fig. 7-81 a Rototrol control generator. When a sufficient speed range can be obtained by field control, the control generator may control the input to the field of the motor instead of the input to the armature, as is shown in the figures. In the circuits of both Fig. 7-80 and Fig. 7-81, the conversion element, or pickup, is the pilot generator, which is coupled to the main dc drive motor. In the circuit of Fig. 7-80, the voltage of the pilot generator is impressed upon the comparison circuit so as to oppose the reference voltage. In the circuit of Fig. 7-81, the voltage of the pilot generator energizes the pilot field of the Rototrol. The comparison circuit of Fig. 7-81 consists of the combination of the pilot- and pattern-field circuits. In the circuits of both figures, the activation circuit is the armature circuit of the control generator. When the signal voltage from the pilot generator is not sufficient to actuate the field circuit of the control generator directly, an electronic amplifier is inserted between the pickup circuit and the field of the control generator. In both the circuits shown in the figures, the circuit is adjusted for the desired speed by means of the adjusting rheostat.

In the amplidyne control circuit of Fig. 7-80, if the speed of the motor rises above the value corresponding to the setting of the adjusting rheostat, the greater opposing voltage of the pilot generator weakens the field of the amplidyne. This will lower the voltage output of the amplidyne and thereby weaken the field of the main generator. The voltage

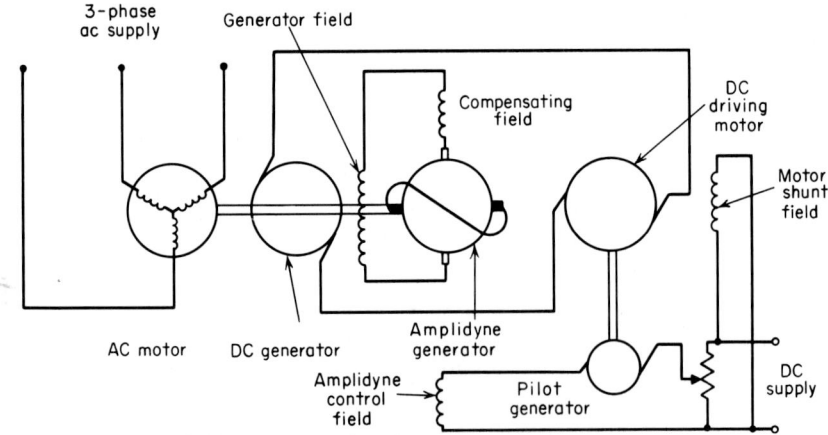

**Fig. 7-80**   Speed control using an amplidyne generator.

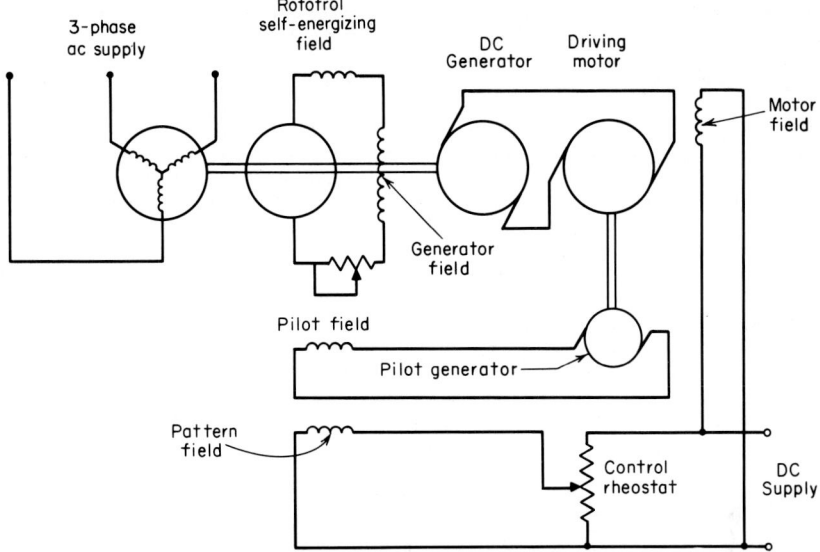

**Fig. 7-81**  Speed control using a Rototrol.

output of the main generator is thus reduced. Therefore, the input and consequently the speed of the motor are reduced until the speed has been brought back to the value determined by the setting of the adjusting rheostat.

In the Rototrol circuit of Fig. 7-81, if the speed of the motor rises above the value corresponding to the setting of the adjusting rheostat, the strength of the pilot field of the Rototrol is increased. Since the pilot field opposes the constant pattern field, the field flux of the Rototrol will be reduced. This will lower the voltage of the Rototrol and thereby weaken the field of the main generator. The voltage output of the main generator is thus reduced. Therefore, the input and consequently the speed of the motor are reduced until the speed has been brought back to the value determined by the setting of the adjusting rheostat.

Circuits for constant-current control of a dc motor by means of electrodynamic control systems are shown in Figs. 7-82 and 7-83. The pickup element consists of a resistor in

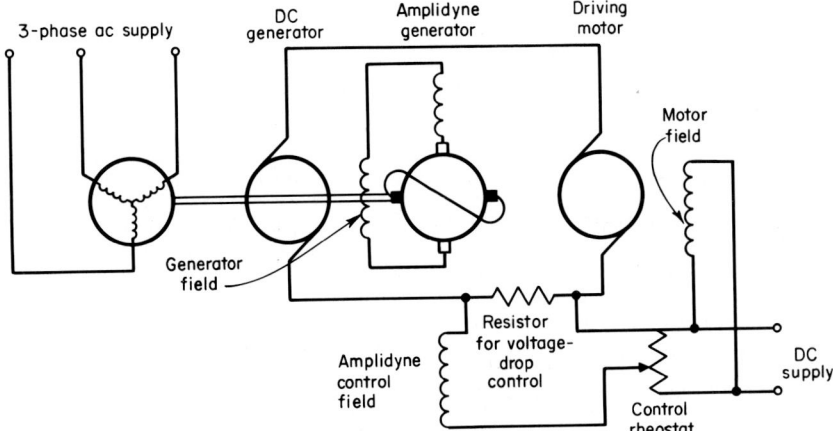

**Fig. 7-82**  Constant-current control using an amplidyne generator.

series with the motor armature circuit. Any deviation in the armature current will result in a proportionate change in voltage across this resistor. The voltage across the resistor activates the comparison circuit in the manner discussed for constant-speed controls.

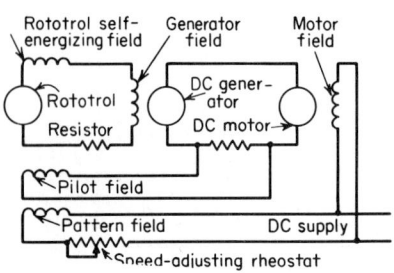

**Fig. 7-83**  Constant-current control using a Rototrol control generator.

**142. Position control.**  It is often desirable to have a control system that functions to control the position of a driven object or rotating machine. The position of a driven object may be indicated by a phototube scanning the driven object. The angular position of a rotating machine may be indicated by the error voltage of a synchro circuit (see Sec. 177). The phototube or synchro pickup and conversion element can actuate either an electronic or an electrodynamic control system so as to control the process or driven machine in accordance with the position desired.

**143.  Approximate Efficiencies of DC Motors: Continuous Rated, 40°C Rise**

| hp | Percentage efficiency | | | hp | Percentage efficiency | | |
|---|---|---|---|---|---|---|---|
| | One-half load | Three-fourths load | Full load | | One-half load | Three-fourths load | Full load |
| ½ | 62.0 | 65.0 | 68.0 | 20 | 82.0 | 85.0 | 86.5 |
| ¾ | 65.0 | 70.0 | 72.0 | 25 | 83.0 | 86.0 | 87.0 |
| 1 | 70.0 | 73.0 | 75.0 | 30 | 84.0 | 87.0 | 88.0 |
| 1½ | 72.0 | 76.0 | 78.0 | 40 | 85.0 | 87.5 | 88.5 |
| 2 | 72.0 | 77.5 | 81.0 | 50 | 85.5 | 88.0 | 89.5 |
| 3 | 72.0 | 77.0 | 79.0 | 60 | 85.5 | 88.0 | 90.0 |
| 5 | 77.0 | 81.0 | 81.5 | 75 | 86.0 | 89.0 | 90.5 |
| 7½ | 79.0 | 82.0 | 83.5 | 100 | 86.0 | 89.0 | 90.5 |
| 10 | 81.0 | 83.0 | 85.0 | 125 | 86.0 | 89.0 | 90.5 |
| 15 | 81.5 | 84.5 | 86.0 | 150 | 87.0 | 90.0 | 91.0 |
| | | | | 200 | 89.0 | 91.5 | 92.0 |

## ALTERNATING-CURRENT MOTORS

**144. General construction of polyphase induction motors.**  Polyphase induction motors have two windings, one on the stationary part of the machine, or stator, and one mounted on the revolving part, or rotor. The stator winding is embedded in slots in the inner surface of the frame of the machine and is similar to the armature winding of a revolving-field type of ac generator. The rotor winding may be one of two types, squirrel-cage or wound-rotor. In a *squirrel-cage machine* the rotor winding consists simply of copper or aluminum bars embedded in slots in the iron core of the rotor and connected together at each end by a copper or aluminum ring. The rotor winding thus forms a complete closed circuit in itself. The rotor winding of the *wound-rotor machine* is similar to the armature winding of a revolving-armature type of ac generator. The free ends of the winding are brought out to slip rings. The rotor circuit is not closed until the slip rings are connected either directly together or through some resistances external to the machine.

**145. Principle of operation of the induction motor.**  The induction motor differs from other types of motors in that there is no electrical connection from the rotor winding to any source of electrical energy. The necessary voltage and current in the rotor circuit are produced by induction from the stator winding. The operation of the induction motor depends upon the production of a revolving magnetic field. The stator winding of a simple two-phase induction motor connected to a simple two-phase alternator is shown in Fig. 7-84.

Windings of the types shown in the illustration are not used in commercial machines, but the general theory involved is the same as with commercial windings. The revolving field of the generator, in turning in the direction shown by the arrow, generates a two-phase current, which is transmitted to the motor. The current in conductors of one phase magnetizes poles $A$ and $B$, and that in the other phase poles $C$ and $D$. The winding is so arranged that a current entering at $A$ will produce a south pole at $A$ and a north pole at $B$.

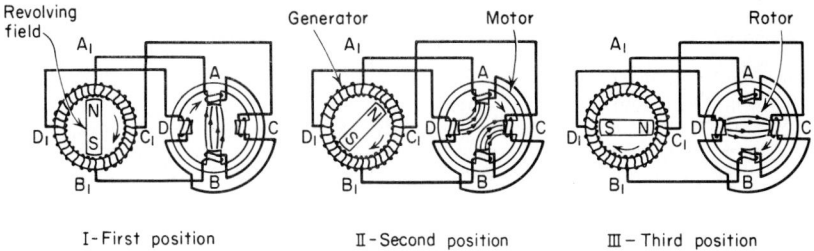

I-First position    II-Second position    III-Third position

**Fig. 7-84** Principle of the induction motor.

At the instant shown at I, motor poles $A$ and $B$ are magnetized while poles $C$ and $D$ are not, because it is a property of a two-phase circuit that when the current in one of the phases is at a maximum value, the current in the other phase is at a zero value. Hence, at this instant a magnetic field will be set up in the machine in a vertical direction, with a north pole at $B$ and a south pole at $A$.

At another later instant, represented at II, the currents in both of the phases are equal and in the same direction. The motor poles will be magnetized as shown. The currents in changing from their values in I to their values in II have moved the magnetic field through an angle of 45° toward the right. At the instant illustrated at III, because of the properties of two-phase currents there is no current in the phase of the conductors which are wound on poles $A$ and $B$, but the current in the phase of the conductors which magnetize poles $C$ and $D$ is at maximum. Hence, the magnetic field produced in the machine at instant III is in a horizontal direction with a south pole at $C$ and a north pole at $D$. The magnetic field has been moved around another 45°. Similar action occurs during successive instants, and the magnetic field is caused to rotate continually in the same direction within the motor frame as long as the two-phase current is applied to the stator terminals. Thus, when the stator winding of a polyphase motor is connected to a source of polyphase power, a revolving field is produced inside the machine.

If a closed electric circuit such as the rotor winding of an induction motor is inserted in the machine, the revolving magnetic field will cut across the conductors of this winding and in that way produce a voltage in them. This voltage acting upon the closed rotor circuit will send a current through the different conductors of the rotor winding. The rotor conductors then will be carrying a current and lying in a magnetic field. Such a condition will produce a force upon the conductors of the rotor winding, tending to turn them in the same direction as that in which the magnetic field is revolving.

If the rotor winding were revolved at the same rate of speed as that at which the magnetic field revolves, there would be no relative motion between the field and the conductors of the rotor winding. Since there would be no relative motion, there would be no cutting of the magnetic field across the conductors and therefore no voltage or current produced in the rotor winding. If there is no current in the rotor winding, no torque can be produced, which tends to keep the machine revolving. As the machine slows down, however, the magnetic field will revolve at a higher speed than the rotor, producing a relative motion between the magnetic field and the conductors of the rotor winding. This relative motion produces a cutting of the conductors by the flux, with the resultant voltage and current necessary to develop a turning effort upon the rotor. The operation of an induction motor depends upon a difference in speed between the revolving magnetic field and the actual speed of the machine or rotor. This difference in speed is called the *slip of the motor.* The rotor is said to slip a certain number of revolutions behind the revolving magnetic field. The greater the load on an induction motor, the more the rotor must slip behind the revolving magnetic field to produce a greater cutting of the flux by the rotor conductors, with the resulting greater current in the rotor and greater turning effort to carry the greater load. Even at no load an induction motor cannot revolve as fast as the magnetic field revolves, since enough torque must be developed to overcome the friction and other losses in the motor.

**146. The synchronous speed of an induction motor** is the speed at which its magnetic field revolves. It depends upon the number of poles for which the stator is wound and the frequency of the supply voltage.

$$\text{Synchronous speed} = \frac{120f}{P} \tag{4}$$

where $f$ = frequency of supply, $P$ = the number of poles per phase for which the stator is wound, and synchronous speed is in revolutions per minute.

**147. The slip of an induction motor** is the ratio of the difference between the rotating magnetic-field speed (rpm or angular velocity) and the actual rotor speed to the rotating magnetic-field speed. The speed of the rotating magnetic field is equivalent to the synchronous speed of the machine (see Table 97), which is determined by the frequency of the current and the number of poles of the machine (Sec. 146). Then

$$\text{Slip in rpm} = \text{synchronous speed} - \text{actual speed} \tag{5a}$$

$$\text{Percent slip} = \frac{\text{synchronous speed} - \text{actual speed}}{\text{synchronous speed}} \times 100 \tag{5b}$$

When there is no load on a motor, the slip is very small; i.e., the rotor speed is practically equal to the synchronous speed. Slip varies with the design of a motor and may vary from 4.0 to 8.5 percent at full load in motors of from 1 to 75 hp of ordinary design.

**example**  What is the slip at full load of a four-pole 60-Hz induction motor which has a full-load speed of 1700 rpm?

**solution**  From Table 97 or Sec. 146 the speed of the rotating field or the synchronous speed of a four-pole 60-Hz motor is 1800 rpm. Then, substituting in the above formula,

$$\text{Percent slip} = \frac{\text{synchronous speed} - \text{actual speed}}{\text{synchronous speed}} \times 100 = \frac{1800 - 1700}{1800} \times 100$$

$$= \frac{100}{1800} \times 100 = 5.5 \text{ percent}$$

Therefore the slip is 5.5 percent. The voltage of the motor and whether it is single-phase, two-phase, or three-phase are not factors in the problem.

**148. Breakdown torque of an induction motor.**  Most induction motors will "break down" at a certain torque if they are overloaded. The breakdown limit, the maximum running torque that can be developed, is the point at which a further increase in load will cause the motor speed to decrease rapidly and then to stop. This point is usually between 2 and 4 times the full-load rated torque, depending on the design and the capacity of the motor. In a few cases there is no breakdown torque, the maximum torque occurring at standstill. Under these conditions there will be no sudden decrease in the speed as the load is increased. Instead, the speed will gradually decrease as the load is increased until standstill is reached.

**149. General operating characteristics of induction motors.**  The following rules, though approximate, are accurate enough for practical use under the normal range of operation of an induction motor from no load to 150 percent of rated load:

1. The slip of an induction motor varies directly with the load. Doubling the load will double the slip.

2. For any given load the slip varies directly with the resistance of the rotor circuit. If the resistance of the rotor circuit is doubled, the slip for the same load will be twice its original value.

3. For any load the slip varies inversely as the square of the voltage impressed on the stator winding. If the voltage is reduced to one-half of its normal value, the slip is increased to 4 times its normal value for any load.

4. The torque developed by an induction motor varies directly with the slip.

5. The torque developed at any particular slip varies inversely with the resistance of the rotor circuit.

6. The torque for any particular slip varies directly as the square of the voltage impressed on the stator.

7. The maximum torque is not affected by the resistance of the rotor circuit. A large value of resistance for the rotor circuit simply makes the maximum torque occur at a high slip.

8. The maximum torque varies directly with the square of the voltage impressed on the stator circuit.

**150. Starting characteristics of induction motors.**    The starting torque varies directly with the square of the voltage impressed on the stator. If three-fourths of normal voltage is impressed at starting, the starting torque will equal only nine-sixteenths of the torque developed with normal voltage impressed at starting.

The value of the starting torque of an induction motor also depends upon the value of the resistance of the rotor circuit. As the resistance of the rotor circuit is increased, the starting torque is increased until the resistance of the rotor circuit is of just the right value to make the maximum torque occur at starting. If the resistance of the rotor circuit is still further increased, the starting torque will be decreased. Thus, a certain value of rotor resistance will produce maximum starting torque, and either a greater or a smaller value of rotor resistance than this amount of resistance produces less starting torque (see Fig. 7-87).

The current taken at starting by an induction motor depends upon the voltage impressed on the stator and the resistance of the rotor circuit. For a fixed rotor resistance the starting current varies directly with the impressed voltage. For a particular impressed voltage, increasing the resistance of the rotor circuit decreases the value of the starting current.

**151. Characteristic curves of the induction motor.**    The curves of Fig. 7-85 are fairly typical of the average normal-torque, normal-starting-current, commercial induction motor. Note that the normal rating of the motor is taken at such a point that both the power factor and the efficiency are the highest possible. The motor could be designed so that either the power factor or the efficiency, but not both, could be higher than shown at normal load, but the design of an induction motor is a compromise between the leading factors that results in the best efficiency and power factor obtainable with suitable overload and starting characteristics.

The characteristics of a polyphase induction motor running single-phase are shown in Fig. 7-86. In Fig. 7-88 is shown the speed-torque characteristic of a polyphase induction motor running single-phase. Note that the starting torque when operated single-phase is zero. A polyphase induction motor will not start when connected to a single-phase supply. Once started, however, it will operate from a single-phase supply and carry approximately half the rated load without overheating.

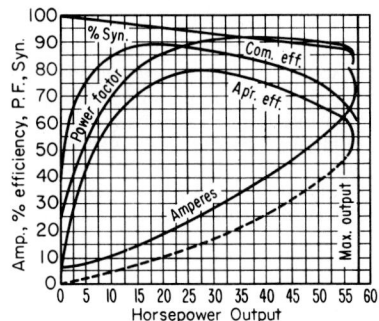
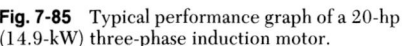

**Fig. 7-85**    Typical performance graph of a 20-hp (14.9-kW) three-phase induction motor.

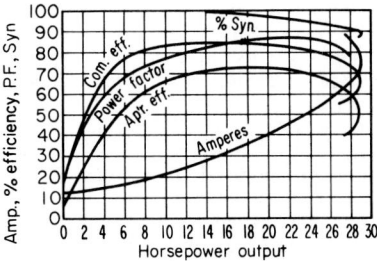

**Fig. 7-86**    Performance graph of the motor shown in Fig. 7-85 when running single-phase.

**151A. High-efficiency induction motors.**    Efficiency is a measure of the effectiveness with which a motor converts electrical energy into mechanical energy; in other words, watts in versus watts out. In the conversion process, watts are lost, transformed into heat which is dissipated by the motor-cooling system, such as through the motor frame, or lost through friction and windage. High-efficiency motors available from many manufacturers reduce the watts loss in one or all of several areas over standard industrial motor designs, in some cases by as much as 50 percent. General techniques for watts-loss reduction include the following:

| Watts-Loss Area | Efficiency Improvements |
|---|---|
| 1. Iron | Iron losses are essentially independent of load. Use of thinner-gage, lower-loss core steel reduces eddy-current losses, and adding iron by making the motor core longer reduces losses due to lower operating flux densities. |
| 2. Stator $I^2R$ | Use of more copper and larger conductors increases the cross-sectional area of stator windings, lowering resistance $R$ and reducing losses due to current flow $I$. $I^2R$ losses are load-dependent. |
| 3. Rotor $I^2R$ | Use of larger rotor conductor bars increases the size of the cross section, lowering resistance $R$ and reducing losses due to current flow $I$. |
| 4. Friction and windage | Friction and windage losses are independent of load. Use of low-loss fan designs (on fan-cooled motors) reduces losses due to air movement. Bearing design can reduce friction losses. |
| 5. Stray load loss | Use of optimized design and improved quality-control procedures minimize stray load losses, such as those due to nonuniform current distribution in stator and rotor conductors, air gap, and so forth. These losses combined may be 10 to 15 percent of total motor losses and tend to increase with load. |

Generally, high-efficiency motors are more expensive than standard industrial motors; however, the additional initial investment may be more than justified by the reduction in the cost of electricity used. There are a number of ways to evaluate the economic impact of motor efficiency. An example of a simple payback analysis is given here. Other methods include a present-worth life-cycle analysis and a cash-flow and payback analysis. Details of the various methods of analysis can be found in financial-management texts.

The simple payback method gives the number of years required to recover the differential investment for a higher-efficiency motor to determine the payback period. The premium for the higher-efficiency motor is divided by the annual savings. First, the annual savings must be determined by using the following formula:

$$S = 0.746 \times \text{hp} \times L \times C \times N \left( \frac{100}{E_B} - \frac{100}{E_A} \right) \tag{5c}$$

where $S$ = annual dollar savings
  hp = horsepower
  $L$ = percentage load $\div$ 100
  $C$ = energy cost, dollars/kWh
  $N$ = annual hours of operation
  $E_B$ = lower-motor percent efficiency
  $E_A$ = higher-motor percent efficiency

The efficiencies for the percent load at which the motor will be operated must be used, since motor efficiency varies with load, as shown in Fig. 7-85. Then, the motor-cost premium is divided by the annual savings to compute the payback period for the higher-efficiency motor. If motor $A$ costs $300 more than motor $B$ and yields annual savings of $100, the simple payback period is $300/$100, or 3 years. The payback method is easy to apply but ignores savings occurring after the end of the payback period and does not consider the time value of money or changes in energy costs.

See Sec. 172 for motor-efficiency marking and minimum efficiencies of "energy-efficient" motors.

**152. The torque curves of an induction motor with a wound rotor,** from rest to synchronism, running both three-phase and single-phase with external resistance and without external resistance, are shown in Fig. 7-87. The curves are plotted with torque in foot pounds at 1-ft (0.3-m) radius. Curve A shows the torque from rest to synchronism without resistance in the rotor circuit. If the proper value of resistance is inserted, curve B is obtained, and the starting torque is 440 ft·lb (596.6 J) against 170 ft·lb (230.5 J) without resistance,

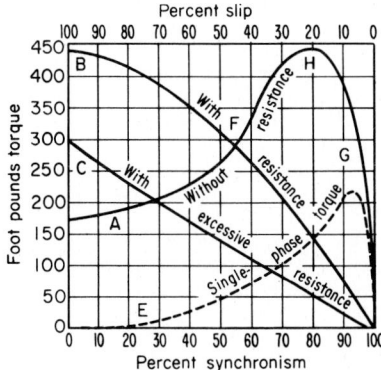

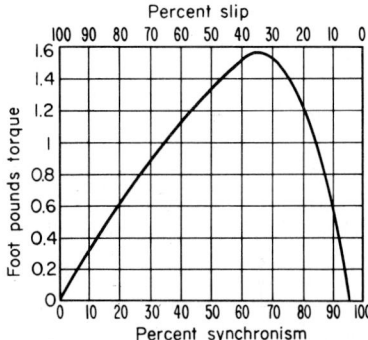

**Fig. 7-87**  Torque graph of a 30-hp (22.4-kW) induction motor.

**Fig. 7-88**  Torque graph of a 1-hp (745.7-W) three-phase induction motor running single-phase.

Curve C indicates the torque when too much resistance is used in the rotor. Curve E illustrates the torque for single-phase, which is zero at starting. A wound-rotor induction motor starts as shown on curve B until it reaches point F, when the resistance is cut out and the motor adjusts itself to its operating position at G. Thus, if the torque required of the motor for which the curve is shown is greater than 440 ft·lb, shown at H, the motor will break down and come to rest. With the resistance in the rotor, a starting torque of 440 ft·lb is available, but this load cannot be brought up to normal speed. The motor can bring only the torque represented by point F, or 290 ft·lb (393.2 J), to normal speed.

In Fig. 7-87, note that the torque of a three-phase motor rises to a maximum and then drops, reaching zero at synchronism. This means that an induction motor never runs at synchronous speed.

**153. Speed control of induction motors.**  It is possible to control the speed of an induction motor by any one of the following methods:

1. Changing the impressed voltage
2. Changing the number of poles for which the machine is wound
3. Changing the frequency of the supply
4. Changing the resistance of the rotor circuit
5. Introducing a foreign voltage into the secondary circuit
6. Operating two or more motors connected in cascade

**154. Reducing the impressed voltage** will increase the slip of an induction motor proportionally to the square of the voltage. At the same time, however, the breakdown torque of the motor is also reduced as the square of the voltage, so that the motor is in danger of stalling under load. For this reason, changing the impressed voltage alone is employed for the speed control of induction motors only in emergencies or other special cases.

**155. Speed control of induction motors by changing the number of poles.**  The actual speed of an induction motor depends upon the speed of the revolving magnetic field produced by the stator winding (synchronous speed). The synchronous speed of an induction motor is inversely proportional to the number of poles for which the stator winding is wound. Thus, on a 60-Hz circuit a two-pole induction motor has a synchronous speed of 3600 rpm; a four-pole motor, 1800 rpm; an eight-pole motor, 900 rpm, etc. It is therefore possible to alter the speed of an induction motor by changing the number of its poles.

This can be accomplished by using two or more separate primary windings, each having a different number of poles, or by using a single winding which can be connected to form different numbers of poles. This method is, of course, not applicable to the ordinary type of induction motor but requires a motor with a number of leads brought out from the primary winding (see Sec. 171). Gradual speed control cannot be obtained by this method. Only a certain number of definite speeds, four at the most, can be obtained.

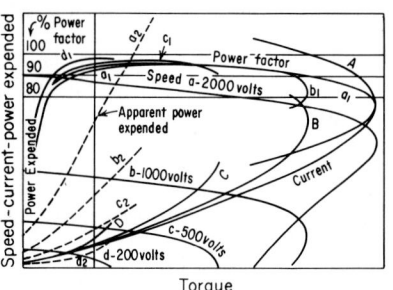

Fig. 7-89   Performance graph of a polyphase induction motor with different applied frequencies and different applied voltages.

**156. Speed control of a polyphase motor by adjusting the frequency of the primary current.** Since the synchronous speed of an induction motor is equal to the alternations per minute of the supply circuit divided by the number of poles in each phase circuit, a change in speed can be effected by changing the frequency of the supply circuit.

Figure 7-89 shows the speed-torque and other curves of a 60-Hz motor when operated at 60, 30, 15, and 6 Hz, or at 100, 50, 25, and 10 percent of the normal frequency. The speed-torque curves corresponding to the above frequencies are $a$, $b$, $c$, and $d$. The current curves are $A$, $B$, $C$, and $D$. This figure shows that for the rated torque $T$ the current is practically constant for all speeds, but the emf varies with the frequency. Consequently, the apparent power supplied, represented by the product of the current by emf, varies with the speed of the motor and is practically proportionate to the power developed.

When only one motor is operated, the generator speed can be varied. If the generator is driven by a waterwheel, its speed can be varied over a wide range and the motor speed will also vary. If the generator field is held at practically constant strength, the motor speed can be varied from zero to a maximum at constant torque with a practically constant current.

Controlling the speed of an induction motor by changing the frequency of the supply requires a separate generator for each motor or an electronic adjustable-speed drive for each motor.

**157. Speed control of induction motors by adjusting the resistance of the secondary circuit.** Inserting resistance in series with the rotor of an induction motor will decrease the speed of the motor. The amount of decrease in speed produced will depend upon the load on the motor. It is not possible to produce much change in speed by this method at light loads. This method of speed control is, of course, applicable only to wound-rotor motors. Speed control by adjusting the resistance of the rotor circuit has two drawbacks. After the speed has been adjusted for any load, it does not remain constant with a change in load. With external resistance in series with the rotor, the motor has a varying-speed characteristic. Considerable power is wasted in the external resistance required for changing the speed. The amount of power wasted depends upon the value of the external resistance. The efficiency of the motor is thus decreased approximately proportionally to the reduction in speed.

Speed control by means of adjustable secondary resistance is, however, very useful when constant speeds are not essential, for example, in operating cranes, hoists, elevators, and dredges and also for service in which the torque remains constant at each speed, as in driving fans, blowers, and centrifugal pumps. In service for which reduced speeds are required only occasionally and speed variation is not objectionable, this method of control can also be used to good advantage. On account of energy loss in the resistors, efficiency is reduced when the motor is operated at reduced speeds, the reduction being greatest at the slowest speeds. The circuits are essentially the same as for starting by varying resistance in the rotor circuit. The rotor usually has a Y-connected winding to which is connected, in series in each phase, an external resistance. If the adjustable arm is moved, the amount of resistance in series in each phase can be varied from a maximum to zero and the speed varied gradually from the highest speed to the lowest.

**158. Speed control of induction motors by introduction of a foreign voltage into the secondary circuit.** If a foreign voltage is introduced into the secondary circuit of an induction motor, it will alter the speed of the motor. If this foreign voltage acts in a direction opposite to that of the voltage induced in the secondary circuit, the speed of the motor will be reduced. If the foreign voltage acts in the same direction as the voltage induced in the secondary circuit, the speed of the motor will be increased. With this method of speed control it is possible to obtain speeds both above and below the synchronous speed of the motor. Speed control by the introduction of a foreign voltage into the secondary circuit has two advantages. No power is wasted in resistance, and the speed is practically constant for all loads with the same value of foreign voltage introduced.

The brush-shifting type of polyphase induction motors, as explained in Sec. 176, employs the foreign-voltage method of speed control. When the operating characteristics of these motors are applicable, the motors provide a very efficient means of obtaining adjustable speed for ac installations.

An electronic method of foreign-voltage speed control is feasible for medium-size and small motors. The diagram of the circuit for such a method is shown in Fig. 7-90. The armature of a dc motor is connected through a three-phase full-wave ignitron or thyratron rectifier to the rotor slip rings of the induction motor. This connection introduces the counter-emf of the dc motor into the rotor circuit of the induction motor. The value of the counter-emf of the dc motor is adjusted by adjustment of the field rheostat of the dc motor. The speed of the induction motor is therefore controlled through adjustment of the field rheostat of the dc motor. The dc motor may be coupled mechanically to the induction motor so that it helps to drive the load, or the dc motor may drive an ac generator so that the greater portion of the power taken by the dc motor is returned to the ac supply system.

There are other methods of obtaining speed control by the foreign-voltage method, but all require additional revolving machines besides the motor itself. Their use is not feasible for ordinary applications, and they are employed in special cases for the speed control of large motors.

**159. Speed control of polyphase motors by operating two or more motors connected in cascade** offers, under some conditions of service, the most convenient and economical method of speed variation. In this arrangement all the rotors are mounted on one shaft, or the several shafts are rigidly connected. The primary of the first motor is connected to the line; its secondary, which must be of the phase-wound slip-ring type, is connected to the primary of the second motor, and so on. The secondary of the last motor can be of either the squirrel-cage or the phase-wound type. In practice more than two motors are rarely used. The arrangement is shown in Fig. 7-91.

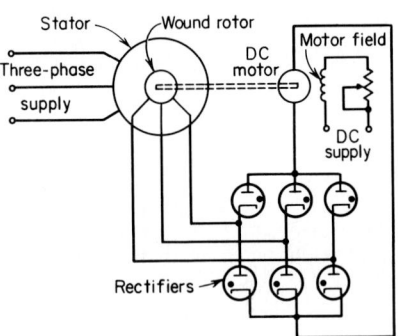

**Fig. 7-90**  Control of a wound-motor induction motor using electronic means.

Speed changes are obtained by varying the connections of the motors. The following combinations are possible with two motors: each motor can be operated separately at its normal speed with its primary connected to the line, the other motor running idle; the motors can be connected in cascade so that the rotors tend to start in the same direction (direct concatenation); or the motors can be connected so that the rotors tend to start in

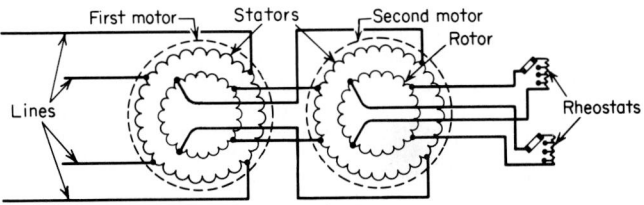

**Fig. 7-91**  Two polyphase motors connected in cascade.

opposite directions (differential concatenation). If the first motor has 12 poles and the second 4, the following synchronous speeds can be obtained on a 25-Hz circuit:

1. Motor 2 (4 poles) running single, 750 rpm
2. Motors in differential concatenation (equivalent of 8 poles), 375 rpm
3. Motor 1 (12 poles) running single, 250 rpm
4. Motors in direct concatenation (equivalent of 16 poles), 187.5 rpm

By the use of adjustable resistance in the secondary circuits, changes from one speed to the next can be made with uniform gradations.

A great number of speed combinations are possible with this method; the control is simple and safe, as few leads are required and main circuits are not opened for most of the speeds. The rotors can be made with smaller diameters than is possible with other multispeed motors; hence the flywheel effect is reduced to a minimum. In general, a cascade set is applicable when speed changes must be made frequently with high horsepower output and primary voltage and when speed ratios are other than 1:2.

**160. Application of the methods of speed control of induction motors.**  From the preceding sections, it is apparent that changing the speed of an induction motor by simple reduction of the impressed voltage (Sec. 154), changing the number of poles (Sec. 155), changing the frequency of the primary current by changing generator speed (Sec. 156), adjusting the resistance of the secondary circuit (Sec. 157), introducing a foreign voltage into the secondary circuit (Sec. 158), and operating motors in cascade (Sec. 159) have very limited application. In recent years, however, rapid changes in the technology of solid-state electronics have resulted in the availability of electronic drive systems for ac induction motors that make speed control of such motors feasible for an increasing number of applications. Known as electronic adjustable-speed drives (ASDs), these systems are available in virtually any horsepower for all kinds of applications from many manufacturers. In all cases, the motor manufacturer should be consulted concerning the use of an ASD and the limits of ASD output characteristics that can be used.

In addition to controlling the speed of a motor at load, ASDs are available to control the motor during starting and stopping periods, providing controlled acceleration and deceleration, thus limiting the starting current and providing gentle start-ups and gradual shutdowns, to increase equipment life and reduce mechanical breakdowns. By matching motor speed to the requirements of the load, ASDs can save energy that mechanical devices, such as valves and vanes, would waste.

ASDs may change the input power from single-phase to three-phase, for the speed control of small three-phase motors from single-phase power supplies, or may control the pulse width of the single-phase supply, typically used for the speed control of ceiling-mounted paddle fans. Frequency control, within limits, may be combined with voltage control, with the lower frequency at low voltage and increasing frequency with increasing voltage.

Commonly, ASDs also include motor-overload protection, although additional external overload protection may be specified to meet Code requirements. Other functions, such as adjustable-load limitation, phase-failure protection, ASD component-failure protection, etc., may also be included.

For further information on the speed control of induction motors, refer to Secs. 289 and 321.

**161. Classification of squirrel-cage induction medium motors.**  Polyphase squirrel-cage induction medium motors rated more than 1 hp were formerly called "integral horsepower" motors. Polyphase squirrel-cage induction medium motors are classified by NEMA standards into four design types as follows:

A *Design A motor* is a squirrel-cage motor designed to withstand full-voltage starting and developing locked-rotor and pull-up torque as shown in Sec. 163 and breakdown torque as shown in Sec. 164, with locked-rotor current higher than the values shown in Sec. 162 and having a slip at rated load of less than 5 percent.[1]

A *Design B motor* is a squirrel-cage motor designed to withstand full-voltage starting, developing locked-rotor, breakdown, and pull-up torques adequate for general application as specified in Secs. 163 and 164, drawing locked-rotor current not to exceed the values shown in Sec. 162 and having a slip at rated load of less than 5 percent.[1]

A *Design C motor* is a squirrel-cage motor designed to withstand full-voltage starting, developing locked-rotor torque for special high-torque application up to the values

---

[1]Motors with 10 and more poles may have slip slightly greater than 5 percent.

shown in Sec. 163, pull-up torque as shown in Sec. 163, and break-down torque up to the values shown in Sec. 164, with locked-rotor current not to exceed the values shown in Sec. 162 and having a slip at rated load of less than 5 percent.

A *Design D motor* is a squirrel-cage motor designed to withstand full-voltage starting, developing high locked-rotor torque as shown in Sec. 163, with locked-rotor current not greater than shown in Sec. 162 and having a slip at rated load of 5 percent or more.

NEMA standards formerly recognized a fifth design, designated *Design F*. A Design F motor is a squirrel-cage motor designed to withstand full-voltage starting developing low locked-rotor torque as shown in Sec. 163, with locked-rotor current not to exceed the values shown in Sec. 162 and having a slip at rated load of less than 5 percent.

Typical speed-torque curves for the five design types are given in Fig. 7-92.

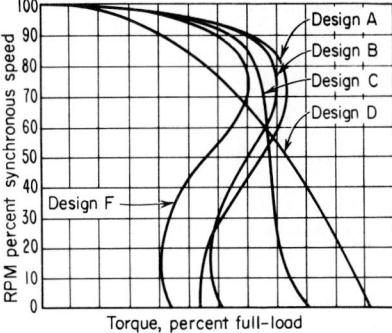

**Fig. 7-92**  Typical speed-torque characteristics for squirrel-cage induction motors. [Westinghouse Electric Corp.]

### 162. Locked-Rotor Current of Three-Phase 60-Hz Medium Squirrel-Cage Induction Motors Rated at 230 V
(NEMA standard)[a]

The locked-rotor current of single-speed, three-phase, constant-speed induction motors, when measured with rated voltage and frequency inpressed and with rotor locked, shall not exceed the following values:

| hp | Locked-rotor current, A[b] | Design letters | hp | Locked-rotor current, A[b] | Design letters |
|----|----|----|----|----|----|
| ½ | 20 | B, D | 50 | 725 | B, C, D |
| ¾ | 25 | B, D | 60 | 870 | B, C, D |
| 1 | 30 | B, D | 75 | 1085 | B, C, D |
| 1½ | 40 | B, D | 100 | 1450 | B, C, D |
| 2 | 50 | B, D | 125 | 1815 | B, C, D |
| 3 | 64 | B, C, D | 150 | 2170 | B, C, D |
| 5 | 92 | B, C, D | 200 | 2900 | B, C |
| 7½ | 127 | B, C, D | 250 | 3650 | B |
| 10 | 162 | B, C, D | 300 | 4400 | B |
| 15 | 232 | B, C, D | 350 | 5100 | B |
| 20 | 290 | B, C, D | 400 | 5800 | B |
| 25 | 365 | B, C, D | 450 | 6500 | B |
| 30 | 435 | B, C, D | 500 | 7250 | B |
| 40 | 580 | B, C, D | | | |

[a]This material is reproduced with the permission of the National Electrical Manufacturers Association from NEMA Standards Publication MG 1-1987, copyright © 1987 by NEMA.

[b]Locked-rotor current of motors designed for voltages other than 230 V shall be inversely proportional to the voltages.

### 163. Locked-Rotor Torque of Single-Speed Polyphase Squirrel-Cage Medium Motors with Continuous Ratings

(NEMA standard)[a]

1. The locked-rotor torque of Designs A and B, 60- and 50-Hz, single-speed, polyphase squirrel-cage medium motors, with rated voltage and frequency applied, shall be not less than the following values, which are expressed in percent of full-load torque requirements. For applications involving higher torque requirements, see the locked-rotor–torque values for Design C and D motors.

| | | Synchronous speeds, rpm | | | | | | |
|---|---|---|---|---|---|---|---|---|
| | 60 Hz | 3600 | 1800 | 1200 | 900 | 720 | 600 | 514 |
| hp | 50 Hz | 3000 | 1500 | 1000 | 750 | ... | ... | ... |
| ½ | | ... | ... | ... | 140 | 140 | 115 | 110 |
| ¾ | | ... | ... | 175 | 135 | 135 | 115 | 110 |
| 1 | | ... | 275 | 170 | 135 | 135 | 115 | 110 |
| 1½ | | 175 | 250 | 165 | 130 | 130 | 115 | 110 |
| 2 | | 170 | 235 | 160 | 130 | 125 | 115 | 110 |
| 3 | | 160 | 215 | 155 | 130 | 125 | 115 | 110 |
| 5 | | 150 | 185 | 150 | 130 | 125 | 115 | 110 |
| 7½ | | 140 | 175 | 150 | 125 | 120 | 115 | 110 |
| 10 | | 135 | 165 | 150 | 125 | 120 | 115 | 110 |
| 15 | | 130 | 160 | 140 | 125 | 120 | 115 | 110 |
| 20 | | 130 | 150 | 135 | 125 | 120 | 115 | 110 |
| 25 | | 130 | 150 | 135 | 125 | 120 | 115 | 110 |
| 30 | | 130 | 150 | 135 | 125 | 120 | 115 | 110 |
| 40 | | 125 | 140 | 135 | 125 | 120 | 115 | 110 |
| 50 | | 120 | 140 | 135 | 125 | 120 | 115 | 110 |
| 60 | | 120 | 140 | 135 | 125 | 120 | 115 | 110 |
| 75 | | 105 | 140 | 135 | 125 | 120 | 115 | 110 |
| 100 | | 105 | 125 | 125 | 125 | 120 | 115 | 110 |
| 125 | | 100 | 110 | 125 | 120 | 115 | 115 | 110 |
| 150 | | 100 | 110 | 120 | 120 | 115 | 115 | |
| 200 | | 100 | 100 | 120 | 120 | 115 | | |
| 250 | | 70 | 80 | 100 | 100 | | | |
| 300 | | 70 | 80 | 100 | | | | |
| 350 | | 70 | 80 | 100 | | | | |
| 400 | | 70 | 80 | | | | | |
| 450 | | 70 | 80 | | | | | |
| 500 | | 70 | 80 | | | | | |

2. The locked-rotor torque of Design C, 60- and 50-Hz, single-speed, polyphase squirrel-cage medium motors, with rated voltage and frequency applied, shall be not less than the following values, which are expressed in percent of full-load torque.

| | | Synchronous speeds, rpm | | |
|---|---|---|---|---|
| | 60 Hz | 1800 | 1200 | 900 |
| hp | 50 Hz | 1500 | 1000 | 750 |
| 3 | | ... | 250 | 225 |
| 5 | | 250 | 250 | 225 |
| 7–1/2 | | 250 | 225 | 200 |
| 10 | | 250 | 225 | 200 |
| 15 | | 225 | 200 | 200 |
| 20 to 200, inclusive | | 200 | 200 | 200 |

3. The locked-rotor torque of Design D, 60- and 50-Hz, single-speed, polyphase squirrel-cage medium motors rated 150 hp and smaller, with rated voltage and frequency applied, shall be not less than 275 percent, expressed in percent of full-load torque.

4. The pull-up torque of Designs A and B, single-speed, polyphase squirrel-cage medium motors, with rated voltage and frequency applied, shall be not less than the following:

| Col. 1<br>Locked-Rotor Torque<br>from Item 1 Above | Col. 2<br>Pull-Up Torque,<br>Percent |
|---|---|
| 110 percent or less | 90 percent of col. 1 |
| Greater than 110 percent<br>but less than 145 percent | 100 percent of full-load torque |
| 145 percent or more | 70 percent of col. 1 |

The pull-up torque of Design C motors, with rated voltage and frequency applied, shall be not less than 70 percent of the locked-rotor torque from item 2 above.

[a]This material is reproduced with the permission of the National Electrical Manufacturers Association from NEMA Standards Publication MG 1-1987, copyright © 1987 by NEMA.

### 164. Breakdown Torque of Single-Speed Polyphase Squirrel-Cage Medium Motors with Continuous Ratings

(NEMA standard)[a]

1. The breakdown torque of Designs A and B, 60- and 50-Hz, single-speed, polyphase squirrel-cage medium motors, with rated voltage and frequency applied, shall be not less than the following values, which are expressed in percent of full-load torque.

| | Synchronous speeds, rpm | | | | | | |
|---|---|---|---|---|---|---|---|
| 60 Hz | 3600 | 1800 | 1200 | 900 | 720 | 600 | 514 |
| hp  50 Hz | 3000 | 1500 | 1000 | 750 | ... | ... | ... |
| ½ | ... | ... | ... | 225 | 200 | 200 | 200 |
| ¾ | ... | ... | 275 | 220 | 200 | 200 | 200 |
| 1 | ... | 300 | 265 | 215 | 200 | 200 | 200 |
| 1½ | 250 | 280 | 250 | 210 | 200 | 200 | 200 |
| 2 | 240 | 270 | 240 | 210 | 200 | 200 | 200 |
| 3 | 230 | 250 | 230 | 205 | 200 | 200 | 200 |
| 5 | 215 | 225 | 215 | 205 | 200 | 200 | 200 |
| 7½ | 200 | 215 | 205 | 200 | 200 | 200 | 200 |
| 10–125, inclusive | 200 | 200 | 200 | 200 | 200 | 200 | 200 |
| 150 | 200 | 200 | 200 | 200 | 200 | 200 | |
| 200 | 200 | 200 | 200 | 200 | 200 | | |
| 250 | 175 | 175 | 175 | 175 | | | |
| 300–350 | 175 | 175 | 175 | | | | |
| 400–500, inclusive | 175 | 175 | | | | | |

2. The breakdown torque of Design C, 60- and 50-Hz, single-speed, polyphase squirrel-cage medium motors, with rated voltage and frequency applied, shall be not less than the following values, which are expressed in percent of full-load torque.

| | Synchronous speeds, rpm | | |
|---|---|---|---|
| 60 Hz | 1800 | 1200 | 900 |
| hp  50 Hz | 1500 | 1000 | 750 |
| 3 | ... | 225 | 200 |
| 5 | 200 | 200 | 200 |
| 7½–200, inclusive | 190 | 190 | 190 |

[a]This material is reproduced with the permission of the National Electrical Manufacturers Association from NEMA Standards Publication MG 1-1987, copyright © 1987 by NEMA.

**165. The Design A (normal-torque, normal-starting-current) motor** is the old standard type of squirrel-cage induction motor. It is a constant-speed machine with a slip of from 2 to 5 percent at full load. It is not adapted for speed-control service. The starting torques with rated voltage applied are at least as great as given in Sec. 163.

The locked-rotor starting currents with rated voltage applied vary from 500 to 1000 percent of full-load current, the smaller values applying to motors with the larger number of poles.

The full-voltage starting currents of motors of 5 hp and smaller will be within the limitations of most power companies, so they are generally started directly across the line. Motors of larger size than 5 hp will have full-voltage starting currents that will not be within the limitations of some power companies and in such cases are started with reduced voltage.

The breakdown torques are from 175 to 300 percent of full-load torque.

**166. The Design B (normal-torque, low-starting-current) squirrel-cage motor** is so designed that its rotor has a high reactance and resistance at start but nearly normal reactance and resistance under running conditions. This enables this type of motor to produce approximately the same starting torque for a given voltage as the normal-torque, normal-starting-current motor but with a much smaller value of starting current.

The running performance will be very nearly but not quite so good as that of the normal-torque, normal-starting-current type, the efficiency, power factor, and breakdown torques being slightly less and the speed regulation slightly larger. The breakdown torques will be at least as great as those given in Sec. 164. The locked-rotor starting current with rated voltage applied will be not greater than that given in Sec. 162. The full-voltage starting currents of motors of 30 hp and smaller will be within the limitations of most power companies, so the motors are generally started directly across the line. Motors of larger size than 30 hp will have full-voltage starting currents that will not be within the limitations of many power companies and therefore frequently must be started with reduced voltage.

**167. Design C (high-torque, low-starting-current) squirrel-cage induction motors** are usually constructed with two squirrel-cage windings on the rotor, located one above the other in the rotor slots. During the starting period most of the current is carried by the top winding, which has a high resistance. This enables the motor to produce a larger starting torque with about the same value of starting current as the normal-torque, low-starting-current type of motor. After the motor has come up to speed, the two windings are equally effective in carrying the rotor current, and its operating characteristics are similar to those of the normal-torque, low-starting-current type of motor.

The locked-rotor starting currents with rated voltage applied will be not greater than those given in Sec. 162. The full-voltage starting current of motors of 30 hp and smaller will be within the limitations of most power companies, so they are generally started directly across the line. Motors of larger size than 30 hp will have full-voltage starting currents that will not be within the limitations of many power companies and therefore frequently must be started with reduced voltage. The starting torques with rated voltage impressed will be at least as great as those given in Sec. 163, Part 2. The breakdown torques will be at least as great as those given in Sec. 164.

**168. The Design D (high-resistance–rotor) squirrel-cage motor,** owing to the high resistance of its rotor winding, produces a high starting torque with a low value of starting current.

These motors are made in two general types, which may be classified as medium-slip and high-slip. The medium-slip motors have full-load slips from 7 to 11 percent, while the high-slip machines have full-load slips from 12 to 17 percent. The medium-slip motors are suitable for driving punch presses, shears, bulldozers, and other heavy-inertia machinery which operates under heavy, fluctuating-load conditions and which is either provided with flywheels or has a flywheel effect. These motors have full-voltage starting torques of at least 275 percent of full-load torque and full-voltage starting currents which will be not greater than those given in Sec. 162. The maximum torque of the motor occurs at standstill, so there is no noticeable breakdown torque. The full-voltage starting currents are within the limitations of most power companies. Reduced-voltage starting would not as a rule be satisfactory because of the nature of the application of these motors.

The high-slip motors are designed for meeting the requirements of elevator drive and small cranes and hoists. These motors have full-voltage starting torques of at least 275 percent of full-load torque and full-voltage starting currents which will be not greater than those given in Sec. 162. Their full-voltage starting currents meet the limitations of practically all power companies. Like the medium-slip motors, they have no breakdown torque. Owing to the nature of their application, reduced-voltage starting would not be satisfactory.

**169. Design F (low-torque, low-starting-current) squirrel-cage motors** are obsolete under the standards of NEMA. They were made by some manufacturers in sizes of 30 hp and larger. Above 30 hp the full-voltage starting current of the normal-torque, low-starting-current type of motor exceeds the limitations of many power companies. By designing a motor for a lower starting torque, however, it is possible to build motors of larger size that will meet these starting-current limitations. The full-voltage starting torque of these motors will be at least 125 percent of full-load torque. The full-voltage starting currents will vary from 350 to 550 percent of the full-load current. The breakdown torques will be at least 135 percent of full-load torque. The full-voltage starting currents of these motors will be within the limitations of most power companies. The motors would not be used if reduced-voltage starting were required. They should be used only for drives which are started with light loads, such as motor-generator sets, fans, and centrifugal pumps.

**170. Automatic-start induction motors** are so arranged that they start with a high-resistance rotor and, after coming up to speed, are automatically changed to have a low-resistance rotor. At start approximately one-third of the rotor copper is in service.

These motors are made in two types. In one type the machine is of generally normal construction but with two rotor windings, one of high resistance and the other of low resistance. At standstill the low-resistance rotor winding is open-circuited so that only the high-resistance winding is in service. As the motor speeds up, a centrifugal device short-circuits the low-resistance winding so that both windings are in service.

In the other type of motor the primary winding is located on the rotor. There are two secondary windings located on the stator, one of high resistance and the other of low resistance. The functioning of the two windings is controlled by magnetic switches so that there is no centrifugal device. The low-resistance winding is of the wound type, and the high-resistance one is a squirrel-cage winding. In starting, a magnetic contactor connects the primary to the line, with the low-resistance winding open-circuited. After a certain lapse of time, a second contactor closes, short-circuiting the low-resistance winding.

These automatic-start motors are started on full voltage and have starting torques from 225 to 250 percent of full-load torque, with starting currents from 350 to 375 percent of full-load current. Their breakdown torques vary from 200 to 250 percent of full-load torque, and they have pull-in torques of from 150 to 200 percent of full-load torque. They are suitable for high-starting-torque duty when starting periods are not too frequent or of too long duration, as in compressors and air-conditioning–equipment service. They can be used in many places where power companies will not allow the starting currents drawn by high-torque, low-starting-current motors when thrown directly across the line.

**171. Multispeed induction motors** are squirrel-cage induction motors constructed so that the number of poles for which the stator is wound can be altered. Since the speed of an induction motor depends upon the number of poles for which the stator is wound, changing the number will alter the speed of the motor. The change in the number of poles is produced either by means of entirely separate windings or by a change in the interconnections of the different parts of a single winding. In either case the necessary leads are brought out from the stator winding to a drum controller through which the necessary change in connections is made. A gradual speed control is not possible with the multispeed motor. Motors of this type which can be operated at two, three, or four different speeds can be obtained. Four speeds are the maximum for which it is feasible to construct multispeed induction motors. To design and construct a machine for more than four speeds would require too many complications in the wiring.

Multispeed induction motors have a practically constant speed from no load to full load for each one of their rated speed connections. They are available in all the following types:

    1. Normal-torque, normal-starting-current
    2. Normal-torque, low-starting-current
    3. High-torque, low-starting-current
    4. Low-torque, low-starting-current
    5. High-resistance, medium- or high-slip

Multispeed induction motors may be designed and rated for constant-torque service, constant-horsepower service, or varying-torque, varying-horsepower service. The allowable rating of constant-torque motors varies directly with the speed. Constant-horsepower motors have the same rated horsepower for all speeds. With varying-torque, varying-horsepower motors the full-load rated torque varies directly with the speed. The horsepower rating of these machines therefore varies as the square of the speed.

**172. Efficiency of squirrel-cage induction motors.**[1]   The nominal full-load efficiency (see Sec. 151A) of Design A and B single-speed polyphase squirrel-cage medium motors in the range of 1 through 125 hp for NEMA frame-size motors, and equivalent Design C motors, is marked on the motor nameplate, identified as "NEMA Nominal Efficiency" or "NEMA Nom. Eff." This marked nominal efficiency is not greater than the average efficiency of a large population of motors of the same design, and is the efficiency value that should be used to compute the motor energy consumption. For the motor to be classified as "energy-efficient," NEMA standards require that the marked nominal efficiency be equal to or exceed the values in the following table.

**Full-Load Efficiencies of "Energy-Efficient" Motors**

| | Open motors | | | | | | | |
|---|---|---|---|---|---|---|---|---|
| | 2 Pole | | 4 Pole | | 6 Pole | | 8 Pole | |
| Hp | Nominal efficiency | Minimum efficiency | Nominal efficiency | Minimum efficiency | Nominal efficiency | Minimum efficiency | Nominal efficiency | Minimum efficiency |
| 1.0 | — | — | 82.5 | 80.0 | 77.0 | 74.0 | 72.0 | 68.0 |
| 1.5 | 80.0 | 77.0 | 82.5 | 80.0 | 82.5 | 80.0 | 75.5 | 72.0 |
| 2.0 | 82.5 | 80.0 | 82.5 | 80.0 | 84.0 | 81.5 | 85.5 | 82.5 |
| 3.0 | 82.5 | 80.0 | 86.5 | 84.0 | 85.5 | 82.5 | 86.5 | 84.0 |
| 5.0 | 85.5 | 82.5 | 86.5 | 84.0 | 86.5 | 84.0 | 87.5 | 85.5 |
| 7.5 | 85.5 | 82.5 | 88.5 | 86.5 | 88.5 | 86.5 | 88.5 | 86.5 |
| 10.0 | 87.5 | 85.5 | 88.5 | 86.5 | 90.2 | 88.5 | 89.5 | 87.5 |
| 15.0 | 89.5 | 87.5 | 90.2 | 88.5 | 89.5 | 87.5 | 89.5 | 87.5 |
| 20.0 | 90.2 | 88.5 | 91.0 | 89.5 | 90.2 | 88.5 | 90.2 | 88.5 |
| 25.0 | 91.0 | 89.5 | 91.7 | 90.2 | 91.0 | 89.5 | 90.2 | 88.5 |
| 30.0 | 91.0 | 89.5 | 91.7 | 90.2 | 91.7 | 90.2 | 91.0 | 89.5 |
| 40.0 | 91.7 | 90.2 | 92.4 | 91.0 | 91.7 | 90.2 | 90.2 | 88.5 |
| 50.0 | 91.7 | 90.2 | 92.4 | 91.0 | 91.7 | 90.2 | 91.7 | 90.2 |
| 60.0 | 93.0 | 91.7 | 93.0 | 91.7 | 92.4 | 91.0 | 92.4 | 91.0 |
| 75.0 | 93.0 | 91.7 | 93.6 | 92.4 | 93.0 | 91.7 | 93.6 | 92.4 |
| 100.0 | 93.0 | 91.7 | 93.6 | 92.4 | 93.6 | 92.4 | 93.6 | 92.4 |
| 125.0 | 93.0 | 91.7 | 93.6 | 92.4 | 93.6 | 92.4 | 93.6 | 92.4 |
| 150.0 | 93.6 | 92.4 | 94.1 | 93.0 | 93.6 | 92.4 | 93.6 | 92.4 |
| 200.0 | 93.6 | 92.4 | 94.1 | 93.0 | 94.1 | 93.0 | 93.6 | 92.4 |

| | Enclosed motors | | | | | | | |
|---|---|---|---|---|---|---|---|---|
| | 2 Pole | | 4 Pole | | 6 Pole | | 8 Pole | |
| Hp | Nominal efficiency | Minimum efficiency | Nominal efficiency | Minimum efficiency | Nominal efficiency | Minimum efficiency | Nominal efficiency | Minimum efficiency |
| 1.0 | — | — | 80.0 | 77.0 | 75.5 | 72.0 | 72.0 | 68.0 |
| 1.5 | 78.5 | 75.5 | 81.5 | 78.5 | 82.5 | 80.0 | 75.5 | 72.0 |
| 2.0 | 81.5 | 78.5 | 82.5 | 80.0 | 82.5 | 80.0 | 82.5 | 80.0 |
| 3.0 | 82.5 | 80.0 | 84.0 | 81.5 | 84.0 | 81.5 | 81.5 | 78.5 |
| 5.0 | 85.5 | 82.5 | 85.5 | 82.5 | 85.5 | 82.5 | 84.0 | 81.5 |
| 7.5 | 85.5 | 82.5 | 87.5 | 85.5 | 87.5 | 85.5 | 85.5 | 82.5 |
| 10.0 | 87.5 | 85.5 | 87.5 | 85.5 | 87.5 | 85.5 | 87.5 | 85.5 |
| 15.0 | 87.5 | 85.5 | 88.5 | 86.5 | 89.5 | 87.5 | 88.5 | 86.5 |
| 20.0 | 88.5 | 86.5 | 90.2 | 88.5 | 89.5 | 87.5 | 89.5 | 87.5 |

[1]This material is reproduced with the permission of the National Electrical Manufacturers Association from NEMA Standards Publication MG 1-1987, copyright © 1987 by NEMA.

**Full-Load Efficiencies of "Energy-Efficient" Motors** (*Continued*)

| | Enclosed motors | | | | | | | |
|---|---|---|---|---|---|---|---|---|
| | 2 Pole | | 4 Pole | | 6 Pole | | 8 Pole | |
| Hp | Nominal efficiency | Minimum efficiency | Nominal efficiency | Minimum efficiency | Nominal efficiency | Minimum efficiency | Nominal efficiency | Minimum efficiency |
| 25.0 | 89.5 | 87.5 | 91.0 | 89.5 | 90.2 | 88.5 | 89.5 | 87.5 |
| 30.0 | 89.5 | 87.5 | 91.0 | 89.5 | 91.0 | 89.5 | 90.2 | 88.5 |
| 40.0 | 90.2 | 88.5 | 91.7 | 90.2 | 91.7 | 90.2 | 90.2 | 88.5 |
| 50.0 | 90.2 | 88.5 | 92.4 | 91.0 | 91.7 | 90.2 | 91.0 | 89.5 |
| 60.0 | 91.7 | 90.2 | 93.0 | 91.7 | 91.7 | 90.2 | 91.7 | 90.2 |
| 75.0 | 92.4 | 91.0 | 93.0 | 91.7 | 93.0 | 91.7 | 93.0 | 91.7 |
| 100.0 | 93.0 | 91.7 | 93.6 | 92.4 | 93.0 | 91.7 | 93.0 | 91.7 |
| 125.0 | 93.0 | 91.7 | 93.6 | 92.4 | 93.0 | 91.7 | 93.6 | 92.4 |
| 150.0 | 93.0 | 91.7 | 94.1 | 93.0 | 94.1 | 93.0 | 93.6 | 92.4 |
| 200.0 | 94.1 | 93.0 | 94.5 | 93.6 | 94.1 | 93.0 | 94.1 | 93.0 |

**173. Wound-rotor induction motors,** when operated without external resistance in the rotor circuit and with the slip rings short-circuited, have operating characteristics similar to those of the normal-torque, normal-starting-current squirrel-cage machine. If external resistance is inserted in the rotor circuit, the speed of the motor can be controlled. With load on the motor the speed can be controlled over a wide range. At no load and with light loads on the motor, it is possible to obtain only slight changes in the speed. When the speed is adjusted for any particular load, the speed does not stay constant as the load varies but decreases considerably as the load increases. The amount that the speed drops off from no load to full load depends upon the amount of external resistance in the rotor circuit; the greater the resistance, the greater the speed change. The characteristics of the wound-rotor motor operating with external resistance in the rotor circuit are similar to those for a dc motor with armature control. Considerable energy is lost in the external resistance, the efficiency of the motor being reduced approximately in proportion to the reduction in speed produced. Wound-rotor motors are started by throwing the stator directly across the line with external resistance in the rotor circuit. The value of the starting torque developed and the corresponding starting current depends upon the amount of external resistance in the rotor circuit (see Secs. 150 and 152). The breakdown torque of a wound-rotor motor will be at least as large as the figures given in Sec. 175. With the proper amount of external resistance in the rotor to produce a maximum torque at standstill the starting current will be from 250 to 300 percent of the full-load current. The maximum value of the starting torque will be from 200 to 275 percent of full-load torque.

Wound-rotor motors with speed control are constant-torque machines, the horsepower decreasing in proportion to the decrease in speed.

**174. Average Efficiencies and Power Factors for Polyphase Wound-Rotor Induction Motors (with Rotor Short-Circuited)**

| Hp | Efficiency | | | Power factor | | |
|---|---|---|---|---|---|---|
| | One-half load | Three-fourths load | Full load | One-half load | Three-fourths load | Full load |
| 2 | 71.0 | 75.0 | 75.5 | 40 | 52 | 60 |
| 3 | 75.0 | 78.0 | 79.0 | 50 | 68 | 70 |
| 5 | 78.0 | 80.0 | 80.5 | 60 | 75 | 77 |
| 7½ | 78.5 | 81.0 | 81.5 | 65 | 75 | 78 |
| 10 | 80.0 | 82.0 | 82.5 | 65 | 75 | 79 |
| 15 | 80.0 | 82.0 | 82.5 | 65 | 75 | 79 |
| 20 | 84.0 | 85.5 | 86.0 | 70 | 76.5 | 81 |
| 25 | 86.0 | 88.0 | 88.0 | 70 | 77 | 81 |
| 30 | 86.0 | 88.0 | 88.0 | 72 | 78 | 82 |
| 40 | 86.0 | 88.0 | 88.0 | 72 | 78 | 82 |
| 50 | 86.0 | 88.0 | 88.0 | 78 | 84 | 86 |
| 60 | 86.5 | 88.5 | 88.5 | 79 | 85 | 88 |
| 75 | 88.0 | 88.0 | 88.8 | 82 | 87 | 89 |

### 175. Breakdown Torque of Polyphase Wound-Rotor Medium Motors with Continuous Ratings

(NEMA standard)[a]

The breakdown torques of 60- and 50-Hz polyphase wound-rotor medium motors, with rated voltage and frequency applied, shall be not less than the following values, which are expressed in percent of full-load torque.

| Hp | Breakdown torque, per cent of full-load torque | | |
|---|---|---|---|
| | Synchronous speed, rpm | | |
| | 1,800 | 1,200 | 900 |
| 1 | ... | ... | 250 |
| 1½ | ... | ... | 250 |
| 2 | 275 | 275 | 250 |
| 3 | 275 | 275 | 250 |
| 5 | 275 | 275 | 250 |
| 7½ | 275 | 250 | 225 |
| 10 | 275 | 250 | 225 |
| 15 | 250 | 225 | 225 |
| 20–200, incl. | 225 | 225 | 225 |

[a]This material is reproduced with the permission of the National Electrical Manufacturers Association from NEMA Standards Publication MG 1-1987, copyright © 1987 by NEMA.

**176. In the commutator brush-shifting type of polyphase induction motor** the relative location of the two windings is reversed from that of the ordinary induction motor. The primary winding, the winding that is connected to the supply, is located on the rotor. The secondary winding is placed in the stator slots. In addition to the primary winding, the rotor carries a second winding which is connected to a commutator in the same manner as in a dc motor. The two ends of each phase of the secondary winding (the stationary winding) are connected to two brushes bearing upon the commutator. The arrangement and connection of the windings are indicated in Fig. 7-93. The position of the two brushes of each phase can be shifted by means of a handwheel. When the handwheel is adjusted so that the brushes of each phase are on the same commutator segment, the secondary winding is short-circuited and the motor operates like a squirrel-cage machine. As the brushes are moved apart, a section of the adjusting winding is placed in series with the secondary winding. This impresses upon the secondary an additional voltage to the one induced in it by the revolving magnetic field. If the brushes are moved in one direction, the voltage impressed by the adjusting winding will oppose the normal induced voltage. Such a shift of the brushes will therefore reduce the speed of the machine. A shifting of the brushes in the opposite direction causes the voltage from the adjusting winding to aid the normal induced voltage in the secondary and thus increases the speed of the motor. These machines are designed to give a speed range of 3:1. The highest speed is approximately 40 percent above synchronous speed, and the lowest speed is approximately 60 percent below synchronous speed. The decrease in speed from no load to full load is from 5 to 10 percent at the high speeds and from 15 to 25 percent at the low speeds. Very low, creeping speeds can be obtained by inserting resistance in the rotor circuit.

These brush-shifting motors are rated on a constant-torque basis, so that the horsepower rating decreases with the speed. The breakdown torque at low speeds is

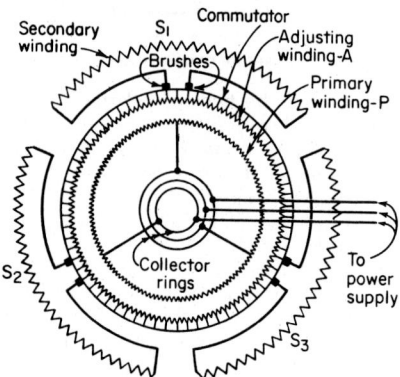

**Fig. 7-93**  Commutator type of polyphase induction motor. [General Electric Co.]

from 140 to 250 percent of full-load torque. The breakdown torque increases somewhat with the speed, so that at high speed it is from 300 to 400 percent of normal torque.

These motors are started by connecting the primary directly to the line with the brushes in the low-speed position. In this way they will develop a starting torque of from 140 to 150 percent of full-load torque with a starting current of from 125 to 175 percent of full-load current for the maximum horsepower rating. In some special cases, resistance is connected in series with the secondary for starting.

**177. Synchros are special ac motors employed for:**

1. Remote indication of position, such as indicating the position of generator rheostats, waterwheel governors, water-reservoir levels, gates or valves, and turntables

2. Remote signaling systems, such as signaling from switchboard to generator room or from steel-mill furnace to blower room and marine signals between engine room and bridge

3. Automatic or remote position control, such as system frequency, synchronizing of incoming generators, waterwheel governors, gates or valves, color screens on lights in theaters, and motor drives (see Sec. 142)

4. Operation of two machines so as to maintain a definite time-position relation or a definite speed relation, as in lift bridges, hoists, kiln drives, elevators, unit printing presses, and conveyors

Synchros are known by various trade names such as Selsyn, Synchrotie, Autosyn, and Telegon. Units are available in single-phase and three-phase types. The polyphase units conform in appearance and general characteristics to a three-phase wound-rotor induction motor. The single-phase units have a three-phase wound secondary and a single-phase primary. Some single-phase units are constructed with the primary located on the stationary member of the machine, and others with the primary on the movable member. Depending upon the torque developed by the machine, synchros are classified as indicating synchros or power synchros. Power synchros are always of three-phase construction. Most of the indicating synchros are constructed single-phase. Indicating synchros are manufactured in a general-purpose type and a high-accuracy type. The general-purpose type will indicate angular position within an accuracy of plus or minus 5°, and the high-accuracy type provides an accuracy of indication within plus or minus 1°.

A synchro system requires the use of at least two synchro machines. The connections for two single-phase units are shown in Fig. 7-94, and those for two three-phase units in Fig. 7-95. With both the single-phase and the polyphase units, if the rotors of the two

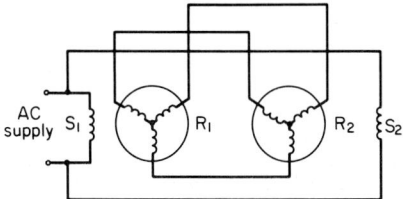
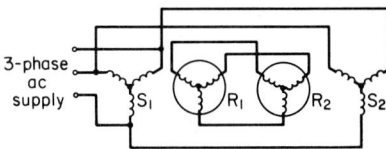

**Fig. 7-94**   Circuit for two single-phase synchros.     **Fig. 7-95**   Circuit for two three-phase synchros.

machines are in corresponding positions, the voltages of the two secondaries at each instant of time will be equal in magnitude and opposite in direction. Therefore, no current will flow in the rotor circuits, no torque will be developed in either machine, and the machines will be in equilibrium. When the two rotors are not in corresponding positions, the voltages of the two secondaries will not neutralize each other and a current will be produced in the two secondaries. This current will produce torque in the machines, which will act upon the rotors, tending to move them into corresponding positions.

In signaling or indicating applications the synchro located at the sending end is called the transmitter, and the synchro at the receiving end is called the receiver. If desired, a number of receivers can be connected in parallel to a single transmitter. The torque exerted upon each receiver will be reduced from that available when only one receiver is used. The torque can be calculated from the following formula:

$$T_r = T_1 \times \frac{2}{N + 1} \qquad (5d)$$

where $T_r$ is the torque available at each receiver, $T_1$ is the torque available if only one receiver is used, and $N$ is the number of receivers.

When the position taken by the receiver is to differ from that of the transmitter, a three-phase synchro is introduced between the transmitter and the receiver. This intermediate synchro is called a *differential synchro*. With such a system, the receiver will take up a position that will be either the sum or the difference of the angles applied to the transmitter and the differential. Conversely, if two synchros are connected through a differential synchro and each is turned through an angle, the differential synchro will indicate the difference between the two angles.

The fundamental synchro circuit generally used for control purposes is shown in Fig. 7-96. When the positions of the rotors coincide, a maximum voltage will be induced in stator $S_2$. When the rotors are displaced 90 electrical degrees from each other, no resultant voltage will be produced between the terminals of $S_2$. This is true since the axis of the winding $S_2$ will be located 90 electrical degrees from the axis of the field produced in machine 2. The system is adjusted so that when the desired condition exists, the rotors will be displaced 90 electrical degrees from each other. This is the equilibrium position. The circuit therefore functions to produce an error voltage in the stator of machine 2 whenever there is a displacement of the shafts from the equilibrium position. The relative polarity of the error voltage depends upon the direction of the displacement. The error voltage can be employed to instigate the functioning of other devices, which will operate to correct the condition causing the displacement. These correction devices will function until the synchros are brought into the equilibrium position, when absence of voltage in $S_2$ will stop their functioning. Refer to Sec. 142 for the application of synchros in position drive control.

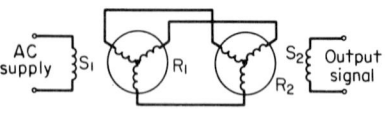

**Fig. 7-96** Fundamental synchro control circuit for the production of an error voltage.

**178. Synchronous motors.** The general construction of a synchronous motor is the same as that of synchronous ac generators with the addition of an auxiliary winding similar to either the squirrel-cage or the wound-rotor winding of an induction motor embedded in the pole faces. The purpose of the auxiliary winding is to produce starting torque so that the motor will be self-starting. A synchronous motor will not develop starting torque through synchronous-motor action. A synchronous motor is started without dc field excitation by induction-motor action through its auxiliary winding. As the motor comes up to speed and dc excitation voltage is applied to its field winding, a pull-in torque is developed, tending to pull the motor into step. If the inertia of the load is not too great, this pull-in torque will accelerate the motor to synchronous speed, and it will continue to operate as a synchronous motor. As long as the motor runs at synchronous speed, the auxiliary winding will be inactive, since no voltage will be produced in the auxiliary winding under this condition. If the motor should instantaneously tend to run at a speed slightly greater or less than synchronous, a voltage is produced in the auxiliary winding, causing a flow of current. This current will produce a torque which will act to bring the motor back to synchronous speed. The auxiliary winding, in addition to furnishing a means of making the motor self-starting, therefore provides a dampening effect tending to eliminate instantaneous fluctuations of the speed.

A synchronous motor has two very important characteristics. It operates at constant speed, and its power factor is under the control of the operator. If the average speed of a synchronous motor deviates from synchronism, the motor pulls out of step and ceases to function as a synchronous motor. The power factor, unlike that of an induction motor, is not a fixed value dependent upon the design and load. It depends upon the value of the direct field current and can be adjusted by the operator to a leading, unity, or lagging power-factor value. For any particular load a certain field current will make the motor operate at unity power factor. This value of field current is called *normal excitation*. A field current less than normal excitation will produce a lagging power factor, and a field current more than normal will produce a leading power factor.

**179. A synchronous motor can be used to correct low power factor** of the system that feeds it in addition to driving a mechanical load provided it has sufficient capacity. This characteristic is often of considerable importance. It is well known that the power factor of the induction motor, even under full-load conditions, is seldom greater than 95 percent, and it often falls as low as 50 or 60 percent at light load. The result is that an ac generator driving a considerable number of induction motors ordinarily operates at a comparatively low power factor. If this alternator is loaded to its full kilowatt capacity at such a low power factor, overheating will result.

If the alternator is not loaded beyond its normal current capacity, it operates at a low energy load but with the same heating losses as at full load on account of the reduced power factor. The advantage of the synchronous motor on such a system is that by proper adjustment of its field current it can be made to draw from the line a current which is leading with respect to the voltage and which will neutralize the lagging current taken by the induction motors. The current in the ac generator can thereby be brought into phase with the voltage, and the generator will operate under its normal conditions. When used in this manner as a compensator for lagging current, the synchronous motor must be of larger size than required by its power output on account of the excess current which it draws from the line.

A *synchronous condenser* is a synchronous motor that operates only to correct the power factor and does not pull any mechanical load.

**180. Classification of synchronous motors**

1. According to power-factor rating
    *a.*  1.0 (unity)-power-factor
    *b.*  0.8 (leading)-power-factor
2. According to speed
    *a.*  High-speed
    *b.*  Low-speed
3. According to service
    *a.*  General-purpose
    *b.*  Special-purpose

The horsepower ratings of synchronous motors are based on unity or 0.8-pf operation. A unity power-factor motor is designed to carry its rated horsepower load when operating at unity power factor. An 0.8-pf motor is designed to carry its rated horsepower load when operating at 0.8 pf leading. Either a unity- or an 0.8-pf–rated motor can be operated at power factors below their rated amount provided the horsepower output is reduced below the rated value by a sufficient amount so that the machine will not be overheated in operation.

Synchronous motors are divided into two groups according to the value of their rated speed. Motors with speeds of 500 rpm or greater are called *high-speed motors,* and those with speeds below 500 rpm are classed as *low-speed motors.*

**181. General-purpose synchronous motors** are rated on a 40°C temperature-rise basis with a service factor of 1.15 (see Sec. 128). When started with full voltage, these motors will have starting kVA inrushes of from 550 to 700 percent of the motor's normal kVA rating. Normal torque values for these motors are given in Sec. 183. The pull-in torques as given in Sec. 183 are based on the standardized normal $Wk^2$ of load as given in Sec. 185.

**182. Operation at other than rated power factors** (NEMA standard)[1]

1. For an 0.8-pf motor which is to operate at 1.0 pf with normal 0.8-pf armature current and with field excitation reduced to correspond to that armature current at 1.0 pf, multiply the rated horsepower and torque values of the motor by the following constants to obtain the horsepower at 1.0 pf and the torques in terms of the 1.0-pf horsepower rating:

| | |
|---|---|
| Horsepower | 1.25 |
| Locked-rotor torques | 0.8 |
| Pull-in torque | 0.8 |
| Pullout torque (approximate) | 0.6 |

For example, consider a 1000-hp, 0.8-pf motor which has a locked-rotor torque of 100 percent, a pull-in torque of 100 percent, and a pullout torque of 200 percent and which is to be operated at 1.0 pf. In accordance with the foregoing, this motor could be operated at 1250 hp, 1.0 pf, 80 percent locked-rotor torque (based upon 1250 hp), 80 percent pull-in torque (based upon 1250 hp), and a pullout torque of approximately 120 percent (based upon 1250 hp).

2. For a 1.0-pf motor which is to operate at 0.8 pf with normal 1.0-pf field excitation and

---

[1]This material is reproduced with the permission of the National Electrical Manufacturers Association from NEMA Standards Publication MG 1-1987, copyright © 1987 by NEMA.

the armature current reduced to correspond to that excitation, multiply the rated horsepower and torque values of the motor by the following constants to obtain the horsepower at 0.8 pf and the torques in terms of the 0.8-pf horsepower rating:

| | |
|---|---|
| Horsepower | 0.35 |
| Locked-rotor torque | 2.85 |
| Pull-in torque | 2.85 |
| Pullout torque (approximate) | 2.85 |

For example, consider a 1000-hp, 1.0-pf motor which has a locked-rotor torque of 100 percent, a pull-in torque of 100 percent, and a pullout torque of 200 percent and which is to be operated at 0.8 pf. In accordance with the foregoing, this motor could be operated at 350 hp, 0.8 pf, 285 percent locked-rotor torque (based upon 350 hp), 285 percent pull-in torque (based upon 350 hp), and a pullout torque of approximately 570 percent (based upon 350 hp).

### 183.    Minimum-Torque Values for Synchronous Motors

(NEMA standard)[a]
The locked-rotor, pull-in, and pullout torques, with rated voltage and frequency applied, shall be not less than the following:

| | | | Torques, percent of rated full-load torque | | |
|---|---|---|---|---|---|
| Speed, rpm | hp | Power factor | Locked-rotor | Pull-in (based on normal $Wk^2$ of load)[b,c] | Pullout[c] |
| 500–1800 | 200 and below | 1.0 | 100 | 100 | 150 |
| | 150 and below | 0.8 | 100 | 100 | 175 |
| | 250–1000 | 1.0 | 60 | 60 | 150 |
| | 200–1000 | 0.8 | 60 | 60 | 175 |
| | 1250 and larger | 1.0 | 40 | 60 | 150 |
| | | 0.8 | 40 | 60 | 175 |
| 450 and below | All ratings | 1.0 | 40 | 30 | 150 |
| | | 0.8 | 40 | 30 | 200 |

[a]This material is reproduced with the permission of the National Electrical Manufacturers Association from NEMA Standards Publication MG 1-1987, copyright © 1987 by NEMA.
[b]Values of normal $Wk^2$ are given in Sec. 185..
[c]With rated excitation current applied.

**184. Effect of load $Wk^2$.**    The inertia of the load (load $Wk^2$) is very important in the selection and operation of synchronous motors. The load inertia that a synchronous motor can accelerate is limited by the thermal capacity of its amortisseur winding. Smaller synchronous motors can accelerate relatively higher load $Wk^2$ than larger motors can. NEMA standards for the load $Wk^2$ which can be accelerated by standard synchronous motors are given in Sec. 185.

## 185. Normal $WK^2$ of Load, Lb·Ft$^2$
(NEMA standard)[a]

| hp | 1800 | 1200 | 900 | 720 | 600 | 514 | 450 | 400 | 360 | 327 |
|---|---|---|---|---|---|---|---|---|---|---|
| 20 | 3.63 | 8.16 | 14.51 | 22.7 | 32.7 | 44.4 | 58.0 | 73.5 | 90.7 | 109.7 |
| 25 | 4.69 | 10.55 | 18.76 | 29.3 | 42.2 | 57.4 | 75.0 | 95.0 | 117.2 | 141.9 |
| 30 | 5.78 | 13.01 | 23.1 | 36.1 | 52.0 | 70.8 | 92.5 | 117.1 | 144.6 | 174.9 |
| 40 | 8.05 | 18.11 | 32.2 | 50.3 | 75.5 | 98.6 | 123.8 | 163.0 | 201 | 244 |
| 50 | 10.41 | 23.4 | 41.6 | 65.0 | 93.7 | 127.5 | 166.5 | 211 | 260 | 315 |
| 60 | 12.83 | 28.9 | 51.3 | 80.2 | 115.5 | 157.2 | 205 | 260 | 321 | 388 |
| 75 | 16.59 | 37.3 | 66.4 | 103.7 | 149.3 | 203 | 265 | 336 | 415 | 502 |
| 100 | 23.1 | 52.0 | 92.4 | 144.3 | 208 | 283 | 369 | 468 | 577 | 699 |
| 125 | 29.8 | 67.2 | 119.3 | 186.6 | 269 | 366 | 478 | 604 | 746 | 903 |
| 150 | 36.8 | 82.8 | 147.2 | 230 | 331 | 451 | 589 | 745 | 920 | 1,114 |
| 200 | 51.2 | 115.3 | 205 | 320 | 461 | 628 | 820 | 1,038 | 1,281 | 1,550 |
| 250 | 66.2 | 149.0 | 265 | 414 | 596 | 811 | 1,060 | 1,341 | 1,656 | 2,000 |
| 300 | 81.7 | 183.8 | 327 | 511 | 735 | 1,001 | 1,307 | 1,654 | 2,040 | 2,470 |
| 350 | 97.5 | 219 | 390 | 610 | 878 | 1,195 | 1,561 | 1,975 | 2,440 | 2,950 |
| 400 | 113.7 | 256 | 455 | 711 | 1,024 | 1,393 | 1,820 | 2,300 | 2,840 | 3,440 |
| 450 | 130.2 | 293 | 521 | 814 | 1,172 | 1,595 | 2,080 | 2,640 | 3,260 | 3,940 |
| 500 | 147.0 | 331 | 588 | 919 | 1,323 | 1,801 | 2,350 | 2,980 | 3,670 | 4,450 |
| 600 | 181.3 | 408 | 725 | 1,133 | 1,632 | 2,220 | 2,900 | 3,670 | 4,530 | 5,480 |
| 700 | 216 | 487 | 866 | 1,353 | 1,948 | 2,650 | 3,460 | 4,380 | 5,410 | 6,550 |
| 800 | 252 | 568 | 1,009 | 1,577 | 2,270 | 3,090 | 4,040 | 5,110 | 6,310 | 7,630 |
| 900 | 289 | 650 | 1,156 | 1,806 | 2,600 | 3,540 | 4,620 | 5,850 | 7,220 | 8,740 |
| 1,000 | 326 | 734 | 1,305 | 2,040 | 2,940 | 4,000 | 5,220 | 6,610 | 8,160 | 9,870 |
| 1,250 | 422 | 949 | 1,687 | 2,640 | 3,790 | 5,160 | 6,750 | 8,540 | 10,540 | 12,750 |
| 1,500 | 520 | 1,170 | 2,080 | 3,250 | 4,680 | 6,370 | 8,320 | 10,530 | 13,000 | 15,730 |
| 1,750 | 621 | 1,397 | 2,480 | 3,880 | 5,590 | 7,610 | 9,930 | 12,570 | 15,520 | 18,780 |
| 2,000 | 724 | 1,629 | 2,900 | 4,520 | 6,510 | 8,870 | 11,580 | 14,660 | 18,100 | 21,900 |
| 2,250 | 829 | 1,865 | 3,320 | 5,180 | 7,460 | 10,150 | 13,260 | 16,780 | 20,700 | 25,100 |
| 2,500 | 936 | 2,110 | 3,740 | 5,850 | 8,420 | 11,460 | 14,970 | 18,950 | 23,400 | 28,300 |

Speed, rpm

**185. Normal WK² of Load, Lb·Ft² (Continued)**
(NEMA standard)[a]

| hp | Speed, rpm | | | | | | | | | |
|---|---|---|---|---|---|---|---|---|---|---|
| | 1800 | 1200 | 900 | 720 | 600 | 514 | 450 | 400 | 360 | 327 |
| 3,000 | 1,154 | 2,600 | 4,620 | 7,210 | 10,390 | 14,140 | 18,460 | 23,400 | 28,800 | 34,900 |
| 3,500 | 1,378 | 3,100 | 5,510 | 8,610 | 12,400 | 16,880 | 22,000 | 27,900 | 34,400 | 41,700 |
| 4,000 | 1,606 | 3,610 | 6,430 | 10,040 | 14,460 | 19,680 | 25,700 | 32,500 | 40,200 | 48,600 |
| 4,500 | 1,839 | 4,140 | 7,360 | 11,500 | 16,550 | 22,500 | 29,400 | 37,200 | 46,000 | 55,600 |
| 5,000 | 2,080 | 4,670 | 8,310 | 12,980 | 18,690 | 25,400 | 33,200 | 42,000 | 51,900 | 62,800 |
| 5,500 | 2,320 | 5,210 | 9,270 | 14,480 | 20,900 | 28,400 | 37,100 | 46,900 | 57,900 | 70,100 |
| 6,000 | 2,560 | 5,760 | 10,240 | 16,000 | 23,000 | 31,400 | 41,000 | 51,900 | 64,000 | 77,500 |
| 7,000 | 3,060 | 6,880 | 12,230 | 19,110 | 27,500 | 37,500 | 48,900 | 61,900 | 76,400 | 92,500 |
| 8,000 | 3,560 | 8,020 | 14,260 | 22,300 | 32,100 | 43,700 | 57,000 | 72,200 | 89,100 | 107,800 |
| 9,000 | 4,080 | 9,180 | 16,330 | 25,500 | 36,700 | 50,000 | 65,300 | 82,700 | 102,000 | 123,500 |
| 10,000 | 4,610 | 10,370 | 18,430 | 28,800 | 41,500 | 56,400 | 73,700 | 93,300 | 115,200 | 139,400 |

| hp | Speed, rpm | | | | | | | | |
|---|---|---|---|---|---|---|---|---|---|
| | 300 | 277 | 257 | 240 | 225 | 200 | 180 | 164 | 150 |
| 20 | 130.6 | 153.3 | 177.8 | 204 | 232 | 294 | 363 | 439 | 522 |
| 25 | 168.8 | 198.1 | 230 | 264 | 300 | 380 | 469 | 567 | 675 |
| 30 | 208 | 244 | 283 | 325 | 370 | 468 | 578 | 700 | 833 |
| 40 | 290 | 340 | 395 | 453 | 515 | 652 | 805 | 974 | 1,159 |
| 50 | 375 | 440 | 510 | 585 | 666 | 843 | 1,041 | 1,259 | 1,499 |
| 60 | 462 | 542 | 629 | 721 | 821 | 1,040 | 1,283 | 1,553 | 1,848 |
| 75 | 597 | 701 | 813 | 933 | 1,062 | 1,344 | 1,659 | 2,010 | 2,390 |
| 100 | 831 | 976 | 1,132 | 1,299 | 1,478 | 1,871 | 2,310 | 2,790 | 3,330 |
| 125 | 1,075 | 1,261 | 1,463 | 1,679 | 1,910 | 2,420 | 2,980 | 3,610 | 4,300 |
| 150 | 1,325 | 1,555 | 1,804 | 2,070 | 2,360 | 2,980 | 3,680 | 4,450 | 5,300 |

| | | | | | | | | | |
|---|---|---|---|---|---|---|---|---|---|
| 200 | 1,845 | 2,170 | 2,510 | 2,880 | 3,280 | 4,150 | 5,120 | 6,200 | 7,380 |
| 250 | 2,380 | 2,800 | 3,250 | 3,730 | 4,240 | 5,370 | 6,620 | 8,010 | 9,540 |
| 300 | 2,940 | 3,450 | 4,000 | 4,600 | 5,230 | 6,620 | 8,170 | 9,880 | 11,760 |
| 350 | 3,510 | 4,120 | 4,780 | 5,490 | 6,240 | 7,900 | 9,750 | 11,800 | 14,050 |
| 400 | 4,090 | 4,800 | 5,570 | 6,400 | 7,280 | 9,210 | 11,370 | 13,760 | 16,380 |
| 450 | 4,690 | 5,500 | 6,380 | 7,320 | 8,330 | 10,550 | 13,020 | 15,760 | 18,750 |
| 500 | 5,290 | 6,210 | 7,200 | 8,270 | 9,410 | 11,910 | 14,700 | 17,790 | 21,200 |
| 600 | 6,530 | 7,660 | 8,880 | 10,200 | 11,600 | 14,680 | 18,130 | 21,900 | 26,100 |
| 700 | 7,790 | 9,140 | 10,610 | 12,180 | 13,850 | 17,530 | 21,600 | 26,200 | 31,200 |
| 800 | 9,090 | 10,660 | 12,370 | 14,200 | 16,150 | 20,400 | 25,200 | 30,500 | 36,300 |
| 900 | 10,400 | 12,210 | 14,160 | 16,260 | 18,490 | 23,400 | 28,900 | 35,000 | 41,600 |
| 1,000 | 11,740 | 13,780 | 15,980 | 18,350 | 20,900 | 26,400 | 32,600 | 39,500 | 47,000 |
| 1,250 | 15,180 | 17,810 | 20,700 | 23,700 | 27,000 | 34,200 | 42,200 | 51,000 | 60,700 |
| 1,500 | 18,720 | 22,000 | 25,500 | 29,200 | 33,300 | 42,100 | 52,000 | 62,900 | 74,900 |
| 1,750 | 22,400 | 26,200 | 30,400 | 34,900 | 39,700 | 50,300 | 62,100 | 75,100 | 89,400 |
| 2,000 | 26,100 | 30,600 | 35,500 | 40,700 | 46,300 | 58,600 | 72,400 | 87,600 | 104,200 |
| 2,250 | 29,800 | 35,000 | 40,600 | 46,600 | 53,000 | 67,100 | 82,900 | 100,300 | 119,400 |
| 2,500 | 33,700 | 39,500 | 45,800 | 52,600 | 59,900 | 75,800 | 93,600 | 113,200 | 134,700 |
| 3,000 | 41,500 | 48,800 | 56,500 | 64,900 | 73,900 | 93,500 | 115,400 | 139,600 | 166,200 |
| 3,500 | 49,600 | 58,200 | 67,500 | 77,500 | 88,200 | 111,600 | 137,800 | 166,700 | 198,400 |
| 4,000 | 57,800 | 67,900 | 78,700 | 90,400 | 102,800 | 130,100 | 160,600 | 194,400 | 231,000 |
| 4,500 | 66,200 | 77,700 | 90,100 | 103,500 | 117,700 | 149,000 | 183,900 | 223,000 | 265,000 |
| 5,000 | 74,700 | 87,700 | 101,700 | 116,800 | 132,900 | 168,200 | 208,000 | 251,000 | 299,000 |
| 5,500 | 83,400 | 97,900 | 113,500 | 130,300 | 148,300 | 187,700 | 232,000 | 280,000 | 334,000 |
| 6,000 | 92,200 | 108,200 | 125,500 | 144,000 | 163,900 | 207,000 | 256,000 | 310,000 | 369,000 |
| 7,000 | 110,100 | 129,200 | 149,800 | 172,000 | 195,700 | 248,000 | 306,000 | 370,000 | 440,000 |
| 8,000 | 128,300 | 150,600 | 174,700 | 201,000 | 228,000 | 289,000 | 356,000 | 431,000 | 513,000 |
| 9,000 | 146,900 | 172,500 | 200,000 | 230,000 | 261,000 | 331,000 | 408,000 | 494,000 | 588,000 |
| 10,000 | 165,900 | 194,700 | 226,000 | 259,000 | 295,000 | 373,000 | 461,000 | 558,000 | 664,000 |

**185. Normal $WK^2$ of Load, Lb·Ft$^2$** (Continued)
(NEMA standard)[a]

| hp | Speed, rpm | | | | | | | | |
|---|---|---|---|---|---|---|---|---|---|
| | 138 | 129 | 120 | 109 | 100 | 95 | 90 | 86 | 80 |
| 20 | 613 | 711 | 816 | 988 | 1,175 | 1,310 | 1,451 | 1,600 | 1,837 |
| 25 | 793 | 919 | 1,055 | 1,277 | 1,519 | 1,693 | 1,876 | 2,070 | 2,370 |
| 30 | 977 | 1,134 | 1,301 | 1,575 | 1,874 | 2,090 | 2,310 | 2,550 | 2,930 |
| 40 | 1,361 | 1,578 | 1,811 | 2,190 | 2,610 | 2,910 | 3,220 | 3,550 | 4,080 |
| 50 | 1,759 | 2,040 | 2,340 | 2,830 | 3,370 | 3,760 | 4,160 | 4,590 | 5,270 |
| 60 | 2,170 | 2,520 | 2,890 | 3,490 | 4,160 | 4,630 | 5,130 | 5,660 | 6,500 |
| 75 | 2,800 | 3,250 | 3,760 | 4,520 | 5,370 | 5,990 | 6,640 | 7,320 | 8,400 |
| 100 | 3,900 | 4,530 | 5,200 | 6,290 | 7,480 | 8,340 | 9,240 | 10,180 | 11,690 |
| 125 | 5,040 | 5,850 | 6,720 | 8,130 | 9,670 | 10,780 | 11,940 | 13,160 | 15,110 |
| 150 | 6,220 | 7,220 | 8,280 | 10,020 | 11,930 | 13,290 | 14,720 | 16,230 | 18,640 |
| 200 | 8,660 | 10,040 | 11,530 | 13,950 | 16,600 | 18,500 | 20,500 | 22,600 | 25,900 |
| 250 | 11,190 | 12,980 | 14,900 | 18,030 | 21,500 | 23,900 | 26,500 | 29,200 | 33,500 |
| 300 | 13,810 | 16,010 | 18,380 | 22,200 | 26,500 | 29,500 | 32,700 | 36,000 | 41,400 |
| 350 | 16,480 | 19,120 | 21,900 | 26,600 | 31,600 | 35,200 | 39,000 | 43,000 | 49,400 |
| 400 | 19,220 | 22,300 | 25,600 | 31,000 | 36,800 | 41,100 | 45,500 | 50,200 | 57,600 |
| 450 | 22,000 | 25,500 | 29,300 | 35,500 | 42,200 | 47,000 | 52,100 | 57,400 | 65,900 |
| 500 | 24,800 | 28,800 | 33,100 | 40,000 | 47,600 | 53,100 | 58,800 | 64,800 | 74,400 |
| 600 | 30,600 | 35,500 | 40,800 | 49,400 | 58,700 | 65,400 | 72,500 | 79,900 | 91,800 |

| | | | | | | | | | |
|---|---|---|---|---|---|---|---|---|---|
| 700 | 36,600 | 42,400 | 48,700 | 58,900 | 70,100 | 78,100 | 86,600 | 95,500 | 109,600 |
| 800 | 42,700 | 49,500 | 56,800 | 68,700 | 81,800 | 91,100 | 100,900 | 111,300 | 127,800 |
| 900 | 48,800 | 56,600 | 65,000 | 78,700 | 93,600 | 104,300 | 115,600 | 127,400 | 146,300 |
| 1,000 | 55,100 | 63,900 | 73,400 | 88,800 | 105,700 | 117,800 | 130,500 | 143,900 | 165,100 |
| 1,250 | 71,300 | 82,600 | 94,900 | 114,800 | 136,600 | 152,200 | 168,700 | 185,900 | 213,000 |
| 1,500 | 87,900 | 101,900 | 117,000 | 141,600 | 168,500 | 187,700 | 208,000 | 229,000 | 263,000 |
| 1,750 | 104,900 | 121,700 | 139,700 | 169,000 | 201,000 | 224,000 | 248,000 | 274,000 | 314,000 |
| 2,000 | 122,300 | 141,900 | 162,900 | 197,100 | 235,000 | 261,000 | 290,000 | 319,000 | 366,000 |
| 2,250 | 140,100 | 162,500 | 186,500 | 226,000 | 269,000 | 299,000 | 332,000 | 366,000 | 420,000 |
| 2,500 | 158,100 | 183,400 | 211,000 | 255,000 | 303,000 | 338,000 | 374,000 | 413,000 | 474,000 |
| 3,000 | 195,000 | 226,000 | 260,000 | 314,000 | 374,000 | 417,000 | 462,000 | 509,000 | 584,000 |
| 3,500 | 233,000 | 270,000 | 310,000 | 375,000 | 446,000 | 497,000 | 551,000 | 608,000 | 697,000 |
| 4,000 | 271,000 | 315,000 | 361,000 | 437,000 | 520,000 | 580,000 | 643,000 | 708,000 | 813,000 |
| 4,500 | 311,000 | 361,000 | 414,000 | 501,000 | 596,000 | 664,000 | 736,000 | 811,000 | 931,000 |
| 5,000 | 351,000 | 407,000 | 467,000 | 565,000 | 673,000 | 750,000 | 831,000 | 916,000 | 1,051,000 |
| 5,500 | 392,000 | 454,000 | 521,000 | 631,000 | 751,000 | 836,000 | 927,000 | 1,022,000 | 1,173,000 |
| 6,000 | 433,000 | 502,000 | 576,000 | 697,000 | 830,000 | 924,000 | 1,024,000 | 1,129,000 | 1,296,000 |
| 7,000 | 517,000 | 599,000 | 688,000 | 832,000 | 991,000 | 1,104,000 | 1,223,000 | 1,348,000 | 1,548,000 |
| 8,000 | 602,000 | 699,000 | 802,000 | 971,000 | 1,155,000 | 1,287,000 | 1,426,000 | 1,572,000 | 1,805,000 |
| 9,000 | 690,000 | 800,000 | 918,000 | 1,111,000 | 1,323,000 | 1,474,000 | 1,633,000 | 1,800,000 | 2,070,000 |
| 10,000 | 779,000 | 903,000 | 1,037,000 | 1,254,000 | 1,493,000 | 1,663,000 | 1,843,000 | 2,030,000 | 2,330,000 |

[a]This material is reproduced with the permission of the National Electrical Manufacturers Association from NEMA Standards Publication MG 1-1987, copyright © 1987 by NEMA.

**186. Special-purpose synchronous motors.**   For applications in which the characteristics of general-purpose motors do not fit the operating requirements, the manufacturer fits the design of the auxiliary windings and motor proportions to meet specific applications. These custom-made motors are built with the following types of auxiliary windings:

1. Conventional squirrel-cage winding
2. Conventional squirrel-cage winding plus single-phase wound-rotor winding
3. Three-phase wound-rotor winding

The fundamental connections and functioning of the different types of auxiliary windings are illustrated in Fig. 7-97.

By means of proper design of a conventional squirrel-cage winding the torque characteristics can be varied over the limits given in Sec. 183. If higher starting torques with low starting kVA are required, a wound-rotor auxiliary winding is employed (see Fig. 7-97). For applications requiring a fairly high starting torque with a fairly low starting-inrush kVA, a single-phase wound-rotor winding is added to the conventional squirrel cage. For applications requiring exceptionally high starting torques with low starting kVA, a three-phase wound-rotor winding is employed. Motors with a wound-rotor auxiliary winding are started under full voltage with external resistance inserted in the auxiliary winding as in starting wound-rotor induction motors.

**187. The repulsion motor** (Fig. 7-98) has an armature which is like that of a dc machine. It is provided with two brushes $E_1$ and $E_2$, which are inclined to the axis of the stator winding. These brushes are connected by a low-resistance conductor $D$, which short-circuits the armature. A machine of this type has a high starting torque with a small starting current and a rapidly decreasing speed with increasing load, being similar in this respect to a dc series motor. The power factor of the repulsion motor increases with the speed, and near synchronous speed it attains a value much higher than is generally obtained in induction motors. These motors are used principally for constant-torque applications such as printing-press drives, fans, and blowers. The direction of rotation of a repulsion motor can be reversed by shifting the brushes to the reverse side of the neutral.

**188. The compensated repulsion motor** (Fig. 7-99) is similar to the repulsion motor, except that it has a compensating winding and two compensating brushes $C_1$ and $C_2$, which the repulsion motor does not have. The compensating winding is wound on the stator. The compensating brushes are connected in series with the compensating winding. These motors have a starting torque equal to about 2½ to 3 times full-load torque with approximately twice full-load current. Owing to the corrective action of the compensating winding, the power factor is very high at all loads. However, the efficiency of this machine is lower than that of the induction motor. Single-phase motors of this type are well adapted for loads involving heavy starting torque and sudden overloads.

**189. Brush-shifting type of single-phase motor.**   The brush-shifting single-phase repulsion motor, as manufactured by the General Electric Co., is a compensated repulsion motor. It is an adjustable varying-speed type of motor. Speed is adjusted by shifting the position of the brushes on the rotor. The speed may be controlled directly at the motor by means of a brush-shifting handle attached to the brush rigging. If the speed is to be controlled from some point remote from the motor, the brush-shifting handle can be connected to a pedestal or foot controller by means of a rocker-shaft assembly (bracket, rocker shaft, and levers). These motors can be obtained in ¼- to 3-hp sizes for either a reversing or a nonreversing service. The motor is reversed simply by shifting the brushes in the opposite direction from the neutral point. The brush-shifting motor is a constant-torque motor. The rated horsepower is the allowable horsepower for the highest speed. As the speed is reduced, the allowable horsepower is reduced in proportion. A speed reduction of 3:1 is possible when operating against a constant torque of 75 percent of the rated value. If operating against 100 percent normal torque, a speed reduction of 4:1 can be obtained. The no-load speed is approximately 60 percent above synchronous speed. At no load or with light loads very little speed reduction is possible. The starting torque of these motors with full voltage varies from 150 to 300 percent of full-load rated torque for four-pole machines and from 125 to 250 percent for six-pole motors, depending upon the brush position.

The breakdown torque is more than 400 percent of full-load rated torque when the brushes are in the maximum-speed position. As the brushes are shifted for the lower speeds, the breakdown torque decreases. These motors are generally started with full voltage.

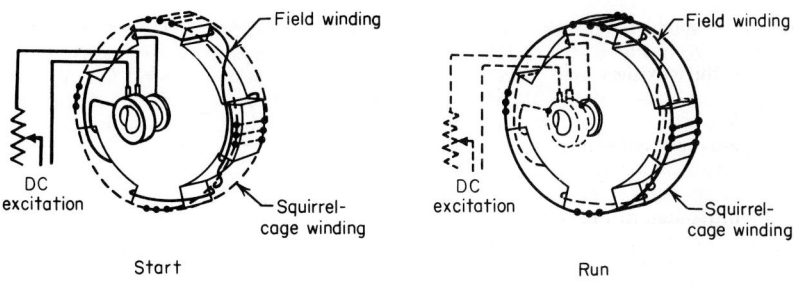

Start                                    Run

*Conventional design. Start:* Motors start like standard induction motors, by means of squirrel-cage windings embedded in the pole faces. *Run:* The motor comes up to speed and after a predetermined definite time, the dc field excitation is applied. This causes the motor to shift from induction to synchronous operation.

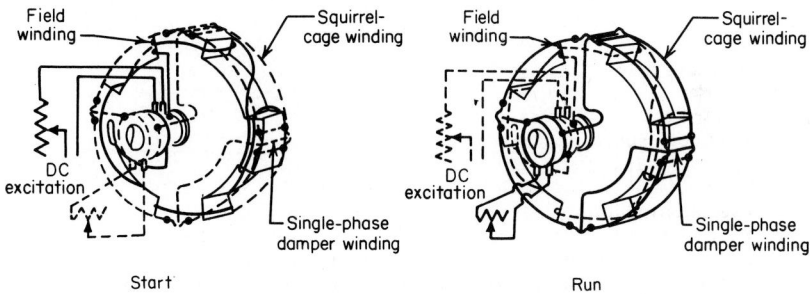

Start                                    Run

*High torque; low starting kVA. Start:* Motor gets high torques at start with low-kVA inrush, by a standard squirrel-cage winding plus a separate series winding on the pole faces connected to an external resistor. *Run:* The motor comes up to speed and after a predetermined definite time, the dc field excitation is applied. This causes the motor to shift from induction to synchronous operation.

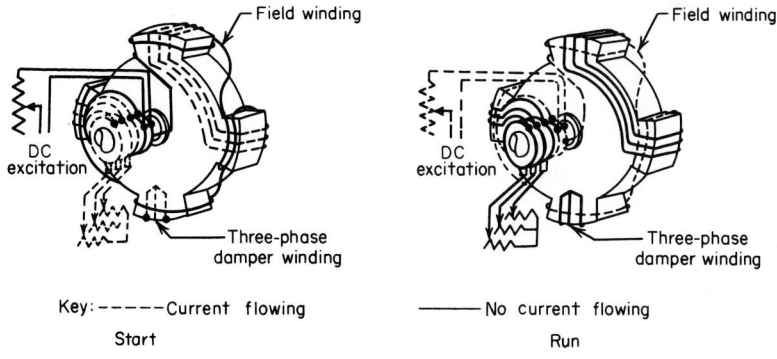

Key: - - - -Current flowing            ——— No current flowing

Start                                    Run

*High torque; starting kVA 50 percent or less of normal. Start:* Motor starts like any ordinary three-phase wound-rotor motor. The external resistor makes possible variable high torque together with extremely low starting kVA. *Run:* The motor comes up to speed and after a predetermined definite time, the dc field excitation is applied. This causes the motor to shift from induction to synchronous operation.

**Fig. 7-97**  Types of auxiliary windings for synchronous motors. [Westinghouse Electric Corp.]

**190. A single-phase induction motor,** when its rotor is not revolving, has no starting torque. After the rotor commences revolving, a continuous turning effort is exerted by a certain interaction of magnetic fields. Single-phase induction motors are provided, therefore, with auxiliary means for making them self-starting. Single-phase induction motors can be divided into the following groups, depending upon the method employed for starting:

1. Shading-coil
2. Inductively split-phase
3. Capacitor split-phase
4. Repulsion-start, induction-run
5. Repulsion-induction

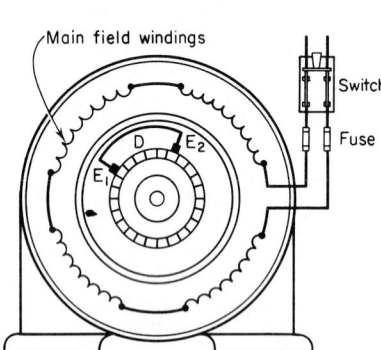

**Fig. 7-98**   Repulsion motor.

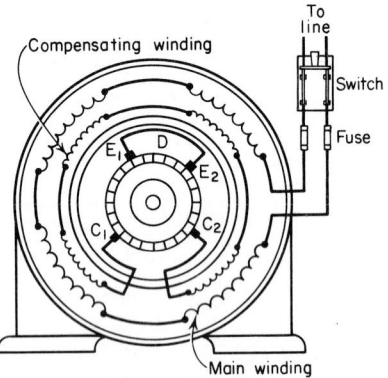

**Fig. 7-99**   Compensated repulsion motor.

**191. Shading-coil single-phase induction motors.** Some fractional-horsepower motors, principally those used on small ventilating fans, are made self-starting by the *shading-coil method.* The stator winding of these motors is not distributed around the frame, as in most induction motors, but is wound on projecting pole pieces, as in dc machines. In

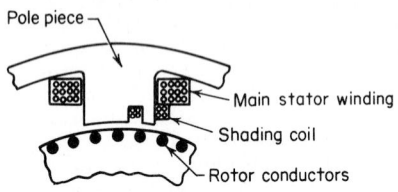

**Fig. 7-100**   Construction of a shading-coil motor.

addition to the main stator winding a short-circuited coil, or shading coil, as it is called, is placed around only a portion of the pole piece (Fig. 7-100). This shading coil makes the flux in the portion of the pole piece surrounded by it lag behind the flux in the other portion of the pole. The shifting of the flux thus produced will develop sufficient torque to start small motors under very light load conditions. These motors are made in sizes of from $\frac{1}{250}$ to $\frac{1}{40}$ hp and are used for driving small fans.

**192. The inductively split-phase single-phase induction motor** is provided with an auxiliary starting winding in addition to the main stator winding. This winding is located in stator slots so that it is displaced 90 electrical degrees from the main stator winding, as shown in Fig. 7-101. If two currents sufficiently out of phase with each other are passed through the main and auxiliary windings, conditions similar to those existing in a two-phase induction motor will be produced. The necessary two currents, which are from 30 to 45 degrees out of phase with each other, are obtained from the single-phase supply by constructing one of the circuits with a relatively high inductance and the other with a relatively high resistance. The main or running winding is normally of high inductance at the time of starting. The auxiliary winding is wound with finer wire, giving it higher resistance than the main winding. The running winding remains in the circuit at all times of operation, while the starting winding remains in the circuit only until the motor has reached 50 to 80 percent of synchronous speed. When the speed at which the starting winding should be cut out is reached, an automatic centrifugal switch operates and opens the starting circuit. This type of single-phase induction motor thus starts as a two-phase machine until the starting winding is cut out, when it continues to operate, by virtue of the running winding only, as a single-phase motor.

These motors are constant-speed machines, not adapted for any speed control. They are made in sizes up to ⅓ hp and in two types, the general-purpose type and the high-torque type. The general-purpose type has the following characteristics: full-voltage starting torque 75 to 175 percent of full-load torque, full-voltage starting current within NEMA

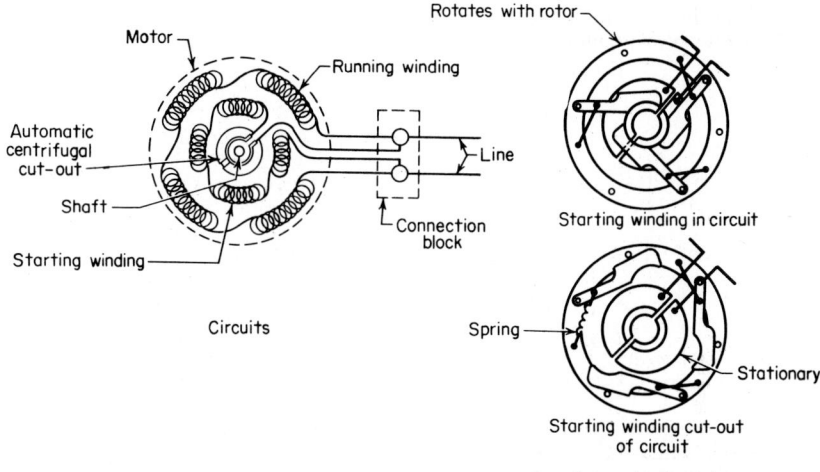

**Fig. 7-101**   Single-phase motor.

specifications of Sec. 200, breakdown torque 175 to 225 percent of full-load torque, and pull-up torque 75 to 200 percent of full-load torque. They are suitable for full-voltage starting on any power system.

The high-torque type of motor has the following characteristics: full-voltage starting torque 150 to 275 percent of full-load torque, full-voltage starting current approximately 850 percent of full-load current, breakdown torque 225 to 350 percent of full-load torque, and pull-up torque 225 to 325 percent of full-load torque. The full-voltage starting currents exceed NEMA specifications of Sec. 200 and may not be permissible for some installations.

**193. Capacitor motors are single-phase induction motors** which have two stator windings displaced 90 electrical degrees from each other just as the inductively split-phase motor does. Therefore their starting torque is developed by two-phase action. In the capacitor motor the necessary phase displacement between the currents of the two stator windings is produced by placing a capacitor in series with the auxiliary winding. By this means the phase displacement can be made more nearly 90 degrees, which results in better starting torque with lower starting current than can be obtained with the inductively split-phase–started motor. Capacitor motors are available in the following types:

  1. Capacitor-start, induction-run
  2. Capacitor-start, capacitor-run
    *a.* Two-value capacitor (high-torque)
    *b.* Permanent-split capacitor (low-torque)
  3. Multispeed, capacitor-start, capacitor-run

Capacitor motors have the advantages of quiet operation, high power factors, and reduction in radio interference.

**194. The capacitor-start, induction-run** motor starts by two-phase action. After the motor has reached approximately 75 percent of full-load speed, a centrifugal switch opens the auxiliary circuit, and the motor continues to run as an ordinary single-phase induction motor. A diagram for a motor of this type is shown in Fig. 7-102. These motors develop starting torques of 275 to 400 percent of rated full-load torque. The breakdown torque is from 200 to 300 percent of rated full-load torque. Starting currents are sufficiently low so

that the motors are started with full voltage. Motors of this type are made in sizes of from
⅛ to ¾ hp.

**195. Capacitor-start, capacitor-run** single-phase induction motors utilize both stator windings at all times. The motor secures its power from a single-phase supply but really operates when both starting and running as a two-phase motor. For the *two-value capacitor motor* a larger value of capacity is used during the starting period than is employed for running conditions. This larger capacity can be obtained by adding a capacitor (Figs. 7-103 and 7-104) or by changing the connections to an autotransformer which supplies the power to a single capacitor (Fig. 7-105). For small motors the change in the connections between starting and running is performed by a centrifugal switch (Fig. 7-103), while for larger motors a definite time relay is employed (Fig. 7-104). The two-value capacitor motor produces a high starting torque of 275 to 450 percent of rated full-load torque with starting currents that in most cases are sufficiently low so that the motors can be started with full voltage. Breakdown torques will be from 200 to 300 percent of rated full-load torque. These motors are available in sizes of from ⅛ to 10 hp.

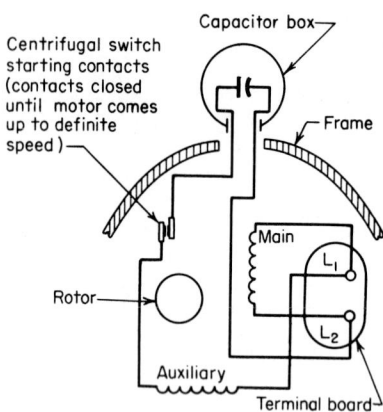

**Fig. 7-102**   Connection diagram of a capacitor-start, induction-run motor. [Wagner Electric Corp.]

The permanent-split-capacitor motor utilizes a single value of capacity that is the same for both starting and running (Fig. 7-106). There is no change in connections between the starting and running conditions. These motors have low starting torques of a value from 40 to 60 percent of rated full-load torque. The breakdown torques will be from 150 to 200 percent of rated full-load torque. The starting current will generally be sufficiently low for full-voltage starting. These motors are made in sizes of from ⅙ to 10 hp.

The principal advantage of capacitor-start, capacitor-run motors over capacitor-start, induction-run motors is that they operate at a higher power factor.

**196. Multispeed capacitor motors** are typically permanent-split-capacitor motors used for

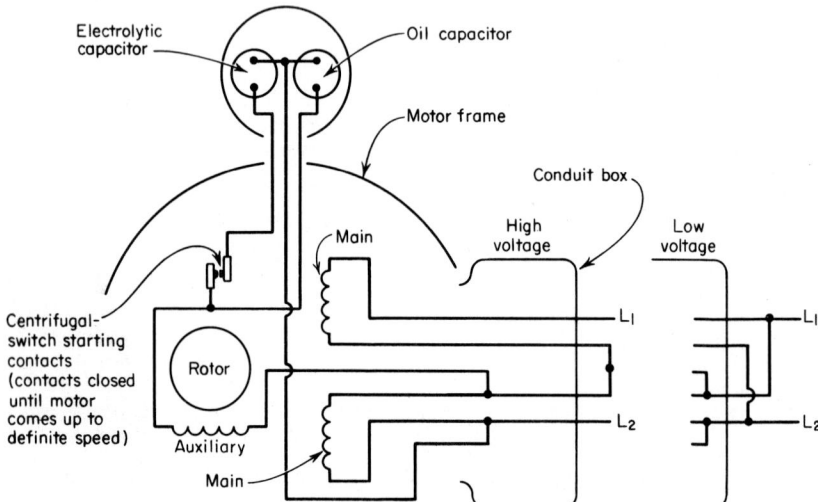

**Fig. 7-103**   Connection diagram for a two-value capacitor motor (capacitor-start, capacitor-run) with centrifugal switch. [Wagner Electric Corp.]

fan service in appliances. These motors have a tapped main winding operated at full voltage and are not intended for general use. Multispeed capacitor motors may also be single-value capacitor-start, capacitor-run motors equipped with a selector switch and an autotransformer for supplying different voltages to the main winding. They are suitable only for special classes of services, since the breakdown torque is reduced in proportion to the square of the voltage impressed on the main winding. It is possible to provide the motor with a centrifugal switch so that when properly connected the motor can be started only with full voltage impressed upon the main winding.

**197. Repulsion-start, induction-run single-phase motors.** Single-phase induction motors which are started by repulsion-motor action are constructed similarly to the straight-repulsion motor with the addition of a centrifugal device which shorts the commutator segments after the motor has come up to speed. The motor starts as a repulsion motor, but after it has come up to the required speed so that the commutator segments are short-circuited, it operates as a straight-induction motor. These motors are constant-speed machines not capable of speed control. The starting torques of different motors vary from 300 to 500 percent of full-load torque. The breakdown torque is at least 175 percent of full-load torque. Starting currents of these motors vary from 200 to 350 percent of full-load current, so generally they can be started directly across the line. When it is necessary or desirable to limit the starting current to smaller values, a resistance type of starter is employed. These motors are made in sizes of from ⅙ to 15 hp.

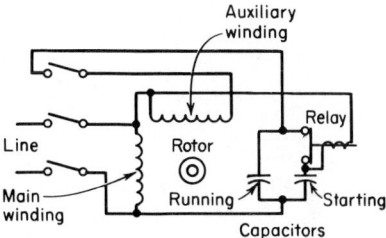

**Fig. 7-104** Connection diagram for a two-value capacitor motor (capacitor-start, capacitor-run) with acceleration relay. [General Electric Co.]

**198. The repulsion-induction motor** has two rotor windings. One of these windings is an ordinary induction rotor winding, and the other is a repulsion winding connected to a commutator with short-circuited brushes. There is frequently a compensating winding on the stator connected to brushes bearing on the commutator, as discussed in Sec. 188. This motor starts by repulsion-motor action. Since there is no short-circuiting device, after the machine has come up to speed, both the repulsion and the induction windings are active.

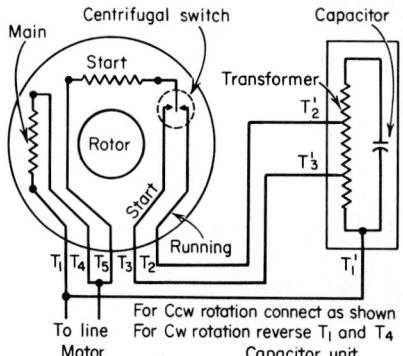
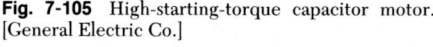

**Fig. 7-105** High-starting-torque capacitor motor. [General Electric Co.]

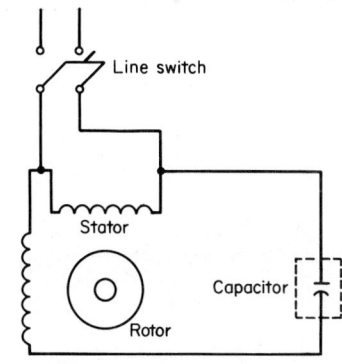

**Fig. 7-106** Permanent-split-capacitor motor. [General Electric Co.]

Under running conditions the motor functions as a combined induction and repulsion motor. The starting torque with rated voltage impressed for different sizes and speeds varies from 225 to 300 percent of full-load torque. There is no definite breakdown point, maximum torque occurring at standstill. The starting currents of these motors, when thrown directly across the line, are within the requirements of most power companies. When starting currents are to be limited to smaller values, a resistance type of starter is employed. The repulsion-induction motor is a constant-speed machine with a speed variation from no load to full load of about 6 percent. Its speed cannot be adjusted. These motors are made in sizes of from ½ to 10 hp.

**199. The single-phase series, or universal, motor** (Fig. 7-107) comprises an armature of the dc type connected in series with the field. It will operate successfully on either ac or dc circuits. The single-phase series motor finds its widest application in fractional-horsepower services such as those for fans, appliance motors, and electric tools. These machines have approximately the same speed-torque characteristics as dc series machines. In the larger capacities, single-phase series motors are equipped with windings which compensate for the armature reaction. The compensating winding is placed in the pole face and must be so connected that at any instant the current will flow in the same direction in all the conductors under any one pole. The winding can be connected in series with the armature. When so connected, it is called a conductively compensated winding. The compensating winding can be short-circuited on itself. When thus arranged, it receives its current by induction, and the machine is said to be inductively compensated.

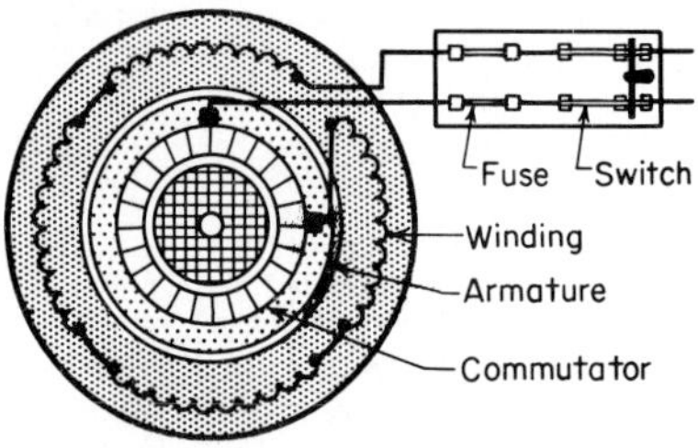


**Fig. 7-107** Series or universal motor.

**200. Small single-phase motors.**[1] Small single-phase motors (1 hp or less) were formerly called "fractional-horsepower" motors. NEMA standardization rules specify that the minimum locked-rotor torque of single-phase, general-purpose small motors, with rated voltage and frequency applied, shall be not less than the following:

| | Minimum locked-rotor, oz·ft | | | | | |
|---|---|---|---|---|---|---|
| | 60-Hz speed, rpm | | | 50-Hz speed, rpm | | |
| hp | 3600<br>3450 | 1800<br>1725 | 1200<br>1140 | 3000<br>2850 | 1500<br>1425 | 1000<br>950 |
| ⅛ | . . . | 24 | 32 | . . . | 29 | 39 |
| ⅙ | 15 | 33 | 43 | 18 | 39 | 51 |
| ¼ | 21 | 46 | 59 | 25 | 55 | 70 |
| ⅓ | 26 | 57 | 73 | 31 | 69 | 88 |
| ½ | 37 | 85 | 100 | 44 | 102 | 120 |
| ¾ | 50 | 119 | . . . | 60 | 143 | . . . |
| 1 | 61 | . . . | . . . | 73 | . . . | . . . |

NEMA standards also specify that the locked-rotor current of 60-Hz single-phase, general-purpose small motors shall not exceed the values given in the following table.

**Two-, Four-, Six-, and Eight-Pole, 60-Hz Motors, Single-Phase**

| | Locked-rotor currents, A | | | |
|---|---|---|---|---|
| | 115 V | | 230 V | |
| hp | Design O | Design N | Design O | Design N |
| ⅙ and smaller | 50 | 20 | 25 | 12 |
| ¼ | 50 | 26 | 25 | 15 |
| ⅓ | 50 | 31 | 25 | 18 |
| ½ | 50 | 45 | 25 | 25 |
| ¾ | . . . | 61 | . . . | 35 |
| 1 | . . . | 80 | . . . | 45 |

[1]This material is reproduced with the permission of the National Electrical Manufacturers Association from NEMA Standards Publication MG 1-1987, copyright © 1987 by NEMA.

**201. Single-phase medium motors.**[1]    Single-phase medium motors rated more than 1 hp were formerly called "integral-horsepower" motors. Single-phase medium motors are classified by NEMA into two design types, Design L and Design M. The distinction is made on the basis of the value of the locked-rotor current of the motor. NEMA standards specify that the locked-rotor current of single-phase, 60-Hz Design L and M motors of all types, when measured with rated voltage and frequency impressed and with the rotor locked, shall not exceed the following values:

| | Locked-rotor currents, A | | |
|---|---|---|---|
| | Design L motors | | Design M motors, 230 V |
| hp | 115 V | 230 V | |
| ½ | 45 | 25 | . . . |
| ¾ | 61 | 35 | . . . |
| 1 | 80 | 45 | . . . |
| 1½ | . . . | 50 | 40 |
| 2 | . . . | 65 | 50 |
| 3 | . . . | 90 | 70 |
| 5 | . . . | 135 | 100 |
| 7½ | . . . | 200 | 150 |
| 10 | . . . | 260 | 200 |

In accordance with NEMA standards the locked-rotor torque of single-phase, 60-Hz, general-purpose medium motors, with rated voltage and frequency applied, shall be not less than the following:

| | Minimum locked-rotor torque, lb·ft | | |
|---|---|---|---|
| | rpm | | |
| hp | 3600 | 1800 | 1200 |
| ¾ | . . . | . . . | 8.0 |
| 1 | . . . | 9.0 | 9.5 |
| 1½ | 4.5 | 12.5 | 13.0 |
| 2 | 5.5 | 16.0 | 16.0 |
| 3 | 7.5 | 22.0 | 23.0 |
| 5 | 11.0 | 33.0 | . . . |
| 7½ | 16.0 | 45.0 | . . . |
| 10 | 21.0 | 52.0 | . . . |

**202. Methods of reversing motors.**    In the following paragraphs instructions are given for reversing the different types of motors.

DIRECT-CURRENT SHUNT OR SERIES MOTORS    Interchange the connections of either the field or the armature winding. Reversing the line leads will not change the direction of rotation. If the motor is not equipped with commutating poles, the brushes should be shifted to the same position on the opposite side of the neutral axis.

DIRECT-CURRENT COMPOUND MOTORS

*Method 1.* Interchange the connections of the two armature leads.

*Method 2.* Interchange the connections of the two shunt-field leads and also the connections of the two series-field leads. Interchanging the connections of just the two shunt-field leads would change the direction of rotation, but the motor would operate as a differential compound motor instead of cumulatively. If the motor is not equipped with commutating poles, the brushes should be shifted to the same position on the opposite side of the neutral axis.

ANY FOUR-WIRE TWO-PHASE INDUCTION MOTOR    Interchange the connections to the line of the two leads of either phase.

[1]This material is reproduced with the permission of the National Electrical Manufacturers Association from NEMA Standards Publication MG 1-1987, copyright © 1987 by NEMA.

ANY THREE-WIRE TWO-PHASE INDUCTION MOTOR    Interchange the connections to the line of the two outside wires.

ANY THREE-PHASE INDUCTION MOTOR    Interchange the connections to the line of any two leads. In reversing the commutator type of polyphase induction motor the brush-shifting mechanism must be adjusted to the proper position for the reversed rotation in addition to reversing the connections of two of the line leads.

SYNCHRONOUS MOTORS    Ordinary synchronous motors are reversed in the same manner as induction motors.

SINGLE-PHASE INDUCTION, SHADING-POLE STARTING    The direction of rotation cannot be reversed except in special cases. If the pole pieces of the machine can be removed, the direction of rotation can be reversed by revolving the pole pieces 180° on their own axes. This will place the shading coils on the opposite sides of the pole pieces. Some special motors are built with shading coils on both sides of the pole pieces. With these machines, each shading-coil circuit must be provided with a switch. The direction of rotation depends on which shading-coil circuit is closed.

SINGLE-PHASE INDUCTION, INDUCTIVELY SPLIT-PHASE STARTING    Interchange the connections of either the main or the auxiliary winding.

SINGLE-PHASE CAPACITOR MOTOR    Interchange the connections of either the main or the auxiliary winding.

SINGLE-PHASE INDUCTION, STARTED BY REPULSION ACTION    Shift the brushes to the same position on the opposite side of the neutral axis.

SINGLE-PHASE REPULSION MOTOR    Shift the brushes to the same position on the opposite side of the neutral axis.

SINGLE-PHASE SERIES MOTOR    It is reversed in the same manner as a dc series motor.

**202A. Electric-motor braking.**    Braking of electric motors can be performed by friction braking, plugging, dynamic braking, or regenerative braking.

*Friction braking* of electric motors consists of retarding the motor speed or of completely stopping the motor by a solenoid-operated friction brake of the shoe or disk type.

*Plugging* is the method of retarding the motor speed and completely stopping the motor or of reversing the motor by applying electric power in the reverse direction so that the motor develops torque in the direction opposite to that of rotation. Plugging will quickly bring a motor and its driven load to a stop. If the reverse power is not removed when the motor comes to a stop, the motor will reverse and accelerate in the opposite direction.

*Dynamic braking* is the method of retarding the motor speed or of completely stopping the motor through making the motor act as a generator with a resistance load.

*Regenerative braking* is the method of retarding the motor speed by making the motor, functioning as a generator, feed its generated power back into the supply line. Regenerative braking will not stop the motor. It is effective only for the braking of overhauling loads. Overhauling loads are those which attain a speed greater than that of the motor speed-load characteristic and under which, therefore, the load tends to drive the motor.

**203. Plugging of motors.**    To employ plugging retardation of an electric motor, a reversing switch or contactor must be provided in the control equipment. To stop or retard the speed of the motor by plugging, the connections to the power supply are simply reversed in the proper manner so that the motor develops torque in the opposite direction to that in which it is rotating. For squirrel-cage induction motors, the braking action can be cushioned to reduce the shock to the machine and power system by the use of resistance connected in series with the reversed connections to the supply. For wound-rotor induction motors, the plugging torque can be controlled by adjusting the resistance in the motor secondary circuit. For synchronous motors, the dc field circuit is deenergized during the plugging operation, and the motor is plugged in the same manner as a squirrel-cage induction motor. For dc motors, additional resistance must be used in series with the reversed power connections. When the motor must come to a stop without reversal, a zero-speed switch or a timing relay must be used to open the motor circuit when the machine comes to a stop.

ADVANTAGES

    1.  Simple arrangement and installation

    2.  Fast operating performance

    3.  Freedom to rotate after stopping in contrast to braking with a mechanically set brake

    4.  No friction parts and freedom from maintenance

    5.  Suitability for large motors and severe duty

DISADVANTAGES

1. Reversing controller required
2. Zero-speed switch or consistent timing relay required if reversing is to be prevented
3. Availability of power required (plugging impossible if power fails)
4. Power system capable of supplying the peak plugging current required
5. Special bracing possibly required for motor windings

If motor and its load must positively be held stationary when the motor is deenergized, an additional spring-set or other type of mechanical brake is necessary.

**204. In dynamic braking,** the armature of the motor is disconnected from the power supply, and the load will drive the motor as a generator. The generator action is loaded either through resistance banks or through part of the motor winding. For dc shunt or compound motors, the armature is disconnected from the line and connected to a resistor bank; the shunt field is left connected to the supply line. For dc series motors, the machine is disconnected from the supply, and the armature and series-field circuits are segregated from each other. The field winding is connected in series with the proper limiting resistance to the supply line. The armature is connected to a resistor bank load. Emergency dynamic braking of series motors can be accomplished by disconnecting the motor from the line, reversing the interconnection between the series field and the armature, and then connecting the two in series to a resistor bank load.

For dynamic braking of squirrel-cage induction motors, a source of excitation must be provided. This excitation may be provided (1) from a separate source of direct current, (2) from the ac supply through a rectifier, or (3) by means of capacitors. In the first method, dynamic braking is accomplished by disconnecting the motor from the supply and connecting one phase of the motor primary to a separate source of direct current. The motor will then be driven by the load as a generator and will load itself through the induced current in the motor secondary flowing through the squirrel-cage winding. The dc excitation must be removed at the end of the braking period; otherwise the windings will overheat. In the second method, dynamic braking is accomplished in the same manner as in the first method except that the dc supply for the excitation is obtained from the ac supply through a single-phase rectifier. The rectifier is usually of the contact, or barrier-layer, type. A diagram of the connections required is shown in Fig. 7-108. In the third method, a capacitor is permanently connected across the motor terminals. In braking, the motor is disconnected from the supply. The capacitor will supply the motor with excitation so that it will function as an induction generator. The braking action is produced by the power losses inside the machine. The braking action can be improved by connecting a loading resistor in parallel with the capacitor. The larger the capacitor, the greater the braking action. All braking action ceases with this method at about one-third of synchronous speed.

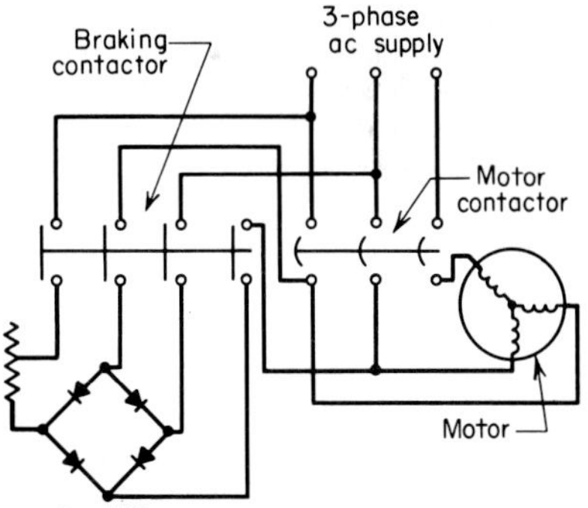

For wound-rotor induction motors, dynamic braking could be accomplished by all three of the methods just described for squirrel-cage motors. However, the first method, using a separate source of direct current for excitation, is usually the only feasible one. The amount of the braking action can be controlled by adjusting the secondary

**Fig. 7-108**  Dynamic braking of an induction motor with rectified direct current.

resistance. For synchronous motors, dynamic braking is accomplished by disconnecting the motor armature from the ac supply and connecting the armature to a resistor bank load. The field is left excited from direct current.

ADVANTAGES

1. The motor cannot be reversed. No zero-speed switch or other antireversing device is required.
2. There is smooth, positive retardation lacking the shock of a mechanically set brake and less severe than plugging.
3. The braking torque is easily adjustable by varying the dc excitation or, as in wound-rotor induction-motor applications, by varying the motor secondary resistors.

4. In dynamic braking of small ac motors by the use of capacitors, the braking is independent of the power supply; furthermore, the capacitor furnishes a power-factor correction while the motor is operating.

DISADVANTAGES

1. A source of dc excitation is required during the braking period except in capacitor braking.

2. Braking is not available if the exciting source fails.

3. This method is more expensive and more complicated than plugging. If the source of excitation is not already available and additional equipment must be installed, the extra cost may be prohibitive.

4. Control must provide means for disconnecting the excitation at the expiration of the braking period.

5. Dynamic braking is useless as a holding force after motor has come to rest and is deenergized.

6. Capacitor dynamic braking is ineffective and ceases to exist below approximately one-third speed; therefore, it is of little value except on drives involving considerable friction. The cost of capacitor braking is prohibitive with large motors.

7. Rectifier-excited dynamic braking is impracticable and expensive in the case of large motors.

**205. Regenerative braking** cannot be used for stopping a motor. It will produce braking action only when the load overhauls the motor and drives the motor at a speed above the speed-load characteristic of the motor. When a load overhauls its drive motor (except for series dc motors), the voltage generated in the machine will be greater than the supply-line voltage. Consequently the motor will function as a generator and feed power back into the supply line. This is an inherent characteristic of the motor and does not require any changing of connections or control equipment. If a series dc motor is overhauled by its load, the motor loses its excitation and becomes incapable of regenerative braking unless the connections are changed. Therefore, to use regenerative braking on a series motor the series field and armature must be segregated from each other and connected individually to the supply. Sufficient resistance must be connected in series with the series field so that the current will be held to normal value.

ADVANTAGES

No extra equipment is required except for series-wound dc motors. This method can be made effective with dc motors at reduced speeds by strengthening the motor field, i.e., by lowering the speed at which the motor will regenerate.

DISADVANTAGES

This method is effective only at overhauling speeds; it is useless as a means of stopping. If dc motors must be made to regenerate at low speeds, the control becomes involved and expensive. The motor may require special fields.

## CONVERSION EQUIPMENT

**206. The conversion of alternating current to direct current** is usually effected with a motor-generator set, a synchronous converter, an electronic rectifier, a contact (barrier-layer) rectifier, or solid-state semiconductor rectifiers.

**207. A motor-generator set** consists of a motor mechanically coupled to and driving one or more generators. The motor is usually alternating-current and is mounted on the same bedplate and on the same shaft with the generator or generators which it drives. A motor-generator set can be employed to convert electrical energy from one voltage or frequency to another voltage or frequency, to convert alternating current to direct current, or to convert direct current to alternating current. When an ac motor is used, it may be of either the induction or the synchronous type. The induction type has the advantages of low first cost and rugged construction. It may also be wound for voltages as high as 13,500 V. This will, in many cases, eliminate the necessity for transformers. This type has, however, poorer speed regulation and a lower power factor. A synchronous-motor drive is probably more frequently used for large-capacity installations than is the induction motor. The synchronous motor is inherently a constant-speed motor. Therefore a more nearly constant voltage can be maintained at the generator terminals than by the use of an induction motor. The field of the synchronous motor may be overexcited. The motor then acts as a synchronous condenser, thus improving the power factor.

**208. A synchronous converter** (sometimes called a *rotary converter*) for converting alternating to direct current or vice versa consists (Fig. 7-109) of a dc motor equipped with slip rings. Taps are brought out from equidistant points on the armature winding and connected to these rings. As ordinarily used, the machine is then driven as a synchronous motor, by an alternating emf which is impressed on the armature through the slip rings. As the armature rotates, an alternating emf is induced in the armature winding, as in any dc generator. This alternating induced emf is rectified by the commutator and leaves the brushes as a pulsating or direct current. If a synchronous converter is utilized for converting alternating current to direct current, it is called a *direct converter,* or merely a converter. If it converts direct current to alternating current, it is called an *inverted converter.* If the main-pole flux distribution is such that the induced emf has a sine waveform (see Table 209 for values),

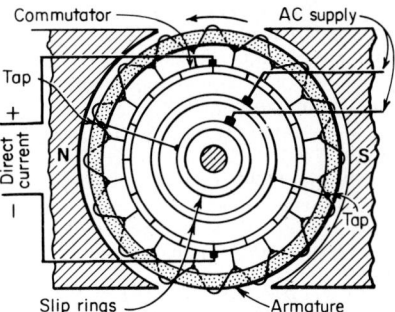

**Fig. 7-109**  Two-pole single-phase synchronous converter.

$$\frac{E_{ac}}{E_{dc}} = 0.707 \left( \sin \frac{180}{n} \right) \qquad (6)$$

where $E_{ac}$ = the effective alternating-current voltage, in volts between adjacent slip rings; $E_{dc}$ = the direct-current voltage, in volts; and $n$ = the number of slip rings for a bipolar machine or the number of taps per pair of poles for a multipolar machine.

**209.  Ratio of the Effective AC to DC Voltage with Different Numbers of Taps**

| Number of taps per pair of poles | Number of phases | $\dfrac{E_{ac}}{E_{dc}} \left( \begin{matrix} \text{between ad-} \\ \text{jacent taps} \end{matrix} \right)$ | $\dfrac{E_{ac}}{E_{dc}}$ (line voltage) |
|:---:|:---:|:---:|:---|
| 2 | 1 | 0.707 | 0.707 |
| 3 | 3 | 0.612 | 0.612 |
| 4 | 4 or 2 | 0.500 | 0.707 |
| 6 | 6 | 0.354 | $\begin{cases} 0.707 \text{ (diametrical)} \\ 0.612 \text{ (double } \Delta) \end{cases}$ |
| 12 | 12 | 0.183 | 0.682 (double chord) |

Note.  Synchronous converters are subject to hunting.  The cause of hunting and the remedy are the same as for a synchronous motor.

**210. The methods of starting a synchronous converter** are (1) from its dc side as a shunt motor, (2) by means of an auxiliary motor mounted on its shaft, and (3), if polyphase, from its ac side as an induction motor. Whatever method is used, the armature must be brought up almost to synchronous speed before the load is applied to the generator.

**211.  Usual Relation among Number of Poles, Phases, Taps, and Slip Rings of a Synchronous Converter**

| Phases | 2 pole | | 4 pole | | 6 pole | | 8 pole | | 12 pole | |
|:---:|:---:|:---:|:---:|:---:|:---:|:---:|:---:|:---:|:---:|:---:|
| | Slip rings | Taps | Slip rings | Taps | Slip rings | Taps | Slip rings | Taps | Slip rings | Taps |
| 1 | 2 | 2 | 2 | 4 | 2 | 6 | 2 | 8 | 2 | 12 |
| 2 | 4 | 4 | 4 | 8 | 4 | 12 | 4 | 16 | 4 | 24 |
| 3 | 3 | 3 | 3 | 6 | 3 | 9 | 3 | 12 | 3 | 18 |
| 6 | 6 | 6 | 6 | 12 | 6 | 18 | 6 | 24 | 6 | 36 |

Note.  The tendency of modern practice in large three-phase conversion installations is to use a six-phase converter.  It is connected to the three-phase line through a set of transformers.

**212. The relation between the values of direct and alternating current of any synchronous converter** is

$$I_{dc} = 0.354 \times I_{ac} \times n \times \text{power factor} \times \text{efficiency} \qquad (7)$$

where $I_{dc}$ = direct current, in amperes; $I_{ac}$ = effective alternating current, in amperes; and $n$ is as given in Eq. (6).

**213. A frequency-changer motor generator** consists of an ac generator or generators driven by an induction or a synchronous motor or motors. The frequency of the generator is different from that of the motor. Since the speed regulation of the induction motor is inherently poorer, a synchronous driving motor is usually applied. The output frequency of a set which is driven by a synchronous motor will vary only with the input frequency and not with the load. The synchronous motor can also be utilized for improving the power factor of the energy-supplying system. Since the motor and the generator are mounted on the same shaft and hence rotate at the same speed, the desired frequency change must be obtained by a proper ratio between the number of motor poles and the number of generator poles. The ratio between the frequency applied to the motor and the frequency which the generator will supply is exactly equal to the ratio between the number of poles on the motor and the number of poles on the generator. That is,

$$\frac{\text{Number of poles on motor}}{\text{Number of poles on generator}} = \frac{\text{frequency applied to motor}}{\text{frequency supplied by generator}} \qquad (8)$$

For 60- to 25-Hz conversion this practically limits the speed to 300 rpm with 24 poles on the 60-Hz machine and 10 poles on the 25-Hz machine.

## TROUBLES OF AC MOTORS AND GENERATORS: THEIR LOCALIZATION AND CORRECTION

**214. Troubles of ac machinery.**  Much of the material under this heading is based on that in the book *Motor Troubles,* by E. B. Raymond.

**215. Induction-motor troubles** (H. M. Nichols, *Power and the Engineer*). The author asserts that the unsatisfactory operation of an induction motor may be due to either external or internal conditions. The voltage or the frequency may be wrong, or there may be an overload on the machine. Low voltage is the most frequent cause of trouble. The starting current sometimes amounts to twice the running current, with the result that the voltage is particularly low at starting. The best remedy for this disorder is larger transformers and larger motor leads, one or both. The troubles that occur most frequently within the motor itself are caused by faulty insulation and by an uneven air gap due to the springing of the motor shaft or to excessive wear in the bearings. If a wound-rotor machine refuses to start, the trouble may be due to an open circuit in the rotor winding. Most wound-rotor motors designed to employ a starting resistance will not start at all if the resistance is left out of the secondary circuit with the secondary circuit open. A short-circuited coil in the motor will make its existence known by local heating in the latter.

**216. Troubles of ac generators** *(Westinghouse Instruction Book).*  The following causes may prevent ac generators from developing their normal emf:

1. The speed of the generator may be below normal.

2. The switchboard instruments may be incorrect, and the voltage may be higher than that indicated, or the current may be greater than is shown by the readings.

3. The voltage of the exciter may be low because its speed is below normal, its series field is reversed, or part of its shunt field is reversed or short-circuited.

4. The brushes of the exciter may be incorrectly set.

5. A part of the field rheostat or other unnecessary resistance may be in the field circuit.

6. The power factor of the load may be abnormally low.

**217. Causes of shutdowns of induction motors.**  Sometimes there is trouble from blowing fuses. Or possibly, and more serious, the fuses do not blow, and the motor, perhaps humming loudly, comes to a standstill. Under these conditions the current may be 10 times normal, so that the heating effect, being increased as the square of the current or a hundredfold, causes the machine to burn out its insulation.

Since the torque, or turning power, of an induction motor is proportional to the square of the applied voltage (one-half voltage produces only one-quarter torque), it is evident that lowering the voltage has a decided effect upon the ability of the motor to carry load and may be the cause of its stopping. Another cause may be that the load on the motor more than equals its maximum output.

The bearings may have become worn, so that the air gap, which ordinarily is not much over 0.040 in (1 mm) and on small motors may be as small as 0.015 in (0.38 mm), has been gradually reduced at the lower side of the rotor to practically zero. The rotor commences to rub on the stator. The friction soon becomes so great that it is more than the motor can carry, and the motor shuts down.

A shutdown may be due to the bearings' introducing excessive friction. Hot bearings, in turn, may be due to excess of belt tension, dirt in the oil, oil rings not turning, or improper alignment of the motor to the machine that it drives. Under such conditions, ascertain whether the voltage has been normal, whether the air gap is such that the armature is free from the field, and whether the load imposed upon the motor is more than that for which it was designed. In any installation a system whereby an inspector will examine the gap, bearings, etc., periodically should be arranged.

Rarely, shutting down may be due to the working out of the starting switch, which may be located within the armature. Such a switch is operated by a lever engaging a collar that bears on contacts which, as they move inward, cut out the resistance that is in series with the rotor winding and is located within it.

If the short-circuiting brushes work back, introducing resistance into the armature circuit while the machine is trying to carry load, the machine will at once slow down and probably stop, usually burning out the starting resistance. Of course, this can occur only from faulty construction. The remedy is to fit the brushes properly so that they will not work out. It is well to inspect them at the time of air-gap inspection.

**218. Low torque while starting induction motors.** Even when the circuit to the motor is closed, it sometimes does not start. The same general laws of voltage, etc., apply to the motor at starting as when it is running. Hence, the points mentioned in Sec. 217 should be investigated and, if necessary, corrected. The resistance, which is frequently inserted in the armature, may be short-circuited, thus giving a low starting torque. To obtain a proper starting torque with a reasonable current with a wound-rotor motor, a resistance must be inserted in the rotor circuit. The resistance not only limits the current, which would be large with the motor standing still, but causes the current of the armature to assume a more effective phase relation, so that a far larger torque is obtained with the same current. A partial or complete short circuit of the resistance will reduce the starting torque.

**219. Low maximum output of induction motors.** The maximum load which a motor can carry may be less than desired or less than the nameplate indicates. This may be due to improper interconnection of the motor windings or to low line voltage. If the line voltage is considerably below the rated voltage of the motor, the maximum torque that the motor can develop may be so reduced that the motor will stall on small overloads. Many motors are designed for operation at two voltages, for example, 115/230 or 230/460. This is accomplished by dividing the stator winding into two parts. For operation on the low-voltage rating, the two parts of the winding should be connected in parallel, while for operation on the high-voltage rating the two parts should be connected in series. Sometimes, when a motor is put in service on a supply which has a voltage corresponding to the low-voltage rating of the motor, the windings are connected in series. With this connection the motor would be operating at half rated voltage. This wrong connection will be indicated by the motor's starting very slowly or failing to start or by its stalling without overheating and without tripping the overcurrent devices.

**220. Winding faults of induction motors.** It sometimes happens that when an attempt is made to operate a new induction motor, although it will start, the currents are excessive and unbalanced, undue heating appears, or a peculiar noise is emitted, accompanied possibly by dimming of the lights on the same circuit and the lowering of speed, with perhaps an actual shutdown of other induction motors thereon. If, after examination, no difficulty is found with the air gap, belt tension, starting resistance, or bearings, the probabilities are that the coils of the motor have been wrongly connected or that the winding has been damaged during transportation. Certain indications of these conditions are shown by instrument readings. The winding faults in a three-phase motor may be:

1. One coil of the rotor may be open-circuited. The armature or the rotor may have a defective winding, just as may the stator.

2. Two coils or phases of the armature may be open-circuited.

3. The armature may be connected properly, but the stator coil or phase may be reversed.

4. Part of the stator may be short-circuited.

5. One phase of the stator may be open-circuited.

**221. With an open circuit in the field or stator** in a three-phase motor, the current would flow only in two legs. There would be no current in the other leg, and the motor would not start from rest with all switches closed. However, a three-phase or two-phase motor will run and do work single-phase if it is assisted in starting. The starting torque is zero, but as the speed increases, the torque increases.

With a small motor, giving a pull on the belt will introduce enough torque so that the motor will pick up its load. Therefore while an open circuit in the field winding should be found and repaired, if there is not time for repairs, the motor can be operated single-phase to about two-thirds of normal load. The power-factor conditions and effects on the rest of the circuit are practically no worse than when the motor is running three-phase. The torque of a 1-hp three-phase induction motor from rest to synchronism, when running single-phase, is indicated in Fig. 7-88.

**222. Squirrel-cage armature or rotor troubles.** Unusual operation due to reversals of phase, phases being open-circuited, and other causes occur with squirrel-cage armatures as well as with wound armatures. Pool soldering of the armature bars may be the cause. Sometimes a solder flux that will ensure proper operation for a while can be used, but time will develop poor electrical contacts because of chemical action at the joints. If the resistances of all the squirrel-cage joints are uniformly high, the effect is simply like that of an armature having a high resistance, which causes a lowering of the speed and local heating at the joints. If some of the joints are perfect but some bad, the motor may not have the ability to come up to speed, and there will be unbalanced currents.

**223. Effects of unbalanced voltages on induction motors.** The maximum output of a polyphase induction motor may be materially decreased if the voltages impressed on the different phases are unequal. On a three-phase system the three voltages between the legs 1–2, 2–3, and 1–3 should be approximately equal. Also on a two-phase system the voltage 1–2 should equal the voltage 3–4. If these voltages impressed on the induction motor are not equal, the maximum output of the motor as well as the current in the various legs is proportionately affected.

For example, with a two-phase motor, if the voltages in the two legs differ by 20 percent, a condition sometimes met in normal practice, the output of the motor may be reduced by 25 percent. Then, instead of being able to give its maximum output of, say, 150 percent for a few moments, it will give only 112 percent. The varying loads which the motor may have to carry may shut it down. In cases of low maximum output, the relative voltages on the various legs should always be investigated. If they vary, the trouble may be due to this variation.

In addition to the effect on the maximum output, the unequal distribution of current in a two-phase motor under such conditions may be quite serious. Consider a specific case of a 15-hp, six-pole, 1200-rpm, 230-V motor with the voltage on one leg 230 and the voltage on the other leg 180; current in leg 1 was 60 A and in leg 2 was 35 A at full load. The normal current at full load was 35 A. Thus the fuse might blow in the phase carrying the high current, causing the motor to run single-phase. If an attempt were made to start the motor, the blown fuse not being noticed, there would be no starting torque.

When unbalanced voltages are detected, make a careful check for a blown fuse. Often a service or feeder fuse will open and go undetected except for slight unbalanced line-to-line voltages. This occurs on installations such as sewer treatment plants, where some large pump, vacuum, or blower motors run continuously. If a "main" fuse opens while one of these motors is running, the motor itself generates the "third" phase, and other motors will start and run at reduced efficiency. If three-phase motors are protected by three overload relays, the large line current in one motor lead will cause the circuit to open. Accordingly, check for open fuses in the feeder or service switch if properly sized motor-overload devices trip out frequently.

If only two overload protective devices are provided for three-phase motors, the excessive current could be in the unprotected leg and cause a motor-winding failure.

If no fuses have been detected and unbalanced line-to-line voltages exist, check to see whether single-phase loads connected to the three-phase system are properly balanced.

After checking for open fuses and balanced single-phase loads, call the electric power company if unbalanced voltages appear on the line side of the main service switch.

**224. Induction-motor starting-compensator troubles.** Sometimes a mistake is made in the connections to the compensator, so that full voltage is used at starting and lesser voltage after throwing over the switch. Then the motor at starting takes excessive current, and since the maximum output is in proportion to the square of the voltage, the motor capacity is much reduced when it is apparently running in the operating position. Such action, therefore, can usually be accounted for by a wrong connection in the compensator. Sometimes a motor connected to a compensator takes more current at starting than it should; under this condition a lower tap should be tried. Compensators are usually supplied with various taps, and the one which produces the least disturbance on the line, giving at the same time the desired starting torque on the motor, should be selected.

When a motor, having been connected to a compensator, will not start, the cause may be entirely in the compensator. The compensator may have become open-circuited owing to a flash within. The switch may have become deranged so that it will not close, or a connection within the compensator may have loosened. Possibly, when a motor will not start when it is connected to a compensator just installed, one secondary coil of the compensator may be improperly interconnected.

**225. Induction-motor collector-ring troubles.**  It is essential that the contact of the brushes on the collector rings be good; otherwise contact resistance will be so great as to slow the motor down and to cause heating of the collector itself. This effect is particularly noticeable when carbon brushes are used. The contact resistance of a carbon brush under normal operation pressure and carrying its usual density of current (40 A/in², or 6.2 A/cm²) is 0.04 Ω/in² (0.258 Ω/cm²). Thus, under normal conditions, the drop is 0.04 × 40, which equals 1.6 V. If the contact is only one-quarter of the surface, this drop would be 6.4 V and might materially affect the speed of the motor. Thus, if the speed is farther below synchronous speed than it should be, an investigation of the fit of the brush upon the collector may show up the trouble.

If copper brushes are used, this trouble is much less liable to occur, since the drop of the voltage due to contact resistance when running at normal density (150 A/in², or 23.25 A/cm²) is only one-tenth that of carbon. The same trouble may occur owing to the pigtail, which is generally used with carbon brushes, making poor contact with the carbon, which gives the same effect as a poor contact with the collector itself.

**226. Improper end play in induction motors.**  Induction motors are so designed that the revolving parts will play endwise in the bearing, $\frac{1}{16}$ in (1.6 mm) or so. If in setting up the machine the bearings so limit this end action that the rotor does not lie exactly in the middle of the stator, a strong magnetic pull tends to center the rotor. If the bearings will not permit this centering, the thrust collars must take the extra thrust, which in an induction motor is considerable. If in addition to the magnetic thrust the belt pull is such as also to draw in the same direction, the trouble is increased. The end force may be such as to heat the bearing excessively and to cause cutting, soon rendering the motor inoperative.

In case of trouble with bearings, the end play should be tested by pushing against the shaft with a small piece of wood, placed on the shaft center. With the machine operating under normal conditions there should be no particular difficulty in pushing the shaft first one way from one side and then the other way from the other side. If it is found that the revolving part is hugging closely against one side, the trouble can be corrected either by pressing the spider along the shaft in a direction toward which the hugging is occurring or by driving the tops of the lamination teeth in the same direction. With a wooden wedge the tops of the teeth can often be driven over $\frac{1}{8}$ to $\frac{3}{16}$ in (3.2 to 4.8 mm) without any difficulty. This movement will usually correct the trouble. Driving the teeth of the stator $\frac{1}{8}$ in or so in the opposite direction to that of the end thrust will usually accomplish the same result. It is best to choose the teeth (stator or rotor) which are most easily driven over. The thin, long ones move more easily than do the short, broad ones.

**227. Oil leakage of induction motors.**  Sometimes a bearing will permit oil to be drawn out, perhaps a very little at a time. Ultimately enough will accumulate to show on the outside or on the windings of the machine. Although a motor will run for a period with its windings wet with ordinary lubricating oil without apparently being injured, insulation soaked with oil will deteriorate and eventually fail.

One of the principal causes is a suction of the oil due to the drafts of air from the rotor, and one of the best methods of stopping the trouble, under ordinary conditions, is to cut grooves as shown in Fig. 7-110 at $B$ and $D$. These grooves on a 50-hp motor may be $\frac{1}{8}$ in (3.2 mm) deep and $\frac{3}{16}$ in (4.8 mm) wide. Each groove has three holes drilled through the bearing shell to convey the oil collected by the grooves into the oil well. These grooves are just as effective with a split as with a solid bearing. It is impossible here to discuss the various causes of oil leakage. The grooves as suggested are a general remedy and cover many cases.

**228. General summary of synchronous-motor troubles.**  Failure of a synchronous motor to start is often due to faulty connections in the auxiliary apparatus. This should be carefully inspected for open circuits or poor connections. An open circuit in one phase of the motor

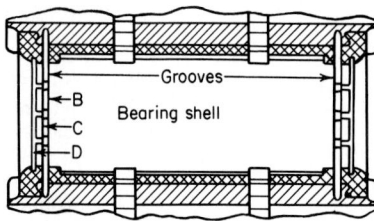

**Fig. 7-110**   Grooves to prevent oil leakage.

itself or a short circuit will prevent the motor from starting. Most synchronous motors are provided with an ammeter in each phase so that the last two causes can be determined from their indications: no current in one phase in case of an open circuit and excessive current in case of a short circuit. Either condition will usually be accompanied by a decided buzzing noise, and a short-circuited coil will often be quickly burned out. The effect of a short circuit is sometimes caused by two grounds on the machine.

Starting troubles should never be assumed until there has been a trial to start the motor light, i.e., with no load except its own friction. It may be that the starting load is too great for the motor.

If the motor starts but fails to develop sufficient torque to carry its load when the field circuit has been closed, the trouble will usually be found in the field circuit. First, determine whether or not the exciter is giving its normal voltage. If the exciter voltage is assumed to be correct, the trouble will probably be due to one of the following causes: (1) open circuit in the field winding or rheostat or (2) short circuit or reversal of one or more of the field spools. An open circuit can often be located by inspection or by use of a magneto.

Most field troubles are caused by excessive induced voltage at the start or by a break in the field circuit. Excessive voltage may break down the insulation between field winding and frame or between turns on any one field spool, thus short-circuiting one or more turns, or it may even burn the field conductor off, causing an open circuit.

Causes of overheating in synchronous motors are about the same as those in ac generators. Probably the most common cause of overheating is excessive armature current due to an attempt to make the motor carry its rated load and at the same time compensate for a power factor lower than that for which it was designed. If the motor is not correcting a low power factor but doing mechanical work only, the field current should be adjusted so that the armature current is mimimum for the average load that the motor carries.

**229. Difficulties in starting synchronous motors.**  A synchronous motor starts as an induction motor. The starting torque, as in an induction motor, is proportional to the square of the applied voltage. For example, if the voltage is halved, the starting effort is quartered. When a synchronous motor will not start, the cause may be that the voltage on the line has been pulled below the value necessary for starting. In general, at least half voltage is required to start a synchronous motor.

Difficulty in starting may also be caused by an open circuit in one of the lines to the motor. Assume the motor to be three-phase. If one of the lines is open, the motor becomes single-phase, and no single-phase synchronous motor, as such, is self-starting. The motor therefore will not start and will soon get hot. The same condition is true of a two-phase motor if one of the phases is open-circuited.

Difficulty in starting may be due to a rather slight increase in static friction. It may be that the bearings are too tight, perhaps from cutting during the previous run. Excessive belt tension, if the synchronous motor is belted to its load, or any cause which increases starting friction will probably give trouble. Difficulty in starting may be due to field excitation on the motor. After excitation exceeds one-quarter of normal value, the starting torque is influenced. With full field on, most synchronous motors will not start at all. The field should be short-circuited through a proper resistance during the starting period.

Usually compensators are used for starting synchronous motors. If there is a reversed phase in a compensator or if the windings of the armature of the synchronous motor are connected incorrectly, there will be little starting torque. Incorrect connection can be located by noting the unbalanced entering current. Readings to determine this unbalancing should be taken with the armature revolving slowly. The revolving can be effected by any mechanical means. While the motor is standing still, even with correct connections, the armature currents of the three phases usually differ somewhat. This is due to the position of the poles in relation to the armature, but when the armature is revolving slowly, the currents should average up. If the rotor cannot be revolved mechanically, similar points on each phase of the armature must be found. Then when the rotor is set successively at these points, the currents at each setting should be the same. Each phase, when located in a certain specific position as related to a pole, should, with right connections, take a certain specific current. With wrong connections the currents will not be the same.

**230. Open circuit in the field of a synchronous motor.**   If in the operation of a synchronous motor the field current breaks for any reason, the armature current will appreciably increase, causing either a shutdown or excessive heat. It becomes important therefore in synchronous motors to have the field circuit permanently established.

**231. Short circuit in an armature coil of a synchronous motor.**   A short circuit in an armature coil of a synchronous motor burns it out completely, charring it down to the bare copper. When this occurs, the symptoms are so evident that there is no difficulty in identifying the trouble. Under ordinary circumstances such a coil may be cut out and operation continued. In an induction motor, the current in the short-circuited coil rises only to a certain value but heats it many times more than normal. It is not necessarily burned out immediately, and perhaps it may not be burned out at all.

**232. Hunting of synchronous motors.**   Synchronous motors served by certain primary sources of energy tend to hunt. The periodicity of the swinging is determined by properties of the armature and the circuit. It may reach a certain magnitude and there stick, or the swinging may increase until finally the motor breaks down altogether. This trouble usually occurs on long lines having considerable resistance between the source of energy and the synchronous motor. Sometimes it occurs under the most favorable conditions. Irregular rotation of a prime mover, such as a single-cylinder steam engine, is often responsible for the trouble. The usual remedy is to apply to the poles bridges of copper or brass in which currents are induced by the wavering of the armature. These currents tend to stop the motion. Different companies use different forms of bridges. When hunting or pulsating occurs and the motor is not already equipped with bridges, it is best to consult the manufacturer. In general, the weaker the field on a synchronous motor, the less the pulsation. Sometimes pulsations may be so reduced that no trouble results, simply by running with a somewhat weaker field current.

**233. Improper armature connections in synchronous motors.**   This trouble usually manifests itself by unbalanced entering currents and by a negligible or very low starting torque. The circuits should be traced out and the connections remade until the three entering currents for three-phase or the two entering currents for two-phase are approximately equal. These currents will not be equal even with correct connection when the armature is standing still.

**234. Bearing troubles of synchronous motors** are similar to those of induction motors. A difference is that with a synchronous motor the air gap between the revolving element and the poles is relatively large, so that the wearing of the bearing, which throws the armature out of center, is not so serious as with an induction motor. End play should be treated in the same way as with an induction motor.

**235. Collector rings and brushes.**   One of the most common sources of trouble with synchronous motors is sparking at the collector rings. Much of this trouble is due to the lack of proper care. The collector rings are one of the most important parts of the machine, and they should be frequently inspected.

Any black spots that appear on the surface of the collector should be removed by rubbing lightly with fine sandpaper the first time that the machine is shut down. It is very important that this be done because although these spots are not serious in themselves, they will lead to pitting of the rings and the necessity of regrinding. However, no harm is done to the rings if the condition is corrected at once.

Sometimes an imprint of the brushes will be found on the surface of the collectors. This may occur on a machine which is subjected to moisture or acid fumes which will act on the surface of the ring. In case the machine is shut down for any length of time, the fumes will act upon the surface of the ring except where it is in contact with the brushes. The surface of the ring will be slightly high at this point. This causes the brush to jump slightly and arc and thus burn its imprint on the ring.

In such an installation it is good practice for the operator, when shutting down the machine, to apply a little paraffin to the rings as the motor comes to a stop and while they are still warm. In general, rings give better service when the motor is run continuously than when it is shut down part of the time.

Another cause of brush imprint on the rings is a slight imbalance in the rotor, which causes a jerk or movement in the rings once every revolution. This results in jumping of the brush, with a consequent small arc, which in time burns an imprint of the brush on the ring.

Brush imprints due to fumes will occur at any point where the motor happens to stop. Imprints due to a rotor imbalance will always occur at the same place on the ring relative to the rotor.

Since there is always an electrolytic action on the surface of an iron ring, the collector operation is improved by occasionally reversing the polarity of the rings. Sometimes trouble will occur on one ring only, and by reversing the polarity every day or so the trouble will entirely disappear.

Occasionally ring trouble will arise from a ring not being of uniform hardness, so that it wears unevenly. Such a ring should be replaced. Small pinholes in the surface of a cast-iron ring will not cause trouble.

**236. Bearing troubles of motors and generators.**    Modern generators and motors have self-oiling bearings. These should be filled to such a height that the rings will carry sufficient oil upon the shaft. If the bearings are too full, oil will be thrown out along the shaft. Watch the bearings carefully from the time at which the machine is first started until the bearings are warmed up; then note the oil level. The expansion of the oil due to heat and foaming raises the level considerably during that time. The oil should be renewed about once in 6 months or oftener if it becomes dirty or causes the bearings to heat.

The bearings must be kept clean and free from dirt. They should be examined frequently to see that the oil supply is properly maintained and that the oil rings do not stick. Use only the best quality of oil. New oil should be run through a strainer if it appears to contain any foreign substances. If the oil is used a second time, it should first be filtered and, if warm, allowed to cool. If a bearing becomes hot, first feed heavy lubricant copiously, loosen the nuts on the bearing cap, and then, if the machine is belt-connected, slacken the belt. If no relief is afforded by these means, shut down, keeping the machine running slowly until the shaft is cool so that the bearing may not "freeze." Renew the oil supply before starting again. A new machine should always be run at a slow speed for an hour or so to see that it operates properly. The bearings should be inspected at regular intervals to ensure that they always remain in good condition. The higher the speed, the greater the care to be taken in this regard.

A warm bearing, or "hot box," is probably due to one of the following causes: (1) excessive belt tension, (2) failure of the oil rings to revolve with the shaft, (3) rough bearing surface, and (4) improper lining up of bearings or fitting of the journal boxes.


**CARE OF MOTORS**[1]

**237. Modern methods of design and construction** have made the electric motor one of the least complicated and most dependable forms of machinery in existence and have thereby made the matter of its maintenance one of comparative simplicity. This statement should not, however, be taken to mean that proper maintenance is not important; on the contrary, it must be given careful consideration if the best performance and longest life are to be expected from the motor. The two major features, from the standpoint of their effect upon the general performance of the motor, are those of proper lubrication and the care given to insulation, because they concern the most vital, and probably the most vulnerable, parts of the machine.

**238. Lubrication.**    The designs of bearings and bearing housings of motors have been remarkably improved. The bearings of modern motors, whether sleeve, ball, or roller, require infrequent attention. This advance in the art is not yet fully appreciated, for although there may have been some necessity for more frequent attention in the case of older designs with housings less tight than on modern machines, oiling and greasing of new motors are quite often entrusted to uninformed and careless attendants, with the

[1]Courtesy of General Electric Co.

result that oil or grease is copiously and frequently applied to the outside as well as the inside of bearing housings. Some of the excess lubricant is carried into the machine and lodges on the windings, where it catches dirt and thereby hastens the ultimate failure of the insulation.

Modern designs provide for a plentiful supply of oil or grease being held in dusttight and oiltight housings. If the proper amount of a suitable lubricant is applied before starting, there should be no need to refill the housings for several months even in dusty places. Infrequent though periodic and reasonable attention to modern bearings of any type will tend to ensure longer life of both bearings and insulation.

**239. Greasing ball bearings.**  Only a high grade of grease, having the following general characteristics, should be used for ball-bearing lubrication:

1. Consistency a little stiffer than that of Vaseline, maintained over the operating-temperature range

2. Melting point preferably over 150°C

3. Freedom from separation of oil and soap under operating and storage conditions

4. Freedom from abrasive matter, acid, and alkali

In greasing a motor, care must be taken not to add too large a quantity of grease, as this will cause too high an operating temperature with resulting expansion and leaking of the grease, especially with large bearings operated at slow speeds.

1. PRESSURE-RELIEF SYSTEMS  The following procedure is recommended for greasing ball-bearing motors equipped with a pressure-relief greasing system:

Make sure, by wiping clean the pressure-gun fitting, bearing housing, and relief plug, that no dirt gets into the bearing with the grease. Always remove the relief plug from the bottom of the bearing before using the grease gun. This prevents putting excessive pressure inside the bearing housing. Excessive pressure might rupture the bearing seals.

With a clean screwdriver or similar tool, free the relief hole of any hardened grease, so that any excess grease will run freely from the bearing. With the motor running, add grease with a hand-operated pressure gun until it begins to flow from the relief hole. This tends to purge the housing of old grease. If it might prove dangerous to lubricate the motor while it is running, follow the procedure with the motor at a standstill.

Allow the motor to run long enough after adding grease to permit the rotating parts of the bearing to expel all excess grease from the housing. This very important step prevents overgreasing the bearing. Stop the motor and replace the relief plug tightly with a wrench.

2. OTHER TYPES  A motor that is not equipped with the pressure-gun fitting and relief plug on the bearing housing cannot be greased by the procedures described. Under average operating conditions, the grease with which the bearing housings of these motors were packed before leaving the factory is sufficient to last approximately 1 year. When the first year of service has elapsed and once a year thereafter (or oftener if conditions warrant), the old grease should be removed and the bearings supplied with new grease. To do this, disassemble the bearing housings and clean the inside of the housings and housing plates or caps and the bearings with a suitable solvent. When they have been thoroughly cleansed of old grease, reassemble all parts except the outer plates or caps.

Apply new grease, either by hand or from a tube, over and between the balls. The amount of grease to be used varies with the type and frame size of the particular motor. The instruction sheet that accompanied the motor should be consulted for this information.

The addition of the correct amount of grease fills the bearing housing one-third to one-half full. More than the amount specified must not be used. When the motor is reassembled, any V grooves that are found in the housing lip should be refilled with grease (preferably a fibrous, high-temperature–sealing grease) which will act as an additional protective seal against the entrance of dirt or foreign particles.

**240. The technique for greasing motors equipped with roller bearings** is very similar to that used for ball bearings. Specific instructions for the individual design should be followed, however, because more frequent greasing or slight changes in technique may sometimes be necessary.

**241. Oiling sleeve bearings.**  The oil level in sleeve-bearing housings should be checked periodically with the motor stopped. If the motor is equipped with an oil-filler gage, the gage should be approximately three-quarters full at all times.

If the oil is dirty, drain it off by removing the drain plug, which is usually located in the bottom or side of the bearing housing. Then flush the bearing with clean oil until the outcoming oil is clean.

In fractional-horsepower motors, there may be no means of checking oil level, as all the oil may be held in the waste packing. In such cases, a good general rule for normal motor service is to add 30 to 70 drops of oil at the end of the first year and to reoil at the end of each subsequent 1000 h of motor operation.

Most fractional-horsepower motors built today require lubrication about once a year. Small fan and agitator motors will often require more frequent lubrication, with 3-month intervals between oilings.

### 242. Cleaning bearings

SLEEVE BEARINGS    Sleeve-bearing housings are provided with liberal settling chambers into which dust, dirt, and oil sludge collect. The only cleaning necessary is to remove the drain plug and drain the oil, which will flush out most of the settled material with it.

Whenever the motor is disassembled for general cleaning, the bearing housing should be washed out with a solvent. Before being assembled, the bearing lining should be dried and the shaft covered with a film of oil.

BALL BEARINGS    The pressure-relief method of greasing motors, described above, tends to purge the bearing housing of used grease. Complete cleaning of bearings, therefore, is required at infrequent intervals only. For a thorough and convenient flushing when the bearings are not disassembled, the following method is recommended.

1. Wipe clean the housing, pressure-gun, and relief fittings, and then remove the fittings. Every care should be taken to keep dirt out of the bearings. Once in a bearing, a bit of abrasive may not be removed even with the most thorough cleaning. Afterward, it may become dislodged and get between the bearing surfaces with subsequent serious results.

2. With a clean screwdriver or a similar tool, free the pressure-fitting hole in the top of the bearing housing of hardened grease. Also free the relief-plug hole in the bottom of the housing of old grease to permit easy expulsion of the old grease during the cleaning process after the solvent has been added.

3. Fill a syringe with any suitable modern grease solvent, and inject some of it into the bearing housing through the pressure-fitting hole while the motor is running. As the grease becomes thinned by the solvent, it will drain out through the relief hole. Continue to add solvent until it drains out quite clear.

4. Replace the relief plug and inject solvent until it can be seen splashing in the filling hole. Allow the solvent to churn for a few minutes, and then remove the relief plug and drain off the solvent. Repeat the churning operation until the solvent runs clean.

5. When using any type of solvent for flushing, replace the relief plug and inject a small amount of light lubricating oil. Allow it to churn for a minute or two before draining off. This will flush out the solvent. To complete the job, grease the bearing, using the method previously described.

This method permits the cleaning of all standard motors operating at an angle not exceeding 15° from the horizontal except totally enclosed, fan-cooled motors. For these motors, the bearing at the pulley end can be flushed as described. To clean the fan-end bearing, first remove the fan cover and fan to make the drain plug at the bottom of the housing accessible.

### 243. Care of insulation.

Motors should be stored in a dry, clean place until ready for installation. Heat should be supplied, especially for larger high-voltage machines, to protect them against alternate freezing and thawing. This advice is equally applicable to spare coils.

Motors that have been long in transit in a moist atmosphere or have been idle for an extended period without heat to prevent the accumulation of moisture should be thoroughly dried out before being placed in service. Machines may also become wet by accident, or they may sweat as a result of a difference in their temperature and that of the surrounding air, just as cold water pipes sweat in a warm, humid atmosphere. This condition is, of course, very injurious and should be prevented, particularly in the case of large or important motors, by keeping them slightly warm at all times. Current at a low voltage can be passed through the windings, electric heaters can be used, or even steam pipes can be utilized for protective purposes. During extended idle periods, tarpaulins can be stretched over the motor, and a small heater put inside to maintain the proper temperature.

### 244. Drying out.

If a motor has become wet from any cause whatever, it should be dried out thoroughly before being operated again. The most effective method is to pass current through the windings, using a voltage low enough to be safe for the winding in its moist condition. For 2300-V motors, 230 V is usually satisfactory for circulating this drying-out current. Thermometers should be placed on the windings to see that they are being heated uniformly. Temperatures should not exceed 90°C (Class A insulation). Applying the heat internally in this manner drives out all moisture and is particularly effective on high-voltage motors, in which the insulation is comparatively thick.

Heat can be applied externally by placing heating units around or in the machine, covering the whole with canvas or other covering and leaving a vent at the top to permit the escape of moisture. In doing this it is essential that warm air circulate over all the surfaces to be dried. The air should be allowed to escape as soon as it has absorbed moisture. Therefore, the heaters should be so placed and baffles so arranged as to get a natural draft, or small fans can be used to force circulation. Twelve-inch (0.3-m) fans set to blow air across the fronts of glow heaters and then into the lower part of a machine from opposite sides, and so on up around the windings and out the top, will produce excellent results. The temperature of the winding should not be allowed to exceed 100°C for Class A–insulated motors. Smaller machines can conveniently be placed in ovens, the same temperature limits being observed.

**245. Insulation-resistance tests.**    The time required for complete drying out depends considerably on the size and voltage of the motor. Insulation-resistance measurements should be taken at intervals of 4 or 5 h until a fairly constant value is reached. This value should at least equal the recommended Institute of Electrical and Electronic Engineers standard, which is

$$\text{Megohms} = \frac{\text{voltage}}{\text{kVA}/100 + 1000} \tag{9}$$

The insulation resistance of dry motors in good condition is considerably higher than this value.

The more convenient way to measure this resistance is through the use of a megohm-meter, although if a 500-V dc source is available, readings can be taken with a voltmeter. The ungrounded side of the system should be connected to all the motor terminals through the voltmeter, the opposite or grounded side being connected directly to the motor frame. The insulation resistance is found by the method of Sec. 177 of Div. 1.

When the voltmeter method is used, the connection to the frame should always be made through a fuse of not more than 10 A in size. The circuit should be tested, and the side showing a complete or partial ground then connected to the frame through the fuse.

Obviously, the insulation resistance varies over a wide range, depending upon moisture, temperature, cleanliness, etc., but it is a good indication of the general condition of the insulation and its ability to stand the operating voltage. Such readings should be taken before a high-potential test to determine whether the insulation is ready for such a test and afterward to make sure that the high potential has not injured the insulation.

**246. High-potential tests.**    High-potential tests should be made after drying out or after repairs to determine the dielectric strength of the insulation. New windings should successfully stand a high-potential 50–60 Hz test of twice normal voltage plus 1000. There is some disagreement as to the proper value to use for motors that have been in operation for some time, but it is reasonable to assume that after thorough cleaning and drying the winding of a used motor should stand 150 percent of normal voltage for 1 min. Small high-potential testing sets are available for such work and are of such capacity that very little damage will result from a breakdown during a test.

**247. Periodic inspection.**    A systematic and periodic inspection of motors is necessary to ensure best operation. Of course, some machines are installed where conditions are ideal and dust, dirt, and moisture are not present to an appreciable degree, but most motors are located where some sort of dirt accumulates in the windings, lowering the insulation resistance and cutting down creepage distance. Steel-mill dusts are usually highly conductive, if not abrasive, and lessen creepage distances. Other dusts are highly abrasive and actually cut the insulation while being carried through by the ventilating air. Fine cast-iron dust quickly penetrates most insulating materials. Hence the desirability of cleaning motors periodically. If conditions are extremely severe, open motors might require a certain amount of cleaning each day. For less severe conditions, weekly inspection and partial cleaning are desirable. Most machines require a complete overhauling and thorough cleaning about once a year.

For the weekly cleaning, the motor should be blown out, using moderate-pressure, dry, compressed air of about 25 to 30 psi (172,369 to 206,843 Pa) pressure. If conducting and abrasive dusts are present, even lower pressure may be necessary, and suction is to be preferred, as damage can easily be caused by blowing the dust and metal chips into the insulation. On most dc motors and large ac motors the windings are usually fairly accessible, and the air can be properly directed to prevent such damage. On the larger ac machines the air ducts should be blown out so that the ventilating air can pass through as intended.

On large machines insulation-resistance readings should be taken in the manner indicated above. As long as the readings are consistent, the condition of the insulation would ordinarily be considered good. Low readings would indicate increased current leakage to ground or to other conductors, owing to one of perhaps several causes, such as deteriorated insulation, moisture, dirty or corroded terminals, etc.

Inspection and servicing should be systematic. Frequency of inspection and degree of thoroughness may vary and must be determined by the maintenance engineer. They will be governed by (1) the importance of the motors in the production scheme (If the motor fails, will production be slowed seriously?), (2) percentage of the day in which the motor operates, (3) the nature of the service, and (4) the environment.

An inspection schedule must therefore be elastic and adapted to the needs of each plant. The following schedule, covering both ac and dc motors, is based on average conditions insofar as dust and dirt are concerned.

EVERY WEEK

1. Examine commutator and brushes.
2. Check oil level in bearings.
3. See that oil rings turn with the shaft.
4. See that the shaft is free of oil and grease from bearings.
5. Examine starter, switch, fuses, and other controls.
6. Start motor and see that it is brought up to speed in normal time.

EVERY SIX MONTHS

1. Clean motor thoroughly, blowing out dirt from windings, and wipe commutator and brushes.
2. Inspect commutator clamping ring.
3. Check brushes and renew any that are more than half worn.
4. Examine brush holders and clean them if dirty. Make sure that brushes ride free in the holders.
5. Check brush pressure.
6. Check brush position.
7. Drain, wash out, and renew oil in sleeve bearings.
8. Check grease in ball or roller bearings.
9. Check operating speed or speeds.
10. See that the end play of the shaft is normal.
11. Inspect and tighten connections on the motor and control.
12. Check current input and compare with normal input.
13. Run the motor and examine the drive critically for smooth running, absence of vibration, and worn gears, chains, or belts.
14. Check motor foot bolts, end-shield bolts, pulley, coupling, gear and journal setscrews, and keys.
15. See that all covers, belt, and gear guards are in good order, in place, and securely fastened.

ONCE A YEAR

1. Clean out and renew grease in ball- or roller-bearing housings.
2. Test insulation by a megohmmeter.
3. Check the air gap.
4. Clean out magnetic dirt that may be hanging on poles.
5. Check clearance between the shaft and journal boxes of sleeve-bearing motors to prevent operation with worn bearings.
6. Clean out undercut slots in the commutator.
7. Examine connections of the commutator and armature coils.
8. Inspect armature bands.

**247A. Records.**   Maintenance persons should have a record card for every motor in the plant. All repair work, with the cost, and every inspection can be entered on the record.

In this way, excessive attention or expense will show up, and the causes can be determined and corrected.

Inspection records will also serve as a guide to tell when motors should be replaced because of the high cost to keep them operating. Misapplications, poor drive engineering, and the like will also be disclosed.

**248. Brush inspection.** The first essential for satisfactory operation of brushes is free movement of the brushes in their holders. Uniform brush pressure is necessary to assure equal current distribution. Adjustment of brush holders should be set so that the face of the holder is approximately ⅛ in (3.175 mm) up from the commutator; any distance greater than this will cause brushes to wedge, resulting in chattering and excessive sparking.

Check the brushes to make sure that they will *not* wear down too far before the next inspection. Keep an extra set of brushes available so that replacements can be made when needed. Sand in new brushes, and run the motor without load as long as possible.

It is false economy to use brushes down to the absolute minimum length before replacement. There have been cases in which brushes wore down until the metal where the pigtail connects to the brush touched the commutator. This, of course, caused severe damage to the commutator.

Make sure that each brush surface in contact with the commutator has the polished finish that indicates good contact and that the polish covers all this surface of the brush. Check the freedom of motion of each brush in the brush holder.

When replacing a brush, be sure to put it in the same brush holder and in its original position. It has been found helpful to scratch a mark on one side of the brush when removing it so that it will be replaced properly.

Check the springs that hold the brushes against the commutator. Improper spring pressure may lead to commutator wear and excessive sparking. Excessive heating may have annealed the springs, in which case they should be replaced and the cause of heating corrected. Larger motors have means for adjusting the spring pressure, which should be 2 to 2½ psi (13,790 to 17,237 Pa) of area of brush contact with the commutator.

**249. Commutator inspection.** Inspect the commutator for color and condition. It should be clean, smooth, and a polished brown color where the brushes ride on it. A bluish color indicates overheating of the commutator.

Roughness of the commutator should be removed by sandpapering or stoning. Never use emery cloth or an emery stone. For this operation, run the motor without load. If sandpaper is used, wrap it partly around a wood block. The stone is essentially a piece of grindstone, known to the trade as a commutator stone. Press the stone or sandpaper against the commutator with moderate pressure with the motor running without load, and move it back and forth across the commutator surface. *Use care not to come in contact with live parts.*

If the armature is very rough, it should be taken out and the commutator turned down in a lathe. When this is done, it is usually necessary to cut back the insulation between the commutator bars slightly. After the commutator has been turned down, the brushes should be sanded and run in as described previously. This is not necessary after light sandpapering or stoning.

**250. Cleaning.** About once a year or oftener if conditions warrant, motors should be cleaned thoroughly. Smaller motors, the windings of which are not particularly accessible, should be taken apart.

First, the heavy dirt and grease should be removed with a heavy, stiff brush, wooden or fiber scrapers, and cloths. Rifle-cleaning bristle brushes can be used in the air ducts. Dry dust and dirt may be blown off, using dry compressed air at moderate pressure, for example, 25 to 50 psi (172,369 to 344,738 Pa) pressure at the point of application, taking care to blow the dirt out from the winding. As stated before, if the dirt and dust are metallic, conducting, or abrasive, air pressure may drive the material into the insulation and damage it. Hence, for such conditions pressure is not so satisfactory as a suction system. If compressed air at low pressure is used, care must be taken to direct it properly so that the dust will not cause damage and will not be pocketed in the various corners.

Grease, oil, and sticky dirt are easily removed by applying cleaning liquids specifically designed for the purpose. Any of these liquids evaporate quickly and, if not applied too generously, will not soak or injure the insulation.

If one of the other liquids must be used, it should be applied outdoors or in a well-ventilated room. It must be remembered that gasoline or naphtha vapor is heavier than air and will flow into pits, basements, etc., and may remain there for hours or even days. The casual smoker or a spark from a hammer or chisel or even from a shoe nail may cause a serious explosion. Therefore, proper ventilation of the room is essential and may require specially piped ventilating fans. The use of carbon tetrachloride obviates the explosion hazard, but some ventilation is required to remove the vapor, which might affect the safety and comfort of the workers.

There are several good methods of applying the cleansing liquid. A cloth, saturated in the liquid, can be used to wipe the coils. A paintbrush, dipped in the liquid container, is handy to get into the corners and crevices and between small coils. Care should be taken not to soak the insulation, as would be the case if coils or small machines were dipped into the liquid.

Probably the best method of applying the liquid is to spray it on. A spray gun, paint-spraying appliance, or an ordinary blowtorch is often used with good results, although the last device is likely to give a heavier spray than desirable. An atomizer will give excellent results using a pressure of about 80 psi (551,581 Pa) if the insulation is in good condition or 40 to 50 psi (275,790 to 344,738 Pa) if the insulation is old. The atomizer should be held not more than 5 or 6 in (127 or 152 mm) away from the coils.

Although the insulation will dry quickly at ordinary room temperature after such cleaning methods, it is highly desirable to heat it to drive off all moisture before applying varnish. This heating or drying-out process has already been discussed and therefore need only be mentioned here.

If the motor can be spared from service long enough, the insulation should be dried out by heating to from 90 to 100°C. While the motor is warm, a high-grade insulating varnish should be applied. For severe acid, alkali, or moisture conditions, a black plastic baking varnish or epoxy resin is best. If oil or dusts are present, a clear or yellow varnish or epoxy resin should be used.

The varnish may be sprayed or brushed on. For small stators or rotors, it is best to dip the windings into the varnish, cleaning off the adjacent metal parts afterward by using a solvent of the varnish. After applying the varnish, the best results are obtained by baking for 6 to 7 h at about 100°C. Experience with particular conditions of operation or the condition of the insulation may indicate the desirability of applying a second coat of the same varnish, followed again by 6 to 7 h of baking at 100°C.

If the machine must be put back in service quickly or if facilities are not available for baking, fairly good results will be obtained by applying one of the quick-drying black or clear varnishes which dry in a few hours at ordinary room temperature.

Insulation-resistance readings should be taken, as explained previously, to determine whether the winding is in satisfactory condition for applying a high-potential test. After this test, it is good practice to run the motor without load for a short time to make certain that everything has been connected, assembled, and adjusted properly.

Motors should generally be given an overhauling at intervals of 5 years or so normally or, if the service is more severe, more frequently. This practice is beneficial in avoiding breakdowns and in extending the useful life of the equipment. When periodic overhauling is practiced, the following advice may be helpful:

1. Check the motor air gap, between stator and rotor, with feelers for uniformity. Too little clearance at the bottom may indicate worn bearings.

2. Take the motor apart and inspect it thoroughly. Measurement of the bearings and journals may disclose a need for new bearing linings. Remove the waste from waste-packed bearings and rearrange or replace it so that any glaze on the wool is removed from its point of contact with the shaft. Any gummy deposit on the wool should be replaced. All lubricant should be cleaned out of the bearings, and a fresh supply put in when the motor is reassembled.

3. The rotors should be cleaned with a solvent to remove any accumulated dirt, after which any rust should be removed with fine sandpaper (not emery paper). When clean and dry, the rotors should be coated with a good grade of clear varnish or lacquer to protect them from moisture. To prevent injury to the bearings, they should be completely protected with a clean rag when the motor is disassembled.

4. The rotors of wound-rotor motors should be given the same treatment as the stators. In addition, soldered joints and binding cords should be inspected and any weakness remedied.

5. The stator bore should be cleaned of dirt with a solvent, and any rust should be removed with fine sandpaper (not emery paper). Care should be taken during this operation not to damage the top sticks or end turns of the stator winding. When the stator bore has dried, any remaining dirt in the bore should be wiped out with a cloth or brushed out with a soft brush. A hand bellows or dry compressed air at low pressure may also be employed.

## CONTROL EQUIPMENT FOR MOTORS

### 251. Electric-controller terminology

*Ambient temperature.*   The temperature of the medium such as air, water, or earth into which the heat of the equipment is dissipated. For self-ventilated equipment, the ambient temperature is the average[1] temperature of the air in the immediate neighborhood of the equipment. For air- or gas-cooled equipment with forced ventilation or secondary water cooling, the ambient temperature is taken as that of the ingoing air or cooling gas. For self-ventilated enclosed (including oil-immersed) equipment considered as a complete unit, the ambient temperature is the average[1] temperature of the air outside the enclosure in the immediate neighborhood of the equipment.

*Apparatus.*   A term said to describe the enclosure, the enclosed equipment, and attached appurtenances.

*Auxiliary contacts of a switching device.*   Those contacts which in addition to the main-circuit contacts function with the movement of the latter.

*Combination controller.*   A full magnetic or semimagnetic controller with additional externally operable disconnecting means contained in a common enclosure. The disconnecting means may be a circuit breaker or a disconnect switch. Combination starters are specific forms of combination controllers.

*Contactor.*   A two-state (on-off) device for repeatedly establishing and interrupting an electric power circuit. Interruption is obtained by introducing a gap or a very large impedance.

*Continuous duty.*   A duty that demands operation at substantially constant load for an indefinitely long time.

*Control circuit of a control apparatus or system.*   That circuit which carries the electric signals directing the performance of the controller but does not carry the main power circuit.

*Drum controller.*   An electric controller that utilizes a drum switch as the main switching element.

*Drum switch.*   A switch in which the electric contacts are made by segments or surfaces on the periphery of a rotating cylinder or sector or by the operation of a rotating cam.

*Duty (of a controller).*   A requirement of service that defines the degree of regularity of the load. The frequency and length of time of operation and rest must be specified.

*Electric controller.*   A device or group of devices which serves to govern in some predetermined manner the electric power delivered to the apparatus to which it is connected. The term *controller* does not apply to a simple disconnecting means whose primary purpose in a motor circuit is to disconnect the circuit, together with the motor and its controller, from the source of power.

*Full-magnetic controller.*   An electric controller having all its basic functions performed by devices that are operated by electromagnets.

*General-use switch.*   A switch which is intended for use in general distribution and branch circuits. It is rated in amperes and is capable of interrupting the rated current at the rated voltage.

*Intermittent duty.*   A requirement of service that demands operation for alternate periods of (1) load and no load, (2) load and rest, or (3) load, no load, and rest, such alternate intervals being definitely specified.

*Isolating switch.*   A switch intended for isolating an electric circuit from the source of power. It has no interrupting rating, and it is intended to be operated only after the circuit has been opened by some other means.

---

[1]The average of temperature readings at several locations.

*Magnetic contactor.*    A contactor actuated by electromagnetic means.

*Magnetic-control relay.*    A relay which is actuated by electromagnetic means. When not otherwise qualified, the term refers to a relay intended to be operated by the opening and closing of its coil circuit and having contacts designated for energizing and/or deenergizing the coils of magnetic contactors or other magnetically operated device.

*Manual controller.*    An electric controller having all its basic functions performed by devices that are operated by hand; for example, a toggle switch or a drum-type controller.

*Master switch.*    A switch dominates the operation of contactors, relays, or other remotely operated devices. A master switch may be automatic, such as a float switch or a pressure regulator, or it may be manually operated, such as a drum, pushbutton, or knife switch.

*Motor-circuit switch.*    A switch intended for use in a motor branch circuit. It is rated in horsepower, and it is capable of interrupting the maximum operating overload (see "Operating overload") current of a motor of the same rating at the rated voltage.

*Open-phase protection (phase-failure protection).*    A form of protection that operates to disconnect the protected equipment on the loss of current in one phase conductor of a polyphase circuit or to prevent the application of power to the protected equipment in the absence of one or more phase voltages of a polyphase system.

*Open-phase relay.*    A polyphase relay designed to operate when one or more input phases of a polyphase circuit are open.

*Operating overload.*    The overcurrent to which electric apparatus is subjected in the course of the normal operating conditions that it may encounter. For example, those currents in excess of running current which occur for a short time as a motor is started or jogged are considered normal operating overloads for control apparatus.

*Overcurrent relay.*    A relay that operates when the current through the relay during its operating period is equal to or greater than its setting.

*Overload protection.*    The effect of a device operative on excessive current, but not necessarily on short circuit, to cause and maintain the interruption of current to the device governed. When it is a function of a controller for a motor, the device employed provides for interrupting any operating overloads but is not required to interrupt short circuits.

*Overload relay.*    A relay that responds to electric load and operates at a preset value of overload. Overload relays are usually current relays, but they may be power, temperature, or other relays.

*Part-winding starter.*    A starter which applies voltage successively to the partial sections of the primary winding of an alternating-current motor.

*Periodic duty.*    A type of intermittent duty in which the load conditions are regularly recurrent.

*Phase-sequence–reversal protection (phase-reversal protection).*    A form of protection that prevents energization of the protected equipment on reversal of the phase sequence in a polyphase circuit.

*Pilot-duty rating.*    A generic term formerly used to indicate the ability of a control device to control other devices.

*Relay.*    An electric device that is designed to interpret input conditions in a prescribed manner and after specified conditions have been met to respond to cause contact operation or similar abrupt change in associated electric control circuits. Inputs are usually electric but may be mechanical, thermal, or of other quantities. Limit switches and similar simple devices are not relays. A relay may consist of several units, each responsive to specified inputs, the combination providing the desired performance characteristic.

*Semimagnetic controller.*    An electric controller having only part of its basic function performed by devices that are operated by electromagnets.

*Snap action.*    A rapid motion of contacts from one position to another position or their return. This action is relatively independent of the rate of travel of the actuator.

*Solid-state contactor.*    A contactor whose function is performed by a semiconductor device or devices.

*Starter.*    An electric controller for accelerating a motor from rest to normal speed, to protect, and to stop the motor. (A device designed for starting a motor in either direction of rotation includes the additional function of reversing and should be designated a *reversing starter.*)

*Three-wire control.* A control function which utilizes a momentary-contact pilot device and a holding circuit contact to provide undervoltage protection.

*Two-wire control.* A control function which utilizes a maintained-contact type of pilot device to provide undervoltage release.

*Undervoltage protection.* The effect of a device, operative on the reduction or failure of voltage, to cause and maintain the interruption of power to the main circuit.

*Undervoltage release.* The effect of a device, operative on the reduction or failure of voltage, to cause the interruption of power to the main circuit but not to prevent the reestablishment of the main circuit on return of voltage.

**252. Enclosure of controllers.** Except when the control equipment is mounted on switchboard panels under the control of experienced operators, the control equipment of motors should be enclosed in a metal or approved nonmetallic case. Different types of enclosures are available to meet the requirements of any surrounding conditions. Most general-purpose controllers are enclosed in a sheet-metal case. If a controller must be placed in a hazardous location in an explosive atmosphere, the advice of the experts of the control manufacturers should be obtained. The standard types of nonventilated enclosures are given in Sec. 253.

**253. Classification of standard types of nonventilated enclosures for electric controllers (National Electrical Manufacturers Association)** is as follows:[1]

TYPE 1    Type 1 enclosures are intended for indoor use primarily to provide a degree of protection against contact with the enclosed equipment, in locations where unusual service conditions do not exist. Type 1 enclosures may be of the nonventilated or the ventilated type.

TYPE 2    Type 2 enclosures are intended for indoor use primarily to provide a degree of protection against limited amounts of falling water and dirt. They are not intended to provide protection against conditions such as dust or internal condensation.

TYPE 3    Type 3 enclosures are intended for outdoor use primarily to provide a degree of protection against windblown dust, rain, and sleet; and to be undamaged by the formation of ice on the enclosure. They are not intended to provide protection against conditions such as internal condensation or internal icing.

TYPE 3R    Type 3R enclosures are intended for outdoor use primarily to provide a degree of protection against falling rain, and to be undamaged by the formation of ice on the enclosure. They are not intended to provide protection against conditions such as dust, internal condensation, or internal icing. Type 3R enclosures may be of the nonventilated or the ventilated type.

TYPE 3S    Type 3S enclosures are intended for outdoor use primarily to provide a degree of protection against windblown dust, rain, and sleet, and to provide for the operation of external mechanisms when ice laden. They are not intended to provide protection against conditions such as internal condensation or internal icing.

TYPE 4    Type 4 enclosures are intended for indoor or outdoor use primarily to provide a degree of protection against windblown dust and rain, splashing water, and hose-directed water; and to be undamaged by the formation of ice on the enclosure. They are not intended to provide protection against conditions such as internal condensation or internal icing.

TYPE 4X    Type 4X enclosures have the same provisions as Type 4 enclosures and, in addition, are corrosion-resistant.

TYPE 5    The Type 5 enclosure is obsolete for control equipment. It has been superseded by Type 12.

TYPE 6    Type 6 enclosures are intended for indoor or outdoor use primarily to provide a degree of protection against the entry of water during temporary submersion at a limited depth, and to be undamaged by the formation of ice on the enclosure. They are not intended to provide protection against conditions such as internal condensation, internal icing, or corrosive environments.

TYPE 6P    Type 6P enclosures are intended for indoor or outdoor use primarily to provide a degree of protection against the entry of water during prolonged submersion at a limited depth, and to be undamaged by the formation of ice on the enclosure. They are

---

[1]This material is reproduced with the permission of the National Electrical Manufacturers Association from NEMA Standards Publication NEMA 250-1985, copyright © 1985 by NEMA.

not intended to provide protection against conditions such as internal condensation or internal icing.

Type 7    Type 7 enclosures are intended for use indoors in locations classified as Class I, Group A, B, C, or D, as defined in the National Electrical Code. The letters A, B, C, and D indicate the gas or vapor atmospheres in the hazardous location. The enclosures are marked with the appropriate Class and Group(s) for which they have been qualified.

These enclosures are capable of withstanding the pressures resulting from an internal explosion of the specified gases and containing such an explosion sufficiently that an explosive gas-air mixture existing in the atmosphere surrounding the enclosure will not be ignited. Enclosed heat-generating devices do not cause external surfaces to reach temperatures capable of igniting explosive gas-air mixtures in the surrounding atmosphere.

Type 8    Type 8 enclosures are for indoor or outdoor use in locations classified as Class I, Group A, B, C, or D, as defined in the National Electrical Code. The letters A, B, C, and D indicate the gas or vapor atmospheres in the hazardous location. The enclosures are marked with the appropriate Class and Group(s) for which they have been qualified.

These enclosures and enclosed devices are arranged so that all arcing contacts, connections, and so forth are immersed in oil. Arcing is confined under the oil such that it will not ignite an explosive mixture of the specified gases in internal spaces above the oil or in the atmosphere surrounding the enclosure. Enclosed heat-generating devices do not cause external surfaces to reach temperatures capable of igniting explosive gas-air mixtures in the surrounding atmosphere.

Type 9    Type 9 enclosures are intended for indoor use in locations classified as Class II, Group E or G, as defined in the National Electrical Code. The letters E and G indicate the dust atmospheres in the hazardous location. The enclosures are marked with the appropriate Class and Group(s) for which they have been qualified.

These enclosures are capable of preventing the entrance of dust. Enclosed heat-generating devices do not cause external surfaces to reach temperatures capable of igniting or discoloring dusts on the enclosure or igniting dust-air mixtures in the surrounding atmosphere.

Type 10    Type 10 enclosures are designed to meet the requirements of the Mine Safety and Health Administration, U.S. Department of the Interior, which may be in effect from time to time. They are for equipment to be used in mines with atmospheres containing methane or natural gas with or without coal dust.

Type 11    Type 11 enclosures are intended for indoor use primarily to provide a degree of protection against the corrosive effects of liquids and gases. In addition, they protect the enclosed equipment against the corrosive effects of fumes and gases by providing for immersion of the equipment in oil.

Type 12    Type 12 enclosures are intended for indoor use primarily to provide a degree of protection against dust, falling dirt, and dripping noncorrosive liquids. They are not intended to provide protection against conditions such as internal condensation.

Type 12K    Type 12K enclosures with knockouts are intended for indoor use primarily to provide a degree of protection against dust, falling dirt, and dripping noncorrosive liquids other than at the knockouts. After installation of the enclosure, the knockout areas shall meet the environmental conditions listed above. They are not intended to provide protection against conditions such as internal condensation.

Type 13    Type 13 enclosures are intended for indoor use primarily to provide a degree of protection against dust, spraying of water, oil, and noncorrosive liquids. They are not intended to provide protection against conditions such as internal condensation.

**253A. The following tables provide comparisons of specific applications of enclosures for indoor and outdoor nonhazardous locations (NEMA standard).[1]**

**Comparison of Specific Applications of Enclosures for Indoor Nonhazardous Locations**

| Provides a degree of protection against the following environmental conditions | Type of enclosure | | | | | | | | | | |
|---|---|---|---|---|---|---|---|---|---|---|---|
| | 1[a] | 2[a] | 4 | 4X | 5 | 6 | 6P | 11 | 12 | 12K | 13 |
| Incidental contact with the enclosed equipment | X | X | X | X | X | X | X | X | X | X | X |
| Falling dirt | X | X | X | X | X | X | X | X | X | X | X |
| Falling liquids and light splashing | ... | X | X | X | ... | X | X | X | X | X | X |
| Dust, lint, fibers, and flyings[b] | ... | ... | X | X | X | X | X | ... | X | X | X |
| Hosedown and splashing water | ... | ... | X | X | ... | X | X | ... | ... | ... | ... |
| Oil and coolant seepage | ... | ... | ... | ... | ... | ... | ... | ... | X | X | X |
| Oil or coolant spraying and splashing | ... | ... | ... | ... | ... | ... | ... | ... | ... | ... | X |
| Corrosive agents | ... | ... | ... | X | ... | ... | X | X | ... | ... | ... |
| Occasional temporary submersion | ... | ... | ... | ... | ... | X | X | ... | ... | ... | ... |
| Occasional prolonged submersion | ... | ... | ... | ... | ... | ... | X | ... | ... | ... | ... |

[a]These enclosures may be ventilated. However, Type 1 may not provide protection against small particles of falling dirt when ventilation is provided in the enclosure top. Consult the manufacturer.

[b]These fibers and flyings are nonhazardous materials and are not considered the Class III–type ignitable fibers or combustible flyings. For Class III–type ignitable fibers or combustible flyings see the National Electrical Code, Article 500.

**Comparison of Specific Applications of Enclosures for Outdoor Nonhazardous Locations**

| Provides a degree of protection against the following environmental conditions | Type of enclosure | | | | | | |
|---|---|---|---|---|---|---|---|
| | 3 | 3R[a] | 3S | 4 | 4X | 6 | 6P |
| Incidental contact with the enclosed equipment | X | X | X | X | X | X | X |
| Rain, snow, and sleet[b] | X | X | X | X | X | X | X |
| Sleet[c] | ... | ... | X | ... | ... | ... | ... |
| Windblown dust | X | ... | X | X | X | X | X |
| Hosedown | ... | ... | ... | X | X | X | X |
| Corrosive agents | ... | ... | ... | ... | X | ... | X |
| Occasional temporary submersion | ... | ... | ... | ... | ... | X | X |
| Occasional prolonged submersion | ... | ... | ... | ... | ... | ... | X |

[a]These enclosures may be ventilated.
[b]External operating mechanisms are not required to be operable when the enclosure is ice-covered.
[c]External operating mechanisms are operable when the enclosure is ice-covered.

**254. Motor controllers[2]** are classified according to size and to the method of performing the necessary operations, shown below.
1. Manual
2. Magnetic
    a. Full magnetic with automatic master switch
    b. Full magnetic with manual master switch
    c. Semimagnetic

The following tables list the NEMA controller sizes for manual and magnetic single- and three-phase motor controllers. There are no NEMA size designations for semimagnetic controllers.

[1]This material is reproduced with the permission of the National Electrical Manufacturers Association from NEMA Standards Publication NEMA 250-1985, copyright © 1985 by NEMA.

[2]This material is reproduced with the permission of the National Electrical Manufacturers Association from NEMA Standards Publication ICS 2-1988, copyright © 1988 by NEMA.

**Manual Motor Controllers: Maximum Ratings**

| NEMA size | Single-phase hp | | Three-phase hp | |
|---|---|---|---|---|
| | 115 V, 60 Hz | 230 V, 60 Hz | 200/230 V, 60 Hz | 460/575 V, 60 Hz |
| M-0 | 1 | 2 | 3 | 5 |
| M-1 | 2 | 5 | 7½ | 10 |
| M-1P | 3 | 5 | | |

**Ratings for Three-Phase Single-Speed Full-Voltage Magnetic Controllers for Nonplugging and Nonjogging Duty**

| Size of controller | Continuous current rating, A | Horsepower[a] at | | | | Service-limit current rating, A |
|---|---|---|---|---|---|---|
| | | 60 Hz | | 50 Hz | 60 Hz | |
| | | 200 V | 230 V | 380 V | 460 or 575 V | |
| 00 | 9 | 1½ | 1½ | 1½ | 2 | 11 |
| 0 | 18 | 3 | 3 | 5 | 5 | 21 |
| 1 | 27 | 7½ | 7½ | 10 | 10 | 32 |
| 2 | 45 | 10 | 15 | 25 | 25 | 52 |
| 3 | 90 | 25 | 30 | 50 | 50 | 104 |
| 4 | 135 | 40 | 50 | 75 | 100 | 156 |
| 5 | 270 | 75 | 100 | 150 | 200 | 311 |
| 6 | 540 | 150 | 200 | 300 | 400 | 621 |
| 7 | 810 | — | 300 | — | 600 | 932 |
| 8 | 1215 | — | 450 | — | 900 | 1400 |
| 9 | 2250 | — | 800 | — | 1600 | 2590 |

[a]These horsepower ratings are based on the locked-rotor current ratings given in Table 47, Div. 11. For motors having higher locked-rotor currents, a larger controller should be used so that its locked-rotor current rating is not exceeded.

**Ratings for Single-Phase Full-Voltage Magnetic Controllers for Nonplugging and Nonjogging Duty**

| Size of controller | Continuous current rating, A | Single-phase horsepower at | | Service-limit current rating, A |
|---|---|---|---|---|
| | | 115 V | 230 V | |
| 00 | 9 | ⅓ | 1 | 11 |
| 0 | 18 | 1 | 2 | 21 |
| 1 | 27 | 2 | 3 | 32 |
| 1P | 36 | 3 | 5 | 42 |
| 2 | 45 | 3 | 7½ | 52 |

In a manual controller all the basic operations, such as the closing of switches and the movement of rheostat handles, are performed by hand.

In a magnetic controller the basic functions of the closing of switches or the movement of rheostat handles are performed by magnetic contactors (see Sec. 255). A full-magnetic controller is one which automatically performs all its operations in the proper sequence after the closure of a master switch. The master switch for a full-magnetic controller, such as a float switch, a pressure switch or regulator, or a time switch, is frequently automatically operated. In other cases, the master switch, such as a pushbutton, is manually operated. A semimagnetic controller operates with a combination of manual and magnetic performance. In certain cases, some of the basic operations are performed by hand and the rest by means of magnetic contactors. In other types of semimagnetic controllers, all the basic operations are performed with magnetic contactors, but the time and sequence of the operation of the contactors are controlled by hand.

**255. A magnetic contactor** is a switch operated by an electromagnet. One form of magnetic contactor is shown in Fig. 7-111, which illustrates the basic parts. It consists of a stationary electric magnet, as at $a$ in the figure, and a movable iron armature $b$, on which are mounted an electrical contact $c$ and a stationary electrical contact $d$. When no current is flowing through the operating coil $e$ of the electric magnet $a$, the armature is held away from the magnet by means of gravitational force or a spring. Under this condition the contacts $c$ and $d$ are open. When the operating coil is energized, the magnet attracts the armature and closes the electrical contacts $c$ and $d$.

Magnetic contactors may be of the shunt or the series type, depending upon the connection of the operating coil. In the shunt contactor the operating coil is connected across the terminals of the circuit or portion of the circuit that the contactor is controlling. In some cases the operating coil of a shunt contactor might be connected to a separate source of supply from the circuit controlled by the contactor. A series contactor has its operating coil connected in series with the circuit controlled by the contactor.

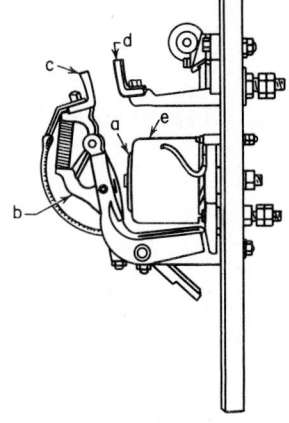

**Fig. 7-111** Magnetic contactor. [Square D Co.]

**256. Series contactors** for motor-control application are constructed so that the contactor will not close until the current of the circuit has dropped below a definite value. They are made in two general forms, *single-coil* and *two-coil*. The single-coil type is referred to simply as a series contactor, while the two-coil type is called a series lockout contactor. Both types, however, possess the lockout feature of the contactor; i.e., they are held open until the current has dropped to some definite value.

The general principle of operation of the single-coil type can be understood from a study of Fig. 7-112. The current passing through the operating coil produces an mmf which acts upon two magnetic paths. One path is through the core of the coil, a section of restricted area $A$, the armature, and main air gap back to the core of the coil. The other path is through the core of the coil, adjustable air gap, tail of armature $B$, armature, and

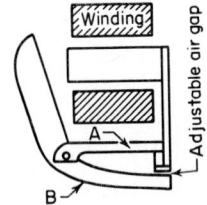

**Fig. 7-112** A series-contactor operating magnet.

main air gap back to the core of the coil. When no current is passing through the coil, the armature is in the open position. When a large current is passing through the operating coil, the mmf is so strong that part $A$ of restricted area becomes saturated so that the flux through this part cannot exceed a certain amount. The magnetic flux through the adjustable air gap and part $B$ will vary with the current through the operating coil, and this path will not become saturated. Any flux passing across the adjustable air gap produces an attracting force on the tail of armature $B$ which tends to hold the armature open against the closing force exerted between the core and the main part of the armature. As the current is reduced in value, the flux across the adjustable air gap is reduced and thus reduces the holding-out force on tail $B$. After the current has been reduced to a certain value, which depends upon the adjustment of the adjustable air gap, the holding-out force on tail $B$ is not sufficient to overcome the closing force exerted by the core upon the main part of the armature, and the contactor closes.

In the two-coil type of series contactor (Fig. 7-113), two coils are connected in series with each other and the circuit to be controlled. The armature is pivoted so that one coil, the larger of the two, acts upon one side of the armature, exerting a force tending to close the contactor. The smaller coil acts upon the other side of the armature and exerts a force tending to hold the contactor open. The areas of the magnetic circuits are so designed that the force tending to close the contactor remains practically constant regardless of the value of the current, while the force tending to hold the contactor open varies with the current. Therefore, after the current has decreased to a certain value, the holding-out force is not sufficient to overcome the closing force and the contactor closes. The air gap of the locking-out circuit can be adjusted so that the current above which the contactor holds open can be adjusted over a wide range.

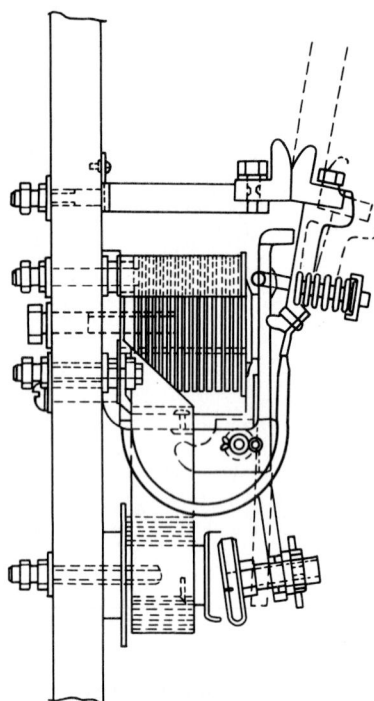

**Fig. 7-113** Magnetic lockout contactor.

**257. Overload protective devices.** Most motor controllers have incorporated in them as an integral part some device for protecting the motor against overloads harmful to it. These overload protective devices are made in many different types and forms. Those made by reliable manufacturers are Listed by the Underwriters Laboratories Inc. Before purchasing control equipment, take care to make sure that the equipment has been Listed by them. All the following types of overload protective devices are used in motor-control equipment.

1. Non-time-delay fuses
2. Time-delay fuses
3. Circuit breakers
4. Thermal sensors
5. Thermal overload relays
6. Magnetic overload relays
7. Solid-state overload relays

For a description of these devices see Div. 4. For selection of overload protective devices with respect to both size and type refer to Secs. 430, 431, and 432.

**258. Undervoltage devices.** Most motor starters are provided with either an undervoltage protective device or an undervoltage release device.

An *undervoltage protective device* is one which opens the supply circuit to the motor upon failure or reduction of voltage and which maintains the circuit open until it is again closed by the operator.

An *undervoltage release device* is one which operates on a reduction or failure of voltage, thereby opening the main circuit, but it does not prevent the automatic reclosing of the main circuit upon the return of normal voltage. It is applied in conjunction with some automatic starters. In certain applications, such as fans, blowers, and pumps, an automatic starter provided with an undervoltage release may be advantageously used. Its operation on a reduction or failure of voltage will stop the motor. Then the reestablishment of voltage causes the automatic starter to restart the motor, thus minimizing the shutdown period. An undervoltage release should not be employed in applications in which an unexpected return of power would endanger an operator.

**259. Methods of controlling the acceleration of motors** with the type of power system on which they may be used are given below. The simplest method is one in which the motor is thrown directly across the line by means of a manual or magnetic switch. The speed of acceleration of the motor and the magnitude and duration of the starting current depend solely upon the characteristics of the motor and its connected load.

With manual control of acceleration the speed of acceleration depends upon the judgment of the operator. Manual control can be obtained through the use of manual controllers or semimagnetic controllers. In semimagnetic controllers of this type the time and sequence of closing of the magnetic contactors are controlled by a manually operated master switch.

With automatic acceleration the only part of the starting operation that may depend upon the operator is the initiation of the sequence of events (such as pushing a START button). When this is done, the resulting sequence of operations is automatically performed and controlled by contactors alone or by contactors controlled by relays.

| Acceleration methods | Applicable to |
|---|---|
| A. No acceleration control (across-the-line starting) | AC and small dc motors |
| B. Manual | AC and dc |
| C. Automatic | |
|    1. Time | |
|       *a.* Solenoid with action retarded by an adjustable dashpot | DC |
|       *b.* Pneumatic timer | AC and dc |
|       *c.* Timing relay | AC and dc |
|       *d.* Inductive time limit | AC and dc |
|       *e.* Electronic timer | AC |
|    2. Current | AC and dc |
|    3. Counter-emf | DC |
|    4. Time-current (magnetic relay) | DC |
|    5. Time-current (electronic) | AC |
|    6. Frequency | AC |

**260. In time-limit acceleration** the closing of the respective contactor contacts is controlled by a timing device so that the starting resistance is shorted out of the circuit or connections to the autotransformer are changed in a certain definite time. Most modern automatic starters are of the time-limit type. They are used for all general-purpose applications. The other types of automatic acceleration are used for special applications in which they have advantages over the time-limit type. The timing of the closing of the respective contactors for time-limit acceleration can be accomplished in several ways, as listed in Sec. 259.

**261. In time-limit acceleration provided by a solenoid with action retarded by an adjustable dashpot,** the time of closing the magnetic accelerating contactors is controlled by a fluid dashpot attached to the lower part of the operating rod of a solenoid. In some starters the dashpot-controlled solenoid is a timing relay (Sec. 265) which controls the time of closing the operating coil of a magnetic accelerating contactor (Sec. 262).

**262. A fluid-dashpot timing relay** is illustrated in Fig. 7-114. The motion of the plunger of the solenoid is retarded by an adjustable fluid dashpot attached to the lower end of the solenoid plunger. The delayed time for closing the relay contacts can be adjusted from 2 to 30 s. The dashpot is filled with a silicone fluid whose viscosity is affected only very slightly by changes in temperature. Such relays will operate satisfactorily in ambient temperatures from $+120$ to $-30°$F ($+49$ to $-35°$C).

**263. Solid-state reduced-voltage starters,** as shown in Fig. 7-115, are available from several manufacturers in ratings up to at least 800 hp for standard three-phase, ac induction motors. These include such features as:

1. Adjustable starting current, from 100 to 400 percent of motor full-load current, to match breakaway torque to applied load.

2. Adjustable current limit, from 100 to 400 percent of motor full-load current, providing stepless acceleration at reduced current and torque.

3. Adjustable current ramp, from 2 to 30 s, for very gradual acceleration, such as to take up slack in the drive system even after the fixed turn-on ramp has expired.

4. Adjustable pulse start, from 0 to 2 s, providing an initial 400 percent current pulse to the motor to break loose high-friction, high-inertia loads.

5. Adjustable current trip, from 50 to 400 percent of motor full-load current, with contacts to initiate audible or visual alarm signaling when the motor reaches the trip setting.

6. Power-saving feature which reduces voltage to the motor when it is running at no load or is lightly loaded, to reduce energy consumption.

It should be noted that solid-state controllers employ silicon-controlled rectifiers (SCRs) which do not provide an air break when off. When the branch circuit disconnect is closed, a solid-state controller has potentially lethal voltages at the motor terminals as well as at the line terminals, even in the OFF state. An isolation contactor or magnetic contactor provides the necessary isolation.

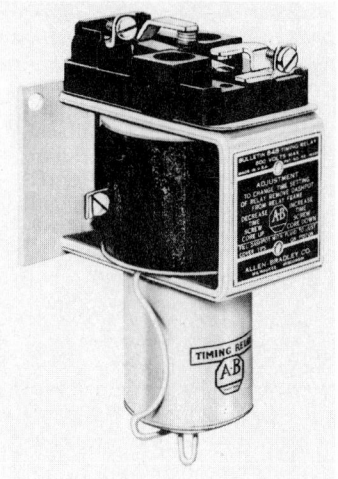

**Fig. 7-114** Fluid-dashpot timing relay. [Allen-Bradley Co.]

**Fig. 7-115** Solid-state reduced-voltage starter. [Eaton Corp., Cutler-Hammer Products]

**264. Contactors equipped with pneumatic timers** provide a means for the time control of acceleration. The timer of Fig. 7-116 consists of an air chamber and an adjustable vent which controls the time delay of the operation of the contactors. The timer is actuated by a lever attached to the arm of a contactor.

The principle of controlling acceleration by a mercury timer is shown in Fig. 7-117. All contactors except the last acceleration contactor are equipped with a pneumatic timer. When the START button is pressed, current flows from $L1$, through STOP and START buttons to 10, through the operating coil ③, through contacts $\overset{O.L.}{\multimap\!\!\multimap}$ , and back to $L2$. This energizes the operating coil ③ of the main-line contactor $\overset{3}{+\!\mid}$ . Contactor $\overset{3}{+\!\mid}$ closes, connecting the motor across the line in series with all the starting resistance. When contactor $\overset{3}{+\!\mid}$ closes, it closes its auxiliary contacts $\overset{3}{\multimap\!\!\multimap}$, which provides a means of ener-

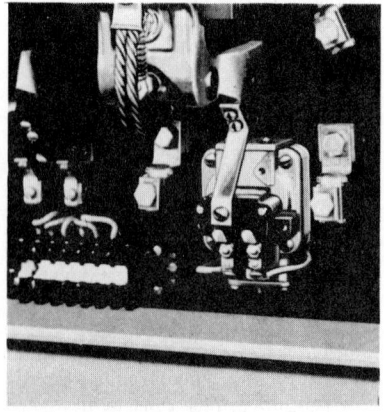

**Fig. 7-116** Pneumatic timer. [Square D Co.]

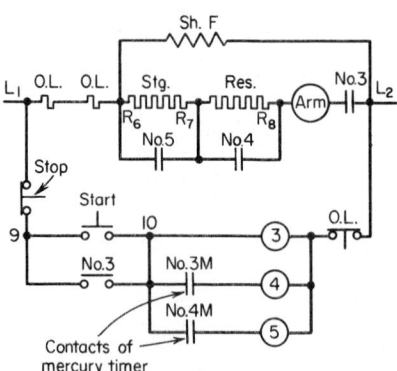

**Fig. 7-117** Principle of time-limit acceleration by means of pneumatic timers.

gizing the operating coils of the contactors when the START button is released. The closing of contactor 3 causes the pneumatic timer $3M$ to start to function. After a certain definite time required for the air to pass through the orifice of the plunger, pneumatic timer $3M$ closes its contact $\overset{3M}{+\,\vdash}$. This energizes the operating coil $(4)$ of accelerating contactor 4, which closes its contacts $\overset{4}{+\,\vdash}$. This shorts the resistance $R_7$–$R_8$. The closing of contactor 4 causes the pneumatic timer $4M$, attached to it, to start to function. After a certain definite time this timer closes its contacts $\overset{4M}{+\,\vdash}$, energizing the operating coil $(5)$ of accelerating contactor 5. Contactor 5 operates, closing its contacts $\overset{5}{+\,\vdash}$ and shorting out resistance $R_6$–$R_7$. The motor is now connected directly across the line.

**265. Timing relays** are the most commonly used devices to obtain automatic control of motor acceleration. A timing relay is a device which, when energized, will operate to close or open its contacts after a definite lapse of time from the instant that the relay is energized. Time-limit control of motor acceleration is accomplished with these relays through their time control of the closing or opening of the operating circuits of the motor-accelerating contactors.

The more common types are pneumatic timing relays, fluid-dashpot timing relays, and synchronous-motor timing relays. The fluid-dashpot and pneumatic relays are operated by a magnetic coil which, when energized, exerts a pull upon the central movable member, or plunger. The time taken for the movable member to close or open the contacts of the relay is governed by the retarding action of the dashpot or air bellows.

Dashpot timing relays (Fig. 7-114) are equipped with an adjustable fluid dashpot. The fluid used should be one whose viscosity is not affected by changes in ambient temperature. The timing range of these relays can be adjusted from about 2 to 30 s.

Pneumatic relays (Fig. 7-118) are equipped with an air chamber and adjustable vent which controls the time delay in the operation of the relay. The timing range of these relays can be adjusted from about 0.2 to 200 s.

Synchronous-motor timer relays (Fig. 7-119) consist of a set of contacts driven from the

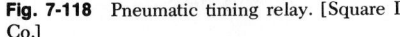

**Fig. 7-118**   Pneumatic timing relay. [Square D Co.]

**Fig. 7-119**   Synchronous-motor timer relay. [General Electric Co.]

shaft of a very small synchronous motor so that there is a definite time lag between the starting of the motor and the closing of the contacts. A controller utilizing such a relay is described in Sec. 312.

**266. Acceleration controlled by inductive time-limit contactors or relays** makes use of the inductive time-lag effect of the decay of current in a heavily inductive circuit. This method of timing is often called *magnetic timing*. In controllers employing this method, each accelerating contactor may be of the inductive time-limit type or of the ordinary shunt type, each being controlled by an inductive time relay. Inductive time-limit contactors may be of the normally open-contact (holdout) type or of the normally closed type.

**267. An inductive time-limit accelerating contactor of the normally open-contact type** (Fig. 7-120) consists of a pivoted armature acted upon by two coils located on opposite sides of the pivot point. The main operating coil acts to close the contactor, while the auxiliary coil acts to hold the contactor open. In the diagram of the controller of Fig. 7-121 the coils marked $2R$, $3R$, $4R$, and $5R$ are the main operating coils of the respective inductive time-limit contactors $2R$, $3R$, $4R$, and $5R$. The coils marked $HC2$, $HC3$, $HC4$, and $HC5$ are the holdout coils of the contactors $2R$, $3R$, $4R$, and $5R$, respectively.

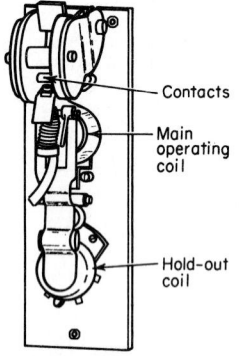

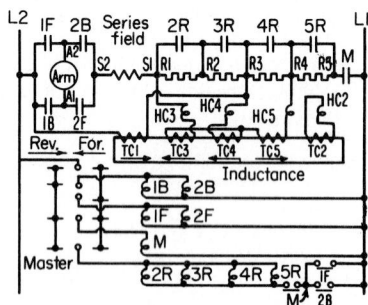

**Fig. 7-120** Inductive time-limit contactor of the normally open-contact type.

**Fig. 7-121** Connections for inductive time-limit acceleration. [Eaton Corp., Cutler-Hammer Products]

**268. The principle of controlling acceleration by inductive time-limit accelerating contactors of the normally open-contact type** is illustrated by the typical control diagram of Fig. 7-121. In addition to the contactors, an iron-core inductive unit is required. This inductance consists of a set of coils $TC1$, $TC2$, $TC3$, $TC4$, and $TC5$ wound on a common iron core. In starting the motor from rest, if the operator moves the master switch to the running position for forward direction of rotation, the circuits of operating coils $1F$, $2F$, and $M$ are closed. These coils operate their respective contactors, causing contacts $1F$, $2F$, and $M$ to close. This connects the motor across the line in series with all the starting resistance. Voltage will be impressed upon coils $TC1$, $TC3$, $TC4$, and $TC5$, causing these coils to be energized and to hold out their respective contactors. But the coils are so connected that the mmf's of the coils just neutralize each other, so that no flux is produced in the core. There is therefore no current in the closed circuit of coils $HC2$ and $TC2$, and contactor $2R$ is free to close when its operating coil is energized. As soon as contactors $1F$ and $M$ have closed, auxiliary contacts on these contactors close the circuits of the operating coils $2R$, $3R$, $4R$, and $5R$. Contactor $2R$ immediately closes, shorting out the section of resistance $R_1$–$R_2$ and shorting coils $HC3$ through $TC3$. This unbalances the mmf's of the inductance, and a definite flux is produced in the core. This change in flux and the inductance of coils $HC3$ and $TC3$ oppose the delay of current in $HC3$ and $TC3$. After a certain lapse of time the current in $HC3$ will drop to nearly zero. The holdout coil $HC3$ of the contactor $3R$, being now deenergized while its main coil $3R$ is energized, causes contactor $3R$ to close. The closing of contactor $3R$, shorting coils $HC4$ and $TC4$, again unbalances the flux. After a definite time lapse the current in $HC4$ will drop to nearly zero, and contactor $4R$ will close. Closing of $4R$ shorts coils $HC5$ and $TC5$ and once again unbalances the flux. After another definite time interval the current $HC5$ drops to nearly zero, and contactor $5R$ is allowed to close. The motor is then connected directly to the line.

When plugging, i.e., throwing from full speed in one direction to full speed in the other, one should have a time delay in the closing of contactor $2R$. Coil $TC1$ is connected across the armature and resistance $R_1$–$R_2$. The current through coil $TC$ therefore depends upon the sum of the voltage across the armature and the resistor $R_1$–$R_2$. The voltage across the armature when starting from rest is very low, but in plugging, it is normal line voltage. In plugging, therefore, the current through $TC1$ is much higher than it is when starting from rest. This larger current through $TC1$ unbalances the mmf's of the inductance and produces a flux which induces a current in shorted coils $HC2$ and $TC2$. This induced current will hold out contactor $2R$ until the current has had time to be reduced to nearly zero.

**269. The principle of time-limit acceleration by inductive time-limit contactors of the normally closed-circuit type** is illustrated in Fig. 7-123. A contactor of this type is shown in Fig. 7-122. A copper tube encircles the magnetic core of the contactor, over which two operating coils are wound. These two coils act in opposition to each other. When the main winding is open-circuited after having been carrying current, the magnetic flux decreases. This decrease in flux induces a heavy current in the enclosed copper tube. The induced current opposes the decay of the flux, so that there is a considerable lapse of time after the circuit of the main coil is opened before the flux decays sufficiently to allow the armature to be released. The length of time delay will depend upon the original value of the flux, which can be controlled by adjusting the current in the neutralizing winding.

Acceleration is automatically performed in the following manner. When the START button is pressed, current flows from $L1$, through STOP and START contacts to 7, through contacts 5, operating coil (11/M), contacts O.L., and back to line $L2$. The armature of contactor 11 is therefore attracted, opening its main contacts 11 and closing its auxiliary contacts 11. The closing of the two sets of auxiliary contacts 11 energizes the main operating coil (12/M) and neutralizing coils (11/N) and (12/N). The armature of contactor 12 is therefore attracted, opening its main contacts and closing its auxiliary contacts 12. This energizes the operating coil (5), which closes the main-line contactor 5. The motor is then connected to the line in series with all the starting resistance. When contactor 5 closes, it

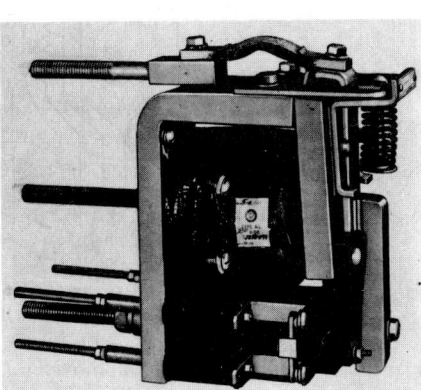

**Fig. 7-122** Inductive time contactor of the normally closed-contact type. [Westinghouse Electric Corp.]

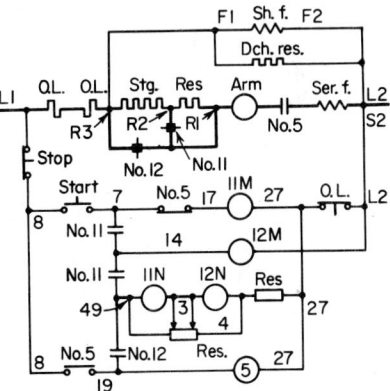

● Main contacts of inductive time-limit accelerating contactors (contacts normally closed)

╪ Auxiliary contacts of inductive time-limit accelerating contactors

**Fig. 7-123** Principle of time-limit acceleration by means of inductive-time-limit contactors of the normally closed-contact type.

opens its auxiliary contacts $\overset{5}{\multimap\!\!\multimap}$ and closes its auxiliary contacts $\overset{5}{\underset{\circ\ \circ}{\rule{0.3cm}{0.4pt}}}$ . The closing of contacts $\overset{5}{\underset{\circ\ \circ}{\rule{0.3cm}{0.4pt}}}$ keeps the operating circuits of the contactors energized when the START button is released. The opening of contacts $\overset{5}{\multimap\!\!\multimap}$ opens the main operating coil circuit $-\!\!\bigcirc\!\!\!\!\tiny{M}\!\!-$ so that after a definite time, determined by the adjustment of the resistances controlling $-\!\!\bigcirc\!\!\!\!\tiny{\substack{11\\N}}\!\!-$ and $-\!\!\bigcirc\!\!\!\!\tiny{\substack{12\\N}}\!\!-$, contactor 11 drops out and thereby closes its main contacts, shorting out the resistance $R_2$–$R_2$. At the same time, auxiliary contacts $\pm$ 11 are opened, which opens the circuit of the main operating coil $-\!\!\bigcirc\!\!\!\!\tiny{\substack{12\\M}}\!\!-$. After a definite time the flux of contactor 12 decreases a sufficient amount so that contactor 12 drops out, closing its main contacts and shorting out resistance $R_3$–$R_2$. The motor is now connected directly across the line. As contactor 12 drops out, its auxiliary contact $\pm$ 12 opens, which disconnects all the operating coils of the accelerating contactors. In case of an overload, the operating coils or bimetallic heater elements open the contacts $\overset{O.L.}{\multimap\!\!\multimap}$, deenergizing the operating coil $\bigcirc\!\!\!\tiny{5}$ of the main-line contactor so that the motor is disconnected from the line and will not start up again until pressing the START button again starts the above sequence of starting operations. To stop the motor, the STOP button is pressed, which opens the circuit of the operating coil $\bigcirc\!\!\!\tiny{5}$ of the main-line contactor.

**270. Acceleration by means of inductive time relays** requires the use, for each step of resistance, of an ordinary shunt contactor in conjunction with a magnetic time relay. A magnetic time (inductive-time) relay can be constructed in just the same manner as the inductive-time accelerating contactors of the normally closed-contact type described in Sec. 269. The only differences are that it is smaller in size and that its main contacts are not capable of carrying as heavy currents. The principle of operation in obtaining acceleration by this means is shown in Fig. 7-124. The same general system of symbols is used as given in Sec. 269 and Fig. 7-123. The main contacts of the line contactor are $\overset{5}{\dashv\vdash}$. The

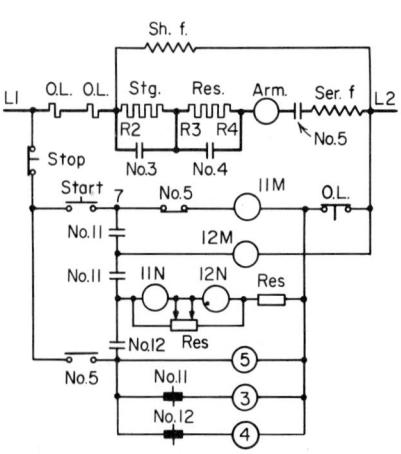

contacts of the regular shunt-coil accelerating contactors are $\overset{3}{\dashv\vdash}$ and $\overset{4}{\dashv\vdash}$. The main contacts of the inductive-time-limit relays are $\blacktriangleright\!\!\dashv$ 11 and $\blacktriangleright\!\!\dashv$ 12 . When the START button is pressed, current flows from $L1$ through the STOP and START button to 7, through contacts $\overset{5}{\multimap\!\!\multimap}$, operating coil $-\!\!\bigcirc\!\!\!\!\tiny{\substack{11\\M}}\!\!-$, contacts $\overset{O.L.}{\multimap\!\!\multimap}$ , and back to line $L2$. The armature of relay 11 is therefore attracted, opening its main contacts $\blacktriangleright\!\!\dashv$ 11, which opens the circuit operating coil $\bigcirc\!\!\!\tiny{3}$. At the same time auxiliary contacts $\overset{11}{\dashv\vdash}$ close, thus energizing operating coil $-\!\!\bigcirc\!\!\!\!\tiny{\substack{12\\M}}\!\!-$ and neutralizing coils $-\!\!\bigcirc\!\!\!\!\tiny{\substack{11\\N}}\!\!-$ and $-\!\!\bigcirc\!\!\!\!\tiny{\substack{12\\N}}\!\!-$. The armature of relay 12 is therefore attracted, opening its main con-

**Fig. 7-124** Principle of operation of time-limit acceleration by means of inductive-time-limit relays.

tacts ✦12 , which open the circuit of operating coil ④. At the same time auxiliary con-

tacts ⊣⊢ close, energizing operating coil ⑤, which closes the main-line contactor 5. The

motor is then connected to the line in series with all the starting resistance. When contac-

tor 5 closes, it closes its auxiliary contact ⟍⟍5⟋⟋ and opens its auxiliary contact ₒ5ₒ. The

closing of ⟍⟍5⟋⟋ makes power available for the operating coils when the START button is

released. Opening of contact ₒ5ₒ opens the main operating-coil circuit ⊸⊛⊷ so that after

a definite time determined by the adjustment of the resistance controlling the neutralizing

coils ⊸⊛⊷ and ⊸⊛⊷ relay 11 drops out and thereby closes its main contacts ✦11. Clos-

ing of contacts ✦11 energizes the operating coil ③, which operates contactor 3, shorting

out the resistance $R_2$–$R_3$. As relay 11 drops out, it opens its auxiliary contacts ⊣⊢, which

open the circuit of the main operating coil ⊸⊛⊷ of relay 12. After a definite time, the flux

of relay 12 decreases a sufficient amount so that relay 12 drops out, closing its main con-

tacts ✦12 and energizing the operating coil ④. Operating coil ④ closes the contacts of

contactor 4, which short out the resistance $R_3$–$R_4$. The motor is now connected directly

across the line. As relay 12 drops out, its auxiliary contacts ⊣⊢ open, which disconnects all

the operating coils of the accelerating relays.

**271. Time-limit acceleration controlled by pneumatic timing relays attached to contactors** (see
Fig. 7-125). When the START button is pressed, current flows from + through the START
and STOP buttons, through the overload relay *OL* contacts, through the main contactor

operating coil ⊸Ⓜ⊷ , and back to −. This energizes the operating coil ⊸Ⓜ⊷ of the main-

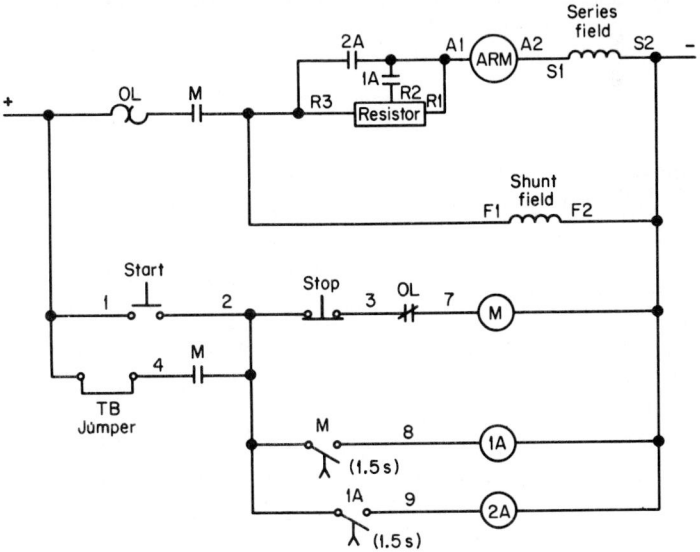

**Fig. 7-125** Principle of time-limit acceleration by means of the pneumatic timing-relay method
applied to a dc motor. [Square D Co.]

line contactor, which closes its contact $\underset{M}{\dashv\vdash}$ and connects the motor across the line in series with all the starting resistance. When the contactor $M$ closes, it also closes its auxiliary contact $\underset{M}{\dashv\vdash}$ , which provides a current path around the START push button, which can now be released. A pneumatic timer attached to the $M$ contactor also begins its time delay when the $M$ contactor closes. After the preset delay, the timed contact $\underset{M}{\multimap\!\!\vdash}$ closes and energizes the operating coil of the first acceleration contactor $(IA)$ . When $(IA)$ is energized, acceleration contact $\underset{IA}{\dashv\vdash}$ closes and bypasses a portion of the resistance in series with the motor. A pneumatic timer attached to the 1A contactor also begins its time delay when the 1A contactor closes. After the preset delay, the timed contact $\underset{IA}{\multimap\!\!\vdash}$ closes and energizes the operating coil of the second acceleration contactor $(2A)$ . When $(2A)$ is energized, acceleration contact $\underset{2A}{\dashv\vdash}$ closes and bypasses the remainder of the resistance in series with the motor. The motor is now running across the line.

Primary-resistor starting of squirrel-cage motors is another effective method (see Fig. 7-126). Operation of the pilot device initiates the timing relay $TR$ and closes its two normally open auxiliary contacts $\underset{TR}{\dashv\vdash}$. This energizes the coil $(M)$ of the main-line contactor and closes its contacts $\underset{M}{\dashv\vdash}$. Since the voltage drop in the resistors causes a decreased voltage at the motor terminals, the motor starts and accelerates at reduced voltage, with a consequent reduction in current and torque. After a certain time delay, the timing relay $TR$ times out, and the timed contact closes and energizes the operating coil $(A)$ of the accelerating contactor. When coil $(A)$ is energized, the contacts $\underset{A}{\dashv\vdash}$ are closed, causing resistors to be shorted out and thereby transferring the motor to line-voltage operation. The motor remains energized during the transfer from the reduced-voltage start to line-voltage operation. Note the elementary wiring diagram for this control in Fig. 7-126.

**272. With current-limit acceleration,** cutting out the various steps of starting resistance or changing taps on an autotransformer is controlled by the decay of the starting current as the motor accelerates. The rapidity of closing the successive contactors depends upon the load of the motor. With light loads the motor will accelerate on each step more rapidly, so that the current decreases more rapidly and causes the successive contactors to function in a shorter time. Total time for the starting period therefore depends upon the load.

Current-limit acceleration is performed by means of series contactors of the types described in Sec. 256. Acceleration can be accomplished by using series contactors of the single-coil type as shown in Fig. 7-127. Pressing the START button energizes the operating coil $(3)$ of the main-line contactor 3 so that line contacts $\underset{3}{\dashv\vdash}$ close, connecting the motor to the line in series with the starting resistance. When the starting current has decreased to the value for which the series contactor 4 is adjusted, operating coil $(4)$ closes its contactor 4. The closing of contacts $\underset{4}{\dashv\vdash}$ shorts out the section of resistance $R_7$–$R_8$ and allows the starting current to pass through operating coil $(5)$ of the second series accelerating contactor. When the starting current has decreased to the value for which the series contactor 5 is adjusted, operating coil $(5)$ closes its contactor 5. Closing of contacts $\overset{6}{\phantom{x}}$ shorts out all the starting resistance and also the main operating coils of contactors 4 and

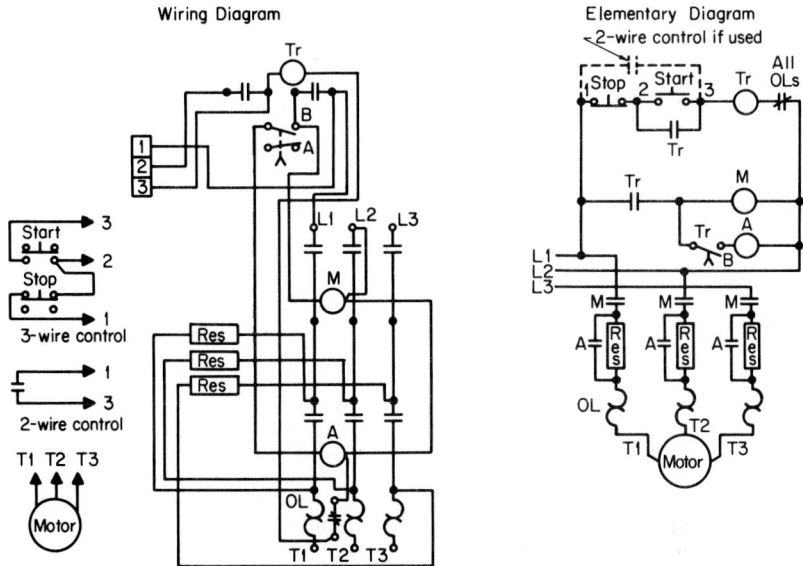

Sizes 3 and 4, three phase primary resistor starter

**Fig. 7-126**   Principle of primary-resistor starting of squirrel-cage motors. [Square D Co.]

5, so that there will be no danger of these contactors dropping out under light loads. Contactor 5 has an auxiliary shunt holding coil $\overset{5}{\sim}$ which holds this contactor closed after it has once been closed by its main operating coil ⑤ . When contactor 5 closes, it also closes its auxiliary contacts $\overset{5}{\circ\ \circ}$, thus energizing the auxiliary holding-coil circuit for the contactor. The motor is now connected directly to the line.

The method of controlling acceleration by means of two-coil series lockout contactors (Sec. 256) is shown in Fig. 7-128. Pressing the START button energizes the operating coil ③ of the main-line contactor 3 so that line contacts $\overset{3}{+\!\vdash}$ close, connecting the motor to the line in series with the starting resistance. When the starting current has decreased to the value below that for which the lockout coil $-④L-$ of contactor 4 will hold this contactor open, contactor 4 closes its contacts $\overset{4}{+\!\vdash}$ . This shorts out the section of starting resistance

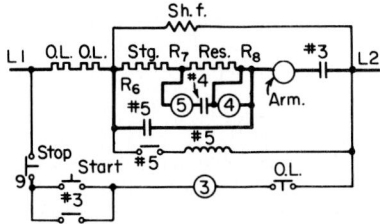

**Fig. 7-127**   Principle of current-limit acceleration by means of series contactors of the single-coil type.

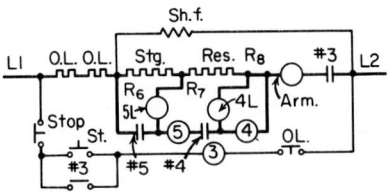

**Fig. 7-128**   Principle of current-limit acceleration by means of series contactors of the two-coil type.

$R_7$–$R_8$ and the lockout coil ④L of contactor 4 so that the contactor will not drop out under light loads. The current now passes through the operating coil ⑤ and lockout coil ⑤L of contactor 5, which locks open until the current has dropped in value to the amount for which the contactor is adjusted to operate. When the current has dropped to this value, contactor 5 closes its contacts ⑤ , which short out the resistance $R_6$–$R_7$ and the lockout coil ⑤L . With the lockout coils shorted out, these contactors do not need

any auxiliary shunt holding coil, since the series operating coils will hold the contactors closed with only about 5 percent of their normal current. The motor is now connected directly across the line.

**273. Counter-emf acceleration** is really a special method of current-limit acceleration, since the current that flows through the motor depends upon the value of the counter-emf. It is applicable only to dc motors. The time required for the starting period will depend upon the load on the motor in just the same way as for the other current-limit acceleration methods.

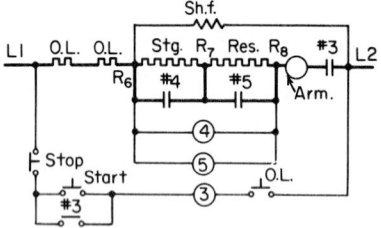

**Fig. 7-129**    Principle of counter-emf acceleration.

When acceleration is controlled by the counter-emf method, the operating coils of the magnetic contactors are connected across the armature winding (Fig. 7-129). At standstill no counter-emf is produced in the armature winding. As soon as the motor starts to revolve, a counter-emf is generated in the armature winding. The counter-emf is directly proportional to the speed so that as the motor speeds up, the counter-emf increases. Since the operating coils of the magnetic contactors are connected across the armature, their operation depends upon the value of the counter-emf. The magnetic contactors are adjusted so that a definite value of counter-emf is required to operate each one, a higher value of counter-emf being required to operate each successive contactor. When the speed has reached a value at which the counter-emf is sufficient to operate one of the contactors, the contactor closes and shorts out a section of the starting resistance.

**274. Time-current control of acceleration,** as the name implies, is a combination of time and current control of the accelerating period. With time control the time of acceleration is entirely independent of the load on the motor, while with current control the time of acceleration depends entirely upon the load on the motor. With time-current control the time of acceleration is modified by the load on the motor but is not entirely dependent upon the load.

One method of obtaining time-current control of acceleration is through the use of the time-current acceleration relay of Fig. 7-130. This relay consists of a series operating coil $A$, a steel core $C$, a core support plate $B$, an air gap $F$, and a movable circular inductor tube $E$, carrying the movable contacts $G$ and a set of stationary contacts $H$. When the circuit of the operating coil is closed, the current changes in a very short time from zero to some definite value, depending upon the circuit in series with which the coil is connected. A magnetic flux is built up rapidly in the iron core $C$ and air gap $F$. This flux also links with the inductor tube $E$, which forms a closed single-turn conductor. The rapid change of flux linked with the induction tube $E$ causes considerable current to be induced in the induction tube. The current through the operating coil and the induced current in the inductor tube are in such directions that they produce a repelling force between the coil and the tube. The tube is therefore rapidly propelled upward upon the closing of the operating-coil circuit. The parts are so designed that considerable force is exerted upon the tube, and it would travel 1 ft (0.3 m) or more if it were not for the fact that as the tube starts to move, it passes through the field $F$, which now causes eddy currents to be set up in the tube. These eddy currents cause a retarding action, just as a wattmeter vane is retarded as it is forced through a magnetic field, and this, combined with the force of gravity, stops the tube, which then starts to fall under the force of gravity. Inasmuch as current still flows through $A$, causing flux in $F$, the inductor tube is likewise retarded by eddy currents in the downward direction just as it is retarded in the upward direction. This retarding action therefore provides a time delay, which may be as much as 1.45 s, in the closing of contacts G and H. The length of time will depend upon the current A and the type of inductor tube used. Since the current in A determines the strength of the flux in F, C, and E, a lower current in A will cause the inductor tube to jump less distance,

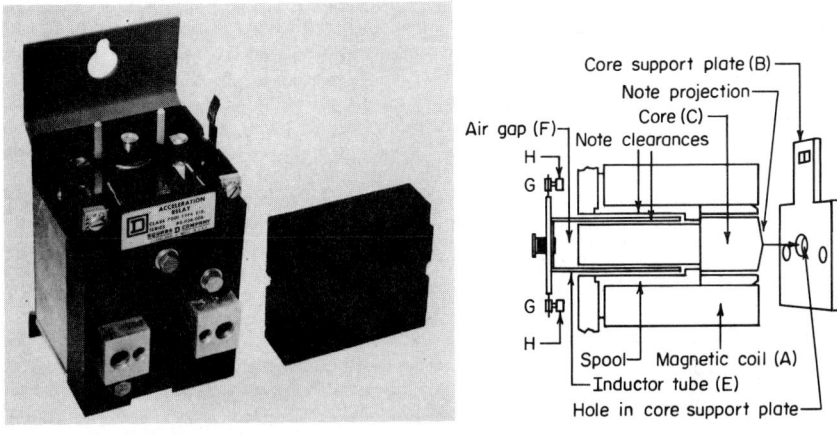

I. Complete relay.                    II. Cross-section view.

**Fig. 7-130**    Time-current relay. [Square D Co.]

and the retarding action in the downward direction will be less, so the time also will be less. In practice, the current in A is the same as the motor current, so that when heavy current flows in the motor owing to a heavy load, the time will be longer, and when the motor current is less as on light loads, the time will be less. Hence we have time-current control, i.e., a system of control in which the time is modified by the current or load on the motor.

The principle of time-current control of acceleration with these time-current relays is illustrated by the elementary control circuit in Fig. 7-131. When the START button is pressed, the operating coil ③ of the main-line contactor 3 is energized. Contacts $\overset{3}{-||-}$ close and connect the motor to the line in series with the starting resistance. Upon the closing of contactor 3, current passes through the operating coil ⑥ of time-current relay 6, and the auxiliary contacts $\overset{3}{-\circ\,\circ-}$ of contactor 3 close. Current will start to flow through operating coil ④, but before its contactor 4 has time to close its contacts $\overset{4}{-||-}$, the inductor-tube relay 6 has been raised and opens the contacts $\overset{6}{-\circ\,\circ-}$ of this relay, thereby opening the circuit of operating coil ④. After a time delay depending upon the setting of relay 6 and the load on the motor, the inductor tube of relay 6 drops and recloses contacts $\overset{6}{-\circ\,\circ-}$. The operating coil ④ of contactor 4 is now ener-

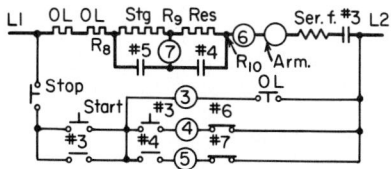

**Fig. 7-131**    Principle of time-current accelera-tion by means of a magnetic time-current relay.

gized, and contactor 4 operates, closing its contacts $\overset{4}{-||-}$. This shorts out the resistance $R_9$–$R_{10}$, closes auxiliary contacts $\overset{4}{-\circ\,\circ-}$, and energizes the operating coil ⑦ of time-current relay 7. Current will start to flow through the operating coil ⑤ of contactor 5, but before

the contactor can close, the inductor tube of relay 7 has jumped up and opened the con-

tacts $\frac{7}{\circ\!-\!\circ}$ . After a time delay, the inductor tube of relay 7 drops, reclosing the contacts

$\frac{7}{\circ\,\,\circ}$ of this relay and energizing the operating coil ⑤ of contactor 5. Contactor 5 now

operates and closes its contacts $\overset{5}{-\!|\!|\!-}$. The motor is now connected directly to the line.

**275. AC static, stepless crane control** for wound-rotor motors is a development of the Square D Co. This type of control is used with a three-phase, ac wound-rotor motor. The control consists of thyristor assemblies on the primary side and fixed secondary resistors. Stepless, regulated hoisting and lowering speeds are obtained by using a thyristor step-less master switch and a permanent-magnet type of tachometer. Indicating lights mounted on control modules help the troubleshooting of modules (Fig. 7-132).

Five pairs of thyristor assemblies are used in the motor primary circuit to provide static, stepless control of the phase sequence and magnitude of motor voltage. The power-limit stop is arranged to disconnect the motor and directly open the brake circuit when the switch trips. Balanced three-phase motor power for lowering out of the tripped limit stop is also provided. A fixed resistor in each phase of the motor secondary is used to alter the basic speed-torque characteristic for speed-regulated operation. The application of this type of control is shown in Fig. 7-132.

Reduced speed is obtained by adjusting the degree of conduction of the power thyris-tors. The solid-state control modules produce properly synchronized firing pulses to turn on the thyristors in response to a speed-control signal.

The master-switch inductor unit produces an ac output voltage which is proportional to the amount of deflection of the master-switch handle from the off point. The signal con-verter changes this ac command signal to a dc signal. A permanent-magnet type of tach-ometer generator, mechanically coupled to a rotating unit of the hoist drive, provides the feedback signal used to obtain speed regulation. The signal comparator module compares these signals. Its output speed-control signal is used to actuate the thyristor control mod-ules, which energize either the hoist or lower thyristors to produce proper motor operation.

The tachometer continuity module is used to apply an ac voltage to the tachometer circuit and its collectors to ensure continuity. This ac voltage operates the undervoltage relay. Overspeed protection is provided by a mechanical centrifugal overspeed switch in the tachometer assembly.

**276. Frequency control of acceleration** is a development of the EC&M Division of the Square D Co. for application to wound-rotor induction motors. The operation of the fre-quency relays employed in this method depends upon the action of resonant circuits. An ac circuit which contains both inductance and capacity will be resonant at some definite frequency, depending upon the value of the inductance and the capacity. At the resonant frequency, such a circuit will pass a very much higher value of current than at any other frequency. The current of such circuits will increase with frequency up to the res-onant frequency. The operating coils of the frequency relays, which have a certain value of inductance, will produce a resonant circuit when connected in series with an external condenser. If condensers of different capacities are used with a set of relays, the respec-tive relays will become resonant at different frequencies (Fig. 7-133). The higher the capacity, the lower the resonant frequency.

The application of this method of acceleration to a wound-rotor induction motor is shown in Fig. 7-134. The relays *PR, 1AR, 2AR,* and *3AR* are frequency relays. Each relay operating coil in series with its capacitor is connected to a potentiometer rheostat so that it can be adjusted to the proper current. The relays are supplied with power from the secondary resistor so as to receive a maximum voltage of 90 V, on 60-Hz circuits, when the motor primary is first connected to the line. The relays have normally closed contacts. After a relay has attracted its armature, the armature will not drop out until the current of the operating coil has decreased to a certain value. The frequency at which the current reaches the dropout value will vary with the resonant frequency of the circuit. The higher the resonant frequency, the higher the frequency at which the dropout current will be reached. The relays are connected in series with progressively larger values of capacity, so that the resonant frequencies of the relays progressively decrease. The dropout cur-rents will occur at progressively lower frequencies. At standstill the frequency of the

**Fig. 7-132** AC static, stepless thyristor control for crane ac wound-rotor motors. [Square D Co.]

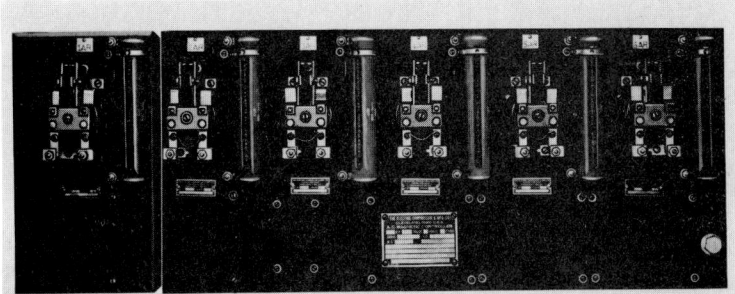

**Fig. 7-133** Frequency relays for the control of acceleration of a wound-rotor induction motor. [Square D Co.]

Elementary Diagram for Bridge or Trolley Control

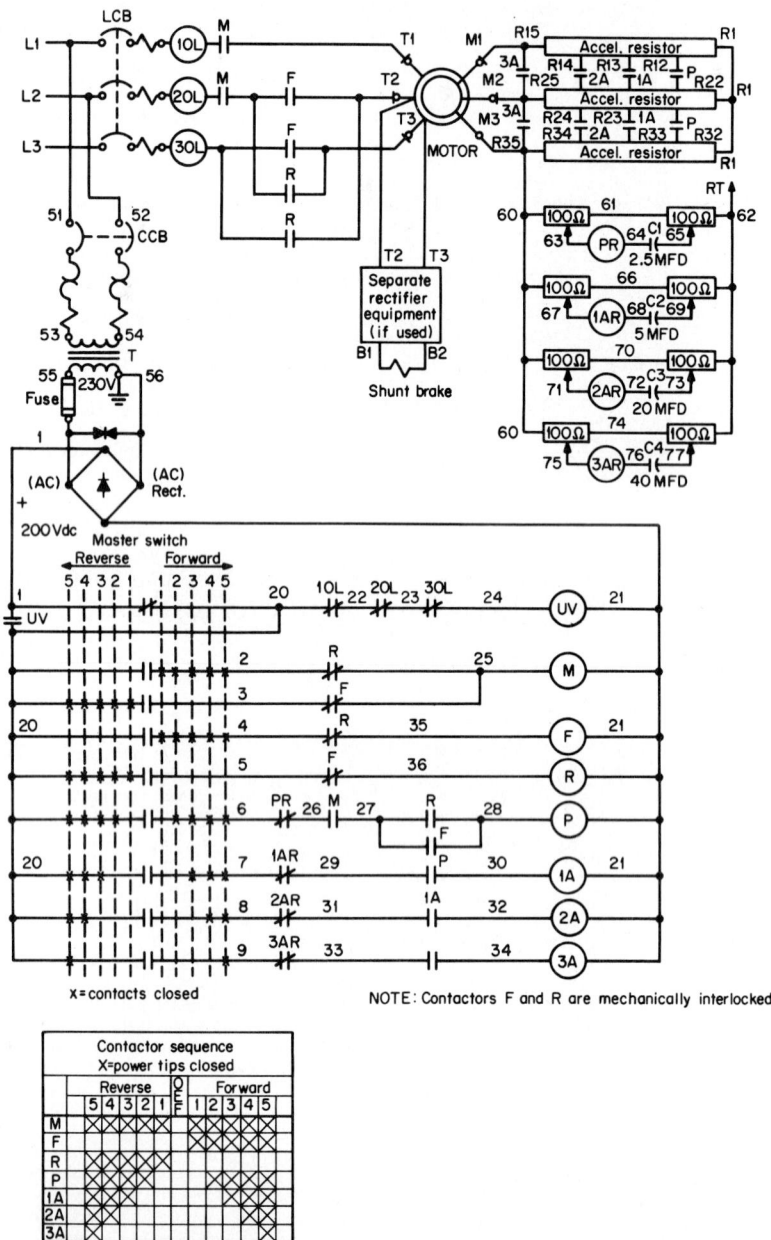

**Fig. 7-134**   Principle of frequency acceleration.

secondary circuit of an induction motor is the same as that of the primary supply circuit and decreases as the speed of the motor increases. It is approximately zero at full speed. The frequency relays will therefore drop out progressively as the motor accelerates. The operation of the motor illustrated can be controlled manually by moving the master controller to successive positions, or the acceleration can be made fully automatic by throwing the master to the last position. If the master is thrown to the last position forward, when the motor is at rest, the master contacts $MF$, $F$, $P$, $1A$, $2A$, and $3A$ will all be closed.

The motor will then be automatically accelerated as follows. Operating coils -(M)- and -(F)- are first energized; they operate contactors $F$ and $M$ and close contacts $\overset{F}{\dashv\vdash}$ and $\overset{M}{\dashv\vdash}$.

Closing contactor $F$ closes its auxiliary contacts $\overset{F}{\underset{\circ\ \circ}{\quad}}$ . The motor is now connected to the line with resistance in its rotor circuit. The starting current in the rotor circuit energizes the operating coils $PR$, $1AR$, $2AR$ and $3AR$ of all the frequency relays. The relay $PR$, which is the plugging relay and has only $2\frac{1}{2}\ \mu\mathrm{F}$ in series with its coil, will not pass enough current to operate. Since all the frequency relays have normally closed contacts, the contacts $\overset{PR}{\underset{\circ\ \circ}{\quad}}$ of relay $PR$ will stay closed, and operating coil -(P)- of contactor $P$ will be energized. Thus contactor $P$ will close its contacts $\overset{P}{\dashv\vdash}$ immediately. The frequency relays $1AR$, $2AR$, and $3AR$ will operate immediately upon passage of current through the motor and therefore will open their contacts $\overset{1AR}{\underset{\circ\ \circ}{\quad}}$ , $\overset{2AR}{\underset{\circ\ \circ}{\quad}}$, and $\overset{3AR}{\underset{\circ\ \circ}{\quad}}$. Closure of contactor $P$ closes its auxiliary contacts $\overset{P}{\underset{\circ\ \circ}{\quad}}$ , but operating coil -(1A)- will not be energized until relay $1AR$ drops out and closes its contacts $\overset{1AR}{\underset{\circ\ \circ}{\quad}}$. Relay $1AR$ will drop out after the motor has accelerated to a sufficient speed so that the frequency of the rotor is such as to cause the current of operating coil -(1AR)- to decrease the dropout value. When relay $1AR$ drops out, it closes its contacts $\overset{1AR}{\underset{\circ\ \circ}{\quad}}$ and energizes the operating coil -(1A)- of contactor $1A$. Contactor $1A$ then operates and closes its contacts $\overset{1A}{\dashv\vdash}$, shorting out some of the resistance. In a similar manner as the motor accelerates, frequency relays $2AR$ and $3AR$ drop out at progressively higher speeds and cause contactors $2A$ and $3A$ to close.

**277. Methods of starting dc motors.**   As discussed in Sec. 134, dc motors are almost always started with resistance in series with the armature winding. Both manual and automatic starters are used.

**278. Types of controllers for dc motors**

| *Controller Type* | *Remarks* |
|---|---|
| 1. Manual starters | |
| *a.* Across-the-line | Available in nonreversing and reversing types without dynamic braking |
| *b.* Starting rheostats (dial-type starters) | Obsolete; suitable only for nonreversing service without dynamic braking |
| *c.* Drum | Suitable for nonreversing and reversing service with or without dynamic braking |
| *d.* Multiple-switch | Suitable only for nonreversing service without dynamic braking |
| 2. Manual-starting and speed-regulating controllers | |
| *a.* Starting and speed-regulating rheostats (dial-type controllers) | Obsolete; suitable only for nonreversing service without dynamic braking |
| *b.* Drum | Suitable for nonreversing and reversing service with or without dynamic braking |
| 3. Magnetic starters | |
| *a.* Time-limit acceleration | |
| *b.* Current-limit acceleration | Available in nonreversing and reversing types with or |
| *c.* Counter-emf–limit acceleration | without dynamic braking |
| *d.* Time-current acceleration | |

**279. Across-the-line manual starters for dc motors** are made as standard equipment in sizes up to and including 2 hp. They consist of a manually operated switch in a suitable enclosure. They can be obtained with or without thermal overload relays. Those of the smaller types are sometimes called *auxiliary* or *secondary breakers* (see Div. 4). Most of the overload relay devices can be reset by throwing the handle to the extreme off and then to the on position. The release mechanism generally is constructed on the trip-free principle so that the switch cannot be held closed against a dangerous overload. These starters cannot be provided with undervoltage release or protective devices. Typical starters of these types are shown in Figs. 7-135, 7-136A, 7-136B, 7-136C, and 7-137.

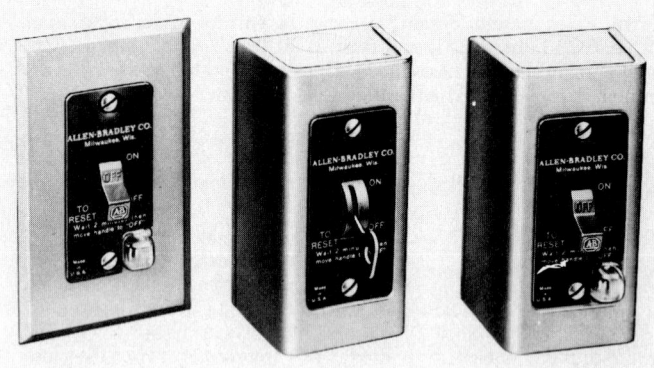

I. Flush-plate type for switch-box mounting.

II. Enclosed key-operated toggle-switch type.

III. Enclosed toggle-switch type.

**Fig. 7-135** Manual across-the-line fractional-horsepower starters for ac or dc motors. [Allen-Bradley Co.]

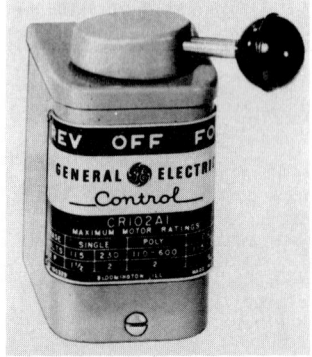

**Fig. 7-136A** Toggle-switch manual across-the-line starter for a dc motor. [Westinghouse Electric Corp.]

**Fig. 7-136B** Drum-type manual across-the-line starting switch for a dc motor. [General Electric Co.]

**280. Starting rheostats for dc motors,** sometimes call *dial* or *faceplate starters*, are obsolete and no longer available. They will be encountered only on very old installations. They consist of a series of contacts mounted on an insulating panel and connected to different points in a resistance unit (see Fig. 7-138). An arm or a dial bearing a contact is arranged so that it can be moved across the stationary row of contacts. A starting rheostat is shown in Fig. 7-139, and wiring diagrams of similar starters in Fig. 7-138. Moving the arm to the first contact closes the armature circuit in series with the starting resistance. As the

**Fig. 7-136C** Toggle-switch manual across-the-line starters for ac or dc motors: left, general-purpose; center, dusttight and watertight; right, flush mount. [General Electric Co.]

**Fig. 7-137** Manual across-the-line fractional-horsepower starters for ac or dc motors: left, toggle-switch type; right, pushbutton type. [Eaton Corp., Cutler-Hammer Products]

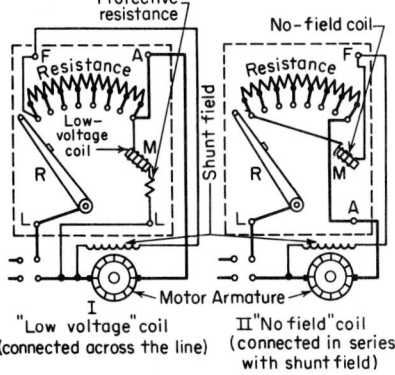

**Fig. 7-138** Low-voltage protection provided by a low-voltage coil and by a no-field coil.

**Fig. 7-139** Dial-type starting rheostat for a dc motor. [Eaton Corp., Cutler-Hammer Products]

arm is moved across the buttons, sections of the starting resistance are successively cut out of the circuit until the last contact is reached. Then all the resistance has been cut out of the circuit, and the armature is connected directly across the line. These starting rheostats are not suitable for reversing or dynamic braking service. They can be used for starting series, shunt, or compound-wound motors.

**281. Undervoltage protective devices for starting rheostats** will also be encountered only on very old installations. Replacements are not available. These can be operated by either a low-voltage coil or a no-field coil (Fig. 7-138). On a manually operated starter, each device usually consists of an electromagnet M, which under conditions of normal voltage holds the operating arm R in the running position, and a spring (not shown in the diagram) which returns the operating arm to the off position upon failure or decrease of voltage. Starters are sometimes so constructed that instead of the electromagnet being made strong enough to hold the arm in the running position against the force of the spring, a catch is provided to engage and hold the arm after it has been manually moved to the running position. Then, if the voltage drops below a predetermined value, the electromagnet disengages the catch. This releases the arm, which is returned to the off position by the spring. The magnet coil may be connected directly across the line (Fig. 7-138, I) with a protective resistance in series, in which case it is called a *low-voltage coil*, or it may be connected in series with the shunt field (Fig. 7-138, II), in which case it is called a *no-field coil*. Both arrangements provide low-voltage protection. If the coil is connected across the line, the current carried by the coil, and therefore the coil losses, will be less than if it is connected in series with the shunt field. The difference between the maximum and minimum field currents of wide-speed-range adjustable-speed motors renders it impractical to design a series-connected holding coil which will operate satisfactorily throughout the range.

**282. In starting a dc motor using a starting rheostat** (see Fig. 7-138), close the line switch and move the operating arm of the rheostat step by step over the contacts, waiting a few seconds on each contact for the motor to accelerate. If this process is performed too quickly, the motor may be damaged by excessive current; if too slowly, the rheostat may be damaged. If the motor fails to start on the first step, move promptly to the second step and if necessary to the third, but no farther. If the motor doesn't start when the third step is reached, open the line switch at once, allow the starter handle to return to the off position, and look for faulty connections, overload, etc. The time of starting a motor with full-load torque should not, as a general rule, exceed 15 s for rheostats for motors of 5 hp and lesser output and 30 s for those of greater output.

**283. To stop a dc motor using a starting rheostat,** open the line switch. The arm will return automatically to the off position. Never force the operating arm of any starting rheostat back to the off position.

**284. Overload protection is not generally provided in starting rheostats.** Overload protection must be provided at some other point in the motor circuit by means of fuses, circuit breakers, or magnetic contactors controlled by overload relays.

**285. Drum-switch controllers** are made in a variety of types. They can be obtained for both reversing and nonreversing service and with or without contacts for dynamic braking. They are used for series, shunt, and compound motors. The essential features of a drum controller are a central movable drum upon which are mounted a set of contact segments and a set of stationary contact fingers. A typical drum controller is illustrated in Fig. 7-140. The movement of the drum is controlled by means of a handle on top of the case. As this handle is moved from the off position, the armature circuit is closed in series with the starting resistance by means of the first set of movable segments on the drum making contact with the first set of stationary fingers. As the handle is moved around, the resistance is short-circuited step by step by the contacts located farther around on the drum. Small controllers of this type are self-contained, the starting resistance being mounted in a ventilated box on the rear of the drum switch. In the larger sizes the resistance is mounted separately from the drum switch. Drum controllers do not provide any overload or undervoltage protection. Short-circuit protection may be provided by fuses or a circuit breaker mounted separately from the controller. A magnetic contactor with an overload relay will provide both overload and undervoltage protection. Such a protective unit has the advantage that it may be electrically interlocked with the drum switch so that the line contactor cannot be closed unless the drum switch is in the off position. This provision safeguards against the possibility of the motor being thrown directly across the line. Connections for a typical magnetic-contactor protective panel are shown in Fig. 7-141.

**286. The manually operated multiple-switch controller for dc motors** (Fig. 7-142) comprises a number of switches $S$, which are manually closed in sequence. Thereby the resistances $R$ are shunted out of circuit as the armature accelerates. The switches are mechanically interlocked so that their being closed in any but the proper sequence is prevented. When the last switch $S_5$ is closed, all the resistance is short-circuited. The switches, after being closed, are held in the closed position by a catch. This catch is in turn held by a low-voltage protective magnet $M$.

Upon the failure of voltage, the catch is released and all the switches are thrown to the off position by gravity or by a spring. A switch of this type is especially adapted for starting large motors. A sufficient time period will elapse in the ordinary closing of the switches to provide a protection against too rapid acceleration. The arrangement of a number of switches in parallel tends to minimize heating of the contacts, particularly of the contact of the last switch $S_5$. One of the contact blocks of each switch is often made of laminated copper strips and the other of carbon. The laminated brushes and the copper-to-carbon arcing contacts render a switch of this type exceptionally adaptable for use with large motors. Overload protection must be provided outside the controller. These controllers are suitable for series, shunt, or compound motors in nonreversing service without dynamic braking.

**287. Apparatus for field control of shunt or compound motors.** The variable resistance in series with the shunt field, necessary for the speed control of shunt or compound motors by the field-control method, may be provided by means of a field rheostat separate from the starting rheostat, or the starting rheostat and the field rheostat may be combined into a single piece of equipment forming a combined starting and speed-adjusting controller. Such combined starting and speed-adjustment controllers are available in both dial and drum types of construction.

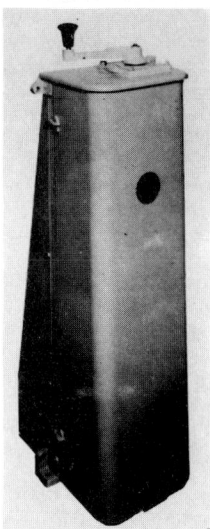

**Fig. 7-140** Nonreversing drum controller. [Westinghouse Electric Corp.]

Drum controllers for combined starting and speed adjustment (field control) are provided with two sets of contacts, one for control of the starting resistance and the other for control of the speed of the motor through variation of shunt-field resistance. Although controllers of this type are most frequently applied on machine tools, they are desirable for any service in which the work is severe and the expense of a controller is justified.

**288. Motor-generator variable-voltage systems,** formerly called Ward-Leonard systems, consist of an adjustable-voltage dc generator driven at a constant speed by an ac motor and a source of direct current to excite the generator separately. The variable-output voltage of the dc generator serves as the input to one or more dc motors whose speed will vary with the voltage supplied. The motor-generator combination acts as an amplifier, since a small current in the generator exciter circuit can control the generator output

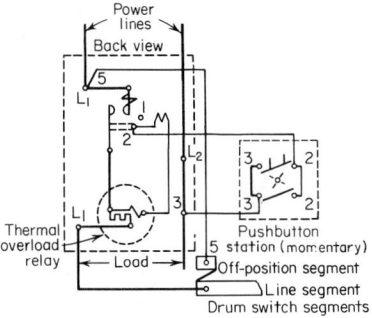

**Fig. 7-141** Connections for a thermal overload protective unit used with a drum starter.

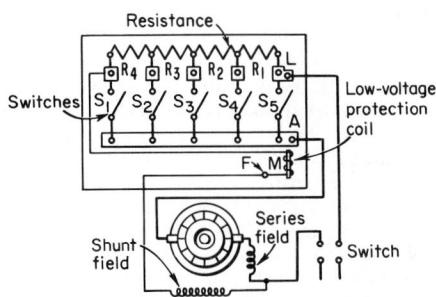

**Fig. 7-142** Connections for a multiple-switch dc controller.

voltage and the armature current of one or more driven dc motors. These systems are regenerative; e.g., the descending load on a hoist driven by a dc motor supplied by the motor-generator system will return energy to the system.

**289. Solid-state variable-voltage systems** supply direct current as the output of a bank of thyristors called a converter. A converter can be described as equipment that will change electrical energy from one form to another. This change is accomplished in two different modes of operation. The first mode is rectification, in which the converter changes alternating current or voltage to direct current or voltage. The second mode is known as inversion, in which the converter changes direct current or voltage to alternating current or voltage. During rectification, power flow is from alternating to direct current. Conversion is accomplished with as few as three or six thyristors, although multiples of these frequently are connected in parallel to increase capacity.

Although the converter and its controls are relatively simple, dc variable-speed drives provide satisfactory performance in the most demanding applications. These applications range from simple, general-purpose nonreversing drives with a fixed shunt field and a 5 percent voltage regulator to sophisticated steel-mill reversing drives with shunt-field controls and high-performance 0.1 percent speed regulators. Generally, the latter also provide inversion capabilities for fast deceleration via power pumpback into the ac line.

Solid-state converters, through proven performance, provide a high degree of reliability. Modern forms of modular construction and packaging techniques allow quick replacement of components or assemblies to minimize downtime if a fault occurs. The compactness and light weight of modern converters eliminate the need for elaborate foundations and require a minimum amount of floor space. In many instances, the converters can be wall-mounted or packaged with other system assemblies to utilize floor space efficiently.

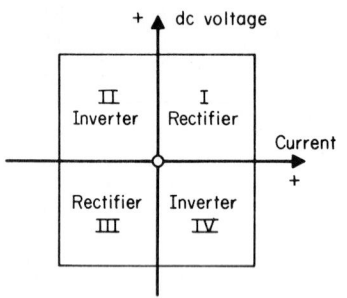

Fig. 7-143  Operating quadrants of thyristor converter units.

FORM CLASSIFICATIONS   Standard designations are used to describe converter characteristics. An operating-quadrant diagram (Fig. 7-143) is used for this purpose. The designations are intended to describe the functional characteristics of converters but not necessarily the circuits of components used. Converters are identified by "form" letters to classify their functional capabilities.

*Form A.*  A single converter capable of unidirectional voltage and current flow. It operates as a rectifier in Quadrant I only.

*Form B.*  A double converter capable of providing a positive voltage and current or a negative voltage and current. It has no inversion capability and operates as a rectifier in Quadrants I and III.

*Form C.*  A single converter capable of bipolarity voltage and unidirectional current flow. It operates as a rectifier in Quadrant I and as an inverter in Quadrant IV.

*Form D.*  A double converter capable of bipolarity voltage and bidirectional current flow. It operates as a rectifier in Quadrants I and III and as an inverter in Quadrants II and IV. Its performance is identical to that of a Ward-Leonard drive.

These forms refer to the converter capabilities only and not to the motor. Motor direction and power flow may be reversed by armature switching or field reversal without the use of an elaborate Form D converter. For example, a motor can be reversed with a Form A converter and a reversing switch or a reversing field. Similar combinations may be obtained for regeneration (reverse power flow).

**290. Apparatus for armature speed control of dc motors.**  When armature speed control is employed, the same piece of equipment serves for both starting and speed control. An ordinary starting controller should not be used for armature speed control, however, as it is not designed for continuous duty but only for intermittent service, as in starting. In employing a controller for armature speed control, make sure that it is designed for that type of service and not simply for starting duty. Controllers suitable for armature control are available in both the rheostat and the drum types.

**291. Rheostat-type armature-control speed regulators** (Fig. 7-144) are used for speed reduction with shunt, compound, or series motors in nonreversing service when the torque required decreases with the speed but remains constant at any given speed, as with fans, blowers, and centrifugal pumps. They can also be used in applications in which the torque is independent of the speed, as in printing presses. However, this method of speed control is not suitable for such applications if operation continues for long periods at reduced speed, since such operation is not economical. If the torque varies, constant speed cannot be obtained with these controllers.

In the regulator shown, the low-voltage release consists of an electromagnet enclosed in an iron shell, a sector on the pivot end of the operating arm, and a strong spring which

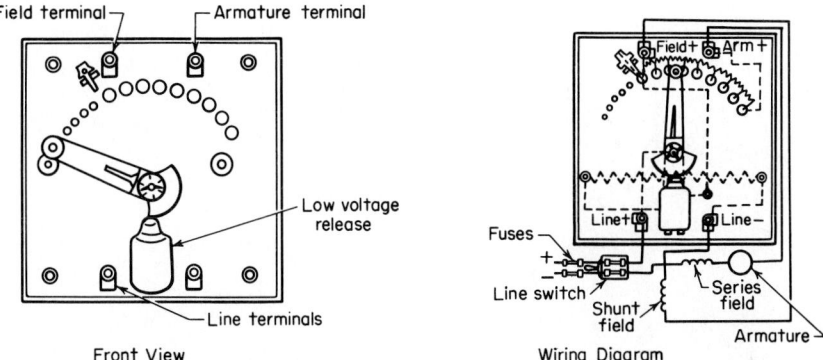

Front View                                   Wiring Diagram

**Fig. 7-144**   Armature-control speed regulator, removed from enclosure.

tends to return the arm to the off position. The magnet is mounted directly below the pivot of the arm, and its coil is connected in shunt across the line in series with a protecting resistance. When the magnet is energized, its plunger rises and forces a steel ball into one of a series of depressions in the sector on the arm with a sufficient force to hold the arm against the action of the spring; each depression corresponds to a contact. The arm can be easily moved by the operator, however, as the ball rolls when the arm is turned. When the voltage fails, the magnet plunger falls and the spring throws the operating arm to the off position. Standard commercial rheostats of this type are designed to give about 50 percent speed reduction on the first notch.

**292. Operation of armature-contol speed regulators** (Fig. 7-144).   Continuous motion of the operating arm starts the motor and brings it gradually to maximum speed. Moving the arm over the first few contact buttons increases the shunt-field strength if the motor is shunt or compound. The movement over the succeeding buttons cuts out armature resistance and permits the motor to speed up.

**293. Drum controllers suitable for armature speed control** are widely used for crane and hoist motors and in railway applications. Their general appearance and construction are the same as for the drum controller shown in Fig. 7-140. Crane controllers are arranged for reversing service and usually for dynamic braking.

**294. Magnetic contactors and starters for dc motors** are available in a number of different types for reversing and nonreversing duty, with or without dynamic braking, and for inching and jogging service. They can be used with constant-speed or adjustable-speed shunt or compound motors and with series motors. When they are used with adjustable-speed motors, speed is controlled by a field rheostat mounted separately from the magnetic controller.

When armature speed control is required, it is obtained by controlling the functioning of the contactors by a master controller. The operator adjusts the master controller to the position for the desired speed, and this action initiates the functioning of the contactors, which adjusts the resistance to the corresponding value.

Magnetic starters consist of the necessary magnetic contactors and relays mounted on a panel (open type) or enclosed in a sheet-metal cabinet (enclosed type). Magnetic starters and controllers are equipped with some form of overload relay which opens the circuit of the operating coil of the line contactor upon the occurrence of an overload on the motor. The master switch for initiating the operation of the controller may consist of an automatic float or pressure switch, a pushbutton switch, or a drum or rotary switch. The automatic switches and the pushbutton masters are used for simple control such as starting and stopping or starting, stopping, and reversing, while the drum or rotary switch is employed when full speed control by the operator is required, as for the control of cranes, hoists, etc. Undervoltage protection or release is provided through the operating coil of the main-line contactors, since upon failure or abnormal reduction of the line voltage, the operating coil of the contactor will be deenergized and the contactor will open the motor circuit. The principles of the different methods of automatically controlling the acceleration are discussed in Secs. 259 to 275. The appearance of typical dc magnetic controllers is shown in Figs. 7-145 to 7-149 inclusive.

**Fig. 7-145** Fluid-dashpot magnetic relay for use with a dc motor. [Allen-Bradley Co.]

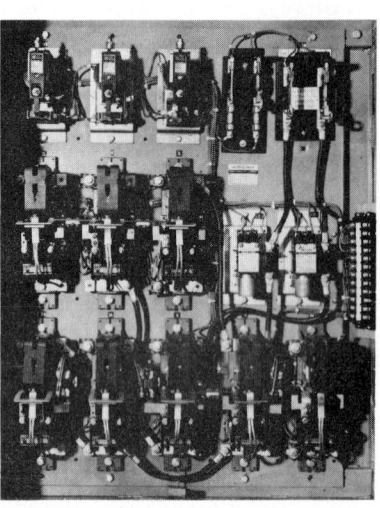

**Fig. 7-146** Pneumatic-relay type of time-limit controller for a dc motor. [Square D Co.]

**Fig. 7-147** Inductive-time-limit controller for a dc motor. [Westinghouse Electric Corp.]

**Fig. 7-148** Pneumatic-timing relay-type time-limit acceleration controller for a dc motor. [Square D Co.]

I. Current-limit type. [Ward-Leonard Electric Co.]

II. Counter-emf type. [Ward-Leonard Electric Co.]

III. Time-current type in NEMA Type 4 (watertight) cabinet. [Square D Co.]

**Fig. 7-149**   Typical magnetic controllers for dc motors.

**295. Field relays** must be used with magnetic starters when field speed control is employed. This is necessary to ensure that the motor will be started with full field. These relays are called *vibrating* or *fluttering relays,* since during the acceleration or deceleration of the motor their armatures flutter or vibrate between the open and closed positions, alternately shorting and cutting in the field resistance. The operating coil of the relay is connected in series with the armature of the motor, and the contacts are connected across the shunt-field rheostat. During the starting period, whenever the armature current exceeds the value for which the relay is set, the relay will close, short-circuiting the field resistance so that the field strength is maximum. After the starting resistance in series with the armature has been cut out of the circuit, the field relay will hold in until the motor has accelerated sufficiently to reduce the armature current to the value which will allow the field relay to drop out. As it drops out, the field resistance will be cut into the field circuit, thus weakening the field strength. This weakening of the field reduces the counter-emf and therefore will cause the armature current to increase and further accelerate the motor. The increase in armature current causes the field relay to close again, short-circuiting the field resistance. This action is rapidly repeated, the relay vibrating between open and closed positions until the motor has been accelerated to the speed for which the field rheostat is adjusted. The armature current is then at its normal value and is not sufficient to operate the field relay, which remains open with the field resistance cut into the field circuit.

IV. Time-current type. [Square D Co.]

**Fig. 7-149**   (Continued)

**Fig. 7-150**   Direct-current motor disconnect switch. [Square D Co.]

**296. DC safety disconnect switches,** as shown in Fig. 7-150, are specially designed to provide in a very satisfactory manner for the disconnect switch required by the **National Electrical Code** for crane installations. The switch in Fig. 7-150 is arranged for lever operation but is not difficult to close, in contrast to most large-size manually operated disconnect switches. Pushing the operating handle up closes the double-pole control-circuit contacts and energizes the magnetic coil to close the switch and connect power to the crane. Pulling the operating handle down breaks both sides of the control circuit.

**297. Methods of starting polyphase squirrel-cage induction motors**
1. Full-voltage (across-the-line)
2. Reduced-voltage
   a. By means of autotransformer (compensator)
   b. By means of series resistor
   c. By means of series reactor
   d. By means of part-winding
   e. By means of star-delta
   f. By means of solid-state reduced-voltage

Any polyphase induction motor can be safely started by applying full voltage at starting without harm to the motor, but as discussed in Sec. 130, the heavy starting current drawn from the line by this method may produce too great power and voltage fluctuations in the power supply system. When the starting current must be reduced, one of the methods listed in item 2 above is employed for reducing the voltage impressed on the motor during starting. In Method a the motor is connected at the start to the secondary of an autotransformer. After the machine has accelerated sufficiently, the motor is switched over to full line voltage. With this method there are just two steps in the sequence of starting operations: reduced voltage of some definite value and, directly, full line voltage. Starters employing the transformer method of reducing the voltage are called compensators or autotransformer starters. In Method b, resistance is connected in series with the motor leads from the supply. One or more steps of resistance can be employed. In most cases only one step is used, so that the motor is first connected to the line with some definite value of resistance in series, and then, after the machine has accelerated sufficiently, it is thrown over to full line voltage. When starting conditions are more severe or the starting current must be more closely regulated, additional steps are employed for cutting out resistance. Method *c* is similar to the series-resistance method except that reactance is inserted in series with the motor during starting instead of resistance. With this method only one step of reactance is used.

For the same value of starting torque developed, the autotransformer method of starting will limit the first value of peak starting current to the lowest value. The amount of power consumed during the starting period will be considerably less than with resistance starting. A disadvantage of this method is that the motor is generally entirely disconnected from the supply during the transition period from reduced to full line voltage. This causes a second peak current to be drawn from the line at the instant that full voltage is applied. Some power companies will not allow the use of autotransformer starters for this reason.

With resistance starting, the motor is connected to the supply continuously throughout the starting period, the motor impressed voltage being increased by shorting out the starting resistance. Although the line current will increase whenever some or all of the starting resistance is short-circuited, the magnitude of such peaks will not be nearly so great as the peak current involved at the instant when full voltage is applied with autotransformer starting. The resistance-starting method has the advantage that it produces very smooth starting conditions with respect both to strains imposed upon the motor and load and to the power and voltage fluctuations produced in the supply circuit during the starting period. Resistance starting is used when the strains imposed upon the load with compensator starting would be detrimental or when this method is required by the regulations of power supply companies. The resistance method is not generally satisfactory unless a starting torque of 50 percent or less of full-load torque is sufficient.

With series-reactance starting, just as in

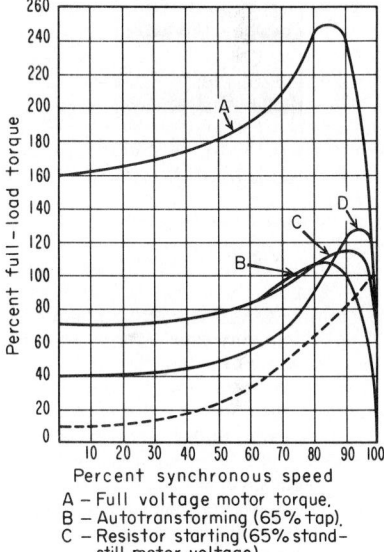

Fig. 7-152  Accelerating torque characteristics of a squirrel-cage induction motor. [Allis-Chalmers Corporation]

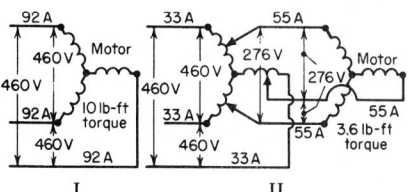

**Fig. 7-151**  Starting with (I) and without (II) compensator.

resistance starting, the motor connections need not be broken during the starting period. The reactor is short-circuited when the motor attains its maximum speed on reduced voltage. For the same current drawn from the line, reactance starting will develop the same starting torque as resistance starting. Reactance starting has the advantages over resistance starting that less power is consumed during the starting period and that the transition peaks of current will be lower because of the opposing effect of the inductance to sudden changes in the current. For the same number of steps reactance starting will produce the smoothest starting conditions of any of the three starting methods. However, in the smaller sizes control equipment of this type is more expensive than either of the other two types and for all sizes is more expensive than resistance starting. It is not practicable to make reactance starters for more than two steps of starting control, so that, by using a greater number of steps, smoother starting can be obtained with resistance starting. For many large motors, a reactance starter will be cheaper than a transformer starter, since only two sets of switching contacts are required for a one-step starter as compared with three sets for a transformer starter. For these reasons, reactance starting is utilized only with very large motors when one or two steps of starting will be satisfactory.

On account of the lower cost of across-the-line starters the general practice is to employ this type of starting whenever it is possible to use motors whose starting currents are within the allowable limits.

Refer to Fig. 7-152 and Secs. 299, 300, and 301 for further comparison.

In method $d$, the motor is provided with two windings, and 12 leads are generally brought out. On initial start-up, only one winding is energized, and the second winding is energized after a short predetermined time delay. Two contactors, rated for part-winding start, are required, and overload protection must be provided for each winding. Part-winding start motors are often dual-voltage motors, such as 230/460 V, with each winding designed for 230 V. When used on 230-V service, the windings are connected in parallel. When used for 460-V service, they are connected in series. In 230-V service, part-winding start (two-step) or full-winding start is available. In 460-V service the motor must be used with full-winding start. See Sec. 312.

In method $e$, the motor has six leads, two for each winding, and is connected in star for the initial start-up, then reconnected in delta for normal running. The transition of connection may be initiated by a timer or a current-sensing device. The transition period from star to delta may be "open," in which case the motor is entirely disconnected from the supply during the transition period, or "closed" transition may be used, which involves additional contactors and resistors to maintain current during the transition period. Mechanical interlocking of contactors, to prevent them from closing a short circuit across the line, is necessary. The overload protection should be connected into only one phase of the winding and selected for 58 percent of the motor full-load current. Starters for star-delta start motors, with mechanically interlocked contactors, are available. See Sec. 313.

In method $f$, solid-state reduced-voltage starters are used, as described in Sec. 263.

**298. Starting with and without compensators.**    The starting current taken by a squirrel-cage induction motor at the instant of starting is equal to the applied emf divided by the impedance of the motor. Only the duration of this current, and not its value, is affected by the torque against which the motor is required to start. The effect of starting without and with a compensator is illustrated by Diagrams I and II in Fig. 7-151. Motor I is thrown directly on a 460-V line. The impedance of the motor is 5 Ω per phase; the starting torque, 10 lb at 1-ft radius; and the current taken, 92 A. In Diagram II a compensator is inserted, stepping down the line voltage from 460 to 60 percent of 460, or 276 V. This reduces the starting current of the motor to 60 percent of 92, or about 55 A, and the starting torque becomes $0.6^2 \times 10$, or 3.6 lb at 1-ft radius. The current in the line is reduced to 60 percent of 55, or 33 A.

Thus when a compensator is used, the starting torque of the motor can be reduced to approximately the value required by the load, and the current taken from the line is correspondingly decreased.

**299.   General Comparison of Starting Conditions among Three Types of Reduced-Voltage Starters**
(Allis-Chalmers Corporation)

| Characteristic | Autotransformer type | Primary resistor type | Reactor type |
|---|---|---|---|
| Starting line current at same motor terminal voltage | Least | More than autotransformer type | |
| Starting power factor | Low | High | Low |
| Quantity of power drawn from the line during starting | Less | More than autotransformer type | |
| Torque | Torque increases slightly with speed | Torque increases greatly with speed | |
| Torque efficiency | The autotransformer-type starter provides the highest motor starting torque of any reduced-voltage starter for each ampere drawn from the line during the starting period | Low | Low |
| Smoothness of acceleration | Motor is momentarily disconnected from the line during transfer from start to run | Smoother. As the motor gains speed, the current decreases. Voltage drop across resistor or reactor decreases and motor terminal voltage increases. Transfer is made with little change in motor terminal voltage | |
| Cost | Equal | Lower in small sizes, otherwise equal | Equal |
| Ease of control | Equal | Equal | Equal |
| Maintenance | Equal | Equal | Equal |
| Safety | Equal | Equal | Equal |
| Reliability | Equal | Equal | Equal |
| Line disturbance | Varies with line conditions and type of load | | |

**300.   Comparison of Current Torque Starting Characteristics for Squirrel-Cage Motor**
(Allis-Chalmers Corporation)
Refer to Fig. 7-152.

| Method of starting | Starting current drawn from the line as a percentage of full-load current | Starting torque as a percentage of full-load torque |
|---|---|---|
| Connecting motor directly to the line-full potential ........ | 470 | 160 |
| Autotransformer 80 % tap.............................. | 335 | 105 |
| Resistor or reactor starter to give 80 % applied voltage ..... | 375 | 105 |
| Autotransformer 65 % tap.............................. | 225 | 67 |
| Resistor or reactor starter to give 65 % applied voltage..... | 305 | 67 |
| Resistor or reactor starter to give 58 % applied voltage..... | 273 | 54 |
| Autotransformer 50 % tap.............................. | 140 | 43 |
| Resistor or reactor starter to give 50 % applied voltage..... | 233 | 43 |

**301. Comparison of autotransformer and resistance for decreasing voltage in starting squirrel-cage motor.** The motor in Fig. 7-153 is supposed to require 100 A to start it, i.e., to provide the energy needed to produce the necessary starting torque. At I, where an autotransformer is used to lower the voltage to 115, a current of 100 A is produced in the motor primary with a current in the line of 50 A. This condition is due to the transformer action of the autotransformer. At II are shown the running conditions at which the autotransformer is entirely disconnected from the circuit. At III are illustrated the conditions that would obtain if the voltage were lowered for starting by inserting resistance in series with the line. Obviously 100 A must flow in all portions of the line even though the resistance of 1.15 Ω reduces the line voltage of 230 to a voltage of 115, which is impressed on the motor. There is a loss of energy in the resistance. For the same line starting current the

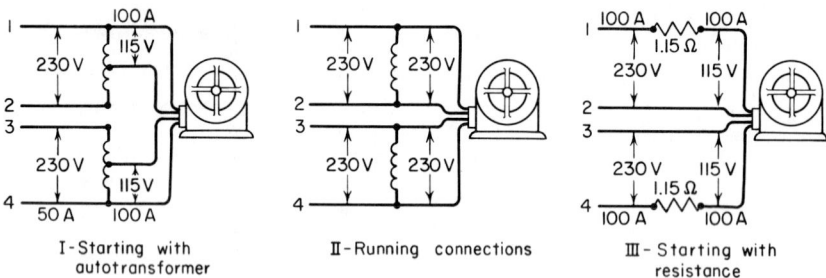

I – Starting with autotransformer    II – Running connections    III – Starting with resistance

**Fig. 7-153**   Starting with resistance and with compensator.

starting torque available with the resistance method will be considerably less than that developed when a compensator is employed. As stated in Sec. 297, the resistance-starting method has the advantage that it produces very smooth starting conditions with respect to both the strains imposed upon the motor and load and the voltage fluctuations produced in the supply circuit during the starting period. A comparison of starting torques and currents when reducing the voltage in the two different ways is given in Table 300.

**302. Across-the-line ac motor controllers** are made in a variety of types. The most common types are:

1. Manual motor-starting switch with no overload protection (toggle and drum switches)
2. Manual motor-starting switch with thermal overload protection
3. Magnetic switch (contactor) without overload protection
4. Magnetic switch with thermal overload protection

**303. The manual motor-starting switch** is simply a tumbler, rotary, or lever switch without overload protection constructed in an especially rugged manner. Some designs provide undervoltage protection. A single switch is used for nonreversing duty and an interlocked pair for reversing duty up to 2 hp. Drum-switch types are more commonly used for reversing applications. Some representative types are illustrated in Figs. 7-154 and 7-155.

**304. Manual across-the-line starting switches with thermal overload protection** are shown in Figs. 7-156, 7-157, and 7-158. The overload protection is obtained by means of a thermal overload relay. (For more comprehensive discussion of the principles of thermal overload relays refer to Div. 4.)

**305. Magnetic across-the-line motor controllers** consist of a magnetically operated contactor controlled from a pushbutton station. Starters are provided with thermal overload and undervoltage protection. For the switch and circuit shown in Fig. 7-159, pressing the START button closes the circuit of the operating coil $M$ of the magnetic contactor. This allows a current to flow from one side of the line, through the START button, through the operating coil, and back to the other side of the line. The operating coil, being energized, closes the magnetic contactor. The closing also closes an auxiliary contact so that when the START button is released, the operating coil is still energized. To stop the motor the STOP button is pressed. This opens the circuit of the operating coil, allowing the switch to reopen. In case of an overload the thermal overload relays operate and open their contacts. This deenergizes the operating coil and allows the contactor to open. If the voltage

**Fig. 7-154**  Toggle-type motor-starting switch without overload protection for polyphase squirrel-cage motors. [Westinghouse Electric Corp.]

**Fig. 7-155**  Drum-switch type of motor-starting switch without overload protection for polyphase squirrel-cage motors. [Eaton Corp., Cutler-Hammer Products]

**Fig. 7-156**  Manual across-the-line starters with thermal overload protection for polyphase squirrel-cage motors. [General Electric Co.]

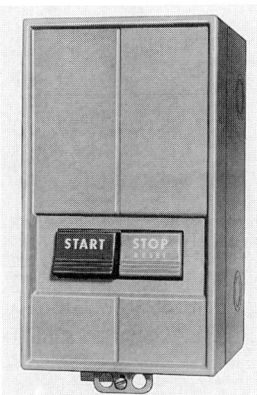

I. [Eaton Corp., Cutler-Hammer Products]          II. [Square D Co.]

**Fig. 7-157**  Manual across-the-line starters with thermal overload protection for squirrel-cage motors.

**Fig. 7-158**  Manual across-the-line starter with overload protection for a squirrel-cage motor. [Allen-Bradley Co.]

I. Size 1 polyphase-motor starter.

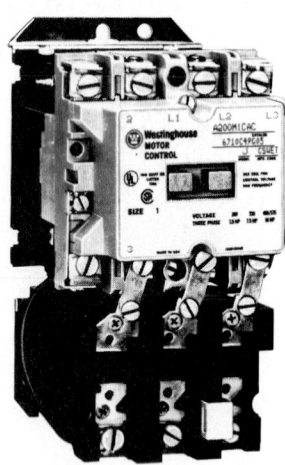

II. Three-wire control of a magnetic across-the-line starter.

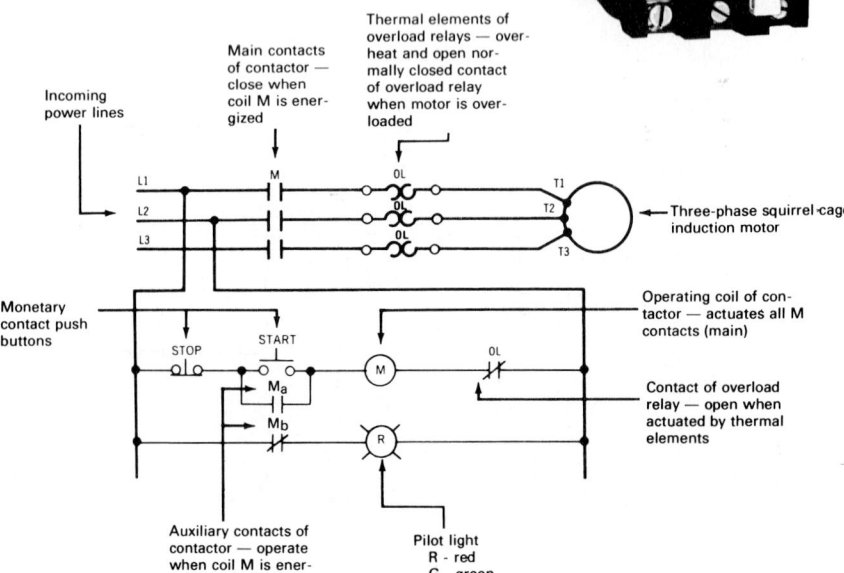

**Fig. 7-159**  Magnetic across-the-line starting. [Westinghouse Electric Corp.]

of the line fails or falls to a low value, the operating coil will not be energized sufficiently to hold the contacts of the contactor closed. As the contactor drops out, the contacts are opened and the motor will not start upon return of normal voltage until the START button is pressed.

Magnetic across-the-line starting control-lers are made in a variety of types for both nonreversing and reversing duty. The reversing types consist of two magnetic contactors electrically interlocked so as to prevent the closing of both switches at the same time (Fig. 7-160).

**306. Combination magnetic across-the-line starters** are the magnetic across-the-line starters which include in the same metal enclosure the motor branch-circuit short-circuit protective device. The following different combinations are available as standard equipment:

1. Magnetic starter and motor-circuit protector (Fig. 7-161)

2. Magnetic starter and fused motor-circuit switch (Fig. 7-162)

3. Magnetic starter and inverse-time circuit breaker

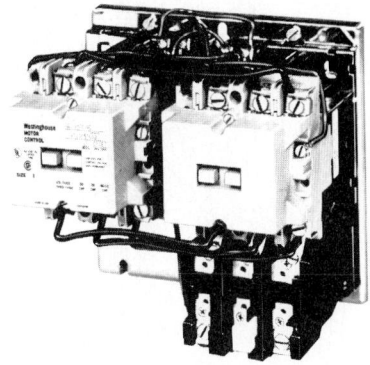

**Fig. 7-160** Reversing-type magnetic across-the-line starter for a squirrel-cage motor. [Westinghouse Electric Corp.]

National Electrical Code rules require the installation of short-circuit protection and a motor-circuit disconnect means on the line side of each magnetic starter. Combination starters make a very compact and neat installation to meet these requirements. Almost any desired combination for nonreversing or reversing service can be obtained (Fig. 7-163).

**307. Motor-circuit protection and isolation.** The disconnect means required by the National Electrical Code provides a way to isolate the motor and other portions of the motor branch circuit from the power lines to permit safe servicing and replacement of fuses.

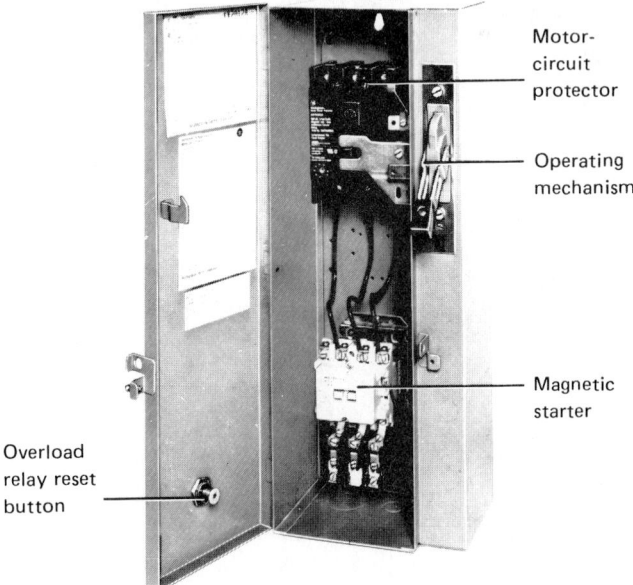

Motor-
circuit
protector

Operating
mechanism

Magnetic
starter

Overload
relay reset
button

**Fig. 7-161** Combination starter with adjustable-trip-only circuit breaker. [Westinghouse Electric Corp.]

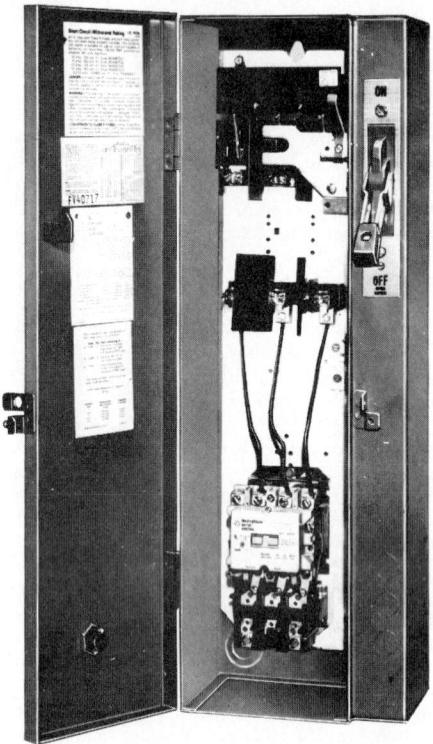

**Fig. 7-162**   Combination starter with fusible motor-circuit switch. [Westinghouse Electric Corp.]

This disconnect means must be a circuit breaker, a motor-circuit switch (rated in either motor horsepower or motor full-load current and motor locked-rotor current), or a properly rated general-use switch. General-use switches are used primarily with small motors. The motor branch circuit must also contain short-circuit protection for the circuit components. This protection may be in the form of either time-delay or non-time-delay fuses selected in accordance with the rules of the National Electrical Code, motor short-circuit protectors (fuse-type devices called MSCPs), or a circuit breaker.

For small motors, adjustable magnetic-trip-only circuit breakers, called motor-circuit

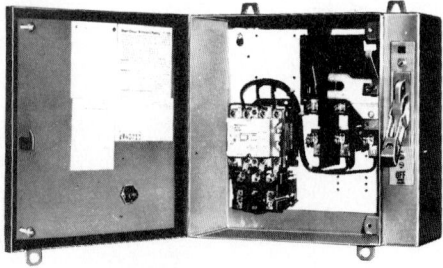

I. Type 1 enclosure with the door closed.    II. Type 12 enclosure with the door open.

**Fig. 7-163**   Across-the-line combination starters with the long axis of the controller horizontal. [Westinghouse Electric Corp.]

protectors, offer close short-circuit protection because the magnetic-trip element of these devices can be adjusted to the lowest possible value that permits the motor to start. On the other hand, molded-case inverse-time circuit breakers in the smaller frame sizes (up to 150 A) have no provision for adjusting their magnetic-trip setting, and the inverse-time characteristic of these circuit breakers is slower for a given overload than the inverse-time characteristic of the thermal overload relays in the circuit. Hence, the latter operate first and open the circuit when an overload or low-level fault occurs. The magnetic-trip portion of an inverse-time circuit breaker is the only portion that ever causes the circuit breaker to operate in a properly designed motor branch circuit containing overload relays.

**308. Reduced-voltage starters.** From the viewpoint of controlling current inrush, the solid-state reduced-voltage starter provides a smoother start for squirrel-cage motors than any other type of starting equipment. Stepless primary resistance starting also provides a smooth start. When the power company severely limits the motor inrush current, these are the ideal automatic motor starters. The primary resistance starter is, however, the most expensive method of reduced-voltage starting and should never be used when an appreciable breakaway torque is required.

The automatic autotransformer starter (Fig. 7-164) is the most effective method of reduced-voltage starting when the application requires a reduction in the current drawn from the line coupled with the greatest possible starting torque. Closed-circuit transition, an option afforded by the autotransformer starter, is particularly advantageous since the high transient currents associated with open-circuit transition starters are eliminated.

Primary-resistance starters are usually employed when the starting current must be reduced and a lower starting torque as compared with that of the autotransformer starter is of little consequence. Primary-reactance starters are similar to the primary-resistance type but are usually applied only to high-voltage starters or on starters for very large motors.

One of the least expensive methods of obtaining reduced–inrush-current starting involves a part-winding starter. It uses two smaller-size contactors as compared with the contactor size employed for across-the-line starting. Voltage-reducing equipment such as

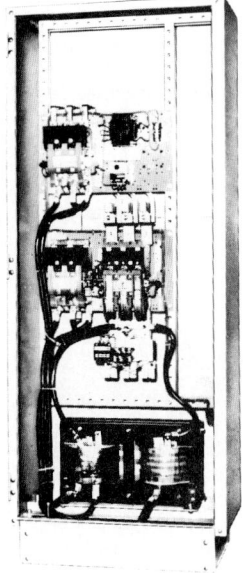

**Fig. 7-164** Autotransformer-type reduced-voltage motor starter, ac, polyphase. [Westinghouse Electric Corp.]

**Fig.    7-165** Primary-resistance-type reduced-voltage motor starter, ac, polyphase. [Allen-Bradley Co.]

transformers, resistors, or reactors is not required. Transition from start to run is inherently closed-circuit transition, and therefore high transient currents during the transition period are nonexistent. Part-winding starting does have some disadvantages. It has a very low and erratic starting torque and can be applied only to induction motors which have been designed for part-winding service. Almost invariably, this type of starting is increment because of the low starting torque. Part-winding starting is not recommended for drives that are frequently started or for high-inertia drives. Its main application is in air-conditioning and ventilating installations.

Star-delta starting is similar to part-winding starting, in that accessory equipment for voltage reduction is not necessary. The stator winding of the motor must be connected in delta with six leads brought out. Starting current and torque are about one-third normal. Star-delta starting is more efficient than part-winding starting and is somewhat more expensive.

**309. Primary-resistance starters for squirrel-cage induction motors** are made in both manual (semimagnetic) and magnetic types. The magnetic types generally provide time-limit acceleration. They are similar to starters for dc motors, except that they are provided with resistors and switching equipment for each phase, all phases being switched simultaneously. In some types all the resistance is cut out in one step, while in others the starting resistance is cut out in several steps. The manual starters may be of the dial or the drum type. With the manual starters, overload and undervoltage (if desired) protection must be provided separately from the starter. A magnetic starter is shown in Fig. 7-165. Magnetic starters are provided with overload protection and with undervoltage protection or release as desired. Primary-resistance starters of the drum and magnetic types can be obtained for reversing or nonreversing service.

**310. Primary-reactance starters.**  It is possible to reduce the motor terminal voltage as required by using series reactors rather than resistors. However, this method is not generally employed except in high-voltage or high-current installations, or both, in which the heating and physical construction of resistors is a problem. The use of reactors results in an exceptionally low starting power factor and somewhat aggravates the already poor line regulation. Then, too, owing to the relative difficulty of changing the reactance values in the field, reactor starting is more expensive and less flexible than resistor starting on all but the largest machines. The added cost has a very real basis, in that the reactors must be carefully designed and applied, since any saturation would lead to inrushes equal to full-voltage starting current. Such a condition would, of course, make the use of the reactors pointless.

Resistor- and reactor-starting methods have one advantage in common: in both cases, the motor terminal voltage is a function of the current taken from the line. It can therefore be assumed that during acceleration the motor voltage will rise as the line current drops, resulting in greater accelerating energy at the higher speeds and a correspondingly less severe disturbance when the transition to full voltage occurs. This advantage is generally overestimated, first, because the line current does not decrease significantly from locked value until rather late in the accelerating period, and, second, because the difference between the line voltage and the motor terminal voltage is a vector and not a scalar difference.

**311. Autotransformer starting.**  One of the most effective methods of reduced-voltage starting uses the autotransformer. It is preferred to the primary-resistance type when the starting current drawn from the line must be held to a minimum value and maximum starting torque per line ampere is needed.

The autotransformer starter has distinctive characteristics which are often overlooked. First, it is plain that the motor terminal voltage is not a function of load current and remains substantially constant during the accelerating period. Of course, there is some voltage variation in the transformers, but this is negligible, and, from the standpoint of overall power factor, so is the magnetizing current.

Second, note that because of the turns-ratio advantage, the ratio of line current to torque is the same for autotransformer starting as for full-voltage starting. Here we must distinguish between motor current and line current, which are not equal, as they are in resistor and reactor starting.

Autotransformers, also known as compensators or autostarters, consist of an autotransformer, a double-throw switching device, and overload and undervoltage protective devices, all compactly mounted in a sheet-metal case. They can be obtained for either manual (semimagnetic) or magnetic operation. The manual type is provided with a lever

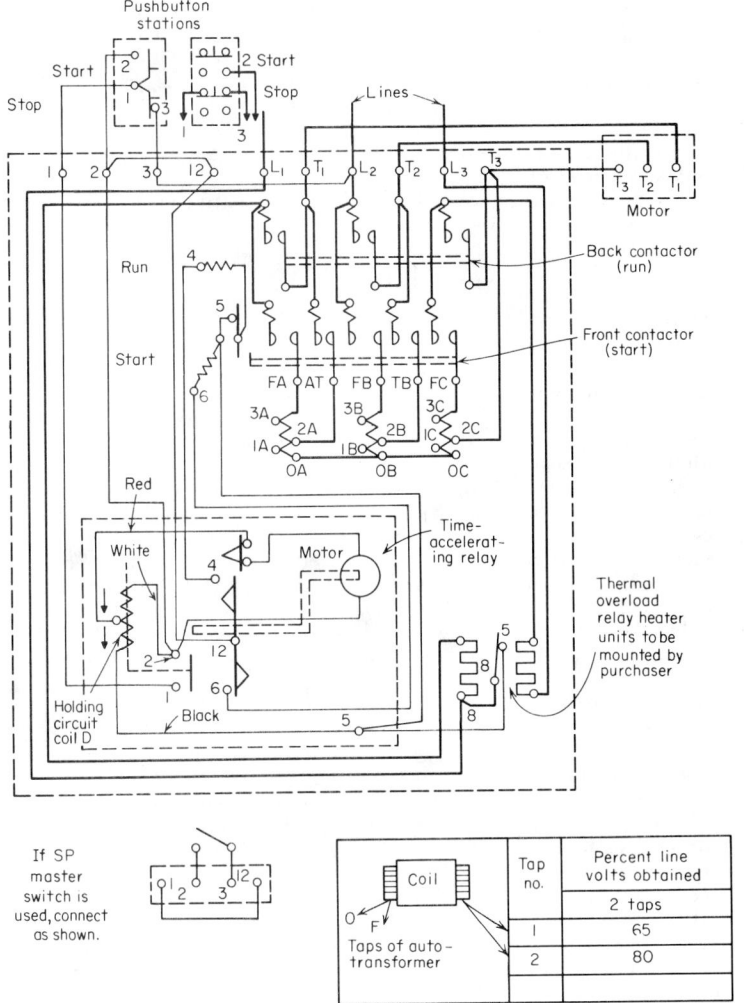

**Fig. 7-166**  Connections for a magnetic compensator (autotransformer starter). [General Electric Co.]

outside the case which controls the double-throw-switch mechanism. When the handle is in the off position, the motor and autotransformer are both disconnected from the line. Throwing the lever to the START position connects the autotransformer to the line and the motor to the secondary taps of the autotransformer applying the reduced voltage for starting. When the handle is thrown to the RUN position, the motor is connected directly to the line and the autotransformer is entirely disconnected from the circuit. The switching mechanism is so designed that it cannot be left in the START position but will automatically return to the off position unless thrown to the RUN position after the start.

Most manual compensators are provided with a STOP button so interconnected with the overload protective device that, if pressed, the button will actuate the device and open the compensator contacts.

In the automatic compensator the same sequence of operations as in the hand-operated type is performed automatically after the operator has pressed the START button. The acceleration may be time-limit- or current-limit-controlled. Most automatic compensator starters employ the time-limit method. The closing of the START button closes the oper-

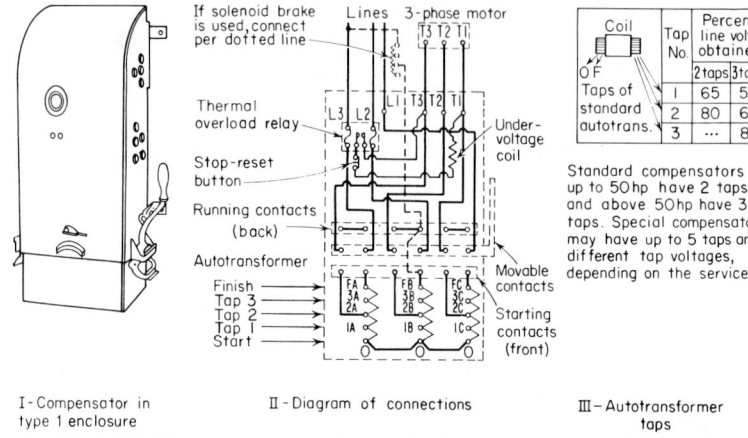

I – Compensator in type 1 enclosure    II – Diagram of connections    III – Autotransformer taps

**Fig. 7-167**    Manual compensator for a squirrel-cage motor. [General Electric Co.]

ating coil of a magnetic contactor which connects the motor to the line through the autotransformer. With time-limit control, after a sufficient time has elapsed, a time-delay relay closes its contacts, completing the operating circuit of a second magnetic contactor. The closing of this second contactor disconnects the autotransformer from the circuit and throws the motor directly across the line. With the current-limit control, a current-limit relay closes the operating-coil circuit of the second contactor after the current has been reduced to the proper value.

To stop the motor it is necessary simply to depress the STOP button, which opens the operating coil of the magnetic contactor, allowing the contacts to open and disconnect the motor from the line.

Compensators are provided with two or more sets of secondary taps on the autotransformers so that the voltage best suited to the required starting conditions may be used. Typical compensators are shown in Fig. 7-164 (magnetic) and Fig. 7-167 (manual).

The standard compensator does not provide for reversing duty. For such service an additional reversing switch of either the manual or the magnetic type is required.

Overload protection on autotransformer starters is provided by thermal or magnetic overload relays. In manual compensators the overload relay opens the holding-coil circuit, allowing the switch of the compensator to return to the off position. In magnetic compensators the overload relay breaks the circuit of the operating coil of the magnetic contactor so that the contactor opens. Typical connections for overload protection with thermal relays are shown in Figs. 7-166 and 7-167. The relays generally are arranged so that they can be adjusted to operate at different currents. Thermal relays inherently have inverse-time characteristics. When the magnetic relays are employed, they generally are equipped with devices to give the inverse-time feature.

Undervoltage protection is almost always provided on autotransformer starters. For manual compensators the holding coil of the compensator switch acts as the undervoltage coil (see Fig. 7-167). The holding coil is connected across the line when the switch is in the running position. If the voltage of the line drops sufficiently, the coil is deenergized to such an extent that the switch is released and returned to its off position by the spring. For magnetic compensators the coil acts as the undervoltage coil (see Fig. 7-166). If the voltage of the line drops sufficiently, this operating coil is deenergized to such an extent that it releases the holding circuit interlock wired in parallel with the START button.

The magnetic compensator is diagrammed in Fig. 7-166. It consists of a run contactor, a start contactor, an autotransformer, an accelerating relay, and an overload relay. In the compensator illustrated, acceleration is time-controlled. Figure 7-166 functions as follows. When the START button is pressed, a circuit is closed for $L_2$, through the STOP and START buttons, to 2, to 12, through the accelerating relay contact to 6, through the operating coil of the start contactor to 5, through overload relay contacts to 8, and back to $L_1$. The operating coil of the start contactor is thus energized, and the start contactor closes, applying reduced voltage to the motor through the transformer. At the same time a circuit is closed from $L_2$, through STOP and START buttons to 2, through the motor of the accelerating relay to 5, through the overload relay to 8, and back to $L_1$. Current through coil $D$ closes contact 1–2 and allows current to flow from $L_2$ through the STOP button to 1, through contact 1–2 to 12, through 12–6 to the operating coil of the start contactor, and back to $L_1$, even though the START button is opened. After a certain definite time the timing motor opens contacts 12–6 and closes contacts 12–4. The opening of contacts 12–6 breaks the circuit of the operating coil of the start contactor, allowing the contactor to open, and closes contact 5. The closing of contacts 12–4 and 5 allows current to flow from $L_2$ through the STOP button to 1, through contact 1–2 to 12, through contacts 12–4, through the operating coil of the run contactor, through contact 5, through the overload relay to 8, and back to $L_1$. The operating coil of the run contactor is thus energized, and this contactor closes, applying full voltage to the motor. Opening the overload relay or pressing the STOP button will open the circuit of coil $D$. This deenergizing of $D$ will allow contact 1–2 to open, which breaks the circuit to the operating coil of the run contactor, allowing this contactor to open and disconnect the motor from the line.

Two types of autotransformer connections of control schemes are in common use today. One is designated *closed-circuit transition or Korndorfer connection,* named after its inventor, and the other *open-circuit transition.* Figure 7-168 is a typical diagram of a closed-circuit transition starter. The taps of the autotransformer are permanently connected to the motor. Three contacts are used. The control sequence is as follows. When the START button is pressed, contactors 1S and 2S close, connecting the motor through the transformer taps to the power line. At the instant that contactor 2S is closed, a timing mechanism is activated. After a predetermined time, the timer drops out 1S, and for a short interval, until $R$ closes, part of the transformer windings functions as a reactor. When contactor $R$ closes, it bypasses the transformer and connects the motor to full line voltage. A late-break auxiliary contact on $R$ drops out 2S.

Figure 7-169 is a diagram of a typical open-circuit transition autotransformer starter. It contains two contactors. A pilot device closes the circuit to contactor S, which connects the motor to the reduced-voltage taps of the autotransformer and the transformer to the line. Simultaneously, a time-delay relay $TR$ is energized. After a time interval, $TR$ opens S, disconnecting the motor and transformer from the line. Contactor $R$ then closes to connect the motor across the line. This is an open-circuit device since contacts $R$ must not be closed until contacts $S$ are open. Any overlap will cause dangerously large currents to flow through part of the transformer windings. The main objection to open-circuit transition is the flow of high transient currents when the motor is reconnected to full-voltage operation.

**312. Part-winding starting.**    If the stator windings of an induction machine are divided into two or more equal parts, with all the terminals for each section available for external connection, the machine may lend itself to what is known as *part-winding starting.* Part-

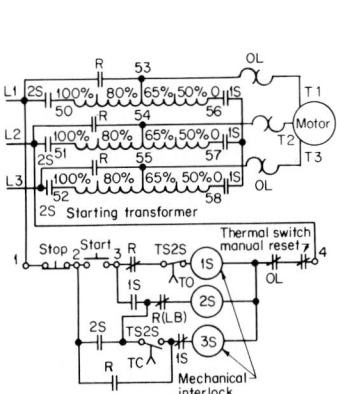

**Fig. 7-168** Typical closed-circuit transition autotransformer starter. [Allen-Bradley Co.]

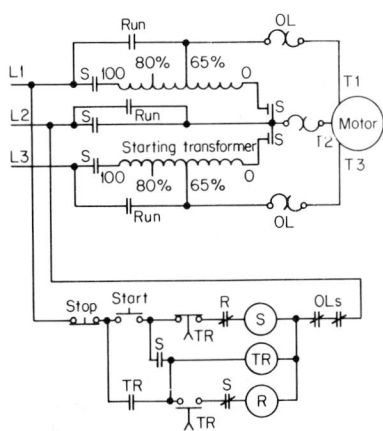

**Fig. 7-169** Typical open-circuit transition autotransformer starter. [Allen-Bradley Co.]

winding starters are designed for use with squirrel-cage motors having two separate parallel stator windings.

Part-winding starters are available in either a two- or a three-step construction. The two-step starter is designed so that when the control circuit is energized, one winding of the motor is connected directly to the line. This winding then draws about 65 percent of normal locked-rotor current and develops approximately 45 percent of normal motor torque. After about 1 s, the second winding is connected in parallel with the first; the motor is then fully across the line and develops its normal torque.

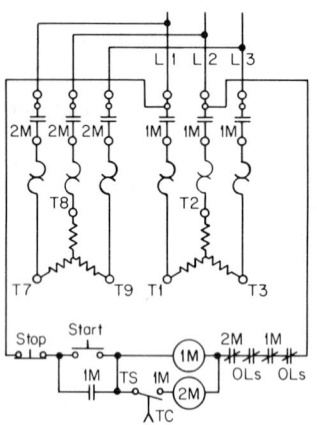

**Fig. 7-170**  Part-winding starter. [Allen-Bradley Co.]

The three-step starter has resistance in series with the motor on the first step. Initially, one winding is connected to the line in series with a resistor. After a short time interval (about 1 s), the resistor is shorted out, and the first winding is connected directly to the line. The second winding is connected to the line in parallel with the first winding 1 s later. The motor then draws full line current and develops normal torque.

By far the most common number of winding sections or parts is two; the stator then consists of two windings, with six leads brought out, and the motor is intended to deliver its rated output with the windings in parallel (see Fig. 7-170). These windings may or may not be identical, but in any case they cannot occupy the same slot simultaneously. Applications of the line voltage to one of the winding sections will cause the motor to draw from 65 to 80 percent of normal starting current, with a torque output of about 40 percent of normal locked torque.

Torque efficiency is roughly equivalent to that of resistor starting, in that the ratio of starting current to starting torque is fairly poor. The similarity ends there, however, since the part-winding method allows no adjustment of these values. Part-winding schemes almost inevitably lead to increment starting, for the starting torque is so low that the motor rarely starts on the first step. At the start, one winding is carrying up to 50 percent more current than it would on a full-voltage start, and heating increases correspondingly. If this characteristic is unacceptable, it is sometimes the practice to use resistors in combination with part-winding starters to reduce further the starting current and torque. However, since part-winding starting is usually adopted for economic reasons, considerable thought should be given before it is modified with such embellishments. It may well be that another form of conventional starting will perform the task more economically and with more satisfactory starting performance.

Part-winding starting has certain obvious advantages. It is less expensive than most other methods, because it requires no voltage-reducing elements such as transformers, resistors, or reactors, and it uses only two half-size contactors. Furthermore, its transition is inherently closed-circuit.

But it also has several disadvantages. In the first place, the starting torque is poor, and it is fixed; the starter is almost always an increment-start device. Second, not all motors should be part-winding–started, and it is important that the motor manufacturer be consulted before this type of starting is applied. Some motors are wound sectionally with part-winding starting in mind, but indiscriminate application to any dual-voltage motor (for example, a 230/460-V motor which is to run at 230 V) can lead to excessive noise and vibration during start, to overheating, and to extremely high transient currents upon switching. Then, too, since the current requirements in part windings allow the use of smaller contactors and coincidentally smaller overload elements, the fuses must be proportionately small to protect these elements. Time-delay fuses are therefore a virtual necessity.

**313. Star-delta starting (also called Y-delta).**  Probably no other method of starting has been so thoroughly investigated as the European technique of star-delta starting. The first prerequisite of this scheme is, of course, that the motor be wound to run with its stator windings connected in delta and with all leads brought out. Delta-wound motors have not been common in the United States, and it was primarily the interest shown by manufacturers of large centrifugal air-conditioning units which prompted their reappraisal for starting purposes.

The fully automatic version of the star-delta starter consists of a full-size, full-voltage magnetic starter, two contactors for switching the star and delta connections, and a timing unit. When the operator presses the START button, the line and star contactors close and the motor starts. The motor is then star-connected and operates on approximately 58 percent of normal voltage. After the starting period, which is determined by the timer setting, the star contactor opens, and an instant later the delta contactor closes. The line contactor remains closed during this automatic transition.

The star-delta starter is also available in a manual style which allows the operator to perform the switching from star to delta. The components of the manual starter are the same as those in the automatic starter, except that the timing unit is omitted and the operator utilizes a START-RUN-STOP button instead of the START-STOP button found on the automatic starter.

As shown in Fig. 7-171, the motor windings are first connected to the line in Y or star. Starting current and torque are consequently about one-third normal. In the basic method, the transition to delta must be preceded by opening the circuit to the motor. The star-delta method therefore is inherently one of open-circuit transition.

The appeal of this method, as with the part-winding technique, lies largely in its lack of accessory voltage-reducing equipment. While it has the disadvantge of open-circuit transition, a star-delta starter does give a higher starting torque per line ampere than a part-winding starter. Star-delta is almost always an increment method, because the low starting torque (33 percent) rarely allows the motor to start on the first step. The time delay allows the sagging voltage caused by motor inrush current to recover. Noise and vibration are conspicuously absent, and except for the normal transients associated with open-circuit transition the current inrushes are not excessive at any point.

Figure 7-172 shows a modification using resistors to maintain continuity and thus avoid the difficulties associated with the open-circuit form of transition. The resistors must be carefully proportioned, both in ohmic value and in watts dissipation, for the application in question.

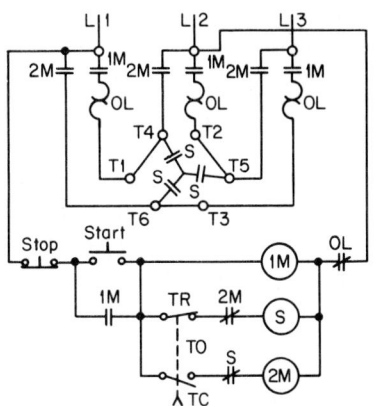

**Fig. 7-171**   Star-delta starter. [Allen-Bradley Co.]

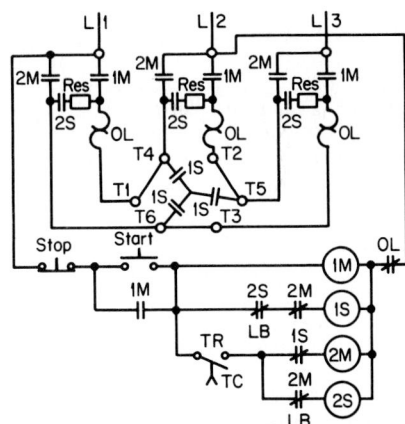

**Fig. 7-172**   Modified star-delta starter. [Allen-Bradley Co.]

**314. Control apparatus for multispeed induction motors.** Controllers for multispeed squirrel-cage motors are predominantly two-speed starters, consisting of two contactors, mechanically and electrically interlocked, and two three-pole overload relays, one for each speed. Various contactor-pole configurations are available. Selection is dependent upon the motor winding pattern.

Two-speed motors with a single winding (called consequent-pole motors) have tap points on their windings connected differently for each of the two speeds. Controllers for consequent-pole motors consist of a three-pole contactor interlocked with a five-pole contactor (called a 3 × 5 controller), where two poles of the latter device are used to tie three tap points together to make a star (Y) connection. The ratio of low to high speed is fixed for such motors.

Two-speed motors with two windings can be made with any desired ratio of low to high speed. The windings can be:

1.  Y-Y            (uses 3 × 3 controller)
2.  Y-delta        (uses 3 × 4 controller)
3.  Delta-Y        (uses 4 × 3 controller)
4.  Delta-delta    (uses 4 × 4 controller)

Two-speed controllers for these motors are various combinations of three- and four-pole contactors as shown above. A four-pole configuration is required for each delta winding so that the delta winding not in use can be opened when the other motor winding is connected to the line. If the unused delta winding is not opened, induced currents in this winding will cause the motor to overheat. See Fig. 7-173 for connection diagrams.

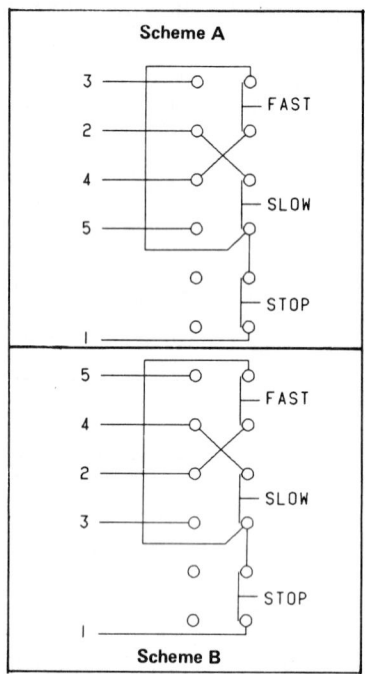

I. Control-station diagram.

II. Connection diagram for consequent-pole motors.

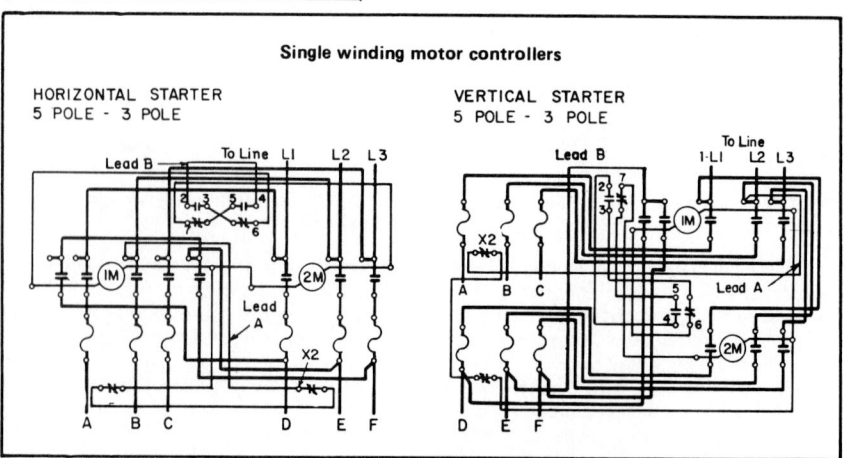

**Fig. 7-173**  Two-speed controllers. [Westinghouse Electric Corp.]

III. Connection diagram for separate-winding motors.

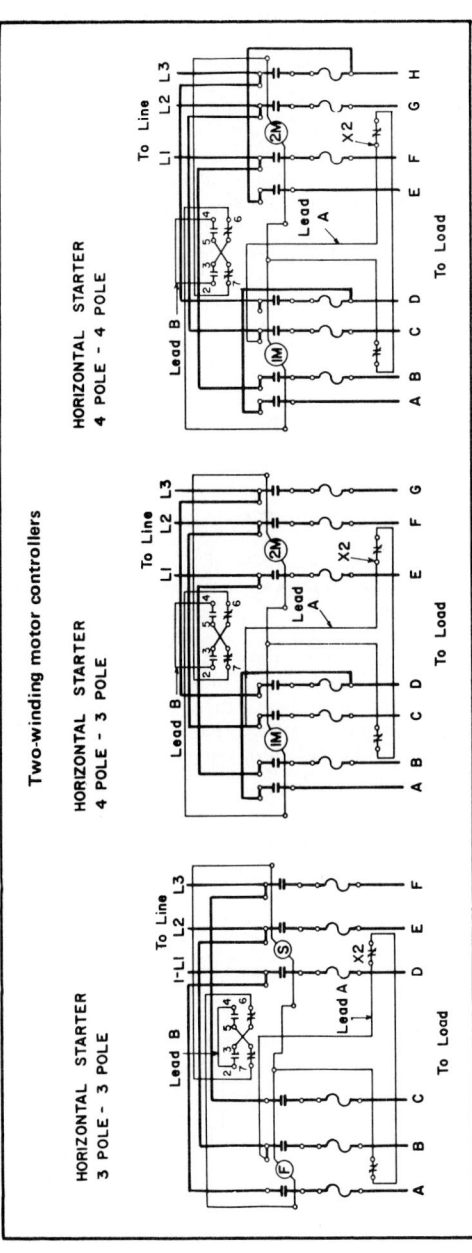

**Two-winding motor controllers**

IV. Motor-to-controller connections.

| Motor type | | Low-speed (slow) connections | High-speed (fast) connections | Contactor poles | Control-station diagram |
|---|---|---|---|---|---|
| One-winding motor | Constant torque | T1-D, T2-E, T3-F, (T7-F) | T4-B, T6-A | 5 x 3 | Scheme A |
| | Variable torque | T1-D, T2-E, T3-F | T4-B, T5-C, T6-A | 5 x 3 | Scheme A |
| | Constant torque | T1-A, T2-B, T3-C | T4-E, T5-F, T6-D, (T7-F) | 5 x 3 | Scheme B |
| Two-winding motor | Y-Y | T1-D, T2-E, T3-F | T11-A, T12-B, T13-C | 3 x 3 | Scheme A |
| | Y-delta | T1-E, T2-F, T3-G | T11-B, T12-C, T13-D, T17-A | 4 x 3 | Scheme A |
| | Delta-Y | T1-B, T2-C, T3-D, T7-A | T11-E, T12-F, T13-G | 4 x 3 | Scheme B |
| | Delta-delta | T1-B, T2-C, T3-D, T7-A | T11-E, T12-F, T13-G, T17-H | 4 x 4 | Scheme B |

**Fig. 7-173** (*Continued*)

**315. Controllers for three-speed and four-speed motors** have a contactor and a three-pole overload relay for each speed, with the contactors and control stations suitably interlocked. The motor full-load current at each speed must be considered when selecting overload relay elements just as for single-speed and two-speed starters. Often the low speed of a multispeed starter is also used to clear a jam, and in those cases the phase rotation within the motor must be reversed. A reversing contactor is required in addition to the multispeed starter for such applications, and the operator must select direction (forward or reverse) in addition to speed. Control for direction is often in the form of a maintained-contact selector switch.

**316. Controllers for wound-rotor motors** are manufactured in dial, drum, and magnetic types. The same piece of equipment is utilized for both starting and speed control when speed control is desired. The same precautions must be taken in using a controller for speed control, however, as are outlined in Sec. 290 for armature speed controllers for dc motors.

**Fig. 7-174** Adjustable frequency controller to provide adjustable speed and acceleration control. [Square D Co.]

**Fig. 7-175** Solid-state reduced-voltage starter. [Allen-Bradley Co.]

**317. Dial controllers for wound-rotor ac motors** are similar to dial controllers for dc motors, except that they have two or three sets of resistances controlled by means of two or three contact arms. These controllers control the resistance in the rotor (secondary) circuit only, so a separate switch and overload protection must be provided in the stator circuit. It is best to have this primary line switch of the magnetic type so that it can be electrically interlocked with the secondary controller. With such an arrangement the motor cannot be started unless the resistance is cut into the rotor circuit. For reversing service a reversing switch is provided in the stator circuit.

**318. Drum controllers for wound-rotor ac motors** are similar in appearance and general construction to those employed for dc motors (see Fig. 7-140), except that the movable contacts control two or three sets of resistances instead of only one set. In addition to these contacts for controlling the rotor resistance, some drum controllers are provided with contacts for opening and closing the stator circuit. If the drum is not so designed, it is best to have the line switch of the magnetic type so interlocked with the secondary controller that the switch cannot be closed unless the resistance is cut into the rotor circuit.

Controllers of this type are suitable for nonreversing or reversing service. Overload protection and undervoltage protection, if desired, must be provided in the stator circuit.

**319. Magnetic controllers for wound-rotor induction motors** consist of a primary contactor and one or more secondary contactors, depending upon the number of steps of secondary resistance employed. They can be obtained for reversing or nonreversing duty. Reversing starters require two electrically interlocked primary contactors. Time acceleration is employed for most of these controllers, but current limit or frequency limit is used for some applications.

A multicircuit-escapement type of timing relay controls the sequence and time of operation of the successive contactors. A controller using frequency control of acceleration is shown in Fig. 7-174.

**320. The use of silicon-controlled rectifiers (SCRs)** is one of the newer methods of reduced-voltage starting. These starters provide smooth, stepless acceleration of standard squirrel-cage motors in applications such as starting conveyors, compressors, and pumps. A controller with vertical open construction is shown in Fig. 7-175, and a typical simplified wiring diagram in Fig. 7-176. The unique feature of this solid-state starter is the use of start and run contactors. Operation is as follows:

1. When the START pushbutton is operated after the disconnect switch has been closed, the start contactor is energized.

2. When the start contactor contacts are closed, the SCRs start to fire. The motor is brought up to speed either with *constant-current acceleration* or with *linear-timed acceleration.*

3. When the sensing circuit of the starter detects a current approaching the full-load current of the motor, the run contactor will be energized.

4. When the run contacts are closed, the SCRs are turned off.

5. After the SCRs stop conducting, the start contactor is deenergized, and the motor runs under the control of the run contactor.

6. When the STOP pushbutton is energized, the run contactor is deenergized, and the motor is isolated from the line.

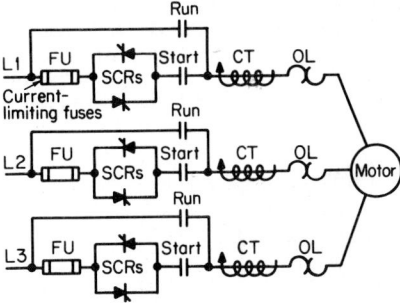

**Fig. 7-176** Simplified wiring diagram of an SCR starter. [Allen-Bradley Co.]

**321. Adjustable-frequency motor controllers**

are designed to drive squirrel-cage induction motors over a range of speed that typically varies from one-thirtieth to twice normal rated speed. The virtue of these controllers is their ability to use standard off-the-shelf ac motors that are lightweight, rugged, and easy to maintain and yet provide speed control almost equivalent to that of dc motors.

In an adjustable-frequency motor controller, the incoming ac supply voltage is rectified to a variable-voltage dc level. The dc voltage is then inverted to an adjustable-frequency output by a microprocessor that creates a new sine wave with properly selected pieces of direct current at the desired frequency. The voltage is generally adjusted proportionately to provide a constant-volts-per-hertz excitation to the motor terminals up to normal rated frequency (50 or 60 Hz). Above normal rated frequency, the voltage supplied to the motor remains at rated motor voltage to obtain energy savings.

Selecting the proper size of motor to drive a given load over a 6:1 ratio of speeds when supplied by an adjustable-frequency motor controller is more complex than for a single- or two-speed application. First, the load torque-versus-speed requirements must be determined, and then a single motor that can cover the range must be selected.

Figure 7-177 illustrates the torque-speed characteristic of a typical induction motor when controlled from an inverter over the frequency range of 2 to 120 Hz. From 2 Hz up to the nominal rated frequency of 60 Hz, the motor can supply constant torque $T_n$. Within this range, the magnitude of the motor slip speed remains the same, in revolutions per minute, for a constant-torque load (Region 1). At 60 Hz, the output voltage reaches

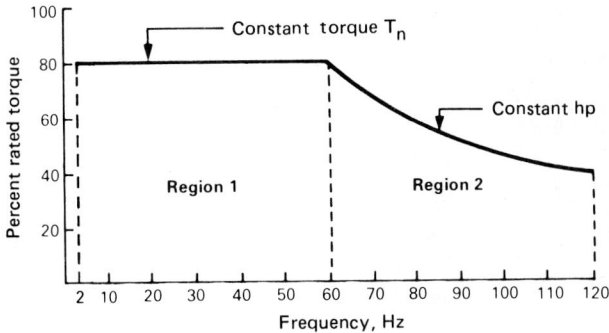

**Fig. 7-177** Induction-motor torque-speed characteristic, 2–120 Hz, Class B temperature rise. [Westinghouse Electric Corp.]

its maximum value. As a result, the motor operates in the weakened field range from 60 to 120 Hz yielding a constant-horsepower characteristic (Region 2). Within this range, the motor's slip speed varies as a constant percentage of its set synchronous speed.

As can be noted from Fig. 7-177, the percentage rated torque shown is 80 percent of the motor nameplate rating. This is to compensate for the additional heating which occurs in a motor owing to the nonsinusoidal output of the inverter.

Figure 7-177 is directly applicable to most separately ventilated and totally enclosed nonventilated Class B rise motors.

**322. The starting equipment for the commutator type of polyphase induction motor** generally consists simply of some type of across-the-line starter, as described in Secs. 302 to 307. In the special cases in which resistance is connected in series with the secondary for starting, the control would be similar to that employed for wound-rotor motors. The speed control can be regulated by shifting the brushes in any one of the following three ways:

1. By means of a handwheel mounted on the motor

2. By means of a handwheel located conveniently for the operator and connected to the brush-operating shaft by a chain drive

3. By means of a pilot motor and reduction-gear mechanism which is mounted on the main motor frame and controlled by push buttons located at the most convenient point for the operator

Typical connections for a magnetic controller used when motor speed is hand-adjusted are shown in Fig. 7-178. The operating coil of the contactor is connected in series with a limit switch on the motor so that the contactor will not operate unless the brushes are adjusted to the low-speed position. When a secondary resistor is employed for special starting conditions or for obtaining creeping speeds, a magnetic controller is always employed. This controller will consist of a primary contactor electrically interlocked with a secondary contactor.

**323. The control equipment for synchronous motors** of conventional design is the same as that for squirrel-cage induction motors with the addition of the necessary field-control equipment. Controllers for synchronous motors may be reduced-voltage (Fig. 7-179) or full-voltage.

The essential parts of a fully automatic controller for reduced-voltage transformer starting are a line contactor, an accelerating contactor, a time-delay relay, a field-actuating relay and contactor, an auto-transformer, a field rheostat, and overload and undervoltage protective devices. (See Fig. 7-180.) When the controller is in the off position, the field circuit is short-circuited through the proper value of resistance. Pushing the start button energizes the accelerating contactor, which connects the autotransformer to the line and the motor to the secondary taps of the auto-transformer. At the same time the time-

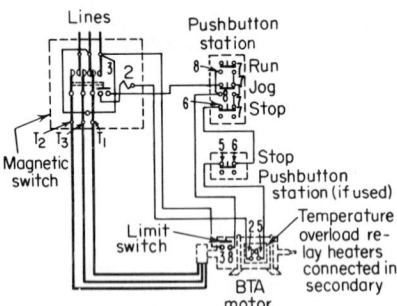

**Fig. 7-178** Connections for control of a commutator-type polyphase induction motor when speed is adjusted by hand. [General Electric Co.]

**Fig. 7-179** Low-voltage autotransformer reduced-voltage synchronous motor starter. [Allen-Bradley Co.]

delay relay is energized. After the time interval set, which corresponds to the setting of the time-delay relay, the accelerating contactor opens and the line contactor closes. The opening of the accelerating contactor disconnects the autotransformer from the line and disconnects the motor from the autotransformer. The closing of the line contactor connects the motor directly to the line and energizes the field-actuating relay. When the

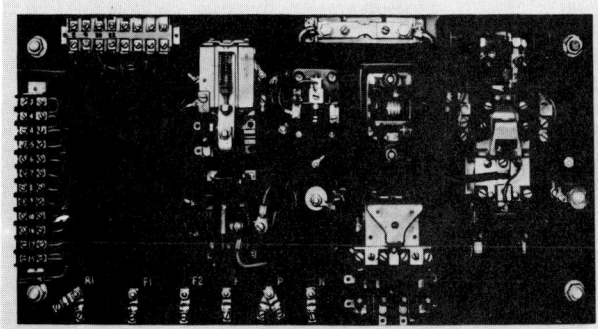

I. Control panel removed from case.

II. Complete starter in case.

**Fig. 7-180**   Automatic synchronous-motor starter. [Westinghouse Electric Corp.]

motor reaches approximately synchronous speed, the field relay closes and energizes the field contactor, which closes and applies excitation to the field circuit.

An automatic starter for full-voltage starting of synchronous motors (Fig. 7-181) consists of a line contactor, a field-actuating relay and contactor, a field rheostat, and overload and undervoltage protective devices. Pushing the START button energizes the line contactor and connects the motor directly across the line with the field short-circuited through the proper value of resistance. When the motor reaches nearly synchronous speed, the contacts of the field-actuating relay automatically close. After a definite time the field contactor closes, applying excitation to the field circuit.

Resistance starters for synchronous motors of conventional design are generally fully automatic. They are similar to primary-resistance starters for squirrel-cage induc-motors with, as mentioned above, the necessary relays and contactors for field control.

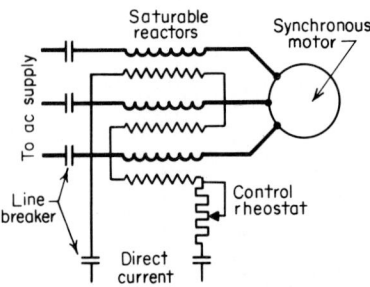

A saturable-reactor type of controller is employed sometimes for very large motors when a large number of starting steps are required. The principle of this method of acceleration is illustrated in Fig. 7-181. Each phase is supplied with a saturable reactor, which is built on a three-leg iron core. Coils on the outside legs of the reactors are connected in series with the motor, and coils on the middle leg are connected to a source of direct current in series with a rheostat. In starting the motor the dc coils are at first disconnected from their supply. The reactor then has a high impedance to alternating current and limits the first inrush of current. The dc coils are then energized with all the resistance of their control rheostat cut into the circuit.

**Figure 7-181** Connections for starting a synchronous motor with saturable reactors.

The current in the dc coils decreases the impedance of the ac coils. As the rheostat cuts resistance out of the dc circuit, the direct current increases still further, reducing the impedance of the ac coils until the reactor core is saturated and the impedance of the ac coils is practically zero. A large number of steps is thus easily available for starting the motor. An actual controller would generally be automatically operated and would include field-control equipment for the motor.

**The control equipment for synchronous motors with wound-rotor secondaries** will generally consist of a fully automatic magnetic controller of the same general type as that used for wound-rotor induction motors with the addition of relays and contactors for field control.

**324. Vacuum contactors for ac applications.** In recent years, metallurgy and vacuum technology have progressed to the point where motor controllers using contacts that operate in a vacuum have become more common. In a vacuum contactor, the main contacts are sealed inside ceramic tubes from which all air has been evacuated; i.e., the contacts are in vacuum. No arc boxes are required because any arc formed between opening contacts in a vacuum has no ionized air to sustain it. The arc simply stops when the current goes through zero as it alternates at line frequency. The arc usually does not survive beyond the first half cycle after the contacts begin to separate. The ceramic tube with the moving and stationary contacts enclosed is called a vacuum interrupter or a bottle, and there is one such bottle for each pole of the contactor. A two-pole contactor has two vacuum bottles, and a three-pole contactor has three vacuum bottles. A metal bellows (like a small, circular accordion) allows the moving contact to be closed and pulled open from the outside without letting air into the vacuum chamber of the bottle. Both the bellows and the metal-to-ceramic seals of modern bottles have been improved to the point that loss of vacuum is no longer cause for undue concern.

The moving contacts are driven by a crossbar supported by two bearings that are clamped in alignment.

The contacts in an unmounted bottle (vacuum interrupter; see Fig. 7-182) are normally closed because the outside air pressure pushes against the flexible bellows. For contactor duty, the contacts must be "normally open" when the operating magnet is not energized. Therefore, the contacts of the vacuum bottles must be held open by a kickout spring. The kickout-spring system pulls against the moving armature and crossbar and thereby forces the bottles into the open position. In the open position, the crossbar is pulling the moving contacts to hold them open.

**325. Control equipment for single-phase motors.** Single-phase motors are almost always started by throwing directly across the line. Some type of across-the-line starter similar to those discussed in Secs. 302 to 307 for polyphase motors is employed. Some of the manual types with thermal overload protection are frequently called *secondary breakers*. They are not really circuit breakers, for they are not designed to interrupt short circuits. Typical single-phase manual across-the-line starters are shown in Fig. 7-183.

When the starting current of the larger sizes is greater than that allowed by the utility, a dial or drum starting rheostat similar to those for dc motors is generally employed.

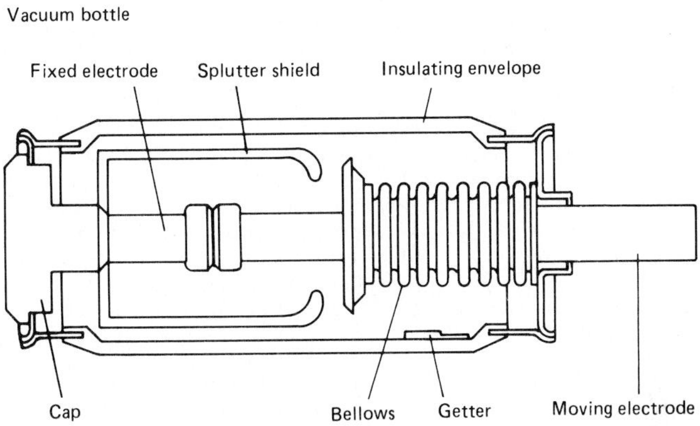

Vacuum bottle

Fixed electrode    Splutter shield    Insulating envelope

Cap        Bellows    Getter    Moving electrode

**Fig. 7-182**    Cross-section view of a vacuum interrupter.

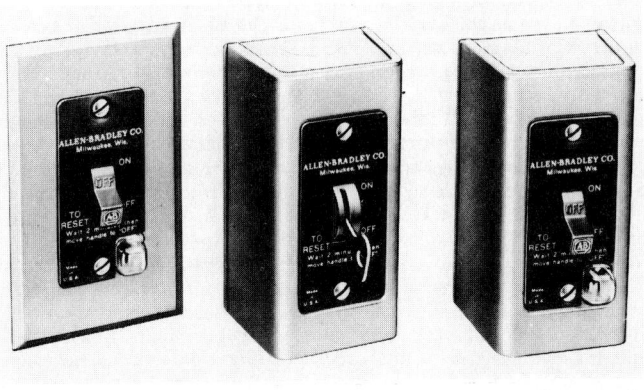

I.  Flush-mounted,    II.  Key-operated.    III.  Toggle-operated.
    toggle-operated.

**Fig. 7-183**    Manual across-the-line starters for single-phase motors. [Allen-Bradley Co.]

## MOTOR DRIVES AND APPLICATION

**326. Design of motor drives.**  The proper design of motor drives for machines is very important. It involves not only the selection of a type and rating of motor but also the selection of a method of connecting the motor to the driven machine and the selection of control equipment. It is affected by power supply, driven-machine requirements, space limitations, safety and working conditions for operators, initial cost and operating costs, production and quality of product, and surrounding conditions. The objective is to obtain a final installation which will be the best when all considerations are taken into account. The manner of attacking and the complexity of the solution of the problem depend upon the particular case. In some cases the limitations are so definitely fixed that they dictate only one answer. For instance, a machine requires a speed of 1600 rpm and full-load power of 24 hp with a very occasional peak of 28 hp and an average load of 20 hp. It is not necessary to maintain an absolutely fixed ratio of speed between driving and driven shaft. No speed control is required, but the speed should not vary by more than 10 percent with load changes. The starting-torque requirements are very light, but the power company will allow only exceptionally low starting currents. The power supply is 60 Hz, 230 V, three-phase. The machine is operated in a dry and clean location with normal temperatures. In this case there is no difficulty in making the proper selection. A normal-torque, normal-starting-current, or a normal-torque, low-starting-current, 25-hp, four-pole, 230-V, three-phase, 1800-rpm, open squirrel-cage induction motor with reduced-voltage starting equipment would be used. The motor would be belt-connected to the machine.

In many cases the problem is not so simple, and it cannot be split into several simple isolated problems that can be answered individually, such as the selection of the type and rating of motor, the method of connection, etc. The selection of the type of motor and the method of connection are interdependent, so that the whole problem must be considered as a unit, carefully weighing all factors involved. Generally several different plans can be used. Each one should be worked out with respect to initial cost, operating costs, and maintenance; then the possible solutions should be weighed against each other, taking into account, in addition to the above costs, the effect of the proposed drives upon the safety and working conditions of operators and upon the production and quality of the product.

It is evident that to design proper motor-drive installations it is necessary to know the different methods of mechanical connections and their advantages and limitations and all the other important factors that will affect the correct selection of motors. In the following sections, the important points that must be considered in the selection of motors and mechanical connections are discussed, followed by tables which, it is believed, will be helpful guides to proper selection. In the latter part of this division (Secs. 377 to 420) is given the procedure for the design of the different types of drives.

**327. Methods of driving machines.**   Two methods can be employed for driving the individual machines of a shop. Each machine may be equipped with its own individual motor. This is called individual drive. In other cases a single motor may be employed to drive a group of machines by means of shafting and belts. This is called group drive. Group drive at one time was widely used. Individual drive, however, has proved to have so many advantages that it is used to drive practically all modern machines.

**328. Motors can be mechanically connected to machines** in several ways. When the speed of the motor is the same as the speed desired for the machine, direct coupling of the motor shaft to the machine shaft is used. When the speed of the motor is different from the speed of the machine shaft, a connection which will give the proper speed ratio between the motor and the machine must be used. A speed ratio may be secured by using a flat belt, V belts, a chain, or a set of gears. With belts the speed ratio is inversely proportional to the effective diameter of the two pulleys. With a chain or gears the speed ratio is inversely proportional to the number of teeth on the two sprockets or gears. In some cases a variation in speed can be obtained by using variable-diameter pulleys or more than one set of gears, which can be shifted.

**329. The factors to be considered in selecting the method of connection of the motor to the machine** are the speed ratio desired, the peak loads to be transmitted, the positiveness of the drive, the space required, the noise, the maintenance, the appearance, and the first cost of the motor and the method of connection. Since the horsepower of a motor is given by the following formula,

$$\text{hp} = \frac{6.28 \times \text{rpm} \times T}{33,000} \text{ (see Sec. 68, Div. 1)} \tag{10}$$

it follows that for a given horsepower, if the speed of the motor is increased, the torque needed is decreased. To produce torque in a motor costs money for windings to create the necessary magnetic flux and for steel to provide a path for the flux. Therefore, the smaller the torque and, consequently, the higher the speed, the cheaper will be the motor. For very high speeds, however, the extra strength which must be built into the rotor to enable it to withstand the high centrifugal force more than offsets the saving due to the lower torque required.

The 1800-rpm motor is usually the cheapest motor, higher-speed motors being more expensive on account of the higher centrifugal force and lower-speed motors being more expensive on account of the greater flux and torque required. Therefore, one of the first factors to consider in selecting the mechanical connection of the motor to the machine is to see if an 1800-rpm motor can be used. The speed of the shaft on the machine to which the motor is to be connected varies with the type of machine and is frequently much lower than 1800 rpm. For example, on forging machines using large flywheels the shaft speed may be as low as 50 to 60 rpm; on machine tools, such as lathes, drills, millers, etc., it may be between 200 and 300 rpm. Speeds as high as 1000 to 2000 rpm occur on grinders and woodworking machines. In most cases, therefore, some reduction in speed is required between the motor and the machine. Direct connection of the motor to the machine shaft with a flexible or rigid coupling is possible only when the speed of the shaft can be the same as the speed of the motor. Belts, gears, or chains must be used in all other cases to effect a change of speed between the motor and the shaft.

For a speed of 1200 rpm, a 1200-rpm motor directly coupled to the shaft will usually be more practical than an 1800-rpm motor with a belt, gear, or chain connection to the 1200-rpm shaft. For speeds between 600 and 13 rpm, gear motors can be used. These motors have a rotor speed of 1800 rpm and gears built into the end frame of the machine to give the output shaft speed desired. These motors cost about twice as much as a high-speed motor with belt drive but are frequently used on account of the saving in space and elimination of troublesome belts. They are available in three classes of construction according to the severity of service (see Sec. 122).

**330. Belt drive** (refer to Secs. 377 to 403) is the least expensive method of connection when speed ratio is required. The factors to be considered when applying a belt drive are speed ratio, pulley sizes, belt speeds, distance between pulley centers, type of belt, size of belt, number of belts, number of pulleys, use of idler pulleys, and slippage. The maximum speed ratio which can be obtained with belt drive is limited by the practical sizes of pulleys obtainable, the arc of contact of the belt on the smaller pulley, and the practical distance between the pulleys. The maximum practical speed ratio may be as high as 10:1 for fractional-horsepower motors but is about 6:1 for ordinary sizes and not more than 4:1 for motors of 50 hp and up. As the diameter of a minimum size of a motor pulley is reduced, the strains on the motor bearings and shaft are increased. A minimum size of pulley is therefore specified by NEMA, as given in Sec. 388. The maximum diameter of a motor pulley is in nearly all cases limited by the belt speed, which should not exceed 5000 ft/min (25.4 m/s). This will occur only for machines which operate at a higher speed than the motor. Flat belts are slightly cheaper than V belts but have a shorter life and are noisier and more dangerous. Flat belts with a high speed ratio usually require idler pulleys, and space must be available for an idler pulley when required. There is always some slippage between the belt and the pulleys, and a belt drive should be used only when a positive drive is not required.

**331. Maximum Horsepower Limits for Two-Bearing Motors for Belt Drive**
(General Electric Co.)

| Motor speed rpm | Flat-belt drive, maximum hp | V-belt drive, maximum hp | Motor speed, rpm | Flat-belt drive, maximum hp | V-belt drive, maximum hp |
|---|---|---|---|---|---|
| 1,700–1,800 | 40 | 60; ball-bearing, 75 | 680–720 | 200 | 300 |
| 1,440–1,500 | 40 | 60; ball-bearing, 75 | 560–600 | 200 | 300 |
| 1,150–1,200 | 75 | 100; ball-bearing, 125 | 500–514 | 150 | 300 |
| 850– 900 | 125 | 200 | 440–450 | 150 | 250 |

**332. Chain drive** (refer to Sec. 419) is a more positive drive than belt drive, and although it is more expensive than belt drive, it is used when slippage under varying loads is not desired. The factors to be considered when applying a chain drive are the speed ratio, the size of sprockets, the width of chain, chain speed, and noise. The maximum speed ratio for a chain drive is about 8:1 with a maximum chain speed of 1500 ft/min (7.6 m/s). For quietness and smoothness a speed of 600 ft/min (3 m/s) or less is desirable. A chain drive is noisier than belt drive.

**333.   Maximum Horsepower Limits for Two-Bearing Motors for Chain Drive**
(General Electric Co.)

| Motor speed, rpm | Maximum hp | Motor speed, rpm | Maximum hp | Motor speed, rpm | Maximum hp |
|---|---|---|---|---|---|
| 1,700–1,800 | 5 | 1,150–1,200 | 25 | 720–750 | 75 |
| 1,440–1,500 | 10 | 850– 900 | 50 | | |

**334. Gear drive** (refer to Sec. 404) is an absolutely positive drive but is the most expensive method of connection. The factors to be considered when applying a gear drive are speed ratio, gear sizes, pitch-line spêed, space, and noise. The maximum speed ratio is about 10:1 for a single set of gears, but any speed ratio is obtainable by using several sets of gears. When the pinion of ordinary spur gearing is mounted on the motor shaft, two-bearing motors should be limited to the maximum horsepower values given for chain drive in Sec. 333. The maximum pitch-line speed with steel pinions is about 1300 ft/min (6.6 m/s). Gear drive usually is compact and takes up only a small space but is noisier than the other methods. It should be used on any application in which slippage or breaking of the belt or chain would be seriously dangerous. It can also be used for much larger speed ratios than are practicable with belts or chains.

### 335. Speed Limitations for Belt, Gear, and Chain Drives
(Allis-Chalmers Corporation)

This table represents good practice, under normal operating conditions for the use of these drives on motors and generators which are not provided with outboard bearings.

| Full-load rpm of motor or generator | | Max hp rating of motor | Max kw rating of generator |
|---|---|---|---|
| Above | Including | | |
| Flat-belt drive | | | |
| 2,400 | 3,600 | 20 | 15 |
| 1,800 | 2,400 | 30 | 20 |
| 1,200 | 1,800 | 40 | 30 |
| 900 | 1,200 | 75 | 50 |
| 750 | 900 | 125 | 75 |
| 720 | 750 | 150 | 100 |
| 560 | 720 | 200 | 150 |
| V-Belt drive[a] | | | |
| 2,400 | 3,600 | 20 | 15 |
| 1,800 | 2,400 | 40 | 30 |
| 1,200 | 1,800 | 75 | 50 |
| 900 | 1,200 | 125 | 75 |
| 750 | 900 | 200 | 100 |
| 720 | 750 | 250 | 150 |
| 560 | 720 | 300 | 200 |
| Gear drive[b,c] | | | |
| 1,500 | 1,800 | 7½ | 5 |
| 1,200 | 1,500 | 15 | 10 |
| 900 | 1,200 | 25 | 15 |
| 750 | 900 | 50 | 30 |
| 560 | 750 | 75 | 50 |
| Chain drive[d] | | | |
| 2,400 | 3,600 | 20 | 15 |
| 1,800 | 2,400 | 40 | 30 |
| 1,200 | 1,800 | 75 | 40 |
| 900 | 1,200 | 125 | 75 |
| 750 | 900 | 200 | 125 |
| 720 | 750 | 250 | 150 |
| 560 | 720 | 300 | 200 |

NOTES  The limitations are based on the use of pulleys, etc. They will be less than those given when motors are belted to low-speed drives such as countershafts.

The values are not intended to establish a definite dividing line below which the use of outboard bearings is not standard but rather to establish a dividing line which will indicate to the motor user what the manufacturers consider to be good practice in general service.

The use of outboard bearings is approved and recommended for belted motors in frame sizes of 250 hp, 575 to 600 rpm and larger.

When an outboard bearing is specified, it is assumed that the necessary three-bearing type of baseplate and slide rails, if required, are included with the motor.

[a] Limiting dimensions of V-belt sheaves for general-purpose motors in frames 505 and smaller are given in NEMA MG1-14.42. Limiting dimensions for V-belt sheaves for motors in frames larger than 505 have not been standardized; they are specified by the motor manufacturer.

[b] These values are based on the use of steel pinions.

[c] In general, for quiet operation and freedom from severe vibration peripheral speed of cut-steel gearing at the pitch diameter should not exceed 1400 ft/min (7.1 m/s).

[d] Limiting dimensions of chain-drive sprockets for general-purpose motors in frames 505 and smaller are given in NEMA MG1-14.07. Limiting dimensions of chain-drive sprockets for motors in frames larger than 505 have not been standardized; they are specified by the motor manufacturer.

**336.  Comparison of Methods of Connection for Adjustable-Speed Drive**

| Type of motor | Type of drive | Space required | Speed variation ratio | First cost |
|---|---|---|---|---|
| Constant speed.......... | Variable pitch motor pulley | Same as V-belt connection | 1¼:1 | Low |
| Constant speed.......... | Separate unit with variable-diameter pulleys | Considerable | 3½:1 | Medium |
| Constant speed.......... | Variable-diameter pulley unit built on motor housing | Moderate | 3½:1 | Medium |
| Constant speed.......... | Variable-position rollers | Small | 10:1 | High |
| Adjustable speed........ | Direct coupling | Small | 4:1 | High |

**337.  Comparison of Methods of Connection for Constant-Speed Drive**

| Type of drive | Relative noise | Space required | Speed ratio obtainable | Peak load transmitted | First cost |
|---|---|---|---|---|---|
| 1. Direct coupling, standard motor | 0 | Small; motor must extend out from end of drive shaft | 0 | Limited only by motor | 1 (considered as 100 per cent) |
| 2. Direct coupling, general-purpose gear motor | Slight | Same as 1. Angle drive obtainable in small motors | Any | Full load | 2 times the standard motor |
| 3. Direct coupling, special-purpose gear motor | Slight | Same as 2 | Any | 150 per cent of full load | 2½ times the standard motor |
| 4. Direct coupling, heavy-duty gear motor | Slight | Same as 2 | Any | 200 per cent of full load | 3 times the standard motor |
| 5. V belt.......... | 0 | Moderate | 10:1 in fractional-horsepower sizes; down to 4:1 for 50 hp and up | Any by multiplying motor horsepower by a service factor[1] | 1½ times the standard motor |
| 6. Flat belt........ | Considerable | Considerable | Same as 5 | Same as 5 | Slightly higher than for standard motor |
| 7. Chain.......... | Considerable | Moderate | 8:1 | Same as 5 | 1½ times the standard motor |
| 8. Gear............ | High | Small | 10:1, single reduction; 100:1, double reduction | Same as 5 | 1½ times the standard motor |

[1] Refer to Sec. **382.**

**338. The important factors to be considered in selecting a motor are:**
1. Power supply
   a. Type of system: dc or ac; frequency; number of phases
   b. Voltage and frequency conditions
   c. Effect of motors on power supply system
   d. Power rates
2. Requirements of driven machine
   a. Speed
   b. Mechanical arrangement
   c. Power and torque
   d. Special conditions
3. Surrounding conditions
   a. Ambient temperature
   b. Atmospheres
   c. Water
   d. Altitude
   e. Quietness
4. Codes, standards, and ordinances
A summary of motor characteristics is given in Sec. 361.

**339. The type of power supply system available** in most cases answers the question of whether ac or dc motors will be used as well as the frequency of the motors for ac. For most applications either ac or dc motors will be satisfactory. In some cases special requirements of extra-wide speed range, large number of adjustable-speed machines, or severe accelerating or reversing duty may justify the conversion of alternating to direct current.

If a new power plant is to be installed, the operating conditions may sometimes affect the choice of current. Even in this case the characteristics of the new plant generally should agree with those of the nearest central station in order to obtain breakdown service and to operate economically with central-station energy on reduced loads. For certain applications, dc motors are preferable—for example, in adjustable-speed service, as in machine-tool operation, in service when frequent starts must be made with very high torque, or in reversing service, as in the operation of cranes, hoists, etc. The voltage of ac circuits can be so readily transformed up or down that such energy is more economical for distribution over considerable areas. For plants extending over a considerable area or distributing energy to distances, say, of ¼ mi (⅖ km) or more, alternating current is nearly always more economical.

For ac motors, three-phase, 60 Hz is most desirable except for small fractional-horse-power motors. The demand for two-phase motors is small, and they are therefore not generally available in stock for quick shipment. If a multispeed motor is desired when the power supply is two-phase, it is best to transform to three-phase for this machine, as multispeed motors are difficult to wind for two-phase. Standard 60-Hz motors are usually suitable for operation on 50-Hz systems with some modification in their ratings and characteristics. The demand for 40- or 25-Hz motors is so small that they are not generally available in stock. For special high-speed applications in the woodworking industry and for portable tools, motors of higher frequencies than 60 Hz are available and are sometimes advantageous. The power supply can be obtained by means of an induction frequency changer.

**340. The conditions of the supply voltage and frequency,** if alternating current, with respect to value and regulation must be known to select the proper motors and control. With ac systems the voltage can, of course, be changed through transformers whenever it is advisable. Table 341 gives a guide for good practice in the maximum and minimum horsepower limits for the different standard motor voltages.

Motors will operate successfully under some conditions of voltage and frequency variations but not necessarily in accordance with the standards established for operation at normal rating (see Sec. 128). When the variations in voltage and frequency exceed those limits, special motors and control will probably be required.

The voltage regulation of the supply should be known to select motors which will deliver sufficient torque, even with the probable drop in voltage, to start and carry the load. All induction-motor torques and synchronous-motor starting torque and pull-in torque vary as the voltage squared.

The speed of ac motors will be affected by the frequency of the supply voltage as discussed in previous sections for each type of motor.

**341.   Guide to Horsepower Limits of Individual Motors for Different Systems**

| Voltage | Min hp | Max hp | Voltage | Min hp | Max hp |
|---|---|---|---|---|---|
| D-c | | | A-c, two- and three-phase | | |
| 115 | None | 30 | 110–115–120 | None | 15 |
| 230 | None | 200 | 220–230–240 | None | 200 |
| 550–600 | ½ | None | 440–550 | None | 500 |
| A-c, one-phase | | | 2,200 | 40 | None |
| | | | 4,000 | 75 | None |
| 110–115–120 | None | 1½ | 6,600 | 400 | None |
| 220–230–240 | None | 10 | | | |
| 440–550 | 5 | 10 | | | |

**342. The effects of motors and their control on the power supply system** will affect the selection of the proper type of motor.

The effect of motor-starting currents upon the supply and permissible limits for starting currents are discussed in Sec. 130.

With fluctuating loads, care should be taken to see that the following limits are not exceeded:

1. Frequencies of current pulsation in the range of 250 to 600 per minute may cause objectionable light flicker even if the voltage variation is as small as 0.5 percent.

2. The usual limit for current pulsation is 66 percent of motor full-load current, although in some cases the system regulation may require a limit of 30 or 40 percent to avoid light flicker.

**343. Effect of power rates on motor selection.**  When speed adjustment is required, the energy charge in the power rates should be considered in the selection of the type of motor and in choosing between mechanical adjustable-speed drive and an adjustable-speed motor.

The power rates are generally based on the maximum demand of the consumers. The choice of motor and use of flywheels may smooth out the power demand. This will keep the maximum demand down and lower the rate.

A power-factor clause is incorporated in many rate schedules. These adjust the rate according to the power factor of the load or involve a penalty or bonus for power factors that are below or above certain values. In such cases, consideration should be given to the use of synchronous motors or power capacitors.

**344. Effect of speed requirements on motor selection.**  Since there is a great difference in the speed characteristics of different types of motors, the maximum allowable speed regulation (variation in speed with load) on the driven machine is an important factor in the selection of the proper motor. The actual operating speed, number of speeds, and the speed range required are factors which must be considered in the joint selection of type of motor and mechanical connection. See Secs. 328 to 337 for description and comparison of different methods of mechanical connection. When speed control of the driven machine is required, it can be obtained through speed control of a motor of the proper type or by means of a motor coupled to the machine through a mechanical speed-adjusting connection.

Close speed regulation is desirable for many applications such as machine tools, textile machinery, and similar work.

When large peak loads occur, wide speed regulation is desirable. This will allow the motor to slow down under the peak loads and reduce the horsepower carried at these moments. If the motor has a small speed regulation, it may be greatly overloaded at the peak loads unless the machine is equipped with an exceptionally large motor.

For peak loads occurring periodically, when the peaks occur less than 25 times per minute, flywheels should be employed with motors having a wide speed regulation. This will allow the use of a smaller motor and will tend to even out the peaks of power drawn from the supply.

**345. Mechanical-arrangement features of the driven machine** should be taken into account in motor selection. The General Electric Co. makes the following recommendations:

1. Horizontal or Vertical Shaft   Arrangement of the driven machine usually determines whether a horizontal or a vertical motor is needed. Horizontal motors are more generally available and less expensive: most grease-lubricated ball-bearing motors will operate in either position. Fractional-horsepower, waste-packed sleeve-bearing motors are satisfactory for short periods of vertical operation when no thrust is involved.

2. Tilted Shaft   If momentary, this will require special construction of bearing housings for oil-ring–lubricated sleeve-bearing motors to avoid loss of lubricant. For long periods of tilted operation, bearings suitable for end thrust may be necessary. Ball-bearing motors with grease lubrication are suitable for tilted operation.

3. Iverted Operation and Rolling

   a. When the driven machine requires mounting the motor with the feet above or to one side of the shaft, rearrangement of the end shields of most motors makes them suitable.

   b. If the driven machine requires operation of the motor at a changing angle from the horizontal (shaft remaining horizontal) of more than 10 or 12°, ball-bearing motors will usually be required. Within this angle, sleeve-bearing motors with modified oil gages are applicable.

4. Portable Machines   A portable type of driven machine may require motors and control of greater compactness and less weight than standard and may necessitate special bearing construction for ring-oiled sleeve-bearing motors.

**346. The horsepower and torque requirements of the driven machine** are main factors in determining the proper motor rating, type of motor, and motor-control equipment. The greatest care should be exercised in determining as accurately as possible the horsepower required and both the starting- and running-torque requirements. The motor not only must have sufficient horsepower and torque capacity under running conditions but must be able to develop sufficient torque at starting to start the load satisfactorily under the most severe starting conditions that may be encountered in the particular application. Sometimes all these requirements will be known from past experience. In all cases, when it is possible, it is best to conduct tests so that the power and torque requirements can be accurately determined. Money spent in such determinations is wisely expended even if it requires the rental of a motor for conducting the tests. In performing tests it should be remembered that the test-motor losses must be subtracted from the power input to the motor, since the motor to be purchased is rated on horsepower output.

If the power and torque requirements cannot be determined from past experience or test, the information can generally be obtained from the manufacturer of the machine to be driven or from the motor experts of motor manufacturers. Whenever there seems to be any question about the accuracy of such information, it should be carefully checked, since machine-tool manufacturers often overestimate the horsepower required to be on the safe side. The result is that the motors run only partially loaded at considerably reduced efficiency. The electrical losses and interest and depreciation on the unnecessary extra investment may be considerable in a large installation.

Whenever the frequency of the starting periods is great (more than 4 to 6 times per hour), it is best to consult the motor manufacturers. Such frequent starting may require special motors and controls.

Many machines work on a definite duty cycle that repeats at regular intervals.[1] Often this horsepower value occurs during the cycle when the values of power required and the length of their duration are known, and the rating of the motor required can be calculated by the rms method.

Multiply the square of the horsepower required for each part of the cycle by the time in seconds necessary to complete that part of the cycle. Divide the sum of these results by the effective time in seconds to complete the whole cycle. Extract the square root of this last result. This gives the rms horsepower. If the motor is stopped for part of the cycle, only one-third of the rest period should be used in determining the effective time for open motors (enclosed motors use one-half of the rest period). This is due to the reduction in cooling effect when the motor is at rest.

**example**  Assume a machining operation in which an open motor operates at 8-hp load for 4 min, 6-hp load for 50 s, and 10-hp load for 3 min and the motor is at rest for 6 min, after which the cycle is repeated.

$$\text{rms hp} = \sqrt{\frac{(8^2 \times 240) + (6^2 \times 50) + (10^2 \times 180)}{240 + 50 + 180 + \dfrac{360}{3}}} = \sqrt{59.5} = 7.7 \text{ hp}$$

Use a 7.5-hp motor.

**NOTE**  For fast-repeating cycles involving reversals and deceleration by plugging, the additional heating due to reversing and external $WR^2$ loads must be taken into consideration, and a more elaborate duty-cycle analysis than shown above is required. It is necessary to know the torque, time, and motor speed for each portion of the duty cycle, such as acceleration, running with load, running without load, deceleration, and at rest.

For such detailed duty cycles it is sometimes more convenient to calculate on the basis of torque required rather than on the horsepower basis. The following formula for the accelerating time for constant torque may be helpful:

$$\text{Time, s} = \frac{wk^2 \times \text{change in rpm}}{308 \times \text{torque (lb·ft) available from motor}} \tag{11}$$

This formula can be used when the accelerating torque is substantially constant. If the accelerating torque varies considerably, the accelerating time should be calculated in increments; the average accelerating torque during the increment should be used. The size of the increment used depends on the accuracy required. Fair accuracy is obtained if the minimum accelerating torque is not less than 50 percent of the maximum accelerating torque during the increment.

[1]Courtesy of Westinghouse Electric Corp.

In addition to determining the horsepower rating with regard to heat capacity, a check should be made to assure that the peak loads imposed by the driven machine, even though of short duration, do not exceed the maximum power that the motor will develop without stalling. From tables of locked-rotor and maximum running-torque values, determine the maximum torque in terms of full-load torque. This value should always exceed peak torque imposed by the duty cycle. Preferably, this should be at least 25 percent greater to compensate for unusual conditions, such as low voltage, high friction, etc.

**347. Horsepower rating of motor.** After having determined the horsepower requirements of the load, select a motor with sufficient horsepower rating to take care of these requirements. See Sec. 128 for a discussion of motor ratings.

**348. Any special conditions of operation of the driven machine** must be carefully considered. Some machines such as propeller-type fans and deep-well pumps impose an axial thrust on the motor bearings. In some cases standard ball-bearing motors will furnish sufficient thrust capacity. For extremely heavy thrusts, special plate-type oil-lubricated bearings may be required.

For some tools such as precision grinders, to ensure high quality of the finished product specially balanced motors are required to give very small amplitudes of vibration.

Sometimes the driven machine requires limited or even zero end play. This will necessitate the use of preloaded ball bearings or sleeve bearings with end-play–limiting devices.

**349. The surrounding conditions of temperature, atmospheres, and water** must be carefully determined in the application of motors. Refer to Sec. 126 for a discussion of the different standard types of enclosures available and a guide to the surrounding conditions under which their applications will be satisfactory. For a discussion of rating of motors and effect of room temperatures upon operation, refer to Sec. 128. The manufacturers should be consulted when the motor and control will be subjected to any unusual risk such as the following:

1. Exposure to chemical fumes
2. Operation in damp places
3. Operation at speeds in excess of specified overspeed
4. Exposure to combustible or explosive dust
5. Exposure to gritty or conducting dust
6. Exposure to lint
7. Exposure to steam
8. Operation in poorly ventilated rooms
9. Operation in pits or where entirely enclosed in boxes
10. Exposure to flammable or explosive gases
11. Exposure to temperatures above 40°C or below 10°C
12. Exposure to oil vapor
13. Exposure to salt air
14. Exposure to abnormal shock or vibration from external sources

**350. Water** (General Electric Co.). If operation for long periods underwater is required, special enclosed equipment, with counter air pressure to prevent the entrance of water, is usually necessary. For occasional flooding or submersion, extremely tight, totally enclosed nonventilated motors equipped with stuffing boxes may suffice. If hosing down or splashing of water is involved, splashproof motors will usually meet all but extreme conditions, which will require enclosed motors.

**351. Altitude** (General Electric Co.). No change in design is considered necessary for most electric machines for altitudes not exceeding 3300 ft (1006 m) above sea level. For higher altitudes, the temperature rise of electric machinery generally will increase approximately 1 percent for each 330-ft (101-m) increase in altitude above 3300 ft.

**352. Quietness** (General Electric Co.). Installations in residences and in public buildings, such as theaters, hospitals, and schools, may require the use of quiet motors and sound-isolating bases. In many cases, special building layout and construction may be needed to ensure the necessary quietness.

**353. Codes, standards, and ordinances** (General Electric Co.). Motors and control must conform to local and national standards such as those shown below to allow connection of power, satisfy safety and fire requirements, and permit lowest insurance rates:

1. NEMA standards, which specify mounting dimensions for induction motors and, in general, minimum performance characteristics for all types of motors and control

2. IEEE standards, which specify temperature limits for insulation materials and prescribe the methods of rating and testing apparatus

3. The **National Electrical Code**, which is the general guide of inspectors in determining acceptability of enclosures, protection, and installation of motors

4. State laws, which stress safety and reduction of fire hazards

5. City ordinances, which may specify particular construction considered necessary locally to avoid fires and accidents

**354. Standard ratings of motors** (NEMA standards).[1] The standard horsepower ratings of motors are 1, 1.5, 2, 3, 5, 7.5, 10, 15, 25, and 35 mhp and $\frac{1}{20}$, $\frac{1}{12}$, $\frac{1}{8}$, $\frac{1}{6}$, $\frac{1}{4}$, $\frac{1}{3}$, $\frac{1}{2}$, $\frac{3}{4}$, 1, 1½, 2, 3, 5, 7½, 10, 15, 20, 25, 30, 40, 50, 60, 75, 100, 125, 150, 200, 250, 300, 350, 400, 450, 500, 600, 700, 800, 900, 1000, 1250, 1500, 1750, 2000, 2250, 2500, 3000, 3500, 4000, 4500, 5000, 5500, 6000, 7000, 8000, 9000, 10,000, 11,000, 12,000, 13,000, 14,000, 15,000, 16,000, 17,000, 18,000, 19,000, 20,000, 22,500, 25,000, 27,500, 30,000 32,500, 35,000, 37,500, 40,000, 45,000, 50,000, 55,000, 60,000, 65,000, 70,000, 75,000, 80,000 90,000, and 100,000 hp.

The standard full-load speed ratings for constant-speed, integral-horsepower dc motors are 3500, 2500, 1750, 1150, 850, 650, 500, 450, 400, 350, 300, 250, 225, 200, 175, 150, 110, 100, 90, 80, 70, 65, 60, 55, and 50 rpm.

The standard full-load ratings for small constant-speed dc motors from $\frac{1}{20}$ to 1 hp are 3450, 2500, 1725, and 1140 rpm.

The standard speeds for 60-Hz synchronous motors are 3600, 1800, 1200, 900, 720, 600, 514, 450, 400, 360, 327, 300, 277, 257, 240, 225, 200, 180, 164, 150, 138, 129, 120, 109, 100, 95, 90, 86, and 80 rpm.

The standard synchronous speeds for 60-Hz, polyphase and single-phase induction motors, except permanent-split-capacitor motors, are 3600, 1800, 1200, 900, 720, 600, 514, 450, 400, 360, 327, 300, 277, 257, 240, and 225 rpm.

The standard synchronous speeds for 60-Hz, single-phase permanent-split-capacitor motors are 3600, 1800, and 1200 rpm.

The approximate full-load speeds of induction motors can be obtained from the synchronous speeds by means of the approximate full-load slips (see Sec. 361).

The standard voltages are as follows:

| | |
|---|---|
| 1. Direct-current fractional-horsepower motors intended for use on adjustable voltage rectifier power supplies[a] | |
|     *a.* Single-phase primary power source | 75, 90, 150, or 180 V |
|     *b.* Three-phase primary power source | 240 V |
| 2. Industrial direct current[a] | 180, 240, 250, 500, 550, or 700 V |
| 3. Universal and single-phase | |
|     *a.* 60 Hz | 115 or 230 V |
|     *b.* 50 Hz | 110 or 220 V |
| 4. Polyphase | |
|     *a.* 60 Hz | 115,[b] 200, 230, 460, 575, 2300, 4000, 4600, or 6600 V |
|     *b.* 50 Hz, three-phase | 220 or 380 V |

[a]Armature voltage.
[b]15 hp and smaller.

NOTE    It is not practical to build motors of all horsepower ratings for all of the standard voltages.

The standard frequencies are as follows:
1. *Alternating-current motors.* The standard frequency shall be 50 or 60 Hz.
2. *Universal motors.* The standard frequency shall be 60 Hz ac.

NOTE    Universal motors will operate successfully on all frequencies below 60 Hz and on direct current.

---

[1]This material is reproduced with the permission of the National Electrical Manufacturers Association from NEMA Standards Publication MG 1-1987, copyright © 1987 by NEMA.

**355. Service factors.** A general-purpose ac motor or, when applicable, any ac motor having a rated ambient temperature of 40°C, continuous duty, is suitable for continuous operation at rated load under normal conditions. When the voltage and frequency are maintained at the values specified on the nameplate, the motor may be overloaded up to the horsepower obtained by multiplying the rated horsepower by the service factor shown on the nameplate. When the motor is operated at any service factor greater than 1, it will have a higher temperature rise and may have different efficiency, power factor, and speed than at rated load, while the locked-rotor torque and current and breakdown torque will remain unchanged. Usual service factors are described in the accompanying table.

**Service Factors**
(NEMA standard)[a]

| Horse-power | Synchronous speed, rpm | | | | | | |
|---|---|---|---|---|---|---|---|
| | 3,600 | 1,800 | 1,200 | 900 | 720 | 600 | 514 |
| 1/20 | 1.4 | 1.4 | 1.4 | 1.4 | | | |
| 1/12 | 1.4 | 1.4 | 1.4 | 1.4 | | | |
| 1/8 | 1.4 | 1.4 | 1.4 | 1.4 | | | |
| 1/6 | 1.35 | 1.35 | 1.35 | 1.35 | | | |
| 1/4 | 1.35 | 1.35 | 1.35 | 1.35 | | | |
| 1/3 | 1.35 | 1.35 | 1.35 | 1.35 | | | |
| 1/2 | 1.25 | 1.25 | 1.25 | 1.15[b] | | | |
| 3/4 | 1.25 | 1.25 | 1.15[b] | 1.15[b] | | | |
| 1 | 1.25 | 1.15[b] | 1.15[b] | 1.15[b] | | | |
| 1½–125[b] | 1.15 | 1.15 | 1.15 | 1.15 | 1.15 | 1.15 | 1.15 |
| 150[b] | 1.15 | 1.15 | 1.15 | 1.15 | 1.15 | 1.15 | |
| 200[b] | 1.15 | 1.15 | 1.15 | 1.15 | 1.15 | | |

[a]This material is reproduced with the permission of the National Electrical Manufacturers Association from NEMA Standards Publication MG 1-1987, copyright © 1987 by NEMA.
[b]For polyphase squirrel-cage medium motors these service factors apply only to designs A, B, and C motors.

**356. Effects of variation of voltage and frequency upon the performance of induction motors** (NEMA standards)[1]

1. Induction motors are at times operated on circuits of voltage or frequency other than those for which the motors are rated. Under such conditions, the performance of the motor will vary from the rating. The following are some of the operating results caused by small variations of voltage and frequency and are indicative of the general character of changes produced by such variation in operating conditions.

2. With a 10-percent increase or decrease in voltage from that given on the nameplate, the heating at rated horsepower load may increase. Such operation for extended periods of time may accelerate the deterioration of the insulation system.

3. In a motor of normal characteristics at full rated-horsepower load, a 10-percent increase of voltage above that given on the nameplate would usually result in a decided lowering in power factor. A 10-percent decrease of voltage below that given on the nameplate would usually give an increase in power factor.

4. The locked-rotor and breakdown torque will be proportional to the square of the voltage applied.

5. An increase of 10 percent in voltage will result in a decrease of slip of about 17 percent, while a reduction of 10 percent will result in an increase of slip of about 21 percent. Thus, if the slip at rated voltage were 5 percent, it would be increased to 6.05 percent if the voltage were reduced 10 percent.

6. A frequency higher than the rated frequency usually improves the power factor but decreases locked-rotor torque and increases the speed and friction and windage loss. At a frequency lower than the rated frequency, the speed is decreased, locked-rotor torque is increased, and power factor is decreased. For certain kinds of motor load, such as in textile mills, close frequency regulation is essential.

7. If variations in both voltage and frequency occur simultaneously, the effects will

[1]This material is reproduced with the permission of the National Electrical Manufacturers Association from NEMA Standards Publication MG 1-1987, copyright © 1987 by NEMA.

be superimposed. Thus, if the voltage is high and the frequency low, the locked-rotor torque will be very greatly increased, but the power factor will be decreased and the temperature rise increased with normal load.

8. The foregoing facts apply particularly to general-purpose motors. They may not always be true with special motors, built for a particular purpose, or as applied to very small motors.

**357. Operation of 230-V general-purpose induction motors on 208-V network systems** (NEMA standard). A 230-V general-purpose induction motor is not recommended for operation on a 208-V system. A motor for such a system should be rated 208 or 200 V. This is so because utilization voltages commonly encountered on 208-V systems are below the $-10$ percent tolerance on the 230-V rating for which the 230-V motor is designed. Such operation will generally cause excessive overheating and serious reduction in torque.

**358. Characteristics of polyphase two-, four-, six-, and eight-pole, 60-Hz medium induction motors operated on 50 Hz**

1. SPEEDS   The synchronous speeds of 60-Hz motors when operated on 50 Hz will be five-sixths of the 60-Hz synchronous speed. The full-load speeds of 60-Hz motors when operated on 50 Hz at the rated 60-Hz voltage will be approximately five-sixths of the 60-Hz full-load speed. When voltages less than the 60-Hz rated voltage are applied, the slip will be increased approximately inversely as the square of the ratios of voltages. For example, when a 60-Hz, 460-V motor is operated at 420 V, 50 Hz, the slip will be increased by approximately 20 percent over the slip at 460 V, 50 Hz. When operated at 380 V, 50 Hz, the slip will be increased by approximately 46 percent over the slip at 460 V, 50 Hz.

2. TORQUES   The full-load torques in pound-feet of 60-Hz motors operated at 50 Hz will be approximately six-fifths of the 60-Hz full-load torques in pound-feet.

The breakdown torques in pound-feet of 60-Hz motors when operated on 50 Hz at the rated 60-Hz voltage will be approximately 135 percent of the 60-Hz breakdown torque in pound-feet. Therefore, the 50-Hz breakdown torque expressed in a percentage of the 50-Hz full-load torque will be approximately 112 percent of the 60-Hz breakdown torque expressed in a percentage of the 60-Hz full-load torque. When voltages less than the rated 60-Hz voltages are applied, the breakdown torques will be decreased as the square of the voltages. For example, when a 460-V, 60-Hz motor is operated at 420 V, 50 Hz, the breakdown torque will be reduced by approximately 17 percent below the torque at 460 V, 60 Hz; when operated at 380 V, 50 Hz, the breakdown torque will be reduced by approximately 32 percent below the torque at 460 V, 50 Hz. It may thus be seen that a 60-Hz, 460-V motor having 200 percent breakdown torque will also have a 200 percent breakdown torque as a 50-Hz motor operated at approximately 435 V, and when operated on 50 Hz at 380 V, the breakdown torque will be approximately 153 percent of the full-load 50-Hz torque.

The locked-rotor torques of 60-Hz motors operated on 50 Hz will vary in the same manner as the breakdown torques described above. For example, when a 460-V, 60-Hz motor having a 135 percent locked-rotor torque expressed in a percentage of 60-Hz full-load torque is operated on 50 Hz, 380 V, it may have as low as 103 percent locked-rotor torque expressed in a percentage of 50-Hz full-load torque.

3. LOCKED-ROTOR CURRENT   The locked-rotor current of 60-Hz motors, when operated on 50 Hz at the rated 60-Hz voltage, will be approximately 15 percent higher in amperes than the 60-Hz values. The 50-Hz value of locked-rotor current will vary in direct proportion to any slight variation in voltage from the rated 60-Hz voltage.

For example, a 60-Hz, 460-V motor with a locked-rotor current of 290 A will have approximately 334 A on 50 Hz at 460 V and approximately 319 A at 440 V and 50 Hz. The code letter appearing on the motor nameplate to indicate locked-rotor kilovolt-amperes per horsepower applies only to the 60-Hz rating of the motor.

4. TEMPERATURE RISE   The temperature rises of 60-Hz motors when operated on 50 Hz will be higher than those attained on 60-Hz operation. When 60-Hz general-purpose open motors having a service factor of 1.15 and an ambient-temperature rating of 40°C are

NOTES   Sixty-Hz motors equipped with speed-responsive switching devices may require mechanical modification for satisfactory operation at 50 Hz.

The information given above is based on motors rated 60 Hz, 460 V. The same relationship will apply to motors of other standard voltage ratings.

suitable for operation on 50-Hz circuits and are operated on 50 Hz at five-sixths of the 60-Hz voltage and horsepower rating, they will have no service factor. In general, the temperature rise of 460-V, 60-Hz 40°C motors when operated at 380 V, 50 Hz, will be approximately the same if the load is five-sixths of the rated load.

**359. Operation of small dc motors on rectified alternating current** (NEMA standard).[1]

1. GENERAL. When direct-current small motors intended for use on adjustable-voltage electronic power supplies are operated from rectified power sources, the pulsating voltage and current wave forms affect the motor performance characteristics. Because of this, the motors should be designed or specially selected to suit this type of operation.

A motor may be used with any power supply if the combination results in a form factor at rated load equal to or less than the motor rated form factor.

A combination of a power supply and a motor which results in a form factor at rated load greater than the motor rated form factor will cause overheating of the motor and will have an adverse effect on commutation.

There are many types of power supplies that can be used; some are:
 *a.* Single-phase, half-wave.
 *b.* Single-phase, half-wave, back rectifier.
 *c.* Single-phase, half-wave, alternating-current voltage controlled.
 *d.* Single-phase, full-wave, firing angle controlled.
 *e.* Single-phase, full-wave, firing angle controlled, back rectifier.
 *f.* Three-phase, half-wave, voltage controlled.
 *g.* Three-phase, half-wave, firing angle controlled.

It is impractical to design a motor or to list a standard motor for each type of power supply. The combination of power supply and motor must be considered. The resulting form factor of the combination is a measure of the effect of the rectified voltage on the motor current as it influences the motor performance characteristics, such as commutation and heating.

2. FORM FACTOR. The form factor of the current is the ratio of the root-mean-square value of the current to the average value of the current.

Armature current form factor of a motor-rectifier circuit may be determined by measuring the rms armature current (using an electrothermic instrument, electrodynamic instrument, or other true rms responding instrument) and the average armature current (using a permanent-magnet moving-coil instrument). The armature current form factor will vary with changes in load, speed, and circuit adjustment.

Armature current form factor of a motor-rectifier circuit may be determined by calculation. For this purpose, the inductance of the motor armature circuit should be known or estimated, including the inductance of any components in the power supply which are in series with the motor armature. The value of the motor inductance will depend upon the horsepower, speed, and voltage ratings and the enclosure of the motor and should be obtained from the motor manufacturer. The method of calculation of the armature current form factor should take into account the parameters of the circuit, such as the number of phases, the firing angle, half-wave, with or without back rectifier, etc., and whether or not the current is continuous or discontinuous.

Ranges of armature current form factors on some commonly used motor-rectifier circuits and recommended rated form factors of motors associated with these ranges are shown in the table on p. 7-189.

**359A. Operation of medium dc motors on rectified alternating current** (NEMA standard). When a direct-current medium motor is operated from a rectified alternating-current supply, its performance may differ materially from that of the same motor when operated from a low-ripple direct-current source of supply, such as a generator of a battery. The pulsating voltage and current wave forms may increase temperature rise and noise and adversely affect commutation and efficiency. Because of these effects, it is necessary that direct-current motors be designed or specially selected to operate on the particular type of rectified supply to be used.

**360. Operation of dc motors by reduced armature voltage** (NEMA standard). When a dc motor is operated below base speed by reduced armature voltage, it may be necessary to reduce its torque load below rated full-load torque to avoid overheating the motor.

---

[1]This material is reproduced with the permission of the National Electrical Manufacturers Association from NEMA Standards Publication MG 1-1987, copyright © 1987 by NEMA.

| Typical combination of power source and rectifier type | Range of armature current form factors[a] | Recommended rated form factors of motors |
|---|---|---|
| Single-phase thyristor (SCR) or thyratron with or without back rectifiers: | | |
| Half-wave | 1.86–2 | 2 |
| Half-wave | 1.71–1.85 | 1.85 |
| Half-wave or full-wave | 1.51–1.7 | 1.7 |
| Full-wave | 1.41–1.5 | 1.5 |
| Full-wave | 1.31–1.4 | 1.4 |
| Full-wave | 1.21–1.3 | 1.3 |
| Three-phase thyristor (SCR) or thyratron with or without back rectifiers: | | |
| Half-wave | 1.11–1.2 | 1.2 |
| Full-wave | 1.0 –1.1 | 1.1 |

[a]The armature current form factor may be reduced by filters or other circuit means which will allow the use of a motor with a lower-rated form factor.

## 361. Summary of Motor Characteristics

| Type of motor | Standard hp ratings | Speed characteristics | Speed control | Locked-rotor starting currents with rated impressed voltage in percentage of full-load current | Starting torque in percentage of full load rated torque | Max. running torque in percentage of full-load rated torque | General application suited for |
|---|---|---|---|---|---|---|---|
| 1. Direct current, shunt, constant speed | Up to 200 | Constant; not more than 10 per cent change in speed from no load to full load | Speed may be increased 25 per cent above normal by field control; speed may be reduced any amount below normal by armature control but speed then varies widely with load | Very high; normally reduced to 150 per cent of rated full-load current by series starting resistance | Maximum with full voltage, 250 to 300; when started in normal manner, average, 150 | Would occur at standstill but limited by commutation to 200 | Constant- or slightly adjustable-speed service with light or medium starting duty |
| 2. Direct current, shunt, adjustable speed | Up to 200 | Nearly constant for any field adjustment, not more than 15 per cent change in speed from no load to full load | Speed range of 4:1 for all loads by field control. Speed may be reduced by armature control but speed then varies widely with load | Same as 1 | Same as 1 | Same as 1 | Adjustable-speed service with close speed regulation and either light or medium starting duty. Motors available for either constant-torque or constant-horsepower service |
| 3. Direct current, compound, 20 per cent series, 80 per cent shunt | Up to 200 | Slightly varying speed, changes approximately 25 per cent from no load to full load | Same as 1 | Same as 1 | Maximum with full voltage; 300 to 350; when started in normal manner, —average 170 | Would occur at standstill but limited by commutation to 300 | Heavy starting duty or intermittent peak-load service or combination of the two duties |
| 4. Direct current, compound, 50 per cent series, 50 per cent shunt | 3 to 200 | Varying | Same as 1 | Same as 1 | Maximum with full voltage, 400; when started in normal manner, average 200 | Would occur at standstill but limited by commutation to 350 | Heavy intermittent starting duty or heavy running loads for short periods with alternate light load periods; adjustable varying-speed service |

| | | | | | | | |
|---|---|---|---|---|---|---|---|
| **5.** Direct current series | 3 to 200 | Varying | Same as 1 | Same as 1 | Maximum with full voltage, 450; when started in normal manner, 225 to 300 | Would occur at standstill but limited by commutation to 400 | Heavy intermittent starting duty or heavy running loads for short periods of time; adjustable, varying-speed service. Motor should at all times be loaded or under control of operator |
| **6.** Polyphase, squirrel cage, normal torque, normal starting current | $\frac{1}{10}$ to 400 | Constant; speed changes approximately 3 to 5 per cent from no load to full load | None | 500 to 1,000. May not meet power companies' limitations. Up to 5 hp, generally started with full voltage; above 5 hp, generally started with reduced voltage. Current drawn from line proportional to square of voltage for transformer starting and directly proportional to voltage for resistance starting | Varies with square of voltage applied to motor; 105 to 150 for full voltage | 200 to 250 | Constant-speed service with no speed control; light-starting-duty service unless it is permissible to start with full voltage |
| **7.** Polyphase, squirrel cage, normal torque, low starting current | 7½ to 200 | Same as 6 | None | 500 to 550. Within the limitations of most power companies up to and including 30-hp size. Above 30-hp size may not be within power companies' limitations. Current depends upon voltage same as for 6 | Same as 6 | 200 to 225 | Constant-speed service with no speed control; medium starting duty. Used in place of motor 6 when its use will eliminate employment of reduced-voltage starting |
| **8.** Polyphase, squirrel cage, high torque, low starting current | 1½ to 150 | Same as 6 | None | Same as 7 | Varies with square of voltage applied to motor; 220 to 275 for full voltage | 200 to 250 | Constant-speed service with no speed control; heavy starting duty if at not too frequent intervals and if starting period is of not too long duration |
| **9.** Polyphase, squirrel cage, low torque, low-starting current | 40 to 100 | Same as 6 | None | 350 to 550. Within the limitations of most power companies, current depends upon voltage same as for 6 | Varies with square of voltage applied to motor, 50 to 100 for full voltage | 125 to 175 | Constant-speed service with no speed control, light starting duty only |

**361. Summary of Motor Characteristics** (*Continued*)

| Type of motor | Standard hp ratings | Speed characteristics | Speed control | Locked-rotor starting currents with rated impressed voltage in percentage of full-load current | Starting torque in percentage of full load rated torque | Max. running torque in percentage of full-load rated torque | General application suited for |
|---|---|---|---|---|---|---|---|
| 10. Polyphase, squirrel cage, high-resistance rotor, medium slip | ½ to 150 | Slightly varying speed; speed changes from 7 to 12 per cent from no load to full load | None | 400 to 800. **Within the limitations of most power companies. Reduced-voltage starting not satisfactory owing to nature of applications** | 300 to 400 | 300 to 400. **Occurs at standstill** | Service with intermittent peak loads, when peak loads occur less than 25 times per minute; heavy starting duty if at not too frequent intervals and if starting period is of not too long duration |
| 11. Polyphase, squirrel cage, high resistance rotor, high slip | 1 to 50 | Medium varying speed; speed changes from 12 to 17 per cent from no load to full load | None | 300 to 500. **Within the limitations of most power companies. Reduced-voltage starting not satisfactory owing to nature of applications** | 225 to 250 | 225 to 400. **Occurs at standstill** | Frequent heavy starting duty, intermittent loads of not too long duration where **varying speed is satisfactory** |
| 12. Polyphase, squirrel cage, multispeed | ½ to 125 | Multispeed 2, 3, or 4 definite speeds available; speed changes approximately 5 per cent from no load to full load at highest speed and approximately 25 per cent at lowest speed | Speed controlled by changing number of poles of stator winding; maximum of 4 definite speeds; no speed control between steps | These motors may be obtained in any one of the types of items 6, 7, 8, or 9 | | | Adjustable-speed service where 2, 3, or 4 definite speeds will be satisfactory. Whether motor with characteristics of 6, 7, 8, or 9 is selected depends upon requirements of service **as given under these types. Available in designs for constant-torque, constant-hp or variable-torque, variable-hp service** |
| 13. Polyphase, squirrel cage, automatic start | ¼ to 75 | Constant: speed changes from 4 to 6 per cent from no load to full load | None | 350 to 375. **Within the limitations of practically all power companies** | 225 to 250. Pull-in torques of from 150 to 200 | 200 to 250 | Constant-speed service with no speed control; heavy starting duty if at not too frequent intervals and if starting period is of not too long duration |

| | Hp | Speed regulation | Speed control | Starting current, % | | | Typical applications |
|---|---|---|---|---|---|---|---|
| 14. Polyphase, wound rotor | ½ to 1,000 | Depends upon the resistance in the rotor circuit; constant (4 to 6 per cent change from no load to full load) with rotor short-circuited; varying with resistance in rotor circuit (greater resistance, greater variation in speed from no load to full load) | Speed can be reduced to 50 per cent below normal at full load by resistance in rotor circuit; only small reduction in speed can be obtained at light loads | 250 to 300. Within limitations of all power companies | 200 to 275 | 200 to 275 | Constant-speed service with no speed control for very heavy or frequent starting duty or where starting period is of long duration (resistance in rotor circuit during starting period only); adjustable varying-speed service; service with peak loads for short periods; different types of service require resistors of different ratings |
| 15. Polyphase, commutator type, brush shifting | 5 to 50 | Adjustable speed; changes from 5 to 25 per cent from no load to full load depending upon speed setting | Speed range of 3:1 by shifting brushes; very low speeds obtained by inserting resistance in secondary | 125 to 175. Within limitations of all power companies | 140 to 250 | 140 to 250 at low speeds; 300 to 400 at high speeds | Adjustable-speed service of the constant-torque type; heavy starting-duty service if at not too frequent intervals and if starting period is of not too long duration |
| 16. Synchronous, general purpose, high speed unity power factor | 20 to 5,000 | Constant: no speed change from no load to full load | None | 400 to 800. May not be within limitations of power company. Current depends upon voltage same as for item 6 | 100 to 125 for full voltage for sizes up to 500 hp; 80 to 100 for full voltage for sizes above 500 hp. Varies with square of voltage applied to motor | 150 to 175 | Constant-speed service with no speed control and light starting duty; fairly steady load; for power-factor correction |

**361. Summary of Motor Characteristics** (*Continued*)

| Type of motor | Standard hp ratings | Speed characteristics | Speed control | Locked-rotor starting currents with rated impressed voltage in percentage of full-load current | Starting torque in percentage of full load rated torque | Max. running torque in percentage of full-load rated torque | General application suited for |
|---|---|---|---|---|---|---|---|
| 17. Synchronous, general purpose, high speed, 80 per cent power factor | 20 to 5,000 | Same as 16 | None | Same as 16 | 125 to 200 for full voltage for sizes up to 500 hp; 100 to 150 for full voltage for sizes above 500 hp. Varies with square of voltage applied to motor | 200 to 250 | Constant-speed service with no speed control where greater power-factor correction is desired than item 16 will provide, or where starting duty is more severe or where peak loads occur. Heavy starting duty if started with full voltage; medium starting duty if started with reduced voltage |
| 18. Synchronous, special purpose, high speed, low torque, low starting current, unity or 80 per cent power factor | 20 to 5,000 | Same as 16 | None | May be designed for starting currents with rated impressed voltage of from 300 to 500 per cent of full-load current | 50 to 100 for full voltage depending upon design; lower values of starting torque correspond to lower values of starting currents. Varies with square of voltage applied to motor | 150 to 300 depending upon design | Constant-speed service with no speed control, light starting duty, for power-factor correction. Used in place of 16 in order to limit starting current without reduce ing voltage |
| 19. Synchronous, special purpose, high speed, high torque, low starting current | 20 to 5,000 | Same as 16 | None | 450 to 500 for single-phase wound rotor; 200 to 300 for polyphase wound rotor | 150 to 200. Always started with rated voltage applied to motor and external resistance in rotor circuit | 150 to 250 | Constant-speed service with no speed control for heavy starting duty, for power-factor correction |

| | | | | | | | |
|---|---|---|---|---|---|---|---|
| 20. Synchronous, low-speed, engine and compressor types, unity and 80 per cent power factor | 20 to 5,000 | Same as 16 | None | 225 to 350 for full voltage. Within limitations of most power companies. Varies with voltage same as 6 | 40 to 50 for full voltage. Varies with square of applied voltage | 140 to 175 for compressor types; 150 to 200 for unity-power-factor-engine type; 200 to 250 for 80 per cent power-factor-engine type | Constant-speed service with no speed control for direct connection to slow-speed machines such as compressors; light starting duty, for power-factor correction |
| 21. Synchronous, low speed special service conventional design, unity or 80 per cent power factor | 20 to 5,000 | Same as 16 | None | 325 to 700 for full voltage. May not be within limitations of power company. Varies with voltage same as 6 | 50 to 200 for full voltage depending upon design; lower values of starting torque correspond to lower values of starting current. Varies with square of applied voltage | 150 to 250 depending upon design | Constant-speed service with no speed control, light or medium starting duty, for power-factor correction. Used in place of 20 for heavier starting duty |
| 22. Synchronous, low speed, special service, wound rotor, unity or 80 per cent power factor | 20 to 5,000 | Same as 16 | None | 450 to 550 for single-phase wound rotor; 265 to 365 for polyphase wound rotor | 150 to 200. Always started with rated voltage applied to motor and external resistance in rotor | 175 to 250 | Constant-speed service with no speed control, heavy starting duty, for power-factor correction |
| 23. Single-phase shading-coil type | $\frac{1}{250}$ to $\frac{1}{4}$ | Constant; speed changes approximately 6 per cent from no load to full load | None | High, but owing to small size always satisfactory for full-voltage starting | Very low | Approximately 150 | Constant-speed service with no speed control; for very light starting duty |
| 24. Single phase, split phase starting (inductive), general purpose | $\frac{1}{60}$ to $\frac{1}{2}$ | Constant; speed changes approximately 6 per cent from no load to full load | None | High, but owing to small size will be within limitations of power companies if manufactured so as to meet NEMA specifications | 75 to 175 for full voltage; varies with square of impressed voltage | 175 to 225 | Constant-speed service with no speed control; for light starting duty |

**361. Summary of Motor Characteristics** (*Continued*)

| Type of motor | Stand-ard hp ratings | Speed characteristics | Speed control | Locked-rotor starting currents with rated impressed voltage in percentage of full-load current | Starting torque in percentage of full load rated torque | Max. running torque in percentage of full-load rated torque | General application suited for |
|---|---|---|---|---|---|---|---|
| 25. Single phase, split-phase starting (inductive), high torque | 1/6 to 1/3 | Same as 23 | None | High; approximately 850; exceeds NEMA specifications | 150 to 275 for full voltage; varies with square of impressed voltage | 225 to 350 | Constant-speed service with no speed control; for heavy starting duty |
| 26. Single phase, capacitor start, induction run | 1/6 to 3/4 | Same as 23 | None | Lower than 24. Always satisfactory to start with full voltage | 275 to 400 | 200 to 300 | Constant-speed service with no speed control; for heavy starting duty |
| 27. Single phase, capacitor start, capacitor run, two-value capacitor, high torque | 1/3 to 10 | Same as 23 | None | Within the limitations of most power companies. Always satisfactory to start smaller sizes with full voltage | 275 to 450 | 200 to 300 | Constant-speed service with no speed control for heavy starting duty. Have better power factors than item 26 |
| 28. Single phase, permanent split-capacitor, low torque | 1/2 to 10 | Same as 23 | None | Within the limitations of most power companies; lower than item 27; generally started with full voltage | 40 to 60 | 150 to 200 | Constant-speed service with no speed control for light starting duty |

| | hp | Speed regulation | Speed control | Starting | | | Remarks |
|---|---|---|---|---|---|---|---|
| 29. Single phase, multispeed capacitor | ½₀ to ½ | Practically constant for high speed; varies considerably for lower speeds | Speed control by changing impressed voltage; 2 or 3 definite speeds, no speed control between steps | Always satisfactory to start with full voltage | Approximately 60–75 for highest voltage setting; varies with square of applied voltage | Approximately 175 for highest voltage and speed setting. Varies as square of voltage for other setting | Adjustable-speed service where 2 or 3 definite speeds are satisfactory; very light starting duty; may not be satisfactory for belt drives |
| 30. Single phase, repulsion start, induction run | ⅛ to 15 | Same as 23 | None | 200 to 350: in the larger sizes may not be within limitation of power company | 300 to 500 for full voltage; varies with square of impressed voltage | 175 to 225 | Constant-speed service with no speed control for light or heavy starting duty |
| 31. Single phase, repulsion induction | ½ to 10 | Same as 23 | None | Within limitations of most power companies | 225 to 300 for full voltage | Occurs at standstill | Constant-speed service with no speed control for either light or heavy starting duty. Will not take care of as heavy starting duty as 30 |
| 32. Single phase, repulsion brush shifting | ¼ to 3 | Varying | Speed range 4:1 for full load. Very little speed control possible at light loads. 3:1 range for 75 per cent load | Within limitations of most power companies | 125 to 300 for full voltage | Above 400 for max. speed; decreases with speed reduction | Adjustable varying speed service; heavy starting duty; service with peak loads for short periods |
| 33. Single phase, series, universal for operation on a-c or d-c | ½₅₀ to 1 | Varying | Speed can be reduced any amount by resistance in series with motor | High, but owing to small size will be within limitations of most power companies. Sizes from ½ to 1 hp may require reduced voltage | 275 to 400 for a-c full voltage; 300 to 500 for d-c full voltage | Would occur at standstill but maximum safe running limited by commutation | Service requiring speed control where large variation in speed with load is not objectionable; heavy starting duty. Must be directly connected to load |

**362. Guide to Selection of Type of Motor**

| Load requirements | | Types of motors suitable | | |
|---|---|---|---|---|
| Running | Starting | Direct current | Polyphase alternating current | Single-phase alternating current |
| All classes of loads requiring constant speed with no speed control. | Very light starting duty | Shunt, constant-speed type | 1. Normal torque, normal starting current, squirrel-cage induction<br>2. Normal torque, low starting current, squirrel-cage induction<br>3. Low torque, low starting current, squirrel-cage induction<br>4. Synchronous, general purpose, high speed<br>5. Synchronous, special purpose, high speed, low torque<br>6. Synchronous, low speed, engine compressor, and special service conventional types | 1. Split phase (inductive)<br>2. Low-torque capacitor<br>3. Shading coil (only for extremely light starting duty) |
| | Light starting duty (starting torque not greater than 100 per cent of full-load torque) | Shunt, constant-speed type | 1. Normal torque, normal starting current, squirrel-cage induction<br>2. Normal torque, low starting current, squirrel-cage induction<br>3. Synchronous general purpose, high speed<br>4. Synchronous, high speed, special service, low torque<br>5. Synchronous, low speed, special service, conventional type | 1. Split-phase (inductive)<br>2. High-torque capacitor or capacitor start induction run |
| | Medium starting duty (starting torque from 100 to 150 per cent of full-load torque) | Shunt, constant-speed type | 1. Normal torque, low starting current, squirrel-cage induction in sizes up to 30 hp inclusive<br>2. High torque, low starting current, squirrel-cage induction<br>3. Synchronous, general purpose, high speed<br>4. Synchronous, special purpose, high speed, high torque<br>5. Synchronous, low speed, special service, conventional or wound-rotor types | 1. Repulsion induction<br>2. Repulsion start, induction run<br>3. High-torque capacitor or capacitor start, induction run |

| | Starting-duty requirement | Recommended type | Direct-current motor | Alternating-current motor |
|---|---|---|---|---|
| | Heavy starting duty if at not too frequent intervals and if starting period is of not too long duration (starting torque from 150 to 200 per cent of full-load torque) | Shunt, constant-speed type | 1. High-torque, low starting-current, squirrel-cage induction in sizes up to 30 hp. incl. <br> 2. Automatic-start induction <br> 3. Wound-rotor induction with resistance in rotor only during starting period <br> 4. Synchronous, general purpose, high speed <br> 5. Synchronous, special purpose, high speed, high torque <br> 6. Synchronous, low speed, special service, conventional or wound-rotor types | 1. Repulsion induction <br> 2. Repulsion start, induction run <br> 3. High-torque capacitor or capacitor start, induction run |
| | Very heavy starting duty requiring starting torques above 200 per cent of full-load torque or very frequent starting or starting periods of long duration | Compound, with series field cut-out after starting | Wound rotor with resistance in rotor only during starting period | 1. Repulsion induction <br> 2. Repulsion start, induction run <br> 3. High-torque capacitor or capacitor start, induction run |
| **All classes of loads requiring adjustable speed with not much change in speed from no load to full load for any speed setting** | Light starting duty (starting torque not greater than 100 per cent of full-load torque) | Shunt, adjustable-speed type for either constant-horsepower or constant-torque service | 1. Multispeed squirrel-cage induction for either constant horsepower or constant-torque service <br> 2. Commutator, brush-shifting type of induction motor for constant-torque service | Multispeed capacitor |
| | Medium starting duty (starting torque from 100 to 150 per cent of full-load torque) | Shunt, adjustable-speed type for either constant-horsepower or constant-torque service | 1. Multispeed squirrel-cage induction if it is permissible to start directly across the line for either constant-horsepower or constant-torque service <br> 2. High-torque, multispeed, squirrel-cage induction <br> 3. Commutator, brush-shifting type of induction motor for constant-torque service | |
| | Heavy starting duty, if at not too frequent intervals and if starting periods are of not too long duration (starting torque from 150 to 200 per cent of full-load torque) | Shunt, adjustable-speed type for either constant-horsepower or constant-torque service | 1. Special high-torque, multispeed, squirrel-cage induction for either constant-horsepower or constant-torque service <br> 2. Commutator, brush-shifting type of induction motor for constant-torque service | |

**362. Guide to Selection of Type of Motor** (*Continued*)

| Load requirements | | Types of motors suitable | | |
|---|---|---|---|---|
| Running | Starting | Direct current | Polyphase alternating current | Single-phase alternating current |
| Intermittent peak loads where it is desirable for motor to have slightly varying-speed characteristics in order to prevent excess motor overloads (flywheel machines); peak loads should occur less than 25 times per minute; any service where a slightly varying-speed characteristic is not objectionable or is desirable | Very heavy starting duty requiring starting torque above 200 per cent of full-load torque, very frequent starting or starting periods of long duration | Compound wound with series field cut out after starting | Commutator, brush-shifting type of induction with resistance in secondary during starting period | |
| | Light or heavy starting duty if at not too frequent intervals and if starting periods are of not too long duration | Compound, 20 per cent series, 80 per cent shunt | High-resistance rotor, squirrel-cage induction (medium slip) | |
| | Very heavy starting duty or starting at very frequent intervals or starting periods of long duration | Compound, 20 per cent series, 80 per cent shunt | 1. Wound-rotor induction with some resistance always in rotor circuit 2. High-resistance rotor, squirrel-cage induction (high slip) | |
| Varying or adjustable varying-speed service; any service where it is desirable or not objectionable to have speed decrease with increasing load; service requiring heavy torque for short periods with some speed control during light loads | Any type of starting duty | 1. Compound, 50 per cent series, 50 per cent shunt 2. Series if motor is always under control of operator or if there is no danger of motor losing its load or if motor is protected by an overspeed device | Wound rotor | 1. Series 2. Repulsion, brush-shifting type 3. Repulsion |

**363. Speed-Control Application Chart: Adjustable-Speed Motors and Drives**

(Westinghouse Electric Corp.)

| Primary power | Type of service | Speed range, Maximum | | | Standard speed ranges available | Speed control adjustable under load | | | | | Speed regulation | | | Hp range | Braking available | Type of motor or drive |
|---|---|---|---|---|---|---|---|---|---|---|---|---|---|---|---|---|
| | | Constant hp | Constant torque | In combination | | By remote operation | For preset jogging speed | For creeping | For presetting | For holding on overhauling load | At max speed | At base speed | At min speed | | | |
| **Motors** | | | | | | | | | | | | | | | | |
| A-C | Constant hp or constant torque | 4:1 | 4:1 | 4:1 | 4 fixed speeds | Standard | Optional[c] | Not available | Standard | Yes | 3-5% | 3-5% | 3-5% | 1-200 | Regenerative or dynamic[a] (special) | A-C multispeed induction motor |
| | Constant torque | ...... | 2:1[a] | ...... | 2:1[a] | Special[b] | Optional | Special | Optional | No | 3-5% | ...... | 50% | 1-200 | Dynamic (special) | A-C wound-rotor induction motor |
| D-C | Constant hp | 8:1 | ...... | ...... | 4:1, 6:1, 8:1 | Special[b] | Optional | Optional | Optional | No | 15-25% | 5-15% | ...... | 1-200 | Dynamic (optional) | D-C standard adjustable-speed motor |
| **Drives** | | | | | | | | | | | | | | | | |
| A-C | Combination constant hp and constant torque | 4:1 | 8:1 | 16:1 | 8:1, 12:1, 16:1 | Standard | Optional | Optional | Standard | Within 20% | 20-30% | 10-15% | 20-40% | 1-200 | Regenerative (standard) or dynamic (optional) | Motor-generator set adjustable-voltage d-c drives |
| | | 4:1 | 50:1 | 200:1 | 5:1, 20:1, 50:1 | Standard | Optional | Optional | Standard | No | 2-8%[d] | 2-8%[d] | 2-8%[d] | ½-30 | Dynamic (standard) | Electronic adjustable-voltage drive |
| | | 20:1 | 1½:1 | 30:1 | 30:1, 120:1 (available) | Not available | Standard | Not available | Standard | No | 1% | 1% | 1% | 10-100 | Regenerative | Planer type variable voltage drive |
| D-C | Constant hp | 4:1, 6:1 | ...... | ...... | 4:1, 6:1 | Not available | Standard | Not available | Standard | No | 15-25% | 15% | ...... | 5-100 | Dynamic (standard) on inching, dynamic and plugging on automatic | Planer-type constant-voltage drive |

[a] Speed ranges up to 4:1 can be obtained if the motor is oversize or if the torque load becomes smaller at reduced speeds (such as on fan, blower, centrifugal pump).

[b] Motor-operated drum or rheostat.

[c] Motor runs at full speed if jog button is held down.

[d] Speed can be set practically flat at minimum speed and will not exceed 2 at 8 per cent at any other speed.

[e] From any speed down to lowest synchronous speed.

### 364.   Representative Applications of DC Motors

| Shunt | Compound | Series |
|---|---|---|
| Agitators | Balers | Bailing presses |
| Blowers | Bending rolls | Bridges |
| Conveyors | Bulldozers | Cranes |
| Fans | Metal drawers | Car retarders and pullers |
| Line shafts | Punch presses | Coke-oven machinery |
| Printing presses | Shears | Hoists |
| Pumps | | Hydraulic gages |
| Mixers | | Lorry cars |
| Most woodworking machines | | Rotary car dumpers |
| Most machine tools | | Yard locomotives |

**365. Representative Applications of Squirrel-Cage Induction Motors**

| Normal-torque, normal-starting-current, and normal-torque, low-starting-current types | High-torque, low-starting current type | Low-torque, low-starting-current type | High-resistance rotor type | Multispeed type |
|---|---|---|---|---|
| Blowers<br>Boring mills<br>Drilling machines<br>Pumps<br>Line-shaft drive<br>Most woodworking machines<br>Most machine tools | Agitators<br>Baking machinery<br>Candy machinery<br>Conveyors<br>Crushers<br>Canning machines<br>Elevators<br>Grinders<br>Flour-mill machinery<br>Milling machines<br>Mixers<br>Pulverizers<br>Reciprocating pumps and compressors<br>Refrigerators<br>Revolving screens | Blowers<br>Centrifugal pumps<br>Fans<br>Motor generators | Balers<br>Bulldozers<br>Metal drawing machines<br>Punch presses<br>Shears<br>Elevators<br>Cranes<br>Hoists<br>Dumbwaiters | Centrifugal transformers<br>Dough mixers<br>Elevators<br>Feed mechanisms<br>Ironers<br>Some machine tools<br>Oil-well pumping and pulling<br>Printing presses<br>Pumps and fans<br>Tumblers<br>Stokers<br>Conveyors |

**366.  Representative Applications of Wound-Rotor, Brush-Shifting, and Synchronous Motors**

| Wound rotor | Polyphase brush-shifting, commutator motor | Synchronous motors |
|---|---|---|
| Bascule lift and swing bridges | Aluminum-foil machines | *High speed* |
| Cranes | Bakery machinery | Band saws |
| Coke pushers | Boilerhouse fans | Blowers |
| Hoists | Centrifugal compressors and pumps | Compressors |
| Hydraulic gates | Cable-stranding machines | Cotton gins |
| Lorry cars | Cloth-printing machines | Fans |
| Main roll drives in iron and steel mills | Cotton-mill calenders | Grinding and crushing machinery |
| Rotary car dumpers | Linoleum calenders | Jordans and beaters |
| Yard locomotives | Paper winders | Metal rolling mills |
| | Printing presses | Mill line shafts |
| | Reciprocating pumps | Pumps |
| | Oil-refinery machinery | *Low speed* |
| | Some machine tools | Air and gas compressors |
| | Some steel-mill main roll drivers | Band-saw mills |
| | Stokers | Jordans |
| | Table drive for plate-glass grinders | Pulp grinders |
| | Tandem drive for textile mills | Refrigerating machinery |
| | Tube machines and calenders in rubber-goods factories | |

**367. Typical Torque Requirements of Synchronous-Motor Applications**

The following, taken from NEMA MG1-21.88, represent typical locked-rotor and pull-in and pullout torque requirements of various synchronous-motor applications. The locked-rotor and pull-in torque values are based on rated voltage being applied at the motor terminals during the starting period, and upon the selection of a motor whose rating is such that the normal running load does not exceed the rated horsepower. The pull-out torque values take into account the peak loads typical of the application and include an allowance for the usual variations in line voltage. In individual cases lower values may be satisfactory or higher values may be necessary, depending upon the characteristics of the particular machine and the effect of locked-rotor kVA on the line voltage.

| Application | Torques in percent of motor full-load torque | | | Ratio of $Wk^2$ of load to normal $Wk^2$ of load |
| | Locked-rotor | Pull-in | Pullout | |
| --- | --- | --- | --- | --- |
| Attrition mills (for grain processing); starting unloaded | 100 | 60 | 175 | 3–15 |
| Ball mills (for rock and coal) | 140 | 110 | 175 | 2–4 |
| Ball mills (for ore) | 150 | 110 | 175 | 1.5–4 |
| Banbury mixers | 125 | 125 | 250 | 0.2–1 |
| Band mills | 40 | 40 | 250 | 50–110 |
| Beaters, Breaker | 125 | 100 | 200 | 3–15 |
| Beaters, Standard | 125 | 100 | 150 | 3–15 |
| Blowers, Centrifugal; starting with | | | | |
|   Inlet or discharge valve closed | 30 | 40–60[a] | 150 | 3–30 |
|   Inlet and discharge valve open | 30 | 100 | 150 | 3–30 |
| Blowers, Positive-displacement, rotary; bypassed for starting | 30 | 25 | 150 | 3–8 |
| Bowl mills (coal pulverizers); starting unloaded | | | | |
|   Common motor for mill and exhaust fan | 90 | 80 | 150 | 5–15 |
|   Individual motor for mill | 140 | 50 | 150 | 4–10 |
| Chippers; starting empty | 60 | 50 | 250 | 10–100 |
| Compressors, Centrifugal; starting with | | | | |
|   Inlet or discharge valve closed | 30 | 40–60[a] | 150 | 3–30 |
|   Inlet and discharge valve open | 30 | 100 | 150 | 3–30 |
| Compressors, Fuller Company | | | | |
|   Starting unloaded (bypass open) | 60 | 60 | 150 | 0.5–2 |
|   Starting loaded (bypass closed) | 60 | 100 | 150 | 0.5–2 |
| Compressors, Nash-Hytor; starting unloaded | 40 | 60 | 150 | 2–4 |
| Compressors, Reciprocating; starting unloaded | | | | |
|   Air and gas | 30 | 25 | 150 | 0.2–15 |
|   Ammonia (discharge pressure, 100–250 psi) | 30 | 25 | 150 | 0.2–15 |
|   Freon | 30 | 40 | 150 | 0.2–15 |
| Crushers, Bradley-Hercules; starting unloaded | 100 | 100 | 250 | 2–4 |
| Crushers, Cone; starting unloaded | 100 | 100 | 250 | 1–2 |
| Crushers, Gyratory; starting unloaded | 100 | 100 | 250 | 1–2 |
| Crushers, Jaw; starting unloaded | 150 | 100 | 250 | 10–50 |
| Crushers, Roll; starting unloaded | 150 | 100 | 250 | 2–3 |
| Defibrators (see Beaters, Standard) | | | | |
| Disintegrators, Pulp (see Beaters, Standard) | | | | |
| Edgers | 40 | 40 | 250 | 5–10 |
| Fans, Centrifugal (except sintering fans); starting with | | | | |
|   Inlet or discharge valve closed | 30 | 40–60[a] | 150 | 5–60 |
|   Inlet and discharge valve open | 30 | 100 | 150 | 5–60 |
| Fans, Centrifugal sintering; starting with inlet gates closed | 40 | 100 | 150 | 5–60 |
| Fans, Propeller-type; starting with discharge valve open | 30 | 100 | 150 | 5–60 |
| Generators, Alternating-current | 20 | 10 | 150 | 2–15 |
| Generators, Direct-current (except electroplating) | | | | |
|   150 kW and smaller | 20 | 10 | 150 | 2–3 |
|   Over 150 kW | 20 | 10 | 200 | 2–3 |
| Generators, Electroplating | 20 | 10 | 150 | 2–3 |
| Grinders, Pulp, single, long-magazine–type; starting unloaded | 50 | 40 | 150 | 2–5 |
| Grinders, Pulp, except single, long-magazine–type; starting unloaded | 40 | 30 | 150 | 1–5 |
| Hammer mills; starting unloaded | 100 | 80 | 250 | 30–60 |
| Hydrapulpers, Continuous-type | 125 | 125 | 150 | 5–15 |
| Jordans (see Refiners, Conical) | | | | |
| Line shafts, Flour-mill | 175 | 100 | 150 | 5–15 |
| Line shafts, Rubber-mill | 125 | 110 | 225 | 0.5–1 |
| Plasticators | 125 | 125 | 250 | 0.5–1 |

7-205

| Application | Torques in percent of motor full-load torque | | | Ratio of $Wk^2$ of load to normal $Wk^2$ of load |
|---|---|---|---|---|
| | Locked-rotor | Pull-in | Pullout | |
| Pulverizers, B&W; starting unloaded | | | | |
|   Common motor for mill and exhaust fan | 105 | 100 | 175 | 20–60 |
|   Individual motor for mill | 175 | 100 | 175 | 4–10 |
| Pumps, Axial-flow, adjustable-blade; starting with | | | | |
|   Casing dry | 5–40[b] | 15 | 150 | 0.2–2 |
|   Casing filled, blades feathered | 5–40[b] | 40 | 150 | 0.2–2 |
| Pumps, Axial-flow, fixed-blade; starting with | | | | |
|   Casing dry | 5–40[b] | 15 | 150 | 0.2–2 |
|   Casing filled, discharge closed | 5–40[b] | 175–250[a] | 150 | 0.2–2 |
|   Casing filled, discharge open | 5–40[b] | 100 | 150 | 0.2–2 |
| Pumps, Centrifugal, Francis-impeller; starting with | | | | |
|   Casing dry | 5–40[b] | 15 | 150 | 0.2–2 |
|   Casing filled, discharge closed | 5–40[b] | 60–80[a] | 150 | 0.2–2 |
|   Casing filled, discharge open | 5–40[b] | 100 | 150 | 0.2–2 |
| Pumps, Centrifugal, radial-impeller; starting with | | | | |
|   Casing dry | 5–40[b] | 15 | 150 | 0.2–2 |
|   Casing filled, discharge closed | 5–40[b] | 40–60[a] | 150 | 0.2–2 |
|   Casing filled, discharge open | 5–40[b] | 100 | 150 | 0.2–2 |
| Pumps, Mixed-flow; starting with | | | | |
|   Casing dry | 5–40[b] | 15 | 150 | 0.2–2 |
|   Casing filled, discharge closed | 5–40[b] | 80–125[a] | 150 | 0.2–2 |
|   Casing filled, discharge open | 5–40[b] | 100 | 150 | 0.2–2 |
| Pumps, Reciprocating; starting with | | | | |
|   Cylinders dry | 40 | 30 | 150 | 0.2–15 |
|   Bypass open | 40 | 40 | 150 | 0.2–15 |
|   No bypass (three-cylinder) | 150 | 100 | 150 | 0.2–15 |
| Refiners, Conical (Jordans, Hydrafiners, Clafins, Mordens); starting with plug out | 50 | 50–100[c] | 150 | 2–20 |
| Refiners, Disc-type; starting unloaded | 50 | 50 | 150 | 1–20 |
| Rod mills (for ore grinding) | 160 | 120 | 175 | 1.5–4 |
| Rolling mills | | | | |
|   Structural and rail roughing mills | 40 | 30 | 300–400[d] | 0.5–1 |
|   Structural and rail finishing mills | 40 | 30 | 250 | 0.5–1 |
|   Plate mills | 40 | 30 | 300–400[d] | 0.5–1 |
|   Merchant mill trains | 60 | 40 | 250 | 0.5–1 |
|   Billet, skelp, and sheet-bar mills, continuous, with lay-shaft drive | 60 | 40 | 250 | 0.5–1 |
|   Rod mills, continuous with lay-shaft drive | 100 | 60 | 250 | 0.5–1 |
|   Hot strip mills, continuous, individual-drive roughing stands | 50 | 40 | 250 | 0.5–1 |
|   Tube-piercing and -expanding mills | 60 | 40 | 300–400[d] | 0.5–1 |
|   Tube-rolling (plug) mills | 60 | 40 | 250 | 0.5–1 |
|   Tube-reeling mills | 60 | 40 | 250 | 0.5–1 |
|   Brass- and copper-roughing mills | 50 | 40 | 250 | 0.5–1 |
|   Brass- and copper-finishing mills | 150 | 125 | 250 | 0.5–1 |
| Rubber mills, individual-drive | 125 | 125 | 250 | 0.5–1 |
| Saws, Band (see Band mills) | | | | |
| Saws, Edger (see Edgers) | | | | |
| Saws, Trimmer | 40 | 40 | 250 | 5–10 |
| Tube mills (see Ball mills) | | | | |
| Vacuum pumps, Hytor | | | | |
|   With unloader | 40 | 30 | 150 | 2–4 |
|   Without unloader | 60 | 100 | 150 | 2–4 |
| Vacuum pumps, Reciprocating; starting unloaded | 40 | 60 | 150 | 0.2–15 |
| Wood hogs | 60 | 50 | 250 | 30–100 |

[a]The pull-in torque varies with the design and operating conditions. The machinery manufacturer should be consulted.

[b]For horizontal-shaft pumps and vertical-shaft pumps having no thrust bearing (entire thrust load carried by the motor), the locked-rotor torque required is usually between 5 and 20 percent, while for vertical-shaft machines having their own thrust bearing a locked-rotor torque as high as 40 percent is sometimes required.

[c]The pull-in torque required required varies with the design of the refiner. The machinery manufacturer should be consulted. Furthermore, even though 50 percent pull-in torque is adequate with the plug out, it is sometimes considered desirable to specify 100 percent to cover the possibility that a start will be attempted without complete retraction of the plug.

[d]The pullout torque varies with the rolling schedule.

**368. Synchronous-Motor Application Data**
(Westinghouse Electric Corp.)
Unusual operating conditions may modify the requirements. For type of motor to produce the required starting torque see Table 183.

| Application | Method of connecting motor to load | Starting conditions | Per cent | | | Remarks |
|---|---|---|---|---|---|---|
| | | | Starting torque | Pull-in torque | Pull-out torque | |
| **Cement, Rock Products, and Metal Mining** | | | | | | |
| Ball mills, mining......... | Coupled to mill pinion | Mills loaded at start | 175–200 | 110–120 | 175 | Full-voltage starting or resistance control for wound-rotor types. Indicating wattmeter on control |
| Crushers, gyratory......... | Coupled or belted | Loaded | 150–200 | 100–125 | 200 | Full-voltage starter required. Continuous operation—varying load |
| Hammer mill............... | Coupled | Loaded | 150 | 115 | 250 | Full-voltage starting recommended |
| Jaw........................ | Coupled or belted | Loaded | 150–225 | 100–125 | 250 | Full-voltage starter required. Continuous load extremely irregular with high peaks |
| Roll....................... | Coupled or belted | Loaded | 225 | 100 | 225 | Full-voltage starter |
| Bradley-Hercules........... | Coupled | Unloaded | 80 | 100 | 175 | Full-voltage starting. Continuous load |
| Hercules mills, cement...... | Coupled to mill pinion shaft by resilient flexible coupling; good for severe misalignment and vibration | Loaded | 150–200 | 100–110 | 175 | Special full-voltage starting. Liners may be installed under pedestals to provide compensation for mill settling. Motors should be provided with dustproof bearings |
| Rod mills, mining........... | Coupled or belted (Lenix drive) to gear | Loaded | 200 | 120 | 175 | Special full-voltage starter required |
| Tube mills................. | Coupled by flexible coupling to reducing gear | Loaded | 190 | 120 | 175 | Special full-voltage starter required |
| **Compressors and Blowers** | | | | | | |
| Blowers, cycloidal, positive... | Coupled or engine type | Unloaded | 40–60 | 40–60 | 150 | Standard Control. Full-voltage starting recommended. Two-speed motors sometimes used |
| Turbo, high-speed.......... | Direct connected or step-up gear | Unloaded, discharge closed | 40 | 60 | 150 | Special reduced-voltage starting |
| Blowing engine reciprocating. | Engine type | Unloaded | 40 | 40–60 | 150 | Standard control. Full-voltage starting recommended |

**368. Synchronous-Motor Application Data** (*Continued*)

| Application | Method of connecting motor to load | Starting conditions | Per cent | | | Remarks |
|---|---|---|---|---|---|---|
| | | | Starting torque | Pull-in torque | Pull-out torque | |
| Compressors, air......... | Engine type | Unloaded | 40 | 40 | 150 | Standard control. Full-voltage starting recommended. Flywheel effect important |
| Ammonia and ammonia booster | High speed, belted; low speed, usually engine type, occasionally coupled | Unloaded by-pass | 40 | 40 | 150 | Standard control. Full-voltage starting recommended. Flywheel effect important |
| Freon......... | High speed, belted; low speed, usually engine type, occasionally coupled | Unloaded by-pass | 45 | 60 | 150 | Standard control. Full-voltage starting recommended. Flywheel effect important |
| Gas, reciprocating......... | High speed, belted; low speed, engine type | Unloaded by-pass | 40 | 40 | 150 | Standard control except on magnetic control a gastight push button is recommended. Full-voltage starting usually used. Flywheel effect important. Control and exciting equipment usually installed in separate room |
| **Fans** | | | | | | |
| Exhaust and ventilating...... | Coupled or belted | Usually loaded | 50 | 60–125 | 150 | Reduced- or full-voltage starting. $WR^2$ of fan must be considered |
| **Flour Mills** | | | | | | |
| Line shaft......... | Coupled or belted | Loaded | 175–225 | 100–125 | 175 | Full-voltage starting required |
| **Lumber** | | | | | | |
| Band saws and resaws...... | Coupled or belted | Unloaded. Torques depend on $WR^2$ of saw | 80–125 | 75–115 | 200–250 | Full-voltage starting. Accelerating time, 45 to 90 sec. Special motor for starting high inertia load |

| | | | | | | |
|---|---|---|---|---|---|---|
| Jordans | Coupled with flexible coupling | Unloaded | 50 | 50 | 150 | Standard control. Full-voltage starting recommended |
| Beaters | Belted | Loaded | 125 | 100 | 175 | Full-voltage starting recommended |
| Pulp grinders | Coupled by flexible coupling | Unloaded | 50 | 50 | 150 | Standard control. Full-voltage type recommended |
| Wood hogs and chippers | Coupled or engine type | Unloaded. Torques depend on $WR^2$ of machine | 83–125 | 80–110 | 225 | Full-voltage starting recommended |

Pumps

| | | | | | | |
|---|---|---|---|---|---|---|
| Centrifugal, water, oil, sewage | Coupled by flexible coupling | Valves either open or closed. Usually closed | Valves open or closed, 35–20 50–75 | Valves closed, 35–50 valves open, 100–120 | 150 | Standard control, full or reduced voltage |
| Reciprocating | Coupled by flexible coupling to pump pinion shaft | Unloaded or by-passed | 30–50 | 25–50 | 150 | Use standard-price-book control. Full-voltage type recommended |
| Reciprocating, vacuum | Belted or direct-connected engine type | Unloaded | 30–50 | 60–85 | 150 | Use standard-price-book control. Full-voltage type recommended. Horsepower should be based on peak-load conditions |
| Rotating vacuum | Belted or direct-connected engine type | Loaded | 40 | 55–100 | 150 | Standard control. Horsepower should be based on peak-load conditions |

Rubber

| | | | | | | |
|---|---|---|---|---|---|---|
| Banbury mixers and plasticators | Direct drive or through reduction gear | Usually unloaded. May reverse under load in emergency | 125–150 | 100–110 | 250 | Special reversing control required. Usually full-voltage starting. Occasionally two-speed motors |
| Mill lines, rubber | Low speed, direct coupled; high-speed, coupled to reducing gear | Usually loaded. May reverse under load in emergency | 110–150 | 110–125 | 175–250 | Full-voltage dynamic braking control with or without reversing recommended. Semimagnetic or magnetic reduced voltage may be used in some cases |

**368. Synchronous-Motor Application Data** (*Continued*)

| Application | Method of connecting motor to load | Starting conditions | Per cent | | | Remarks |
| --- | --- | --- | --- | --- | --- | --- |
| | | | Starting torque | Pull-in torque | Pull-out torque | |
| **Steel** | | | | | | |
| Piercing mills.............. | Direct-connected by flexible coupling; sometimes coupled to reducing gears | Unloaded. May be reversed under load in emergency | 80–100 | 50–60 | 300–350 | Special control required. Full-voltage type recommended |
| Continuous mills; steel rolling, sheet, bar, billet, skelp, strip mills | Direct-connected by flexible coupling, usual practice. Sometimes coupled to reducing gear using higher speed motors | Unloaded. May be reversed under load in emergency | 50–100 | 50–100 | 250–350 | Special control required. Full-voltage type recommended |
| Rod mills.............. | Direct-connected by flexible coupling | Unloaded. May reverse under load in emergency | 60–100 | 50–60 | 225–275 | Special control required. Full-voltage type recommended |
| Sizing and reducing mills...... | Direct-connected by flexible coupling; sometimes coupled to reducing gears | Unloaded. May reverse under load in emergency | 80–100 | 50–60 | 300–350 | Special control required. Full-voltage type recommended |

### 369. Operating Speeds of Various Machine Tools

| | |
|---|---|
| Saws, circular, wood............................. | 9,000 ft per min at rim |
| Band, wood ..................................... | 4,000 ft per min at rim |
| Band, hot iron and steel.................... | 200–300 ft per min at rim |
| Grindstones...................................... | 800 ft per min at rim |
| Emery wheels.................................... | 5,000 ft per min at rim |
| Drills, for wrought iron........................ | 12 ft per min outer edge |
| For cast iron ................................... | 8 ft per min outer edge |
| Milling cutters, for brass ..................... | 120 ft per min outer edge |
| For cast-iron................................... | 60 ft per min outer edge |
| For wrought iron............................. | 50 ft per min outer edge |
| Wrought steel ................................. | 35 ft per min outer edge |
| Screw cutting, gun metal, etc............... | 30 ft per min at circum |
| Steel............................................. | 8 ft per min at circum |
| Boring, cast-iron................................ | 10 ft per min at circum |
| Sawing | |
| Brass.............................................. | 70 ft per min at circum |
| Gun metal ....................................... | 30 ft per min at circum |
| Steel.............................................. | 25 ft per min at circum |
| Wrought iron .................................. | 30 ft per min at circum |
| Cast iron ........................................ | 20 ft per min at circum |

**370. Motor-driven pumps** (Westinghouse Electric Corp.). Either dc or ac motors are satisfactory. For most cases, shunt-wound dc or squirrel-cage ac motors and synchronous motors above 25 hp are suitable, but when starting conditions are severe, as when the pump must be started against a full discharge pipe, compound-wound dc and wound-rotor ac motors are preferable. For small centrifugal pumps, squirrel-cage motors are suitable. On large units, squirrel-cage or synchronous motors are used. In many government buildings wound-rotor motors are specified for their low starting currents.

$$\text{Water hp} = \frac{\text{gpm} \times \text{tdh (in ft)}}{3960} = \frac{\text{gpm} \times \text{tdh (in lb)}}{1720}$$

where gpm = gallons per minute and tdh = total dynamic head.

$$\text{bhp} = \frac{\text{gpm} \times \text{tdh (in ft)}}{3960 \times \text{pump eff}}$$

where bhp = brake horsepower.

$$\text{kW input} = \frac{\text{gpm} \times \text{tdh (in ft)}}{5308 \times \text{pump eff} \times \text{motor eff}}$$

$$\text{kWh per million gallons pumped} = \frac{\text{tdh (in ft)} \times 3.1456}{\text{pump eff} \times \text{motor eff}}$$

$$\text{Velocity} = \frac{\text{gpm}}{2.45 \times D^2} \qquad \text{Velocity head} = \frac{V^2}{64.4}$$

NOTE   When the head is expressed in feet, formulas are correct only for liquids of 1.0 specific gravity.

On plunger pumps high-starting-torque squirrel-cage motors are recommended to take care of the full pump load plus the accelerating load. For boiler-feed service and other applications requiring variable-speed motors, a wound-rotor motor with continuous-duty resistors should be used.

**371. Motor-driven fans and blowers** (Westinghouse Electric Corp.). The approximate horsepower required to drive fans and blowers may be obtained from the curve in Fig. 7-184 when the cubic feet per minute and water-gage pressure are known. Horsepower for pressures and quantities not shown on the curves may be figured from the following formula (the average ventilating fan has an efficiency of 60 percent):

$$\text{hp} = \frac{\text{ft}^3/\text{min} \times \text{pressure in inches water gage}}{6.346 \times \text{fan efficiency}} \qquad (12)$$

The horsepower required by a fan varies approximately as the cube of the speed; the volume, directly as the speed. Motor horsepower should exceed the rated horsepower of a forward-curved fan by about 20 percent but can be near the rated horsepower of backward-curved fans with nonoverload characteristics. If fan speed will be increased by adjustable pitch sheaves or sheave changes, base the motor rating on the highest speed.

Standard shunt-wound direct-current or normal-torque ac motors have sufficient starting torque. Synchronous motors are used when power-factor correction is desired.

Multispeed variable-torque squirrel-cage motors are recommended when the volume of air must be varied. Any combination of 1800-, 1200-, 900-, and 600-rpm speeds can be supplied. If the volume variation obtained with fixed motor speeds is not suitable, a wound-rotor induction motor with the proper control will allow speed adjustment between 100 and 50 percent of the motor speed. Shunt-wound dc motors with suitable control give variable-speed adjustment over the same range as wound-rotor motors.

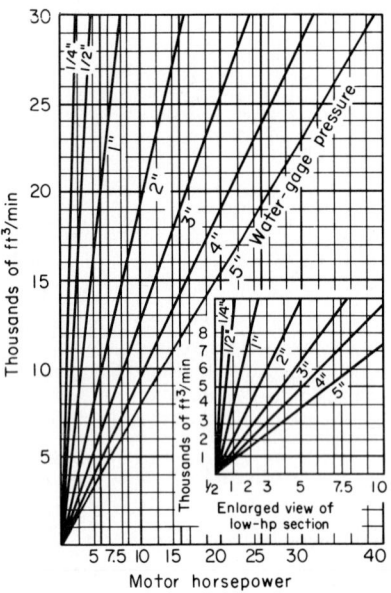

**Fig. 7-184** Curves for horsepower required to drive fans and blowers. [Westinghouse Electric Corp.]

**372. Motor-driven compressors.** Compressors equipped with unloaders can be started by normal-starting-torque ac motors of NEMA Design B. Compressors which start loaded require high-starting-torque ac motors of NEMA Design C.

Most air-conditioning compressors, especially hermetic refrigerant motor-compressors and centrifugal compressors, start unloaded. If the compressor starts loaded, it will require motor-starting torques of 200 to 220 percent of full-load torque. High-starting-torque Type A motors (Design C) will develop these torques when started across the line. With across-the-line starting, the total current inrush is the locked-rotor current of the motor. When a low-starting-current, high-starting-torque motor will not meet the inrush or torque requirements, a wound-rotor motor and Class 13-100 magnetic starter are recommended.

Large reciprocating air compressors are generally driven by high-speed belted or low-speed direct-connected engine-type synchronous motors.

The horsepower required can be determined by using the following instructions applied to the curves in Fig. 7-185.

AIR COMPRESSORS AT ANY LEVEL   Curve A can be used when the discharge pressure and cubic feet per minute are known. The curves are lines of constant horsepower. Dotted lines are for two-stage water-cooled units.

AIR COMPRESSORS AT ANY ALTITUDE   First select the equivalent sea-level pressure from curve B. This will take care of the increase in compression ratio above sea-level ratio. Then from curve A determine the total horsepower for the pressure corresponding to the equivalent sea-level pressure and free air capacity you desire. Multiply this horsepower by the correction factor from curve C.

AIR-CONDITIONING AND REFRIGERATION   The vast majority of compressors used in air-conditioning and refrigeration equipment are hermetic refrigerant motor-compressors (see Sec. 446), which employ special low-starting-torque motors as an integral part of the compressor and must be started unloaded. To estimate accurately the total electrical starting and running load of a hermetic refrigerant motor-compressor and the air-conditioning or refrigerating equipment of which it is a part, it is necessary to know the electrical ratings of the specific motor-compressor or equipment. As detailed in Sec. 446, a hermetic refrigerant motor-compressor is not rated in horsepower. Air-conditioning and refrigerating equipment is rated for its cooling capacity in British thermal units per hour or in tons of cooling capacity. One ton of cooling capacity equals 12,000 Btu/h.

The electrical input per ton of cooling depends not only on the efficiency of the compressor motor but also on the efficiency of the compressor in its system. The efficiency of the system is dependent on many factors, including the cooling temperature (high temperature for air conditioning, medium temperature for unfrozen-food preservation, low temperature for frozen-food storage and food freezing). Other factors affecting the overall efficiency of the system include the system condensing medium, such as water-cooled, evaporatively cooled, or air-cooled, the kind of refrigerant, the number and type (and efficiency) of auxiliary motors, such as fan motors, etc. Even approximate conversion from tons of cooling to electrical-input power is feasible only when there is adequate knowledge of the equipment to be installed.

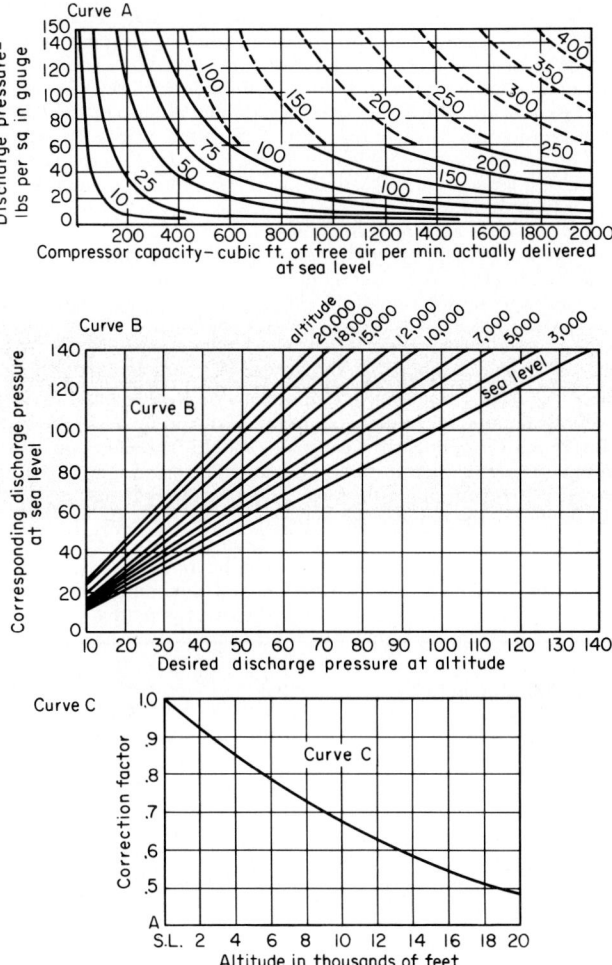

**Fig. 7-185** Curves for determining horsepower required to drive compressors. [Westinghouse Electric Corp.]

It is traditional to convert tons of cooling to motor horsepower by estimating 1 hp/ton, but this can result in a 100 percent error on the high side for some high-efficiency equipment and can be as much as 25 percent on the low side for other equipment. It is best to consult the mechanical contractor or the equipment manufacturer to obtain accurate information.

**373. Foundations are necessary** to support and maintain in alignment generators and other electrical machines of any considerable size. They are made of masonry. Brick or stone set in mortar (preferably cement mortar) will do, but concrete is almost universally used because it is usually the cheapest. A 1:3:6 mixture of concrete (1 part cement, 3 parts sand, and 6 parts crushed stone or gravel by bulk) or even a 1:3:7 mixture will give excellent results. Brick or stone for foundations can be set in mortar consisting of 1 part cement and 3 parts sand.

**374. The size of a foundation** is determined by the size of the machine supported and by the stresses imposed by the machine. The base area of any foundation must be great enough so that its weight and the weight of the machine supported will not cause it to sink into the soil. If a machine is not subjected to any external forces (i.e., if it is self-contained), the only requirement of the foundation (provided the machine is not one that vibrates excessively) is to keep it from sinking into the ground, and the lightest foundation that will do this is satisfactory. Therefore motor generators and rotary converters do not require foundations heavy enough to resist the tendency of the external apparatus to tip or to displace the foundation. No rule can be given for determining the proper weight for a foundation in such a case. However, with a solid foundation it is usually true that if the foundation is large enough to include all the foundation bolts of the machine and to extend to good bottom, it will be sufficiently heavy. Experience is required to enable one to design the lightest foundation that will do, so it is well for the beginner to be sure that the foundation is heavy enough.

**375. Foundation for a gear motor** (Westinghouse Electric Corp.). A foundation or mounting which provides rigidity and prevents weaving or flexing, with resultant misalignment of the shafts, is essential to the successful operation of a gear drive. A concrete foundation should be used whenever possible and should be carefully prepared to conform with data regarding bolt spacing and physical measurements contained in the identified dimension leaflet.

Gear motors are designed with a tolerance of $+0$ and $-\frac{1}{32}$ in $(-0.794 \text{ mm})$ between the shaft center and the base. Therefore, shimming may be required. Shims of various thicknesses, slotted to slide around the foundation bolts, should be used. After alignment has been secured through shimming, the gear motor should be bolted down and grouted in. When the unit must be installed on structural foundations, a supporting baseplate of steel should be provided to obtain proper rigidity. This baseplate should be of a thickness not less than the diameter of the holding-down bolts.

**376. Alignment of motors.** After the motor has been properly located and the right kind of mounting has been built, the next step is to align the motor properly with its drive. The following procedure is recommended by the General Electric Co. Belt drive is the most common form of mechanical transmission, but the principles involved and the methods used in aligning generally apply equally well for chain drive. The tools usually used for aligning motors or line shafting are the square, plumb bob, and level. Tool manufacturers also build combination squares and levels and similar tools that aid in quickly and accurately aligning machinery.

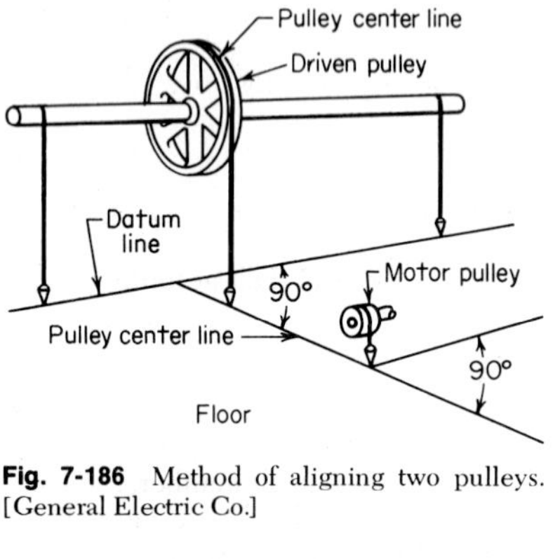

The first step in aligning two machines is to see if they are level. It is possible for a machine to be out of line in more than one plane. If by placing a level or plumb on the machine it is found to be out by a certain amount, the motor must be mounted so that it will be out by a like amount.

A simple and easy method of aligning a motor pulley with the driven pulley is shown in Fig. 7-186. First, the crown or centerline of the pulleys must be on the same centerline; second, the motor shaft must be parallel to the driven shaft.

**Fig. 7-186** Method of aligning two pulleys. [General Electric Co.]

By using a plumb bob and drawing a datum line on the floor, a base of operation is established. Next, drop a plumb line from the center of the driven pulley to the floor. With a square, draw a line perpendicular to the datum line. Next, drop a plumb line from the center of the motor pulley and move the motor up or back until the plumb bob rests on the centerline of the driven pulley. From the pulley centerline, perpendiculars can be drawn through the centers of the holes in the motor feet. A level should be used to see that the line shafting is level. If it is not, the motor feet must be shimmed up so that the motor shaft and the line shaft will be out of level by the same amount. Chain drive can be aligned in a similar manner.

With belt drive, a sliding base is nearly always used to allow for belt adjustment. Another method, therefore, is to use the following procedure when aligning two pulleys:

1. Place the motor on the base so that there will be an equal amount of adjustment in either direction, and firmly fasten the motor to the base by the four holding-down bolts.

2. Mount the motor pulley on the motor shaft.

3. Locate the base and motor in approximately the final position, as determined by the length of belt.

4. Stretch a string from the face of the driven pulley toward the face of the motor pulley.

5. Parallel the face of the motor pulley with this string.

6. Using a scratch pin, mark the end positions of the sliding base.

7. Extend these lines.

8. Move the base and motor away from the string by an amount equal to one-half the difference in face width of the two pulleys. Use the two extended lines as a guide to keep the base in its proper position.

9. The belt should now be placed on the pulleys to see if it operates satisfactorily. If it does not operate properly, the base can be shifted slightly until proper operation is obtained.

10. Firmly fasten the base to the floor, ceiling, or sidewalls by lag screws or bolts.

**377. In designing a V-belt drive** the following factors should be considered: speed-reduction factor, pitch diameters of pulleys, belt speed, distance between pulley centers, cross-sectional size of belt, and number of belts to use.

**378. The method of designing V-belt drives** for satisfactory operation is outlined as follows through the courtesy of the Allis-Chalmers Corporation:

1. Determine the speed ratio = motor rpm/machine rpm.

2. From Table 379 determine the section of V belt to use according to the horsepower and synchronous-speed rating of the motor. When two sections are given, the design should be made for both and at the end a decision made as to which should be used.

3. From Table 381 and the speed ratio determined in 1, select pitch diameters for the sheaves. The pitch diameter for the machine sheave should equal the pitch diameter for the motor sheave multiplied by the speed ratio. The motor sheave should not be too small, as the horsepower transmitted increases with the increase in size of the small sheave. The machinery sheave must not be too large, however. Consideration must be given to the size of the machine and the available space for the sheave.

4. From Table 382 determine the service factor for the drive according to the type of motor and type of machine.

5. From Table 383 determine the arc of contact of the V belt on the small sheave and the correction factor for belt horsepower.

6. Compute the horsepower required for the belts by multiplying the rated motor horsepower by the service factor obtained from 4.

7. Compute the velocity of the belt = $\pi dn/12$, where $d$ = diameter of motor sheave in inches and $n$ = approximate full-load speed of motor.

8. From Table 379 determine the horsepower rating of each belt according to the belt velocity and the pitch diameter of the motor sheave.

9. Determine the number of belts required =

$$\frac{\text{horsepower required (from 6)}}{\text{horsepower rating of belt (from 8)} \times \text{service factor (from 4)}} \tag{13}$$

Use the next larger whole number. This should preferably be a small number, such as 4 or 5, although in the case of large horsepower ratings or low speeds a larger number may be necessary. In case of doubt try the next larger belt section and recalculate.

### 379.  V-Belt Sections to Use

| Hp rating of motor | Syn. speed of motor | V-belt section | Hp rating of motor | Syn. speed of motor | V-belt section |
|---|---|---|---|---|---|
| 1½ and less | All | *A* | 60 | 1,800 | *C* |
|  |  |  | 60 | 1,200–900 | *C* or *D* |
| 2–7½ | All | *A* or *B* | 60 | 720 and less | *D* |
| 7½ | 900 and less | *B* or *C* | 75 | 1,800 | *C* |
|  |  |  | 75 | 1,200–900 | *C* or *D* |
| 10 | 1,800 | *A* or *B* | 75 | 720 and less | *D* or *E* |
| 10 | 1,200–900 | *B* |  |  |  |
| 10 | 720 and less | *B* or *C* | 100 | 1,800 | *C* |
|  |  |  | 100 | 1,200 and less | *D* |
| 15 | 1,800 | *B* |  |  |  |
| 15 | 1,200–900 | *B* or *C* | 125 | 1,200–900 | *D* |
| 15 | 720 and less | *C* | 125 | 720 and less | *D* or *E* |
| 20 | 1,800 | *B* or *C* | 150 | 1,200–900 | *D* |
| 20 | 1,200 and less | *C* | 150 | 720 and less | *E* |
| 25–30 | All | *C* | 200 | 1,200–900 | *D* |
|  |  |  | 200 | 720 and less | *E* |
| 40 | 1,800 | *C* |  |  |  |
| 40 | 1,200–900 | *C* or *D* | 250 and over | All | *E* |
| 40 | 720 and less | *D* |  |  |  |
| 50 | 1,800–900 | *C* or *D* |  |  |  |
| 50 | 720 and less | *D* |  |  |  |

### 380.  Standard Sections of V Belts

| Section | Greatest width, in. | Thickness, in. | Section | Greatest width, in. | Thickness, in. |
|---|---|---|---|---|---|
| *A* | ½ | $\frac{11}{32}$ | *D* | 1¼ | ¾ |
| *B* | $\frac{21}{32}$ | $\frac{7}{16}$ | *E* | 1½ | 1 |
| *C* | ⅞ | $\frac{17}{32}$ |  |  |  |

## 381.  Standard Pitch Diameters for Sheaves for V-Belt Drives

(Allis-Chalmers Corporation)

| Section A | Section B | Section C | Section D | Section E | Section A | Section B | Section C | Section D | Section E |
|---|---|---|---|---|---|---|---|---|---|
| Pressed steel | | Cast iron | | | Cast iron | | | | |
| 3.0 | 5.4 | 9.0 | 13 | 20 | 12 | 15 | 38 | 48 | 72 |
| 3.2 | 5.6 | 9.2 | 13.4 | 21 | 13 | 16 | 40 | 50 | 74 |
| 3.4 | 5.8 | 9.4 | 13.8 | 22 | 14 | 17 | 42 | 52 | 76 |
| 3.6 | 6.0 | 9.6 | 14 | 23 | 15 | 18 | 44 | 54 | 78 |
| 3.8 | 6.2 | 9.8 | 14.2 | 24 | 16 | 19 | 46 | 56 | 80 |
| 4.0 | 6.4 | 10.0 | 14.6 | 25 | 17 | 20 | 48 | 58 | 82 |
| 4.2 | 6.6 | 10.2 | 15 | 26 | 18 | 21 | 50 | 60 | 84 |
| 4.4 | 6.8 | 10.6 | 15.4 | 27 | 19 | 22 | 52 | 62 | 86 |
| 4.6 | 7.4 | 11 | 16 | 28 | 20 | 23 | 54 | 64 | 88 |
| 4.8 | 8.6 | 12 | 17 | 29 | 21 | 24 | 56 | 66 | 90 |
| 5.0 | 9.4 | 13 | 18 | 30 | 22 | 25 | 58 | 68 | 92 |
| 5.2 | 11.0 | 14 | 19 | 31 | 23 | 26 | 60 | 70 | 94 |
| 5.4 | 12.4 | 15 | 20 | 32 | 24 | 27 | 62 | 72 | 96 |
| 5.6 | 15.4 | 16 | 21 | 33 | 25 | 28 | 64 | 74 | |
| 5.8 | 18.4 | 17 | 22 | 34 | 26 | 29 | 66 | 76 | |
| 6.0 | Cast iron | 18 | 23 | 35 | 27 | 30 | 68 | 78 | |
| 6.2 | 5.4 | 19 | 24 | 36 | 28 | 31 | 70 | 80 | |
| 6.4 | 5.6 | 20 | 25 | 38 | 29 | 32 | 72 | 82 | |
| 7.0 | 5.8 | 21 | 26 | 40 | 30 | 33 | | 84 | |
| 8.2 | 6.0 | 22 | 27 | 42 | 31 | 34 | | 86 | |
| 9.0 | 6.2 | 23 | 28 | 44 | 32 | 35 | | 88 | |
| 10.6 | 6.4 | 24 | 29 | 46 | 33 | 36 | | 90 | |
| 12.0 | 6.6 | 25 | 30 | 48 | 34 | 38 | | 92 | |
| 15.0 | 6.8 | 26 | 31 | 50 | 35 | 40 | | 94 | |
| 18.0 | 7.0 | 27 | 32 | 52 | 36 | 42 | | 96 | |
| | 8.0 | 28 | 33 | 54 | 38 | 44 | | | |
| Cast iron | 8.6 | 29 | 34 | 56 | 40 | 46 | | | |
| 4 | 9.0 | 30 | 35 | 58 | | 48 | | | |
| 5 | 10.0 | 31 | 36 | 60 | | 50 | | | |
| 6 | 11.0 | 32 | 38 | 62 | | 52 | | | |
| 7 | 12.0 | 33 | 40 | 64 | | 54 | | | |
| 8 | 13.0 | 34 | 42 | 66 | | 56 | | | |
| 9 | 13.6 | 35 | 44 | 68 | | 58 | | | |
| 10 | 14 | 36 | 46 | 70 | | 60 | | | |
| 11 | | | | | | | | | |

### 382.  Service Factors for Horsepower Loads on V-Belt Drives
(Allis-Chalmers Corporation)

| Applications | Electric motors | | | | | | | | | | Engines | | | |
|---|---|---|---|---|---|---|---|---|---|---|---|---|---|---|
| | A-C | | | | | | | | D-C | | Gas and diesel | | | |
| | Squirrel cage | | | | Synchronous | | Single phase | | | | | | | |
| | Normal torque, line start | Normal torque, compensator start | High torque | Wound rotor (slip-ring) | Normal torque | High torque | Repulsion and split-phase | Capacitor | Shunt wound | Compound wound | 4 or more cyl.; above 700 rpm | 4 or more cyl.; below 700 rpm | Steam | Line shafts and clutch starting |
| **Agitators, paddle-propeller:** | | | | | | | | | | | | | | |
| Liquid | 1.0 | 1.0 | 1.2 | | | | | | | | | | | |
| Semiliquid | 1.2 | 1.0 | 1.4 | 1.2 | | | | | | | | | | |
| **Brick and clay machinery:** | | | | | | | | | | | | | | |
| Auger machines | ... | 1.2 | 1.4 | 1.4 | ... | ... | ... | ... | 1.4 | ... | ... | ... | ... | 2.0 |
| De-airing machines | ... | 1.2 | 1.4 | 1.4 | ... | ... | ... | ... | 1.4 | ... | ... | ... | ... | 2.0 |
| Cutting table | ... | 1.2 | 1.4 | 1.4 | ... | ... | ... | ... | ... | ... | ... | ... | ... | 2.0 |
| Pug mill | 1.5 | 1.3 | 1.8 | 1.5 | | | | | | | | | | |
| Mixer | ... | 1.2 | 1.6 | 1.4 | | | | | | | | | | |
| Granulator | ... | 1.2 | 1.4 | 1.4 | | | | | | | | | | |
| Dry press | ... | 1.2 | 1.6 | 1.4 | | | | | | | | | | |
| Rolls | ... | 1.2 | 1.4 | 1.4 | | | | | | | | | | |
| **Bakery machinery:** | | | | | | | | | | | | | | |
| Dough mixer | 1.2 | ... | ... | ... | ... | ... | 1.2 | 1.0 | | | | | | |
| **Compressors:** | | | | | | | | | | | | | | |
| Centrifugal | 1.2 | 1.2 | ... | 1.4 | 1.4 | ... | ... | ... | 1.2 | ... | 1.2 | | | |
| Rotary | 1.2 | 1.2 | ... | 1.4 | 1.4 | ... | 1.2 | 1.2 | 1.2 | ... | 1.2 | | | |
| Reciprocating: | | | | | | | | | | | | | | |
| 3 or more cylinders | 1.2 | 1.2 | ... | 1.4 | 1.4 | ... | ... | ... | 1.2 | | | | | |
| 1 or 2 cylinders | 1.4 | 1.4 | ... | 1.5 | 1.5 | ... | ... | ... | 1.2 | | | | | |
| **Conveyors:** | | | | | | | | | | | | | | |
| Apron | ... | 1.4 | 1.6 | ... | ... | ... | ... | ... | 1.4 | ... | ... | ... | ... | 1.6 |
| Belt (ore, coal, sand) | ... | 1.2 | 1.4 | ... | ... | ... | ... | ... | 1.2 | ... | ... | ... | ... | 1.4 |
| Belt (light package) | ... | 1.0 | 1.1 | ... | ... | ... | ... | ... | 1.0 | ... | ... | ... | ... | 1.2 |
| Oven | ... | 1.0 | 1.1 | ... | ... | ... | ... | ... | 1.0 | ... | ... | ... | ... | 1.2 |
| Screw | ... | 1.6 | 1.8 | ... | ... | ... | ... | ... | 1.6 | ... | ... | ... | ... | 1.8 |
| Bucket | ... | 1.4 | 1.6 | ... | ... | ... | ... | ... | 1.4 | ... | ... | ... | ... | 1.6 |
| Pan | ... | 1.4 | 1.6 | ... | ... | ... | ... | ... | 1.4 | ... | ... | ... | ... | 1.6 |
| Flight | ... | 1.6 | 1.8 | ... | ... | ... | ... | ... | 1.6 | ... | ... | ... | ... | 1.8 |
| Elevator | ... | 1.4 | 1.6 | ... | ... | ... | ... | ... | 1.4 | ... | ... | ... | ... | 1.6 |
| **Crushing machinery:** | | | | | | | | | | | | | | |
| Jaw crushers | ... | 1.4 | 1.6 | 1.4 | ... | ... | ... | ... | ... | 1.4 | 1.4 | ... | ... | 1.6 |
| Gyratory crushers | ... | 1.4 | 1.6 | 1.4 | 1.4 | 1.6 | ... | ... | ... | 1.4 | 1.4 | ... | ... | 1.6 |
| Cone crushers | ... | 1.4 | 1.6 | 1.4 | ... | ... | ... | ... | ... | 1.6 | 1.4 | ... | ... | 1.6 |
| Crushing rolls | ... | 1.4 | 1.6 | 1.4 | ... | ... | ... | ... | ... | 1.4 | 1.4 | ... | ... | 1.6 |
| Ball, pebble and | ... | 1.4 | 1.6 | 1.4 | 1.4 | 1.6 | ... | ... | ... | 1.4 | 1.6 | ... | ... | 1.6 |
| tube mills | ... | 1.4 | 1.6 | 1.4 | 1.4 | ... | ... | ... | ... | 1.4 | ... | ... | ... | 1.6 |
| **Fans and blowers:** | | | | | | | | | | | | | | |
| Centrifugal | 1.2 | 1.2 | ... | 1.4 | ... | ... | ... | ... | 1.2 | ... | 1.2 | | | |
| Propeller | 1.4 | 1.4 | 2.0 | 1.6 | ... | 2.0 | ... | ... | 1.4 | ... | 1.4 | | | |
| Induced draft | 1.2 | 1.2 | ... | 1.4 | ... | ... | ... | ... | 1.4 | ... | 1.4 | | | |
| Mine fans | 1.6 | 1.4 | 2.0 | ... | ... | 2.0 | ... | ... | ... | ... | 1.6 | | | |
| Positive blowers | 1.6 | 1.6 | ... | 2.0 | 2.0 | 2.0 | ... | ... | ... | ... | 1.6 | | | |
| Exhausters | 1.2 | 1.2 | ... | 1.4 | ... | ... | ... | ... | 1.4 | ... | ... | ... | 1.5 | 1.5 |
| **Flour-, feed-, cereal-mill machinery:** | | | | | | | | | | | | | | |
| Bolters and sifters | ... | 1.0 | | | | | | | | | | | | |
| Grinders and hammer mills | ... | 1.4 | ... | ... | ... | ... | ... | ... | ... | ... | 1.6 | | | |
| Purifiers and reels | 1.2 | 1.4 | | | | | | | | | | | | |
| Main-line shaft drive | 1.4 | 1.4 | 1.6 | 1.4 | 1.4 | ... | ... | ... | ... | ... | 1.8 | | | |
| Separator | 1.0 | 1.0 | | | | | | | | | | | | |
| Roller mills | ... | 1.4 | | | | | | | | | | | | |
| Generator and exciters | 1.2 | ... | ... | ... | ... | ... | ... | ... | 1.2 | ... | 2.0 | ... | 1.4 | 1.4 |

**382.   Service Factors for Horsepower Loads on V-Belt Drives**

| Applications | Electric motors | | | | | | | | | | Engines | | | |
| --- | --- | --- | --- | --- | --- | --- | --- | --- | --- | --- | --- | --- | --- | --- |
| | A-C | | | | | | | | D-C | | Gas and diesel | | | |
| | Squirrel cage | | | | Syn-chronous | | Single phase | | | | | | | |
| | Normal torque, line start | Normal torque, compensator start | High torque | Wound rotor (slip-ring) | Normal torque | High torque | Repulsion and split-phase | Capacitor | Shunt wound | Compound wound | 4 or more cyl.; above 700 rpm | 4 or more cyl.; below 700 rpm | Steam | Line shafts and clutch starting |
| **Laundry machinery:** | | | | | | | | | | | | | | |
| Washers | 1.2 | | | | | | | | | 1.2 | | | | |
| Extractors | 1.2 | | | | | | | | | 1.2 | | | | |
| Tumblers | 1.2 | | | | | | | | | 1.2 | | | | |
| Dampeners | 1.2 | | | | | | | | | 1.2 | | | | |
| Flatwork ironers | 1.2 | | | | | | | | | 1.2 | | | | |
| Line shafts | 1.4 | 1.4 | | 1.4 | 1.4 | 2.0 | 1.4 | 1.4 | 1.4 | 1.4 | 1.6 | | 1.6 | 1.6 |
| **Machine tools:** | | | | | | | | | | | | | | |
| Grinders | 1.2 | | | 1.4 | | | 1.2 | 1.0 | 1.2 | 1.2 | | | | |
| Boring mills | 1.2 | | | 1.4 | | | | | 1.2 | 1.2 | | | | |
| Lathes | 1.0 | | | 1.2 | | | 1.0 | 1.0 | 1.0 | 1.0 | | | | |
| Milling machines | 1.2 | | | 1.4 | | | | | 1.2 | 1.2 | | | | |
| Screw machines | 1.0 | | | 1.0 | | | 1.0 | 1.0 | 1.0 | 1.0 | | | | |
| Cam cutters | 1.0 | | | 1.0 | | | | | 1.0 | 1.0 | | | | |
| Planers | 1.2 | | | 1.4 | | | 1.2 | 1.0 | 1.2 | 1.2 | | | | |
| Shapers | 1.0 | | | 1.0 | | | 1.0 | 1.0 | 1.0 | 1.0 | | | | |
| Drill press | 1.0 | | | 1.0 | | | 1.0 | 1.0 | 1.0 | 1.0 | | | | |
| Drop hammers | 1.0 | | | 1.0 | | | 1.0 | 1.0 | 1.0 | 1.0 | | | | |
| Shears | 1.2 | | | 1.4 | | | 1.2 | 1.2 | 1.2 | 1.0 | | | | |
| **Mills:** | | | | | | | | | | | | | | |
| Pebble | | 1.4 | 1.6 | 1.4 | | | | | | 1.4 | | | | 1.6 |
| Rod | | 1.4 | 1.6 | 1.4 | | | | | | 1.4 | | | | 1.6 |
| Ball | | 1.4 | 1.6 | 1.4 | | | | | | 1.4 | | | | 1.6 |
| Roller mills | | 1.4 | 1.6 | 1.4 | | | | | | 1.4 | | | | 1.6 |
| Flaking mills | | 1.6 | 1.6 | 1.4 | | | | | | 1.4 | | | | 1.6 |
| Tumbling barrels | | 1.6 | 1.6 | 1.4 | | | | | | 1.4 | | | | 1.6 |
| **Oil-field machinery:**[a] | | | | | | | | | | | | | | |
| Slush pumps | | | | | | | | | | 1.4 | 1.4 | 1.6 | 1.4 | 1.4 |
| Pumping units | 1.2 | 1.2 | 1.4 | | | | | | | 1.4 | | | | 1.6 |
| Pipe-line pumps, centrifugal | 1.2 | 1.2 | 1.4 | | | | | | 1.4 | 1.4 | | | | |
| Draw works (intermittent) | | | | | | | | | | 1.3 | 1.0 | 1.0 | | |
| **Paper machinery:** | | | | | | | | | | | | | | |
| Jordan engines | 1.5 | 1.3 | 1.8 | 1.5 | 1.6 | 1.8 | | | 1.5 | 1.5 | | | | |
| Beaters | 1.4 | 1.4 | | 1.4 | | | | | 1.4 | 1.4 | | | | 1.8 |
| Calenders | 1.2 | 1.2 | | 1.2 | | | | | 1.2 | 1.2 | | | | |
| Agitators | 1.2 | 1.0 | 1.4 | 1.2 | | | | | 1.2 | 1.2 | | | | 1.6 |
| Driers | 1.2 | 1.2 | | 1.2 | | | | | 1.2 | 1.2 | | | | |
| Paper machines | 1.4 | 1.4 | | 1.5 | | | | | 1.5 | 1.5 | | | | 1.6 |
| **Printing machinery:** | | | | | | | | | | | | | | |
| Rotary press | 1.2 | 1.2 | | 1.2 | | | | | 1.2 | | | | | |
| Embossing press | 1.2 | 1.2 | | 1.2 | | | | | 1.2 | 1.2 | | | | |
| Folders | 1.2 | 1.2 | | 1.2 | | | | | 1.2 | | | | | |
| Paper cutters | 1.2 | 1.2 | | 1.2 | | | | | 1.2 | 1.2 | | | | |
| Linotype machines | 1.2 | 1.2 | | | | | | | 1.2 | | | | | |
| Flat-bed press | 1.2 | 1.2 | | 1.2 | | | | | 1.2 | | | | | |
| **Pumps:** | | | | | | | | | | | | | | |
| Centrifugal | 1.2 | 1.2 | 1.4 | 1.4 | | | 1.2 | 1.2 | | | | | | |
| Gear | 1.2 | 1.2 | 1.4 | 1.4 | | | 1.2 | 1.2 | | | | | | |
| Rotary | 1.2 | 1.2 | 1.4 | 1.4 | | | 1.2 | 1.2 | 1.2 | | 1.2 | | | |

[a] Hoisting service factor based on total engine horsepower (continuous rating).   Electric drive factor based on continuous rating of motor.

### 382.   Service Factors for Horsepower Loads on V-Belt Drives (*Continued*)

| Applications | Electric motors | | | | | | | | | D-C | | Engines | | | Line shafts and clutch starting |
|---|---|---|---|---|---|---|---|---|---|---|---|---|---|---|---|
| | A-C | | | | | | | | | | | Gas and diesel | | | |
| | Squirrel cage | | | Wound rotor (slip-ring) | Syn-chronous | | Single phase | | | | | | | | |
| | Normal torque, line start | Normal torque, compensator start | High torque | | Normal torque | High torque | Repulsion and split-phase | Capacitor | Shunt wound | Compound wound | 4 or more cyl.; above 700 rpm | 4 or more cyl.; below 700 rpm | Steam | |
| **Pumps:—(*Continued*)** | | | | | | | | | | | | | | | |
| Reciprocating: | | | | | | | | | | | | | | | |
| 3 or more cylinders | 1.2 | 1.2 | ... | 1.4 | 1.4 | 1.6 | ... | ... | | | 1.8 | ... | 1.8 | |
| 1 or 2 cylinders | 1.4 | 1.4 | ... | 1.6 | 1.6 | 1.8 | ... | ... | | | 2.0 | ... | 2.0 | |
| Dredge pumps | 1.4 | 1.4 | ... | 1.4 | ... | ... | ... | ... | | | 2.0 | ... | 2.0 | |
| Rubber-plant machinery: | | | | | | | | | | | | | | | |
| Calenders | 1.4 | 1.4 | 1.4 | 1.4 | ... | 1.8 | ... | ... | | | ... | ... | 2.0 | |
| Banbury mills | 1.4 | 1.4 | 1.4 | 1.4 | ... | 1.8 | ... | ... | | | ... | ... | 2.0 | |
| Mixers | 1.4 | 1.4 | 1.4 | 1.4 | ... | 1.8 | ... | ... | | | ... | ... | 2.0 | |
| Screens: | | | | | | | | | | | | | | | |
| Vibrating | 1.2 | 1.2 | 1.4 | | | | | | | | | | | |
| Conical | 1.2 | 1.2 | | | | | | | | | | | | |
| Revolving | 1.2 | 1.2 | | | | | | | | | | | | |
| Textile machinery: | | | | | | | | | | | | | | | |
| Spinning frames | 1.6 | ... | 1.8 | | | | | | | | | | | |
| Twisters | 1.6 | ... | 1.8 | | | | | | | | | | | |
| Looms | 1.2 | | | | | | | | | | | | | |
| Warpers | 1.2 | | | | | | | | | | | | | |
| Reels | 1.2 | | | | | | | | | | | | | |

### 383.   Correction Factors for Horsepower Rating of V Belts According to Arc of Contact on Small Sheave

| Arc of contact,ª deg | V-belt sheaves on both shafts | V-belt sheave to flat pulley | Arc of contact,ª deg | V-belt sheaves on both shafts | V-belt sheave to flat pulley |
|---|---|---|---|---|---|
| | Correction factor | | | Correction factor | |
| 180 | 1.00 | 0.74 | 130 | 0.86 | 0.86 |
| 170 | 0.98 | 0.77 | 120 | 0.83 | 0.83 |
| 160 | 0.95 | 0.79 | 110 | 0.79 | 0.79 |
| 150 | 0.92 | 0.82 | 100 | 0.74 | 0.74 |
| 140 | 0.89 | 0.84 | 90 | 0.69 | 0.69 |

ª Arc of contact $= 180 - (D - d)60/C$, where $D$ = pitch diameter of large pulley (for flat pulleys add the thickness of belt to the pulley diameter); $d$ = pitch diameter of small pulley; and $C$ = center distance.

**384.  Horsepower Ratings for V Belts**

"A" Section

| Velocity, ft per min | Pitch diam, in. | | | | | | |
|---|---|---|---|---|---|---|---|
| | 2.6 | 3.0 | 3.4 | 3.8 | 4.2 | 4.6 | 5.0 and larger |
| 1,000 | 0.5 | 0.7 | 0.8 | 0.9 | 0.9 | 1.0 | 1.0 |
| 1,100 | 0.6 | 0.7 | 0.9 | 1.0 | 1.0 | 1.1 | 1.1 |
| 1,200 | 0.6 | 0.8 | 1.0 | 1.0 | 1.1 | 1.2 | 1.2 |
| 1,300 | 0.7 | 0.9 | 1.0 | 1.1 | 1.2 | 1.3 | 1.3 |
| 1,400 | 0.7 | 0.9 | 1.1 | 1.2 | 1.3 | 1.4 | 1.4 |
| 1,500 | 0.8 | 1.0 | 1.1 | 1.3 | 1.4 | 1.4 | 1.5 |
| 1,600 | 0.8 | 1.0 | 1.2 | 1.3 | 1.4 | 1.5 | 1.6 |
| 1,700 | 0.9 | 1.1 | 1.3 | 1.4 | 1.5 | 1.6 | 1.7 |
| 1,800 | 0.9 | 1.2 | 1.4 | 1.5 | 1.6 | 1.7 | 1.8 |
| 1,900 | 0.9 | 1.2 | 1.4 | 1.6 | 1.7 | 1.8 | 1.9 |
| 2,000 | 1.0 | 1.3 | 1.5 | 1.6 | 1.8 | 1.9 | 2.0 |
| 2,100 | 1.0 | 1.3 | 1.5 | 1.7 | 1.9 | 2.0 | 2.1 |
| 2,200 | 1.1 | 1.4 | 1.6 | 1.8 | 1.9 | 2.0 | 2.2 |
| 2,300 | 1.1 | 1.4 | 1.6 | 1.8 | 2.0 | 2.1 | 2.3 |
| 2,400 | 1.1 | 1.4 | 1.7 | 1.9 | 2.0 | 2.2 | 2.3 |
| 2,500 | 1.1 | 1.5 | 1.7 | 2.0 | 2.1 | 2.3 | 2.4 |
| 2,600 | ... | 1.6 | 1.8 | 2.0 | 2.2 | 2.3 | 2.5 |
| 2,700 | ... | 1.6 | 1.8 | 2.1 | 2.3 | 2.4 | 2.6 |
| 2,800 | ... | 1.6 | 1.9 | 2.1 | 2.3 | 2.5 | 2.6 |
| 2,900 | ... | 1.6 | 1.9 | 2.2 | 2.4 | 2.6 | 2.7 |
| 3,000 | ... | ... | 2.0 | 2.2 | 2.4 | 2.6 | 2.7 |
| 3,100 | ... | ... | 2.0 | 2.3 | 2.5 | 2.6 | 2.8 |
| 3,200 | ... | ... | 2.0 | 2.3 | 2.5 | 2.7 | 2.9 |
| 3,300 | ... | ... | 2.1 | 2.4 | 2.6 | 2.7 | 2.9 |
| 3,400 | ... | ... | ... | 2.4 | 2.7 | 2.8 | 3.0 |
| 3,500 | ... | ... | ... | 2.5 | 2.7 | 2.8 | 3.0 |
| 3,600 | ... | ... | ... | 2.5 | 2.7 | 2.9 | 3.1 |
| 3,700 | ... | ... | ... | 2.5 | 2.8 | 2.9 | 3.1 |
| 3,800 | ... | ... | ... | ... | 2.8 | 3.0 | 3.2 |
| 3,900 | ... | ... | ... | ... | 2.8 | 3.0 | 3.2 |
| 4,000 | ... | ... | ... | ... | 2.8 | 3.0 | 3.3 |
| 4,100 | ... | ... | ... | ... | 2.9 | 3.1 | 3.3 |
| 4,200 | ... | ... | ... | ... | ... | 3.1 | 3.3 |
| 4,300 | ... | ... | ... | ... | ... | 3.1 | 3.3 |
| 4,400 | ... | ... | ... | ... | ... | 3.2 | 3.4 |
| 4,500 | ... | ... | ... | ... | ... | 3.2 | 3.4 |
| 4,600 | ... | ... | ... | ... | ... | ... | 3.4 |
| 4,700 | ... | ... | ... | ... | ... | ... | 3.4 |
| 4,800 | ... | ... | ... | ... | ... | ... | 3.4 |
| 4,900 | ... | ... | ... | ... | ... | ... | 3.4 |
| 5,000 | ... | ... | ... | ... | ... | ... | 3.4 |

**384.  Horsepower Ratings for V Belts** (*Continued*)

"B" Section

| Velocity, ft per min | Pitch diam, in. | | | | | |
|---|---|---|---|---|---|---|
| | 5.0 | 5.4 | 5.8 | 6.2 | 6.6 | 7.0 and larger |
| 1,000 | 1.3 | 1.4 | 1.5 | 1.6 | 1.7 | 1.8 |
| 1,100 | 1.4 | 1.5 | 1.6 | 1.7 | 1.8 | 1.9 |
| 1,200 | 1.5 | 1.6 | 1.7 | 1.8 | 1.9 | 2.0 |
| 1,300 | 1.6 | 1.8 | 1.9 | 2.0 | 2.1 | 2.2 |
| 1,400 | 1.7 | 1.9 | 2.0 | 2.1 | 2.2 | 2.3 |
| 1,500 | 1.8 | 2.0 | 2.1 | 2.3 | 2.4 | 2.5 |
| 1,600 | 1.9 | 2.1 | 2.3 | 2.4 | 2.5 | 2.6 |
| 1,700 | 2.0 | 2.2 | 2.4 | 2.5 | 2.7 | 2.8 |
| 1,800 | 2.2 | 2.4 | 2.5 | 2.7 | 2.8 | 2.9 |
| 1,900 | 2.3 | 2.5 | 2.6 | 2.8 | 3.0 | 3.1 |
| 2,000 | 2.4 | 2.6 | 2.8 | 2.9 | 3.1 | 3.2 |
| 2,100 | 2.4 | 2.7 | 2.9 | 3.1 | 3.2 | 3.4 |
| 2,200 | 2.5 | 2.8 | 3.0 | 3.2 | 3.4 | 3.5 |
| 2,300 | 2.6 | 2.9 | 3.1 | 3.3 | 3.5 | 3.6 |
| 2,400 | 2.7 | 3.0 | 3.2 | 3.4 | 3.6 | 3.7 |
| 2,500 | 2.8 | 3.1 | 3.3 | 3.5 | 3.7 | 3.9 |
| 2,600 | 2.9 | 3.2 | 3.4 | 3.6 | 3.8 | 4.0 |
| 2,700 | 3.0 | 3.3 | 3.5 | 3.7 | 3.9 | 4.1 |
| 2,800 | 3.0 | 3.3 | 3.6 | 3.8 | 4.1 | 4.3 |
| 2,900 | 3.1 | 3.4 | 3.7 | 3.9 | 4.2 | 4.4 |
| 3,000 | 3.2 | 3.5 | 3.8 | 4.0 | 4.3 | 4.5 |
| 3,100 | 3.2 | 3.6 | 3.9 | 4.1 | 4.4 | 4.6 |
| 3,200 | 3.3 | 3.6 | 4.0 | 4.2 | 4.5 | 4.7 |
| 3,300 | 3.3 | 3.7 | 4.0 | 4.3 | 4.6 | 4.8 |
| 3,400 | 3.4 | 3.7 | 4.1 | 4.4 | 4.7 | 4.9 |
| 3,500 | 3.4 | 3.8 | 4.1 | 4.4 | 4.7 | 4.9 |
| 3,600 | 3.4 | 3.8 | 4.2 | 4.5 | 4.8 | 5.0 |
| 3,700 | 3.5 | 3.9 | 4.3 | 4.5 | 4.8 | 5.1 |
| 3,800 | 3.5 | 3.9 | 4.3 | 4.6 | 4.9 | 5.2 |
| 3,900 | 3.5 | 3.9 | 4.3 | 4.6 | 4.9 | 5.2 |
| 4,000 | 3.5 | 4.0 | 4.4 | 4.7 | 5.0 | 5.3 |
| 4,100 | 3.5 | 4.0 | 4.4 | 4.7 | 5.0 | 5.3 |
| 4,200 | 3.5 | 4.0 | 4.4 | 4.7 | 5.1 | 5.4 |
| 4,300 | 3.5 | 4.0 | 4.4 | 4.7 | 5.1 | 5.4 |
| 4,400 | 3.5 | 4.0 | 4.4 | 4.8 | 5.1 | 5.4 |
| 4,500 | 3.5 | 4.0 | 4.4 | 4.8 | 5.1 | 5.4 |
| 4,600 | 3.4 | 4.0 | 4.4 | 4.8 | 5.1 | 5.5 |
| 4,700 | 3.4 | 3.9 | 4.4 | 4.8 | 5.2 | 5.5 |
| 4,800 | 3.4 | 3.9 | 4.4 | 4.8 | 5.2 | 5.5 |
| 4,900 | 3.3 | 3.9 | 4.3 | 4.8 | 5.1 | 5.5 |
| 5,000 | 3.3 | 3.8 | 4.3 | 4.7 | 5.1 | 5.5 |

**384. Horsepower Ratings for V Belts**
"C" Section

| Velocity, ft per min | Pitch diam, in. | | | | | |
|---|---|---|---|---|---|---|
| | 7.0 | 8.0 | 9.0 | 10.0 | 11.0 | 12.0 and larger |
| 1,000 | 2.0 | 2.5 | 2.8 | 3.1 | 3.4 | 3.6 |
| 1,100 | 2.2 | 2.7 | 3.1 | 3.5 | 3.7 | 3.9 |
| 1,200 | 2.4 | 2.9 | 3.4 | 3.8 | 4.0 | 4.3 |
| 1,300 | 2.6 | 3.2 | 3.7 | 4.1 | 4.4 | 4.6 |
| 1,400 | 2.7 | 3.4 | 3.9 | 4.4 | 4.7 | 5.0 |
| 1,500 | 2.9 | 3.6 | 4.2 | 4.6 | 5.0 | 5.3 |
| 1,600 | 3.1 | 3.9 | 4.5 | 4.9 | 5.3 | 5.7 |
| 1,700 | 3.3 | 4.1 | 4.7 | 5.3 | 5.6 | 6.0 |
| 1,800 | 3.4 | 4.3 | 5.0 | 5.5 | 5.9 | 6.3 |
| 1,900 | 3.6 | 4.5 | 5.2 | 5.8 | 6.2 | 6.6 |
| 2,000 | 3.7 | 4.7 | 5.5 | 6.1 | 6.5 | 7.0 |
| 2,100 | 3.9 | 4.9 | 5.7 | 6.3 | 6.8 | 7.3 |
| 2,200 | 4.0 | 5.1 | 5.9 | 6.6 | 7.1 | 7.6 |
| 2,300 | 4.2 | 5.3 | 6.2 | 6.8 | 7.4 | 7.9 |
| 2,400 | 4.3 | 5.5 | 6.4 | 7.1 | 7.7 | 8.2 |
| 2,500 | 4.5 | 5.7 | 6.6 | 7.3 | 8.0 | 8.5 |
| 2,600 | 4.6 | 5.8 | 6.8 | 7.6 | 8.2 | 8.8 |
| 2,700 | 4.7 | 6.0 | 7.0 | 7.8 | 8.4 | 9.0 |
| 2,800 | 4.8 | 6.2 | 7.2 | 8.0 | 8.7 | 9.3 |
| 2,900 | 4.9 | 6.3 | 7.4 | 8.3 | 9.0 | 9.6 |
| 3,000 | 5.0 | 6.5 | 7.6 | 8.5 | 9.2 | 9.8 |
| 3,100 | 5.1 | 6.6 | 7.8 | 8.7 | 9.5 | 10.1 |
| 3,200 | 5.2 | 6.7 | 7.9 | 8.9 | 9.7 | 10.3 |
| 3,300 | 5.3 | 6.9 | 8.1 | 9.1 | 9.9 | 10.6 |
| 3,400 | ... | 7.0 | 8.3 | 9.3 | 10.1 | 10.8 |
| 3,500 | ... | 7.1 | 8.4 | 9.4 | 10.3 | 11.0 |
| 3,600 | ... | 7.2 | 8.5 | 9.6 | 10.5 | 11.2 |
| 3,700 | ... | 7.3 | 8.7 | 9.8 | 10.7 | 11.4 |
| 3,800 | ... | 7.4 | 8.8 | 9.9 | 10.9 | 11.6 |
| 3,900 | ... | ... | 8.9 | 10.1 | 11.0 | 11.8 |
| 4,000 | ... | ... | 9.0 | 10.2 | 11.2 | 12.0 |
| 4,100 | ... | ... | 9.1 | 10.3 | 11.3 | 12.2 |
| 4,200 | ... | ... | 9.2 | 10.4 | 11.5 | 12.3 |
| 4,300 | ... | ... | 9.3 | 10.5 | 11.6 | 12.5 |
| 4,400 | ... | ... | ... | 10.6 | 11.7 | 12.6 |
| 4,500 | ... | ... | ... | 10.7 | 11.8 | 12.7 |
| 4,600 | ... | ... | ... | 10.8 | 11.9 | 12.9 |
| 4,700 | ... | ... | ... | 10.8 | 12.0 | 13.0 |
| 4,800 | ... | ... | ... | 10.9 | 12.1 | 13.3 |
| 4,900 | ... | ... | ... | .... | 12.1 | 13.2 |
| 5,000 | ... | ... | ... | .... | 12.2 | 13.2 |

**384. Horsepower Ratings for V Belts** (*Continued*)

"D" Section

| Velocity, ft per min | Pitch diam, in. | | | | | |
|---|---|---|---|---|---|---|
| | 12.0 | 13.0 | 14.0 | 15.0 | 16.0 | 17.0 and larger |
| 1,000 | 4.5 | 5.1 | 5.6 | 6.1 | 6.5 | 6.8 |
| 1,100 | 4.9 | 5.8 | 6.2 | 6.7 | 7.2 | 7.5 |
| 1,200 | 5.3 | 6.1 | 6.7 | 7.3 | 7.7 | 8.2 |
| 1,300 | 5.7 | 6.5 | 7.2 | 7.8 | 8.4 | 8.8 |
| 1,400 | 6.1 | 7.0 | 7.8 | 8.4 | 9.0 | 9.4 |
| 1,500 | 6.5 | 7.5 | 8.2 | 9.0 | 9.6 | 10.1 |
| 1,600 | 6.9 | 7.9 | 8.8 | 9.5 | 10.2 | 10.7 |
| 1,700 | 7.3 | 8.4 | 9.3 | 10.1 | 10.8 | 11.4 |
| 1,800 | 7.7 | 8.8 | 9.8 | 10.6 | 11.3 | 12.0 |
| 1,900 | 8.0 | 9.2 | 10.2 | 11.1 | 11.9 | 12.6 |
| 2,000 | 8.4 | 9.7 | 10.7 | 11.7 | 12.5 | 13.2 |
| 2,100 | 8.8 | 10.1 | 11.2 | 12.2 | 13.0 | 13.8 |
| 2,200 | 9.1 | 10.5 | 11.7 | 12.7 | 13.6 | 14.4 |
| 2,300 | 9.5 | 10.9 | 12.1 | 13.2 | 14.1 | 14.9 |
| 2,400 | 9.8 | 11.3 | 12.6 | 13.7 | 14.6 | 15.5 |
| 2,500 | 10.1 | 11.6 | 12.9 | 14.1 | 15.1 | 16.0 |
| 2,600 | 10.4 | 12.0 | 13.4 | 14.6 | 15.6 | 16.6 |
| 2,700 | 10.6 | 12.3 | 13.8 | 15.0 | 16.1 | 17.1 |
| 2,800 | 10.9 | 12.7 | 14.1 | 15.5 | 16.6 | 17.6 |
| 2,900 | 11.2 | 13.0 | 14.5 | 15.9 | 17.0 | 18.1 |
| 3,000 | 11.4 | 13.3 | 14.9 | 16.3 | 17.5 | 18.6 |
| 3,100 | 11.7 | 13.6 | 15.2 | 16.7 | 17.9 | 19.0 |
| 3,200 | 11.9 | 13.9 | 15.6 | 17.1 | 18.4 | 19.5 |
| 3,300 | 12.1 | 14.1 | 15.9 | 17.4 | 18.9 | 19.9 |
| 3,400 | 12.2 | 14.4 | 16.2 | 17.7 | 19.1 | 20.3 |
| 3,500 | 12.4 | 14.6 | 16.5 | 18.1 | 19.5 | 20.7 |
| 3,600 | 12.6 | 14.8 | 16.7 | 18.4 | 19.8 | 21.1 |
| 3,700 | 12.7 | 15.0 | 17.0 | 18.7 | 20.2 | 21.5 |
| 3,800 | 12.8 | 15.2 | 17.2 | 18.9 | 20.4 | 22.2 |
| 3,900 | .... | 15.3 | 17.4 | 19.2 | 20.7 | 22.3 |
| 4,000 | .... | 15.5 | 17.6 | 19.4 | 21.0 | 22.5 |
| 4,100 | .... | 15.6 | 17.8 | 19.6 | 21.3 | 22.8 |
| 4,200 | .... | .... | 17.9 | 19.8 | 21.5 | 23.0 |
| 4,300 | .... | .... | 18.0 | 20.0 | 21.8 | 23.0 |
| 4,400 | .... | .... | 18.1 | 20.1 | 21.9 | 23.3 |
| 4,500 | .... | .... | .... | 20.3 | 22.1 | 23.5 |
| 4,600 | .... | .... | .... | 20.4 | 22.3 | 23.9 |
| 4,700 | .... | .... | .... | 20.5 | 22.4 | 24.1 |
| 4,800 | .... | .... | .... | .... | 22.5 | 24.2 |
| 4,900 | .... | .... | .... | .... | 22.6 | 24.3 |
| 5,000 | .... | .... | .... | .... | 22.6 | 24.4 |

### 384. Horsepower Ratings for V Belts
"E" Section

| Velocity, ft per min | Pitch diam, in. | | | | | | | | |
|---|---|---|---|---|---|---|---|---|---|
| | 20 | 21 | 22 | 23 | 24 | 25 | 26 | 27 | 28 and larger |
| 1,000 | 8.0 | 8.5 | 9.0 | 9.4 | 9.8 | 10.2 | 10.5 | 10.9 | 11.1 |
| 1,100 | 8.7 | 9.3 | 9.8 | 10.3 | 10.8 | 11.2 | 11.6 | 11.9 | 12.2 |
| 1,200 | 9.5 | 10.1 | 10.7 | 11.2 | 11.7 | 12.2 | 12.6 | 13.0 | 13.3 |
| 1,300 | 10.2 | 10.9 | 11.6 | 12.1 | 12.7 | 13.1 | 13.6 | 14.0 | 14.4 |
| 1,400 | 11.0 | 11.7 | 12.4 | 13.0 | 13.6 | 14.1 | 14.6 | 15.0 | 15.4 |
| 1,500 | 11.7 | 12.5 | 13.2 | 13.9 | 14.5 | 15.1 | 15.6 | 16.1 | 16.5 |
| 1,600 | 12.4 | 13.3 | 14.1 | 14.8 | 15.4 | 16.0 | 16.6 | 17.1 | 17.5 |
| 1,700 | 13.1 | 14.0 | 14.9 | 15.6 | 16.3 | 16.9 | 17.5 | 18.1 | 18.6 |
| 1,800 | 13.8 | 14.8 | 15.7 | 16.5 | 17.2 | 17.9 | 18.5 | 19.0 | 19.6 |
| 1,900 | 14.5 | 15.5 | 16.4 | 17.3 | 18.0 | 18.8 | 19.4 | 20.0 | 20.6 |
| 2,000 | 15.2 | 16.2 | 17.2 | 18.1 | 18.9 | 19.6 | 20.3 | 21.0 | 21.6 |
| 2,100 | 15.8 | 17.0 | 18.0 | 18.9 | 19.7 | 20.5 | 21.2 | 21.9 | 22.5 |
| 2,200 | 16.5 | 17.6 | 18.7 | 19.7 | 20.6 | 21.4 | 22.1 | 22.8 | 23.5 |
| 2,300 | 17.1 | 18.3 | 19.4 | 20.4 | 21.4 | 22.2 | 23.0 | 23.7 | 24.4 |
| 2,400 | 17.7 | 18.9 | 20.1 | 21.2 | 22.2 | 23.1 | 23.9 | 24.6 | 25.3 |
| 2,500 | 18.3 | 19.6 | 20.8 | 21.9 | 23.0 | 23.9 | 24.7 | 25.5 | 26.3 |
| 2,600 | 18.9 | 20.2 | 21.5 | 22.6 | 23.7 | 24.7 | 25.5 | 26.4 | 27.1 |
| 2,700 | 19.4 | 20.8 | 22.1 | 23.3 | 24.4 | 25.4 | 26.4 | 27.2 | 28.0 |
| 2,800 | 19.9 | 21.4 | 22.8 | 24.0 | 25.1 | 26.2 | 27.1 | 28.0 | 28.9 |
| 2,900 | 20.4 | 22.0 | 23.4 | 24.7 | 25.8 | 26.9 | 27.9 | 28.8 | 29.7 |
| 3,000 | 20.9 | 22.7 | 24.0 | 25.3 | 26.5 | 27.6 | 28.6 | 29.6 | 30.5 |
| 3,100 | 21.4 | 23.0 | 24.5 | 26.0 | 27.2 | 28.3 | 29.4 | 30.4 | 31.3 |
| 3,200 | 21.8 | 23.5 | 25.1 | 26.5 | 27.8 | 29.0 | 30.1 | 31.1 | 32.0 |
| 3,300 | 22.3 | 24.0 | 25.6 | 27.1 | 28.4 | 29.6 | 30.7 | 31.8 | 32.8 |
| 3,400 | 22.6 | 24.5 | 26.1 | 27.6 | 29.0 | 30.2 | 31.4 | 32.5 | 33.5 |
| 3,500 | 23.0 | 24.9 | 26.6 | 28.1 | 29.5 | 30.8 | 32.0 | 33.1 | 34.2 |
| 3,600 | 23.4 | 25.3 | 27.0 | 28.6 | 30.0 | 31.4 | 32.6 | 33.8 | 34.8 |
| 3,700 | 23.7 | 25.6 | 27.4 | 29.1 | 30.6 | 31.9 | 33.2 | 34.4 | 35.5 |
| 3,800 | 24.0 | 26.0 | 27.8 | 29.5 | 31.0 | 32.4 | 33.7 | 34.9 | 36.1 |
| 3,900 | 24.2 | 26.3 | 28.2 | 29.9 | 31.5 | 32.9 | 34.3 | 35.5 | 36.6 |
| 4,000 | 24.5 | 26.6 | 28.5 | 30.3 | 31.9 | 33.4 | 34.7 | 36.0 | 37.2 |
| 4,100 | .... | 26.8 | 28.8 | 30.6 | 32.3 | 33.8 | 35.2 | 36.5 | 37.7 |
| 4,200 | .... | 27.1 | 29.1 | 30.9 | 32.6 | 34.2 | 35.6 | 37.0 | 38.2 |
| 4,300 | .... | .... | 29.3 | 31.2 | 33.0 | 34.6 | 36.0 | 37.4 | 38.7 |
| 4,400 | .... | .... | 29.5 | 31.5 | 33.3 | 34.9 | 36.4 | 37.8 | 39.1 |
| 4,500 | .... | .... | .... | 31.7 | 33.5 | 35.2 | 36.7 | 38.2 | 39.5 |
| 4,600 | .... | .... | .... | 31.9 | 33.7 | 35.5 | 37.0 | 38.5 | 39.9 |
| 4,700 | .... | .... | .... | .... | 33.9 | 35.7 | 37.3 | 38.8 | 40.2 |
| 4,800 | .... | .... | .... | .... | 34.1 | 35.9 | 37.5 | 39.1 | 40.5 |
| 4,900 | .... | .... | .... | .... | .... | 36.0 | 37.7 | 39.3 | 40.7 |
| 5,000 | .... | .... | .... | .... | ....: | 36.2 | 37.9 | 39.5 | 40.9 |

**385. Procedure in planning a flat-belt drive** (General Electric Co.)

1. Check reduction ratio of speed required against limits of Sec. 330.

2. Determine pulley diameters from Secs. 388 and 390.

3. Check available distances between centers of shafts against limits of Secs. 387 and 391.

4. Determine belt speed from Secs. 389 and 392.

5. Determine horsepower that will be transmitted per inch of width of belt from Secs. 393 and 394.

6. Determine width of belt required from Sec. 395.

7. Determine required width of pulleys from Sec. 396 and check against standard motor-pulley widths from Sec. 388.

**386. Belt slip.**   The effect of creepage and slip is to reduce the speed of the driven pulley and, consequently, the power transmitted. For commercial applications it will be close enough to multiply the calculated speed of the driven shaft by 98 percent (a 2 percent slip is assumed). The diameter of the driven pulley should be slightly smaller than that calculated on the basis of no slip in order to get a certain speed at the driven shaft. In other words, the speed of the driven shaft is reduced by slip unless slip is compensated for by reducing the driven-pulley diameter by a proportionate amount.

A small amount of belt slip is not harmful, but excessive slip is. Severe slippage burns the belt, quickly destroying its usefulness, while at best the belt surface is polished so that the grip on the pulley is materially reduced. Belt slip can be detected by noting the condition of the pulley surface. When the belt is slipping, the pulley will have a very shiny appearance, as contrasted with the desired smooth but rather dull appearance.

Excessive slip is caused by poorly designed drives in which the driving pulley is too small or the load too great, by running the belt too loosely, or by not giving proper attention to the care of the belt.

**387. Arc of contact for flat-belt drives.**   One of the factors affecting the power that can be transmitted by a belt of given width is the arc of contact (the distance the belt wraps around the pulley). As this arc of contact decreases, the belt width must be increased to transmit the same horsepower with 2 percent slip. Table 394, however, allows a factor of safety sufficient to take care of installations which are in accordance with the recommendations for belt ratios and center distances given in this subdivision. If the limits given must be exceeded, it may be necessary to use a belt tightener.

**388. Application of flat-belt pulley dimensions and belt widths to general-purpose motors.** General-purpose motors having continuous time rating with the frame sizes, horsepower, and speed ratings given in the following paragraphs are designed to operate with flat-belt pulleys and belts within the limiting dimensions listed.

### A. Direct-current Medium Motors

| Frame number | Horsepower at, rpm | | | | | | Standard pulley | | | Minimum pulley | | | Pulley bore, in. |
|---|---|---|---|---|---|---|---|---|---|---|---|---|---|
| | 1,750 | 1,150 | 850 | 690 | 575 | 500 | Pulley diam, in. | Pulley width, in. | Belt width, in. | Min pulley diam, in. | Max pulley width, in. | Belt width, in. | |
| 203 | 1 | .... | .... | ..... | ..... | ....... | 3 | 3 | 2½ | 2½ | 3 | 2½ | ¾ |
| 204 | 1½ | 1 | .... | ..... | ..... | ....... | 3 | 3 | 2½ | 2½ | 3 | 2½ | ¾ |
| 224 | 2 | 1½ | ¾ | ½ | ..... | ....... | 4 | 3½ | 3 | 2½ | 3½ | 3 | 1 |
| 225 | 3 | 2 | 1 | ¾ | ½ | ....... | 4 | 3½ | 3 | 3 | 3½ | 3 | 1 |
| 254 | 5 | 3 | 2–1½ | 1½–1 | 1–¾ | ....... | 4½ | 4½ | 4 | 3½ | 4½ | 4 | 1½ |
| 284 | 7½ | 5 | 3 | 2 | 2–1½ | 1½–1¾ | 5 | 4½ | 4 | 4 | 5½ | 5 | 1¼ |
| | 10 | 7½ | 5 | ..... | ..... | ....... | 6 | 5½ | 5 | 5 | 6¾ | 6 | |
| | 15 | 10 | 7½ | ..... | ..... | ....... | 7 | 6¾ | 6 | 6 | 7¾ | 7 | |
| | 20 | 15 | 10 | ..... | ..... | ....... | 8 | 6¾ | 6 | 6 | 9¾ | 9 | |
| | 25 | .... | ..... | ..... | ..... | ....... | 9 | 7¾ | 7 | 6 | 9¾ | 9 | |
| | 30 | 20 | 15 | ..... | ..... | ....... | 9 | 7¾ | 7 | 7 | 11 | 10 | |
| | 40 | 25 | 20 | ..... | ..... | ....... | 10 | 7¾ | 7 | 7 | 11 | 10 | |
| | | 30 | 25 | ..... | ..... | ....... | 10 | 7¾ | 7 | 9 | 12 | 11 | |
| | | 40 | 30 | ..... | ..... | ....... | 11 | 11 | 10 | 9 | 13 | 12 | |
| | | 50 | ..... | ..... | ..... | ....... | 12 | 12 | 11 | 11 | 13 | 12 | |
| | | 60 | ..... | ..... | ..... | ....... | 12 | 12 | 11 | 11 | 17 | 16 | |
| | | .... | 40 | ..... | ..... | ....... | 12 | 12 | 11 | 11 | 13 | 12 | |
| | | 75 | ..... | ..... | ..... | ....... | 14 | 13 | 12 | 12 | 17 | 16 | |
| | | 50 | ..... | ..... | ..... | ....... | 14 | 13 | 12 | 11 | 17 | 16 | |
| | | 60 | ..... | ..... | ..... | ....... | 14 | 13 | 12 | 12 | 17 | 16 | |
| | | 75 | ..... | ..... | ..... | ....... | 15 | 15 | 14 | | | | |

### B. Single-phase Medium Motors

| Frame number | Horsepower at, rpm | | | | Standard pulley | | | Minimum pulley | | | Pulley bore, in. |
|---|---|---|---|---|---|---|---|---|---|---|---|
| | 3,600 | 1,800 | 1,200 | 900 | Pulley diam, in. | Pulley width, in. | Belt width, in. | Min pulley diam, in. | Max pulley width, in. | Belt width, in. | |
| 203 | 1½ | 1 | .... | ...... | 3 | 3 | 2½ | 2½ | 3 | 2½ | ¾ |
| 204 | 2 | 1½ | ¾ | . .... | 3 | 3 | 2½ | 2½ | 3 | 2½ | ¾ |
| 224 | 3 | 2 | 1 | ½ | 4 | 3½ | 3 | 2½ | 3½ | 3 | 1 |
| 225 | 5 | 3 | 1½ | ¾ | 4 | 3½ | 3 | 3 | 3½ | 3 | 1 |
| 254 | 7½ | 5 | 2 | 1½–1 | 4½ | 4½ | 4 | 3½ | 4½ | 4 | 1⅛ |
| | | .... | 3 | 2 | 5 | 4½ | 4 | 3½ | 4½ | 4 | |
| | | 7½ | 5 | 3 | 5 | 4½ | 4 | 4 | 5½ | 5 | |
| | | 10 | 7½ | 5 | 6 | 5½ | 5 | 5 | 6¾ | 6 | |
| | | 15 | 10 | 7½ | 7 | 6¾ | 6 | 6 | 7¾ | 7 | |
| | | 20 | 15 | 10 | 8 | 6¾ | 6 | 6 | 9¾ | 9 | |
| | | .... | 20 | 15 | 9 | 7¾ | 7 | 7 | 11 | 10 | |
| | | 25 | .... | ...... | 9 | 7¾ | 7 | 6 | 9¾ | 9 | |
| | | | 25 | 20 | 10 | 7¾ | 7 | 7 | 11 | 10 | |
| | | | | 25 | 10 | 7¾ | 7 | 9 | 12 | 11 | |

C. Medium Motors—Polyphase Induction

| Frame number | Horsepower at, rpm | | | | | | | | Standard pulley | | | Minimum pulley | | | Pulley bore, in. |
|---|---|---|---|---|---|---|---|---|---|---|---|---|---|---|---|
| | 3,600 | 1,800 | 1,200 | 900 | 720 | 600 | 514 | 450 | Pulley diam, in. | Pulley width, in. | Belt width, in. | Min pulley diam, in. | Max pulley width, in. | Belt width, in. | |
| **110, 208, 220, 440 and 550 Volts** | | | | | | | | | | | | | | | |
| 203 | 1½ | 1 | ¾ | .... | .... | .... | .... | .... | 3 | 3 | 2½ | 2½ | 3 | 2½ | ¾ |
| 204 | 2 | 1½ | 1 | ½ | .... | .... | .... | .... | 3 | 3 | 2½ | 2½ | 3 | 2½ | ¾ |
| 224 | 3 | 2 | 1½ | ¾ | ½ | .... | .... | .... | 4 | 3½ | 3 | 2½ | 3½ | 3 | 1 |
| 225 | 5 | 3 | 2 | 1 | ¾ | ½ | .... | .... | 4 | 3½ | 3 | 3 | 3½ | 3 | 1 |
| 254 | 7½ | 5 | 3 | 2–1½ | 1½–1 | 1–¾ | .... | .... | 4½ | 4½ | 4 | 3½ | 4½ | 4 | 1⅛ |
| 284 | 10 | 7½ | 5 | 3 | 2 | 1½ | 1–¾ | ¾ | 5 | 4½ | 4 | 4 | 5½ | 5 | 1¼ |
| 324 | 15 | .... | .... | .... | .... | .... | .... | .... | 5 | 6¾ | 6 | 5 | 6¾ | 6 | 1⅝ |
| 324 | .... | 10 | 7½ | 5 | 3 | 2 | 1½ | 1 | 6 | 5½ | 5 | 5 | 6¾ | 6 | 1⅝ |
| 326 | 20 | .... | .... | .... | .... | .... | .... | .... | 5 | 6¾ | 6 | 5 | 6¾ | 6 | 1⅝ |
| 326 | .... | 15 | 10 | 7½ | 5 | 3 | 2 | 1½ | 8 | 6¾ | 6 | 6 | 7¾ | 7 | 1⅝ |
| 364 | .... | 25–20 | 15 | 10 | 7½ | 5 | .... | .... | 9 | 7¾ | 7 | 6 | 9¾ | 9 | 1⅞ |
| 365 | .... | 30 | 20 | 15 | 10 | 7½ | 3 | 2 | 9 | 7¾ | 7 | 7 | 11 | 10 | 1⅞ |
| 404 | .... | 40 | 25 | 20 | 15 | 10 | 5 | 3 | 10 | 8¾ | 8 | 7 | 11 | 10 | 2⅛ |
| 405 | .... | .... | 30 | 25 | 20 | 15 | 7½ | 5 | 10 | 8¾ | 8 | 9 | 12 | 11 | 2⅛ |
| 444 | .... | .... | 40 | 30 | 25 | 20 | 10 | 7½ | 12 | 11 | 10 | 9 | 13 | 12 | 2⅜ |
| 445 | .... | .... | 50 | 40 | 30 | 25 | 15 | 10 | 12 | 11 | 10 | 11 | 13 | 12 | 2⅜ |
| 504U | .... | .... | 60 | 50 | 40 | 30 | 20 | 15 | 15 | 13 | 12 | 11 | 17 | 16 | 2⅞ |
| 505 | .... | .... | 75 | 60 | 50 | 40 | 25 | 20 | 15 | 13 | 12 | 12 | 17 | 16 | 2⅞ |
| **2300 Volts, Three Phase** | | | | | | | | | | | | | | | |
| 445 | .... | .... | 40 | .... | .... | .... | .... | .... | 12 | 11 | 10 | 11 | 13 | 12 | 2⅜ |
| 504U | .... | .... | 50 | 40 | .... | .... | .... | .... | 15 | 13 | 12 | 11 | 17 | 16 | 2⅞ |
| 505 | .... | .... | 60 | 50 | .... | .... | .... | .... | 15 | 13 | 12 | 12 | 17 | 16 | 2⅞ |
| | | | 75 | .... | .... | .... | .... | .... | 15 | 13 | 12 | 12 | 17 | 16 | 2⅞ |

## 389. Belt Speeds for Flat Belts

| Full-load speed of motor, rpm | Pulley diam, in. | | | | | | | | | | |
|---|---|---|---|---|---|---|---|---|---|---|---|
| | 3½ | 4 | 4½ | 5 | 5½ | 6 | 7 | 8 | 9 | 10 | 11 |
| | Belt speed, ft per min | | | | | | | | | | |
| 1,750 | 1,605 | 1,835 | 2,060 | 2,290 | 2,520 | 2,750 | 3,205 | 3,670 | 4,125 | 4,580 | 5,040 |
| 1,450 | 1,330 | 1,520 | 1,710 | 1,900 | 2,090 | 2,275 | 2,660 | 3,040 | 3,420 | 3,800 | 4,170 |
| 1,150 | 1,055 | 1,205 | 1,355 | 1,505 | 1,655 | 1,810 | 2,110 | 2,410 | 2,710 | 3,010 | 3,315 |
| 860 | 790 | 900 | 1,015 | 1,130 | 1,240 | 1,350 | 1,575 | 1,800 | 2,030 | 2,255 | 2,480 |
| 690 | ..... | 725 | 815 | 905 | 995 | 1,085 | 1,265 | 1,445 | 1,625 | 1,810 | 1,990 |
| 575 | ..... | ..... | 680 | 755 | 830 | 905 | 1,055 | 1,205 | 1,355 | 1,505 | 1,655 |

**389.   Belt Speeds for Flat Belts** (*Continued*)

| Full-load speed of motor, rpm | Pulley diam, in. | | | | | | | | | | |
|---|---|---|---|---|---|---|---|---|---|---|---|
| | 12 | 13 | 14 | 15 | 16 | 17 | 18 | 19 | 20 | 21 | 22 |
| | Belt speed, ft per min | | | | | | | | | | |
| 1,450 | 4,560 | 4,930 | | | | | | | | | |
| 1,150 | 3,615 | 3,915 | 4,220 | 4,520 | 4,820 | 5,120 | | | | | |
| 860 | 2,705 | 2,930 | 3,160 | 3,380 | 3,610 | 3,835 | 4,060 | 4,280 | 4,510 | 4,740 | 4,960 |
| 690 | 2,170 | 2,350 | 2,530 | 2,710 | 2,890 | 3,075 | 3,250 | 3,435 | 3,615 | 3,800 | 3,980 |
| 575 | 1,805 | 1,955 | 2,105 | 2,260 | 2,410 | 2,560 | 2,710 | 2,860 | 3,010 | 3,160 | 3,310 |

| Full-load speed of motor, rpm | Pulley diam, in. | | | | | | | | | | |
|---|---|---|---|---|---|---|---|---|---|---|---|
| | 23 | 24 | 25 | 26 | 27 | 28 | 29 | 30 | 31 | 32 | 33 |
| | Belt speed, ft per min | | | | | | | | | | |
| 690 | 4,160 | 4,340 | 4,520 | 4,700 | 4,880 | 5,060 | | | | | |
| 575 | 3,460 | 3,610 | 3,760 | 3,915 | 4,060 | 4,210 | 4,360 | 4,515 | 4,665 | 4,815 | 4,965 |

Table 389 is based on the formula

$$\text{Belt speed in feet per minute} = \frac{3.1416 \times \text{pulley diameter in inches} \times \text{rpm}}{12} \qquad (14)$$

**390. Determining pulley diameters.**   The following rules are given for determining pulley diameters and speeds:

1. Diameter of driven pulley =

$$\frac{\text{diameter of motor pulley} \times \text{full-load motor speed in rpm}}{\text{rpm of driven pulley}}$$

2. Diameter of motor pulley =

$$\frac{\text{diameter of driven pulley} \times \text{rpm of driven pulley}}{\text{full-load motor speed in rpm}}$$

3. rpm of driven pulley =

$$\frac{\text{diameter of motor pulley} \times \text{full-load motor speed}}{\text{diameter of driven pulley}}$$

4. rpm of motor =

$$\frac{\text{diameter of driven pulley} \times \text{speed of driven pulley}}{\text{diameter of motor pulley}}$$

The above formulas can be applied to any belt drive by substituting the terms *driver* and *rpm of driver* for *motor pulley* and *motor speed.*

*Caution.* The desired driven speed should be increased by 2 percent in the formulas to allow for belt slip.

It is desirable to utilize the standard pulleys furnished with the motors whenever possible. The use of special pulleys on electric motors is limited by mechanical dimensions and by the permissible bearing pressure, which, for the transmission of a certain horsepower, increases as the pulley diameter decreases. The maximum pulley diameter is determined by mechanical dimensions. These are in addition to the usual requirements of proper belt speed and width to transmit the power involved.

**391. Center distances for flat-belt drives.**   A short center distance is bad for two reasons: (1) it means a shorter belt, so that flexure occurs at each cross section more frequently;

and (2) the arc of contact on the smaller pulley frequently becomes so small that power can no longer be effectively transmitted. The minimum center distance should be 2½ times the diameter of the larger pulley, with from 3 to 5 times a better figure.

**392. Flat-belt speed.** Belts should not be run at speeds above 5000 ft/min (25.4 m/s). Above this speed there is not sufficient friction contact between belt and pulley because of the effect of centrifugal force which tends to throw the belt.

Table 389 gives the belt speed in feet per minute with various motor speeds and pulley diameters.

**393. Horsepower transmitted by flat belts.** The pulley must be large enough to transmit the horsepower, which ordinarily will not be the motor-horsepower rating but a somewhat larger figure to take care of possible overloads and peak loads. In ordinary applications, 125 percent of the motor rating may be allowed for a safety factor in the usual sizes of motors.

In general, the product of the diameter and belt width of a *special* pulley should at least equal the product of the corresponding dimensions of the *standard* pulley recommended for that size and speed of motor. If the diameters are assumed to have been tentatively chosen as best for the desired purpose, the horsepower that can be transmitted will depend upon the pulley material, the belt width, its quality and number of plies, the distance between centers, and the relative angular location of the driving and driven pulley.

Table 394 shows the horsepower per inches of width transmitted by good leather belting running over paper pulleys with 180° arc of contact and no slip. For iron pulleys, multiply figures given by 0.62; for wood pulleys, multiply by 0.41 (for belt speeds see Table 389).

The figures shown in Table 394 are for regular single-ply and double-ply belts. The values for horsepower transmitted by light belts will be 80 percent of values shown; by heavy belts, 115 percent of values shown.

**394. Horsepower Transmitted per Inch of Width of Flat Leather Belts**

To obtain the horsepower transmitted by belts of any width, multiply the figure shown for the given belt speed by the width of the belt used.

| Belt speed, ft per min | Hp per in. of width | Belt speed, ft per min | Hp per in. of width | Belt speed, ft per min | Hp per in. of width | Belt speed, ft per min | Hp per in. of width |
|---|---|---|---|---|---|---|---|
| | | | Single Belting | | | | |
| 900 | 1.25 | 2,000 | 2.78 | 3,100 | 4.31 | 4,200 | 5.84 |
| 1,000 | 1.39 | 2,100 | 2.92 | 3,200 | 4.45 | 4,300 | 5.98 |
| 1,100 | 1.53 | 2,200 | 3.06 | 3,300 | 4.59 | 4,400 | 6.12 |
| 1,200 | 1.67 | 2,300 | 3.20 | 3,400 | 4.72 | 4,500 | 6.26 |
| 1,300 | 1.81 | 2,400 | 3.33 | 3,500 | 4.86 | 4,600 | 6.40 |
| 1,400 | 1.95 | 2,500 | 3.47 | 3,600 | 5.00 | 4,700 | 6.54 |
| 1,500 | 2.09 | 2,600 | 3.61 | 3,700 | 5.14 | 4,800 | 6.68 |
| 1,600 | 2.22 | 2,700 | 3.75 | 3,800 | 5.28 | 4,900 | 6.82 |
| 1,700 | 2.36 | 2,800 | 3.89 | 3,900 | 5.42 | 5,000 | 6.96 |
| 1,800 | 2.50 | 2,900 | 4.03 | 4,000 | 5.56 | | |
| 1,900 | 2.64 | 3,000 | 4.17 | 4,100 | 5.70 | | |
| | | | Double Belting | | | | |
| 900 | 1.79 | 2,000 | 3.97 | 3,100 | 6.16 | 4,200 | 8.34 |
| 1,000 | 1.99 | 2,100 | 4.17 | 3,200 | 6.36 | 4,300 | 8.54 |
| 1,100 | 2.19 | 2,200 | 4.37 | 3,300 | 6.55 | 4,400 | 8.74 |
| 1,200 | 2.39 | 2,300 | 4.57 | 3,400 | 6.75 | 4,500 | 8.94 |
| 1,300 | 2.58 | 2,400 | 4.77 | 3,500 | 6.95 | 4,600 | 9.13 |
| 1,400 | 2.78 | 2,500 | 4.97 | 3,600 | 7.15 | 4,700 | 9.33 |
| 1,500 | 2.98 | 2,600 | 5.16 | 3,700 | 7.35 | 4,800 | 9.53 |
| 1,600 | 3.18 | 2,700 | 5.36 | 3,800 | 7.55 | 4,900 | 9.73 |
| 1,700 | 3.38 | 2,800 | 5.56 | 3,900 | 7.74 | 5,000 | 9.93 |
| 1,800 | 3.58 | 2,900 | 5.76 | 4,000 | 7.94 | | |
| 1,900 | 3.78 | 3,000 | 5.96 | 4,100 | 8.14 | | |

**395. Belt width for flat belts.** For a given diameter, the minimum belt width is determined by the horsepower to be transmitted; the maximum width is limited by the possible overhang and the safe strain which may be placed on the bearing.

To obtain the belt width necessary to transmit a given horsepower, divide the horsepower which is to be transmitted by the figures given in Table 394 if paper pulleys are used. For iron pulleys, multiply the quotient thus obtained by 1.6; for wood pulleys, by 2.4.

Stock sizes of belts increase in width by the following increments:

| Belt Width, In. | Increment, In. |
|:---:|:---:|
| ½–1 | ⅛ |
| 1 –4 | ¼ |
| 4 –7 | ½ |
| 7 –30 | 1 |
| 30 –56 | 2 |
| Above 56 | 4 |

The foregoing represents average practice; stock sizes of different manufacturers may vary slightly from the figures given.

NOTES   A good rule to remember is: A single-ply belt 1 in wide and running at 1000 ft/min (5.08 m/s) will deliver approximately 1 hp (up to about 3000 rpm).

Round belts of 0.25- and 0.5-in (6.35- and 12.7-mm) diameter are fully equal to single belts of 1 and 3 in (25.4 and 76.2 mm), respectively.

**396. Pulley widths should exceed belt widths** by the following amounts:

| Pulley diam, in. | Iron pulley, in. | Paper pulley, in. |
|:---:|:---:|:---:|
| Under 2 | ¼ | ¼ |
| 2– 5 | ½ | ½ |
| 5–10 | ½ | ¾ |
| 10–20 | ½ | 1 |
| 20–24 | ¾ | 1 |
| 24–36 | ¾ | 1½ |
| Above 36 | ¾ | 2 |

**397. Arrangement of belts and pulleys**   A typical belt-drive layout is shown in Fig. 7-187. The best arrangement of belts and pulleys is with the centerline through the pulleys in a horizontal plane. This arrangement is called *horizontal drive*. The drive should be arranged so that the lower side of the belt is driving in order that the sag of the upper side of the belt will tend to increase the arc of contact. This is illustrated in Fig. 7-188. The sag should be about 1½ in (38 mm) for every 10 ft (3 m) of center distance between shafts. If it is too loose, the belt will have an unsteady flapping motion which will injure both belt and machinery; if it is too tight, the bearings will be worn and the belt quickly destroyed.

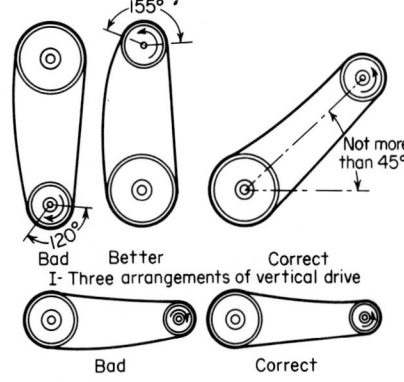

**Fig. 7-188**   Arrangement of belts. [General Electric Co.]

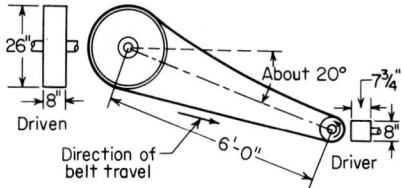

**Fig. 7-187**   Belt layout. [General Electric Co.]

Vertical drives (the mounting of one pulley above another) should be avoided. This is particularly true when the lower pulley is the smaller. The effective tension and arc of contact are reduced in vertical drives so that the normal load cannot be carried. It is better if the angle of the belt with the floor does not exceed 45° (see Fig. 7-188).

When several belts transmit power from a line shaft, it is advantageous, if possible, to locate the line shaft so that the bearing pressures can be equalized and reduced by alternating the direction of drive, first on one side and then on the other.

**398. Rule for finding length of belts.** When it is not feasible to measure with a tapeline the length required, the following rule, which gives an accurate result when the pulleys are of the same diameter and an approximately accurate result when the pulleys are of different diameters, can be used.

Add the diameters of the two pulleys ($D$ and $d$, Fig. 7-189), divide the result by 2, and multiply the quotient by $3\frac{1}{7}$; add the product to twice the distance $L$ between the centers

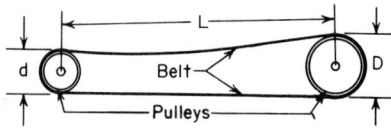

**Fig. 7-189**   Notation for belt-length formula.          **Fig. 7-190**   Example of finding belt length.

of the shafts, and the result is the length required. All values should be expressed either in feet or in inches. Expressed as a formula, using the notation of Fig. 7-189, the rule becomes

$$\text{Length of belt} = (D + d)1.57 + 2L \qquad (15)$$

**example**   What is the length of the belt required for the two pulleys of Fig. 7-190? Diameters of pulleys are 16 and 18 in. Distance between centers is 10 ft, or 120 in.
**solution**   Substitute in the formula

$$\text{Length of belt} = (18 + 16)1.57 + 2 \times 120 = (34 \times 1.57) + 240 = 293.4 \text{ in} = \frac{293.4}{12} = 24.4 \text{ ft}$$

If one pulley is considerably larger than the other, a little extra allowance should be made, because the distance between the points of tangency of the belt on the two pulleys is somewhat greater than the exact distance between the centers of the shafts.

**399. Rule for measuring belts in the roll.** Add to the diameter of the roll in inches the diameter of the hole in the center of the roll. Multiply this sum by the number of coils in the roll, and multiply this product by 1.32. The three figures on the left indicate the number of feet in the roll.

**example**   A roll of 5-in single leather belt measures 37⅝ in, outside diameter; the hole is 4⅝ in in diameter; and the number of coils in the roll is 84. How long is the belt?
**solution**   Using the above rule,

$$(37\tfrac{5}{8} + 4\tfrac{5}{8}) \times 84 = 42\tfrac{1}{4} \times 84 = 3549 \times 1.32 = 4684.68$$

The first three figures on the left indicate that the roll contains 468½ ft. By actual measurement the roll is found to contain 469 ft (Page Belting Co.).

**400. Splicing belts** (Page Belting Co.).   When possible, the ends of the belt should be fastened together by splicing and cementing. If belts are to be laced or fastened otherwise than with cement, cut off the ends perfectly true, using a try square. Punch the holes exactly opposite one another in the two ends as in Fig. 7-191. The grain side of the belt should be run next to the pulley, and the belt should be run off, not against, the laps. Undoubtedly, exclusive of cementing, lacing is the best method for fastening belt ends together, as the lacing is as flexible as the belt and runs noiselessly over the pulleys. The best lacing is in the end the cheapest. Cheap lacing is very expensive in the long run.

Use a small lace so that the holes will be small. For belts 1 to 2¼ in wide, use ¼-in lacing; 2½ to 4½ in wide, ⁵⁄₁₆-in lacing; and 5 to 12 in wide, ⅜-in lacing. For wider belts use wider lacing in proportion. Avoid thick lacing. Light, strong lacing is the best.

In punching a belt for lacing, it is desirable to use an oval punch, the longer diameter of the punch lying parallel with the length of the belt, so that a minimum amount of leather across the belt will be cut out. There should be in each end of the belt two rows of holes, placed zigzag. Make the holes the smallest possible that will admit the lace. In a 2-in belt there should be 3 holes in each end; in a 2½-in belt, 4 holes; in a 3-in belt, 5 holes; in a 4-in belt, 7 holes; in a 5-in belt, 9 holes; in a 6-in belt, 11 holes; in an 8-in belt, 15 holes; in a 10-in belt, 19 holes; in a 12-in belt, 23 holes.

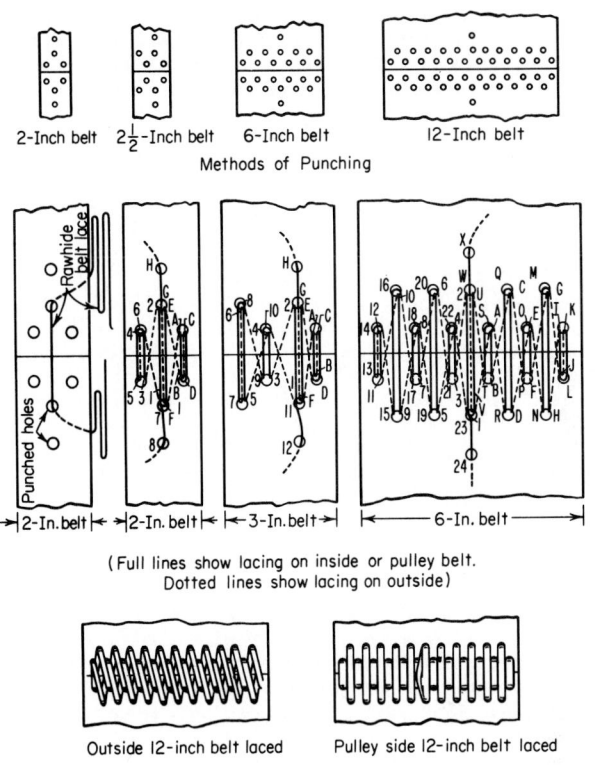

2-Inch belt    2½-Inch belt    6-Inch belt    12-Inch belt

Methods of Punching

(Full lines show lacing on inside or pulley belt. Dotted lines show lacing on outside)

Outside 12-inch belt laced    Pulley side 12-inch belt laced

Finished Joint

**Fig. 7-191**  Method of lacing belts.

The center of no hole should come nearer to the side of the belt than ⅝ in or nearer the end than ⅞ in. The second row should be at least 1¾ in from the end. On wide belts these distances should be a little greater.

Begin to lace in the center of the belt, and take great care to keep the ends exactly in line and to lace both sides with equal tightness. The lacing must not be crossed on the side of the belt that runs next to the pulley.

**401. In putting on new belts** a common rule is to draw them up and stretch them ⅛ in for every foot (3.175 mm for 0.3 m) in length of belt.

The strongest part of belt leather is near the flesh side, about one-third of the way through from that side. It is therefore desirable to run the grain (hair) side on the pulley so that the strongest part of the belt may be subject to the least wear. The flesh side is not so apt to crack as is the grain side when the belt is old; hence it is better to crimp the grain than to stretch it. Leather belts that are run with the grain side to the pulley will drive 30 percent more than if run with the flesh side. The belt as well as the pulley adheres best when smooth, and the grain side adheres best because it is smoother.

**402. A belt adheres much better and is less apt to slip when run at a high speed than at a low speed.** Therefore, it is better to gear a mill with small pulleys and run them at high velocity than to have large pulleys and run them more slowly. A mill thus geared costs less and has a much neater appearance than with large, heavy pulleys.

**403. Belt troubles.**   The belt on any belt-connected machine should be tight enough to run without slipping, but the tension should not be too great, or the bearings will heat. The crowns of driving and driven pulleys should be alike, as belts sometimes wobble because of pulleys having unlike crowns. If this condition is caused by bad joints, they should be broken and cemented over again. A wave motion or flapping is usually caused by slippage between the belt and pulley, resulting from grease spots, etc. It may, however, be a warning of an excessive overload. This fault may sometimes be corrected by increasing the tension, but a better remedy is to clean the belt. A back-and-forth movement on the pulley is caused by unequal stretching of the edges of the belt. If this does not cure itself shortly, examine the joints. If they are evenly made and remain so, the belt is bad and should be discarded.

**404. Gear drive.**   Gearing is the most positive form of power transmission and is frequently employed when the motor can be mounted directly on a machine. Gear motors, with integral gearing, are available in a large variety of motors and output speeds. The points to be considered on an external gear drive are (1) speed reduction, (2) pitch of the gears, (3) number of teeth on the gears (pinion and gear), (4) face of the gear, (5) pitch-line speed, (6) distance between centers, (7) use of idler gears, and (8) mounting of the motor.

The speed reduction is the same as for the belt drive. Each motor rating has a minimum pinion to limit stresses to safe values. The pitch, number of teeth, and face for motor pinions have been standardized for back-geared motors, and the best practice when gearing a motor directly to machines is to use these motor pinions if possible. Table 415 gives the standard motor ratings and other valuable gearing information. (The information on gear drives given herein is largely from an article in the *American Machinist* by A. G. Popcke.)

**405. The method of gearing** depends largely upon the distance between centers and the space available for the motors. In all cases the pinion selected must not be smaller than the minimum specified in Table 415. The pitch-line speed must not exceed the limits given. Two general cases cover the mounting of a motor to drive a machine through gears:

1. The dimension of the motor or machine limits the distance between the centers of the motor shaft and the driven shaft.

2. The limitation of case 1 does not exist.

The first case occurs when a motor is mounted on the top, side, or bottom of a machine, as shown in Fig. 7-192. The dimension causing limitations is indicated by *A* in these illustrations. The proper distance can be obtained by using large enough gears; the limit is pitch-line speed. An intermediate idler gear frequently overcomes the difficulties experienced here.

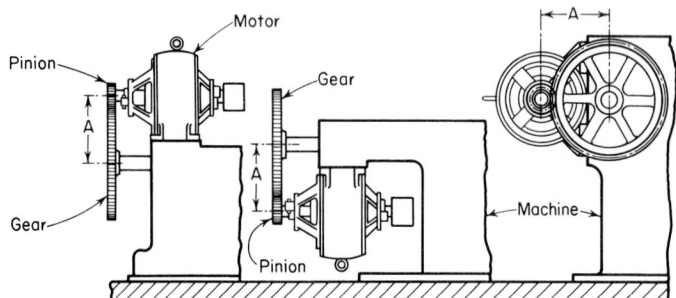

**Fig. 7-192.**   Motor mountings for gear drive (distances between gear centers limited).

The relation of the motor and machine in the second case is shown in Fig. 7-193. Here the motor can be mounted on a base, and the motor pinion can mesh with the gear on the machine in any convenient position.

If reductions greater than 7:1 are required, it is usually necessary to obtain the reduction by the use of two sets of gears. The back-geared motors discussed in Sec. 404 can be used to furnish one set of gears in these cases. Thus if a reduction of 10:1 is desired, a back-geared motor with a standard 6:1 reduction, with a further reduction from the countershaft of the motor to the machine of $1\%_6$:1 (1.66:1), will fulfill the requirements.

**example** The speed of the driven shaft of the machine is 210 rpm, the hp is 10, the motor is mounted on the machine, and the limiting distance between centers is 12 in. What are the sizes of gear and pinion to be used? The machine is a punch and shear.

**solution** In this case a pitch-line speed of approximately 1000 ft/min will be employed. Table 415 shows that a 10-hp motor at 850 rpm is the highest-speed motor that can be used for this pitch-line speed. The ratio of reduction is then

$$\frac{850}{210} = 4.05 \text{ (use 4:1)}$$

The distance between centers for any set of gears is determined by the formula

$$a = \frac{b}{2P} \text{ in} \qquad\qquad (16)$$

where $a$ = the distance between centers in inches, $b$ = the sum of the number of teeth in both gears, and $P$ = the diametral pitch. In this case

$$a, \text{ or } 12 = \frac{b}{2 \times 5} \qquad b = 120$$

The number of teeth in the pinion is

$$\frac{b}{\text{Ratio of reduction plus 1}} = \frac{120}{5} = 24 = \text{number of teeth}$$

The number of teeth in the gear is $4 \times 24 = 96$.

Table 415 shows that the pitch-line speed for the motor with 20 teeth is 890 ft/min. The pitch-line speed with 24 teeth is

$$\frac{24}{20} \times 890 = 1068 \text{ ft/min}$$

If quiet operation is desired, a cloth or rawhide pinion should be used with a 3¾-in face. Thus the gears are specified as follows:

Motor pinion—rawhide, $P = 5$, face 3 ¾ in, 24 teeth. Machine gear—steel, $P = 5$, face 3 in, 96 teeth.

The bore for each is determined by the diameter of the shaft to which it is connected. The pinion is wider than the gears, so that the rawhide engages only with the gear. If it were the same width, the brass end plates of the rawhide pinion would engage with the gear, causing noise.

**406. In the selection or specification of pinions for motors** (C. W. Drake, *Electric Journal*) for geared applications, three dimensions must be determined, namely, the face, the diameter, and the pitch. These dimensions vary symmetrically according to the strength required, or, in other words, according to the torque exerted in transmitting power. Because the horsepower and speed of the motor in any case determine the torque, it is evident that these factors determine the proper dimensions of a pinion for the motor. A line of pinions with dimensions increasing symmetrically with the torque will therefore answer the purpose for all combinations of horsepower and speed. Every geared application requires special consideration, since the nature of the service, the shaft diameter, etc., may affect the dimensions of the pinion.

In gear drive the pinion is subject to most rapid wear owing to its smaller diameter. It is as important to have a pinion of good wearing qualities as it is to have one of sufficient strength, and for a pinion of a given material the ability to withstand wear depends mainly, if not wholly, on the width of the face. With a steel pinion and a cast-iron gear, the former is usually the limiting factor of life and the latter the limiting factor of strength.

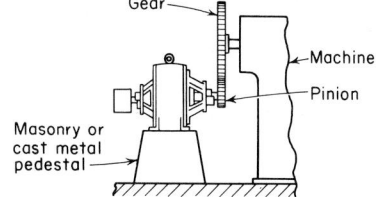

**Fig. 7-193** Motor mounting for gear drive (no limitation to center distance).

**407. Cast-steel gears** are about twice as strong as cast-iron and should be used when the face of a corresponding cast-iron gear would be 4.5 in (114 mm) or more, although the cost is approximately double that of cast-iron gears. With continuous contact along the length of the teeth, the strength of the gear is approximately proportional to the face and the square of the circular pitch, but gear teeth seldom make such contact until they have been worn down in service. With new gears the whole pressure is brought to bear on the high spots, and stripping may occur before they are worn down; hence the necessity of using the stronger material.

**408. Bronze and rawhide pinions.**   For equal strength the working face of rawhide pinions must be about 25 percent wider than corresponding steel pinions. For quiet operation, only the rawhide should be in contact with the gear, although for high-torque motors and other severe service the gears may be widened to cover the entire pinion, thus making use of the metal flanges. If steel pinions would make objectionable noise, rawhide pinions should be used if the stresses permit, since the pitch-line speed with a rawhide pinion is limited more by the rapid wear of the pinion than by noise. A pitch-line speed of 2000 ft/min (10.2 m/s) is considered a fair average limit for rawhide, but speeds of 2500 to 3000 ft/min (12.7 to 15.2 m/s) can be used under especially favorable conditions regarding attendance, lubrication, absence of moisture, or high temperature and for intermittent service or cases in which the life of the pinion is not important.

The wear and noise of bronze pinions are intermediate between those of rawhide and steel. Bronze pinions are particularly adapted to conditions in which heat and moisture prohibit the use of rawhide. Their cost is about the same as that of rawhide.

**409. Noise of gears and pitch-line speed limits.**   Spur gears ordinarily begin to make a noticeable noise at pitch-line speed of about 600 ft/min (3 m/s) but under average conditions may not become disagreeably noisy with pitch-line speeds under 1200 ft/min (6.1 m/s). The amount of noise allowable depends on the noise made by surrounding machinery, the character of the workers, and the nature of the work in the vicinity. A noise that would be unnoticeable in a boiler shop might be exceedingly disagreeable in a shop that was otherwise comparatively quiet. When noise is not a limiting feature, there is no limit to allowable pitch-line speeds except the increased wear and depreciation of the motor, gears, and driven machine, but depreciation may become a very important factor with high pitch-line speeds, say 2500 ft/min (12.7 m/s) or sometimes less. Tests made to determine a design for gears that will give the least noise and yet have sufficient strength and wearing qualities indicate the following facts, other conditions being the same:

1. Gears having large teeth give forth a relatively greater volume of noise at a low pitch that does not carry far, while gears having smaller teeth give forth a smaller volume of noise at a higher pitch that carries farther.

2. Most of the noise comes from the gear and not from the pinion or the motor.

3. A gear designed so that it will give a dead sound when struck a blow with the hammer will be the least noisy in operation.

**410. Conditions for noiseless operation of gears.**   Rigid and massive supports and close-fitting bearings for both the motor and the driven machine are conducive to a noiseless gear drive, and the pinion should always be placed close to the motor bearing. A gear application with the motor mounted upon the ceiling might be twice as noisy as the same application with the motor mounted on a concrete foundation.

**411. Pinions for high-torque motors.**   For series motors and those heavily compounded, such as bending-roll motors, or for motors subject to very severe service of any kind, select a pinion suitable for a constant-speed motor of the same rated rpm but of about 50 percent higher horsepower.

**412. Selection of ratio for back-geared motors.**   A ratio of about 6:1 is usually standard for back-geared motors and should be selected whenever possible, but smaller ratios down to 3:1 or maximum ratios up to 7:1 can be obtained in certain capacities of motors for service if the conditions of the application warrant the use of such ratios.

**413. Outboard bearings for gear drive.**   Outboard bearings should be used for motors of about 40 hp and above in heavy geared service requiring continuous operation with frequent reversing and overloads and also for all motors of about 100 hp and above in any geared service. The proper use of outboard bearings cannot be emphasized too strongly, since increased expense results in a tendency to omit them even when good engineering demands their use.

**414. How to use the chart for determining gear dimensions** (C. W. Drake, *Electric Journal*).   The dimensions for pinions for average conditions of motor-drive service are given in Fig. 7-194. This chart is useful in making preliminary estimates or selections of pinions for geared motors. It applies without correction to *steel pinions only*. The diameters are considered about standard for the various ratings, although both smaller and larger pinions can generally be used, the limiting size for small pinions being the strength and number of teeth, and for large pinions the pitch-line speed.

**example**  To determine the steel pinion for a 5-hp motor at 1200 rpm, find the intersection of the oblique line marked 5 hp with the horizontal line through 1200 rpm. On the vertical line through this intersection may be found a 21.9-lb torque, a 2.3-in pinion face, a 3.2-in pitch diameter, and a diametral pitch of 4.85. A 2.25-in pinion face is good practice here, since pinion-face dimensions with fractions smaller than 0.25 in are not commonly used. The diametral pitch is also usually a whole number except for very large pinions, for which half pitches are sometimes used, so that a pitch of 5 would probably be used in the above case. Since the number of teeth is the product of the pitch diameter and the diametral pitch, the assumed pitch diameter, 3.2 in, is satisfactory with the 5 pitch, because it gives a whole number of teeth, i.e., 16.

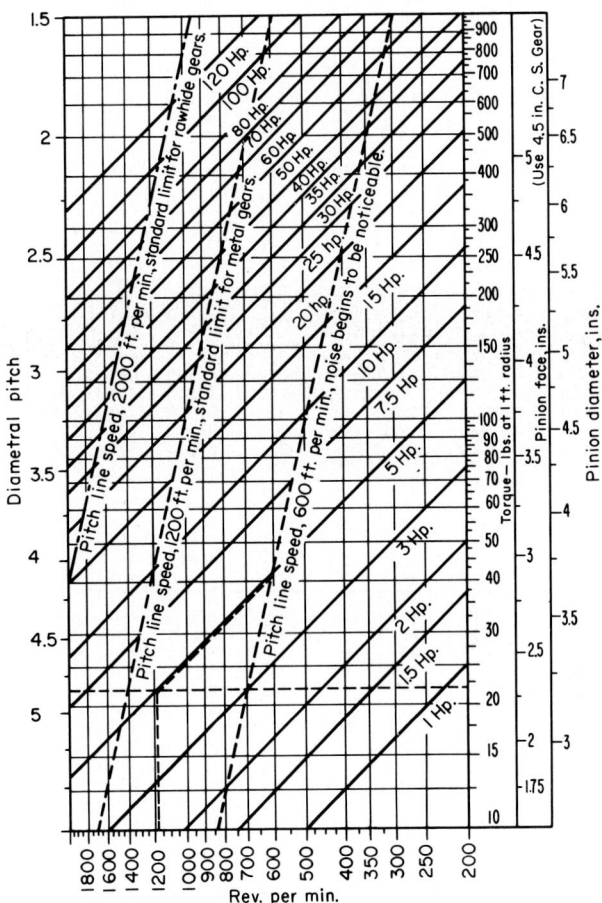

**Fig. 7-194**  Chart for determining approximate commercial values of pinion diameter, pinion face, diametral pitch, pitch-line speed, and torque, with revolutions per minute and horsepower of motor given.

## 415.  Gear Data for Motor Applications
(A. G. Popcke, *American Machinist*)

| Hp | Rpm | Diam. pitch | Number of teeth | | | Face | | Standard pitch-line speed | Min diam | Max no. of teeth for pitch-line speed of | |
|---|---|---|---|---|---|---|---|---|---|---|---|
| | | | Standard pinion | Min rawhide pinion | Min steel pinion | Steel | Rawhide and cloth | | | 1,000 ft per min | 2,000 ft per min |
| 1 | 1,700 | 8 | 17 | 15 | 13 | 1 | 1½ | 940 | 1.63 | 18 | 36 |
| 1 | 1,200 | 8 | 17 | 15 | 13 | 1½ | 2 | 665 | 1.63 | 25 | 50 |
| 2 | 1,700 | 8 | 17 | 15 | 13 | 1½ | 2 | 940 | 1.63 | 18 | 36 |
| 2 | 1,200 | 8 | 22 | 20 | 19 | 1¾ | 2¼ | 870 | 2.38 | 25 | 50 |
| 2 | 850 | 6 | 18 | 21 | 19 | 1¾ | 2¼ | 615 | 2 38 | 36 | 72 |
| 3 | 1,800 | 8 | 22 | 20 | 19 | 1¾ | 2¼ | 1,300 | 2.38 | ... | 34 |
| | 1,150 | 8 | 22 | 21 | 19 | 1¾ | 2¼ | 830 | 2.38 | 26 | 52 |
| | 850 | 6 | 18 | 18 | 18 | 2¾ | 3⅜ | 670 | 3.0 | 27 | 54 |
| 5 | 1,800 | 8 | 22 | 21 | 19 | 1¾ | 2¼ | 1,300 | 2.38 | ... | 34 |
| | 1,200 | 6 | 18 | 18 | 18 | 2¾ | 3⅜ | 940 | 3.0 | 19 | 38 |
| | 850 | 6 | 21 | 19 | 18 | 2¾ | 3⅜ | 990 | 3.0 | 27 | 54 |
| 7½ | 1,700 | 6 | 18 | 18 | 18 | 2¾ | 3⅜ | 1,400 | 3.0 | ... | 26 |
| | 1,150 | 6 | 21 | 19 | 18 | 2¾ | 3⅜ | 1,050 | 3.0 | 20 | 40 |
| | 975 | 5 | 19 | 18 | 18 | 3 | 3¾ | 970 | 3.6 | 19 | 38 |
| | 850 | 5 | 19 | 18 | 18 | 3 | 3¾ | 850 | 3.6 | 22 | 44 |
| | 650 | 5 | 20 | 18 | 18 | 3 | 3¾ | 685 | 3.6 | 29 | 58 |
| 10 | 1,700 | 6 | 21 | 19 | 18 | 2¾ | 3⅜ | 1,420 | 3.0 | ... | 27 |
| | 1,300 | 6 | 22 | 19 | 18 | 2¾ | 3⅜ | 1,250 | 3.0 | ... | 35 |
| | 1,150 | 5 | 19 | 18 | 18 | 3 | 3¾ | 1,150 | 3.6 | ... | 33 |
| | 850 | 5 | 20 | 18 | 18 | 3 | 3¾ | 890 | 3.6 | 22 | 44 |
| | 730 | 5 | 21 | 18 | 18 | 3½ | 4¼ | 805 | 3.6 | 26 | 52 |
| | 600 | 5 | 21 | 19 | 19 | 3½ | 4¼ | 665 | 3.8 | 31 | 62 |
| 15 | 1,700 | 5 | 19 | 18 | 18 | 3 | 3¾ | 1,700 | 3.6 | ... | 22 |
| | 1,250 | 5 | 20 | 18 | 18 | 3 | 3¾ | 1,300 | 3.6 | ... | 30 |
| | 1,150 | 5 | 21 | 18 | 18 | 3½ | 4¼ | 1,210 | 3.6 | ... | 35 |
| | 825 | 5 | 21 | 19 | 19 | 3½ | 4¼ | 910 | 3.8 | 23 | 46 |
| | 675 | 4½ | 22 | 18 | 18 | 4 | 5 | 870 | 4.0 | 25 | 50 |
| | 600 | 4½ | 22 | 19 | 19 | 4 | 5 | 770 | 4.22 | 29 | 58 |
| 20 | 1,700 | 5 | 20 | 18 | 18 | 3 | 3¾ | 1,780 | 3.6 | ... | 22 |
| | 1,100 | 5 | 21 | 19 | 19 | 3½ | 4¼ | 1,220 | 3.8 | ... | 35 |
| | 900 | 4½ | 22 | 18 | 18 | 4 | 5 | 1,150 | 4.0 | 19 | 38 |
| | 750 | 4½ | 22 | 19 | 19 | 4 | 5 | 960 | 4.22 | 23 | 46 |
| | 650 | 4 | 21 | 18 | 18 | 4¼ | 5¼ | 890 | 4.5 | 23 | 46 |
| 25 | 1,400 | 5 | 21 | 19 | 19 | 3½ | 4¼ | 1,550 | 3.8 | ... | 27 |
| | 1,100 | 4½ | 22 | 18 | 18 | 4 | 5 | 1,400 | 4.0 | ... | 31 |
| | 950 | 4½ | 22 | 19 | 19 | 4 | 5 | 1,220 | 4.22 | .. | 36 |
| | 825 | 4 | 21 | 18 | 18 | 4¼ | 5¼ | 1,130 | 4.5 | 18 | 36 |
| | 600 | 4 | 22 | 19 | 18 | 4¼ | 5¼ | 860 | 4.5 | 25 | 50 |
| 30 | 1,700 | 5 | 21 | 19 | 19 | 3½ | 4¼ | 1,880 | 3.8 | ... | 22 |
| | 1,150 | 4½ | 22 | 19 | 19 | 4 | 5 | 1,470 | 4.22 | ... | 30 |
| | 975 | 4 | 21 | 18 | 18 | 4¼ | 5¼ | 1,330 | 4.5 | ... | 31 |
| | 725 | 4 | 22 | 19 | 18 | 4¼ | 5¼ | 1,050 | 4.5 | 21 | 42 |
| | 600 | 3¼ | 20 | ... | 18 | 4½ | ... | 970 | 5.53 | 20 | 40 |
| 35 | 1,700 | 4½ | 22 | 18 | 18 | 4 | 5 | 2,180 | 4.0 | ... | 20 |
| | 1,150 | 4 | 21 | 18 | 18 | 4¼ | 5¼ | 1,580 | 4.5 | ... | 27 |
| | 850 | 4 | 22 | 19 | 18 | 4¼ | 5¼ | 1,220 | 4.5 | 18 | 36 |
| | 675 | 3¼ | 20 | ... | 18 | 4½ | ... | 1,080 | 5.53 | 18 | 36 |
| 40 | 1,700 | 4½ | 22 | 19 | 19 | 4 | 5 | 2,180 | 4.22 | ... | 20 |
| | 950 | 4 | 22 | 19 | 18 | 4¼ | 5¼ | 1,370 | 4.5 | ... | 32 |
| | 775 | 3¼ | 20 | ... | 18 | 4½ | ... | 1,250 | 5.53 | ... | 32 |
| | 600 | 3 | 18 | ... | 15 | 4½ | ... | 940 | 5.0 | 19 | 38 |
| 50 | 1,700 | 4 | 21 | 18 | 18 | 4¼ | 5¼ | 2,340 | 4.5 | ... | 18 |
| | 975 | 3¼ | 20 | ... | 18 | 4½ | ... | 1,580 | 5.53 | ... | 25 |
| | 750 | 3 | 18 | ... | 15 | 4½ | ... | 1,170 | 5.0 | 15 | 30 |
| | 565 | 3 | 20 | ... | 18 | 4½ | ... | 990 | 6.0 | 20 | 40 |

**416.   Adjustable-Speed Motor Ratings and Pulley and Gear Data for Use When Connecting Adjustable-Speed Motors to Drive Machinery**

(A. G. Popcke, *American Machinist*)

| Hp | Rpm Min | Rpm Max | Smallest pulley Diam | Smallest pulley Face | Gear data Pitch | Face Steel | Face Raw-hide | Min teeth | Min diam | Pitch-line speed at min diam Min speed | Max speed | Max teeth not to exceed 2,000 ft per min at max speed |
|---|---|---|---|---|---|---|---|---|---|---|---|---|
| 1 | 740 | 2,200 | 3 | 3 | 8 | 1¾ | 2¼ | 19 | 2.38 | 460 | 1,380 | 27 |
|   | 600 | 1,800 | 3 | 3 | 8 | 1¾ | 2¼ | 19 | 2.38 | 375 | 825 | 46 |
|   | 450 | 1,800 | 3 | 4 | 8 | 1¾ | 2¼ | 19 | 2.38 | 280 | 1,120 | 34 |
| 2 | 1,100 | 2,200 | 3 | 3 | 8 | 1¾ | 2¼ | 19 | 2.38 | 690 | 1,380 | 27 |
|   | 740 | 2,200 | 3 | 4 | 8 | 1¾ | 2¼ | 19 | 2.38 | 460 | 1,380 | 27 |
|   | 450 | 1,800 | 4 | 4½ | 6 | 2¾ | 3⅜ | 18 | 3.0 | 355 | 1,420 | 25 |
| 3 | 1,000 | 2,000 | 3 | 4 | 8 | 1¾ | 2¼ | 19 | 2.38 | 625 | 1,250 | 30 |
|   | 660 | 2,000 | 4 | 4½ | 6 | 2¾ | 3⅜ | 13 | 3.0 | 520 | 1,560 | 23 |
|   | 450 | 1,800 | 4½ | 5 | 6 | 2¾ | 3⅜ | 18 | 3.0 | 355 | 1,422 | 25 |
|   | 375 | 1,500 | 5 | 6 | 6 | 2¾ | 3⅜ | 18 | 3.0 | 294 | 1,176 | 30 |
| 5 | 1,000 | 2,000 | 4 | 4½ | 6 | 2¾ | 3⅜ | 18 | 3.0 | 790 | 1,580 | 23 |
|   | 750 | 1,500 | 4½ | 5 | 6 | 2¾ | 3⅜ | 18 | 3.0 | 590 | 1,180 | 30 |
|   | 600 | 1,800 | 5 | 6 | 6 | 2¾ | 3⅜ | 18 | 3.0 | 470 | 1,410 | 25 |
|   | 450 | 1,800 | 6 | 7 | 5 | 3 | 3¾ | 18 | 3.6 | 425 | 1,700 | 21 |
|   | 375 | 1,500 | 6 | 7½ | 5 | 3½ | 4¼ | 18 | 3.6 | 355 | 1,420 | 25 |
| 7½ | 900 | 1,800 | 5 | 6 | 6 | 2¾ | 3⅜ | 18 | 3.0 | 705 | 1,410 | 25 |
|   | 800 | 1,600 | 5 | 6 | 5 | 3 | 3¾ | 18 | 3.6 | 755 | 1,510 | 24 |
|   | 600 | 1,800 | 6 | 7 | 5 | 3 | 3¾ | 18 | 3.6 | 570 | 1,710 | 21 |
|   | 500 | 1,500 | 6 | 7½ | 5 | 3½ | 4¼ | 18 | 3.6 | 475 | 1,425 | 25 |
|   | 450 | 1,800 | 6½ | 9 | 5 | 3½ | 4¼ | 19 | 3.8 | 450 | 1,800 | 21 |
|   | 350 | 1,400 | 6½ | 9 | 5 | 3½ | 4¼ | 19 | 3.8 | 350 | 1,400 | 27 |
| 10 | 850 | 1,700 | 6 | 7 | 5 | 3 | 3¾ | 18 | 3.6 | 800 | 1,600 | 22 |
|   | 750 | 1,500 | 6 | 7 | 5 | 3 | 3¾ | 18 | 3.6 | 710 | 1,420 | 25 |
|   | 600 | 1,800 | 6 | 7½ | 5 | 3½ | 4¼ | 18 | 3.6 | 570 | 1,710 | 21 |
|   | 500 | 1,500 | 6½ | 9 | 5 | 3½ | 4¼ | 19 | 3.8 | 500 | 1,500 | 25 |
|   | 450 | 1,800 | 6½ | 9 | 5 | 3½ | 4¼ | 19 | 3.8 | 450 | 1,800 | 21 |
|   | 375 | 1,500 | 7 | 8 | 4½ | 4 | 5 | 18 | 4.0 | 390 | 1,560 | 23 |
| 15 | 780 | 1,560 | 6½ | 9 | 5 | 3½ | 4¼ | 19 | 3.8 | 780 | 1,560 | 24 |
|   | 600 | 1,200 | 7 | 8 | 4½ | 4 | 5 | 18 | 4.0 | 630 | 1,260 | 28 |
|   | 500 | 1,500 | 7½ | 9½ | 4½ | 4 | 5 | 19 | 4.22 | 555 | 1,665 | 23 |
|   | 400 | 1,200 | 8 | 9½ | 4 | 4¼ | 5¼ | 18 | 4.5 | 470 | 1,410 | 25 |
|   | 375 | 1,500 | 9 | 10½ | 4 | 4¼ | 5¼ | 18 | 4.5 | 440 | 1,760 | 20 |
| 20 | 650 | 1,300 | 7½ | 9½ | 4½ | 4 | 5 | 19 | 4.22 | 720 | 1,440 | 26 |
|   | 550 | 1,100 | 8 | 9½ | 4 | 4¼ | 5¼ | 18 | 4.5 | 645 | 1,290 | 28 |
|   | 500 | 1,500 | 9 | 10½ | 4 | 4¼ | 5¼ | 18 | 4.5 | 590 | 1,770 | 20 |
|   | 400 | 1,200 | 10 | 11 | 3¼ | 4½ | ... | 18 | 5.53 | 580 | 1,740 | 21 |
|   | 300 | 1,200 | 12 | 13 | 3 | 4½ | ... | 15 | 5.0 | 390 | 1,560 | 19 |
| 25 | 550 | 1,100 | 9 | 10½ | 4 | 4¼ | 5¼ | 18 | 4.5 | 645 | 1,290 | 28 |
|   | 400 | 1,200 | 12 | 13 | 3 | 4½ | ... | 15 | 5.0 | 525 | 1,575 | 19 |
|   | 300 | 1,200 | 12½ | 15 | 3 | 4½ | ... | 18 | 6.0 | 470 | 1,880 | 19 |
| 30 | 550 | 1,100 | 10 | 11 | 3¼ | 4½ | ... | 18 | 5.53 | 800 | 1,600 | 23 |
|   | 350 | 1,050 | 12½ | 15 | 3 | 4½ | ... | 18 | 6.0 | 550 | 1,650 | 22 |
|   | 250 | 1,000 | 14 | 18 | 3 | 4½ | ... | 18 | 6.0 | 390 | 1,560 | 23 |
| 40 | 550 | 1,100 | 12 | 13 | 3 | 4½ | ... | 15 | 5.0 | 720 | 1,440 | 21 |
|   | 350 | 1,050 | 12½ | 15 | 3 | 4½ | ... | 18 | 6.0 | 550 | 1,650 | 22 |
|   | 250 | 1,000 | 16 | 21 | 3 | 4½ | ... | 19 | 6.33 | 415 | 1,660 | 23 |
| 50 | 500 | 1,000 | 12½ | 15 | 3 | 4½ | ... | 18 | 6.0 | 790 | 1,580 | 23 |
|   | 325 | 975 | 16 | 21 | 3 | 4½ | ... | 19 | 6.33 | 540 | 1,620 | 23 |

**417. Gearing definitions and formulas.** A circle whose circumference passes through the point of contact on each tooth of a gear or pinion when this point is on the line connecting the centers of the two wheels is called the *pitch circle*. The diameter of this is the *pitch diameter,* and its circumference is the *pitch line.*

*Diameter,* when applied to gears, is always understood to mean pitch diameter.

*Diametral pitch* is the number of teeth to each inch of the pitch diameter. To illustrate, if a pinion has 18 teeth and the pitch diameter is 3 in, there are 6 teeth to each inch of the pitch diameter, and the diametral pitch is 6.

*Circular pitch* is the distance from the center of one tooth to the center of the next, measured along the pitch line.

In the following formulas, for use in gear problems,

$d_1$ = pitch diameter of pinion  $D^1$ = pitch diameter of gear
$d$ = outside diameter of pinion  $D$ = outside diameter of gear
$p$ = circular pitch       $n$ = number of teeth on pinion
$p^1$ = diametral pitch      $N$ = number of teeth on gear

$S$ = distance between centers  $r$ = gear ratio $= \dfrac{N}{n} = \dfrac{D^1}{d^1} = \dfrac{D}{d}$
  = $\frac{1}{2}(D^1 + d^1)$

$$\pi = 3.1416$$

$$p = \frac{\pi}{p^1} = \frac{\pi d}{n + 2} \qquad (17)$$

$$p^1 = \frac{\pi}{p} = \frac{n + 2}{d} \qquad (18)$$

$$n = d^1 p^1 = \frac{\pi d^1}{p} = dp^1 - 2 \quad (19)$$

$$p^1 = \frac{n}{d^1} \qquad (20)$$

$$S = \frac{N + n}{2p^1} \qquad (21)$$

$$d^1 = \frac{2S}{r + 1} \qquad (22)$$

$$D^1 = \frac{2S\,r}{r + 1} \qquad (23)$$

$$n = \frac{2Sp^1}{r + 1} \qquad (24)$$

$$N = \frac{3Sp^1 r}{r + 1} \qquad (25)$$

$$S = \frac{d^1(r + 1)}{2} \qquad (26)$$

**418. Gear and belt drives for adjustable-speed motors.**  Table 415 deals with constant-speed motors. Adjustable-speed–motor problems are solved similarly. The belt speeds and pitch-line speeds must be carefully considered at the maximum speeds of the motors. The minimum pulleys and pinions are determined by the minimum speeds of the motors. Table 416 gives the ratings commonly used and pulley and gear information.

**419. The design of chains** is more complicated than that of belts and gears, and it is therefore best to let the various chain manufacturers specify the chain, giving them the necessary information. The minimum sprocket to be used on a motor as specified by NEMA standards is given in Sec. 420. Chain speed should not exceed 1200 to 1600 ft/min (6.1 to 8.1 m/s). The best practice does not exceed 1000 ft/min (5.1 m/s). For factors that must be considered see Sec. 332.

**420. Application of sprocket dimensions to general-purpose motors.**  General-purpose motors having continuous-time rating with the frame sizes, horsepower, and speed ratings given in the following paragraphs are designed to operate with chain drives within the limiting sprocket dimensions listed. Selection of sprocket dimensions is made by the chain-drive vendor and the motor purchaser, but to assure satisfactory motor operation the selected pitch diameter shall not be smaller than the dimensions listed.

**A. Direct-current Medium Motors**

| Frame number | Horsepower at, rpm | | | | | | | Chain sprocket min pitch diam, in. |
|---|---|---|---|---|---|---|---|---|
| | 3,500 | 1,750 | 1,150 | 850 | 690 | 575 | 500 | |
| 203 | 1½ | 1 | ¾ | ..... | ..... | .... | .... | 2.04 |
| 204 | 2 | 1½ | 1 | ½ | ..... | .... | .... | 2.04 |
| 224 | 3 | 2 | 1½ | ¾ | ½ | .... | .... | 2.04 |
| 225 | 5 | 3 | 2 | 1 | ¾ | ½ | .... | 2.04 |
| 254 | 7½ | 5 | 3 | 2–1½ | 1½–1 | 1–¾ | .... | 2.16 |
| 284 | 10 | 7½ | 5 | 3 | 2 | .... | 1–¾ | 2.28 |

B. Single-phase Medium Motors

| Frame number | Horsepower at, rpm | | | | Chain sprocket min pitch diam, in. |
|---|---|---|---|---|---|
| | 3,600 | 1,800 | 1,200 | 900 | |
| 203 | 1½ | 1 | ... | ..... | 2.04 |
| 204 | 2 | 1½ | ¾ | ..... | 2.04 |
| 224 | 3 | 2 | 1 | ½ | 2.04 |
| 225 | 5 | 3 | 1½ | ¾ | 2.04 |
| 254 | 7½ | 5 | 2 | 1½–1 | 2.16 |

C. Medium Motors—Polyphase Induction

| Frame number | Horsepower at, rpm | | | | | | | | Chain sprocket min pitch diam, in. |
|---|---|---|---|---|---|---|---|---|---|
| | 3,600 | 1,800 | 1,200 | 900 | 720 | 600 | 514 | 450 | |
| 110, 208, 220, 440, and 550 Volts | | | | | | | | | |
| 203 | 1½ | 1 | ¾ | ..... | ...... | .... | .... | .... | 2.04 |
| 204 | 2 | 1½ | 1 | ½ | ...... | .... | .... | .... | 2.04 |
| 224 | 3 | 2 | 1½ | ¾ | ½ | .... | .... | .... | 2.04 |
| 225 | 5 | 3 | 2 | 1 | ¾ | ½ | .... | .... | 2.04 |
| 254 | 7½ | 5 | 3 | 2–1½ | 1½–1 | 1–¾ | .... | .... | 2.16 |
| 284 | 10 | 7½ | 5 | 3 | 2 | 1½ | 1–¾ | ¾ | 2.28 |
| 324 | 15 | 10 | 7½ | 5 | 3 | 2 | 1½ | 1 | 2.75 |
| 326 | 20 | 15 | 10 | 7½ | 5 | 3 | 2 | 1½ | 3.11 |
| 364 | .... | 20–25 | 15 | 10 | 7½ | 5 | .... | .... | 3.11 |
| 365 | .... | 30 | 20 | 15 | 10 | 7½ | 3 | 2 | 3.67 |
| 404 | .... | 40 | 25 | 20 | 15 | 10 | 5 | 3 | 3.67 |
| 405 | .... | 50 | 30 | 25 | 20 | 15 | 7½ | 5 | 4.56 |
| 444 | .... | 60 | 40 | 30 | 25 | 20 | 10 | 7½ | 4.56 |
| 445 | .... | 75 | 50 | 40 | 30 | 25 | 15 | 10 | 5.51 |
| 504U | .... | ..... | 60 | 50 | 40 | 30 | 20 | 15 | 5.51 |
| 505 | .... | ..... | 75 | 60 | 50 | 40 | 25 | 20 | 6.10 |
| 2,300 Volts, Three Phase | | | | | | | | | |
| 445 | .... | 50–60 | 40 | ..... | ...... | .... | .... | .... | 5.51 |
| 504U | .... | 75 | 50 | 40 | ...... | .... | .... | .... | 5.51 |
| 505 | .... | ..... | 60 | 50 | ...... | .... | .... | .... | 6.10 |

## MOTOR CIRCUITS

**421.** National Electrical Code **rules definitely specify certain requirements** with respect to the size and location of the component parts of motor branch circuits. The following tables and rules are in accordance with the 1990 edition of the Code. Certain cities have adopted motor wiring tables and rules which may be more exacting than those required by the Code. In such communities these local requirements should be followed in preference to the tables presented herein.

Sections 421 through 445 are the rules for conventional motors and motor circuits. Sections 446 through 454 are the rules for air-conditioning and refrigerating equipment which includes a hermetic refrigerant motor-compressor. See Sec. 446.

In the following sections the terminology of the National Electrical Code with respect to *within sight* is adhered to. A distance of more than 50 ft (15.2 m) is considered equivalent to being *out of sight*, even though there may be no obstructions between the two points.

FULL-LOAD MOTOR CURRENTS   According to the National Electrical Code, whenever the current rating of a motor is used to determine the ampacity (current-carrying capacity) of conductors or the ampere ratings of switches, branch-circuit overcurrent devices, etc., the values given in Tables 7 to 10 of Div. 11, including footnotes, should be used in lieu of the actual current rating marked on a motor nameplate. Motor-overload protection should be based on the motor-nameplate current rating. If a motor is marked in amperes but not in horsepower, the horsepower rating should be assumed to be that corresponding to the value given in Tables 7 to 10 of Div. 11, prorated if necessary.

*Exceptions*

*1. For multispeed motors, the selection of conductors on the line side of the controller shall be based on the highest of the full-load current ratings shown on the motor nameplate; selection of conductors between the controller and the motor, which are energized for that particular speed, shall be based on the motor-nameplate current rating for that speed.*

*2. For torque motors, in which the rated current is the locked-rotor current, this nameplate current shall be used.*

*3. For ac motors used on adjustable-voltage, variable-torque drive systems, the maximum operating current marked on the motor and/or control nameplate shall be used. If the maximum operating current does not appear on the nameplate, the ampacity determination shall be based on 150 percent of the values given in Tables 7 to 10 of Div. 11.*

*4. The incoming branch circuit or feeder to power conversion equipment included as part of an adjustable-speed drive system shall be based on the rated input to the power conversion equipment.*

**421A. Phase converters.** The National Electrical Code defines a phase converter as an electric device that converts single-phase electrical power for the operation of equipment that normally operates from a three-phase electrical supply. Such phase converters are of two types: (1) static with no moving parts, and (2) rotary with an internal rotor that is required to be rotating before load is applied. Single- to three-phase converters are typically used in rural areas with only single-phase utility power and where three-phase motor use is desirable because of the lower cost and greater efficiency of three-phase motors. The National Electrical Code includes rules for three-phase motor circuits operated using phase converters. These rules specify that three-phase motor loads be converted to single-phase for circuits and equipment on the input side of the phase converter by multiplying the three-phase current value specified to be used in computations by the square root of 3 (1.73). For example, where Sec. 437 specifies that the wires of a branch circuit have an ampacity of at least 125 percent of the full-load rated current of a motor, the single-phase branch circuit on the input side of a phase converter supplying only that one three-phase motor would be required to have an ampacity of at least 2.16 times the full-load rated current (1.25 × 1.73 = 2.16). Three-phase horsepower ratings, as in Table 10 of Div. 11, are to be used as single-phase horsepower ratings for the input side of the phase converter, as in Table 8 of Div. 11, because power conversions at the same voltage between single- and three-phase do not require a conversion factor.

**422. Types of motor branch circuits.** Branch circuits supplying motors as required by the National Electrical Code may be divided into three types, as follows: (1) a branch circuit supplying only a single motor, (2) a branch circuit supplying several motors but no other equipment except that directly associated with the motors, and (3) a branch circuit supplying one or more motors, lamps, receptacles, and/or electric heaters, etc. (combination load).

**423. Rating or setting for an individual motor circuit.** The motor branch-circuit short-circuit and ground-fault protective device shall be capable of carrying the starting current of the motor. For motor circuits of 600 V or less, a protective device having a rating or setting not exceeding the values given in Table 43 of Div. 11 shall be permitted. An instantaneous-trip circuit breaker (without time delay) shall be used only if it is adjustable and is part of a listed combination controller having motor overload and also short-circuit and ground-fault protection in each conductor. A motor short-circuit protector shall be permitted in lieu of devices listed in Table 43 of Div. 11 if the motor short-circuit protector is part of a listed combination controller having both motor overload protection and short-circuit and ground-fault protection in each conductor if it will operate at not more than 1300 percent of full-load motor current. An instantaneous-trip circuit breaker or motor short-circuit protector shall be used only as part of a listed combination motor controller which provides coordinated motor branch-circuit overload and short-circuit and ground-fault protection.

When the values for branch-circuit protective devices determined by Table 43 of Div.

11 do not correspond to the standard sizes or ratings of fuses, nonadjustable circuit break-ers, or thermal protective devices or the possible settings of adjustable circuit breakers adequate to carry the load, the next higher size, rating, or setting shall be permitted.

*Exception. When the values for branch-circuit short-circuit and ground-fault protective devices specified in the tables are not sufficient for the starting current of the motor:*

*1. The rating of a non-time-delay fuse not exceeding 600 A shall be permitted to be increased but shall in no case exceed 400 percent of the full-load current.*

*2. The rating of the time-delay (dual-element) fuse shall be permitted to be increased but shall in no case exceed 225 percent of the full-load current.*

*3. The rating of an inverse-time circuit breaker shall be permitted to be increased but in no case shall exceed (a) 400 percent for full-load currents of 100 A or less or (b) 300 percent for full-load currents greater than 100 A.*

*4. Torque motor branch circuits shall be protected at the motor-nameplate current rating.*

*5. The rating of a fuse of 601–6000-A classification shall be permitted to be increased but shall in no case exceed 300 percent of the full-load current.*

For a multispeed motor, a single short-circuit and ground-fault protective device shall be permitted for one or more windings of the motor provided the rating of the protective device does not exceed the above applicable percentage of the nameplate rating of the smallest winding protected.

*Exception: For a multispeed motor, a single short-circuit and ground-fault protective device shall be permitted to be used and sized according to the full-load current of the highest current winding, provided that each winding is equipped with individual overload protection sized according to its full-load current and that the branch-circuit conductors feeding each winding are sized according to the full-load current of the highest full-load current winding.*

When maximum branch-circuit protective-device ratings are shown in the manufactur-er's overload-relay table for use with a motor controller or are otherwise marked on equipment, they shall not be exceeded even if higher values are allowed as shown above.

**424. Several motors or loads on one branch circuit.** Two or more motors or one or more motors and other loads shall be permitted to be connected to the same branch circuit under the conditions of 1, 2, or 3 below:

1. *Not more than 1 hp.* Several motors each not exceeding 1 hp in rating shall be permitted on a branch circuit protected at not more than 20 A at 125 V or less, or 15 A at more than 125 V but not more than 600 V if all the following conditions are met:

   *a.* The full-load rating of each motor does not exceed 6 A.

   *b.* The rating of the branch-circuit protective device marked on any of the controllers is not exceeded.

   *c.* Individual running overload protection conforms to Sec. 430.

2. *If the smallest motor is protected.* If the branch-circuit protective device is selected not to exceed that allowed by Sec. 423 for the motor of the smallest rated motor, two or more motors or one or more motors and other load or loads, with each motor having individual overload protection, shall be permitted to be connected to a branch circuit when it can be determined that the branch-circuit protective device will not open under the most severe normal conditions of service that might be encountered.

3. *Other group installation.* Two or more motors of any rating or one or more motors and other load or loads, with each motor having individual overload protection, shall be permitted to be connected to one branch circuit when the motor controller or controllers and overload device or devices are (*a*) installed as a listed factory assembly and the motor branch-circuit short-circuit and ground-fault protective device is either factory-provided as part of the assembly or specified by a marking on the assembly, or (*b*) the motor branch-circuit short-circuit and ground-fault protective device, the motor controller or controllers, and the overload device or devices are field-installed as separate assemblies listed for such use and provided with manufacturers' instructions for use with each other, or (*c*) all the following conditions are met:

   *a.* Each motor-running overload device is listed for group installation with a speci-fied maximum rating for fuse and/or inverse-time circuit breaker.

   *b.* Each motor controller is listed for group installation with a specified maximum rating of fuse and/or circuit breaker.

*c.* Each circuit breaker is of the inverse-time type and is listed for group installation.

*d.* The branch circuit shall be protected by fuses or inverse-time circuit breakers having a rating not exceeding that specified in Sec. 423 for the highest rated motor connected to the branch circuit plus an amount equal to the sum of the full-load current ratings of all other motors and the ratings of other loads connected to the circuit. When this calculation results in a rating less than the ampacity of the supply conductors, it shall be permitted to increase the maximum rating of the fuses or circuit breaker to a value not exceeding the next standard size higher than the ampacity of the conductors if this rating does not exceed 800 A.

*e.* The branch-circuit fuses or inverse-time circuit breakers are not larger than allowed by condition 12 of Sec. 430 for the overload relay protecting the smallest rated motor of the group.

*4. Single-motor taps.* For group installations described in 1, 2, and 3 above, the conductors of any tap supplying a single motor shall not be required to have an individual branch-circuit protective device provided it complies with either of the following: (*a*) no conductor to the motor shall have an ampacity less than that of the branch-circuit conductors; or (*b*) no conductor to the motor shall have an ampacity less than one-third that of the branch-circuit conductors, with a minimum in accordance with Sec. 437, the conductors to the motor-running overload protective device being not more than 25 ft (7.6 m) long and being protected from physical damage.

**425. Motors on general-purpose branch circuits.**    Overload protection for motors used on general-purpose branch circuits shall be provided as specified in 1, 2, 3, or 4 below.

*1. Not more than 1 hp.* One or more motors without individual overload protection shall be permitted to be connected to general-purpose branch circuits only when the installation complies with the limiting conditions specified in conditions 1*a* and 1*b* of Sec. 424.

*2. More than 1 hp.* Motors of larger ratings than specified in condition 1 of Sec. 424 may be connected to general-purpose branch circuits only when each motor is protected by overload protection selected to protect the motor as specified in Sec. 430. Both the controller and the motor-running overload device shall be approved for group installation with the short-circuit and ground-fault protective device selected in accordance with Sec. 424.

*3. Cord- and plug-connected.* Where a motor is connected to a branch circuit by means of an attachment plug and receptacle and individual overload protection is omitted as provided in condition 1 above, the rating of the attachment plug and receptacle shall not exceed 15 A at 125 V or 10 A at 250 V. Where individual overload protection is required as provided in condition 2 for a motor or motor-operated appliance that is attached to the branch circuit with an attachment plug and receptacle, the overload device shall be an integral part of the motor or of the appliance. The rating of the attachment plug and receptacle shall determine the rating of the circuit to which the motor may be connected, as is provided in Article 210 of the National Electrical Code.

*4. Time delay.* The branch-circuit short-circuit and ground-fault protective device protecting a branch circuit to which a motor or motor-operated appliance is connected shall have sufficient time delay to permit the motor to start and accelerate its load.

**426. Multimotor and combination load equipment.**    The rating of the branch-circuit short-circuit and ground-fault protective device for multimotor and combination load equipment shall not exceed the rating marked on the equipment.

**427. The component parts of a motor branch circuit for an individual motor are shown in Fig. 7-195.**

$P_1$: *Overcurrent protective device in feeder or main.* It may consist of fuses or a circuit breaker. Refer to Sec. 440.

$S_B$: *Branch-circuit switch.* It is not required by National Electrical Code rules. It may consist of a switch or a circuit breaker. It must have an ampere rating at least 115 percent of the full-load current rating of the motor. For ac motors of 100 hp or less and dc motors of 40 hp or less, the switch must be of the motor-circuit type. If a switch is used, it must have fuse clips large enough to accommodate fuses required at $P_2$. If a circuit breaker is used as a branch-circuit short-circuit and ground-fault protective device, it will serve the double purpose of a branch-circuit switch and a circuit protective device. Refer to Sec. 436.

$P_2$: *Branch-circuit short-circuit and ground-fault protective device.* It may consist of fuses or a circuit breaker. If a circuit breaker is used, it will serve the double purpose of a branch-circuit switch and an overload protective device. Refer to Sec. 435.

$S_M$: *Motor switch.* It may consist of a switch or a circuit breaker. It must have a rating and be of a type as described for $S_B$. If the switch is fused, the fuse clips must be large enough to accommodate the size of fuses required at $P_3$. Switch $S_M$ must be located within sight of motor controller $C$. If switch $S_M$ is of the two-throw or two-position type (Diagram II) for cutting the motor overload protective device out of the circuit during the starting period, the switch must be so arranged that it cannot be left in the starting position. It must be within sight from the location of controller $C$, and it must also be within sight from the motor and driven machinery location unless it is capable of being locked in the open position.

$P_3$. *Motor-overload protective device.* It may consist of fuses, thermal relays, magnetic relays, circuit breakers, or a thermal protector in the motor. It generally is not required for intermittent-duty motors. Except for a thermal protector in the motor, the device shall be rated or selected to trip at not more than 125 percent of the motor full-load current rating for motors marked with a service factor not less than 1.15 or marked to have a temperature rise not over 40°C and at not more than 115 percent for all other types of motors. This value may be modified as permitted by Sec. 430, condition 6. The motor-overload protective device is generally incorporated in controller $C$ and mounted inside the controller case. Refer to Secs. 430 to 432.

$C$: *Motor controller.* Refer to Sec. 429.

$W_1$: *Branch-circuit tap wires from point of connection to feeder to branch-circuit overload protective device.* If a branch-circuit short-circuit and ground-fault protective device is located at the point of connection to the feeder, the ampacity of the tap wires must be at least 125 percent of the motor full-load current. If the length $L$ of the wires from the point of connection to the feeder to the branch-circuit short circuit and ground-fault protective device is 25 ft (7.6 m) or less, the ampacity of the tap wires must be at least one-third of the ampacity of the feeder. If the length $L$ of the wires from the point of connection to the feeder to the branch-circuit short-circuit and ground-fault protective device is greater than 25 ft, the tap wires must be of the same size as the feeder, except in high-bay manufacturing buildings over 35 ft (10.67m) high at walls, where the length $L$ is permitted to be not over 25 ft long horizontally and not over 100 ft (30.48 m) in total length under special conditions detailed in the exception to Sec. 430-28 of the 1990 National Electrical Code.

$W_2$: *Branch-circuit wires from motor-overload overload protective device to motor.* They must have an ampacity of at least 125 percent of the motor full-load current for continuous-duty motors. Refer to Sec. 437.

**428. The component parts of a motor branch circuit supplying several motors are shown in Fig. 7-196.**

$P_1$: *Short-circuit and ground-fault protective device in branch circuit.* It may consist of fuses or an inverse-time circuit breaker listed for group installation. If the motor-overload protective device consists of overload relays, the rating of the fuses or the setting of the circuit breaker must not be greater than is permitted in condition 12 of Sec. 430 unless the motor-overload protective device $P_3$ and controller $C$ are listed for group installation.

If the motor-overload protective devices are cut out during the starting period, $P_1$ must consist of fuses or an inverse-time circuit breaker rated not more than 400 percent of the full-load current of the smallest motor so protected. Refer to Sec. 435.

$S_M$: *Combined motor and branch-circuit switch.* It

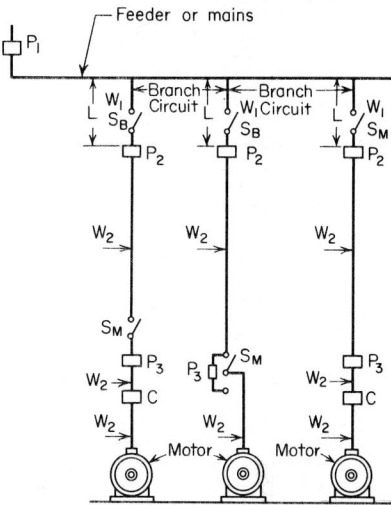

**Fig.   7-195** Single-line diagrams giving requirements for component parts of branch motor circuits with branch-circuit short-circuit and ground-fault protection.

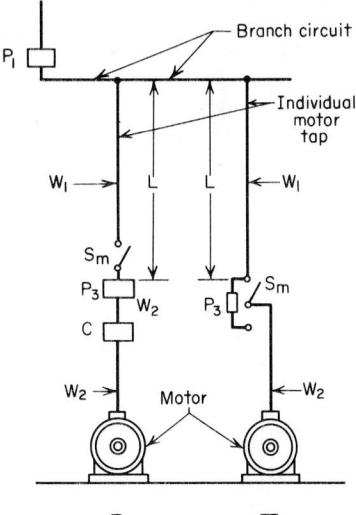

**Fig. 7-196** Single-line diagrams giving requirements for motor circuits when a branch-circuit short-circuit and ground-fault protective device is used for more than one motor.

may consist of a switch or a circuit breaker. It must have a continuous-duty rating of at least 115 percent of the motor full-load current. For ac motors of 100 hp or less and dc motors of 40 hp or less and for all cases in which a switch is used for shutting down the motor, the switch must be of the motor-circuit type. If the switch is fused, the fuse clips must be large enough to accommodate the size of fuses required at $P_3$. If the switch is used for shutting down the motor, it should be located as close as possible to the motor and its driven machinery and in the most accessible location for the operator. If the switch is not used for shutting down the motor, it may be remote from the motor provided that it is located within sight from the controller $C$ and is either located within sight of the motor and its driven machinery or so arranged that it can be locked in the open position. If the switch is of the two-position or double-throw type (Diagram II) for cutting the motor overload protective device out of the circuit during the starting period, it must be so arranged that it cannot be left in the starting position. If $S_M$ is a fused switch or circuit breaker, the supply from $P_1$ is a feeder. Refer to Sec. 436.

$P_3$. *Motor-overload protective device.* It may consist of fuses, thermal relays, magnetic relays, a circuit breaker, or a thermal protector in the motor. Except for a thermal protector in the motor, the device shall be rated or selected to trip at not more than 125 percent of the motor full-load current rating for motors marked with a service factor not less than 1.15 or marked to have a temperature rise not over 40°C and at not more than 115 percent for all other types of motors. This value may be modified as permitted by Sec. 430, condition 6. The motor-overload protective device is generally incorporated in the controller $C$ and mounted inside the controller case. Refer to Secs. 424 and 430 to 432.

$C$: *Motor controller.* Refer to Secs. 424 and 429.

$W_1$: *Individual motor tap wires from points of connection to branch circuit to motor-overload protective device.* If the motor-overload protective device is located at the point of connection to the branch circuit, the ampacity of the wires must be at least 125 percent of the motor full-load current. If the length $L$ of the wires is 25 ft (7.6 m) or less, the ampacity of the wires must be at least one-third of the ampacity of the branch-circuit wires. If the length $L$ of the wires is greater than 25 ft, the wires must be of the same size as the branch-circuit wires, except in high-bay manufacturing buildings as described in Sec. 427.

**429. Motor controllers.** Each motor must be provided with a suitable controller that can start and stop the motor and perform any other control functions required for the satisfactory operation of the motor, such as speed control. The different types of motor-control equipment are discussed in Secs. 251 to 325. A controller must be capable of interrupting the current of the motor under stalled-rotor conditions. Each motor must have its own individual controller except that, for motors of 600 V or less, a single controller may serve a group of motors under any one of the following conditions:

1. If a number of motors drive several parts of a single machine or a piece of apparatus such as metalworking and woodworking machines, cranes, hoists, and similar apparatus.

2. If a group of motors is under the protection of one overcurrent device as permitted in condition 1 for motor branch circuits as given in Secs. 424 and 425.

3. If a group of motors is located in a single room within sight of the controller location. A distance of more than 50 ft (15.2 m) is considered equivalent to being out of sight.

The controller must have a horsepower rating that is not less than that of the motor, except as follows:

1. *Stationary motor of ⅛ hp or less.* For a stationary motor rated at ⅛ hp or less that is normally left running and is so constructed that it cannot be damaged by overload or failure to start, such as clock motors and the like, the branch-circuit protective device may serve as the controller.

2. *Stationary motor of 2 hp or less.* For a stationary motor rated at 2 hp or less and 300 V or less, the controller may be a general-use switch having an ampere rating at least twice the full-load current rating of the motor. On ac circuits, general-use snap switches suitable for use only on ac (not general-use ac, dc snap switches) may be employed to control a motor having a full-load current rating not exceeding 80 percent of the ampere rating of the switch.

3. *Portable motor of ⅓ hp or less.* For a portable motor rated ⅓ hp or less, the controller may be an attachment plug and receptacle.

4. *Circuit breaker as controller.* A branch-circuit circuit breaker, rated in amperes only, may be used as a controller. When this circuit breaker is also used for short-circuit and ground-fault and/or overload protection, it shall conform to the appropriate provisions of this section governing the type of protection afforded.

A controller need not open all conductors to the motor unless it serves as a disconnecting means or unless one pole of the controller is placed in a permanently grounded conductor, in which case the controller must be designed so that the pole in the grounded conductor cannot be opened without simultaneously opening all conductors of the circuit.

Machines of the following types shall be provided with speed-limiting devices or other speed-limiting means unless the inherent characteristics of the machines, the system, or the load and the mechanical connection thereto are such as to limit the speed safely or unless the machine is always under the manual control of a qualified operator:

1. Separately excited dc motors
2. Series motors
3. Motor-generators and converters that can be driven at excessive speed from the dc end, as by a reversal of current or a decrease in load

Adjustable-speed motors, if controlled by field regulation, shall be so equipped and connected that they cannot be started under a weakened field unless the motors are designed for such starting.

The controller for a motor should be as near as possible to the motor to make the length of the wires between the controller and the motor as short as possible. For motor circuits 600 V or less, a disconnecting means must be within sight from the controller location, and must also be within sight from the location of the motor and its driven machinery unless the controller disconnecting means is capable of being locked in the open position. For motor circuits over 600 V, the controller disconnecting means is permitted to be out of sight from the controller, provided the controller is marked with a warning label giving the location and identification of the disconnecting means to be locked in the open position. See Sec. 433 for type and rating of the disconnecting means.

The control point should be located so that it will be in the most accessible location for the operator of the machine driven by the motor. In the case of manual controllers the location of the control point will be the same as the location of the controller. For magnetic controllers the location of the control point will be the point of location of the START and STOP pushbuttons, which may or may not be the same as the location of the controller.

For motors rated over 600 V, the controller must have a continuous ampere rating not less than the current at which the motor-circuit overload protective device is selected to trip. See Sec. 435.

**430. Overload protection of motors.** Each continuous-duty motor must be protected against excessive overloads under running conditions by some approved protective device. This protective device, except for motors rated at more than 600 V, may consist of thermal relays, thermal release devices, or magnetic relays in connection with the motor controller, fuses, an inverse-time circuit breaker, or a thermal protector built into the motor.

Except for a thermal protector in the motor, the **National Electrical Code** requires that the protection be as follows:

1. MORE THAN 1 HP   Each continuous-duty motor rated more than 1 hp shall be protected against overload by one of the following means:

*a.* A separate overload device that is responsive to motor current. This device shall be rated or selected to trip at no more than the following percentage of the motor-nameplate full-load current rating:

|  | *Percent* |
|---|---|
| Motors with a marked service factor not less than 1.15 | 125 |
| Motors with a marked temperature rise not over 40°C | 125 |
| All other motors | 115 |

For a multispeed motor, each winding connection shall be considered separately. This value may be modified as permitted by Par. 6.

When a separate motor-running overload device is so connected that it does not carry the total current designated on the motor nameplate, as for Y-delta starting, the proper percentage of nameplate current applying to the selection or setting of the overload device shall be clearly designated on the equipment, or the manufacturer's selection table shall take this into account.

*b.* A thermal protector integral with the motor, approved for use with the motor which it protects on the basis that it will prevent dangerous overheating of the motor due to overload and failure to start. The **National Electrical Code** specifies the level of protection to be provided by the motor manufacturer. If the motor current-interrupting device is separated from the motor and its control circuit is operated by a protective device integral with the motor, it shall be so arranged that the opening of the control circuit will result in an interruption of current to the motor.

*c.* A protective device integral with the motor that will protect the motor against damage due to failure to start shall be permitted if the motor is part of an approved assembly that does not normally subject the motor to overloads.

*d.* For motors larger than 1500 hp, a protective device employing embedded temperature detectors which cause current to the motor to be interrupted when the motor attains a temperature rise greater than marked on the nameplate in an ambient of 40°C.

2. 1 HP OR LESS, NONAUTOMATICALLY STARTED

*a.* Each continuous-duty motor rated at 1 hp or less that is not permanently installed, is nonautomatically started, and is within sight from the controller location shall be permitted to be protected against overload by the branch-circuit short-circuit and ground-fault protective device. This branch-circuit protective device shall not be larger than that specified in Table 44 of Div. 11, except that any such motor may be used at 125 V or less on a branch circuit protected at not over 20 A.

*b.* Any such motor that is not in sight from the controller location shall be protected as specified in Par. 3. Any motor rated at 1 hp or less that is permanently installed shall be protected in accordance with Par. 3.

3. 1 HP (745.7 W) OR LESS, AUTOMATICALLY STARTED   Any motor of 1 hp or less that is started automatically shall be protected against overload by the use of one of the following means:

*a.* A separate overload device that is responsive to motor current. This device shall be selected to trip or be rated at no more than the following percentage of the motor-name-plate full-load current rating:

|  | *Percent* |
|---|---|
| Motors with a marked service factor not less than 1.15 | 125 |
| Motors with a marked temperature rise not over 40°C | 125 |
| All other motors | 115 |

For a multispeed motor, each winding connection shall be considered separately. This value may be modified as permitted by Par. 6.

*b.* A thermal protector integral with the motor, approved for use with the motor which it protects on the basis that it will prevent dangerous overheating of the motor due to overload and failure to start. When the motor current-interrupting device is separated from the motor and its control circuit is operated by a protective device integral with the motor, it shall be so arranged that the opening of the control circuit will result in interruption of current to the motor.

*c.* A protective device integral with the motor that will protect the motor against damage due to failure to start shall be permitted (1) if the motor is part of an approved assembly that does not normally subject the motor to overloads or (2) if the assembly is also equipped with other safety controls (such as safety combustion controls of a domestic oil burner) that protect the motor against damage due to failure to start. Where the assembly has safety controls that protect the motor, this fact shall be indicated on the nameplate of the assembly where it will be visible after installation.

*d.* When the impedance of the motor windings is sufficient to prevent overheating due to failure to start, the motor may be protected as specified in Par. 2*a* for nonautomatically started motors if the motor is part of an approved assembly in which the motor will limit itself so that it will not be dangerously overheated.

Many ac motors of less than 1/20 hp, such as clock motors, series motors, etc., and also some larger motors such as torque motors come within this classification. It does not include split-phase motors having automatic switches to disconnect the starting windings.

4. WOUND-ROTOR SECONDARIES   The secondary circuits of wound-rotor ac motors, including conductors, controllers, and resistors, shall be considered to be protected against overload by the motor-running overload device.

5. INTERMITTENT AND SIMILAR DUTY   A motor used for service which is inherently short-time, intermittent, periodic, or varying-duty (as illustrated by Table 12 of Div. 11) is considered to be protected against overload by the branch-circuit short-circuit and ground-fault protective device, provided the protective-device rating or setting does not exceed that specified in Table 43 of Div. 11.

Any motor is considered to be for continuous duty unless the nature of the apparatus which it drives is such that the motor cannot operate continuously with load under any condition of use.

6. SELECTION OF SETTING OF OVERLOAD RELAY   When the overload relay selected in accordance with Par. 1a or 3a is not sufficient to start the motor or to carry the load, the next higher size of overload relay shall be permitted to be used. However, it should not exceed 140 percent of the motor full-load current rating of motors marked with a service factor not less than 1.15, and of motors marked to have a temperature rise not over 40°C, and not be higher than 130 percent of the full-load current rating for all other motors. If not shunted during the starting period of the motor, the overload protective device shall have sufficient time delay to permit the motor to start and accelerate its load.

7. SHUNTING DURING STARTING PERIOD   If the motor is manually started (including starting with a magnetic starter having pushbutton control), the running overload protection may be shunted or cut out of the circuit during the starting period of the motor, provided the device by which the overload protection is shunted or cut out cannot be left in the starting position and provided fuses or inverse-time circuit breakers rated or set at not over 400 percent of the full-load current of the motor are so located in the circuit as to be operative during the starting period of the motor. The motor-running overload protection shall not be shunted or cut out during the starting period if the motor is automatically started, except as follows: (a) the motor-starting period exceeds the time delay of available motor-overload protective devices; and (b) listed means are provided to (1) sense motor rotation and automatically prevent shunting or cutout in the event that the motor fails to start, (2) limit the time of overload protection shunting or cutout to less than the locked-rotor time rating of the protected motor, and (3) provide for shutdown and manual restart if the motor-running condition is not reached.

8. FUSES INSERTED IN CONDUCTORS   When fuses are used for motor-running protection, a fuse shall be inserted in each ungrounded conductor. A fuse shall also be inserted in the grounded conductor if the supply system is three-wire, three-phase, ac with one conductor grounded.

9. DEVICES OTHER THAN FUSES INSERTED IN CONDUCTORS   If devices other than fuses are used for motor-running overload protection, Table 41 of Div. 11 shall govern the minimum allowable number and location of overload units such as trip coils or relays.

10. NUMBER OF CONDUCTORS OPENED BY OVERLOAD DEVICE   Motor overload devices other than fuses or thermal protectors shall simultaneously open a sufficient number of underground conductors to interrupt current flow to the motor.

11. MOTOR CONTROLLER AS RUNNING OVERLOAD PROTECTION   A motor controller shall also be permitted to serve as the overload device if the number of overload units complies with Par. 9 and if these units are operative in both the starting and the running positions in the case of a dc motor and in the running position in the case of an ac motor.

12. OVERLOAD RELAYS   Overload relays and other devices for motor-overload protection that are not capable of opening short circuits shall be protected by fuses, circuit breakers with ratings or settings in accordance with Sec. 423, or a motor short-circuit protector, unless listed for group installation, and marked to indicate the maximum size of fuse or inverse-time circuit breaker by which they must be protected.

13. AUTOMATIC RESTARTING   A motor-overload device that can restart a motor automatically after overload tripping shall not be installed unless approved for use with the motor it protects. A motor that can restart automatically after shutdown shall not be installed so that its automatic restarting can result in injury to persons.

**431. Application of fuses for motor-overload protection.** If regular fuses are used for the overload protection of a motor, they must be shunted during the starting period, since a regular fuse having a rating of 125 percent of the motor full-load current would be blown by the starting current. Many cases of dc motor and some wound-rotor–induction-motor installations will be an exception to this rule. Aside from these exceptions it is not common practice to use regular fuses for the overload protection of motors. Time-delay fuses can sometimes be satisfactorily employed for the overload protection of motors, since ones rated at 125 percent of the motor full-load current will not be blown by the starting current for ordinary conditions of motor service. In fact, the manufacturers of these fuses recommend that for ordinary service fuses of a smaller rating than 125 percent of the motor full-load current be employed. Typical heavy-service conditions are:

1. Motors slightly overloaded
2. Motors with severe starting conditions, such as large fans, compressors, etc.
3. Motors which are started frequently or reversed quickly
4. Motors which have the protective time-delay fuses located in places where high ambient temperatures prevail

Even time-delay fuses may not be satisfactory unless they are shunted during the starting period, because the 125 percent value cannot be exceeded and available ratings of fuses may not permit selection close to but not exceeding 125 percent.

Direct-current motors under ordinary conditions of service will not require a starting current of such value that it would blow regular fuses rated at 125 percent of the motor full-load current. Standard manual starters for dc motors are equipped with a starting resistance of a value which will limit the starting current on the first step to 150 percent of the motor full-load current. This current for the normal duration of the starting period would not blow a fuse rated at 125 percent of the full-load current. For ordinary conditions of service, therefore, regular fuses can often be used for the protection of dc motors. In many cases what has been said for dc motors will hold true for wound-rotor induction motors.

**432. Thermally protected motors** are equipped with built-in overload protection mounted directly in the motor. The built-in protector usually consists of a thermostat which when heated above a certain temperature opens two small contacts within the enclosing case of the thermostat. The thermostat is mounted on the windings or the stator iron of a motor. The contacts may open the motor circuit directly. On three-phase motors, the thermostat may be connected to open the center point of a Y-connected motor winding. When the thermostat does not directly open the motor circuit, the contacts are provided with leads so that they may be connected in a control circuit in a manner that will cause the controller to open the motor circuit or give a warning signal when the thermostat operates because of overheating of the motor (see Fig. 7-197).

The thermostat type of thermal protector employs a bimetallic element which warps on a temperature increase, causing the contact to open. A thermal protector may also consist of PTC thermistors, one in each phase and connected in series for three-phase motors (Fig. 7-198).

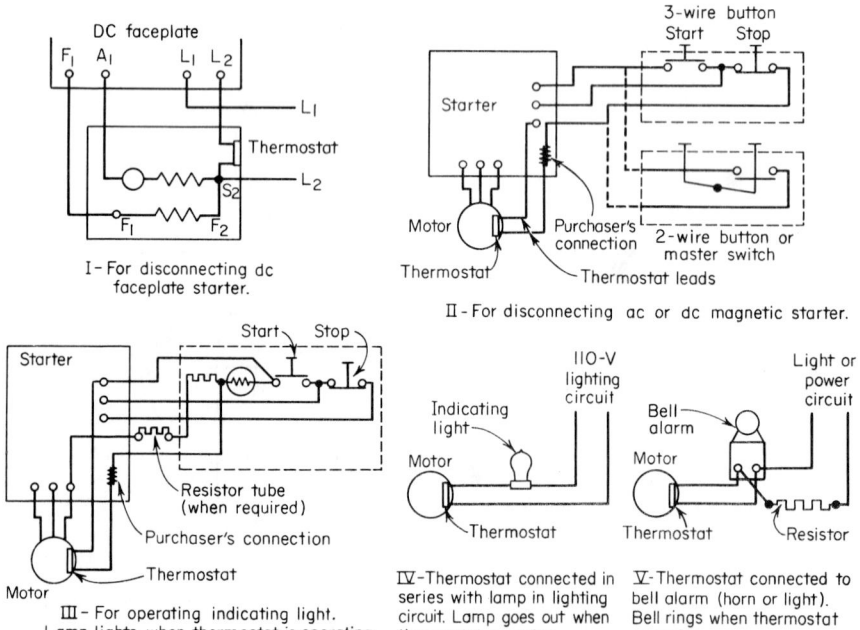

**Fig. 7-197** Connections for thermostats of Thermoguard motors. [Westinghouse Electric Corp.]

At normal winding temperatures, the resistance of the PTC thermistor is low, and it remains virtually constant up to the critical temperature. Beyond this point, a small increase in temperature greatly increases the resistance. When the temperature of any of the PTC thermistors transgresses the critical or switching temperature range, the circuit operates to give a warning or to disconnect the power. When cooling occurs, the circuit returns to normal resistance and the motor can be restarted. Either manual or automatic reset can be used.

On the thermally protected motor designed for warning only, a neon-glow lamp may be mounted on the conduit box for visual warning (see Fig. 7-199, I). This lamp can also be supplied for remote mounting (if specified) for installations in which the motor conduit box would not be visible.

**433. Motor switches.** Each motor with its controller must be provided with some form of approved manual disconnecting means, as at $S_M$ in Figs. 7-195 and 7-196. This disconnecting means, when in the open position, must disconnect both the controller and the motor from all ungrounded supply wires. The switch is also permitted to open a permanently grounded supply wire provided it is so designed that the pole opening the grounded wire cannot be opened without

**Fig. 7-198** Method of locating PTC thermistors for the protection of a three-phase motor. [Westinghouse Electric Corp.]

simultaneously opening all supply wires. It must be a type which will plainly indicate whether it is in the open or the closed position. It may consist of a manually operated air-break switch, a manually operated circuit breaker, a molded case switch (nonautomatic

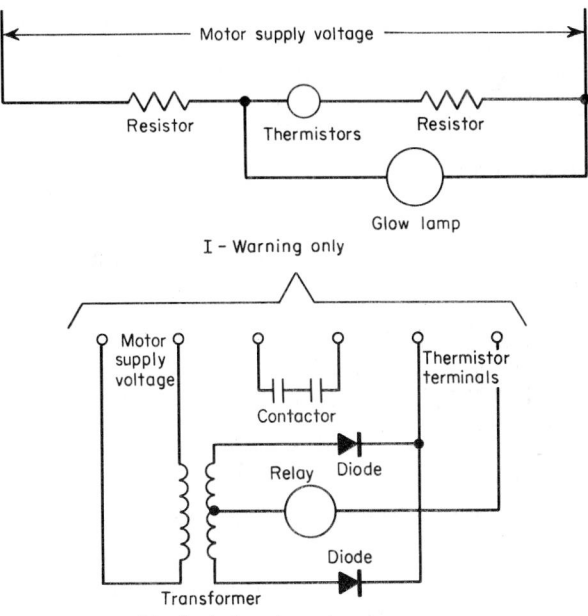

**Fig. 7-199** Basic circuits for Guardistor thermally protected motors. [Westinghouse Electric Corp.]

circuit interrupter), or a manually operated oil switch. The oil switch or circuit breaker may be equipped for electrical operation in addition to the hand mechanism. More than one switch may be used in series, as at $S_B$ and $S_M$ in Fig. 7-195, but at least one of the switches must be readily accessible.

For motors rated more than 600 V, the switch must have a continuous ampere rating not less than the current at which the motor-circuit overload protective device is selected to trip. See Sec. 7-435.

No additional switch is required with most types of manual across-the-line starters, since their switch acts as the disconnecting means. All magnetic types of starters and controllers must be provided with a switch ahead of the starter and within sight of it. All autotransformer starters require a separate disconnecting switch ahead of the starter.

If fuses are used for the overload protection of the motor and the starting current of the motor is greater than the time delay of the fuses, a double-throw or two-position switch will be required to short-circuit the running fuses during the starting period, as at II in Figs. 7-195 and 7-196. This switch must be of such a type that it cannot be left in the starting position.

Every switch in the motor circuit between the point of attachment to the feeder and the point of connection to the motor must be a motor-circuit switch, rated in horsepower, or a circuit breaker except under the following conditions:

1. For stationary motors of ⅛ hp or less, the branch-circuit overcurrent device is permitted to serve as the disconnecting means.

2. For stationary motors of 2 hp or less operating at not more than 300 V, a general-use switch having an ampere rating of at least twice the full-load motor current rating is permitted to be used. On ac circuits general-use snap switches suitable only for use on ac (not general-use ac, dc snap switches) are permitted to be used to disconnect a motor having a full-load current rating not exceeding 80 percent of the ampere rating of the switch.

3. A general-use switch is permitted to be used for motors of a size over 2 hp but not larger than 100 hp which are equipped with a compensator starter, provided that all the following provisions are complied with:

   a. The motor drives a generator provided with overcurrent protection.

   b. The controller is capable of interrupting the stalled-rotor current and is provided with a no-voltage release and with running overload protection not exceeding 125 percent of the motor full-load current rating.

   c. Separate fuses or an inverse-time circuit breaker, rated or set at not more than 150 percent of the motor rated current, is provided in the branch circuit.

4. For stationary motors that are rated at more than 40 hp, dc, or 100 hp, ac, the disconnecting means is permitted to be a general-use switch, or an isolating switch plainly marked "Do not operate under load."

5. For cord- and plug-connected motors, a horsepower-rated attachment plug and receptable having ratings no less than the motor ratings are permitted to serve as the disconnecting means. Horsepower-rated plugs and receptacles are not required for cord- and plug-connected appliances or a portable motor rated ⅓ hp or less.

6. A general-use switch is permitted to serve as the disconnecting means for a torque motor.

A motor-circuit switch is intended for use in a motor branch circuit and is rated in horsepower. It must be capable of interrupting the maximum operating overload current of a motor of the same horsepower as the switch at the rated voltage. It also must have a continuous rating of at least 115 percent of the full-load current of the motor. If a fused switch is used, it must be large enough to accommodate the size of fuses employed.

If the motor switch is used for shutting down the motor, it must be located as close as possible to the motor and its driven machinery and in the most accessible location for the operator. One switch may be employed to serve the double purpose of motor- and branch-circuit switch as shown at $S_M$ in III of Fig. 7-195. If an installation consists of a single motor, the service switch may serve as the disconnecting means provided it meets all the above conditions.

**434. Group switching.** Under any one of the following conditions, the **National Electrical Code** rules will allow a single switch to be provided for a group of motors if the motors are rated at not more than 600 V:

1. When a number of motors drive several parts of a single machine or piece of apparatus, such as metalworking and woodworking machines, cranes, and hoists

2. When a group of motors is under the protection of one set of overcurrent devices as permitted by the Code (see Sec. 430)

3. When a group of motors is in a single room within sight of the location of the disconnecting means

The disconnecting means shall have a horsepower rating not less than is required for a single motor, the rating of which equals the sum of the horsepowers or full-load and locked-rotor currents, whichever results in the largest horsepower rating, of all the motors of the group. See Tables 7 through 10 and Table 47 of Div. 11 for full-load and locked-rotor currents and equivalent horsepower ratings.

**435. Branch-circuit short-circuit and ground-fault protection.** The wires of motor branch circuits must be protected against short circuits and ground faults by a protective device located at the point where the branch-circuit wires are connected to the feeder or main, as at location $P_2$ in Fig. 7-195. National Electrical Code rules, however, allow the size of these branch-circuit protective devices to be considerably greater than the allowable ampacity of the wires. This provision is for the purpose of permitting the passage of the starting current without the employment of excessively large conductors.

With this practice the motor-overload device will protect against overloads of moderate severity, and the branch-circuit protective device will protect against short circuits and ground faults. The recommended maximum rating of fuses and setting of circuit breakers for motor branch circuits supplying an individual motor are given in Tables 42 and 43 of Div. 11.

The National Electrical Manufacturers Association has adopted a standard of identifying code letters that may be marked by the manufacturers on motor nameplates to indicate the motor kilovolt-ampere input with a locked rotor. These code letters with their classification are given in Table 42 of Div. 11. In determining the starting current to employ for circuit calculations, use values from Table 43 of Div. 11.

Although the branch-circuit short-circuit and ground-fault protective device must pass the starting current of the motor, the rating of the protective device need not be as great as the maximum value of the starting current of the motor. Owing to the time-lag characteristics of fuses and inverse-time circuit breakers, the starting current, which lasts for only a short time, will not rupture fuses or operate inverse-time circuit breakers of a considerably smaller rating than the maximum value of the starting current. The percentage of the maximum starting current that fuses must be rated for or inverse-time circuit breakers be set for, so that the starting current will not open the circuit, depends upon the frequency and duration of the starting periods. Under ordinary starting conditions, fuses rated or inverse-time circuit breakers set at from 50 to 67 percent of the maximum starting current will allow the passage of the starting current. See Secs. 423 and 424 for maximum ratings or settings of short-circuit and ground-fault protective devices for motors rated 600 V or less.

Either fuses or circuit breakers may be employed for the protection of motor branch circuits for motors rated at not more than 600 V. When circuit breakers are used, they must have a continuous current rating of not less than 115 percent of the full-load current rating of the motor.

It is often not convenient or practicable to locate the branch-circuit short-circuit and ground-fault protective device directly at the point where the branch-circuit wires are connected to the main. In such cases, the size of the branch-circuit wires between the feeder and the branch-circuit protective device must be the same as that of the mains, unless the length of these wires is not greater than 25 ft (7.6 m). When the length of the branch-circuit wires between the main and the branch-circuit protective device is not greater than 25 ft, Code rules allow the size of these wires to be such that they have an ampacity not less than one-third of the ampacity of the mains (see Fig. 7-196) if they are protected against physical damage.

Each motor and motor branch circuit and feeder of more than 600 V shall be protected by devices coordinated to interrupt automatically the motor-running overcurrent (overload) and fault currents in the motor, the motor-circuit wiring, and the motor controller.

1. *Motor-overload protection.* This protection can consist of one or more thermal protectors integral with the motor or external current-sensing devices, or both. The operation of the overload interrupting device must simultaneously disconnect all ungrounded conductors and should not automatically reset so as to restart the motor after reset unless such automatic restarting cannot be dangerous. The rating of the controller and disconnect switch must be coordinated with the trip current of the overload device.

2. *Fault-current protection.* Protection against fault currents can be obtained by one of the following means:

*a.* A circuit breaker of suitable rating so arranged that it can be serviced without hazard. It must simultaneously disconnect all ungrounded conductors and be able to sense the fault current by integral or external sensing elements.

*b.* Fuses of a suitable type and rating in each ungrounded conductor. Fuses shall be used with suitable disconnecting means, or they shall be of a type that can also serve as the disconnecting means. They shall be so arranged that they cannot be refused or replaced while they are energized.

Differential protection can be employed to provide the protection for both item 1 and item 2. A differential protective system is a combination of two or more sets of current transformers and a relay or relays energized from their interconnected secondaries.

The primaries of the current transformers are connected on both sides of the equipment to be protected, both ends of the motor phase windings being brought out for this purpose. All the apparatus and circuits included between the sets of current transformer primaries constitute the protected zone. The current transformer secondaries and the relay elements are so interconnected that the relay elements respond only to a predetermined difference between the currents entering and leaving the protected zone. When actuated, the relay or relays serve to trip the branch-circuit circuit breaker, thus disconnecting the motor, the control apparatus in the motor circuit, and the branch-circuit conductors from the source of power and, in the case of a synchronous motor, deenergizing its field circuit.

**436. Branch-circuit switches.** National Electrical Code rules do not require a switch to be located on the line side of the short-circuit and ground-fault protective device, as at $S_B$ in Fig. 7-195, unless no switch is located as at $S_M$ in Fig. 7-195, I. When no switch is located at $S_M$, a switch must be located at $S_B$ to serve as a motor switch (see Sec. 433). If a circuit breaker is employed for branch-circuit short-circuit and ground-fault protection, it will serve as both a switch and a protective device. Although not required, a branch-circuit switch is very convenient, providing a means of disconnecting the entire branch circuit from the supply. It is recommended that a branch-circuit switch be provided on all circuits operating at a voltage greater than 230 V unless a safety-fuse type of panel is used (see Sec. 86, Div. 4). The branch-circuit switch must have fuse clips large enough to accommodate the branch-circuit fuses, as required by Sec. 435 and Table 43 of Div. 11.

If a branch-circuit switch is used, it must comply with the requirements for a motor switch. Refer to Sec. 433.

**437. Size of wire for motor branch circuits.** The wires of a branch circuit supplying a single continuous-duty motor must have an ampacity of at least 125 percent of the full-load rated current of the motor. The size of the wires of a motor branch circuit supplying a single motor in short-time–duty service may be different from those for a continuous-duty motor. In most cases, the ampacity of the wires need not be greater than the percentages of the full-load current of the motor given in Table 12 of Div. 11.

The wires between the slip rings and the controller of a wound-rotor induction motor must have an ampacity of at least 125 percent of the full-load secondary current of the motor for continuous-duty motors. For other than continuous-duty motors, the ampacity of the wires must be not less than the percentages of full-load secondary currents as given in Table 12 of Div. 11. If the secondary resistor is separate from the controller, the wires between the secondary controller and the resistance of a wound-rotor induction motor must have an ampacity which is not less than the percentages of the full-load secondary current of the motor given in Table 13 of Div. 11.

The wires of a branch circuit supplying several motors but no other equipment except that directly associated with the motors must have an ampacity of not less than 125 percent of the full-load current rating of the largest motor (highest full-load current rating) plus the sum of the full-load current ratings of the rest of the motors.

For the size of the tap conductors, refer to Secs. 427 and 428.

If the tap between the feeder or main and the branch-circuit protective devices is not over 10 ft (3 m) long and is protected against mechanical injury by a conduit, electrical metallic tubing, or metal gutters, the tap from the feeder or main to the protective devices may be of the same size as the branch-circuit conductors.

**438. Remote-control circuits.** Control circuits shall be so arranged that they will be disconnected from all sources of supply when the disconnecting means is in the open position. The disconnecting means is permitted to consist of two separate devices, one of which disconnects the motor and the controller from the source or sources of power supply for the motor, and the other the control circuit from its power supply. When separate devices are used, they shall be located immediately adjacent to one another, although the National Electrical Code provides exceptions to this rule if more than 12 motor control-circuit conductors must be disconnected and it would be unsafe to open control-circuit conductors because of the nature of the controlled machine or process.

If a transformer or other device is used to obtain a reduced voltage for the control circuit and is located in the controller, such a transformer or other device shall be connected to the load side of the disconnecting means for the control circuit.

The conductors of these control circuits must be protected against overcurrent at their ampacities (7 A for No. 18, 10 A for No. 16).

1. If the conductors do not extend outside the control-equipment enclosure, they are permitted to be protected by the motor branch-circuit short-circuit and ground-fault protective device, provided this device is rated at not more than 400 percent of the 60°C ampacity given in Table 19 of Div. 11 of the control-circuit conductors No. 14 and larger, and at not more than 25 A for No. 18 and 40 A for No. 16.

2. Conductors which extend outside the control-equipment enclosure and are No. 14 or larger are permitted to be protected by the motor branch-circuit short-circuit and ground-fault protective device if this device is rated at not more than 300 percent of the 60°C ampacity given in Table 18 of Div. 11 of the control-circuit conductors.

3. Except as indicated in item 4, when the conductors are in the secondary of a single-phase control-circuit transformer having only a two-wire (single-voltage) secondary, they are permitted to be protected in the primary (supply) side of the transformer by protective devices suitable to protect the transformer if the primary protection does not exceed the ampacity of the secondary conductors or the values given in items 1 and 2 when the value of primary protection is multiplied by the primary-to-secondary transformer ratio.

4. When the conductors are in the secondary of a control transformer and the conditions comply with the requirements of Article 725 of the NEC for a Class 1 power-limited circuit or a Class 2 or Class 3 remote-control circuit.

5. If the opening of the control circuit would create a hazard, as, for example, in the control circuit of fire pump motors, etc.

**439. Size of motor feeders or mains.** For the determination of the size of motor mains or feeders, refer to Div. 3.

**440. Overload protection of motor feeders or mains.** The size of the protective equipment for motor feeders or mains is determined from the so-called starting current of the circuit, as explained in Sec. 54 of Div. 3. Fuses or circuit breakers can be used for the overload protection of motor feeders or mains.

**441. Motor-wiring data** are given in Tables 44 to 49 of Div. 11. These tables give data on the size of the component parts of motor branch circuits in accordance with the 1990 edition of the National Electrical Code (refer to Sec. 421). If the wiring is to be done in a locality which has adopted motor-wiring tables of its own, the tables of that particular district should be adhered to instead of the values given here. The values for the sizes of wires given in the tables are the minimum sizes allowed by the Code. They do not take into account the voltage drop in the circuit. For long circuits the voltage drop should be computed to determine whether the size of wire should be increased to keep the voltage drop within an allowable amount (see Div. 3).

**442. Protection of live parts.** The National Electrical Code requires that live parts shall be protected in a manner judged adequate to the hazard involved. The rules state:

1. *Where required.* Exposed live parts of motor and controllers operating at 50 V or more between terminals, except for stationary motors having commutators, collectors, and brush rigging located inside motor end brackets and not connected conductively to supply circuits operating at more than 150 V to ground, shall be guarded against accidental contact by enclosure, or by location as follows:

    *a.* By installation in a room or enclosure accessible only to qualified persons

    *b.* By installation on a suitable balcony, gallery, or platform so elevated and arranged as to exclude unqualified persons

    *c.* By elevation 8 ft (2.4 m) or more above the floor

2. *Guards for attendants.* If the live parts of motors or controllers operating at more than 150 V to ground are guarded against accidental contact only by location as specified

in Par. 1 and if adjustment or other attendance may be necessary during the operation of the apparatus, suitable insulating mats or platforms shall be provided so that the attendant cannot readily touch live parts without standing on the mats or platforms.

## CONTROL CIRCUITS

**443. Electric control circuits** constitute a broad subject, and the reader is urged to refer to special books devoted to this important phase of motor circuitry. Two such books are *Control of Electric Motors*, by Paisley B. Harwood, published by John Wiley & Sons, Inc., and *Modern Electric Controls*, by J. F. McPartland, published by *Electrical Construction and Maintenance* of McGraw-Hill, Inc. These textbooks provide an excellent insight on how to understand, select, and design control circuits.

**444. Arrangement of control wiring in grounded circuits.** It is a requirement of the National Electrical Code that if one side of a control circuit is *intentionally* grounded, the control circuit shall be so arranged that an accidental ground in the remote-control devices will not *start* the motor. Figure 7-200 shows two methods that satisfy this code rule. The elementary wiring diagram at the top of Fig. 7-200 shows the arrangement of a START-STOP push button connected to the holding coil of a magnetic starter. Note that all control devices are connected to the $L1$ side of coil $C$ and that the other side is connected to the grounded circuit conductor $L2$.

The $L2$ conductor could be the grounded conductor of an end-grounded three-phase delta supply or the grounded conductor of the secondary winding of a control transformer.

**445. Control circuits from ungrounded systems.** It is significant that the Code rule described in Sec. 444 applies only to grounded-type control circuits and is concerned only with preventing a motor from being accidentally started. In ungrounded three-phase 230- or 460-V three-wire systems, there is always a possibility of one or more accidental grounds occurring in a control circuit connected to such systems. If *two* grounds occur in

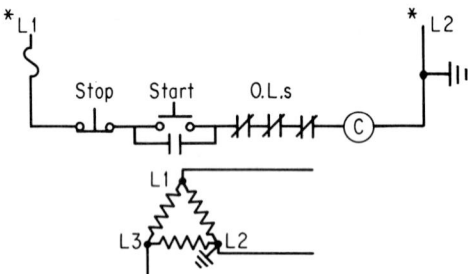

$^{*}$ L1 & L2 from end-grounded delta supply or;

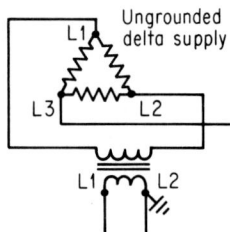

$^{*}$ L1&L2 from control trans.
where secondary L2 is grounded

When one side of the control circuit is grounded circuit shall be so arranged that an accidental gr. in remote-control devices will not <u>start</u> motor. Drawing shows acceptable arrangements

**Fig. 7-200**  Arrangement of control wiring in grounded circuits. [*Electrical Construction and Maintenance*]

the same control-circuit conductor at certain critical points, a motor could accidentally start, or if a second ground develops after a motor has been started, it is possible that the motor will not stop after the STOP button is pressed. There are two ways to prevent this. The first method is illustrated in Fig. 7-200, in which a control transformer provides a grounded secondary and the control circuit is wired as shown.

The second method applies when the control circuit is taken directly from an ungrounded supply. The arrangement illustrated in Fig. 7-201 utilizes double-pole START

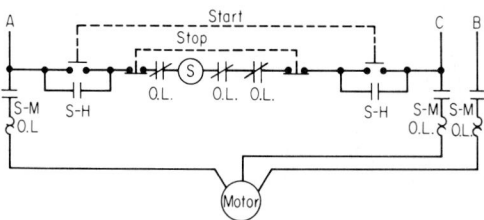

**Fig. 7-201** Ideal control circuit when the supply system or the control circuit is ungrounded. [*Electrical Construction and Maintenance*]

and STOP buttons wired with one pole of each button on opposite sides of coil S. With this arrangement one or more accidental grounds will not start the motor; nor will a ground prevent the motor from being stopped after the motor is in operation and a second ground occurs.

The arrangement shown in Fig. 7-201 is ideal when long control circuits are used because pressing the STOP button opens both lines to coil S and line-to-line capacitive coupling will not keep the coil energized, as it occasionally has done when standard START-STOP control circuits were used.

## AIR-CONDITIONING AND REFRIGERATING EQUIPMENT

**446. The rules in the** National Electrical Code **for air-conditioning and refrigerating equipment incorporating a hermetic refrigerant motor-compressor** and for the circuits supplying such equipment are somewhat different from the rules applying to conventional motors and motor circuits because of the special nature of a hermetic refrigerant motor-compressor. The following sections describe only those rules that differ from those described in Secs. 421 through 445. If a situation is not described in the following sections, the rules of Secs. 421 through 445 apply, except as noted in Sec. 447.

Types of air-conditioning and refrigerating equipment which do not incorporate a hermetic refrigerant motor-compressor and circuits which do not supply such a motor-compressor are not to be installed under the rules given in Secs. 446 through 454. Examples of such equipment are devices that employ a refrigeration compressor driven by a conventional motor, furnaces with air-conditioning evaporator coils installed, fan-coil units, remote forced-air–cooled condensers, and remote commercial refrigerators.

The National Electrical Code defines a hermetic refrigerant motor-compressor as follows:

*Hermetic refrigerant motor-compressor.* A combination consisting of a compressor and motor, both of which are enclosed in the same housing, with no external shaft or shaft seals, the motor operating in the refrigerant.

Because there is no motor external shaft (see Fig. 7-202), the hermetic refrigerant motor-compressor does not have a horsepower rating. The motor is rated instead by the rated-load current and the locked-rotor current. The National Electrical Code defines the rated-load current as follows:

*Rated-load current.* The rated-load current for a hermetic refrigerant motor-compressor is the current resulting when the motor-compressor is operated at the rated load, rated voltage, and rated frequency of the equipment it serves.

The rated-load current is established in the equipment driven by the motor-compressor, rather than being the maximum current for a given motor temperature rise, because the motor winding is cooled by the refrigerant and because the degree of cooling depends on the characteristics of the refrigerant used, its flow rate, temperature, density, etc., and these factors are determined by the equipment. Therefore, the manufacturer of the motor-compressor, not knowing all the equipment in which the motor-compressor will be used, selects the motor-overload protection system to protect the motor and motor-circuit components and wiring at the maximum-load conditions for which the motor-compressor is intended. If the motor-compressor is used in an application in which it is effectively larger than needed for the rating conditions of the equipment, the rated-load current for the motor-compressor marked on the equipment nameplate may be significantly less than its maximum, and the motor-compressor–overload protection system may not protect the motor-circuit wiring, motor controller, and motor-circuit switches against overheating due to overloads.

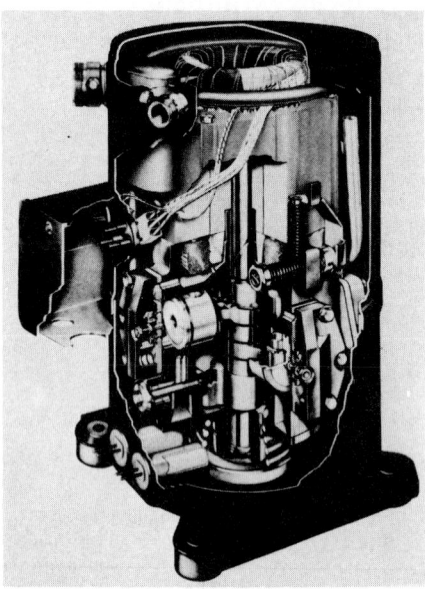

**Fig. 7-202**  Cutaway of a hermetic motor-compressor. [Tecumseh Products Co.]

The National Electrical Code establishes limiting relationships between the maximum continuous current permitted by the motor-compressor–overload protection system and the ampacity of the motor-circuit wiring, the rating of the motor controller, motor-circuit switches, and motor-circuit short-circuit and ground-fault protective devices (see Sec. 449). When this limiting relationship is exceeded for the marked rated-load current of the motor-compressor in the equipment, an additional equipment-rating current must be marked by the equipment manufacturer. This is designated *branch-circuit selection current* and is defined in the National Electrical Code as follows:

*Branch-circuit selection current.* Branch-circuit selection current is the value in amperes to be used instead of the rated-load current in determining the ratings of motor branch-circuit conductors, disconnecting means, controllers, and branch-circuit short-circuit and ground-fault protective devices wherever the running overload protective device permits a sustained current greater than the specified percentage of the rated-load current. The value of branch-circuit selection current will always be equal to or greater than the marked rated-load current.

A significant portion of air-conditioning and refrigerating equipment incorporating a hermetic refrigerant motor-compressor also incorporates conventional motors, such as fan and/or water-pump motors, and/or electric resistance heaters on the same circuit. The National Electrical Code requires such equipment to be marked with (1) the minimum supply-circuit conductor ampacity (usually shortened to "Min. Circuit Amp.") and (2) the maximum rating of the branch-circuit short-circuit and ground-fault protective device. The ratings marked are to be based on the maximum concurrent loading condition, since all motors and other loads may not be operating at the same time. For example, on an air conditioner with electric resistance space heaters the motor-compressor does not usually operate at the same time as the heaters, as it does on a heat pump. The National Electrical Code requires that the supply wires have an ampacity at least as great as the marked minimum and that the branch-circuit short-circuit and ground-fault protective device rating not exceed the marked maximum rating. If the type of branch-circuit short-circuit and ground-fault protective device (fuse or circuit breaker) is marked, the proper type should be used. When these minimum and maximum markings appear, the installer need not calculate the values required for the installation.

Since the marked motor-compressor rated-load current and branch-circuit selection current (if marked) and the marked minimum supply-circuit conductor ampacity and maximum rating of the branch-circuit short-circuit and ground-fault protective device vary somewhat from one make and model of equipment to another, even though the equipment may have the same cooling or heating capacity, the installer must know the marking on the particular equipment to be installed. This information is generally available in the equipment manufacturer's literature, so that the equipment itself need not be available before the supply circuit can be designed and partially installed. However, the equipment marking should be checked before the installation is completed.

**447. Ampacity and rating.** For a hermetic refrigerant motor-compressor, the rated-load current marked on the nameplate of the equipment in which the motor-compressor is employed shall be used in determining the rating or ampacity of the disconnecting means, the branch-circuit conductors, the controllers, the branch-circuit short-circuit and ground-fault protection, and the separate motor-overload protection. If no rated-load current is shown on the equipment nameplate, the rated-load current shown on the compressor nameplate shall be used.

*Exception No. 1. When so marked, the branch-circuit selection current shall be used instead of the rated-load current to determine the rating or ampacity of the disconnecting means, the branch-circuit conductors, the controller, and the branch-circuit short-circuit and ground-fault protection.*

*Exception No. 2. For cord- and plug-connected equipment, as explained under Par. 2 of Sec. 448.*

**448. Branch-circuit short-circuit and ground-fault protection.** The rules of Sec. 423 generally apply except for the rating of the protective device. As noted in Sec. 446, for multimotor and combination-load air-conditioning and refrigerating equipment, the equipment nameplate is required by the National Electrical Code to specify the maximum rating of the supply-circuit short-circuit and ground-fault protective device. When this rating is marked, a calculation by the installer is not necessary, and the marked maximum rating cannot be exceeded.

1. RATING OR SETTING FOR AN INDIVIDUAL MOTOR-COMPRESSOR The motor-compressor branch-circuit short-circuit and ground-fault protective device shall be capable of carrying the starting current of the motor. The required protection shall be considered as being obtained when this device has a rating or setting not exceeding 175 percent of the motor-compressor rated-load current or branch-circuit selection current, whichever is greater (15 A minimum size), provided that if the protection specified is not sufficient for the starting current of the motor, it shall be permitted to be increased but shall not exceed 225 percent of the motor rated-load current or branch-circuit selection current, whichever is greater.

2. RATING OR SETTING FOR EQUIPMENT The equipment branch-circuit short-circuit and ground-fault protective device shall be capable of carrying the starting current of the equipment. When the hermetic refrigerant motor-compressor is the only load on the circuit, the protection shall conform with Par. 1 above. When the equipment incorporates more than one hermetic refrigerant motor-compressor or a hermetic refrigerant motor-compressor and other motors or other loads, the equipment protection shall conform with Sec. 424 and the following:

*a.* When a hermetic refrigerant motor-compressor is the largest load connected to the circuit, the rating or setting of the protective device shall not exceed the value specified in Par. 1 above for the largest motor-compressor plus the sum of the rated-load current or branch-circuit selection current, whichever is greater, of the other motor-compressor or motor-compressors and the ratings of the other loads supplied.

*b.* When a hermetic refrigerant motor-compressor is not the largest load connected to the circuit, the rating or setting of the protective device shall not exceed a value equal to the sum of the rated-load current or branch-circuit selection current, whichever is greater, the rating or ratings for the motor-compressor or motor-compressors, and the value specified in Par. 3*d* of Sec. 424 for other loads supplied.

The following are exceptions to Par. 2 above.

1. Equipment which will start and operate on a 15- or 20-A, 120-V or a 15-A, 208- or 240-V single-phase branch circuit shall be considered as protected by the 15- or 20-A overcurrent device protecting the branch circuit, but if the maximum circuit-protective-device rating marked on the equipment is less than these values, the circuit protective device shall not exceed the value marked on the equipment nameplate.

2. The nameplate marking of cord- or plug-connected equipment rated not greater than 250 V, single-phase, such as some room air conditioners (see Sec. 454), household refrigerators and freezers, drinking-water coolers, and beverage dispensers, shall be used in determining the branch-circuit requirements, and each unit shall be considered a single motor unless the nameplate is marked otherwise.

**449. Overload protection.** Overload-protection provisions in the National Electrical Code for hermetic refrigerant motor-compressors and equipment employing such motor-compressors have the same purpose as overload protection for conventional motor circuits, but there are some significant differences in the rules. Some reasons for this variation are described in Sec. 446. The device that protects the refrigerant-cooled motor against overheating must do so while the motor is being cooled by the refrigerant under running conditions and also under locked-rotor conditions without such refrigerant cooling. The conditions necessary to do this and also to protect the other components and wiring in the motor-compressor branch circuit commonly result in the need for both temperature- and current-sensing devices, or *system protection*. The National Electrical Code therefore treats protection of the motor-compressor and protection of the motor-compressor control apparatus and branch-circuit conductors separately, as described in Pars. 1 and 2 below.

1. PROTECTION OF MOTOR-COMPRESSOR   One of the following means shall be used:

*a.* A separate overload relay which is responsive to the motor-compressor current. This device shall be selected to trip at not more than 140 percent of the motor-compressor rated-load current.

*b.* A thermal protector integral with the motor-compressor, approved for use with the motor-compressor which it protects, on the basis that it will prevent dangerous overheating of the motor-compressor due to overload and failure to start. If the current-interrupting device is separate from the motor-compressor and if its control circuit is operated by a protective device integral with the motor-compressor, it shall be so arranged that the opening of the control circuit will result in an interruption of current to the motor-compressor.

*c.* A fuse or inverse-time circuit breaker responsive to the motor current, which shall also be permitted to serve as the branch-circuit short-circuit and ground-fault protective device. This device shall be rated at not more than 125 percent of the motor-compressor rated-load current. It shall have sufficient time delay to permit the motor-compressor to start and accelerate its load.

*d.* A protective system, furnished or specified and approved for use with the motor-compressor which it protects on the basis that it will prevent dangerous overheating of the motor-compressor due to overload and failure to start. If the current-interrupting device is separate from the motor-compressor and if its control circuit is operated by a protective device which is not integral with the current-interrupting device, it shall be so arranged that the opening of the control circuit will result in an interruption of current to the motor-compressor.

2. PROTECTION OF MOTOR-COMPRESSOR CONTROL APPARATUS AND BRANCH-CIRCUIT CONDUCTORS   Except for motor-compressors and equipment on 15- or 20-A branch circuits as described in Par. 3 below, one of the following means shall be used (the device used may be the same device or system described in Par. 1 above):

*a.* An overload relay selected in accordance with Par. 1 above

*b.* A thermal protector which is applied in accordance with Par. 1 above and which will not permit a continuous current in excess of 156 percent of the marked rated-load current or branch-circuit selection current

*c.* A fuse or inverse-time circuit breaker selected in accordance with Par. 1 above

*d.* A protective system which is in accordance with Par. 1 above and which will not permit a continuous current in excess of 156 percent of the marked rated-load current or branch-circuit selection current

3. MOTOR-COMPRESSORS AND EQUIPMENT ON 15- OR 20-A BRANCH CIRCUITS   Overload protection for motor-compressors and equipment used on 15- or 20-A, 120-V or 15-A, 208- or 240-V single-phase branch circuits shall be permitted as indicated in *a* and *b* below.

*a.* The motor-compressor shall be provided with overload protection selected as specified in Par. 1 above. Both the controller and the motor-overload protective device shall be approved for installation with the short-circuit and ground-fault protective device for the branch circuit to which the equipment is connected.

*b.* The short-circuit and ground-fault protective device protecting the branch circuit shall have sufficient time delay to permit the motor-compressor and other motors to start and accelerate their loads.

*c.* For cord– and attachment-plug–connected equipment, the rating of the attachment plug and receptacle shall not exceed 20 A at 125 V or 15 A at 250 V.

**450. Disconnecting means.** The rules of Secs. 433 and 436 generally apply, but the method of determining the required rating of the switch differs, and a disconnecting means must be located within sight from and be readily accessible from the air-conditioning or refrigerating equipment.

1. SINGLE MOTOR-COMPRESSOR  A disconnecting means serving a hermetic refrigerant motor-compressor shall be selected on the basis of the nameplate rated-load current or branch-circuit selection current, whichever is greater, and locked-rotor current, respectively, of the motor-compressor as follows:

*a.* The ampere rating shall be at least 115 percent of the nameplate rated-load current or branch-circuit selection current, whichever is greater.

*b.* To determine the equivalent horsepower in complying with the requirements of Sec. 433, select the horsepower rating from Tables 7 to 10 of Div. 11 corresponding to the rated-load current or branch-circuit selection current, whichever is greater, and also the horsepower rating from Table 47 of Div. 11 corresponding to the locked-rotor current. If the nameplate rated-load current or the branch-circuit selection current and locked-rotor current do not correspond to the currents shown in Tables 7 to 10 or 47 of Div. 11, select the horsepower rating corresponding to the next higher value. If different horsepower ratings are obtained when applying these tables, select a horsepower rating at least equal to the larger of the values obtained.

2. MULTIMOTOR AND COMBINATION LOADS  When one or more hermetic refrigerant motor-compressors are used together or are used in combination with other motors and/or loads such as resistance heaters and when the combined load may be simultaneous on a single disconnecting means, the rating for the combined load shall be determined as follows.

*a.* Determine the horsepower rating of the disconnecting means from the summation of all currents, including resistance loads, at the rated-load condition and also at the locked-rotor condition. Consider the combined rated-load current and the combined locked-rotor current so obtained as a single motor for the purpose of this requirement as follows:

(1) Select the full-load current equivalent to the horsepower rating of each motor, other than a hermetic refrigerant motor-compressor, from Tables 7 to 10 of Div. 11. Add these full-load currents to the motor-compressor rated-load current or currents or the branch-circuit selection current or currents, whichever is greater, and to the rating in amperes of other loads to obtain an equivalent full-load current for the combined load.

(2) Select the locked-rotor current equivalent to the horsepower rating of each motor, other than a hermetic refrigerant motor-compressor, from Table 47 of Div. 11. Add the locked-rotor currents to the motor-compressor locked-rotor current or currents and to the rating in amperes of other loads to obtain an equivalent locked-rotor current for the combined load. When two or more motors and/or other loads cannot be started simultaneously, select appropriate combinations of locked-rotor and rated-load current or branch-circuit selection current, whichever is greater, to determine the equivalent locked-rotor current for the simultaneous combined load.

Exceptions to (1) and (2) are:

1. The rated full-load current for small shaded-pole and permanent split-capacitor fan and blower motors provided as part of multimotor equipment may be used in the calculation instead of the motor horsepower rating.

2. For small motor-compressors not having the locked-rotor current marked on the nameplate or for small motors not covered by Tables 7 to 10 of Div. 11, assume the locked-rotor current to be 6 times the rated-load current.

3. When the rated-load or locked-rotor current as determined above would indicate a disconnecting means rated in excess of 100 hp, the provisions of Par. 4 of Sec. 433 would apply.

4. For cord-connected equipment such as room air conditioners, household refrigerators and freezers, drinking-water coolers, and beverage dispensers, a separable connector or an attachment plug and receptacle are permitted to serve as the disconnecting means.

5. When part of the concurrent load is a resistance load and the disconnecting means is a switch rated in horsepower and amperes, the horsepower rating of the switch shall be

not less than the combined load to the motor-compressor or motor-compressors and other motor or motors at the locked-rotor condition, and the ampere rating is permitted to be not less than this locked-rotor load plus the resistance load. This alternative calculation may permit a smaller switch than would otherwise be required, especially when the load of the resistance heaters is large compared with the motor load.

The ampere rating of the disconnecting means must be at least 115 percent of the summation of all currents at the rated-load condition determined in accordance with Par. 2*a*(1) above.

**451. Group switching.** The rules of Sec. 434 do not apply to an air-conditioning or refrigeration system incorporating a hermetic refrigerant motor-compressor. Such equipment is considered a single machine and is permitted to have a single switch for a group of motors even if the motors are remote from each other.

**452. Controllers.** The rules for motor controllers for hermetic refrigerant motor-compressors are generally the same as described in Sec. 429 except for the method of determining the rating.

A motor-compressor controller shall have both a continuous-duty full-load current rating and a locked-rotor current rating not less than the nameplate rated-load current or branch-circuit selection current, whichever is greater, and locked-rotor current, respectively, of the motor-compressor. If the motor controller is rated in horsepower but is without one or both of the foregoing current ratings, determine equivalent currents as follows. Use Tables 7 to 10 of Div. 11 to determine the equivalent full-load current rating and Table 47 of Div. 11 to determine the equivalent locked-rotor current rating.

A controller serving more than one motor-compressor or a motor-compressor and other loads shall have a continuous-duty full-load current rating and a locked-rotor current rating not less than the combined load as determined in accordance with Par. 2 of Sec. 450.

**453. Size of wire for branch circuits.** The rules given in Sec. 437 generally apply to branch circuits supplying hermetic refrigerant motor-compressors that are permanently connected, except that the motor-compressor rating is established as described in Sec. 447. As noted in Sec. 446, for multimotor and combination-load air-conditioning and refrigerating equipment, the equipment nameplate is required by the National Electrical Code to specify the minimum supply-circuit conductor ampacity in accordance with the rules in the Code. When this rating is marked, a calculation by the installer is not necessary.

**454. Provisions for room air conditioners.** The National Electrical Code has special provisions for room air conditioners which are (1) rated not over 250 V ac, single-phase, and (2) which are cord– and attachment-plug–connected. The Code defines for this purpose a room air conditioner (with or without provisions for heating) as an appliance of the air-cooled window, console, or in-wall type which is installed in the conditioned room and which incorporates a hermetic refrigerant motor-compressor.

1. Room air conditioners shall be grounded.

2. A room air conditioner shall be considered a single motor unit in determining its branch-circuit requirements when all the following conditions are met:

*a.* It is cord– and attachment-plug–connected.

*b.* Its rating is not more than 40 A and 250 V, single-phase.

*c.* Total rated-load current is shown on the room-air-conditioner nameplate rather than individual motor currents.

*d.* The rating of the branch-circuit short-circuit and ground-fault protective device does not exceed the ampacity of the branch-circuit conductors or the rating of the receptacle, whichever is less.

3. The total marked rating of a cord– and attachment-plug–connected room air conditioner shall not exceed 80 percent of the rating of a branch circuit when no other loads are supplied.

4. The total marked rating of a cord– and attachment-plug–connected room air conditioner shall not exceed 50 percent of the rating of a branch circuit when lighting units or other appliances are also supplied.

5. An attachment plug and receptacle shall be permitted to serve as the disconnecting means for a single-phase room air conditioner rated 250 V or less if (*a*) the manual controls on the room air conditioner are readily accessible and are located within 6 ft (1.8 m) of the floor or (*b*) an approved manually operable switch is installed in the readily accessible location within sight from the room air conditioner.

**ENGINE- AND GAS-TURBINE GENERATORS**

**455. Engine-driven generators** (diesel, gasoline, or gas) are the most economical and practical source of *standby power* in a wide range of applications. They are sized from a few hundred watts to several hundred kilowatts. Control is obtained manually, remotely, or automatically. Units are designed as either stationary or mobile power stations. Cooling is obtained by natural-convection fan-forced air or by circulating water.

Generator engines are generally four-cycle units having one to six cylinders, depending on size and capacity. In larger sets, ignition current is supplied by starting batteries. A typical arrangement is shown in Fig. 7-203.

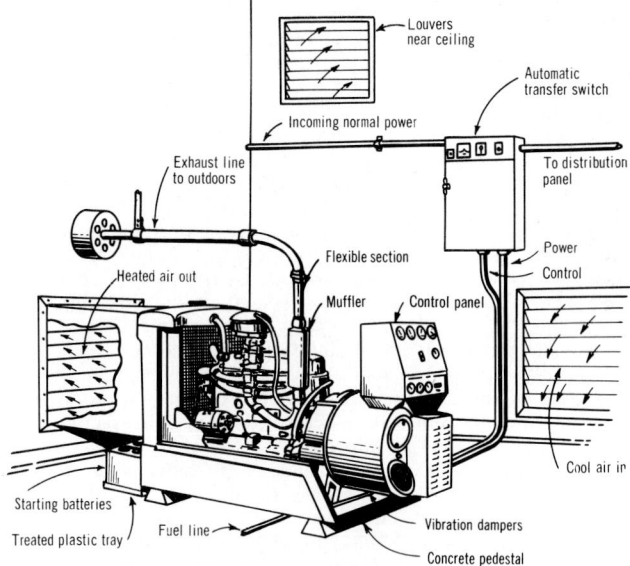

**Fig. 7-203**  Typical installation of an engine-generator for a standby or emergency supply sustem. [*Electrical Construction and Maintenance*]

**456. Location.**   In most instances, the power plant should be close to the normal electric service, in a place where it is warm enough to start easily. Locations where ambient temperatures may be unusually high should be avoided because inefficient cooling may result. On the other hand, in locations where temperatures fall beflow 50°F (10°C), special accessories, such as an electric water jacket or a manifold heater, may be needed to ensure dependable automatic starts.

**457. Mounting.**   Smaller units, of 10 kW and less, should be installed on a concrete foundation equipped with anchor bolts. In some cases, steel-beam sections will make a satisfactory base. Such a base raises the plant off the floor for convenience in servicing and lessens the chance of accidental damage from loose objects, seepage, etc. For larger plants that are equipped with steel skids, a separate foundation may not be required.

Always allow at least 2 ft (0.6 m) of unobstructed space around the power plant on all sides. This permits room for servicing the unit. If this space is not available, position the plant so that the carburetor, distributor, governor, fuel pump, oil fill, etc., are easily accessible.

**458. Battery mounting.**   Batteries should be mounted in a clean, dry location free from falling dirt or dust, and they should be installed on a wood base rather than on a concrete floor. Allow room for easy maintenance of the battery.

**459. Exhaust piping.**   Exhaust pipes should be of wrought iron, cast iron, or steel. Because exhaust fumes are dangerous, care should be taken to prevent leaks in the exhaust system. The exhaust pipe should terminate outside away from doors, windows, or other openings.

**460. Cooling.**  Ventilation openings must always be provided and located so that cool air can be brought in and forced through the engine-cooling system and directly out without recirculating.

Other important installation considerations include fuel-line and fuel-tank installation, muffler requirements, heaters, battery chargers, and automatic controls.

**461. Automatic transfer switches** are being used increasingly wherever continuity of light and power is needed to preserve life and to prevent accidents, theft, panic, or loss of goodwill and revenue. An automatic transfer switch is an emergency device that automatically transfers electrical loads from a normal source, usually a utility, to an emergency source, usually an engine-generator set, when the normal source fails or its voltage is substantially reduced. The transfer switch automatically retransfers the load back to the normal source when it is restored.

CONSTRUCTION CONSIDERATIONS    The automatic transfer switch consists of two major components, namely, an electrically operated, double-throw transfer switch and a control panel. The control panel provides the necessary voltage sensing, time delays, and signals for operating the transfer switch.

Automatic transfer switches of double-throw construction are used primarily for emergency and standby power generation systems rated 600 V and less. These transfer switches do not normally incorporate overcurrent protection and are designed and applied in accordance with the National Electrical Code, particularly Articles 517, 700, 701, and 702. They are available in ratings from 30 to 4000 A. For reliability, most automatic transfer switches rated above 100 A are mechanically held and electrically operated from the power source to which the load is to be transferred. To provide maximum versatility and further reliability, today's control panels utilize solid-state sensing and timing circuitry.

An automatic transfer switch is usually located in the main or secondary distribution system which feeds the branch circuits. Because of its location in the system, the abilities which must be designed into the transfer switch are unqiue and extensive as compared with the design requirements for other electrical devices. For example, special consideration should be given to the following characteristics of an automatic transfer switch: (1) its ability to close against high inrush currents, (2) its ability to carry full-rated current continuously from either normal or emergency sources, (3) its ability to withstand fault currents, (4) its ability to interrupt at least 6 times the full-load currents, and (5) its ability to withstand the stress of two out-of-phase power sources connected to it. In addition to considering each of the above characteristics individually, it is also necessary to consider the effect each has upon the others.

In arrangements to provide protection against failure of the utility service, consideration should also be given to (1) an open circuit within the building area on the load side of the incoming service, (2) overload or fault condition, and (3) electrical or mechanical failure of the electric power distribution system within the building. It is desirable to locate transfer switches close to the load and to have their operation independent of overcurrent protection. It is often desirable to use multiple smaller transfer switches located near the load rather than one large transfer switch at the point of incoming service. As shown in Fig. 7-204, multiple transfer switches can also provide varying degrees of protection, depending upon how critical the loads are.

RATINGS    An automatic transfer switch is unique in the electrical distribution system in that it is one of the few electrical devices that may have two unsynchronized power sources connected to it. This means that the voltages impressed on the insulation may actually be as high as 960 V on a 480-V ac system. A properly designed transfer switch provides sufficient spacings and insulation to meet these increased voltage stresses.

Domestic ac voltage ratings of automatic transfer switches are normally 120, 208, 240, 480, or 600 V, single or polyphase. Standard frequencies are 50 to 60 Hz. Automatic transfer switches can also be supplied for other voltages and frequencies when required.

Today automatic transfer switches are available in continuous-current ratings ranging from 30 through 4000 A. Typical ratings include 30, 70, 100, 150, 260, 400, 600, 800, 1000, 1200, 1600, 2000, 3000, and 4000 A.

Most transfer switches are capable of carrying 100 percent of rated current continuously at an ambient temperature of 40°C. However, some transfer switches, such as those incorporating integral overcurrent protective devices, may be limited to a continuous load current not to exceed 80 percent of the switch rating.

When selecting an automatic transfer switch, it usually is necessary only to determine the maximum continous load current which the transfer switch must carry. Momentary inrushes, which occur when lighting or motor loads are first energized, can be ignored if the transfer switch is rated for total-system loads according to UL 1008. The automatic transfer switch should be sized equal to or greater than the calculated continuous current.

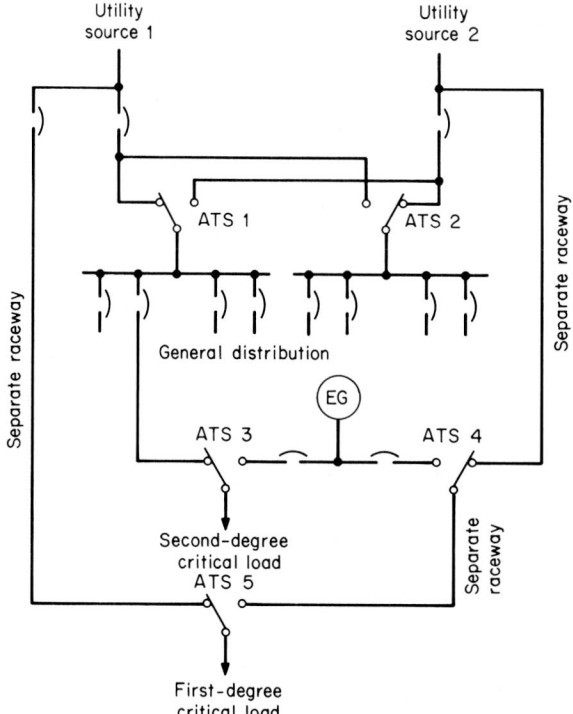

**Fig. 7-204**  Two utility sources combined with an engine-generator set to provide varying degrees of emergency power. *ATS* = automatic transfer switch; *EG* = engine-generator set.

The *withstand current rating* pertains to the ability of an automatic transfer switch to withstand the magnetic and thermal stresses of high-fault currents until the fault has been cleared by an overcurrent protective device. The overcurrent protective device is usually located externally from the transfer switch, although there are some transfer switches that, for purposes of economy, do include integral overcurrent protection. To differentiate between the two, one recognized standard defines the following type designations:

*Transfer switch, Type A* means an automatic transfer switch that does not employ integral overcurrent devices.

*Transfer switch, Type B* means an automatic transfer switch that employs integral overcurrent devices.

In a Type A transfer switch, the main contacts stay closed while the system's coordinated protective devices clear the fault, whereas a Type B transfer switch depends upon trip elements to open the contacts of the transfer switch during a short-circuit fault. The latter arrangement depends upon *interruption rating*, which should not be confused with withstand rating. This discussion addresses itself to the conventional Type A transfer switch as found in most properly engineered and fault-current–coordinated applications.

To evaluate and coordinate transfer switching and overcurrent protection, a short-circuit study should be made on the system to determine the available fault-current magnitude and the $X/R$ ratio of the circuit at each point in the distribution system where a transfer switch is to be located. This study is usually made by the system designer.

SOLID NEUTRAL on AC and DC SYSTEMS    Solid neutrals can be used with the grounding connections made as required in Sec. 250-23 of the 1990 National Electrical Code where automatic ac-to-ac transfer switches are used. Where multiple grounding creates objectionable ground current, the corrective action specified in Sec. 250-21(b) should be taken.

Section 230-95 requires ground-fault protection of equipment. Because the normal source and the emergency source are typically grounded at their locations, the multiple neutral-to-ground connections usually require some additional means or devices to assure proper ground-fault sensing by the ground-fault protection device. Additional means or devices are generally required because the normal alterations to stop objectional current per Sec. 250-21(b) do not apply when the objectionable current is a ground-fault current. [See Sec. 250-21(c).] Rather, solutions such as an overlapping neutral transfer pole or a conventional fourth pole are often added to the transfer switch. However, when switching the neutral conductor with a conventional fourth pole, consideration should be given to possible resulting high-voltage transients. Other solutions are isolation transformers and special ground-fault circuits.

On ac-to-dc automatic transfer switches, a solid-neutral tie between the ac and dc neutrals is not permitted where both sources of supply are exterior distribution systems. Section 250-22 regarding location of grounds for dc exterior systems clearly specifies that the dc system can be grounded only at the supply station.

Where the dc system is an interior isolated system, such as a storage battery, solid-neutral connection between the ac system neutral and the dc source is acceptable.

On an ac-to-dc automatic transfer switch where the neutral must be switched, the size of the neutral switching pole must be considered. A four-pole double-throw switch must be used where a three-phase, four-wire normal source and a two-wire dc emergency source are transferred. Owing to the neutral being switched, a four-pole double-throw transfer switch would be required. In this instance, one pole of the dc emergency source would carry 3 times the current of the other poles.

CLOSE DIFFERENTIAL-VOLTAGE SUPERVISION OF NORMAL SOURCE    Most often the normal source is an electric utility company whose power is transmitted many miles to the point of utilization. The automatic-transfer-switch control panel continuously monitors the voltage of all phases. Because utility frequency is, for all practical purposes, constant, only the voltage need be monitored. For single-phase power systems, the line-to-line voltage is monitored. For three-phase power systems, all three line-to-line voltages should be monitored to provide full-phase protection.

In addition, monitoring protects against operation at reduced voltage, such as brownouts, which can damage loads. Since the voltage sensitivity of loads varies, the pickup (acceptable) voltage setting and dropout (unacceptable) voltage setting of the monitors should be adjustable. The typical range of adjustment for the pickup is 85 to 100 percent of nominal, while the dropout setting, which is a function of the pickup setting, is 75 to 98 percent of the pickup selected. Usual settings for most loads are 95 percent of nominal for pickup and 85 percent of nominal for dropout (approximately 90 percent of pickup).

Consideration must be given to close differential-voltage supervision at installations where the loads are more sensitive to voltage reduction. Starter-type fluorescent lighting is extremely voltage-sensitive, and below 90 percent voltage it becomes uncertain as to whether or not a fluorescent lamp will light. Therefore, close differential sensing is required for this type of lighting.

Electronic-equipment load is frequently voltage-critical. Such installations include patient-care equipment in health care facilities, television stations, microwave communications, telephone communications, computer centers, and similar applications.

Lightly loaded polyphase motors have a tendency to run single-phase, with the loss of voltage in one phase, thus leading to motor burnups. Automatic transfer switches with properly set close differential-voltage levels will help to detect phase outages.

COORDINATION WITH AUTOMATICALLY STARTED ENGINE-GENERATOR SETS    In these installations, the normal source is usually a utility power line, and the emergency source is an automatically started engine-generator set which starts upon failure of the normal source. To ensure proper operation and maximum reliability, a minimum installation as illustrated in Fig. 7-206 should be arranged to:

1. Initiate engine starting of the engine-generator set from a pilot engine-start contact on the automatic-transfer-switch control panel when utility voltage drops to an unacceptable level.

2. Sustain connection of load circuits to the normal source during the starting period to provide utilization of any existing service on the normal source.

3. Measure output voltage and frequency of emergency source through the use of a voltage-frequency–sensitive monitor and effect transfer of the load circuits to the engine-generator set when both voltage and frequency of the power plant are approximately normal. Sensing of the emergency source need only be single-phase since most applications involve an on-site engine-generator with a relatively short line run to the automatic

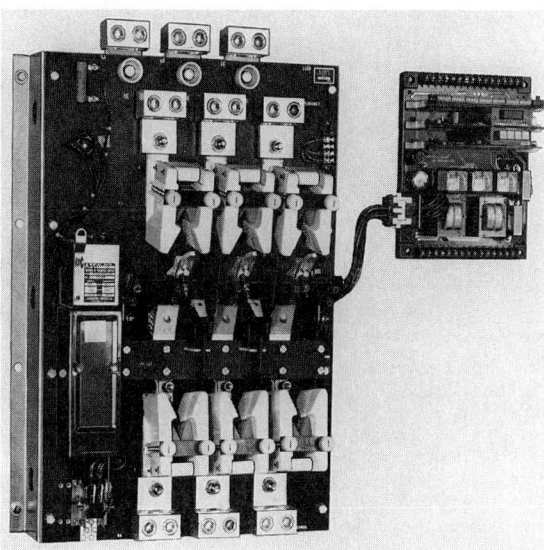

**Fig. 7-205**   Automatic emergency transfer switch and control panel. [Automatic Switch Co.]

transfer switch. In addition to voltage, the emergency source's frequency varies during startup. Combined frequency and voltage monitoring will protect against transferring loads to an engine-generator set with an unacceptable output.

4. Provide visual signal and auxiliary contact for remote indication when the emergency source is feeding the load.

TIME DELAYS   Time delays are provided to program operation of the automatic transfer switch. To avoid unnecessary starting and transfer to the alternative supply, a nominal 1-s time delay, adjustable up to 6 s, can override momentary interruptions and reductions in the normal source voltage but allow starting and transfer if the reduction or outage is sustained. (See Fig. 7-206.) The advantages of this feature are realized in all types of automatic-transfer installations. In standby plant installations the reduced number of false starts is especially important to minimize wear on the starting motor, ring gear, battery, and associated equipment. This delay is generally set at 1 s but may be set higher if reclosers on the high lines take longer to operate or if momentary power dips exceed 1 s. If longer delay settings are used, care must be taken to ensure that sufficient time remains to meet Code 10-s power restoration requirements.

Once the load has been transferred to the alternate source, another timer delays retransfer to the normal source until that source has had time to stabilize. The timer is adjustable from 0 to 30 min and is normally set at 30 min. Another important function of this timer is to assure that the engine-generator set operates under load for a preselected minimum time, thus assuring continued good performance of the engine and its starting system. This delay should be automatically nullified if the engine-generator set fails and the normal source is available.

Engine-generator manufacturers often recommend a cooldown period for their sets which allows them to run unloaded after the load has been retransferred to the normal source. A third time delay, usually 5 min, is provided for this purpose. Running an unloaded engine for more than 5 min is neither necessary nor recommended since it can cause deterioration in engine performance.

It is sometimes desirable to sequence transfer of the loads to the alternate source when more than one automatic transfer switch is connected to the same engine-generator. Utilization of such a sequencing scheme can reduce starting kilovolt-ampere capacity requirements of the generator. A fourth timer, adjustable from 0 to 1 min, will delay transfer to emergency for this and similar requirements.

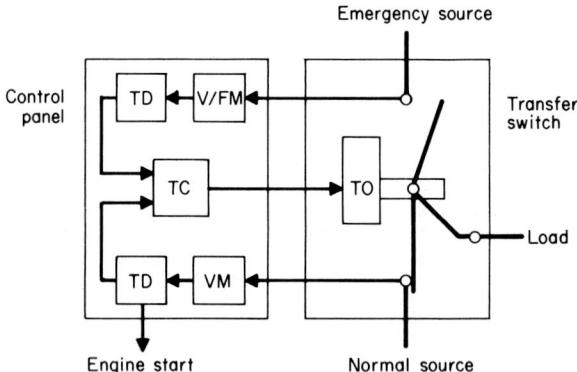

**Fig. 7-206**   Block diagram of a typical automatic transfer switch. *TD* = time delay; *TO* = transfer operator; *TC* = transfer controls; *V/FM* = voltage/frequency monitor; *VM* = voltage monitor.

TRANSFER OF MOTOR LOADS    Automatic transfer switches operate rapidly, with total operating time usually less than 0.5 s, depending upon the manufacturer and the rating of the transfer switch. Therefore, transferring motor loads may require special consideration in that the residual voltage of the motor may be out of phase with that of the power source to which the motor is being transferred. On transfer, this phase difference may cause damage to the motor, and the excessive current drawn by the motor may trip the overcurrent protective device. Motor loads above 50 hp with relatively low load inertia in relation to torque requirements, such as pumps and compressors, may require special controls.

Automatic transfer switches can be provided with accessory controls that disconnect motor controllers prior to transfer and reconnect them after transfer when the residual voltage has been substantially reduced. Automatic transfer switches can also be provided with in-phase monitors that allow transfer between sources when both sources are synchronized. Another approach is to use a three-position transfer switch with accessory controls that allow the switch to pause in the neutral position while the residual voltage decays substantially before completing the transfer.

Of the above approaches, many engineers prefer the use of the in-phase monitor because loads are not intentionally disconnected for a period of time, there is no control interwiring between the automatic transfer switch and the motor controllers, and the transfer switch remains a two-position double-throw device.

MANUALLY OPERATED TRANSFER SWITCHES    Though this discussion pertains primarily to *automatic* transfer switches, there are applications that justify the use of *nonautomatic* transfer switches. Manually operated transfer devices are commonly used in applications where operating personnel are available and the load is not of an emergency nature requiring immediate automatic restoration of power.

Typical applications are found in health-care-facility equipment systems, industrial plants, sewage plants, centers for civil defense control, farms, residences, telephone buildings, and other communication facilities. Double-throw knife switches and safety switches are sometimes used for these applications. However, many of these devices must be operated under "no-load" conditions; i.e., operators must be sure that they have disconnected the load by opening breakers or other load-interrupting devices before they operate the switch. Though some double-throw switches are load-break and may be rated as single-throw switches, they may not be suitable for transferring all classes of load between two unsynchronized power sources. Because these devices are marginal adaptations, there often is a reluctance on the part of personnel to operate them.

There are nonautomatic transfer switches that meet the safety requirements of UL 1008. Such switches are electrically operated by manually actuating a control station. The pole construction of these switches is similar to that of automatic transfer switches, and thus they have the ability to transfer all classes of load between two unsynchronized power sources.

Whether the transfer switch be automatic or nonautomatic, careful consideration should be given to its coordination with upstream overcurrent protective devices. Furthermore, the selection of the transfer switch as well as other emergency power equipment should be based on the required degree of reliability, elimination of unnecessary outages, convenience of operation, and safety of personnel as needed for the application.

**462. Gas-turbine generators.** In the larger sizes, from about 700 kW up, the turbine-driven generator offers a number of advantages as a source of power. These units are small in size and weight when compared with other types of power sources of the same capacity. They can provide much more power capability in the same space required for standby power-using reciprocating engines.

The installation of gas-turbine generators is relatively simple because foundations either are small and light or are not required at all for some skid-mounted units. Vibration is minimal. Cooling water is not required because the unit is normally air-cooled. However, soundproofing of the room or location in which the unit is installed may be required if the high-frequency vibrations emitted are objectionable. Effective equipment for soundproofing is available.

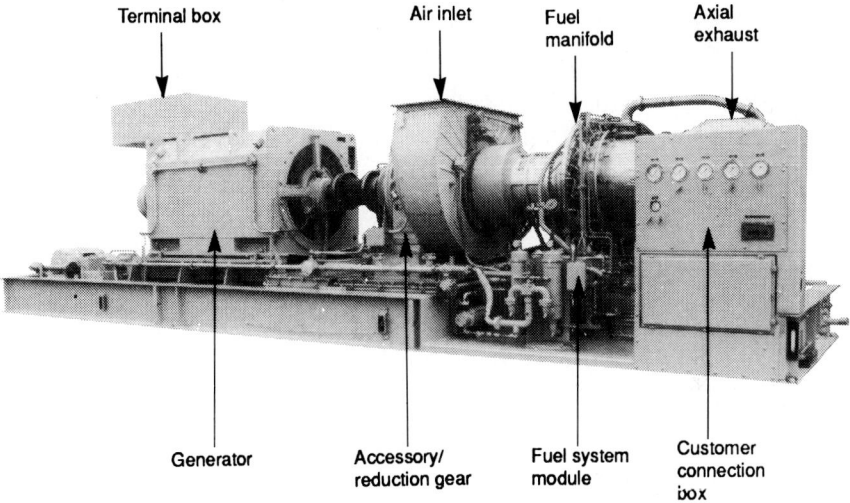

**Fig. 7-207**   4.4-MW gas turbine generator set. [Solar Turbines Inc.]

Other applications in which the gas-turbine generator has been particularly successful include cogeneration systems and utility peak-load service.

Figure 7-207 shows an unclosed gas turbine generator set that provides 4370 kW, 60 Hz (continuous-duty rating at ISO conditions), along with 165,600 lb/h of clean, dry exhaust heat at 927°F for many commercial and industrial applications. The major components are, left to right, the generator, the accessory-reduction gear, and the gas turbine engine. The unit's Centaur Taurus gas turbine is a single-shaft, constant-speed, axial-flow machine with 12 air compressor stages, an annular combustor, and three turbine stages. The unit is mounted on a single skid and measures about 32 ft long, 8 ft wide, and 8.5 ft high. Its weight is approximately 42,500 lb.

# Outside Distribution

## POLE LINES: GENERAL, CONSTRUCTION, AND EQUIPMENT

**1. The reports of the Committee on Overhead Line Construction of the National Electric Light Association** contain what are probably the best and most complete specifications for pole-line construction for lighting and power distribution that have ever been compiled. Some of the material in this division regarding pole lines has been abstracted from those reports. The National Electric Light Association (NELA) has been reorganized and succeeded by the Edison Electric Institute (EEI). As specifications and standards of the NELA organization are revised, they will be known as those of the EEI.

**2. The National Electrical Safety Code,** IEEE ANSI-C2, comprises an exhaustive collection of rules for proper installation, maintenance, and operation of electrical distribution systems. Companies are now, in general, endeavoring to follow these rules. Ultimately, it is believed, their adoption will be universal.

**3. There are three principal types of overhead line construction:**

1. Bare or weatherproof covered wire supported by insulators mounted on or hung from crossarms located on buildings or near the tops of poles.

2. Weatherproof covered wire supported by insulators mounted on racks on the sides of buildings or poles.

3. Insulated aerial cable which either is self-supporting from clamps bolted to sides of poles or crossarms or is supported by messenger cable from clamps bolted to the sides of poles or crossarms. See Div. 2 for descriptions of the types of wire and cable.

**4. There are four main types of poles used for overhead line construction:**

1. Wood poles
2. Steel poles
3. Reinforced-concrete poles
4. Structural-steel fabricated towers

**5. The most common woods for poles are:**

1. Western red cedar, found in the Rocky Mountain regions from southern Alaska to the central part of California

2. Northern white cedar, found in Michigan, Wisconsin, Minnesota, and elsewhere throughout northeastern United States and southeastern Canada

3. Southern yellow pine, found in Virginia, North and South Carolina, Tennessee, and south into Florida and west to Texas

4. Chestnut, found in Virginia, West Virginia, and some of the other southern states, virtually all blight-killed but in a good state of preservation

5. Redwood, found on the Pacific coast

Of these, the yellow pine is most commonly used in the eastern states, red or white cedar in the central states, and redwood on the west coast. According to the *Standard Handbook for Electrical Engineers*, about two-thirds of all poles used are cedar.

**6. Wood poles are classified** into 10 classes according to the circumference at the top of the pole and the circumference at a point 6 ft (1.8 m) from the butt, as given in the following table:

| Class..................... | 1 | 2 | 3 | 4 | 5 | 6 | 7 | 8 | 9 | 10 |
|---|---|---|---|---|---|---|---|---|---|---|
| Min top circumference, in.. | 27 | 25 | 23 | 21 | 19 | 17 | 15 | 18 | 15 | 12 |

| Length of pole, ft | Min circumference 6 ft from butt, in. | | | | | | | | | |
|---|---|---|---|---|---|---|---|---|---|---|
| **Northern White Cedar Poles** | | | | | | | | | | |
| 16 | .... | .... | .... | .... | 26.0 | 24.0 | 22.0 | No butt requirement | | |
| 18 | .... | .... | 32.5 | 30.0 | 28.0 | 25.5 | 23.5 | | | |
| 20 | 39.5 | 37.0 | 34.0 | 31.5 | 29.0 | 27.0 | 25.0 | | | |
| 22 | 41.0 | 38.5 | 36.0 | 33.0 | 30.5 | 28.0 | 26.0 | | | |
| 25 | 43.5 | 41.0 | 38.0 | 35.5 | 32.5 | 30.0 | 28.0 | | | |
| 30 | 47.5 | 44.5 | 41.5 | 38.5 | 35.5 | 33.0 | 30.5 | | | |
| 35 | 50.5 | 47.5 | 44.0 | 41.0 | 38.0 | 35.0 | 32.5 | | | |
| 40 | 53.5 | 50.0 | 46.5 | 43.5 | 40.0 | 37.0 | | | | |
| 45 | 56.0 | 52.5 | 49.0 | 45.5 | 42.0 | | | | | |
| 50 | 58.5 | 55.0 | 51.5 | 47.5 | 44.0 | | | | | |
| 55 | 61.0 | 57.5 | 53.5 | 49.5 | 46.0 | | | | | |
| 60 | 63.5 | 59.5 | 55.5 | 51.5 | | | | | | |
| **Western Red Cedar Poles** | | | | | | | | | | |
| 16 | .... | .... | .... | .... | 23.0 | 21.5 | 19.5 | No butt requirement | | |
| 18 | .... | .... | 28.5 | 26.5 | 24.5 | 22.5 | 21.0 | | | |
| 20 | 34.5 | 32.0 | 30.0 | 28.0 | 25.5 | 23.5 | 22.0 | | | |
| 22 | 36.0 | 33.5 | 31.5 | 29.0 | 27.0 | 25.0 | 23.0 | | | |
| 25 | 38.0 | 35.5 | 33.0 | 30.5 | 28.5 | 26.0 | 24.5 | | | |
| 30 | 41.0 | 38.5 | 35.5 | 33.0 | 30.5 | 28.5 | 26.5 | | | |
| 35 | 43.5 | 41.0 | 38.0 | 35.5 | 32.5 | 30.5 | 28.0 | | | |
| 40 | 46.0 | 43.5 | 40.5 | 37.5 | 34.5 | 32.0 | | | | |
| 45 | 48.5 | 45.5 | 42.5 | 39.5 | 36.5 | | | | | |
| 50 | 50.5 | 47.5 | 44.5 | 41.0 | 38.0 | | | | | |
| 55 | 52.5 | 49.5 | 46.0 | 42.5 | 39.5 | | | | | |
| 60 | 54.5 | 51.0 | 47.5 | 44.0 | | | | | | |
| 65 | 56.0 | 52.5 | 49.0 | 45.5 | | | | | | |
| 70 | 57.5 | 54.0 | 50.5 | 47.0 | | | | | | |
| 75 | 59.5 | 55.5 | 52.0 | 48.5 | | | | | | |
| 80 | 61.0 | 57.0 | 53.5 | 49.5 | | | | | | |
| 85 | 62.5 | 58.5 | 54.5 | | | | | | | |
| 90 | 63.5 | 60.0 | 56.0 | | | | | | | |

| Class | 1 | 2 | 3 | 4 | 5 | 6 | 7 | 8 | 9 | 10 |
|---|---|---|---|---|---|---|---|---|---|---|
| Min top circumference, in.. | 27 | 25 | 23 | 21 | 19 | 17 | 15 | 18 | 15 | 12 |

| Length of pole, ft | Min circumference 6 ft from butt, in. | | | | | | | | | |
|---|---|---|---|---|---|---|---|---|---|---|
| **Creosoted Southern Pine Poles** | | | | | | | | | | |
| 16 | .... | .... | .... | .... | 21.5 | 19.5 | 18.0 | No butt requirement | | |
| 18 | .... | .... | 26.5 | 24.5 | 22.5 | 21.0 | 19.0 | | | |
| 20 | 31.5 | 29.5 | 27.5 | 25.5 | 23.5 | 22.0 | 20.0 | | | |
| 22 | 33.0 | 31.0 | 29.0 | 26.5 | 24.5 | 23.0 | 21.0 | | | |
| 25 | 34.5 | 32.5 | 30.0 | 28.0 | 26.0 | 24.0 | 22.0 | | | |
| 30 | 37.5 | 35.0 | 32.5 | 30.0 | 28.0 | 26.0 | 24.0 | | | |
| 35 | 40.0 | 37.5 | 35.0 | 32.0 | 30.0 | 27.5 | 25.5 | | | |
| 40 | 42.0 | 39.5 | 37.0 | 34.0 | 31.5 | 29.0 | 27.0 | | | |
| 45 | 44.0 | 41.5 | 38.5 | 36.0 | 33.0 | 30.5 | 28.5 | | | |
| 50 | 46.0 | 43.0 | 40.0 | 37.5 | 34.5 | 32.0 | 29.5 | | | |
| 55 | 47.5 | 44.5 | 41.5 | 39.0 | 36.0 | 33.5 | | | | |
| 60 | 49.5 | 46.0 | 43.0 | 40.0 | 37.0 | 34.5 | | | | |
| 65 | 51.0 | 47.5 | 44.5 | 41.5 | 38.5 | | | | | |
| 70 | 52.5 | 49.0 | 46.0 | 42.5 | 39.5 | | | | | |
| 75 | 54.0 | 50.5 | 47.0 | 44.0 | | | | | | |
| 80 | 55.0 | 51.5 | 48.5 | 45.0 | | | | | | |
| 85 | 56.5 | 53.0 | 49.5 | | | | | | | |
| 90 | 57.5 | 54.0 | 50.5 | | | | | | | |
| **Chestnut Poles** | | | | | | | | | | |
| 16 | .... | .... | .... | .... | 22½ | 21 | 19½ | No butt requirement | | |
| 18 | .... | .... | 28 | 26 | 24 | 22 | 20½ | | | |
| 20 | 33½ | 31½ | 29½ | 27 | 25 | 23 | 21½ | | | |
| 22 | 35 | 33 | 30½ | 28½ | 26½ | 24½ | 22½ | | | |
| 25 | 37 | 34½ | 32½ | 30 | 28 | 25½ | 24 | | | |
| 30 | 40 | 37½ | 35 | 32½ | 30 | 28 | 26 | | | |
| 35 | 42½ | 40 | 37½ | 34½ | 32 | 30 | 27½ | | | |
| 40 | 45 | 42½ | 39½ | 36½ | 34 | 31½ | 29½ | | | |
| 45 | 47½ | 44½ | 41½ | 38½ | 36 | 33 | 31 | | | |
| 50 | 49½ | 46½ | 43½ | 40 | 37½ | 34½ | 32 | | | |
| 55 | 51½ | 48½ | 45 | 42 | 39 | 36 | | | | |
| 60 | 53½ | 50 | 46½ | 43½ | | | | | | |
| 65 | 55 | 51½ | 48 | 45 | | | | | | |
| 70 | 56½ | 53 | | | | | | | | |

## 7. Approximate Weights of Wood Poles in Pounds
(The MacGillis & Gibbs Co.)

| Length of pole, ft | Northern white cedar — Class | | | | | | | | | | Western red cedar — Class | | | | | | | | | |
|---|---|---|---|---|---|---|---|---|---|---|---|---|---|---|---|---|---|---|---|---|
| | 1 | 2 | 3 | 4 | 5 | 6 | 7 | 8 | 9 | 10 | 1 | 2 | 3 | 4 | 5 | 6 | 7 | 8 | 9 | 10 |
| 16 | | | | | 230 | 190 | 135 | 135 | 105 | 85 | | | | | | | | | | |
| 18 | | | 420 | 300 | 250 | 210 | 155 | 155 | 125 | 95 | | | | | | | | | | |
| 20 | 720 | 600 | 440 | 350 | 300 | 230 | 190 | 190 | 130 | 100 | 700 | 600 | 500 | 400 | 300 | 225 | 200 | 180 | 135 | 100 |
| 22 | 820 | 780 | 540 | 500 | 400 | 315 | 200 | 225 | 170 | 120 | | | | | | | | | | |
| 25 | 1,020 | 980 | 600 | 515 | 420 | 300 | 225 | 250 | 200 | 150 | 850 | 720 | 600 | 480 | 400 | 320 | 250 | 225 | 200 | 135 |
| 30 | 1,320 | 1,170 | 870 | 630 | 520 | 420 | 350 | 350 | 275 | | 1,000 | 850 | 730 | 610 | 500 | 420 | 350 | 325 | 250 | |
| 35 | 1,620 | 1,320 | 1,060 | 820 | 720 | 510 | 450 | | | | 1,200 | 1,000 | 850 | 750 | 650 | 560 | 470 | 450 | | |
| 40 | 2,040 | 1,675 | 1,280 | 1,020 | 790 | 625 | | | | | 1,500 | 1,300 | 1,100 | 900 | 800 | 700 | | | | |
| 45 | 2,640 | 1,970 | 1,535 | 1,215 | 1,080 | | | | | | 1,800 | 1,550 | 1,300 | 1,150 | 1,000 | | | | | |
| 50 | 3,200 | 2,640 | 1,860 | 1,470 | 1,380 | | | | | | 2,000 | 1,800 | 1,550 | 1,400 | 1,300 | | | | | |
| 55 | 3,800 | 2,960 | 2,260 | 1,620 | 1,560 | | | | | | 2,300 | 2,000 | 1,750 | 1,600 | 1,600 | | | | | |
| 60 | 4,500 | 3,460 | 2,640 | 1,700 | | | | | | | 2,600 | 2,200 | 2,000 | 1,900 | | | | | | |
| 65 | | | | | | | | | | | 3,200 | 2,500 | 2,300 | 2,200 | | | | | | |
| 70 | | | | | | | | | | | 3,600 | 3,000 | 2,700 | 2,600 | | | | | | |
| 75 | | | | | | | | | | | 4,200 | 3,600 | 3,100 | 3,000 | | | | | | |
| 80 | | | | | | | | | | | 5,000 | 4,200 | 3,600 | 3,500 | | | | | | |
| 85 | | | | | | | | | | | 5,500 | 4,500 | 4,000 | | | | | | | |
| 90 | | | | | | | | | | | 6,600 | 5,600 | 4,800 | | | | | | | |

**8. Wood-pole specifications.** The American National Standards Institute (ANSI) has standardized the specifications and dimensions for wood poles. ANSI specifications provide that all poles shall be free from sap rot, cracks, bird holes, plugged holes, injurious checks, and damage by marine borers. All poles shall be free from splits, shakes, hollow, and decay in the tops. Nails, spikes, and other metal shall not be present unless specifically authorized by the purchaser. There are further detailed restrictions regarding blue stain, hollow centers, shakes in the butt surface, splits, grain twist, insect damage, scars, etc., which vary with the different types of poles. Anyone interested in such details is advised to consult the ANSI.

Allowable deviations from straight poles are illustrated in Fig. 8-1. Dimension A should be taken from the following table:

| *A*, ft | 3½ | 4 | 5 | 5½ | 6 | 6½ | 7 | 7½ | 8 | 8½ | 9 | 9½ | 10 | 10½ | 11 |
|---|---|---|---|---|---|---|---|---|---|---|---|---|---|---|---|
| Total length of pole, ft............... | 16–18 | 20–22 | 25 | | 30 | 35–40 | 45 | 50 | 55 | 60 | 65 | 70 | 75 | 80 | 85 | 90 |

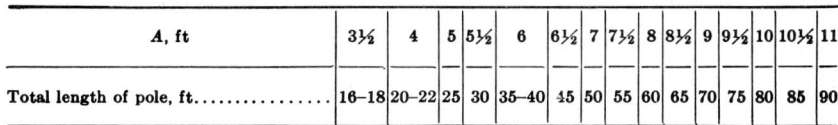

I- Sweep in one plane and one direction

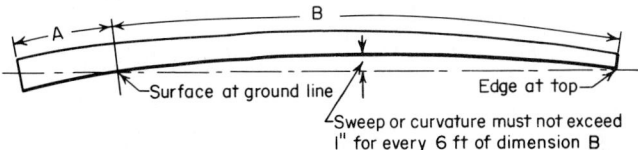

II–Sweep in two planes (Double Sweep) or in two directions in one plane (Reverse Sweep)

(Applies to western red cedar and southern pine poles only) For northern white cedar and chestnut poles the sweep in each plane must not exceed the value in I

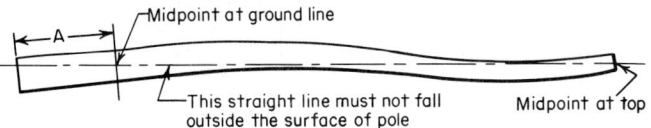

III–Short crook where the reference axes are approximately parallel

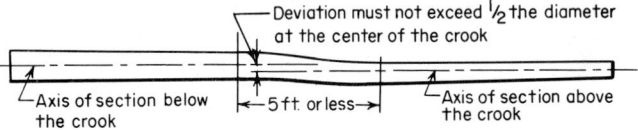

IV –Short crook where axes of sections above and below the crook coincide or are practically coincident

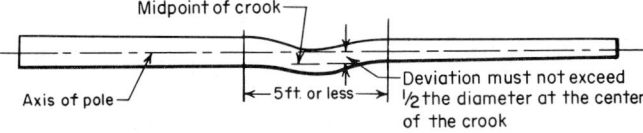

V- Short crook where axis of section above short crook is not parallel or coincident with axis below the crook

**Fig. 8-1**   Allowable sweeps and crooks in wood poles. [American National Standards Institute]

**9. Preservation of poles.**  Owing to the continually increasing scarcity of timber and the labor cost of replacing decayed poles, general practice is to impregnate poles with some substance which resists or retards decay. Southern pine poles generally are pressure-treated for their entire length. Common practice is to treat only the butt of poles of other woods, because since these woods are naturally resistant to decay, practically all decay occurs within the portion 1½ ft (0.46 m) above or below the groundline.

**10. Methods of butt treating** (Naugle Pole & Tie Co.).   The specifications for each of the three methods of treating require that poles be seasoned at least four *seasoning months* before the treatment (see Sec. 11). All fibrous inner bark and foreign substances must be thoroughly removed from that portion of the pole between the points 1½ ft above and 1½ ft below the groundline. In general, the methods of treatment are as follows. *Treatment A* provides for a continuous submersion in hot Carbolineum (a coal-tar derivative) for a minimum interval of 15 min. *Treatment AA* provides for a continuous submersion in hot creosote for a minimum interval of 15 min. *Treatment B* provides for a continuous submersion in hot creosote for a minimum interval of 4 h, to be followed by a continuous submersion in cold creosote for a minimum interval of 2

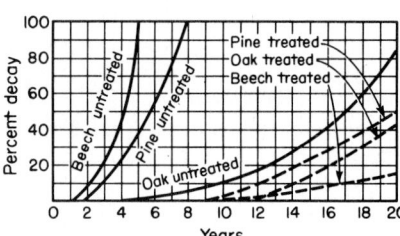

**Fig. 8-2**   Life of treated and untreated poles.

h. In each of these three treatments the poles are placed in upright tanks with the butts continuously submerged to a height of 1½ ft above the groundline.

Figure 8-2 shows, as would be expected, that the softer and more porous woods which suffer most rapid decay are most benefited and have the longest life after treatment. Such woods can absorb the most oil. The cost of treatment varies with the amount of oil injected and with local conditions.

**11. Calendar months are rated in equivalent seasoning months as follows:**

| Calendar month...... | Jan. | Feb. | Mar. | Apr. | May | June | July | Aug. | Sept. | Oct. | Nov. | Dec. |
| --- | --- | --- | --- | --- | --- | --- | --- | --- | --- | --- | --- | --- |
| Seasoning month..... | ⅛ | ⅛ | ¼ | ½ | ¾ | 1 | 1 | 1 | 1 | ¾ | ⅜ | ⅛ |

Poles which have been properly piled for seasoning must be seasoned for a total of 4 seasoning months, counting the seasoning months as given in the above table. For example, poles which are cut in January or February must be seasoned to the end of August before the time will have equaled 4 seasoning months.

**12. Steel poles are made in three general types:** (1) latticed, (2) expanded truss, and (3) tubular. The latticed type is made from structural-steel angles or channels joined and braced by latticed crossbars. The expanded truss is made from structural H sections, the web of which has been sheared by a rotary shear at fixed intervals so as to remove some of the metal. The pole is then heated to a cherry red and expanded by a special machine which grips the flanges and pulls them apart to the desired width (see Fig. 8-3). Latticed and expanded-truss steel poles are used for medium-voltage transmission lines with spans 250 to 350 ft (76.2 to 106.7 m) in length. Their life is 25 to 50 years or more, depending on the adequacy of painting and upkeep. Tubular poles are built up from lengths of steel tubing, using a large-diameter tube for the base and successively smaller tubes for the upper sections. They were extensively used for street-railway and trolleybus wiring in cities, where appearance is important. Their higher cost prevented their use through open country.

**13. Fabricated steel towers** are made from four main structural angles as uprights, cross-braced with smaller structural angles. Steel towers are ordinarily used for high-voltage lines with spans of more than 350 ft (106.7 m).

**14. The design of reinforced-concrete poles** requires considerable skill. When persons unfamiliar with concrete-pole design must build them, they had best accept the proportions of poles that have been built and are giving good service. Figure 8-4 shows some of

the major considerations. The pole is proportioned for a 150-ft (45.7-m) span to withstand
a gale successfully, with the wind at a velocity of 70 mi/h (113 km/hr) and ½ in (12.7 mm)
of ice on the wires. The horizontal load thus
imposed by the wind on all the 18 wires,
tending to overturn the pole, is 2100 lb (9341
N). The sides of the reinforcing bars are 1¼
in (31.75 mm) in from the faces of the pole.
The concrete is a 1:2:4 mixture. It should be
mixed wet, using carefully selected materials with the fine aggregate next to the
forms. Air bubbles should be eliminated by
careful tamping or churning. Corners of the
pole should be chamfered off. The square
reinforcing rods, which are of the mechani-

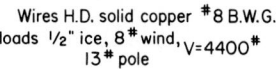

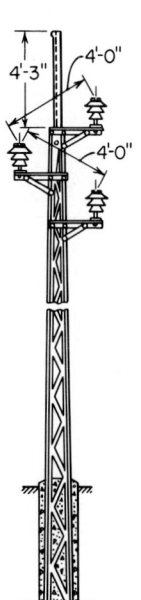

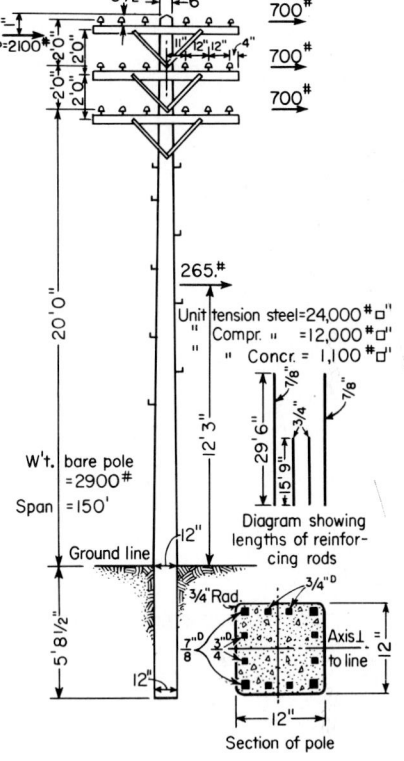

**Fig. 8-3**   Bates expanded steel pole.

**Fig. 8-4**   A 30-ft (9.1-m) reinforced-concrete
pole. [Universal Portland Cement Co.]

cal-bond type, are bound together by a web system not shown in the illustration. The web
system consists of a spiral of No. 12 steel wire wound outside the rods and securely bound
to them. The rods are also secured together with horizontal ties 1 in (25.4 mm) wide and
¼ in (6.35 mm) thick, spaced 3 to 5 ft (0.9 to 1.5 m) apart. The reinforcement thus forms an
independent skeleton which can be assembled and lowered into the forms. It is likely
that it is most economical to cast poles exceeding 35 ft (10.7 m) in height in their final
vertical positions. Shorter poles are erected with a derrick. Gains for crossarms and holes
for bolts are cast in poles. Metal pole steps may be cast in solid also.

**15. Reinforced-concrete poles** (*Standard Handbook*) are the most durable and usually the
most expensive. The life of a properly designed concrete pole is practically unlimited.
The facility with which special purposes can be served with reinforced concrete is also a
great advantage. The exterior form can easily be modified to harmonize with any desired
scheme of decoration. When it is desired to lead wires from the pole top to ground, the
poles can be made hollow, and thus at a slight additional cost the wires are completely
hidden from view and protected from the weather. Concrete poles may fail, but they will
not fall to the ground. The principal drawback to this form of construction has been the

cost and the difficulty of manufacture. The poles are heavy and cumbersome to transport, so that, when possible, it is well to make them in the neighborhood where they are to be used. Both concrete and steel poles can be transported in small packages over mountains and erected on the spot, but in this respect steel is much superior to concrete. Concrete poles are classified according to the horizontal load which can be applied to the pole 2 ft (0.6 m) from the top:

| Class | *A* | *B* | *C* | *D* | *E* |
|---|---|---|---|---|---|
| Horizontal load applied 2 ft from top, lb | 4000 | 3000 | 2000 | 1500 | 1000 |

**16. Depth to set poles in the ground.**   One rule is that poles should, on straight lines, be set in the ground one-sixth of their lengths. The following table indicates good practice for normal soils:

| Pole length, over all, ft | Depth to set in ground, ft | | Pole length, over all, ft | Depth to set in ground, ft | |
|---|---|---|---|---|---|
| | Straight lines | Curves, corners, and points of extra strain | | Straight lines | Curves, corners, and points of extra strain |
| 30 | 5.5 | 6.0 | 55 | 7.0 | 7.5 |
| 35 | 6.0 | 6.5 | 60 | 7.5 | 8.0 |
| 40 | 6.0 | 6.5 | 65 | 8.0 | 8.5 |
| 45 | 6.5 | 7.0 | 70 | 8.0 | 8.5 |
| 50 | 7.0 | 7.5 | 75 | 8.5 | 9.0 |

**17. The size of pole** to use cannot be definitely specified without knowing local conditions. Municipal ordinances sometimes regulate the heights of poles and wires. If there are trees along the line, the wires should be carried entirely above them or through the lower branches, which interfere less than do the higher ones, on which the foliage is thicker. For lines on highways it is usually customary to place the lowest wire at least 18 ft (5.5 m) above the highway, and 21 ft (6.4 m) is better. Railway companies frequently specify 22 ft (6.7 m) between the top of the rail and the lowest crossarm. Wires should be at least 15 ft (4.6 m) above sidewalks. The height of the pole will depend upon the number of crossarms to be carried. It is desirable to avoid abrupt changes in the level of wire. Hence, if the line runs uphill and downdale, the longer poles should be used in the valleys.

Guy wires should, except when otherwise provided by ordinance, be at least 18 ft above a highway and 12 ft (3.7 m) above a sidewalk.

Also in cities it is good practice to use 35-ft (10.7-m) poles to carry either one or two crossarms, 40-ft (12.2-m) poles to carry three or four crossarms, and 45-ft (13.7-m) poles to carry more than four crossarms. For suburban lines 30-ft (9.1-m) poles are often used. For very light lines carrying only three or four wires, 6-in (152-mm) poles 25 ft (7.6 m) long are sometimes used, though so light a pole is inexpedient if the number of wires is likely to increase. The height of a pole is always considered as the total length overall.

**18. Poles should be spaced,** in straight portions of a line, about 125 ft (38.1 m) apart. The usual maximum is 150 ft (45.7 m), and the minimum about 100 ft (30.5 m). In curves and at corners the spans should be about as indicated in Fig. 8-5. For rural lines the span may be lengthened to 200 or 250 ft (61 or 76.2 m). Spans up to 400 to 500 ft (122 or 152.4 m) have been used for rural feeder lines employing aluminum cable, steel-reinforced.

**19. Holes for poles** should be large enough to admit the poles without any slicing or chopping and should be of the same diameter from top to bottom. The diameter of the hole should always be at least large enough so that a tamping bar can be worked on all sides between the pole and the sides of the hole.

**20. Setting poles.**   On straight lines poles should be set perpendicularly. On curves poles should slant slightly so that the tension of the wires will tend to straighten them. In filling a hole after the pole is in it, only one shoveler should be employed and as many more persons as can conveniently work around the pole should tamp in the earth as the shoveler throws it in. Some of the surplus earth should be piled around the butt of the pole so that the water will drain away. Figure 8-6 illustrates the method of setting a pole with pikes.

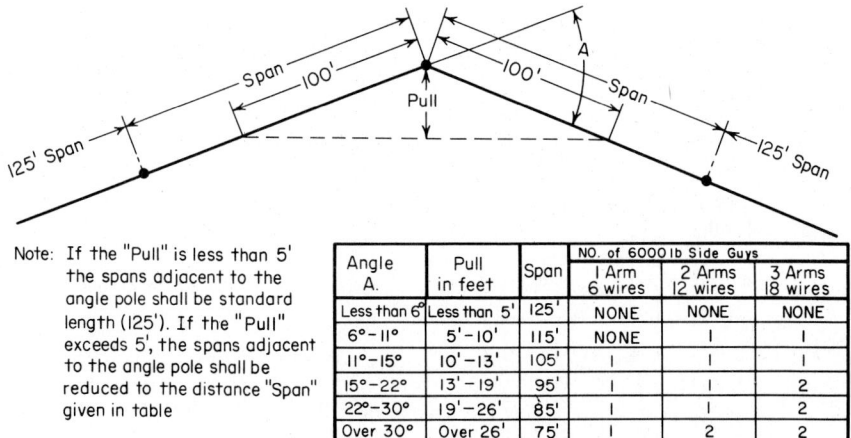

Note: If the "Pull" is less than 5' the spans adjacent to the angle pole shall be standard length (125'). If the "Pull" exceeds 5', the spans adjacent to the angle pole shall be reduced to the distance "Span" given in table

| Angle A. | Pull in feet | Span | NO. of 6000 lb Side Guys | | |
|---|---|---|---|---|---|
| | | | 1 Arm 6 wires | 2 Arms 12 wires | 3 Arms 18 wires |
| Less than 6° | Less than 5' | 125' | NONE | NONE | NONE |
| 6°–11° | 5'–10' | 115' | NONE | 1 | 1 |
| 11°–15° | 10'–13' | 105' | 1 | 1 | 1 |
| 15°–22° | 13'–19' | 95' | 1 | 1 | 2 |
| 22°–30° | 19'–26' | 85' | 1 | 1 | 2 |
| Over 30° | Over 26' | 75' | 1 | 2 | 2 |

**Fig. 8-5**   Pole spacing and side guys on curves. [Edison Electric Institute]

**21. Setting poles with a gin pole.**   A few workers can set a large pole with a gin pole, as suggested in Fig. 8-7. The gin pole can be a short wooden pole or, if the poles to be raised are not too heavy, a length of wrought-iron pipe. The gin need be only one-half as long as the pole to be raised. In setting a pole the gin is first raised to an almost vertical position with its top over the pole hole. It is held in that position by fastening the guy lines. Then the hook of the tackle blocks is engaged in a sling around the pole, and the pole is raised, by workers or by a power unit, by pulling on the free end of the tackle-block line. When high enough so that its lower end can be slipped into it, the pole is dropped into the hole and adjusted to a vertical position with pikes, and the earth is tamped in. In modern practice gin poles are permanently mounted on trucks for transportation and are then called pole derricks. They are great savers of time and money. The methods illustrated in Figs. 8-6 and 8-7 are shown to assist those who install poles only occasionally.

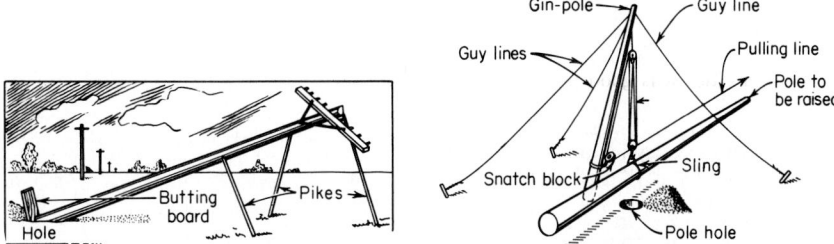

**Fig. 8-6**   Method of setting a pole with pikes.          **Fig. 8-7**   Gin pole for raising poles.

**22. Setting poles in loose and weak soils.**   On important lines it is customary to use a concrete setting for the pole (Fig. 8-8, I). A suitable mixture is 1 part cement, 3 parts sand, and 3 parts broken stone or coarse gravel.

For somewhat less important lines when the soil is fairly firm, the sand barrel (Fig. 8-8, II) is a valuable expedient. This consists of a strong barrel or barrels placed at the bottom of the hole into which the pole is set. The barrel is filled with a firm, substantial soil. By this means the pole is given a larger bearing area. Sometimes a temporary sand barrel, consisting of a special iron cylinder that is placed around the pole, is filled with firm dirt and then hoisted away.

In marshy ground another expedient for less important lines is to build a triangular brace from sound portions of old poles (Fig. 8-8, III and IV) bolted together. This provides

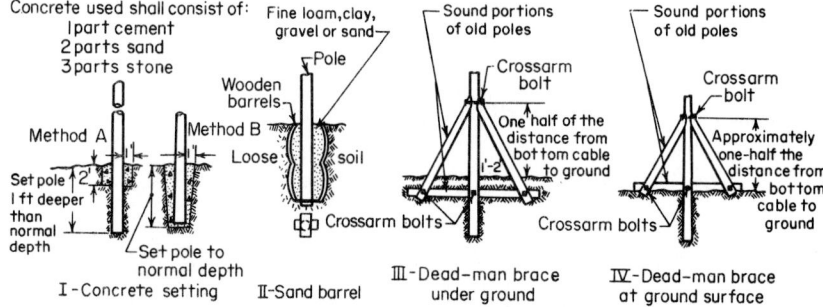

**Fig. 8-8**    Methods of setting poles in poor soils. [General Electric Co.]

a bearing area on or near the top of the soil which resists overturning of the pole. The method of Fig. 8-8, III, is preferable to the one in Fig. 8-8, IV, because the ground member will not rot so fast when it is completely underground. There is, however, somewhat more labor involved in digging the trench for the ground member. These methods are not so permanent as a concrete foundation and are recommended only when the expense of a concrete foundation does not seem warranted.

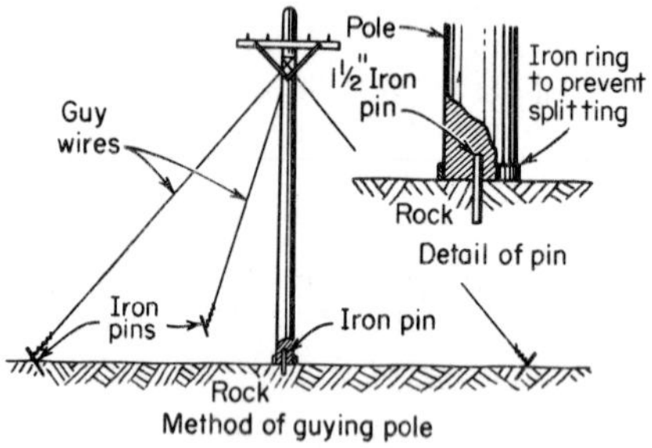

**23. When poles are set in rock,** the hole may be blasted or a hole 1½ in (38.1 mm) in diameter may be drilled in the rock (Fig. 8-9); in it is placed an iron pin that extends about 6 in (152.4 mm) above the surface. A similar hole is drilled in the butt of the pole, and the pole mounted on the pin. It must then be braced by three or four wood struts spiked to the pole 6 ft (1.8 m) from the ground, running diagonally to the rock and formed thereto, or it can be braced by

**Fig. 8-9**    Method of setting a pole on rock.

guy wires made fast to metal pins set in the rock.

**24. Wooden poles which have rotted at the butt can be reinforced** with a short stub pole set in the ground next to the regular pole and extending a few feet above the groundline. Various methods of stubbing are shown in Fig. 8-10, with the required materials in Fig. 8-11. In all cases two bands should be employed with spacings as given below. In Fig. 8-10 only the top band of the two is illustrated in several cases. If a wire wrap is employed (Fig. 8-10C), four wraps should be made around the pole and six wraps around both the pole and stub for poles up to 40 ft (12.2 m) long. For poles longer than 40 ft, 10 wraps should be used around both pole and stub. In employing the method of Type C, Fig. 8-10, the following dimensions should be used. Dimension C should be not less than 30 in (762 mm) for poles up to 45 ft (13.7 m) long or 36 in (914.4 mm) for poles 45 to 50 ft (13.7 to 15.2 m) long; dimension A should be not less than 38 in (965.2 mm) for poles up to 45 ft long or 44 in (1117.6 mm) for poles from 45 to 50 ft long; dimension B should be less than 24 in (609.6 mm) for poles up to 45 ft long or 30 in for poles from 45 to 50 ft long. Poles are generally stubbed so that the stub is across the line. Methods D and E will not be satisfactory if the stub is located along the line. The other methods can be used for either across-the-line or along-the-line stubbing. With Method B the ⅝-in dowels are not required if the poles are stubbed across the line.

**25. Reinforcing old poles with concrete and steel.** Wooden poles usually become unsafe because of butt rot at the groundline. Such poles can be repaired without moving the

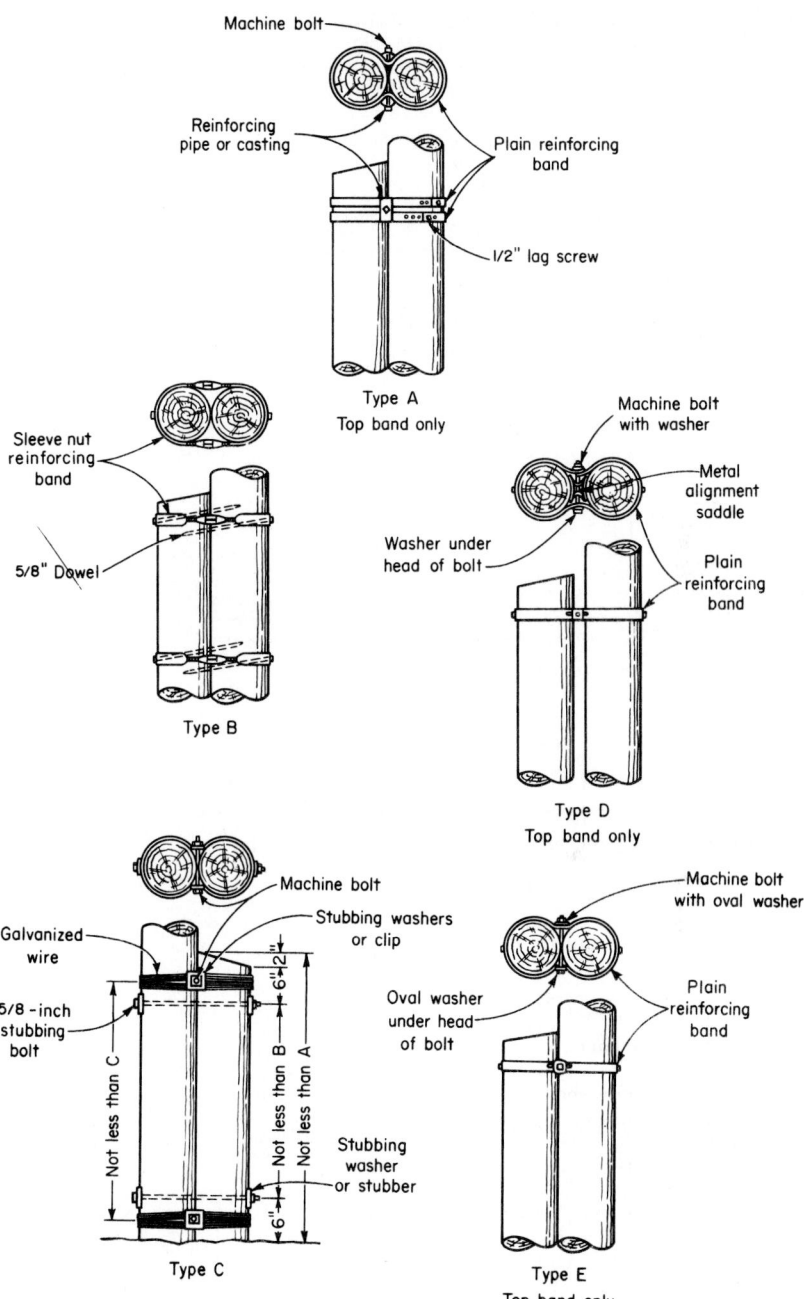

**Fig. 8-10**  Methods of reinforcing with stubs. [Hubbard & Co.]

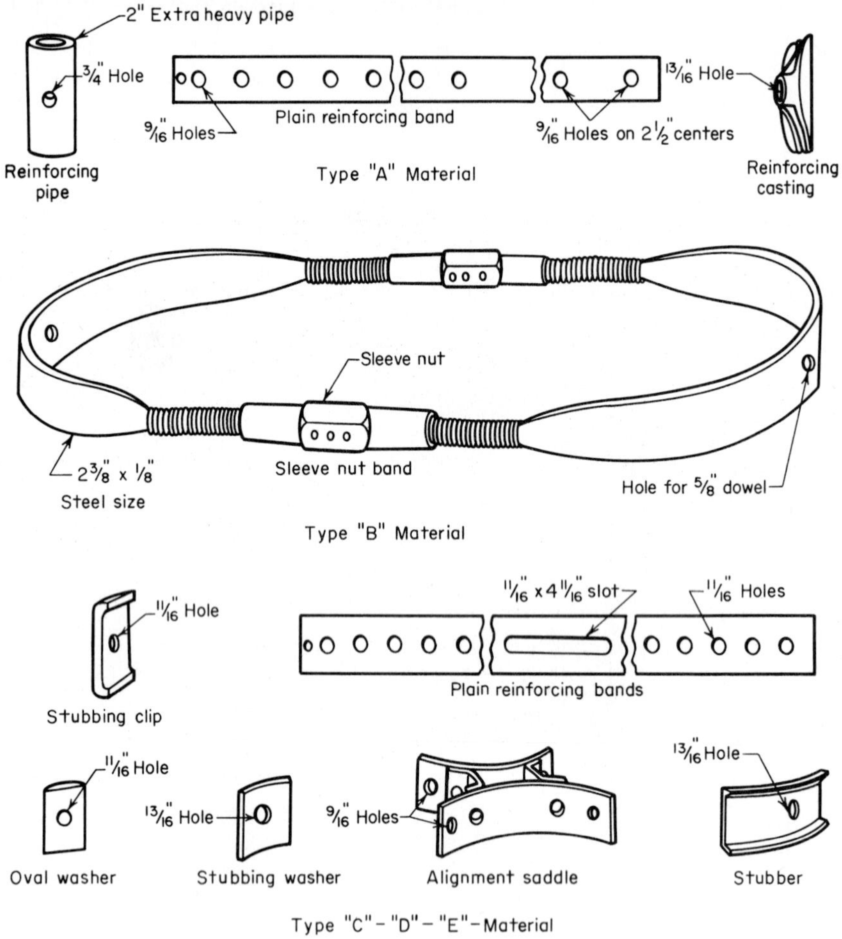

2" Extra heavy pipe
3/4" Hole
Reinforcing pipe
9/16" Holes
Plain reinforcing band
9/16" Holes on 2 1/2" centers
13/16" Hole
Reinforcing casting
Type "A" Material
Sleeve nut
2 3/8" x 1/8"
Steel size
Sleeve nut band
Hole for 5/8" dowel
Type "B" Material
11/16" Hole
Stubbing clip
11/16" x 4 11/16" slot
11/16" Holes
Plain reinforcing bands
11/16" Hole
Oval washer
13/16" Hole
Stubbing washer
9/16" Holes
Alignment saddle
13/16" Hole
Stubber
Type "C" - "D" - "E" - Material


**Fig. 8-11** Materials for reinforcing with stubs. [Hubbard & Co.]

Pole
Reinforcing rod
Pointed
1/2" Mild round steel
4' to 6' long
Pointed
Concrete
Pole
Ground
Rods in pole
Reinforcing rod
Rods and concrete in place


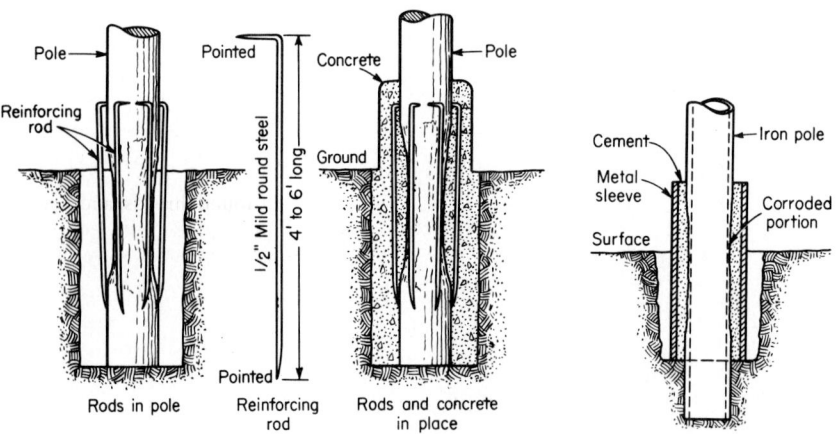

**Fig. 8-12** Reinforcing a pole with concrete.

Cement
Iron pole
Metal sleeve
Corroded portion
Surface


**Fig. 8-13** Repairing a metal pole.

wires they support by reinforcing them with steel and concrete as shown in Fig. 8-12. For ordinary poles and conditions, 10 mild-steel rods ½ in (12.7 mm) in diameter and 4 to 6 ft (1.2 to 1.8 m) long are used for reinforcing. The lower end of each reinforcing rod is pointed and is driven into the portion of the butt that remains in the hole. The other end is bent at right angles, and pointed. It is driven into the pole above the groundline. A 1:2½:5 mixture of concrete is used for the main body, and a richer mixture is used for the portion above the groundline and is molded in a cylindrical sheet-iron form. The concrete extends to about 1½ ft (0.46 m) above the groundline. Poles 15 to 20 years old have been satisfactorily repaired by this method without moving the supported wires.

**26. Repairing steel poles.**   Metal poles sometimes corrode very rapidly at the ground, and when discovered the corrosion is often too far advanced to make any preventive measures effective. A very satisfactory method (Fig. 8-13) of repairing steel poles is to place a loose-fitting metal sleeve around the butt of the pole and fill the space between the two with portland cement.

**27. Crossarm bolts** are used for fastening crossarms to the poles. These bolts are standard ⅝-in galvanized machine bolts (see machine-bolt data in Div. 4 for dimensions). A square washer is used under both head and nut, as shown in Figs. 8-47 and 8-70, I.

**28. Double-arming bolts** (Fig. 8-14) give a more rigid construction and greater economy than the old block-spacer method of tying two crossarms together. The double-arming bolts are furnished with a long rolled thread, galvanized and fitted with four square nuts. A square washer is used under each nut, as shown in Figs. 8-48 and 8-70, II.

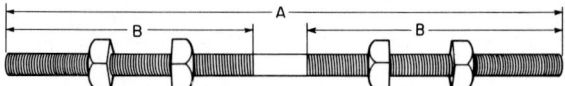

**Fig. 8-14**   Double-arming bolts. [Line Material Co.]

| Diam of bolt, in. | A, in.[a] | B, in.[a] | Diam of bolt, in. | A, in.[a] | B, in.[a] |
|---|---|---|---|---|---|
| ½ | 12 | 6 | ¾ | 12 | 5 |
|   | 14 | 6 |   | 14 | 6 |
|   | 16 | 8 |   | 16 | 8 |
|   | 18 | 8 |   | 18 | 8 |
|   | 20 | 8 |   | 20 | 8 |
|   | 22 | 8 |   | 22 | 8 |
|   | 24 | 8 |   | 24 | 8 |
| ⅝ | 12 | 6 |   |   |   |
|   | 14 | 6 |   |   |   |
|   | 16 | 6 |   |   |   |
|   | 18 | 8 |   |   |   |
|   | 20 | 8 |   |   |   |
|   | 22 | 8 |   |   |   |
|   | 24 | 8 |   |   |   |

[a]Refer to Fig. 8-14.

**29. Eyebolts** (Fig. 8-15) are furnished with a forged oval eye, rolled threads, and a square nut, galvanized. They are used for fastening guys, dead-ending insulators, and suspending insulators to crossarms, as shown in Figs. 8-41, I; 8-87; and 8-88. A square washer is used under nut and eye.

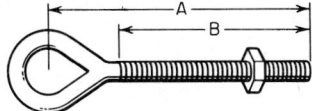

**Fig. 8-15**   Standard eyebolt. [Line Material Co.]

| Diam of bolt, in. | Opening in forged eye, in. | A, in. | B, in. | Diam of bolt, in. | Opening in forged eye, in. | A, in. | B, in. |
|---|---|---|---|---|---|---|---|
| ½ | 1¼ × 1½ | 6 | 3 | ¾ | 1½ × 2 | 6 | 4 |
|   | 1¼ × 1½ | 8 | 4 |   | 1½ × 2 | 8 | 4 |
|   | 1¼ × 1½ | 10 | 4 |   | 1½ × 2 | 10 | 4 |
|   | 1¼ × 1½ | 12 | 4 |   | 1½ × 2 | 12 | 4 |
|   | 1¼ × 1½ | 14 | 6 |   | 1½ × 2 | 14 | 6 |
|   | 1¼ × 1½ | 16 | 6 |   | 1½ × 2 | 16 | 6 |
|   | 1¼ × 1½ | 18 | 6 |   | 1½ × 2 | 18 | 6 |
|   | 1¼ × 1½ | 20 | 6 |   | 1½ × 2 | 20 | 6 |
| ⅝ | 1½ × 2 | 6 | 3 |   |   |   |   |
|   | 1½ × 2 | 8 | 4 |   |   |   |   |
|   | 1½ × 2 | 10 | 4 |   |   |   |   |
|   | 1½ × 2 | 12 | 4 |   |   |   |   |
|   | 1½ × 2 | 14 | 6 |   |   |   |   |
|   | 1½ × 2 | 16 | 6 |   |   |   |   |
|   | 1½ × 2 | 18 | 6 |   |   |   |   |
|   | 1½ × 2 | 20 | 6 |   |   |   |   |
|   | 1½ × 2 | 22 | 6 |   |   |   |   |
|   | 1½ × 2 | 24 | 6 |   |   |   |   |

**30. Double-arming eyebolts** (Fig. 8-16) are furnished with a forged oval eye and three square nuts, galvanized. The opening of the eye is 1½ by 2 in. These eyebolts are used for the same purposes as eyebolts in double-arm construction (see Figs. 8-41, IV; 8-48; and 8-86). Thimble-eye double-arming bolts, which eliminate the use of a separate thimble in attaching guy wires to the eye of the bolt, are also available.

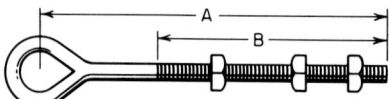

**Fig. 8-16**   Double-arming eyebolt. [Line Material Co.]

| Diam of bolt, in. | A, in. | B, in. | Diam of bolt, in. | A, in. | B, in. |
|---|---|---|---|---|---|
| ⅝ | 16 | 12 | ¾ | 20 | 16 |
|   | 18 | 14 |   | 22 | 18 |
|   | 20 | 16 |   | 24 | 20 |
|   | 22 | 18 |   |   |   |
|   | 24 | 20 |   |   |   |

**31. Eye nuts** (Fig. 8-17) are used on through bolts, eyebolts, double-arming bolts, straight and angle thimble-eye bolts, crossarm bolts, and anchor rods and for other attachments when it is desired to convert a standard threaded bolt to a thimble-eye bolt. They are

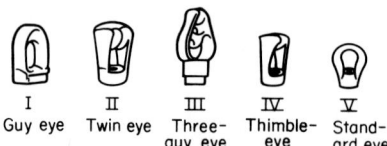

| I | II | III | IV | V |
|---|---|---|---|---|
| Guy eye | Twin eye | Three-guy eye | Thimble-eye | Stand-ard eye |

**Fig. 8-17**   Eye nuts. [Hubbard & Co.]

commonly used for dead-ending a messenger wire or span guy on the threaded end of an angle thimble-eye bolt on the opposite end of which is a down guy.

There are two types of eye nuts: (1) the standard eye (Fig. 8-17, V), which requires the use of a thimble; and (2) the thimble eye, which has a thimble forged in it. The thimble-eye type is also made in a twin style (Fig. 8-17, II) to take two guy strands and in a three-guy style (Fig. 8-17, III) to take three guy strands. The guy eye (Fig. 8-17, I) serves the same purpose as the thimble eye, merely being forged in a slightly different manner. Refer to Fig. 8-71, II, for application.

**32. Common, or buttonhead, carriage bolts** are used for attaching crossarm braces to cross-arms. They are provided with a square shoulder under the head. A round washer is used under the head, as shown in Fig. 8-47.

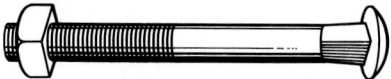

**Fig. 8-18**  Carriage bolt.

| Diam, in. | Length, in. | Length of thread. in. | Approx shipping weight, lb per 100 pieces | Diam, in. | Length, in. | Length of thread, in. | Approx shipping weight, lb per 100 pieces |
|---|---|---|---|---|---|---|---|
| ⅜ | 3 | 1¾ | 14.5 | ½ | 3 | 2½ | 26.7 |
|   | 3½ | 1¾ | 16.5 |   | 3½ | 3 | 29.2 |
|   | 4 | 1¾ | 18.3 |   | 4 | 3 | 33.3 |
|   | 4½ | 1¾ | 20.0 |   | 4½ | 3 | 36.7 |
|   | 5 | 1¾ | 21.1 |   | 5 | 3 | 38.6 |
|   | 5½ | 1¾ | 22.5 |   | 5½ | 3 | 41.2 |
|   | 6 | 1¾ | 23.3 |   | 6 | 3 | 44.0 |
|   |   |   |   |   | 7 | 3 | 50.0 |
|   |   |   |   |   | 8 | 4 | 59.0 |
|   |   |   |   |   | 10 | 4 | 72.0 |
|   |   |   |   |   | 12 | 6 | 85.0 |
|   |   |   |   |   | 14 | 6 | 99.0 |
|   |   |   |   |   | 16 | 6 | 105.0 |

**33. Lag screws** are made in both fetter-drive and gimlet-point types. The ¼- and ⁵⁄₁₆-in screws are generally of the gimlet-point type, while all screws of larger diameter are generally of the fetter-drive type. Lag screws are used for attaching crossarm braces to poles, as shown in Figs. 8-47, 8-50, and 8-60.

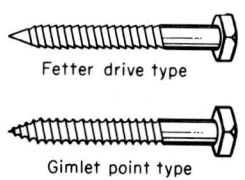

Fetter drive type

Gimlet point type

**Fig. 8-19**  Lag screws.

| Diam, in. | Length, in. | Length of thread, in., and type | Approx shipping weight, lb per 100 pieces | Diam, in. | Length, in. | Length of thread, in., and type | Approx shipping weight, lb. per 100 pieces |
|---|---|---|---|---|---|---|---|
| 1/4 | 1 1/2 | 1 1/8 G.P. | 2.0 | 1/2 | 2 1/2 | 2 F.D. | 18.4 |
| | 2 | 1 5/8 G.P. | 3.5 | | 3 | 2 1/2 F.D. | 20.9 |
| | 2 1/2 | 1 3/4 G.P. | 5.0 | | 3 1/2 | 3 F.D. | 23.4 |
| | 3 | 2 G.P. | 6.5 | | 4 | 2 1/2 F.D. | 26.0 |
| | 4 | 2 1/2 G.P. | 8.0 | | 4 1/2 | 2 7/8 F.D. | 27.8 |
| 5/16 | 2 | 1 3/4 G.P. | 5.2 | | 5 | 3 1/4 F.D. | 32.1 |
| | 2 1/2 | 2 G.P. | 6.2 | | 5 1/2 | 3 F.D. | 33.9 |
| | 3 | 2 1/4 G.P. | 7.5 | | 6 | 3 F.D. | 38.3 |
| | 3 1/2 | 2 1/2 G.P. | 9.7 | | 6 1/2 | 2 7/8 F.D. | 43.2 |
| | 4 | 2 1/2 G.P. | 11.9 | | 7 | 3 F.D. | 46.4 |
| 3/8 | 2 1/4 | 2 F.D. | 8.8 | 5/8 | 4 | 3 F.D. | 42.6 |
| | 2 1/2 | 2 F.D. | 9.7 | | 4 1/2 | 3 F.D. | 46.0 |
| | 3 | 2 F.D. | 11.0 | | 5 | 3 1/2 F.D. | 50.6 |
| | 3 1/2 | 2 1/2 F.D. | 12.8 | | 5 1/2 | 3 F.D. | 55.2 |
| | 4 | 2 7/8 F.D. | 14.6 | | 6 | 2 7/8 F.D. | 60.0 |
| | 4 1/2 | 3 F.D. | 16.4 | 3/4 | 5 | 3 F.D. | 74.5 |
| | 5 | 3 F.D. | 16.9 | | 6 | 3 1/2 F.D. | 84.9 |
| | 6 | 3 F.D. | 19.9 | | 7 | 4 F.D. | 99.4 |
| | | | | | 8 | 4 1/2 F.D. | 112.2 |

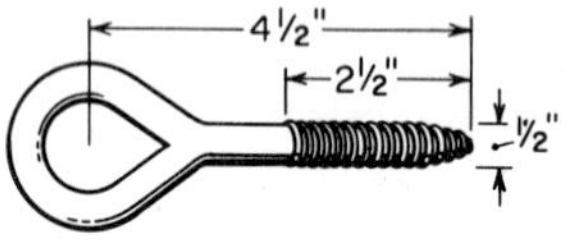


**Figure 8-20** Lag-screw eye. [Line Material Co.]

**34. Lag-screw eyes** (Fig. 8-20) are galvanized and furnished with gimlet-point, lag-screw thread. The forged eye has a 1¼- by 1½-in opening.

**35. Washers** used in overhead line construction are made of galvanized iron in both round and square types. Square washers are used under the heads and nuts of all crossarm bolts. Round washers are used under the heads of carriage bolts which fasten crossarm braces to crossarms.

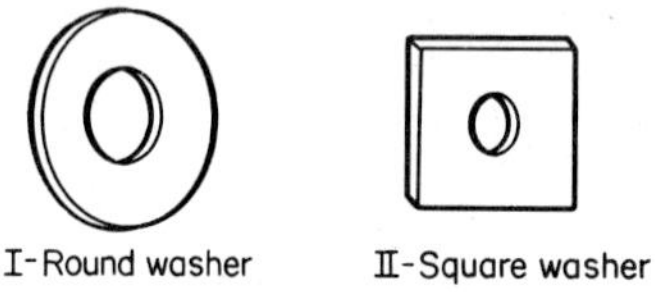


**Fig. 8-21** Washers.

| Outside diam, in. | Size of hole, in. | Thick-ness, in. | Size of bolt, in. | | Dimensions, in. | Diam of hole, in. | Size of bolt, in. |
|---|---|---|---|---|---|---|---|
| | | | Mach | Carriage | | | |
| 1 | 7/16 | 5/64 | 3/8 | | 2 × 2 × 1/8 | 11/16 | 1/2 or 5/8 |
| 1 1/4 | 1/2 | 5/64 | . . . | 3/8 | 2 1/4 × 2 1/4 × 3/16 | 13/16 | 5/8 or 3/4 |
| 1 3/8 | 9/16 | 7/64 | 1/2 | 3/8 | 3 × 3 × 3/16 | 13/16 | 5/8 or 3/4 |
| 1 3/4 | 11/16 | 1/8 | 5/8 | 1/2 | 3 × 3 × 1/4 | 13/16 | 5/8 or 3/4 |
| 2 | 13/16 | 1/8 | 3/4 | 5/8 | 4 × 4 × 3/16 | 13/16 | 5/8 or 3/4 |
| | | | | | 4 × 4 × 1/4 | 15/16 | 3/4 or 7/8 |
| | | | | | 4 × 4 × 1/2 | 1 3/16 | 1 |
| | | | | | 3 1/2 × 3 1/2 × 3/8 | 15/16 | 3/4 or 7/8 |
| | | | | | 6 × 6 × 3/8 | 1 3/16 | 1 |

### 36. Standard Wooden Crossarms
(General Electric Supply Co.)

| Pin holes Spacings, in. Center | Sides | Ends | Size, in. | Center bolt hole, in. | Brace length, in. | Size and length |
|---|---|---|---|---|---|---|
| **Electric Light Arms** | | | | | | $3\frac{1}{4} \times 4\frac{1}{4}$ |
| 28 | .... | 4 | $1\frac{17}{32}$ | $\frac{5}{8}$ | 25 | 3 ft 2 pin |
| 16 | 12 | 4 | $1\frac{17}{32}$ | $\frac{5}{8}$ | 28 | 4 ft 4 pin |
| 18 | 17 | 4 | $1\frac{17}{32}$ | $\frac{5}{8}$ | 28 | 5 ft 4 pin |
| 22 | 21 | 4 | $1\frac{17}{32}$ | $\frac{5}{8}$ | 32 | 6 ft 4 pin |
| 16 | 12 | 4 | $1\frac{17}{32}$ | $\frac{5}{8}$ | 32 | 6 ft 6 pin |
| 18 | $17\frac{1}{2}$ | 4 | $1\frac{17}{32}$ | $\frac{5}{8}$ | 32 | 8 ft 6 pin |
| 16 | 12 | 4 | $1\frac{17}{32}$ | $\frac{5}{8}$ | 32 | 8 ft 8 pin |
| 16 | $9\frac{3}{4}$ | 4 | $1\frac{17}{32}$ | $\frac{5}{8}$ | 32 | $8\frac{1}{2}$ ft 10 pin |
| $17\frac{1}{2}$ | $15\frac{3}{4}$ | 4 | $1\frac{17}{32}$ | $\frac{5}{8}$ | 42 | 10 ft 8 pin |
| 16 | 12 | 4 | $1\frac{17}{32}$ | $\frac{5}{8}$ | 42 | 10 ft 10 pin |
| 16 | $9\frac{5}{8}$ | $3\frac{7}{8}$ | $1\frac{17}{32}$ | $\frac{5}{8}$ | 42 | 10 ft 12 pin |
| **R.S.A. Arms** | | | | | | $3 \times 4\frac{1}{4}$ |
| 20 | 22 | 4 | $\frac{9}{16}$ | $1\frac{1}{16}$ | ... | 6 ft 4 pin |
| 19 | $17\frac{1}{4}$ | 4 | $\frac{9}{16}$ | $1\frac{1}{16}$ | ... | 8 ft 6 pin |
| 19 | $15\frac{1}{2}$ | 4 | $\frac{9}{16}$ | $1\frac{1}{16}$ | ... | 10 ft 8 pin |
| 16 | $12\frac{3}{8}$ | $2\frac{1}{2}$ | $\frac{9}{16}$ | $1\frac{1}{16}$ | ... | 10 ft 10 pin |
| **Western Union Arms** | | | | | | $3 \times 4\frac{1}{4}$ |
| 20 | $11\frac{1}{2}$ | 3 | $\frac{9}{16}$ | $2\frac{1}{32}$ | ... | 6 ft 6 pin |
| 21 | $11\frac{1}{2}$ | 3 | $\frac{9}{16}$ | $2\frac{1}{32}$ | ... | 8 ft 8 pin |
| 22 | $11\frac{1}{2}$ | 3 | $\frac{9}{16}$ | $2\frac{1}{32}$ | ... | 10 ft 10 pin |
| **Pony Telephone Arms** | | | | | | $2\frac{3}{4} \times 3\frac{3}{4}$ |
| 17 | .... | $3\frac{1}{2}$ | $1\frac{9}{32}$ | $\frac{5}{8}$ | ... | 24 in. 2 pin |
| 23 | .... | $3\frac{1}{2}$ | $1\frac{9}{32}$ | $\frac{5}{8}$ | ... | 30 in. 2 pin |
| 29 | .... | $3\frac{1}{2}$ | $1\frac{9}{32}$ | $\frac{5}{8}$ | 25 | 36 in. 2 pin |
| 16 | $9\frac{1}{2}$ | $3\frac{1}{2}$ | $1\frac{9}{32}$ | $\frac{5}{8}$ | 28 | 42 in. 4 pin |
| 16 | $9\frac{3}{4}$ | $3\frac{1}{2}$ | $1\frac{9}{32}$ | $\frac{5}{8}$ | 28 | 62 in. 6 pin |
| 16 | $9\frac{3}{4}$ | $3\frac{3}{4}$ | $1\frac{9}{32}$ | $\frac{5}{8}$ | 28 | 82 in. 8 pin |
| 16 | $9\frac{3}{4}$ | 4 | $1\frac{9}{32}$ | $\frac{5}{8}$ | 28 | 102 in. 10 pin |
| 16 | $9\frac{5}{8}$ | $3\frac{7}{8}$ | $1\frac{9}{32}$ | $\frac{5}{8}$ | 28 | 120 in. 12 pin |
| **N.E.L.A. Arms** | | | | | | $3\frac{1}{2} \times 4\frac{1}{2}$ |
| 30 | .... | 4 | $1\frac{17}{32}$ | $1\frac{1}{16}$ | 28 | 3 ft 2 in. 2 pin |
| 30 | $14\frac{1}{2}$ | 4 | $1\frac{17}{32}$ | $1\frac{1}{16}$ | 38 | 5 ft 7 in. 4 pin |
| 30 | $14\frac{1}{2}$ | 4 | $1\frac{17}{32}$ | $1\frac{1}{16}$ | 38 | 8 ft 6 pin |
| 30 | 12 | 4 | $1\frac{17}{32}$ | $1\frac{1}{16}$ | 38 | 9 ft 2 in. 8 pin |
| **N.E.L.A. (Light) Arms** | | | | | | $3\frac{1}{4} \times 4\frac{1}{4}$ |
| 30 | .... | 4 | $1\frac{17}{32}$ | $1\frac{1}{16}$ | 28 | 3 ft 2 in. 2 pin |
| 30 | $14\frac{1}{2}$ | 4 | $1\frac{17}{32}$ | $1\frac{1}{16}$ | 38 | 5 ft 7 in. 4 pin |
| 30 | $14\frac{1}{2}$ | 4 | $1\frac{17}{32}$ | $1\frac{1}{16}$ | 38 | 8 ft 6 pin |
| 30 | 12 | 4 | $1\frac{17}{32}$ | $1\frac{1}{16}$ | 38 | 9 ft 2 in. 8 pin |
| **New England Arms** | | | | | | $3\frac{1}{4} \times 4\frac{1}{4}$ |
| 30 | .... | 3 | $1\frac{17}{32}$ | $1\frac{1}{16}$ | 33 | 3 ft 2 pin |
| 30 | $13\frac{1}{2}$ | $4\frac{1}{2}$ | $1\frac{17}{32}$ | $1\frac{1}{16}$ | 36 | 5 ft 6 in. 4 pin |
| 30 | $13\frac{1}{2}$ | $4\frac{1}{2}$ | $1\frac{17}{32}$ | $1\frac{1}{16}$ | 36 | 7 ft 9 in. 6 pin |
| 30 | $13\frac{1}{2}$ | $4\frac{1}{2}$ | $1\frac{17}{32}$ | $1\frac{1}{16}$ | 36 | 10 ft 8 pin |
| **New England Power Arms** | | | | | | $3\frac{3}{4} \times 4\frac{3}{4}$ |
| 30 | .... | 3 | $1\frac{17}{32}$ | $1\frac{1}{16}$ | 33 | 3 ft 2 pin |
| 30 | $13\frac{1}{2}$ | $4\frac{1}{2}$ | $1\frac{17}{32}$ | $1\frac{1}{16}$ | 36 | 5 ft 6 in. 4 pin |
| 30 | $13\frac{1}{2}$ | $4\frac{1}{2}$ | $1\frac{17}{32}$ | $1\frac{1}{16}$ | 36 | 7 ft 9 in. 6 pin |
| 30 | $13\frac{1}{2}$ | $4\frac{1}{2}$ | $1\frac{17}{32}$ | $1\frac{1}{16}$ | 36 | 10 ft 8 pin |
| **Pacific Arms** | | | | | | $3\frac{1}{4} \times 4\frac{1}{4}$ |
| 28 | .... | 4 | $1\frac{17}{32}$ | $\frac{5}{8}$ | 32 | 3 ft 2 pin |
| 28 | 12 | 4 | $1\frac{17}{32}$ | $\frac{5}{8}$ | 32 | 5 ft 4 pin |
| 28 | 12 | 4 | $1\frac{17}{32}$ | $\frac{5}{8}$ | 32 | 7 ft 6 pin |
| 28 | 12 | 4 | $1\frac{17}{32}$ | $\frac{5}{8}$ | 42 | 9 ft 8 pin |
| 28 | 12 | 4 | $1\frac{17}{32}$ | $\frac{5}{8}$ | 42 | 11 ft 10 pin |

**37. Wooden crossarms** are generally made of Douglas fir, which, according to tests made by the U.S. Forest Service, is best suited for the purpose. Longleaf yellow pine and Norway pine also are used for crossarms. Crossarm dimensions have not actually been standardized throughout the country. The dimensions of those carried as standard by one manufacturer are given in Sec. 36. Great care should be exercised in ordering crossarms to give full specifications. It is generally best to furnish a sketch (Fig. 8-22) with accompanying dimensions for the required boring.

**Fig. 8-22**   Sketch for boring of crossarm.

Figure 8-23 shows the dimensions of crossarms recommended by the Edison Electric Institute. These arms have a spacing between center pins of 30 in (762 mm), which is believed to provide a safe climbing space. Arms of the EEI cross-sectional dimensions (Fig. 8-23) have not come into extensive use. It is modern practice not to paint crossarms as soon as they are made. They either are treated with a wood preservative or are permitted to season naturally for at least 3 months and are then painted with two coats of green white-lead paint before erection. Crossarms of Southern pine generally are pressure-treated against decay. It is not general practice to treat crossarms of other woods. No crossarm having a spacing of less than 20 in (508 mm) between center pins or 12 in (304.8 mm) between side pins should be used. The six-pin arm (Fig. 8-23) is recommended for general use.

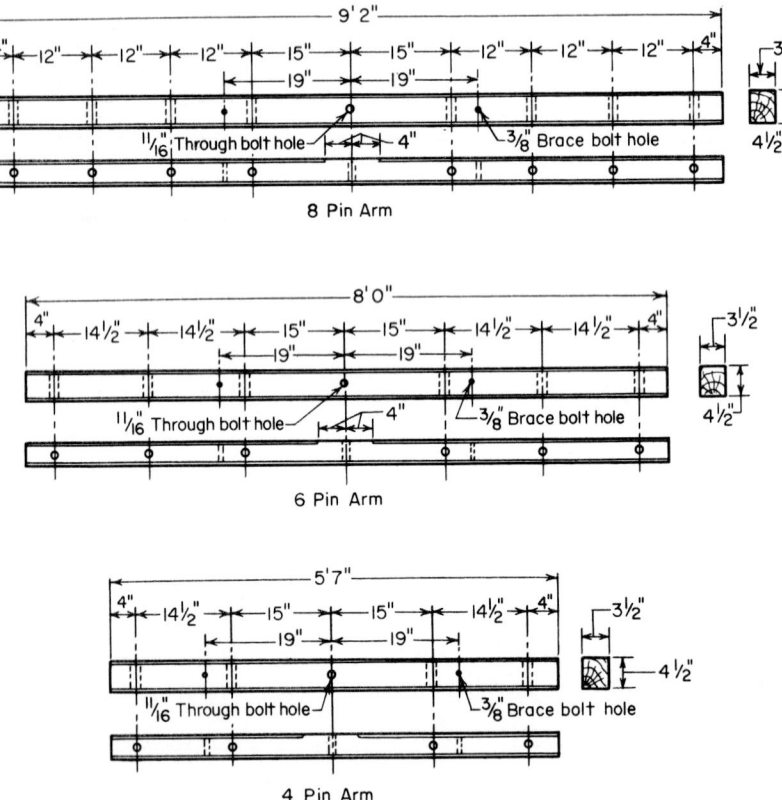

**Fig. 8-23**   Crossarm dimensions recommended by the Edison Electric Institute.

**38. Galvanized-steel crossarms** are made from standard steel angles in three types, as given in Sec. 39. They cost somewhat more than wooden crossarms but are more durable and will carry heavier loads. They are used for installations subjected to heavy strains and for higher-voltage lines.

### 39.  Types of Galvanized-Steel Crossarms
(Hubbard & Co.)

| Number of pins | Length, in. | Brace style | Pin spacing, in. | | Size of angle, in. | Approximate shipping weight, lb per 100 pieces |
|---|---|---|---|---|---|---|
| | | | Pole pins | Side pins | | |
| Telephone Arms—Pinholes, $1\frac{3}{16}$ In., Pole-mounting Hole, $1\frac{1}{16}$ In. | | | | | | |
| 2 | 20 | Flat | 16 | .... | $3 \times 2 \times \frac{3}{16}$ | 575 |
| 4 | 40 | Flat | 16 | 10 | $3 \times 2 \times \frac{3}{16}$ | 1,125 |
| 6 | 60 | Flat | 16 | 10 | $3 \times 3 \times \frac{1}{4}$ | 2,700 |
| 8 | 80 | Flat | 16 | 10 | $3 \times 3 \times \frac{1}{4}$ | 3,600 |
| 10 | 100 | Flat | 16 | 10 | $3 \times 3 \times \frac{1}{4}$ | 4,510 |
| Electric-light Arms—Pinholes, $1\frac{3}{16}$ In., Pole-mounting Hole, $1\frac{1}{16}$ In. | | | | | | |
| 2 | 36 | Flat | 30 | .... | $3 \times 3 \times \frac{1}{4}$ | 1,625 |
| 4 | 65 | Flat | 30 | $14\frac{1}{2}$ | $3 \times 3 \times \frac{1}{4}$ | 2,915 |
| 6 | 94 | Angle | 30 | $14\frac{1}{2}$ | $3\frac{1}{2} \times 3\frac{1}{2} \times \frac{5}{16}$ | 6,215 |
| 8 | $117\frac{3}{4}$ | Angle | 30 | $13\frac{5}{8}$ | $3\frac{1}{2} \times 3\frac{1}{2} \times \frac{5}{16}$ | 7,700 |
| Power transmission Arms—Pinholes, $1\frac{3}{16}$ In., Pole-mounting Hole, $1\frac{1}{16}$ In. | | | | | | |
| 2 | 28 | Flat | 24 | .... | $3 \times 3 \times \frac{1}{4}$ | 1,290 |
| 2 | 40 | Flat | 36 | .... | $3 \times 3 \times \frac{1}{4}$ | 1,840 |
| 2 | 52 | Flat | 48 | .... | $3 \times 3 \times \frac{1}{4}$ | 2,410 |
| 4 | 76 | Angle | 24 | 24 | $3 \times 3 \times \frac{1}{4}$ | 3,490 |
| 2 | 80 | Angle | 74 | .... | $3\frac{1}{2} \times 3\frac{1}{2} \times \frac{5}{16}$ | 5,280 |
| 4 | 116 | Angle | 38 | 36 | $3\frac{1}{2} \times 3\frac{1}{2} \times \frac{5}{16}$ | 7,645 |

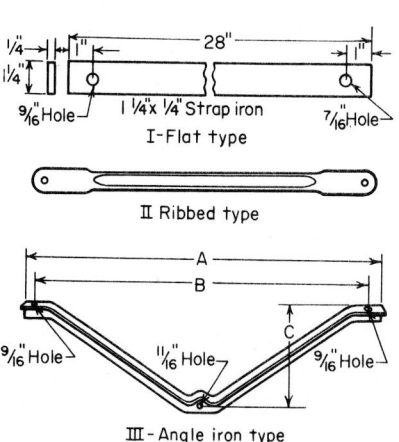

**Fig. 8-24**  Standard crossarm braces. [Hubbard & Co.]

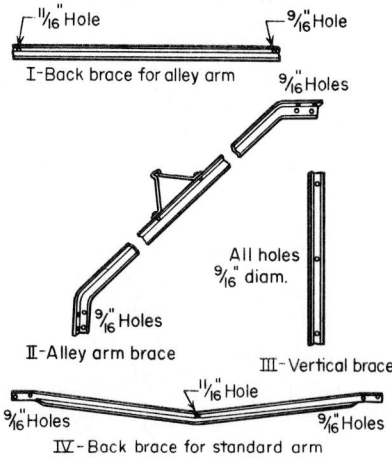

**Fig. 8-25**  Special types of crossarm braces. [Hubbard & Co.]

**40. Crossarm braces** are made from galvanized flat strap iron (Fig. 8-24, I), ribbed strap iron (Fig. 8-24, II), or angle iron (Fig. 8-24, III). Braces are attached to the back of each crossarm by a carriage bolt, with the head of the bolt at the front with a round washer under it (Figs. 8-47 and 8-60). The braces are secured to the pole with a through bolt or a lag screw. The flat type is used for ordinary loads. It is made in lengths varying in 2-in (50.8-mm) increments from 20 to 32 in (508 to 812.8 mm), the 28-in (711.2-mm) length being the most common. The ribbed type, being about 25 percent stronger, is used for heavier loads. The angle-iron type is used for very heavy lines. The standard sizes are given in Table 41, the first four being the Edison Electric Institute standard for power company use. The alley-arm brace (Fig. 8-25, II) is used to support alley-type crossarms (see Sec. 62). The vertical brace (Fig. 8-25, III) supports the upper crossarms from the lowest crossarm in alley-arm construction (Fig. 8-49). The back brace (Fig. 8-25, IV) is fastened at the back of the crossarm and on the back of the pole. It is used as an alternative to double-arm construction at crossings and at abrupt turns in the line. A back brace for alley arms is shown in Fig. 8-25, I.

**41.  Angle Crossarm Braces**[a]
(Hubbard & Co.)

| Dimensions, in. | | | | Approx shipping weight, lb per 100 pieces |
|---|---|---|---|---|
| Angle size | A | B | C | |
| $1\frac{1}{2}$ × $1\frac{1}{2}$ × $\frac{3}{16}$ | 45 | 42 | 12 | 858 |
| $1\frac{1}{2}$ × $1\frac{1}{2}$ × $\frac{3}{16}$ | 51 | 48 | 18 | 1,067 |
| $1\frac{1}{2}$ × $1\frac{1}{2}$ × $\frac{3}{16}$ | 63 | 60 | 18 | 1,210 |
| $1\frac{3}{4}$ × $1\frac{3}{4}$ × $\frac{3}{16}$ | 75 | 72 | 22 | 1,716 |
| $1\frac{1}{2}$ × $1\frac{1}{2}$ × $\frac{3}{16}$ | 51 | 48 | 14 | 974 |
| $1\frac{1}{2}$ × $1\frac{1}{2}$ × $\frac{3}{16}$ | 40 | 37 | 12 | 781 |
| $1\frac{1}{2}$ × $1\frac{1}{2}$ × $\frac{3}{16}$ | 51 | 48 | $14\frac{3}{4}$ | 979 |
| $1\frac{3}{4}$ × $1\frac{3}{4}$ × $\frac{3}{16}$ | 63 | 60 | 18 | 1,408 |
| $1\frac{3}{4}$ × $1\frac{3}{4}$ × $\frac{3}{16}$ | 69 | 66 | 20 | 1,551 |
| $1\frac{3}{4}$ × $1\frac{3}{4}$ × $\frac{3}{16}$ | 75 | 72 | 18 | 1,639 |
| 2    × 2    × $\frac{3}{16}$ | 75 | 72 | 22 | 1,958 |

[a]Refer to Fig. 8-24, III.

**42.  Crossarm Back Braces** (Fig. 8-25, IV)

| Angle size, in. | Length over-all, in. | Approx shipping weight, lb per 100 pieces | Angle size, in. | Length over-all, in. | Approx shipping weight, lb per 100 pieces |
|---|---|---|---|---|---|
| $1\frac{1}{2}$ × $1\frac{1}{2}$ × $\frac{3}{16}$ | 48 | 550 | $1\frac{3}{4}$ × $1\frac{3}{4}$ × $\frac{3}{16}$ | 94 | 1,540 |
| $1\frac{1}{2}$ × $1\frac{1}{2}$ × $\frac{3}{16}$ | 60 | 825 | $1\frac{3}{4}$ × $1\frac{3}{4}$ × $\frac{3}{16}$ | 109 | 2,204 |
| $1\frac{1}{2}$ × $1\frac{1}{2}$ × $\frac{3}{16}$ | 72 | 1,200 | | | |

**43. Insulators** are used for insulating the electrical conduction wires and guy wires from the pole structure. The different types may be classified as pin, suspension, spool, strain, and wire-holder insulators. Either pin-type (Fig. 8-26) or suspension-type insulators (Fig. 8-30) are used for supporting the wires of the main electrical circuits. Spool insulators (Fig. 8-32) are used principally in the assembly of insulator racks (Fig. 8-33) for supporting low-voltage distribution mains from the sides of poles or buildings. Combined with a clevis, they are sometimes used for dead-ending low-voltage main electrical circuits. Strain insulators are used principally for insulating guy wires from the pole structure so that contact with the lower portion of the guy will not be dangerous. One type of porcelain strain insulator is frequently used for dead-ending low-voltage main electrical circuits or distribution wires. Wire-holder insulators are used in various assemblies for supporting service wires on buildings and for supporting, on the pole structure, taps for certain street-lighting luminaires. Insulators are made of either glass or porcelain, the choice for any particular size and voltage being determined by weighing the electrical and mechanical characteristics against first cost. Glass has been most commonly used for pin insulators (Fig. 8-26, I) for voltages up to 5000 V. Porcelain has been most commonly used for pin insulators for higher voltages and for the other types of insulators.

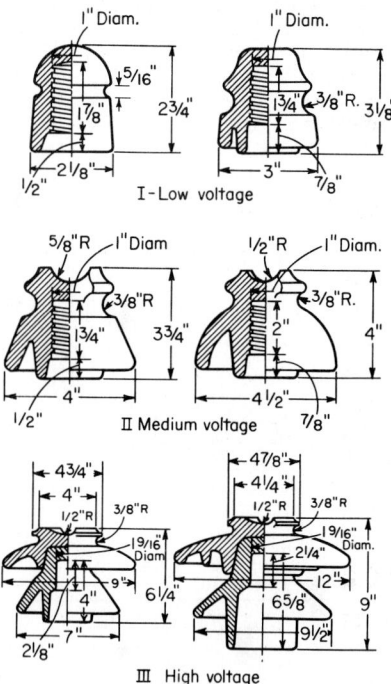

**Fig. 8-26**  Typical pin-type insulators. [Locke Insulators, Inc.]

**44. Pin-type insulators** are molded with a central threaded hole so that the insulator is supported from threaded insulator pins mounted on the crossarms. These insulators are made in various shapes and with varying contours to meet the insulating requirements for different voltage classifications. For low-voltage work the insulators are generally grooved on the side for support of the wires. Medium-voltage insulators frequently have wire grooves on both top and sides, and the higher-voltage ones are made with only top grooves. Typical pin-type insulators are shown in Fig. 8-26.

**45. Insulator pins.**  Pin insulators of the type shown in Fig. 8-26 are fastened on cross-arms with insulator pins which thread into the cast thread inside the insulator. Insulator pins may be wood, steel, steel and lead, or steel and wood (Fig. 8-27). The wood pins (Fig. 8-27, IV) are made from oak or locust, the latter being preferred as less apt to crack. Wood pins are used on communications circuits and low-voltage distribution circuits on account of their low cost. Steel pins (Fig. 8-27, I and II) are used on the higher-voltage, longer-span lines, where the side loads on the pin are greater. Combination wood-cob-with-steel-bolt pins (Fig. 8-27, III) provide stronger construction than the plain wood at lower cost than all-steel pins. Lead adapters (Fig. 8-27, V) can be used with 1-in steel pins to fit 1⅜-in insulator threads.

**46. Wooden insulator pins.**  All standard insulator pins of 1- and 1⅜-in top diameter have four threads to the inch and a tapering diameter of $\frac{1}{16}$-in increase for each inch in length.

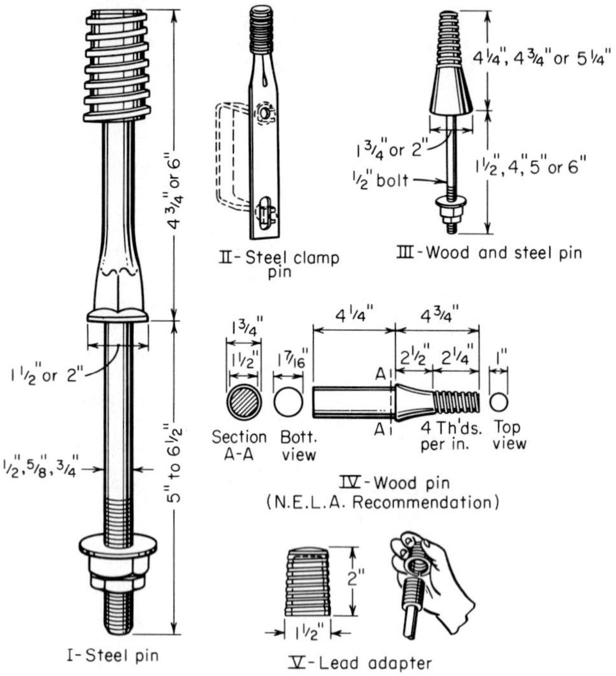


**Fig. 8-27** Insulator pins. [Line Material Co.]

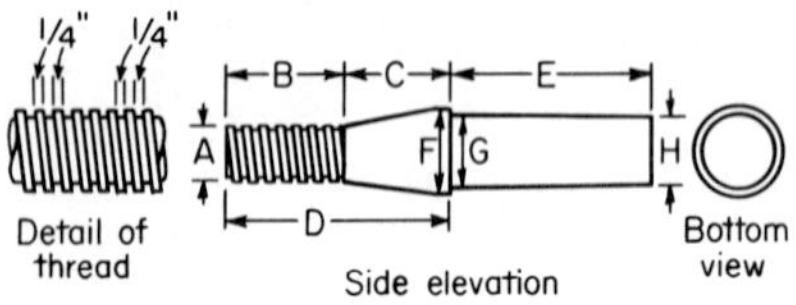


**Fig. 8-28** Wooden insulator pin.

| Dimensions, In. | | | | | | | | Shipping weight per 1,000, lb |
|---|---|---|---|---|---|---|---|---|
| *A* | *B* | *C* | *D* | *E* | *F* | *G* | *H* | |
| 1 | 2½ | 2¼ | 4¾ | 4¼ | 1¾ | 1½ | 1 7/16 | 400 |
| 1 | 2½ | 4¾ | 7¼ | 4¼ | 2 | 1½ | 1 7/16 | 510 |
| 1 | 2½ | 4¾ | 7¼ | 4¼ | 2¼ | 1¾ | 1 11/16 | 700 |
| 1 | 2½ | 4 | 6½ | 5 | 2½ | 2 | 1 15/16 | 930 |
| 1⅜ | 2¼ | 2½ | 4¾ | 4¼ | 1⅞ | 1½ | 1 7/16 | 400 |
| 1⅜ | 2¼ | 5 | 7¼ | 4¼ | 2 | 1½ | 1 7/16 | 510 |
| 1⅜ | 2¼ | 5 | 7¼ | 4¼ | 2¼ | 1¾ | 1 11/16 | 700 |
| 1⅜ | 2¼ | 4¼ | 6½ | 5 | 2½ | 2 | 1 15/16 | 930 |
| 1⅜ | 2¼ | 7¼ | 9½ | 5 | 3 | 2 | 1 15/16 | 1,160 |
| 1⅜ | 2¼ | 8¾ | 11 | 5 | 3½ | 2 | 1 15/16 | 1,280 |
| 1⅜ | 2¼ | 9¾ | 12 | 5 | 3½ | 2 | 1 15/16 | 1,360 |

**47. Wood pins are held in crossarms** with a sixpenny nail, as shown in Fig. 8-29. The nail should not be driven entirely in. Enough of its length should extend so that the cutting jaws of a pair of pliers can be forced under the head and the nail thereby withdrawn. If this suggestion is followed, a pin can readily be removed as required.

**48. Steel pins** are either clamped over the crossarm as in Fig. 8-27, II, or the rod projects through a hole in the crossarm and is held in place with a washer, square-head nut, and pal nut (Fig. 8-27, I). The clamp type does not weaken the crossarm by the drilling of the holes but is more expensive.

**49. Suspension insulators** are used to support wires from under crossarms. They are used on high-voltage lines, where pin insulators would become large, expensive, and

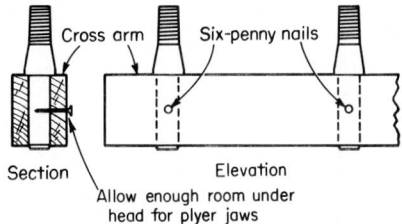

**Fig. 8-29**    Fastening pin in crossarm.

unwieldy, and are sometimes used on lower-voltage lines also. They are made in a number of different styles varying in hardware and petticoat diameters. There are three general types of hardware: hook-and-clevis, ball-and-socket, and the tongue-and-clevis, as illustrated in Fig. 8-30. The hook type (Fig. 8-30, I) is used principally to support street-lighting units. Whether to use the ball-and-socket type (Fig. 8-30, II) or the tongue-and-clevis type (Fig. 8-30, III) on power circuits is mostly a matter of individual choice. With the tongue-and-clevis type a round pin is used to hold the tongue of one unit in the clevis of the other. A cotter pin slips through a hole in the end of the round pin to keep the pin from slipping out. With ball-and-socket hardware the round pin is eliminated. The insulators are assembled by sliding the ball of one insulator into the socket of the next one from the side. A cotter pin is slipped in from the back of the socket, taking up sufficient space in the socket so that the ball cannot slide out. The socket clevis (Fig. 8-30, IV) is used when it is desired to support other hardware from the bottom of the insulator.

Suspension insulators are made up in strings, as shown in Fig. 8-31. A clevis is used to support them from steel crossarms as shown in I. With wood crossarms a hook bolt through the crossarm can be used as in II, or a clevis bolt as shown in III. The insulators are supported from each other with no additional hardware except the pins. The wire is supported from the bottom insulator with a wire clamp which is made in various types

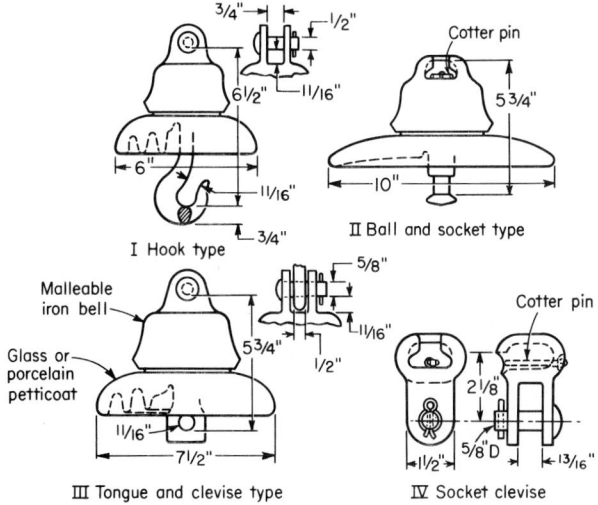

**Fig. 8-30**    Suspension insulators. [Locke Insulators, Inc.]

depending on the size of the wire. The voltage rating of a complete assembly is determined by the manufacturer from flashover tests. The diameter and shape of the petticoat and the number of units in the string affect the voltage rating. Small-diameter petticoats are preferred for locations where the insulators are subject to malicious damage from stones or bullets. The larger-diameter units are ordinarily used on the longer strings on high-voltage lines, where the wires are at a greater height from the ground.

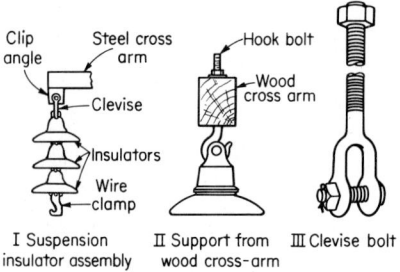

I Suspension    II Support from    III Clevise bolt
insulator assembly    wood cross-arm

**Fig. 8-31** Details of suspension insulator supports.

**50. Spool- and wire-holder–type** insulators are used in single-wire or rack assemblies for supporting low-voltage circuits on the sides of poles and buildings and for dead-ending service wires on buildings, poles, and crossarms. In the rack assemblies (Fig. 8-33) the insulators are removable from the rack for easy replacement in case of breakage. The racks are made in two-, three-, and four-wire assemblies. Typical spool and wire-holder insulator assemblies are shown in Figs. 8-32 to 8-35.

**51. Strain insulators** are used in guys and are also used in line wires at dead-ending points. Composition, porcelain, and wooden strain insulators are made. Wooden strain insulators (Fig. 8-36) were once popular with some companies and afford excellent insulation but have the objection that if an insulator burns, the wire that it supports falls. Composition and porcelain strain insulators can be made so that even if the insulating material fails, the supported wires will not fall. Figures 8-37

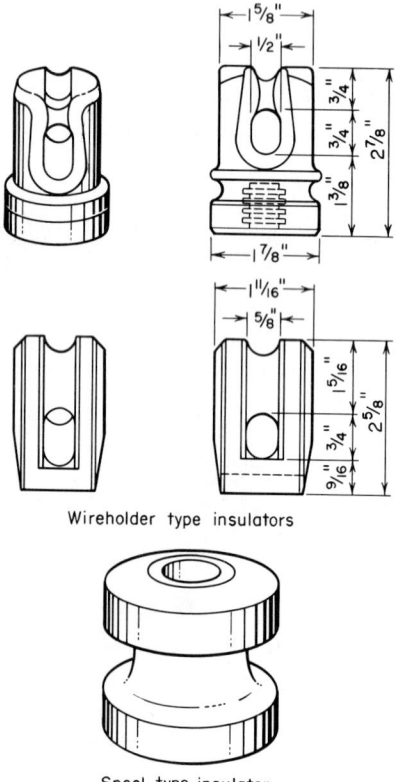

Wireholder type insulators

Spool type insulator

**Fig. 8-32** Typical spool- and wire-holder-type insulators. [General Electric Co.]

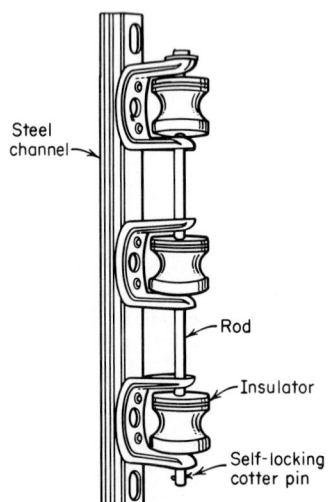

**Fig. 8-33** Low-voltage secondary rack. [Line Material Co.]

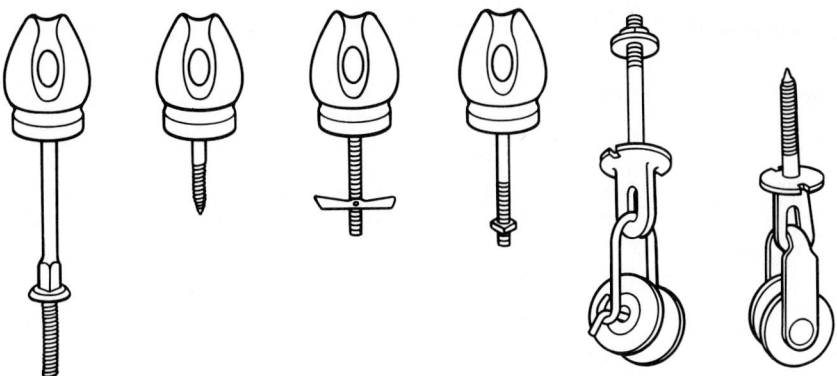

**Fig. 8-34**  Typical single-wire insulator assemblies. [General Electric Co.]

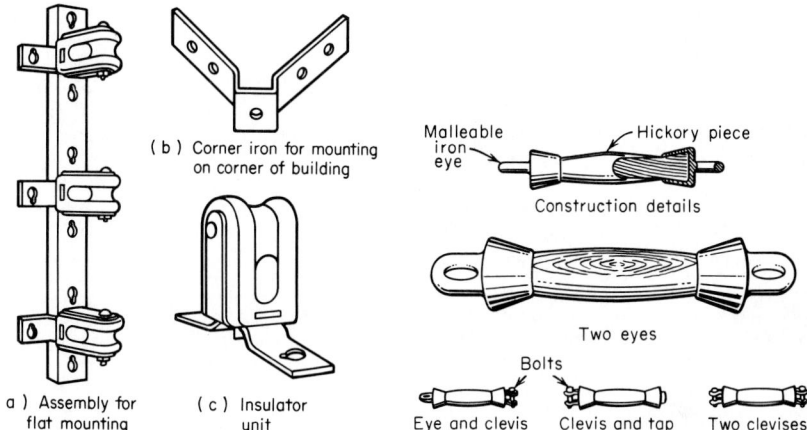

( b )  Corner iron for mounting
on corner of building

( a )  Assembly for
flat mounting

( c )  Insulator
unit

**Fig. 8-35**  Typical wire-holder rack assembly.
[General Electric Co.]

Malleable
iron
eye
Hickory piece

Construction details

Two eyes

Bolts

Eye and clevis    Clevis and tap    Two clevises

**Fig. 8-36**  Wooden strain insulators.

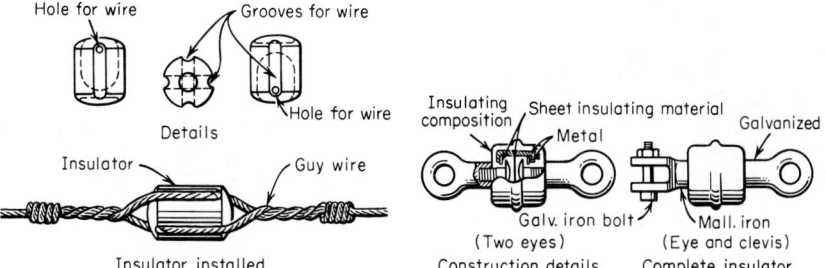

Hole for wire    Grooves for wire

Details
Hole for wire

Insulator    Guy wire

Insulator installed

**Fig. 8-37**  A porcelain strain insulator.

Insulating    Sheet insulating material
composition                                   Galvanized
                            Metal

Galv. iron bolt    Mall. iron
(Two eyes)         (Eye and clevis)
Construction details    Complete insulator

**Fig. 8-38**  Composition strain insulators.

and 8-38 show types meeting these requirements. The strain insulator of Fig. 8-37 is inexpensive and satisfactory and has become very popular in electric-lighting-line construction.

**52. Emergency strain insulators** can be made by knocking the end out of common glass line-wire insulators, as illustrated in Fig. 8-39. To break out the end, hold the insulator in one hand and strike the inside of the top a sharp blow with the handle of a pair of pliers or a pair of connectors or with a screwdriver held in the other hand. If one emergency insulator will not give sufficient insulation, two or more can be used in series. Emergency strain insulators thus made are not strong enough for heavy guy wires but are more suitable for insertion in line wires.

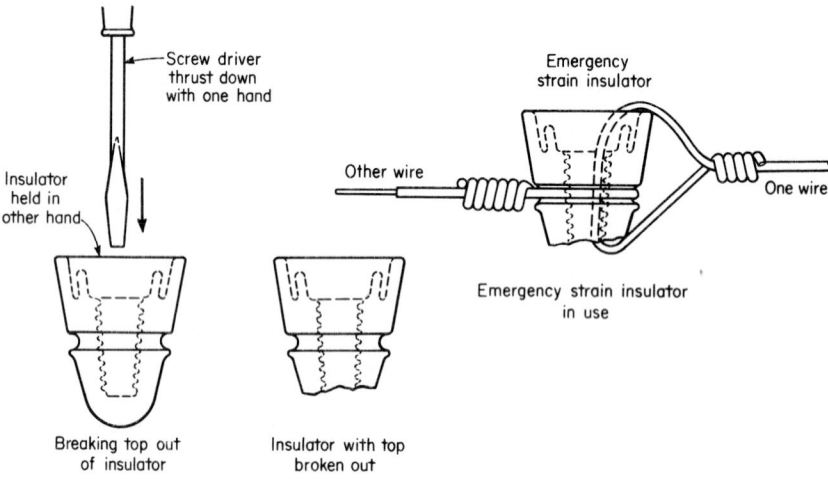

**Fig. 8-39**   Emergency strain insulators.

**53. Clevises** are used in making up assemblies connecting insulators, straps, strands, etc., together. Figure 8-40 illustrates some types of clevises. The forged clevises (Fig. 8-40, I) and bent strap clevises (Fig. 8-40, II) are used to connect dead-end insulators to crossarms (Fig. 8-41, I and II). The eye clevis (Fig. 8-40, III) will connect a dead-end insulator to another piece of hardware having a pin connection. The thimble clevis (Fig. 8-40, IV) will connect a stranded cable to an insulator assembly (Fig. 8-41, III). The strain-insulator clevis (Fig. 8-40, V) is used to connect a strain insulator of the type in Fig. 8-37 to an eyebolt or eye nut (Fig. 8-41, IV).

**54. Guy anchor rods** are used between the anchor in the ground and the guy strand. Figure 8-42, I, illustrates a roller type which allows the strand to turn on the rod and prevents twisting of the guy strand. Figure 8-42, II, is used to attach to a log anchor. Figure 8-42, III, is used when anchoring to solid rock or masonry. Guy anchor rods are made in $\frac{1}{2}$-, $\frac{5}{8}$-, $\frac{3}{4}$-, and 1-in diameters.

**55. In connecting a guy strand to the pole,** the strand may be wrapped around the pole or may be connected to a special bolt. When the strand is wrapped around the pole, guy hooks (Fig. 8-43, I) are used to prevent the strand from slipping down the pole, and guy plates (Fig. 8-44, II) or shims (Fig. 8-44, I) are placed under the guy strand to prevent its cutting into the pole. The molding strain plate of Fig. 8-44, III, is used under the guy-wire wrap on poles when a ground wire runs down the pole. The plate will fit over EEI standard 1-in ground-wire molding and prevent the guy strand from cutting or crushing the ground-wire molding.

An alternative to this method is to connect the guy to a thimble-eye bolt. The straight thimble-eye type (Fig. 8-43, II) is used when guying between a main pole and a stub pole. The angle thimble-eye type (Fig. 8-43, III) is used when guying into the ground. A J bolt (Fig. 8-43, IV) can also be used for attaching a guy wire to a pole, but this requires the use of a thimble (Fig. 8-43, V) in addition to the bolt. A stubbing washer should be used under the nut of the thimble-eye or J bolt. Under the eye of the bolt a stubbing washer (Fig. 8-44, V) is used for a 90° guy and a guy plate of the type shown in Fig. 8-44, IV, for an angle guy. Guying construction is discussed in Sec. 85.

**56. When a guy strand is connected to a bolt or rod of small diameter,** the strand should be wrapped around a thimble (Fig. 8-43, V) to prevent its being cut or broken from too sharp a bend.

**57. Guy clamps** (Fig. 8-45) are used for securing the ends of guy-wire strands. They are made in two-, three-, and four-bolt types.

**58. Guy protector guards** are installed over guy rods, at locations where the guy might be damaged by traffic, to conceal the connection of the guy strand to the eye of the guy rod, to make the guy more visible, and to minimize accidents. They may be made of a wood

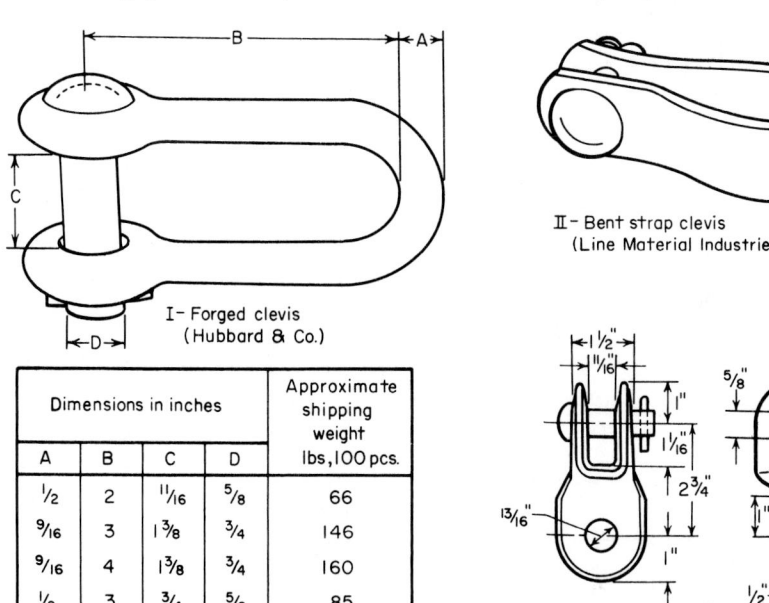

| Dimensions in inches | | | | Approximate shipping weight |
|---|---|---|---|---|
| A | B | C | D | lbs,100 pcs. |
| 1/2 | 2 | 11/16 | 5/8 | 66 |
| 9/16 | 3 | 1 3/8 | 3/4 | 146 |
| 9/16 | 4 | 1 3/8 | 3/4 | 160 |
| 1/2 | 3 | 3/4 | 5/8 | 85 |

I– Forged clevis
(Hubbard & Co.)

II– Bent strap clevis
(Line Material Industries)

III–Clevis eye
(Corning Glass Co.)

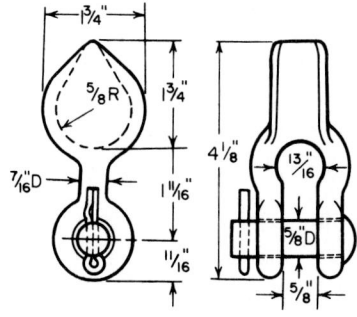

IV– Thimble clevis
(Ohio Brass Co.)

V– Strain insulator clevis
(Hubbard & Co.)

**Fig. 8-40** Clevises.

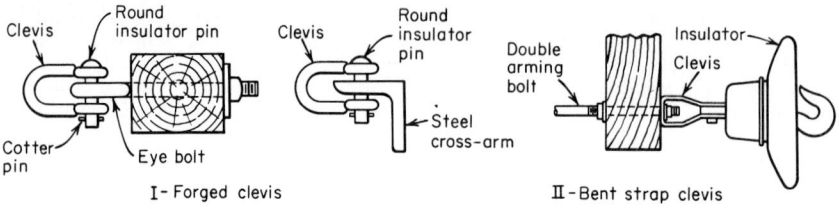

I– Forged clevis

II– Bent strap clevis

III– Thimble clevis connection to dead end insulator

IV– Strain insulator clevis

**Fig. 8-41**   Applications of clevises to crossarm connections. [Hubbard & Co.]

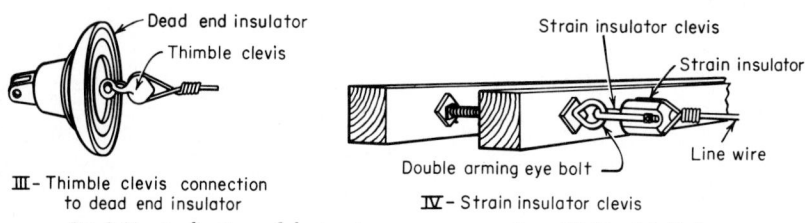

I– Roller type

II– Log type

III– Rock anchor

I– Guy hook

II– Straight thimble-eye bolt

III– Angle thimble-eye bolt

IV– J-bolt

V– Thimble

**Fig. 8-42**   Guy anchor rods. [Line Material Co.]

**Fig. 8-43**   Guy hook, thimble-eye bolts, and J bolt. [Line Material Co.]

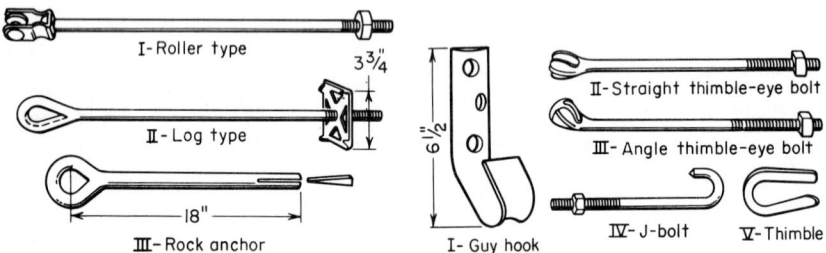

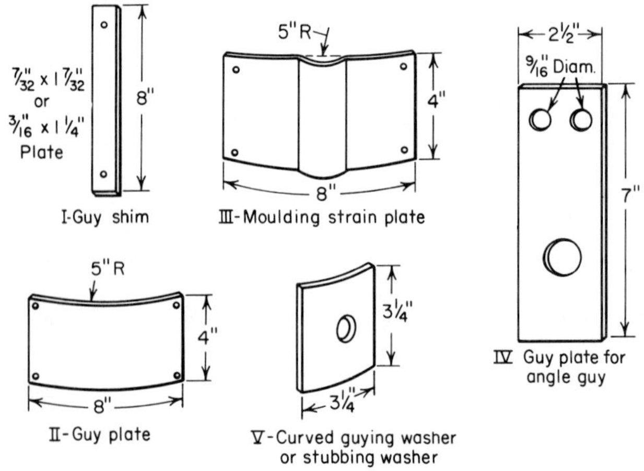

I-Guy shim

II-Guy plate

III-Moulding strain plate

IV Guy plate for angle guy

V-Curved guying washer or stubbing washer

**Fig. 8-44**   Guy plates and shims. [Locke Insulators, Inc., and Hubbard & Co.]

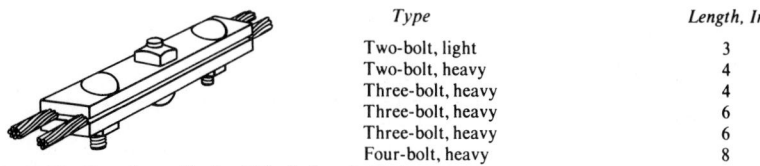

| Type | Length, In |
|---|---|
| Two-bolt, light | 3 |
| Two-bolt, light | 4 |
| Three-bolt, heavy | 4 |
| Three-bolt, heavy | 6 |
| Three-bolt, heavy | 6 |
| Four-bolt, heavy | 8 |

**Fig. 8-45**   Guy clamp. [Joslyn Mfg. & Supply Co.]

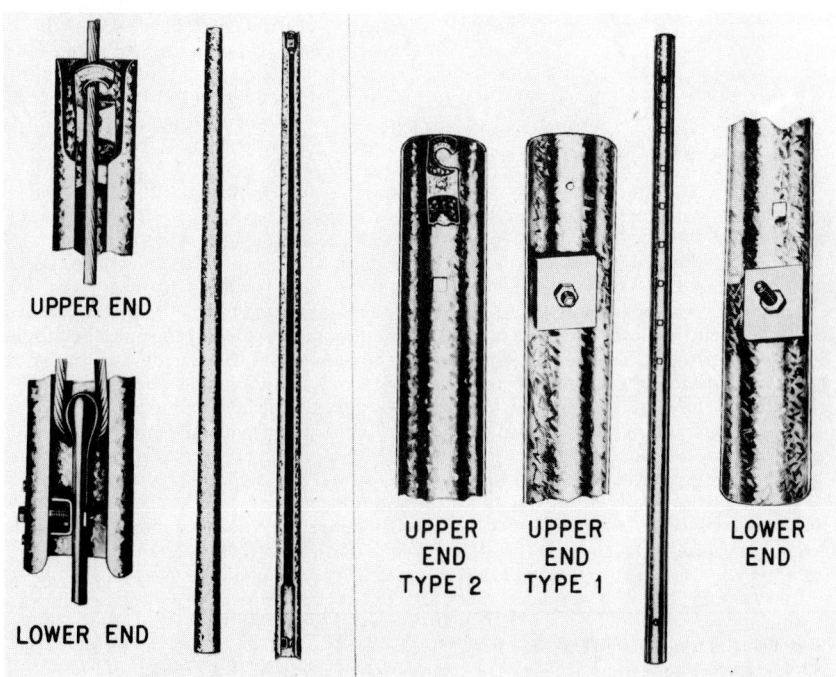

UPPER END

LOWER END

UPPER END TYPE 2

UPPER END TYPE 1

LOWER END

I. Full-round type.                    II. Half-round type.

**Fig. 8-46**   Guy-wire protectors. [Line Material Co.]

strip or of steel. One type of guard (Fig. 8-46) consists of a steel plate rolled into cylindrical form, leaving a split on one side. Special clamps are fastened to the strand, and the guard is slipped over the strand and held by the clamps. The split is turned down toward the ground so that it is not ordinarily noticeable.

**59. For guy wire,** seven-strand galvanized-steel cable is used. The National Electrical Safety Code recommends that for Grades A and B construction the maximum load in the guy should not exceed 50 percent of the ultimate strength; for Grade C construction 75 percent is allowable. The following table gives the strand sizes, weights, and ultimate strength of guy wire:

| Diam, in | Approx weight, lb per 1,000 ft | Approx strength, lb | Diam, in. | Approx weight, lb per 1,000 ft | Approx strength, lb |
|---|---|---|---|---|---|
| $\frac{3}{4}$ | 1,200 | 16,700 | $\frac{3}{8}$ | 273 | 4,250 |
| $\frac{5}{8}$ | 813 | 11,600 | $\frac{5}{16}$ | 205 | 3,200 |
| $\frac{9}{16}$ | 671 | 9,600 | $\frac{9}{32}$ | 164 | 2,570 |
| $\frac{1}{2}$ | 517 | 7,400 | $\frac{1}{4}$ | 121 | 1,900 |
| $\frac{7}{16}$ | 399 | 5,700 | | | |

## POLE-LINE CONSTRUCTION

**60. A single crossarm assembly** for a two-wire single-phase circuit is shown in Fig. 8-47. Additional insulators for three- or four-wire lines can be added to the arm 14½ in (368.3 mm) closer to the pole from the insulators shown. This type cf arm is used for normal spans and loads when there is a straight line with no obstructions.

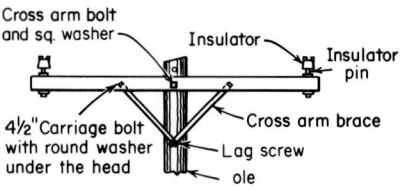

Fig. 8-47  Single crossarm assembly.

**61. Double arms** are used whenever the stress on the arms is unusually severe or when every precaution is necessary to ensure safety. Double arms are often used on the poles at each side of a street intersection, at each side of railroad crossings, and at corners or other points where the direction of a line changes. Figures 8-48, 8-51, 8-52, 8-70, 8-86, 8-87, 8-89, and 8-90 show examples of double-armed poles.

The two crossarms can be separated by wooden spacing blocks (Fig. 8-51), by spacing bolts or double-arming bolts (Fig. 8-48A), or by spacing nipples (Fig. 8-48B).

The spacing blocks can be sawed from a crossarm. An ¹¹⁄₁₆-in (17.462-mm) hole bored through the block and crossarm accommodates a ⅝-in bolt. A washer is used under bolt head and nut. Spacing bolts or double-arming bolts can be used instead of spacing blocks (Fig. 8-48A); the bolts are threaded for most of their length. A galvanized-iron pipe nipple (Fig. 8-48B) can also be used to separate the crossarms. A ¾-in spacing bolt probably provides the preferable and most economical method; spacing blocks rot readily, and the bolts through them corrode rapidly. When an arm guy is to be attached to the crossarm, a double-arming eyebolt can be used as shown. Single-armed poles are now often used, particularly on junction poles, as shown in the accompanying illustrations, in locations where double arms were formerly thought necessary. The single arms are preferable in that they allow greater climbing space for lineworkers.

**62. Alley arms** (Figs. 8-49 and 8-50) are used in alleys and other locations where it is necessary to clear obstructions by offsetting the crossarm toward one side of the pole. The crossarms will be special arms, but standard material can generally be used for the rest of the construction. Types of standard braces for side-arm construction are discussed in Sec. 40. Side guys or crib braces (see Fig. 8-49) are used when the line wires are heavy to counteract the tendency of the pole to tip.

**63. Reverse or buck arms** are used at corners (see Figs. 8-51, 8-52, 8-86, and 8-87). In placing buck arms, ample room must be provided through which a lineworker can reach the top of the pole. A 20-in (508-mm) square space is the minimum. Crossarm braces on buck-arm poles should be attached to the arms at a point 23½ in (596.9 mm) from the center of arm instead of at 19 in (482.6 mm), the standard distance for ordinary framing. The ⅜-in (9.525-mm) holes for the brace (carriage) bolts must be specially bored, at the

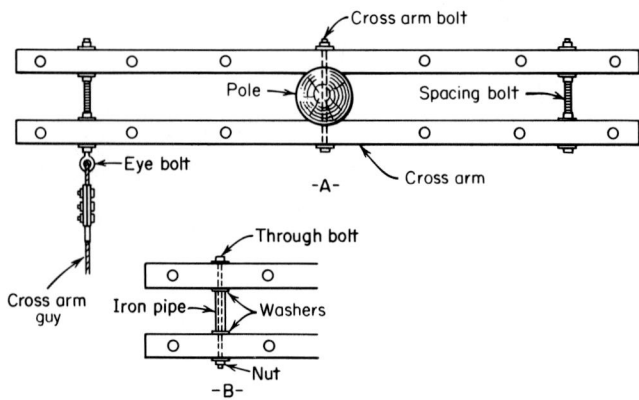

**Fig. 8-48**  Method of arranging double arms.

above spacing, in buck arms. The 23-in (584.2-mm) spacing is correct for 28-in (711.2-mm) braces.

**64. A gain** is a flat bearing surface on the pole at the point of attachment of the crossarm used so that the crossarm will be rigid. The gain is usually formed by cutting a notch in the side of the pole, as shown in Fig. 8-53. The gaining template of Fig. 8-54 is convenient in laying out the gains. The gain should be exactly the width of the crossarm, ensuring a snug fit. In good work, gains should always be spaced on 24-in (609.6-mm) centers and should be ½ in (12.7 mm) deep. Nothing is accomplished by making them deeper. In using the gaining template (Fig. 8-53, II) the point of the "roof" of the template is placed exactly over the point of the roof of the pole and the template is shifted until its centerline (which should be marked thereon) lies exactly under a cord stretched from the roof point to the center of the pole at the groundline. Then the positions of the crossarm gains are indicated by knife scratches made along the sides of the crosspieces on the template.

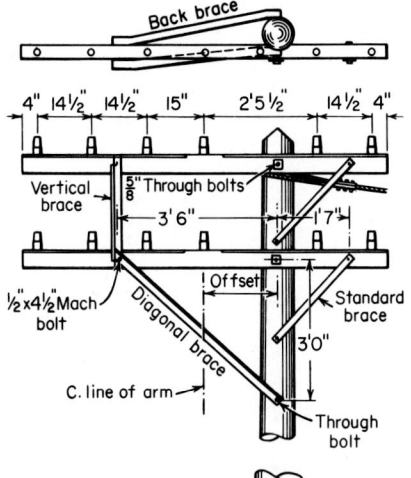

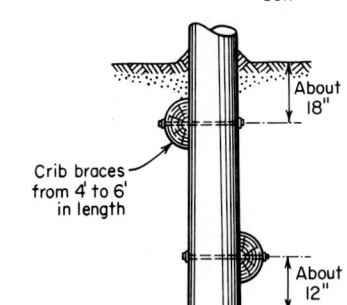

**Fig. 8-49**    A side-arm pole.

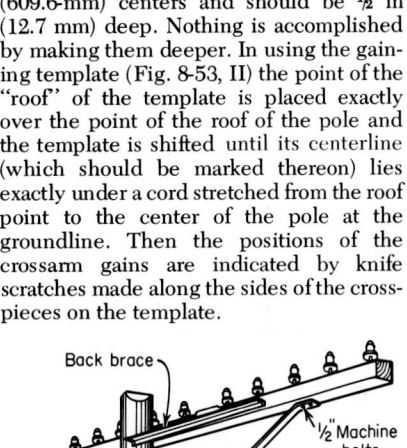

**Fig. 8-50**    Bracing for alley arms. [Hubbard & Co.]

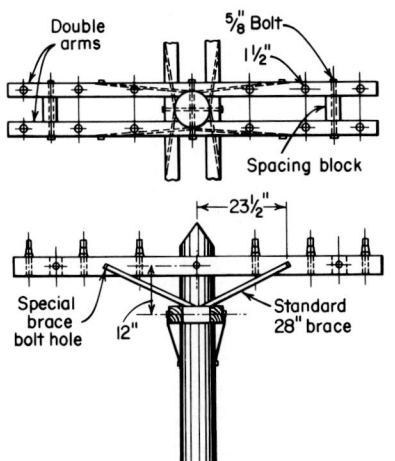

**Fig. 8-51**    A buck-armed pole, double arms.

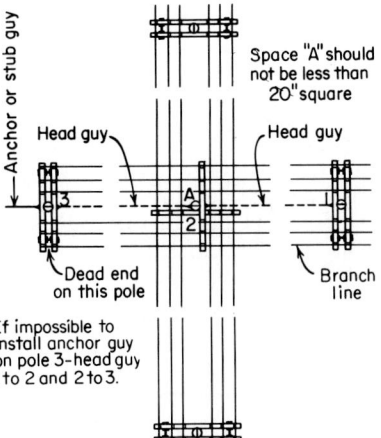

**Fig. 8-52**    Junction pole without double arms.

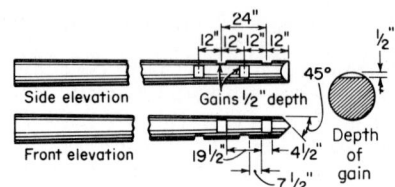

**Fig. 8-53, I**   Gains in pole.

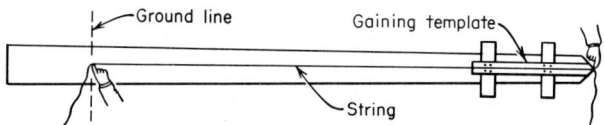

**Fig. 8-53, II**   Use of gaining template.

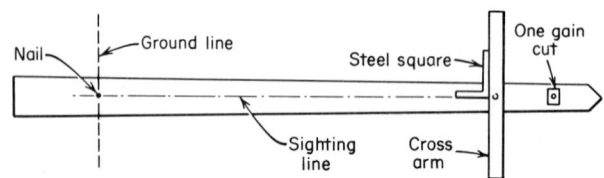

**Fig. 8-53, III**   Locating a gain with steel square and crossarm.

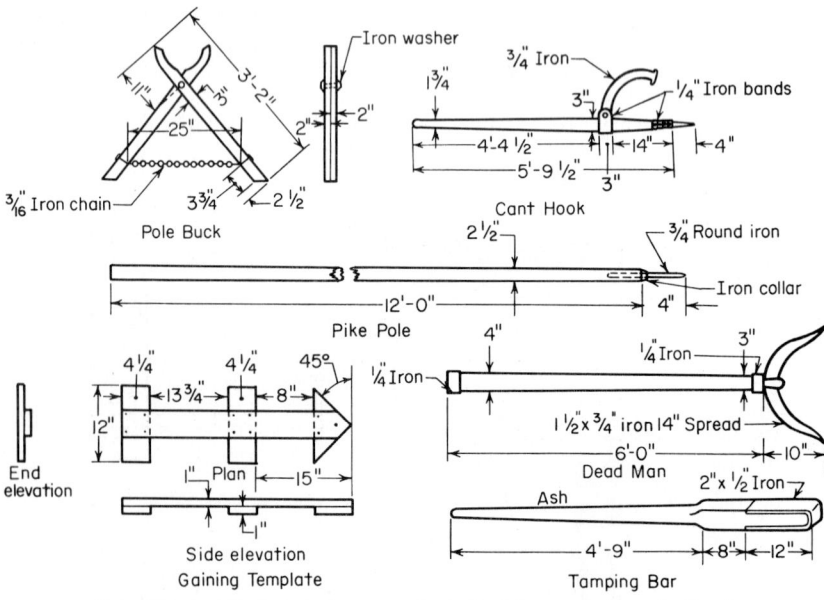

Note: The tool illustrated above as a Cant–Hook is properly termed a Peavie.
A peavie has a spike in the end of its handle, while a cant-hook has no spike.

**Fig. 8-54**   Some line-construction tools.

If a gaining template is not available, a crossarm (Fig. 8-53, III) can be laid on the pole with a steel square held against its lower face, the outer edge of the short limb of the square lying at the center of the pole and the center of the crossarm. Rotate the crossarm in a horizontal plane until, by sighting, it is evident that the edge of the square coincides with an imaginary centerline to a nail in the center of the butt at the groundline. Indicate the gain location by knife scratches along the crossarm sides.

**65. Metal gains** (Fig. 8-55, I) can be used in attaching the crossarm to the pole instead of notching the pole. By spacing the crossarm about ½ in (12.7 mm) away from the pole metal gains allow for complete drainage of water from the joint and prevent rotting of the pole and crossarm. Also, at double-arming locations, the use of metal gains spaces cross-arms 2 in (50.8 mm) farther apart, which allows more room between the pin insulators on the two arms and more room for tie wires, an important factor on the higher-voltage lines. Metal gains are a great convenience when installing an additional crossarm to an existing pole which did not have the gain cut in it. In using metal gains a square washer can be employed under the head of the through bolt (Fig. 8-55, III). A better construction is to use a reinforcing plate (Fig. 8-55, II) located on the outside of the cross-arm (Fig. 8-55, IV). This makes for a more rigid construction and helps to prevent checking and splitting of the crossarm at the attachment point.

**66. Pole steps** (Fig. 8-56, I) can be located on the pole, as shown in Fig. 8-57. A ⅜-in (9.525-mm) hole 4 in (101.6 mm) deep is bored for the iron step. It is driven into the hole until it projects 6 in (152. 4 mm) from the pole and is then turned with a wrench until the hook end is vertical. The wooden step is held to the pole with one 40-penny and one 20-penny nail or with cut spikes. Often the wooden steps are omitted. The lowest iron step should be at least 6½ ft (2 m) from the ground or other readily accessible place. One wood block (on a private right-of-way, more than one) may be placed on poles carrying communications cables or conductors below supply conductors, but the lowest block is not to be less than

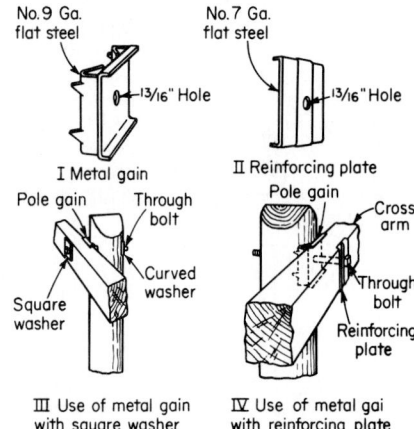

Fig. 8-55 Metal gains. [Hubbard & Co.]

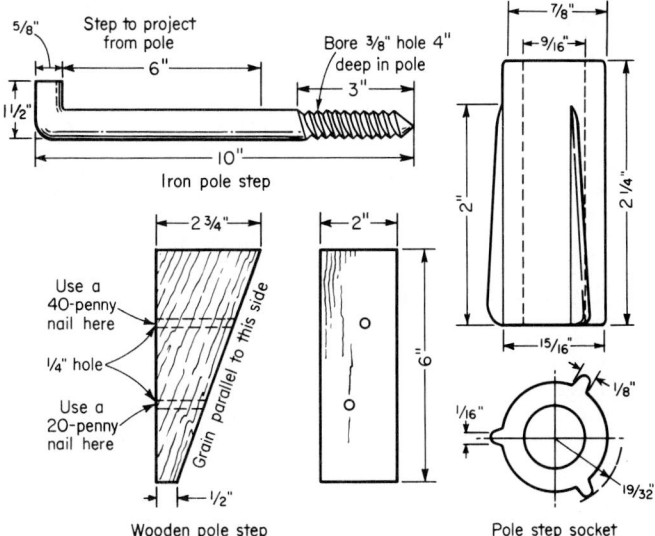

**Fig. 8-56, I** Pole steps and socket.

3½ ft (1.1 m) from the ground or other readily accessible place. On poles carrying only communications conductors, additional wood blocks may be used.

Detachable pole steps are available. The one shown in Fig. 8-56, II, slips over the head of the special ⁹⁄₁₆- by4-in lag screw. Turning is prevented by a punched projection. The steps can be removed to prevent unauthorized climbing.

*Pole-step sockets* (Fig. 8-56, I) are sometimes substituted for the wooden steps. The sockets drive into a ⅞-in (22.225-mm) hole. To climb the pole a lineworker can temporarily insert bolts or similar pieces of metal into the sockets.

**67. Secondary racks** are mounted on the sides of poles or buildings to carry low-voltage mains, usually of 440 V or less, to locations where building services are tapped off. Figure 8-58, II, illustrates the rack used as a dead end with a head guy to take the tension of the wires. The building service in this case is tapped with open wires. The higher-voltage wires are shown mounted on the crossarm at the top of the pole. Figure 8-58, I, shows two secondary lines crossing; Fig. 8-58, III, shows a corner pole; Fig. 8-58, IV, shows the secondary wires supported on the side of the pole with an open-wire service tap; and Fig. 8-58, V, shows a similar construction with taps to a service cable.

**Fig. 8-56, II**   Detachable pole step. [Joslyn Mfg. & Supply Co.]

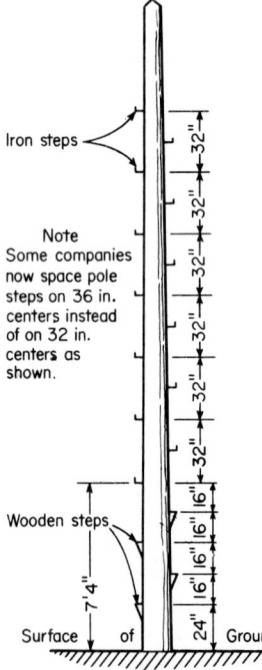

**Fig. 8-57**   Location of pole steps on pole.

Iron steps

Note
Some companies now space pole steps on 36 in. centers instead of on 32 in. centers as shown.

Wooden steps

Surface    of    Ground

**68. Distribution transformers** reduce the primary-line voltage to a low secondary voltage suitable for customers' use. They are mounted on the side of poles or on a platform between poles. Figure 8-59, I, shows such a transformer mounted on special crossarms in the space between the high-voltage wires at the top of the pole and the low-voltage wires on the secondary rack. Taps connect the primary of the transformer, through fuses mounted on the extra crossarm, to the high-voltage wires. The secondary of the transformer is connected directly to the secondary wires at the secondary rack. Attention should be given to making these connections in a neat manner, using right-angle bends and running the several wires of a circuit parallel to each other. Whenever there is danger of a wire touching another wire or a crossarm or brace or coming too close to a pole, wire-holder insulators (Fig. 8-32) or pin insulators should be installed on the sides of crossarms to keep the connecting wires in their proper position. Figure 8-59, II, shows a similar arrangement when the secondary wires are closer to the primary wires and the transformer must be mounted below the secondary wires. The insulators on the transformer crossarm keep the primary taps at a safe distance from the secondary wires. A compact installation is shown in Fig. 8-59, III. Such an arrangement is quite satisfactory when there are only two primary wires spaced sufficiently far apart to give plenty of climbing and working space. A three-phase bank consisting of three single-phase transformers can be mounted on a platform supported from two poles located fairly close together. This is common practice for supplying an industrial plant. For three-phase distribution along city streets the three transformers can be mounted singly on adjacent poles, each one similar to Fig. 8-59, I. This practice is preferable when the transformer platform would be conspicuous and unsightly. Additional details for the mounting of transformers are given in Div. 5.

**69. In locating circuits on poles,** the through wires or trunk lines should be carried on the upper crossarms, and the local wires, those which are tapped frequently, should be carried on the lower crossarms. The two or three wires of a circuit should always be carried on adjacent pins. This is of particular importance with ac circuits, as the inductance, and consequently the inductive voltage drop, is increased as the distance between the wires of a circuit is increased. Wires of a circuit should always take the same pin positions on all poles to facilitate trouble hunting. Series circuits which do not operate during the daytime may often be carried on the pole pins of a crossarm. High-tension multiple circuits that are "hot" continuously are well placed at the ends of the arms out of the way of lineworkers. The neutral wire should always be in the center of a three-wire circuit. Figure 8-60 shows one good arrangement on a pole with two four-pin arms.

**70. Joint use of poles for circuits of different power companies or for circuits of a power company and a telephone and telegraph company** is frequently advantageous if the lines run along the same route (Fig. 8-61). In the rules for location of conductors and clearances, the power company circuits are frequently called simply supply wires; the telephone and telegraph company circuits are called communications wires. The National Electrical Safety Code provides that when supply circuits of different companies are located on the same pole, the higher-voltage line should be carried at the higher level except that the lines of each company may be grouped together and always carried in the same relative position. For each company the lines of higher voltage should be located in the higher position. Whenever a lower-voltage line is carried above a higher-voltage one, it must be on the opposite side of the pole. A vertical clearance of at least 4 ft (1.2 m) must be maintained between wires of different power companies, and a clearance of 6 ft (1.8 m) must be used if required by the rules of the NESC. Supply wires must be carried above communications wires except that minor extensions may be made to existing systems

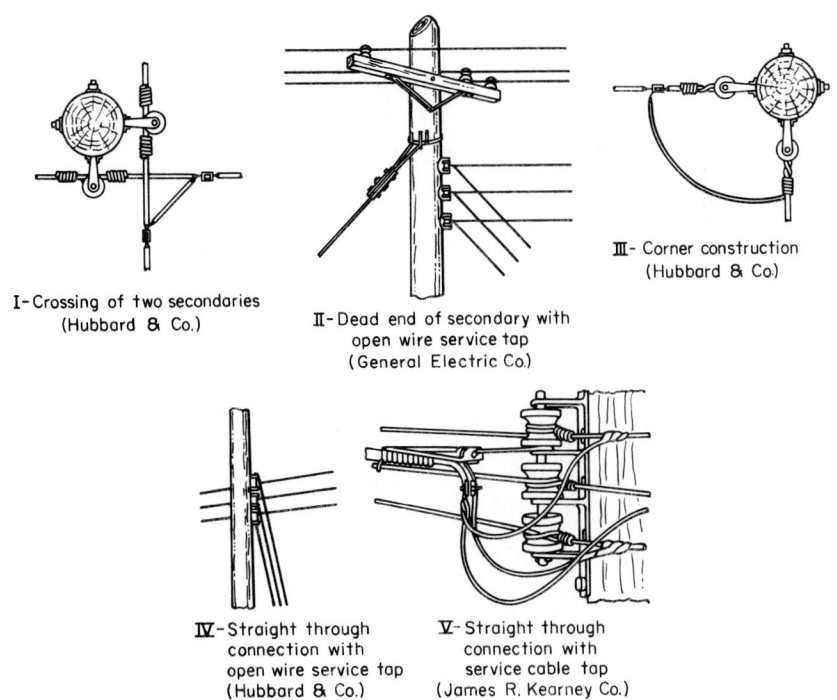

I- Crossing of two secondaries
(Hubbard & Co.)

II- Dead end of secondary with
open wire service tap
(General Electric Co.)

III- Corner construction
(Hubbard & Co.)

IV- Straight through
connection with
open wire service tap
(Hubbard & Co.)

V- Straight through
connection with
service cable tap
(James R. Kearney Co.)

**Fig. 8-58** Ways of using secondary racks.

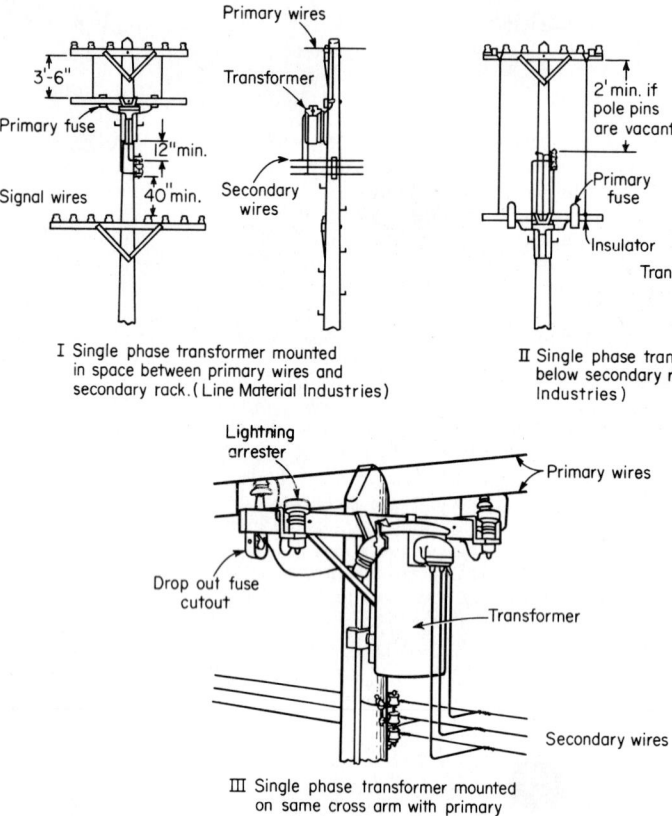

I Single phase transformer mounted in space between primary wires and secondary rack.(Line Material Industries)

II Single phase transformer mounted below secondary rack.(Line Material Industries)

III Single phase transformer mounted on same cross arm with primary supply wires.(General Electric Co.)

**Fig. 8-59**    Mounting and wiring of distribution transformers.

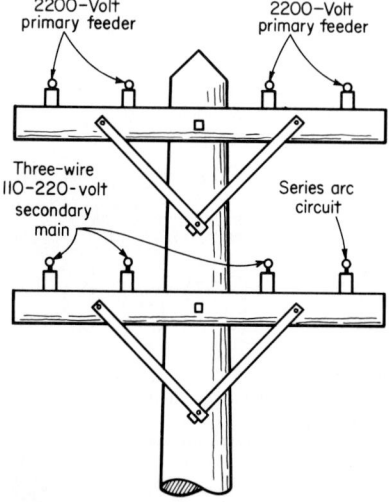

**Fig. 8-60**    Location of circuits on a two-arm pole.

which were installed with the communications wires above. Trolley feeder wires may be carried at the approximate level of the trolley wires irrespective of whether that location brings them above or below the communications wires.

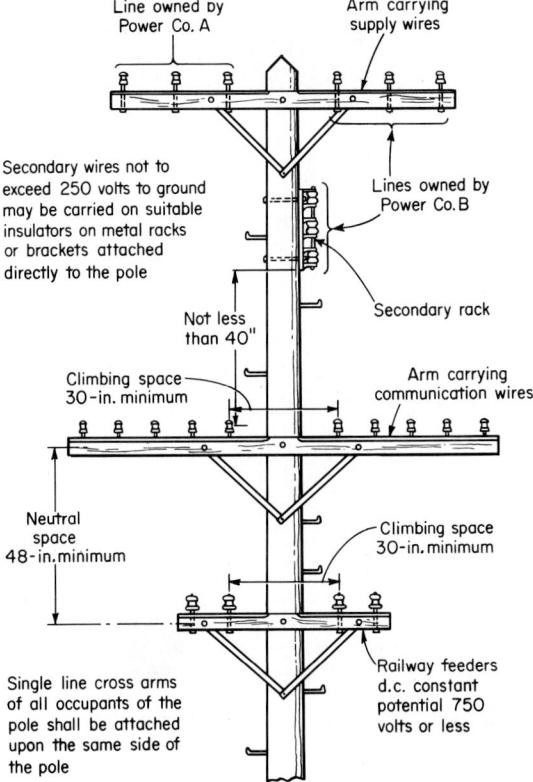

**Fig. 8-61**   Construction of jointly used wood-pole lines carrying supply circuits and signal circuits, showing arrangement of attachments and clearances. [Hubbard & Co.]

**71. Outside wiring as specified by the** National Electrical Code.   The article on outside wiring applies to electrical equipment and wiring for the supply of utilization equipment located on or attached to the outside of public and private buildings or run between buildings, structures, or poles on other premises served, but it shall not apply to equipment or wiring of an electric or communications utility used in the exercise of its function as a utility.

1. APPLICATION OF OTHER ARTICLES   Application of other articles, including additional requirements to specific cases of equipment and conductors, is as follows:

Article 200: polarity identification
Article 210: branch circuits
Article 215: feeders
Article 230: services
Article 250: grounding
Article 500: hazardous locations—general
Article 510: hazardous locations—specific
Article 600: signs and outline lighting
Article 710: circuits and equipment operating at more than 600 V
Article 725: remote-control, signal circuits
Article 800: communications circuits and power-limited
Article 810: radio and television circuits

2. Calculation of Load

*a. Branch circuits.* The load on every outdoor branch circuit is to be determined by the applicable provisions of Article 220, as given in Div. 3 of this book.

*b. Feeders.* The load to be expected on every outdoor feeder is to be determined by the procedure specified in Article 220, as given in Div. 3 of this book.

3. Conductor Covering  Where within 10 ft (3 m) of any building or structure, open conductors supported on insulators shall be insulated or covered. Conductors in cables or raceways, except Type MI cable, shall be of the rubber-covered or thermoplastic type and in wet locations shall comply with Sec. 310-8. Conductors for festoon lighting shall be of the rubber-covered or thermoplastic type.

4. Size of Conductors  The ampacity of outdoor branch circuits and feeder conductors shall be according to the rating in Tables 18 to 21 of Div. 11.

5. Minimum Size of Conductor

*a. Overhead spans.* Overhead conductors shall not be smaller than No. 10 for spans up to 50 ft (15.2 m) in length and not smaller than No. 8 for longer spans.

*b. Festoon lighting.* Overhead conductors for festoon lighting shall not be smaller than No. 12 unless supported by messenger wires (see Sec. 255-6).

*Definition.* Festoon lighting is a string of outdoor lights suspended between two points more than 15 ft (4.57 m) apart.

*c. Over 600 V.* Overhead conductors operating at more than 600 V shall not be smaller than No. 6 when open individual conductors or smaller than No. 8 when in cable.

6. Lighting Equipment on Poles or Other Structures

*a.* For the supply of lighting equipment installed on a single pole or structure, the branch circuits shall comply with the requirements of Article 210 and Par. *c* below.

*b.* A common neutral may be used for a multiwire branch circuit. The ampacity of the neutral conductor shall be not less than the calculated sum of the currents in all ungrounded conductors connected to any one phase of the circuit.

*c.* 227 V to ground may be used to supply lighting fixtures of industrial establishments, office buildings, schools, stores, and other commercial or public buildings where a 3-ft clearance to windows, platforms, fire escapes, and the like is maintained.

*d.* Branch circuits exceeding 227 V for permanently installed electric discharge lighting fixtures for area illumination mounted on poles clear of buildings or on other structures may operate at not more than 600 V between conductors, provided fixtures on poles are mounted at a height of 22 ft (6.71 m) or more and on other structures at a height of 18 ft (5.49 m) or more.

7. Disconnection

*a.* For branch circuits as required in Article 210, as given in Div. 9 of this book.

*b.* For feeders as required in Article 215, as given in Div. 9 of this book. (At each building supplied by a feeder see Sec. 230-84 and Div. 9 of this book.)

8. Overcurrent Protection

*a.* For branch circuits as required in Article 210, as given in Div. 9 of this book.

*b.* For feeders as required in Article 215, as given in Div. 9 of this book.

9. Wiring on Buildings  Outside wiring on surfaces of buildings may be installed for circuits when not in excess of 600 V as open conductors on insulating supports, as multiconductor cable, as MC cable or MI cable, as messenger supported wiring, in rigid metal conduit, in intermediate metal conduit, in rigid nonmetallic conduit, in cable trays, as cablebus, in wireways, in auxiliary gutters, in flexible metal conduit, in liquidtight flexible metal conduits, in liquidtight flexible nonmetallic conduit, in busways, or in electrical metallic tubing. Circuits of more than 600 V shall be installed as provided for services in Sec. 710-3. Circuits for signs and outline lighting shall be installed as provided in Article 600.

10. Circuit Exits and Entrances  Where outside branch and feeder circuits exit from or enter buildings, the installation shall comply with the requirements of Article 230 which apply to service-entrance conductors.

11. Open-Conductor Supports  Open conductors shall be supported on glass or porcelain knobs, racks, brackets, or strain insulators.

12. Festoon Supports  In spans exceeding 40 ft (12.2 m) the conductors shall be supported by messenger wire supported by strain insulators. Conductors or messenger wires shall not be attached to any fire escape, downspout, or plumbing equipment.

13. Open-Conductor Spacings  Conductors shall conform to the following spacings:

*a. Open conductors exposed to weather.* 600 V or less, as provided in Sec. 230-51(c).

*b. Over 600 V.* As provided in Sec. 710, Part D.

*c. Separation from other circuits.* Open conductors shall be separated from open conductors of other circuits or systems by not less than 4 in (102 mm).

*d. Conductors on poles.* Conductors on poles shall have a separation of not less than 1 ft (305 mm) where not placed on racks or brackets. Conductors supported on poles shall provide a horizontal climbing space not less than the following:

| | |
|---|---|
| Power conductors, below communications conductors | 30 in |
| Power conductors alone or above communications conductors: | |
|    300 V or less | 24 in |
|    Over 300 V | 30 in |
| Communications conductors below power conductors | Same as power conductors |
| Communications conductors alone | No requirement |

14. SUPPORTS OVER BUILDINGS  See Sec. 230-29.

15. POINT OF ATTACHMENT TO BUILDINGS  See Sec. 230-26.

16. MEANS OF ATTACHMENT TO BUILDINGS  See Sec. 230-27.

17. CLEARANCE FROM GROUND  Open conductors of not over 600 V shall conform to the llowing:

. / ft: above finished grade, sidewalks, or any platform or projection from which they might be ·eached where the supply conductors are limited to 150 V to ground and accessible to pedestrians only

1 2 ft: over residential driveways and those commercial areas not subject to truck traffic where the voltage is limited to 300 V to ground

1 5 ft: for those areas listed in the 12-ft classification where the voltage exceeds 300 V to ground

18 ft: over public streets, alleys, roads, parking areas subject to truck traffic, driveways on other than residential property, and other land traversed by vehicles such as cultivated, grazing, forest, and orchard land

18. CLEARANCES OVER BUILDINGS FOR CONDUCTORS NOT IN EXCESS of 600 V  See Sec. 225-19 of the NEC.

    NOTE  For clearance of conductors of over 600 V, consult National Electrical Safety Code.

19. MECHANICAL PROTECTION OF CONDUCTORS  Mechanical protection of conductors on buildings, structures, or poles shall be as provided for services (Sec. 230-50).

20. MULTICONDUCTOR CABLES ON EXTERIOR SURFACES OF BUILDINGS  Multiconductor cables on exterior surfaces of buildings shall be as provided for service cable (Sec. 230-51)

21. RACEWAYS ON EXTERIOR SURFACES OF BUILDINGS  Raceways on exterior surfaces of buildings shall be made raintight and be suitably drained.

22. UNDERGROUND CIRCUITS  Underground circuits shall meet the requirements of Sec. 300-5.

23. OUTDOOR LIGHTING EQUIPMENT: LAMPHOLDERS  Where lampholders are attached as pendants, they shall have the connections to the circuit wires staggered. If lampholders have terminals of a type that puncture the insulation and make contact with the conductors, they shall be attached only to conductors of the stranded type.

24. OUTDOOR LIGHTING EQUIPMENT: LOCATION OF LAMPS  Location of lamps for outdoor lighting shall be below all energized conductors, transformers, or other electrical equipment, unless clearances or other safeguards are provided for relamping operations or unless the installation is controlled by a disconnecting means which can be locked in the open position.

**72. Tying in wires.**  Normally the wires should rest in the insulator grooves as shown in Fig. 8-62*A*, but if there is a side stress, the wires should be so arranged that the pull comes against the insulators rather than away from them (Fig. 8-62*B*). A single tie for the smaller

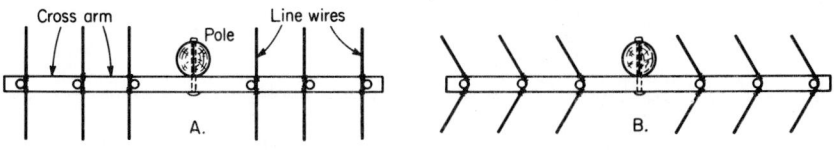

Arrangement on straight lines            Arrangement where there is a side strain

**Fig. 8-62**  Positions of wire in insulator grooves.

wires is shown in Fig. 8-63, I, and a back tie for the larger wires is shown in II. The single tie wire is about 12 in (304.8 mm) long, and the back tie about 18 in (457.2 mm) long. The back tie is made as follows. Bend the tie around the insulator under the line wire, with 4 in (101.6 mm) on one side of the insulator and the balance on the other side. Wrap the short end 3 times around the line wire, leaving a space equal to the diameter of the tie wire between successive wraps. Now wrap the long end of the wire 3 times completely

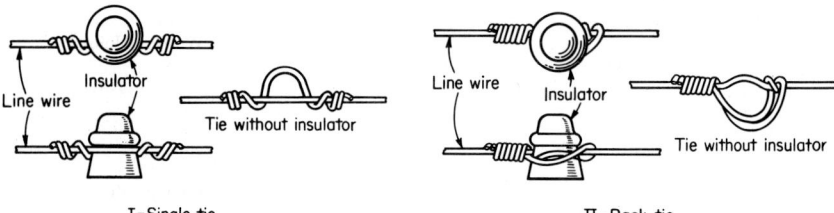

I—Single tie                    II—Back tie

**Fig. 8-63**  Methods of tying. Preferably the tie wires should not be cut off flush as shown above. Instead, their ends should be permitted to stick out or to lie along the line wires. This procedure facilitates their removal.

around the line wire, then back around the insulator, and wrap it in the spaces left between the wraps of the other end of the tie wire. Refer to Fig. 8-63 for instructions regarding cutting of tie wires.

**73. Different types of insulator ties** which have been standardized by a western power company are shown in Fig. 8-64 and Table 74. The company uses medium-hard-drawn line wire. Although good ties can be made with a medium-hard-drawn weatherproof tie on a weatherproof line wire, medium-hard-drawn ties are not desirable on bare wires, as the bending radius is too small for these stiff wires. Hence, annealed bare-wire ties are used for bare line wires. Annealed ties can be cut from purchased annealed copper. Do not use a tie wire twice, as, after once being bent and strained, it will be brittle. The following table of sizes has been recommended for ordinary stresses. For very heavy stresses heavier tie wires should be used.

| Line wire, size | Tie wire, size | Kind of tie | Line wire, size | Tie wire, size | Kind of tie |
|:---:|:---:|:---:|:---:|:---:|:---:|
| 6 | 6 | Single | 1 | 4 | Back |
| 4 | 6 | Single | 2/0 and larger | 2 | Back |
| 2 | 4 | Single | | | |

Tie wire should usually have the same kind of insulation as the line wire which it confines. The **National Electrical Code** requires this on indoor work, and it is also good practice on outdoor work.

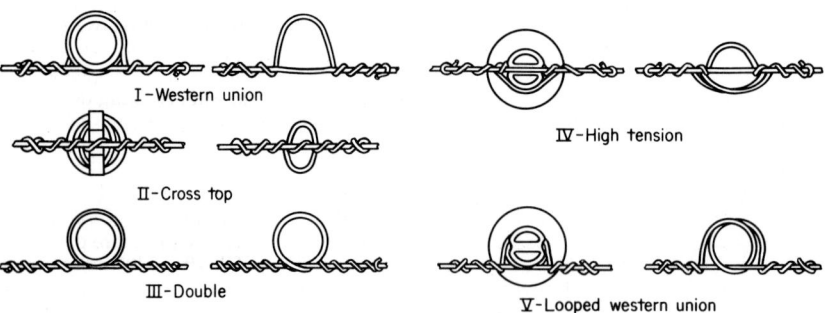

I—Western union

II—Cross top

III—Double

IV—High tension

V—Looped western union

**Fig. 8-64**  Ties of the different types used in ordinary distribution systems (see Table 89).

**74. Types of Insulator Ties for Different Services**

| Size of line wire, gage or in | Insulator | Top or side | Tie material | Tie length, in | Type of tie | Fig. 8–64 |
|---|---|---|---|---|---|---|
| | | | Copper | | | |
| 6 or 4 WP | DGDP | Side | 6 WP mhd | 25 | Western Union | I |
| 2 WP | DGDP | Side | 4 WP mhd | 27 | Western Union | I |
| 1/0 to 4/0 WP | Cable | Top | 4 WP mhd | 36 | Crosstop | II |
| 1/0 to 4/0 WP | Cable | Side | 4 WP mhd | 30 | Western Union | I |
| 6 or 4 bare | DGDP | Side | 6 bare annealed | 27 | Double | III |
| 6 or 4 bare | 11kV | Top | 6 bare annealed | 40 | High-tension | IV |
| 2 bare | 11 kV | Top | 4 bare annealed | 40 | High-tension | IV |
| 1/0 to 4/0 bare | 11 kV | Top | 2 bare annealed | 44 | High-tension | IV |
| 6 to 4 bare | 11 kV | Side | 6 bare annealed | 40 | Looped WU | V |
| 2 bare | 11 kV | Side | 4 bare annealed | 40 | Looped WU | V |
| 1/0 to 4/0 bare | 11 kV | Side | 2 bare annealed | 48 | Looped WU | V |
| | | | Steel | | | |
| ¼ and ⁵⁄₁₆ | DGDP | Side | 1 strand, ⁵⁄₁₆ in | 27 | Double | III |
| ¼ and ⁵⁄₁₆ | 11 kV | Top | 1 strand, ⁵⁄₁₆ in | 44 | High-tension | IV |
| ⅜ | 11 kV | Top | 1 strand, ⅜ in | 44 | High-tension | IV |
| ¼ and ⁵⁄₁₆ | 11 kV | Side | 1 strand, ⁵⁄₁₆ in | 48 | Looped WU | V |
| ⅜ | 11 kV | Side | 1 strand, ⅜ in | 48 | Looped WU | V |

**75. The bare tie wires for uninsulated line wires** should always be of the same metal as the line wire. If dissimilar metals are employed, electrolytic corrosion and a consequent weakening and possible breakage of the line wire at the tying point may result. That is, bare copper wires should always be used on bare copper line wires and galvanized-iron tie wires on galvanized-iron line wires.

**76. In making up line-wire ties,** the tie wire should not be drawn too tightly around the line wire. The tendency of the beginner is to do this, with the result that the line wire is weakened and will probably sever when it is first subjected to an unusual stress such as that which is likely to occur when a low temperature prevails.

**77. In making ties on ACSR wire** (see Div. 2), the conductor must first be reinforced by wrapping with armor rods twisted around the conductor with the aid of two special wrenches.

Armor rods are shipped in sets (bundles) with the ends taped and with the center marked. Remove the tape from one end of the set and thread the rods through the two armor-rod wrenches, making sure that both the wrenches open between the *same two rods*. Leaving the set taped at one end keeps the rods from sliding while the assembly is being sent aloft and keeps ends even while making the first twist.

Center the armor rods on the conductor over the point of support. Open the wrenches and then close them with the rods around the conductor (Fig. 8-65, I). Shift the wrenches on the rods to such a position that the middle third of the armor rods will be twisted with the first turn of the wrenches. Rotate the wrenches counterclockwise to make the lay of the rods in the *same direction* as the lay of the cable. Twist the rods to give them a permanent set, so that the wrenches do not tend to spring back. After twisting the middle third, cut the remaining tape and slide the wrenches halfway to the ends. Turn the wrenches and twist the rods as before. Finally slide the wrenches as close as possible to the ends, twisting this last section the same as for previous settings.

The rods should be twisted snugly with a smooth lay. If they are twisted too tightly, there is a tendency to force one or more rods out of the layer, and the rods will stand away from the cable so that the assembly can be slipped along the conductor before the armor-rod clips are applied. There is a stopping point, which can be felt after a little practice.

The armor-rod (V-bolt) should be applied about 1½ in (38.1 mm) from the ends of the assembly, which can best be accomplished by installing and tightening the V bolts while the armor-rod wrenches are still in their final twisting position (Fig. 8-65, II). This procedure leaves the ends of the rods slightly flared away from the conductor. If the wrenches are removed prior to tightening the armor-rod clips, the ends of the rods may have to be flared as a separate operation.

After installation of the armor rods the ties are made as shown in Fig. 8-66. Figure 8-66, I, shows a top tie with the conductor on top of the insulator, and Fig. 8-66, II, shows the conductor tied on the side groove of the insulator. Above each view is shown a sketch of the manner in which the tie wire is wrapped and looped.

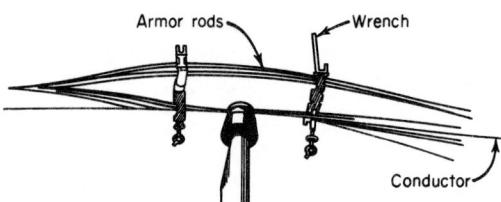

I – Initial step. Wrenches ready to be closed around conductor leaving one end of rods taped.

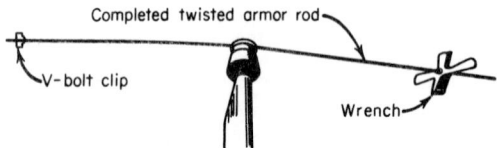

II – Final step. Wrench in final position ready for application of second V-bolt clip.

**Fig. 8-65**   Installing armor rods at insulator support for ACSR wire. [Aluminum Co. of America]

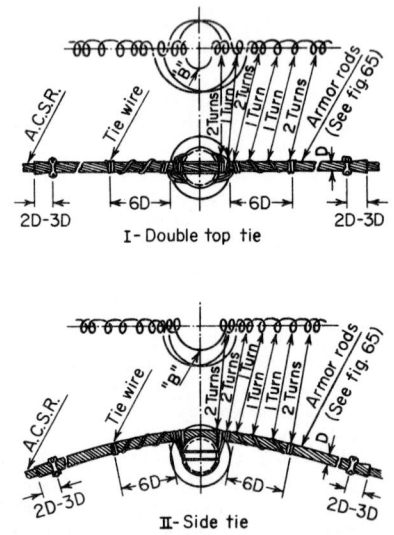

I – Double top tie

II – Side tie

**Fig. 8-66**   Ties for ACSR wire. [Aluminum Co. of America]

**78. For ties on Copperweld wire** use annealed-copper tie wires of the size given in the following table as recommended by the Copperweld Corp.

| Size of Conductor | Size of[1] Tie Wire | Length of Tie Wire[2] | |
|---|---|---|---|
| | | Top Tie | Side Tie |
| **3-Wire Stranded Conductors** | | | |
| 3 No.  6 AWG | 6 AWG | 54 in. | 60 in. |
| 3 No.  7 AWG | 6 AWG | 52 in. | 58 in. |
| 3 No.  8 AWG | 6 AWG | 50 in. | 56 in. |
| 3 No.  9 AWG | 6 AWG | 48 in. | 54 in. |
| 3 No. 10 AWG | 8 AWG | 44 in. | 50 in. |
| 3 No. 12 AWG | 8 AWG | 44 in. | 50 in. |
| **Type F Copperweld-copper Conductors** | | | |
| 1/0F | 6 AWG | 56 in. | 62 in. |
| 1F | 6 AWG | 54 in. | 60 in. |
| 2F | 6 AWG | 54 in. | 60 in. |
| **Copperweld-copper 3-wire Stranded Conductors** | | | |
| 2A | 6 AWG | 56 in. | 62 in. |
| 3A | 6 AWG | 54 in. | 60 in. |
| 4A | 6 AWG | 52 in. | 58 in. |
| 5A | 6 AWG | 50 in. | 56 in. |
| 6A | 8 AWG | 46 in. | 52 in. |
| 7A | 8 AWG | 44 in. | 50 in. |
| 8A | 8 AWG | 44 in. | 50 in. |
| 8C | 8 AWG | 44 in. | 50 in. |
| 8D | 8 AWG | 44 in. | 50 in. |
| 9½D | 8 AWG | 44 in. | 50 in. |

[1]Sizes listed apply to annealed copper tie wires. When using annealed Copperweld for tie wire, size should be two gage sizes smaller than listed for annealed copper.

[2]Includes 4 in. additional length on each end for convenience in applying tie.

For top ties on Copperweld wire, the type shown in Fig. 8-67*B* is recommended. For side ties, the one shown in Fig. 8-67*A* with three loops around the insulator is used.

NOTES ON TYING PRACTICE

1. Use only fully annealed tie wire.

2. Use a size of tie wire which can be readily handled, yet one which will provide adequate strength.

3. Use a length of tie wire sufficient for making the complete tie, including an end allowance for gripping with the hands. The extra length should be cut from each end after the tie is completed.

4. A good tie should

*a.* Provide a secure binding between line wire, insulator, and tie wire.

*b.* Have positive contacts between line wire and tie wire to avoid any chafing contacts.

*c.* Reinforce line wire in the vicinity of insulator.

5. The ties illustrated can and should be applied without the use of pliers.

6. Avoid nicking the line wire.

7. Do not use a tie wire which has been previously used.

8. Do not use hard-drawn wires for tying or fire-burned wire, which is usually either only partially annealed or injured by overheating.

9. Careful workmanship is essential.

**79. For dead-ending of line wires** at the end of a run or at corners, various types of construction may be used. For low-voltage, secondary rack construction, a rack may be located on the front of the pole and the tension taken by bolts through the pole (Fig. 8-68). The wires are looped around the rack insulators, and the end of the wire is wrapped around the line conductor. Other low-voltage dead ends may be made up as in Fig. 8-69. In many current installations a single-wire bracket, similar to the one shown in Fig. 8-69,

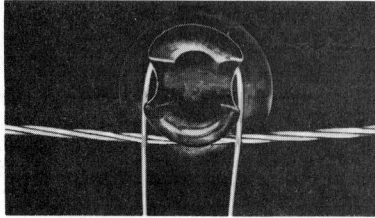

I. Bend tie wire around insulator above conductor to form a U. The two legs of the tie wire should be of equal length after bending.

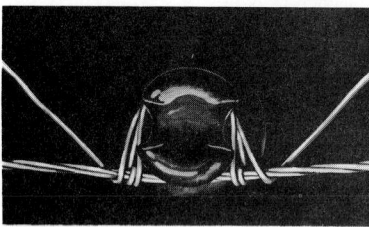

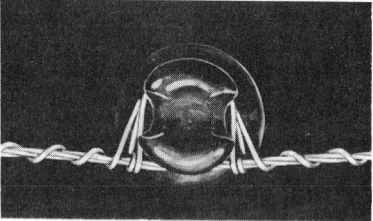

II. Holding the tie wire tightly against the insulator, throw two tight close wraps around conductor on each side of the insulator; then cross legs of tie wire around insulator.

III. Finish in same way as top tie with six tight 45° spiral wraps on each side of the insulator. Bend back ends and cut off extra length of tie wire close to conductor.

**Fig. 8-67A** Side tie for Copperweld® wire. [Copperweld Corp.]

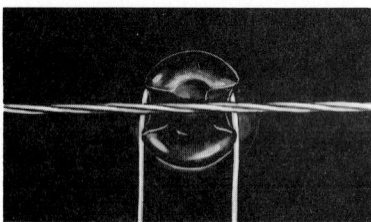

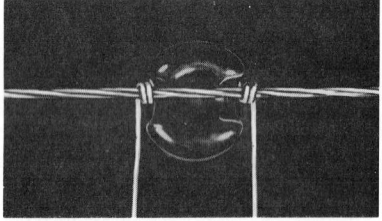

I. Bend annealed tie wire of proper size and length around insulator under the conductor, forming a U. Both legs of tie wire should be of equal length after bending.

II. Holding the tie wire tightly against the insulator, throw two tight close wraps around the conductor on each side of the insulator, keeping these wraps snugly against the insulator.

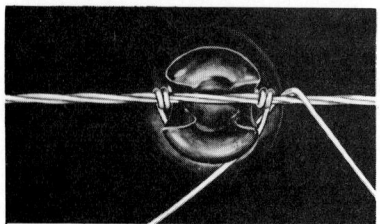

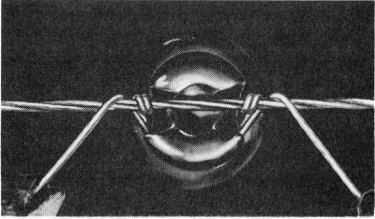

III. Cross the legs of the tie wire around the insulator, right to left and left to right.

IV. With both legs of the tie wire crossed, tightly wrap each leg spirally around the conductor at an angle of 45°.

**Fig. 8-67B** Top tie for Copperweld® wire. [Copperweld Corp.]

 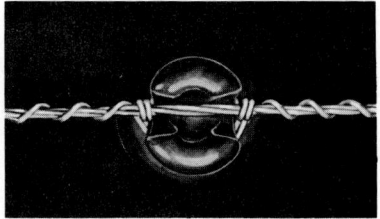

V. Complete six spiral 45° wraps on each side of the insulator, bending back the ends and cutting them off short.

VI. The finished top tie when made properly and tightly will greatly reduce any possibility of the conductor's chafing at the insulator.

**Fig. 8-67B**   (*Continued*)

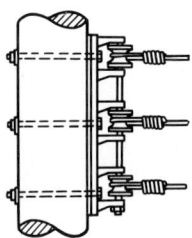

On secondary rack

**Fig. 8-68**    Low-voltage dead ends. [Hubbard & Co.]

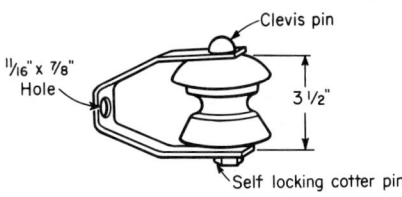

I Single wire type

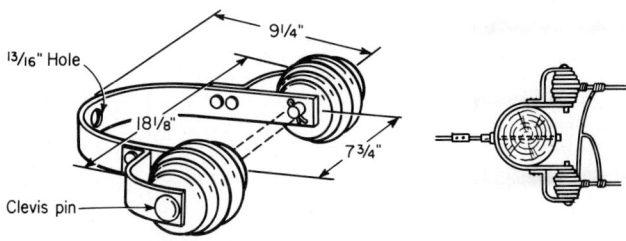

II Double wire type

**Fig. 8-69**   Dead-end insulators. [Line Material Co.]

I, is used to attach triplex or four-plex cables, consisting of two or three insulated cables which are spiral-wrapped around a bare messenger wire (usually the neutral of the supply). For heavier loads a low-voltage dead end on a crossarm can be made by using a strain nsulator, as shown in Fig. 8-70, I. For higher voltages one or more suspension insulators with a clamp to grip the conductor, as shown in Fig. 8-70, II and III, can be used. Instead

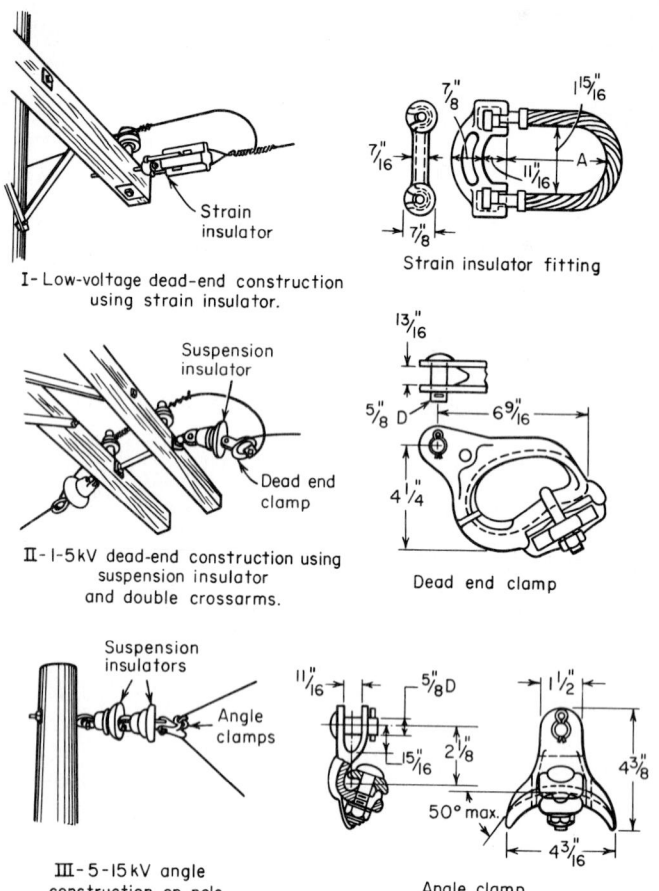

I- Low-voltage dead-end construction using strain insulator.

Strain insulator fitting

II-1-5kV dead-end construction using suspension insulator and double crossarms.

Dead end clamp

III- 5-15 kV angle construction on pole.

Angle clamp

**Fig. 8-70** Types of clamps and dead-end construction. [Ohio Brass Company]

of the conductor clamp, the conductor can be looped around a dead-end thimble (Fig. 8-71).

**80. When wires are carried though trees** so that they would rub against branches, special tree wire (Fig. 8-72, I) should be used. However, wires through trees should be avoided if at all possible. Tree wire has a tougher insulation which will resist abrasion from the branches better than weatherproof wire. If the tree branches are more or less continuous, tree wire should be used exclusively. If there is a span or more of clearing, it will be economical to splice in weatherproof wire to run through the clear space. For especially severe conditions a length of tree-wire molding (Fig. 8-72, II, III, and IV) should be added to the wire where the wire passes close to heavy limbs or tree trunks. In some cases a tree insulator (Fig. 8-73) can be screwed to the limb or trunk of the tree. This insulator has a removable pin and roller (Fig. 8-73, II) so that the wire can be inserted without cutting and the pin replaced. The roller allows the wire to move freely without

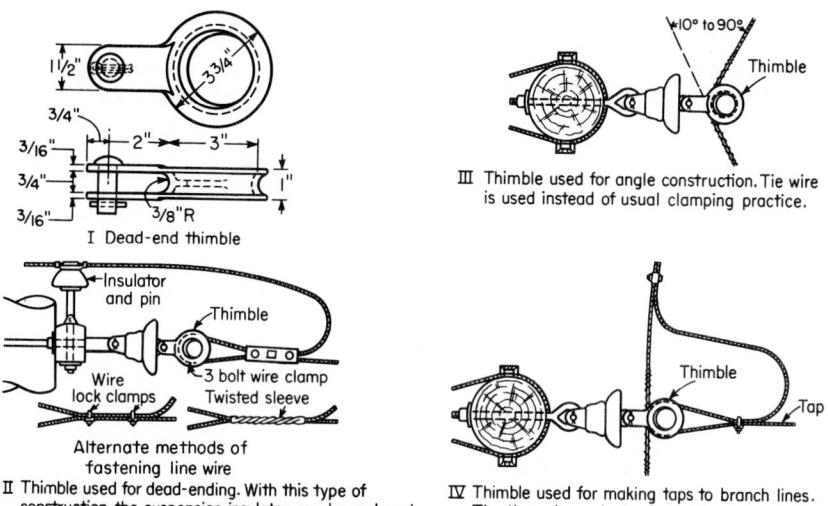

I Dead-end thimble

III Thimble used for angle construction. Tie wire is used instead of usual clamping practice.

Alternate methods of fastening line wire

II Thimble used for dead-ending. With this type of construction the suspension insulator can be replaced without cutting the jumper and making a new dead-end.

IV Thimble used for making taps to branch lines. The through conductor is tied in at the clevis

**Fig. 8-71**   Dead-end construction with dead-end thimble. [Ohio Brass Company]

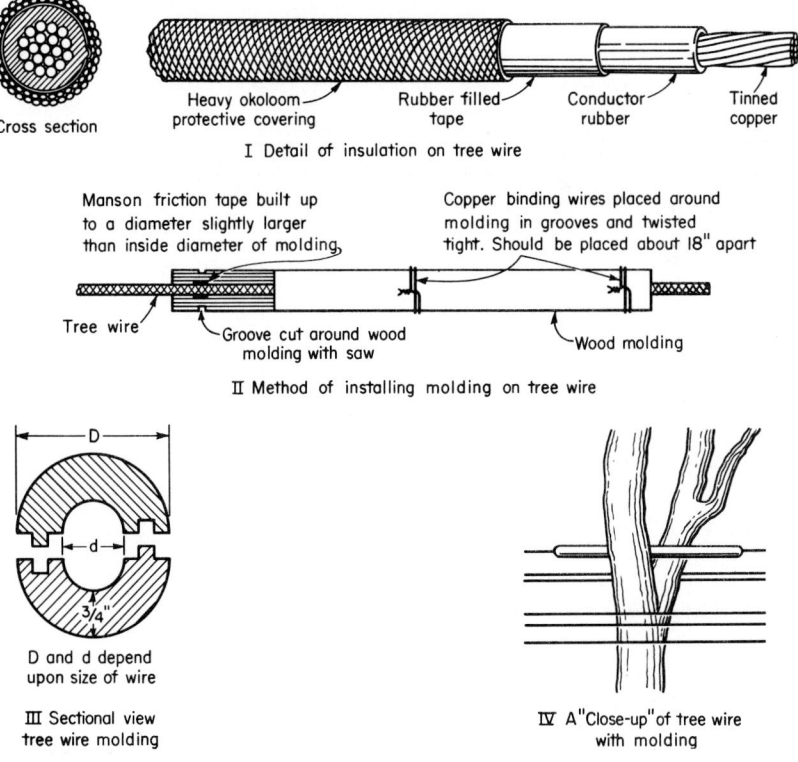

I Detail of insulation on tree wire

II Method of installing molding on tree wire

III Sectional view tree wire molding

IV A"Close-up" of tree wire with molding

**Fig. 8-72**   Tree wiring. [The Okonite Co.]

being cut. Wires should never be attached rigidly to trees because the swaying of the tree might break the wires.

**81. Wire and wire sizes for electric light and power lines.**   No wire smaller than No. 6 is used in good construction for line wire. No. 8 is sometimes used for services not more than 75 ft (22.9 m) long that do not cross a street. Solid wires are often used for sizes up to and including No. 2/0, and cable (stranded wire) is used for larger conductors. Triple-braid weatherproof is the standard insulation of aerial line wires. Annealed or soft-drawn wire is preferable to hard-drawn for services and secondary distribution lines because it is more readily handled and, in the sizes used, has ample tensile strength. Hard-drawn copper wire or steel-reinforced aluminum cable is preferable for transmission lines and feeders.

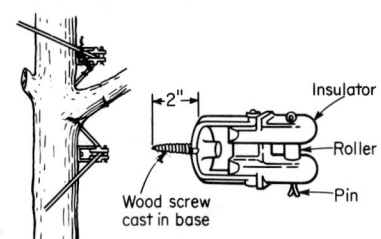

I Insulator mounted on tree       II Detail of tree insulator

**Fig.  8-73**  Tree  insulator.  [Line  Material Co.]

**82. Stringing wire.**   There are two methods, the choice depending on local conditions and the size and length of the circuit. By one method a reel of wire is set at one end of the line and a rope carried 1500 or 2000 ft (457 or 610 m) over the crossarms and attached to the wire, which is then drawn over the arms. The other way is to place the reel on a cart, and after the end of the wire has been secured to the last pole, the cart is started and the wire paid out till the second pole is reached, and then the wire is hoisted up and laid on the arm. Wire should always be paid out from the coil, the coil revolving, so that the wire will not be twisted. If wire is not received on reels, it should be placed on them before paying out.

## POLE-LINE GUYING

**83. Guying.**   Probably there are not so many guys on pole lines as there should be to ensure continuity of service and minimum maintenance expense. Lines should be guyed not for normal conditions but for the more severe conditions that are apt to obtain. The guys should be frequent and heavy enough to sustain the line after the heaviest snow-storm or during the worst possible windstorm. A guy should be used on every pole where the tension of the wires tends to pull the pole from its normal position.

Terminal poles should always be head-guyed, and on lines carrying three or more cross-arms the two poles next to the terminal pole should also be head-guyed to distribute the stress.

Line guys are installed on straight pole lines to reinforce them against the excess stresses introduced by storms. It is good practice to install head line guys, as shown in Fig. 8-74, at about every twentieth pole. This recommendation applies only to lines carrying more than one crossarm.

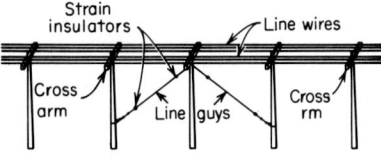

**Fig. 8-74**  Line guys.

The installation of additional side guys, arranged at right angles to the line, to trees, stubs, or anchors is recommended. The side guys are attached to the same pole as the head guys.

Side guys should be installed at curves, the guys taking the direction of radii of the curves. Figure 8-5 shows a table that gives the number of side guys required for a given line with a given *pull*. The pull is the distance from the pole to a line joining two points in the line 100 ft (30.5 m) on either side of the pole (see Fig. 8-5).

Poles on either side of a long span should be head-guyed as shown in Fig. 8-75, I.

Lines on steep hills should be guyed as shown in Fig. 8-75, II, with head guys.

Poles at each side of a railroad crossing, important highway crossing, or a crossing over another wire line should be head-guyed away from the crossing as in Fig. 8-75, I.

Guy wires should be kept more than 8 ft. (2.4 m) from the ground where possible to prevent persons from making accidental contact with them. A clearance of at least 3 ft (0.9 m) should be maintained between guys and electric wires; otherwise changes in temperature may cause the wires to come in contact.

**84. A guy assembly** consists of the following parts:

1. Attachment to pole being guyed
2. Guy wire containing one or two strain insulators
3. Attachment to stub pole or to lower part of another pole or ground anchor, anchor rod, attachment of strand, and guard

**85. Some of the methods of** making up a guy on a pole are shown in Fig. 8-76. The method of Fig. 8-76, I, is easy when some of the other materials are not at hand. The methods of Fig. 8-76, II and III, make a neater-looking job when the hardware is available. The method of Fig. 8-76, IV, is used at a stub pole. Refer to Sec. 55 for a discussion of the hardware required. When there is only one guy per pole, it is made around the pole just under the top crossarm as shown in Figs. 8-77 and 8-78. If there are two guys to the pole, one is made up just under the top crossarm and one just under the bottom crossarm. If more than two guys are used, the additional ones should be spaced as equally as possible between the top and the bottom guys that are made up under the top and bottom crossarms, respectively. If there are two or more guys to a pole, each should be independent of the others. The different guy wires should not cross one another. If two or more guys support a pole, a turnbuckle should be inserted in all but one to equalize the stresses.

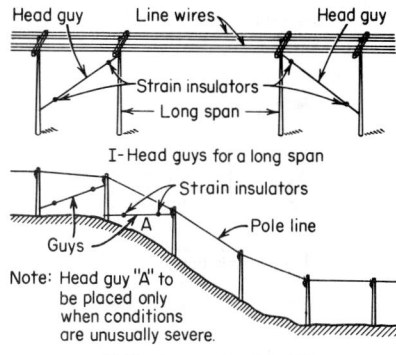

**Fig. 8-75**   Head-guy locations.

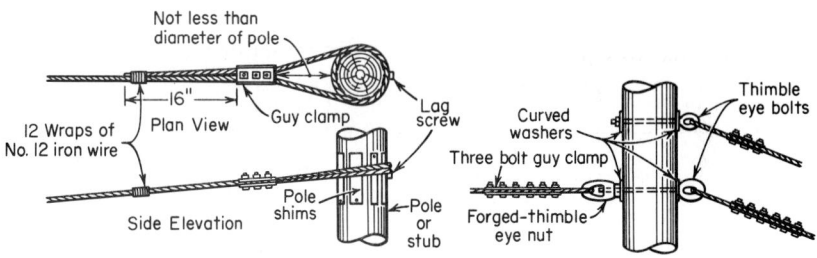

I-With strand wrapped around pole, using guy shims

II-With straight thimble eye bolt

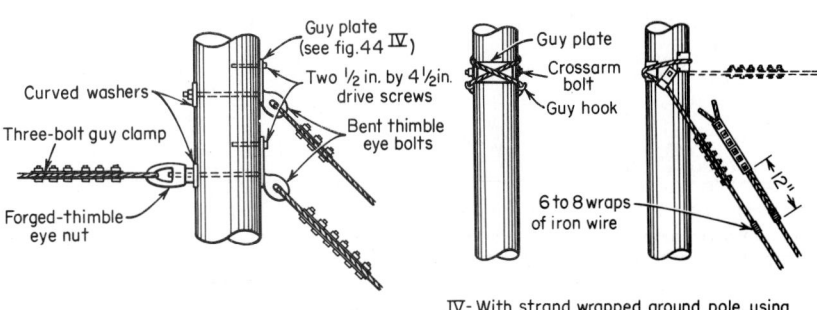

III-With bent thimble eye bolt

IV- With strand wrapped around pole, using guy plates and hooks

**Fig. 8-76**   Methods of attaching guys to pole. [General Electric Co.]

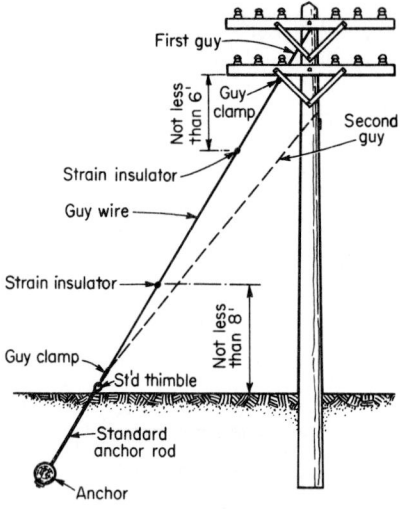

First guy

Guy clamp

Not less than 6'

Second guy

Strain insulator

Guy wire

Strain insulator

Not less than 8'

Guy clamp

St'd thimble

Standard anchor rod

Anchor

**Fig. 8-77**   An anchor-guyed pole.

**86. Guy-wire insulation.** Strain insulators should be inserted in all guy wires to poles carrying electric lighting or power wires. Two insulators should be inserted in each guy. One is located at least 6 ft (1.8 m) from the pole itself or 6 ft below the lowest line wire. The other is located at least 6 ft from the lower end of the guy and at least 8 ft (2.4 m) from the ground. The two strain insulators are sometimes coupled in series in short guys.

The assembly makeup of a strain insulator for a guy is shown in Fig. 8-79. The two ends of the guy strand are looped through the separate holes in the insulator. The guy-wire clamps hold the wire from slipping, and the No. 12 wrapping wire secures the end of the strand to the guy wire and provides a neat appearance for the assembly. Suitable clamps can be used in lieu of the No. 12 wrapping wire at guy-wire ends.

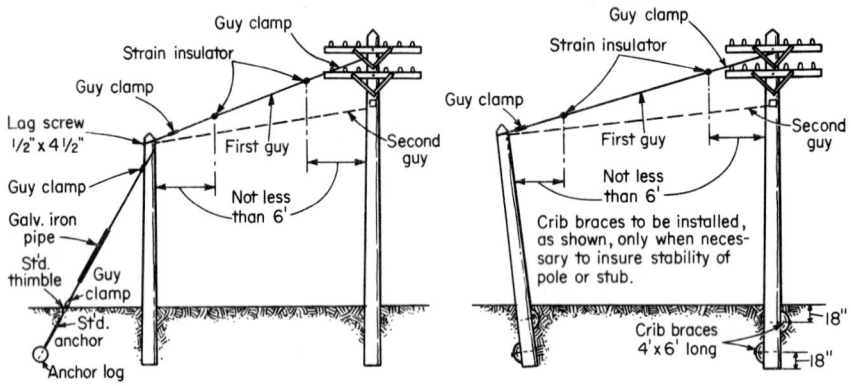

**Fig. 8-78**   Poles guyed to stubs.

**87. The number and size of clamps** to use for each looped-back connection on a guy strand are as follows:

| Breaking strength of guy strand, lb | No. of clamps | Type of clamp | Breaking strength of guy strand, lb | No. of clamps | Type of clamp |
|---|---|---|---|---|---|
| 4,000 | 1 | Two bolt | 10,000 | 2 | Three bolt |
| 6,000 | 1 | Three bolt | 16,000 | 3 | Three bolt |

**88. The attachment of the guy strand to the anchor rod** is shown in Fig. 8-80. Refer to Sec. 54 and Fig. 8-42 for description of anchor rods. The eye of the guy rod should extend about 3 ft (0.9 m) above the surface of the earth. Guy rods should not be installed where they will interfere with traffic. A guy protector guard (see Sec. 58) extending to about 8 ft (2.4 m) above the ground should be placed over anchor guy wires near roadways, as

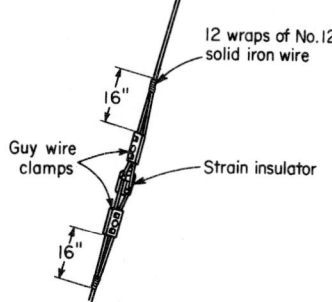

**Fig. 8-79**   Method of making up a strain insulator in a guy strand. [Hubbard & Co.]

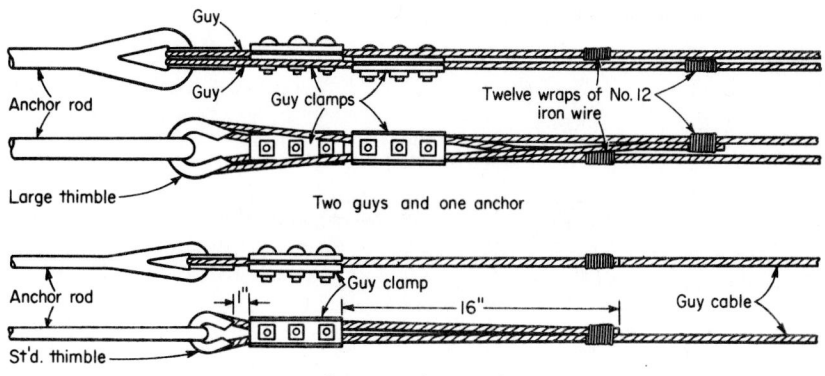

**Fig. 8-80**   Methods of making up guys on guy rods.

shown in Fig. 8-78. The foot of an anchor guy should be as far away from the foot of the pole as possible—at least a distance equal to one-fourth the height of the pole.

**89. Anchors** can be made from logs cut from the tops of old poles (Fig. 8-81), or patented metal or concrete anchors can be used. Figure 8-82 shows a variety of modern guy anchors which provide an effective support for guy rods. Table 90 gives the minimum size for anchor logs and rods and the depth of setting for anchors.

**90.   Size of Anchor Logs and Depth of Setting for Anchors**

| Size and number of guys | Min size of round logs for good soil | Length of rod below ground, ft | Size of guy rod |
|---|---|---|---|
| One No. 6 Wire or one 4,000 lb | 3 ft × 6 in. | 4 | 7 ft × ½ in. |
| One  6,000 lb | 3 ft × 7 in. | 4½ | 8 ft × ⅝ in. |
| One 10,000 lb | 4 ft × 8 in. | 5½ | 9 ft × ¾ in. |
| One 16,000 lb | 5 ft × 10 in. | 6½ | 10 ft × 1  in. |
| Two  6,000 lb | 5 ft × 8 in. | 5½ | 9 ft × 1  in. |
| Two 10,000 lb | 5 ft × 10 in. | 6 or 6½ | 10 ft × 1¼ in. |
| Two 16,000 lb | 6 ft × 12 in. | 6½ or 7 | 11 ft × 1½ in. |

For the depth and holding strength of patented anchors the catalogs of the particular manufacturer must be consulted.

**91. Patented guy anchors** can be used in certain kinds of soil very effectively. Figure 8-83 shows one kind of anchor that is screwed into the earth with a wrench. The resistance of this sort of anchor to withdrawal from compact homogeneous soils is not determined by the weight of a column of earth the diameter of the anchor screw, but is determined by that of a cone with sides slating at 45° and having the point of the anchor as an apex.

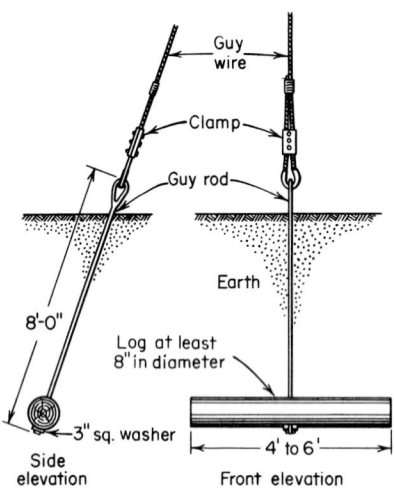

**Fig. 8-81**    A guy rod and anchor.

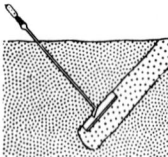

**PLATE ANCHORS** are used for general anchoring; are made of malleable iron or steel plates. They provide excellent holding power because pull is against undisturbed earth.

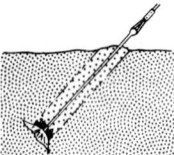

**EXPANDING ANCHORS** are used for general anchoring. Size of anchors in folded position ranges from 6 to 12 in. dia. Expanded sizes range from 50 to 300 sq. in.

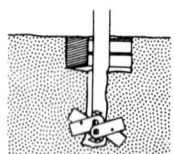

**KEY ANCHORS,** sometimes called pole keys, are used where standard guying is impossible because of sidewalks, property restrictions, etc. Additional keying is necessary at ground level.

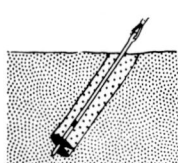

**SCREW TYPE ANCHORS** can be selected for general anchoring in gravel, sand, clay, fill, silt, and with extension, swamp soils. The corkscrew-like blade is made of drop-forged steel. No anchor hole is required; anchor is screwed into the ground with a wrench or bar.

**ROCK ANCHORS** are used for anchoring in rock, rocky soil or masonry. The expanding rock anchor is made of malleable iron, develops tremendous holding power through a wedging self-action.

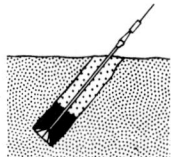

**CONE ANCHORS** are made of steel and are used in bed rock, laminated rock, slate, sandstone, shale, hardpan, gravel, and medium-firm clay. Sizes range from 6 to 23 in. dia. It develops adequate holding power in class 2 soil through wedging action.

**Fig. 8-82**    Types of modern guy anchors.

Figure 8-84 illustrates this and shows the formula used for computing the withdrawal resistance. Figure 8-84 shows the anchor inserted perpendicularly, but an anchor should always be inserted at the same angle that the guy wire assumes so that the rod of the anchor will be in a direct line with the guy wire. Section 92 gives the actual resistances to withdrawal of the anchors.

**92. The actual holding power of patented guy anchors** is an uncertain quantity, except in hard clays (adobe, hardpan, or gumbo) for which Carpenter's formula (Fig. 8-84) can be used. The safest procedure is to test the anchor's resistance to withdraw in the soil in which it is to be used. A dynamometer can be employed for measuring the pull. Anchor manufacturers will gladly ship anchors on approval for test and 30 days' trial. In general, the 5-in Matthews Scrulix anchor (Fig. 8-83) should, in other than the hard clays, be used only for the lightest strains, say 1100 to 2500 lb, the 6-in for 1500 to 2500 lb, the 7-in for 2500 to 5000 lb, the 8-in for 5000 to 7000 lb, the 10-in for 7000 to 10,000 lb, and the 12-in for

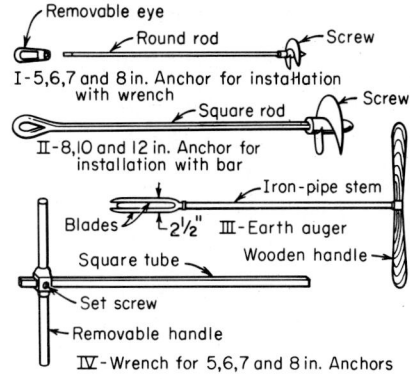

**Fig. 8-83** Matthews Scrulix anchors and wrenches.

10,000 to 12,000 lb. These holding powers are based on the anchor being screwed into the earth for a depth of at least 5 ft (1.5 m). For anchorages under shallow strands of water or in swampy or caving ground, the 8-, 10-, and 12-in anchors having the square rods are very convenient.

**93. Directions for installing Matthews guy anchors**

5-, 6-, 7-, AND 8-IN WITH RODS   Remove the eye of the anchor; pass the rod through the wrench and replace the eye, which will serve to hold the wrench rigidly to the anchor; then screw the anchor in, at the same angle as the guy wire is to run, as far as ground conditions will permit. When it is in as far as possible, remove the eye and pull out the wrench. Then replace the eye, thus making anchor ready for guy wires. The handlebars of the wrench are adjustable and held in place with a setscrew. They can be moved back as the anchor screws in.

8-, 10-, 12-IN WITH RODS   Place bar or other lever in eye of anchor and screw it in as far as ground conditions will permit, always at the angle that the guy wire is to run. Time will be saved and the anchor start more easily if a few spadefuls of earth are removed before starting anchor. When the anchor is set, attach the guy strand to the eye. Always pull anchor back as far as possible before finally tying the guy wire.

IN DRY, HARD GROUND   In setting all anchors in hard ground, the work will be much easier if a small hole is first made with an earth auger (Fig. 8-83), a crowbar, or a wood

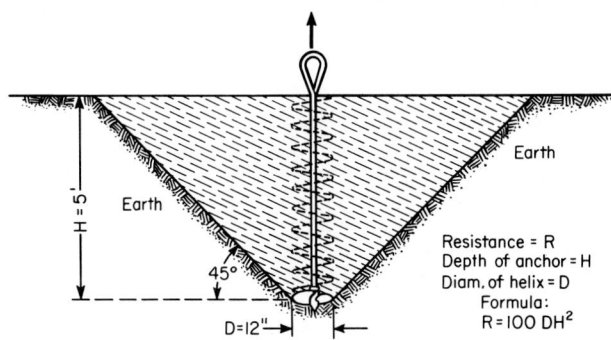

$H = 5'. D = 12''.$ Then $R = 100\,DH^2$ or $R = 100 \times 12 \times 25$ or $R = 30,000$ lbs.

**Fig. 8-84**   Carpenter's formula for resistance of Matthews anchor to withdrawal.

auger with a long shank. This renders the insertion of the anchor easier. A little water poured down this hole before starting the anchor will help considerably if the ground is hard and dry. In installing 8-, 10-, and 12-in anchors in very hard ground, clamp a lever to the rod by means of a chain a foot or so above the ground. As the anchor is screwed down, the lever can be moved up. The anchor will start more easily if a few spadefuls of earth are removed at the angle desired to set the guy. If someone stands on the helix of the anchor when starting until the point bites the ground, it will be helpful.

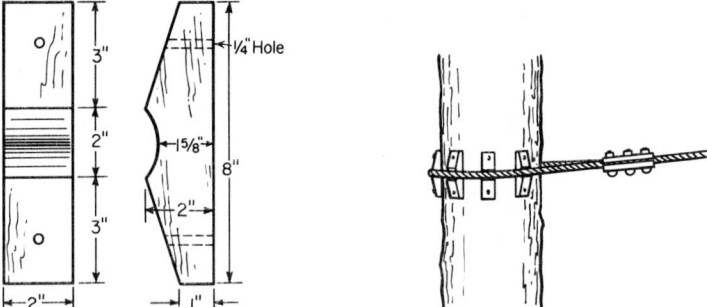

**Fig. 8-85-I**   Details of tree blocks.          **Fig. 8-85-II**   Guy wire fastened to tree.

In localities where loose gravel or small flat rock occurs, drill a hole with an earth auger, a digging bar, or a crowbar as suggested above. If a small rock is encountered, it can be broken by the bar. If a large rock is "discovered," the bar can be removed and the hole drilled in another place. This will allow the use of anchors in many places where otherwise it would seem impossible to install them.

**94. The methods of installing stub guys** are shown in Fig. 8-78. The stubs should be long enough so that the guy wires will clear roadways by at least 18 ft (5.5 m), sidewalks by 15 ft (4.6 m), and electric wires by 3 ft (0.9 m). Stubs are used only when a line cannot be guyed properly to trees or poles. Stubs should satisfy the specification for poles as given in Sec. 8.

**95. In guying to a tree,** tree blocks (Fig. 8-85) should be used, and the wire should pass

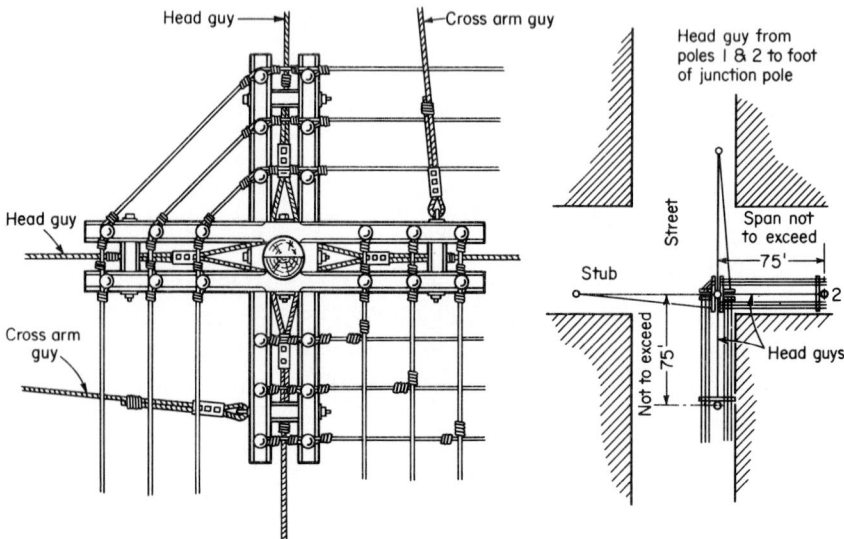

**Fig. 8-86**   Method of turning corner with one pole.

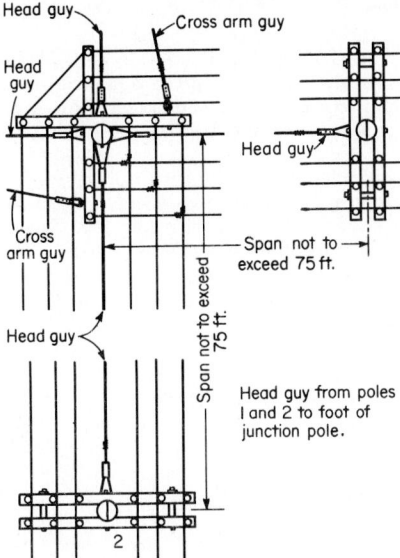

**Fig. 8-87**  Corner pole without double arms.

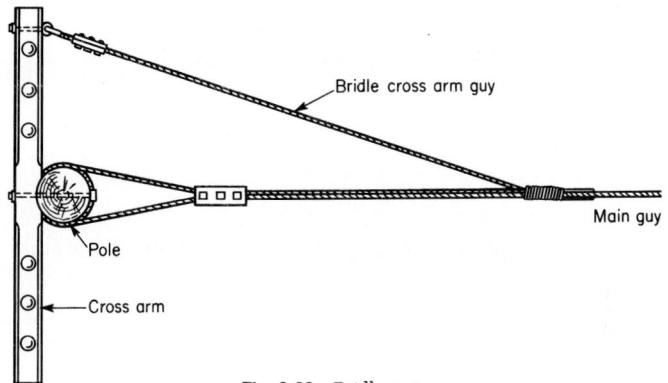

**Fig. 8-88**  Bridle guy.

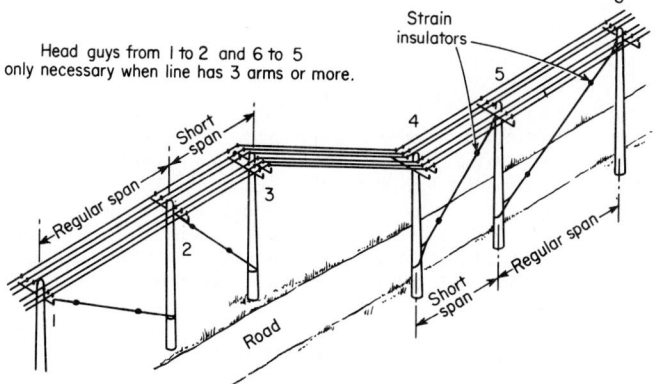

**Fig. 8-89**  Guying at road crossing without side guys.

only once around the tree. Tree guying is undesirable and should not be done unless absolutely necessary. Guys should preferably be attached to trunks or to limbs that are not less than 8 in (203 mm) in diameter. Do not attach a guy wire to a limb that will swing with the wind and sway the pole. Enough tree blocks should be used so that the guy wire cannot touch the tree.

**96. Crossarm guys** are used when the pull on a crossarm is unbalanced. Figures 8-48, 8-86, and 8-87 show examples. Crossarm guys usually extend from the arm to a pole or stub, but sometimes for light strains the Y, or bridle, guy (Fig. 8-88) is used.

**97. A line must be thoroughly guyed where it crosses a road.** Figures 8-89 and 8-90 show two methods of holding a line at such a point. The method involving the use of side guys is preferable, but the other one will give good service when side guys cannot be installed.

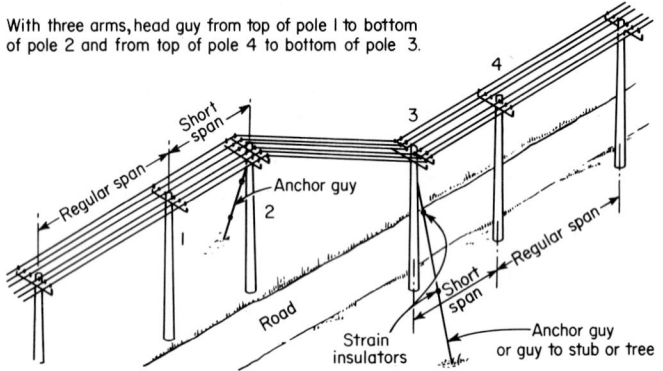

**Fig. 8-90**   Guying at road crossing with side guys.

**98. When it is impossible to secure guying privileges** along the desired route for a line or obstructions make guying difficult, the problem is generally solved by the use of self-supporting poles, a construction commonly referred to as *hog guying.* This method of strengthening a pole is especially useful on corners, around curves, and at points where space is too limited for ordinary guying. The method of hog guying is illustrated in Fig. 8-91.

**99. Pole braces** are used when guying is not feasible. They cost more than equivalent guys. Figure 8-92 shows methods of bracing poles. The upper end of each brace fits in a notch cut in the pole and is bolted thereto.

## UNDERGROUND WIRING

**100. Underground systems** are of three main types:
1. Duct line
2. Cable buried directly in the ground
3. Conduits located in tunnels

A duct line consists of one or more conduits spaced close together with or without a concrete casing. The duct line is laid in a trench with the top at least 2 ft (0.6 m) underground and covered over with earth; it should be covered with pavement if it is located under a street. The duct lines terminate in underground vaults, called manholes, where transformers and protective equipment are located and where cable splices are made. The cables are drawn into the ducts from one manhole to the next. This system is ordinarily used for general underground distribution work when more than one cable is located in the same run or when the cable may have to be increased in size or replaced. Nonmetallic conduits must be encased in not less than 2 in (50.8 mm) of concrete, unless listed by a nationally recognized testing agency as suitable for direct burial without encasement, if contained conductors operate in excess of 600 V.

Under the system in which the cable is buried directly in the ground, a trench is dug, and the cable, which is of a type to resist the corrosive effects of the ground (called parkway or nonmetallic-armored cable; see Div. 2), is laid in the trench without any protection other than that afforded by the cable insulation and earth covering (refer to Sec. 133 for further details). This method is used for single-circuit runs on which the load is apt to be steady and definite, as in street or parkway lighting systems. This method is also used for underwater lines, where the construction of a duct line would be difficult or impossible.

For groups of buildings in an industrial plant or institution, where tunnels between buildings must be constructed for steam and water lines, it is quite common to locate the electrical circuits in conduits or cable trays on the walls of the same tunnels.

**101. Commercial conduit duct materials** are fiber, vitrified tile, iron, polyvinyl chloride (PVC), polyethylene (PE), styrene, and monolithic concrete.

**102. Fiber conduit** is widely used for ducts in underground construction. Fiber conduit is made of wood pulp which is treated and thoroughly saturated with a bituminous compound. The compound contains about 6 percent of creosote to prevent rotting. The

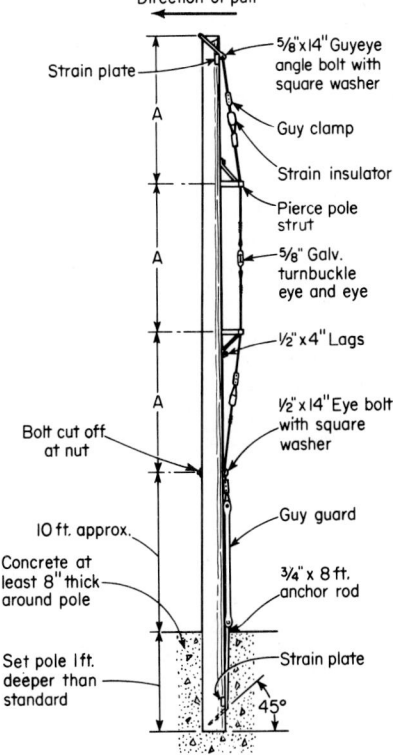

**Fig. 8-91** Details of self-supporting pole.

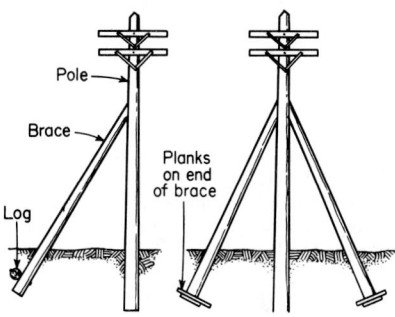

**Fig. 8-92** Bracing poles.

different sections of conduit are joined by means of self-aligning joints. They can be laid in the trench with great rapidity. The cost of laying fiber conduit is considerably less than for tile duct owing to its lightness and to the greater length of sections, which require fewer joints. Fiber conduit permits cables to be drawn into it readily because of its smooth oily interior. The ducts are laid in concrete with walls 3 in (76.2 mm) thick on all sides. Generally, a spacing of 3 in is employed between the different ducts of a run.

The standard sizes (nominal inside diameters) are 2, 3, 4, 4½, 5, and 6 in. Fiber conduit is made in standard lengths of 5 and 8 ft (1.5 and 2.4 m).

**103. Plastic conduits,** such as polyvinyl chloride, polyethylene, and styrene, described in Divs. 4 and 9, are being widely used for underground installation because of low cost, ease of installation, and minimum use of labor. Such conduits are available in lengths up to 30 ft (9.1 m), thereby reducing the number of couplings required in a given run.

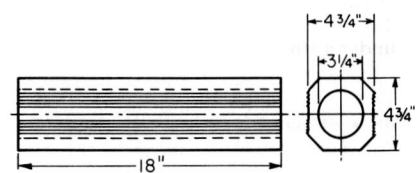

**Fig. 8-93** A piece of single duct.

**104. Vitrified-clay single duct** or hollow brick is the most popular for communications cables and is sometimes used for low-voltage power work. Figure 8-93 shows a typical length. The inside diameters of the round ducts are 3¼, 3½, 4¼, and 4½ in. A 3½-in-square size is also available. The single duct is preferred because its walls are thick and, in laying, every joint is broken, eliminating the possibility of the arc of a short circuit on one cable affecting cables in other ducts in the same run. Single ducts can also be more readily laid around obstructions such as pipes, and, furthermore, curves can be more readily formed with them than with multiple ducts.

**105. Vitrified-clay multiple duct** is frequently used for communications cables on account of the ease of assembly. The four-way multiple duct is the most popular size (Fig. 8-94), although the six-way duct is frequently used. Nine-, twelve-, and sixteen-way multiple ducts can be manufactured, but they are seldom used because of their excessive weight and liability to breakage. The dimensions of the multiple ducts of a certain nominal size made by different manufacturers vary. Those shown in Fig. 8-94 for a four-way duct are typical. A 3¼-in-square size is also available. Two- and three-duct sections are 24 in (609.6 mm) long; the other, 36 in (914.4 mm).

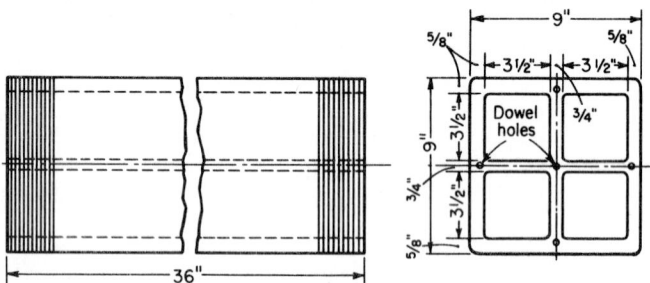

**Fig. 8-94**   A piece of multiple duct.

**106. Iron conduit** is sometimes used for duct lines, but since it costs more than other materials, it has not had general acceptance for this purpose. One important advantage is that it gives the cables better protection from damage during excavation. Wrought iron is preferable to steel conduit, which is often sold for wrought iron, because wrought iron resists corrosion much more effectively than does steel. See Index for dimensions of wrought-iron conduit.

Wrought-iron conduit is frequently used to carry parkway lighting circuits if they are laid in filled ground which might be injurious to cable laid directly in the ground. The principal use for wrought-iron conduit in underground work is for laterals (see Sec. 116).

**107. Cable-in-duct.**   A new popular system is cable-in-duct, which consists of a semirigid nonmetallic conduit and factory-installed conductors. Shipped in large reels, this system is easy to install, offset fittings and couplings are eliminated, and conductors can be withdrawn if necessary. Typical applications are for parkway, highway, and street-lighting underground systems.

**108. Monolithic concrete ducts** can be built in two ways, by using flexible Para rubber tubing as a core and by using a duct-forming machine called a boat. In the rubber-tubing method, thick rubber tubes which are flexible but not readily collapsible are laid on a concrete base and separated with terra-cotta spacers. Concrete is then poured around them as in the other methods. When the concrete has set for a few hours, the tubings are withdrawn, leaving a very smooth concrete duct. In the boat method a track is built at the bottom of the trench. On this track runs the boat, which is really a die containing as many steel tubes as the number of ducts wanted. Concrete is fed into a hopper at the top and is tamped down around the tubes. The boat is moved slowly along the track either by hand or by power from an electric motor, leaving the finished duct extruding from the rear. The concrete is of a special mixture which is sufficiently solid to hold the form of the ducts as the boat moves along.

The rubber-tubing method is especially advantageous when the duct line must be threaded around other underground structures and pipes. It can be used to advantage in bringing duct lines into power stations and substations, as the duct can be formed in one continuous run from the street manhole to the switching-gallery floor. The main advantage of the boat method as well as the rubber-tubing method is that there is no material in the duct to decay. Both methods have proved more expensive than using fiber conduit, so that for normal straight runs they cannot compete economically. For special irregularly shaped curves a section of the monolithic concrete duct formed by the rubber-tubing method may well be inserted even if the main run is made with one of the other materials.

**109. In installing any kind of duct line** a trench is excavated to such a depth that the top conduit will be 2 to 3 ft (0.6 to 0.9 m) below the surface of the ground and so that the grade of the duct line will pitch toward the manholes about 1 ft in 100 ft (0.3 m in 30.4 m) so as to ensure effective drainage. For power cables the ducts should be arranged

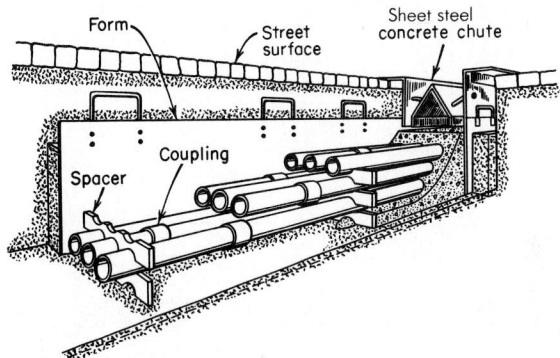

**Fig. 8-95**    Construction of duct line [Manville Corp.]

either two conduits wide or two conduits deep so that every conduit will have earth on at least one side of it. This is necessary for conduction into the cooler ground of the heat generated by the $I^2R$ losses in the cables. For communications cables the heat loss is low, and the ducts can be arranged in any convenient way. There should be from 1 to 3 in (25.4 to 76.2 mm) of earth or concrete between conduits for power systems to ensure that flame and heat from a short circuit in one conduit will not affect the adjacent cables. The 3-in separation should be used on the higher-voltage lines. Conduits are held in position during construction of the duct line by plastic spacers (Fig. 8-95). The width of the trench will depend on the working space required and on whether concrete and concrete forms are used or not. If the soil is firm, the trench may be merely 1 or 2 in (25.4 or 50.8 mm) wider than the space required by the conduits. Where concrete is required, it should be about 3 in thick outside the conduits. In good, firm soils the trench can be dug to exactly this width and forms omitted, the concrete occupying all the space between the conduits and the earth wall. In ordinary and poor soils the trench should be made an additional 3 in wider on each side to provide space for installing the forms.

The bottom of the trench should be carefully rammed until solid so that undue stress will not be placed on the conduits. With concrete construction the 3-in bed of concrete is then laid, over which the lower tier of conduits is placed with spacers to ensure the correct separation between conduits. Then the concrete is tamped around them, and thus each level is built up until the duct is completed with 3 in of concrete on top. Sometimes the spacers are removed after the conduits are held firmly in position by the backfill and before the spacers are completely covered over. At other times the spacers are left in to form a permanent part of the duct line.

**110. Installing single-duct vitrified-tile conduit.** See Fig. 8-96 for sections of duct runs.

After the bed is set, the duct is laid in cement mortar. A mandrel (Figs. 8-97 and 8-98) is used to keep the successive pieces in line. It is customary to enclose the conduit in a continuous concrete encasement 3 to 4 in (76.2 to 101.6 mm) thick.

The mandrel is pulled through with a long hook as the conduit progresses to align the ducts. The leather washer scrapes away any mortar that has oozed through between joints and leaves the duct quite clean. The end of a No. 12 galvanized-iron wire is frequently attached to the inner end of the mandrel and is pulled into the conduit as its construction progresses. The wire is employed to pull in the drawing-in rope which is used to pull in the cable.

The single ducts furnished by some manufacturers are provided with male and female ends, which assist in aligning the ducts.

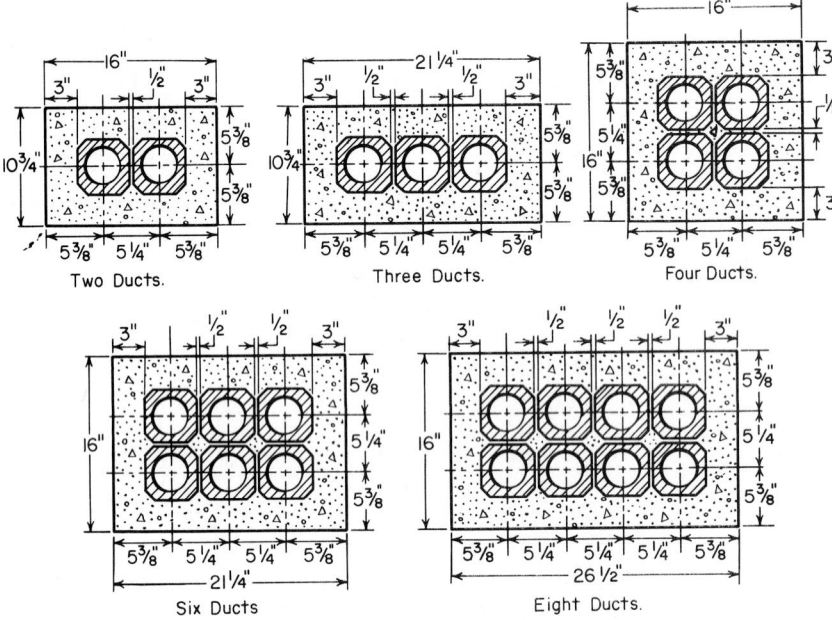

Two Ducts.          Three Ducts.          Four Ducts.

Six Ducts          Eight Ducts.

**Fig. 8-96**   Arrangement of single-duct subway.

**111. Laying multiple duct.** Multiple duct is laid in about the same way as single duct. Figure 8-99 shows sections of multiple-duct runs. The pieces are laid end to end, and the joints, if there is more than one tier in the subway, are broken. The pieces are maintained in alignment by iron dowel pins or keys (Fig. 8-100), which fit in holes in the pieces. All joints are wrapped with pieces of burlap or coarse muslin, 4 in (101.6 mm) wide and 3 ft (0.9 m) long for four-duct tile, which are moistened to make them stick. They are then coated with cement. The cloth prevents the entrance of cement or concrete into the ducts.

Sometimes a mandrel (Fig. 8-97) ¼ to ½ in (6.35 to 12.7 mm) smaller than the hole is drawn through as the construction progresses, as suggested in Fig. 8-98, to ensure alignment of the pieces. The handle on the mandrel should be long enough to reach back two joints so that one can be sure that the last three pieces set align, as they may have become displaced in setting.

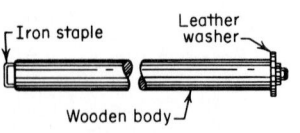

**Fig. 8-97** Mandrel for aligning conduit.

After the pieces are set, be careful that they are not displaced before the concrete jacket is deposited. The ducts should be cleaned out after the concrete jacket has been placed by drawing a wire brush or flue cleaner (Fig. 8-101) through them. The brush is somewhat bigger than the duct hole. Sometimes a metal scraper (Fig. 8-101) also is used.

Multiple duct has been laid without any concrete casing but with merely a concrete bed, as suggested in Fig. 8-102. This construction is economical in first cost but is apt to give trouble through settling of the earth or displacement due to future excavation. Creosoted boards are sometimes laid on top of a conduit run to protect against laborers' picks. Experience has shown that average laborers will stop when their picks strike a

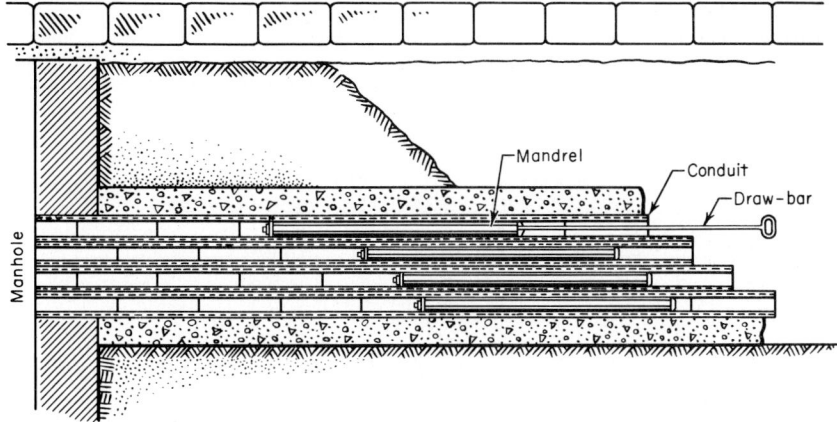

**Fig. 8-98**   Use of a mandrel.

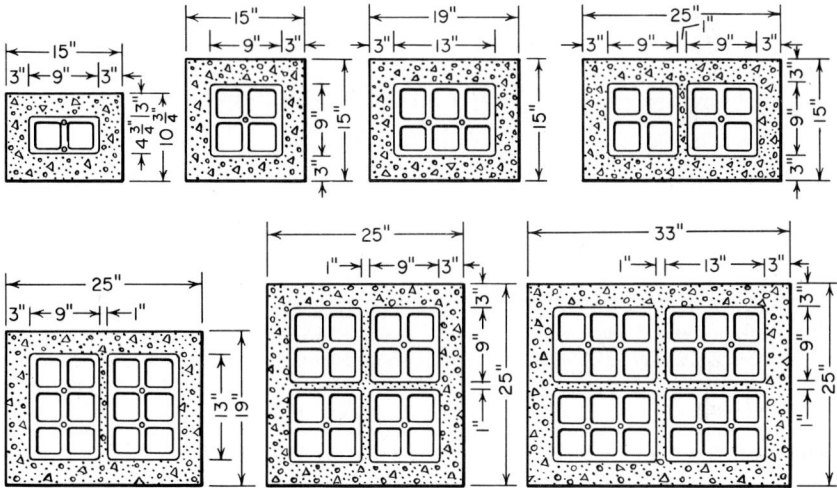

**Fig. 8-99**   Sections of multiple duct.

board but that they will pick their way through concrete or duct material. Multiple-duct conduit can be carried around obstructions, as shown in Fig. 8-103, by beveling the ends of the pieces. If the turn is too short, it may be difficult or impossible to "rod" the duct and to pull the cable in.

**112. To cut vitrified conduit** a groove is chipped completely around the piece on the line at which it is desired to cut it. A hammer and cold chisel are used for chipping the groove.

Usually it will break off on the chipped line after continued chipping, but it may not. Some experience is required before one becomes skillful at this work. Short lengths can be furnished by the conduit manufacturers, and their use is recommended.

**113. In installing wrought-iron conduit** no concrete casing is necessary if only one duct is involved. If there are several ducts in the run, the ducts are sometimes laid on a 3-in

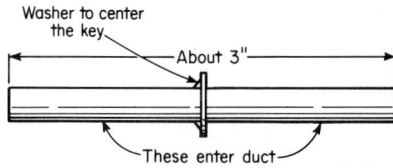

**Fig. 8-100**   Steel key for multiple duct.

(76.2-mm) foundation of concrete, and concrete is tamped between and around the ducts as shown in Fig. 8-96 for single vitrified duct. When the ducts will not be exposed to the dangers of future excavation, the cost of the concrete is probably not justified. Metal conduits can be obtained in 20-ft (6.1-m) lengths. Joints between adjacent lengths are made with conventional couplings. The burrs at the ends of each conduit must be care-

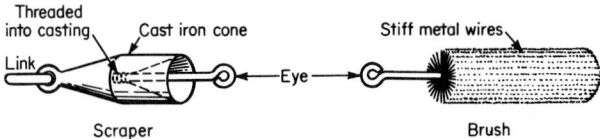

**Fig. 8-101**   Scraper and brush for conduit.

fully removed to prevent damage to the cable. When it is inconvenient to use a coupling, a union can be used instead. Metal conduit should be painted on the outside with asphalt or similar coating material to minimize corrosion.

**114. Installing fiber conduit.**   There are two methods of installing fiber conduit. In the *tier-by-tier method*, the lower tier of ducts is placed upon a concrete foundation in the trench and then covered with fresh concrete. The second tier of ducts is then put in place on top of the concrete, covering the first tier of ducts. This process is repeated with each tier of ducts until the required number of ducts have been installed.

In the *built-up method*, all the ducts are installed in a framework of spacers. Then side forms are put in place, and the concrete is poured over and around the ducts, thereby forming a monolithic concrete block.

When installing heavy-wall fiber conduit, it is recommended that after the line gradient is established, the trench surface be examined and all sharp or hard objects, such as stones, be eliminated. The floor of the trench is then shaped to receive the lower one-quarter of the outside surface of the conduit.

Where the ground is too solid or rocky, 3 in (76.2 mm) of selected backfill should be used under the conduit. The conduit should then be covered with a 3-in fine fill and tamped. Care should be taken not to tamp the conduit directly. A power tamper may be used on the final fill. If more than one bank of ducts is being installed, there should be a 3-in fill between banks and at least a 2-in (50.8-m) fill between duct rows.

Each length of conduit is provided with a taper sleeve joint coupling which provides a means of quick jointing. No cement or mastic is required to produce watertight joints. A block of wood and an ordinary hammer are all that are required to drive the taper sleeve joint onto the machined end of the conduit. The friction thus generated causes a bonding of the material which requires over 200 lb (889.6 N) of straight-line pull to separate one from the other.

Many fittings are available to meet the varied needs of installation. These include:

1. Angle couplings to provide for offsets to avoid obstructions

2. Adapters for connecting fiber conduit to threaded metal pipe of the same or different size

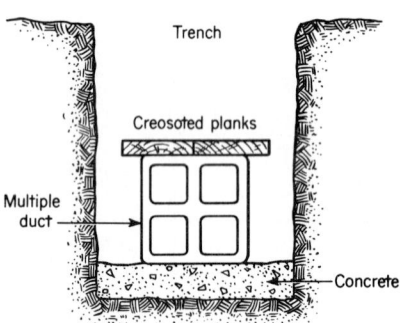

**Fig. 8-102**   Protective planks over a multiple duct.

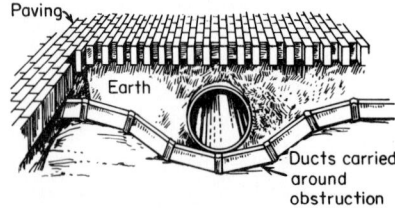

**Fig. 8-103**   Breaking around an obstruction with vitrified duct. If possible, the bend should pass over rather than under the obstruction. When it passes under the obstruction, water will collect in the pocket.

3.  45° and 90° elbows and bends
4.  S bends
5.  Watertight caps for sealing duct ends
6.  Expansion joints for use when expansion and contraction are likely to occur
7.  Bell ends for use at conduit terminals in manholes
8.  Reducers for connecting one size of fiber conduit to another size of fiber conduit

**115. Concrete for conduit work** should be clean; i.e., foreign substances should not be permitted to enter its composition. If the surface on which it is to be mixed is not smooth and clean, mixing boards or pans should be used. Foreign material impairs the strength of concrete, and it becomes porous and leaky. The concrete should be mixed from portland cement, clean sand, and gravel or broken stone in the proportions, by volume, of 1 part cement to 3 parts of sand and 5 parts of gravel or stone. Just sufficient water should be used to wet the mixture thoroughly and to permit a small amount of water to come to the surface when the concrete is tamped into final position. The cement, sand, and stone should be "turned" on the mixing board at least 3 times dry and at least twice after wetting. The concrete should be placed immediately after mixing. When the concrete has been placed in the trench, several hours should be allowed for it to take its initial set before the trench is filled in. This is necessary to prevent throwing the ducts out of alignment or fracturing the "green" concrete.

**116. For laterals which carry the service cables** from the duct lines to the buildings served, concrete construction is usually not justified. Iron conduit has been commonly employed for this purpose. Iron conduit can be bent at will, and it is often necessary to thread a network of underground structures with a lateral, especially if it must cross the street. Usually wrought-iron conduit is used because this material resists the corrosive effect of the ground. An application of wrought-iron conduit from overhead to underground construction is illustrated in Fig. 8-104.

**117. Manholes** are necessary in a subway system to permit the installation, removal, splicing, and rearrangement of the cables. A manhole is merely a subterranean vault or masonry chamber of sufficient size to permit proper manipulation of the cables. The conduits enter the vault, and on its sides are arranged devices whereby the cables within the manhole can be supported.

The location of manholes is determined largely by the layout of the district that is to be supplied with power. Whenever a branch of lateral extends from the main subway, there must be a manhole, and there must be manholes at intersections of

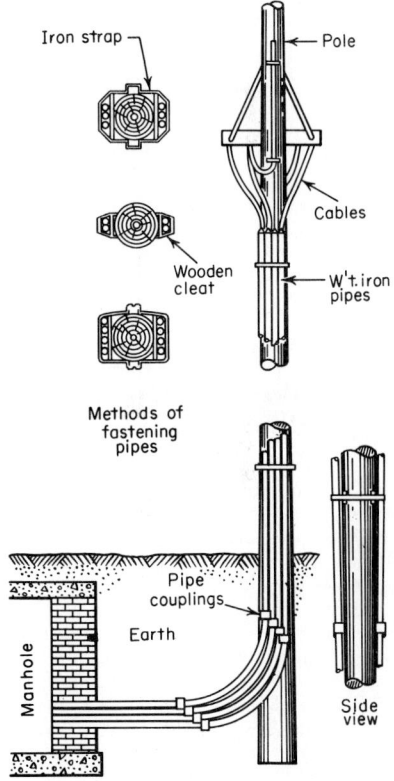

**Fig. 8-104**  Wrought-iron pipe laterals.

subways. In general, cables are not made in lengths exceeding 400 to 600 ft (122 to 183 m), and as it is necesary to locate splices in manholes, the distance between manholes cannot exceed these values. Furthermore, it is not advisable to pull in very long lengths of cable, because the mechanical strain on the conductors and sheath may then become too great during the pulling-in process. It is recommended that manholes be located not more than 500 ft (152 m) apart. The lines should preferably be run straight between manholes. If curves are necessary, the offsets from a straight line and the distance between manholes should not exceed the values given in the following tables:

| Single curve (see Fig. 8-105, I) | | | | Reverse curve (see Fig. 8-105, II) | | | |
|---|---|---|---|---|---|---|---|
| *A* Curve offset | *B* Min. length curve, ft | Max. length *L* between manholes, ft | | *A* Curve offset | *B* Min. length curve, ft | Max. length *L* between manholes, ft | |
| | | 4 ducts | 6–16 ducts | | | 4 ducts | 6–16 ducts |
| 0′ 1″ | 5 | 800 | 700 | 0′ 1″ | 5 | 800 | 700 |
| 0′ 3″ | 10 | 740 | 650 | 0′ 3″ | 15 | 740 | 650 |
| 0′ 6″ | 15 | 685 | 600 | 0′ 6″ | 25 | 655 | 575 |
| 0′ 10″ | 20 | 640 | 560 | 1′ 0″ | 35 | 570 | 500 |
| 1′ 3″ | 25 | 600 | 525 | 1′ 6″ | 40 | 515 | 450 |
| 1′ 9″ | 30 | 570 | 500 | 2′ 0″ | 45 | 485 | 425 |
| 2′ 4″ | 35 | 545 | 475 | 2′ 6″ | 50 | 455 | 400 |
| 3′ 0″ | 40 | 515 | 450 | 3′ 0″ | 55 | 435 | 380 |
| 3′ 9″ | 45 | 485 | 425 | 3′ 6″ | 60 | 420 | 365 |
| 4′ 7″ | 50 | 470 | 410 | 4′ 0″ | 65 | 400 | 350 |
| 5′ 6″ | 55 | 445 | 390 | 4′ 6″ | 70 | 385 | 335 |
| 6′ 6″ | 60 | 425 | 370 | 5′ 0″ | 75 | 365 | 320 |
| 7′ 6″ | 65 | 400 | 350 | 6′ 0″ | 80 | 345 | 300 |
| 8′ 9″ | 70 | 370 | 325 | 7′ 0″ | 85 | 320 | 280 |
| 10′ 0″ | 75 | 345 | 300 | 7′ 6″ | 90 | 310 | 270 |

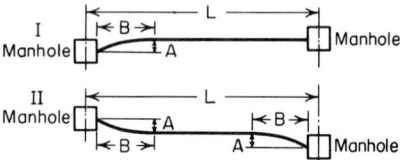

**Fig. 8-105**   Offsets for main duct line: (I) single curve; (II) reverse curve.

**118. Manholes are made in many shapes and sizes** to meet the ideas of the designer and to satisfy local conditions. It is established, however, that the form shown in Fig. 8-106 is best for the average condition. If there are obstacles about the point where a manhole is to be located, the form of the manhole must be modified to avoid them. The form approximating an ellipse (Fig. 8-108, I) is used so that the cables will not be abruptly bent in training them around the manhole. When the rectangular type of manhole is used (Fig. 8-108, II), care must be taken not to bend the cables too sharply.

The size of manholes will vary with the number of cables to be accommodated, but in any case there must be sufficient room to work in the manhole. A 5- by 7-ft (1.5- by 2.1-m) manhole (Fig. 8-106) is probably as large as will be required in isolated plant work, while a 3- by 4-ft (0.9- by 1.2-m) manhole (Fig. 8-107) is about as small as should be used. When transformers are located in a manhole, the size should be increased to allow for working space around the transformer and for ventilation. About 2 of 3 ft³ (0.057 to 0.085 m³) of volume should be allowed per kilovolt-ampere of transformer rating.

Manholes are built of either brick or concrete or of both of these materials. When many manholes are to be built of one size and there are no subterranean obstructions, concrete is usually the cheapest and best material. But when only a few are to be constructed or when there are many obstructions, a manhole with a concrete bottom, brick sides, and a concrete top is probably the best. Such a manhole can be constructed without having to wait for concrete to set before forms can be removed. There is a growing use of precast concrete manholes, shipped directly to the site.

**119. A concrete manhole is built** by first depositing the concrete floor (Fig. 8-106, II) and then erecting the form for the sides on this floor. In a self-supporting soil the sides of the hole constitute the form for the outside of the manhole. If the soil is not self-supporting, there must be an outer form of rough planks, which is usually left in the ground. Steel reinforcing (old rails are good) must be placed in the concrete top of a large manhole. In

a small manhole the manhole head or cover will extend over the sidewalks, and no reinforcing or manhole roof, for that matter, is required. All reinforcing steel should be completely encased in concrete to prevent corrosion.

**120. A manhole with brick walls is built** (Fig. 8-106, I) by first depositing the concrete floor
and then building up the brick walls thereon. If the manhole is large, the roof can be

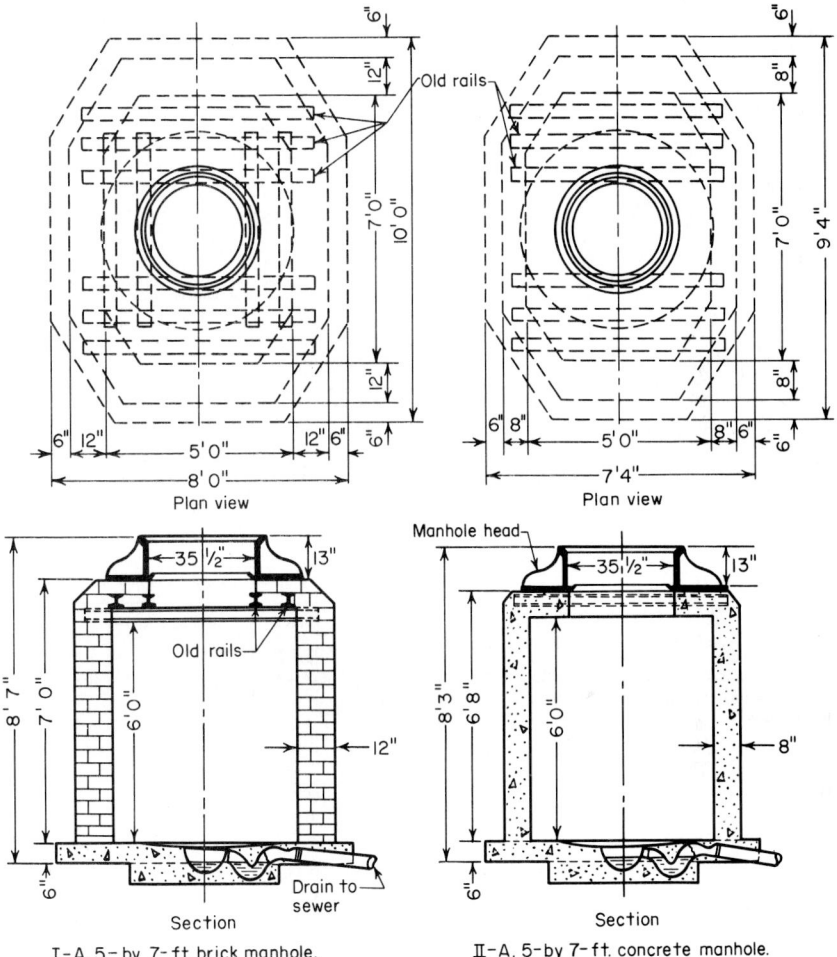

Plan view                    Plan view

Manhole head

Section                      Section

I–A. 5– by 7–ft. brick manhole.    II–A. 5–by 7–ft. concrete manhole.

**Fig. 8-106**   Typical manholes.

either of steel-reinforced concrete or of brick set between rails. Probably for installations
in which only a few manholes are to be built, the brick-between-rails method is the best.
For a small manhole no masonry roof is necessary, as the cast-steel manhole head forms
the roof.

Cement mortar for building brick manholes or for conduit construction can be made
by mixing together 1 part of cement, 3 parts of sand, and about ⅙ part of water, all by
volume.

**121. Distribution or service boxes,** so called, which are really small manholes, often serve
the purpose of and can be used instead of larger vaults in industrial and isolated plant
installations. A design for a concrete service box is shown in Fig. 8-109. A brick box
would be of approximately the same dimensions. The depressions in the sidewalls are

sleeve pockets. The splicing sleeves on the cables lie partially in these, after installation, and therefore less of the valuable working space of the box is occupied by them. In spite of the fact that a square manhole cover can fall into the hole, heads with square covers are often used for distribution boxes to provide an orifice giving maximum working room.

**122. Manhole heads** are frequently made of cast iron, but cast steel is better; the cover should always be of cast steel. Figure 8-110 shows a design for cast steel for a large man-

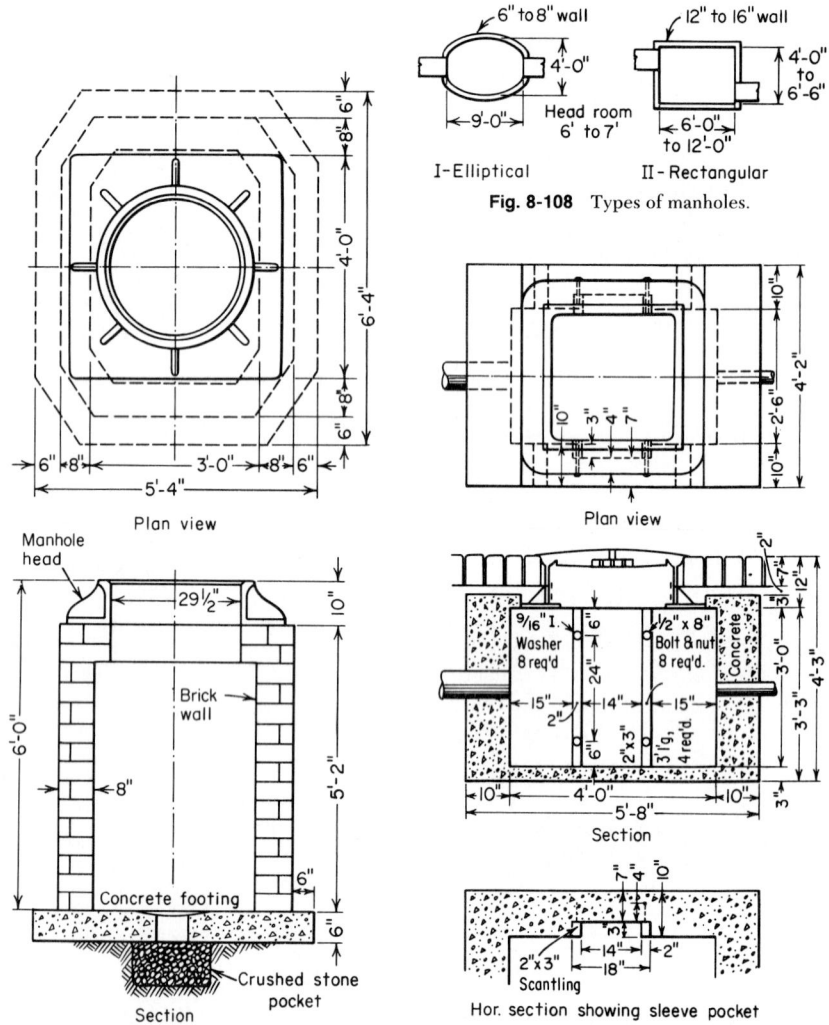

Fig. 8-108  Types of manholes.

Fig. 8-107  A 3- by 4-ft (0.9- by 1.2-m) manhole.

Fig. 8-109  A concrete service box.

hole, and Fig. 8-111 one for a smaller manhole. Covers should be round so that they cannot drop into the hole accidentally.

So-called watertight covers are now seldom used, as it is not feasible to make a satisfactory watertight cover at reasonable expense and water gets into the manholes in any event. Covers should not be fastened down because if they are and accumulated gas in a manhole explodes, the vault and cover will be shattered. A ventilated cover should be used to allow the escape of gas. The newer types have ventilating slots over approximately 50 percent of their area. Dirt and water will get into the hole, but the dirt can be cleaned out, and the water will drain out and no harm will result. If ventilation is not provided, an explosion of gas may occur and do great damage.

**123. Draining manholes.** When feasible, a sewer connection should lead from the bottom of every manhole (see Fig. 8-106). The mouth of the trap should be protected by a strainer, made of noncorrodible wire, such as that used for leader pipes. If a sewer connection cannot be made, there should be a hole in the manhole floor so that water can

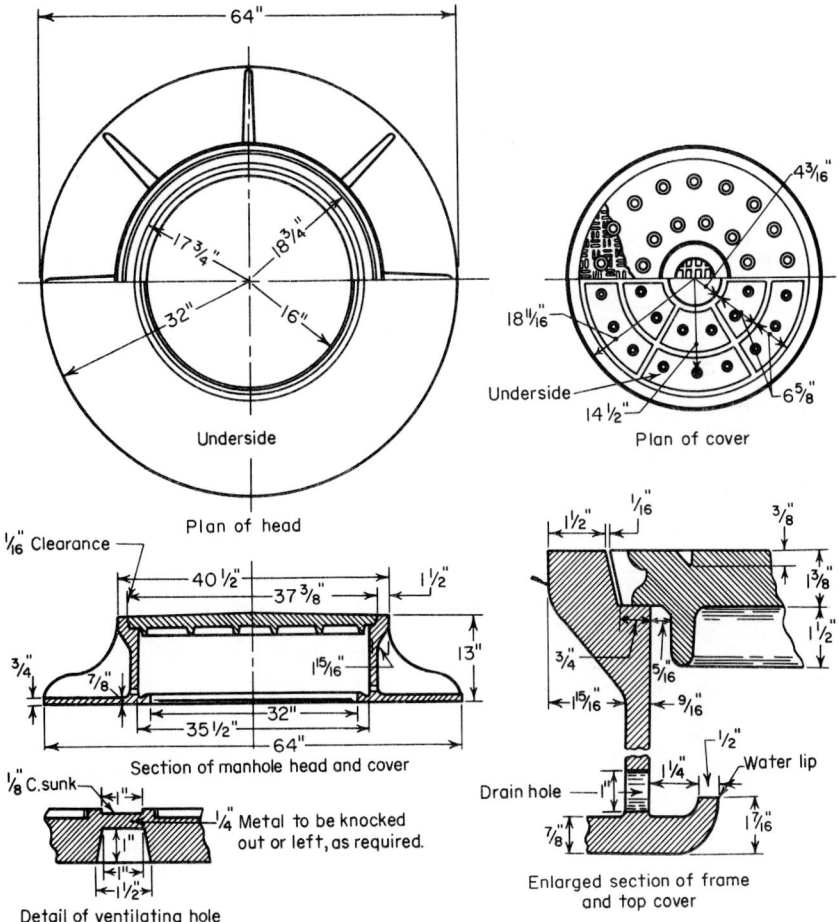

**Fig. 8-110** Head for large manholes.

drain out. A pocket under the manhole filled with broken rock will promote effective drainage (see Fig. 8-107).

**124. At the point where a conduit line** enters a manhole, the walls should be chamfered off as shown in Fig. 8-112 to prevent the damage that might occur if a cable is bent over a sharp corner.

A precast conduit entrance can be made up on forms on the surface of the ground and then lowered into position in the side of the manhole. One type using porcelain duct bells is shown in Fig. 8-113. The smooth, curved surface of the bell prevents the cable sheath from being cut where it starts to bend coming out of the duct line.

**125. A manhole hook,** a convenient tool for removing manhole heads, is shown in Fig. 8-114. A common pick can be used, but the tool shown is much more convenient.

**126. In installing cables in conduit,** if a pull-in wire was not installed at the time the ducts were placed, the conduit is rodded, a pull-in wire or the drawing-in cable is drawn through, cleaners are pulled through, and then the cable is drawn in.

**127. Rodding.** Rods are pieces of round hickory about ¾ in (19.05 mm) in diameter and 3 ft (0.9 m) long (Fig. 8-115). The ends of the rods are equipped with brass knuckle-joint fittings so that the rods can be readily joined together and disjoined. In rodding, a rod is

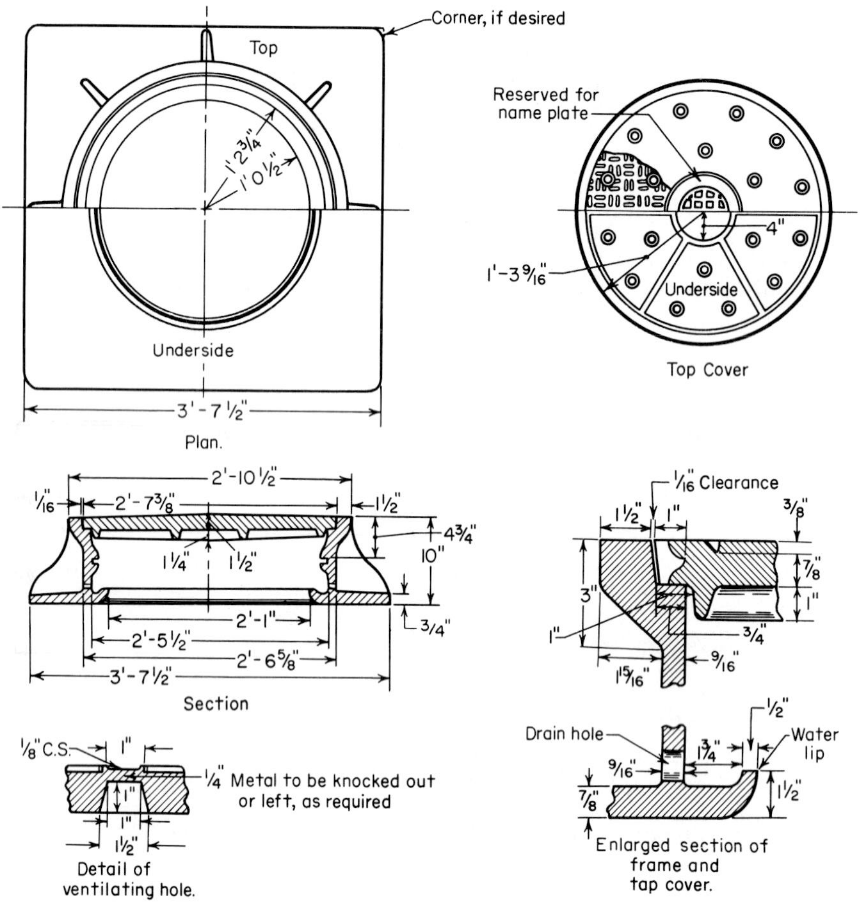

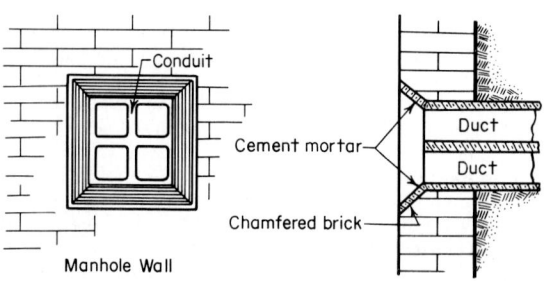

**Fig. 8-111**   Head for small manholes.

**Fig. 8-112**   Chamfered wall at conduit entrance.

pushed into the duct, and a second rod is coupled to it. The two are pushed into the duct, a third rod joined on, and the process is repeated until the rods extend from manhole to manhole. A galvanized-iron wire is attached to the last rod, and the wire is drawn into the duct.

A rope or flexible steel cable to which are attached a scraper and a brush (Fig. 8-101) is drawn through to ensure that the duct is clear and clean. To the end of this rope or cable is attached another which is used to pull in the electrical conductor cable.

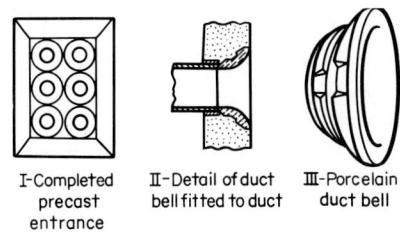

I–Completed precast entrance    II–Detail of duct bell fitted to duct    III–Porcelain duct bell

**Fig. 8-113**   Manhole duct entrance using porcelain bells. [Line Material Co.]

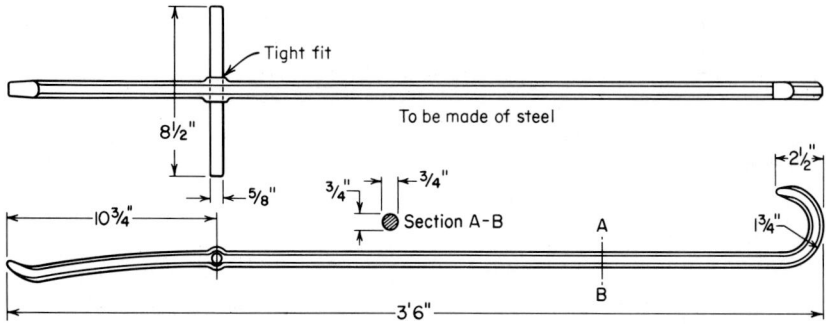

**Fig. 8-114**   A manhole hook.

**Fig. 8-115**   Rods for conduit

If the conduit is short, a steel fish wire or ribbon, like that used by electricians in wrought-iron conduit work, can be inserted instead of the rods.

**128. Pulling in cable.** The cable can be attached to the pulling-in wire by any one of several methods. Fig. 8-116C shows one that was formerly much used. Probably the best methods are those illustrated in Fig. 8-116A and B. At A, a galvanized-iron wire is laced around the cable in such a way that the harder it is pulled, the tighter it grips. At B is shown a *grip* spirally laced from flexible steel strands. It slips over the cable sheath readily, but when tension is applied, it effectively grips the cable. A swivel should always be inserted in any pulling-in line to prevent the untwisting of the drawing-in line under tension from twisting the cable.

After the cable has been fastened to the pulling-in line, a *protector* is placed in the mouth of the duct to prevent abrasion of the cable. Metal protectors can be purchased, but a good one can be formed from a piece of sole leather.

The cable is bent, as shown in Fig. 8-117, from the cable reel to the mouth of the duct, and the pulling in commences. In the far manhole, sheaves are arranged over which the pulling-in line passes (see Fig. 8-117). If eye-bolts are built in the manhole sides, the sheaves (snatch blocks) can be fastened to them. Otherwise a guide-sheave rack (see Fig. 8-118 for detail and Fig. 8-117 for application) can be set up in the manhole.

A winch, truck, or tractor can be used for the pulling. Workers can pull a cable in if the run is not too long. The cable sheath should be greased as it is drawn in to ensure

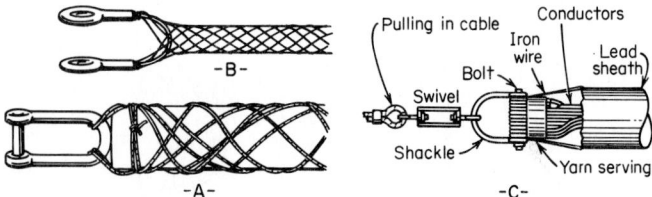

**Fig. 8-116**   Methods of attaching cable to drawing-in line.

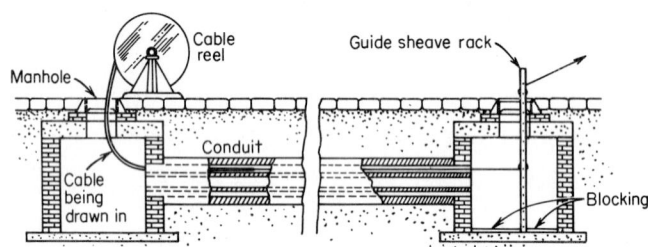

**Fig. 8-117**   Drawing in cable with a winch.

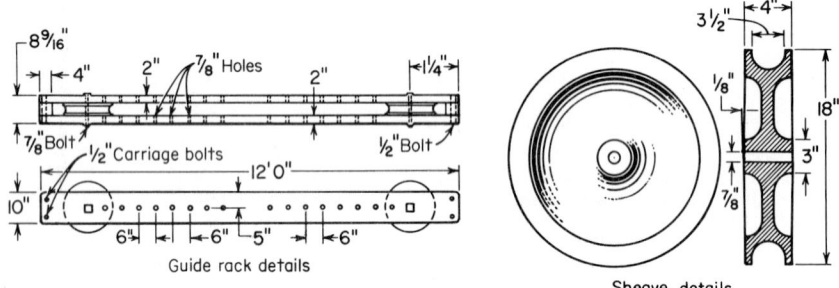

**Fig. 8-118**   Guide-sheave rack and sheave.

easy pulling. When a length of cable longer than the distance between manholes is to be pulled through, it can pass over the sheaves on the guide-sheave rack provided they are large enough in diameter. A manhole capstan (Fig. 8-119) is sometimes used instead of a winch on the surface of the ground. Enough cable should be pulled into the manhole to allow for forming it around the hole and splicing it. Do not permit a cable to hang over the sharp edge of a duct. Support it in the rack.

**129. Supporting cables in manholes.**   Some provision must be made for supporting cables. Creosoted planks (Fig. 8-120) are sometimes bolted to the manhole sides, and the cables are held to the cleats with pipe straps. In other cases metal supports, several forms of which are on the market, are used. One that can be readily made is shown in Fig. 8-121. A rack made of porcelain arms on a steel framework is shown in Fig. 8-122.

**130. Eyebolts or stirrups should be set in manhole walls to provide** means of attachment for the tackle used in pulling in cable (Fig. 8-120). An eyebolt or stirrup should be set opposite the point of entrance of each subway. Figure 8-123 shows the dimensions of a suitable stirrup.

**131. Several cables should not be placed in one duct.**   Experience has shown that although it is easy enough to install cables under such conditions and mechanically easy to withdraw them, the removal almost invariably ruins the cable, because after long lying in a duct the cables become so impacted with dust and grit that when one is drawn out, the sheath is stripped either from the cable itself or from one of its companions. Consequently, conduits are now almost exclusively built by arranging a sufficient number of ducts so that each cable can have its own exclusive compartment.

**132. In manholes and ducts, cables should be so arranged** that there will be a minimum of crossing and recrossing. An underground cable system should be carefully planned, and the ducts should be so chosen for the cables that, as far as feasible, a cable will take the duct in the same relative position throughout the subway.

**133. Installation of cable buried directly in the ground.**   Cable buried directly in the ground finds a wide field of application for installations of single circuits for which the cost of duct construction would be prohibitive. Some of the more common applications are:

1.   Street-lighting circuits, especially in cities whose outlying sections are without ducts

2.   Lighting and power circuits in parks, on estates, in industrial plants, on parkways, etc.

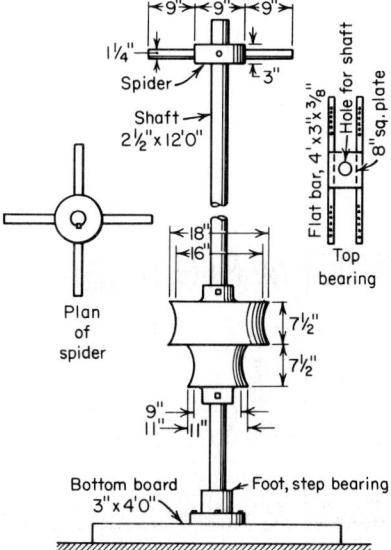

**Fig. 8-119**   Manhole capstan.

3.   Connecting residences to mains, especially in subdivisions and for underground service in alleys

4.   Connecting houses, garages, and other buildings

5.   Railroad-yard and airport lighting

6.   Lighting and power circuits for amusement parks, baseball, football, and general sports

7.   High-wall power cable for strip mines

8.   Railway-signal circuits

9.   Crossings under small lakes and streams

10.   Horizontal runs in mines for power, lighting, and telephone circuits

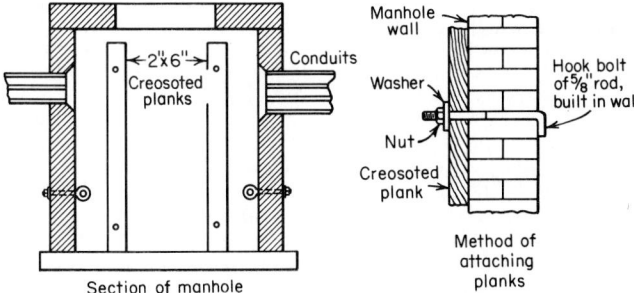

**Fig. 8-120**   Creosoted-plank cable supports.

Both nonmetallic-armored cable and metallic-armored cable (parkway cable) are used for direct burial in the earth. The nonmetallic-armored types are lighter in weight, more flexible, and easier to splice and are not subject to rust, crystallization, induced sheath power loss, or trouble from stray currents. On the other hand, they do not give such good protection against mechanical injury.

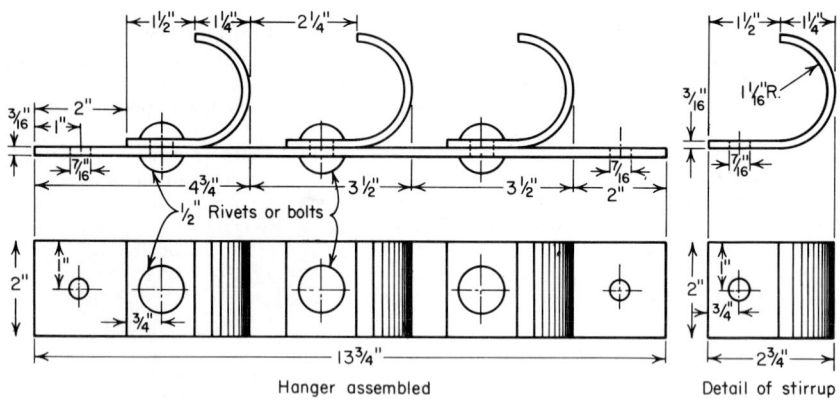

Hanger assembled                     Detail of stirrup

**Fig. 8-121**   Iron cable support.

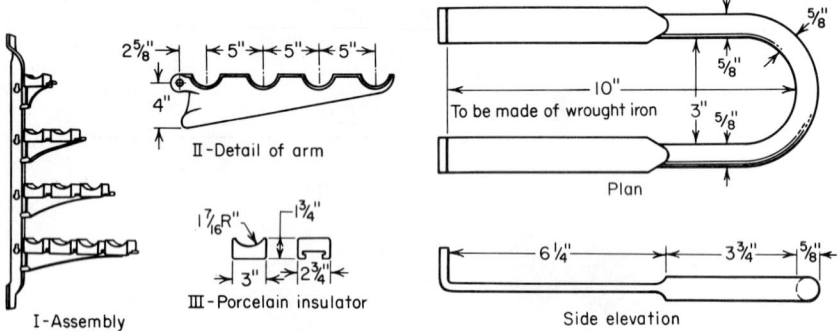

I-Assembly

**Fig. 8-122**   Cable rack. [Line Material Co.]     **Fig. 8-123**   Stirrup for manhole wall.

The cables are installed either by digging a shallow open trench, laying in the cables, and refilling or by means of a gasoline-powered trencher–cable layer. When a trench is dug, it needs to be made only wide enough for convenient handling of a cable. The cables (600 V or less) need not be laid more than 18 in (457.2 mm) under the surface except when subject to extreme mechanical hazard, such as at street intersections or under roadways. At such points it is advisable to bury the cable at least 30 in (762 mm) deep and to cover it with protective pads.

**134. Types UF and USE cables** are Code-designated single-conductor or multiconductor cables suitable for direct burial in earth. These cables are described in Divs. 4 and 9, where typical applications at 600 V or less are discussed.

**135. Underground conductors over 600 V.**   The National Electrical Code includes rules for the protection of underground conductors when the supply voltage exceeds 600 V. These rules were introduced to minimize the hazards of "dig-ins," and Sec. 710-3(*b*) of the Code covers such rules.

## GROUNDING OF SYSTEMS

**136. Grounding of distribution circuits.**  The National Electrical Code requires that on systems supplying interior-wiring circuits one wire of the circuit shall be grounded, provided that the voltage from any other conductor to ground will not exceed 300 V on dc systems or 150 V on ac systems. Grounding helps to prevent accidents to persons and damage by fire to property in case of lightning, breakdown between primary and secondary windings of the transformers, or accidental contact between high-voltage wires and low-voltage wires. If some point on the low-voltage circuit is grounded, (1) lightning striking the wires will be conducted into the ground, and (2) breakdown of the transformer insulation between primary and secondary coils will reveal itself through blowing of the primary transformer fuses if one wire of the primary circuit comes in contact with one of the secondary wires. See Sec. 150 for the importance of a low-impedance ground-path return.

**137. The value of groundings** is illustrated in Fig. 8-124. One of the primary circuit wires is grounded at the transformer. The secondary is grounded at the customer's water pipe. An accidental cross between ungrounded primary and secondary wires is shown by the dotted line. Current will flow through the cross-connection, the secondary of the transformer, and the customer's water-pipe ground and up the ground connection to the other primary line. The voltage of the secondary wires to ground will be increased by the $IZ$ drop through the water-pipe ground. A current of 100 A times a ground resistance of 2 $\Omega$ (there is practically no reactance) giving 200 V would represent very severe conditions. This means that the secondary voltage to ground would be the normal 110 V plus the 200 V, giving 310 V, which would probably not do any damage before the primary protective devices would open and break the circuit. This type of accidental cross should open the substation-feeder circuit breaker. If the cross were in the transformer itself, the primary fuse would open. If, however, there were no ground connection, the primary 2300 V would be impressed on the secondary system, which is insulated for only 250 to 600 V, and numerous insulation breakdowns and considerable damage to equipment

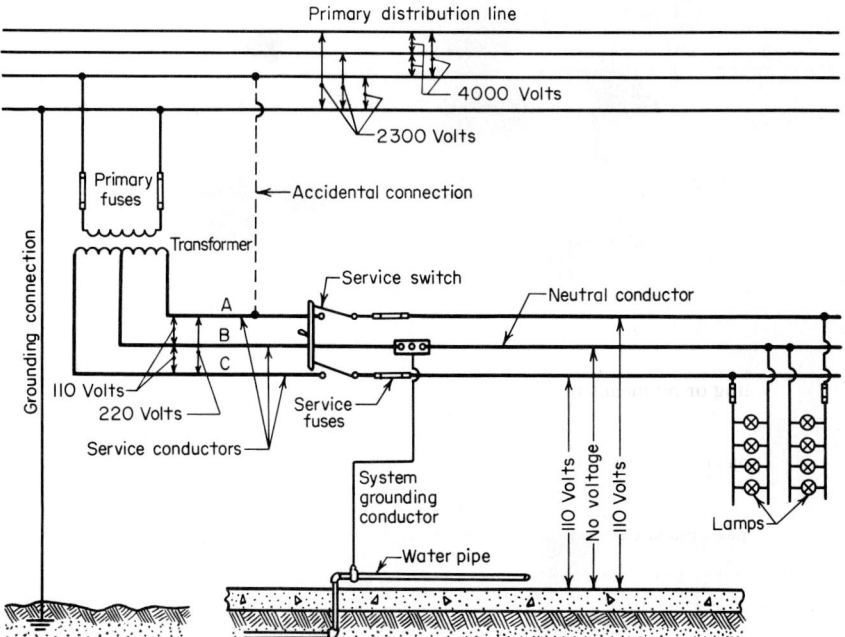

**Fig. 8-124**  Ground connections for interior-wiring systems.

would occur. Also, there would be danger to human life for anyone touching a fixture at such a time.

**138. High-voltage transmission systems are also frequently grounded** to reduce the voltage to ground and consequently the line insulation required. For example, if a 220-kV three-phase circuit has the windings of the supply transformer connected in Y, with the mid-point of the Y grounded, the voltage to ground will be $220 \div \sqrt{3} = 127$ kV. This reduction of the voltage to ground is of great value in insulating the transformer windings and in the number of insulators required for the line supports. At the same time, the advantage of the 220 kV in reducing the line current and line losses is retained.

**139. Ground-wire connections to transformer secondaries** should be made to the neutral point or wire if one is accessible. If no neutral point is accessible, one side of the secondary circuit may be grounded. Figure 8-125 shows theoretical diagrams of ground connections to transformer secondaries, and Fig. 8-126 illustrates how some of these connections are arranged with commercial transformers. The neutral point of each transformer feeding a two-phase, four-wire secondary should be grounded unless the motors taking energy from the secondary have interconnected windings. If they are inter-

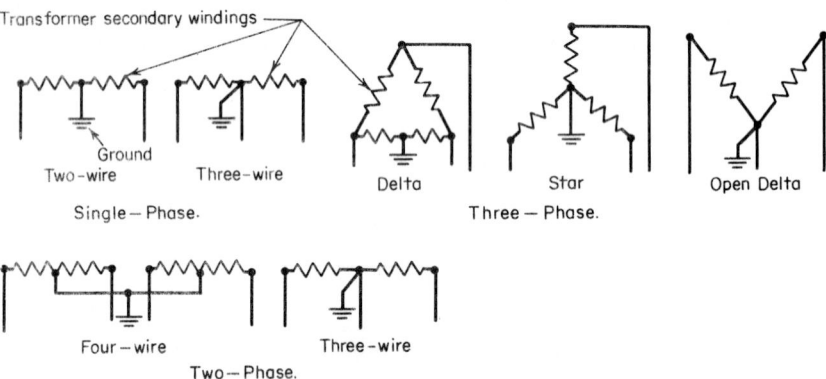

**Fig. 8-125** Theoretical diagrams of secondary-ground connections.

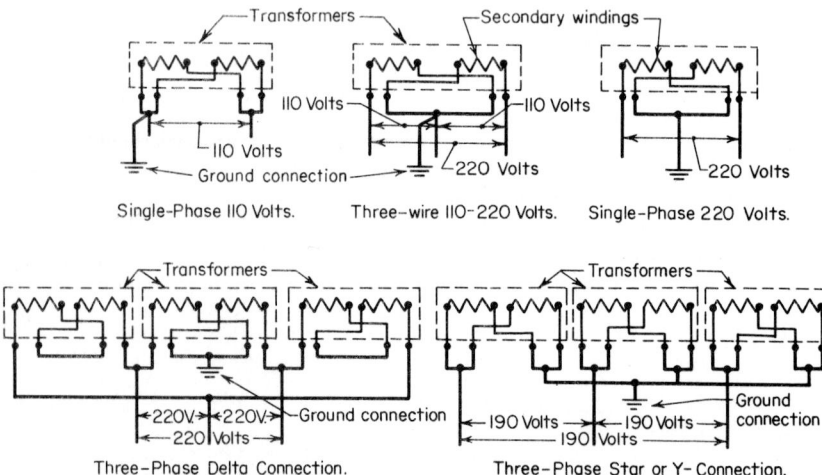

**Fig. 8-126** Ground connections to secondaries of commercial transformers.

connected, the center or neutral point of only one transformer is grounded. No primary windings are shown in Figs. 8-125 and 8-126. In Fig. 8-126 the secondary wind-ing of each transformer is shown divided into two sections, as it is in commercial trans-formers.

**140. Ground rods** made of steel pipe with copper-welded exterior surface (pointed at the bottom; Fig. 8-127) are extensively used. The rod is driven into the ground, and the ground wire fastened by a clamp near the top of the rod. The copper coating provides for a good electrical connection and affords resistance to corrosion from the weather and earth.

**141. Ground wires** should be encased by wooden molding for a distance of at least 8 ft (2.4 m) from the surface to protect against shocks to passersby. Under certain conditions of soil moisture, a shock can be received from a ground wire by a person standing on the earth's surface. The ground pipe extends about 1 ft (0.3 m) above ground and is not usually protected. Some companies encase the entire length of the ground wire in molding to protect the lineworkers.

No copper wire smaller than No. 8 should be used for a ground wire, and some companies use nothing smaller than No. 4. Copper wire is preferable. Bare wire is sat-isfactory and should be attached to the poles with cleats or straps. Staples, although used, should not be. The National Electrical Code requirements with respect to the size of ground wires are given in Div. 9. There should be a ground for each transformer or group of transformers, and when transformers feed a network with a neutral wire, there should, in addition, be a ground at least every 500 ft (152.4 m).

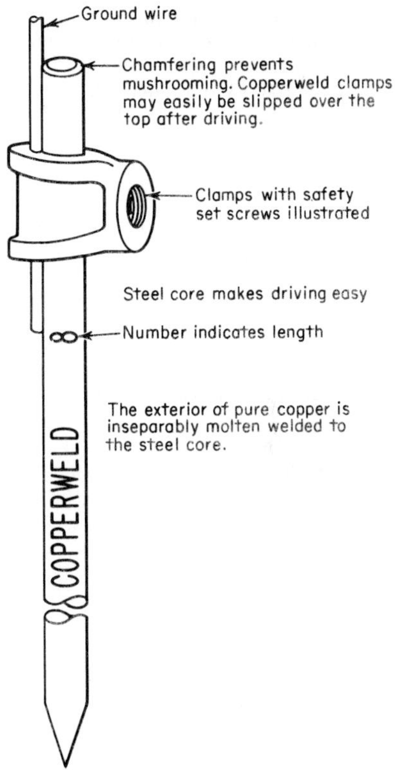

**Fig. 8-127**   Ground rod and clamp. [Copperweld Corp.]

The National Electrical Safety Code spec-ifies that all ground pipes shall be galva-nized and have a minimum nominal diameter of ½ in. Steel or iron rods shall have a min-imum cross-sectional dimension of ⅝ in.

**142. The National Electrical Safety Code implies that the ground resistance** should be mea-sured at the time of installing the ground and that the ground resistance must not exceed 25 Ω for artificial (buried or driven) grounds.

**143. To measure the ground resistance,** two additional temporary grounds, consisting of short rods 2 or 3 ft (610 or 914 mm) long, must be driven in the ground at least 20 ft (6.1 m) away from the ground being tested (Fig. 8-128, I). A direct-reading instrument called a groundohmer or groundometer can then be connected to the three grounds by means of insulated leads. A magneto or a battery in the instrument furnishes the necessary power. The instrument reads the ground resistance directly in ohms. If a source of direct or alternating current is available, the resistance can be measured by the fall-of-potential method with a voltmeter and ammeter, connections for which are shown in Fig. 8-128, II. The resistance is then calculated from $R_g = E/I$.

**144. When testing ground resistance in city streets** there may no convenient place to drive the auxiliary grounds. In such a case a fire hydrant can be used for the auxiliary ground, or a pad made of several thicknesses of heavy cloth about 1 ft (0.3 m) square and saturated with salt water can be placed on the sidewalk (Fig. 8-129).

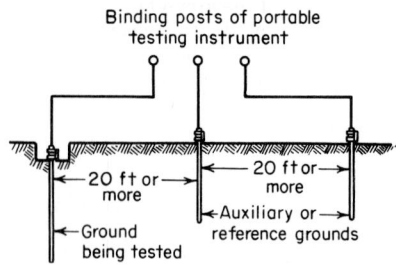

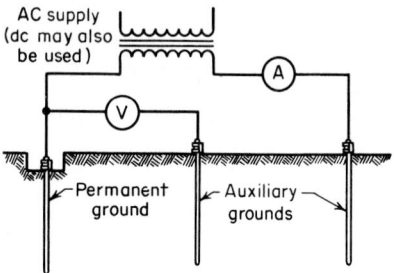

I – Location of reference grounds for direct-reading, single-test method of measuring the resistance of a ground connection.

II – Connections for the fall-of-potential method of measuring ground resistance.

**Fig. 8-128**   Circuits for measurement of ground resistance. [Copperweld Corp.]

**145. Reduction of ground resistance** when it exceeds the allowable value of 25 Ω can be accomplished in several ways:

1. Use of a larger-diameter ground rod
2. Use of a longer ground rod
3. Putting two, three, or more rods in parallel
4. Chemical treatment of the soil

**146. The effect of the use of a larger-diameter rod** is shown in Fig. 8-130. It will be noted that increasing the diameter of the ground rod does not effect much reduction in resistance, and use of ground rods larger than ¾ in (19.05 mm) is not recommended. The diameter of the rod should be only large enough to permit it to be driven into the soil without bending.

**147. Long ground rods** made in sections coupled together are very effective where there is considerable depth of dry sand and good moist soil is many feet underground. Although the 8-ft (2.4-m) minimum depth required by the National Electrical Code is sufficient for normal soil conditions, grounds as deep as 50 ft (15.2 m) have been used in very poor soil. The calculated effect of the depth of rod on ground resistance is shown in Fig. 8-131. The curve is for uniform soil at all depths. In practice the soil is always firmer and more moist the deeper the ground, so that the decrease of resistance with increase of depth is usually greater than is shown by the curve.

**148. Using several rods in multiple** is a very effective means of lowering ground resistance, as shown in Fig. 8-132. The recommended distance between rods is at least 6 ft (1.8 m), so that this method is more applicable to large steel-tower lines and substation structures than to wood-pole lines. Spacing rods as close together as the opposite sides of a wood pole is not recommended.

**149. Chemical treatment of the soil** is recommended for reducing ground resistance when the other methods of Secs. 146 to 148 are not practical.

There are two practical ways of accomplishing this result, as shown in Fig. 8-133. In the method of Fig. 8-133, I, a length of tile pipe is sunk in the ground a few inches

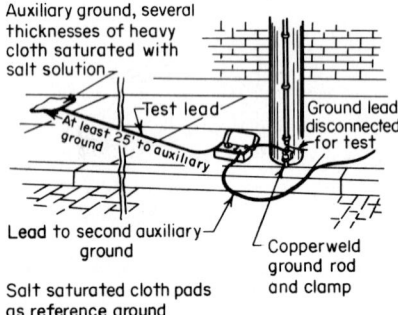

**Fig. 8-129**   Testing ground resistances in paved streets.

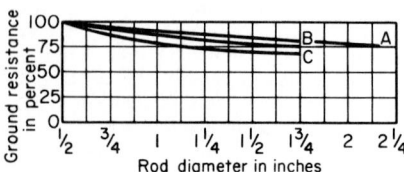

**Fig. 8-130**   Effect of diameter of ground rod on ground resistance. [Copperweld Corp.]

from the ground rod and filled to within 1 ft (0.3 m) or so of the ground level with magnesium sulfate, copper sulfate, or common rock salt. The magnesium sulfate is the most common, as it combines low cost with high electrical conductivity and low corrosive effect on the rod. This method is effective where there is limited space for soil treating, such as on city streets. The method of Fig. 8-133, II, is applicable where a circular or semicircular trench can be dug around the ground rod to hold the chemical. The chemical must be kept several inches away from direct contact with the ground rod to avoid corrosion of the rod. The first treatment requires 40 to 90 lb (18.1 to 40.8 kg)

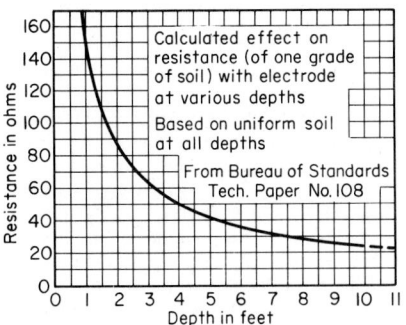

**Fig. 8-131** Effect of depth of ground rod on ground resistance. [Copperweld Corp.]

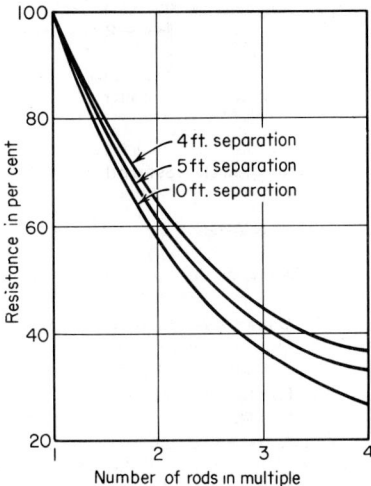

**Fig. 8-132** Effect on ground resistance of placing ground rods in multiple. [Copperweld Corp.]

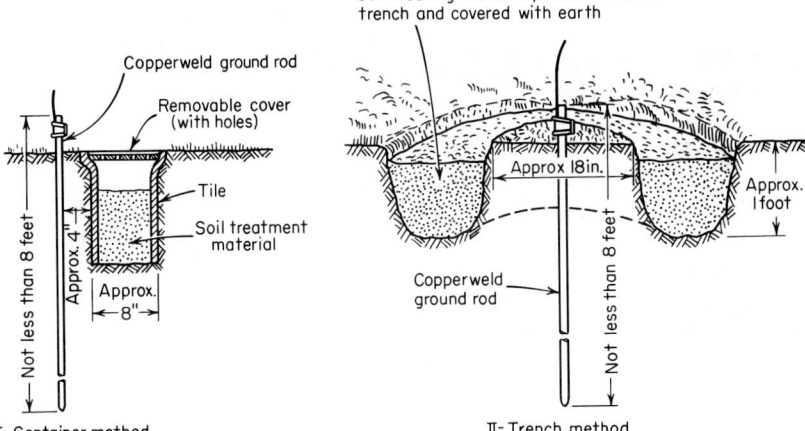

Soil treating material placed in circular trench and covered with earth

Copperweld ground rod
Removable cover (with holes)
Tile
Soil treatment material
Approx. 4"
Approx. 8"
Not less than 8 feet

Approx 18in.
Approx. I foot
Copperweld ground rod
Not less than 8 feet

I- Container method

II- Trench method

**Fig. 8-133** Methods of soil treatment for lowering of ground resistance. [Cooperweld Corp.]

of chemical and will retain its effectiveness for 2 to 3 years. Each replenishment of the chemical extends the effectiveness for a longer period, so that future retreating becomes less and less frequent. When the chemical is first installed, it is well to flood the treated area with water so that the chemical will diffuse through the soil. Thereafter normal rainfall will usually provide the necessary water for carrying the solution into the earth.

**150. A low-impedance ground-path return** is absolutely essential if line-to-ground faults in grounded systems are to open overcurrent devices in buildings. It is an important rule in the National Electrical Code (Sec. 250-23b), which requires a grounded conductor of a grounded utility system to extend to every building served. Such grounded conductors are then bonded to the metallic enclosures of the service equipment, and the grounded conductor is connected to an electrode on the premises by a grounding electrode conductor which usually extends from a neutral bar in the service-equipment enclosure to the electrode. See Div. 9 for detailed Code rules to size grounded and grounding conductors.

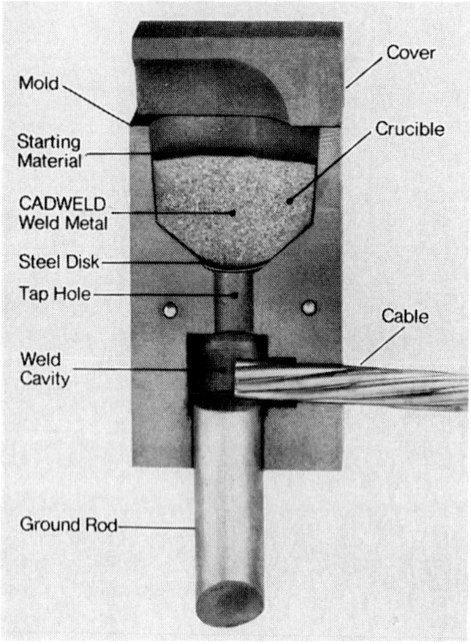

When a grounded conductor is properly bonded to the metallic service equipment, any line-to-ground fault from an ungrounded conductor to grounded equipment (raceways, cables, boxes, etc.) on the load side of overcurrent devices will be able to follow a low-impedance path to the grounded conductor at the supply transformer, thus permitting an overcurrent device to open quickly. All ground faults must be cleared as fast as possible to avoid extended exposure of energized metallic parts, which provides a serious shock hazard. It cannot be stressed too strongly that proper bonding of the grounded conductor of the supply system to the service-entrance equipment is essential for the proper operation of overcurrent devices in the event of line-to-ground faults. For splices of the grounding electrode conductor, see Fig. 8-134. For low-magnitude ground faults, see information on ground-fault circuit interrupters in Div. 4.

**Fig. 8-134**  Cadweld® is an exothermic welding process suitable for grounding electrode connections and for extending the grounding electrode conductor at commercial and industrial buildings. [Erico Products, Inc.]

# Division 9

# Interior Wiring

## GENERAL

**1. The** (NEC) **rules,** which are the recommendations of the National Fire Protection Association (NFPA), should be followed in installing all interior wiring. These rules are revised every 3 years, and so it is inadvisable to include them in entirety in this book. A copy of the rules can be obtained from any local inspection bureau or purchased from the NFPA. All statements are made in accordance with the 1990 edition of the Code. The

Occupational Safety and Health Act (OSHA) adopted the 1971 National Electrical Code (initially the 1968 edition) as the electrical standard for employee safety in the workplace. NFPA has written a standard titled "Electrical Safety Requirements for Employee Workplaces" that is intended to serve as an enforcement document by compliance officials of the Department of Labor.

**2. There are local regulations covering the installation of wiring,** in force in many localities, which have been enacted by city or state governments. Sometimes these differ from the Code regulations, and so it is always well to be familiar with all the regulations in force before starting any work. The city and state rules are in reality laws and therefore take precedence over the NEC rules, which of themselves have no legal status.

**3. Definitions** (National Electrical Code). Definitions that duplicate those in ANSI/IEEE-100 *Standard Definitions of Electrical and Electronic Terms, C42,* are marked with an asterisk (°). Those not so marked either differ from or are not found in the latest ANSI *Standard* or the latest revision thereof.

°ACCESSIBLE (AS APPLIED TO WIRING METHODS)    Capable of being removed or exposed without damaging the building structure or finish, or not permanently closed in by the structure or finish of the building. SEE also CONCEALED; EXPOSED.

°ACCESSIBLE (AS APPLIED TO EQUIPMENT)    Admitting close approach because not guarded by locked doors, elevation, or other effective means. See also READILY ACCESSIBLE.

AMPACITY    The current in amperes that a conductor can carry continuously under the conditions of use without exceeding its temperature rating.

APPLIANCE    Utilization equipment, generally other than industrial, normally built in standardized sizes or types, which is installed or connected as a unit to perform one or more functions such as clothes, washing or drying, air conditioning, food mixing, deep frying, etc.

APPROVED    Acceptable to the authority having jurisdiction.

*ATTACHMENT PLUG (PLUG CAP; CAP)    A device which, by insertion in a receptacle, establishes a connection between the conductors of the attached flexible cord and the conductors connected permanently to the receptacle.

*AUTOMATIC    Self-acting; operating by its own mechanism when actuated by some impersonal influence, as, for example, a change in current strength, pressure, temperature, or mechanical configuration. See also NONAUTOMATIC.

BONDING JUMPER    A reliable conductor to assure the required electric conductivity between metal parts required to be electrically connected.

BONDING JUMPER, CIRCUIT    The connection between portions of a conductor in a circuit to maintain the required ampacity of the circuit.

BONDING JUMPER, EQUIPMENT    The connection between two or more portions of the equipment-grounding conductor.

BONDING JUMPER, MAIN    The connection between the grounded circuit conductor and the equipment-grounding conductor at the service.

BRANCH CIRCUIT    The circuit conductors between the final overcurrent device protecting the circuit and the outlet or outlets. See Sec. 240-9 of the NEC for thermal relays, supplementary overcurrent protection, and other devices.

°BRANCH CIRCUIT, APPLIANCE    A branch circuit supplying energy to one or more outlets to which appliances are to be connected, such a circuit to have no permanently connected lighting fixtures not a part of an appliance.

*BRANCH CIRCUIT, GENERAL-PURPOSE    A branch circuit that supplies a number of outlets for lighting and appliances.

°BRANCH CIRCUIT, INDIVIDUAL    A branch circuit that supplies only one utilization equipment.

°BRANCH CIRCUIT, MULTIWARE    A branch circuit consisting of two or more ungrounded conductors having a potential difference between them and a grounded conductor which has equal potential difference between it and each ungrounded conductor of the circuit and which is connected to the neutral (grounded) conductor of the system.

*BUILDING    A structure that stands alone or is cut off from adjoining structures by fire walls with all openings therein protected by approved fire doors.

°CABINET    An enclosure designed for either surface or flush mounting and provided with a frame, mat, or trim in which a swinging door or doors are or may be hung.

°CIRCUIT BREAKER    A device designed to open and close a circuit by nonautomatic means and to open the circuit automatically on a predetermined overcurrent without damage to itself when properly applied within its rating.

NOTE  The automatic opening means can be integral, direct acting with the circuit breaker, or remote from the circuit breaker. See definition of "Switching Devices" in Part B of Art. 100 of the **NEC** for definition applying to circuits and equipment over 600 V, nominal.

*Adjustable (as applied to circuit breakers).* A qualifying term indicating that the circuit breaker can be set to trip at various values of current and/or time within a predetermined range.

*Instantaneous-trip (as applied to circuit breakers).* A qualifying term indicating that no delay is purposely introduced in the tripping action of the circuit breaker.

*Inverse-time (as applied to circuit breakers).* A qualifying term indicating a delay is purposely introduced in the tripping action of the circuit breaker, which delay decreases as the magnitude of the current increases.

*Nonadjustable (as applied to circuit breakers).* A qualifying term indicating that the circuit breaker does not have any adjustment to alter the value of current at which it will trip or the time required for its operation.

*Setting (of a circuit breaker).* The value of current and/or time at which an adjustable circuit breaker is set to trip.

°CONCEALED  Rendered inaccessible by the structure or finish of the building. Wires in concealed raceways are considered concealed even though they may become accessible by withdrawing them. See also ACCESSIBLE (AS APPLIED TO WIRING METHODS).

°CONDUCTOR, BARE  A conductor having no covering or electrical insulation whatsoever. See also CONDUCTOR, COVERED.

°CONDUCTOR, COVERED  A conductor encased within material of a composition or a thickness that is not recognized by the **Code** as electrical insulation. See also CONDUCTOR, BARE.

CONDUCTOR, INSULATED  A conductor encased within material of composition and thickness that is recognized by the **Code** as electrical insulation.

CONNECTOR, PRESSURE (SOLDERLESS)  A device that establishes a connection between two or more conductors or between one or more conductors and a terminal by means of mechanical pressure and without the use of solder.

°CONTINUOUS LOAD  A load where the maximum current is expected to continue for 3 h or more.

CONTROLLER  A device or group of devices which serves to govern, in some predetermined manner, the electric power delivered to the apparatus to which it is connected. See also Sec. 430-81(a) of the **NEC**.

COOKING UNIT, COUNTER-MOUNTED  A cooking appliance designed for mounting in or on a counter and consisting of one or more heating elements, internal wiring, and built-in or separately mountable controls. See also OVEN, WALL-MOUNTED.

*CUTOUT BOX  An enclosure designed for surface mounting and having swinging doors or covers secured directly to and telescoping with the walls of the box proper. See also CABINET.

°DEMAND FACTOR  The ratio of the maximum demand of a system or part of a system to the total connected load of a system or the part of the system under consideration.

°DEVICE  A unit of an electric system that is intended to carry but not utilize electric energy.

*DISCONNECTING MEANS  A device, or group of devices, or other means whereby the conductors of a circuit can be disconnected from their source of supply.

DRY  See LOCATION: *Dry Location.*

*DUSTPROOF  So constructed or protected that dust will not interfere with successful operation.

°DUSTTIGHT  So constructed that dust will not enter the enclosing case under specified test conditions.

DUTY

*Continuous duty.*  Operation at a substantially constant load for an indefinitely long time.

*Intermittent duty.*  Operation for alternate intervals of (1) load and no load, or (2) load and rest, or (3) load, no load, and rest.

*Periodic duty.*  Intermittent operation in which the load conditions are regularly recurrent.

*Short-time duty.*  Operation at a substantially constant load for a short and definitely specified time.

*Varying duty.*  Operation at loads, and for intervals of time, both of which may be subject to wide variation.

See Table 430–22(a), Exception, of the NEC for illustration of various types of duty.

DWELLING

*Dwelling unit.*   One or more rooms for the use of one or more persons as a housekeeping unit with space for eating, living, and sleeping and permanent provisions for cooking and sanitation.

*Multifamily dwelling.*   A building containing three or more dwelling units.

*One-family dwelling.*   A building consisting solely of one dwelling unit.

*Two-family dwelling.*   A building consisting solely of two dwelling units.

ELECTRIC SIGN   A fixed, stationary or portable, self-contained, electrically illuminated utilization equipment with words or symbols designed to convey information or attract attention.

ENCLOSED   Surrounded by a case, housing, fence, or walls which will prevent persons from accidentally contacting energized parts.

EQUIPMENT   A general term including material, fittings, devices, appliances, fixtures, apparatus, and the like used as a part of, or in connection with, an electrical installation.

°EXPLOSIONPROOF APPARATUS   Apparatus enclosed in a case which is capable of withstanding an explosion of a specified gas or vapor that may occur within it and of preventing the ignition of a specified gas or vapor surrounding the enclosure by sparks, flashes, or explosion of the gas or vapor within and which operates at such an external temperature that a surrounding flammable atmosphere will not be ignited thereby.

°EXPOSED (AS APPLIED TO LIVE PARTS)   Capable of being inadvertently touched or approached nearer than a safe distance by a person. The term is applied to parts not suitably guarded, isolated, or insulated. See also ACCESSIBLE; CONCEALED.

EXPOSED (AS APPLIED TO WIRING METHODS)   On or attached to the surface or behind panels designed to allow access. See also ACCESSIBLE (AS APPLIED TO WIRING METHODS).

EXTERNALLY OPERABLE   Capable of being operated without exposing the operator to contact with live parts.

FEEDER   Any circuit conductor between the service equipment or the source of a separately derived system and the final branch-circuit overcurrent device.

*FITTING   An accessory such as a locknut, bushing, or other part of a wiring system that is intended primarily to perform a mechanical rather than an electrical function.

GARAGE   A building or portion of a building in which one or more self-propelled vehicles carrying volatile, flammable liquid for fuel or power are kept for use, sale, storage, rental, repair, exhibition, or demonstrating purposes and all that portion of a building which is on or below the floor or floors in which such vehicles are kept and which is not separated therefrom by suitable cutoffs.

GROUND   A conducting connection, whether intentional or accidental, between an electric circuit or equipment and earth or to some conducting body that serves in place of the earth.

GROUND-FAULT CIRCUIT INTERRUPTER   A device intended for the protection of personnel that functions to deenergize a circuit or portion thereof within an established period of time when a current to ground exceeds some predetermined value that is less than that required to operate the overcurrent protective device of the supply circuit.

GROUND-FAULT PROTECTION OF EQUIPMENT   A system intended to provide protection of equipment from damaging line-to-ground fault currents by operating to cause a disconnecting means to open all ungrounded conductors of the faulted circuit. This protection is provided at current levels less than those required to protect conductors from damage through the operation of a supply-circuit overcurrent device.

GROUNDED   Connected to earth or to some conducting body that serves in place of the earth.

GROUNDED CONDUCTOR   A system or circuit conductor which is intentionally grounded.

GROUNDING CONDUCTOR   A conductor used to connect equipment or the grounded circuit of a wiring system to a grounding electrode or electrodes.

GROUNDING CONDUCTOR, EQUIPMENT   The conductor used to connect the non-current-carrying metal parts of equipment, raceways, and other enclosures to the system grounded conductor and/or the grounding electrode conductor at the service equipment or at the source of a separately derived system.

GROUNDING ELECTRODE CONDUCTOR   The conductor used to connect the grounding electrode to the equipment-grounding conductor and/or to the grounded conductor of the circuit at the service equipment or at the source of a separately derived system.

GUARDED   Covered, shielded, fenced, enclosed, or otherwise protected by means of suitable covers, casings, barriers, rails, screens, mats, or platforms to remove the likelihood of approach or contact by persons or objects to a point of danger.

HOISTWAY   Any shaftway, hatchway, wellhole, or other vertical opening or space in which an elevator or dumbwaiter is designed to operate.

*ISOLATED   Not readily accessible to persons unless special means for access are used.

LABELED   Equipment or materials to which has been attached a label, symbol, or other identifying mark of an organization acceptable to the authority having jurisdiction and concerned with product evaluation that maintains periodic inspection of production of labeled equipment or materials and by whose labeling the manufacturer indicates compliance with appropriate standards or performance in a specified manner.

LISTED   Equipment or materials included in a list published by an organization acceptable to the authority having jurisdiction and concerned with product evaluation that maintains periodic inspection of production of listed equipment or materials and whose listing states that the equipment or material either meets appropriate standards or has been tested and found suitable for use in a specified manner.

> NOTE   The means for identifying listed equipment may vary for each of the organizations concerned with product evaluation, some of which do not recognize equipment as listed unless it is also labeled. The authority having jurisdiction should utilize the system employed by the listing organization to identify a listed product.

*LIGHTING OUTLET   An outlet intended for the direct connection of a lampholder, a lighting fixture, or a pendant cord terminating in a lampholder.

LOCATION

*Damp location.*   A parially protected location under a canopy, marquee, roofed open porch, and like location and an interior location subject to moderate degrees of moisture, such as some basements, some barns, and some cold-storage warehouses.

*\*Dry location.*   A location not normally subject to dampness or wetness. A location classified as dry may be temporarily subject to dampness or wetness, as in the case of a building under construction.

*Wet location.*   An installation underground or in concrete slabs or masonry in direct contact with the earth and a location subject to saturation with water or other liquids, such as vehicle-washing areas and locations exposed to weather and unprotected.

°MULTIOUTLET ASSEMBLY   A type of surface or flush raceway designed to hold conductors and attachment plug receptacles assembled in the field or at the factory.

°NONAUTOMATIC   Action requiring personal intervention for its control (see also AUTOMATIC). As applied to an electric controller, nonautomatic control does not necessarily imply a manual controller but implies only that personal intervention is necessary.

*OUTLET   A point on the wiring system at which current is taken to supply utilization equipment.

OUTLINE LIGHTING   An arrangement of incandescent lamps or gaseous tubes to outline and call attention to certain features such as the shape of a building or the decoration of a window.

OVEN, WALL-MOUNTED   An oven for cooking purposes designed for mounting in or on a wall or other surface and consisting of one or more heating elements, internal wiring, and built-in or separately mountable controls. See also COOKING UNIT, COUNTER-MOUNTED.

OVERCURRENT   Any current in excess of the rated current of equipment or the ampacity of a conductor. It may result from overload, short circuit, or ground fault (see also OVERLOAD). A current in excess of rating may be accommodated by certain equipment and conductors for a given set of conditions. Hence the rules for overcurrent protection are specific for particular situations.

OVERLOAD   Operation of equipment in excess of normal full-load rating or of a conductor in excess of rated ampacity which, when it persists for a sufficient length of time, would cause damage or dangerous overheating. A fault, such as a short circuit or ground fault, is not an overload (see also OVERCURRENT). For motor apparatus applications see Sec. 430-31 of the NEC.

°PANELBOARD   A single panel or group of panel units designed for assembly in the form of a single panel, including buses and automatic overcurrent devices, with or without switches for the control of light, heat, or power circuits; designed to be placed in a cabinet or cutout box placed in or against a wall or partition and accessible only from the front. See also SWITCHBOARD.

POWER OUTLET   An enclosed assembly which may include receptacles, circuit breakers, fuseholders, fuse switches, buses, and watthour-meter mounting means; intended to supply and control power to mobile homes, recreational vehicles, or boats or to serve as a means for distributing power required to operate mobile or temporarily installed equipment.

PREMISES WIRING (SYSTEM)   That interior and exterior wiring, including power, lighting, control, and signal circuit wiring together with all its associated hardware, fittings, and wiring devices, both permanently and temporarily installed, which extends from the load end of the service drop or the load end of the service lateral conductors or source of a separately derived system to the outlet(s). Such wiring does not include wiring internal to appliances, fixtures, motors, controllers, motor-control centers, and similar equipment.

QUALIFIED PERSON   One familiar with the construction and operation of the equipment and the hazards involved.

RACEWAY   An enclosed channel designed expressly for holding wires, cables, or busbars, with additional functions as permitted in the NEC. Raceways may be of metal or insulating material, and the term includes rigid metal conduit, rigid nonmetallic conduit, intermediate metal conduit, liquidtight flexible conduit, flexible metallic tubing, flexible metal conduit, electrical nonmetallic tubing, electrical metallic tubing, underfloor raceways, cellular-concrete-floor raceways, cellular-metal-floor raceways, surface raceways, wireways, and busways.

RAINPROOF   So constructed, protected, or treated as to prevent rain from interfering with the successful operation of the apparatus under specfied test conditions.

RAINTIGHT   So constrcuted or protected that exposure to a beating rain will not result in the entrance of water under specified test conditions.

°READILY ACCESSIBLE   Capable of being reached quickly, for operation, renewal, or inspections, without requiring those to whom ready access is requisite to climb over or remove obstacles or to resort to portable ladders, chairs, etc. See also ACCESSIBLE.

RECEPTACLE   A contact device installed at the outlet for the connection of a single attachment plug. A single receptacle is a single contact device with no other contact device on the same yoke. A multiple receptacle is a single device containing two or more receptacles.

*RECEPTACLE OUTLET   An outlet in which one or more receptacles are installed.

*REMOTE-CONTROL CIRCUIT   Any electric circuit that controls any other circuit through a relay or an equivalent device.

SEALABLE EQUIPMENT   Equipment enclosed in a case or cabinet that is provided with means of sealing or locking so that live parts cannot be made accessible without opening the enclosure. The equipment may or may not be operable without opening the enclosure.

SERVICE   The conductors and equipment for delivering energy from the electricity supply system to the wiring system of the premises served.

*SERVICE CABLE   Service conductors made up in the form of a cable.

SERVICE CONDUCTORS   The supply conductors that extend from the street main or from transformers to the service equipment of the premises supplied.

SERVICE DROP   The overhead service conductors from the last pole or other aerial support to and including the splices, if any, connecting to the service-entrance conductors at the building or other structure.

*SERVICE-ENTRANCE CONDUCTORS (OVERHEAD SYSTEM)   The service conductors between the terminals of the service equipment and a point usually outside the building, clear of building walls, where joined by tap or splice to the service drop.

*SERVICE-ENTRANCE CONDUCTORS (UNDERGROUND SYSTEM)   The service conductors between the terminals of the service equipment and the point of connection to the service lateral. Where service equipment is located outside the building walls, there may be no service-entrance conductors, or they may be entirely outside the building.

SERVICE EQUIPMENT   The necessary equipment, usually consisting of a circuit breaker or switch and fuses and their accessories, located near the point of entrance of supply conductors to a building or other structure or an otherwise defined area and intended to constitute the main control and means of cutoff of the supply.

SERVICE LATERAL   The underground service conductors between the street main, including any risers at a pole or other structure or from transformers, and the first point of connection to the service-entrance conductors in a terminal box or meter or other enclosure with adequate space, inside or outside the building wall. If there is no terminal box, meter, or other enclosure with adequate space, the point of connection shall be considered to be the point of entrance of the service conductors into the building.

SETTING (OF A CIRCUIT BREAKER)   The value of the current and/or time at which the adjustable circuit breaker is set to trip.

SHOW WINDOW   Any window used or designed to be used for the display of goods or advertising material, whether it is fully or partly enclosed or entirely open at the rear and whether or not it has a platform raised higher than the street-floor level.

SIGN   See ELECTRIC SIGN.

SIGNALING CIRCUIT   Any electric circuit that energizes signaling equipment.

SPECIAL PERMISSION   The written consent of the authority having jurisdiction.

SWITCHES

*General-use switch.*   A switch intended for use in general-distribution and branch circuits. It is rated in amperes, and it is capable of interrupting its rated current at its rated voltage.

*General-use snap switch.*   A form of general-use switch so constructed that it can be installed in flush device boxes or on outlet-box covers or otherwise used in conjunction with wiring systems recognized by the NEC. For the ac general-use snap switch, see Sec. 380-14(a) of the NEC; for the ac, dc general-use snap switch, see Sec. 380-14(b) of the NEC.

*Isolating switch.*   A switch intended for isolating an electric circuit from the source of power. It has no interrupting rating, and it is intended to be operated only after the circuit has been opened by some other means.

*Motor-circuit switch.*   A switch, rated in horsepower, capable of interrupting the maximum operating overload current of a motor of the same horsepower rating as the switch at the rated voltage.

*SWITCHBOARD   A large single panel, frame, or assembly of panels on which are mounted, on the face or back or both, switches, overcurrent and other protective devices, buses, and, usually, instruments. Switchboards are generally accessible from the rear as well as from the front and are not intended to be installed in cabinets. See also PANELBOARD.

THERMAL PROTECTOR (AS APPLIED TO MOTORS)   A protective device for assembly as an integral part of a motor or motor-compressor and which, when properly applied, protects the motor against dangerous overheating due to overload and failure to start. The thermal protector may consist of one or more sensing elements integral with the motor or motor-compressor and an external control device.

THERMALLY PROTECTED (AS APPLIED TO MOTORS)   The words *thermally protected* appearing on the nameplate of a motor or motor-compressor indicate that the motor is provided with a thermal protector.

UTILIZATION EQUIPMENT   Equipment that utilizes electric energy for mechanical, chemical, heating, lighting, or similar purposes.

*VENTILATED   Provided with a means to permit circulation of air sufficient to remove an excess of heat, fumes, or vapors.

VOLTAGE (OF A CIRCUIT)   The greatest root-mean-square (effective) difference of potential between any two conductors of the circuit concerned. Some systems, such as three-phase four-wire, single-phase three-wire, and three-wire direct-current, may have various circuits of various voltages.

VOLTAGE, NOMINAL   A nominal value assigned to a circuit or system for the purpose of conveniently designating its voltage class (as 120/240, 480Y/277, 600, etc.). The actual voltage at which a circuit operates can vary from the nominal within a range that permits satisfactory operation of equipment. See *Voltage Ratings for Electric Power Systems and Equipment (60 Hz)*, ANSI C84.1-1982.

VOLTAGE TO GROUND   For grounded circuits, the voltage between the given conductor and that point or conductor of the circuit that is grounded; for ungrounded circuits, the greatest voltage between the given conductor and any other conductor of the ungrounded circuit.

°WATERTIGHT   So constructed that moisture will not enter the enclosure under specified test conditions.

°WEATHERPROOF   So constructed or protected that exposure to the weather will not interfere with successful operation. Rainproof, raintight, or watertight equipment can fulfill the requirements for weatherproof where varying weather conditions other than wetness, such as snow, ice, dust, or temperature extremes, are not a factor.

### 4. Methods of installing interior wiring

1. Open wiring on insulators
2. Concealed knob-and-tube wiring
3. Rigid metal, intermediate metal, or nonmetallic conduit
4. Flexible metal conduit
5. Liquidtight flexible conduit
6. Metal-clad cable
7. Armored cable
8. Surface raceways
9. Electrical metallic tubing
10. Nonmetallic-sheathed cable
11. Mineral-insulated metal-sheathed cable
12. Underground-feeder and branch-circuit cable
13. Service-entrance cable
14. Nonmetallic extensions
15. Underfloor raceways
16. Underplaster extensions
17. Wireways
18. Busways
19. Cellular-metal-floor raceways
20. Cellular-concrete-floor raceways
21. Flexible metallic tubing
22. Multioutlet assemblies
23. Cable tray
24. Cablebus
25. Flat conductor cable
26. Electrical nonmetallic tubing

**5. In the open-wiring method** of installing interior wiring the wires are run either concealed or exposed on the ceiling, roof structure, or walls. They are supported on porcelain insulators of either the knob (Figs. 4-139, 4-140, and 4-141) or the cleat type (Figs. 4-143 and 4-144) so as to provide the necessary clearance between the wires and the surface being wired over.

**6. Concealed knob-and-tube wiring** was one of the earliest methods of installing wiring. The wires were mounted on knobs and run through tubes where they pierced structural members of the building. They were then concealed by the wall and ceiling surface. This method is still allowed by the **NEC** but is not used very often because of the large amount of labor required to install the wiring.

**7. In the rigid-metal-conduit or intermediate-metal-conduit method** of installing interior wiring the wires are supported and protected from mechanical injury by being installed in ferrous or nonferrous types of rigid metal conduits. Ferrous types are of wrought iron or steel with coatings such as black enamel, electrogalvanizing, hot-dip galvanizing, or similar material. Nonferrous types include aluminum or brass (silicon bronze) conduits. Various types of ferrous and nonferrous metal conduits are available with outer plastic coatings to provide optimum protection from corrosion.

Rigid or intermediate metal conduit may be run exposed, supported directly on walls, ceilings, or roof structures or on suitable hanger assemblies. It may also be concealed in partitions, ceilings, or floors. Provision for connection to the circuit at outlet and switch points is made by the insertion of sheet-steel, cast-metal boxes (steel or aluminum) or special conduit fittings in the conduit run. The special conduit fittings, LB, T, LL, LR, X, etc., are called by various trade names, such as Condulets or Unilets. Although rigid metal conduits are usually employed with threaded fittings and connections, many threadless fittings are available for use without threading the conduit.

Conduit, fittings, and boxes are installed complete without wires. Then the wires are pulled into the conduit from fitting to fitting, box to box, or fitting to box. In hazardous locations only threaded connections (five full threads) are permitted.

**8. Rigid nonmetallic conduit and fittings** are constructed of a suitable nonmetallic material that is resistant to moisture and chemical atmospheres. For use aboveground the conduit must be flame-retardant and resistant to impact and crushing, to distortion due to heat under conditions likely to be encountered in service, and to low-temperature or sunlight effects. The only type of nonmetallic conduit that is suitable for aboveground applications is the heavy-wall (Schedule 40 or Schedule 80) type constructed of polyvinyl chloride (PVC). Such conduits may be embedded in concrete in buildings or installed in wet locations such as laundries or dairies. They may also be buried directly in earth.

Other types of nonmetallic conduits, recognized solely for underground use, are those constructed of fiber, asbestos cement, soapstone, fiberglass epoxy, and high-density polyethylene. Depending upon construction and Underwriters Laboratories listings, such conduits may or may not require concrete encasement. Complete lines of nonmetallic fittings and boxes are available. For underground application refer to Div. 8.

**9. The flexible-metal-conduit method** of installing wiring is used to a limited degree in frame buildings or similar applications in which rigid raceways would be difficult to install and a pull-in, pullout conduit system is desirable. (Flexible metal conduit is often referred to as Greenfield.) The runs of conduit, boxes, and fittings are installed first as a complete system. Then the wires are installed in the same manner as in rigid metal conduit. The NEC states that flexible metal conduit may be used as a grounding means in 6-ft (1.8-m) lengths or if both conduit and fittings are listed for grounding and the circuit conductors are protected by a 20-A overcurrent device or less. If they are not so approved, each run of flexible metal conduit must contain a bare or insulated grounding conductor, and this grounding conductor must be attached to each box or other equipment supplied by such conduit.

**10. Liquidtight flexible metal conduit** has an outer liquidtight jacket. It is not intended for general-use wiring. Wiring of this kind may be used only for the connection of motors or portable or stationary equipment when flexibility of connections is required. When used, it must be provided with suitable terminal connectors approved for the purpose. It cannot be used in sizes larger than 2 in, where subject to physical damage, where subject to temperatures above its rating, or in any hazardous location except for limited flexibility, and motor connection in Div. 2 areas. See Tables 21 through 36 in Div. 11 for maximum sizes of conductors allowed.

**11. Type MC metal-clad cable** is a factory assembly of one or more conductors, each individually insulated and enclosed in a metallic sheath of interlocking tape or in a smooth or corrugated tube. The metallic covering may be a smooth metallic sheath, a welded and corrugated metallic sheath, or interlocking metal-tape armor. This cable is widely used for concealed and exposed wiring when flexibility and ease of installation are important factors. Type MC cable is not permitted to be used when exposed to destructive corrosive conditions, such as direct burial in earth or in concrete or when exposed to cinder fills, strong chlorides, caustic alkalies, or vapors of chlorine or of hydrochloric acids except when the metallic sheath is suitable for the conditions or is protected by material suitable for the conditions.

**12. Armored cable** includes Types AC, ACT, and ACL. Type AC contains insulated conductors of a type accepted for general wiring applications in the NEC. Type ACT conductors are individually covered by a moisture-resistant fibrous covering. Type ACL cable contains lead-covered conductors and is used in locations exposed to the weather or to continuous moisture; for underground runs in raceways and embedded in masonry, concrete, or fill in buildings in course of construction; or in locations exposed to oil or in which other conditions have a deteriorating effect on the insulation. Type AC cable, which is commonly called BX, is widely used for exposed and concealed work when flexibility and ease of installation are desirable.

**13. Surface raceways** include both metallic and nonmetallic types. In the metallic type wires are inserted into a thin sheet-metal casing. The raceway is installed exposed on interior building surfaces. The metal molding is made with a flattened-oval or rectangular cross section. Numerous fittings, adapters, and boxes, specially designed for this system, are readily available from major manufacturers of this wiring system. The NEC recognizes the nonmetallic type of raceway, and the installation, boxes, and fittings are similar to those in the metallic system.

**14. In the electrical-metallic-tubing (EMT) method** of installing wiring the wires are installed in a thin-walled metallic tube or thin-walled conduit, as it is sometimes called. Electrical metallic tubing is similar to rigid conduit except that it is constructed of much thinner material. Provisions for connecting outlets and switches to the circuit are made by means of special metallic-tubing fittings or by means of regular rigid-conduit fittings provided with an adapter. After the tubing and fittings have been installed as a complete system, the wires are pulled in from fitting to fitting as in the rigid-conduit method. Connectors and couplings are threadless types such as setscrew, compression, indenter, or tap-on.

**15. Nonmetallic-sheathed cable** is made in two types, Type NM and Type NMC. Type NM consists of two or more rubber- or thermoplastic-insulated wires bound together and protected by a plastic jacket or by a cotton-bound paper jacket with an outer heavy cotton braid treated with a moisture- and heat-resisting compound. Type NMC cables are protected by an outer flame-retardent, moisture-resistant, fungus-resistant, and corrosion-resistant sheath which contains no cotton or paper. Nonmetallic cable may be run exposed on walls and ceilings or concealed in the hollow spaces between partitions or between floors and ceilings. The cable is fastened directly to the surface of the walls, ceilings, or structural members with approved supports. Provision for outlets and switches is made by running the cable into outlet boxes. In most cases such cables contain an equipment-grounding conductor.

**16. Mineral-insulated metal-sheathed cable (Type MI)** consists of one or more electrical conductors insulated and separated by highly compressed refractory mineral insulation and protected from injury by enclosure in a liquidtight and gastight copper or alloy steel sheath. The wires, insulation, and protective sheathing are manufactured as a unit. The method of installation is similar to that of metal-clad cable. Special approved fittings are used for terminating and connecting the cable to boxes and other equipment. At all points of termination of the cable an approved moisture seal must be provided.

**17. Underground-feeder and branch-circuit cable (Type UF)** resembles Type USE service-entrance cable in general appearance. The insulation employed may consist of a special synthetic plastic compound. When single-conductor cables are used, all conductors of the circuit must be installed in the same trench or raceway. For multiconductor cable installations all conductors of the circuit together with the protective covering of the cable are installed as a unit, just as in metal-clad or nonmetallic-sheathed–cable installations. Provision for outlets is made by running the cable into suitable outlet boxes. Underground-feeder and branch-circuit cable wiring provides a convenient approved method for interior wiring in wet or corrosive locations. It may be installed underground in raceways or be buried directly in the earth. Multiconductor cables usually contain an equipment-grounding conductor.

**18. Service-entrance cable** (see Div. 2) normally is used for the circuit from the point of attachment of the service conductors on the outside of the building to the service switch, load center, or meter cabinet just inside the building. It may, however, also be employed for the interior wiring on the load side of the service switch. It is not commonly used for general interior wiring but is used for certain special portions of the wiring such as to supply large appliances like an electric range, as a feeder to a distribution panel elsewhere in the building, or as a service cable to other buildings. For unrestricted use all the conductors must be rubber- or thermoplastic-insulated. If an uninsulated neutral is used, the cable must have a final nonmetallic outer covering, the voltage to ground must not exceed 150 V, and it may be used only to supply a range, wall-mounted oven, counter-mounted cooking equipment, or clothes dryer and the branch circuit originates at the service equipment or, from a master service cabinet, to supply other buildings. The cable is fastened to the surface with straps spaced not over 4 ft 6 in (1.37 m) apart.

**19. In the underfloor-raceway method** of wiring the wires are installed in a sheet-metal or fiber casing which is embedded in the concrete or cement fill of the floor. Generally, the ducts or raceways are laid out in the floor to form a network. Provision is made at the intersections of the ducts for the pulling in and splicing of wires by means of special floor junction boxes. Sheet-metal–type raceways are provided with outlet openings spaced at regular intervals along the raceway. These outlet openings are either plugged or equipped with special floor-outlet fittings. With the fiber-duct type of raceway the raceway is first installed in the floor without any outlet openings. After the floor has been finished, outlets may be provided at any desired point along the duct runs simply by cutting a hole in the floor and duct and inserting a special floor-outlet fitting. In the under-floor-raceway systems the wires are pulled through the ducts in the same manner as for rigid-conduit work.

**20. Underplaster extensions** are wiring systems laid in a channel cut in the plaster surface of a wall down to the face of masonry or other wall material. Rigid, intermediate, or flexible metal conduit, Type AC cable, electrical metallic tubing, Type MI cable, or metal raceways approved for the purpose are laid in the channel and secured to the surface of the wall or ceiling structure. Provision for outlets is made by inserting shallow outlet boxes in the run of conduit, cable, tubing, or raceway. After the system has been secured in the channel, it is plastered over flush with the wall or ceiling surface.

**21. A nonmetallic extension** is an assembly of two insulated conductors within a nonmetallic jacket or an extruded thermoplastic covering. The classification includes both surface extensions, intended for mounting directly on the surface of walls or ceilings, and aerial cable, containing a supporting messenger cable as an integral part of the cable assembly. Nonmetallic extensions are permitted only if (1) the extension is from an existing outlet on a 15- or 20-A branch circuit and (2) the extension is run exposed and in a dry location. Nonmetallic *surface* extensions are limited to residential or office buildings, and *aerial* cable is limited to industrial buildings when a highly flexible means for connecting equipment is required.

**22. In the wireway method** of wiring the wires are supported and protected in a sheet-metal trough. The trough is installed exposed, being mounted on the ceiling or walls or supported from the roof structure. One side of the trough is fitted with a sheet-metal cover so that access can be had to the interior sections of the trough throughout its length. After the wireway has been installed as a complete system, the wires are laid in the trough and the cover is installed. Mains and feeders are run in the wireways with taps for branches taken off at the most convenient point through rigid or flexible metal conduit, connected to the wireway through knockouts provided on three sides of the trough. This system provides a flexible distribution system.

**23. Busway systems** of wiring consist of bare or insulated busbars insulated from each other and supported and protected by a sheet-metal housing. The housing and busbars are assembled as a unit in 10-ft (3-m) or longer sections. The system is installed by supporting the housing from the ceiling, walls, or roof structure. Busway systems are extensively used for the feeder and main circuits in industrial plants. They are available in three general types: without provision for outlets, with provision for fixed outlets, and with provision for continuously movable outlets for portable tools and electric-welding work. With the type which has provision for fixed outlets, openings are provided every few inches for branch taps. The connection of a tap to the busbars is accomplished through the insertion of a special power attachment plug in these openings. For the wiring of the branch circuit from the power plug to the apparatus to be supplied with power, flexible conduit, rigid conduit, or metal-clad cable may be employed. This method of wiring provides a very flexible system which meets in an economical manner the requirements of plants in which production conditions necessitate the frequent shifting or replacement of machinery.

**24. In cellular-metal-floor–raceway wiring,** the floor is constructed of special metal structural members containing hollow spaces or cells which form raceways for the wires. Provision for pulling in and splicing wires is made by locating special junction boxes in the floor between sections of the floor members. Outlet openings and fittings may be originally provided at regular intervals along the raceway formed by the cells, or outlets may be provided at any time after completion of the building construction simply by cutting a hole in the floor and cell wall and inserting a special floor-outlet fitting. After the complete floor structure has been erected and finished over, the wires are pulled through the raceway formed by the floor cell structure and the junction boxes.

**25. In cellular-concrete-floor–raceway wiring,** the floor is constructed of precast reinforced-concrete members. These precast members are provided with hollow voids which form smooth, round cells. The cells form raceways for the wires. Connections to the cells from a distribution center can be made by metal header ducts run horizontally across the precast slabs and embedded in the concrete fill over the slabs. Connection from these headers to the cells is made through handhole metal junction boxes. An outlet can be located at any point along a cell. An opening into the cell at the desired point is formed by drilling a hole through the concrete floor slab. The hole is then fitted with the proper outlet fitting, and the wires fished from a handhole junction box to the outlet.

**26. Flexible metallic tubing,** a recent innovation, is a flexible, liquidtight, metallic raceway intended to be used in protected locations such as suspended ceilings. Flexible metallic tubing is restricted to 6-ft (1.8-m) lengths and was developed for use in ducts or plenums through which the excursion of smoke or gases from the raceway could be controlled.

**27. Multioutlet assemblies** are a special form of surface raceway with built-in single receptacle outlets every few inches. The assembly is supported on the surface of walls or in or on top of baseboards. It is used for wiring in dwellings and commercial buildings to provide very convenient and adequate convenience outlets. The assembly is supplied with power through a cable or conduit run concealed in the walls and brought into the back or end of the assembly.

**28. A cablebus system** is an approved assembly of insulated conductors mounted in spaced relationship in a ventilated metal protective supporting structure, including fittings and conductor terminations. Cablebus may be used at any voltage up to 35 kV or current for which the spaced conductors are rated. It is ordinarily assembled at the point of installation from components specified by the manufacturer. First, the supporting structure is installed in the same manner as in a cable-tray support system. Next, the insulated conductors, not less than 1/0 with insulation ratings of 75°C or higher, are inserted. After this, insulating supports are installed so that the conductors are properly separated and supported at intervals of not less than 3 ft (0.9 m) for horizontal runs and 1½ ft (0.46 m) for vertical runs. Conductor ampacity is based on the values for open conductors.

**29. Cable tray** is a unit or an assembly of units or sections and associated fittings made of metal or other noncombustible materials forming a continuous rigid structure used to support cables. The support system includes ladders, troughs, channels, and similar support systems. Cable tray is not a wiring method and may be used only as the mechanical support for approved raceways or multiconductor cable wiring methods or specially approved multiconductor cables designed for use in cable trays; or 1/0 or larger single conductors in industrial establishments where conditions of maintenance and supervision assure that only qualified persons will service the cable-tray system. First, the cable tray is installed, and then the cables or raceways are installed and secured to the cable tray. This system has particular merit in industrial applications or similar uses for which many power, control, or signal cables are required and flexibility is a major consideration.

**30. Medium-voltage (MV) cable** is a single or multiconductor cable with solid dielectric insulated conductors rated 2001 V or higher. Type MV cables are permitted to be used on power systems up to 35,000 V in wet or dry locations, in raceways, or directly buried when installed in accordance with Sec. 710-3(b) of the NEC. Type MV cables are not permitted in locations exposed to direct sunlight or in cable trays.

**31. Flat conductor cable** is a field-installed wiring system for branch circuits incorporating Type FCC cable and associated accessories for installation under carpet squares no larger than 36 in (914 mm) square. FCC systems shall not be used outdoors or in other wet locations, where subject to corrosive vapors, in any hazardous location, or in residential, school, and hospital buildings.

**31A. Electrical nonmetallic tubing** is a pliable, corrugated raceway of circular cross section with integral or associated couplings, connectors, and fittings listed for the installation of electric conductors. It is composed of a material that is resistant to moisture and chemical atmospheres and is flame-retardant.

## OPEN WIRING ON INSULATORS

**32. Open wiring on insulators** is one of the cheapest methods of installing wiring. It finds application in factories and mills for the installation of feeder circuits in places where appearance is of little consequence. The wires may contain rubber or thermoplastic insulation. The wires must be supported at least every 4½ ft (1.37 m) except in mill buildings, where a support on each beam may be approved for wires No. 8 and larger if they are separated by at least 6 in (152 mm). Also, circuits using No. 8 or larger wire, if not likely to be disturbed, may be run across open spaces and supported at distances not greater than 15 ft (4.57 m) if approved noncombustible, nonabsorptive insulating separators providing not less than 2½-in (63.5-mm) separation between conductors are installed at intervals of not over 4½ ft (1.37 m).

**33. Knobs and methods of supporting conductors on them.** Knobs for supporting conductors in interior work are of porcelain. Split knobs or cleats may be used for supporting conductors smaller than No. 8 American wire gage. Some methods of securing wires to knobs are shown in Fig. 9-1. The line tie of I is made by winding the conductor once around the knob, so both ends of the wire must be under tension to hold the wire in position. A tie wire is used at II; in making up the tie wire the slack can be drawn out of the conductor. A dead end, or termination, is shown at III. If it is necessary to change the direction of a run to get the conductor to an outlet or for any other reason, tap ties IV, V, and VI are used. It is not practicable to tie large conductors, so they may be supported as at VII.

For a discussion of types of knobs see Div. 4.

Nails used to support knobs must be not smaller than tenpenny and must be provided with cushion washers. When screws are used, they must penetrate the supporting wood to a depth equal to at least one-half the height of the knob or to at least the thickness of the cleat used.

**34. Tie wires** must have an insulation equal to that of the conductors they confine and may be used in connection with solid knobs for the support of wires of size No. 8 or larger.

**35. The method of dead-ending on a cleat** at the end of a run is illustrated in Fig. 9-2. After the wire has been passed through the groove, the free end is given several short turns

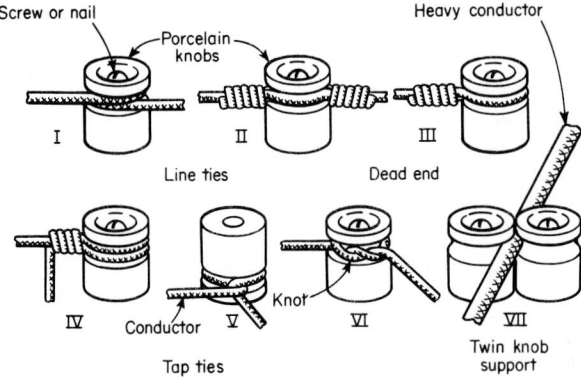

**Fig. 9-1**  Methods of attaching conductors to knobs.

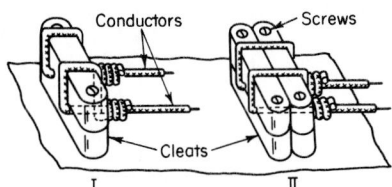

**Fig. 9-2**  Dead-ending on cleats.

around the line. When a long run is dead-ended, it is often advisable to fasten two sets of cleats in such a way that one bears against the other so that both will assume the strain as shown at II. For types and dimensions of cleats see Div. 4.

**36. Methods of terminating heavy conductors.** At the ends of all important open-wire runs of wires larger than, say, No. 8, strain insulators engaging in some wire-tightening device should be used. Figure 9-3 illustrates some methods. Either tightening bolts or turnbuckles can be used. The insulator may be of the type extensively used in trolley-line construction as in I, II, and III, or it may be a heavy knob (IV) held to the tightening device with stout wire. When a run changes direction, a cable clamp can often be used with economy, particularly with large conductors. If a cable clamp is used, it is unnecessary to cut the conductor to change its direction, and the need to make up turns about the line wire as in I and II is eliminated.

**37. Different approved methods of exposed surface wiring** are illustrated in Fig. 9-4. Which method should be used in any particular case is determined largely by the size of the wire involved and other local conditions.

**38. Physical protection of exposed surface wiring.**  Wires must be protected on sidewalls from physical damage and, when crossing timbers where they might be disturbed, must be protected by guard strips or running board and guard strips as shown in Figs. 9-5 and 9-6. The guard strips must be at least 1 in (25.4 mm) thick and at least as high as the insulators. If a running board is used, it must be at least ½ in (12.7 mm) thick and must extend at least 1 in (25.4 mm) outside the conductors but not more than 2 in (50.8 mm). The wires should also be protected by porcelain tubes when passing over pipes (Fig. 9-7) or any other members. Tie wires may be used in lieu of tape.

Conductors within 7 ft (2.1 m) from the floor shall be considered exposed to physical damage. Suitable protection on sidewalls therefore should extend not less than 7 ft from the floor (Fig. 9-8). This protection may consist of substantial boxing, providing an air space of 1 in (25.4 mm) around the conductors, closed at the top (the wire passing through porcelain-bushed holes), or of a conduit with fittings on each end providing individual bushed holes for each wire (Fig. 9-8, III).

**39. Methods of carrying exposed wiring around and through beams** are illustrated in Fig. 9-

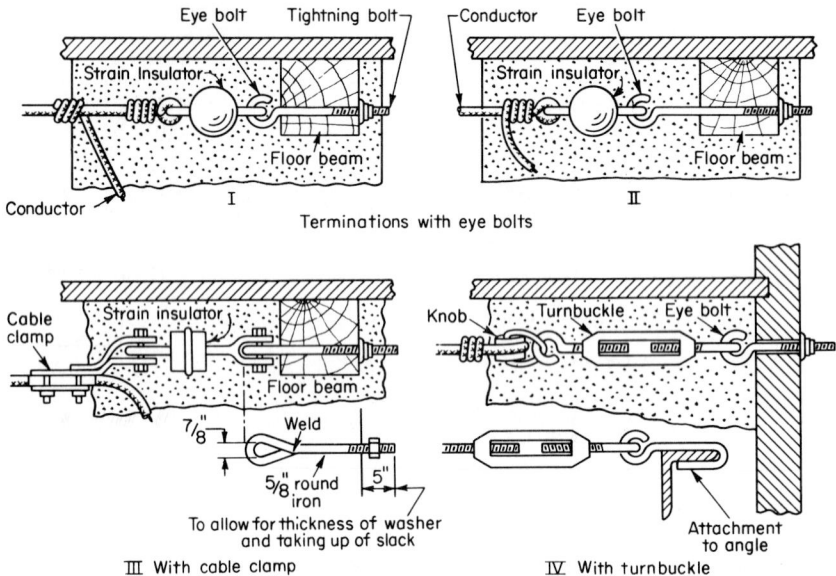

**Fig. 9-3**    Methods of terminating conductors.

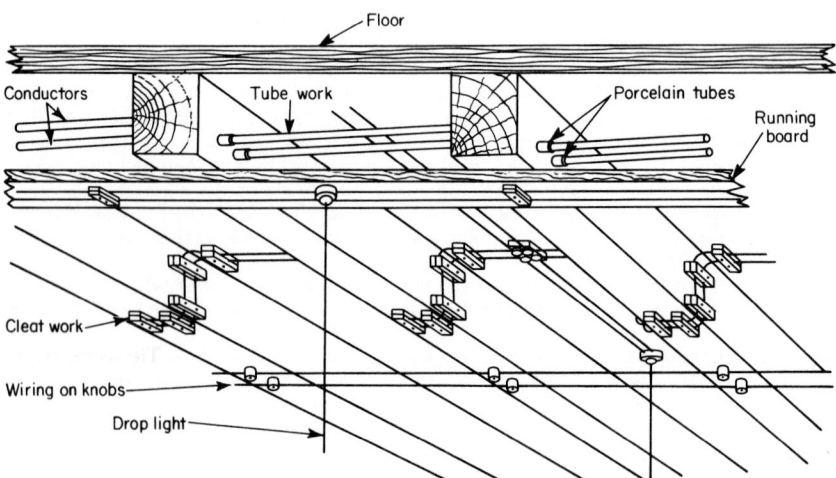

**Fig. 9-4**    Openwork wiring in a mill building.

4, which shows tube-and-cleat arrangements. In Fig. 9-9 are shown methods that can be used when wires are supported on knobs.

**40. In steel-mill buildings heavy conductors may be carried** on the lower chords of the roof trusses (Fig. 9-10). This is a good location because the conductors are out of the way and not apt to be disturbed. At each truss the conductors can be supported by one of the methods illustrated in Fig. 9-11. With the method of Fig. 9-11, I, the conductor merely rests in the insulator, and the entire longitudinal strain is taken by strain insulators

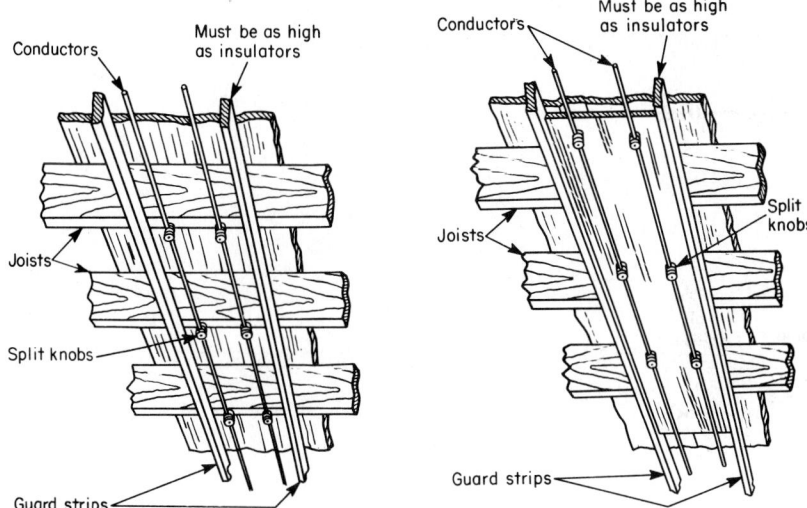

**Fig. 9-5** Guard strips to protect open wiring crossing ceiling joists.

**Fig. 9-6** Running board with guard strips to protect open wiring crossing ceiling joists.

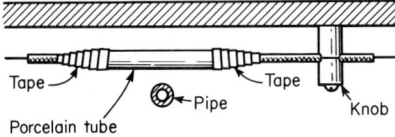

**Fig. 9-7** Protection of conductor passing over pipe.

attached to tightening bolts or turnbuckles at the ends of the run. This method has the disadvantage that if the conductor breaks at any point or is burnt in two, it will fall to the floor. The tie-wire method of II is seldom used, though it is satisfactory if cleats are not obtainable. (Split knobs or cleats must be used for conductors smaller than No. 8.) The cleat and through-bolt method of III is probably the best, all things considered. After the conductor has been drawn taut with the tightening bolts at the ends of the run, the cleat bolts are tightened and each cleat then assumes its share of the stress. Tie wires, which are unreliable and may cut into the insulation of the conductor, are unnecessary. Leather washers should be used between the insulator and the bolt to prevent breakage.

**41. For supporting conductors on steel angles** the Universal Insulator Support (Figs. 9-12 and 9-13 and Table 42) is a convenient fitting. It is of malleable iron and can be clamped on the flanges of steel beams, angles, channels, and Z bars and on round, square, and flat bars. It can also be attached to gas and water pipes and to the edges of plates and tanks.

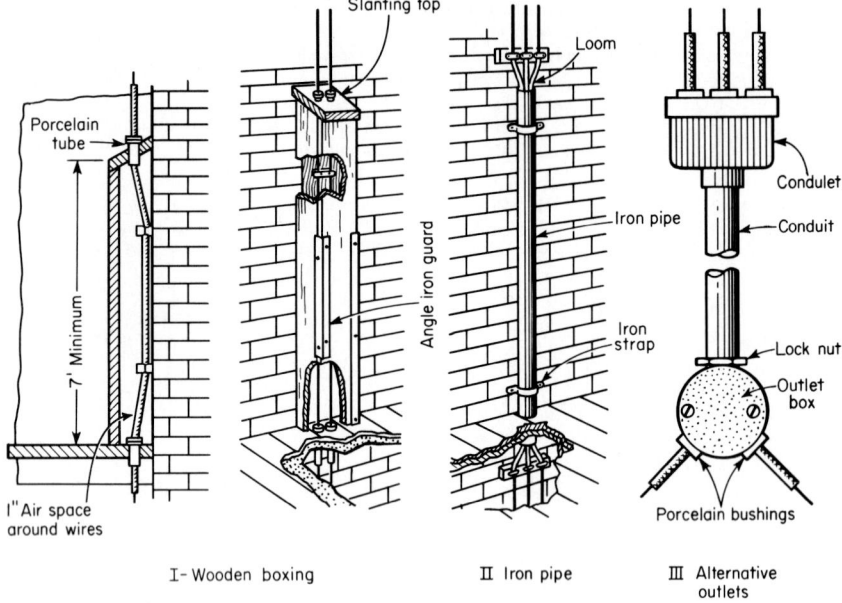

**Fig. 9-8**   Protection of conductors on sidewalls.

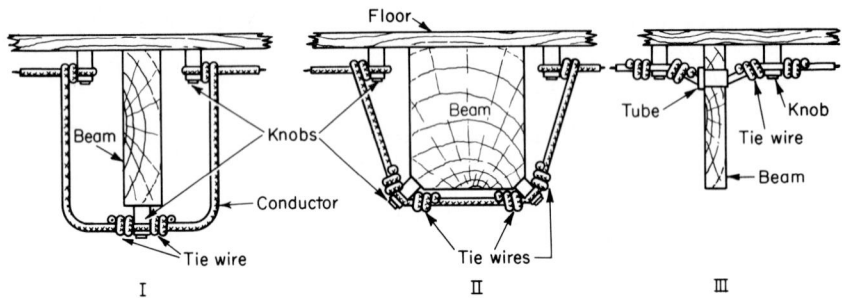

**Fig. 9-9**   Openwork wiring with knobs.

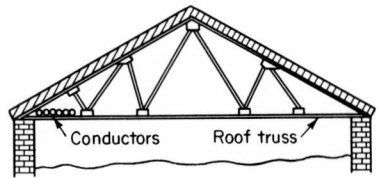

**Fig. 9-10**   Conductors carried on roof truss.

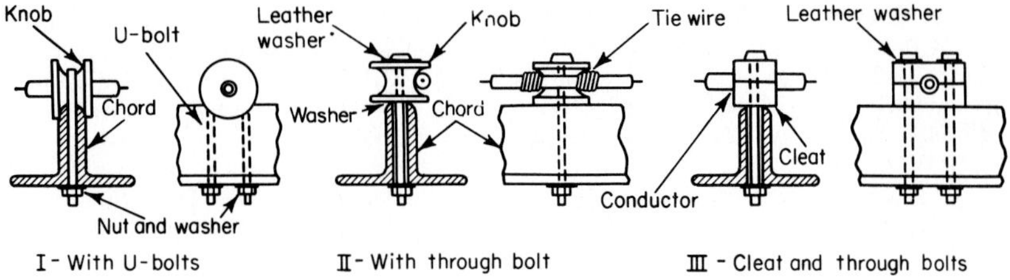
Knob
U-bolt
Leather washer
Knob
Tie wire
Leather washer
Chord
Washer
Chord
Cleat
Conductor
Nut and washer
I - With U-bolts
II - With through bolt
III - Cleat and through bolts

**Fig. 9-11** Attaching knobs to truss chords.

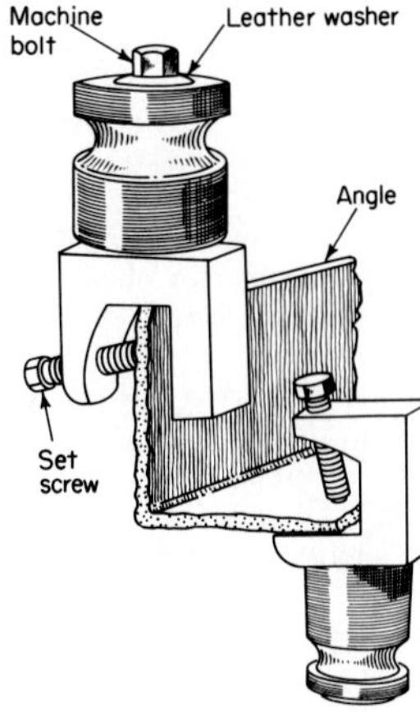
Machine bolt
Leather washer
Angle
Set screw

**Fig. 9-12** Universal Insulator Supports on an angle.

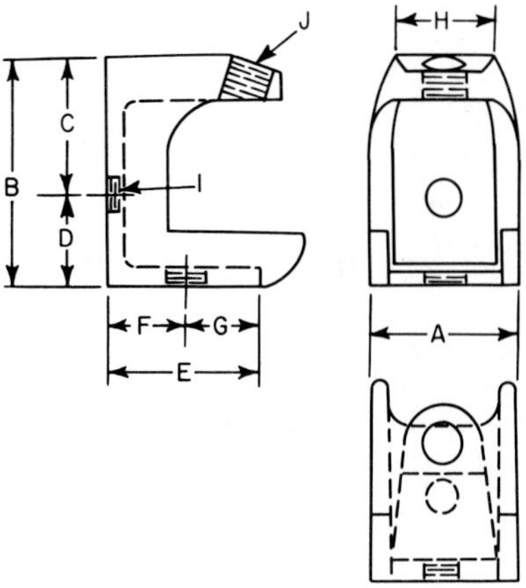
J
H
C
B
I
D
F
G
E
A

**Fig. 9-13** The Universal Insulator Support.

Two insulators can be fastened to each support when necessary. Cup-pointed, case-hardened setscrews are used. Leather washers should be used under the bolts that hold the insulators.

**42. Dimensions of Universal Insulator Supports**
(Steel City Division, Midland-Ross Corp., Pittsburgh)

| *A*, size, in. | For insulators, Nos. | *B*, in. | *C*, in. | *D*, in. | *E*, in. | *F*, in. | *G*, in. | *H*, in. | *I*, diam of tapped hole | *J*, diam of setscrew furnished |
|---|---|---|---|---|---|---|---|---|---|---|
| 1 | 5, 5½ | 1½ | ⅞ | ⅝ | 1¹⁄₁₆ | ⁹⁄₁₆ | ½ | ¾ | ¼ | ⁵⁄₁₆ |
| 1½ | 10, 4, 4½ | 1¾ | 1¾ | ⅝ | 1½ | ⁹⁄₁₆ | ¾ | ¾ | ⁵⁄₁₆ | ⁷⁄₁₆ |
| 2 | 1, 3, 3 W.G., 3½, 24 | 2 | 1 | 1 | 2 | 1 | 1 | ¾ | ⅜ | ½ |
| 2½ | 25, 29, 34 | 2½ | 1¼ | 1¼ | 2½ | 1¼ | 1¼ | ¾ | ½ | ⅝ |

**43. For supporting conductors on steel columns** a wooden baseboard for the cleats clamped to the column with hook bolts (Fig. 9-14) is a good arrangement. The board must be cut out in back for the rivet heads in the column. Strap-iron cleats through which the hook bolts pass prevent warping and splitting.

**44. Wire racks** are used to support conductors, principally heavy ones, when there are many conductors in the run. The conductors should have flameproof or slow-burning covering. A wire rack can be made of wood fashioned into a framework somewhat along the lines of the steel ones of Figs. 9-15 and 9-16. The cleats insulating the conductors are held to the frame with wood screws or, preferably, with machine or stove bolts. A commercial wire rack with a cast-iron base that can be bolted to any surface is shown in Fig. 9-17. Generally a steel-frame rack is preferable to a wooden one. The rack of steel angles of Fig. 9-15 was designed for installation in the top of a pipe tunnel. The insulators are held to the cross angles with bolts with a leather washer under the head of each. The structural-steel rack of Fig. 9-16, III, is arranged for supporting from a ceiling. Angle crossarms can be used as at II, or the crossarms can each be formed of two iron straps as at I. With the two-strap method, drilling for the cleat bolts is unnecessary, and the cleats can be shifted along the arm into any desired position and there clamped fast. Strain insulators engaging in turnbuckles or tightening bolts should be used at the ends of each straight run to assume the strain and to provide for tightening, or else the arms and cleats at the run ends should be reinforced to assume the stress that will come on them.

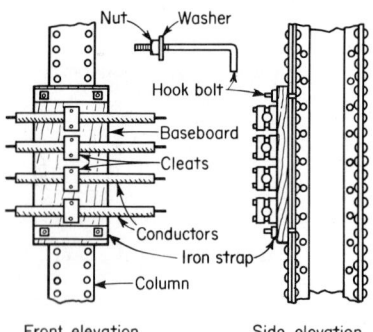

**Fig. 9-14**  Attachment of wiring board to column.

**45. A method of supporting open wiring in concrete buildings** is shown in Fig. 9-18. A round groove of ⅜-in (9.525-mm) radius is cast in the faces of the beams, by having ¾-in (19.05-mm) half-round molding nailed in the forms. Wrought-iron yokes are bent to fit the grooves as shown, and ½-in (12.7-mm) bolts clamp them in position. Wooden blocks or sections of strap or angle iron can be bolted to the yokes for supporting the wire cleats.

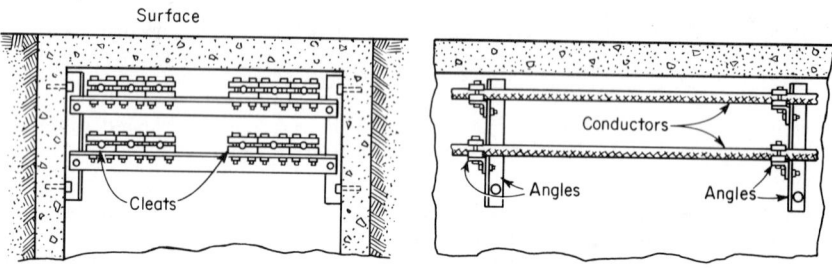

**Fig. 9-15**  Angle-iron rack for conductors.

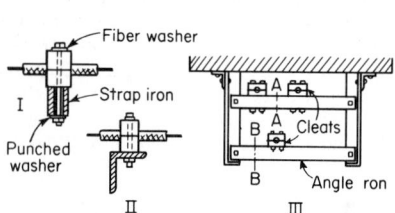

**Fig. 9-16**  Rack composed of angles and strap iron: I, section A–A; II, section B–B.

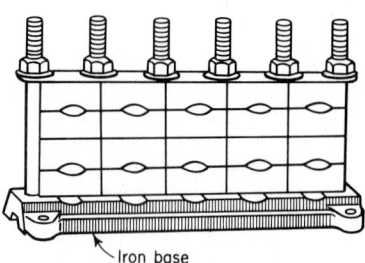

**Fig. 9-17**  A commercial insulator rack.

**46. When conductors pass through floors, walls, or partitions,** they must always be protected. Openwork wires can be protected with porcelain tubes (Fig. 9-19). The tube or bushing must be long enough to bush the entire length of the hole in one continuous piece, or else the hole must first be bushed by a continuous waterproof sleeve of noninductive material and the conductor so installed that it will be absolutely out of contact with the sleeve.

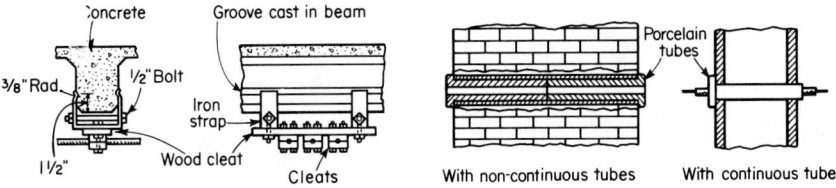

**Fig. 9-18**  Supports for open wiring on a concrete beam.

**Fig. 9-19**  Protection through walls and partitions.

**47. A tube for protecting a wire where it crosses another wire** should always be so placed that the tube will not force the unprotected wire against the surface supporting the conductors.

**48. Flexible tubing, or "loom"** (Fig. 9-20), is used mostly in connection with open wiring and concealed knob-and-tube wiring. When metal outlet boxes or switch boxes are used, flexible tubing is required from the last porcelain support, extending into the outlet box at least 1 in (25.4 mm), and held in place by an approved fitting such as a universal bushing or a clamp.

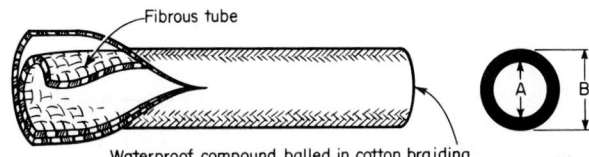

Waterproof compound balled in cotton braiding

**Fig. 9-20**  Flexible tubing.

Another application for flexible tubing is in completed buildings when the wires are fished in between the walls and ceilings. The tubing is used as a covering on such wires separately encased. In concealed knob-and-tube work it is frequently impracticable to place wires 3 in (76.2 mm) apart and 1 in (25.4 mm) from the surface wired over as required by the **National Electrical Code**, and in such cases the wires must be separately encased in flexible tubing. In open wiring when the amount of separation required by the **Code** from the surface wired over cannot be maintained, the wires may also be encased in flexible tubing.

The following is a list of places where flexible tubing is applicable: in openwork where wires are exposed nearer each other than 2½ in (63.5 mm); on wires crossing other wires; on wires crossing gas pipes, water pipes, iron beams, woodwork, brick, or stone; and at distributing centers or when space is limited and the 3-in (76.2-mm) separation required cannot be maintained. Each wire must be separately encased in a continuous length of flexible tubing.

Only continuous (unspliced) lengths of flexible tubing can be used for wire protection at outlets.

### 49.    Properties of Flexible Tubing or Loom

| $A$, inside diameter, in | $B$, outside diameter, in | | Ft per coil | Largest wire, AWG or cmil | Weight, lb/1000 ft |
|---|---|---|---|---|---|
| | In decimals | To nearest 64 in | | | |
| 7/32 | 0.46 | 15/32 | 250 | No. 14 | 50 |
| 1/4 | 0.50 | 1/2 | 250 | No. 14 | 58 |
| 3/8 | 0.65 | 21/32 | 250 | No. 12 | 75 |
| 1/2 | 0.78 | 25/32 | 200 | No. 8 | 90 |
| 5/8 | 0.98 | 63/64 | 200 | No. 4 | 120 |
| 3/4 | 1.06 | 1 1/16 | 150 | No. 2 | 196 |
| 1 | 1.31 | 1 5/16 | 100 | No. 2/0 | 250 |
| 1 1/4 | 1.63 | 1 41/64 | 100 | 200,000 | 400 |
| 1 1/2 | 1.88 | 1 57/64 | Odd lengths | 400,000 | 480 |
| 1 3/4 | 2.25 | 2 1/4 | Odd lengths | 600,000 | 590 |
| 2 | 2.63 | 2 41/64 | Odd lengths | 800,000 | 800 |
| 2 1/4 | 2.81 | 2 13/16 | Odd lengths | 1,100,000 | 810 |

**50. When conductors cross damp pipes,** they should be carried over rather than under so that drippings will not strike the wires. Porcelain tubes, securely taped to the conductors, should be placed on the conductors over the point where they cross.

**51. Rosettes for open surface wiring** are used to connect drop cords for incandescent lamps to the branch circuits. A rosette with protected (concealed) contact lugs is preferable to one with exposed lugs. Figure 9-21 shows one good type. Another good method of supporting drop cords, particularly if there is vibration, is with the ceiling button illustrated in Fig. 9-22.

**52. Switches can be supported in exposed surface wiring** as shown in Fig. 9-23. The switch can be mounted on a commercial porcelain switch block.

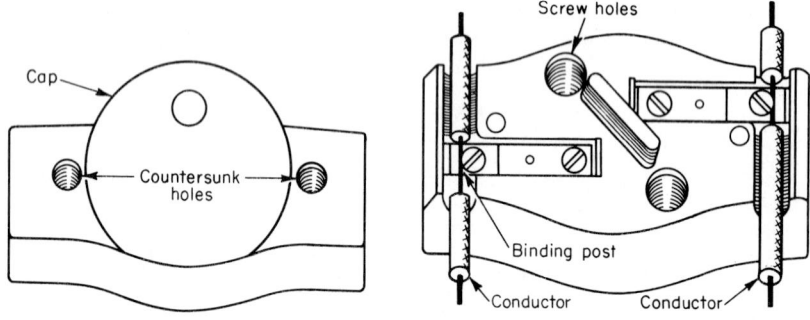

**Fig. 9-21**  A cleat rosette.

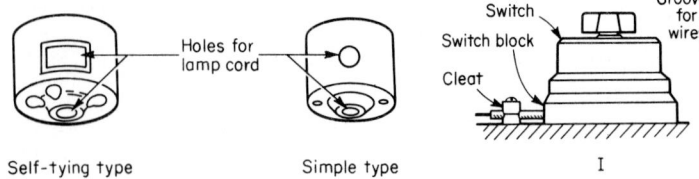

**Fig. 9-22**  Ceiling buttons.

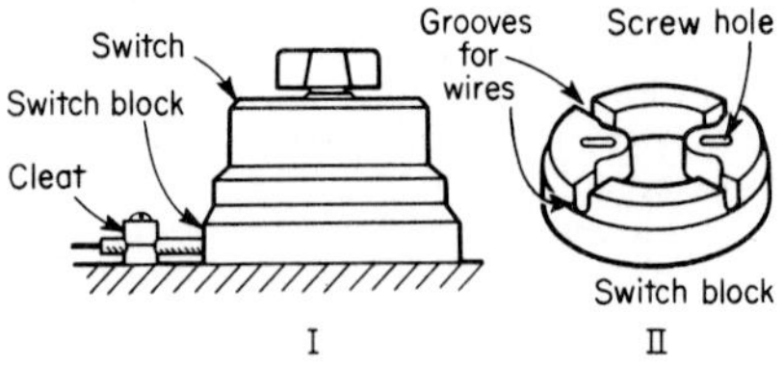

**Fig. 9-23**  Support switches in exposed surface wiring.

**53. Entering damp or wet locations** (National Electrical Code). Conductors entering or leaving damp or wet locations shall have drip loops formed on them and shall then pass upward and inward from the outside of buildings or from the damp or wet location through noncombustible, nonabsorptive tubes.

## CONCEALED KNOB-AND-TUBE WIRING

**54. Concealed knob-and-tube wiring** was used extensively at one time for the wiring of frame buildings when a low-cost installation was essential. It is still allowed by the National Electrical Code but is rarely used because more labor is required than with common cable systems. It has been superseded for low-cost installations by nonmetallic-sheathed cable.

**55. When changing from open wiring or concealed knob-and-tube wiring** to conduit, electrical-metallic-tubing, nonmetallic-sheathed–cable, armored-cable, or surface-raceway wiring, a fitting or outlet box having a separately bushed hole for each conductor must be used.

## RIGID-METAL-CONDUIT AND INTERMEDIATE-METAL-CONDUIT WIRING

**56. Rigid-metal-conduit wiring** is approved for both exposed and concealed work and for use in nearly all classes of buildings. For ordinary conditions, wiring in metal conduit is probably the best, although it is the most expensive. The advantages of metal conduit are that (1) wires can be inserted and removed, (2) it provides a good ground path, (3) it is strong enough mechanically so that nails cannot be driven through it and so that it is not readily deformed by blows or by wheelbarrows being run over it, and (4) it successfully resists the normal action of cement when embedded in the partitions or walls of concrete buildings.

A recent innovation is intermediate metal conduit, which has a thinner wall and is acceptable whenever rigid metal conduit is permitted.

**57. When to use rigid-metal-conduit or intermediate-metal-conduit wiring.**  In general, conduit wiring should be used whenever the job will stand the cost. Rigid conduit protects the conductors it contains and provides a smooth raceway, permitting ready insertion or removal.

Conduits and fittings exposed to severe corrosive influences shall be of corrosion-resistant material suitable for the conditions. If practicable, the use of dissimilar metals in contact anywhere in the system shall be avoided to eliminate the possibility of galvanic action.

Meat-packing plants, tanneries, hide cellars, casing rooms, glue houses, fertilizer rooms, salt storage, some chemical works, metal refineries, pulp and paper mills, sugar mills, roundhouses, textile bleacheries, plants producing synthetic staples, some stables, and similar locations are judged to be occupancies where severe corrosive conditions are likely to be present.

Cinder Fill   Conduit, unless of corrosion-resistant material suitable for the purpose, shall not be used in or under cinder fill where subject to permanent moisture unless protected on all sides by a layer of noncinder concrete at least 2 in (50.8 mm) thick or unless the conduit is at least 18 in (457.2 mm) under the fill.

Wet Locations   In portions of dairies, laundries, canneries, and other wet locations and in locations where walls are frequently washed, the entire conduit system, including all boxes and fittings used therewith, shall be so installed and equipped as to prevent water from entering the conduit, and the conduit shall be mounted so that there is at least ¼-in (6.35-mm) air space between the conduit and the wall or other supporting surface.

All supports, bolts, straps, screws, etc., shall be of corrosion-resistant materials or be protected against corrosion by approved corrosion-resistant materials.

**58. Corrosion-resistant rigid conduit** is available in three general types: (1) aluminum, (2) silicon bronze alloy, and (3) plastic-coated.

Aluminum rigid conduit is useful in installations where there are present certain chemical fumes or vapors which have little effect upon aluminum but have a severe corrosive effect upon steel. It is widely used because its light weight reduces installation labor costs. Uncoated or unprotected aluminum conduit should not be embedded in concrete or buried in earth, particularly if soluble chlorides are present.

A conduit of silicon bronze alloy known as Everdur electrical conduit has special corrosive-resistance characteristics. Its use is advantageous for installations exposed to the weather, as on bridges, piers, and dry docks and along the seacoast; in oil refineries and chemical plants; in sewage-disposal works; underground for water-supply works; and in wiring to underwater lighting fixtures for swimming pools.

Plastic-coated rigid conduit consists of standard galvanized-steel conduit over the surface of which seamless coatings of polyvinyl chloride plastic have been extruded. Thus a uniform nonporous protective coating is formed over the entire length of the raceway. All couplings and fittings should be covered with plastic sleeves or carefully wrapped with three thicknesses of standard vinyl electric insulating tape. This plastic-coated conduit is flame-retardant and highly resistant to the action of oils, grease, acids, alkalies, and moisture; does not oxidize, deteriorate, or shrink when exposed to sunlight and weather; and provides excellent resistance to abrasion, impact, and other mechanical wear. Typical applications are meat-packing plants, malt-liquor industries, paper and allied industries, production plants for industrial organic and inorganic products, soap and related-products industries, tanneries and leather-finishing plants, canneries, food-processing industries, dairies, fertilizer plants, and petroleum industries.

**59. Dimensions of corrosion-resistant rigid conduit.** The internal and external diameters for aluminum conduit are the same as those for rigid steel conduit. The internal diameter of plastic-coated rigid conduit, of course, is the same as that of standard rigid steel conduit. The outside diameter of the plastic-coated is slightly greater, as given in Table 61. The external diameter of conduit of silicon bronze alloy is practically the same as that of rigid steel conduit, but its internal diameter is slightly greater, as shown in Table 60A.

**60A.  Everdur Rigid Conduit**
(American Brass, Anaconda Co.)

| Nominal size, in. | Nominal dimensions, in. | | | Unit length without couplings, ft.-in. | Min weight of 10 unit lengths of conduit with couplings attached, lb |
|---|---|---|---|---|---|
| | Diameter | | Wall thickness | | |
| | Outside | Inside | | | |
| ¼ | 0.540 | 0.382 | 0.079 | 9-11½ | 40 |
| ⅜ | 0.675 | 0.503 | 0.086 | 9-11½ | 55 |
| ½ | 0.840 | 0.636 | 0.102 | 9-11¼ | 86 |
| ¾ | 1.050 | 0.834 | 0.108 | 9-11¼ | 115 |
| 1 | 1.315 | 1.075 | 0.120 | 9-11 | 164 |
| 1¼ | 1.660 | 1.382 | 0.139 | 9-11 | 242 |
| 1½ | 1.900 | 1.614 | 0.143 | 9-11 | 288 |
| 2 | 2.375 | 2.077 | 0.149 | 9-11 | 380 |
| 2½ | 2.875 | 2.519 | 0.178 | 9-10½ | 553 |
| 3 | 3.500 | 3.084 | 0.208 | 9-10½ | 790 |
| 3½ | 4.000 | 3.548 | 0.226 | 9-10¼ | 980 |
| 4 | 4.500 | 4.026 | 0.237 | 9-10¼ | 1,160 |

### 60B.  Intermediate Metal Conduit
(Allied Tube & Conduit Corp.)

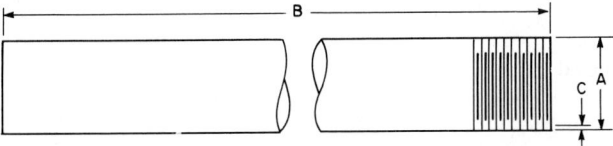

| Trade size, in | $A$, Outside diameter, in | | | $B$, Length of finished conduit without coupling, in | $C$, Wall thickness, in[a] | Nominal weight per 100, lb[b] |
|---|---|---|---|---|---|---|
| | Nominal | Minimum (−.005) | Maximum (+.005) | | | |
| ½ | 0.815 | 0.810 | 0.820 | 119¾ | 0.070 | 57 |
| ¾ | 1.029 | 1.024 | 1.034 | 119¾ | 0.075 | 78 |
| 1 | 1.290 | 1.285 | 1.295 | 119 | 0.085 | 112 |
| 1¼ | 1.638 | 1.630 | 1.645 | 119 | 0.085 | 144 |
| 1½ | 1.883 | 1.875 | 1.890 | 119 | 0.090 | 176 |
| 2 | 2.360 | 2.352 | 2.367 | 119 | 0.095 | 235 |
| 2½ | 2.857 | 2.847 | 2.867 | 118½ | 0.130 | 393 |
| 3 | 3.476 | 3.466 | 3.486 | 118½ | 0.130 | 483 |
| 3½ | 3.971 | 3.961 | 3.981 | 118¼ | 0.130 | 561 |
| 4 | 4.466 | 4.456 | 4.476 | 118¼ | 0.130 | 625 |

[a]The wall thickness is + .015 and −.000 for IMC.

[b]Because of the specification of tube dimensions, weight specifications are not part of Underwriters Laboratories (UL) Standard 1242 but have been included for comparative purposes. At regular production intervals, UL inspects wall thicknesses of all trade sizes with a micrometer to check consistency with the specified standard tolerance.

### 61.  Plastic-Coated Rigid Conduit
(General Electric Co.)

| Size, in | Weight, lb/ 1000 ft | Ft per bundle | External diameter, in |
|---|---|---|---|
| ½ | 855 | 100 | 0.900 |
| ¾ | 1119 | 50 | 1.110 |
| 1 | 1701 | 50 | 1.375 |
| 1¼ | 2247 | 30 | 1.720 |
| 1½ | 2688 | 10 | 1.960 |
| 2 | 3601 | 10 | 2.435 |
| 2½ | 6132 | 10 | 2.935 |
| 3 | 7860 | 10 | 3.560 |
| 3½ | 9433 | 10 | 4.060 |

**62. Rigid steel conduit is made in three types:** white (galvanized), black (enameled), and green (sheradized). The white galvanized conduit is the type to be used where exposed to the weather, when embedded in concrete, or when installed in wet locations. The black conduit, which is coated with black enamel, is the type commonly used for general interior wiring. The green (sheradized) conduit has a special corrosion-resistant coating. Except for special conditions no size smaller than ½ in is allowed.

## 63. Standard Rigid Wrought-Iron or Steel Conduit and Couplings

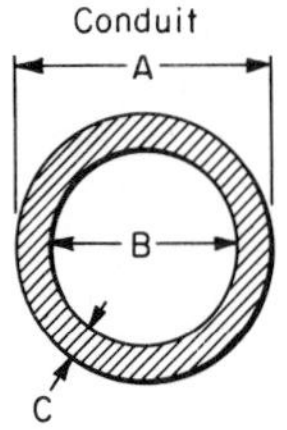


**Fig. 9-24** Section of conduit.

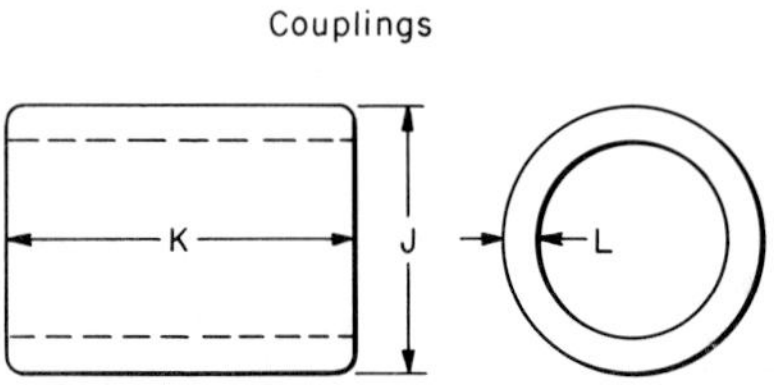


**Fig. 9-25** Conduit couplings.

| Rigid steel conduit | | | | | | Couplings | | |
|---|---|---|---|---|---|---|---|---|
| Trade size, in. | *A*, Approx. OD, in. | *B*, Approx. ID, in. | *C*, Approx. wall thickness, in. | Threads per in. | Approx wt per 1,000 ft, lb | *J*, OD, in. | *K*, Length, in. | Approx. wt per 100, lb |
| 1/2 | 0.840 | 0.622 | 0.109 | 14 | 820 | 1.010 | 1 9/16 | 14 |
| 3/4 | 1.050 | 0.824 | 0.113 | 14 | 1,120 | 1.250 | 1 5/8 | 24 |
| 1 | 1.315 | 1.049 | 0.133 | 11 1/2 | 1,600 | 1.525 | 2 | 39 |
| 1 1/4 | 1.660 | 1.380 | 0.140 | 11 1/2 | 2,160 | 1.869 | 2 1/16 | 47 |
| 1 1/2 | 1.900 | 1.610 | 0.145 | 11 1/2 | 2,680 | 2.155 | 2 1/16 | 66 |
| 2 | 2.375 | 2.067 | 0.154 | 11 1/2 | 3,500 | 2.730 | 2 1/8 | 105 |
| 2 1/2 | 2.875 | 2.469 | 0.203 | 8 | 5,600 | 3.250 | 3 1/8 | 180 |
| 3 | 3.500 | 3.068 | 0.216 | 8 | 7,120 | 4.000 | 3 1/4 | 300 |
| 3 1/2 | 4.000 | 3.548 | 0.226 | 8 | 8,520 | 4.500 | 3 3/8 | 390 |
| 4 | 4.500 | 4.026 | 0.237 | 8 | 10,300 | 5.000 | 3 1/2 | 400 |
| 5 | 5.563 | 5.047 | 0.258 | 8 | 13,910 | 6.296 | 3 3/4 | 760 |
| 6 | 6.625 | 6.065 | 0.280 | 8 | 18,500 | 7.390 | 4 | 1075 |

NOTE All tubes are 10 ft long, threaded both ends, with coupling on one end.

**64. Data on conduit couplings and elbows** are given in Secs. 63, 66A, and 71. The weight columns are convenient for estimating transportation charges. The internal area is useful in determining the size of conduit for unusual combinations of conductors, as explained in Sec. 79. Dimensions of elbows and couplings are often used in laying out work on the drawing board or in estimating clearances in advance. Table 67 gives the dimensions of standard conduit threads.

Running threads are not allowed on conduit for connection at couplings.

**65. Rigid aluminum conduit.** The use of rigid aluminum conduit has gained wide acceptance because of its light weight, excellent grounding conductivity, ease of threading, bending, and installation, resistance to corrosion, and low losses for installed ac circuits. Installations of rigid aluminum conduit require no maintenance, painting, or protective treatment in most applications. Because of its high resistance to corrosion this conduit should be used in many severely corrosive industrial environments such as sewerage plants, water treatment stations, filtration plants, many chemical plants, and installations around salt water.

When aluminum conduit is buried in concrete or mortar, a limited chemical reaction on the conduit surface forms a self-stopping coating. This prevents significant corrosion for the life of the structure. However, calcium chloride or similar soluble chlorides sometimes are used to speed concrete setting. In limiting the use of such speeding agents, the American Concrete Institute and most building codes recognize that embedded metals can be damaged by chlorides. As a result, if aluminum conduit is to be buried in concrete, the installer should be absolutely sure that the concrete will contain no chlorides. If there is any doubt, rigid steel conduit should be used, because even though chlorides can damage steel conduit to some degree, this will not lead to cracking or spalling of concrete.

Although aluminum conduit can be buried safely in many soils, precautions are recommended because of unpredictable moisture, stray electric current, and chemical variations in almost every soil. To assure protection when buried directly in earth, aluminum or steel conduit should be coated with bituminous or asphalt paint, wrapped with plastic tape, or encased in a chloride-free concrete envelope.

Being a nonmagnetic metal, aluminum conduit reduces voltage drop in installed copper or aluminum wire up to 20 percent of a corresponding steel-conduit installation when ac circuits are involved.

Tables 66 and 66A list important data on rigid aluminum conduit.

**66.  Aluminum Rigid Conduit: Nominal 10-Ft Lengths**

| Trade size | Lb per 100 ft [a] | No. per bundle | Lb per bundle [a] | Master shipping package | | | |
|---|---|---|---|---|---|---|---|
| | | | | Bundles | Pieces | Feet | Lb [a] |
| 1/2 | 29.8 | 10 | 29.80 | 20 | 200 | 2,000 | 596 |
| 3/4 | 39.8 | 10 | 39.80 | 20 | 200 | 2,000 | 796 |
| 1 | 58.9 | 10 | 58.90 | 10 | 100 | 1,000 | 589 |
| 1 1/4 | 79.8 | 5 | 39.90 | 10 | 50 | 500 | 399 |
| 1 1/2 | 95.6 | 5 | 47.80 | 10 | 50 | 500 | 478 |
| 2 | 128.8 | 5 | 64.40 | 9 | 45 | 450 | 580 |
| 2 1/2 | 204.7 | 1 | 20.47 | Loose | 30 | 300 | 614 |
| 3 | 268.0 | 1 | 26.80 | Loose | 20 | 200 | 536 |
| 3 1/2 | 321.3 | 1 | 32.13 | Loose | 20 | 200 | 642 |
| 4 | 382.1 | 1 | 38.21 | Loose | 20 | 200 | 764 |
| 5 | 521.5 | 1 | 52.15 | Loose | 8 | 80 | 417 |
| 6 | 677.5 | 1 | 67.75 | Loose | 6 | 60 | 406 |

[a] Nominal shipping weights.

**66A.  Aluminum Rigid Conduit Couplings**

| Trade size, inches | Outside diameter, inches | Length, inches | Nominal weight per 100 pieces, lb |
|---|---|---|---|
| 1/2 | 1 5/64 | 1.56 | 6.1 |
| 3/4 | 1 21/64 | 1.62 | 9.1 |
| 1 | 1 9/16 | 2.00 | 12.5 |
| 1 1/4 | 1 0 1/64 | 2.06 | 18.9 |
| 1 1/2 | 2 7/32 | 2.06 | 23.3 |
| 2 | 2 3/4 | 2.12 | 34.6 |
| 2 1/2 | 3 9/32 | 3.12 | 68.3 |
| 3 | 3 15/16 | 3.25 | 91.4 |
| 3 1/2 | 4 7/16 | 3.37 | 108.0 |
| 4 | 5 | 3.50 | 142.0 |
| 5 | 6 7/32 | 3.75 | 241.9 |
| 6 | 7 5/16 | 4.00 | 321.0 |

### 67. Standard Conduit: Dimensions of Threads

| Trade size of conduit, inches | Number of threads per inch | Total length $L_4$ of threads, inches[a] | Effective length $L_2$ of thread, inches | Pitch diameter $E_0$ at end of conduit, inches[b] |
|---|---|---|---|---|
| ¼ | 18 | 0.59 | 0.40 | 0.477 |
| ⅜ | 18 | 0.60 | 0.41 | 0.612 |
| ½ | 14 | 0.78 | 0.53 | 0.758 |
| ¾ | 14 | 0.79 | 0.55 | 0.968 |
| 1 | 11½ | 0.98 | 0.68 | 1.214 |
| 1¼ | 11½ | 1.01 | 0.71 | 1.557 |
| 1½ | 11½ | 1.03 | 0.72 | 1.796 |
| 2 | 11½ | 1.06 | 0.76 | 2.269 |
| 2½ | 8 | 1.57 | 1.14 | 2.720 |
| 3 | 8 | 1.63 | 1.20 | 3.341 |
| 3½ | 8 | 1.68 | 1.25 | 3.838 |
| 4 | 8 | 1.73 | 1.30 | 4.334 |
| 4½ | 8 | 1.78 | 1.35 | 4.831 |
| 5 | 8 | 1.84 | 1.41 | 5.391 |
| 6 | 8 | 1.95 | 1.51 | 6.446 |

[a] A minus tolerance of one thread applies to the total length of threads.
[b] A tolerance of ±0.005 in. applies to the pitch diameter.

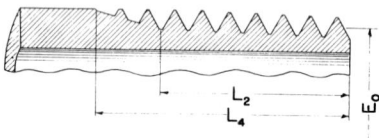

**Fig. 9-26**  Conduit threads. The taper of threads shall be ¾ in/ft, and the perfect thread shall be tapered for its entire length.

**68. Bushings are required** to protect wires from abrasion where a conduit enters a box or other fitting unless the design of the box or fitting is such as to afford equivalent protection.

Connections of rigid conduit to enclosures can be made by a single locknut (outside) and a single bushing (inside) if the bushing is tightly fitted against the wall of the enclosure. Double locknuts (one inside and one outside) and a bushing always make a better connection than the single-locknut–bushing method. The **NEC** requires the double-locknut method if the circuit voltage exceeds 250 V to ground or if wholly insulated bushings are utilized.

When ungrounded conductors of No. 4 and larger size enter a raceway in a cabinet, pull box, junction box, or auxiliary gutter, the conductors shall be protected by a substantial bushing providing a smoothly rounded insulating surface. Figures 9-27 and 9-28 show two types of bushings that provide insulation protection. The bushing in Fig. 9-27 is wholly insulated and constructed of a tough phenolic with reinforced fibers. The bushing in Fig. 9-28 is a metal grounding bushing with provisions for installing a bonding jumper from the bushing lug to a metal enclosure. The throat, or collar, is of insulating material to protect wires.

**Fig. 9-27**  Wholly insulated conduit bushing.
[Union Insulating Co.]

**Fig. 9-28**  Metal grounding bushing
with insulated throat. [Midwest Electric Mfg. Corp.]

Tables 69 and 73 are useful when cutting knockouts into enclosures. The spacings
listed in Table 69 allow enough room between adjacent knockouts for locknuts and bushings to be attached.

### 69. Spacings for Conduit with Given Clearances and Punched-Steel Conduit Locknuts

(All dimensions in inches)

The *clearance*, *C*, values are based on the *D* values given at the top of the next page.

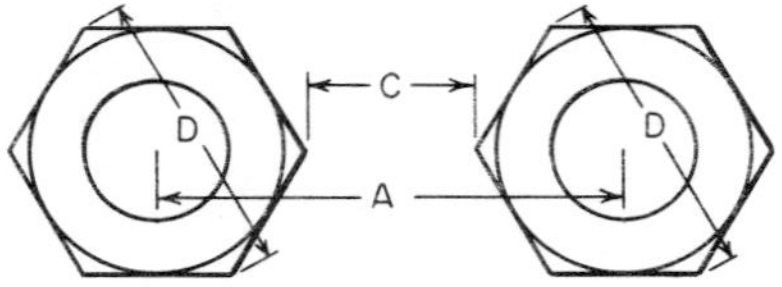


**Fig. 9-29** End view.

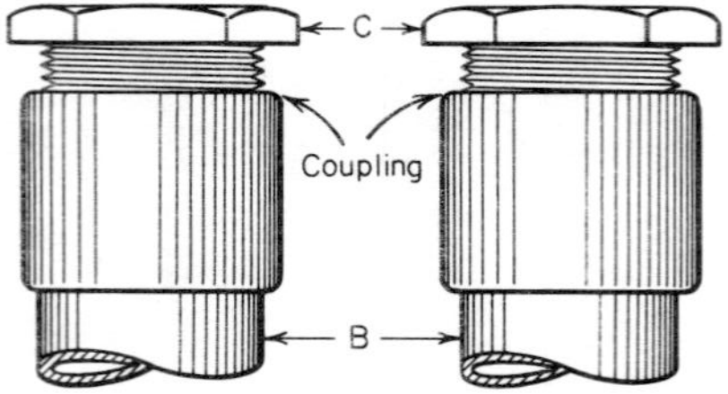


**Fig. 9-30** Elevation.

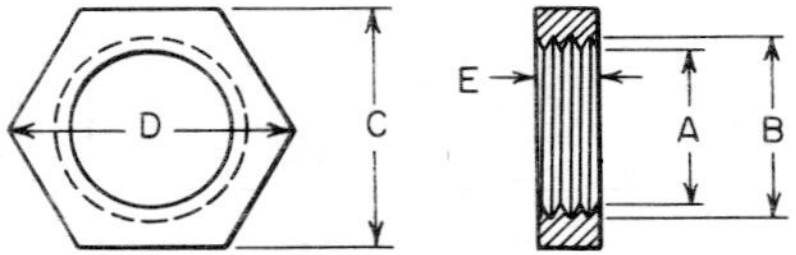


**Fig. 9-31** Conduit locknut.

**Spacings**

*C* = 1/4 in.

| Size conduit | | 1/2 | 3/4 | 1 | 1 1/4 | 1 1/2 | 2 | 2 1/2 | 3 |
|---|---|---|---|---|---|---|---|---|---|
| 1/2 | *A* | 1 3/8 | 1 1/2 | 1 5/8 | 1 13/16 | 1 15/16 | 2 1/4 | 2 9/16 | 2 7/8 |
| | *B* | 0.53 | 0.55 | 0.54 | 0.55 | 0.56 | 0.64 | 0.70 | 0.70 |
| 3/4 | *A* | 1 1/2 | 1 5/8 | 1 3/4 | 1 15/16 | 2 1/16 | 2 3/8 | 2 11/16 | 3 |
| | *B* | 0.55 | 0.57 | 0.56 | 0.58 | 0.58 | 0.66 | 0.72 | 0.72 |
| 1 | *A* | 1 3/8 | 1 3/4 | 1 7/8 | 2 1/16 | 2 3/16 | 2 1/2 | 2 13/16 | 3 1/8 |
| | *B* | 0.55 | 0.57 | 0.56 | 0.58 | 0.58 | 0.66 | 0.72 | 0.72 |
| 1 1/4 | *A* | 1 13/16 | 1 15/16 | 2 1/16 | 2 1/4 | 2 3/8 | 2 11/16 | 3 | 3 3/8 |
| | *B* | 0.55 | 0.58 | 0.58 | 0.58 | 0.59 | 0.67 | 0.73 | 0.79 |
| 1 1/2 | *A* | 1 15/16 | 2 1/16 | 2 3/16 | 2 3/8 | 2 1/2 | 2 13/16 | 3 1/8 | 3 1/2 |
| | *B* | 0.56 | 0.58 | 0.58 | 0.59 | 0.59 | 0.68 | 0.74 | 0.79 |
| 2 | *A* | 2 1/4 | 2 3/8 | 2 1/2 | 2 11/16 | 2 13/16 | 3 1/8 | 3 7/16 | 3 3/4 |
| | *B* | 0.64 | 0.66 | 0.65 | 0.67 | 0.68 | 0.75 | 0.81 | 0.81 |
| 2 1/2 | *A* | 2 9/16 | 2 11/16 | 2 13/16 | 3 | 3 1/8 | 3 7/16 | 3 3/4 | 4 1/16 |
| | *B* | 0.70 | 0.72 | 0.72 | 0.73 | 0.74 | 0.81 | 0.87 | 0.88 |
| 3 | *A* | 2 7/8 | 3 | 3 1/8 | 3 3/8 | 3 1/2 | 3 3/4 | 4 1/16 | 4 3/16 |
| | *B* | 0.70 | 0.72 | 0.72 | 0.79 | 0.79 | 0.81 | 0.88 | 1.06 |
| 3 1/2 | *A* | 3 1/4 | 3 3/8 | 3 1/2 | 3 11/16 | 3 13/16 | 4 1/8 | 4 7/16 | 4 7/8 |
| | *B* | 0.83 | 0.86 | 0.85 | 0.86 | 0.87 | 0.94 | 1.00 | 1.12 |

*C* = 3/8 in.

| Size conduit | | 1/2 | 3/4 | 1 | 1 1/4 | 1 1/2 | 2 | 2 1/2 | 3 |
|---|---|---|---|---|---|---|---|---|---|
| 1/2 | *A* | 1 1/2 | 1 5/8 | 1 3/4 | 1 15/16 | 2 1/16 | 2 3/8 | 2 11/16 | 3 |
| | *B* | 0.66 | 0.68 | 0.67 | 0.68 | 0.69 | 0.77 | 0.83 | 0.83 |
| 3/4 | *A* | 1 5/8 | 1 3/4 | 1 7/8 | 2 1/16 | 2 3/16 | 2 1/2 | 2 13/16 | 3 1/8 |
| | *B* | 0.68 | 0.70 | 0.69 | 0.71 | 0.71 | 0.79 | 0.85 | 0.85 |
| 1 | *A* | 1 3/4 | 1 7/8 | 2 | 2 3/16 | 2 5/16 | 2 5/8 | 2 15/16 | 3 1/4 |
| | *B* | 0.68 | 0.69 | 0.69 | 0.71 | 0.71 | 0.78 | 0.85 | 0.85 |
| 1 1/4 | *A* | 1 15/16 | 2 1/16 | 2 3/16 | 2 3/8 | 2 1/2 | 2 3/16 | 3 1/8 | 3 1/2 |
| | *B* | 0.68 | 0.71 | 0.71 | 0.71 | 0.72 | 0.80 | 0.87 | 0.92 |
| 1 1/2 | *A* | 2 1/16 | 2 3/16 | 2 5/16 | 2 1/2 | 2 5/8 | 2 15/16 | 3 1/4 | 3 5/8 |
| | *B* | 0.69 | 0.71 | 0.71 | 0.72 | 0.72 | 0.81 | 0.87 | 0.92 |
| 2 | *A* | 2 3/8 | 2 1/2 | 2 5/8 | 2 13/16 | 2 15/16 | 3 1/4 | 3 7/16 | 3 7/8 |
| | *B* | 0.77 | 0.79 | 0.78 | 0.80 | 0.81 | 0.88 | 0.88 | 0.94 |
| 2 1/2 | *A* | 2 11/16 | 2 13/16 | 2 15/16 | 3 1/8 | 3 1/4 | 3 9/16 | 3 7/8 | 4 3/16 |
| | *B* | 0.83 | 0.85 | 0.85 | 0.86 | 0.87 | 0.94 | 1.01 | 1.01 |
| 3 | *A* | 3 | 3 1/8 | 3 1/4 | 3 1/2 | 3 5/8 | 3 7/8 | 4 3/16 | 4 11/16 |
| | *B* | 0.83 | 0.85 | 0.85 | 0.92 | 0.92 | 0.94 | 0.99 | 0.99 |
| 3 1/2 | *A* | 3 3/8 | 3 1/2 | 3 5/8 | 3 13/16 | 3 15/16 | 4 1/4 | 4 9/16 | 5 |
| | *B* | 0.96 | 0.98 | 0.98 | 0.99 | 0.99 | 1.07 | 1.13 | 1.25 |

**Lock-nut dimensions**

| Nominal size of conduit, in. | Threads per in. | *A*, in. |
|---|---|---|
| 1/2 | 14 | 0.701 |
| 3/4 | 14 | 0.911 |
| 1 | 11 1/2 | 1.144 |
| 1 1/4 | 11 1/2 | 1.488 |
| 1 1/2 | 11 1/2 | 1.727 |
| 2 | 11 1/2 | 2.223 |
| 2 1/2 | 8 | 2.620 |
| 3[a] | 8 | 3.241 |

[a] The 3-in. lock nut is octagonal instead of hexagonal

| Nominal size of conduit, in. | *B*, in. | *C*, in. |
|---|---|---|
| | | oct. hex. |
| 1/2 | 0.815 | 1–1 1/16 |
| 3/4 | 1.025 | 1 1/4 |
| 1 | 1.283 | 1 1/2 |
| 1 1/4 | 1.627 | 2 1/32 |
| 1 1/2 | 1.866 | 2 1/4 |
| 2 | 2.339 | 2 23/32 |
| 2 1/2 | 2.820 | 3 1/4 |
| 3 | 3.441 | 4 11/32–3 11/16 |

## 69. Spacings for Conduit with Given Clearances and Punched-Steel Conduit Locknuts *(Continued)*

**C = 1/2 in.**

| Size conduit | | 1/2 | 3/4 | 1 | 1 1/4 | 1 1/2 | 2 | 2 1/2 | 3 |
|---|---|---|---|---|---|---|---|---|---|
| 1/2 | A | 1 5/8 | 1 3/4 | 1 7/8 | 2 1/16 | 2 3/16 | 2 1/2 | 2 13/16 | 3 1/8 |
| | B | 0.78 | 0.80 | 0.79 | 0.80 | 0.81 | 0.89 | 0.95 | 0.95 |
| 3/4 | A | 1 3/4 | 1 7/8 | 2 | 2 3/16 | 2 5/16 | 2 5/8 | 2 15/16 | 3 1/4 |
| | B | 0.80 | 0.82 | 0.81 | 0.83 | 0.83 | 0.91 | 0.97 | 0.97 |
| 1 | A | 1 7/8 | 2 | 2 1/8 | 2 5/16 | 2 7/16 | 2 3/4 | 3 1/16 | 3 3/8 |
| | B | 0.80 | 0.81 | 0.81 | 0.83 | 0.83 | 0.90 | 0.97 | 0.97 |
| 1 1/4 | A | 2 1/16 | 2 3/16 | 2 5/16 | 2 1/2 | 2 5/8 | 2 15/16 | 3 1/4 | 3 5/8 |
| | B | 0.80 | 0.83 | 0.83 | 0.83 | 0.84 | 0.92 | 0.99 | 1.04 |
| 1 1/2 | A | 2 3/16 | 2 5/16 | 2 7/16 | 2 5/8 | 2 3/4 | 3 1/16 | 3 3/8 | 3 3/4 |
| | B | 0.81 | 0.83 | 0.83 | 0.84 | 0.84 | 0.93 | 0.98 | 1.04 |
| 2 | A | 2 1/2 | 2 5/8 | 2 3/4 | 2 15/16 | 3 1/16 | 3 3/8 | 3 11/16 | 4 |
| | B | 0.89 | 0.91 | 0.90 | 0.92 | 0.93 | 1.00 | 1.00 | 1.06 |
| 2 1/2 | A | 2 13/16 | 2 15/16 | 3 1/16 | 3 1/4 | 3 3/8 | 3 11/16 | 4 | 4 5/16 |
| | B | 0.95 | 0.97 | 0.97 | 0.98 | 0.09 | 1.06 | 1.13 | 1.13 |
| 3 | A | 3 1/8 | 3 1/4 | 3 3/8 | 3 5/8 | 3 3/4 | 4 | 4 5/16 | 4 13/16 |
| | B | 0.95 | 0.97 | 0.97 | 1.04 | 1.04 | 1.06 | 1.14 | 1.14 |
| 3 1/2 | A | 3 1/2 | 3 5/8 | 3 3/4 | 3 15/16 | 4 1/16 | 4 1/8 | 4 11/16 | 5 5/16 |
| | B | 1.01 | 1.04 | 1.03 | 1.04 | 1.11 | 1.13 | 1.31 | 1.31 |

**C = 5/8 in.**

| Size conduit | | 1/2 | 3/4 | 1 | 1 1/4 | 1 1/2 | 2 | 2 1/2 | 3 |
|---|---|---|---|---|---|---|---|---|---|
| 1/2 | A | 1 3/4 | 1 7/8 | 2 | 2 3/16 | 2 5/16 | 2 5/8 | 2 15/16 | 3 1/4 |
| | B | 0.91 | 0.93 | 0.92 | 0.93 | 0.94 | 1.02 | 1.08 | 1.08 |
| 3/4 | A | 1 7/8 | 2 | 2 1/8 | 2 5/16 | 2 7/16 | 2 3/4 | 3 1/16 | 3 3/8 |
| | B | 0.93 | 0.95 | 0.94 | 0.96 | 0.96 | 1.04 | 1.10 | 1.10 |
| 1 | A | 2 | 2 1/8 | 2 1/4 | 2 7/16 | 2 9/16 | 2 7/8 | 3 3/16 | 3 1/2 |
| | B | 0.92 | 0.94 | 0.94 | 0.96 | 0.96 | 1.03 | 1.10 | 1.10 |
| 1 1/4 | A | 2 3/16 | 2 5/16 | 2 7/16 | 2 5/8 | 2 3/4 | 3 1/16 | 3 3/8 | 3 3/4 |
| | B | 0.93 | 0.96 | 0.96 | 0.96 | 0.97 | 1.05 | 1.12 | 1.17 |
| 1 1/2 | A | 2 5/16 | 2 7/16 | 2 9/16 | 2 3/4 | 2 7/8 | 3 3/16 | 3 1/2 | 4 |
| | B | 0.94 | 0.96 | 0.96 | 0.97 | 0.97 | 1.06 | 1.11 | 1.17 |
| 2 | A | 2 5/8 | 2 3/4 | 2 7/8 | 3 1/16 | 3 3/16 | 3 1/2 | 3 13/16 | 4 1/4 |
| | B | 1.02 | 1.04 | 1.03 | 1.05 | 1.05 | 1.13 | 1.13 | 1.19 |
| 2 1/2 | A | 2 15/16 | 3 1/16 | 3 3/16 | 3 3/8 | 3 1/2 | 3 13/16 | 4 1/8 | 4 1/2 |
| | B | 1.08 | 1.10 | 1.10 | 1.11 | 1.12 | 1.19 | 1.26 | 1.26 |
| 2 | A | 3 1/4 | 3 3/8 | 3 1/2 | 3 3/4 | 4 | 4 1/4 | 4 1/2 | 4 15/16 |
| | B | 1.08 | 1.10 | 1.10 | 1.17 | 1.17 | 1.19 | 1.24 | 1.24 |
| 3 1/2 | A | 3 5/8 | 3 3/4 | 3 7/8 | 4 1/16 | 4 3/16 | 4 1/2 | 4 13/16 | 5 1/4 |
| | B | 1.14 | 1.17 | 1.16 | 1.24 | 1.24 | 1.32 | 1.39 | 1.44 |

| Nominal size of conduit, in. | D, in. | E, in. |
|---|---|---|
| | oct. hex. | oct. hex. |
| 1/2 | 1 1/16–1 7/32 | 5/32–1/8 |
| 3/4 | 1 7/16 | 5/32 |
| 1 | 1 23/32 | 3/16 |
| 1 1/4 | 2 5/16 | 3/16 |
| 1 1/2 | 2 9/16 | 3/16 |
| 2 | 3 1/8 | 7/32 |
| 2 1/2 | 3 3/4 | 1/4 |
| 3 | 4 11/16–4 | 11/32–17/32 |

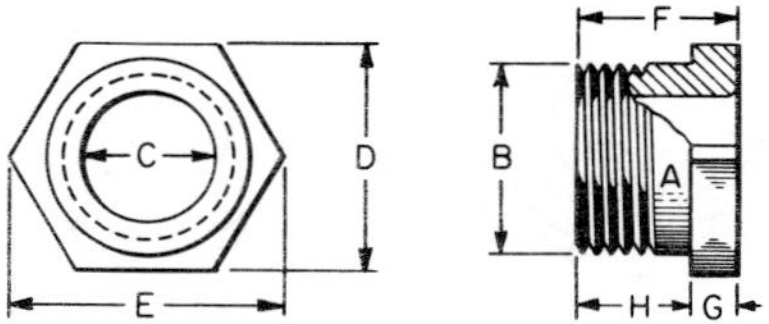


**Fig. 9-32** Conduit chase nipple.

## 70. Conduit Chase Nipples

| Size of conduit, in. | A, threads per in. | B, diam of threads, in. | C, in. | D, in. | E, in. | F, in. | G, in. | H, in. |
|---|---|---|---|---|---|---|---|---|
| 1/2 | 14.0 | 0.82 | 0.62 | 1.00 | 1.15 | 0.62 | 0.12 | 0.50 |
| 3/4 | 14.0 | 1.02 | 0.82 | 1.25 | 1.44 | 0.81 | 0.19 | 0.62 |
| 1 | 11.5 | 1.28 | 1.04 | 1.37 | 1.59 | 0.94 | 0.25 | 0.69 |
| 1 1/4 | 11.5 | 1.63 | 1.38 | 1.75 | 2.02 | 1.06 | 0.25 | 0.81 |
| 1 1/2 | 11.5 | 1.87 | 1.61 | 2.00 | 2.31 | 1.12 | 0.31 | 0.81 |
| 2 | 11.5 | 2.34 | 2.06 | 2.50 | 2.89 | 1.31 | 0.31 | 1.00 |
| 2 1/2 | 8.0 | 2.82 | 2.46 | 3.00 | 3.46 | 1.44 | 0.37 | 1.06 |
| 3 | 8.0 | 3.44 | 3.06 | 3.75 | 4.33 | 1.50 | 0.37 | 1.12 |
| 3 1/2 | 8.0 | 3.94 | 3.54 | 4.25 | 4.91 | 1.62 | 0.44 | 1.19 |

### 71.  Rigid Conduit Elbows

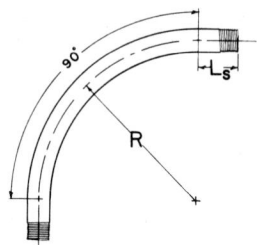

**Fig. 9-33**  Conduit elbows: minimum radius for factory ells and ells made by a one-shot bender.

**Standard-Radius Conduit Elbows**

| Trade size of conduit, inches | Minimum radius to center of tube, inches | Minimum straight length $L_S$ at each end, inches |
|:---:|:---:|:---:|
| ½ | 4 | 1½ |
| ¾ | 4½ | 1½ |
| 1 | 5¾ | 1⅞ |
| 1¼ | 7¼ | 2 |
| 1½ | 8¼ | 2 |
| 2 | 9½ | 2 |
| 2½ | 10½ | 3 |
| 3 | 13 | 3⅛ |
| 3½ | 15 | 3¼ |
| 4 | 16 | 3⅜ |
| 5 | 24 | 3⅝ |
| 6 | 30 | 3¾ |

### 72.  Conduit Nipples

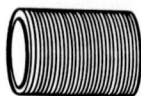  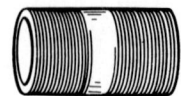

a. Close Nipples          b. Meter Service Nipples          c. Short Nipples

**Fig. 9-34**  Conduit nipples.

| Size nipple, in. | Close | | Short | | 2 in. long | 3 in. long | 4 in. long | 5 in. long | 6 in. long | 8 in. long | 10 in. long | 12 in. long |
|:---:|:---:|:---:|:---:|:---:|:---:|:---:|:---:|:---:|:---:|:---:|:---:|:---:|
| | Length | Wt, lb per 100 pieces | Length | Wt, lb per 100 pieces | Wt, lb per 100 pieces | Wt, lb per 100 pieces | Wt, lb per 100 pieces | Wt, lb per 100 pieces | Wt, lb per 100 pieces | Wt, lb per 100 pieces | Wt, lb per 100 pieces | Wt, lb per 100 pieces |
| ½ | 1⅛ | 7 | 1½ | 9 | 12 | 19 | 25 | 32 | 38 | 51 | 64 | 77 |
| ¾ | 1⅜ | 11 | 2 | 17 | … | 25 | 34 | 43 | 51 | 68 | 85 | 103 |
| 1 | 1½ | 19 | 2 | 25 | … | 38 | 51 | 63 | 76 | 102 | 127 | 153 |
| 1¼ | 1⅝ | 28 | 2½ | 43 | … | 51 | 69 | 86 | 103 | 138 | 172 | 207 |
| 1½ | 1¾ | 36 | 2½ | 51 | … | 62 | 82 | 103 | 124 | 165 | 206 | 248 |
| 2 | 2 | 55 | 2½ | 69 | … | 83 | 111 | 139 | 167 | 228 | 278 | 334 |
| 2½ | 2½ | 110 | 3 | 132 | … | … | 176 | 220 | 264 | 353 | 441 | 529 |
| 3 | 2⅝ | 151 | 3 | 173 | … | … | 231 | 288 | 346 | 462 | 577 | 692 |
| 3½ | … | … | … | … | … | … | … | … | 418 | 558 | 697 | 837 |

### 73. Malleable-Iron Conduit Bushings

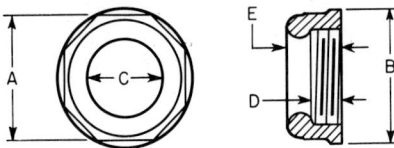

**Fig. 9-35**   Conduit bushing.

| Nominal size of conduit, in. | A, in. | B, in. | C, in. | D, in. | E, in. |
|---|---|---|---|---|---|
| $\frac{3}{8}$ | $\frac{25}{32}$ | $\frac{27}{32}$ | $\frac{15}{32}$ | $\frac{3}{16}$ | $1\frac{1}{32}$ |
| $\frac{1}{2}$ | $1\frac{5}{16}$ | $1\frac{1}{32}$ | $1\frac{9}{32}$ | $\frac{1}{4}$ | $1\frac{3}{32}$ |
| $\frac{3}{4}$ | $1\frac{5}{32}$ | $1\frac{1}{4}$ | $\frac{3}{4}$ | $\frac{1}{4}$ | $1\frac{5}{32}$ |
| 1 | $1\frac{13}{32}$ | $1\frac{17}{32}$ | $8\frac{1}{32}$ | $\frac{5}{16}$ | $1\frac{7}{32}$ |
| $1\frac{1}{4}$ | $1\frac{11}{16}$ | $1\frac{29}{32}$ | $1\frac{5}{16}$ | $\frac{3}{8}$ | $1\frac{9}{32}$ |
| $1\frac{1}{2}$ | $2\frac{1}{16}$ | $2\frac{5}{32}$ | $1\frac{17}{32}$ | $1\frac{3}{32}$ | $\frac{5}{8}$ |
| 2 | $2\frac{9}{16}$ | $2\frac{11}{16}$ | $1\frac{5}{16}$ | $1\frac{3}{32}$ | $2\frac{1}{32}$ |
| $2\frac{1}{2}$ | $3\frac{1}{16}$ | $3\frac{5}{32}$ | $2\frac{1}{32}$ | $1\frac{5}{32}$ | $\frac{3}{4}$ |
| 3 | $3\frac{11}{16}$ | $3\frac{25}{32}$ | $2\frac{27}{32}$ | $1\frac{7}{32}$ | $\frac{7}{8}$ |

**74. Table 70, giving the dimensions of conduit chase nipples,** is, since the function of the chase nipple is about the same as that of the bushing, used for the same purposes as the bushing table. The chase nipple screws into a coupling (see Fig. 9-30), while the bushing screws onto the threaded end of a length of conduit. The chase nipple is more compact than the bushing and hence is preferable for some work.

**75. Punched-steel locknuts are shown in Table 69.** Locknuts are used on conduit on the outside of the box whenever the conduit enters an outlet box, and their dimensions must often be known in laying out panel or outlet boxes so that proper turning clearances can be provided for the nuts.

**76. Table 69 of conduit spacings for different clearances between conduits and their locknuts or nipples** is exceedingly valuable to someone who is designing or erecting conduit work. From it the designer can determine directly just what the distance between centers of conduits should be for given clearances between nipples or conduit. These data are indispensable when laying out the centers of a row of holes through which conduit is to enter a panel box or in laying out the supports for a multiple-conduit run.

**77. Bends.** The NEC specifies that bends of rigid conduit shall be so made that the conduit will not be injured and that the internal diameter of the conduit will not be effectively reduced. The radius of the curve of the inner edge of any field bend shall not be less than shown in Table 78. See Table 71 for the minimum radius permitted for field bends made by one-shot benders.

The NEC also specifies that a run of conduit between outlet and outlet, between fitting and fitting, or between outlet and fitting shall not contain more than the equivalent of four quarter bends (360° total), including bends located immediately at the outlet or fitting.

### 78.  Radius of Conduit Field Bends, In

| Size of conduit, in. | Conductors without lead sheath | Conductors with lead sheath |
|:---:|:---:|:---:|
| 1/2 | 4 | 6 |
| 3/4 | 5 | 8 |
| 1 | 6 | 11 |
| 1¼ | 8 | 14 |
| 1½ | 10 | 16 |
| 2 | 12 | 21 |
| 2½ | 15 | 25 |
| 3 | 18 | 31 |
| 3½ | 21 | 36 |
| 4 | 24 | 40 |
| 5 | 30 | 50 |
| 6 | 36 | 61 |

**79. The NEC specifies the size of conduit to use,** depending upon the number, type, and size of conductors to be installed therein. The proper size of conduit can be determined from the following tables in Div. 11 (wiring at 600 V or less):

| | |
|---|---|
| 1.  For new work | Tables 30 to 33 |
| 2.  For lead-covered cables | Tables 34 and 36 |
| 3.  For combinations of conductors | Tables 34 and 35 |
| 4.  Dimensions and areas of conduits | Table 34 |
| 5.  Dimensions of rubber-covered and thermoplastic-covered conductors | Table 35 |

If bare conductors are to be installed in conduit, the dimensions of such conductors, as given in Table 61 of Div. 11, may be used in sizing a conduit.

**80. Outlets for current-consuming equipment and wiring devices** require the use of outlet boxes for concealed or exposed work, whereas conduit bodies (Fig. 9-38) are limited to exposed work.

**81. Outlet boxes** that are used for conduit wiring are of sheet steel, preferably coated with zinc. They not only hold the conduit ends firmly in position and form a pocket for enclosing wire joints but also constitute electrical connectors between the elements of the conduit system, all of which must be in good electrical contact. Each conduit run in an installation must terminate in an accessible outlet box. Outlet boxes are made in many different forms. For complete data and descriptions of outlet boxes see Div. 4. The required size of boxes is given in Sec. 320.

**82. All boxes should be installed** so that the outer edge of the box or the cover mounted on the box will come flush with the surface of the plaster. The top of an outlet or junction box must never be concealed, as the wire splices inside must be accessible for inspection. The conduit must be fastened securely to the box by using a locknut on the outside and a bushing on the inside of the box (see Sec. 68).

**83. Conduit junction boxes,** which in reality are nothing more than pull boxes on a large scale, are often very convenient at points where several conduit lines intersect, as for instance over a switchboard (Fig. 9-36) from which conduit lines radiate. The junction box is usually supported from the ceiling and is best made of sheet iron on an angle-iron frame. The sides should be held on with machine screws turning into tapped holes in the frame so that they can be readily removed. Round holes can be cut in the sheet-iron sides for the conduits, or, often preferably, slots can be provided instead. The conductors within the box can be carried from conduit outlet to conduit outlet in any direction desired, and the use of elbows and troublesome conduit crossings can thereby be avoided.

**84. Pull boxes can often be advantageously substituted for elbows** (Fig. 9-37).  Large elbows are expensive. If there are three or more right-angle turns in a run, a pull box should be inserted in any event. One pull box may be substituted for several elbows. Wire can be pulled in more readily when there are pull boxes. Pull boxes are generally made of sheet steel (Fig. 9-37, I). Boxes should be made and drilled in the shop, where proper tools are available, rather than on the job. The required size of pull boxes is given in Sec. 324.

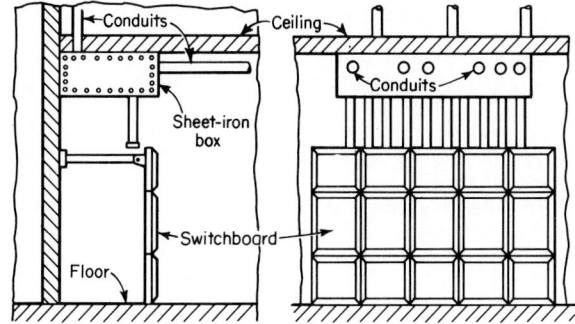

**Fig. 9-36**   A sheet-iron conduit junction box.

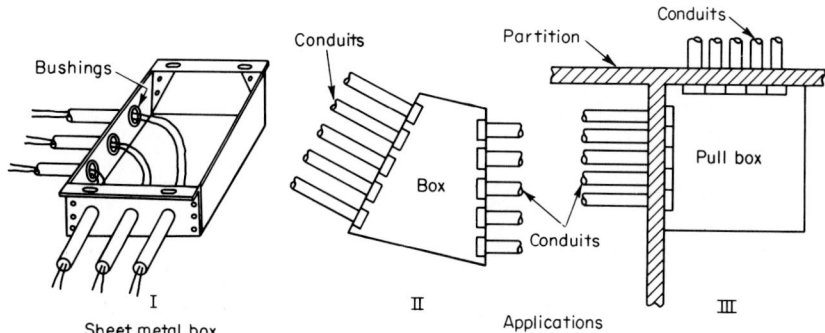

**Fig. 9-37**   Conduit pull boxes.

**85. Conduit bodies** are fittings used to adapt conduit to different situations and conditions. Figure 9-38 shows some popular fittings, the applications of which are obvious. The **NEC** specifies that every conduit outlet must be equipped with an outlet box or approved fitting. A fitting like that of I or II placed on a conduit end fulfills this requirement. Elbows, crosses, and tees, as in III, V, and VI, respectively, can be obtained fitted with either metal or composition or porcelain wire-hole covers or with outlet or other devices. The weather head of IV is used on the end of an outdoor piece of conduit into which wires enter. Its shape is such that wires must enter upwardly, preventing the entrance of water. The wires can be pulled into the conduit and the fitting slipped over them and attached to the flange without the necessity to turn the fitting. The fittings of VII and VIII can be used as pull boxes, to support switches, or for a number of other purposes. Everyone interested in wiring should have the catalogs of the fitting manufacturers. These illustrate a great number of fitting combinations and applications.

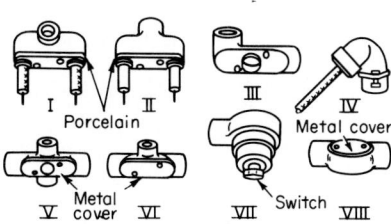

**Fig. 9-38**   Popular conduit fittings.

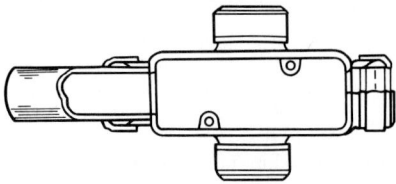

**Fig. 9-39**   No-thread conduit body. [Appleton Electric Co.]

Conduit fittings are made with their hubs for attachment to the conduit of either threaded or threadless types.

**86. Threadless fittings** (Fig. 9-39) can be used with unthreaded conduit. Tightening glands or setscrews clamp the conduit within the fitting. Threadless fittings materially reduce the labor cost of conduit installation in many instances.

The NEC requires that threadless fittings be made tight and that, if installed in wet places or if buried in masonry, concrete, or fill, they be of a type to prevent water from entering the conduit.

**87. Conduit should run as straight and direct as possible.** There should never be more than the equivalent of four right-angle bends between outlets or fittings.

**88. In installing exposed conduit runs** where there are several conduits in the run, it is usually better to carry the erection of all of them together rather than to complete one line before starting the others. If all are carried together, it is easier to keep all the raceways parallel, particularly at turns, and the chances are that the job will look better.

**89.  Galvanized-Iron Pipe Straps of the Two-Hole Type**

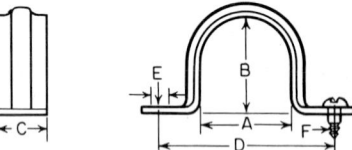

**Fig. 9-40**   Pipe strap.

| Nominal size of conduit, in. | *A*, width of opening, in. | *B*, height of opening, in. | *C*, width of strap, in. | *D*, distance between centers of screw holes, in. | *E*, diam of screw hole, in. | *F*, size of wood screw to use | Approximate number per lb |
|---|---|---|---|---|---|---|---|
| | | | | | | No.    in. | |
| 1/4 | 9/16 | 1 7/32 | 5/8 | 1 9/16 | 0.20 | 8 × 5/8 | 75 |
| 3/8 | 1 1/16 | 2 1/32 | 5/8 | 1 3/8 | 0.20 | 8 × 3/4 | 72 |
| 1/2 | 7/8 | 2 5/32 | 5/8 | 1 5/8 | 0.20 | 8 × 3/4 | 40 |
| 3/4 | 1 1/8 | 1 | 3/4 | 2 1/8 | 0.22 | 10 × 3/4 | 29 |
| 1 | 1 3/8 | 1 11/32 | 3/4 | 2 3/8 | 0.22 | 10 × 7/8 | 21 |
| 1 1/4 | 1 3/4 | 1 5/8 | 1 3/16 | 2 3/4 | 0.22 | 10 × 1 | 18 |
| 1 1/2 | 2 | 1 7/8 | 1 3/16 | 3 | 0.22 | 10 × 1 | 14 |
| 2 | 2 1/2 | 2 5/16 | 1 | 3 3/4 | 0.22 | 10 × 1 1/4 | 12 |
| 2 1/2 | 2 3/4 | 2 15/16 | 7/8 | 4 3/8 | 0.25 | 11 × 1 1/4 | 6 |

**90. Conduit can be supported** on surfaces with pipe straps made in two-hole (Fig. 9-40) and one-hole (Fig. 9-41) types. On wooden surfaces, wood screws secure the straps in position. On masonry surfaces, machine screws turning into lead expansion anchors can be used. (Refer to Div. 4 for types of expansion anchors.) Wooden plugs should never be used because no matter how well seasoned a plug appears to be, it will usually dry out to some extent and loosen in the hole. The dimensions in Table 89 are valuable, when laying out multiple-conduit runs, to determine the spacings necessary between the conduits to permit proper plac-

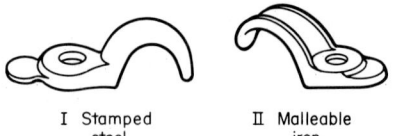

I  Stamped        II  Malleable
    steel                   iron

**Fig. 9-41**  One-hole pipe straps. [General Electric Supply Co.]

ing of the straps. The screw-hole dimensions enable one to order, in advance, screws of the proper diameters to support the straps.

**91. Location of conduit supports.** The NEC states that rigid metal conduit shall be firmly secured within 3 ft (0.9 m) of each outlet box, junction box, cabinet, or fitting. Such conduit shall be supported at least every 10 ft (3 m) except that straight runs of conduit made up with approved threaded couplings may be secured as indicated in Table 92, provided such fastening prevents transmission of stresses to terminations when conduit is deflected between supports.

**92.   Spacing of Rigid-Metal-Conduit Supports**

| Conduit Size, In | Maximum Distance between Rigid-Metal-Conduit Supports, Ft |
|---|---|
| ½ and ¾ | 10 |
| 1 | 12 |
| 1¼ and 1½ | 14 |
| 2 and 2½ | 16 |
| 3 and larger | 20 |

**93. Some commercial I-beam conduit hangers** are shown in Figs. 9-42 and 9-43. The one at I is an I-beam clamp formed from wrought-iron strap. The hanger or clamp (the part

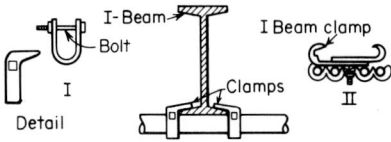

**Fig. 9-42**   Commercial conduit clamps.

that grips the beam) of that at II can be purchased of either stamped steel or malleable iron. The support (the yoke in which the conduits rest) is of malleable iron or spring steel and can be purchased to accommodate one or several conduits of different sizes.

**94. Conduit hangers and supports.** A variety of conduit hangers and supports and several applications are shown in Fig. 9-43. U-channel supports are ideal for supporting several runs of conduits. In laying out these supports consideration should be given to future conduit runs as well as those to be installed initially. It is a simple matter to provide U channels or trapeze hangers with additional space for future conduits. This procedure greatly reduces the cost of installing new conduit at a later date. With the U-channel system, as shown in Fig. 9-43, special clamps are slipped into the channel slot, and the top bolt of the clamp securely fastens the conduit to the U channel.

The U channel can be directly fastened to a wall or ceiling, or it can be attached to bolted threaded rod hangers, suspended from ceilings, roof structures, or similar members.

Another excellent application for the U channel is in suspended ceilings which contain lift-out ceiling panels. In modern construction these lift-out panels provide ready access to mechanical and electrical equipment within the suspended-ceiling area. Accordingly, it is important that conduits installed in such an area do not prevent the removal of panels or access to the area. Rod-suspended U channels provide the solution to conduit wiring in such areas.

Sections of U channel and associated fittings are available in aluminum or steel types. Another type of material that can be used for supports is slotted-angle–steel units. Numerous prepunched slots allow installers to bolt on rods, straps, and similar material without drilling holes. Slotted steel has unlimited applications in forming special structures, racks, braces, or similar items.

**95. Conduit in concrete buildings** (much of it, at any rate) can be installed while the building is being erected. The outlets should be attached to the forms, and the conduits between outlets should be attached to reinforcing steel with metal tie wires so that the conduit can be poured around them (Fig. 9-44). When several conduits pass through a wall, partition, or floor, a plugged sheet-metal tube (Fig. 9-45, I) should be set in the forms to provide a hole for them in the concrete. When a single conduit is to pass through, a nipple (Fig. 9-45, II) or a plugged sheet-metal tube can be set in the forms.

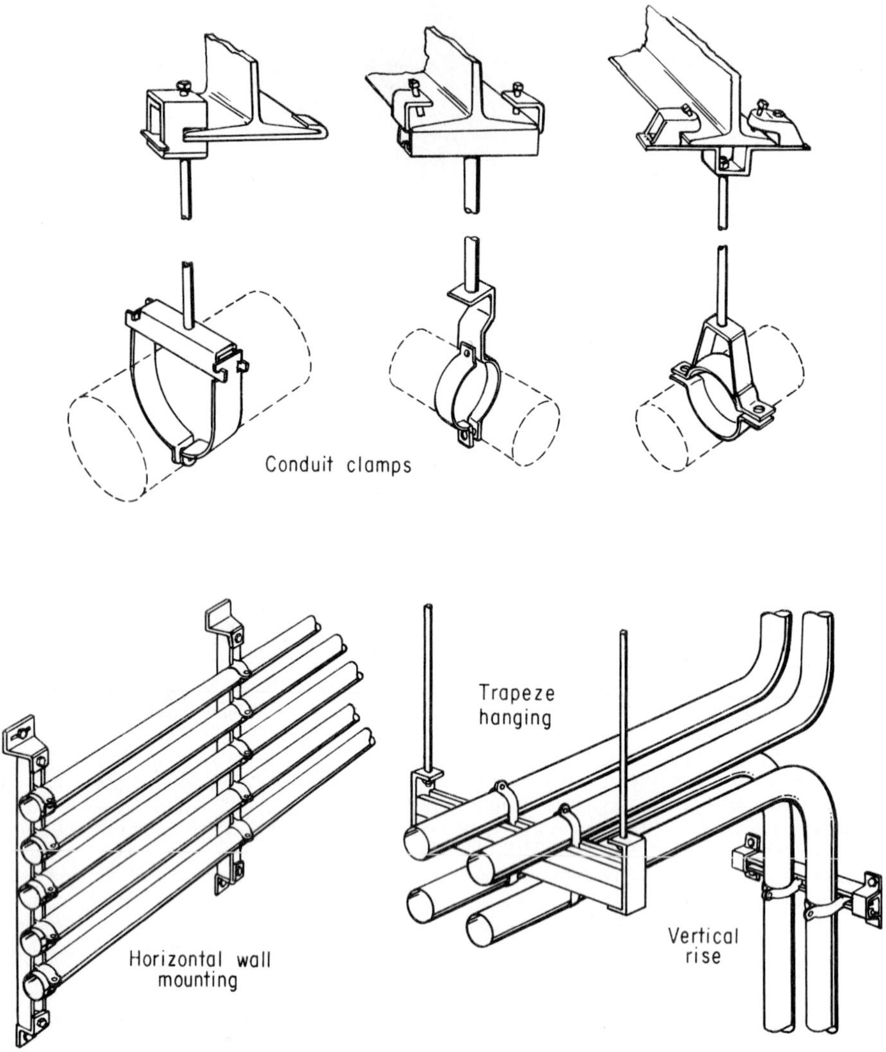

Conduit clamps

Trapeze
hanging

Horizontal wall
mounting

Vertical
rise

**Fig. 9-43** Beam clamps, rod hangers, and U-channel supports.

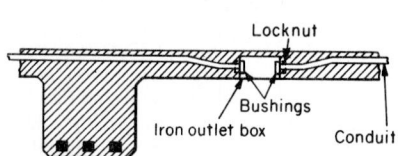

**Fig. 9-44** Conduit and outlet box in concrete.

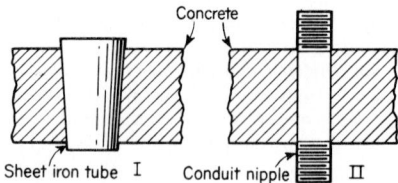

**Fig. 9-45** Methods of providing passage through concrete.

**96. Conductors in vertical conduits must be supported** within the conduit system as indicated in the following table.

| Conductor size | Conductors, ft | |
|---|---|---|
| | Aluminum or copper-clad aluminum | Copper |
| | Not greater than | |
| No. 18 through No. 8 | 100 | 100 |
| No. 6 through No. 0 | 200 | 100 |
| No. 00 through No. 0000 | 180 | 80 |
| 211,601 through 350,000 cmil | 135 | 60 |
| 350,001 through 500,000 cmil | 120 | 50 |
| 500,001 through 750,000 cmil | 95 | 40 |
| Above 750,000 cmil | 85 | 35 |

The following methods of supporting cables will satisfy NEC requirements:

1. Approved clamping devices are constructed of or employ insulated wedges inserted in the ends of the conduits (Fig. 9-46). With cables having varnished-cambric insulation, it may also be necessary to clamp the conductor.

2. Junction boxes may be inserted in the conduit system at the required intervals. Insulating supports of an approved type must be installed in them and secured in a satisfactory manner so as to withstand the weight of the conductors attached thereto. The boxes must be provided with proper covers.

3. The cables may be supported in junction boxes by deflecting them (Fig. 9-

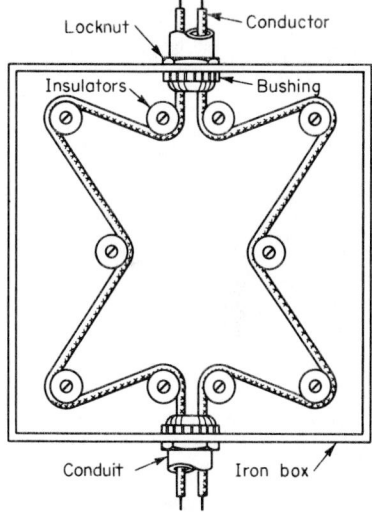

**Fig. 9-47** Supporting conductors in a vertical conductor run.

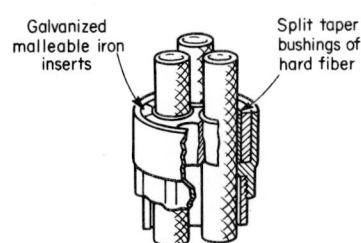

**Fig. 9-46** Cable support screwed on the end of a conduit.

47) not less than 90° and carrying them horizontally to a distance not less than twice the diameter of the cable, the cables being carried on two or more insulating supports and additionally secured thereto by tie wires if desired. When this method is used, the cables shall be supported at intervals not greater than 20 percent of those mentioned in the preceding tabulation.

**97. Properly bent conduit turns look better than elbows** and are therefore preferable for exposed work (see Fig. 9-48). If bends are formed to a chalk line, drawn as suggested in Sec. 98, the conduits can be made to lie parallel at a turn in a multiple run as shown in Fig. 9-48, II. If standard elbows are used, it is impossible to make them lie parallel at the turns. They will have an appearance similar to that shown in I.

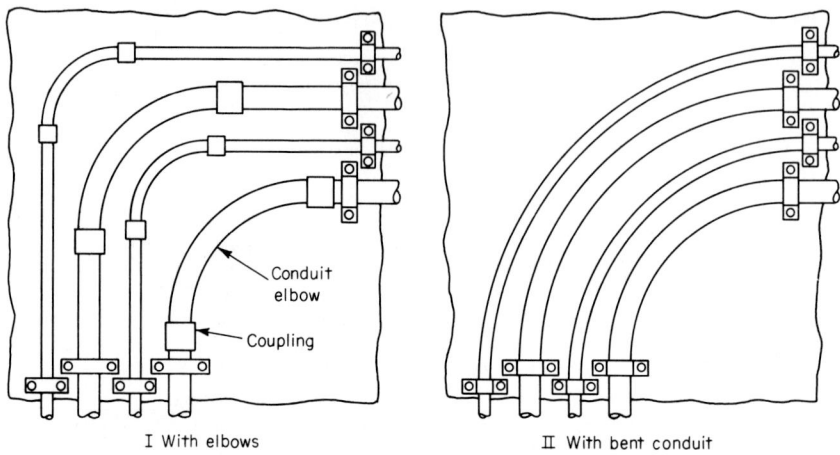

I With elbows          II With bent conduit

**Fig. 9-48**   Right-angle turns with elbows and with bent conduit.

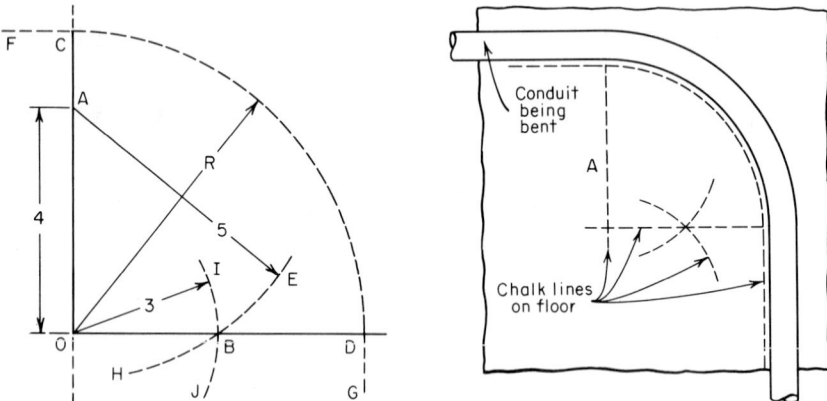

**Fig. 9-49**   Laying out a right angle.          **Fig. 9-50**   Forming a conduit to chalk lines.

**98. Laying out a right-angle conduit bend.**   Draw a chalk-line diagram of the contour of the bend on the floor as follows (see Fig. 9-49): Draw a base line *ÇO* of any length. Lay off *AO* 4 units long. (The units may be of any dimensions.) With a cord and a piece of chalk with *O* as a center and a radius of 3 units describe the arc *IJ*. With *A* as a center and a radius of 5 units describe the arc *EH*. The line *OD* drawn from *O* through *B*, the intersection of the two arcs, will be at right angles with *CO*.

*CO* and *OD* can now be prolonged for any desired distance. The arc *CD* is drawn with the cord and chalk with any required radius *R*. The conduit bend should lie parallel to this arc when the bend is laid on the floor for inspection as shown in Fig. 9-50. The radius of curvature of the inner edge of any field bend is required by the **NEC** to be not less than shown in Tables 71 and 78.

**99. Conduit hickeys and benders** are used to hand-bend smaller sizes of rigid steel conduit, rigid aluminum conduit, and the new Type IMC (intermediate metal conduit) in ½-, ¾-, and 1-in sizes, as well as in a 1¼-in size for small offsets. Benders and hickeys for thin-wall electrical metallic tubing (EMT) are widely used for all types of bends in sizes from ½ to 1¼ in inclusive. A typical thin-wall conduit bender, shown in

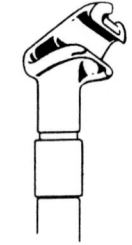

**Fig. 9-51**   The standard conduit hickey.

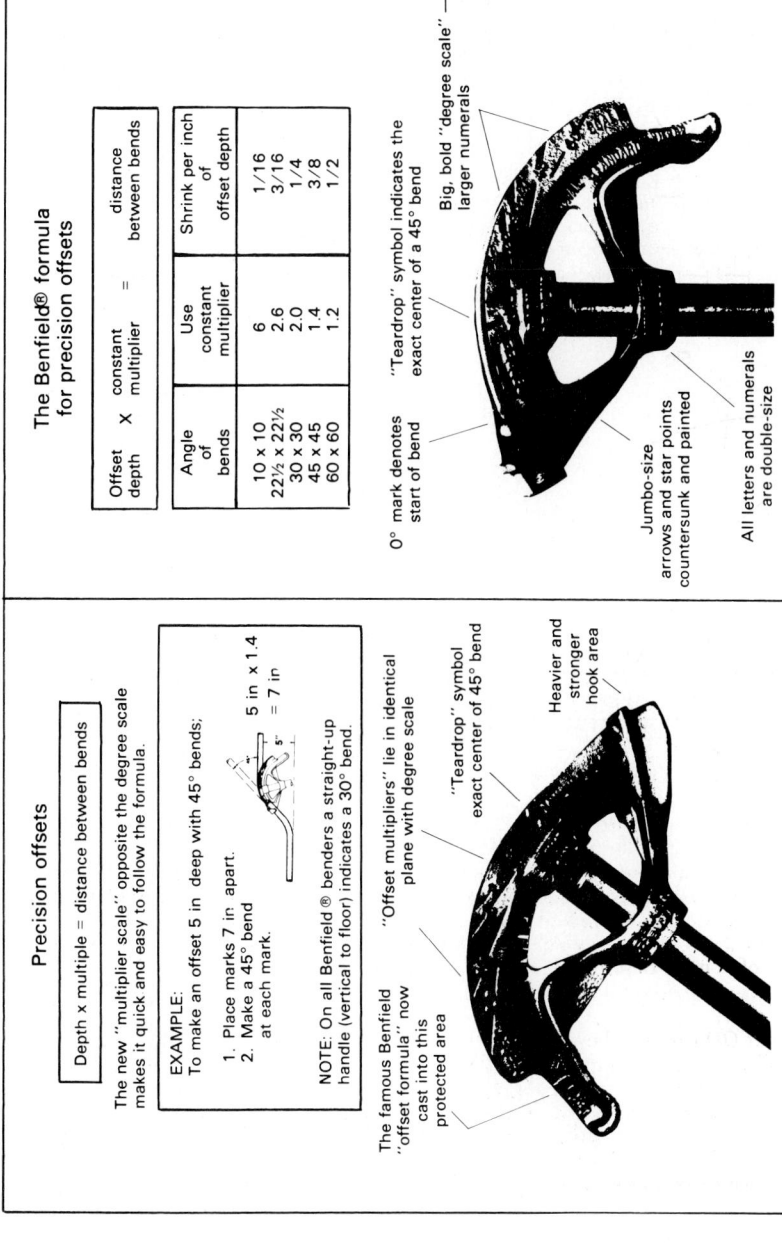

## The Benfield® formula for precision offsets

Offset depth × constant multiplier = distance between bends

| Angle of bends | Use constant multiplier | Shrink per inch of offset depth |
| --- | --- | --- |
| 10 × 10 | 6 | 1/16 |
| 22½ × 22½ | 2.6 | 3/16 |
| 30 × 30 | 2.0 | 1/4 |
| 45 × 45 | 1.4 | 3/8 |
| 60 × 60 | 1.2 | 1/2 |

"Teardrop" symbol indicates the exact center of a 45° bend

Big, bold "degree scale" — larger numerals

0° mark denotes start of bend

Jumbo-size arrows and star points countersunk and painted

All letters and numerals are double-size

## Precision offsets

Depth × multiple = distance between bends

The new "multiplier scale" opposite the degree scale makes it quick and easy to follow the formula.

EXAMPLE:
To make an offset 5 in deep with 45° bends;

1. Place marks 7 in. apart.
2. Make a 45° bend at each mark.

5 in × 1.4 = 7 in

NOTE: On all Benfield® benders a straight-up handle (vertical to floor) indicates a 30° bend.

The famous Benfield "offset formula" now cast into this protected area

"Offset multipliers" lie in identical plane with degree scale

"Teardrop" symbol exact center of 45° bend

Heavier and stronger hook area

**Fig. 9-52A** No. 1 Benfield bender for ½-in electrical metallic tubing (EMT). This tool makes the minimum National Electrical Code radius. [Jack Benfield, patent holder]

Fig. 9-52*A*, was invented by Jack Benfield of Fort Lauderdale, Florida. Listed below are the features of this bender that have greatly improved the art of precision-bending conduit and EMT over the years:

1. Bold cast-in symbols not only are recessed but also are paint-filled to serve as permanent bench marks or guide points for making accurate stub lengths, back-to-back bends, saddles, and offsets.

2. The Benfield formula for making precision offsets is printed in bold letters cast into the metal as a constant reference for the operator at point of use. The formula reads:

### OFFSET DEPTH X MULTIPLIER = DIST. BTWN BENDS

A cast-in degree scale on one side and a cooperating cast-in multiplier scale on the opposite side of the tool guide the operator in this precision offset technique.

3. A patented inner-hook contour prevents the tool from slipping. It will not chew into or nick galvanized surfaces or gouge aluminum conduits.

4. The hook area also includes a back-pusher contour which allows the bender to grab on in reverse to remove an overbend or to shift the bend from side to side.

5. The back of the hook has a double-width outrigger ground flat (like a T square in relation to the bender groove). This flat shoulder provides stability and maintains floor contact long enough to set the course of the bend in a plane vertical to the working surface.

6. For purposes of safety the socket for a handle has recessed threads. The entire tool is cast of tough pearlitic-alloy malleable iron with heavy sections and guard points to protect the groove from damage.

7. The extended nonskid foot pedal invites foot pressure and increases leverage by 75 percent.

8. The Benfield No. 1 bender (Fig. 9-52*A*) is designed for bending ½-in EMT only. A similar model, No. 2, is used for ¾-in EMT, and No. 3 is for bending 1-in EMT. All three tools bend to the minimum **NEC** radius. If larger radii are not objectionable, the No. 2 bender will bend ½-in rigid conduit. The No. 3 bender will bend ¾-in rigid conduit.

The Benfield No. 6 Powr-Jack® bender, shown in Fig. 9-52*B*, has every feature of the No. 2 bender plus an extended two-position power step that provides 200 percent more leverage, thereby extending the range for hand benders to include 1-in rigid or IMC conduit and 1¼-in EMT.

The Benfield No. 9 double-grooved bender (Fig. 9-52*D*) is shown along with a special extra-high-strength-steel handle. The tool is a one-piece malleable-iron casting with no moving parts. One groove bends ½-in EMT only; the other groove is for ¾-in EMT or ½-in rigid conduit. The No. 9 Benfield makes shorter radii than standard models, i.e., 3-in inside radius for ½-in EMT and a 4-in radius for ¾-in EMT and ½-in rigid conduit or IMC. Handles are now available in IPS ¾-, 1-, and 1¼-in sizes and are factory-engineered ready for use. They are marketed complete with an applied mylar-coated weatherproof bumper-sticker–type Zip Guide® to give the operator basic bending allowances at the point of use.

**Fig. 9-52*B***   No. 6 Benfield Powr-Jack® bender for 1-in rigid conduit or 1¼-in EMT. [Jack Benfield, patent holder]

**Fig. 9-52*C***   No. RH7 Benfield hickey for ½- or ¾-in rigid conduit. [Jack Benfield, patent holder]

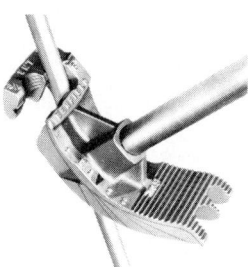

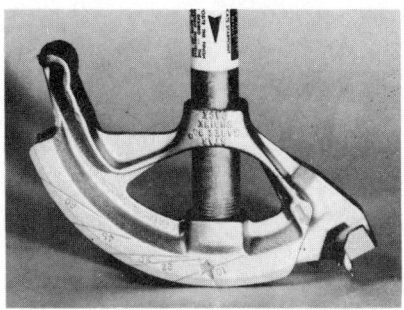

**Fig. 9-52D** No. 9 Benfield double-grooved bender for ½-in EMT and ½-in rigid conduit. Note that the handle has a weatherproof Zip Guide® plus an expanded end that permits its use as a stub straightener. [Jack Benfield, patent holder]

**Fig. 9-52E** Model No. 16 series especially designed for rigid steel conduit and Type IMC (intermediate metal conduit). It is heavier throughout than EMT benders and is made to bend to the minimum radius that the National Electrical Code allows. [Jack Benfield, patent holder]

A conduit new to the industry is Type IMC (intermediate metal conduit). To provide the right NEC radii for IMC, a new series of Benfield benders is being marketed. The new models bend all three types: rigid steel conduit, rigid aluminum conduit, and IMC intermediate metal conduit. Figure 9-52E, Benfield Model No. 16, shows an extra-heavy beefed-up hook plus extra-rugged sections throughout to withstand the rigors of bending rigid conduit and Type IMC conduit to the correct NEC radius. The No. 16 bends three conduit types: ½-in rigid steel, ½-in IMC, and ½-in aluminum rigid conduit. Larger models bend both rigid conduit and IMC in the ¾- and 1-in sizes. The No. 16 series tools are not suitable for bending EMT.

Figure 9-52F, the Benfield Zip Guide®, is an on-the-handle reference that speeds the work. This precalculated table for spacing between bends is accurate for any make of bender, whether hand-type, mechanical, or hydraulic. Zip Guides® are a definite aid for the electrician.

The Benfield No. 7 one-shot hickey (Fig. 9-52G) bends ½- and ¾-in rigid or IMC conduit. Model No. 8 bends 1- or 1¼-in rigid or IMC conduit. The unbreakable plumb bob

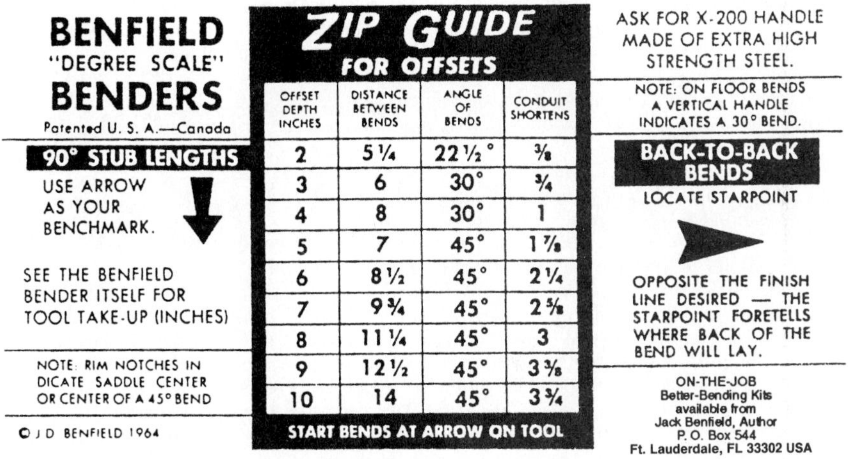

**BENFIELD**
"DEGREE SCALE"
**BENDERS**
Patented U. S. A.—Canada

**ZIP GUIDE**
**FOR OFFSETS**

ASK FOR X-200 HANDLE MADE OF EXTRA HIGH STRENGTH STEEL.

| | OFFSET DEPTH INCHES | DISTANCE BETWEEN BENDS | ANGLE OF BENDS | CONDUIT SHORTENS |
|---|---|---|---|---|
| **90° STUB LENGTHS** | 2 | 5¼ | 22½° | ⅜ |
| USE ARROW AS YOUR BENCHMARK. | 3 | 6 | 30° | ¾ |
| | 4 | 8 | 30° | 1 |
| | 5 | 7 | 45° | 1⅞ |
| SEE THE BENFIELD BENDER ITSELF FOR TOOL TAKE-UP (INCHES) | 6 | 8½ | 45° | 2¼ |
| | 7 | 9¾ | 45° | 2⅝ |
| | 8 | 11¼ | 45° | 3 |
| NOTE: RIM NOTCHES IN DICATE SADDLE CENTER OR CENTER OF A 45° BEND | 9 | 12½ | 45° | 3⅜ |
| | 10 | 14 | 45° | 3¾ |

© J D BENFIELD 1964

**START BENDS AT ARROW ON TOOL**

NOTE: ON FLOOR BENDS A VERTICAL HANDLE INDICATES A 30° BEND.

**BACK-TO-BACK BENDS**
LOCATE STARPOINT

OPPOSITE THE FINISH LINE DESIRED — THE STARPOINT FORETELLS WHERE BACK OF THE BEND WILL LAY.

ON-THE-JOB Better-Bending Kits available from Jack Benfield, Author P.O. Box 544 Ft. Lauderdale, FL 33302 USA

**Fig. 9-52F** A weatherproof Zip Guide® label for any bender or hickey handle invites precision bending at the point of use. [Jack Benfield, patent holder]

swings free on a stainless-steel bolt as a constant degree-of-angle indicator. (Two shots are required for the No. 8 model.) Multiple shots may be used on both models for larger radii and concentric bending.

The Benfield No. 10 series stubber-hickey line (Fig. 9-52*H*) is unique. These tools are inch-along short-radius hickeys that bend *both* EMT and rigid or IMC conduit: No. 10 for ½-in EMT only, No. 11 for ½-in rigid IMC or ¾-in EMT, No. 12 for ¾-in rigid IMC or 1-in EMT, and No. 13 for 1-in rigid IMC or 1¼-in EMT.

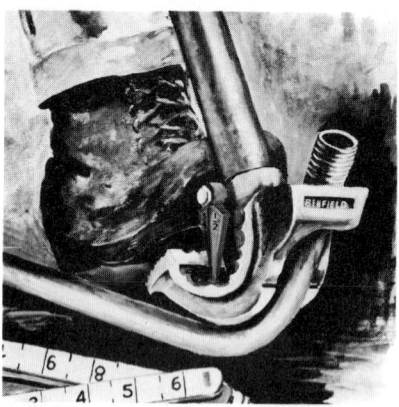

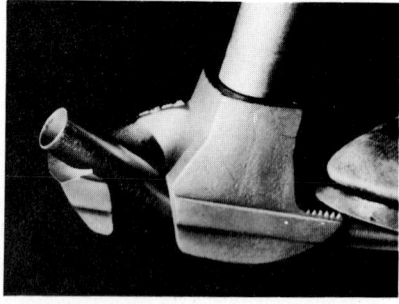

**Fig. 9-52***G*  Action photograph of the Benfield one-shot Model No. 7 hickey for rigid or IMC conduit. It is not suitable for EMT. [Jack Benfield, patent holder]

**Fig. 9-52***H*  Benfield inch-along short-radius hickey. Model No. 10 series. It has a pedal and arrow symbol to control the distance between bites and thereby varies the bending radius for concentric conduit runs. [Jack Benfield, patent holder]

These Benfield tools are available from wholesale electrical distributors throughout the United States and Canada. A 22-page instruction booklet and basic training guide for any make of bender or hickey (whether hand-type or mechanically or electrically powered) is available from Benfield Benders, P.O. Box 544, Fort Lauderdale, Florida 33302. This guide is provided in kit form and includes the booklet, a Zip Guide® 9-52*E* for the bender handle, and a wallet-size offset formula card. The combination is called the Benfield Better Bending Kit. It is refreshingly nontechnical, replete with easy-to-grasp diagrams and simple formulas for making stub-ups, back-to-back bends, offsets, concentric bends, and crossover saddles.

Benfield benders are produced in the United States by Klein Tools, Inc., of Chicago, Illinois. Klein Tools, Inc., franchises other manufacturers to market Benfield benders as follows:

1. American Electric Company, St. Louis, Missouri 63114
2. Appleton Electric Company, Chicago, Illinois 60657
3. Klein Tools, Inc., Chicago, Illinois 60645
4. The Wiremold Company, West Hartford, Connecticut 06110
5. Appleton Electric Ltd., Cambridge, Ontario, Canada N3H 2N1

SURFACE METAL MOLDING BENDERS  The Wiremold Benfield catalog No. 600-B (Fig. 9-52*I*) is a one-sweep bender with no moving parts. It makes 90° bends, offsets, and saddles in both No. 500 and No. 700 Wiremold surface metal raceway. The bender makes smooth sweeps flange-in or flange-out. An adapter that snaps into the bender groove is used to bend No. 200 Wiremold. NOTE  This tool will not bend this rectangular raceway *sideways*. To make sideway turns fittings must be used. Two separate degree scales are cast into the tool, one scale for No. 500 and the other for No. 700 Wiremold. Painted symbols serve as bench marks to enable the operator to make precision bends as with EMT benders (Fig. 9-52*I*). The 600-B comes complete with a two-piece handle of ¾-in IPS coupled in the center. Applied to the handle is a weatherproof 3-in-by-6-in Zip-Guide® instruction label, a basic reference at point of use.

**Fig. 9-52***I*   Action view.

**Fig. 9-52***J*   Top view.                    **Fig. 9-52***K*   Side view with two-piece handle

**100. Power conduit benders.**   For larger EMT or conduit sizes, hydraulic benders similar to the one shown in Fig. 9-53 are used in modern practice. Such benders are operated by a hand pump or a motor-driven pump. In general, two types are available. One type utilizes the *progression-bend method,* in which bends are laid out in marked spacings. Then every few degrees the conduit is shifted to the next mark, and so on, until the bend is completed. The second type is called a *one-shot bender,* in which the bend

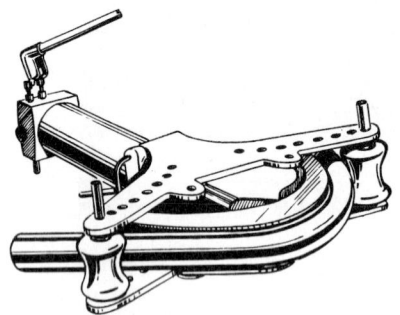

**Fig. 9-53**   Hydraulic conduit bender.

**Fig. 9-54A**   Condumatic power bender for ½-, ¾-, and 1-in rigid metal conduit. [The Ridge Tool Co.]

can be made without shifting the conduit. A 90° bend made with a one-shot bender provides a radius about the size of a standard factory elbow. Various shoes are available for different EMT and conduit sizes and for different sweep sizes. Information on the proper use of hydraulic benders is available from manufacturers of such equipment, and their instruction books should be followed to obtain precision bends.

Figure 9-54A shows a Condumatic power bender for making precision bends with ½-, ¾-, and 1-in rigid conduit. This motor-operated tool is extremely useful on large construction jobs for which many repetitive bends are required. The Greenlee power bender shown in Fig. 9-54B bends ½- through 2-in conduit or EMT. The photo shows four ¾-in conduits being bent at one time.

**101. A hydraulic pipe pusher** is used for underground installations of rigid metal conduit. The pusher eliminates tedious trenching operations. The conduit, which must be equipped with a cap or point, is driven through the ground laterally. A common application is under streets or roadways.

Pipe pushers can be operated at different pressures, depending on the conduit size and the type of ground through which the conduit is driven. Standard pipe pushers permit pushes up to 7 ft (2.1 m) without the pipe clamp being changed.

**102. Cable pullers and wire reels** (Fig. 9-55) are used for the installation of larger conductor sizes. Both are extremely useful on larger installations and save a great deal of time and labor. Multiple wire reels are also useful when a large number of small conductors are to be pulled into a single conduit. Figure 9-56 shows an electric-motor–operated cable puller which is also helpful in pulling in conductors with minimum effort.

**103. Threading conduit.**   Dies for threading conduit are designed to produce a taper of ¾ in (19 mm)/ft as described in Sec. 67. It is usual practice when a lot of conduit is received to rethread all the ends, which may have become filled with paint or dirt or distorted by blows. Rethreading will save more than its cost in that it ensures rapid erection. Always ream conduit after cutting a thread on it. Pipe-threading machines for threading conduit, preferably those operated by motors, should be used on big jobs, as they will soon pay for themselves in the time that they save.

**Fig. 9-54*B*** A power bender provides 90° bends for four conduits at the same time. [Greenlee Tool, a division of Greenlee Bros. & Co.]

**104. Wrenches for turning conduit.** The form of wrench shown in Fig. 9-57, I, appears to be the most popular for turning conduit. Chain wrenches (II) are not as yet much used for conduit work, but when they have been tried, they have proved very satisfactory. Their advantages are that they can be used with one hand after the chain is around the conduit and that they can be used in confined places and close to walls where a Stillson wrench could not be utilized.

**105. Conduit ends should always be reamed.** A reamer like that of Fig. 9-58 which can be turned by a bit brace is a good tool for small and medium-size conduit. For conduit of the larger sizes, reamers which have long handles attached, giving the needed leverage, can be obtained. When conduit is received and after it is cut, the ends are frequently turned in (Fig. 9-59, I) and, when screwed together in a coupling, form a knifelike edge which will abrade insulation. When the ends are properly reamed, they appear as shown in Fig. 9-59, II, but if they are screwed together too tightly, they may turn up as at I, defeating the purpose that reaming should accomplish. If no other tool is available, conduit can be reamed by hand with a half-round file.

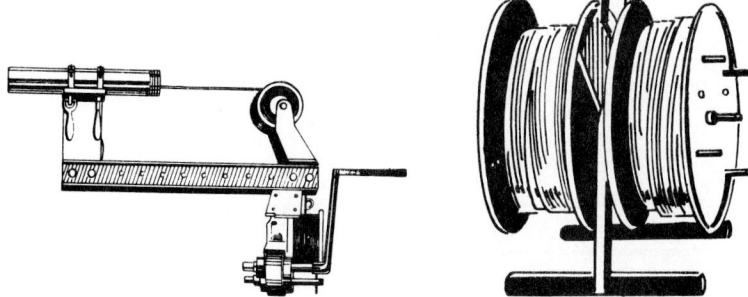

**Fig. 9-55**   Hand-crank–operated cable puller and wire reel.

**106. The usual tool for cutting conduit** is a hacksaw. Pipe cutters should not be used because they leave a large burr on the inside of the conduit, which takes time to ream out. While the conduit is being cut, it should be held in a vise. On jobs with a great deal of conduit to cut, the installation of a motor-driven cold-cutoff saw, such as is used for cutting structural steel and rails, will prove economical. A rapidly rotating steel disk without teeth cuts the pipe. Water must be sprayed on the disk to keep it cool. Other types of power saws are available.

**Fig. 9-56** Motor-operated cable puller. [Emerpac]

**107. Fishing wire** is tempered-steel wire of rectangular cross section. It can be readily obtained at electrical supply houses. A fishing wire is sometimes termed a *snake* (see Table 111). So that a fishing wire will slide readily past small obstructions, hooks should be bent in its ends as shown in Fig. 9-60. Before being bent, the ends should be annealed by heating them to a red heat and allowing them to cool slowly. If a small brass knob is riveted to the end of a fishing wire (Fig. 9-61) instead of a hook, the wire can be pushed through conduit more easily. When fishing is difficult, it is sometimes necessary to push two snakes with hook ends into the wireway, one from each of the outlets, as shown in Fig. 9-60. The wires must be worked back and forth and twisted

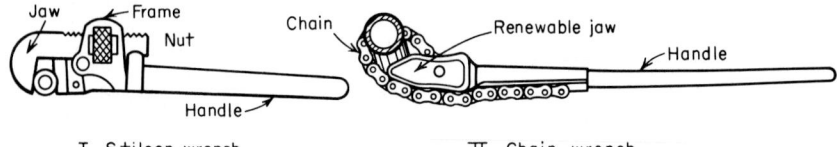

I  Stilson wrench          II  Chain wrench

**Fig. 9-57**  Wrenches for conduit.

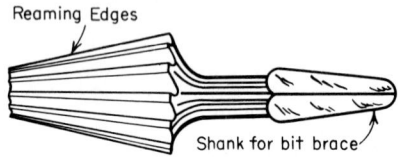

**Fig. 9-58**  A bit-brace reamer.

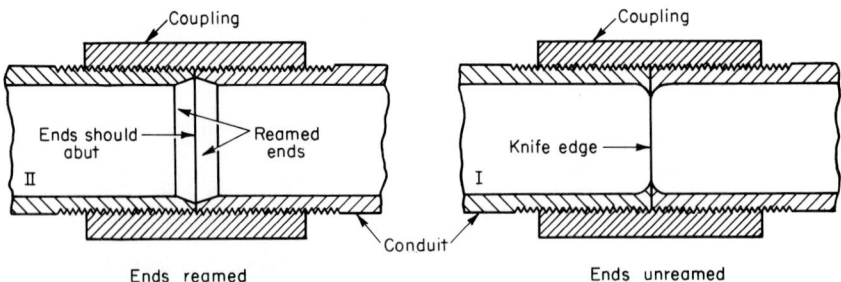

Ends reamed                Ends unreamed

**Fig. 9-59**  Reamed and unreamed conduit ends.

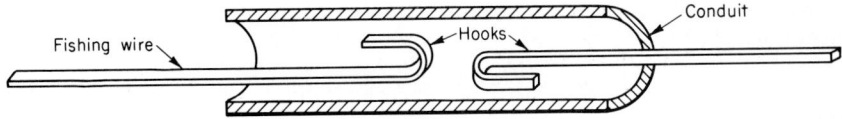

**Fig. 9-60**   Hooks bent in fishing-wire ends.

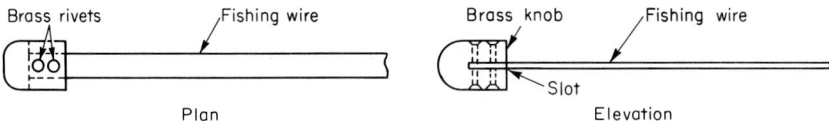

Plan                                    Elevation

**Fig. 9-61**   Knob on the end of fishing wire.

around until the two hooked ends engage. Then one wire can be pulled into the raceway with the other.

**108. When fishing from two ends,** as in Fig. 9-60, it is often advisable to tie a loop of cord, possibly 1 ft (0.3 m) long, in the hook of one wire and bend down the hook (Fig. 9-62).

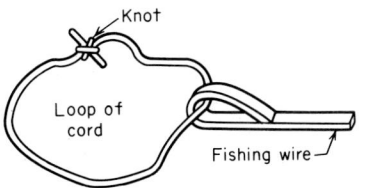

**Fig. 9-62**   Cord loop on the end of fishing wire.

The other wire has an open hook which can be made to engage in the cord loop quite readily. It has been found that a fish wire will go through conduit more readily if prepared as in Fig. 9-63, by loosely winding the end with small wire or cord, so that the wire or cord cannot pull off.

**109. Chain is used for vertical fishing.** A small chain can be made to drop down a vertical wireway with little difficulty. With a partition, the noise made by the lower end of a chain that is jiggled up and down will disclose its location almost exactly.

**110. Galvanized-steel wire can be used for fishing.** Any size from No. 14 up to, possibly, No. 6 may be utilized, but in nearly every case the flat steel ribbon wire is preferable.

**111.   Dimensions of Steel Fish Wire**
Wire ¼ in wide is the size most frequently used. The wire is usually put up in coils of 50, 75, 100, 150, and 200 ft (15.2, 22.9, 30.5, 45.7, and 61 m).

| $W$, width, in | $T$, thickness, in | Weight per 100 ft |
|---|---|---|
| ⅟₁₆ | 0.015 | 11 oz |
| ⅛ | 0.030 | 1 lb  4 oz |
| ³⁄₁₆ | 0.030 | 1 lb 14 oz |
| ¼ | 0.030 | 2 lb  8 oz |
| ⁵⁄₁₆ | 0.035 | 3 lb  8 oz |
| ⅜ | 0.035 | 3 lb 12 oz |

**112. In drawing wire into conduit** it is a mistake to use so much force that the wire cannot be withdrawn. Conduits should be big enough so that excessive force is not necessary. Small conductors (Nos. 14 and 12) can be pulled in with the fish wire. Figure 9-65 shows how they can be attached to the fish wire, and Fig. 9-66 shows how the attachment should be served with tape to render pulling easy. It spoils any fishing wire to draw in with it conductors that pull hard.

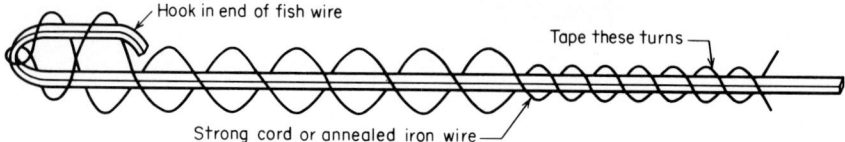

**Fig. 9-63**   Fish-wire end prepared with cord.

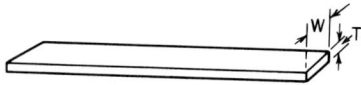

**Fig. 9-64**   Steel fishing wire.

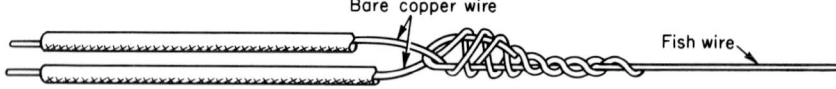

**Fig. 9-65**   Attachment of stranded conductor to fishing line.

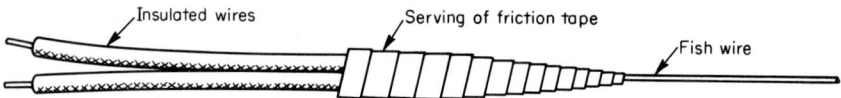

**Fig. 9-66**   Attachment taped over.

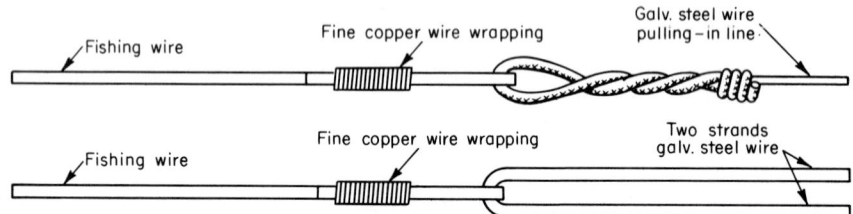

**Fig. 9-67**   Attachment of pulling-in line to fish wire.

For conductors larger than No. 12 and No. 14, unless the pulls are very easy, a pulling-in line which is drawn into the conduit with the fish wire should be used for hauling in the conductors. Number 10 or 12 Birmingham wire gage galvanized-steel wire makes a good pulling-in line and is probably better for heavy work than rope. Two strands can be used if necessary. Braided cord is better than twisted rope for a pulling-in line, because when tension is applied, rope tends to untwist. Sash cord is satisfactory for light work. Galvanized-steel pulling-in wire can be attached to fish wire as in Fig. 9-67. Nylon rope, blown into conduits by a Jet-Line tool, is also recommended.

Three or four links from a chain of no greater diameter than the line should be made up in the end of a rope or cord pulling-in line. Wires to be drawn in or a fish line can be attached to the links (Fig. 9-68). One stranded conductor can be attached to a pulling-in

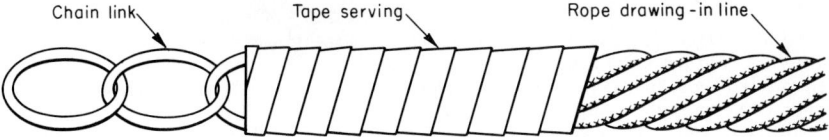

**Fig. 9-68**    Chain links in the end of pulling-in line.

line as in Fig. 9-69. The attachment should be taped over to make it smooth. If two conductors are to be pulled in, one of them is made up into a loop in the end of the pulling-in line, as shown in Fig. 9-69. The insulation is trimmed from this conductor for 6 in (152.4 mm) or more, and the bared end of the other conductor is made up about the first one, forming a long, tapering connection. If a hard pull is expected, it is advisable to solder the connection. The whole should be served with tape as in Fig. 9-66. If three wires are to be drawn in instead of two, the attachment is the same, with the addition that the bared end of the third conductor is made up around the other two. The diameter at any section of the attachment must not exceed the overall diameter of the wires, and the attachment

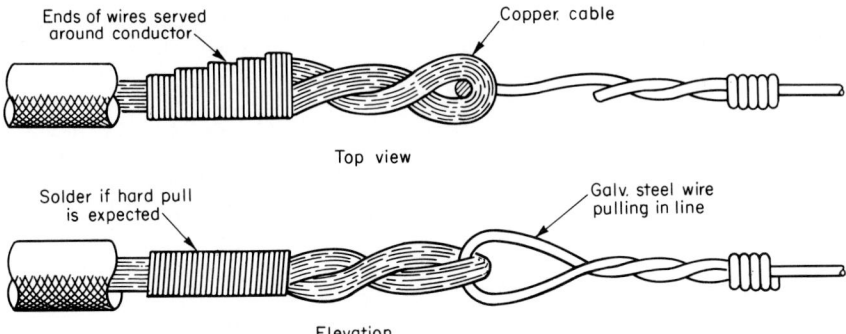

**Fig. 9-69**    Attachment of stranded conductor to pull-in line.

**Fig. 9-70**    Metal basket grip.

should be in the form of a conical wedge. It is sometimes necessary to cut off possibly half of the strands of the bared ends and make up only the strands that remain to ensure that the attachment will be of sufficiently small diameter. A flexible-metal basket grip (Fig. 9-70), available in many sizes, is also recommended. The weblike grip slips over the ends of cables, and when tension is applied, the grip tightens and the conductors are firmly secured.

**113. Force for pulling-in wires** is, in the case of easy pulls, supplied by the worker. Tackle blocks are permissible for heavy pulls. Cranes can often be used very effectively. Snatch blocks can be used to guide the pulling-in line to some point at which it can be pulled either with the crane hook or by the crane in traveling along its runway. A lever can be used for hard pulls either by repeatedly fastening the pulling-in line to the lever or by gripping it with a pair of pliers and then prying against the pliers with the lever. Only a short pull is possible with each setting, and a succession of settings and pries is necessary to draw the conductors in.

**114. In feeding conductors into conduit,** care must be taken that they go in symmetrically and without lapping or twisting. If one conductor crosses another, it may make a hump that will wedge in the conduit. Powdered soapstone blown into conduit renders pulling easier. One convenient way to get the soapstone into the raceways is to place it in an elbow, place the elbow to the conduit, and then blow on the elbow. Powdered soapstone should be rubbed on the conductors as they are drawn into conduits when the pulls are hard. Liquid paste, commonly known as *wire lube*, is widely used in modern cable-pulling operations. Although in a paste form during cable pulls, the substance dries into a powder form, which aids in withdrawing conductors later on.

**115. Calculations for determining tension.** When conductors are installed with pulling eyes attached, maximum tension should not exceed 0.008 cmil. Therefore

$$T = 0.008 \times N \times \text{cmil}$$

where $T$ (tension) is determined in pounds, $N$ is the number of conductors being pulled in, and cmil is the circular-mil area per conductor.

When cables are pulled with basket grips around sheaths, maximum strain on sheaths should not exceed 1500 lb/in³ (10.3 million Pa) if sheaths are of lead. Therefore

$$T = 4712t(D - t)$$

where $T$ (tension) is determined in pounds per square inch; $t$ is thickness, in inches, of the sheath; and $D$ is outside diameter of cable, also in inches.

When basket grips are used to pull in nonleaded cables, maximum strain should not exceed 1000 lb (6.9 million Pa) or the limitation determined by the pulling-eye method.

When cables are pulled into straight raceway sections, pulling tension depends on total cable weight plus the coefficient of friction. Therefore

$$T = LWF$$

where $T$ (tension) is total pulling tension in pounds; $L$ is length, in feet, of the raceway run; $W$ is weight of the cable in pounds per square foot; and $F$, the coefficient, is 0.5 when raceways are well made.

## INTERIOR OR ABOVEGROUND WIRING WITH RIGID NONMETALLIC CONDUIT

**116. Rigid nonmetallic conduit** of the Schedule 40 polyvinyl chloride (PVC) type may be used in walls, floors, and ceilings and for:
1. Portions of dairies, laundries, canneries, or other wet locations
2. Locations where walls are frequently washed
3. Locations subject to severe corrosive influences
4. Damp or dry locations not prohibited by Sec. 117
5. Locations subject to chemicals for which the PVC material is specifically approved

It may also be used for exposed work when it is approved for the purpose and is not exposed to physical damage. Schedule 80 may be used aboveground on poles.

**117. Heavy-wall PVC conduit must not be used:**
1. In hazardous locations, except as permitted in Secs. 514-8, 515-5, and the exception to 501-4(b) of the NEC.
2. Where exposed to physical damage unless identified for such use.
3. For the support of fixtures or other equipment.
4. Where subject to ambient temperatures exceeding those for which the conduit is approved.
5. PVC Schedule 40 and Schedule 80 are suitable for use aboveground and underground and for direct burial without encasement in concrete.

**Underwriters' Laboratories, Inc.**
**LISTED**

**Underwriters' Laboratories, Inc.**
**LISTED**

RIGID NONMETALLIC CONDUIT
ABOVEGROUND AND UNDERGROUND
(Schedule 40)

RIGID NONMETALLIC CONDUIT
ABOVEGROUND AND UNDERGROUND
EXTRA HEAVY WALL
(Schedule 80)

**Fig. 9-71**    Identification of aboveground PVC conduits.

**118. Underwriters Laboratories** lists rigid nonmetallic PVC conduit for aboveground applications with two different labels, as shown in Fig. 9-71. The conduits are manufactured in sizes from ½ to 6 in. The extra-heavy-wall–type Schedule 80 conduit is utilized for special applications which require additional physical protection, such as pole risers or similar uses when conduit is subject to physical damage.

**119. Manufacturers' classifications** for rigid nonmetallic PVC conduit are:

1. *Type A.*    Thin-wall for underground, encased in concrete
2. *Type 40.*    Heavy-wall for direct burial in earth and limited aboveground installations
3. *Type 80.*    Extra-heavy-wall for direct burial in earth and aboveground installations for general applications and when subject to physical damage as indicated in Sec. 118

From the classifications indicated in Secs. 117 and 118 it should be noted that only heavy-wall (Type 40) and extra-heavy-wall (Type 80) PVC conduits are suitable for aboveground applications. The Type A PVC conduit is limited to underground installations when encased in concrete, and such installations should be installed as indicated for underground work in Div. 8.

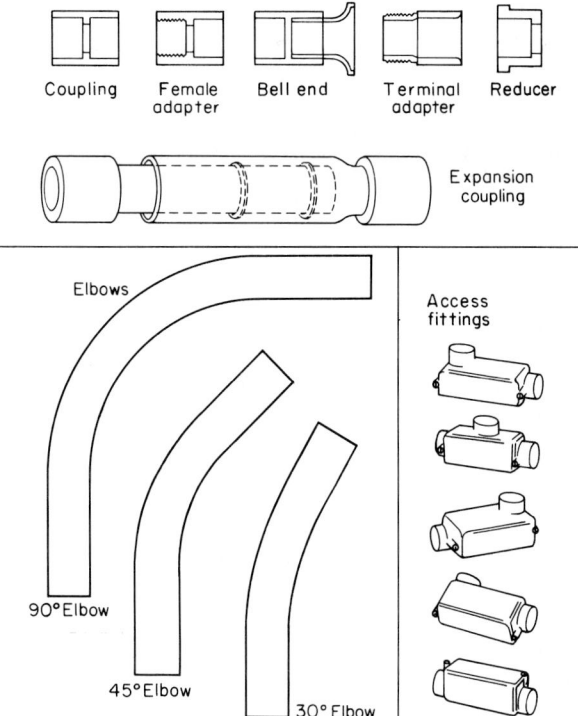

Coupling    Female adapter    Bell end    Terminal adapter    Reducer

Expansion coupling

Elbows

90° Elbow

45° Elbow

30° Elbow

Access fittings

**Fig. 9-72**    Fittings and accessories for PVC conduit.

**120. PVC conduit fittings** are shown in Fig. 9-72. Also available are a wide variety of outlet boxes constructed of polyvinyl chloride or fiber-reinforced Bakelite.

**121. Cutting PVC conduit.** This conduit can be easily cut with a fine-tooth handsaw. For sizes from 2 through 6 in a miter box or similar saw guide should be used to keep the material steady and assure a square cut. After cutting, deburr the conduit ends and wipe off dust, dirt, and plastic shavings. Deburring can be readily accomplished with a pocketknife.

**122. Joining PVC conduit.** The conduit should be wiped clean and dry. Then apply a coat of cement (use a natural-bristle brush or aerosol-spray cement) to the end of the conduit including the length of the fitting socket to be attached. Finally, push conduit and fittings firmly together, and then rotate the conduit in the fitting about a half turn to distribute the cement evenly. The cementing and joining operation will not require more than 1 min. Many manufacturers supply aerosol-spray cans or brush-top cans of PVC solvent cement, which may be used as timesaving expedients. In adapting PVC conduit to existing or metal equipment, a complete range of threaded adapters is available. Standard metallic locknuts are used at these termination points, and bushings normally are not necessary.

**123. Bending PVC conduit.** Although a complete line of factory elbows (90, 45, or 30°) are available (Fig. 9-72), heat-bending of PVC conduit in the field can be easily accomplished. The most accepted method for sizes from ½ through 2 in is with an electric *hot box*, which consists of a 115-V, 15-A heating element and a 2-ft (0.6-m) slot into which the section of conduit to be bent is placed.

Under normal conditions a complete bend can be made in less than 1 min. When exact measurements are required, preheat a length of PVC conduit and then form it in place. This is possible because PVC conduit will remain pliable for at least 3 min after it has been heated. Never use flame-type devices as a heat application to PVC conduit, because this excessive heat can ruin the conduit.

For installations of PVC larger than 2 in, factory elbows as shown in Fig. 9-72 should be obtained.

Other methods of bending PVC conduit include heater blankets and infrared heaters. The primary function of a heating device is to supply uniform heat over the entire area without scorching the conduit.

**124. Expansion and contraction.** In applications in which the conduit installation will be subject to constantly changing temperatures and the runs are long, consideration should be given to expansion and contraction of PVC conduit. In such instances an expansion coupling, such as shown in Fig. 9-72, should be installed near the fixed end of the run to take up any expansion or contraction that may occur. The coupling shown in Fig. 9-72 has a normal expansion range of 6 in (152.4 mm). The coefficient of linear expansion of PVC conduit can be obtained from manufacturers' data.

Expansion couplings may be used in either underground or exposed applications. In underground or slab applications such couplings are seldom used, because expansion and contraction may generally be controlled by bowing the conduit slightly or by immediate burial. After the conduit has been buried, expansion and contraction cease to be a factor. Care should be taken, however, to assure that contraction does not take place in a buried installation if the conduit should be left exposed for an extended period of time during widely variable temperature conditions.

**125. Support of rigid nonmetallic conduit** should be not less than specified in the following NEC table:

| Conduit Size, In | Maximum Spacing between Supports, Ft |
|---|---|
| ½–1 | 3 |
| 1–2 | 5 |
| 2½ and 3 | 6 |
| 3½ to 5 | 7 |
| 6 | 8 |

In addition, rigid nonmetallic conduit shall be supported within 3 ft (0.9 m) of each box, cabinet, or other conduit termination.

**126. Number of conductors.**    The number of conductors permitted in a single nonmetallic conduit is determined in the manner described in Sec. 79. In many instances nonmetallic conduits will contain equipment-grounding conductors to ground metal-enclosed equipment or grounding-type receptacles supplied by such conduits. If *bare* equipment conductors are used, the cross-sectional areas of these conductors, as listed in Table 41 of Div. 11, may be used in sizing the conduit. Accordingly, smaller conduit sizes may be permitted if bare instead of insulated grounding conductors are used.

It should be noted that extra-heavy-wall PVC conduit (Type 80) has a reduced cross-sectional area available for wiring space. The actual cross-sectional area is prominently marked on such conduits, and the conductor fill may be determined accordingly.

**127. The number of bends** in any run of conduit between outlet and outlet, fitting and fitting (LB, T, X), or outlet and fitting shall not contain more than the equivalent of four quarter bends (360°), including bends located immediately at the outlet or fitting.

**128. Nonmetallic conduit lengths** are normally available in 10-ft (3-m) lengths. However, lengths up to 30 ft (9.1 m) may be obtained for special applications. The longer lengths require fewer coupling connections and save time and labor.

**129.  Weight of Heavy-Wall (Type 40) Rigid Nonmetallic PVC Conduit**

| Trade Size, In | Weight, Lb/100 Ft |
|---|---|
| ½ | 16 |
| ¾ | 21 |
| 1 | 31 |
| 1¼ | 42 |
| 1½ | 50 |
| 2 | 67 |
| 2½ | 106 |
| 3 | 139 |
| 3½ | 168 |
| 4 | 197 |
| 5 | 268 |
| 6 | 348 |

**130. PVC is a nonmagnetic conduit,** and being nonmetallic, it provides less voltage drop and fewer reactance losses than are possible with metallic conduits with corresponding conductors used on ac circuits.

**131. Other types of rigid nonmetallic conduits,** such as thin-wall PVC, fiber, asbestos cement, soapstone, fiberglass epoxy, and high-density polyethylene, are designed solely for use in underground work. They are not intended for interior and aboveground applications. These conduits and installation procedures are described in the underground-wiring sections of Div. 8.

For systems of 600 V or less all types of rigid nonmetallic conduits must be 18 in (457 mm) below the surface if buried directly in earth. Lesser depths are permitted if the conduit is run below 2 in (50.8 mm) of concrete. Refer to Sec. 135 of Div. 8 for minimum burial depths for high voltages.

## FLEXIBLE-METAL-CONDUIT WIRING

**132. Flexible metal conduit** can be used for many kinds of wiring, being in some cases preferable to rigid conduit. Its installation is much easier and quicker than the installation of rigid conduit, the latter coming in short pipe lengths, whereas flexible conduit may be had in lengths of 25 to 250 ft (7.6 to 76.2 m), depending on the size of the conduit. The rules governing the insulation on the wires are the same as for rigid conduit: outlet or switch boxes must be installed at all outlets or switches, and the conduit must be continuous from outlet to outlet and be securely fastened to the boxes.

Flexible metal conduit shall not be used (1) in wet locations, unless conductors are of the lead-covered type or of another type specially approved for the conditions; (2) in hoistways, except as provided in Sec. 620-21 of the NEC; (3) in storage-battery rooms; (4) in any hazardous location except as permitted by Sec. 501-4(b) and 504-20 of the Code; (5) where rubber-covered conductors are exposed to oil, gasoline, or other materials having a deteriorating effect on rubber; or (6) underground or embedded in poured concrete or aggregate.

Flexible metal conduit consists of a single strip of aluminum or galvanized steel, spirally wound on itself and interlocked so as to provide a round cross section of high mechanical strength and great flexibility.

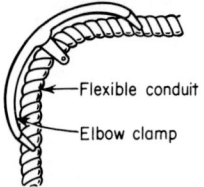

Its flexibility together with the continuous length procurable makes its use practicable when rigid conduit would be out of the question. For this reason it can be employed to advantage in finished frame buildings in place of the other forms of wiring so largely employed in these structures. The conduit is easily fished and requires no elbow fittings. These can be made with the conduit itself, but care should be exercised in properly fastening the conduit at elbows (Fig. 9-73). There are on the market fittings whereby changes from other forms of wiring can be easily made to flexible-conduit wiring. Iron plates should be used to protect the conduit from nails where it passes through slots in floor beams or studding.

**Fig. 9-73**  Elbow clamp.

### 133.  Standard Sizes of Flexible Steel Conduit

**Fig. 9-74**  Flexible steel conduit.

| B, nominal inside diameter, in | A, minimum OD, in | Bending radius in | Approximate ft per coil | Minimum weight, lb/1000 ft |
|---|---|---|---|---|
| ⁵⁄₁₆ | 0.470 | 1¾ | 250 | 150 |
| ⅜ | 0.560 | 2 | 250 | 255 |
| ½ | 0.860 | 3 | 100 | 470 |
| ¾ | 1.045 | 4 | 50 | 595 |
| 1 | 1.300 | 5 | 50 | 1020 |
| 1¼ | 1.550 | 6¼ | 50 | 1250 |
| 1½ | 1.850 | 7½ | 25 | 1625 |
| 2 | 2.350 | 10 | 25 | 2125 |
| 2½ | 2.860 | 12½ | 25 | 2630 |
| 3 | 3.360 | 15 | 25 | 3130 |

**134. Installation of flexible metal conduit.**  If exposed, flexible metal conduit can be fastened to the surface with single-hole or two-hole straps (Figs. 9-40 and 9-41). The conduit must be secured at intervals not exceeding 4½ ft (1.37 m) and within 12 in (304.8 mm) from every outlet box or fitting. If concealed, the conduit can be fished into place just as any concealed conductors are fished in.

Exposed runs of the conduit shall closely follow the finish of the building finish or of running boards to which the conduit is fastened, except for:

1. Lengths of not more than 36 in (914.4 mm) at terminals where flexibility is necessary

2. Locations in accessible attics and roof spaces

3. The underside of floor joists in basements where the conduit is supported at each joist and is so located as not to be subject to physical damage.

In making bends, care must be exercised that the conduit is not damaged.

**135. The number and size of conductors to be installed in flexible metal conduits** is the same as for rigid metal conduit (refer to Sec. 79).

**136. The minimum size of flexible conduit** that is allowed by the NEC is ½ in except as allowed by the Code for underplaster extensions, for motor wiring, or for a connection which is not over 72 in (1828.8 mm) in length, for taps to lighting fixtures or for utilization equipment.

**137. Flexible metal conduit is joined** with couplings as illustrated in Fig. 9-75. Short lengths can be coupled to longer pieces with the clamp of I and waste thereby be prevented. The clamp at II is used for coupling rigid to flexible conduit.

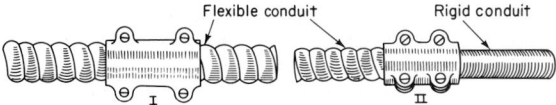

Flexible conduit    Rigid conduit

**Fig. 9-75**   Couplings for flexible steel conduit.

**138. When elbows are formed in flexible conduit,** some provision must be made to prevent the conduit from straightening out and thereby preventing the withdrawal of the conductors. Pipe straps can be used in some cases, and in others an elbow clamp (Fig. 9-73) can be applied.

**139. Flexible metal conduit can be connected into steel boxes** with the connector illustrated in Fig. 9-76. It is clamped to the armor with a bolt which provides a connection between armor and steel box. Figure 9-77 shows flexible metal conduit connected into an outlet box.

**140. To cut flexible metal conduit** a fine hacksaw should be used. Special vises have a slot across their jaws to guide the saw blade, the conduit being held between grooves in the jaws. Always cut flexible conduit straight across to ensure a proper fit into connectors.

**141. For reaming flexible metal conduit** a special reamer (Fig. 9-78) has been made, but inasmuch as the burr resulting from the hacksaw cut is very small, it can be readily removed with a three-cornered scraper made from a three-cornered file, or it can be removed with an ordinary file. The reamer illustrated is for conduit with a diameter of 1¼ in or less.

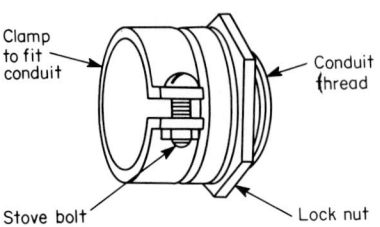

Clamp to fit conduit

Conduit thread

Stove bolt

Lock nut

**Fig. 9-76**   Connector for attaching flexible steel conduit to an outlet box.

**142. Fish plugs for pulling in flexible conduit** (Fig. 9-79) are furnished by the conduit manufacturer for ⅜-, ½-, and ¾-in flexible conduit. After the conduit has been cut off square in a vise with a hacksaw, the fish plug is screwed into the tube and the fish or drawing-in wire attached to the plug for pulling in.

**143. Pulling in conductors.** Conductors are pulled into flexible conduits in much the same manner as in rigid conduits (refer to Sec. 107) with fish tapes or similar pulling means. It should be emphasized that the number of bends in any one run of flexible conduit should never exceed the equiva-

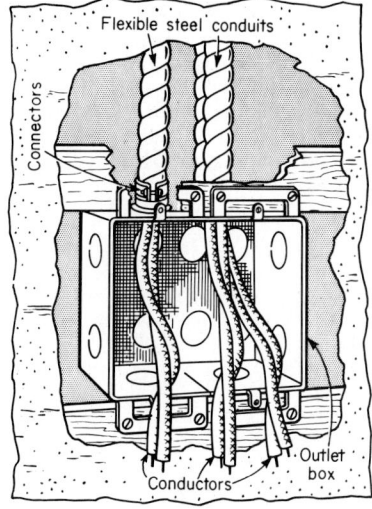

Flexible steel conduits

Connectors

Conductors

Outlet box

**Fig. 9-77**   Flexible-steel-conduit outlet box.

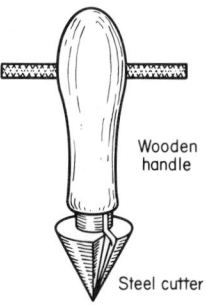

Wooden handle

Steel cutter

**Fig. 9-78**   Reamer for flexible conduit.

lent of four quarter bends (360°), and preferably there should not be more than two quarter bends in a run. A fish tape which is to be inserted into flexible conduit to pull in conductors should contain a ball or oval end instead of a hook. A rounded tip permits the tape to be inserted without hanging onto convolutions of the flexible conduit.

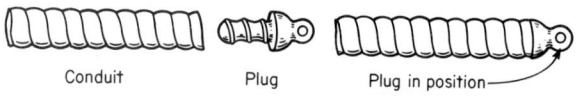

Conduit          Plug          Plug in position

**Fig. 9-79**  Plug for pulling in flexible conduit.

**144. Flexible metal conduit and fittings** cannot be used as a grounding means unless they are specifically listed for the purpose. When not so approved, the conduit must contain a bare or insulated equipment-grounding conductor, which must be attached to all equipment supplied and connected to the flexible metal conduit. Flexible metal conduit with a total ground return path of 6 ft may be used for grounding if the circuit conductors are protected at 20 A or less and the fittings are listed for grounding.

### LIQUIDTIGHT FLEXIBLE-METAL-CONDUIT WIRING

**145. Liquidtight flexible metal conduit** is similar to regular flexible metal conduit except that it is covered with a liquidtight plastic sheath (Fig. 9-80). It is not intended for general-purpose wiring but has definite advantages in many cases for the wiring of machines and portable equipment. Its use is restricted by the NEC as follows:

1. It is allowed for connections of motors or portable and stationary equipment when flexibility of connection is required.

2. Its use is not allowed (a) when subject to physical damage; (b) under conditions such that its operating temperature exceeds the rating of the material; and (c) in any hazardous location except as described in Secs. 501-4(b), 502-4, 503-3, and 504-20 of Article 500 of the Code unless it is specially approved for such use.

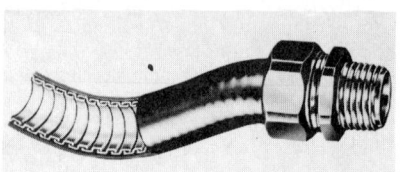

**Fig. 9-80**  Liquidtight flexible metal conduit. [Electri-Flex Co.]

Care must be exercised in its installation to employ only suitable terminal fittings that are approved for the purpose. Liquidtight flexible metal conduit in sizes 1½ in and larger shall be bonded in accordance with Sec. 250-79 of the NEC. Trade sizes of 1¼ in and smaller are suitable for grounding in any ground return path of 6 ft or less if ⅜ in and ½ in are protected at 20A or less, and 60A or less for ¾ in through 1¼ in.

The Code specifies that the maximum size used must not be larger than the 4-in electrical trade size.

**146.   Dimensions of Liquidtight Flexible Metal Conduit**

| Nominal size electrical conduit, in | Actual | | | |
|---|---|---|---|---|
| | ID, in | | OD, in | |
| | Mini-mum | Maxi-mum | Mini-mum | Maxi-mum |
| ⅜ | 0.484 | 0.504 | 0.690 | 0.710 |
| ½ | 0.622 | 0.642 | 0.820 | 0.840 |
| ¾ | 0.820 | 0.840 | 1.030 | 1.050 |
| 1 | 1.041 | 1.066 | 1.290 | 1.315 |
| 1¼ | 1.380 | 1.410 | 1.630 | 1.660 |
| 1½ | 1.560 | 1.590 | 1.870 | 1.900 |
| 2 | 2.005 | 2.045 | 2.335 | 2.375 |
| 2½ | 2.449 | 2.489 | 2.855 | 2.895 |
| 3 | 3.048 | 3.088 | 3.480 | 3.520 |

**147. Conductor size.** The NEC specifies that the maximum number of conductors installed in liquidtight flexible metal conduit shall not exceed those given in Table 29, Div. 11, for raceways ½ in and larger and Sec. 350-3 of the NEC for ⅜-in raceways.

## METAL-CLAD–CABLE WIRING: TYPES AC AND MC

**148. Type AC cable** is a fabricated assembly of insulated conductors in a flexible metallic enclosure.

**149. Type AC cables** (generally referred to as *armored cable*) are branch-circuit and feeder cables with armor or flexible metal tape (Fig. 9-81). Cables of the AC type, except ACL (lead-covered conductors), must have an internal bonding strip of copper or aluminum in intimate contact with the armor for its entire length. Most manufacturers use a flat aluminum ribbon as the bonding strip. At terminations, the bond wire is simply folded back without connection to the enclosure or other such conductors. Being in intimate contact with the metal armor, the bond wire is intended to reduce the overall resistance of the cable. For cables of the AC type, insulated conductors must be one of the types listed in Sec. 128 of Div. 2. In addition, the conductors must have an overall moisture-resistant and fire-retardant fibrous covering.

For ACT (thermoplastic conductors), a moisture-resistant fibrous covering is required only on the individual conductors.

**Fig. 9-81**   Type AC cable with bond wire.

When not subject to physical damage, Type AC cable may be installed in both exposed and concealed work. Type AC cable may be used in dry locations and for underplaster extensions and be embedded in plaster finish on brick or other masonry except in damp or wet locations. This cable may be run or fished in the air voids of masonry-block or tile walls; when such walls are exposed or subject to excessive moisture or dampness or are below the grade line, Type ACL cable must be used. Also, Type ACL must be used if the cable is exposed to the weather or to continuous moisture; run underground; embedded in masonry, concrete, or fill in buildings in course of construction; or exposed to oil or other conditions having a deteriorating effect on the insulation.

Type AC cable must not be used where expressly prohibited in the NEC including (1) in theaters or assembly halls, except where permitted by Article 518, Places of Assembly, of the NEC; (2) in motion-picture studios; (3) in any hazardous location; (4) where exposed to corrosive fumes or vapors; (5) on cranes or hoists, except as provided in Sec. 610-11, Exception No. 3 of the Code; (6) in storage-battery rooms; (7) in hoistways or on elevators, except as provided in Sec. 620-21 of the Code; or (8) commercial garages as prohibited in Article 511 of the Code. Type AC cable shall not be used for direct burial in earth.

**150. Manipulating Type AC cable.**   Type AC cable must never be spliced except at outlets. Many methods of fastening cable in outlet boxes are in vogue. Clamps or box connectors can be used. Outlet and switch boxes specially adapted for flexible metal-clad cable should be used. Special cutters are available.

**151. For cutting the cable to length,** a shear, bolt cutter, or hacksaw can be used. At each outlet or junction box the armor must also be cut back about 8 in (203.2 mm) in order to leave sufficient free ends of wire for splicing. This cutting is done with a hacksaw held so as to cut across the armor at an angle of about 60° with the axis of the cable (Fig. 9-82). Great care must be taken not to cut into the insulation on the conductors.

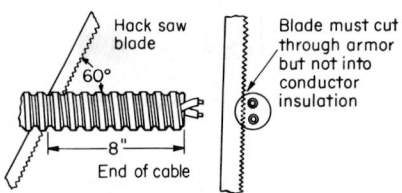

**Fig. 9-82**  Method of cutting armor back from the end of cable.

Greenlee Textron, Inc., has developed a Flex Splitter rotary tool that is far superior for cutting Type AC cable. An adjustment knob varies the depth of the cut and a rotary crank easily cuts the sheath.

**152. The wires at the ends of Type AC cable** should be protected by fiber bushings, although these are not required with lead-covered cable. The fiber bushings employed for this purpose are called *antishort insulating bushings* (Fig. 9-83). The split bushing is pinched together and slipped over the wires inside the armor, as illustrated in Fig. 9-84. It protects the insulation of the wires from the rough edges of the cut ends of the armor.

**153. For fastening Type AC cable to outlet boxes** a connector (Fig. 9-85) can be slipped over the end of the cable. The setscrew is tightened down against the armor. The threaded end of the connector slips into standard conduit knockouts, and the locknut is then put on, holding the connector tight to the box. Another method is to use outlet boxes of the type shown in Fig. 9-86, which contain built-in cable clamps in which the clamp tightens down against the armor and holds it firmly in place.

**154. Type AC cable for old-building wiring.**   Type AC cable can be used to great advantage in this work. It can be run with almost utter disregard of its contact with pipes or other materials and can be fished for long distances. Its own weight is sufficient to carry it down

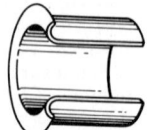

**Fig. 9-83**  Antishort fiber bushing.

**Fig. 9-84**  Method of protecting wires with antishort bushings.

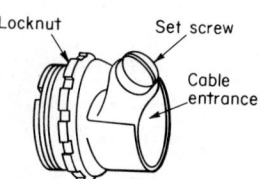

**Fig. 9-85**  Box connector for Type AC armor-clad cable. [General Electric Supply Co.]

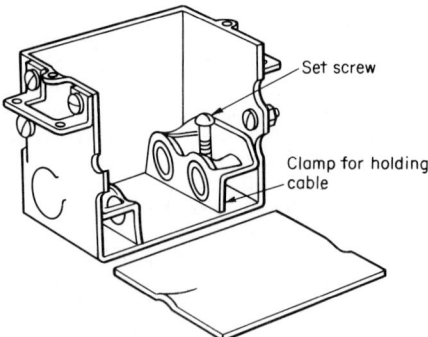

**Fig. 9-86**  Outlet box with clamps for Type AC armor-clad cable. [General Electric Supply Co.]

partitions, and it is stiff enough to fish between joists without the use of a fish wire. It or nonmetallic-sheathed cable can also be installed more quickly and with less cutting of walls, floors, etc., than can wires in flexible tubing.

**155. The loop system of wiring** is nearly always used for concealed work in order to avoid the use of junction boxes. The cables loop from one outlet to the next outlet, and splices are made only at outlet boxes. All splices must be accessible for inspection, and if a junction box is used, the cover must be left accessible.

**156. Type MC cables** are power cables (Fig. 9-87) limited in size to conductors of No. 18 AWG and larger for copper and No. 12 AWG and larger for aluminum. The metal enclosure shall be either a smooth metallic sheath, welded and corrugated metallic sheath, or interlocking metal-tape

**Fig. 9-87** Type MC cable with aluminum sheath and aluminum conductors. [Kaiser Aluminum & Chemical Corp.]

armor. Supplemental protection of an outer covering of corrosion-resistant material must be provided when such protection is required by Sec. 300-6 of the NEC. The cables must provide an adequate path for grounding purposes. For cables of the MC type, insulated conductors must be of a type listed in Sec. 128 of Div. 2, and No. 18 and No. 16 may employ a fixture-wire insulation.

Except as otherwise specified in the Code and where not subject to physical damage, circuits, services, or feeders in both concealed and exposed work as follows.

The cable may be used in partially protected areas, as in cable trays and the like, in dry locations; and when any of the following conditions are met, it may be used in wet locations:

1. The metallic covering is impervious to moisture.

2. A lead sheath or moisture-impervious jacket is provided under the metal covering.

3. The insulated conductors under the metallic covering are approved for use in wet locations.

At all points where Type MC cable terminates, suitable fittings designed for use with the particular wiring cable and the conditions of service must be used (see Fig. 9-88).

**157. Bends.** All bends shall be so made that the cable will not be damaged; and the radius of the curve of the inner bend of any bend must not be less than 7 times the diameter of Type MC cable with an interlocked or corrugated sheath, 10 times the diameter of a smooth sheath of a size not more than ¾ in, 12 times the diameter for 1-in but not less than 1½ in, and 15 times the diameter for cable more than 1½ in.

**158. Supports.** Metal-clad cables must be secured by approved staples, straps, hangers, or similar fittings so designed and installed as not to injure the cable. The cable may be installed on the underside of floor joists in basements when it is supported at each joist and is so located as not to be subject to physical damage.

Type AC cable must be secured at intervals not exceeding 4½ ft (1.37 m) and within 12 in (305 mm) from every outlet box or fitting, except when the cable is fished for lengths of not over 24 in (610 mm) at terminals when flexibility is necessary, and lengths of not more than 6 ft (610 mm) from an outlet for connections within an accessible ceiling to lighting fixtures or equipment. Staples or straps are used to support the cables to wired-over surfaces. The cable may be installed on metal racks, trays, troughs, or cable trays, grounded as required in Article 250 of the NEC. The cables must be separated from each other by a distance of not less than one cable diameter.

**Fig. 9-88** Connectors for Type MC cables.

**159. Through studs, joists, and rafters.** Metal-clad cable can be run through bored holes in studs, joists, or rafters. The holes must be bored at the approximate centers of the wood members or at least 1¼ in (31.75 mm) from the nearest edge when practical. If there is no objection because of weakening of the structure, the Code will allow metal-clad cable to be laid in notches in the studding or joists, provided that the cable is protected against the driving of nails into it by having the notch covered with a steel plate at least 1⁄16 in (1.588 mm) thick before the building finish is applied.

**160. In accessible attics.** Type AC cables in accessible attics shall be installed as follows:

1. If run across the top of floor joists, or within 7 ft (2.1 m) of the floor or floor joists across the face of rafters or studs, in attics or roof spaces which are accessible, the cable shall be protected by substantial guard strips which are at least as high as the cable. If this space is not accessible by permanent stairs or ladders, protection only will be required only within 6 ft (1.8 m) of the nearest edge of the scuttle hole or attic entrance.

2. If cable is carried along the sides of rafters, studs, or floor joists, neither guard strips nor running boards shall be required.

## SURFACE-RACEWAY WIRING

**161. Wiring in surface metal raceway** can be used for exposed work for branch-circuit or feeder conductors. It may be installed in dry locations. It shall not be used (1) when concealed, except that metal raceways approved for the purpose may be used for under-plaster extensions; (2) when subject to severe physical damage unless approved for the purpose; (3) when the voltage is 300 V or more between conductors unless the metal has a thickness of not less than 0.040 in (1.016 mm); (4) when subject to corrosive vapors; (5) in hoistways; or (6) in any hazardous location, except in a Class I, Div. 2, location as permitted by Sec. 501-4(b), exception. The number of conductors installed in any race-way shall be no greater than the number for which the raceway is designed. Metal race-way must be continuous from outlet to outlet, junction box, or approved fittings designed especially for use with metal raceway. All outlets must be provided with approved terminal fittings which will protect the insulation of conductors from abrasion unless such protection is afforded by the construction of the boxes or fittings. Metal raceway should not be used in damp places.

When combination raceways are used for the installation of signal, power, and lighting circuits, each system shall be run in a separate compartment of the raceway.

Raceways may be extended through dry walls, dry partitions, and dry floors if in unbroken lengths where passing through.

**162. Surface nonmetallic raceway** is intended as a raceway for baseboard receptacles and other circuit supplies. It also serves a dual function as an attractive baseboard trim and replaces conventional baseboards.

Basically, the baseboard system contains a plastic backplate which is secured directly to the wall. Where receptacles are to be located, a metal bracket is attached through the backplate into the wall with wood screws or other fastening devices. Wires are then pulled in, and receptacles are mounted in place and wired. The wires are held in place along the raceway by a retaining clip strip, which is fastened to the wall of the raceway about every 5 ft (1.5 m). This holds the wiring in place. Finally, the external baseboard cover and receptacle cover are snapped into place. Covers over receptacles are doubly anchored by means of a flange and a 6-32 screw into the outlet body.

The material used in this system is the same as is used with rigid PVC conduit. End caps and inside or outside corners are available. One side of each of these fittings is cemented to the baseboard cover.

National Electric Code rules state that surface nonmetallic raceway and fittings must be of suitable nonmetallic material which is resistant to moisture and chemical atmospheres. It must also be flame-retardant and resistant to impact and crushing, distortion due to heat under conditions likely to be encountered in service, and resistant to low-temperature effects.

Surface nonmetallic raceways may be installed in dry locations. They must not be used (1) when concealed, (2) when subject to severe physical damage, (3) when the voltage is 300 V or more between conductors, (4) in hoistways, (5) in any hazardous location, except Class I, Div. 2, locations as permitted by Sec. 501-4(b), (6) when subject to ambient temperatures exceeding the listing, or (7) for conductors whose insulation exceeds those for which the nonmetallic raceway is listed.

Other Code rules such as the number and size of conductors are the same as for surface metal raceways.

**163. Walker Parkersburg surface raceway,** called SNAP CAP®, is manufactured by the Walker Parkersburg Division of Textron Inc. Two-piece construction, which eliminates wire fishing, consists of the base and the cap. The parts and assembly are shown in Fig. 9-89. SNAP CAP surface raceway is made in 15 types:

1. 111 surface raceway
2. 222 surface raceway
3. 333 surface raceway
4. 888 surface raceway
5. 711 Florduct
6. 733 Florduct
7. 1700 Surfaceduct
8. 3400 Twinduct
9. 5100 Tripleduct
10. 6800 Superduct
11. 2GW Series plug-in strip
12. 2SG Series plug-in strip
13. 4GW Series plug-in strip
14. 4SG Series plug-in strip
15. 4GWC Series plug-in strip

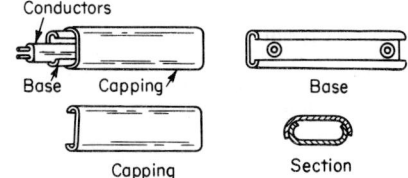

**Fig. 9-89**  Parts and assembly of SNAP CAP© surface raceway. [Walker Parkersburg Division of Textron Inc.]

The general installation procedure for all SNAP CAP surface raceway is the same, consisting of three basic steps:

1. Mount the base to the supporting structure.

2. Lay in the wires. Wires are held in place by retaining clips. Four clips are furnished with each standard length of raceway.

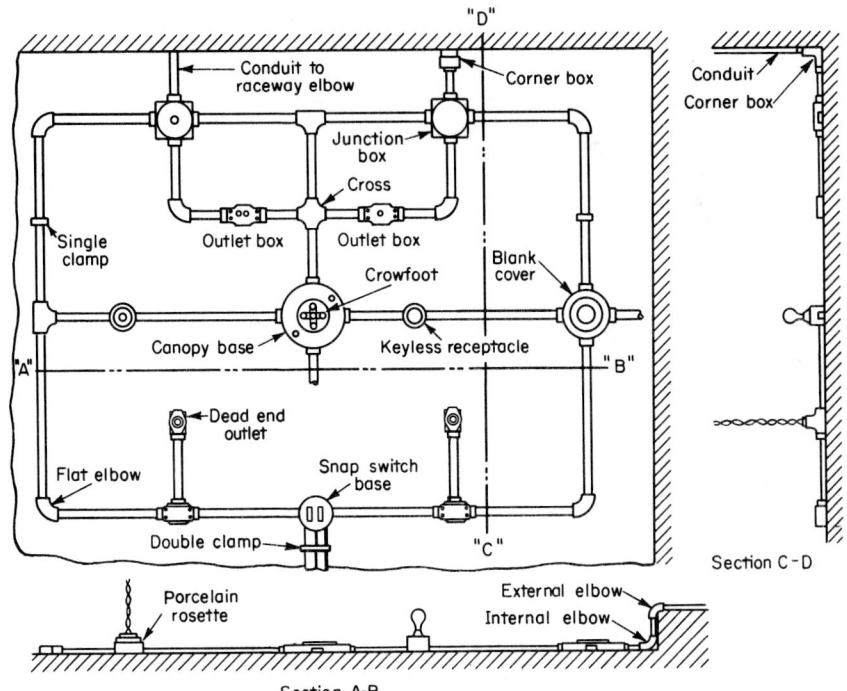

**Fig. 9-90**  Application of surface raceway and fittings.

3. Snap the capping over the base at one end, and then press the capping into place by using the palm and shifting hands progressively toward the other end.

These raceways are supported in place through holes in the back of the base by wood screws, toggle bolts, expansion plugs, or Rawl-Drives.

A complete line of fittings, including elbows, tees, crosses, terminations, junction boxes, receptacle and switch boxes, and rosettes, is available. The application of raceway and fittings is illustrated in Fig. 9-90, which shows an imaginary layout to indicate how the material may be used.

**164. No. 111 surface raceway,** the smallest-size SNAP CAP surface raceway, comes in standard 5-ft (1.5-m) lengths. No. 222 surface raceway is manufactured in standard 10-ft (3.05-m) lengths.

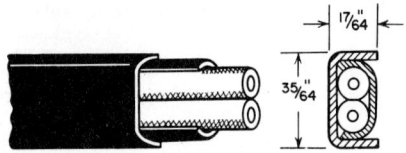

**Fig. 9-91**   No. 111 surface raceway. [Walker Parkersburg Division of Textron Inc.]

**165. No. 333 surface raceway** is manufactured in standard 10-ft (3.05-m) lengths.

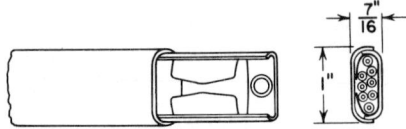

**Fig. 9-92**   No. 333 surface raceway. [Walker Parkersburg Division of Textron Inc.]

**166. No. 888 surface raceway** is manufactured in standard 10-ft (3.05-m) lengths.

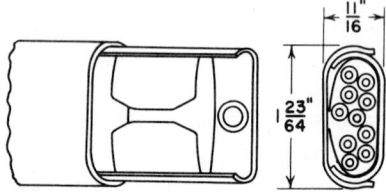

**Fig. 9-93**   No. 888 surface raceway. [Walker Parkersburg Division of Textron Inc.]

**167. SNAP CAP® 711 Florduct,** the smaller-size Florduct, is manufactured in standard 5-ft (1.5-m) lengths.

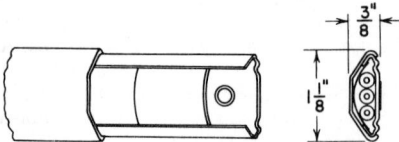

**Fig. 9-94**   No. 711 Florduct. [Walker Parkersburg Division of Textron Inc.]

**168. SNAP CAP® 733 Florduct,** the larger-size Florduct, is manufactured in standard 5-ft (1.5-m) lengths.

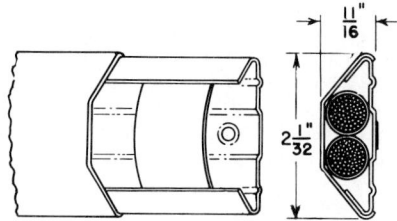

**Fig. 9-95**   No. 733 Florduct. [Walker Parkersburg Division of Textron Inc.]

**169. To install Florduct** on wood floors, use No. 4 wood screws for the smaller-size duct and No. 8 wood screws for the larger-size duct. For stone or concrete floors, drill a hole about 1 in (25.4 mm) deep with a ³⁄₁₆-in drill. A Rawl-Drive (Fig. 9-96) is then driven into the hole with a hammer and firmly grips the sides of the hole in the masonry. After the base of the duct has been fastened in this way, the wires are laid in, and the capping is snapped over the base as with the other raceways. Figure 9-97, representing a completed installation of Florduct, shows the method of attachment to wall-mounted raceway and some of the fittings.

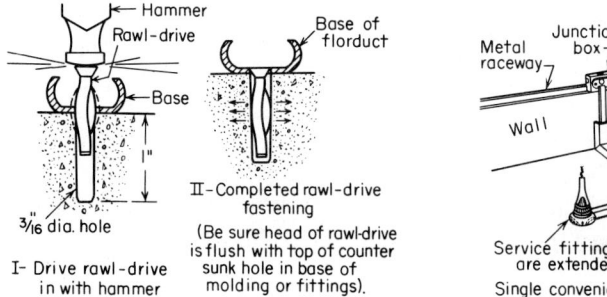

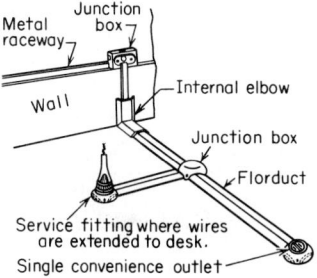

**Fig. 9-96**  Method of fastening Florduct on masonry floor. [Walker Parkersburg Division of Textron Inc.]

**Fig. 9-97**  Installation of Florduct and outlets. [Walker Parkersburg Division of Textron Inc.]

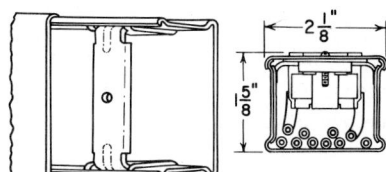

**Fig. 9-98**   SNAP CAP© 1700 surface raceway. [Walker Parkersburg Division of Textron Inc.]

**170. 1700 Surfaceduct** is a SNAP CAP surface raceway designed for commercial and industrial use. All types of service up to 60 A can be supplied from it. Device covers are available to adapt any manufacturer's approved outlet devices. A Surfaceduct system is quickly accessible for future changes or relocation of service points. One run of Surfaceduct can be used as a feeder duct for lighting and to furnish power to bench tools and other devices, as well as to provide heavier service for motor-driven equipment. The varied applications of Surfaceduct are illustrated in Fig. 9-99. It is made in lengths of 10 ft (3.05 m). The base is provided with ½- and ¾-in knockouts.

## SNAP CAP® Wire Fill
(Walker Parkersburg Division of Textron Inc.)

| | Wire size gage no. | Number of wires in one raceway | | | | |
|---|---|---|---|---|---|---|
| Type of raceway | | Types RHH, RHW | Type RH | Type THW | Type TW | Types THHN, THWN |
| No. 111 | 12 | . . . | . . . | . . . | 2 | 3 |
| | 14 | . . . | . . . | . . . | 2 | 3 |
| No. 222 | 8 | . . . | . . . | . . . | 2 | 2 |
| | 10 | 2 | . . . | 2 | 4 | 4 |
| | 12 | 2 | . . . | 3 | 5 | 7 |
| | 14 | 3 | . . . | 4 | 6 | 10 |
| No. 333 | 7 | . . . | . . . | . . . | . . . | 2 |
| | 8 | . . . | . . . | 2 | 2 | 3 |
| | 10 | 2 | 2 | 3 | 5 | 6 |
| | 12 | 2 | 4 | 4 | 6 | 9 |
| | 14 | 3 | 4 | 5 | 8 | 12 |
| No. 888 | 6 | 2 | 2 | 3 | 3 | 5 |
| | 8 | 3 | 3 | 5 | 7 | 9 |
| | 10 | 6 | 6 | 9 | 13 | 16 |
| | 12 | 7 | 10 | 11 | 17 | 25 |
| | 14 | 9 | 12 | 14 | 22 | 34 |
| No. 711 | 8 | . . . | . . . | . . . | 2 | 2 |
| | 10 | . . . | . . . | 2 | 3 | 4 |
| | 12 | 2 | 2 | 3 | 4 | 7 |
| | 14 | 2 | 3 | 4 | 6 | 9 |
| No. 733 | 6 | 3 | 3 | 4 | 4 | 7 |
| | 8 | 4 | 4 | 7 | 9 | 11 |
| | 10 | 8 | 8 | 11 | 16 | 20 |
| | 12 | 9 | 13 | 14 | 21 | 31 |
| | 14 | 11 | 16 | 17 | 27 | 42 |

Raceway dimensions:
- No. 111: 17/64 in. × 35/64 in.
- No. 222: 3/8 in. × 7/8 in.
- No. 333: 7/16 in. × 1 in.
- No. 888: 11/16 in. × 1-23/64 in.
- No. 711: 3/8 in. × 1-1/8 in.
- No. 733: 11/16 in. × 2-1/32 in.

**SNAP CAP© Wire Fill** (*Continued*)
(Walker Parkersburg Division of Textron Inc.)

| Type of raceway | Wire size gage no. | Types RHH, RHW | | Type RH | | Type THW | | Type TW | | Types THHN, THWN | |
|---|---|---|---|---|---|---|---|---|---|---|---|
| 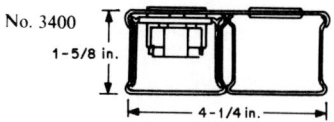 No. 1700[a] 1-5/8 in ← 2-1/8 in. → | 6 | 3[b] | 18 | 3[b] | 18 | 5[b] | 13 | 5[b] | 13 | 8[b] | 21 |
| | 8 | 5[b] | 14 | 5[b] | 14 | 8[b] | 21 | 10[b] | 27 | 13[b] | 34 |
| | 10 | 9[b] | 24 | 9[b] | 24 | 13[b] | 35 | 19[b] | 49 | 23[b] | 60 |
| | 12 | 11[b] | 28 | 15[b] | 39 | 17[b] | 43 | 24[b] | 64 | 36[b] | 94 |
| | 14 | 13[b] | 33 | 18[b] | 48 | 20[b] | 53 | 31[b] | 81 | 49[b] | 126 |

No. 3400
1-5/8 in.
← 4-1/4 in. →

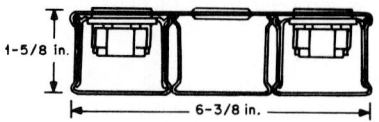

Catalog No. 3400 is a raceway consisting of two No. 1700 housings in a common cover. Each channel has the same wire fill as No. 1700.

No. 5100
1-5/8 in.
← 6-3/8 in. →

Catalog No. 5100 is a raceway consisting of three No. 1700 housings in a common cover. Each channel has the same wire fill as No. 1700.

| No. 6800[a] | 2/0 | 6[b] | 8 | 6[b] | 8 | 8[b] | 11 | 8[b] | 11 | 10[b] | 14 |
|---|---|---|---|---|---|---|---|---|---|---|---|
| 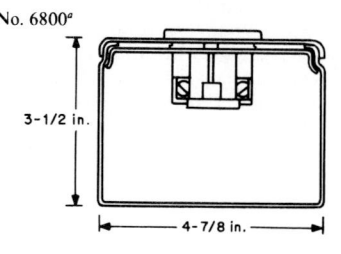 3-1/2 in. ← 4-7/8 in. → | 1/0 | 7[b] | 10 | 7[b] | 10 | 10[b] | 13 | 10[b] | 13 | 13[b] | 16 |
| | 1 | 9[b] | 11 | 9[b] | 11 | 12[b] | 15 | 12[b] | 15 | 15[b] | 20 |
| | 2 | 11[b] | 15 | 11[b] | 15 | 16[b] | 21 | 16[b] | 21 | 20[b] | 26 |
| | 3 | 13[b] | 17 | 13[b] | 17 | 19[b] | 25 | 19[b] | 25 | 25[b] | 31 |
| | 4 | 15[b] | 19 | 15[b] | 19 | 22[b] | 29 | 22[b] | 29 | 29[b] | 37 |
| | 6 | 20[b] | 25 | 20[b] | 25 | 30[b] | 38 | 30[b] | 38 | 47[b] | 61 |
| | 8 | 32[b] | 41 | 32[b] | 41 | 47[b] | 60 | 60[b] | 77 | 78[b] | 100 |
| | 10 | 53[b] | 69 | 53[b] | 69 | 79[b] | 102 | 110[b] | 141 | 134[b] | 172 |
| | 12 | 64[b] | 82 | 89[b] | 114 | 98[b] | 126 | 143[b] | 185 | 211[b] | 271 |
| | 14 | 75[b] | 97 | 107[b] | 138 | 120[b] | 154 | 183[b] | 235 | 284[b] | 365 |

[a]Figures are without devices except as noted.
[b]With devices.

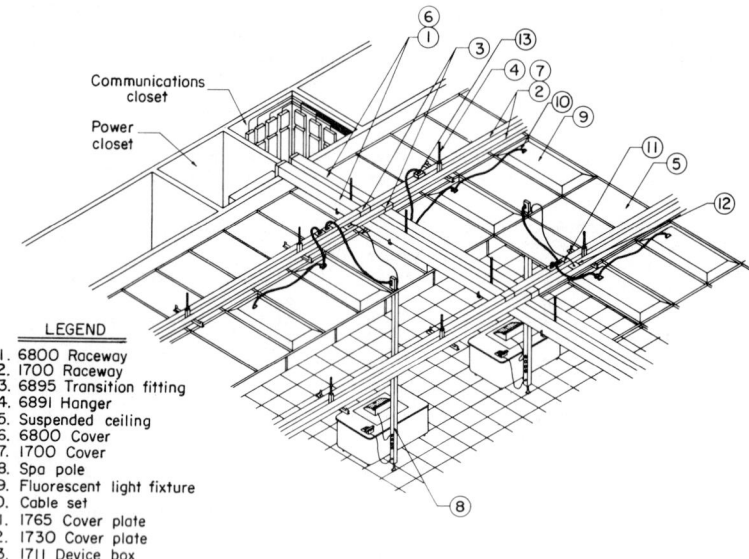

Communications closet

Power closet

LEGEND
1. 6800 Raceway
2. 1700 Raceway
3. 6895 Transition fitting
4. 6891 Hanger
5. Suspended ceiling
6. 6800 Cover
7. 1700 Cover
8. Spa pole
9. Fluorescent light fixture
10. Cable set
11. 1765 Cover plate
12. 1730 Cover plate
13. 1711 Device box

**Fig. 9-99**   No. 1700 Surfaceduct application used as Walkertrak ™ overhead raceway system. [Walker Parkersburg Division of Textron Inc.]

**171. SNAP CAP 3400 Twinduct** surface raceway provides a combination high- and low-potential raceway. Required devices and terminal blocks are mounted and wired prior to the mounting of the screw-supported cover. Both raceways are readily accessible for maintenance, wiring changes, or additions to systems by the removal of the single cover. The low-potential raceway accommodates a 50-pair telephone cable or several 25-pair telephone cables. Covers are available for telephone jacks. No. 3400 Twinduct is manufactured in standard 10-ft (3.05-m) base length with two 5-ft (1.5-m) caps. The wire fill for the high-potential (upper) raceway is the same as that given for No. 1700 Surfaceduct.

SNAP CAP 5100 Tripleduct surface raceway provides a combination of high- and low-potential raceway and signal application under one cover. Required devices and terminal blocks are mounted and wired prior to the mounting of the screw-supported cover. All three raceways are readily accessible by the removal of the single cover. No. 5100 Tri-

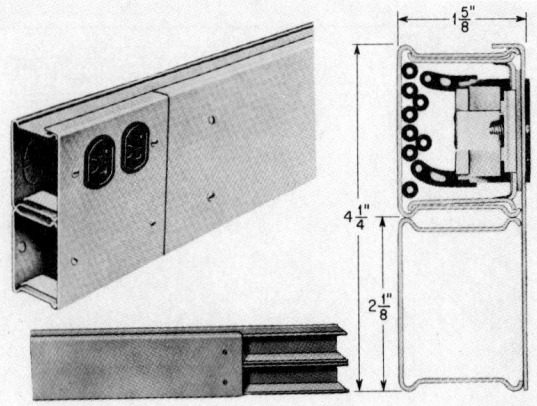

**Fig. 9-100**   SNAP CAP© Twinduct surface raceway. [Walker Parkersburg Division of Textron Inc.]

pleduct is manufactured in standard 10-ft (3.05-m) base length with two 5-ft (1.5-m) caps. The wire fill and telephone cable data for each of the three ducts are identical to those of No. 1700 Surfaceduct.

SNAP CAP 6800 Superduct provides capacity for large jobs. However, like No. 1700, it has no isolation and must be used for either high or low potential. It is manufactured in standard 5-ft lengths.

**172. SNAP CAP plug-in strips** are factory prewired multioutlet assemblies with a varying number of receptacles per piece, depending on how they are ordered. Plug-in strips come in five different series:

1. *No. 2GW; wire ground.* Available in 3-, 6-, or 10-ft (0.9-, 1.8-, or 3.05-m) lengths. Receptacles are the gray, 15-A, 125-V, three-wire type.

2. *No. 2SG; system ground.* Available in 3-, 6-, or 10-ft lengths. Receptacles are the gray, 15-A, 125-V, two-wire type.

3. *No. 4GW; wire ground.* Same as No. 2GW but larger to accommodate additional circuit wiring.

4. *No. 4SG; system ground.* Same as No. 2SG but larger to accommodate additional circuit wiring.

5. *No. 4GWC; receptacles wired alternately.* Same as No. 4GW but already wired for two or three circuits.

**173. Installing SNAP CAP® surface raceway.**   Reasonable care should be exercised in separating the backing and capping preparatory to installation. As the quickest, most satisfactory method, hooking one of the punched holes in the backing over a convenient nail or screw and drawing the capping off is recommended.

**174. Cutting raceway.**   Except in cases in which the backing of the raceway passes through under the fittings and is not cut, backing and capping should always be cut before being separated.

**175. Bending metal raceway.**   The raceway is readily bent and with reasonable care can be worked to any radius down to one of 4½ in (114.3 mm). Bends must be made in all cases before backing and capping are separated.

**176. Supporting raceway.**  The backing is punched and countersunk every 2 ft (0.6 m) for the supporting screws or bolts. The support so afforded will usually be found more than ample, but further support can be secured through additional punching with a special punch or through the use of a metal raceway clamp. Figure 9-101 shows a toggle-bolt support for a metal raceway. When the metal raceway is installed on uneven surfaces, such as the ceilings of old buildings, the capping has a tendency to spring away from the backing. This tendency can be overcome by the use of two or three straps fastened over each length.

**177. Loose raceway capping.**  If the capping of the raceway is loose, it should be removed from the backing and tightened by tapping it with a mallet at points 8 in (203.2 mm) apart on one edge only.

**178. Wiremold metal raceways** are made in three general types by the Wiremold Co. These types are designated as (1) Wiremold surface raceway, used for general surface-raceway wiring systems; (2) Pancake overfloor raceway, used for overfloor surface-raceway wiring; and (3) Plugmold raceway, used primarily for multioutlet-assemblies wiring (see Sec. 278). Plugmold can be used without receptacles as a general high-capacity surface-wiring raceway. The Wiremold surface raceway is a unit assembly, while the other two types consist of two separable parts, the base and the cover.

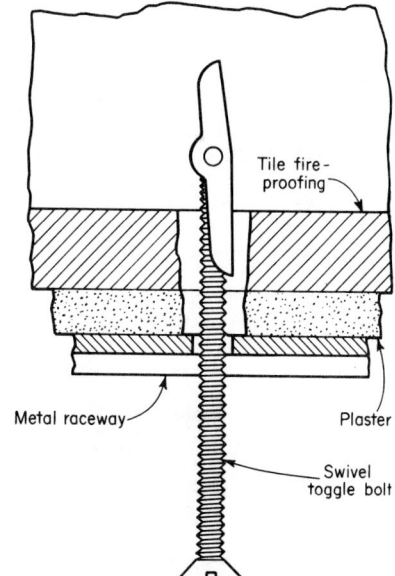

**Fig. 9-101**  Toggle bolt being inserted.

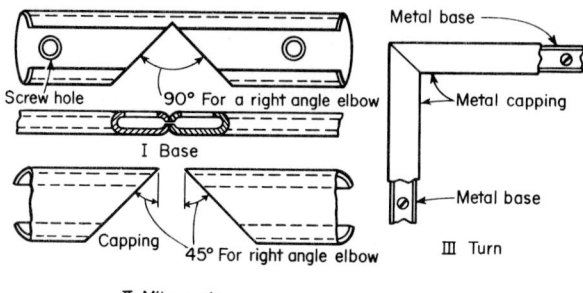

**Fig. 9-102**   Mitered turn.

The appearance and dimensions of the different types and sizes are shown in Fig. 9-104. The wire capacities are given in Secs. 179 to 181A inclusive. Wiremold No. 200 and Plugmold Nos. 2000 and 2200 are made in 5-ft (1.5-m) lengths. The standard length for the other types is 10 ft (3.05 m), although some types are available also in 5-ft lengths.

Wiremold surface metal raceway consists of a metal capping with its edges crimped over a slightly curved base. The capping is crimped on the base at the factory so that capping and base are installed as a unit. The wires are pulled through the raceway from fitting to fitting in a manner similar to that employed in conduit work. The lengths are fastened together with a coupling (Fig. 9-105, I) which is fastened to the wall surface with a wood screw or toggle bolt. The base of the raceway is slipped under the end of the coupling before the screw or bolt is drawn up tight. A connection cover (Fig. 9-105, II) snaps over the ends of the raceway, closing the joint. Each section of raceway is secured at one or two intermediate points between the couplings by means of either one- or two-

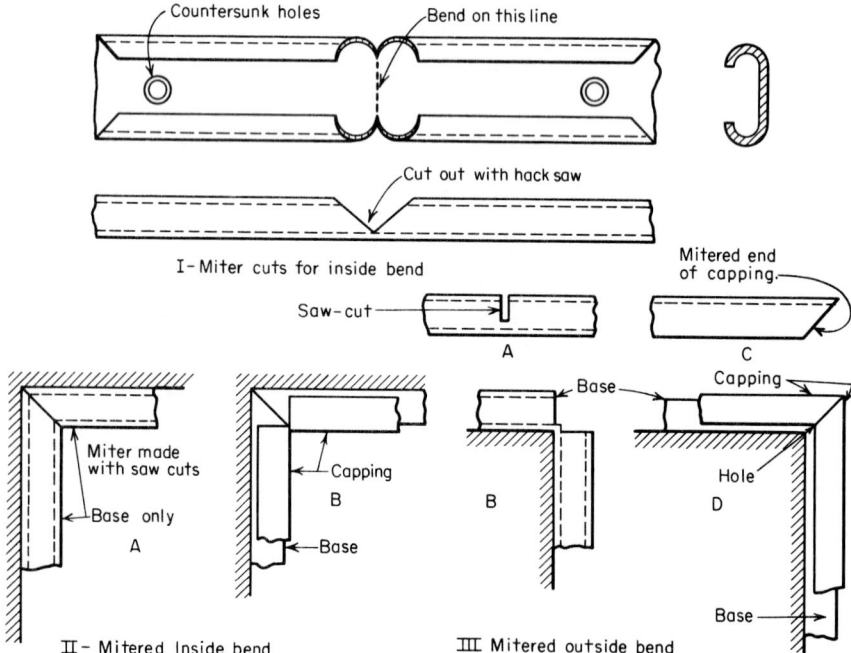

**Fig. 9-103**   Mitered elbows.

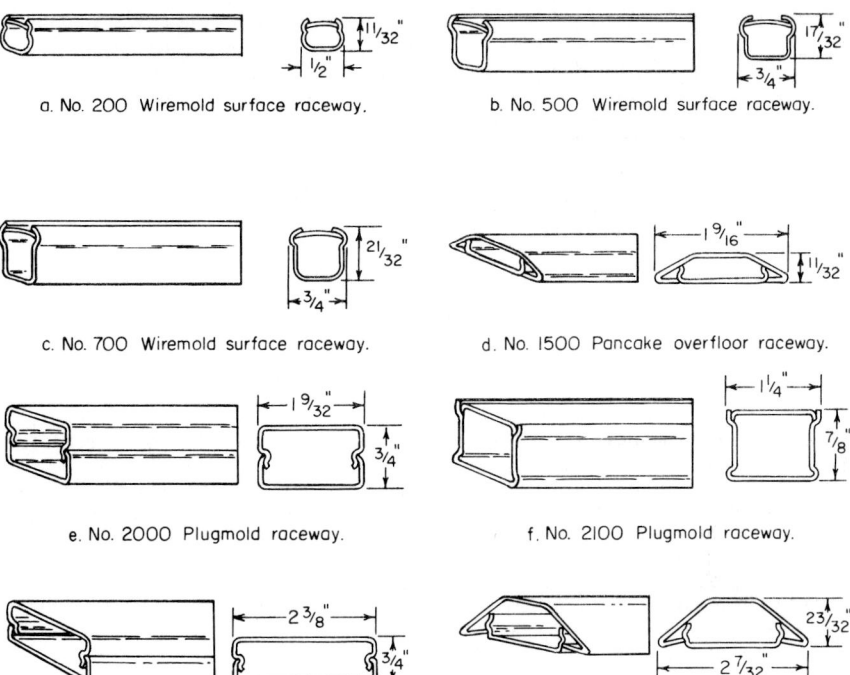

a. No. 200  Wiremold surface raceway.

b. No. 500  Wiremold surface raceway.

c. No. 700  Wiremold surface raceway.

d. No. 1500  Pancake overfloor raceway.

e. No. 2000  Plugmold raceway.

f. No. 2100  Plugmold raceway.

g. No. 2200  Plugmold baseboard raceway.

h. No. 2600  Pancake overfloor raceway.

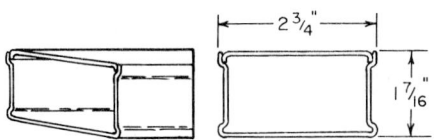

i. No. G-3000  surface raceway.

j. No. G-4000 surface raceway.

k. No. G-6000  surface raceway.

**Fig. 9-104**  Types and dimensions of Wiremold raceways. [The Wiremold Co.]

hole straps (Fig. 9-105, III and IV) screwed or bolted to the wall surface. The fittings and capping of Wiremold are finished in a neutral brown tint which blends well with most surroundings. The base of the raceway and the baseplates of all fittings are given a special zinc treatment to prevent corrosion.

Pancake overfloor raceway consists of two steel parts, a base and a cover. The parts have a natural galvanized finish. When this raceway is installed, the base is fastened to the floor by the appropriate means of wood screws, expansion shields, toggle bolts, etc. The con-

I - Coupling          II-Connection cover          III-Two hole strap          IV - One hole strap

**Fig. 9-105**   Wiremold metal raceway and support details.

ductors are then laid flat in the base of the raceway. After the connections have been completed, the cover is snapped in place over the base.

Plugmold raceway consists of two steel parts, a channel base and a snap-on cover. Both base and cover are available in the standard brown Wiremold finish or in a gray finish. When this raceway is installed, the channel base is fastened to the supporting structure with the appropriate means. The conductors are laid in the channel base, the necessary connections made, and the installation completed by snapping the cover over the base.

A very complete line of fittings, outlets, and receptacles is available for the different Wiremold raceways to meet the varied needs of installation.

**179.  Wire Capacities for Nos. 200 to 3000 Wiremold Raceways**

(For not more than the number of wires of sizes and types indicated when no wiring devices are mounted on or in raceway)

| Catalog numbers | | Wire size gage no. | Wire fill | | | |
|---|---|---|---|---|---|---|
| | | | Types RHH, RHW | Type THW | Type TW | Types THHN, THWN |
| No. 200 | 11/32 in. — 1/2 in. | 12 | ... | 2 | 3 | 3 |
| | | 14 | ... | 2 | 3 | 3 |
| No. 500 | 17/32 in. ← 3/4 in. → | 8 | | | 2 | 2 |
| | | 10 | 2 | 2 | 3 | 4 |
| | | 12 | 2 | 3 | 4 | 7 |
| | | 14 | 2 | 4 | 6 | 9 |
| No. 700 | 21/32 in. ← 3/4 in. → | 6 | ... | ... | ... | 2 |
| | | 8 | ... | 2 | 2 | 3 |
| | | 10 | 2 | 3 | 4 | 5 |
| | | 12 | 2 | 4 | 6 | 8 |
| | | 14 | 3 | 5 | 7 | 11 |
| No. 2000 | ← 1- 9/32 in. → 3/4 in. | 12 | ... | ... | 3 | 3 |
| | | 14 | ... | ... | 3 | 3 |
| No. 2100 | 7/8 in. ← 1-1/4 in. → | 6 | 2 | 4 | 4 | 6 |
| | | 8 | 4 | 6 | 8 | 10 |
| | | 10 | 7 | 10 | 14 | 17 |
| | | 12 | 8 | 13 | 19 | 28 |
| | | 14 | 10 | 15 | 24 | 37 |
| No. 2200 | ← 2-3/8 in. → 3/4 in. | 6 | 5 | 7 | 7 | 11 |
| | | 8 | 8 | 11 | 14 | 19 |
| | | 10 | 13 | 19 | 26 | 32 |
| | | 12 | 15 | 23 | 34 | 51 |
| | | 14 | 18 | 29 | 44 | 69 |
| No. G3000 | 1-7/16 in. ← 2-3/4 in. → | 6 | 11 | 19 | 17 | 27 |
| | | 8 | 18 | 26 | 34 | 44 |
| | | 10 | 30 | 45 | 62 | 76 |
| | | 12 | 36 | 55 | 81 | 119 |
| | | 14 | 42 | 67 | 103 | 160 |

### 180.   Wire Capacities for Nos. 1500 and 2600 Overfloor Wiremold

(For use on floors in dry locations not subject to excessive mopping or scrubbing; raceway to be used with not more than the number of wires of sizes and types indicated)

| Type of raceway | Wire size gage no. | Number of wires | | | |
|---|---|---|---|---|---|
| | | Types RHH, RHW | Type THW | Type TW | Types THHN, THWN |
| No. 1500 — 1-9/16 in. — 11/32 in. | 6 | . . . | . . . | . . . | 2 |
| | 8 | . . . | . . . | 2 | 3 |
| | 10 | 2 | 3 | 4 | 5 |
| | 12 | 2 | 3 | 5 | 7 |
| | 14 | 2 | 4 | 6 | 10 |
| No. 2600 — 23/32 in. — 2-7/32 in | 6 | 2 | 3 | 3 | 5 |
| | 8 | 4 | 5 | 7 | 9 |
| | 10 | 6 | 9 | 12 | 15 |
| | 12 | 7 | 11 | 16 | 24 |
| | 14 | 9 | 14 | 21 | 33 |

**181. Wire Capacities for Wiremold Nos. 2000 to 3000 Plugmold**

(For use with Wiremold attachment-plug receptacles designed for mounting in raceway; raceway to be used with not more than the number of wires of sizes and types indicated)

### Catalog No. 2000 wire capacity

|  |  | Single-conductor | |
|---|---|---|---|
| Type of wire | | No. 12 | No. 14 |
| With receptacles | RHH, RHW, TW, THW, THHN, THWN | 7 | 7 |

### Catalog No. 2200 wire capacity

|  |  | Single-conductor | | | |
|---|---|---|---|---|---|
|  | Type of wire | No. 6 | No. 8 | No. 10 | No. 12 | No. 14 |
| With receptacles | TW | 3 | 7 | 10 | 10 | 10 |

### Plugmold No. 2100: wire capacities with devices

| Device no. | Description | Types TW. | | Types R, RH | | Types THHN, THWN | |
|---|---|---|---|---|---|---|---|
|  |  | No. 12 | No. 14 | No. 12 | No. 14 | No. 12 | No. 14 |
| 2126 | Keyless socket | 12 | 14 | 6 | 6 | 14 | 18 |
| 2126P | Keyless socket | 12 | 14 | 6 | 6 | 14 | 18 |
| 2127DC | Lumiline single | 10 | 10 | 6 | 6 | 10 | 14 |
| 2127G | Grounding | 4 | 6 | 3 | 3 | 6 | 8 |
| 2727G3 | With 6-in ground wire | 4 | 6 | 3 | 3 | 6 | 8 |
| 2127GA | Three-wire grounding | 8 | 10 | 6 | 6 | 10 | 10 |
| 2127GB | Three-wire grounding | 8 | 10 | 6 | 6 | 10 | 10 |
| 2127GT | Three-wire grounding | 8 | 10 | 6 | 6 | 10 | 10 |
| 2127PG | Three-wire polarized grounding | 8 | 10 | 6 | 6 | 10 | 10 |
| 2140 | Single-pole switch | 8 | 10 | 10 | 10 | 10 | 14 |

(For use with standard flush-mounted attachment-plug receptacle of type not having pilot lights; raceway to be used with not more than the number of wires of sizes and types indicated)

### Catalog No. G-3000 wire capacity

| Single-conductor | Wire types | No. 6 | No. 8 | No. 10 | No. 12 | No. 14 |
|---|---|---|---|---|---|---|
| With devices | RHH, RHW | 4 | 6 | 10 | 14 | 16 |
|  | THHN, THWN, TW, THW | 6 | 8 | 10 | 18 | 26 |

## 181A.  Wire Capacities for Nos. 4000 and 6000 Wiremold Raceways
(For use with not more than the number of wires of sizes and types indicated)

### Catalog No. G-4000 wire capacities

| Single-conductor | Without devices | | | | | | | | With devices | | | | |
|---|---|---|---|---|---|---|---|---|---|---|---|---|---|
| Wire types | No. 2 | No. 3 | No. 4 | No. 6 | No. 8 | No. 10 | No. 12 | No. 14 | No. 6 | No. 8 | No. 10 | No. 12 | No. 14 |
| With divider: Power compartment only | | | | | | | | | | | | | |
| TW | 10 | 11 | 13 | 18 | 36 | 66 | 86 | 110 | 4 | 7 | 11 | 15 | 17 |
| THW | 10 | 11 | 13 | 18 | 28 | 48 | 59 | 72 | 4 | 7 | 11 | 15 | 17 |
| THWN, THHN | 12 | 15 | 17 | 28 | 47 | 81 | 128 | 171 | 7 | 8 | 15 | 24 | 32 |
| RHH, RHW | 7 | 8 | 9 | 12 | 19 | 32 | 39 | 45 | 4 | 7 | 11 | 15 | 17 |
| Without divider: | | | | | | | | | | | | | |
| TW | 20 | 23 | 27 | 36 | 78 | 133 | 174 | 222 | 8 | 10 | 12 | 16 | 17 |
| THW | 20 | 23 | 27 | 36 | 57 | 96 | 119 | 145 | 8 | 10 | 15 | 21 | 21 |
| THWN, THHN | 25 | 30 | 35 | 57 | 94 | 163 | 256 | 344 | 10 | 15 | 18 | 34 | 34 |
| RHH, RHW | 14 | 16 | 18 | 24 | 39 | 65 | 78 | 91 | 8 | 10 | 15 | 21 | 21 |

### Communications and signal wiring, with divider[a]

| Telephone cable | 5-Pair | 12-Pair | 16-Pair | 25-Pair | 50-Pair | 75-Pair | 100-Pair |
|---|---|---|---|---|---|---|---|
| No. of cables | 19 | 12 | 9 | 7 | 4 | 3 | 2 |

| Wire type | | JKT-2 | JKT-3 | JKT-4 |
|---|---|---|---|---|
| Without devices | | 70 | 60 | 50 |
| With Jack Nos. 493A and 541A; connecting block Nos. 47B and 47D; and KS-type plug connector | | 35 | 30 | 25 |

[a]Without divider: Double the capacity for raceway with divider.

### Catalog No. G-6000 wire capacities

| Wire types | Number and conductor size (AWG) | | | | | | | | | | |
|---|---|---|---|---|---|---|---|---|---|---|---|
| | 2/0 | 1/0 | 1 | 2 | 3 | 4 | 6 | 8 | 10 | 12 | 14 |
| Without devices: | | | | | | | | | | | |
| TW | 22 | 26 | 31 | 43 | 50 | 58 | 77 | 134 | 282 | 368 | 469 |
| THW | 22 | 26 | 31 | 43 | 50 | 58 | 77 | 106 | 203 | 252 | 307 |
| THWN, THHN | 27 | 33 | 39 | 53 | 63 | 74 | 122 | 169 | 343 | 540 | 726 |
| RHH, RHW | 17 | 20 | 23 | 30 | 34 | 39 | 51 | 74 | 137 | 164 | 193 |
| With devices: | | | | | | | | | | | |
| TW | 12 | 14 | 17 | 23 | 27 | 32 | 42 | 73 | 154 | 200 | 255 |
| THW | 12 | 14 | 17 | 23 | 27 | 32 | 42 | 57 | 111 | 137 | 167 |
| THWN, THHN | 15 | 18 | 21 | 29 | 34 | 40 | 66 | 92 | 187 | 295 | 396 |
| RHH, RHW | 10 | 11 | 12 | 16 | 19 | 21 | 27 | 40 | 75 | 90 | 105 |

| Telephone cable | 25-Pair | 50-Pair | 75-Pair | 100-Pair |
|---|---|---|---|---|
| No. of cables | 39 | 20 | 14 | 10 |

**182.  Cross-Sectional Areas of Wiremold Raceways**

| Raceway | Area, in² | Raceway | Area, in² |
|---------|-----------|---------|-----------|
| 200 | 0.11 | | |
| 500 | 0.20 | 2600 | .72 |
| 700 | 0.25 | | |
| | | G-3000 | 3.51 |
| | | G-400 | 7.50 (undivided) |
| | | | 3.70 (each compartment when |
| 1500 | 0.22 | | raceway is divided) |
| 2000 | 0.80 | | |
| 2100 | 0.81 | G-6000 | 15.82 |
| 2200 | 1.50 | | |

## ELECTRICAL-METALLIC-TUBING WIRING

**183. Wiring employing electrical metallic tubing** (EMT) is widely used for commercial and industrial wiring systems. The NEC allows the installation of EMT, either concealed or exposed, when either during construction or afterward it will not be subject to severe physical damage. Electrical metallic tubing may be supported in buildings of fire-resistive construction on the face of masonry or other material of which walls or ceilings are composed. It may then be buried in concrete or plaster finish. Connections between lengths of tubing and between tubing and any fitting shall provide adequate mechanical strength and electrical continuity.

In installations having cinder concrete or fill, EMT must be protected on all sides by a layer of noncinder concrete at least 2 in (50.8 mm) thick unless the tubing is at least 18 in (457.2 mm) under the fill.

In installations in which corrosive fumes or vapors may exist, as in meat-packing plants, tanneries, hide cellars, casing rooms, glue houses, fertilizer rooms, salt storage, some chemical works, metal refineries, pulp mills, sugar mills, roundhouses, some stables, and similar locations, special tubing and fittings made of corrosion-resistant material approved for the purpose must be used (refer to Sec. 192).

When installed in wet locations, such as dairies, laundries, and canneries, and in locations where the walls are frequently washed, the entire tubing system, including all its boxes and other fittings, shall be installed and equipped so as to prevent water from entering the tubing system, and the tubing shall be mounted so that there is at least a ¼-in (6.35-mm) air space between it and the wall or other supporting surface. For these installations all the supporting materials, such as straps, bolts, and screws, must be of corrosion-resistant material or be protected against corrosion by approved materials.

The rules regarding bends, minimum size, and allowable number of conductors are the same as for rigid metal conduit. The maximum allowable size of tubing is 4 in.

**184. Table 185 gives the sizes and weights of electrical metallic tubing.** Since the inside diameters are the same as those for rigid metal conduit, the same tables of number of allowable wires in a conduit apply to the tubing. The wall is thinner, however, and the outside diameters of tubing are less than those for conduit. The regular metallic tubing has a round cross section and is made of galvanized steel. However, sizes over 2 in have the same outside diameters as corresponding sizes of rigid metal conduit. Accordingly,

**185.  Electrical Metallic Tubing**

| Size, inches | Approximate weight per 1,000 ft, lb | Diameter, inches | | Wall thickness, inch |
|--------------|-------------------------------------|------------------|------------------|----------------------|
| | | Inside | Outside | |
| ½ | 295 | 0.622 | 0.706 | 0.042 |
| ¾ | 445 | 0.824 | 0.922 | 0.049 |
| 1 | 650 | 1.049 | 1.163 | 0.057 |
| 1¼ | 960 | 1.380 | 1.510 | 0.065 |
| 1½ | 1,110 | 1.610 | 1.740 | 0.065 |
| 2 | 1,410 | 2.067 | 2.197 | 0.065 |
| 2½ | 2,300 | 2.731 | 2.875 | 0.072 |
| 3 | 2,700 | 3.356 | 3.500 | 0.072 |
| 4 | 4,000 | 4.334 | 4.500 | 0.083 |

the internal diameters of these larger sizes are slightly greater than those of other types of conduit.

**186. All outlet boxes and fittings** for use with EMT must be of the threadless type, since owing to the thin wall it is not permissible to thread the tubing. Special threadless fittings similar to those employed for standard rigid conduit are manufactured for tubing. Standard rigid conduit fittings can be used with tubing if an adapter is employed. Sections of tubing can be joined by means of threadless couplings, which are available in steel and malleable-iron construction. Tubing is connected to outlet boxes by means of connectors. One end of the connector is made for threadless attachment to the tubing, and the other end is threaded and supplied with a locknut for attachment to the box through a knockout. Raintight connections can be made to the tubing with some couplings and connectors. Standard 90° elbows (Fig. 9-106) are attached to the tubing by means of the above-described couplings. The dimensions of standard elbows are given in Sec. 187. Raintight elbows and angle-box connectors are available with the ends fitted with threadless fittings.

### 187.  Standard 90° Elbows for EMT

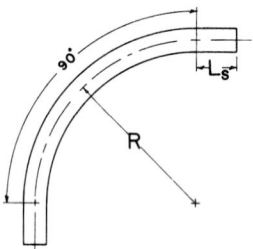

**Fig. 9-106**   Standard radius, 90° elbow.

(All dimensions in inches)

| Trade size of tubing | Smallest acceptable radius $R$ to center line of tubing | Shortest acceptable length $Ls$ of each straight end portion of tubing |
|:---:|:---:|:---:|
| ½ | 4 | 1½ |
| ¾ | 4½ | 1½ |
| 1 | 5¾ | 1⅞ |
| 1¼ | 7¼ | 2 |
| 1½ | 8¼ | 2 |
| 2 | 9½ | 2 |
| 2½ | 10½ | 3 |
| 3 | 13 | 3⅛ |
| 4 | 16 | 3⅜ |

**188. Threadless couplings and connectors** for EMT are of setscrew, compression, indenter, tap-on, two-piece, or squeeze types. Such fittings are classified as concrete-tight only, or raintight and concrete-tight. Accordingly, only fittings that are classified as being raintight can be used outdoors exposed to the weather or in wet locations. Except for indenter-type fittings, EMT couplings and connectors can be installed with common hand tools. Indenter-type fittings require a special tool, and these fittings are made only up to 1 in in size.

**189. Supports for EMT.**   The NEC requires EMT to be securely fastened in place at least every 10 ft (3.05 m) and within 3 ft (914 mm) of each outlet box, junction box, cabinet, or fitting. Methods of support are similar to those described for rigid metal conduit. Figure 9-107 shows the use of spring clips to secure EMT to U-channel supports.

Typical Tubing Installation

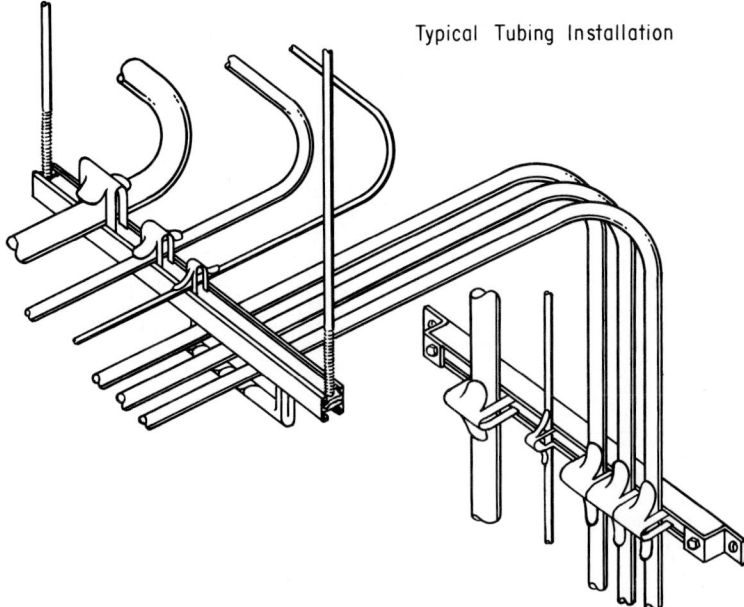

**Fig. 9-107**　Spring clips secure EMT to U-channel supports.

**190. Bending EMT.** Bends in EMT must be so made that the tubing will not be injured and the internal diameter of the tubing will not be effectively reduced. The radius of the curve of the inner bend or any field bend shall not be less than shown in Secs. 71 and 78. For ½-, ¾-, and 1-in EMT a hand bender, as shown in Fig. 9-108, I, is used. The EMT hickey shown in Fig. 9-108, II, is only for making small offsets in 1¼-, 1½-, or 2-in EMT. Almost all bends in sizes larger than 1 in are made with hydraulic benders, and many rigid-metal-conduit hand *benders* are suitable for EMT. Refer to discussions in Secs. 99 and 100.

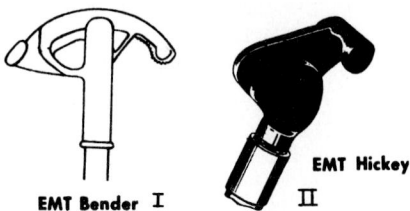

EMT Bender  I　　EMT Hickey  II

**Fig. 9-108**　Typical EMT bender and hickey.

**191. Installation methods.** The techniques of installation described in the sections for rigid metal conduit will apply, in general, to EMT. Such items as junction or pull boxes, number of conductors permitted in a single raceway, and support hardware apply to both these wiring methods.

**192. Corrosion-resistant metallic tubing** is available in two types, copper alloy and plastic-coated. The principles of construction and the characteristics of both these types are the same as for the corresponding types of corrosion-resistant rigid conduit, as discussed in Sec. 58. The dimensions for silicon bronze EMT, called Everdur EMT, are given in Sec. 193. They are practically the same as for steel EMT.

**193.    Data for Everdur Silicon Bronze Electrical Metallic Tubing**

(American Brass, Anaconda Co.)

| Nominal size, in. | Nominal dimensions, in. | | | Unit length, ft | Min weight, lb per 100 ft |
|---|---|---|---|---|---|
| | Diameter | | Wall thickness | | |
| | Outside | Inside | | | |
| ⅜ | 0.577 | 0.493 | 0.042 | 10 | 25½ |
| ½ | 0.706 | 0.622 | 0.042 | 10 | 32 |
| ¾ | 0.922 | 0.824 | 0.049 | 10 | 48½ |
| 1 | 1.163 | 1.049 | 0.057 | 10 | 71½ |
| 1¼ | 1.510 | 1.380 | 0.065 | 10 | 106 |
| 1½ | 1.740 | 1.610 | 0.065 | 10 | 123 |
| 2 | 2.197 | 2.067 | 0.065 | 10 | 156 |

## NONMETALLIC-SHEATHED–CABLE WIRING

**194. Nonmetallic-sheathed cable is used** for the wiring of new and old residences and for small offices and stores. It is permitted in one- and two-family dwellings or in multifamily dwellings or other structures not exceeding three floors above grade. This cable is frequently employed for extensions to existing installations and is commonly used in small apartment houses. Nonmetallic-sheathed cable can be run either exposed on ceilings and walls or concealed in hollow partitions or between floors and ceilings. It provides a reasonably safe and reliable wiring installation of low first cost. It is made in two types, Type NM and Type NMC. Conductors are copper or aluminum.

Type NM may be used only in normally dry locations which are free from corrosive vapors or fumes. It may be run or fished in air voids in masonry-block or tile walls where not exposed or subject to excessive moisture or dampness. Type NM must not be embedded in masonry, concrete, fill, or plaster or run in a shallow chase in masonry or concrete and covered with plaster or similar finish.

Type NMC is moisture- and corrosion-resistant, so that its use is allowed in dry, moist, damp, or corrosive locations and in outside and inside walls of masonry block or tile. If embedded in plaster or run in a shallow chase in masonry walls and covered with plaster, it shall be protected against damage from nails by a cover of corrosion-resistant coated steel at least 1⁄16 in (1.59 mm) thick under the final surface finish.

Neither Type NM nor Type NMC is allowed to be used (1) as service-entrance cable, (2) in commercial garages, (3) in theaters and similar locations except as provided in Article 518 of the NEC, (4) in motion-picture studios, (5) in storage-battery rooms, (6) in hoistways, (7) in any hazardous location except as permitted by Secs. 501-4(b), Ex., and 504-20, or (8) embedded in poured cement, concrete, or aggregate.

**195. Nonmetallic-sheathed cable, with or without grounding wires, is manufactured** in both two- and three-conductor cables. Each wire is insulated with rubber or thermoplastic of the same thickness as required on single wires, except that the cable may have an approved size of covered or bare conductor for grounding purposes only. In Type NM cable the two or three conductors have an outer flame-retardant and moisture-resistant coverings. Type NMC cable is similar in construction to Type NM cable except that it is protected by an outer covering that is flame-retardant and moisture-, fungus-, and corrosion-resistant.

**196. When nonmetallic-sheathed cable is installed exposed,** it is supported directly on the walls or ceiling with straps. For three-wire cable, ordinary pipe straps are used. Standard ⅜-in pipe straps fit Nos. 14(3), 12(3), and 10(3) sizes of sheathed cable; ½-in pipe straps are required for No. 8(3); ¾-in straps, for No. 6(3); and 1-in straps, for No. 4(3). Two-wire cable requires special flat straps, which are available from various manufacturers. The distance between supports must not be greater than 4½ ft (1.37 m), and there must be a support within 12 in (305 mm) of each outlet box or fitting. The cable must follow the surface of the wall or ceiling and not break from beam to beam. Whenever nonmetallic-sheathed cable does not follow the building finish, it must be protected from physical damage. This protection may consist of a wooden or metal protecting strip or a piece of conduit or EMT. It is preferable to use either conduit or EMT, especially where the cable passes through floors.

**197. When nonmetallic-sheathed cable is installed concealed** in new buildings, the cable is fastened to the joists or beams by means of staples. The distance between supports should not be greater than 4½ ft (1.37 m). In wiring finished buildings, when it is impracticable to support the cable between outlets, it can be fished from outlet to outlet in the same manner as Type AC metal-clad cable. Where the cable is run through holes bored in studs, joists, etc., no protection is required. The holes should be bored as near to the center of the timber as possible and not less than 1¼ in (31.8 mm) from the nearest edge when practical. An approved method of installing nonmetallic-sheathed cable in a new building is shown in Fig. 9-109.

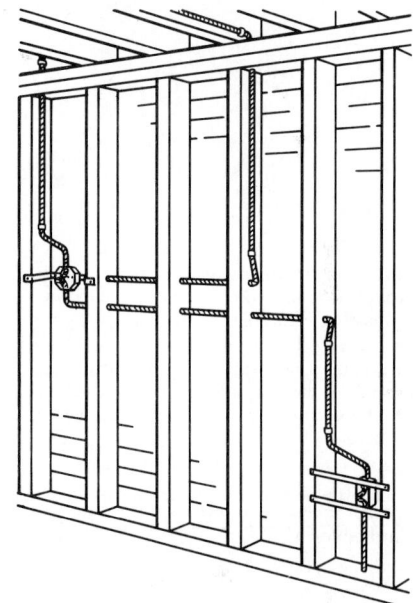

**198. No joints or splices** should be made in nonmetallic-sheathed–cable wiring except in outlet or junction boxes. This rule applies to both exposed and concealed installations. The loop system is employed generally for concealed work. The cable is looped from outlet to outlet, and the use of junction boxes is avoided.

**199. Bends in cable** (National Electrical Code) shall be so made and other handling shall be such that the protective coverings of the cable will not be injured, and no bend shall have a radius less than 5 times the diameter of the cable.

**200. An approved outlet box or fitting** must be provided at all outlets or switch locations. The cable must be attached to the box with some form of approved device which will substantially close the opening in the box. This connection of the cable to the box can be accomplished

**Fig. 9-109** Method of installing nonmetallic-sheathed cable concealed in new buildings.

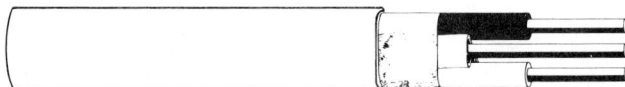

**Fig. 9-110** Type NM cable marked NM-B has insulation rated 90°C with a cable ampacity of 60°C.

by means of a box with a built-in clamp, or a clamp connector (Fig. 9-111) or a squeeze connector (Fig. 9-112) can be employed with a box having conduit knockouts. For data on metal boxes, see Sec. 115 of Div. 4.

**201. When NEC rules require that outlet boxes be grounded** or when grounding-type receptacles are used, nonmetallic-sheathed cable can be obtained with a grounding wire built into the cable under the outer sheath. This grounding wire should be ripped back out of the cable for 6 or 8 in (152 or 203 mm) before the cable is secured to the outlet box. If outlet boxes with built-in clamps are employed, the required contact between the box and the grounding wire is made by means of a small grounding clip. The connection is made by fastening the clip over the grounding wire and box (Fig. 9-113). These copper grounding clips are commonly called G clips.

**202. Boxes of insulating material** (Fig. 4-156 of Div. 4) can be used with nonmetallic-sheathed cable instead of metal boxes. With these boxes, grounding is limited, and it is not necessary to connect each box to a grounding conductor. The boxes shown in Fig. 4-156 of Div. 4 are made of plastic. They are used in the same way as metal boxes except that the cable need not be fastened to the box but must be fastened to the building structure within 8 in (203 mm) of the box. Ceiling outlets for fixtures must employ cable clamps. Thin sections of material are knocked out to provide openings to bring the cable into the box. Nonmetallic covers should be used with devices in these boxes.

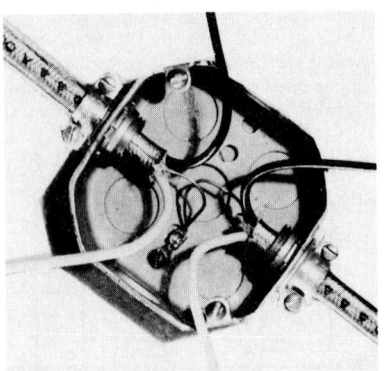

**Fig.** **9-111** Nonmetallic-sheathed cable attached to outlet box by means of clamp connectors.

**203. A line of outlets for surface mounting** with exposed nonmetallic-sheathed-cable wiring is available (Fig. 9-114). They are made of porcelain or plastic; a clamp inside the case holds the cable. There are terminal screws near each end so that cables can be brought in from both ends. NEC requirements for these devices are given in Sec. 204.

**204. Devices of insulating material.** Switch, outlet, and tap devices of insulating material may be used without boxes in exposed cable wiring and for concealed work for rewiring in existing buildings when the cable is concealed and fished. Openings in such devices shall form a close fit around the outer covering of the cable, and the device shall fully enclose that part of the cable from which any part of the covering has been removed.

If connections to conductors are by binding screw terminals, there shall be available as many terminals as conductors, unless cables are clamped within the structure and terminals are of a type approved for multiple conductors.

**205. Methods of installing nonmetallic-sheathed cable in accessible attics or roof spaces** are shown in Fig. 9-115. When the cable is carried across the top of floor beams or across the face of rafters within 7 ft (2.13 m) of the floor, the cable should be protected by guard strips, as shown at *B*. If the attic is accessible only through a scuttle hole without permanent stairs or ladders, the protection is required only within 6 ft (1.83 m) of the edge of the scuttle hole. The cable may be supported along the sides of rafters, studs, or floor joists without any additional protection, as at *D*.

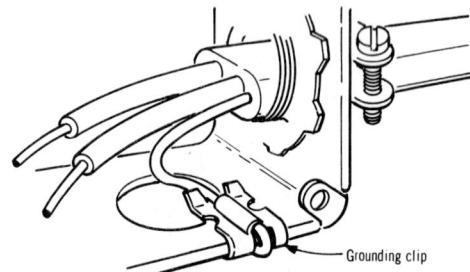

Grounding clip

**Fig. 9-112** Squeeze connector for attaching nonmetallic-sheathed cable to outlet box.

**Fig. 9-113** Grounding clip attached to metal box. [*Electrical Construction and Maintenance*]·

**206. Methods of installing nonmetallic-sheathed cable in unfinished cellars or basements** are shown in Fig. 9-116. The cable may be run through bored holes in the floor joists and along the sides or faces of joists without any additional protection such as running boards or guard strips, as shown at *A* and *B*. When the cable is run at angles across the bottom faces of floor joists, the cable must be supported on the underside of running boards as shown in *C*. Assemblies containing not smaller than two No. 6 or three No. 8 conductors may be secured directly on the bottom of the joists.

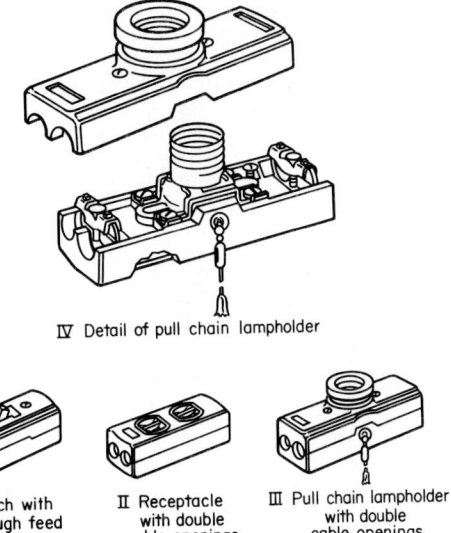

IV Detail of pull chain lampholder

I Switch with
through feed

II Receptacle
with double
cable openings

III Pull chain lampholder
with double
cable openings

**Fig. 9-114**  Surface outlets for use with nonmetallic-sheathed cable.

## MINERAL-INSULATED METAL-SHEATHED–CABLE WIRING

**207. Mineral-insulated metal-sheathed cable (Type MI)** was developed to fill the need for an electrical cable which would have high heat and water resistance and would not be bulky or difficult to install. It provides a general wiring system which is suitable for a great variety of applications for systems with voltages up to and including 600 V. Type MI is approved for use in almost any type of installation. The **NEC** states that MI cable may be used for services, feeders, and branch circuits in both exposed and concealed work, in

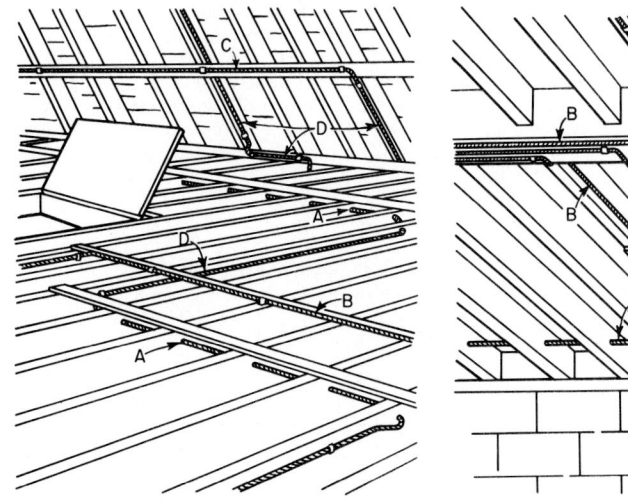

**Fig. 9-115**  Methods of installing nonmetallic-sheathed cable in accessible attics or roof spaces.

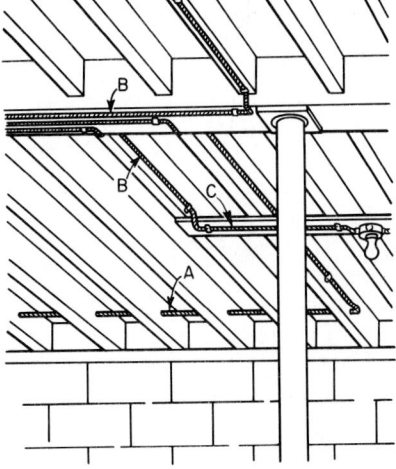

**Fig. 9-116**  Methods of installing nonmetallic-sheathed cable in unfinished cellars or basements.

dry or wet locations, for underplaster extensions as provided in Article 344, and embedded in plaster finish on brick or other masonry. It may be used where exposed to weather or continuous moisture, for underground runs and embedded in masonry, concrete, or fill, in buildings in the course of construction or where exposed to oil, gasoline, or other conditions not having a deteriorating effect on the metal sheath. The sheath of mineral-insulated metal-sheathed cable exposed to destructive corrosive conditions, such as some types of cinder fill, must be protected by materials suitable for those conditions. In addition, MI cable is approved for installation in all hazardous locations.

**Fig. 9-117**   Type MI cable [General Cable Co.]

Type MI cable is composed of only two materials, which are assembled as three components: the conductor, the insulation, and the outer sheath. The conductors are fabricated from high-conductivity copper. The insulation is a specially selected, highly compressed pure magnesium oxide which is processed to maintain permanently the excellent electrical characteristics inherent in this material. The sheath is a round, seamless copper tube which provides the requirements of heat and flame resistance, mechanical protection, and ductility. The wires, insulation, and protective copper or alloy steel sheath are manufactured as a unit, as shown in Fig. 9-117. MI cable is available in single-, two-, three-, four-, and seven-conductor assemblies. Refer to Sec. 129 of Div. 2 for data. The cable possesses exceptional ability to withstand severe mechanical abuse, as shown in Sec. 208 by the summary of mechanical properties as determined from tests.

Both ends of each coil of cable are factory-sealed to prevent ingress of moisture into the cable. It is advisable to reseal unused cable in coils that will stand around for any length of time and to apply permanent seals to each cable run as promptly as possible. Should moisture enter cable ends during field handling and installation, the affected insulation will normally be discarded in the process of stripping back and terminating the cable. If necessary, the cable can be further dried by careful application of a torch over approximately a 2-ft (0.6-m) section of cable adjacent to the end. Insulation-resistance measurements checked by an insulation tester should indicate adequate dryness.

The general method of installation is similar to that of metal-clad cable. Standard junction or outlet boxes may be used at each point of termination or when two lengths of cable are to be joined. The tap size of these units should be the same as listed under "Corresponding Conduit Size," as given in Sec. 125 of Div. 2.

**208.   Summary of Mechanical Tests on Mineral-Insulated Cable**

|  | Type MI | Metal-clad cable | EMT Type T | Rigid conduit Type T |
|---|---|---|---|---|
| Impact............ | B | D | C | A |
| Crushing......... | A | D | C | B |

NOTES.   Performance is indicated by letter "grades"; A signifies best performance.   EMT is electrical metallic tubing.   Type T refers to Underwriters' Code Grade thermoplastic-insulated conductors.   In the impact test, the cable was laid over an anvil hardy having a smoothly rounded edge with a ¼ in. radius.   Weights of 50, 20, and 10 lb were dropped on the cable, and the length of drop determined which produced not more than two failures in 10 separate blows.   In the crushing test the cable was crushed against the same anvil hardy used in the impact test by a flat steel plate in a compression-testing machine.   The force required to crush the cable to 50 per cent of its original diameter was determined.

**209. Installation of MI cable.**    It is suggested that the MI be trained into position where the installation allows. The cable can also be snaked and fished into and through congested areas. In some cases, a simple rotating X frame has been used to advantage, facilitating paying off the cable as it is snaked through barriers. After training to position, the cable can be secured with one-hole clips, as offered by many manufacturers of conduit clips, multicable gang straps, channel and spring-set clip-supporting arrangements, cable trays, conduit hangers, and spacing bars and straps.

The cable can be trained into final position by hand in most cases. True alignment can be ensured by tapping the cable lightly along its axis with a wooden mallet. True round sweeps can be made in the larger-sized cable runs with a hand hickey, but this usually is not necessary.

NOTE  NEC  requires that the radius of bends (at the inner edge of the bend) shall be not less than five cable diameters and that cable (except where fished) be supported at intervals not exceeding 6 ft (1.8 m).

The copper itself being in an annealed condition, MI is easily worked. Keeping the cable straight when unwinding the coil facilitates final training. The cable is cut after measurement, allowing for conductor tails (stretching steel wire over the proposed run is an ideal method of measuring). MI can be terminated before or after it is laid in position. If undue stressing and sharp bending are avoided during preliminary handling and roughing in, work hardening of the cable will be minimized, resulting in a more readily installed system.

In portions of dairies, laundries, canneries, and other wet locations and in locations where walls are frequently washed, the entire wiring system, including all boxes and fittings used therewith, shall be made watertight, and the cable shall be mounted so that there is at least ¼-in (6.35-mm) air space between it and the wall or other supporting surface.

The Code states that at all points where mineral-insulated metal-sheathed cable terminates, an approved seal shall be provided immediately after stripping to prevent the entrance of moisture into the mineral insulation and that, in addition, the conductors extending beyond the sheath shall be insulated with an approved insulating material. Also, when Type MI cable is connected to boxes or equipment, the fittings shall be approved for the conditions of service. Safety MI cable terminal fittings (Fig. 9-118) are designed to seal the cable ends hermetically to prevent the entrance of moisture into the cable as well as secure mechanically the cable to the outlet box, motor, or other apparatus. The complete terminal consists of an end seal and a threaded gland.

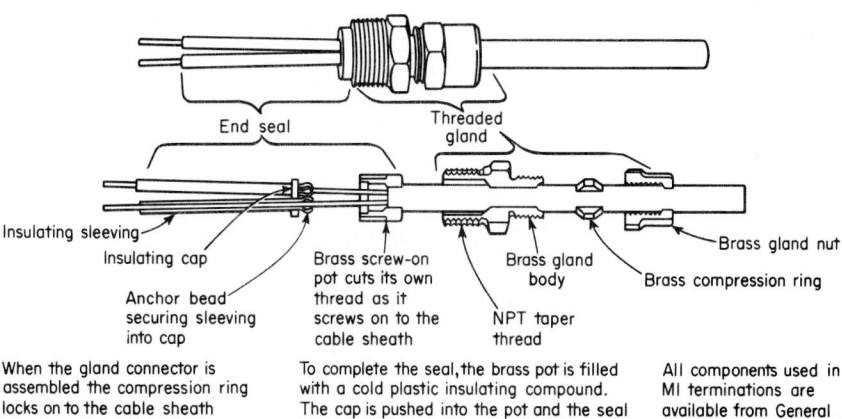

When the gland connector is assembled the compression ring locks on to the cable sheath providing a watertight joint.

To complete the seal, the brass pot is filled with a cold plastic insulating compound. The cap is pushed into the pot and the seal is completed with a compressing tool.

All components used in MI terminations are available from General Cable.

**Fig. 9-118**  Standard termination for mineral-insulated cables suitable for use at temperatures up to 85°C. [General Cable Co.]

### 210.  Termination Dimensions

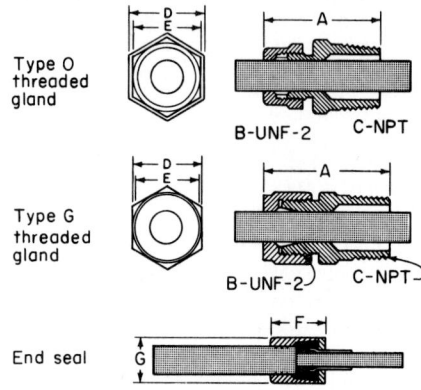

Type O
threaded
gland

B-UNF-2    C-NPT

Type G
threaded
gland

B-UNF-2    C-NPT

End seal

(Dimensions in inches)

| Group size | Pot | | | | | | |
|---|---|---|---|---|---|---|---|
| | A | B | $C^a$ | D | E | F | G |
| **Type O Gland** | | | | | | | |
| $^3/_8$ | $1^5/_{16}$ | $^1/_2$–20 | $^3/_8$–18 | $^{11}/_{16}$ | $^5/_8$ | $^9/_{16}$ | $^{15}/_{32}$ |
| $^1/_2$ | $1^9/_{16}$ | $^5/_8$–18 | $^1/_2$–14 | $^7/_8$ | $^3/_4$ | $^{27}/_{32}$ | $^{19}/_{32}$ |
| $^3/_4$ | $1^3/_4$ | $^7/_8$–14 | $^3/_4$–14 | $1^1/_{16}$ | 1 | $1^1/_8$ | $^{25}/_{32}$ |
| 1 | $1^{15}/_{16}{}^b$ | 1–14 | $1$–$11^1/_2$ | $1^5/_{16}$ | $1^1/_8$ | $1^3/_8$ | $^{31}/_{32}$ |
| **Type G Gland** | | | | | | | |
| $^1/_2$ | $1^3/_4$ | $^3/_4$–16 | $^1/_2$–14 | $^7/_8$ | $^7/_8$ | $^{27}/_{32}$ | $^{19}/_{32}$ |
| $^3/_4$ | $2^1/_8$ | $^7/_8$–14 | $^3/_4$–14 | $1^1/_{16}$ | 1 | $1^1/_8$ | $^{25}/_{32}$ |
| 1 | $2^1/_2$ | $1^1/_8$–12 | $1$–$11^1/_2$ | $1^5/_{16}$ | $1^1/_4$ | $1^3/_8$ | $^{31}/_{32}$ |

$^a$ STD electrical conduit pipe thread.
$^b$ This dimension is $2^5/_{16}$ on sizes 637 and 684.

### 211.  Termination Selection Chart

| Application | Seals | | | Glands |
|---|---|---|---|---|
| | Designation | Sleeving | Sealant | |
| General use up to 85°C, wet and dry locations— UL listed | Type O | PVC | Type O plastic | Type O |
| Hazardous locations, wet and dry up to 85°C – UL listed | Type H | PVC | Type H epoxy | Type G |
| Low temperatures (below −10°C) | Low temperature | Silicone rubber (Fiberglas reinforced) | Type O plastic | Type O |
| High temperatures (up to 125°C) | High temperature | Silicone rubber (Fiberglas reinforced) | Type O epoxy | Type O |

Ceramic-to-metal seals are available for temperatures in excess of 125°C.
NOTES.
1. Type G glands may be used as an alternative for Type O glands in any application.
2. Silicone rubber Fiberglas-reinforced sleeving may be used as an alternative for PVC sleeving in any general-use application.
3. Type H epoxy may be used as an alternative sealant in any application.
4. Type O plastic and Type O epoxy may be used as alternative sealants in Class I, Group B, and Class II, Groups E, F, and G, hazardous locations.

### 212. Termination procedure for MI cable

CABLE END SEAL    Basic parts of the cable-seal termination and the complete seal are shown in Fig. 9-119. The following parts are required for making the seal.

1. Screw-on pot, a self-thread brass cylinder
2. Silicone rubber or PVC tubing, for insulating the exposed conductor ends
3. Insulating cap
4. Sealing compound, a soft plastic material or two-part epoxy compound
5. Crimping and stripping tools

The cable-seal termination is made by removing the copper sheath with the stripping tool to expose the length of conductors needed for the connections, screwing the self-threading screw-on pot onto the end of the cable sheath, filling the pot with sealing compound, slipping the insulating tubing and cap assembly over the exposed conductors and pressing the cap down into the end of the screw-on pot, and, finally, securing in place by crimping the edge of the pot down over the insulating cap. This insulates the end of each exposed conductor and provides a moisture seal over the insulation.

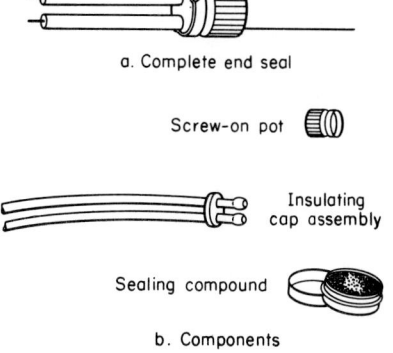

a. Complete end seal

Screw-on pot

Insulating cap assembly

Sealing compound

b. Components

**Fig. 9-119**    End seal for MI cable. [General Cable Co.]

THREADED GLAND    Figure 9-120 shows the parts of the threaded gland, with the gland in place before tightening and, finally, with the gland locked in the final position. The following parts are included:

1. Gland body
2. Compression ring
3. Gland nut

These gland parts are slipped over the end of the cable before the termination is applied, in the order of first the gland nut, then the compression nut, and finally the gland body. When the gland nut is screwed into the gland body over the compression ring, the

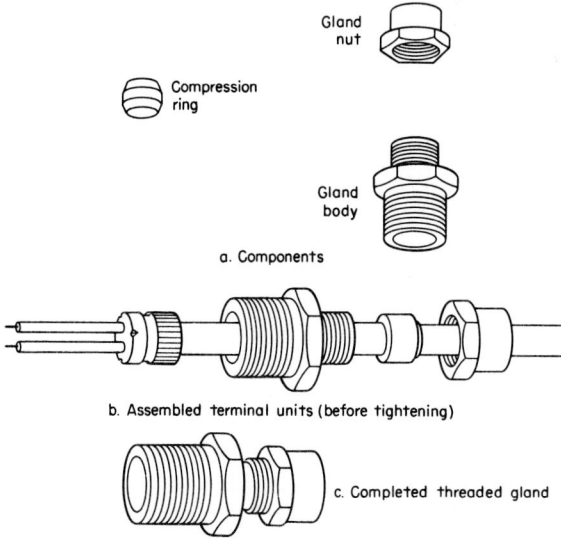

Gland nut

Compression ring

Gland body

a. Components

b. Assembled terminal units (before tightening)

c. Completed threaded gland

**Fig. 9-120**    Threaded gland for MI cable. [General Cable Co.]

ring is compressed onto the cable sheath, which anchors it into position. This can be applied at any point on the cable, but usually it is brought clear up to the end with the cable-seal termination seated into the counterbored end of the gland body. The outer end of the gland body is threaded for locknuts.

ASSEMBLY TOOLS
1. One hand vise
2. One screwdriver or nut driver
3. One copper tube cutter
4. One diagonal cutting pliers
5. Two pipe wrenches
6. One crimping and compressing tool
7. One stripping tool

**213.   Details of Termination Procedure for MI Cable**
(General Cable Co.)

l. Mark sheath only with a pencil at point which will expose desired conductor length

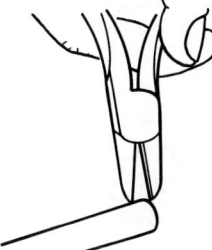

2. Use diagonal cutters to start the rip.

3. Use wrist motion to tear up "tag."

4. Engage the tag in the slot of the rod and twist it around the cable. Wrist motion is continued as rod revolves about cable axis.

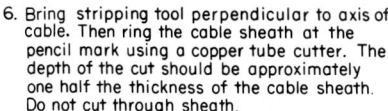

5. Keep rod at about 45° to the line of the cable. Keep bare rod against sheath and 45° will be regulated. Do not force tearing.

6. Bring stripping tool perpendicular to axis of cable. Then ring the cable sheath at the pencil mark using a copper tube cutter. The depth of the cut should be approximately one half the thickness of the cable sheath. Do not cut through sheath.

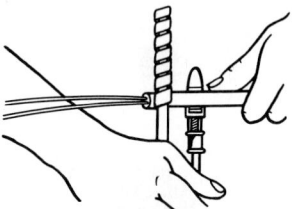

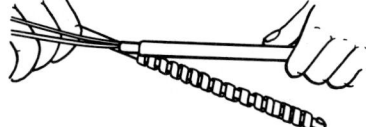

7. After ringing, continue stripping as shown in Fig. 5 and tear off squarely at the ring. Remove any burrs at the end of the cable sheath with diagonal cutters.

### 213. Details of Termination Procedure for MI Cable (*Continued*)

8. Slip the gland connector parts on to the cable in this order: gland nut, compression ring, gland body.

9. Engage the self-threading pot finger tight. See that pot is square. Then screw on with pipe wrench engaging all threads. Examine inside of pot for cleanliness and metallic slivers or dust. Test for alignment by bringing gland body up to enclose pot.

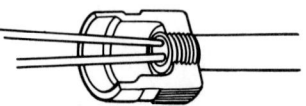

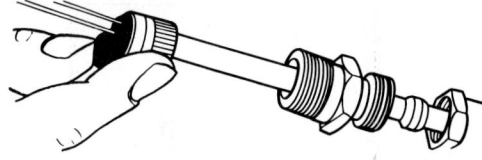

10. Press plastic compound into sealing pot until it is packed tightly. Important: Be sure the hands are free from metal dust or filings while feeding compound into pot.

11. Slip cap and sleeving sub assembly into position

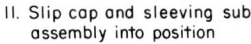

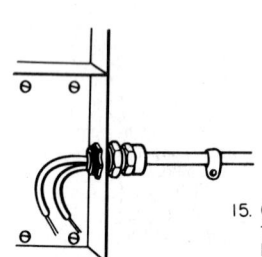

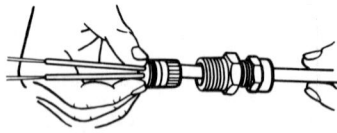

12. Force the insulating cap assembly into position on top of compound.

13. Slip the compression and crimping tool over the insulating leads

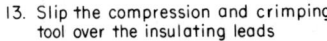

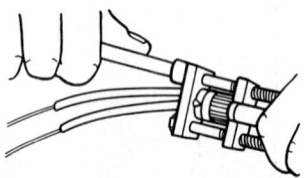

14. After positioning crimping tool on pot, compress the insulating washer into place flush with the top of the pot by tightening the two slotted cap screws. This operation compresses the compound in the pot and crimps the edge of the pot into the insulating cap, holding it in place. Remove compression tool.

15. Completed termination in box utilizing standard lock nut.

## MEDIUM-VOLTAGE–CABLE WIRING

**214. Type MV cable** is a single-conductor or multiconductor cable with extruded solid dielectric insulation. The cable may have an extruded jacket, a metallic covering, or a combination of both over the single conductors or the cabled conductors.

**215. Type MV cables are rated** from 2001 V up to 35,000 V. The ampacities are governed by Sec. 310-15 of the National Electrical Code.

**216. The construction of Type MV cable** shall be in accordance with Article 310 of the NEC and may have aluminum, copper, or copper-clad aluminum conductors.

**217. Power systems** up to 35,000 V use Type MV cable, which is permitted in wet or dry locations, in raceways, in cable trays as specified in Sec. 318-3(b) of the NEC, in messenger-supported wiring, or directly buried, installed in accordance with Articles 300 and 710 of the NEC.

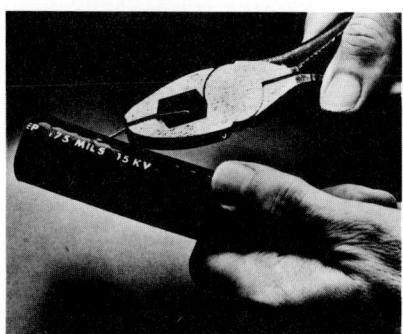

**Fig. 9-121** UniShield® EP power cable, 5 kV to 35 kV. [Anaconda Company]

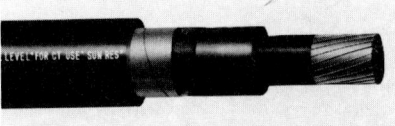

**Fig. 9-122** UniBlend® EP shielded power cable, Type MV 90 copper conductor, 15,000 V, 90°C. [Anaconda Company]

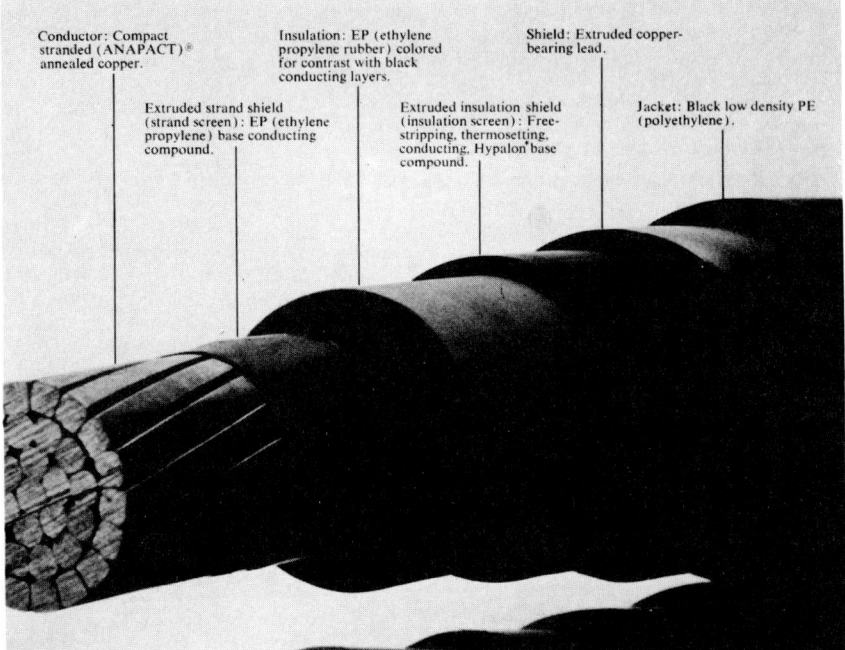

Conductor: Compact stranded (ANAPACT)® annealed copper.

Insulation: EP (ethylene propylene rubber) colored for contrast with black conducting layers.

Shield: Extruded copper-bearing lead.

Extruded strand shield (strand screen): EP (ethylene propylene) base conducting compound.

Extruded insulation shield (insulation screen): Free-stripping, thermosetting, conducting, Hypalon base compound.

Jacket: Black low density PE (polyethylene).

**Fig. 9-123A** Compact stranded medium-voltage cable. [Anaconda Company]

### UNDERGROUND-FEEDER AND BRANCH-CIRCUIT–CABLE WIRING

**218. Underground-feeder and branch-circuit cable** provides an economical wiring system for wet and corrosive installations. Single-conductor Type UF cable resembles Type USE service-entrance cable in general appearance. The insulation employed consists of a plastic compound. NEC statements with respect to its use are: Underground-feeder and branch-circuit cable may be used underground, including direct burial in the earth, as feeder or branch-circuit cable when provided with overcurrent protection not in excess of the rated current-carrying capacity of the individual conductors. If single-conductor cables are installed, all cables of the feeder circuit, subfeeder circuit, or branch circuit, including the neutral and equipment grounding conductor, if any, shall be run together in the same trench or raceway. If the cable is buried directly in the earth, the minimum

**Fig. 9-123***B*   Type UF two-conductor cable suitable for direct burial.

burial depth permitted is 24 in (610 mm) if the cable is unprotected and 18 in (457 mm) when a supplemental covering, such as a 2-in (50.8-mm) concrete pad, metal raceway, pipe, or other suitable protection, is provided. Type UF cable may be used for interior wiring in wet, dry, or corrosive locations under the recognized wiring methods of the Code, and when installed as nonmetallic-sheathed cable it shall conform with the provisions of Article 336 and be of a multiconductor type. It must also be of a multiconductor type if installed in cable tray.

This type of cable shall not be used (1) as service-entrance cable, (2) in commercial garages, (3) in theaters, (4) in motion-picture studios, (5) in storage-battery rooms, (6) in hoistways, (7) in any hazardous location, (8) embedded in poured cement, concrete, or aggregate except as provided in Article 424 of the Code and (9) where exposed to direct sunrays, unless identified as sunlight-resistant (Fig. 9-123*B*).

### INTERIOR WIRING WITH SERVICE-ENTRANCE CABLE

**219. Service-entrance cable of Type SE or Type USE** may be employed for interior wiring, installed either concealed or exposed. The construction of these cables is described in Sec. 221. Cables with all the conductors insulated may be used for general interior wiring. Those without insulation on the grounded conductor may be used only on systems not exceeding 150 V to ground, for a range, wall-mounted oven, counter-mounted cooking unit, and clothes-dryer circuit, or as feeders to supply other buildings from the one containing the master service-entrance equipment. The cables with uninsulated grounding conductors must have a final nonmetallic outer covering. Cables which are used for interior wiring must be installed in accordance with the rules for nonmetallic-sheathed cable discussed in Secs. 194 to 206.

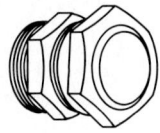   

I  Watertight  Screw  Connector.          II Nonwatertight  Connector.

**Fig. 9-124**   Box connectors for service-entrance cable. [General Electric Co.]

**220. The principal use of service-entrance cable for interior wiring** is for branch circuits supplying ranges or water heaters and in farm wiring as feeders from the master service to supply the other buildings that are wired for electricity. The cable is supported on the building structure by means of single-hole or two-hole cable straps made in different sizes to accommodate the size of cable. They are similar in appearance to those used for rigid conduit. Connection to switch or outlet boxes is made by means of a connector (Fig. 9-124), which is made in watertight and nonwatertight types. A typical range installation is shown in Fig. 9-125.

**221. Types of cables.** Two common types of cables are shown in Figs. 9-126 and 9-127. The three-conductor Type SE cable in Fig. 9-126 consists of two insulated conductors and one concentric-stranded insulated conductor (generally used as the grounded-neutral conductor). This cable is limited to services that do not exceed 150 V to ground, such as a 120/240-V single-phase supply. If used as a three-wire 120/240-V branch circuit to a range, wall-mounted oven, counter-mounted cooking unit, or clothes dryer, this cable must originate at the service equipment. In such instances, the uninsulated neutral conductor may be used to ground the metal frame of the equipment supplied.

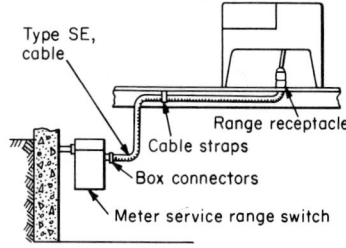

Fig. 9-125  Installation of electric range wired with service-entrance cable.

Figure 9-127 shows a single-conductor 4/0 aluminum Type USE cable, which is also suitable for use as a Type RHW (in wet locations) and a type RHH (in dry locations). When used in direct-burial applications, such a cable has a 75°C rating. The ampacity ratings of conductors are listed in Tables 19 through 23 of Div. 11. For underground applications Type USE cable should be buried not less than the depths specified for Type UF cables in Sec. 218. For over 600 V refer to Sec. 135 of Div. 8.

**Fig. 9-126**  Four-conductor Type SE cables uses three insulated conductors and a bare equipment ground.

**Fig. 9-127**  Single-conductor Type USE cable suitable for direct burial.

## UNDERFLOOR-RACEWAY WIRING

**222. Underfloor-raceway systems** are used principally in office buildings for the installation of the wiring for telephone and signal systems and for convenience outlets for electrically operated office machinery. They provide a flexible system by which the location of outlets may be easily changed to accommodate the rearrangement of furniture and partitions. The NEC allows their use when embedded in the concrete or in the concrete fill of floors. Their installation is allowed only in locations which are free from corrosive or hazardous conditions. No wires larger than the maximum size approved for the particular raceway shall be installed. The voltage of the system must not exceed 600 V. The total cross-sectional area of all conductors in a duct must not be greater than 40 percent of the interior cross-sectional area of the duct.

**223. An underfloor-raceway system** consists of ducts laid below the surface of the floor and interconnected by means of special cast-iron floor junction boxes. The ducts for underfloor-raceway systems are made of either fiber or steel. Fiber ducts are made in two types, the open-bottom type and the completely enclosing type. Steel ducts are always of the completely enclosing type, usually having a rectangular cross section. In the underfloor-raceway system, provision is made for outlets by means of specially designed floor-

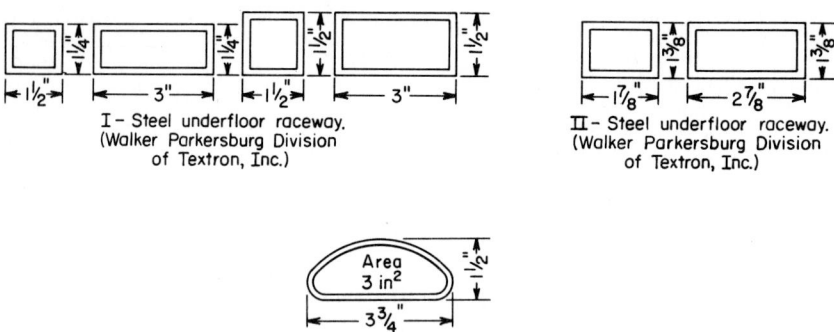

I – Steel underfloor raceway.
(Walker Parkersburg Division
of Textron, Inc.)

II – Steel underfloor raceway.
(Walker Parkersburg Division
of Textron, Inc.)

III – Nonmetallic raceway.

**Fig. 9-128**  Dimensions of ducts for underfloor raceways.

outlet fittings which are screwed into the walls of the ducts. When fiber ducts are employed, the duct system is laid in the floor with or without openings or inserts for outlets. After the floor has been poured and finished as desired, the outlet fittings are installed into inserts or at any points along the ducts at which outlets are required. The method of installing outlet fittings is described in Sec. 237. When steel ducts are employed, provision for the outlet fittings must be made at the time that the ducts are laid, before the floor or floor fill is poured. The steel ducts are manufactured with threaded openings for outlet connections at regularly spaced intervals along the duct. During the installation of the raceway and the floor these outlet openings are closed with specially constructed plugs whose height can be adjusted to suit the floor level. The shapes and dimensions of the different types of ducts for underfloor raceways are shown in Fig. 9-128. For telephone and similar circuits, much wider ducts can be obtained.

**224. General rules for the installation of underfloor raceways.**  In general, underfloor raceways should be installed so that there is at least ¾ in (19 mm) of concrete or wood over the highest point of the ducts (see Figs. 9-129 and 9-130). However, in offices approved raceways may be laid flush with

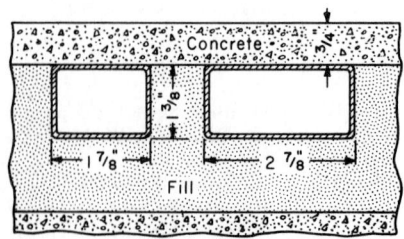

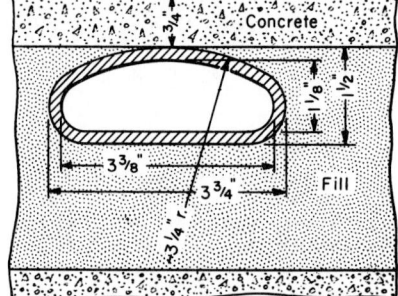

**Fig. 9-129**  Any flat-top raceway 4 in (101.6 mm) or less in width must be covered with wood or concrete at least ¾ in (19.05 mm) in thickness except in office occupancies.

**Fig. 9-130**  Any fiber raceway 4 in (101.6 mm) or less in width must be covered with wood or concrete at least ¾ in (19.05 mm) in thickness.

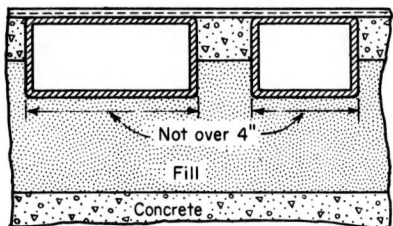

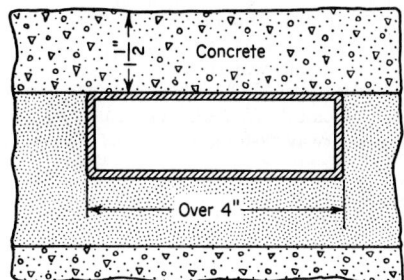

**Fig. 9-131**  Approved raceways flush with concrete and covered with linoleum in an office occupancy.

**Fig. 9-132**  Any flat-top raceway over 4 in (101.6 mm) wide must be covered with concrete at least 1½ in (38.1 mm) in thickness if raceways are spaced less than 1 in (25.4 mm) apart.

the concrete if covered with linoleum or equivalent floor covering (see Fig. 9-131). When two or three raceways are installed flush with the concrete, they must either be contiguous with each other and joined to form a rigid assembly. Flat-top ducts over 4 in (102 mm) wide but not over 8 in (203 mm), spaced less than 1 in (25.4 mm), must be covered with at least 1½ in (38 mm) of concrete (see Fig. 9-132). It is standard practice to allow ¾-in (19-mm) clearance between ducts run side by side. The centerline of the ducts should form a straight line between junction boxes. If the spacing between raceways is 1 in or more, the raceway may be covered with only 1 in of concrete. All joints in the raceway between sections of ducts and at junction boxes should be made waterproof and with good electrical contact so that the raceways will be electrically continuous. Metal raceways must be properly grounded.

**225. Wires of signal or telephone systems** must not be run in the same duct as wires of lighting systems. Combination junction boxes accommodating the two or three ducts of multiple-duct systems may be employed, provided separate compartments are furnished in the boxes for each system. It is best to keep the same relative location of compartments for the respective systems throughout the installation.

All joints in or taps to the conductors must be made in the junction boxes. No joints or taps should be made in the ducts of the raceway or at outlet insert points.

**226. Junction boxes for use with underfloor raceways** are specially constructed cast-iron boxes. They are available in single-compartment types and in combination boxes to accommodate multiple-duct systems having two or three raceways run side by side. The general construction of the junction boxes is similar for both fiber- and steel-duct systems. A typical single-compartment box is illustrated in Fig. 9-133, and a two-compartment box in Fig. 9-134. Most boxes are equipped with leveling screws so that they can be leveled up with the duct runs. The

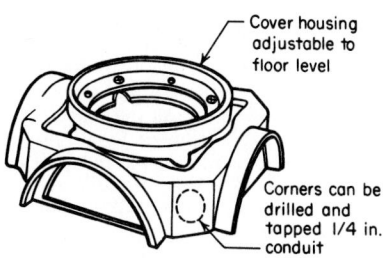

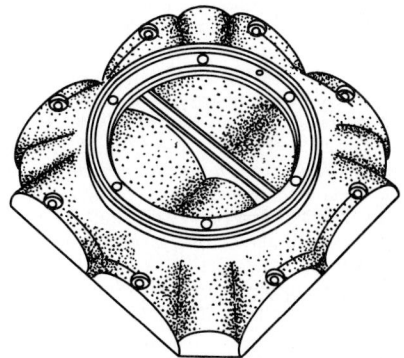

**Fig. 9-133**  Single-compartment floor junction box.

**Fig. 9-134**  Two-compartment floor junction box for accommodating double-duct systems.

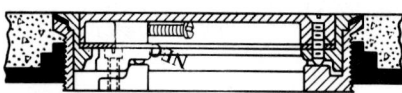

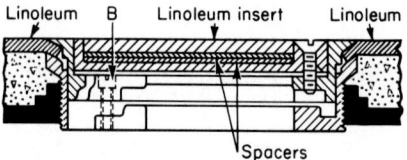

**Fig. 9-135**  Floor junction-box cover for cement-finished floors.

**Fig. 9-136**  Floor junction-box cover for lino-leum-covered floors.

covers of the boxes also are adjustable with respect to height so that they can be accurately leveled to agree with the surface of the finished floor. Different types of covers that will be suitable for the different types of floor finishes are available. With plain cement-finished floors a plain brass cover set flush with the floor (Fig. 9-135) is sometimes used. In other cases a recessed cover filled with cement is employed. For linoleum- or cork-covered floors a cover with a recess for a cork or linoleum insert is provided, as illustrated in Fig. 9-136. For marble or hardwood floors a recessed cover with a marble insert (Fig. 9-137) may be employed. With the recessed covers only a thin ring of brass shows in the finished floor, marking the location of the junction box.

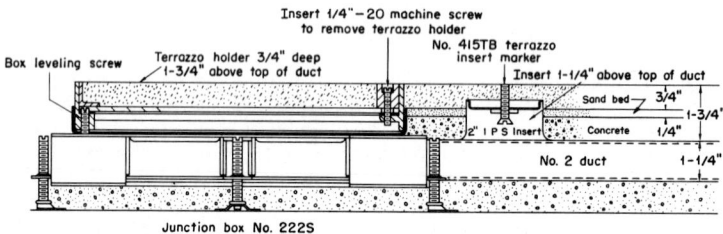

**Fig. 9-137**  Floor junction-box cover for terrazzo floors. [Walker Parkersburg Division of Textron Inc.]

**227. Fittings must be installed** at every end of a line of raceway with a marker extending through the floor to show the line of the duct. All dead ends of raceway must be closed with a suitable fitting.

**228. In installing raceways** every effort should be made to avoid low points which could form traps for water.

All connections to a raceway except those made through the inserts provided in the raceway shall be made by means of rigid conduit, intermediate metal conduit, EMT, or a special approved fitting (refer to Fig. 9-138). Where a metallic underfloor raceway system provides for the termination of an equipment grounding conductor, rigid nonmetallic conduit, electrical nonmetallic tubing, or liquidtight flexible nonmetallic conduit where not installed in concrete may be used.

**229. Inserts** (National Electrical Code) shall be leveled to the floor grade and sealed against the entrance of water. Inserts used with metal raceways shall be metal and shall be electrically continuous with the raceway. Inserts set in or on fiber raceways before the floor has been laid shall be mechanically secured to the raceway. Inserts set in fiber raceways after the floor has been laid shall be screwed into the raceway. In cutting through the raceway wall and setting inserts, chips and other dirt shall not be allowed to fall into the raceway, and tools shall be so designed as to prevent them from entering the raceway and injuring conductors that may be in place.

When an outlet is discontinued, the conductors supplying the outlet shall be removed from the raceway.

**230. Steel-duct raceways** are made for two types of underfloor electrical distribution systems:

1. The two-level system, recommended for use when electrical growth and expansion of facilities are anticipated

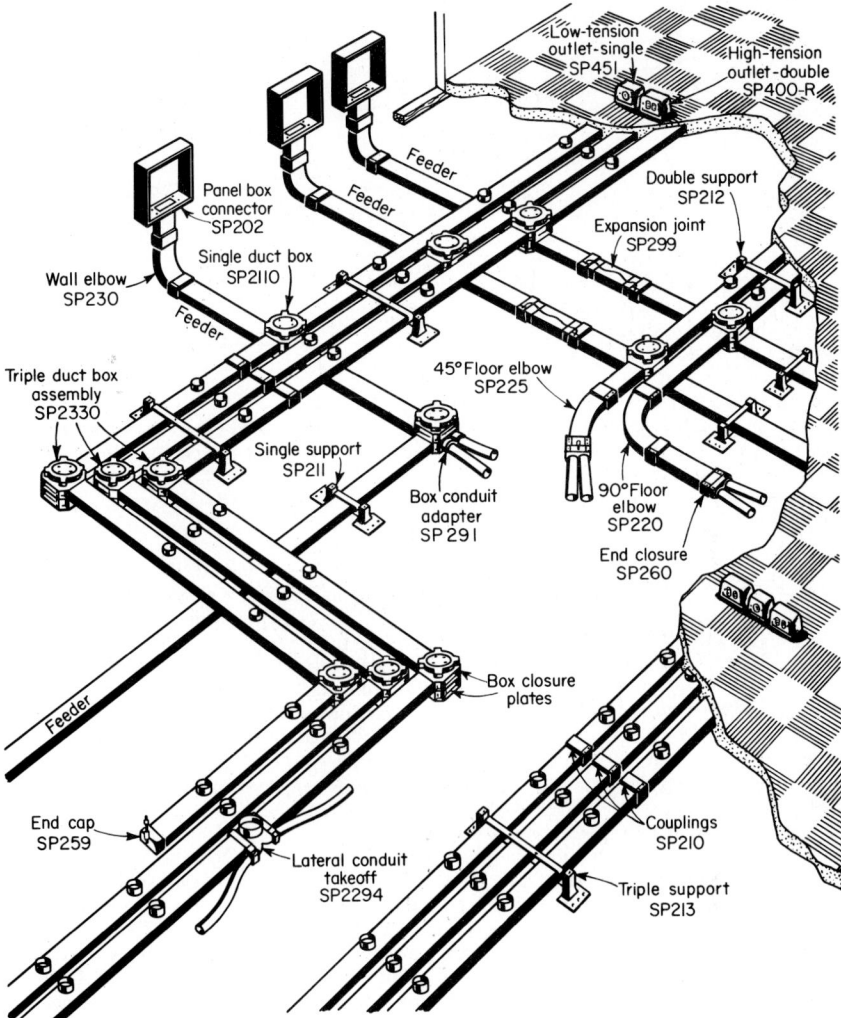

Low-tension
outlet-single
SP451

High-tension
outlet-double
SP400-R

Feeder

Double support
SP212

Panel box
connector
SP202

Feeder

Expansion joint
SP299

Single duct box
SP2110

Wall elbow
SP230

Feeder

45° Floor elbow
SP225

Triple duct box
assembly
SP2330

Single support
SP211

Box conduit
adapter
SP291

90° Floor
elbow
SP220

End closure
SP260

Box closure
plates

Feeder

End cap
SP259

Couplings
SP210

Lateral conduit
takeoff
SP2294

Triple support
SP213

**Fig. 9-138**   Component parts and method of installation for two-level steel underfloor raceways. [General Electric Co.]

2. The single-level system, recommended for underfloor wiring installations when cost and design limitations are a factor

The features of a two-level system showing raceways, junction boxes, typical supports, and other fittings are given in Fig. 9-138. Duct details are shown in Fig. 9-139.

The features of a one-level system showing raceways, junctions boxes, typical supports, and other fittings are given in Fig. 9-140. Duct details for the raceway of one manufacturer are given in Fig. 9-141. For dimensions of ducts available from other manufacturers refer to Fig. 9-128 and the catalogs of such manufacturers.

**231. Installation of steel-duct raceways.** The method of installing Nepcoduct, the steel-raceway system manufactured by the Walker Parkersburg Division of Textron Inc., in the concrete fill of floors is illustrated in Fig. 9-142. The ducts are supported by saddle supports, which consist, as shown in Fig. 9-143, of two detachable parts, a base and a saddle. These supports are available for single-duct runs and in various combinations for supporting multiple-duct runs. Supports should be located not more than 5 ft (1.5 m) apart.

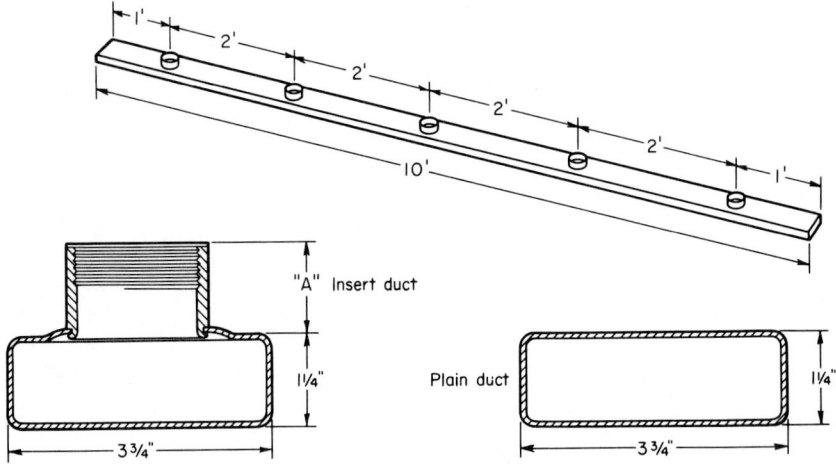

**Fig. 9-139**    Duct and duct insert for a two-level raceway system. [General Electric Co.]

After the locations of junction boxes and duct runs have been marked out, the bases of the saddle supports are fastened to the floor by expansion bolts for concrete construction or by special toggle wires for hollow-tile work (see Fig. 9-142). The saddle is then assembled to the base, and the ducts fastened to the saddle by the tie wires. The complete raceway is then leveled and adjusted to the proper height by means of the height-adjusting bolts of the saddles and the leveling screws of the couplings and junction boxes. The adjacent sections of duct are fastened together by duct couplings (Fig. 9-144). The pointed set-screws of the couplings fasten the duct to the coupling and at the same time provide an electrical ground connection between the duct and the coupling. Special fittings are available for connecting ducts and junction boxes to conduit, for connecting ducts directly to distribution cabinets, for blanking off the ends of ducts, etc.

**232. The outlet openings in the ducts** are closed by means of outlet plugs whose height can be adjusted to coincide with the floor level. A marker screw can be inserted in one plug of each 10-ft (3-m) length of duct if it is desired to facilitate the location of the outlets. The appearance of a duct after the installation of the floor fill with an outlet opening closed by means of an outlet plug is as shown in Fig. 9-145. When the floor is covered with linoleum, rubber, cork, etc., the outlets are not extended through the covering until needed. It is good practice to indicate the location of the outlet inserts by means of an escutcheon marking screw in the plugs on each side of the junction boxes, as indicated in Fig. 9-146. Wood or marble floors can be laid over the outlet inserts (Fig. 9-147), and the outlets are not extended through the finished floor until actually needed.

**233. When the raceway is installed in the concrete of the floor,** the ducts are supported from the forms by saddle supports, and the whole system is accurately leveled and adjusted for the proper height, by means of the leveling screws and bolts on the fittings and saddle supports, before the floor is poured. In concrete-joist construction when building codes will not allow the saddle support to be mounted directly on the concrete pans, a saddle bridge must be used to support the duct saddles. This method of installation is shown in Fig. 9-148.

**234. Installation of outlet fittings in steel-duct–raceway systems.**    When a duct outlet is to be used for service, the small amount of concrete is removed from the top depression of the outlet plug. The plug is removed by unscrewing it from the duct with a plug-removal wrench. The removal of the plug forms a neat preformed passage through the concrete to the duct-outlet opening. The necessary wires are pulled in from the nearest junction box and threaded through the service fitting. The service fitting is then screwed into the duct-outlet opening, and the floor flange adjusted.

When it is desired to locate an outlet at a point in a raceway where there is no preset insert plug, an afterset insert for the outlet can be made by the use of special tools availa-

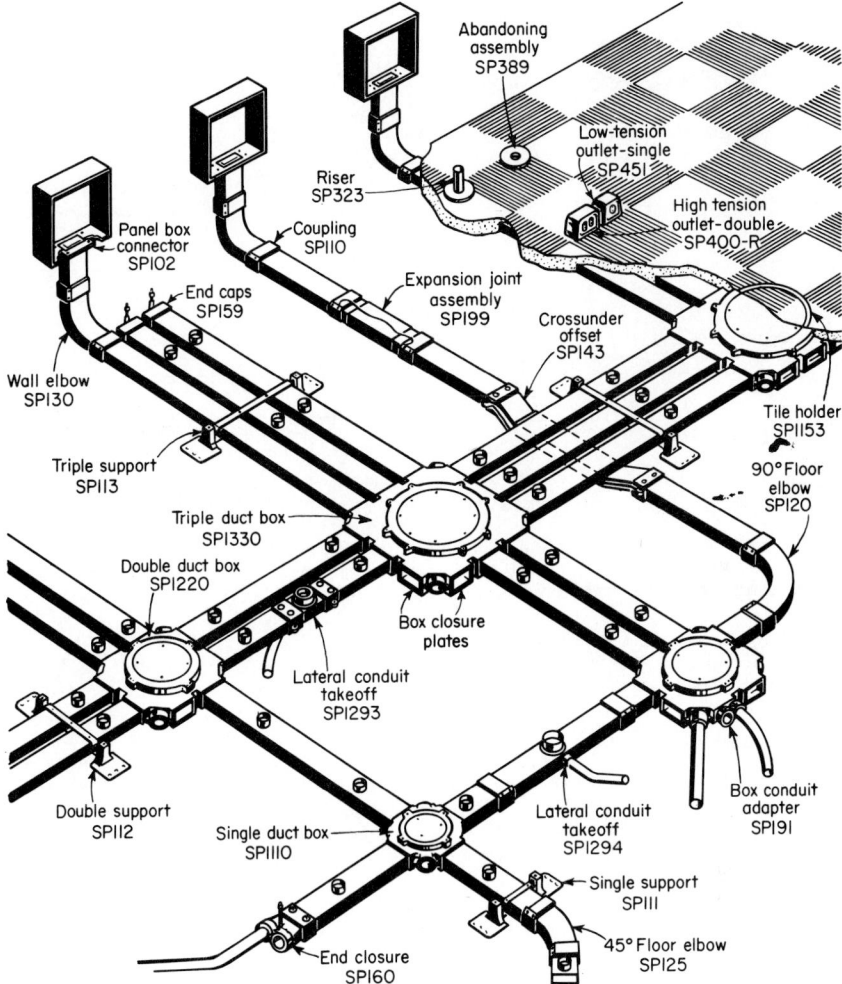

**Fig. 9-140**  Component parts and method of installation for a single-level steel underfloor raceway. [General Electric Co.]

ble from the manufacturer. The method of making an afterset insert is illustrated in Fig. 9-149.

**235. Trenchduct.**  Several makes of steel underfloor raceways are suitable for use with Trenchduct. Figure 9-150 shows a section of Trenchduct supplying two runs of underfloor steel raceways. Trenchduct is available in standard 6-ft (1.8-m) lengths at a depth of 2⅛ in (53.975 mm). Other depths are available on special order. Ducts are shipped complete with three 18-in- (457.2-mm-) long cover sections. Cover plates are ¼-in- (6.35-mm-) thick roller-leveled steel. The Trenchduct is used to extend branch circuits from a panelboard to the underfloor raceways, and the Trenchduct shown in Fig. 9-150 has a cross-sectional area of 22½ in² (145.15cm²) which allows many more conductors than a conduit feed.

Cover plates for Trenchducts are adjustable upward to ⅜ in (9.525 mm) and downward to ⅜ in from a standard position by means of leveling screws flush with the top surface cover. Adjustment to finished-floor height is made by a screwdriver from the top.

Side-feed connectors provide bushed openings for feeds into the side of the Trench-

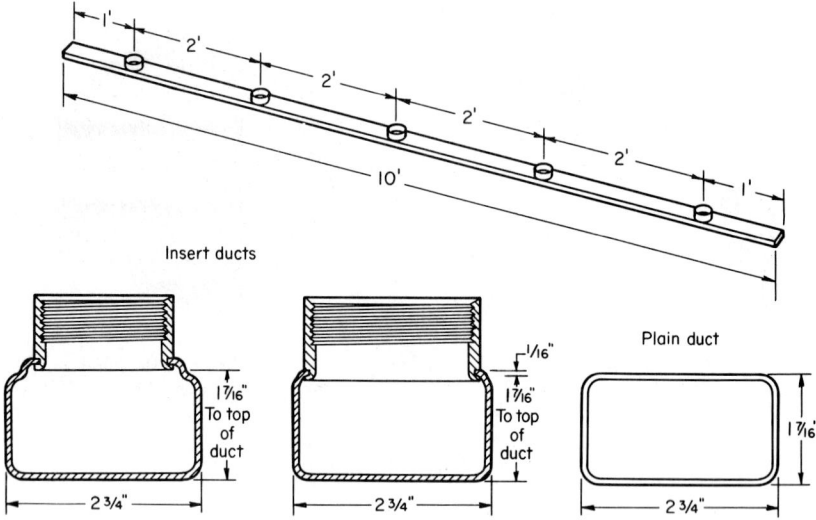

Insert ducts

Plain duct

**Fig. 9-141**    Duct and duct inserts for single-level raceway system. [General Electric Co.]

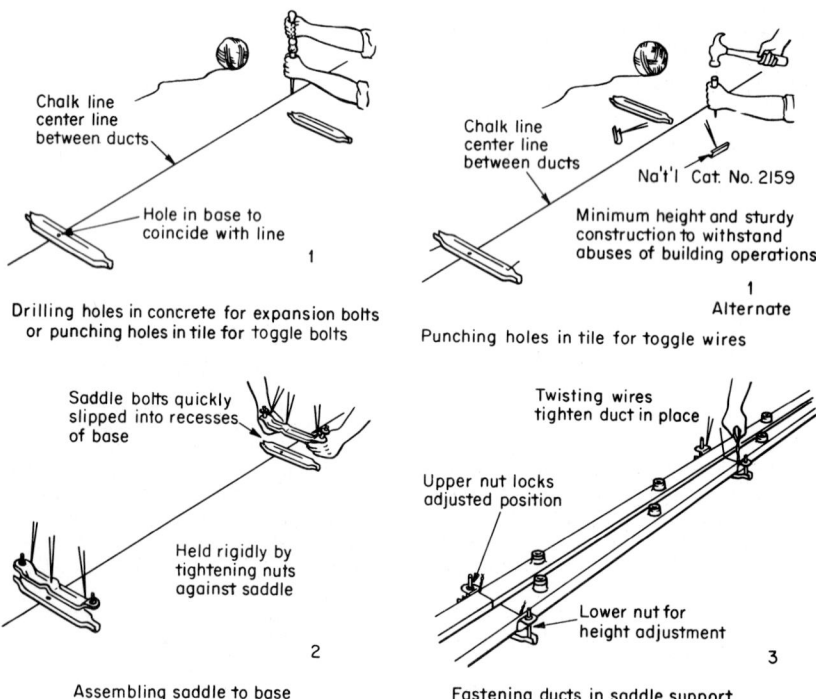

Drilling holes in concrete for expansion bolts or punching holes in tile for toggle bolts

Punching holes in tile for toggle wires

Assembling saddle to base

Fastening ducts in saddle support

**Fig. 9-142**    Method of installing a steel underfloor raceway. [Walker Parkersburg Division of Textron Inc.]

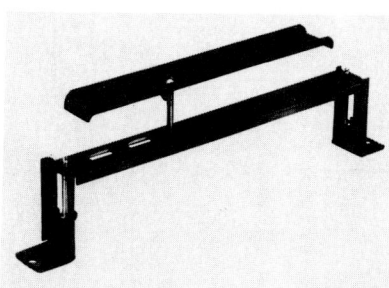

**Fig. 9-143**  Saddle supports for one small and one large steel underfloor duct. [Walker Parkersburg Division of Textron Inc.]

**Fig. 9-144**  Coupling for a steel underfloor raceway. [Walker Parkersburg Division of Textron Inc.]

**Fig. 9-145**  Appearance of outlet opening in a steel underfloor raceway with opening closed by means of an outlet plug. [Walker Parkersburg Division of Textron Inc.]

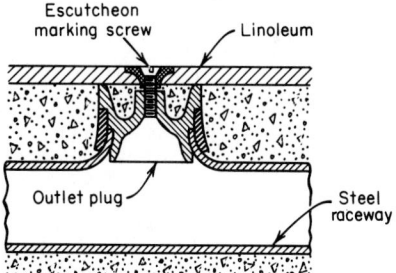

**Fig. 9-146**  Outlet opening in a steel underfloor raceway closed with an outlet plug and concealed with linoleum floor covering.

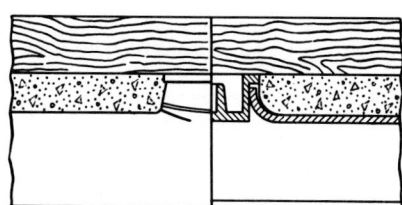

**Fig. 9-147**  Outlet opening in a steel underfloor raceway concealed by a wood floor until outlet is needed. [Walker Parkersburg Division of Textron Inc.]

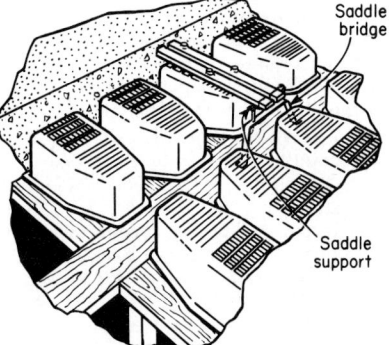

**Fig. 9-148**  Method of supporting a steel underfloor raceway in concrete-joist construction. [Walker Parkersburg Division of Textron Inc.]

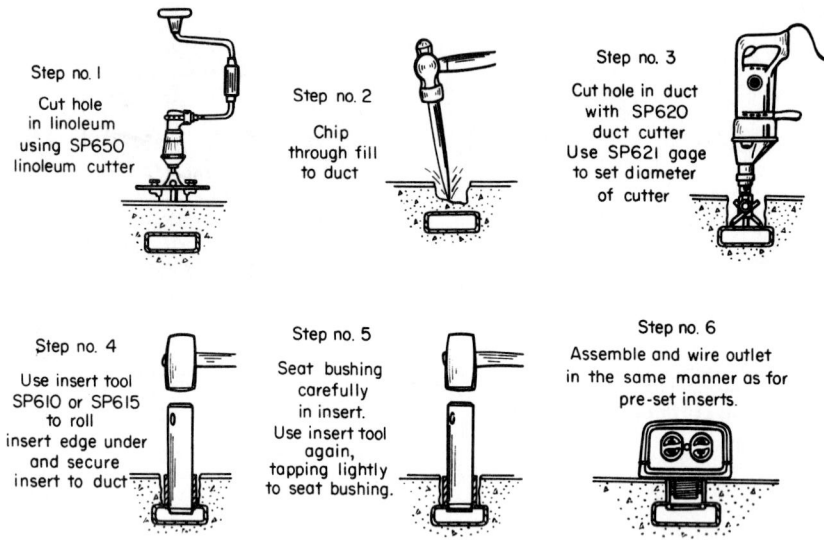

**Fig. 9-149**   Method of making an outlet with an afterset insert. [General Electric Co.]

duct. When specified, openings in Trenchduct are precut by the manufacturer. The Trenchduct is, in effect, a continuous junction box, because the cover plates are flush with the finished floor and one or more sections can be easily removed as required.

Vertical elbows turn the Trenchduct system up to a panelboard with full duct capacity. Figure 9-150 shows typical components of the system.

**236. Installation of completely enclosing fiber-duct raceways.**   Completely enclosing fiber-duct underfloor raceways may be installed in either the concrete or the concrete fill of floors in a manner similar to that described for steel-duct raceways in Sec. 231. With fiber-duct raceways there are preinserted provisions for outlets, or they can be provided without any openings for outlets. After the finished floor has been laid, an outlet may be provided at any preset insert or at any point desired along the duct runs, as described in Sec. 237. The main component parts of a fiber-duct underfloor-raceway system are illustrated in Figs. 9-151 and 9-152. The different sections of duct are clamped together by means of a combination steel coupling and support provided with leveling screws for leveling the runs and adjusting to the proper height.

**237. Installation of outlet fittings in fiber-duct–raceway systems.**   To establish outlets in a preset system after the finish is in place, it is necessary to determine the location of the insert. This can be done by means of an insert finder. Then the flooring is chipped down to expose the insert cap. The cap is removed and a hole cut in the duct so that wires can be fished through and connected to the receptacle.

The procedures shown in Fig. 9-153 should be employed in installing an outlet fitting at any point in a fiber underfloor raceway of the completely enclosing type. The special tools provided by the manufacturers for this purpose should be used to ensure satisfactory workmanship.

**238. Layout of underfloor-raceway systems.**   A very simple layout for an underfloor-raceway system, called the single-runaround layout, is shown in Fig. 9-154. Such a layout provides only the minimum of flexibility in the location of outlets. It is satisfactory for certain types of buildings when the installation cost must be limited and desks will be located only around the sidewalls. This layout will provide outlets only for telephone and signal systems. A similar layout is shown in Fig. 9-155. This layout, having two separate raceways, provides outlets for electrically operated office machines from one raceway and outlets for telephones and signal systems from the other raceway.

**239. To provide greater flexibility** in the possible location of outlets over the entire area than is accomplished with the runaround layouts, grid layouts are employed. A typical

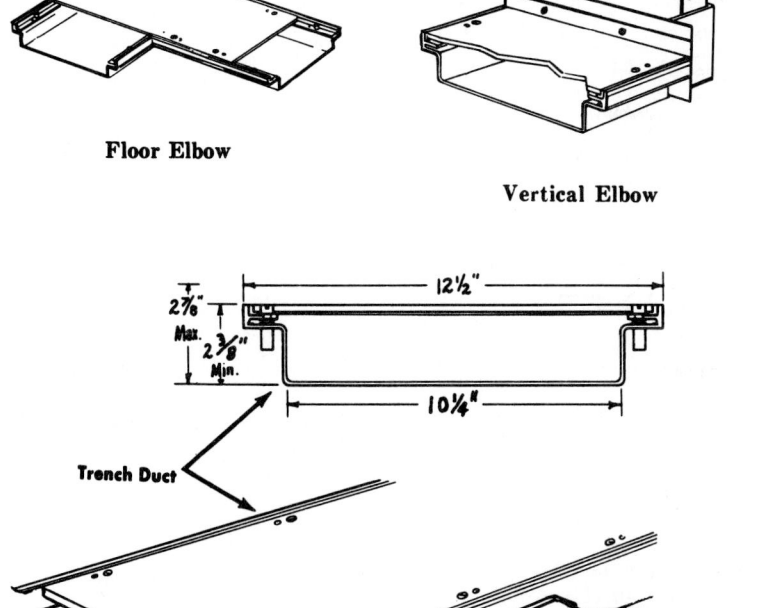

**Floor Elbow**

**Vertical Elbow**

**Fig. 9-150**   Typical components of a Trenchduct used in connection with an underfloor steel raceway. [Triangle Conduit & Cable Co., Inc., subsidiary of Triangle Industries, Inc.]

single-grid layout is shown in Fig. 9-156, and a double-grid layout in Fig. 9-157. The flexibility provided for the location of outlets depends upon the distance between parallel duct lines. For maximum flexibility the parallel lines of ducts should be located approximately 5 ft (1.5 m) apart with the outside runs located 3 ft (0.9 m) from the outside walls. The distance between rows of junction boxes should be from 20 to 60 ft (from 6.1 to 18.3 m), depending upon the estimated number of outlets that will be required at any one time between junction boxes.

## WIREWAY WIRING

**240. Wireways** may be installed only for exposed work, except as permitted in Sec. 640-4, Ex., of the NEC. They must not be used where they will be subjected to severe physical damage or to corrosive vapors or in any hazardous location, except for Class I, Div. 2, and Class II, Div. 2, as permitted by Secs. 501-4(b), Ex., 502-4(b), and 504-20 of the NEC. Wireways for outdoor use shall be of approved raintight construction. Their chief application is for the installation of mains and feeders in industrial plants or other locations where a flexible means of supplying branch circuits is advantageous or where many wires must be located in one run. They permit access to the wires and cables at all points

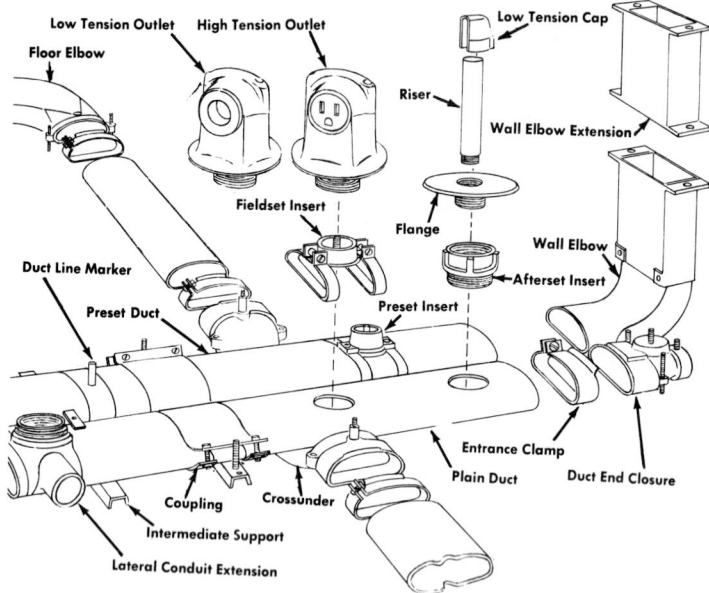

**Fig. 9-151**  Component parts of a completely enclosed fiber underfloor-raceway installation.

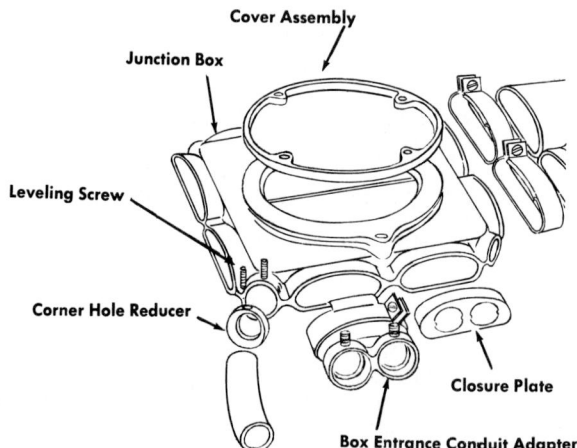

**Fig. 9-152**  Junction box and associated fittings for a completely enclosed fiber underfloor-raceway system.

**Step No. 1—** locate duct line

**Step No. 2—**cut hole in linoleum

**Step No. 3—** chip hole down to duct

**Step No. 4—** cut hole in duct

**Step No. 5—** screw insert into duct

**Step No. 6—**anchor insert with grouting compound

**Step No. 7—** screw outlet into insert

**Fig. 9-153**   Method of installing afterset inserts into fiber underfloor raceways.

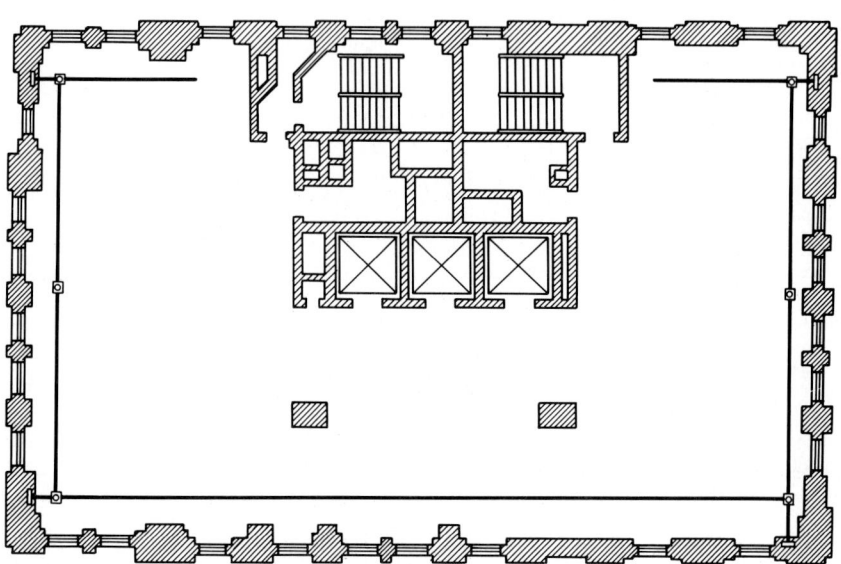

**Fig. 9-154**   Single-runaround layout for an underfloor raceway.

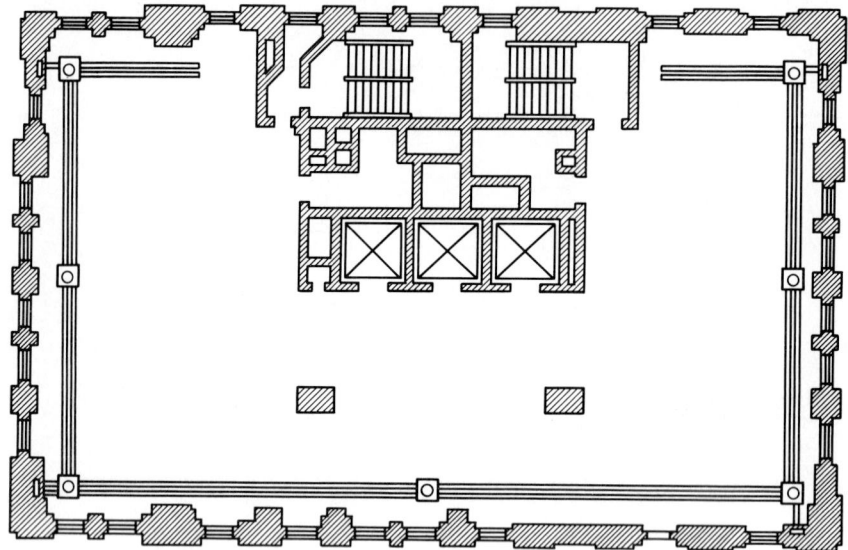

**Fig. 9-155**   Double-runaround layout for an underfloor raceway.

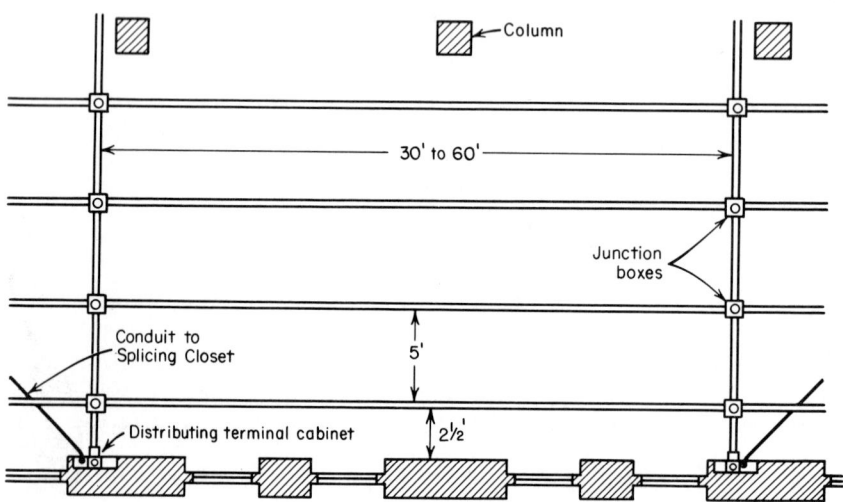

**Fig. 9-156**   Single-grid layout for an underfloor raceway.

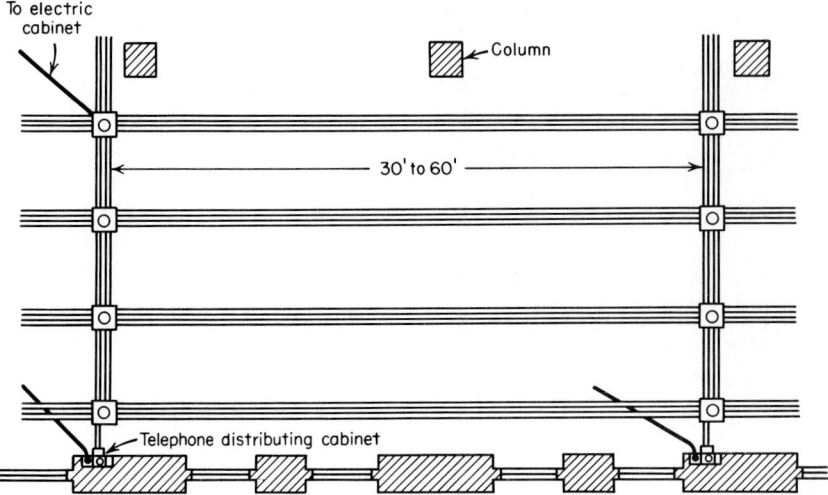

**Fig. 9-157**   Double-grid layout for an underfloor raceway.

system for tapping, splicing, rerouting, inspection, installation of additional wiring, or other changes, all without disturbing the existing work. A typical section of wireway is shown in Fig. 9-158. Three sides are provided with closely spaced knockouts for conduit or metal-clad–cable attachment. The fourth side consists of a hinged cover which can be opened for installing wires and making splices. The lengths are coupled together with nuts and bolts which secure a sheet-metal hanger (Fig. 9-159, I) between the flanged ends of the duct sections for supporting the duct. Mains and feeders are run in the wireway with taps for branches taken off at the most convenient point through rigid or flexible conduit or metal-clad cable, connected to the wireway through the knockouts. Square plates are obtainable (Fig. 9-159, IV) for closing the ends of the duct. Junctions of wireways are formed by using a junction box or a tee (Fig. 9-159, II and III). Wireways are available in four sizes: 2½, 4, 6, and 8 in (63.5, 101.6, 152.4, and 203.2 mm) square. No conductor larger than that for which the wireway is designed shall be installed in wireways, provided that the sum of the cross-sectional areas of all the conductors at any cross section does not exceed 20 percent of the interior area of the wireway. Splices and taps may be located within the wireway, provided the cover is left accessible. The conductors including the splices must not fill the wireway to more than 75 percent of its area. The wireway must not contain more than 30 conductors at any point except for signaling circuits or for control conductors between a motor and its starter, used only for starting duty. More than 30 *current-carrying* conductors may be installed in a wireway, at a 20 percent maximum fill, if the reduction factors specified in Note 8 of Sec. 17, Div. 11, are applied. Horizontal runs of wireways must be supported every 5 ft (1.52 m) unless specifically approved for supports at greater intervals, but in no case shall the distance between supports exceed 10 ft (3.05 m). Vertical runs of wireway shall be securely supported at intervals not exceeding 15 ft (4.57 m) and shall not have more than one joint between supports. Adjoining wireway sections shall be securely fastened together to provide a rigid joint. An unbroken length of wireway may extend transversely through walls.

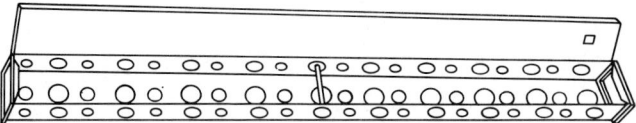

**Fig. 9-158**   Length of wireway with a hinged cover.

**Fig. 9-159**    Fittings for a square duct. [Square D Co.]

### 241.    Cross-Sectional Areas of Wireways

| Wireway size, in | Cross-sectional areas, in$^2$ | | |
|---|---|---|---|
| | 100% | 75% | 20% |
| 2½ × 2½ | 6.25 | 4.68 | 1.25 |
| 4 × 4 | 16.00 | 12.00 | 3.20 |
| 6 × 6 | 36.00 | 27.00 | 7.20 |
| 8 × 8 | 64.00 | 48.00 | 12.80 |

**242. In computing the proper size of wireway,** refer to Secs. 240 and 241. Then determine the total cross-sectional area of the conductors to be installed from Tables 24 to 29 of Div. 11.

#### BUSWAY WIRING (NEMA)

**243. Busway** is a modern system of power distribution. It provides a flexible means for distributing power and light in industrial and commercial buildings and/or connecting branch motor circuits to the mains so that the location of the branch taps can be readily and economically changed to suit the conditions of rearrangement of machines. This system finds wide use in industrial plants when the layout of machines must be continually

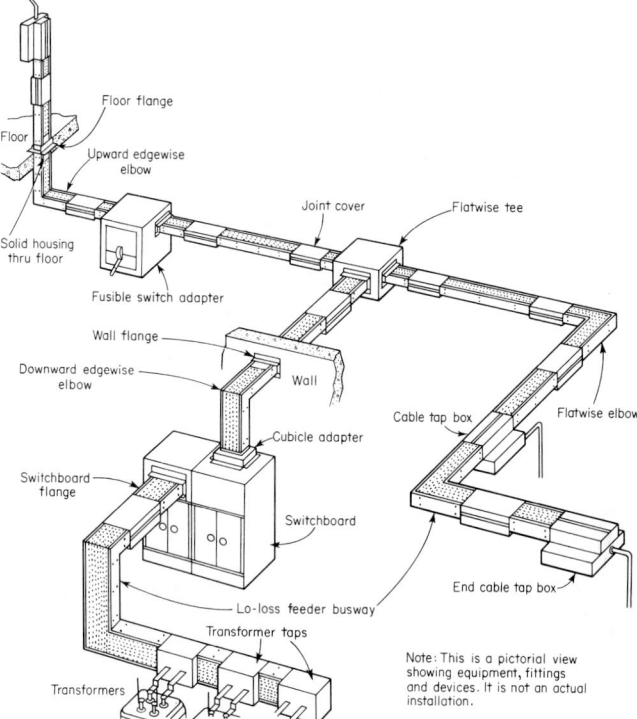

**Fig. 9-160A**   Low-voltage-drop feeder busway system.

changed to meet changing manufacturing conditions because busway's flexibility in accommodating such change is its greatest advantage. (See Fig. 9-160A.)

By definition (NEMA standard 11-3-1977) a busway is a prefabricated electric distribution system consisting of busbars in a protective enclosure, including straight lengths, fittings, devices, and accessories. Busways are of the following types:

1. *Feeder busway.* A feeder busway is a busway without plug-in openings which is intended primarily for conducting electric power from the source of supply to centers of distribution but which can have provisions for bolt-on devices (see Fig. 9-160B).

**Fig. 9-160B**   Feeder busway.

2. *Plug-in busway.* A plug-in busway is a busway having plug-in openings on one or both sides at spaced intervals, offering means for electrical connection of plug-in or bolt-on devices to the busbars. Such plug-in or bolt-on devices (bus plugs) may incorporate switches, circuit breakers, transformers, motor controllers, or other auxiliary equipment (see Fig. 9-160C).

**244. Major requirements related to the application and testing of low-voltage busway.**   This material is intended to assist in understanding the application and testing of low-voltage busway by summarizing the principal codes and standards applicable to the product. Since all codes and standards are revised from time to time and requirements vary from locality to locality, contact your local inspection authority and the manufacturer for the latest requirements applicable to your installation.

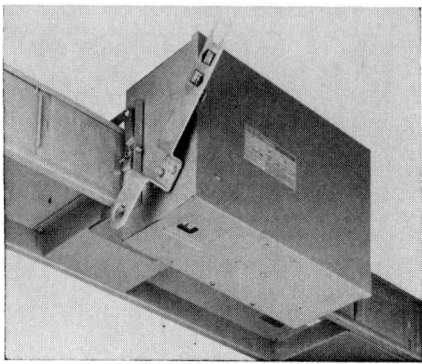

**Fig. 9-160C**  Fusible bus plug installed on a plug-in busway.

Groups which maintain standards enumerating busway construction and installation practices include:

1. Underwriters Laboratories Inc.
2. The National Electrical Manufacturers Association
3. The National Fire Protection Association (The National Electrical Code)
4. State and local electrical codes
5. Power companies, which frequently specify minimum conductor sizes and spacing for connection to their equipment

UNDERWRITERS LABORATORIES REQUIREMENTS  Busway bearing the Underwriters Laboratories listing mark must pass a series of performance tests (see UL 857). Among the tests which may be required are:

*Heating (temperature-rise) test.* Busway with suitably plated joint connections must carry its rated current continuously without any part showing a temperature rise above ambient of more than 55°C. Tests are conducted on three 10-ft (3-m) sections bolted together, at any convenient voltage, at design frequency, with rated current flowing.

*Resistance-to-impact test.* The resistance-to-impact test will ordinarily not be required for nonventilated busway if its construction conforms to the specifications spelled out in Table A. However, all other busway must be tested to ensure that no hazardous condition will exist after any side of the enclosure has been subjected to a blow from a 50-lb weight suspended from a 30-in- (762-mm-) long pendulum arm going through a 60° angle from the vertical.

*Dielectric test.* A busway shall withstand for 1 min without breakdown the application of a 60-Hz potential of 1000 V plus twice the maximum rated voltage between live parts and the enclosure and between live parts of opposite polarity.

*Resistance-to-bending test.* Busway intended for a spacing of not more than 5 ft (1.5 m) between mounting supports will ordinarily be accepted without a bending test. However,

**Table A    Thickness of Sheet Metal**

| Maximum inside width of widest surface | | Minimum acceptable thickness of sheet metal | | | | | |
|---|---|---|---|---|---|---|---|
| | | Steel | | | | Aluminum | |
| | | Zinc-coated | | Uncoated | | | |
| in | mm | in | mm | in | mm | in | mm |
| 12[a] | 305 | 0.056 | 1.42 | 0.053 | 1.35 | 0.075 | 1.91 |
| 18 | 457 | 0.070 | 1.78 | 0.067 | 1.70 | 0.095 | 2.41 |
| 30 | 762 | 0.097 | 2.46 | 0.093 | 2.36 | 0.122 | 3.1 |
| More than 30 | More than 762 | 0.126 | 3.20 | 0.123 | 3.12 | 0.153 | 3.89 |

[a]18 in if enclosure does not support the weight of the busbar assembly.

to qualify for horizontal mounting on a 10-ft (3-m) support basis or on 16-ft (4.9-m) centers in vertical risers, a busway must withstand for 5 min without rupture of the joint, appreciable permanent distortion, or short-circuiting or grounding of the phase or neutral busbars a bending moment equivalent to the weight of a 10-ft section plus an additional 100-lb (45.36-kg) force applied at the end of the 10-ft section.

*Resistance-to-crushing test.* Busway intended to employ a spacing of not more than 5 ft (1.5 m) between mounting supports and whose construction is in accordance with Table A will ordinarily be accepted without a crushing test. However, if the busway does not meet these spacing or construction requirements, it shall be tested to ensure that it will be able to withstand a crushing force of 200 lb (90.7 kg) plus 3 times the weight of a 10-ft (3-m) section.

*Busbar pullout test.* Busway intended for mounting in a vertical run shall be capable of supporting a downward force equal to 4 times the weight of an individual busbar in a 10-ft (3-m) section applied simultaneously to each of the busbars without rupture or permanent distortion of busbar supports.

*Insulation-resistance and dielectric-strength test after exposure to rain.* Outdoor busway must withstand a simulated rain for 8 h per day for 5 successive days without dielectric breakdown.

*Short-circuit test.* All busway and fittings rated 100 A or less and having a short-circuit current rating of more than 5000 A shall be tested and marked with their respective rating(s). Busway and fittings rated over 100 A and having a short-circuit current rating over 10,000 A shall be tested and marked with their respective ratings. To verify these rating(s), representative samples of the busway shall be subjected to phase-to-phase, phase-to-enclosure, and phase-to-ground conductor (when applicable) short-circuit tests. Tests may be conducted either with or without fuse or circuit-breaker protection ahead of the busway being tested, and the ratings and markings will so indicate.

*Tests when busway enclosure is used as an equipment-grounding conductor.* For the busway enclosure to be accepted as an equipment-grounding conductor, the busway shall be subjected to the short-circuit tests described above. In addition, two other conditions shall be met: (1) the busway enclosure shall comply with minimum cross-sectional–area requirements, and (2) a maximum resistance of 0.005 $\Omega$ is allowed when passing a direct current of 30 A between adjacent busway sections.

NATIONAL ELECTRICAL MANUFACTURERS ASSOCIATION STANDARDS   In addition to the Underwriters Laboratories tests listed above, the NEMA standard for busways specifies the following:

*Short-circuit current ratings.* There are no minimum or maximum standards on short-circuit current ratings specified by Underwriters Laboratories. The minimum rating depends upon the maximum short-circuit current available at the line and of the installed busway run. However, NEMA recommends the following minimum ratings:

| Continuous current rating of busway, A | Short-circuit current ratings, symmetrical amperes | |
| --- | --- | --- |
| | Plug-in busway | Feeder busway |
| 100 | 10,000 | |
| 225 | 14,000 | |
| 400 | 22,000 | |
| 600 | 22,000 | 42,000 |
| 800 | 22,000 | 42,000 |
| 1000 | 42,000 | 75,000 |
| 1200 | 42,000 | 75,000 |
| 1350 | 42,000 | 75,000 |
| 1600 | 65,000 | 100,000 |
| 2000 | 65,000 | 100,000 |
| 2500 | 65,000 | 150,000 |
| 3000 | 85,000 | 150,000 |
| 4000 | 85,000 | 200,000 |
| 5000 | . . . . . | 200,000 |

*Voltage drop.* There are no minimum or maximum standards on voltage drop since these values may vary considerably depending upon several factors, including (1) the design, (2) the conductor material, (3) the continuous-current rating, (4) the current load, (5) the power factor of the load, (6) the operating temperatures, (7) the length of the run, and (8) the distribution of the load. Actual values of voltage drop in given situations can best be determined by obtaining the necessary data from a busway manufacturer and applying it to the specific situation. In general, the line-to-line voltage drop of fully loaded, three-phase busway will usually be less than 6 V/100 ft. Voltage drops for plug-in busway are based upon an evenly distributed load along the run and are one-half that of a load concentrated at one end.

*Resistance, reactance, and impedance.* There are no minimum or maximum values of resistance, reactance, or impedance since these values vary considerably with different ratings and busway constructions. Busway manufacturers can supply these data for their particular products. These data are used for calculating the voltage drop for a specific application.

*Unusual service conditions.* For applications in which the ambient temperature is higher than 40°C, the busway current rating should be derated in accordance with the manufacturer's recommendations, if furnished, or the following:

**Table B**

|  | Ambient-temperature limits | NEMA standards publication no. |
|---|---|---|
| Busway lengths and fittings | −30°C through +40°C | NEMA BU 1-1983 |
| Low-voltage-power circuit breakers | Not to exceed +40°C | NEMA SG 3-1981 |
| Molded-case circuit breakers | 0°C through + 40°C | NEMA AB 1-1975 (R1981) |
| Enclosed switches | −30°C through +40°C | NEMA KS 1-1983 |
| Low-voltage cartridge fuses | See applicable UL standards. | NEMA ICS 1-1978 (R1983), |
| Electromagnetic and manual motor controls at 6000 ft and less | 0°C through + 40° C maximum | ICS 2-1978 (R1983), and ICS 3-1983 |

**245. National Electrical Code requirements.**    The National Electrical Code, which is sponsored by the National Fire Protection Association, outlines the general guidelines which are necessary for the safe application of electrical equipment. Article 364 of the Code specifically applies to busways, although other articles in the Code are applicable in certain situations. Thorough familiarization with the NEC requirements for busways is recommended before attempting to apply busways to specific applications.

Some of the major requirements for safe application of buways as outlined in the 1984 NEC are as follows:

*364-4. Use*

*a. Use permitted.* Busways shall be installed only where located in the open and are visible.

*Exception. Busways shall be permitted to be installed behind panels if means of access are provided and if all the following conditions are met.*

   *a. No overcurrent devices are installed on the busway other than for an individual fixture.*
   *b. The space behind the access panels is not used for air-handling purposes.*
   *c. The busway is the totally enclosed, nonventilating type.*
   *d. Busway is so installed that the joints between sections and fittings are accessible for maintenance purposes.*

*b. Use prohibited.* Busways shall not be installed (1) where subject to severe physical damage or corrosive vapors; (2) in hoistways; (3) in any hazardous (classified) location, unless specifically approved for such use [see Sec. 501-4 (b) of the NEC]; or (4) outdoors or in wet or damp locations unless identified for such use.

*364-5. Support.* Busways shall be securely supported at intervals not exceeding 5 ft (1.52 m) unless otherwise designed and marked.

*364-6. Through walls and floors.* It shall be permissible to extend unbroken lengths of busway through drywalls. It shall be permissible to extend busways vertically through dry floors if totally enclosed (unventilated) where passing through and for a minimum distance of 6 ft (1.83 m) above the floor to provide adequate protection from physical damage.

*300-21. Spread of fire or products of combustion.* Electrical installations in hollow spaces, vertical shafts, and ventilation or air-handling ducts shall be so made that the possible spread of fire or products of combustion will not be substantially increased. Openings around electrical penetrations through fire-resistance–rated walls, partitions, floors, or ceilings shall be fire-stopped by using approved methods to maintain the fire-resistance rating. NOTE   Fire-stops may require decreasing the allowable load on the conductors.

*364-11. Reduction in size of busway.* Omission of overcurrent protection shall be permitted at points where busways are reduced in size, provided that the smaller busway does not extend more than 50 ft (15.2 m) and has a current rating at least equal to one-third of the rating or setting of the overcurrent device next back on the line and provided further that such busway is free from contact with combustible material.

STATE AND LOCAL ELECTRICAL CODES   Many states and cities have electrical codes which supplement the National Electrical Code. Before attempting to apply or install busway as well as any other type of electrical equipment, contact your local electrical inspection authority for a copy of the latest requirements applicable to the location of the installation.

ELECTRIC-POWER-COMPANY REQUIREMENTS   Busways are commonly used for the main services of buildings, in which case they are usually connected to one or more distribution transformers owned by the local electric power company. This connection may be made by either cable or busbars, depending upon local practices. Since the various electric power companies throughout the United States often prefer different methods of connection to busway services, it is recommended that the local company supplying the power be contacted before applying or installing a service-entrance busway run.

CONTACTING BUSWAY MANUFACTURERS   Since the practical applications and uses of busway are nearly unlimited, it is not possible to outline here all the technical information available pertaining to busway. For this reason, it is advisable to contact one or more of the manufacturers of busway for more specific information whenever a possible application for busway is contemplated. All companies in the industry have technical experts who will assist in properly selecting and applying the busway system best suited to the specific requirements.

**246. Busway layout.**  Present-day busway systems are designed for flexibility of layout and can be applied on applications such as a simple straight run of plug-in busway fed from a cable-tap box or a complex run with multiple feeders and plug-in devices for a high-rise building. However simple or complex, the system must be planned in advance.

The purpose of this section is to show how to plan the layout in an easy, step-by-step approach. A number of examples and details of how the busway is applied, as well as specific details on how to coordinate the installation of busway with other products, are included.

This section should be used as a reference guide only. The final decision on any system must be dictated by the particular details of the application.

When busway manufacturers receive orders for busway, an approval drawing is generally required. Experience has shown that a complete, detailed approval drawing as shown in Fig. 9-161 expedites release for manufacture. It is therefore important that the information sent to the manufacturer be as nearly complete as possible. If it is incomplete, a great deal of time can be wasted in obtaining missing information.

The following is some of the information typically missing from sketches or drawings received by busway manufacturers:
1. Front location of switchgear, switchboards, transformers, or other cubicles
2. Floor heights and floor thicknesses on risers
3. Location of busway risers in relation to closet walls
4. Location of plugs on a riser in relation to closet walls
5. Location of walls and wall thickness

The approval drawing that is sent back to the field for approval may require additional dimensions for the complete busway layout. The electrician will be required to fill in these dimensions after field-measuring the job, approve the drawing for manufacture, and submit it through channels to the busway manufacturer.

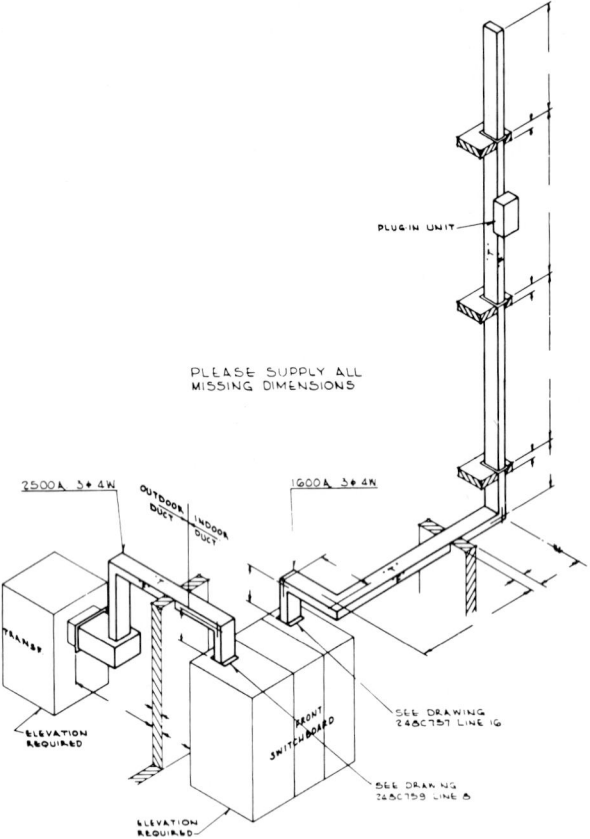

**Fig. 9-161**  Typical customer-approval drawing.

GUIDELINES FOR GETTING BASIC ON-SITE MEASUREMENTS   Before even looking at a busway job, you should have these basic materials for measuring: a sketch pad, a 50- or 100-ft (15- or 30-m) measuring tape, an 8- or 10-ft (2.4- or 3-m) folding wood rule, chalk and chalk line, plumb bob and string, and a yellow lumber crayon. Armed with these tools, you are prepared to lay out a busway system successfully.

First, it is always good practice to walk through the entire facility to get an idea of general routing. While walking through, watch for obstructions, both vertical and horizontal. At the same time, try to establish one elevation so that offsets are not required. To do so, you must check floor elevations between rooms or various parts of the building. Otherwise, the busway installed at a specific elevation in one part of the building might be too high or too low at another point.

After the walk-through, you are ready to begin measuring the system. For reference points, use structures already available in the building, such as walls or columns. To save time, don't attempt to break the run down into specific lengths of busway. A busway manufacturer's engineer will do this and at the same time prepare an erection drawing to show the breakdown.

Actual measurement should start in the area where the switchboard or switchgear is located. If the switchborad has not been installed, chalk in the specified location and the point of entry for the busway. At this time, if the service entrance run is part of the job, you can measure its location.

One of the most important steps in planning a large busway layout is coordination between the trades. This is a good time for the electrician to meet with plumbing, heating, ventilating, and sprinkler-system trades to establish a definite right of way throughout the construction area. Once this has been established, hours of frustrating rerouting can be saved. Immediately after the right of way has been determined, electricians will often install busway hangers to claim and identify their areas.

The next step is to establish the busway elevation, then start horizontal measurements using the following examples as a guide. The examples indicate the dimensions required (see Figs. 9-162A through 9-162J).

**247. Busway-riser applications.**     Several conditions will dictate the number of plug-in units that will fit physically on each floor: floor height, size of plugs, maximum height of plug-operating handle, other equipment or walls in close proximity to the busway riser, and orientation and phasing. Let's consider each of these conditions in turn.

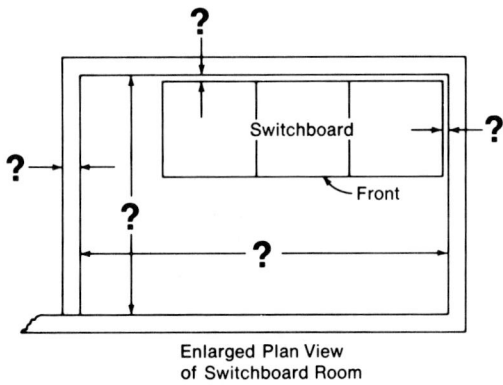

Enlarged Plan View
of Switchboard Room

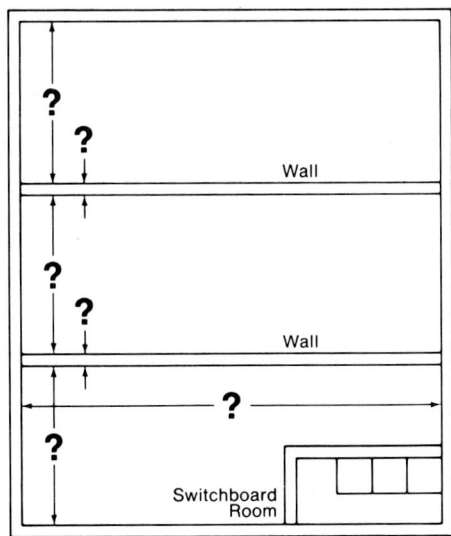

**Fig. 9-162A**     Basic site dimensions required before starting busway layout.

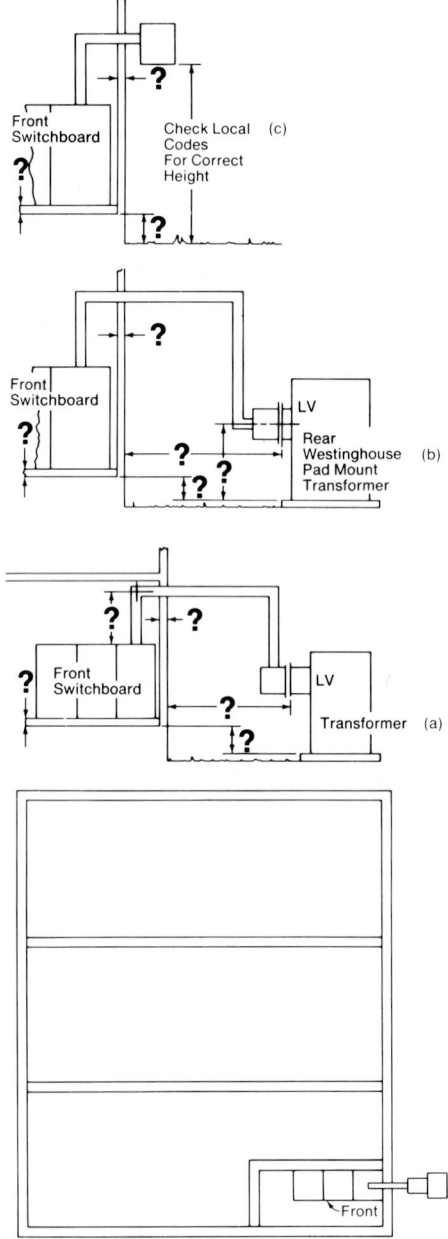

**Fig. 9-162B**  Service entrance run from power-center transformer, pad-mount transformer and weatherhead.

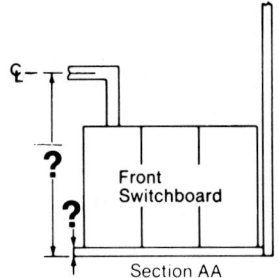

Section AA

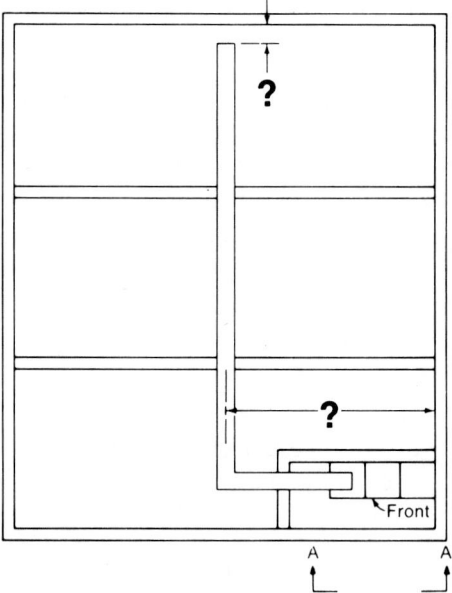

**Fig. 9-162C**  Elevation.

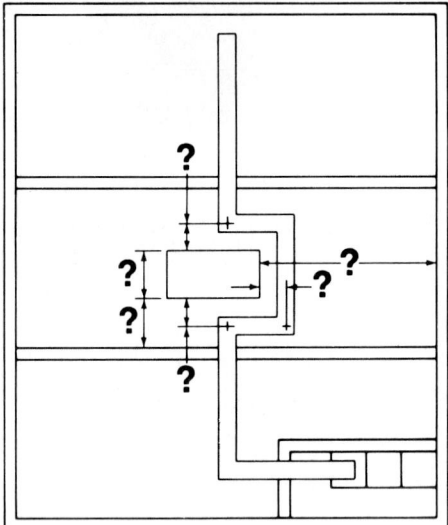

**Fig. 9-162D**   Horizontal offset.

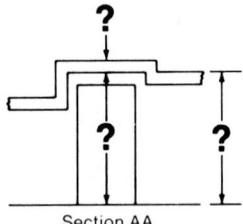

Section AA

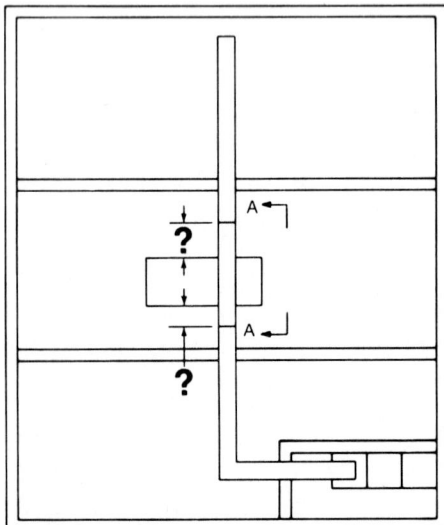

**Fig. 9-162E**   Vertical offset.

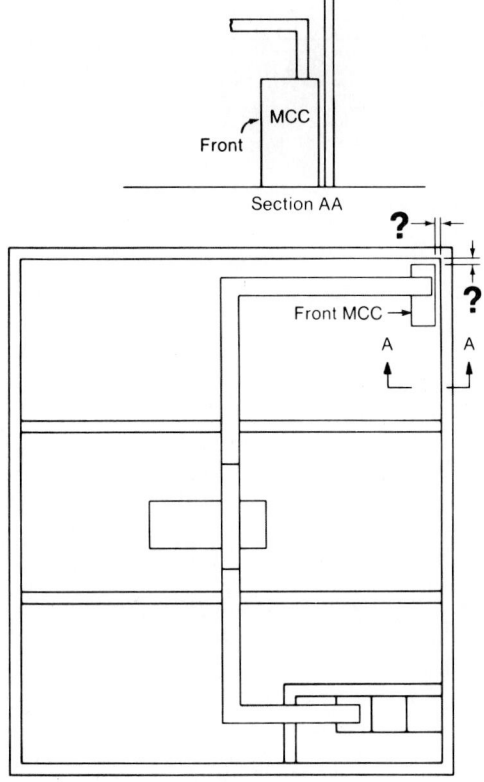

**Fig. 9-162F** Vertical drop into an MCC.

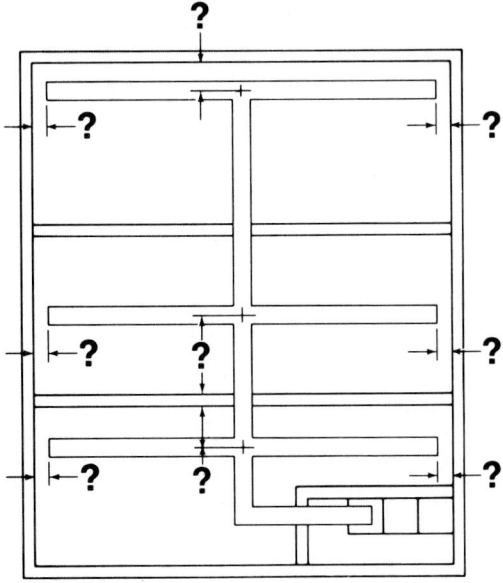

**Fig. 9-162G** Single-run distribution system.

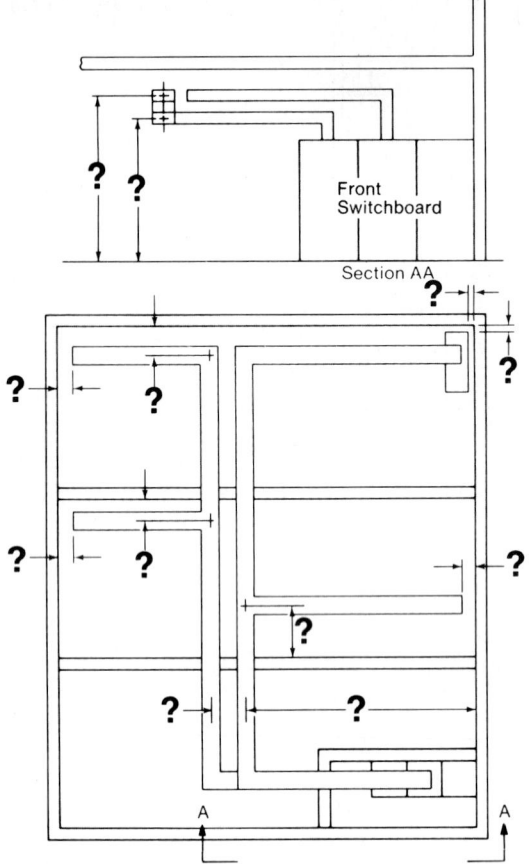

**Fig. 9-162H** Double-run distribution system.

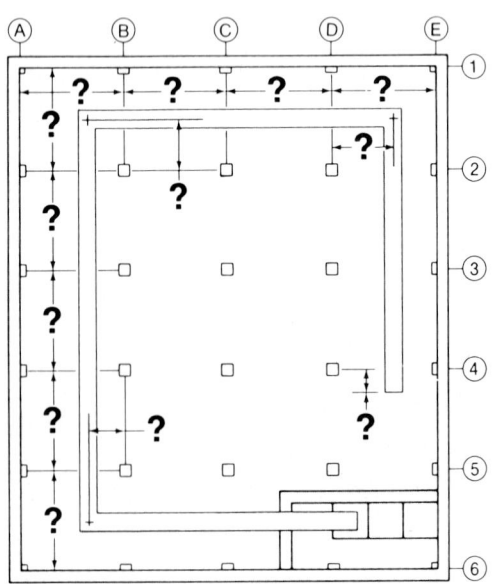

**Fig. 9-162I** Using columns as reference points.

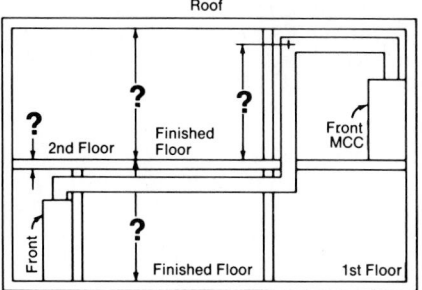

West Elevation View

Plan View 2nd Floor

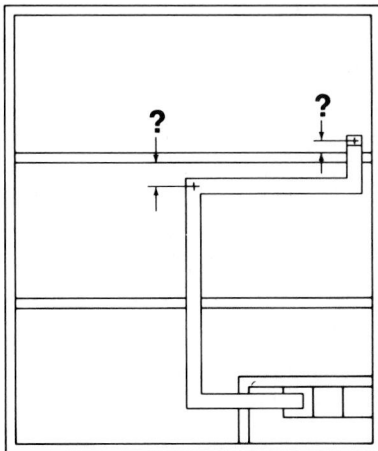

Plan View 1st Floor

**Fig. 9-162J** Multifloor.

FLOOR-HEIGHT MEASUREMENTS    The basic dimensions (height from finished floor to finished floor plus floor thickness) required for every busway riser are shown in Fig. 9-162K. Do not assume that all floors are the same height. NOTE    If a curb is to be supplied around the floor opening, specify the height of the curb.

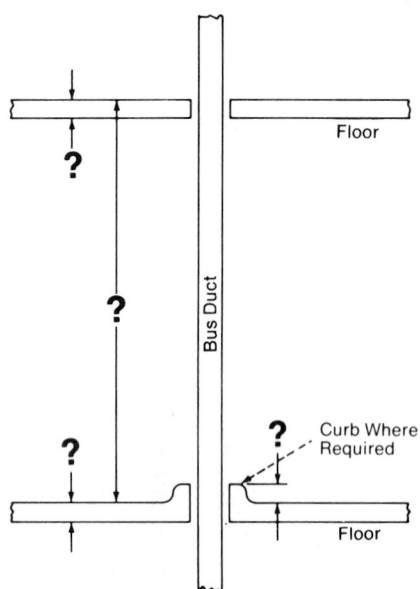

**Fig. 9-162K**   Basic height measurements required for risers.

FLOOR HEIGHT    The individual planning the riser should realize that clearance above the floor must be maintained to install vertical mounting hangers and joint-cover plates. Generally a distance of 15 to 24 in (381 to 609.6 mm) above the floor to the centerline of the joint is required.

**248. Size of plug-in devices.**   The floor height will dictate the number of plug-in openings physically possible for each riser on each floor. If the floor height is less that 12½ ft (3.8 m) and floor thickness remains at 6 in (152.4 mm), it is impossible to use a 10-ft (3.05-m) straight length of plug-in busway. Instead, a shorter straight length would be used.

Using the manufacturer's plug-in unit dimensions, you can experiment with plugs to be supplied and determine the number that can fit on each floor.

MAXIMUM HEIGHT OF PLUG-OPERATING HANDLES    Generally, there is no maximum height for plug-in units when a suitable means of operating such as a hook stick, rope, or chain is provided (see Sec. 364-12, 1984 NEC). However, since requirements vary from locality to locality, contact the local inspection authority and busway manufacturer for the latest requirements applicable to your installation.

OTHER EQUIPMENT OR WALLS IN CLOSE PROXIMITY TO THE BUSWAY RISER    Applying busway with plug-in or bolt-on units in a vertical riser requires preplanning. See Figs. 9-162L, 9-162M, and 9-162N for examples of busway risers in an electrical closet. Busway with plug-in or bolt-on units must be arranged to clear other equipment and walls in the wiring closet. In particular, make sure that the plug-in or bolt-on hinged cover can be opened fully for maintenance.

ORIENTATION AND PHASING    To ensure that plug-in units are properly located, consult the manufacturer for coordination of busway orientation and phasing.

BUSWAY CONNECTIONS FROM SWITCHGEAR THROUGH WALL AND FLOOR OPENINGS    Wall and floor openings for busway should be at least 1 in (25.4 mm) larger than the outside dimensions of the busway. In some areas where there is doubt about the accuracy of the measurement,

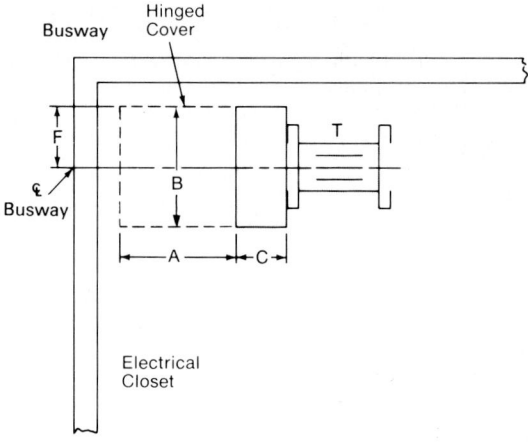

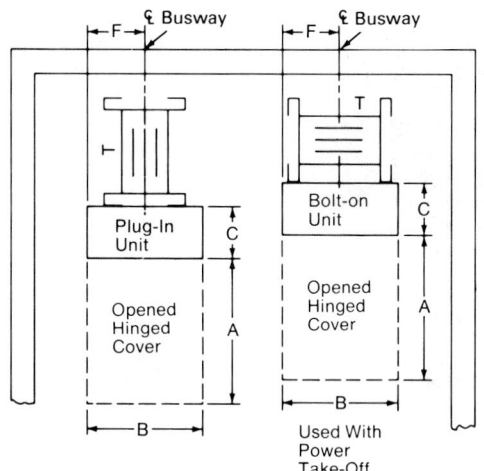

**Fig. 9-162L**   Examples of busway risers in an electrical closet.

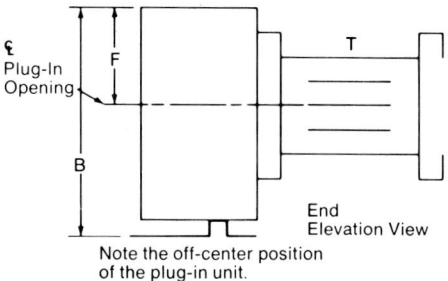

**Fig. 9-162M**   Plug-in position on busing.

a flange must be installed through a wall, the hole must be large enough for the largest dimension.

SPECIAL CONDITIONS    Certain conditions require preplanning to circumvent installation problems. Figure 9-162Q is an example. If the *X* or *Y* dimensions are below the minimum needed to permit a joint between the elbow and the walls or floor, the joints must fall as shown. It then follows that openings in the walls or floors must be made large enough to install the busway. (See Fig. 9-162R.) Wall thickness must also be considered.

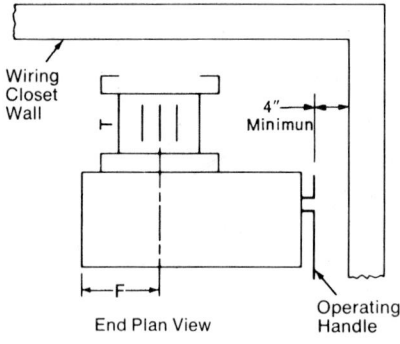

**Fig. 9-162N**  Busway riser in an electrical closet.

BUSWAY PLUG-IN APPLICATION    When plug-in units are to be installed, consideration must be given to providing adquate clearance from the plug-in busway to obstructions to allow the plug-in unit to swing in place. Consult the manufacturer for specific requirements. (See Fig. 9-162S.)

CUSTOMER ERECTION DRAWINGS    Prior to shipment of busway from the factory, the installer will receive a customer drawing. A sample appears in Fig. 9-162T. This drawing contains a complete layout of the entire installation and a bill of material that includes (1) an item number which can be correlated with the drawing and a description of each piece and (2) the identification number and quantity of each piece.

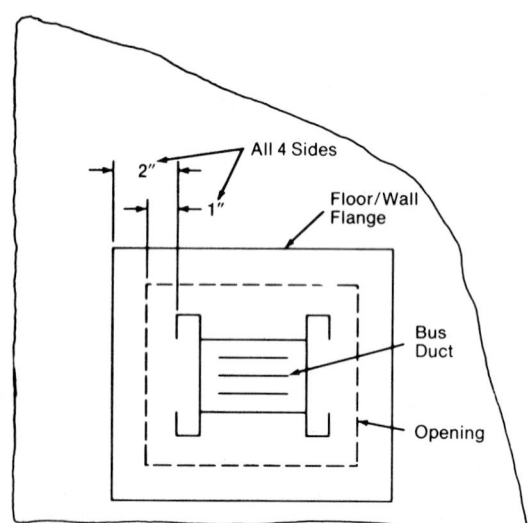

**Fig. 9-162P**  Minimum clearances for wall and floor openings.

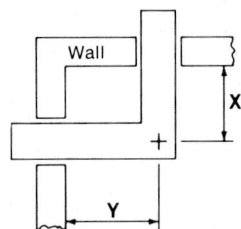

**Fig. 9-162Q**  Example of an installation problem.

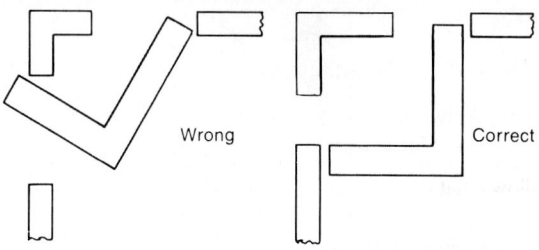

Wrong                                    Correct

**Fig. 9-162R** Solution to an installation problem.

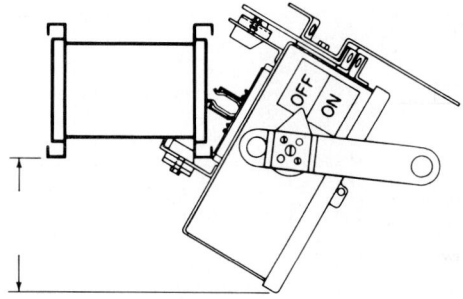

**Fig. 9-162S** Clearance required to mount plug-in units on a busway.

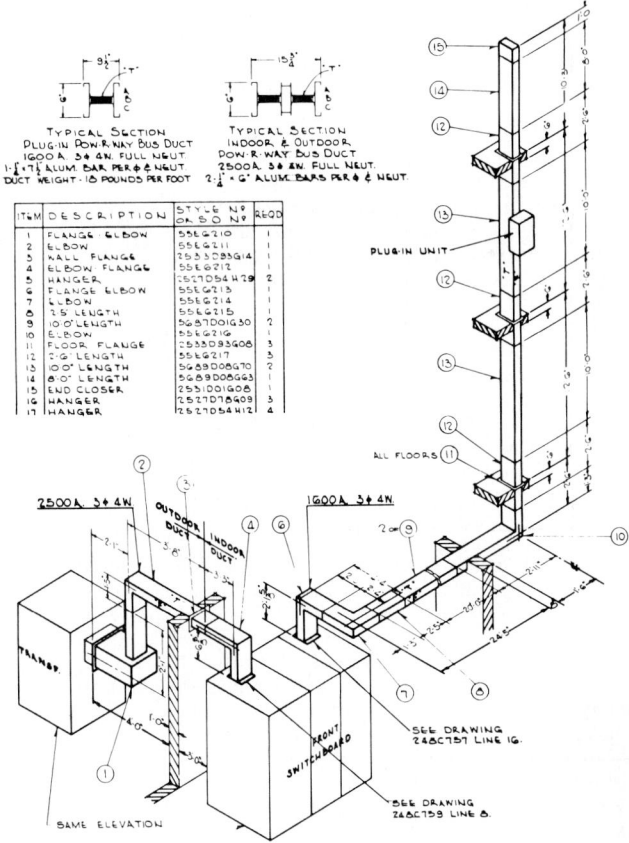

**Fig. 9-162T** Typical drawing sent by a busway manufacturer to an installer.

### 249. Busway installation (see Fig. 9-163*A*)

GENERAL    To ensure maximum reliability and long life for a busway system, proper installation is imperative. Manufacturers supply installation drawings on all but the simplest busway layouts. Study these drawings carefully. If drawings are not supplied, make· your own. Verify actual components on hand against those shown on installation drawings to ensure that there are no missing items. Manufacturers' drawings identify components by catalog number and specify their location in the installation. Catalog numbers appear on nameplates and carton labels. Read the manufacturers' instructions for installation of the individual components. If you are still in doubt, ask for more information; *never guess.* Almost all busway sections are built with a busway joint on one end. Refer to the installation drawing to orient each component properly. Pay particular attention to "Top" labels and other orientation marks where applicable.

**Fig. 9-163*A***    Installing busway.

STORAGE AND HANDLING    Busway sections and fittings which are not to be installed and energized immediately upon receipt should be stored in a clean, dry place that has a uniform temperature to prevent condensation. Preferably, they should be stored in a heated building having adequate air circulation and protected from dirt, fumes, water, and physical damage. Because of moisture it is extremely risky to store busway outdoors. If busway must be stored outdoors, it must be securely covered for protection from weather and dirt. Temporary electrical heating should be installed beneath the cover to prevent condensation.

Indoor busway must always be protected from exposure to moisture, plaster, and other types of contaminants. This is extremely important during initial installation when the busway may be exposed to higher levels of dust and moisture.

Outdoor busway must be treated exactly the same as indoor busway until it is installed. It is not weather-resistant until it is completely and properly installed. Care must be used in unpacking. Band cutters should be used on all banding securing the package. If a crane is employed to install busway, use nylon straps and distribute, or balance, the weight of each lift. If cables are employed, spurs should be used to avoid damage to the metal housing. Do not drag busway across the floor. When installing vertical sections, it is easier to lower the busway from the floor above where it will be installed. Vertical sections are often stored on the floor above their final location to facilitate lowering them into position.

SUPPORTS AND HANGERS    Use the hangers or the hanging recommendations supplied by the manufacturer in the manner prescribed to afford a secure installation and proper hanger spacing. A busway which is mounted horizontally should be supported at 5-ft (1.5-m) intervals and not more than 10-ft (3-m) intervals if it is marked as being suitable for such intervals. Consider sway bracing for plug-in busway. Caution must be emphasized when all plug-in units are used on one side of a busway or when there are more plugs on one side than the other. Large plug-in units are cubicles which have considerable weight and when mounted on a horizontal run of busway should be separately supported from the building structure to prevent twisting or deformation of the busway. A busway which is to be mounted vertically should be marked to indicate that it is suitable for that purpose. When the distance between floor supports is more than that for which the busway is marked as suitable, the manufacturers' recommendation regarding intermediate supports should be followed; however, support intervals should not exceed 16 ft (4.9 m). Busway cubicles and bus plugs should be separately supported by the building structure when so recommended by the manufacturer. (See Figs. 9-163*B* and 9-163*C*.)

TESTING    Be sure to recheck all steps to ensure that you have followed the manufacturer's recommendations for installation and that you have not forgotten anything. At this point the busway installation should almost be complete. Before energizing it, the com-

**Fig. 9-163B**  Installing busway on hangers.

**Fig. 9-163C**  Riser busway section with floor support.

plete installation should be properly tested. The testing procedures should first verify that the proper phase relationships exist between the busway and associated equipment. This phasing and continuity test can be performed in the same manner as similar tests on other pieces of electrical equipment on the job. All busway installations should then be tested with a megohmmeter or high-potential-voltage tester to be sure that excessive leakage paths between phases and ground do not exist. Megohmmeter values depend on busway construction, type of installation, size and length of busway run, and atmospheric conditions. Acceptable values for a particular busway should be obtained from the manufacturer.

ENERGIZING   When the equipment is energized for the first time, qualified electrical personnel should be present. Energizing a run of busway for the first time is potentially dangerous. If faults caused by damage or poor installation practices have not been detected in the checkout procedure and corrected, serious damage can result when the power is turned on. There should be no load on the busway when it is energized. Since busway typically extends through several rooms and floor levels, care should be taken to see that all devices fed from the busway are switched to the off position. The equipment should be energized in sequence by starting at the line end of the system and working toward the load end. In other words, energize the main devices, then the feeder devices, and then the branch-circuit devices. Turn the devices on with a firm, positive motion. Protective devices which are not quick-acting should not be teased into the closed position. After all the devices have been closed, loads such as lighting circuits, contactors, heaters, and motors may be turned on. Busway, when operating properly, will have a moderate 60-Hz hum. However, excessive noise may be an indication of hardware that has not been tightened or of metal parts that have been improperly assembled. Occurrence of sparking at any joint along the busway is an abnormal condition. The busway should be deenergized until the sparking condition has been corrected.

**250. Busway-system inspection and maintenance**
*Warning.* Turn off power ahead of the busway before performing any of the following maintenance operations. Check incoming line terminals with a voltmeter prior to proceeding with work. *Do not attempt to work on energized busway.*

**251. Busway should be inspected** once each year or after any severe electrical faults. Look for any moisture or signs of previous wetness or dripping onto the busway or onto connection boxes from leaky roofs, pipes, sprinklers, or other sources of moisture. Look for any recent changes in sprinklers or other plumbing that might now be a source of trouble to the busway. Eliminate the source of any liquids dripping onto the busway and any other source of moisture (indoor busway). Replace or thoroughly dry and clean any insulating material which is damp or wet or shows accumulation of deposited material from previous wettings. If there is an appreciable accumulation of dust and dirt, clean it off by using a brush, vacuum cleaner, or clean, lint-free rags. To avoid blowing dust into busway joints, circuit breakers, or other equipment do not use a blower or compressed air. Carefully inspect all visible electrical joints and terminals, checking for tightness of bolts, nuts, etc.,

in accordance with the manufacturers' instructions (see Fig. 9-164A). If joints or terminations appear to be badly discolored, corroded, or pitted or show evidence of having
been subjected to high temperatures, the parts should be disassembled and replaced or
cleaned. Do not remove plating on aluminum parts, joints, or terminations. Damaged aluminum parts should normally be replaced.

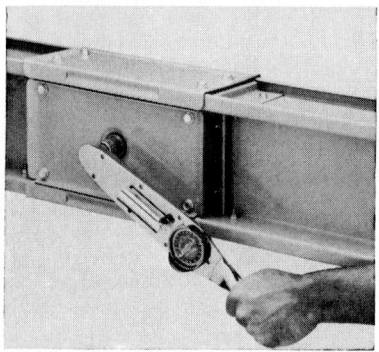

Check the insulation resistance prior to
reenergizing the busway. A permanent record
should be kept of resistance readings. If readings decrease appreciably with time, deterioration is taking place. If resistance readings are
below recommended manufacturer's values,
contact the busway manufacturer for the
appropriate corrective action.

**252. Busway plugs and protective devices.**
Check the operation of all mechanical components. Check all switch-operator mechanisms
and external operators of circuit breakers.
Make sure that each operator mechanism
quickly and positively throws contacts fully on

**Fig. 9-164A** Tightening a joint bolt by using a torque wrench.

and fully off. Check the mechanisms of all electrical and mechanical interlocks and padlock-
ing means. Check for missing or broken parts,
proper spring tension, free movement, rusting
or corrosion, dirt, and excessive wear. Examine
all readily accessible arc chutes and insulating parts for cracks or breakage and for arc
spatter, sooty deposits, oil, or tracking. Clean off arc spatter, oil, and sooty deposits.
Replace the device if appreciable material has burned away or if parts are charred or
cracked. Closely examine the fuse clips. If there is any sign of overheating or looseness,
check the spring pressure, tightness of clamps, etc. Replace the fuse clips if the spring
pressure compares unfavorably with that of identical fuse clips and other equipment. Look
for any signs of deterioration in insulating material or melting of the sealing compound.
Replace such insulating parts and assemblies where the sealing compound has melted. Be
sure that the condition which caused the overheating has been corrected. Lubricate the
operating parts of switch mechanisms, etc., according to the manufacturer's instructions.
Use a clean nonmetallic light grease or oil as instructed. Do not oil or grease parts of
molded-case circuit breakers (see Fig. 9-164B). If no instructions are given on the devices,
sliding copper contacts, operating mechanisms, and air locks may be lubricated with
clean, light grease. Operate the switch or circuit breakers several times to make sure that
all mechanisms are free and in proper working order. Retighten all wire connections.
Check fuses to make sure that they have the proper ampere rating and interrupting rating.
Make sure that non-current-limiting fuses are not used as replacements for current-limiting fuses. If the bus plug has been removed
from the busway, check the insulation resistance of the device prior to its reinstallation.

**253. Specific-purpose busway.** Lighting busway and trolley busway are considered to be
special-purpose busway. *Lighting busway*
provides a low-cost, flexible power-distribution system for supplying low-capacity loads.
It provides maximum flexibility for these
low-capacity loads, since a power tap-off can
be located at any point along the busway. It
is used effectively as a power supply system
for lighting fixtures, small power tools, and
machines in industrial and assembly plants,
office and commercial buildings, department
stores, vocational schools, garages, warehouses, truck and rail terminals, shipping
docks, museums, and galleries. *Trolley busway* provides a completely mobile and flexi-

**Fig. 9-164B** Power tap-off cubicle with a molded-case circuit breaker.

ble power-distribution system for such applications as cranes, hoists, monorails, conveyors, assembly lines, observatories, rotating restaurants, industrial trolley cars, and portable tools.

*Lighting busway* (Fig. 9-165A) provides the greatest flexibility in the location of outlets for low-current-capacity loads. Several ampere ratings (20 A, 30 A, 35 A, 50 A, and 60 A) are available in two-, three-, and four-wire systems, with a maximum voltage rating of 300 V, ac or dc, to ground. Lighting busway may be used as a branch-circuit or feeder circuit as defined in Articles 364-2, 210-2, and 210-23 of the National Electrical Code. It may be utilized in exposed areas except in unsuitable locations where subject to mechanical injury, corrosive vapors, hoistways, battery rooms, outdoors, or in wet or hazardous locations (see Section 364-4 of the NEC).

Standard straight sections are typically available in 4- and 10-ft (1.2- and 3-m) lengths, while nonstandard sections are typically available in lengths of 1, 2, 3, 5, 6, 7, 8, and 9 ft (0.3, 0.6, 0.9, 1.5, 1.8, 2.1, 2.4, and 2.7 m). In this type of busway, the steel housing serves as the ground path. A continuous narrow opening or slot along the bottom of the housing provides power tap-off capability along the entire length of the busway run by allowing the insertion of plugs at any point. If required, a rubber slot closure which can be field-cut to length is available to close off openings between plugs to prevent the entrance of dust or other objects.

Standard components, along with straight sections, include coupling for joining section to section, feed-in devices for supplying power to individual runs, and caps to terminate a run, various types of hangers and heavy-gage–steel channels to suspend and support lighting runs with fixtures attached, and a variety of tap-off devices in both terminal and receptacle types with or without circuit protectors (fuses or circuit breakers).

*Couplings* (Fig. 9-165B) serve three functions: first, to connect electrically the circuit conductors between sections; and, second, to join section housing rigidly and securely to section housing. A third function is available with a trolley-entrance coupling which allows the insertion of a trolley-type tap-off device for movable-load applications.

*Feed-in boxes* serve to power lighting busway circuits and are designed to attach at the end (Fig. 9-165C) or in the center of a run (Fig. 9-166).

*End caps* serve to terminate or seal off the end of a run (Fig. 9-167). Similar in design to couplings, they snap into place and can be easily removed and reused when extending a lighting busway run.

METHODS OF SUSPENSION   (Figs. 9-168 and 9-169) Lighting busway can be suspended by various methods. Manufacturers provide a variety of hangers for flush or surface mounting, drop-rod suspension, or the messenger-cable method. The distance between points

**Fig. 9-165B**   Couplings.

**Fig. 9-165A**   Example of a lighting busway.

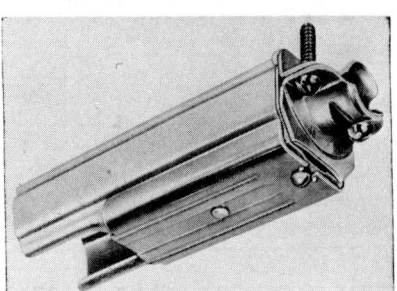

**Fig. 9-165C**   Feed-in boxes.

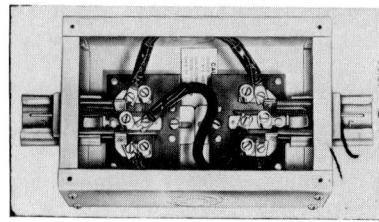

**Fig. 9-166**  Typical center feed-in box.

**Fig. 9-167**  Typical end cap.

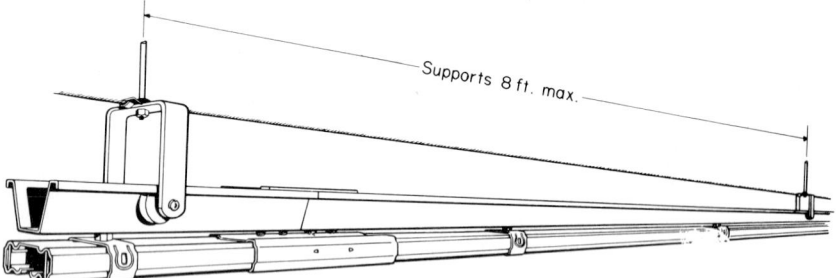

**Fig. 9-168**  Hat section, suspension-fixed.

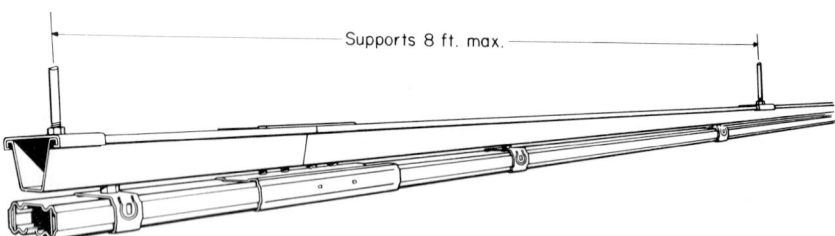

**Fig. 9-169**  Hat section, suspension–roll-in.

of suspension depends on the weight loads directly supported by the busway, building truss or beam locations, and manufacturers' allowable limits. Hangers can be positioned anywhere along the busway housing. When heavy loads (such as high-bay lighting fix-

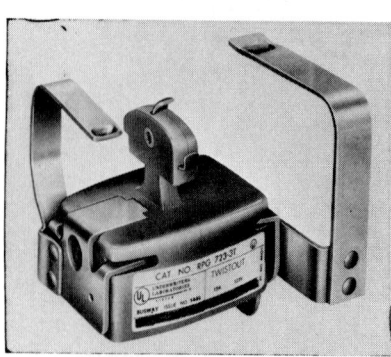

**Fig. 9-170**  Twistout.

tures) are to be supported or when building trusses are spaced beyond standard allowable limits, a supplementary steel support channel (hat section) can be provided by busway manufacturers. By using the support channel and roller-type rod hangers, lighting busway can be rolled into position, with lighting fixtures prepositioned (installed on the busway and electrically connected), from one end of a building.

**254. Power tap-off devices** (twistouts and trolleys; Figs. 9-170, 9-171, and 9-172) are typically available as follows:

   1. *Without circuit protection.* Polarized.
   2. *Receptacle twistouts.* Rated 15 A at 125 V, ac or dc; for three-wire (line, neutral, and ground) convenience gaps.

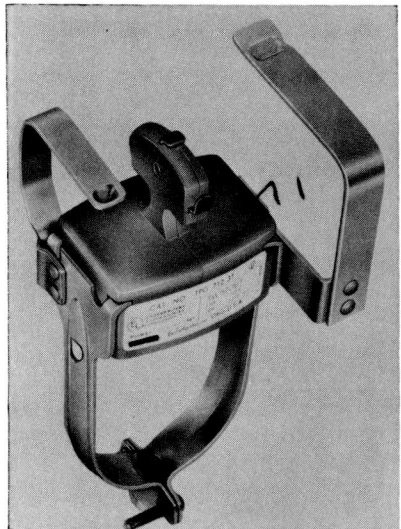

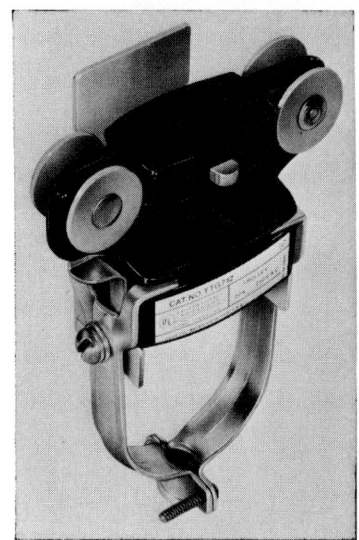

**Fig. 9-171**  Twistout.                    **Fig. 9-172**  Trolley.

3. *Terminal twistouts.* Rated 20 A at 125 V, ac or dc, and 15 A at 300 V ac; for three-wire (line, neutral, and ground) permanent connection to electrical devices.

4. *Terminal trolleys.* Rated 20 A at 300 V, ac, for three-wire (line, neutral, and ground) permanent connection to electrical devices.

Terminal twistouts and trolleys are equipped with cord strain-relief clamps or can be furnished with outlet boxes with a standard convenience three-wire receptacle installed.

5. *With circuit protection* (Figs. 9-173 and 9-174). Tap-off devices (twistout and trolley types) can be provided with outlet boxes containing circuit protection devices and accessories prewired to line side terminals.

APPLICATION DATA  Lighting busway can be furnished in two-, three-, or four-pole configuration. All are polarized, use conductors rated at 300 V, ac or dc, with four available

**Fig. 9-173**  Twist-type tap-off device with circuit protection.

**Fig. 9-174**  Trolley-type tap-off device with circuit protection.

ampere ratings: 20 A, 35 A, 50 A, and 60 A. Neutrals are sized at 100 percent cross-sectional area of the conductor bars.

For single-circuit loads, two-pole lighting busway is normally used for low-initial-cost installations. It can be used for 120-, 240-, or 277-V ac systems or dc systems up to 300 V.

Three-pole lighting busway is used for most applications, as it offers two-circuit capability at only slightly higher installed cost and provides greater load-carrying capacity. While a two-pole system rated at 277 V, 60 A can carry 16.6 kVA, a three-pole system of the same rating can carry 33.2 kVA.

Four-pole lighting busway may be used where there is a need to serve both single-phase and three-phase loads.

Three-pole systems are normally operated at either 120/240 or 120/208 V, ac or dc, up to 300 V. They can also be connected to a 480Y/277-V system provided the ungrounded legs are all connected to the same phase or that the maximum voltage between any two conductors never exceeds 300 V, the maximum rating of lighting busway.

In effect, three-pole lighting busway provides two independent circuits in the same housing, thus providing greater switching capability, extra circuits at low cost, and greater service continuity.

Independently controlled circuits are desirable whenever it is necessary to control separate loads or separate portions of the same load. For example, a department store may require two different lighting circuits, depending on the material displayed, and an industrial plant may require one circuit for work periods and another for maintenance crews. By using single-pole switching in each of the hot legs with a grounded neutral, it is easy to obtain maximum flexibility with three-pole lighting busway.

In most distribution systems, normal day-lighting circuits are paralleled by other branch circuits. These additional circuits supply night lighting, convenience outlets, small offices or washrooms, heaters, fans, portable tools, etc. If two-pole busway is used, other loads must be supplied by other independent circuits, which may be either lighting busway or wire and conduit. In many instances, a considerable saving can be effected by using three-wire lighting busway to supply two, three, or more of these circuits.

If service continuity is essential, as in emergency lighting systems, it is desirable to split the loads in a three-pole, two-circuit single housing. Then, if a temporary overload should occur on one circuit, the single-pole breaker protecting that circuit may trip but the other circuit will be left in operation. Thus, only a portion of the load will be shut down, and all production will not be lost.

**255. Typical installation procedures for lighting busway.** After the suspension method has been put into place and hangers located as required, the installer may select any desired location along the designated run to install the first section of lighting busway. A coupling can first be installed by snapping it onto the end of a section. A tab on the coupling engages a square slot on the top of the section housing which locks it into place. As no tools are required, couplings can be attached after a section has been installed on the hangers.

To install lighting busway, simply slide each section through the hanger(s), snap the sections together at the coupling, and repeat the process until the run has been completed. Lighting fixtures and tap-off devices, as well as end or center feed-ins and end caps, can be installed at the same time.

Lighting busway is designed for simplicity of installation and reusability. It can be removed and reinstalled, affording flexibility and cost savings for future changes.

TROLLEY BUSWAY (Fig. 9-175)   This is a modern enclosed, mobile power-distribution system designed to provide moving-current tap-off for applications such as cranes, hoists, maintenance hanger doors, and rotating platforms. Automobile plants, for example, use trolley busway to provide mobile power for portable assembly tools and parts carrying monorail systems. Direct-current trolley busway is also used for testing the vehicle electrical system as it moves down the assembly line. Appliance manufacturers use trolley busway for assembly and testing of their products. When compared with a bare bar or cable used in the same application, trolley busway offers not only lower installed cost but greatly reduced maintenance in addition to safety, as the conductor bars are enclosed and protected in a single steel housing.

Two- and three-pole trolley busway (Fig. 9-176) is available in continuous ampere ratings of 100, 225, 375, 400, 500, 600, and 800 A. All have 150 percent intermittent

**Fig. 9-175** Typical example of a trolley busway.

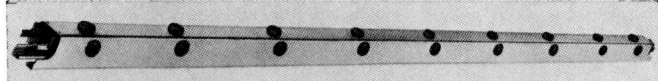

**Fig. 9-176** Typical section of a trolley busway.

ratings based on 1-min-on, 1-min-off duty cycles and can be used on system voltages up to 600 V ac (50, 600, 400, or 800 Hz) or 250 V dc. Four-pole trolley busway is rated 100/150 A at 480 V ac or 250 V dc. All use copper conductor bars. In this type of busway the steel housing serves as the ground path. Trolley busway is usually furnished in standard 10-ft (3-m) straight sections. However, section lengths from 1 ft 2 in to 10 ft (0.35 to 3 m), in any increments, are available as well as horizontal or vertical curved sections (Fig. 9-177). For exceptionally long straight runs or where rigidly supported on the building structure or at both ends of the run or where a run crosses a building expansion joint, a trolley busway expansion section should be included.

Polarized dropout sections, which provide trolley access points, may be built into a standard section and may be installed at any convenient location along the run. Sections can also be sectionalized (for circuit isolation) or be provided busless for trolley return on closed-loop applications. When switching trolleys from one run to another run, insu-

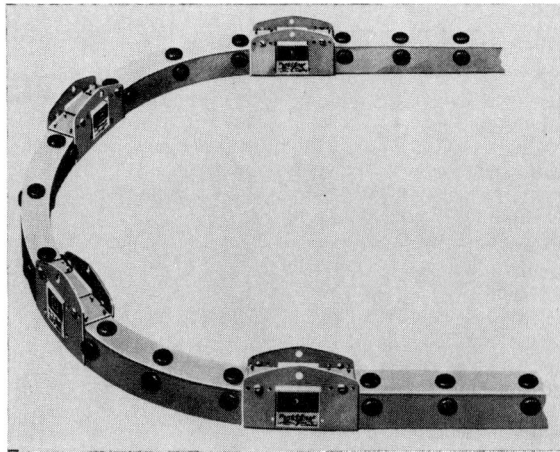

**Fig. 9-177** Horizontal curved section.

lated flared ends are installed on the end of each facing section. These serve to align trolleys when transferring from one run to the other.

Accessories include couplings, hangers, and plates, feed-ins, and a variety of trolleys.

*Couplings* are used to join section to section together mechanically and electrically (Fig. 9-178) and also serve as the hanger. Intermediate hangers are used on long-radius curves or straight sections subject to heavy hanging loads between couplings. [Trolleys can support up to 25-lb (11.3-kg) loads.]

*End plates* close off or terminate a trolley-busway run and prevent trolleys from traveling beyond the end of a run.

*Feed-in adapters* attach to the top of a coupling and provide means to power a run at the ends or at any coupling location in between. This permits feeding a run in several locations to control voltage drop or to maintain or increase system capacity, as, for example, on assembly test lines.

*Trolleys* serve as the power tap-off and transverse inside the trolley-busway housing. Contacts on the trolley move along the copper busbars for power collection. Trolleys are available in several types and should be selected in accordance with the application and tap-off–ampere requirement. Contacts (button, roller, and shoe types) are typically made of an alleged graphite-impregnated bronze and are replaceable, as they are intended to wear instead of the copper busbars. Contact life varies depending on the severity of the duty and working atmospheric conditions. To prevent incorrect phasing, trolleys are polarized. The number and type of contacts determine trolley tap-off capacities. Ampere ratings range from 30 A to 200 A, 600 V or less, ac or dc. Trolleys can be close-coupled and interconnected for higher-rating applications (Fig. 9-179).

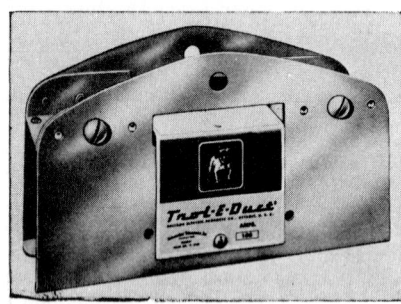

**Fig. 9-178** Coupling.

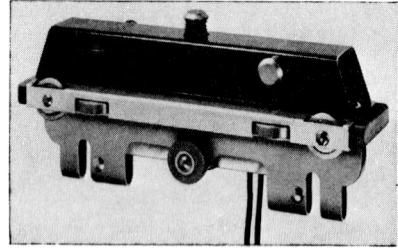

**Fig.9-179** Typical busway trolley.

The steel carriages on trolleys are equipped with sealed ball-bearing wheels which support and guide the trolley inside the trolley-busway housing. Side wheels, which are of copper-impregnated steel, provide positive ground contact between trolley and busway housing. Undercarriage wheels prevent the trolley from riding up and down when being pushed or towed in either direction.

Accessory boxes, which attach directly to trolley carriages, can be equipped with fuses, circuit breakers, receptacles, contactors, starters, timers, pilot lights, and push buttons (Fig. 9-180). A tool hanger with wiring box is also available (Fig. 9-181).

Nylon-brush cleaning trolleys are used to clean the copper conductor bars when a trolley busway is subject to dusty conditions. An abrasive cleaning trolley can be used intermittently to resurface the copper conductor bars if they become rough or coated from severe operating conditions.

Trolley busway, being manufactured in bolt-together sections, can easily be disassembled and moved to a new location within a facility or a new operation.

### 256. Glossary

ACCESSORY  Non-current-carrying component of a busway system used to mount or adapt the busway to the building structure (NEMA standard 11-3-1977).

ADAPTER  Fitting which permits the joining together of lengths and fittings of different shapes or designs (NEMA standard 11-3-1977).

**Fig. 9-180**  Busway trolley with accessory box.

**Fig. 9-181**  Busway trolley with wiring box.

AMBIENT TEMPERATURE   Temperature of the surrounding air that comes in contact with the outside of the busway enclosure, device, or fitting (NEMA standard 9-15-1983).

ASYMMETRICAL CURRENT   Alternating current having a waveform which is offset with respect to the zero axis. The offset occurs at the initiation of a short circuit or other change in current. It usually decays quickly until steady-state conditions are reached and the current becomes symmetrical. Asymmetrical current is composed of the alternating symmetrical component. It is expressed in rms asymmetrical amperes at a specific time (normally one-half cycle) after initiation of a short circuit or other change in current. Peak current refers to the maximum instantaneous amperes within this first half cycle (NEMA standard 11-3-1977).

AVAILABLE SHORT-CIRCUIT CURRENT   Maximum current which a circuit is capable of delivering at the system terminals ahead of the apparatus being supplied. Available short-circuit current may be expressed in either rms symmetrical amperes or rms asymmetrical amperes (NEMA standard 9-15-1983).

BOLT-ON DEVICE   Power takeoff means which can be bolted to the busways at a joint between lengths and fittings, at the end of a run, or at any predetermined location, and which makes electrical connection to the busbars (NEMA standard 11-3-1977).

BUSBARS   Conductors which carry current through busway lengths and fittings and which have one of the following arrangements: (1) single-bar arrangement, which refers to a busway having one conductor per pole; and (2) multibar arrangement, which refers to a busway having two or more conductors for one or more of its poles (NEMA standard 11-3-1977).

BUSWAY   Prefabricated electric distribution system consisting of busbars in a protective enclosure, including straight lengths, fittings, devices, and accessories. Busways are of the following types: (1) feeder busway, a busway that has no plug-in openings and is intended primarily for conducting electric power from the source of supply to centers of distribution but can have provisions for bolt-on devices; and (2) plug-in busway, a busway having plug-in openings on one or both sides at spaced intervals, offering means for electrical connection of plug-in or bolt-in devices to the busbars (NEMA standard 11-3-1977).

CENTER CABLE TAP BOX   Fitting or device which provides for the attachment of cables at a location other than the end of a busway run (NEMA standard 11-3-1977).

CIRCUIT BREAKER   Device designed to open and close a circuit by nonautomatic means and to open the circuit automatically on a predetermined overcurrent without injury to itself when properly applied within its rating (NEMA standard 11-3-1977).

CONTINUOUS-CURRENT RATING    Designated maximum direct current or alternating current in rms amperes at rated frequency which a busway can carry continuously without exceeding its temperature-rise limits when subjected to specified heating tests (NEMA standard 11-3-1977).

CROSS    Fitting suitable for connection in four directions (NEMA standard 11-3-1977).

CUBICLE    Enclosure attached to a length or fitting for the purpose of enclosing electrical components (NEMA standard 11-3-1977).

DEVICE    Enclosed component used on rather than in a run of a busway system. The device may carry current from the busway system to supply a load circuit or be a non-load-supplying unit, such a ground detector (NEMA standard 11-3-1977).

ELBOW    Angular fitting (NEMA standard 11-3-1977).

END CABLE-TAP BOX    Fitting which provides for the attachment of cables and conduits at the end of the busway run (NEMA standard 11-3-1977).

END CLOSURE    Fitting which terminates and closes the end of a busway run (NEMA standard 11-3-1977).

EQUIPMENT-GROUNDING CONDUCTOR    Conductor which is used to connect non-current-carrying metal parts of the busway (NEMA standard 11-3-1977).

EXPANSION FITTING    Fitting which accommodates expansion and contraction of the busway or building (NEMA standard 11-3-1977).

FITTING (BUSWAY)    Component in a run of a busway system, other than a straight length. It may be either current-carrying, such as an elbow, tee, or cross, or non-current-carrying, such as an end closure (NEMA standard 9-15-1983).

FLANGED END (OR SWITCHBOARD STUB)    Fitting which provides means for mechanically and electrically connecting a busway run to other apparatus (NEMA standard 11-3-1977).

FLOOR FLANGE    Accessory on the outside of a busway enclosure that provides means for the installer to cover the floor opening penetrated by the busway (NEMA standard 11-3-1977).

INDOOR    Suitable for installation within a building which protects the busway from exposure to the weather (NEMA standard 11-3-1977).

NEUTRAL CONDUCTOR    Conductor which is connected to the midpoint of a three-wire single-phase system, center point of a Y-connected three-phase system, or midpoint of one side of a delta-connected three-phase system (NEMA standard 11-3-1977).

OFFSET    Fitting providing two or more angles (NEMA standard 11-3-1977).

OUTDOOR    So constructed that exposure to the weather will not interfere with successful operation. NOTE    Outdoor busway is not suitable for outdoor use until completely and properly installed as recommended by the manufacturer (NEMA standard 9-15-1983).

PLUG-IN CIRCUIT BREAKER    Plug-in device containing an externally operable circuit breaker (NEMA standard 11-3-1977).

PLUG-IN DEVICE (OR BUS PLUG)    Power takeoff means which can be plugged into a plug-in busway and which makes electrical connection to the busbars (NEMA standard 11-3-1977).

PLUG-IN FUSIBLE SWITCH    Plug-in device containing an externally operable fusible switch (NEMA standard 11-3-1977. A *handle-operated switch* is a switch which is externally opeated without opening or closing the cover; a *cover-operated switch* is a switch which is operated by opening and closing the cover (NEMA standard 11-3-1977).

PLUG-IN GROUND DETECTOR    Plug-in device which indicates a ground on any of the normally ungrounded busway conductors (NEMA standard 11-3-1977).

PLUG-IN POTENTIALIZER (OR NEUTRALIZER)    Plug-in device having means for limiting the potential difference between a busway enclosure and the busbars (NEMA standard 11-3-1977).

RATING    Designated limit(s) of the rated operating characteristic(s) of a busway length, fitting, or device. NOTE    Such operating characteristics as current, voltage, frequency, etc., may be given in the rating (NEMA standard 11-3-1977).

REDUCER    Fitting designed for connection between lengths and fittings of different ampere ratings (NEMA standard 11-3-1977).

ROOF FLANGE    Accessory on the outside of a busway enclosure that provides means for the installer to weatherproof the roof opening penetrated by the busway (NEMA standard 11-3-1977).

STRAIGHT LENGTH   Straight section of a busway system (NEMA standard 11-3-1977).

SWITCH   Device for opening and closing or for changing the connection of a circuit (NEMA standard 11-3-1977).

SYMMETRICAL CURRENT   Alternating current having no offset or transient component and therefore having a waveform essentially symmetrical about the zero axis. Symmetrical current is expressed in rms amperes (NEMA standard 9-17-1972).

TEE   Fitting suitable for connection in three directions (NEMA standard 11-3-1977).

TOTALLY ENCLOSED   So constructed as to prevent the free exchange of air between the inside and outside of a housing but not sufficiently enclosed to be termed *airtight* (NEMA standard 11-3-1977).

TRANSFORMER FITTING   Fitting wherein busbar positions are interchanged to equalize the impedance of the different phases or to change the phase relationship (NEMA standard 11-3-1977).

TRANSFORMER TAP   Fitting having busbars extended through its housing or end barrier for connections by open cable to terminals of the transformer(s) (NEMA standard 11-3-1977).

TRANSFORMER THROAT   Fitting which provides enclosed busbar connections to transformer terminals (NEMA standard 11-3-1977).

TRANSVERSE BARRIER   Dividing barrier or other means of restricting the free flow of either air or moisture through the inside of a busway (NEMA standard 11-3-1977).

VAULT TERMINATION   Fitting having open busbars extending from one end which provides for connections to vault equipment (NEMA standard 11-3-1977).

VENTILATED   So constructed as to provide for the circulation of external air through an enclosure to remove heat (NEMA standard 11-3-1977).

VOLTAGE DROP   Arithmetical difference between voltages at the load and supply ends (NEMA standard 7-54-1954).

WALL FLANGE   Accessory on the outside of a busway enclosure that provides means for the installer to cover the wall opening penetrated by the busway (NEMA standard 11-3-1977).

**257. Useful references**   The following publications may be useful in the application, maintenance, and testing of busway:

AMERICAN NATIONAL STANDARDS INSTITUTE
ANSI C37.16-1980, *Preferred Ratings, Related Requirements, and Application Recommendations for Low-Voltage Power Circuit Breakers and AC Power Circuit Protectors*
ANSI/UL 857-1982, *Standard for Busways and Associated Fittings*
ANSI/ASTM D1535-80, *Method of Specifying Color by the Munsell System*
INSTITUTE OF ELECTRICAL AND ELECTRONIC ENGINEERS
IEEE Standard 141-1976, *Recommended Practices for Electric Power Distribution for Industrial Plans*

NATIONAL ELECTRICAL MANUFACTURERS ASSOCIATION
NEMA AB1-1975(R1981), *Standards Publication for Molded Case Circuit Breakers*
NEMA BU1-1983, *Standards Publication for Busways*
NEMA BU1-1-1981, *Instructions for Safe Handling, Installation, Operation and Maintenance of Busway and Associated Fittings Rated 600 Volts or Less*
NEMA ICS1-1978, *Standards Publication for General Standards for Industrial Control Systems*
NEMA KS1-1983, *Standards Publication for Enclosed Switches*
NEMA PB1-1984, *Standards Publications for Panelboards*
NEMA SG3-1975, *Standards Publication for Low-Voltage Power Circuit Breakers*

NATIONAL FIRE PROTECTION ASSOCIATION
NFPA 70-1984, The National Electrical Code
Copies of the above standards may be purchased from the following:

American National Standards Institute
1430 Broadway
New York, New York 10018

Institute of Electrical and Electronic Engineers Service Center
Publication Sales Department
445 Hoes Lane
Piscataway, New Jersey 08854

National Electrical Manufacturers
  Association
Sales Office
2101 L Street, NW
Washington, D.C. 20037

National Fire Protection Association
Publications Sales Department
Batterymarch Park
Quincy, Massachusetts 02269

Underwriters Laboratories Inc.
Publications Stock
333 Pfingsten Road
Northbrook, Illinois 60062

## CELLULAR-METAL-FLOOR–RACEWAY WIRING

**258. Cellular-metal-floor raceways** are formed by employing for the floor construction of the building structural members made of corrugated sheet steel. Seven types of cellular steel floor, as made by the H. H. Robertson Co. of Pittsburgh, are shown in Fig. 9-182. Cross-sectional views of completed floors and their adaptation for the installation of the electric wiring system are given in Figs. 9-183 and 9-184.

**259. The cells** which are not filled with concrete form the raceways for wiring and piping. In all types the cells are 6 in (152.4 mm) apart on centers, and the members are manufactured four cells wide with varying lengths to suit the span between beams. The cells are cross-connected by headers formed over the beams in the space left between adjacent lengths of flooring. When headers connected to alternate cells are used, one raceway system can be formed for signal wires and another for power wires, with both systems kept entirely separate from each other. If any cells are used for piping, those cells must not be interconnected to the electrical raceway system.

**260. In forming these floors, close cooperation is required between the structural and the electrical workers.** The structural workers lay the floor members across the spans from beam to beam, fastening them to the beams and leaving about 4½ in (114.3 mm) of space on top of the beams between the abutting ends of the floor members for formation of the header.

The raceway headers are formed by the electrical workers in the space between members on top of the floor beams. A line of cells which is to form a raceway is interconnected at the ends of the sections by locating an access unit (Fig. 9-185) between the open ends of the aligning cells of adjacent sections.

The other cells are closed off with connecting units (Fig. 9-186) so as to form a continuous header across the beam to the raceway line of cells. When one raceway system is to be built for power wiring and another for telephone wiring, the header for one system is built on one junction of floor beams and the header for the other system on another junction (Fig. 9-187). The wires for one system will pass through the lower portion of its cells under the header for the other system. The joints at the edges of the header units are sealed with a cold-flowing asphaltic compound.

**261. The connection from a header to a panelboard** is made from special fittings as shown in Fig. 9-188. Connection to the header is made with an adapter (Fig. 9-189, II) to which is connected the vertical elbow (Fig. 9-189, I). Straight sections of duct 1 ft (0.3 m) long (Fig. 9-189, IV) coupled together with couplings (Fig. 9-189, V) form the vertical run up the wall. The final section to the panelboard is cut to length with a hacksaw. The box connector (Fig. 9-189, III) holds the duct to the panelboard. It requires an opening 1⅞ by 5¾ in (47.625 by 146.05 mm) in the bottom of the panelboard box.

**262. Method of installing floor outlets.** A hole of 1⅝-in (41.275-mm) diameter is cut in the center of a cell with a saw-type drill (Fig. 9-190, I). A forming tool (Fig. 9-190, II) is turned into the hole to prepare the hole for reception of the outlet tap. For floors with a concrete fill of 2 to 3¼ in (50.8 to 82.55 mm) a standard adjustable tap (Fig. 9-191, I) is then screwed into the hole. For fills of 1 to 2 in (25.4 to 50.8 mm) a special tap is available. For fills of less than 1 in a shallow tap (Fig. 9-191, IV) is set in the hole and held in place with three No. 12 self-tapping screws. Holes for the screws must be drilled in the cell with a No. 9 drill. The top of the standard tap can be adjusted in height to the floor level, and a blanking cover (Fig. 9-191, II) installed until after the floor is poured. When a floor outlet is installed, the blanking cover is removed, and an extension (Fig. 9-191, III) is screwed into the tap. The floor outlet is then screwed onto the extension, using a Warnock wrench (Fig. 9-190, III) or other type of fabric wrench to avoid marring the fitting. With the shallow tap (Fig. 9-191, IV) the fitting screws onto the tap directly without the use of the extension. The complete outlet assembly is shown in Fig. 9-192. A duplex convenience outlet is shown in Fig. 9-193.

**263. Cell markers** should be located on the center of each cell near the opposite side of the room from the header to facilitate future location of cells when outlets are not installed initially. The access units in the header will identify the cell locations at the header. A straight line between the center of an access unit and a cell marker will then identify the cell so that an outlet can be located accurately at any time in the future. Several types of cell markers are shown in Fig. 9-194. The double-slot screw (Fig. 9-194, I) can be used to identify power-wire cells, and the single-slot one (Fig. 9-194, II) to identify telephone-

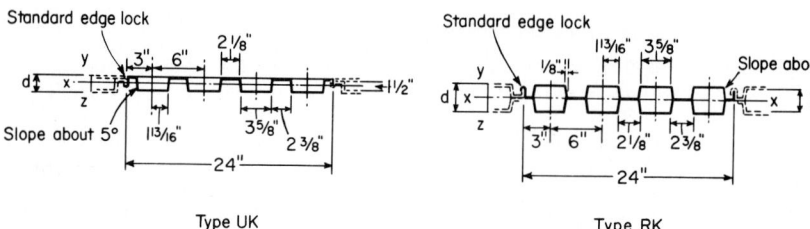

Type UK

\*Not recommended for use with concrete fill

Type RK

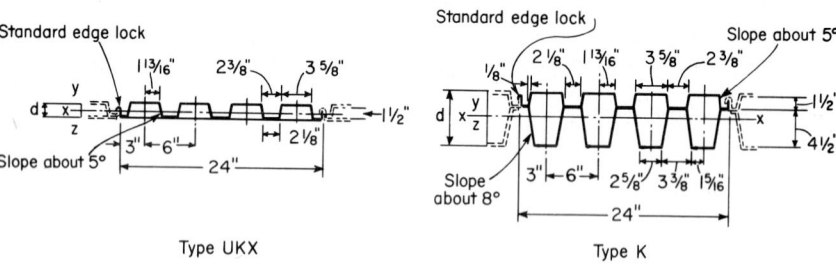

Type UKX

Type K

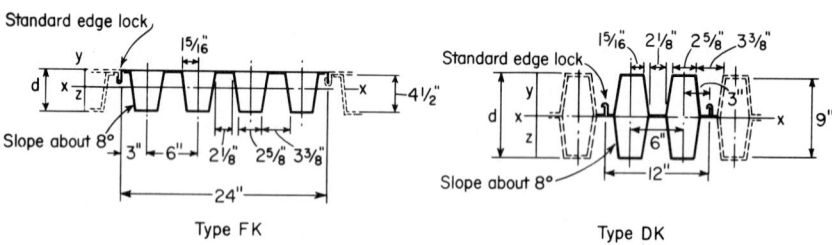

Type FK

\*Not recommended for use with concrete fill

Type DK

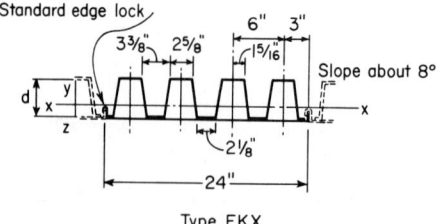

Type FKX

**Fig. 9-182**   Types of cellular-steel floors. [H. H. Robertson Co.]

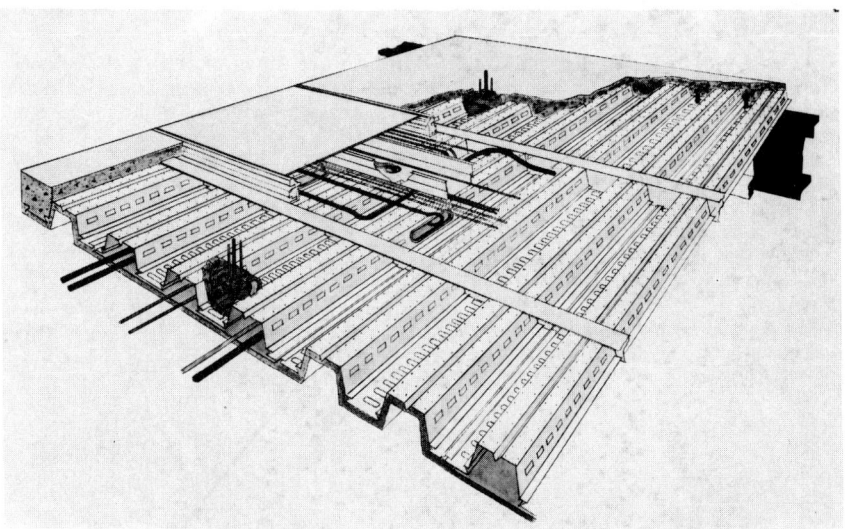

**Fig. 9-183**   Model illustrating cellular-metal-floor–raceway wiring. [H. H. Robertson Co.]

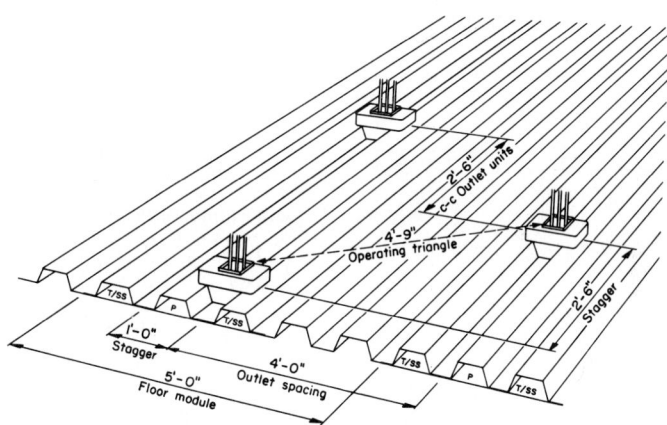

**Fig. 9-184**   Model illustrating typical 5-ft (1.5-m) staggered module of cellular-metal-floor raceway. [H. H. Robertson Co.]

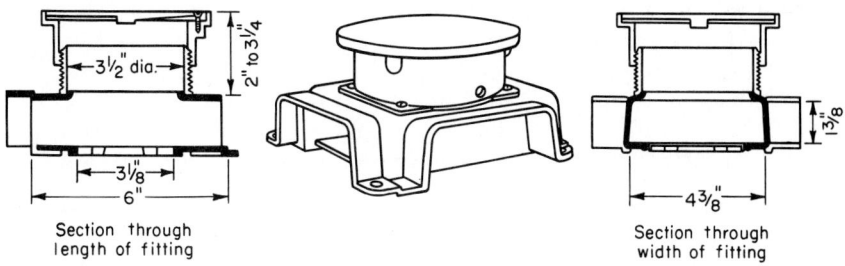

**Fig. 9-185**   Access unit for cellular-metal floors which have concrete fill. [H. H. Robertson Co.]

wire cells. In installing these markers (Fig. 9-195) a hole is drilled and tapped in the top of the cell for a 12-24 screw before the floor fill is laid. The brass marker with a headless screw is then fastened in place and adjusted in height to the level of the finished floor. The marker should then be grouted in place. After the floor has been poured and the finished floor covering laid, the headless screw is replaced with one of the grommeted brass screws (Fig. 9-194, I or II).

**264. Underside (ceiling) headers** (Fig. 9-196) are generally used with Types FK inverted, UK inverted, RK, and UK floors of Fig. 9-182 because there is not sufficient room for the other headers in the shallow floor fill. Two sizes are available:

    1.  2½ in (63.5 mm) deep and 5 in (127 mm) wide

    2.  4 in (101.6 mm) deep and 6 in (152.4 mm) wide

A ceiling header mounted in the corner

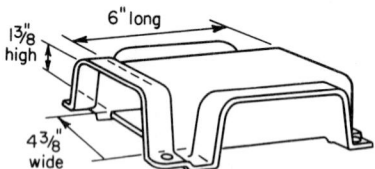

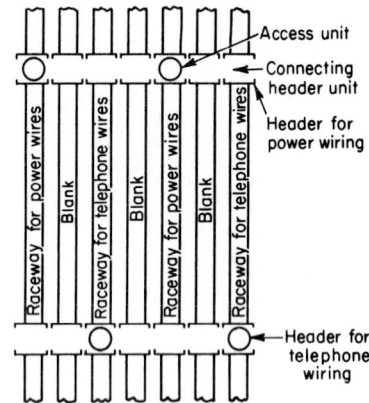

**Fig. 9-186**  Connecting header unit.

**Fig. 9-187**  Method of forming separate headers for power and for telephone wiring.

formed by the ceiling and wall so as to be as inconspicuous as possible is shown in Fig. 9-197. It is held in place with No. 12 self-tapping screws which pass through holes drilled in the bottom of the raceways. Connection to a panelboard is made with a special ell (Fig. 9-198, II), as shown in Fig. 9-199. Connection between the cells and the header is made by drilling a 2-in (50.8-mm) hole in the cell to line up with the pilot opening in the header. A grommet (Fig. 9-198, I) with the locknut removed is then inserted in the hole diagonally, as shown in Fig. 9-200.

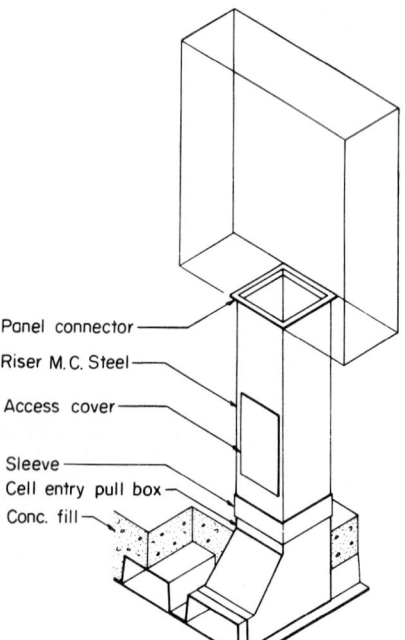

**Fig. 9-188**  Connection of cellular-metal-floor raceways to panelboards. [H. H. Robertson Co.]

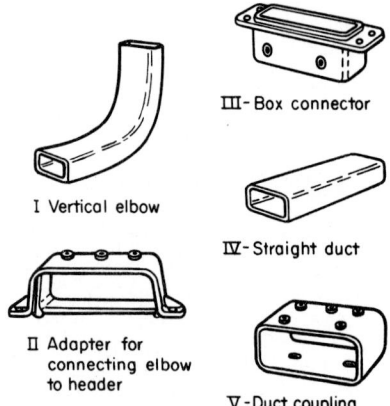

**Fig. 9-189**  Fittings for connecting cellular-metal-floor header to wall panelboard. [H. H. Robertson Co.]

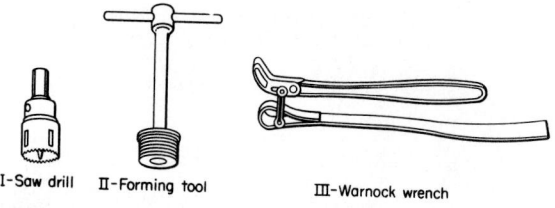

I-Saw drill    II-Forming tool    III-Warnock wrench

**Fig. 9-190**   Tools for installing outlets in cellular-metal-floor raceways. [H. H. Robertson Co.]

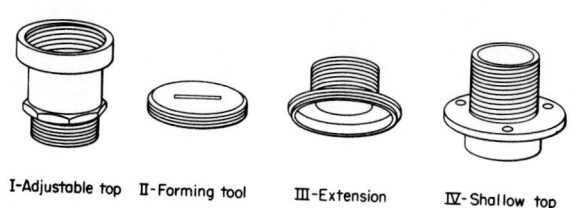

I-Adjustable top    II-Forming tool    III-Extension    IV-Shallow top

**Fig. 9-191**   Fittings for cellular-metal-floor raceways. [H. H. Robertson Co.]

Extension
Base of outlet
2"-3¼" Floor fill as shown
1"-2" Floor fill when special tap is used
Putty
Saw 1⅝" dia. hole
Floor tap

I- For floors with 1" to 3¼" fill

Secure with four self-tapping screws
Base of outlet
Shallow floor tap
Saw 1⅝" dia. hole
Putty
Floor fill less than 1"

II- For floors with shallow fill

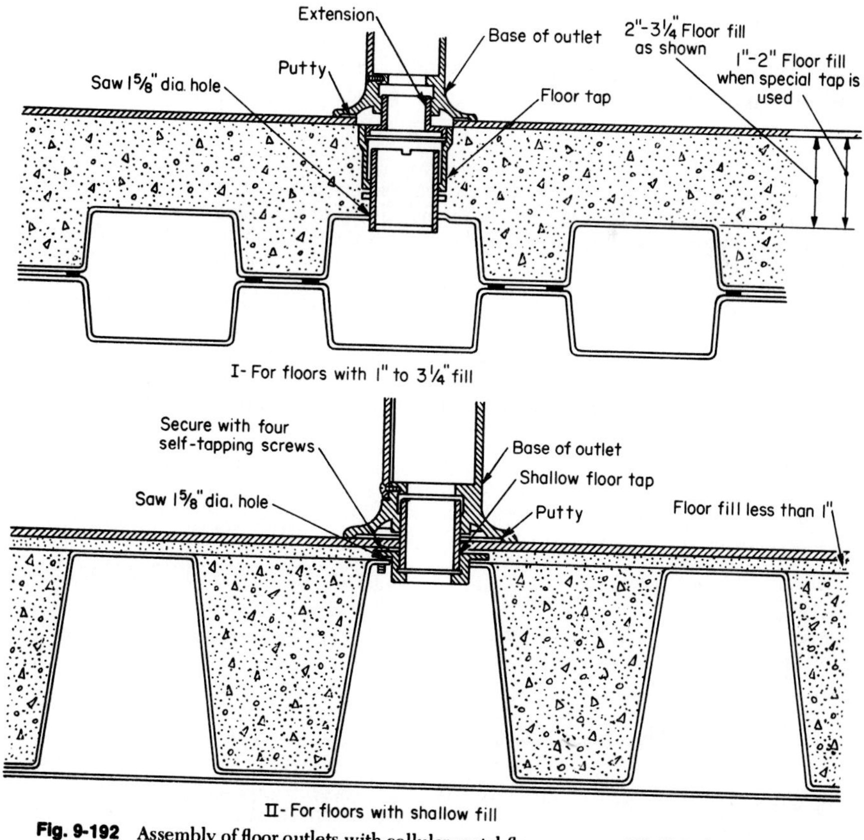

**Fig. 9-192**   Assembly of floor outlets with cellular-metal-floor raceway. [H. H. Robertson Co.]

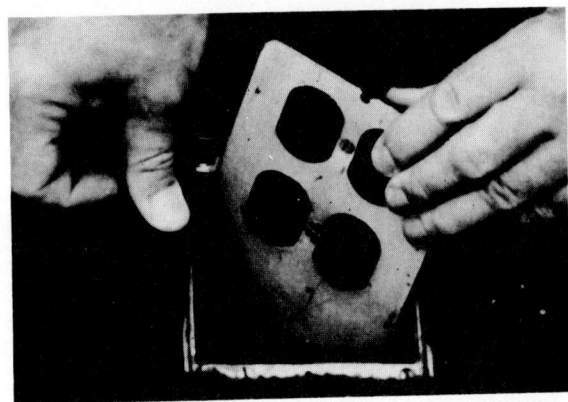

**Fig. 9-193**  Outlets for use with cellular-metal-floor raceways. [H. H. Robertson Co.]

I-Double slot
marker screw

II- Single slot
marker screw

III  Round
marker

IV- Hexagonal
marker

Headless
screw in
marker

V-Square
marker

**Fig. 9-194**  Markers for cellular-metal-floor raceways. [H. H. Robertson Co.]

After the grommet has been adjusted to its proper position, the locknut is screwed on and holds the grommet firmly in place. At least two markers, as explained in Sec. 263, should be located on top of the floor for each cell which does not have outlets installed initially.

**265. Ceiling lighting fixtures** can be hung and wired from cellular-metal-floor raceways as shown in Fig. 9-201. At I is shown a fixture-stud mounting using the stud illustrated in Fig. 9-202, I. The fixture stud is inserted through a 2-in (50.8-mm) hole cut in the cell similar to the grommet assembly of Fig. 9-200. A mounting for fixtures which fasten on the cover of a 4-in (101.6-mm) box is shown in Fig. 9-201, II. The box (Fig. 9-202, II) is held in place with a grommet. The knockouts in the box can be used for surface conduit wiring if desired. A conduit-hung fixture is supported from the cell by means of a hanger as shown in Fig. 9-201, III. The hanger (Fig. 9-202, III) is held in place with a locknut in the same way as the grommet.

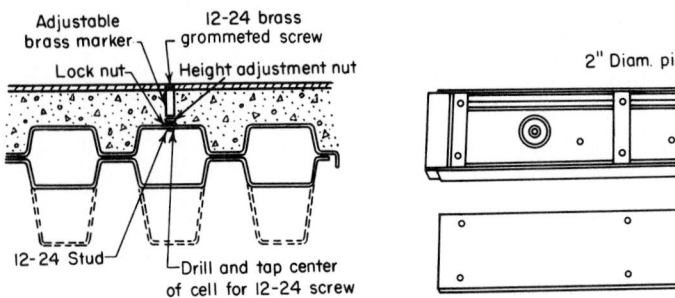

**Fig. 9-195**  Detail showing cell marker. [H. H. Robertson Co.]

**Fig. 9-196**  Underside header with cover removed. [H. H. Robertson Co.]

**266. Trench header duct.** A trench header duct can be used in lieu of the standard headers shown in Figs. 9-183 and 9-184. A typical trench-header-duct installation is shown in Fig. 9-203. It is run across the floor cells at right angles, and access to each cell is made through a grommet connection. The trench header duct shown in Fig. 9-203 contains barriers so that feeders, branch circuits, and telephone circuits are separated. The cover is flush with the floor, and flush screws secure the cover to the header duct. Each cover section is 3 to 6 ft (0.9 to 1.8 m) long.

Trench header duct is made in widths of 9 to 36 in (228.6 to 914.4 mm) and arranged to

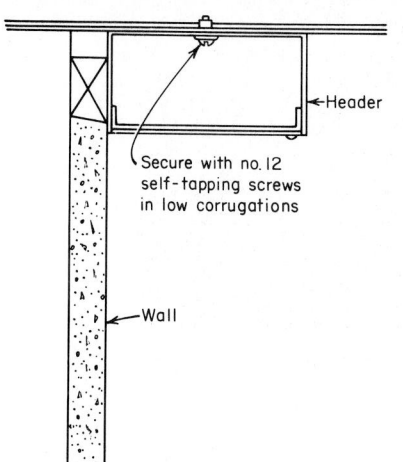

**Fig. 9-197** Location and mounting of underside header. [H. H. Robertson Co.]

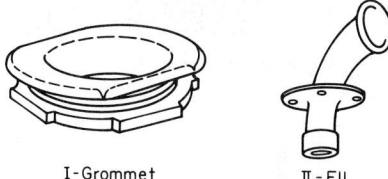

I-Grommet        II-Ell

**Fig. 9-198** Fittings for underside header. [H. H. Robertson Co.]

accept standard floor tiles or similar floor coverings. This header system makes a neat installation, and access openings to the header duct are practically unnoticeable after the covers have been installed. One or more cover sections can be easily removed when access to the trench duct or floor cells is necessary.

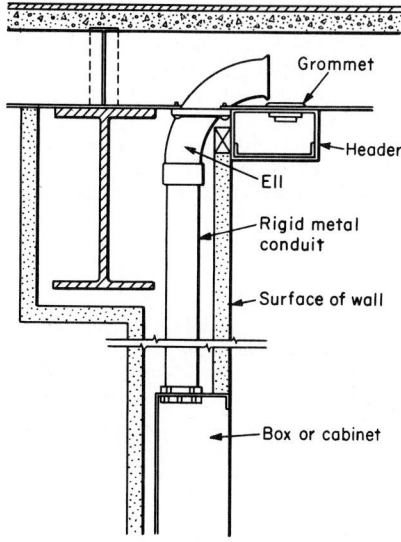

**Fig. 9-199** Method of connecting between underside header and panel. [H. H. Robertson Co.]

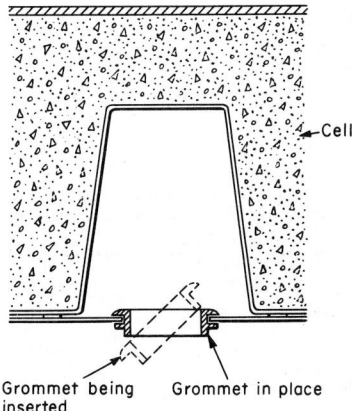

**Fig. 9-200** Method of inserting grommet into hole in cell. [H. H. Robertson Co.]

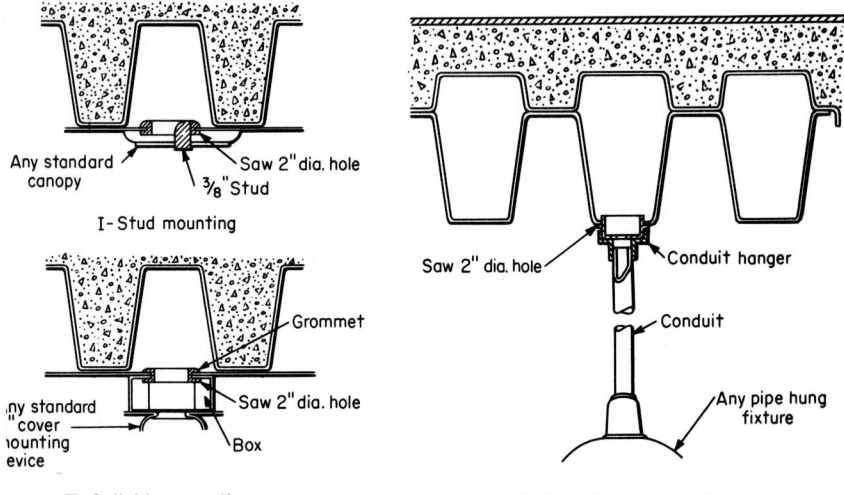

**Fig. 9-201**    Methods of mounting ceiling fixtures from cellular-metal-floor raceways. [H. H. Robertson Co.]

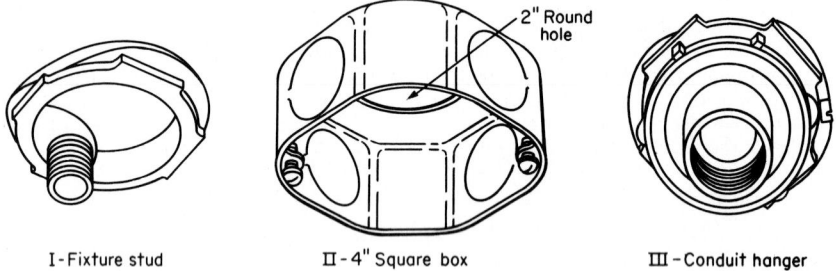

**Fig. 9-202**    Fittings for ceiling fixtures. [H. H. Robertson Co.]

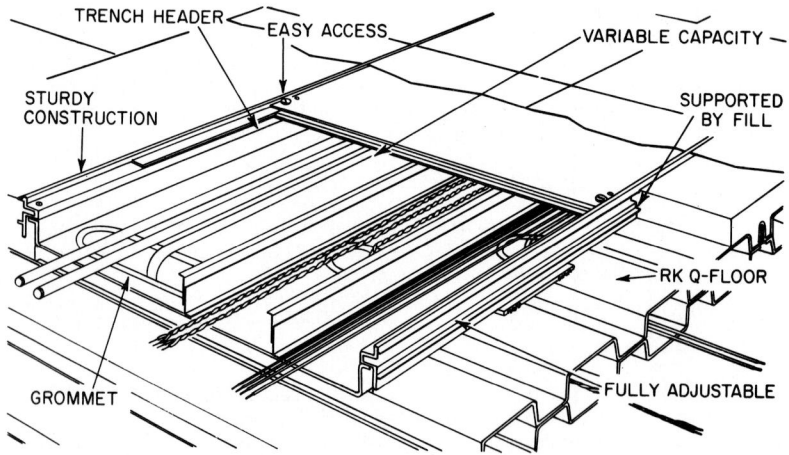

**Fig. 9-203**    Trench header duct used for cellular-metal-floor raceways. [H. H. Robertson Co.]

**267. The cellular raceways are wired** by fishing the wires from the access units to the outlets. No conductor larger than No. 1/0 may be installed except by special permission. The number of conductors in a cell is not limited as long as the total cross-sectional area of the conductors and cables does not exceed 40 percent of the internal area of the header or cell. All splices and taps must be made in the trench header duct or at junction boxes. So-called loop wiring (continuous, unbroken conductors) is not considered a splice or tap. The internal areas of the raceways are as follows:

|  | Entire area, sq in. | 40 per cent |
|---|---|---|
| K cells......................... | 20.25 | 8.10 |
| RK cells....................... | 11.25 | 4.50 |
| FK, FK inverted cells........... | 14.62 | 5.85 |
| UK, UK inverted cells........... | 5.62 | 2.24 |
| DK cells....................... | 29.24 | 11.70 |

All junction boxes and inserts must be leveled to the floor grade and sealed against the entrance of water. They must be made of metal and so installed that they will be electrically continuous with the raceway. In cutting a cell and in setting inserts great care must be exercised so that chips or other dirt will not fall into the raceway. Only tools which are specially designed so that they will not enter the cell and injure the conductors should be used.

The NEC requires that when an outlet is discontinued, the conductors supplying it must be removed from the raceway and that connections to cabinets and extensions from cells to outlets unless made by approved fittings must be made by means of rigid, intermediate, or flexible metal conduit or electrical metallic tubing. Where there are provisions for the termination of an equipment grounding conductor, nonmetallic conduit, electrical nonmetallic tubing, or liquidtight flexible nonmetallic conduit where not installed in concrete may be used.

**268. Cellular-metal-floor–raceway wiring is not allowed** in any hazardous location except Class I, Div. 2 locations as permitted by the exceptions to Secs. 501-4(b) and 504-20 of the NEC or where subject to corrosive vapors. It is not allowed in commercial garages except for supplying ceiling outlets or extensions to the area below the floor but not above. Any cell or header that is used for electrical conductors may not be used for any other service, such as steam, water, air, gas, and drainage.

**269. Cellular-metal-floor raceways are used** principally for office buildings in which the needs of the tenants for wiring outlets are not known definitely in advance and in which a high quality of adequacy and convenience of outlets and interior appearance without exposed wiring is desired. The position of the raceways may be varied in 6-in (152.4-mm) increments, and a very flexible system of outlets for telephones, business machines, and desk lamps provided, with all outlets located under the desks out of sight. The cost may be lower than would be incurred if a standard floor with a large number of underfloor raceways were installed. The wiring for ceiling-lighting outlets is also simplified by use of a type of floor in which the underside of the raceways serves as the ceiling.

## CELLULAR-CONCRETE-FLOOR–RACEWAY WIRING

**270. Cellular-concrete-floor raceways** provide a complete underfloor electrical distribution system. For this system the floor is constructed of precast reinforced-concrete members (Fig. 9-204). These precast members are provided with hollow voids which form smooth, round cells. The cells form raceways for the electrical wires. Connections to the cells from a distribution center are made by means of metal header ducts which are run horizontally across the precast slabs and embedded in the concrete fill over the slabs. Connection from these headers to the cells is made through handhole metal junction boxes. An outlet can be located at any point along a cell. An opening into the cell at the desired point is formed by drilling a hole through the concrete floor slab. The hole is then fitted with the proper outlet fitting, and the wires fished from a handhole junction box to the outlet.

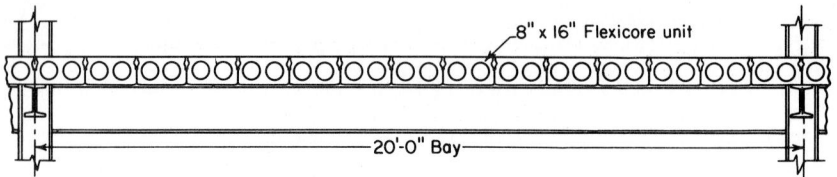

**Fig. 9-204**    Typical cross section of a cellular-concrete floor. [The Flexicore Co., Inc.]

**271. The** National Electrical Code **makes the following definitions:** For the purpose of this article *precast cellular concrete floor raceways* shall be defined as the hollow spaces in floors constructed of precast cellular-concrete slabs, together with suitable metal fittings designed to provide access to the floor cells in an approved manner. A *cell* shall be defined as a single, enclosed tubular space in a floor made of precast cellular-concrete slabs, the direction of the cell being parallel to the direction of the floor member. *Header ducts* shall be defined as transverse metal raceways for electrical conductors, furnishing access to predetermined cells of a precast cellular-concrete floor, thus providing for the installation of electrical conductors from a distribution center to the floor cells.

**271A. Use of cellular-concrete-floor–raceway wiring** (National Electrical Code).    Conductors shall not be installed in precast-cellular-concrete-floor raceways (1) where subject to corrosive vapor; (2) in hazardous locations except Class I, Div. 2, locations as permitted by

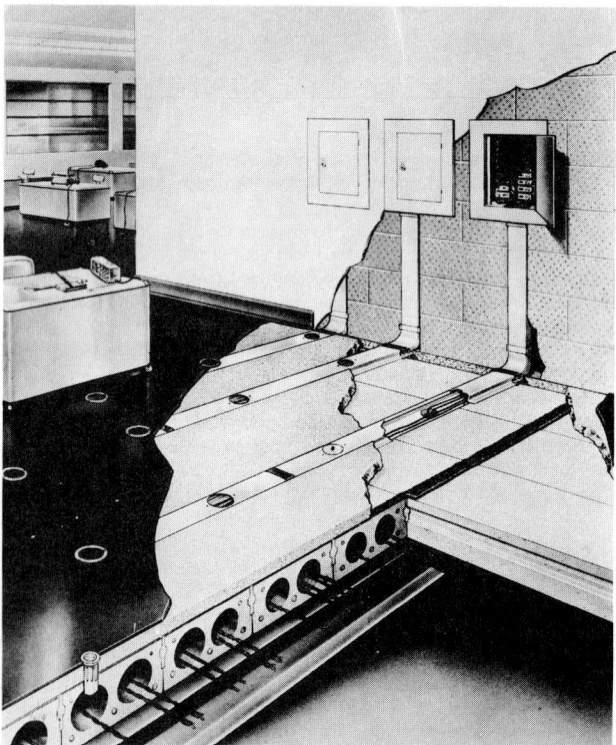

**Fig. 9-205**    Electrical distribution with cellular-concrete-floor–raceway wiring. [The Flexicore Co., Inc.]

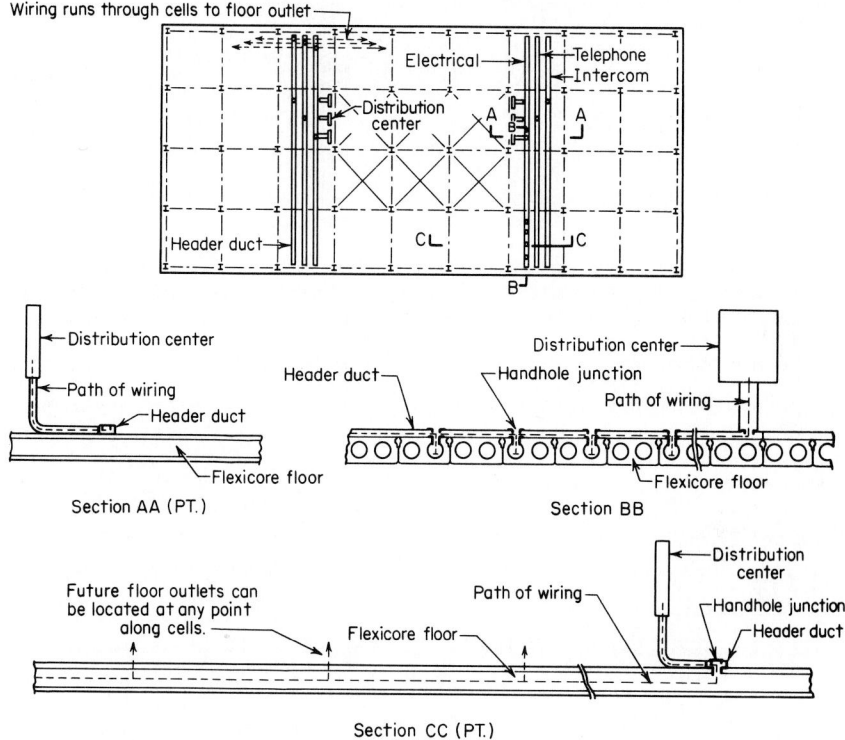

**Fig. 9-206**  General principles of electrical distribution by means of cellular-concrete-floor raceways. [The Flexicore Co., Inc.]

the exceptions to Secs. 501-4(b) and 504-20 of the NEC; or (3) in commercial garages except for supplying ceiling outlets or extensions to the area below the floor but not above. No electrical conductors shall be installed in any cell or header which contains a pipe for steam, water, air, gas, drainage, or any service other than electrical.

**272. Electrical distribution by means of cellular-concrete-floor raceways** is illustrated in Figs. 9-205, 9-206, and 9-207.

The precast slabs are available in two sizes as shown in Fig. 9-208. Flexicore Hi-Stress Deck is available in five sizes, as shown in Fig. 9-209.

Header ducts are available in different sizes.

The **NEC** makes the following specifications for header ducts: The header duct shall be installed in a straight line, at right angles to the cells. It shall be mechanically secured to the top of the precast cellular-concrete floor, and the top surface leveled to the floor finish. The end joints shall be closed by metallic closure fittings and sealed against the penetration of water. The header duct shall be electrically continuous throughout its length and be electrically bonded to the enclosure of the distribution center.

The method of making connection to cabinets or other enclosures is shown in Figs. 9-205, 9-206, and 9-210.

When the panel box or cabinet is located at right angles to the header (Types A and B, Fig. 9-210), the connection is made by an elbow and riser duct to the panel box. This is simply a continuation of the header up to the panel box.

When the panel box is parallel to the header (Type C), the connection is made with a T junction box, a short piece of header duct, an elbow, and a riser.

When parallel to the header, telephone and intercom cabinets are usually connected to

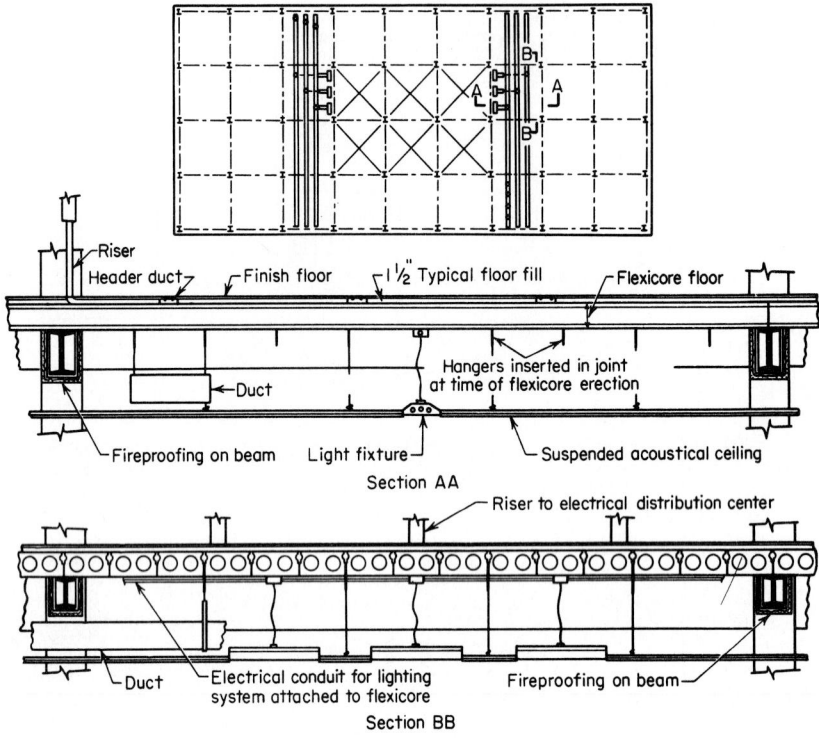

**Section AA**

**Section BB**

**Fig. 9-207**   Details of installation of a cellular-concrete-floor–raceway wiring system. [The Flexicore Co., Inc.]

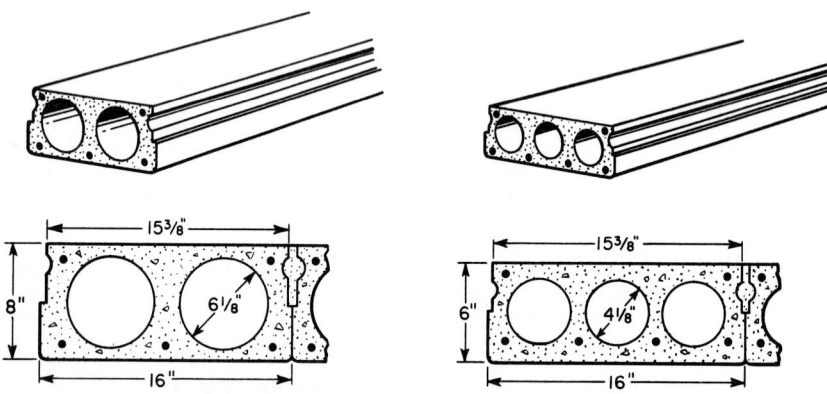

**Fig. 9-208**   Cellular-concrete floor slabs. [The Flexicore Co., Inc.]

the header as shown by Type D. The wire drops from the header through a handhole junction into a cell, through the cell, up through a jack junction box, and then to the cabinet.

The NEC requires that these connections be made by means of metal raceways and approved fittings.

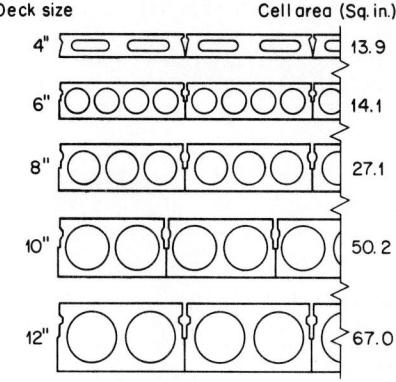

| Deck size | | Cell area (Sq. in.) |
|---|---|---|
| 4" | | 13.9 |
| 6" | | 14.1 |
| 8" | | 27.1 |
| 10" | | 50.2 |
| 12" | | 67.0 |

**Fig. 9-209**    Flexicore Hi-Stress Deck. [The Flexicore Co., Inc.]

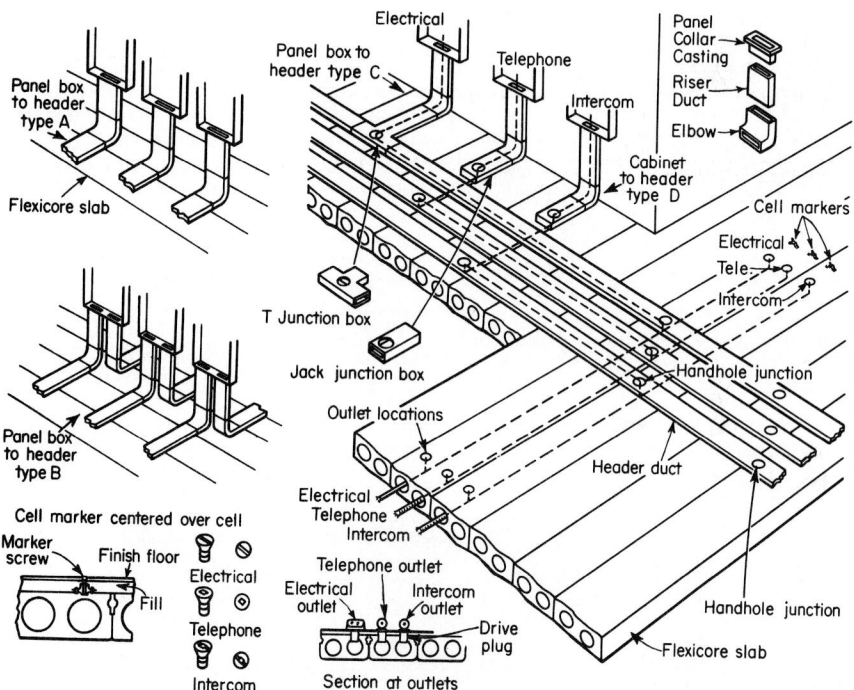

**Fig. 9-210**    Details of installation of a cellular-concrete-floor raceway. [The Flexicore Co., Inc.]

Handhole junction boxes and their relation to the header duct and concrete cells are shown in Fig. 9-210. The NEC requires that junction boxes be leveled to the floor grade and sealed against the free entrance of water. Junction boxes shall be of metal and be mechanically and electrically continuous with the header ducts.

Markers (Figs. 9-210 and 9-211) are installed along each cell for the determination of its location and to mark the location of all hidden access points between a header and a cell

**Fig. 9-211** Installation of outlets and markers. [The Flexicore Co., Inc.]

**Fig. 9-212** Outlet drive plug. [The Flexicore Co., Inc.]

which is intended for future use. These markers must extend through the floor covering, and a suitable number must be used to locate the cells and provide for system identification.

**273. High-capacity system.** The system shown in Fig. 9-214 is designed to handle eighty-two 100-pair plus one hundred thirty-seven 25-pair telephone cables, or the equivalent in other sizes. The cables feed from the panel through the channel slab, transversely through the 24-in (609.6-mm) trench header duct, then down into the cells of the Flexicore deck. Wiring can run in either direction to telephone floor outlets located at any point along the cells. Every second cell is assigned to telephones, providing lines of availability only 16 in (406.4 mm) apart.

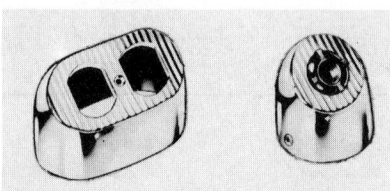

**Fig. 9-213** Outlet fittings: *left*, electric-outlet fitting; *right*, telephone or intercom outlet. [The Flexicore Co., Inc.]

Electrical distribution is handled through the high-capacity trench-header-duct system shown. Many variations of both systems are possible.

**274. Inserts for the location of outlets** can be located at any point along a cell. Typical outlets and their installation are shown in Figs. 9-210, 9-211, 9-212, and 9-213. The procedure in installing an outlet is as follows:

1. Mark location of outlet. Use cell markers to line up center of cell.

2. Drill hole for outlet using 1⅞-in (47.625-mm) core bit.

3. Drive in outlet drive plug.

4. Screw in floor-outlet nipple.

5.  Fish wire to outlet location.

[The Flexicore Co., Inc.]

6.  Install outlet box. Floor outlets can be installed at any time during the life of the building.

**275.** NEC **requirements for inserts** are as follows: Inserts shall be leveled to the floor grade and sealed against the entrance of concrete. They shall be of metal and be fitted with receptacles of the grounded type. A grounding conductor shall connect the insert receptacles to a positive ground connection provided on the header duct. In cutting through the cell wall for setting inserts or other purposes (such as providing access openings between header duct and cells), chips and other dirt shall not be allowed to fall into the raceway, and the tool used shall be so designed as to prevent it from entering the cell and damaging the conductors.

**276. Discontinued outlets** (National Electrical Code).  When an outlet is discontinued, the conductors supplying the outlet shall be removed from the header and cell. No splices or reinsulated conductors such as abandoned loop wiring shall be allowed in raceways.

**277. Conductors** (National Electrical Code)  No conductor larger than No. 1/0 shall be installed except by special permission. The total cross-sectional area of all conductors or cables in a header or in an individual cell shall not exceed 40 percent of the interior cross-sectional area of the header or cell.

Splices and taps shall be made only in header-duct access units or junction boxes.

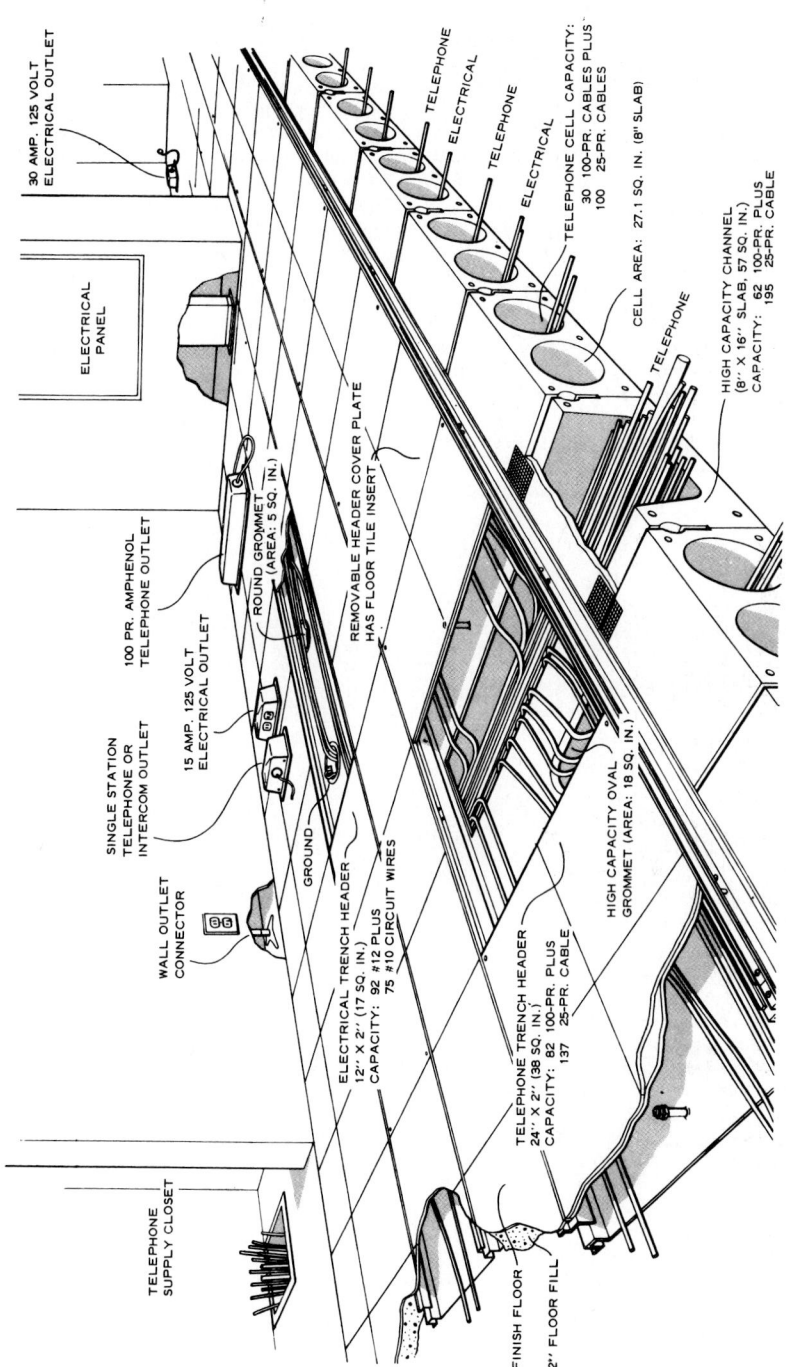

**Fig. 9-214** High-capacity system with trench header ducts. [The Flexicore Co., Inc.]

30 AMP. 125 VOLT ELECTRICAL OUTLET

TELEPHONE

ELECTRICAL

TELEPHONE

ELECTRICAL

TELEPHONE

ELECTRICAL

TELEPHONE CELL CAPACITY: 30 100-PR. CABLES PLUS 100 25-PR. CABLES

TELEPHONE

CELL AREA: 27.1 SQ. IN. (8" SLAB)

HIGH CAPACITY CHANNEL (8" X 16" SLAB, 57 SQ. IN.) CAPACITY: 62 100-PR. PLUS 195 25-PR. CABLE

ELECTRICAL PANEL

ROUND GROMMET (AREA: 5 SQ. IN.)

REMOVABLE HEADER COVER PLATE HAS FLOOR TILE INSERT

100 PR. AMPHENOL TELEPHONE OUTLET

SINGLE STATION TELEPHONE OR INTERCOM OUTLET

15 AMP. 125 VOLT ELECTRICAL OUTLET

WALL OUTLET CONNECTOR

GROUND

ELECTRICAL TRENCH HEADER 12" X 2" (17 SQ. IN.) CAPACITY: 92 #12 PLUS 75 #10 CIRCUIT WIRES

TELEPHONE TRENCH HEADER 24" X 2" (38 SQ. IN.) CAPACITY: 82 100-PR. PLUS 137 25-PR. CABLE

HIGH CAPACITY OVAL GROMMET (AREA: 18 SQ. IN.)

TELEPHONE SUPPLY CLOSET

FINISH FLOOR

2" FLOOR FILL

## WIRING WITH MULTIOUTLET ASSEMBLIES

**278. Multioutlet assemblies** consist of a metallic or nonmetallic assembly with convenience outlets built in every few inches. In one type made by the Wiremold Co. (Fig. 9-215) the outlets may be spaced as desired along the raceway and connected by wires laid in the raceway. In another type (Fig. 9-216) the outlets are built in at fixed intervals and connected by copper strips which terminate in terminal screws at each end of each length of raceway. This type is made in three styles according to the position in which it is to be used: (1) the chair-rail style for flat-surface mounting, (2) the baseboard-cap style for mounting on the top of baseboards or in the corners of mantels and cabinets, and (3) the baseboard-insert style for recessing in the baseboard or in a plastered wall. Blank sections through which standard-type wires may be run to join plug-in sections are available for use behind radiators or for other inaccessible places where the expensive plug-in strip is not needed. Multioutlet assemblies are used to provide adequate plug receptacles for the utmost convenience in the use of appliances, in dwelling-type use, and for industrial and commercial applications.

NEC specifications for the use of multioutlet assemblies are as follows: A multioutlet assembly may be used in dry locations. It shall not be used (1) when concealed, except that the back and sides of a metal multioutlet assembly may be surrounded by the building finish and a nonmetallic multioutlet assembly may be recessed in the baseboard; (2)

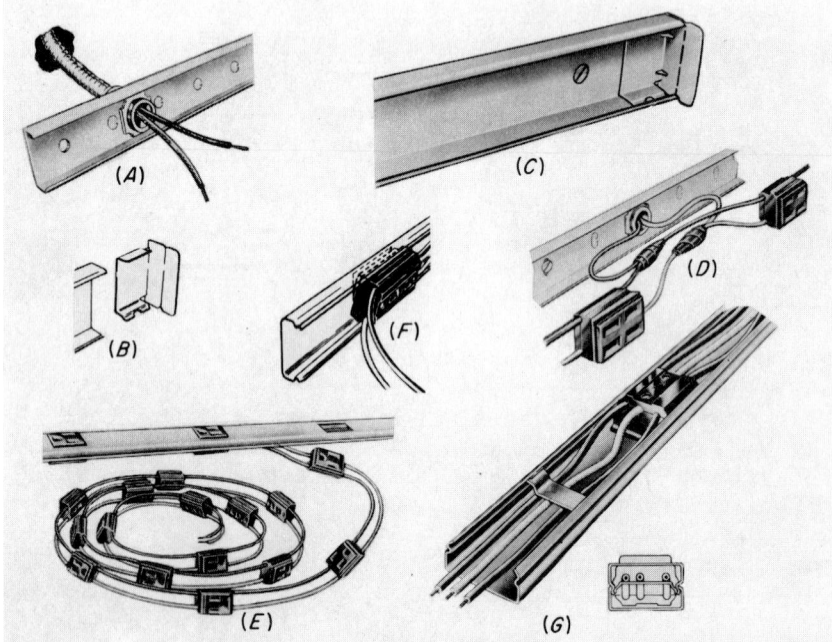

**Fig. 9-215** Installation of multioutlet assembly of the Plugmold-Snapicoil type: (1) Remove knockout (from inside out) for feed connection and fasten ½-in connector into base with a locknut A. No special fittings are needed; there is plenty of room to spin the locknut. (2) At the end of the run, install a No. 2010B blank-end fitting in the base, B and C. (3) Mount base to surface through screw piercings or knockouts, using No. 6 flathead screws. (Some types of surfaces require other methods of fastening.) (4) Lay out Snapicoil, and cover sections to avoid having outlet directly in a corner or over a feed connection. (5) Splice feed wires onto Snapicoil conductors D. NOTE   In a three-wire harness, black wire with yellow lettering should be used for the switch leg. (6) Snap Snapicoil receptacles into 2000C Holecut cover, E and F: a wire clip can be used to cover G or base to keep wires inside raceways. (7) Snap cover, complete with receptacles, into base. [The Wiremold Co.]

when subject to severe physical damage; (3) when the voltage is 300 V or more between conductors unless the assembly is of metal having a thickness of not less than 0.040 in (1.02 mm); (4) when subject to corrosive vapors; (5) in hoistways; or (6) in hazardous locations, except Class I, Div. 2, locations as permitted by the exception of Sec. 501-4(b) of the NEC. A metal multioutlet assembly may be extended through (not run within) dry partitions providing arrangements are made for removing the cap or cover on all exposed portions and no outlet falls within the partitions.

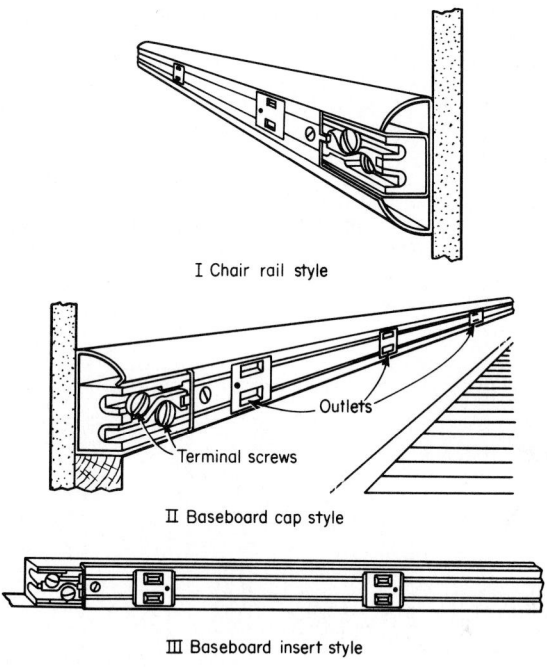

I Chair rail style

II Baseboard cap style

III Baseboard insert style

**Fig. 9-216**    Plug-in strip.

**279. Various ways of mounting the chair-rail style** on plastered walls are shown in Fig. 9-217. Toggle bolts are used unless a wood strip is available, as in III, when wood screws can be used. The chair-rail style is intended for mounting on flat surfaces at a considerable height above the baseboard. The curved surfaces on the top and bottom give the installation a neat appearance.

**280. The baseboard-cap type** has a flat surface on one edge and is curved on the other edge. It is used as illustrated in Fig. 9-218, when there is a corner against which the flat edge can be placed. The curved edge gives a finished appearance to the corner. In this type of mounting, there is usually a wood background so that wood screws can be used for fastening.

**281. The baseboard type** is set into the wall so that the surface of the strip is flush with the wall surface. When finished buildings are wired with this type, the molding on top of the baseboard is removed, the plug-in strip is placed on top of the baseboard, and the molding is placed against the plug-in strip (Fig. 9-219, III). For new installations a channel can be left in tile, glass, or marble surfaces as shown in Fig. 9-219, I and II, or the plaster can be channeled as shown in Fig. 9-219, IV.

**282. Plug-in strip is made in various lengths** from 6 in to 9 ft (from 152.4 mm to 2.7 m). The lengths are joined together as shown in Fig. 9-220. Copper links connect the terminal screws on two adjoining lengths of plug-in strip (Fig. 9-220, I). Then insulating plates are

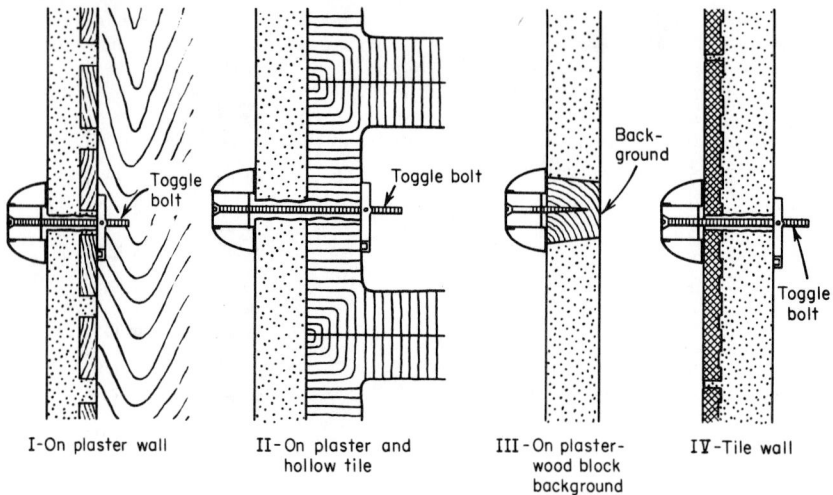

I-On plaster wall   II-On plaster and   III-On plaster-   IV-Tile wall
                    hollow tile        wood block
                                       background

**Fig. 9-217**   Methods of mounting chair-rail style of plug-in strip.

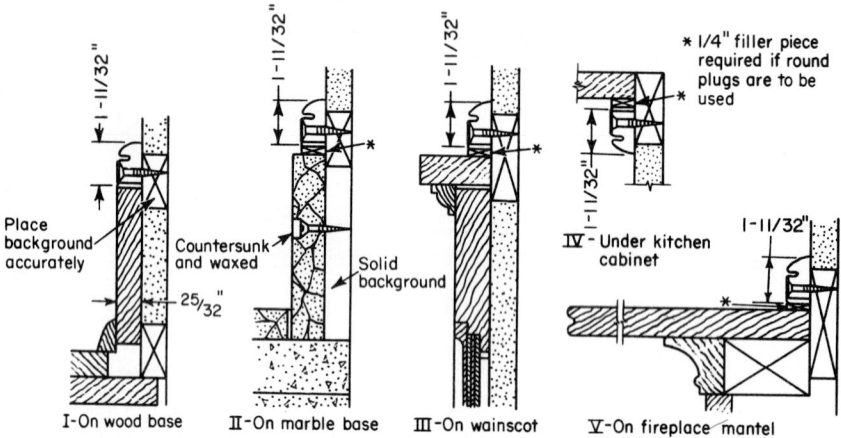

I-On wood base   II-On marble base   III-On wainscot   V-On fireplace mantel

**Fig. 9-218**   Methods of mounting baseboard-cap style of plug-in strip.

placed over the terminals (Fig. 9-220, II), and the joint is covered with a coupling of the same exterior finish as the plug-in strip.

**283. The method of bringing in** the supply circuit is shown in Fig. 9-221. Junction boxes are available for either metal-clad cable or rigid conduit. The junction box has a hub in the back with a setscrew for clamping metal-clad cable or threaded for ½-in conduit connection (Fig. 9-221, I). The circuit wires are stripped as shown, and two short lengths of building wire are spliced to them so that the circuit will feed both ways from the junction box. The junction box is then placed against the wall, and the wires are connected to the terminal screws in the adjoining plug-in strips (Fig. 9-221, II). The middle section of the junction box is closed by means of a cover, held in place by two screws. Over each pair of terminals an insulating plate is placed and then covered with a coupling plate.

**284. Heavier-duty multioutlet assemblies.**   The types of multioutlet assemblies discussed in the preceding sections are of the light-duty type. Heavier-duty multioutlet assemblies are available in the following types:

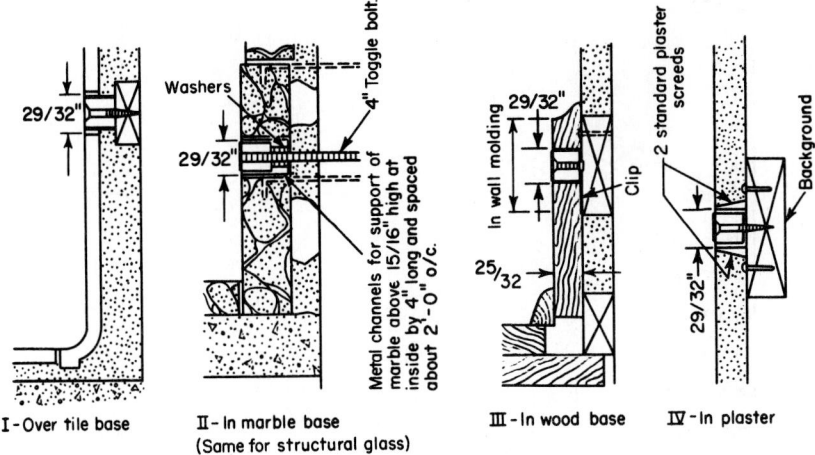

**Fig. 9-219**   Methods of mounting baseboard-insert style of plug-in strip.

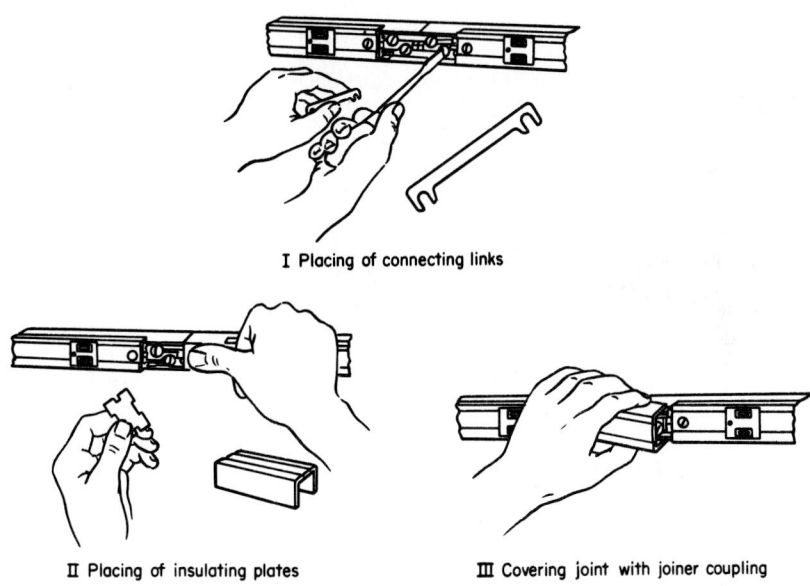

**Fig. 9-220**   Method of coupling plug-in strips together.

1. Two-wire with grounding strip for grounding of equipment
2. 20-A receptacle type
3. 30-A receptacle type
4. 50-A receptacle type
5. Combination raceway and multioutlet assembly (Fig. 9-222)

Baseboard types of combination multioutlet assembly and surface raceway are available. Two types are shown in Figs. 9-223 and 9-224.

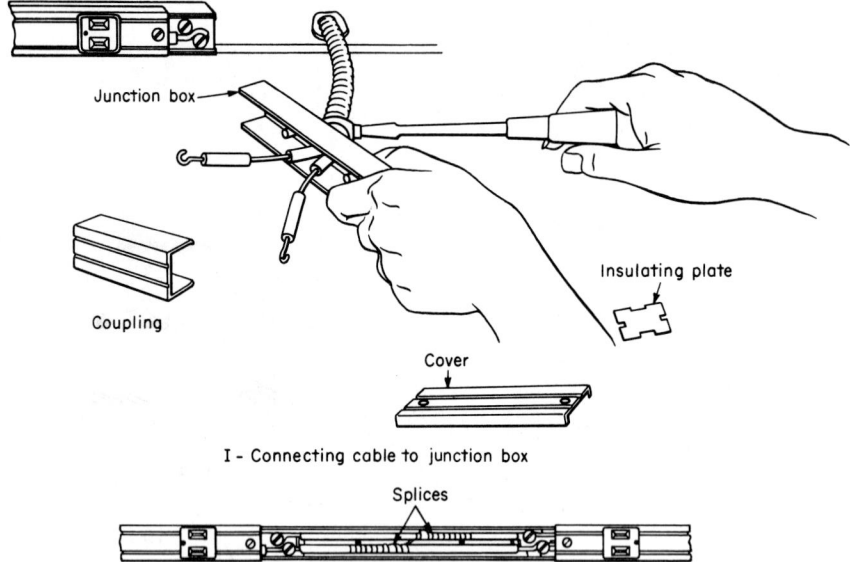

I - Connecting cable to junction box

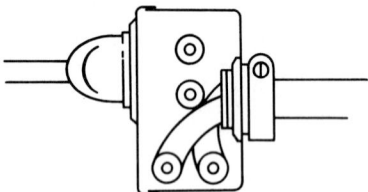

II - With splices and wiring completed

**Fig. 9-221**  Method of bringing supply circuit into plug-in strip.

**Fig. 9-222**  Combination raceway and multioutlet assembly. Up to three No. 12 Type TW electric conductors can be installed in the raceway space located beneath the standard plug-in–strip assembly. The additional conductors may be wired either three-phase or single-phase.

## CABLEBUS WIRING

**285. Cablebus is an approved assembly** of insulated conductors mounted in a spaced relationship in a ventilated metal protective supporting structure including fittings and conductors. Typical components are shown in Fig. 9-225, which also includes some of the major NEC rules for cablebus.

The original manufacturer of the cablebus wiring system was Husky Products, Inc., subsidiary of Burndy Corp. The system is highly desirable in industrial applications for larger feeders or branch circuits at 600 V or less and in higher distribution voltages of 5 or 15 kV. It can also be used for primary feeders in high-rise buildings or commercial buildings in which load-center unit substations are to be used. Cablebus can be used indoors or outdoors in wet or dry locations. The entire system is factory-fabricated with bottom support blocks (Fig. 9-225) in place.

The size of cablebus structures will vary according to the number of conductors. Figure 9-226 shows the four sizes of insulator blocks that are presently available: 3-conductor, 6-conductor, 12-conductor, and 18-conductor. Phase arrangements for three-phase circuits are also shown in Fig. 9-226 to provide a closely balanced impedance. Because of care-

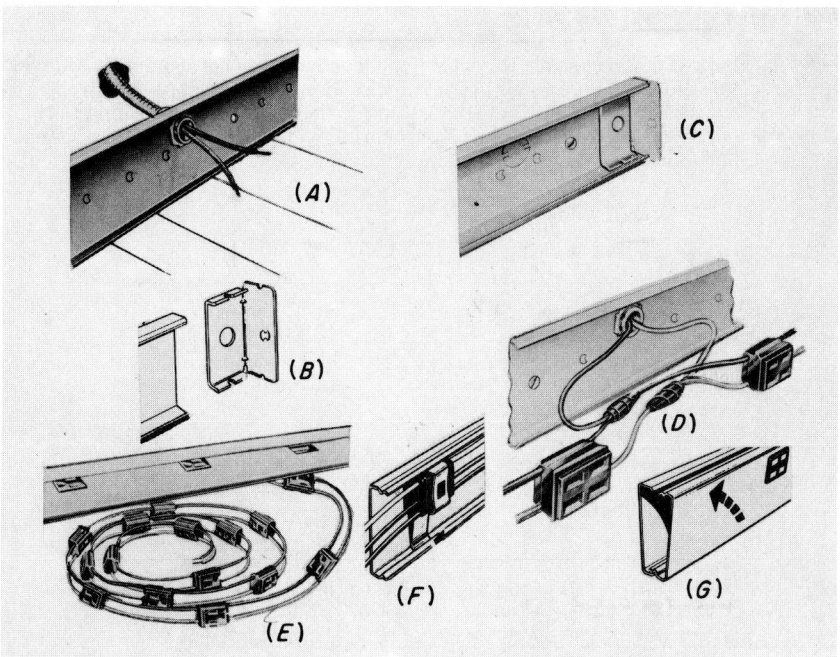

**Fig. 9-223** Installation of Plugmold baseboard: (1) Remove knockout (from inside out) for feed connection and fasten ½-in connector into base with a locknut *A*. No special fittings are needed. (2) At the end of the run, install No. 2210B blank-end fitting in the base, *B* and *C*. (3) Mount base to surface through screw piercings or knockouts, using No. 6 flathead screws. (Some types of surfaces require other methods of fastening.) (4) Lay out Snapicoil, and cover sections to avoid having outlet directly in a corner or over a feed connection. (5) Splice feed wires onto Snapicoil conductors *D*. NOTE In a three-wire harness, black wire with yellow lettering should be used for the switch leg. (6) Insert Snapicoil receptacles into 2200C Holecut cover, and snap device clip in place, *E* and *F*. The clip can also be used in the base to keep wires inside raceways. (7) Snap cover, complete with receptacles, into 2200B base *G*. [The Wiremold Co.]

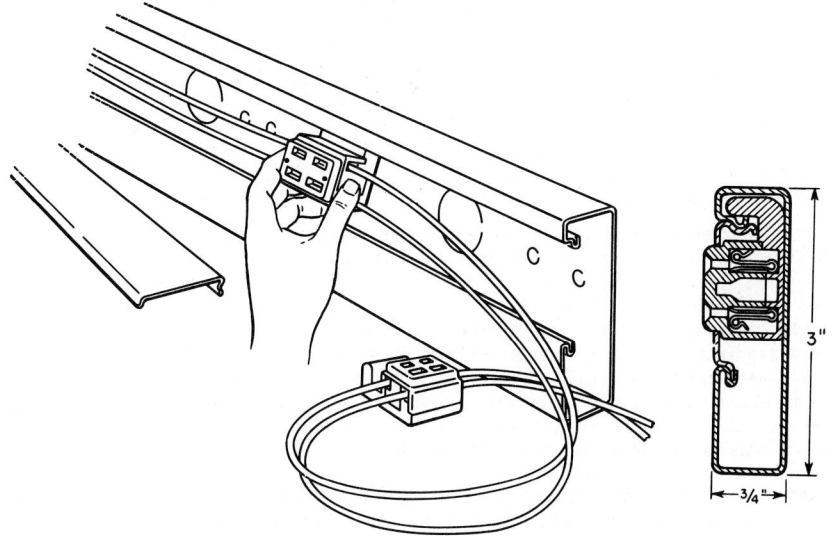

**Fig. 9-224** Base duct.

fully selected conductor spacing, transposition of conductors is not required. Controlled spacing also assures low impedance and low voltage drop. The cablebus framework is an all-welded construction for maximum strength and 12-ft (3.7-m) spacing of supports. High-pressure splice joints between framework sections provide an excellent path or ground, and the installed framework serves as a grounding conductor.

Article 365: Cablebus

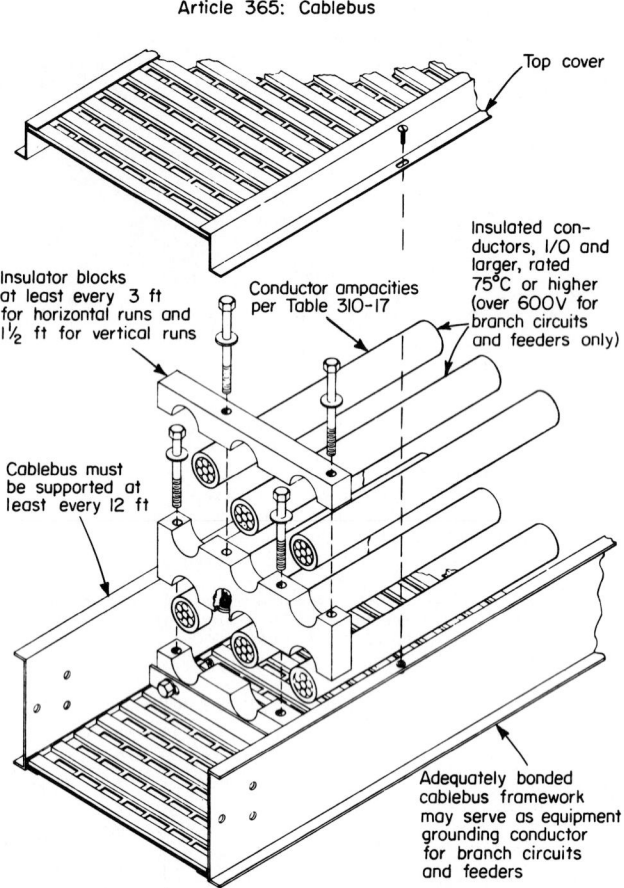

Top cover

Insulated con-
ductors, I/O and
larger, rated
75°C or higher
(over 600V for
branch circuits
and feeders only)

Insulator blocks
at least every 3 ft
for horizontal runs and
1½ ft for vertical runs

Conductor ampacities
per Table 310-17

Cablebus must
be supported at
least every 12 ft

Adequately bonded
cablebus framework
may serve as equipment
grounding conductor
for branch circuits
and feeders

**Fig. 9-225**   Elements of a typical cablebus system and major National Electrical Code requirements. [*Electrical Construction and Maintenance*]

**286. Types of conductor.**   The manufacturer of cablebus recommends that only thermoplastic or thermosetting conductors be used in cablebus. At the present time, combining the insulation's electrical and physical qualities with the cost, the preferred insulation is cross-linked polyethylene. Since cablebus is a ventilated enclosure and conductors are separated by insulating blocks, conductor ampacities may be determined on the basis of single conductors in free air, as in Tables 19 and 22 of Div. II. Generally, 90°C-rated conductors are recommended.

**287. Installation procedures.**   The basic installation of the cablebus framework is similar to that of a cable-tray system (Sec. 289). Typical arrangements are shown in Fig. 9-227, in which cablebus is used for primary and secondary feeders and circuits. The conductor-pulling tools are the same as those shown in Fig. 9-233 for cable tray. Conductors are pulled in after the basic framework has been completely installed. After conductors have been pulled in and arranged in proper phase sequence, they are secured to the insulating blocks. The final step is the installation of top covers.

**288.** National Electrical Code **rules.**   The following Code rules apply to cablebus:

1. Cablebus shall be installed only for exposed work. It may be installed outdoors or in corrosive, wet, or damp locations if identified for such use.

2. Cablebus shall not be installed in hoistways or hazardous locations unless specifically approved for such use.

3. If adequately bonded, cablebus framework may be used as the equipment-grounding conductor for branch circuits and feeders.

4. Cablebus may be used for systems in excess of 600 V, nominal.

5. The current-carrying conductors in cablebus shall have an insulation rating of at least 75°C and be of an approved type.

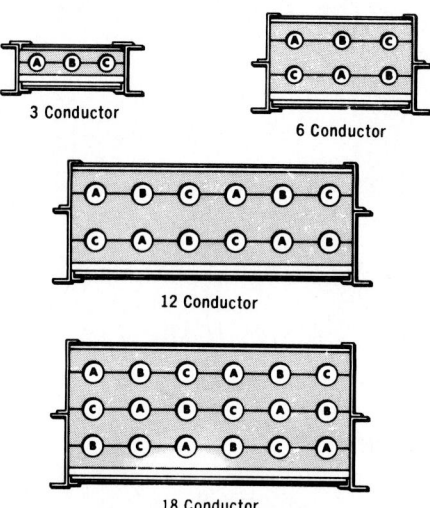

**Fig. 9-226**   Sizes of insulating blocks and three-phase conductor arrangements. [Husky Products, Inc., subsidiary of Burndy Corp.]

6. The size and number of conductors shall be that for which the cablebus is designed, and in no case less than 1/0.

7. Insulated conductors shall be supported on blocks or other mounting means designed for the purpose.

8. The individual conductors in a cablebus shall be supported at intervals not greater than 3 ft (914 mm) for horizontal runs and 1½ ft (457 mm) for vertical runs. Vertical and horizontal spacing between conductors shall be not less than one conductor diameter at the points of support.

9. When the ampacity of cablebus conductors does not correspond to a standard rating of the overcurrent device, the next-higher-rated overcurrent device may be used.

10. Cablebus shall be securely supported at intervals not exceeding 12 ft (3.66 m). Where spans greater than 12 ft are required, the structure shall be specifically designed for the required span length.

11. Cablebus may extend transversely through partitions or walls, other than fire walls, provided the section within the wall is continuous, protected against physicial damage, and unventilated.

12. Except when fire stops are required, cablebus may extend vertically through floors and platforms in wet locations where (*a*) there are curbs or other suitable means to prevent water flow through the floor or platform opening and (*b*) the cablebus is totally enclosed at the point where it passes through the floor or platform and for a distance of 6 ft (1.83 m) above the floor or platform.

13. Except when fire stops are required, cablebus may extend vertically through dry

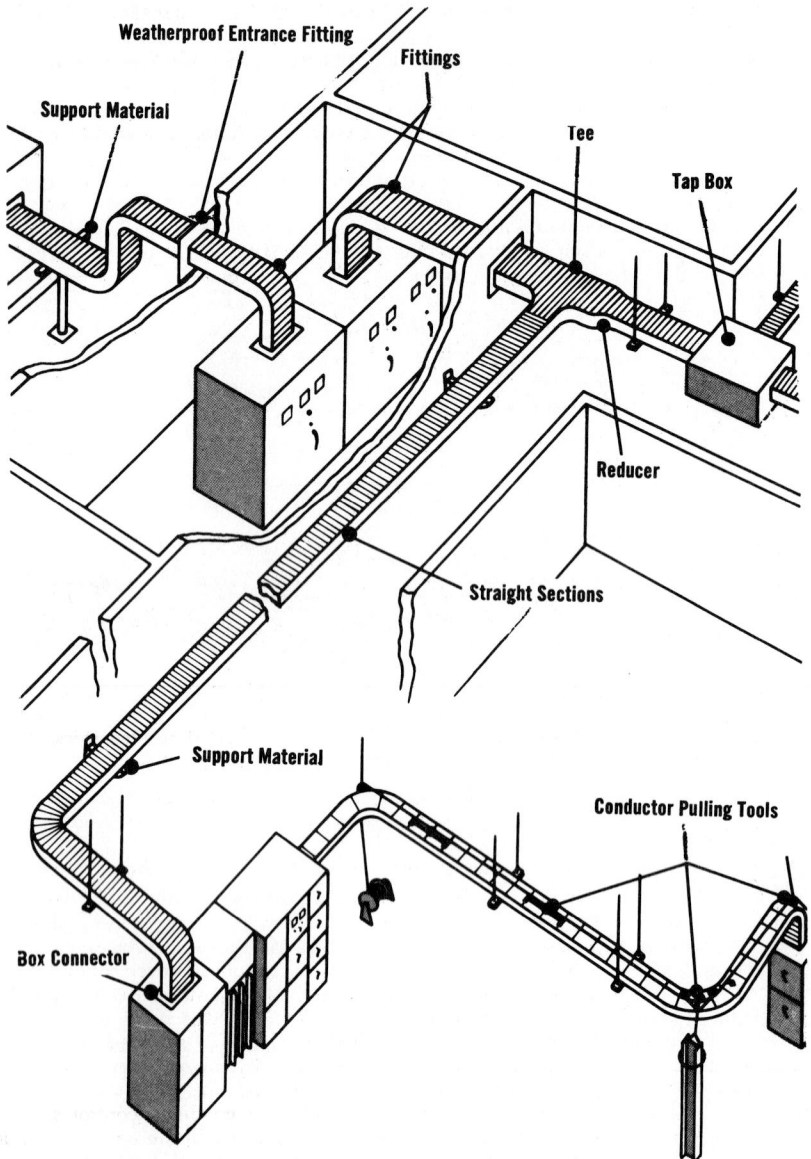

**Fig. 9-227**  Typical cablebus runs for primary and secondary distribution. [Husky Products, Inc., subsidiary of Burndy Corp.]

floors and platforms, provided the cablebus is totally enclosed at the point where it passes through the floor or platform and for a distance of 6 ft (1.83 m) above the floor or platform.

14. Cablebus shall utilize approved fittings for changes in the horizontal or vertical direction of the run, dead ends, terminations in or on connected apparatus or equipment or the enclosures for such equipment, and additional physical protection when required, such as guards for severe mechanical protection.

15. Approved terminating means shall be used for connections to cablebus connections.

16. Sections of cablebus shall be electrically bonded either by inherent design of the mechanical joints or by applied bonding means.

17. Each section of cablebus shall be marked with the manufacturer's name or trade designation and the maximum diameter, number, voltage rating, and ampacity of conductors to be installed. Markings shall be so located as to be visible after installation.

**Fig. 9-228** A 600-V power tray cable for use in cable trays. [Plastic Wire & Cable Corp., subsidiary of Triangle Industries, Inc.]

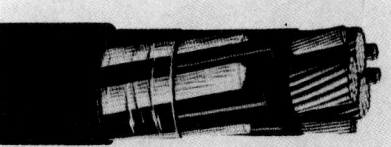

**Fig. 9-229** A 600-V 19-conductor tray cable suitable for power-control circuits and for use in cable trays. [Plastic Wire & Cable Corp., subsidiary of Triangle Industries, Inc.]

**Fig. 9-230** A 600-V power and control tray cable for use in cable trays. [Plastic Wire & Cable Corp., subsidiary of Triangle Industries, Inc.]

## CABLE TRAYS

**289. A cable-tray system** is a unit or an assembly of units or sections, with associated fittings, forming a continuous rigid structural system used to support cables. Cable trays include ladders, troughs, channels, solid-bottom trays, and similar structures.

The National Electrical Code allows cable tray to contain any approved conduit or raceway with its contained conductors; mineral-insulated metal-sheathed cable (Type MI); armored cable (Type AC); metal-clad cable (Type MC); power-limited tray cable (Type PLTC); nonmetallic-sheathed cable (Types NM and NMC); shielded nonmetallic-sheathed cable (Type SNM); multiconductor service-entrance cable (Types SE and USE); multiconductor underground-feeder and branch-circuit cable (Type UF); power and control tray cables (Type TC); and other factory-assembled multiconductor control, signal, or power cables which are specifically approved for installation in cable trays. In practice, the most common types used in cable trays are Type MC metal-clad cable (Fig. 9-87) and Type TC tray cables (Figs. 9-228, 9-229, and 9-230) listed by Underwriters Laboratories Inc. for cable-tray systems. The cable shown in Fig. 9-228 is a polyvinyl chloride- and nylon-insulated 600-V power cable with three 2/0 conductors and a grounding conductor. The cable shown in Fig. 9-229 has the same insulation and contains 19 No. 12 stranded conductors, suitable for 600-V power or control circuits. Figure 9-230 also has the same insulation, and contains three 4/0 power conductors and four No. 12 control conductors, all suitable for 600 V. Cables similar to these can be obtained with a wide variety of conductor sizes. These nonmetallic cables and Type MC metal-clad cables can be installed in cable trays much more easily than other types of cables.

**290. Types of cable tray.** There are several manufacturers of cable tray, and the types shown in Figs. 9-231 and 9-232 are manufactured by the Globe Division of United States

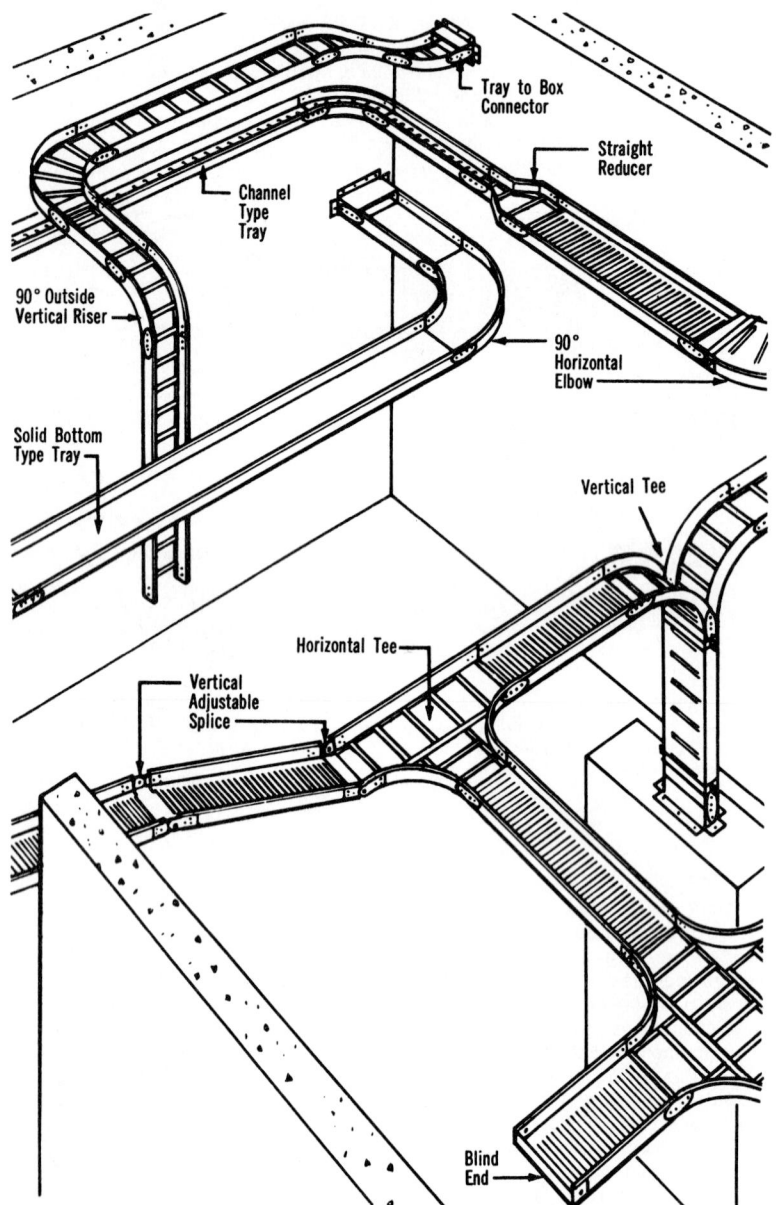

**Fig. 9-231** Typical components of a CABLE-STRUT system. [Globe Division, United States Gypsum Co.]

Gypsum. These drawings show various types of cable tray (channel type, solid-cover type, louvered-cover type, ladder type, and solid-bottom type). Also illustrated are a variety of fittings used with the support system, such as elbows or offsets, blind ends, tees, crosses, reducers, dividers, and box connectors. As can be seen in Figs. 9-231 and 9-232, cable trays provide a highly flexible system in any application in which a number of power and/

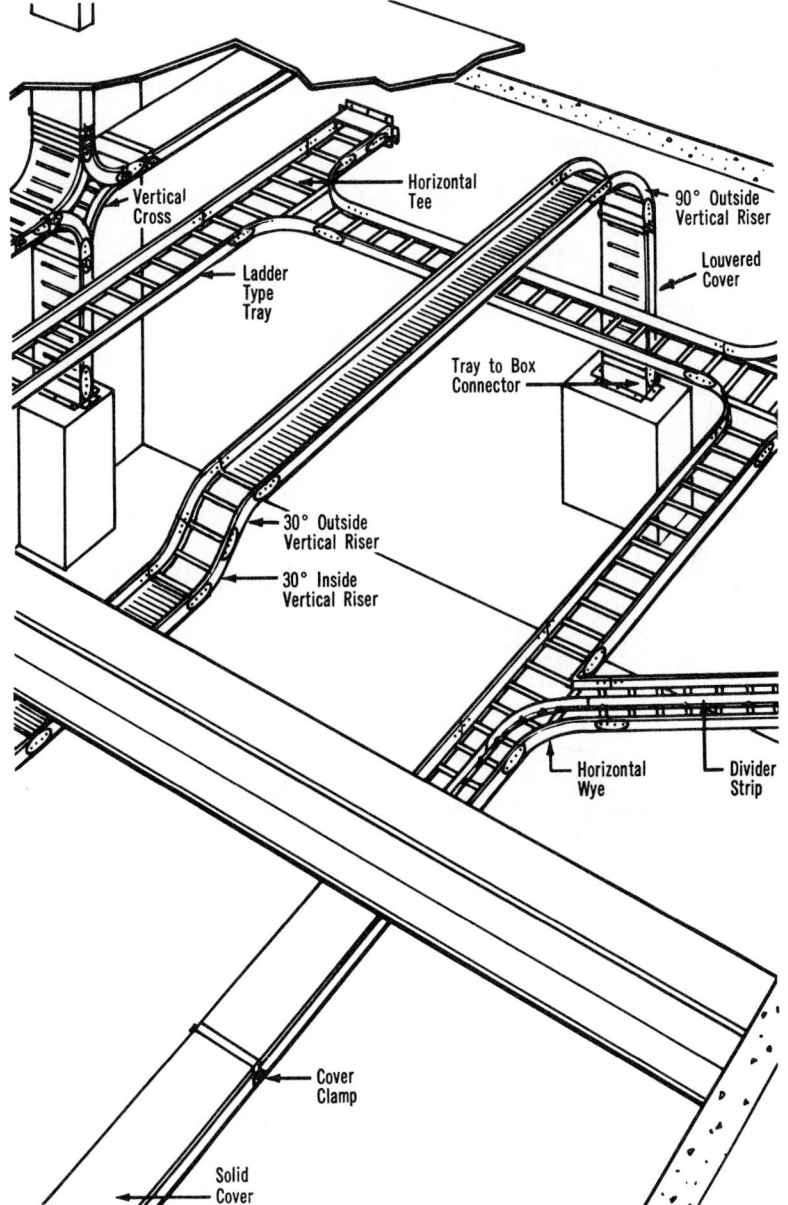

Vertical
Cross

Horizontal
Tee

90° Outside
Vertical Riser

Ladder
Type
Tray

Louvered
Cover

Tray to Box
Connector

30° Outside
Vertical Riser

30° Inside
Vertical Riser

Horizontal
Wye

Divider
Strip

Cover
Clamp

Solid
Cover

**Fig. 9-232** Typical components of a CABLE-STRUT system. [Globe Division, United States Gypsum Co.]

or control cables are required. Cable trays are secured in a manner similar to that described in the sections on rigid metal conduit and busways.

An excellent feature of the cable-tray system is that cables can be added, revamped, or removed with minimum effort. When laying out such systems, it is well to plan for future needs by obtaining trays larger than necessary for the initially installed cables. Manufacturers offer many different widths and lengths for each section of cable-tray system, and it is recommended that representatives of these firms be approached when such installations are contemplated.

**291. Installation aids.**   The tools shown in Fig. 9-233 to simplify the pulling of cables into cable trays are typical of those offered by manufacturers of these systems. Locating rollers and pulleys at the proper locations in the system speeds up the installation of

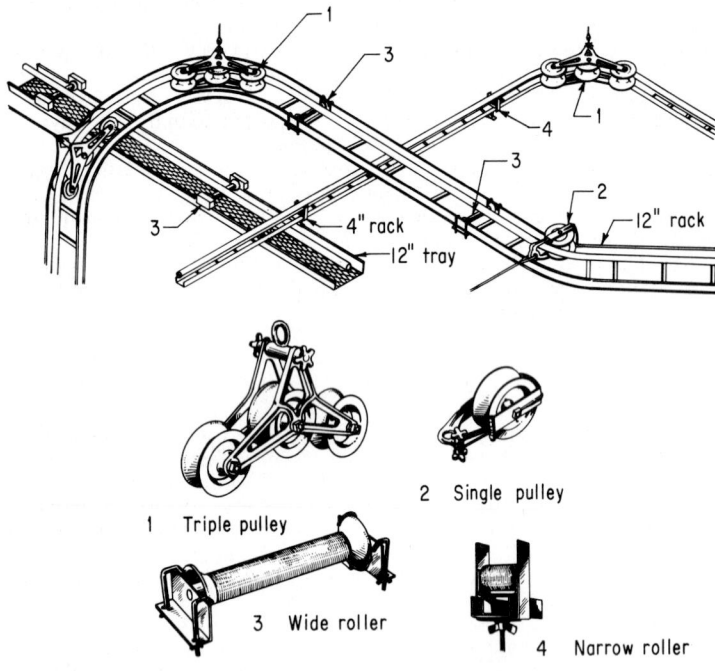

Installation aids available

**Fig. 9-233**   Installation aids available to simplify cable pulling in cable trays. [*Electrical Construction and Maintenance*]

cables and prevents damage to them during installation. After cables have been installed, they should be spaced and supported as indicated in Sec. 292.

**292.** National Electrical Code **rules** for cable trays are as follows:

1. They shall have suitable strength and rigidity to provide adequate support for all contained wiring.

2. They shall not present sharp edges, burrs, or projections injurious to the insulation or jackets of the wiring.

3. If made of metal, they shall be adequately protected against corrosion or shall be made of corrosion-resistant material.

4. They shall have side rails or equivalent structural members.

5. They shall include fittings or other suitable means for changes in direction and elevation of runs.

6. Nonmetallic cable trays shall be made of fire-retardant material.

7. Cable trays shall be installed as a complete system. Field bends or modifications shall be so made that the electrical continuity of the cable-tray system and support for the cables shall be maintained.

8. Each run of cable tray shall be completed before the installation of cables.

9. Supports shall be provided to prevent stress on cables where they enter another raceway or enclosure from cable-tray systems.

10. In portions of runs where additional protection is required, covers or enclosures providing the required protection shall be of a material compatible with the cable tray.

11. Multiconductor cables rated 600 V or less shall be permitted to be installed in the same cable tray.

12. Cables rated over 600 V shall not be installed in the same cable tray with cables rated 600 V or less except where separated by solid fixed barriers or where cables over 600 V are Type MC.

13. Cable trays shall be permitted to extend transversely through partitions and walls or vertically through platforms and floors in wet or dry locations where the installations, complete with installed cables, are made in accordance with the requirements of Sec. 300-21 of the Code.

14. Cable trays shall be exposed and accessible except as permitted by Par. 13.

15. Sufficient space shall be provided and maintained around cable trays to permit adequate access for installing and maintaining the cables.

16. Metallic cable trays which support electrical conductors shall be grounded as required for conductor enclosures in Article 250 of the Code.

17. When steel or aluminum cable-tray systems are used as equipment-grounding conductors, all the following provisions shall be complied with:

   *a.* The cable-tray sections and fittings shall be identified for grounding purposes.

   *b.* The minimum cross-sectional area of cable trays shall conform to the requirements in Table 318-7(b)(2) of the NEC.

   *c.* All cable-tray sections and fittings shall be legibly and durably marked to show the cross-sectional area of metal in channel-type cable trays or cable trays of one-piece construction and the total cross-sectional area of both side rails for ladder or trough-type cable trays.

   *d.* Cable-tray sections, fittings, and connected raceways shall be bonded in accordance with Sec. 250-75 of the NEC, using bolted mechanical connectors or bonding jumpers sized and installed in accordance with Sec. 250-79 of the Code.

18. Cable splices made and insulated by approved methods shall be permitted to be located within a cable tray provided they are accessible and do not project above the side rails.

19. In other than horizontal runs, the cables shall be fastened securely to transverse members of the cable trays.

20. A box shall not be required where cables or conductors are installed in bushed conduit or tubing used for support or for protection against physical damage.

21. Where single-conductor cables comprising each phase or neutral of a circuit are connected in parallel as permitted in Sec. 310-4 of the NEC, the conductors shall be installed in groups consisting of not more than one conductor per phase or neutral in order to prevent current unbalance in the paralleled conductors due to inductive reactance.

Single conductors shall be securely bound in circuit groups to prevent excessive movement due to fault-current magnetic forces.

*Exception. When single conductors are twisted together, as in triplexed assemblies.*

## GENERAL REQUIREMENTS FOR WIRING INSTALLATIONS

CONDUCTOR REQUIREMENTS

**293. Size of conductors.** No wire smaller than No. 14 is allowed for general interior-wiring work. Wires of No. 18 or No. 16 may be used for fixture wiring, flexible cords, and some other special applications as given in Sec. 32 of Div. 3. Although No. 14 wire is allowed for branch circuits, it is good practice not to use wire smaller than No. 12. Number 12 wire costs only about 20 percent more than No. 14, can usually be placed in the same-size conduit or raceway, will have greater mechanical strength, and will reduce the voltage drop and the $I^2R$ heating loss in the wires by approximately 40 percent. Solid conductors of larger size than No. 10 are not allowed to be used in conduit or raceways. For ease in handling, it is good practice to use stranded conductors in the larger sizes for open wiring also. The size of wire required, according to the amount of current to be carried by the circuit, is discussed in Sec. 31 of Div. 3, and the ampere current-carrying capacities are given in Tables 12 to 23 of Div. 11.

**294. All ungrounded conductors must be protected by fuses or circuit breakers** rated in accordance with the allowable ampacity of the conductors as given in Sec. 295 and located at the point where the conductor receives its supply, with the following exceptions:

1. For building-service conductors the overcurrent protection is located at the same point as the other service-entrance equipment, which may be at the load end of the service-entrance conductors or at the outer end.

2. The overcurrent protection may be omitted provided the overcurrent unit farther back on the circuit is rated not greater than the ampacity of the smaller conductor.

3. Taps to individual outlets and circuit conductors supplying a single household electric range shall be considered to be protected by the branch-circuit overcurrent devices when in accordance with the requirements of Secs. 210-19, 210-20, and 210-24 of the NEC.

4. If a feeder tap conductor is not over 10 ft (3.05 m) long, does not extend beyond the switchboard, panelboard, or control devices which it supplies, is enclosed in a raceway, and has an ampacity not less than the sum of the computed loads of the circuits which it supplies, not less than the rating of the device supplied by the tap conductor, or not less than the rating of the overcurrent protective device at the termination of the tap conductors, it need not be separately protected.

5. If the feeder tap conductor is not over 25 ft (7.62 m) long, is protected from physical damage, has an ampacity not less than one-third that of the conductor from which it is supplied, and terminates in a single circuit breaker or set of fuses, the circuit breaker and fuses may be located at the load end of the conductor.

6. For individual motor branch circuits, refer to Sec. 435, Div. 7.

7. For feeders or main conductors supplying more than one motor, refer to Sec. 440, Div. 7.

8. *Protection of fixture wires and cords.* No. 18 fixture wire up to 50 ft (15.2 m) or flexible cord, size No. 16 or No. 18, and tinsel cord shall be considered to be protected by 20-A overcurrent devices.

Flexible cord or a tinsel cord approved for use with specific listed appliances or portable lamps shall be considered to be protected by the overcurrent device of the branch circuit of Article 210 of the NEC when conforming to the following:

20-A circuits: Tinsel cord or No. 18 cord and larger
30-A circuits: Cord of size No. 16 and larger
40-A circuits: Cord of 20-A capacity and over
50-A circuits: Cord of 20-A capacity and over

Fixture wire shall be considered to be protected by the overcurrent device of the branch circuit of Article 210 of the NEC when conforming to the following:

20-A circuits: No. 18 up to 50 ft (15.2 m)
20-A circuits: No. 16 up to 100 ft (30.5 m)
30-A circuits: No. 14 and larger
40-A circuits: No. 12 and larger
50-A circuits: No. 12 and larger

**295. Rating of the overcurrent protective devices.**    The rating of the overcurrent protective device must be as given below except for the exceptions referred to in Pars. 6 to 8 of Sec. 294. Fuses or circuit breakers shall have ratings not greater than the allowable ampacity of the conductor which they protect, except that when the standard ampere ratings and settings of overcurrent devices do not correspond with the allowable ampacity of conductors, the next-higher standard rating and setting may be used in ratings of 800 A or less.

The effect of the temperature on the operation of thermally controlled circuit breakers should be taken into consideration in the application of such circuit breakers when they are subjected to extremely low or extremely high temperatures.

The allowable ampacity of wires is given in Tables 18 to 23 of Div. 11, and the common sizes of fuses and circuit breakers in Table 38 of Div. 11 and Sec. 240-6 of the NEC.

**296. Grounded conductors must not be fused and must not have a circuit breaker** in them unless the circuit breaker opens all the wires of the circuit at the same time. The size of a grounded conductor may be changed whenever the sizes of the other conductors are changed, provided the other conductors are protected as outlined in Secs. 294 and 295 (see Div. 8, Sec. 136, for an explanation of the reason for grounding one conductor of a wiring system).

**297. Overcurrent devices in multiple.** Except for fuses or circuit breakers which are factory-assembled and approved as a single unit, overcurrent devices must not be arranged or installed in parallel.

**298. Enclosures for protective devices** (National Electrical Code).  Overcurrent devices shall be enclosed in cutout boxes or cabinets unless they are a part of a specially approved assembly which affords equivalent protection or unless they are mounted on switchboards, panelboards, or controllers; or when mounted as open-type assemblies, they shall be located in rooms or enclosures free from easily ignitible material and dampness and accessible only to qualified personnel. The operating handle of a circuit breaker may be accessible without opening a door or cover.

DAMP OR WET LOCATIONS  Enclosures for overcurrent devices in damp or wet locations shall be of a type identified for use in such locations and shall be mounted so there is at least a ¼-in (6.35-mm) air space between the enclosure and the wall or other supporting surface.

VERTICAL POSITION  Enclosures for overcurrent devices shall be mounted in a vertical position unless in individual instances this is shown to be impracticable.

DISCONNECTION OF FUSES BEFORE HANDLING  Disconnecting means shall be provided on the supply side of all fuses in circuits of more than 150 V to ground and cartridge fuses in circuits of any voltage, if accessible to other than qualified persons, so that each individual circuit containing fuses can be independently disconnected from the source of electrical energy, except that a single disconnecting means may be used to control a group of circuits each protected by fuses under the conditions described in Secs. 430-112 and 424-22 of the NEC.

ARCING OR SUDDENLY MOVING PARTS  Arcing or suddenly moving parts shall comply with the following:

1. *Location.* Fuses and circuit breakers shall be so located or shielded that persons will not be burned or otherwise injured by their operation.

2. *Suddenly moving parts.* Handles or levers of circuit breakers and similar parts which may move suddenly in such a way that persons in the vicinity are likely to be injured by being struck by them shall be guarded or isolated.

**299. Flexible cords are used for wiring fixtures,** for the connection of pendants or portable lamps or appliances, for elevator cables, for the wiring of cranes and hoists, for the connection of approved stationary equipment to facilitate their frequent interchange, to prevent transmission of noise or vibration, or to facilitate the removal of fixed or stationary appliances for maintenance or repair, for data processing cables, and connection of moving parts, or for temporary wiring as permitted by Article 305 of the NEC. (See Div. 2 for a complete listing of flexible cords.) If the load connected to them is not greater than their carrying capacity, they are considered to be sufficiently protected by the branch-circuit overcurrent device of the branch circuit from which they are tapped. For the ampacity of fixture wire see Table 47, Div. 11. Cords must be connected to devices so that tension in a cord will not be taken by the terminal screws. This can be accomplished by knotting the cord inside the device (Fig. 9-234), by winding it with tape, or by using a clamp fitting. When flexible cords pass through the cover of an outlet box, the hole must have a smooth, well-rounded surface; when they enter lampholders, they must be protected by insulating bushings. Flexible cords shall be installed in continuous lengths without splice or tap. For voltages between 300 and 600, cords of No. 10 and smaller shall have at least 45 mils (1.143 mm) of thermoset or thermoplastic insulation on each conductor unless Type S, SO, SE, SEO, SOO, STO, STOO, or ST cord is used.

In the wiring of fixtures, use conductors having insulation suitable for the current, voltage, and temperature to which they will be subjected. The splice between the fixture conductors and the circuit conductors must be easily accessible for inspection without requiring disconnection of the wiring. This is accomplished by making the canopy so that it will slide down the fixture stem (Fig. 9-236).

**300. Splicing of conductors.** Conductors in conduit or raceway systems must be continuous from outlet to outlet and must not be spliced or tapped within the raceway or conduit itself except in wireways or auxiliary gutters. Instructions for the proper splicing of conductors are given in Div. 2.

**301. Conductors may be placed in parallel** in sizes 1/0 and larger provided thay are of the same length and size and have the same type of insulation and conductor material. The conductors shall be arranged and terminated at each end in such a manner as to ensure equal division of the total current between all the parallel conductors. If the parallel circuits should be carried in different conduits or raceways, at least one wire of each phase must be carried in each conduit or raceway.

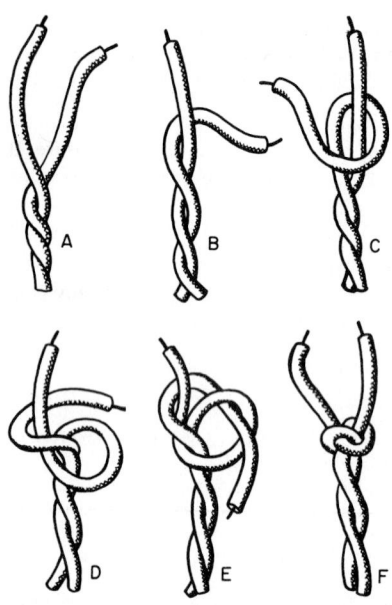

**Fig. 9-234** Method of tying a supporting knot in flexible cord (see Sec. 299 for description).

**302. Induced currents in metal enclosures** (National Electrical Code). In metal enclosures, the conductors of circuits operating on alternating current shall be so arranged as to avoid overheating of the metal by induction. When the capacity of a circuit is such that it is impracticable to run all conductors in one enclosure, the circuit may be divided and two or more enclosures may be used, provided each phase conductor of the circuit and the neutral conductor, when one is used, are installed in each enclosure. The conductors of such an installation must conform to the provisions of Sec. 301 for paralleled conductors.

Induced currents in an enclosure can be avoided by so grouping the conductors in one enclosure that the current in one direction will be substantially equal to the current in the opposite direction.

In the case of circuits supplying vacuum or electric discharge lighting systems or signs or x-ray apparatus and underplaster extensions permitted by Secs. 344-1 to 344-5 inclusive of the NEC, the currents carried by the conductors are so small that a single conductor may be placed in a metal enclosure or pass through metal without causing trouble from induction.

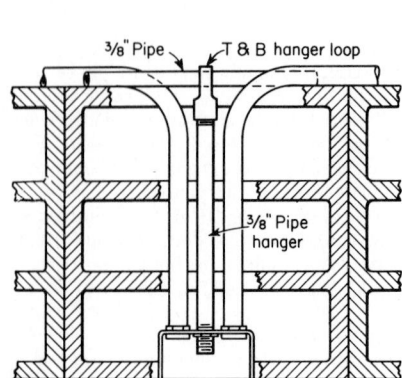

**Fig. 9-235** Support for an outlet box in a terracotta ceiling.

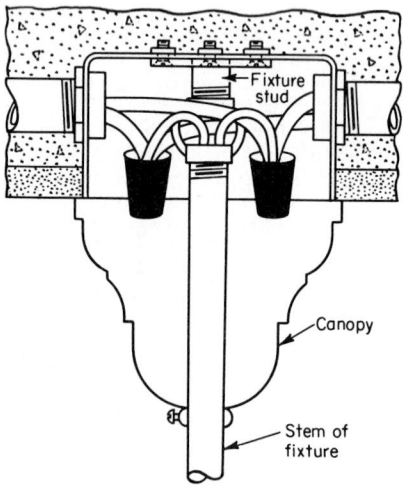

**Fig. 9-236** Method of supporting a fixture from a fixture stud.

When the conductors of a circuit pass through individual holes in the wall of a metal cabinet, the effect of induction may be minimized by (1) cutting slots in the metal between the individual holes through which the conductors of the circuit pass or (2) passing all the conductors in the circuit through an aluminum wall or insulating block used to cover a hole in the metal cabinet sufficiently large for all the conductors of the circuit and providing individual holes for the separate conductors.

**303. At outlets and switch locations at least 6 in (152 mm) of free conductor** must be left for the making of joints and the connection of wiring devices except when conductors are intended to loop without joints through boxes and fittings.

**304. When ungrounded conductors of No. 4 and larger enter a cabinet,** junction, or pull box, they shall be protected by a smoothly rounded insulating bushing unless they are separated from the raceway fitting by substantial insulating material securely fastened in place. If conduit bushings are constructed wholly of insulating material, a locknut to which the conduit is attached shall be provided both inside and outside the enclosure. If the conductors enter the cabinet vertically, a wiring space must be provided with a width as specified in Sec. 82, Div. 4.

**305. Conductors of different systems**

1. Conductors of light and power systems of 600 V or less may occupy the same enclosure, without regard to whether the individual circuits are alternating-current or direct-current, only if all conductors are insulated for the maximum voltage of any conductor within the enclosure.

2. Conductors of light and power systems of over 600 V shall not occupy the same enclosure with conductors of light and power systems of 600 V or less.

3. Secondary wiring to electric discharge lamps of 1000 V or less may occupy the same enclosure as the branch-circuit conductors.

4. Control, relay, and ammeter conductors used in connection with any motor or starter may occupy the same enclosure as the motor-circuit conductors.

5. Primary leads of electric discharge lamp ballasts, insulated for the primary voltage of the ballast, when contained within the individual wiring enclosure may occupy the same fixture enclosure as the branch-circuit conductors.

**306. When conductors or cables are run through bored holes in studs and joists,** the holes should be bored at the approximate centers of wood members or at least 1¼ (31.8 mm) from the nearest edge when practical. If there is no objection because of weakening the building structure, cables or raceways may be laid in notches in the studding or joists if the cable at those points is protected against the driving of nails into it by having the notch covered with a steel plate at least ⅟₁₆ in (1.59 mm) thick before the building finish is applied.

**307. Conductors in vertical raceways (any metallic enclosure for conductors)** must be supported at intervals not greater than those given in Sec. 96.

**308. Wiring in ducts, plenums, and other air-handling spaces** (National Electrical Code). No wiring systems of any type shall be installed in ducts used to transport dust, loose stock, or flammable vapors; nor shall any wiring system of any type be installed in any duct, or shaft containing only such ducts, used for vapor removal or ventilation of commercial-type cooking equipment.

Only wiring methods consisting of mineral-insulated metal-sheathed cable, Type MC cable employing a smooth or corrugated impervious metal sheath without an overall nonmetallic covering, electrical metallic tubing, flexible metallic tubing, intermediate metal conduit, or rigid metal conduit shall be installed in ducts or plenums specifically fabricated to transport environmental air. Flexible metal conduit and liquidtight flexible metal conduit shall be permitted, in lengths not to exceed 4 ft (1.22 m), to connect physically adjustable equipment and devices permitted to be in these ducts and plenum chambers. The connectors used with flexible metal conduit shall effectively close any openings in the connection. Equipment and devices shall be permitted within such ducts or plenum chambers only if necessary for their direct action upon, or sensing of, the contained air. Where equipment or devices are installed and illumination is necessary to facilitate maintenance and repair, enclosed gasketed-type fixtures shall be permitted.

Only totally enclosed, nonventilated insulated busways having no provisions for plug-in connections and wiring methods consisting of mineral-insulated metal-sheathed cable, Type MC cable without an overall nonmetallic covering, and Type AC cable and other factory-assembled multiconductor control or power cable which is specifically listed for the use shall be used for wiring in systems installed in other space used for environmental air. Other types of cables and conductors shall be installed in electrical metallic tubing, flexible metallic tubing, intermediate metal conduit, rigid metal conduit, metal surface raceway or wireway with metal covers if accessible, or flexible metal conduit.

Electric equipment with a metal enclosure or with a nonmetallic enclosure listed for the use and having adequate fire-resistant and low-smoke–producing characteristics, and associated wiring material suitable for the ambient temperature may be installed in such other space unless prohibited elsewhere in the **NEC**.

In ducts and plenums or other spaces used for environmental air, liquidtight flexible metal conduit in single lengths not exceeding 6 ft (1.83 m) is permitted. Habitable rooms or areas of buildings the prime purpose of which is not air handling are not considered ducts or plenums.

NFPA *Standard for the Installation of Air Conditioning and Ventilating Systems*, No. 90A, sets forth requirements of building used for ducts and plenums.

The wiring systems used for data processing systems and located within air-handling areas created by raised floors shall conform to Article 645 of the **NEC**.

**309. The approved types of insulation of conductors** for different installations are given in Sec. 31.

**310. Insulation resistance.**   To provide a reasonable factor of safety against short circuits and grounds, completed wiring systems should undergo an insulation-resistance test. The proper instrument for making this test is a megohmmeter. The **NEC** formerly contained recommended values of resistance for various circuit sizes. However, these values were deleted from the **Code** because they were misleading. Cable and conductor installations provide a wide variation of conditions with respect to the resistance of insulation. These variations are due to the many types of insulating materials used, insulation thickness, voltage rating, and the length of the circuit. Long circuits may be subject to wide variations in temperature, and this can influence the insulation-resistance values in a test.

While insulation-resistance readings are quantitative, they are also relative and comparable. Accordingly, such readings can be used to indicate the presence of moisture, dirt, and deterioration. Although the operation of a megohmmeter insulation tester is fairly simple, one must know how to interpret the results, and instructions provided for such instruments should be carefully followed.

Because a discussion of all the factors involved with these instruments and various test methods is beyond the scope of this book, anyone interested in this subject should review an instruction manual for insulation testers. Such a manual, *Instruction Manual for Megger Insulation Testers*, can be purchased from the James G. Biddle Co., Plymouth Meeting, Pennsylvania. This manual includes instructions for connecting and operating Megger insulation-testing instruments, directions for making insulation-resistance tests on various types of electrical equipment, supplementary instructions and explanations to assist the person who makes the test, and valuable material on how to interpret insulation-resistance readings.

**311. Grounded conductor.**   Every interior-wiring system must have a grounded conductor except as allowed by the **Code** for certain specific conditions. The grounded conductor must be continuously identified throughout the system.

If a neutral (grounded) conductor is run to a lampholder or any other screw-shell device, the neutral (grounded) conductor must be connected to the screw shell. This requirement does not apply to screw shells that serve as fuseholders.

GENERAL INSTALLATION REQUIREMENTS

**312. Working space about electric equipment, 600 V or less** (National Electrical Code). Sufficient access and working space shall be provided and maintained around all electric equipment to permit ready and safe operation and maintenance of such equipment.

1. WORKING CLEARANCES   Except as elsewhere required or permitted in the **NEC** the dimension of the working space in the direction of access to living parts operating at 600 V or less and likely to require examination, adjustment, servicing, or maintenance while energized shall not be less than indicated in Table 312A. In addition to the dimensions shown in Table 312A, the work space shall not be less than 30 in (762 mm) wide in front of the electric equipment. Distances shall be measured from the live parts if such are

exposed or from the enclosure front or opening if such are enclosed. Concrete, brick, or tile walls shall be considered as grounded.

**Table 312A. Working Clearances[a]**
(Table 110-16(a), 1990 NEC)

| Voltage to ground | Minimum clear distance, ft | | |
|---|---|---|---|
| | Condition 1 | Condition 2 | Condition 3 |
| 0–150 | 3 | 3 | 3 |
| 151–600 | 3 | 3½ | 4 |

[a]Reprinted with permission from NFPA 70-1990. National Electrical Code®. Copyright © 1990. National Fire Protection Association. Quincy, Massachusetts 02269. This reprinted material is not the complete and official position of the NFPA on the referenced subject, which is represented only by the standard in its entirety.

The conditions are as follows:

*Condition 1.* Exposed live parts on one side and no live or grounded parts on the other side of the working space, or exposed live parts on both sides effectively guarded by suitable wood or other insulating materials. Insulated wire or insulated busbars operating at not over 300 V shall not be considered live parts.

*Condition 2.* Exposed live parts on one side and grounded parts on the other side.

*Condition 3.* Exposed live parts on both sides of the work space (not guarded as provided in Condition 1) with the operator between.

*Exception No. 1. Working space shall not be required in back of assemblies such as dead-front switchboards or motor-control centers where there are no renewable or adjustable parts such as fuses or switches on the back and where all connections are accessible from locations other than the back.*

*Exception No. 2. By special permission smaller spaces may be permitted where it is judged that the particular arrangement of the installation will provide adequate accessibility or the voltages are less than 30 V rms or 42 V dc.*

2. CLEAR SPACES  Working space required by this section shall not be used for storage. When normally enclosed live parts are exposed for inspection or servicing, the working space, if in a passageway or general open space, shall be suitably guarded.

3. ACCESS AND ENTRANCE TO WORKING SPACE  At least one entrance of sufficient area shall be provided to give access to the working space about electric equipment. For equipment rated 1200 A or more and over 6 ft (1.83 m) wide, containing overcurrent devices, switching devices, or control devices, there shall be one entrance not less than 24 in (610 mm) wide and 6½ ft (1.98 m) high at each end.

*Exception No. 1. Where the equipment location permits a continuous and unobstructed way of exit travel.*

*Exception No. 2. Where the work space required by Sec. 110-16(a) of the NEC is doubled, only one entrance to the working space is required.*

*Working space with one entrance provided shall be so located that the edge of the entrance nearest the equipment is the minimum clear distance given in Table 312A away from such equipment.*

4. FRONT WORKING SPACE  In all cases when there are live parts normally exposed on the front of switchboards of motor-control centers, the working space in front of such equipment shall not be less than 3 ft (914 mm).

5. ILLUMINATION  Illumination shall be provided for all working spaces about service equipment, switchboards, panelboards, or motor control centers installed indoors.

*Exception. Service equipment or panelboards in dwelling units that do not exceed 200 A.*

6. HEADROOM  The minimum headroom of working spaces about service equipment, switchboards, panelboards, or motor control centers shall be 6¼ ft (1.91 m).

*Exception. Service equipment or panelboards in dwelling units that do not exceed 200 A.*

**313. All cables and raceways** must be continuous from outlet to outlet and from fitting to fitting and must be securely fastened in place. Fittings must be mechanically secured to conduits, raceways, and boxes; and the entire metallic enclosure system must form an electrically continuous conductor. If different portions of a conduit or raceway system are exposed to widely different temperatures, as in refrigerating or cold-storage plants, provision must be made to prevent circulation of air through the raceway from a warmer to a colder section.

**314. Protection against corrosion.** Metal raceways, cable armor, boxes, cable sheathing, cabinets, elbows, couplings, fittings, supports, and support hardware shall be of materials suitable for the environment in which they are to be installed.

1. Ferrous raceways, cable armor, boxes, cable sheathing, cabinets, metal elbows, couplings, fittings, supports, and support hardware shall be suitably protected against corrosion inside and outside (except threads at joints) by a coating of approved corrosion-resistant material, such as zinc, cadmium, or enamel. If protected from corrosion solely by enamel, they shall not be used outdoors or in wet locations as described in Par. 3. When boxes or cabinets have an approved system of organic coatings and are marked "Raintight," "Rainproof," or "Outdoor type," they may be used outdoors.

2. Unless made of materials judged suitable for the condition or unless corrosion protection approved for the condition is provided, ferrous or nonferrous metal raceways, cable armor, boxes, cable sheathing, cabinets, elbows, couplings, fittings, supports, and support hardware shall not be installed in concrete or in direct contact with the earth or in areas subject to severe corrosive influences.

3. In portions of dairies, laundries, canneries, and other indoor wet locations, and in locations where walls are frequently washed or where there are surfaces of absorbent materials, such as damp paper or wood, the entire wiring system, including all boxes, fittings, conduits, and cable used therewith, shall be mounted so that there is at least a ¼-in (6.35-mm) air space between it and the wall or supporting surface.

Meat-packing plants, tanneries, hide cellars, casing rooms, glue houses, fertilizer rooms, salt storage, some chemical works, metal refineries, pulp mills, sugar mills, roundhouses, some stables, and similar locations are judged to be occupancies in which severe corrosive conditions are likely to be present.

**315. Outlet boxes** (see Div. 4) must be provided at each outlet, switch, or junction point for all wiring systems except open wiring. The boxes must be made of metallic material, except that for open wiring on insulators, concealed knob-and-tube wiring, nonmetallic-sheathed cable, or nonmetallic raceways the boxes may be of nonmetallic construction. Whenever a change is made from open or concealed knob-and-tube wiring to one of the other types of wiring, a box or terminal fitting which has a separately bushed hole for each conductor must be used. In terminating conduit wiring at switchboards and control-apparatus locations an insulating bushing instead of a box may be used, and the conductors bunched, taped, and painted with insulating paint. If conductors are lead-covered, the bushing need not be insulated.

Care must be exercised in installing boxes and fittings in damp or wet locations so that they are either placed or equipped to prevent moisture or water from entering and accumulating within the box or fitting. In wet locations the boxes or fittings must be of the weatherproof type. The NEC recommends that boxes of nonconductive material be used with nonmetallic-sheathed cable when such cable is used in locations where occasional moisture is likely to be present, as in dairy barns.

Round outlet boxes are not allowed when conduits or connectors requiring the use of locknuts or bushings are to be connected to the side of the box.

**316. Supports.** Boxes shall be securely and rigidly fastened to the surface upon which they are mounted or securely and rigidly embedded in concrete or masonry. Except as otherwise provided in this section, boxes shall be supported from a structural member of the building either directly or by using a substantial and approved metal or wooden brace. If of wood, the brace shall not be less than nominal 1-in (25.4-mm) by 2-in (50.8 mm) thickness. If of metal, it shall be corrosion-resistant and shall be not less than 0.020 in (508 $\mu$m) uncoated.

When mounted in new walls in which no structural members are provided or in existing walls in previously occupied buildings, boxes shall be affixed with approved anchors or clamps so as to provide a rigid and secure installation.

Enclosures less than 100 in$^3$ in size which do not contain devices, receptacles, or switches and do not support fixtures may be considered adequately supported if two or more conduits are threaded into the box wrenchtight and are supported within 3 ft (914 mm) of the box on two or more sides; or device boxes or conduit bodies may be supported by the connecting conduit when at least two connecting conduits are threaded into the box and rigidly supported within 18 in (457 mm) of the junction box.

**317. Ceiling outlet boxes may be supported** with wood screws projecting through the holes in the bottom of the box, in frame construction when a wood background is available. If boxes must be supported between wooden beams, bar hangers as shown in Sec. 119 of Div. 4 are used. The hangers are fastened with nails or wood screws to the beams. The box is fastened to the hanger by means of the fixture stud or with two bolts. With concealed conduit wiring in reinforced-concrete ceilings the conduits and the edges of the backing plate of the concrete boxes, bearing against the concrete, support the box. With terra-cotta ceilings a tee made of ⅜-in pipe substantially bearing on the top of the terra-cotta should be used (Fig. 9-235). The pipe projects through the fixture-stud knockout in the back of the box and forms the fixture stud. For exposed wiring on concrete ceilings, screws and lead expansion sleeves should be used. For exposed wiring on steelwork the box should be bolted or clamped to the steel. For exposed wiring on plastered ceilings the box may be fastened on the surface with toggle bolts.

**318. Wall outlet boxes** in frame construction are fastened to the studs with side hangers which form part of the box (see Sec. 117, Div. 4). With concealed wiring in concrete or brick walls the masonry, being built tight around the box, supports it. The conduit prevents the box from coming loose from the wall. If a pocket has been cut in a brick wall to receive an outlet box and the wall has been channeled for the conduit, the box should be fastened to the wall by two expansion sleeves. With concealed wiring in concrete construction, for the support of a fixture which is too heavy to be safely hung on an ordinary ⅜-in fixture stud, the outlet box should be supported by a pipe support similar to that of Fig. 9-235, embedded in the concrete. For exposed wiring on concrete and brick walls the box may be fastened with wood screws in lead expansion sleeves.

**319. Lighting fixtures are usually supported** from the fixture stud in the outlet box (Fig. 9-236). A hickey is screwed onto the fixture stud, and the fixture stem is screwed into the hickey. The fixture wires are brought out of the fixture stem through the holes in the hickey and spliced to the circuit wires. The canopy is then slipped up against the ceiling and fastened to the fixture stem with a setscrew. In many cases it is preferable to use outlet boxes which have the fixture stud made integral with the back of the box. In case that type of box is not available, a separate fixture stud (Fig. 9-237) may be installed. The no-bolt type (Fig. 9-237, I) is used for supporting the box from a bar hanger. The body of the stud is on the back of the hanger with the threaded portion projecting through the hanger and through a hole in the center of the box. The nut is screwed on the stud inside the box and holds the box to the hanger. The bolted type (Fig. 9-237, II) either is bolted to the back of the box or, when the box is supported from a wooden member, is held in place with wood screws. The NEC requires that a fixture which weighs more than 6 lb (2.72 kg) or exceeds 16 in (406 mm) in any dimension must not be supported by the screw shell of the lampholder. Small lighting fixtures consisting of a lampholder and a small shade may be supported from a pendant cord. In this case a box cover with a round hole in the center is used. A knot is tied in the cord inside the box (Fig. 9-234) so that the knot will bear against the cover and support the weight of the fixture. With exposed

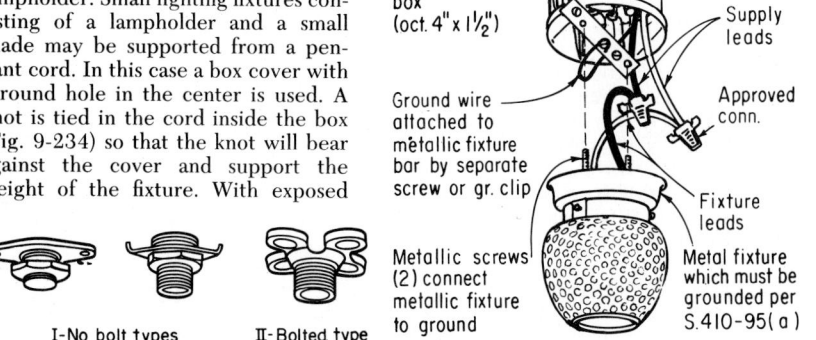

Metallic fixture bar

Nonmetallic box (oct. 4" x 1½")

Ground wire attached to metallic fixture bar by separate screw or gr. clip

Metallic screws (2) connect metallic fixture to ground

NM cable 12/2 w/No.12 gr. wire

Supply leads

Approved conn.

Fixture leads

Metal fixture which must be grounded per S.410-95(a)

**Fig. 9-238** Method of securing and connecting a lightweight fixture to a nonmetallic box.

I-No bolt types    II-Bolted type

**Fig. 9-237** Fixture studs.

conduit wiring, elbow or tee conduit fittings are frequently used. The fixture is supported with a short length of conduit which threads into the fixture and into the hub which projects down from the fitting. Sometimes the upper part of the fixture is designed to be fastened as the cover on a 3¼- or 4-in outlet box. A fixture which weighs more than 50 lb (22.7 kg) must be supported independently of the outlet box. Figure 9-238 shows a typical method of attaching a lightweight fixture to a nonmetallic box.

**320. Number of conductors in switch, outlet, receptacle, device, and junction boxes** (National Electric Code). Boxes shall be of sufficient size to provide free space for all conductors enclosed in the box. The provisions of this section shall not apply to terminal housings supplied with motors. Boxes and conduit bodies containing conductors of size No. 4 or larger shall also comply with the provisions of Sec. 370-18 of the NEC.

1. STANDARD BOXES  The maximum number of conductors permitted in standard boxes shall be as is listed in Table 370-6(a) of the NEC. See Sec. 370-18 of the NEC if boxes or conduit bodies are used as junction or pull boxes.

a.  Table 370-6(a) of the NEC shall apply where no fittings or devices, such as fixture studs, cable clamps, hickeys, switches, or receptacles, are contained in the box and where no grounding conductors are part of the wiring within the box. Where one or more of these types of devices such as fixture studs, cable clamps, or hickeys are contained in the box, the number of conductors shall be one less than shown in the tables for each type of device; an additional deduction of two conductors shall be made for each strap containing one or more devices; and a further deduction of one conductor shall be made for one or more grounding conductors entering the box. A conductor running through the box shall be counted as one conductor, and each conductor originating outside the box and terminating inside the box is counted as one conductor. Conductors no part of which leaves the box shall not be counted. The volume of a wiring enclosure (box) shall be the total volume of the assembled sections and, where used, the space provided by plaster rings, domed covers, extension rings, etc., that are marked with their volume in cubic inches or are made from boxes listed in Table 370-6(a) of the NEC.

b.  For combinations of conductor sizes shown in Table 370-6(a) of the NEC the volume per conductor listed in Table 370-6(b) of the NEC shall apply. The maximum number and size of conductors listed in Table 370-6(a) of the NEC shall not be exceeded.

2. OTHER BOXES  Boxes 100 in³ (1640 cm³) or less other than those described in Table 370-6(a) of the NEC, conduit bodies having provision for more than two conduit entries, and nonmetallic boxes shall be durably and legibly marked by the manufacturer with their cubic-inch capacity, and the maximum number of conductors permitted shall be computed by using the volume per conductor listed in Table 370-6(b) of the NEC and the deductions provided for in Par. 1, Subsec. a. Boxes described in Table 370-6(a) of the NEC that have a larger cubic-inch capacity than is designated in the table shall be permitted to have their cubic-inch capacity marked as required by this section, and the maximum number of conductors permitted shall be computed by using the volume per conductor listed in Table 370-6(b) of the NEC.

Where No. 6 conductors are installed, the minimum wire-bending space required in Table 373-6(a) of the NEC shall be provided.

3. CONDUIT BODIES  Conduit bodies enclosing No. 6 conductors or smaller shall have a cross-sectional area not less than twice the cross-sectional area of the largest conduit to which it is attached. The maximum number of conductors permitted shall be the maximum number permitted by Table 29, Div. 11, for the conduit to which it is attached.

Conduit bodies having provisions for fewer than three conduit entries shall not contain splices, taps, or devices unless they comply with the provisions of Par. 2 and are supported in a rigid and secure manner.

**321. Conductors entering boxes or fittings** (National Electrical Code).  Conductors entering boxes or fittings shall be protected from abrasion and conform to the following:

1. *Openings to be closed.* Openings through which conductors enter shall be adequately closed.

2. *Metal boxes, conduit bodies, and fittings.* If metal outlet boxes, conduit bodies, or fittings are used with open wiring or concealed knob-and-tube work, conductors shall enter through insulating bushings or, in dry places, through flexible tubing extending from the last insulating support and firmly secured to the box, conduit body, or fitting. Where raceway or cable is used with metal outlet boxes, conduit bodies, or fittings, the raceway or cable shall be secured to such boxes, conduit bodies, and fittings. Fittings such as capped elbows and service-entrance elbows shall not contain splices, taps, or devices

and shall be of sufficient size to provide free space for all conductors enclosed in the fitting.

NONMETALLIC BOXES   Nonmetallic boxes shall be suitable for the lowest-temperature–rated conductor entering the box. Where nonmetallic boxes are used with open wiring or concealed knob-and-tube wiring, the conductors shall enter the box through individual holes. When flexible tubing is used to encase the conductors, the tubing shall extend from the last insulating support to no less than ¼ in (6.35 mm) inside the box. Where nonmetallic-sheathed cable is used, the cable assembly, including the sheath, shall extend into the box no less than ¼ in through a nonmetallic-sheathed cable knockout opening. Where nonmetallic-sheathed cable is used with 2¼- by 4-in boxes mounted in walls and the cable is fastened within 8 in (203.2 mm) of the box measured along the sheath and when the sheath extends into the box no less than ¼ in, securing the cable to the box shall not be required. In all instances all permitted wiring methods shall be secured to nonmetallic boxes.

**322. Location.**   Conduit bodies and junction, pull, and outlet boxes shall be so installed that the wiring contained in them can be rendered accessible without removing any part of the building or, in underground circuits, without excavating sidewalks, paving earth, or other substance that is to be used to establish the finished grade.

*Exception. Listed boxes shall be permitted where covered by gravel, light aggregate, or noncohesive granulated soil if their location is effectively identified and accessible for excavation.*

In walls or ceilings of concrete, tile, or other noncombustible material (**National Electrical Code**), boxes and fittings shall be so installed that the front edge of the box or fitting will not set back of the finished surface more than ¼ in (6.35 mm). In walls and ceilings constructed of wood or other combustible material, outlet boxes and fittings shall be flush with the finished surface or project therefrom.

**323. Covers and canopies** (National Electrical Code).   In completed installations each outlet box shall be provided with a cover, faceplate, or fixture canopy.

1. Nonmetallic or metal covers and plates may be used with nonmetallic outlet boxes. Where metal covers or plates are used, they are subject to the grounding requirements of the NEC.

2. If a fixture canopy or pan is used, any combustible wall or ceiling finish exposed between the edge of the canopy or pan and the outlet box shall be covered with noncombustible material.

3. Covers of outlet boxes and conduit bodies having holes through which flexible-cord pendants pass shall be provided with bushings designed for the purpose or shall have smooth, well-rounded surfaces on which the cords may bear. So-called hard-rubber or composition bushings shall not be used.

**324. Pull and junction boxes** (National Electrical Code).   Pull boxes, junction boxes, and conduit bodies shall conform to the following:

1. MINIMUM SIZE   For raceways of ¾-in trade size and larger, containing conductors of No. 4 size or larger, and for cables[1] containing conductors of No. 4 size or larger, the minimum dimensions of a pull or junction box installed in a raceway or cable run shall conform to the following:

*a. Straight pulls.*   In straight pulls the length of the box shall be not less than 8 times the trade diameter of the largest raceway.

*b. Angle or U pulls.*   Where angle or U pulls are made, the distance between each raceway entry inside the box and the opposite wall of the box shall not be less than 6 times the trade diameter of the largest raceway in a row. This distance shall be increased for additional entries by the amount of the sum of the diameters of all other raceway entries in the same row on the same wall of the box. Each row shall be calculated individually, and the single row that provides the maximum distance shall be used. The distance between raceway entries enclosing the same conductor shall not be less than 6 times the trade diameter of the larger raceway.

*Exception. Where a raceway or cable entry is in the wall of a box or conduit body opposite a removable cover and where the distance from that wall to the cover is in conformance with the column for one wire per terminal in Table 373-6(a) at the NEC.*

*c.* Boxes of lesser dimensions than those required in Subpars. *a* and *b* of this section

---

[1] When transposing cable size into raceway size in Subpars. *a* and *b*, the minimum trade-size raceway required for the number and size of conductors in the cable shall be used.

may be used for installations of combinations of conductors that are less than the maximum conduit fill (of conduits being used) provided that the box has been approved for and is permanently marked with the maximum number of conductors and the maximum AWG size permitted.

*Exception. Terminal housings supplied with motors which shall comply with the provisions of Sec. 430-12 of the* Code.

2. CONDUCTORS IN PULL OR JUNCTION BOXES  In pull boxes or junction boxes having any dimension over 6 ft (1.83 m), all conductors shall be cabled or racked up in an approved manner.

3. COVERS  All pull boxes, junction boxes, conduit bodies, and fittings shall be provided with covers suitable for the enclosure. Where metal covers are used, they shall comply with the grounding requirements of Sec. 250-42 of the NEC.

4. PERMANENT BARRIERS  Where permanent barriers are installed in a box, each section shall be considered as a separate box.

**325. Cabinets and cutout boxes.**  The requirements of the NEC with respect to cabinets and cutout boxes are as follows:

1. DAMP OR WET LOCATIONS  In damp or wet locations, cabinets and cutout boxes of the surface type shall be so placed or equipped as to prevent moisture or water from entering and accumulating within the cabinet or cutout box and shall be mounted so there is at least a ¼-in (6.35-mm) air space between the enclosure and the wall or other supporting surface. Cabinets or cutout boxes installed in wet locations shall be weatherproof.

2. POSITION IN WALL  In walls of concrete, tile, or other noncombustible material, cabinets shall be so installed that the front edge of the cabinet will not set back of the finished surface more than ¼ in (6.35 mm). In walls constructed of wood or other combustible material, cabinets shall be flush with the finished surface or project therefrom.

3. UNUSED OPENINGS  Unused openings in cabinet or cutout boxes shall be effectively closed to afford protection substantially equivalent to that of the wall of the cabinet or cutout box. If metal plugs or plates are used with nonmetallic cabinets or cutout boxes, they shall be recessed at least ¼ in (6.35 mm) from the outer surface.

4. CONDUCTORS ENTERING CABINETS OR CUTOUT BOXES  Conductors entering cabinets or cutout boxes shall be protected from abrasion and shall conform to the following:

*a. Openings to be closed.* Openings through which conductors enter shall be adequately closed.

*b. Metal cabinets and cutout boxes.* If metal cabinets or cutout boxes are used with open wiring or concealed knob-and-tube wiring, conductors shall enter through insulating bushings or, in dry places, through flexible tubing extending from the last insulating support and firmly secured to the cabinet or cutout box. Where cable is used, each cable shall be secured to the cabinet or cutout box.

5. DEFLECTION OF CONDUCTORS  Conductors entering or leaving cabinets or cutout boxes and the like shall conform to the following:

*a. Width of gutters.* Conductors in parallel shall be judged on the basis of the number of conductors in parallel. Conductors shall not be deflected within a cabinet unless a gutter having a width in accordance with Table 373-6(a) of the NEC is provided.

*b. Conductors entering or leaving the enclosure through the wall opposite the terminal.* They shall have additional wire-bending space specified by Table 373-6(b) of the NEC.

*c. Insulated fittings.* Where ungrounded conductors of No. 4 or larger size enter a raceway in a cabinet, pull box, junction box, or auxiliary gutter, they shall be protected by a substantial bushing providing a smoothly rounded insulating surface unless the conductors are separated from the raceway fitting by substantial insulating material securely fastened in place. If conduit bushings are constructed wholly of insulating material, a locknut shall be provided both inside and outside the enclosure to which the conduit is attached. The insulating bushing or insulating material shall have a temperature rating not less than the insulation temperature rating of the conductor.

*Exception. Where threaded hubs or bosses that are an integral part of an enclosure provide a smoothly rounded or flared entry for conductors.*

6. SPACE IN ENCLOSURES  Cabinets and cutout boxes shall conform to the following:

*a. To accommodate conductors.* Cabinets and cutout boxes which have sufficient space to accommodate all conductors installed in them without crowding shall be selected.

*b. Used as junction boxes.*   Switch or overcurrent enclosures shall not be used as junction boxes, auxiliary gutters, or raceways for conductors feeding through or tapping off to other switches or overcurrent devices unless special designs are employed to provide adequate space for this purpose.

7. SIDE- OR BACK-WIRING SPACES OR GUTTERS   Cabinets and cutout boxes shall be provided with back-wiring spaces, gutters, or wiring compartments as required by Pars. *c* and *d* of Sec. 373-11 of the **Code**.

**326. Auxiliary gutters** (National Electrical Code).   Auxiliary gutters, used to supplement wiring spaces at meter centers, distribution centers, switchboards, and similar points of wiring systems, may enclose conductors or busbars but shall not be used to enclose switches, overcurrent devices, appliances, or other similar equipment.

1. EXTENSION BEYOND EQUIPMENT   An auxiliary gutter shall not extend a greater distance than 30 ft (9.14 m) beyond the equipment which it supplements, except in elevator work. Any further extension shall comply with the rules for wireways or busways.

2. SUPPORTS   Gutters shall be supported throughout their length at intervals not exceeding 5 ft (1.52 m).

3. COVERS   Covers shall be securely fastened to the gutter.

4. NUMBER OF CONDUCTORS   Auxiliary gutters shall not contain more than 30 conductors at any cross section unless the conductors are for elevators or signaling circuits or are controller conductors between a motor and its starter and are used only for starting duty and the conductors are derated in accordance with Note 8 to Tables 310-16 through 310-19 of the **NEC**. The sum of the cross-sectional areas of all contained conductors at any cross section of an auxiliary gutter shall not exceed 20 percent of the interior cross-sectional area of the gutter.

5. AMPACITY OF COPPER BARS   The current carried continuously in bare conductors in auxiliary gutters shall not exceed 1000 A/in$^2$ of cross section of the conductor. For bare aluminum bars, the current carried continuously is limited to 700A/in$^2$.

6. CLEARANCE OF BARE LIVE PARTS   Bare conductors shall be securely and rigidly supported so that the minimum clearance between bare current-carrying metal parts of opposite polarities mounted on the same surface shall be not less than 2 in (50.8 mm) or less than 1 in (25.4 mm) for parts that are held free in the air. A spacing not less than 1 in shall be secured between bare current-carrying metal parts and any metal surface. Adequate provision shall be made for expansion of busbars.

7. SPLICES AND TAPS   Splices and taps shall conform to the following:

*a.* Splices or taps, made and insulated by approved methods, may be located within gutters if they are accessible by means of removable covers or doors. The conductors, including splices and taps, shall not fill the gutter to more than 75 percent of its area.

*b.* Taps from bare conductors shall leave the gutter opposite their terminal connections, and conductors shall not be brought in contact with uninsulated current-carrying parts of opposite polarity.

*c.* All taps shall be suitably identified at the gutter as to the circuit or equipment which they supply.

*d.* Tap connections from conductors in auxiliary gutters shall be provided with overcurrent protection in conformity with the provisions of Secs. 294 and 295.

8. CONSTRUCTION AND INSTALLATION   Auxiliary gutters shall comply with the following:

*a.* Gutters shall be so constructed and installed that adequate electrical and mechanical continuity of the complete raceway system will be secured.

*b.* Gutters shall be of substantial construction and shall provide a complete enclosure for the contained conductors. All surfaces, both interior and exterior, shall be suitably protected from corrosion. Corner joints shall be made tight, and if the assembly is held together by rivets or bolts, these shall be spaced not more than 12 in (305 mm) apart.

*c.* Suitable bushings, shields, or fittings having smooth, rounded edges shall be provided when conductors pass between gutters, through partitions, around bends, between gutters and cabinets or junction boxes, and at other locations where necessary to prevent abrasion of the insulation of the conductors.

*d.* Where insulated conductors are deflected within the auxiliary gutter, either at the ends or where conduits, fittings, or other raceways or cables enter or leave the gutter, or where the direction of the gutter is deflected more than 30°, dimensions correspond to those required for cabinets and cutout boxes as given in Sec. 373-6 of the **Code**.

*e.* Auxiliary gutters installed in wet locations shall be of raintight construction.

GROUNDING

### 327. There are two types of grounds in interior-wiring systems:

1. System grounds (grounding electrode conductors)
2. Equipment and conductor-enclosure grounds

A system ground refers to the condition of having one wire of a circuit connected to ground. The reason for doing this is explained in Secs. 136 and 137, Div. 8.

An equipment or conductor-enclosure ground refers to connecting the non-current-carrying metal parts of the wiring system or equipment to ground. This is done so that the metal parts with which a person might come in contact are always at or near ground potential. With this condition there is less danger that a person touching the equipment or conductor enclosure will receive a shock. Also metal conduit, raceways, and boxes may be in contact with metal parts of the building at several points. If an accidental contact occurs between an ungrounded conductor and its metal enclosure, a current may flow to ground through a stray path made up of sections of metal lath, metal partitions, piping, or other similar conductors. If the equipment is grounded, the resistance of the path through the grounding conductor will usually be much less than the resistance through the stray path, and not much current will flow through the stray path. Sufficient current will usually flow through the grounded path to blow the circuit fuse or trip the circuit breaker and thus open the circuit until repairs can be made. On the other hand, if the equipment is not grounded, sufficient current may flow through the stray path to heat up some section to a sufficient temperature to ignite wood or other flammable material with which it is in contact. The current may not be sufficient to operate the fuse or circuit breaker, expecially if the contact occurs on large-size feeders.

### 328. The following items of equipment and conductor enclosures must be grounded:

1. METAL CONDUCTOR ENCLOSURES    All these enclosures, except in runs of less than 25 ft (7.62 m), which are free from probable contact with ground, grounded metal, metal lath, or conductive thermal insulation and which, if within reach from grounded surfaces, are guarded against contact by persons.

2. FIXED EQUIPMENT: GENERAL    Under any of the following conditions, exposed non-current-carrying metal parts of fixed equipment, which are likely to become energized, shall be grounded.

a. Where equipment is supplied by means of metallic wiring methods.

b. Where equipment is located in a wet location and is not isolated.

c. Where equipment is located within 8 ft (2.44 m) vertically and 5 ft (1.52 m) horizontally of grounded objects and ground and within reach of a person who can make contact with any grounded surface or object.

d. Where equipment is in a hazardous location.

e. Where equipment is in electrical contact with metal.

f. Where equipment operates with any terminal at more than 150 V to ground, except as follows:

(1) Enclosures for switches or circuit breakers used for other than service equipment and accessible to qualified persons only

(2) Metal frames of electrically heated devices, exempted by special permission, in which case these metal frames shall be permanently and effectively insulated from ground

(3) Transformers and capacitor cases mounted on wooden poles at a height of more than 8 ft (2.44 m) from the ground

(4) Listed equipment protected by a system of double insulation, or its equivalent, is not required to be grounded. Where such a system is used, the equipment shall be distinctively marked.

3. FIXED EQUIPMENT: SPECIFIC    Exposed non-current-carrying metal parts of the following kinds of equipment, regardless of voltage, shall be grounded:

a. Frames of motors as specified in Sec. 430-142 of the Code

b. Controller cases for motors, except lined covers of snap switches and enclosures attached to ungrounded portable equipment

c. Electric equipment of elevators and cranes

d. Electric equipment in garages, theaters, and motion-picture studios, except pendant lampholders on circuits of not more than 150 V to ground

e. Motion-picture projection equipment

f. Electric signs and associated equipment

*g.* Generator and motor frames in an electrically operated pipe organ

*h.* Switchboard frames and structures supporting switching equipment, except that frames of 2-wire dc switchboards need not be grounded where effectively insulated

*i.* Equipment supplied by Class 1, Class 2, and Class 3 remote-control and signaling circuits and by fire-protective signaling circuits when Part B of Article 250 of the Code requires system grounding of those circuits

*j.* Lighting fixtures as provided in Part E of Article 410 of the NEC

*k.* Motor-operated water pumps including the submersible type

4. NONELECTRICAL EQUIPMENT   The following metal parts shall be grounded:

*a.* Frames and tracks of electrically operated cranes

*b.* The metal frame of a nonelectrically driven elevator car to which electric conductors are attached

*c.* Hand-operated metal shifting ropes or cables of electric elevators

*d.* Metal enclosures such as partitions and grillwork around equipment carrying voltages in excess of 1 kV between conductors, unless in substations or vaults under the sole control of the supply company

*e.* Mobile homes and recreational vehicles as required to be grounded by Articles 550 and 551 of the NEC

If extensive metal in or on buildings may become energized and is subject to personal contact, adequate bonding and grounding will provide additional safety.

5. EQUIPMENT CONNECTED BY CORD AND PLUG   Under any of the following conditions, exposed non-current-carrying metal parts of cord- and plug-connected equipment which are likely to become energized shall be grounded:

*a.* In hazardous locations (see Articles 500 to 517 of the Code).

*b.* Where operated at more than 150 V to ground, except:

(1) Motors, where guarded

(2) Metal frames of electrically heated appliances exempted by special permission, in which case the frames shall be insulated from ground

(3) Listed equipment protected by a system of double insulation or its equivalent, shall not be required to be grounded.

*c.* In residential occupancies, (1) refrigerators, freezers, and air conditioners; (2) clothes-washing, clothes-drying, and dish-washing machines, sump pumps, and electrical aquarium equipment; and (3) hand-held, motor-operated tools and motor-operated appliances of the following types: hedge clippers, lawn mowers, snowblowers, and wet scrubbers; and (4) portable hand lamps.

*Exception. Listed tools and listed appliances protected by a system of double insulation or its equivalent need not be grounded. Where such a system is employed, the equipment shall be distinctively marked.*

Portable tools or appliances are not intended to be used in damp, wet, or conductive locations unless they are grounded, double-insulated, or supplied by an isolation transformer.

*d.* In other than residential occupancies, (1) refrigerators, freezers, and air conditioners; (2) clothes-washing, clothes-drying, and dish-washing machines, electronic computer/data processing equipment, sump pumps, and electrical aquarium equipment; (3) hand-held, motor-operated tools and motor-operated appliances of the following types: hedge clippers, lawn mowers, snowblowers, and wet scrubbers; (4) cord- and plug-connected appliances used in damp or wet locations or by persons standing on the ground or on metal floors or working inside metal tanks or boilers; (5) tools that are likely to be used in wet and conductive locations; and (6) portable hand lamps.

*Exception No. 1. Tools and portable hand lamps that are likely to be used in wet and conductive locations need not be grounded when supplied through an isolating transformer with an ungrounded secondary of not over 50 V.*

*Exception No. 2. Listed portable tools and listed appliances protected by an approved system of double insulation or its equivalent need not be grounded. Where such a system is employed, the equipment shall be distinctively marked.*

**329. The types of wiring systems required to have a system ground** are as follows:

1. Two-wire dc systems of not more than 300 V supplying premises wiring unless the system supplies industrial equipment in limited areas and is equipped with a ground detector as described in Div. 1. A system operating at 50 V or less between conductors need not be grounded.

Rectifier-derived dc systems supplied by an ac system complying with Sec. 250-5 of the Code and dc fire-protective signaling circuits having a maximum current of 0.030 A need not be grounded.

2. Three-wire dc systems supplying premises wiring.

3. AC systems, provided the grounded conductor is so arranged that the maximum voltage to ground is between 50 and 150 V. Grounding is also recommended when the voltage to ground will be between 150 and 300 V and is allowable in some cases on higher voltages. Any uninsulated service conductor must be grounded.

4. AC systems of less than 50 V if supplied by transformers from systems of more than 150 V to ground or from ungrounded systems or if run overhead outside buildings.

5. AC systems of 50 to 1000 V supplying premises wiring of the following types are not required to be grounded:

*a.* Electric systems used exclusively to supply industrial electric furnaces for melting, refining, tempering, and the like.

*b.* Separately derived systems used exclusively for rectifiers supplying only adjustable-speed industrial drives.

*c.* Separately derived systems supplied by transformers that have a primary voltage rating less than 1000 V, provided that all the following conditions are met:

(1) The system is used exclusively for control circuits.

(2) The conditions of maintenance and supervision assure that only qualified persons will service the installation.

(3) Continuity of control power is required.

(4) Ground detectors are installed on the control system.

*d.* Isolated systems as permitted in Articles 517 and 668 of the NEC.

*e.* High-impedance grounded neutral systems in which a grounding impedance, usually a resistor, limits the ground-fault current to a low value. High-impedance grounded neutral systems are permitted for three-phase ac systems of 480 to 1000 V where all of the following conditions are met:

(1) The conditions of maintenance and supervision assure that only qualified persons will service the installation.

(2) Continuity of power is required.

(3) Ground detectors are installed on the system.

(4) Line-to-neutral loads are not served.

**330. Circuits which must not be grounded.** Circuits for electric cranes operating over combustible fibers in Class III hazardous locations (refer to Sec. 383 for description of hazardous locations) and isolated circuits in health care facilities.

**331. Location of the system ground** (Fig. 9-239)

1. For dc systems the ground must be at the supply station and nowhere else.

2. For ac systems there must be a ground at the secondary of the transformer supplying the system. There must be another ground at each building service to the grounded service conductor at an accessible point from the point of connection of the service drop or service lateral conductor to the service disconnecting means.

3. For two or more buildings served by a master service the ground must be located at each building if it utilizes two or more branch circuits, if the building houses livestock, or if equipment is required to be grounded, except that livestock buildings may utilize an equipment-grounding conductor run with the circuit.

4. For a separately derived system which is not connected to an exterior distribution

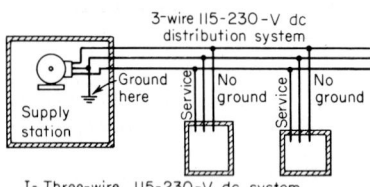

I- Three-wire  115-230-V dc system

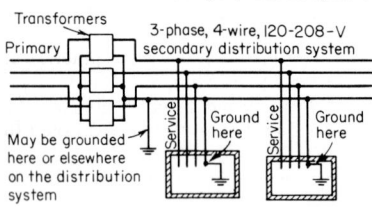

II-Three-wire 115-230-V single-phase ac system.

III-Four-wire 120-208-V three-phase ac system.

**Fig. 9-239**  Location of system grounds.

system the common main grounding conductor shall be made at the transformer, generator, or other source of supply or at the switchboard on the supply side of the first switch controlling the system.

**332. The conductor to be grounded.** For alternating-current premises-wiring systems the conductor to be grounded shall be as follows:

1. Single-phase two-wire: the grounded-circuit conductor.

2. Single-phase three-wire: the grounded-circuit (neutral) conductor.

3. Multiphase systems having one wire common to all phases: the grounded-circuit common conductor.

4. Multiphase systems having one phase grounded: the grounded-circuit conductor.

5. Multiphase systems in which one phase is used as in Par. 2: the grounded-circuit (neutral) conductor. One phase only can be grounded.

The grounded-circuit conductor is commonly known as the *white wire.*

**333. The non-current-carrying metallic parts of fixed equipment are considered to be satisfactorily grounded** under any one of the following conditions (fixed equipment includes boxes, cabinets, and fittings):

1. They are metallically connected to grounded cable armor or the grounded metal enclosure of conductors. This is the common and preferable method.

2. By a grounding conductor run with circuit conductors, this conductor may be uninsulated, but if an individual covering is provided for this conductor, it shall be finished to show a green color.

3. DC electric equipment secured to and in contact with the grounded structural metal frame of a building shall be deemed to be grounded.

4. Metal car frames supported by metal hoisting cables attached to or running over metal sheaves or drums of elevator machines shall be deemed to be grounded if the machine is grounded in accordance with the Code.

**334. Portable and/or cord- and plug-connected equipment.** The non-current-carrying metal parts of cord- and plug-connected equipment required to be grounded may be grounded in any one of the following ways:

1. By means of the metal enclosure of the conductors feeding such equipment, provided an approved grounding-type attachment plug is used, one fixed contacting member being for the purpose of grounding the metal enclosure, and provided, further, that the metal enclosure of the conductors is attached to the attachment plug and to the equipment by approved connectors.

*Exception. The grounding contacting member of grounding-type attachment plugs on the power supply cord of portable hand-held, hand-guided, or hand-supported tools or appliances may be of the movable self-restoring type.*

2. By means of a grounding conductor run with the power supply conductors in a cable assembly or flexible cord that is properly terminated in a grounding-type attachment plug having a fixed grounding contacting member. The grounding conductor in a cable assembly may be uninsulated, but if an individual covering is provided for such conductors, it shall be finished in a continuous green color or a continuous green color with one or more yellow stripes.

*Exception. The grounding contacting member of grounding-type attachment plugs on the power supply cord of portable tools or portable appliances may be of the movable self-restoring type.*

3. A separate flexible wire or strap, insulated or bare, protected as well as practicable against physical damage may be used where part of the equipment.

**335. Bonding at service equipment.** The electrical continuity of the grounding circuit for the following equipment and enclosures shall be assured by one of the means given in Sec. 336.

1. The service raceways, cable trays, or service cable armor or sheath, except as provided in Sec. 250-55 of the Code

2. All service equipment enclosures containing service-entrance conductors, including meter fittings, boxes, or the like, interposed in the service raceways or armor

3. Any metallic raceway or armor enclosing a grounding electrode conductor

**336. Continuity at service equipment.** Electrical continuity at service equipment shall be assured by one of the following means:

1. Bonding equipment to the grounded service conductor in a manner provided in Sec. 250-113 of the Code.

2. Threaded couplings and threaded bosses on enclosures with joints made up wrenchtight when rigid metal conduit or intermediate metal conduit is involved.

3. Threadless couplings and connector made up tight for rigid and intermediate metal conduit and electrical metallic tubing. Standard locknuts or bushings shall not be used for the bonding requirements of this section.

4. Bonding jumpers meeting the other requirements of this section. Bonding jumpers shall be used around concentric or eccentric knockouts which are punched or otherwise formed so as to impair the electric connection to ground.

5. Other approved devices such as bonding locknuts and bushings.

**337. The equipment and conductor-enclosure ground** should be located as close to the service entrance or source of supply as practicable. This ground must be so located that no conductor enclosure is grounded through a grounding conductor smaller than required by Table 338A. A common method is to run the equipment ground from the service-switch enclosure. As far as practicable, all other equipment is then connected to ground by electrically continuous conductor enclosures from the service switch. Bonding jumpers or ground wires are used to connect isolated equipment to grounded equipment. Figure 9-240 shows a single common grounding conductor which grounds a grounded system and the metallic equipment of both an ungrounded system and the grounded system.

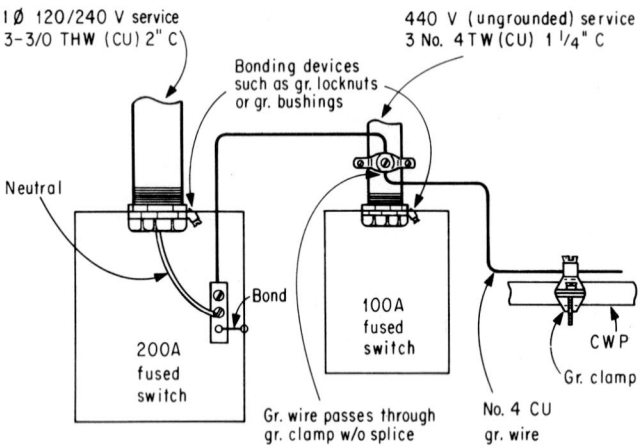

**Fig. 9-240**    A single grounding conductor for grounded and ungrounded systems.

**338. Size of system grounding conductor**

1. For dc systems: not smaller than the largest conductor supplied by the system and in no case smaller than No. 8. An exception to this rule is a system supplied by a balancer set. In such installations, if the grounded circuit conductor is a neutral derived from a balancer winding or a balancer set protected in an approved manner, the size of the grounding electrode conductor shall not be less than that of the neutral conductor but in no case smaller than No. 8.

2. For ac systems and service equipment as follows:

*a.* The size of the grounding electrode conductor for an ac system shall not be less than is given in Table 338A, except that when connected to made electrodes the conductor need not be larger than No. 6 copper wire or No. 4 aluminum.

*b.* If the wiring system is not grounded at the premises, the size of a grounding conductor for a service raceway, for the metal sheath or armor of a service cable, and for service equipment shall be not less than is given in Table 338A, except that when connected to made electrodes the conductor need not be larger than No. 6 copper or No. 4 aluminum.

**Table 338A. Grounding Electrode Conductor for AC Systems**[a]
(Table 250-94, 1990 NEC)

| Size of largest service-entrance conductor or equivalent for parallel conductors, AWG or 1000 cmil | | Size of grounding electrode conductor, AWG or 1000 cmil | |
|---|---|---|---|
| Copper | Aluminum or copper-clad aluminum | Copper | Aluminum or copper-clad aluminum[b] |
| 2 or smaller | 1/0 or smaller | 8 | 6 |
| 1 or 1/0 | 2/0 or 3/0 | 6 | 4 |
| 2/0 or 3/0 | 4/0 or 250 kcmil | 4 | 2 |
| Over 3/0 through 350 kcmil | Over 250 kcmil through 500 kcmil | 2 | 1/0 |
| Over 350 kcmil through 600 kcmil | Over 500 kcmil through 900 kcmil | 1/0 | 3/0 |
| Over 600 kcmil through 1100 kcmil | Over 900 kcmil through 1750 kcmil | 2/0 | 4/0 |
| Over 1100 kcmil | Over 1750 kcmil | 3/0 | 250 kcmil |

[a]Reprinted with permission from NFPA 70-1990 National Electrical Code®, Copyright © 1990, National Fire Protection Association, Quincy, Massachusetts 02269. This reprinted material is not the complete and official position of the NFPA on the referenced subject, which is represented only by the standard in its entirety.

[b]See installation restrictions in Sec. 250-92(a) of the NEC.

NOTE Where there are no service-entrance conductors, the size of the grounding electrode conductor shall be determined by the equivalent size of the largest service-entrance conductor required for the load to be served. Where multiple sets of service-entrance conductors are used as permitted in Sec. 230-40, Exception No. 2, of the NEC, the equivalent size of the largest service-entrance conductor shall be determined by the largest sum of the areas of the corresponding conductors of each set.

**339.  Size of Equipment Grounding Conductors for Grounding Interior Raceway and Equipment**[a]
(Table 250-95, 1990 NEC)

| Rating or setting of automatic overcurrent device in circuit ahead of equipment, conduit, etc., A not exceeding: | Size of wire, AWG or 1000 cmil | |
|---|---|---|
| | Copper | Aluminum or copper-clad aluminum[b] |
| 15 | 14 | 12 |
| 20 | 12 | 10 |
| 30 | 10 | 8 |
| 40 | 10 | 8 |
| 60 | 10 | 8 |
| 100 | 8 | 6 |
| 200 | 6 | 4 |
| 400 | 3 | 1 |
| 600 | 1 | 2/0 |
| 800 | 1/0 | 3/0 |
| 1000 | 2/0 | 4/0 |
| 1200 | 3/0 | 250 kcmil |
| 1600 | 4/0 | 350 kcmil |
| 2000 | 250 kcmil | 400 kcmil |
| 2500 | 350 kcmil | 600 kcmil |
| 3000 | 400 kcmil | 600 kcmil |
| 4000 | 500 kcmil | 800 kcmil |
| 5000 | 700 kcmil | 1200 kcmil |
| 6000 | 800 kcmil | 1200 kcmil |

[a]Reprinted with permission from NFPA 70-1990, National Electrical Code®, Copyright © 1990, National Fire Protection Association. Quincy, Massachusetts 02269. This reprinted material is not the complete and official position of the NFPA on the referenced subject, which is represented only by the standard in its entirety.

[b]See installation restrictions of Sec. 250-92(a) of the NEC.

**340. NEC specifications for the material and installation of grounding conductors** are as follows:

1. MATERIAL

*a. For grounding electrode conductor.*  The grounding conductor of a wiring system shall be of copper or other corrosion-resistant material. The conductor may be solid or stranded, insulated or bare. Except in cases of up to six separate service enclosures utilizing tap conductors, exothermic welding to extend the grounding electrode conductor in commercial and industrial locations, and busbars, the grounding conductor shall be without joint or splice throughout its length.

*b. For conductor enclosures and equipment only.*  The grounding conductor for equipment and for conduit and other metal raceways or enclosures for conductors may be a conductor of copper or other corrosion-resistant material, stranded or solid, insulated or bare, a busbar, or a rigid metal conduit, intermediate metal conduit, electrical metallic tubing or flexible metal conduit and fittings both listed for grounding purposes, the armor of Type AC armored cable, Type MI, or MC cable or the combined metal sheath and grounding conductors of Type MC cable; cable trays as permitted in Secs. 318-3(c) and 318-7 of the NEC, cablebus framework, and other electrically continuous metal raceways listed for grounding. Under conditions favorable to corrosion, a suitable corrosion-resistant material shall be used. All bolted and threaded connections at joints and fittings shall be made tight by the use of suitable tools.

2. INSTALLATION

*a. Grounding electrode conductor.* A grounding electrode conductor, No. 4 or larger, may be attached to the surface on which it is carried without the use of knobs, tubes, or insulators. It need not have protection unless exposed to severe physical damage. A No. 6 grounding conductor, which is free from exposure to physical damage, may be run along the surface of the building construction without metal covering or protection when it is rigidly stapled to the construction; otherwise it shall be in rigid metal conduit, intermediate metal conduit, electrical metallic tubing, or cable armor. Grounding conductors smaller than No. 6 shall be in rigid metal conduit, intermediate metal conduit, electrical metallic tubing, or cable armor. Metal enclosures for grounding conductors shall be electrically continuous from the point of attachment to cabinets or equipment to the grounding electrode and shall be securely fastened to the ground clamp or fitting. Metal enclosures which are not physically continuous from cabinet or equipment to the grounding electrode can be made electrically continuous by bonding each end to the grounding conductor. Where a raceway is used as protection for a grounding conductor, the installation shall comply with the appropriate raceway article of the Code. Aluminum or copper-clad aluminum grounding conductors shall not be used in direct contact with masonry or the earth or when subject to corrosive conditions. If used outside, aluminum or copper-clad aluminum grounding conductors shall not be installed within 18 in (457 mm) of the earth.

*b. Conductor enclosures and equipment only.* A grounding conductor for conductor enclosures and equipment only shall meet the requirements of Par. *a*, except that when smaller than No. 6, as permitted by Sec. 339, it need not be armored or installed in a raceway if run through the hollow spaces of a wall or partition or otherwise run so as not to be subject to mechanical injury.

**341. Cord-connected and pendant equipment** (National Electrical Code).    For grounding listed cord-connected or pendant equipment, the conductors of which are protected by fuses or circuit breakers rated or set at not exceeding 20 A, No. 18 copper wire may be used. Conductors of No. 16 or 18 copper which are used for grounding cord-connected equipment shall be part of a listed flexible-cord assembly. For grounding cord-connected or pendant equipment protected at more than 20 A, the table in Sec. 339 shall be followed.

**342. Grounding electrode system.**    If available on the premises at each building or structure served, each item Nos. 1 through 4 below shall be bonded together to form the grounding electrode system. The bonding jumper shall be sized in accordance with Sec. 250-79(d) of the NEC and shall be connected in the manner specified in Sec. 250-115 of the NEC.

1. A metal underground water pipe in direct contact with the earth for 10 ft (3.05 m) or more (including any metal well casing effectively bonded to the pipe) and electrically continuous (or made electrically continuous by bonding around insulating joints or sections or insulating pipe) to the points of connection of the grounding electrode conductor and the bonding conductors. Continuity of the grounding path or the bonding connection to interior piping shall not rely on water meters. A metal underground water pipe shall be supplemented by an additional electrode of a type specified in Secs. 2 through 9.

2. The metal frame of the building, when effectively grounded.

3. An electrode encased by at least 2 in (50.8 mm) of concrete, located within and near the bottom of a concrete foundation or footing that is in direct contact with the earth, consisting of at least 20 ft (6.1 m) of one or more steel reinforcing bars or rods of not less than ½-in (12.7-mm) diameter or consisting of at least 20 ft of bare copper conductor not smaller than No. 4 AWG.

4. A ground ring encircling the building or structure, in direct contact with the earth at a depth below the earth surface not less than 2½ ft (762 mm), consisting of at least 20 ft (6.1 m) of bare copper conductor not smaller than No. 2 AWG.

5. *Made and other electrodes.* If none of the electrodes specified in Pars. 1 through 4 is available, one or more of the electrodes specified in Pars. 6 through 9 below shall be used. Where practicable, made electrodes shall be embedded below permanent-moisture level. Made electrodes shall be free from nonconductive coatings, such as paint or enamel. If more than one electrode system is used (including systems used for lightning rods), each electrode of one system shall not be less than 6 ft (1.83 m) from any other electrode of another grounding system. Two or more electrodes that are effectively bonded together are to be treated as a single electrode system in this sense.

6. A metal underground gas-piping system shall not be used as a grounding electrode.

7. Other local metal underground systems or structures, such as piping systems and underground tanks.

8. *Rod and pipe electrodes.* Rod and pipe electrodes shall not be less than 8 ft (2.44 m) in length, shall consist of the following materials, and shall be installed in the following manner:

*a.* Electrodes of pipe or conduit shall not be smaller than ¾-in trade size and, when of iron or steel, shall have the outer surface galvanized or otherwise metal-coated for corrosion protection.

*b.* Electrodes of rods of steel or iron shall be at least ⅝ in (15.87 mm) in diameter. Nonferrous or stainless steel rods or their equivalent less than ⅝ in (15.87 mm) in diameter shall be listed and shall be not less than ½ in (12.7 mm) in diameter.

*c.* If rock bottom is not encountered, the electrode shall be driven to a depth of 8 ft (2.44 m). If rock bottom is encountered, the rod may be driven at an oblique angle not exceeding 45° or shall be buried in a trench at least 2½ ft (762 mm) deep. The upper end of the electrode shall be flush with or below ground level unless the above-ground end and the grounding electrode conductor attachment are protected against physical damage as specified in Sec. 250-117 of the NEC.

9. *Plate electrodes.* Each plate electrode shall expose not less than 2 ft$^2$ (0.186 m$^2$) of surface to exterior soil. Electrodes of iron or steel plates shall be at least ¼ in (6.35 mm) in thickness. Electrodes of nonferrous metal shall be at least 0.06 in (1.52 mm) in thickness.

10. *Resistance.* Made electrodes shall, if practicable, have a resistance to ground not to exceed 25 Ω. If the resistance is not as low as 25 Ω, an additional electrode connected in parallel shall be used.

Continuous metallic underground water systems in general have a resistance to ground of less than 3 Ω. Metal frames of buildings and local metallic underground piping systems, metal well casings, and the like have, in general, a resistance substantially below 25 Ω.

**343. Location of connection to grounding electrode** (National Electrical Code)

1. To WATER PIPES  System or common grounding conductors shall be attached to a water-piping system on the street side of the water meter or on a cold-water pipe of adequate ampacity as near as practicable to the water service entrance to the building. If the source of the water supply is a driven well in the basement of the premises, the connection shall be made as near as practicable to the well. The point of attachment shall be accessible. If the point of attachment is not on the street side of the water meter, the water-piping system shall be made electrically continuous by bonding together all parts between the attachment and the street side of the water meter or the pipe entrance which are liable to become disconnected, as at meters, valves, and service unions.

2. To OTHER ELECTRODES  The grounding conductor shall be attached to other electrodes permitted in Sec. 342 at a point which will assure a permanent ground.

**344. Attachment of grounding conductor** (National Electrical Code)

ATTACHMENT TO CIRCUITS AND EQUIPMENT  The equipment-grounding conductor, bond, or bonding jumper shall be attached to circuits, conduits, cabinets, equipment, and the like which are to be grounded by means of exothermic welding, listed pressure connectors, listed clamps, or other listed means, except that connections which depend solely upon solder shall not be used.

ATTACHMENT TO ELECTRODES  The grounding electrode conductor shall be attached to the grounding electrode by means of (1) a listed bolted clamp of cast bronze or brass or of plain or malleable cast iron, (2) a pipe fitting, plug, or other approved device, screwed into the pipe or into the fitting, (3) a listed sheet-metal-strap-type ground clamp having a rigid metal base that seats on the electrode and having a strap of such material and dimensions that it is not likely to stretch during or after installation, or (4) other equally substantial approved means. The grounding electrode conductor shall be attached to the grounding fitting by means of exothermic welding, listed lugs, listed pressure connectors, listed clamps, or other listed means, except that connections which depend upon solder shall not be used. Not more than one conductor shall be connected to the grounding electrode by a single clamp or fitting unless the clamp or fitting is of a type listed for multiple conductors.

GROUND CLAMPS  For the grounding conductor of a wiring system the sheet-metal–strap type of ground clamp is not considered adequate unless the strap is attached to a rigid metal base which, when installed, is seated on the water pipe or other electrode and the strap is of such material and dimensions that it is not liable to stretch during or after installation.

Ground clamps for use on copper water tubing and copper, brass, or lead pipe should preferably be of copper, and those for use on galvanized or iron pipe should preferably be of galvanized iron and so designed as to avoid physical damage to pipe. Ground clamps used with aluminum grounding conductors should be listed for the purpose.

PROTECTION OF ATTACHMENT    Ground clamps or other fittings shall be approved for general use without protection and shall be protected from ordinary physical damage.

Figure 9-241 shows some types of grounding clamps. For the smaller ground wires, which must be enclosed in conduit, the type in Fig. 9-241, I, is used. For bare ground wire connected to a water or gas pipe, use the type in Fig. 9-241, II. The type in Fig. 9-241, III, is for a bare wire to be attached to a ground rod.

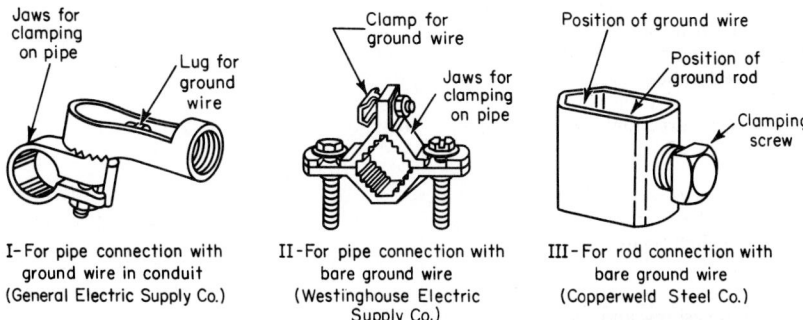

I–For pipe connection with ground wire in conduit (General Electric Supply Co.)    II–For pipe connection with bare ground wire (Westinghouse Electric Supply Co.)    III–For rod connection with bare ground wire (Copperweld Steel Co.)

**Fig. 9-241**    Ground-wire clamps.

**345. On grounded wiring systems the conductor which is connected to ground must be identified throughout the system.**

1. Insulated conductors of No. 6 size or smaller, except conductors of Type MI cable and multiconductor cables covered under Exception 3 to Sec. 200-6(a) of the NEC, fixture wires and multiconductor varnished-cloth-insulated cables, shall have an outer identification of white or natural-gray color along its entire length. The grounded conductors of Type MI cable shall be identified by distinctive marking at the terminals during the process of installation.

2. Insulated conductors larger than No. 6 shall have an outer identification of white or natural-gray color or shall be identified by distinctive white marking at terminals during the process of installation, except for multiconductor cables in occupancies employing qualified persons. Multiconductor flat cable No. 4 or larger may employ an external ridge on the grounded conductor.

3. If, on a four-wire delta-connected secondary, the midpoint of one phase is grounded to supply lighting and similar loads, that phase conductor having the higher voltage to ground shall be indicated by the color orange or by tagging or other effective means at any point at which a connection is to be made if the neutral conductor is present.

All ungrounded conductors must be finished in black or red or any color contrasting to white. At the terminals of wiring devices the grounded conductor must be connected to the nickel- or zinc-colored terminal. On screw-shell lampholders the grounded conductor must be connected to the screw shell itself. No single-pole switch or circuit breaker and no fuse shall be used in a grounded conductor. Where more than one nominal voltage system exists in a building, each ungrounded system conductor shall be identified by phase and system. The means of identification shall be permanently posted at each branch circuit panelboard.

SERVICE-ENTRANCE REQUIREMENTS AND INSTALLATION

**346. Service entrance for the electrical system of a building** may be made by means of overhead or underground service conductors. A service consists of the service conductors and the service equipment. The service conductors are the conductors which supply a building with electrical energy from an electric power system outside the building. They consist of that portion of the supply conductors which extend from the street distribution main or distribution transformer to the service equipment of the building. The service conductors may be located overhead or underground. For overhead conductors the service conductors include the conductors from the last line pole to the service equipment. The service conductors of an overhead service are further classified into two parts: the service-drop conductors and the service-entrance conductors. The *service-drop conductors* are the overhead service conductors from the last line pole or other aerial support to and including the splices, if any, connecting to the service-entrance conductors at the building or other structure. The *service-entrance conductors* of an overhead service are that portion of the service conductors between the terminals of service equipment and a point usually outside the building, clear of the building walls, where they are joined by tap or splice to the service drop.

**347. The service equipment** must consist of an externally operated disconnecting device and overcurrent protective device. The disconnecting device must provide a readily accessible means for disconnecting all the service conductors from the source of supply. It may consist of a single switch or manually operated circuit breaker or not more than six switches or six manually operated circuit breakers. In Fig. 9-242, if there had been only

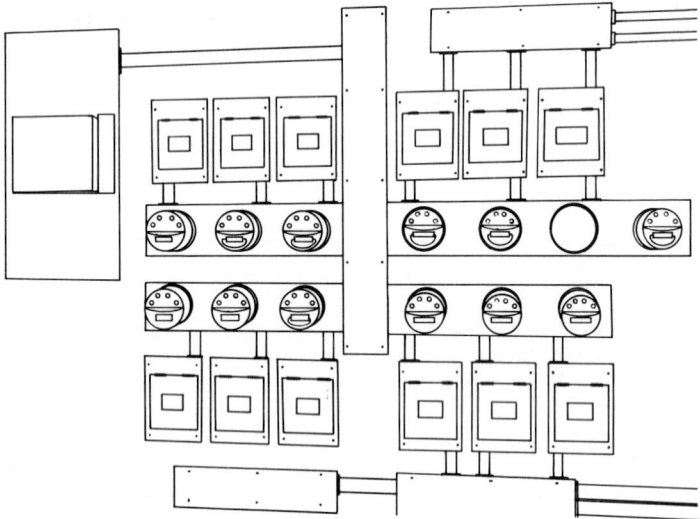

**Fig. 9-242**  Arrangement of meters and service for a multitenant building.

six occupants in the building, the master service switch would not have been required. In multiple-occupancy buildings each occupant must have access to his or her disconnecting means. Circuit breakers may be equipped with electrical remote control, provided that the breaker has in addition a manually operable handle. The disconnecting means must be of a type which will plainly indicate whether it is in the open or the closed position. The service switch or breaker must be located as close as is feasible to the point of entrance of the service conductors to the building. It may be located either outside or inside the building. If the switch or breaker does not open the grounded conductor, other means must be provided at the service entrance for disconnecting the grounded conductor of the supply from the interior wiring. The connection to a neutral terminal bar in the service switch will constitute a satisfactory disconnecting means.

A service switch must have a rating which is not less than the fuseholder in series with it, and the rating must be at least as great as the minimum allowable load (refer to Sec. 52 of Div. 3). Except in special cases of very small occupancies, it is not good practice to install a service switch or circuit breaker that has a rating less than 60 A. A 30-A switch should not be used for a two-wire service supplying more than two 15-A branch circuits. If a single-family residence has an initial load of 10 kW or more computed in accordance with Code regulations or if the initial installation has more than five two-wire branch circuits (refer to Div. 3), the service equipment shall have a rating that is not less than 100 A.

Each ungrounded service conductor must be provided with overcurrent protection. The protection may consist of a single set of fuses or circuit breakers or not more than six sets of fuses or six circuit breakers. For the rating of fuses or the setting of circuit breakers, refer to Secs. 294 and 295. No overcurrent device shall be inserted in a grounded conductor except a circuit breaker which simultaneously opens all conductors of the service. The protective devices must be located at the same point as the disconnecting means unless located at the outer end of the service raceway.

In addition to the disconnecting means and the protective devices, when the building is supplied directly from public-utility mains, the service equipment will include the kilowatthour meter for the measurement of power consumption. For small or multiple-occupancy buildings, the branch-circuit distribution equipment frequently is grouped along with the service equipment (see Fig. 9-242). If the service overcurrent protective devices are locked or sealed or otherwise not readily accessible, the branch-circuit overcurrent devices must be installed as close as is practicable to the load side of the service equipment. They must be in an accessible location and be of a lower rating than the service overcurrent device.

The construction and types of service equipment are discussed in Div. 4.

**348. There are six possible different arrangements or sequences for service equipment consisting of switch, fuses, and meter,** as follows:

1. Switch-fuse-meter
2. Switch-meter-fuses
3. Fuses-switch-meter
4. Fuses-meter-switch
5. Meter-switch-fuses
6. Meter-fuses-switch

When circuit breakers are employed, there are, of course, only two possible sequences: meter–circuit breaker or circuit breaker–meter.

The proper order for placing the switch, fuses, and meter in the circuit has been the subject for much dispute and not enough standardization, each power company reserving for itself the decision on the proper method. Each method has its advantages, and it has been a matter for local decision as to which advantage seems most important to the individual power company.

**349. A master service** is a service which supplies more than one building under single management. If the buildings are supplied from a private distribution system, the services should be installed in accordance with the general requirements for services from a public utility. In installations which include properties with their own generating plant, the conductors running from one building to another are not considered service conductors.

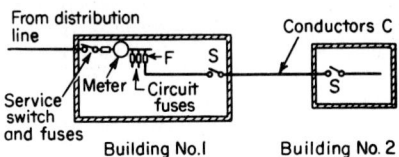

**Fig. 9-243** Two buildings under single management. No service fuses are required in building 2 if fuses F are of proper rating to protect conductors C. A switch S must be installed to control conductors C.

The supply to each building must be separately controlled and properly protected by overcurrent devices. The control should consist of an externally operable enclosed safety switch or a circuit breaker. Garages and outbuildings of residences and farms come under this class of installation. Methods of meeting these requirements are illustrated in Figs. 9-243 and 9-244. The protection for each building should have a lower rating or setting than the protection at F in the figures, so that in case of trouble the protection of any one building will function without operating the device at F.

No service should supply one building from another unless they are under the same management.

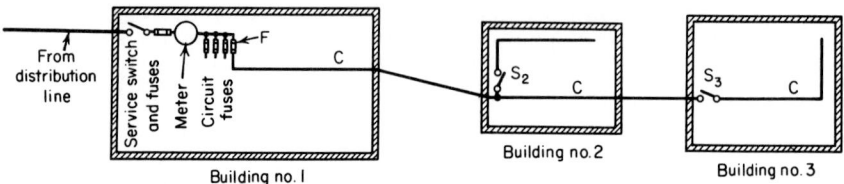

**Fig. 9-244**  Three buildings under single management. If properly protected by fuses $F$, conductors $C$ require no other fuse protection. A switch $S_2$ must be provided to control the wiring in building 2, and a switch $S_3$ to control the wiring in building 3.

**350. No building should be supplied by more than one set of service conductors from the same secondary distribution system or from the same transformer except as follows:**

1. FIRE PUMPS   If a separate service is required for fire pumps.

2. EMERGENCY LIGHTING   If a separate service is required for emergency, legally required standby, optional standby, or parallel power production systems.

3. CAPACITY REQUIREMENTS   If capacity requirements make multiple services desirable.

4. BUILDINGS OF LARGE AREA   By special permission, if more than one service drop is necessary owing to the area over which a single building extends.

5. MULTIPLE-OCCUPANCY BUILDINGS   By special permission, in multiple-occupancy buildings where there is no available space for service equipment accessible to all the occupants.

6. BUILDINGS OF MULTIPLE OCCUPANCY   These buildings may have two or more separate sets of service-entrance conductors which are tapped from one service drop (Fig. 9-245), or two or more subsets of service-entrance conductors may be tapped from a single set of main service conductors (Fig. 9-246).

7. WHEN ADDITIONAL SERVICES ARE REQUIRED FOR DIFFERENT CLASSES OF USE   These include needs for different voltage, frequency, or phase or different rate schedules for different types of use of electric power.

8. Underground conductors 1/0 or larger serving two to six disconnecting means.

Separate services may be brought into a single building for power and light, respectively, when the lighting is supplied from a separate secondary distribution main, as shown in Fig. 9-247. Separate services may be brought into a single building for power and light, respectively, if each service is supplied through a separate transformer (Fig. 9-248) or if the lighting service is supplied from one transformer of a polyphase bank of transformers supplying the power.

**351. Size of service conductors.**  Service conductors must have an ampacity sufficient to supply the load of the building (refer to Div. 3) and regardless of the actual load must have an ampacity not less than that required by the minimum load requirements of Div. 3. Also, regardless of the load, no wire should be smaller than the following minimum sizes:

| *Conductors* | *Minimum Allowable Size* |
| --- | --- |
| Service entrance for all installations, except ones supplying a single branch circuit | No. 8 if supplying only two two-wire branch circuits or by special permission for loads limited by demand or by the source of supply |
| Service entrance supplying a single branch circuit | No. 12 (but in no case smaller than size of branch circuit supplied) |
| Service drop | No. 12 hard-drawn copper |
| Service of over 600 V | No. 6, except that in multiconductor cables No. 8 may be used |

For single-family residences with an initial load of 10 kVA or more computed in accordance with the Code; or if the initial installation has more than five two-wire branch circuits, the service shall be a minimum of 100 A, three-wire.

It is recommended that a minimum of 100-A three-wire service be provided for all individual residences.

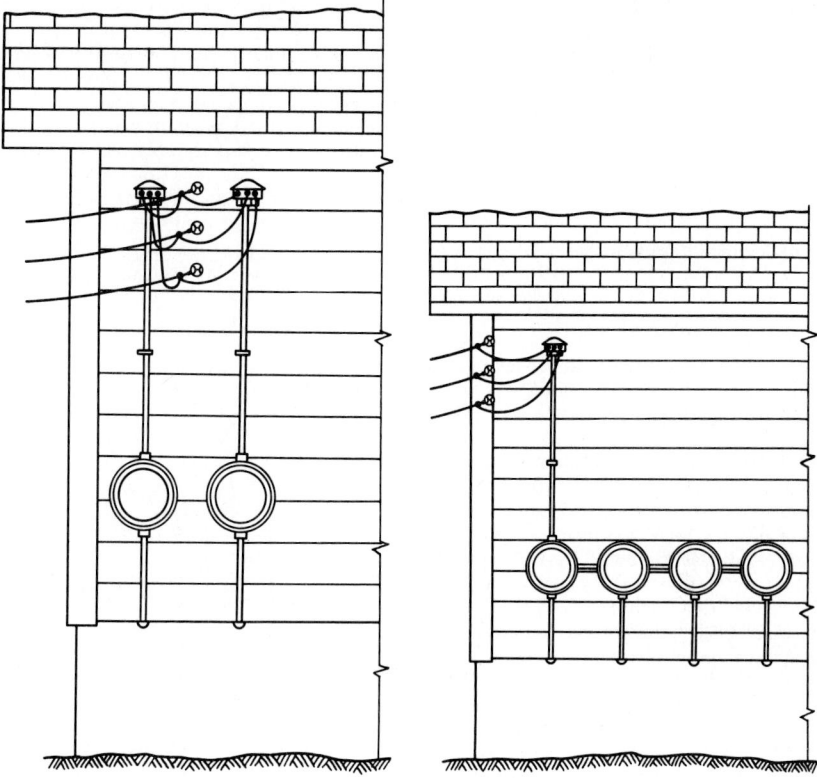

**Fig. 9-245** Two sets of service-entrance conductors tapped from one service drop.

**Fig. 9-246** Four sets of service-entrance conductors tapped from one set of main service conductors.

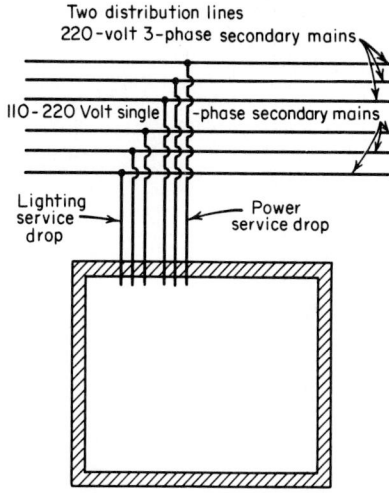

Two distribution lines
220-volt 3-phase secondary mains

110-220 Volt single -phase secondary mains

Lighting service drop

Power service drop

**Fig. 9-247** Separate services for power and light.

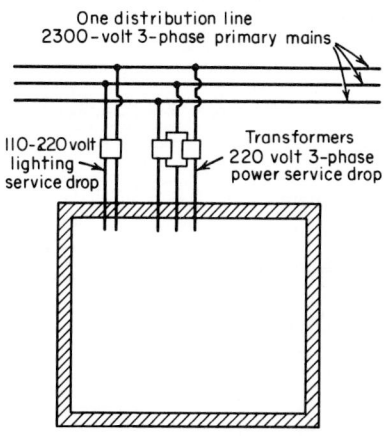

One distribution line
2300-volt 3-phase primary mains

110-220 volt lighting service drop

Transformers
220 volt 3-phase power service drop

**Fig. 9-248** Separate services for power and light.

**352. Splicing of service conductors.**    Service conductors shall be without splices except as follows:

1.  At point of junction of service-drop and service-entrance conductors.

2.  Clamped or bolted connections in a meter enclosure are permitted.

3.  Taps to main service conductors are permitted as provided in Subpar. 6 of Sec. 350.

4.  A connection is permitted, if properly enclosed, when an underground service conductor enters a building and is to be extended to the service equipment or meter in another form of approved service raceway or service cable.

5.  A connection is permitted when service conductors are extended from a service drop to an outside meter location and returned to connect to the service-entrance conductors of an existing installation.

6.  Where busway sections are used.

**353. Terminating raceway at service equipment** (National Electrical Code).  If conduit, electrical metallic tubing, or service cable is used for service conductors, the inner end shall enter a terminal box or cabinet or be made up directly to an equivalent fitting, enclosing all live metal parts, except that if the service disconnecting means is mounted on a switchboard having exposed busbars on the back, the raceway may be equipped with a bushing which shall be of the insulating type unless lead-covered conductors are used.

**354. Typical service installations of proper and improper construction** are shown in Fig. 9-249. Overhead services to occupancies having relatively small loads, such as residences and small stores, are generally made with one of the installations shown in Fig. 9-249, II and III. The installation in Fig. 9-249, I, violates Code rules because the service head is not located above the point of attachment of the service-drop

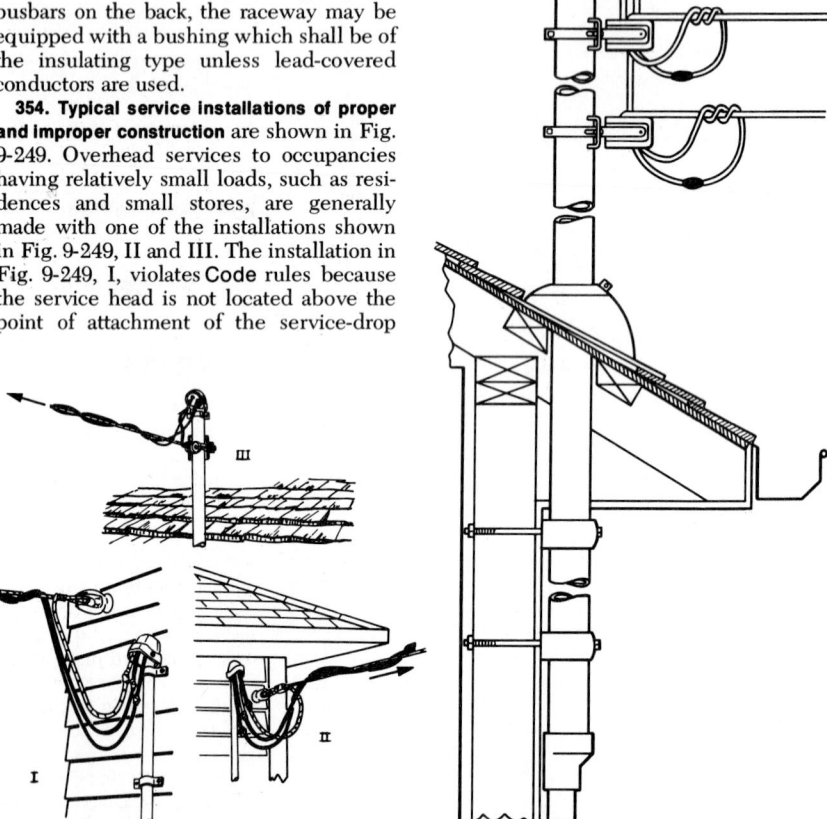

**Fig. 9-249**  Various types of service installations. Drawing I violates the **National Electrical Code.** [*Electrical Construction and Maintenance*]

**Fig.  9-250**  Low-roof  service  installation. [Hubbard & Co.]

conductors to the building. The present trend is toward installations of the types shown in Figs. 9-249, II and III, and 9-250. A good service installation for a low-roof building is shown in Fig. 9-250. Occasionally the service switch is also mounted outside the building along with the meter.

In congested city residential districts an underground service using underground service-entrance cable is frequently run from overhead lines when a change to underground distribution is planned at a later date. This procedure facilitates the changeover, as the part of the cable which is aboveground may in the future be taken down and placed underground and run directly into the underground distribution box. In cities where the distribution is all underground, direct-burial cable or nonmetallic raceway is the most common method of running services. The advantage of the conduit system is that in case the service needs to be replaced for any reason, the cable may be pulled out of the conduit and another one installed without having to dig up the pavement or sidewalk. For apartment-house service a meter trough in which are mounted the meter sockets, one for each apartment, in one or more horizontal rows (Fig. 9-242), is installed. Directly under each meter is a branch-circuit distribution panel containing fuses or circuit breakers for the branch-circuit protection for the apartment. In some cases a service switch is provided for each apartment, and a feeder is run to the apartment with the branch-circuit distribution panel located in the apartment.

**355. For commercial or industrial services** the maximum demand is usually metered as well as the kilowatthours. When the demand is of the order of a few kilowatts, this may be done with a single meter which has a demand-indicating pointer as well as the usual meter dials (Fig. 9-251). For larger installations a separate indicating-demand meter is installed (Fig. 9-252). A recording type of demand meter may also be employed for installations of more than about 150 A; current transformers are usually required to reduce the current which the meter has to handle. For an underground service from an

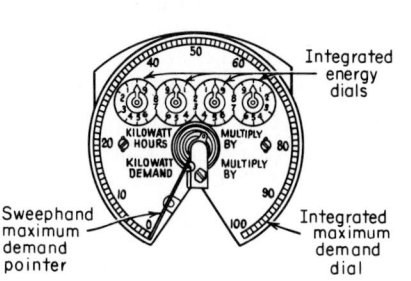

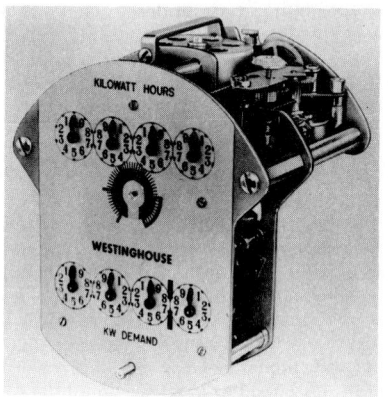

**Fig. 9-251** Dial for a watthour meter with demand-indicating pointer. [Westinghouse Electric Corp.]

**Fig. 9-252** Indicating-demand meter. [Westinghouse Electric Corp.]

urban network system, the service equipment consists of a main fused switch or circuit breaker, a set of current transformers in a sheet-metal cabinet, and meters as outlined above. If the supply comes from a radial feeder system, a transformer bank with an oil circuit breaker and disconnect switches is installed at the building ahead of the above service equipment. For an overhead service a transformer bank is installed just outside the building, on a concrete foundation on the ground or on an elevated platform between two poles. The bank will contain high-voltage fuses and lightning arresters in addition to the transformers. The service conductors are then brought into the building in conduit or as individual wires to the service equipment. Sometimes the metering is installed so as to measure the input to the transformers so that the customer pays directly for the transformer losses. When this is done, high-voltage current and potential transformers are installed in the leads to the primary of the transformers. The meters may then be located in a weatherproof cabinet outside the building or located inside the building wall, as before, with a conduit to carry the secondary leads from the instrument transformers.

**356. Connections ahead of disconnecting means.** The National Electrical Code allows meters, high-impedance shunt circuits (such as potential coils of meters, etc.), supply conductors for load management devices, meters, surge-protective devices, instrument transformers, surge arresters and circuits for emergency systems, fire-pump equipment, and fire and sprinkler alarms as provided in Sec. 230-94 of the Code to be connected on the supply side of the disconnecting means.

## CRANE WIRING

**357. Wiring for cranes consists** of the following elements:
1. Collector switch
2. Main collector contact conductors
3. Cab main-line switch
4. Bridge collector contact conductors
5. Motor-circuit wiring

**358. A collector switch or circuit breaker** must be installed in every circuit feeding crane wires. A fused switch is most frequently used to provide overload protection, although in important installations, where reliability is essential (where equipment must be placed back in service after an accident, in minimum time), circuit breakers are used. From an electrical standpoint the best location for the switch is at the center of the run.

The switch must be readily accessible and operable from the ground, be within view of the crane or hoist and the collector conductors, and be of a type which can be locked in the open position. It shall open all ungrounded conductors simultaneously.

**359. The main collector contact conductors** are bare copper trolley wires or structural-steel conductors that are erected parallel to the crane runway. The location of the conductors is determined by conditions. Sometimes the crane builder specifies that the conductors must be located in a certain position, and in other cases the purchaser selects and specifies the position of the conductors, and the crane manufacturer arranges the current collectors on the crane accordingly.

**360. Trolley wire for cranes** is hard-drawn copper, the same as used in electric traction. Hard-drawn wire must be used to prevent excessive stretching. Round wire is erected when the method of support adopted does not involve the use of trolley ears for holding it at intermediate points between the ends of the run. When trolley ears are used, figure-eight or, preferably, grooved trolley wire should be used, because it can be readily held in screw-clamp trolley ears. Round wire can be used with trolley ears, but the ear flanges of these must be hammered down around the wire, a time-consuming operation requiring some skill; also, a round-wire ear introduces a hump on the wire and makes the trolley wheel jump and draw an arc when the wheel passes over the raised place. Either 1/0, 2/0, 3/0, or 4/0 wire is usually required. The wire size required is ordinarily specified by the crane builder, but in any case it should be large enough so that the voltage drop in it will not much exceed 3 or 4 percent of the line voltage with all the crane motors operating at full load. In no case should the wires be smaller than the minimum allowable sizes given in Sec. 371 for bridge collector conductors.

**361. Trolley rails of structural steel** are being used to some extent instead of copper trolley wire to supply current to cranes and other moving electrical machinery. The steel rails are made sufficiently heavy so that they cannot break and fall, as copper wires occasionally do. Sometimes strap-steel bars are used, but more frequently a section such as an angle or a tee, which has considerable stiffness in all directions, is adopted. Steel conductors should be painted, except on the contact edge or face, to prevent corrosion. Either a shoe or a trolley wheel can be used to collect current from a steel conductor rail. A shoe or spoon, which makes a rubbing contact, is probably preferable for the average application. Trolley-wheel collectors that travel at high speeds are not successful for current collection from steel conductors, because the wheel tends to bounce and jump from the rigid rail at joints and uneven places.

**362. The main collector contact conductors** must be isolated by elevation or provided with suitable guards so that persons cannot inadvertently touch the live parts while in contact with the ground or with conducting material connected to the ground. If the crane travels over easily ignitible combustible fibers and materials, the collector conductors must be protected by barriers so arranged as to prevent the escape of sparks or hot particles. Probably the best location for the collector conductors on a bridge-crane runway is between the flanges of and parallel to one of the crane girders. Here the conductors are out of the way and well protected and can be readily supported. It is not often that they are erected in any other position. Occasionally the trolley wires can be supported from the roof trusses above the crane runway and are installed similarly to the trolley wires for trolley cars. A pole collector with a wheel at its upper end, exactly like a trolley-car pole but much shorter, is used. When the spacing between roof trusses is very great, this method may not give good results because of the wire swinging and the trolley coming off. The main contact conductors, when carried along runways, should normally be supported on insulating supports at intervals not exceeding 20 ft (6.1 m) and separated from each other at least 6 in (152 mm), except that a spacing of not less than 3 in (76 mm) may be used for monorail hoists. If wires are used, they should be secured at the ends with approved strain insulators and so mounted on intermediate insulators that the extreme displacement of the wires will not bring them closer than 1½ in (38 mm) from the surface wired over. If necessary, the intervals between supports may be increased up to 40 ft (12.2 m), pro-

vided that the separation between conductors is increased proportionally. All sections of the conductors must be so joined that a continuous electrical connection is provided. The collector conductors must not supply any equipment other than that of the crane or cranes which they are designed to serve.

**363. Methods of supporting crane trolley wires** differ with conditions. If the crane is provided with hook-shoe collectors (Fig. 9-253), which slide along under and carry the weight of the trolley wire, the wire is rigidly held only at the terminations at the ends of the run and is kept from sagging by intermediate insulating brackets like those shown in Fig. 9-253. If trolley-wheel current collectors are used, the tension in the wire is taken by the terminations, and the wire is also rigidly held by trolley ears at intermediate points.

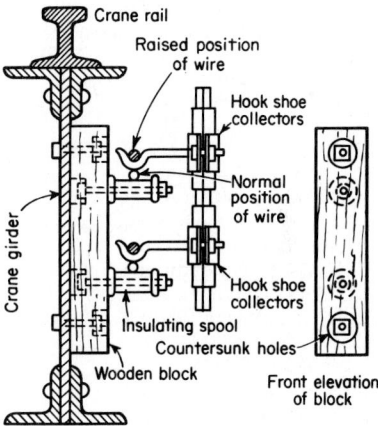

**Fig. 9-253**  Intermediate bracket for trolley wire.

**364. Terminations** are made as shown in Fig. 9-254. Strain insulators separate the trolley wire electrically from the building members, and either a turnbuckle or an eyebolt with a long thread can be used for pulling the slack from the wire and adjusting its tension. The terminations should be depended upon to assume the entire tension of the wire. The intermediate supports are placed merely to prevent excessive sagging. The members which take the stresses of the eyebolts at the terminations should be very substantial or thoroughly braced, because on them depends the reliability of the entire installation. The eyebolts, turnbuckles, and strain insulators should be not smaller than the ⅝-in size.

**365. Intermediate supports for crane trolley wires,** the supports installed between the terminations to prevent sagging, can be arranged as shown in Figs. 9-253, 9-255, and 9-256. The bracket of Fig. 9-253 is, as above explained, applicable only if the crane has hook-

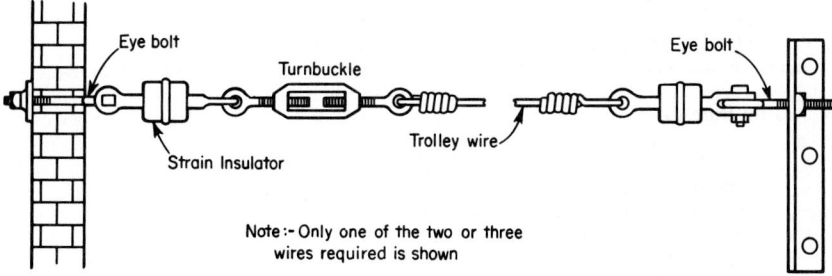

**Fig. 9-254**  Trolley supports at ends of run.

shoe collectors. The block of wood that supports the insulating spools should be thoroughly dry and treated with an insulating paint. The bolt holes in it should be deeply countersunk to eliminate the possibility of grounds. The spools should be of porcelain. Porcelain tubes with porcelain knobs at the ends to form flanges will do. The length of the insulator between flanges should be at least 4 in (101.6 mm). Spools turned from fiber are sometimes used. The spools should be so arranged that there will be at least a ½-in (12.7-mm) clearance between them and the hook shoe when it passes along.

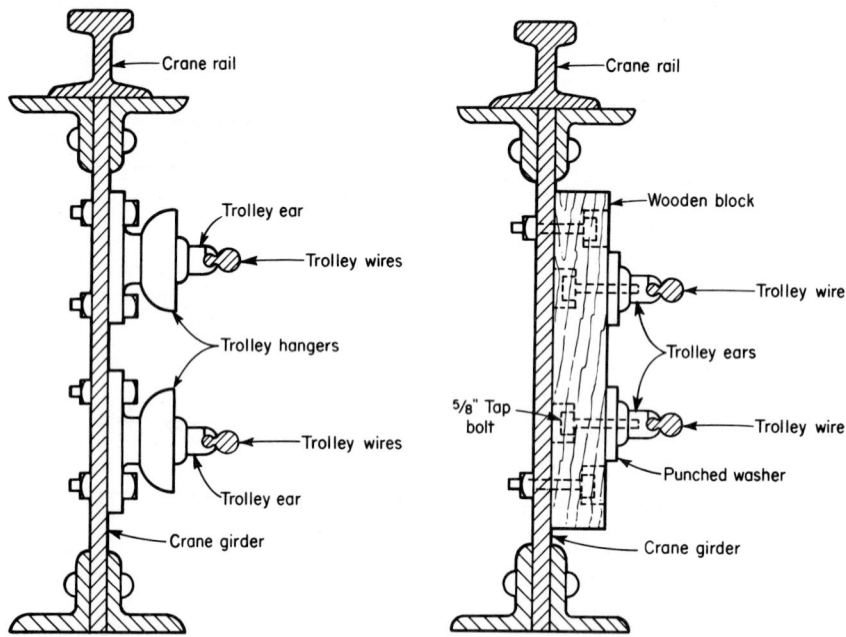

**Fig. 9-255**  Trolley hanger support.          **Fig. 9-256**  Trolley ear supports.

In a fireproof building, when the crane has trolley-wheel collectors, the wires can be supported by trolley ears as in Fig. 9-256. The wooden supporting block must be thoroughly painted, and the bolt holes in it deeply countersunk to prevent the possibility of grounds. Tap bolts, screwing in from the rear, support the screw-clamp trolley ears which seat against washers. Figure-eight or, preferably, grooved trolley wire is used. The wire should be drawn tightly at the terminations, and the ears should be installed every 8 or 10 ft (2.4 or 3 m).

For an outdoor crane, as well as for indoor applications, the wires can be supported in some cases as shown in Fig. 9-255. Standard street-railway–type trolley hangers and ears, which provide excellent insulation, are used. The hangers can, provided a proper location of the trolley wires results, be bolted directly to the crane girder.

**366. There are almost numberless ways in which steel conductor rails can be arranged and supported.** The arrangement of a structural-steel tee conductor or trolley rail is shown in Fig. 9-257. Although the arrangement illustrated was developed for serving monorail cranes, which travel on the lower flanges of I beams, only minor modifications in the supporting forging would be required to adapt it for serving bridge cranes or other similar traveling electrical machines. Note that a feature of the method is that no drilling or close fitting is required in the field. The only piece that is different for different jobs is the supporting forging, but this can be formed and drilled in the shop. The only tools required to erect the rail are a hacksaw for cutting the tee, which is purchased in 30-ft (9.1-m) lengths, and a wrench for setting the bolts. No bolt smaller than ⅝-in diameter is used,

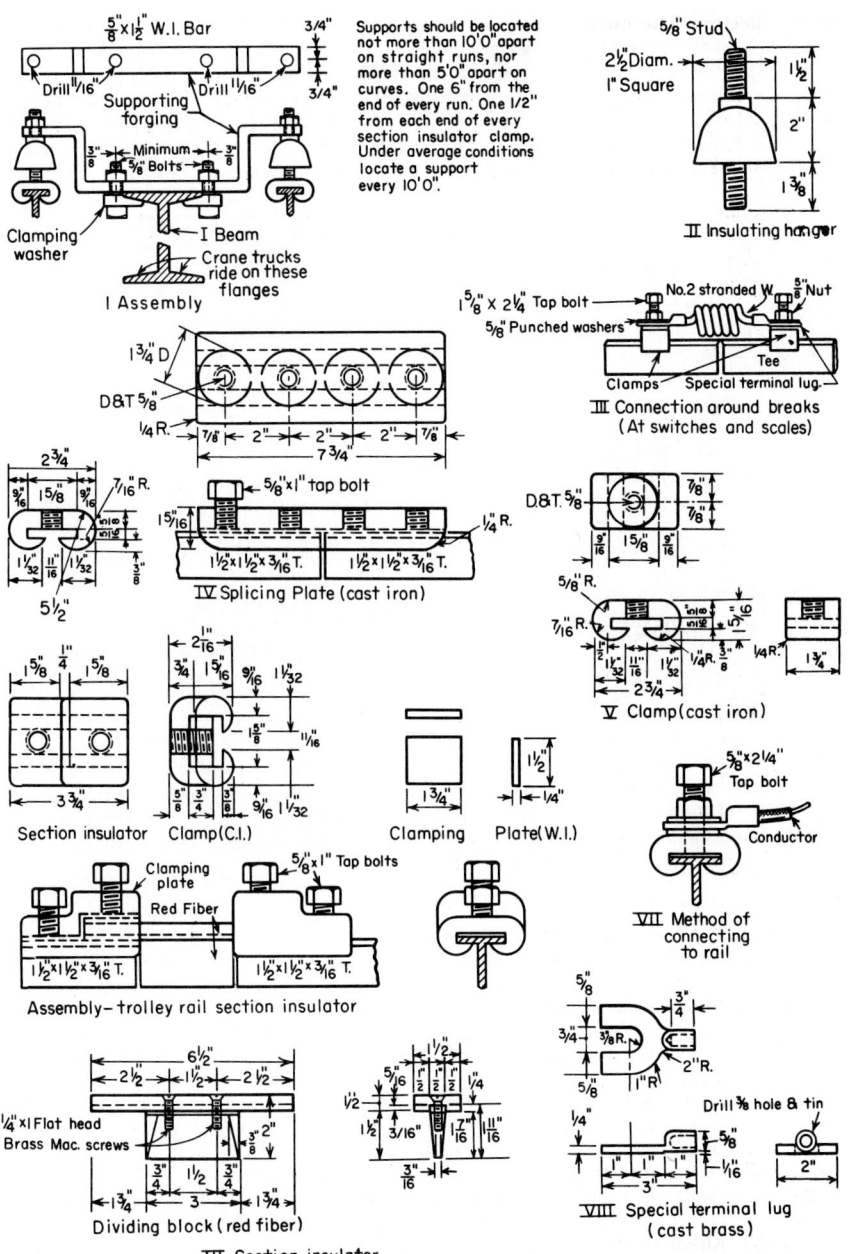

Supports should be located not more than 10'0" apart on straight runs, nor more than 5'0" apart on curves. One 6" from the end of every run. One 1/2" from each end of every section insulator clamp. Under average conditions locate a support every 10'0".

I Assembly

II Insulating hanger

III Connection around breaks (At switches and scales)

IV Splicing Plate (cast iron)

V Clamp (cast iron)

Section insulator    Clamp (C.I.)    Clamping    Plate (W.I.)

VII Method of connecting to rail

Assembly–trolley rail section insulator

Dividing block (red fiber)

VI Section insulator

VIII Special terminal lug (cast brass)

**Fig. 9-257**   Steel trolley rail.

because smaller ones than this can be twisted asunder too easily. A 1½- by 1½- by ³⁄₁₆-in T bar was selected, because this is about the smallest size that is rigid enough to sustain itself effectively between supports. A T bar of this size has a conductivity equivalent to that of a 109,800-cmil copper conductor, that is, a copper conductor between Nos. 1/0 and 2/0 in size.

The insulating hanger (Fig. 9-257, II) is similar to a trolley hanger but smaller. A malleable-iron bell encloses the molded material that supports and insulates the hanger stud. The Manville Corp. makes the insulator to special order, and its mold number is 4689-B. The splicing plate (IV) and the clamp (V) are castings, preferably of malleable iron, and the only machine work on them is the drilling and tapping of the holes. The section insulator (VI) consists of two castings, a fiber dividing block, and two wrought-iron clamping plates. Directions for spacing the insulating supports when erecting the conductor are given on the illustration. The terminal lug (VIII) is forked instead of annular so that it can be readily disconnected for isolating circuits for testing without taking out a bolt.

Rigid collector conductors must have insulating supports spaced at intervals of not more than 80 times the vertical dimension of the conductor, but in no case greater than 15 ft (4.57 m), and spaced apart sufficiently to give a clear electrical separation of conductors or adjacent collectors of not less than 1 in (25.4 mm).

**367. Track as circuit conductor** (National Electrical Code).   Monorail, tramrail, or crane-runway tracks may be used as a conductor of current for one phase of a three-phase ac system furnishing power to the carrier, crane, or trolley, provided the following conditions are fulfilled:

1. The conductors for supplying the other two phases of the power supply shall be insulated.

2. The power for all phases shall be obtained from an insulating transformer.

3. The voltage shall not exceed 300 V.

4. The rail serving as a conductor shall be effectively grounded at the transformer and may also be grounded by the fittings used for the suspension or attachment of the rail to a building or structure.

**368. In computing the resistance of steel trolley rails** the area in square inches of the section involved can be taken from one of the steel companies' handbooks, such as those issued by the Cambria and Carnegie steel companies; the area in circular mils can then be obtained by using the rule given below. By dividing this area by 6.14, which is the approximate ratio of the resistance of mild steel to that of copper, the equivalent copper area of the steel conductor results. Then by using the standard formula for the resistance of a copper conductor, the actual resistance of the steel is obtained.

**example**   What is the resistance of 160 ft of 1½- by 1½- by 1½- by ³⁄₁₆-in steel angle? (See Fig. 9-258 for a picture of the section.)

**solution**   By referring to a handbook, it will be noted that the area of 1½- by 1½- by ³⁄₁₆-in steel angle is 0.53 in². Then, to find its area in circular mils,

$$\text{Cmil} = \frac{\text{area in in}^2}{0.000,000,785,4} = \frac{0.53}{0.000,000,785,4}$$
$$= 674,800 \text{ cmil}$$

$$\text{Equivalent in copper} = \frac{\text{cmil area of steel}}{6.14}$$
$$= \frac{674,800}{6.14} = 109,800 \text{ cmil}$$

Then the resistance of the 160-ft length will be

$$\text{Resistance (for copper)} = \frac{11 \times \text{ft}}{\text{cmil}} = \frac{11 \times 160}{109,800} = 0.016 \ \Omega$$

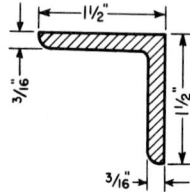

**Fig. 9-258**   Section of 1½- by 1½- by ³⁄₁₆-in steel angle.

The resistance, therefore, of 160 ft of 1½- by 1½- by ³⁄₁₆-in steel angle is 0.016 Ω. It is evident from the equivalent copper area of the steel (109,800 cmil) that the conductivity of the steel section will lie between the conductivities of No. 1/0 (105,500 cmil) and No. 2/0 (133,100 cmil) copper wire.

The equivalent copper area in circular mils can be used in any of the wiring formulas for computing drop in a steel conductor, just as the actual area of a copper conductor is used in the same formulas, and the result will be a correct one for the steel section. Obviously the above method is approximate because the constants are approximate, but it is quite accurate enough for wiring computations, which always involve necessarily inaccurate assumptions.

**369. Switch in cab** (National Electrical Code). When a crane is operated from a cage or cab, a motor-circuit switch or circuit breaker shall be provided in the leads from the runway contact conductors. The switch or circuit breaker shall be in the cage or cab or mounted on the bridge and operable from the cage or cab when the trolley is at one end of the bridge, or means shall be provided to open the power circuits from the operating station.

On both ac and dc crane protective panels, the continuous ampere rating of the main-line switch or circuit breaker and main-line contactors shall be not less than 50 percent of the combined short-time ampere ratings of the motors nor less than 75 percent of the short-time ampere rating of the motors required for any single crane motion.

**370. Contact conductors** are required across the crane bridge for carrying the circuit to the carriage and hoist motors. They are usually made of bare copper wires supported in the general way described in Secs. 362 to 365, except that bridge collector conductors shall be kept at least 2½ in (64 mm) apart, and if the span exceeds 80 ft (24.4 m), insulating saddles shall be placed at intervals not exceeding 50 ft (15.2 m).

It is recommended that the distance between wires be greater than 2½ in where practicable.

**371. Contact conductors.** The National Electrical Code requires that the size of contact wires shall be not less than the following:

| Distance between End Strain Insulators or Clamp-type Intermediate Supports, Ft | Size of Wire |
|---|---|
| 0–30 | No. 6 |
| 31–60 | No. 4 |
| Over 60 | No. 2 |

**372. The motor-circuit wiring** must be of the types listed in Table 610-14(a) of the NEC. Type MI cable is allowed in both dry and wet locations. Conductors exposed to external heat or connected to resistors shall have an insulation approved by the Code for the temperature and location. If conductors not having a flame-resistant outer covering are grouped together, the group shall be covered with a flame-resistant tape.

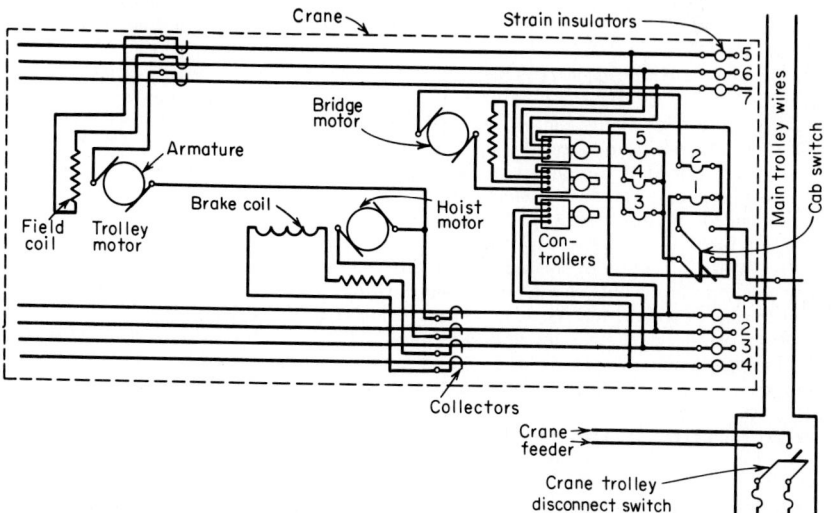

**Fig. 9-259**  Direct-current crane-wiring diagram.

The conductors must be of a size as given in Table 610-14(a) of the NEC. The wiring must be enclosed in a raceway or be Type MI or Type MC cable. When flexible connections to motors or controllers are necessary, flexible metal raceways, metal-clad cable, multiple-conductor cable, or an approved nonmetallic enclosure may be used. Short runs for control equipment in the cab or on the bridge may be enclosed in auxiliary gutters. Short lengths of conductors at resistors and collectors may be left open. A common-return conductor may be used for the several motors of a single crane or hoist.

**373. Motor control and protective equipment.** The hoist of cranes must be provided with a limit switch which will control the upper limit of travel. It is best practice to arrange the control levers of all traveling cranes, in any one installation, in the same relative position in all the cages of the different cranes. The NEC allows two motors with their leads, which operate a single hoist, carriage, truck, or bridge and are controlled as a unit by one controller, to be protected by a single overload device. The device should be located in the cab if there is one.

**374. A crane-wiring diagram** is given in Fig. 9-259, which is typical for a dc, three-motor, traveling bridge crane. Variations in crane wiring and control schemes are practically numberless. For dc cranes, series motors are almost invariably used, while for ac cranes, wound-rotor motors are used.

## WIRING FOR CIRCUITS OF OVER 600 VOLTS

**375. For wiring circuits operating at more than 600 V,** the conductors must have an insulation sufficient for the particular voltage employed. The wires must be installed in rigid or intermediate metal conduit, in rigid nonmetallic conduit, in cable trays, as busways, as cablebus, in other identified raceways, or in open runs of metal-clad cable suitable for the use and purpose, except that in locations accessible to qualified persons only open runs of Type MV cable, bare conductors, and bare busbars may be used. Open runs of braid-covered insulated conductors must have flame-retardant braid, or the braid covering must be treated with a flame-retardant saturant after installation. The braid covering must be stripped back at conductor terminals to a safe distance according to the operating voltage, and is not less than 1 in (25.4 mm) for each kilovolt of the conductor-to-ground voltage.

The metallic and semiconducting insulation shielding components of shielded cables shall be stripped back to a safe distance according to the circuit voltage at all terminations of the shielding, as in potheads and joints. At such points, suitable methods such as the use of potheads, terminators, stress cones, or similar devices shall be employed for stress reduction, and the metallic shielding tape and semiconducting components shall be grounded.

**376. Circuit breakers** (National Electrical Code). Indoor installations shall consist of metal-enclosed units or fire-resistant cell-mounted units except that open mounting of circuit breakers is permissible in locations accessible to qualified persons only.

Circuit breakers used to control oil-filled transformers should be located outside the transformer vault or be capable of operation from outside the vault.

Circuit breakers shall have a means of indicating the open and closed positions of the breakers at the point or points from which they may be operated.

Circuit breakers shall be trip-free in all positions. In every installation the circuit-breaker rating in respect to closing, carrying, or interrupting capabilities shall not be less than the short-circuit duty at the point of application.

Oil circuit breakers shall be so arranged or located that adjacent readily combustible structures or materials are safeguarded in an approved manner. Adequate space separation, fire-resistant barriers or enclosures, trenches containing sufficient coarse crushed stone, and properly drained oil enclosures such as dikes or basins are recognized as suitable for this purpose.

**377. Fuseholders and Fuses** (National Electrical Code)

1. Fuses which expel flame in opening the circuit shall be so designed or arranged that they will function properly without hazard to persons or property.

2. Fuseholders shall be designed so that they will be deenergized while replacing a fuse unless the fuse and fuseholder are designed to permit fuse replacement by qualified persons using equipment designed for the purpose without deenergizing the fuseholder.

3. When high-voltage fused cutouts are installed in a building or a transformer vault, they shall be of a type designed for use in buildings. If such cutouts are not suitable to interrupt the circuit manually while carrying a full load, an approved means which is capable of interrupting the entire load shall be provided. Unless the fused cutouts are interlocked with the switch to prevent opening of the cutouts under load, a conspicuous sign shall be placed at such cutouts reading: "Warning: do not open under load."

The cutouts shall be so located that they may be readily and safely operated and re-fused. Fuses shall be accessible from a clear floor space.

4. Load-interrupter switches may be used, provided suitable fuses or circuit breakers are applied in conjunction with these devices to interrupt fault currents. Where these devices are used in combination, they shall be so coordinated electrically that they will safely withstand the effects of closing, carrying, or interrupting all possible currents up to the assigned maximum short-circuit rating.

**378. Isolating means** (National Electrical Code).   Means shall be provided to isolate an item of equipment completely. The use of isolating switches is not necessary if there are other ways of deenergizing the equipment for inspection and repairs. Isolating switches should be interlocked with the associated circuit-interrupting device to prevent their being opened under load; otherwise signs warning against opening them under load shall be provided. A fuseholder and fuse designed for the purpose may be used as an isolating switch.

**379. Space separation.**   The NEC requires that a minimum air separation in field fabricated installations between live conductors and between such conductors and adjacent grounded surfaces be not less than the values given in Table 710-33 of the NEC.

**380. Work space and guarding.**   The National Electrical Code requires the following protection:

1. Working Space   The minimum clear working space in front of electric equipment such as switchboards, control panels, switches, circuit breakers, motor controllers, relays, and similar equipment shall be not less than set forth in Table 110-34(a) of the NEC unless otherwise specified in the Code. Distances shall be measured from the live parts if such are exposed, or from the enclosure front or opening if such are enclosed.

The conditions are as follows:

*a.* Exposed live parts on one side and no live or grounded parts on the other side of the working space or exposed live parts on both sides effectively guarded by suitable wood or other insulating materials. Insulated wire or insulated busbars operating at not over 300 V shall not be considered live parts.

*b.* Exposed live parts on one side and grounded parts on the other side. Concrete, brick, or tile walls will be considered grounded surfaces.

*c.* Exposed live parts on both sides of the work space (not guarded as provided in Condition 1) with the operator between.

*Exception. Working space is not required in back of equipment such as dead-front switchboards or control assemblies where there are no renewable or adjustable parts (such as fuses or switches) on the back and where all connections are accessible from locations other than the back. Where rear access is required to work on deenergized parts on the back of enclosed equipment, a minimum working space of 30 in (762 mm) horizontally shall be provided.*

2. Separation from Low-Voltage Equipment   Where switches, cutouts, or other equipment operating at 600 V or less are installed in a room or enclosure where there are exposed live parts or exposed wiring operating at more than 600 V, the high-potential equipment shall be effectively separated from the space occupied by the low-potential equipment by a suitable partition, fence, or screen.

*Exception. Switches or other equipment operating at 600 V or less and serving only equipment within the high-voltage vault, room, or enclosure may be installed in the high-voltage enclosure, room, or vault if accessible to qualified persons only.*

3. Locked Rooms or Enclosures   The entrances to all buildings, rooms, or enclosures containing exposed live parts or exposed conductors operating in excess of 600 V shall be kept locked, except where such entrances are at all times under the observation of a qualified person.

If the voltage exceeds 600 V, permanent and conspicuous warning signs shall be provided, reading substantially as follows: "Warning—high voltage—keep out."

4. Illumination   Adequate illumination shall be provided for all working spaces

around electric equipment. The light outlets shall be so arranged that persons changing lamps or making repairs on the lighting system will not be endangered by live parts or other equipment. The points of control shall be so located that persons are not likely to come into contact with any live part or moving part of the equipment while turning on the lights.

**381. Isolation by elevation.** The National Electrical Code requires that the distances from working spaces to unguarded live parts be not less than those specified in Table 110-34(e).

## WIRING FOR CIRCUITS OF LESS THAN 50 VOLTS

**382. For wiring circuits of less than 50 V,** conductors not smaller than No. 12 must be used. If the conductors supply more than one appliance or appliance receptacle, they must not be smaller than No. 10. Receptacles must be rated at not less than 15 A; in kitchens, laundries, and other locations where portable appliances are likely to be used, the receptacles must be rated at not less than 20 A.

## WIRING FOR HAZARDOUS LOCATIONS

**383. Hazardous locations** are divided by the National Electrical Code into three classes. Each of these is further subdivided into Divs. 1 and 2. In Div. 1 the hazardous material is more or less freely present in the air in connection with manufacture. In Div. 2 the hazardous material is confined in containers, and explosive mixtures with air will occur only in case of accident or through failure of ventilating systems to operate properly.

CLASS I   Locations in which flammable gases or vapors are or may be present in the air in quantities sufficient to produce explosive or ignitible mixtures.

CLASS II   Locations which are hazardous because of the presence of combustible dust.

CLASS III   Locations which are hazardous because of the presence of easily ignitible fibers or flyings but in which such fibers or flyings are not likely to be in suspension in air in quantities sufficient to produce ignitible mixtures.

**384. Equipment for installation in hazardous locations** must be tested and approved for use, according to the following classification of the hazard involved:

*Group A.*   Atmospheres containing acetylene.

*Group B.*   Atmospheres containing hydrogen, fuel, and combustible process gases containing more than 30 percent hydrogen by volume or gases or vapors of equivalent hazard such as butadiene, ethylene oxide, propylene oxide, and acrolein.

*Group C.*   Atmospheres containing cyclopropane, ethyl ether, ethylene, or gases or vapors of equivalent hazard.

*Group D.*   Atmospheres such as acetone, ammonia, benzene, butane, ethanol, gasoline, hexane, methanol, methane, natural gas, naptha, propane, or gases or vapors of equivalent hazard.

*Group E.*   Atmospheres containing combustible metal dusts regardless of resistivity or other combustible dusts of similarly hazardous characteristics having resistivity of less than $10^2\ \Omega \cdot \mathrm{cm}$.

*Group F.*   Atmospheres containing carbon black, charcoal, coal, or coke dusts which have more than 8 percent total volatile material (coal or coke dusts per ASTM 3175-82), or atmospheres containing these dusts sensitized by other materials so that they present an explosion hazard and having resistivity greater than $10^2\ \Omega \cdot \mathrm{cm}$ but equal to or less than $10^8\ \Omega \cdot \mathrm{cm}$.

*Group G.*   Atmospheres containing combustible dusts having resistivity of $10^8\ \Omega \cdot \mathrm{cm}$ or greater.

**385. In Class I, Div. 1, hazardous locations all wiring** must be threaded rigid metal conduit or intermediate metal conduit with explosionproof fittings or Type PLTC, MI, MC, MV, TC, or SNM cable, except extra-hard usage flexible cord can be used at motor terminals. Division 2 locations may use flexible metal fittings, flexible metal or nonmetallic raceway with approved fittings, or flexible cord approved for extra-hard usage when necessary, as at motor terminals. All equipment such as circuit breakers, fuses, motors, generators, and controllers must be totally enclosed in explosionproof housings.

**386. In Class II, Div. 1, hazardous locations the wiring** must be in threaded rigid metal conduit, intermediate metal conduit, or Type MI cable, with liquidtight flexible metal or nonmetallic conduit or extra-hard usage cord where necessary, as at motor terminals, and pendant fixtures, and must have threaded fittings. All equipment must be in dustproof cabinets with motors and generators in totally enclosed, fan-cooled housings. In Div. 2 locations electrical metallic tubing, dusttight wireways, and MI, MC, or SNM cables with approved termination fittings may also be used.

**387. In Class III, Div. 1, hazardous locations the wiring** must be of the same type as in Class II, and the use of electrical metallic tubing, dusttight wireways, flexible cord approved for extra-hard usage, and Type MC or SNM cables is also permitted. In Div. 2 locations open wiring is permitted under certain conditions. Motors or generators must be totally enclosed. For further special rules for hazardous locations in regard to receptacles, plugs, lighting fixtures, portable cords, cranes, control equipment, service equipment, panelboards, etc., consult the NEC.

## INSTALLATION OF APPLIANCES

**388. Connection to circuit and protection.**   A portable appliance may be connected only to a receptacle having a rating at least as great as the appliance. The standard duplex convenience-outlet receptacle is rated at 15 A and may supply a single 15-A fixed appliance or a 12-A portable appliance if used on a 15-A branch circuit. Heavy-duty receptacles rated at 20, 30, and 50 A may be obtained for higher-current appliances. On new installations all 15- and 20-A receptacles must be of the grounding type. (Refer to Div. 4 for a description of receptacles.) Most household appliances, such as toasters, hot plates, percolators, flatirons, waffle irons, refrigerators, radiant heaters, roasters, portable ovens, etc., are rated at less than 12 A, so that they may be used in the standard outlet on a 15-A circuit.

Appliances other than motor-operated ones are not generally required to have individual overload protection.

**389. In wiring appliances** one of the approved types of cords listed in Div. 2 should be used. If the appliance is rated at more than 50 W and temperatures of more than 121°C (250°F) are produced on surfaces with which the cord is apt to be in contact, one of the types of heater cords must be used. Every heating appliance which is intended to be applied to combustible material as a smoothing iron must be equipped with a stand, which may be separate or a part of the appliance. Each heating appliance intended to be located in a fixed position must have ample protection provided between the appliance and adjacent combustible material.

**390. Disconnecting means** (National Electrical Code).   Each appliance shall be provided with a means for disconnection from all ungrounded conductors as follows:

1. PORTABLE APPLIANCES   For portable appliances (including household ranges and clothes dryers) a separable connector or an attachment plug and receptacle may serve as the disconnecting means. The rating of a receptacle or of a separable connector shall not be less than the rating of any appliance connected thereto, except that demand factors authorized elsewhere in the NEC may be applied. Attachment-plug caps and connectors shall conform to the following:

*a. Live parts.*   They shall be so constructed and installed as to guard against inadvertent contact with live parts.

*b. Interrupting capacity.*   They shall be capable of interrupting their rated current without hazard to the operator.

*c. Interchangeability.*   They shall be so designed that they will not fit into receptacles of lesser rating.

For household electric ranges, a plug-and-receptacle connection at the rear base of a range, if it is accessible from the front by removal of a drawer, is considered as meeting the intent of this rule.

2. PERMANENTLY CONNECTED APPLIANCES   For permanently connected appliances rated at not over 300 VA or ⅛ hp, the branch-circuit overcurrent device may serve as the disconnecting means. For permanently connected appliances of greater rating the branch-circuit switch or circuit breaker within sight of the user of the appliance, and capable of being locked in the open position, may serve as the disconnecting means.

3. UNIT SWITCHES WITH A MARKED "OFF" POSITION   Switches that are a part of an appliance and that disconnect all ungrounded conductors shall not be considered as taking the place

of the single disconnecting means required by this section unless there are other means for disconnection as follows:

*a. Multifamily dwellings.* In multifamily (more than two families) dwellings, the disconnecting means shall be within the apartment or on the same floor as the apartment in which the appliance is installed and may control lamps and other appliances.

*b. Two-family dwellings.* In two-family dwellings, the disconnecting means may be outside the apartment in which the appliance is installed. This will permit an individual switch for the apartment to be used and shall also be permitted to control lamps and other appliances.

*c. One-family dwellings.* In one-family dwellings, the service disconnecting means may be used.

*d. Other occupancies.* In other occupancies, the branch-circuit switch or circuit breaker, if readily accessible to the user of the appliance, may be used for this purpose.

4. SWITCH OR CIRCUIT BREAKER TO BE INDICATING    Switches or circuit breakers used as disconnecting means shall be of the indicating type.

5. MOTOR-DRIVEN APPLIANCES    A switch or circuit breaker which serves as the disconnecting means for a permanently connected motor-driven appliance of more than ⅛ hp shall be located within sight of the motor controller.

*Exception. A switch or circuit breaker that serves as the other disconnecting means as required in Par. 3 shall be permitted to be out of sight from the motor controller of an appliance which is provided with a unit switch or switches with a marked-off position and which disconnects all ungrounded conductors.*

**391. Portable appliances need to be grounded as specified in Sec. 250-45 of the** NEC .

**392. Signals and temperature-limiting devices for heated appliances.** In other than dwelling-type occupancies, each electrically heated appliance or group of electrically heated appliances intended to be applied to combustible material shall be provided with a signal unless the appliance is provided with an integral temperature-limiting device.

**393. The electric range** consists of a number of heater elements each rated at 1000, 1200, 1500, or 2000 W, so that it must usually be supplied with an individual three-wire branch circuit. The heater elements usually contain two units controlled by connecting the units in various combinations of series and parallel between the two lines and the neutral. A range, hot plate, or similar appliance with surface heating elements, having a maximum demand of more than 60 A, as calculated in accordance with Table 2 of Div. II, shall have the main circuit subdivided into two or more circuits, each provided with overcurrent protection rated at not more than 50 A. Infrared-lamp heating appliances shall have overcurrent protection not exceeding 50 A.

Many modern residential installations utilize counter-mounted cooking units and wall ovens.

## ELECTRIC COMFORT CONDITIONING

**394. Introduction.** As a practical guide to electric comfort conditioning, these sections are designed to provide a general understanding of when to use electric heating and cooling systems, how to estimate heat loss, what type of equipment is available, and where comfort conditioning equipment should be located.

**395. When to use electric heating and cooling.** Electric heating and cooling systems are used in all types of buildings, and there are outdoor spot-heating systems in addition to snow-melting and deicing systems.

**396. Residential buildings.** Electric heating and cooling systems for single-family dwellings and many multifamily dwellings have become accepted throughout the United States. Whether such a system is most desirable economically, a personal preference, or a luxury depends upon its usage by the occupants, the construction of the building, and the local electric utility rates (cost per kilowatthour of electricity for heating).

1. ADVANTAGES OF ELECTRIC HEATING    include individual room control, long life of the equipment, space-saving features, cleanliness, and safety. However, the user can abuse these advantages by leaving doors, windows, or fireplace chimney flues open; overheating; or improper building construction or insulation. Improper usage by the occupant is usually the reason for excessive utility bills.

2. INSULATION    If electric heat is to be operated economically, it is important to consider

the value of insulation. A well-insulated residence requires less equipment and incurs lower operating costs. Savings in equipment cost and annual operating costs generally pay for the increased cost of added insulation, storm windows, and storm doors.

3. SELECTING INSULATION   It is important to consider the effectiveness of insulation in terms of resistance to heat transfer and not just thickness. The effectiveness of insulation (for ceilings, walls, and floors) is judged in terms of its R (resistance) value. The insulation manufacturer will indicate the R value for a given thickness. For buildings with electric comfort conditioning, the National Mineral Wool Insulation Association specifies a minimum resistance as follows: ceilings = R19; walls = R11; floors = R13.

4. CONSTRUCTION OF THE WALL OR CEILING ROOF   The construction is not very significant if the minimum R value is provided. Storm windows (and doors) or double-glazed glass more than double the R value of the glass area.

5. HOUSE WITH A HEATED BASEMENT   Such a house need not have insulation under the floor, but there should be insulation under a floor which is over an exposed crawl space. The most effective insulation for concrete slabs is around the exposed outside edge. Perimeter-slab insulation can be extended down vertically about 24 in (609.6 mm) or can be installed around the edge of the slab at about the same distance.

6. INSULATION PROTECTION   To protect insulation from the negative effect of moisture from within the house, a vapor barrier on the inside face of the insulation is recommended. The vapor barrier must be protected from damage, and any tears or holes must be repaired.

7. WELL-INSULATED BUILDINGS   A well-insulated building may become an excessively humid building. To avoid excessive humidity buildup, a *humidistat* should be used to control one or more exhaust fans or a fresh-air ventilator. The humidistat closes the circuit to the fans when the relative humidity exceeds the preselected humidistat setting.

The operation of a humidistat is similar to that of a thermostat except that the sensing element responds to humidity instead of heat. Also, the element in a humidistat most commonly consists of a series of human hairs, which respond to changes in the relative humidity. In a home the relative humidity rises because of the use of dishwashers, dryers, washing machines, showers, sinks, and similar sources when use of water or production of moisture occurs.

8. HEAT LOSS   A residential building, insulated to the previously indicated specification, will have a heat loss of about 10 W/ft$^2$ or 1 W/ft$^3$.

9. HEAT-LOSS SURVEY   Before actual load requirements are determined, a heat-loss survey of all exposed rooms should be made. If, for example, in the New York City area, with an outside design temperature of 0°F (− 18°C) and an inside design temperature of 70°F (18°C), the heat loss is to be determined for a frame residence, the loss for each exposed room can be calculated in the same manner as in the following example for a 12- by 10- by 8-ft (3.7- by 3- by 2.4-m) corner bedroom (bedroom A).

| | Sq ft | Factor | | | Heat loss, watts |
|---|---|---|---|---|---|
| Gross outside wall windows..... | 176 | − | | | − |
| (and doors)......................... | 24 | × | 12 | = | 288 |
| Net outside walls.................... | 152 | × | 1.6 | = | 243 |
| Ceiling.................................. | 120 | × | 1.0 | = | 120 |
| Floor..................................... | 120 | × | 2.0 | = | 240 |
| Infiltration............................. | 960 | × | 0.4 | = | 384 |
| Total heat loss....................... | | | | | 1,275 |

10. STEPS IN CALCULATING HEAT LOSS

*a.* Determine heat-loss factor in watts per square foot by the R value. Refer to the *NEMA Electric Heating Guide, The Electric Heating and Cooling Handbook* (published by Edison Electric Institute), or the ASHRAE *Guide and Data Books.*

*b.* Multiply the total square-foot area for all exposed areas by this factor. (Deduct exposed-glass and door area from the gross wall area to determine the net wall area.)

*c.* Multiply the cubic-foot volume by the infiltration or air-change factor.

*d.* Add all products for the total heat loss of the room.

*e.* Add the heat loss of all the rooms to determine the total heat loss of the house. See the following example:

| Rooms | Heat Loss, Each Room, Watts |
|---|---|
| Living room | 2,350 |
| Bedroom A | 1,275 |
| Bedroom B | 980 |
| Bathroom | 610 |
| Kitchen–dining | 1,420 |
| Hall | 735 |
| Total heat loss | 7,370 |

11. DETERMINING ANNUAL OPERATING COST    Check with the local utility for the conversion factor for kilowatthours. If the heat loss is 7370 W and the conversion factor is 1.4, multiply the heat loss of 7370 W by 1.4, which equals 10,318, the total number of kilowatthours required to provide a 70°F (18°C) indoor temperature during the heating period. A similar heat-gain study can be made to determine the cooling load. Once the kilowatthour consumption is estimated, multiply 10,318 kilowatthours by 0.0458 (4.58 cents), the 1985 base rate for electric heating in Chicago. The product is $472.56, which is the estimated annual operating cost for electric heating. On the basis of an 8-month heating season, the average operating cost per month is $59.07. Heat-loss calculations of this type may also be made for apartment buildings, small commercial buildings, and other small buildings. For large commercial or institutional buildings, other factors, particularly that of fresh-air changes, must be considered.

**397. Commercial or industrial buildings.**    Many factors are involved in considering the use of electric heating in a commercial, industrial, or institutional building. Usually, a consulting engineer or heating specialist will perform a feasibility study which compares electric heating with other types of heating systems. The study will compare not only the cost of fuel but also the annual capital costs and maintenance costs over a period of years. The selection of electric heating, however, will generally depend upon the cost of electricity.

**398. Types of equipment.**    There are many types of electric heaters, varieties of housings, heating elements, and methods of distributing the heat into the heated area.

1. ELECTRIC HEATERS    There are convector, fan-forced, or radiant types of heaters. Each heater contains a resistance element which heats air or water. All such heaters provide both radiant and convection heat. The ratio depends upon the type of heater. Radiant heaters provide the highest radiation-to-convection ratios.

2. RESIDENTIAL BUILDINGS    A centralized system may be an electric furnace, boiler, or heat-pump system. A decentralized system could include the use, entirely or in combination, of baseboard heaters, wall heaters, floor–drop-in heaters, ceiling heaters, or ceiling panels or heating cables. The decentralized systems permit individual room control. Central systems have central control but often have the advantage of combining heating and cooling in the same duct system . The conversion of a fossil-fuel–fired hot-air or hot-water system can be readily accomplished by replacing the existing furnace or boiler with an *electric* furnace or hot-water boiler.

3. COMMERCIAL BUILDINGS    Central systems are available with large banks of electric duct heaters in conjunction with air handlers. Electric hot-water or steam boilers are available in large kilowatt capacities. Central heating and cooling systems can be used in conjunction with perimeter-baseboard–type heaters.

Perimeter heaters will counteract external-wall heat loss and allow control of the heaters in modular sections.

4. UNITARY SYSTEMS    In commercial buildings unitary systems permit combinations of heating and cooling with control of each room or area.

5. THROUGH-THE-WALL HEATING AND AIR-CONDITIONING UNITS    Particularly in school buildings, such units help to reduce engineering and maintenance costs. Unitary-equipment breakdown affects only a single room or area instead of the entire building or large section that is affected when a central system becomes defective.

6. ROOFTOP EQUIPMENT    Such equipment saves building space in low-ceiling commercial buildings while providing large-capacity electric heat and cooling into a ceiling plenum or duct system.

7. ELECTRICALLY HEATED BUILDINGS    Most such buildings will use electricity for domestic hot water. All-electric kitchens complete the application of electricity in the *total-electric* concept.

8. PARTIAL-BUILDING HEAT    This is another common application for unitary heaters.

9. INFRARED HEATERS    These provide spot heating for indoor or outdoor heating applications. Electric-heating cables or mats for snow melting, deicing, or pipeline heating offer convenience and reduction of costs unavailable in other systems.

**399. Equipment capacities.**    The following equipment is described for various applications in Secs. 400 to 406.

**400. Electric furnaces** (Fig. 9-260) are made in capacities of 4 to 35 kW at 240 V, single-phase or three-phase.

**Fig. 9-260**    A 20-kW electric furnace for homes. [Square D Co.]

**Fig. 9-261**    Central electric furnace for homes with heating elements, controls, blower motor, and ac cooling coil. [Square D Co.]

This heating assembly is a compact package heating system for new buildings or conversion applications in homes, apartments, or small commercial buildings. The furnace can be positioned in the ducts so that it can be used for upflow, downflow, or horizontal discharge in basements, attics, utility rooms, or utility closets.

Heating elements can be internally wired to be energized in 5-kW stages, at 60-s intervals, to prevent power surges. As part of the electric-furnace system, summer air conditioning or an electronic air cleaner may be added at any time without changing the blower or duct system. Some units are designed to be used with a split-system air conditioner. Low-voltage wall-mounted controls are usually installed in a central part of the building.

**401. Hydronic boilers** have capacities from 6 to 30 kW at 240 V, single-phase. These wall-mounted hot-water boilers are compact in size (between 2 and 8 ft$^3$, or between 0.057 and 0.227 m$^3$) and weigh about 120 lb (54.4 kg). A typical boiler has incremental stages (5 kW or less), which sequence at 30- to 45-s intervals to eliminate a large power surge. Centrally located low-voltage wall-mounted controls operate the boiler.

**402. Heat pump.**    Capacities range from 23,000 to 236,000 Btu for heating and 2 to 20 tons for cooling, single-phase or three-phase, at several voltage ratings.

The residential heat pump is a boxlike unit which is installed inside or outside the building, although it is usually outside with air-to-air heat transfer. The heat pump designed for outdoor installation is roof-mounted, through-the-wall, or remote-mounted. In cold climates (below 20°F, or −7°C, outdoor design) supplementary space-heating equipment is used in conjunction with the heat pump. An accessory kit is available for low-ambient operation.

The heat pump is regarded as a low–operating-cost heating and cooling system with a centrally mounted combination heating-cooling thermostat. Large-capacity heat pumps are sometimes used for commercial or industrial buildings.

**403. Residential unitary systems** include several types of electric heaters or combination units:

1. BASEBOARD HEATERS   Figure 9-262 shows a typical unit of these commonly used heaters. Capacities range from 375 to 3500 W at 120, 208, or 240 V.

These heaters are usually surface-mounted convector units with finned-tubular or cast-aluminum (Fig. 9-263) heating elements. Ratings are in terms of watt density. *Low watt density* is about 175 W/ft, *medium watt density* is about 225 W/ft, and *high watt density* is about 275 W/ft.

**Fig. 9-262**   Baseboard heating section. [Square D Co.]

**Fig. 9-263**   A 3-ton residential central air-conditioning condensing unit. [Square D Co.]

Heater lengths range from 2 to 12 ft (from 0.6 to 3.7 m), and heights are usually from 6 to 9 in (from 152.4 to 228.6 mm). The heaters are controlled by integral line-voltage thermostats, thermostat sections, line-voltage wall thermostats, or low-voltage wall thermostats.

These heaters are most commonly used to provide perimeter heating, particularly under the windows of outside walls. They can be easily installed in both new or existing buildings. Many baseboard heaters provide a full-length wireway. They have knockouts in the bottom, side, and back for electrical-supply connections. A plastic seal can be provided on the back of the heater to maintain a neat appearance when the heater is installed against an uneven wall. An air deflector directs heated air away from the wall. A free-floating element reduces expansion and contraction noises.

Heater accessories include blank sections which can be cut on the job for wall-to-wall installations. A thermostat section is available with or without duplex receptacles, although a receptacle section itself is available with some heaters. A control section, with or without a thermostat, may be used to switch the same source of power from an off position to a heat position or to a cool position. A room air conditioner is plugged into the top of this section. Also available is a section that contains a relay or transformer-relay for use with a low-voltage thermostat.

2. RADIANT WALL HEATERS    These heaters are also widely used. Capacities range from 500 to 6000 W at 120, 208, or 240 V.

Residential units of this type are for surface or recessed mounting. Small-capacity (500- to 1500-W) heaters are used in bathrooms or other small rooms. Large-capacity (2- to 6-kW) heaters are designed for add-on rooms, new construction, or existing room areas. The radiant heater may have a resistance element or glass panel with embedded resistance material. A typical glass unit, rated at 750 W, is shown in Fig. 9-264.

This heater features a built-in thermostat and shallow depth for recessed applications (2 in, or 50.8 mm, in some cases). For ease of installation, there are knockouts on the back, bottom, or side.

3. FORCED-AIR WALL HEATERS    Such a heater is shown in Fig. 9-265. Capacities range from 660 to 4000 W at 120, 208, 240, or 277 V.

These residential units can be recessed, semirecessed, or surface-mounted. Most commonly, this type of heater has a built-in thermostat with a positive off position. The lower-capacity heaters (660 to 1500 W) are used in bathrooms and other small rooms. The higher-capacity heaters (2 to 4 kW) are commonly used in basements, recreation rooms, or add-on rooms. The larger heaters offer several options. Some units have a fan-delay switch, a built-in safety switch, a fuse-disconnect, or a low-voltage transformer. A self-lubricating motor is common in the larger heaters. Also, a permanent type of filter is available in large-capacity heaters.

**Fig. 9-264**    A 750-W recessed heater with a glass heating panel. [Berko Electric Mfg. Corp., division of Weil-McLain]

**Fig. 9-265**    A recessed forced-air wall heater with built-in thermostat and fan switch. [Square D Co.]

Some of these heaters utilize the counterflow principle of heat and air distribution. They may be controlled by a separate line- or low-voltage thermostat in conjunction with an integral or remote relay. A fan switch permits summer operation.

4. CONVECTOR WALL HEATERS    Such a heater is shown in Fig. 9-266. Capacities range from 500 to 5000 W at 120, 208, or 240 V.

**Fig. 9-266**  Convector wall heater listed for Class I, Group D, hazardous locations. [Emerson Chromalox]

These convector units can be surface- or recess-mounted with integral or remote controls. Low-wattage heaters can be used in shops, corridors, breezeways, and basements. Larger-wattage convectors can be used to heat basements, attics, or any area where limited wall space prevents the use of baseboard heaters.

5. FAN-FORCED KICK-SPACE OR BASEBOARD HEATERS.    Such a heater is shown in Fig. 9-267. Capacities range from 750 to 2000 W at 208 or 240 V.

Such heaters can be surface- or recess-mounted. They are well suited to locations where baseboard radiation is desired but only limited space is available. The kick-space heater has a quiet tangential blower and is particularly suitable for kitchens or bathrooms where adequate wall space is not available. It is recessed into the kick-space area under a cabinet or counter. Control is by a separate wall-mounted line-voltage thermostat.

6. WALL-MOUNTED ELECTRIC FIREPLACES    Such a fireplace is shown in Fig. 9-268. Capacities range from 2000 to 5000 W at 240 V.

**Fig. 9-267**  Fan-forced recessed kick-space heater. [Emerson Chromalox]

This unit is a forced-air surface-mounted heater designed to look like a fireplace. It has a glowing grate with realistic logs in a steel cabinet which is finished in matte black or copper-tone enamel. It is ideal for dens, beach and mountain cabins, offices, motels, restaurants, and lounges. The electric fireplace can have a built-in thermostat and fan-delay control; it is also available with a remote wall-mounted low-voltage thermostat.

7. RADIANT-HEAT CEILING UNITS   Such a unit is shown in Fig. 9-269. The capacity of the heater is 1500 W at 120 V. Lighting consists of two 60-W incandescent lamps, and the small exhaust fan is rated at 60 ft³/min (0.028 m³/s).

This unit is recessed in the ceiling of bathrooms. A wall-mounted switch or time-control switch provides individual control of the heater and lamps.

8. RADIANT AND FORCED-AIR CEILING HEATERS   These heaters are available in capacities of 600 to 1250 W at 120 V.

A typical unit is attached directly to a flush 3½- or 4-in ceiling outlet box. This heater is commonly used in small bathrooms, playrooms, and powder rooms. A small fan is used to force heat away from the ceiling; the primary heat is radiant. In some heaters, a center-mounted 100-W incandescent lamp is available for lighting.

9. FORCED-AIR CEILING HEATERS   Such heaters range in capacities from 1000 to 1525 W at 120, 208, or 240 V.

Forced-air units can be surface-mounted on a standard flush 3½- or 4-in ceiling outlet box. Commonly used in bathrooms or powder rooms, the fan-forced heater is approved for standard 60°C supply wires, such as Type TW. It is available with adjustable mounting brackets and an automatic plug-in connection. Control is by a wall-mounted thermostat.

10. RADIANT CEILING HEATERS   These range in capacities from 600 to 1500 W at 120, 208, or 240 V.

The ceiling-mounted residential radiant heater can be recessed or surface-mounted on a standard flush 3½- or 4-in ceiling outlet box. Available with a center-located 100-W incandescent lamp and plug-in connections, the heaters are approved for 60°C supply wires. The radiant ceiling heater is commonly used in bathrooms, powder rooms, or other small rooms. Control is by a wall-mounted thermostat.

**Fig. 9-268**   Wall-mounted electric fireplace.

**Fig. 9-269**   Radiant heater with fan circulation for ceiling mounting. [Emerson Chromalox]

11. RADIANT CEILING PANEL HEATERS   These heaters range in capacities from 235 to 1000 W at 120, 208, or 240 V.

The residential radiant ceiling panel is constructed of laminated glass, gypsum drywall, or radiant metal. It can be surface-mounted, recessed, or used in a T-bar construction module, usually in panels 2 to 4 ft (0.6 to 1.2 m) wide and 3 to 12 ft (0.9 to 3.7 m) long. The depth of the heater is ¾ to 1 in (19.05 to 25.4 mm). The surface finish can be painted. A typical unit contains a 4-in terminal box and a 3-ft (0.9-m) flexible metal conduit. Controlled by a separate wall-mounted thermostat, the ceiling panel may be used for small rooms, spot heating, or complete area heating.

12. CEILING HEATING CABLE   A typical run of this cable is shown in Fig. 9-270. Capacities

range from 200 to 5000 W at 120, 208, or 240 V, and lengths are from 75 to 1800 ft (from 22.9 to 548.6 m).

Ceiling cable is embedded in plaster or laminated ceiling construction for individual rooms or complete house heating. Color-coded cables are electrically insulated and resistant to high temperature, water absorption, aging effects, and chemical action. The cables are usually rated at 2¾ W/ft. They must not be cut in the field, and identification labels must not be removed.

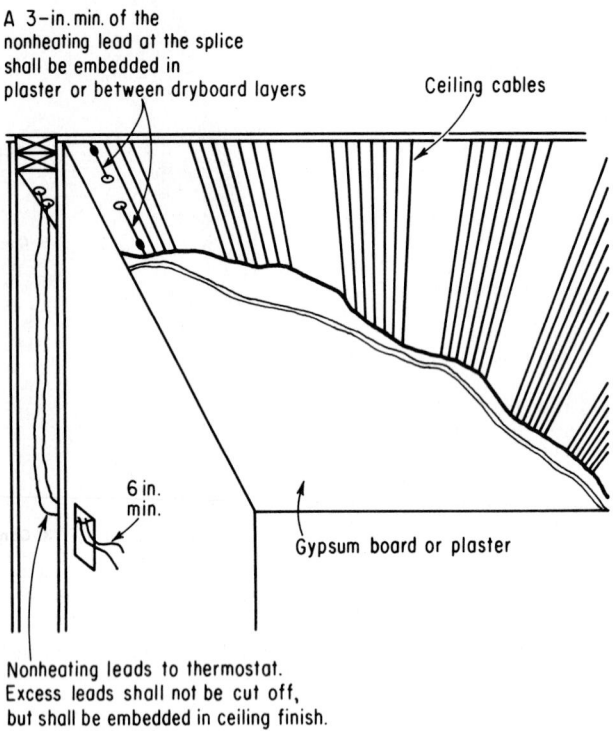

A 3-in. min. of the
nonheating lead at the splice
shall be embedded in
plaster or between dryboard layers

Ceiling cables

6 in.
min.

Gypsum board or plaster

Nonheating leads to thermostat.
Excess leads shall not be cut off,
but shall be embedded in ceiling finish.

**Fig. 9-270**  Installation of ceiling heating cable. Adjacent cable runs must be spaced not less than 1½ in (38 mm) on centers. [*Electrical Construction and Maintenance*]

Nonheating lead wires, 7 ft (2.1 m) or longer, extend to the thermostat or a junction box. For installation (Fig. 9-271), the cable is stapled to a nonmetallic fire-resistant surface at least every 16 in (406 mm). After the first plaster coat has been applied, cables should be tested for continuity and for insulation resistance (at least 100,000 Ω measured to ground).

Minimum spacing between cable runs is 1½ in (38 mm) on centers with the closest spacing near the outside walls. Cables should clear lighting fixtures by at least 8 in (203 mm). General electric wiring (for lights, etc.) should be run over the thermal insulation and at least 2 in (50.8 mm) over the heated ceiling surface.

**13. Floor Heating Cable**  Such cable has unlimited heat capacities which vary with the conductor resistance, type of cable, and length of cable. Ratings are from 120 to 600 V. In residential occupancies ratings are 120, 208, or 240 V.

Heating cables are used in floors to form radiant panel systems. Cables most commonly used are polyvinyl chloride–sheathed cables or mineral-insulated (MI) copper-sheathed cables. In concrete, cables must be of a material that is resistant to any chemical action in the concrete. Cables may be interwoven as part of a properly grounded prefabricated mat. They may be of predesigned lengths which are spaced to provide a desired radiant-heat output. Cables such as Type MI are available at several different resistances. Single- or two-conductor cables at specified lengths will have varied heat capacities, depending upon the applied voltage. Heating cables may be installed in one or two pours with adjacent runs not less than 1 in (25.4 mm) on centers and with adequate spacing between cable runs and metallic bodies embedded in the floor. Nonheating leads must be protected when they leave the floor by rigid metal conduit, IMC, or EMT or, with MI cable, by approved current-carrying nonheating cable sections. Control of floor heating systems is by a thermostat with a capillary and sensing element attached. The sensing bulb is embedded in the slab so that it will respond to typical temperature changes.

14. FLOOR DROP-IN HEATERS   Such a heater is shown in Fig. 9-272. Capacities range from 300 to 2000 W at 120, 208, 240, or 277 V.

Residential floor drop-in heaters are resistance convector units, fan-forced units, or units that have a chamber with antifreeze water sealed in and heated electrically. They are best applied to counteract downdrafts from floor-to-ceiling windows or sliding glass doors. They can be installed in wood or concrete floors. A fan-forced heater may have an integral thermostat, but the control for most units is by a line- or low-voltage wall thermostat.

**404. Central systems for commercial-type occupancies** are of the types described in the following paragraphs.

1. OUTDOOR ROOF-MOUNT HEATING AND COOLING UNITS   Such a unit is shown in Fig. 9-273. Electric-heating capacities are custom-designed for a particular installation.

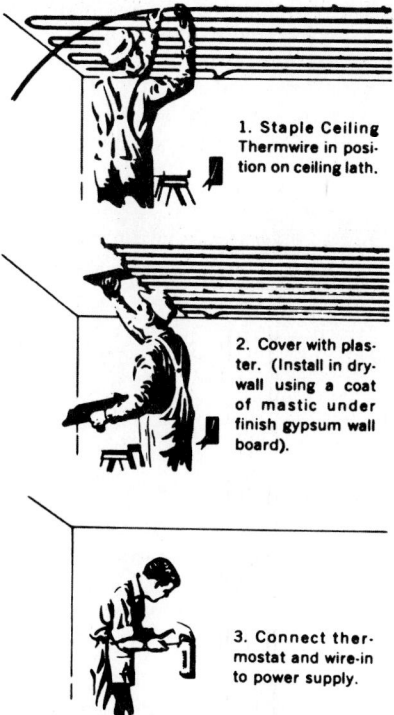

1. Staple Ceiling Thermwire in position on ceiling lath.

2. Cover with plaster. (Install in drywall using a coat of mastic under finish gypsum wall board).

3. Connect thermostat and wire-in to power supply.

**Fig. 9-271**   Three steps in installing ceiling heating cable. [Emerson Chromalox]

**Fig. 9-272**   Drop-in floor heater. [Emerson Chromalox]

Cooling capacities range from 2 to 200 tons at 208 V, single- or three-phase; 277 V, single-phase; and 480 V, single- or three-phase.

All-electric heating and cooling machines are designed for roof-mount outdoor conditions. Custom fabrication provides machines with job-specified heating and cooling capacities. Multizone staging from a remote monitoring panel can establish the roof-

**Fig. 9-273**    Large-capacity electric heating-cooling central-system unit mounted on roof.

mount machine as the heating and cooling plant for a one- or two-story building. Supermarkets, office buildings, laboratories, warehouses, machine shops, and aircraft plants are good applications for roof-mount equipment. Space-saving economy is a major feature of such equipment. Staged or step control for the heating section, with fresh-air damper control, provides a well-balanced automated system.

2. COMMERCIAL STEAM OR HOT-WATER BOILERS (Fig. 9-274) These boilers are available in capacities up to 3000 kW at 208, 240, and 480 V, single- or three-phase.

Commercial steam or hot-water boilers are designed for conversion of electric energy to heat in the form of steam or hot water. Hot-water boilers are used as a replacement for fuel-fired boilers; as the heating component for a chilled water-cooling system; as a safe method of heating a hazardous area; or as a side-arm–faucet water heater. A typical boiler is provided with a low-water cutoff, a safety relief valve, temperature control, a water gage, a high-temperature limit switch, contactors, internal fusing, full insulation, and an enameled-steel jacket. The electric boiler occupies a much smaller floor area than other types of boilers and does not require retubing, flue cleaning, burner cleaning, damper adjusting, or chemical additives.

Steam boilers are available with ½ to 30 boiler hp (4.9 to 294.3 W), 1 to 250 lb/in², and are built to ASME and Underwriters Laboratories standards. Boilers include a

**Fig. 9-274**    Commercial-type electric boiler.

safety valve, a boiler steam pressure gage, a pressure regulator, a low-water cutoff, an automatic-feed control, a control switch, a pilot light, outlet valves, and drain valves. A condensate tank is recommended to reuse the condensed steam.

**405. Commercial unitary systems** are described in the following paragraphs.

1. INFRARED HEATERS   Such a heater is shown in Fig. 9-275. Capacities are 500 to 7600 W at 120, 208, 240, 277, 440, 480, and 600 V.

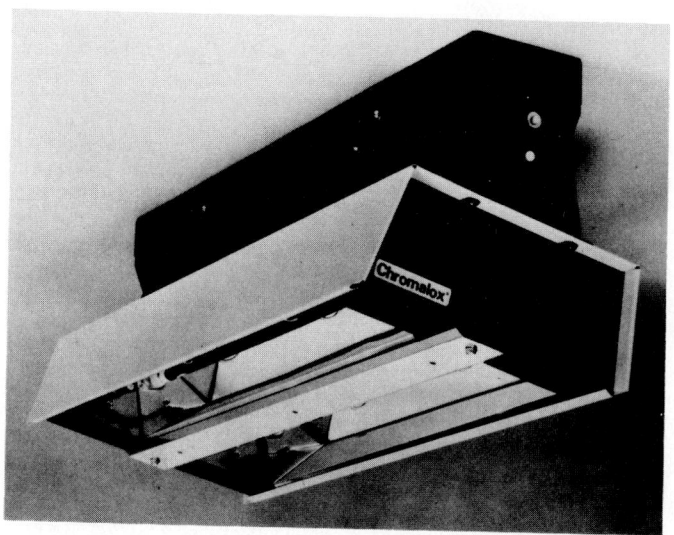

**Fig. 9-275**   An infrared heater with quartz lamps. [Emerson Chromalox]

The heating elements consist of metal-sheathed units, quartz tubes, or quartz lamps. The following table describes the characteristics of these three infrared sources:

**Characteristics of Three Infrared Sources**

| Characteristic | Quartz Lamp | Quartz Tube | Metal Sheath |
|---|---|---|---|
| Relative heat intensity | Very high | Low to medium high | Medium to high |
| Splash resistance | Very good | Very good | Very good |
| Color blindness | Fair | Very good | Very good |
| Life expectancy | 5,000 hr or more | 5,000 hr or more | Consult manufacturers |
| Visible light | High | Low | Low |
| Heat-up and cool-down time | A few seconds | 2 min or more | About 2 min |
| Relative cost per kw | Medium high | Medium | Medium |
| Filament or source temperature | About 4,000°F | About 1,200 to 1,800°F | About 1,200 to 1,800°F |
| Vibration resistance | Medium | Medium | Very good |

Infrared heaters are used for complete indoor heating, spot heating, outdoor heating, and outdoor snow melting. For outdoor heating and snow melting it is preferable to use the quartz lamp and to shield the element and fixture from strong winds as much as possible. Rapid cooling of the element diminishes radiation effectiveness. For spot heating, use two fixtures aimed at the center of the target area at about a 45° angle. Do not irradiate outside walls or surfaces over 6 ft (1.8 m) above the floor. For complete comfort heating, install ½ W for each degree of operational temperature desired above the minimum ambient ever reached in a specific area. For spot heating provide 1 W/ft²/degree of temperature difference. For outdoor heating provide 2 W/ft²/degree of temperature difference. If, for example, the area to be heated is a storefront sidewalk 100 ft (30.5 m) long by 10 ft (3 m) wide in New York City, proceed as follows (assuming an outside design of 0°F and a desired temperature of 50°F):

$$100 \times 10 = 1000 \text{ ft}^2$$
$$\underline{\times\ 2 \text{ W/ft}^2}$$
$$2000$$
$$\underline{\times 50}$$
$$100,000 \text{ W, or } 100 \text{ kW}$$

Select fixtures in accordance with the manufacturers' recommendations. The radiation patterns are determined by the reflector design and the resulting beam pattern. Beam patterns range from a 30° to a 90° angle. They may be symmetrical or asymmetrical. Fixtures may have one, two, or three elements. Marquee-type heaters are recessed into an overhead masonry canopy. For snow-melting applications, twice the suggested outdoor capacity is recommended. Provide 150 to 200 W/ft² of area for most applications. Control of infrared heaters is by switch, input control, ambient-temperature thermostat, or any combination of these devices. A high-limit thermostat is recommended for most applications.

2. HORIZONTAL UNIT HEATERS  Such a heater is shown in Fig. 9-276. Capacities range from 1.5 to 50 kW at 120 or 208 V, single- and three-phase; 240 V, single- and three-phase; 277 V, single-phase; and 480 or 600 V, single- and three-phase.

Horizontal unit heaters can be installed on the wall or hung from the ceiling with bracket or swivel adjustments. They are ideal for use in warehouses, factories, or entranceways. Unit heaters are normally located along the outside wall with the direction of airflow in a circular fashion. Large-capacity heaters can be provided with duct collars or deflectors. Unit heaters are commonly controlled by a remote thermostat and may have remote or built-in contactors. A fan-delay switch can be provided to prevent the fan from operating until the heating elements have reached the desired temperature. An explosionproof model is available to conform with Class I, Group D, requirements.

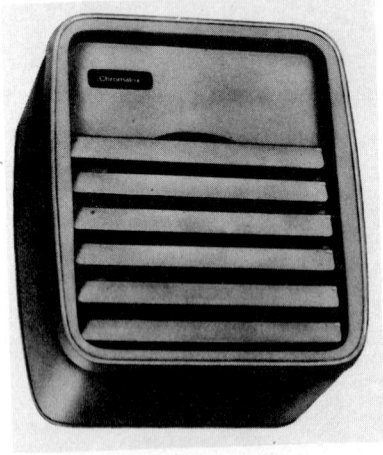

3. VERTICAL UNIT HEATERS  Several types of these heaters are shown in Fig. 9-277. Capacities are 10 to 50 kW at 208 and 480 V, single- or three-phase. The ceiling-suspended vertical unit heater has integral relays and one- or two-stage remote control. It can be surface-mounted or recessed with air discharge through four optional louver arrangements which permit virtually any desired air pattern at the floor level. Diffusers are of the radial, cone, anemostat, or louver type. Internally mounted automatic-control relays and overheat and motor-overload protection are usually provided. These heavy-duty units can be used in factories, garages, and warehouses, with high-bay mounting up to 33 ft (10.1 m) above the floor.

**Fig. 9-276** Horizontal unit heaters can be mounted on wall or ceiling. [Emerson Chromalox]

4. AUDITORIUM UNIT VENTILATORS  Such a ventilator is shown in Fig. 9-278. Capacities are 12 to 400 kW; 1200 to 15,000 ft³/min (0.57 to 7.08 m³/s), and 208, 240, or 480 V, three-phase. This unit ventilator provides the proper heating, ventilating, and natural cooling required in school auditoriums, gymnasiums, libraries, and other areas where large-capacity systems and quiet operation are essential. It may be installed for horizontal, upright, or inverted operation on the ceiling, wall, or floor; directly in the area to be served; or in an adjacent location and connected by ductwork. Such a unit features a special attenuator and an enclosed motor and drive designed for quiet operation. Step control is frequently used.

5. WALL CONVECTOR UNITS  A typical unit is shown in Fig. 9-279. Capacities are 750 to 4000 W at 208, 240, and 480 V, single- and three-phase. Commercial wall convectors

can be surface-mounted or recessed with front or top discharge. These units are used in commercial, industrial, or institutional buildings in offices, corridors, and entrances. Cabinet convectors are available with integral or remote controls.

6. COMMERCIAL FORCED-AIR WALL HEATERS    This type of heater is similar to the heater shown

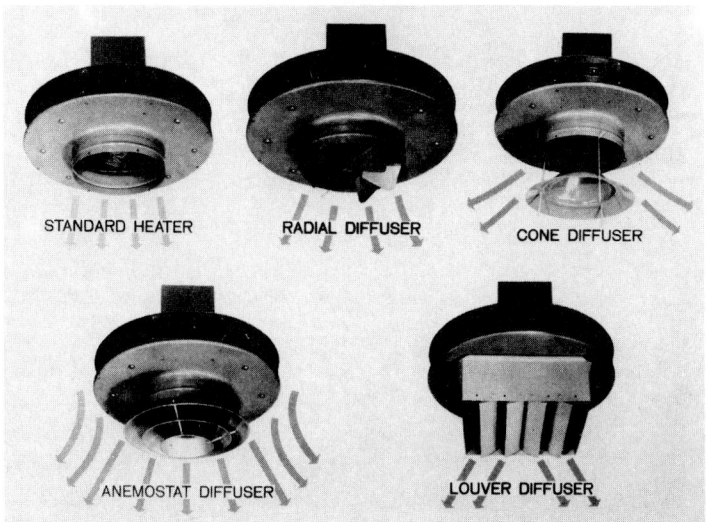

STANDARD HEATER    RADIAL DIFFUSER    CONE DIFFUSER

ANEMOSTAT DIFFUSER    LOUVER DIFFUSER

**Fig. 9-277**    Types of vertical unit heaters. [Emerson Chromalox]

**Fig. 9-278**    Auditorium unit ventilator, 15 kW. [Square D Co.]

in Fig. 9-265. Capacities are 1.5 to 4 kW at 120, 208, 240, or 277 V, single-phase. The commercial forced-air heater can be surface-mounted or recessed. Tamperproof construction makes this unit ideal for entryways, lobbies, corridors, stairwells, and rest rooms in all types of commercial buildings. Fan and discharge louvers direct air downward, keeping the floor warm and dry.

7. SILL-LINE CONVECTOR UNITS    Such a unit is shown in Fig. 9-280. Capacities are 120 to 750 W/ft at 208, 240, 277, and 480 V, three-phase, in one- or two-stage control. Commercial sill-line heaters are available at different heights and many different wattage outputs. Lengths vary from 2 to 8 ft (0.6 to 2.4 m) to provide modular or wall-to-wall design. Units

of this type are available with blank sections, filler sections, matching sleeves, end caps, inside and outside corners, and conduit covers. Sill-line heaters are surface-mounted with front or top discharge when perimeter radiation is desired. They should be mounted at least 2 in (50.8 mm) above the floor. Built-in controls may be integral to the unit, or they can be provided in a filler section. The filler section can contain a relay for a 120-V remote-control circuit, a disconnect switch, or a circuit breaker.

**Fig. 9-279**   Wall convector unit for surface mounting. [Emerson Chromalox]

**Fig. 9-280**   Sill-line convector heater. [Emerson Chromalox]

8. PEDESTAL CONVECTOR HEATERS   Such a heater is shown in Fig. 9-281. Capacities range from 500 to 3500 W at 208, 240, and 277 V, single-phase. This heater is a convector which is used to heat floor-to-ceiling glass areas. It can be mounted on a new concrete floor or be bolted to the surface of an existing floor. Pedestal-type heaters can be installed individually or end to end in continuous runs. They provide a basic perimeter system designed especially for corridors, lobbies, foyers, and entryways, particularly for hard-to-

heat areas where floor-to-ceiling glass creates cold downdraft problems and prohibits the use of wall-mounted equipment. This special downdraft heater is controlled by a separate wall-mounted thermostat with an optional integral relay for remote control by a time clock.

9. DRAFT-BARRIER WALL HEATERS   Such a heater is shown in Fig. 9-282. Capacities are 100 to 200 W/ft at 120, 208, 240, and 277 V, single-phase. The draft-barrier wall heaters are low-silhouette units (3 to 6 in, or 25.4 to 76.2 mm, high) designed to counteract downdraft at window areas. They are mounted under the sill, from wall to wall at floor level, or within modular confines in commercial office buildings. Electric connections can be made from either end on some units or through the back. Built-in thermal relays and thermostats are available.

**Fig. 9-281**   Pedestal convector heater. [Emerson Chromalox]

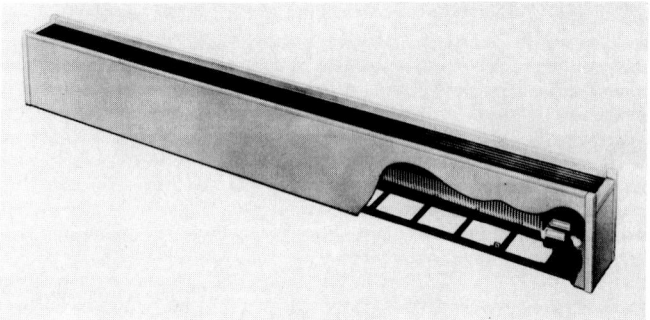

**Fig. 9-282**   An architectural electric convection heater provides a draft barrier when installed along glass curtain walls. [Emerson Chromalox]

10. COMMERCIAL BASEBOARD HEATERS   These heaters are similar to the unit shown in Fig. 9-262. Capacities are 300 to 1250 W at 120, 208, 240, and 277 V, single-phase. Commercial-grade baseboard heaters are available in lengths of 2, 5, and 8 ft (0.6, 1.5, and 2.4 m). They are surface-mounted 2 in (50.8 mm) off the floor. These units can be controlled by an attached thermostat section or by a wall-mounted line- or low-voltage thermostat. End caps, blank sections, and inside and outside corners for wall-to-wall installation are available as required.

11. CABINET UNIT HEATERS   Such a heater is shown in Fig. 9-283. Capacities are 1.5 to 30 kW at 208, 240, 277, and 480 V, single-phase or three-phase.

The commercial cabinet unit heater can be used freestanding, wall-mounted, or ceiling-mounted. Surface or recessed installation is applicable to wall- or ceiling-type heaters. An inverted model is available for unusual installations. The cabinet unit heater can be provided with tamperproof integral or remote wall-mounted controls. A two-or three-speed fan motor, providing 200 to 1250 ft³/min (0.094 to 0.59 m³/s), may have a fan selector switch to provide quiet operation. A fan-delay thermostat is used to prevent circulation of unheated air. This type of heater is designed primarily for entranceways, corridors, and lobbies in offices, airport terminals, schools, and retail stores. There are some units which can provide a choice of 25 or 100 percent outside air with dampers for ventilation cooling and air filters that can be cleaned.

12. UNIT VENTILATORS AND AIR CONDITIONERS    Such a ventilator and air conditioner is shown in Fig. 9-284. Heating capacities are 1.3 to 36 kW at 208, 240, 277, and 480 V, single- or three-phase. Cooling capacities range from 8000 to 27,000 Btu.

The packaged electric unit ventilator and its companion unit, the unitary air conditioner, are through-the-wall machines. The electric unit ventilator is installed under win-

**Fig. 9-283**    Cabinet unit heater with top discharge and front inlet. [Emerson Chromalox]

**Fig. 9-284**    Combination heating-cooling ventilator.

dows, has a wall sleeve and louvered opening in the wall, and is designed for use in school classrooms, auditoriums, gymnasiums, offices, libraries, or other high-occupancy areas. It heats, cools, and ventilates with outdoor air (600 to 1500 ft³/min, or 0.28 to 7.08 m³/s). Usually, there is internal wiring for step-control switching. An optional built-in electric demand limiter is available. Controls may be integral or remote; electric, electronic, or pneumatic. The controls of one unit (the master) may be used in the same area to control other units (slaves). The unit ventilator is usually installed against an outside wall, though it may also be ceiling-mounted. The through-the-wall heating and cooling unit may have an integral cooling (compressor) section, or it may be remote-mounted.

13. DUCT HEATERS A typical heater is shown in Fig. 9-285. Wattage combinations are virtually unlimited with single- and three-phase connections and multistage control at 120

**Fig. 9-285**  Duct heater section with six elements.

to 600 V. Electric duct heaters are used in ducts, plenums, or air handlers for primary heat, terminal booster heat, or air-conditioning reheat. Open-coil or metal-sheathed finned tubular heaters (Fig. 9-285) are available for insertion into an existing duct or, with a flange, as part of a new duct. To conform to duct sizes, heaters are manufactured in different horizontal and vertical dimensions for mounting on the side or bottom of the duct. Integral controls may include automatic reset of thermal cutouts, built-in relays, a transformer, fuses or circuit breakers, a fan interlock relay, pilot switches, and pilot lights. The NEC specifies a maximum 60-A branch circuit for overcurrent protection with a maximum connected load of 48 A, and zero clearance, automatic reset of thermal overload, and separate manual reset for a backup thermal overload system. Duct heaters are controlled by duct stats or remote wall controls. For large-capacity heaters a step controller is used.

**406. Electric snow-melting systems** consist of resistance-type wire or cable embedded in concrete or asphalt. The design of snow-melting systems depends upon the area, geographical conditions, wind conditions, and expectations of the user (see Table 30 of Div. 11). Generally, for melting snow at temperatures between 32 and 0°F (between 0 and −16°C), the system will be designed at 40 to 60 W/ft². Depending on wind factor, temperature, and the rate of snowfall, the area will be kept free and clear of any snow formation if the system is energized about 30 min prior to a snowfall.

The heat-transfer environment may be asphalt or concrete, and the heating element may be in the form of a mat or a heating cable. In asphalt, install the mats 12 in (304.8 mm) from the edge with the largest mats on a straight run. Mats may be installed on an existing slab with a 2-in (50.8-mm) blacktop surface cover. The surface asphalt should be fine-grain with no stones larger than ⅜-in (9.525-mm) diameter.

In concrete, the mats or cable should be located within expansion joints generally no farther than 20 ft (6.1 m) apart. The concrete slab should contain crushed-rock aggregate (not river gravel) and should be at least 4 in (101.6 mm) thick. There should be a well-prepared base with adequate drainage and suitable reinforcement.

Junction boxes for connection of feeders to cables or mat lead wires should be located out of the area or in a nearby wall. They should not be installed flush with the snow-melting–surface area. Mats or cables should be installed from 1½ to 2 in (from 38.1 to 50.8 mm) below the surface. If they are installed in two pours of concrete, a binder or bonding agent should be used between pours. Cables should be tested before installation, during installation, and on completion of installation.

Snow-melting mats for walkways, driveways, or steps (Fig. 9-286) are made of resistance wires embedded in a polyvinyl chloride sheath, interwoven in chicken wire or plastic webbing. The wire may have a copper overbraid for positive grounding.

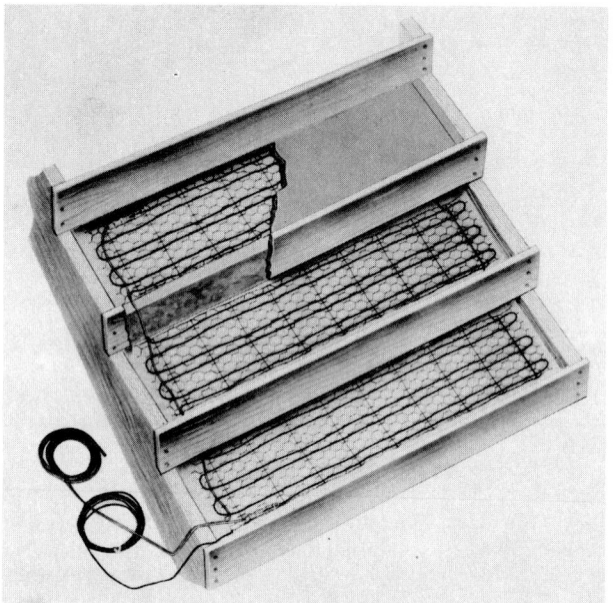

**Fig. 9-286**  Snow-melting cable for steps.

Single- or two-conductor Type MI (mineral-insulated) cable with varied resistances is factory-fabricated at specified lengths of heating leads and nonheating leads. Hot-to-cold wire connectors and end caps are factory-brazed. Type MI cable is semirigid and can be hand-bent at the construction site to conform to desired spacing (usually 3 to 9 in, or 76.2 to 228.6 mm). It has the advantage of being durable in different asphalt or concrete environments.

Heating cables may be controlled by switch or by automatic snow control (Fig. 9-287). It is recommended that a high-limit thermostat be installed in the slab to prevent the system from operating at higher temperatures (50°F, or 10°C, or more).

**407. Electric-heating controls.**  Electric space-heating equipment may be specified with a built-in thermostat and relay or with a wall thermostat. Commercial installations often require the use of a contactor. Step controllers or solid-state controls may be used with electric comfort heating equipment. For safety, most heaters are provided with spot or linear overheat protection.

**408. Sizing room air conditioners.**  As a general rule, room air conditioners are sized on the basis of 30 Btu/h/ft² of space in a given room. Room air conditioners are rated in Btu (British thermal units) of cooling capacity. Units are available at 120-V ratings up to 12,000 Btu in most cases. Current ratings range from 7½ to 12 A for 120-V units. In planning outlets, however, it is important to know whether the room air conditioner (window or through-the-wall type) will be 120 or 240 V, because some manufacturers provide 240-V ratings for units of 9000 Btu or more.

**409. Code rules for fixed space-heating equipment.**  Article 424 of the NEC covers installations of fixed electric space-heating equipment. The following paragraphs describe the major provisions of this article.

1. Branch Circuits  All circuits for space-heating, snow-melting, and deicing equipment are considered continuous loads. As such, the load on any single or multioutlet branch circuit cannot exceed 80 percent of the branch-circuit rating (12 A on a 15-A circuit, 16 A on a 20-A circuit, etc.). Generally, space-heating equipment can be connected to 15-, 20-, or 30-A multioutlet branch circuits. Infrared or snow-melting systems can be connected to 15-, 20-, 30-, 40-, or 50-A multioutlet branch circuits. Refer to Div. 3 for feeder or service calculations for a group of heaters.

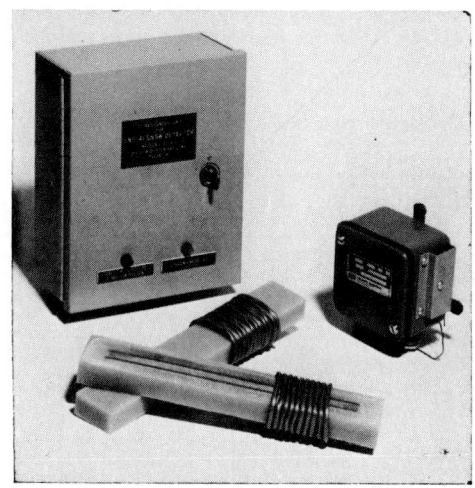

**Fig. 9-287**  Automatic snow-control device. [Nelson Electric]

2. Controllers and Disconnecting Means
   *a.* Thermostats and thermostatically controlled switching devices which indicate an off position and which interrupt line current shall open all ungrounded conductors when the control device is in this off position.
   *b.* Thermostats and thermostatically controlled switching devices which do not have an off position are not required to open all ungrounded conductors.
   *c.* Remote-control thermostats need not meet the requirements of Pars. *a* and *b*. These devices shall not serve as the disconnecting means.
   *d.* Switching devices consisting of combined thermostats and manually controlled switches that serve both as controllers and disconnecting means shall open all ungrounded conductors when manually placed in the off position or be so designed that the circuit cannot be energized automatically after the device has been manually placed in the off position.

3. Duct Heaters
   *a.* Means shall be provided to assure uniform and adequate airflow over the heater.
   *b.* Means should be provided to ensure that the fan circuit is energized when the first heater circuit is energized.
   *c.* Each duct heater shall be provided with an integral approved automatic-reset temperature-limiting control or controllers to deenergize the circuit (or circuits). In addition, an integral, independent, supplemental control or controllers shall be provided in each duct heater which will disconnect enough conductors to interrupt current flow. This control shall be manually resettable or replaceable.
   *d.* Duct-heater controller equipment shall be accessible with the disconnecting means installed at or within sight of the controller except as permitted by Sec. 424-19 (a) of the Code.

**410. Other reference material.** With the continued development of the electric-heating market, reference should be made to the latest NEMA standards and publications of the Electric Heating Association (EHA).

### WIRING FOR ELECTRIC SIGNS AND OUTLINE LIGHTING

**411. Types of signs.** Electric signs or outline lighting may consist of incandescent lamps or gaseous-discharge tubing (commonly called neon signs). Refer to Div. 10 for a discussion of lamps and tubing.

**412. Methods of wiring incandescent electric signs** (National Electric Light Association, *Data on Electric Signs*). Lamps burning in multiple may be connected either two-wire or three-wire, as shown in Fig. 9-288. In series wiring, lamps may be connected in either

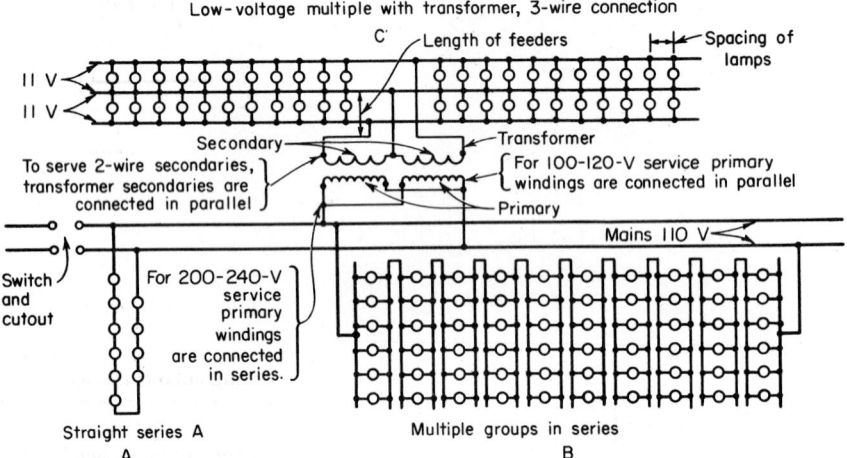

NOTE: Diagram C for alternating current only; others may be used on ac or dc.

**Fig. 9-288** Methods of connecting sign lamps.

straight series or multiple series, as shown. If transformers (see Sec. 60 of Div. 5 for information on sign transformers) are used to obtain low voltage, lamps may be connected either two-wire or three-wire as in standard multiple wiring, the transformer reducing the voltage from the regular 110- or 220-V circuits to the voltage required by the lamp. The ordinary multiple wiring can be changed to straight series wiring by merely clipping the alternate connections between lamps (Fig. 9-289).

In a large sign any combination of series or multiple series may be used. With straight series wiring, should one lamp in the series burn out, all the lamps in that series will be out. If the lamps are connected in multiple series, the failure of one lamp does not cause any of the other lamps to go out. However, there should not be less than 8 to 10 lamps in each multiple group, or the failure of one lamp will cause too much current to flow through the other lamps of the same group, thus shortening their lives.

**413. Different effects of flashing or motion** can be obtained with incandescent signs by the use of a sign flasher. Cams or a drum are mounted on a shaft that is rotated by a small electric motor (Fig. 9-

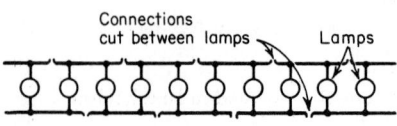

**Fig. 9-289** Method of changing sign wiring from multiple to series.

290). The circumferences of the cams or drums are so cut that, in brush-type flashers, the brushes will make contact only during certain predetermined portions of a revolution and thereby complete the electric circuit through the sign lamps only during that period. In the carbon-type flashers the cams, instead of carrying current and making and breaking the contacts directly, operate to open and close carbon-break contact switches which control the sign lamps. The possible variations in arrangement of cams and drums for produc-

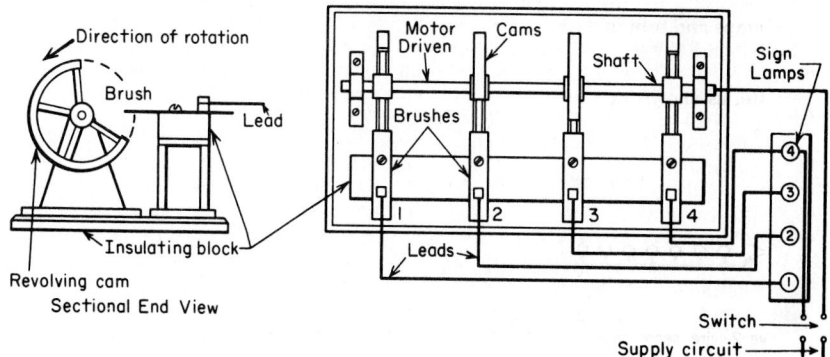

Plan and Wiring Diagram

**Fig. 9-290**   The elementary sign flasher.

ing different effects are almost numberless. Flashers are available for high- or low-voltage circuits.

**414. Method of wiring neon-lamp signs.**   Neon lamps used for signs are of the high-voltage type (see Div. 10). The lamps are supplied from ordinary lighting circuits through small step-up transformers. The voltage required depends upon the length of neon-lamp tubing used in the sign. The maximum voltage employed on the lamps is 15,000. When the length of tubing is greater than can be illuminated with 15,000 V, two or more transformers are employed as required. Practically always, neon signs are completely wired and installed by the manufacturers, the purchaser simply providing the necessary outlets from the general-lighting system. Also, the servicing and maintenance of the signs generally are handled by specialists.

**415. The installation of all signs or outline lighting,** whether of the incandescent or the gaseous-discharge type, must meet the following requirements: All equipment and devices used with signs must be enclosed in metal boxes. Except for portable incandescent signs, the boxes must be grounded unless they are insulated from ground and other conducting surfaces and are inaccessible to unauthorized persons. Each sign, except a portable one, must be controlled by an enclosed, externally operable switch or breaker located within sight of the sign, except signs operated by electronic or electromechanical controllers located external to the sign shall have the disconnect located within sight of the controller location and shall be capable of being locked in the open position. All devices, such as switches, flashers, and similar devices controlling transformers shall be either rated for controlling inductive loads or must have a current rating which is not less than twice the current rating of the transformer supplying the sign. The circuits should be so arranged that none will have a load of more than 15 A.

**416. The installation of signs (600 V or less)** must be in accordance with the following NEC requirements. Conductors shall be installed as follows:

1. METHOD OF WIRING   Any wiring method in Chapter 3 of the NEC suitable for the conditions.

2. INSULATION AND SIZE   Conductors shall be of a type approved for general use and, except in portable signs and for short leads permanently attached to lampholders or ballasts, shall be not smaller than No. 14.

3. CONDUCTORS EXPOSED TO THE WEATHER   Conductors in raceways, metal-clad cable, or enclosures exposed to the weather shall be of a type listed for the condition.

4. NUMBER OF CONDUCTORS IN RACEWAY    For signs the number of conductors in conduit or tubing may be in accordance with Table 29 of Div. 11.

5. OPEN CONDUCTORS    Open conductors on insulators shall comply with the provisions of the Code as given in Secs. 32 to 53 and, if outdoors, with the requirements given in Div. 8, except that the separation between conductors concealed indoors need be only 1½ in (38 mm) for voltages above 10,000 and not more than 1 in (25.4 mm) for voltages of 10,000 or less.

6. CONDUCTORS SOLDERED TO TERMINALS    If the conductors are fastened to lampholders other than of the pin type, they shall be soldered to the terminals or made with wire connectors, and the exposed parts of conductors and terminals shall be treated to prevent corrosion. If the conductors are fastened to pin-type lampholders which protect the terminals from the entrance of water and which have been found acceptable for sign use, the conductors shall be of the stranded type but need not be soldered to the terminals.

7. LAMPHOLDERS    Lampholders shall be of the unswitched type with bodies made of suitable insulating material and shall be so constructed and installed as to prevent turning. The screw-shell of all sign lampholders shall be connected to the grounded conductor of the circuit. Lampholders for outdoor signs shall be suitable for the application.

**417. The installation of signs and outline lighting exceeding 600 V** must be in accordance with the following NEC requirements (Secs. 418 to 423).

**418. Installation of conductors.**    Conductors shall be installed as follows:

1. WIRING METHODS    Conductors shall be installed as concealed conductors on insulators or in rigid metal conduit, intermediate metal conduit, rigid nonmetallic conduit, liquid-tight flexible metal conduit, Type MC cable, flexible metal conduit, or electrical metallic tubing, except that liquidtight flexible nonmetallic conduit may be used where requirements for flexibility exist and where weather conditions require corrosion protection. Conductors may be run from the ends of tubing to the grounded midpoint of transformers specifically designed for the purpose and provided with terminals at the midpoint. When such connections are made to the transformer-grounded midpoint, the connections between the high-voltage terminals of the transformer and the line ends of the tubing shall be as short as possible.

2. INSULATION AND SIZE    Conductors shall be of a type identified for the voltage of the circuit and shall not be smaller than No. 14, except conductors not smaller than No. 18 shall be permitted as leads not more than 8 ft (2.44 m) long permanently attached to electric-discharge lampholders or electric-discharge ballasts if the leads are enclosed in wiring channels, and in showcase displays or small portable signs, as leads not more than 8 ft (2.44 m) long.

3. BENDS IN CONDUCTORS    Sharp bends in the conductors shall be avoided.

4. CONCEALED CONDUCTORS ON INSULATORS: INDOORS    Concealed conductors on insulators shall be separated from each other and from all objects other than the insulators on which they are mounted by a spacing of not less than 1½ in (38 mm) for voltages above 10,000 V, and not less than 1 in (25.4 mm) for voltages of 10,000 or less. They shall be installed in channels lined with noncombustible material and used for no other purpose, except that the primary-circuit conductors may be in the same channel. The insulators shall be of noncombustible, nonabsorptive material. Concealed conductors or insulators shall not be allowed outside the enclosure.

5. SHOW WINDOWS AND SIMILAR LOCATIONS    If conductors hang freely in the air, away from combustible material, and are not subject to physical damage, as in some show-window displays, they need not be otherwise protected.

6. CONDUCTORS IN RACEWAYS    If conductors are covered with metal sheathing, the covering shall extend beyond the end of the raceway, and the surface of the cable shall not be damaged where the covering terminates. In damp or wet locations, the insulation on all conductors shall extend beyond the metal covering or raceway at least 4 in (102 mm) for voltages over 10,000, 3 in (76 mm) for voltages over 5000 but not exceeding 10,000, and 2 in (50.8 mm) for voltages of 5000 or less. In dry locations, the insulation shall extend beyond the end of the metal covering or raceways not less than 2½ in (64 mm) for voltages over 10,000, 2 in for voltages over 5000 but not exceeding 10,000, and 1½ in (38 mm) for voltages of 5000 or less. For conductors at grounded midpoint terminals, no spacing is required. Not more than 20 ft (6.1 m) of single conductor from one secondary terminal shall be run in a metal raceway.

**419. Transformers.**   Transformers shall comply with the following:

1. VOLTAGE   The transformer secondary open-circuit voltage shall not exceed 15,000 V with an allowance on test of 1000 V additional. In end-grounded transformers the secondary open-circuit voltage shall not exceed 7500 V with an allowance on test of 500 V additional.

2. TYPE   Transformers shall be of a type identified for use with electric-discharge tubing and shall be limited in rating to a maximum of 4500 VA. Open core-and-coil-type transformers shall be limited to 5000 V with an allowance on test of 500 V and to indoor applications in small portable signs. Transformers for outline-lighting installations shall have secondary-current ratings not in excess of 30 mA unless they and all wiring connected to them are installed in accordance with the provisions of Article 410 of the Code for electric discharge lighting of the same voltage.

3. TRANSFORMERS EXPOSED TO WEATHER   Transformers used outdoors shall be of the weatherproof type or shall be protected from the weather by enclosure in the sign body or in a separate metal box.

4. TRANSFORMER SECONDARY CONNECTIONS   The secondary windings of transformers shall not be connected in parallel or in series, except that two transformers each having one end of its high-voltage winding connected to the metal enclosure may have their high-voltage windings connected in series to form the equivalent of a midpoint-grounded transformer. The grounded ends shall be connected by insulated conductors not smaller than No. 14, except that transformers for small portable signs, show windows, and similar locations that are equipped with leads permanently attached to the secondary winding within the transformer enclosure and that do not extend more than 8 ft (2.44 m) beyond the enclosure for attaching to the line ends of the tubing shall not be smaller than No. 18.

5. ACCESSIBILITY   Transformers shall be accessible and securely fastened in place.

6. ATTIC LOCATIONS   Transformers may be located in attics provided there is a passageway at least 3 ft (914 mm) in height and at least 2 ft (610 mm) in width, provided with a suitable permanent fixed walkway or catwalk at least 12 in (305 mm) in width extending from the point of entry into the attic to each transformer.

7. WORKING SPACE   A workspace at least 3 ft (914 mm) high and measuring at least 3 ft (914 mm) by 3 ft (914 mm) horizontally shall be provided about each transformer or its enclosure where not installed in a sign.

**420. Tubing.**   Electric discharge tubing shall conform to the following:

1. DESIGN   The tubing shall be of such length and design as not to cause a continuous overvoltage on the transformer.

2. SUPPORT   Tubing shall be adequately supported on noncombustible, nonabsorptive supports. Tubing supports should, if practicable, be adjustable.

3. CONTACT WITH FLAMMABLE MATERIAL AND OTHER SURFACES   The tubing shall be free from contact with flammable material and shall be located where not normally exposed to physical damage. If operating in excess of 7500 V, the tubing shall be supported on noncombustible, nonabsorptive insulating supports which maintain a spacing of not less than ¼ in (6.35 mm) between the tubing and the nearest surface.

**421. Terminals and receptacles for electric discharge tubing** shall comply with the following:

1. TERMINALS   The terminals of the tubing shall be inaccessible to unqualified persons and isolated from combustible material and grounded metal or shall be enclosed. If enclosed, they shall be separated from grounded metal and combustible material by nonabsorbent, noncombustible insulating material or by 1½ in (38 mm) of air. Terminals shall be relieved from stress by the independent support of the tubing.

2. TUBE CONNECTIONS OTHER THAN WITH RECEPTACLES   If tubes do not terminate in receptacles designed for the purpose, all live parts of tube terminals and conductors shall be so supported as to maintain a separation of at least 1½ in (38 mm) between conductors or between conductors and any grounded metal.

3. RECEPTACLES   Electrode receptacles for gas tubing shall be of noncombustible, nonabsorbent insulating material.

4. BUSHINGS   If electrodes enter the enclosure of outdoor signs or of an indoor sign operating at a voltage in excess of 7500 V, bushings shall be used unless receptacles are provided. Electrode terminal assemblies shall be supported not more than 6 in (152 mm) from the electrode terminals. All electric-discharge tubing terminals shall be insulated

with a material suitable for the voltages and environmental conditions expected during use.

5. SHOW WINDOWS    In the exposed type of show-window signs, terminals shall be enclosed by receptacles.

6. RECEPTACLES AND BUSHING SEALS    A flexible, nonconducting seal may be used to close the opening between the tubing and the receptacle or bushing against the entrance of dust or moisture. This seal shall not be in contact with grounded conductive material and shall not be depended upon for the insulation of the tubing.

7. ENCLOSURES OF METAL    Metal enclosures for electrodes shall be of not less than No. 24 MSG sheet metal.

8. ENCLOSURES OF INSULATING MATERIAL    Enclosures of insulating material shall be noncombustible, nonabsorbent, and suitable for the voltage of the circuit.

9. ENERGIZED PARTS    Energized parts shall be enclosed or suitably guarded to prevent contact.

**422. Switches on doors.**    Door or covers giving access to uninsulated parts of indoor signs or outline lighting exceeding 600 V and accessible to unqualified persons either shall be provided with interlock switches which on the opening of the doors or covers disconnect the primary circuit or shall be so fastened that the use of other than ordinary tools will be necessary to open them.

**423. Enclosures.**    Enclosures for signs and outline lighting shall conform to the following:

1. CONDUCTORS AND TERMINALS    Sign boxes, cabinets, and outline troughs shall have conductors and terminals, except supply leads, enclosed in metal or other noncombustible material.

2. CUTOUTS, FLASHERS, ETC.    Cutouts, flashers, and similar devices shall be enclosed in metal boxes, the doors of which shall be arranged so that they can be opened without removing obstructions or finished parts of the enclosure.

3. ENCLOSURES EXPOSED TO WEATHER    Enclosures for outdoor use shall be weatherproof and shall have at least two drain holes, each not larger than ½ in (12.7 mm) or smaller than ¼ in (6.35 mm).

4. MATERIAL    Except for portable signs of the indoor type, signs and outline lighting shall be constructed of metal or other noncombustible material. Wood may be used for external decoration if placed at least 2 in (50.8 mm) from the nearest lampholder or current-carrying part.

5. STRENGTH    Enclosures shall have ample strength and rigidity.

6. THICKNESS OF METAL    Sheet copper shall be at least 20 oz (0.028 in, or 71 μm). Sheet steel may be of No. 28 MSG, except that for outline lighting and for electric discharge signs, sheet steel shall be of No. 24 MSG unless ribbed, corrugated, or embossed over its entire surface, when it may be of No. 26 MSG.

7. PROTECTION OF METAL    All steel parts of enclosures shall be galvanized or otherwise protected from corrosion.

**424. Current-carrying capacities of flashers.**    Double-pole flashers are made in four sizes that will carry respectively 15, 30, 45, or 60 A per switch. Single-pole carbon flashers that will carry 5 A per switch are made. Brush-type flashers are rated at from 2 to 5 A on each brush and are not reliable for greater currents.

## REMOTE-CONTROL, SIGNALING, AND POWER-LIMITED CIRCUITS

**425. A remote-control circuit** is any electric circuit that controls any other circuit through a relay or an equivalent device.

**426. Low-voltage power circuits.**    Circuits which are neither remote-control nor signal circuits but which operate at not more than 30 V when the current is not limited in accordance with the requirements of Class 2 and Class 3 circuits and which are supplied from a source not exceeding 1000 VA are considered Class 1 remote-control circuits with respect to NEC requirements for their installation. These circuits shall be protected at not more than 167 percent of their VA ratings divided by their rated voltage.

**427. A signal circuit** is any electric circuit that energizes signaling equipment. Such circuits include circuits for doorbells, buzzers, code-calling systems, signal lights, and the like.

**428. Safety-control devices.**  Remote-control circuits to safety-control devices, the failure of operation of which would introduce a direct fire or life hazard, shall be considered Class 1 circuits.

Room thermostats, service hot-water temperature-regulating devices, and similar controls used in conjunction with electrically controlled domestic heating equipment are not considered to be safety-control devices.

**429. Remote-control, power-limited, and signal circuits in communication cables.**  Remote-control, power-limited, and signal circuits, which use conductors in the same cable with communication circuits, shall, for the purpose of this discussion, be classified as communication circuits and meet the requirements of such circuits (refer to Article 800 of the NEC).

**430. Hazardous locations.**  Any remote-control or signal circuit or any circuit considered such by the Code which is installed in a hazardous location must meet the Code requirements for hazardous-location installation in addition to the requirements for remote-control circuits. Cables installed in hazardous locations shall be Type PLTC.

**431. Remote-control, power-limited, and signal circuits are divided** by the NEC into two classes as follows:

1. CLASS 1    Systems in which the voltage and power is limited in accordance with Sec. 725-11 of the NEC.

2. CLASS 2 AND CLASS 3    Systems in which the voltage and power are limited in accordance with Sec. 725-31 of the NEC.


CLASS 1 SYSTEMS

**432. The installation of Class 1 systems** must be in accordance with general NEC requirements for the installation of electrical circuits except as stated in the special Code specifications given in Secs. 433 to 442 inclusive.

**433. Conductor sizes.**  Numbers 18- and 16-gage conductors may be used if installed in a raceway, an approved enclosure, or a listed cable. Flexible cords shall comply with the provisions of Article 400 of the Code.

**434. Conductor insulation.**  Conductors larger than No. 16 may be any of the conductors listed in Table 310-13 of the NEC for general wiring. Conductors of Nos. 18 and 16 gage shall have an insulation of a type listed in Sec. 725-16(b) of the NEC.

**435. Number of conductors in raceways.**  The number of conductors of remote-control or signal circuits in a raceway must be determined according to Table 29, Div. 11; Note 8 of Sec. 17, Div. 11, applies only to continuous loads (3 h or more) in excess of 10 percent of the ampacity of the conductor.

**436. Conductors of different systems.**  Conductors of two or more Class 1 remote-control and/or signal circuits may occupy the same cable, enclosure or raceway without regard to whether the individual systems or circuits are ac or dc, provided all conductors are insulated for the maximum voltage of any conductor in the cable, enclosure or raceway.

**437. Physical protection of remote-control circuits.**  If damage to a remote-control circuit would introduce a hazard as covered in Sec. 428, all conductors of such remote-control circuits shall be installed in rigid metal conduit, intermediate metal conduit, rigid nonmetallic conduit, electrical metallic tubing, Type MI cable, or Type MC cable or be otherwise suitably protected from physical damage.

**438. Overcurrent protection.**  Conductors shall be protected against overcurrent in accordance with the general Code requirements except for conductors of Nos. 18 and 16 gage, which shall not exceed 7 A for No. 18 and 10 A for No. 16 except when other articles of the Code have other protection requirements; except for No. 14 and larger Class 1 power-limited circuits that have main overcurrent protection and the rating is not over 300 percent of the ampacity of the Class 1 circuit conductor; and except for two-wire (single-voltage) secondary transformers that may be protected by the primary overcurrent device.

**439. Location of overcurrent protection.**  Overcurrent devices shall be located at the point where the conductor to be protected receives its supply unless the overcurrent device protecting the larger conductor also protects the smaller conductor in accordance with the current-carrying–capacity tables of Div. 11; or the overcurrent protection is provided in accordance with Sec. 725-12, Exception 2, of the NEC.

**440. Circuits extending beyond one building.**  Class 1 circuits which extend aerially beyond one building shall also meet the requirements of Article 225 of the Code.

**441. Grounding.**  Class 1 remote-control and signal circuits shall be grounded in accordance with Article 250 of the Code.

**442. Transformers.**  Transformers used to supply power-limited Class 1 circuits shall comply with Article 450 of the NEC and the power limitations of Sec. 725-11 of the NEC.

CLASS 2 AND CLASS 3 SYSTEMS

**443. NEC requirements for the installation of Class 2 and Class 3 systems** are given in the following Secs. 444 to 449, inclusive.

**444. Overcurrent protection and mounting.**  If current is limited in Class 2 and Class 3 systems by means of overcurrent protection, such overcurrent protection devices shall not be interchangeable with protection of a higher rating. The overcurrent protection may be an integral part of the power supply.

**445. Power limitations.**  Protection for power-limited circuits shall comply with Tables 725-31(a) for ac circuits and 725-31(b) for dc circuits in the NEC.

**446. On the supply side of overcurrent protection, transformer, or current-limiting devices.**  Conductors and equipment on the supply side of overcurrent protection, transformers, or current-limiting devices shall be installed in accordance with the appropriate requirements of Chap. 3 of the Code. Transformers or other devices supplied from electric-light and power circuits shall be protected by an overcurrent device with a rating or setting not exceeding 20 A, except the input leads of a transformer or other power source supplying Class 2 and Class 3 circuits may be smaller than No. 14, but not smaller than No. 18 if they are not over 12 in (305 mm) long and if they have insulation that complies with Sec. 725-16(b) of the NEC.

**447. On the load side of overcurrent protection, transformer, or current-limiting devices.**  Conductors on the load side of overcurrent protection, transformer, or current-limiting devices shall be insulated and shall comply with Sec. 725-38 and not less than the requirements of Sec. 725-50 of the Code.

**448. Circuits extending beyond one building.**  Class 2 and Class 3 circuits which extend beyond one building and are so run as to be subject to accidental contact with light or power conductors operating at a potential exceeding 300 V shall also meet the requirements of Secs. 800-10, 800-12, 800-13, 800-30, 800-31, and 800-32 for communication circuits in the Code.

**449. Grounding.**  Class 2 and Class 3 circuits and equipment shall be grounded in accordance with Article 250 of the Code.

**WIRING FOR SPECIAL OCCUPANCIES**

**450. The NEC should be consulted for special rules on installations in the following special occupancies** or for the following special equipment or conditions:

Garages
Aircraft hangars
Gasoline-dispensing and service stations
Bulk-storage plants
Finishing processes
Health care facilities
Places of assembly
Theaters and similar locations
Motion-picture and television studios and similar locations
Motion-picture projectors
Manufactured buildings
Agricultural buildings
Elevators, dumbwaiters, and escalators
Electric welders
Sound-recording and similar equipment
Electrolytic cells
Organs
X-ray equipment
Induction and dielectric heating equipment

Industrial machinery
Electrically driven or controlled irrigation machines
Emergency systems
Radio and television equipment
Mobile homes
Recreational vehicles
Standby power generation systems
Swimming pools, fountains, and similar installations
Marina and boatyard wiring
Electronic computer/data processing systems
Solar photovoltaic systems
Floating buildings

## DESIGN OF INTERIOR-WIRING INSTALLATIONS

**451. Factors affecting wiring layouts** *(Standard Handbook).*  In conduit work the space available often dictates that feeders be split into two or more feeder lines so as to guard against complete shutdown should anything happen to one feeder.

In reinforced-concrete floors a cross section of the floor must be studied to see that there will be sufficient space for the reinforcing rods, conduit or raceways, and necessary thickness of concrete. This is especially true when conduits cross. The local building code should be consulted to see what restrictions are placed on floor installations. A minimum thickness of 1 in (25.4 mm) of concrete over conduits and raceways should be used to prevent cracking.

> **example**  The building code may allow floors 4 in (101.6) thick with ¾-in reinforcing rods located with 1 in (25.4 mm) of concrete covering the underside (Fig. 9-291). This would mean that there is a minimum space of 1¼ in (31.75 mm) between the top of the reinforcing rods and the top of any conduit. For this installation, ¾-in conduit, which has an outside diameter of 1.05 in (Sec. 63), would be the largest size allowable. There could be no crosses of conduit because even two ½-in conduits would take up 2 × 0.,84 = 1.68 of space, which is more than the 1¼ in available.

In brick walls there must be two thicknesses of brick to conceal conduits, and the distance between them must be studied to determine the maximum size for vertical runs of conduit. These are only a few of the possible examples in which space for wiring must be considered. Very often the mistake of installing feeders just large enough to carry the present load is made, and when additions are called for, the feeders are overloaded and additional feeders must be installed at great expense. Conduits for feeders and mains should be of sufficient size to permit the installation of feeders or mains of a carrying capacity of 150 percent of the present connected load. Also, spare conduits can be installed.

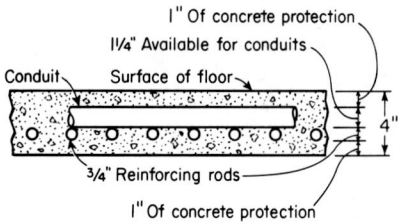

**Fig. 9-291**  Study of space available for conduits in reinforced-concrete floors.

**452. Wiring methods to be used are determined** by a consideration of national and local code requirements, reliability, appearance, and cost. Those planning a wiring installation should decide which methods of wiring are allowable for the particular occupancy and electric system being considered. Then the cost of each allowable method should be weighed against the appearance, the degree of protection afforded, reliability, and probable years of service that will be realized. Both cost of material and cost of labor must be considered.

1. RESIDENCES  Type AC (commonly referred to as Bx) cable concealed in partitions is very practical, giving good reliability at moderate cost. If a less expensive installation is desired, nonmetallic-sheathed cable may be used. Sometimes this is used for a few large-size circuits such as a circuit to the electric range if the difference in cost is considerable, even if the rest of the wiring is metal-clad cable. The expense of metal raceways is usually not justified except in very large residences of the highest grade of construction. When the highest degree of adequacy for convenience outlets is desired, multioutlet assemblies should be employed for this part of the wiring.

2. Rewiring Old Buildings of Residential or Commercial Occupancy    Surface raceways, Type AC cable, nonmetallic-sheathed cable, and electrical metallic tubing are the usual expedients in rewiring. Surface raceways give the best appearance if the wiring must be exposed. Type AC cable or nonmetallic-sheathed cable is the most practical for that part of the wiring which can be fished through the partitions.

3. Small Commercial Buildings and Apartment Buildings    For buildings of frame or brick and frame construction, Type AC cable, nonmetallic-sheathed cable, or electrical metallic tubing concealed in the partitions is generally the most practical.

4. Office and Public Buildings    For buildings of steel, concrete, terra-cotta, or similar construction, the cost of wiring is only a small percentage of the total cost of the building. To keep the offices rented to business firms, the electric system should provide the very best in appearance and adequacy of service. Therefore the wiring should be concealed, and rigid metal conduit, intermediate metal conduit, or electrical metallic tubing be employed for ceiling lights and feeders, and underfloor raceways or cellular-floor raceways for desk outlets for power and signaling circuits.

5. Industrial Buildings    Metal raceways are generally used for branches, with busways, metal-clad cables, or conduits for feeders.

6. Large-Sized Main Feeders    The method of wiring should be considered very carefully because the high unit costs may make a considerable difference in the total cost of the installation.

Very often, combinations of wiring methods are used in the same building, each method being used where its advantages make it desirable.

**453. In rewiring buildings using existing conduits,** the use of new insulations with thinner insulations permits more conduit fill, depending on the type and size of conductors (see Tables 29 through 33 of Div. 11).

**454. The branch circuits used in interior wiring may be classified** as follows:

Type 1    This is an individual branch circuit that supplies only one utilization equipment.

Type 2    This is a branch circuit that supplies two or more outlets. Such branch circuits are further classified by the NEC according to the rating of the fuse or circuit breaker that protects the circuit. The Code recognizes five types of these multioutlet branch circuits. They are 15-, 20-, 30-, 40-, and 50-A branch circuits respectively. The Code provides certain restrictions on the rating of outlets, size of wire, and equipment to be supplied, as given in Sec. 457.

**455. Maximum loads for branch circuits.**    Type 1 branch circuits may supply loads of any capacity. If the load consists of a motor-operated appliance, the load must not exceed 80 percent of the branch-circuit wires. Also the load must not exceed 80 percent of the rating of the branch circuit when any load will continue for 3 h or more. For details of the requirements for motor branch circuits refer to Div. 7.

Branch circuits are required by the Code to have maximum loads which conform to the following:

1. Appliances Consisting of Motors and Other Loads    When a circuit supplies only motor-operated appliance loads, Article 430 of the Code shall apply (refer to Div. 7). For other than a portable appliance, the branch-circuit size shall be calculated on the basis of 125 percent of motor load if the motor is larger than ⅛ hp plus the sum of the other loads.

2. Other Loads    The total load shall not exceed the branch-circuit rating and shall not exceed 80 percent of the rating when the load will continue for 3 h or more, as for store lighting and similar loads. In computing the load of lighting units that employ ballasts, transformers, or autotransformers, the load shall be based on the total of the ampere rating of such units and not on the wattage of the lamps.

*Exception: range loads. See Note 4 of Table 2 in Div. 11.*

**456. Permissible loads** for Type 2 branch circuits are specified by the NEC as follows:

1. 15- and 20-A Branch Circuits    Lighting units and/or appliances: the rating of any one portable appliance shall not exceed 80 percent of the branch-circuit rating; the total rating of fixed appliances shall not exceed 50 percent of the branch-circuit rating if lighting units or portable appliances are also supplied.

2. 30-A Branch Circuits    Fixed lighting units with heavy-duty lampholders in other than dwelling occupancies or utilization equipment in any occupancy: the rating of any one portable appliance shall not exceed 24 A.

3. 40- AND 50-A BRANCH CIRCUITS   Fixed lighting units with heavy-duty lampholders or infrared heating appliances or utilization equipment in other than dwelling occupancies or fixed cooking appliances in any occupancy.

4. Branch circuits larger than 50 A shall supply only nonlighting loads.

The term *fixed* as used in this section recognizes cord connections when otherwise permitted.

**457.   Lighting and Appliance Branch Circuits**[a]

(Table 210-24, 1990 NEC)

(Types FEP, FEPB, SA, TW, RH, RHW, RHH, THHN, THW, THWN, and XHHW conductors in raceway or cable)

| Circuit rating, A | 15 | 20 | 30 | 40 | 50 |
|---|---|---|---|---|---|
| Conductors (minimum size): | | | | | |
| Circuit wires[b] | 14 | 12 | 10 | 8 | 6 |
| Taps | 14 | 14 | 14 | 12 | 12 |
| Fixture wires and cords | Refer to Sec. 240-4 of Code. | | | | |
| | | | | | |
| Overcurrent protection, A | 15 | 20 | 30 | 40 | 50 |
| Outlet devices: | | | | | |
| Lampholders permitted | Any type | Any type | Heavy-duty | Heavy-duty | Heavy-duty |
| Receptacle rating, A[c] | 15 maximum | 15 or 20 | 30 | 40 or 50 | 50 |
| Maximum load, A | 15 | 20 | 30 | 40 | 50 |
| Permissible load | ... [d] | ... [d] | ... [e] | ... [f] | ... [f] |

[a]Reprinted with permission from NFPA 70-1990. National Electrical Code® Copyright © 1990, National Fire Protection Association, Quincy, Massachusetts 02269. This reprinted material is not the complete and official position of the NFPA on the referenced subject, which is represented only by the standard in its entirety.

[b]These gages are for copper conductors.

[c]Refer to Sec. 410-30(c) of the NEC for receptacle rating of cord-connected electric discharge fixtures.

[d]Refer to Sec. 210-23(a).

[e]Refer to Sec. 210-23(b).

[f]Refer to Sec. 210-23(c).

**458. The minimum required loads** for which branch circuits and feeders must be supplied shall meet the Code requirements as determined from Table 5, Div. 11. If the maximum load of a branch circuit will continue for 3 h or more, as with store lighting and similar loads, the minimum unit loads specified in Table 5 of Div. 11 shall be increased by 25 percent so that the wiring system may have sufficient branch-circuit and feeder capacity to ensure safe operation.

**459. Branch circuits required.**   The NEC requires that branch circuits shall be installed as follows:

1. LIGHTING AND APPLIANCE CIRCUITS   For lighting and for appliances, including motor-operated appliances, not specifically provided for in Par. 2, branch circuits shall be provided for a computed load not less than that determined by Sec. 220-3 of the Code. (Refer to Table 5, Div. 11.)

The number of circuits shall be not less than that determined from the total computed load and the capacity of circuits to be used, but in every case the number shall be sufficient for the actual load to be served, and the branch-circuit loads shall not exceed the maximum loads specified in Secs. 455 to 457.

If the load is computed on a watt-per-square-foot basis, the total load, insofar as practical, shall be evenly proportioned among the branch circuits according to their capacity.

2. SMALL-APPLIANCE BRANCH CIRCUITS: DWELLING OCCUPANCIES   For the small-appliance load, including refrigeration equipment, in kitchen, pantry, dining room, and breakfast room or similar area of dwelling occupancies, two or more 20-A appliance branch circuits in addition to the branch circuits specified in Par. 1 of this section shall be provided for all receptacle outlets in these rooms, and such circuits shall have no other outlets except clock outlets or outdoor receptacles. In addition to the required receptacles, switched receptacles are permitted in lieu of lighting outlets in other than the kitchen.

Receptacle outlets supplied by at least two appliance-receptacle branch circuits shall be installed in the kitchen to serve countertop surfaces.

At least one 20-A branch circuit shall be provided for a laundry receptacle (or receptacles) required in Sec. 210-52(f) of the Code.

Receptacle outlets installed solely for the support of and the power supply for electric clocks may be installed on lighting branch circuits or small appliance circuits.

A three-wire 120/240-V branch circuit is the equivalent of two 120-V receptacle branch circuits.

When a grounding receptacle is required as in Sec. 210-7 of the Code, the branch circuit shall include or provide an equipment grounding conductor to which the grounding contacts of the grounding receptacle shall be connected. The metal armor of Type AC cable, the sheath of Type MI or Type MC cable, or a metallic raceway shall be acceptable as a grounding conductor.

3. OTHER CIRCUITS    For specific loads not otherwise provided for in Pars. 1 or 2, branch circuits shall be as required by other sections of the Code.

**460. Feeder loads should be determined** in accordance with instructions given in Div. 3. Instructions for determining the minimum allowable loads required by the Code are given in Table 5, Div. 11.

**461. The size of conductors required** with respect to both carrying capacity and voltage drop is discussed in Div. 3. For Type 2 branch circuits also refer to Sec. 457 of this division.

**462. Overcurrent protection of circuits.**  Except for the branch circuit to an individual motor, the rating of the overcurrent protective devices of a circuit shall not be in excess of the carrying capacity of the circuit conductor. If a branch circuit supplies only a single non-motor-operated appliance, it shall not exceed that marked on the appliance; or if not marked and rated over 13.3 A, it shall not exceed 150 percent of the appliance rated current; or if not marked and rated 13.3 A or less, it may not exceed 20-A branch circuit protection. For branch circuits supplying two or more outlets refer to Sec. 456. For motor branch circuits refer to Div. 7.

**463. The voltage to ground** on branch circuits supplying lampholders, fixtures, or standard receptacles of 15-A or lower rating shall not exceed 120 V between conductors, except for railway properties, infrared industrial heating appliances, and lighting in industrial establishments as permitted by Sec. 210-6 of the NEC. The branch circuits supplying the ballasts for electric discharge lamps mounted in permanently installed fixtures on poles for the illumination of areas such as highways, roads, bridges, athletic fields, parking lots, at a height not less than 22 ft (6.71 m), or on other structures such as tunnels at a height not less than 18 ft (5.49 m) shall not exceed 600 V between conductors.

**464. In dwelling occupancies,** the voltage between conductors supplying lighting fixtures, and cord- and plug-connected loads 1440 VA, or less, or less than ¼ hp shall not exceed 120 V between conductors.


### WIRING FOR RESIDENTIAL OUTDOOR LIGHTING

**465. General considerations.**  The following recommendations for wiring for residential outdoor lighting have been taken from a booklet on the subject published by the Construction Materials Division of the General Electric Co.

Factors that must be considered before making any outdoor wiring installation are family habits and the facilities that the family wishes to enjoy, the style and construction of the house, the size and topography of the property, and, most important of all, whether the service entrance is large enough to handle the additional load, whether spare circuits are available at the panelboard or new circuits must be added, and the length of the circuits required for the various outdoor lighting and outlet installations.

In general, it will probably be found that permanently installed outdoor wiring to be used beyond the house should be run underground from the point where it emerges from the house to the point where the equipment is to be used. There are, of course, exceptions. When the house is multistoried, overhead wiring may be used. In almost all other cases, from the viewpoint of both safety and aesthetics, underground circuits will be preferable. Sometimes a combination of overhead and underground wiring will be most satisfactory.

Any circuit installed in an existing dwelling for outdoor use should be a new circuit originating at the panelboard rather than an extension of an existing circuit. In all cases circuits for outdoor use should be restricted to this purpose only.

In laying out new outdoor wiring, it should be recognized that even though the home-owner considers that the primary purpose of the new circuits is to provide lighting only, this cannot be taken for granted. Once the circuits are installed and outdoor outlets are available, there will probably be a heavy demand from this source for power for appliances and tools. Not less, therefore, than two 20-A circuits or one three-wire 120/240-V circuit should be installed.

**466. Running the circuits from panelboard to the outside.**  The style and construction of the house will determine how the circuits are run from the panelboard to the outside. Most installations will be either (1) from a basement when the walls are concrete or cement-block or (2) from above ground level when the panelboard is located either in the garage or in another interior location at first-floor level.

1. WHEN THE PANELBOARD IS IN THE BASEMENT  In this situation, wiring is generally brought to the outside either through the masonry wall of the basement or through the sill on top of the foundation. Factors to be considered are whether the installation is to be underground or overhead. If it is underground, going directly through the basement wall and coming out below ground level (Fig. 9-292) on the outside is the least conspicuous. If convenient shrubbery will hide the conduit, it may be easier to come out through the sill (Fig. 9-293).

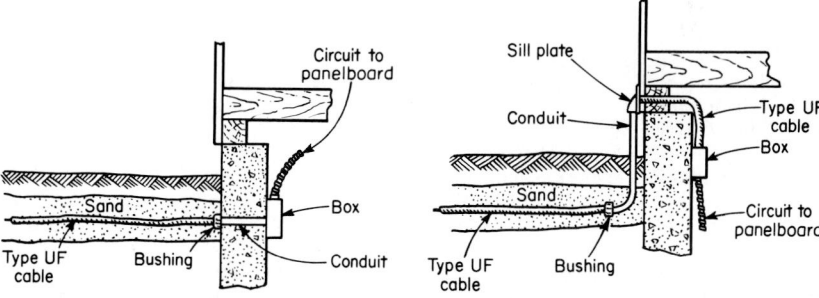

**Fig. 9-292**  Running the circuit from the panel-board through masonry.

**Fig. 9-293**  Running the circuit from the panel-board through the sill.

Either of these methods may be chosen when the wiring is to be overhead. In this case, since a length of conduit will have to run up the side of the house, it is well to come out of the basement at a point where shrubbery will partially conceal the installation. Occasionally, in frame construction, the circuit for overhead wiring can be fished up through the walls to a point of departure at the desired elevation of the house.

2. WHEN THE PANELBOARD IS IN THE GARAGE OR THE HOUSE  In many newer houses, especially those without basements, the panelboard is installed in the attached garage or in some other handy but inconspicuous part of the house. In such cases, the circuit is run from the panelboard through the wall of the garage or house (Fig. 9-294) at the most convenient point. Thought should be given to the point selected so that the installation is as unobtrusive as possible.

**467. Underground circuits.**  Circuits run underground offer a number of advantages to the homeowner. They are not subject to damage from storms or exposed to mechanical damage. Most important, such circuits are invisible and do not, therefore, mar the appearance of the grounds and gardens.

The most economical underground installation for both feeders and branch circuits is Type UF (underground-feeder) cable. This cable is approved by the NEC for direct burial in the ground without additional protection.

Underground cable is laid in a trench at least 20 in (508 mm) deep to prevent possible damage from normal spading. The bottom of the trench should be free from stones. This is easily accomplished by using a layer of sand or sifted dirt in the bottom of the trench

(Fig. 9-295). Cable is laid directly on top of this layer. When cable enters the building or leaves the ground, slack should be provided in the form of an S curve to permit expansion with extreme changes in temperature. After the cable has been installed, fill all openings through the foundation with sealing compound so that water from rain or melting snow cannot follow the cable into the building.

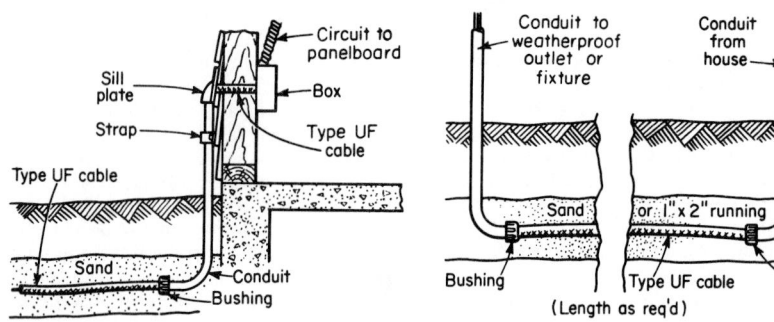

**Fig. 9-294**  Running the circuit from the panelboard through the garage wall.

**Fig. 9-295**  Type UF cable used for underground circuits.

The NEC requires that underground-feeder cables be installed in continuous lengths from outlet to outlet and from fitting to fitting. Splices can be made only within a proper enclosure.

In some areas of a yard, when it is felt that digging might accidentally occur to the depth of the cable, a concrete pad or a 1- by 2-in (25.4- by 50.8-mm) running board can be laid over the cable before the trench is filled (Fig. 9-295). Under driveways or roadways or where heavy loading might occur, it is also desirable to use a board, concrete pad, or conduit.

For outdoor wiring in sections of the United States that are known to be termite-infested or subject to attack by rodents, or whenever extra protection to the Type UF underground circuits is required, rigid galvanized-steel conduit may be used for mechanical protection. When rigid conduit for the complete circuit run is used, feeders and branch circuits may be of any approved type of moisture-resistant wire such as Types TW, RHW, or THWN.

**468. Multiple-family buildings or apartments.**  The recommendations for residences can be applied to individual apartments for lighting, convenience, and switch outlets. If central heating is supplied, it is unlikely that an electric heater will be used in the bathroom or that many heavy-duty appliances will be used at one time. Therefore two 20-A branch circuits supplying convenience outlets in the dining room and kitchen generally will be sufficient in addition to a 15-A branch circuit for every 500 ft² (46.5 m²) of floor area. It is recommended that a three-wire, 120/240-V, No. 8 range circuit be installed for each apartment.

For nonrentable areas under the control of the building management the following rules from the *Handbook of Interior Wiring Design,* prepared by the Industry Committee on Interior Wiring Design, represent good practice:

1. Entrances

   a. Sufficient outlets shall be installed to provide electrical service consistent with the general architectural treatment, but in no case shall there be less than one outlet for an outside lantern or two for outside wall brackets.

   b. At least one ceiling outlet shall be installed for each 150 ft² (13.9 m²) of vestibule space.

   c. At least one wall outlet shall be installed over each group of letter boxes covering a lineal distance of 8 ft (2.4 m) or less.

2. Lobby or Main Hallway

*a.* At least one ceiling outlet per 125 ft² (11.6 m²) of floor area shall be installed provided that:

(1) No area has outlets spaced more than 15 ft (4.6 m) apart.

(2) No elevator door or stairwell is more than 6 ft (1.8 m) from the nearest ceiling outlet.

*b.* At least one convenience outlet shall be installed for every 12 ft (3.7 m) of wall space for portable lamps and general-utility purposes (vacuum cleaners, floor polishers, and so on). Hallways 10 ft or longer require a minimum of one convenience outlet.

*c.* Provision shall be made for additional outlets to furnish such illumination as is demanded by the architectural treatment. This may be in the form of decorative wall brackets, urns, pedestals, built-in units, etc.

3. UPPER HALLWAYS

*a.* At least one ceiling outlet shall be installed per 150 ft² (13.9 m²) of floor area, provided that:

(1) No area has ceiling outlets spaced more than 15 ft (4.6 m) apart if they are arranged on alternate circuits or 30 ft (9.1 m) if adjacent outlets are on the same circuit.

(2) No elevator door or staircase is located more than 6 ft (1.8 m) from the nearest ceiling outlet.

(3) No section of a hallway not in a straight line from another shall be without at least one ceiling outlet.

*b.* At least one convenience outlet shall be installed per floor, with additional outlets spaced so that no point in a hallway is located more than 25 ft (7.6 m) from a convenience outlet, and a minimum of one for hallways 10 ft or longer.

4. HALL SWITCHES

*a.* All ceiling outlets plus the outlets for entrance lighting shall be controlled by switches.

*b.* When all ceiling outlets on any one floor are controlled by a single switch, this switch shall be of a type or shall be located so as to be inaccessible to the general public.

*c.* When more than one switch controls the ceiling outlets on a particular floor, at least one switch shall be inaccessible to the general public. This switch shall control outlets installed under the following provisions:

(1) It must control the outlet nearest the stairwell or the elevator door.

(2) It must control outlets so that the spacing between any two is not greater than 30 ft (9.1 m).

(3) It must control an outlet in every space not in a direct line from the outlet near the stairwell or the elevator door.

5. STAIRWELLS

*a.* When enclosed stairwells are installed (as in masonry construction), there shall be an outlet for lighting on each landing.

*b.* If exit lights are required, they shall be installed in accordance with all rules and regulations covering their use.

*c.* Interior stairways with six steps or more require a wall switch at each floor level.

6. BASEMENT AREAS   Lighting outlets shall be provided in basement areas on the basis of at least one ceiling outlet per 200 ft² (18.6 m²), plus additional outlets and necessary switch control if such basement areas are available to tenants for laundry, storage, or other purposes.

7. OIL BURNERS, AIR-CONDITIONING EQUIPMENT, ETC.   Circuits shall be individually provided for each such piece of equipment in accordance with the size and rating of each motor or heater.

8. ELEVATORS   Power and light feeders for elevator service shall be of the size required by the elevator manufacturer.

**469. The arrangement of the service and feeders** will depend on the system of metering employed. When each apartment is to be metered and billed for all its power, a meter service switch and meter should be provided for each apartment and one or two for the building management. These should be located in rows along the walls in the basement. Each apartment must be on an individual feeder from the meter service switch. The load on the feeder should be computed in accordance with the instructions given in Div. 3. If an electric range is not supplied by the feeder, a two-wire feeder is usually sufficient. If an electric range is supplied, three-wire feeders should be used. When the power company will bill the building management on a lower rate for power than it does for lighting, two meters with separate circuits for each should be provided. If there are more than six meter service switches, the service must be brought in through a single circuit breaker or fused service switch, the size to be determined in accordance with the instructions of Div. 3. When the power is to be billed to the building management and included in the rent, the circuit is brought in through a single circuit breaker or fused service switch as before.

Circuits for either one or two meters, depending on whether the power company bills the power on a separate rate from the lighting, should be provided. A distribution center with a safety enclosed fused knife switch or circuit breaker for each feeder should be installed. The feeders can then supply distribution panelboards in the halls on each floor, or one feeder can be installed for each three apartments. The latter plan is preferred, as it facilitates a change to the other metering system if a different policy should prove advisable in the future. Frequently the building employs a combination of these two systems in which the management is billed for and provides certain definite services, such as the electric range, refrigerator, or air-conditioning equipment, the probable cost of which is relatively easy to determine and much cheaper when paid for on a wholesale rate that takes advantage of the low demand factor which applies to a large number of customers. Then the customer has a separate service through his or her individual meter for lighting and small appliances. The method of wiring is with Type AC cable or nonmetallic-sheathed cable for small buildings of frame construction. Conduit wiring is usually employed for larger buildings of masonry construction.

**470. With any of the above plans each apartment should have its own distribution panelboard** containing fuses or circuit breakers for the branch circuits. Fuse panels are cheaper, but circuit breakers will reduce labor in calls for service due to short circuits in portable cords and appliances.

## WIRING FOR COMMERCIAL AND INDUSTRIAL OCCUPANCIES

**471. Electrical distribution systems for commercial and industrial use** (Joseph F. McPartland, *Electrical Systems Design*). In any electrical system, the distribution system involves the methods and equipment used to carry power from the service equipment to the overcurrent devices protecting the branch circuits. The distribution system carries power to lighting panelboards, power panelboards, and motor-control centers and to the branch-circuit protective devices for individual motor or power loads. Depending upon the type of building, the size and nature of the total load, various economic factors, and local conditions, a distribution system may operate at a single voltage level or may involve one or more transformations of voltage. A distribution system might also incorporate a change in frequency of ac power or rectification from ac to dc power.

Design of a distribution system, therefore, is a matter of selecting circuit layouts and equipment to accomplish electrical actions and operations necessary for the conditions of voltage, current, and frequency. This means relating such factors as service voltage, distribution voltage or voltages, conductors, transformers, converters, switches, protective devices, regulators, and power-factor corrective means to economy, load conditions, continuity of service, operating efficiency, and future power requirements. Of course, the factors of capacity, accessibility, flexibility, and safety must be carefully included in design considerations for distribution.

Basically, distribution systems are classified according to the voltage level used to carry the power either directly to the branch circuits or to load-center transformers or substations at which feeders to branch circuits originate. The most common types of distribution systems based on voltage are discussed in Div. 3.

**472. Types of electrical distribution systems.** The descriptions and discussions of the different types of electrical distribution systems given in this section and Secs. 473 to 484 inclusive are reproduced from *Westinghouse Architects' and Engineers' Electrical Data Book* through the courtesy of the Westinghouse Electric Corporation.

In the great majority of cases, power is supplied to a building at the utilization voltage. In practically all these cases, the distribution of power within the building is achieved through the use of a simple radial distribution system. This system is the first type described on the following pages from the low-voltage bus to the loads.

When service is available at the building at some voltage higher than the utilization voltage to be used, the system design engineer has a choice of a number of types of systems. This discussion covers seven major types of distribution systems and practical modifications of several of them.

Simple radial
Loop-primary radial
Banked-secondary radial
Primary-selective radial
Secondary-selective radial
Simple network
Primary-selective network

**473. Simple radial system.**   The conventional simple radial system receives power at the utility supply voltage at a single substation and steps the voltage down to the utilization level. Low-voltage feeder circuits run from the substation bus to switchgear assemblies and panelboards that are located with respect to the loads as shown in Fig. 9-296. Each feeder is connected to the substation bus through a circuit breaker. Relatively small circuits are used to distribute power to the loads from the switchgear assemblies and panelboards.

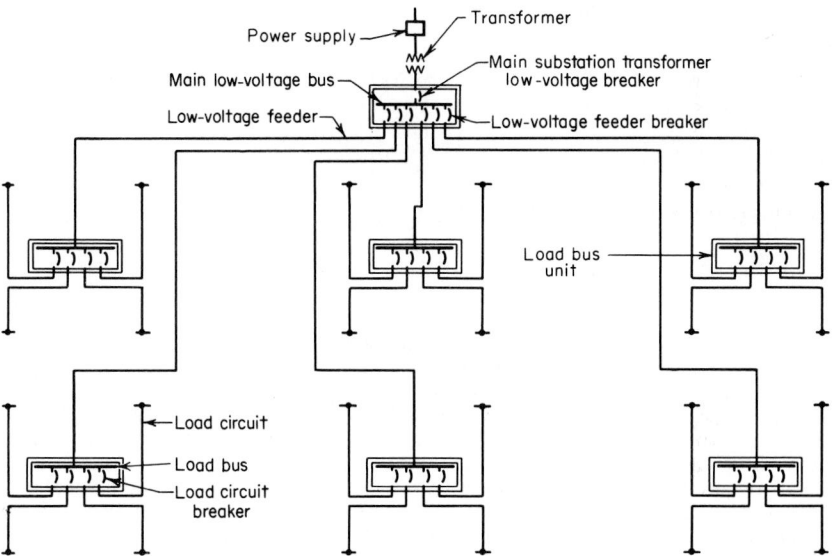

**Fig. 9-296**   Conventional simple radial system.

Since the entire load is served from a single source, full advantage can be taken of the diversity among the loads. This makes it possible to minimize the installed transformer capacity. However, the voltage regulation and efficiency of this system are poor because of the low-voltage feeders and single source. The costs of the low-voltage feeder circuits and their associated circuit breakers are high when the feeders are long and the peak demand is above 1000 kVA.

A fault on the substation bus or in the substation transformer will interrupt service to all loads. Service cannot be restored until the necessary repairs have been made. A low-voltage feeder-circuit fault will interrupt service to all loads supplied over that feeder circuit.

A modern and improved form of the conventional simple radial system distributes power at a primary voltage. The voltage is stepped down to utilization level in the several load areas within the building through power-center transformers. The transformers are usually connected to their associated load bus through a circuit breaker, as shown in Fig. 9-297. Each power center is a factory-assembled unit substation consisting of a three-phase liquid-filled or air-cooled transformer, an integrally mounted primary switch, and low-voltage metal-enclosed circuit breakers. Circuits are run to the loads from these circuit breakers.

Since each transformer is located within a specific load area, it must have sufficient capacity to carry the peak load of that area. Consequently, if any diversity exists among the load areas, this modified system requires a greater transformer capacity than the basic form of the simple radial system. However, because power is distributed to the load areas at a primary voltage, losses are reduced, voltage regulation is improved, feeder-circuit

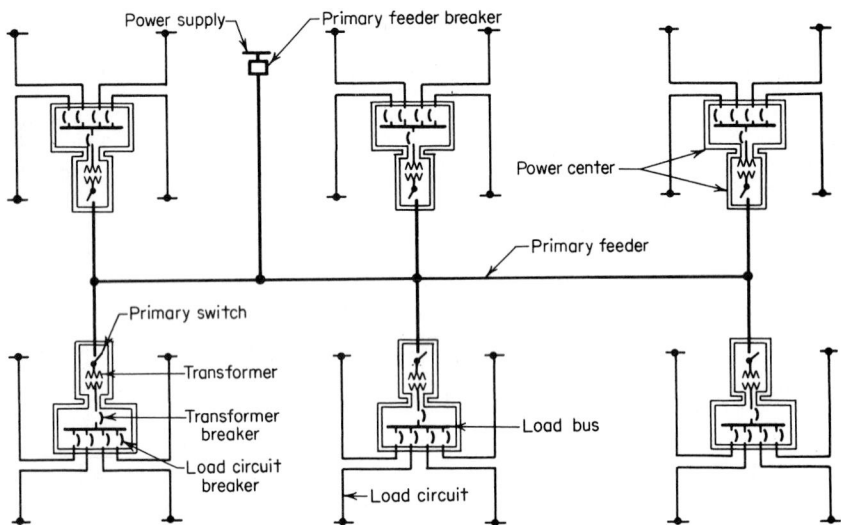

**Fig. 9-297**   Modern simple radial system.

costs are reduced substantially, and the large low-voltage feeder circuit breakers are eliminated. In many cases the interrupting duty imposed on the load circuit breakers is reduced.

This modern form of the simple radial system will usually be lower in initial investment than any other type of distribution system for buildings in which the peak load is above 1000 kVA. It is poor, however, from the standpoints of service continuity and flexibility. A fault on the primary feeder circuit or in one transformer will cause an outage to all loads. In the case of the feeder fault, service is interrupted to all loads until the trouble has been eliminated. When a transformer fault occurs, service can be restored to all loads except those served by the faulty transformer by opening the transformer primary switch and reclosing the primary feeder breaker.

Reducing the number of transformers per feeder by adding primary feeder circuits will improve the flexibility and service continuity of this system. This, of course, increases the investment in the system but minimizes the extent of an outage due to a transformer or feeder fault.

If one primary feeder is used per transformer, as in Fig. 9-298, the system is comparable in service continuity to the single substation form of the simple radial system. This sytem arrangement will usually be very expensive. The cost can be materially reduced, however, by replacing the automatic feeder circuit breakers with manually operated breakers or load-break switches and by backing up a number of these switches with one automatic circuit breaker.

In this modified form of the simple radial system, a primary feeder fault interrupts service to all loads. Service can be restored to all loads except those associated with the faulted system element by opening the load-break switch on the faulted circuit and closing the circuit breaker. Service can be restored to the remaining loads when the necessary repairs are made.

**474. Loop-primary radial system.**    This system is similar in principle to the modern form of simple radial system. It provides for quick restoration of service when a primary feeder or transformer fault occurs, as does the modified simple radial system of Fig. 9-298, and at a lower investment. A sectionalized primary loop controlled by a single primary feeder breaker as shown in Fig. 9-299 is used rather than a radial primary feeder.

Manually operated load-break switches are installed in the feeder for sectionalizing

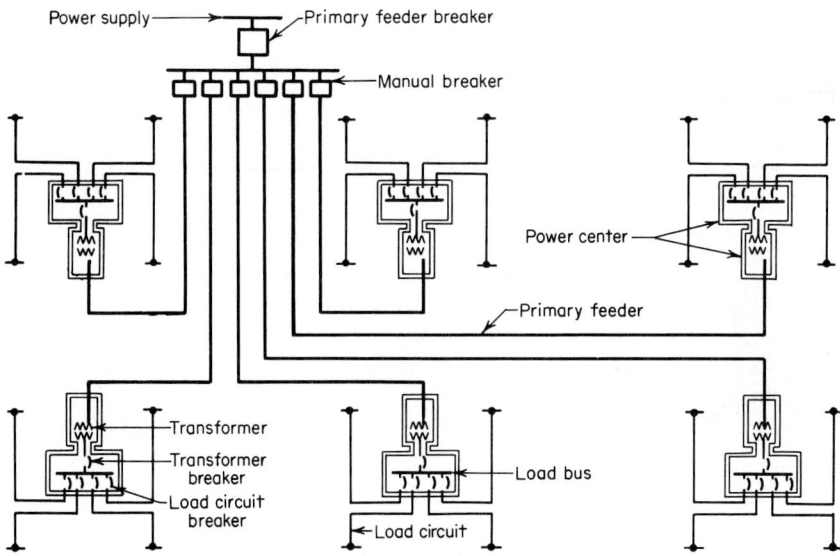

**Fig. 9-298**    Modified modern simple radial system

purposes. Two are located where the feeder branches to form the loop, and the others are located at the transformers. This makes it possible to disconnect any transformer and its associated section of loop from the rest of the system.

When a transformer or primary-feeder fault occurs, the main feeder breaker opens and interrupts service to all loads. Increasing the number of primary loop feeders will reduce the extent of the outage from a fault but will also increase the system investment. The faulty section of loop and its associated transformer can be disconnected from the system by the load-break switches. Service can then be restored to all but the disconnected portion of the system by closing the primary feeder breaker. Service cannot, however, be restored to the loads in the area where the transformer or primary loop section is in trouble until the necessary repairs have been made.

The cost of this system as illustrated is only slightly more than that of the modern simple radial system of Fig. 9-297. The primary-cable cost is a little greater, and there are two additional load-break switches in the system. The load-break switches associated with the transformers will cost little more than the disconnecting switches in the primary leads of the transformers of the simple radial system.

**475. Banked-secondary radial system.**    This system permits quick restoration of service to all loads following a primary-feeder or transformer fault. It uses a secondary loop, as shown in Fig. 9-300, to provide an emergency supply when a fault occurs in a transformer or a section of the primary loop. The primary-circuit arrangement is the same as that in the loop-primary radial system of Fig. 9-299.

The primary-feeder arrangement of the simple radial system of Fig. 9-298 can be used to give the same results, but it usually will be more costly.

A primary-feeder or transformer fault causes the main breaker to operate and drop all load. Service can be restored by opening the two load-break switches and the low-voltage breaker of the transformer adjacent to the fault and closing the primary feeder breaker. When the primary feeder breaker is reclosed, service is restored to all loads. Those loads connected to the load bus associated with the disconnected transformer are supplied over the secondary loop from adjacent load buses.

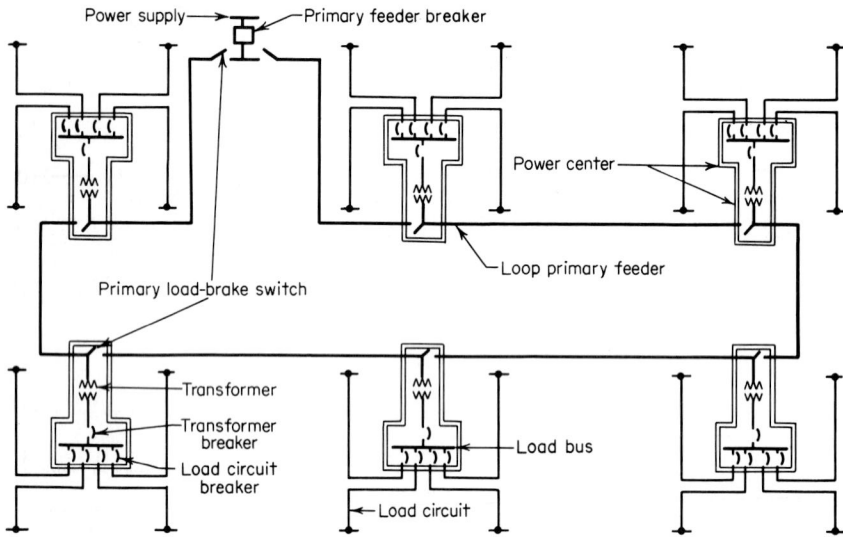

**Fig. 9-299** Loop-primary radial system.

The secondary loop gives a number of important advantages other than providing an emergency supply to restore service. It helps equalize the loads on all transformers and thus makes it unnecessary to match the transformer capacity at each load center to the load connected to each load bus.

Balancing transformer capacity and load in each load area may be a real problem on all other types of radial systems except the radial system of Fig. 9-296. The secondary loop permits one kVA rating of transformer to be used throughout the system, and load circuits can be run directly from the nearest bus to the loads.

Advantage is taken of the diversity among the loads because all transformers are connected in parallel. As a result, there usually is a saving in transformer capacity over that required in any other form of radial system except the simple radial system of Fig. 9-296. Also, this system is better adapted for across-the-line starting of relatively large motors than is any other type of radial system because the starting current is supplied through several transformers in parallel rather than through a single transformer. Thus, if the starting of large motors is involved, the use of this system may result in a saving in the cost of motor-starting equipment. Likewise, it is the most satisfactory type of radial system for combined light and power secondary circuits.

In the banked-secondary radial system, fault current flows not only through the transformer associated with a faulted load bus or circuit but also through other transformers and over the secondary loop to the fault. The resulting increase in fault current may require the use of load circuit breakers having a higher interrupting rating than needed in other types of modern radial systems. This possible increase in cost may or may not be offset by savings in transformer capacity and secondary conductors.

It is difficult to make a general cost comparison between this system and the loop-primary radial system. In some buildings the additional cost of the secondary loop will be offset by the saving in transformer capacity and secondary conductors. In other buildings, because of lack of diversity and a relatively uniform distribution of load, these savings are very small, and the system will cost more than the loop-primary radial system by about the cost of the secondary loop. However, the advantages of quick restoration of service to all loads, greater flexibility, higher efficiency, and better voltage conditions may justify the extra cost.

**476. Primary-selective radial system.** The primary-selective radial system, as shown in Fig. 9-301, differs from the systems previously described in that it employs at least two primary-feeder circuits in each load area. It is designed so that when one primary circuit is out of service, the remaining feeder or feeders have sufficient capacity to carry the total

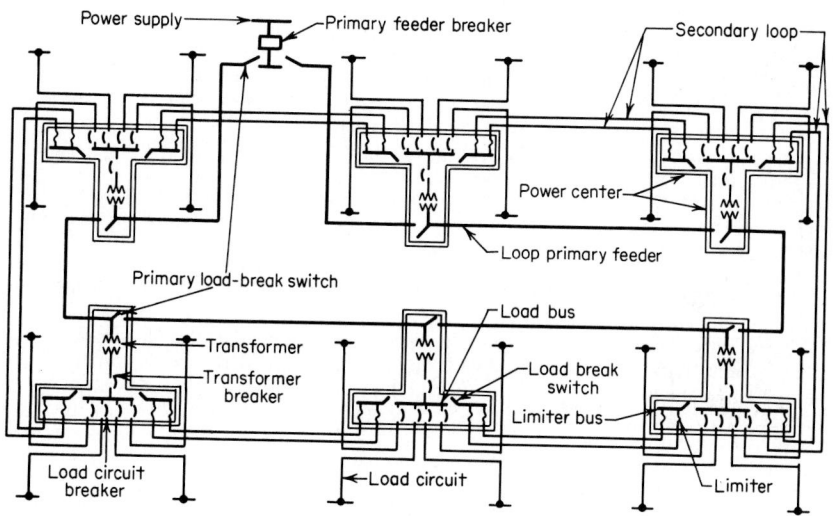

**Fig. 9-300** Banked-secondary radial system.

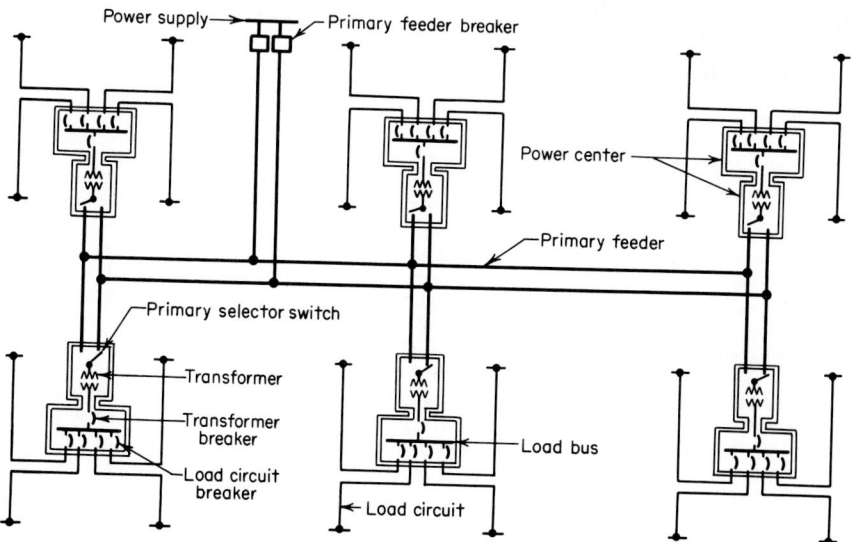

**Fig. 9-301** Primary-selective radial system.

load. While three or more primary feeders may be used, usually only two feeders are employed. Half of the transformers are normally connected to each of the two feeders. When a fault occurs on one of the primary feeders, only half of the load in the building is dropped. In the systems discussed previously a primary-feeder fault causes an outage to all loads in the building.

Double-throw or primary selector switches are used with all transformers so that when a primary-feeder fault occurs, the transformers normally supplied from the faulted feeder can be switched to the good feeder and restore service to all loads.

The short-circuit duty on the load circuit breakers is less than in the banked-secondary radial system and is about the same as for all other types of radial systems using transformers within the load areas.

If a fault occurs in one transformer, the associated primary-feeder breaker opens and interrupts service to half of the load in the building, just as when a primary feeder fails. The faulted transformer is disconnected by moving its primary selector switch to the open position. Service is then restored to all loads except those normally supplied by the defective transformer by closing the primary-feeder breaker.

Service cannot be restored to the loads normally served by the faulted transformer until the transformer is repaired or replaced.

Cost of the primary-selective radial system is greater than that of the simple radial system of Fig. 9-297 because of the additional primary-feeder breaker, the use of primary selector switches, and the greater amount of primary-feeder cable required. The benefits derived from the reduction in the amount of load dropped when a primary-feeder or transformer fault occurs, plus the quick restoration of service to all or most of the loads, may more than offset the greater cost.

The primary-selective radial system, however, is not so good as the banked-secondary radial system from the standpoint of restoring service to the loads after a primary-feeder or transformer fault. It is only a little better than the loop-primary radial system in this respect. In some cases, particularly in fairly large buildings with medium- or light-load density, about the same quality of service can be rendered at less cost by using the loop-primary radial system employing two separate loops instead of one. In this case, half of the transformers are connected to each loop of the system.

**477. Secondary-selective radial system.**  This system uses the same principle of duplicate feed from the power supply point as the primary-selective radial system. In this system, however, the duplication is carried all the way to each load bus on the secondary side of the transformers instead of just to the primary terminals of the transformers. This arrangement permits quick restoration of service to all loads when a primary-feeder or transformer fault occurs, as does the banked-secondary radial system.

The usual form of secondary-selective radial system is shown in Fig. 9-302. Each load area in the building is supplied over two primary feeders and through two transformers. The capacity of each of these transformers must be such that it can safely carry the entire load in the area served by both transformers. Each transformer is connected through a transformer breaker to a secondary bus section to which half of the radial load circuits are connected.

A bus-tie breaker is provided for connecting the two secondary or load-bus sections at each load center. Normally this bus-tie breaker is open, and the system operates as two parallel systems entirely independent of each other beyond the power supply points. The bus-tie breaker is interlocked with the two transformer breakers so that it cannot be closed unless one of the transformer breakers is open. This is done to keep the interrupting duty imposed on the load circuit breakers to a minimum.

A primary-feeder fault causes half of the load in the building to be dropped, as with the primary-selective radial system. Service can be restored by opening the transformer breakers associated with the faulted feeder and closing all bus-tie breakers. When this is done, the entire load is fed over one primary feeder and through half of the transformers. Therefore, both primary feeders must be capable of carrying the entire load, as with the primary-selective radial system, and a much greater transformer capacity is required.

A fault in a transformer causes the associated primary feeder breaker to trip and interrupts service to half of the building load. Service can be restored by opening the disconnecting switch on the primary and the circuit breaker on the secondary of the faulted transformer, closing the associated bus-tie breaker, and reclosing the primary-feeder breaker. This manual switching restores the system to normal operating conditions, except that the faulted transformer is deenergized and its load is being supplied through the adjacent transformer.

Cost of the secondary-selective radial system usually will be considerably more than that of the radial systems previously discussed with the exception of the system shown in Fig. 9-296, which uses one large substation and low-voltage feeders, and the banked-

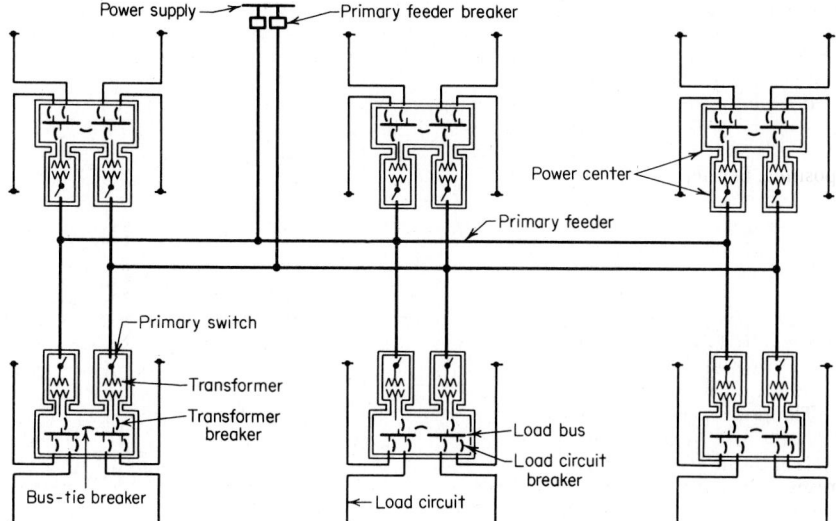

**Fig. 9-302**   Secondary-selective radial system.

secondary radial system of Fig. 9-300. This is due chiefly to the large amount of transformer capacity required to provide complete duplicate power supply to the secondary load buses. Because of this spare transformer capacity, the regulation provided by the secondary-selective radial system under normal conditions is better than that of the systems previously discussed, with the possible exception of the banked-secondary radial system.

From the standpoint of voltage fluctuation when a load such as a relatively large motor is thrown on the system, this system is not so good as the banked-secondary radial system. The advantage that the secondary-selective radial system offers over the banked-secondary radial system is that a primary-feeder or transformer fault causes only one-half of the load to be dropped instead of all the load in the area. Both systems permit the quick restoration of service to all loads when a primary-feeder or transformer fault occurs.

The banked-secondary radial system has greater flexibility and can be more readily adapted to changing load conditions than can the secondary-selective radial system.

A modified form of secondary-selective radial system that will often be less costly than the common form previously described is shown in Fig. 9-303. In this system there is only one transformer at each load center instead of two. Pairs of adjacent load buses are connected by secondary cables or bus and thus permit picking up the load of any bus when a primary-feeder or transformer fault occurs.

Each tie circuit is connected to each of its two load buses through a secondary-tie circuit breaker. Each secondary-tie breaker is interlocked so that it cannot be closed unless one of the two transformer breakers is open.

There are two ways in which this single-transformer-substation form of secondary-selective radial system can be handled. One is to use the same number and rating of transformers as in the usual form of the system. Essentially, this results in twice as many load areas with half as much load in each area. This arrangement has the advantage of reducing the average length of the radial load circuits and thereby lowering the cost of the secondary conductors. When compared with the other types of distribution systems, both this and the usual form of secondary-selective radial system introduce the problem of balancing the loads on twice the usual number of load buses.

The other way in which the system can be handled is to use the same number of load areas as in Fig. 9-302. With one transformer of twice the kVA rating at each load center,

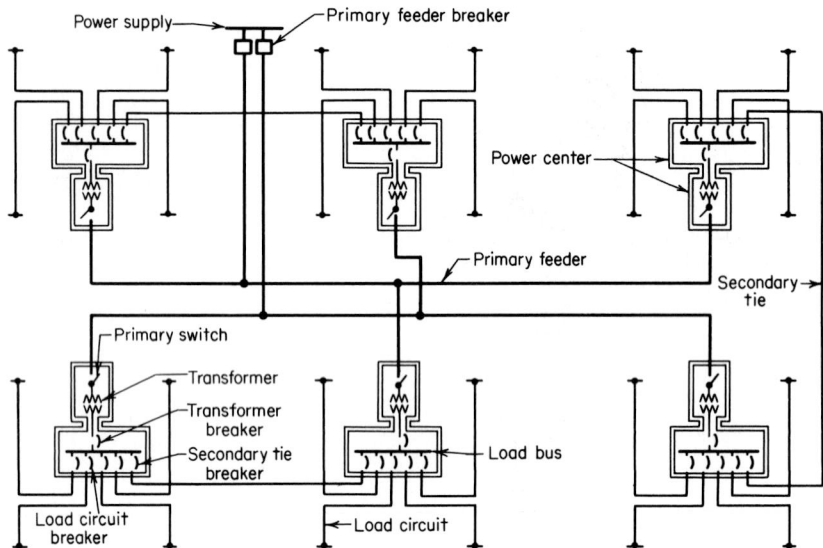

**Fig. 9-303**   Modified secondary-selective radial system.

the result is half of the number of transformers but the same total transformer capacity. Because of the smaller number of transformers of larger rating, this will often be the lowest-cost form of secondary-selective radial system.

Basically, the single-transformer-substation form of secondary-selective radial system, regardless of the number and size of the transformers used, functions as described in connection with the more usual form of this system.

**478. Simple network system.** The ac secondary network system is the system that has been used for many years to distribute electric power in the high-density, downtown areas of cities. Modifications of this type of system make it applicable to serve loads within buildings.

The best-known advantage of the secondary network system is continuity of service. No single fault anywhere on the system will interrupt service to more than a small part of the system load. Most faults will be automatically cleared without interrupting service to any load. Another outstanding advantage that the network system offers is its flexibility to meet changing and growing load conditions at minimum cost and with minimum interruption of service to the other loads. In addition to flexibility and service reliability, the secondary network system prov des exceptionally uniform and good voltage regulation, and its high efficiency materially reduces the cost of losses.

Three major differences between the network system and the simple radial system account for the outstanding advantages of the network. First, a network protector is connected in the secondary leads of each network transformer in place of the usual transformer secondary breaker, as shown in Fig. 9-304. Second, the secondaries of all transformers are connected together by a ring bus or secondary loop from which the loads are supplied over short radial circuits. Third, the primary supply has sufficient capacity to carry the entire building load without overloading when any one primary feeder is out of service.

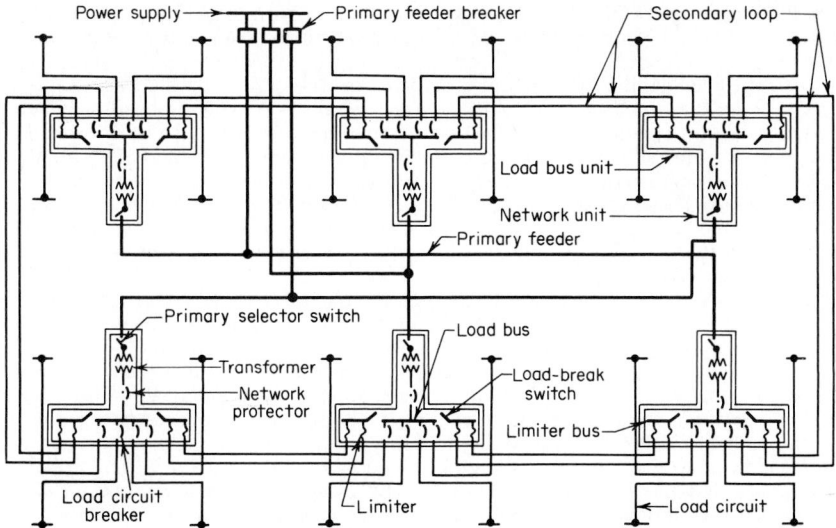

**Fig. 9-304**   Simple network system.

A network protector is a specially designed, motor-closed, and shunt-tripped air circuit breaker controlled by network relays. Tbe network relays function to close the breaker automatically only when voltage conditions are such that the associated transformer will supply load to the secondary loop and to open the breaker automatically when power flows from the secondary loop to the network transformer. The purpose of the network protector is to protect the secondary loop and the loads served from it against transformer and primary-feeder faults by disconnecting the defective feeder-transformer unit from the loop when a back feed occurs.

The chief purpose of the secondary loop is to provide an alternative supply to any load bus when the primary feeder which normally supplies it is deenergized, so as to prevent any service interruption when a transformer or feeder fault occurs. The use of the loop, however, gives at least four other important advantages. First, it saves transformer capacity. It permits lightly loaded transformers to help those which are more heavily loaded and thus tends to equalize the load on all transformers. Second, it saves secondary load circuit conductors and conduit. The secondary loop makes it unnecessary to match the load supplied from any one load bus with the transformer capacity at that point. Loads can be served from the loop at load buses located between network transformers, and the load circuits can be run directly to the loads from the nearest load bus. This permits the use of very short radial circuits and results in a considerable saving in secondary cable and conduit when compared with the load circuits of a simple radial system. Third, the use of the loop gives lower system losses and greatly improved voltage conditions. The voltage regulation on a network system is such that both lights and power can be fed from the same load bus. Much larger motors can be started across the line than on a simple radial system. This factor often results in greatly simplified motor control and permits the use of relatively large low-voltage motors with their less expensive control.

The fourth important advantage resulting from the use of the secondary loop is the much greater system flexibility it provides. New or relocated loads can be supplied directly from the nearest load bus without the necessity of reconnecting other loads to keep within the transformer capacity of the secondary substation to which the load is to be added. Complete network power centers or load-bus units can be relocated in the loop, or new units can be added to the loop to meet changing load conditions. The loop capacity requires no change when additional network power centers are connected to it if each new transformer is of no larger rating than those originally used. The loop permits these system changes to be made without interrupting service to existing loads. This greater flexibility of the network system permits the system to be changed more easily, more quickly, and at less cost to meet changing load conditions than can be done with a radial system.

Since the length of the secondary loop (which is in essence an extended bus) is considerably longer than the buses used with the usual radial systems, the probability of faults on it is greater. It therefore becomes necessary to protect the system against faults on this bus or loop so that a fault on it will be automatically disconnected without interrupting service. To accomplish this, the secondary loop should consist of a minimum of two three-phase circuits. These separate circuits should be bused at each point where power is supplied to the loop through a transformer and at each point where a radial load circuit is served from the loop. Each individual loop circuit should be connected to these bus points through limiters.

A limiter is a device for disconnecting a faulted cable from a distribution system and for protecting the unfaulted portions of that cable against serious thermal damage. This is accomplished by means of a circuit-opening member which is heated and destroyed by the current passing through it. The fusible circuit-opening member is completely enclosed so that there is no visible flame or smoke when the limiter operates. Limiters for building networks are suitable for use on circuits of 600 V and below and have an interrupting rating of 50,000 A. Limiters are designed for use with various sizes of cable and are rated in cable size. The largest limiter is rated at 600,000 cmil. The minimum fusing current of a limiter is about 3 to 3½ times the NEC continuous-current rating of the cable with which it is designed to be used. This value of fusing current is necessary to ensure positive selectivity among the limiters in the secondary loop. If limiters having appreciably lower values of fusing current were used, those on both the good conductors and the faulted conductor would function on the minimum definite time portion of their time-current curves for high fault currents. This means that not only the limiters in the faulted loop conductor but a number of those in the good conductors would blow almost instantly and in some cases cause unnecessary interruptions of service. In addition to preventing unnecessary limiter blowings, the high fusing-current value of limiters also has the advantage of keeping the normal temperature of the limiter terminals relatively low so that the associated cable can safely carry its rated current.

If loads are to be supplied from the secondary loop at points other than those at which the network transformers are connected to the loop, the current rating of the loop should be equal to the full-load current of one of the transformers supplying the loop. The current rating of the loop can be safely reduced to 67 percent of the above value if loads are supplied from the loop only at transformer locations. If more than one kVA rating of network transformer is used, the current rating of the loop should be based on the full-load current of the largest network transformer connected to the loop. When loads are served from the loop at points other than transformer locations, the maximum amount of load which should be supplied from the loop between any two adjacent transformers should not exceed the full-load rating of the larger of the two transformers.

To obtain satisfactory selectivity between limiters under all fault conditions, a minimum of two similar single-conductor cables per phase should be used in the secondary loop. The size and number of conductors used in any case should be selected to give the lowest installed loop cost. However, the conductors used should not be larger than about 600,000-cmil cable.

Loads are served from the secondary loop at conveniently located load buses over relatively short radial circuits. These load circuits are taken off the load buses through air circuit breakers. The breakers should provide adequate overload protection for their associated circuits and should be capable of interrupting the fault current available at that point on the system. The average interrupting duty on the load breakers in a network system will often be higher than the duty on the load breakers in a radial system having the same connected transformer capacity.

The optimum size and number of primary feeders can be used in the secondary network system because the loss of any primary feeder and its associated transformers does not result in the loss of any load. In spite of the spare capacity required in primary feeders, this will often result in a saving in primary-feeder and switchgear cost when compared with a radial system. This saving is due to the fact that in many radial systems more and smaller feeders are often used to keep down the extent of the outage when a primary-feeder fault occurs.

When a fault occurs on a primary feeder or in a transformer, the fault is isolated from the system through the automatic tripping of the primary feeder circuit breaker and all the network protectors associated with that feeder circuit. This operation does not interrupt service to any loads. When the necessary repairs have been made, the system is restored to normal operating conditions by closing the feeder circuit breaker. All network protectors associated with that feeder will close automatically.

The simple network system sometimes takes a form commonly known as a spot-network

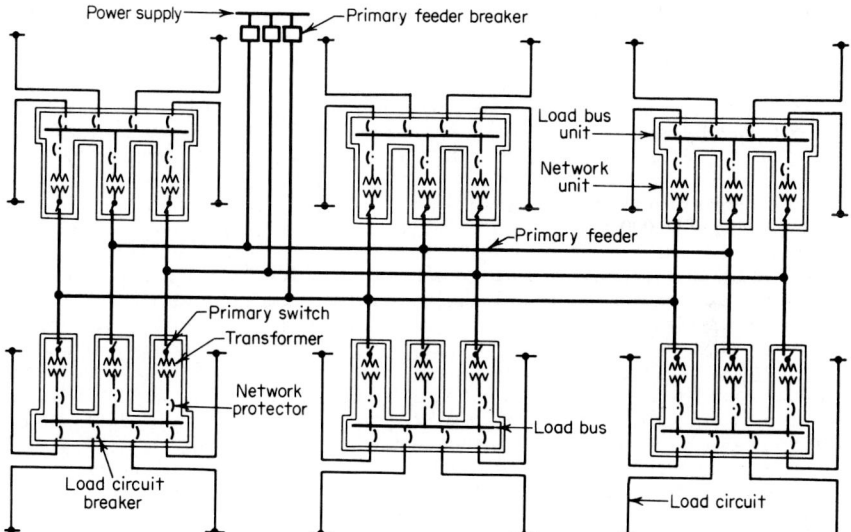

**Fig. 9-305**   Simple spot-network system.

system. With this system, emergency capacity to prevent a service interruption when a primary-feeder or transformer fault occurs is provided by making a spare primary cable and transformer available at each load center instead of tying the load areas together by a secondary loop, as in the usual simple network.

The simple spot-network system resembles the secondary-selective radial system in that each load area is supplied over two or more primary feeders through two or more transformers. In this system, however, the transformers are connected through network protectors to a single load bus, as shown in Fig. 9-305. Since the transformers at each load bus operate in parallel, a primary-feeder or transformer fault does not cause any service interruption. The transformers supplying each load bus will normally carry equal loads, whereas equal loading of the two transformers supplying a load center in the secondary-selective radial system is difficult to obtain. The interrupting duty imposed on the load circuit breakers will be greater with the simple spot-network system than with the secondary-selective radial system.

Simple spot-network systems are more economical than the other forms of network systems for buildings in which there are heavy concentrations of load covering small areas, with considerable distances between these concentrations and very little load in the areas between them. The chief disadvantage of spot-network systems is that they are not nearly so flexible for growing and shifting loads as are the other forms of network systems. Simple spot-network systems are used when a high degree of service continuity is required, flexibility is of minor importance, and their cost is less than that of the network systems using a secondary loop. The simple spot-network system is most economical when three or more primary feeders are required. This is true because supplying each load bus through three or more transformers greatly reduces the spare cable and transformer capacity that is required.

**479. Primary-selective network.** This is the most generally applicable and widely used form of industrial secondary network system. It is similar in principle to the simple network system, offers almost all its advantages, and is the lowest-cost network system for buildings or load areas requiring only two or three primary feeders. Saving in first cost when compared with the simple network system is accomplished by using the principle of duplicate or alternate feed to each transformer, as with the primary-selective radial system. This procedure practically eliminates the necessity for spare transformer capacity

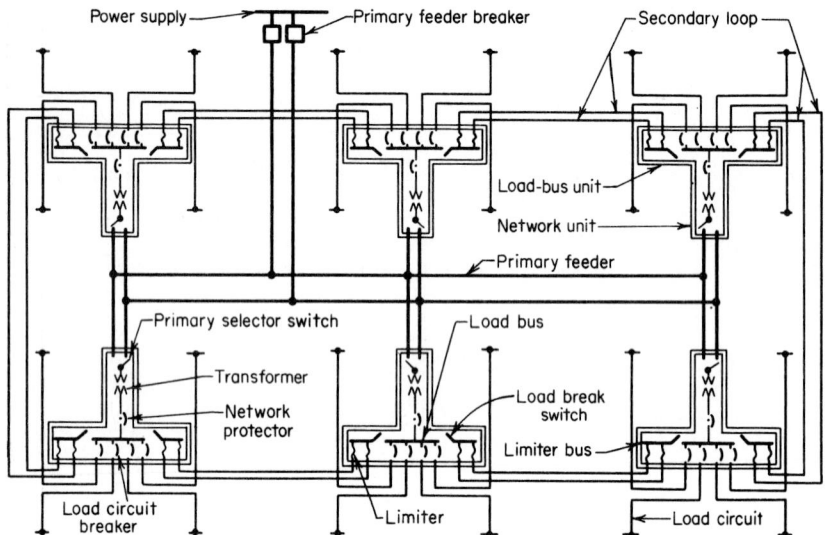

**Fig. 9-306**    Primary-selective network system.

and reduces the interrupting duty imposed on the load circuit breakers but not the spare primary-feeder capacity required.

Each transformer in the primary-selective network system is equipped with a primary selector switch. Two primary feeders are run to each transformer, as shown in Fig. 9-306. In a building requiring two primary feeders, each feeder must be capable of supplying the entire load in the building. Each transformer is connected to a load bus through a network protector. The radial circuits serving the loads are connected to the load bus through circuit breakers. A secondary loop, as is used in the banked-secondary radial and the simple network systems, connects each load bus to the two adjacent load buses. One-half of the network transformers are normally connected to each primary feeder.

A primary-feeder fault causes the faulty feeder and its transformers to be disconnected from the system by the automatic operation of the primary-feeder breaker and associated network protectors. The entire building load is then carried over the remaining primary feeder and one-half of the network transformers. The transformers associated with the faulty feeder can be connected to the good feeder by manually operating their selector switches. This relieves the overload on the transformers normally associated with the good feeder.

A transformer fault is isolated from the system by the operation of the primary-feeder breaker and the network protectors associated with the primary feeder. The defective transformer can be disconnected from the primary feeder by opening its selector switch. The feeder and its good transformers can be immediately returned to service. A primary-feeder or transformer fault will not cause any interruption of service to the load.

Cost of the primary-selective network system compares very favorably with that of the secondary-selective radial system. In buildings in which the load is quite uniform and there is little diversity among load centers, the secondary-selective radial system usually will be lower in initial investment than the network system. However, in buildings in which the load is nonuniform and there is appreciable diversity among load centers, the primary-selective network system often will be lower in first cost than the secondary-selective radial system.

It is common practice to supply the radial circuits only from the load buses to which the network transformers are connected. At times a reduction in system cost can be made by serving loads from the loop at points between transformers. When this is done, the loop

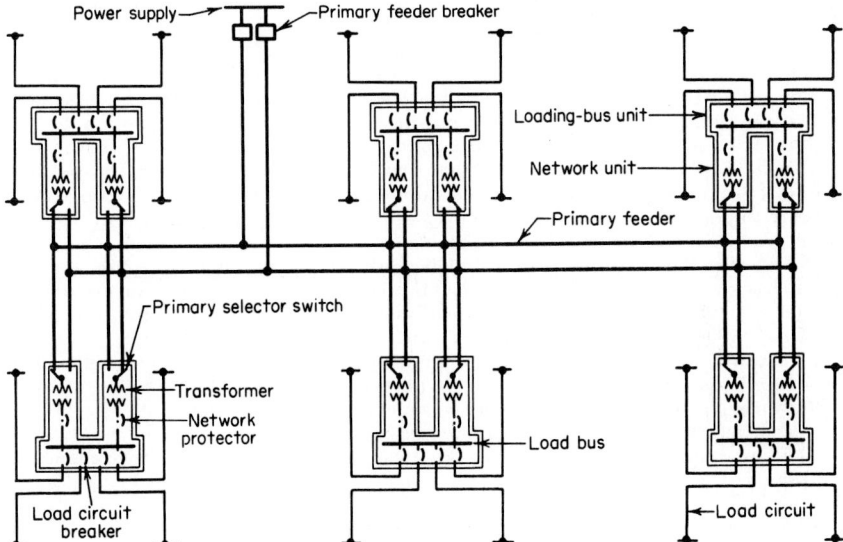

**Fig. 9-307**   Primary-selective spot-network system.

cables should preferably be paralleled and the loads taken off through load-bus units. Each load-bus unit should be similar to those used when network transformers connect to the loop; it consists of a load bus, two limiter buses with load-break switches and limiters, and load circuit breakers. Connecting loads to the secondary loop in this manner reduces the average length of the secondary load circuits.

By tapping all loads from the secondary loop at the nearest point, the amount of secondary load circuit conductors can be reduced to a minimum. This cannot be done practically when using a cable secondary loop if limiters are installed in the loop conductors at each load takeoff point. Such a form of network system is practical, however, if two or more parallel runs of plug-in bus duct are used for the secondary loop.

In this system, the transformers connect to a bus through network protectors, as in Fig. 9-306, except that there usually will be no load circuit breakers in the bus units. All loads usually are supplied through the load-break switches to the limiter buses and then through the limiters into the bus ducts that tie the adjacent bus units together.

This form of network system requires larger conductors in the secondary loop than those using the cable loop previously described. It also sacrifices something in service continuity when compared with network systems in which the loads are fed from load-bus units connected in a cable loop. This is true because a fault in a section of the bus-duct loop will cause an outage to the loads served from that section.

As with the simple network system, the primary-selective network may also take the form of a spot-network system. Each transformer in the primary-selective spot-network system is equipped with a primary selector switch and arranged for connection to either of two primary feeders, as shown in Fig. 9-307. This largely eliminates the necessity for spare transformer capacity, and the system is ordinarily designed without providing for such capacity.

Primary-feeder and transformer faults on this system are cleared without any interruption of service to the loads as in the simple spot-network system. When a primary feeder is deenergized because of a feeder fault, the transformers associated with the good feeder will carry as high as 160 to 200 percent of their rated kVA until the transformers normally associated with the defective feeder are switched to the good feeder by operation of their primary selector switches. After this switching operation has been completed, all the overloads on the system are relieved.

When a spot-network system of this design is used, a transformer failure requires a reduction in the load served from its load bus to prevent serious overloading of the remaining transformer connected to that bus. A transformer failure will occur so infrequently, however, that the cost of providing spare transformer capacity throughout the system cannot be justified to eliminate the remote probability of having to reduce load on one bus. If the possibility of having to reduce load is objectionable, it will cost less to use the simple spot-network system than to provide the necessary spare capacity in the primary-selective spot-network system. The primary-selective spot-network system, when designed without spare transformer capacity, will cost less than the simple spot-network system if two primary feeders and two transformers per load bus are used. If three or more primary feeders are required, the simple spot-network system will be lower in cost than the primary-selective spot-network system.

**480. Influence of buildings on design.**   Building distribution systems are too often selected on the basis of using the system having the lowest initial investment. This practice usually results in the installation of a system which is neither the best nor the most economical for the building when all the factors are considered and properly evaluated.

All the major characteristics of the different types of distribution systems should be evaluated in the light of the building load requirements, in order to select the best system. The six major characteristics are initial investment, flexibility, service continuity, voltage regulation, efficiency, and cost of operation and maintenance.

The best system is that which gives the greatest value per dollar of investment and provides adequate electric service. This is rarely the system with the lowest *initial investment.*

*Flexibility* is the system's ability to be adapted to changing load conditions with the minimum of service interruptions and minimum cost of necessary changes.

*Service continuity* is measured in terms of the amount of load dropped with faults at different locations in the system, taking into account the likelihood of faults in the various locations.

Under *voltage regulation,* the change in voltage with load variations, the uniformity of this change at the various load points throughout the system, and the change in voltage resulting from suddenly applied loads should be considered.

The overall *efficiency* of the system from the power supply point or points to the utilization devices should be taken into account for both light-load and heavy-load conditions of operation.

*Operating and maintenance costs* are influenced by the nature and amount of equipment in the system and the ability to inspect, test, and maintain the equipment with minimum interruption of service.

The general arrangement and choice of the type of system depend upon the type and purpose of the building. In this discussion buildings shall be considered as belonging to one of these five classifications:

1. Large vertical buildings less than about 12 stories high
2. Tall (high-rise) commercial buildings
3. Large horizontal buildings such as shopping centers and large one-floor offices
4. Groups of small buildings such as hospitals and educational institutions
5. Industrial plants

The quality of service required by the loads varies somewhat within each of these classifications, depending upon the purposes of the buildings and the activities carried on within them.

**481. Large vertical buildings.**   Large vertical buildings, such as office buildings, stores, apartments, hospitals, and hotels, require a high quality of electric service because the operation of essential equipment in the buildings depends directly on a reliable supply of electric power. Network systems are recognized as the best means of assuring satisfactory electric service in large buildings. However, in some cases, such as warehouses, interruptions of service for inspection and maintenance or as a result of faults may not interfere with the function of the building or cause hazards. In these cases, simple radial, primary-selective radial, or secondary-selective radial systems are used, depending upon the extent and duration of interruptions that can be tolerated and the importance of small savings in initial cost.

The network system is generally accepted as the most desirable form of distribution system for large buildings because it is the most economical method of providing reliability and good voltage characteristics. It is essential to have these characteristics in the distribution system for most large buildings because the activities in the buildings usually depend on a continuous supply of electric power with good voltage characteristics. Reliability is important because a failure of power that may stop elevators, pumps, and ventilating systems not only halts the activities in the buildings but may endanger the occupants. Obviously, good voltage regulation is essential to obtain the most effective output from lamps and other utilization devices. Good voltage regulation becomes particularly important when the same system supplies both the lighting and the power loads.

The relatively high load demands in large buildings encourage the use of large undesirable concentrations of transformer capacity in secondary substations with the result that very high vault currents are available in many cases. This requires expensive switchgear equipment to give adequate protection to the system against all possible faults in low-voltage circuits within the system. The power supply in buildings should be carefully designed to limit the fault currents to reasonable values so that less expensive equipment can be used. Several small substations instead of a few big substations can be used to reduce interrupting duties.

The interrupting duty on the low-voltage breakers in the substations directly affects the cost of the distribution system. High-interrupting-capacity breakers are more expensive than low-interrupting-capacity breakers. In general, the cost of the distribution system will materially increase if the interrupting duty exceeds 50,000 A. Therefore, it is desirable to plan the distribution system so that the interrupting duty does not exceed this value. In buildings in which the total load is not more than a few hundred kVA, it is not difficult to keep the interrupting duty below this value, and frequently it is possible to limit the interrupting duty to 25,000 A. The difference in cost between a substation using 25,000-A breakers and one using 50,000-A breakers is a relatively small part of the total cost of the substation. Therefore, it is not economical to make extensive special arrangements to reduce the interrupting duty when it is between 25,000 and 50,000 A. On the other hand, when it is necessary to install switchgear having an interrupting capacity higher than 50,000 A, the switchgear may become a large part of the total cost, and considerable expenditures to reduce the interrupting duty to 50,000 A may be justified. Relatively small substations distributed throughout the building generally mean that the low-voltage feeders will be short and that the voltage regulation will be better. Furthermore, the small substations can be installed in smaller spaces. Frequently, it will be easier to find several small spaces for substations than to find or provide one large space in the building for one substation. Another advantage of using a number of small substations is that it is easier to find space to carry the low-voltage feeder conduits away from the substation.

The form of the network system most frequently used in large buildings is the simple spot-network system. However, a simple network system having several interconnected secondary substations is used in some cases.

When the distribution system in a building is continuously supervised by capable operators, the total transformer capacity sometimes can be reduced by using high-voltage selector switches instead of disconnecting switches on the network transformers. When four or more primary feeders are available, it is generally as economical to use disconnect switches and provide the necessary reserve transformer capacity as to use the more expensive selector switches. Selector switches should not be used whenever it is likely that they may not be operated promptly in an emergency.

The distributed type of simple network system is primarily applicable to very large buildings in which the low-voltage feeders can be materially shorter with this type of network than if a spot network were used. In such applications, the extra cost of the secondary ties may be more than offset by savings in low-voltage feeder circuits because the substations are located closer to the load centers. This form of network system may also provide better voltage conditions at the load centers than could be provided by a spot network and relatively long low-voltage feeder circuits.

The spot network is particularly applicable to medium-size buildings having only a few floors. In these cases, the savings in low-voltage feeders that could be obtained by using two or more small substations in a distributed form of network usually do not justify the expense of the secondary ties and their protective equipment. The spot network, of course, gives the same degree of reliability as does the distributed form of network. Obviously, two or more spot networks can be used in larger buildings with the same resulting savings in low-voltage feeder circuits. However, this means that there would be a greater number of smaller transformers and a correspondingly greater number of network protectors. The smaller units generally have a higher cost per kVA than the larger units, so that a number of small spot networks may cost more than a distributed network in an extensive system.

**482. Tall buildings.** Tall buildings are generally built in the densely loaded areas of cities. Almost without exception, the loads in these areas are served by secondary network systems. This makes it practical and economical to use some form of network system in a tall building. The nature of such a building practically requires a network system. Access to the major part of the building is accomplished almost entirely by elevators. Hence, it is imperative that power be continuously available for the operation of the elevators to avoid the hazard of panic or serious inconvenience resulting from a power outage. Furthermore, fire protection, water supply, heating, and ventilating depend for their operation on the availability, at all times, of an electric power supply. These considerations practically limit the choice of a distribution system to some form of secondary network. Usually three or more primary feeders are available within the area to supply the building, so that either a simple spot-network system or a simple network system is feasible. Various arrangements of these two systems are used, depending on the height of the building, the location of the major loads within the building, and the total load demand.

In most modern tall commercial buildings there is likely to be a large air-conditioning load. When this load and the other power loads such as elevators, ventilating and heating equipment, fire pumps, water pumps, and related auxiliaries are taken into account, the power demand is likely to be more than the lighting load and may be as much as 3 or 4 times the lighting load. This proportion of power and lighting loads should be considered in the choice of utilization voltage. The power load will usually be the principal factor in the choice of locations for the distribution equipment.

Generally speaking, the practical maximum height that 208Y/120-V circuits can be carried in a tall building is about 12 stories. This limitation depends to a considerable extent on the relative cost of space for the distribution equipment in various parts of the building. The limitation of about 12 stories is based on including in the cost of the distribution system a charge for all the space used above the basements equal to the value of the same amount of commercially usable space. On this basis there is a thumb rule that major distribution centers will occur about every 25 floors in a tall office building. This is a very general rule, and the locations of the elevator equipment and major air-conditioning loads and other power loads in the building may make other locations of the distribution equipment more economical. The locations of the secondary substations involved in the distribution system in the building should be very carefully considered. Some specific comparisons have indicated that a much closer spacing of the secondary substations than 25 stories is economical. A closer spacing of the units improves the voltage conditions within the building and reduces losses in the system.

The simplest form of network system for a tall building is the simple spot-network system. In this system, a spot-network substation comprising the necessary number of network transformers and the associated low-voltage switchgear equipment is located at or near the major loads in the building. Therefore, the spot-network substation in the basement is generally the biggest of the various secondary substations. The locations of the power loads in the basement area or a logical segregation of the load may make two or more spot-network substations feasible. In this form of the network system, several primary-feeder circuits are carried up through the building to the various spot-network substation locations where the corresponding network transformers are connected to the primary feeders.

In many locations where tall buildings are built, it is practical to operate the building distribution system as part of the low-voltage secondary network in the area around the building. This involves the use of a simple network system within the building with secondary-tie circuits between the various network substations and the secondary-main system in the area around the building. The ties between the secondary mains of the utility network system and the building distribution system permit interchange of power between the two systems. This augments the reliability of the network system within the building and to some extent may improve the voltage conditions within the building. Since the same primary feeders that supply the building may also supply network units in the area around the building, there is some advantage in interconnecting the two systems to make sure that all the network equipment functions in the same manner. The desirability of interconnecting the building network system with a surrounding local network system may offset some of the advantages of a 460-V system within the building. The network system in the area around the building will be a 208Y/120-V system in most cases.

The simple network system involves some variations of the secondary ties between the network substations within the building. In their simplest form, these ties are purely tie circuits and do not supply loads at any time except by interchange of power from one substation to another. When this form of secondary tie is used, it generally will have only about two-thirds as much carrying capacity as the rated current of the largest of the associated network transformers. This provides enough capacity to make up for the outage of one transformer at one of the network substations. Generally, the carrying capacity of the tie circuits is practically independent of the number of network units at any load bus. The secondary ties between the building system and the surrounding network system usually have about the same capacity as the ties between network substations within the building.

The capacity of the secondary ties can be increased to provide for tapping loads directly from the secondary ties between network substations. Basically, such a tie can be treated in two general ways. One way is to form load buses at every point along the tie where loads are tapped, as indicated in Fig. 9-308. If this is done, the capacity of the tie circuit will depend on the amount of load tapped off the circuit and the size of the network units associated with the tie. The capacity should be enough to carry all the loads tapped from the circuit, from one end of the circuit, as well as a reasonable amount of power interchanged between the network substations. In a simple case in which single-transformer network substations are interconnected by a tie circuit of this type, the total capacity of the circuit should be about equal to the rated current of one of the associated network transformers.

If load buses are not established in the tie circuit and the loads are tapped from one or another of the parallel branches of the tie circuit, the total carrying capacity of the tie circuit should be correspondingly increased. The reason for this is that it is unlikely that the load can be equally divided among the parallel branches of the tie circuit. Even if an equal division of connected load were possible, the diversity among the loads would mean that the load on one branch would not always be the same as that on another. Therefore, in this type of circuit the total carrying capacity should be at least 1½ times the rated current of one of the associated transformers when there is only one transformer in each of the interconnected network substations.

The forms of the secondary-tie circuit illustrated in Fig. 9-308 consider only the use of paralleled three-phase circuits to form the tie circuit because practical and convenient sizes of conductors can be used and overcurrent protection for the circuits can be economically provided. It is possible, of course, to use a single conductor per phase, such as a heavy bus, to form the tie circuit. To get the necessary carrying capacity in this way is likely to be expensive, and providing overcurrent protection for such a circuit is expensive and requires heavy protective equipment.

Another form of tie circuit that has been used is a secondary bus running the full height of the building. This form of tie circuit in effect is an extended spot network with the spot-network bus extending from the basement of the building to the uppermost network unit. In this arrangement the load circuits at the various floors are connected directly to the riser bus through protective devices or through a switchboard connected to the riser bus through a suitable protective device. The major disadvantage of this arrangement is the length of the spot-network bus. It is necessary to install this bus in such a way that there

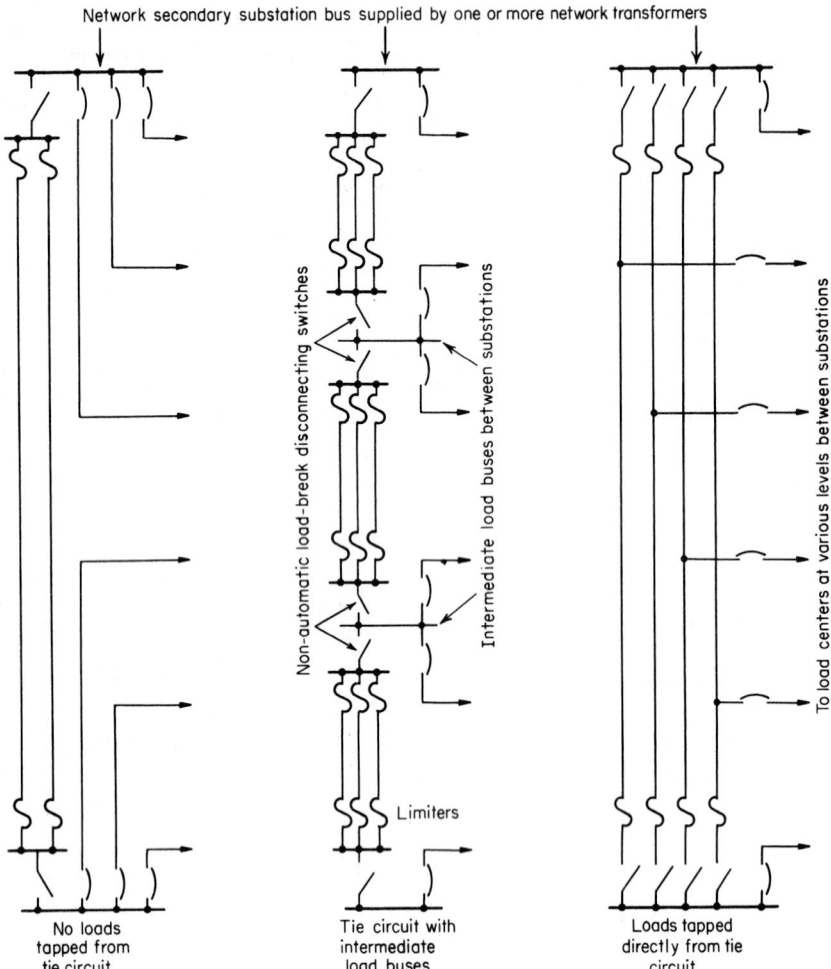

**Fig. 9-308**   Three arrangements of secondary-tie circuits between network substations in tall commercial buildings.

is practically no possibility of a fault on it. This can be accomplished by segregating the phases by barriers between the buses. However, such an installation is expensive. The capacity of such a bus should be somewhat more than the total load connected to the bus between adjacent network transformers. Some excess above this total load value would be required for the interchange of power between network units when one of the primary-feeder circuits is out of service.

The distribution of power within a tall building necessarily involves the use of transformers within the building. It is very important to choose transformers that minimize the possibility of fire or other damage to the building or injury to people in the building resulting from a transformer failure. For this reason, ventilated or sealed dry-type transformers are used in buildings. These types of transformers eliminate the hazard of fire and explosion damage resulting from a transformer failure.

**483. Large horizontal buildings.**   Under the classification of large horizontal buildings, probably the most important with respect to numbers, size, and magnitude of load are shopping centers.

Shopping centers may be classified into three types with regard to size. They are the neighborhood shopping center that is designed to serve 5000 to 20,000 people, the district shopping center that serves 20,000 to 100,000 people, and the regional shopping center that serves a market of over 100,000 people.

Shopping centers will usually take one of five general physical arrangements. The simplest type, usually associated with the neighborhood shopping center, is the strip layout. A common variation of this simple layout is the L-shape layout. The mall, the court, and the ring are three layouts that are used mostly in the district shopping center. The fifth type, the group or cluster of buildings, is generally used for large regional shopping centers.

The physical layout of a shopping center will influence the design of the electric distribution system. For instance, a system utilizing a loop-secondary circuit can be used to advantage in the mall-, ring-, or group-type shopping centers but may not be economical for the strip or court arrangement. The large loads in a shopping center will usually dictate where the power-center transformers must be located. If power centers must be located at these large loads, the loads in between them may be served over radial circuits from the power centers. However, if the distance between the power centers becomes great, using a system that permits the tapping of services off a secondary circuit running between power centers would be advantageous.

The selection of the utilization voltage or voltages to be used in shopping centers should not be limited to the choice of 120/240 V single-phase, 440 V three-phase, or 208Y/120 V three-phase. The magnitude, locations, and characteristics of the loads in shopping centers make it appear that the use of a three-phase, four-wire, grounded-Y, 460Y/265-V system would prove to be more economical in many cases.

Table 484 compares the distribution systems discussed previously for several magnitudes of load in each of the five types of shopping centers. The comparisons are made on a column basis, and it is not intended that the ratings in different columns should be compared.

The ratings take into account initial investment, service continuity, flexibility to handle new or changing loads, operation and maintenance costs, voltage regulation, and efficiency. Each of these factors has been weighted in proportion to its relative importance for each load magnitude and type of shopping center. The A ratings indicate that analysis of the system has shown it to rate at the top of the 12 systems compared in each column. The other quality ratings are given in terms of percentage of value of the A system:

| Rating | Percent |
|:------:|:-------:|
| B | 81–90 |
| C | 71–80 |
| D | 61–70 |
| E | 51–60 |
| F | 41–50 |

**484. Comparison of Distribution Systems for Several Types of Shopping Centers**

(Type of shopping center, loads in kVA)

| | Strip or L | | | Mall | | | Court | | | Ring | | | Group or cluster | | |
|---|---|---|---|---|---|---|---|---|---|---|---|---|---|---|---|
| | 100–1,000 | 1,000–2,500 | 2,500 and up | 100–1,000 | 1,000–2,500 | 2,500 and up | 100–1,000 | 1,000–2,500 | 2,500 and up | 100–1,000 | 1,000–2,500 | 2,500 and up | 1,000–2,500 | 2,500–5,000 | 5,000 and up |
| Conv. simple-radial | D | D | E | E | E | F+ | D | D | E | E | E | F+ | E | F+ | F |
| Modern simple-radial | A | A | B | A— | B— | C | C | A | B | A— | B— | C | B— | C | D |
| Modified modern simple-radial | C | C | C | C | D | D— | C | C | C | C | D | D— | D | D— | E |
| Loop-primary radial | A | A | B+ | B+ | B— | C | C | A | B+ | B+ | B— | C | B— | C | D |
| Banked-secondary radial | C | C | B | B | C | D+ | B | C | B | B | C | D+ | C | D+ | C |
| Primary-selective radial | B | B | C | D | D | D— | C— | B | C | D | D | D— | D | D— | D |
| Secondary-selective radial | C— | C | C | D | D | D— | C— | C | C | D | D | D— | D | D— | E |
| Modified secondary-selective radial | C— | C | A— | B+ | A— | A | D | C | A— | B+ | A— | A | A— | A | E |
| Simple network | D | D | B | B | B | B | D | D | B | B | B | B | B | B | A |
| Simple spot-network | C | B | A | A | A | A | C | B | A | A | A | A | A | A | B |
| Primary-selective network | D | C— | A | B | A | B | D | C— | A | B | A | B | A | A | A |
| Primary-selective spot-network | C | B | B | B | B | B | C | B | B | B | B | B | B | B | B |

**485. Methods of supplying power to individual motor loads.** The layout of the last stage of a motor distribution system will depend upon the type of building, construction characteristics of the building, motor size, types of motors to be served, and the flexibility desired in order to take care of changing location of loads. In one type of layout the branch circuit to each motor is run from a power panel. In another type the branch circuits to individual motors are tapped from a feeder. The features of the different types of motor distribution layouts are shown in Fig. 9-309.

The following discussion of the application of these methods is reproduced from *Electrical Systems Design*, by Joseph F. McPartland.

Miscellaneous motor loads in most commercial and institutional buildings and in many industrial buildings are usually circuited from power panels to which feeders deliver power. This type of distribution is a standard method of motor circuiting, generally limited to handling a number of motors of small integral-horsepower or fractional-horsepower sizes, located in a relatively small area such as a fan room or a pump room. The feeder to a power panel may be a riser in a multistory commercial or institutional building, run from a basement switchboard. In a one-level industrial area, the feeder to a power panel may be run from a main or load-center switchboard or from a load-center substation in a high-voltage distribution system. In commercial and industrial buildings, motor feeders may be run from a main switchboard or a load-center substation to a motor-control center serving a large group of motors in a machine room or in any compact area where the motors are relatively close together and close to the control-center assembly. The motor-control center contains all the control, protection, and disconnecting means for the motor circuits supplied.

In industrial buildings, in which a large number of motors are used over a large area,

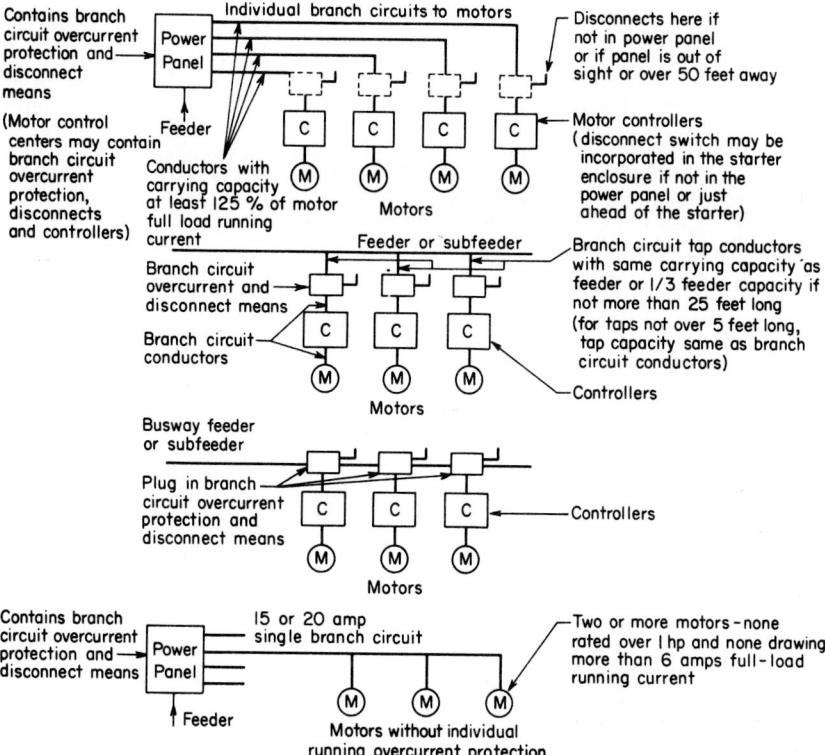

**Fig. 9-309** Methods of power distribution to individual motors. [*Electrical Systems Design*]

distribution of power to the individual motor loads generally follows the method of tapping motor branch circuits from the feeder. Such circuits may be tapped in several ways, with various combinations of branch-circuit overcurrent protection and motor control and disconnect means. The feeder in such an arrangement may originate from a main switchboard or from a load-center substation. In multistory buildings, such as office buildings or institutions, taps to motor branch circuits on a particular floor may be taken from a subfeeder originating in a load-center switchboard which is fed by a riser from the main switchboard in the basement. The feeder-tap method of supplying motor branch circuits is, of course, the basis for plug-in busway distribution to spread-out motor loads. In the system, the plug-in busway may be a feeder or a subfeeder, depending upon its origin. From a main switchboard, it would be a feeder; from a load-center switchboard, it would be a subfeeder. However, from a load-center substation, the plug-in busway would be a feeder—a secondary feeder derived from the transformation down from the primary feeder to the substation.

**486. Adequate wiring for lighting and convenience outlets in commercial, public, and industrial structures.** It is very important in designing the interior wiring for a building to make adequate provision not only for present demands but also for probable future demands. The levels of illumination demanded by industry and commerce are being constantly increased. In order that a building may keep pace with this advancement without excessive expense in the reconstruction of its wiring, the wiring must be more than adequate to provide merely for present needs. Double carrying capacity can be provided in the original wiring installation for less than 25 percent additional cost. To provide a guide for adequate wiring, the following recommendations from the *Handbook of Interior Wiring Design* are given:

1. Outlet Location: Commercial and Public Occupancies

    *a. Ceiling outlets for general illumination.* Whenever possible, ceiling outlets should be uniformly spaced and located so that the spacing will be approximately the same in both directions. In *offices* and *schoolrooms*, the spacing between adjacent outlets shall not exceed the distance from floor to ceiling. In other interiors, the space may be extended to 1¼ times the ceiling height. The space between an outlet and a wall should not exceed one-half of the average distance between outlets. In *halls* and *corridors*, spacing should not exceed 20 ft (6.1 m).

    In locating outlets on any floor plan, construction features should be considered, especially ceiling beams. These are not always shown on the floor plans, yet they may seriously affect a lighting layout. This is particularly true for indirect lighting.

    *b. Stores: convenience outlets.* There shall be installed at least one convenience outlet for each 400 ft² (37.2 m²) of floor area or major part thereof, these outlets to be uniformly distributed over the entire area. Preference shall be given to the placing of these outlets in supporting columns and sidewalls, but no part of the floor area shall be more than 15 ft (4.6 m) from an outlet.

    *c. Stores: outlets for window illumination.* Provision for show-window illumination shall be made by (1) a junction box on a wall or column at a suitable height above the transom bar or possible false ceiling to which all circuits of show-window reflectors or spotlights can be connected or (2) individual outlets spaced as recommended below.

    For medium-base lampholders, outlets for show-window reflectors should be located 12 to 18 in (304.8 to 457.2 mm) apart. For mogul-base lampholders, outlets for show-window reflectors should be located 15 to 24 in (381 to 609.6 mm) apart; for fluorescent lamps, as required by the lighting layout. The specific layout is to be governed by the standard loads of Sec. 487. Whenever more than one circuit per window is necessary, adjacent outlets should be on different circuits.

    For spot or floodlights, at least two outlets per window shall be symmetrically placed, with an additional outlet for every 8 ft (2.4 m) or major fraction in excess of 16 ft (4.9 m).

    *d. Stores: convenience outlets in show windows.* In or near the floor of each window, there shall be installed at least one convenience outlet for each 50 ft² (4.6 m²) of platform or floor area used for window display, but in no case shall there be fewer than two convenience outlets per window.

    *e. Stores: outlets for case lighting.* Outlets shall be provided in the floor or wall for termination of the circuits for show-case and wall-case lighting. Such outlets shall be suitably located for connection to the lighting equipment and shall be governed by the standard loads as listed in Table 487.

*f. Offices and schools: convenience outlets.* In each separate office room with 400 ft² (37.2 m²) or less of floor area, there shall be installed at least one convenience outlet for each 20 ft (6.1 m) of wall space. In each separate office with more than 400 ft² of floor area, there shall be installed at least four convenience outlets for the first 400 ft² of floor area and at least two outlets for each additional 400 ft² or major fraction thereof. Outlets should be placed at suitable locations to serve all parts of the office space. Certain offices, for example, professional and display offices, require many more outlets than are specified above, and the wiring needs must be individually studied if such probable occupancy is known.

For typical schoolrooms at least one convenience outlet shall be placed along the front wall and at least one along the rear wall.

*g. Offices and stores: fans.* Unless made unnecessary by provision for complete air conditioning, outlets for fans shall be installed on the basis of at least two for each 400 ft² (37.2 m²) or major portion thereof. Such outlets shall be placed approximately 7 ft (2.1 m) above the floor and shall be of a type from which the fan may be mechanically suspended.

*h. Stores: signs.* If no other equivalent provision is made for sign lighting, a rigid metal conduit or electrical metallic tubing, not smaller than 1 in, shall be run to the front face of the building for each intended or probable individual store occupancy. Such a raceway shall terminate outside the building at a point suitable for connection to the sign and shall terminate inside the building at a cabinet. Feeder capacity to this cabinet and service capacity shall be based upon a sign load of not less than 50 W/ft of store frontage, only the frontage on the principal street being considered if the store faces more than one street. Space provision shall be allowed for the connection and placement of a time switch to the sign circuit.

2. OUTLET LOCATION: INDUSTRIAL OCCUPANCIES

*a. Ceiling outlets for general illumination.* The many special considerations of industrial lighting make it necessary to develop the outlet locations around the lighting layout. Local lighting units frequently supplement general illumination; special directional overhead sources are often necessary. In planning outlet locations, these must be considered as part of the complete installation. In halls and corridors lighted from a single row of outlets, the spacing between outlets shall not exceed 20 ft (6.1 m).

*b. Convenience outlets.* There should be at least one convenience outlet in each bay in both manufacturing and storage spaces.

3. STANDARD LOADS FOR GENERAL ILLUMINATION   The number of branch circuits and the capacities of the feeders and service shall be based upon the loads specified in Tables 487 and 488 as a minimum. If a known load that will exceed the loading here specified is to be supplied, the circuits, feeders, and service shall be based upon such known load.

The tabulated values represent the present acceptable standards of Illuminating Engineering Society foot-candle values. From these, the watts per square foot for each occupancy were determined by considering average conditions found in such occupancy, together with the accepted methods and equipment used in providing such illumination levels. In certain cases the standard loads include allowance for convenience outlets as well as for general illumination. The calculated wattage standards were based on the use of fluorescent and mercury-vapor equipment for large areas and incandescent equipment for local and supplementary lighting.

4. BRANCH CIRCUITS[1]   The minimum number of branch circuits shall be based on the present connected load as follows:

*a.* For two-wire 15-A circuits, the load per circuit shall not exceed 1000 W.

*b.* For multiwire 15-A circuits, the load shall not exceed 1000 W between each outside wire and the neutral.

*c.* For circuits designed for heavy-duty lampholders, the load per circuit shall not exceed 1500 W for No. 10 wire, 2500 W for No. 8, and 3000 W for No. 6.

The following considerations of wire size and circuit runs shall apply:

*d.* No wire smaller than No. 12 shall be used for any circuit.

*e.* When the run from a panelboard to the first outlet of a lighting branch circuit exceeds

---

[1]These standards are based upon 15-A protection of branch circuits. If local regulations limit the load on branch circuits to less than these requirements, the wattage and area limits herein specified must be proportionately reduced.

50 ft (15.2 m), the size of wire used shall be at least one size larger than that determined by any of the above considerations. The calculated size may be used between outlets.

*f.* When the run from a panelboard to the first outlet of a convenience-outlet circuit exceeds 100 ft (30.5 m), No. 10 wire or larger shall be used for such run.

*g.* No runs longer than 100 ft (30.5 m) between the panelboard and the first outlet of a lighting branch circuit shall be made unless the intended load is so small that the voltage drop can be restricted to 2 percent between the panelboard and any outlet on that circuit. To avoid this condition, panelboards should be relocated or additional panelboards installed.

The following considerations in regard to combinations of outlets shall apply:

*h.* No convenience outlet shall be supplied by the same branch circuit that supplies ceiling or show-window lighting outlets.

*i.* Outlets for show-window spotlights shall be on separate circuits from general show-window outlets.

*j.* The number of convenience outlets included on one circuit shall be based on the listing in the following table.

### Outlets per Circuit

| Location | Maximum Number |
|---|---|
| Barbershops, beauty parlors, etc. | 2 |
| Medical, dental, and similar offices | 2 |
| Store show windows (for spolights) | 3 |
| Store show windows (at floor) | 6 |
| Display areas in retail stores | 6 |
| School classrooms | 6 |
| Manufacturing spaces | 6 |
| Office spaces | 8 |
| Storage spaces | 10 |

5. CIRCUIT CONTROL   Suitable provision shall be made for the control of all circuits except those supplying convenience outlets only, and control of the latter circuits is recommended. Circuits may be controlled individually, in groups, or by a combination of both methods.

In a retail store, individual control at the panelboards by means of circuit switches or circuit breakers is usually preferable. In most other occupancies, control should be provided by means of local switches or circuit breakers. Frequently it is necessary to provide two or more switches for the local control of outlets supplied by a single branch circuit.

Provision shall be made for control of all circuits supplying a single sign by means of an external manually operable switch or circuit breaker, and in each case a suitable time switch shall also be provided to effect the same control.

In a retail store, each circuit supplying outlets in one or more show windows shall be individually controlled, and provision shall be made for the installation of a time switch for group control of all such circuits.

Group control of circuits for general illumination or for special decorative effects, by means of remote-control switches, is desirable in certain large spaces such as large offices, large reading rooms, museums, art galleries, and ballrooms.

6. PANELBOARDS   The number and location of panelboards shall be based on the number of branch circuits and the distance or runs as specified in Par. 4 of these adequacy standards.

On each panelboard one spare circuit shall be provided for each five circuits utilized in the initial installation. When flush-type cabinets are used, provision shall be made for bringing a corresponding number of circuit conductors to the ceiling of the story served, the ceiling of the story immediately below, or both points. Such provision may consist of circuit conductors or of empty raceways terminating in boxes suitably located for future extensions.

Provision shall be made for control of show-window lighting by such grouping of circuits to a panelboard as to enable a time switch to be suitably installed.

7. FEEDERS

*a. Carrying capacity.* The carrying capacity of each feeder or subfeeder shall be based on the number of separate circuits which it supplies, computed as follows:

(1) *Overhead lighting circuits.* These are assumed as having 1000 W for each 15-A circuit and 1500 W for each 20-A circuit.

(2) *Convenience-outlet circuits.* The assumed load is 1000 W per circuit.

(3) *Spare panelboard circuits.* The assumed load is 500 W.

(4) *Nonitemized and heavy-duty additional circuits.* The capacity is based on the specific load for which they are designed.

To the total of these four, such demand factors as permitted by the NEC may be applied. Refer to Div. 3 for additional information.

*b. Provision for the future.* In all determinations of feeder size, consideration should be given to increasing the capacity of the initial system by 50 percent to provide for future growth at a minimum of unwarranted expense. If an ultimate feeder size does not exceed No. 4 wire, the excess capacity should be installed immediately. In other cases, one of the following methods should be made a part of the original layout:

(1) The installation of oversize raceways, so that the conductors installed can be withdrawn at any time and replaced by conductors of a suitable larger size

(2) The installation or arrangement of the installed equipment so that additional feeders can be installed later at minimum expense to furnish the ultimate capacity

(3) The installation of feeders of excess size

*c. Voltage drop.* Feeders and subfeeders shall be of such size that the total voltage drop from any panelboard to the point where connection is made to the lines of the utility supplying services shall not exceed 1 percent (see Div. 3 for method of calculation). To compute such a drop, the ultimate demand as calculated above shall be used whenever the wire size based on such demand is installed immediately. If provision for the future is made in some other way (oversize raceways, etc.), the voltage drop shall be computed on the basis of the carrying capacity of the wire used.

8. FEEDER DISTRIBUTION CENTER   A feeder distribution center in the form of a panelboard, switchboard, or group of enclosed switches or circuit breakers shall be provided for the control and protection of each feeder.

Except when oversize feeders are provided in the original installation as recommended in the preceding section, provision shall be made at the feeder distribution center for the connection and suitable protection of feeders of increased size or of supplemental feeders. This can be accomplished by providing additional protective devices as part of the original installation, or by so designing the original equipment that space, bus capacity, and facilities for making connections will be available.

9. SERVICE-ENTRANCE ADEQUACY STANDARDS (75°C RATED COPPER CONDUCTORS)

| Initial load, amp | Service switch, amp | Conductor size, gage No. or cir mils | Initial load, amp | Service switch, amp | Conductor size, gage No. or cir mils |
|---|---|---|---|---|---|
| 1– 23 | 60 | 8 | 118–133 | 200 | 4/0 |
| 24– 33 | 60 | 6 | 134–150 | 400 | 4/0 |
| 34– 47 | 100 | 4 | | | |
| 48– 60 | 100 | 2 | 151–167 | 400 | 250.000 |
| 61– 67 | 100 | 1 | 168–183 | 400 | 300,000 |
| | | | 184–200 | 400 | 350,000 |
| 68– 83 | 200 | 1/0 | 201–217 | 400 | 400,000 |
| 84–100 | 200 | 2/0 | 218–267 | 400 | 500,000 |
| 101–117 | 200 | 3/0 | | | |

10. SERVICE CONDUCTORS AND EQUIPMENT   The minimum capacity of the service required for the initial load depends on the total number of feeders supplied by it, whose individual loads are determined as specified in Par. 7. When applicable, demand factors as permitted by the NEC may be used (see Sec. 44 of Div. 3). To determine the probable required ultimate capacity of the service conductors and equipment, 50 percent should be added to the calculated initial load.

When the calculated initial load does not exceed 267 A, service conductors and equipment having the capacity needed for the ultimate load should be included as part of the original installation. The recommended equipment for various initial loads is tabulated in Par. 9. When the initial load exceeds 267 A, a study should be made of each individual case to determine what provision should be made for a future increase in the load.

### 487. Standard Loads for Lighting in Commercial Buildings

| Occupancy | Watts per sq ft | Occupancy | Watts per sq ft |
|---|---|---|---|
| Armories: drill sheds and exhibition halls. (This does not include lighting circuits for demonstration booths, special exhibit spaces, etc.) | 5 | Minor surgeries—1,500 watts per area This and the above figure include allowance for directional control. Special wiring for emergency systems must also be considered | |
| Art galleries: | | Laboratories | 5 |
| General | 3 | Hotels: | |
| On paintings—50 watts per running foot of usable wall area | | Lobby, not including provision for conventions, exhibits | 5 |
| Auditoriums | 4 | Dining room | 4 |
| Automobile showrooms | 6 | Kitchen | 5 |
| Banks: | | Bedrooms, including allowance for convenience outlets | 3 |
| Lobby | 4 | Corridors—20 watts per running foot | |
| Counters—50 watts per running foot including service for signs and small motor applications, etc. | | Writing room, including allowance for convenience outlets | 5 |
| Offices and cages | 5 | Library: | |
| Barber shop and beauty parlors. (This does not include circuits for special equipment) | 5 | Reading rooms. This includes allowance for convenience outlets | 6 |
| Billiards: | | Stack room—12 watts per running foot of facing stacks | |
| General | 3 | Motion-picture houses and theaters: | |
| Tables—450 watts per table. | | Auditoriums | 2 |
| Bowling: | | Foyer | 3 |
| Alley runway and seats | 5 | Lobby | 5 |
| Pins—300 watts per set of pins | | Museums: | |
| Churches: | | General | 3 |
| Auditoriums | 2 | Special exhibits—supplementary lighting | 5 |
| Sunday-school rooms | 5 | Office buildings: | |
| Pulpit or rostrum | 5 | Private offices, no close work | 5 |
| Club rooms: | | Private offices, with close work | 7 |
| Lounge | 2 | General offices, no close work | 5 |
| Reading rooms | 5 | General offices, with close work | 7 |
| The above two uses are so often combined that the higher figure is advisable. It includes provision for convenience outlets | | File room, vault, etc. | 3 |
| | | Reception room | 2 |
| Courtrooms | 5 | Post office: | |
| Dance halls. (No allowance has been included for spectacular lighting, spots, etc.) | 2 | Lobby | 3 |
| | | Sorting, mailing, etc. | 5 |
| | | Storage, file room, etc. | 3 |
| Drafting rooms | 7 | Professional offices: | |
| Fire-engine houses | 2 | Waiting rooms | 3 |
| Gymnasiums: | | Consultation rooms | 5 |
| Main floor | 5 | Operating offices | 7 |
| Shower rooms | 2 | Dental chairs—600 watts per chair | |
| Locker rooms | 2 | Railway: | |
| Fencing, boxing, etc. | 5 | Depot—waiting room | 3 |
| Handball, squash, etc. | 5 | Ticket room—general | 5 |
| Halls and interior passageways—20 watts per running foot | | On counters 50 watts per running foot | |
| Hospitals: | | Rest room, smoking room | 3 |
| Lobby, reception room | 3 | Baggage, checking office | 3 |
| Corridors—20 watts per running foot | | Baggage storage | 2 |
| Wards, including allowance for convenience outlets for local illumination | 3 | Concourse | 2 |
| | | Train platform | 2 |
| Private rooms, including allowance for convenience outlets for local illumination | 5 | Restaurants, lunchrooms, and cafeterias: | |
| | | Dining areas | 3 |
| Operating room | 5 | Food displays—50 watts per running foot of counter (including service aisle) | |
| Operating tables or chairs: | | Schools: | |
| Major surgeries—3,000 watts per area | | Auditoriums | 3 |
| | | If to be used as a study hall—5 watts per sq ft | |

**487. Standard Loads for Lighting in Commercial Buildings** (Continued)

| Occupancy | Watts per sq ft | Occupancy | Watts per sq ft |
|---|---|---|---|
| Schools (Continued): | | Medium cities:[a] | |
| Class and study rooms.............. | 5 | Brightly lighted district—500 watts per running foot of glass | |
| Drawing room..................... | 7 | Neighborhood stores—250 watts per running foot of glass frontage | |
| Laboratories...................... | 5 | Small cities and towns—300 watts per running foot of glass frontage[a] | |
| Manual training.................. | 5 | Lighting to reduce daylight window reflections—1,000 watts per running foot of glass | |
| Sewing room...................... | 7 | | |
| Sight-saving classes.............. | 7 | | |
| Showcases—25 watts per running foot | | Stores, large department and specialty: | |
| Show windows: | | Main floor......................... | 6 |
| Large cities:[a] | | Other floors....................... | 6 |
| Brightly lighted district—700 watts per running foot of glass | | Stores in outlying districts............. | 5 |
| Secondary business locations—500 watts per running foot of glass | | Wall cases—25 watts per running foot | |
| Neighborhood stores—250 watts per running foot of glass | | | |

NOTE. Figures based on use of fluorescent equipment for large-area application, incandescent for local or supplementary lighting.

[a] Wattages shown are for white light with incandescent filament lamps. Where color is to be used, wattages should be doubled.

**488. Standard Loads for General Lighting in Industrial Occupancies**

| Occupancy | Watts per sq ft | Occupancy | Watts per sq ft |
|---|---|---|---|
| Aisles, stairways, passageways 10 watts per running foot | | Cloth products: | |
| Assembly: | | Cutting, inspecting, sewing: | |
| Rough............................ | 3.0 | Light goods...................... | 4.5 |
| Medium.......................... | 4.5 | Dark goods...................... | *4.5 |
| Fine............................. | *4.5 | Pressing, cloth treating (oil cloth, etc.): | |
| Extra fine....................... | *4.5 | Light goods...................... | 3.0 |
| Automobile manufacturing: | | Dark goods...................... | 6.0 |
| Assembly line.................... | *4.5 | Coal breaking, washing, screening....... | 2.0 |
| Frame assembly.................. | 3.0 | Dairy products...................... | 4.0 |
| Body assembly................... | 4.5 | Engraving........................... | *4.5 |
| Body finishing and inspecting........ | *4.5 | Forge shops, welding................. | 2.0 |
| Bakeries............................ | 4.0 | Foundries: | |
| Book binding: | | Charging floor, tumbling, cleaning, pouring, shaking out.............. | 2.0 |
| Folding, assembling, pasting.......... | 3.0 | Rough molding and core making...... | 2.0 |
| Cutting, punching, stitching, embossing | 4.0 | Fine molding and core making........ | 4.0 |
| Breweries: | | Garages: | |
| Brewhouse........................ | 3.0 | Storage........................... | 2.0 |
| Boiling, keg washing, etc............ | 3.0 | Repair and washing................. | *3.0 |
| Bottling.......................... | 4.0 | Glassworks: | |
| Candy making...................... | 4.0 | Mixing and furnace rooms, pressing and Lehr glass-blowing machines.... | 3.0 |
| Canning and preserving............... | 4.0 | Grinding, cutting glass to size, silvering. | 4.5 |
| Chemical works: | | Fine grinding, polishing, beveling, etching, inspecting, etc................. | *4.5 |
| Hand furnaces, stationary driers and crystallizers...................... | 2.0 | Glove manufacturing: | |
| Mechanical driers and crystallizers, filtrations, evaporators, bleaching... | 2.0 | Light goods: | |
| Tanks for cooking, extractors, percolators, nitrators, electrolytic cells..... | 3.0 | Cutting, pressing, knitting, sorting.. | 4.5 |
| Clay products and cements | | Stitching, trimming, inspecting..... | 4.5 |
| Grinding, filter presses, kiln rooms..... | 2.0 | Dark goods: | |
| Moldings, pressing, cleaning, trimming. | 2.0 | Cutting, pressing, etc.............. | *4.5 |
| Enameling......................... | 3.0 | Stitching, trimming, etc........... | *4.5 |
| Glazing........................... | 4.0 | Hangars—aeroplane: | |
| | | Storage—live..................... | 2.0 |

## 488.    Standard Loads for General Lighting in Industrial Occupancies *(Continued)*

| Occupancy | Watts per sq ft | Occupancy | Watts per sq ft |
|---|---|---|---|
| Hangars (Continued): | | Fine hand painting and finishing...... | *3.0 |
| Repair department................ | *3.0 | Extra-fine hand painting and finishing | |
| Hat manufacturing: | | (automobile bodies, piano cases, etc.). | *3.0 |
| Dyeing, stiffening, braiding, cleaning, | | Paper-box manufacturing: | |
| and refining: | | Light........................... | 3.0 |
| Light........................... | 2.0 | Dark............................ | 4.0 |
| Dark............................ | 4.5 | Storage of stock.................... | 2.0 |
| Forming, sizing, pouncing, flanging, | | Paper manufacturing: | |
| finishing, and ironing: | | Beaters, grinding, calendering......... | 2.0 |
| Light........................... | 3.0 | Finishing, cutting, trimming.......... | 4.5 |
| Dark............................ | 6.0 | Plating............................ | 2.0 |
| Sewing: | | Polishing and burnishing............. | 3.0 |
| Light........................... | 4.5 | Power plants, engine rooms, boilers: | |
| Dark............................ | *4.5 | Boilers, coal and ash handling, storage- | |
| Ice making, engine and compressor room. | 2.0 | battery rooms.................... | 2.0 |
| Inspection: | | Auxiliary equipment, oil switches and | |
| Rough........................... | 3.0 | transformers.................... | 2.0 |
| Medium.......................... | 4.5 | Switchboard, engines, generators, blow- | |
| Fine............................ | *4.5 | ers, compressors................. | 3.0 |
| Extra fine....................... | *4.5 | Printing industries: | |
| Leather manufacturing: | | Matrixing and casting.............. | 2.0 |
| Vats............................ | 2.0 | Miscellaneous machines............. | 3.0 |
| Cleaning, tanning, and stretching..... | 2.0 | Presses and electrotyping............ | 4.5 |
| Cutting, fleshing, and stuffing........ | 3.0 | Lithographing...................... | *4.5 |
| Finishing and scarfing.............. | 4.5 | Linotype, monotype, typesetting, im- | |
| Leatherworking: | | posing stone, engraving............ | *4.5 |
| Pressing, winding, and glazing: | | Proof reading..................... | *4.5 |
| Light........................... | 2.0 | Receiving and shipping................ | 2.0 |
| Dark............................ | 4.5 | Rubber manufacturing and products: | |
| Grading, matching, cutting, scarfing, | | Calenders, compounding mills, fabric | |
| sewing: | | preparation, stock cutting, tubing | |
| Light........................... | 4.5 | machines, solid tire operations, me- | |
| Dark............................ | *4.5 | chanical goods, building, vulcanizing. | 3.0 |
| Locker rooms...................... | 2.0 | Bead building, pneumatic-tire building | |
| Machine shops: | | and finishing, inner-tube operation, | |
| Rough bench- and machine work...... | 3.0 | mechanical-goods trimming, treading | 4.5 |
| Medium bench- and machine work, | | Sheet-metal works: | |
| ordinary automatic machines, rough | | Miscellaneous machines, ordinary | |
| grinding, medium buffing and polish- | | bench work...................... | 3.0 |
| ing.............................. | 4.5 | Punches, presses, shears, stamps, | |
| Fine bench- and machine work, fine | | welders, spinning, medium bench- | |
| automatic machines, medium grind- | | work............................ | 4.5 |
| ing, fine buffing and polishing...... | *4.5 | Tin-plate inspection................ | *4.5 |
| Extra-fine bench- and machine work, | | Shoe manufacturing: | |
| grinding, fine work............... | *4.5 | Hand turning, miscellaneous bench- | |
| Meat packing: | | and machine work................ | 2.0 |
| Slaughtering....................... | 2.0 | Inspecting and sorting raw material, | |
| Cleaning, cutting, cooking, grinding, | | cutting, and stitching: | |
| canning, packing.................. | 4.5 | Light........................... | 4.5 |
| Milling—grain foods: | | Dark............................ | *4.5 |
| Cleaning, grinding, and rolling........ | 2.0 | Lasting and welting................. | 4.5 |
| Baking or roasting.................. | 4.5 | Soap manufacturing: | |
| Flour grading...................... | 4.5 | Kettle houses, cutting, soap chip and | |
| Offices: | | powder......................... | 3.0 |
| Private and general: | | Stamping, wrapping and packing, fill- | |
| No close work..................... | 5.0 | ing, and packing soap powder....... | 4.5 |
| Close work....................... | 7.0 | Steel and iron mills: bar, sheet and wire | |
| Drafting rooms.................... | 7.0 | products: | |
| Packing and boxing................. | 3.0 | Soaking pits and reheating furnaces... | 2.0 |
| Paint manufacturing................. | 3.0 | Charging and casting floors.......... | 2.0 |
| Paint shops: | | Muck and heavy rolling, shearing | |
| Dipping, spraying, firing, rubbing, | | (rough by gage), pickling, and clean- | |
| ordinary hand painting and finishing. | 3.0 | ing............................. | 2.0 |
| | | Plate inspection, chipping........... | *4.5 |

**488. Standard Loads for General Lighting in Industrial Occupancies** *(Continued)*

| Occupancy | Watts per sq ft | Occupancy | Watts per sq ft |
|---|---|---|---|
| Steel and iron mills (Continued): | | Silk: | |
| Automatic machines, light and cold rolling, wire drawing, shearing (fine by line)............................ | 4.5 | Winding, throwing, dyeing......... | 4.5 |
| | | Quilling, warping, weaving, finishing: | |
| Stone crushing and screening: | | Light goods.................... | 4.5 |
| Belt-conveyor tubes, main-line shafting spaces, chute rooms, inside of bins............................. | 2.0 | Dark goods.................... | 6.0 |
| | | Woolen: | |
| Primary breaker room, auxiliary breakers under bins................... | 2.0 | Carding, picking, washing, combing.. | 3.0 |
| | | Twisting, dyeing................. | 3.0 |
| Screens........................... | 3.0 | Drawing-in, warping: | |
| Storage-battery manufacturing, molding of grids........................... | 3.0 | Light goods.................... | 4.5 |
| | | Dark goods.................... | 6.0 |
| Store- and stock rooms: | | Weaving: | |
| Rough bulky material............... | 2.0 | Light goods.................... | 4.5 |
| Medium or fine material requiring care. | 3.0 | Dark goods.................... | 6.0 |
| Structural-steel fabrication............ | 3.0 | Knitting machines............... | 4.5 |
| Sugar grading........................ | 5.0 | Tobacco products: | |
| Testing: | | Drying, stripping, general........... | 3.0 |
| Rough............................ | 3.0 | Grading and sorting.............. | *4.5 |
| Fine.............................. | 4.5 | Toilet and wash rooms............... | 2.0 |
| Extra-fine instruments, scales, etc..... | *4.5 | Upholstering, automobile, coach, furniture............................. | 4.5 |
| Textile mills: | | Warehouse......................... | 2.0 |
| Cotton: | | Woodworking: | |
| Opening and lapping, carding, drawing, roving, dyeing............... | 3.0 | Rough sawing and benchwork........ | 2.0 |
| | | Sizing, planing, rough sanding, medium machine and bench-work, gluing, veneering, cooperage.............. | 4.5 |
| Spooling, spinning, drawing, warping, weaving, quilling, inspecting, knitting, slashing (over beam end). | 4.5 | Fine bench- and machine work, fine sanding and finishing............. | 6.0 |

The figures given in this table are average design loads for general lighting. In those cases marked with an asterisk (*), the load values provide only for large-area lighting applications. Local lighting must then be provided as an additional load.

The figures given are based on the use of fluorescent and mercury-vapor equipment of standard design. Adjustments can be made for use of higher or lower efficiency equipment. For equal lighting intensities from incandescent units, the figures must be at least doubled.

Use of these figures in computing number and loading of circuits and feeders should always be checked against the conditions and requirements of the particular area. The figures are not substitutes for lighting design.

In any case, need for particular color quality of light, special intensities, or control of light must be determined as part of the lighting design, and wattages and circuit requirements must be provided accordingly.

**489. Allowance for growth in feeders** *(Electrical Systems Design)* should begin with the spare capacity designed into the branch circuits. For all circuits loaded to 50 percent of capacity, it can be assumed that there is an allowance for growth of each circuit load by an amount equal to 30 percent of the circuit capacity. This allowance is based on Code limitation of 80 percent load on circuits which are in operation for 3 h or longer or on circuits which supply motor-operated appliances in addition to other appliances and/or lighting. The 30 percent growth allowance in each circuit should be converted to a total watts figure and added to the total load on the feeder as calculated above. This grand total then represents the required feeder capacity to handle the full circuit load on the panelboard. The next step is to provide capacity in the feeder for anticipated load growth (plus some amount for possible unforeseen future requirements).

From experience, modern design practice dictates the sizing of feeders to allow an increase of at least 50 percent in load on a feeder when analysis reveals any load-growth possibilities. Such an analysis depends upon the type of building, the work performed, the plans or expectations of management with respect to expansion of facilities or growth of business, the type of distribution system used, locations of centers of loads, permanence of various load conditions, and particular economic conditions. Depending upon a thorough study of all these factors, the advisability of spare capacity can be determined for each feeder in a distribution system. But if study demands extra capacity in a feeder, substantial growth allowance (50 percent) is generally essential to realize the sought-after economy of future electrical expansion. Skimpy upsizing of feeder conductors or raceways has proved a major shortcoming of past electrical design work. This is particularly true in tall office buildings, apartment houses, and other commercial buildings in which the elimination of riser bottlenecks represents a large part of the total modernization cost.

Spare capacity in feeders may be provided in one or more of several ways. If the anticipated increase in feeder load is to be made in the near future, the extra capacity should usually be included in the conductor size and installed as part of the initial electrical system. In many small office or commercial buildings and apartment houses, extra capacity should automatically be included in the initial size of the feeder conductors, provided the size is not greater than No. 2 after adding the 50 percent of extra capacity. Other methods of providing for growth in feeder load are as follows:

1. Selection of raceways larger than required by the initial size and number of feeder conductors. With this provision, the conductors can be replaced with larger sizes if required at a later date. The advisability of this step should be carefully determined in the case of risers or underground or concealed feeder raceway runs. In some cases, a compromise can be made between providing spare capacity in the feeder conductor size and upsizing the feeder raceway.

2. Including spare raceways in which conductors can be installed at a later date to obtain capacity for load growth. Such arrangements of multiple raceways must be carefully laid out and related to the existing overall system. The types of feeder-distribution centers or main switchboards and the layouts of local branch-circuit panelboards must be able to accommodate future expansion of distribution capacity based on the use of spare raceways, with modification and regrouping of feeder loads.

**490. Control of groups of receivers (other than hall or night lights) from the main switchboard.** If it is desirable to control a group of receivers from the main switchboard in the basement, a separate feeder must be carried from it to each group to be so controlled. Usually the feeder system can be laid out without regard to the control of the room lights, because, as a rule, they need not be controlled from the switchboard. It is usually advisable to have each of the lower floors, up to and including the ground floor, on a separate switch, as these floors often require light when the others do not. Special lighting appliances such as sign, clock-dial, and outside dome lights require separate feeders from the switchboard, because they are turned on and off at set times from the switchboard. Certain motors may require similar control. In hotels the feeder switches are never opened except in case of accident, so from a control standpoint only it is not necessary to subdivide hotel feeders. When tenants of portions of buildings pay for the light they use, it is often desirable to carry a separate feeder from the switchboard to each tenant's suite so that all meters can be located together at the switchboard. Suites can be metered separately by cutting meters in the mains at the suites, but this may be undesirable.

**491. The control of hall lights from the main switchboard** is an important consideration. In private dwellings it does not usually pay to install a separate feeder for the hall lights, and it may not be necessary in a hotel, where attendants are constantly passing in the halls. In a majority of public buildings, however, separate control of the hall lights is very desirable, if not necessary. The usual problem, then, is whether there shall be one or two sets of hall-light feeders. With two sets of feeders for hall lights, local switches can be eliminated and control effected entirely from the main switchboard. Two sets of hall feeders increase the cost of installation, but the saving in energy usually justifies them. By arranging two sets of feeders, one set serving, say, one-third of the hall lights and the other the remaining two-thirds, the smaller group can be used for dark days and for an all-night circuit, and a saving in energy will result. When there are two sets of hall lights thus controlled, the wiring of outlets should be such that there will be a uniform distribution of light whichever set is lighted. When tenants pay for the energy used in their suites, a separate feeder for the hall lights is indispensable.

## FARM WIRING

**492. Wiring.** The use of electric power on the farm is continually increasing. So that the farm electrical system may be adequate and give the best results in expenditure of money and satisfactory service, it is essential that the system be properly and wisely planned, that good materials and equipment be employed throughout, and that good workmanship be employed in its installation. In most cases the best installation will be achieved by the use of a power pole, as shown in Fig. 9-310. The power pole will be the central distribution point of the electrical system. The power company will run its supply lines from its main pole line to the power pole of the farm and down the pole to its meter. From the power pole, the consumer will distribute the electric power by overhead or underground lines to the different buildings. In rare instances, it may not be possible to erect a power

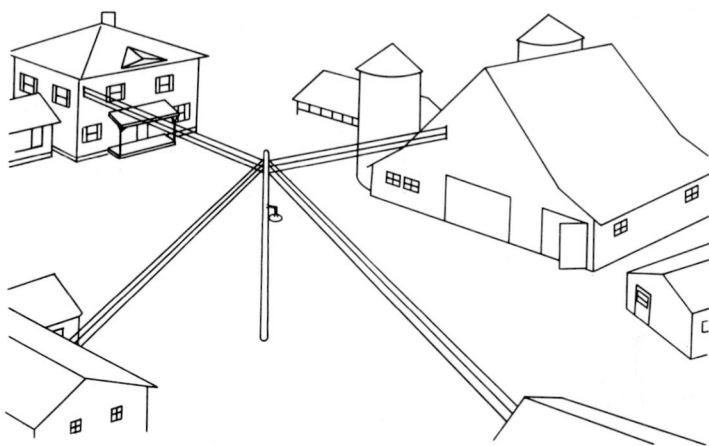

**Fig. 9-310**   Power pole for farm-building distribution feeders. [General Electric Co.]

pole. In such cases the power lines from the public utility will terminate at a building, probably the farmhouse. The meter will then be installed at this point. This method introduces some complications and is recommended only when the erection of the power pole is very difficult or impossible.

The first step in planning the farm electric system is to determine the power demands for each building. The diversity of farm work makes it impossible to set down definite load requirements to fit all farms. Every kind of farming has its own needs insofar as power, machinery, and appliances are concerned. However, a number of wiring requirements are fundamental to the various farm buildings. No matter where the farm is located or how the buildings are constructed, certain minimum standards of wiring must be followed to obtain satisfactory electrical service. The following sections specify proper wiring for average conditions. They take into consideration the correct number of outlets, the proper switching, modern lighting, and adequate wire sizes for the machinery and appliances that are normally used in specific farm buildings. They are intended as a basic guide. If more equipment is to be installed than is included in these specifications or if the farming operations are of a special nature, the specifications must be varied to suit the special needs.

**493.    Typical Computed Loads and Probable Maximum Demands**

| Building | Computed load | Probable demand |
|---|---|---|
| Farm residence........................... | 19   kw | 13   kw |
| Dairy barn, medium size, including milkhouse and one 5-hp motor........................... | 20   kw | 12   kw |
| Dairy barn, large size, including milkhouse and one 7½-hp motor........................... | 27   kw | 15   kw |
| Same, two 7½-hp motors.................... | 36   kw | 24   kw |
| Milking barn with milkhouse................ | 7.5 kw | 5.8 kw |
| Milkhouse only........................... | 6.0  kw | 4.8 kw |
| Beef-cattle, horse, sheep, and hog barns, medium size, with 5-hp motor..................... | 14   kw | 10   kw |
| Same, large size, with 7½-hp motor.......... | 20   kw | 15   kw |
| Poultry laying house: | | |
| 1,000 sq ft........................... | 1.8  kw | 1.8 kw |
| 4,000 sq ft........................... | 8.7  kw | 6.4 kw |
| 8,000 sq ft........................... | 15.2 kw | 11.7 kw |
| Brooder house, per brooder: | | |
| Infrared............................ | 2.25 kw | 2   kw |
| Standard brooder.................... | 1.25 kw | 1   kw |
| Farm shop, with welder.................... | 17   kw | 10   kw |
| Sweet-potato curing and storage.............. | 1.5  watts per cu ft or 4  watts per bu | 1.5 watts per cu ft or 4  watts per bu |
| Machinery sheds, stock shelters, and miscellaneous buildings........................... | 1  watt per sq ft floor area | 1  watt per sq ft floor area |

SOURCE: "Farmstead Wiring Handbook."

**494. Other loads** (*Design Manual on Steel Electrical Raceways*, published by American Iron and Steel Institute). Over and beyond the electrical requirements of individual buildings are a number of other loads for which provision must be made. These include such diverse items as water pumps (water system and irrigation), immersion heaters, water warmers, hay dryers and hoists, forage handlers, silo unloaders, and other miscellaneous equipment. Some idea of typical loads encountered is given by the following selected examples:

```
Forage handling:
   Blower type         20-hp motor
   Inclined elevator   2- to 3-hp motor
Silo unloaders         3- to 5-hp motor
Bulk-milk coolers      Varies with size and type; 1 to 5 hp for one or more motors
```

Such additional loads should be kept in mind when surveying a farm operation for initial wiring or rewiring, and provisions made in the system design to handle such equipment.

**495. Specifications for dairy barn**

LIGHTS   In back of cows there should be one lighting outlet every 12 ft (3.7 m), and in front of cows one approximately every 20 ft (6.1 m). There should be one or two lights on top of the silo, so installed that they will light the silo and chute. Only general illumination is needed in the haymow. Floodlights above the hay line on opposite walls of the barn will therefore give the required lighting.

CONVENIENCE OUTLETS   Outlets should be placed approximately 15 ft (4.6 m) apart in a row in back of the cows. If the cows face in, outlets should be on the wall, up high enough so that cattle cannot come in contact with them. If the cows face out, outlets should be suspended from the ceiling on heavy-duty cords. They must be hung so that they can be reached easily yet high enough not to obstruct traffic.

In open spaces a convenience outlet should be installed for each 600-ft² (55.7-m²) area.

SWITCHES   It is best to install separate switches for each row of lights in the barn. In small barns, however, all lights in back of the cows may be on one switch. When two or more entrances are used frequently, at least one row of lights should be controlled by three-way switches. The switch for the lights in the haymow should be mounted at a convenient location on the barn floor and also be equipped with a pilot light. To have a visible indication when the light is burning in the silo, a switch with a pilot light mounted at the entrance of the silo chute is recommended.

CIRCUITS  The service-entrance panel for the barn must be adequate to take care of the load. Since some of the equipment needs special circuits, it is important to have an entrance panel of the proper capacity. The following number of circuits is recommended as a minimum. However, additional equipment or a change in the barn layout may make it necessary to deviate from this recommendation.

| | |
|---|---|
| Lights | 2 circuits (depending on size of barn) |
| Milking machine | 1 circuit (230 V) |
| Feed grinder | 1 circuit(230 V) |
| Utility motor | 1 circuit (230 V) |
| Convenience outlets | 1 circuit (depending on size of barn) |
| Fans | 1 or more (depending on size of barn) |
| Hay dryer | 1 circuit (230 V) |

### 496. Specifications for milk house

LIGHTS  In most cases a center light will be sufficient unless the milk house is unusually large. Then additional lights will be necessary over the work areas, especially where the utensils are washed.

CONVENIENCE OUTLETS  At least one convenience outlet for general use should be installed at a convenient location in the milk house. It is advisable to put this outlet on a heavy-duty 20-A circuit.

SWITCHES  All lights should be controlled by a switch at the entrance.

CIRCUITS  When the milk house is connected to the barn, its circuits may be included in the barn entrance panel. When the milk house is separate, a special entrance panel must be installed.

The following circuits are recommended as a minimum:

| | |
|---|---|
| Lights and outlets | 1 circuit |
| Milk cooler | 1 circuit |
| Water heater | 1 circuit (230 V) |
| Sterilizer | 1 circuit (230 V) |
| Ventilating fan | 1 circuit (if fan is small, it may be connected to light and outlet circuit) |

### 497. Specifications for workshop, garage, and machine shed.
Frequently the workshop, garage, and machine shed are one building, subdivided to serve all three purposes. The wiring for this building must therefore be arranged to suit this special condition.

LIGHTS

*Workshop.* A well-lighted workshop is absolutely necessary for doing good work. Two types of lighting are needed: general lighting for the entire workshop and localized lighting over the workbench and near permanently located equipment.

For this reason, a ceiling light must be installed for every 200 ft (61 m) of floor area, at least one and possibly two lights over the workbench, and conveniently located lights over the stationary tools such as the drill press, plane, forge, or saw.

*Garage.* One light over the front of the cars and another one at the rear will permit making minor repairs to the car and other equipment.

*Machine shed.* One or two ceiling lights, depending on the size of the shed, give sufficient general illumination. Lights must be properly placed for best results.

CONVENIENCE OUTLETS

*Workshop.* Convenience outlets in the workshop must serve many different purposes. There must be a sufficient number, properly located, to facilitate the work. Install at least two convenience outlets over the workbench and other outlets throughout the workshop for the connection of the electric tools. A heavy-duty outlet must be installed for the portable welder.

*Garage.* A convenience outlet at the rear wall of the garage permits the use of portable tools, a battery charger, and extension lights.

*Machine shed.* Since the welder may frequently be used in the machine shed, a heavy-duty outlet should be provided. Other convenience outlets should be considered for the connection of portable tools such as drills and soldering irons.

SWITCHES

*Workshop.* The ceiling lights in the workshop should be controlled by a single-pole switch near the entrance. The lights over the workbench and fixed equipment can be controlled by pull-chain switches inside the fixture.

*Garage.* A single-pole switch near the entrance of the garage is needed for the control of the ceiling lights.

*Machine shed.* The same type of switch control should be used here as in the garage.

CIRCUITS   As in the barn and milk house, the service-entrance panel for the workshop must have the correct capacity for the load. Since all the equipment will never be used at the same time, a reasonable diversity factor may be taken into consideration.

The number of circuits depends, of course, on the equipment and machinery installed. However, the following minimum should be carefully considered:

| | |
|---|---|
| Lights | 1 circuit |
| Convenience outlet | 1 circuit |
| Wood saw and planer | 1 circuit (No. 12 wire) |
| Portable welder | 1 circuit (230 V) |

### 498. Specifications for poultry house

LIGHTS   Lighting in the poultry house serves to increase egg production in addition to providing light for seeing. One light should be installed for each 200 ft² (18.6 m²) of floor area, but if pens are smaller than 200 ft², at least one light per pen must be provided for. The same general-lighting arrangement should also be kept in the poultry-house work-room with additional local lighting for egg cleaning, grading, poultry scalding, waxing, etc.

Since ultraviolet light promotes healthier, stronger birds, sunlamps should be installed over feeding troughs. It is also recommended that at least one germicidal fixture for each 100 ft² (9.3 m²) of floor area of laying pens be installed. These lamps should be on a separate switch so that they can be turned off when someone is working in the pen.

CONVENIENCE OUTLETS   To connect the water warmer in the laying pens, at least one convenience outlet should be installed for every 200 ft² (18.6 m²) of floor area. The outlets are best located at the ceiling and should be equipped with the locking type of receptacle.

In the brooder house, the same arrangement of receptacles is recommended to serve brooders and water warmers. In the workroom, convenience outlets must be installed for the egg cleaner, candler, grader, poultry scalder, and waxer.

The feed-grinder and feed-mixer motors must have special heavy-duty outlets on a 230-V circuit.

SWITCHES   Lights for all the laying pens should be connected to a regular switch as well as to an automatic time switch. The time switch will make morning lights to control egg production possible.

The ceiling lights in the workroom should be on a separate switch from the lights in the laying pens. The local lighting over the work areas can be controlled by wall switches or by pull-chain switches in the fixtures. Germicidal lamps, being on at all times, do not need a switch.

CIRCUITS   The following circuits are recommended for the poultry house as a minimum:

| | |
|---|---|
| Lights | 1 circuit (depending on size of poultry house) |
| Ultraviolet lamps | 1 circuit (depending on size of poultry house) |
| Germicidal lamps | 1 circuit (depending on size of poultry house) |
| Convenience outlets | 1 circuit |
| Brooder outlets | 1 circuit each brooder |
| Feed mixer | 1 circuit (230 V) |
| Feed grinder | 1 circuit (230 V) |

**499. Specifications for hog house or sheep shed.**  For general illumination, a lighting outlet every 20 ft (6.1 m) along the passageways should be sufficient. These lights should be controlled by a single-pole switch near the entrance. Convenience outlets are needed in the farrowing pen for the connection of the pig brooders. This outlet should be about 3 ft (0.9 m) above the floor and in a corner of the pen.

**500. Specifications for granary and corncrib.**  Lights over the center of the drive spaced 20 ft (6.1 m) apart are needed in the granary and corncrib. These lights should be controlled with a switch at the entrance. Upstairs, a light is needed for approximately every 200 ft² (18.6 m²) of storage area with a switch and pilot light installed downstairs near the stairs. At least one power outlet is needed downstairs for the connection of the grain elevator.

**501. Specifications for farmyard.**  In the farmyard, electricity serves many important functions. Lights properly switched help in performing the evening chores. Outlets are needed throughout the yard for connection of the portable utility motor. Therefore, a plan for the wiring of a farm should include proper lighting and outlets outdoors. In general, the following simple rules will be sufficient:

LIGHTS   Install at least three lights in the yard: one at the house, one at the barn, and one at the workshop or garage. They should be at least 15 ft (4.6 m) above the ground to allow a wide spread of light.

OUTLETS   Heavy-duty weatherproof outlets are needed for the connection of the utility motor. One should be mounted at the barn near the silo, and others at all places where farm chores will employ the utility motor. If the pump house is located adjacent to the yard, a separate circuit to the pump motor should be provided.

SWITCHES   To make it possible to turn the lights on or off from several different points, three- and four-way switches must be installed. A switch at the house, one at the barn, and a third one at another convenient place are recommended for maximum efficiency.

**502. Planning the farm wiring system.**  After the required loads have been determined, the wiring system can be planned. The system will be considered in three parts: (1) the power pole, (2) distribution feeders, and (3) interior wiring of each building.

**503. Power pole.**  To be of best service, the power pole (Fig. 9-310) should be erected in a spot as central to the different electrical loads as possible. Often this is somewhere halfway between the house and the barn.

Normally a main switch and fuse or circuit breaker will not be required. However, it is good practice to install one on the power pole. A raintight main switch on the power pole will permit the power to be cut off from one centrally located point. A feeder of the required current-carrying capacity for the entire farm load must be run from the meter to the main switch if used and up to the top of the pole. The distribution feeders will be tapped off from this main feeder at the top of the pole. Regardless of the present load, the main feeder should not be less than No. 2 wire.

The entire wiring system must be grounded at the power pole. This is usually done with driven ground rods (refer to Sec. 342). The grounding is a very important safety measure, and the advice of the power supplier or local inspection bureau should be obtained with respect to the proper method to employ.

**504. Distribution feeders.**  An individual feeder from the power pole to each building is best. However, the location of the buildings may make it necessary to connect one building from another. Similarly, to buildings with very small electrical loads, it may be equally good and less expensive to extend feeders from a nearby building. This feeder, when tapped on the outside of the building, without going through the service-entrance panel of the first building, permits additional load and future extensions in the second building. Additional wiring can be installed without affecting the load in the first building. No matter how small the load in any one building may be, this method is recommended.

The length of the feeder from pole to building and the electrical load inside determine the size of the feeder. For mechanical strength, no feeder can be smaller than No. 10 wire when the span is 50 ft (15.2 m) or less. A span longer than 50 ft needs wire No. 8 or larger. The actual size of the feeder can be calculated by the methods given in Div. 3. The voltage drop for any feeder should not exceed 2 percent.

If one or more buildings are connected from another, feeder sizes from the power pole to the first building must be increased. Use 100 percent of the building having the largest load plus 50 percent of the combined load of the other buildings. No one can foretell how quickly the use of electricity will be increased. It is therefore recommended that feeders at least one size larger than the load figures indicate should be used. Refer to Sec. 5, Part B, of Div. 11 for NEC feeder demand factors on farm properties.

It is recommended that all feeders shall be three-wire 115/230 V. However, for a small load in a particular building, a two-wire 115-V feeder may be satisfactory. Feeders to buildings having a load of 3450 W or more or to buildings having a motor of ½ hp must be three-wire.

**505. Wiring of buildings.**  Normally the power-distribution feeder for the power pole to a building is secured to the building with an insulator bracket (see Div. 8). Brackets should be mounted high enough so that the power feeders are never suspended lower than 18 ft (5.5 m) over driveways, for clearance of loaded wagons, and 10 ft (3 m) over footwalks.

From the insulator bracket service-entrance conductors are run down the side of the building to a point where they enter the building and connect to the service-entrance panel. Service-entrance cable is recommended for this purpose. This cable with its tough nonmetallic outer covering protects cattle from coming in contact with metallic parts of the wiring system. The table in Sec. 506 will give an idea of the size of service-entrance panels required (refer to Secs. 346 to 356 for information on service entrances).

It is recommended that armored cable or nonmetallic-sheathed cable be employed for the interior wiring of the buildings (refer to preceding sections of this division for information on wiring, and refer to Secs. 495 to 501 of this division for information on requirements for lights, convenience outlets, switches, and circuits).

At each building the wiring system must be grounded. This provision is in addition to the ground at the power pole. Grounds must be established at each point of entrance to each building, and, if possible, all these grounds should be tied together on driven grounds. Also, for added safety, the farm water system should be tied at each building to the driven ground for that building. This is important. A well-grounded wiring system adds to the safety of the entire installation.

| Type of building | Installation | Service-entrance conductors, amp (All services are 3-wire, 115/230-volt, unless otherwise specified) |
|---|---|---|
| **Dairy barn (complete barn, including milk room) up to 2,500 sq ft** | Normal amount of lighting, ventilation, milker, milk cooler, water pump, dairy water heater, milkhouse heater, gutter cleaner (3 hp), hay drier, or utility motor (5 hp) | 100 |
| **Same, above 2,500 sq ft** | Normal amount of lighting, ventilation, milker, milk cooler, water pump, dairy water heater, milkhouse heater, gutter cleaner (5 hp), hay drier, or utility motor (7 hp) | 125 |
| **Same, above 2,500 sq ft** | Same as above, but two 7½-hp motors | 195 |
| **Milking barn or "milking parlor"—includes milk room—(cows not housed)** | Normal amount of lighting, milker, milk cooler, dairy water heater, milkhouse heater, ventilation | 55 |
| **Milkhouse (when separate from barn)** | Normal amount of lighting, milk cooler, dairy water heater, milkhouse heater | 55 |
| **Horse, beef-cattle, sheep, and hog barns up to 2,500 sq ft** | Normal amount of lighting, ventilation, pig or lamb brooders, water pump, antifreezing protection on plumbing, utility motor (5 hp) | 70 |
| **Same, above 2,500 sq ft** | Normal amount of lighting, ventilation, pig or lamb brooders, water pump, antifreezing protection on plumbing, utility motor (7½ hp) | 100 |
| **Poultry laying house (including feed room) up to 4,000 sq ft pen floor area** | Normal amount of lighting, ventilation, water warmers, antifreezing protection on plumbing, automatic feeders, feed grinder (up to 1 hp) | 55 |
| **Same, above 4,000 sq ft pen floor area** | Normal amount of lighting, ventilation, water warmers, antifreezing protection on plumbing, automatic feeders, feed grinder (up to 2 hp) | 70 |
| **Poultry brooder house, portable type** | Light and outlet for single brooder and water warmer | 20 (2-wire, 115-volt) |
| **Poultry brooder house, permanent type, up to six pens** | Normal amount of lighting, ventilation, brooders, water warmers, antifreezing protection on plumbing | 55 |
| **Poultry brooder house, permanent type, above six pens** | Normal amount of lighting, ventilation, brooders, water warmers, antifreezing protection on plumbing | 70 minimum; estimate 2,000 watts per pen |
| **Poultry-cleaning and -dressing house** | Normal amount of lighting, poultry scalder, waxer, wax reclaimer, picker, and refrigeration | 55 |
| **Farm shop** | Normal amount of lighting, bench tools (tool grinder, drill, soldering iron, etc.), farm welder, air compressor, saw, and battery charger | 70 |
| **General farm buildings** | Minimum capacity where lighting, motors and miscellaneous 115- and 230-volt equipment will be used | 50 |
| **Storage sheds, stock shelters** | Lighting and small portable equipment—not over two 2-wire bench circuits | 40[a] (2-wire, 115-volt) |

[a]Ordinarily such buildings are supplied by a feeder or circuit from another building. In such cases, this is not a service as defined by the **National Electrical Code**. If a brooder house or similar structure is supplied directly from the power source, the minimum permissible size of service-entrance conductors would be No. 8 for two two-wire branch circuits.

If the building is served by a feeder from another building the minimum size of service-entrance conductors would be No. 10 for two two-wire branch circuits. For a single branch-circuit source, the minimum permissible size of service-entrance conductors would be No. 8 for two two-wire branch circuits.

NOTE  All the capacities listed assume service for the individual building only. If other buildings are supplied from any of these services, the capacity of the service would have to be increased accordingly.

# Electric Lighting

## PRINCIPLES AND UNITS

**1. Explanation of light.** Light *(Standard Handbook for Electrical Engineers)* may be defined as radiant energy of those wavelengths to which the human eye is sensitive. Figure 10-1 shows the complete radiant-energy spectrum of electromagnetic waves, which travel through space at the velocity of approximately 186,000 mi/s. The longer waves are the ones used in radio communications; the shortest ones are the x-rays and cosmic rays. The waves to which the eye is sensitive are those near the middle of the spectrum, of a length of about 0.0004 to 0.0008 mm. An enlarged section of this part of the spectrum is shown in the figure.

The effect of light upon the eye gives us the sensation of sight. The impression of color depends upon the wavelength of the light falling upon the eye. There are three primary colors of light: red, green, and violet. Violet light has the shortest wavelength of the radiant energy to which the eye is sensitive, red the longest, and green an intermediate wavelength between those of violet and red. These three colors are called the primary colors, because light of any one of them cannot be produced by combining light of any other colors. Light of any other color than these three can be produced by combining in the proper proportions light of two or all three of the primary colors.

**2. Propagation of light.** Rays of light travel in straight lines unless interfered with by some medium that absorbs or deflects them. Whenever a light wave strikes a different medium from that through which it has been passing, there are three fundamental phenomena that may occur: absorption, reflection, or refraction. Whenever light waves strike any object, a portion of their energy is absorbed, the amount depending upon the nature of the substance. This absorbed energy is dissipated in the form of heat. The remaining portion of the light may be all transmitted through the substance, all reflected back from the surface, or part transmitted and part reflected, depending upon the nature of the substance and the angle at which the light impinges upon the surface of the object. If the light strikes the object perpendicularly to the surface, it is either transmitted in a straight line through the substance or reflected back from its surface in the same direction in which it impinged upon the surface. If light strikes an object at an angle other than 90° to its surface, then either the light is transmitted through the object but in an altered direc-

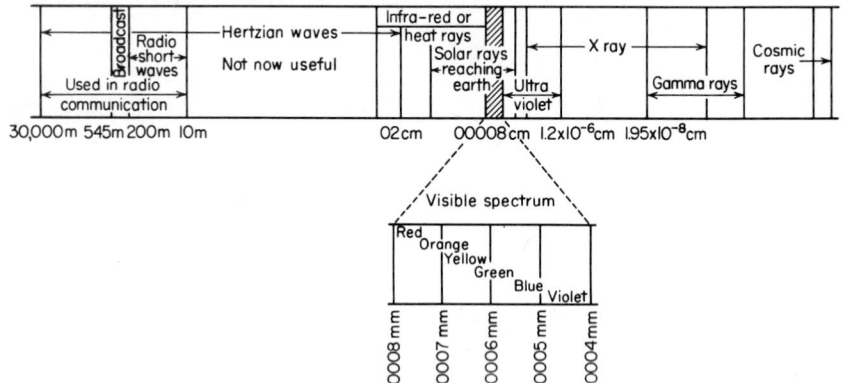

**Fig. 10-1**   The radiant-energy spectrum.

tion (refraction) or the light is reflected back from the object but in a different direction from that in which it impinged upon the object (reflection). With most objects all three of the phenomena occur, some of the light impinging upon them being absorbed, some transmitted through (refracted), and some reflected back from the surface.

**3. Absorption.**   Although some of the energy of a ray of light is always absorbed whenever a light ray impinges upon an object, the amount absorbed varies over wide limits, depending upon the nature of the object, the molecular construction, the wavelength or color of the incident light, and the angle at which the light strikes the surface. All objects do not absorb light of different wavelengths in the same proportion. It is this phenomenon which accounts for the characteristic color of objects (see Sec. 13). Since objects do not absorb the same proportion of the incident light of different colors, the amount of light absorbed by an object depends upon the color or wavelength of the light impinging upon the object. Tables 4 and 12 give the percentage of incident white light that is absorbed by various types of surfaces.

**4.   Coefficients (Percent) of Absorption of Lighting Materials**

| Material | Absorption, per cent | Material | Absorption, per cent |
|---|---|---|---|
| Clear glass globes.................... | 5–12 | Medium opalescent globes............ | 25–40 |
| Light sandblasted globes.............. | 10–20 | Amber glass......................... | 40–60 |
| Alabaster globes..................... | 10–20 | Heavy opalescent globes............. | 30–60 |
| Canary-colored globes................ | 15–20 | Flame-glass globes................... | 30–60 |
| Light-blue alabaster globes........... | 15–25 | Enameled glass...................... | 60–70 |
| Heavy blue alabaster globes.......... | 15–30 | White diffuse plastic................. | 65–90 |
| Ribbed glass globes.................. | 15–30 | Signal-green globes.................. | 80–90 |
| Clear plastic globes.................. | 20–40 | Ruby-glass globes................... | 85–90 |
| Opaline glass globes................. | 15–40 | Cobalt-blue globes.................. | 90–95 |
| Ground-glass globes................. | 20–30 | | |

**5. Absorption is the loss of light flux** which occurs when the flux is reflected by a reflecting surface or when it passes through a translucent or so-called transparent material.

$$\text{Percent absorption} = 100 \text{ percent} - \text{percent reflection} - \text{percent transmitted} \qquad (1)$$

**6. Reflection of light.**   (Fig. 10-2) is the redirecting of light rays by a reflecting surface. Whenever light energy strikes an opaque object or surface, part is absorbed by the surface and part is reflected. Light-colored surfaces reflect (Table 9) a larger part of the light thrown on them than do dark-colored surfaces, whereas dark surfaces absorb a larger part of the light and black surfaces absorb nearly all the light which reaches them.

NOTE  Consider first a smooth surface $AB$ (Fig. 10-2, I), on which a ray of light $L$ falls. This ray will be so reflected in the direction $R$ that the angle $i$ is exactly equal to the angle $r$. Consider now the

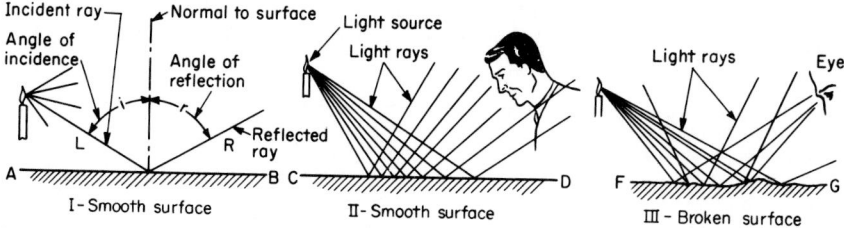

**Fig. 10-2** The reflection of light. Note that the angle of incidence $i$ always equals the angle of reflection $r$.

effect of a number of rays falling on a smooth surface $CD$ (Fig. 10-2, II). Each ray will be reflected in such a way that it leaves the surface at the same angle at which it strikes it. The eye if held as shown would perceive only the light reflected into it. Consider now a broken surface such as $FG$ (Fig. 10-2, III). Each ray of light is reflected from that portion of the surface on which it falls, just as though that point were on a smooth surface. The result is that the light is scattered, and if the surface is irregular enough, the eye placed at any point will receive reflections from many points of the surface. All opaque surfaces except polished surfaces have innumerable minute irregularities like the surface in Fig. 10-2, III. This alone enables them to be seen.

**7. The different kinds of reflection** will now be considered. *Regular reflection* is that (Fig. 10-3A, I, and 10-3B, I) in which the angle of incidence $i$ is equal to the angle of reflection $r$. This kind of reflection is obtained from mirrored glass, prismatic glass, and polished metal surfaces. *Spread reflection* (Fig. 10-3A, II, and 10-3B, II) is that in which the maxi-

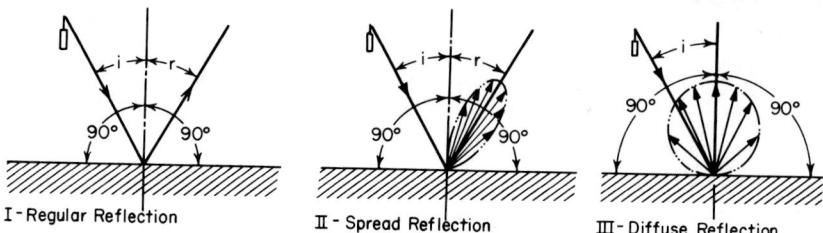

**Fig. 10-3A**  Classifications of reflection.

mum intensity of the reflected light follows the law of regular reflection, except that a part of the light is scattered slightly out of this line. Spread reflection is obtained from etched prismatic glass and from rough metallic surfaces. *Diffuse reflection* (Fig. 10-3A, III) is that in which the maximum intensity of the reflected light is normal to the reflecting surface. This holds over a large range of the angle of incidence. This kind of reflection is usually caused by reflection from particles beneath the surface (see Fig. 10-3B, III). Diffuse

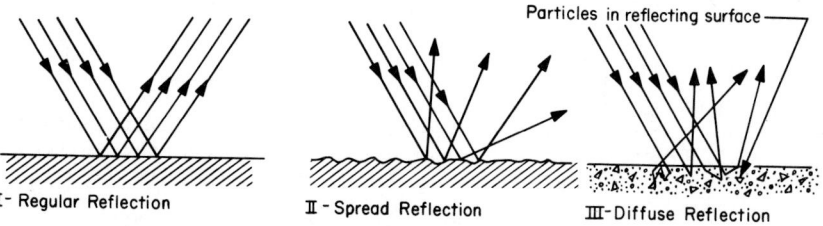

**Fig. 10-3B**  Magnified view of Fig. 10-3A.

reflection may be obtained from opal glass, porcelain enamel, paint enamel, and paint finishes commonly used for interior decoration of walls and ceilings.

**8. Reflecting power of surfaces.**  Different surfaces reflect different percentages of the light falling upon them. The illumination of a small room having poorly reflecting walls may sometimes be improved by changing the wall coverings. If the room is large or if reflectors are used to throw the light downward so that not much light reaches the walls, a change in the wall covering will have little effect on the general illumination.

**9. The following table of reflection coefficients** is useful in showing the relative value of wall coverings in rooms.

| Material | Reflection, % | Material | Reflection, % |
|---|---|---|---|
| Highly polished silver | 92 | Porcelain enamel | 70–80 |
| White plaster | 90 | Polished aluminum | 67 |
| White paint | 75–90 | Chrome-yellow paper | 62 |
| Optical mirrors silvered on surface | 75–85 | Yellow paper | 40 |
| Highly polished brass | 70–75 | Light-pink paper | 36 |
| Highly polished copper | 60–70 | Blue paper | 25 |
| Highly polished steel | 60 | Dark-brown paper | 13 |
| Speculum metal | 60–80 | Vermilion paper | 12 |
| Limestone | 35–65 | Blue-green paper | 12 |
| Brushed aluminum | 55 | Cobalt blue | 12 |
| Polished gold | 50–55 | Glossy black paper | 5 |
| Burnished copper | 40–50 | Deep chocolate paper | 4 |
| White paper | 80 | Black cloth | 1.2 |

**10. Refraction.**  Whenever a light ray passes from one medium into another of greater or less density, the direction of the ray is altered. This is called refraction. Refraction may be one of three types: regular, irregular or spread, or diffuse, depending upon the nature of the construction of the substance and the character of its surfaces.

*Regular refraction* occurs with plain glass or glass prisms, as shown in Fig. 10-4. Light in passing through a substance goes through two refractions, one upon entering the substance and one upon leaving the substance. If the surfaces of the object are parallel, as in a piece of glass (Fig. 10-4, I), the direction of the light leaving the object is parallel to the direction of the light impinging upon the object. If the surfaces of the object are not parallel, as in the

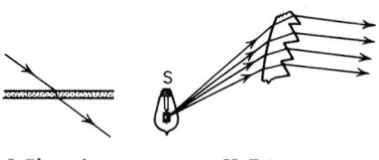

I. Plain glass.         II. Prisms.

**Fig. 10-4**   Regular refraction.

prism of Fig. 10-4, II, the light leaving the object will not be in a direction parallel to the incident light. A prism can be constructed to refract the light from its different surfaces so that light is not transmitted through the prism but is reflected back as shown in Fig. 10-5.

*Irregular, or spread, refraction* occurs with light transmitted through glass with a rough surface such as etched or frosted glass, as shown in Fig. 10-6. Such a surface can be considered as consisting of a great number of very small smooth surfaces making slight angles with each other. The individual rays of light being emitted from such a surface are refracted at slightly different angles but all in the same general direction. Thus the light transmitted through a substance with such a surface is refracted in the same general direction but with the beam spread somewhat from what it would be for regular refraction.

The composition of opal glass is such that it contains a number of minute opaque particles throughout its structure. Light striking such an object travels through the glass until it strikes one of these opaque particles, from which it is either reflected back or transmitted through the glass to the other surface. The total beam of light striking the object is thus split up by the innumerable small opaque particles, part being reflected back in all directions, and part being refracted through the glass in all directions. The portion that is transmitted (refracted) through the glass is *diffusely refracted.* An idea of how diffuse refraction takes place can be gained from Fig. 10-7. In Fig. 10-8 the diffuse reflection and refraction of a ray of light impinging upon a piece of opal glass are indicated.

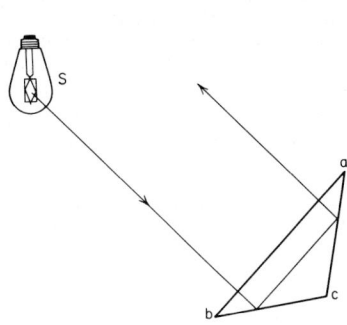

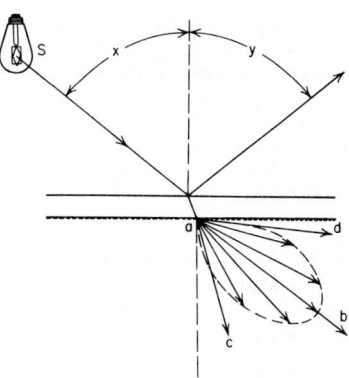

**Fig. 10-5** Total reflection with prism. [General Electric Co.]

**Fig. 10-6** Spread refraction with etched glass. [General Electric Co.]

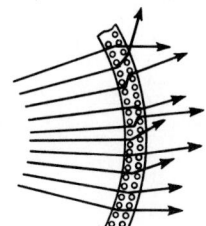

**Fig. 10-7** Diffuse refraction.

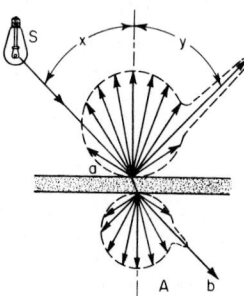

**Fig. 10-8** Diffuse reflection and refraction with opal glass. [General Electric Co.]

## 11. Most Frequently Used Lighting Units, Abbreviations, and Symbols and the Corresponding Hydraulic Analogies

| Photometric quantity | Name of unit | Abbreviation | Symbol | Hydraulic analogy |
|---|---|---|---|---|
| Luminous flux | lumen | lm | $F$ | gal/min |
| Luminous intensity | candela | cd | $I$ | pressure, $lbf/in^2$ |
| Illumination | footcandle | fc | $E$ | incident $gal/(ft^2 \cdot min)$ |
| Luminance | lambert | lambert | $L$ | issuing $gal/(ft^2 \cdot min)$ |

**12. Reflection, Transmission, and Absorption Characteristics of Materials**
(General Electric Co.)

| Material | Light reflected | | | Light transmitted | | | Light absorbed |
|---|---|---|---|---|---|---|---|
| | In concentrated beam | In spread beam | Diffused in all directions | In concentrated beam | In spread beam | Diffused in all directions | |
| **Crystal glass:** | | | | | | | |
| Clear | 8-10[a] | | | 80-85 | 70-85 | | 5-10 |
| Frosted or pebbled[b] | 4-5 | 5-10 | | | 72-87 | | 5-15 |
| Frosted or pebbled[c] | | 8-12 | | | | | 5-15 |
| **White glass:** | | | | | | | |
| Very light density[b] | 4-5[a] | | 10-20 | 5-20 | 5-20 | 50-55 | 8-12 |
| Very light density[c] | | 3-4 | 10-20 | | | 50-55 | 10-15 |
| Heavy density | 4-5[a] | | 40-70 | | | 10-45 | 10-20 |
| Mirrored glass | 82-88 | | | | | | 12-18 |
| **Polished metal:** | | | | | | | |
| Silver | 92 | | | | | | 8 |
| Chromium | 65 | | | | | | 35 |
| Aluminum | 62 | | | | | | 38 |
| Alzak aluminum | 75-85 | 70-80 | | | | | 15-30 |
| Nickel | 55 | | | | | | 45 |
| Tin | 63 | | | | | | 37 |
| Steel | 60 | | | | | | 40 |
| Porcelain enamel steel | 4-5[a] | | 60-70 | | | | 25-35 |
| **Mat-finished metal:** | | | | | | | |
| Aluminum | | 62 | | | | | 38 |
| White oxidized aluminum | | 70-75 | | | | | 25-30 |
| Aluminum paint | | 60-65 | | | | | 35-40 |
| **Mat surfaces:** | | | | | | | |
| White plaster | | | 90-95 | | | | 5-10 |
| White blotting paper | | | 80-85 | | | | 15-20 |
| White paper (calendered) | 4-5[a] | | 75-80 | | | | 15-20 |
| White paint (dull) | | | 75-80 | | | | 20-25 |
| White paint (semimat) | | 2-4 | 70-75 | | | | 20-25 |
| White paint (gloss) | 4-5[a] | | 70-75 | | | | 20-25 |
| Black paint (gloss) | 4-5[a] | | 3-5 | | | | 85-92 |
| Black paint (dull) | | | 3-5 | | | | 95-97 |
| Magnesium carbonate | | | 98-99 | | | | 1-2 |
| **Plastic:** | | | | | | | |
| Clear | | | | | 60-80 | | 40-20 |
| White diffuse | | | | | | 5-35 | 65-95 |

[a] For angles up to 45°; for angles greater than 45°, this value rises considerably; angle of incidence as X, Fig. 10-6.
[b] Smooth side toward light source.
[c] Roughed on side toward light source.

**13. Color of objects.**   Our impression of the color of objects is due to the color of the light that the object reflects or transmits to our eyes. In Sec. 2 it was stated that all objects absorb a certain portion of the light that falls upon them but that all objects do not absorb the same proportion of light of the different wavelengths. It is through this different absorption that different objects have unlike colors. If all objects absorbed light in exactly the same manner, all objects would appear to have the same color. In order that things may appear in their true colors, they must be observed under white light, i.e., light containing all three primary colors in the right proportion. An article which absorbs no light or which absorbs light of the three primary colors in the same proportion as they are combined to produce white light will transmit from its surface light in a condition unchanged from that in which the light fell upon the object. Such an object would appear white in color. An object which absorbs all or nearly all the light which falls upon it will have no color or, in other words, will be black. An object which absorbs all the green and violet rays will be red in color, since it will transmit to our eyes only red light. In order that an object may appear in its true color, the light falling upon it must contain light of the wavelength that the object reflects or transmits. Thus, light falling upon a red object must contain red light if the object is to appear in its true color. A red object viewed under a light which contains only green and violet rays will appear black, since the object is capable of reflecting only red rays.

**14. Light-source color.**   The color of light sources has two basic characteristics: *chromaticity* and *color rendering*. Chromaticity, or color temperature, defines its "whiteness," its blueness or yellowness, its warmth or coolness. It does not describe how colors will appear when lighted by the source. Chromaticity (color temperature) is a term sometimes used to describe the color of the light from a source by comparing it with the color of a *blackbody*, a theoretical *complete radiator* which absorbs all radiation that falls on it and in turn radiates a maximum amount of energy in all parts of the spectrum. A blackbody, like any other incandescent body, changes color as its temperature is raised. The light from a warm white fluorescent lamp is similar in color to the light from a blackbody at a temperature of approximately 3000 K,[1] and the lamp is accordingly said to have a color temperature of 3000 K. The light from a cool white fluorescent lamp is bluer, and the blackbody must be raised to 4200 K to match it. Hence the cool white lamp has a color temperature of 4200 K. A daylight fluorescent lamp has a color temperature of 6200 K.

Color temperature is not a measure of the *actual temperature* of an object. It defines *color only*. Some light sources, such as a sodium-vapor lamp or a green or pink fluorescent lamp, will not match the color of a blackbody at any temperature, and therefore no color temperatures can be assigned to them.

**Color Temperatures, K**
(Approximate values)

| | |
|---|---|
| Blue sky | 10,000–30,000 |
| Overcast sky | 7000 |
| Noon sunlight | 5250 |
| Fluorescent lamps | |
| Daylight | 6200 |
| Cool white | 4200 |
| White | 3500 |
| Warm white | 3000 |
| Clear mercury lamps | 5700 |
| Deluxe mercury lamps | 3900 |
| Clear metal halide lamps | 4100 |
| High-pressure sodium lamps | 2100 |
| General-service incandescent lamps | 2500–3050 |
| Candle flame | 1800 |

The other characteristic, the color-rendering index (CRI), attempts to describe how colors will appear when illuminated by a light source of specified chromaticity. The CRI rating system is an international system that mathematically compares how a light source causes eight selected colors to appear compared with a reference source. However, there are limitations to its use which should be recognized. Although the system provides for ratings up to 100, it can only be used to compare sources of approximately the same chro-

[1]Kelvin is a temperature scale which has its zero point at $-273°C$.

maticity. Comparison of the CRI of light sources with widely different color temperatures such as a cool light source and a warm light source can be misleading. Light sources with the same CRI may render some colors differently, and light sources with CRIs as much as 5 points different may cause colors to look the same. Owing to their spectral distribution, HID sources may actually look better than their CRIs would indicate.

The CRI system is the best presently available to describe the relative appearance of colors under different sources, and it can be useful when applied within the limits of the system.

**15. Luminous flux** (which, as is explained later, is measured in lumens) is a flow of light, i.e., light energy or light waves. Luminous flux always originates from some source of light, such as the sun, a candle, or an incandescent lamp. But luminous flux can be redirected by reflecting surfaces. The luminous flux which emanates directly or is reflected from objects to the human eye is the medium whereby the objects are seen. Although there is no actual flow of anything material in a flux of light, there is a flow of light waves.

**16. A true point source of light** (Fig. 10-9) is a luminous mathematical point which emits luminous flux uniformly in all directions. It is a theoretical concept which cannot actually exist. It is, however, used (and is necessary) for the development of the quantities and units employed in lighting computations.

NOTE   An actual light source may be considered a point source without prohibitive error if the distance between the source and the location at which the source is viewed or examined is at least 10 times the greatest dimension of the source.

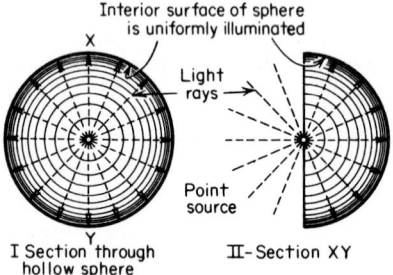

Interior surface of sphere is uniformly illuminated

Light rays →

Point source

I Section through hollow sphere        II-Section XY

**Fig. 10-9**   Diagrammatical illustration of light rays emitted by a point source of light.

**17. The luminous intensity** of a given source of light in a certain specified direction is a measure of the ability of the source to project light in that direction. Luminous intensity is measured in a unit which is called the candela, formerly the candle.

NOTE   The luminous intensity of actual light sources is generally greater in certain directions than in others. Thus, in Fig. 10-10, the luminous intensity of the lamp is greatest in the horizontal direction.

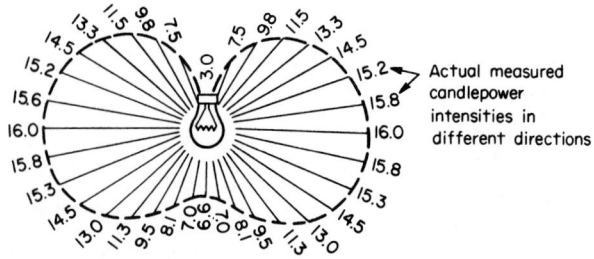

Actual measured candlepower intensities in different directions

**Fig. 10-10**   Luminous intensities in different directions around an incandescent lamp.

**18. The candela (cd)** is defined as the luminous intensity or light-producing power in the horizontal direction of a standard lamp which is made and used in accordance with U.S. Bureau of Standards specifications.

NOTE   The luminous intensity of an ordinary sperm candle (Fig. 10-11) in the horizontal direction is about 1 candle (cd). Thus was derived the name of the unit, candela.

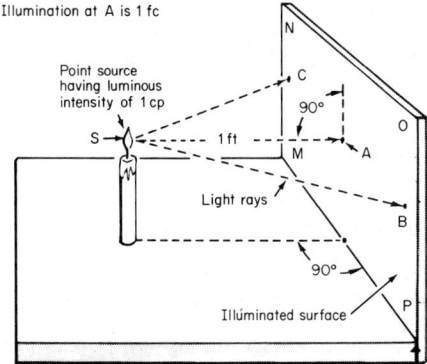

**Fig. 10-11**    An intensity of illumination of 1 fc (10.76 lx).

**19. The true luminous intensity or candlepower** of a light source can be obtained only when the source is a true point source (Sec. 16). But a true point source does not exist. It is merely a mathematical concept. Now, the luminous intensity of an actual light source, in a certain direction (Fig. 10-12), is due to the combined effects of a number of point sources in the surface of the actual light source. Hence, the luminous intensity of an actual light source, in a given direction, as determined with a photometer (Sec. 27), is not the true candlepower (luminous intensity) but is the apparent luminous intensity (apparent candlepower) in that direction.

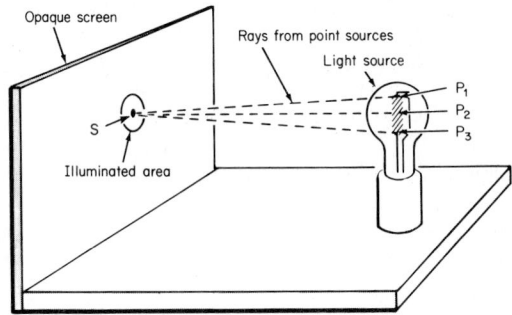

**Fig. 10-12**    The illumination at point S is due to the combined effects of an infinite number of point sources $P_1$, $P_2$, $P_3$, etc.

**20. Illumination is measured in footcandles (fc).**    The footcandle is defined as that illumination which is produced (Fig. 10-11) by a 1-cd point source (or its equivalent) on a surface which is exactly 1 ft (0.3048 m) distant from the point source.

EXPLANATION   If, in Fig. 10-11, the light source S is assumed to be a point source of luminous flux, then the illumination at point A, which is exactly 1 ft from S, is (by definition) 1 fc. Since the illuminated surface MNOP is a plane, point A is the only point on the surface which has an illumination of 1 fc. The illumination at any other point on the surface, such as B or C, is less than 1 fc because it is farther away from S than is A. If, however, the sphere of Fig. 10-9 has an internal radius of 1 ft and the true point source (Sec. 16) has a luminous intensity of 1 cd, then every point on the interior surface of the sphere will have an illumination of 1 fc (10.76 lx).

**21. Illumination is really the density of the luminous flux** which impinges on the surface of an illuminated object. The average density of anything on a surface may be numerically represented by the number of things on the whole surface divided by the number of unit areas in the surface. Thus, as will be further explained in Sec. 25, if the luminous flux, in lumens, which impinges on a surface is divided by the area of that surface in square feet, the average illumination over the surface will be the result.

**22. Luminous flux is measured in lumens (lm).**  A lumen is defined as that quantity of incident luminous flux which will, when uniformly distributed over a surface having an area of 1 ft² (0.0929 m²), produce an illumination of 1 fc on every point of the surface.

NOTE   When luminous flux impinges nonuniformly on a surface, then a lumen is the quantity of luminous flux which will, on 1-ft² area of the surface, produce an average illumination of 1 fc.

**23. A point source of light of 1-cd luminous intensity emits 12.57 lm.**   It has been shown (Sec. 20 and Fig. 10-9) that a 1-cd point source of light located at the center of a hollow sphere of 1-ft radius will produce an illumination of 1 fc on every point of the interior surface of the sphere. Now the superficial area of a sphere = 4 × 3.1416 × $r^2$. Hence, this 1-ft-radius sphere will have an area of

$$4 \times 3.1416 \times 1 \times 1 = 12.57 \text{ ft}^2$$

Since every point on the surface of this sphere has an illumination of 1 fc, there must, to satisfy the definition of the lumen (Sec. 22), be as many lumens emitted by the 1-cd point source as there are square feet in the surface of the sphere. The sphere has an area of 12.57 ft². Therefore every 1-cd point source of light emits 12.57 lm.

**24. To obtain the output of a light source in lumens when its mean spherical candlepower is known,** substitute in the following formula. This formula is strictly accurate only for a true point source (Sec. 16) of light, but it gives results sufficiently accurate for all practical purposes if the mean spherical candlepower (Sec. 29) is substituted for $I$.

$$F = 12.57 \times I \text{ lm} \qquad (2)$$

where $F$ = total luminous flux emitted by the light source, in lumens, and $I$ = luminous intensity or candlepower of a point source, in candelas, or, with sufficient accuracy for most practical purposes, the mean spherical candlepower of an actual light source.

example   The mean spherical candlepower of a light source is 68.8. What quantity of luminous flux is emitted in lumens?
solution   By For. (2), luminous flux $F$ = 12.57 × $I$ = 12.57 × 68.8 = 865 lm.

**25. Illumination in footcandles is really lumens per square foot.**   From the definition of the lumen in Sec. 22, 1 lm of flux spread out over an area of 1 ft² produces 1 fc average illumination over that surface; 2 lm would produce 2 fc, etc. Similarly, 10 lm spread out over an area of 5 ft² would produce an average illumination of 10 ÷ 5 = 2 fc. Likewise, 1 lm spread out over an area of 2 ft² would produce an illumination of 1 ÷ 2 = 0.5 fc. Thus, it is evident that lumens ÷ ft² area = average footcandles illumination.

example   A room which has an area of 600 ft² has 3300 lm of luminous flux impinging on the working plane. What is the average illumination, in footcandles, on the working plane?
solution   By the equation given above, average footcandles illumination = lumens ÷ ft² area = 3300 ÷ 600 = 5.5 fc.

**26. The illumination, in footcandles, which from a light source impinges on a surface varies inversely as the square of the distance from the source.**  This is true absolutely for a true point source (Sec. 16). It is approximately true if the distance is at least 10 times the largest dimension of the light source.

EXPLANATION   Assume (Fig. 10-13) that the 1-cd point source $P$ is placed at the center of a hollow spherical shell which has an internal radius $D$ of 1 ft. Then, a section $A$ of the shell, having an area of 1 ft², will have on its inner surface an illumination of 1 fc. This follows from the preceding discussions. Furthermore, from the definition of the lumen (Sec. 22), just 1 lm of flux is lighting this surface. Now, if the 1-ft-radius sphere be removed and the 1-cd source be placed at the center of a 2-ft-radius sphere, i.e., one of double the radius, then the same luminous flux will be spread out over an area $B$ of 4 ft². (This follows because of the geometric fact that similar areas vary directly as the square of similar dimensions.) The illumination on $B$ will be 1 lm ÷ 4 ft² = 0.25 fc. Thus by doubling the distance from the source, the illumination has been quartered. Similarly, it can be shown for $C$ that with the distance $D$ increased 3 times the illumination provided by 1 lm of flux is one-ninth of what it was with the distance of 1 ft. Thus, illumination varies inversely as the square of the distance from the source.

**27. The photometer** is an instrument which is used for determining the luminous intensity of a light source. Any determination made with a photometer gives apparent candlepower. However (Sec. 26), if the location at which the intensity is measured is at a sufficient distance from the source, then the source may for all practical purposes be consid-

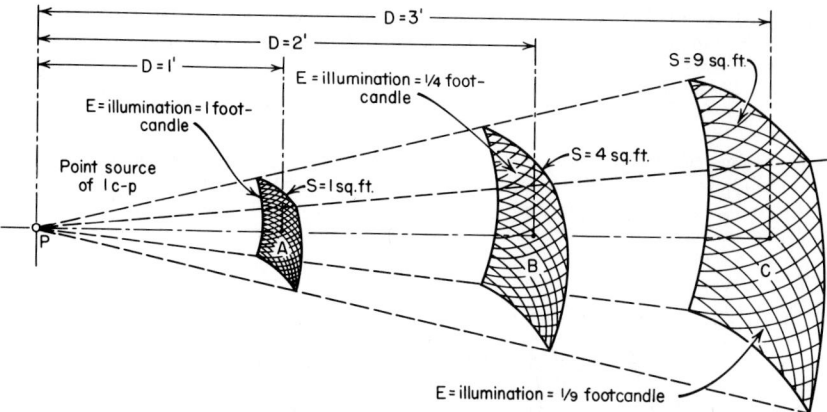

**Fig. 10-13** The inverse-square law.

ered a point source, and the value obtained for the unknown candlepower will be accurate well within the limits of error of experimental observation.

Photometric measurements are obtained from visual comparison of an unknown light source compared with a known source or with direct-reading photoelectric instruments which have been calibrated from a known source. The latter method is the most widely used today in the measurement of lamps and lighting fixtures.

**28. Mean horizontal candlepower** is the average of the candlepowers of a lamp in all directions in a horizontal plane. This term is now applied only to special lamps for laboratory test work.

**29. Mean spherical candlepower** is the average of the candlepowers of a lamp in all directions. It is measured by putting the lamp in the center of a sphere photometer (Fig. 10-14). The sphere has a small window of milk glass which is shielded from the direct rays of the lamp by a small opaque screen. The inner surface of the sphere is painted flat white for good reflection of the light. The candlepower of the window is compared with the horizontal candlepower of a standard lamp. This candlepower must be multiplied by a constant for the particular sphere to take account of the loss of light absorbed on the inner

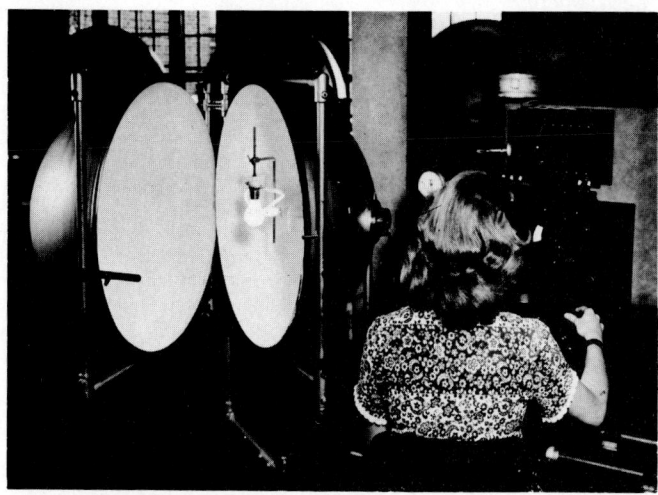

**Fig. 10-14** Sphere photometer. [General Electric Co.]

surface of the sphere and in the glass window. The mean spherical candlepower is used principally with the equation of Sec. 24 to obtain the lumen output in which the lamp is rated.

**30. The efficacy (also referred to as efficiency)** of an electric-light source is stated in *lumens per watt*. This term is obtained by dividing the lumen output of the source by the watts input.

**31. Candlepower distribution curves.**   Since the common light-giving sources either alone or in conjunction with the reflecting equipment used with them do not have the same light-giving power or candlepower in all directions, photometric graphs are employed to indicate the candlepower of the source in all directions. Curves giving this information for a light source are called candlepower distribution curves or simply distribution curves. These curves are obtained from a photometer which measures the luminous intensity of a light source in all directions (Fig. 10-15).

**Fig. 10-15**   Mirror photometer used to measure the luminous intensity of a light source or luminaire in all directions. [General Electric Co.]

Many lamps or lamps with their reflectors have the same candlepower in all directions in any one horizontal plane. This fact enables the candlepower in any direction from such a source to be determined from a single distribution curve which gives the candlepowers in all directions in a vertical plane through the center of the light source.

**32. How to read a photometric graph.**   In the photometric graph of Fig. 10-16, I, the candlepower directly downward is indicated by measuring it off on the vertical to a given scale. Thus, *XA* represents the candlepower directly below the light. Similarly, the distances *XB, XC, XD, XE, XF,* and *XG* represent to scale candlepowers around the light at angles above the vertical of 15, 30, 45, 60, 75, and 90°. Similarly, the candlepowers above 90° can be measured off to the given scale along lines at their respective angles. These points are then joined by a continuous line *GFED*, etc., and this line, completed for 360°, is called the *photometric distribution graph* of the light source. Figure 10-16, I, shows such a completed photometric curve, but in practice it is customary to use circular lines, as indicated on Fig. 10-16, II, to show the scale to which the candlepowers are plotted. The candlepower of the light unit can be measured at as few or as many angles as necessary, the accuracy of the resultant graph being largely determined by the number of angles taken.

**33. The area of the distribution graph is not proportional to the amount of light** given off. *B* (Fig. 10-78) represents a smaller total flux, by an amount equal to the absorption in the reflector, than does Graph *A*, though it has a larger area. Such a graph as *B* is useful for determining the intensity of light at any given angle and for determining the total luminous output, as explained in Sec. 34. These data may be required for any one of a number of practical operations.

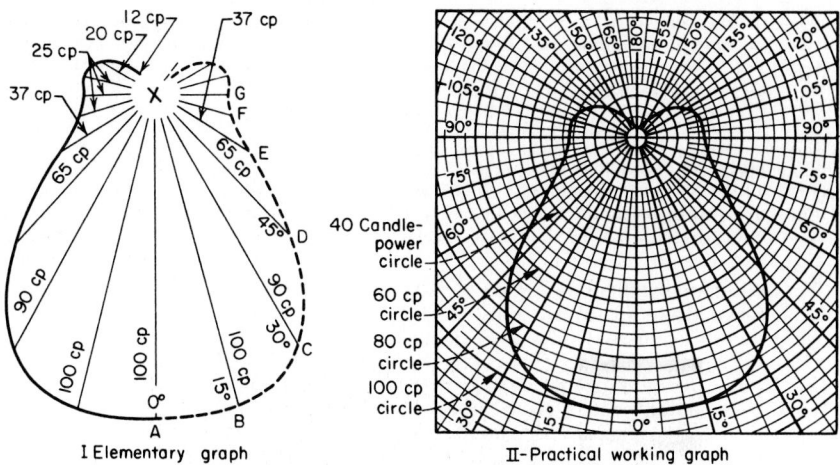

**Fig. 10-16** Photometric graphs.

**34. The method of computing from its distribution graph the total flux, in lumems,** emitted by a symmetrical light source is this: From the distribution graph of any light unit, as Fig. 10-16, take the candlepower at 5° and multiply it by the 0-10 factor as given in Table 36. This gives the lumens in the 0-10° zone. Similarly, to obtain the lumens in the 10-20° zone, take the candlepower at 15° and multiply it by the 10-20° factor as given in Table 36. The total lumens emitted in any large zone is obtained by adding the lumens of all the 10° zones contained in the large zone. If the sum total of the lumens is thus obtained for the 0-180° zone, the result is the total flux, in lumens, emitted by the source.

**example** What is the total flux emitted by a light source having a distribution graph as shown in Fig. 10-16?
**solution**

| Degrees | Candlepower | Zone factor | Lumens |
|---|---|---|---|
| (1) | (2) | (3) | (4) |
| 5 | 100 | 0.095 | 9.50 |
| 15 | 98 | 0.283 | 27.45 |
| 25 | 94 | 0.463 | 43.50 |
| 35 | 84 | 0.628 | 52.75 |
| 45 | 66 | 0.774 | 51.10 |
| 55 | 46 | 0.897 | 41.25 |
| 65 | 33 | 0.992 | 32.71 |
| 75 | 27 | 1.058 | 28.55 |
| 85 | 26 | 1.091 | 28.37 |
| 95 | 26 | 1.091 | 28.37 |
| 105 | 25 | 1.058 | 26.48 |
| 115 | 24 | 0.992 | 23.80 |
| 125 | 20 | 0.897 | 17.94 |
| 135 | 15 | 0.774 | 11.60 |
| 145 | 12 | 0.628 | 7.54 |
| 155 | 8 | 0.463 | 3.70 |
| 165 | 4 | 0.283 | 1.13 |
| 175 | 4 | 0.095 | 0.38 |
| | | Total = | 436.12 |

Total flux emitted = 436.12 lm.
Column 2 is obtained from the graph (Fig. 10-16). Column 3 is obtained from Table 36. Column 4 is obtained by multiplying the values in col. 2 by the corresponding values of col. 3.

NOTE    The total flux in lumens emitted by a light source can be computed graphically as follows. On the candlepower distribution graph (Fig. 10-16) measure the horizontal distance between the vertical axis (0–180° line) and the point where the candlepower graph crosses the 5° line. Then lay off this distance on the candlepower scale to which the distribution graph is plotted. Multiply the value thus obtained by 1.1, and the result is the quantity of luminous flux, in lumens, emitted by the light source in the 0–10° zone. To determine the flux in any 10° zone, it is only necessary to measure the horizontal distance between the vertical axis and the point where the candlepower graph crosses the center of the 10° zone under consideration and then proceed as above. To obtain the total lumens emitted in any large zone lay off the horizontal distances between the vertical axis and the point where the candlepower graph cuts the center of each 10° zone contained within the large zone successively along the edge of a strip of paper. Then lay off the total length on the candlepower scale and multiply the result by 1.1.

For the determination of the lumen output of nonsymmetric or asymmetric luminaires, such as conventional fluorescent units, readings of candlepower must be taken in a number of planes. From these readings a weighted average candlepower is obtained for each zone.

For fluorescent lighting units, candlepower readings often are taken in five planes, at 0, 22½, 45, 67½, and 90° from a plane through the luminaire axis. Candlepower values are measured in each of these planes at 10° intervals (5, 15, 25°, etc.). If the candlepower readings in the five planes (0, 22½, 45, 67½, and 90°) for any one zone are designated respectively as $A$, $B$, $C$, $D$, and $E$, then their weighted average for that zone is obtained by the formula

$$cd = \frac{A + 2B + 2C + 2D + E}{8}$$

In some laboratories for similar tests, candlepower readings are taken in three planes only (0, 45, and 90°), as in Fig. 10-17. The candlepower values in $A$ (crosswise) and $B$ (lengthwise) plus twice the values in $C$ (45°) are added. The sum divided by 4 equals the average candlepower.

Similarly, nonsymmetric or asymmetric luminaires for filament lamps have such wide variations in candlepower at a given angle about the vertical that an average reading from which to compute zonal lumens cannot be obtained by rotating the unit. Candlepower distribution curves for such equipment are prepared from data obtained in specific planes, and in interpreting such curves one must be careful to observe tbe planes they represent (see Fig. 10-18).

**35. The mean spherical candlepower** (Sec. 29) of a light source can be determined (1) directly, by means of the sphere photometer (Sec. 29 and Fig. 10-14), and (2) indirectly, by computing the total lumens from the distribution graph, as explained in the preceding paragraph, and dividing the result by 12.57.

**example**    What is the mean spherical candlepower of the light source of which the distribution graph is shown in Fig. 10-16?
**solution**    By the example in Sec. 34, the total flux of the source = 436.12 lm. The mean spherical candlepower = 436.12 ÷ 12.57 = 34.6.

**36. Factors to obtain the lumens in the 10° zones around a light source.**    To obtain the lumens in any zone, multiply the candlepower at the center of the zone by the factor for that zone.

| Degree zones | | Factor | Degree zones | | Factor |
|---|---|---|---|---|---|
| 0–10 | 170–180 | 0.095 | 50–60 | 120–130 | 0.897 |
| 10–20 | 160–170 | 0.283 | 60–70 | 110–120 | 0.992 |
| 20–30 | 150–160 | 0.463 | 70–80 | 100–110 | 1.058 |
| 30–40 | 140–150 | 0.628 | 80–90 | 90–100 | 1.091 |
| 40–50 | 130–140 | 0.774 | | | |

**37. The footcandle meter** (also called the light meter or luminance meter) is an instrument for measuring illumination directly in footcandles. It consists of one or two photovoltaic cells connected to a microammeter. When light falls on the photovoltaic cells, a voltage is generated as current flow through the meter approximately in proportion to the footcandle illumination on the cell. The small pocket type of footcandle meter is commonly

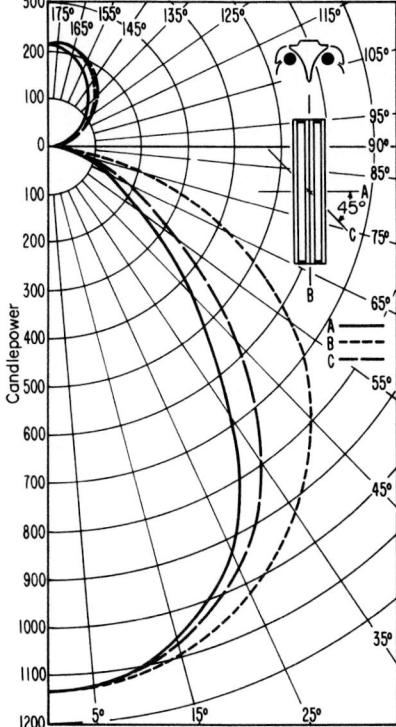

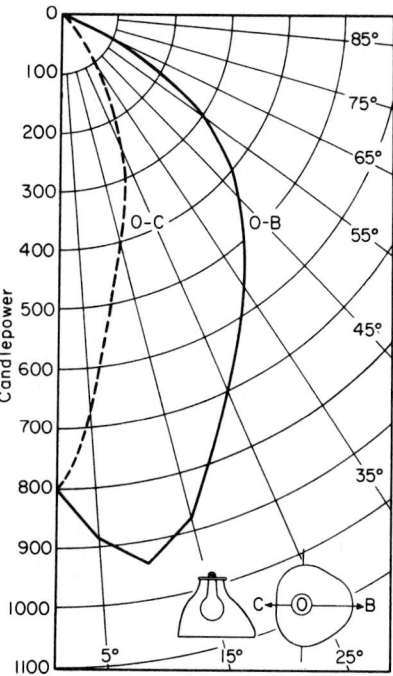

**Fig. 10-17** The average candlepower multiplied by the zone constant gives the zone lumens. Candlepower values at a given angle in curves A and B are added to twice the value in curve C at the same angle. This sum is divided by 4 to get the weighted average candlepower.

**Fig. 10-18** Candlepower distribution curves of nonsymmetrical sources such as show-window reflectors vary widely, and their interpretation depends on the specific planes in which they are taken.

referred to as a light meter (Fig. 10-19). It has a single photovoltaic cell and is calibrated in footcandles. The meter is placed with the window of the photovoltaic cell parallel to the plane in which it is desired to measure the illumination. Higher levels of illumination (such as those of daylight intensity) are measured via multiple scales on the same meter which are chosen through a selector switch.

Larger and more accurate meters (Fig. 10-20) have additional cells connected in par-

**Fig. 10-19** The light meter. [General Electric Co.]

**Fig. 10-20** Light meter for more accurate readings. [Minolta Corporation]

allel and utilize solid-state electronic circuits to amplify the output of the photovoltaic cells and maintain excellent accuracy through a wide range of footcandle levels.

Since the spectral response of photovoltaic cells is not the same as that of the human eye, it is necessary to incorporate filters into the light meter for the meter to read lighting levels properly under all different light sources. Another factor that requires correction within the meter is the proper measurement of light from various angles. If no correction is included, light at high incident angles will be reflected off the photovoltaic-cell surfaces with a resulting low reading of lighting level. Most footcandle meters today include a means of correcting the meter response to include light from all angles. This is referred to as *cosine correction.*

**38. Glare** (Fig. 10-21) is defined as any brightness within the field of vision of such a character as to cause discomfort, annoyance, interference with vision, or eye fatigue.

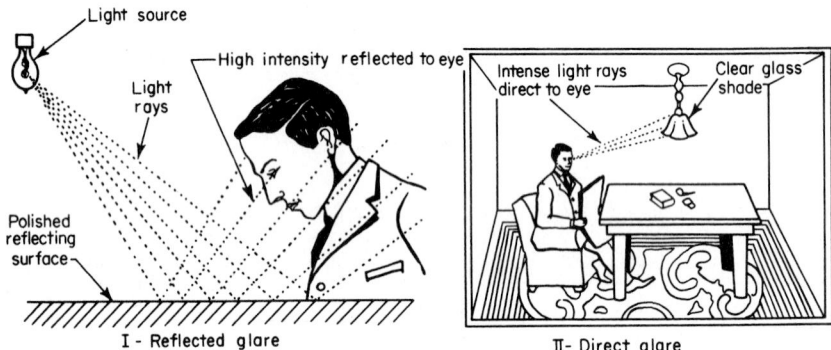

**Fig. 10-21**   Methods of causing glare.

NOTE    Glare may be subdivided into three different classes: (1) Direct glare (Fig. 10-21, II), which results when one looks directly at a brilliant light source. (2) Contrast glare, which is caused by brightness contrast. An example of this is the visual discomfort produced by a brilliant automobile headlight shining in the eyes on a dark night. The same headlight would cause no discomfort in the daytime. The annoyance experienced at night is due solely to the contrast between the bright headlight and the dark surroundings. (3) Reflected glare (Fig. 10-21, I), which is produced by light being reflected directly into the eyes from a polished, a white, or a light-colored surface. Glass desk plates, highly polished furniture, glazed paper, and mirrors may occasion reflected glare.

**39. Brightness** is the property of lighted objects which enables them to be seen by virtue of the light issuing from them. Technically, brightness is the density of luminous flux issuing or projected from a light source or from some illuminated surface.

NOTE    The relation between brightness and illumination: Illumination is the density of the luminous flux impinging on an illuminated surface. Just as illumination measures incident–luminous-flux density, so brightness measures issuing–luminous-flux density. It is due solely to the light issuing or reflected from things that we see. There might be great illumination (incident–luminous-flux density) on a dead-black object in a room with dead-black walls, and yet that object would be invisible, because the black would absorb (see Table 9) all the light which impinged on it and would reflect none to the eye.

**40. The units of brightness** are the *lambert,* the *millilambert,* the *footlambert,* and the *candela per unit of area.* The lambert (L) is by definition the brightness of a perfectly diffusing surface which radiates or reflects 1 lm/cm², while the millilambert (mL) is 0.001 of this value. The footlambert (fL) is the brightness of a surface which radiates or reflects 1 lm/ft². As its name implies, the candela per square inch is the brightness of a surface which radiates 1 cd/in². The lambert and the candela per square inch are commonly used for high brightness such as light sources, while the millilambert and the footlambert are used for designating ordinary illuminated surfaces.

**41. The working equations for brightness computations are**

$$b = 2.05 \, L \qquad (3)$$
$$F = 6.45 \, SL \qquad (4)$$

where $b$ = brightness in candelas per square inch, $L$ = brightness in lamberts, $F$ = total luminous flux in lumens, and $S$ = area of the surface in square inches.

**example** A light source has a brightness of 500 cd/in². What is its brightness expressed in lamberts?
**solution** By For. (3),

$$L = \frac{b}{2.05} = \frac{500}{2.05} = 243.5 \, L$$

**example** An incandescent-lamp filament has a surface area of 0.4 in² and emits 1580 lm. What is its brightness in lamberts?
**solution** By For. (4),

$$L = \frac{F}{6.45S} = \frac{1580}{6.45 \times 0.4} = 612 \, L$$

**42. The relation between the illumination incident on and the brightness reflected from an illuminated surface** can be understood from a consideration of the following facts. No surface, however smooth, is a perfect reflector, since some of the luminous flux impinging on the surface will be absorbed thereby. Therefore, when the surface brightness resulting from the reflection of the impinging light rays or illumination is computed, the coefficient of reflection (Table 9) of the material must be considered. The brightness equals the illumination times the coefficient of reflection.

$$mL = 1.076 \times E \times m = mL \qquad (5)$$

where $mL$ = average surface brightness over the surface under consideration in millilamberts; $E$ = the incident illumination on the surface in footcandles; and $m$ = the absolute coefficient of reflection (Table 9) of the surface material.

**example** A surface of white plaster has illumination of 5 fc. What is its brightness?
**solution** The coefficient of reflection of white plaster (Table 9) is 0.90. By For. (5),

$$\text{Brightness} = 1.076 \times E \times m = 1.076 \times 5 \times 0.90 = 4.84 \, mL$$

**43.  Conversion Factors for Various Units of Brightness**

| Values in units in this column × conversion factor = values in units at top of column | Candelas/in² | Lamberts | Millilamberts | Footlamberts |
|---|---|---|---|---|
| Candelas/in² | 1 | 0.487 | 487 | 452 |
| Lamberts | 2.054 | 1 | 1000 | 929 |
| Millilambert | 0.00205 | 0.001 | 1 | 0.929 |
| Footlambert | 0.00221 | 0.00108 | 1.076 | 1 |

**44. The maximum brightness** beyond which glare will result (see Sec. 38) from a lighting unit usually should not exceed from 1 to 1.56 (2 to 3 cd/in² of projected area) in the central portion of the visual field. Higher values of brightness will produce a sensation of glare. The value depends on the size of the source, the position in the line of vision, and the contrast between the source and its surrounding. If the eyes are not to be fatigued when the unit is viewed continually, the brightness should not exceed 0.25 L. It should be noted that the maximum permissible brightness is somewhat influenced by the darkness of the surroundings (see Sec. 38).

### 44A.   Brightness of Light Sources

| Source | Approximate Lamberts | | |
|---|---|---|---|
| Incandescent lamp with opal globe | 0.24 | – | 1.46 |
| Incandescent lamp, inside-frosted globe | 7 | – | 15 |
| Fluorescent lamp, white and daylight | 1.2 | – | 2.6 |
| Fluorescent lamp, green | 2.4 | – | 3.6 |
| Fluorescent lamp, red | 0.12 | – | 0.23 |
| Fluorescent lamp, pink, blue, or gold | 0.83 | – | 1.8 |
| Candle flame | 1.46 | – | 1.95 |
| Acetylene-burner flame | 29 | – | 50 |
| Vacuum–incandescent-lamp filament | 460 | – | 700 |
| Gas-filled–incandescent-lamp filament | 3,000 | –4,000 | |
| Neon tube, red | 0.24 | | |
| Neon-mercury tube, green or blue | 0.05 | – | 0.1 |
| Mercury arc (quartz tube) | 290 | – | 490 |
| Metal halide arc | 1,100 | –1,700 | |
| High-pressure sodium arc | 1,100 | –1,900 | |
| Sky | 0.7 | – | 2.0 |
| Sun, on horizon | 1,000 | | |
| Sun, 30° above horizon | 243,000 | | |
| Sun, at noon | 450,000 | | |
| Ceiling over indirect-lighting fixture | 0.01 | – | 0.1 |
| Walls when lighted by diffused daylight | 0.001 | – | 0.05 |

**45. Brightness (or luminance)** is usually measured by portable meters which operate on the same basic principle as the footcandle meter. Brightness meters use a photoreceptor such as a photomultiplier tube or photovoltaic cell which creates an electrical signal in proportion to the brightness of an object or surface. Some form of optical system is designed into the meter to permit the operator to focus on the meter or the object or surface being measured. The electrical signal is sent through an amplification circuit to the meter, which may be either analog or digital. With proper calibration, a direct reading of brightness is obtained. A typical portable brightness meter is shown in Fig. 10-22.

**46. Fiber optics** is a relatively new optical technology that refers to optical systems consisting of thin cylindrical glass or plastic fibers with excellent optical properties. Light enters one end of a fiber and is transmitted along the fiber core to the other end by internal reflections off the clad or fiber wall (Fig. 10-23). Large quantities of fibers are placed

**Fig. 10-22**   Portable brightness meter. [Minolta Corporation]

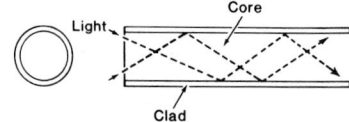

**Fig. 10-23**   Light transmission through optical fibers.

together to form a bundle. Each fiber is insulated with a special glass coating to prevent light from transforming one fiber to another. The bundle has external tubing to protect the fibers, and the ends of the bundle are bonded and polished. Initial use of fiber optics was in medical applications to observe conditions inside the human body. The most significant application has been in optical communications where voice signals are encoded and transmitted at high speed over the fibers. Fiber-optic systems have two advantages over conventional cable transmission systems: they are free from cross talk, and they are unaffected by random electrical disturbances such as lightning.

In lighting applications, fiber optics are being used in some automobile models to monitor information on signal-light operation on the dashboard as well as in optical sensors in varying applications. They have also been used in signs and other advertising and decorative applications to create unique effects of colors and/or motion.

## ELECTRIC-LIGHT SOURCES

**47. Electric-light sources may be classified as follows:**

    *A.* Visible-light sources
        1. Incandescent (filament)
        2. Fluorescent
        3. High-intensity–discharge (mercury, metal halide, and high-pressure sodium)
        4. Other gaseous-discharge
            *a.* Low-pressure sodium
            *b.* Neon
            *c.* Glow (argon and neon)
    *B.* Ultraviolet-light sources
        1. Sunlight lamps
        2. Black-light lamps
        3. Germ-killing lamps
        4. Photochemical lamps
    *C.* Infrared heating lamps

## INCANDESCENT (FILAMENT) LAMPS

**48. The incandescent lamp** consists of a filament which is a highly refractory conductor mounted in a transparent or translucent glass bulb and provided with a suitable electrically connecting base. The filament is heated by the passage of an electric current through it to such a high temperature that it becomes incandescent and emits light. In incandescent lamps of the older types the air was, insofar as practicable, exhausted from the space within the bulb and surrounding the conductor (filament), forming a partial vacuum. But in many of the modern lamps this space is filled with an inert transparent gas such as nitrogen. The conductor must have a high melting point or a high vaporizing temperature and a high resistance; it must be hard and not become plastic when heated. In vacuum-type lamps the vacuum must be good, not only to prevent the oxidation of the filament but also to prevent the loss of heat, which would reduce the efficiency. In non-vacuum-type lamps (gas-filled lamps) the gas used must be inert so as not to combine chemically with the filament material. The bulb must be transparent or translucent to permit the passage of light, not porous, so that it will retain the vacuum or inert gas, and strong to withstand handling and use.

**49. Classification of incandescent lamps.** Incandescent lamps may be classified in six different ways, according to:

    1. The class of lamp           4. The type of the base
    2. The shape of the bulb       5. The type of filament
    3. The finish of the bulb       6. The type of service

**50. Class of lamp.** Incandescent lamps are classified as Type B or Type C. The Type B lamp is one in which the filament operates in a vacuum. The Type C lamp is one which is gas-filled. Gas-filled lamps are the most widely used types.

**51. Classification according to shape of bulb** (Fig. 10-24). The standard-line shape is employed for general-lighting-service lamps up to and including the 100-W size. Lamps of 150 W and larger for general lighting service are made in the pear shape. The other shapes are utilized for lamps designed for special classes of service. Lumiline lamps are included in the tubular classification.

| Shape of bulb | Designating letter | Shape of bulb | Designating letter |
|---|---|---|---|
| Standard-line | A | Pear-shaped | P or PS |
| Cone-shaped | C | Reflector | R |
| Flame-shaped | F | Straight-side | S |
| Globular | G | Tubular | T |
| Parabolic | PAR | | |

Lamps are designated by a letter and figure such as PS-30. The letter indicates the shape of bulb and the figure the greatest diameter of the bulb in eighths of an inch. Thus, a PS-30 lamp is a lamp with a pear-shaped bulb with a diameter of $^{30}\!/_8$, or 3¾ in.

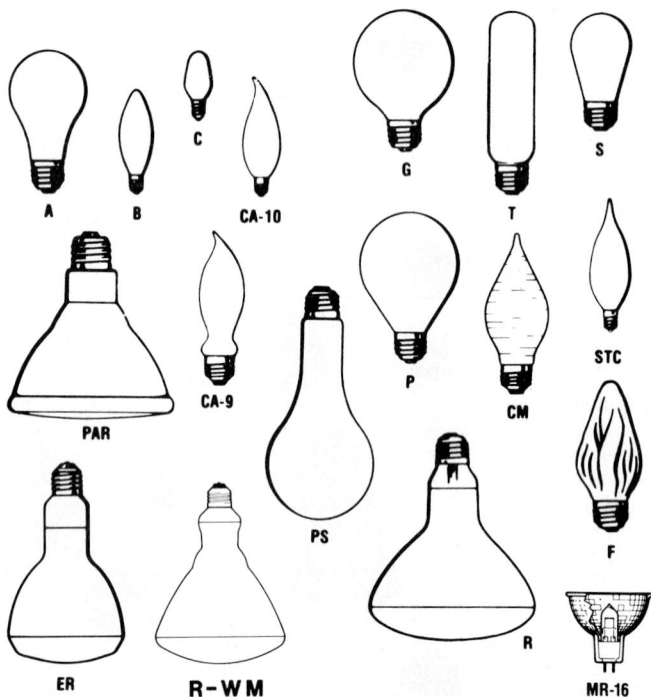

**Fig. 10-24**  Bulb designations of incandescent lamps. [General Electric Co.]

### 52. Classification according to finish of bulb
1. Clear
2. Inside-frosted
3. Silvered-bowl
4. White
5. Daylight
6. Inside-colored
7. Outside-colored
8. Colored-glass
9. Outside-coating

**53. Finish of bulbs.**    Incandescent lamps can be obtained with the bulbs finished in several different ways as listed in Sec. 52. With the clear lamps the bulb is made of clear glass which leaves the filament exposed to view. Clear-bulb lamps are used with reflecting equipment which completely conceals the lamp from view. They are employed with open-bottom types of reflecting equipment in some cases when the units are mounted so high that the lamps are not in the line of vision. Inside-frosted lamps have the entire inside surface of the bulb coated with a frosting which leaves the exterior surface perfectly smooth. This finish conceals the bright filament and diffuses the light emitted from the lamp. Inside-frosted lamps are used with open-bottom types of reflecting equipment and in places where no reflecting equipment is employed. Silvered-bowl lamps (Fig. 10-25) have a coating of mirror silver on the lower half of the bowl, which shields the brilliant filament and forms a highly efficient reflecting surface for indirect lighting. The upper part of the bulb is inside-frosted to eliminate streaks and shadows of fixture supports. Silvered-bowl lamps should be used only in fixtures designed for them, because the silvering directs the heat toward the socket assembly, which will therefore tend to operate at a higher temperature than with clear lamps. White lamps have the entire interior surface of the lamp covered with a fine coating of silica. This coating gives a high degree of diffusion which softens shadows and reduces shiny reflections. Since bulb blackening is not apparent through the diffusing coating, the lamps appear clean and white throughout their life.

**54. Daylight lamps** have a blue bulb made of a special blue-green glass to approximate noon sunlight. The ordinary incandescent lamp produces light in which the red rays predominate. The blue-green glass of the bulb of daylight lamps absorbs part of the reddish rays emitted by the filament of the lamp, giving a light which approaches the whiteness of noon sunlight. The glass absorbs about one-third of the total light emitted by the filament.

**55. Colored bulbs.** Colored light is obtained from filament lamps by the subtractive method, that is, by means of a separate filter or a bulb so processed that light of colors other than that desired is largely absorbed. Colored bulbs used in lamp manufacture are of natural-colored glass or of clear glass having a coating applied to either the inner or the outer surface of the bulb by one of several different processes. Natural-colored bulbs, wherein chemicals are added to the ingredients of the glass to produce the desired color, are regularly available in daylight blue, blue, amber, green, and ruby. Natural-colored bulbs produce light of purer colors than coated bulbs and are often used in preference to the latter for theatrical and photographic lighting.

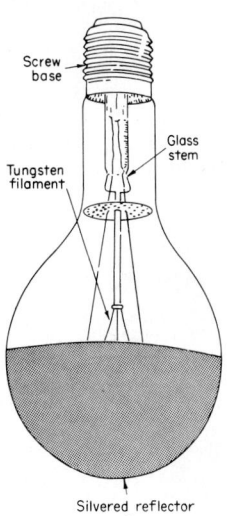

Fig.    **10-25**   Silvered-bowl incandescent lamp.

Coated colored lamps are made by spraying either the inside or the outside of the bulb or by applying a fused enamel (ceramic) or acrylic coating to the outside of the bulb. The colors in most common use are red, blue, green, yellow, orange, and white.

When decorative or display lighting is involved, coated colored lamps are to be preferred to natural-colored lamps because of their lower cost. Enameled, acrylic-coated, and inside-sprayed bulbs are satisfactory for either indoor or outdoor use.

**56. Tungsten-halogen lamps** are incandescent (filament) lamps but are significantly different from conventional lamps in size and design. They represent the practical application of the halogen regenerative cycle in filament lamps. All tungsten-halogen lamps are filled with a gas of the halogen family. Iodine and bromine are the most commonly used. As the lamp burns, the halogen gas combines with the tungsten that is evaporated from the filament. As the gas circulates inside the bulb, the tungsten is deposited back on the filament rather than on the inside bulb wall. This keeps the bulb wall clean and allows the lamp to deliver essentially its initial light output throughout life. Owing to the lamp temperatures, quartz rather than glass is used for the lamp bulb.

Four types of tungsten-halogen lamps are available (Fig. 10-26). The most widely used designs are the small tubular double-ended lamps. A coiled filament extends from one end of the lamp to the other. Lamp wattages range from 200 to 1500 W. A second type utilizes a base at only one end of the quartz bulb. In the third type, the quartz filament tubes are sealed into outer bulbs such as PAR lamps for good optical control.

The last type of tungsten-halogen lamp for lighting is an adaptation of the small photographic halogen projection lamp. Small 12-V quartz filament tubes are mounted in small multifaceted glass ellipsoidal reflectors with infrared transmitting reflectors which reflect light but transmit much of the infrared heat out the back of the lamp. These lamps are extensively used in display-lighting applications.

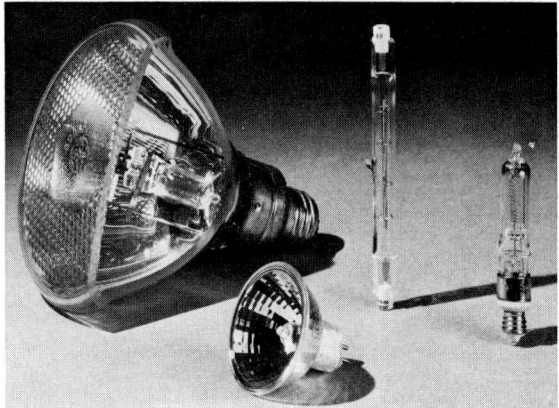

**Fig. 10-26** Tungsten-halogen lamps. The PAR design is cut away to show the internal filament tube.

The average life of most general-lighting tungsten-halogen lamps is about twice the life of regular general-service incandescent lamps, or 2000 h. Although for the same wattage the initial light output of these lamps is about the same, at the end of 1000 h the tungsten-halogen lamp produces about 13 percent more light than the standard general-service lamp.

These lamps offer high light output from compact lighting equipment. For example, the reflector for the 1500-W T-3 tungsten-halogen lamp is only 1 ft (0.3 m) long and 6 to 7 in (152 to 177 mm) wide.

Tungsten-halogen PAR lamps combine the excellent efficiency and long life of quartz lamps with the light control of a PAR bulb. The quartz tube is mounted at the focal point of the PAR bulb's reflector for accurate beam control. At the end of life (4000 h), 500-W tungsten-halogen PAR lamps provide 40 percent more light than standard 500-W PAR lamps rated at 2000 h.

Tungsten-halogen lamps are widely used in general lighting and floodlighting. Special tungsten-halogen designs are also used in specialty applications such as stage, film, and TV lighting, copying machines, and optical devices. For data on tungsten-halogen lamps refer to Sec. 100.

**56A. Classification according to type of base (Fig. 10-27):**

| | | |
|---|---|---|
| Bayonet | Medium prefocus | Extended mogul end-prong |
| Candelabra | Mogul prefocus | Medium two-pin |
| Intermediate | Medium bipost | Medium side-prong |
| Medium | Mogul bipost | Candelabra prefocus |
| Three-contact medium | Medium skirted | Screw terminal |
| Admedium | Minicandelabra | Lug sleeve |
| Mogul | Recessed single-contact | Two-pin |
| Three-contact mogul | Mogul end-prong | Double-contact medium |
| Disc | | |

**57. Lamp bases.** A number of different types of bases for incandescent lamps (Fig. 10-27) are in use. Of these, the bayonet, candelabra, and intermediate base are used on small-size (miniature) lamps. The medium base, used on general-service lamps of 300 W and less, is the most common type. The mogul base is used on sizes of 300 W and up. The admedium is slightly larger in diameter than the medium and is used on some mercury lamps. The three-contact base is used with a three-lite type of lamp. The disc base is used on Lumiline lamps.

The medium and mogul prefocused bases are used on certain types of concentrated-filament lamps, such as those for picture projection and aviation service, for which it is desirable to have the light source accurately located. The medium bipin base is for fluorescent lamps. The medium bipost base is made in 500-, 750-, and 1000-W lamp sizes for use principally with indirect fixtures, for which it allows a better design of fixture and better radiation of heat than are obtainable with the mogul-base type. For the very large-size lamps of 1500 W and up for floodlighting service the mogul bipost is the standard.

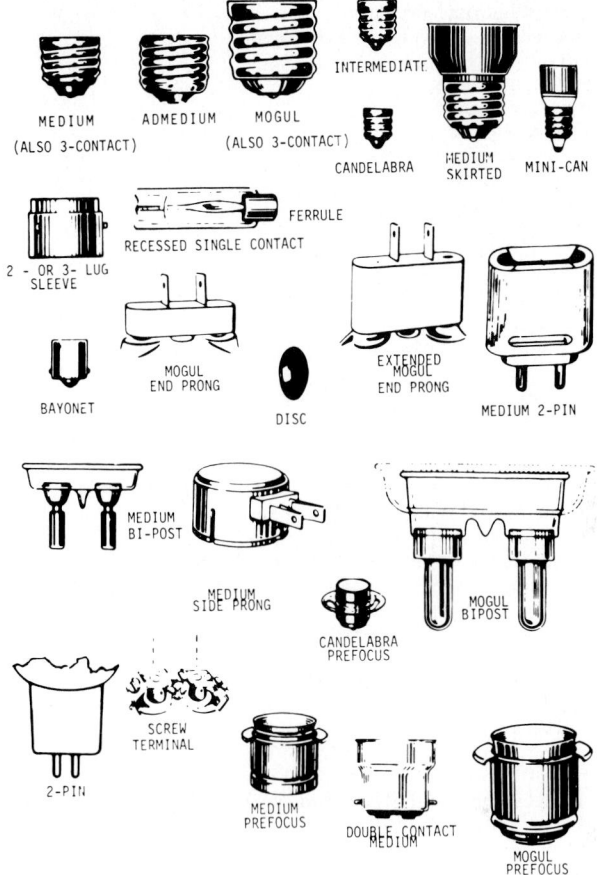

**Fig. 10-27**    Bases for incandescent lamps. [General Electric Co.]

**58. Classification according to type of filament.**    Several different types of filament structure are used. The filament structure is designated by a letter or letters to indicate whether
the wire is straight or coiled and by an arbitrary number sometimes followed by a letter
to indicate the arrangement of the filament on the supports. Prefix letters include S
(straight; wire is straight or slightly corrugated), C (coil; wire is wound into a helical coil,
or it may be deeply fluted), and CC (coiled coils; wire is wound into a helical coil, and
this coiled wire is again wound into a helical coil).

**59. Classification of incandescent lamps according to type of service:**
1. General lighting service
    *a.* Clear bulb
    *b.* Inside-frosted bulb
    *c.* Silvered-bowl bulb
    *d.* White bulb
2. Special lighting service
    *a.* Daylight lamps
    *b.* Decorative lamps
    *c.* Rough-service lamps
    *d.* Three-lite lamps
    *e.* Tubular lamps
    *f.* Vibration lamps

3. Miscellaneous lighting service
   *a.* Applicance- and indicator-service lamps
   *b.* Aviation-service lamps
   *c.* High-voltage lamps
   *d.* Low-voltage lamps
   *e.* Marine-service lamps
   *f.* Mine-service lamps
   *g.* Optical-service lamps
   *h.* Photographic lamps
   *i.* Photoservice lamps
   *j.* Projection lamps
   *k.* Projector and reflector lamps
   *l.* Sign lamps
   *m.* Spotlight- and floodlight-service lamps
   *n.* Street-lighting–service lamps
   *o.* Traffic-signal lamps
   *p.* Train- and locomotive-service lamps

**60. General-lighting-service lamps** are those of 120- or 130-V rating for ordinary uses in homes, stores, offices, schools, factories, etc.

**61. Special-lighting-service lamps** are for use in similar locations, but they have a special design feature, such as shape or color of bulb, or other special features.

For an explanation of daylight lamps refer to Sec. 54.

Decorative lamps for general and special lighting are colored lamps that are available in a number of different types (see Sec. 55). They can be used to provide special effects in homes, theaters, shops, restaurants, and lobbies and foyers of public buildings.

Special yellow enameled lamps, which often are called "bug" lamps, are available. Although these lamps are excellent for decorative lighting, they are designed primarily for outdoor lighting during the season of night-flying insects. They have less attraction for insects than lamps of other colors. These lamps are used in open porches, outdoor recreation areas, filling stations, camps, roadside stands, and any other place where people enjoy outside activities under lights.

Rough-service lamps are specially constructed so that the filament can withstand sudden bumps and other forms of rough treatment. They are used principally in extension-cord service in garages, industrial plants, and similar applications in which they will be subjected to excessive shock in service.

For an explanation of three-lite lamps refer to Sec. 62.

Tubular incandescent lamps for special general-lighting service are available in the Lumiline type and the showcase type.

The Lumiline lamps with their long bulb, 1 in (25.4 mm) in diameter, give a continuous line of light which is well suited to places where space is limited, as in displays, niches, small coves, signs, mirrors, paintings, and luminous panels. These lamps have contact

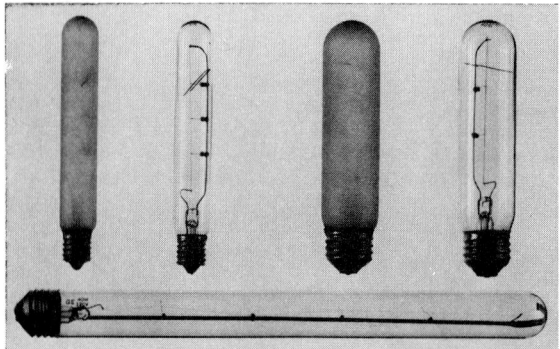

**Fig. 10-28**   Showcase lamps. [General Electric Co.]

caps of the disk-base type at each end of the bulb. Specially designed sockets or lamp-holders are required. Lumiline lamps are available with clear, inside-frosted, white, or colored-glass tubes.

Showcase lamps (Fig. 10-28) are tubular lamps with conventional screw bases, which are designed primarily for the lighting of showcases but which also are used for the lighting of shallow-depth displays and other special applications requiring small trough-type reflectors. These lamps are available in clear, inside-frosted, and special showcase-reflector types. The showcase-reflector type is made with a tubular bulb with the upper half inside-aluminized so that it can be used in showcases, shelves, speakers' stands, etc., in an ordinary socket without any reflector. A spring contact on the base allows the lamp to be adjusted in the socket to throw the light rays in any desired direction.

Vibration lamps are designed particularly for use on or near rotating machinery and other places where relatively high-frequency vibration exists. Certain of these lamps are equipped with a special type of filament wire designed to operate suitably under vibration conditions.

**62. The three-lite lamp** has two separate filaments in one bulb (Fig. 10-29). One filament consumes twice the wattage of the other, and the filaments can be lighted separately or together to produce three different levels of illumination, such as 50/100/150 W or 100/200/300 W. These lamps are particularly applicable to study lamps, reading lamps, and indirect floor lamps so that the user, with a single lamp, can adjust the illumination from that for decorative and casual use to full brilliancy for close seeing. They can also be used in stores so that daylight can be supplemented with an economical use of electric light and the illumination can be varied to suit different occasions.

**63. Appliance- and indicator-service lamps** are specially designed for use with appliances and equipment for home and commercial use. These lamps, when properly used, provide effective illumination of equipment exteriors and interiors and also give clear indications of operations in progress.

**64. Aviation-service (airport) lamps** are designed to suit the special conditions encountered in the vital lighting for safety at airport landing fields. Several types are required to provide satisfactory approach, beacon, running, taxiway, etc., lighting.

**65. High-voltage–service lamps,** rated at 230 and 250 V, are available for use in locations where only the higher voltage is available. These lamps are less rugged and less efficient than the general-lighting-service type. The reason for this is that the filament must be of longer and finer wire to have sufficient resistance to limit the current to approximately one-half the value for the same-wattage general-service lamp. For this reason use in any new installation in which the lower voltage could be made available is not encouraged.

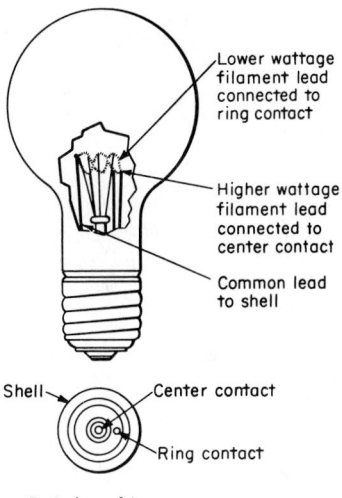

Fig. 10-29 Three-lite lamp. [General Electric Co.]

**66. Low-voltage–service lamps,** rated at 6, 12, 30, 32, 60, and 64 V respectively, are available for use on electrical systems such as those of automobiles, boats, garden lighting, miniature interior fixtures or desk lamps, underwater swimming-pool fixtures, battery-generator sets, and trains.

**67. Marine lamps** are specially designed to take care of the special maritime illumination requirements. These lamps are used on shipboard to outline and identify vessels for seaway safety and to signal between ships. On land they provide a source for lighthouse beacons. Underwater, they illuminate areas where divers must work.

**68. Mine lamps** are specially designed to meet the conditions encountered in the general illumination of mines and mine equipment.

**69. Lamps for optical devices** are available in a great variety of types to meet the special requirements of devices used in the optical field of science, industry, and education.

**70. Photographic lamps** are spotlight lamps designed with concentrated filaments for maximum light output in the controlled beams of spotlights used in theaters, television studios, and motion-picture and other photographic studios. For best lighting results, the filaments of these lamps must be accurately positioned, and the lamps should include mounting characteristics that will properly locate the filament in relation to the spotlight optical system. These lamps operate at a higher color temperature and have a shorter life than lamps for general service.

**71. Photoservice lamps** are made in two types:

1. Photoflash lamps
2. Photoflood lamps

The *photoflash lamp* (Fig. 10-30) consists of a bulb containing flammable material such as zirconium in an atmosphere of oxygen. When the lamp is connected to a source of voltage, the foil burns with a brilliant flash lasting about $\frac{1}{50}$ s.

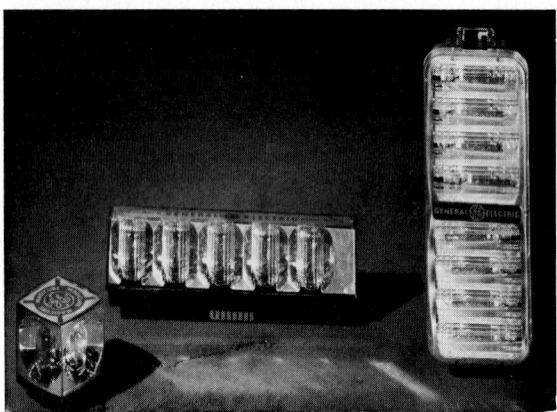

**Fig. 10-30**   Photoflash lamps. [General Electric Co.]

The *photoflood lamp* is similar to the regular inside-frosted incandescent lamp except that the filament is designed to operate at a higher temperature. This lamp emits much more light than the general-service lamp for the same wattage, but its life is much shorter, for example, 2 h for the smallest size to 10 h for the largest size.

The *photoflash lamp* is used to illuminate scenes for the taking of photographs, whereas the photoflood lamp is used for continuous illumination in the taking of moving pictures. The photoflood lamp is also used by commercial photographers for studio portrait work and by amateur photographers for interior photographs.

**72. Projection lamps** (Fig. 10-31) are used in slide and motion-picture projections. The lamps utilize carefully positioned concentrated filaments and a base type permitting accurate positioning of the filaments in the projectors. The filaments operate at high temperatures, resulting in higher efficiency and thus shorter life, usually from 25 to 50 h.

**73. Reflector and projector lamps** (Fig. 10-32) are made with a parabolic-shaped bulb having a mirrored surface on the inside of the neck and a fixed-focus filament. They can be used with a plain socket to form a highly efficient spotlight. The projector type has a lens built into the face of the lamp for better control of the light. This lamp is made of heat-resisting glass so that it can be used either indoors or in locations exposed to the weather. The reflector type has an inside-frosted glass bulb which is not weatherproof and does not provide so accurate a control of the light but costs only about two-thirds as much as the projector type. Both types can be obtained in either a concentrating-spotlight or a wide-floodlight version. These bulbs are available with integral quartz-halogen lamps.

**74. Sign and decorative lamps** are lamps designed especially for outdoor signs, Christmas and other decorations, carnivals, fairs, and festoon lighting. They are used also for many interior applications. While many standard gas-filled lamps are used in enclosed and other

**Fig. 10-31**   Projection lamps. [General Electric Co.]

types of electric signs, those designated particularly as sign lamps are mostly of the vacuum type. Lamps of this type are best adapted for exposed sign and festoon service because the lower bulb temperature of vacuum lamps minimizes thermal cracks due to the contact of rain and snow. Some low-wattage sign lamps, however, are gas-filled for use in flashing signs. This causes more rapid cooling of the filament and reduces the trailing effect which slow-cooling vacuum lamps might produce. Bulb temperatures of these low-wattage gas-filled lamps are sufficiently low to permit exposed outdoor use.

**75. Spotlight and floodlight lamps** have concentrated filaments, accurately positioned with respect to the base. They are used in equipment which produces accurately controlled beams of light. There are several companion listings of spotlight and floodlight lamps having the same dimensions but differing in life design. Floodlight lamps are used when burning hours are long, as in building floodlighting and show-window lighting. Spotlight lamps are used for applications in which burning hours are short and higher light output is needed, particularly in the blue and green portions of the visible spectrum. The T-12 and T-14 lamps are for use in ellipsoidal projectors when employed for show windows and interior displays. The floodlight lamps are used also for underwater units. A typical lamp is shown in Fig. 10-33.

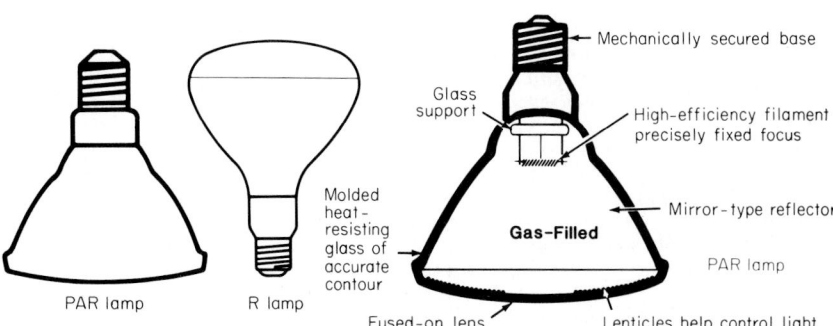

**Fig. 10-32**   Reflector and projector lamps. [General Electric Co.]

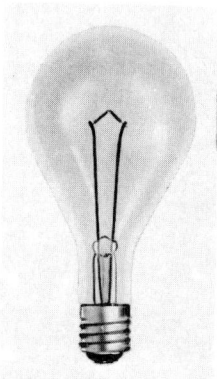

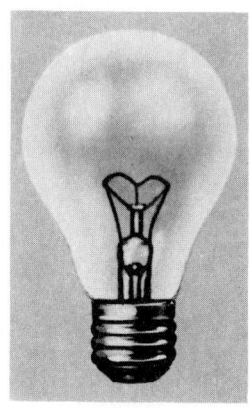

**Fig.    10-33** Floodlight lamp. [General Electric Co.]

**Fig.    10-34** Street-lighting–service lamp. [General Electric Co.]

**Fig.        10-35** Traffic-signal lamp. [General Electric Co.]

**76. Street-lighting–service lamps** are lamps made specifically for use in street-lighting luminaires. They are made for series or multiple operation. A typical lamp is shown in Fig. 10-34. The series lamps are for operation on constant-current series circuits. They are made for 6.6- and 20-A circuits.

**77. Traffic-signal lamps** (Fig 10-35) are clear lamps which are designed with a short light-center length for focusing the rays to give a signal indication of required brightness for traffic-signal use.

**78. Train and locomotive lamps** are specially designed to withstand the intense vibrations and shocks encountered in this service. They are available in voltages from 30 to 75 V for regular train-lighting service and for locomotive-cab and headlighting service.

**79. Burning position.** With a few exceptions general-lighting-service lamps can be burned in any position. However, the lumen maintenance of Type C lamps is best when they are burned base up. This is because the tungsten blackening is conducted upward by the gas and is always deposited above the filament. When the lamp is burned base up, the blackening collects in the area of the bulb adjacent to the base, where the light is already partially intercepted by the base, socket, and luminaire husk. If the lamp is burned base down, the blackening collects in the bowl of the bulb, where it causes a much greater reduction in light output. Lamps burned in a horizontal position are affected similarly.

Certain types of lamps, particularly projection, spotlight, floodlight, and some street series lamps, are not designed for universal burning and should always be used in the position designated by the manufacturer's published data. The reason for this may be the construction of the filament, which would be likely to sag or short-circuit if burned in a position other than that for which it is designed. Operation in an incorrect position sometimes places the filament directly under a glass part which might be softened by the heat. If a lamp containing a collector grid is burned in any other position than with the grid directly above the filament, the special construction will not be effective in controlling blackening.

**80. Lamp temperatures.** Operation of lamps under conditions which cause excessive bulb and base temperatures may result in melting of the bulb, softening of the base cement, and loosening of the base or, in extreme cases, damage to the socket and adjacent wiring. Most fixtures are properly designed to dissipate the heat generated by the lamps, but severe conditions such as might be induced by overvoltage operation or the use of lamps of higher wattage than the manufacturer's rating may give rise to difficulty. If metal parts of shades, reflectors, or fixtures are allowed to come in contact with the bulb of a gas-filled lamp, the local cooling effect may result in glass cracks which will cause lamp failure (sometimes violent). Maximum safe operating temperatures for best lamp performance are shown in the following table.

Gas-filled lamps, because of the convection currents within the bulb, have higher bulb

**Maximum Safe Operating Temperatures**
(Approximate figures)

| | |
|---|---|
| Soft-glass bulb | 370°C |
| Hard-glass bulb | 450–510°C |
| Cemented base | |
|   Regular | 170°C |
|     Special Hi-Temp | 200°C |
| Mechanical base | 210°C |
| Bipost base | 285°C |

temperatures than vacuum lamps. Therefore vacuum lamps are preferable for use in exposed outdoor locations where snow or rain may strike the hot bulb. Gas-filled lamps exposed to the elements should have hard or heat-resisting glass bulbs.

**81. Vibration and shock.**   Tungsten wire heated to incandescence becomes somewhat soft and pliable, and filament coils may be distorted or broken if the lamp is subjected to shock or vibration while burning. Vibration, especially of the low-amplitude high-frequency variety, is an insidious enemy of satisfactory lamp performance and should be guarded against whenever possible. There are available a number of commercial sockets and fixtures designed to protect the lamps by absorbing vibration.

If vibration cannot be eliminated by these means, special *vibration-service lamps*, provided with extra filament supports, should be used. The construction of vibration-service lamps is such that they will give most satisfactory performance if burned in a vertical position, either base up or base down. They should never be used where they are likely to receive extreme shocks.

For use on extension cords and in any service where excessive shocks may be encountered, there are available *rough-service lamps* with a special type of shock-resisting filament construction. Rough-service lamps are designed to operate in any burning position and can be used in place of vibration-service lamps when it is necessary to burn lamps horizontally. Both vibration- and rough-service lamps sacrifice some efficiency to strength of construction and are more expensive than standard lamps; they should, therefore, be used only where required by service conditions.

**82. The life of incandescent lamps** is defined as the average life of a large group of lamps burned under specified conditions. The range of typical mortality curves for incandescent lamps is shown in Fig. 10-36.

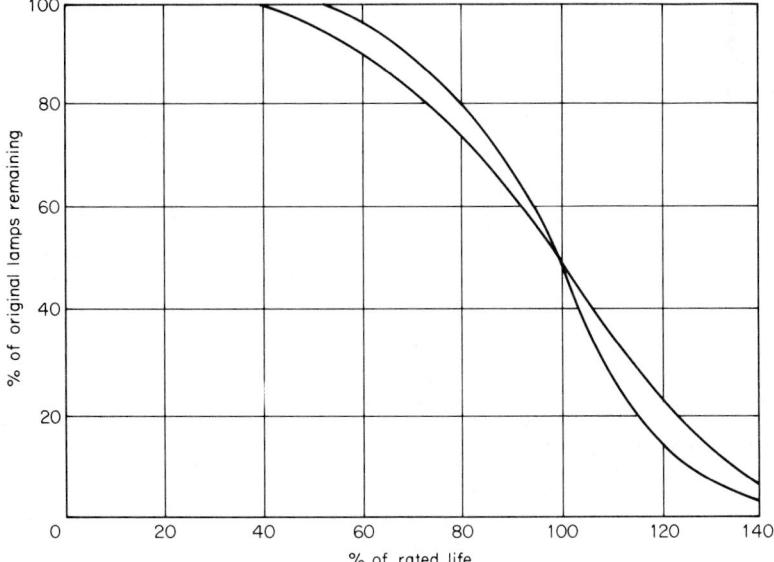

**Fig. 10-36**   Incandescent-lamp mortality curves. [General Electric Co.]

The average life of standard lamps for general lighting purposes varies from 750 h for some lamps and 1000 h for others to 2500 h for certain sizes and types. Approximate hours of life are specified by the manufacturer. The life of a lamp is materially affected by the voltage impressed upon it. Operating a lamp at less than its rated voltage will prolong the life of the lamp while operating at higher than its rated voltage will shorten its life.

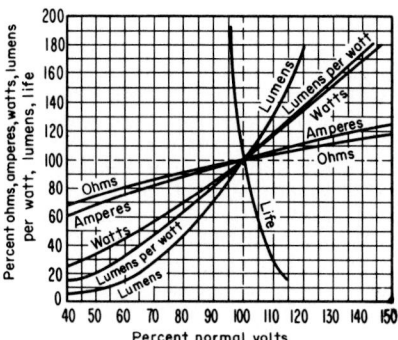

**Fig. 10-37**   Variation in characteristics of incandescent lamps with voltage. [General Electric Co.]

This does not mean that it is good practice to operate lamps at less than their rated voltage. Decreasing the voltage on a lamp decreases the light given out by the lamp, but the percentage of decrease in light is much greater than the percentage of decrease in voltage, so that the efficiency of operation is much poorer. Figure 10-37 shows percentage variation in characteristics of lamps with variation from normal voltage. Consider a 120-V lamp operating at 114 V. The voltage impressed on the lamp is 95 percent of rated voltage. From Fig. 10-37 the lumens emitted by the lamp will be reduced to 84 percent of the rated value. Thus a 5 percent reduction in voltage results in a 16 percent reduction in light output.

**82. Clear-Bulb Incandescent Lamps for General Lighting Service, 120 V**
(General Electric Co.)

| Lamp-ordering code | Watts | Bulb | Base | Class | Filament | Approximate life, h | Approximate initial lumens | Light-center length, in | Maximum overall length, in |
|---|---|---|---|---|---|---|---|---|---|
| 25ACL | 25 | A-19 | Medium | B | CC-6 | 2500 | 190 | 2⅜ | 3⅞ |
| 40ACL | 40 | A-19 | Medium | C | CC-6 | 1500 | 480 | 3⅛ | 4 7/16 |
| 60ACL | 60 | A-19 | Medium | C | CC-6 | 1000 | 870 | 3⅛ | 4 7/16 |
| 60A/52WM/CL | 60 | A-19 | Medium | C | CC-6 | 1000 | 800 | 3⅛ | 4 7/16 |
| 75ACL | 75 | A-19 | Medium | C | CC-6 | 750 | 1,190 | 3⅛ | 4 7/16 |
| 75A/67WM/CL | 67 | A-19 | Medium | C | CC-8 | 750 | 1,130 | 3⅛ | 4 7/16 |
| 100A/90WM/CL | 90 | A-19 | Medium | C | CC-8 | 750 | ..... | 3⅛ | 4 7/16 |
| 100ACL | 100 | A-19 | Medium | C | CC-8 | 750 | 1,750 | 3⅛ | 4 7/16 |
| 150/135WM/CL | 135 | A-21 | Medium | C | CC-8 | 750 | 2,580 | 4 1/16 | 5⅝ |
| 150CL | 150 | PS-25 | Medium | C | C-9 | 750 | 2,680 | 5 3/16 | 6 |
| 200A/CL | 250 | A-23 | Medium | C | CC-8 | 750 | 4,010 | 4⅝ | 6 5/16 |
| 200 | 200 | PS-30 | Medium | C | C-9 | 750 | 3,710 | 6 | 8 1/16 |
| 200PS30/12 | 200 | PS-30 | Mogul | C | C-9 | 750 | ..... | 6⅜ | 8 7/16 |
| 300M | 300 | PS-25 | Medium | C | CC-8 | 750 | 6,360 | 5 3/16 | 6 15/16 |
| 300 | 300 | PS-35 | Mogul | C | C-9 | 1000 | 5,820 | 7 | 9⅜ |
| 500 | 500 | PS-35 | Mogul | C | CC-8 | 1000 | 10,850 | 7 | 9⅜ |
| 500PS40 | 500 | PS-40 | Mogul | C | C-9 | 1000 | 9,900 | 7 | 9¾ |
| 750 | 750 | PS-52 | Mogul | C | CC-8 | 1000 | 17,040 | 9½ | 13 |
| 1000 | 1000 | PS-52 | Mogul | C | CC-8 | 1000 | 23,740 | 9½ | 13 |

**83. Inside-Frosted–Bulb Incandescent Lamps for General Lighting Service, 120 V**
(General Electric Co.)

| Lamp-ordering code | Watts | Bulb | Base | Class | Filament | Approximate life, h | Approximate initial lumens | Light-center length, in | Maximum overall length, in |
|---|---|---|---|---|---|---|---|---|---|
| 15A15 | 15 | A-15 | Medium | B | C-9 | 2500 | 125 | 2⅜ | 3½ |
| 25A | 25 | A-19 | Medium | B | CC-6 | 2500 | 190 | 2⅜ | 3⅞ |
| 40A | 40 | A-19 | Medium | C | CC-6 | 1000 | 505 | 3⅛ | 4 7/16 |
| 40A/34WM | 34 | A-19 | Medium | C | CC-6 | 1500 | 410 | 3⅛ | 4 7/16 |
| 60A | 60 | A-19 | Medium | C | CC-6 | 1000 | 870 | 3⅛ | 4 7/16 |
| 60A/52WM | 52 | A-19 | Medium | C | CC-8 | 1000 | 800 | 3⅛ | 4 7/16 |
| 75A | 75 | A-19 | Medium | C | CC-6 | 750 | 1,190 | 3⅛ | 4 7/16 |
| 75A/67WM | 67 | A-19 | Medium | C | CC-8 | 750 | 1,130 | 3⅛ | 4 7/16 |
| 100A | 100 | A-19 | Medium | C | CC-8 | 750 | 1,750 | 3⅛ | 4 7/16 |
| 100A/90WM | 90 | A-19 | Medium | C | CC-8 | 750 | 1,620 | 3⅛ | 4 7/16 |
| 100A21 | 100 | A-21 | Medium | C | CC-6 | 750 | 1,690 | 3⅞ | 5¼ |
| 150A23 | 150 | A-23 | Medium | C | CC-8 | 750 | 2,780 | 4⅝ | 6 3/16 |
| 150A | 150 | A-21 | Medium | C | CC-8 | 750 | 2,850 | 4 1/16 | 5⅜ |
| 150A/135WM | 135 | A-21 | Medium | C | CC-8 | 750 | 2,580 | 4 1/16 | 5⅜ |
| 150/IF | 150 | PS-25 | Medium | C | C-9 | 750 | 2,680 | 5 3/16 | 6 15/16 |
| 200A | 200 | A-23 | Medium | C | CC-8 | 750 | 4,010 | 4⅝ | 6 3/16 |
| 200/IF | 200 | PS-30 | Medium | C | C-9 | 750 | 3,710 | 6 | 8 1/16 |
| 300M/IF | 300 | PS-25 | Medium | C | CC-8 | 750 | 6,360 | 5 3/16 | 6 15/16 |
| 300M/PS30/IF | 300 | PS-30 | Medium | C | C-9 | 750 | 6,110 | 6 | 8 1/16 |
| 300/IF | 300 | PS-35 | Mogul | C | C-9 | 1000 | 5,820 | 7 | 9⅜ |
| 500/IF | 500 | PS-35 | Mogul | C | CC-8 | 1000 | 10,850 | 7 | 9⅜ |
| 750/IF | 750 | PS-52 | Mogul | C | CC-8 | 1000 | 17,040 | 9½ | 13 |
| 1000/IF | 1000 | PS-52 | Mogul | C | CC-8 | 1000 | 23,740 | 9½ | 13 |

**84. Extended-Service Incandescent Lamps for General Lighting Service**
(General Electric Co.)

| Lamp-ordering code | Watts | Bulb | Description | Base | Volts | Approximate life, h | Approximate initial lumens | Light-center length, in | Maximum overall length, in |
|---|---|---|---|---|---|---|---|---|---|
| 40A/34WMP/99 | 34 | A-19 | Inside-frosted | Medium | 120 | 2500 | 365 | 3⅛ | 4 7/16 |
| 60A/52WMP/99 | 52 | A-19 | Inside-frosted | Medium | 120 | 2500 | 700 | 3⅛ | 4 7/16 |
| 75A/67WMP/99 | 67 | A-19 | Inside-frosted | Medium | 120 | 2500 | 930 | 3⅛ | 4 7/16 |
| 100A/90WMP/99 | 90 | A-19 | Inside-frosted | Medium | 120 | 2500 | 1,360 | 3⅛ | 4 7/16 |
| 100A21/99 | 100 | A-21 | Inside-frosted | Medium | 120 | 2500 | 1,440 | 3⅞ | 5¼ |
| 150A23/99 | 150 | A-23 | Inside-frosted | Medium | 120 | 2500 | 2,310 | 4⅝ | 6 3/16 |
| 150A/135WMP/99 | 135 | A-21 | Inside-frosted | Medium | 120 | 2500 | 2,145 | 4 1/16 | 5⅜ |
| 150A/135WMP/CL | 135 | A-21 | Clear | Medium | 120 | 2500 | 2,145 | 4 1/16 | 5⅜ |
| 150/99 | 150 | PS-25 | Inside-frosted | Medium | 120 | 2500 | 2,300 | 5 3/16 | 6 15/16 |
| 150/99CL | 150 | PS-25 | Clear | Medium | 130 | 2500 | 2,300 | 5 3/16 | 6 15/16 |
| 200A/99 | 200 | A-23 | Inside-frosted | Medium | 120 | 2500 | 3,410 | 4⅝ | 6 3/16 |
| 200/99 | 200 | PS-30 | Clear | Medium | 120 | 2500 | 3,260 | 6 | 8 1/16 |
| 200/99IF | 200 | PS-30 | Inside-frosted | Medium | 120 | 2500 | ..... | 6 | 8 1/16 |
| 300M/99 | 300 | PS-30 | Clear | Medium | 120 | 2500 | 5,190 | 6 | 8 1/16 |
| 300M/99IF | 300 | PS-30 | Inside-frosted | Medium | 120 | 2500 | ..... | 6 | 8 1/16 |
| 300/99 | 300 | PS-35 | Clear | Mogul | 120 | 2500 | 5,190 | 7 | 9⅜ |
| 300/99IF | 300 | PS-35 | Inside-frosted | Mogul | 120 | 2500 | ..... | 7 | 9⅜ |
| 500/99 | 500 | PS-40 | Clear | Mogul | 120 | 2500 | 9,070 | 7 | 9¾ |
| 500/99IF | 500 | PS-40 | Inside-frosted | Mogul | 120 | 2500 | 9,070 | 7 | 9¾ |
| 750/99 | 750 | PS-52 | Clear | Mogul | 120 | 2500 | 14,200 | 9½ | 13 |
| 1000/99 | 1000 | PS-52 | Clear | Mogul | 120 | 2500 | 19,800 | 9½ | 13 |
| 1500/99 | 1500 | PS-52 | Clear | Mogul | 130 | 2500 | ..... | 9½ | 13 |

**85. White Incandescent Lamps for General Lighting Service, 120 V**
(General Electric Co.)

| Lamp-ordering code | Watts | Bulb | Base | Class | Filament | Approximate life, h | Approximate initial lumens | Light-center length, in | Maximum overall length, in |
|---|---|---|---|---|---|---|---|---|---|
| 15A/W | 15 | A-15 | Medium | C | C-9 | 2500 | 120 | . . | 3½ |
| 25A/W | 25 | A-19 | Medium | C | C-6 | 2500 | 190 | . . | 4¼ |
| 40A/W | 40 | A-19 | Medium | C | CC-6 | 1000 | 490 | 3⅛ | 4 7/16 |
| 55A/MI/LL | 55 | A-19 | Medium | C | CC-8 | 1000 | 810 | 3⅛ | 4 7/16 |
| 60A/W | 60 | A-19 | Medium | C | CC-6 | 1000 | 855 | 3⅛ | 4 7/16 |
| 70A/MI/LL | 70 | A-19 | Medium | C | CC-8 | 750 | 1140 | 3⅛ | 4 7/16 |
| 75A/W | 75 | A-19 | Medium | C | CC-6 | 750 | 1170 | 3⅛ | 4 7/16 |
| 95A/MI/LL | 95 | A-19 | Medium | C | CC-8 | 750 | 1620 | 3⅛ | 4 7/16 |
| 100A/W | 100 | A-19 | Medium | C | CC-8 | 750 | 1710 | 3⅛ | 4 7/16 |
| 150A/W | 150 | A-21 | Medium | C | CC-8 | 750 | 2780 | 4 1/16 | 5⅜ |
| 170A/RL/SW | 170 | A-23 | Medium | C | CC-8 | 750 | 3100 | — | 5 15/16 |
| 250A/RL/SW | 250 | A-23 | Medium | C | CC-8 | 750 | 4500 | — | 5 15/16 |

**86. Protection-Coated Incandescent Lamps, 120 V**
(General Electric Co.)

| Lamp-ordering code | Watts | Bulb | Description | Base | Class | Filament | Approximate life, h | Maximum overall length, in |
|---|---|---|---|---|---|---|---|---|
| 25A/CVG | 25 | A-19 | Inside-frosted | Medium | B | C-9 | 2500 | 3⅞ |
| 40A/CVG | 40 | A-19 | Inside-frosted | Medium | C | C-9 | 1500 | 4¼ |
| 50A/RS/CVG | 50 | A-19 | Inside-frosted, rough-service | Medium | B | C-22 | 1000 | 3⅞ |
| 60A/CVG | 60 | A-19 | Inside-frosted | Medium | C | CC-6 | 1000 | 4⁹⁄₁₆ |
| 75A/RS/CVG | 75 | A-21 | Inside-frosted, rough-service | Medium | C | C-17 | 1000 | 5¼ |
| 75R30/FL/CVG | 75 | R-30 | Reflector | Medium | C | CC-6 | 2000 | 5⅝ |
| 100A/RS/CVG | 100 | A-21 | Inside-frosted, rough-service | Medium | C | C-17 | 1000 | 5¼ |
| 100A21/CVG | 100 | A-21 | Inside-frosted | Medium | C | CC-6 | 750 | 5¼ |
| 150PS25/CVG | 150 | PS-25 | Inside-frosted | Medium | C | C-9 | 1000 | 6¹⁵⁄₁₆ |
| 150/RS/CVG | 150 | PS-25 | Inside-frosted, rough-service | Medium | C | C-17 | 1000 | 6⁷⁄₁₆ |
| 150R/FL/CVG[a] | 150 | R-40 | Reflector | Medium | C | CC-6 | 2000 | 6⁹⁄₁₆ |
| 200PS30/CVG | 200 | PS-30 | Inside-frosted | Medium | C | C-9 | 1000 | 8³⁄₁₆ |
| 200PS30/RS/CVG | 200 | PS-30 | Inside-frosted, rough-service | Medium | C | C-9 | 1000 | 8³⁄₁₆ |

[a] 130-V design.
NOTE  All lamps are coated with Teflon.® Teflon is a registered trademark of E. I. du Pont de Nemours & Co.

**87. Silvered-Bowl Incandescent Lamps for General Lighting Service, 120 V**
(General Electric Co.)

| Lamp-ordering code | Watts | Bulb | Base | Class | Filament | Approximate life, h | Approximate initial lumens | Light-center length, in | Maximum overall length, in |
|---|---|---|---|---|---|---|---|---|---|
| 60/A/SB | 60 | A-19 | Medium | C | CC-6 | 1000 | 740 | 3⅛ | 4 7/16 |
| 100A21/SB[a] | 100 | A-21 | Medium | C | CC-6 | 1000 | ..... | 3⅞ | 5¼ |
| 100A/1SBIF | 100 | A-21 | Medium | C | CC-6 | 1000 | ..... | 3⅞ | 5¼ |
| 100A/SB | 100 | A-23 | Medium | C | CC-6 | 750 | 1,470 | 4 7/16 | 5 15/16 |
| 150/SB | 150 | PS-25 | Medium | C | C-9 | 1000 | 2,370 | 5 5/16 | 6 15/16 |
| 200/SBIF | 200 | PS-30 | Medium | C | C-9 | 1000 | 3,320 | 6 | 8 1/16 |
| 300/SBIF | 300 | PS-35 | Mogul | C | C-9 | 1000 | 5,410 | 7 | 9⅜ |
| 500/SBIF | 500 | PS-40 | Mogul | C | C-9 | 1000 | 9,530 | 7 | 9¾ |
| 750/SBIF | 750 | PS-52 | Mogul | C | C-7A | 1000 | ..... | 9½ | 13 |
| 1000/SBIF | 1000 | PS-52 | Mogul | C | C-7A | 1000 | 20,400 | 9½ | 13 |

[a]Clear-silvered. All other lamps in this table are inside-frosted–silvered.

**88. Daylight Incandescent Lamps, 120 V**
(General Electric Co.)

| Lamp-ordering code | Watts | Bulb | Base | Description | Class | Filament | Approximate life, h | Approximate initial lumens | Light-center length, in | Maximum overall length, in |
|---|---|---|---|---|---|---|---|---|---|---|
| 60A/D | 60 | A-19 | Medium | Inside-frosted | C | CC-6 | 1000 | 500 | 3⅜ | 4 7⁄16 |
| 100A/D | 100 | A-23 | Medium | Inside-frosted | C | CC-6 | 750 | 910 | 4⅜ | 6 7⁄16 |
| 150/D | 150 | PS-25 | Medium | Inside-frosted | C | C-9 | 1000 | — | 5¼ | 6 15⁄16 |

**89. General Decorative Incandescent Lamps, 120 V**
(General Electric Co.)

| Lamp-ordering code | Watts | Bulb | Description | Base | Filament | Approximate life, h | Maximum overall length, in |
|---|---|---|---|---|---|---|---|
| 2BC/FL | 2 | B-10 | Clear-flicker | Candelabra | ..... | 2500 | 3⅞ |
| 15BC | 15 | B-10 | Clear–bent-tip | Candelabra | C-7A | 1500 | 3¾ |
| 15CAC | 15 | CA-8 | Clear–bent-tip | Candelabra | C-7A | 1500 | 3 9/16 |
| 15CAC/F | 15 | CA-8 | Frost–bent-tip | Candelabra | C-7A | 1500 | 3 9/16 |
| 15CAC/L | 15 | CA-8 | Clear–bent-tip | Candelabra | C-7A | 4000 | 3 9/16 |
| 15FC | 15 | F-10 | Clear-flame | Candelabra | C-7A | 1500 | 3⅛ |
| 15FC/L | 15 | F-10 | Clear-flame | Candelabra | C-7A | 4000 | 3⅛ |
| 15FC/A | 15 | F-10 | Amber-flame | Candelabra | C-7A | 1500 | 3⅛ |
| 15FC/A/L | 15 | F-10 | Amber-flame | Candelabra | C-7A | 4000 | 3⅛ |
| 15FC/AU | 15 | F-10 | Auradescent-flame | Candelabra | C-7A | 1500 | 3⅛ |
| 15GC | 15 | G-16½ | Clear-globe | Candelabra | C-7A | 1500 | 3 |
| 15GC/W | 15 | G-16½ | White-globe | Candelabra | C-7A | 1500 | 3 |
| 15GC/L | 15 | G-16½ | Clear-globe | Candelabra | C-7A | 4000 | 3 |
| 15GC/W/L | 15 | G-16½ | White-globe | Candelabra | C-7A | 4000 | 3 |
| 25BC | 25 | B-10 | Clear–blunt-tip | Candelabra | C-7A | 1500 | 3¾ |
| 25BC/W | 25 | B-10 | White–blunt-tip | Candelabra | C-7A | 1500 | 3¾ |
| 25BM | 25 | B-13 | Clear–blunt-tip | Medium | C-9 | 1500 | 4⅝ |
| 25BM/W | 25 | B-13 | White–blunt-tip | Medium | C-9 | 1500 | 4⅝ |
| 25CAM/L | 25 | CA-9 | Clear–bent-tip | Medium | CC-2V | 4000 | 4 9/16 |
| 25CAM/F/L | 25 | CA-9 | Frost–bent-tip | Medium | CC-2V | 4000 | 4 9/16 |
| 25CAC | 25 | CA-10 | Clear–bent-tip | Candelabra | CC-2V | 1500 | 4 5/16 |
| 25CAC/F | 25 | CA-10 | Frost–bent-tip | Candelabra | CC-2V | 1500 | 4 5/16 |
| 25CAC/L | 25 | CA-10 | Clear–bent-tip | Candelabra | CC-2V | 4000 | 4 5/16 |
| 25CAC/F/L | 25 | CA-10 | Frost–bent-tip | Candelabra | CC-2V | 4000 | 4 5/16 |
| 25CRC | 25 | HX-10½ | Clear-faceted | Candelabra | C-7A | 1500 | 4⅝ |
| 25CC/FW | 25 | C-11 | Frost | Candelabra | C-7A | 1500 | 4 |
| 25STC | 25 | ST-9½ | Clear–straight-tip | Candelabra | CC-2V | 1500 | 4 |
| 25STC/F | 25 | ST-9½ | Frost–straight-tip | Candelabra | CC-2V | 1500 | 4 |
| 25FM | 25 | F-15 | Clear-flame | Medium | C-9 | 1500 | 4⅜ |
| 25FM/W | 25 | F-15 | White-flame | Medium | C-9 | 1500 | 4⅜ |

| | | | | | | | |
|---|---|---|---|---|---|---|---|
| 25FM/A | 25 | F-15 | Amber-flame | Medium | C-9 | 1500 | 4⅜ |
| 25FM/A/L | 25 | F-15 | Amber-flame | Medium | C-9 | 4000 | 4⅜ |
| 25FM/AU | 25 | F-15 | Auradescent-flame | Medium | C-9 | 1500 | 4⅜ |
| 25FC | 25 | F-10 | Clear-flame | Candelabra | C-7A | 1500 | 3⅛ |
| 25GC | 25 | G-16½ | Clear-globe | Candelabra | C-7A | 1500 | 3 |
| 25GC/W | 25 | G-16½ | White-globe | Candelabra | C-7A | 1500 | 3 |
| 25GC/MW | 25 | G-16½ | Matte–white-globe | Candelabra | C-7A | 1500 | 3 |
| 25GC/L | 25 | G-16½ | Clear-globe | Candelabra | C-7A | 4000 | 3 |
| 25GC/W/L | 25 | G-16½ | White-globe | Candelabra | C-7A | 4000 | 3 |
| 25G 181/2/W | 25 | G-18½ | White-globe | Medium | C-9 | 1500 | 3 9/16 |
| 25G25 | 25 | G-25 | Clear-globe | Medium | C-9 | 1500 | 4⅝ |
| 25G25/W | 25 | G-25 | White-globe | Medium | C-9 | 1500 | 4⅝ |
| 25G25/L | 25 | G-25 | Clear-globe | Medium | C-9 | 4000 | 4⅝ |
| 25G25/W/L | 25 | G-25 | White-globe | Medium | C-9 | 4000 | 4⅝ |
| 25G40 | 25 | G-40 | Clear-globe | Medium | C-9 | 2500 | 6 15/16 |
| 25G4C/W | 25 | G-40 | White-globe | Medium | C-9 | 2500 | 6 15/16 |
| 40BC | 40 | B-10 | Clear–bent-tip | Candelabra | C-9 | 1500 | 3¾ |
| 40BC/W | 40 | B-10 | White–bent-tip | Candelabra | C-9 | 1500 | 3¾ |
| 40BC/L | 40 | B-10 | Clear–bent-tip | Candelabra | C-9 | 4000 | 3¾ |
| 40BM | 40 | B-13 | Clear–bent-tip | Medium | C-9 | 1500 | 4⅝ |
| 40BM/W | 40 | B-13 | White–bent-tip | Medium | C-9 | 1500 | 4⅝ |
| 40CAM | 40 | CA-9 | Clear–bent-tip | Medium | CC-2V | 1500 | 4 9/16 |
| 40CAM/F | 40 | CA-9 | Frost–bent-tip | Medium | CC-2V | 1500 | 4 9/16 |
| 40CAM/L | 40 | CA-9 | Clear–bent-tip | Medium | CC-2V | 4000 | 4 9/16 |
| 40CAC | 40 | CA-10 | Clear–bent-tip | Candelabra | CC-2V | 1500 | 4 5/16 |
| 40CAC/F | 40 | CA-10 | Frost–bent-tip | Candelabra | CC-2V | 1500 | 4 5/16 |
| 40CAC/L | 40 | CA-10 | Clear–bent-tip | Candelabra | CC-2V | 4000 | 4 5/16 |
| 40CAC/F/L | 40 | CA-10 | Frost–bent-tip | Candelabra | CC-2V | 4000 | 4 5/16 |
| 40CC/FW | 40 | C-11 | Frost | Candelabra | C-7A | 1500 | 4 |
| 40CRC | 40 | HX-10½ | Clear-faceted | Candelabra | C-7A | 1500 | 4⅝ |
| 40FM | 40 | F-15 | Clear-flame | Medium | C-9 | 1500 | 4⅜ |
| 40FM/L | 40 | F-15 | Clear-flame | Medium | C-9 | 4000 | 4⅜ |
| 40FM/W | 40 | F-15 | White-flame | Medium | C-9 | 1500 | 4⅜ |
| 40FM/AU | 40 | F-15 | Auradescent-flame | Medium | C-9 | 1500 | 4⅜ |
| 40GC | 40 | G-16½ | Clear-globe | Candelabra | C-9 | 1500 | 3 |

**89. General Decorative Incandescent Lamps, 120 V** (*Continued*)

| Lamp-ordering code | Watts | Bulb | Description | Base | Filament | Approximate life, h | Maximum overall length, in |
|---|---|---|---|---|---|---|---|
| 40GC/W | 40 | G-16½ | White-globe | Candelabra | C-9 | 1500 | 3 |
| 40GC/L | 40 | G-16½ | Clear-globe | Candelabra | C-9 | 4000 | 3 |
| 40GC/W/L | 40 | G-16½ | White-globe | Candelabra | C-9 | 4000 | 3 |
| 40G25 | 40 | G-25 | Clear-globe | Medium | C-9 | 1500 | 4 9/16 |
| 40G25/W | 40 | G-25 | White-globe | Medium | C-9 | 1500 | 4 9/16 |
| 40G25/L | 40 | G-25 | Clear-globe | Medium | C-9 | 4000 | 4 9/16 |
| 40G25/W/L | 40 | G-25 | White-globe | Medium | C-9 | 4000 | 4 9/16 |
| 40G40 | 40 | G-40 | Clear-globe | Medium | C-9 | 2500 | 6 15/16 |
| 40G40/W | 40 | G-40 | White-globe | Medium | C-9 | 2500 | 6 15/16 |
| 40G40/L | 40 | G-40 | Clear-globe | Medium | C-9 | 4000 | 6 15/16 |
| 60BC | 60 | B-10 | Clear–bent-tip | Candelabra | CC-2V | 1500 | 3¾ |
| 60CAM | 60 | CA-9 | Clear–bent-tip | Medium | CC-2V | 1500 | 4 9/16 |
| 60CAM/F | 60 | CA-9 | Frost–bent-tip | Medium | CC-2V | 1500 | 4 9/16 |
| 60CAM/L | 60 | CA-9 | Clear–bent-tip | Medium | CC-2V | 4000 | 4 9/16 |
| 60CAC | 60 | CA-10 | Clear–bent-tip | Candelabra | CC-2V | 1500 | 4 9/16 |
| 60CAC/F | 60 | CA-10 | Frost–bent-tip | Candelabra | CC-2V | 1500 | 4 9/16 |
| 60CAC/L | 60 | CA-10 | Clear–bent-tip | Candelabra | CC-2V | 4000 | 4 9/16 |
| 60G25 | 60 | G-25 | Clear-globe | Medium | CC-9 | 1500 | 4 9/16 |
| 60G25/W | 60 | G-25 | White-globe | Medium | CC-9 | 1500 | 4 9/16 |
| 60G30/W | 60 | G-30 | White-globe | Medium | CC-9 | 2500 | 5 |
| 60G40 | 60 | G-40 | Clear-globe | Medium | CC-6 | 2500 | 6 15/16 |
| 60G40/W | 60 | G-40 | White-globe | Medium | CC-6 | 2500 | 6 15/16 |
| 60G40/W/L | 60 | G-40 | White-globe | Medium | CC-6 | 4000 | 6 15/16 |
| 75G40 | 75 | G-40 | Clear-globe | Medium | CC-6 | 2500 | 6 15/16 |
| 75G40/W | 75 | G-40 | White-globe | Medium | CC-6 | 2500 | 6 15/16 |
| 100G40 | 100 | G-40 | Clear-globe | Medium | CC-6 | 2500 | 6 15/16 |
| 100G40/W | 100 | G-40 | White-globe | Medium | CC-6 | 2500 | 6 15/16 |
| 100G40/W/L | 100 | G-40 | White-globe | Medium | CC-6 | 4000 | 6 15/16 |
| 150G40/W | 150 | G-40 | White-globe | Medium | CC-6 | 2500 | 6 15/16 |

**90. Yellow Bug-Lite Incandescent Lamps, 120 V**
(General Electric Co.)

| Lamp-ordering code | Watts | Bulb | Base | Class | Filament | Approximate life, h | Approximate initial lumens | Maximum overall length, in |
|---|---|---|---|---|---|---|---|---|
| 40A/Y | 40 | A-19 | Medium | C | CC-6 | 1000 | 320 | 4 7/16 |
| 60A/Y | 60 | A-19 | Medium | C | CC-6 | 1000 | 550 | 4 7/16 |
| 100A/Y | 100 | A-19 | Medium | C | CC-8 | 1000 | 1010 | 4 7/16 |
| 150PS25/Y | 150 | PS-25 | Medium | C | C-9 | 1000 | 1560 | 6 15/16 |

## 91. Rough-Service and Vibration-Service Incandescent Lamps, 120 V
(General Electric Co.)

| Lamp-ordering code | Watts | Bulb | Description | Class | Base | Filament | Approximate life, h | Approximate initial lumens | Light-center length, in | Maximum overall length, in |
|---|---|---|---|---|---|---|---|---|---|---|
| 25A/RS | 25 | A-19 | Rough-service, inside-frosted | B | Medium | C-17 | 1000 | 190 | 2 9/16 | 3 7/8 |
| 50A/RS | 50 | A-19 | Rough-service, inside-frosted | B | Medium | C-22 | 1000 | 480 | 2 9/16 | 3 7/8 |
| 50A/RS/CVG | 50 | A-19 | Rough-service, inside-frosted, Teflon®-coated[a] | B | Medium | C-22 | 1000 | 480 | . . . | 3 7/8 |
| 50A19/5 | 50 | A-19 | Rough-service, inside-frosted | B | Medium | C-22 | 1000 | . . . . . | 2 9/16 | 3 7/8 |
| 50A/VS | 50 | A-19 | Vibration-service, inside-frosted | B | Medium | C-9 | 1000 | 550 | 2 3/8 | 3 7/8 |
| 75A/RS | 75 | A-21 | Rough-service, inside-frosted | C | Medium | C-17 | 1000 | 750 | 3 13/16 | 5 1/4 |
| 75A/RS/CVG | 75 | A-21 | Rough-service, inside-frosted, Teflon-coated | C | Medium | C-17 | 1000 | 740 | . . . | 5 1/4 |
| 100A/RS | 100 | A-21 | Rough-service, inside-frosted | C | Medium | C-17 | 1000 | 1260 | 3 13/16 | 5 1/4 |
| 100A/RS/CVG | 100 | A-21 | Rough-service, inside-frosted, Teflon-coated | C | Medium | C-17 | 1000 | 1230 | . . . | 5 1/4 |
| 100A23/VS | 100 | A-23 | Vibration-service, inside-frosted | C | Medium | C-9 | 1000 | 1340 | 4 7/16 | 5 15/16 |
| 150A21/RS | 150 | A-21 | Rough-service, inside-frosted | C | Medium | CC-17 | 1000 | . . . . . | 3 13/16 | 5 1/4 |
| 150/RS | 150 | PS-25 | Rough-service, inside-frosted | C | Medium | C-17 | 1000 | 2160 | 5 3/16 | 6 15/16 |
| 150/RS/CVG | 150 | PS-25 | Rough-service, inside-frosted, Teflon-coated | C | Medium | C-17 | 1000 | 2120 | . . . | 6 15/16 |
| 150/VS | 150 | PS-25 | Vibration-service, inside-frosted | C | Medium | C-9 | 1000 | 2400 | 5 3/16 | 6 15/16 |
| 200PS30/24 | 200 | PS-30 | Rough-service, clear | C | Medium | C-9 | 1000 | 3400 | 6 | 8 1/16 |
| 200PS30/23 | 200 | PS-30 | Rough-service, inside-frosted | C | Medium | C-9 | 1000 | 3400 | 6 | 8 1/16 |
| 200PS30/RS/CVG | 200 | PS-30 | Rough-service, inside-frosted, Teflon-coated | C | Medium | C-9 | 1000 | 3330 | . . . | 8 1/16 |
| 300/RS | 300 | PS-35 | Rough-service, inside-frosted | C | Mogul | C-9 | 1000 | 5340 | 7 | 9 3/8 |
| 500/RS | 500 | PS-40 | Rough-service, inside-frosted | C | Mogul | C-9 | 1000 | . . . . . | 7 | 9 3/4 |

[a]Teflon is a registered trademark of E. I. Du Pont de Nemours & Co.

## 92. White-Three-Way Incandescent Lamps, 120 V
(General Electric Co.)

| Lamp-ordering code | Watts | Bulb | Base | Filament | Approximate life, h | Approximate initial lumens | Light-center length, in | Maximum overall length, in |
|---|---|---|---|---|---|---|---|---|
| 15/150 | 15<br>135<br>150 | A-21 | Three-contact medium | C-2R<br>CC-8 | 1500<br>1200<br>1200 | 80<br>2330<br>2140 | 3⅝ | 5¾ |
| 50/185 | 50<br>135<br>185 | A-21 | Three-contact medium | C-2R<br>CC-8 | 1500<br>1200<br>1200 | 580<br>2330<br>2910 | 3⅝ | 4⁹⁄₁₆ |
| 30/100 | 30<br>70<br>100 | A-21 | Three-contact medium | C-2R<br>CC-8 | 1500<br>1200<br>1200 | 300<br>1035<br>1315 | ... | 5¾ |
| 50/150 | 50<br>100<br>150 | A-21 | Three-contact medium | C-2R<br>CC-8 | 1500<br>1200<br>1200 | 580<br>1640<br>2220 | 3⅝ | 5¾ |
| 50/250 | 50<br>200<br>250 | A-23 | Three-contact medium | C-2R<br>CC-8 | 1500<br>1200<br>1200 | 580<br>3660<br>4240 | 4⁹⁄₁₆ | 5¾ |
| 80/250A/RL/SW | 80<br>170<br>250 | A-23 | Three-contact medium | CC-8 | 1000<br>1000<br>1000 | 1100<br>2800<br>3900 | ... | 5¹³⁄₁₆ |
| 100/300 | 100<br>200<br>300 | PS-25 | Three-contact mogul | C-2R<br>CC-8 | 1500<br>1200<br>1200 | 1320<br>3300<br>4620 | ... | 6¹¹⁄₁₆ |

**93. Lumiline Incandescent Lamps, 120 V**

(General Electric Co.)

| Lamp-ordering code | Watts | Bulb | Base | Description | Class | Filament | Approximate life, h | Approximate initial lumens | Maximum overall length, in |
|---|---|---|---|---|---|---|---|---|---|
| L30 | 30 | T-8 | Disc | Clear | B | C-8 | 1500 | — | 17¾ |
| L30/W | 30 | T-8 | Disc | White | B | C-8 | 1500 | 210 | 17¾ |
| L40 | 40 | T-8 | Disc | Clear | B | C-8 | 1500 | — | 11¾ |
| L40/W | 40 | T-8 | Disc | White | B | C-8 | 1500 | 295 | 11¾ |
| L60 | 60 | T-8 | Disc | Clear | B | C-8 | 1500 | — | 17¾ |
| L60/W | 60 | T-8 | Disc | White | B | C-8 | 1500 | 450 | 17¾ |

## 94. Showcase Incandescent Lamps
(General Electric Co.)

| Lamp-ordering code | Watts | Bulb | Description | Volts | Base | Class | Filament | Approximate life, h | Approximate initial lumens | Maximum overall length, in |
|---|---|---|---|---|---|---|---|---|---|---|
| 25T6 1/2 | 25 | T-6 1/2 | Clear | 120 | Intermediate | B | C-8 | 1000 | 244 | 5% |
| 25T6 1/2 | 25 | T-6 1/2 | Clear | 130 | Intermediate | B | C-8 | 1000 | 244 | 5% |
| 25T6 1/2/F | 25 | T-6 1/2 | Inside-frosted | 120 | Intermediate | B | C-8 | 1000 | 240 | 5% |
| 25T6 1/2/F | 25 | T-6 1/2 | Inside-frosted | 130 | Intermediate | B | C-8 | 1000 | 240 | 5% |
| 25T10 | 25 | T-10 | Clear | 120 | Medium | B | C-8 | 1000 | 248 | 5% |
| 25T10 | 25 | T-10 | Clear | 130 | Medium | B | C-8 | 1000 | 248 | 5% |
| 25T10/F | 25 | T-10 | Inside-frosted | 120 | Medium | B | C-8 | 1000 | 244 | 5% |
| 25T10/F | 25 | T-10 | Inside-frosted | 130 | Medium | B | C-8 | 1000 | 244 | 5% |
| 40T8 | 40 | T-8 | Clear | 120 | Medium | B | C-23 | 1000 | 430 | 11% |
| 40T8 | 40 | T-8 | Clear | 130 | Medium | B | C-23 | 1000 | 430 | 11% |
| 40T8/F | 40 | T-8 | Inside-frosted | 120 | Medium | B | C-23 | 1000 | 425 | 11% |
| 40T10 | 40 | T-10 | Clear | 120 | Medium | B | C-8 | 1000 | 420 | 5% |
| 40T10 | 40 | T-10 | Clear | 130 | Medium | B | C-8 | 1000 | 420 | 5% |
| 40T10/F | 40 | T-10 | Inside-frosted | 120 | Medium | B | C-8 | 1000 | 415 | 5% |
| 40T10/F | 40 | T-10 | Inside-frosted | 130 | Medium | B | C-8 | 1000 | 415 | 5% |
| 60T10/F | 60 | T-10 | Inside-frosted | 120 | Medium | C | C-8 | 1000 | ... | 5% |
| 60T10/64 | 60 | T-10 | Clear | 120 | Medium | C | C-8 | 1000 | 745 | 5% |
| 75T10/1 | 75 | T-10 | Inside-frosted | 120 | Medium | B | C-23 | 1000 | ... | 11% |
| 75T10/45 | 75 | T-10 | Clear | 120 | Medium | B | C-23 | 1000 | ... | 11% |

## 95. Higher-Voltage Incandescent Lamps for General Lighting Service
(General Electric Co.)

| Lamp-ordering code | Watts | Bulb | Description | Volts | Base | Class | Filament | Approximate life, h | Approximate initial lumens | Light-center length, in | Maximum overall length, in |
|---|---|---|---|---|---|---|---|---|---|---|---|
| 15A | 15 | A-17 | Inside-frosted | 230 | Medium | B | C-7A | 1000 | 120 | $2\frac{3}{8}$ | $3\frac{5}{8}$ |
| 25A | 25 | A-19 | Inside-frosted | 230 | Medium | B | C-17A | 1000 | 220 | $2\frac{9}{16}$ | $3\frac{7}{8}$ |
| 25A | 25 | A-19 | Inside-frosted | 250 | Medium | B | C-17A | 1000 | 220 | $2\frac{9}{16}$ | $3\frac{7}{8}$ |
| 25A | 25 | A-19 | Inside-frosted | 300 | Medium | B | C-17 | 1000 | ..... | $2\frac{9}{16}$ | $3\frac{7}{8}$ |
| 50A | 50 | A-19 | Inside-frosted | 230 | Medium | B | C-17A | 1000 | 490 | $2\frac{9}{16}$ | $3\frac{7}{8}$ |
| 50A | 50 | A-19 | Inside-frosted | 250 | Medium | B | C-17A | 1000 | 190 | $2\frac{9}{16}$ | $3\frac{7}{8}$ |
| 50A/RS | 50 | A-19 | Inside-frosted, rough-service | 250 | Medium | B | C-22 | 1000 | 490 | $2\frac{9}{16}$ | $3\frac{7}{8}$ |
| 50A19 | 50 | A-19 | Inside-frosted | 300 | Medium | B | C-17A | 1000 | ..... | $2\frac{9}{16}$ | $3\frac{7}{8}$ |
| 50A19/35 | 50 | A-19 | Clear | 300 | Medium | B | C-17A | 1000 | ..... | $2\frac{9}{16}$ | $3\frac{7}{8}$ |
| 60A21 | 60 | A-21 | Inside-frosted | 230 | Medium | B | C-17A | 1000 | ..... | $2\frac{3}{4}$ | $4\frac{3}{8}$ |
| 75A21 | 75 | A-21 | Inside-frosted | 230 | Medium | C | C-7A | 1000 | ..... | $3\frac{13}{16}$ | $5\frac{1}{4}$ |
| 100A | 100 | A-21 | Inside-frosted | 230 | Medium | C | C-7A | 750 | 1280 | $3\frac{13}{16}$ | $5\frac{1}{4}$ |
| 100A | 100 | A-21 | Inside-frosted | 250 | Medium | C | C-7A | 750 | 1280 | $3\frac{13}{16}$ | $5\frac{1}{4}$ |
| 100A | 100 | A-21 | Inside-frosted | 277 | Medium | C | C-7A | 750 | ..... | $3\frac{13}{16}$ | $5\frac{1}{4}$ |
| 100A | 100 | A-21 | Inside-frosted | 300 | Medium | C | C-7A | 1000 | ..... | $3\frac{13}{16}$ | $5\frac{1}{4}$ |
| 100A/RS | 100 | A-21 | Inside-frosted, rough-service | 250 | Medium | C | C-17 | 1000 | 960 | $3\frac{13}{16}$ | $5\frac{1}{4}$ |
| 150PS25/IF | 150 | PS-25 | Inside-frosted | 250 | Medium | C | C-7A | 1000 | ..... | $5\frac{3}{16}$ | $6\frac{15}{16}$ |
| 200 | 200 | PS-30 | Clear | 250 | Medium | C | C-9 | 1000 | 2,980 | 6 | $8\frac{1}{16}$ |
| 200 | 200 | PS-30 | Clear | 300 | Medium | C | C-9 | 1000 | ..... | 6 | $8\frac{1}{16}$ |
| 200/IF | 200 | PS-30 | Inside-frosted | 250 | Medium | C | C-9 | 1000 | 2,980 | 6 | $8\frac{1}{16}$ |
| 200PS30/RS | 200 | PS-30 | Inside-frosted, rough-service | 250 | Medium | C | C-9 | 1000 | ..... | 6 | $8\frac{1}{16}$ |
| 300 | 300 | PS-35 | Clear | 277 | Mogul | C | C-7A | 1000 | ..... | 7 | $9\frac{3}{8}$ |
| 500 | 500 | PS-40 | Clear | 250 | Mogul | C | C-7A | 1000 | 9,270 | 7 | $9\frac{3}{4}$ |
| 500 | 500 | PS-40 | Clear | 277 | Mogul | C | C-7A | 1000 | ..... | 7 | $9\frac{3}{4}$ |
| 750 | 750 | PS-52 | Clear | 250 | Mogul | C | C-7A | 2000 | ..... | $9\frac{1}{2}$ | 13 |
| 1000 | 1000 | PS-52 | Clear | 250 | Mogul | C | C-7A | 2000 | 17,700 | $9\frac{1}{2}$ | 13 |
| 1000 | 1000 | PS-52 | Clear | 277 | Mogul | C | C-7A | 2000 | — | $9\frac{1}{2}$ | 13 |

## 96. PAR Reflector Incandescent Lamps
(General Electric Co.)

| Lamp-ordering code | Watts | Bulb | Base | Volts | Beam type | Approximate life, h | Approximate total lumens | Approximate center-beam candlepower | Approximate beam spread, °[a] | Maximum overall length, in |
|---|---|---|---|---|---|---|---|---|---|---|
| 55PAR/FL | 55 | PAR-38 | Medium skirted | 120 | Flood | 3000 | 560 | 1,210 | 37 | 5 5/16 |
| 55PAR/SP | 55 | PAR-38 | Medium skirted | 120 | Spot | 3000 | 560 | 3,840 | 17 | 5 5/16 |
| 75PAR/FL/65WM[b] | 65 | PAR-38 | Medium skirted | 120 | Flood | 2000 | 675 | 1,670 | 30 | 5 5/16 |
| 75PAR/SP/65WM[b] | 65 | PAR-38 | Medium skirted | 120 | Spot | 2000 | 675 | 5,620 | 14 | 5 5/16 |
| 75PAR/3FL/65WM[b] | 65 | PAR-38 | Medium side-prong | 120 | Flood | 2000 | 675 | 1,670 | 30 | 4 5/16 |
| 75PAR/3SP/65WM[b] | 65 | PAR-38 | Medium side-prong | 120 | Spot | 2000 | 675 | 5,620 | 14 | 4 5/16 |
| 75PAR38/FL | 75 | PAR-38 | Medium skirted | 120 | Flood | 2000 | 765 | 1,480 | 32 | 5 5/16 |
| 75PAR38/SP | 75 | PAR-38 | Medium skirted | 120 | Spot | 2000 | 765 | 3,900 | 16 | 5 5/16 |
| 100PAR/FL/85WM[b,c] | 85 | PAR-38 | Medium skirted | 120 | Flood | 2000 | 930 | 2,010 | 37 | 5 5/16 |
| 100PAR/SP/85WM[b] | 85 | PAR-38 | Medium skirted | 120 | Spot | 2000 | 930 | 6,450 | 15 | 5 5/16 |
| 150PAR/FL[d] | 150 | PAR-38 | Medium skirted | 120 | Flood | 2000 | 2,000 | 3,060 | 39 | 5 5/16 |
| 150PAR/SP | 150 | PAR-38 | Medium skirted | 120 | Spot | 2000 | 2,000 | 11,500 | 17 | 5 5/16 |
| 150PAR/WFL | 150 | PAR-38 | Medium skirted | 120 | Wide flood | 2000 | 1,740 | 1,730 | 54 | 5 5/16 |
| 150PAR/FL/120WM[b] | 120 | PAR-38 | Medium skirted | 120 | Flood | 2000 | 1,500 | 3,600 | 31 | 5 5/16 |
| 150PAR/SP/120WM[b] | 120 | PAR-38 | Medium skirted | 120 | Spot | 2000 | 1,500 | 7,730 | 18 | 5 5/16 |
| 150PAR/3FL/120WM[b] | 120 | PAR-38 | Medium side-prong | 120 | Flood | 2000 | 1,500 | 3,600 | 31 | 4 5/16 |
| 150PAR/3SP/120WM[b] | 120 | PAR-38 | Medium side-prong | 120 | Spot | 2000 | 1,500 | 7,730 | 18 | 4 5/16 |
| 150PAR46/3MFL | 150 | PAR-46 | Medium side-prong | 120 | Medium flood | 2000 | 1,500 |  |  | 4 |
| 150PAR46/3NSP | 150 | PAR-46 | Medium side-prong | 120 | Narrow spot | 2000 | 1,500 | 13,200 | 14×15 | 4 |
| 200PAR46/3NSP | 200 | PAR-46 | Medium side-prong | 120 | Narrow spot | 2000 | 2,270 | 21,700 | 13×14 | 4 |
| 200PAR46/3MFL | 200 | PAR-46 | Medium side-prong | 120 | Medium flood | 2000 | 2,270 | 7,100 | 20×31 | 4 |
| 200PAR56/MFL | 200 | PAR-56 | Mogul end-prong | 120 | Medium flood | 2000 |  |  |  | 5 |
| 300PAR56/NSP | 300 | PAR-56 | Mogul end-prong | 120 | Narrow spot | 2000 | 3,840 | 51,400 | 10×13 | 5 |
| 300PAR56/MFL | 300 | PAR-56 | Mogul end-prong | 120 | Medium flood | 2000 | 3,840 | 22,600 | 14×24 | 5 |
| 300PAR56/WFL | 300 | PAR-56 | Mogul end-prong | 120 | Wide flood | 2000 | 3,840 | 7,320 | 25×43 | 5 |
| 500PAR64/NSP | 500 | PAR-64 | Extended mogul end-prong | 120 | Narrow spot | 2000 | 6,500 | 94,300 | 9×13 | 5 |
| 500PAR64/MFL | 500 | PAR-64 | Extended mogul end-prong | 120 | Medium flood | 2000 | 6,500 | 26,500 | 15×28 | 6 |
| 500PAR64/WFL | 500 | PAR-64 | Extended mogul end-prong | 120 | Wide flood | 2000 | 6,500 | 15,300 | 21×40 | 6 |

## 96. PAR Reflector Incandescent Lamps (*Continued*)

| Lamp-ordering code | Watts | Bulb | Base | Volts | Beam type | Approximate life, h | Approximate total lumens | Approximate center-beam candlepower | Approximate beam spread, °[a] | Maximum overall length, in |
|---|---|---|---|---|---|---|---|---|---|---|
| Tungsten-halogen PAR lamps; inner filament tube | | | | | | | | | | |
| 50PAR20/NSP/H | 50 | PAR-20 | Medium | 120 | Narrow spot | 2000 | 560 | 4,500 | 13 | 3⅛ |
| 50PAR20/NFL/H | 50 | PAR-20 | Medium | 120 | Narrow flood | 2000 | 560 | 1,500 | 28 | 3⅛ |
| 50PAR30/FL/H | 50 | PAR-30 | Medium | 120 | Flood | 2000 | 670 | 1,450 | 38 | 3⅝ |
| 50PAR30/NFL/H | 50 | PAR-30 | Medium | 120 | Narrow flood | 2000 | 670 | 2,300 | 28 | 3⅝ |
| 50PAR30/NSP/H | 50 | PAR-30 | Medium | 120 | Narrow spot | 2000 | 670 | 6,800 | 13 | 3⅝ |
| 60PARFL/HIR[e] | 60 | PAR-38 | Medium skirted | 120 | Flood | 2000 | — | 3,200 | 32 | 55/16 |
| 60PARSP/HIR[e] | 60 | PAR-38 | Medium skirted | 120 | Spot | 2000 | — | 13,000 | 13 | 55/16 |
| 75PAR30/FL/H | 75 | PAR-30 | Medium | 120 | Flood | 2000 | 1,150 | 2,400 | 38 | 3⅝ |
| 75PAR30/NFL/H | 75 | PAR-30 | Medium | 120 | Narrow flood | 2000 | 1,150 | 3,500 | 28 | 3⅝ |
| 75PAR30/NSP/H | 75 | PAR-30 | Medium | 120 | Narrow spot | 2000 | 1,150 | 11,000 | 13 | 3⅝ |
| 100PARFL/HIR[e] | 100 | PAR-38 | Medium skirted | 120 | Flood | 2000 | — | 5,000 | 32 | 55/16 |
| 100PARSP/HIR[e] | 100 | PAR-38 | Medium skirted | 120 | Spot | 2000 | — | 25,000 | 13 | 55/16 |
| Q150PAR38/FL | 150 | PAR-38 | Medium skirted | 120 | Flood | 4000 | 2,000 | 4,260 | 34 | 55/16 |
| Q150PAR38/SP | 150 | PAR-38 | Medium skirted | 120 | Spot | 4000 | 2,000 | 12,300 | 14 | 55/16 |
| Q250PAR38/FL | 250 | PAR-38 | Medium skirted | 120 | Flood | 6000 | 3,500 | 6,500 | 39 | 55/16 |
| Q250PAR38/SP | 250 | PAR-38 | Medium skirted | 120 | Spot | 6000 | 3,500 | 15,800 | 17 | 55/16 |
| Q500PAR56MFL | 500 | PAR-56 | Mogul end-prong | 120 | Medium flood | 4000 | 8,000 | 41,100 | 14×29 | 5 |
| Q500PAR56NSP | 500 | PAR-56 | Mogul end-prong | 120 | Narrow spot | 4000 | 8,000 | 88,800 | 9×17 | 5 |
| Q500PAR56WFL | 500 | PAR-56 | Mogul end-prong | 120 | Wide flood | 4000 | 8,000 | 14,500 | 25×47 | 5 |
| Q1000PAR64MFL | 1000 | PAR-64 | Extended mogul end-prong | 120 | Medium flood | 4000 | 19,400 | 73,700 | 14×32 | 6 |
| Q1000PAR64WFL | 1000 | PAR-64 | Extended mogul end-prong | 120 | Wide flood | 4000 | 19,400 | 36,200 | 24×44 | 6 |
| Q1000PAR64NSP | 1000 | PAR-64 | Extended mogul end-prong | 120 | Narrow spot | 4000 | 19,400 | 157,000 | 10×19 | 6 |

[a]To 50% of maximum beam candlepower.
[b]Energy-saving type.
[c]Available in amber, blue, blue-white, green, pink, red, and yellow colors with a silicone filter.
[d]Available in amber, blue, green, red, and yellow colors with a dichroic filter.
[e]Infrared reflecting film on inner filament tube.

NOTE 75-W PAR-38 flood, 150-W PAR-38 spot and flood, and 300-W PAR-56 narrow-spot, medium flood, and wide flood are available with an infrared transmitting reflector to reduce the heat in the beam.

**97. Reflector-Bulb Incandescent Lamps**
(General Electric Co.)

| Lamp-ordering code | Watts | Bulb | Base | Volts | Beam type | Approximate life, h | Approximate total lumens | Approximate beam spread, °[a] | Maximum overall length, in |
|---|---|---|---|---|---|---|---|---|---|
| 25R14N | 25 | R-14 | Intermediate | 120 | Flood | 1500 | 190 | 67 | 2 9/16 |
| 27R20/ML | 27 | R-20 | Medium | 120 | Flood | 2000 | 190 | 43 | 3 15/16 |
| 30R20 | 30 | R-20 | Medium | 120 | Flood | 2000 | 210 | 43 | 3 15/16 |
| 50ER30 | 50 | ER-30 | Medium | 120 | . . . [b] | 2000 | 525 | 43 | 6 1/4 |
| 50R20 | 50 | R-20 | Medium | 120 | Flood | 2000 | 440 | 50 | 3 15/16 |
| 50R30/FL | 50 | R-30 | Medium | 120 | Flood | 2000 | 525 | 66 | 5 3/8 |
| 75R30/FL/65WM | 65 | R-30 | Medium | 120 | Flood | 2000 | 770 | 64 | 5 3/8 |
| 75R30/SP/65WM | 65 | R-30 | Medium | 120 | Spot | 2000 | 770 | 29 | 5 3/8 |
| 75ER30 | 75 | ER-30 | Medium | 120 | . . . [b] | 2000 | 850 | 42 | 6 1/4 |
| 75R30/SP | 75 | R-30 | Medium | 120 | Spot | 2000 | 900 | 31 | 5 3/8 |
| 75R30/FL | 75 | R-30 | Medium | 120 | Flood | 2000 | 900 | 72 | 5 3/8 |
| 75R/FL | 75 | R-40 | Medium | 120 | Flood | 2000 | 890 | 65 | 6 9/16 |
| 100R/FL | 100 | R-40 | Medium | 120 | Flood | 2000 | 1,190 | 66 | 6 9/16 |
| 100R/SP | 100 | R-40 | Medium | 120 | Spot | 2000 | 1,190 | 24 | 6 9/16 |
| 120ER40 | 120 | ER-40 | Medium | 120 | . . . [b] | 2000 | 1,475 | 45 | 7 3/8 |
| 150R/SP | 150 | R-40 | Medium | 120 | Spot | 2000 | 1,900 | 25 | 6 9/16 |
| 150R/FL | 150 | R-40 | Medium | 120 | Flood | 2000 | 1,900 | 59 | 6 9/16 |
| 150R/SP/120WM[c] | 120 | R-40 | Medium | 120 | Spot | 2000 | 1,600 | 21 | 6 9/16 |
| 150R/FL/120WM[c] | 120 | R-40 | Medium | 120 | Flood | 2000 | 1,600 | 55 | 6 9/16 |
| 300R/SP | 300 | R-40 | Medium | 120 | Spot | 2000 | 3,750 | . . | 6 9/16 |
| 300R/FL | 300 | R-40 | Medium | 120 | Flood | 2000 | 3,750 | 67 | 6 9/16 |
| 300R/3FL | 300 | R-40 | Mogul | 120 | Flood | 2000 | 3,750 | . . | 7 1/4 |
| 500R/3SP | 500 | R-40 | Mogul | 120 | Spot | 2000 | 6,500 | . . | 7 1/4 |
| 500R/3FL | 500 | R-40 | Mogul | 120 | Flood | 2000 | 6,500 | . . | 7 1/4 |
| 500R52 | 500 | R-52 | Mogul | 120 | . . . . | 2000 | 7,600 | . . | 11 3/4 |
| 750R52 | 750 | R-52 | Mogul | 120 | . . . . | 2000 | 13,000 | . . | 11 3/4 |

[a]To 50% of maxium beam candlepower.
[b]Elliptical reflector bulb.
[c]Energy-saving type.

**98. Sign Incandescent Lamps**
(General Electric Co.)

| Lamp-ordering code[a] | Watts | Bulb | Base | Volts | Approximate life, h | Light-center length, in | Maximum overall length, in | Available finishes[b] |
|---|---|---|---|---|---|---|---|---|
| 10S11N | 10 | S-11 | Intermediate | 130 | 1500 | $1\frac{5}{8}$ | $2\frac{5}{16}$ | Clear, CB, CG, CR, CW, CY, TB, TG, TR, TY |
| 10S14 | 10 | S-14 | Medium | 120 | 1500 | $2\frac{1}{2}$ | $3\frac{1}{2}$ | Clear, CB, CG, CO, CR, CW, CY |
| 10S14/IF | 10 | S-14 | Medium | 120 | 1500 | $2\frac{1}{2}$ | $3\frac{1}{2}$ | Inside-frosted |
| 11S14 | 11 | S-14 | Medium | 130 | 3000 | $2\frac{1}{2}$ | $3\frac{1}{2}$ | Clear, inside-frosted, B, G, O, R, W, Y, TB, TG, TO, TR, TY |
| 15A | 15 | A-15 | Medium | 120 | 2500 | . . . | $3\frac{1}{2}$ | B, G, O, R, Y |
| 15S14GR | 15 | S-14 | Medium | 130 | 3000 | $2\frac{1}{2}$ | $3\frac{1}{2}$ | Clear, inside-frosted |
| 25A15/RFL | 25 | A-15 | Medium | 130 | 3000 | $2\frac{3}{8}$ | $3\frac{5}{8}$ | Light, inside-frosted, reflector |
| 25A19/GR | 25 | A-19 | Medium | 130 | 3000 | $2\frac{3}{8}$ | $3\frac{7}{8}$ | Clear, inside-frosted, TB, TG, TO, TR, TY |
| 30A15 | 30 | A-15 | Medium | 130 | 5000 | $2\frac{3}{8}$ | $3\frac{1}{2}$ | Inside-frosted, changing message |
| 30A15/CL | 30 | A-15 | Medium | 130 | 5000 | $2\frac{3}{8}$ | $3\frac{1}{2}$ | Clear, changing message |
| 33A19/5 | 33 | A-19 | Medium | 130 | 3000 | . . . | $3\frac{7}{8}$ | Clear, changing message |
| 40A21/GR | 40 | A-21 | Medium | 130 | 3000 | $2\frac{15}{16}$ | $4\frac{3}{8}$ | Clear, inside-frosted, TB, TG, TR, TY |

[a]Basic lamp code; finish not included.
[b]Colors are blue (B), green (G), red (R), yellow (Y), orange (O), and white (W). Transparent colors are indicated with a T prefix to the color designation. Ceramic colors are indicated with a C prefix to the color designation.

## 99. Incandescent Lamps for Display, Architectural, and Theatrical Spotlights

(General Electric Co.)

| Lamp-ordering code | ANSI code | Watts | Bulb | Base | Volts | Burning position[a] | Filament | Approximate life, h | Approximate initial lumens | Light-center length, in | Maximum overall length, in |
|---|---|---|---|---|---|---|---|---|---|---|---|
| 100G16½/29DC | .... | 100 | G-16½ | Double-contact bayonet | 120 | BDTH | CC-13 | 200 | 1,660 | 1⅝ | 3 |
| 250G/FL | .... | 250 | G-30 | Medium screw | 120 | BDTH | C-5 | 800 | 3,650 | 3 | 5⅛ |
| 250G/SP | .... | 250 | G-30 | Medium screw | 120 | BDTH | C-5 | 200 | 4,500 | 3 | 5⅛ |
| 400G/FL | .... | 400 | G-30 | Medium screw | 120 | BDTH | C-5 | 800 | 6,800 | 3 | 5⅛ |
| 400G/SP | .... | 400 | G-30 | Medium screw | 120 | BDTH | C-5 | 200 | 8,400 | 3 | 5⅛ |
| 500G/FL | .... | 500 | G-40 | Mogul screw | 120 | BDTH | C-5 | 800 | 9,300 | 4¼ | 7¹⁄₁₆ |
| 500T12/8 | DEB | 500 | T-12 | Medium prefocus | 120 | BU30 | C-13D | 800 | 9,000 | 3½ | 6⅛ |
| 500T12/9 | DNS | 500 | T-12 | Medium prefocus | 120 | BU30 | C-13D | 200 | 11,000 | 8½ | 6⅛ |
| 500T14/7 | .... | 500 | T-14 | Medium bipost | 120 | BU30 | C-13D[b] | 800 | 9,000 | 4 | 6⁷⁄₁₆ |
| 500T20/64 | DNW | 500 | T-20 | Medium prefocus | 120 | BDTH | C-13 | 500 | 10,000 | 2³⁄₁₆ | 5¾ |
| 750T12/9 | DNT | 750 | T-12 | Medium prefocus | 120 | BU30 | C-13D | 200 | 17,000 | 3½ | 6⅛ |
| 750T14 | .... | 750 | T-14 | Medium bipost | 120 | BU30 | C-13D[b] | 200 | 17,000 | 4 | 6⁷⁄₁₆ |
| 750T20P/SP | BFE | 750 | T-20 | Medium prefocus | 120 | BD45 | C-13 | 200 | 17,000 | 2³⁄₁₆ | 5¾ |
| 1MG40FL | .... | 1000 | G-40 | Mogul screw | 120 | BDTH | C-5 | 800 | 20,000 | 5¼ | 7⅞ |
| 1500G48/6 | .... | 1500 | G-48 | Mogul screw | 120 | BDTH | C-5 | 800 | 32,200 | 5¼ | 8⁵⁄₁₆ |
| **Tungsten-halogen lamps** | | | | | | | | | | | |
| Q500CL/P | EGE | 500 | T-4 | Medium prefocus | 120 | Any | CC-8[b] | 2000 | 10,450 | 3½ | 6 |
| Q500CL/TP | EHD | 500 | T-4 | Medium two-pin | 120 | Any | CC-8[b] | 2000 | 10,400 | 2⅜ | 4 |
| Q500T6/CL/P | BTL | 500 | T-6 | Medium prefocus | 120 | Any | C-13D[b] | 500 | 11,000 | 2³⁄₁₆ | 4½ |
| Q750CL/P | EGG | 750 | T-6 | Medium prefocus | 120 | Any | CC-8[b] | 2000 | 15,750 | 3½ | 6 |
| Q750CL/TP | EHG | 750 | T-6 | Medium two-pin | 120 | Any | CC-8[b] | 2000 | 15,400 | 2⅜ | 4⅛ |
| Q750/4CL | EHF | 750 | T-6 | Medium two-pin | 120 | Any | CC-8[b] | 300 | 20,400 | 2⅜ | 4⅛ |
| Q750/4CL/P | EGF | 750 | T-6 | Medium prefocus | 120 | Any | CC-8[b] | 500 | 20,400 | 3½ | 6 |
| Q750T7/CL/2P | BTN | 750 | T-7 | Medium prefocus | 120 | Any | C-13D[b] | 500 | 17,600 | 2³⁄₁₆ | 4½ |
| Q1000/4CL | FEL | 1000 | T-6 | Medium two-pin | 120 | Any | CC-8[b] | 300 | 27,500 | 2⅜ | 4⅛ |
| Q1000CL/P | EGM | 1000 | T-6 | Medium prefocus | 120 | Any | CC-8[b] | 2000 | 21,500 | 3½ | 6 |
| Q1000T7/CL/MP | BVT | 1000 | T-7 | Mogul prefocus | 120 | Any | C-13D[b] | 500 | 24,500 | 3¹⁵⁄₁₆ | 7¼ |

[a] BDTH, base down to horizontal; BU30, within 30° of vertical base up; BD45, within 45° of vertical base down.
[b] Low-noise construction.

**100. Tungsten-Halogen Incandescent Lamps for General-Lighting Service**
(General Electric Co.)

| Lamp-ordering code | Watts | Bulb | Base | Volts | Filament | Approximate life, h | Approximate initial lumens | Lighted length or light-center length[a] | Maximum overall length, in |
|---|---|---|---|---|---|---|---|---|---|
| Q75CL | 75 | T-3 | Miniature screw | 28 | CC-8 | 2000 | 1,400 | $1\frac{3}{16}^{\circ}$ | $2\frac{1}{2}$ |
| Q100MC[b] | 100 | T-4 | Minicandelabra | 120 | CC-8 | 2000 | 1,550 | $1\frac{3}{8}$ | $2\frac{13}{16}$ |
| Q100CL/MC | 100 | T-4 | Minicandelabra | 120 | CC-8 | 2000 | 1,600 | $1\frac{3}{8}$ | $2\frac{13}{16}$ |
| Q150CL/DC | 150 | T-4 | Double-contact bayonet | 120 | CC-8 | 2000 | 2,800 | $1\frac{3}{8}^{\circ}$ | $2\frac{7}{16}$ |
| Q150DC[b] | 150 | T-4 | Double-contact bayonet | 120 | CC-8 | 2000 | 2,700 | $1\frac{3}{8}^{\circ}$ | $2\frac{7}{16}$ |
| Q150MC[b] | 150 | T-4 | Minicandelabra | 120 | CC-8 | 2000 | 2,700 | $1\frac{3}{8}^{\circ}$ | $2\frac{13}{16}$ |
| Q150CL/MC | 150 | T-4 | Minicandelabra | 120 | CC-8 | 2000 | 2,800 | $1\frac{3}{8}^{\circ}$ | $2\frac{13}{16}$ |
| Q200T3/CL | 200 | T-3 | Recessed single-contact | 120 | CC-8 | 1500 | 3,460 | $\frac{15}{16}$ | $3\frac{1}{8}$ |
| Q200T3[b] | 200 | T-3 | Recessed single-contact | 120 | CC-8 | 1500 | 3,350 | $\frac{15}{16}$ | $3\frac{1}{8}$ |
| Q250CL/DC | 250 | T-4 | Double-contact bayonet | 120 | CC-8 | 2000 | 5,000 | $1\frac{5}{8}^{\circ}$ | 3 |
| Q250CL/MC | 250 | T-4 | Minicandelabra | 120 | CC-8 | 2000 | 5,000 | $1\frac{5}{8}^{\circ}$ | $3\frac{5}{32}$ |
| Q250DC[b] | 250 | T-4 | Double-contact bayonet | 120 | CC-8 | 2000 | 4,850 | $1\frac{5}{8}^{\circ}$ | 3 |
| Q250MC[b] | 250 | T-4 | Minicandelabra | 120 | CC-8 | 2000 | 4,850 | $1\frac{5}{8}^{\circ}$ | $3\frac{5}{32}$ |
| Q300T3[b] | 300 | T-3 | Recessed single-contact | 120 | C-8 | 2000 | 5,900 | $2\frac{5}{16}$ | $4\frac{11}{16}$ |
| Q300T3/CL | 300 | T-3 | Recessed single-contact | 120 | C-8 | 2000 | 5,950 | $2\frac{5}{16}$ | $4\frac{11}{16}$ |
| Q300T4/CL | 300 | T-4 | Recessed single-contact | 120 | CC-8 | 2000 | 5,650 | $\frac{11}{16}$ | $3\frac{1}{8}$ |
| Q400MC[b] | 400 | T-4 | Minicandelabra | 120 | CC-8 | 2000 | 7,850 | $2^{\circ}$ | $3\frac{5}{8}$ |
| Q400CL/MC | 400 | T-4 | Minicandelabra | 120 | CC-8 | 2000 | 8,250 | $2^{\circ}$ | $3\frac{5}{8}$ |
| Q400T4/CL | 400 | T-4 | Recessed single-contact | 120 | CC-8 | 2000 | 7,750 | $\frac{13}{16}$ | $3\frac{1}{8}$ |
| Q425T3/CL | 425 | T-3 | Recessed single-contact | 120 | C-8 | 2000 | 8,900 | $2\frac{7}{32}$ | $4\frac{11}{16}$ |

| | Watts | Bulb | Base | Volts | Filament | Rated life, h | Lumens | Dim. A, in | Dim. B, in |
|---|---|---|---|---|---|---|---|---|---|
| Q500/350WM[c] | 350 | T-3 | Recessed single-contact | 120 | C-8 | 2000 | 10,000 | 2 3/16 | 4 11/16 |
| Q500CL/DC | 500 | T-4 | Double-contact bayonet | 120 | CC-8 | 2000 | 10,450 | 2 1/8* | 3 5/8 |
| Q500DC[b] | 500 | T-4 | Double-contact bayonet | 120 | CC-8 | 2000 | 10,100 | 2 1/8* | 3 5/8 |
| Q500T3[b] | 500 | T-3 | Recessed single-contact | 120 | C-8 | 2000 | 10,300 | 2 1/4 | 4 11/16 |
| Q500T3/CL | 500 | T-3 | Recessed single-contact | 120 | C-8 | 2000 | 10,450 | 2 1/4 | 4 11/16 |
| Q500T3/CL/6[d] | 500 | T-3 | Recessed single-contact | 120 | C-8 | 1500 | 10,950 | 2 3/8 | 4 11/16 |
| Q1000T3/CL | 1000 | T-3 | Recessed single-contact | 220 | C-8 | 2000 | 21,500 | 6 1/8 | 10 5/16 |
| Q1000T3/CL | 1000 | T-3 | Recessed single-contact | 240 | C-8 | 2000 | 21,500 | 6 1/8 | 10 5/16 |
| Q1000T6/CL | 1000 | T-6 | Recessed single-contact | 120 | CC-8 | 2000 | 23,400 | 1 | 5 5/8 |
| Q1500/ 900WM[c] | 900 | T-3 | Recessed single-contact | 240 | C-8 | 2000 | 32,000 | 6 5/16 | 10 5/16 |
| Q1500/ 900WM[c] | 900 | T-3 | Recessed single-contact | 277 | C-8 | 2000 | 31,000 | 6 5/16 | 10 5/16 |
| Q1500T3/CL | 1500 | T-3 | Recessed single-contact | 208 | C-8 | 2000 | 35,800 | 6 5/16 | 10 5/16 |
| Q1500T3/CL | 1500 | T-3 | Recessed single-contact | 220 | C-8 | 2000 | 35,800 | 6 5/16 | 10 5/16 |
| Q1500T3/CL | 1500 | T-3 | Recessed single-contact | 240 | C-8 | 2000 | 35,800 | 6 5/16 | 10 5/16 |
| Q1500T3/CL | 1500 | T-3 | Recessed single-contact | 277 | C-8 | 2000 | 34,400 | 6 5/16 | 10 5/16 |
| Q1500T3/CL/6[d] | 1500 | T-3 | Recessed single-contact | 240 | C-8 | 1000 | 35,800 | 6 3/4 | 10 5/16 |

[a] Length shown with an asterisk is light-center length.
[b] Frosted design.
[c] Energy-saving design with dichroic coating on quartz tube.
[d] Rough-service design.

**101. Low-Voltage Display Incandescent Lamps**
(General Electric Co.)

| Lamp-ordering code | Watts | Bulb | Base | Volts | Beam type | Approximate life, h | Approximate center-beam candlepower | Approximate beam spread,[a] | Maximum overall length, in |
|---|---|---|---|---|---|---|---|---|---|
| 25PAR36 | 25 | PAR-36 | Screw terminal | 5.5 | Very narrow spot | 1000 | 24,000 | 4.6×5.4 | 2¾ |
| 25PAR36NSP | 25 | PAR-36 | Screw terminal | 12 | Narrow spot | 2000 | 24,000 | 7.5×9.3 | 2¾ |
| 25PAR36WFL | 25 | PAR-36 | Screw terminal | 12 | Wide flood | 2000 | 4,000 | 25×37 | 2¾ |
| 25PAR36VWFL | 25 | PAR-36 | Screw terminal | 12 | Very wide flood | 2000 | 158 | 54×56 | 2¾ |
| 25PAR46 | 25 | PAR-46 | Screw terminal | 5.5 | Very narrow spot | 1000 | 18,000 | 5×6 | 3¾ |
| 50PAR36NSP | 50 | PAR-36 | Screw terminal | 12 | Narrow spot | 2000 | 8,800 | 9.3×10 | 2¾ |
| 50PAR36VNSP | 50 | PAR-36 | Screw terminal | 12 | Very narrow spot | 2000 | 19,300 | 6×9 | 2¾ |
| 50PAR36WFL | 50 | PAR-36 | Screw terminal | 12 | Wide flood | 2000 | 870 | 27×39 | 2¾ |
| 50PAR36WFL/4 | 50 | PAR-36 | Screw terminal | 12 | Wide flood | 4000 | 620 | 30×40 | 2¾ |
| 50PAR36VWFL | 50 | PAR-36 | Screw terminal | 12 | Very wide flood | 2000 | 235 | 69×69 | 2¾ |
| 120PAR56VNSP | 120 | PAR-56 | Screw terminal | 12 | Very narrow spot | 2000 | 53,400 | 6.8×9.3 | 4½ |
| 120PAR56MFL | 120 | PAR-56 | Screw terminal | 12 | Medium flood | 2000 | 15,500 | 12×20 | 4½ |
| 120PAR56WFL | 120 | PAR-56 | Screw terminal | 12 | Wide flood | 2000 | 3,080 | 25×44 | 4½ |
| 240PAR56VNSP | 240 | PAR-56 | Screw terminal | 12 | Very narrow spot | 2000 | 74,200 | 7.5×10 | 4½ |
| 240PAR56MFL | 240 | PAR-56 | Screw terminal | 12 | Medium flood | 2000 | 36,500 | 14×20 | 4½ |
| 240PAR56WFL | 240 | PAR-56 | Screw terminal | 12 | Wide flood | 2000 | 7,000 | 27×45 | 4½ |

**Faceted 2-in reflector; infrared transmitting reflector**

| | | | | | | | | | |
|---|---|---|---|---|---|---|---|---|---|
| Q20MR16/FL(BAB)[b] | 20 | MR-16 | Two-pin | 12 | Flood | 4000 | 525 | 40 | 1% |
| Q20MR16/NSP (ESX)[b] | 20 | MR-16 | Two-pin | 12 | Narrow spot | 3000 | 3,850 | 13 | 1% |
| Q20MR16/VNSP (EZX) | 20 | MR-16 | Two-pin | 12 | Very narrow spot | 3000 | 8,200 | 7 | 1% |
| Q42MR16/NFL (EYS) | 42 | MR-16 | Two-pin | 12 | Narrow flood | 3500 | 2,100 | 27 | 1% |
| Q42MR16/VNSP (EZY) | 42 | MR-16 | Two-pin | 12 | Very narrow spot | 3500 | 13,100 | 9 | 1% |
| Q50MR16/FL (EXN)[b] | 50 | MR-16 | Two-pin | 12 | Flood | 4000 | 1,725 | 40 | 1% |
| Q50MR16/NSP (EXT)[b] | 50 | MR-16 | Two-pin | 12 | Narrow spot | 4000 | 10,200 | 14 | 1% |
| Q50MR16/NFL (EXZ)[b] | 50 | MR-16 | Two-pin | 12 | Narrow flood | 4000 | 2,900 | 27 | 1% |
| Q50MR16/NFL/1 (FNY)[b] | 50 | MR-16 | Two-pin | 12 | Narrow flood | 4000 | 2,450 | 32 | 1% |
| Q75MR16/FL (EYC)[b] | 75 | MR-16 | Two-pin | 12 | Flood | 4000 | 2,100 | 42 | 1% |
| Q75MR16/NSP (EYF)[b] | 75 | MR-16 | Two-pin | 12 | Narrow spot | 4000 | 12,300 | 14 | 1% |
| Q75MR16/NFL (EYJ)[b] | 75 | MR-16 | Two-pin | 12 | Narrow flood | 4000 | 4,600 | 25 | 1% |

**Faceted 1%-in reflector; infrared transmitting reflector**

| | | | | | | | | | |
|---|---|---|---|---|---|---|---|---|---|
| Q20MR11/NSP | 20 | MR-11 | Two-pin | 12 | Narrow spot | 2000 | 3,900 | 10 | 1% |
| Q20MR11/SP | 20 | MR-11 | Two-pin | 12 | Spot | 2000 | 1,550 | 20 | 1% |
| Q20MR11/NFL | 20 | MR-11 | Two-pin | 12 | Narrow flood | 2000 | 600 | 30 | 1% |
| Q35MR11/NSP | 35 | MR-11 | Two-pin | 12 | Narrow spot | 2000 | 5,850 | 10 | 1% |
| Q35MR11/SP | 35 | MR-11 | Two-pin | 12 | Spot | 2000 | 2,750 | 20 | 1% |
| Q35MR11/NFL | 35 | MR-11 | Two-pin | 12 | Narrow flood | 2000 | 1,300 | 30 | 1% |

[a] To 50% of center-beam candlepower.
[b] Also available with an integral cover glass.

**102. Street-Lighting Incandescent Lamps**
(General Electric Co.)

| Lamp-ordering code | Nominal lumens | Watts | Bulb | Base | Volts or amperes[a] | Class | Filament | Approximate life, h | Light-center length, in | Maximum overall length, in |
|---|---|---|---|---|---|---|---|---|---|---|
| Multiple street-lighting lamps | | | | | | | | | | |
| 92A23/49 | 1,000 | 92 | A-23 | Medium | 120 | C | C-9 | 3000 | 4 7/16 | 5 15/16 |
| 189PS25/64 | 2,500 | 189 | PS-25 | Medium | 120 | C | C-9 | 3000 | 5 1/4 | 6 15/16 |
| 189PS25/64 | 2,500 | 189 | PS-25 | Medium | 130 | C | C-9 | 3000 | 5 1/4 | 6 15/16 |
| 202PS25 | 2,500 | 202 | PS-25 | Medium | 125 | C | C-9 | 6000 | 5 1/4 | 6 15/16 |
| 295PS35/58 | 4,000 | 295 | PS-35 | Mogul | 120 | C | C-9 | 3000 | 7 | 9 3/8 |
| 295PS35/58 | 4,000 | 295 | PS-35 | Mogul | 125 | C | C-9 | 3000 | 7 | 9 3/8 |
| 327PS35 | 4,000 | 327 | PS-35 | Mogul | 120 | C | C-9 | 6000 | 7 | 9 3/8 |
| 327PS35 | 4,000 | 327 | PS-35 | Mogul | 125 | C | C-9 | 6000 | 7 | 9 3/8 |
| 405PS40/54 | 6,000 | 405 | PS-40 | Mogul | 120 | C | C-9 | 3000 | 7 | 9 3/4 |
| 448PS40 | 6,000 | 448 | PS-40 | Mogul | 125 | C | C-9 | 6000 | 7 | 9 3/4 |
| Series street-lighting lamps | | | | | | | | | | |
| 2500/66R | 2,500 | . . . | PS-35 | Mogul | 6.6 | C | C-2V | 3000 | 7 | 9 3/8 |
| 2500/66G | 2,500 | . . . | PS-35 | Mogul | 6.6 | C | C-2V | 6000 | 7 | 9 3/8 |
| 4M/66G | 4,000 | . . . | PS-35 | Mogul | 6.6 | C | C-2V | 6000 | 7 | 9 3/8 |
| 4M/66R | 4,000 | . . . | PS-35 | Mogul | 6.6 | C | C-2V | 3000 | 7 | 9 3/8 |
| 6M/66 | 6,000 | . . . | PS-40 | Mogul | 6.6 | C | C-2V | 2000 | 7 | 9 3/4 |
| 6M/66G | 6,000 | . . . | PS-40 | Mogul | 6.6 | C | C-2V | 6000 | 7 | 9 3/4 |

[a]Multiple lamps are shown with design voltage. Series lamps are shown with design current in amperes.
[b]Burn base up only.

## FLUORESCENT LAMPS

**103. The fluorescent lamp** (Fig. 10-38) is an electronic device. It functions through conduction in a gas. The lamp consists of a long straight or circular tube containing a drop of mercury and a small amount of argon gas with electrodes sealed into each end. Both electrodes are so constructed that they can function as cathodes (emitters of electrons into the enclosure). The lamp will therefore conduct in either direction and pass alternating cur-

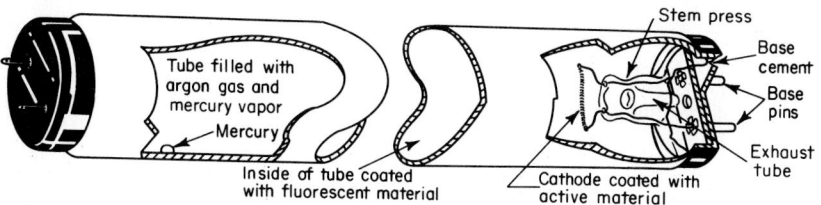

**Fig. 10-38**   Fluorescent-lamp construction. [General Electric Co.]

rent. The inside surface of the tube is coated with a fluorescent chemical (Sec. 106). Fluorescent lamps can be constructed with hot or cold cathodes (see Fig. 10-39). Hot cathodes are constructed of coated tungsten filaments wound in coil form. Cold cathodes are made of iron, shaped in thimblelike form to give a large emitting area. Cold cathodes require a higher voltage drop across the lamp for the production of the emission. Lamps with cold cathodes will consequently not have such high efficiency, i.e., lumens per watt. However, because of their inherently long life, cold-cathode lamps are advantageous for special custom-built shapes and patterns and for applications in which it is difficult to replace the lamp. Most general-purpose lamps are operated with hot cathodes. Some of the general-purpose lamps are started with cold cathodes by applying a sufficiently high voltage to produce electric-field emission. However, after the lamp has conducted for only a fraction of a second, the impinging arc will heat a few of the segments of the cathode filament wire to a red-hot temperature. The cathode then produces thermionic emission, and the lamp continues to function under hot-cathode operation.

**104. Operation of fluorescent lamps.** Although all general-purpose fluorescent lamps operate with hot cathodes, they are started with either hot-cathode or cold-cathode action. Hot-cathode starting is called *preheat* or *rapid-start starting,* and cold-cathode starting is called *instant starting,* since the lamps light almost instantaneously upon the closing of the lamp circuit. The preheat-start lamps are usually constructed with bipin bases at each end, as shown in Fig. 10-38. The instant-start lamps are constructed with either bipin or single-pin bases at each end. In either construction the filament of instant-start lamps is short-circuited in the base. The only purpose in making some of the instant-start lamps with bipin bases is that these lamps can be used in the conventional type of bipin lampholder which is employed for preheat-start lamps. However, the bipin instant-start lamps will not operate on the same ballast circuits as those employed for preheat or rapid-start lamps.

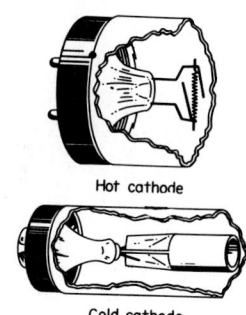

Hot cathode

Cold cathode

**Fig. 10-39**   Cathodes for fluorescent lamps.

The principles of operation of a preheat-start fluorescent lamp can be understood from a study of the schematic diagram of Fig. 10-40. This figure shows the simplest connections that it is possible to employ for operation of these lamps. A reactor, connected in series with the lamp, is always required for starting and stabilizing the operation. Switch S is an automatic starting switch, or starter, as it is called. This switch is in its closed position when no current is passing through the circuit. When the circuit switch is closed, the two filaments are connected in series through switch S to the supply voltage. No current will flow through the lamp between the filaments, since this path is short-circuited by switch

S and also the argon gas is not in a conducting state when the filaments are cold. Current will flow through the filaments, heating them to the proper temperature and vaporizing the mercury. After a few seconds, switch S automatically opens, breaking the circuit and causing the reactor to produce a high voltage between the two filaments. This strikes an arc through the argon gas and the mercury vapor, the purpose of the argon being to facilitate the starting of the arc. Refer to Sec. 126 for the operating characteristics of various types of starters. The mercury-vapor arc produces a large amount of ultraviolet-light radiation, which impinges on the fluorescent chemicals on the inside of the tube, causing them to fluoresce and emit a brilliant light, the color of which may be different tones of white or colors such as pink, gold, green, or blue, depending upon the chemical used (see Table 106). By combining chemicals, a white light can be obtained.

The simplest circuit that is possible to employ for an instant-start lamp is shown in Fig. 10-41. A ballast reactor is always required, as with the preheat-start lamp. Since the lamp starts with cold cathodes, no starter switch is required. However, the circuit must be so

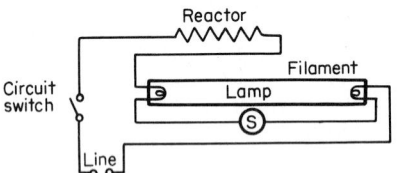

**Fig. 10-40**   Circuit for fluorescent lamp and simple reactor.

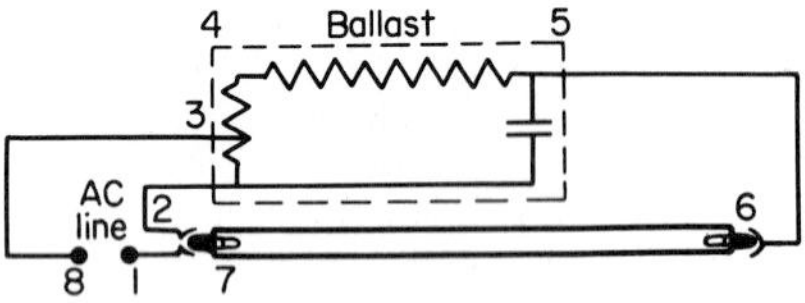

**Fig. 10-41**   Circuit for instant-start fluorescent lamp.

arranged that (1) a high voltage will be impressed across the lamp when the lamp circuit is closed and (2) the voltage across the lamp is reduced to its normal operating value as soon as conduction takes place and the lamp is started. The simple circuit of Fig. 10-41 will perform these two functions. At the instant that the circuit to the lamp is closed, there is no conduction through the lamp and terminals 6 and 7 are open-circuited. The reactor 2-3-4 will act as a step-up autotransformer and produce from 6 to 7 a voltage of a magnitude considerably greater than the line voltage from 1 to 8. This high voltage will strike an arc through the argon gas and start conduction. As soon as the lamp conducts, the self-inductance of the ballast reactor 4-5 reduces the voltage across the lamp to normal operating voltage. A fraction of a second after the arc has been struck, the filaments are heated sufficiently by the arc to produce thermionic emission. The production of visible light in the instant-start lamp is the same as that described in the preceding paragraph for the preheat-start lamp.

**105. There are five principal methods of operation.**   Some lamps may be operated by only one method; others may be operated by more than one. The following describes these five operating methods.

1. Preheat or Switch Starting (with Starters or Manual Starting Switches)   If fluorescent-lamp cathodes are preheated before the lamp is started, relatively inexpensive ballasts can be used. Such preheating is readily accomplished by means of manual switches (used in desk lamps and portable lamps) or by automatic starters (with which fixtures are controlled from a wall switch). Starters are available in either standard or No-Blink types. The latter are obtainable in either the manual reset (Watch-Dog) or automatic reset designs and in the range of sizes needed for the different lamps. The Watch-Dog is recommended in most instances because it eliminates flashing or blinking at the end of lamp life, saves ballast wear, and lasts much longer.

2. Trigger Start (No Starters)   This method permits operation of some smaller preheat-start fluorescent lamps without starters, yet gives practically instant starting. Although lamp life is a little shorter and thus lamp cost a little higher, maintenance is greatly simplified and convenience of use much improved. No special lamp is required, but the lighting fixture must be equipped with the proper size of trigger-start ballast. This automatically provides cathode preheat without starters. Trigger-start ballasts are cur-

rently available for 14-, 15-, and 20-W general-line fluorescent lamps and for 6-, 8-, and 12-in Circline lamps.

3. RAPID START (NO STARTERS)   This system, used with rapid-start, high-output, 1500-mA T-10, T-12, or PG-17 lamps, combines the simplicity of trigger start with the low cost of conventional switch starting. It requires the use of special low-loss cathodes to reduce cathode heating losses and is coated with a silicone film to assure rapid starting even under adverse conditions. Rapid-start lamps will give good performance in fixtures employing glow-type starters. The lamps should be used with rapid-start ballasts designed to provide adequate preheat automatically with low losses. Lamps glow as soon as they are turned on and come up to uniform full brightness in approximately 2 s.

While lamp and ballast prices are slightly higher, these are offset by elimination of the starter and starter-maintenance costs. Rapid-start lamps are available in the 30- and 40-W T-12 size, 16-in Circline sizes, high-output, and 1500-mA T-10, T-12, and noncircular PG-17 lamps designed for greater current to secure higher light output.

4. INSTANT START (NO STARTERS)   Through the use of higher-voltage ballasts, these lamps can be started without preheat. They are equipped with cathodes that afford in general the same long life obtained from the popular sizes of general-line switch-start lamps. While they look just like switch-start lamps of the same wattage, instant-start lamps are not electrically interchangeable with them, for the cathode leads are short-circuited inside the lamp base to ensure safety in use. Therefore, instant-start lamps cannot be preheated in starter-type circuits. Furthermore, general-line lamps should not be used on instant-start ballasts; otherwise, much shorter lamp life will result.

Instant-start lamps are available in 40-W T-12 and 40-W T-17 sizes.

5. SLIMLINE (INSTANT START WITHOUT STARTERS)   Slimline lamps combine all the advantages of the instant-start lamp with much greater convenience in handling and easier maintenance. The lamps are equipped with extra-strong single-pin bases that fit easily and solidly in rugged push-pull sockets. This combination makes lamp installation fast and easy.

In the 8-ft (2.4-m) sizes, slimlines are among the most efficient lamps made. In addition to increased efficiency, the longer length reduces the number of lamps and fixtures required in a given installation. This, together with the elimination of starters, reduces the amount of maintenance required in a fluorescent lighting system.

These advantages, together with the long trouble-free life offered by slimline lamps, assure continuing growth in popularity.

Slimline fluorescent lamps are available in 42- and 64-in (1067- and 1626-mm) lengths in the T-6 bulb size, in 72- and 96-in (1829- and 2438-mm) lengths in T-8, and in 24-in (610-mm) to 96-in (2438-mm) lengths in the most popular T-12 diameter.

**106.  Data on Fluorescent Phosphors[a]**
(In addition to the phosphor, manganese is usually present as an activator.)

| Phosphor | Lamp color | Wavelength, nm | | | |
| --- | --- | --- | --- | --- | --- |
| | | Exciting range | Sensitivity peak | Emitted range | Emitted peak |
| Barium silicate | Black light | 180–280 | 200–240 | 310–400 | 346 |
| Barium-strontium-magnesium silicate | Black light | 180–280 | 200–250 | 310–450 | 360 |
| Cadmium borate | Pink | 220–360 | 250 | 520–750 | 615 |
| Calcium halophosphate | White | 180–320 | 250 | 350–750 | 580 |
| Calcium tungstate | Blue | 220–300 | 272 | 310–700 | 440 |
| Magnesium tungstate | Blue-white | 220–320 | 285 | 360–720 | 480 |
| Strontium halophosphate | Greenish blue | 180–300 | 230 | 400–700 | 500 |
| Strontium ortho phosphate | Orange | 180–320 | 210 | 450–750 | 610 |
| Yttrium oxide | Orange | 180–300 | 220–280 | 550–650 | 611 |
| Zinc silicate | Green | 220–296 | 253.7 | 460–640 | 525 |

[a]All values are given in nanometers. The nanometer, used to measure wavelength of radiation, is 1/100,000,000 m in length.

**107. Fluorescent-lamp colors.**   Fluorescent lamps are available in a range of strong colors and in several different whites. The saturated colors—red, pink, gold, green, and blue—are used for decorative effects, while the whites serve for both decorative and general lighting purposes. All fluorescent lamps except gold and red are white when unlighted. Different phosphors produce the different colors when lamps are lighted.

White fluorescent lamps are designed to combine three elements important in lighting effects: (1) efficiency—most light per dollar, (2) color-rendering properties—the ability to bring out the beauty of colored materials and objects, and (3) whiteness—their appearance in relation to either natural outdoor daylight or the traditional artificial illumination such as filament lamps.

The choice among fluorescent whites always involves compromise among these three elements. Obtaining the best color-rendering properties necessitates reduction in efficiency. Choice of whiteness affects both efficiency and color-rendering properties. The descriptions below outline the effects obtained from the most popular whites.

Cool white combines high efficiency with reasonably good color rendition. It is the most widely used fluorescent-lamp color in factories, offices, and schools. It blends well with natural daylight.

Warm white provides the highest efficiency in white fluorescent lamps. It emphasizes orange, yellow, and yellow green at the expense of other colors. It is generally used where highest efficiency is more important than color rendition.

Deluxe cool white most closely simulates the appearance and color-rendering properties of natural daylight. It is widely used in stores such as supermarkets, florist shops, menswear shops, and other places where excellent color rendition of natural daylight is needed. It is also used in factory and office installations in which best appearance of colors is important.

Deluxe warm white simulates the warm, friendly effects of filament lighting in both whiteness and color rendering. It is usually first choice in residences, restaurants, beauty parlors, department stores, bakeries, and other places where homelike lighting effects are wanted. For types used in the home, deluxe warm white is also called soft white.

Deluxe cool and warm lamps provide about 30 percent less light than regular cool and warm white lamps. New phosphors have been developed which provide excellent color rendition in warm and cool tones with no loss in light output compared with cool white and warm white lamps.

Special fluorescent-lamp colors have been developed for applications such as outdoor plastic signs, color checking and matching in the textile and printing industries, the lighting of aquariums, and the growing of indoor plants.

Daylight, soft white, and white are still available for replacement purposes in existing installations and for new installations in which their appearance or color-rendering properties are particularly suitable.

**108. Bases for fluorescent lamps** (See Fig. 10–42). Lamps incorporating preheat or rapid-start cathodes require four electrical contacts, which in the standard line of lamps take the form of a bipin base at either end. There are three standard types of bipin bases: miniature bipin, medium bipin, and mogul bipin. In Circline lamps the contacts are brought

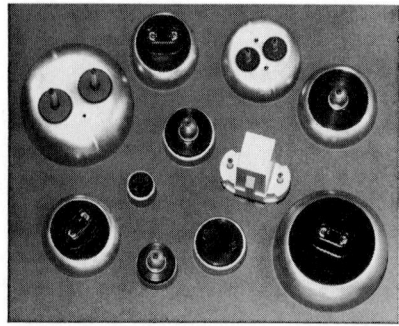

**Fig. 10-42**   Bases for fluorescent lamps. [General Electric Co.]

together in a four-pin base located between the two cathodes where the ends of the lamp adjoin. The high-output rapid-start lamps have recessed double-contact bases.

Lamps of the instant-start type require only two contacts, slimline lamps having single-pin bases. The 40-W instant-start hot-cathode lamp has a medium-bipin base but contains an electrical bridging member between the pair of contacts at each end, producing in effect a single contact for each cathode. Because of this construction, these lamps cannot be operated in circuits containing starters. Cold-cathode lamps are also instant-start lamps and have a single contact at each end, usually in the form of a cap base.

### 109. Types of visible-light fluorescent lamps

1. General-line
   - a. T-5 lamps (for use with starters)
   - b. T-8 lamps (for use with starters)
   - c. T-12 lamps (for use with starters)
   - d. T-12 lamp—instant-start (no starter used)
   - e. T-17 lamps (for use with starters)
   - f. T-17 lamp—instant-start (no starter used)
2. Rapid-start (no starter used)
3. High-output—rapid-start (no starter used)
4. 1500-mA rapid-start (no starter used)
5. Slimline—instant-start
6. Circline—rapid-start (no starter used)
7. Compact fluorescent (internal starter)

### 110. General-line lamps.
The 4-, 6-, 8-, and 13-W T-5 fluorescent lamps are generally used when space for lamps is limited and where the inherent cool light and color quality of fluorescent lamps are desired. In stores they are applied in niches, showcases, and shelving to enhance the function and appearance of miniature displays, signs, and models. In industrial plants they supply light locally for machine work, fine assembly, inspection, and other supplementary lighting applications. In offices they are built into business machines and similar devices for increased visibility of dials, scales, and keyboards.

The 14-W T-8, 14-W T-12, and 15-W T-12 lamps are used in homes for kitchens and mirror lighting. They are also used in stores for showcases, niches, and signs, in industrial plants for local lighting at work stations, and in portable lamps. The 15-W T-12 has a lower brightness than the 15-W T-8 owing to its larger diameter and is preferred if used without shielding.

The 20-W T-12 lamp is one of the most widely used fluorescent lamps. It is employed in home fixtures for lighting in kitchens, bathrooms, basements, and recreation rooms. It is used in window valances and under shelving and cupboards for decorative and utilitarian lighting. It can be used to light closets, washrooms, and small areas. It is also employed for supplementary lighting in offices and factories. In stores it lights fitting mirrors, niches, and wall-case displays. It can be operated by trigger-start ballasts.

Several sizes of T-8 and T-12 lamps are available in lengths from 22 to 33 in (559 to 838 mm) for use in appliance and home applications. The 33-in 25-W T-12 is the longest lamp which can be generated from 120 V ac with a simple reactor ballast.

The 30-W T-8 [36 in (914 mm) long] is used in stores for showcase, wall-case, and perimeter lighting where its small diameter provides opportunity for good light control.

The 40-W T-12 rapid-start lamp is used today in place of the former 40-W T-12 preheat lamp. The 40-W T-12 and the 40-W T-17 instant-start lamp and 90-W T-17 preheat lamp are used only for replacement in older installations. Newer fluorescent systems provide lower lighting costs in applications for these lamps.

### 111. Rapid-start 40-W T-12 fluorescent lamps
are the most widely used fluorescent lamps. They simplify lighting maintenance for the user and give, in effect, instant starting at costs comparable to those of the 40-W preheat lamp. Starters are eliminated from the electrical circuit. This is accomplished with a cathode design in the lamp somewhat different from that of the preheat lamp and with a ballast having low-voltage windings which apply heating to the cathodes at starting and during operation. Rated lamp life and light output are the same as for the preheat.

Forty-W rapid-start lamps with increased light output and efficiency are available. These lamps are also available as U-shaped lamps in two sizes, 3½-in (89-mm) leg spacing and 6-in (152-mm) leg spacing. Both 40-W increased-output and U-shaped lamps operate on standard 40-W ballasts.

The 30-W T-12 rapid-start lamp is used in applications where a 36-in (914-mm) lamp is required to fit the dimensions of the installation.

DIMMING   The 40-W T-12 rapid-start lamps can be dimmed from full brightness to nearly full blackout by means of ballasts and dimmers especially designed for this service.

FLASHING   Special ballasts similar to the dimming ballasts, but providing somewhat higher cathode-heating current, have been designed for flashing rapid-start and high-output lamps.

**112. High-output rapid-start lamps.**   The high-output line of T-12 lamps (24- to 96-in, or 610- to 2438-mm) operates at 800 to 1000 mA. Since the lamps are of rapid-start design, two electrical contacts are required at each base. The recessed double-contact base was developed to meet this requirement and, at the same time, to eliminate any hazard from electrical shock.

The high-output rapid-start lamp gives about 40 percent more light than the 96-in (2438-mm) T-12 slimline or 40-W lamp. Because of the higher current load and thus higher

**Fig. 10-43**   1500-mA fluorescent lamps. [General Electric Co.]

bulb-wall temperature, this lamp performs best in ventilated fixtures. Typical open-top fixtures that allow substantial amounts of upward light provide excellent ventilation. Efficient surface-mounted and recessed fixtures have also been developed.

Because of the higher bulb-wall temperature, high-output lamps perform better in low-temperature applications than 430-mA lamps. They are widely used in outdoor plastic-sign applications.

The highest-output fluorescent lamps operate at 1500 mA. These lamps are available in T-10, T-12, and a noncircular PG-17 bulb (Fig. 10-43). The PG-17 is the most efficient 1500-mA lamp for indoor applications. This lamp has a U- or crescent-shaped cross section. The exciting ultraviolet radiation produced within the bulb has a shorter distance to travel before striking the fluorescent material or phosphor than it would have from the center of a corresponding bulb of circular cross section. The full benefit of the greater amount of ultraviolet radiation generated is obtained with the U-shaped construction. Less opportunity is provided for reabsorption of this radiation by the mercury vapor before it strikes the phosphor. The rails along the grooves serve to keep the mercury pressure inside the bulb near the optimum value by providing cool spots, which condense out excessive mercury vapor. The bridges between the grooves assure adequate bulb strength.

The distribution of light from this lamp differs from that of conventional sources. The total light in the 0 to 90° zone is equal to the total in the 90 to 180° zone, but in the 0 to 30° and 150 to 180° zones the relative light output is approximately 12 percent more than that from a tubular cross section of equal total light output. In most fixture designs this directional effect can be used to advantage.

Like the high-output lamp, 1500-mA lamps maintain their light output well at low temperatures. Again, enclosed fixtures will provide maximum output in most low-temperature applications.

For exposed lamps, outdoor applications, or indoor applications such as refrigerated rooms for meat storage, jacketed T-10 and T-12 1500-mA lamps are available.

**113. Slimline lamps** are a family of fluorescent lamps which are instant-starting, with single-pin bases. In general, they have advantages in wiring installation and maintenance over other fluorescent lamps. The rugged single-pin base combined with push-pull insertion in sockets has become very popular. Slimline lamps are available in a range of lengths and diameters to fit general and supplementary lighting needs in nearly all fields of application.

The 96T12 is the most prominent slimline lamp for general-lighting service. The maintenance advantages of the 8-ft (2.4-m) lamp result from fewer parts in the lighting systems than preheat types require for the same amount of light—lamps, sockets, ballasts, starters, etc., being considered.

The 72T12 and 48T12 are 6- and 4-ft (1.8- and 1.2-m) companions, respectively, of the 96T12 which permit finishing out the ends of continuous rows in which the 8-ft lamp is used. These lamps are also used for general lighting when shorter fixture lengths are desired for scale or to fit an architectural module. The 24-, 36-, 42-, 60-, 64-, and 84-in (585-, 914-, 1067-, 1524-, 1626-, and 2134-mm) T-12 lamps are used primarily in outdoor plastic-sign applications.

The 96T8 and 72T8 are used primarily as replacements in existing installations. The smaller diameter of these lamps makes them suited for use in coves or other restricted areas.

The 42T6 and 64T6 are designed to fit standard 4- and 6-ft (1.2- and 1.8-m) store showcases. Their ¾-in diameter means minimum visual obstruction for all types of displays. The smaller diameter also permits accurate control with polished, concentrating reflectors for wall-case, show-window, cove, wall, or mural lighting and many other specialized applications.

**114. Reduced-wattage fluorescent lamps** in most of the widely used types and sizes have been introduced to minimize energy consumption. Depending on the lamp type, lamp wattage is reduced by 13 to 20 percent with minimal loss in light output. These lamps use krypton in place of argon as the fill gas, which reduces the voltage across the lamps. Reduced-wattage lamps are used on the same ballasts as standard lamps. They are available for use in lighting fixtures intended for 30-W and 40-W rapid-start, 90-W T-17 preheat, 96-in (2438-mm) T-8 slimline, 48-in (1219-mm) and 96-in T-12 slimline, 48-in and 96-in high-output, and 48-in and 96-in 1500-mA fluorescent lamps.

Reduced-wattage lamps are intended for use where lamp ambient temperatures are 60°F (16°C) or higher. Lamp flickering may occur where the lamp ambient temperature is below 60°F or where strong air drafts blow directly on bare bulbs.

The 40-W rapid-start lamps are available with an internal switch which removes the cathode-heating current during lamp operation after the lamp has started to reduce energy use further.

**115. Circline lamps.** Fluorescent lamps are also made in the form of a circle. These are known as Circline lamps. Circline fluorescent lamps (Fig. 10-44) are now available in four

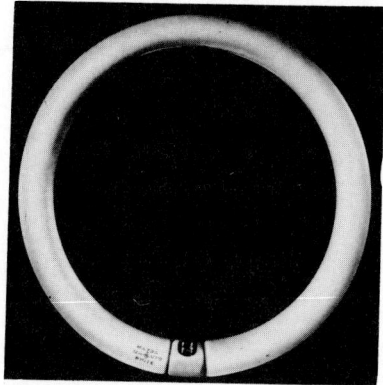

**Fig. 10-44**  Circline fluorescent lamp. [General Electric Co.]

diameters, 6, 8, 12, and 16 in. They are used in home lighting fixtures and portable lamps. They are also used for decorative lighting in restaurants, theaters, lobbies, lounges, and other commercial areas. They are adapted for some inspection processes in industry. The 6-, 8-, and 12-in-diameter lamps have a preheat type of cathode and can be operated on trigger-start ballasts. The 16-in-diameter lamp is a rapid-start lamp. See Table 125 for lamp data.

Adapters for Circline lamps, including a ballast, which permit their installation in incandescent medium-base lampholders are available. They are used in portable lamps as well as ceiling sockets for incandescent lamps.

**116. Compact fluorescent lamps.** The newest types of fluorescent lamps are compact lamps ranging in wattage from 5 to 40 W. Utilizing T-4 or T-5 tubing, the lamps are formed in a biaxial shape with a single-end base. In addition, other types employ a double-

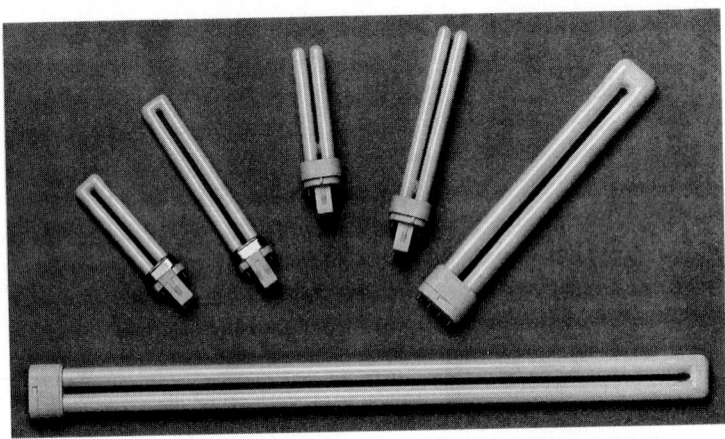

**Fig. 10-45** Compact fluorescent lamps. [General Electric Co.]

biaxial design with four tubular sections. This results in a more compact design for a specific lamp wattage. The 5- to 13-W biaxial lamps and 13- to 26-W double-biaxial lamps are preheat designs with internal starters in the lamp base. The higher lumen biaxial lamps are rapid-start types. All compact fluorescent lamps use new rare-earth phosphors which provide both high lamp efficacy and excellent color rendition.

Adapters with built-in ballasts are available for the installation of the 7- to 13-W biaxial and 13-W double-biaxial lamps in incandescent fixtures. When replacing incandescent lamps with compact fluorescent lamps and adaptors, care must be taken to ensure that the optical and thermal characteristics of the fixture will result in satisfactory performance.

Higher wattage rapid-start compact lamps provide the opportunity for efficient and smaller fluorescent luminaires for general lighting applications. See Sec. 126 for design specifications.

## 117. T-5 Fluorescent Lamps
(General Electric Co.)
(For use with starters)

| Lamp-ordering code | Nominal lamp watts | Bulb | Length, in[a] | Base | Description | Approximate life, h[b] | Approximate initial lumens[c] | Approximate % lumens at 40% rated average life |
|---|---|---|---|---|---|---|---|---|
| F4T5/CW | 4 | T-5 | 6 | Miniature bipin | Cool white | 6000 | 135 | 67 |
| F4T5/D | 4 | T-5 | 6 | Miniature bipin | Daylight | 6000 | 115 | 67 |
| F6T5/CW | 6 | T-5 | 9 | Miniature bipin | Cool white | 7500 | 295 | 75 |
| F6T5/CWX | 6 | T-5 | 9 | Miniature bipin | Deluxe cool white | 7500 | 205 | 75 |
| F6T5/D | 6 | T-5 | 9 | Miniature bipin | Daylight | 7500 | 230 | 75 |
| F8T5/CW | 8 | T-5 | 12 | Miniature bipin | Cool white | 7500 | 400 | 75 |
| F8T5/WW | 8 | T-5 | 12 | Miniature bipin | Warm white | 7500 | 410 | 75 |
| F8T5/CWX | 8 | T-5 | 12 | Miniature bipin | Deluxe cool white | 7500 | 275 | 75 |
| F8T5/D | 8 | T-5 | 12 | Miniature bipin | Daylight | 7500 | 330 | 75 |
| F13T5/CW | 13 | T-5 | 21 | Miniature bipin | Cool white | 7500 | 850 | 75 |

[a]Including standard lampholders.
[b]Life when operated at 3 h per start on ballasts meeting published specifications.
[c]After 100 h of operation.

## 118. T-8 Fluorescent Lamps
(General Electric Co.)
(For use with starters)

| Lamp-ordering code | Nominal lamp watts | Bulb | Length, in[a] | Base | Description | Approximate life, h[b] | Approximate initial lumens[c] | Approximate % lumens at 40% rated average life |
|---|---|---|---|---|---|---|---|---|
| F13T8/CW | 13 | T-8 | 12 | Medium bipin | Cool white | 7500 | 575 | 85 |
| F14T8/CW | 14 | T-8 | 15 | Medium bipin | Cool white | 7500 | 700 | 85 |
| F14T8/D | 14 | T-8 | 18 | Medium bipin | Daylight | 7500 | 575 | 85 |
| F15T8/CW | 15 | T-8 | 18 | Medium bipin | Cool white | 7500 | 825 | 88 |
| F15T8/WW | 15 | T-8 | 18 | Medium bipin | Warm white | 7500 | 860 | 88 |
| F15T8/SW | 15 | T-8 | 18 | Medium bipin | Soft white (deluxe warm white) | 7500 | 580 | 85 |
| F15T8/CWX | 15 | T-8 | 18 | Medium bipin | Deluxe cool white | 7500 | 610 | 85 |
| F15T8/D | 15 | T-8 | 18 | Medium bipin | Daylight | 7500 | 700 | 88 |
| F15T8/W | 15 | T-8 | 18 | Medium bipin | White | 7500 | 850 | 88 |
| F15T8/C50 | 15 | T-8 | 18 | Medium bipin | Chroma-50 | 7500 | 620 | 85 |
| F15T8/N | 15 | T-8 | 18 | Medium bipin | Natural | 7500 | 580 | 85 |
| F15T8/PL/AQ | 15 | T-8 | 18 | Medium bipin | Plant and aquarium | 7500 | 510 | — |

| | | | | | | | |
|---|---|---|---|---|---|---|---|
| F15T8/PL | 15 | T-8 | 18 | Medium bipin | Plant lite | 7500 | 215 | — |
| F15T8/R | 15 | T-8 | 18 | Medium bipin | Red | 7500 | 40 | — |
| F30T8/CW | 30 | T-8 | 36 | Medium bipin | Cool white | 7500 | 2175 | 91 |
| F30T8/WW | 30 | T-8 | 36 | Medium bipin | Warm white | 7500 | 2250 | 91 |
| F30T8/SW | 30 | T-8 | 36 | Medium bipin | Soft white (deluxe warm white) | 7500 | 1500 | 88 |
| F30T8/CWX | 30 | T-8 | 36 | Medium bipin | Deluxe cool white | 7500 | 1550 | 88 |
| F30T8/D | 30 | T-8 | 36 | Medium bipin | Daylight | 7500 | 1850 | 91 |
| F30T8/N | 30 | T-8 | 36 | Medium bipin | Natural | 7500 | 1500 | 88 |
| F30T8/PL | 30 | T-8 | 36 | Medium bipin | Plant lite | 7500 | 575 | — |
| F22″T8/D/4 | 18 | T-8 | 22 | Medium bipin | Daylight | 7500 | 925 | 88 |
| F24″T8/CW/4 | 18 | T-8 | 24 | Medium bipin | Cool white | 7500 | 1250 | 88 |
| F26″T8/CW/4 | 19 | T-8 | 26 | Medium bipin | Cool white | 7500 | 1300 | 88 |
| F28″T8/CW/4 | 19 | T-8 | 28 | Medium bipin | Cool white | 7500 | 1375 | 88 |
| F30″T8/CW/4 | 19 | T-8 | 30 | Medium bipin | Cool white | 7500 | 1400 | 88 |

[a] Including standard lampholders.
[b] Life when operated at 3 h per start on ballasts meeting published specifications.
[c] After 100 h of operation.

## 119. T-12 Fluorescent Lamps
(General Electric Co.)
(For use with starters)

| Lamp-ordering code | Nominal lamp watts | Bulb | Length, in[a] | Base | Description | Approximate life, h[b] | Approximate initial lumens[c] | Approximate % lumens at 40% rated average life |
|---|---|---|---|---|---|---|---|---|
| F14T12/CW | 14 | T-12 | 15 | Medium bipin | Cool white | 9000 | 650 | 88 |
| F14T12/SW | 14 | T-12 | 15 | Medium bipin | Soft white (deluxe warm white) | 9000 | 460 | 88 |
| F14T12/D | 14 | T-12 | 15 | Medium bipin | Daylight | 9000 | 575 | 88 |
| F15T12/CW | 15 | T-12 | 18 | Medium bipin | Cool white | 9000 | 760 | 88 |
| F15T12/WW | 15 | T-12 | 18 | Medium bipin | Warm white | 9000 | 780 | 88 |
| F15T12/SW | 15 | T-12 | 18 | Medium bipin | Soft white (deluxe warm white) | 9000 | 525 | 88 |
| F15T12/D | 15 | T-12 | 18 | Medium bipin | Daylight | 9000 | 640 | 88 |
| F20T12/CW | 20 | T-12 | 24 | Medium bipin | Cool white | 9000 | 1200 | 92 |
| F20T12/CW/1 | 20 | T-12 | 24 | Medium bipin | Cool white; dc operation | 7500 | 960 | 92 |
| F20T12/WW | 20 | T-12 | 24 | Medium bipin | Warm white | 9000 | 1250 | 92 |
| F20T12/SP30 | 20 | T-12 | 24 | Medium bipin | SP30-3000K | 9000 | 1275 | 94 |
| F20T12/SP35 | 20 | T-12 | 24 | Medium bipin | SP35-3500K | 9000 | 1275 | 94 |
| F20T12/SP41 | 20 | T-12 | 24 | Medium bipin | SP41-4100K | 9000 | 1275 | 94 |
| F20T12/SPX30 | 20 | T-12 | 24 | Medium bipin | SPX30-3000K | 9000 | 1300 | 94 |
| F20T12/SPX35 | 20 | T-12 | 24 | Medium bipin | SPX35-3500K | 9000 | 1300 | 94 |

| Lamp | | Bulb | | Base | Color | Life | Lumens | CRI |
|---|---|---|---|---|---|---|---|---|
| F20T12/SW | 20 | T-12 | 24 | Medium bipin | Soft white | 9000 | 875 | 90 |
| F20T12/CWX | 20 | T-12 | 24 | Medium bipin | Deluxe cool white | 9000 | 890 | 90 |
| F20T12/WWX | 20 | T-12 | 24 | Medium bipin | Deluxe warm white | 9000 | 875 | 90 |
| F20T12/D | 20 | T-12 | 24 | Medium bipin | Daylight | 9000 | 1025 | 82 |
| F20T12/W | 20 | T-12 | 24 | Medium bipin | White | 9000 | 1250 | 92 |
| F20T12/C50 | 20 | T-12 | 24 | Medium bipin | Chroma 50 | 9000 | 875 | 90 |
| F20T12/N | 20 | T-12 | 24 | Medium bipin | Natural | 9000 | 850 | 90 |
| F20T12/B | 20 | T-12 | 24 | Medium bipin | Blue | 9000 | 450 | — |
| F20T12/G | 20 | T-12 | 24 | Medium bipin | Green | 9000 | 1575 | — |
| F20T12/GO | 20 | T-12 | 24 | Medium bipin | Gold | 9000 | 1000 | — |
| F20T12/R | 20 | T-12 | 24 | Medium bipin | Red | 9000 | 80 | — |
| F20T12/CW/26 | 20 | T-12 | 26 | Medium bipin | Cool white | 7500 | 1200 | 92 |
| F25T12/CW/28 | 25 | T-12 | 28 | Medium bipin | Cool white | 7500 | 1700 | 92 |
| F25T12/D/28 | 25 | T-12 | 28 | Medium bipin | Daylight | 7500 | 1450 | 92 |
| F30″T12/CW | 21 | T-12 | 30 | Medium bipin | Cool white | 7500 | 1350 | 92 |
| F25T12/CW/33 | 25 | T-12 | 33 | Medium bipin | Cool white | 7500 | 1900 | 92 |
| F25T12/WW/33 | 25 | T-12 | 33 | Medium bipin | Warm white | 7500 | 1950 | 92 |
| F25T12/D/33 | 25 | T-12 | 33 | Medium bipin | Daylight | 7500 | 1600 | 92 |

[a] Including standard lampholders.
[b] Life when operated at 3 h per start on ballasts meeting published specifications.
[c] After 100 h of operation.

**120. T-17 Fluorescent Lamps**
(General Electric Co.)

| Lamp-ordering code | Nominal lamp watts | Bulb | Length, in[a] | Base | Description | Approximate life, h[b] | Approximate initial lumens[c] | Approximate % lumens at 40% rated average life |
|---|---|---|---|---|---|---|---|---|
| | | | | Instant-start lamp; no starters used | | | | |
| F40T17/CW/IS | 40 | T-17 | 60 | Mogul bipin | Cool white | 7500 | 2850 | 90 |
| | | | | For use with starters | | | | |
| F90T17/CW | 90 | T-17 | 60 | Mogul bipin | Cool white | 9000 | 6000 | 90 |
| F90T17/CW/WM[d] | 82 | T-17 | 60 | Mogul bipin | Cool white | 9000 | 5750 | 90 |
| F90T17/D | 90 | T-17 | 60 | Mogul bipin | Daylight | 9000 | 5150 | 90 |

[a]Including standard lampholders.
[b]Life when operated at 3 h per start on ballasts meeting published specifications.
[c]After 100 h of operation.
[d]Energy-saving lamp.

**121. T-12 Rapid-Start Fluorescent Lamps**
(General Electric Co.)

| Lamp-ordering code | Nominal lamp watts | Bulb | Length, in[a] | Base | Description | Approximate life, h[b] | Approximate initial lumens[c] | Approximate % lumens at 40% rated average life |
|---|---|---|---|---|---|---|---|---|
| | | | | 36-in fluorescent lamps[d] | | | | |
| F30T12/CW/RS | 30 | T-12 | 36 | Medium bipin | Cool white | 18,000 | 2225 | 86 |
| F30T12/WW/RS | 30 | T-12 | 36 | Medium bipin | Warm white | 18,000 | 2275 | 86 |
| F30T12/SP30/RS | 30 | T-12 | 36 | Medium bipin | SP30-3000K | 18,000 | 2375 | 89 |
| F30T12/SP35/RS | 30 | T-12 | 36 | Medium bipin | SP35-3500K | 18,000 | 2350 | 89 |
| F30T12/SP41/RS | 30 | T-12 | 36 | Medium bipin | SP41-4100K | 18,000 | 2350 | 89 |
| F30T12/SPX30/RS | 30 | T-12 | 36 | Medium bipin | SPX30-3000K | 18,000 | 2375 | 89 |
| F30T12/SPX35/RS | 30 | T-12 | 36 | Medium bipin | SPX35-3500K | 18,000 | 2375 | 89 |
| F30T12/CWX/RS | 30 | T-12 | 36 | Medium bipin | Deluxe cool white | 18,000 | 1600 | 82 |
| F30T12/WWX/RS | 30 | T-12 | 36 | Medium bipin | Deluxe warm white | 18,000 | 1575 | 82 |
| F30T12/D/RS | 30 | T-12 | 36 | Medium bipin | Daylight | 18,000 | 1900 | 86 |
| F30T12/W/RS | 30 | T-12 | 36 | Medium bipin | White | 18,000 | 2250 | 86 |
| F30T12/C50/RS | 30 | T-12 | 36 | Medium bipin | Chroma 50 | 18,000 | 1650 | 82 |
| | | | | Energy-saving 36-in lamps | | | | |
| F30T12/CW/RS/WM | 25 | T-12 | 36 | Medium bipin | Cool white | 18,000 | 1925 | 86 |
| F30T12/WW/RS/WM | 25 | T-12 | 36 | Medium bipin | Warm white | 18,000 | 1975 | 86 |
| F30T12/SP30/RS/WM | 25 | T-12 | 36 | Medium bipin | SP30-3000K | 18,000 | 2025 | 89 |
| F30T12/SP35/RS/WM | 25 | T-12 | 36 | Medium bipin | SP35-3500K | 18,000 | 2025 | 89 |
| F30T12/SP41/RS/WM | 25 | T-12 | 36 | Medium bipin | SP41-4100K | 18,000 | 2025 | 89 |
| | | | | 48-in fluorescent lamps[d] | | | | |
| F40CW | 40 | T-12 | 48 | Medium bipin | Cool white | 20,000+ | 3050 | 88 |
| F40WW | 40 | T-12 | 48 | Medium bipin | Warm white | 20,000+ | 3150 | 88 |
| F40SP30 | 40 | T-12 | 48 | Medium bipin | SP30-3000K | 20,000+ | 3200 | 90 |
| F40SP35 | 40 | T-12 | 48 | Medium bipin | SP35-3500K | 20,000+ | 3200 | 90 |
| F40SP41 | 40 | T-12 | 48 | Medium bipin | SP41-4100K | 20,000+ | 3200 | 90 |

**121. T-12 Rapid-Start Fluorescent Lamps** (*Continued*)

| Lamp-ordering code | Nominal lamp watts | Bulb | Length, in[a] | Base | Description | Approximate life, h[b] | Approximate initial lumens[c] | Approximate % lumens at 40% rated average life |
|---|---|---|---|---|---|---|---|---|
| | | | | | 48-in fluorescent lamps[d] (*Continued*) | | | |
| F40SPX30 | 40 | T-12 | 48 | Medium bipin | SPX30-3000K | 20,000+ | 3250 | 90 |
| F40SPX35 | 40 | T-12 | 48 | Medium bipin | SPX35-3500K | 20,000+ | 3250 | 90 |
| F40SPX41 | 40 | T-12 | 48 | Medium bipin | SPX41-4100K | 20,000+ | 3250 | 90 |
| F40SW | 40 | T-12 | 48 | Medium bipin | Soft white (deluxe warm white) | 20,000+ | 2200 | 82 |
| F40CWX | 40 | T-12 | 48 | Medium bipin | Deluxe cool white | 20,000+ | 2250 | 82 |
| F40WWX | 40 | T-12 | 48 | Medium bipin | Deluxe warm white | 20,000+ | 2200 | 82 |
| F40D | 40 | T-12 | 48 | Medium bipin | Daylight | 20,000+ | 2600 | 88 |
| F40W | 40 | T-12 | 48 | Medium bipin | White | 20,000+ | 3050 | 88 |
| F40SGN | 40 | T-12 | 48 | Medium bipin | Sign white | 20,000+ | 2300 | 82 |
| F40/C50 | 40 | T-12 | 48 | Medium bipin | Chroma 50 | 20,000+ | 2250 | 82 |
| F40/C75 | 40 | T-12 | 48 | Medium bipin | Chroma 75 | 20,000+ | 1950 | 82 |
| F40N | 40 | T-12 | 48 | Medium bipin | Natural | 20,000+ | 2100 | 82 |
| F40PL/AQ | 40 | T-12 | 48 | Medium bipin | Plant and aquarium | 20,000+ | 1900 | — |
| F40PL | 40[e] | T-12 | 48 | Medium bipin | Plant lite | 20,000+ | 800 | — |
| F40CG | 40[e] | T-12 | 48 | Medium bipin | Cool green | 20,000+ | 2850 | 88 |
| F40B | 40[e] | T-12 | 48 | Medium bipin | Blue | 20,000+ | 1200 | — |
| F40G | 40[e] | T-12 | 48 | Medium bipin | Green | 20,000+ | 4000 | — |
| F40GO | 40[e] | T-12 | 48 | Medium bipin | Gold | 20,000+ | 2550 | — |
| F40PK | 40[e] | T-12 | 48 | Medium bipin | Pink | 20,000+ | 1175 | — |
| F40R | 40[e] | T-12 | 48 | Medium bipin | Red | 20,000+ | 200 | — |
| FR40CW | 40[e] | T-12 | 48 | Medium bipin | Cool white reflector, 135° window | 20,000+ | 2600 | 88 |

| Designation | Watts | Bulb | Length (in) | Base | Color | Life (h) | Lumens | CRI |
|---|---|---|---|---|---|---|---|---|
| F40CW/RS/WM | 34 | T-12 | 48 | Medium bipin | Cool white | 20,000 | 2650 | 88 |
| F40LW/RS/WMII | 34 | T-12 | 48 | Medium bipin | Lite white | 20,000 | 2825 | 88 |
| F40WW/RS/WM | 34 | T-12 | 48 | Medium bipin | Warm white | 20,000 | 2750 | 88 |
| F40SP30/RS/WM | 34 | T-12 | 48 | Medium bipin | SP30-3000K | 20,000 | 2800 | 90 |
| F40SP35/RS/WM | 34 | T-12 | 48 | Medium bipin | SP35-3500K | 20,000 | 2800 | 90 |
| F40SP41/RS/WM | 34 | T-12 | 48 | Medium bipin | SP41-4100K | 20,000 | 2800 | 90 |
| F40CWX/RS/WM | 34 | T-12 | 48 | Medium bipin | Deluxe cool white | 20,000 | 1950 | 82 |
| F40WWX/RS/WM | 34 | T-12 | 48 | Medium bipin | Deluxe warm white | 20,000 | 1925 | 82 |
| F40D/RS/WM | 34 | T-12 | 48 | Medium bipin | Daylight | 20,000 | 2225 | 88 |
| F40W/RS/WM | 34 | T-12 | 48 | Medium bipin | White | 20,000 | 2650 | 88 |
| F40SPX30/RS/WM | 34 | T-12 | 48 | Medium bipin | SPX30-3000K | 20,000 | 2850 | 90 |
| F40SPX35/RS/WM | 34 | T-12 | 48 | Medium bipin | SPX35-3500K | 20,000 | 2850 | 90 |
| F40CW/RS/WMP[f] | 32 | T-12 | 48 | Medium bipin | Cool white | 20,000[e] | 2525 | 88 |
| F40WW/RS/WMP[f] | 32 | T-12 | 48 | Medium bipin | Warm white | 20,000[e] | 2625 | 88 |
| F40SP30/RS/WMP[f] | 32 | T-12 | 48 | Medium bipin | SP30-3000K | 20,000[e] | 2650 | 90 |
| F40SP35/RS/WMP[f] | 32 | T-12 | 48 | Medium bipin | SP35-3500K | 20,000[e] | 2650 | 90 |
| F40SP41/RS/WMP[f] | 32 | T-12 | 48 | Medium bipin | SP41-4100K | 20,000[e] | 2650 | 90 |
| F40LW/RS/WMP[f] | 32 | T-12 | 48 | Medium bipin | Lite white | 20,000[e] | 2700 | 88 |
| F40SPX41/RS/WM | 34 | T-12 | 48 | Medium bipin | SPX41-4100K | 20,000 | 2850 | 90 |

Increased-light-output 48-in lamps

| Designation | Watts | Bulb | Length (in) | Base | Color | Life (h) | Lumens | CRI |
|---|---|---|---|---|---|---|---|---|
| F40CW/SXL/SP30 | 40 | T-12 | 48 | Medium bipin | SP30-3000K | 24,000 | 3400 | 90 |
| F40WW/SXL/SP35 | 40 | T-12 | 48 | Medium bipin | SP35-3500K | 24,000 | 3400 | 90 |
| F40SP30/SXL/SP41 | 40 | T-12 | 48 | Medium bipin | SP41-4100K | 24,000 | 3400 | 90 |
| F40SP35/SXL/SPX30 | 40 | T-12 | 48 | Medium bipin | SPX30-3000K | 24,000 | 3500 | 90 |
| F40SP41/SXL/SPX35 | 40 | T-12 | 48 | Medium bipin | SPX35-3500K | 24,000 | 3500 | 90 |
| F40D/SXL/SPX41 | 40 | T-12 | 48 | Medium bipin | SPX41-4100K | 24,000 | 3500 | 90 |

**121. T-12 Rapid-Start Fluorescent Lamps** (*Continued*)

| Lamp-ordering code | Nominal lamp watts | Bulb | Length, in[a] | Base | Description | Approximate life, h[b] | Approximate initial lumens[c] | Approximate % lumens at 40% rated average life |
|---|---|---|---|---|---|---|---|---|
| | | | | | **U-shaped lamps, 3⅝-in leg spacing** | | | |
| F40CW/U/3 | 40 | T-12 | 24 | Medium bipin | Cool white | 12,000 | 2725 | 88 |
| F40WW/U/3 | 40 | T-12 | 24 | Medium bipin | Warm white | 12,000 | 2825 | 88 |
| F40SP35/U/3 | 40 | T-12 | 24 | Medium bipin | SP35-3500K | 12,000 | 2925 | 91 |
| F40CWX/U/3 | 40 | T-12 | 24 | Medium bipin | Deluxe cool white | 12,000 | 2025 | 85 |
| F40WWX/U/3 | 40 | T-12 | 24 | Medium bipin | Deluxe warm white | 12,000 | 2000 | 85 |
| F40W/U/3 | 40 | T-12 | 24 | Medium bipin | White | 12,000 | 2725 | 88 |
| | | | | | **Energy-saving U-shaped lamps, 3⅝-in leg spacing** | | | |
| F40CW/U/3/WM | 35 | T-12 | 24 | Medium bipin | Cool white | 12,000 | 2350 | 88 |
| F40LW/U/3/WM | 35 | T-12 | 24 | Medium bipin | Lite white | 12,000 | 2500 | 88 |
| F40WW/U/3/WM | 35 | T-12 | 24 | Medium bipin | Warm white | 12,000 | 2425 | 88 |
| | | | | | **U-shaped lamps, 6-in leg spacing** | | | |
| F40CW/U/6 | 40 | T-12 | 24 | Medium bipin | Cool white | 12,000 | 2800 | 88 |
| F40WW/U/6 | 40 | T-12 | 24 | Medium bipin | Warm white | 12,000 | 2900 | 88 |
| F40SP30/U/6 | 40 | T-12 | 24 | Medium bipin | SP30-3000K | 12,000 | 3050 | 91 |
| F40SP35/U/6 | 40 | T-12 | 24 | Medium bipin | SP35-3500K | 12,000 | 3050 | 91 |
| F40SP41/U/6 | 40 | T-12 | 24 | Medium bipin | SP41-4100K | 12,000 | 3050 | 91 |
| F40SPX30/U/6 | 40 | T-12 | 24 | Medium bipin | SPX30-3000K | 12,000 | 3100 | 91 |
| F40SPX35/U/6 | 40 | T-12 | 24 | Medium bipin | SPX35-3500K | 12,000 | 3100 | 91 |
| F40CWX/U/6 | 40 | T-12 | 24 | Medium bipin | Deluxe cool white | 12,000 | 2050 | 85 |
| F40WWX/U/6 | 40 | T-12 | 24 | Medium bipin | Deluxe warm white | 12,000 | 1975 | 85 |
| F40W/U/6 | 40 | T-12 | 24 | Medium bipin | White | 12,000 | 2800 | 88 |

Energy-saving U-shaped lamps, 6-in leg spacing

| | | | | | | | | |
|---|---|---|---|---|---|---|---|---|
| F40CW/U/6/WM | 35 | T-12 | 24 | Medium bipin | Cool white | 12,000 | 2400 | 88 |
| F40LW/U/6/WM | 35 | T-12 | 24 | Medium bipin | Lite white | 12,000 | 2550 | 88 |
| F40WW/U/6/WM | 35 | T-12 | 24 | Medium bipin | Warm white | 12,000 | 2500 | 88 |
| F40W/U/6/WM | 35 | T-12 | 24 | Medium bipin | White | 12,000 | 2400 | 88 |

T-8 rapid-start lamps

| | | | | | | | | |
|---|---|---|---|---|---|---|---|---|
| F17T8/SP30/RS | 17 | T-8 | 24 | Medium bipin | SP30-3000K | 20,000[g] | 1325 | 90 |
| F17T8/SP35/RS | 17 | T-8 | 24 | Medium bipin | SP35-3500K | 20,000[g] | 1325 | 90 |
| F17T8/SP41/RS | 17 | T-8 | 24 | Medium bipin | SP41-4100K | 20,000[g] | 1325 | 90 |
| F25T8/SP30/RS | 25 | T-8 | 36 | Medium bipin | SP30-3000K | 20,000[g] | 2125 | 90 |
| F25T8/SP35/RS | 25 | T-8 | 36 | Medium bipin | SP35-3500K | 20,000[g] | 2125 | 90 |
| F25T8/SP41/RS | 25 | T-8 | 36 | Medium bipin | SP41-4100K | 20,000[g] | 2125 | 90 |
| F32T8/SP30/RS | 32 | T-8 | 48 | Medium bipin | SP30-3000K | 20,000[g] | 2850 | 90 |
| F32T8/SP35/RS | 32 | T-8 | 48 | Medium bipin | SP35-3500K | 20,000[g] | 2850 | 90 |
| F32T8/SP41/RS | 32 | T-8 | 48 | Medium bipin | SP41-4100K | 20,000[g] | 2850 | 90 |
| F40T8/SP30/RS | 40 | T-8 | 60 | Medium bipin | SP30-3000K | 20,000[g] | 3600 | 90 |
| F40T8/SP35/RS | 40 | T-8 | 60 | Medium bipin | SP35-3500K | 20,000[g] | 3600 | 90 |
| F40T8/SP41/RS | 40 | T-8 | 60 | Medium bipin | SP41-4100K | 20,000[g] | 3600 | 90 |

[a]Including standard lampholders.
[b]Life when operated at 3 h per start on ballasts meeting published specifications.
[c]After 100 h of operation.
[d]Average life on preheat circuits is 15,000 h at 3 h per start.
[e]Life when operated at 12 h per start on ballasts meeting published specifications.
[f]Internal cathode cutout switch.
[g]Average life on instant start circuits is 15,000 h at 3 h per start.

**122. High-Output 800-mA Rapid-Start Fluorescent Lamps**
(General Electric Co.)

| Lamp-ordering code | Nominal lamp watts | Bulb | Length, in[a] | Base | Description | Approximate life, h[b] | Approximate initial lumens[c] | Approximate % lumens at 40% rated average life |
|---|---|---|---|---|---|---|---|---|
| F18T12/CW/HO | 25 | T-12 | 18 | Recessed double-contact | Cool white | 9,000 | 1025 | 87 |
| F24T12/CW/HO | 35 | T-12 | 24 | Recessed double-contact | Cool white | 9,000 | 1650 | 87 |
| F24T12/D/HO | 35 | T-12 | 24 | Recessed double-contact | Daylight | 9,000 | 1400 | 87 |
| F36T12/CW/HO | 45 | T-12 | 36 | Recessed double-contact | Cool white | 9,000 | 2800 | 87 |
| F36T12/D/HO | 45 | T-12 | 36 | Recessed double-contact | Daylight | 9,000 | 2350 | 87 |
| F36T12/SGN/HO | 45 | T-12 | 36 | Recessed double-contact | Sign white | 9,000 | 2150 | 87 |
| F42T12/CW/HO | 55 | T-12 | 42 | Recessed double-contact | Cool white | 9,000 | 3400 | 87 |
| F42T12/D/HO | 55 | T-12 | 42 | Recessed double-contact | Daylight | 9,000 | 2900 | 87 |
| F42T12/SGN/HO | 55 | T-12 | 42 | Recessed double-contact | Sign white | 9,000 | 2600 | 87 |
| F48T12/CW/HO | 60 | T-12 | 48 | Recessed double-contact | Cool white | 12,000 | 4050 | 87 |
| F48T12/WW/HO | 60 | T-12 | 48 | Recessed double-contact | Warm white | 12,000 | 4200 | 87 |
| F48T12/SPX30/HO | 60 | T-12 | 48 | Recessed double-contact | SPX30-3000K | 12,000 | 4350 | 90 |
| F48T12/CWX/HO | 60 | T-12 | 48 | Recessed double-contact | Deluxe cool white | 12,000 | 2900 | 84 |
| F48T12/D/HO | 60 | T-12 | 48 | Recessed double-contact | Daylight | 12,000 | 3400 | 87 |
| F48T12/SGN/HO | 60 | T-12 | 48 | Recessed double-contact | Sign white | 12,000 | 3100 | 87 |
| F48T12/LW/HO/WM[d] | 55 | T-12 | 48 | Recessed double-contact | Lite white | 12,000 | 3900 | 87 |
| F48T12/SP30/HO/WM | 55 | T-12 | 48 | Recessed double-contact | SP30-3000K | 12,000 | 3850 | 90 |
| F48T12/SP35/HO/WM | 55 | T-12 | 48 | Recessed double-contact | SP35-3500K | 12,000 | 3850 | 90 |
| F48T12/SP41/HO/WM[d] | 55 | T-12 | 48 | Recessed double-contact | SP41-4100K | 12,000 | 3850 | 90 |
| F60T12/CW/HO | 75 | T-12 | 60 | Recessed double-contact | Cool white | 12,000 | 5150 | 87 |
| F60T12/D/HO | 75 | T-12 | 60 | Recessed double-contact | Daylight | 12,000 | 4400 | 87 |
| F60T12/SGN/HO | 75 | T-12 | 60 | Recessed double-contact | Sign white | 12,000 | 4000 | 87 |
| F64T12/CW/HO | 80 | T-12 | 64 | Recessed double-contact | Cool white | 12,000 | 5600 | 87 |
| F64T12/D/HO | 80 | T-12 | 64 | Recessed double-contact | Daylight | 12,000 | 4750 | 87 |
| F64T12/SGN/HO | 80 | T-12 | 64 | Recessed double-contact | Sign white | 12,000 | 4300 | 87 |

| Lamp | Watts | Bulb | Length | Base | Color | Life (h) | Lumens | CRI |
|---|---|---|---|---|---|---|---|---|
| F72T12/CW/HO | 85 | T-12 | 72 | Recessed double-contact | Cool white | 12,000 | 6350 | 87 |
| F72T12/WW/HO | 85 | T-12 | 72 | Recessed double-contact | Warm white | 12,000 | 6650 | 87 |
| F72T12/SP30/HO | 85 | T-12 | 72 | Recessed double-contact | SP30-3000K | 12,000 | 6650 | 90 |
| F72T12/SP35/HO | 85 | T-12 | 72 | Recessed double-contact | SP35-3500K | 12,000 | 6650 | 90 |
| F72T12/SP41/HO | 85 | T-12 | 72 | Recessed double-contact | SP41-4100K | 12,000 | 6650 | 90 |
| F72T12/SPX30/HO | 85 | T-12 | 72 | Recessed double-contact | SPX30-3000K | 12,000 | 6800 | 90 |
| F72T12/SPX35/HO | 85 | T-12 | 72 | Recessed double-contact | SPX35-3500K | 12,000 | 6800 | 90 |
| F72T12/SPX41/HO | 85 | T-12 | 72 | Recessed double-contact | SPX41-4100K | 12,000 | 6800 | 90 |
| F72T12/CWX/HO | 85 | T-12 | 72 | Recessed double-contact | Deluxe cool white | 12,000 | 4600 | 84 |
| F72T12/D/HO | 85 | T-12 | 72 | Recessed double-contact | Daylight | 12,000 | 5350 | 87 |
| F72T12/SGN/HO | 85 | T-12 | 72 | Recessed double-contact | Sign white | 12,000 | 4900 | 87 |
| F72T12/N/HO | 85 | T-12 | 72 | Recessed double-contact | Natural | 12,000 | 4300 | 87 |
| F84T12/CW/HO | 100 | T-12 | 84 | Recessed double-contact | Cool white | 12,000 | 7700 | 87 |
| F84T12/D/HO | 100 | T-12 | 84 | Recessed double-contact | Daylight | 12,000 | 6500 | 87 |
| F96T12/CW/HO | 110 | T-12 | 96 | Recessed double-contact | Cool white | 12,000 | 8900 | 87 |
| F96T12/WW/HO | 110 | T-12 | 96 | Recessed double-contact | Warm white | 12,000 | 9150 | 87 |
| F96T12/SP30/HO | 110 | T-12 | 96 | Recessed double-contact | SP30-3000K | 12,000 | 9200 | 90 |
| F96T12/SP35/HO | 110 | T-12 | 96 | Recessed double-contact | SP35-3500K | 12,000 | 9200 | 90 |
| F96T12/SP41/HO | 110 | T-12 | 96 | Recessed double-contact | SP41-4100K | 12,000 | 9200 | 90 |
| F96T12/SPX30/HO | 110 | T-12 | 96 | Recessed double-contact | SPX30-3000K | 12,000 | 9350 | 90 |
| F96T12/SPX35/HO | 110 | T-12 | 96 | Recessed double-contact | SPX35-3500K | 12,000 | 9350 | 90 |
| F96T12/SPX41/HO | 110 | T-12 | 96 | Recessed double-contact | SPX41-4100K | 12,000 | 9350 | 90 |
| F96T12/CWX/HO | 110 | T-12 | 96 | Recessed double-contact | Deluxe cool white | 12,000 | 6600 | 84 |
| F96T12/WWX/HO | 110 | T-12 | 96 | Recessed double-contact | Deluxe warm white | 12,000 | 6400 | 84 |
| F96T12/D/HO | 110 | T-12 | 96 | Recessed double-contact | Daylight | 12,000 | 7600 | 87 |
| F96T12/W/HO | 110 | T-12 | 96 | Recessed double-contact | White | 12,000 | 8950 | 87 |
| F96T12/SGN/HO | 110 | T-12 | 96 | Recessed double-contact | Sign white | 12,000 | 6900 | 87 |
| F96T12/C50/HO | 110 | T-12 | 96 | Recessed double-contact | Chroma 50 | 12,000 | 6750 | 84 |
| F96T12/N/HO | 110 | T-12 | 96 | Recessed double-contact | Natural | 12,000 | 6200 | 84 |
| F96T12/GO/HO | 110 | T-12 | 96 | Recessed double-contact | Gold | 12,000 | 7500 | — |

**122. High-Output 800-mA Rapid-Start Fluorescent Lamps** (*Continued*)

| Lamp-ordering code | Nominal lamp watts | Bulb | Length, in[a] | Base | Description | Approximate life, h[b] | Approximate initial lumens[c] | Approximate % lumens at 40% rated average life |
|---|---|---|---|---|---|---|---|---|
| F96T12/CW/HO/WM[d] | 110 | T-12 | 96 | Recessed double-contact | Cool white | 12,000 | 8000 | 87 |
| F96T12/LW/HO/WM[d] | 95 | T-12 | 96 | Recessed double-contact | Lite white | 12,000 | 8500 | 87 |
| F96T12/WW/HO/WM[d] | 95 | T-12 | 96 | Recessed double-contact | Warm white | 12,000 | 8200 | 87 |
| F96T12/SP30/HO/WM | 95 | T-12 | 96 | Recessed double-contact | SP30-3000K | 12,000 | 8350 | 90 |
| F96T12/SP35/HO/WM[d] | 95 | T-12 | 96 | Recessed double-contact | SP35-3500K | 12,000 | 8350 | 90 |
| F96T12/SP41/HO/WM[d] | 95 | T-12 | 96 | Recessed double-contact | SP41-4100K | 12,000 | 8350 | 90 |
| F96T12/SPX30/HO/WM | 95 | T-12 | 96 | Recessed double-contact | SPX30-3000K | 12,000 | 8500 | 90 |
| F96T12/SPX35/HO/WM | 95 | T-12 | 96 | Recessed double-contact | SPX35-3500K | 12,000 | 8500 | 90 |
| F96T12/CWX/HO/WM[d] | 95 | T-12 | 96 | Recessed double-contact | Deluxe cool white | 12,000 | 5900 | 84 |
| F96T12/W/HO/WM[d] | 95 | T-12 | 96 | Recessed double-contact | White | 12,000 | 8000 | 87 |

[a]Including standard lampholders.
[b]Life when operated at 3 h per start on ballasts meeting published specifications.
[c]After 100 h of operation.
[d]Energy-saving design.

**123. 1500-mA Rapid-Start Fluorescent Lamps**
(General Electric Co.)

| Lamp-ordering code | Nominal lamp watts | Bulb | Length, in[a] | Base | Description | Approximate life, h[b] | Approximate initial lumens[c] | Approximate % lumens at 40% rated average life |
|---|---|---|---|---|---|---|---|---|
| | | | | | **T-12 lamps** | | | |
| FR24T12/CW/1500/HT | 70 | T-12 | 24 | Recessed double-contact | Cool white reflector; 135° reflector | 9,000 | 2,500 | — |
| F48T12/CW/1500 | 110 | T-12 | 48 | Recessed double-contact | Cool white | 10,000 | 6,200 | 70 |
| F48T12/WW/1500 | 110 | T-12 | 48 | Recessed double-contact | Warm white | 10,000 | 6,400 | 70 |
| F48T12/CW/1500/0[d] | 110 | T-12 | 48 | Recessed double-contact | Cool white—low-temperature | 9,000 | 7,000 | — |
| F72T12/CW/1500 | 165 | T-12 | 72 | Recessed double-contact | Cool white | 10,000 | 9,700 | 70 |
| F72T12/WW/1500 | 165 | T-12 | 72 | Recessed double-contact | Warm white | 10,000 | 10,000 | 70 |
| F72T12/CW/1500/0[d] | 170 | T-12 | 72 | Recessed double-contact | Cool white—low-temperature | 9,000 | 10,800 | — |
| F96T12/CW/1500 | 215 | T-12 | 96 | Recessed double-contact | Cool white | 10,000 | 13,500 | 70 |
| F96T12/WW/1500 | 215 | T-12 | 96 | Recessed double-contact | Warm white | 10,000 | 14,000 | 70 |
| F96T12/CWX/1500 | 215 | T-12 | 96 | Recessed double-contact | Deluxe cool white | 10,000 | 10,000 | 67 |
| F96T12/D/1500 | 215 | T-12 | 96 | Recessed double-contact | Daylight | 10,000 | 11,500 | 70 |
| F96T12/CW/1500/0[d] | 220 | T-12 | 96 | Recessed double-contact | Cool white—low-temperature | 9,000 | 14,400 | — |
| FR96T12/CW/1500 | 215 | T-12 | 96 | Recessed double-contact | Cool white reflector; 135° window | 10,000 | 9,700 | — |
| F96T12/CW/1500/WM[e] | 185 | T-12 | 96 | Recessed double-contact | Cool white | 9,000 | 12,500 | 70 |
| F96T12/LW/1500/WM[e] | 185 | T-12 | 96 | Recessed double-contact | Lite white | 9,000 | 13,250 | 70 |

**123. 1500-mA Rapid-Start Fluorescent Lamps** (*Continued*)

| Lamp-ordering code | Nominal lamp watts | Bulb | Length, in[a] | Base | Description | Approximate life, h[b] | Approximate initial lumens[c] | Approximate % lumens at 40% rated average life |
|---|---|---|---|---|---|---|---|---|
| | | | | PG-17 lamps | | | | |
| F48PG17/CW | 110 | PG-17 | 48 | Recessed double-contact | Cool white | 12,000 | 6,900 | 75 |
| F48PG17/D | 110 | PG-17 | 48 | Recessed double-contact | Daylight | 12,000 | 6,000 | 75 |
| F48PG17/CW/WM[e] | 95 | PG-17 | 48 | Recessed double-contact | Cool white | 12,000 | 6,000 | 75 |
| F72PG17/CW | 165 | PG-17 | 72 | Recessed double-contact | Cool white | 12,000 | 11,500 | 75 |
| F72PG17/D | 165 | PG-17 | 72 | Recessed double-contact | Daylight | 12,000 | 9,000 | — |
| F96PG17/CW | 215 | PG-17 | 96 | Recessed double-contact | Cool white | 12,000 | 15,300 | — |
| F96PG17/WW | 215 | PG-17 | 96 | Recessed double-contact | Warm white | 12,000 | 15,700 | — |
| F96PG17/CWX | 215 | PG-17 | 96 | Recessed double-contact | Deluxe cool white | 12,000 | 11,000 | — |
| F96PG17/D | 215 | PG-17 | 96 | Recessed double-contact | Daylight | 12,000 | 12,700 | — |
| F96PG17/CW/WM[e] | 185 | PG-17 | 96 | Recessed double-contact | Cool white | 12,000 | 13,500 | — |
| F96PG17/LW/WM[e] | 185 | PG-17 | 96 | Recessed double-contact | Lite white | 12,000 | 14,100 | — |

| | Watts | Bulb | Length | Base | Color | | | |
|---|---|---|---|---|---|---|---|---|
| F48T10/CW[d] | 110 | T-10 | 48 | Recessed double-contact | Cool white | 9,000 | 6,200 | — |
| F48T10/CWX[d] | 110 | T-10 | 48 | Recessed double-contact | Deluxe cool white | 9,000 | 4,500 | — |
| F60T10/CW[d] | 135 | T-10 | 60 | Recessed double-contact | Cool white | 9,000 | 8,200 | — |
| F72T10/CW[d] | 160 | T-10 | 72 | Recessed double-contact | Cool white | 9,000 | 9,700 | — |
| F72T10/CWX[d] | 160 | T-10 | 72 | Recessed double-contact | Deluxe cool white | 9,000 | 7,550 | — |
| F96T10/CW[d] | 205 | T-10 | 96 | Recessed double-contact | Cool white | 9,000 | 13,500 | — |
| **Jacketed T-12 lamps; 1⅟₁₆-in clear outer jacket** | | | | | | | | |
| F48T12J/CW/1500/0[d] | 110 | T-12J | 48 | Recessed double-contact | Cool white | 10,000 | 6,800 | — |
| F72T12J/CW/1500/0[d] | 170 | T-12J | 72 | Recessed double-contact | Cool white | 10,000 | 10,500 | — |
| F96T12J/CW/1500/0[d] | 220 | T-12J | 96 | Recessed double-contact | Cool white | 10,000 | 14,000 | — |
| **Jacketed T-10 lamps; 1⅟₁₆-in clear outer jacket** | | | | | | | | |
| F48T10J/CW[d] | 110 | T-10J | 48 | Recessed double-contact | Cool white | 9,000 | 6,000 | — |
| F48T10J/CWX[d] | 110 | T-10J | 48 | Recessed double-contact | Deluxe cool white | 9,000 | 4,300 | — |
| F60T10J/CW[d] | 135 | T-10J | 60 | Recessed double-contact | Cool white | 9,000 | 8,000 | — |
| F72T10J/CW[d] | 160 | T-10J | 72 | Recessed double-contact | Cool white | 9,000 | 9,600 | — |
| F96T10J/CW[d] | 205 | T-10J | 96 | Recessed double-contact | Cool white | 9,000 | 13,300 | — |

[a] Including standard lampholders.
[b] Life when operated at 3 h per start on ballasts meeting published specifications.
[c] After 100 h of operation.
[d] Nominal lamp watts and initial lumens are peak values; actual wattage and lumen values in service depend on ambient-temperature conditions.
[e] Energy-saving design.

**124. Slimline Instant-Start Fluorescent Lamps**
(General Electric Co.)

| Lamp-ordering code | Nominal lamp watts | Bulb | Length, in[a] | Base | Description | Approximate life, h[b] | Approximate initial lumens[c] | Approximate % lumens at 40% rated average life |
|---|---|---|---|---|---|---|---|---|
| F42T6/CW[d] | 25 | T-6 | 42 | Single-pin | Cool white | 7,500 | 1750 | 91 |
| F42T6/WW[d] | 25 | T-6 | 42 | Single-pin | Warm white | 7,500 | 1825 | 91 |
| F42T6/CWX[d] | 25 | T-6 | 42 | Single-pin | Deluxe cool white | 7,500 | 1300 | 86 |
| F64T6/CW[d] | 40 | T-6 | 64 | Single-pin | Cool white | 7,500 | 2800 | 91 |
| F64T6/WW[d] | 40 | T-6 | 64 | Single-pin | Warm white | 7,500 | 2900 | 91 |
| F64T6/CWX[d] | 40 | T-6 | 64 | Single-pin | Deluxe cool white | 7,500 | 2050 | 86 |
| F72T8/CW[d] | 35 | T-8 | 72 | Single-pin | Cool white | 7,500 | 3000 | 91 |
| F72T8/WW[d] | 35 | T-8 | 72 | Single-pin | Warm white | 7,500 | 3100 | 91 |
| F96T8/CW[d] | 50 | T-8 | 96 | Single-pin | Cool white | 7,500 | 4050 | 91 |
| F96T8/CW/WM[d,e] | 40 | T-8 | 96 | Single-pin | Cool white | 7,500 | 3450 | 91 |
| F24T12/CW | 20 | T-12 | 24 | Single-pin | Cool white | 7,500 | 1200 | 91 |
| F36T12/CW | 30 | T-12 | 36 | Single-pin | Cool white | 7,500 | 2000 | 91 |
| F36T12/WW | 30 | T-12 | 36 | Single-pin | Warm white | 7,500 | 2050 | 91 |
| F42T12/CW | 35 | T-12 | 42 | Single-pin | Cool white | 7,500 | 2400 | 91 |
| F48T12/CW | 40 | T-12 | 48 | Single-pin | Cool white | 9,000 | 2875 | 91 |
| F48T12/WW | 40 | T-12 | 48 | Single-pin | Warm white | 9,000 | 2975 | 91 |
| F48T12/SPX30 | 40 | T-12 | 48 | Single-pin | SPX30-3000K | 9,000 | 3050 | 94 |
| F48T12/CWX | 40 | T-12 | 48 | Single-pin | Deluxe cool white | 9,000 | 2100 | 86 |
| F48T12/D | 40 | T-12 | 48 | Single-pin | Daylight | 9,000 | 2450 | 91 |
| F48T12/W | 40 | T-12 | 48 | Single-pin | White | 9,000 | 2875 | 91 |
| F48T12/N | 40 | T-12 | 48 | Single-pin | Natural | 9,000 | 1950 | 91 |
| F48T12/CW/WM[e] | 30 | T-12 | 48 | Single-pin | Cool white | 9,000 | 2475 | 91 |
| F48T12/LW/WM[e] | 30 | T-12 | 48 | Single-pin | Lite white | 9,000 | 2650 | 91 |
| F48T12/SP30 | 40 | T-12 | 48 | Single-pin | SP30-3000K | 9,000 | 3000 | 94 |
| F48T12/SP35 | 40 | T-12 | 48 | Single-pin | SP35-3500K | 9,000 | 3000 | 94 |

| Lamp | Watts | Bulb | Length | Base | Color | | | CRI |
|------|-------|------|--------|------|-------|------|------|-----|
| F48T12/SPX35 | 40 | T-12 | 48 | Single-pin | SPX35-3500K | 9,000 | 3050 | 94 |
| F48T12/SP30/WM[e] | 30 | T-12 | 48 | Single-pin | SP30-3000K | 9,000 | 2575 | 94 |
| F48T12/SP35/WM[e] | 30 | T-12 | 48 | Single-pin | SP35-3500K | 9,000 | 2575 | 94 |
| F48T12/SP41/WM[e] | 30 | T-12 | 48 | Single-pin | SP41-4100K | 9,000 | 2575 | 94 |
| F48T12/WW/WM[e] | 30 | T-12 | 48 | Single-pin | Warm white | 9,000 | 2550 | 91 |
| F60T12/CW | 50 | T-12 | 60 | Single-pin | Cool white | 12,000 | 3600 | 91 |
| F60T12/D | 50 | T-12 | 60 | Single-pin | Daylight | 12,000 | 3000 | 91 |
| F64T12/CW | 50 | T-12 | 64 | Single-pin | Cool white | 12,000 | 3850 | 91 |
| F64T12/D | 50 | T-12 | 64 | Single-pin | Daylight | 12,000 | 3300 | 91 |
| F72T12/CW | 55 | T-12 | 72 | Single-pin | Cool white | 12,000 | 4500 | 91 |
| F72T12/WW | 55 | T-12 | 72 | Single-pin | Warm white | 12,000 | 4650 | 91 |
| F72T12/CWX | 55 | T-12 | 72 | Single-pin | Deluxe cool white | 12,000 | 3300 | 86 |
| F72T12/SP30 | 55 | T-12 | 72 | Single-pin | SP30-3000K | 12,000 | 4700 | 94 |
| F72T12/SP35 | 55 | T-12 | 72 | Single-pin | SP35-3500K | 12,000 | 4700 | 94 |
| F72T12/SP41 | 55 | T-12 | 72 | Single-pin | SP41-4100K | 12,000 | 4700 | 94 |
| F72T12/SPX30 | 55 | T-12 | 72 | Single-pin | SPX30-3000K | 12,000 | 4800 | 94 |
| F72T12/SPX35 | 55 | T-12 | 72 | Single-pin | SPX35-3500K | 12,000 | 4800 | 94 |
| F72T12/D | 55 | T-12 | 72 | Single-pin | Daylight | 12,000 | 3800 | 91 |
| F72T12/N | 55 | T-12 | 72 | Single-pin | Natural | 12,000 | 3000 | 91 |
| F72T12/GO | 55 | T-12 | 72 | Single-pin | Gold | 12,000 | 3800 | — |
| F84T12/CW | 65 | T-12 | 84 | Single-pin | Cool white | 12,000 | 5300 | 91 |
| F84T12/D | 65 | T-12 | 84 | Single-pin | Daylight | 12,000 | 4450 | 91 |
| F96T12/CW | 75 | T-12 | 96 | Single-pin | Cool white | 12,000 | 6150 | 91 |
| F96T12/WW | 75 | T-12 | 96 | Single-pin | Warm white | 12,000 | 6400 | 91 |
| F96T12/SP30 | 75 | T-12 | 96 | Single-pin | SP30-3000K | 12,000 | 5750 | 94 |
| F96T12/SP35 | 75 | T-12 | 96 | Single-pin | SP35-3500K | 12,000 | 6425 | 94 |
| F96T12/SP41 | 75 | T-12 | 96 | Single-pin | SP41-4100K | 12,000 | 5750 | 94 |
| F96T12/SPX30 | 75 | T-12 | 96 | Single-pin | SPX30-3000K | 12,000 | 6550 | 94 |
| F96T12/SPX35 | 75 | T-12 | 96 | Single-pin | SPX35-3500K | 12,000 | 6550 | 94 |
| F96T12/SPX41 | 75 | T-12 | 96 | Single-pin | SPX41-4100K | 12,000 | 6550 | 94 |
| F96T12/CWX | 75 | T-12 | 96 | Single-pin | Deluxe cool white | 12,000 | 4500 | 86 |
| F96T12/WWX | 75 | T-12 | 96 | Single-pin | Deluxe warm white | 12,000 | 4400 | 86 |
| F96T12/D | 75 | T-12 | 96 | Single-pin | Daylight | 12,000 | 5250 | 91 |
| F96T12/W | 75 | T-12 | 96 | Single-pin | White | 12,000 | 6150 | 91 |
| F96T12/C50 | 75 | T-12 | 96 | Single-pin | Chroma 50 | 12,000 | 4600 | 86 |

**124. Slimline Instant-Start Fluorescent Lamps** (*Continued*)

| Lamp-ordering code | Nominal lamp watts | Bulb | Length, in[a] | Base | Description | Approximate life, h[b] | Approximate initial lumens[c] | Approximate % lumens at 40% rated average life |
|---|---|---|---|---|---|---|---|---|
| F96T12/N | 75 | T-12 | 96 | Single-pin | Natural | 12,000 | 4250 | 91 |
| F96T12/GO | 75 | T-12 | 96 | Single-pin | Gold | 12,000 | 5150 | — |
| F96T12/R | 75 | T-12 | 96 | Single-pin | Red | 12,000 | 400 | — |
| F96T12/CW/WM[e] | 60 | T-12 | 96 | Single-pin | Cool white | 12,000 | 5500 | 91 |
| F96T12/LW/WM[e] | 60 | T-12 | 96 | Single-pin | Lite white | 12,000 | 5800 | 91 |
| F96T12/WW/WM[e] | 60 | T-12 | 96 | Single-pin | Warm white | 12,000 | 5700 | 91 |
| F96T12/SP30/WM[e] | 60 | T-12 | 96 | Single-pin | SP30-3000K | 12,000 | 5750 | 94 |
| F96T12/SP35/WM[e] | 60 | T-12 | 96 | Single-pin | SP35-3500K | 12,000 | 5750 | 94 |
| F96T12/SP41/WM[e] | 60 | T-12 | 96 | Single-pin | SP41-4100K | 12,000 | 5750 | 94 |
| F96T12/SPX30/WM[e] | 60 | T-12 | 96 | Single-pin | SPX30-3000K | 12,000 | 5900 | 94 |
| F96T12/SPX35/WM[e] | 60 | T-12 | 96 | Single-pin | SPX35-3500K | 12,000 | 5900 | 94 |
| F96T12/SPX41/WM[e] | 60 | T-12 | 96 | Single-pin | SPX41-4100K | 12,000 | 5900 | 94 |
| F96T12/CWX/WM[e] | 60 | T-12 | 96 | Single-pin | Deluxe cool white | 12,000 | 4000 | 86 |
| F96T12/WWX/WM[e] | 60 | T-12 | 96 | Single-pin | Deluxe warm white | 12,000 | 3900 | 86 |
| F96T12/D/WM[e] | 60 | T-12 | 96 | Single-pin | Daylight | 12,000 | 4650 | 91 |
| F96T12/W/WM[e] | 60 | T-12 | 96 | Single-pin | White | 12,000 | 5200 | 91 |
| F96T12/C50/WM[e] | 60 | T-12 | 96 | Single-pin | Chroma 50 | 12,000 | 4000 | 86 |

[a]Including standard lampholders.
[b]Life when operated at 3 h per start on ballasts meeting published specifications.
[c]After 100 h of operation.
[d]Nominal lamp watts; life and lumen ratings are for 200-mA operation.
[e]Energy-saving design.

**125. Circline Fluorescent Rapid-Start Lamps**
(General Electric Co.)

| Lamp-ordering code | Nominal lamp watts | Bulb | Outside diameter, in | Base | Description | Approximate life, h[a] | Approximate initial lumens[b] | Approximate % lumens at 40% rated average life |
|---|---|---|---|---|---|---|---|---|
| FC6T9/CW | 20 | T-9 | 6¾ | Four-pin | Cool white | 12,000 | 800 | — |
| FC6T9/WW | 20 | T-9 | 6¾ | Four-pin | Warm white | 12,000 | 825 | — |
| FC8T9/CW | 22 | T-9 | 8¼ | Four-pin | Cool white | 12,000 | 1000 | — |
| FC8T9/WW | 22 | T-9 | 8¼ | Four-pin | Warm white | 12,000 | 1050 | — |
| FC8T9/SW | 22 | T-9 | 8¼ | Four-pin | Soft white (deluxe warm white) | 12,000 | 825 | — |
| FC8T9/CWX | 22 | T-9 | 8¼ | Four-pin | Deluxe cool white | 12,000 | 875 | — |
| FC8T9/D | 22 | T-9 | 8¼ | Four-pin | Daylight | 12,000 | 875 | — |
| FC12T9/CW | 32 | T-9 | 12 | Four-pin | Cool white | 12,000 | 1825 | — |
| FC12T9/WW | 32 | T-9 | 12 | Four-pin | Warm white | 12,000 | 1875 | — |
| FC12T9/SW | 32 | T-9 | 12 | Four-pin | Soft white (deluxe warm white) | 12,000 | 1500 | — |
| FC12T9/CWX | 32 | T-9 | 12 | Four-pin | Deluxe cool white | 12,000 | 1550 | — |
| FC12T9/D | 32 | T-9 | 12 | Four-pin | Daylight | 12,000 | 1550 | — |
| FC16T9/CW | 40 | T-9 | 16 | Four-pin | Cool white | 12,000 | 2700 | — |
| FC16T9/WW | 40 | T-9 | 16 | Four-pin | Warm white | 12,000 | 2800 | — |
| FC16T9/SW | 40 | T-9 | 16 | Four-pin | Soft white (deluxe warm white) | 12,000 | 1950 | — |
| FC16T9/CWX | 40 | T-9 | 16 | Four-pin | Deluxe cool white | 12,000 | 2000 | — |
| FC16T9/D | 40 | T-9 | 16 | Four-pin | Daylight | 12,000 | 2250 | — |

[a]Life when operated at 3 h per start on ballasts meeting published specifications. Life will also be attained on preheat circuits for earlier Circline lamps of the same wattage.
[b]After 100 h of operation.

10-85

## 126. Compact Fluorescent Lamps

(General Electric Co.)

| Lamp-ordering code | Nominal lamp watts | Bulb | Length, in[a] | Base | Description | Approximate life, h[b] | Approximate initial lumens | Approximate % lumens at 40% rated average life |
|---|---|---|---|---|---|---|---|---|
| Twin-tube preheat lamps: starter in the lamp base | | | | | | | | |
| F5BX/SPX27 | 5 | T-4 | 4$\frac{5}{32}$ | G23 | SPX27-2700K | 10,000 | 250 | 84 |
| F7BX/SPX27 | 7 | T-4 | 5$\frac{5}{16}$ | G23 | SPX27-2700K | 10,000 | 400 | 84 |
| F7BX/SPX35 | 7 | T-4 | 5$\frac{5}{16}$ | G23 | SPX35-3500K | 10,000 | 400 | 84 |
| F9BX/SPX27 | 9 | T-4 | 6$\frac{5}{16}$ | G23 | SPX27-2700K | 10,000 | 600 | 84 |
| F9BX/SPX30 | 9 | T-4 | 6$\frac{5}{16}$ | G23 | SPX35-3500K | 10,000 | 600 | 84 |
| F13BX/SPX27 | 13 | T-4 | 7$\frac{1}{2}$ | GX23 | SPX27-2700K | 10,000 | 825 | 84 |
| F13BX/SPX30 | 13 | T-4 | 7$\frac{1}{2}$ | GX23 | SPX30-3000K | 10,000 | 900 | 84 |
| F13BX/SPX35 | 13 | T-4 | 7$\frac{1}{2}$ | GX23 | SPX35-3500K | 10,000 | 825 | 84 |
| Twin-tube preheat lamps: external starter | | | | | | | | |
| F18BX/SPX30 | 18 | T-5 | 9 | 2G11 | SPX30-3000K | 10,000 | 1200 | 90 |
| F18BX/SPX35 | 18 | T-5 | 9 | 2G11 | SPX35-3500K | 10,000 | 1200 | 90 |
| F18BX/SPX41 | 18 | T-5 | 9 | 2G11 | SPX41-4100K | 10,000 | 1250 | 90 |
| Twin-tube rapid start lamps | | | | | | | | |
| F18BX/ SPX30/RS | 18 | T-5 | 10$\frac{1}{2}$ | 2G11 | SPX30-3000K | 20,000 | 1250 | 90 |
| F18BX/ SPX35/RS | 18 | T-5 | 10$\frac{1}{2}$ | 2G11 | SPX35-3500K | 20,000 | 1250 | 90 |
| F18BX/ SPX41/RS | 18 | T-5 | 10$\frac{1}{2}$ | 2G11 | SPX41-4100K | 20,000 | 1250 | 90 |
| F27BX/SPX30 | 27 | T-5 | 12$\frac{27}{32}$ | 2G11 | SPX30-3000K | 12,000 | 1800 | 90 |
| F27BX/SPX35 | 27 | T-5 | 12$\frac{27}{32}$ | 2G11 | SPX35-3500K | 12,000 | 1800 | 90 |
| F27BX/SPX41 | 27 | T-5 | 12$\frac{27}{32}$ | 2G11 | SPX41-4100K | 12,000 | 1800 | 90 |
| F39BX/SPX30 | 39 | T-5 | 16$\frac{1}{2}$ | 2G11 | SPX30-3000K | 12,000 | 2850 | 90 |
| F39BX/SPX35 | 39 | T-5 | 16$\frac{1}{2}$ | 2G11 | SPX35-3500K | 12,000 | 2850 | 90 |
| F39BX/SPX41 | 39 | T-5 | 16$\frac{1}{2}$ | 2G11 | SPX41-4100K | 12,000 | 2900 | 90 |
| F40BX/SPX30 | 40 | T-5 | 22$\frac{1}{2}$ | 2G11 | SPX30-3000K | 20,000 | 3150 | 90 |
| F40BX/SPX35 | 40 | T-5 | 22$\frac{1}{2}$ | 2G11 | SPX35-3500K | 20,000 | 3150 | 90 |
| F40BX/SPX41 | 40 | T-5 | 22$\frac{1}{2}$ | 2G11 | SPX41-4100K | 20,000 | 3150 | 90 |
| Quad-tube preheat lamps: starter in the lamp base | | | | | | | | |
| F13DBX23T4/ SPX27 | 13 | T-4 | 4$\frac{29}{32}$ | GX23 | SPX27-2700K | 10,000 | 900 | — |
| F13DBX23T4/ SPX35 | 13 | T-4 | 4$\frac{29}{32}$ | GX23 | SPX35-3500K | 10,000 | 860 | — |
| F13DBXT4/ SPX27 | 13 | T-4 | 6 | G24D1 | SPX27-2700K | 10,000 | 900 | — |
| F13DBXT4/ SPX35 | 13 | T-4 | 6 | G24D1 | SPX35-3500K | 10,000 | 900 | — |
| F18DBXT4/ SPX27 | 18 | T-4 | 6$\frac{29}{32}$ | G24D2 | SPX27-2700K | 10,000 | 1200 | — |
| F18DBXT4/ SPX30 | 18 | T-4 | 6$\frac{29}{32}$ | G24D2 | SPX35-3500K | 10,000 | 1200 | — |
| F26DBXT4/ SPX27 | 26 | T-4 | 7$\frac{25}{32}$ | G24D3 | SPX27-2700K | 10,000 | 1800 | — |
| F26DBXT4/ SPX30 | 26 | T-4 | 7$\frac{25}{32}$ | G24D3 | SPX35-3500K | 10,000 | 1800 | — |

[a] Overall length including base.
[b] Life when operated at 3 h per start in ballasts meeting published specifications.

**127. Auxilliary equipment.**  The operation of fluorescent lamps requires certain auxiliary equipment. Each fluorescent lamp requires a ballast assembly, generally called a *ballast* (Fig. 10-46). The mercury arc in the tube has a varying unstable resistance, which requires a high voltage to start the arc. Each lamp must therefore be equipped with a ballast reactor, which performs a twofold purpose. It provides a high induced starting

**Fig. 10-46**   Cases for fluorescent-lamp ballasts. [General Electric Co.]

voltage for striking the arc and, after conduction has started, stabilizes the operating impedance of the circuit to maintain the operating current at a steady value. Figure 10-47 shows the wiring diagrams for the more common ballasts.

Since the ballast reactor used with the lamps gives an overall lagging power factor of only 0.45 to 0.50, the installation of any considerable number of fluorescent lamps with simply a ballast reactor would require an uneconomical size of wire and distribution transformer ratings in the circuits supplying the lamps. The power factor of the lamp circuit can be raised to a high value by connecting a capacitor of the proper rating across the lamp circuit. The use of power-factor-correcting capacitors is recommended. Many power companies have adopted rules, and several states have passed laws, that limit the minimum power factor permissible in fluorescent-lamp installations. The power-factor-correcting capacitor is usually included as an integral part of the ballast assembly, or a large capacitor may be connected across a circuit supplying several lamps.

For lamps started by the preheat-start method a starting switch is required. The different types of starting switches are explained in Sec. 129.

Trigger-, rapid-, and instant-start lamps require only a ballast; no starting switch is used. The ballasts, however, must be properly designed for the particular type of starting employed.

Ballasts with reduced internal losses and thus increased lighting efficiency are available for the widely used fluorescent-lamp types. The lower losses are achieved through the use of more efficient materials and designs. In some 40-W rapid-start–ballast designs, additional efficiency gains are obtained through an electronic switch in the ballast that opens the cathode-heating circuit after the lamp has started.

Some states require the use of these more efficient ballasts both in new installations and as replacement ballasts.

**128. Ballast assemblies.**  Certain parts of the auxiliary equipment required for fluorescent lamps are made up by the manufacturers and assembled inside a single metal case (Fig. 10-46). The complete assembly is generally termed the *ballast*. Single-ballast assemblies are available for single-lamp, two-lamp (tulamp), three-lamp, and four-lamp luminaire installations. The automatic starting switch required for the preheat-start lamps is not combined with the ballast unit but is made as a separate unit.

Data for fluorescent ballasts may be obtained from manufacturers of these devices. Such manufacturers offer catalogs which include tables describing the characteristics of each type of ballast. Information includes nominal lamp watts, circuit voltage, sound rating, dimensions, weight, watts loss, power factor, line current, and type of case. Since ballast designs are frequently revised or new types are added, only the lastest catalogs should be used, thus assuring proper information.

**129. The starting switches** required for preheat-start lamps must perform at least the following functions:

1. Closing the circuit between the two filaments when the lamp circuit is energized

2. Opening the circuit between the two filaments after a sufficient lapse of time for the filaments to be heated to the proper temperature

Starting switches are available in four types: (1) manual starting switches, (2) automatic glow-switch starters, (3) automatic Watch-Dog or No-Blink starters, and (4) automatic thermal-switch starters. The manual starting switch is the simplest starter switch. It consists of a single pushbutton switch, as indicated in Fig. 10-48. When the lamp is started, the push button is held down for a second or two and then released. Manual starting switches of this kind are often used for desk-type fluorescent lamps.

The automatic *glow-switch starter* (Fig. 10-49) consists of a glass bulb containing neon or argon gas and two electrodes. One electrode is a U-shaped bimetallic strip, and the other is a fixed rod. On starting, when there is practically no voltage drop at the ballast,

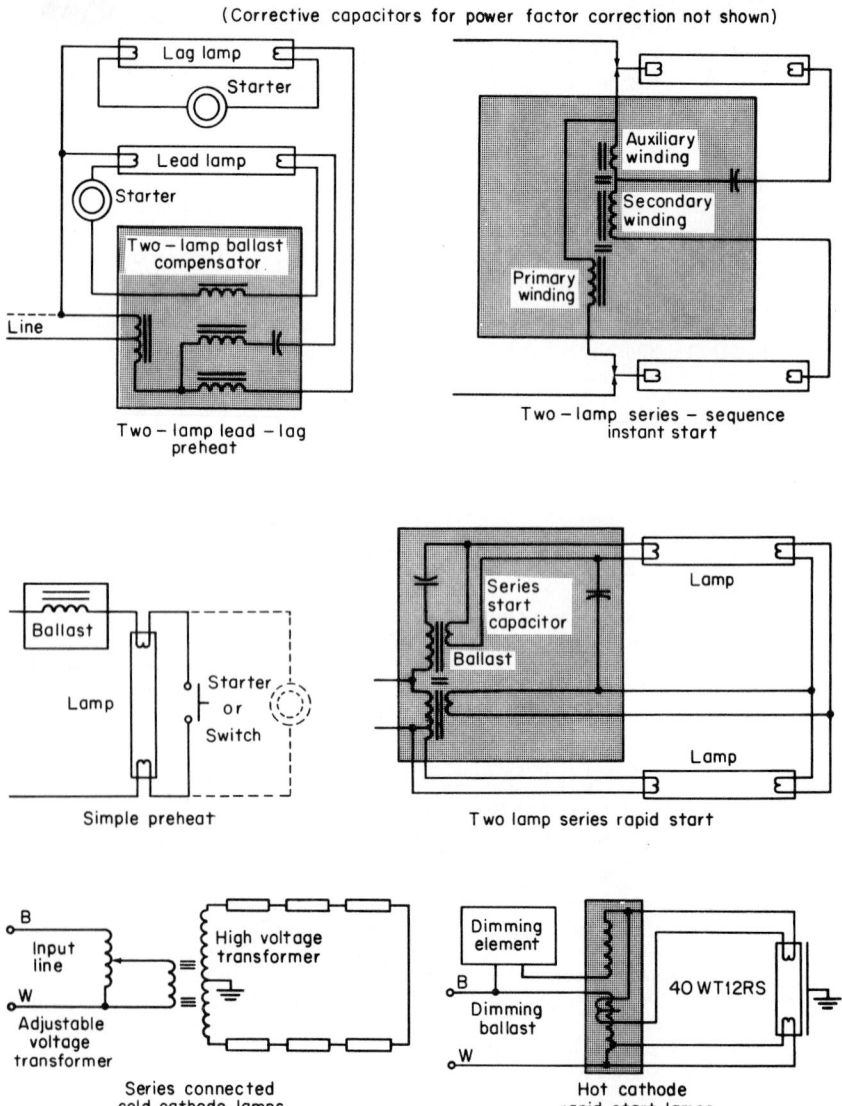

**Fig. 10-47**   Wiring diagrams for fluorescent lamps. [*IES Lighting Handbook*]

the voltage at the starter is sufficient to produce a glow discharge between the U-shaped strip and the fixed-contact or center electrode (see Fig. 10-50A). The heat from the glow actuates the strip, the contacts close, and cathode preheating begins (see Fig. 10-50B). This shorts out the glow discharge so that the bimetal cools, and in a very short time the contacts open (see Fig. 10-50C). The inductive voltage kick from the ballast is then suf-

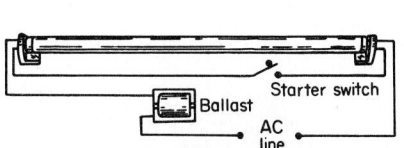

**Fig. 10-48** Fluorescent-lamp circuit with push-button starting switch.

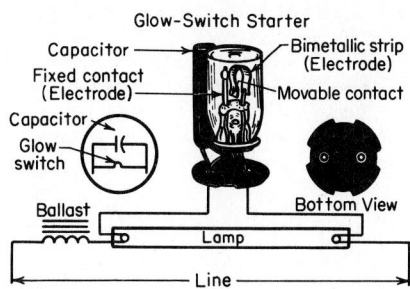

**Fig. 10-49** Fluorescent-lamp circuit with glow-switch starter.

ficient to start the lamp. During normal operation, there is not enough voltage across the lamp to produce further starter glow; the contacts therefore remain open, and the starter consumes no energy.

The *Watch-Dog starter* (one manufacturer's trade name: Fig. 10-51) contains a glow switch, which during normal starting functions in the manner described. This starter has an added feature, which consists of a wire-coil heater element actuating a bimetallic arm that serves as a latch to hold a second switch in a normal closed position. When a lamp is deactivated or for some reason will not start after repeated attempts have been made by blinking it on and off, enough heat is developed by the intermittent flow of cathode-pre-heating current so that the latcg pulls away and releases a spring-operated switch in the starter circuit. This occurs after 15 to 20 s at rated line voltage, thus removing the annoyance of blinking and conserving the life of the starter. When the lamp is replaced, the starter is reset to operating position by pushing down on the reset button at the top of the starter. No-Blink is a manufacturer's trade name for a starter that performs the same function as the Watch-Dog starter and operates in a similar way.

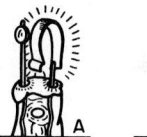

**Fig. 10-50** Glow-switch starter, showing operating positions of contacts.

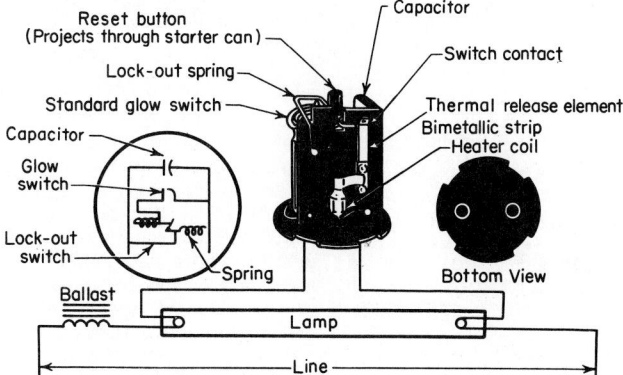

**Fig. 10-51** Fluorescent-lamp circuit with Watch-Dog starter.

The *thermal-switch starter* consists of a bimetallic switch element heated by a separate heater element. The arrangement and connections for such a starter are shown in Fig. 10-52. When the lamp is cold, the switch element is closed. When the lamp circuit is energized, the switch element connects the ballast, starter heating element, and lamp cathodes in series across the supply line. The cathode-preheating current, in passing through the starter heating element, heats the bimetallic switch element. After the proper lapse of time, the heating of the switch element opens its contacts and breaks the circuit through the lamp filaments. The inductive kick then starts the lamp. The normal operating current passes through the starter heating element. The bimetallic switch element is

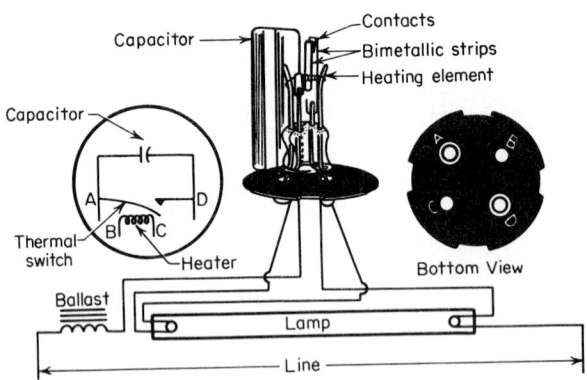

**Fig. 10-52**    Circuit for fluorescent lamp with thermal-switch starter.

therefore held open as long as the lamp continues to conduct. During operation, the thermal-switch starters consume some energy, ½ W for the smaller size and 1½ W for the larger size. These starters probably give the best all-around performance. They provide an adequate preheating period and a high induced starting voltage and are inherently less susceptible to line-voltage variations.

A small capacitor is connected across the switch contacts in many types of automatic starting switches. It aids in the starting but is primarily useful to shunt out line-lead harmonics that may cause radio interference.

The general appearance of automatic starting switches is shown in Fig. 10-53.

Automatic starting switches are mounted in special starter sockets that are available as a separate unit or as a unit combined with a fluorescent lampholder.

**130. Wiring arrangements.**  A typical wiring arrangement for a fluorescent lamp and its auxiliaries is shown in Fig. 10-54. The sockets, or *lampholders*, as they are called for fluorescent lamps, are entirely different from those for incandescent lamps. Care must be exercised to obtain the proper lampholder for the lamp being used.

**131. Ballast life and temperatures.**  The conventional ballast is enclosed in a container filled with a heavy impregnating compound which congeals around the choke coils and condenser. This serves to radiate heat and, by its compactness, to eliminate or minimize noise or hum. Since the wattage loss in the ballast creates heat, suitable ventilation must be provided in fixture design or in places where ballasts are installed to keep the temperature within safe limits. If the temperature rises too high, the capacitor will fail and cause excessive heating and eventual failure of the internal windings.

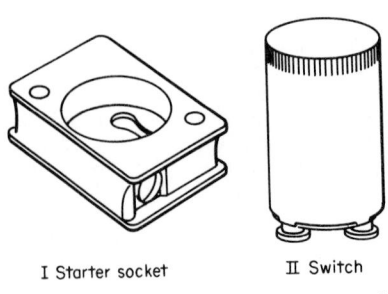

I Starter socket        II Switch

**Fig. 10-53**  Fluorescent-lamp starter switch and socket.

Excessive ballast temperatures may be the result of one of the following:

1. Shorted starter in preheat circuits.

2. Burned-out lamp.
3. Rectifying lamp (near the end of lamp life).
4. Line-voltage fluctuation. Certain types of ballasts are sensitive to line-voltage fluctuation and therefore overheat at higher line voltages.
5. Improper design of fixture so that ballast heat cannot be dissipated efficiently.

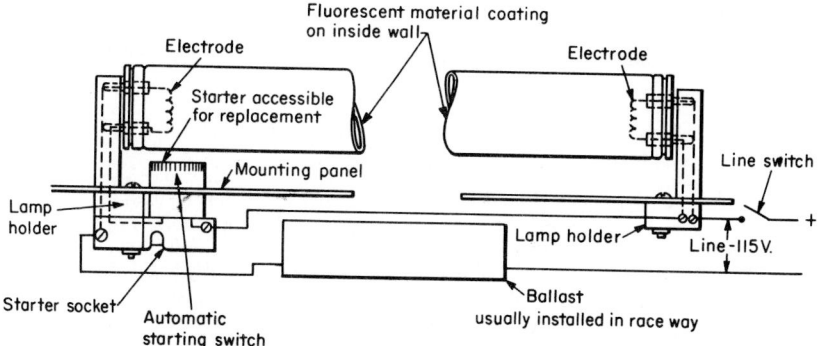

**Fig. 10-54**   Assembly of fluorescent lamp, starter, and reactor.

6. Improper location of fixture which prevents proper heat dissipation.
7. Improper selection of ballast for the type of lamps used.

At a maximum temperature of 105°C (221°F) *within the winding* or 90°C (194°F) *measured on the case* and a maximum capacitor temperature of 70°C (158°F), the expected ballast life is about 12 to 15 years, depending on the operating hours each year. An increase of 10°C (18°F) beyond these limits will cut ballast life by as much as 50 percent.

The NEC requires that all fluorescent fixtures used indoors incorporate ballast protection other than that for simple reactor ballasts, as shown in Figs. 10–48 and 10–49. In response to this Code rule Underwriters Laboratories has a standard that requires the use of Class P (protected) ballasts for indoor fixtures. Class P ballasts contain inherent *thermal devices* or *thermal fuses*, or both, which open automatically if the maximum temperature of the ballast is exceeded. Only Class P ballasts should be used as replacements in fixtures where ballasts other than reactor types are installed.

Although external fusing of a ballast (consisting of small quick-blow glass-tube fuses rated close to the normal ballast current rating) provides some degree of ballast protection, such fuses respond only to line current and not to temperatures inside the ballast. Since some internal ballast faults will cause excessive temperatures without an increase in line current, the fuse will not sense such faults. Accordingly, all present Class P ballasts contain internal thermal devices or fuses which respond to internal temperatures. External fusing will, however, provide a backup protection for Class P ballasts, and they will afford a high degree of protection for other types of ballasts if sized according to the recommendations of fuse and ballast manufacturers.

**132. Noise ratings of ballasts.**   Ballasts are noise-rated according to laboratory standards. Such a rating is only a guide for installation practice and does not attempt to specify one ballast type over another for specific installations of fluorescent lamps.

Noise ratings are designated by letter, from A, the quietest, through F. The A rating is best, for example, for home applications, in which the surrounding and competitive noise level may be at a minimum. In an industrial plant with attendant operation noises, ballast hum may be of no importance. Not all ballasts, regardless of their general noise rating, produce annoyable hum or noise. Chances are that the noise potential of different ballast designs is significant only in exceptionally quiet places. Individual ballasts of any noise rating may by some chance become offensive. Likewise, the ballast hum or noise vibration may be induced into the wiring channel or fixture and may be minimized or emphasized by the method used in mounting and clamping the ballast within the fixture.

**133. The life of a fluorescent lamp** is affected not only by the voltage and current supplied to it but also by the number of times it is started. Electron emission material is sputtered off from the electrodes continuously during the operation of the lamp and in larger quantities each time the lamp starts. Since life normally ends when the emission material is completely consumed from one of the electrodes, the greater the number of burning hours per start, the longer the life of the lamp. When the emission material is exhausted, lamps on a preheat type of circuit will blink on and off as the electrodes heat but the arc fails to strike. Lamps designed for instant or rapid start will simply fail to operate. Blinking lamps should be removed from the circuit promptly to protect both the starter and the ballast from overheating.

The rated average life of a fluorescent lamp in burning hours is based upon the average life of large representative groups of lamps measured in the laboratory under specified test conditions. Many fluorescent lamps have a rated average life of 12,000 to 20,000 h at 3 burning hours per start.

With the proper ballast operating voltage within the line-voltage limits shown on the ballast label, rated lamp life should be obtained.

**134. The temperature of the bulb** has a decided effect on the efficiency of a fluorescent lamp (Fig. 10-55). The best efficiency occurs at 100 to 120°F (38 to 49°C), which is the operating temperature corresponding to an ambient room temperature of 70 to 80°F (21 to 27°C). Efficiency decreases slowly as the temperature is increased above normal but decreases very rapidly as the temperature is decreased below normal. For this reason the fluorescent lamp is not satisfactory for locations where it will be subject to wide variations in temperature. This reduction in efficiency with low surrounding air temperature can be minimized by enclosing the lamp within a clear-glass tube or by using higher-current special fluorescent lamps designed for exposed low-temperature application (Sec. 112) so that the lamp will operate at more nearly its desirable temperature (Fig. 10-56).

**135. Variations in line voltage** produce less effect on fluorescent lamps than on incandescent lamps, the phase relationship among line voltage, lamp voltage, and auxiliary voltage being such that lamp voltage varies inversely as the square root of the line voltage (Fig. 10-57). For this reason, most fluorescent lamp ballasts are rated for a nominal circuit voltage of 120 V with an acceptable range of 110 to 125 V. Line voltage less than 110 may make starting of the arc uncertain; a voltage over 125 may overheat the ballast. Line voltage less than 110 or greater than 125 can decrease the life of the lamp. Ballasts are also available for widely employed lamp types for use on 227-V systems with an acceptable range of 254 to 289 V. In addition, ballasts are available for some lamp types for 208-, 240-, and 480-V systems.

**136. Radio interference.**   The performance of the mercury arc of a fluorescent lamp at the electrodes is associated with an electrical instability that sets up a continuous series of radio waves. There are three ways in which these waves may reach the radio and interfere with reception:

  1.  Direct radiation from the bulb to the radio aerial circuit

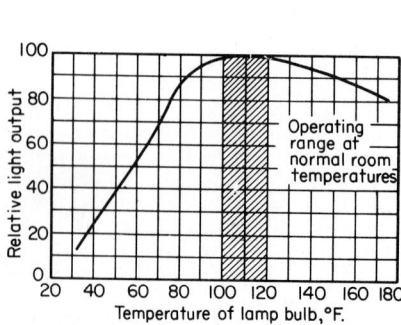

**Fig. 10-55**   Effect of bulb-wall temperature on the efficiency of the fluorescent lamp.

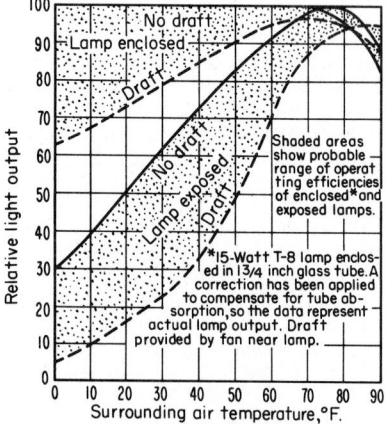

**Fig. 10-56**   Effect of surrounding air temperature on the efficiency of the fluorescent lamp.

2. Direct radiation from the electric supply line to the aerial circuit
3. Line feedback from the lamp through the power line to the radio

The direct radiation from the bulb diminishes rapidly as the radio is separated from a lamp, and this effect can be controlled by proper positioning of the radio and its aerial.

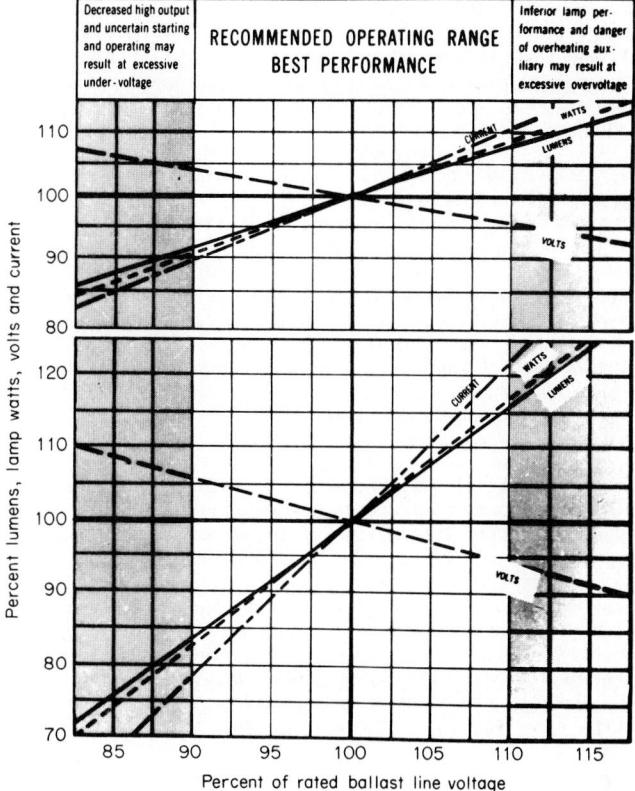

**Fig. 10-57**  Effect of variation of line voltage on the operating characteristics of fluorescent lamps. [General Electric Co.]

The following table shows the recommended distance of popular fluorescent lamps from an AM radio antenna to minimize the effect of direct bulb radiation.

**Recommended Distance from Lamp to AM Radio Antenna**

| Lamp Size | Distance, Ft |
|---|---|
| 14-, 15-, and 20-W | 4 |
| 32-W Circline | 5 |
| 30-W | 6 |
| 40-W | 8 |
| 90-W, 72- and 96-in slimline | 10 |

If the radio must remain within the bulb-radiation range, it will be necessary to take the following precautions.

1. Connect the aerial to the radio by means of a shielded lead-in wire with the shield grounded, or install a doublet type of aerial with twisted pair leads.

2. Provide a good ground for the radio.

3. The aerial proper must be out of bulb- and line-radiation range. The use of a correct antenna system will usually help reduce radio interference by providing a better station-signal strength.

The use of fixture-shielding material such as electrically conductive glass or special louver materials can aid in reducing radiated interference.

Interference from line radiation and line feedback can best be minimized by the proper application of line filters at each lamp or fixture. A simple form of filter is a three-section capacitor unit. One such unit per fixture, or for each 8 ft (2.4 m) of lamps in a cove, will reduce line noise by approximately 75 percent. If it is desirable to eliminate line noise completely, the inductive-capacitor type is recommended. This filter has a current-carrying capacity of 2 A, which is, for example, about the load of four 40-W lamps.

When only one or two radios are located near a fluorescent installation and the aerial circuit has been properly shielded from bulb and line radiation, a single line filter located at the radio power outlet will suffice.

When radios located in buildings adjoining the fluorescent installation are receiving line-feedback type of interference, it is practical to install a single filter such as the three-section capacitor unit at each panel box feeding fluorescent-lamp circuits.

When it is necessary to filter each lamp or fixture, the filter should be located as close to the lamps as possible. This precaution should be taken because of line radiation between lamp and filter.

The improper spacing of lampholders, resulting in poor contact with lamp base pins, can also generate interference, as can improperly grounded fluorescent fixtures. Failure to ground the neutral of branch circuits (as required in the **National Electrical Code**) is an additional cause. If the service lines are not properly grounded, filters will be much less effective.

Fluorescent equipment destined for homes or other places where radios are likely to be present should have the proper radio-interference filter in each fixture.

**137. Direct-current operation.** If fluorescent lamps are used on dc circuits, special auxiliaries and proper series resistance must be employed. However, dc operation is inferior to normal ac operation in several respects: (1) The overall efficiency of light production is much less on dc circuits because the series resistance consumes about as much wattage as the lamp itself. (2) Rated average life is usually less than the ac value. (3) Suitable line reversing switches to change polarity with each lamp start should be provided for lamps longer than 24 in (610 mm). (4) Lumen maintenance may be somewhat less favorable. On the other hand, the color of the light and the total light output compare favorably with normal ac operation, and problems of power factor and stroboscopic effect are, of course, eliminated.

**138. High-frequency operation.** While 60 Hz is the standard frequency for ac distribution, higher frequencies have a number of advantages in the operation of fluorescent lamps. As the frequency increases, lamp efficacy is increased. The increase in efficacy depends on the lamp type and the frequency of power delivered to the lamp. The gain in efficacy for 40-W T-12 rapid-start lamps is shown in Fig. 10-58. In addition, ballast losses are reduced, further improving system efficiency, and ballasts can be smaller and lighter-weight.

Earlier high-frequency systems utilized central rotary or static inverters providing 360- to 400-Hz power with small capacitance- and inductance-type ballasts. These systems required separate electrical distribution systems for lighting.

The recent advances in solid-state devices together with the availability of improved integrated-circuit functions have resulted in the development of commercially available electronic ballasts. These ballasts operate on standard 60-Hz power systems but operate these lamps at approximately 25,000 Hz. Electronic ballasts are available for the popular lamp types: 30-W and 40-W rapid-start and 96-in (2438-mm) slimline lamps. They provide high lamp efficacy and reduced system wattage. The conversion to higher frequency within the ballast eliminates the need for a separate electrical distribution system for lighting.

Special 40-W lamps are available for improved performance on high-frequency electronic ballasts.

High-frequency systems have been used on aircraft to reduce weight with 400-Hz generators and small capacitative ballasts. In motor vehicles, lamps are operated at variable frequencies up to 500 Hz.

**139. Swirling and spiraling** (General Electric Co.). This phrase refers to all conditions in which the lighted lamp appears to fluctuate in brightness from end to end. The cause is principally particles of materials loosened from the cathode and floating in the arc stream. Such particles usually settle to the bulb when the current is turned off for a few minutes. This temporary swirling may occur in new lamps and is not serious. If swirling persists, some operating condition may be the cause of the particles being cast into the arc stream.

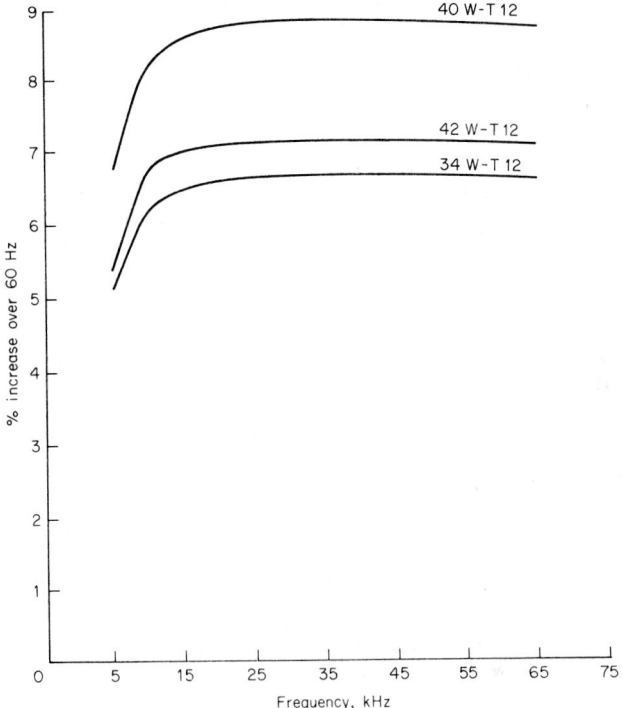

**Fig. 10-58** Effect of frequency on 40-W T-12 fluorescent-lamp efficacy. [General Electric Co.]

High-impact starting voltage may jar loose such material; this high-voltage starting may be due to (1) a ballast of inadequate design, (2) a lack of proper starting aids in trigger-start and rapid-start systems, (3) a starter in a preheat circuit that is not working properly to give adequate preheat, or (4) any other circuit condition such as low supply voltage in which neither starter nor ballast functions normally yet in which lamps may occasionally start by high-voltage impact alone without preheat.

Instant-start lamps have cathodes designed to withstand high-voltage starting. When new lamps of this type swirl persistently, the lamp may be at fault and should be replaced. At the end of normal lamp life, when no electron-emissive material remains on the cathode, the high starting voltage disintegrates even the metallic parts of the cathode, which then invade the arc stream. This is the cause of the excessive swirling that characterizes the end-of-life period of instant-start lamps.

**140. Operating suggestions** (General Electric Co.). To secure the best performance of fluorescent lamps, it is important that users understand how to maintain their fluorescent installations properly. Many factors that affect the performance of these lamps were never encountered with incandescent filament lamps, and users must realize that some of these elements of satisfactory service are within their control.

For example, if a filament lamp does not light when current is applied, the one single, positive conclusion is that the lamp is burned out or defective. No such conclusion should be made in the case of the fluorescent lamp. This lamp, though perfect in all respects, may not start or operate properly through no fault of the design or manufacture of the lamp.

AVERAGE LIFE    Since mortality laws apply to fluorescent as well as to filament lamps, it is not unusual for some fluorescent lamps to fail before rated life while life will last for a longer period than their rated average life to balance out the normal early failures.

NORMAL FAILURE    The electroemissive material on fluorescent-lamp electrodes is used up during the life of the lamp, the sputtering off being more rapid during starting. Therefore, a longer life will be obtained if lamps are allowed to operate continuously instead of being turned on and off frequently. Average rated life should be obtained if lamps are operated for normal periods of 3 or 4 h with each start.

When the active material on the electrodes is used up, the lamp will no longer operate. On preheat circuits, it may continue to blink on and off as the starter attempts to start the lamp. On instant-start circuits, the lamp may spiral at the end of normal life.

NORMAL DEPRECIATION    The lighted output at 100 h is used for rating purposes because the loss during this period may amount to as much as 10 percent. The average light output during life is approximately 90 percent of the 100-h rating value.

BLACKENING    A fluorescent lamp darkens rather uniformly throughout the length of the tube during life, though this is not noticed unless an old lamp is compared with a new one in front of a light source. At the end of life, the lamp usually shows a dense blackening either at one end or at both. Also, there may be dark rings, slightly brownish in color, at one end or both. On the average, there should be little indication of either blackening or rings during the first 500 h of operation. If the lamp has not given a proper length of service when this occurs, this may be due to improper starting, frequent starting with short operating periods, improper ballast equipment, unusually high or low voltage, improper wiring, or a defect in the lamp.

End blackening should not be confused with a mercury deposit which sometimes condenses around the bulb at the ends. This mercury condensation appears to be more common with the 1-in- (25.4-mm-) diameter lamps than with the 1½-in (38.1-mm) sizes. It is occasionally visible on new lamps but should evaporate after the lamp has been in operation for some time. However, it may reappear later when the lamp cools. Frequently, dark streaks appear lengthwise of the tube as small globules of mercury cool on the lower (cooler) part of the lamp. Mercury condensation is quite common on lamps in louvered units owing to the cooling effects of air circulation around the louvers.

Mercury may condense at any place on the tube if a cold object is allowed to lie against it for a short period. Such spots near the center section may not again evaporate. When condensation occurs in this manner, rotating the lamp 180° in the lampholder may give a more favorable position for evaporation or may place the spot in a less conspicuous place from an appearance standpoint.

Near the end of life, some lamps may develop a very dense spot about ½ in (12.7 mm) wide and extending almost halfway around the bulb, centering about 1 in (25.4 mm) from the base. This is quite normal, but a spot developing early in life is an indication of excessive starting or operating current. This may be due to a ballast off-rating or to an unusually high circuit voltage.

Occasionally a lamp may develop a ring or gray band at one end or at both. Such rings are usually located about 2 in (50.8 mm) from either base. These rings have no effect on the lamp performance and are no indication that a lamp is near failure and must soon be replaced.

CIRCUIT VOLTAGE    The circuit voltage should be within the ballast rating, although 110- to 125-V ballasts *may in some cases* give satisfactory lamp performance on circuits as low as 105 or as high as 130 V. Ballasts designed for 265 to 277 V may sometimes give satisfactory results on line voltages as low as 260 and as high as 285 V. Excessive undervoltage or overvoltage is injurious to the lamp.

BALLAST HUM    Characteristic transformer hum is inherent in lamp ballast equipment, although the noise may come and go, varying considerably with individual ballasts. If ballasts are mounted on soft rubber, neoprene, or similar nonrigid mountings, the vibrations due to the magnetic action in the chokes will not be transferred to supporting members, and disconcerting auxiliary hum will be reduced to a minimum. Such vibration insulation, however, will inhibit heat radiation and transfer to such a point that ballast ambient-temperature limits will be exceeded.

LOW-TEMPERATURE OPERATION   In most cases, satisfactory starting and performance can be expected at temperatures considerably below 50°F (10°C) by (1) keeping the voltage in the upper half of the ballast rating, (2) conserving lamp heat by enclosure or by other suitable means, and (3) using thermal starting switches.

STARTING DIFFICULTY   Starting difficulties may be due to a number of causes other than the starter itself. In general, any difficulty in starting may result in premature end blackening and short lamp life.

A LAMP THAT MAKES NO EFFORT TO START   First the lamp should be checked to make sure that it is properly seated in the sockets. If this fails to correct the trouble, the lamp should be checked in another circuit, and, if necessary, the voltage can be checked at the sockets. If no indication of power is found at the sockets, the circuit connections are incorrect or the ballast is probably defective. In a preheat circuit, the starter should be checked. It may have reached the end of life and should be replaced.

A LAMP THAT IS SLOW IN STARTING   Low line voltage, low ballast rating, or in preheat circuits a sluggish starter can also result in slow starting.

A LAMP THE ENDS OF WHICH REMAIN LIGHTED   This condition indicates a short circuit in the starter, which should be replaced. Starters that have been in service for some time frequently fail in this manner. If the ends of a lamp in a new installation remain lighted, it is possible that the wiring is incorrect. Of course, if short-circuiting types of No-Blink or Watch-Dog starters are used, the lamp ends will remain lighted in a normally failed lamp; this condition automatically corrects itself when the failed lamp is replaced with a new lamp.

A LAMP THAT BLINKS ON AND OFF IN PREHEAT CIRCUITS   Blinking is the usual indication of a normal failure of a lamp. It will usually be found more convenient to put in a new starter first and then, if blinking continues, to renew the lamp. Low circuit voltage, low ballast rating, low temperature, and cold drafts may individually cause difficulties of this nature, or several of these may be contributing factors. It is also possible for improper circuit connections to cause such blinking. The annoyance of blinking lamps can be avoided by using No-Blink starters in installations employing lamps for which such special types of starters are available.

If the ends of a lamp remain lighted or the lamp blinks on and off, either the trouble should immediately be corrected or the lamp or starter should be removed from the circuit. A blinking lamp can shortly ruin both lamp and starter. Ordinary starters can be replaced with No-Blink starters (where these starters are available), which remove the starter element from the circuit if lamps fail to light after a reasonable number of attempts have been made.

EARLY END BLACKENING   Heavy end blackening early in life indicates that the active material on the electrodes is being sputtered off too rapidly. It may be due to:

1. High or low voltage. For best results, the circuit voltage should be within ballast rating.

2. Loose contacts (most likely at the lampholder). Remedy: lampholders should be rigidly mounted and properly spaced. See that lamps are securely seated in lampholders.

3. Improperly designed ballasts or ballasts outside specification limits. Remedy: the use of ballasts approved by the Electrical Testing Laboratories will usually eliminate trouble of this nature.

4. In preheat circuits a defective or worn-out starter, causing the lamp to blink on and off, prolonged flashing of the lamp at each start, or the ends of the lamps remaining lighted over a period of time. Remedy: replace with a new starter.

5. Improper wiring of units.

**141. Damaged lamps.**   Fluorescent lamps are of rugged construction and will withstand a considerable amount of rough handling. However, a severe side blow may jar the cathode loose or break the stem. If a lamp or a package of lamps is dropped or severely bumped on the end, the shock may break the base insulation, and, more important, the concealed exhaust tube may be broken; this will admit air into the bulb. The breakage may not be revealed by casual inspection.

Cathodes can be examined by viewing the end of the bulb against a pinhole of light which casts a shadow of the cathode on the bulb wall. If cathodes appear to be intact in a lamp that has leaked air, no end glow will take place and the lamp will not start; if only a minute quantity of air has seeped in, the end glow may appear deep violet or pinkish in color. If large numbers of lamps are to be tested, a spark-coil tester brushed over the base pins is the quickest way to test for glow, the lack of which identifies defective or broken lamps.

With preheat bipin lamps a simple test will show whether or not cathodes are intact. The filament cathodes are designed for preheating at relative low voltage, and if 115 V is applied directly to the base pins, the cathode filament will instantly burn off, generally at both leads with little fusing of the metal. The stem may also crack, resulting in an air leak. To test the cathode circuit, test each end separately by connecting the two base pins in series with an ordinary 115- or 120-V incandescent lamp on a 115- or 120-V circuit. Use a 25-W lamp for miniature-bipin-based lamps, a 60-W lamp for 14- to 40-W lamps, and a 200-W size for testing 90- and 100-W fluorescent-lamp cathodes. Instant-start cathodes cannot be tested in this manner since they are internally shorted within the base.

## GASEOUS-DISCHARGE LAMPS

**142. Light is produced by gaseous-conduction methods** when proper energy transitions result from electron displacement within the atomic structure of the gas involved. Applied voltage at the electrodes accelerates free electrons, which in the course of their travel strike atoms and displace electrons from their normal atomic positions. Radiations of a particular wavelength result as the displaced electrons return to their normal position in the atomic structure; this wavelength depends on the gas used and the degree of electron displacement.

Once the discharge begins, the enclosed arc becomes a light source with one electrode acting as a cathode and the other as an anode. The electrodes will exchange functions as the supply changes polarity. This principle is employed in high-intensity discharge and neon lamps.

**143. High-intensity–discharge lamps** (also commonly referred to as HID lamps) for general-lighting applications consists of three groups of lamps: mercury lamps, metal halide lamps, and high-pressure sodium lamps. All high-intensity–discharge lamps produce light from an arc tube which is usually contained in an outer glass bulb.

**144. The high-pressure mercury lamp** consists of a quartz arc tube sealed within an outer glass jacket or bulb. The inner arc tube is made of quartz to withstand the high temperatures resulting when the lamp builds up to normal wattage. Two main electron-emissive electrodes are located at opposite ends of the tube; these are made of coiled tungsten wire. Near the upper main electrode is a third, or starting, electrode in series with a ballasting resistor and connected to the lower main-electrode lead wire.

The arc tube in the mercury lamp contains a small amount of pure argon gas which is used as a conducting medium to facilitate the starting of the arc before the mercury is vaporized. When voltage is applied, an electric field is set up between the starting electrode and the adjacent main electrode. This ionizing potential causes current to flow, and as the main arc strikes, the heat generated gradually vaporizes the mercury. When the arc tube is filled with mercury vapor, it creates a low-resistance path for current to flow between the main electrodes. When this takes place, the starting electrode and its high-resistance path become automatically inactive.

Once the discharge begins, the enclosed arc becomes a light source with one electrode acting as a cathode and the other as an anode. The electrodes will exchange functions as the ac supply changes polarity.

The quantity of mercury in the arc tube is very carefully measured to maintain quite an exact vapor pressure under design conditions of operation. This pressure differs with wattage sizes, depending on arc-tube dimensions, voltage-current relationships, and various other design factors.

Efficient operation requires the maintenance of a high temperature of the arc tube. For this reason the arc tube is enclosed in an outer bulb made of heat-resistant glass, which makes the arc tube less subject to surrounding temperature or cooling by air circulation. About half an atmosphere of nitrogen is introduced in the space between the arc tube and the outer bulb. The operating pressure for most mercury lamps is in the range of 2 to 4 times atmospheric pressure. Lamps can operate in any position. However, light output is reduced when burned in positions other than vertical. Mercury lamps for lighting applications range in wattage from 40 to 1000 W. The 175- and 400-W types are the most popular. Mercury lamps are used in street lighting, industrial lighting, and outdoor area lighting. In new installations today, mercury lamps are being replaced with more efficient metal halide or high-pressure sodium systems. A typical clear mercury lamp is shown in Fig. 10-59. Some special mercury lamps are available for use in ultraviolet applications.

**145. Phosphor-coated mercury lamps** are the most widely used mercury lamps. Within the visible region the mercury-lamp spectrum, or energy radiated, consists of principal wavelengths. The lack of radiation in the red end results in a greenish-blue light. When one looks at a lighted mercury lamp, the source itself appears to emit a daylight white. The absence of red radiation causes most colored objects to appear distorted in color value. Blue, green, and yellow colors of objects are emphasized, while orange and red objects appear brownish or black.

Color improvement has been incorporated in the design of mercury lamps by the use of special phosphors on the inside of the outer bulb (Fig. 10-60). The function of the phosphor is to convert the invisible ultraviolet energy into visible light in the same manner as light is produced in fluorescent lamps. The phosphor produces light principally in the red region, where the light from the arc tube is deficient. The original color-improved phosphors provided better color appearance with a small loss in lamp efficacy. Present phosphor-coated mercury lamps utilize a vanadate phosphor which not only significantly improves color appearance but increases lamp efficacy over clear mercury lamps by 7 to 11 percent. Most phosphor-coated mercury lamps provide a cool tone of light output somewhat similar to cool-white fluorescent lamps. Lamps are also available with a phosphor that creates a warmer tone of light output close to that of incandescent lamps.

See Sec. 153 for data on clear and phosphor-coated mercury lamps.

**146. Metal halide lamps** are similar to mercury lamps in construction, in that the lamp consists basically of a quartz arc tube mounted within an outer glass bulb. However, in addition to mercury, the arc tubes contain halide salts, usually sodium and scandium iodide. During lamp operation, the heat from the arc discharge evaporates the iodides along with the mercury. The result is an increase in efficacy approximately 50 percent higher than that of a mercury lamp of the same wattage together with excellent color quality from the arc.

The amount of the iodides vaporized determines lamp efficacy and color and is temperature-dependent. Metal halide arc tubes have carefully controlled seal shapes to maintain temperature consistency between lamps. In addition, one or both ends of the arc tube are coated to maintain the desired arc-tube temperature. There is some color variation

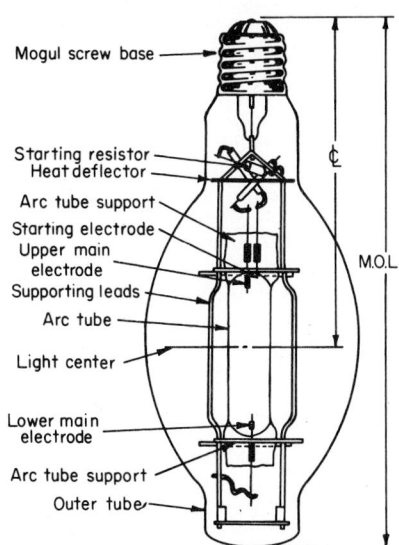

**Fig. 10-59** Mercury lamp.

**Fig. 10-60** Phosphor-coated mercury lamp. [General Electric Co.]

between individual metal halide lamps owing to differences in the characteristics of each lamp.

Metal halide lamps utilize a starting electrode at one end of the arc tube which operates in the same manner as the starting electrode in a mercury lamp. A bimetal shorting switch is placed between the starting electrode and the adjacent main electrode. This switch closes during lamp operation and prevents a small voltage from developing between the two electrodes, which in the presence of the halides could cause arc-tube seal failure. A typical metal halide lamp is shown in Fig. 10-61.

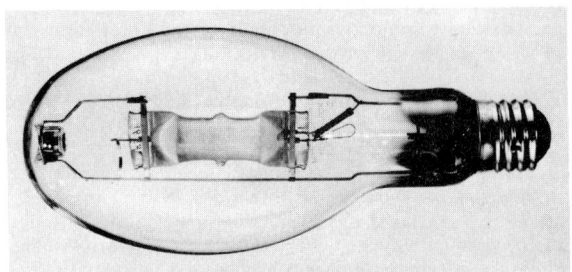

**Fig. 10-61**    Metal halide lamp. [General Electric Co.]

Metal halide lamps are available in 175-, 250-, 400-, 1000-, and 1500-W sizes. There are also special metal halide lamps with limitations in burning positions which have higher lamp efficacy. While many metal halide lamps can be operated in any burning position, initial light output, lumen maintenance, and life ratings change with burning position.

Although metal halide ballasts are designed to the specific starting and operating requirements for these lamps, there are several metal halide lamps which are designed to operate on certain 400- or 1000-W mercury-lamp ballasts. They provide substantially higher light output over life than the mercury lamps they have replaced.

Phosphor-coated metal halide lamps have the same external appearance as phosphor-coated mercury lamps. They are available in 175-, 250-, 400-, and 1000-W types. Although the light from the metal halide arc tube provides good color rendition, the phosphor, similarly to phosphors used in mercury lamps, adds further to the color rendition. It also provides a diffuse light source for lighting fixtures when optical characteristics are better suited to this type of lamp.

Data for metal halide lamps are shown in Sec. 154.

**147. The ultraviolet energy** produced by mercury and metal halide arc tubes is absorbed by the outer bulb in normal operation. However, in the event that the outer bulb is broken by being struck by an external object such as a basketball in a gymnasium, the arc tube, if not damaged, may continue to operate for many hours. With sufficient exposure, the ultraviolet energy from an arc tube operating without the protection of the outer bulb can cause temporary sunburn of the skin or eyes, similar to an excess exposure to the sun.

For lighting applications in open fixtures where lamps maybe subject to accidental or intentional breakage, special self-extinguishing lamps are available. These lamps are designed with an internal fuse in series with the arc tube, which opens within 15 min when exposed to air, causing the arc to be extinguished.

**148. The high-pressure sodium** (commonly referred to as HPS) lamp has the highest light-producing efficacy of any commerical source of white light (Fig. 10-62). Like most other high-intensity–discharge lamps, high-pressure sodium lamps consist of an arc tube enclosed within an outer glass bulb. The arc operates in a sodium vapor at a temperature and pressure which provide a warm color with light in all portions of the visible spectrum at a high efficacy. Owing to the chemical activity of hot sodium, quartz cannot be used as the arc-tube material. Instead, high-pressure sodium arc tubes are made of an alumina ceramic (polycrystalline alumina oxide) which can withstand the corrosive effects of hot sodium vapor.

There are coated-tungsten electrodes sealed at each end of the arc tube. The sodium is placed in the arc tube in the form of a sodium-mercury amalgam which is chemically inactive. The arc tube is filled with xenon gas to aid in starting.

High-pressure sodium lamps are available in sizes from 35 to 1000 W. They can be operated in any burning position and have the best lumen-maintenance characteristic of the three types of HID lamps. Except for the 35-W lamp, most high-pressure sodium lamps have rated lives of more than 24,000 h. The 35-W lamp has a rated life of 16,000 h. The 50-, 70-, 100-, and 150-W sizes are available in both a mogul-base and a medium-base design.

See Sec. 155 for data on high-pressure sodium lamps.

Diffuse-coated high-pressure sodium lamps are available in several wattages for use in lighting fixtures where a larger effective light-source size is desired, such as units for low mounting height and decorative applications. The coating does not change the lamp color and reduces lamp efficacy by 5 to 7 percent.

Improved-color high-pressure sodium lamps are available in a few types with significantly better color appearance. This is achieved through a change in arc-tube pressure. The light output and life of improved-color HPS lamps are less than with the standard high-pressure sodium lamp.

Special high-pressure sodium lamps have been designed to operate on specified mercury ballasts. Lamp design is similar to that of standard high-pressure sodium lamps except that a neon-argon starting gas is used and an internal starting aid is located in the lamp to permit starting on approved mercury ballasts. These lamps can only be used on reactor or autotransformer lag-type mercury ballasts. Lamp life and efficacy are reduced for these special HPS lamps.

**149. HID spectral-distribution** mercury, metal halide, and high-pressure sodium lamps have significantly different color characteristics in the light produced by each type. All HID lamps produce light in definite lines or bands rather than the continuous spectrum of an incandescent lamp. The typical spectral-power-distribution curves for HID lamps are shown in Fig. 10-63.

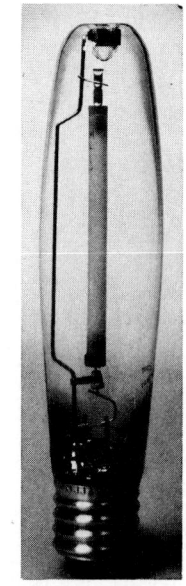

**Fig.    10-62** High-pressure sodium lamp. [General Electric Co.]

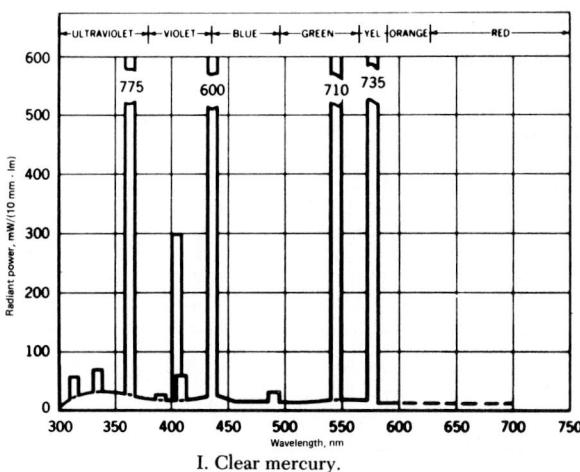

I. Clear mercury.

**Fig. 10-63**  Spectral characteristics of high-intensity–discharge lamps. [General Electric Co.]

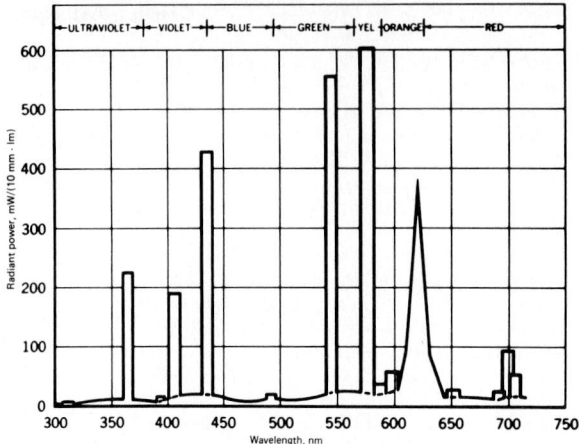

II. Phosphor-coated mercury.

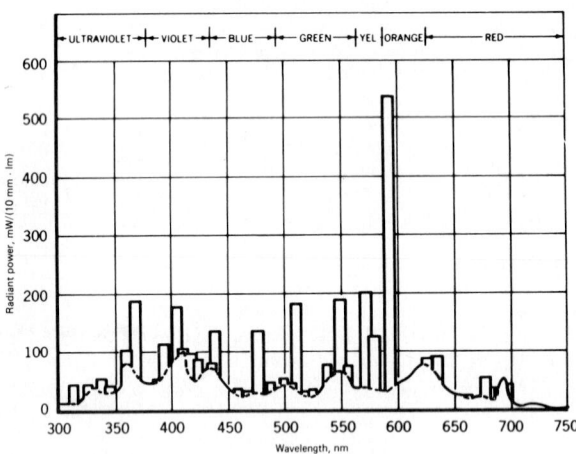

III. Clear metal halide.

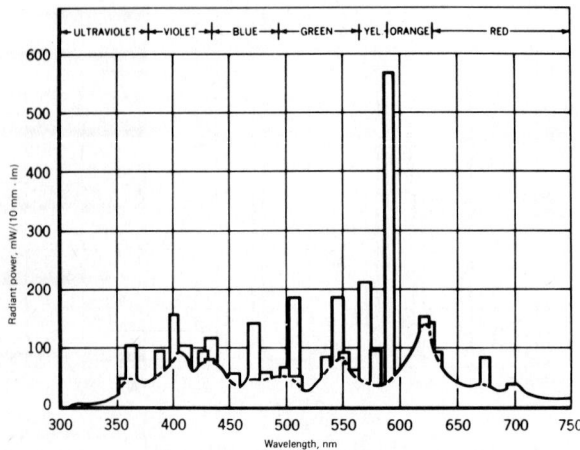

IV. Phosphor-coated metal halide.

**Fig. 10-63**  *(Continued)*

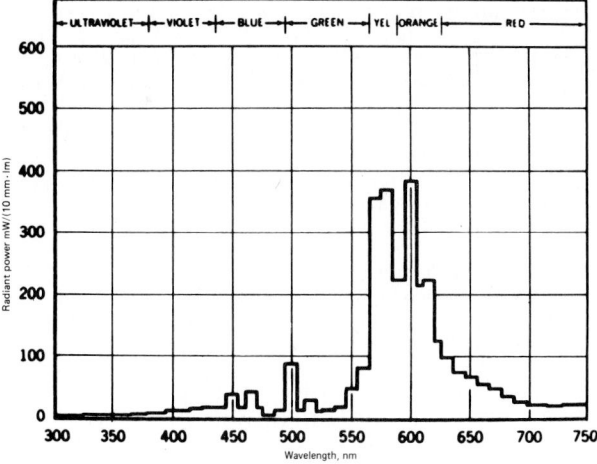

V. High-pressure sodium.

**Fig. 10-63**    *(Continued)*

**150. Application of HID lamps** for general lighting. HID lamps are widely used in flood-lighting and roadway, industrial, and sports lighting and to an increasing extent in indoor commercial applications. Owing to their greater efficacy, metal halide and high-pressure sodium lamps have replaced mercury lamps in new HID installations. In addition, many mercury installations are being changed to metal halide or HPS systems. Mercury lamps are used today primarily for replacement purposes.

HPS lamps are used predominantly in applications where efficacy and lowest lighting-

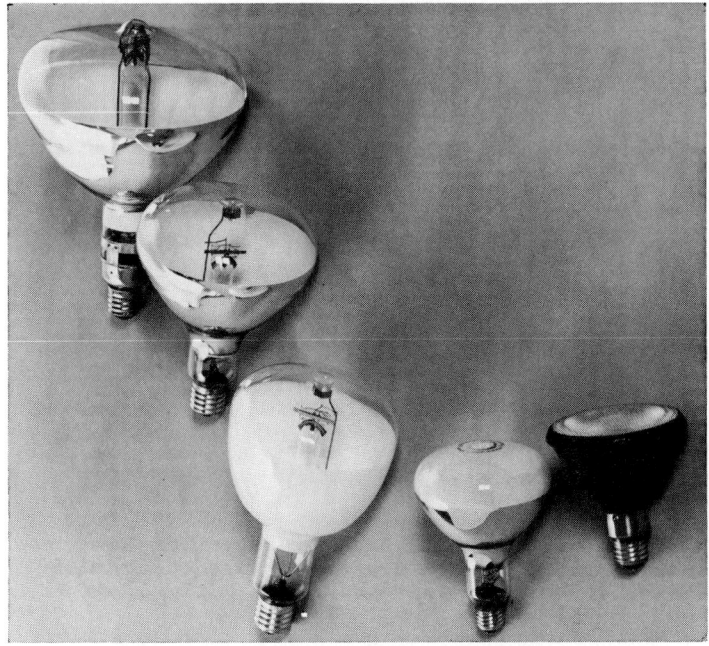

**Fig. 10-64**    Reflector-type mercury lamps. [General Electric Co.]

system operating costs are the primary factors affecting system choice. Metal halide systems are used primarily in installations where color appearance together with high efficiency are important, such as store applications or sports lighting for television.

**151. Reflector-type mercury lamps** are available in several wattages for use in applications where it is desirable to provide optical control of the light within the lamp. As in the case of incandescent reflector-type lamps, the sealed-in reflecting surface does not deteriorate during the life of the lamp. Reflector-type lamps are available in 100-, 175-, 250-, 400-, and 1000-W sizes. They are used primarily in floodlighting applications. Typical types are shown in Fig. 10-64. Data for reflector mercury lamps are shown in Sec. 156.

**152. Self-ballasted mercury lamps.** All mercury lamps require a ballast to provide proper starting voltage and control the current in the lamp. Self-ballasted mercury lamps are lamps which will start with the available line voltage and contain an internal tungsten filament in series with the arc tube to limit the lamp current. The filament ballast increases the losses in the lamp. Thus, the lamp efficacy of self-ballasted mercury lamps is much lower than that of other mercury lamps. Since there is no ballast in the circuit, self-ballasted mercury lamps, like incandescent lamps, must be matched to the voltage at the socket. Data on self-ballasted mercury lamps are shown in Sec. 157.

**153. Mercury Lamps**
(General Electric Co.)

| Lamp-ordering code | ANSI code | Nominal lamp watts | Bulb | Base | Rated average life,[a] | Approximate initial lumens[b] | Approximate mean lumens | Light-center length, in | Maximum overall length, in |
|---|---|---|---|---|---|---|---|---|---|
| | | | | Clear lamps | | | | | |
| HR100A38 | H38HT-R100 | 100 | E-23½ | Mogul | 24,000+ | 3,850 | 2,800 | 5 | 7½ |
| HR100A-38/A23 | H38LL-R100 | 100 | A-23 | Medium | 18,000 | 3,700 | 3,000 | 3½ | 5 7/16 |
| HR175A39 | H39KB-R175 | 175 | E-28 | Mogul | 24,000+ | 7,950 | 7,000 | 5 | 8¼ |
| HR250A37 | H37KB-R250 | 250 | E-28 | Mogul | 24,000+ | 11,200 | 9,850 | 5 | 8¼ |
| HR400A33 | H33CD-R400 | 400 | E-37 | Mogul | 24,000+ | 21,000 | 18,200 | 7 | 11 15/16 |
| HR1000A36[c] | H36GV-R1000 | 1000 | BT-56 | Mogul | 24,000+ | 57,000 | 44,450 | 9½ | 15 1/16 |
| | | | | Phosphor-coated lamps | | | | | |
| HR40/50DX45-46 | H45AY-R40/50/DX | 40/50 | E-17 | Medium | 16,000 | 40-W-1140<br>50-W-1575 | 910<br>1,250 | 3⅛ | 5⅛ |
| HR75DX43 | H46AV-R75/DX | 75 | E-17 | Medium | 16,000 | 2,800 | 2,250 | 3½ | 5 7/16 |
| HR100DX38 | H38JA-R100/DX | 100 | E-23½ | Mogul | 24,000+ | 4,200 | 3,200 | 5 | 7½ |
| HR100WDX38[d] | H38JA-R100/WDX | 100 | E-23½ | Mogul | 24,000+ | 3,400 | 2,600 | 5 | 7½ |
| HR100DX38/A23 | H38MP-R100/DX | 100 | A-23 | Medium | 18,000 | 4,000 | 3,050 | 3½ | 5 7/16 |
| HR175DX39 | H39KC-R175/DX | 175 | E-28 | Mogul | 24,000+ | 8,600 | 7,200 | 5 | 8¼ |
| HR175WDX39[d] | H38KC-R175/WDX | 175 | E-28 | Mogul | 24,000+ | 7,000 | 5,400 | 5 | 8¼ |
| HR250DX37 | H37KC-R250/DX | 250 | E-28 | Mogul | 24,000+ | 12,100 | 9,800 | 5 | 8¼ |
| HR250WDX37[d] | H37KC-R250/WDX | 250 | E-28 | Mogul | 24,000+ | 10,000 | 7,200 | 5 | 8¼ |
| HR400DX33 | H33GL-R400/DX | 400 | E-37 | Mogul | 24,000+ | 22,500 | 17,500 | 5 | 8¼ |
| HR400WDX33[d] | H33GL-R400/WDX | 400 | E-37 | Mogul | 24,000+ | 19,500 | 14,450 | 7 | 11 5/16 |
| HR700DX35 | H35ND-R700/DX | 700 | BT-46 | Mogul | 24,000+ | 42,000 | 30,250 | 9½ | 14 5/16 |
| HR1000DX36[c] | H36GW-R1000/DX | 1000 | BT-56 | Mogul | 24,000+ | 63,000 | 40,000 | 9½ | 15 1/16 |
| HR1000DX34[c] | H34GW-R1000/DX | 1000 | BT-56 | Mogul | 16,000 | 62,000 | 47,700 | 9⅜ | 15 1/16 |

**153. Mercury Lamps** (*Continued*)

| Lamp-ordering code | ANSI code | Nominal lamp watts | Bulb | Base | Rated average life,[a] | Approximate initial lumens[b] | Approximate mean lumens | Light-center length, in | Maximum overall length, in |
|---|---|---|---|---|---|---|---|---|---|
| | | | | | Self-extinguishing phosphor-coated lamps[e] | | | | |
| HT100DX38 | H38JA-T100/DX | 100 | E-23½ | Mogul | 16,000 | 3,700 | 3,100 | 5 | 7⅞ |
| HT175DX39 | H39KC-T100/DX | 175 | E-28 | Mogul | 16,000 | 7,700 | 6,450 | 5 | 8¾ |
| HT250DX37 | H37KC-T250/DX | 250 | E-28 | Mogul | 24,000 | 11,200 | 9,050 | 5 | 8¾ |
| HT400DX33 | H33GL-T400/DX | 400 | E-37 | Mogul | 24,000 | 20,000 | 11,400 | 7 | 11¹⁵⁄₁₆ |
| HT1000DX36[c] | H36GW-T1000/DX | 1000 | BT-56 | Mogul | 24,000 | 55,450 | 44,900 | 9½ | 15⅝₁₆ |

[a]Life when operated at 10 or more h per start on ballasts meeting published specifications.
[b]After 100 h of operation.
[c]The two 1000-W mercury types are not interchangeable on their respective ballasts. Check to ensure that the lamp is matched to the ballast.
[d]Warm phosphor, approximately 3300 K.
[e]Self-extinguishing lamps will cease operating within 15 min if the outer bulb is broken or punctured.

## 154. Metal Halide Lamps
(General Electric Co.)

| Lamp-ordering code | ANSI code | Nominal lamp watts | Bulb | Base | Burning position | Rated average life, h[a] | Approximate initial lumens,[dc] | Approximate mean lumens,[c] | Light-center length, in | Maximum overall length, in |
|---|---|---|---|---|---|---|---|---|---|---|
| | | | | | Clear lamps | | | | | |
| MKR100/BU/BD | MT90TW-100/BU-BD | 100 | E-17 | Medium | Vert. ±15° | 10,000 | 9,000 | 6,800 | 3 7/16 | 5 7/16 |
| MVR175/U | M57PE-R175/U | 175 | E-28 | Mogul | Any | 10,000 | 14,000 | 10,350 | 5 | 8 1/4 |
| MXR175/BU[d] | M57PE-R175/XBU | 175 | E-23 1/2 | Mogul | BU ±15° | 10,000 | 16,600 | 13,300 | 5 | 7 3/4 |
| MXR175/BD[d] | M57PE-R175/XBD | 175 | E-23 1/2 | Mogul | BD ±15° | 10,000 | 16,600 | 13,300 | 5 | 7 3/4 |
| MVR175/HOR | M57PE-R175HOR | 175 | ED-28 | Posit. mogul | Hor. ±15° | 10,000 | 15,000 | 12,000 | 5 | 8 1/4 |
| MVR250/U | M58PG-R250/U | 250 | E-28 | Mogul | Any | 10,000 | 20,500 | 17,000 | 5 | 8 1/4 |
| MVR250/HOR | M58PG-R250HOR | 250 | ED-28 | Posit. mogul | Hor. ±15° | 10,000 | 23,000 | 18,000 | 5 | 8 1/4 |
| MVR400/U | M59PJ-R400/U | 400 | E-37 | Mogul | Any | 20,000 | 36,000 | 28,800 | 7 | 11 5/16 |
| MVR400/VBU | M59PJ-R400/VBU | 400 | E-37 | Mogul | BU ±15° | 20,000 | 40,000 | 32,000 | 7 | 11 5/16 |
| MVR400/VBD | M59PJ-R400/VBD | 400 | E-37 | Mogul | BD ±15° | 20,000 | 40,000 | 32,000 | 7 | 11 5/16 |
| MVR400/HOR | M59PJ-R400HOR | 400 | ED-37 | Posit. mogul | Hor. ±15° | 20,000 | 40,000 | 32,000 | 7 | 11 1/2 |
| MPR400/VBU[k] | — | 400 | E-37 | Mogul | BU ±15° | 20,000 | 38,000 | 30,500 | 7 | 11 5/16 |
| MVR1000/U | M47PA-R1000/U | 1000 | BT-56 | Mogul | Any | 12,000 | 110,000 | 88,000 | 9 1/2 | 15 1/16 |
| MVR1000/VBU | M47PA-R1000/VBU | 1000 | BT-56 | Mogul | BU ±15° | 12,000 | 115,000 | 92,000 | 9 1/2 | 15 1/16 |
| MVR1000/VBD | M47PA-R1000/VBD | 1000 | BT-56 | Mogul | BD ±15° | 12,000 | 115,000 | 92,000 | 9 1/2 | 15 1/16 |
| MVR1500/HBU/E | M48PC-R1500/HBU | 1500 | BT-56 | Mogul | BU-HOR[e] | 3,000[f] | 155,000 | 140,000 | 9 1/2 | 15 1/16 |
| MVR1500/HBD/E | M48PC-R1500/HBD | 1500 | BT-56 | Mogul | BD-HOR[g] | 3,000[f] | 155,000 | 140,000 | 9 1/2 | 15 1/16 |
| | | | | | Phosphor-coated lamps | | | | | |
| MXR32/C/VBU | — | 32 | E-17 | Medium | BU ±15° | 10,000 | 2,500 | 1,900 | 3 7/16 | 5 7/16 |
| MXR100/C/BU/BD | MT90TX-100/BU-BD | 100 | E-17 | Medium | Vert. ±15° | 10,000 | 8,000 | 6,400 | 3 7/16 | 5 7/16 |
| MVR175/C/U | M57PF-R175/U | 175 | E-28 | Mogul | Any | 10,000 | 14,000 | 9,950 | 5 | 8 1/4 |
| MXR175/C/BU[d] | M57PF-R175/XBU | 175 | E-23 1/2 | Mogul | BU ±15° | 10,000 | 15,750 | 12,150 | 5 | 7 3/4 |
| MXR175/C/BD[d] | M57PF-R175/XBD | 175 | E-23 1/2 | Mogul | BD ±15° | 10,000 | 15,750 | 12,150 | 5 | 7 3/4 |

## 154. Metal Halide Lamps (*Continued*)

| Lamp-ordering code | ANSI code | Nominal lamp watts | Bulb | Base | Burning position | Rated average life, h[a] | Approximate initial lumens,[dc] | Approximate mean lumens,[c] | Light-center length, in | Maximum overall length, in |
|---|---|---|---|---|---|---|---|---|---|---|
| | | | | | Phosphor-coated lamps | | | | | |
| MVR175/C/HOR | M57PF-R175HOR | 175 | ED-28 | Posit. mogul | Hor. ± 15° | 10,000 | 15,000 | 11,300 | 5 | 8¼ |
| MVR250/C/U | M58PH-R250/U | 250 | E-28 | Mogul | Any | 10,000 | 20,500 | 16,000 | 5 | 8¼ |
| MVR250/C/HOR | M58PH-R250HOR | 250 | ED-28 | Posit. mogul | Hor. ± 15° | 10,000 | 23,000 | 17,000 | 5 | 8¼ |
| MVR400/C/U | M59PK-R400/U | 400 | E-37 | Mogul | Any | 20,000 | 36,000 | 27,700 | 7 | 11 5/16 |
| MVR400/C/VBU | M59PK-R400/VBU | 400 | E-37 | Mogul | BU ± 15° | 20,000 | 40,000 | 31,000 | 7 | 11 5/16 |
| MVR400/C/VBD | M59PK-R400/VBD | 400 | E-37 | Mogul | BD ± 15° | 20,000 | 40,000 | 31,000 | 7 | 11 5/16 |
| MVR400/C/HOR | M57PK-R400HOR | 400 | ED-37 | Posit. mogul | Hor. ± 15° | 20,000 | 40,000 | 31,000 | 7 | 11½ |
| MVR1000/C/U | M47PB-R1000/U | 1000 | BT-56 | Mogul | Any | 12,000 | 105,000 | 79,800 | 9½ | 15 1/16 |
| | | | | | Clear interchangeable metal halide lamps for mercury ballasts[h] | | | | | |
| MVR325/I/U/WM[i] | . . . . . . . . . . . . . . . . . . | 325 | E-37 | Mogul | Any | 20,000 | 28,000 | 18,200 | 7 | 11 5/16 |
| MVR400/I/U | M59PJ-R400/I/U | 400 | E-37 | Mogul | Any | 15,000 | 36,000 | 21,600 | 7 | 11 5/16 |
| MVR950/I/VBU | . . . . . . . . . . . . . . . . . . | 950 | BT-56 | Mogul | BU ± 15° | 12,000 | 107,000 | 85,600 | 9½ | 15 1/16 |
| MVR950/I/VBD | . . . . . . . . . . . . . . . . . . | 950 | BT-56 | Mogul | BD ± 15° | 12,000 | 107,000 | 85,600 | 9½ | 15 1/16 |
| | | | | | Phosphor-coated interchangeable metal halide lamps for mercury ballasts[h] | | | | | |
| MVR325/C/I/U/WM[i] | . . . . . . . . . . . . . . . . . . | 325 | E-37 | Mogul | Any | 20,000 | 28,000 | 17,600 | 7 | 11 5/16 |
| MVR400/C/I/U | M59PK-R400/I/U | 400 | E-37 | Mogul | Any | 15,000 | 36,000 | 20,900 | 7 | 11 5/16 |

| | | | | | | | | | | |
|---|---|---|---|---|---|---|---|---|---|---|
| | | | | Clear self-extinguishing metal halide lamps[j] | | | | | | |
| MVT325/I/U/WM[i] | .................. | 325 | E-37 | Mogul | Any | 15,000 | 26,700 | 17,350 | 7 | $11\frac{5}{16}$ |
| MVT400/I/U | M59PJ-T400/I/U | 400 | E-37 | Mogul | Any | 15,000 | 34,300 | 20,600 | 7 | $11\frac{5}{16}$ |
| MVT400/VBU | M59PJ-T400/VBU | 400 | E-37 | Mogul | BU ± 15° | 20,000 | 37,400 | 29,900 | 7 | $11\frac{5}{16}$ |
| MVT1000/VBU | M47PA-T1000/VBU | 1000 | BT-56 | Mogul | BU ± 15° | 10,000 | 105,000 | 83,000 | $9\frac{1}{2}$ | $15\frac{1}{16}$ |
| | | | | Phosphor-coated self-extinguishing metal halide lamps[j] | | | | | | |
| MVT325/C/I/U/WM[i] | .................. | 325 | E-37 | Mogul | Any | 15,000 | 26,700 | 16,800 | 7 | $11\frac{5}{16}$ |
| MVT400//C/I/U | M59PK-T400/I/U | 400 | E-37 | Mogul | Any | 15,000 | 34,300 | 19,750 | 7 | $11\frac{5}{16}$ |
| MVT400/VBU | M59PK-T400/VBU | 400 | E-37 | Mogul | BU ± 15° | 20,000 | 37,400 | 28,800 | 7 | $11\frac{5}{16}$ |
| | | | | Double-ended metal halide lamps[l] | | | | | | |
| MQI/70/T6/30 | M85-PX70 | 70 | T-6 | Recessed single contact | Hor. ± 45° | 6,000 | 5,000 | 4,200 | 0.3 | 4.7 |
| MQI/150/T7/43 | M81-PS150 | 150 | T-7 | Recessed single contact | Hor. ± 45° | 6,000 | 11,250 | 10,000 | 0.7 | 5.4 |

[a]Life when operated in a vertical burning position at 10 or more h per start (except 1500-W lamps) on ballasts meeting published specifications. See manufacturer's data for life ratings in other burning positions.
[b]After 100 h of operation.
[c]Lumen ratings are given for vertical operation. See manufacturer's data for ratings in other burning positions.
[d]Improved-efficiency, warm-color 3100 K design
[e]Base up to 15° below horizontal.
[f]Life when operated at 5 or more h per start on ballasts meeting published specifications.
[g]Base down to 15° above horizontal.
[h]For use only on specified mercury ballasts. See manufacturer's data for ballast requirements.
[i]Do not use on 400-W metal halide ballasts.
[j]Self-extinguishing lamps will cease operating within 15 min if the outer bulb is broken or punctured.
[k]Protective shroud around arc tube.
[l]Operate only in enclosed fixtures with UV-absorbing glass.

## 155. High-Pressure Sodium Lamps
(General Electric Co.)

| Lamp-ordering code | ANSI code | Nominal lamp watts | Bulb | Base | Rated average life, h[a] | Approximate initial lumens[b] | Approximate mean lumens | Light-center length, in | Maximum overall length, in |
|---|---|---|---|---|---|---|---|---|---|
| | | | | Clear lamps | | | | | |
| LU35/MED | S76HA-35 | 35 | E-17 | Medium | 24,000 | 2,250 | 2,025 | 3 7/16 | 5 7/16 |
| LU50 | S68MS-50 | 50 | E-23 1/2 | Mogul | 24,000+ | 4,000 | 3,600 | 5 | 7 3/4 |
| LU50/MED | S68XX-50 | 50 | E-17 | Medium | 24,000+ | 4,000 | 3,600 | 3 7/16 | 5 7/16 |
| LU70 | S62ME-70 | 70 | E-23 1/2 | Mogul | 24,000+ | 6,400 | 5,450 | 5 | 7 3/4 |
| LU70/MED | S62LG-70 | 70 | E-17 | Medium | 24,000+ | 6,400 | 5,450 | 3 7/16 | 5 7/16 |
| LU70DX/MED | S62LG-70DX | 70 | E-17 | Medium | 10,000 | 3,800 | 3,040 | 3 1/2 | 5 1/2 |
| LU70/SBY[i] | — | 70 | E-23 | Mogul | 24,000+ | 5,600 | 5,050 | 5 | 7 1/2 |
| LU100 | S54SB-100 | 100 | E-23 1/2 | Mogul | 24,000+ | 9,500 | 8,550 | 5 | 7 3/4 |
| LU100/SBY[i] | — | 100 | E-23 | Mogul | 24,000+ | 9,100 | 8,190 | 5 | 7 1/2 |
| LU100/MED | S54SG-100 | 100 | E-17 | Medium | 24,000+ | 9,500 | 8,550 | 3 7/16 | 5 7/16 |
| LU150/55[c] | S55SC-150 | 150 | E-23 1/2 | Mogul | 24,000+ | 16,000 | 14,400 | 5 | 7 3/4 |
| LU150/55DX | S55SC-150DX | 150 | E-23 1/2 | Mogul | 10,000 | 10,500 | 9,135 | 5 | 7 3/4 |
| LU150/55/SBY[i] | — | 150 | E-23 | Mogul | 24,000+ | 15,600 | 14,000 | 5 | 7 1/2 |
| LU150/55/MED | ............ | 150 | E-17 | Medium | 24,000+ | 16,000 | 14,400 | 3 1/2 | 5 3/4 |
| LU150/55DX/MED | S55RN-150DX | 150 | E-17 | Medium | 10,000 | 10,500 | 9,135 | 3 1/2 | 5 1/2 |
| LU150/100[c] | S56SD-150 | 150 | E-28 | Mogul | 24,000+ | 15,000 | 13,500 | 5 | 8 5/16 |
| LU200 | S66MN-200 | 200 | E-18 | Mogul | 24,000+ | 22,000 | 19,800 | 5 3/4 | 9 3/4 |
| LU250 | S50VA-250 | 250 | E-18 | Mogul | 24,000+ | 27,500 | 24,750 | 5 3/4 | 9 3/4 |
| LU250/S[d] | S50VA-250/S | 250 | E-18 | Mogul | 24,000+ | 30,000 | 27,000 | 5 3/4 | 9 3/4 |
| LU250/DX[e] | S50VA-250/DX | 250 | E-18 | Mogul | 10,000 | 22,500 | 20,700 | 5 3/4 | 9 3/4 |
| LU250/SBY[i] | — | 250 | E-18 | Mogul | 24,000+ | 27,500 | 24,750 | 5 3/4 | 9 3/4 |
| LU310 | S67MR-310 | 310 | E-18 | Mogul | 24,000+ | 37,000 | 33,300 | 5 3/4 | 9 3/4 |
| LU400 | S51WA-400 | 400 | E-18 | Mogul | 24,000+ | 50,000 | 45,000 | 5 3/4 | 9 3/4 |
| LU400/DX | — | 400 | E-28 | Mogul | 10,000 | 37,400 | 34,400 | 5 1/4 | 9 |
| LU400/SBY[i] | — | 400 | E-18 | Mogul | 24,000+ | 50,000 | 45,000 | 5 3/4 | 9 3/4 |
| LU1000 | S52XB-1000 | 1000 | E-25 | Mogul | 24,000+ | 140,000 | 126,000 | 8 3/4 | 15 1/16 |

Diffuse-coated lamps

| Lamp | Order code | Watts | Base | Socket | Life (h) | Lumens | Lumens | Dim. (in) | Dim. (in) |
|---|---|---|---|---|---|---|---|---|---|
| LU35/D/MED | S76HB-35 | 35 | E-17 | Medium | 24,000 | 2,150 | 1,935 | 3 3/16 | 5 5/16 |
| LU50/D | S68MT-50 | 50 | E-23½ | Mogul | 24,000+ | 3,800 | 3,420 | 5 | 7¼ |
| LU50/D/MED | S68YY-50 | 50 | E-17 | Medium | 24,000+ | 3,800 | 3,420 | 3 3/16 | 5 5/16 |
| LU70/D | S62MF-70 | 70 | E-23½ | Mogul | 24,000+ | 5,950 | 5,050 | 5 | 7¼ |
| LU70/D/MED | S62LG-70 | 70 | E-17 | Medium | 24,000+ | 5,950 | 5,050 | 3 3/16 | 5 5/16 |
| LU70DX/D/MED | — | 70 | E-17 | Medium | 10,000 | 3,600 | 2,880 | 3 3/16 | 5 5/16 |
| LU100/D | S54MC-100 | 100 | E-23½ | Mogul | 24,000+ | 8,800 | 7,920 | 5 | 7¼ |
| LU100/D/MED | S54SH-100 | 100 | E-17 | Medium | 24,000+ | 8,800 | 7,920 | 3 3/16 | 5 5/16 |
| LU150/55/Dᶜ | S55MD-150 | 150 | E-23½ | Mogul | 24,000+ | 15,000 | 13,500 | 5 | 7¼ |
| LU150/D/MED | ·········· | 150 | E-17 | Medium | 24,000+ | 15,000 | 13,500 | 3½ | 5¾ |
| LU150DX/D/MED | — | 150 | E-17 | Medium | 10,000 | 9,900 | 8,600 | 3 3/16 | 5 5/16 |
| LU250/D | S50VC-250 | 250 | E-28 | Mogul | 24,000+ | 26,000 | 23,400 | 5 | 9 |
| LU250DX/D | — | 250 | E-28 | Mogul | 10,000 | 20,000 | 18,400 | 5 | 9 |
| LU400/D | S51WB-400 | 400 | E-37 | Mogul | 24,000+ | 47,500 | 42,750 | 7 | 11 5/16 |
| LU400DX/D | — | 400 | E-28 | Mogul | 10,000 | 35,500 | 32,600 | 5 | 9 |

Interchangeable HPS lamps for mercury ballasts

| Lamp | Order code | Watts | Base | Socket | Life (h) | Lumens | Lumens | Dim. (in) | Dim. (in) |
|---|---|---|---|---|---|---|---|---|---|
| LUH150/EZᶠ | S63MG-150 | 150 | E-28 | Mogul | 12,000 | 12,000 | 10,800 | 5 | 9 |
| LUH150/D/EZᶠ·ᵍ | S63HG-150 | 150 | E-28 | Mogul | 12,000 | 11,000 | 9,900 | 5 | 9 |
| LUH215/EZʰ | S66ML-215 | 215 | E-28 | Mogul | 12,000 | 19,000 | 17,100 | 5 | 9 |

[a] Life when operated at 10 or more h per start on ballasts meeting published specifications.
[b] After 100 h of operation.
[c] The two 150-W high-pressure sodium types are not interchangeable on their respective ballasts. Check to ensure that the lamp is matched to the ballast.
[d] Improved-light-output design.
[e] Improved-color-rendition design.
[f] For use only on specified 175-W mercury ballasts. See manufacturer's data for ballast requirements.
[g] Diffuse-coated lamp.
[h] For use only on specified 250-W mercury ballasts. See manufacturer's data for ballast requirements.
[i] Dual arc tube design for standby hot restrike.

**156. Reflector Mercury Lamps**
(General Electric Co.)

| Lamp-ordering code | ANSI code | Nominal lamp watts | Bulb | Base | Description | Rated average life, h[a] | Approximate initial lumens[b] | Approximate mean lumens | Maximum overall length, in |
|---|---|---|---|---|---|---|---|---|---|
| HR75/100PFL43-44 | H43/44-R75/100PFL | 75/100 | PAR-38 | Medium side-prong | Clear flood | 12,000 | 75-W—1,600; 100-W—2,450 | 1,100; 1,700 | 4 11/32 |
| HR100RFL38 | H38BM-R100 | 100 | R-40 | Medium | Clear flood | 24,000+ | 2,850 | 2,000 | 7 |
| HR100RDXFL38 | H38BP-R100/DX | 100 | R-40 | Medium | Phosphor flood | 24,000+ | 2,850 | 2,050 | 7 |
| HR100PFL44 | H44JM-R100 | 100 | PAR-38 | Admedium | Clear flood | 12,000 | 2,450 | 1,700 | 5 3/16 |
| HR100PSP44 | H44GS-R100 | 100 | PAR-38 | Admedium | Clear spot | 12,000 | 2,450 | 1,700 | 5 3/16 |
| HR175RFL39 | H39BM-R175 | 175 | R-40 | Medium | Clear flood | 24,000+ | 5,700 | 4,800 | 7 |
| HR175RFL39/M | H39BS-R175 | 175 | R-40 | Mogul | Clear flood | 24,000+ | 5,700 | 4,800 | 7 1/2 |
| HR175RDXFL39 | H39BP-R175/DX | 175 | R-40 | Medium | Phosphor flood | 24,000+ | 5,700 | 4,350 | 7 |
| HR400RDX33 | H33DN-R400DX | 400 | R-52 | Mogul | Phosphor high-bay | 24,000+ | 22,000 | 17,100 | 11 3/4 |
| HR400R33 | H33FY-R400 | 400 | R-52 | Mogul | Inside-frosted high-bay | 24,000+ | 18,300 | 15,900 | 11 3/4 |
| HR400RDXFL33 | H33FS-R400/DX | 400 | R-60 | Mogul | Phosphor flood | 24,000+ | 15,500 | 8,950 | 10 3/4 |
| HR400RSP33 | H33FP-R400 | 400 | R-60 | Mogul | Clear spot | 24,000+ | 15,300 | 11,050 | 10 3/8 |
| HR1000RDX36 | H36KY-R1000/DX | 1000 | BT-56 | Mogul | Phosphor semireflector | 24,000+ | 58,500 | 32,750 | 15 3/16 |
| HR1000RFL36 | . . . . . . . . . . . . . . . | 1000 | R-80 | Mogul | Clear flood | 24,000+ | 42,000 | 23,500 | 13 5/8 |
| HR1000RSDX36 | . . . . . . . . . . . . . . . | 1000 | R-80 | Mogul | Phosphor medium flood | 24,000+ | 43,000 | 20,200 | 13 5/8 |

[a]Life when operated at 10 or more h per start on ballasts meeting published specifications.
[b]After 100 h of operation.

**157. Self-Ballasted Phosphor-Coated Mercury Lamps**
(General Electric Co.)

| Lamp-ordering code | Nominal lamp watts | Bulb | Base | Voltage | Rated average life, h[a] | Approximate initial lumens | Approximate mean lumens | Light-center length, in | Maximum overall length, in |
|---|---|---|---|---|---|---|---|---|---|
| HSB160/M | 160 | PS-30 | Medium | 120 | 12,000 | 2,300 | 1,600 | 5⅞ | 8⅝ |
| HSB250 | 250 | E-28/BT-28 | Mogul | 120 | 12,000 | 5,000 | 3,750 | 5⁵⁄₁₆ | 8⅜ |
| HSB250/M | 250 | E-28/BT-28 | Medium | 120 | 12,000 | 5,000 | 3,750 | 5⁵⁄₁₆ | 8⅜ |
| HSB450 | 450 | BT-37 | Mogul | 120 | 16,000 | 9,100 | 8,280 | 7⅜ | 11¹³⁄₃₂ |
| HSB750R/120 | 750 | R-57 | Mogul | 120 | 16,000 | 14,000 | 11,200 | 8⅜ | 12¹³⁄₃₂ |
| HSB750R/240 | 750 | R-57 | Mogul | 240 | 16,000 | 17,300 | 15,400 | 8⅜ | 12¹³⁄₃₂ |

[a]Life when operated at 10 or more h per start.

**158. High-intensity–discharge lamp ballasts.** All HID lamps have a negative-resistance characteristic. As a result, unless a current-limiting device is used, the lamp current will increase until the lamp is destroyed. Ballasts for HID lamps provide three basic functions: to control lamp current to the proper value, to provide sufficient voltage to start the lamp, and to match the lamp voltage to the line voltage. Ballasts are designed to provide proper electrical characteristics to the lamp over the range of primary voltage stated for each ballast design. Most HID ballasts are designed into the luminaire. Typical ballasts are shown in Fig. 10-65.

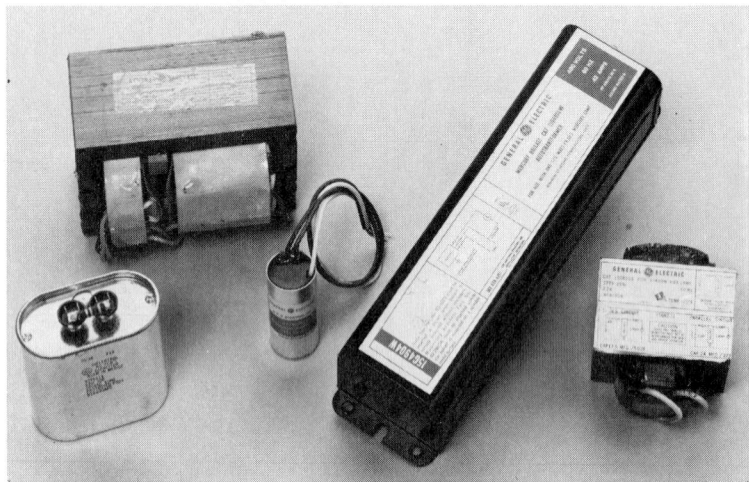

**Fig. 10-65**   Ballasts for high-intensity–discharge lamps. [General Electric Co.]

Ballasts are classified into three major categories depending on the basic circuit involved: nonregulating, lead-type regulating, and lag-type regulating. Each type has different operating characteristics.

Circuits for mercury-lamp ballasts are shown in Fig. 10-66. Reactor ballasts can be used if sufficient line voltage is available to start and operate the mercury lamp. These ballasts have a low power factor unless corrected by a capacitor. The lag ballast consists of an autotransformer plus a reactor combined in a single structure. Performance is similar to that of a reactor, and the ballast is used where normal line voltage is lower than the required lamp-starting voltage. This ballast also has a low power factor unless corrected by a capacitor. Lag and reactor ballasts should be used when line-voltage regulation is good, since lamp wattage will vary $\pm$ 10 percent with the $\pm$ 5 percent in line voltage.

The regulator ballast, also referred to as the constant-wattage (CW) or stabilized ballast, has primary and secondary windings electrically isolated from each other. A capacitor is used together with the magnetic portion of the ballast to control the lamp current. Owing to the capacitor, regulator ballasts have a high power factor. The basic advantage of regulator ballasts is their excellent regulation of lamp wattage with changes in line voltage. Lamp watts vary only $\pm$ 2 to 3 percent with line-voltage changes of $\pm$ 13 percent.

Autoregulator ballasts combine an autotransformer with the regulator circuit. As a result, these ballasts are smaller and less costly than the regulator design and have lower losses. They are also called constant-wattage–autotransformer (CWA) ballasts. The regulation of lamp wattage with line voltage is very good although somewhat less than with the regulator design: approximately a $\pm$ 5 percent change in lamp wattage with a $\pm$ 10 percent change in line voltage. Autoregulator ballasts have a high power factor. Multiple primary-voltage ratings are available by using a series of taps in the transformer. Ballasts designed for 120/208/240/277 V are available.

Two-lamp mercury-lamp ballasts are available. The circuit is basically the same as that of the single-lamp regulator ballast except that two lamps are operated in series.

Metal halide ballasts use the same circuit as mercury autoregulator ballasts. The magnetic design is different to provide the required peaked open-circiut voltage for satisfactory lamp starting and operation. Metal halide ballasts provide a regulation of lamp wattage between those of an autoregulator and a reactor or lag ballast. With a 10 percent change in low voltage, lamp wattage will vary by about 10 percent. All-metal-halide ballasts have a high power factor.

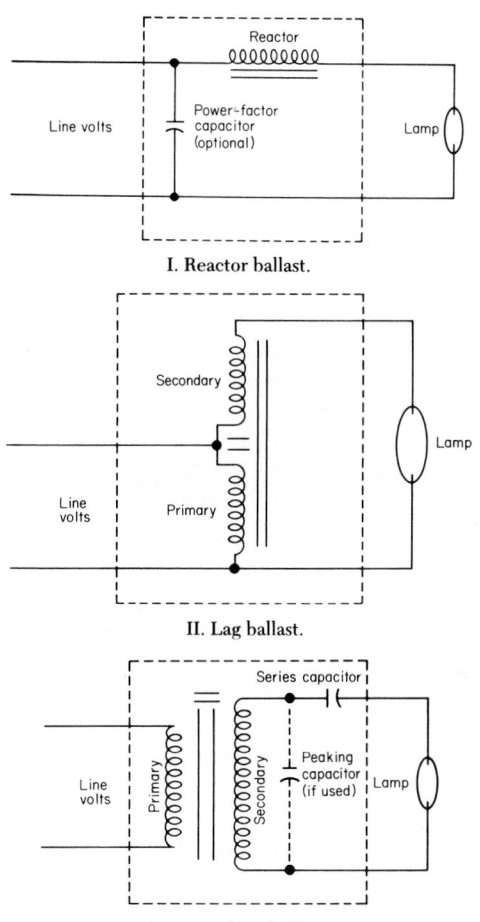

I. Reactor ballast.

II. Lag ballast.

III. Regulator ballast.

**Fig. 10-66**   Mercury-lamp ballast circuits. [General Electric Co.]

Ballasts for high-pressure sodium lamps, in addition to providing the basic functions of all ballasts, have an additional requirement. In contrast to mercury and metal halide lamps, the voltage of high-pressure sodium lamps increases significantly during lamp life. The ballast must be able to handle this change and operate the lamp within specified limits. The voltage required to start HPS lamps is much greater than that required for mercury and metal halide types. This necessitates an auxiliary starting circuit which supplies a low-energy, high-voltage pulse of several thousand volts. This starting aid is a separate device from the magnetic and capacitor portions of the ballast. As the result of the high-voltage starting pulse required by high-pressure sodium lamps, the lampholder for these lamps must be rated to handle this voltage.

The circuits of high-pressure sodium ballasts are shown in Fig. 10-67.

As with mercury reactor ballasts, reactor ballasts for high-pressure sodium lamps can be used only when there is sufficient line voltage for proper lamp operation. They have a low power factor, unless corrected, and have poor regulation of lamp wattage with line-voltage variation. The required starting aid is part of the circuit.

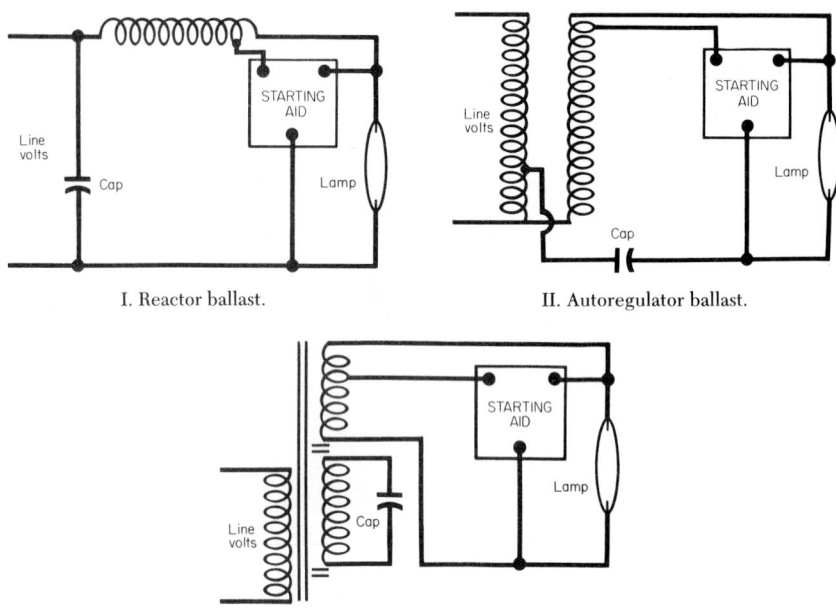

I. Reactor ballast.          II. Autoregulator ballast.

III. Magnetic regulator ballast.

**Fig. 10-67**   Circuits of high-pressure sodium lamp ballasts. [General Electric Co.]

Autoregulator and magnetic-regulator HPS ballasts are similar to mercury autoregulator and regulator ballasts, respectively, with the magnetic design changed to provide the proper lamp characteristics and the addition of the starting aid. They have regulation characteristics similar to those of mercury ballasts and have a high power factor.

As indicated with the various ballast designs, the ability of the ballast to control lamp wattage and thus light output as the line voltage changes is referred to as lamp-wattage regulation. The effect of line-voltage variation on lamp wattage for the various HID ballast designs is shown in Fig. 10-68.

Data on HID ballasts can be obtained from ballast manufacturers.

**159. HID lamp warm-up and restrike.**   All high-intensity lamps require time for the arc to stabilize at its operating value and achieve full light output. This requires several minutes and is controlled by the ballast type. During the warm-up period, ballast line current and wattage will change continuously until they reach the stable operating point. Figure 10-69 shows the characteristics for a typical lag autotransformer or reactor ballast and a regulator ballast.

**160. Low voltage-dip tolerance.**   The ability of an HID lamp to tolerate a dip in line voltage depends on ballast design. Ballasts are usually rated on their ability to tolerate dips of up to 4 s while keeping a lamp operating. Generally, nonregulating ballasts will tolerate dips of only 10 to 20 percent, while regulating ballasts will sustain lamp operation during dips in line voltage of as much as 30 to 50 percent for lag-type regulator ballasts and 20 to 40 percent for lead-type regulator ballasts. However, in HPS systems ballast dip tolerance changes over time owing to the increase in lamp voltage. After midlife of the lamp, regulating and lead types of regulator ballasts become susceptible to lamp dropout with more than 10 to 15 percent dips in line voltage. Dip tolerance of lag-type regulator HPS ballasts also changes but remains above the tolerance for lead-type regulator ballasts.

When HID lamps are extinguished owing to a voltage dip or turning the fixture off, there is a delay in restrike time until the arc tube can cool and internal pressure can drop to the level where the arc will restrike with the available voltage from the ballast. This time varies with different HID lamps. Mercury lamps will restrike in 3 to 7 min depending on the type. Metal halide lamps can take as long as 15 min to restrike, while high-pressure sodium lamps will restrike in the shortest time, usually within 1 min.

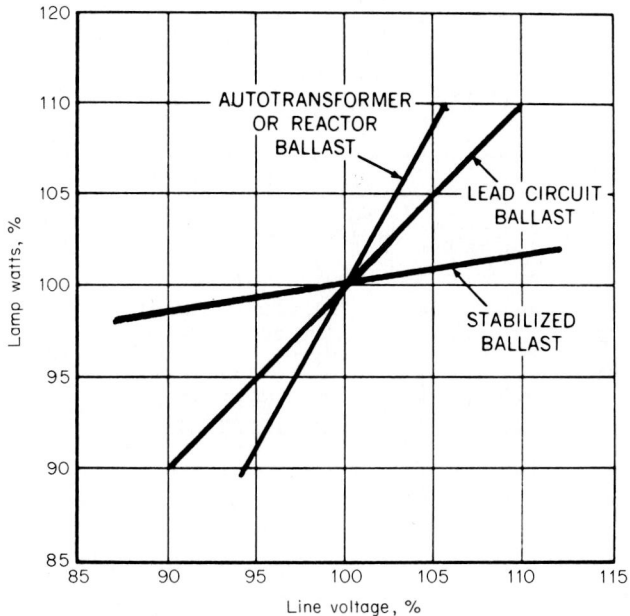

**Fig. 10-68**   Effect of line-voltage variation on HID lamps. [General Electric Co.]

**161. Line fusing of HID ballasts.**   There are times when line fusing of an HID ballast may be desirable, as in high-bay or other relatively inaccessible installations. The principal reason for fusing is to remove a faulty luminaire from a circuit before it causes a branch-circuit outage. A fused ballast aids in troubleshooting when a faulty ballast is encountered. Fuse ratings should meet ballast manufacturers' recommendations. Slow-blow types are desirable when available to minimize nonfault opening.

**162. Remote installation of HID ballasts.**   Although most HID lighting fixtures have the ballast installed in the fixture, in occasional applications it is desirable to install the ballast remote from the fixture. With mercury and metal halide systems, correctly sized wire is required to maintain proper ballast voltage at the lamp socket as the primary design factor. In HPS systems, the maximum distance is limited by the characteristics of the starting aid which provides the low-energy starting pulse. The ballast manufacturer should be consulted for the remote-mounting distance limitation of HPS ballasts.

**163. Effect of temperature.**   Excessive temperatures can result in lamp failure or unsatisfactory performance. Most HID lamps have a maximum bulb-temperature limit of 400°C (752°F) and a maximum base-temperature limit of 210°C (410°F) for mogul-base types and 190°C (374°F) for medium-base designs.

Most HID ballasts have sufficient voltage to start lamps at −29°C (−20°F). HID ballasts have maximum operating-temperature limits for reliable life performance. Ballasts designed for mounting within fixtures have a coil-temperature limit of 165°C (329°F). Enclosed ballasts are generally designed for ambient temperatures of 40°C (104°F) or less. Special high-temperature enclosed-ballast designs are available for some lamp types for ambient temperatures as high as 60°C (140°F).

**164. High-intensity–discharge system troubleshooting.** HID lighting systems include the power supply system (wiring, circuit breakers, and switches), lighting fixture (socket, reflector, refractor or lens, and housing), ballast, lamp, and, in outdoor fixtures, frequently a photoelectric cell to turn on the fixture at dusk. When an HID system does not operate as expected, the source of the problem can be in any part of the total system.

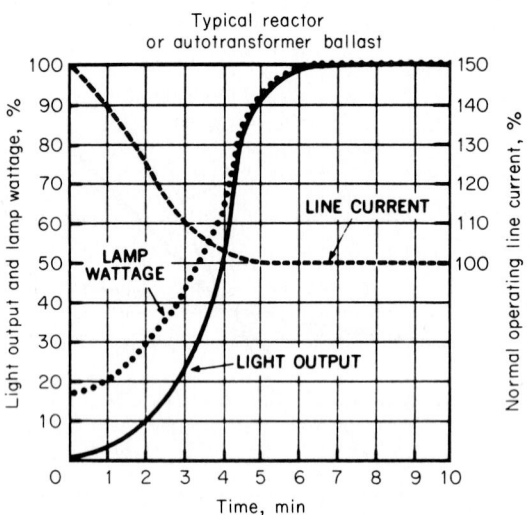

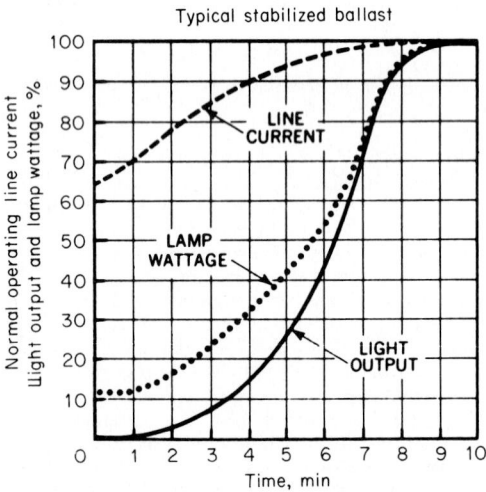

**Fig. 10-69**  HID-lamp light output, lamp wattage, and line current during the lamp warm-up period.

It is important to understand normal lamp-failure characteristics to determine whether or not operation is abnormal. All HID lamps have expected lamp-failure patterns over life; these are published by lamp manufacturers. Rated life represents the expected failure point for one-third to one-half of the lamps, depending on the lamp type and the lamp manufacturer's rating.

End-of-life characteristics vary for the different HID lamp types:

1. *Mercury.* Normal end of life is a nonstart condition or very low light output resulting from blackening of the arc tube owing to electrode deterioration during life.

2. *Metal halide.* Normal end of life is a nonstart condition resulting from a change in the electrical characteristic so that the ballast can no longer sustain the lamp. Lamp color at the end of life will usually be warmer (pinker) than that of a new lamp owing to arc-tube blackening during life that changes thermal balance in the arc tube. Metal halide lamps can have a nonpassive failure mode at the end of life. This varies with lamp type and burning position. The lamp manufacturer's recommendations regarding metal-halide lamp enclosure should be reviewed.

3. *High-pressure sodium.* Normal end of life is on-off cycling. This results when an aging lamp requires more voltage to stabilize and operate than the ballast is able to provide. When the lamp's normally rising voltage exceeds the ballast output voltage, the lamp is extinguished. Then, after a cool-down period of about 1 min, the arc will restrike and the cycle is repeated. This cycle starts slowly at first and increases in frequency, if the lamp is not replaced, until ultimately the lamp fails owing to overheating of the arc-tube seal.

There are four basic visual variations in the lamp of an HID lighting system which indicates that a problem may exist: (1) the lamp does not start, (2) the lamp cycle is on and off or is unstable, (3) the lamp is extra bright, or (4) the lamp is dim. The following table indicates the most likely possible causes for each of these system conditions.

| HID-system conditions | Possible causes | |
| --- | --- | --- |
| | Other than lamp | Lamp |
| Lamp does not start. | Ballast failure<br>Incorrect or loose wiring<br>Low supply voltage<br>Low ambient<br>  temperature<br>Circuit breakers tripped<br>Inoperative photocell<br>Starting-aid failure<br>  (HPS) | Lamp loose in socket<br>Incorrect lamp<br>Normal end of life<br>Lamp internal structure<br>  broken |
| Lamp cycle is on and<br>  off or is unstable. | Low supply voltage<br>Incorrect ballast<br>High supply voltage<br>  (HPS)<br>Ballast voltage low<br>System voltage dipping<br>Fixture concentrating<br>  energy on lamp (HPS) | Normal end of life (HPS)<br>Lamp operating voltage<br>  too high (HPS)<br>Lamp arc tube unstable |
| Lamp is extra bright. | Shorted or partially<br>  shorted ballast or<br>  capacitor<br>Overwattage operation | Incorrect lamp<br>High lamp voltage |
| Lamp is dim. | Low supply voltage<br>Incorrect ballast<br>Low ballast voltage to<br>  lamp<br>Dirt accumulation<br>Ballast capacitor shorted<br>Corroded connection in<br>  fixture | Incorrect lamp<br>Low lamp voltage<br>Lamp a hard starter |

**165. Low-pressure sodium lamps** (Fig. 10-70) utilize sodium vapor to conduct the arc. They provide very high efficacy with light that is almost totally yellow in color. The lamps

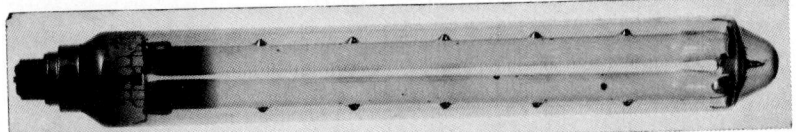

**Fig. 10-70**  Low-pressure sodium lamp. [North American Philips Lighting Corp.]

are constructed with two glass envelopes. The inner arc tube is usually U-shaped and is made of a special sodium-resistant glass. Glass can be used in low-pressure sodium lamps because the sodium vapor operates at temperatures and pressures substantially lower than those of a high-pressure sodium lamp. Electrodes are sealed into the ends of the arc tube. The arc tube is filled with a starting gas of neon with a small amount of argon. Sodium metal is placed in the arc tube. To reduce heat loss, the outer jacket is evacuated and the inside of the outer bulb is coated with an indium oxide coating that transmits light but reflects infrared energy. Low-pressure sodium lamps are available in wattages from 18 through 185 W. As with any other discharge lamp, low-pressure sodium lamps must be operated on a ballast designed to meet the lamp-starting and -operating requirements. Owing to the monochromatic yellow color of the light from low-pressure sodium lamps, they are generally used in applications where the appearance of people and colors is not important.

## NEON LAMPS

**166. Neon lamps may be classified into two types:** (1) high-voltage and (2) low-voltage.

**167. High-voltage neon lamps** (Fig. 10-71) consist of two terminals or electrodes set into the opposite ends of a glass tube which contains neon, helium, or argon gas, with or without mercury, at low vapor pressure.

As in all gaseous-discharge lamps a higher voltage is required to start the arc than is required for continuous operation. These lamps are supplied from a high-reactance trans-

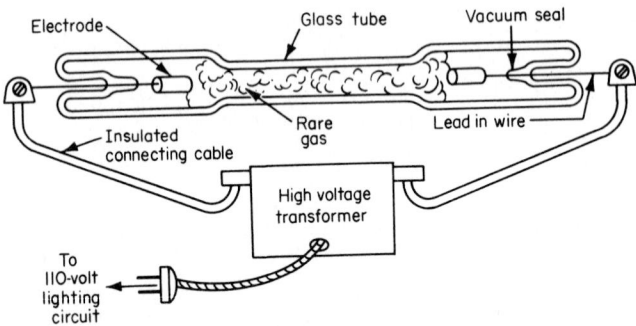

**Fig. 10-71**  Schematic diagram of gas-filled tube and transformer. [S. C. Miller and D. G. Fink, *Neon Signs*]

former in which the secondary voltage falls rapidly as the current drain is increased. This feature provides the higher voltage required to start the lamp and tends to make the arc circuit stable. Even if the secondary terminals should become short-circuited, no more than full-load current will flow, so that the transformer will not be damaged. A high voltage, 2000 to 15,000 V depending on the length and diameter of the tubing and the gas used, is applied to the electrodes at the ends of the tubing by means of this transformer.

The tubing can be made to produce a number of different colors, depending on the gas used, the color of the glass tubing, and whether or not mercury is added. Table 168 gives the colors available.

**168.  Colors Obtainable from Neon Tubes**
(From S. C. Miller and D. G. Fink, *Neon Signs*)

| Color produced | Gas used | Mercury used | Color of glass tubing used | Gas pressure, mm of mercury |
|---|---|---|---|---|
| | | Standard colors | | |
| Red | Neon | No | Clear | 10 and over |
| Dark red | Neon | No | Soft red | 10 and over |
| Gold | Helium | No | Soft yellow (noviol) | 3 |
| White | Helium | No | Clear | 3 |
| Light green | Argon | Yes | Soft canary (uranium oxide) | 8 |
| Medium green | Argon | Yes | Soft yellow (noviol) | 8 |
| Light blue | Argon | Yes | Clear | 8 |
| Dark blue | Argon | Yes | Soft blue | 8 |
| | | Colors available but not widely used | | |
| Soft red | Neon | No | Soft opal | 10 and over |
| Orange | Neon | No | Soft yellow (noviol) | 10 and over |
| Soft white | Helium | No | Soft opal | 3 |
| Dark green | Argon | Yes | Soft medium amber | 8 |
| Red lavender | Neon | No | Soft dark purple | 10 and over |

**169. The gas pressure** as given in Table 170 must be determined quite accurately during manufacture. If the pressure is too low, the resistance of the tubing and the cathode voltage drop will be high and the tendency to sputter will be great. Sputtering means that the metal of the electrode evaporates and deposits on the tube, causing the ends of the tube to blacken. This process lowers the gas pressure still further until the tube flickers and finally fails to light. If the pressure is too high, the light will not be brilliant. The gas pressure can be tested by holding a special type of spark coil (Fig. 10-72) against either the glass or the electrode wire.

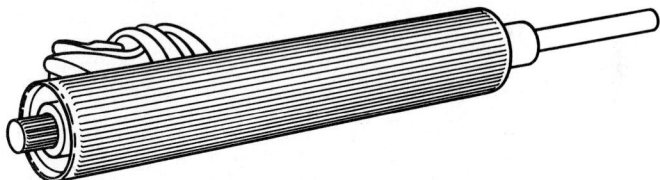

**Fig. 10-72**  A high-tension spark coil for testing the vacuum system. The knob at the end is used to control the intensity of the spark which appears at the metal tip of the device. [S. C. Miller and D. G. Fink, *Neon Signs*]

**170.  Gas Pressure Determined by Color, Using Spark-Coil Tester**
(S. C. Miller and D. G. Fink, *Neon Signs*)

| Coil held against | Color observed | Gas pressure, mm of mercury | Remarks |
|---|---|---|---|
| Electrode wire | Purple | 200 | Color visible only at edge of electrode shell |
| Electrode wire | Purple | 150 | Faint glow in tube |
| Electrode wire | Purple | 50 | Fair glow in tube |
| Electrode wire | Blue purple | 15 | Dark glow |
| Electrode wire | Red purple | 4 | |
| Electrode wire | Lavender | 2 | |
| Electrode wire | Light lavender | 1 | |
| Glass | Dark blue | 0.25 | |
| Glass | Light blue | 0.10 | |
| Glass | Very light blue | 0.01 | |
| Glass | Blue disappears | 0.005 | No glow in tube, blue haze near glass wall |

**171. The principal application of neon lamps is for sign lighting.** They are particularly adapted for this class of illumination owing to their bright colors and the fact that the tubes lend themselves readily to the forming of letters and figures. The brilliant single-color red, blue, or green light has an eye appeal in outdoor advertising which white light cannot offer. The orange-red light of the neon tube penetrates great distances, making neon signs stand out with brilliance and sparkle even on rainy nights. The flexibility of these thin glass tubes in the forming of trademarks, special figures, and animated designs also adds to their desirability to the advertiser. Finally, the relatively low wattage (about 4 to 6 W/ft of tubing) makes the operating cost relatively low.

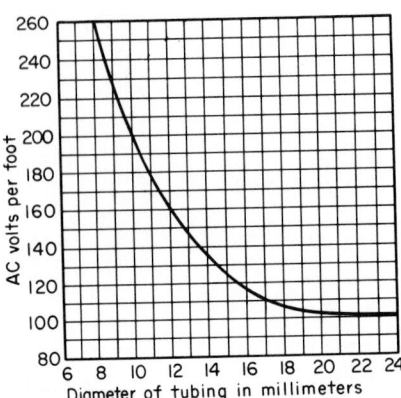

**Fig. 10-73** The voltage per foot required to operate tubing of various diameters filled with neon gas. [S. C. Miller and D. G. Fink, *Neon Signs*]

**172. The diameter of tubing used** is from 7 to 15 mm (%₂ to 19⁄₂ in). These sizes are convenient to work with and thoroughly practical for almost all applications. The voltage required per foot of tubing varies with the diameter of the tubing, as given in Fig. 10-73. The smaller-size tubes give more brilliant light but require the greatest voltage per foot of tube.

**173. The transformers** are rated according to secondary voltage and short-circuit current in milliamperes. Table 175 gives a list of standard transformers with the length of tubing in feet which can be supplied by each. This length varies with the diameter of the tubing and the gas used. When it is desired to combine two different sizes and colors of tubing in series on one transformer, the chart of Fig. 10-74 can be used. Any combination of lengths which are side by side in one column in the chart can be used on the transformer whose rating is given above the chart.

For example, a 9000-V, 30-mA transformer may supply (in the first column of the chart) 20 ft (6.1 m) of 15-mm blue tubing and 16 ft (4.9 m) of 15-mm red or (in the third column) 20 ft of 15-mm red and 11 ft (3.4 m) of 12-mm blue, making a total of 36 ft (11 m) in the first case of the larger tubing and 31 ft (9.4 m) in the second case of the combination of the larger and the medium-size tubing. Note that as the diameter of the tubing is decreased, the number of feet which can be supplied decreases on account of the rise in the voltage per foot, as shown in Fig. 10-73.

**174. In the makeup of signs** one section of tubing usually forms two or three letters. The different sections of tubes are then connected in series with wire jumpers until as many feet of tubing have been assembled as can be handled by the transformer to be used, as determined from Table 175. In large signs several transformers, each connected to its own section of tubing, can be used. The crossovers of tubing between letters can be blocked out by winding the tubing with tape and covering it with a waterproof varnish, or the tubes can be painted with a nonmetallic opaque paint. The glass should be made perfectly clean before painting by rubbing it with a wet cloth and drying. Metallic paint (with a lead or copper base) should never be used on the tubing, as it will conduct electricity and may cause a corona discharge between the tube and the housing which will attack the glass.

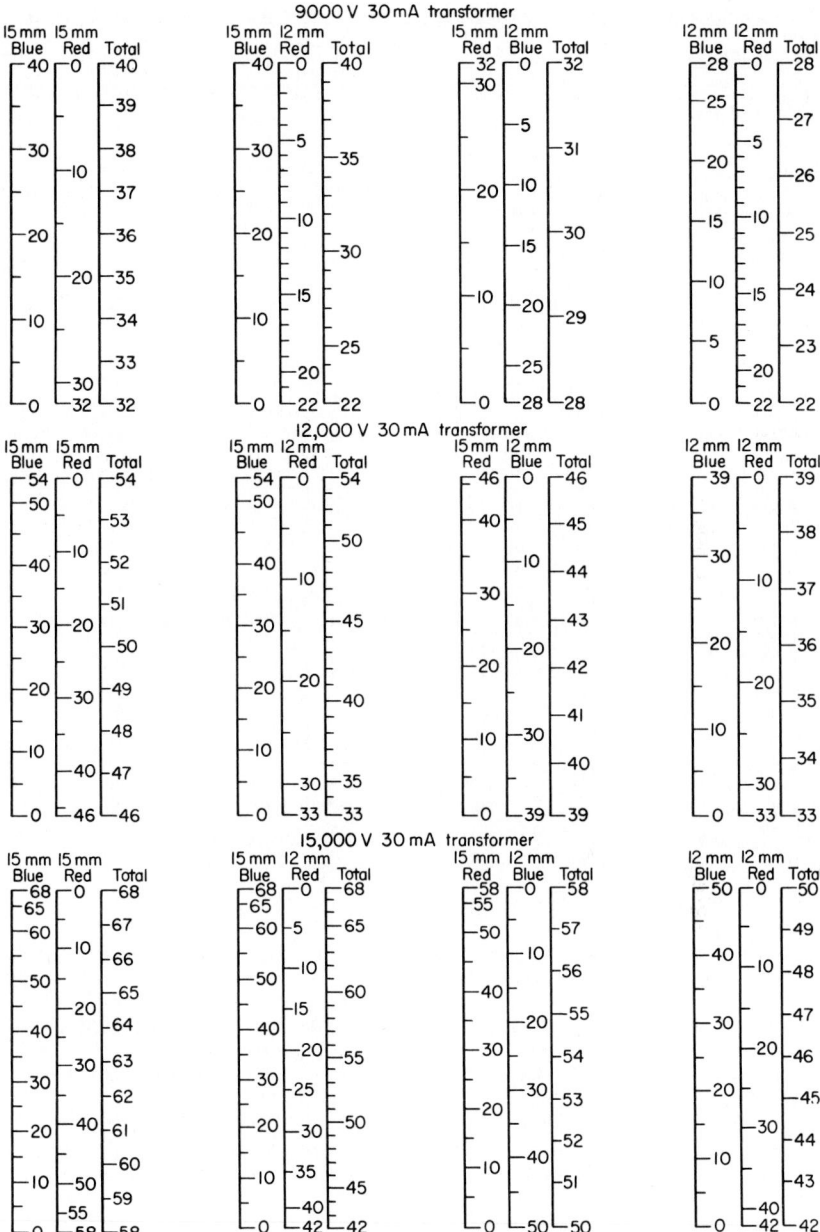

**Fig. 10-74**   Chart for determining the footage of combinations of red and mercury tubing which may be run from a single transformer. [S. C. Miller and D. G. Fink, *Neon Signs*]

**175.   Transformer Chart, Showing Maximum Number of Feet of Tubing a Given Transformer Will Carry**[a]

(S. C. Miller and D. G. Fink, *Neon Signs*)

| Transformer specifications | | | | | Color of tubing | | | | | | | | | | | | | | | |
| --- | --- | --- | --- | --- | --- | --- | --- | --- | --- | --- | --- | --- | --- | --- | --- | --- | --- | --- | --- | --- |
| | | | | | Red | | | | | | Blue | | | | | | White or gold | | | |
| Secondary volts | Secondary milliamperes short-circuited | Volt-amperes | Operating primary watts | Primary amperes, open | Diameter of tubing, mm | | | | | | | | | | | | | | | |
| | | | | | 7 | 9 | 10 | 11 | 12 | 15 | 7 | 9 | 10 | 11 | 12 | 15 | 9 | 10 | 11 | 12 |
| 15,000 | 60 | 875 | 400 | 8.0 | ... | 28 | 32 | 36 | 43 | 60 | ... | 34 | 38 | 44 | 54 | 70 | .. | 13 | 18 | 23 |
| 15,000 | 30 | 450 | 210 | 4.0 | ... | 27 | 31 | 34 | 42 | 58 | ... | 32 | 36 | 42 | 50 | 68 | 9 | 12 | 16 | 22 |
| 12,000 | 30 | 350 | 175 | 3.2 | ... | 21 | 24 | 28 | 33 | 46 | ... | 26 | 29 | 34 | 39 | 54 | 7 | 9 | 12 | 17 |
| 12,000 | 25 | 280 | 145 | 2.5 | ... | 18 | 23 | 25 | 30 | 40 | ... | 22 | 27 | 31 | 36 | 49 | .. | 7 | 10 | 15 |
| 9,000 | 30 | 280 | 130 | 2.5 | ... | 14 | 17 | 19 | 22 | 32 | ... | 18 | 20 | 23 | 28 | 40 | .. | 6 | 9 | 12 |
| 9,000 | 18 | 200 | 80 | 1.8 | 9 | 12 | 15 | 17 | 20 | 28 | 11 | 15 | 18 | 20 | 24 | 34 | | | | |
| 7,500 | 30 | 245 | 100 | 2.2 | ... | 9 | 12 | 15 | 17 | 21 | ... | 12 | 15 | 18 | 21 | 26 | | | | |
| 7,500 | 20 | 150 | 72 | 1.3 | 7 | 9 | 12 | 15 | 17 | .. | 8 | 12 | 15 | 18 | 21 | | | | | |
| 7,500 | 18 | 140 | 70 | 1.2 | 7 | 9 | 12 | 15 | 17 | .. | 8 | 12 | 15 | 18 | 21 | | | | | |
| 6,000 | 30 | 170 | 88 | 1.5 | ... | ... | 9 | 11 | 13 | 17 | ... | ... | 11 | 13 | 15 | 22 | | | | |
| 6,000 | 20 | 130 | 60 | 1.1 | 6 | 7 | 9 | 11 | 13 | 17 | 7 | 9 | 11 | 13 | 15 | 22 | | | | |
| 5,000 | 30 | 160 | 70 | 1.4 | ... | ... | 8 | 9 | 11 | 15 | ... | ... | 9 | 11 | 13 | 18 | | | | |
| 5,000 | 20 | 100 | 50 | 0.9 | ... | 6 | 8 | 9 | 11 | 15 | ... | ... | 9 | 11 | 13 | 18 | | | | |
| 5,000 | 18 | 95 | 48 | 0.8 | 5 | 6 | 8 | 9 | 11 | 15 | 6 | 8 | 9 | 11 | 13 | 17 | | | | |
| 4,000 | 30 | 130 | 60 | 1.1 | ... | 5 | 6 | 7 | 8 | 12 | ... | ... | 8 | 9 | 10 | 14 | | | | |
| 4,000 | 20 | 75 | 38 | 0.6 | ... | 5 | 6 | 7 | 8 | .. | ... | 6 | 8 | 9 | 10 | | | | | |
| 4,000 | 18 | 70 | 36 | 0.6 | ... | 5 | 6 | 7 | 8 | .. | ... | 6 | 8 | 9 | 10 | | | | | |
| 3,500 | 18 | 70 | 35 | 0.6 | 2 | 3 | 4 | 5 | 6 | .. | 3 | 4 | 5 | 6 | 7 | | | | | |
| 3,000 | 20 | 55 | 30 | 0.5 | 1.5 | 2 | 3 | 4 | 5 | .. | 2 | 3 | 4 | 5 | 6 | | | | | |
| 2,000 | 20 | 45 | 22 | 0.4 | 1 | 1.5 | 2 | 3 | 4 | .. | 1.5 | 2.5 | 3 | 4 | 5 | | | | | |

[a]This chart was compiled from an average taken from various charts listed by leading transformer companies in the United States. Values given are conservative, to allow for variations in glass diameters.

**176. Low-voltage neon-glow lamps** (Fig. 10-75) are operated on 105- to 125-V dc or ac systems. They contain two plates to form electrodes spaced with their abutting edges about 1/16 to 1/8 in (1.6 to 3.2 mm) apart. The bulb is filled with neon gas. A high-resistance coil in the base or external is connected in series with the electrodes and serves to limit the current, no auxiliary devices being necessary. When the circuit is closed, the 110 to 125 V causes the small air space between the plates to become ionized. A glow discharge takes place between the plates and then spreads so as virtually to cover the surface of the plates. The plates are very rugged and long-lived, failure usually being due to the burning out of the series resistance. When these lamps are used on direct current, only one electrode glows, whereas on alternating current both electrodes glow. They can therefore be

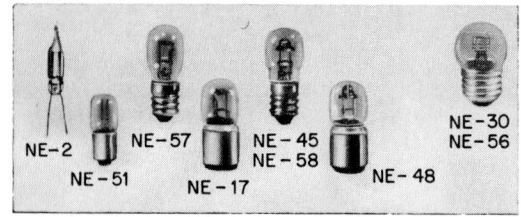

**Fig. 10-75**   Low-voltage neon-glow lamps. [General Electric Co.]

used as an ac, dc detector device. They may be used as pilot and indicator lamps in industry, for stroboscopic lamps in the laboratory, and as all-night lights in homes. Their low wattage, long life, and dependability make them ideal for such uses. The characteristics of neon-glow lamps are given in Table 177.

**177.  Glow Lamps, 105–125 V**
(General Electric Co.)

| Lamp-ordering code | Nominal current, mA | Bulb | Base | Maximum overall length, in | Starting voltage | | Series resistance, Ω |
|---|---|---|---|---|---|---|---|
| | | | | | ac | dc | |
| B1A (NE-51) | 0.3 | T-3¼ | Miniature bayonet | 1³⁄₁₆ | 65 | 90 | 220,000 |
| A1A (NE-2) | 0.7 | T-2 | Wire leads | 1 | 65 | 90 | 100,000 |
| C7A (NE-2D) | 0.7 | T-2 | Single-contact midget flange | ¹⁵⁄₁₆ | 65 | 90 | 100,000 |
| A1H | 1.2 | T-2 | Single-contact midget flange | ⅝ | 95 | 135 | 47,000 |
| C9A (NE-2J) | 1.9 | T-2 | Single-contact midget flange | ¹⁵⁄₁₆ | 95 | 135 | 30,000 |
| B5A (NE-17) | 2.0 | T-4½ | Double-contact bayonet candelabra | 1½ | 65 | 90 | 30,000 |
| B7A (NE-45) | 2.0 | T-4½ | Candelabra screw | 1¹⁷⁄₃₂ | 65 | 90 | 30,000 |
| B8A (NE-47) | 2.0 | T-4½ | Single-contact bayonet | 1½ | 65 | 90 | 30,000[a] |
| B9A (NE-48) | 2.0 | T-4½ | Double-contact bayonet candelabra | 1½ | 65 | 90 | 30,000 |
| F3A (NE-57) | 2.0 | T-4½ | Candelabra screw | 1¹⁷⁄₃₂ | 65 | 90 | 30,000[a] |
| F4A (NE-58)[b] | 2.0 | T-4½ | Candelabra screw | 1¹⁷⁄₃₂ | 65 | 90 | 100,000[a] |
| J9A (NE-56) | 5.0 | S-11 | Medium screw | 2³⁄₁₆ | 65 | 90 | 33,000[a] |
| J5A (NE-30) | 12.0 | S-11 | Medium screw | 2³⁄₁₆ | 65 | 90 | 7,500[a] |

[a]Internal resistor.
[b]For 210- to 250-V applications.

**178. Argon-glow lamps,** which consist of a mixture of gases, radiate mainly blue and violet and in the near-ultraviolet region. The negative glow appears blue violet. The fact that there is strong radiation in the near-ultraviolet region can be demonstrated by the fluorescent effects produced on uranium glass and many phosphorescent and fluorescent substances. Commercially, therefore, argon-glow lamps are used to some extent as convenient ultraviolet sources.

## ULTRAVIOLET-LIGHT SOURCES

**179. Ultraviolet-light sources** are lamps which are designed primarily to produce light of wavelengths in the ultraviolet part of the spectrum (see Fig. 10-1). It should be understood that not all ultraviolet light is the same. There are at present four recognized bands of ultraviolet light. The band in which the light from any source falls depends upon the wavelength of the light emitted. Each of the recognized bands differs widely from the others in its characteristics, usefulness, and field of application. The four bands are as follows: the near-ultraviolet or fluorescent region, the erythemic region, the abiotic or bactericidal region, and the Schumann region. They are listed in order of decreasing wavelength of the light, the wavelength of the near-ultraviolet being closest to that of visible light. The light from the ordinary fluorescent lamp originates at its source as light in the near-ultraviolet band. This nonvisible light is transformed into visible light by the fluorescence of the coating on the lamp tube under the influence of the near-ultraviolet light. The light output from the lamp therefore becomes visible light, so that the ordinary fluorescent lamp is not an ultraviolet-light source. However, by the use of special phosphors for the coating of fluorescent lamps, fluorescent lamps can be made to be sources of ultraviolet light.

The common methods of producing ultraviolet radiation are (1) by carbon and tungsten arcs, (2) by tungsten filaments operated at higher temperatures than in ordinary lamps, and (3) by gaseous-discharge lamps of proper design. Ultraviolet light can be used for healthful radiation as in the case of the sunlight lamps, for photographic printing, for illuminating fluorescent materials for analysis or for theatrical effects, or for the killing of germs as in the case of the germicidal lamps.

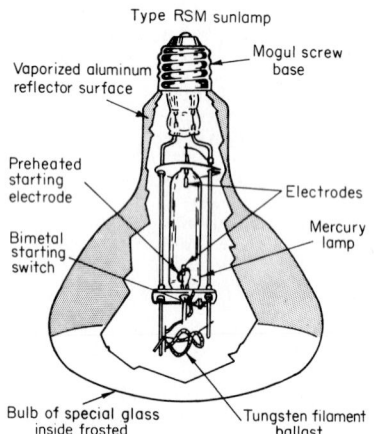

**Fig. 10-76**   RSM mercury-vapor sunlamp. [General Electric Co.]

**180. Sunlight lamps** (General Electric Co.). All mercury-vapor lamps generate ultraviolet rays which cause sunburn or suntan. These rays are transmitted only when special glass bulbs are used. The middle ultraviolet radiation between 2800 and 3300 Å is effective, but the peak of effectiveness comes at 2967 Å, which is one of the important mercury lines. Shorter wavelengths in the germicidal region cause quick sunburn or skin reddening (erythema), but they do not penetrate deeply into the skin to cause tanning. The longer wavelengths in the above range penetrate deeper and cause tanning.

Various types of sunlamps, including fluorescent lamps, have been developed. The self-contained RSM reflector sunlamp with built-in starting switch and filament ballast is the most widely used type because of its simplicity in use.

The RSM sunlamp (Fig. 10-76) is a self-contained unit with a 100-W mercury discharge element and a 175-W tungsten-filament resistance ballast incorporated in an ultraviolet-transmitting reflector-type bulb. The internal bimetallic starting switch first allows current to pass through the filament ballast and an auxiliary starting electrode which brings the latter up to emitting temperature. The switch is so placed that in several seconds the heat from the filament ballast causes the bimetallic switch to open, and the full line voltage starts the mercury element. The temperature of the filament and the resultant light output are then reduced because the filament is in series with the mercury arc, and this change is noticeable both in light intensity and in color quality as a starting characteristic. During operation the heat from the filament ballast is sufficient to hold the starting switch open. The starting switch is similar in operating characteristics to thermal-type starters used for fluorescent lamps.

As presently designed, the lamp will operate satisfactorily only on either 50- or 60-Hz current on 110- to 125-V ac circuits. Design voltage is 118 V, and the ultraviolet output rated at 35,000 E-Vitons varies about 1 percent for each percentage variation from this median line voltage.

The RSM lamp takes about 3 min to reach full ultraviolet output on starting and approximately 5 min for restarting. Part of this time is required for the bimetallic switch to cool down and close, and no light is produced in this interval. Direct-current operation is not recommended because arcing at the contacts of the built-in thermal switch results in short life. The life of the RSM lamp on normal ac circuits in household use has been estimated at approximately 1200 applications of 10 min per application or 1000 h when operated at 5 h per start.

Since the sun's rays reaching the earth contain no rays shorter than 2800 Å, the bulbs of sunlamps are made of special glass which, like the atmosphere, absorbs the shorter rays. Quartz-bulb lamps generate and transmit much erythemal energy in the far ultraviolet region not encountered in natural sunshine and therefore are not well adapted for popular sunlamp uses.

**181. Black light** (General Electric Co.) is the popular name for near-ultraviolet radiant energy in the 320- to 380-mm range. The label is not a very precise one. Whether or not this energy *looks* black is debatable. Certainly it is *not* light, because the human eye is insensitive to it.

The interesting thing about black light is that when it falls on certain materials, it makes them fluoresce, i.e., emit visible light. What actually occurs is a conversion of energy: the black light that falls upon the fluorescent surface is absorbed and then reradiated at longer wavelengths—wavelengths to which the eye is sensitive. So surfaces that contain or are treated with fluorescent chemicals glow, white or in color, when irradiated by black light. If the surroundings are kept dark, as they often are, the effect is that of self-luminous elements floating in space.

The weirdly beautiful and dramatic effects that black light makes possible account for its popularity in decoration, display, and advertising. In its many and rapidly expanding workaday uses, black light makes it possible to see things that otherwise could not be seen.

The most important artificial sources of black light are black-light fluorescent and mercury lamps. All mercury lamps used for black-light applications require external filters. Filament lamps are weak and inefficient sources of black light, and argon-glow lamps produce it in only very small quantities. Carbon arcs produce useful amounts of black light. Radiant energy from the sun and sky contains a strong black-light component.

Fluorescent black-light lamps are used in insect control. Many night-flying insects are particularly sensitive to near-ultraviolet and blue light. Black-light lamps are used in insect traps to attract the insects into the trap to be exterminated by an electric grid or trapped.

Data for black-light lamps are given in Table 182.

## 182.   Black-Light Lamps
(General Electric Co.)

| Lamp-ordering code | Nominal lamp watts | Bulb | Base | Length, in[a] | Related average life, h[b] | Approximate relative black-light energy[c] |
|---|---|---|---|---|---|---|
| *Fluorescent lamps* | | | | | | |
| F4T5/BL | 4 | T-5 | Miniature bipin | 6 | 6,000 | 4 |
| F4T5/BLB[d] | 4 | T-5 | Miniature bipin | 6 | 6,000 | 3 |
| F6T5/BLB[d] | 6 | T-5 | Miniature bipin | 9 | 7,500 | 6 |
| F8T5/BLB[d] | 8 | T-5 | Miniature bipin | 12 | 7,500 | 8 |
| F15T8/BL | 15 | T-8 | Medium bipin | 18 | 7,500 | 25 |
| F15T8/BLB[d] | 15 | T-8 | Medium bipin | 18 | 7,500 | 20 |
| F20T12/BL | 20 | T-12 | Medium bipin | 24 | 9,000 | 42 |
| F20T12/BLB[d] | 20 | T-12 | Medium bipin | 24 | 9,000 | 31 |
| F25T8/BL | 25 | T-8 | Medium bipin | 18 | 7,500 | 40 |
| F20T12/BL/U/3 | 20/25 | T-12 | Medium bipin | 12 | 9,000 | 38–48 |
| F35T12/BL/U/3 | 35/40 | T-12 | Medium bipin | 12 | 7,500 | 65–75 |
| F40BL/U/3 | 40 | T-12 | Medium bipin | 24 | 12,000 | 90 |
| F40BL | 40 | T-12 | Medium bipin | 48 | 20,000 | 100 |
| F40BLB[d] | 40 | T-12 | Medium bipin | 48 | 20,000 | 81 |
| F72T12/BL/HO | 85 | T-12 | Recessed double-contact | 72 | 12,000 | 190 |
| *Mercury lamps* | | | | | | |
| HR100A38 | 100 | E-23½ | Mogul | 7½ | 24,000+ | 68 |
| HR100A38/A23 | 100 | A-23 | Medium | 5⁷⁄₁₆ | 18,000 | 68 |
| HR100PFL44 | 100 | PAR-38 | Admedium | 5⁷⁄₁₆ | 12,000 | 18 |
| HR100PSP44 | 100 | PAR-38 | Admedium | 5⁷⁄₁₆ | 12,000 | 18 |
| HR175A39 | 175 | E-28 | Mogul | 8¼ | 24,000+ | 120 |
| HR175RFL39 | 175 | R-40 | Medium | 7 | 24,000+ | 30 |
| HR175RFL39/M | 175 | R-40 | Mogul | 7½ | 24,000+ | 30 |
| HR250A37 | 250 | E-28 | Mogul | 8¼ | 24,000+ | 165 |
| HR400A33 | 400 | E-37 | Mogul | 11⁵⁄₁₆ | 24,000+ | 270 |

[a]Length of fluorescent lamps includes standard lampholders. Length of mercury lamps is the maximum overall length.

[b]Rated average life for fluorescent lamps is the life when operated at 3 h per start on ballasts meeting published specifications. Rated average life for mercury lamps is the life when operated at 10 or more h per start on ballasts meeting published specifications.

[c]Relative value of 100 = 8100 fluorers.

[d]Integral-filter lamp.

**183. Filters**   (General Electric Co.). Most sources of black light produce visible light along with the ultraviolet. In the great majority of applications this visible light is undesirable. If it is not screened out by filters, it illuminates not only the luminous areas but also their surroundings. This, of course, reduces brightness contrasts and robs the display (or whatever the application may be) of its effectiveness. So almost always some sort of light-absorbing filter is placed between the black-light source and the irradiated surface. In the case of the BLB fluorescent lamps, the filter is an integral part of the lamp.

Filters having a wide range of characteristics are used for black-light applications. At one extreme are filters that pass virtually no visible light and not very much ultraviolet; at the other extreme are deep-blue sheet glasses that transmit appreciable visible light and a great deal of near-ultraviolet. Scientifically designed black-light filters are regularly supplied in molded squares or roundels or in sheet glass. They are used in all black-light applications that require a high degree of absorption of the visible light. Some of the deep-blue sheet glasses being used as black-light filters were not originally intended for this service. But since their near-ultraviolet transmission is high, their cost is low, and they are easily cut to any desired size, they are used in many applications in which some visible light can be tolerated.

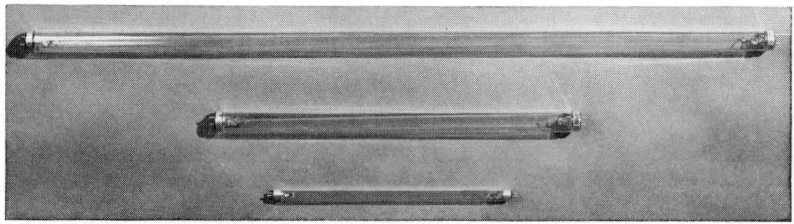

**Fig. 10-77**   Germicidal lamps. [General Electric Co.]

**184. Applications of black light**
   1.  Decorative purposes in the lighting of murals and show-window displays
   2.  Insect control
   3.  Advertising: producing effects in outdoor signboards
   4.  Inspection of cast and machined parts
   5.  Detection of leaks in hydraulic systems
   6.  Diagnostic use in medicine and biology
   7.  Inspection of food
   8.  Inspection of textiles
   9.  Sorting of laundry via invisible markings
  10.  Checking cleanliness in restaurants, dairies, etc.
  11.  Photoproductive processes
  12.  Location of mineral deposits

**185. Germ-killing or bactericidal lamps,** called germicidal lamps, have been developed for use in many places for sterilizing and prevention of mold. The germ-killing lamp is a gaseous-discharge lamp consisting of a long tube containing mercury vapor and inert gases. It gives out ultraviolet radiation in the bactericidal band which kills bacteria of many kinds. Germicidal lamps are discharge lamps similar in operation to fluorescent lamps and thus require the correct ballast for proper operation. Many germicidal lamps are designed to operate on standard fluorescent-lamp ballasts.

**186. The uses for germicidal lamps** are:
   1.  Storage and aging of meat. The meat when exposed to ultraviolet radiation can be kept in a warm, high-humidity atmosphere where it will age and become tender much more rapidly with no formation of mold or slime and with no loss of weight due to drying out.
   2.  Water sterilization.
   3.  Irradiation of air to prevent the spread of disease.
   4.  Sterilizing drinking glasses, hospital instruments, toilet seats, etc., to prevent the spread of disease.

**187. Photochemical lamps** (General Electric Co.) are a family of mercury lamps designed for specialized applications of ultraviolet radiation. The lamps with tubular quartz bulbs transmit the full range of generated radiation. Since these emit shortwave germicidal radiation, they must be used with utmost caution. Bare tubes must not be exposed to the eyes or skin. When lamps are used exposed, shortwave-absorbing glass filters should surround the lamp.

Ultraviolet radiation from quartz lamps is extremely useful as a catalyst in many industrial chemical processes such as polymerization, halogenation, chlorination, and oxidization. Similarly, the lamps are highly effective for ultraviolet therapy.

Quartz lamps are used by manufacturers for weathering tests on paints, dyes, and other finishes. Since a high concentration of energy from lamps is possible, a few hours' exposure may be equivalent to many days' exposure to natural sunlight. These lamps have also found widespread use in the sterilization of water. This method is particularly desirable when the taste and odor from chemical sterilization are objectionable.

Photochemical lamps are used for black-and-white printing, blueprinting, copyboard lighting, diazo printing, and vacuum-frame printing. Some lamps use an ultraviolet-transmitting glass which does not transmit far ultraviolet. These are excellent photographic lamps, since many of their strongest emission lines lie close to the maximum sensitivity of most photosensitive materials.

## INFRARED HEATING LAMPS

**188. Heating or drying lamps** are incandescent lamps with filaments which operate at a lower color temperature (2500 instead of about 3000 K) so that most of the radiation occurs in the infrared part of the spectrum with wavelengths longer than those of visible light.

Infrared lamps have many uses in commercial and industrial applications for heating and drying and on farms for brooding of poultry and other animals. Important features of these lamps include rapid heat transfer, efficient operation, simple oven construction, low oven first cost, adaptability to conveyor-line production, cleanliness, and low maintenance cost. The several wattages in each bulb size permit a wide range of temperatures.

The T-3 infrared quartz lamps are capable of delivering several times the energy concentration provided by the R-40 or G-30 types of lamps. They can be used in compact trough reflectors for concentrated radiation.

## 189. Infrared Heating Lamps
(General Electric Co.)

| Lamp-ordering code | Watts | Bulb | Volts | Base | Filament | Light-center length or lighted length, in[a] | Maximum overall length, in | Description |
|---|---|---|---|---|---|---|---|---|
| | | | | Reflector or G-bulb lamps | | | | |
| 125R40 | 125 | R-40 | 115 | Medium skirted | C-9 | ... | 7 7/8 | Reflector, light inside-frosted |
| 125R40/1 | 125 | R-40 | 120 | Medium | C-9 | ... | 6 9/16 | Reflector, clear |
| 250G30 | 250 | G-30 | 120 | Medium skirted | C-7A | 5 | 7 7/16 | Reflector, clear |
| 250R40/5 | 250 | R-40 | 120 | Medium skirted | C-9 | ... | 7 1/2 | Reflector, clear |
| 250R40/4 | 250 | R-40 | 120 | Medium skirted | C-9 | ... | 7 7/8 | Reflector, light inside-frosted |
| 250R40/1 | 250 | R-40 | 120 | Medium | C-9 | ... | 6 9/16 | Reflector, clear |
| 250R40/10 | 250 | R-40 | 120 | Medium | C-9 | ... | 6 3/4 | Reflector, red bowl |
| 375G30 | 375 | G-30 | 115 | Medium skirted | C-7A | 5 | 7 9/16 | Reflector, light inside-frosted |
| 375R40 | 375 | R-40 | 115 | Medium skirted | C-9 | ... | 7 7/8 | Reflector, clear |
| 375R40/1 | 375 | R-40 | 115 | Medium skirted | C-9 | ... | 7 1/2 | Reflector, clear |
| 375R40/10 | 375 | R-40 | 115 | Medium skirted | C-9 | ... | 7 1/2 | Reflector, red bowl |
| 500G30 | 500 | G-30 | 115 | Medium skirted | C-7A | 5 | 7 9/16 | Reflector, clear |
| | | | | Tubular quartz lamps | | | | |
| QH300T3 | 300 | T-3 | 115–125 | Sleeve leads | C-8 | 4 5/16 | 8 5/32 | |
| QH375T3 | 375 | T-3 | 115–125 | Sleeve leads | C-8 | 5 5/16 | 8 13/16 | |
| QH375T3/7 | 375 | T-3 | 115–125 | Recessed single-contact | C-8 | 5 5/16 | 8 11/16 | |
| QH500T3 | 500 | T-3 | 115–125 | Sleeve leads | C-8 | 4 13/16 | 8 13/16 | |
| QH500T3/7 | 500 | T-3 | 115–125 | Recessed single contact | C-8 | 4 5/16 | 8 7/8 | |
| QH500T3/CL | 500 | T-3 | 115–125 | Sleeve leads | C-8 | 4 5/16 | 8 13/16 | Clear |
| QH1000T3 | 1000 | T-3 | 200–220 | Sleeve leads | C-8 | 10 | 13 13/16 | Burn horizontal |
| QH1000T3 | 1000 | T-3 | 230–250 | Sleeve leads | C-8 | 10 | 13 13/16 | Burn horizontal |
| QH1000T3/CL | 1000 | T-3 | 230–250 | Sleeve leads | C-8 | 10 5/8 | 11 7/8 | Clear, burn horizontal |
| QH1000T3/2CL/HT | 1000 | T-3 | 230–250 | Sleeve leads | C-8 | 10 | 13 13/16 | Clear, high-temperature, burn horizontal |

**189. Infrared Heating Lamps** (*Continued*)

| Lamp-ordering code | Watts | Bulb | Volts | Base | Filament | Light-center length or lighted length, in[a] | Maximum overall length, in | Description |
|---|---|---|---|---|---|---|---|---|
| | | | | Tubular quartz lamps | | | | |
| QH1200T3/CL | 1200 | T-3 | 144 | Sleeve leads | C-8 | 6 | 8 5/16 | Clear, burn horizontal |
| QH1600T3 | 1600 | T-3 | 200–220 | Sleeve leads | C-8 | 16 | 19 13/16 | Burn horizontal |
| QH1600T3 | 1600 | T-3 | 230–250 | Sleeve leads | C-8 | 16 | 19 13/16 | Burn horizontal |
| QH1600T3 | 1600 | T-3 | 277 | Sleeve leads | C-8 | 16 | 19 13/16 | Burn horizontal |
| QH1600T3/7 | 1600 | T-3 | 200–220 | Recessed single-contact | C-8 | 16 | 19 7/8 | Burn horizontal |
| QH1600T3/7 | 1600 | T-3 | 230–250 | Recessed single-contact | C-8 | 18 | 19 7/8 | Burn horizontal |
| QH1600T3/CL | 1600 | T-3 | 230–250 | Recessed single-contact | C-8 | 16 | 19 13/16 | Clear, burn horizontal |
| QH2M/T3/CL/HT | 2000 | T-3 | 230–250 | Sleeve leads | C-8 | 10 | 13 7/8 | Clear, high-temperature, burn horizontal |
| QH2M/T3/1CL/HT | 2000 | T-3 | 230–250 | Sleeve leads | C-8 | 9 3/4 | 11 13/16 | Clear, high-temperature, burn horizontal |
| QH2500T3 | 2500 | T-3 | 460–500 | Sleeve leads | C-8 | 25 | 28 3/16 | Burn horizontal |
| QH2500T3 | 2500 | T-3 | 575–625 | Sleeve leads | C-8 | 25 | 28 3/16 | Burn horizontal |
| QH2500T3/7 | 2500 | T-3 | 460–500 | Recessed single-contact | C-8 | 25 | 28 7/8 | Burn horizontal |
| QH2500T3/CL | 2500 | T-3 | 460–500 | Sleeve leads | C-8 | 25 | 28 13/16 | Clear, burn horizontal |
| QH2500T3/VB | 2500 | T-3 | 460–500 | Sleeve leads | C-8 | 25 | 28 3/16 | Burn in any position |
| QH3650T3/5 | 3650 | T-3 | 480 | Special | C-8 | 38 | 41 7/8 | Burn horizontal |
| QH3800T3 | 3800 | T-3 | 550–600 | Sleeve leads | C-8 | 38 | 41 13/16 | Burn horizontal |
| QH3800T3/VB | 3800 | T-3 | 550–600 | Sleeve leads | C-8 | 38 | 41 13/16 | Burn in any position |
| QH5M/T3/CL | 5000 | T-3 | 920–1000 | Sleeve leads | C-8 | 50 | 53 13/16 | Clear, burn horizontal |
| QH5M/T3/1CL/HT | 5000 | T-3 | 575–625 | Sleeve leads | C-8 | 25 | 28 3/16 | Clear, high-temperature, burn horizontal |

[a] Light-center length shown for G-bulb lamps; lighted length shown for tubular quartz lamps.
NOTE  All lamps are rated at 5000+ h of life.

# LUMINAIRES

**190. Purpose of luminaires.** A luminaire is a device which directs, diffuses, or modifies the light given out by the illuminating source in such a manner as to make its use more economical, effective, and safe to the eye. The luminaire includes the reflector, lamp sockets, enclosing materials, ballasts in fluorescent and HID units, and stems and canopies where used. Since the light from a bare lamp is given off approximately equally in all directions, to use the light economically some accessory is required to direct the light to the desired areas. As most lamps have a high brightness, it is desirable in producing satisfactory illumination that the eye be shielded from the source in order to reduce direct glare. In many cases the appearance of the illuminating system is of great importance, so that the luminaire must possess decorative features as well as those necessary for satisfactory illumination.

**191. Distribution graphs of luminaires.** The effect of a luminaire in changing the direction and distribution of the light given out by a light source is best expressed by a distribution graph. Figure 10-78 shows such a graph for a bare lamp and for the same lamp with a reflector. The graph represents the light in a single vertical plane through the center of the light unit, and it is assumed that the light in all similar vertical planes is similarly distributed. See Sec. 32, "How to Read a Photometric Graph."

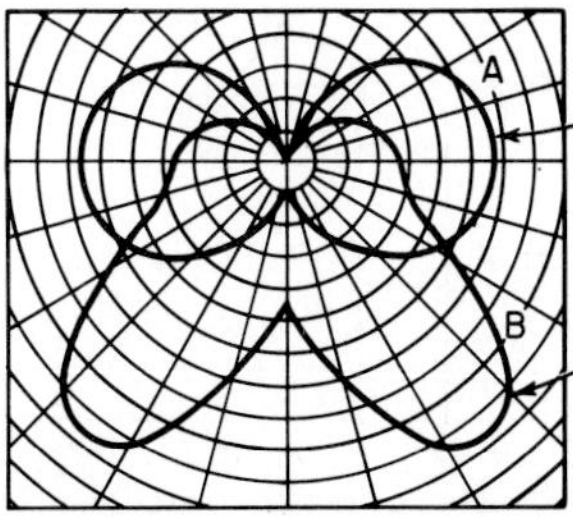

**Fig. 10-78**   Comparison between the distribution curve of a bare incandescent lamp and that of the same lamp equipped with a suitable reflector.

Figure 10-79 shows a sample photometric test report for a 400-W HPS luminaire. A report of this type is used in comparing luminaires with one another as to relative distribution of light, type of illumination, lumen-output efficiency, and brightness.

**192. Classification of luminaires.**   Luminaires can be classified by several methods:

1. According to system of illumination produced
    - *a.* Direct
    - *b.* Semidirect
    - *c.* General diffuse or direct-indirect
    - *d.* Semi-indirect
    - *e.* Indirect
2. According to the amount that the luminaire encloses the lamp
    - *a.* Open
    - *b.* Enclosed
3. According to class of service
    - *a.* Industrial
    - *b.* Commercial and institutional
    - *c.* Residential
    - *d.* Street lighting
    - *e.* Floodlighting
4. According to the material used for reflection or transmission of the light
    - *a.* Steel
    - *b.* Aluminum
    - *c.* Opal glass
    - *d.* Prismatic glass
    - *e.* Glass and metal
    - *f.* Plastic
    - *g.* Plastic and metal
5. According to method of mounting
    - *a.* Suspended
    - *b.* Surface-mounted
    - *c.* Recessed or built-in
6. According to light source
    - *a.* Incandescent lamp

# PHOTOMETRIC DATA
## GENERAL ⓖⓔ ELECTRIC

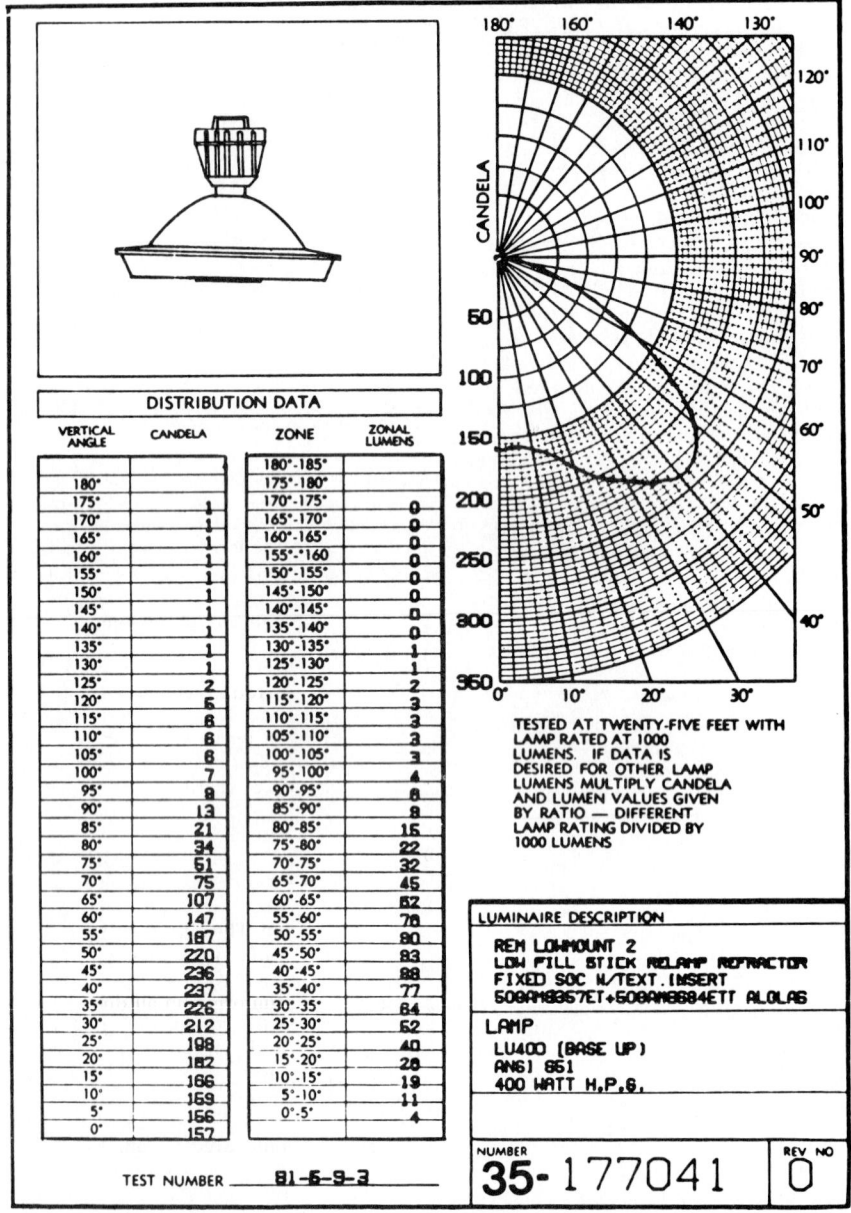

### DISTRIBUTION DATA

| VERTICAL ANGLE | CANDELA | ZONE | ZONAL LUMENS |
|---|---|---|---|
| | | 180°-185° | |
| 180° | | 175°-180° | |
| 175° | 1 | 170°-175° | 0 |
| 170° | 1 | 165°-170° | 0 |
| 165° | 1 | 160°-165° | 0 |
| 160° | 1 | 155°-160 | 0 |
| 155° | 1 | 150°-155° | 0 |
| 150° | 1 | 145°-150° | 0 |
| 145° | 1 | 140°-145° | 0 |
| 140° | 1 | 135°-140° | 0 |
| 135° | 1 | 130°-135° | 1 |
| 130° | 1 | 125°-130° | 1 |
| 125° | 2 | 120°-125° | 2 |
| 120° | 5 | 115°-120° | 3 |
| 115° | 6 | 110°-115° | 3 |
| 110° | 6 | 105°-110° | 3 |
| 105° | 6 | 100°-105° | 3 |
| 100° | 7 | 95°-100° | 4 |
| 95° | 9 | 90°-95° | 6 |
| 90° | 13 | 85°-90° | 8 |
| 85° | 21 | 80°-85° | 15 |
| 80° | 34 | 75°-80° | 22 |
| 75° | 51 | 70°-75° | 32 |
| 70° | 75 | 65°-70° | 45 |
| 65° | 107 | 60°-65° | 62 |
| 60° | 147 | 55°-60° | 78 |
| 55° | 187 | 50°-55° | 80 |
| 50° | 220 | 45°-50° | 83 |
| 45° | 236 | 40°-45° | 88 |
| 40° | 237 | 35°-40° | 77 |
| 35° | 226 | 30°-35° | 64 |
| 30° | 212 | 25°-30° | 52 |
| 25° | 198 | 20°-25° | 40 |
| 20° | 182 | 15°-20° | 28 |
| 15° | 166 | 10°-15° | 19 |
| 10° | 169 | 5°-10° | 11 |
| 5° | 166 | 0°-5° | 4 |
| 0° | 157 | | |

TESTED AT TWENTY-FIVE FEET WITH LAMP RATED AT 1000 LUMENS. IF DATA IS DESIRED FOR OTHER LAMP LUMENS MULTIPLY CANDELA AND LUMEN VALUES GIVEN BY RATIO — DIFFERENT LAMP RATING DIVIDED BY 1000 LUMENS

### LUMINAIRE DESCRIPTION

REM LOWMOUNT 2
LOW FILL STICK RELAMP REFRACTOR
FIXED SOC W/TEXT.INSERT
508AM8357ET+508AM8684ETT ALGLAS

**LAMP**

LU400 (BASE UP)
ANSI 851
400 WATT H.P.S.

TEST NUMBER ___81-6-9-3___

NUMBER  **35-** 177041

REV NO  0

Fig. 10-79   Photometric test report. [General Electric Co.]

    *b.* High-intensity–discharge lamp
    *c.* Fluorescent lamp

**193. Direct lighting** is defined as a lighting system in which practically all (90 to 100 per-cent) of the light of the luminaries is directed in angles below the horizontal directly toward the usual working areas (Fig. 10-80, VI). This type of lighting is produced by lumi-naires, which range from industrial-type steel or aluminum reflectors to fixtures mounted above large light-source areas such as glass or plastic panels and skylights. Although, in general, such a system provides illumination on working surfaces most efficiently, this achievement may be at the expense of other factors such as excessive contrasts of the light source with the surroundings, troublesome shadows, or direct and reflected glare.

    **194. Indirect lighting.** From 90 to 100 percent of the light output of the luminaire is directed toward the ceiling at angles above the horizontal (Fig. 10-80, I). Practically all the light effective at the working plane is redirected downward by the ceiling and to a lesser extent by the sidewalls. Since the ceiling is in effect the light source, the illumina-tion produced is quite diffuse in character. While indirect lighting is not as efficient as some of the other systems on a purely quantitative basis, the even distribution and absence of shadows and reflected glare frequently make it the most desirable type of installation for offices, schools, and similar applications. Because room finishes play such an important part in redirecting the light, it is particularly important that they be as light in color as possible and carefully maintained in good condition. The ceilings should always have a matte finish if reflected images of the light source are to be avoided.

    Glass or plastic luminaires in this classification are known as *luminous indirect,* while metal luminaires which transmit no light are *totally indirect.* The translucent type is sometimes more desirable than the totally indirect because a luminous fixture is less sharply silhouetted against the relatively bright ceiling. Indirect illumination may also be provided by means of architectural coves. Luminaire suspension length, or cove propor-tions, must be carefully selected to provide uniform ceiling coverage, where desired, and to prevent excessive ceiling brightness.

    **195. Semi-indirect lighting.** From 60 to 90 percent of the light output of the luminaire is directed toward the ceiling at angles above the horizontal (Fig. 10-80, II), while the bal-ance is directed downward. Semi-indirect lighting has most of the advantages of the indi-rect system but is slightly more efficient and is sometimes preferred to achieve a desirable brightness ratio between ceiling and luminaire in high-level installations. The diffusing medium employed in these luminaires is glass or plastic of a lower density than that employed in indirect equipment.

    **196. General diffuse or direct-indirect lighting.** From 40 to 60 percent of the light is directed downward at angles below the horizontal. The major portion of the illumination produced on ordinary working planes is a result of the light coming directly from the luminaire. However, a substantial portion of the light is directed to the ceiling and side-walls. If these are light in color, the upward light provides a brighter background against which to view the luminaire, in addition to supplying a substantial indirect component which adds materially to the diffuse character of the illumination. The difference between the *general diffuse* (Fig. 10-80, III) and *direct-indirect* (Fig. 10-80, IV) classifications is in the amount of light produced in a horizontal direction. The general diffuse type is exem-plified by the enclosing globe which distributes light nearly uniformly in all directions, while the direct-indirect luminaire produces very little light in a horizontal direction, owing to the density of its side panels. Glass, plastic, or louvered bottoms are commonly used with the latter type of luminaire to provide lamp shielding.

    **197. Semidirect lighting.** From 60 to 90 percent of the light is directed downward at angles below the horizontal (Fig. 10-80, V). The footcandles effective under this system at normal working planes are primarily a result of the light coming directly from the lumi-naire. The portion of the light directed to the ceiling results in a relatively small indirect component, the greatest value of which is that it brightens the ceiling area around the luminaire, with a resultant lowering of brightness contrasts. Equipment of this type is exemplified by the suspended plastic-sided fluorescent luminaire or the open-bottom glass shade for incandescent lamps.

    **198. An open luminaire** is one which only partially encloses the lamp. Open direct-lighting luminaires have the lamp exposed to view from at least one direction. The luminaires of Figs. 10-83, 10-84, 10-86, I, and 10-87, III, and IV are open-type direct-lighting units.

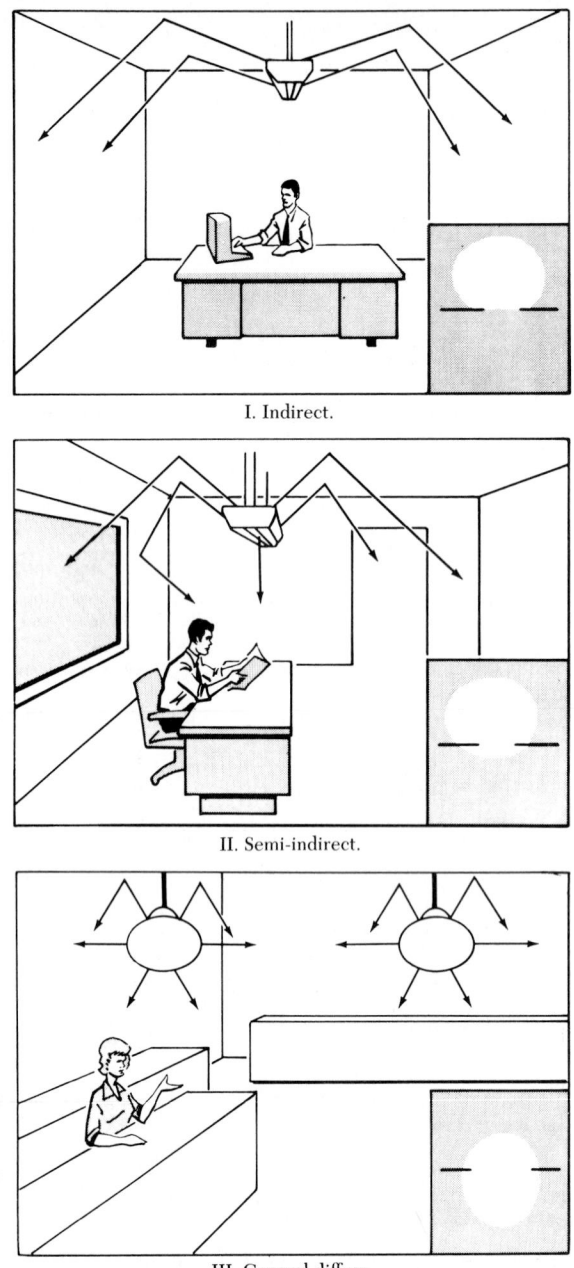

I. Indirect.

II. Semi-indirect.

III. General diffuse.

**Fig. 10-80**   Systems of illumination.

**199. An enclosed luminaire** completely encloses the lamp from view from any direction. The ones shown in Figs 10-87, I and II, and 10-88 are enclosed direct-lighting luminaires.

**200. Classification of luminaires according to class of service.**   Industrial service refers to use in factories and manufacturing establishments. *Commercial service* refers to use in offices, stores, and public buildings. *Residential use* means use in houses and apartments;

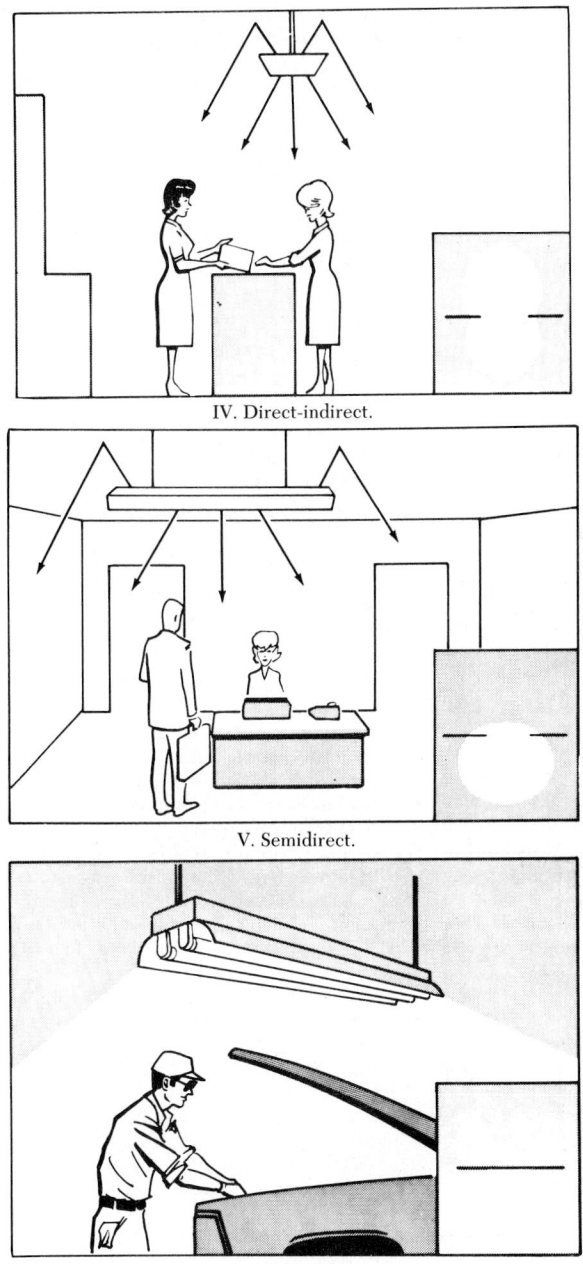

IV. Direct-indirect.

V. Semidirect.

VI. Direct.

**Fig. 10-80**   *(Continued)*

*street lighting* means the illumination of streets, roads, and highways; and *floodlighting* means the illumination of outdoor areas such as painted signs, recreational grounds, automobile parking spaces, and building exteriors.

**201. Steel reflectors** have a sheet-steel base coated with porcelain enamel, paint enamel, or aluminum paint to form the reflecting surface. The best types of steel reflectors are those coated with porcelain enamel, since their reflecting surface is very efficient, durable, and easily cleaned. Paint-enamel reflectors when new are quite efficient, but the reflecting surface deteriorates very rapidly. Aluminum-painted reflectors are slightly better than paint-enamel ones but not so durable as porcelain-enamel ones.

Steel reflectors are used in the manufacture of many luminaires of the direct and indirect types.

**202. Aluminum reflectors** are used in many different types of luminaires. The reflecting surface is usually given an electrochemical treatment called Alzak which removes the impurities from the aluminum and endows it with a hard, durable surface which has a high reflection efficiency of approximately 90 percent. Like steel reflectors, aluminum reflectors are used in luminaires for both direct and indirect lighting.

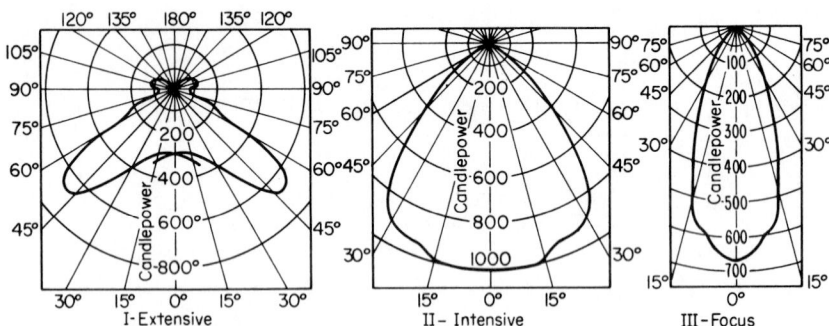

**Fig. 10-81**   Distribution characteristic curves for direct prismatic-glass luminaires. [Manville Special Products Group, Holophane Division]

**203. Opal glass** is a white diffusing glass (Sec. 10) which transmits light with a loss of only about 15 to 35 percent, the amount depending on the density of the glass. It breaks up the light rays so that the lamp is not visible and glare is kept at a minumum. Opal glass is used for enclosed diffusing luminaires for incandescent and HID lamps and for side-panel shields in some commercial fluorescent luminaires.

**204. Prismatic glass** consists of clear glass which is molded into scientifically designed prisms. Each prism is designed with reference to the position of the light source. By proper design, distribution curves of the extensive, intensive, or focusing types are obtained. The extensive reflector or globe refractor distributes the light over a wide angle below the horizontal (Fig. 10-81, I), the maximum intensity being at an angle of about 45° with the vertical.

Intensive reflectors, globes, and lenses (Fig. 10-81, II) throw the light directly downward, the maximum intensity occurring between 0 and 25° with the vertical.

Focusing globes, lenses, and reflectors (Fig. 10-81, III) concentrate the light to a small area, producing the greatest intensity of illumination along the axis of the reflector. Focusing units give an end-on candlepower approximately 3½ times as great as the rated horizontal candlepower of the lamp. The area intensely illuminated is a circle, the diameter of which should be one-half the height of the lamp above the plane of illumination; outside this limit the intensity falls rapidly but not so abruptly as to give the effect of a spot of light. Generally focusing units give their maximum candlepowers at about 10° from the vertical.

Prismatic-glass units are used in many direct, semidirect, and semi-indirect luminaires. The different types of prismatic units employed are reflectors, globes, and lenses. Some luminaires employ only a single type of unit, and others employ a combination of types of units. Luminaires employing prismatic-glass units are available for incandescent, fluorescent, and HID lamps.

**205. Combination glass-and-metal luminaires** are used principally for enclosed direct-lighting units. The metal part of the unit usually directs the light, while the glass is used for diffusing purposes or simply for the enclosure of the unit so that it will be dusttight or vaportight.

**206. Plastics** are used to some extent in luminaires for incandescent lamps and HID lamps and to quite a great extent for fluorescent lamps. The incandescent luminaires employing plastics are of the semi-indirect type (very close to indirect). With fluorescent units plastic is used in many direct and direct-indirect luminaires for side-panel shielding. It is also used quite extensively for bottom and side enclosures of direct and semidirect fluorescent units.

**207. A combination of plastic and metal** is used in some fluorescent luminaires. In some units the plastic is used for side-panel shielding. In others it is used to enclose the bottom of the unit for diffusing purposes or for making the unit dusttight and vaportight.

**208. Method of mounting.** *Suspended mounting* refers to hanging the luminaire from the ceiling with a chain, tube, conduit, or cord-pendant suspension. *Surface mounting* refers to placing the body of the luminaire directly against the ceiling. *Recessed or built-in mounting* refers to placing the luminaire in recesses in the ceiling or on the walls or structural members, so that it forms a part of the architectural treatment of the building interior.

**209. Luminaires for use with incandescent lamps** are available for all classes of service. For a discussion of street-lighting and floodlighting equipment refer to Secs. 261 and 269. Incandescent luminaires for residential use are available in a great variety of types, which are too numerous to be discussed here.

**210. Industrial luminaires for incandescent lamps** are of either the metal or the glass-metal type. The more common types are discussed in the following sections.

**211. Enameled-steel–reflector luminaires for incandescent lamps** are made in several different types of assemblies. One type has a reflector and hood pressed in one piece. When the nut is loosened at the top, the reflector will slide up on the stem, exposing the socket for wiring. This is a weatherproof type for use outdoors and for indoor installations where interchangeability of reflectors and ease of cleaning are not the important considerations. With the turn-lock type of reflector, the entire reflector and socket can be removed from the supporting cap by a quarter turn and taken down for cleaning. The prongs of the turn-locking device, in addition to supporting the reflector and socket, provide the electrical connection from the socket to the circuit terminated in the cap. This type is especially advantageous in locations where the cleaning of reflectors in position is going to be difficult. The threaded-reflector type has a reflector which can be removed from the socket and hood by unscrewing. Several shapes of reflectors will fit the same hoods. Hoods are available with a threaded connection for fixture-stem suspension or with a flange having holes for mounting as the cover of a 4-in (102-mm) outlet box. With the threaded type the reflectors can be taken down for cleaning, and different reflectors inserted in the same hood, depending on the distribution curve of light desired. The snap-on type is designed so that it can be fastened on pendant sockets. The reflector holder snaps on the reflector, and then the whole assembly screws onto the threads on the outside of an ordinary brass socket. The vaportight type (Fig. 10-82) is for interior use in wet atmospheres. It requires a cast outlet box with a threaded opening. The reflector is bolted to a ring which threads on the outside of the box opening. A threaded glass globe screws into the inside of the box opening against a gasket, which seals the box against entrance of moisture. A metal guard is used for low mounting heights or other locations where it is necessary to protect the globe from mechanical injury. The guard screws on the outside of the box opening.

These enameled-steel–reflector assemblies are all available in several shapes for specific purposes as shown in Figs. 10-83 and 10-84. The flat cone (Fig. 10-83, I) and the shallow bowl (Fig. 10-83, II) are used for yard and aisle lighting for wide distribution of light when the quality of the light is unimportant and glare from the partially exposed lamp is not objectionable. The RLM dome (Fig. 10-83, III) is a standard dome reflector which must comply with standard specifications of the Reflector and Lamp Manufacturers as regards contour and quality of the porcelain reflecting surface. This type of reflector is commonly used for general interior industrial lighting. The deep-bowl type (Fig. 10-83, IV) is used frequently when units must be hung very low, as over workbenches. Angle reflectors (Fig. 10-84) are used in industrial lighting to supplement overhead units in very high interiors and for special types of service which require very good illumination on vertical planes. They are used also for the lighting of billboards and painted signs.

Specially designed deep-bowl units also are available for high-bay mounting.

The use of silvered-bowl lamps in steel reflectors will provide well-diffused illumination.

Glass covers are available for enclosing the bottom of steel reflectors to make them dust-tight.

**212. The Glassteel diffuser luminaire for incandescent lighting** is a combination glass-and-steel direct-lighting unit which is employed for high-quality industrial lighting. It consists of a completely enclosing opal-glass globe (Fig. 10-85) mounted inside a dome-shaped, porcelain-enamel reflector. The porcelain reflector has the same general construction as the standard RLM dome reflector. Small openings in the top of the dome allow a small portion of the light to be directed to the ceiling. This illumination of the ceiling produces

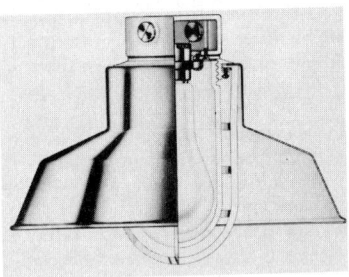

I. With junction box.

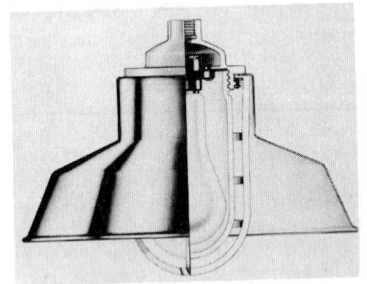

II. For pendant mounting.          III. For outlet-box mounting.

**Fig. 10-82**   Vaportight units. [Miller Lighting, Hubbell Lighting Division]

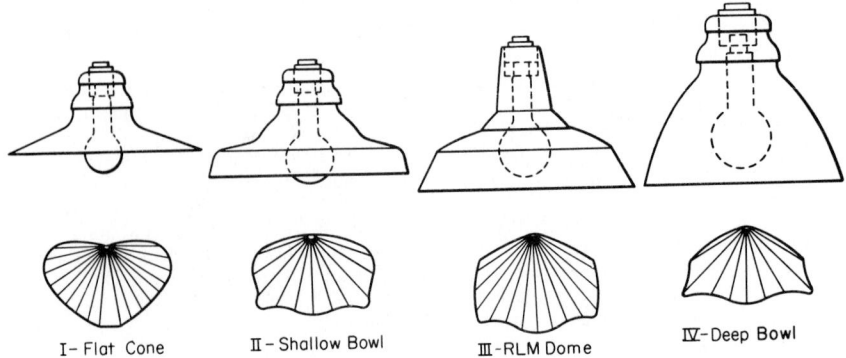

I– Flat Cone          II – Shallow Bowl          III – RLM Dome          IV – Deep Bowl

**Fig. 10-83**   Contour outlines of typical enameled-steel reflectors with their chracteristic distribution curves.

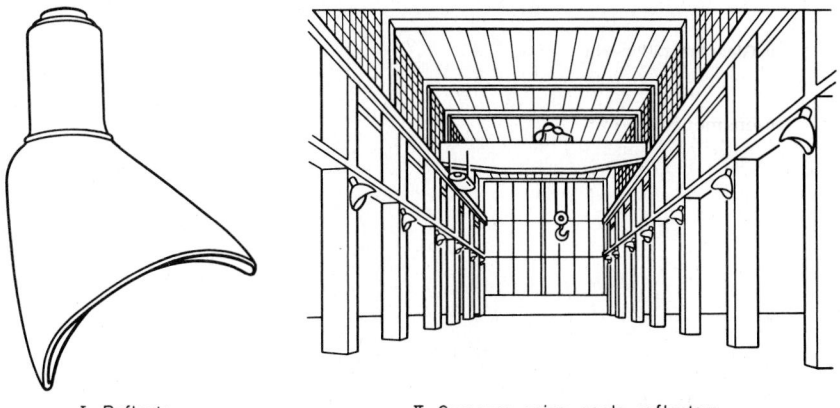

I -Reflector                    II- Craneway using angle reflectors

**Fig. 10-84**  Elliptical-angle steel reflector. [Benjamin Division, Thomas Industries, Inc.]

natural and cheerful working conditions, whereas with the RLM dome reflectors the ceiling is left in darkness. The enclosing globe of the Glassteel diffuser shields the lamp from view and diffuses the light, reducing the glare effect. The Glassteel diffuser is made in all the types of assemblies explained above under the discussion of enameled-steel luminaires.

**213. Aluminum luminaires for incandescent lamps** are chiefly of the high-bay type (Fig. 10-86). They provide more accurate control of light than is obtainable with the enameled-steel reflectors and are recommended when the mounting height is high compared with the width of the area to be lighted. A concentrating type is used when the units are sufficiently high so that it is best to locate them closer together than the mounting height and for high-

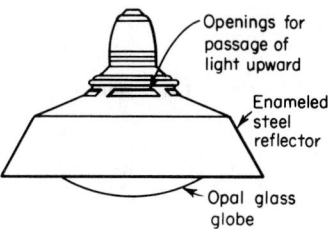

**Fig. 10-85**  Glassteel diffuser.

mounted units in narrow interiors. A spread type gives better illumination on vertical surfaces and can be used when the spacing is up to 1½ times the mounting height. These luminaires are made in the one-piece and turn-lock types as described in Sec. 211.

Steel-clad units, which consist of an Alzak aluminum reflector protected by a porcelain-enamel steel housing, are available for severe service conditions. These units may be provided with glass covers for protection against moisture and noncombustible dust.

**214. Prismatic-glass luminaires for industrial use with incandescent and HID lamps** are available

I. Open type.

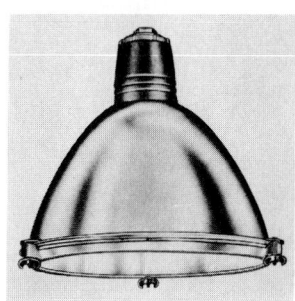

II. Vaportight type.

**Fig. 10-86**  Aluminum high-bay luminaires. [Miller Lighting, Hubbell Lighting Division]

in low-bay (Fig. 10-87, I, II, and IV) and high-bay (Fig. 10-87, III) types. The basic unit is provided with an open-bottom prismatic-glass reflector which produces semidirect illumination. When upward light is not required but protection against moisture (condensate, leakage, etc.) from the ceiling is desirable, an aluminum cover can be used over the prismatic reflector without materially affecting the downward distribution of light. Guards of either wire- or louver-type construction are available for protection of the lamp. These units, in addition to their industrial applications, are widely used in certain commercial

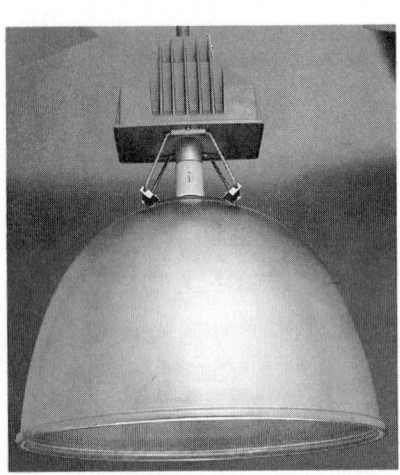

I. Low-bay enclosed industrial luminaire suitable for damp locations.

II. Low-bay general-area lighting or hazardous-area lighting; high-pressure sodium, metal halide, or mercury lamp unit.

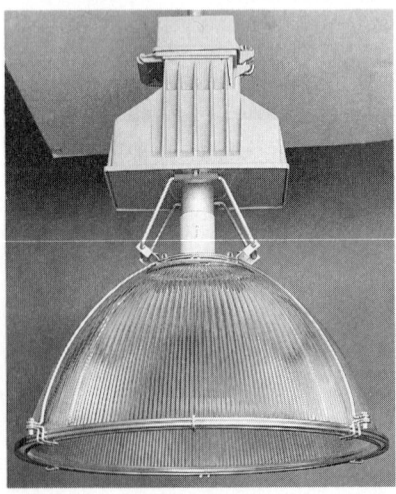

III. High-bay metal halide, mercury, or high-pressure sodium lamp unit.

IV. Low-bay high-pressure sodium, metal halide, or mercury lamp unit.

**Fig. 10-87**  Industrial-type prismatic-glass HID luminaires. [Manville Special Products Group, Holophane Division]

types of buildings, such as gymnasiums and supermarkets.

Another prismatic-glass luminaire for industrial use with incandescent lamps is an all-purpose vaportight and dusttight unit. This luminaire consists of a prismatic-glass reflector-refractor and a cast-metal filter with stainless-steel supporting bails assembled to normally closed-type cam latches. This unit is shown in Fig. 10-87, II. It is used in chemical, petroleum, ordnance, generating-station, textile, woodworking, food-processing, and similar industries.

**215. Luminaires for high-intensity–discharge lamps** are generally intended for high-bay mounting in industrial interiors. Luminaires are available in enameled steel, Alzak aluminum, and prismatic-glass types. The metal units are of the same general type of construction as similar luminaires used for lighting with incandescent lamps. Refer to Secs. 211 and 213.

Prismatic-glass units are available for both high-bay and low-bay mounting (refer to Fig. 10-87). Both the high-bay and the low-bay units consist of a prismatic-glass reflector with the outside of the reflector protected by a sealed and permanently spun-on metal cover.

Other high-bay luminaires utilize reflectors with a coated-glass surface for high efficiency and easy maintenance. Enclosed high-bay fixtures are also available with internal

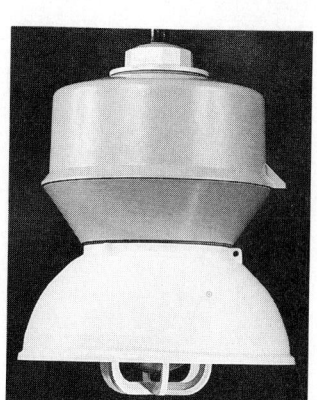

Hazardous location.

High bay.

Low bay.

**Fig. 10-88**    HID industrial luminaires. [GE Lighting Systems]

**Fig. 10-89**    Partially open-end type of steel fluorescent luminaire. [Miller Lighting, Hubbell Lighting Division]

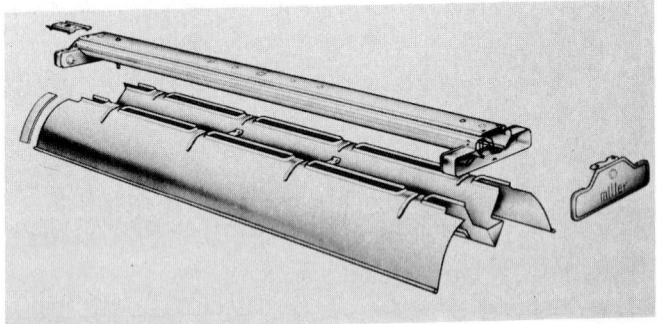

**Fig. 10-90**    Construction of an open-type fluorescent luminaire, showing longitudinal shield. [Miller Lighting, Hubbell Lighting Division]

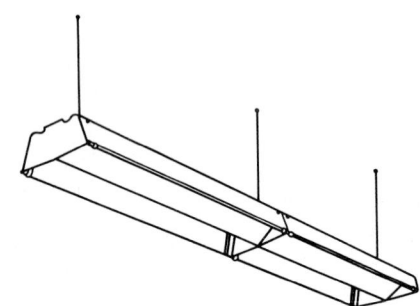

**Fig. 10-91**    Closed-end type of steel fluorescent luminaire.

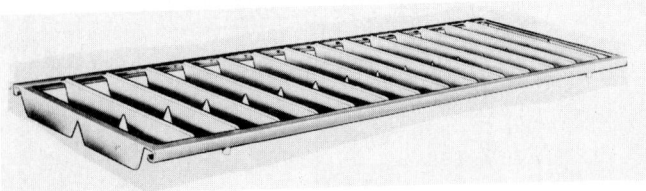

**Fig. 10-92**    Louver for use with an open-type fluorescent luminaire. [Miller Lighting, Hubbell Lighting Division]

filter systems which greatly reduce light-output depreciation due to dirt. Low-bay units are available with acrylic and polycarbonate plastic refractors. (See Fig. 10-88).

**216. Industrial luminaires for fluorescent lamps** generally employ porcelain-enamel steel reflectors. They provide direct illumination. Both open and enclosed units are available for two- to four-lamp assemblies. The open units may be of the open- or closed-end type of construction. The tops of the reflectors may be solid or be provided with apertures (refer to Figs. 10-89, 10-90, and 10-91.) These open units may have a plain reflector, or the reflector may be fitted with a longitudinal shield or with louvers (refer to Fig. 10-92). The shield is positioned between the lamps and provides a crosswise shielding angle of approximately 27° from the horizontal. Closed units are provided with glass or plastic covers across the bottom of the luminaire. Closed units with sealed covers are available for hazardous locations (see Fig. 10-93).

Different mounting methods are shown in Fig. 10-94.

**217. The ballast for fluorescent lamps is mounted** in the supporting channel or base of the luminaire. The construction to accommodate the ballast for one type of luminaire is shown in Fig. 10-95.

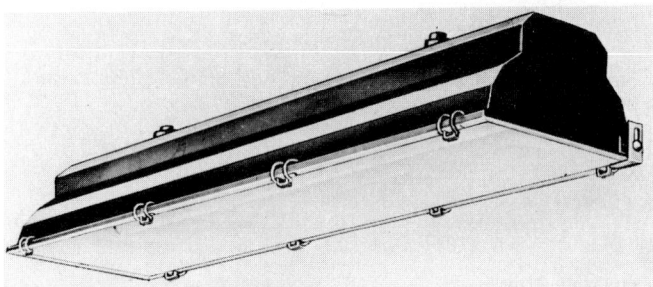

**Fig. 10-93** Closed-type fluorescent luminaire for hazardous locations. [Benjamin Division, Thomas Industries, Inc.]

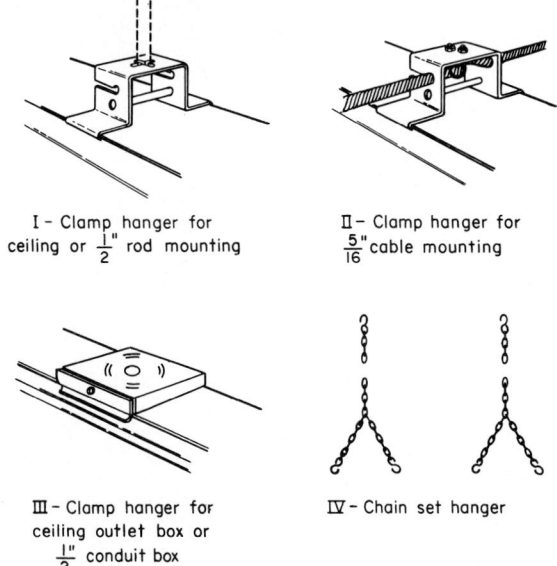

I - Clamp hanger for ceiling or $\frac{1}{2}$" rod mounting

II - Clamp hanger for $\frac{5}{16}$" cable mounting

III - Clamp hanger for ceiling outlet box or $\frac{1}{2}$" conduit box

IV - Chain set hanger

**Fig. 10-94** Methods of suspension for industrial fluorescent luminaires. [Miller Lighting, Hubbell Lighting Division]

**218. Fluorescent luminaires for commercial lighting** are made in a great variety of types. They can be obtained to produce illumination of any one of the different systems of illumination listed in Sec. 192, item 1. All the materials listed in Sec. 192, item 4, are employed for the reflection and diffusion of the light in fluorescent luminaires of the different constructions available. Probably the best method of obtaining an appreciation

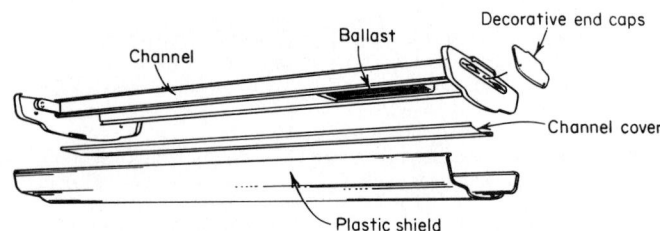

**Fig. 10-95**   Construction of a fluorescent luminaire.

of the various types available is through a study of the latest catalogs of fixture manufacturers.

Commercial fluorescent luminaires may be classified as follows:

1. Large-area ceiling units
2. Troffers
3. Convertional suspended or surface units

Large-area ceiling units (refer to Figs. 10-96, 10-97, and 10-98) are available in 2- by 2-, 2- by 4-, and 4- by 4-ft (0.6- by 0.6-, 0.6- by 1.2-, and 1.2- by 1.1-m) panels which

**Fig. 10-96**   Luminaires utilizing mercury, metal halide, or high-pressure sodium lamps. [Manville Special Products Group, Holophane Division]

**Fig. 10-97**  Illumination with large-area fluorescent ceiling units.

can be mounted to achieve almost any geometrical design. The luminous surface of the panels may be made of glass or plastic. Units are available for surface or recess mounting.

Translucent wall-to-wall lighting (Fig. 10-99) is illumination by means of one large luminous area. It does not employ luminaires unless the whole luminous ceiling is considered one large luminaire. The diffusing panels for wall-to-wall lighting may be of plastic or louver construction.

**Fig. 10-98**  Large-area surface-mounted fluorescent luminaire. [Day-Brite Lighting Division, Emerson Electric Co.]

Troffers are fluorescent luminaires which are mounted in recesses in the ceiling so that the surface of the unit is practically flush with the ceiling. A typical troffer luminaire is shown in Fig. 10-100, and the method of installing troffers in a suspended plaster ceiling in Fig. 10-101. Troffers are available for installation in almost any type of ceiling. Troffers (Fig. 10-102) may be of the open type, louvered, or covered with a glass, prismatic-glass, or plastic cover.

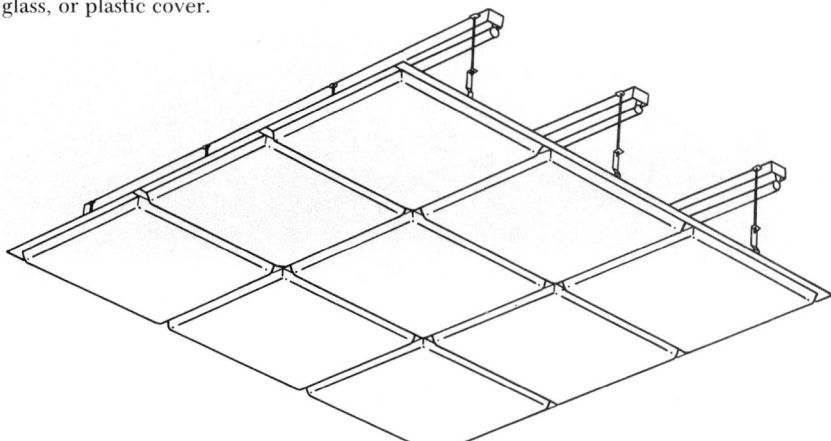

**Fig. 10-99**  Wall-to-wall fluorescent lighting.

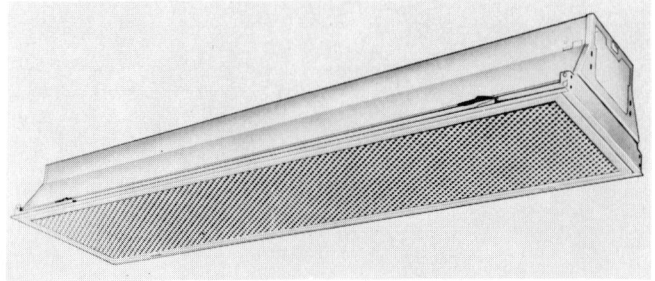

**Fig. 10-100**   Troffer fluorescent luminaire. [Miller Lighting, Hubbell Lighting Division]

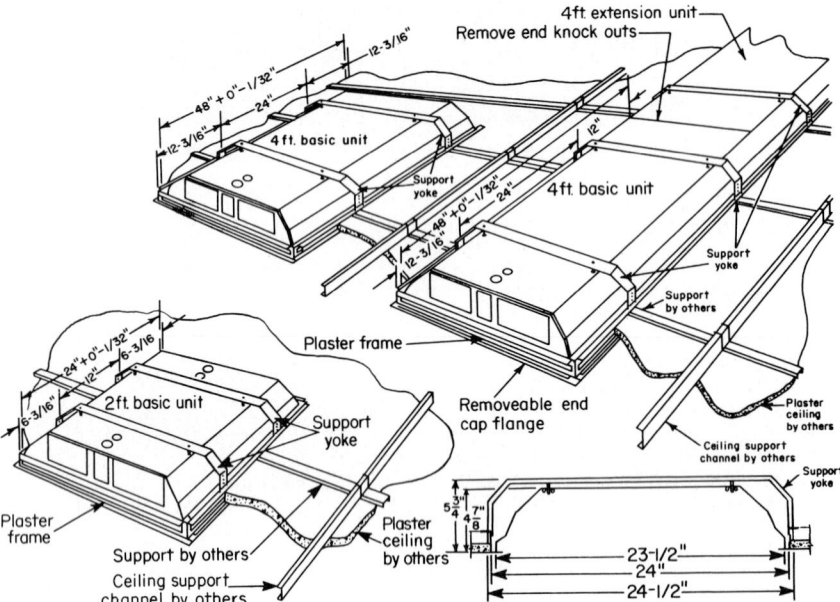

**Fig. 10-101**   Installation of flanged troffers in a suspended plaster ceiling.

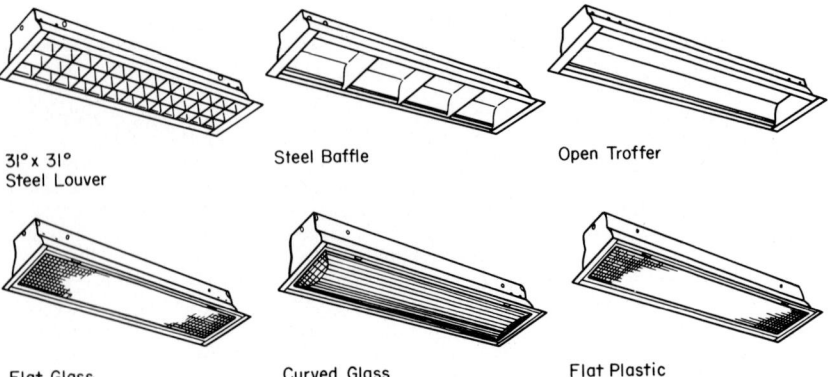

**Fig. 10-102**   Covers for troffer luminaires. [Benjamin Division, Thomas Industries, Inc.]

Conventional suspended or surface luminaires (Figs. 10-103 and 10-104) are available in units for producing illumination of any one of the systems.

**219. The size of luminaire for an incandescent lamp** should be such that its wattage rating will correspond to the wattage rating of the lamp to be used with it. This is necessary for the proper distribution of light, for adequate dissipation of the heat produced by the lamp, and for prevention of excessive brightness of the unit. An exception is the use of a smaller-size lamp if an adapter is employed in the socket to bring the lamp bowl into the proper position in the luminaire. When replacing lamps, always use the same size that was used previously unless careful investigation has shown a different size to be satisfactory.

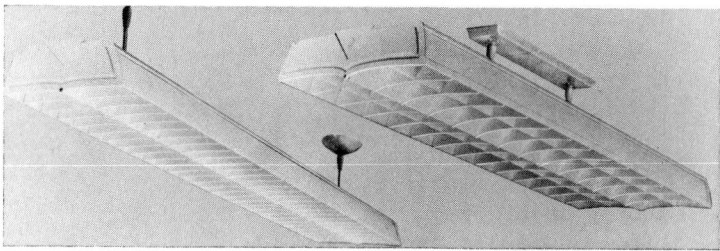

**Fig. 10-103** Direct-indirect fluorescent luminaire. [Day-Brite LIghting Division, Emerson Electric Co.]

**Fig. 10-104** Semi-indirect fluorescent luminaire. [Miller Lighting, Hubbell Lighting Division]

## PRINCIPLES OF LIGHTING-INSTALLATION DESIGN

**220. The general purpose of artificial lighting** is to enable things to be seen readily. Since things are seen by the light which is reflected from them (see "Brightness," Sec. 39) into the eye, in an effective lighting installation it is necessary that the number and arrangement of lighting units render most easily seen those things which it is desired to see. To accomplish this, recognition must be made of the effect of light on the human eye.

**221. Physiological features of artificial lighting.** To appreciate properly the principles of scientific lighting, it is necessary to understand the mechanism of the eye. Figure 10-105 (from *Primer of Illumination,* copyright by the Illuminating Engineering Society) shows the parts of the eye as they would appear if it were cut through from back to front vertically. In the process of seeing, the light passes through the cornea, pupil, and lens of the eye to the retina, just as in a camera light passes through the lens to the sensitized film. The picture is formed on the retina, which is a layer made up of the ends of nerve fibers that gather into the optic nerve and go directly to the brain.

**222. The optic nerve sends the picture to the brain.** The lens of the eye, unlike that of the camera, automatically changes in thickness to focus or make a clear image on the retina for seeing at different distances. This focusing action is called the accommodation of the eye, and when the light is dim or bad, the focusing muscle vainly hunts for some focus which may make objects look clear and gets tired in trying to do it. The muscles which move the eye about also get tired in the same way, and the result is eyestrain, which stirs up pain and headache just as any other overtired muscles of the body may set up an ache.

**223. The iris,** which gives the eye its color, serves to regulate the amount of light that reaches the eye. In very dim light it opens out, making the pupil big, as shown in Fig. 10-106), and in very bright light it shuts up as shown and thus keeps out a flood of brilliant light which might hurt the retina. The protective action of the pupil is fairly good but by no means complete, for it seldom becomes smaller than shown in the illustration, however bright the light. From a study of Fig. 10-106 we may deduce:

1. When trying to see any object, do not allow a light to shine into the eyes, and do not face a brightly lighted area. In addition to tiring the retina, the superfluous light causes the pupil to contract, so that less light from the lighted object reaches the retina. An object which would seem well lighted in a room with dark walls and with no light shining in the eyes will appear poorly lighted in a bright room with light walls or when a light is shining

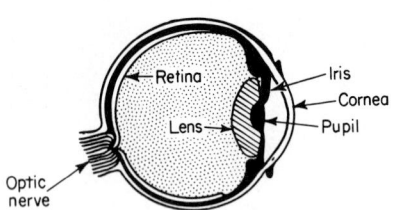

**Fig. 10-105**  Essential parts of the eye shown in section.

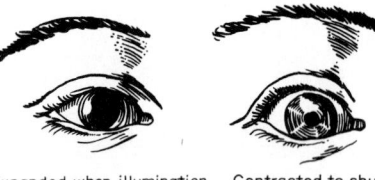

Expanded when illumination is dim        Contracted to shut out excessive light

**Fig. 10-106**  Expansion and contraction of the pupil of the eye.

in the eyes, simply because the pupil is smaller. This also explains why a higher light intensity is necessary in the daytime than at night. It is generally easier to read with the same light source in a room having dark walls than if the walls are light in color, though the total illumination on the page will probably be less. Reflected light from glossy paper produces the same effect as light surroundings. The effect produced by a light shining directly into the eyes is termed *glare* (see Sec. 38).

2. A fluctuating light causes the pupil to be constantly changing. This is very tiring to the muscles controlling the iris and if long continued may cause a permanent injury.

3. The lens of the eye is not corrected, as is a photographic lens, for color variations. It cannot focus sharply red and blue light from the same object simultaneously, although ordinarily this is not noticed. As white light is composed of all colors, it follows that we can see more clearly (i.e., objects appear sharper and more distinct) by a monochromatic light (light of only one color) than even by daylight. The light from mercury-vapor lamps closely approximates this condition.

4. Illumination should be uniform; otherwise the eye, in continually attempting to adapt itself to the unequal conditions, becomes tired as with a fluctuating light.

Correct lighting enables one to see clearly with minimum tiring of the eyes. To secure this, all the above conditions must be satisfied.

**224. The requirements of a satisfactory lighting installation** (Ward Harrison, *Electric Lighting*) are (1) sufficient light of unvarying intensity on all principal surfaces, whether horizontal, vertical, or oblique planes; (2) a comparable intensity of light on adjacent areas and on the walls; (3) light of a color and spectral character suited to the purpose for which it is employed; (4) freedom from glare and from glaring reflections; (5) light so directed and diffused as to prevent objectionable shadows or contrasts of intensity; (6) a lighting effect appropriate for the location and lighting units which are in harmony with their surroundings, whether lighted or unlighted; and (7) a system which is simple, reliable, easy to maintain, and in initial and operating cost not out of proportion to the results attained. Neglect of any of these may result in an unsatisfactory installation.

**225. Intensity of illumination.**  Although the eye is capable of adjusting itself to perceive objects over a wide range of intensities, the speed of this perception and the ability to distinguish fineness of detail are improved as the intensity increases, until the intensity becomes so great that a blinding effect is produced. An intensity of illumination that will be so high as to produce a blinding effect is, however, far above the range of intensities utilized in artificial illumination. The effect of the intensity of illumination upon the length of time required for the perception of objects is shown in Fig. 10-107A. The effect of the intensity of illumination upon the length of time required for the discrimination of detail is shown in Fig. 10-107B. From a study of the curve of Fig. 10-107A it is seen that the speed of perception increases approximately proportionally to the intensity of illumination, until an illumination of 5 fc (53.8 lx) is reached. Above this point the increase in speed of perception for a given increase in intensity gradually decreases. From these facts it was at one time erroneously considered that intensities of illumination above 5 fc

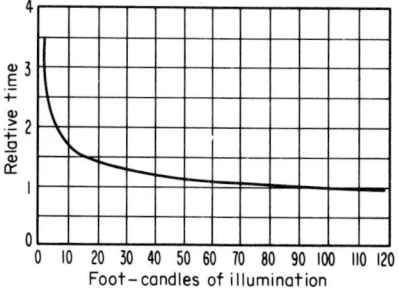

**Fig. 10-107A** Effect of intensity of illumination upon time for perception. [General Electric Co.]

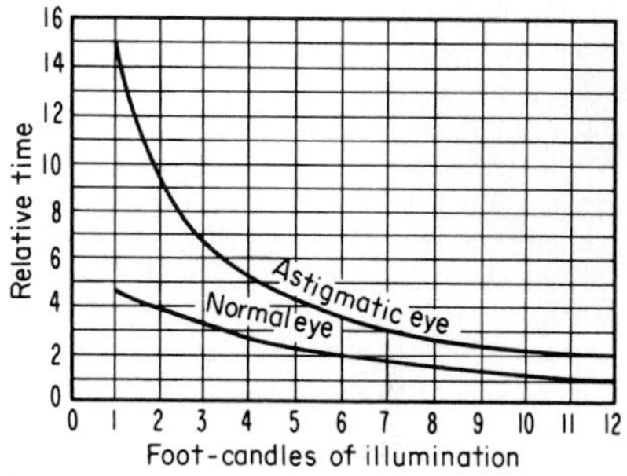

**Fig. 10-107B** Effect of intensity of illumination upon time required for discrimination of detail. [General Electric Co.]

were of little practical value in making things more clearly seen. That this is not true has been definitely proved not only by laboratory tests but also by practical tests in actual working areas. High levels of illumination have been found to result in an actual saving in dollars and cents owing to the increased production and decrease in spoilage produced by their use. Recommended values of intensities are given in Table 248.

**226. Blink test.** To have some definite and scientific way of measuring eyestrain and fatigue, the General Electric Co. devised the *blink test*. In this test the number of blinks per minute which the eye makes involuntarily are counted. This is done when the eyes are in a rested condition and then again after an hour of work at the various usual tasks of the individual under different levels of illumination. The rate of increase in blinking is taken as a measure of eye and nerve fatigue. The rate of blinking always increases somewhat with working or reading, but the higher the illumination intensity, up to at least 75 or 100 fc (807.3 or 1076.4 lx), the lower the rate of blinking, especially when attention to detail is required. Results of these tests have been applied to the values for Table 248.

**227. Elimination of glare.** In designing a lighting installation the greatest care should be exercised to avoid the presence of any one of the three types of glare described in Sec. 38. So that a lighting unit in the central portion of the visual field may not produce glare, its brightness should not exceed from 2 to 3 cd/in² of apparent area. If the eyes are not to be fatigued when the unit is viewed continually, the brightness should not exceed ½ cd/in² of apparent area. The maximum permissible brightness is, of course, influenced by the darkness of the surroundings.

No less important than the elimination of direct glare is the avoidance of reflected glare of specular reflection. Its effects are often more harmful than those due to direct glare. The use of highly polished furniture, plate-glass desktops, and glossy finishes on walls should be avoided. A large expanse of too-light-finished wall surface will cause harmful glare in offices, schoolrooms, and other locations where the occupants of the room must sit facing the wall for long periods of time. Although it is always desirable to have walls finished in light colors such as cream or buff in order to utilize as effectively as possible the light emitted by the source, the reflection factor of the lower walls should not be so high as to cause eye fatigue. For rooms in which occupants must face the walls for long periods, the reflection factor of the wall below eye level should not exceed 50 percent.

**228. Production of shadows.** The character of the shadow cast by an object depends upon the uniformity of the illumination in the area and upon the direction of the light falling upon the object. If the light falling upon an object is coming from one direction only, a sharp, dense shadow will be cast. As the number of directions in which light is impinging upon an object is increased, the sharpness and denseness of the shadow cast will be reduced. A shadow will still be cast even if the object is illuminated from all directions unless the intensity of the illumination is the same in all directions. Therefore, to eliminate shadows completely the illumination must be uniform and completely diffused; i.e., illumination of any point will be produced by light from all directions. The less uniform the illumination is or the fewer the directions of light falling upon an object, the more dense and sharp will be the shadows cast.

The desirability of shadows in an illuminated area depends upon the nature of the installation. In no installation is it desirable to have complete uniformity of illumination, i.e., illumination with no shadows or contrast in the intensity of illumination. Such illumination would produce a cold and flat effect tiresome to the eye. Nor could the shape and contour of objects be discerned. On the other hand, it is equally bad to have illumination which results in very dark, dense shadows and excessive contrasts in the intensity of illumination of different areas. Such illumination would be very hard on the eyes, owing to the continual change in the size of the pupil as the field of vision varies. When shadows are too dark, an object cannot be distinguished from its shadow; the shape of object is discernible from its shadow and without great contrast in intensity of illumination.

Shadows of the proper quality are of great value in observing objects in their three dimensions, in determining the shape of an object, and in determining irregularities in surfaces. They are of no value in the observation of plane surfaces. Shadows should never be dark and dense. If shadows are essential, they should be soft and luminous so that the object is clearly discernible from its shadow and so that there is no great marked contrast in intensity of illumination.

**229. Uniformity of illumination** is usually expressed as a maximum deviation from the average or mean intensity of illumination. It is not necessary to have the illumination of an area exactly uniform in order that the contrasts in the illumination of different portions will be unobservable by the eye or that the shadows will be too dense or sharp. The degree of uniformity of illumination obtained depends upon the ratio of the spacing distance of the lighting units to their mounting heights. For any reflector there is a maximum spacing of the units for a given mounting height that should not be exceeded if satisfactory uniformity of illumination is to be obtained. The effect of the spacing of the units upon the uniformity of illumination is illustrated in Fig. 10-108. Although this maximum spacing distance varies somewhat for reflectors distributing differently the light emitted by them, it is a good general rule to make the spacing distance equal to the mounting height. A greater spacing distance is likely to result in sharp and too dense shadows. One need not hesitate to use a closer spacing distance than that equal to the mounting height if desired, since the closer the spacing, the more uniform will be the illumination of the area. For the average installation it is best to keep the spacing distance of the units within the values recommended for specific luminaires. If the ceiling is unusually high, closer spacing

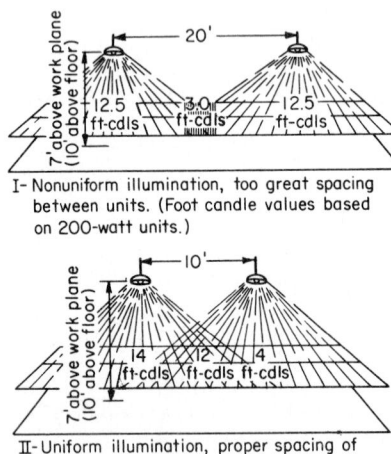

I- Nonuniform illumination, too great spacing between units. (Foot candle values based on 200-watt units.)

II-Uniform illumination, proper spacing of units. (Foot candle values based on 200-watt units.)

**Fig. 10-108** Effect of spacing of outlets on uniformity of illumination.

distances may be advisable to reduce the density of the shadows. In the illumination of certain areas such as storage spaces, it is permissible to space the units slightly farther apart.

**230. Methods of illumination.**  With respect to the arrangement of the light sources the methods of illumination may be divided into four classes as follows:

1. Localized
2. General
3. Group or localized general
4. Combination of general and localized

In the localized method an individual lighting unit of small wattage is supplied for each worker, machine, or workbench; the units are supported at a low level to bring the light close to the work. There is no attempt to produce uniform illumination over the total area.

With the general illumination method the lighting units are supported fairly close to the ceiling or at least at a considerable distance above the working plane. They are spaced uniformly throughout the area without special regard to the location of machinery, furniture, etc., and at such a distance from one another as to give nearly uniform illumination.

The group method is somewhat of a compromise between the localized and general methods. The lighting units are mounted close to the ceiling or at a considerable distance above the working plane. The units are not spaced uniformly but are located with respect to the machinery, position of operators, etc., so as to give sufficient illumination for work at each machine. The illumination of the area is not uniform.

In the combination of general and localized methods uniform illumination is supplied over the complete area by means of units located according to the general method. This illumination is supplemented with local lights for certain operations that require a higher intensity than that required for the main area.

**231. Comparison of methods of illumination.**  The localized arrangement of lighting units in industrial and most commercial applications should be a thing of the past and should never be tolerated. It produces very dense shadows and such low intensities in the general area that it is likely to cause very serious accidents. Such a method will prove to be very harmful to the occupants of the room, owing to direct glare from the low-hung lamps, and may so blind the operators of machines that they can be severely injured.

For the great majority of installations the general method of arranging light sources is the most satisfactory. It is the method most commonly employed. For some classes of work which require a very high intensity of illumination at the working point, it would not be economical to light the whole area to this high intensity. These conditions can be satisfactorily met by means of the combination of the general and localized methods, provided the local lamps are well shielded to prevent the possibility of glare. With the combination method the ratio of local to general illumination intensity should not be greater than 10:1. In other words, if some operations require 75 fc (807.3 lx), the general illumination should be not less than 7½ fc (80.7 lx).

In places where there are machines with large overhanging parts and in rooms with high obstructions, the group method will direct the light upon the working parts better than the general arrangement. The group method is frequently a good arrangement for rooms containing a number of similar machines arranged in rows.

**232. Effect of proportions of a room.**  Of the light which strikes the walls of a room, only part is reflected back into the room to become useful upon the working plane. Consequently, the greater the proportion of the total light that is directed to the walls, the lower the intensity of illumination on the working plane, or, stated in another way, the greater the proportion of the total light that is directed to the walls, the larger must the lamps be to produce a given intensity on the working plane. The proportion of the light that strikes the walls depends upon the size of the room and the height at which the units are mounted. In a large room the ratio between the wall area and the floor area is less than for a small room. Thus, for the same mounting height a smaller proportion of the total light will strike the walls than in a small room. In large rooms, therefore, less light is absorbed by the walls than in small rooms.

In a room of given dimensions, if the mounting height of the units is increased, more light will fall upon the walls. One should be careful not to get the impression that it is best to mount the units as low as possible. Although less light is absorbed by the walls with low-hung units, there are other more important factors that are affected by the mounting height. As the mounting height of the units is reduced, the spacing between them must also be reduced so that uniform illumination is produced and the shadows are not too dense. The cost of installing a large number of small units is greater than that of installing a smaller number of larger units. Also, the larger-size lamps are much more efficient with respect to the number of lumens produced per watt. As a result, the saving from the use of large units mounted high more than offsets the increased absorption of light by the walls. For practically all installations it is best to mount the units as high as possible. There is less probability of glare with high-mounted units than with low-hung ones, since the units are not so likely to be in the field of view.

**233. Room cavity ratios.** A room cavity ratio is used as the measure of the effect of room proportions upon the useful utilization of the light given out by the luminaire. Values of room cavity ratios are given in Table 243A. The room cavity ratio serves as a reference index for use with Tables 245, 246, and 247 in determining the coefficient of utilization.

**234. Effect of character of finish of walls.** Of the light that falls upon the walls or ceiling of a room, the percentage that is reflected back into the room depends upon the color of the wall and ceilings and the type of paint employed. The lighter the color, the greater the coefficient of reflection. The reflection factor of two freshly painted walls of practically the same color may differ, however, by several percent, depending upon the grade of paint employed. The reflection factor of a surface decreases as time elapses after painting. The amount of this depreciation in reflecting power varies widely with different grades of paint. It is important to have the walls and ceiling light in color and to employ a good grade of paint which will have a high initial reflection factor that does not depreciate an excessive amount with age. The walls, however, should not be made so bright as to be a source of glare (see Sec. 227).

**235. Location of lighting equipment.** The lighting units for general illumination should be located symmetrically throughout the area to be illuminated. When the room is divided into bays by means of columns, roof trusses, or girders or when the ceiling is divided into panels, the units should be arranged symmetrically, for the sake of appearance, with respect to such architectural divisions, provided the arrangement will not interfere with the uniformity of illumination. Fluorescent luminaires generally are arranged in rows or in some other symmetrical pattern which will fit in with the architectural arrangement of the area and the utilization of the space. Incandescent and mercury-vapor units generally are arranged in the form of squares or rectangles. Typical layouts are shown in Fig. 10-109.

It is important that the units be placed at the centers of squares and not at the corners. Figure 10-110 shows a method of locating outlets which is undesirable because it gives a very low intensity of illumination near the walls compared with that at the center of the room. Figure 10-111 shows the correct method of locating outlets in the centers of the squares. In certain cases, notably in office lighting and in rooms with benches located along the walls, it may, to minimize shadows, be desirable to place the outer rows of outlets somewhat nearer the sidewalls of the room than they would be if symmetrically arranged as shown in Fig. 10-111.

In laying out a lighting installation the maximum permissible spacing distance for the production of uniform illumination and satisfactory density of shadows should be determined first. The maximum permissible spacing for direct, semidirect, or general diffusing luminaires depends upon the candlepower-distribution characteristics of the luminaire.

After the maximum spacing has been determined, locate the units on a plan of the area so that they will be positioned symmetrically with respect to architectural conditions and at the same time will not exceed the maximum permissible spacing distance.

When the outlets are located above traveling cranes, the staggered system (Fig. 10-109e) is preferable so that as the crane moves along, it cuts off the light from only one unit at a time. If it is not desired to use the staggered system, additional lighting outlets, with special shock-absorbing sockets, should be located on the underside of the crane truss so that as the crane moves along, the light from those units replaces the light from the regular outlets which are cut off by the crane.

In buildings of mill-type construction the units are usually supported on the lower chord of the roof truss or suspended from the roof purlins. When the units are mounted on the lower chord of the roof truss, the illumination near the walls at each end of the

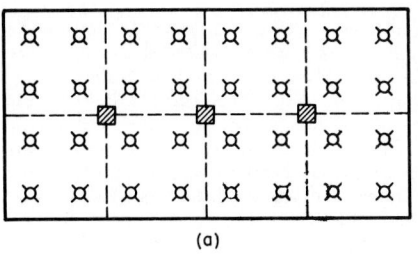

(a)

Four units per bay — This is the most common system for the square bay of usual dimensions.

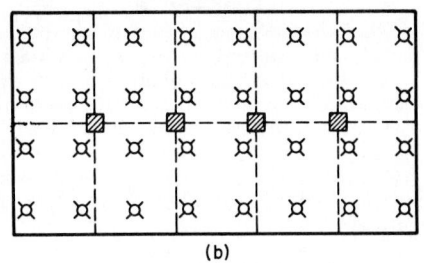

(b)

Four - two system — This is equivalent to three units per bay and is an alternative to four per bay where spacing allows.

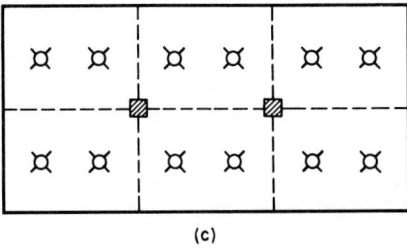

(c)

Two units per bay — Usually applicable only in narrow bays where the width is less than two-thirds the length.

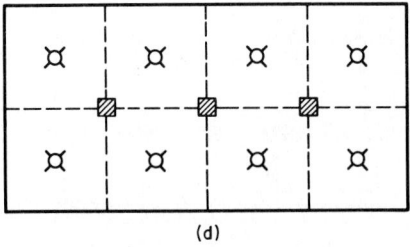

(d)

One unit per bay — A very common practice, but satisfactory only where bay size is no greater than the maximum permissible spacing — an unusual condition.

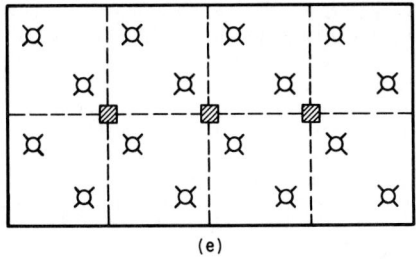

(e)

Staggered system — A recourse where one unit per bay is unsatisfactory and where four per bay is unnecessary. Less favorable appearance, and certain areas near walls may be inadequately lighted. Often expensive to wire.

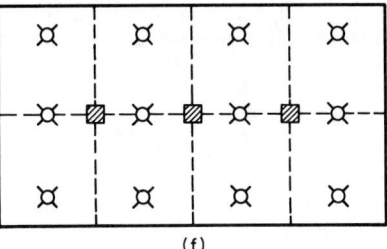

(f)

Interspaced system — Applicable in rectangular bays where one unit per bay would exceed the permissible spacing in one direction, and where center row will not interfere with future structural arrangements, such as added office partitions.

**Fig. 10-109**   Layouts of lighting units for symmetrical spacing. [General Electric Co.]

building is low. When work must be carried on in these areas, it is better to support the units from the purlins or to locate a row of angle units along each end wall.

High mounting is desirable because then the lamps are out of the way of cranes and are less apt to be broken, glare is reduced to a minimum, and in the case of a light ceiling there is more reflection and better diffusion of light. The lamps should be lowered in locations where there is horizontal overhead belting, to the level of the bottom of the belting; otherwise, a portion of the light is ineffective. It may be necessary, for the same reason, to install two or three units in an area where conditions would otherwise warrant only one unit.

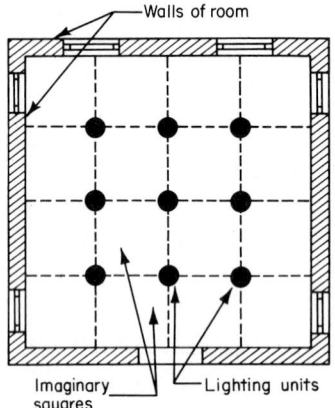
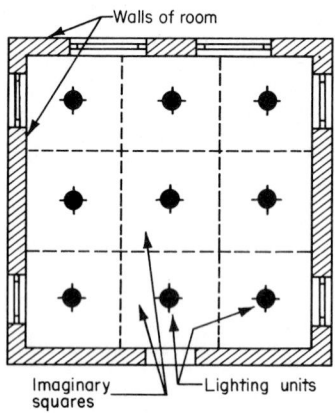

**Fig. 10-110**  Incorrect arrangement of lighting units.

**Fig. 10-111**  Correct arrangement of lighting units.

**236. Reflection factor** is the percentage of light reflected from a surface, such as a wall or ceiling, to the total light falling on the surface. The difference between the reflection factor and 100 percent represents the percentage of light which is absorbed by the surface and lost. Reflection factors for some of the common colors are given in Table 243D. The reflection factor varies with the roughness of the surface as well as with the darkness of the color.

**237. The coeffecient of utilization** is the overall efficiency of the lighting installation: the ratio of the useful lumens which get down to the working plane to the total lumens generated by the lamp. Tables 245, 246, and 247 give values for the coefficient of utilization for different types of luminaires, subdivided according to room cavity ratios and reflection factors for walls and for the ceiling. For more complete data refer to the *IES Lighting Handbook*.

**238. Maintenance of lighting equipment.**  The intensity of illumination produced by a lighting installation will be somewhat less after the system has been in use for some time than it was when the system was first installed. This depreciation in the lighting system with time is due to the decrease in efficiency of the lamp in use and to the decrease in reflecting efficiency of the reflecting equipment, walls, and ceiling, which is the result of the natural deterioration of the surface with age and of the collection of dust and dirt.

In the design of lighting installations, a maintenance factor is used to take into account this depreciation. It has values less than unity, so that when the initial illumination is multiplied by the maintenance factor, the actual illumination after a period of several months' use will be obtained.

The amount of depreciation will vary from 25 to 45 percent, corresponding to a maintenance factor of 0.75 to 0.55, depending upon the type and material of the reflecting equipment, the system of illumination employed, the local conditions of dust and dirt, and the frequency of cleaning of equipment and repainting of walls. All lighting units should be adequately cleaned at regular intervals to prevent a waste of energy and low intensity of illumination. The frequency of the cleaning periods depends upon the degree of prevalence of dust and dirt and upon the type of luminaire.

**239. Four principal methods are used in calculating illumination:**
1. Point-by-point method
2. Lumen-per-foot method
3. Zonal-cavity method (general lighting)
4. Beam-lumens method (floodlighting)

**240. The point-by-point method** is based on the inverse-square law that the intensity of light flux varies inversely as the square of the distance from the light source to the point of measurement (see Sec. 26 and Fig. 10-112). The illumination on any plane perpendicular to the light rays is given by the following formula:

$$\text{Footcandles (on perpendicular plane)} = \frac{\text{cd}}{D^2} \tag{6}$$

where cd = the candlepower of the light source in the direction in which the distance $D$ is taken and $D$ = the perpendicular distance from the light source to the illuminated plane in feet.

Since the illumination of the horizontal plane is the value usually desired, the formula must be multiplied by the ratio $H/D$ (see Fig. 10-112), or

$$\text{Footcandles (on horizontal plane)} = \frac{\text{cd} \times H}{D^3} \tag{7}$$

If the illumination on a vertical plane is desired, the basic formula (6) must be multiplied by the ratio $X/D$ (see Fig. 10-112), or

$$\text{Footcandles (on vertical plane)} = \frac{\text{cd} \times X}{D^3} \tag{8}$$

The illumination at any point is obtained by adding the footcandles due to each light source which is sending rays of light to the point.

It is obvious that the point-by-point method is practical for use only with the direct type of lighting, for with any other system of illumi-

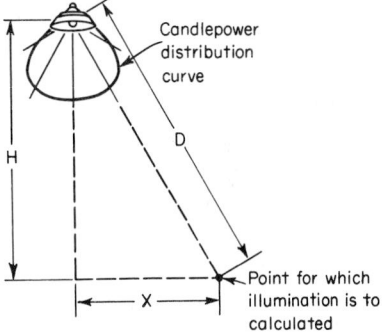

**Fig. 10-112**  Point-by-point method of lighting calculations.

nation the number of light sources which would have to be considered would make the calculations prohibitively tedious. Even with the direct system it is usually necessary to calculate the footcandles from several different light sources. This method is especially applicable to calculating localized lighting when only a single light source needs to be considered.

**241. Lumen-per-foot method** (General Electric Co.). The predetermining of lighting levels for supplementary lighting systems in which continuous linear sources are used can be accomplished from empirical data based on the lumens per foot of the source. These data are adaptable for relatively short distances between the light source and the work or display to which the inverse-square law obviously does not apply. The lumen-per-foot method employs the following formulas:

Where the lamp and reflector have been selected,

$$\text{Footcandles} = K \times \text{lumens per foot of source} \tag{9}$$

Where a footcandle level is desired,

$$\text{Necessary lumens per foot} = \frac{\text{footcandles desired}}{K} \tag{10}$$

In formulas (9) and (10) the constant $K$ refers, respectively, to either one of two constants $K_H$ and $K_V$ as given in Tables 241A, 241B, 241C, and 241D for horizontal and vertical illumination.

Two types of luminaires are specified: broad distribution as from a unit with a mat-finish reflector and narrow distribution as from a polished metal reflector. The beam in both cases is presumed to be aimed at a plane 4 ft (1.2 m) from the source through point A.

The importance of the reflector is evident from a comparison of Tables 241C and 241D. The average $K$ value at 4 ft in 236C is 0.009; in 236D, 0.021. This difference is 133 percent.

The actual values from which the tables were compiled are readings taken at the midpoint of luminaires 12 ft (3.7 m) in length. For conventional lighting systems, the footcandle levels would normally drop at the ends of rows unless additional lamps or lamps of higher output were provided.

**Table 241A.    Horizontal Illumination**
(Horizontal fc = $K_H$ × lamp lumens per foot)
(Broad distribution—white-enamel reflector)

| Distance between lamp center and plane of measurement, ft | $K_H$ values (distance from centerline of unit), ft | | | |
|---|---|---|---|---|
| | 0 | 1 | 2 | 3 |
| 1 | 0.438 | 0.127 | 0.008 | 0.001 |
| 2 | 0.223 | 0.150 | 0.061 | 0.017 |
| 3 | 0.145 | 0.120 | 0.077 | 0.041 |
| 4 | 0.106 | 0.095 | 0.072 | 0.048 |

**Table 241B.    Horizontal Illumination**
(Horizontal fc = $K_H$ × lamp lumens per foot)
(Narrow distribution—polished aluminum reflector)

| Distance between lamp center and plane of measurement, ft | $K_H$ values (distance from centerline of unit), ft | | | |
|---|---|---|---|---|
| | 0 | 1 | 2 | 3 |
| 1 | 0.753 | 0.079 | | |
| 2 | 0.330 | 0.165 | 0.035 | 0.006 |
| 3 | 0.212 | 0.161 | 0.066 | 0.022 |
| 4 | 0.153 | 0.131 | 0.086 | 0.038 |

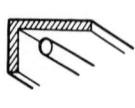

**Table 241C.    Vertical Illumination**
(Vertical fc = $K_V$ × lamp lumens per foot)
(Broad distribution—white-painted cornice; no reflector)

| Distance between lamp center and plane of measurement, ft | $K_V$ values (distance from centerline of unit), in | | | | |
|---|---|---|---|---|---|
| | 3 | 6 | 9 | 12 | 18 |
| 1 | 0.185 | 0.159 | 0.175 | 0.165 | 0.129 |
| 2 | 0.011 | 0.028 | 0.044 | 0.057 | 0.068 |
| 3 | 0.004 | 0.010 | 0.017 | 0.023 | 0.032 |
| 4 | 0.002 | 0.005 | 0.008 | 0.012 | 0.018 |

**Table 241D.   Vertical Illumination**
(Vertical fc = $K_V$ × lamp lumens per foot)
(Narrow distribution—polished aluminum reflector)

| Distance between lamp center and plane of measurement, ft | $K_V$ values (distance from centerline of unit), in | | | | |
|---|---|---|---|---|---|
| | 3 | 6 | 9 | 12 | 18 |
| 1 | 0.121 | 0.125 | 0.135 | 0.096 | 0.080 |
| 2 | 0.028 | 0.056 | 0.077 | 0.086 | 0.090 |
| 3 | 0.010 | 0.028 | 0.036 | 0.044 | 0.059 |
| 4 | 0.006 | 0.013 | 0.020 | 0.031 | 0.037 |

**242.  Lumens per Foot (Initial Lumens per Nominal Length) for Fluorescent Lamps (Standard Cool White)[a]**
(General Electric Co.)

**General-line lamps**

| Lamp designation | Lumens per foot | Lumens (initial) | Nominal length, ft |
|---|---|---|---|
| 4WT-5 | 270 | 135 | 6 |
| 6WT-5 | 393 | 295 | 9 |
| 8WT-5 | 400 | 400 | 12 |
| 13WT-5 | 486 | 850 | 21 |
| 14WT-12 | 520 | 650 | 15 |
| 15WT-8 | 550 | 825 | 18 |
| 15WT-12 | 507 | 760 | 18 |
| 20WT-12 | 600 | 1200 | 24 |
| 30WT-8 | 725 | 2175 | 36 |
| 30WT-12 | 742 | 2225 | 36 |
| 40WT-12 | 762 | 3050 | 48 |

**Slimline lamps**

| Lamp designation | Lumens per foot | Lumens (initial) | Nominal length, in | Watts (nominal) |
|---|---|---|---|---|
| 42″ T-6 | 500 | 1750 | 42 | 25.0 |
| 64″ T-6 | 525 | 2800 | 64 | 40.0 |
| 72″ T-8 | 500 | 3000 | 72 | 35.0 |
| 96″ T-8 | 506 | 4050 | 96 | 50.0 |
| 48″ T-12 | 600 | 2400 | 48 | 40.0 |
| 72″ T-12 | 750 | 4500 | 72 | 55.0 |
| 96″ T-12 | 769 | 6150 | 96 | 75.0 |

**High-output rapid-start lamps**

| | | | | |
|---|---|---|---|---|
| 48″ T-12 | 1012 | 4050 | 48 | 60.0 |
| 72″ T-12 | 1058 | 6350 | 72 | 85.0 |
| 96″ T-17 | 1112, | 8900 | 96 | 110.0 |

[a]For lumens per foot of other colors use the following multiplying factors: deluxe cool white, 0.71; deluxe warm white, 0.71; daylight, 0.93; soft white, 0.68; green, 1.2; gold, 0.6; blue, 0.45; and pink, 0.45.

**243. The zonal-cavity method.**    The lumen method of calculating illumination, which has been in use for many years, is based on the theory that average illumination is equal to lumens divided by the work area over which they are distributed. Newer methods of analysis of lighting distributions, taking into account the concept of interreflection of light, have led to more accurate coefficient-of-utilization data and have been adopted in the Illuminating Engineering Society (IES) approved method of lighting calculations called the *zonal-cavity method.* This method provides increased flexibility in lighting calculations, including greater accuracy, but still adheres to the basic concept that footcandles are equal to light flux over an area (the lumen method).

The zonal-cavity method assumes that an area to be lighted will be made up of a series of cavities which have effective reflectances with respect to one another and to the work plane. Basically, the area is divided into three spaces, or cavities, a *ceiling cavity,* a *room cavity,* and a *floor cavity.* These are shown graphically in Fig. 10-113.

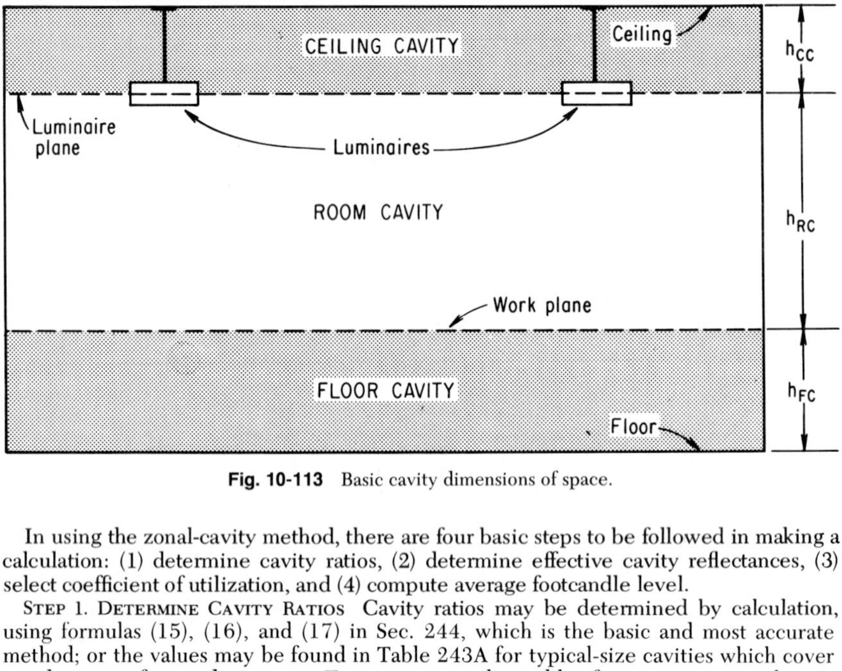

**Fig. 10-113**    Basic cavity dimensions of space.

In using the zonal-cavity method, there are four basic steps to be followed in making a calculation: (1) determine cavity ratios, (2) determine effective cavity reflectances, (3) select coefficient of utilization, and (4) compute average footcandle level.

STEP 1. DETERMINE CAVITY RATIOS    Cavity ratios may be determined by calculation, using formulas (15), (16), and (17) in Sec. 244, which is the basic and most accurate method; or the values may be found in Table 243A for typical-size cavities which cover a wide range of room dimensions. For a more complete table of cavity ratios, see the *IES Lighting Handbook.*

STEP 2. DETERMINE EFFECTIVE CAVITY REFLECTANCES    Table 243B provides a means of converting the combination of actual wall and ceiling or actual wall and floor reflectances into a single effective ceiling cavity reflectance $\rho cc$ and a single effective floor cavity reflectance $\rho fc$. Note that if the luminaire is recessed or surface-mounted or if the floor is the work plane, the ceiling cavity ratio or floor cavity ratio will be 0, and then the actual reflectance of the ceiling or floor will also be the effective reflectance. When using reflectance values of room surfaces, the expected maintained values should be used for calculation of maintained footcandles, or if initial footcandle values are desired, the initial reflectance values should be used.

STEP 3. SELECT COEFFICIENT OF UTILIZATION (CU)    Using the $\rho cc$, $\rho fc$, and $\rho W$ (wall reflectance) determined in step 2 and knowing the room cavity ratio previously calculated in step 1, formula (16), in Sec. 244, refer to the coefficient of utilization in the CU table for the luminaire under consideration. Tables showing CU values for currently typical, popular types of luminaires are given in Tables 245, 246, and 247. For more complete data, see the *IES Lighting Handbook.* Manufacturers of lighting equipment will supply CU data for their own luminaires upon request. Note that since the CU tables are linear,

linear interpolations can be made for exact cavity ratios or reflectance combinations. Most CU tables are based on an effective floor cavity reflectance of 20 percent. Table 243C can be used to revise the CU for other effective floor cavity reflectances.

In determining the coefficient of utilization for a fluorescent luminaire, it is important that the specific lamp and ballast combination planned for an installation be identified and checked with the fixture manufacturer for any effect on the published CU. Fluorescent-lamp rated light output is established at 25°C (77°F). The light output from all fluorescent lamps varies with temperature (see Sec. 134). Different lamp designs, such as energy-saving types, do not operate at the same temperature in a specific luminaire, thus will differ in relative light output. Different ballast designs operate at different wattages and vary in losses, both of which affect the temperature in the luminaire and thus the light output.

Fluorescent-lamp ballasts usually do not operate lamps at rated wattage and light output. The ratio of lamp light output on a ballast compared with rated light output is called the *ballast factor*. Published coefficients of utilization assume rated lamp light output. Thus the CU for fluorescent luminaires must be multiplied by the ballast factor for the lamp-ballast combination being used in order to obtain accurate design calculations. The lamp type and ballast type must be identified to establish the specific ballast factor. Ballast factors can be obtained from ballast manufacturers.

STEP 4. COMPUTE AVERAGE FOOTCANDLE LEVEL  Use the standard lumen method [formula (13) in Sec. 244] to compute the footcandle level that will be obtained. When maintained illumination levels are to be calculated, the maintenance factor (MF) should include lamp lumen depreciation (LLD) and luminaire dirt depreciation (LDD). Light-

**Table 243A.  Cavity Ratios**
(From *IES Lighting Handbook*, 4th edition)

| Room dimensions, feet | | Cavity depth | | | | | | | | | | |
|---|---|---|---|---|---|---|---|---|---|---|---|---|
| Width | Length | 1.0 | 1.5 | 2.0 | 2.5 | 3.0 | 4.0 | 6 | 8 | 10 | 12 | 16 |
| 8 | 10 | 1.1 | 1.7 | 2.2 | 2.8 | 3.4 | 4.5 | 6.7 | 9.0 | 11.3 | | |
| | 14 | 1.0 | 1.5 | 2.0 | 2.5 | 3.0 | 3.9 | 5.9 | 7.8 | 9.7 | 11.7 | |
| | 20 | 0.9 | 1.3 | 1.7 | 2.2 | 2.6 | 3.5 | 5.2 | 7.0 | 8.8 | 10.5 | |
| 10 | 10 | 1.0 | 1.5 | 2.0 | 2.5 | 3.0 | 4.0 | 6.0 | 8.0 | 10.0 | 12.0 | |
| | 14 | 0.9 | 1.3 | 1.7 | 2.1 | 2.6 | 3.4 | 5.1 | 6.9 | 8.6 | 10.4 | |
| | 20 | 0.7 | 1.1 | 1.5 | 1.9 | 2.3 | 3.0 | 4.5 | 6.0 | 7.5 | 9.0 | 12.0 |
| | 40 | 0.6 | 0.9 | 1.2 | 1.6 | 1.9 | 2.5 | 3.7 | 5.0 | 6.2 | 7.5 | 10.0 |
| 12 | 12 | 0.8 | 1.2 | 1.7 | 2.1 | 2.5 | 3.3 | 5.0 | 6.7 | 8.4 | 10.0 | |
| | 16 | 0.7 | 1.1 | 1.5 | 1.8 | 2.2 | 2.9 | 4.4 | 5.8 | 7.2 | 8.7 | 11.6 |
| | 36 | 0.6 | 0.8 | 1.1 | 1.4 | 1.7 | 2.2 | 3.3 | 4.4 | 5.5 | 6.6 | 8.8 |
| | 50 | 0.5 | 0.8 | 1.0 | 1.3 | 1.5 | 2.1 | 3.1 | 4.1 | 5.1 | 6.2 | 8.2 |
| 14 | 14 | 0.7 | 1.1 | 1.4 | 1.8 | 2.1 | 2.9 | 4.3 | 5.7 | 7.1 | 8.5 | 11.4 |
| | 20 | 0.6 | 0.9 | 1.2 | 1.5 | 1.8 | 2.4 | 3.6 | 4.9 | 6.1 | 7.3 | 9.8 |
| | 42 | 0.5 | 0.7 | 1.0 | 1.2 | 1.4 | 1.9 | 2.9 | 3.8 | 4.7 | 5.7 | 7.6 |
| 20 | 20 | 0.5 | 0.7 | 1.0 | 1.2 | 1.5 | 2.0 | 3.0 | 4.0 | 5.0 | 6.0 | 8.0 |
| | 45 | 0.4 | 0.5 | 0.7 | 0.9 | 1.1 | 1.4 | 2.2 | 2.9 | 3.6 | 4.3 | 5.8 |
| | 90 | 0.3 | 0.5 | 0.6 | 0.8 | 0.9 | 1.2 | 1.8 | 2.4 | 3.0 | 3.6 | 4.8 |
| 30 | 30 | 0.3 | 0.5 | 0.7 | 0.8 | 1.0 | 1.3 | 2.0 | 2.7 | 3.3 | 4.0 | 3.4 |
| | 60 | 0.3 | 0.4 | 0.5 | 0.6 | 0.7 | 1.0 | 1.5 | 2.0 | 2.5 | 3.0 | 4.0 |
| | 90 | 0.2 | 0.3 | 0.4 | 0.6 | 0.7 | 0.9 | 1.3 | 1.8 | 2.2 | 2.7 | 3.6 |
| 42 | 42 | 0.2 | 0.4 | 0.5 | 0.6 | 0.7 | 1.0 | 1.4 | 1.9 | 2.4 | 2.8 | 3.8 |
| | 90 | 0.2 | 0.3 | 0.3 | 0.4 | 0.5 | 0.7 | 1.0 | 1.4 | 1.7 | 2.1 | 2.8 |
| | 200 | 0.1 | 0.2 | 0.3 | 0.4 | 0.4 | 0.6 | 0.9 | 1.1 | 1.4 | 1.7 | 2.3 |
| 60 | 60 | 0.2 | 0.2 | 0.3 | 0.4 | 0.5 | 0.7 | 1.0 | 1.3 | 1.7 | 2.0 | 2.7 |
| | 100 | 0.1 | 0.2 | 0.3 | 0.3 | 0.4 | 0.5 | 0.8 | 1.1 | 1.3 | 1.6 | 2.1 |
| | 300 | 0.1 | 0.1 | 0.2 | 0.2 | 0.3 | 0.4 | 0.6 | 0.8 | 1.0 | 1.2 | 1.6 |
| 100 | 100 | 0.1 | 0.1 | 0.2 | 0.2 | 0.3 | 0.4 | 0.6 | 0.8 | 1.0 | 1.2 | 1.6 |
| | 300 | 0.1 | 0.1 | 0.1 | 0.2 | 0.2 | 0.3 | 0.4 | 0.5 | 0.7 | 0.8 | 1.1 |

source manufacturers can supply the LLD factor for any type of lamps; the factor should be based upon the luminaire's dirt attraction or dirt retention characteristics and the degree of dirtiness of the areas where it is to be used. Procedures for making these calculations are given in the *IES Lighting Handbook*.

In making lighting calculations for proposed projects, lighting designers should have all the facts required for use in their calculations. Included would be the end-use application of the space; the physical dimensions; the color and reflectance values of the ceiling, walls, and floor; the layout of furniture or machinery, including colors of the various furnishings, etc.; and a knowledge of the degree of cleanliness (or dirtiness) of the area which will affect maintenance of the lighting levels. The degree of accuracy of the light-

**Table 243B.  Percent Effective Ceiling or Floor Cavity Reflectance for Various Reflectance Combinations**
(From *IES Lighting Handbook*, 4th ed.)

| Ceiling or floor cavity ratio | % ceiling or floor reflectance | | | | | | | | | | | | | | |
|---|---|---|---|---|---|---|---|---|---|---|---|---|---|---|---|
| | 80 | | | 70 | | | 50 | | | 30 | | | 10 | | |
| | % wall reflectance | | | | | | | | | | | | | | |
| | 70 | 50 | 30 | 70 | 50 | 30 | 70 | 50 | 30 | 50 | 30 | 10 | 50 | 30 | 10 |
| 0 | 80 | 80 | 80 | 70 | 70 | 70 | 50 | 50 | 50 | 30 | 30 | 30 | 10 | 10 | 10 |
| 0.1 | 79 | 78 | 78 | 69 | 69 | 68 | 49 | 49 | 48 | 30 | 29 | 29 | 10 | 10 | 10 |
| 0.3 | 77 | 75 | 74 | 68 | 66 | 64 | 49 | 47 | 46 | 29 | 28 | 27 | 10 | 10 | 9 |
| 0.5 | 75 | 73 | 70 | 66 | 64 | 61 | 48 | 46 | 44 | 28 | 27 | 25 | 11 | 10 | 9 |
| 0.6 | 75 | 71 | 68 | 65 | 62 | 59 | 47 | 45 | 43 | 28 | 26 | 25 | 11 | 10 | 9 |
| 0.8 | 73 | 69 | 65 | 64 | 60 | 56 | 47 | 43 | 41 | 27 | 25 | 23 | 11 | 10 | 8 |
| 1.0 | 71 | 66 | 61 | 63 | 58 | 53 | 46 | 42 | 39 | 27 | 24 | 22 | 11 | 9 | 8 |
| 1.2 | 70 | 64 | 58 | 61 | 56 | 50 | 45 | 41 | 37 | 26 | 23 | 20 | 12 | 9 | 7 |
| 1.4 | 68 | 62 | 55 | 60 | 54 | 48 | 45 | 40 | 35 | 26 | 22 | 19 | 12 | 9 | 7 |
| 1.6 | 67 | 60 | 53 | 59 | 52 | 45 | 44 | 39 | 33 | 25 | 21 | 18 | 12 | 9 | 7 |
| 1.8 | 65 | 58 | 50 | 57 | 50 | 43 | 43 | 37 | 32 | 25 | 21 | 17 | 12 | 9 | 6 |
| 2.0 | 64 | 56 | 48 | 56 | 48 | 41 | 43 | 37 | 30 | 24 | 20 | 16 | 12 | 9 | 6 |
| 2.2 | 63 | 54 | 45 | 55 | 46 | 39 | 42 | 36 | 29 | 24 | 19 | 15 | 13 | 9 | 6 |
| 2.4 | 61 | 52 | 43 | 54 | 45 | 37 | 42 | 35 | 27 | 24 | 19 | 14 | 13 | 9 | 6 |
| 2.6 | 60 | 50 | 41 | 53 | 43 | 35 | 41 | 34 | 26 | 23 | 18 | 13 | 13 | 9 | 5 |
| 2.8 | 59 | 48 | 39 | 52 | 42 | 33 | 41 | 33 | 25 | 23 | 18 | 13 | 13 | 9 | 5 |
| 3.0 | 58 | 47 | 38 | 51 | 40 | 32 | 40 | 32 | 24 | 22 | 17 | 12 | 13 | 8 | 5 |
| 3.3 | 56 | 44 | 35 | 49 | 39 | 30 | 39 | 31 | 23 | 22 | 16 | 11 | 13 | 8 | 5 |
| 3.6 | 54 | 42 | 33 | 48 | 37 | 28 | 39 | 30 | 21 | 21 | 15 | 10 | 13 | 8 | 5 |
| 3.9 | 53 | 40 | 30 | 47 | 36 | 26 | 38 | 29 | 20 | 21 | 15 | 10 | 13 | 8 | 4 |
| 4.2 | 51 | 39 | 29 | 46 | 34 | 25 | 37 | 28 | 19 | 20 | 14 | 9 | 13 | 8 | 4 |
| 4.5 | 50 | 37 | 27 | 45 | 33 | 24 | 37 | 27 | 19 | 20 | 14 | 8 | 14 | 8 | 4 |
| 4.8 | 49 | 36 | 25 | 44 | 32 | 23 | 36 | 26 | 18 | 19 | 13 | 8 | 14 | 8 | 4 |
| 5.0 | 48 | 35 | 25 | 43 | 32 | 22 | 36 | 26 | 17 | 19 | 13 | 7 | 14 | 8 | 4 |

**Table 243C.  Factors for Effective Floor Cavity Reflectance Other Than 20 Percent[a]**
(From *IES Lighting Handbook*, 4th ed.)

| Room cavity ratio | % effective ceiling cavity reflectance, $\rho cc$ | | | | | | | |
|---|---|---|---|---|---|---|---|---|
| | 80 | | 70 | | 50 | | 10 | |
| | % wall reflectance, $\rho W$ | | | | | | | |
| | 50 | 30 | 50 | 30 | 50 | 30 | 50 | 30 |
| 1 | 1.08 | 1.08 | 1.07 | 1.06 | 1.05 | 1.04 | 1.01 | 1.01 |
| 3 | 1.05 | 1.04 | 1.05 | 1.04 | 1.03 | 1.03 | 1.01 | 1.01 |
| 5 | 1.04 | 1.03 | 1.03 | 1.02 | 1.02 | 1.02 | 1.01 | 1.01 |
| 6 | 1.03 | 1.02 | 1.03 | 1.02 | 1.02 | 1.02 | 1.01 | 1.01 |
| 8 | 1.03 | 1.02 | 1.02 | 1.02 | 1.02 | 1.01 | 1.01 | 1.01 |
| 10 | 1.02 | 1.01 | 1.02 | 1.01 | 1.02 | 1.01 | 1.01 | 1.01 |

[a]For 30% effective floor cavity reflectance, multiply by appropriate factor above. For 10% effective floor cavity reflectance, divide by appropriate factor above.

**243D.  Reflection Factors of Colored Surfaces**

| Color | Reflection factor, % | Color | Reflection factor, % |
|---|---|---|---|
| Flat white | 75–85 | Light green | 40–50 |
| Ivory | 70–75 | Gray | 30–50 |
| Buff | 60–70 | Blue | 25–35 |
| Yellow | 55–65 | Red | 15–20 |
| Light tan | 45–55 | Dark brown | 10–15 |

ing calculations is influenced directly by the degree of accuracy involved in these various factors. The zonal-cavity method of making lighting calculations offers a high degree of accuracy in the calculated results, but these results cannot be any more accurate than the data used to make the calculations, including estimates of maintenance factors.

**244. Lighting design formulas[1]**

LUMEN METHOD

$$\text{Footcandles} = \frac{\text{lumens}}{\text{area, ft}^2} \tag{11}$$

$$\text{Initial fc} = \frac{\text{LL} \times \text{CU}}{\text{area, ft}^2} \tag{12}$$

$$\text{Maintained fc} = \frac{\text{LL} \times \text{CU} \times \text{MF}}{\text{area, ft}^2} \tag{13}$$

$$\text{MF} = \text{LLD} \times \text{LDD} \tag{14}$$

where fc = footcandles
  LL = lamp lumens
  CU = coefficient of utilization
  MF = maintenance factor
  LLD = lamp lumen depreciation factor
  LDD = luminaire dirt depreciation factor

IES ZONAL-CAVITY METHOD

Ceiling cavity ratio:

$$\text{CCR} = \frac{5h_{cc}(L + W)}{L \times W} \tag{15}$$

Room cavity ratio:

$$\text{RCR} = \frac{5h_{rc}(L + W)}{L \times W} \tag{16}$$

Floor cavity ratio:

$$\text{FCR} = \frac{5h_{fc}(L + W)}{L \times W} \tag{17}$$

where $h_{cc}$ = distance in feet from luminaire to ceiling
  $h_{rc}$ = distance in feet from luminaire to work plane
  $h_{fc}$ = distance in feet from work plane to floor
  $L$ = length of room in feet
  $W$ = width of room in feet

[1]From *IES Lighting Handbook,* 4th ed.

## TABLES FOR INTERIOR ILLUMINATION DESIGN

### 245.  Incandescent Lighting

(Lighting manufacturers' published reference material)
(Coefficients for utilization for typical luminaires; $\times$ 0.01)
(Effective floor cavity reflectance 20 percent; $\rho$fc = 0.20)

| Room cavity ratio, RCR | Effective ceiling cavity reflectance, $\rho$cc | | | | | | Typical luminaires, description |
|---|---|---|---|---|---|---|---|
| | 80 | | 70 | | 50 | | |
| | Wall reflectance, $\rho$W | | | | | | |
| | 50 | 30 | 50 | 30 | 50 | 30 | |
| | | | | | | | PAR-38 open unit |
| 1 | 89 | 88 | 88 | 87 | 85 | 84 | |
| 2 | 83 | 81 | 82 | 80 | 81 | 79 | |
| 5 | 79 | 77 | 79 | 76 | 78 | 75 | |
| 6 | 78 | 75 | 77 | 75 | 77 | 75 | |
| 8 | 74 | 71 | 73 | 71 | 73 | 71 | |
| 10 | 71 | 69 | 71 | 69 | 71 | 69 | |
| | | | | | | | PS-25 Fresnelens |
| 1 | 69 | 68 | 69 | 68 | 66 | 65 | |
| 3 | 60 | 57 | 59 | 56 | 57 | 55 | |
| 5 | 52 | 48 | 51 | 47 | 50 | 47 | |
| 6 | 49 | 44 | 48 | 44 | 47 | 44 | |
| 8 | 42 | 38 | 41 | 37 | 40 | 37 | |
| 10 | 36 | 32 | 36 | 32 | 35 | 32 | |
| | | | | | | | A-23 low brightness |
| 1 | 59 | 58 | 58 | 57 | 56 | 55 | |
| 3 | 52 | 50 | 52 | 50 | 50 | 49 | |
| 5 | 47 | 44 | 46 | 44 | 46 | 43 | |
| 6 | 45 | 42 | 44 | 42 | 43 | 41 | |
| 8 | 39 | 37 | 39 | 36 | 39 | 36 | |
| 10 | 35 | 32 | 35 | 32 | 34 | 32 | |
| | | | | | | | PAR-38 microgroove |
| 1 | 72 | 72 | 70 | 70 | 68 | 68 | |
| 3 | 67 | 65 | 66 | 64 | 65 | 63 | |
| 5 | 63 | 60 | 62 | 60 | 61 | 59 | |
| 6 | 61 | 59 | 61 | 59 | 60 | 58 | |
| 8 | 58 | 56 | 58 | 56 | 57 | 56 | |
| 10 | 56 | 54 | 56 | 54 | 55 | 53 | |
| | | | | | | | PAR-38 ellipsoidal unit |
| 1 | 49 | 48 | 48 | 48 | 47 | 46 | |
| 3 | 45 | 44 | 45 | 43 | 44 | 42 | |
| 5 | 42 | 40 | 41 | 40 | 41 | 39 | |
| 6 | 40 | 38 | 40 | 38 | 39 | 38 | |
| 8 | 37 | 35 | 37 | 35 | 37 | 35 | |
| 10 | 35 | 33 | 35 | 33 | 34 | 33 | |

## 246.  High-Intensity–Discharge Lighting
(*IES Lighting Handbook*, 4th ed., and lighting manufacturers' published reference material)
(Coefficients of utilization for typical luminaires; × 0.01)
(Effective floor cavity reflectance 20 percent; $\rho fc = 0.20$)

| Room cavity ratio, RCR | Coefficients of utilization | | | | | | Typical luminaires, description |
|---|---|---|---|---|---|---|---|
| Effective ceiling cavity reflectance, $\rho cc$ | 80 | | 50 | | 10 | | |
| Wall reflectance, $\rho W$ | 50 | 30 | 50 | 30 | 50 | 30 | |
| | | | | | | | **400W aluminum** |
| 1 | 89 | 87 | 81 | 79 | 72 | 71 | |
| 3 | 77 | 72 | 71 | 68 | 65 | 63 | |
| 5 | 65 | 60 | 62 | 58 | 57 | 54 | |
| 6 | 61 | 56 | 57 | 53 | 54 | 50 | |
| 8 | 52 | 46 | 49 | 45 | 46 | 43 | |
| 10 | 42 | 36 | 39 | 35 | 37 | 37 | |
| $\rho cc$ | 80 | | 70 | | 50 | | |
| $\rho W$ | 50 | 30 | 50 | 30 | 50 | 30 | |
| | | | | | | | **400W H.P. sodium** |
| 1 | 88 | 86 | 86 | 84 | 82 | 81 | |
| 3 | 77 | 73 | 76 | 72 | 73 | 70 | |
| 5 | 67 | 63 | 66 | 62 | 65 | 61 | |
| 6 | 63 | 58 | 62 | 58 | 61 | 57 | |
| 8 | 56 | 51 | 54 | 51 | 54 | 50 | |
| 10 | 48 | 44 | 46 | 44 | 47 | 43 | |
| $\rho cc$ | 70 | | 50 | | 30 | | |
| $\rho W$ | 50 | 30 | 50 | 30 | 50 | 30 | |
| | | | | | | | **400W R-60 mercury** |
| 1 | 103 | 100 | 99 | 96 | 93 | 91 | |
| 3 | 82 | 75 | 79 | 73 | 72 | 68 | |
| 5 | 64 | 56 | 62 | 56 | 54 | 50 | |
| 6 | 54 | 49 | 53 | 48 | 47 | 43 | |
| 8 | 40 | 36 | 39 | 36 | 35 | 33 | |
| 10 | 27 | 25 | 27 | 25 | 24 | 23 | |
| | | | | | | | **400W prismatic reflector** |
| 1 | 77 | 76 | 73 | 71 | 68 | 67 | |
| 3 | 66 | 63 | 63 | 60 | 59 | 57 | |
| 5 | 56 | 51 | 53 | 49 | 51 | 47 | |
| 6 | 52 | 47 | 49 | 45 | 47 | 43 | |
| 8 | 43 | 38 | 41 | 37 | 40 | 36 | |
| 10 | 35 | 30 | 33 | 29 | 32 | 28 | |

### 247. Fluorescent Lighting

(*IES Lighting Handbook* and manufacturers' catalogs)
(Coefficients of utilization for typical luminaires; × 0.01)
(Effective floor cavity reflectance 20 percent; $\rho fc = 0.20$)

| Room cavity ratio, RCR | Effective ceiling cavity reflectance, $\rho cc$ | | | | | | Typical luminaires, description |
|---|---|---|---|---|---|---|---|
| | 80 | | 70 | | 50 | | |
| | Wall reflectance, $\rho W$ | | | | | | |
| | 50 | 30 | 50 | 30 | 50 | 30 | |
| | | | | | | | **12″-wide troffer** |
| 1 | 73 | 70 | 70 | 68 | 66 | 64 | |
| 3 | 59 | 54 | 57 | 53 | 54 | 50 | |
| 5 | 47 | 42 | 46 | 41 | 44 | 39 | |
| 6 | 43 | 38 | 42 | 37 | 40 | 35 | |
| 8 | 35 | 30 | 34 | 29 | 32 | 28 | |
| 10 | 29 | 23 | 28 | 23 | 27 | 22 | |
| | | | | | | | **2′ × 4′ white troffer** |
| 1 | 69 | 66 | 67 | 65 | 65 | 63 | |
| 3 | 53 | 49 | 52 | 48 | 50 | 47 | |
| 5 | 42 | 36 | 41 | 36 | 40 | 34 | |
| 6 | 38 | 31 | 37 | 31 | 36 | 30 | |
| 8 | 30 | 25 | 30 | 25 | 29 | 24 | |
| 10 | 25 | 19 | 24 | 19 | 24 | 18 | |
| | | | | | | | **2′ × 2′ troffer (U-tubes)** |
| 1 | 71 | 69 | 69 | 67 | 67 | 65 | |
| 3 | 57 | 53 | 56 | 52 | 54 | 51 | |
| 5 | 46 | 41 | 46 | 41 | 44 | 40 | |
| 6 | 42 | 36 | 41 | 36 | 40 | 36 | 40W U-Tubes |
| 8 | 34 | 29 | 33 | 28 | 33 | 28 | |
| 10 | 28 | 23 | 27 | 22 | 27 | 22 | |
| | | | | | | | **4-lamp 2′ × 4′ troffer** |
| 1 | 58 | 56 | 57 | 55 | 55 | 53 | |
| 3 | 47 | 43 | 46 | 43 | 45 | 42 | |
| 5 | 38 | 34 | 37 | 33 | 37 | 33 | |
| 6 | 34 | 30 | 34 | 30 | 33 | 29 | |
| 8 | 27 | 23 | 27 | 23 | 26 | 23 | |
| 10 | 23 | 19 | 23 | 19 | 22 | 19 | |
| | | | | | | | **Prismatic wrap-around** |
| 1 | 70 | 68 | 67 | 65 | 63 | 61 | |
| 3 | 56 | 51 | 54 | 50 | 51 | 47 | |
| 5 | 45 | 40 | 44 | 39 | 41 | 37 | |
| 6 | 41 | 35 | 39 | 34 | 37 | 33 | |
| 8 | 33 | 27 | 32 | 27 | 30 | 26 | |
| 10 | 26 | 21 | 26 | 21 | 24 | 20 | |
| | | | | | | | **2-lamp porcelain** |
| 1 | 86 | 83 | 82 | 79 | 75 | 72 | |
| 3 | 68 | 62 | 65 | 60 | 60 | 56 | |
| 5 | 54 | 47 | 52 | 46 | 48 | 43 | |
| 6 | 49 | 42 | 47 | 40 | 43 | 38 | |
| 8 | 39 | 32 | 38 | 31 | 35 | 30 | |
| 10 | 32 | 26 | 31 | 25 | 29 | 23 | |
| | | | | | | | **4-lamp Alzak unit** |
| 1 | 83 | 80 | — | — | 74 | 72 | |
| 3 | 67 | 61 | — | — | 60 | 56 | |
| 5 | 53 | 47 | — | — | 48 | 44 | |
| 6 | 48 | 42 | — | — | 43 | 39 | |
| 8 | 37 | 32 | — | — | 34 | 30 | |
| 10 | 29 | 24 | — | — | 26 | 22 | |
| | | | | | | | **2/40W-3′ × 3′ unit** |
| 1 | 66 | 64 | — | — | 62 | 60 | |
| 3 | 53 | 48 | — | — | 50 | 46 | |
| 5 | 42 | 37 | — | — | 40 | 36 | |
| 6 | 38 | 33 | — | — | 37 | 32 | |
| 8 | 31 | 26 | — | — | 30 | 25 | |
| 10 | 25 | 20 | — | — | 24 | 20 | |

**248. Levels of illumination.** The present *Recommended Levels of Illumination*, RP-15, utilizes a new approach to lighting that has been introduced by the IES. The new approach is intended to provide flexibility in establishing lighting levels so that lighting systems can be set to meet specific needs. The procedure is fully described in the *IES Lighting Handbook: Application Volume.*

## INTERIOR-LIGHTING SUGGESTIONS

**249. Residence lighting.** In the lighting of the home, the decorative element should predominate. The lighting must, however, comply with the general rules of lighting (Sec. 224) concerning color, shadows, glare, and illumination. A room in which yellow or red is the predominant color gives a warm, cheerful impression, whereas a room furnished in blue tends to produce the opposite sensation. Shade and softened shadows are preferable to sharp shadows or to no shadows. It is especially important that any glare be minimized. Luminous ceilings can be installed in such rooms as kitchens and bathrooms.

Detailed recommendations for residence lighting are given in a booklet prepared by the Committee on Residence Lighting of the Illuminating Engineering Society and titled *Design Criteria for Lighting Interior Spaces.*

KITCHEN   The kitchen requires plenty of well-diffused light for timely preparation of food without accidents. In most kitchens, general lighting from two 40-W fluorescent lamps or two 150-W to four 100-W incandescent lamps will provide satisfactory general illumination. For very large kitchens, 80 to 120 W of fluorescent lighting or two 100-W incandescent lamps per 60 ft$^2$ will be needed. Unless the room is very small with light-colored walls and ceilings, the general illumination should be augmented by additional localized ceiling or bracket units. A single light in the center of the kitchen usually compels the cook to work entirely in his or her own shadow, whether at the range, the sink, or the kitchen cabinet or table. A single fluorescent ceiling unit over the center of the sink or a bracket at each side of the sink is nearly always necessary to eliminate shadows and provide adequate illumination at the sink. Often it is advisable to locate similar units at the range and counter work areas.

BEDROOM   The illumination requirements of the bedroom are somewhat similar to those of the kitchen. General illumination should be provided by means of a centrally located ceiling unit of low brightness so that it will not be uncomfortable to the eyes of one lying in bed. Additional illumination should be provided at the dresser or bureau by means of either brackets or a portable lamp on each side of the mirror. A bracket or wall-mounted fluorescent unit should be provided at the head of each bed unless localized table luminaires are used for providing illumination for reading at these locations. Better illumination for this purpose is generally obtained by a wall-mounted bracket or fluorescent units.

LIVING ROOM   Although ceiling fixtures have been eliminated in many living rooms, an attractive, properly shielded fixture is a very effective method of obtaining the general level of illumination needed for festive occasions. For a quiet evening at home a lower level of general illumination can be obtained by means of the indirect type of floor and table lamps. The chief function of bracket lamps is ornamentation, and they should not be relied upon to provide necessary light for useful purposes.

DINING ROOM   There are several satisfactory methods of illuminating the dining room, the selection depending upon personal taste. General illumination may be provided by means of a centrally located chandelier. It should be mounted 30 in (762 mm) above the table in an 8-ft- (2.4-m-) ceiling room. In rooms with higher ceilings, the chandelier should be raised 3 in (76 mm) for each foot. Recessed downlights can be used with chandeliers for dramatic lighting of table settings. Cove lighting by means of fluorescent lamps concealed in a trough a few inches down from the ceiling may be used. Wall brackets for decorative effects are also common.

BATHROOM   In the bathroom illumination of the mirror requires the greatest consideration. For the best illumination of the mirror three luminaires, one at each side of the mirror and one mounted overhead, are needed. Fairly good results can be obtained with only two units located one at each side of the mirror. If only one unit is used, it should be mounted overhead and centered with respect to the mirror. For small bathrooms the mirror wall brackets will supply sufficient general illumination for the rest of the room. In rooms of any appreciable size a central enclosing unit should be used in addition to the bracket lamps. A ceiling-mounted heat or sun lamp will provide healthful ultraviolet light for the winter months.

**250. Store lighting.** The object of lighting in a store is twofold. Primarily, sufficient illumination must be provided to enable articles for sale to be seen plainly, but of almost equal importance is the advertising value. The lighting units should be so selected as to give a pleasing and cheerful appearance to the store as a whole, without glare. Stores may be divided into three classes: (1) the small store, in which efficiency is of first importance; (2) the large store, such as a department store, in which efficiency is necessary on account of the large areas to be lighted but in which it must be balanced by artistic appearance, the result being a compromise between the two; and (3) shops, large or small, in which the articles for sale are of a special type and the profits large enough so that even the most inefficient system can be afforded if it is sufficiently attractive to appeal to customers. The general requirements which must be satisfied are outlined separately in the following sections.

GENERAL FEATURES OF STORE LIGHTING  General lighting best meets the requirements for the lighting of stores. The lamps should be arranged symmetrically with respect to bays or pillars. Direct glare must be very carefully avoided. The position of the counters need not be considered in spacing the lamps. Local accurate-color–matching units are advisable at certain points in the store, such as counters for the sale of ribbon and dress goods.

The intensity of illumination must be varied with the articles which are to be sold. Furniture requires well-diffused lighting of relatively low intensity. Colored dress goods, men's clothing, rugs and carpets, etc., require a high intensity. In many installations, side light is necessary and should be given special attention in selecting types of units and reflectors. Crystal and jewelry should be so lighted as to sparkle and glitter. Bare lamps and mirrored reflectors may be used in ornamental-type luminaires for this purpose. In this case glare is to a certain extent unavoidable, but light units can usually be so located as to be out of the range of vision of customers when they are inspecting the wares on display.

Pictures require a high intensity, with the light units at such an angle that light will not be reflected from the surface of the painting or from the glass directly into the observer's eyes. Individual units or mirrored-trough reflectors, with fluorescent or tubular tungsten lamps, are ordinarily used.

High-color-rendition fluorescent lamps should be used to illuminate all items in which color is important, such as women's dresses, coats, and furs, men's suits and coats, draperies and curtains, tapestries, and neckties. Concealed fluorescent units may also be used in wall valances or coves for decorative effects.

SHOWCASES  A showcase interior should have more illumination than the top of the case but not more than about twice as much. If the illumination on top is in the 30- to 75-fc (323- to 807-lx) range, T-6 or T-8 slimlines inside the case, operated at 200 mA, provide sufficient light; for higher footcandles on the top, an operating current of 300 mA is sometimes recommended. Too great a differential between the illumination inside the case and that on the top, where the merchandise is normally inspected more closely prior to purchase, is not advisable. Many products examined under an illumination level significantly lower than that under which they were displayed lose some of their attraction. This is not a consideration with feature displays, since they are seldom removed for inspection.

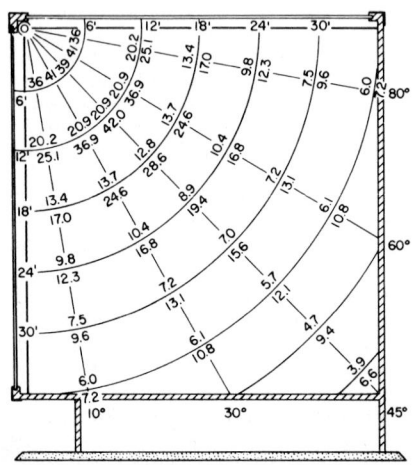

**Fig. 10-114**  Initial footcandles for each 100 rated lamp lm/ft of showcase length.

The plot in Fig. 10-114 indicates the

approximate initial footcandles perpendicular to the source at various angles for each 100 rated lamp lm/ft of case length. The figures above the arcs represent footcandles obtained with a fluorescent lamp in a specular showcase reflector. The figures below the arcs represent footcandles obtained with standard filament reflector showcase lamps and differ from the fluorescent values because of beam control.

WALL CASES    Wall cases generally should be illuminated to approximately the same level as counter display areas. However, higher levels are desirable when the brightness of the wall case and the adjoining sales space is to be used partly to pull traffic into the area. One of the most important features of the lighting of wall cases for wearing apparel is that it be designed to light the entire case from top to bottom rather than just the shoulders of the garments. The illumination at the bottom should be at least one-tenth of that at the top.

The accompanying table lists the approximate illumination in footcandles produced by fluorescent lamps on a vertical surface for each 100 rated lamp lm/ft of case length. The specular symmetrical reflector was so adjusted that its maximum candlepower was directed at a point 42 in (1067 mm) below the lamp. When no reflector was used, the entire inner surface of the valance was painted white.

| | Horizontal distance—$H$, in | | | | | | | |
|---|---|---|---|---|---|---|---|---|
| Vertical distance— $V$, in | Values with symmetrical reflector, fc | | | | Values without reflector, fc | | | |
| | 6 | 12 | 18 | 24 | 6 | 12 | 18 | 24 |
| 6 | 21.4 | 15.7 | 12.1 | 9.8 | 30.6 | 22.8 | 16.5 | 13.5 |
| 12 | 12.0 | 12.8 | 12.1 | 10.2 | 11.3 | 14.7 | 13.6 | 12.2 |
| 18 | 7.4 | 9.6 | 10.0 | 9.1 | 5.2 | 8.8 | 10.2 | 9.6 |
| 24 | 4.4 | 6.9 | 8.8 | 7.7 | 2.6 | 5.0 | 6.6 | 7.2 |
| 30 | 2.6 | 4.8 | 6.9 | 6.8 | 1.4 | 3.3 | 4.4 | 5.2 |
| 36 | 2.0 | 3.8 | 5.2 | 6.1 | 1.0 | 2.2 | 3.2 | 4.1 |
| 42 | 1.4 | 2.8 | 4.1 | 5.0 | 0.6 | 1.4 | 2.2 | 3.0 |
| 48 | 0.8 | 1.9 | 2.8 | 4.1 | 0.4 | 1.0 | 1.6 | 2.2 |
| 54 | 0.8 | 1.6 | 2.4 | 3.4 | 0.4 | 0.8 | 1.4 | 1.6 |
| 60 | 0.6 | 1.2 | 1.9 | 2.6 | 0.3 | 0.8 | 1.0 | 1.4 |

**251. Show-window lighting.**    Since the primary object of a show window is to attract attention, it should be illuminated to a sufficient intensity so that it will stand out from its surroundings. Several factors need to be considered in designing show-window lighting systems such as type of merchandise, location of the store, time-of-day usage, enclosed or open back design, and the size and shape of the show window. Great care should be exercised in designing show-window lighting so that the lamps will be concealed from view. The lighting equipment should be concealed either by recessing it in the ceiling or by employing a valance between the reflector and the front glass of the window. A good way to blind prospective customers so that they cannot see the goods on display in your window is to put exposed lamps around the window borders, suspend them from chandeliers, or so install them in the top of the window that their eyes cannot escape them.

Shadows are necessary in window lighting to produce the desired effects, but they should not be too sharply defined. The light should come from in front of the goods. If the lamps are placed in the middle of the show-window ceiling, the front of the goods displayed in the front of the window will be in darkness because of shadows they cast.

If the back of the display window is not boxed in or if the window back is of glass, a valance or a curtain should be provided to screen the window lamps from the store and to prevent back reflection to an observer on the sidewalk.

Customers' attention should be attracted by illuminating selected articles with spotlights. If decorative effects are desired, colored spotlights may be used, and motor-operated dimmers can add motion and dramatic effects. Fluorescent units bring out the beauty of floral displays and emit very little heat. Daylight lamps bring out the beauty of polished aluminum and chromium-plated articles.

**252. Show-window lighting equipment.**  Incandescent lamps are extensively used in show-window applications owing to the wide variety of wattages, sizes, reflectorized types, and colored lamps. Lighting units with special reflectors for show-window lighting are available in many shapes to meet the varied requirements of different types and sizes of windows. Reflectorized (reflector and projector) lamps are widely used owing to the built-in light control within the lamp. Low-voltage PAR and MR-16 reflector lamps provide excellent beam control for lighting special areas of a show window to a higher intensity. Colored reflector and PAR lamps or the use of colored gelatin screen or glass roundels can provide desired effects with color. No general rules for shadows and color of light to be used can be given. An infinite number of effects can be obtained. By experiment, the display manager should determine the best effect for each display to be lighted.

**253. School and office lighting.**  As school and office work is usually performed during daylight hours, any artificial illumination is usually in conjunction with available daylight. Since practically all the work done in an office or school is on plane surfaces such as papers and books, shadows not only are unnecessary but are objectionable. Also, since the persons must use artificial illumination for long periods of time, the glare should be minimal. These requirements render indirect and semi-indirect units especially suitable for office and school illumination. The units should, as a general rule, be more closely spaced than for other classes of service. This spacing will produce a more gradual shading of any shadows and will also afford a greater flexibility in desk and partition arrangement. The extensive use of video-display units in offices and classrooms has established the need to design lighting systems to avoid reflected brightness in VDU screens.

**254. Lighting codes and legislation.**  Some states have enacted statutes or lighting codes regulating lighting in industrial plants. Since it is their intention only to protect the welfare of workers against undue accident hazard, the illuminations required by the statutes are usually much too low for best efficiency. Any lighting installation planned according to the principles set forth in this division will in all probability satisfy any state code. However, before a lighting installation is made in a factory or industrial plant, designers should fully acquaint themselves with any pertinent legislation.

## HEAT WITH LIGHT FOR BUILDING SPACES

**255. Luminaires for environmental control.**  Not too many years ago architects designed buildings primarily as space enclosures. The end use for which the space enclosures were intended usually dictated the type and design of the buildings and the division and arrangement of the space. Comfort conditioning of the space for people normally consisted of a heating system for use in cold weather and a ventilating or air-conditioning system for warm weather.

Progress in building design and construction and in technological developments in space conditioning has changed the architects' objective in the design of buildings. Present emphasis is not only on space enclosure tailored to the needs of occupants but also on environment and environmental control of lighting, heating, cooling, acoustics, space flexibility, and appearance (pleasing colors, etc.).

Visual and thermal conditions have become two of the most important considerations in planned interior environment. Visual comfort is controlled primarily by quality and quantity of illumination. And adequate illumination of proper quality for high visual performance and high visual comfort increases electric-energy loads, which also means higher heat gain. Thus, lighting systems that provide adequate quality illumination may have a considerable effect on thermal conditions within buildings. Thermal comfort is controlled through a proper balance in temperature, relative humidity, and air motion. Accordingly, light- and heat-producing characteristics of the lighting system have become dominant factors in the thermal equation.

This expanded use of lighting heat heralds a new era in environmental design  It provides the opportunity to supply all or a substantial part of a building's heating requirements from lighting systems with accompanying benefits, and control of lighting heat is necessary for the most effective utilization of this energy.

**256. Integrated lighting-heating-cooling systems.**  There has been a growing coordinated activity between electrical and mechanical engineers since 1958, when the Illuminating

Engineering Society adopted new and higher recommended levels of illumination for most seeing tasks. This activity has resulted in many new techniques for handling lighting heat loads, which are based on the integration and correlation of lighting, heating, and cooling systems. Although many approaches to this problem are possible, the basic one is to divert all possible lighting heat gain from the occupied spaces in buildings to keep the capacity of the *cooling* system as low as possible. Another objective is to recapture some or all of this lighting heat gain to keep the capacity of the *heating* system as low as possible. This heat may be removed from interior areas of the building which need cooling, for example, and redirected to perimeter areas which need heating at the same time.

**257. Air-handling troffers.** In the typical integrated system, heat from the lighting system is removed by passing return air from the air-conditioned space through vents integral with the luminaires over the warm surfaces of lamps, ballasts, and luminaires, into the plenum or return air duct, and back through the return side of the air-conditioning system. After tempering, the supply air is redistributed, entering the conditioned spaces through a different set of vents, which are also integral with the luminaires. Luminaires designed for this purpose are called *air-handling troffers.* There is also available an *air-water luminaire,* in which circulating water through the luminaire removes the lighting heat.

By combining lighting, heating, and cooling services in a single outlet element (the luminaire), the conflict for space both above and on the ceiling is reduced. This integration forces coordinated design and produces as its visual result an architecturally clean, uncluttered ceiling.

Many lighting manufacturers now have available air-handling troffers in a range of sizes and for a variety of air-handling requirements: (1) static (non-air-handling), (2) air supply, (3) air return, and (4) a combination of air supply and air return. These luminaires may be integrated with either cooling systems or heating systems, or both. Installation methods conform generally to those for standard fluorescent troffers.

Also available are several complete ceiling systems combining both air-supply and air-return provisions, some of which use conventional fluorescent luminaires and some of which incorporate special air-handling luminaires that are an integral part of the ceiling system. Typical examples are shown in Figs. 10-115, 10-116, and 10-117. The air-handling troffer in Fig. 10-115 supplies cool air to the occupied space below the ceiling and returns warm air from the occupied space through the troffer to the open plenum above. Heat from lamps and ballasts is also removed. Warm plenum air may be discarded or mixed with fresh air and recirculated for heating specific occupied spaces.

The unified ceiling system in Fig. 10-116 provides complete environmental control through a modular-size integrated package. Each module combines lighting, heating, cooling, sound control, partition tracks for flexibility, shielding of the lighted portion of a luminaire in the line of vision, and good appearance.

The recessed incandescent downlight (Fig. 10-117) is integrated with heating and cooling systems to the extent that lighting heat may be exhausted to the plenum space and discharged as waste or reused.

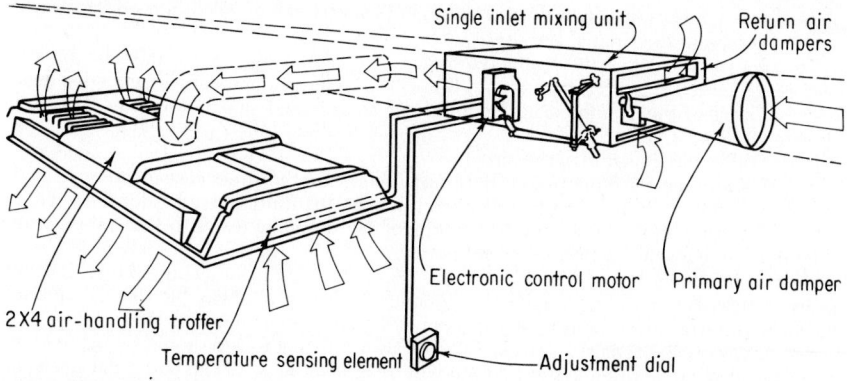

**Fig. 10-115** Air-handling troffer. [Day-Brite Lighting Division, Emerson Electric Co., *Electrical Construction and Maintenance*]

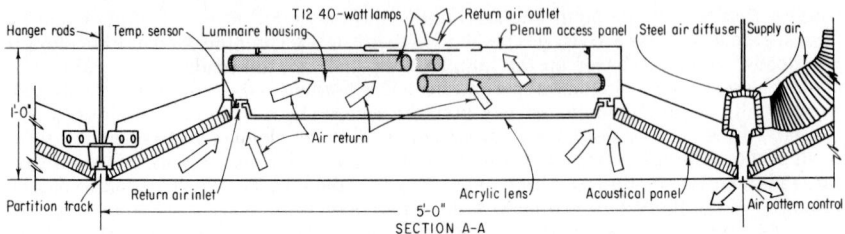

**Fig. 10-116**   Unified ceiling system. [Day-Brite Lighting Division, Emerson Electric Co.]

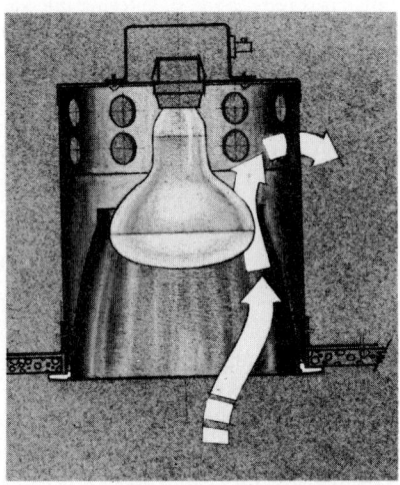

**Fig. 10-117**   Recessed incandescent downlights integrated with heating and cooling systems.

## STREET LIGHTING

**258. General requirements for street lighting.**   Good roadway lighting, often termed *traffic safety lighting*, not only promotes safer conditions for drivers but provides greater safety for pedestrians as well. It enhances the community value of a street by its attractive appearance, which is usually reflected in higher property values. Well-lighted streets also act as a deterrent to criminal activity.

To achieve truly effective street lighting it is essential that the installation be well planned. Planned street lighting should follow the American Standard Practice for Roadway Lighting of the Illuminating Engineering Society and will involve the following considerations:

    1. Traffic classification of the street

    2. Determination of the proper lighting intensity for the street classification

    3. Selection of luminaires according to the light distribution needed for the street

    4. Determination of the mounting height of the luminaire above the road surface and the proper linear spacing between luminaires

Each step follows the preceding one in logical order, and the following information will assist in the planning of an installation. Manufacturers of roadway lighting equipment have computerized programs to assist in detailed planning of an installation.

**259. Roadway classification.**   A traffic classification relating to use should be made of all streets so that the lighting-system design will be in keeping with the particular needs of each street or highway. It is also recommended that roadways be classified by the area of location to establish the volume of pedestrian traffic to be considered. Typical roadway and area classifications are defined here (General Electric Co.).

ROADWAY CLASS

1. *Freeway.* A divided **major** roadway with full control of access and with no crossings at grade. This definition applies to toll as well as to nontoll roads.

2. *Expressway.* A divided **major** roadway for through traffic with partial control of access and generally with interchanges at major crossroads. Expressways for noncommercial traffic within parks and parklike areas are generally known as parkways.

3. *Major roadways.* The part of the roadway system that serves as the principal network for through-traffic flow. The routes connect areas of principal traffic generation and important rural highways entering the city.

4. *Collector roadways.* The distributor and collector roadways serving traffic between major and local roadways. These are roadways used mainly for traffic movements within residential, commercial, and industrial areas.

5. *Local roadways.* Roadways used primarily for direct access to residential, commercial, industrial, or other abutting property. They do not include roadways carrying through traffic. Long local roadways will generally be divided into short sections by collector roadway systems.

AREA CLASS

1. *Commercial.* That portion of a municipality in a business development where ordinarily there are large numbers of pedestrians and heavy demand for parking space during periods of peak traffic or sustained high pedestrian volume and continuously heavy demand for off-street parking during business hours. This definition applies to densely developed business areas outside, as well as those that are within, the central part of a municipality.

2. *Intermediate.* That portion of a municipality which is outside a downtown area but generally within the zone of influence of a business or industrial development, characterized often by moderately heavy nighttime pedestrian traffic and somewhat lower parking turnover than is found in a commercial area. This definition includes densely developed apartment areas, hospitals, public libraries, and neighborhood recreational centers.

3. *Residential.* A residential development, or a mixture of residential and commercial establishments, characterized by few pedestrians and lower parking demand or turnover at night. This definition includes areas with single-family homes, townhouses, and/ or small apartments. Regional parks, cemeteries, and vacant lands are also included.

**260. Lighting intensity.** The recommended illumination level is dependent both on the area and on the roadway classification. Recommended minimum average maintained levels and recommended average-to-minimum uniformity ratios for several classifications are shown in the following table.

**ANSI-IES Recommendations**

| Roadway classification | Minimum average maintained footcandles | Uniformity, average footcandles/ minimum footcandles |
|---|---|---|
| Residential areas | | |
| Local | 0.4 | 6:1 |
| Collector | 0.6 | 3:1 |
| Major | 1.0 | 3:1 |
| Intermediate areas | | |
| Local | 0.6 | 3:1 |
| Collector | 0.9 | 3:1 |
| Major | 1.4 | 3:1 |
| Commercial areas | | |
| Collector | 1.2 | 3:1 |
| Major | 2.0 | 3:1 |

**261. Selection of luminaire.** Luminaires should be selected according to their pattern of light distribution so as to conform not only to the light intensity required but to the physical characteristics of the street to be lighted. Typical lateral candlepower distribution curves for several types of luminaires are shown. Lateral beam width is measured to half maximum candlepower.

TYPE I LUMINAIRE    Type I is a two-way lateral distribution having a preferred lateral width of 15° on each side of the reference line (total, 30°), with an acceptable range of from 10 to less than 20°. It projects two beams in opposite directions along the roadway parallel to the curbline. The candlepower distribution is similar on both sides of the vertical plane of maximum candlepower. Luminaires with this type of distribution are generally applicable to locations near the center of a roadway where the mounting height is approximately equal to the roadway width.

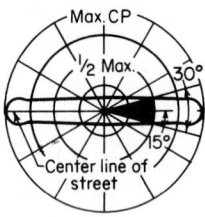

FOUR-WAY TYPE I LUMINAIRE    Four-way Type I is a distribution having four principal concentrations at lateral angles of approximately 90° to one another, each with a width of from 20 to less than 40° as in Type I. This distribution is generally applicable to luminaires located over or near the center of a right-angle intersection.

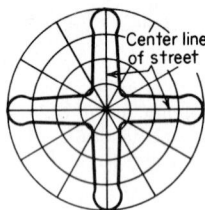

TYPE II LUMINAIRE    Type II distributions have a preferred lateral width of 25° with an acceptable range of from 20 to less than 30°. This distribution is generally applicable to luminaires located at or near the side of relatively narrow roadways when the width of the roadway does not exceed 1.6 times the mounting height.

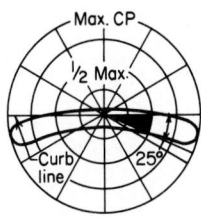

FOUR-WAY TYPE II LUMINAIRE    Four-way Type II distributions have four principal light concentrations, each with a width of from 20 to less than 30° as in Type II. This distribution is generally applicable to luminaire locations near one corner of a right-angle intersection.

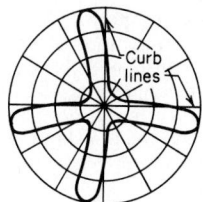

TYPE III LUMINAIRE     Type III distributions have a preferred lateral width of 40° with an acceptable range of from 30 to less than 50°. This type of distribution is intended for luminaires mounted at or near the side of medium-width roadways when the width of the roadway does not exceed 2.7 times the mounting height.

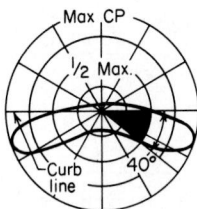

TYPE IV LUMINAIRE     Type IV distributions have a preferred lateral width of 60° with an acceptable width of 50° or more. This type of distribution is intended for side-of-road mounting and is generally used on wide roadways (width up to 3.7 times the mounting height).

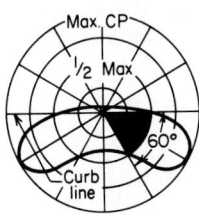

TYPE V LUMINAIRE     Type V distribution is essentially circular, with equal candlepower at all lateral angles. This type of distribution is intended for luminaire mounting at or near the center of a roadway, in the center islands of parkways, and at intersections.

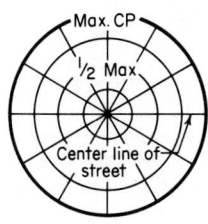

**262. Mounting height and spacing of luminaires.** Two considerations are of paramount importance in determining optimum mounting height: the desirability of minimizing direct glare from the luminaire and the need for a reasonably uniform distribution of illumination on the street surface. The higher the luminaire is mounted, the farther it is above the normal line of vision and the less glare it creates. However, the attainment of uniform illumination requires a certain relationship between mounting height, luminaire spacing, and the vertical angle of maximum candlepower for the specific luminaire (usually between 70 and 80°).

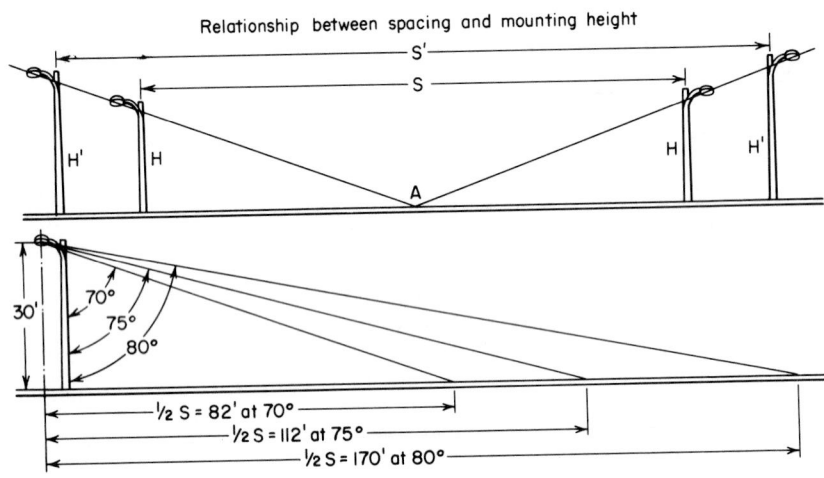

Relationship between spacing and mounting height

For a given luminaire the ratio of pole spacing to mounting height should be low enough so that the light at the angle of maximum vertical candlepower will strike the street at least halfway to the adjacent pole. To provide greater uniformity on busy streets the spacing is often reduced by as much as 50 percent, which provides a 100 percent overlap of vertical beams.

The mounting heights recommended by the Illuminating Engineering Society's Roadway Lighting Committee take into account both the objective of minimum glare and that of maximum uniformity. Greater mounting heights may often be preferable, but heights less than those recommended should be avoided.

## FLOODLIGHTING

**263. Modern floodlighting** meets many utilitarian requirements as well as many applications concerned with decoration, aesthetics, or advertising values. Protecting property after nightfall, completing a construction job within the time allotted, illuminating a dangerous traffic intersection, and prolonging the hours of play in recreational areas are only a few of the almost infinite applications of utilitarian floodlighting.

As an advertising medium that compels attention without detracting from the beauty or dignity of a building, floodlighting offers its best proof by the many excellent examples to be found in almost every city. The natural beauty of churches, civic buildings, monuments, and gardens is often enhanced by skillfully applied floodlighting.

**264. Floodlighting design.** The type of area to be lighted, the possible location of equipment, and the variation in surrounding conditions impose problems in design which tend to make floodlighting-design standardization difficult. Manufacturers of floodlighting equipment can be consulted for detailed design information. There are, however, certain basic rules which may be followed in installation design.

**265. Step 1: Determine the level of illumination.** In Table 248 are listed the illumination levels for many floodlighting applications. These levels are designated as *footcandles in service*, and allowance must be made for reasonable depreciation in the original design.

In lighting buildings, monuments, etc., the reflection factor of the object and the brightness of the surroundings must be considered to determine the amount of light necessary (Table 266).

**266.   Recommended Illumination Levels for Floodlighting**
*(IES Lighting Handbook)*

| Surface material | Reflec- tance, % | Bright | Dark |
|---|---|---|---|
| | | Surround | |
| | | Recommended level, fc | |
| Light marble, white or cream terra- cotta, white plaster | 70–85 | 15 | 5 |
| Concrete, tinted stucco, light gray and buff limestone, buff face brick | 45–70 | 20 | 10 |
| Medium gray limestone, common tan brick, sandstone | 20–45 | 30 | 15 |
| Common red brick, brownstone, stained wood shingles, dark gray brick | 10–20[a] | 50 | 20 |

[a]Buildings or areas of materials having a reflectance of less than 20% usually cannot be floodlighted economically unless they carry a large amount of high-reflectance trim.

## 267    Recommendations for Sports Lighting

| Application | Foot-candles maintained in service | | Application | Foot-candles maintained in service | |
|---|---|---|---|---|---|
| **Sports lighting** | | | Club........................... | 20 | |
| Badminton: | | | Recreational.................. | 10 | |
| Tournament.................. | 30 | | Horseshoe pitching: | | |
| Club.......................... | 20 | | Tournament.................. | 10 | |
| Recreational................. | 10 | | Recreational.................. | 5 | |
| Baseball: | | | | | |
| Seats: | | | Playgrounds................... | 5 | |
| During game.............. | 2 | | | | |
| Before and after game....... | 5 | | Race tracks: | | |
| | In-field | Out-field | Seats......................... | 2 | |
| Major League.................. | 150 | 100 | Track......................... | 20 | |
| AAA and AA League........... | 75 | 50 | | | |
| A and B League............... | 50 | 30 | Rifle range: | | |
| C and D League............... | 30 | 20 | | Out-door | In-door |
| Semipro and municipal......... | 20 | 15 | On target..................... | 30[a] | 50[a] |
| | | | Firing point................... | 10 | 10 |
| Basketball: | | | Range......................... | ... | 5 |
| College and Professional........ | 50 | | | | |
| High school................... | 30 | | Shuffleboard: | | |
| Recreational.................. | 10 | | Tournament................... | 10 | |
| | | | Recreational.................. | 5 | |
| Bathing beaches (surf).......... | 3[a] | | | | |
| | | | Skating: | | |
| Billiards: | | | Rink (indoor or outdoor)........ | 5 | |
| | Gen-eral | On Table | Pond or flooded area........... | 1 | |
| Tournament.................. | 10 | 50 | | | |
| Recreational................. | 10 | 30 | Skeet shooting, target surface at 60 ft.......................... | 30[a] | |
| Bowling: | | | | | |
| | Gen-eral | On Pins | Ski slope, practice................ | 0.5 | |
| Tournament.................. | 20 | 50[a] | Soccer: | | |
| Recreational................. | 10 | 30[a] | College and professional........ | 30 | |
| | | | High school.................... | 20 | |
| Boxing or wrestling: | | | Recreational.................. | 10 | |
| Seats: | | | | | |
| During bout................. | 2 | | Softball: | | |
| Before and after bout......... | 5 | | | In-field | Out-field |
| Ring: | | | Pro and championship.......... | 50 | 30 |
| Championship.............. | 500 | | Semipro...................... | 30 | 20 |
| Professional................. | 200 | | Industrial league............... | 20 | 10 |
| Amateur.................... | 100 | | Recreational.................. | 10 | 5 |
| Croquet: | | | Squash: | | |
| Tournament.................. | 10 | | Tournament................... | 30 | |
| Recreational................. | 5 | | Club.......................... | 20 | |
| | | | Recreational.................. | 10 | |
| Football: | | | | | |
| Class I....................... | 100 | | Swimming pools ............. | 10 | |
| Class II...................... | 50 | | | | |
| Class III..................... | 30 | | Tennis: | | |
| Class IV..................... | 20 | | | Table | Lawn |
| Class V...................... | 10 | | Tournament.................... | 50 | 30 |
| | | | Club.......................... | 30 | 20 |
| Gymnasium: | | | Recreational.................. | 20 | 10 |
| Locker and shower rooms....... | 10 | | | | |
| Exercise rooms, fencing, boxing, wrestling, basketball, volleyball, softball, and general exercise..................... | 20 | | Trapshooting: | | |
| | | | Target surface at 150 ft......... | 30[a] | |
| | | | Firing point—general........... | 10 | |
| Exhibition games and matches... | 30 | | | | |
| | | | Volleyball: | | |
| Handball: | | | Tournament................... | 20 | |
| Tournament.................. | 30 | | Recreational.................. | 10 | |

[a] Vertical.

**268. Step 2: Determine the type and location of floodlights.** Floodlighting equipment may be of either the *general-purpose* or the *heavy-duty* class. The general-purpose floodlight is one in which the inner surface of the housing serves as the reflecting surface. The heavy-duty floodlight is more rugged, since its aluminum or glass reflector is protected by a metal housing.

The finish of the reflector has an important influence on the beam spread. Most enclosed units are available with either a narrow-beam (specular-finish) or a wide-beam (diffuse-finish) reflector. Essentially all floodlighting equipment is enclosed today for improved maintenance.

**269. Data on Typical Floodlight Equipment**

| Description | Advantages | Lamp types | Beam-spread NEMA type[a] |
|---|---|---|---|
| Group A<br><br>Heavy-duty or general-purpose<br><br> | Good light control for distance<br>Weathertight<br>Useful in a variety of applications, mostly sports lighting | 1000- to 1500-W metal halide<br>400- to 1000-W high-pressure sodium | 1–5 |
| Group B<br><br>Heavy-duty high-wattage HID<br><br> | Good light control<br>Weathertight<br>Easily serviced<br>High fixture efficiency | 250-, 400-, and 1000-W metal halide<br>250-, 400-, and 1000-W high-pressure sodium | 6 × 2<br>6 × 5<br>6 × 6<br>6 × 7 |
| Group C<br><br>Heavy-duty low-wattage HID<br><br> | Good light control<br>Weathertight<br>Easily serviced<br>High fixture efficiency<br>Use at lower mounting or where lower lighting levels are satisfactory | 175-W metal halide<br>35- through 250-W high-pressure sodium | 6 × 6<br>7 × 6<br>7 × 7 |
| Group D<br><br>Heavy-duty tungsten-halogen<br><br> | Good light control<br>Weathertight<br>Easily serviced<br>Low cost | 300-, 500-, 1000-W, and 1500-W tungsten halogen | 6 × 4<br>6 × 5 |

[a]Asymmetrical beam spreads have a horizontal and a vertical designation. The horizontal is stated first; i.e., a 6 × 2 beam spread is a NEMA Type 6 in the horizontal dimension and a NEMA Type 2 in the vertical dimension.

**270. To simplify the specification of equipment** the National Electrical Manufacturers Association has introduced a series of type numbers specifying beam spread and group letters specifying floodlight construction. To meet NEMA standards a floodlight must conform to certain construction requirements as well as have a certain minimum efficiency for a given beam spread.

Although the choice of beam spread for a particular application depends upon individual circumstances, the following general principles apply:

1. The greater the distance from the floodlight to the area to be lighted, the narrower the beam spread desired.

2. Since by definition the candlepower at the edge of a floodlight beam is 10 percent of the candlepower near the center of the beam, the illumination level at the edge of the beam is one-tenth or less of that at the center. To obtain reasonable uniformity of illumination the beams of individual floodlights must overlap each other as well as the edge of the surface to be lighted.

3. The percentage of beam lumens falling outside the area to be lighted is usually lower with narrow-beam units than with wide-beam units. Thus narrow-beam floodlights are preferable when they will provide the necessary degree of uniformity of illumination and the proper footcandle level.

### Outdoor Floodlighting Designations

| Beam Spread,° | NEMA Type |
|---|---|
| 10–18 | 1 |
| 18–29 | 2 |
| 29–46 | 3 |
| 46–70 | 4 |
| 70–100 | 5 |
| 100–130 | 6 |
| 130 and up | 7 |

**271.  Typical Floodlighting Applications**

| Application | Location of equipment | Class of equipment |
|---|---|---|
| | At edge of area and mounted as high as possible. Spacing not to exceed 4 times mounting height. | General-purpose enclosed or heavy-duty Type 4, 5, or 6 |
| | Immediately inside and below concealing parapet— maximum distance from face of building to be lighted. | Heavy-duty or general-purpose Type 3, 4, or 5 |
| | In banks at suitable location. One bank normally should cover an area whose height and length are not greater than the distance from the units to the face of the building. | Heavy-duty or general-purpose Type 1, 2, 3, or 4 |
| | At edge of area where they will not hinder traffic. Minimum mounting height 20 ft. | Heavy-duty Type 5 or 6 |
| | On ground 5 to 25 ft from vertical surface, concealed by hedge, low structure, or natural elevation. | Heavy-duty or general purpose Type 3, 4, 5, or 6 |

The location of floodlighting equipment is usually dictated by the type of application and the physical surroundings. If the area is large, individual towers or poles spaced at regular intervals may be required to light it evenly; smaller areas may require only one tower with all equipment concentrated on it, or adjacent buildings may be utilized as floodlight locations. The accompanying chart will aid in the selection of the right equipment and its proper location for a number of typical floodlighting applications.

In planning any floodlighting system it is important that the light be properly controlled. Strong light directed parallel to a highway or a railroad track can be a dangerous source of glare to oncoming traffic, and light thrown indiscriminately on adjacent property may be a serious nuisance.

**272. Step 3: Determine the coefficient of beam utilization.**  To determine the number of floodlights that will be required to produce a specified level of illumination in a given situation, it is necessary to know the number of lumens in the beam of the floodlights and the percentage of the beam lumens striking the area to be lighted. The beam lumens can be obtained from the manufacturer's catalog. The ratio of the lumens striking the floodlighted surface to the beam lumens is called the *coefficient of beam utilization* (CBU). If an area is uniformly lighted, the average CBU of the floodlights in the installation is always less than 1.0

The coefficient of beam utilization for any individual floodlight will depend on its location, the point at which it is aimed, and the distribution of light within its beam. In general, it may be said that the average CBU of all the floodlights in an installation should fall within the range of 0.60 to 0.90. Utilization of less than 60 percent of the beam lumens is a definite indication that a more economical lighting plan should be possible by using different locations or narrower-beam floodlights. On the other hand, if the CBU is over 0.90, the beam spread selected probably is too narrow and the resultant illumination will be spotty. Accurate determination of the CBU is possible only after the aiming points have been selected. However, an estimated CBU can be determined by experience or by making calculations for several potential aiming points and using the average figure thus obtained.

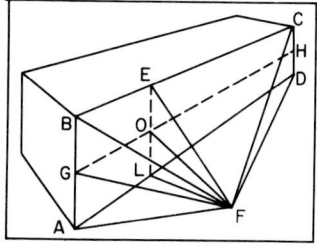

If:     EO = OL = 25
        AL = FL = 40
        LD = 80

Then:  angle  LFO = 32°
              EFO = 19°
              BFE = 32°
              GFO = 41°
              AFL = 45°
              CFE = 51°
              HFO = 60°
              DFL = 64°

The CBU of the floodlight at F is about .81

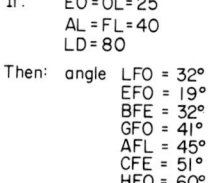

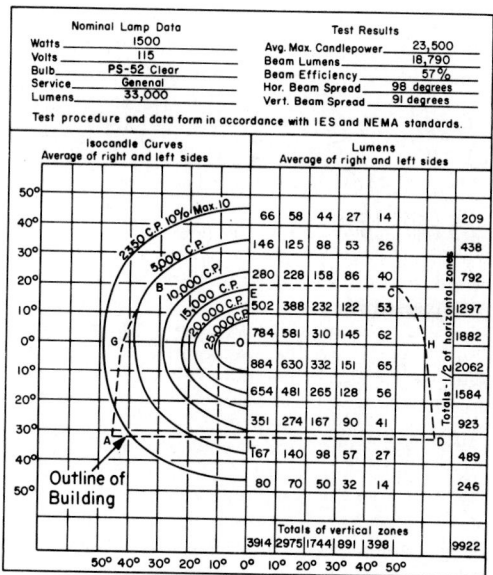

To make such calculations the floodlighted area is superimposed on the photometric grid and the ratio of the lumens inside this area to the total beam lumens is determined. All horizontal lines on a building (or straight lines on a ground area which are parallel to a line perpendicular to the beam axis) appear as straight horizontal lines on the grid if the floodlight is so aimed that its beam axis is perpendicular to a horizontal line on the face of the building. All vertical lines except the one through the beam axis appear slightly curved.

**273. Step 4: Estimate the maintenance factor.** Lighting efficiency is seriously impaired by blackened lamps and by dirt on the reflecting and transmitting surfaces of the equipment. To compensate for the gradual depreciation of illumination on the floodlighted area, a maintenance factor must be applied in the calculations to make allowance for the following:

1. Loss of light output due to dirt on the lamp, reflector, and cover glass.

2. Loss in light output of the lamp with life. Because some of the light must pass through the bulb more than once before finally leaving the floodlight, bulb blackening also lowers floodlight efficiency, the reduction in beam lumens being about double the reduction in bare-lamp output.

A maintenance factor of 0.75 has been widely used for floodlights.

However, if the atmosphere is not clean, if the floodlights are cleaned infrequently, or if lamps are replaced only on burnout, a realistic appraisal of in-service conditions will require the use of considerably lower maintenance factors. Differences in lumen maintenance among lamp types and sizes should also be considered. With incandescent lamps the higher the wattage for a given bulb size or the smaller the bulb for a given wattage, the denser the bulb blackening and the greater the depreciation in light output.

**Approximate Average Lamp Lumens throughout Life**

| Lamp Type | % of Initial Lumens |
|---|---|
| Standard incandescent | 83 |
| Tungsten-halogen incandescent | 98 |
| Mercury | 60 |
| Metal halide | 80 |
| High-pressure sodium | 90 |

With narrow-beam floodlights, dirt on the reflector and cover glass tends to widen the beam spread, reducing the maximum candlepower more than the total light output. Thus for a small lighted area utilizing only the central part of a beam (e.g., a 4-ft, or 1.2-m) archery target at 100 yd, or 91.4 m), a smaller percentage of the beam lumens will strike the target after the floodlight has become dirty. Therefore the depreciation in footcandle intensity will be greater than the depreciation in total light output, and it will be necessary to consider this in selecting a maintenance factor.

**274. Step 5: Determine the number of floodlights required.**

$$\text{Number of floodlights} = \frac{\text{area} \times \text{footcandles}}{\text{beam lumens} \times \text{CBU} \times \text{MF}} \tag{18}$$

where area is the surface area to be lighted in square feet and footcandles are as selected from Table 266. For beam lumens, refer to the manufacturer's catalog for the equipment under consideration. If the lamps are to be burned at other than rated voltage, the beam lumens, and hence the number of floodlights required, is altered. The increase in lumen output for 5 and 10 percent overvoltage operation is indicated in Sec. 277. For the CBU (coefficient of beam utilization), refer to step 3; and for the MF (maintenance factor), to step 4.

**275. Step 6: Check for coverage and uniformity.** After a tentative layout has been made (steps 1 to 5), uniformity can be checked by calculating the intensity of illumination at a few points on the floodlighted surface. This may be done by the point-by-point method described in Sec. 240, using either a candlepower-distribution curve or an isocandle diagram. If uniformity is found to be unsatisfactory, a larger number of units may have to be used.

**276. Application notes**

TYPE OF LIGHT SOURCE    Most floodlighting installations today utilize high-intensity–discharge lamps owing to their higher efficacy and long lamp life. Where incandescent lamps are used, they are generally the tungsten-halogen types.

BUILDINGS AND MONUMENTS    The floodlighting of a building or a monument is primarily a problem in aesthetics, and each installation must be studied individually. Under some circumstances, particularly with small utilitarian buildings or larger buildings which lack special architectural features, uniform illumination is desirable. To create an appearance of uniform brightness over the entire facade of a building, it is usually necessary to increase the actual brightness appreciably toward the top. Higher brightness at the top of a building also increases its apparent height.

With buildings of classical design or special architectural character uniform illumination often defeats the purpose of the lighting, which should aim to preserve and emphasize the architectural form. Buildings are designed primarily for daytime appearance, when the light comes from above. This effect is almost impossible to duplicate with floodlights, which must be mounted in nearby locations and usually at a height no greater than the elevation of the building. However, it is often possible to achieve a result that is interesting and pleasing, although quite different from the daytime appearance. Shadows are essential to relief, and contrasts in brightness levels or sometimes in color can be used advantageously to bring out important details and to suppress others. Sculpture or architectural details require particularly careful treatment to avoid flatness or grotesque shadows that may entirely distort the appearance as it has been conceived by the artist or architect.

EXCAVATION AND CONSTRUCTION    Approximately one 1500-W tungsten-halogen, one 400-W metal halide or HPS, or two 250-W metal halide or HPS units will be required for each 5000 ft² (465 m²) of excavation area or for each 1000 to 2000 ft² (93 to 186 m²) of construction area. It is usually most satisfactory to mount floodlights in groups of two or more on wood poles or towers 40 to 70 ft (12.2 to 21.3 m) above ground. A minimum of two poles should be used, with enough poles on larger jobs to provide coverage at any working point from two or three floodlight banks. Fixed-pole spacing may be from 1½ to 3 times mounting height and as much as 5 times mounting height on large projects. Occasionally it is practical to provide one or two portable poles mounted on timber-sled bases. If a mechanical shovel or crane is used, it is advisable to mount an automatic leveling floodlight on the boom.

DIRECTIONAL APPLICATION SUCH AS RAILWAY YARDS AND FREIGHT TERMINALS    There are two general types of railway-yard lighting, the unidirectional system and the parallel opposing system. The first is applicable only to tracks on which the traffic is all in one direction, the light being directed with the traffic flow. Glare is entirely eliminated, but seeing is entirely by direct light without the advantages of silhouette effect. The parallel opposing system is used when the traffic flows in both directions. Here seeing is accomplished by direct light from the tower behind the observer and by the silhouette of cars and glint from the rails produced by light from the tower ahead of the observer.

Tower locations are determined by track layouts and operating methods, but in general spacings should not exceed 1000 to 2000 ft (305 to 610 m) for the parallel opposing system or 800 to 1000 ft (245 to 305 m) for the unidirectional system. The first tower in the classification yard should be near the ladder tracks but on the approach side of the hump, so that spill light or a separate floodlight, if necessary, illuminates the hump area. Narrow yards can be lighted by single towers located in the center. Wide yards should have pairs of towers opposite each other, each about one-fourth of the yard width from the edge. Except for spacings well below the values noted above, 90 ft (36 m) should be considered a minimum mounting height.

The lighting of outdoor freight terminals without covered platforms is similar to that of railroad yards except that the intensities must be much higher. Poles must be placed at the ends of the platforms to prevent interference with vehicular traffic and in line with the platforms to avoid shadows thrown by cars standing on the tracks. Mounting heights of from 60 to 80 ft (18 to 24 m), depending on the average length of throw, are required. Heavy-duty floodlights are recommended for all railroad service.

COLOR    Color is provided in floodlight installations usually by colored cover glass, which is generally available for enclosed floodlights to replace regular lenses. When smaller amounts of colored light are needed, colored R or PAR incandescent lamps may be used. Any color of filter absorbs a large amount of light, and this loss must be considered in designing the installation. The following approximate transmission values are found in typical commercial color filters: amber, 40 to 60 percent; red, 10 to 20 percent; green, 5 to 20 percent; and blue, 3 to 10 percent. Although less colored light than white light is usually required for equal advertising or decorative effectiveness, it is desirable to determine by experiment the exact amount of colored light necessary for any particular application.

The color of light from high-intensity–discharge light sources is particularly desirable in specific decorative floodlighting applications. Metal halide and phosphor-coated mercury lamps are used where a cooler color of light is preferred, while high-pressure sodium lamps are applied in installations better suited to a warmer tone. Clear mercury lamps are very effective in floodlighting trees and landscaping owing to the greenish-blue character of the light from these lamps.

**277. Sports lighting.** The level of illumination required depends upon several factors, among which are the speed of the game, the skill of the players, and the number of spectators and their distance from the field of play. If television coverage is involved, this also affects the lighting level. Economic considerations also are important. High intensities of illumination are desirable in almost all sports, but since lower levels are acceptable, a variety of layouts are suggested.

HID lighting systems with metal halide and high-pressure sodium lamps are extensively used in sports lighting owing to their high efficacy and longer lamp life. The warm-up and restrike characteristics of these light sources should be recognized in designing the lighting system. A standby incandescent system may be desired in spectator areas.

Incandescent lighting systems are still used in sports-lighting applications owing to the lower system cost when the annual hours of usage are limited. In sports installations in which the lighting is operated during fewer than about 500 h per year, it may be economical to use short-life incandescent lamps or to burn standard lamps at higher than rated voltages. This reduces the power cost and the required number of floodlights. For fewer than 200 h per season, general-service incandescent lamps at 10 percent overvoltage are generally used. If the system is used between 200 and 500 h per year, operation of general-service lamps at 5 percent overvoltage is preferable. The following table shows the effect of overvoltage operation on the life, wattage, and light output of 1000- and 1500-W incandescent lamps.

**Approximate Performance of 1000- and 1500-W Incandescent Lamps**

| Voltage | Light, % | Watts, % | Lamp life, % |
|---|---|---|---|
| Rated | 100 | 100 | 100 |
| 5% over rated | 117 | 108 | 50 |
| 10% over rated | 135 | 116 | 30 |

The layouts that follow are guidelines to obtain the recommended levels of illumination if high-quality floodlights are mounted and operated as indicated. Some latitude in beam-spread requirements is acceptable when several floodlights are mounted on a single pole, provided the resulting average beam spread is approximately the same as that specified.

NOTE    For current recommended practice on sports lighting, consult Illuminating Engineering Society, *IES Sports Lighting*, RP-6.

**278. Archery**

The floodlight provides visibility of the arrow throughout flight.

| | |
|---|---|
| Floodlights | Per target |
| | One Type 2, Group A |
| | One Type 6 × 5, Group B |
| Lamps | For enclosed floodlight |
| | Distance up to 30 yd: 175-W metal halide |
| | 30 to 50 yd: 250-W metal halide |
| | 50 to 100 yd: 400-W metal halide |
| Mounting height | Enclosed floodlight 15 ft above ground |
| | Flood on same pole 12 ft above ground |
| Poles | One per target |

### 279. Badminton

Indoor courts may be lighted with industrial diffusing units along the sidelines.

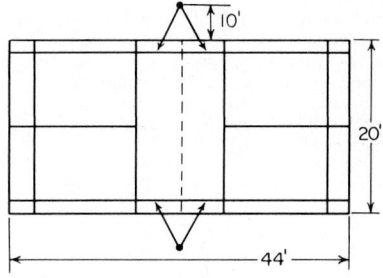

| | | | Floodlights | | | |
|---|---|---|---|---|---|---|
| Class | Type | Group | No. per pole | Total no. | Lamp watts | Total load, kW |
| Recreational | 6 × 5 | D | 2 | 4 | 500 | 2 |

| | |
|---|---|
| Lamps | 500-W tungsten-halogen |
| Mounting height | 20 to 25 ft above court |
| Poles | Two per court |

### 280. Baseball

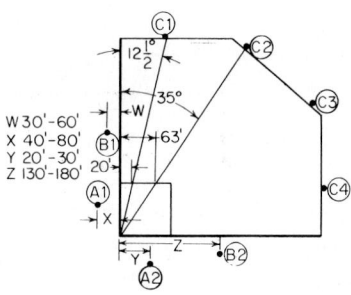

| | | | Floodlights | | | | | |
|---|---|---|---|---|---|---|---|---|
| | | | Poles | | | | Total load, kW | Minimum mounting height, ft |
| Class | Type | Group | A's | B's | C's° | Total no. | | |
| Major league | 2 | A | ...ᵃ | ...ᵃ | ...ᵃ | ...ᵃ | ...ᵃ | 120 |
| AAA and AA | 2, 3 | A | 12 | 24 | 18 | 144 | 235 | 110 |
| A and B | 2, 3 | A | 10 | 16 | 8 | 84 | 137 | 90 |
| C and D | 3, 5 | A | 6 | 8 | 6 | 52 | 85 | 70 |
| Semipro and municipal | 5 | A | 4 | 6 | 4 | 36 | 59 | 60 |

ᵃLuminaire quantities based on field and stadium dimensions.

| | |
|---|---|
| Lamps | 1500-W metal halide |

### 281. Basketball: indoor

If ceiling is lower than 20 ft, more units on closer spacing should be used and recessed in the ceiling if possible. Luminaires should be rigidly mounted, and lamps should be protected from the ball.

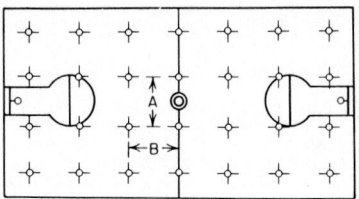

| Class | Fixture | Lamp watts | Spacing, ft A | B |
|---|---|---|---|---|
| | | Metal halide | | |
| College or pro | High-bay industrial | 250 | 14.2 | 16.8 |
| High school | High-bay industrial | 175 | 15.9 | 21.0 |
| Recreational | High-bay industrial | 175 | 26.0 | 28.0 |
| | | High-pressure sodium | | |
| College or pro | High-bay industrial | 250 | 17.0 | 16.0 |
| High school | High-bay industrial | 150 | 17.0 | 16.0 |
| Recreational | High-bay industrial | 70 | 17.5 | 19.2 |

| Mounting height | On ceiling |
|---|---|

### 282. Basketball: Outdoor

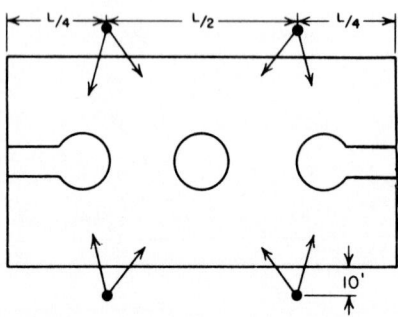

Wide-beam floodlights are required to illuminate the ball when it is a considerable distance above the court.

| | Floodlights | | | | Total load, kW | |
|---|---|---|---|---|---|---|
| Class | Type | Group | No. per pole | Total no. | | Lamp type |
| Recreational | 6 × 5 | B | 2 | 8 | 3.6 | 400-W metal halide |
| | 6 × 5 | B | 2 | 8 | 2.5 | 250-W HPS |

| Mounting height | 30 ft above court |
|---|---|
| Poles | Four per court |

### 283. Billards

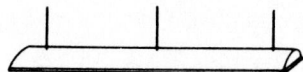

In large commercial parlors where a number of tables are installed, general illumination of high intensity proves more satisfactory than individual luminaires over each table.

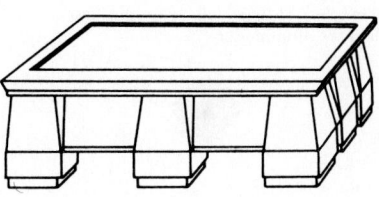

| | |
|---|---|
| Equipment | Tournament: two two-lamp fluorescent fixtures with louvers |
| Lamps | Recreational: one two-lamp fluorescent fixture with louvers |
| | 40-WT-12 fluorescent |
| Load | 200 or 100 W |
| Mounting height | 7 ft above floor |

### 284. Bowling

Luminaires should be shielded by false ceiling beams or baffles unless they are of the asymmetric type. They should be positioned so as to provide even illumination along the alley with higher intensity on the pins. Behind the foul line any type of general area-lighting equipment may be employed.

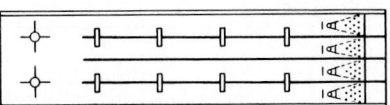

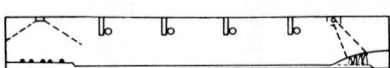

| | |
|---|---|
| Equipment | Four two-lamp 40-W direct fluorescent fixtures per pair of alleys |
| | One 20-W fluorescent unit per alley over the pins |
| Lamp | 40-W T-12 fluorescent |
| | 20-W T-12 fluorescent |
| Mounting height | 9 ft minimum |

### 285. Boxing

The class of bout will determine the level of illumination.

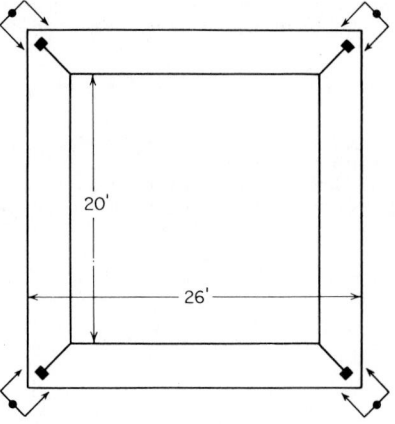

| Class | Floodlights | | | | Lamp watts | Total load, kW |
|---|---|---|---|---|---|---|
| | Type | Group | No. per pole | Total no. | | |
| Championship | 5 | A | 2 | 8 | 1000 | 8.7 |
| Professional | 5 | A | 1 | 4 | 1000 | 4.3 |
| Amateur | 6 | A | 2 | 8 | 400 | 3.6 |

Lamps            Metal halide
Mounting height  20 to 25 ft above ring

### 286. Croquet

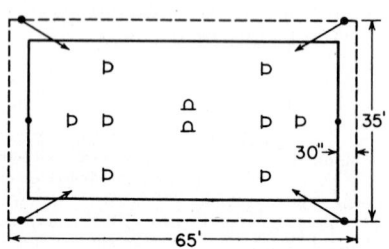

| Class | Floodlights | | | | Lamp watts | Total load, kW | Lamp type |
|---|---|---|---|---|---|---|---|
| | Type | Group | No. per pole | Total no. | | | |
| Tournament | 6 × 5 | D | 1 | 4 | 250 | 1.2 | Metal halide |
| | 6 × 5 | D | 1 | 4 | 250 | 1.2 | HPS |
| Recreational | 6 × 5 | D | 1 | 4 | 175 | 0.8 | Metal halide |
| | 6 × 5 | D | 1 | 4 | 100 | 0.5 | HPS |

Lamps            Metal halide or high-pressure sodium
Mounting height  20 to 25 ft above court
Poles            Four per court

## 287. Football

The distance between the farthest row of spectators and the nearest sideline (see table) determines the lighting requirements, but the seating capacity should also be considered. Either of the pole plans shown or any intermediate longitudinal spacing is considered good practice. Local conditions determine the exact pole locations.

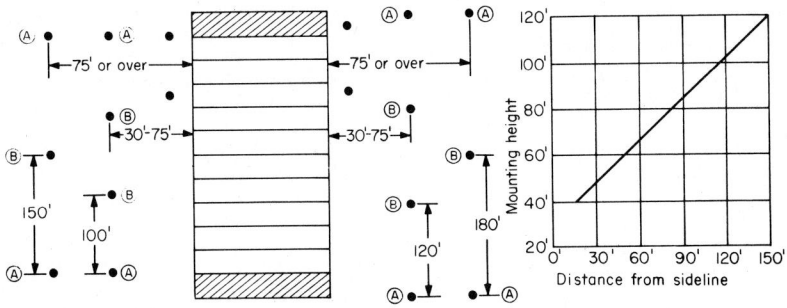

| Class | Recommended footcandles | Distance— poles to sideline, ft | No. of poles | Floodlights | | | | | Total load, kW |
|---|---|---|---|---|---|---|---|---|---|
| | | | | Type | Group | No. per pole | Total no. | | |
| I | 100 | Over 140 | 6 | 2 | A | 24 | 144 | | 234 |
| | | 100–140 | 6 | 2 | A | 22 | 132 | | 222 |
| II | 50 | 75–100 | 6 | 2 | A | 12 | 72 | | 117 |
| | | 50–75 | 6 | 3 | A | 10 | 60 | | 98 |
| III | 30 | 30–50 | 6 | 3, 4, 5 | A | 6A, 8B | 40 | | 65 |
| IV | 20 | 15–30 | 8 | 4 | A | 3 | 24 | | 39 |

| Class | Distance farthest spectators to field, ft | Seating capacity | | |
|---|---|---|---|---|
| I | Over 100 | Over 30,000 | Lamps | 1500-W metal halide |
| II | 50–100 | 10,000–30,000 | | |
| III | 30–50 | 5,000–10,000 | | |
| IV | Under 30 | 5,000 | | |

### 288. Golf Driving Range

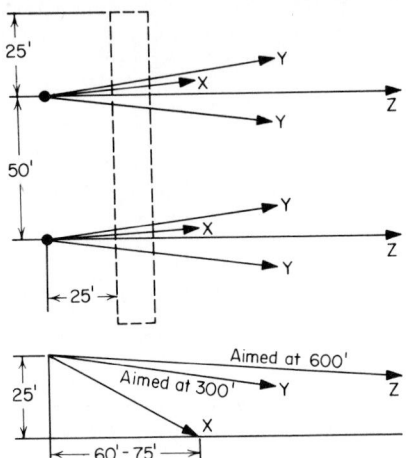

The floodlights should be directed so as to provide illumination on the ball throughout its flight.

| Floodlights | | | | | Load per pole, kW |
|---|---|---|---|---|---|
| Aiming point | Type | Group | No. per pole | Lamp | |
| X | 5 | B | 1 | 1000-W metal halide | 1.1 |
| Y | 2 | A | 2 | 1000-W metal halide | 2.2 |
| Z | 2 | A | 1 | 1000-W metal halide | 1.1 |

| | |
|---|---|
| Mounting height | 25 to 30 ft above tees |
| Poles | One for every 50 ft of range width; minimum: two poles |

### 289. Handball: Outdoor

Glare is largely eliminated by locating the floodlights behind the players.

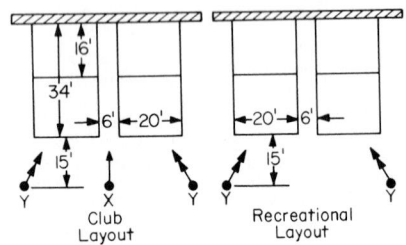

| | | | Floodlights | | | | |
|---|---|---|---|---|---|---|---|
| | | | No. per pole | | Total no. | Lamp watts | Total load, kW |
| Class | Type | Group | X | Y | | | |
| Club | 6 × 5 | D | 1 | 1 | 3 | 1500 | 4.5 |
| Recreational | 6 × 5 | D | ... | 1 | 2 | 1500 | 3.0 |

| | |
|---|---|
| Lamps | 1500-W tungsten-halogen |
| Mounting height | At least 25 ft above court |
| Poles | For club play, three per pair of courts |
| | For recreational play, two per pair of courts |

### 290. Horseshoe Pitching

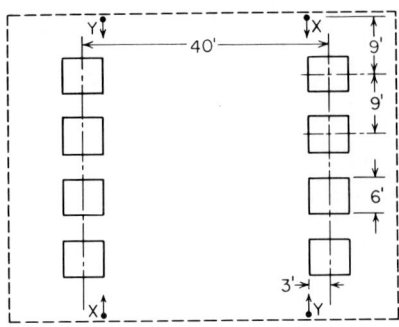

Floodlights should be directed across courts to prevent direct glare.

| Class | No. of courts | Floodlights Type | Group | No. per pole X | No. per pole Y | Total no. | Lamp watts | Total load, kW | Lamp type |
|---|---|---|---|---|---|---|---|---|---|
| Tournament | 4–6 | 6 × 5 | B | 1 | 1 | 4 | 400 | 1.84 | Metal halide |
|  | 1–3 | 6 × 5 | B | ⋯ | 1 | 2 | 400 | 0.92 | Metal halide |
| Recreational | 4–6 | 6 × 5 | B | 1 | 1 | 4 | 250 | 1.24 | HPS |
|  | 1–3 | 6 × 5 | B | ⋯ | 1 | 2 | 250 | 0.62 | HPS |

| | |
|---|---|
| Lamps | Metal halide or high-pressure sodium |
| Mounting height | At least 20 ft above court |
| Poles | Four for 4–6 court layout, two for 1–3 court layout |

### 291. Ice Skating: Outdoor

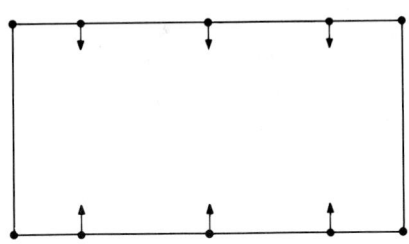

The design suggested produces satisfactory illumination for recreational skating.

| Area | Floodlights Type | Group | W/ft² | Lamp type |
|---|---|---|---|---|
| Rink | 6 × 5 | B | 0.10 | HPS |
|  | 6 × 5 | B | 0.16 | Metal halide |
| Pond | 6 × 5 | B | 0.02 | HPS |
|  | 6 × 5 | B | 0.03 | Metal halide |

The size of the area determines the number and wattage of the floodlights.

| | |
|---|---|
| Lamps | High-pressure sodium or metal halide |
| Mounting height | At least 20 ft above ice |
| Pole spacing | Not to exceed 4 times mounting height |

**292. Racetracks**

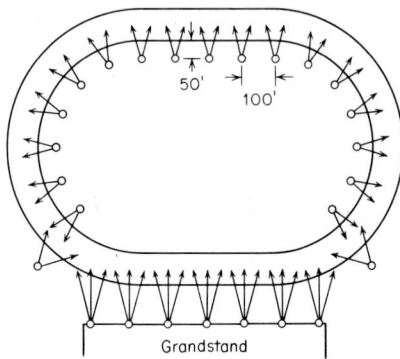

The lighting equipment should be so positioned and directed as to keep glare and shadows at a minimum.

| | |
|---|---|
| Floodlights | Types A and B |
| Lamps | Metal halide |
| Load | Varies with track size |
| Mounting data | 50 to 80 ft high and 50 ft inside inner edge of track |

**293. Shuffleboard**

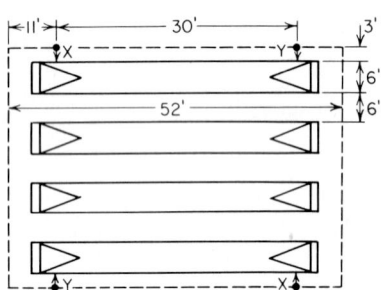

Floodlights should be directed across the court to prevent glare.

| Class | No. of courts | Type | Group | No. per pole X | No. per pole Y | Total no. | Lamp watts | Total load, kW | Lamp type |
|---|---|---|---|---|---|---|---|---|---|
| Tournament | 4–6 | 6 × 5 | D | 1 | 1 | 4 | 400 | 1.84 | Metal halide |
| | 1–3 | 6 × 5 | D | · · · | 1 | 2 | 400 | 0.92 | Metal halide |
| Recreational | 4–6 | 6 × 5 | D | 1 | 1 | 4 | 250 | 1.24 | HPS |
| | 1–3 | 6 × 5 | D | · · · | 1 | 2 | 250 | 0.62 | HPS |

| | |
|---|---|
| Lamps | Metal halide or high-pressure sodium |
| Mounting height | At least 20 ft above court |
| Poles | Four for 4–6 court layout, two for 1–3 court layout |

**294. Skeet Shooting**

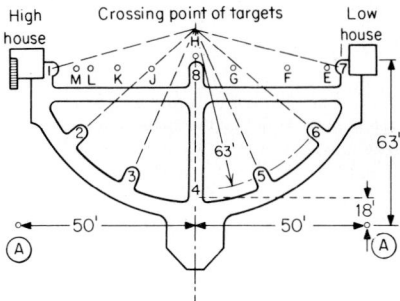

| | | Floodlights | | | | | |
|---|---|---|---|---|---|---|---|
| Location | Group | Type | No. per pole | Total no. | Mounting height, ft | Lamp watts | Total load, kW |
| A | B | 6 × 5 | 2 | 4 | 25 | 1000 | 4.3 |

| Lamps | 1000-W metal halide |
|---|---|

**295. Soccer**

For larger or smaller fields, the number of floodlights and the pole spacings should be altered in proportion to the area of the field.

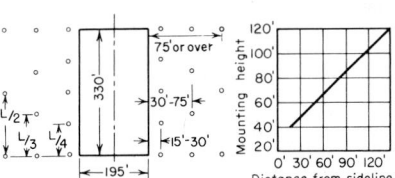

| | | | Floodlights | | | | Total |
|---|---|---|---|---|---|---|---|
| Class | Distance poles to sideline, ft | No. of poles | Type | Group | No. per pole | Total no. | load, kW |
| Professional and college | Over 140 | 6 | 2 | A | 24 | 144 | 232 |
| | 100–140 | 6 | 2 | A | 23 | 138 | 222 |
| | 75–100 | 6 | 2 | A | 20 | 120 | 193 |
| High school | 30–50 | 6 | 3 | A | 7 | 42 | 68.7 |
| | 15–30 | 6 | 5 | A | 3 | 18 | 29.4 |
| Recreational | 15–30 | 8 | 6 × 5 | B | 2 | 16 | 17.3[a] |

[a]1000-W metal halide.

| Lamps | 1500-W metal halide |
|---|---|

### 296. Softball: Professional and Industrial League

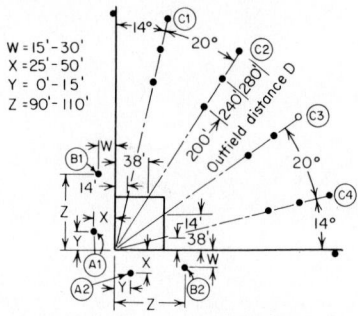

W = 15'-30'
X = 25'-50'
Y = 0'-15'
Z = 90'-110'

The distance to the poles in the outfield is determined by the size of the field.

| Class | Outfield distance $D$, ft | Group A floodlights | | | | | Minimum mounting height, ft | |
| | | No. per pole | | | | | Poles | |
| | | Poles | | | Total no. | Total load, kW | 1–4 | 5–8 |
| | | A's | B's | C's | | | | |
|---|---|---|---|---|---|---|---|---|
| Pro and championship | 280 | 4 | 8 | 4 | 40 | 65.2 | 50 | 60 |
| | 240 | 4 | 6 | 3 | 32 | 52.2 | 50 | 55 |
| Semipro | 280 | 3 | 4 | 3 | 26 | 42.4 | 40 | 55 |
| Industrial League | 280 | 2 | 4 | 2 | 20 | 32.6 | 35 | 50 |
| | 240 | 2 | 3 | 2 | 18 | 29.3 | 35 | 45 |

Lamps          1500-W metal halide

### 297. Softball: recreational

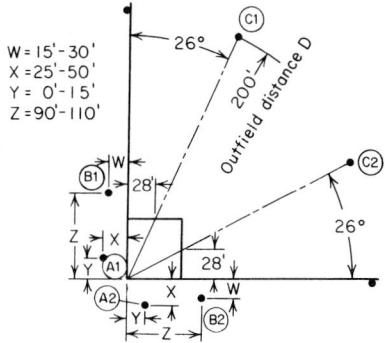

W = 15'– 30'
X = 25'– 50'
Y = 0'– 15'
Z = 90'– 110'

| Class | Outfield distance D, ft | Layout for Type 4 or 5 Group A floodlights | | | | Layout for Type 6 × 5 Group B floodlights | | | | Miminum mounting height, ft | |
|---|---|---|---|---|---|---|---|---|---|---|---|
| | | No. per pole | | | | No. per pole | | | | Poles | |
| | | Poles | | | Total no. | Poles | | | Total no. | A's & B's | C's |
| | | A's | B's | C's | | A's | B's | C's | | | |
| Recreational | 200 | 2 | 3 | 3 | 16 | 3 | 3 | 3 | 18 | 35 | 40 |

|  |  |
|---|---|
| Lamps | 1000-W metal halide |

### 298. Swimming pool: underwater floodlights

Especially designed equipment is mounted in niches in the walls of the pool.

| Location of pool | W/ft² | |
|---|---|---|
| | Good practice | Minimum |
| Outdoors | 0.5 | 0.3 |
| Indoors | 2.0 | 1.6 |

| Lamp watts | B maximum ft, where D is over 5 ft | B' maximum ft, where D is less than 5 ft | E, in below waterline |
|---|---|---|---|
| 250–400 | 4 | 5 | 18–24 |

| | |
|---|---|
| Lamps | Metal halide Consult fixture manufacturer if 12. V units are to be used. |

### 299. Swimming pool: overhead floodlights

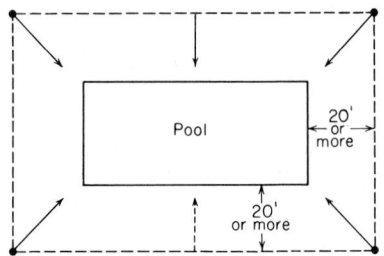

The number of floodlights and the lamp size are determined by the size of the area and the type of equipment.

| | |
|---|---|
| Floodlights | Type 6 × 5, Group B |
| Lamps | Metal halide or high-pressure sodium |
| Load | 0.32 W/ft² for metal halide, 0.20 W/ft² for HPS (Both pool and surrounding area to be lighted) |
| Mounting height | At least 20 ft above water |
| Pole spacing | Not to exceed 4 times mounting height |

### 300. Tennis: single court

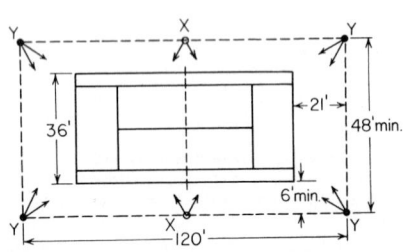

Floodlights must be directed sufficiently high to provide even illumination on the ball during flight.

| | | Floodlights | | | | | | |
|---|---|---|---|---|---|---|---|---|
| | | | | No. per pole | | | | Total |
| Class | No. of poles | Type | Group | X | Y | Total no. | Lamp watts | load, kW |
| Tournament | 6 | 5 | A | 2 | 2 | 12 | 1000 | 13.1 |
| Club | 6 | 5 | A | 2 | 1 | 8 | 1000 | 8.7 |
| Recreational | 4 | 5 | A | ··· | 1 | 4 | 1000 | 4.4 |

| | |
|---|---|
| Lamps | 1000-W metal halide |
| Mounting height | 30 ft above court |

## 301. Tennis: Two Courts

Floodlights must be directed sufficiently high to provide even illumination on the ball during flight.

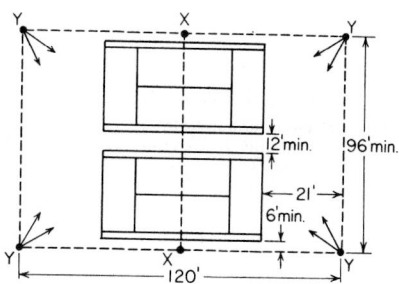

| | | | Floodlights | | | | |
|---|---|---|---|---|---|---|---|
| | | | | No. per pole | | | |
| Class | No. of poles | Type | Group | X | Y | Total no. | Total load, kW |
| Club | 6 | 5 | A | 2 | 2 | 12 | 13 |
| Recreational | 6 | 6 × 5 | B | 1 | 1 | 6 | 6.5 |

| | |
|---|---|
| Lamps | 1000-W metal halide |
| Mounting height | 35 to 40 ft above the court |

## 302. Table tennis

When louvered fluorescent luminaires are used, they may be mounted
either lengthwise or crosswise of the
table.

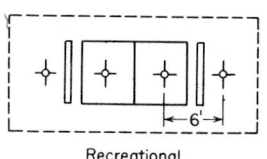

Recreational

| | Fixtures | | | |
|---|---|---|---|---|
| Class | No. | Type | Lamp watts | Total load, kW |
| Recreational | 2 | Fluorescent | 40 | 0.16 |
| | 2 or 4[a] | Deep-bowl[b] | 150 | 0.3 or 0.6 |

[a]For skilled play four luminaires mounted 4½ to 6 ft above the table should be used.
[b]Louvered fluorescent luminaires for two 40-W lamps may also be used.

| | |
|---|---|
| Lamps | 40-W T-12 fluorescent |
| | 150-W incandescent |

### 303. Trapshooting

Uniform illumination will prevent apparent variation in bird speed.

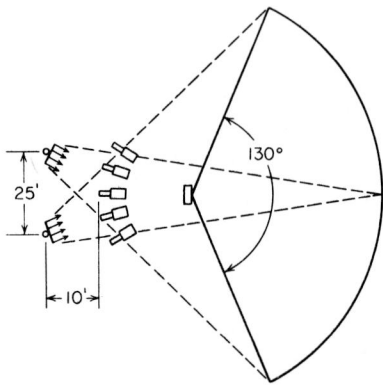

| Floodlights | Four 6 × 5, Group B |
|---|---|
| Lamps | 1000-W metal halide |
| Load | 4.3 kW |
| Mounting height | 20 ft above ground |
| Poles | Two |

### 304. Volleyball: outdoor

Wide-beam floodlights are necessary to provide uniform illumination.

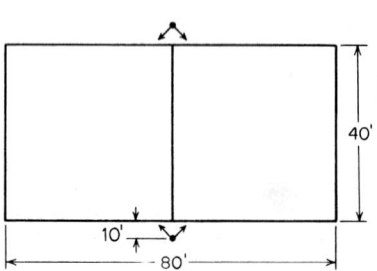

| | | Floodlights | | | | |
|---|---|---|---|---|---|---|
| Class | Type | Group | No. per pole | Total no. | Total load, kW | Lamp type |
| Tournament | 6 × 5 | B | 3 | 6 | 2.7 | 400-W metal halide |
| | 6 × 5 | B | 3 | 6 | 2.8 | 400-W HPS |
| Recreational | 6 × 5 | B | 2 | 4 | 1.8 | 400-W metal halide |
| | 6 × 5 | B | 2 | 4 | 1.2 | 250-W HPS |

| | |
|---|---|
| Mounting height | 20 to 25 ft above court |
| Poles | Two per court |

# Wiring and
# Design Tables

## 1. Standard Sizes of Lamps for General Illumination, in Watts[a]

| General incandescent | Tungsten-halogen incandescent | Fluorescent (preheat) | Fluorescent (rapid-start, standard) | Fluorescent (slimline) | Fluorescent (rapid-start, high-output) | Fluorescent (rapid-start, 1500 mA) | Mercury | Metal halide | High-pressure sodium |
|---|---|---|---|---|---|---|---|---|---|
| 10 | 75 | 4 | 30/25 | 20 | 35 | 110/95 | 40 | 32 | 35 |
| 15 | 100 | 6 | 30 | 25 | 45 | 110 | 50 | 70 | 50 |
| 25 | 150 | 8 | 40/34 | 30 | 55 | 165 | 75 | 100 | 70 |
| 40 | 200 | 13 | 40 | 40/30 | 60/55 | 215/185 | 100 | 150 | 100 |
| 50 | 250 | 14 | | 35 | 60 | 215 | 175 | 175 | 110 |
| 60/52 | 300 | 15 | | 40 | 75 | | 250 | 250 | 150 |
| 75/67 | 400 | 20 | | 50 | 85 | | 400 | 400 | 200 |
| 75 | 500 | 25 | | 55 | 110/95 | | 1000 | 1000 | 250 |
| 100/90 | 1000 | 30 | | 75/60 | 100 | | | 1500 | 310 |
| 100 | 1250 | 40 | | 65 | 110 | | | | 400 |
| 150 | 1500 | 90 | | 75 | | | | | 1000 |
| 200 | | | | | | | | | |
| 300 | | | | | | | | | |
| 500 | | | | | | | | | |
| 750 | | | | | | | | | |
| 1000 | | | | | | | | | |
| 1500 | | | | | | | | | |

[a]For more complete data on lamps refer to Div. 10.

## 2. Demand Loads for Household Electric Ranges, Wall-Mounted Ovens, Counter-Mounted Cooking Units, and Other Household Cooking Appliances over 1¾-kW Rating[a]

(Table 220-19, 1990 NEC)
Column *A* to be used in all cases except as otherwise permitted in Note 3.

| Number of appliances | Maximum demand, kW (see Notes) | Demand factors (see Note 3), percent | |
| --- | --- | --- | --- |
| | Col. *A* (not over 12 kW rating) | Col. *B* (less than 3½ kW rating) | Col. *C* (3½- to 8¾- kW rating) |
| 1 | 8 | 80 | 80 |
| 2 | 11 | 75 | 65 |
| 3 | 14 | 70 | 55 |
| 4 | 17 | 66 | 50 |
| 5 | 20 | 62 | 45 |
| 6 | 21 | 59 | 43 |
| 7 | 22 | 56 | 40 |
| 8 | 23 | 53 | 36 |
| 9 | 24 | 51 | 35 |
| 10 | 25 | 49 | 34 |
| 11 | 26 | 47 | 32 |
| 12 | 27 | 45 | 32 |
| 13 | 28 | 43 | 32 |
| 14 | 29 | 41 | 32 |
| 15 | 30 | 40 | 32 |
| 16 | 31 | 39 | 28 |
| 17 | 32 | 38 | 28 |
| 18 | 33 | 37 | 28 |
| 19 | 34 | 36 | 28 |
| 20 | 35 | 35 | 28 |
| 21 | 36 | 34 | 26 |
| 22 | 37 | 33 | 26 |
| 23 | 38 | 32 | 26 |
| 24 | 39 | 31 | 26 |
| 25 | 40 | 30 | 26 |
| 26–30 | 15 kW plus 1 kW for each range | 30 | 24 |
| 31–40 | | 30 | 22 |
| 41–50 | 25 kW plus ¾ kW for each range | 30 | 20 |
| 51–60 | | 30 | 18 |
| 61 and over | | 30 | 16 |

[a]Reprinted with permission from NFPA 70-1990. National Electrical Code®, Copyright © 1990, National Fire Protection Association. Quincy, Massachusetts 02269. This reprinted material is not the complete and official position of the NFPA on the referenced subject, which is represented only by the standard in its entirety.

NOTE 1   *Over-12- to 27-kW ranges all of the same rating.* For ranges individually rated more than 12 kW but not more than 27 kW, the maximum demand in col. *A* shall be increased by 5 percent for each additional kilowatt of rating or major fraction thereof by which the rating of individual ranges exceeds 12 kW.

NOTE 2   *Over-8¾- to 27-kW ranges of unequal ratings.* For ranges individually rated more than 8¾ kW and of different ratings but none exceeding 27 kW, an average value of rating shall be calculated by adding together the ratings of all ranges to obtain the total connected load (using 12 kW for any range rated less than 12 kW) and dividing by the total number of ranges. Then the maximum demand in col. *A* shall be increased by 5 percent for each kilowatt or major fraction thereof by which this average value exceeds 12 kW.

NOTE 3   *Over 1¾ kW through 8¾ kW.* In lieu of the method provided in col. *A*, it shall be permissible to add the nameplate ratings of all ranges rated more than 1¾ kW but not more than 8¾ kW and to multiply the sum by the demand factors specified in col. *B* or *C* for the given number of appliances.

NOTE 4   *Branch-circuit load.* It shall be permissible to compute the branch-circuit load for one range in accordance with Table 2. The branch-circuit load for one wall-mounted oven or one counter-mounted cooking unit shall be the nameplate rating of the appliance. The branch-circuit load for a counter-mounted cooking unit and not more than two wall-mounted ovens, all supplied from a single branch circuit and located in the same room, shall be computed by adding the nameplate rating of the individual appliances and treating this total as equivalent to one range.

NOTE 5   This table also applies to household cooking appliances rated over 1¾ kW and used in instructional programs.

### 3.  Demand Factors for Household Electric Clothes Dryers[a]
(Table 220-18, 1990 NEC)

| Number of Dryers | Demand Factors, Percent |
|:---:|:---:|
| 1 | 100 |
| 2 | 100 |
| 3 | 100 |
| 4 | 100 |
| 5 | 80 |
| 6 | 70 |
| 7 | 65 |
| 8 | 60 |
| 9 | 55 |
| 10 | 50 |
| 11–13 | 45 |
| 14–19 | 40 |
| 20–24 | 35 |
| 25–29 | 32.5 |
| 30–34 | 30 |
| 35–39 | 27.5 |
| 40 up | 25 |

[a]Reprinted with permission from NFPA 70-1990, National Electrical Code®, Copyright © 1990, National Fire Protection Association. Quincy, Massachusetts 02269. This reprinted material is not the complete and official position of the NFPA on the referenced subject, which is represented only by the standard in its entirety.

### 3A.  Feeder Demand Factors for Kitchen Equipment Other Than Dwelling Units[a]
(Table 220-20, 1990 NEC)

| Number of Units of Equipment | Demand Factors, Percent |
|:---:|:---:|
| 1 | 100 |
| 2 | 100 |
| 3 | 80 |
| 4 | 80 |
| 5 | 70 |
| 6 and over | 65 |

[a]Reprinted with permission from NFPA 70-1990, National Electrical Code®, Copyright © 1990, National Fire Protection Association, Quincy, Massachusetts 02269. This reprinted material is not the complete and official position of the NFPA on the referenced subject, which is represented only by the standard in its entirety.

### 4.  Demand Factors for General Lighting Loads[a]
(Table 220-3(b), 1990 NEC)

| Type of Occupancy | Unit Load per Ft², Voltamperes |
|:---|:---:|
| Armories and auditoriums | 1 |
| Banks | 3½[b] |
| Barbershops and beauty parlors | 3 |
| Churches | 1 |
| Clubs | 2 |
| Courtrooms | 2 |
| Dwelling units[c] | 3 |
| Garages, Commercial (storage) | ½ |
| Hospitals | 2 |

**4.   Demand Factors for Lighting Loads**[a] *(Continued)*

| Type of Occupancy | Unit Load per Ft², Voltamperes |
|---|---|
| Hotels and motels, including apartment houses without provisions for cooking by tenants[c] | 2 |
| Industrial commercial (loft) buildings | 2 |
| Lodge rooms | 1½ |
| Office buildings | 3½[b] |
| Restaurants | 2 |
| Schools | 3 |
| Stores | 3 |
| Warehouses (storage) | ¼ |
| In any of the above occupancies except one-family dwellings and individual dwelling units of multifamily dwellings: | |
| Assembly halls and auditoriums | 1 |
| Halls, corridors, and closets | ½ |
| Storage spaces | ¼ |

[a]Reprinted with permission from NFPA 70-1990, National Electrical Code®, Copyright © 1990, National Fire Protection Association, Quincy, Massachusetts 02269. This reprinted material is not the complete and official position of the NFPA on the referenced subject, which is represented only by the standard in its entirety.

[b]In addition a unit load of 1 W/ft² shall be included for general-purpose receptacle outlets when the actual number of general-purpose receptacle outlets is unknown.

[c]All general-use receptacle outlets of 20 A or less rating in one-family, two-family, and multifamily dwellings and in guest rooms of hotels and motels (except those connected to the receptacle circuits specified in Secs. 220-4(b) and (c) of the NEC) shall be considered outlets for general illumination, and no additional load calculations need be included for such outlets.

**5.   Data for determining** National Electrical Code **minimum allowable lighting and appliance loads: calculation of feeder loads.**   The computed load of a feeder shall be not less than the sum of all branch-circuit loads supplied by the feeder, subject to the following provisions:

1. GENERAL LIGHTING   The demand factors listed in Table A shall apply to that portion of the total branch-circuit load computed for general illumination. They shall not be applied in determining the number of branch circuits for general illumination supplied by the feeders. See Pars. 7 and 8.

These demand factors are based on minimum load conditions and 100 percent power factor and in specific instances may not provide sufficient capacity for the installation

**A.   Lighting-Load Feeder Demand Factors**

| Type of occupancy | Portion of lighting load to which demand factor applies (wattage) | Demand factor, percent |
|---|---|---|
| Dwelling units | First 3000 or less at | 100 |
| | From 3001 to 120,000 at | 35 |
| | Remainder over 120,000 at | 25 |
| Hospitals[a] | First 50,000 or less at | 40 |
| | Remainder over 50,000 at | 20 |
| Hotels and motels including apartment houses without provision for cooking by tenants[a] | First 20,000 or less at | 50 |
| | From 20,001 to 100,000 at | 40 |
| | Remainder over 100,000 at | 30 |
| Warehouses (storage) | First 12,500 or less at | 100 |
| | Remainder over 12,500 at | 50 |
| All others | Total voltamperes | 100 |

[a]The demand factors of this table shall not apply to the computed load of feeders to areas in hospitals, hotels, and motels where the entire lighting is likely to be used at one time, as in operating rooms, ballrooms, or dining rooms.

contemplated. In view of the trend toward higher-intensity lighting systems and increased loads because of the more general use of fixed and portable appliances, each installation should be considered with regard to the load likely to be imposed and the capacity increased to ensure safe operation. When electric discharge lighting systems are to be installed, a high-power-factor type should be used; otherwise the conductor capacity may need to be increased.

2. SHOW-WINDOW LIGHTING   For show-window lighting a load of not less than 200 VA shall be included for each linear foot of show window, measured horizontally along its base.

3. MOTORS   For motors, a load computed according to the provisions of Secs. 6 to 14 shall be included.

4. NEUTRAL FEEDER LOAD   The neutral feeder load shall be the maximum imbalance of the load determined by Sec. 4. The maximum unbalanced load shall be the maximum connected load between the neutral and any one ungrounded conductor, except that the load thus obtained shall be multiplied by 140 percent for three-wire two-phase or five-wire two-phase systems. For a feeder supplying household electric ranges, wall-mounted ovens, and counter-mounted cooking units, the maximum unbalanced load shall be considered to be 70 percent of the load on the ungrounded conductors, as determined in accordance with Sec. 2. For three-wire dc or single-phase ac, four-wire three-phase, three-wire two-phase, or five-wire two-phase systems, a further demand factor of 70 percent may be applied to that portion of the unbalanced load in excess of 200 A. There shall be no reduction of the neutral capacity for that portion of the load which consists of electric-discharge lighting, electronic computer/data processing, or similar equipment and supplied from a four-wire wye-connected three-phase system.

5. FIXED ELECTRICAL SPACE HEATING   The computed load of a feeder supplying fixed electrical space-heating equipment shall be the total connected load on all branch circuits.

*Exception No. 1. When reduced loading of the conductors results from units operating on duty cycle, intermittently, or from all units not operating at one time, the authority enforcing the* Code *may grant permission for feeder conductors to be of a capacity less than 100 percent, provided the conductors are of sufficient capacity for the load so determined.*

*Exception No. 2. Paragraph 5 does not apply when feeder capacity is calculated in accordance with the optional method in Sec. 46 of Div. 3 for one-family dwellings.*

6. NONCOINCIDENT LOAD   In adding the branch-circuit loads to determine the feeder load, the smaller of two dissimilar loads may be omitted from the total if it is unlikely that both loads will be served simultaneously.

7. SMALL APPLIANCES   The computed branch-circuit load for receptacle outlets in other than dwelling occupancies, for which the allowance is not more than 180 VA per outlet, may be included with the general-lighting load and subject to the demand factors in Par. 1.

### DWELLING UNITS

A dwelling unit is defined as one or more rooms for the use of one or more persons as a housekeeping unit with space for eating, living, and sleeping and permanent provisions for cooking and sanitation. The requirements in Pars. 8 to 11 apply to dwelling-type occupancies and are supplemental to Pars. 1 to 7.

8. SMALL APPLIANCES
 *a. Dwelling units.*   In dwelling units a feeder load of not less than 1500 VA for each two-wire circuit shall be included for small appliances (portable appliances supplied from receptacles of 15- or 20-A rating) in pantry and breakfast room, dining room, and kitchen. When the load is subdivided through two or more feeders, the computed load for each shall include not less than 1500 VA for each two-wire circuit for small appliances. These loads may be included with the general-lighting load and subject to the demand factors in Par. 1.

 *b. Laundry circuit.*   A feeder load of not less than 1500 VA shall be included for each two-wire laundry circuit. This load may be included with the general-lighting load and subject to the demand factors in Par. 1.

9. ELECTRIC RANGES   The feeder load for household electric ranges and other cooking appliances, individually rated more than 1¾ kW, may be calculated in accordance with Table 2. To provide for possible future installations of ranges or higher ratings, it is recommended that when ranges of less than 8¾-kW ratings or wall-mounted ovens and

counter-mounted cooking units are to be installed, the feeder capacity be not less than the maximum demand value specified in col. A of Table 2.

When a number of single-phase ranges are supplied by a three-phase four-wire feeder, the current shall be computed on the basis of the demand of twice the maximum number of ranges connected between any two-phase wires. See example in Sec. 50 of Div. 3.

10. FASTENED-IN-PLACE APPLIANCES (OTHER THAN RANGES, CLOTHES DRYERS, AIR-CONDITIONING EQUIP-MENT, OR SPACE-HEATING EQUIPMENT) When four or more fastened-in-place appliances other than electric ranges, clothes dryers, air-conditioning equipment, or space-heating equipment are connected to the same feeder in a one-family, two-family, or multifamily dwelling unit, a demand factor of 75 percent may be applied to the appliance load.

11. SPACE HEATING AND AIR COOLING  In adding branch-circuit loads for space heating and air cooling in dwelling units, the smaller of the two loads may be omitted from the total if it is unlikely that both loads will be served simultaneously.

12. FARM BUILDINGS  Feeders supplying farm buildings (excluding dwellings) or loads consisting of two or more branch circuits shall have minimum capacity computed in accordance with Table B.

## B. Method of Computing Farm Loads for Other Than Dwelling Units

| *Load in Amperes at 240 V* | *Percent of Connected Load* |
|---|---|
| Loads expected to operate without diversity, but not less than 125 percent full-load current of the largest motor and not less than first 60 A of load | 100 |
| Next 60 A of all other loads | 50 |
| Remainder of other load | 25 |

NOTE 1    For services to farm dwellings, see Pars. 8 to 11.
NOTE 2    For service at the main point of delivery to the farmstead, see Par. 13.

13. FARM SERVICES

*a.* Service-equipment and service-entrance conductors for individual farm buildings (excluding dwellings) shall have a minimum capacity computed in accordance with Par. 12.

*b.* Minimum capacity of service conductors and service equipment, if any, at the main point of delivery to farms (including dwellings) shall be determined in accordance with the following formula:

100 percent of the largest demand computed in accordance with Par. 12
75 percent of the second largest demand computed in accordance with Par. 12
65 percent of the third largest demand computed in accordance with Par. 12
50 percent of the demands of remaining loads computed in accordance with Par. 12

NOTE 1    Consider as a single computed demand the total of the computed demands of all buildings or loads having the same function.
NOTE 2    The demand of the farm dwelling, if included in the demands of this formula, should be computed in accordance with Note 1 of Table B.

## 6. Average Demand Factors for Motor Loads[a]

| Number of motors | Character of load | Demand factor |
|---|---|---|
| 1– 5 | Individual drives—tools, etc. | 1.00 |
| 6– 10 | Individual drives—tools, etc. | 0.75 |
| 10– 15 | Individual drives—tools, etc. | 0.70 |
| 15– 20 | Individual drives—tools, etc. | 0.65 |
| 20– 30 | Individual drives—tools, etc. | 0.60 |
| 30– 50 | Individual drives—tools, etc. | 0.50 |
| 50– 75 | Individual drives—tools, etc. | 0.45 |
| 75–100 | Individual drives—tools, etc. | 0.40 |
| Above 100 | Individual drives—tools, etc. | 0.40 |
| 6– 10 | Group drives | 0.85 |
| Above 10 | Group drives | 0.70–0.45 |
|  | Fans, compressors, pumps, etc. | 1.00–0.85 |

[a]Permission for the use of a demand factor should be obtained from the authority enforcing the National Electrical Code.

### 7.  Full-Load Current in Amperes, Direct-Current Motors[a]

(Table 430-147, 1990 NEC)

The following values of full-load currents[b] are for motors running at base speed.

| | Armature voltage rating[b] | | | | | |
|---|---|---|---|---|---|---|
| hp | 90 V | 120 V | 180 V | 240 V | 500 V | 550 V |
| ¼ | 4.0 | 3.1 | 2.0 | 1.6 | | |
| ⅓ | 5.2 | 4.1 | 2.6 | 2.0 | | |
| ½ | 6.8 | 5.4 | 3.4 | 2.7 | | |
| ¾ | 9.6 | 7.6 | 4.8 | 3.8 | | |
| 1 | 12.2 | 9.5 | 6.1 | 4.7 | | |
| 1½ | ... | 13.2 | 8.3 | 6.6 | | |
| 2 | ... | 17 | 10.8 | 8.5 | | |
| 3 | ... | 25 | 16 | 12.2 | | |
| 5 | ... | 40 | 27 | 20 | | |
| 7½ | ... | 58 | ... | 29 | 13.6 | 12.2 |
| 10 | ... | 76 | ... | 38 | 18 | 16 |
| 15 | ... | ... | ... | 55 | 27 | 24 |
| 20 | ... | ... | ... | 72 | 34 | 31 |
| 25 | ... | ... | ... | 89 | 43 | 38 |
| 30 | ... | ... | ... | 106 | 51 | 46 |
| 40 | ... | ... | ... | 140 | 67 | 61 |
| 50 | ... | ... | ... | 173 | 83 | 75 |
| 60 | ... | ... | ... | 206 | 99 | 90 |
| 75 | ... | ... | ... | 255 | 123 | 111 |
| 100 | ... | ... | ... | 341 | 164 | 148 |
| 125 | ... | ... | ... | 425 | 205 | 185 |
| 150 | ... | ... | ... | 506 | 246 | 222 |
| 200 | ... | ... | ... | 675 | 330 | 294 |

[a]Reprinted with permission from NFPA 70-1990, National Electrical Code®, Copyright © 1990, National Fire Protection Association, Quincy, Massachusetts 02269. This reprinted material is not the complete and official position of the NFPA on the referenced subject, which is represented only by the standard in its entirety.

[b]These are average direct-current quantities.

**8.  Full-Load Currents in Amperes, Single-Phase Alternating-Current Motors**[a]
(Table 430-148, 1990 NEC)

| hp | 115 V | 200 V | 208 V | 230 V |
|----|-------|-------|-------|-------|
| ⅙ | 4.4 | 2.5 | 2.4 | 2.2 |
| ¼ | 5.8 | 3.3 | 3.2 | 2.9 |
| ⅓ | 7.2 | 4.1 | 4.0 | 3.6 |
| ½ | 9.8 | 5.6 | 5.4 | 4.9 |
| ¾ | 13.8 | 7.9 | 7.6 | 6.9 |
| 1 | 16 | 9.2 | 8.8 | 8 |
| 1½ | 20 | 11.5 | 11 | 10 |
| 2 | 24 | 13.8 | 13.2 | 12 |
| 3 | 34 | 19.6 | 18.7 | 17 |
| 5 | 56 | 32.2 | 30.8 | 28 |
| 7½ | 80 | 46 | 44 | 40 |
| 10 | 100 | 57.5 | 55 | 50 |

[a]Reprinted with permission from NFPA 70-1990, National Electrical Code®, Copyright © 1990, National Fire Protection Association, Quincy, Massachusetts 02269. This reprinted material is not the complete and official position of the NFPA on the referenced subject, which is represented only by the standard in its entirety.

These values of full-load currents are for motors running at usual speeds and motors with normal torque characteristics. Motors built for especially low speeds or high torques may have higher full-load currents, and multispeed motors will have full-load currents varying with speed, in which case the nameplate current ratings shall be used.

The voltages listed are rated motor voltages. The currents listed shall be permitted for system voltage ranges of 110 to 120 and 220 to 240.

### 9.  Average Full-Load Currents of Two-Phase Four-Wire Motors[a]
(Table 430-149, 1990 NEC)

| hp | Induction-type squirrel-cage and wound-rotor, A | | | | |
|---|---|---|---|---|---|
| | 115 V | 230 V | 460 V | 575 V | 2,300 V |
| ½ | 4 | 2 | 1 | 0.8 | |
| ¾ | 4.8 | 2.4 | 1.2 | 1.0 | |
| 1 | 6.4 | 3.2 | 1.6 | 1.3 | |
| 1½ | 9 | 4.5 | 2.3 | 1.8 | |
| 2 | 11.8 | 5.9 | 3 | 2.4 | |
| 3 | ....... | 8.3 | 4.2 | 3.3 | |
| 5 | ....... | 13.2 | 6.6 | 5.3 | |
| 7½ | ....... | 19 | 9 | 8 | |
| 10 | ....... | 24 | 12 | 10 | |
| 15 | ....... | 36 | 18 | 14 | |
| 20 | ....... | 47 | 23 | 19 | |
| 25 | ....... | 59 | 29 | 24 | |
| 30 | ....... | 69 | 35 | 28 | |
| 40 | ....... | 90 | 45 | 36 | |
| 50 | ....... | 113 | 56 | 45 | |
| 60 | ....... | 133 | 67 | 53 | 14 |
| 75 | ....... | 166 | 83 | 66 | 18 |
| 100 | ....... | 218 | 109 | 87 | 23 |
| 125 | ....... | 270 | 135 | 108 | 28 |
| 150 | ....... | 312 | 156 | 125 | 32 |
| 200 | ....... | 416 | 208 | 167 | 43 |

[a]Reprinted with permission from NFPA 70-1990. National Electrical Code®, Copyright © 1990, National Fire Protection Association, Quincy, Massachusetts 02269. This reprinted material is not the complete and official position of the NFPA on the referenced subject, which is represented only by the standard in its entirety.

These values of full-load current are for motors running at speeds usual for belted motors and motors with normal torque characteristics. Motors built for especially low speeds or high torques may require more running current, and multispeed motors will have full-load current varying with speed, in which case the nameplate current rating shall be used. Current in common conductor of two-phase three-wire system will be 1.41 times value given.

The voltages listed are rated motor voltages. Corresponding nominal system voltages are 110 to 120, 220 to 240, 440 to 480, and 550 to 600 V.

**10. Full-Load-Current Three-Phase Alternating-Current Motors**[a]
(Table 430-150, 1990 NEC)

| hp | Induction-type squirrel-cage and wound-rotor, A | | | | | | | Synchronous-type unity power factor, A[b] | | | |
|---|---|---|---|---|---|---|---|---|---|---|---|
| | 115 V | 200 V | 208 V | 230 V | 460 V | 575 V | 2300 V | 230 V | 460 V | 575 V | 2300 V |
| ½ | 4 | 2.3 | 2.2 | 2 | 1 | .8 | | | | | |
| ¾ | 5.6 | 3.2 | 3.1 | 2.8 | 1.4 | 1.1 | | | | | |
| 1 | 7.2 | 4.1 | 4.0 | 3.6 | 1.8 | 1.4 | | | | | |
| 1½ | 10.4 | 6.0 | 5.7 | 5.2 | 2.6 | 2.1 | | | | | |
| 2 | 13.6 | 7.8 | 7.5 | 6.8 | 3.4 | 2.7 | | | | | |
| 3 | | 11.0 | 10.6 | 9.6 | 4.8 | 3.9 | | | | | |
| 5 | | 17.5 | 16.7 | 15.2 | 7.6 | 6.1 | | | | | |
| 7½ | | 25.3 | 24.2 | 22 | 11 | 9 | | | | | |
| 10 | | 32.2 | 30.8 | 28 | 14 | 11 | | | | | |
| 15 | | 48.3 | 46.2 | 42 | 21 | 17 | | | | | |
| 20 | | 62.1 | 59.4 | 54 | 27 | 22 | | | | | |
| 25 | | 78.2 | 74.8 | 68 | 34 | 27 | | 53 | 26 | 21 | |
| 30 | | 92 | 88 | 80 | 40 | 32 | | 63 | 32 | 26 | |
| 40 | | 119.6 | 114.4 | 104 | 52 | 41 | | 83 | 41 | 33 | |
| 50 | | 149.5 | 143.0 | 130 | 65 | 52 | | 104 | 52 | 42 | |
| 60 | | 177.1 | 169.4 | 154 | 77 | 62 | 16 | 123 | 61 | 49 | 12 |
| 75 | | 220.8 | 211.2 | 192 | 96 | 77 | 20 | 155 | 78 | 62 | 15 |
| 100 | | 285.2 | 272.8 | 248 | 124 | 99 | 26 | 202 | 101 | 81 | 20 |
| 125 | | 358.8 | 343.2 | 312 | 156 | 125 | 31 | 253 | 126 | 101 | 25 |
| 150 | | 414 | 396.0 | 360 | 180 | 144 | 37 | 302 | 151 | 121 | 30 |
| 200 | | 552 | 528.0 | 480 | 240 | 192 | 49 | 400 | 201 | 161 | 40 |

[a]Reprinted with permission from NFPA 70-1990, National Electrical Code®, Copyright © 1990, National Fire Protection Association, Quincy, Massachusetts 02269. This reprinted material is not the complete and official position of the NFPA on the referenced subject, which is represented only by the standard in its entirety.

These values of full-load currents are for motors running at speeds usual for belted motors and motors with normal torque characteristics. Motors built for especially low speeds or high torques may require more running current, and multispeed motors will have full-load currents varying with speed, in which case the nameplate current rating shall be used.

[b]For 90 and 80 percent power factor the above figures shall be multiplied by 1.1 and 1.25, respectively.

The voltages listed are rated motor voltages. The currents listed shall be permitted for system-voltage ranges of 110 to 120, 220 to 240, 440 to 480, and 550 to 600 V.

## 11. Capacitor Ratings for Use with Three-Phase 60-Hz Motors[a]

| Induction motor horse-power rating | Nominal motor speed, rpm | | | | | | | | | | | |
|---|---|---|---|---|---|---|---|---|---|---|---|---|
| | 3,600 | | 1,800 | | 1,200 | | 900 | | 720 | | 600 | |
| | Capacitor rating, kvar | Line current reduction, % | Capacitor rating, kvar | Line current reduction, % | Capacitor rating, kvar | Line current reduction, % | Capacitor rating, kvar | Line current reduction, % | Capacitor rating, kvar | Line current reduction, % | Capacitor rating, kvar | Line current reduction, % |
| 3 | 1.5 | 14 | 1.5 | 15 | 1.5 | 20 | 2 | 27 | 2.5 | 35 | 3.5 | 41 |
| 5 | 2 | 12 | 2 | 13 | 2 | 17 | 3 | 25 | 4 | 32 | 4.5 | 37 |
| 7½ | 2.5 | 11 | 2.5 | 12 | 3 | 15 | 4 | 22 | 5.5 | 30 | 6 | 34 |
| 10 | 3 | 10 | 3 | 11 | 3.5 | 14 | 5 | 21 | 6.5 | 27 | 7.5 | 31 |
| 15 | 4 | 9 | 4 | 10 | 5 | 13 | 6.5 | 18 | 8 | 23 | 9.5 | 27 |
| 20 | 5 | 9 | 5 | 10 | 6.5 | 12 | 7.5 | 16 | 9 | 21 | 12 | 25 |
| 25 | 6 | 9 | 6 | 10 | 7.5 | 11 | 9 | 15 | 11 | 20 | 14 | 23 |
| 30 | 7 | 8 | 7 | 9 | 9 | 11 | 10 | 14 | 12 | 18 | 16 | 22 |
| 40 | 9 | 8 | 9 | 9 | 11 | 10 | 12 | 13 | 15 | 16 | 20 | 20 |
| 50 | 12 | 8 | 11 | 9 | 13 | 10 | 15 | 12 | 19 | 15 | 24 | 19 |
| 60 | 14 | 8 | 14 | 8 | 15 | 10 | 18 | 11 | 22 | 15 | 27 | 19 |
| 75 | 17 | 8 | 16 | 8 | 18 | 10 | 21 | 10 | 26 | 14 | 32.5 | 18 |
| 100 | 22 | 8 | 21 | 8 | 25 | 9 | 27 | 10 | 32.5 | 13 | 40 | 17 |
| 125 | 27 | 8 | 26 | 8 | 30 | 9 | 32.5 | 10 | 40 | 13 | 47.5 | 16 |
| 150 | 32.5 | 8 | 30 | 8 | 35 | 9 | 37.5 | 10 | 47.5 | 12 | 52.5 | 15 |
| 200 | 40 | 8 | 37.5 | 8 | 42.5 | 9 | 47.5 | 10 | 60 | 12 | 65 | 14 |
| 250 | 50 | 8 | 45 | 7 | 52.5 | 8 | 57.5 | 9 | 70 | 11 | 77.5 | 13 |
| 300 | 57.5 | 8 | 52.5 | 7 | 60 | 8 | 65 | 9 | 80 | 11 | 87.5 | 12 |
| 350 | 65 | 8 | 60 | 7 | 67.5 | 8 | 75 | 9 | 87.5 | 10 | 95 | 11 |
| 400 | 70 | 8 | 65 | 6 | 75 | 8 | 85 | 9 | 95 | 10 | 105 | 11 |
| 450 | 75 | 8 | 67.5 | 6 | 80 | 8 | 92.5 | ... | 100 | 9 | 110 | 11 |
| 500 | 77.5 | 8 | 72.5 | 6 | 82.5 | 8 | 97.5 | 9 | 107.5 | 9 | 115 | 10 |

[a]For use with three-phase 60-Hz NEMA Classification B motors to raise full-load power factor to approximately 95 percent.

## 12. Minimum Ampacities for Conductors Supplying Motors Used for Short-Time, Intermittent, Periodic, or Varying Duty[a]
(Table 430-22(a); Exception, 1990 NEC)

| Classification of service | Percentages of nameplate current rating | | | |
|---|---|---|---|---|
| | 5-min rating | 15-min rating | 30- and 60-min rating | Continuous rating |
| Short-time duty: operating valves, raising or lowering rolls, etc. | 110 | 120 | 150 | |
| Intermittent duty: freight and passenger elevators, tool heads, pumps, drawbridges, turntables, etc. | 85 | 85 | 90 | 140 |
| Periodic duty: rolls, ore- and coal-handling machines, etc. | 85 | 90 | 95 | 140 |
| Varying duty | 110 | 120 | 150 | 200 |

[a]Reprinted with permission from NFPA 70-1990, National Electrical Code®, Copyright © 1990, National Fire Protection Association, Quincy, Massachusetts 02269. This reprinted material is not the complete and official position of the NFPA on the referenced subject, which is represented only by the standard in its entirety.

NOTE  Any motor application shall be considered continuous-duty unless the nature of the apparatus it drives is such that the motor will not operate continuously with load under any condition of use.

### 13. Ampacities of Wires between Secondary Controller of Wound-Rotor Induction Motors and Resistors

| Resistor-Duty Classification (NEMA Apparatus Division) | Ampacity of Wire, Percent of Full-Load Secondary Current |
|---|---|
| Light starting duty | 35 |
| Heavy starting duty | 45 |
| Extra-heavy starting duty | 55 |
| Light intermittent duty | 65 |
| Medium intermittent duty | 75 |
| Heavy intermittent duty | 85 |
| Continuous duty | 110 |

### 14. Ampacities of Insulated Copper Conductors Used with Short-Time Rated Crane and Hoist Motors, Based on Ambient Temperature of 30°C (86°F), Up to Four Conductors in Raceway or Cable,[a] Up to 3 ac[b] or 4 dc[a] Conductors in Raceway or Cable
(Table 610-14(a), 1990 NEC)

| Maximum operating temperature | 75°C (167°F) | | 90°C (194°F) | | 125°C (257°F) | | Maximum operating temperature |
|---|---|---|---|---|---|---|---|
| | Types MTW, RH, RHW, THW, THWN, XHHW, USE, ZW | | Types TA, TBS, SA, SIS, PFA, FEP, FEPB, RHH, THHN, XHHW, Z, ZW | | Types FEP, FEPB, PFA, PFAH, SA, TFE, Z, ZW | | |
| Size AWG, kcmil | 60 min | 30 min | 60 min | 30 min | 60 min | 30 min | Size AWG, kcmil |
| 16 | 10 | 12 | · · · | · · · | · · · | · · · | 16 |
| 14 | 25 | 26 | 31 | 32 | 38 | 40 | 14 |
| 12 | 30 | 33 | 36 | 40 | 45 | 50 | 12 |
| 10 | 40 | 43 | 49 | 52 | 60 | 65 | 10 |
| 8 | 55 | 60 | 63 | 69 | 73 | 80 | 8 |
| 6 | 76 | 86 | 83 | 94 | 101 | 119 | 6 |
| 5 | 85 | 95 | 95 | 106 | 115 | 134 | 5 |
| 4 | 100 | 117 | 111 | 130 | 133 | 157 | 4 |
| 3 | 120 | 141 | 131 | 153 | 153 | 183 | 3 |
| 2 | 137 | 160 | 148 | 173 | 178 | 214 | 2 |
| 1 | 143 | 175 | 158 | 192 | 210 | 253 | 1 |
| 1/0 | 190 | 233 | 211 | 259 | 253 | 304 | 1/0 |
| 2/0 | 222 | 267 | 245 | 294 | 303 | 369 | 2/0 |
| 3/0 | 280 | 341 | 305 | 372 | 370 | 452 | 3/0 |
| 4/0 | 300 | 369 | 319 | 399 | 451 | 555 | 4/0 |
| 250 | 364 | 420 | 400 | 461 | 510 | 635 | 250 |
| 300 | 455 | 582 | 497 | 636 | 587 | 737 | 300 |
| 350 | 486 | 646 | 542 | 716 | 663 | 837 | 350 |
| 400 | 538 | 688 | 593 | 760 | 742 | 941 | 400 |
| 450 | 600 | 765 | 660 | 836 | 818 | 1042 | 450 |
| 500 | 660 | 847 | 726 | 914 | 896 | 1143 | 500 |

**14.  Ampacities of Insulated Copper Conductors Used with Short-Time Rated Crane and Hoist Motors, Based on Ambient Temperature of 30°C (86°F), Up to Four Conductors in Raceway or Cable,[a] Up to 3 ac[b] or 4 dc[a] Conductors in Raceway or Cable** (*Continued*)

| | Types MTW, RH, RHW, THW, THWN, XHHW, USE, ZW | | Types TA, TBS, SA, SIS, PFA, FEP, FEPB, RHH, THHN, XHHW, Z, ZW | | Types FEP, FEPB, PFA, PFAH, SA, TFE, Z, ZW | | |
|---|---|---|---|---|---|---|---|
| | 60 min | 30 min | 60 min | 30 min | 60 min | 30 min | |

Ampacity correction factors

| Ambient temp., °C | For ambient temperatures other than 30°C (86°F), multiply the ampacities shown above by the appropriate factor shown below | | | | | | Ambient temp., °F |
|---|---|---|---|---|---|---|---|
| 21–25 | 1.05 | 1.05 | 1.04 | 1.04 | 1.02 | 1.02 | 70–77 |
| 26–30 | 1.00 | 1.00 | 1.00 | 1.00 | 1.00 | 1.00 | 79–86 |
| 31–35 | .94 | .94 | .96 | .96 | .97 | .97 | 88–95 |
| 36–40 | .88 | .88 | .91 | .91 | .95 | .95 | 97–104 |
| 41–45 | .82 | .82 | .87 | .87 | .92 | .92 | 106–113 |
| 46–50 | .75 | .75 | .82 | .82 | .89 | .89 | 115–122 |
| 51–55 | .67 | .67 | .76 | .76 | .86 | .86 | 124–131 |
| 56–60 | .58 | .58 | .71 | .71 | .83 | .83 | 133–140 |
| 61–70 | .33 | .33 | .58 | .58 | .76 | .76 | 142–158 |
| 71–80 | — | — | .41 | .41 | .69 | .69 | 160–176 |
| 81–90 | — | — | — | — | .61 | .61 | 177–194 |
| 91–100 | — | — | — | — | .51 | .51 | 195–212 |
| 101–120 | — | — | — | — | .40 | .40 | 213–248 |

[a]For 5 to 8 simultaneously energized power conductors in raceway or cable, the ampacity of each power conductor shall be reduced to a value of 80 percent of that shown in the table.

[b]For 4 to 6 simultaneously energized 125°C (257°F) ac power conductors in raceway or cable, the ampacity of each power conductor shall be reduced to a value of 80 percent of that shown in the table.

NOTE    Other insulations for general-wiring conductors approved for the temperature and location shall be permitted to be substituted for those shown in Table 14. The allowable ampacities of conductors used with 15-min motors shall be the 30-min ratings increased by 12 percent. Reprinted with permission from NFPA 70-1990, National Electrical Code, Copyright © 1990, National Fire Protection Association, Quincy, Massachusetts 02269. This reprinted material is not the complete and official position of the NFPA on the referenced subject, which is represented only by the standard in its entirety.

CONTACT CONDUCTORS    The size of contact wires shall be not less than the following:

| Distance between End Strain Insulators or Clamp-Type Intermediate Supports, Ft | Size of Wire |
|---|---|
| 0–30 | No. 6 |
| 31–60 | No. 4 |
| Over 60 | No. 2 |

**15.  Curve for Determining Ampacities at 60 Hz of Bare Concentric-Stranded Copper Conductors for 30°C Rise over 40°C Ambient**

For outdoor service, a 25 percent increase in ampacities is permissible.

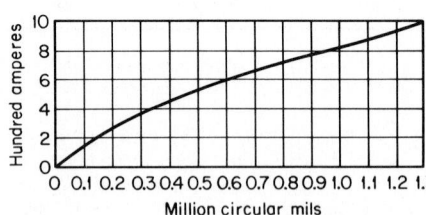

Million circular mils

**16.   Curve for Determining Ampacities for Weatherproof-Covered Concentric-Stranded Copper Conductors Installed outside Buildings**

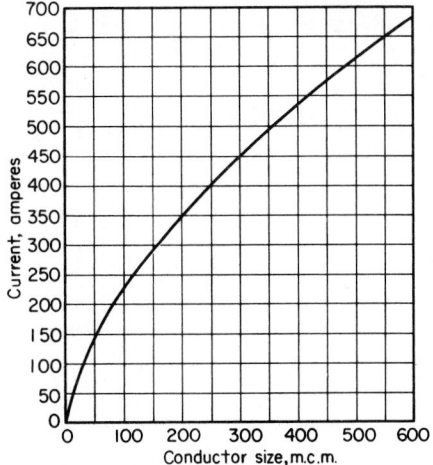

**17.** National Electrical Code **allowable ampacities of conductors for all interior wiring** (Tables 18 to 22, inclusive)

1. EXPLANATION OF TABLES   For an explanation of type letters and for the recognized size of conductors for the various conductor insulations, see Sec. 310-13 in the NEC. For installation requirements, see Sec. 310-1 through 310-10 and the various articles of NEC. For flexible cords, see Table 24.

3. THREE-WIRE SINGLE-PHASE DWELLING SERVICE   In dwelling units conductors, as listed below, shall be permitted to be utilized as three-wire, single-phase, service-entrance and feeder conductors.

**Conductor Types and Sizes**
**RH-RHH-RHW-THW-THWN-THHN-XHHW**

| Copper | Aluminum and copper-clad aluminum | Service rating, A |
|---|---|---|
| AWG | AWG | . . . |
| 4 | 2 | 100 |
| 3 | 1 | 110 |
| 2 | 1/0 | 125 |
| 1 | 2/0 | 150 |
| 1/0 | 3/0 | 175 |
| 2/0 | 4/0 | 200 |
| 3/0 | 250 kcmil | 225 |
| 4/0 | 300 kcmil | 250 |
| 250 kcmil | 350 kcmil | 300 |
| 350 kcmil | 500 kcmil | 350 |
| 400 kcmil | 600 kcmil | 400 |

5. BARE CONDUCTORS   Where bare conductors are used with insulated conductors, their allowable ampacities shall be limited to that permitted for the insulated conductors of the same size.

6. MINERAL-INSULATED METAL-SHEATHED CABLE   The temperature limitation on which the ampacities of mineral-insulated, metal-sheathed cable are based is determined by the insulating materials used in the end seal. Termination fittings incorporating unimpregnated organic insulating materials are limited to 90°C operation.

7. TYPE MTW MACHINE-TOOL WIRE    The ampacities of Type MTW wire are specified in Table 11-1(b) of the electrical standard for *Electrical Metalworking Machine Tools and Plastics Machinery,* 1977 (NFPA Publication No. 79).

8. MORE THAN THREE CONDUCTORS IN A RACEWAY OR CABLE    When the number of conductors in a raceway or cable exceeds three, the ampacities given in Tables 18 and 21 shall be reduced as shown in the following table:

| Number of conductors | Column A<br>Percent of values in tables as adjusted for ambient temperature if necessary | Number of conductors | Column B[a]<br>Percent of values in tables as adjusted for ambient temperature if necessary |
|---|---|---|---|
| 4 through 6 | 80 | 4 through 6 | 80 |
| 7 through 9 | 70 | 7 through 9 | 70 |
| 10 through 24[b] | 70 | 10 through 20 | 50 |
| 25 through 42[b] | 60 | 21 through 30 | 45 |
| 43 and above[b] | 50 | 31 through 40 | 40 |
| | | 41 through 60 | 35 |

[a]No diversity.
[b]These factors include the effects of a load diversity of 50 percent.

When single conductors or multiconductor cables are stacked or bundled longer than 24 in (604 mm) without maintaining spacing and are not installed in raceways, the ampacity of each conductor shall be reduced as shown in the above table.

*Exception No. 1. When conductors of different systems, as provided in Sec. 300-3 of the NEC, are installed in a common raceway or cable, the derating factors shown above shall apply to the number of power and lighting (Articles 210, 215, 220, and 230 of the NEC) conductors only.*

*Exception No. 2. For conductors installed in cable trays, the provisions of Section 318-11 of the NEC shall apply.*

*Exception No. 3. Derating factors shall not apply to conductors in nipples having a length not exceeding 24 in (610 mm).*

*Exception No. 4. Derating factors shall not apply to underground conductors entering or leaving an outdoor trench if those conductors have physical protection in the form of rigid metal conduit, intermediate metal conduit, or rigid nonmetalic conduit having a length not exceeding 10 ft (3.05 m) above grade and the number of conductors does not exceed 4.*

9. OVERCURRENT PROTECTION    When the standard ratings and settings of overcurrent devices do not correspond with the ratings and settings allowed for conductors, the next higher standard rating and setting shall be permitted.

*Exception: As limited in Section 240-3 of the NEC.*

10. NEUTRAL CONDUCTOR

*a.* A neutral conductor which carries only the unbalanced current from other conductors, as in the case of normally balanced circuits of three or more conductors, shall not be counted in determining ampacities as provided for in Par. 8.

*b.* In a three-wire circuit consisting of two-phase wires and the neutral of a four-wire, three-phase Y-connected system, a common conductor carries approximately the same current as the other conductors and shall be counted in determining ampacities as provided in Par. 8.

*c.* On a four wire, three-phase Y circuit where the major portion of the load consists of electric discharge lighting, data processing, or similar equipment there are harmonic currents present in the neutral conductor, and the neutral shall be considered a current-carrying conductor.

11. GROUNDING CONDUCTOR    A grounding conductor shall not be counted in determining ampacities as provided for in Par. 8.

**18. Ampacities of Insulated Conductors Rated 0–2000 V, 60 to 90°C[a]**
(Table 310-16; 1990 NEC)
Not more than three conductors in raceway or cable or earth (directly buried), based on ambient temperature of 30°C (86°F).

Temperature rating of conductor (see Table 310-13 of NEC)

| Size, AWG or 1000 cmil | Copper | | | | Aluminum or copper-clad aluminum | | | | Size, AWG or 1000 cmil |
|---|---|---|---|---|---|---|---|---|---|
| | 60°C (140°F) | 75°C (167°F) | 85°C (185°F) | 90°C (194°F) | 60°C (140°F) | 75°C (167°F) | 85°C (185°F) | 90°C (194°F) | |
| | Types TW,[b] UF[b] | Types FEPW,[b] RH,[b] RHW,[b] THHW,[b] THW,[b] THWN,[b] XHHW,[b] USE,[b] ZW[b] | Type V | Types TA, TBS, SA, SIS, FEP,[b] FEPB,[b] RHH,[b] THHN,[b] THHW,[b] XHHW[b] | Types TW,[b] UF[b] | Types RH,[b] RHW,[b] THHW,[b] THW,[b] THWN,[b] XHHW,[b] USE[b] | Type V | Types TA, TBS, SA, SIS, RHH,[b] THHN,[b] THHW,[b] XHHW[b] | |
| 18 | ... | ... | ... | 14 | | | | | |
| 16 | ... | ... | 18 | 18 | | | | | |
| 14 | 20[b] | 20[b] | 25 | 25[b] | | | | | |
| 12 | 25[b] | 25[b] | 30 | 30[b] | 20[b] | 20[b] | 25 | 25[b] | 12 AWG |
| 10 | 30 | 35[b] | 40 | 40[b] | 25 | 30[b] | 30 | 35[b] | 10 |
| 8 | 40 | 50 | 55 | 55 | 30 | 40 | 40 | 45 | 8 |
| 6 | 55 | 65 | 70 | 75 | 40 | 50 | 55 | 60 | 6 |
| 4 | 70 | 85 | 95 | 95 | 55 | 65 | 75 | 75 | 4 |
| 3 | 85 | 100 | 110 | 110 | 65 | 75 | 85 | 85 | 3 |
| 2 | 95 | 115 | 125 | 130 | 75 | 90 | 100 | 100 | 2 |
| 1 | 110 | 130 | 145 | 150 | 85 | 100 | 110 | 115 | 1 |
| 1/0 | 125 | 150 | 165 | 170 | 100 | 120 | 130 | 135 | 1/0 |
| 2/0 | 145 | 175 | 190 | 195 | 115 | 135 | 145 | 150 | 2/0 |
| 3/0 | 165 | 200 | 215 | 225 | 130 | 155 | 170 | 175 | 3/0 |
| 4/0 | 195 | 230 | 250 | 260 | 150 | 180 | 195 | 205 | 4/0 |

## 18. Ampacities of Insulated Conductors Rated 0–2000 V, 60 to 90°C[a] (Continued)

Temperature rating of conductor (see Table 310-13 of NEC)

**Copper**

| Size AWG or 1000 cmil | 60°C (140°F) Types TW,[b] UF[b] | 75°C (167°F) Types FEPW,[b] RH,[b] RHW,[b] THHW,[b] THW,[b] THWN,[b] XHHW,[b] USE,[b] ZW[b] | 85°C (185°F) Type V | 90°C (194°F) Types TA, TBS, SA, SIS, FEP,[b] FEPB,[b] RHH,[b] THHN,[b] THHW,[b] XHHW[b] |
|---|---|---|---|---|
| 250 kcmil | 215 | 255 | 275 | 290 |
| 300 | 240 | 285 | 310 | 320 |
| 350 | 260 | 310 | 340 | 350 |
| 400 | 280 | 335 | 365 | 380 |
| 500 | 320 | 380 | 415 | 430 |
| 600 | 355 | 420 | 460 | 475 |
| 700 | 385 | 460 | 500 | 520 |
| 750 | 400 | 475 | 515 | 535 |
| 800 | 410 | 490 | 535 | 555 |
| 900 | 435 | 520 | 565 | 585 |
| 1000 | 455 | 545 | 590 | 615 |
| 1250 | 495 | 590 | 640 | 665 |
| 1500 | 520 | 625 | 680 | 705 |
| 1750 | 545 | 650 | 705 | 735 |
| 2000 | 560 | 665 | 725 | 750 |

**Aluminum or copper-clad aluminum**

| 60°C (140°F) Types TW,[b] UF[b] | 75°C (167°F) Types RH,[b] RHW,[b] THHW,[b] THW,[b] THWN,[b] XHHW,[b] USE[b] | 85°C (185°F) Type V | 90°C (194°F) Types TA, TBS, SA, SIS, RHH,[b] THHN,[b] THHW,[b] XHHW[b] | Size, AWG or 1000 cmil |
|---|---|---|---|---|
| 170 | 205 | 220 | 230 | 250 kcmil |
| 190 | 230 | 250 | 255 | 300 |
| 210 | 250 | 270 | 280 | 350 |
| 225 | 270 | 295 | 305 | 400 |
| 260 | 310 | 335 | 350 | 500 |
| 285 | 340 | 370 | 385 | 600 |
| 310 | 375 | 405 | 420 | 700 |
| 320 | 385 | 420 | 435 | 750 |
| 330 | 395 | 430 | 450 | 800 |
| 355 | 425 | 465 | 480 | 900 |
| 375 | 445 | 485 | 500 | 1000 |
| 405 | 485 | 525 | 545 | 1250 |
| 435 | 520 | 565 | 585 | 1500 |
| 455 | 545 | 595 | 615 | 1750 |
| 470 | 560 | 610 | 630 | 2000 |

Ampacity correction factors

For ambient temperatures other than 30°C (86°F), multiply the ampacities shown above by the appropriate factor shown below

| Ambient temp., °C | | | | | | | | | Ambient temp., °F |
|---|---|---|---|---|---|---|---|---|---|
| 21–25 | 1.08 | 1.05 | 1.04 | 1.04 | 1.08 | 1.05 | 1.04 | 1.04 | 70–77 |
| 26–30 | 1.00 | 1.00 | 1.00 | 1.00 | 1.00 | 1.00 | 1.00 | 1.00 | 79–86 |
| 31–35 | .91 | .94 | .95 | .96 | .91 | .94 | .95 | .96 | 88–95 |
| 36–40 | .82 | .88 | .90 | .91 | .82 | .88 | .90 | .91 | 97–104 |
| 41–45 | .71 | .82 | .85 | .87 | .71 | .82 | .85 | .87 | 106–113 |
| 46–50 | .58 | .75 | .80 | .82 | .58 | .75 | .80 | .82 | 115–122 |
| 51–55 | .41 | .67 | .74 | .76 | .41 | .67 | .74 | .76 | 124–131 |
| 56–60 | — | .58 | .67 | .71 | — | .58 | .67 | .71 | 133–140 |
| 61–70 | — | .33 | .52 | .58 | — | .33 | .52 | .58 | 142–158 |
| 71–80 | — | — | .30 | .41 | — | — | .30 | .41 | 160–176 |

[a]Reprinted with permission from NFPA 70-1990, National Electrical Code®, Copyright © 1990, National Fire Protection Association, Quincy, Massachusetts 02269. This reprinted material is not the complete and official position of the NFPA on the referenced subject, which is represented only by the standard in its entirety.

[b]Unless otherwise specifically permitted elsewhere in the Code, the overcurrent protection for this conductor type shall not exceed 15 A for 14 AWG, 20 A for 12 AWG, and 30 A for 10 AWG copper or 15 A for 12 AWG and 25 A for 10 AWG aluminum and copper-clad aluminum after any correction factors for ambient temperatures and number of conductors have been applied.

**19. Ampacities of Single Insulated Conductors Rated 0–2000 V, 60 to 90°C, in Free Air[a]**
(Table 310-17, 1990 NEC)
Based on ambient temperature of 30°C (86°F).

**Copper**

| Size, AWG or 1000 cmil | 60°C (140°F) Types TW[b], UF[b] | 75°C (167°F) Types FEPW[b], RH[b], RHW[b], THHW[b], THW[b], THWN[b], XHHW[b], ZW[b] | 85°C (185°F) Type V | 90°C (194°) Types TA, TBS, SA, SIS, FEP[b], FEPB[b], RHH[b], THHN[b], THHW[b], XHHW[b], MI |
|---|---|---|---|---|
| 18 | · · · | · · · | · · · | 18 |
| 16 | · · · | · · · | 23 | 24 |
| 14 | 25[b] | 30[b] | 30 | 35[b] |
| 12 | 30[b] | 35[b] | 40 | 40[b] |
| 10 | 40[b] | 50[b] | 55 | 55[b] |
| 8 | 60 | 70 | 75 | 80 |
| 6 | 80 | 95 | 100 | 105 |
| 4 | 105 | 125 | 135 | 140 |
| 3 | 120 | 145 | 160 | 165 |
| 2 | 140 | 170 | 185 | 190 |
| 1 | 165 | 195 | 215 | 220 |
| 1/0 | 195 | 230 | 250 | 260 |
| 2/0 | 225 | 265 | 290 | 300 |
| 3/0 | 260 | 310 | 335 | 350 |
| 4/0 | 300 | 360 | 390 | 405 |

**Aluminum or copper-clad aluminum**

| Size, AWG or 1000 cmil | 60°C (140°F) Types TW[b], UF[b] | 75°C (167°F) Types RH[b], RHW[b], THHW[b], THW[b], THWN[b], XHHW[b] | 85°C (185°F) Type V | 90°C (194°F) Types TA, TBS, SA, SIS, RHH[b], THHW[b], THHN[b], XHHW[b], MI |
|---|---|---|---|---|
| 12 AWG | 25[b] | 30[b] | 30 | 35[b] |
| 10 | 35[b] | 40[b] | 40 | 40[b] |
| 8 | 45 | 55 | 60 | 60 |
| 6 | 60 | 75 | 80 | 80 |
| 4 | 80 | 100 | 105 | 110 |
| 3 | 95 | 115 | 125 | 130 |
| 2 | 110 | 135 | 145 | 150 |
| 1 | 130 | 155 | 165 | 175 |
| 1/0 | 150 | 180 | 195 | 205 |
| 2/0 | 175 | 210 | 225 | 235 |
| 3/0 | 200 | 240 | 265 | 275 |
| 4/0 | 235 | 280 | 305 | 315 |

| 250 MCM | | | | | | | | |
|---|---|---|---|---|---|---|---|---|
| 250 | 340 | 405 | 440 | 455 | 265 | 315 | 345 | 355 |
| 300 | 375 | 445 | 485 | 505 | 290 | 350 | 380 | 395 |
| 350 | 420 | 505 | 550 | 570 | 330 | 395 | 430 | 445 |
| 400 | 455 | 545 | 595 | 615 | 355 | 425 | 465 | 480 |
| 500 | 515 | 620 | 675 | 700 | 405 | 485 | 525 | 545 |
| 600 | 575 | 690 | 750 | 780 | 455 | 540 | 595 | 615 |
| 700 | 630 | 755 | 825 | 855 | 500 | 595 | 650 | 675 |
| 750 | 655 | 785 | 855 | 885 | 515 | 620 | 675 | 700 |
| 800 | 680 | 815 | 885 | 920 | 535 | 645 | 700 | 725 |
| 900 | 730 | 870 | 950 | 985 | 580 | 700 | 760 | 785 |
| 1000 | 780 | 935 | 1020 | 1055 | 625 | 750 | 815 | 845 |
| 1250 | 890 | 1065 | 1160 | 1200 | 710 | 855 | 930 | 960 |
| 1500 | 980 | 1175 | 1275 | 1325 | 795 | 950 | 1035 | 1075 |
| 1750 | 1070 | 1280 | 1395 | 1445 | 875 | 1050 | 1145 | 1185 |
| 2000 | 1155 | 1385 | 1505 | 1560 | 960 | 1150 | 1250 | 1335 |

## Ampacity correction factors

For ambient temperatures other than 30°C (86°F), multiply the ampacities shown above by the appropriate factor shown below

| Ambient temp., °C | | | | | | | | | Ambient temp., °F |
|---|---|---|---|---|---|---|---|---|---|
| 21–25 | 1.08 | 1.05 | 1.04 | 1.04 | 1.08 | 1.05 | 1.04 | 1.04 | 70–77 |
| 26–30 | 1.00 | 1.00 | 1.00 | 1.00 | 1.00 | 1.00 | 1.00 | 1.00 | 79–86 |
| 31–35 | .91 | .94 | .95 | .96 | .91 | .94 | .95 | .96 | 88–95 |
| 36–40 | .82 | .88 | .90 | .91 | .82 | .88 | .90 | .91 | 97–104 |
| 41–45 | .71 | .82 | .85 | .87 | .71 | .82 | .85 | .87 | 106–113 |
| 46–50 | .58 | .75 | .80 | .82 | .58 | .75 | .80 | .82 | 115–122 |
| 51–55 | .41 | .67 | .74 | .76 | .41 | .67 | .74 | .76 | 124–131 |
| 56–60 | — | .58 | .67 | .71 | — | .58 | .67 | .71 | 133–140 |
| 61–70 | — | .33 | .52 | .58 | — | .33 | .52 | .58 | 142–158 |
| 71–80 | — | — | .30 | .41 | — | — | .30 | .41 | 160–176 |

[a] Reprinted with permission from NFPA 70-1990, National Electrical Code®, Copyright © 1990, National Fire Protection Association, Quincy, Massachusetts 02269. This reprinted material is not the complete and official position of the NFPA on the referenced subject, which is represented only by the standard in its entirety.

[b] Unless otherwise specifically permitted elsewhere in the Code, the overcurrent protection for this conduction type shall not exceed 20 A for 14 AWG, 25 A for 12 AWG, and 10 A for 10 AWG copper or 20 A for 12 AWG and 30 A for 10 AWG aluminum and copper-clad aluminum after any correction factors for ambient temperatures have been applied.

## 20. Ampacities of Insulated Copper Conductors of Various Deratings[a]

In raceways or cables at room temperatures not over 30°C (86°F).

| Size, AWG or 1000 cmil | Ampacity | | | | | | | | | | | | | | | |
| --- | --- | --- | --- | --- | --- | --- | --- | --- | --- | --- | --- | --- | --- | --- | --- | --- |
| | 100% | | | 80% | | | 70% | | | 60% | | | 50% | | | |
| | 60°C | 75°C | 90°C | 60°C | 75°C | 90°C | 60°C | 75°C | 90°C | 60°C | 75°C | 90°C | 60°C | 75°C | 90°C |
| **AWG** | | | | | | | | | | | | | | | |
| 14[a] | 20 | 20 | 25 | 16 | 16 | 20 | 14 | 14 | 18 | 12 | 12 | 15 | 10 | 10 | 13 |
| 12[a] | 25 | 25 | 30 | 20 | 20 | 24 | 18 | 18 | 21 | 15 | 15 | 18 | 13 | 13 | 15 |
| 10[a] | 30 | 35 | 40 | 24 | 28 | 32 | 21 | 25 | 28 | 18 | 21 | 24 | 15 | 18 | 20 |
| 8 | 40 | 50 | 55 | 32 | 40 | 44 | 28 | 35 | 39 | 24 | 30 | 33 | 20 | 25 | 28 |
| 6 | 55 | 65 | 75 | 44 | 52 | 60 | 38 | 45 | 53 | 33 | 39 | 45 | 27 | 32 | 38 |
| 4 | 70 | 85 | 95 | 56 | 68 | 76 | 49 | 59 | 67 | 42 | 51 | 57 | 35 | 42 | 48 |
| 3 | 85 | 100 | 110 | 68 | 80 | 88 | 60 | 70 | 77 | 51 | 60 | 66 | 43 | 50 | 55 |
| 2 | 95 | 115 | 130 | 76 | 92 | 104 | 66 | 80 | 91 | 57 | 69 | 78 | 47 | 57 | 65 |
| 1 | 110 | 130 | 150 | 88 | 104 | 120 | 77 | 91 | 105 | 66 | 78 | 90 | 55 | 65 | 75 |
| 1/0 | 125 | 150 | 170 | 100 | 120 | 136 | 87 | 105 | 119 | 75 | 90 | 102 | 62 | 75 | 85 |
| 2/0 | 145 | 175 | 195 | 116 | 140 | 156 | 101 | 122 | 137 | 87 | 105 | 117 | 72 | 87 | 98 |
| 3/0 | 165 | 200 | 225 | 132 | 160 | 180 | 115 | 140 | 158 | 99 | 120 | 135 | 82 | 100 | 113 |
| 4/0 | 195 | 230 | 260 | 156 | 184 | 200 | 136 | 161 | 175 | 117 | 138 | 150 | 97 | 115 | 125 |
| 250 kcmil | 215 | 255 | 290 | 172 | 204 | 232 | 150 | 178 | 203 | 129 | 153 | 174 | 107 | 127 | 145 |
| 300 | 240 | 285 | 320 | 192 | 228 | 256 | 168 | 199 | 224 | 144 | 171 | 192 | 120 | 142 | 160 |
| 350 | 260 | 310 | 350 | 208 | 248 | 280 | 182 | 217 | 245 | 156 | 186 | 210 | 130 | 155 | 175 |
| 400 | 280 | 335 | 380 | 224 | 268 | 304 | 196 | 234 | 266 | 168 | 201 | 228 | 140 | 167 | 190 |
| 500 | 320 | 380 | 430 | 256 | 304 | 344 | 224 | 266 | 301 | 192 | 228 | 258 | 160 | 190 | 215 |

[a]The overcurrent protection shall not exceed 15 A for 14 AWG, 20 A for 12 AWG, and 30 A for 10 AWG after any correction factors for ambient temperatures and number of conductors have been applied.

NOTE  60°C conductors: TW, UF; 75°C conductors: FEPW, RH, RHW, THHW, THW, THWN, XHHW (wet locations), USE, ZW; 90°C conductors: THHN, THHW, RHH, FEP, FEPB, XHHW (dry or damp locations).

a

**21. Ampacities for Three Single Insulated Conductors Rated 0–2000 V, 150 to 250°C, in Raceway or Cable**[a]
(Table 310-18, 1990 NEC)
Based on ambient temperature of 40°C (104°F).

| Size, AWG, kcmil | Temperature rating of conductor | | | | Size, AWG, kcmil |
| | 150°C (302°F) | 200°C (392°F) | 250°C (482°F) | 150°C (302°F) | |
| | Type Z | Types FEP, FEBP, PFA | Types PFAH, TFE | Type Z | |
| | Copper | | Nickel or nickel-coated copper | Aluminum or copper-clad aluminum | |
|------|------|------|------|------|------|
| 14 | 34 | 36 | 39 | . . . . | 14 |
| 12 | 43 | 45 | 54 | 30 | 12 |
| 10 | 55 | 60 | 73 | 44 | 10 |
| 8 | 76 | 83 | 93 | 57 | 8 |
| 6 | 96 | 110 | 117 | 75 | 6 |
| 4 | 120 | 125 | 148 | 94 | 4 |
| 3 | 143 | 152 | 166 | 109 | 3 |
| 2 | 160 | 171 | 191 | 124 | 2 |
| 1 | 186 | 197 | 215 | 145 | 1 |
| 1/0 | 215 | 229 | 244 | 169 | 1/0 |
| 2/0 | 251 | 260 | 273 | 198 | 2/0 |
| 3/0 | 288 | 297 | 308 | 227 | 3/0 |
| 4/0 | 332 | 346 | 361 | 260 | 4/0 |
| 250 | . . . . | . . . . | . . . . | . . . . | 250 |
| 300 | . . . . | . . . . | . . . . | . . . . | 300 |
| 350 | . . . . | . . . . | . . . . | . . . . | 350 |
| 400 | . . . . | . . . . | . . . . | . . . . | 400 |
| 500 | . . . . | . . . . | . . . . | . . . . | 500 |
| 600 | . . . . | . . . . | . . . . | . . . . | 600 |
| 700 | . . . . | . . . . | . . . . | . . . . | 700 |
| 750 | . . . . | . . . . | . . . . | . . . . | 750 |
| 800 | . . . . | . . . . | . . . . | . . . . | 800 |
| 1000 | . . . . | . . . . | . . . . | . . . . | 1000 |
| 1500 | . . . . | . . . . | . . . . | . . . . | 1500 |
| 2000 | . . . . | . . . . | . . . . | . . . . | 2000 |

**Ampacity correction factors**

| Ambient temp., °C | For ambient temperatures other than 40°C (104°F), multiply the ampacities shown above by the appropriate factor shown below | | | | Ambient temp., °F |
|------|------|------|------|------|------|
| 41–50 | .95 | .97 | .98 | .95 | 106–122 |
| 51–60 | .90 | .94 | .95 | .90 | 124–140 |
| 61–70 | .85 | .90 | .93 | .85 | 142–158 |
| 71–80 | .80 | .87 | .90 | .80 | 160–176 |
| 81–90 | .74 | .83 | .87 | .74 | 177–194 |
| 91–100 | .67 | .79 | .85 | .67 | 195–212 |
| 101–120 | .52 | .71 | .79 | .52 | 213–248 |
| 121–140 | .30 | .61 | .72 | .30 | 249–284 |
| 141–160 | . . . . | .50 | .65 | . . . . | 285–320 |
| 161–180 | . . . . | .35 | .58 | . . . . | 321–356 |
| 181–200 | . . . . | . . . . | .49 | . . . . | 357–392 |
| 201–225 | . . . . | . . . . | .35 | . . . . | 393–437 |

[a]Reprinted with permission from NFPA 70-1990, National Electrical Code®, Copyright © 1990, National Fire Protection Association, Quincy, Massachusetts 02269. This reprinted material is not the complete and official position of the NFPA on the referenced subject, which is represented only by the standard in its entirety.

### 22. Ampacities for Single Insulated Conductors Rated 0–2000 V, 150 to 250°C, in Free Air Based on Ambient Temperature of 40°C (104°F)[a]
(Table 310-19, 1990 NEC)

| | Temperature rating of conductor | | | | | | |
| | 150°C (302°F) | 200°C (392°F) | | 250°C (482°F) | 150°C (302°F) | | |
| | Type Z | Types FEP, FEPB, PFA | Bare or covered conductors | Types PFAH, TFE | Type Z | Bare or covered conductors | |
| Size, AWG, kcmil | | Copper | | Nickel or nickel-coated copper | Aluminum or copper-clad aluminum | | Size, AWG, kcmil |
|---|---|---|---|---|---|---|---|
| 14 | 46 | 54 | 30 | 59 | .... | 25 | 14 |
| 12 | 60 | 68 | 35 | 78 | 47 | 30 | 12 |
| 10 | 80 | 90 | 50 | 107 | 63 | 35 | 10 |
| 8 | 106 | 124 | 70 | 142 | 83 | 55 | 8 |
| 6 | 155 | 165 | 95 | 205 | 112 | 75 | 6 |
| 4 | 190 | 220 | 125 | 278 | 148 | 100 | 4 |
| 3 | 214 | 252 | 150 | 327 | 170 | 120 | 3 |
| 2 | 255 | 293 | 175 | 381 | 198 | 135 | 2 |
| 1 | 293 | 344 | 200 | 440 | 228 | 160 | 1 |
| 1/0 | 339 | 399 | 235 | 532 | 263 | 185 | 1/0 |
| 2/0 | 390 | 467 | 275 | 591 | 305 | 215 | 2/0 |
| 3/0 | 451 | 546 | 320 | 708 | 351 | 250 | 3/0 |
| 4/0 | 529 | 629 | 370 | 830 | 411 | 285 | 4/0 |
| 250 | .... | .... | 415 | .... | .... | 325 | 250 |
| 300 | .... | .... | 460 | .... | .... | 360 | 300 |
| 350 | .... | .... | 520 | .... | .... | 405 | 350 |
| 400 | .... | .... | 560 | .... | .... | 435 | 400 |
| 500 | .... | .... | 635 | .... | .... | 495 | 500 |
| 600 | .... | .... | 710 | .... | .... | 560 | 600 |
| 700 | .... | .... | 780 | .... | .... | 615 | 700 |
| 750 | .... | .... | 805 | .... | .... | 635 | 750 |
| 800 | .... | .... | 835 | .... | .... | 660 | 800 |
| 900 | .... | .... | 865 | .... | .... | 715 | 900 |
| 1000 | .... | .... | 895 | .... | .... | 770 | 1000 |
| 1500 | .... | .... | 1205 | .... | .... | 980 | 1500 |
| 2000 | .... | .... | 1420 | .... | .... | 1215 | 2000 |

#### Ampacity correction factors

| Ambient temp., °C | For ambient temperatures other than 40°C (104°F), multiply the ampacities shown above by the appropriate factor shown below | | | | | | Ambient temp., °F |
|---|---|---|---|---|---|---|---|
| 41–50 | .95 | .97 | .... | .98 | .95 | .... | 106–122 |
| 51–60 | .90 | .94 | .... | .95 | .90 | .... | 124–140 |
| 61–70 | .85 | .90 | .... | .93 | .85 | .... | 142–158 |
| 71–80 | .80 | .87 | .... | .90 | .80 | .... | 160–176 |
| 81–90 | .74 | .83 | .... | .87 | .74 | .... | 177–194 |
| 91–100 | .67 | .79 | .... | .85 | .67 | .... | 195–212 |
| 101–120 | .52 | .71 | .... | .79 | .52 | .... | 213–248 |
| 121–140 | .30 | .61 | .... | .72 | .30 | .... | 249–284 |
| 141–160 | .... | .50 | .... | .65 | .... | .... | 285–320 |
| 161–180 | .... | .35 | .... | .58 | .... | .... | 321–356 |
| 181–200 | .... | .... | .... | .49 | .... | .... | 357–392 |
| 201–225 | .... | .... | .... | .35 | .... | .... | 393–437 |

**23. Ampacities of Insulated Aluminum or Copper-Clad Aluminum Conductors at Various Deratings**
In raceways or cables at room temperatures not over 30°C (86°F).

| Size, AWG or 1000 cmil | Ampacity | | | | | | | | | | | | | | |
|---|---|---|---|---|---|---|---|---|---|---|---|---|---|---|---|
| | 100% | | | 80% | | | 70% | | | 60% | | | 50% | | |
| | 60°C | 75°C | 90°C | 60°C | 75°C | 90°C | 60°C | 75°C | 90°C | 60°C | 75°C | 90°C | 60°C | 75°C | 90°C |
| **AWG** | | | | | | | | | | | | | | | |
| 12[a] | 20 | 20 | 25 | 16 | 16 | 20 | 14 | 14 | 18 | 12 | 12 | 15 | 10 | 10 | 13 |
| 10[a] | 25 | 30 | 35 | 20 | 24 | 28 | 17 | 21 | 25 | 15 | 18 | 21 | 12 | 15 | 18 |
| 8 | 30 | 40 | 45 | 24 | 32 | 36 | 21 | 28 | 32 | 18 | 24 | 27 | 15 | 20 | 23 |
| 6 | 40 | 50 | 60 | 32 | 40 | 48 | 28 | 35 | 42 | 24 | 30 | 36 | 20 | 25 | 30 |
| 4 | 55 | 65 | 75 | 44 | 52 | 60 | 38 | 45 | 53 | 33 | 39 | 45 | 27 | 32 | 38 |
| 3 | 65 | 75 | 85 | 52 | 60 | 68 | 45 | 52 | 60 | 39 | 45 | 51 | 32 | 37 | 43 |
| 2 | 75 | 90 | 100 | 60 | 72 | 80 | 52 | 63 | 70 | 45 | 54 | 60 | 37 | 45 | 50 |
| 1 | 85 | 100 | 115 | 68 | 80 | 92 | 59 | 70 | 81 | 51 | 60 | 69 | 42 | 50 | 58 |
| 1/0 | 100 | 120 | 135 | 80 | 96 | 108 | 70 | 84 | 95 | 60 | 72 | 81 | 50 | 60 | 68 |
| 2/0 | 115 | 135 | 150 | 92 | 108 | 120 | 80 | 94 | 105 | 69 | 81 | 90 | 57 | 67 | 75 |
| 3/0 | 130 | 155 | 175 | 104 | 124 | 140 | 91 | 108 | 123 | 78 | 93 | 105 | 65 | 77 | 88 |
| 4/0 | 150 | 180 | 205 | 120 | 144 | 164 | 105 | 126 | 144 | 90 | 108 | 123 | 75 | 90 | 103 |
| 250 kcmil | 170 | 205 | 230 | 136 | 164 | 184 | 119 | 143 | 161 | 102 | 123 | 138 | 85 | 102 | 115 |
| 300 | 190 | 230 | 255 | 152 | 184 | 204 | 133 | 161 | 179 | 114 | 138 | 153 | 95 | 115 | 128 |
| 350 | 210 | 250 | 280 | 168 | 200 | 224 | 147 | 175 | 196 | 126 | 150 | 168 | 105 | 125 | 140 |
| 400 | 225 | 270 | 305 | 180 | 216 | 244 | 157 | 189 | 214 | 135 | 162 | 183 | 112 | 135 | 153 |
| 500 | 260 | 310 | 350 | 208 | 248 | 280 | 182 | 217 | 245 | 156 | 186 | 210 | 130 | 155 | 175 |

[a]The overcurrent shall not exceed 15 A for 12 AWG and 25 A for 10 AWG after any correction factors for ambient temperatures and number of conductors have been applied.

NOTE  60°C conductors: TW, UF; 75°C conductors: RH, RHW, THHW, THW, THWN, XHHW (wet locations), USE; 90°C conductors: RHH, THHN, THHW, THW, XHHW (dry or damp locations).

**24. Ampacity of flexible cords and cables.**  Tables 400-5A and 400-5B of the NEC (Sec. 47) give the allowable ampacity for not more than three current-carrying copper conductors in a cord. If the number of current-carrying conductors in a cord exceeds three, the allowable ampacity of each conductor shall be reduced as shown in the following table.

| Column A | | Column B[a] | |
|---|---|---|---|
| Number of conductors | Percent of values in Tables 400-5(A) and 400-5(B) | Number of conductors | Percent of values in Tables 400-5(A) and 400-5(B) |
| 4 through 6 | 80 | 4 through 6 | 80 |
| 7 through 9 | 70 | 7 through 9 | 70 |
| 10 through 24[b] | 70 | 10 through 20 | 50 |
| 25 through 42[b] | 60 | 21 through 30 | 45 |
| 43 and above[b] | 50 | 31 through 40 | 40 |
| | | 41 through 60 | 35 |

[a]No diversity.
[b]These factors include the effects of a load diversity of 50 percent.

A conductor used for equipment grounding and a neutral conductor which carries only the unbalanced current from other conductors, as in the case of normally balanced circuits of three or more conductors, shall not be considered current-carrying conductors.

Where a single conductor is used for both equipment grounding and to carry unbalanced current from other conductors, as provided in Sec. 250-60 of the NEC for electric ranges and electric clothes dryers, it shall be considered as a current-carrying conductor.

**Ampacity of Flexible Cords and Cables**[a]
(Table 400-5(A), 1990 NEC)
Based on ambient temperature of 30°C (86°F). See Section 400-13 and Table 400-4 in the NEC.

| Size, AWG | Thermoset types TS Thermoplastic types TPT, TST | Thermoset types C, E, DO, PD, S, SJ, SJO, SJOO, SO, SOO, SP-1, SP-2, SP-3, SRD, SV, SVO, SVOO Thermoplastic types ET, ETT, ETLB, SE, SEO, SJE, SJEO, SJT, SJTO, SJTOO, SPE-1, SPE-2, SPE-3, SPT-1, SPT-2, SPT-3, ST, STO, STOO, SRDE, SRDT, SVE, SVEO, SVT, SVTO, SVTOO | | Types AFS, AFSJ, HPD, HPN, HS, HSJ, HSJO, HSO | Asbestos types AFC[b] AFPD[b] AFPO |
|---|---|---|---|---|---|
| | | $A^c$ | $B^c$ | | |
| 27[d] | 0.5 | | | | |
| 20 | ... | 5[e] | 7[e] | | |
| 18 | ... | 7 | 10 | 10 | 6 |
| 17 | ... | ... | 12 | | |
| 16 | ... | 10 | 13 | 15 | 8 |
| 15 | ... | ... | ... | 17 | |
| 14 | ... | 15 | 18 | 20 | 17 |
| 12 | ... | 20 | 25 | 30 | 23 |
| 10 | ... | 25 | 30 | 35 | 28 |
| 8 | ... | 35 | 40 | | |
| 6 | ... | 45 | 55 | | |
| 4 | ... | 60 | 70 | | |
| 2 | ... | 80 | 95 | | |

[a]Reprinted with permission from NFPA 70-1990, National Electrical Code®, Copyright © 1990, National Fire Protection Association. Quincy, Massachusetts 02269. This reprinted material is not the complete and official position of the NFPA on the referenced subject, which is represented only by the standard in its entirety.

NOTE   *Ultimate insulation temperature.* In no case shall conductors be associated together in such a way with respect to the kind of circuit, wiring method used, or number of conductors that the limiting temperature of the conductors will be exceeded.

[b]These types are used almost exclusively in fixtures where they are exposed to high temperatures, and ampere ratings are assigned accordingly.

[c]The ampacities under subheading *A* apply to three-conductor cords and other multiconductor cords connected to utilization equipment so that only three conductors are current-carrying. The ampacities under subheading *B* apply to two-conductor cords and other multiconductor cords connected to utilization equipment so that only two conductors are current-carrying.

[d]Tinsel cord.

[e]Elevator cables only.

**Ampacity of Cables Types W and G**

(Table 400-5(B), 1990 NEC)

Based on ambient temperature of 30°C (86°F). See Table 400-4 in the NEC.

| Size, AWG, kcmil | Temperature rating of cable | | | | | | | | |
|---|---|---|---|---|---|---|---|---|---|
| | 60°C (140°F) | | | 75°C (167°F) | | | 90°C (194°F) | | |
| | D | E | F | D | E | F | D | E | F |
| 8 | 60 | 55 | 48 | 70 | 65 | 57 | 80 | 74 | 65 |
| 6 | 80 | 72 | 63 | 95 | 88 | 77 | 105 | 99 | 87 |
| 4 | 105 | 96 | 84 | 125 | 115 | 101 | 140 | 130 | 114 |
| 3 | 120 | 113 | 99 | 145 | 135 | 118 | 165 | 152 | 133 |
| 2 | 140 | 128 | 112 | 170 | 152 | 133 | 190 | 174 | 152 |
| 1 | 165 | 150 | 131 | 195 | 178 | 156 | 220 | 202 | 177 |
| 1/0 | 195 | 173 | 151 | 230 | 207 | 181 | 260 | 234 | 205 |
| 2/0 | 225 | 199 | 174 | 265 | 238 | 208 | 300 | 271 | 237 |
| 3/0 | 260 | 230 | 201 | 310 | 275 | 241 | 350 | 313 | 274 |
| 4/0 | 300 | 265 | 232 | 360 | 317 | 277 | 405 | 361 | 316 |
| 250 | 340 | 296 | 259 | 405 | 354 | 310 | 455 | 402 | 352 |
| 300 | 375 | 330 | 289 | 445 | 395 | 346 | 505 | 449 | 393 |
| 350 | 420 | 363 | 318 | 505 | 435 | 381 | 570 | 495 | 433 |
| 400 | 455 | 392 | 343 | 545 | 469 | 410 | 615 | 535 | 468 |
| 500 | 515 | 448 | 392 | 620 | 537 | 470 | 700 | 613 | 536 |

The ampacities under subheading D are the allowable ampacities for single-conductor Types W cable only where the individual conductors are not installed in raceways and are not in physical contact with each other except in lengths not to exceed 24 in (610 mm) where passing through the wall of an enclosure.

The ampacities under subheading E apply to two-conductor cables and other multiconductor cables connected to utilization equipment so that only two conductors are current carrying. The ampacities under subheading F apply to three-conductor cables and other multiconductor cables connected to utilization equipment so that only three conductors are current carrying.

NOTE    *Ultimate insulation temperature.* In no case shall conductors be associated together in such a way with respect to the kind of circuit, the wiring method used, or the number of conductors that the limiting temperature of the conductors will be exceeded.

**25. Ampacity of fixture wires** (Table 402-5, 1990 NEC).[a] The ampacity of fixture wire shall not exceed the following:

| Size, AWG | Ampacity |
|---|---|
| 18 | 6 |
| 16 | 8 |
| 14 | 17 |
| 12 | 23 |
| 10 | 28 |

[a]Reprinted with permission from NFPA 70-1990, National Electrical Code®, Copyright © 1990, National Fire Protection Association, Quincy, Massachusetts 02269. This reprinted material is not the complete and official position of the NFPA on the referenced subject, which is represented only by the standard in its entirety.

**26. Curves for Determining Ampacity of Aluminum Cable, Steel-Reinforced**
(Aluminum Co. of America)

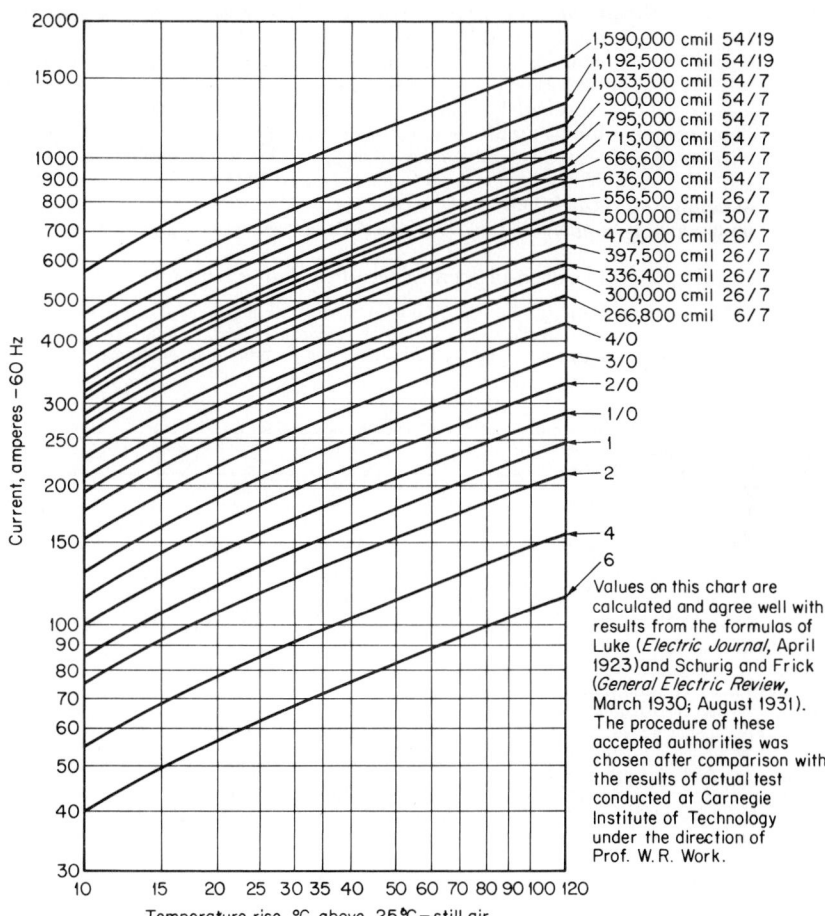

**1,590,000 cmil 54/19**
**1,192,500 cmil 54/19**
**1,033,500 cmil 54/7**
**900,000 cmil 54/7**
**795,000 cmil 54/7**
**715,000 cmil 54/7**
**666,600 cmil 54/7**
**636,000 cmil 54/7**
**556,500 cmil 26/7**
**500,000 cmil 30/7**
**477,000 cmil 26/7**
**397,500 cmil 26/7**
**336,400 cmil 26/7**
**300,000 cmil 26/7**
**266,800 cmil 6/7**
**4/0**
**3/0**
**2/0**
**1/0**
**1**
**2**
**4**
**6**

Values on this chart are calculated and agree well with results from the formulas of Luke (*Electric Journal*, April 1923) and Schurig and Frick (*General Electric Review*, March 1930; August 1931). The procedure of these accepted authorities was chosen after comparison with the results of actual test conducted at Carnegie Institute of Technology under the direction of Prof. W. R. Work.

Current, amperes – 60 Hz

Temperature rise, °C above 25 °C – still air

**27. Ampacities of Parkway Cables Buried Directly in Ground**[a]

| Conductor size, AWG or 1000 cmil | Maximum allowable ampacity per conductor (copper) | | Conductor size, AWG or 1000 cmil | Maximum allowable ampacity per conductor (copper) | |
|---|---|---|---|---|---|
| | Single-conductor cable, A | Three-conductor cable, A | | Single-conductor cable, A | Three-conductor cable, A |
| 8 AWG | 85 | 60 | 3/0 AWG | 335 | 225 |
| 6 | 115 | 80 | 4/0 | 385 | 255 |
| 4 | 150 | 100 | 250 MCM | 445 | 285 |
| 2 | 195 | 130 | 300 | 470 | 315 |
| 1 | 220 | 150 | 350 | 515 | 345 |
| 1/0 | 255 | 170 | 400 | 555 | 375 |
| 2/0 | 295 | 195 | 500 | 635 | 425 |

[a]Table based on earth temperature of 20°C (68°F). Rubber-insulated conductors in accordance with ICEA specification. Average soil condition: sandy loam, 10 percent moisture. Average depth of burial 18 in. Other cables sufficiently far away to prevent mutual heating. Rated voltages up to 5 kV, grounded neutral.

### 28. Notes to Tables 29 through 36

1. Tables 31, 32, and 33 apply only to complete conduit or tubing systems and are not intended to apply to short sections of conduit or to tubing used to protect exposed wiring from physical damage.

2. Equipment-grounding or bonding conductors, when installed, shall be included when calculating conduit or tubing fill. The actual dimensions of the equipment-grounding or bonding conductor (insulated or bare) shall be used in the calculation.

3. When conduit or tubing nipples having a maximum length not to exceed 24 in are installed between boxes, cabinets, and similar enclosures, the nipple shall be permitted to be filled to 60 percent of its total cross-sectional area, and Note 8 of Tables 18 through 22 does not apply to this condition.

4. For conductors not included in Tables 30 through 36, actual dimensions shall be used.

5. See Table 29 for the allowable percentage of conduit or tubing fill.

Table 29 is based on common conditions of proper cabling and alignment of conductors where the length of the pull and the number of bends are within reasonable limits. It should be recognized that for certain conditions a larger size conduit or a lesser conduit fill should be considered.

### 29. Percentage of Cross Section of Conduit and Tubing for Conductors
See Table 30 for fixture wires.

| Number of conductors | 1 | 2 | 3 | 4 | Over 4 |
|---|---|---|---|---|---|
| All conductor types except lead-covered | 53 | 31 | 40 | 40 | 40 |
| Lead-covered conductors | 55 | 30 | 40 | 38 | 35 |

NOTE 1   See Tables 31, 32, and 33 for number of conductors all of the same size in trade sizes of conduit or tubing ½-in through 6-in.

NOTE 2   For conductors larger than 750,000 cmil or for combinations of conductors of different sizes, use Tables 34 through 36 for the dimensions of conductors, conduit, and tubing.

NOTE 3   When the calculated number of conductors, all of the same size, includes a decimal fraction, the next higher whole number shall be used if this decimal is 0.8 or larger.

NOTE 4   When bare conductors are permitted by other sections of the NEC, the dimensions for bare conductors in Table 61 shall be permitted.

NOTE 5   A multiconductor cable of two or more conductors shall be treated as a single-conductor cable for calculating the percentage of conduit fill area. For cables that have an elliptical cross section, the cross-sectional–area calculation shall be based on using the major diameter of the ellipse as a circle diameter.

## 30. Maximum Number of Fixture Wires in Trade Sizes of Conduit or Tubing[a]
(Table 2, Chap. 9, 1990 NEC)

| Conduit trade size, in | ½ | | | | | ¾ | | | | | 1 | | | | | 1¼ | | | | | 1½ | | | | | 2 | | | | |
|---|---|---|---|---|---|---|---|---|---|---|---|---|---|---|---|---|---|---|---|---|---|---|---|---|---|---|---|---|---|---|
| Wire types | 18 | 16 | 14 | 12 | 10 | 18 | 16 | 14 | 12 | 10 | 18 | 16 | 14 | 12 | 10 | 18 | 16 | 14 | 12 | 10 | 18 | 16 | 14 | 12 | 10 | 18 | 16 | 14 | 12 | 10 |
| PTF, PTFF, PGFF, PGF, PFF, PF, PAF, PAFF, ZF, ZFF | | | | | | | | | | | | | | | | | | | | | | | | | | | | | | |
| ZFF | 23 | 18 | 14 | | | 40 | 31 | 24 | | | 65 | 50 | 39 | | | 115 | 90 | 70 | | | 157 | 122 | 95 | | | 257 | 200 | 156 | | |
| TFFN, TFN | 19 | 15 | | | | 34 | 26 | | | | 55 | 43 | | | | 97 | 76 | | | | 132 | 104 | | | | 216 | 169 | | | |
| SF-1 | 16 | | | | | 29 | | | | | 47 | | | | | 83 | | | | | 114 | | | | | 186 | | | | |
| SFF-1 | 15 | | | | | 26 | | | | | 43 | | | | | 76 | | | | | 104 | | | | | 169 | | | | |
| CF | 13 | 10 | 8 | | | 23 | 18 | 14 | 7 | 6 | 38 | 30 | 23 | 12 | 9 | 66 | 53 | 40 | 21 | 16 | 91 | 72 | 55 | 29 | 22 | 149 | 118 | 90 | 48 | 37 |
| TF | 11 | 10 | | | | 20 | 18 | | | | 32 | 30 | | | | 57 | 53 | | | | 79 | 72 | | | | 129 | 118 | | | |
| RFH-1 | 11 | | | | | 20 | | | | | 32 | | | | | 57 | | | | | 79 | | | | | 129 | | | | |
| TFF | 11 | 10 | 7 | | | 20 | 17 | 12 | | | 32 | 27 | 20 | | | 56 | 49 | 36 | | | 77 | 66 | 49 | | | 126 | 109 | 81 | | |
| AF | 11 | 9 | 7 | 4 | 3 | 19 | 16 | 12 | 7 | 5 | 31 | 26 | 20 | 11 | 8 | 55 | 46 | 36 | 19 | 15 | 75 | 63 | 49 | 27 | 20 | 123 | 104 | 81 | 44 | 34 |
| SFF-2 | 9 | 7 | 6 | | | 16 | 12 | 10 | | | 27 | 20 | 17 | | | 47 | 36 | 30 | | | 65 | 49 | 42 | | | 106 | 81 | 68 | | |
| SF-2 | 9 | 8 | 6 | | | 16 | 14 | 11 | | | 27 | 23 | 18 | | | 47 | 40 | 32 | | | 65 | 55 | 43 | | | 106 | 90 | 71 | | |
| FFH-2 | 9 | 7 | | | | 15 | 12 | | | | 25 | 19 | | | | 44 | 34 | | | | 60 | 46 | | | | 99 | 75 | | | |
| RFH-2 | 7 | 5 | | | | 12 | 10 | | | | 20 | 16 | | | | 36 | 28 | | | | 49 | 38 | | | | 80 | 62 | | | |
| KF-1, KFF-1, KF-2, KFF-2 | 36 | 32 | 22 | 14 | 9 | 64 | 55 | 39 | 25 | 17 | 103 | 89 | 63 | 41 | 28 | 182 | 158 | 111 | 73 | 49 | 248 | 216 | 152 | 100 | 67 | 406 | 353 | 248 | 163 | 110 |

[a] 40 percent fill, based on individual diameters. Reprinted with permission from NFPA 70-1990, National Electrical Code®, Copyright © 1990, National Fire Protection Association, Quincy, Massachusetts 02269. This reprinted material is not the complete and official position of the NFPA on the referenced subject, which is represented only by the standard in its entirety.

## 31. Maximum Number of Conductors in Trade Sizes of Conduit or Tubing[a]
(Table 3A, Chap. 9, 1990 NEC)

| Type letters | Conductor size, AWG or 1000 cmil | ½ | ¾ | 1 | 1¼ | 1½ | 2 | 2½ | 3 | 3½ | 4 | 5 | 6 |
|---|---|---|---|---|---|---|---|---|---|---|---|---|---|
| TW, XHHW (14 through 8) | 14 AWG | 9 | 15 | 25 | 44 | 60 | 99 | 142 | 171 | | | | |
| | 12 | 7 | 12 | 19 | 35 | 47 | 78 | 111 | 131 | 176 | | | |
| | 10 | 5 | 9 | 15 | 26 | 36 | 60 | 85 | | | | | |
| | 8 | 2 | 4 | 7 | 12 | 17 | 28 | 40 | 62 | 84 | 108 | | |
| RHW and RHH (without outer covering), THW | 14 | 6 | 10 | 16 | 29 | 40 | 65 | 93 | 143 | 192 | | | |
| | 12 | 4 | 8 | 13 | 24 | 32 | 53 | 76 | 117 | 157 | | | |
| | 10 | 4 | 6 | 11 | 19 | 26 | 43 | 61 | 95 | 127 | 163 | | |
| | 8 | 1 | 3 | 5 | 10 | 13 | 22 | 32 | 49 | 66 | 85 | 133 | |
| TW, THW, FEPB (6 through 2), RHW and RHH (without outer covering) | 6 | 1 | 2 | 4 | 7 | 10 | 16 | 23 | 36 | 48 | 62 | 97 | 141 |
| | 4 | 1 | 1 | 3 | 5 | 7 | 12 | 17 | 27 | 36 | 47 | 73 | 106 |
| | 3 | 1 | 1 | 2 | 4 | 6 | 10 | 15 | 23 | 31 | 40 | 63 | 91 |
| | 2 | 1 | 1 | 2 | 4 | 5 | 9 | 13 | 20 | 27 | 34 | 54 | 78 |
| | 1 | | 1 | 1 | 3 | 4 | 6 | 9 | 14 | 19 | 25 | 39 | 57 |
| | 1/0 | ... | ... | 1 | 2 | 3 | 5 | 8 | 12 | 16 | 21 | 33 | 49 |
| | 2/0 | ... | ... | 1 | 1 | 3 | 5 | 7 | 10 | 14 | 18 | 29 | 41 |
| | 3/0 | ... | ... | 1 | 1 | 2 | 4 | 6 | 9 | 12 | 15 | 24 | 35 |
| | 4/0 | ... | ... | 1 | 1 | 1 | 3 | 5 | 7 | 10 | 13 | 20 | 29 |
| | 250 MCM | ... | ... | ... | 1 | 1 | 2 | 4 | 6 | 8 | 10 | 16 | 23 |
| | 300 | ... | ... | ... | 1 | 1 | 2 | 3 | 5 | 7 | 9 | 14 | 20 |
| | 350 | ... | ... | ... | 1 | 1 | 1 | 3 | 4 | 6 | 8 | 12 | 18 |
| | 400 | ... | ... | ... | 1 | 1 | 1 | 2 | 4 | 5 | 7 | 11 | 16 |
| | 500 | ... | ... | ... | ... | 1 | 1 | 1 | 3 | 4 | 6 | 9 | 14 |
| | 600 | ... | ... | ... | ... | 1 | 1 | 1 | 3 | 4 | 5 | 7 | 11 |
| | 700 | ... | ... | ... | ... | 1 | 1 | 1 | 2 | 3 | 4 | 7 | 10 |
| | 750 | ... | ... | ... | ... | 1 | 1 | 2 | 2 | 3 | 4 | 6 | 9 |

## 32. Maximum Number of Conductors in Trade Sizes of Conduit or Tubing[a]
(Table 3B, Chap. 9, 1990 NEC)

| Type letters | Conductor size, AWG or 1000 cmil | ½ | ¾ | 1 | 1¼ | 1½ | 2 | 2½ | 3 | 3½ | 4 | 5 | 6 |
|---|---|---|---|---|---|---|---|---|---|---|---|---|---|
| THWN, THHN, FEP (14 through 2), FEPB (14 through 8), PFA (14 through 4/0), PFAH (14 through 4/0), Z (14 through 4/0), XHHW (4 through 500 kcmil) | 14 AWG | 13 | 24 | 39 | 69 | 94 | 154 | | | | | | |
| | 12 | 10 | 18 | 29 | 51 | 70 | 114 | 164 | | | | | |
| | 10 | 6 | 11 | 18 | 32 | 44 | 73 | 104 | 160 | | | | |
| | 8 | 3 | 5 | 9 | 16 | 22 | 36 | 51 | 79 | 106 | 136 | | |
| | 6 | 1 | 4 | 6 | 11 | 15 | 26 | 37 | 57 | 76 | 98 | 154 | |
| | 4 | 1 | 2 | 4 | 7 | 9 | 16 | 22 | 35 | 47 | 60 | 94 | 137 |
| | 3 | 1 | 1 | 3 | 6 | 8 | 13 | 19 | 29 | 39 | 51 | 80 | 116 |
| | 2 | … | 1 | 3 | 5 | 7 | 11 | 16 | 25 | 33 | 43 | 67 | 97 |
| | 1 | … | 1 | 1 | 3 | 5 | 8 | 12 | 18 | 25 | 32 | 50 | 72 |
| | 1/0 | … | 1 | 1 | 3 | 4 | 7 | 10 | 15 | 21 | 27 | 42 | 61 |
| | 2/0 | … | 1 | 1 | 2 | 3 | 6 | 8 | 13 | 17 | 22 | 35 | 51 |
| | 3/0 | … | … | 1 | 1 | 3 | 5 | 7 | 11 | 14 | 18 | 29 | 42 |
| | 4/0 | … | … | 1 | 1 | 2 | 4 | 6 | 9 | 12 | 15 | 24 | 35 |
| | 250 MCM | … | … | … | 1 | 1 | 3 | 4 | 7 | 10 | 12 | 20 | 28 |
| | 300 | … | … | … | 1 | 1 | 3 | 4 | 6 | 8 | 11 | 17 | 24 |
| | 350 | … | … | … | 1 | 1 | 2 | 3 | 5 | 7 | 9 | 15 | 21 |
| | 400 | … | … | … | 1 | 1 | 1 | 3 | 5 | 6 | 8 | 13 | 19 |
| | 500 | … | … | … | … | 1 | 1 | 2 | 4 | 5 | 7 | 11 | 16 |
| | 600 | … | … | … | … | 1 | 1 | 1 | 3 | 4 | 5 | 9 | 13 |
| | 700 | … | … | … | … | 1 | 1 | 1 | 3 | 4 | 5 | 8 | 11 |
| | 750 | … | … | … | … | 1 | 1 | 1 | 2 | 3 | 4 | 7 | 11 |
| **XHHW** | 6 AWG | 1 | 3 | 5 | 9 | 13 | 21 | 30 | 47 | 63 | 81 | 128 | 185 |
| | 600 MCM | … | … | … | 1 | 1 | 1 | 1 | 3 | 4 | 5 | 9 | 13 |
| | 700 | … | … | … | … | 1 | 1 | 1 | 3 | 4 | 5 | 7 | 11 |
| | 750 | … | … | … | … | 1 | 1 | 1 | 2 | 3 | 4 | 7 | 10 |

[a]Based on Table 29. Reprinted with permission from NFPA 70-1990, National Electrical Code®, Copyright © 1990, National Fire Protection Association, Quincy, Massachusetts 02269. This reprinted material is not the complete and official position of the NFPA on the referenced subject, which is represented only by the standard in its entirety.

NOTE  This table is for concentric stranded conductors only. For cables with compact conductors, the dimensions in Table 5B NEC, (Sec. 35) shall be used.

**33. Maximum Number of Conductors in Trade Sizes of Conduit or Tubing**[a]
(Table 3C, Chap. 9, 1990 NEC)

| Conductor size, AWG or 1000 cmil | ½ | ¾ | 1 | 1¼ | 1½ | 2 | 2½ | 3 | 3½ | 4 | 5 | 6 |
|---|---|---|---|---|---|---|---|---|---|---|---|---|
| **Type letters** | | | | | | | | | | | | |
| **RHW, RHH (with outer covering)** | | | | | | | | | | | | |
| 14 AWG | 3 | 6 | 10 | 18 | 25 | 41 | 58 | 90 | 121 | 155 | | |
| 12 | 3 | 5 | 9 | 15 | 21 | 35 | 50 | 77 | 103 | 132 | | |
| 10 | 2 | 4 | 7 | 13 | 18 | 29 | 41 | 64 | 86 | 110 | | |
| 8 | 1 | 2 | 4 | 7 | 9 | 16 | 22 | 35 | 47 | 60 | 94 | 137 |
| 6 | 1 | 1 | 2 | 5 | 6 | 11 | 15 | 24 | 32 | 41 | 64 | 93 |
| 4 | 1 | 1 | 1 | 3 | 5 | 8 | 12 | 18 | 24 | 31 | 50 | 72 |
| 3 | 1 | 1 | 1 | 3 | 4 | 7 | 10 | 16 | 22 | 28 | 44 | 63 |
| 2 | 1 | 1 | 1 | 3 | 4 | 6 | 9 | 14 | 19 | 24 | 38 | 56 |
| 1 | | 1 | 1 | 1 | 3 | 5 | 7 | 11 | 14 | 18 | 29 | 42 |
| 1/0 | | 1 | 1 | 1 | 2 | 4 | 6 | 9 | 12 | 16 | 25 | 37 |
| 2/0 | | | 1 | 1 | 1 | 3 | 5 | 8 | 11 | 14 | 22 | 32 |
| 3/0 | | | 1 | 1 | 1 | 3 | 4 | 7 | 9 | 12 | 19 | 28 |
| 4/0 | | | 1 | 1 | 1 | 2 | 4 | 6 | 8 | 10 | 16 | 24 |
| 250 MCM | | | | | | 1 | 3 | 5 | 6 | 8 | 13 | 19 |
| 300 | | | | | | 1 | 3 | 4 | 5 | 7 | 11 | 17 |
| 350 | | | | | | 1 | 2 | 4 | 5 | 6 | 10 | 15 |
| 400 | | | | | | 1 | 1 | 3 | 4 | 6 | 9 | 14 |
| 500 | | | | | | 1 | 1 | 3 | 4 | 5 | 8 | 11 |
| 600 | | | | | | 1 | 1 | 2 | 3 | 4 | 6 | 9 |
| 700 | | | | | | 1 | 1 | 1 | 3 | 3 | 6 | 8 |
| 750 | | | | | | 1 | 1 | 1 | 3 | 3 | 5 | 8 |

[a]Based on Table 29. Reprinted with permission from NFPA 70-1990, National Electrical Code®, Copyright © 1990, National Fire Protection Association, Quincy, Massachusetts 02269. This reprinted material is not the complete and official position of the NFPA on the referenced subject, which is represented only by the standard in its entirety.

NOTE This table is for concentric stranded conductors only. For cables with compact conductors, the dimensions in Table 5B NEC (Sec. 35) shall be used.

**34. Dimensions and Percent Area of Conduit and of Tubing**[a]
(Table 4, Chap. 9, 1990 NEC)
Areas of conduit or tubing for the combinations of wires permitted in Table 29.

| Trade size | Internal diameter, in | Total, 100% | Not lead-covered | | | One conductor, 55% | Lead-covered | | | |
|---|---|---|---|---|---|---|---|---|---|---|
| | | | Two conductors, 31% | Over two conductors, 40% | One conductor, 53% | | Two conductors, 30% | Three conductors, 40% | Four conductors, 38% | Over four conductors, 35% |
| ½ | .622 | .30 | .09 | .12 | .16 | .17 | .09 | .12 | .11 | .11 |
| ¾ | .824 | .53 | .16 | .21 | .28 | .29 | .16 | .21 | .20 | .19 |
| 1 | 1.049 | .86 | .27 | .34 | .46 | .47 | .26 | .34 | .33 | .30 |
| 1¼ | 1.380 | 1.50 | .47 | .60 | .80 | .83 | .45 | .60 | .57 | .53 |
| 1½ | 1.610 | 2.04 | .63 | .82 | 1.08 | 1.12 | .61 | .82 | .78 | .71 |
| 2 | 2.067 | 3.36 | 1.04 | 1.34 | 1.78 | 1.85 | 1.01 | 1.34 | 1.28 | 1.18 |
| 2½ | 2.469 | 4.79 | 1.48 | 1.92 | 2.54 | 2.63 | 1.44 | 1.92 | 1.82 | 1.68 |
| 3 | 3.068 | 7.38 | 2.29 | 2.95 | 3.91 | 4.06 | 2.21 | 2.95 | 2.80 | 2.58 |
| 3½ | 3.548 | 9.90 | 3.07 | 3.96 | 5.25 | 5.44 | 2.97 | 3.96 | 3.76 | 3.47 |
| 4 | 4.026 | 12.72 | 3.94 | 5.09 | 6.74 | 7.00 | 3.82 | 5.09 | 4.83 | 4.45 |
| 5 | 5.047 | 20.00 | 6.20 | 8.00 | 10.60 | 11.00 | 6.00 | 8.00 | 7.60 | 7.00 |
| 6 | 6.065 | 28.89 | 8.96 | 11.56 | 15.31 | 15.89 | 8.67 | 11.56 | 10.98 | 10.11 |

[a]Reprinted with permission from NFPA 70-1990, National Electrical Code®, Copyright © 1990, National Fire Protection Association, Quincy, Massachusetts 02269. This reprinted material is not the complete and official position of the NFPA on the referenced subject, which is represented only by the standard in its entirety.

**35. Dimensions of Rubber-Covered and Thermoplastic-Covered Conductors[a]**
(Table 5, Chap. 9, 1990 NEC)

| Size, AWG or 1000 cmil (1) | Types RFH-2, RH, RHH,[b] RHW,[b] SF-2 | | Types TF, THW,[c] TW | | Types TFN, THHN, THWN | | Types[e] FEP, FEPB, FEPW, TFE, PF, PFA, PFAH, PGF, PTF, Z, ZF, ZFF | | Type XHHW, ZW[f] | | Types KF-1, KF-2, KFF-1, KFF-2 | |
|---|---|---|---|---|---|---|---|---|---|---|---|---|
| | Approximate diameter, in (2) | Approximate area, in² (3) | Approximate diameter, in (4) | Approximate area, in² (5) | Approximate diameter, in (6) | Approximate area, in² (7) | Approximate diameter, in (8) | Approximate area, in² (9) | Approximate diameter, in (10) | Approximate area, in² (11) | Approximate diameter, in (12) | Approximate area, in² (13) |
| 18 AWG | .146 | .0167 | .106 | .0088 | .089 | .0062 | .081 | .0052 | .... | .... | .065 | .0033 |
| 16 | .158 | .0196 | .118 | .0109 | .100 | .0079 | .092 | .0066 | .... | .... | .070 | .0038 |
| 14 | 30 mil .171 | .0230 | .131 | .0135 | .105 | .0087 | .105 .105 | .0087 .0087 | | | .083 | .0054 |
| 14 | 45 mil .204[g] | .0327[g] | | | | | | | | | | |
| 14 | | | .162[c] | .0206[c] | | | | | .129 | .0131 | | |
| 12 | 30 mil .188 | .0278 | .148 | .0172 | .122 | .0117 | .121 .121 | .0115 .0115 | | | .102 | .0082 |
| 12 | 45 mil .221[g] | .0384[g] | .179[c] | .0252[c] | | | | | .146 | .0167 | | |
| 10 | .242 | .0460 | .168 | .0222 | .153 | .0184 | .142 .142 | .0158 .0158 | | | .124 | .0121 |
| 10 | | | .199[c] | .0311[c] | | | | | .166 | .0216 | | |
| 8 | .328 | .0845 | .245 | .0471 | .218 | .0373 | .206 .186 | .0333 .0272 | | | | |
| 8 | | | .276[c] | .0598[c] | | | | | .241 | .0456 | | |
| 6 | .397 | .1238 | .323 | .0819 | .257 | .0519 | .244 .302 | .0468 .0716 | .282 | .0625 | | |
| 4 | .452 | .1605 | .372 | .1087 | .328 | .0845 | .292 .350 | .0670 .0962 | .328 | .0845 | | |
| 3 | .481 | .1817 | .401 | .1263 | .356 | .0995 | .320 .378 | .0804 .1122 | .356 | .0995 | | |
| 2 | .513 | .2067 | .433 | .1473 | .388 | .1182 | .352 .410 | .0973 .1320 | .388 | .1182 | | |
| 1 | .588 | .2715 | .508 | .2027 | .450 | .1590 | .420 .... | .1385 .... | .450 | .1590 | | |

| | | | | | | | | | |
|---|---|---|---|---|---|---|---|---|---|
| 1/0 | .629 | .3107 | .549 | .2367 | .491 | .1893 | .462 ..... | .1676 ..... | .491 | .1893 |
| 2/0 | .675 | .3578 | .595 | .2781 | .537 | .2265 | .498 ..... | .1948 ..... | .537 | .2265 |
| 3/0 | .727 | .4151 | .647 | .3288 | .588 | .2715 | .560 ..... | .2463 ..... | .588 | .2715 |
| 4/0 | .785 | .4840 | .705 | .3904 | .646 | .3278 | .618 ..... | .3000 ..... | .646 | .3278 |
| 250 MCM | .868 | .5917 | .788 | .4877 | .716 | .4026 | ..... | ..... | .716 | .4026 |
| 300 | .933 | .6837 | .843 | .5581 | .771 | .4669 | ..... | ..... | .771 | .4669 |
| 350 | .985 | .7620 | .895 | .6291 | .822 | .5307 | ..... | ..... | .822 | .5307 |
| 400 | 1.032 | .8365 | .942 | .6969 | .869 | .5931 | ..... | ..... | .869 | .5931 |
| 500 | 1.119 | .9834 | 1.029 | .8316 | .955 | .7163 | ..... | ..... | .955 | .7163 |
| 600 | 1.233 | 1.1940 | 1.143 | 1.0261 | 1.058 | .8791 | ..... | ..... | 1.073 | .9043 |
| 700 | 1.304 | 1.3355 | 1.214 | 1.1575 | 1.129 | 1.0011 | ..... | ..... | 1.145 | 1.0297 |
| 750 | 1.339 | 1.4082 | 1.249 | 1.2252 | 1.163 | 1.0623 | ..... | ..... | 1.180 | 1.0936 |
| 800 | 1.372 | 1.4784 | 1.282 | 1.2908 | 1.196 | 1.1234 | ..... | ..... | 1.210 | 1.1499 |
| 900 | 1.435 | 1.6173 | 1.345 | 1.4208 | 1.259 | 1.2449 | ..... | ..... | 1.270 | 1.2668 |
| 1000 | 1.494 | 1.7530 | 1.404 | 1.5482 | 1.317 | 1.3623 | ..... | ..... | 1.330 | 1.3893 |
| 1250 | 1.676 | 2.2062 | 1.577 | 1.9532 | ..... | ..... | ..... | ..... | 1.500 | 1.7671 |
| 1500 | 1.801 | 2.5475 | 1.702 | | ..... | ..... | ..... | ..... | 1.620 | 2.0612 |
| 1750 | 1.916 | 2.8832 | 1.817 | 2.5930 | ..... | ..... | ..... | ..... | 1.740 | 2.3779 |
| 2000 | 2.021 | 3.2079 | 1.922 | 2.9013 | ..... | ..... | ..... | ..... | 1.840 | 2.6590 |

[a]Reprinted with permission from NFPA 70-1990, National Electrical Code®, Copyright © 1990, National Fire Protection Association, Quincy, Massachusetts 02269. This reprinted material is not the complete and official position of the NFPA on the referenced subject, which is represented only by the standard in its entirety.

[b]Dimensions of RHH and RHW.
[c]Dimensions of THW in sizes 14 to 8. No. 6 THW and larger have the same dimensions as TW.
[d]No. 14 to No. 2.
[e]In cols. 8 and 9 the values shown for sizes No. 1 through 4/0 are for TFE and Z only. The right-hand values in cols. 8 and 9 are for FEPB, Z, ZF, and ZFF only.
[f]No. 14 to No. 2.
[g]Dimensions of Types RHH and RHW.

**Maximum Number of Compact Conductors in Trade Sizes of Conduit or Tubing**
(Table 5B, 1990 NEC)

Insulation type / Conduit trade size

| Conductor size, AWG or kcmil | 1 in THW | 1 in THHN THHW | 1 in XHHW | 1¼ in THW | 1¼ in THHN THHW | 1¼ in XHHW | 1½ in THW | 1½ in THHN THHW | 1½ in XHHW | 2 in THW | 2 in THHN THHW | 2 in XHHW | 2½ in THW | 2½ in THHN THHW | 2½ in XHHW | 3 in THW | 3 in THHN THHW | 3 in XHHW | 3½ in THW | 3½ in THHN THHW | 3½ in XHHW | 4 in THW | 4 in THHN THHW | 4 in XHHW | Conductor size, AWG or kcmil |
|---|---|---|---|---|---|---|---|---|---|---|---|---|---|---|---|---|---|---|---|---|---|---|---|---|---|
| 6 | 5 | 7 | 6 | 9 | 13 | 11 | 12 | 18 | 15 | 15 | 18 | 18 | | | | | | | | | | | | | 6 |
| 4 | 4 | 4 | 4 | 7 | 8 | 8 | 9 | 11 | 11 | 11 | 13 | 13 | 11 | 14 | 14 | | | | | | | | | | 4 |
| 2 | 3 | 3 | 3 | 5 | 6 | 6 | 7 | 8 | 8 | 8 | 10 | 10 | 9 | 12 | 12 | | | | | | | | | | 2 |
| 1 | | | | 3 | 4 | 4 | 5 | 6 | 6 | 7 | 8 | 8 | 8 | 10 | 10 | | | | | | | | | | 1 |
| 1/0 | | | | 3 | 3 | 3 | 4 | 5 | 5 | 5 | 7 | 7 | 7 | 8 | 8 | 12 | 15 | 15 | | | | | | | 1/0 |
| 2/0 | | | | 3 | 3 | 3 | 3 | 4 | 4 | 5 | 6 | 6 | 6 | 7 | 7 | 10 | 13 | 13 | | | | | | | 2/0 |
| 3/0 | | | | | | | 3 | 3 | 3 | 4 | 5 | 5 | 5 | 6 | 6 | 9 | 10 | 10 | | | | | | | 3/0 |
| 4/0 | | | | | | | 3 | 3 | 3 | 4 | 5 | 5 | 5 | 6 | 6 | 9 | 10 | 10 | 12 | 14 | 14 | | | | 4/0 |
| 250 | | | | | | | | | | 3 | 4 | 4 | 4 | 5 | 5 | 7 | 8 | 8 | 9 | 11 | 11 | 10 | 12 | 12 | 250 |
| 300 | | | | | | | | | | 3 | 3 | 3 | 4 | 4 | 4 | 6 | 7 | 7 | 8 | 9 | 10 | 9 | 11 | 11 | 300 |
| 350 | | | | | | | | | | 3 | 3 | 3 | 3 | 4 | 4 | 5 | 6 | 6 | 7 | 8 | 8 | 8 | 9 | 11 | 350 |
| 400 | | | | | | | | | | | | | 3 | 3 | 4 | 5 | 5 | 6 | 6 | 7 | 8 | 8 | 9 | 10 | 400 |
| 500 | | | | | | | | | | | | | | 3 | 3 | 4 | 4 | 5 | 5 | 6 | 6 | 7 | 8 | 8 | 500 |
| 600 | | | | | | | | | | | | | | 3 | 3 | 3 | 4 | 4 | 4 | 5 | 5 | 6 | 6 | 7 | 600 |
| 700 | | | | | | | | | | | | | | 3 | 3 | 3 | 3 | 3 | 4 | 4 | 4 | 5 | 6 | 6 | 700 |
| 750 | | | | | | | | | | | | | | | | 3 | 3 | 3 | 4 | 4 | 4 | 5 | 5 | 5 | 750 |
| 1000 | | | | | | | | | | | | | | | | | | | 3 | 3 | 3 | 4 | 4 | 4 | 1000 |

**36.   Dimensions of Lead-Covered Conductors: Types RL and RHL**[a]
(Table 6, 1990 NEC)

| Size, AWG or 1000 cmil | Single-conductor | | Two-conductor | | Three-conductor | |
|---|---|---|---|---|---|---|
| | Diameter, in | Area, in$^2$ | Diameter, in | Area, in$^2$ | Diameter, in | Area, in$^2$ |
| 14 AWG | 0.28 | 0.062 | 0.28 × 0.47 | 0.115 | 0.59 | 0.273 |
| 12 | 0.29 | 0.066 | 0.31 × 0.54 | 0.146 | 0.62 | 0.301 |
| 10 | 0.35 | 0.096 | 0.35 × 0.59 | 0.180 | 0.68 | 0.363 |
| 8 solid | 0.41 | 0.132 | 0.41 × 0.71 | 0.255 | 0.82 | 0.528 |
| 8 stranded | 0.43 | 0.145 | 0.43 × 0.75 | 0.282 | 0.86 | 0.581 |
| 6 | 0.49 | 0.188 | 0.49 × 0.86 | 0.369 | 0.97 | 0.738 |
| 4 | 0.55 | 0.237 | 0.54 × 0.96 | 0.457 | 1.08 | 0.916 |
| 2 | 0.60 | 0.283 | 0.61 × 1.08 | 0.578 | 1.21 | 1.146 |
| 1 | 0.67 | 0.352 | 0.70 × 1.23 | 0.756 | 1.38 | 1.49 |
| 1/0 | 0.71 | 0.396 | 0.74 × 1.32 | 0.859 | 1.47 | 1.70 |
| 2/0 | 0.76 | 0.454 | 0.79 × 1.41 | 0.980 | 1.57 | 1.94 |
| 3/0 | 0.81 | 0.515 | 0.84 × 1.52 | 1.123 | 1.69 | 2.24 |
| 4/0 | 0.87 | 0.593 | 0.90 × 1.64 | 1.302 | 1.85 | 2.68 |
| 250 MCM | 0.98 | 0.754 | . . . . . . . . . | . . . . | 2.02 | 3.20 |
| 300 | 1.04 | 0.85 | . . . . . . . . . | . . . . | 2.15 | 3.62 |
| 350 | 1.10 | 0.95 | . . . . . . . . . | . . . . | 2.26 | 4.02 |
| 400 | 1.14 | 1.02 | . . . . . . . . . | . . . . | 2.40 | 4.52 |
| 500 | 1.23 | 1.18 | . . . . . . . . . | . . . . | 2.59 | 5.28 |

[a]Reprinted with permission from NFPA 70-1990, National Electrical Code®, Copyright © 1990, National Fire Protection Association, Quincy, Massachusetts 02269. This reprinted material is not the complete and official position of the NFPA on the referenced subject, which is represented only by the standard in its entirety.

NOTES   No. 14 to No. 10, solid conductors; No. 8, solid or stranded conductors; No. 6 and larger, stranded conductors.

The above cables are limited to straight runs or with nominal offsets equivalent to not more than two quarter bends.

## 37. Elevation of Unguarded Energized Parts above Working Space[a]
(Table 110-34(e), 1990 NEC)

| *Nominal Voltage between Phases* | *Elevation* |
|---|---|
| 601–7500 V | 8 ft 6 in |
| 7501–35,000 V | 9 ft |
| Over 35,000 V | 9 ft 0.37 in/kV above 35 V |

[a]Reprinted with permission from NFPA 70-1990, National Electrical Code®, Copyright © 1990, National Fire Protection Association, Quincy, Massachusetts 02269. This reprinted material is not the complete and official position of the NFPA on the referenced subject, which is represented only by the standard in its entirety.

## 38. Ordinary Ratings of Overload Protective Devices, in Amperes

| Plug fuses | Cartridge fuses[a] | Large-sized magnetic breakers Rating | Large-sized magnetic breakers Range of setting | Small magnetic breakers | Thermal breakers[b] | Time-lag fuses Screw-base type | Time-lag fuses Cartridge type | Overload relays and releases for use with manual or magnetic switches |
|---|---|---|---|---|---|---|---|---|
| | 1 | 3 | 3– 6 | Any ampere or frac-tional-ampere rating from 50 mA to 35 A | 10 | 1.0 | 1 | ½ |
| 3 | 3 | 6 | 6– 12 | | 15 | 1.25 | 1.25 | ¾ |
| 6 | 6 | 10 | 10– 20 | | 20 | 1.6 | 1.6 | 1 |
| 10 | 10 | 15 | 15– 30 | | 25 | 2.0 | 2.0 | 1½ |
| 15 | 15 | 20 | 20– 40 | | 35 | 2.5 | 2.5 | 2 |
| 20 | 20 | 25 | 25– 50 | | | | | |
| 25 | 25 | 35 | 35– 70 | | 50 | 3.2 | 3.2 | 2½ |
| 30 | 30 | 50 | 50– 100 | | 70 | 4.0 | 4.0 | 3 |
| ... | 35 | 70 | 70– 140 | | 90 | 5.0 | 5.0 | 3½ |
| ... | 40 | 90 | 90– 180 | | 100 | 6.25 | 6.25 | 4 |
| ... | 45 | 100 | 100– 200 | | 125 | 8.0 | 8 | 4½ |
| ... | 50 | 125 | 125– 250 | | 150 | 10.0 | 10 | 5 |
| ... | 60 | 150 | 150– 300 | | 175 | 12.0 | 12 | 6 |
| ... | 70 | 175 | 175– 350 | | 200 | 14.0 | 15 | 7 |
| ... | 80 | 200 | 200– 400 | | 225 | 15.0 | 20 | 8 |
| ... | 90 | 225 | 225– 450 | | 250 | 20.0 | 25 | 9 |
| ... | 100 | 250 | 250– 500 | | 275 | 25.0 | 30 | 10 |
| ... | 110 | 275 | 275– 550 | | 300 | 30.0 | 35 | 12 |
| ... | 125 | 300 | 300– 600 | | 325 | ..... | 40 | 14 |
| ... | 150 | 325 | 325– 650 | | 350 | ..... | 45 | 16 |
| ... | 175 | 350 | 350– 700 | | 400 | ..... | 50 | 18 |
| ... | 200 | 400 | 400– 800 | | 450 | ..... | 60 | 20 |
| ... | 225 | 500 | 500– 1,000 | | 500 | ..... | 70 | 22 |
| ... | 250 | 600 | 600– 1,200 | | 550 | ..... | 80 | 24 |
| ... | 300 | 800 | 800– 1,600 | | 600 | ..... | 90 | 28 |
| ... | 350 | 1,000 | 1,000– 2,000 | | ... | ..... | 100 | 32 |
| ... | 400 | 1,200 | 1,200– 2,400 | | ... | ..... | 110 | 36 |
| ... | 450 | 1,600 | 1,600– 3,200 | | ... | ..... | 125 | 40 |
| ... | 500 | 2,000 | 2,000– 4,000 | | ... | ..... | 150 | 45 |
| ... | 600 | 2,500 | 2,500– 5,000 | | ... | ..... | 175 | 50 |
| ... | ... | 3,000 | 3,000– 6,000 | | ... | ..... | 200 | 55 |
| ... | ... | 4,000 | 4,000– 8,000 | | ... | ..... | 225 | 60 |
| ... | ... | 5,000 | 5,000–10,000 | | ... | ..... | 250 | 65 |
| ... | ... | 6,000 | 6,000–12,000 | | ... | ..... | 300 | 70 |
| ... | ... | 8,000 | 8,000–16,000 | | ... | ..... | 350 | 75 |
| ... | ... | 10,000 | 10,000–20,000 | | ... | ..... | 400 | 80 |
| ... | ... | ...... | ............ | | ... | ..... | 450 | |
| ... | ... | ...... | ............ | | ... | ..... | 500 | |
| ... | ... | ...... | ............ | | ... | ..... | 600 | |

[a]Class H, J, or K fuses. Class G fuses are 15, 20, 25, 30, 40, 45, 50, and 60 A.
[b]Nonadjustable thermal-magnetic breakers.

### 39.  Nonrenewable Cartridge Fuses
(Underwriters Laboratories Inc.)

Fuses designated as Class K1, K5, K9 (0–600 A, 250 V, ac, or 600 V, ac) are classified as to interrupting capacity and in terms of maximum clearing amperes squared–seconds and maximum peak let-through current. They incorporate dimensional features equivalent to, and are interchangeable with, other nonrenewable cartridge fuses intended for installation in conventional designs of equipment recognized by the National Electrical Code for branch-circuit, service, and motor-overload use. They are not marked "current-limiting."

Fuses designated as Class RK1 and RK5 (0–600 A, 250 V, ac, or 600 V, ac) are high-interrupting-capacity types and are marked "current-limiting." They have maximum peak let-through currents and maximum clearing amperes squared–seconds at 50,000, 100,000, and 200,000 rms symmetrical amperes. They incorporate features which permit their insertion into holders for other nonrenewable cartridge fuses intended for installation in conventional designs of equipment recognized by the NEC for branch-circuit, service, and motor-overload use. They are provided with a feature which allows their insertion into rejection-type fuseholders designed to accept only Class RK1 or RK5 fuses. Rejection-type fuseholders are used in equipment as covered in the NEC.

Fuses designated as Class G (0–60 A, 300 V, ac) are high-interrupting-capacity types and are marked "current-limiting." They are not interchangeable with other fuses mentioned above and below.

Fuses designated as Class J (0–600 A, 600 V, ac) or Class L (601–6000 A, 600 V, ac) are high-interrupting-capacity types and are marked "current-limiting." They are not interchangeable with other fuses such as those mentioned above.

Fuses designated as Class T (0–600 A, 250 and 600 V, ac) are high-interrupting-capacity types and are marked "current-limiting." They are not interchangeable with other fuses mentioned above.

The term *current-limiting* indicates that a fuse, when tested on a circuit capable of delivering a specific short-circuit current (rms symmetrical amperes) at rated voltage, will start to melt within 90 electrical degrees and will clear the circuit within 180 electrical degrees (½ cycle).

Because the time required for a fuse to melt depends on the available current of the circuit, a fuse which may be current-limiting when subjected to a specific short-circuit current (rms symmetrical amperes) may not be current-limiting on a circuit of lower maximum available current.

The performance of a fuse as it is determined by the ability of the fuse to open and clear a circuit is indicated by the following maximum permissible let-through values which are obtained when fuses of the Class K designs are connected to circuits having an available current of up to 100,000 A maximum (rms symmetrical) but not greater than the marked interrupting rating of the fuse if it is less than 100,000 A, when fuses of the Class G design are connected to circuits having an available current of 100,000 A maximum (rms symmetrical), and when fuses of the Class J, L, RK, and T designs are connected to circuits having available currents of 50,000, 100,000, and 200,000 A maximum (rms symmetrical).

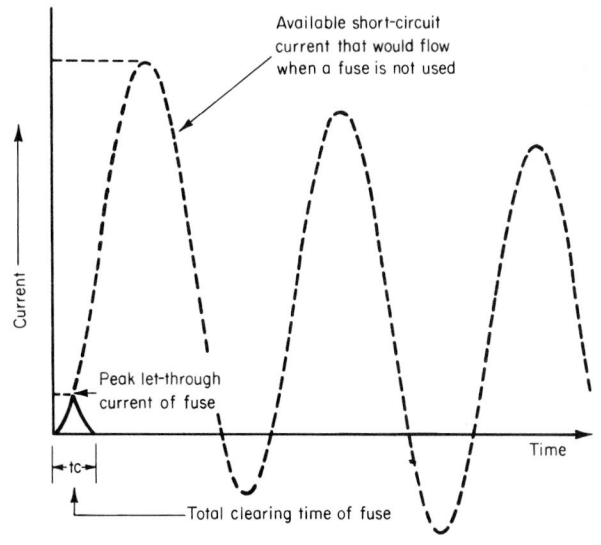

Current-Limiting Effect of Fuse

**39. Nonrenewable Cartridge Fuses** (*Continued*)

### Class G fuses

| Cartridge size, A | Maximum peak let-through current, A $\times 10^3$ | Maximum clearing $I^2t$ (A$^2$ − s $\times 10^3$) |
|---|---|---|
| 0–15 | 3 | 2 |
| 16–20 | 4 | 3 |
| 21–30 | 6 | 5 |
| 31–60 | 10 | 25 |

### Class J and T fuses

Maximum peak let-through current ($I_p$) and clearing $I^2t$ between threshold and

| Cartridge size, A | 50 kA | | 100 kA | | 200 kA | |
|---|---|---|---|---|---|---|
| | $I^2t \times 10^3$ | $I_p \times 10^3$ | $I^2t \times 10^3$ | $I_p \times 10^3$ | $I^2t \times 10^3$ | $I_p \times 10^3$ |
| 0–30 | 7 | 6 | 7 | 8 | 7 | 12 |
| 31–60 | 30 | 8 | 30 | 10 | 30 | 16 |
| 61–100 | 60 | 12 | 80 | 14 | 80 | 20 |
| 101–200 | 200 | 16 | 300 | 20 | 300 | 30 |
| 201–400 | 1000 | 25 | 1100 | 30 | 1100 | 45 |
| 401–600 | 2500 | 35 | 2500 | 45 | 2500 | 70 |

### Class K fuses

| Cartridge size, A | Maximum peak let-through current, A $\times 10^3$ | | | Maximum clearing $I^2t$ (A$^2$ − s $\times 10^3$) | | |
|---|---|---|---|---|---|---|
| | K1 | K5 | K9 | K1 | K5 | K9 |
| 30 | 10 | 11 | 14 | 10 | 50 | 50 |
| 60 | 12 | 21 | 28 | 40 | 200 | 250 |
| 100 | 16 | 25 | 35 | 100 | 500 | 650 |
| 200 | 22 | 40 | 60 | 400 | 1,600 | 3,500 |
| 400 | 35 | 60 | 80 | 1,200 | 5,000 | 15,000 |
| 600 | 50 | 80 | 130 | 3,000 | 10,000 | 40,000 |

### Class L fuses

Maximum peak let-through current ($I_p$) and clearing $I^2t$

| Cartridge size, A | 50 kA or threshold current, whichever is greater | | 100 kA | | 200 kA | |
|---|---|---|---|---|---|---|
| | $I^2t \times 10^6$ | $I_p \times 10^3$ | $I^2t \times 10^6$ | $I_p \times 10^3$ | $I^2t \times 10^6$ | $I_p \times 10^3$ |
| 601–800 | 10 | 80 | 10 | 80 | 10 | 80 |
| 801–1200 | 12 | 80 | 12 | 80 | 15 | 120 |
| 1201–1600 | 22 | 100 | 22 | 100 | 30 | 150 |
| 1601–2000 | 35 | 110 | 35 | 120 | 40 | 165 |
| 2001–2500 | ... | ... | 75 | 165 | 75 | 180 |
| 2501–3000 | ... | ... | 100 | 175 | 100 | 200 |
| 3001–4000 | ... | ... | 150 | 220 | 150 | 250 |
| 4001–5000 | ... | ... | 350 | ... | 350 | 300 |
| 5001–6000 | ... | ... | 350 | ... | 500 | 350 |

**39.  Nonrenewable Cartridge Fuses** (*Continued*)

Class RK1 fuses

Maximum peak let-through current ($I_p$) and clearing $I^2t$ between threshold and

| Cartridge size, A | 50 kA | | 100 kA | | 200 kA | |
|---|---|---|---|---|---|---|
| | $I^2t \times 10^3$ | $I_p \times 10^3$ | $I^2t \times 10^3$ | $I_p \times 10^3$ | $I^2t \times 10^3$ | $I_p \times 10^3$ |
| 0–30 | 10 | 6 | 10 | 10 | 11 | 12 |
| 31–60 | 40 | 10 | 40 | 12 | 50 | 16 |
| 61–100 | 100 | 14 | 100 | 16 | 100 | 20 |
| 101–200 | 400 | 18 | 400 | 22 | 400 | 30 |
| 201–400 | 1200 | 33 | 1200 | 35 | 1600 | 50 |
| 401–600 | 3000 | 43 | 3000 | 50 | 4000 | 70 |

Class RK5 fuses

Maximum peak let-through current ($I_p$) and clearing $I^2t$ between threshold and

| Cartridge size, A | 50 kA | | 100 kA | | 200 kA | |
|---|---|---|---|---|---|---|
| | $I^2t \times 10^3$ | $I_p \times 10^3$ | $I^2t \times 10^3$ | $I_p \times 10^3$ | $I^2t \times 10^3$ | $I_p \times 10^3$ |
| 0–30 | 50 | 11 | 50 | 11 | 50 | 14 |
| 31–60 | 200 | 20 | 200 | 21 | 200 | 26 |
| 61–100 | 500 | 22 | 500 | 25 | 500 | 32 |
| 101–200 | 1600 | 32 | 1600 | 40 | 2000 | 50 |
| 201–400 | 5000 | 50 | 5000 | 60 | 6000 | 75 |
| 401–600 | 10000 | 65 | 10000 | 80 | 12000 | 100 |

Class K-1, K-5, and K-9 fuses are marked, in addition to their regular voltage and current ratings, with an interrupting rating of 200,000, 100,000, or 50,000 A (rms symmetrical).

Class RK1, RK5, J, L, and T fuses are marked, in addition to their regular voltage and current ratings, with an interrupting rating of 200,000 A (rms symmetrical).

Equipment (switches, motor starters, panelboards, etc.) which has been investigated and found suitable for use with these fuses is marked with the class of fuse intended to be used in the equipment and an available current rating applicable to that piece of equipment. Equipment when so marked, with these fuses installed, is considered suitable for use on circuits which can deliver currents under short-circuit conditions up to the available current rating of the equipment or the interrupting rating of the fuse, whichever is lower. An interrupting rating on a fuse included in a piece of equipment does not automatically qualify the equipment in which the fuses are installed for use on circuits with higher available currents than the rating of the equipment itself.

Class L fuses are designed for use in equipment to which line and load connections are made by means of solid busbars. For this reason temperature rises on Class L fuse blades may exceed those observed in connection with other cartridge-fuse designs. Terminal connections for wires in such equipment must be designed to avoid excessive temperatures on the wire insulation.

Some manufacturers are in a position to provide fuses which are advertised and marked to indicate that they have *time-delay* characteristics. In the case of Class G, H, K, and RK fuses, time-delay characteristics (minimum blowing time) have been investigated. Class G fuses, which can carry 200 percent of rated current for 12 s or more, and Class H, K, or RK fuses, which can carry 500 percent of rated current for 10 s or more, may be marked with "D," "time-delay," or some equivalent designation. Class L fuses are permitted to be marked "time-delay" but have not been evaluated for such performance. Class J and T fuses are not permitted to be marked "time-delay."

DEFINITIONS

*Case size.* Fuses of all ratings for which the size of the cartridge is the same; for example, 31–60 A, 61–100 A.

*Clearing time.* The total time measured from the beginning of the specified overcurrent condition until the interruption of the circuit, at rated voltage. It is equal to the sum of the melting time and the arcing time.

*Current-limiting fuse.* A fuse which safely interrupts all available currents within its interrupting rating and, within its current-limiting range, limits the clearing time at rated voltage to an interval equal to or less than the first major or symmetrical current-loop duration and limits peak let-through current to a value less than the peak current that would be possible if the fuse were replaced by a solid conductor of the same impedance.

*Current-limiting range.* The range of rms symmetrical available currents equal to or less than the interrupting rating in which the total clearing time at rated voltage and frequency is less than ½ cycle.

*Interrupting rating.* The rating specifying the highest rms symmetrical, alternating current which the fuse is required to interrupt under the conditions of these requirements.

*$I^2t$ (amperes squared–seconds).* An expression of the energy available as a result of current flow. The term is normally applied during the melting, arcing, or total clearing time of the fuse under which, in the latter instance, it is often designated as "clearing $I^2t$."

*Maximum energy.* The test condition under which in a specified maximum time (½ cycle), the maximum amount of heat possible is generated in the fuse before clearing.

*Melting time.* The time required for the current to melt the current-responsive element.

*Normal-frequency recovery voltage.* The normal-frequency rms voltage impressed upon the fuse after the circuit has been interrupted and after high-frequency transients have subsided.

*Peak let-through current ($I_p$).* The maximum instantaneous current through the fuse during the total clearing time.

*Rating tests.* Carrying-capacity, temperature, and clearing-time–current tests.

*Recovery voltage.* The voltage impressed upon the fuse after a circuit has been interrupted.

*Threshold current.* The rms symmetrical available current at the threshold of the current-limiting range, when melting of the current-responsive element has occurred at approximately 90° on the symmetrical current wave and total clearing time is less than ½ cycle at rated voltage and rated frequency and with a circuit power factor of 20 percent or less.

*Threshold ratio.* The ratio obtained by dividing the threshold current by the continuous-current rating of the fuse.

**40. Number of Overcurrent Units, Such as Trip Coils or Relays, for Protection of Circuits[a]**

| Systems | Number and Location of Overcurrent Units[b] |
|---|---|
| Two-wire single-phase ac or dc, ungrounded | Two (one in each conductor). Diagram 1 of Fig. 11-1. |
| Two-wire single-phase ac or dc, one wire grounded | One (in ungrounded conductor). Diagram 2 of Fig. 11-1. |
| Two-wire single-phase ac or dc, midpoint grounded | Two (one in each conductor). Diagram 3 of Fig. 11-1. |
| Two-wire single-phase ac derived from three-phase, with ungrounded neutral | Two (one in each conductor). Diagram 4 of Fig. 11-1. |
| Two-wire single-phase derived from three-phase, grounded neutral system by using outside wires of three-phase circuit | Two (one in each conductor). Diagram 5 of Fig. 11-1. |
| Three-wire single-phase ac or dc, ungrounded neutral | Three (one in each conductor). Diagram 6 of Fig. 11-1. |
| Three-wire single-phase ac or dc, grounded neutral | Two (one in each conductor except neutral conductor). Diagram 7 of Fig. 11-1. |
| Three-wire two-phase, ac, common wire, ungrounded | Three (one in each conductor). Diagram 8 of Fig. 11-1. |
| Three-wire two-phase, ac, common wire, grounded | Two (one in each conductor except common conductor). Diagram 9 of Fig. 11-1. |
| Four-wire two-phase, ungrounded, phases separate | Four (one in each conductor). Diagram 10 of Fig. 11-1. |
| Four-wire two-phase, grounded neutral or five-wire two-phase, grounded neutral | Four (one in each conductor except neutral conductor). Diagrams 11 and 12 of Fig. 11-1. |
| Three-wire three-phase, ungrounded | Three (one in each conductor). Diagram 13 of Fig. 11-1.[c] |
| Three-wire three-phase, one wire grounded | Two (one in each ungrounded conductor). Diagram 14 of Fig. 11-1. |
| Three-wire three-phase, grounded neutral | Three (one in each conductor). Diagram 15 of Fig. 11-1.[c] |
| Three-wire three-phase, midpoint of one phase grounded | Three (one in each conductor). Diagram 17 of Fig. 11-1.[c] |
| Four-wire three-phase, grounded neutral | Three (one in each ungrounded conductor). Diagram 18 of Fig. 11-1.[c] |
| Four-wire three-phase, ungrounded neutral | Four (one in each conductor). Diagram 19 of Fig. 11-1. |

[a]Refer to Sec. 294 of Div. 9 and Fig. 11-1. When standard devices are not available with three or four overcurrent units as required in the table, it is permissible to substitute two overcurrent units and one fuse when three overcurrent units are called for and two overcurrent units and two fuses when four overcurrent units are called for. The fuse or fuses are to be placed in the conductors not containing an overcurrent unit. However, this practice of substituting fuses for overcurrent units is to be discouraged for obvious reasons.

[b]An overcurrent unit may consist of a series overcurrent tripping device or the combination of a current transformer and a secondary overcurrent tripping device. Either two or three secondary overcurrent tripping devices may be used with three current transformers on a three-phase system similar to those shown in Diagrams 15 and 18 in Fig. 11-1.

[c]When three current transformers are used instead of three series overcurrent tripping devices shown in Diagrams 13, 15, 17, and 18 in Fig. 11-1, the secondary tripping devices may consist of three secondary overcurrent tripping devices or two secondary overcurrent tripping devices with a residual current tripping device of a lower range. See Diagram 16 of Fig. 11-1.

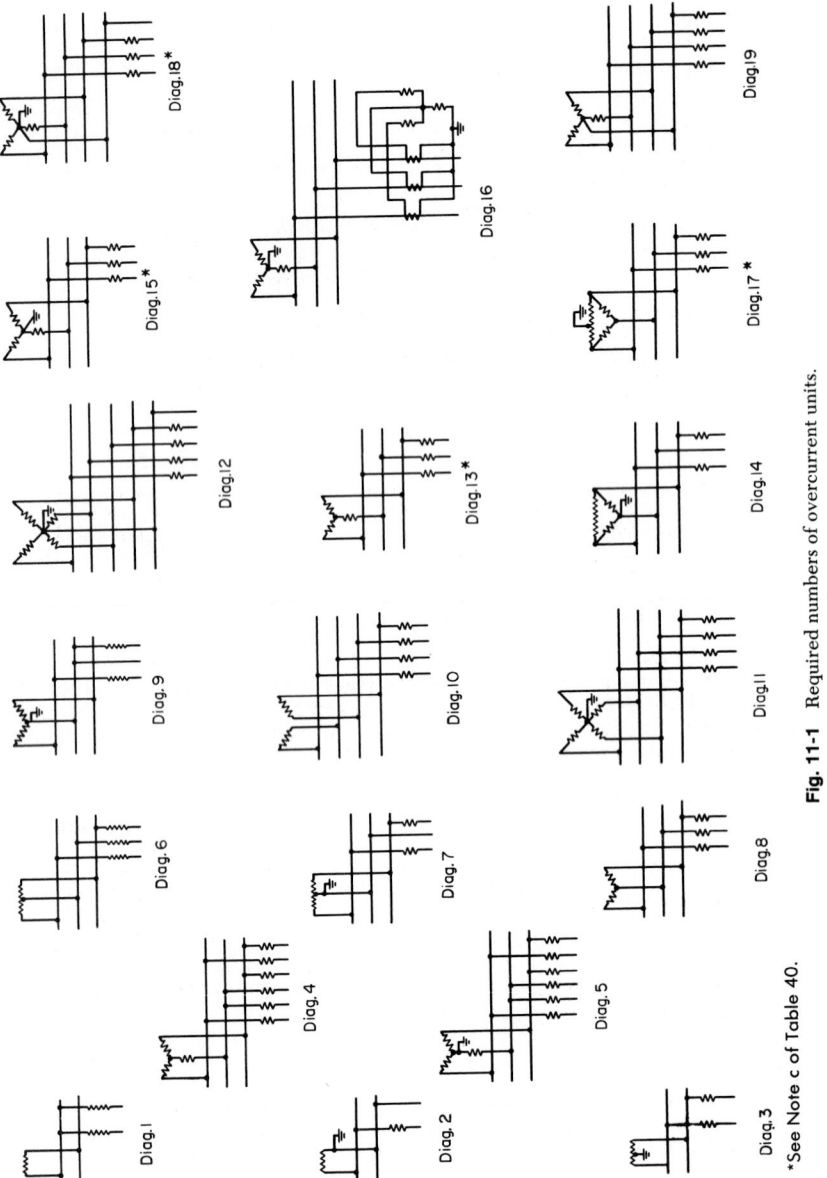

**Fig. 11-1**  Required numbers of overcurrent units.

*See Note c of Table 40.

**41.  Number and Location of Overload Units Other Than Fuses for Protection of Motors[a]**
(Table 430-37, 1990 NEC)

| Kind of motor | Supply system | Number and location of overload units such as trip coils, relays, or thermal cutouts |
|---|---|---|
| One-phase, ac or dc | Two-wire, one-phase, ac or dc, ungrounded | One in either conductor |
| One-phase, ac or dc | Two-wire, one-phase, ac or dc, one conductor grounded | One in ungrounded conductor |
| One-phase, ac or dc | Three-wire, one-phase, ac or dc, grounded neutral | One in either ungrounded conductor |
| Two-phase ac | Three-wire, two-phase, ac, ungrounded | Two, one in each phase |
| Two-phase ac | Three-wire, two-phase, ac, one conductor grounded | Two in ungrounded conductors |
| Two-phase ac | Four-wire, two-phase ac, grounded or ungrounded | Two, one per phase in ungrounded conductors |
| Two-phase ac | Five-wire, two-phase ac, grounded neutral or ungrounded | Two, one per phase in any ungrounded phase wire |
| Three-phase ac | Any three-phase | Three, one in each phase[b] |

[a]Reprinted with permission from NFPA 70-1990, National Electrical Code®, Copyright © 1990, National Fire Protection Association, Quincy, Massachusetts 02269. This reprinted material is not the complete and official position of the NFPA on the referenced subject, which is represented only in the standard in its entirety.

[b]*Exception. When protected by other approved means.*

**42. Motor code letters and locked-rotor kilovolt-amperes.**  The National Electrical Manufacturers Association has adopted a standard of identifying code letters that may be marked by the manufacturer on motor nameplates to show the motor kilovolt-ampere input with a locked rotor. The meaning of the code letters is as follows:

| Letter designation | Kva per hp with locked rotor | Letter designation | Kva per hp with locked rotor |
|---|---|---|---|
| A | 0–3.15 | K | 8.0–9.0 |
| B | 3.15–3.55 | L | 9.0–10.0 |
| C | 3.55–4.0 | M | 10.0–11.2 |
| D | 4.0 –4.5 | N | 11.2–12.5 |
| E | 4.5 –5.0 | P | 12.5–14.0 |
| F | 5.0 –5.6 | R | 14.0–16.0 |
| G | 5.6 –6.3 | S | 16.0–18.0 |
| H | 6.3 –7.1 | T | 18.0–20.0 |
| J | 7.1 –8.0 | U | 20.0–22.4 |
|  |  | V | 22.4 and up |

NOTE  Locked kilovolt-ampere per horsepower range includes the lower figure up to but not including the higher figure. For example, 3.14 is designated by letter A and 3.15 by letter B.

**43.  Maximum Rating or Setting of Motor Branch-Circuit Short-Circuit and Ground-Fault Protective Devices**[a]
(Table 430-152, 1990 NEC)

| Type of motor | Percent of full-load current | | | |
|---|---|---|---|---|
| | Non-time-delay fuse | Dual-element (time-delay) fuse | Instantaneous-trip breaker | Inverse-time breaker[b] |
| Single-phase, all types | | | | |
| No code letter | 300 | 175 | 700 | 250 |
| All ac single-phase and polyphase squirrel-cage and synchronous motors[c] with full-voltage resistor or reactor starting | | | | |
| No code letter | 300 | 175 | 700 | 250 |
| Code letter F to V | 300 | 175 | 700 | 250 |
| Code letter B to E | 250 | 175 | 700 | 200 |
| Code letter A | 150 | 150 | 700 | 150 |
| All ac squirrel-cage and synchronous motors[c] with autotransformer starting | | | | |
| Not more than 30 A | | | | |
| No code letter | 250 | 175 | 700 | 200 |
| More than 30 A | | | | |
| No code letter | 200 | 175 | 700 | 200 |
| Code letter F to V | 250 | 175 | 700 | 200 |
| Code letter B to E | 200 | 175 | 700 | 200 |
| Code letter A | 150 | 150 | 700 | 150 |
| High-reactance squirrel-cage motors | | | | |
| Not more than 30 A | | | | |
| No code letter | 250 | 175 | 700 | 250 |
| More than 30 A | | | | |
| No code letter | 200 | 175 | 700 | 200 |
| Wound-rotor motors | | | | |
| No code letter | 150 | 150 | 700 | 150 |
| Direct-current (constant-voltage) motors | | | | |
| No more than 50 hp | | | | |
| No code letter | 150 | 150 | 250 | 150 |
| More than 50 hp | | | | |
| No code letter | 150 | 150 | 175 | 150 |

NOTE   For an explanation of code-letter marking, see Table 42. For certain exceptions to the values specified, see Secs. 430-52 through 430-54 of the NEC.

[a]Reprinted with permission from NFPA 70-1990, National Electrical Code®, Copyright © 1990, National Fire Protection Association, Quincy, Massachusetts 02269. This reprinted material is not the complete and official position of the NFPA on the referenced subject, which is represented only by the standard in its entirety.

[b]The values given in the last column also cover the ratings of nonadjustable inverse-time types of circuit breakers that may be modified as in Sec. 430-52 of the NEC.

[c]Synchronous motors of the low-torque, low-speed type (usually 450 rpm or lower), such as are used to drive reciprocating compressors, pumps, etc., that start unloaded, do not require a fuse rating or circuit-breaker setting in excess of 200 percent of full-load current.

## 44. Overcurrent and Overload Protection for Motors

See Table 43. The current values in col. 1 are to be taken from Tables 7 to 10, including footnotes, but the values shown for overload protection in cols. 2 and 3 must be modified if nameplate full-load–current values are different, as provided in Sec. 430-6 of the National Electrical Code (refer to Sec. 421, Div. 7). The current values shown in cols. 5 and 6 must be reduced by 8 percent for all motors other than open-type motors marked to have a temperature rise not over 40°C. For certain exceptions to the values in cols. 4, 5, 6, and 7, see Sec. 423 of Div. 7. See Sec. 424 of Div. 7 for values to be used for several motors on one branch circuit.

| | For overload protection of motors[a] | | Maximum allowable rating or setting of branch-circuit short-circuit and ground-fault protective devices | | | | | | | |
| Full-load-current rating of motor, amp | Max rating of nonadjustable protective devices, amp | Max setting of adjustable protective devices, amp | With code letters: Single-phase, squirrel-cage, and synchronous. Full voltage, resistor or reactor starting, Code letters F to V inclusive. Without code letters: same as above | | With code letters: Single-phase, squirrel-cage, and synchronous. Full voltage, resistor or reactor start, Code letters B to E inclusive. Autotransformer start, Code letters F to V inclusive. Without code letters (not more than 30 amp): Squirrel-cage and synchronous, autotransformer start, high-reactance squirrel-cage[b] | | With code letters: Squirrel-cage and synchronous autotransformer start, Code letters B to E inclusive. Without code letters (more than 30 amp): Squirrel-cage and synchronous, autotransformer start, high-reactance squirrel-cage[b] | | With code letters: All motors Code letter A Without code letters: D-c and wound-rotor motors | |
| | | | Fuses | Circuit breakers (nonadjustable overload trip) | Fuses | Circuit breakers (nonadjustable overload trip) | Fuses | Circuit breakers (nonadjustable overload trip) | Fuses | Circuit breakers (nonadjustable overload trip) |
| (1) | (2) | (3) | (4) | | (5) | | (6) | | (7) | |
| 1[c] | 2[d] | 1.25[d] | 15 | 15 | 15 | 15 | 15 | 15 | 15 | 15 |
| 2[c] | 3[d] | 2.50[d] | 15 | 15 | 15 | 15 | 15 | 15 | 15 | 15 |
| 3[c] | 4[d] | 3.75[d] | 15 | 15 | 15 | 15 | 15 | 15 | 15 | 15 |
| 4[c] | 6[d] | 5.0[d] | 15 | 15 | 15 | 15 | 15 | 15 | 15 | 15 |
| 5[c] | 8[d] | 6.25[d] | 15 | 15 | 15 | 15 | 15 | 15 | 15 | 15 |
| 6[d] | 8[d] | 7.50[d] | 20 | 15 | 15 | 15 | 15 | 15 | 15 | 15 |
| 7[d] | 10[d] | 8.75[d] | 25 | 20 | 20 | 15 | 15 | 15 | 15 | 15 |
| 8 | 10[d] | 10.0[d] | 25 | 20 | 20 | 20 | 20 | 20 | 15 | 15 |
| 9 | 12[d] | 11.25[d] | 30 | 30 | 25 | 20 | 20 | 20 | 15 | 15 |
| 10 | 15[d] | 12.50[d] | 30 | 30 | 25 | 20 | 20 | 20 | 15 | 15 |
| 11 | 15[d] | 13.75 | 35 | 30 | 30 | 30 | 25 | 30 | 20 | 20 |
| 12 | 15 | 15.00 | 40 | 30 | 30 | 30 | 25 | 30 | 20 | 20 |
| 13 | 20 | 16.25 | 40 | 40 | 35 | 30 | 30 | 30 | 20 | 20 |
| 14 | 20 | 17.50 | 45 | 40 | 35 | 30 | 30 | 30 | 25 | 30 |
| 15 | 20 | 18.75 | 45 | 40 | 40 | 30 | 30 | 30 | 25 | 30 |

**44. Overcurrent and Overload Protection for Motors** (*Continued*)

Maximum allowable rating or setting of branch-circuit short-circuit and ground-fault protective devices

| Full-load-current rating of motor, amp | For overload protection of motors[a] | | With code letters: Single-phase, squirrel-cage, and synchronous. Full voltage, resistor or reactor starting, Code letters F to V, inclusive. Without code letters: same as above | | With code letters: Single-phase, squirrel-cage, and synchronous. Full voltage, resistor or reactor start, Code letters B to E inclusive. Autotransformer start, Code letters F to V inclusive. Without code letters (not more than 30 amp): Squirrel-cage and synchronous, autotransformer start, high-reactance squirrel-cage[b] | | With code letters: Squirrel-cage and synchronous autotransformer start, Code letters B to E inclusive. Without code letters (more than 30 amp): Squirrel-cage and synchronous, autotransformer start, high-reactance squirrel-cage[b] | | With code letters: All motors Code letter A. Without code letters: D-c and wound-rotor motors | |
|---|---|---|---|---|---|---|---|---|---|---|
| | Max rating of nonadjustable protective devices, amp | Max setting of adjustable protective devices, amp | Fuses | Circuit breakers (nonadjustable overload trip) | Fuses | Circuit breakers (nonadjustable overload trip) | Fuses | Circuit breakers (nonadjustable overload trip) | Fuses | Circuit breakers (nonadjustable overload trip) |
| (1) | (2) | (3) | (4) | | (5) | | (6) | | (7) | |
| 16 | 20 | 20.00 | 50 | 40 | 40 | 40 | 35 | 40 | 25 | 30 |
| 17 | 25 | 21.25 | 60 | 50 | 45 | 40 | 35 | 40 | 30 | 30 |
| 18 | 25 | 22.50 | 60 | 50 | 45 | 40 | 40 | 40 | 30 | 30 |
| 19 | 25 | 23.75 | 60 | 50 | 50 | 40 | 40 | 40 | 30 | 30 |
| 20 | 25 | 25.00 | 60 | 50 | 50 | 40 | 40 | 40 | 30 | 30 |
| 22 | 30 | 27.50 | 70 | 70 | 60 | 50 | 45 | 50 | 35 | 40 |
| 24 | 30 | 30.00 | 80 | 70 | 60 | 50 | 50 | 50 | 40 | 40 |
| 26 | 35 | 32.50 | 80 | 70 | 70 | 70 | 60 | 70 | 40 | 40 |
| 28 | 35 | 35.00 | 90 | 70 | 70 | 70 | 60 | 70 | 45 | 50 |
| 30 | 40 | 37.50 | 90 | 100 | 80 | 70 | 60 | 70 | 45 | 50 |

| (1) | (2) | (3) | (4) | (5) | (6) | (7) | (8) | (9) | (10) | (11) |
|---|---|---|---|---|---|---|---|---|---|---|
| 32 | 40 | 40.00 | 100 | 100 | 80 | 70 | 70 | 70 | 50 | 50 |
| 34 | 45 | 42.50 | 110 | 100 | 90 | 70 | 70 | 70 | 60 | 70 |
| 36 | 45 | 45.00 | 110 | 100 | 90 | 100 | 80 | 100 | 60 | 70 |
| 38 | 50 | 47.50 | 125 | 100 | 100 | 100 | 80 | 100 | 60 | 70 |
| 40 | 50 | 50.00 | 125 | 100 | 100 | 100 | 80 | 100 | 60 | 70 |
| 42 | 50 | 52.50 | 125 | 125 | 110 | 100 | 90 | 100 | 70 | 70 |
| 44 | 60 | 55.00 | 125 | 125 | 110 | 100 | 90 | 100 | 70 | 70 |
| 46 | 60 | 57.50 | 150 | 125 | 125 | 100 | 100 | 100 | 70 | 70 |
| 48 | 60 | 60.00 | 150 | 125 | 125 | 100 | 100 | 100 | 80 | 100 |
| 50 | 60 | 62.50 | 150 | 125 | 125 | 100 | 100 | 100 | 80 | 100 |
| 52 | 70 | 65.00 | 175 | 150 | 150 | 125 | 110 | 125 | 80 | 100 |
| 54 | 70 | 67.50 | 175 | 150 | 150 | 125 | 110 | 125 | 90 | 100 |
| 56 | 70 | 70.00 | 175 | 150 | 150 | 125 | 125 | 125 | 90 | 100 |
| 58 | 80 | 72.50 | 200 | 150 | 150 | 125 | 125 | 125 | 90 | 100 |
| 60 | 80 | 75.00 | 200 | 150 | 150 | 125 | 125 | 125 | 90 | 100 |
| 62 | 80 | 77.50 | 200 | 175 | 175 | 150 | 125 | 125 | 100 | 100 |
| 64 | 80 | 80.00 | 200 | 175 | 175 | 150 | 125 | 150 | 100 | 100 |
| 66 | 80 | 82.50 | 225 | 175 | 175 | 175 | 150 | 150 | 100 | 125 |
| 68 | 90 | 85.00 | 225 | 175 | 175 | 175 | 150 | 150 | 110 | 125 |
| 70 | 90 | 87.50 | 225 | 175 | 175 | 175 | 150 | 150 | 110 | 125 |
| 72 | 90 | 90.00 | 250 | 200 | 200 | 175 | 150 | 150 | 110 | 125 |
| 74 | 90 | 92.50 | 250 | 200 | 200 | 175 | 150 | 150 | 125 | 125 |
| 76 | 100 | 95.00 | 250 | 200 | 200 | 175 | 175 | 175 | 125 | 125 |
| 78 | 100 | 97.50 | 250 | 200 | 200 | 175 | 175 | 175 | 125 | 125 |
| 80 | 100 | 100.00 | 250 | 200 | 200 | 175 | 175 | 175 | 125 | 125 |
| 82 | 110 | 102.50 | 250 | 225 | 225 | 175 | 175 | 175 | 125 | 125 |
| 84 | 110 | 105.00 | 300 | 225 | 225 | 200 | 175 | 175 | 150 | 150 |
| 86 | 110 | 107.50 | 300 | 225 | 225 | 200 | 175 | 175 | 150 | 150 |
| 88 | 110 | 110.00 | 300 | 225 | 225 | 200 | 200 | 200 | 150 | 150 |
| 90 | 110 | 112.50 | 300 | 225 | 225 | 200 | 200 | 200 | 150 | 150 |
| 92 | 125 | 115.00 | 300 | 250 | 250 | 200 | 200 | 200 | 150 | 150 |
| 94 | 125 | 117.50 | 300 | 250 | 250 | 200 | 200 | 200 | 150 | 150 |
| 96 | 125 | 120.00 | 300 | 250 | 250 | 200 | 200 | 200 | 150 | 150 |
| 98 | 125 | 122.50 | 300 | 250 | 250 | 200 | 200 | 200 | 150 | 150 |
| 100 | 125 | 125.00 | 300 | 250 | 250 | 200 | 200 | 200 | 150 | 150 |

## 44. Overcurrent and Overload Protection for Motors (Continued)

| Full-load-current rating of motor, amp | For overload protection of motors[a] | | Maximum allowable rating or setting of branch-circuit short-circuit and ground-fault protective devices | | | | | | | |
|---|---|---|---|---|---|---|---|---|---|---|
| | Max rating of nonadjustable protective devices, amp | Max setting of adjustable protective devices, amp | With code letters: Single-phase, squirrel-cage, and synchronous. Full voltage, resistor or reactor starting, Code letters F to V inclusive. Without code letters: same as above | | With code letters: Single-phase, squirrel-cage, and synchronous. Full voltage, resistor or reactor start, Code letters B to E inclusive. Autotransformer start, Code letters F to V inclusive. Without code letters (not more than 30 amp): Squirrel-cage and synchronous, autotransformer start, high-reactance squirrel-cage[b] | | With code letters: Squirrel-cage and synchronous autotransformer start, Code letters B to E inclusive. Without code letters (more than 30 amp): Squirrel-cage and synchronous, autotransformer start, high-reactance squirrel-cage[b] | | With code letters: All motors Code letter A. Without code letters: D-c and wound-rotor motors | |
| | | | Fuses | Circuit breakers (nonadjustable overload trip) | Fuses | Circuit breakers (nonadjustable overload trip) | Fuses | Circuit breakers (nonadjustable overload trip) | Fuses | Circuit breakers (nonadjustable overload trip) |
| (1) | (2) | (3) | (4) | | (5) | | (6) | | (7) | |
| 105 | 150 | 131.50 | 350 | 300 | 300 | 225 | 225 | 225 | 175 | 175 |
| 110 | 150 | 137.50 | 350 | 300 | 300 | 225 | 225 | 225 | 175 | 175 |
| 115 | 150 | 144.00 | 350 | 300 | 300 | 250 | 250 | 250 | 175 | 175 |
| 120 | 150 | 150.50 | 400 | 300 | 300 | 250 | 250 | 250 | 200 | 200 |
| 125 | 175 | 156.50 | 400 | 350 | 350 | 250 | 250 | 250 | 200 | 200 |
| 130 | 175 | 162.50 | 400 | 350 | 350 | 300 | 300 | 300 | 200 | 200 |
| 135 | 175 | 169.00 | 450 | 350 | 350 | 300 | 300 | 300 | 225 | 225 |
| 140 | 175 | 175.00 | 450 | 350 | 350 | 300 | 300 | 300 | 225 | 225 |
| 145 | 200 | 181.50 | 450 | 400 | 400 | 300 | 300 | 300 | 225 | 225 |
| 150 | 200 | 187.50 | 450 | 400 | 400 | 300 | 300 | 300 | 225 | 225 |
| 155 | 200 | 194.00 | 500 | 400 | 400 | 350 | 350 | 350 | 250 | 250 |
| 160 | 200 | 200.00 | 500 | 400 | 400 | 350 | 350 | 350 | 250 | 250 |
| 165 | 225 | 206.00 | 500 | 500 | 450 | 350 | 350 | 350 | 250 | 250 |
| 170 | 225 | 213.00 | 500 | 500 | 450 | 350 | 350 | 350 | 300 | 300 |
| 175 | 225 | 219.00 | 600 | 500 | 450 | 350 | 350 | 350 | 300 | 300 |

| | | | | | | | | | | |
|---|---|---|---|---|---|---|---|---|---|---|
| 180 | 225 | 225.00 | 600 | 500 | 450 | 400 | 400 | 400 | 300 | 300 |
| 185 | 250 | 231.00 | 600 | 500 | 500 | 400 | 400 | 400 | 300 | 300 |
| 190 | 250 | 238.00 | 600 | 500 | 500 | 400 | 400 | 400 | 300 | 300 |
| 195 | 250 | 244.00 | 600 | 500 | 500 | 400 | 400 | 400 | 300 | 300 |
| 200 | 250 | 250.00 | 600 | 500 | 500 | 400 | 400 | 400 | 300 | 300 |
| 210 | 250 | 263.00 | 800 | 600 | 600 | 500 | 450 | 500 | 350 | 350 |
| 220 | 300 | 275.00 | 800 | 600 | 600 | 500 | 450 | 500 | 350 | 350 |
| 230 | 300 | 288.00 | 800 | 600 | 600 | 500 | 500 | 500 | 350 | 350 |
| 240 | 300 | 300.00 | 800 | 600 | 600 | 500 | 500 | 500 | 400 | 400 |
| 250 | 300 | 313.00 | 800 | 700 | 800 | 500 | 500 | 500 | 400 | 400 |
| 260 | 350 | 325.00 | 800 | 700 | 800 | 600 | 600 | 600 | 400 | 400 |
| 270 | 350 | 338.00 | 1000 | 700 | 800 | 600 | 600 | 600 | 450 | 500 |
| 280 | 350 | 350.00 | 1000 | 700 | 800 | 600 | 600 | 600 | 450 | 500 |
| 290 | 350 | 363.00 | 1000 | 800 | 800 | 600 | 600 | 600 | 450 | 500 |
| 300 | 400 | 375.00 | 1000 | 800 | 800 | 600 | 600 | 600 | 450 | 500 |
| 320 | 400 | 400.00 | 1000 | 800 | 800 | 700 | 800 | 700 | 500 | 500 |
| 340 | 450 | 425.00 | 1200 | : | 1000 | 700 | 800 | 700 | 600 | 600 |
| 360 | 450 | 450.00 | 1200 | : | 1000 | 800 | 800 | 800 | 600 | 600 |
| 380 | 500 | 475.00 | 1200 | : | 1000 | 800 | 800 | 800 | 600 | 600 |
| 400 | 500 | 500.00 | 1200 | : | 1000 | 800 | 800 | 800 | 600 | 600 |
| 420 | 600 | 525.00 | 1600 | : | 1200 | : | 1000 | : | 800 | 700 |
| 440 | 600 | 550.00 | 1600 | : | 1200 | : | 1000 | : | 800 | 700 |
| 460 | 600 | 575.00 | 1600 | : | 1200 | : | 1000 | : | 800 | 700 |
| 480 | 600 | 600.00 | 1600 | : | 1200 | : | 1000 | : | 800 | 800 |
| 500 | : | 625.00 | 1600 | : | 1600 | : | 1000 | : | 800 | 800 |

[a] For the overload protection of motors, see Sec. 430 of Div. 7.
[b] High-reactance squirrel-cage motors are those designed to limit the starting current by means of deep-slot secondaries or double-wound secondaries and are generally started on full voltage.
[c] For the grouping of small motors under the protection of a single set of fuses, see Sec. 424 of Div. 7.
[d] For overload protection of motors of 1 hp or less, see Sec. 430 of Div. 7. For setting of branch-circuit breakers, see Table 43.

## 45. Condensed Three-Phase Motor Table

Induction-type, full-voltage starting 230/160 V; code letters F to V and no code letters.

| | | Minimum AWG or 1000 cmil $\frac{\text{wire}}{\text{conduit}}$ sizes | | Overload protection | | Branch-circuit short-circuit and ground-fault protection | |
|---|---|---|---|---|---|---|---|
| hp | Full-load current, A | CU[a] | AL[b] | Nonadjustable types | Adjustable types | Switch-fuse (maximum size for standard fuses) | CB (maximum size) |
| ½ | 2/1 | $\frac{14}{1/2} / \frac{14}{1/2}$ | $\frac{12}{1/2} / \frac{12}{1/2}$ | 3/2 | 2.5/1.25 | 30–15/30–15 | 15/15 |
| ¾ | 2.8/1.4 | $\frac{14}{1/2} / \frac{14}{1/2}$ | $\frac{12}{1/2} / \frac{12}{1/2}$ | 4/2 | 3.5/1.75 | 30–15/30–15 | 15/15 |
| 1 | 3.6/1.8 | $\frac{14}{1/2} / \frac{14}{1/2}$ | $\frac{12}{1/2} / \frac{12}{1/2}$ | 5/4 | 4.5/3.25 | 30–15/30–15 | 15/15 |
| 1½ | 5.2/2.6 | $\frac{14}{1/2} / \frac{14}{1/2}$ | $\frac{12}{1/2} / \frac{12}{1/2}$ | 8/4 | 6.5/3.25 | 30–15/30–15 | 15/15 |
| 2 | 6.3/3.4 | $\frac{14}{1/2} / \frac{14}{1/2}$ | $\frac{12}{1/12} / \frac{12}{1/2}$ | 8/5 | 7.87/4.25 | 30–20/30–15 | 20/15 |
| 3 | 9.6/4.8 | $\frac{14}{1/2} / \frac{14}{1/2}$ | $\frac{12}{1/2} / \frac{12}{1/2}$ | 12/6 | 12.0/6.0 | 30–30/30–15 | 25/15 |
| 5 | 15.2/7.6 | $\frac{12}{1/2} / \frac{14}{1/2}$ | $\frac{10}{1/2} / \frac{12}{1/2}$ | 20/10 | 19.0/9.5 | 60–45/30–25 | 40/20 |
| 7½ | 22/11 | $\frac{10}{1/2} / \frac{14}{1/2}$ | $\frac{8}{3/4} / \frac{12}{1/2}$ | 30/15 | 27.5/13.75 | 100–70/60–35 | 70/30 |
| 10 | 28/14 | $\frac{8}{3/4} / \frac{12}{1/2}$ | $\frac{8}{3/4} / \frac{10}{1/2}$ | 35/20 | 35.0/17.5 | 100–90/60–45 | 70/40 |

| | | | | | | | |
|---|---|---|---|---|---|---|---|
| 15 | 42/21 | $\frac{6}{1} \big/ \frac{10}{\frac{1}{2}}$ | $\frac{4}{1} \big/ \frac{8}{\frac{3}{4}}$ | 50/25 | 52.5/26.25 | 200–125/60–60 | 100/60 |
| 20 | 54/27 | $\frac{4}{1} \big/ \frac{8}{\frac{3}{4}}$ | $\frac{3}{1} \big/ \frac{8}{\frac{3}{4}}$ | 70/35 | 67.5/33.75 | 200–175/100–80 | 150/70 |
| 25 | 68/34 | $\frac{4}{1} \big/ \frac{6}{1}$ | $\frac{2}{1\frac{1}{4}} \big/ \frac{6}{1}$ | 80/40 | 85/42.5 | 200–225/100–125 | 175/90 |
| 30 | 80/40 | $\frac{3}{1\frac{1}{4}} \big/ \frac{6}{1}$ | $\frac{1}{1\frac{1}{4}} \big/ \frac{6}{1}$ | 100/50 | 100/50 | 400–250/200–125 | 200/100 |
| 40 | 104/52 | $\frac{1}{1\frac{1}{4}} \big/ \frac{6}{1}$ | $\frac{2/0}{1\frac{1}{2}} \big/ \frac{4}{1}$ | 125/70 | 130/65 | 400–350/200–175 | 300/150 |
| 50 | 130/65 | $\frac{2/0}{1\frac{1}{2}} \big/ \frac{4}{1}$ | $\frac{4/0}{2} \big/ \frac{2}{1\frac{1}{4}}$ | 150/80 | 162.5/81.25 | 400–400/200–200 | 350/175 |
| 60 | 154/77 | $\frac{3/0}{2} \big/ \frac{3}{1\frac{1}{4}}$ | $\frac{250}{2\frac{1}{2}} \big/ \frac{1}{1\frac{1}{4}}$ | 200/100 | 192.5/96.25 | 600–500/400–250 | 400/200 |
| 75 | 192/96 | $\frac{250}{2\frac{1}{2}} \big/ \frac{1}{1\frac{1}{4}}$ | $\frac{350}{2\frac{1}{2}} \big/ \frac{1/0}{1\frac{1}{2}}$ | 250/125 | 240/120 | 600–600/400–300 | 500/250 |
| 100 | 248/124 | $\frac{350}{2\frac{1}{2}} \big/ \frac{00}{2\frac{1}{2}}$ | $\frac{500}{3} \big/ \frac{3/0}{2}$ | 300/150 | 310/155 | 800–800/400–400 | 700/350 |

NOTE 1: In the dual values shown, the figures to the left in each column are for motors connected at 230 V; the figures to the right are for 460-V motors.
NOTE 2: Conduit sizes are based on Tables 31, 32, and 33.
[a]60°C wire: No. 14 to No. 8; 75°C wire: No. 6 and larger.
[b]60°C wire: No. 12 to No. 10; 75°C wire: No. 8 and larger.

### 46. Standard General-Purpose-Switch Sizes
All sizes in amperes.

| Regular open-knife blade | Open Klamp-tite | Open toggle | Enclosed safety | Regular open-knife blade | Open Klamp-tite | Open toggle | Enclosed safety |
|---|---|---|---|---|---|---|---|
| 30  | 400  | 300  | 30  | 1200  | 4000  | 2500  | 1200 |
| 60  | 600  | 500  | 60  | ..... | 5000  | 3000  | 1600 |
| 100 | 800  | 600  | 100 | ..... | ..... | 4000  | 2000 |
| 200 | 1000 | 800  | 200 | ..... | ..... | 5000  | 2500 |
| 400 | 1200 | 1000 | 400 | ..... | ..... | 6000  | 3000 |
| 600 | 1600 | 1200 | 600 | ..... | ..... | 8000  | 4000 |
| 800 | 2000 | 1600 | 800 | ..... | ..... | 10000 | 5000 |
| ... | 3000 | 2000 | ... | ..... | ..... | ...... | 6000 |

### 47. Conversion Table of Locked-Rotor Currents for Selection of Disconnecting Means and Controllers as Determined from Horsepower and Voltage Rating[a]
(Table 430-151, 1990 NEC)
For use only with Secs. 430-110, 440-12, and 440-41 of the NEC.

| Motor locked-rotor current amperes [b] | | | | | | | |
|---|---|---|---|---|---|---|---|
| Single-phase | | Two- or three-phase | | | | | Maximum hp rating |
| 115 V | 230 V | 115 V | 200 V | 230 V | 460 V | 575 V | |
| 58.8 | 29.4 | 24 | 18.8 | 12 | 6 | 4.8 | ½ |
| 82.8 | 41.4 | 33.6 | 19.3 | 16.8 | 8.4 | 6.6 | ¾ |
| 96 | 48 | 43.2 | 24.8 | 21.6 | 10.8 | 8.4 | 1 |
| 120 | 60 | 62 | 35.9 | 31.2 | 15.6 | 12.6 | 1½ |
| 144 | 72 | 81 | 46.9 | 40.8 | 20.4 | 16.2 | 2 |
| 204 | 102 | ... | 66 | 58 | 26.8 | 23.4 | 3 |
| 336 | 168 | ... | 105 | 91 | 45.6 | 36.6 | 5 |
| 480 | 240 | ... | 152 | 132 | 66 | 54 | 7½ |
| 600 | 300 | ... | 193 | 168 | 84 | 66 | 10 |
| ... | ... | ... | 290 | 252 | 126 | 102 | 15 |
| ... | ... | ... | 373 | 324 | 162 | 132 | 20 |
| ... | ... | ... | 469 | 408 | 204 | 162 | 25 |
| ... | ... | ... | 552 | 480 | 240 | 192 | 30 |
| ... | ... | ... | 718 | 624 | 312 | 246 | 40 |
| ... | ... | ... | 897 | 780 | 390 | 312 | 50 |
| ... | ... | ... | 1063 | 924 | 462 | 372 | 60 |
| ... | ... | ... | 1325 | 1152 | 576 | 462 | 75 |
| ... | ... | ... | 1711 | 1488 | 744 | 594 | 100 |
| ... | ... | ... | 2153 | 1872 | 936 | 750 | 125 |
| ... | ... | ... | 2484 | 2160 | 1080 | 864 | 150 |
| ... | ... | ... | 3312 | 2880 | 1440 | 1152 | 200 |

[a]Reprinted with permission from NFPA 70-1990, National Electrical Code®, Copyright © 1990, National Fire Protection Association, Quincy, Massachusetts 02269. This reprinted material is not the complete and official position of the NFPA on the referenced subject, which is represented only by the standard in its entirety.

[b]These values of motor locked-rotor current are approximately 6 times the full-load current values given in Tables 430-148 and 430-150 of the NEC.

Refer to Secs. 450 and 452 of Div. 7 for hermetic refrigerant motor-compressors.

**48. Horsepower Ratings of Fused Switches**
(Underwriters Laboratories, Inc.—Standard UL98)

| Rating of fuseholders, A | Switch rating, V | Two-pole, single-phase | Two-pole, dc | Three-pole, three-phase | Four-pole, two-phase |
|---|---|---|---|---|---|
| | | **"Standard" and "maximum" horsepower ratings** | | | |
| 30 | 120 ac | ¾ / 2° | | 1½ / 3° | 2 / 3° |
| 60 | | 1½ / 3° | | 3 / 7½° | 3 / 10° |
| 30 | 125 dc | | 2 / 3° | | |
| 60 | | | 5 | | |
| 30 | 240 ac | 1½ / 3° | | 3 / 7½° | 3 / 10° |
| 60 | | 3 / 10° | | 7½ / 15° | 7½ / 20° |
| 100 | | 7½ / 15° | | 15 / 30° | 15 / 30° |
| 200 | | 15 | | 25 / 60° | 30 / 50° |
| 400 | | | | 50 / 125° | 50 |
| 600 | | | | 75 / 200° | |
| 800 | | | | 100 / 250° | |
| 30 | 250 dc | | 5 | | |
| 60 | | | 10 | | |
| 100 | | | 20 | | |
| 200 | | | 40 | | |
| 400 | | | 50 | | |
| 30 | 480 ac | 3 / 7½° | | 5 / 15° | 7½ / 20° |
| 60 | | 5 / 20° | | 15 / 30° | 15 / 40° |
| 100 | | 10 / 30° | | 25 / 60° | 25 / 50° |
| 200 | | 25 / 50° | | 50 / 125° | 50 |
| 400 | | | | 100 / 250° | |
| 600 | | | | 150 / 400° | |
| 800 | | | | 200 / 500° | |

## 48. Horsepower Ratings of Fused Switches (Continued)

| Rating of fuseholders, A | Switch rating, V | "Standard" and "maximum" horsepower ratings | | | | | | | |
|---|---|---|---|---|---|---|---|---|---|
| | | Two-pole, single-phase | | Two-pole, dc | | Three-pole, three-phase | | Four-pole, two-phase | |
| 30 | | 3 | 10* | | | 7½ | 20* | 10 | 25* |
| 60 | | 10 | 20* | | | 15 | 50* | 20 | 50* |
| 100 | | 15 | 40* | | | 30 | 75* | 30 | 50* |
| 200 | 600 ac | 30 | 50* | | | 60 | 150* | 50 | |
| 400 | | | | | | 125 | 350* | | |
| 600 | | | | | | 200 | 500* | | |
| 800 | | | | | | 250 | 500* | | |
| 30 | | | | 10 | 15* | | | | |
| 60 | | | | 25 | 30* | | | | |
| 100 | 600 dc | | | 40 | 50* | | | | |
| 200 | | | | 50 | | | | | |

*The ratings in this table are "standard" ratings except for the "maximum" ratings, which are indicated by the asterisks. "Maximum" horsepower ratings are achieved with the use of time-delay fuses.

## 49.   Minimum Branch-Circuit Sizes for Motors

Based on conductor ampacities when not more than three conductors are installed in a raceway or cable.

| Full-load current rating of motor, A | Wire size, AWG or 1000 cmil | | | |
|---|---|---|---|---|
| | 60°C | | 75°C | |
| | Cu | Al | Cu | Al |
| 12 or less | 14 AWG | 12 AWG | 14 AWG | 12 AWG |
| to   16 | 12 | 10 | 12 | 10 |
| to   20 | 10 | 10 | 10 | 10 |
| to   24 | 10 | 8 | 10 | 8 |
| to   32 | 8 | 6 | 8 | 8 |
| to   36 | 6 | 4 | 8 | 6 |
| to   40 | 6 | 4 | 8 | 6 |
| to   44 | 6 | 4 | 6 | 4 |
| to   52 | 4 | 3 | 6 | 4 |
| to   56 | 4 | 2 | 4 | 3 |
| to   60 | 3 | 2 | 4 | 3 |
| to   64 | 3 | 1 | 4 | 2 |
| to   68 | 3 | 1 | 4 | 2 |
| to   72 | 2 | 1/0 | 3 | 2 |
| to   76 | 2 | 1/0 | 3 | 1 |
| to   80 | 1 | 1/0 | 3 | 1 |
| to   88 | 1 | 2/0 | 2 | 1/0 |
| to   92 | 1/0 | 2/0 | 2 | 1/0 |
| to   96 | 1/0 | 3/0 | 1 | 1/0 |
| to 100 | 1/0 | 3/0 | 1 | 2/0 |
| to 104 | 2/0 | 3/0 | 1 | 2/0 |
| to 108 | 2/0 | 4/0 | 1/0 | 2/0 |
| to 116 | 2/0 | 4/0 | 1/0 | 3/0 |
| to 120 | 3/0 | 4/0 | 1/0 | 3/0 |
| to 124 | 3/0 | 250 kcmil | 2/0 | 3/0 |
| to 132 | 3/0 | 250 | 2/0 | 4/0 |
| to 136 | 4/0 | 250 | 2/0 | 4/0 |
| to 140 | 4/0 | 300 | 2/0 | 4/0 |
| to 144 | 4/0 | 300 | 3/0 | 4/0 |
| to 152 | 4/0 | 300 | 3/0 | 250 kcmil |
| to 156 | 4/0 | 350 | 3/0 | 250 |
| to 160 | 250 kcmil | 350 | 3/0 | 250 |
| to 164 | 250 | 350 | 4/0 | 250 |
| to 168 | 250 | 350 | 4/0 | 300 |
| to 172 | 250 | 400 | 4/0 | 300 |
| to 180 | 300 | 400 | 4/0 | 300 |
| to 184 | 300 | 500 | 4/0 | 300 |
| to 200 | 350 | 500 | 250 kcmil | 350 |
| to 204 | 350 | 500 | 250 | 400 |

## 50.  Electric Snow-Melting–System Design Data

Design heat density installed in slab, watts per square foot of heated area.

| Location | Residential,[a] Class I | | Commercial-Industrial,[b] Class II | | Critical,[c] Class III | |
|---|---|---|---|---|---|---|
| | Theoretical[d] | Common densities actually installed[e] | Theoretical[d] | Common densities actually installed[e] | Theoretical[d] | Common densities actually installed[e] |
| **Arkansas** | | | | | | |
| Ft. Smith | .... | 20 | .... | 40 | ...... | 45 |
| Little Rock | .... | 20 | .... | 30 | ...... | 50 |
| **Colorado** | | | | | | |
| Colorado Springs | 20 | .... | 26 | .... | 120 | .... |
| Denver | .... | 42 | .... | 50 | ...... | 60 |
| Pueblo | .... | .... | .... | 45 | ...... | 60 |
| **Connecticut** | | | | | | |
| Hartford | 47 | 30 | 104 | 50 | 107 | 70 |
| Middletown | .... | 40–60 | .... | 40–60 | ...... | 60–70 |
| New Haven | .... | 40 | .... | 40 | ...... | 60 |
| **Delaware** | | | | | | |
| Wilmington | .... | 30 | .... | 40 | ...... | 50 |
| **District of Columbia** | | | | | | |
| Washington | 48 | 30–40 | 50 | 40–55 | 59 | 55–60 |
| **Idaho** | | | | | | |
| Mt. Home | 21 | .... | 37 | .... | 57 | .... |
| **Illinois** | | | | | | |
| Chicago | 37 | 40 | 68 | 50 | 144 | 60 |
| Peoria | .... | 40 | .... | 45–50 | ...... | 55–60 |
| Rockford | .... | 42 | .... | 40–60 | ...... | .... |
| Springfield | .... | 40 | .... | 45–50 | ...... | 55–60 |
| **Indiana** | | | | | | |
| Elkhart | .... | .... | .... | 42 | ...... | .... |
| Hartford City | .... | .... | .... | 35 | ...... | .... |
| Indianapolis | .... | 40 | .... | 40 | ...... | 40–60 |
| South Bend | .... | .... | .... | 52 | ...... | 50–55 |
| **Iowa** | | | | | | |
| Dubuque | .... | 40 | .... | 40–60 | ...... | .... |
| **Kansas** | | | | | | |
| Kansas City | .... | 40 | .... | 50 | ...... | 60 |
| Salina | 35 | .... | 49 | .... | 94 | .... |
| Topeka | .... | 40 | .... | 40 | ...... | 60 |
| Wichita | .... | 50 | .... | 50 | ...... | 50 |
| **Kentucky** | | | | | | |
| Ashland | .... | .... | .... | 42 | ...... | .... |
| **Maine** | | | | | | |
| Caribou-Limestone | 37 | .... | 57 | .... | 126 | .... |
| Bangor | .... | 40 | .... | 40 | ...... | 60 |
| Portland | .... | 40 | .... | 40 | ...... | 60 |
| **Maryland** | | | | | | |
| Baltimore | 44 | 30–45 | 95 | 45–55 | 105 | 50–70 |
| **Massachusetts** | | | | | | |
| Boston | 44 | 40–45 | 95 | 50–60 | 105 | 60–75 |
| Falmouth | 38 | .... | 59 | .... | 68 | .... |
| Fall River | .... | 40 | .... | 40 | ...... | 60 |
| Springfield | .... | 40 | .... | 40 | ...... | 80 |

**50.  Electric Snow-Melting–System Design Data** (*Continued*)

| Location | Residential,[a] Class I | | Commercial-Industrial,[b] Class II | | Critical,[c] Class III | |
|---|---|---|---|---|---|---|
| | Theoretical[d] | Common densities actually installed[e] | Theoretical[d] | Common densities actually installed[e] | Theoretical[d] | Common densities actually installed[e] |
| **Michigan** | | | | | | |
| Detroit | 28 | 40–60 | 57 | 60 | 105 | 60 |
| Sault Ste. Marie | 21 | .... | 59 | .... | 87 | .... |
| Jackson | .... | 40 | .... | 60 | ...... | 80 |
| **Minnesota** | | | | | | |
| Duluth | 34 | .... | 85 | .... | 153 | .... |
| Minneapolis–St. Paul | 26 | 42–75 | 64 | 60–75 | 104 | 70–75 |
| **Missouri** | | | | | | |
| Kansas City | .... | 42 | .... | 40–50 | ...... | 60–70 |
| St. Joseph | .... | 42 | .... | 42 | ...... | .... |
| St. Louis | 50 | 40–60 | 62 | 40–60 | 81 | 60 |
| **Montana** | | | | | | |
| Great Falls | 34 | .... | 57 | .... | 153 | .... |
| **Nebraska** | | | | | | |
| Lincoln | 26 | 40–50 | 83 | 40–50 | 101 | 60 |
| Omaha | .... | 40–45 | .... | 60 | ...... | 60 |
| **Nevada** | | | | | | |
| Reno | 40 | .... | 63 | .... | 64 | .... |
| **New Hampshire** | | | | | | |
| Concord–Manchester | .... | 50 | .... | 50 | ...... | 75 |
| **New Jersey** | | | | | | |
| Atlantic City | .... | 30 | .... | 40 | ...... | 60 |
| Morristown | .... | 40 | .... | 50 | ...... | 60 |
| **New Mexico** | | | | | | |
| Albuquerque | 29 | .... | 34 | .... | 69 | .... |
| **New York** | | | | | | |
| Buffalo–Niagara Falls | 33 | .... | 79 | 60 | 126 | .... |
| New York City | 50 | 35–50 | 122 | 40–50 | 140 | 50–60 |
| Poughkeepsie | .... | 40 | .... | 60–70 | ...... | 50–80 |
| Syracuse | .... | 40–60 | .... | 60 | ...... | 60 |
| **North Carolina** | | | | | | |
| Charlotte | .... | 42 | .... | 30–42 | | 42 |
| **Ohio** | | | | | | |
| Canton | .... | 30 | .... | 36 | ...... | .... |
| Cincinnati | .... | 40 | .... | 50 | ...... | 60 |
| Cleveland | .... | 40 | .... | .... | ...... | 45–55 |
| Columbus | 21 | 30 | 30 | 40 | 104 | 50 |
| Findlay | .... | .... | .... | 40 | ...... | 60 |
| Ironton | .... | 30 | .... | 40 | ...... | .... |
| Lima | .... | 40 | .... | 40 | ...... | 70 |
| Portsmouth | .... | 30–40 | .... | 40 | ...... | .... |
| Steubenville | .... | 40 | .... | 45 | ...... | .... |
| **Oklahoma** | | | | | | |
| Oklahoma City | 27 | 25 | 33 | 40 | 144 | 45 |
| Tulsa | .... | 20 | .... | 30 | ...... | 40 |
| **Oregon** | | | | | | |
| Portland | 35 | 42 | 40 | 42 | 46 | 60 |

### 50. Electric Snow-Melting–System Design Data (*Continued*)

| Location | Residential,[a] Class I | | Commercial-Industrial,[b] Class II | | Critical,[c] Class III | |
|---|---|---|---|---|---|---|
| | Theoret-ical[d] | Common densities actually installed[e] | Theoret-ical[d] | Common densities actually installed[e] | Theoret-ical[d] | Common densities actually installed[e] |
| **Pennsylvania** | | | | | | |
| Allentown | .... | 40 | .... | 44–55 | ...... | 60 |
| Johnstown | .... | 30 | .... | 40 | ...... | 70 |
| Philadelphia | 40 | 40 | 94 | 40–60 | 108 | 40–100 |
| Pittsburgh | 37 | 30–40 | 65 | 30–60 | 113 | 60 |
| **Rhode Island** | | | | | | |
| Providence | .... | 45 | .... | 65 | ...... | .... |
| **So. Dakota** | | | | | | |
| Rapid City | 24 | .... | 42 | .... | 183 | .... |
| **Tennessee** | | | | | | |
| Memphis | 55 | .... | 59 | .... | 87 | .... |
| Kingsport | .... | .... | .... | 42 | ...... | .... |
| Nashville | .... | 40 | .... | 40 | ...... | 60 |
| **Texas** | | | | | | |
| Amarillo | 40 | .... | 59 | .... | 99 | .... |
| **Utah** | | | | | | |
| Ogden | 40 | .... | 89 | .... | 89 | .... |
| **Vermont** | | | | | | |
| Bennington | .... | 50 | .... | 50 | ...... | 75 |
| Burlington | 37 | 50 | 58 | 50 | 100 | 75 |
| **Virginia** | | | | | | |
| Richmond | .... | 30 | .... | 40 | ...... | .... |
| Abingdon | .... | 30 | .... | 30 | ...... | .... |
| **Washington** | | | | | | |
| Seattle | 38 | .... | 53 | .... | 55 | .... |
| Spokane | 36 | 30–40 | 52 | 30–45 | 78 | .... |
| **West Virginia** | | | | | | |
| Wheeling | .... | 35 | .... | 35–50 | ...... | .... |
| Bluefield | .... | .... | .... | 40 | ...... | 80 |
| Parkersburg | .... | 30 | .... | 45 | ...... | 60 |
| Beckley | .... | 40 | .... | 40 | ...... | .... |
| Charleston | .... | 40 | .... | 45 | ...... | 50 |
| Morgantown | .... | 30 | .... | 45 | ...... | 60 |
| **Wisconsin** | | | | | | |
| Milwaukee | .... | 45 | .... | 50–65 | ...... | 65 |
| **Wyoming** | | | | | | |
| Cheyenne | 34 | .... | 53 | .... | 174 | .... |

[a]*Residential.* Residential walks or driveways and interplant areaways—will allow snow completely to cover area temporarily but will not accumulate for worse conditions of 93 percent of snow frequency. Minimum recommended installed intensity: 30 W/ft$^2$.

[b]*Commercial-industrial.* Commercial sidewalks, steps, and driveways—will allow snow completely to cover area temporarily but will not accumulate for worse conditions of 100 percent of snow frequency. Minimum recommended installed intensity: 40 W/ft$^2$.

[c]*Critical.* Toll plazas of highways and bridges and aprons and loading areas of airports—will melt snow immediately for 98 percent of snow frequency. Minimum recommended intensity: 60 W/ft$^2$. Per. Sec. 426–20(a) of the National Electrical Code, installed heating intensity of embedded cable or wire systems shall not exceed 120 W/ft$^2$.

[d]Based on adjustment for assumed 40 percent heat loss through edge and bottom of slab. Figures from American Society of Heating, Refrigerating and Air-Conditioning Engineers.

[e]Based on survey of actual practice as reported by 66 electric utilities. Survey made during spring, 1966.

## 51.  Outside Design Temperatures and Yearly Degree-Days

The outside design temperatures given in this table are design dry-bulb temperatures in common use. These temperatures and the yearly degree-days are taken from American Society of Heating, Refrigerating and Air-Conditioning Engineers, *Heating, Ventilating, Air Conditioning Guide.* Outside temperatures are based on an occurrence of once in 13 years.

| State and city | Outside design temperature, °F | Yearly degree-days | State and city | Outside design temperature, °F | Yearly degree-days |
|---|---|---|---|---|---|
| **Alabama** | | | Dubuque | −15 | 7,271 (A) |
| Anniston | 12 | 2,820 (A) | Keokuk | −13 | 5,663 |
| Birmingham | 12 | 2,780 (A) | Sioux City | −15 | 7,012 (A) |
| Mobile | 22 | 1,612 (A) | **Kansas** | | |
| Montgomery | 18 | 2,137 (A) | Concordia | −11 | 5,323 |
| **Arizona** | | | Dodge City | −9 | 5,058 (A) |
| Flagstaff | −4 | 7.525 (A) | Topeka | −8 | 5,209 (A) |
| Phoenix | 36 | 1,698 (A) | Wichita | −6 | 4,571 (A) |
| Yuma | 38 | 951 (A) | **Kentucky** | | |
| **Arkansas** | | | Lexington | −2 | 4,979 (A) |
| Fort Smith | 6 | 3,188 (A) | Louisville | −2 | 4,439 (A) |
| Little Rock | 8 | 2,982 (A) | **Louisiana** | | |
| **California** | | | New Orleans | 26 | 1,317 (A) |
| Eureka | 32 | 4,632 | Shreveport | 14 | 2,117 (A) |
| Fresno | 32 | 2,532 (A) | **Maine** | | |
| Los Angeles | 41 | 2,015 (A) | Eastport | −9 | 8,246 |
| Red Bluff | ...... | 2,546 (A) | Greenville | ...... | 9,439 |
| Sacramento | 30 | 2,822 (A) | Portland | −9 | 7,681 (A) |
| San Diego | 43 | 1,574 (A) | **Maryland** | | |
| San Francisco | 37 | 3,421 (A) | Baltimore | 8 | 4,787 (A) |
| San Jose | 38 | 2,410 | **Massachusetts** | | |
| **Colorado** | | | Boston | 0 | 5,791 (A) |
| Denver | −12 | 6,132 (A) | Fitchburg | ...... | 6,743 |
| Durango | −6 | 7,143 | Nantucket | ...... | 6,102 (A) |
| Grand Junction | −3 | 5,796 (A) | **Michigan** | | |
| Pueblo | −14 | 5,709 (A) | Detroit | −4 | 6,404 (A) |
| **Connecticut** | | | Grand Rapids | −4 | 7,075 (A) |
| Hartford | −2 | 6,139 (A) | Lansing | −8 | 6,982 (A) |
| New Haven | 0 | 6,026 (A) | Sault Ste. Marie | −19 | 9,475 (A) |
| **District of Columbia** | | | **Minnesota** | | |
| Washington | 10 | 4,333 (A) | Duluth | −27 | 9,937 (A) |
| **Florida** | | | Minneapolis | −23 | 7,853 (A) |
| Apalachicola | 25 | 1,307 | St. Paul | −23 | 7,804 (A) |
| Jacksonville | 28 | 1,243 (A) | **Mississippi** | | |
| Key West | 53 | 89 (A) | Meridian | 14 | 2,333 (A) |
| Miami | 35 | 178 (A) | Vicksburg | 15 | 2,000 |
| Pensacola | 24 | 1,435 | **Missouri** | | |
| Tampa | 36 | 674 (A) | Columbia | −9 | 5,113 (A) |
| **Georgia** | | | Kansas City | −8 | 4,888 (A) |
| Atlanta | 11 | 2,826 (A) | St. Louis | −5 | 4,699 (A) |
| Augusta | 20 | 2,138 (A) | Springfield | −5 | 4,693 (A) |
| Macon | 20 | 2,049 (A) | **Montana** | | |
| Savannah | 24 | 1,710 (A) | Billings | −31 | 7,106 |
| **Idaho** | | | Havre | −39 | 8,213 |
| Boise | −10 | 5,890 (A) | Helena | −39 | 8,250 (A) |
| Lewiston | −12 | 5,483 (A) | Kalispell | ...... | 8,055 (A) |
| Pocatello | −17 | 6,976 (A) | Miles City | −35 | 7,850 (A) |
| **Illinois** | | | Missoula | ...... | 7,873 (A) |
| Cairo | 0 | 3,756 | **Nebraska** | | |
| Chicago | −11 | 6,310 (A) | Lincoln | −15 | 6,104 (A) |
| Peoria | −13 | 6,087 (A) | North Platte | −15 | 6,546 (A) |
| Springfield | −10 | 5,693 (A) | Omaha | −17 | 6,160 (A) |
| **Indiana** | | | Valentine | −21 | 7,075 |
| Evansville | −4 | 4,360 (A) | **Nevada** | | |
| Fort Wayne | −7 | 6,287 (A) | Reno | 3 | 6,036 (A) |
| Indianapolis | −8 | 5,611 (A) | Winnemucca | −9 | 6,369 (A) |
| Terre Haute | −6 | 5,366 (A) | **New Hampshire** | | |
| **Iowa** | | | Concord | −11 | 7,612 (A) |
| Charles City | −21 | 7,504 | **New Jersey** | | |
| Davenport | −12 | 6,091 | Atlantic City | 8 | 4,741 |
| Des Moines | −13 | 6,446 (A) | Cape May | ...... | 4,870 |

(A)  Temperatures recorded at airport stations.  Other temperatures recorded at city stations.

## 51. Outside Design Temperatures and Yearly Degree-Days (Continued)

| State and city | Outside design temperature, °F | Yearly degree-days | State and city | Outside design temperature, °F | Yearly degree-days |
|---|---|---|---|---|---|
| Newark | 0 | 5,252 (A) | Greenville | 10 | 3,060 (A) |
| Sandy Hook | 0 | 5,369 | South Dakota | | |
| Trenton | 2 | 5,068 | Huron | −21 | 7,902 (A) |
| New Mexico | | | Rapid City | −22 | 7,535 (A) |
| Albuquerque | 8 | 4,389 (A) | Tennessee | | |
| Roswell | 4 | 3,424 (A) | Chattanooga | 8 | 3,384 (A) |
| Sante Fe | 3 | 6,123 | Knoxville | 5 | 3,590 (A) |
| New York | | | Memphis | 6 | 3,137 (A) |
| Albany | −9 | 6,962 (A) | Nashville | 3 | 3,513 (A) |
| Binghamton | −7 | 7,537 (A) | Texas | | |
| Buffalo | −5 | 6,838 (A) | Abilene | 7 | 2,657 (A) |
| Canton | −22 | 8,305 | Amarillo | −2 | 4,345 (A) |
| Ithaca | −4 | 6,914 | Austin | ...... | 1,713 (A) |
| New York–LaGuardia | 5 | 4,989 (A) | Brownsville | 30 | 617 (A) |
| Oswego | −7 | 6,975 | Corpus Christi | 23 | 1,011 (A) |
| Rochester | −4 | 6,863 (A) | Dallas | 8 | 2,272 (A) |
| Syracuse | −10 | 6,520 (A) | Del Rio | ...... | 1,407 (A) |
| North Carolina | | | El Paso | 20 | 2,641 (A) |
| Asheville | 5 | 4,072 | Fort Worth | 8 | 2,361 (A) |
| Charlotte | 14 | 3,205 (A) | Galveston | 23 | 1,233 (A) |
| Raleigh | 14 | 3,369 (A) | Houston | 19 | 1,388 (A) |
| Wilmington | 20 | 2,323 (A) | Palestine | 11 | 1,980 |
| North Dakota | | | Port Arthur | 20 | 1,340 (A) |
| Bismarck | −31 | 9,033 (A) | San Antonio | 19 | 1,579 (A) |
| Devils Lake | −32 | 9,940 | Utah | | |
| Williston | −35 | 9,068 | Modena | −15 | 6,598 |
| Ohio | | | Salt Lake City | −1 | 5,866 (A) |
| Cincinnati | −3 | 5,195 (A) | Vermont | | |
| Cleveland | −5 | 6,006 (A) | Burlington | −17 | 7,865 (A) |
| Columbus | −3 | 5,615 (A) | Virginia | | |
| Dayton | −4 | 5,597 (A) | Cape Henry | 17 | 3,307 |
| Sandusky | −4 | 5,859 | Lynchburg | 11 | 4,153 (A) |
| Toledo | −5 | 6,394 (A) | Norfolk | 15 | 3,454 (A) |
| Oklahoma | | | Richmond | 11 | 3,955 (A) |
| Oklahoma City | −1 | 644 (A) | Washington | | |
| Oregon | | | North Head | 20 | 5,211 |
| Baker | −14 | 7,087 | Seattle | 15 | 4,438 |
| Medford | 5 | 4,547 (A) | Seattle–Tacoma | ...... | 5,275 (A) |
| Portland | 10 | 4,632 (A) | Spokane | −16 | 6,852 (A) |
| Roseburg | 19 | 4,122 | Tacoma | 15 | 4,866 |
| Pennsylvania | | | Walla Walla | −12 | 4,848 |
| Erie | −3 | 6,116 | Yakima | ...... | 5,845 (A) |
| Harrisburg | 4 | 5,258 (A) | West Virginia | | |
| Philadelphia | 6 | 4,866 (A) | Elkins | −4 | 5,773 (A) |
| Pittsburgh | −3 | 5,905 (A) | Parkersburg | −1 | 4,750 |
| Reading | 3 | 5,060 | Wisconsin | | |
| Scranton | −2 | 6,047 | Green Bay | −20 | 8,259 (A) |
| Rhode Island | | | La Crosse | −20 | 7,650 (A) |
| Block Island | 7 | 5,843 (A) | Madison | −19 | 7,417 (A) |
| Providence | 1 | 6,125 (A) | Milwaukee | −17 | 7,205 (A) |
| South Carolina | | | Wyoming | | |
| Charleston | 22 | 1,973 (A) | Cheyenne | −19 | 7,562 (A) |
| Columbia | 19 | 2,435 (A) | Lander | −30 | 8,303 (A) |

(A) Temperatures recorded at airport stations. Other temperatures recorded at city stations.

## 52.  Maximum Allowable Voltage Drop

In accordance with good practice, refer to Sec. 35 of Div. 3.

| Circuits | Voltage drop, percent | System voltages | | | | | | Remarks |
|---|---|---|---|---|---|---|---|---|
| | | 110 | 115 | 220 | 230 | 440 | 550 | |
| | | Voltage drop | | | | | | |
| For total drop of 5 percent | | | | | | | | |
| Lighting | | | | | | | | |
| Branches | 3 | 3.3 | 3.4 | 6.6 | 6.9 | | | In accordance with recommendations of the National Electrical Code |
| Mains and feeders combined | 2 | 2.2 | 2.3 | 4.4 | 4.6 | | | |
| Total | 5 | 5.5 | 5.8 | 11.0 | 11.5 | | | |
| For total drop of 4 percent | | | | | | | | |
| Lighting | | | | | | | | |
| Branches | 2 | 2.2 | 2.3 | 4.4 | 4.6 | | | In accordance with recommendations prepared by Industry Committee on Interior Wiring Design |
| Mains and feeders combined | 2 | 2.2 | 2.3 | 4.4 | 4.6 | | | |
| Total | 4 | 4.4 | 4.6 | 8.8 | 9.2 | | | |
| For total drop of 5 percent | | | | | | | | |
| Power | | | | | | | | |
| Branches | 2 | 2.2 | 2.3 | | | | | In accordance with recommendations prepared by Industry Committee on Interior Wiring Design |
| Mains and feeders combined | 3 | 3.3 | 3.5 | | | | | |
| Total | 5 | 5.5 | 5.8 | | | | | |
| For total drop of 5 percent | | | | | | | | |
| Power | | | | | | | | |
| Branches | 1 | ... | ... | 2.2 | 2.3 | 4.4 | 5.5 | In accordance with recommendations prepared by Industry Committee on Interior Wiring Design |
| Mains and feeders combined | 4 | ... | ... | 8.8 | 9.2 | 17.6 | 22.0 | |
| Total | 5 | ... | ... | 11.0 | 11.5 | 22.0 | 27.5 | |
| For total drop of 5 percent | | | | | | | | |
| Power | | | | | | | | |
| Branches | 3 | ... | ... | 6.6 | 6.9 | 13.2 | 16.5 | In accordance with recommendations of the National Electrical Code |
| Mains and feeders combined | 2 | ... | ... | 4.4 | 4.6 | 8.8 | 11.0 | |
| Total | 5 | ... | ... | 11.0 | 11.5 | 22.0 | 27.5 | |

**53. Graph for Computing Copper-Conductor Sizes for Circuits According to Voltage Drop**
See Sec. 59 of Div. 3.

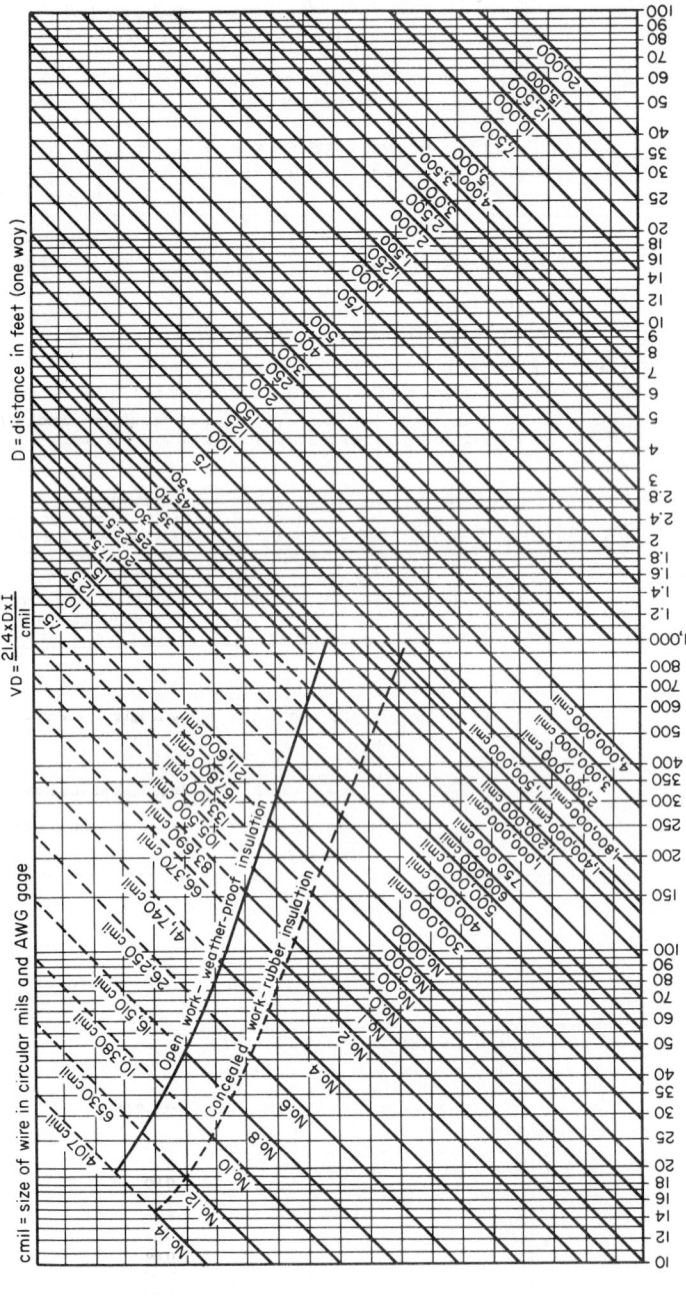

$$VD = \frac{21.4 \times D \times I}{cmil}$$

cmil = size of wire in circular mils and AWG gage

D = distance in feet (one way)

V = drop in volts

I = current in amperes

**54. The effect of inductance can be neglected unless the size of wire exceeds the following values:**

| Type of load | Size of wire with following spacings | | | | | | | | | | |
|---|---|---|---|---|---|---|---|---|---|---|---|
| | In conduit | 2½ in | 4 in | 5 in | 6 in | 8 in | 12 in | 18 in | 24 in | 36 in | 48 in |
| Incandescent lamps: 60 Hz | 4/0 | 2/0 | 1/0 | 1/0 | 1 | 1 | 2 | 2 | 2 | 3 | 3 |
| Incandescent lamps: 25 Hz | 600,000 | 400,000 | 300,000 | 4/0 | 4/0 | 4/0 | 3/0 | 3/0 | 3/0 | 3/0 | 2/0 |
| Motors: 60 Hz | 1 | 3 | 4 | 4 | 5 | 5 | 5 | | | | |
| Motors: 25 Hz | 4/0 | 00 | 0 | 0 | 0 | 1 | 1 | 2 | 2 | 2 | 3 |

**55. Effective Spacings of Wires to Use for Different Systems**

| System | Arrangement of wires | Effective spacing that should be used with Mershon diagram or table of ratio of reactance to resistance |
|---|---|---|
| Single-phase three-wire | Any | Use spacing between outside wires |
| Two-phase, any number of wires | Any | Use a spacing equal to average distance between centers of wires of the same phase |
| Three-phase | | Effective spacing = $\sqrt[3]{ABC}$ when wires are transposed so that each wire occupies each position for approximately one-third of the length of the line |
| | | Effective spacing = $\sqrt[3]{ABC}$ when wires are transposed so that each wire occupies each position for approximately one-third of the length of the line |
| | | Effective spacing = $1.26A$ when wires are transposed so that each wire occupies each position for approximately one-third of the length of the line<br><br>The position of the neutral in a four-wire system does not affect the spacing |

**56. Approximate Power Factors of Apparatus**

| Type of Apparatus | Power Factor |
|---|---|
| Incandescent lamps and heaters | 1.0 |
| Fluorescent lamps and mercury-vapor lamps[a] | 0.5–0.95 |
| Induction motors:[b] | |
|     1 hp | 0.67 |
|     2 hp | 0.73 |
|     3 hp | 0.80 |
|     5 hp | 0.83 |
|     7½–25 hp | 0.86 |
|     30–50 hp | 0.89 |
|     60–125 hp | 0.92 |
|     150–200 hp | 0.93 |
| Synchronous motors | 0.8 to 1.0 leading |

NOTE. All power factors less than unity are lagging except for synchronous motors.

[a] For more accurate information refer to Div. 10.

[b] The values of power factor for induction motors are average values when the motors are operating at 75 per cent of rated load.

## 57. 60-Hz Ratios: Reactance to Resistance

|  |  |  |  | Ratios for distance between wires (see Sec. 55) | | | | | | | | | | | | | | |
|---|---|---|---|---|---|---|---|---|---|---|---|---|---|---|---|---|---|---|
| Type | Size, AWG | Area, cmil | In conduit | 2½ in | 3 in | 4 in | 5 in | 6 in | 8 in | 12 in | 18 in | 2 ft | 3 ft | 4 ft | 5 ft | 6 ft | 7 ft | 8 ft |
|  | ... | 2,000,000 | ... | ... | 7.07 | 8.23 | 9.16 | 9.90 | 11.10 | 12.80 | 14.40 | 15.70 | 17.30 | 18.50 | 19.20 | 20.10 | 20.80 | 21.40 |
|  | ... | 1,800,000 | ... | ... | 6.52 | 7.58 | 8.46 | 9.15 | 10.20 | 11.70 | 13.20 | 14.30 | 15.80 | 17.00 | 17.70 | 18.40 | 19.00 | 19.40 |
|  | ... | 1,700,000 | ... | ... | 6.30 | 7.30 | 8.05 | 8.72 | 9.70 | 11.20 | 12.60 | 13.60 | 15.00 | 16.10 | 16.90 | 17.40 | 18.00 | 18.50 |
|  | ... | 1,600,000 | ... | ... | 5.97 | 6.96 | 7.67 | 8.30 | 9.22 | 10.60 | 11.90 | 12.90 | 14.20 | 15.20 | 16.00 | 16.40 | 17.00 | 17.50 |
|  | ... | 1,500,000 | ... | ... | 5.70 | 6.62 | 7.25 | 7.90 | 8.75 | 9.95 | 11.30 | 12.10 | 13.40 | 14.30 | 15.00 | 15.60 | 16.00 | 16.40 |
|  | ... | 1,400,000 | ... | ... | 5.40 | 6.26 | 6.90 | 7.40 | 8.28 | 9.43 | 10.60 | 11.50 | 12.60 | 13.50 | 14.10 | 14.70 | 15.20 | 15.50 |
|  | ... | 1,200,000 | ... | ... | 4.84 | 5.52 | 6.10 | 6.55 | 7.28 | 8.26 | 9.20 | 10.00 | 11.00 | 11.70 | 12.20 | 12.80 | 13.10 | 13.40 |
|  | ... | 1,100,000 | ... | ... | 4.51 | 5.20 | 5.70 | 6.11 | 6.75 | 7.70 | 8.60 | 9.25 | 10.20 | 10.80 | 11.40 | 11.80 | 12.10 | 12.40 |
|  | ... | 1,000,000 | 2.89 | 3.66 | 4.22 | 4.81 | 5.29 | 5.65 | 6.25 | 7.10 | 7.92 | 8.52 | 9.37 | 9.96 | 10.40 | 10.80 | 11.10 | 11.40 |
|  | ... | 950,000 | ... | ... | 4.05 | 4.60 | 5.05 | 5.42 | 6.00 | 6.80 | 7.55 | 8.12 | 8.95 | 9.50 | 9.95 | 10.30 | 10.60 | 10.90 |
| Stranded | ... | 900,000 | 2.69 | 3.39 | 3.92 | 4.43 | 4.87 | 5.20 | 5.75 | 6.50 | 7.24 | 7.82 | 8.50 | 9.10 | 9.52 | 9.82 | 10.10 | 10.40 |
|  | ... | 850,000 | ... | ... | 3.76 | 4.24 | 4.65 | 4.97 | 5.46 | 6.20 | 6.90 | 7.40 | 8.15 | 8.65 | 9.03 | 9.33 | 9.65 | 9.85 |
|  | ... | 800,000 | 2.49 | 3.12 | 3.55 | 4.03 | 4.42 | 4.70 | 5.17 | 5.83 | 6.55 | 7.00 | 7.65 | 8.15 | 8.50 | 8.80 | 9.10 | 9.30 |
|  | ... | 750,000 | ... | ... | 3.39 | 3.84 | 4.18 | 4.48 | 4.91 | 5.55 | 6.18 | 6.63 | 7.25 | 7.70 | 8.05 | 8.35 | 8.60 | 8.78 |
|  | ... | 700,000 | 2.06 | 2.84 | 3.21 | 3.66 | 3.98 | 4.25 | 4.65 | 5.23 | 5.82 | 6.25 | 6.83 | 7.25 | 7.60 | 7.85 | 8.10 | 8.27 |
|  | ... | 650,000 | ... | ... | 3.06 | 3.42 | 3.73 | 3.98 | 4.35 | 4.90 | 5.46 | 5.85 | 6.40 | 6.77 | 7.10 | 7.32 | 7.55 | 7.73 |
|  | ... | 600,000 | 1.85 | 2.52 | 2.85 | 3.21 | 3.47 | 3.70 | 4.07 | 4.55 | 5.05 | 5.42 | 5.93 | 6.30 | 6.55 | 6.80 | 6.98 | 7.15 |
|  | ... | 550,000 | ... | ... | 2.66 | 2.98 | 3.22 | 3.44 | 3.76 | 4.22 | 4.68 | 5.00 | 5.45 | 5.80 | 6.05 | 6.27 | 6.45 | 6.60 |
|  | ... | 500,000 | 1.75 | 2.30 | 2.48 | 2.78 | 2.99 | 3.20 | 3.51 | 3.92 | 4.35 | 4.63 | 5.05 | 5.35 | 5.60 | 5.78 | 5.90 | 6.08 |
|  | ... | 450,000 | ... | ... | 2.26 | 2.52 | 2.74 | 2.90 | 3.16 | 3.54 | 3.90 | 4.18 | 4.55 | 4.84 | 5.04 | 5.20 | 5.32 | 5.47 |

| | Size | Circular Mils | | | | | | | | | | | | | | | | |
|---|---|---|---|---|---|---|---|---|---|---|---|---|---|---|---|---|---|---|---|
| Stranded | : : : | 400,000 | 1.49 | 1.93 | 2.07 | 2.30 | 2.48 | 2.64 | 2.88 | 3.21 | 3.54 | 3.77 | 4.12 | 4.35 | 4.55 | 4.68 | 4.82 | 4.91 |
| | : : : | 350,000 | | | 1.87 | 2.08 | 2.23 | 2.37 | 2.58 | 2.87 | 3.17 | 3.38 | 3.68 | 3.88 | 4.05 | 4.17 | 4.30 | 4.39 |
| | : : : | 300,000 | 1.01 | 1.54 | 1.63 | 1.81 | 1.95 | 2.06 | 2.23 | 2.49 | 2.74 | 2.92 | 3.16 | 3.34 | 3.48 | 3.60 | 3.70 | 3.77 |
| | : : | 250,000 | | | 1.41 | 1.56 | 1.67 | 1.76 | 1.91 | 2.13 | 2.34 | 2.50 | 2.70 | 2.85 | 2.97 | 3.06 | 3.13 | 3.20 |
| | : 4/0 | 211,600 | 0.76 | 1.17 | 1.23 | 1.36 | 1.46 | 1.54 | 1.66 | 1.84 | 2.02 | 2.15 | 2.33 | 2.46 | 2.55 | 2.63 | 2.69 | 2.76 |
| | 3/0 | 167,772 | 0.64 | 0.97 | 1.02 | 1.11 | 1.20 | 1.26 | 1.36 | 1.50 | 1.64 | 1.74 | 1.88 | 1.98 | 2.06 | 2.12 | 2.16 | 2.22 |
| | 2/0 | 133,079 | 0.54 | 0.80 | 0.84 | 0.91 | 0.98 | 1.03 | 1.11 | 1.22 | 1.33 | 1.41 | 1.52 | 1.60 | 1.66 | 1.71 | 1.75 | 1.79 |
| | 1/0 | 105,560 | 0.38 | 0.66 | 0.69 | 0.75 | 0.80 | 0.84 | 0.90 | 1.00 | 1.08 | 1.15 | 1.23 | 1.30 | 1.35 | 1.39 | 1.42 | 1.45 |
| | 1 | 83,694 | 0.32 | 0.54 | 0.57 | 0.62 | 0.66 | 0.70 | 0.74 | 0.81 | 0.88 | 0.94 | 1.00 | 1.06 | 1.10 | 1.13 | 1.15 | 1.17 |
| | 2 | 66,358 | 0.26 | 0.45 | 0.47 | 0.50 | 0.53 | 0.56 | 0.60 | 0.65 | 0.71 | 0.75 | 0.81 | 0.85 | 0.88 | 0.90 | 0.92 | 0.95 |
| | 3 | 52,624 | 0.22 | 0.37 | 0.38 | 0.41 | 0.44 | 0.45 | 0.48 | 0.53 | 0.58 | 0.60 | 0.65 | 0.68 | 0.70 | 0.72 | 0.74 | 0.75 |
| | 4 | 41,738 | 0.15 | 0.30 | 0.31 | 0.34 | 0.35 | 0.37 | 0.40 | 0.43 | 0.47 | 0.49 | 0.53 | 0.55 | 0.57 | 0.59 | 0.60 | 0.61 |
| | 5 | 33,088 | 0.14 | 0.25 | 0.25 | 0.27 | 0.29 | 0.30 | 0.32 | 0.35 | 0.37 | 0.40 | 0.42 | 0.44 | 0.46 | 0.47 | 0.48 | 0.49 |
| Solid | 4/0 | 211,600 | 0.77 | 1.18 | 1.25 | 1.38 | 1.48 | 1.56 | 1.69 | 1.87 | 2.06 | 2.18 | 2.37 | 2.50 | 2.60 | 2.68 | 2.72 | 2.81 |
| | 3/0 | 167,772 | 0.65 | 0.98 | 1.04 | 1.14 | 1.22 | 1.29 | 1.39 | 1.53 | 1.68 | 1.78 | 1.92 | 2.04 | 2.10 | 2.17 | 2.20 | 2.27 |
| | 2/0 | 133,079 | 0.55 | 0.81 | 0.85 | 0.94 | 1.00 | 1.05 | 1.13 | 1.25 | 1.36 | 1.44 | 1.55 | 1.64 | 1.70 | 1.75 | 1.79 | 1.84 |
| | 1/0 | 105,560 | 0.39 | 0.67 | 0.70 | 0.77 | 0.82 | 0.86 | 0.92 | 1.01 | 1.11 | 1.17 | 1.26 | 1.32 | 1.37 | 1.41 | 1.45 | 1.48 |
| | 1 | 83,694 | 0.33 | 0.55 | 0.59 | 0.63 | 0.67 | 0.71 | 0.76 | 0.83 | 0.90 | 0.95 | 1.03 | 1.08 | 1.12 | 1.15 | 1.17 | 1.20 |
| | 2 | 66,358 | 0.27 | 0.46 | 0.48 | 0.52 | 0.55 | 0.57 | 0.61 | 0.67 | 0.73 | 0.77 | 0.83 | 0.87 | 0.89 | 0.92 | 0.94 | 0.96 |
| | 3 | 52,624 | 0.23 | 0.38 | 0.39 | 0.42 | 0.44 | 0.46 | 0.49 | 0.54 | 0.58 | 0.61 | 0.66 | 0.69 | 0.71 | 0.74 | 0.75 | 0.77 |
| | 4 | 41,738 | 0.16 | 0.31 | 0.32 | 0.34 | 0.36 | 0.38 | 0.40 | 0.44 | 0.48 | 0.50 | 0.53 | 0.56 | 0.58 | 0.60 | 0.61 | 0.62 |
| | 5 | 33,088 | 0.15 | 0.25 | 0.26 | 0.28 | 0.30 | 0.31 | 0.33 | 0.36 | 0.39 | 0.40 | 0.43 | 0.45 | 0.47 | 0.48 | 0.50 | 0.50 |

## 58. 25-Hz Ratios: Reactance to Resistance

| | | | | | | | | | | Distance between wires (see Sec. 55) | | | | | | | | |
| Type | Size, AWG | Area, cmil | In conduit | 2½ in | 3 in | 4 in | 5 in | 6 in | 8 in | 12 in | 18 in | 2 ft | 3 ft | 4 ft | 5 ft | 6 ft | 7 ft | 8 ft |
|---|---|---|---|---|---|---|---|---|---|---|---|---|---|---|---|---|---|---|
| | | | | | | | | | | Ratio | | | | | | | | |
| | ... | 2,000,000 | ... | ... | 2.93 | 3.42 | 3.82 | 4.12 | 4.60 | 5.32 | 5.98 | 6.52 | 7.20 | 7.68 | 8.00 | 8.35 | 8.67 | 8.92 |
| | ... | 1,800,000 | ... | ... | 2.74 | 3.17 | 3.50 | 3.79 | 4.25 | 4.86 | 5.47 | 5.95 | 6.60 | 7.02 | 7.40 | 7.64 | 7.88 | 8.06 |
| | ... | 1,700,000 | ... | ... | 2.62 | 3.02 | 3.34 | 3.64 | 4.04 | 4.65 | 5.24 | 5.67 | 6.25 | 6.70 | 7.02 | 7.27 | 7.50 | 7.70 |
| | ... | 1,600,000 | ... | ... | 2.48 | 2.89 | 3.19 | 3.46 | 3.84 | 4.38 | 4.93 | 5.35 | 5.90 | 6.30 | 6.65 | 6.85 | 7.10 | 7.25 |
| | ... | 1,500,000 | ... | ... | 2.38 | 2.76 | 3.04 | 3.27 | 3.63 | 4.15 | 4.68 | 5.05 | 5.60 | 5.96 | 6.24 | 6.48 | 6.68 | 6.83 |
| | ... | 1,400,000 | ... | ... | 2.27 | 2.60 | 2.87 | 3.08 | 3.44 | 3.92 | 4.42 | 4.77 | 5.25 | 5.60 | 5.88 | 6.10 | 6.30 | 6.43 |
| | ... | 1,200,000 | ... | ... | 2.02 | 2.31 | 2.54 | 2.72 | 3.02 | 3.44 | 3.85 | 4.18 | 4.58 | 4.87 | 5.12 | 5.32 | 5.45 | 5.60 |
| | ... | 1,100,000 | ... | ... | 1.87 | 2.15 | 2.38 | 2.55 | 2.81 | 3.20 | 3.56 | 3.85 | 4.24 | 4.50 | 4.75 | 4.90 | 5.03 | 5.17 |
| | ... | 1,000,000 | 1.20 | 1.53 | 1.78 | 2.00 | 2.20 | 2.36 | 2.60 | 2.95 | 3.31 | 3.55 | 3.91 | 4.16 | 4.35 | 4.48 | 4.63 | 4.76 |
| | ... | 950,000 | ... | ... | 1.70 | 1.91 | 2.10 | 2.25 | 2.49 | 2.83 | 3.16 | 3.38 | 3.70 | 3.96 | 4.14 | 4.27 | 4.40 | 4.53 |
| Stranded | ... | 900,000 | 1.12 | 1.42 | 1.63 | 1.85 | 2.03 | 2.17 | 2.38 | 2.71 | 3.02 | 3.25 | 3.54 | 3.79 | 3.97 | 4.08 | 4.20 | 4.32 |
| | ... | 850,000 | ... | ... | 1.57 | 1.77 | 1.95 | 2.08 | 2.28 | 2.57 | 2.86 | 3.08 | 3.39 | 3.60 | 3.75 | 3.89 | 4.01 | 4.10 |
| | ... | 800,000 | 1.03 | 1.30 | 1.48 | 1.69 | 1.84 | 1.96 | 2.15 | 2.42 | 2.72 | 2.91 | 3.17 | 3.38 | 3.54 | 3.65 | 3.77 | 3.87 |
| | ... | 750,000 | ... | ... | 1.41 | 1.60 | 1.74 | 1.87 | 2.05 | 2.32 | 2.57 | 2.77 | 3.02 | 3.21 | 3.36 | 3.47 | 3.57 | 3.66 |
| | ... | 700,000 | 0.86 | 1.18 | 1.34 | 1.52 | 1.65 | 1.77 | 1.93 | 2.18 | 2.42 | 2.60 | 2.85 | 3.02 | 3.16 | 3.27 | 3.36 | 3.44 |
| | ... | 650,000 | 0.77 | 1.05 | 1.27 | 1.43 | 1.56 | 1.66 | 1.81 | 2.04 | 2.28 | 2.42 | 2.66 | 2.83 | 2.96 | 3.04 | 3.14 | 3.22 |
| | ... | 600,000 | ... | ... | 1.19 | 1.34 | 1.44 | 1.55 | 1.70 | 1.90 | 2.11 | 2.25 | 2.47 | 2.62 | 2.72 | 2.82 | 2.90 | 2.98 |
| | ... | 550,000 | 0.73 | 0.96 | 1.10 | 1.24 | 1.34 | 1.43 | 1.56 | 1.76 | 1.94 | 2.08 | 2.27 | 2.41 | 2.51 | 2.62 | 2.68 | 2.74 |
| | ... | 500,000 | ... | ... | 1.03 | 1.15 | 1.25 | 1.33 | 1.46 | 1.63 | 1.81 | 1.93 | 2.12 | 2.23 | 2.33 | 2.42 | 2.46 | 2.53 |
| | ... | 450,000 | ... | ... | 0.95 | 1.05 | 1.14 | 1.21 | 1.32 | 1.47 | 1.63 | 1.75 | 1.90 | 2.01 | 2.09 | 2.16 | 2.22 | 2.28 |

| Type | Size | CM | | | | | | | | | | | | | | | | |
|---|---|---|---|---|---|---|---|---|---|---|---|---|---|---|---|---|---|---|---|
| Stranded | … | 400,000 | 0.62 | 0.81 | 0.86 | 0.96 | 1.03 | 1.10 | 1.20 | 1.33 | 1.47 | 1.57 | 1.72 | 1.81 | 1.89 | 1.95 | 2.01 | 2.05 |
| | … | 350,000 | … | … | 0.78 | 0.86 | 0.93 | 0.99 | 1.07 | 1.20 | 1.32 | 1.41 | 1.53 | 1.62 | 1.68 | 1.75 | 1.80 | 1.83 |
| | … | 300,000 | 0.42 | 0.64 | 0.68 | 0.76 | 0.81 | 0.86 | 0.93 | 1.04 | 1.15 | 1.22 | 1.32 | 1.40 | 1.45 | 1.50 | 1.54 | 1.57 |
| | … | 250,000 | … | … | 0.59 | 0.65 | 0.69 | 0.74 | 0.80 | 0.89 | 0.97 | 1.04 | 1.12 | 1.19 | 1.23 | 1.27 | 1.30 | 1.34 |
| | 4/0 | 211,600 | 0.32 | 0.49 | 0.51 | 0.57 | 0.61 | 0.64 | 0.69 | 0.77 | 0.84 | 0.90 | 0.97 | 1.02 | 1.06 | 1.09 | 1.12 | 1.15 |
| | 3/0 | 167,772 | 0.27 | 0.40 | 0.42 | 0.46 | 0.50 | 0.52 | 0.56 | 0.62 | 0.68 | 0.73 | 0.79 | 0.83 | 0.86 | 0.88 | 0.90 | 0.93 |
| | 2/0 | 133,079 | 0.23 | 0.33 | 0.35 | 0.38 | 0.41 | 0.43 | 0.46 | 0.51 | 0.55 | 0.59 | 0.63 | 0.66 | 0.69 | 0.71 | 0.73 | 0.75 |
| | 1/0 | 105,560 | 0.16 | 0.28 | 0.29 | 0.31 | 0.33 | 0.35 | 0.38 | 0.41 | 0.45 | 0.48 | 0.51 | 0.54 | 0.56 | 0.58 | 0.59 | 0.60 |
| | 1 | 83,694 | 0.13 | 0.23 | 0.24 | 0.26 | 0.28 | 0.29 | 0.31 | 0.34 | 0.37 | 0.39 | 0.42 | 0.44 | 0.46 | 0.47 | 0.48 | 0.49 |
| | 2 | 66,358 | 0.11 | 0.19 | 0.19 | 0.21 | 0.22 | 0.24 | 0.25 | 0.27 | 0.30 | 0.31 | 0.34 | 0.35 | 0.36 | 0.38 | 0.38 | 0.39 |
| | 3 | 52,624 | 0.09 | 0.15 | 0.16 | 0.17 | 0.18 | 0.19 | 0.20 | 0.22 | 0.24 | 0.25 | 0.27 | 0.28 | 0.29 | 0.30 | 0.31 | 0.31 |
| | 4 | 41,738 | 0.06 | 0.12 | 0.13 | 0.14 | 0.15 | 0.15 | 0.16 | 0.18 | 0.19 | 0.20 | 0.22 | 0.23 | 0.24 | 0.24 | 0.25 | 0.25 |
| | 5 | 33,088 | 0.06 | 0.10 | 0.10 | 0.11 | 0.12 | 0.12 | 0.13 | 0.14 | 0.16 | 0.16 | 0.18 | 0.18 | 0.19 | 0.19 | 0.20 | 0.20 |
| Solid | 4/0 | 211,600 | 0.33 | 0.50 | 0.52 | 0.58 | 0.62 | 0.65 | 0.71 | 0.78 | 0.86 | 0.91 | 0.99 | 1.04 | 1.08 | 1.12 | 1.14 | 1.17 |
| | 3/0 | 167,772 | 0.28 | 0.41 | 0.43 | 0.48 | 0.51 | 0.54 | 0.58 | 0.64 | 0.70 | 0.74 | 0.80 | 0.84 | 0.88 | 0.90 | 0.92 | 0.95 |
| | 2/0 | 133,079 | 0.24 | 0.34 | 0.36 | 0.39 | 0.42 | 0.44 | 0.47 | 0.52 | 0.57 | 0.60 | 0.65 | 0.68 | 0.71 | 0.73 | 0.74 | 0.76 |
| | 1/0 | 105,560 | 0.17 | 0.29 | 0.29 | 0.32 | 0.34 | 0.36 | 0.38 | 0.42 | 0.46 | 0.49 | 0.52 | 0.55 | 0.57 | 0.59 | 0.60 | 0.72 |
| | 1 | 83,694 | 0.14 | 0.24 | 0.24 | 0.26 | 0.28 | 0.29 | 0.31 | 0.35 | 0.38 | 0.40 | 0.43 | 0.45 | 0.47 | 0.48 | 0.49 | 0.50 |
| | 2 | 66,358 | 0.12 | 0.20 | 0.20 | 0.22 | 0.23 | 0.24 | 0.25 | 0.28 | 0.30 | 0.32 | 0.35 | 0.36 | 0.37 | 0.39 | 0.39 | 0.40 |
| | 3 | 52,624 | 0.10 | 0.16 | 0.16 | 0.17 | 0.18 | 0.19 | 0.20 | 0.22 | 0.24 | 0.26 | 0.27 | 0.29 | 0.30 | 0.31 | 0.31 | 0.32 |
| | 4 | 41,738 | 0.07 | 0.13 | 0.13 | 0.14 | 0.15 | 0.16 | 0.17 | 0.18 | 0.20 | 0.21 | 0.22 | 0.23 | 0.24 | 0.25 | 0.25 | 0.26 |
| | 5 | 33,088 | 0.07 | 0.11 | 0.11 | 0.13 | 0.12 | 0.13 | 0.14 | 0.15 | 0.16 | 0.17 | 0.18 | 0.19 | 0.20 | 0.20 | 0.21 | 0.21 |

59.   Alternating-Current Drop Factors

| Ratio of reactance to resistance | Power factor | | | | | | | |
|---|---|---|---|---|---|---|---|---|
| | 1.00 | 0.95 | 0.90 | 0.85 | 0.80 | 0.70 | 0.60 | 0.40 |
| | Drop factor | | | | | | | |
| 0.1 | 1.00 | 1.00 | 1.00 | 0.94 | 0.88 | 0.80 | 0.70 | 0.60 |
| 0.2 | 1.00 | 1.01 | 1.01 | 0.98 | 0.92 | 0.86 | 0.82 | 0.67 |
| 0.3 | 1.00 | 1.05 | 1.05 | 1.02 | 0.99 | 0.93 | 0.89 | 0.74 |
| 0.4 | 1.00 | 1.08 | 1.10 | 1.08 | 1.04 | 1.00 | 0.93 | 0.82 |
| 0.5 | 1.00 | 1.11 | 1.14 | 1.13 | 1.10 | 1.07 | 1.01 | 0.92 |
| 0.6 | 1.01 | 1.15 | 1.18 | 1.19 | 1.15 | 1.14 | 1.09 | 1.01 |
| 0.7 | 1.02 | 1.18 | 1.23 | 1.24 | 1.21 | 1.20 | 1.17 | 1.11 |
| 0.8 | 1.02 | 1.21 | 1.28 | 1.29 | 1.28 | 1.27 | 1.24 | 1.20 |
| 0.9 | 1.03 | 1.25 | 1.33 | 1.34 | 1.34 | 1.35 | 1.32 | 1.29 |
| 1.0 | 1.04 | 1.28 | 1.37 | 1.39 | 1.40 | 1.41 | 1.39 | 1.38 |
| 1.1 | 1.05 | 1.32 | 1.41 | 1.44 | 1.45 | 1.48 | 1.47 | 1.46 |
| 1.2 | 1.06 | 1.35 | 1.46 | 1.30 | 1.51 | 1.55 | 1.54 | 1.55 |
| 1.3 | 1.07 | 1.39 | 1.51 | 1.55 | 1.57 | 1.62 | 1.63 | 1.64 |
| 1.4 | 1.08 | 1.43 | 1.55 | 1.61 | 1.64 | 1.70 | 1.71 | 1.72 |
| 1.5 | 1.10 | 1.47 | 1.60 | 1.67 | 1.70 | 1.77 | 1.80 | 1.81 |
| 1.6 | 1.10 | 1.51 | 1.65 | 1.74 | 1.77 | 1.85 | 1.87 | 1.90 |
| 1.7 | 1.13 | 1.55 | 1.70 | 1.79 | 1.84 | 1.92 | 1.95 | 1.99 |
| 1.8 | 1.15 | 1.59 | 1.76 | 1.85 | 1.91 | 1.99 | 2.04 | 2.08 |
| 1.9 | 1.17 | 1.63 | 1.82 | 1.91 | 1.98 | 2.06 | 2.11 | 2.16 |
| 2.0 | 1.18 | 1.68 | 1.87 | 1.96 | 2.04 | 2.14 | 2.19 | 2.25 |
| 2.1 | 1.20 | 1.72 | 1.92 | 2.03 | 2.10 | 2.21 | 2.28 | 2.35 |
| 2.2 | 1.22 | 1.77 | 1.98 | 2.09 | 2.17 | 2.29 | 2.37 | 2.45 |
| 2.3 | 1.23 | 1.82 | 2.03 | 2.15 | 2.23 | 2.37 | 2.45 | 2.53 |
| 2.4 | 1.25 | 1.87 | 2.09 | 2.22 | 2.30 | 2.44 | 2.53 | 2.62 |
| 2.5 | 1.27 | 1.91 | 2.14 | 2.28 | 2.37 | 2.52 | 2.60 | 2.71 |
| 2.6 | 1.30 | 1.95 | 2.20 | 2.34 | 2.44 | 2.60 | 2.67 | 2.80 |
| 2.7 | 1.32 | 1.99 | 2.26 | 2.41 | 2.51 | 2.68 | 2.74 | 2.98 |
| 2.8 | 1.35 | 2.05 | 2.32 | 2.47 | 2.57 | 2.76 | 2.82 | 3.07 |
| 2.9 | 1.37 | 2.10 | 2.39 | 2.54 | 2.64 | 2.83 | 2.91 | 3.15 |
| 3.0 | 1.40 | 2.15 | 2.45 | 2.60 | 2.72 | 2.90 | 3.00 | 3.23 |
| 3.1 | 1.42 | 2.20 | 2.51 | 2.66 | 2.80 | 2.97 | 3.10 | 3.31 |
| 3.2 | 1.45 | 2.26 | 2.57 | 2.73 | 2.87 | 3.05 | 3.20 | 3.39 |
| 3.3 | 1.48 | 2.31 | 2.63 | 2.80 | 2.93 | 3.12 | 3.30 | 3.47 |
| 3.4 | 1.51 | 2.36 | 2.69 | 2.87 | 3.00 | 3.20 | 3.39 | 3.56 |
| 3.5 | 1.53 | 2.42 | 2.74 | 2.94 | 3.08 | 3.27 | 3.48 | 3.65 |
| 3.6 | 1.57 | 2.47 | 2.80 | 3.00 | 3.15 | 3.35 | 3.56 | 3.75 |
| 3.7 | 1.60 | 2.52 | 2.86 | 3.07 | 3.23 | 3.43 | 3.65 | 3.85 |

**60. Mershon diagram.** See Sec. 72 of Div. 3 for directions as to the diagram's use and application. The side of each small square equals 1 percent; percentage of resistance is measured horizontally, and percentage of reactance vertically.

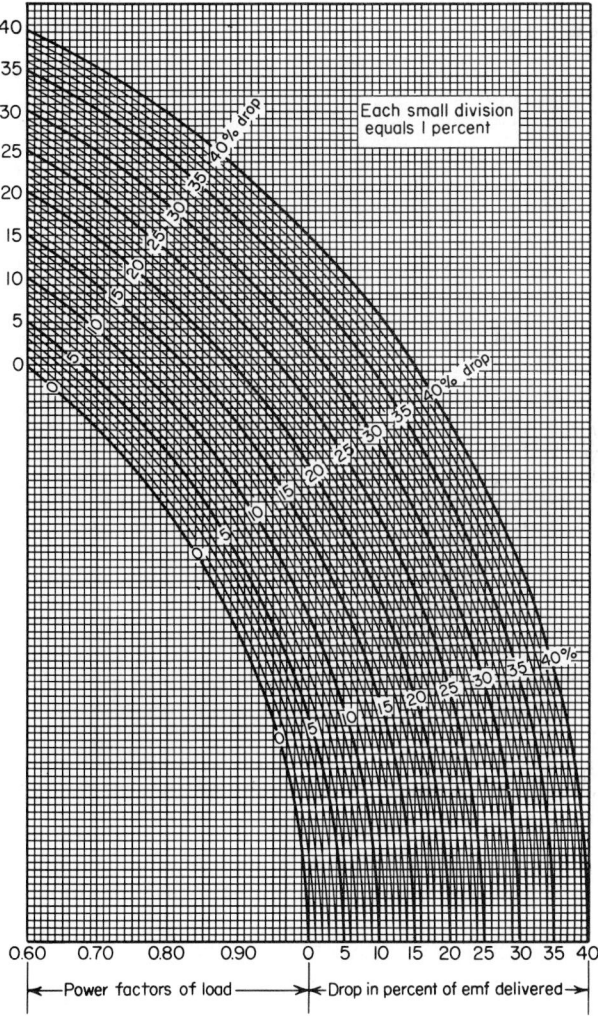

**61. Conductor Properties**[a]
(Table 8, 1990 NEC)

| Size, AWG or 1000 cmil | Area, cmil | DC resistance at 75°C (167°F) | | | | | | |
|---|---|---|---|---|---|---|---|---|
| | | Stranding | | Overall | | Copper | | Aluminum Ω/1000 ft |
| | | Quantity | Diameter, in | Diameter, in | Area, in² | Uncoated, Ω/1000 ft | Coated, Ω/1000 ft | |
| 18 AWG | 1,620 | 1 | .... | 0.040 | 0.001 | 7.77 | 8.08 | 12.8 |
| 18 | 1,620 | 7 | 0.015 | 0.046 | 0.002 | 7.95 | 8.45 | 13.1 |
| 16 | 2,580 | 1 | .... | 0.051 | 0.002 | 4.89 | 5.08 | 8.05 |
| 16 | 2,580 | 7 | 0.019 | 0.058 | 0.003 | 4.99 | 5.29 | 8.21 |
| 14 | 4,110 | 1 | .... | 0.064 | 0.003 | 3.07 | 3.19 | 5.06 |
| 14 | 4,110 | 7 | 0.024 | 0.073 | 0.004 | 3.14 | 3.26 | 5.17 |
| 12 | 6,530 | 1 | .... | 0.081 | 0.005 | 1.93 | 2.01 | 3.18 |
| 12 | 6,530 | 7 | 0.030 | 0.092 | 0.006 | 1.98 | 2.05 | 3.25 |
| 10 | 10,380 | 1 | .... | 0.102 | 0.008 | 1.21 | 1.26 | 2.00 |
| 10 | 10,380 | 7 | 0.038 | 0.116 | 0.011 | 1.24 | 1.29 | 2.04 |
| 8 | 16,510 | 1 | .... | 0.128 | 0.013 | 0.764 | 0.786 | 1.26 |
| 8 | 16,510 | 7 | 0.049 | 0.146 | 0.017 | 0.778 | 0.809 | 1.28 |
| 6 | 26,240 | 7 | 0.061 | 0.184 | 0.027 | 0.491 | 0.510 | 0.808 |
| 4 | 41,740 | 7 | 0.077 | 0.232 | 0.042 | 0.308 | 0.321 | 0.508 |
| 3 | 52,620 | 7 | 0.087 | 0.260 | 0.053 | 0.245 | 0.254 | 0.403 |
| 2 | 66,360 | 7 | 0.097 | 0.292 | 0.067 | 0.194 | 0.201 | 0.319 |
| 1 | 83,690 | 19 | 0.066 | 0.332 | 0.087 | 0.154 | 0.160 | 0.253 |
| 1/0 | 105,600 | 19 | 0.074 | 0.373 | 0.109 | 0.122 | 0.127 | 0.201 |
| 2/0 | 133,100 | 19 | 0.084 | 0.419 | 0.138 | 0.0967 | 0.101 | 0.159 |
| 3/0 | 167,800 | 19 | 0.094 | 0.470 | 0.173 | 0.0766 | 0.0797 | 0.126 |

| | | | | | | | | |
|---|---|---|---|---|---|---|---|---|
| 4/0 | 211,600 | 19 | 0.106 | 0.528 | 0.219 | 0.0608 | 0.0626 | 0.100 |
| 250 MCM | ...... | 37 | 0.082 | 0.575 | 0.260 | 0.0515 | 0.0535 | 0.0847 |
| 300 | ...... | 37 | 0.090 | 0.630 | 0.312 | 0.0429 | 0.0446 | 0.0707 |
| 350 | ...... | 37 | 0.097 | 0.681 | 0.364 | 0.0367 | 0.0382 | 0.0605 |
| 400 | ...... | 37 | 0.104 | 0.728 | 0.416 | 0.0321 | 0.0331 | 0.0529 |
| 500 | ...... | 37 | 0.116 | 0.813 | 0.519 | 0.0258 | 0.0265 | 0.0424 |
| 600 | ...... | 61 | 0.99 | 0.893 | 0.626 | 0.0214 | 0.0223 | 0.0353 |
| 700 | ...... | 61 | 0.107 | 0.964 | 0.730 | 0.0184 | 0.0189 | 0.0303 |
| 750 | ...... | 61 | 0.111 | 0.998 | 0.782 | 0.0171 | 0.0176 | 0.0282 |
| 800 | ...... | 61 | 0.114 | 1.03 | 0.834 | 0.0161 | 0.0166 | 0.0265 |
| 900 | ...... | 61 | 0.122 | 1.09 | 0.940 | 0.0143 | 0.0147 | 0.0235 |
| 1000 | ...... | 61 | 0.128 | 1.15 | 1.04 | 0.0129 | 0.0132 | 0.0212 |
| 1250 | ...... | 91 | 0.117 | 1.29 | 1.30 | 0.0103 | 0.0106 | 0.0169 |
| 1500 | ...... | 91 | 0.128 | 1.41 | 1.57 | 0.00858 | 0.00883 | 0.0141 |
| 1750 | ...... | 127 | 0.117 | 1.52 | 1.83 | 0.00735 | 0.00756 | 0.0121 |
| 2000 | ...... | 127 | 0.126 | 1.63 | 2.09 | 0.00643 | 0.00662 | 0.0106 |

These resistance values are valid *only* for the parameters as given. Using conductors having coated strands, different stranding type, and, especially, other temperatures, changes the resistance.

Formula for temperature change: $R_2 = R_1 [1 + \alpha(T_2 - 75)]$ where $\alpha_{cu} = 0.00323$ and $\alpha_{AL} = 0.00330$.

Class B stranding is listed as well as solid for some sizes. Its overall diameter and area are those of its circumscribing circle. The construction information is per NEMA WC8-1976 (Rev. 5-1980). The resistance is calculated per National Bureau of Standards Handbook 100, dated 1966, and Handbook 109, dated 1972.

Conductors with compact and compressed stranding have about 9% and 3%, respectively, smaller bare conductor diameters than those shown.

IACS conductivities used: bare copper = 100%; aluminum = 61%.

See Table 5B, NEC (Sec. 35) for actual compact cable dimensions.

## 62. AC Resistance and Reactance: 600-V Cables, Three-Phase, 60 Hz, 75°C; Three Single Conductors in Conduit[a]
(Table 9, 1990 NEC)

Ohms to neutral per 1000 ft

| Size, AWG, kcmil | $X_L$ (reactance) for all wires: PVC, al. conduits | $X_L$ (reactance) for all wires: Steel conduit | AC resistance for uncoated copper wires: PVC conduit | AC resistance for uncoated copper wires: Al. conduit | AC resistance for uncoated copper wires: Steel conduit | AC resistance for aluminum wires: PVC conduit | AC resistance for aluminum wires: Al. conduit | AC resistance for aluminum wires: Steel conduit | Effective Z at .85 PF for uncoated copper wires: PVC conduit | Effective Z at .85 PF for uncoated copper wires: Al. conduit | Effective Z at .85 PF for uncoated copper wires: Steel conduit | Effective Z at .85 PF for aluminum wires: PVC conduit | Effective Z at .85 PF for aluminum wires: Al. conduit | Effective Z at .85 PF for aluminum wires: Steel conduit | Size, AWG, kcmil |
|---|---|---|---|---|---|---|---|---|---|---|---|---|---|---|---|
| 14 | .058 | .073 | 3.1 | 3.1 | 3.1 | — | — | — | 2.7 | 2.7 | 2.7 | — | — | — | 14 |
| 12 | .054 | .068 | 2.0 | 2.0 | 2.0 | 3.2 | 3.2 | 3.2 | 1.7 | 1.7 | 1.7 | 2.8 | 2.8 | 2.8 | 12 |
| 10 | .050 | .063 | 1.2 | 1.2 | 1.2 | 2.0 | 2.0 | 2.0 | 1.1 | 1.1 | 1.1 | 1.8 | 1.8 | 1.8 | 10 |
| 8 | .052 | .065 | .78 | .78 | .78 | 1.3 | 1.3 | 1.3 | .69 | .69 | .70 | 1.1 | 1.1 | 1.1 | 8 |
| 6 | .051 | .064 | .49 | .49 | .49 | .81 | .81 | .81 | .44 | .45 | .45 | .71 | .72 | .72 | 6 |
| 4 | .048 | .060 | .31 | .31 | .31 | .51 | .51 | .51 | .29 | .29 | .30 | .46 | .46 | .46 | 4 |
| 3 | .047 | .059 | .25 | .25 | .25 | .40 | .41 | .40 | .23 | .24 | .24 | .37 | .37 | .37 | 3 |
| 2 | .045 | .057 | .19 | .20 | .20 | .32 | .32 | .32 | .19 | .19 | .20 | .30 | .30 | .30 | 2 |
| 1 | .046 | .057 | .15 | .16 | .16 | .25 | .26 | .25 | .16 | .16 | .16 | .24 | .24 | .25 | 1 |
| 1/0 | .044 | .055 | .12 | .13 | .12 | .20 | .21 | .20 | .13 | .13 | .13 | .19 | .20 | .20 | 1/0 |
| 2/0 | .043 | .054 | .10 | .10 | .10 | .16 | .16 | .16 | .11 | .11 | .11 | .16 | .16 | .16 | 2/0 |
| 3/0 | .042 | .052 | .077 | .082 | .079 | .13 | .13 | .13 | .088 | .092 | .094 | .13 | .13 | .14 | 3/0 |
| 4/0 | .041 | .051 | .062 | .067 | .063 | .10 | .11 | .10 | .074 | .078 | .080 | .11 | .11 | .11 | 4/0 |
| 250 | .041 | .052 | .052 | .057 | .054 | .085 | .090 | .086 | .056 | .070 | .073 | .094 | .098 | .10 | 250 |
| 300 | .041 | .051 | .044 | .049 | .045 | .071 | .076 | .072 | .059 | .063 | .065 | .082 | .086 | .088 | 300 |
| 350 | .040 | .050 | .038 | .043 | .039 | .061 | .066 | .063 | .053 | .058 | .060 | .073 | .077 | .080 | 350 |
| 400 | .040 | .049 | .033 | .038 | .035 | .054 | .059 | .055 | .049 | .053 | .056 | .066 | .071 | .073 | 400 |
| 500 | .039 | .048 | .027 | .032 | .029 | .043 | .048 | .045 | .043 | .048 | .050 | .057 | .061 | .064 | 500 |
| 600 | .039 | .048 | .023 | .028 | .025 | .036 | .041 | .038 | .040 | .044 | .047 | .051 | .055 | .058 | 600 |
| 750 | .038 | .048 | .019 | .024 | .021 | .029 | .034 | .031 | .036 | .040 | .043 | .045 | .049 | .052 | 750 |
| 1000 | .037 | .046 | .015 | .019 | .018 | .023 | .027 | .025 | .032 | .036 | .040 | .039 | .042 | .046 | 1000 |

[a]Reprinted with permission from NFPA 70-1990, National Electrical Code®, Copyright © 1990, National Fire Protection Association, Quincy, Massachusetts 02269. This reprinted material is not the complete and official position of the NFPA on the referenced subject, which is represented only by the standard in its entirety.

NOTE 1 These values are based on the following constants: UL-type RHH wires with Class B stranding, in cradled configuration. Wire conductivities are 100% IACS copper, and 61% IACS aluminum, and aluminum conduit is 45% IACS. Capacitive reactance is ignored since it is negligible at these voltages. These resistance values are valid only at 75°C (167°F) and for the parameters as given, but are representative for 600-volt wire types operating at 60 Hz.

NOTE 2 "Effective Z" is defined as R cos(theta) + X sin(theta), where "theta" is the power factor angle of the circuit. Multiplying current by effective impedance gives a good approximation for line to neutral voltage drop. Effective impedance values shown in this table are valid only at .85 power factor. For another circuit power factor (PF), effective impedance (Ze) can be calculated from R and $X_L$ values given in this table as follows:

$$Ze = R \times PF + X_L \sin[\arccos(PF)]$$

**63. Metric practice (ANSI/IEEE 268-1982).**  The International System of Units, developed and maintained by the General Conference on Weights and Measures (abbreviated CGPM from the official French name Conférence Générale des Poids et Mesures), is intended as a basis for worldwide standardization of measurement units. The name International System of Units and the international abbreviation SI were adopted by the Eleventh CGPM in 1960.

CLASSES OF SI UNITS  SI units are divided into three classes: base units, supplementary units, and derived units

BASE UNITS  SI is based on seven well-defined units which by convention are regarded as dimensionally independent:

| Quantity | Unit | Symbol |
|----------|------|--------|
| Length | meter | m |
| Mass | kilogram | kg |
| Time | second | s |
| Electric current | ampere | A |
| Thermodynamic temperature | kelvin | K |
| Amount of substance | mole | mol |
| Luminous intensity | candela | cd |

SUPPLEMENTARY UNITS  The units listed below are called *supplementary units* and may be regarded either as base units or as derived units:

| Quantity | Unit | Symbol |
|----------|------|--------|
| Plane angle | radian | rad |
| Solid angle | steradian | sr |

DERIVED UNITS  Derived units are formed by combining base units, supplementary units, and other derived units according to the algebraic relations linking the corresponding quantities. The symbols for derived units are obtained by means of the mathematical signs for multiplication, division, and use of exponents. For example, the SI unit for velocity is the meter per second (m/s or $m \cdot s^{-1}$), and that for angular velocity is the radian per second (rad/s or $rad \cdot s^{-1}$). Derived SI units which have special names and symbols approved by the CGPM are listed below:

| Quantity | Unit | Symbol | Formula |
|----------|------|--------|---------|
| Frequency (of a periodic phenomenon) | hertz | Hz | $1/s$ |
| Force | newton | N | $(kg \cdot m)/s^2$ |
| Pressure, stress | pascal | Pa | $N/m^2$ |
| Energy, work, quantity of heat | joule | J | $N \cdot m$ |
| Power, radiant flux | watt | W | $J/s$ |
| Quantity of electricity, electric charge | coulomb | C | $A \cdot s$ |
| Electric potential, potential difference, electromotive force | volt | V | $W/A$ |
| Capacitance | farad | F | $C/V$ |
| Electric resistance | ohm | Ω | $V/A$ |
| Conductance | siemens | S | $A/V$ |
| Magnetic flux | weber | Wb | $V \cdot s$ |
| Magnetic flux density | tesla | T | $Wb/m^2$ |
| Inductance | henry | H | $Wb/A$ |
| Luminous flux | lumen | lm | $cd \cdot sr$ |
| Illuminance | lux | lx | $lm/m^2$ |
| Activity (of radionuclides) | becquerel[a] | Bq | $1/s$ |
| Absorbed dose | gray[a] | Gy | $J/kg$ |

[a]Adopted by the CGPM in 1975.

### 64.   Some Common Derived Units of SI

| Quantity | Unit | Symbol |
|---|---|---|
| Acceleration | meter per second squared | $m/s^2$ |
| Angular acceleration | radian per second squared | $rad/s^2$ |
| Angular velocity | radian per second | $rad/s$ |
| Area | square meter | $m^2$ |
| Concentration (of amount of substance) | mole per cubic meter | $mol/m^3$ |
| Current density | ampere per square meter | $A/m^2$ |
| Density, mass | kilogram per cubic meter | $kg/m^3$ |
| Electric charge density | coulomb per cubic meter | $C/m^3$ |
| Electric field strength | volt per meter | $V/m$ |
| Electric flux density | coulomb per square meter | $C/m^2$ |
| Energy density | joule per cubic meter | $J/m^3$ |
| Entropy | joule per kelvin | $J/K$ |
| Heat capacity | joule per kelvin | $J/K$ |
| Heat flux density ⎫<br>Irradiance ⎭ | watt per square meter | $W/m^2$ |
| Luminance | candela per square meter | $cd/m^2$ |
| Magnetic field strength | ampere per meter | $A/m$ |
| Molar energy | joule per mole | $J/mol$ |
| Molar entropy | joule per mole kelvin | $J/(mol \cdot K)$ |
| Molar heat capacity | joule per mole kelvin | $J/(mol \cdot K)$ |
| Moment of force | newton meter | $N \cdot m$ |
| Permeability | henry per meter | $H/m$ |
| Permittivity | farad per meter | $F/m$ |
| Radiance | watt per square meter steradian | $W/(m^2 \cdot sr)$ |
| Radiant intensity | watt per steradian | $W/sr$ |
| Specific heat capacity | joule per kilogram kelvin | $J/(kg \cdot K)$ |
| Specific energy | joule per kilogram | $J/kg$ |
| Specific entropy | joule per kilogram kelvin | $J/(kg \cdot K)$ |
| Specific volume | cubic meter per kilogram | $m^3/kg$ |
| Surface tension | newton per meter | $N/m$ |
| Thermal conductivity | watt per meter kelvin | $W/(m \cdot K)$ |
| Velocity | meter per second | $m/s$ |
| Viscosity, dynamic | pascal second | $Pa \cdot s$ |
| Viscosity, kinematic | square meter per second | $m^2/s$ |
| Volume | cubic meter | $m^3$ |
| Wave number | 1 per meter | $1/m$ |

## 65.  Recommended Pronunciations

| Prefix | Pronunciation (United States)[a] |
|--------|----------------------------------|
| Exa | ex′ a (*a* as in a*b*out) |
| Peta | pet′ a (*e* as in p*e*t, *a* as in a*b*out) |
| Tera | as in *terr*ace |
| Giga | jig′ a (*i* as in j*i*g, *a* as in a*b*out) |
| Mega | as in *mega*phone |
| Kilo | kill′ oh |
| Hecto | heck′ toe |
| Deka | deck′ a (*a* as in a*b*out) |
| Deci | as in *deci*mal |
| Centi | as in *centi*pede |
| Milli | as in *mili*tary |
| Micro | as in *micro*phone |
| Nano | nan′ oh (*an* as in a*n*t) |
| Pico | peek′ oh |
| Femto | fem′ toe (*fem* as in *fem*inine) |
| Atto | as in ana*to*my |

| Selected units | Pronunciation |
|----------------|---------------|
| Candela | can *dell′* a |
| Joule | rhyme with *tool* |
| Kilometer | *kill′* oh meter |
| Pascal | rhyme with *rascal* |
| Siemens | same as *seamen's* |

[a]The first syllable of every prefix is accented to assure that the prefix will retain its identity. Therefore, the preferred pronunciation of *kilometer* places the accent on the first syllable, *not* the second.

## 66.  Inch-Millimeter Equivalents

| | 0 | 1 | 2 | 3 | 4 | 5 | 6 | 7 | 8 | 9 |
|---|---|---|---|---|---|---|---|---|---|---|
| In | | | | | mm | | | | | |
| 0 | 0.0 | 25.4 | 50.8 | 76.2 | 101.6 | 127.0 | 152.4 | 177.8 | 203.2 | 228.6 |
| 10 | 254.0 | 279.4 | 304.8 | 330.2 | 355.6 | 381.0 | 406.4 | 431.8 | 457.2 | 482.6 |
| 20 | 508.0 | 533.4 | 558.8 | 584.2 | 609.6 | 635.0 | 660.4 | 685.8 | 711.2 | 736.6 |
| 30 | 762.0 | 787.4 | 812.8 | 838.2 | 863.6 | 889.0 | 914.4 | 939.8 | 965.2 | 990.6 |
| 40 | 1016.0 | 1041.4 | 1066.8 | 1092.2 | 1117.6 | 1143.0 | 1168.4 | 1193.8 | 1219.2 | 1244.6 |
| 50 | 1270.0 | 1295.4 | 1320.8 | 1346.2 | 1371.6 | 1397.0 | 1422.4 | 1447.8 | 1473.2 | 1498.6 |
| 60 | 1524.0 | 1549.4 | 1574.8 | 1600.2 | 1625.6 | 1651.0 | 1676.4 | 1701.8 | 1727.2 | 1752.6 |
| 70 | 1778.0 | 1803.4 | 1828.8 | 1854.2 | 1879.6 | 1905.0 | 1930.4 | 1955.8 | 1981.2 | 2006.6 |
| 80 | 2032.0 | 2057.4 | 2082.8 | 2108.2 | 2133.6 | 2159.0 | 2184.4 | 2209.8 | 2235.2 | 2260.6 |
| 90 | 2286.0 | 2311.4 | 2336.8 | 2362.2 | 2387.6 | 2413.0 | 2438.4 | 2463.8 | 2489.2 | 2514.6 |
| 100 | 2540.0 | 2565.4 | 2590.8 | 2616.2 | 2641.6 | 2667.0 | 2692.4 | 2717.8 | 2743.2 | 2768.6 |

NOTE  All values in this table are exact, based on the relation 1 in = 25.4 mm. By manipulation of the decimal point any decimal value or multiple of an inch may be converted to its exact equivalent in millimeters.

**67. Inch-Millimeter Equivalents of Decimal and Common Fractions, from ¹⁄₆₄ to 1 in**

| Inch | ½s | ¼s | 8ths | 16ths | 32ds | 64ths | Decimals of an inch[a] | Millimeters |
|---|---|---|---|---|---|---|---|---|
| | | | | | | 1 | 0.015 625 | 0.397 |
| | | | | | 1 | 2 | 0.031 25 | 0.794 |
| | | | | | | 3 | 0.046 875 | 1.191 |
| | | | | 1 | 2 | 4 | 0.062 5 | 1.588 |
| | | | | | | 5 | 0.078 125 | 1.984 |
| | | | | | 3 | 6 | 0.093 75 | 2.381 |
| | | | | | | 7 | 0.109 375 | 2.778 |
| | | 1 | 2 | 4 | 8 | 0.125 0 | 3.175[a] |
| | | | | | | 9 | 0.140 625 | 3.572 |
| | | | | | 5 | 10 | 0.156 25 | 3.969 |
| | | | | | | 11 | 0.171 875 | 4.366 |
| | | | | 3 | 6 | 12 | 0.187 5 | 4.762 |
| | | | | | | 13 | 0.203 125 | 5.159 |
| | | | | | 7 | 14 | 0.218 75 | 5.556 |
| | | | | | | 15 | 0.234 375 | 5.953 |
| | 1 | 2 | 4 | 8 | 16 | 0.250 0 | 6.350[a] |
| | | | | | | 17 | 0.265 625 | 6.747 |
| | | | | | 9 | 18 | 0.281 25 | 7.144 |
| | | | | | | 19 | 0.296 875 | 7.541 |
| | | | | 5 | 10 | 20 | 0.312 5 | 7.938 |
| | | | | | | 21 | 0.328 125 | 8.334 |
| | | | | | 11 | 22 | 0.343 75 | 8.731 |
| | | | | | | 23 | 0.359 375 | 9.128 |
| | | 3 | 6 | 12 | 24 | 0.375 0 | 9.525[a] |
| | | | | | | 25 | 0.390 625 | 9.922 |
| | | | | | 13 | 26 | 0.406 25 | 10.319 |
| | | | | | | 27 | 0.421 875 | 10.716 |
| | | | | 7 | 14 | 28 | 0.437 5 | 11.112 |
| | | | | | | 29 | 0.453 125 | 11.509 |
| | | | | | 15 | 30 | 0.468 75 | 11.906 |
| | | | | | | 31 | 0.484 375 | 12.303 |
| | 1 | 2 | 4 | 8 | 16 | 32 | 0.500 0 | 12.700[a] |
| | | | | | | 33 | 0.515 625 | 13.097 |
| | | | | | 17 | 34 | 0.531 25 | 13.494 |
| | | | | | | 35 | 0.546 875 | 13.891 |
| | | | | 9 | 18 | 36 | 0.562 5 | 14.288 |
| | | | | | | 37 | 0.578 125 | 14.684 |
| | | | | | 19 | 38 | 0.593 75 | 15.081 |
| | | | | | | 39 | 0.609 375 | 15.478 |
| | | 5 | 10 | 20 | 40 | 0.625 0 | 15.875[a] |
| | | | | | | 41 | 0.640 625 | 16.272 |
| | | | | | 21 | 42 | 0.656 25 | 16.669 |
| | | | | | | 43 | 0.671 875 | 17.066 |
| | | | | 11 | 22 | 44 | 0.687 5 | 17.462 |
| | | | | | | 45 | 0.703 125 | 17.859 |
| | | | | | 23 | 46 | 0.718 75 | 18.256 |
| | | | | | | 47 | 0.734 375 | 18.653 |
| | 3 | 6 | 12 | 24 | 48 | 0.750 0 | 19.050[a] |
| | | | | | | 49 | 0.765 625 | 19.447 |
| | | | | | 25 | 50 | 0.781 25 | 19.844 |
| | | | | | | 51 | 0.796 875 | 20.241 |
| | | | | 13 | 26 | 52 | 0.812 5 | 20.638 |
| | | | | | | 53 | 0.828 125 | 21.034 |
| | | | | | 27 | 54 | 0.843 75 | 21.431 |
| | | | | | | 55 | 0.859 375 | 21.828 |
| | | 7 | 14 | 28 | 56 | 0.875 0 | 22.225[a] |
| | | | | | | 57 | 0.890 625 | 22.622 |
| | | | | | 29 | 58 | 0.906 25 | 23.019 |
| | | | | | | 59 | 0.921 875 | 23.416 |
| | | | | 15 | 30 | 60 | 0.937 5 | 23.812 |
| | | | | | | 61 | 0.953 125 | 24.209 |
| | | | | | 31 | 62 | 0.968 75 | 24.606 |
| | | | | | | 63 | 0.984 375 | 25.003 |
| 1 | 2 | 4 | 8 | 16 | 32 | 64 | 1.000 0 | 25.400[a] |

NOTE  *Caution.* Convert only to the number of decimal places consistent with the stated or implied accuracy.

[a]Exact.

## 68.  Definitions of Derived Units of the International System Having Special Names

| Physical Quantity | Unit and Definition |
|---|---|
| Absorbed dose | The *gray* is the energy imparted by ionizing radiation to a mass of matter corresponding to one joule per kilogram. |
| Activity | The *becquerel* is the activity of a radionuclide having one spontaneous nuclear transition per second. |
| Electric capacitance | The *farad* is the capacitance of a capacitor between the plates of which there appears a difference of potential of one volt when it is charged by a quantity of electricity equal to one coulomb. |
| Electric conductance | The *siemens* is the electric conductance of a conductor in which a current of one ampere is produced by an electric potential difference of one volt. |
| Electric inductance | The *henry* is the inductance of a closed circuit in which an electromotive force of one volt is produced when the electric current in the circuit varies uniformly at a rate of one ampere per second. |
| Electric potential difference, electromotive force | The *volt* (unit of electric potential difference and electromotive force) is the difference of electric potential between two points of a conductor carrying a constant current of one ampere when the power dissipated between these points is equal to one watt. |
| Electric resistance | The *ohm* is the electric resistance between two points of a conductor when a constant difference of potential of one volt, applied between these two points, produces in this conductor a current of one ampere, this conductor not being the source of any electromotive force. |
| Energy | The *joule* is the work done when the point of application of a force of one newton is displaced a distance of one meter in the direction of the force. |
| Force | The *newton* is that force which, when applied to a body having a mass of one kilogram, gives it an acceleration of one meter per second squared. |
| Frequency | The *hertz* is the frequency of a periodic phenomenon of which the period is one second. |
| Illuminance | The *lux* is the illuminance produced by a luminous flux of one lumen uniformly distributed over a surface of one square meter. |
| Luminous flux | The *lumen* is the luminous flux emitted in a solid angle of one steradian by a point source having a uniform intensity of one candela. |
| Magnetic flux | The *weber* is the magnetic flux which, linking a circuit of one turn, produces in it an electromotive force of one volt as it is reduced to zero at a uniform rate in one second. |
| Magnetic flux density | The *tesla* is the magnetic flux density given by a magnetic flux of one weber per square meter. |
| Power | The *watt* is the power which gives rise to the production of energy at the rate of one joule per second. |
| Pressure or stress | The *pascal* is the pressure or stress of one newton per square meter. |
| Quantity of electricity | The *coulomb* is the quantity of electricity transported in one second by a current of one ampere. |

## 69.  Alphabetical List of Units
(Symbols of SI units in parentheses)

| To convert from | To | Multiply by |
|---|---|---|
| abampere | ampere (A) | 1.000 000*E+01 |
| abcoulomb | coulomb (C) | 1.000 000*E+01 |
| abfarad | farad (F) | 1.000 000*E+09 |
| abhenry | henry (H) | 1.000 000*E−09 |
| abmho | siemens (S) | 1.000 000*E+09 |
| | | |
| abohm | ohm (Ω) | 1.000 000*E−09 |
| abvolt | volt (V) | 1.000 000*E−08 |
| acre foot (US survey)ᵃ | meter³ (m³) | 1.233 489 E+03 |
| acre (US survey)ᵃ | meter² (m²) | 4.046 873 E+03 |
| ampere hour | coulomb (C) | 3.600 000*E+03 |
| | | |
| are | meter² (m²) | 1.000 000*E+02 |
| angstrom | meter (m) | 1.000 000*E−10 |
| astronomical unitᵇ | meter (m) | 1.495 979 E+11 |
| atmosphere (standard) | pascal (Pa) | 1.013 250*E+05 |
| atmosphere (technical = 1 kgf/cm²) | pascal (Pa) | 9.806 650*E+04 |
| | | |
| bar | pascal (Pa) | 1.000 000*E+05 |
| barn | meter² (m²) | 1.000 000*E−28 |
| barrel (for petroleum, 42 gal) | meter³ (m³) | 1.589 873 E−01 |
| board foot | meter³ (m³) | 2.359 737 E−03 |
| | | |
| British thermal unit (International Table)ᶜ | joule (J) | 1.055 056 E+03 |
| British thermal unit (mean) | joule (J) | 1.055 87  E+03 |
| British thermal unit (thermochemical) | joule (J) | 1.054 350 E+03 |
| British thermal unit (39°F) | joule (J) | 1.059 67  E+03 |
| British thermal unit (59°F) | joule (J) | 1.054 80  E+03 |
| British thermal unit (60°F) | joule (J) | 1.054 68  E+03 |
| | | |
| Btu (International Table)·ft/h·ft²·°F (k, thermal conductivity) | watt per meter kelvin (W/m·K) | 1.730 735 E+00 |
| Btu (thermochemical)·ft/h·ft²·°F (k, thermal conductivity) | watt per meter kelvin (W/m·K) | 1.729 577 E+00 |
| Btu (International Table)·in/h·ft²·°F (k, thermal conductivity) | watt per meter kelvin (W/m·K) | 1.442 279 E−01 |
| Btu (thermochemical)·in/h·ft²·°F (k, thermal conductivity) | watt per meter kelvin (W/m·K) | 1.441 314 E−01 |
| Btu (International Table)·in/s·ft²·°F (k, thermal conductivity) | watt per meter kelvin (W/m·K) | 5.192 204 E+02 |
| Btu (thermochemical)·in/s·ft²·°F (k, thermal conductivity) | watt per meter kelvin (W/m·K) | 5.188 732 E+02 |

ᵃSince 1893 the United States basis of length measurement has been derived from metric standards. In 1959 a small refinement was made in the definition of the yard to resolve discrepancies both in the United States and other countries; this changed its length from 3600/3937 m to 0.9144 m exactly. The change resulted in the new value being shorter by 2 parts in 1 million. At the same time it was decided that any data in feet derived from and published as a result of geodetic surveys within the United States would remain with the old standard (1 ft = 1200/3937 m) until further decision. This foot is named the United States survey foot. As a result all United States land measurements in United States customary units will relate to the meter by the old standard. All the conversion factors in these tables for units referenced to this footnote are based on the United States survey foot rather than on the international foot.

Conversion factors for the land measure given below may be determined from the following relationships:

$$1 \text{ league} = 3 \text{ mi (exactly)}$$
$$1 \text{ rod} = 16\tfrac{1}{2} \text{ ft (exactly)}$$
$$1 \text{ section} = 1 \text{ mi}^2 \text{ (exactly)}$$
$$1 \text{ township} = 36 \text{ mi}^2 \text{ (exactly)}$$

ᵇThis value conflicts with the value printed in NBS 330 (17). The value requires updating in NBS 330.
ᶜThis value was adopted in 1956. Some of the older international tables use the value 1.055 04 E+03. The exact conversion factor is 1.055 055 852 62*E+03.

**69.  Alphabetical List of Units** (*Continued*)

| To convert from | To | Multiply by |
|---|---|---|
| Btu (International Table)/h | watt (W) | 2.930 711  E−01 |
| Btu (thermochemical)/h | watt (W) | 2.928 751  E−01 |
| Btu (thermochemical)/min | watt (W) | 1.757 250  E+01 |
| Btu (thermochemical)/s | watt (W) | 1.054 350  E+03 |
| Btu (International Table)/ft$^2$ | joule per meter$^2$ (J/m$^2$) | 1.135 653  E+04 |
| Btu (thermochemical)/ft$^2$ | joule per meter$^2$ (J/m$^2$) | 1.134 893  E+04 |
| Btu (thermochemical)/ft$^2$·h | watt per meter$^2$ (W/m$^2$) | 3.152 481  E+00 |
| Btu (thermochemical)/ft$^2$·min | watt per meter$^2$ (W/m$^2$) | 1.891 489  E+02 |
| Btu (thermochemical)/ft$^2$·s | watt per meter$^2$ (W/m$^2$) | 1.134 893  E+04 |
| Btu (thermochemical)/in$^2$·s | watt per meter$^2$ (W/m$^2$) | 1.634 246  E+06 |
| Btu (International Table)/h·ft$^2$·°F<br>(*C,* thermal conductance) | watt per meter$^2$ kelvin (W/m$^2$·K) | 5.678 263  E+00 |
| Btu (thermochemical)/h·ft$^2$·°F<br>(*C,* thermal conductance) | watt per meter$^2$ kelvin (W/m$^2$·K) | 5.674 466  E+00 |
| Btu (International Table)/s·ft$^2$·°F | watt per meter$^2$ kelvin (W/m$^2$·K) | 2.044 175  E+04 |
| Btu (thermochemical)/s·ft$^2$·°F | watt per meter$^2$ kelvin (W/m$^2$·K) | 2.042 808  E+04 |
| Btu (International Table)/lb | joule per kilogram (J/kg) | 2.326 000*E+03 |
| Btu (thermochemical)/lb | joule per kilogram (J/kg) | 2.324 444  E+03 |
| Btu (International Table)/lb·°F<br>(*c,* heat capacity) | joule per kilogram kelvin (J/kg·K) | 4.186 800*E+03 |
| Btu (thermochemical)/lb·°F<br>(*c,* heat capacity) | joule per kilogram kelvin (J/kg·K) | 4.184 000  E+03 |
| bushel (US) | meter$^3$ (m$^3$) | 3.523 907  E−02 |
| caliber (inch) | meter (m) | 2.540 000*E−02 |
| calorie (International Table) | joule (J) | 4.186 800*E+00 |
| calorie (mean) | joule (J) | 4.190 02   E+00 |
| calorie (thermochemical) | joule (J) | 4.184 000*E+00 |
| calorie (15°C) | joule (J) | 4.185 80   E+00 |
| calorie (20°C) | joule (J) | 4.181 90   E+00 |
| calorie (kilogram, International Table) | joule (J) | 4.186 800*E+03 |
| calorie (kilogram, mean) | joule (J) | 4.190 02   E+03 |
| calorie (kilogram, thermochemical) | joule (J) | 4.184 000*E+03 |
| cal (thermochemical)/cm$^2$ | joule per meter$^2$ (J/m$^2$) | 4.184 000*E+04 |
| cal (International Table)/g | joule per kilogram (J/kg) | 4.186 800*E+03 |
| cal (thermochemical)/g | joule per kilogram (J/kg) | 4.184 000*E+03 |
| cal (International Table)/g·°C | joule per kilogram kelvin (J/kg·K) | 4.186 800*E+03 |
| cal (thermochemical)/g·°C | joule per kilogram kelvin (J/kg·K) | 4.184 000*E+03 |
| cal (thermochemical)/min | watt (W) | 6.973 333  E−02 |
| cal (thermochemical)/s | watt (W) | 4.184 000*E+00 |
| cal (thermochemical)/cm$^2$·min | watt per meter$^2$ (W/m$^2$) | 6.973 333  E+02 |
| cal (thermochemical)/cm$^2$·s | watt per meter$^2$ (W/m$^2$) | 4.184 000*E+04 |
| cal (thermochemical)/cm·s·°C | watt per meter kelvin (W/m·K) | 4.184 000*E+02 |
| carat (metric) | kilogram (kg) | 2.000 000*E−04 |
| centimeter of mercury (0°C) | pascal (Pa) | 1.333 22   E+03 |
| centimeter of water (4°C) | pascal (Pa) | 9.806 38   E+01 |
| centipoise | pascal second (Pa·s) | 1.000 000*E−03 |
| centistokes | meter$^2$ per second (m$^2$/s) | 1.000 000*E−06 |
| circular mil | meter$^2$ (m$^2$) | 5.067 075  E−10 |
| clo | kelvin meter$^2$ per watt (K·m$^2$/W) | 2.003 712  E−01 |
| cup | meter$^3$ (m$^3$) | 2.365 882  E−04 |
| curie | becquerel (Bq) | 3.700 000*E+10 |

**69.  Alphabetical List of Units** (*Continued*)

| To convert from | To | Multiply by |
|---|---|---|
| day (mean solar) | second (s) | 8.640 000 E+04 |
| day (sidereal) | second (s) | 8.616 409 E+04 |
| degree (angle) | radian (rad) | 1.745 329 E−02 |
| degree Celsius | kelvin (K) | $t_K = t_C + 273.15$ |
| degree centigrade | [see 3.4.2] | |
| degree Fahrenheit | degree Celsius | $t_C = (t_F - 32)/1.8$ |
| degree Fahrenheit | kelvin (K) | $t_K = (t_F + 459.67)/1.8$ |
| degree Rankine | kelvin (K) | $t_K = t_R/1.8$ |
| °F·h·ft²/Btu (International Table)<br>(*R*, thermal resistance) | kelvin meter² per watt (K·m²/W) | 1.761 102 E−01 |
| °F·h·ft²/Btu (thermochemical)<br>(*R*, thermal resistance) | kelvin meter² per watt (K·m²/W) | 1.762 280 E−01 |
| denier | kilogram per meter (kg/m) | 1.111 111 E−07 |
| dyne | newton (N) | 1.000 000*E−05 |
| dyne·cm | newton meter (N·m) | 1.000 000*E−07 |
| dyne/cm² | pascal (Pa) | 1.000 000*E−01 |
| electronvolt | joule (J) | 1.602 19  E−19 |
| EMU of capacitance | farad (F) | 1.000 000*E+09 |
| EMU of current | ampere (A) | 1.000 000*E+01 |
| EMU of electric potential | volt (V) | 1.000 000*E−08 |
| EMU of inductance | henry (H) | 1.000 000*E−09 |
| EMU of resistance | ohm (Ω) | 1.000 000*E−09 |
| ESU of capacitance | farad (F) | 1.112 650 E−12 |
| ESU of current | ampere (A) | 3.335 6   E−10 |
| ESU of electric potential | volt (V) | 2.997 9   E+02 |
| ESU of inductance | henry (H) | 8.987 554 E+11 |
| ESU of resistance | ohm (Ω) | 8.987 554 E+11 |
| erg | joule (J) | 1.000 000*E−07 |
| erg/cm²·s | watt per meter² (W/m²) | 1.000 000*E−03 |
| erg/s | watt (W) | 1.000 000*E−07 |
| faraday (based on carbon-12) | coulomb (C) | 9.648 70  E+04 |
| faraday (chemical) | coulomb (C) | 9.649 57  E+04 |
| faraday (physical) | coulomb (C) | 9.652 19  E+04 |
| fathom | meter (m) | 1.828 8   E+00 |
| fermi (femtometer) | meter (m) | 1.000 000*E−15 |
| fluid ounce (US) | meter³ (m³) | 2.957 353 E−05 |
| foot | meter (m) | 3.048 000*E−01 |
| foot (US survey)[a] | meter (m) | 3.048 006 E−01 |
| foot of water (39.2°F) | pascal (Pa) | 2.988 98  E+03 |
| ft² | meter² (m²) | 9.290 304*E−02 |
| ft²/h (thermal diffusivity) | meter² per second (m²/s) | 2.580 640*E−05 |
| ft²/s | meter² per second (m²/s) | 9.290 304*E−02 |
| ft³ (volume; section modulus) | meter³ (m³) | 2.831 685 E−02 |
| ft³/min | meter³ per second (m³/s) | 4.719 474 E−04 |
| ft³/s | meter³ per second (m³/s) | 2.831 685 E−02 |
| ft⁴ (moment of section)[a] | meter⁴ (m⁴) | 8.630 975 E−03 |
| ft/h | meter per second (m/s) | 8.466 667 E−05 |
| ft/min | meter per second (m/s) | 5.080 000*E−03 |
| ft/s | meter per second (m/s) | 3.048 000*E−01 |
| ft/s² | meter per second² (m/s²) | 3.048 000*E−01 |
| footcandle | lux (lx) | 1.076 391 E+01 |
| footlambert | candela per meter² (cd/m²) | 3.426 259 E+00 |

[a]This is sometimes called the moment of inertia of a plane section about a specified axis.

**69.  Alphabetical List of Units** (*Continued*)

| To convert from | To | Multiply by |
|---|---|---|
| ft·lbf | joule (J) | 1.355 818 E+00 |
| ft·lbf/h | watt (W) | 3.766 161 E−04 |
| ft·lbf/min | watt (W) | 2.259 697 E−02 |
| ft·lbf/s | watt (W) | 1.355 818 E+00 |
| ft·poundal | joule (J) | 4.214 011 E−02 |
| free fall, standard (*g*) | meter per second$^2$ (m/s$^2$) | 9.806 650*E+00 |
| gal | meter per second$^2$ (m/s$^2$) | 1.000 000*E−02 |
| gallon (Canadian liquid) | meter$^3$ (m$^3$) | 4.546 090 E−03 |
| gallon (UK liquid) | meter$^3$ (m$^3$) | 4.546 092 E−03 |
| gallon (US dry) | meter$^3$ (m$^3$) | 4.404 884 E−03 |
| gallon (US liquid) | meter$^3$ (m$^3$) | 3.785 412 E−03 |
| gal (US liquid)/day | meter$^3$ per second (m$^3$/s) | 4.381 264 E−08 |
| gal (US liquid)/min | meter$^3$ per second (m$^3$/s) | 6.309 020 E−05 |
| gal (US liquid)/hp·h <br> (SFC, specific fuel consumption) | meter$^3$ per joule (m$^3$/J) | 1.410 089 E−09 |
| gamma | tesla (T) | 1.000 000*E−09 |
| gauss | tesla (T) | 1.000 000*E−04 |
| gilbert | ampere (A) | 7.957 747 E−01 |
| gill (UK) | meter$^3$ (m$^3$) | 1.420 654 E−04 |
| gill (US) | meter$^3$ (m$^3$) | 1.182 941 E−04 |
| grad | degree (angular) | 9.000 000*E−01 |
| grad | radian (rad) | 1.570 796 E−02 |
| grain (1/7000 lb avoirdupois) | kilogram (kg) | 6.479 891*E−05 |
| grain (lb avoirdupois/7000)/gal <br> (US liquid) | kilogram per meter$^3$ (kg/m$^3$) | 1.711 806 E−02 |
| gram | kilogram (kg) | 1.000 000*E−03 |
| g/cm$^3$ | kilogram per meter$^3$ (kg/m$^3$) | 1.000 000*E+03 |
| gram-force/cm$^2$ | pascal (Pa) | 9.806 650*E+01 |
| hectare | meter$^2$ (m$^2$) | 1.000 000*E+04 |
| horsepower (550 ft·lbf/s) | watt (W) | 7.456 999 E+02 |
| horsepower (boiler) | watt (W) | 9.809 50  E+03 |
| horsepower (electric) | watt (W) | 7.460 000*E+02 |
| horsepower (metric) | watt (W) | 7.354 99  E+02 |
| horsepower (water) | watt (W) | 7.460 43  E+02 |
| horsepower (UK) | watt (W) | 7.457 0   E+02 |
| hour (mean solar) | second (s) | 3.600 000 E+03 |
| hour (sidereal) | second (s) | 3.590 170 E+03 |
| hundredweight (long) | kilogram (kg) | 5.080 235 E+01 |
| hundredweight (short) | kilogram (kg) | 4.535 924 E+01 |
| inch | meter (m) | 2.540 000*E−02 |
| inch of mercury (32°F) | pascal (Pa) | 3.386 38  E+03 |
| inch of mercury (60°F) | pascal (Pa) | 3.376 85  E+03 |
| inch of water (39.2°F) | pascal (Pa) | 2.490 82  E+02 |
| inch of water (60°F) | pascal (Pa) | 2.488 4   E+02 |
| in$^2$ | meter$^2$ (m$^2$) | 6.451 600*E−04 |
| in$^3$ (volume; section modulus)$^c$ | meter$^3$ (m$^3$) | 1.638 706 E−05 |
| in$^3$/min | meter$^3$ per second (m$^3$/s) | 2.731 177 E−07 |
| in$^4$ (moment of section)$^d$ | meter$^4$ (m$^4$) | 4.162 314 E−07 |
| in/s | meter per second (m/s) | 2.540 000*E−02 |
| in/s$^2$ | meter per second$^2$ (m/s$^2$) | 2.540 000*E−02 |
| kayser | 1 per meter (1/m) | 1.000 000*E+02 |
| kelvin | degree Celsius | $t_{·C} = t_K - 273.15$ |

$^c$The exact conversion factor is 1.638 706 4*E−05.

**69.  Alphabetical List of Units** (*Continued*)

| To convert from | To | Multiply by |
|---|---|---|
| kilocalorie (International Table) | joule (J) | 4.186 800*E+03 |
| kilocalorie (mean) | joule (J) | 4.190 02  E+03 |
| kilocalorie (thermochemical) | joule (J) | 4.184 000*E+03 |
| kilocalorie (thermochemical)/min | watt (W) | 6.973 333 E+01 |
| kilocalorie (thermochemical)/s | watt (W) | 4.184 000*E+03 |
| | | |
| kilogram-force (kgf) | newton (N) | 9.806 650*E+00 |
| kgf·m | newton meter (N·m) | 9.806 650*E+00 |
| kgf·s²/m (mass) | kilogram (kg) | 9.806 650*E+00 |
| kgf/cm² | pascal (Pa) | 9.806 650*E+04 |
| kgf/m² | pascal (Pa) | 9.806 650*E+00 |
| kgf/mm² | pascal (Pa) | 9.806 650*E+06 |
| | | |
| km/h | meter per second (m/s) | 2.777 778 E−01 |
| kilopond | newton (N) | 9.806 650*E+00 |
| kW·h | joule (J) | 3.600 000*E+06 |
| kip (1000 lbf) | newton (N) | 4.448 222 E+03 |
| kip/in² (ksi) | pascal (Pa) | 6.894 757 E+06 |
| knot (international) | meter per second (m/s) | 5.144 444 E−01 |
| | | |
| lambert | candela per meter² (cd/m²) | 1/π     *E+04 |
| lambert | candela per meter² (cd/m²) | 3.183 099 E+03 |
| langley | joule per meter² (J/m²) | 4.184 000*E+04 |
| league | meter (m) | ... *a* |
| light year | meter (m) | 9.460 55  E+15 |
| liter*f* | meter³ (m³) | 1.000 000*E−03 |
| | | |
| maxwell | weber (Wb) | 1.000 000*E−08 |
| mho | siemens (S) | 1.000 000*E+00 |
| microinch | meter (m) | 2.540 000*E−08 |
| micrometer | meter (m) | 1.000 000*E−06 |
| mil | meter (m) | 2.540 000*E−05 |
| | | |
| mile (international) | meter (m) | 1.609 344*E+03 |
| mile (statute) | meter (m) | 1.609 3   E+03 |
| mile (US survey)*a* | meter (m) | 1.609 347 E+03 |
| mile (international nautical) | meter (m) | 1.852 000*E+03 |
| mile (UK nautical) | meter (m) | 1.853 184*E+03 |
| mile (US nautical) | meter (m) | 1.852 000*E+03 |
| | | |
| mi² (international) | meter² (m²) | 2.589 988 E+06 |
| mi² (US survey)*a* | meter² (m²) | 2.589 998 E+06 |
| mi/h (international) | meter per second (m/s) | 4.470 400*E−01 |
| mi/h (international) | kilometer per hour (km/h) | 1.609 344*E+00 |
| mi/min (international) | meter per second (m/s) | 2.682 240*E+01 |
| mi/s (international) | meter per second (m/s) | 1.609 344*E+03 |
| | | |
| millibar | pascal (Pa) | 1.000 000*E+02 |
| millimeter of mercury (0°C) | pascal (Pa) | 1.333 22  E+02 |
| minute (angle) | radian (rad) | 2.908 882 E−04 |
| minute (mean solar) | second (s) | 6.000 000 E+01 |
| minute (sidereal) | second (s) | 5.983 617 E+01 |
| month (mean calendar) | second (s) | 2.628 000 E+06 |
| | | |
| oersted | ampere per meter (A/m) | 7.957 747 E+01 |
| ohm centimeter | ohm meter (Ω·m) | 1.000 000*E−02 |
| ohm circular-mil per ft | ohm millimeter² per meter (Ω·mm²/m) | 1.662 426 E−03 |

*f*In 1964 the General Conference on Weights and Measures adopted the name *liter* as a special name for the cubic decimeter. Prior to this decision the liter differed slightly (previous value, 1.000 028 dm³), and in expression of precision volume measurement this fact must be kept in mind.

**69.** **Alphabetical List of Units** (*Continued*)

| To convert from | To | Multiply by |
|---|---|---|
| ounce (avoirdupois) | kilogram (kg) | 2.834 952 E−02 |
| ounce (troy or apothecary) | kilogram (kg) | 3.110 348 E−02 |
| ounce (UK fluid) | meter³ (m³) | 2.841 307 E−05 |
| ounce (US fluid) | meter³ (m³) | 2.957 353 E−05 |
| ounce-force | newton (N) | 2.780 139 E−01 |
| ozf·in | newton meter (N·m) | 7.061 552 E−03 |
| | | |
| oz (avoirdupois)/gal (UK liquid) | kilogram per meter³ (kg/m³) | 6.236 021 E+00 |
| oz (avoirdupois)/gal (US liquid) | kilogram per meter³ (kg/m³) | 7.489 152 E+00 |
| oz (avoirdupois)/in³ | kilogram per meter³ (kg/m³) | 1.729 994 E+03 |
| oz (avoirdupois)/ft² | kilogram per meter² (kg/m²) | 3.051 517 E−01 |
| oz (avoirdupois)/yd² | kilogram per meter² (kg/m²) | 3.390 575 E−02 |
| parsec[b] | meter (m) | 3.085 678 E+16 |
| peck (US) | meter³ (m³) | 8.809 768 E−03 |
| | | |
| pennyweight | kilogram (kg) | 1.555 174 E−03 |
| perm (0°C) | kilogram per pascal second meter² (kg/Pa·s·m²) | 5.721 35  E−11 |
| perm (23°C) | kilogram per pascal second meter² (kg/Pa·s·m²) | 5.745 25  E−11 |
| perm·in (0°C) | kilogram per pascal second meter (kg/Pa·s·m) | 1.453 22  E−12 |
| perm·in (23°C) | kilogram per pascal second meter (kg/Pa·s·m) | 1.459 29  E−12 |
| | | |
| phot | lumen per meter² (lm/m²) | 1.000 000*E+04 |
| pica (printer's) | meter (m) | 4.217 518 E−03 |
| pint (US dry) | meter³ (m³) | 5.506 105 E−04 |
| pint (US liquid) | meter³ (m³) | 4.731 765 E−04 |
| point (printer's) | meter (m) | 3.514 598*E−04 |
| poise (absolute viscosity) | pascal second (Pa·s) | 1.000 000*E−01 |
| | | |
| pound (lb avoirdupois)[g] | kilogram (kg) | 4.535 924 E−01 |
| pound (troy or apothecary) | kilogram (kg) | 3.732 417 E−01 |
| lb·ft² (moment of inertia) | kilogram meter² (kg·m²) | 4.214 011 E−02 |
| lb·in² (moment of inertia) | kilogram meter² (kg·m²) | 2.926 397 E−04 |
| | | |
| lb/ft·h | pascal second (Pa·s) | 4.133 789 E−04 |
| lb/ft·s | pascal second (Pa·s) | 1.488 164 E+00 |
| lb/ft² | kilogram per meter² (kg/m²) | 4.882 428 E+00 |
| lb/ft³ | kilogram per meter³ (kg/m³) | 1.601 846 E+01 |
| lb/gal (UK liquid) | kilogram per meter³ (kg/m³) | 9.977 633 E+01 |
| lb/gal (US liquid) | kilogram per meter³ (kg/m³) | 1.198 264 E+02 |
| | | |
| lb/h | kilogram per second (kg/s) | 1.259 979 E−04 |
| lb/hp·h (SFC, specific fuel consumption) | kilogram per joule (kg/J) | 1.689 659 E−07 |
| lb/in³ | kilogram per meter³ (kg/m³) | 2.767 990 E+04 |
| lb/min | kilogram per second (kg/s) | 7.559 873 E−03 |
| lb/s | kilogram per second (kg/s) | 4.535 924 E−01 |
| lb/yd³ | kilogram per meter³ (kg/m³) | 5.932 764 E−01 |
| | | |
| poundal | newton (N) | 1.382 550 E−01 |
| poundal/ft² | pascal (Pa) | 1.488 164 E+00 |
| poundal·s/ft² | pascal second (Pa·s) | 1.488 164 E+00 |
| | | |
| pound-force (lbf)[h] | newton (N) | 4.448 222 E+00 |
| lbf·ft | newton meter (N·m) | 1.355 818 E+00 |
| lbf·ft/in | newton meter per meter (N·m/m) | 5.337 866 E+01 |

[g]The exact conversion factor is 4.535 923 7*E−01.
[h]The exact conversion factor is 4.448 221 615 260 5*E+00.

**69.    Alphabetical List of Units** (*Continued*)

| To convert from | To | Multiply by |
|---|---|---|
| lbf·in | newton meter (N·m) | 1.129 848 E−01 |
| lbf·in/in | newton meter per meter (N·m/m) | 4.448 222 E+00 |
| lbf·s/ft² | pascal second (Pa·s) | 4.788 026 E+01 |
| | | |
| lbf/ft | newton per meter (N/m) | 1.459 390 E+01 |
| lbf/ft² | pascal (Pa) | 4.788 026 E+01 |
| lbf/in | newton per meter (N/m) | 1.751 268 E+02 |
| lbf/in² (psi) | pascal (Pa) | 6.894 757 E+03 |
| lbf/lb (thrust/weight [mass] ratio) | newton per kilogram (N/kg) | 9.806 650 E+00 |
| | | |
| quart (US dry) | meter³ (m³) | 1.101 221 E−03 |
| quart (US liquid) | meter³ (m³) | 9.463 529 E−04 |
| rad (radiation dose absorbed) | gray (Gy) | 1.000 000*E−02 |
| rhe | 1 per pascal second (1/Pa·s) | 1.000 000*E+01 |
| rod | meter (m) | ...ᵃ |
| roentgen | coulomb per kilogram (C/kg) | 2.58      E−04 |
| | | |
| second (angle) | radian (rad) | 4.848 137 E−06 |
| second (sidereal) | second (s) | 9.972 696 E−01 |
| section | meter² (m²) | ...ᵃ |
| shake | second (s) | 1.000 000*E−08 |
| | | |
| slug | kilogram (kg) | 1.459 390 E+01 |
| slug/ft·s | pascal second (Pa·s) | 4.788 026 E+01 |
| slug/ft³ | kilogram per meter³ (kg/m³) | 5.153 788 E+02 |
| | | |
| statampere | ampere (A) | 3.335 640 E−10 |
| statcoulomb | coulomb (C) | 3.335 640 E−10 |
| statfarad | farad (F) | 1.112 650 E−12 |
| stathenry | henry (H) | 8.987 554 E+11 |
| statmho | siemens (S) | 1.112 650 E−12 |
| | | |
| statohm | ohm (Ω) | 8.987 554 E+11 |
| statvolt | volt (V) | 2.997 925 E+02 |
| stere | meter³ (m³) | 1.000 000*E+00 |
| stilb | candela per meter² (cd/m²) | 1.000 000*E+04 |
| stokes (kinematic viscosity) | meter² per second (m²/s) | 1.000 000*E−04 |
| | | |
| tablespoon | meter³ (m³) | 1.478 676 E−05 |
| teaspoon | meter³ (m³) | 4.928 922 E−06 |
| tex | kilogram per meter (kg/m) | 1.000 000*E−06 |
| therm | joule (J) | 1.055 056 E+08 |
| | | |
| ton (assay) | kilogram (kg) | 2.916 667 E−02 |
| ton (long, 2240 lb) | kilogram (kg) | 1.016 047 E+03 |
| ton (metric) | kilogram (kg) | 1.000 000*E+03 |
| ton (nuclear equivalent of TNT) | joule (J) | 4.184      E+09ᶠ |
| ton (refrigeration) | watt (W) | 3.516 800 E+03 |
| ton (register) | meter³ (m³) | 2.831 685 E+00 |
| | | |
| ton (short, 2000 lb) | kilogram (kg) | 9.071 847 E+02 |
| ton (long)/yd³ | kilogram per meter³ (kg/m³) | 1.328 939 E+03 |
| ton (short)/h | kilogram per second (kg/s) | 2.519 958 E−01 |
| ton-force (2000 lbf) | newton (N) | 8.896 444 E+03 |
| tonne | kilogram (kg) | 1.000 000*E+03 |
| | | |
| torr (mm Hg, 0°C) | pascal (Pa) | 1.333 22   E+02 |
| township | meter² (m²) | ...ᵃ |
| unit pole | weber (Wb) | 1.256 637 E−07 |
| W·h | joule (J) | 3.600 000*E+03 |

ᶠDefined (not measured) value.

**69.  Alphabetical List of Units** (*Continued*)

| To convert from | To | Multiply by |
|---|---|---|
| W·s | joule (J) | 1.000 000*E+00 |
| W/cm$^2$ | watt per meter$^2$ (W/m$^2$) | 1.000 000*E+04 |
| W/in$^2$ | watt per meter$^2$ (W/m$^2$) | 1.550 003 E+03 |
| yard | meter (m) | 9.144 000*E−01 |
| yd$^2$ | meter$^2$ (m$^2$) | 8.361 274 E−01 |
| yd$^3$ | meter$^3$ (m$^3$) | 7.645 549 E−01 |
| yd$^3$/min | meter$^3$ per second (m$^3$/s) | 1.274 258 E−02 |
| year (calendar) | second (s) | 3.153 600 E+07 |
| year (sidereal) | second (s) | 3.155 815 E+07 |
| year (tropical) | second (s) | 3.155 693 E+07 |

## 70. Electrical Symbols for Architectural Plans[a]

**Ceiling  Wall**

**GENERAL OUTLETS**

| | | |
|---|---|---|
| Ⓞ | ─Ⓞ | Outlet |
| Ⓑ | ─Ⓑ | Blanked Outlet |
| Ⓓ | | Drop Cord |
| Ⓔ | ─Ⓔ | Electric Outlet |
| | | For use only when circle used alone might be confused with columns, plumbing symbols, etc. |
| Ⓕ | ─Ⓕ | Fan Outlet |
| Ⓙ | ─Ⓙ | Junction Box |
| Ⓛ | ─Ⓛ | Lamp Holder |
| Ⓛ$_{PS}$ | ─Ⓛ$_{PS}$ | Lamp Holder with Pull Switch |
| Ⓢ | ─Ⓢ | Pull Switch |
| Ⓥ | ─Ⓥ | Outlet for Vapor Discharge Lamp |
| Ⓧ | ─Ⓧ | Exit-light Outlet |
| Ⓒ | ─Ⓒ | Clock Outlet (Specify Voltage) |

**CONVENIENCE OUTLETS**

Duplex Convenience Outlet

Convenience Outlet other than Duplex
1=Single, 3=Triplex, etc.

Weatherproof Convenience Outlet

Range Outlet

Switch and Convenience Outlet

Radio and Convenience Outlet

Special Purpose Outlet (Des. in Spec.)

Floor Outlet

**SWITCH OUTLETS**

| | |
|---|---|
| S | Single-pole Switch |
| S$_2$ | Double-pole Switch |
| S$_3$ | Three-way Switch |
| S$_4$ | Four-way Switch |
| S$_D$ | Automatic Door Switch |
| S$_E$ | Electrolier Switch |
| S$_K$ | Key-operated Switch |
| S$_P$ | Switch and Pilot Lamp |
| S$_{CB}$ | Circuit Breaker |
| S$_{WCB}$ | Weatherproof Circuit Breaker |
| S$_{MC}$ | Momentary Contact Switch |
| S$_{RC}$ | Remote-control Switch |
| S$_{WP}$ | Weatherproof Switch |
| S$_F$ | Fused Switch |
| S$_{WF}$ | Weatherproof Fused Switch |

**SPECIAL OUTLETS**

$\ominus_{a,b,c,etc.}$    S$_{a,b,c,etc.}$

Any standard symbol as given above with the addition of a lower-case subscript letter may be used to designate some special variation of standard equipment of particular interest in a specific set of architectural plans.

When used they must be listed in the Key of Symbols on each drawing and if necessary further described in the specifications.

**AUXILIARY SYSTEMS**

| | |
|---|---|
| ▣ | Pushbutton |
| | Buzzer |
| | Bell |
| ◇ | Annunciator |
| ◄ | Outside Telephone |
| ◁ | Interconnecting Telephone |
| ◁ | Telephone Switchboard |
| Ⓣ | Bell-ringing Transformer |
| D | Electric Door Opener |
| F▷ | Fire-alarm Bell |
| F | Fire-alarm Station |
| ▨ | City Fire-alarm Station |
| FA | Fire-Alarm Central Station |
| FS | Automatic Fire-alarm Device |
| W | Watchman's Station |
| W | Watchman's Central Station |
| H | Horn |
| N | Nurse's Signal Plug |
| M | Maid's Signal Plug |
| R | Radio Outlet |
| SC | Signal Central Station |
| | Interconnection Box |
| ⊣⊢⊣⊢ | Battery |
| ─·─·─ | Auxiliary System Circuits |

Note: Any line without further designation indicates a 2-wire system. For a greater number of wires designate with numerals in manner similar to ─·─
12-No. 18W-3/4"C., or designate by number corresponding to listing in Schedule.

☐$_{a,b,c}$  Special Auxiliary Outlets

Subscript letters refer to notes on plans or detailed description in specifications.

**70.   Electrical Symbols for Architectural Plans**[a] (*Continued*)

PANELS, CIRCUITS, AND MISCELLANEOUS

Lighting Panel

Power Panel

Branch Circuit; Concealed in Ceiling or Wall

Branch Circuit; Concealed in Floor

Branch Circuit; Exposed

Home Run to Panel Board. Indicate number of circuits by number of arrows.
Note: Any circuit without further designation indicates a two-wire circuit. For a greater number of wires indicate as follows: ─╫─ (3 wires)   ─╫╫─ (4 wires), etc.

Feeders. Note: Use heavy lines and designate by number corresponding to listing in Feeder Schedule.

Underfloor Duct and Junction Box-- Triple System
Note: For a double or single systems eliminate one or two lines. This symbol is equally adaptable to auxiliary system layouts.

(G)   Generator

(M)   Motor

(I)   Instrument

(T)   Power Transformer (Or draw to scale.)

Controller

Isolating Switch

[a]Abstracted from ANSI/IEEE 315-1975. It is not the intention of this standard to set the size of the particular conventions as used on the drawing. The size of wire or conduit is not to be designated or implied by the symbol.

The symbols can be modified by upper-case letters and other distinguishing marks placed in the center and by subscript letters or numbers placed at the lower right. Upper-case letters as subscripts refer to standard types, numerals to standard numerical variations; lower-case letters should be explained in the key of symbols accompanying the drawing or drawings.

# Index

The numbers referenced in the Index represent the *division* and *section number*, not the page number. For example, an Index reference to 10-57 means Division 10, Section 57. Since within each division all section numbers are placed in numerical order, there should be no difficulty in locating any section number.

At the beginning of each division is a table of contents which describes the various topics within the division. These are listed by page numbers.

*Note*: Numbers in the Index represent the *division* and *section*, not the page.

*Note*: Numbers in the Index represent the *division* and *section*, not the page.

*Note:* Numbers in the Index represent the *division* and *section*, not the page.

*Note*: Numbers in the Index represent the *division* and *section*, not the page.

*Note*: Numbers in the Index represent the *division* and *section*, not the page.

*Note*: Numbers in the Index represent the *division* and *section*, not the page.

*Note*: Numbers in the Index represent the *division* and *section*, not the page.

Note: Numbers in the Index represent the *division* and *section*, not the page.

*Note:* Numbers in the Index represent the *division* and *section*, not the page.

*Note*: Numbers in the Index represent the *division* and *section*, not the page.

Note: Numbers in the Index represent the *division* and *section*, not the page.

*Note*: Numbers in the Index represent the *division* and *section*, not the page.

*Note:* Numbers in the Index represent the *division* and *section*, not the page.

*Note*: Numbers in the Index represent the *division* and *section*, not the page.

*Note:* Numbers in the Index represent the *division* and *section*, not the page.

*Note*: Numbers in the Index represent the *division* and *section*, not the page.

*Note*: Numbers in the Index represent the *division* and *section*, not the page.

*Note:* Numbers in the Index represent the *division* and *section*, not the page.

Note: Numbers in the Index represent the *division* and *section*, not the page.

*Note:* Numbers in the Index represent the *division* and *section*, not the page.

## ABOUT THE EDITORS

Terrell Croft (deceased) was a consultant in the electrical contracting field.

Wilford I. Summers is former executive director of the International Association of Electrical Inspectors, a nonprofit trade association with over 22,000 members. Well-known through his many lectures and seminars on electricity before major trade groups, he has 40 years' experience in the electrical construction industry, including 10 years as chief electrical inspector for the city of Colorado Springs. He has edited numerous books on industry codes and standards, including the Tenth and Eleventh Editions of the *American Electricians' Handbook*, published by McGraw-Hill.